# BRIDGE REHABILITATION
# AND
# REPLACEMENT

## ( BRIDGE REPAIR PRACTICE )

BY

SUNG H. PARK, PE , LS

S. H. PARK

P. O. BOX 7474
TRENTON, NJ 08628–0474

To Bridge Engineers



ISBN :0-9604440-1-7
Library of Congress Catalog Card Number : 82-90094

Printed in the United States of America

# PREFACE

After the collapse of the Silver Bridge spanning the Ohio River on December 15, 1967, the Congress established a National Bridge Inspection Standard and Bridge Program. The collapse of the I-95 Mianus River Bridge on June 28, 1983 raised fears about the safety of the Nation's bridges. It brought renewed alert that bridges are deteriorating at an alarming rate and several States were able to pass Bridge Bonds to rebuild the crumbling, obsolete and aging bridges to bring them up to current structural and functional standards.

This book provides common causes of bridge defects and deterioration along with countermeasures for repairs, rehabilitations, retrofits, and replacements. It also covers a wide range of bridge related problem areas and exposes potential conflicts in engineering, design, materials, construction and fabrication. Beside engineering and design procedures, this book offers an overview of all areas of funding, environmental requirements, FHWA rules, selection of schemes, permits, and utility involvements.

The most difficult problem faced by bridge engineers today is the tendency involving only in bridge design while bridge projects require more general engineering, administration and communication skills. As bridge engineers, we had overlooked other potential problem areas such as highway geometrics, hydraulics, traffic control plans, construction, public relations and political influence. It is my intention to bring the bridge engineer's attention to other bridge related fields so as to play leading roles in rebuilding our aging bridges economically and effectively with fewer construction problems.

When repair or rehabilitation schemes are developed, the schemes often focuss on the condition of the main structural members but not secondary members and structural details of the main members. The results of such oversights could cause unnecessary delays and higher cost during construction. It is prudent to check whether details are compatible with field conditions or can be built on the fabrication shop or both. Details may appear to be perfect on the drawing board but may not be built because the design does not reflect field conditions or construction realities. Unfortunately, bridge construction is still labor intensive and precise construction is difficult to obtain without higher cost. Therfore simple construction and fabrication methods should be used whenever possible. For example, a level bridge seat for a prestressed box beam bridge with variable deck overlay depth on the beam is much simpler to build than a sloped bridge seat with a uniform depth overlay. If complex and complicated details and methods are required, the details and procedures should be clearly shown or specified on the plans or specifications to achieve desired results.

Like us, each bridge has its own characteristics. On rehabilitation projects, the designer should use engineering judgement to select proper repair details instead of blindly following standard details or design concepts and details from previous projects. Certain details or design approaches may be appropriate at one location but could fail to perform at others.

Current rapid development and use of micro computers and CADD systems open a new door for bridge engineers. Wide use of the computer free us from routine detailing work and lead us to visualize the structural behavior of the bridge in three dimentions instead of conventional two dimentional approach. This helps us to understand the structural need of a particular bridge and to provide better repair schemes and designs.

In selecting a rehabilitation or replacement scheme, some engineers are influenced by the visual condition of some particular members of the bridge instead of the overall condition of the structure. They attempt to save a few good members while replacing others. This approach could be uneconomical due to the additional labor costs and equipment required during demolition and rehabilitation operations. Often, a lower initial cost is the deciding factor in selecting the scheme but caution should be excercized since all other factors such as future maintenace and public inconvience must also be considered and may be more important than "initial cost".

Bridge rehabilitation is much more involved than new bridge construction and requires a great deal of effort to bring all disciplines together to build a cost effective bridge. In the design process, it requires a great deal of engineering judgement as to whether current AASHTO or state standards should be applied. If not, alternates should be evaluated. It may be prudent to design substandard bridge at a much lower cost at certain situation.

This book is to be used only as a guide in developing and selecting rehabilitation schemes, and preparing repair and replacement construction plans. Because the condition and structural requirements of the bridge differ substantially at each location, the designers are required to have engineering knowledge, experience, judgement, and a familiarity of the bridge site.

It is not intended that this book be construed as a substitute for engineering knowledge, experience or judgement. Even though the contents of this book have been developed with about ten years of bridge repair and rehabilitation experience along with many years of previous inspection and design experience in bridge engineering fields with consulting engineering firms and government agencies. Numerous information concerning bridge repair are available from the research reports of Transportation Research Board, FHWA and State Transportation Agencies, various state Bridge Manuals and Guidelines. Author attempts to bring them together for bridge repair practices. Different view points, opinions, details and design approaches are included and final application should be based on each engineer's judgement.

Although the material in this book has been checked, no warranty, expressed or implied, is made by author, as to accuracy of the material in this book nor shall the fact of application of generally accepted engineering practices constitute any such warranty. Author assumes no responsibility for the use of the material contained herein and its user should excercise his own engineering judgement in making any decision and applications.

I wish to express appreciation to friends, colleagues, and my family for their constructive suggestions and contributions in editing, reviewing, and revising manucript.

Acknowledgement is made for the invaluable help received or permission to reproduced the information from following organization: American Association of State Highway Officials (AASHTO), Ameican Welding Society (AWS), Armco Construction Div., Bethlehem Steel Corp., Energy Absorption System, Doublewal Corp., Fox Industries, Hobart Brothers Co., Lincoln Electric Co., NL Chemicals, Portland Cement Association (PCA), Prestressed Concrete Institute (PCI), Reinforced Earth Co., Royston, Maccaferri Gabion, National Corrugated Steel Pipe Association, Wire Reinforcement Institute, Inc., Transportation Research Board (TRB), Offices of Research, Development and Technology, Office of Implementation, Bridge Division and other Divisions of Federal Highway Administration (FHWA), Various State Departments of Transportation or Highways: Arizona, California, Connecticut, Colorado, Florida, Illinois, Iowa, Maine, Maryland, Massachusetts, Michigan, Minnesota, New Jersey, New Mexico, New York, Ohio, Pennsylvania, Texas, Virginia, Washington, West Virginia, Wisconsin.

May 1, 1984 Sung H. Park, PE & LS

# CONTENTS

## 1. GENERAL


1-1 **Introduction** .............................................. 1
1-2 **Bridge Inspection** ....................................... 3
1-3 **Deck Condition Surevey** ................................ 4
1-3.1 Field Survey 4
1-3.2 Category Classification 6
1-3.3 Recommended Restoration Procedure and Guideline 6
1-4 **Structural Inventory & Appraisal Sheet** .................. 8
1-4.1 General 8
1-4.2 Sufficiency Rating 8
1-4.3 Deficient Bridge 9
1-5 **Project Selection** ........................................ 10


## 2. REHABILITATION STUDY


2-1 **Studies** .................................................. 14
2-2 **Field Inspection** ......................................... 14
2-3 **General Consideration** .................................. 15
2-3.1 Maintenance of Traffic 15
2-3.2 Availability of Fund and Resources 16
2-3.3 Functional Requirements 16
2-3.4 Energy Conservation 23
2-3.5 Safety 23
2-3.6 Aesthetics 24
2-3.7 Environmental Consideration 24
2-3.8 Maintenance 27
2-3.9 Permits and Agreements 28
2-3.10 Value Engineering 28
2-4 **Developments of Schemes and Selection Process** .......... 31
2-4.1 Rehabilitation Schemes 32
2-4.2 Replacement Schemes 33
2-5 **Economic Analysis** ....................................... 34
2-5.1 Cost 34
2-5.2 Economic Analysis 34
2-6 **Comprehensive Evaluation** ............................... 37
2-7 **Report** ................................................... 38


## 3. PRELIMINARY DESIGN


3-1 **General** .................................................. 40
3-2 **Design Criteria** ......................................... 41
3-2.1 Design Live Loads 41
3-2.2 Design Method 42
3-2.3 Live Load Deflection 42
3-2.4 Material 42
3-3 **Structure Type Selection** ............................... 43
3-3.1 Selection of Superstructure Type 43
3-3.2 Selection of Abutment 48
3-3.3 Selection of Pier 51

| | | | |
|---|---|---|---|
| | 3-3.4 | Selection of Foundatioin | 51 |
| | 3-3.5 | Deck Drainage | 56 |
| | 3-3.6 | Approach Slab | 57 |
| | 3-3.7 | Railroad Grade Separation Structure | 57 |
| **3-4** | **Permits** ........................................ | | 58 |


## 4. DECK RECONSTRUCTION

| | | | |
|---|---|---|---|
| **4-1** | **General** ........................................ | | 61 |
| **4-2** | **Concrete Deck Repairs** ........................................ | | 62 |
| | 4-2.1 | Patching | 62 |
| | 4-2.2 | Epoxy Injection | 67 |
| **4-3** | **Concrete Deck Protective System** ........................................ | | 70 |
| | 4-3.1 | Low Slump, High Density Concrete | 70 |
| | 4-3.2 | Latex Modified Concrete | 71 |
| | 4-3.3 | Waterproofing Membrane System | 73 |
| | 4-3.4 | Polymer Impregnated and Polymer Concrete | 74 |
| | 4-3.5 | Wax Bead Concrete (Internally Sealed Concrete) | 75 |
| | 4-3.6 | Cathodic Protection System | 75 |
| | 4-3.7 | Chloride Removal | 76 |
| | 4-3.8 | Protective Coating and Sealants | 76 |
| | 4-3.8 | Corrosion Inhibiting Admixture | 77 |
| **4-4** | **Deck Replacement** ........................................ | | 78 |
| | 4-4.1 | Concrete Deck Slab Replacement | 78 |
| | 4-4.2 | Other Factors | 79 |
| **4-5** | **Skid Resistance** ........................................ | | 81 |
| **4-6** | **Deck Design** ........................................ | | 83 |
| | 4-6.1 | Concrete Deck Slab Design | 83 |
| | 4-6.2 | Precast Deck | 83 |
| | 4-6.3 | Steel Grid Deck | 85 |
| | 4-6.4 | Corrugated Metal Deck | 86 |
| | 4-6.5 | Timber Deck | 86 |
| | 4-6.6 | Orthotropic Steel Deck | 87 |


## 5. CONCRETE STRUCTURE

| | | | |
|---|---|---|---|
| **5-1** | **General Defects** ........................................ | | 91 |
| **5-2** | **Investigation** ........................................ | | 92 |
| **5-3** | **Concrete Repairs** ........................................ | | 93 |
| | 5-3.1 | General Repairs | 93 |
| | 5-3.2 | Repair Materials | 95 |
| | 5-3.3 | Repair Methods | 97 |
| | 5-3.4 | Surface Preparation and Reinfocement Steel | 97 |
| | 5-3.5 | Dry Pack Method | 98 |
| | 5-3.6 | Preplaced Aggregate Method (Prepacked Method) | 98 |
| | 5-3.7 | Patching | 100 |
| | 5-3.8 | Shotcrete (Pneumatically Applied Mortar or Gunite) | 101 |
| | 5-3.9 | Pressure Injection | 101 |
| | 5-3.10 | Pressure Grouting | 102 |
| | 5-3.11 | Others | 102 |
| **5-4** | **Reinforcing and Strengthening** ........................................ | | 103 |
| | 5-4.1 | Reinforcing | 103 |
| | 5-4.2 | Strengthening | 105 |
| **5-5** | **Design Details** ........................................ | | 106 |

## 6. PRESTRESSED CONCRETE STRUCTURE


| | | | |
|---|---|---|---|
| **6-1** | **Prestressed Concrete Repairs** ........................... | | 109 |
| | 6-1.1 | General | 109 |
| | 6-1.2 | Damage Evaluation | 109 |
| | 6-1.3 | Countermeasure and repairs | 110 |
| | 6-1.4 | Minor Damage | 111 |
| | 6-1.5 | Moderate Damage | 111 |
| | 6-1.6 | Severe Damage | 114 |
| | 6-1.7 | Preloading | 114 |
| | 6-1.8 | Restoration of Prestress | 115 |
| | 6-1.9 | Girder Replacement | 117 |
| **6-2** | **Prestressed Concrete Bridge** ............................. | | 117 |
| | 6-2.1 | Replacement Bridge | 117 |
| | 6-2.2 | Design Consideration | 118 |
| | 6-2.3 | Materials | 123 |
| **6-3** | **Preloadings and Repairs** ................................ | | 124 |
| | 6-3.1 | Preloading | 124 |
| | 6-3.2 | Repairs | 126 |
| **6-4** | **Segmental Concrete Box Bridge** ........................... | | 128 |
| | 6-4.1 | General | 128 |
| | 6-4.2 | Casting of Segmental Bridge | 129 |
| | 6-4.3 | Erection Methods | 130 |
| | 6-4.4 | Design Consideration | 132 |
| | 6-4.5 | Problems of Segmental Concrete Bridges | 133 |


## 7. STEEL STRUCTURE


| | | | |
|---|---|---|---|
| **7-1** | **Corrosion** ................................................ | | 140 |
| | 7-1.1 | General | 140 |
| | 7-1.2 | Galvanic Corrosion | 141 |
| | 7-1.3 | Differential Environmental Corrosion | 142 |
| | 7-1.4 | Stress Corrosion | 142 |
| | 7-1.5 | Fretting Corrosion | 142 |
| | 7-1.6 | Chemical Corrosion | 142 |
| | 7-1.6 | Stray Current Corrosion | 142 |
| **7-2** | **Corrosion Control and Remedial Measures** ................. | | 142 |
| | 7-2.1 | General | 142 |
| | 7-2.2 | Protective Methods | 145 |
| | | 1. Metallic Coating | 145 |
| | | 2. Organic Coating - painting | 146 |
| | | 3. Inorganic Coating | 150 |
| | | 4. Cathodic Protectiion | 150 |
| **7-3** | **Fracture and Fatigue** .................................... | | 152 |
| | 7-3.1 | Brittle Fracture | 153 |
| | 7-3.2 | Stress Corrosion Crack | 153 |
| | 7-3.3 | Fatigue Cracks | 156 |
| | 7-3.4 | Testing Methods | 156 |
| | 7-3.5 | Welded Cover Plate | 158 |
| | 7-3.6 | Welded Stiffener | 161 |
| | 7-3.7 | Floor Beam to Girder or Stringer Connection | 163 |
| | 7-3.8 | Cantilever Bracket Connection | 168 |
| | 7-3.9 | Longitudinal Welds | 169 |
| | 7-3.10 | Gusset Plate Connection | 170 |
| | 7-3.11 | Geometric Stress Riser and Reenterant Corners | 173 |
| | 7-3.12 | Lamellar Tearing and Lamination | 173 |

| | | | |
|---|---|---|---|
| | 7-3.13 | Eye Bars and Hanger Assembly | 175 |
| | 7-3.14 | Other Defects | 176 |
| **7-4** | **Fracture Control Plan** ............................... | | 177 |
| **7-5** | **Rehabilitation** ............................................ | | 180 |
| | 7-5.1 | Rehabilitation Procedure | 180 |
| | 7-5.2 | Repair Design Consideration | 181 |
| | 7-5.3 | Strengthening | 184 |
| | 7-5.4 | Adding Supplementary Member | 191 |
| | 7-5.5 | Modifying Structural System | 192 |
| | 7-5.6 | Repairs of Collision Damaged Members | 194 |

**8. TIMBER STRUCTURE**

| | | | |
|---|---|---|---|
| **8-1** | **Defects** ................................................ | | 197 |
| | 8-1.1 | Common Defects | 197 |
| | 8-1.2 | Common damages | 198 |
| **8-2** | **Maintenace and Repairs** ................................. | | 199 |
| | 8-2.1 | Replacement | 199 |
| | 8-2.2 | Replacement of Member | 199 |
| | 8-2.3 | Remedial Repairs | 200 |
| | 8-2.4 | Preventive Maintenance | 202 |
| **8-3** | **New Timber Bridge** ...................................... | | 203 |

**9. BEARINGS**

| | | | |
|---|---|---|---|
| **9-1** | **General** ................................................. | | 204 |
| | 9-1.1 | Forces | 204 |
| | 9-1.2 | Movement | 204 |
| **9-2** | **Bearing Types** ........................................... | | 205 |
| | 9-2.1 | Types of Bearing | 205 |
| | 9-2.2 | Sliding Plate Type | 206 |
| | 9-2.3 | Rocker Bearing | 206 |
| | 9-2.4 | Roller Bearing | 206 |
| | 9-2.5 | Elastomeric Bearing | 207 |
| | 9-2.6 | Pot Bearing | 208 |
| | 9-2.7 | Spherical Bearing | 208 |
| **9-3** | **Problems and Repairs** .................................. | | 209 |
| | 9-3.1 | Rusting | 209 |
| | 9-3.2 | Alignment | 210 |
| | 9-3.3 | Girder Bearing | 211 |
| | 9-3.4 | Bearing Surface | 211 |
| | 9-3.5 | Looseness in Bearing | 211 |
| | 9-3.6 | Bearing Seat | 212 |
| | 9-3.7 | Repairs of Bearing | 212 |
| | 9-3.8 | Guideline for New Bearing | 212 |

**10. JOINTS AND DRAINAGE**

| | | | |
|---|---|---|---|
| **10-1** | **Joints** .................................................. | | 214 |
| | 10-1.1 | Expansion Joints | 214 |
| | 10-1.2 | Types of Joints and Problems | 215 |
| **10-2** | **Drainage** ................................................ | | 219 |

## 11. SUBSTRUCTURE


11-1 **Abutment** ........................................ 221
 11-1.1 Cracks in Abutment Wall 221
 11-1.2 Crack Repairs 221
 11-1.3 Bridge Seat 222
 11-1.4 Skew Effects 223
 11-1.5 Other defects 224
 11-1.6 Raising the Bridge Seat 224
 11-1.7 Types of Abutment and Integral Abutment 226
11-2 **Stone Masonry** ........................................ 227
11-3 **Pier** ........................................ 232
 11-3.1 Types of Pier 232
 11-3.2 Defects and Repairs 233


## 12. RETAINING WALL


12-1 **General** ........................................ 234
 Reinforced Earth 235
 Doublewal 236
 Gabion 237
 Tiebacks 240
12-2 **Defects and Repairs** ........................................ 241


## 13. FOUNDATION


13-1 **Foundation** ........................................ 242
 13-1.1 General 242
 13-1.2 Spread Footing 242
 13-1.3 Pile Foundation 243
 13-1.4 Other Types of Footing 245
13-2 **Foundation Failure** ........................................ 245
 13-2.1 General 245
 13-2.2 Settlement 246
 13-2.3 Seepage and Water Table Variation 247
 13-2.4 Expansive Soil 247
13-3 **Monitoring** ........................................ 248
 13-3.1 General 248
 13-3.2 Instrumentation 248
13-4 **Foundation Repairs** ........................................ 249
 13-4.1 Strengthening 250
 13-4.2 Stabilizing Soil 252
 13-4.3 Pile Repairs 257


## 14. WATERWAY


14-1 **General** ........................................ 261
14-2 **Scour** ........................................ 261
14-3 **Degradation and Aggradatiion** ........................................ 262
14-4 **Countermeasures** ........................................ 263
 14-4.1 General 263
 14-4.2 Scour 266
 14-4.3 Degradation 267
 14-4.4 Aggradatioin 269
14-5 **Preventive Measures** ........................................ 269
14-6 **Risk Analysis** ........................................ 270

14-6.1 General 270
14-6.2 Risk Analysis 271
14-6.3 Design Procedure 271
**14-7 Erosion Control** ........................................ 273

## 15. CULVERT

**15-1 General** ........................................ 274
**15-2 Defects and Repairs** ........................................ 274
15-2.1 Structure 274
15-2.2 Hydraulics 277
15-2.3 Culvert Extension 280

## 16. TRAFFIC SAFETY

**16-1 General** ........................................ 281
**16-2 Railing** ........................................ 283
**16-3 Traffic Barrier** ........................................ 286
16-3.1 General 286
16-3.2 Guard Rail 287
**16-4 Impact Attenuater** ........................................ 288
16-4.1 Gore Area on Bridge 288
16-4.2 Impact Attenuater 289
**16-5 Narrow Bridge** ........................................ 290
16-5.1 General 290
16-5.2 Remedial Treatment 291
**16-6 Traffic Control Plan** ........................................ 292
16-6.1 Work Zone Traffic Control Plan 292
16-6.1 Traffic Control for Construction and Operatioin 294

## 17. WIDENING

**17-1 Review of Studies** ........................................ 296
**17-2 Design Consideration** ........................................ 296
**17-3 Design** ........................................ 298

## APPENDIX

**A: General** ........................................ 303
**B: Deck** ........................................ 332
**C: Concrete Structure** ........................................ 369
**D: Prestressed Concrete Structure** ........................................ 401
**E: Steel Structure** ........................................ 486
**F: Timber Structure** ........................................ 574
**G: Deck Joints** ........................................ 582
**H: Deck Drainage** ........................................ 600
**I: Bearing** ........................................ 608
**J: Substructure** ........................................ 634
**K: Wall** ........................................ 667
**L: Foundation** ........................................ 686
**M: Culvert** ........................................ 706
**N: Waterway (Hydraulic)** ........................................ 722
**O: Traffic Safety** ........................................ 742
**P: Widening** ........................................ 753
**Q: Specification and Special Provisions** ........................................ 763
**R: Miscellaneous** ........................................ 804

# 1 GENERAL

## 1-1 INTRODUCTION

National growth and economic expansion are dependent on highway sprawl and the growth of competing forms of transportation. However, as the nation's transportation sectors have almost matured and the importance of energy conservation measures has emerged, we are forced to take a second look into the themes of transportation.

The emphasis has begun to shift from building new roads to improving and maintaining existing facilities on highway programs. As originally conceived, federal funds were to be used solely in the construction of nationwide highway systems and maintenance was to be a responsibility of the State.

**Fig. 1.1 Bridge Failure**

However, because of the financial problems of the states and local governments and substantial increases in the amount of highway mileage which had to be maintained by the states and the local government, maintenance efforts are dwindled and some of our old highway bridges have deteriorated to such a critical point that they had to be closed to assure public safety. Therefore the Federal Highway Administration was forced to assume more responsibility of maintaining and rehabilitating the existing bridges to prevent any catastrophic disasters.

Many county and local bridges are too old, deteriorated, or otherwise inadequate to accommodate today's traffic volume, and vehicular weight. Therefore, these bridges must be restricted for clearance and weight. It is estimated that there are about 558,000 bridges in the States and slightly over 248,500 bridges, almost 45 percent are deficient. Approximately 126,600 of them are structurally deficient (inadequate for heavy loads) and the other 121,900 bridges are functionally obsolete (narrow widths, poor clearance and critical approach roadways).

**Fig. 1.2 Bridge Closing**

There are over 3,400 bridges which are reported as closed to all traffic as of December 31, 1981. Furthermore, about 61 percent of bridges located on Off-System roads and 26 percent of Federal aid system bridges are deficient.

The Federal Government has become the major source of funds through the Bridge Program to rehabilitate or replace deficient bridges. Most state and local governments are heavily depending on these Bridge Replacement and Rehabilitation Funds and 4R (Resurfacing, Restoring, Rehabilitation, or Reconstruction) Funds.

In 1978, the highway bridge replacement and rehabilitation program was expanded to cover Off-System (non-federally aided) bridges, and the federal share of the cost was increased to 80 percent. Some bridges are being replaced or repaired under other Federal-aid programs (Interstate 4R, Primary, Secondary, or Urban) while others are replaced with states and local funds without federal participation. However, at current funding levels, it would take many years to restore the deficient bridges already identified and additional bridges will be added to the list during this period.

Because of the necessity of keeping highways open, most state and local governments have used up their available funds and resources for temporary repairs and strengthening of the bridges. However, with inspection and proper maintenance programs, as well as increased federal participation of these projects, state and local governments will be able to rehabilitate and replace bridges on a systematic basis as they become structurally deficient in load carrying capacity and functionally obsolete in geometric aspects.

In 1982, the Congress passed "Surface Transportation Assistance Act of 1982" and raised the federal gasoline tax from 4 cents to 9 cents per gallon to generate additional $5.5 billion a years over the next four years. In the bridge program, each deficient bridge will be placed into one of following categories:

1) Federal-aid system bridges eligible for replacement
2) Federal-aid system bridges for rehabilitation
3) Off-system bridges for replacement
4) Off-system bridges for rehabilitation.

The apportionment factor of the bridge program will be the total square foot cost of deficient bridges in each category in each State divided by the total cost of the deficient bridges in all states. Maximum share for a state will be 10% and minimum be .25%. Also, Off-system bridges will have maximum 35% and minimum 10% of the bridge funds apportioned to each State.

In 1982, Reconstruction was added to the 3R program (Resurfacing, Restoring, or Rehabilitation) for Primary, Secondary, and Urban system to make 4R program similar to the Interstate system. Beginning with Fiscal Year 1984, a minimum of 40% of Primary, Secondary, or Urban program will be used for the 4R project.

**Fig. 1.3 Temporary Repairs**

In order to restore the structural integrity of a bridge and to correct its major defects which threaten public safety, the following procedure may be used:

1) Inspect the bridge to identify its deficiencies, appraise and rate it for its condition.
2) After inspecting the bridge, complete the Structural Inventory and Appraisal (SI & A) sheet which summarizes its condition, and computes sufficiency rating. This step is necessary because the SI & A sheet must be submitted to the federal government to be eligible for the bridge program. This information can be used to determine whether the condition of the bridge justifies its rehabilitation or replacement.
3) For a structure in poor condition, classification should be made as to whether the bridge is either structurally deficient or functionally obsolete. Generally, this job is performed by the State. Then, from these bridges, each State will make a project selection list based on the condition of the structures in question, on the availability of funds for the action being considered and other requirements.
4) After selecting the eligible projects for rehabilitation or replacement, studies should be made to evaluate feasible alternatives or to evaluate economic schemes for bridge replacement when rehabilitation is not feasible.

5) After adopting a scheme, the preliminary plans should be developed.

In the 1970's, the bankruptcy of railroads in the Northeast and Midwest became a national concern and Congress passed the Railroad Revitalization and Regulatory Reform Act (4R Act) of 1976 authorizing $1.6 billion to help the nation's railroads to overcome deferred maintenance. This act also granted $2.5 billion for capital improvement in the Northeast Corridor Programs. One project which constitutes a large portion of these program, the main line between Washington, D.C. and Boston, is scheduled to be completed by 1984. In 1982 Act, a one cent equivalent highway user revenue is dedicated to Mass Transit Account of Highway Trust Fund but most of the fund will be used for operation subsidies of mass transit.

Major railroad bridge deterioration problems still exist on most marginal branch lines, where railroads do not have enough traffic to generate the extensive capital investments required to correct deferred maintenance. It appears that abandoning infrequently used tracks and consolidating the remaining operations are the most economical choices for the railroad and the tax paying public in general. But these choices however are not feasible in the long run because the opening of rail lines is essential for commuters, practical in anticipating future energy need, and vital to local areas economic growth and interests.

It appears that the program assisting local rail services will continue to support the local branches and the feeder railroad lines with federal assistance and the aim of the railroad bridge rehabilitation and replacement program will be focused on public safety.

## 1—2 BRIDGE INSPECTION

The collapse of the Silver Bridge in West Virginia in 1967 aroused national attention in the inspection and maintenance of bridges. Congress added a section to the Surface Transportation Assistant Act of 1978 concerning bridge rehabilitation and replacement programs. This act requires that all public bridges be inspected and inventoried in accordance with the National Bridge Inspection Standard (NBIS), and that between 15 and 35 percent of the total money apportioned to the States should be used for replacing and rehabilitating Off-System bridges.

**Fig. 1.4 Inspection by Snooper**

Even without the mandatory Federal Act, bridge inspection is necessary for both old and new bridges. Old bridges were designed for light traffic volume and small load capacities and their basic construction materials have deteriorated due to weather and age. For new bridges designed according to current requirements, inspection is important to observe whether each element in the structure is behaving correctly, and to detect whether there are any faults in the design or displays of poor workmanship in the construction details.

In the past, the use of salt on ice slicked decks and the absence of systematic preventative maintenance have caused an alarming number of bridges to deteriorate and fail. Therefore, the primary function of bridge inspection is to assure public safety and confidence; the second aim is to provide timely and economical planning for preventive maintenance and remedial measures which include major rehabilitation and bridge replacement programs.

Accurate and detailed bridge inspection data will not only minimize the repair design efforts in preparing construction plans but will also avoid costly design changes during repairs. It is strongly recommended that design engineers inspect the bridge themselves before preparing any rehabilitation plans. The engineer may use the inspection report as a general guideline or reference, but at least one field visit should be made before any studies are made or any repair schemes initiated.

## 1—3 DECK CONDITION SURVEY

### 1-3.1 FIELD SURVEY

Many bridge decks across the country show some degree of deterioration. Even recently built concrete decks show corroded rebars and high salt concentration reaching about 2 pounds of chloride per cubic yard of concrete. This problem becomes worse every year due to the heavy use of de-icing salt.

Maintenance people face the problems of maintaining a snow and ice free deck and at the same time, protecting the concrete deck from disintegrating and deteriorating.

There are commonly three conditions of deck deterioration: cracking, scaling and spalling. A crack itself is not a serious problem and our design concepts even allow for cracks in the tensile zone of the concrete. Scaling is the disintegration of concrete on the surface evidenced by flaking and loose aggregate. On the other hand, spalls are deep circular depressions revealing top reinforcement. These spalls eventually become pot holes through the deck.

**Fig. 1.5 Crack in Deck**

The bridge deck testing serves the following three important purposes:

1) To establish the initial condition of the deck before a protective system or a deck repair is started.
2) To determine a priority list for bridge deck repair.
3) To define and estimate repair areas prior to bridge deck repair work.

The following bridge deck evaluation techniques may be used to detect the defects and to evaluate the condition of the deck.

**1. Visual Observation**
This is quick, very subjective, and used as a basis for further deck condition surveys. Visual observation gives the degree of area spalled and the percentage of the deteriorated deck area while noting other deficiencies. However, visual observation usually does not reveal the hidden structural problems such as delamination, rebar corrosion, the thickness of the existing concrete cover for the reinforcement and the chloride contents of the concrete.

**2. Delamination Detection**
A delamination is a horizontal plane of separation in the concrete deck slab, usually occurring at the level of the top reinforcement steel.

**a. Chain Drag:** The main problem with this method is that heavy traffic noise will prevent a true reading of the hollow sound created over the delaminated area.

**b. Sounding by Hammer:** This is an effective but time consuming method.

**Fig. 1.6 Delamination Detector**

**c. Delamination Detector:** the concrete delamination detector is a mechanical-electrical instrument for sonically identifying and recording information concerning the location and the extent of subsurface delamination of bridge deck concrete.

### 3. Electrical Corrosion Detection Device

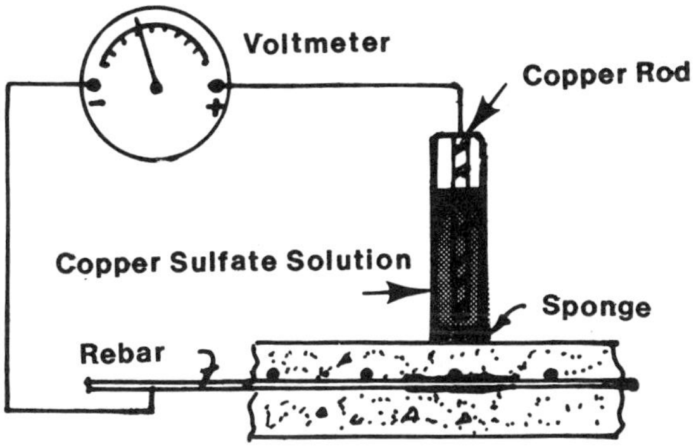


**Fig. 1.7 Half-Cell of Copper/Copper Sulfate**

The active corrosion of reinforcing bars is caused by electric current flowing from one point (anode) to another point (cathode) when an electrical (potential) difference occurs between the anode and the cathode. Copper/copper sulfate half-cells are used to measure these electrical potentials. Potential differences of -.35 volts or more indicate the possibility of active corrosion occuring in the reinforcing steel. However, this does not necessarily indicate that any positive relationship truly exists between the measurement shown by the Copper/Copper sulfate half-cell.

### 4. Pachometer

The pachometer is used to determine the thickness of the concrete cover above the reinforcing steel. This data may be used to determine at what depth the chloride level should be measured and to evaluate the overall condition of the bridge deck. This data also supplements the findings of the other tests.

### 5. Chloride Analysis

A chloride analysis provides a quantitative measure of the chloride ion content of the concrete in the deck. Rotary hammer or coring may be used to obtain chloride samples. Measuring the total chloride content using the Berman technique requires digestion of the sample in dilute nitric acid and titration of the filtered extract with standard silver nitrate solution, using a chloride-ion-specific electrode as an indicator. The level of chloride content needed to initiate corrosion, which is also known as **the threshold of chloride content, is between 1.3 and 2.0 pounds per cubic yard of concrete.**

### 6. Coring

Even though the FHWA does not require a compression test of the existing concrete, it is suggested that a minimum of two corings be taken from the bridge and tested to determine actual compression strength of the concrete.

### 7. Summary

Visual observations are important in revealing the probable overall condition of the deck in general terms. However, caution should be exercised in the measurement of spalled area as a major factor in determining the condition of the deck, because deterioration could extend beyond the area of spalling.

The delamination detector will define the areas of subsurface fracture within the concrete, and the electrical corrosion detection device aids in finding the areas of active corrosion at a given time. However, the delamination detector will not identify whether the concrete is contaminated or sound and the corrosion detection device will not indicate whether the concrete at the passive voltage site is contaminated.

To determine the condition of the contamination and to detect future deterioration potentials, a chloride analysis is necessary at the level of the top mat of reinforcement

If the proper test procedures are followed, the results of a chloride analysis are more accurate and reliable than the results of other tests. However in discovering and deciding the areas of concrete for permanent repairs, the chloride analysis should not be used as sole evaluation method. Thus, the optimum evaluation procedure would be to examine the results of all tests and supplement these results with a visual observation.

### 1-3.2 CATEGORY CLASSIFICATION

The categories are based on the best judgement available nationally. The user will note that Category 2 will in many cases overlap Category 1. In such cases, the State will exercise its best judgement based on engineering, economics and other factors in categorizing a given bridge deck. The three categories of condition ratings for a concrete deck as defined by the FHWA are as follows:

**Category 1: Extensive Active Corrosion**

5% or more of the deck area spalled OR 40% or more of the deck area deteriorated or contaminated as indicated by the summation of any non-duplicating combination of the following: 1) spalls, 2) delaminations, and 3) corrosion potentials over 0.35 volt (CSE) OR 40% of the area of the bridge deck contains greater than 2 lbs. of chloride per cubic yard of concrete at the level of the top reinforcing bars as indicated by random chloride sampling.

**Category 2: Moderate Active Corrosion**

0 to 5% of the deck area spalled OR 5 to 40% of the deck area deteriorated or contaminated as indicated by summation of nonduplicating combination of the following: 1) spalls 2) delaminations and 3) corrosion potential over 0.35 volt(CSE) OR 5 to 40% of deck area indicated by random chloride sample contains greater than 2 lbs. of chloride per cubic yard of concrete at the level of the top reinforcement bars.

**Category 3: Light to No Active Corrosion**

No spalls, OR 0 to 5% of the deck area deteriorated or contaminated as indicated by summation of any non-duplicating combination of the following: 1) delamination, 2) corrosion potentials over 0.35 volt, OR 0 to 5% of the deck area contains greater than 2 lbs. of chloride per cubic yard of concrete at the level of rebars as shown by random chloride sampling.

### 1-3.3 RECOMMENDED RESTORATION PROCEDURE AND GUIDELINE

Based on the categories of the condition of the bridge deck an acceptable reconstruction alternative has been developed for use on Federal Aid projects as shown on the following page. These recommendations are based on the best judgement of FHWA, bridge engineers, maintenance engineers, and research engineers currently involved in the rehabilitation of concrete bridge decks.

In 1983, FHWA revised the rules of concrete bridge deck to allow States the flexibility to select a protective and reconstruction methods. This revision eliminate the strict provision requiring removal of contaminated deck element for permanent reconstruction.

The following are some of the FHWA guidelines for the bridge programs:

> To be eligible for the Bridge Rehabilitation and Replacement program, the sufficiency rating of the bridge should be less than 80 and the bridge should be structurally deficient or functionally obsolete.
>
> The same design standard applies to bridge replacement (BR) and rehabilitation (BH) funded projects as to regular federal aid projects. Off-System projects are processed the same as secondary road projects.
>
> Major rehabilitation consists of reconstruction work at least as extensive as total deck replacement. Minor rehabilitation consists of restoration work less extensive than total deck replacement.
>
> Any rehabilitation must restore structural integrity for a minimum period of 10 years (20 years desirable) and correct major safety defects. Railings must substantially comply with current AASHTO requirements.
>
> Painting of the existing structural steel is eligible for federal funds if the rehabilitation is a major project.
>
> Roadway widths and live load capacity must comply with applicable AASHTO and State standards. For minor rehabilitation, previous criteria may be used as approved. Major rehabilitation should conform to current standards. For new or reconstructed bridges which have over 400 aver-

*Acceptable reconstruction procedures for Federal-aid participation* [1]

| Category | Procedures | Acceptable permanent reconstruction | Experimental cost-effective reconstruction (estimated extended life 10 to 15 yr.) |
|---|---|---|---|
| Structurally inadequate. | .......... | Complete deck replacement (unless restorable). | |
| Extensive deck deterioration. | Required restoration work. | Complete deck replacement .......... | Removal of all deteriorated concrete. Follow the repair procedure approved for the protective system selected. |
| | Testing .......... | Steps 1 through 5 as necessary (probably only steps 1 and 2). | Steps 1 and 2 only, except all steps on the first 3 plus 10 pct of the remaining decks. |
| | Suggested protective systems. | Epoxy coated rebars. Two course-Iowa system or latex modified concrete. Membrane W/ac overlay. Cathodic protection. | Two course-Iowa system or latex modified concrete. Membrane W/ac overlay. Cathodic protection. |
| Moderate deck deterioration.. | .......... | Same as for category 1 above.—or .......... Same as for category 3 below as determined by the State.. | Same as for category 1 above. |
| Light deck deterioration. | Required restoration work. | Removal and replacement of all areas of deterioration and chloride contaminated concrete as determined by corrosion potentials and/or chloride sampling (less than 5 pct of deck area).. | Same as for category 1 above. [2] |
| | Testing .......... | Steps 1 through 5 .......... | Steps 1 and 2 only, except all steps on the first 3 plus 10 pct of the remaining decks. |
| | Suggested protective systems. | Two course-Iowa system or latex modified concrete. Membrane W/ac overlay. Cathodic protection. | Two course-Iowa system or latex modified concrete. Membrane W/ac overlay. Cathodic protection. |

[1]Testing steps: 1 Visual, 2 Delamination, 3 Electrical potential, 4 Rebar Depth, 5 Chloride Content.
[2]For this category of condition permanent restoration is strongly recommended.

**TABLE I Deck Reconstruction Procedure**

age daily traffic, HS 20 live load capacity should be obtained based on inventory rating procedures.

The Ohio Department of Transportation has developed following guidelines for bridge deck rehabilitation and repairs:

**Required Work:**

1. If more than 25% of the deck area is deteriorated, overlay with Low Slump High Density Concrete or Latex Modified Concrete or replace the deck.
2. If the deteriorated area with spalls and delamination is 25% or less, remove all the deteriorated concrete, patch the deteriorated area and waterproof and overlay with 2 1/2" asphalt concrete.
3. If no repairs are required, waterproof and overlay with 2 1/2" asphalt concrete.

**Other Considerations:**

Bridges without curbs must have drip strip along the edges. For bridges with curbs, consideration should be given to providing floor drain with asphalt overlay.

A deck condition survey report is required for each deck. It should be conducted as near to the plan preparation stage as practical. If the time lapse between plan preparation and award of the project for construction is expected to be considerable, the patching quantities should be increased.

Bridges with existing asphalt concrete overlays should be cored if there is any doubt to their condition.

For timber bridge decks, if the timber is in good condition and tight, consideration should be given to water proofing and overlaying with 2 1/2" asphalt concrete.

For bridges with more than 2 1/2" existing asphalt concrete, the preferred treatment is to remove all existing asphalt concrete and treat as outlined in guidelines above. Approaches must be profiled to lower grade. Older decks with excessive asphalt concrete may be "skimmed over" with 3/4" asphalt.

End Dams of joints should be vertically extended and sealed. Bare bridge decks in excellent condition (no cracks) may be sealed with a penetrating sealant.

Bridges on a grade of 4% or more should not be waterproofed but consider using a concrete overlay.

STRUCTURE INVENTORY AND APPRAISAL

37.8 SUFFICIENCY RATING (STRUC DEF)

05/12/80

**IDENTIFICATION**

| | |
|---|---|
| (1) STATE | STATE NAME |
| (2) HIGHWAY DISTRICT | 03 |
| (3) COUNTY 075 (4) CITY/TOWN | 0000 |
| (5) INVENTORY ROUTE | 140000090 |
| (6) FEATURE INTERSECTED SUGAR CREEK | |
| (7) FACILITY CARRIED | CH-9 |
| (8) STRUCTURE NUMBER | 075000090502564 |
| (9) LOCATION | 1.42 MI W OF WHITTIER |
| (10) VERT CLEARANCE | 16 FT 00 IN |
| (11) MILEPOINT | 22.37 |
| (12) ROAD SECTION NO | 0000 |
| (13) DEFENSE BRIDGE LETTER | |
| (14) DEFENSE MILEPOINT | 0.00 |
| (15) DEFENSE SECTION LENGTH | 0.0 |
| (16) LATITUDE 40D 37.8' (17) LONGITUDE 087D 43.4' | |
| (18) PHYSICAL VULNERABILITY | |
| (19) BYPASS DETOUR LENGTH | 12 MI |
| (20) TOLL | ON FREE ROAD |
| (21) CUSTODIAN | S H DEPT |
| (22) OWNER | COMBINATION |
| (23) FEDERAL-AID PROJECT NUMBER | 0000000 |

**CLASSIFICATION**

| | |
|---|---|
| (24) HIGHWAY SYSTEM | FA SECONDARY, STATE |
| (25) ADMINISTRATIVE | STATE |
| (26) FUNCTIONAL CLASS | MAJOR RURAL |

**STRUCTURE DATA**

| | |
|---|---|
| (27) YEAR BUILT | 2200 |
| (28) LANES ON STR 02 UNDER | 00 |
| (29) ADT ON STR | 001100 |
| (30) YEAR OF ADT | 1977 |
| (31) DESIGN LOAD | H-15 |
| (32) APP RDWY WDTH W/SHLD | 30 FT |
| (33) BRIDGE MEDIAN | NONE |
| (34) SKEW | 00 |
| (35) STRUCTURE FLARED | NO |
| (36) TRAFFIC SAFETY FTRS. | 0111 |
| (37) | |
| (38) NAV CONTROL | NO |
| (39) NAV VERT CLEARANCE | 000 FT |
| (40) NAV HORIZ CLEARANCE | 0000 FT |
| (41) OPEN OR CLOSED | POSTED |

| | | |
|---|---|---|
| (42) TYPE SERVICE | HIGHWAY | WATERWAY |
| (43) STRUCTURE TYPE MAIN | STEEL / | TRUSS-THR |
| (44) STRUCTURE TYPE APP | OTHER / | OTHER |
| (45) NO OF SPANS MAIN | | 001 |
| (46) NO OF APPROACH SPANS | | 0000 |
| (47) TOTAL HORIZ CLEARANCE | | 20.0 FT |
| (48) MAX SPAN LENGTH | | 0181 FT |
| (49) STRUCTURE LENGTH | | 000183 FT |
| (50) SIDEWALK | LEFT 0.0 FT RIGHT | 0.0 FT |
| (51) BRIDGE RDWY WIDTH CURB TO CURB | | 19.2 FT |
| (52) DECK WIDTH OUT TO OUT | | 19.2 FT |
| (53) VERT CLEARANCE OVER DECK | | 16 FT 00 IN |
| (54) MIN VERT UNDERCLEARANCE | | 00 FT 00 IN |
| (55) MIN LAT UNDERCLEARANCE RIGHT | | 99.9 FT |
| (56) MIN LAT UNDERCLEARANCE LEFT | | 0.0 FT |
| (57) WEARING SURFACE | | ASPHALT |

| CONDITION | RATING |
|---|---|
| (58) DECK | 4 |
| (59) SUPERSTRUCTURE | 5 |
| (60) SUBSTRUCTURE | 7 |
| (61) CHANNEL & CHANNEL PROTECTION | 8 |
| (62) CULVERT & RETAINING WALLS | N |
| (63) ESTIMATED REMAINING LIFE | 05 YRS |
| (64) OPERATING RATING | 225 HS 14T |
| (65) APPROACH ROADWAY ALIGNMENT | 7 |
| (66) INVENTORY RATING | 217 HS 091 |

| APPRAISAL | RATING |
|---|---|
| (67) STRUCTURE CONDITION | 2 |
| (68) DECK GEOMETRY | 2 |
| (69) UNDERCLEARANCE VERTICAL & LATERAL | N |
| (70) SAFE LOAD CAPACITY | 4 |
| (71) WATERWAY ADEQUACY | 8 |
| (72) APPROACH ROADWAY ALIGNMENT | 6 |

**PROPOSED IMPROVEMENTS**

| | |
|---|---|
| (73) YEAR NEEDED | 1980 |
| (74) TYPE OF SERVICE | HIGHWAY |
| (75) TYPE OF WORK | 311 |
| (76) IMPROVEMENT LENGTH | 000230 FT |
| (77) DESIGN LOADING | HS-20 |
| (78) ROADWAY WIDTH | 0030 FT |
| (79) NUMBER OF LANES | 02 |
| (80) ADT 001050 (81) YEAR | 1980 |
| (82) PROPOSED RDWY IMPROVEMENT YEAR | 1900 |
| (83) APPROACH IMPROVEMENT | NOT APP |
| (84) COST OF IMPROVEMENTS | $460,000 |
| (85) P E COST | $40,000 |
| (86) DEMOLITION COST | $10,000 |
| (87) SUBSTRUCTURE COST | $165,000 |
| (88) SUPERSTRUCTURE COST | $245,000 |
| (89) | |
| (90) DATE OF LAST INSPECTION | 04/10/80 |

**TABLE II Structural Inventory and Appraisal Sheet**

# 1–4 STRUCTURAL INVENTORY & APPRAISAL SHEET

## 1-4.1 GENERAL

To provide a complete and thorough inventory of the nation's bridges and to com-pile a report concerning defense bridges and highway facilities, the FHWA requires an SI & A sheet.

The FHWA also requires this sheet in order to meet National Bridge Inspection Standards and to consider highway bridges for replacement or rehabilitation. Based on this inventory data and their evaluation, a sufficiency rating will be assigned to each bridge inventoried.

## 1-4.2 SUFFICIENCY RATING

This assigned rating must be computed in accordance with AASHTO's Sufficiency Rating Formula which evaluates the factors indicative of the bridge's competence. That is, a method which decides how worthy the bridge is in serving the public in years to come and the method assesses the sufficiency of the bridge and its continued function. If it is an sufficient structure, it is awarded a rating of 100. An entirely insufficient or deficient structure gets a grade of 0. A bridge with a sufficiency rating lower than an 80 is currently eligible for rehabilitation measures. Those with ratings below 50 can be replaced or rehabilitated and others with ratings of 50 through 80 generally can be only rehabilitated. Engineers use the sufficiency rating as a basis to establish priorities for rehabilitation or replacement of the bridges.

Sufficiency ratings are numerical ratings from 0 to 100 percent which reflect a priority for replacement or rehabilitation. The lower the rating, the higher the priority. The sufficiency rating formula is a method used to evaluate factors which indicate the sufficiency of the bridge to remain in service in its current condition. A rating of 100 represents an entirely sufficient bridge which does not require any work. A 0 rating indicates an entirely insufficient or deficient bridge which has many safety hazards and should be closed.

If you consider the ratings as a sole measure of the structural condition, Sufficiency rating may be misleading because in the sufficiency formula, structural adequacy accounts for 55 percent maximum. When two bridges have the same structural conditions, the rating may be substantially different due to other factors involved.

The formula has three general categories with following maximum assigned relative weights:

1. Structural Adequacy and Safety: 55 %
2. Serviceability and Functional Obsolescence: 30 %
3. Essentiality for Public Use: 15 %

Structural adequacy and safety consists of the ratings of the superstructure, substructure, culvert and the inventory ratings. Serviceability and functional obsolescence are made up of items such as the approach roadway alignment, underclearance, deck condition, deck geometry, average daily traffic and the number of lanes on structure. The average daily traffic, the detour length, when the bridge is closed and whether the bridge is on the defense highway would indicate the essentiality for public use.

If the sufficiency ratings are 50 or more, the formula provides for an additional special reduction up to 13 % for long detour lengths, guardrails and bridge railings which do not meet current standards, and structure types such as suspension and movable bridges. These special considerations may qualify some bridges for replacement which otherwise would not be ordinarily eligible.

Within each category, various items are weighted. When the substructure or superstructure is critically weak, the maximum of 55% would be reduced and when the structure is in marginal condition 25% is deducted. No point would be subtracted if the structure is in fair condition. For example, if the deck is somewhat better than minimum adequacy, the sufficiency rating would be dropped 1%, but if deck geometry is basically intolerable, a reduction of 4% would result.

### 1-4.3 DEFICIENT BRIDGE

The Federal Highway Administration (FHWA) defines deficient to include a variety of bridge inadequacies and conditions. The manner in which FHWA defines deficient is important because a bridge must be deficient to be eligible for the bridge programs and it also should have a sufficiency rating of 80 or less to qualify. FHWA considers a bridge structurally deficient if either its deck, superstructure, substructure, or culvert (major elements of a bridge) has deteriorated to the point of marginal condition for major rehabilitation or waterway adequacy has an appraisal rating of basically intolerable conditions requiring replacement. A bridge is functionally obsolete if its deck geometry, underclearances or approach roadway alignment is basically in intolerable condition.

The following is a classification of ratings for deficient bridges:

**Structurally Deficient:**

1. A condition rating of 4 or less for one of following items:
   Item 58: Deck
   Item 59: Superstructure
   Item 60: Substructure
   Item 62: Culvert and retainining wall(*1)
2. An Appraisal rating of 2 or less for one of following:

   Item 67: Structural Condition
   Item 71: Waterway Adequacy(*2)

**Functionally Obsolete**

1. An appraisal rating of 3 or less for one of following:
   Item 68: Deck Geometry
   Item 69: Underclearance (*3)
   Item 72: Approach Rdwy Alignment
2. An Appraisal rating of 3 for
   Item 67: Structural Condition or
   Item 71: Waterway adequacy(*2)

Any bridge classified as structurally deficient is excluded from the functionally obsolete category.

-------------

*1) Item 62 applies only if the last two digits of Item 43 are coded 07 or 19.
*2) Item 71 applies only if the last digit of Item 42 is coded 0, 5, 6, 7, 8 or 9
*3) Item 69 applies only if the last digit of Item 42 is coded 0, 1, 2, 4, 6, 7 or 8.

Those bridges appearing on the select list with a sufficiency rating of less than 50.0 are eligible for replacement or rehabilitation (There are 26,600 bridges on Federal-aid system and 121,600 on Off-system in this category), while those with a sufficiency rating of 80.0 or less are eligible for rehabilitation (34,600 bridges on Federal-aid system and 50,700 On Off-system). Bridges that are to be rehabilitated must have potential for an extended service life. Note that Item 58 the deck condition is figured in determining structural deficiency and a rating code of 4 and less is included.

Since a rehabilitated bridge should not reappear as functionally obsolete bridge, the new deck must be wide enough to rate higher than 3 (Item 58).

Data items 75 and 84 through 88 must be completed for all bridges on the selection list. This is applicable to those bridges entered in the file with only the data listed in the abridged (Short Form) Structure Inventory and Appraisal sheet for Off-system bridges. Since Item 52 is used to compute area, it is included in the Short Form and must be completed for all bridges.

For the actual coding of the SI & A sheet, refer to the Recording and Coding Guide for the Structure Inventory and Appraisal of the Nation's Bridges, January, 1979, Design and Inspection Branch, Bridge Division, Federal Highway Administration, Washington, D.C. 20590.

## 1–5 PROJECT SELECTION

The Federal Highway Administration has broad eligibility criteria and their definition of a deficient bridge includes a wide variety of structural inadequacies and conditions. Under this criteria, deficient bridges with sufficiency rating of 80 or less are eligible although selection of project would require many other factors. Most States are selecting bridges with lower sufficiency ratings for Federal funds. However, they also are selecting some bridges with relatively high ratings after considering other factors, some unique to an individual State.

For bridges whose replacement or rehabilitation cost would be greater than $10 million, the Federal Highway Administration selects the bridges for discretionary funding (subject to final approval by the Secretary of Transportation). Meantime the States select the bridges to be replaced or rehabilitated with the regular bridge program funds. The Federal Highway Administration had given preferential consideration to bridges which have legislative history (such as they are mentioned in congressional committee reports or the Congressional Record as candidates for discretionary funding).

However, in 1982, the Congress provided $200 million dollars per year Discretionary Fund and directed the Secretary of Transportation to develop a selection process which is to include a formula resulting in a rating factor based on criteria including sufficiency rating, average daily traffic, average daily truck traffic, defense highway system status and the State's unobligated balance of funds. In addition, the Secretary is directed to give consideration to bridges that are closed to all traffic or that have a load restriction of 10 tons or less. Those bridges already eligible for discretionary bridge funds will remain eligible. New eligible bridges will be limited to those with a rating factor of 100 or less based on a scale of 0 to infinity.

On May 23, 1983, FHWA issued the following proposed rating factor formula for use in selecting bridges to be funded under the Discretionary Bridge Program:

$$\text{Rating Factor (RF):} = \frac{SR}{ADT} \times \frac{10}{D} \times \frac{TPC}{ADT} \times \left[1 + \frac{\text{Unobligated HBRRP Balance}}{\text{Total HBRRP Funds Received}}\right]$$

SR: Sufficiency Rating computed as illustrated in Appendix A of the Recording and Coding Guide for the Structure Inventory and Appraisal of Nation's Bridges. (If SR is less than 1.0, use SR=1.0)

ADT: ADT is ADT plus ADTT. ADT is average daily traffic in thousands taking the most current value from the national bridge inventory data. ADTT is average daily truck traffic in thousands excluding pick up trucks and light delivery trucks.

D: Defense Highway System Status
D=1 if not on defense highway
D=2 if bridge carries a designated defense highway

TCP: Total Project Cost in million of dollars
HBRRP: Highway Bridge Replacement and Rehabilitation Program

The last term of the rating factor expression includes the State's unobligated balance of funds received under 23 USC 144 at the time the State's application is received by FHWA, and the total funds received under 23 USC 144 for the last four fiscal years ending with the most recent fiscal year of the FHWA7s annual call for discretionary bridge candidate submittals. If unobligated HBRRP balance is less than $10 million, use zero balance. This will limit the effect of the modifier on those States with small apportionments or those who may accumulating funds to finance a major bridge.

The selection process for new discretionary bridge projects will be based upon the rating factor priority ranking. However, although not specifically included in the rating factor formula, special consideration will be given to bridges that are closed to all traffic or that have a load restriction of less than 10 tons. Consideration will also given to bridges with other unique situations, and to bridge candidates in the States which have not previously been allocated discretionary bridge funds.

About 95 percent of all deficient bridges are now eligible for replacement or rehabilitation, or both under the Federal bridge program. About 60 percent have sufficiency rating below 50 and thus qualify for replacement or rehabilitation. About 35 percent have rating of 50 through 80 and generally are eligible for rehabilitation only. FHWA has approved the replacement of some bridges with ratings of 50 through 80 at the request of the States when the cost comparisons of rehabilitation versus replacement showed that it would be more beneficial in the long run to replace them with bridges of a different type of material.

Rehabilitation is generally less expensive than replacement because in rehabilitation, part of the bridge is left in place. During rehabilitation all major defects are corrected and the bridge's life is extended. However this work is done without replacing the whole bridge. For example, only the deck may be replaced or it is widened with much of the bridge left undisturbed.

Due to outside constraints such as the requirement of maintenance of traffic during construction, high cost of relocation or replacement, lengthy involvement of environmental analysis and outcome of public hearings, rehabilitation may be the only alternative for many bridges. If the bridge is a major structure located in the large city or an urban area, rehabilitation will be the choice but if the bridges are small and medium size and located in rural areas, bridge replacement may be cost effective.

Several States use various methods to select bridge project for funding. Some have selected projects on a first-come, first-serve basis while others consider a variety of factors by using systematic analysis computer program. The States normally use the sufficiency ratings to identify eligible projects and as one of the factors in project selection. Besides the sufficiency ratings, factors such as effect on industry and commerce, access for police and firefighters, type of bridge, average daily traffic, school bus routes, structural condition of the bridge and project cost effectiveness, continuity of the route, future potential for an increase in the volume of traffic, continued maintenance costs, and the number of injuries or fatalities should also be considered when bridges are ranked and selected for funding. While one state emphasizes traffic flow, another state focuses primarily on the level of service and risk to the public. Some states use Federal bridge program funds to work on their worst bridges, and they concentrate on deficient bridges with sufficiency ratings below 50 and which had superstructure or substructure ratings of 3 or less or a load-carrying capacity of 10 tons or less.

Generally project selection is focused on bridges which are in the worst condition and are in need of replacement or rehabilitation with some flexibility for State and local governments.

The Wisconsin Department of Transportation has developed a computer model that uses life cycle cost analysis, in addition to information on the structural adequacy and functional obsolescence of the bridges to determine the least cost mix of bridge repair and replacement work. The following are the excerpt from Paper "A Computer Model For Life Cycle Cost Analysis Of Statewide Bridge Repair and Replacement Needs" presented at Transportation Research Board 1983 Annual Meeting:

The following basic principals are made: A structurally inadequate bridge is in immediate need of major repairs, rehabilitation, or replacement, but it does not require replacement. A functionally obsolete bridge may be in excellent condition and have many additional years of useful life even if it is narrow, has substandard clearance, or has a poor alignment. Thus such criteria as structural adequacy and functional obsolescence are not sufficient to determine replacement need by themselves.

A more pertinent issue is whether repair or rehabilitation is more cost effective than replacement at various times during the life cycle of the bridge.

A bridge whose condition is beyond repair and unsafe should be replaced.

If the cost to continue maintaining a bridge over its life and subsequently replacing it is greater than the sum of the cost of constructing and maintaining a new bridge over the same life cycle analysis period, then the bridge should be replaced.

All costs are treated in constant dollars so that cost estimate are not influenced by inflation.

All future costs are discounted.

Three long term policy directions are analyzed: an austerity option, a preventative maintenance option, and a functional obsolescence option.

**1. Austerity Option**

- Revenue Shortage.
- Maximize useful life.
- Assumed life expectancy:
  Concrete slab, reinforced concrete deck girder, and culverts: 60 years
  Steel deck girders: 75 years
  Other steel bridge and prestressed concrete structure: 80 years
- Sacrifice condition to obtain cost savings.
- Types and rate of repairs in future similar to today.
- Tolerate narrow bridges.

**2. Preventive Maintenance Option**

- Adequate revenues.
- Preventive maintenance to achieve longer life expectancies:
  Slab and reinforced concrete deck girders: 60 years
  All other bridges: 80 years
- Frequent and periodic painting of steel structures and concrete overlay.
- Minimum roadway width standards similar today's.

**3. Functional Obsolescence Option**

- Adequate revenue.
- Bridges become functionally obsolete at 57 years on average and life expectancy for all bridge type is 57 years.
- Type and rate of repairs in future is similar to today.
- Minimum roadway width standards greater than today's.

The results have been expressed in average annual levels of work by decade partly to smooth out the extreme peak of the estimated annual replacement needs. Total repair cost will rise noticeably from this decade to the next under all three policy directions.

When total replacement and repair costs are compared through 1999 in 1982 dollars, the austerity option ($371.3 million) proves to be the most inexpensive. The next is the functional obsolescence option ($4467.3 million), and the preventive maintenance alternative ($484.0 million) is the most costly. However, using real costs (1982 dollars) discounted at a 10% rate, the three options ranked from least to most expensive through 1999 are the austerity ($171 million), preventive maintenance ($214 million) and the functional obsolescence option ($221 million).

Also, the model has showed the following:

1. Virtually every bridge is cheaper to repair than replace a bridge (except for widening options) if a bridge has not become

unsafe and beyond repair. The repairs include new deck, concrete and bituminous overlay, painting, joint repairs, and other routine repairs to the superstructure or substructure such as full depth deck patching and bearing replacement.

2. Wisconsin should replace between 27 and 38 bridges per years to the year 2000 out of total 4500 bridges under state responsibility. Nearly all estimated bridge replacement needs occur in any period because the age of a bridge exceeds its life expectancy and its condition appraisal has been forecast to fall below the rating of 2, indicating the bridge is unsafe.

3. Average condition of bridges should be allowed to decline over the next two decades to take advantage of the remaining years of useful life in its bridge reflected in their age distribution.

4. Major repair costs for concrete overlay and new deck on steel girders and prestressed concrete structure built after 1955 will increase substantially from the 1980's to the 1990's. These bridges are reaching their midlife and require new deck or additional overlays.

For many bridge engineers, the life cycle cost analysis and the statistical approach are slight deviation from the general engineering evaluation methods by inspection and judgement but it appears that the computer model is very effective management tools to set the general policy directions for bridge program.

## REFERENCE:


1. AASHTO, "Manual for Maintenance Inspection of Bridge, 1978" AASHTO Washington, D.C.
2. Federal Highway Administration:
   "Bridge Inspector's Training Manual"
   "Federal-Aid Highway Program Manual Vol. 6 - Engineering and Traffic Operation"
   "Third Annual Report to Congress, Highway Bridges Replacement and Rehabilitation Program" Bridge Division, March 1982;
   "Recording and Coding Guide for the Structural Inventory and Appraisal of Nation's Bridges"
   "FCP Annual Progress Report-Cost Effective Rigid Concrete Construction and Rehabilitation in Adverse Environments"
   "Extending the Service Life on Existing Bridges by Increasing Their Load Carrying Capacity" Report No. FHWA-RD-78-133
3. General Accounting Office:
   "Better targeting of federal funds needed to eliminate unsafe bridges" (CED-81-126, Aug. 11, 1981) "Deteriorating highways and lagging revenue" (CED-81-42, March 5, 1981) "Excessive Truck weight: An expensive burden We can no longer support" (CED-79-94, June 27,1980) "Federal assistance to rehabilitate railroads should be reassessed" (CED-80-90, June 27,1980)
4. Ohio Depart. of Transp., "Design Regulation" Bureau of Bridge and Structure Design.
5. Transportation Research Board:
   "Bridges on secondary highways and local roads - Rehabilitation and replacement" National Cooperative Highway Research Program (NCHRP) Report 222
   "Rehabilitation and Replacement of Bridge on Secondary Highways and Local Roads" NCHRP Report 243, Dec. 1981
   "Bridge Deck Repairs" Research Result Digest 85. March 1976
   "Durability of Concrete Bridge Deck" NCHRP Report 54
   "Bridge Engineering Volume 1 & 2 " Transportation Research Record 664 & 665
6. Congressional Record-House, December 21, 1982; Conference Report on HR 6211.
7. "Policy and Procedure Manual" Maryland Depart. of Transp. Division of Bridge Development.
8. AREA "Manual of Railway Engineering" Chapter 7 Steel Structure - Part 7 Existing bridges; Chapter 8 Concrete Structure and Foundation - Part 21 Instruction for the Inspection of Concrete and Masonry Structures. AREA
9. "Manual of Concrete Practice (Part 2)" ACI
10. "Guideline Developing a Bridge Maintenance Program" 1974 American Public Works Assoc.
11. Hyman, William A., Dennis J. Hughe " A Computer Model for Life Cycle Cost Analysis of Statewide Bridge Repair and Replacement Needs" Paper presented at TRB Annual Meeting, January, 1983.
12. Federal Register/ Vol.48, No. 100/ May 23, 1983. Proposed Rules "Bridges, Structures, and Hydraulics; Discretionary Bridge Criteria"
13. "The Infrastructure Crisis: Learning to cope" Paper presented National Bridge Conference, June 1, 1983, Pittsburgh, Pennsylvania.
14. Kutz, Scott A., Warren W. Zeiss "Improved Highway Program Management Through Use of Integrated Information Systems" TRB Record 867, TRB
15. Park, Sung H "Bridge Inspection and Structural Analysis" 1980

# 2 REHABILITATION STUDY

## 2—1 STUDIES

A rehabilitation report includes the processes of gathering, coordinating and analyzing the available information and data, and after developing a few feasible schemes, selecting the one which is the most effective and economical in reconstructing and upgrading the bridge. This scheme must have a minimum adverse impact on the natural environment, must conserve energy, and must have minimum economic impact on the traveling public and minimum economic and social impacts on the local community.

Each State has established project priorities based on the previous bridge inspections, the Structural Inventory and Appraisal (SI & A) computer data submitted for each bridge, and sufficiency rating assigned to it. However, these established project priorities could be revised. For instance, if the engineer submits an SI & A sheet that is revised to show that a particular bridge has developed symptoms of severe deterioration, then the eligibility of that bridge for rehabilitation or replacement would increase comparative to the other bridges on the priority list.

The following five steps to be included in the study are:

1) Review the existing bridge inspection report, perform the bridge and approach roadway inspection survey, conduct a deck condition survey, identify whether the bridge has any structural deficiencies and physical deterioration, and whether it is functionally obsolete.
2) Analyze and appraise the existing condition of the bridge and approach roadway based on the available information. Generally bridge problems can be broken down into six major groups.
   a. Total bridge replacement
   b. Total superstructure replacement
   c. Total deck replacement
   d. Major substructure rehabilitation.
   e. Minor repairs
   f. Preventative measures such as LMC overlay, resurfacing with waterproofing membrane and silane surface treatment.
3) Develop schemes to correct a major portion of these deficiencies. Decide to restore the areas of severe deterioration in the bridge, and determine methods that correct any functional problems occurring in it. Make sure that these procedures have a minimum social and economic impacts.
4) Compare the cost of alternates such as replacing the bridge with that of widening or relocating it (building a new bridge in different location).
5) After reviewing the feasibility study, recommend the most effective scheme either for correcting the faults in deficient bridges or for removing unsafe bridges from service.

For some bridges, the choice of the scheme is limited to one of total replacement due either to the advanced stage of deterioration or to the hazards because of narrow bridge width. In these cases, the project evaluations become routine, and determination can be made easily without extensive studies. For other bridges, the choice may not be as clear without thorough studies involving number of complex factors.

Whereas a large sum of funds and a certain amount of environmental study are required to replace a bridge, only small percentage of this cost may be necessary to repair it. The engineer making the choice should also consider that a bridge's life often can be extended with repairs and it may offer the public an acceptable (although diminishing) level of service without spending large funds.

## 2—2 FIELD INSPECTION

A visual field inspection, regardless of its duration, should be performed by the project manager, who are in charge of the study and design phase of the project. During the visual inspection, he should be able to obtain the required scope of work and form a preliminary conceptual design for future rehabilitation or replacement of the structure. During the study and design phase, he should visit the site as often as time permits to expand on and revise, where necessary, his concept of preliminary structural rehabilitation design.

Especially for a rehabilitation or widening project, each designer should familiarize himself with the current condition of the structure by performing overall visual inspection. This inspection should verify and supplement the previous inspection report regarding the condition, deficiencies, functional problems, type of structure and material composition of the structure. Each defect and problem area should be photographed in detail and in color, with the scale, for future reference.

Also, if the facts gathered in the field indicate that the as-built plans did not record the changes properly or were changed without proper record, the designer should update the plans. Typical revisions should include such items as the span length, the bridge elevation, alignment, and the beam spacing. In addition, the traffic volume, vertical and horizontal alignment, clearance, the superelevation, the approach roadway and surrounding facilities should also be included in the report.

After finishing the field inspection, the engineer should analyze the adequacy of the structure.

Additional information on bridge inspection and rating can be found in the publication "Bridge Inspection and Structural Analysis" by S.H. Park, P.O. Box 7474 Trenton, NJ 08628.

## 2—3 GENERAL CONSIDERATION

Many other factors must be considered besides the deficiencies and the actual or potential functional obsolescence inherent in the design of a bridge. Some of these common factors are listed below.

1. Traffic maintenance; the volume of present and future traffic (20 year projection): peak traffic, hourly rate, pedestrian traffic, average daily traffic, etc. and accident data.
2. Cost and available funds & resources.
3. Requirements regarding the structure's functional capabilities and its remaining service life and alignment of existing roadway and required geometric improvement.
4. The area topography; existing or needed right-of-way for the bridge reconstruction; existing location and type of utilities which may be involved; stream crossing and flood plain; known drainage and hydrology features; airport involvement; railroad crossing and electrified railroad involvement; and functional classification of road system.
5. Safety.
6. Aesthetics.
7. Impact of bridge reconstruction on the surrounding environment and community; historical and archaeological feature likely to cause controversy; air & water qualities and noise level during construction.
8. Maintenance.
9. Conservation of energy.

### 2-3.1 MAINTENANCE OF TRAFFIC

In an urban area, traffic control is the most expensive part of a project because closing a bridge is often impossible, and traffic detour system is not often acceptable to the public. Therefore, engineers should try to maintain and protect the traffic flow surrounding the project site, and plan alternate routes which will adequately handle the vehicles which formerly traveled through, over, and under the project site. If the facilities are owned by other agencies or railroad, the engineer should develop the traffic maintenance scheme in advance so that the agencies or railroads have the information of the intended work. After contacting the owners of the facilities, the engineer assigned to the project must make sure that these agencies have reviewed the proposed schemes for maintaining and protecting the traffic flow and have agreed with the proposed procedures and schemes.

While developing the traffic scheme in accordance with the Uniform Traffic Control Manual,the engineer should decide the best possible traffic detours and should consider whether certain lanes should be closed, which signs should be mounted to ease the new traffic flow and which signs should be altered or covered, how to channel the traffic volume so that they are suited to the facilities as ideally as possible and meet the requirements of FHWA Standard and Guideline Work Zone Traffic Control (Part VI of Manual on Uniform Traffic Control Devices). The working hours should be scheduled for the construction crews so as to cause minimum interference with the traffic rerouting and they should correlate with each stage of construction. The construction barrier, striping, protection and safety features should be temporarily devised and used during the duration of the project.

Also during developing the traffic scheme, the engineer must closely coordinate his plans with those of the traffic bureau of the particular agency which is responsible for meeting the traffic requirements at the location of project site. Such a requirement might include a rule such as, "two traffic lanes with a minimum of ten feet per lane (11 feet preferred), shall be provided in every stage of the construction".

Often, the traffic schemes will not meet the particular agency's strict requirements because of limited funds and resources. Whenever this happens, the engineer should develop a few alternate schemes which will be adaptable.

The project manager should also become familiar with the area's local traffic problems by consulting state and local officials and, if necessary, by conducting actual field investigations. The state and local highway planning organization is a source for past and present traffic volumes, and the engineer should obtain this valuable information. With the aid of this information, the project can be oriented toward the temporary restriction of the traffic movement during construction.

The project manager should also look ahead into the future, while deciding how to relieve the local traffic problems, where by he may be able to promote future growth and development in the local area. The need for future traffic is great when the current traffic is approaching the capacity of highway and decision should be made whether additional lanes be provided.

### 2-3.2 AVAILABILITY OF FUND AND RESOURCES

To transform deteriorated bridges into the structures which are able to meet current safety requirements or handle ever-increasing traffic volumes, most states must solve the common problems of gathering the necessary funds. In the past, this has usually meant that the state had to invest substantial capital. However, the ability of such revenue sources to handle the demand caused by the bridge replacement program has diminished. Even with increased federal participation, it is becoming more difficult for states to continue their current levels of investment in the program. Therefore, whenever a structure needs a major overhaul, engineers should be especially careful to select reconstruction schemes that are most effective and at the same time they are economically feasible. For instance, whenever, a structure could be rehabilitated to provide extended service life and minimum level of service, rehabilitation could be chosen over the alternative entire bridge replacement depending on the overall cost. (If over-all estimated rehabilitation cost is less than half of the replacement cost, it is preferred to rehabilitate the bridge.)

In addition, one of the restrictions in an expanding program is the limited in-house staff to administrate, plan and design these projects and restriction on public employments. Another drawback to the rehabilitation program is the large time lag after the decision to rehabilitate has been made. The minimum time requirement for most rehabilitation design phases is two years and complex projects take four to six years.

### 2-3.3 FUNCTIONAL REQUIREMENTS

In replacement schemes, the proposed type of structure, roadway width, vertical and horizontal clearances and design loadings should satisfy the current functional and hydraulic requirements. However, sometimes fulfilling today's requirements would be costly with minor benefits. Therefore, it is necessary to waive some of the current requirements and to develop alternative schemes to achieve a minimum desirable functional level. For example, a 30 feet minimum horizontal under clearance may not be necessary on most state or county roads and a reduced minimum clearance would meet the functional requirement with a large reduction in construction cost.

For rehabilitation schemes, the functional requirement would depend on the degree of the deficiencies and the severity of the problems to be corrected. Every effort should be made to satisfy the current requirements as much as is economically practical and provide a minimum desirable level of service by improving geometrics, strengthening critical members, jacking up structures or lowering the existing roadways, retrofitting and widening the structure. Also, the remaining estimated service life of the structure after the rehabilitation should be considered and brought to attention. If the life expectancy of the structure is less than 10 to 15 years, permanent reconstruction would be a logical choice for the project. When the bridge is structurally in good condition but it has narrow roadway width and light traffic, the choice may be to tolerate the functional obsolescence and keep the bridge until the condition become critical.

**A. Structure Type, Span and Loading**

For a widening project, the proposed structure shall be of the same type as the existing structure. However, precast or prestressed members may be used when the horizontal or vertical clearances during the construction of cast-in-place type structures become insufficient. For a replacement project, the intended structure type should be based on careful appraisal of the site, foundation and hydraulic condition, geometric and clearance requirements, availability of materials, environmental impact. Cost also should be considered when several structures are to be replaced in one general area. The type of structure should be similar so that the same details can be used again. Also grouping several bridges into one contract will result in substantial savings.

Some specific recommendations should be kept in mind when developing schemes for the type of structure and span are as follows:

1. Consider replacing bridges over streams with following hierarchy of structures if the hydraulic requirements can be satisfied:
   a. Pipe (precast concrete or corrugated metal) or pipe arch (concrete or metal)
   b. Box culvert (single or multi-cell)
   c. Bridge
2. From the safety and aesthetic view point, consider single spans to minimize obstruction and enhance openness under the structure.
3. For a long viaduct over flood plains, consider using medium size span lengths of 60 to 100 feet instead of many short spans.
4. Consider using continuous spans instead of simple spans in longer multiple span bridges to eliminate deck joints and future maintenance at piers as well as reduce structural steel weight.
5. For low traffic volume road, consider using prestressed concrete deck slabs, box beams or tee beams.
6. After reviewing the geographic locations and the foundation conditions of any major highway that is in deep cut or any highway that is crossing over a deep valley, consider using an arch or a frame.
7. Consider using a composite design for steel beams spanning over 40 feet and for prestressed concrete structures.
8. Stub, spill-through, and integral types of abutments in place of cantilever abutments. In most cases, lengthening the bridge by the use of stub abutment is more economical and also has aesthetic and safety advantages.
9. An architecturally pleasing pier is desirable. However, excessive massive concrete piers with various shapes and forms should be avoided. Consider slim functional and economical piers with simple shapes which provide architectural effects. Also special attention should be given to locate the piers such

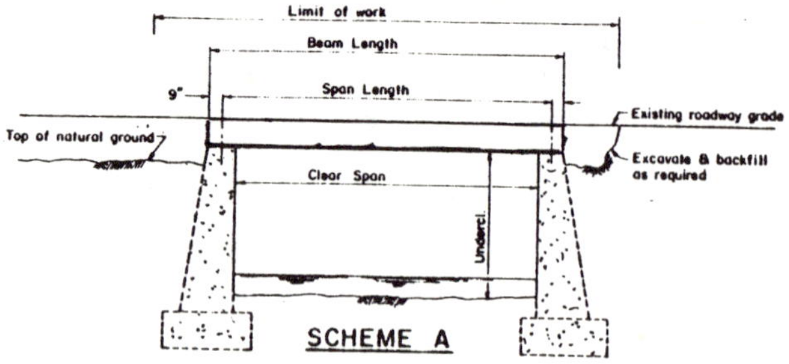


SCHEME A

**Replace superstr. on existing abuts. Modify top of abut. and/or adjust approach as required.**

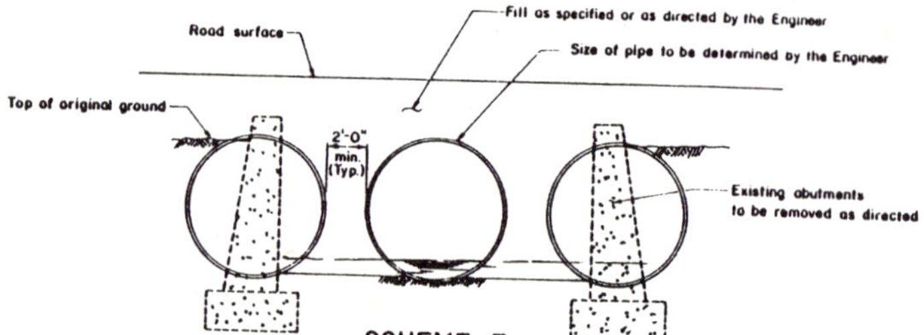


SCHEME E

METAL PIPE CULVERTS

7'ϕ

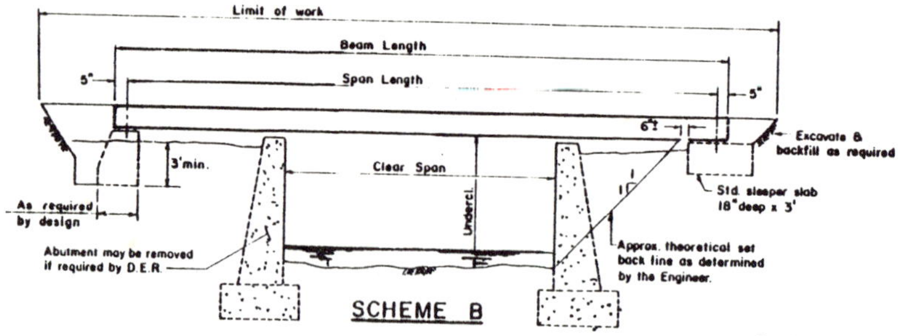


SCHEME B

**Remove superstr. Place new beams on sleeper slabs. Straddle old abuts. which stay in place. Remove top of abut. as required or raise grade and adj. approaches.**

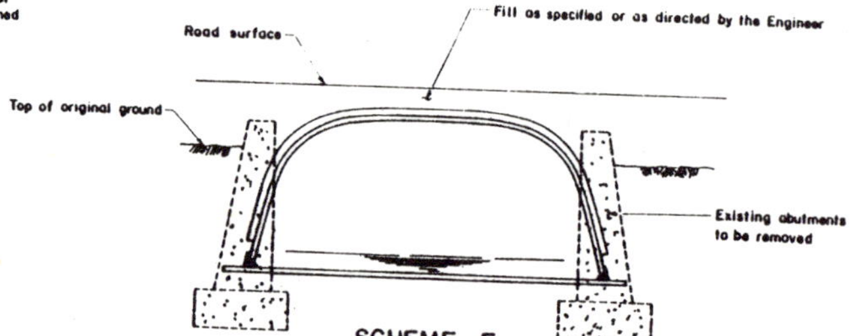


SCHEME F

PREFAB METAL BOX CULVERT

Max. Span 20'-11"

FOR MULTIPLE METAL BOX CULVERT USE CLEAR SPACING AS SHOWN FOR SCHEME E

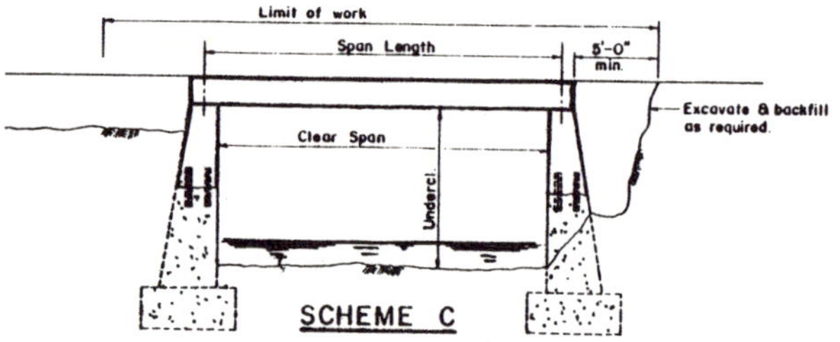


SCHEME C

**Replace superstr. on rebuilt abuts.**

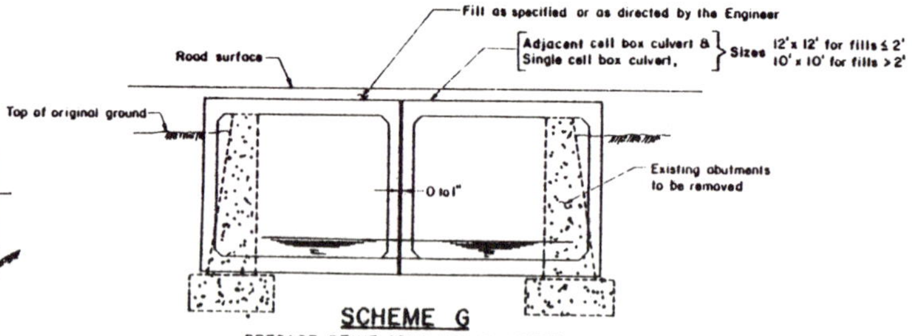


SCHEME G

PRECAST REINF. CONC BOX CULVERTS

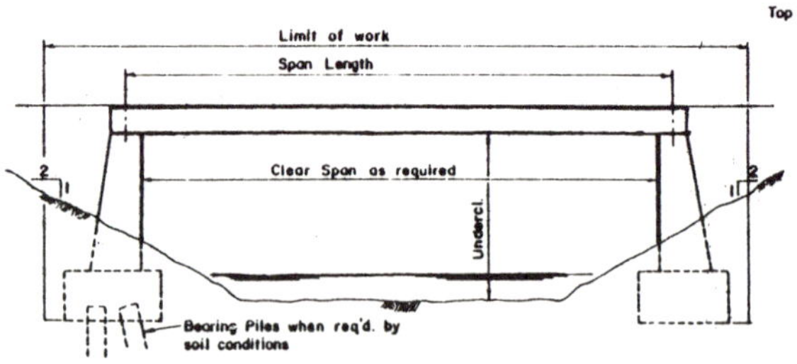


SCHEME D

**Replace entire bridge-Improve channel Reconstruct Abutments(Relocate as specified)**

PENN DOT

## ALTERNATES for LOW COST BRIDGE

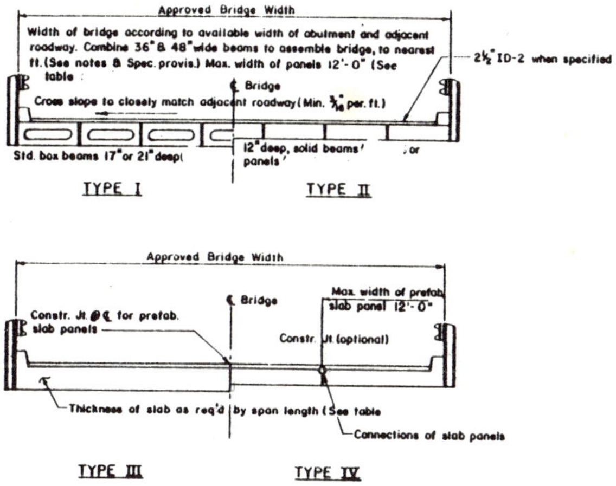


## TYPICAL SECTIONS

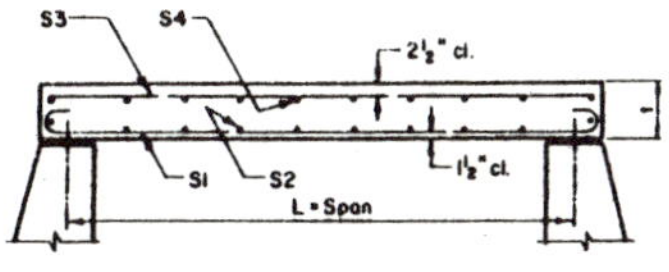


## LONGITUDINAL SECTION
FOR PRECAST R.C. BRIDGES
(Showing Reinf.)

DESIGN TABLE

| PRECAST R.C. SLAB BRIDGES | | | | | |
|---|---|---|---|---|---|
| SPAN L (ft) | DEPTH t (in.) | BOTTOM BARS S1 | BOTTOM BARS S2 | TOP BARS S3 | TOP BARS S4 |
| 8 | 10½" | #8 @ 12" | #6 @ 12" | #5 @ 18" | #4 @ 12" |
| 10 | 11½" | #8 @ 10½" | #6 @ 12" | #5 @ 16" | #4 @ 12" |
| 12 | 12½" | #8 @ 9½" | #6 @ 12" | #5 @ 14" | #4 @ 12" |
| 14 | 13" | #8 @ 8" | #6 @ 12 | #5 @ 12" | #4 @ 12" |
| 16 | 14" | #9 @ 9½" | #6 @ 12" | #5 @ 14" | #4 @ 12" |
| 18 | 15" | #9 @ 8½" | #6 @ 12" | #5 @ 13" | #4 @ 12" |
| 20 | 15½" | #9 @ 7½" | #6 @ 12" | #5 @ 12" | #4 @ 12" |

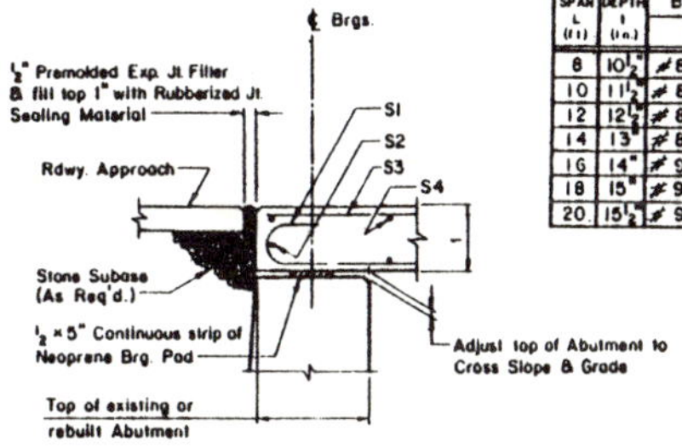


## SECTION AT ABUTMENT (TYP.)
FOR PRECAST R.C. BRIDGES

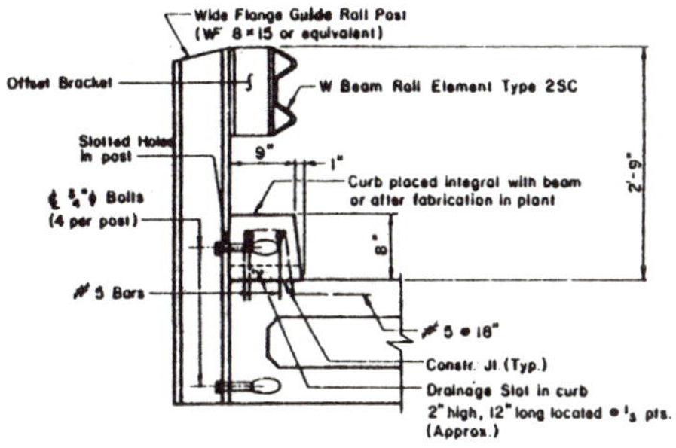


## GUIDE RAIL & POST DETAIL
TYP. FOR 17" & DEEPER BEAM SIZES

DESIGN TABLE

| PRECAST P/S SLAB BRIDGES | | | | | | | | |
|---|---|---|---|---|---|---|---|---|
| SPAN L (ft.) | PLANK DEPTH t | STRAND SIZE ksi | STRAND SIZE ϕ | STRANDS (No.) | INITIAL P/S FORCE (kips) | e (in.) | CAMBER (in) | REMARKS |
| 8 | 8" | 250 | ¼" | 32 | 201.6 | 2.125 | 0.08 | |
| 10 | | | | 32 | 201.6 | 2.125 | 0.12 | |
| 12 | | | | 33 | 207.9 | 2.182 | 0.16 | * |
| 14 | 10" | 250 | ¼" | 40 | 252 | 2.65 | 0.18 | |
| 16 | | | | 40 | 252 | 2.65 | 0.22 | ** |
| 18 | | | | 58 | 365.4 | 3.239 | 0.27 | *** |
| 18 | 12" | 250 | ¼" | 48 | 302.4 | 3.25 | 0.21 | |
| 20 | | | | 48 | 302.4 | 3.168 | 0.24 | |
| 22 | | | | 59 | 371.7 | 3.326 | 0.33 | |

* $M_{UD} > 3.64\% M_{UC}$
** $M_{UD} > 2.95\% M_{UC}$
*** $M_{UD} > 2.58\% M_{UC}$

$M_{UD}$ Denotes Ultimate Design Moment
$M_{UC}$ Denotes Ultimate Moment Capacity

STANDARDIZED SPAN LENGTH

| 90° SKEW | SCHEME B | | | | | | | SCHEMES A, C & D | | | | | | 90° SKEW |
|---|---|---|---|---|---|---|---|---|---|---|---|---|---|---|
| CLEAR SPAN | 8' | 12' | 16' | 20' | 24' | 28' | UNDERCL. | 16' | 20' | 24' | 28' | 32' | 36' | CLEAR SPAN |
| BM. LENGTH | 24'-10" | 28'-10" | 32'-10" | 36'-10" | 40'-10" | 44'-10" | 6' | | | | | | | |
| SPAN LENGTH | 24' | 28' | 32' | 36' | 40' | 44' | | | | | | | | |
| BM. LENGTH | 29.5' | 33.5' | 37.5' | 41.5' | 45.5' | 49.5' | 8' | 19' | 23' | 27' | 31' | 35' | 39' | BM. LENGTH |
| SPAN LENGTH | 28' | 32' | 36' | 40' | 44' | 48' | | 17.5' | 21.5' | 25.5' | 29.5' | 33.5' | 37.5' | SPAN LENGTH |
| BM. LENGTH | 33.5' | 37.5' | 41.5' | 45.5' | 49.5' | 53.5' | 10' | | | | | | | |
| SPAN LENGTH | 32' | 36' | 40' | 44' | 48' | 52' | (SEE NOTES ON SHT. 1) | | | | | | | |

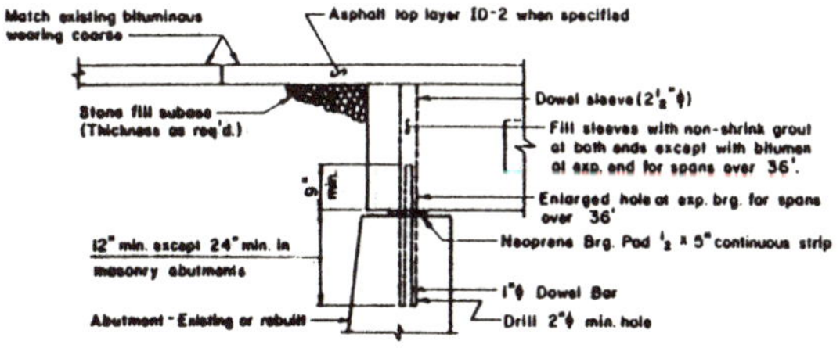


## BEAM ANCHOR DETAIL
DETAIL ON SLEEPER SLAB SIMILAR

| DESIGN TABLE PRESTRESSED BEAMS | | | |
|---|---|---|---|
| SPAN LENGTH | BEAM SIZE | NO. OF STRANDS | C. G. |
| 20' | 48/12* | 50 @ ¼"ϕ | 2.82 in. |
| 25' | 48/12* | 30 @ ⅜"ϕ | 3.02 in. |
| 20' | 36/12* | 40 @ ¼"ϕ | 3.09 in. |
| 25' | 36/12* | 24 @ ⅜"ϕ | 3.05 in. |
| 20 | 48/17* | 20 @ ⅜"ϕ | 2.60 in. |
| 25' | 48/17* | 20 @ ⅜"ϕ | 3.02 in. |
| 30' | 48/17* | 15 @ 7/16"ϕ | 2.01 in. |
| 35' | 48/17* | 21 @ 7/16"ϕ | 2.71 in. |
| 36' | 48/17* | 22 @ 7/16"ϕ | 2.79 in. |
| 40' | 48/21 | 19 @ 7/16"ϕ | 3.15 in. |
| 45' | 48/27 | 18 @ 7/16"ϕ | 3.06 in. |
| 50' | 48/27 | 21 @ 7/16"ϕ | 3.39 in. |

*-Beam width may be increased to 96 × 12, 144 × 12 or 96 × 17 and either P/S or Precast.

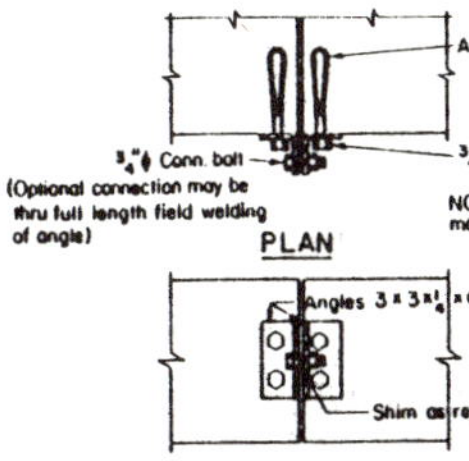


## END OF BEAM CONNECTION
ELEVATION

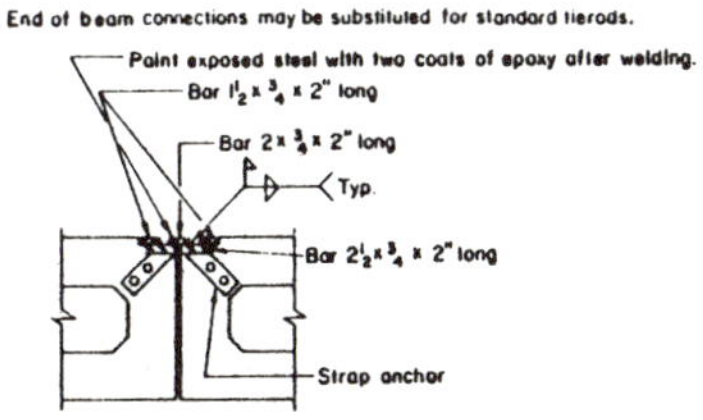


## BEAM CONNECTIONS ✱

✱ Welded connections may be substituted for shear keys.

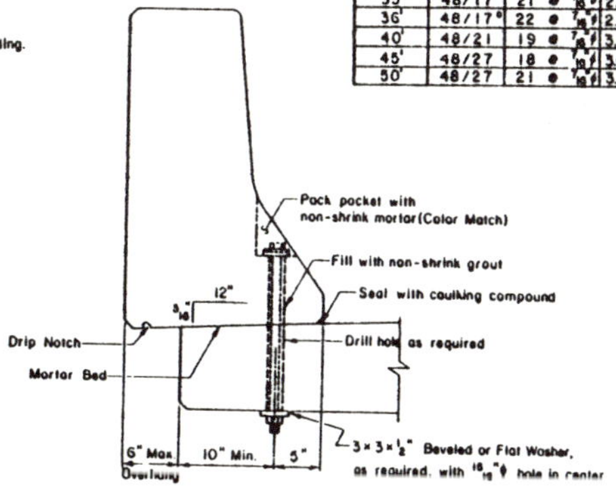


## PRECAST PARAPET
(FOR PRECAST R.C. & P/S SLAB BRIDGES)
(Alternate: Cast-in-place in field or Fabricating Plant)

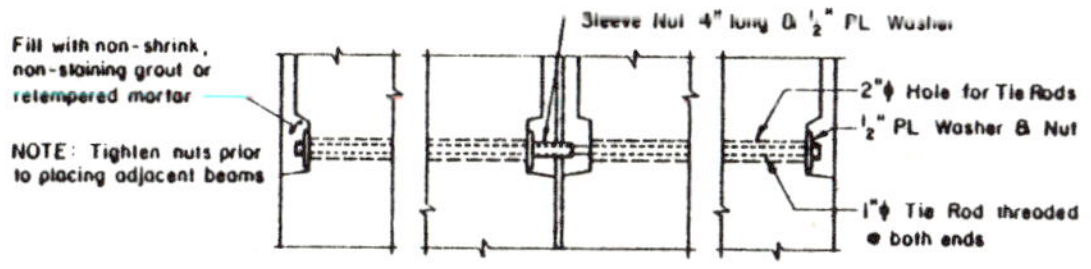


## TIE ROD DETAIL

# DETAILS for LOW COST BRIDGE

a way that it would not gather debris, impede flow or cause scour.

10. Avoid fracture critical members in the main support systems for the bridge.

Design loads for bridge replacement should comply with current AASHTO or the State requirements. However, for minor rehabilitation, the strict compliance of current standards would result in an additional construction cost without substantial benefits. Therefore, it is suggested that consideration be given to apply previous criteria for a minor project where the bridge will remain in place and the traffic volume will not be critical.

AASHTO "Geometric Design Guide for Resurfacing, Restoration, and Rehabilitation (R-R-R) of Highways and Streets, 1977" suggests the following for the bridge to remain in place:

| Traffic | | Design Loading | |
|---|---|---|---|
| ADT | DHV | Desirable Minimum | Minimum |
| 0- 50 | - | H-15 | H-10 |
| 50-250 | - | H-15 | H-15 |
| 250-400 | 100-200 | H-15 | H-15 |
| | 200-400 | H-15 | H-15 |
| | over 400 | H-20 | H-15 |

**B. Bridge Width**

The width of a bridge shall be equal to the full width of the traffic lanes and shoulders on the approaches. In the past, as a cost reducing measure, some major long span bridges have been built with shoulder widths less than approach shoulder widths and these structures should be widened. On multi-lane divided highways, many states require that median 30 feet (some states use 24 feet criteria) or less be closed and decked.

Bridge width is based on the number and width of traffic lanes and full shoulder width. The number of lanes and shoulder widths are related to selected design speed, future traffic volume, and functional classification. Therefore, a determination of bridge width is generally made by roadway engineers in conjunction with an overall geometric and highway capacity review for that particular section of the roadway. Widths must be able to handle the current average daily traffic but in most cases, widths are established based on future ADT or design hourly volume for the following 20 years. When ADT indicates a greater bridge width than the existing, a widening or replacement scheme should be considered.

For a replacement bridge on a local road, the minimum width should be based on the 1971 AASHTO Geometric Design Guide for Local Roads and Streets which is summarized below.

1. Where the approach roadway including shoulder is surfaced, the entire surfaced width should be carried over the bridge.
2.

| Design Speed | Volume (ADT) | Bridge Width (curb to curb) |
|---|---|---|
| over 50 mph | over 750 | Approach Roadway Width (including unpaved shoulder) |
| over 50 mph | under 750 | Pavement width plus 6 feet |
| under 50 mph | over 400 | Pavement width plus 6 feet |
| under 50 mph | under 400 | pavement width plus 4 feet |

For an existing bridge which is to remain in place, but if deck is to be replaced, you may use following AASHTO requirement from "Geometric Design Guide for Resurfacing, Restoration, and Rehabilitation (R-R-R) of Highways and Streets, 1977."

| Traffic | | Roadway Clear Width* | |
|---|---|---|---|
| Current ADT | DHV | Desirable Minimum | Minimum** |
| 0- 50 | - | 24 ft. | 20 ft. |
| 50-250 | - | 26 | 20 |
| 250-400 | - | 28 | 22 |
| 400-750 | 100-200 | 28 | 22 |
| | 200-400 | 32 | 24 |
| | Over 400 | 36 | 28 |

* Clear width between curbs or rails, whichever is the lesser.

** For design speed of 50 MPH or less, minimum clear widths that are two ft. narrower may be used on minor roads with few trucks. In no case shall the minimum clear width be less than the approach traveled way width.

## C. Geometric and Clearance

In addition to horizontal and vertical clearance requirements, geometric design features, such as vertical and horizontal alignment, sight distances, and superelevation, should be analyzed for improvements by considering the character of terrain, type of highway, traffic volume, waterway openings and economic considerations. A highway serving longer trips and carrying a large volume of traffic may justify for the upgrading of highway features along with bridge replacements by extending project limits particularly where the savings in vehicle operation and other costs are sufficient to offset the additional cost of right of way and increased roadway construction. Either the project limits for rehabilitation or in-kind replacement project should be extended for a reasonable and practical distance from the bridge so as to make proper transition to the approach roadway. In developing rehabilitation schemes, it is recommended that the cost effective and practical geometric improvement schemes be included for a review by the the responsible agency.

**Lateral Clearance:**

For safety reasons, fixed objects should be placed as far from the edge of the roadway as is economically feasible and lateral clearance requirements have been increased from a 10 foot to 30 foot setback.

If it is not feasible to provide the minimum clearance, due to structural restrictions and economic reasons, then the following alternatives should be considered:

1. Provide minimum desirable clearance by placing guide rails in front of piers or abutments.
2. Provide minimum clearance by installing concrete barrier curb in front of the substructure or as an integral part of the substructure.

**Vertical Clearance:**

Adequate vertical under clearance is an essential factor for the protection of the superstructure from collision damages. the following criteria are used by most states:

**Interstate Highways & Defense Route:**
16'-6" (Desirable)
16'-0" (Minimum )
**State Routes:**
15'-6" (Desirable)
15'-0" (Minimum )
**County & Local Roads:** 14'-6"
**During Construction :** 14'-6"(Minimum )

Minimum requirements vary from 16'-6" to 14'-6" absolute.

The maximum vehicle height allowed is 14'-0" (actual average height is 13'-10"), therefore, any bridge with less than 14'-0" under-clearance is exposed to critical collision damage. Because of truck bounce, there are potential problems for any bridges with underclearance less than 14'-6".

If an existing underclearance of a bridge is less than 14'-6", a study should be made to investigate the possible alternatives to increase the clearance by raising the structure, lowering the roadway, or combination of both. Generally combination scheme will cost more than others.

## D. Hydraulic Consideration

Structures which cross a stream are influenced by the stream flow, drift, passage of flood debris, ice jamming, scour, channel condition, effect of the pier as an obstruction, bank protection, navigational clearances, and environmental requirements. Hydraulic information should be collected and reviewed and a complete hydraulic analysis report, as necessary, should be made in conjunction with widening or replacement schemes.

In the hydraulic analysis, flood flow should be estimated for a design flood, the 50 year flood, the 100 year flood and the greatest recorded flood if that information is available. The following guideline may be used for selecting the frequency of the design flood:

| Type of Road | Design Freqn. |
|---|---|
| Interstate & Major Routes | The greatest recorded flood or 50 year minimum |
| Local Road with more than 1000 ADT | 25 years |

| | |
|---|---|
| Local Road with less than 1000 ADT | 10 years |
| Highway encroached in flood plain | 100 years (prefer)<br>50 years min. |

However, in all cases, flow for a 50 year and 100 year flood should be computed to check the effect of flood water on the surrounding area. A hydraulic risk analysis should also be made to determine the cost effective design for a flood flow.

A waterway opening for a bridge shall be designed to carry the designed flood water and maintain a desirable free board of 2 feet (where ice jams are expected, the minimum should be 3 feet). It should carry the flood of record or a 100 year flood with reduced freeboard. Each project should be checked with the State Environmental Protection Agency, the U.S. Army Corp of Engineers, and the Federal Insurance Administration of the U.S. Department of Housing and Urban Development (HUD) for clearance or involvement. Also, in areas where local authorities are likely to provide agricultural or other drainage improvements, they should be consulted for the probability of a future lowering of the channel and proper provision should be included in design.

Before establishing new waterway openings, the performances of existing bridges should be carefully evaluated because the area under a bridge may not truly show the effective openings due to obstructions such as dead trees, either upstream or downstream. If so, relief openings, spur dikes, debris deflection, and channel work should be considered to minimize the effect of adverse flood condition.

To establish new waterway opening, it may require to prepare hydraulic and hydrologic studies including a variety of risk analysis and its recommendation.

The following are the steps:

1. Collect data which will indicate potential flood problems including data from such sources as aerial and field surveys; interview; water resources, fish and wildlife agencies; local planning board; newspaper; and flood studies.
2. Collect data on topography, land use, hydrology, existing structures, channel characteristics, environmental data, and irrigation supplies.
3. Conduct field hydraulic survey to obtain necessary bridge and hydraulic data.
4. Perform preliminary hydrologic estimates at probable flood plain encroachment and preliminary hydraulic studies to assess significance of encroachment and outline the impacted area by flood plain encroachment.
5. Prepare site maps and data on existing bridges.
6. Upon receiving review comments for preliminary studies, prepare final hydrology and hydraulic analysis report with recommendations.

### E. Existing Utilities

In urban areas, public and private utilities generally carry their facilities, (such as sewers, water mains, pipes, ducts or electrical lines, underground,) on or over the bridge or in the stream bed at river crossing. When these facilities cannot be relocated, temporary supports or reinstallations should be considered during bridge reconstruction. The location of these facilities may interfere with the most economical span arrangement due to site restrictions and the aesthetics of the bridge and surrounding areas. Large utilities carried by the structure may require a change in the proposed type of structure.

### F. Soil Condition

Poor foundation conditions of the existing structure which caused by soil movements will influence the reconstruction and geometry of the structure. In some cases, it may be more economical to use longer spans and fewer substructure units with pile foundations. It may be necessary to conduct field investigation to obtain samples, boring log and subsurface conditions, and to perform physical testing of soils. Upon receiving test results, prepare the foundation report outlining surface condition and design recommendations including alternate type of foundation (pile foundation and type of pile or spread footing), removal and treatment of unsuitable materials, and cost analysis.

**G. Construction**

Most rehabilitation and replacement projects are built over the existing highway on which traffic should be maintained during construction. On heavily travelling urban highways, any obstruction, such as falsework which imposes undesirable vertical restrictions and removal of existing materials which may result in sudden dropping of the materials underneath, would be hazardous. This is also true for the bridges over railroads. In situations involving restricted vertical clearances during construction, consideration should be given to the use of prefabricated members such as integral deck precast concrete beams, prestressed box beams, and erection of a pair of steel girders with shop installed stay-in-place deck forms.

**2-3.4 ENERGY CONSERVATION**

Since World War II, the economic growth has been fueled by inexpensive, readily available crude oil. However, this situation changed remarkedly in the 1970's as domestic oil resources began to decline and we have rapidly come to rely on imported oil. The price of the crude oil will increase with periods of uncertain supply.

Since automobile alone consume about 40 percent of the petroleum used in the States, increased emphasis should be placed on energy conservation in selecting replacement or rehabilitation scheme to improve operating efficiency. If the bridges on high volume traffic road are to be closed or partially closed for an extensive period, consideration should be given for alternate schemes which will advance the project expeditiously.

In addition to expeditious construction and proper traffic control measures, energy saving would result from gentle geometry, smooth surface, flat grade profile and adequate bridge maintenance. In many cases, such improvements are not possible due to local conditions and restrictions but every effort should be made to reduce vehicle energy consumption by adopting most realistic energy saving measures into the design.

To encourage energy conservation, 1982 Transportation Act include that federal share will be 100% for carpool and van pool project, bicycle, and pedestrian walkway project. Also, for project which make 4R Improvement (Resurfacing, Restoring, Rehabilitation, or Reconstruction) to roads on Federal-aid systems incur a substantial use as result of transportation activities to meet national energy requirements, the federal share will increase to 85% of the cost of such project.

**2-3.5 SAFETY**

Since motor vehicle accidents are a leading cause of death for persons between 15 and 45 years of age, and each year about 50,000 lives are lost on our nation's highways, highway safety is an important aspect of bridge rehabilitation or replacement programs. While traffic safety hinges heavily on the driver's performance, the improved roadway environment will decrease the driver's errors and provide a better chance of survival when an error is made.

Most functional deficiencies consist of narrow traffic lanes, inadequate shoulder widths, inadequate sight, passing and stopping distances, restricted vertical clearances, and improper horizontal alignments.

Other areas of inadequacy are bridge railings, guard rail transition, guard rail terminals, gore areas on bridges and hazardous approach roadways. The FHWA bridge program requires correcting and upgrading the bridge traffic railings and hazardous approach geometrics. Because unprotected sidewalks on highspeed traffic bridges are hazardous to pedestrians and motorist, consideration should be given to install of N.J. type traffic barriers at the curb.

At all bridge undercrossing, It is desirable to provide minimum a 30 feet wide recovery area which is free of the common road hazards. It is also desirable to eliminate side piers from the roadside area. However, on those locations where a 30 foot clearance is impractical, guiderail or barrier curbs should be installed to protect the vehicles from crashing into the substructure. During construction, safety considerations should be given to adjacent property, utilities, workers, and pedestrian.

### 2-3.6 AESTHETICS

A pleasing appearance should also be considered along with safety, economy and other factors considered in developing replacement schemes. In a natural scenic environment, consideration must be given to designing the bridge which is in harmony with adjacent land and development. The following factors should be kept in mind from an aesthetic standpoint:

1. The general alignment and profile of the replacement bridge should fit the character of the area so that unsightly scars of embankments or excavation will be minimized.
2. An agreeable and natural roadside appearance should be preserved. Destruction of valuable trees should be avoided as much as it is practical.
3. The absence of shoulder piers (side pier) or abutments and wider medians would result in a more pleasing appearance.
4. Where the aesthetic value of the area warrants major treatments, the structure type and length should be determined on the basis of its aesthetic value rather than the cost. For example, an arch or a rigid frame structure were used on parkways.

### 2-3.7 ENVIRONMENTAL CONSIDERATION

With the passage of the National Environmental Policy Act (NEPA) in 1969, transportation agencies are required to assess the environmental impact of their projects on the human environment. Since highway projects have major impact on land use and community development, federal law through NEPA requires that the social, economic and environmental impact resulting from existing and proposed projects be formally documented and that appropriate measures be taken to reduce adverse environmental effects.

The environmental consideration also related to legislations such as the Clean Water Act of 1977, the Noise Control Act of 1972, the Coastal Zone Management Act of 1972, the Fish and Wildlife Act of 1956, the Clean Air Act, the National Historic Preservation Act of 1966, the Endangered Species Act, the Wild and Scenic River Act, Archaeological and Historic Preservation Act of 1974, Executive Order 11988 (Flood Plain Management), Executive Order 11990 (Protection of Wetlands), etc.

The environmental consideration could be divided into the following three classes by the differences in their complexity and environmental impact:

**Class I (EIS):**
Action that normally has significant impact on environment and which requires an Environmental Impact Statement (EIS)

**Class II (CE):**
Categorical exclusion (CE) is action that does not individually o: cumulatively have a significant effect on the environment.

**Class III (Environmental Assessment):**
Action in which the significance of impact on environment is not clearly established then the environmental assessment (EA) and a finding of no significant impact (FONSI) will be required.

### CATEGORICAL EXCLUSION:

Normal categorical exclusions are those which do not involve substantial planning, time, resources or expenditures. These projects will not induce significant, foreseeable alterations in land use, planned growth, development patterns, traffic volumes, traffic patterns, natural and cultural resources and thus do not require the environmental impact statement (EIS) or the environmental assessment (EA).

The following examples are ordinarily considered to be categorical exclusions:

1. Modernization of an existing highway by resurfacing, restoration, rehabilitation, widening by less than a single lane width, adding shoulders and adding auxiliary lanes for localized purposes.
2. Lighting, signing, pavement marking, signalization, freeway surveillance and

control system and railroad protective devices.

3. Safety projects such as grooving, glare screens, safety barriers, energy attenuators, etc.
4. Reconstruction or modification of an existing bridge on essentially the same alignment or location by widening less than a single travel lane, adding shoulders or safety lanes, walkways, bikeways, pipe lines or fender works, unless the bridge is on or is eligible for the National Register of Historic Places.
5. Reconstruction or modification of an existing one lane bridge, presently serviced by a two lane road and used for a two way traffic, to a two way bridge on essentially the same alignment or location, except for the bridges with historical significance.
6. Construction of pedestrian and bicycle bridges and pipe line bridges for transporting portable water.
7. Temporary replacement of a highway bridge and facilities, after the occurrence of a natural disaster or catastrophic failure, to restore the highway for the health, welfare, and safety of the public.
8. Approval of utilities installations along or across a highway.
9. Maintenance of track, railbed, bridge structure, electrifications, communications and other existing railroad or transit facilities when carried within existing right-of-way.
10. Rehabilitation or reconstruction of existing rail and facilities where no additional land is required.
11. Construction of bus shelters and bays.
12. Highway landscaping and rest area projects.

Actions ordinarily classified as categorical exclusions which involve significant environmental impact will require preparation of an EIS or EA. Therefore it is strongly recommended that early coordination be initiated with appropriate local, state, and federal agencies to identify and evaluate the social, economic, and environmental effects of bridge reconstruction and replacement, and to develop reasonable measures to reduce adverse impacts.

**ENVIRONMENTAL ASSESSMENT (EA):**

This is not classified as a categorical exclusion. In this case the environmental studies and early coordination indicate that proposed action will not have significant impact on the quality of human environment. After reviewing the environmental assessment and examining the social economic and environmental issues, the appropriate federal agencies will adopt the EA as a finding of no significant impact (FONSI). If there is significant impact, an environmental impact statement is required.

These projects are normally widenings or realignments to upgrade existing roadways or bridges and requiring additional right-of-way with the potential of causing alterations to the local environment such as public use land, historic site, archaeological site, wet lands, productive farm land, and fishable stream. Within this broad range of projects are major widening of bridges, stream crossings with changes in the channel and relocation of the railroad and highway bridges in isolated areas.

**ENVIRONMENTAL IMPACT STATEMENT:**

Environmental Impact Statement is required when the actions have potentially significant environmental effects and they generate a controversy because of their effects on the human environment, abutting property, wetlands, endangered species and cultural resources.

The following are some of the projects which require EIS: New freeway or expressway highway projects that would be expected to cause major shifts in travel patterns and significant changes in land use or development; and relocation of major highways and bridges involving demolition of large numbers of buildings and substantial disruption of local traffic patterns and changes in land use.

The EIS should include the probable effects of the proposed work on the following: Conservation, Economics, Aesthetics, General environmental concerns, Wetlands, Cultural values, Fish and wildlife value, Flood hazards, Flood plains value, Land use, Navigation, Shoreline erosion and control, Recreation, Water supply and conservation, Water quality, Energy needs, Safety, Food production, Needs and welfare of the public and appropriate alternative and their beneficial or detrimental effect, or both in permanent nature.

In developing reconstruction schemes, the lengthy procedure of preparing Environmental Impact Statement should be kept in mind and in-kind replacement or temporary rehabilitations with permanent future relocations should be considered as alternatives.

**Environmental Processing:**
The following steps are generally required to obtain approvals:

**Step 1: Determination of Level** of Action.
Determine whether the project is categorical exclusion(CE) or whether an environmental assessment (EA) or a draft environmental impact statement (DEIS) will be required.

**Step 2: Preparing Environmental Assessment**

1. Compile and analyze all data from rehabilitation report and other sources.
2. Perform required engineering, social, economic and environmental studies.
3. Prepare preliminary Environmental Assessment.
4. Prepare historic preservation document as required.
5. Circulate for review within State Highway Department, receive comments and revise preliminary Environmental Assessment.
6. Send to FHWA for their review and comments.
7. Revise Assessment and circulate for public review.
8. When necessary, hold public hearing and prepare hearing document.

**Step 3: Preparing Finding of No Significant Impact (FONSI).**

1. Revise and update the Environmental Assessment.
2. Review comments from public hearing if held.
3. Prepare preliminary FONSI, indicate alternate selected and reasons. Summarize comments from hearing and responses, or indicate that an opportunity for hearing was offered but no response, and attach updated environment assessment.
4. Send to FHWA for review and comment.
5. Revise FONSI as required.
6. Forward to FHWA for approval and obtain their approval. Otherwise DEIS should be prepared.

**Step 4: Draft Environmental Impact Statement (DEIS).**

1. Revise and evaluate available information and data.
2. Perform engineering, social, economic and environmental studies.
3. Prepare preliminary DEIS containing a summary, table of contents, purpose and need of proposed action, alternative considered, affected environment, environmental consequences, list of preparers, distribution list, index, and appendix.
4. Circulate pre-DEIS within Highway Department for review. Upon receiving comments, send FHWA copies of pre-DEIS and in-house review comments for their review and comments.
5. After FHWA review and incorporating all comments, update the Draft EIS and send to FHWA for approval.
6. Upon receiving signed title page from FHWA, distribute Draft Environmental Impact Statement to all interested parties for comments prior to final EIS.

**Step 5: Public Hearing**

1. Advertise "opportunity for public hearing", evaluate need for hearing and prepare for hearing if required.
2. Hold meeting with public to discuss project alternative, costs, effects, and to give the public opportunity to express their opinions. Record meeting and prepare transcript. Document that hearing requirements have been met.

**Step 6: Selection of Alternate**

1. Receive, summarize, and evaluate comments from hearing or environmental documents, or both.
2. Review and evaluate the additional data and study materials as required.
3. Prepare recommendation including summary of comments, findings, conclusions and recommendations.
4. Obtain approval of selected alternate.

**Step 7: Final Environmental Impact Statement**

1. Revise and update DEIS to include a summary containing recommendation of alternate; hearing certification; list of

reviewing agencies and commenting agencies; list of required State and Federal permits; an analysis and summary of comments received at public hearing: and the resolution of comments received on the DEIS.
2. Upon completion of internal review, send pre-FEIS to FHWA for review and comments.
3. Prepare Final EIS incorporating FHWA comments.
4. Obtain approval of FEIS and prepare Draft Record of Decision (ROD).
5. Send FEIS and draft ROD to FHWA for approval.
6. Upon FHWA approval, notify public of availability of FEIS, distribute to all interested parties and agencies. Prepare any additional documents required by the State, and maintain availability for at least 30 days.

## SECTION 4(f) EVALUATIONS

A public document of evaluation is required for any federally funded transportation project which will use land from a public-owned park, recreation area, wild life and waterfowl refuge, or from historic sites on or eligible for the National Register, and archaeological site. Section 4(f) evaluations which required by Section 4(f) of the Dept. of Transportation Act of 1966 must document the fact that 1). there is no feasible and prudent alternative to the use of land protected under this section and 2). the action includes all possible planning to minimize harm resulting from use of the land. Coordination with certain concerned agencies is also documented.

The section 4(f) evaluation is usually a separate section in an EIS or EA, or a separate document for Categorical Exclusions (CE) with minimal effect on lands covered under the section. The section 4(f) evaluation is submitted to FHWA and FHWA's separately written "Section 4(f) Determination" will conclude their decision regarding to the requirement of the section 4(f) and approve the action to proceed.

## SECTION 106 PROCESS (Cultural Resources):

Under Section 106 of the National Historic Preservation Act of 1966, A Federal Aid project which may affect archaeological, historical, architectural or cultural site of National Register quality require to do following procedures:

Survey a project area for sites on or eligible for inclusion on the National Register.

Prepare a report which analyzes the impacts which the project would have on any (if any) such site.

Investigate options which avoid or minimize impacts.

Coordinate the above studies and determinations with the State Historic Preservation Officer, the Keeper of the National Register, and the Advisory Council on Historic Preservation.

## WETLANDS

Executive Order 1190 of May 24, 1977 requires the protection, preservation and enhancement of wetlands to the fullest extend practical during construction and planning of highway. New construction in wetland should be avoided unless there is no other practical alternative. It is required to document the impacts on and value of the wetland, accumulated impacts, mitigation measures and their cost. The principles of minimizing harm to wetland must be followed for all Federally funded projects and the project requiring the U.S. Army Corps of Engineers Section 404 Permit and the U.S. Coast Guard Permit.

## 2-3.8 MAINTENANCE

All bridges need maintenance during their service lives to preserve the structural and operational characteristics for which they were originally designed. These include safe, smooth, skid-resistant surfaces; proper surface drainage (minimum drainage grade of 0.3% across the bridge is desirable to eliminate small ponds on the deck); adequately functioning bearing devices and substructure elements, and replaced or repaired structural railings and approach guardrail.

The result of limited maintenance spending, increasing inflation, and increasing vehicle usage is causing our bridges to wear out faster than they can be repaired. Therefore, consideration should be given to develop schemes for maintenance free structures.

Concrete and prestressed concrete bridges generally require a minimum maintenance.

Deck expansion joints have been a maintenance problem and bridges with a minimum number of joints are desirable.

Bridge decks carrying heavy traffic loads and receiving greater amount of deicers should be considered for double deck protection systems. A double deck protective design utilizes the use of epoxy coated bars at the top layer of deck reinforcement steel and latex modified concrete or low slump concrete as an overlay topping the deck slab.

Steel girders and beams require periodic painting unless weathering steel is used. Even this steel requires painting when it is exposed to salt water and industrial fumes.

## 2-3.9 PERMITS AND AGREEMENTS

The studies should identify which types of permits and agreements will probably be required. The following is the list of permits and agreements:

- U.S. Army Corps of Engineers: Section 404 Permit for dredging and filling in all water and Section 10 Permit.
- U.S. Coast Guard: Crossing, relocation and encroachment of navigable water.
- Coastal Zone Management Certificate.
- Wetland permit.
- Water quality certificate.
- Stream encroachment or water resource permit.
- Riparian permit or grant.
- Utility (electric, water, gas, sewer, telephone) Agreement.
- Railroad Agreements and Right-of-Entry if required.
- Local jurisdiction agreement.

## 2-3.10 VALUE ENGINEERING

Value engineering is an objective, and systematic method of optimizing the total cost of a facilities or system for a specific number of years. Total cost are overall life cycle cost including the cost of design, construction, maintenance and replacement for a specific life. Using value engineering, this optimization of cost is attempted by the systematic development of alternates covering isolated high cost areas or items. In other words, the goal of value engineering is **better for less.** This means by which the design cost can be increased a small percentage to enable a larger percentage reduction in total cost. Value engineering is devoted to find better ways of getting things done, improving the cost and efficiency of providing needed functions and eliminating the waste. Value engineering is management tool that complements rather replace other cost reducing methods.

Value engineering can be used during any phase of a project. However, the best time to use this technique is during the study phase or early design stage because once the criteria or policy have been established, engineering decisions have the major impact on the total project cost. Also, earlier design phase, the possibility of design change will be easier and the effects on implementation cost and construction schedule will be minimum.

Value engineering can be achieved by job plans and functional analysis. Functional analysis requires identification of the function, cost estimate of that function and development of alternates. Job plan is the systematic application of recognized techniques to solve the problems by identifying problems, developing several alternates, selecting the alternate and executions of selected alternate.

Generally job plans consist of the following procedures:

1. Information gathering
2. Problem definition and investigation. The following should be asked: a) Is there a problems ? b) What is the problem ? c) What is not problem ?
3. Functional analysis: analyze cost and function, and evaluate functional worth by following methods: a) Define basic function by two words. b) List all functions. c) Challenge everything: high cost items, elimination of certain functions, question the need, etc.

4. Speculation: speculate alternates. Develop ways through creative thought processes to achieve the same basic function of the items selected.
5. Evaluate alternates and identify best alternate for possible implementation by comparison.
6. Develop alternate.
7. Execution: presenting and implementing the alternate.

In selecting value engineering project, the following guide lines should be considered:

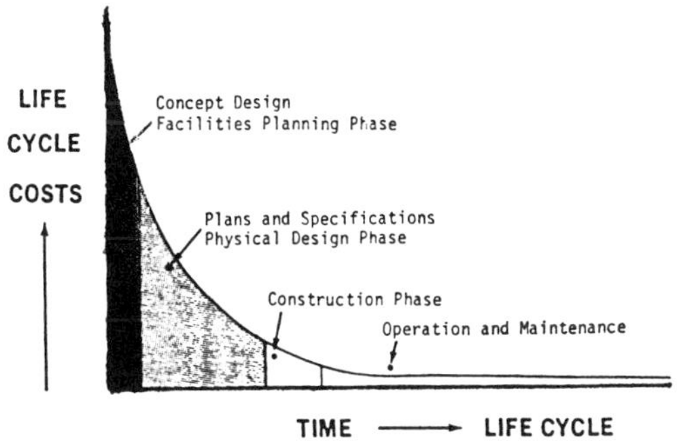


**Fig. 2.1 Major decion makers influence on life-cycle cost.**

1. Select the project in design stage either the conceptual design study phase or at the time the plans and specification are developed. This early action will provide the maximum cost saving. It is recommended to adopt value engineering at early design phase or study stage. Application of value engineering to construction project should be limited the project which are unique, extremely difficult or complex.
2. Select project with higher probability of acceptance and construction.
3. Select project with sufficient high cost (over $2 million) to make it worth the effort.

**Value Engineering during Construction:**
Value engineering incentive clauses in construction specification encourage participation by contractors and subcontractors. The clauses enable a contractor to share in savings resulting from changes they suggest in methods of construction or materials. Generally the incentive clauses should not allow the change of basic structure type and design criteria.

**ALTERNATE DESIGN**

Other form of value engineering will be the use of alternate designs and materials or options. It is better to investigate all possibilities for most economical design during study and preliminary design phase but when local condition warrant competition and cost savings, alternate design and different type of material can be shown on the plans as alternate bid items. The FHWA issued Technical Advisory T 5140.12, Alternate Bridge Design, dated December 4, 1979 which includes the following recommendation:

General:

1. To receive the most economical construction between basic structural materials, consistent with geographic, environmental, ecological or other site restrictions, there should be maximum opportunity for competition between structural steel and concrete.

2. Within environmental, aesthetic, site constraints, etc., the plans and bid documents should show or otherwise indicate what alternate type structures will be allowed or considered. The contractor should be allowed the option to bid any designated alternate design that is consistent with the contractor's expertise, available equipment, etc.

3. Bid documents and the contract plans should clearly indicate the design criteria and what type of alternate designs or contractors options, or both will be acceptable. Determination of practical and economical alternates or contractors options, or both should be developed in the preliminary design.

4. Bid documents should be considered as "open" documents in regard to construction method, erection systems, and prestressing systems.

5. Consistent design criteria should be used for alternates, for example that is, if load factor design is used, it should be used for all alternates.

6. Span length should be identified on the contract plans. However, other than where pier locations are constrained by physical and geological conditions at the site, con-

sideration should be given to allowing a tolerance in pier location to avoid placing a particular alternate at an economic disadvantage. For example, in a typical three span structure, the side span should be approximately 80 percent of the center span for structural steel, 70 percent for conventional cast-in-place concrete on falsework, and 65 to 70 percent in segmental balanced cantilever construction.

7. To avoid an economic disadvantage to a particular superstructure alternate, alternate superstructure design may be required. Limitations on the sub structure, such as allowable axial load and moment, should be clearly identified on the contract plans.

8. Where specific design requirements are not covered by the AASHTO Bridge Specifications, the contractor should be allowed to use other recognized codes and standards where applicable. However, the alternate design should be document where these provisions are to be used, why the AASHTO requirements do not apply, and which articles of the substituted code or standard are to be used. Such provisions should be subject to approval by the engineer and appropriate agencies.

9. Prebid conferences are to be encouraged as a means of communication between the engineer, highway agencies and contractors.

10. In order to allow a contractor adequate time to investigate the various alternatives and prepare plans, it is recommended that advertising time be commensurate with a minimum of 60 days.

11. In order to allow adequate review and checking of the low bidder's proposal, award of contract should be extended commensurate with the size of project.

**Prestressed Concrete Alternate:**

1. To increase the competition in post-tesioned concrete construction, it is recommended that plans and other bid documents allow conventional cast-in-place on falsework, precast prestressed span units, and segmental construction or combinations thereof.

2. Segmental construction should allow the following at the contractors option:

2.1 Precast or cast-in-place segmental construction.

2.2 Any of the post tensioning systems, for example, strands, wire or bar or combinations thereof.

2.3 Any of the following construction methods: balanced cantilever, span-by-span, progressive planning, incremental launching, or combination thereof.

2.4 Exterior dimensions of the cross-section should be fixed. At the contractor's option, the thickness of webs and flanges may be varied to accommodate proposed construction and erection methods and post-tensioning systems, providing that any changes in the dead weight, shear, and others, are accommodated in the design.

3. The contract plans should indicated the maximum and minimum final prestressing force (Pf) and moment (Pf x e) required, after all losses, for the final condition of the structure; for example, dead, live, impact and all superimposed loads. Any increase in prestressing force requirements as a result of the method of construction, erection, type of tendon system should be evaluated at the shop drawing stage.

4. Changes in eccentricity of prestress should be accompanied with appropriate changes in prestress force to produce the same minimum compressive stress due to prestress.

5. The minimum prestress force should be such that under any loading condition, both during and after construction, stresses will be within allowable limits. Consideration should be given to secondary moments due to prestress, redistributed moments due to creep, and stresses resulting from thermal gradient (between the top and bottom of the girder and between the inside and outside of webs).

6. Contractor's revisions to contract plans, with supporting calculations, should be submitted to the engineer for approval.

**Structural Steel Alternate:**

1. Any changes in member sizes or details resulting from erection procedures or methods should be evaluated at the shop drawing stage.

2. To increase the competition in structural steel construction, it is recommended that plans and other bid documents allow the following, where feasible:

2.1 Option in field and shop splice locations.
2.2 In bolted compression splices, the option of transferring 50 percent of the load through milled ends of the members.
2.3 Consideration of options in details to simplify or facilitate fabrication and erection.
2.4 Option of using rolled sections in lie of fabricated members.

On May 12, 1983, FHWA issued "Alternate Designs for Bridge; Policy Statement" and the final FHWA policy is as follows:

1. Preliminary plan development for major bridges should be based on engineering and economic evaluation of acceptable alternate designs. The definition of a major bridge is contained in the Federal-Aid Highway Program Manual (FHPM 6-1-2-1). (Major bridges are bridges estimated to cost more than $5 million or having spans in excess of 150 feet, or viaduct type structures whose total length is in excess of 1,000 feet. This criterion applies to individual units of separated dual bridges.)

2. Alternate designs should consider the utilization of competitive materials and/or structural types.

3. Economic evaluation of preliminary estimates should take into consideration, to the maximum extent possible, the relative accuracy of estimates for state-of-the-art type methods of construction.

4. When comparative economic estimates are reasonably close to each other, two or more complete sets of contract documents should be prepared and advertised.

5. Value engineering at the design stage should be strongly encouraged.

6. Options should be considered for structure components (piling, expansion joints, bearings, prestressing systems, etc.).

7. A determination of FHWA approval and participation in the development of alternate or single project plans, specifications and estimates for major bridges will be made after evaluation of a State's submittal of preliminary bridge plans and cost studies. Recognizing that the States are experienced and have broad expertise in factors pertinent to bridge type selection in their States, the FHWA will review comprehensive State's studies and subsequent recommendations as a primary factor in reaching preliminary bridge plan approval decisions.

At the some time, FHWA approvals will be based upon the need to ensure safe, efficient and cost effective bridge projects which meet the aesthetic and structural requirements of site and are based upon the latest, proven technology and techniques.

8. The submittal of preliminary bridge plans for approval should be accompanied by the State's plan for developing the project to completion (horizontal or vertical stage construction, number of contracts, etc.).

9. This policy shall not be interpreted as deterring any State or FHWA field office from considering alternate designs for any other type or classification of structure which they would consider appropriate.

This policy is written with the intent of taking advantage of evolving state-of-the-art of bridge constructions and fluctuating economic conditions in the market place while not compromising sound engineering, safety, quality control, or aesthetics.

## 2–4 DEVELOPMENT OF SCHEMES AND SELECTION PROCESS

Repair and replacement schemes should be developed after considering every factor. In some cases, the structural damage or deficiency is so critical that the only choice is to replace the structure. In such cases, there is no need of comparison studies between repair and replacement schemes and further study should be focused on the replacement. In other cases, the decision may be more difficult in selection of repairs and replacement and comparison studies may be required.

The following steps may be used in comparison studies:

1. Evaluate the condition of the structure and set goals.
2. Identify the factors involved.
3. Evaluate the effectiveness of repair or replacement to meet the goals and satisfy the requirements of all factors.
4. Estimate the cost of the repairs and replacement.
5. Select the cost effective scheme which meets the objects and requirements or perform a further detailed study.

After selecting the repairs or replacement, detailed repair or replacement schemes should be developed. In many cases, it is practical and economical to develop a few schemes with estimated costs but without going through the selection process.

## 2-4.1 REHABILITATION SCHEMES

The methods of repairing and upgrading bridges vary substantially depending on the degree and the severity of the deficiencies and the problems to be corrected. They may need a protective overlay with minor deck repairs, total deck replacement, strengthening of critical members, replacing bearings, correcting undermining, foundation problems and other defects.

In the FHWA report "Extending the service life of existing bridges by increasing their load carrying capacity" (FHWA-RD-78-133), rehabilitation procedures are classified into the following four groups:

1. Increase live load carrying capacity.
2. Improve geometrics.
3. Correct mechanical deficiencies.
4. Miscellaneous: drainage, concrete, safety.

**Increasing Live Load Carrying Capacity:**

Four general procedures are available to increase the live load carrying capacity of a bridge.

a. **Strengthen critical members** by adding new material to the existing member or replacing the entire member with new material. Often, the connection is the critical part of the member. This deficiency can be corrected by adding additional connectors or by replacing the entire connection.
b. **Add supplemental members** to assist in carrying the load. On structurally inadequate floor systems of trusses or multi-girder bridges, additional stringers could be installed between the existing ones to provide necessary overall capacity.
c. **Reduce the dead load** with a lighter deck system, such as open steel grid, orthotropic deck, etc.
d. **Modify the structural system** in a number of ways to provide additional capacity to support live load. Three modifications are composite action, beam continuity, and king post truss beam reinforcement.

**Improve Geometrics:**

Geometrically deficient bridges can be improved by increasing the vertical clearance, widening the usable roadway and upgrading the horizontal and vertical alignment of the approach roadway.

a. **Vertical Clearance:** Inadequate vertical clearances of through truss bridges could be corrected by reducing the depth of the portal frames and sway frames. On grade-separation structures, vertical clearances could be improved by lowering the roadway or raising the superstructure.
b. **Widening the roadway:** On multi girder systems, the widening could be achieved without any major problems, but on thru-girder and through truss systems, this would be technically difficult and very costly. Replacement of the structure often is the most practical choice.
c. **Improve Approach Alignment:** Sometimes on railroad crossing and stream crossing, the local roads and bridges were built

at right angles and the approach roadways made abrupt turns adjacent to the bridge. Alignments of this nature are extremely difficult to correct due to the need of additional right-of-way and environmental restrictions. Improving geometric generally will requires major corridor improvement, environmental studies, and considerable time and efforts. Therefore consideration should be given to make minor improvements of a temporary nature while initiating major rehabilitation, deck replacements and widenings or replacement projects.

**Correct Mechanical Deficiencies:**

These repairs are associated with deck joints, expansion bearings, hangers and similar devices that permit the structure to expand and contract. The replacement of joints may require the rebuilding of the headers and is normally done during rehabilitation of the deck slab. To minimize the future maintenance, elastomeric type bearings are recommended whenever possible in place of metal plate sliding bearings.

**Miscellaneous Repairs:**

**Drainage:**
An ineffective bridge deck drainage is a major cause of bridge deterioration and should be replaced with an adequate system. Special attention should be given to details that will prevent deck run off from reaching the bearings, superstructure members, piers and abutments. Scuppers should be properly spaced and should be large enough to prevent clogging.

**Concrete Bridge Decks:**
Deck deterioration is a major bridge problem and its repairs vary from minor patching of isolated areas to total replacement.

**Safety and Barriers:**
These include replacement of inadequate bridge railing; alteration of parapets and raking ends where these face oncoming traffic, protection with attenuators at the ends of through girders or through trusses, in gore areas on structures, or in front of piers within recovery zones. Also, adequate protection of pedestrians should be provided.

**Substructure Stabilization:**
For longitudinal movements, tiebacks could be used by anchoring abutment backwalls to the deadman or by utilizing soil anchor systems. Underpinning and soil stabilization procedures can be used for vertical settlement and foundation rotation problems. Where lateral soil pressure causes problems, sheet piles may be used behind the walls or proper drainage systems should be placed to reduce hydrostatic pressure behind the walls.

### 2-4.2 REPLACEMENT SCHEMES

In developing replacement schemes, the following items should be considered:

1. Estimate the cost and the available fund.
2. Level of service provided to meet traffic volume.
3. Anticipated use.
4. Environmental requirements for constructing replacement bridges.
5. Delays caused by permit requirements.

A few replacements should be developed to correct previously identified deficiencies and provide adequate levels of service.

The following are some of the suggestions for developing schemes:

1. If an environmental clearance time and approval are anticipated to be lengthy, develop in-kind replacement scheme to meet the minimum requirements. This scheme may have essentially the same alignment or location and can be classified as a Categorical Exclusion or a moderate project requiring Environmental Assessment (EA) and finding no significant impact (FONSI). Also, develop a scheme to provide a desirable future service level meeting all current requirements. A detailed analysis should be made for comparison of the two schemes.
2. On low volume roads, consider precast prestressed concrete beams with an integral deck. For short to medium span bridges with simple geometries the prestressed concrete bridge has the advantages of low construction cost, minimum traffic interruptions, fast construction time, and few maintenance problems.

3. Consider replacing small stream crossing bridges with pipe or culverts if the culverts can handle the required flood flows.
4. Consider replacing bridges with embankments and roadway by filling in when railroad tracks or roads are abandoned.
5. Consider reducing the overall length of existing bridge when it is not required

## 2—5 ECONOMIC ANALYSIS

### 2-5.1 COST

In developing schemes, cost is a primary consideration in selecting the scheme to be used. Therefore, it is necessary to estimate the true construction costs of the proposed schemes to establish their ranking, feasibility or to compare one scheme to another.

Initially it is hard to estimate accurately the cost of the project due to insufficient information. At this period, the course of the project is determined primarily based on the estimated cost. Adequate information on facts such as foundation condition, environmental restrictions and underground construction may not be available but due care must be exercised in estimating cost. The following are some items which affect the final cost:

1. Maintenance of traffic.
2. Clearing site, Demolition of existing structure.
3. Availability of local material.
4. Labor requirements, government regulations and permits.
5. Local contractor's experience and preference.
6. Capabilities and experience of local fabricator.
7. Total construction time.
8. Stage construction and degree of interference with other phases of construction.
9. Additional Right-of-way cost.
10. Operation and maintenance cost.

Roughly, the estimated cost of bridge is based on the average cost of the bridge per square foot of deck area and the cost of the retaining wall per lineal feet. It is known from several studies that the effects of the span length and type of substructure excluding foundation have relatively little effect on the total estimated construction cost if the total length and width of the structure remains the same. In general, it is estimated that the superstructure cost will be about two third (2/3) of the total structure cost on grade separations and about one half (1/2) of the structure cost on stream crossing structures. The square foot bridge costs have been summarized in the TABLE as a guide. The costs represent the average cost for usual highway and stream crossings and it is strongly recommended to consult with the estimator of the state and local highway agencies to arrive at reasonable cost estimate to reflect various factors involved.

### 2-5.2 ECONOMIC ANALYSIS

In an earlier study, we had established the objective and had determined all of the feasible and realistic alternate schemes for meeting that objective. Occasionally a choice has to be made between several schemes. Reasonable alternate should be sought in order to preserve the value of the analysis. For example, consider the situation in which the following two alternatives have been presented: 1). rehabilitation of a movable bridge 2). replacement in-kind bridge. The replacement scheme is favored over the first because it has same environmental restrictions but lower life-cycle costs (present value). However, a third alternative scheme, a high elevation bridge at a different location, had not been considered because its estimated construction cost was deemed too high. Upon further investigation, it was found that due to the high elevation, operating and maintenance costs will be minimal and thus the third alternative will have the lowest life-cycle cost.

Economical analysis is a comparison of two or more alternatives and deals with future expenditures and thus it has elements of uncertainty. On bridge rehabilitation projects, the deficiencies may not be revealed during inspections and the areas that need repairs may increase substantially during construction.

**1982 Construction Cost**
(Dollars per Sq.Ft.)

| | Low | Average | High |
|---|---|---|---|
| Prestressed Girder | | | |
| Stream Cross w/pile | 40 | 50 | 60 |
| Stream Cross w/o pile | 35 | 40 | 48 |
| Rdwy Cross w/pile | 36 | 45 | 55 |
| Rdwy Cross w/o pile | 30 | 40 | 50 |
| Prestressed Slab | 40 | 50 | 60 |
| Prestressed Box | 35 | 50 | 60 |
| Concrete T Beam | | | |
| Rdwy Cross w/pile | 50 | 60 | 70 |
| Rdwy Cross w/o pile | 45 | 50 | 60 |
| Reinforced Concrete Box Girder | | | |
| Stream Cross w/pile | 60 | 68 | 75 |
| Stream Cross w/o pile | 57 | 60 | 67 |
| Rdwy Cross w/pile | 40 | 50 | 60 |
| Rdwy Cross w/o pile | 40 | 48 | 55 |
| Post-tensioned Concrete Box Girder | | | |
| Stream Cross w/pile | 65 | 70 | 80 |
| Stream Cross w/o pile | 60 | 65 | 70 |
| Rdwy Cross w/pile | 40 | 50 | 60 |
| Rdwy Cross w/o pile | 40 | 48 | 55 |
| Reinforced Concrete Slab | 35 | 40 | 50 |
| Steel Structure (Rdwy Crossing) | | | |
| Rolled beam | 50 | 60 | 80 |
| Plate girder | 50 | 65 | 90 |
| Truss | | 120 | |
| Arch | | 130 | |
| Treated Timber Str. | 30 | 35 | 45 |

**Reconstruction and Repairs**

| | Low | Average | High |
|---|---|---|---|
| • Conc. Deck Replacement inc. removal: | 20 | 30 | 50 |
| • Conc. Deck Repairs: | 10 | 20 | 40 |
| • Superstr. Replacement. | | | |
| Steel grid deck : | 20 | 27 | 35 |
| Deck & steel girder: | 60 | 80 | 100 |
| • Widening Exist. Str.: | 60 | 75 | 90 |
| • LMC overlay (sq. ft): | 3 | 4 | 7 |
| • Asphalt overlay with water proofing (s.f): | 1.5 | 2 | 3 |
| • Silane Surf. Treat. : | .8 | 1 | 1.5 |
| • Crack repairs (ft.) : | 15 | 20 | 30 |
| • Repack joint (ft.) : | 8 | 10 | 15 |
| • Conc. Repairs*(c.y) | 300 | 600 | 900 |
| • Steel Repairs(lbs.) | 1 | 2 | 4 |

*price include removal and per cubic yard.

Certain assumptions have to be made in order to proceed with the analysis. These assumptions should be clearly identified for what they are and must be justified. Frequently, formulation of the assumptions could be made during earlier stages of the studies.

Each alternative under consideration must be evaluated in terms of all the costs and benefits occurring during the entire project life cycle, such as the costs of engineering and administration, construction, right-of-way, maintenance and operation. Costs, although difficult to estimate, are easy to measure in terms of dollars. Benefits, on the other hand, are very difficult to measure explicitly. This is especially true for the bridge project alone since it is hard to assess the benefits associated with each of the alternatives involving travel time, accident cost, car running cost, volume to capacity ratio, community growth, safety, social and aesthetic considerations. Therefore, these factors will be considered at scheme selection process, not at cost comparison.

In comparing costs and ranking the alternative, we will assume them to have equal benefits, although the benefits for different alternatives are not truly equal. The commonly used methods of analyzing these alternatives are based on the annual cost and the present worth.

**Economic Life:**

In general, the economic life of an alternative scheme is a period during which it provides the service to the travelling public. Economic life starts only when the bridge opens to the general public. The estimated economic life may vary depending on the type of structure and maintenance.

The following guidelines are to be used only in the absence of other specific information.

---

For more realistic cost estimate, the quantities of major items should be computed then multiplied by the current unit price to obtain the raw estimated cost. It is suggested to add 20% over the raw cost for miscellaneous items.

- Concrete bridge 50 years
- Steel bridge 50 years
- Timber bridge 25 years
- Rehabilitated bridge 25 years
- Major deck replacement 35 years
- Asphalt overlay with waterproofing membrane 10 years
- LMC overlay 15 years

**Equivalent Uniform Annual Cost:**

Since some of the alternative schemes considered may not have equal life span, equivalent uniform annual cost method may be used to compare the alternatives. In this method, all costs are expressed in the same terms of uniformly recurring costs per year for an equivalent level of total benefit by converting each alternative into an equivalent hypothetical alternative. In reality, we cannot expect the annual recurring cost to be uniform. However, this method will permit us a direct comparison of alternatives.

For example, when a precast deck slab bridge and a timber bridge are under consideration as alternatives. Their costs and physical lives are as follows:

| | Precast | Timber |
|---|---|---|
| Physical Life | 50 Years | 25 Years |
| Capital Invest. | $100,000 | $ 60,000 |
| Annual Maint. Cost | $ 1,000 | $ 3,000 |
| Restoration Cost | $ 20,000 | $ 20,000 |
| resurf. overlay | at 25 yrs. | at 15 yrs. |
| Salvage Value | 0 | 0 |
| Interest Rate 10% | | |
| Unif. Annual Cost | $ 11,289 | $ 10,082 |

From 10% Interest Table (Appendix).

Precast Bridge
Capital Recovery Factor:
(CRF N=50)=0.10086
Sinking Fund Factor:
(SFF N=25)=0.01017

| | |
|---|---|
| 100,000 X 0.10086= | 10,086 |
| 20,000 X 0.01017= | 203 |
| Annual Maint. | 1,000 |
| Total Annual Cost | $11,289 |

Timber Bridge
(CRF N=25)=0.11017
(SFF N=15)=0.03147

| | |
|---|---|
| 60,000 X 0.11017 = | 6,610 |
| 15,000 X 0.03147 = | 472 |
| Annual Maint. | 3,000 |
| Total Annual Cost | $10,082 |

**Present Worth Method:**

The present worth method is also known as the capitalized cost method. The annual costs for operation and maintenance are converted into capital costs by multiplying them by the present worth factor. In the previous example, the present worth of the precast bridge would be as follows:

| | |
|---|---|
| Initial cost: | = $100,000 |
| Capitalized Maintenance cost | |
| 1,000 x 9.915(PWF, N=50)= | $ 9,915 |
| 20,000 x.0923(CAF, N=25)= | $ 1,846 |
| | $111,761 |

The present worth method should be used only if the following two conditions are satisfied: 1). All alternatives have the same physical lives. 2). All alternatives provide roughly the same level of benefits over their common physical lives. In selecting schemes, the one with the lowest present worth should be selected.

**Sensitivity of the Economic Analysis:**

It should be recognized that certain factors of economic analysis are highly sensitive and significantly affect the outcome of the analysis. The sensitivity analysis is simple process involving changing variables, one at a time, to determine the relative effect of the variable change on the estimated cost in order to select optimum scheme. Initially, the input of variables in the analysis can be selected from any reasonable estimate, however as experience is gained, the most sensitive variables can be identified. In selecting variable, it requires judgement and experience. Particular attention should be given to interest rates, terminal values and service life span. The key to realistic results is in the selection of realistic values and consistency in their selection.

In the selection of interest rates, it is logical to select the rate at which the agency can borrow for capital expenditures. Long-term improvements are affected substantially by a radical departure in borrowing from either higher or lower than the prevailing rate. Current high interest rates make annual cost comparison insensitive and original capital cost is heavily weighted.

The sensitivity of annual cost and service life should be recognized as important considerations in the economic analysis. The service life is the period of time during which the improvement can reasonably be expected to provide adequate level of service. Fifty years is usually the maximum time for major changes in bridges. The useful life is the most sensitive to analysis, and care should be exercised in its selection.

Both the costs and the benefits of improvements should be calculated for a period of time designated for the expected service life. The analysis period should not extend beyond this time. For some bridges, the annual cost for maintenance will be relatively small, and it can be ignored in the economic analysis.

## 2—6 COMPREHENSIVE EVALUATION

In economic analysis, we could rank each alternative based on its life-cycle project cost. In selecting alternative scheme, the following may be used as guide lines:

If the rehabilitation cost for providing the required level of service is less than 30 percent of the replacement cost, the rehabilitation scheme should be selected. On the other hand, if the rehabilitation cost is over 60 percent of the replacement cost, the replacement is preferred. Additional deficiencies will be discovered during repairs and the change order will eventually increase the total construction cost from 30 to 50 percent of the original bid price for a rehabilitation project. Thus the cost effective threshold figure would be 60 percent of the replacement cost. The value analysis technique may be used in the selection of the alternate when the repair costs are between 30 and 60 percent of the replacement cost.

The following steps could be used to evaluate the alternatives and to aid in the selection of the best scheme.

1. Determine criteria and objectives
2. Weigh the criteria and objectives
3. Rate the alternatives
4. Compute numerical ratings
5. Rank alternatives
6. Select the best alternative

**1. Determination of Objectives and Criteria**
Whenever a new project is going to be undertaken, first determine its primary objective which may be the improving of safety, the reducing of traffic congestion, the increasing of its structural capacity, or improvement of other factors. In order to assess the objective, select criteria on which to base the decision. These are as follows:

1. Equivalent uniform annual cost or life cycle project cost
2. Traffic maintenance and detour
3. Availability of materials, labor and equipment
4. Aesthetics
5. Impacts on local community
6. Environmental restrictions and permit requirements
7. Functional requirements

**2. Weighing the Criteria and Objectives**
Objectives or criteria under consideration seldom have equal importance and some of these have greater influence on the final decision than the others. Therefore, we should determine weight factors which will be applied to each criteria or objective indicating how important each factor is to reach a decision. The intention of this process is ranking and rating the factors according to their importance. A weight factor can be introduced as shown in the blocks along the top of the form in figure. Here weights from 1 to 5 or 1 to 10 can be used, with the highest number being given to the factor with the most importance. In assigning weight factors, the matrix method is generally used. It has been described in National Cooperative Highway Research Program Report 222 "Bridge on Secondary Highways and Local Roads Rehabilitation and Replacement".

**3. Rating Alternative**
Each alternative is rated within the category by using 4 for excellent, 3 for good, 2 for fair, and 1 for poor. It is recommended that the rating be done from top to bottom, rather than left to right.

RANKING ANALYSIS

**WEIGH ALTERNATIVES**

4 Excellent
3 Good
2 Fair
1 Poor

CONTRIBUTION TO: OBJECTIVES

| ALTERNATIVES | WEIGHT | Initial Cost | Implemenability | Maintenance Cost | Aesthetic | Public Acceptance | | | | TOTAL | RANKING |
|---|---|---|---|---|---|---|---|---|---|---|---|
| | | 10 | 10 | 3 | 7 | 7 | | | | | |
| I Reinforced Concrete Retaining Wall | | 2 / 20 | 4 / 40 | 4 / 12 | 2 / 14 | 2 / 14 | | | | 100 | 3 |
| II Reinforced Earth | | 3 / 30 | 2 / 20 | 3 / 9 | 3 / 21 | 3 / 21 | | | | 101 | 2 |
| III Gabion | | 4 / 40 | 4 / 40 | 2 / 6 | 4 / 28 | 4 / 28 | | | | 142 | 1 |
| | | | | | | | | | | | |

**TABLE I Ranking and Evaluation**

**4. Compute Numerical Rating**

This could be done by multiplying the weight factor of each criteria by the rating of each alternative in that category. The result is inserted in the lower right hand portion of each square. Then summed horizontally and the totals are posted in the right hand column.

**5. Rank the Alternative with the Highest Aggregate Score**

The results obtained by this method heavily depend on the judgement and experience of the engineer. However, this technique is helpful in selection of the best practical scheme.

For example, the existing highway is to be widened to provide an additional traffic lane along a five(5) miles stretch by side of the river. To retain the earth, the following three alternative schemes are considered for length of a 1,500 feet wall: 1. Reinforced Concrete Wall, 2. Reinforced Earth 3. Gabion Wall.

The economic analysis indicates that the gabion is favored. also ranking analysis by value engineering is made to select the best scheme to satisfy the various conditions.

| | I. REINFOCED CONCRETE WALL | II.REINFORCED EARTH | III: GABION |
|---|---|---|---|
| Capital Invest<br>Annal Mainten.<br>Physical Life | $600,000<br>$1,000<br>50 yrs. | $400,000<br>$2,000<br>50 yrs. | $300,000<br>$2,000<br>30 yrs. |
| Remarks | Conventioinal Costruction method and little maintenance. | Local contractor may not have construction experience. Aesthetically acceptable. | Lower initial cost. Aesthetically pleasing & blend in to nature. Material available locally. |
| Annual Cost | $61,516 | $42,344 | $33,824 |

**TABLE II Economic Analysis:**

Based on the ranking analysis shown on above, Scheme III Gabion is a definite choice for retaining earth for widening project

## 2–7 REPORT

The rehabilitation study report could have the following parts:

**1. Summary.** A brief description of the structure and a summary of its condition, identifying major problems, proposed alternative schemes with cost estimates, conclusion, and recommendations.

**2. Finding.** The summary of findings of bridge inspection and deck condition survey.

**3. Purpose and Need.** A statement of the objective of the project and its basic functional requirement. Identify and describe the problem and explain the need of improvement such as structural condition of the existing bridge, traffic capacity, high accident rate, social demand and economic development.

**4. Analysis.** A statement for developing alternatives to meet the requirement of physical and geometric problems of the structure and social, economic and environmental impacts associated with the project and statements of the comparison of schemes, economic analysis, evaluations and selection of the scheme. The selection of the scheme should include a discussion of how the reasonable scheme were selected and why other alternative were eliminated.

**5. Conclusion and Recommendations**
Concise presentation of the finding, selection of the best schemes, and of the recommendations.

## REFERENCE


1. AASHTO:
   "Standard Specifications for Highway Bridges, 1977" and current interims.
   "AASHTO Manual for Bridge Maintenance, 1976";
   "Geometric Design Standards for Highway Other Than Freeway, 1969";
   "Geometric Design Guide for Resurfacing, Restoration, and Rehabilitation (R-R-R) of Highways and Streets, 1977";
   "A Guide for Highway Landscape and Environmental Design, 1970"
2. American Society of Civil Engineer "Engineering Economy" ASCE NY 1970
3. Berger, Roland "Extending the Service Life of Existing Bridges by Increasing Their Load Carrying Capacity" FHWA-RD-78-133, Washington D.C.: FHWA June 1978
4. General Accounting Office:
   "Excessive Truck Weight" CD-ED-79-94 Washington D.C.: GAO, July 1979.
   "Value engineering has the potential to reduce mass transit construction cost" RCED-83-34 Dec. 29. 1982
   "Energy Conservation and the Federal-aid highway program, June 1981"
   "Preconstruction Engineering Management System Manual-Draft" NHI
   "Highway Engineering Economy, 1980"
   "Federal-aid Highway Program Manual Vol. 6 and Vol. 7"
   "Third Annual Report to Congress-Highway bridge replacement and rehabilitation program" FHWA Bridge Division, March, 1982
   " Value Engineering for Highway" FHWA, 1980
   "Alternate Bridge Design" FHWA Technical Advisory T 5140.12, Dec. 4, 1979
6. Federal Register: "National Environmental Policy Act (NEPA) Implementation Procedure, Dept. of Transportation.": Vol. 44, No.200 : Oct. 15, 1979; Vol. 45, No. 98 : May 19, 1980; Vol. 45, No.117 : June 16, 1980; Corp of Engineer, Dept. of Army:Vol. 45, No. 3 : Jan.4, 1980;Vol. 45, No.184 : Sep. 19, 1980 Vol. 42, No.138 : July 19, 1977
7. Dept. of the Navy, Naval Facilities Engineering Command: "Economic Analysis Handbook, NAVFAC P-442" 1975
8. Tokerud, Roy " Precast Prestressed Concrete Bridge for Low-Volume Roads" PCI Journal, July/August 1979 pp42-56
9. Transportation Research Board
   "Bridge Engineering Vol. I & II" Transportation Research Report 664 & 665, TRB Conference, Sept. 25-27, 1978. St. Louis, Missouri, National Academy of Science, Washington D.C.
   "Bridge on Secondary Highways and Local Roads - Rehabilitation and Replacement" NCHRP Report 222, 1980
   "Rehabilitation and Replacement of Bridges on Secondary Highway and Local Roads" NCHRP Report 243, 1981
   "Value Engineering in Preconstruction and Construction" NCHRP Practice 78, Sep. 1981
10. Florida Dept. of Transportation
    "Manual for Bridge Maintenance Planning and Repair Methods"
11. "The New York State Department of Transportation Bridge Inspection and Rehabilitation Design Program" Paper presented 1983 TRB Annual Meeting
12. "Environmental Digest" NJ DOT, Feb., 1981
13. Federal Register/ Vol. 48 No. 93/ May 12, 1983
    "Alternate Design for Bridges; Policy Statement" FHWA pp21409

# 3 PRELIMINARY DESIGN

## 3–1 GENERAL

Preliminary plans should reflect a definite design for the structure's type, size, and locations. Thus developing the plans is known as type, size, and locations (TS & L) design and the plan submittal for FHWA approval is called TS & L submission. Before developing preliminary plans, the rehabilitation study report should be reviewed for it's validity and for the changes which occurred after the completion of the final report. If the structural concept, geometric, structure type and repair method have been agreed, the preliminary plans should be developed based on the selected scheme. The format of the plans will vary depending upon the type of work, which can be either repair, widening, or replacement.

For repairs and minor widening projects, the wash-off photo copies with light lines of the original General Plan and Elevation Sheet may be used. These wash-off plans show the required repairs and their locations and notes highlighted by the use of darker lines and lettering. By doing this, the sections of the existing structure will not be duplicated.

The preliminary bridge plans generally include the following:

1. A general plan and elevation of the structure show repair areas, the grades of the structure and immediate approaches, the span length, the skew, and the controlling horizontal and vertical clearances. Also, the plans show the type of replacement structure, typical superstructure section, substructure locations, and the location of the expansion and fixed bearings. If the substructure is to be replaced, then the typical abutment and/or pier sections, the proposed foundation type and the bottom elevation of the footing shall be indicated.
2. A typical section of replacement bridge should show the following: The type and arrangement of beams, the width of the structure including width of the median, travelling lanes, shoulders, curbs, sidewalks, etc. and type of railing.
3. Also, a typical section of approach roadway should be shown.
4. The plan should include information on design loading, allowable stress and specifications under which the structure is to be widened, rehabilitated or replaced.
5. If a bridge is to be relocated, the plan should show the relocation site with the soil boring log identification numbers.
6. If architectural treatments are proposed, sketches of the same should be furnished.
7. A roadway plan showing the physical features of the approach roadway at each side of the bridge for minimum of 500 feet. This should show topography, roads, ramps, railroads, streams and any other features.
8. The location of subsurface utilities and the proposed relocation of utilities in the structure should be shown if information is available at the time of plan submission. Overhead utilities should also be included in the plans.
9. Hydraulic and hydrologic data shall be noted on the plans for a waterway structure.
10. If the bridge crosses a railroad, the existing tracts, the profile of the tracks, the proposed horizontal and vertical clearances and topography along the railroad must be shown.
11. The proposed foundation type and design soil pressure should be shown on the plans.
12. The schematic bridge deck plan should outline the anticipated limits of each deck or superstructure replacement with respect to the available traffic lanes. Also, a section of the required stage construction and available traffic lanes should be shown.
13. Plans should show general areas of major repairs or replacement.

14. The estimated cost and traffic data should also be included.

More information furnished on the study report will result in a more expeditious review and afford a mutually acceptable preliminary design. It also reduces time and money which may be expended on detailed design and plan preparation later on. Also, an agreement reached during a preliminary review will minimize the possibility of costly and time consuming changes of the plans at a later date.

If the preliminary submission involves new or unusual features of design, construction or specifications out of the AASHTO Specifications or previously approved Standards, then supporting information to justify their use should be supplied along with the submission of preliminary plans.

Upon the approval of preliminary plans from FHWA, the construction plans can be developed. After completing the design computations and all drawings, the plans should be submitted along with special provisions of the standard specification and estimated cost for review. At this time, the plans are essentially completed and are subject only to minor modifications. Upon receiving comments and incorporating the comments into the plans, the revised plans, special provisions and estimated cost are submitted to FHWA for review and comments as Pre-PS & E submission. It is required to have a Right-of-way availability certificate, Utilities clearance letter, the signed agreements and the permits at this submission. After receiving FHWA comments, revise and update the PS & E packages and prepare proposal booklet and bid documents then submit the final PS & E packages to the FHWA for approval.

## 3–2 DESIGN CRITERIA

### 3-2.1 DESIGN LIVE LOADS

The majority of highway bridges are designed according to the AASHTO Standard Specifications for the Highway Bridges. The heaviest loadings, designated HS 20, were adopted in 1944 and have been supplemented by alternate military loadings. These loadings are not as heavy as those used in Europe where engineers tend to design for longer life by taking into account heavy military and special loading conditions.

In 1975, the Federal government increased the truck gross weight limits from 73,280 pounds to 80,000 pounds and the single and tandem axle weight limits by 2,000 pounds to save fuel. Since then, most States have raised their legal limits to the Federal ceilings while keeping their bridge design loads the same 72,000 pounds which are 8,000 pounds below the legal limits. This discrepancy and the increased heavy truck traffic are among the principle causes of bridge deterioration.

In 1982, the Congress made mandatory the previous permissive maximum weight limits of 80,000 pounds gross, 20,000 pounds single axle, and 34,000 pounds tandem axles on Interstate system with reasonable access to be permitted from the Interstate road to terminals. Also, the Congress requires states to allow twin trailer combination trucks on any segment of the Interstate system and designated Federal-aid primary highways. Minimum trailer length is set 28 feet for double and 48 feet for single trailer combinations with no length restriction permitted on the tractor or overall configuration. Also, maximum 102 inches motor vehicle width is established.

Although a five-axle tractor trailer loaded to the current 80,000 pound of federal weight limits weighs about the same as twenty automobiles, its impact on a highway is dramatically higher. It is estimated that one tractor trailer causes as much damage as at least 9,600 automobiles. For example, the parkway bridges where trucks are prohibited, show no major deterioration whereas the Interstate bridges of the same age need extensive deck repairs or replacement.

In order to offset the increase of truck weight and to extend the service life of structures, some states require that all new bridges shall be designed for the AASHTO HS-25 live loads and New Jersey requires that all new bridges shall be designed for HS-20 loadings supplemented with an additional 10%, while others adhere to HS-20 loading. Since HS 25 loading is not included in the

AASHTO Specifications, its configurations and values may be based on those of the HS-20 by proportional adjustment. The additional construction cost of a bridge designed for HS-25 loads is estimated to be only 3 to 5% more than that of a similar bridge designed for HS-20 loads.

The highway routes which are subjected to substantial heavy truck traffic, such as coal or log hauling route adjacent to a coal mine or the Forest Service roads or highway routes leading to grain elevators, petroleum depots, or muti-year construction projects should receive special consideration for the design of the increased live loads for the bridges on the routes.

The following are suggested guidelines for design live loads:

**Replacement Bridge:**

1. Interstate and other highway with heavy trucks: HS-25
2. Other Highway with ADT greater than 400 : HS-20
3. ADT less than 400: H-20

It is suggested that a widening section of a bridge should meet the live loads of the current design requirement instead of original design loads.

**Rehabilitation:**
The design load to be used for the rehabilitation of the existing bridges should be determined in each individual case by considering the inherent strength of the structure and the cost involved in providing additional load carrying capacity without making major changes to the structural members. In the case of low volume local roads, the rehabilitated structure should at least meet the AASHTO requirements. (Geometric Design Guide for Resurfacing, Restoration, and Rehabilitation of Highways and Streets, 1977". That is H-15 loads for highway with ADT 400-750).

## 3-2.2 DESIGN METHOD

The current AASHTO bridge specifications give the option of using either the Service Load Method (Working Stress Design Method) or the Strength Method (Load Factor Design). Most State Highway Agencies have adopted the Strength Design Method to achieve the material and cost savings which it provides. However, it is suggested that designs involving the rehabilitation, in-kind widening, or extension of the existing bridges should use the same design method that was used in the existing bridges.

## 3-2.3 LIVE LOAD DEFLECTION

Live load deflection preferably shall be not more than L/1200 for highway bridges but in no case more than L/800 of AASHTO requirement. ("L" represents the span length of the bridge).

The live load deflection may be computed by assuming that all beams act together and have equal deflections. All lanes may be loaded simultaneously subject to the reduction of intensity for improbable coincidental maximum loading. For bridges with longitudinal joints, the live load deflection should be computed as a separate transverse unit.

## 3-2.4 MATERIAL

**Steel:**
The type of steel commonly used in steel girders are Structural Carbon Steel (ASTM A-36, Fy=36 ksi) and High Strength Low Alloy Structural Steel (ASTM A441, A572, and A588, Fy=50 ksi). Generally, A-588 steel reduces the weight and total cost of material but the main reason it is preferred is its maintenance free nature.

It is recommended that bare A-588 steel should not be used in a corrosive industrial atmosphere, where it is subjected to recurrent wetting by salt water, accumulated dirt and debris, or where it is exposed to continued moisture. When bare A-588 steel is used, all steel for a distance of three times of the beam depth or 10 feet each way from the deck joints should be given two shop coats of paint to avoid rust stains on the substructure.

Currently more attention is being focused on the ability of structural steel to withstand fracture. Some high strength structural steels such as ASTM A-514 and A-517 are considered to have low toughness and these steels should be used with care even if the thickness is less than or equal to 2 inches.

**Concrete:**
In service design, the allowable stress for deck slabs is normally 1,200 psi based on the 3,000 psi ultimate strength while bridge decks designed by the Load Factor Design method require a concrete with an ultimate strength of 4,000 psi.

**Reinforcement Bars:**
Previously, reinforcement steel of Grade 40 was used in designing reinforced concrete structures and its allowable design stress was 20 ksi by the Working Stress method and 40 ksi by the Load Factor Method. But currently, certain sizes of Grade 40 bars are not available and most states specify the use of Grade 60 bars with allowable stress of 24 ksi for the Working Stress Method and 60 ksi for the Load Factor Method.

## 3–3 STRUCTURE TYPE SELECTION

In selecting structure type for widening and paralleling of the existing bridges, it is recommended that a choice be made of the existing type of structure even if the cost of it is more than some other type. This is done to maintain the compatibility and aesthetic value of the structure.

### 3-3.1 SELECTION OF SUPERSTRUCTURE TYPE

The selection of the type of the proposed replacement structure should be determined by evaluation and recommendation of the rehabilitation studies with the supplemental information of current estimated construction cost and the preference based on past experience and judgement.

In selecting the most economical structure, the problems of fabrication and erection should be considered along with the general features of terrain, highway geometrics, soil conditions, and geographical location. The environmental impacts have become a most important subject and the proposed replacement structure should blend with the existing site in a manner which will minimize detrimental impacts on the surroundings. The engineer should therefore make every effort to pick an architecturally pleasing structure consistent with structural requirement, economy, and surrounding locations.

In selecting type, shape and location of the structure, the aesthetic value should be considered and the following are some of the suggestion: On a multi-span bridge, the end span generally become shorter than the interior spans and require shallow depth of beams. The abrupt changes in fascia depth at the end beam should be avoided and uniform or smooth tapered depth of beam is desirable for the appearance of the structure. Also, longer spans are preferred to multiple short spans so as to provide openness in view. A short span should not be placed between longer spans. An architectural treatment of wall or abutment should be considered by scoring, surface texture or installing stone facing or precast panels. The pier shape should be balanced so that it should be structurally functional, relatively economical and also have a good appearance.

In deciding span lengths, if the substructure become relatively expensive, it may be more economical to select longer span lengths available for the proposed structure type. For grade separation bridges providing four or more traffic lanes underneath, two span bridges with semi-stub abutment are preferred but four span bridges with end span cantilevered from the side piers and integral type diaphragm abutment or stub abutment may be used as an alternates. For two span bridges, the use of the same structure length is recommended if the difference in the structure lengths is within 5 feet. Also, for a multi-span structure, the span lengths should be equal if the theoretical differences between the spans are within 5 feet. This additional length would increase the material cost but the easy fabrication would compensate the for extra material cost.

In selecting skew angles, it is desirable to avoid between 24 and 27 degrees because snow plow blades which use this angle would likely damage the joints by dropping the plow blade into the deck joint. If skew angle is used, it is preferred to select nearest

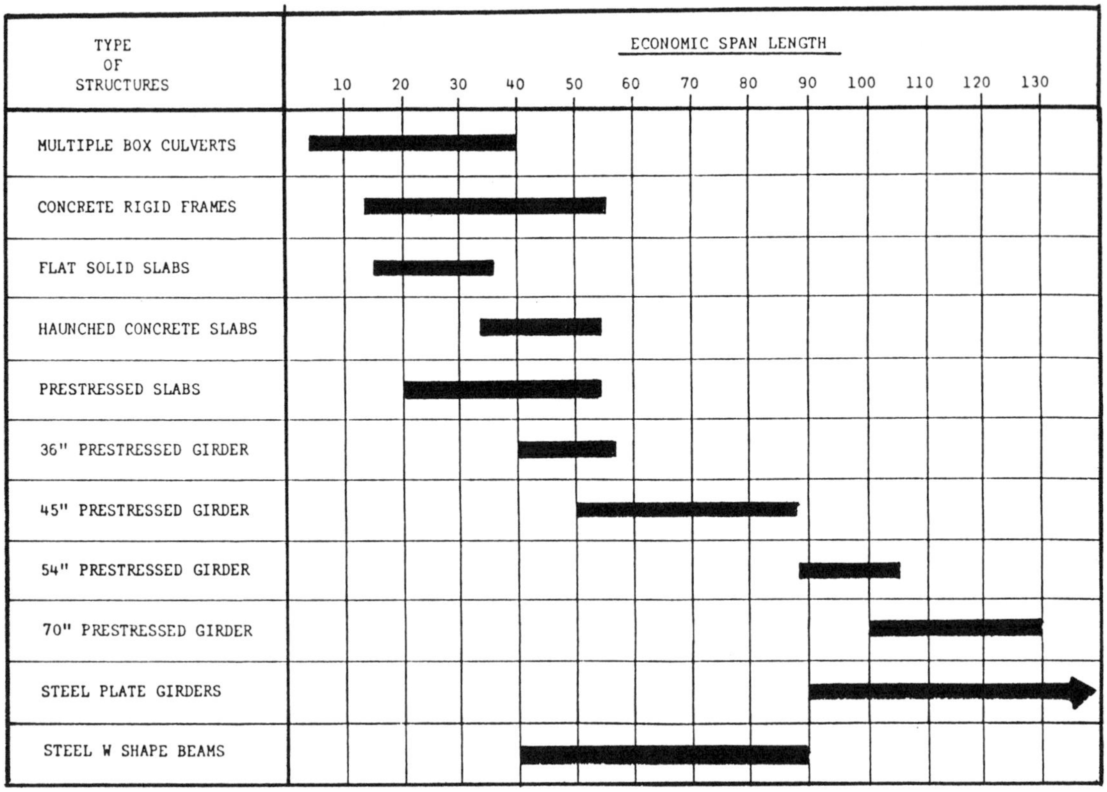


**TABLE I Economical Type of Structure** (WI DOT)

degree by rounding off the minutes. If the roadway skew angle is less than two (2) degrees, consideration should be given to construct the bridge without skew angle and lengthen the span to accommodate the effect of the skew on the roadway underneath the bridge.

Generally, the required number of beam is depend on their cost. It is more economical to use fewer numbers of heavy girders by spacing them wider and using thicker deck slabs because the increased cos, and not be used in an urban area. The inlet should be placed directly above down spouts attached to the substructure. Mins of the deck where the discharge would fall on the roadway, walkway, railroad, unprotected embankment and erodible areating wider spacing of the beams (more than 9 feet).

A spacing of 7 to 9 feet of a beam is considered to be economical, but it may be necessary to use closer spacings when the vertical clearance requires a shallower superstructure.

The following are the typical type of superstructures:

## 1. CONCRETE SLAB BRIDGE

Some slab bridges are composed of one or more simple spans, but continuous spans are preferred where good condition in foundation can be provided. With continuity, the cost of joint construction and maintenance is reduced substantially and a smoother riding surface is secured due to decreased deflections.

Slab bridges are usually suitable for bridges with end spans of up to about 40 feet and interior spans proportionally longer. The practical ranges in span length can be increased by using the haunched or voided slab. A concrete slab bridge is limited to sites requiring a skew angle of 30 degrees or less.

The shallow superstructure inherent in a slab bridge helps by solving the problems of vertical clearance and minimum grade difference which result in a overall economy because the height of the embankments is reduced.

The following are the generally used depth-span ratios:

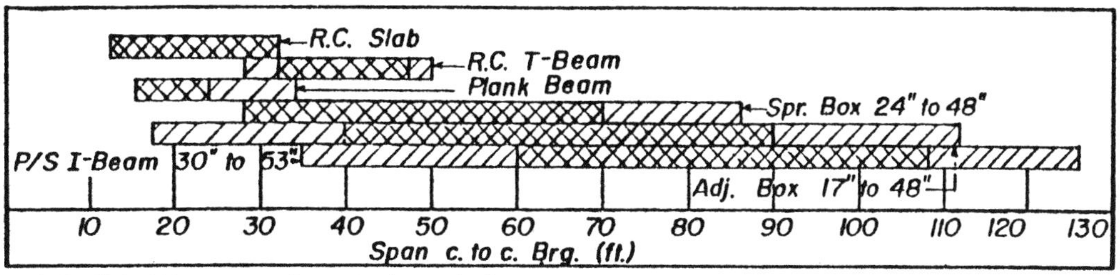


TABLE II Economical Comparison of Concrete Beam (PennDOT)

a. **Standard slab bridge**
Simple Span : 1/22
Continuous Span : 1/25

b. **Haunched slab bridge**
The ratio should be adjusted to account for change in relative stiffenes of positive and negative sections.

## 2. CONCRETE "T" BEAM

Currently, the concrete T-beam has not been used widely in most states. The T-beam is suitable to roadway with sharp horizontal curvature due to its cast-in-place nature to adopt any alignment. The main problems are maintaining of vertical clearance below the depth of false work during construction and deck replacements after many years of service. The deck replacement of a T-beam structure requires extensive engineering analysis to be compatible with the monolithically constructed existing bridge system.

## 3. CONCRETE BOX GIRDER

The concrete box girder bridge is structurally desirable and aesthetically suitable for urban locations where roadways have a high degree of horizontal curvature as well as large skew angles. This type of bridge is often used in multi-level interchanges where horizontal clearance is limited. The pier cap is an integral part of the deck and may impose some vertical and horizontal clearance problems if conventional pier design is used.

The disadvantages of concrete box girders are initial high cost, cracks over the piers, and problems of deck replacements requiring shoring and extensive design analysis. Because of these problems, the concrete box girder is not desirable as a replacement bridge. The generally used depth-span ratio are 1/18 and 1/20 for simple and continuous spans respectively.

## 4. PRESTRESSED CONCRETE SLAB

The prestressed concrete slab bridge is widely used as a replacement bridge for a short span because of its advantages of elimination of false work during construction of superstructure and rapid erection. This type of bridge has a particular advantage when vertical clearances are critical. Also, the shallow structure depth permits minimum grade profile difference between intersecting roadways which results in savings. The optimum depth span ratio is 1/25.

## 5. PRESTERESSED CONCRETE GIRDER

Since the 1950's, prestressed concrete girder bridges have been built in most states. Prestressed girder bridges are competitive from a initial construction cost standpoint and they require little maintenance.

Plant fabrication of girders usually give better quality controlled products and easy erection enables the contractor to bid prestressed girders as an alternate to the other type of structure. The optimum depth-span ratio is 1/18 to 1/20.

## 6. OTHER PRESTRESSED BRIDGE

**Standard T-girder:**
T-girders are heavier than I-girders with the closured section poured in place and the bulb T-girder has the most efficient section. Optimum depth-span ratio is 1/20.

**Channel Girder:**
Channel girders would cost more than slabs and no standard forms are available from most fabricators. This girder is difficult to built but it is more stable than I-girders for handling and placing. They require detailed design and the optimum depth-span ratio is 1/18.

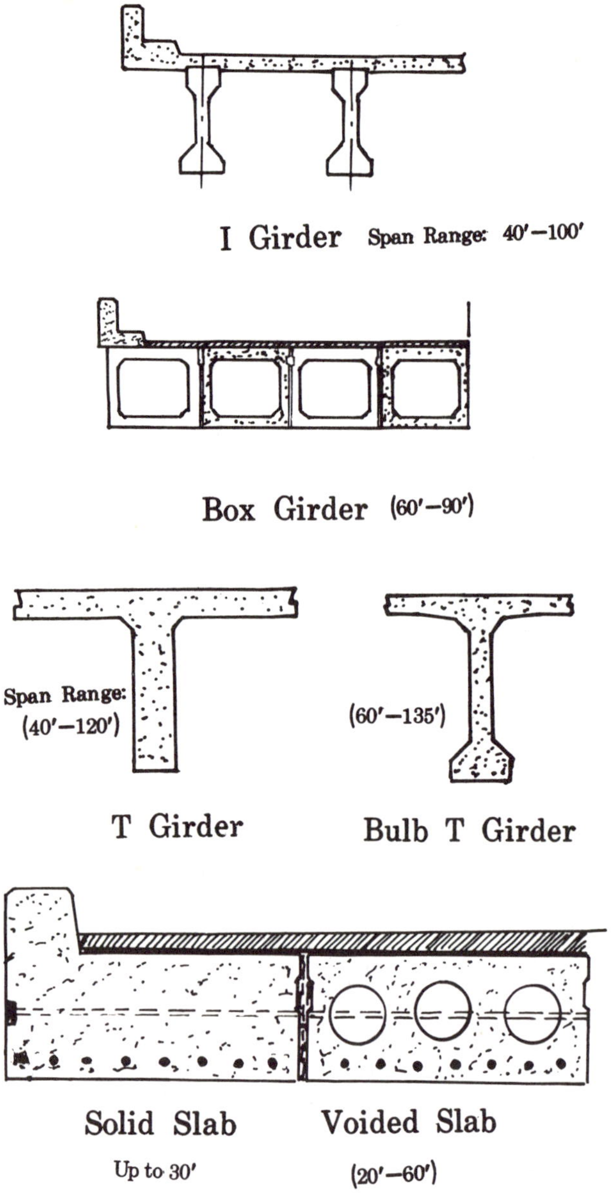


**Fig. 3.1 Typical Prestressed Section**

**Standard Box Beams:**
This type of beam is used for span length between 60 to 100 feet and it is popular as a replacement bridge for short and simple spans. It is considered economical and desirable alternate where maintenance of traffic is critical. The main problem associated with the box beam is reflectional longitudinal cracks in the wearing surface over joints between the beams.

**Other Considerations:**
The suitability of a prestressed concrete bridge will depend on the simplicity, availability and reusability of formwork, ease in placing concrete, functional and aesthetical requirements, and theoretical considerations. It is preferred to have constant depth of beams in multi-span bridges. Asphalt wearing surface over the adjacent muti-beam type slab, box beams and channel girders are having longitudinal cracks over the joints and deteriorations therefore it is suggested the use of non-shrink grout at the joints and concrete wearing surface of minimum 3.5 inches depth over the beams. Reinforcement steel (#3 at 12" or welded fabrics) should be epoxy coated and have minimum 1.5 inch cover.

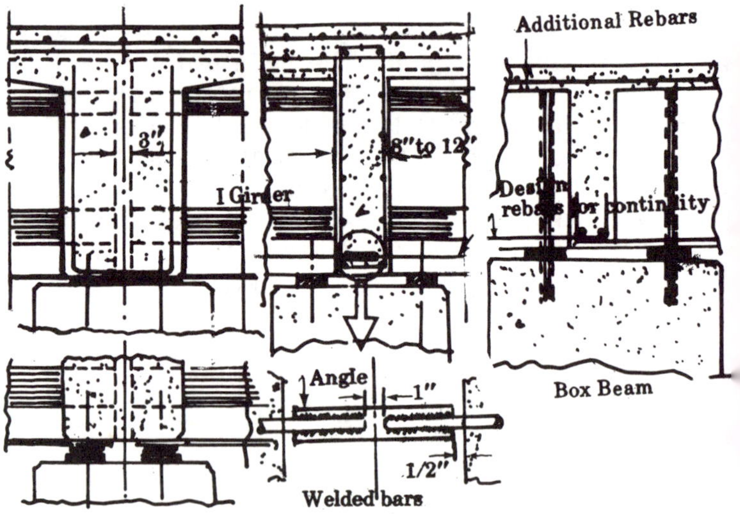


**Fig. 3.2 Continuity for Live Load**

Where the foundation of the piers is resting on rock or piles, multi-span box beams, slabs, and girders are designed for continuity for live loads so that the joint over the pier is eliminated. In order to provide the continuity for live load, it is preferred to have 5" to 6" depth concrete reinforced with #4 at 12" or fabric reinforcement (zero gage 6"x6") for a positive moments section of continuous spans and simple span bridge. For the negative moment area of continuous spans, the bars should be designed to develop the negative moments for live load in conjunction with prestressed members. All bars should be epoxy coated and have 2.5 inch clear covers. The shear key between prestressed member should be filled with non-shrink grout to avoid longitudinal cracks.

The rectangular shape is easier to form but uneconomical materialwise. The T-girder is suitable where there is little danger of over stressing at transfer and where the concrete is effectively concentrated at the compressive flange. The I-girder and box beam have more concrete near the extreme fibers and are efficient both at transfer and under unlimited loads, but they have weaker webs and require more complicated forming.

From the standpoint of economy and aesthetics, higher depth-span ratios are always used for prestressed concrete than for reinforced concrete. The use of a higher ratio is possible for prestressed concrete bridges because deflection can be better controlled than for reinforced concrete bridges.

On the other hand, when these ratios become high, camber and deflection become quite sensitive to variations in loadings, properties of materials, magnitude and location of prestressing, and temperature. Furthermore, the effects of vibration become pronounced with higher depth-span ratio. Care should be exercised in designing cantilever beams because they are particularly sensitive to deflection and vibration.

If the existing girders are reused and designed as composite beam, significant shrinkage stress is developed by the restrained shrinkage of the concrete deck slab over the existing girders. The differential shrinkage stress would be significant in a short span bridge. The shrinkage stress of normal highway bridge with less than 100 feet span would be about 10% of total dead load stress but may increase to 50% in shorter bridges.

In continuous post tensioned box girder bridges, temperature difference between top slab and box can be such significant magnitude, temperature induced stresses could contribute the cracks in the deck and the damage of the bridge. It is recommend to consider temperature gradient in the design of the continuous post tensioned girders or box beams.

## 7. CONCRETE RIGID FRAME

The concrete rigid frame is more expensive than other superstructures but the concrete rigid frame bridge is aesthetically pleasing. Therefore, it was used mostly in parks, urban areas, and parkways where the shorter span length was needed.

## 8. STEEL GIRDER

The emphasis of highway safety and sight distance requirements have favored the use of steel girders due to their ability to support longer span lengths.

Welded plate girders are used for structures requiring longer span lengths than the span which could be extended by use of rolled beams. Generally, rolled beams are cheaper than plate girders per weight, but the maximum available depth of a rolled beam is limited to 36". Welded plate girders are also preferred over rolled beams with cover plates because of the inefficient use of the section by the reduced allowable fatigue design stress and susceptibility of the fatigue failures. Welded plate girder give greater choice of sections by allowing variable web thickness, depth, variable flange width and thickness.

If there is no differential settlement problems, continuous span bridges are preferred over multiple simple span bridges.

For cost comparison of different type of steel girders, the Bethlehem Steel Corporation offer free preliminary girder bridge studies which include various girder spacing; a variety of span length; a number of AASHTO steel grades; simple and continuous spans; composite and non-composite analysis; combination of transverse and longitudinal stiffeners. For more information contact Bethlehem Steel Corp. Bethlehem, PA 18016. Also, United Steel Corp. offer similar program called SIMON for a steel girder structure.

**Non-composite Rolled Beam:**
This type uses more steel than the composite rolled beam, but it requires practically no shop work. Therefore, net cost may not be greater than composite beams. It is competitive in short spans ranging 30 to 50 feet. The optimum depth-span ratios are 1/15 in simple spans and 1/18 in continuous spans.

**Composite Rolled Beam:**
In this type more steel is wasted than compared in composite welded girders but it requires minimum shop work and net cost may be competitive with welded girders for short spans. This type of structure also has low dead loads which may be of value when foundation conditions are poor and it could be used for a span length of less than 90 feet. The optimum depth-span ratio is 1/20.

**Composite Welded Girder:**
This is economical for spans of 60 to 150 feet and it is the most widely used in the eastern States. The optimum depth-span ratio is 1/20 to 1/25.

## 3-3.2 SELECTION OF ABUTMENT

Abutments are used at the ends of bridges to retain the embankment and carry the vertical and horizontal forces from the superstructure. They could be designed as piers or retaining walls and they should be able to withstand against overturning and sliding. The footings should be designed to avoid differential settlement and excessive horizontal movements.

Several types of abutment are available for a particular bridge site and the selection of an abutment is made basically by the economics. For roadway crossings, the minimum horizontal distance and sight distance should be considered while the minimum waterway opening and channel area is critical for stream crossings in selecting a type of an abutment. Also, aesthetic factors sometimes play a key role in the selection process.

### 1. FULL HEIGHT ABUTMENT

A full height abutment is constructed at the lower level roadway and should support the entire embankment. This abutment is costly and is generally used in urban and metropolitan area where right-of-way is critical. In some cases the use of a full height abutment would reduce the total structure cost by shortening the span length and thus reducing the superstructure cost. It is suggested that cost comparison with other types of structure be made before selecting a full height abutment.

**Fig. 3.3 Full Height Abutment**

If both abutments are connected by fixed bearings to the superstructure as the rigid frame is constructed monolithically with superstructure, the wing walls should be joined to the abutment breast wall by an expansion joint. For a non-skewed abutment, this would allow the abutment wall to rotate freely about its footing and permit contraction and expansion of the superstructure. It also allows differential settlement between the wingwall footing and the abutment footing which is caused by the different loading conditions of the wingwalls and the breast wall of the abutment.

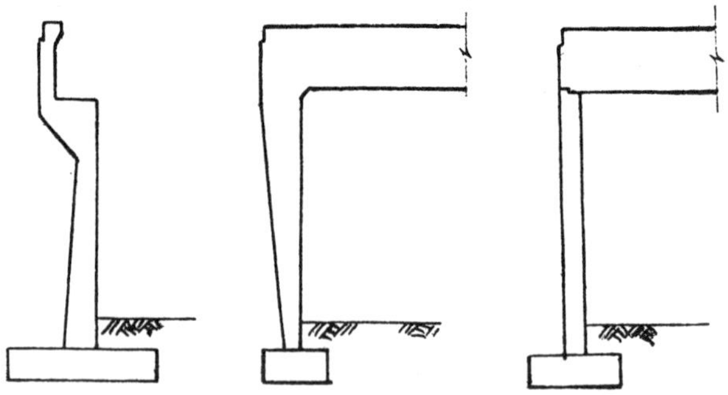

**Fig 3.4 Types of Full Height Abutment**

In a full height abutment there are difficulties of placing and compacting backfill materials against the abutment wall. It is possible that the wall may be pushed out of the vertical alignment if heavy equipment is used near the walls. The fill and compaction of the embankment after construction of an abutment would cause foundation settlements,

| ABUTMENT ARRANGEMENTS<br>L = Length of continuous superstructure between abutments | SUPERSTRUCTURES<br>CONCRETE SLAB SPANS | PRESTRESSED GIRDERS | STEEL GIRDERS |
|---|---|---|---|
| | L = Length and S = Skew , AL = Abutment Length | | |
| (1) TYPE A1 WITH FIXED SEAT 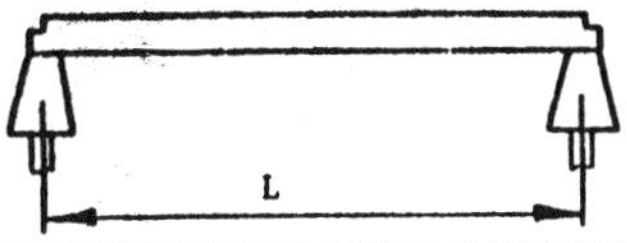 TYPE A1 WITH FIXED SEAT | L ≤ 300'<br>S ≤ 30°<br>AL ≤ 50' | L ≤ 300'<br>S ≤ 15°<br>AL ≤ 50' | L ≤ 150'<br>S ≤ 15°<br>AL ≤ 50' |
| (2) TYPE A1 WITH SEMI-EXP. SEAT 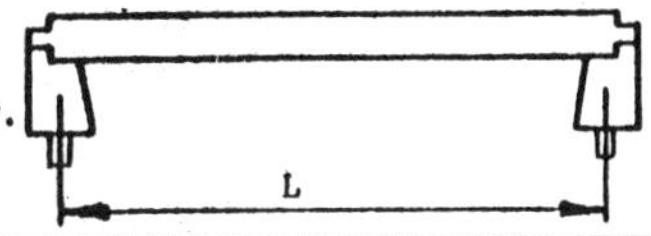 TYPE A1 WITH SEMI-EXP. SEAT | L ≤ 300'<br>S ≤ 30°<br>AL > 50' | L ≤ 300'<br>S ≤ 40° | L ≤ 200'<br>S ≤ 40° |
| (3) TYPE A2 WITH FIXED SEAT 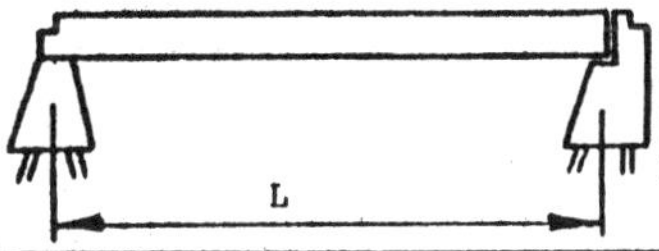 TYPE A3 WITH EXPANSION BEARING | L > 300'<br>with flexible Piers and<br>S ≤ 30°<br>AL ≤ 50' | 300' < L ≤ 400'<br>AL ≤ 50'<br>and with flexible Piers | NOT USED |
| (4) TYPE A3 WITH FIXED BEARING 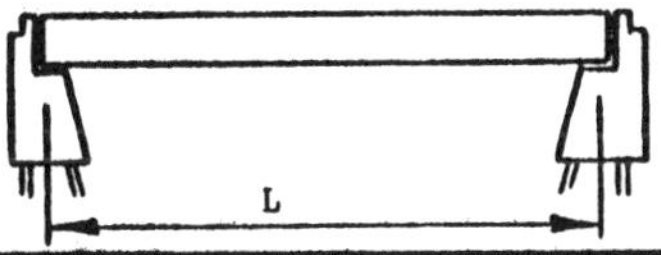 TYPE A3 WITH EXPANSION BEARING | NOT USED | NOT USED | L ≤ 200'<br>S > 40° |
| (5) TYPE A3 WITH EXPANSION BEARING 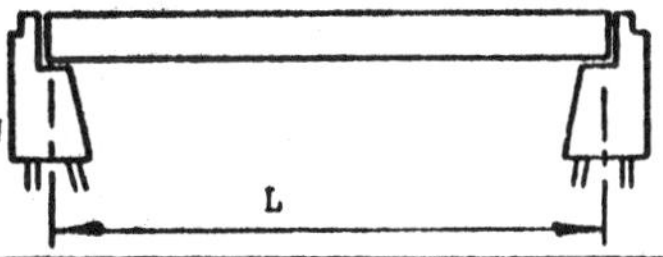 TYPE A3 WITH EXPANSION BEARING | (a.) L > 300'<br>with rigid Piers and S ≤ 30° | Exceeds above Criteria for (1) and (2) and (3) | L > 200' |
| (6) TYPE A4 WITH EXPANSION BEARING  TYPE A4 WITH EXPANSION BEARING | NOT USED | (b.) Based on Geometry and Economics | (c.) Based on Geometry and Economics |

**TABLE III Abutment Types** (WI DOT)

a. Consider the flexibility of the piers when choosing these abutment types. Only one expansion bearing is needed if the structure is capable of expanding easily in one direction. With rigid piers, symmetry is important for getting equal expansion movements and minimizing the forces on substructures.

b. For two span prestressed girder bridges, it may be necessary to investigate both sill type and semi-retaining type abutments. For example, semi-retaining abutments with 45 inch girders may be more economical than sill abutments with 54 inch girders. The sill abutment is usually more economical than a semi-retaining, if both are feasible for a given girder depth.

c. For two-span steel structures with long spans, the semi-retaining abutments are usually more economical than sill abutments due to the shorter bridge lengths.

therefore it is recommended that a roadway embankment be constructed before constructing the abutment and backfilling of the abutment should not be started until the superstructure is in place. Other problems associated with full height abutments are minimum horizontal clearance, sight distance, a collision safety hazard and footing settlements.

## 2. SEMI-STUB ABUTMENT (Semi-Retaining)

A semi-stub abutment is constructed somewhere between the top and bottom of an embankment and its height is between the height of full and stub abutment. Since this type of abutment is placed on the embankment, it has less chance of collision hazard and provide more horizontal clearance & sight distance.

**Fig. 3.5 Semi-stub Abutment**

It is generally used in a two (2) span bridge to eliminate an intermediate shoulder pier and stub abutment of a four (4) span bridge.

### 3. STUB ABUTMENT (Sill)

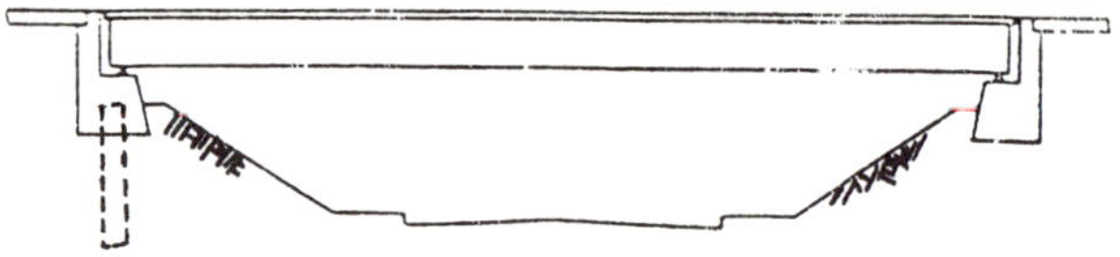

**Fig. 3-6 Stub Abutment**

A stub abutment is built at the top of embankment slope and is considered as the best type of abutment for avoiding rough and settled approach pavements. It can easily obtain adequate compaction behind the wall and thus the differential settlement of abutment and embankment under the approach roadway could be minimized.

Generally, a berm is built in front of the abutment wall to prevent settlement of the embankment by bulging up the slope in front of the abutment.

A stub abutment is economical but gives a longer span length and a higher superstructure cost. However, the overall structure would be less costly than other alternate and many structure have this type of abutment. The stub abutment has a problem of spillage of earth and water on the bearing seats and erosion of the slope of embankment.

The stub abutment has a height of 5 to 8 feet and it could be an integral part of the superstructure if constructed monolithically with a concrete deck slab, thus eliminating the problems at joints.

### 4. OPEN OR SPILL-THROUGH ABUTMENT

This is used where an additional span will be added at a later date. It is essentially a pier functioning as an abutment. It is important to provide provisions for easy removal of the wing walls by placing joints between them and the abutment without crossing the reinforcement through the joints. In case of a tensioned section, bolts or reinforcement with threaded inserts could be used for future removal.

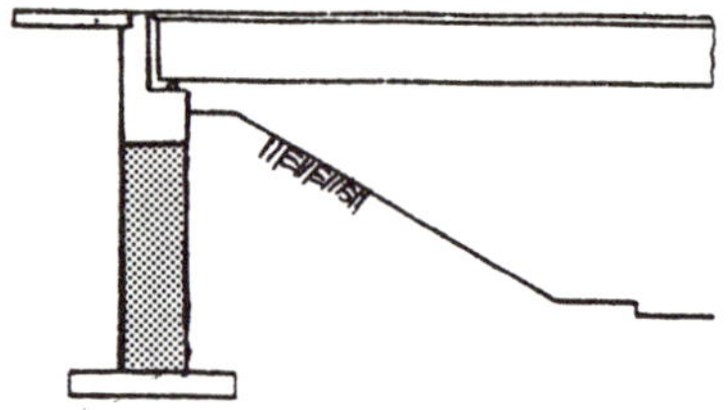

**Fig. 3.7 Spill-thru Abutment**

The main problems with this type of abutment are compaction of the embankment around the abutment, early settlement and erosion.

### 5. OTHER TYPES OF ABUTMENTS

**a. Open End Type Abutment**

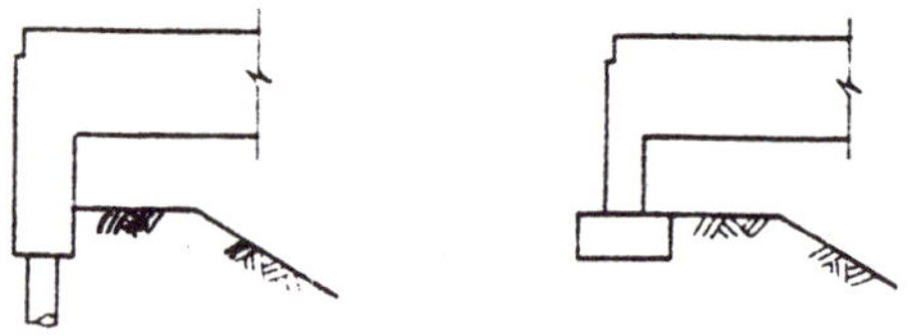

**Fig. 3.8 Diaphragm Abutment**

**Diaphragm abutment** is integral with the end of beams or girders and functions as a backwall and diaphragm. It eliminates expansion bearings and deck slab joints. However, It only allows limited thermal or shrinkage movement and it should not be for used for structures more than 300 feet in length.

**Pile bent abutment** consists of a backwall and a pile cap supported on the piles and it is used on timber structures.

**b. Closed End Vaulted or Cellular Abut.**

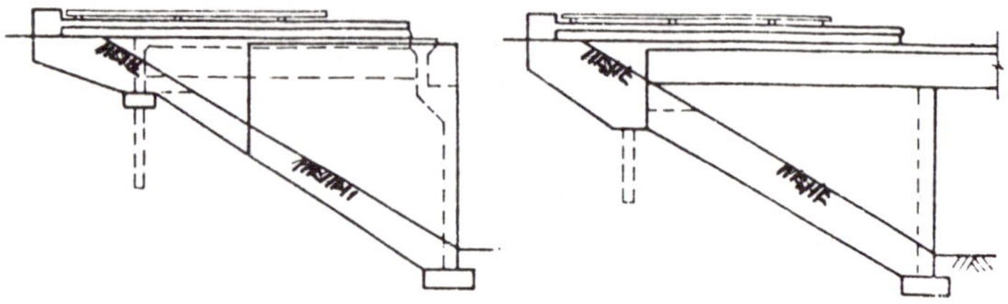

Bin Closure

**Fig. 3.9 Closed End and Cellular**

Another special abutment is the cellular or closed end vaulted abutment which is designed to reduce weight and pressure both vertically and horizontally by eliminating the approach embankment. This abutment is really a transition section between the bridge and the embankment. The cellular type is preferred over the backfilled type abutment because it will provide smoother approach to the bridge and usually is more economical than backfilled abutment.

### 3-3.3 SELECTION OF PIER

The piers generally in use for highways are column with cap, solid wall and hammerhead type piers. The column piers are predominant with 3 feet diameter columns with or without cap. This type is used generally for a pier height of less than 30 feet. The T-type pier or wall piers are normally used when the bridge is over a stream or railroad.

**Column Pier:**
The column pier which consists of a pier cap with columns supporting an individual or a continuous footing, is widely used on grade separation structures. On widely spaced girders, the cap could be omitted and the girders set directly on the columns. This has advantages for future widening of the structures, appearance, and cost. On longer spans, large rectangular columns have been used. The column pier for railroad crossing bridges requires a crashwall but if the pier is located more than 25 feet from the centerline of the adjacent tracks, a crashwall is not required for protection.

**Solid Wall Pier:**
A wall pier is used for most stream crossings to avoid collecting of debris and floating ices between columns. Solid wall piers are useful on structures with deep haunches and on stream crossings where a hammerhead pier cap is close to the water. A wall pier is also used on railroad bridges where high crashwalls are required.

**Hammerhead Pier:**
The hammerhead type T pier is generally cheaper than the solid wall pier because it requires less quantities of materials and a smaller cofferdam to construct it on stream crossings. On a hammerhead pier, the junction between the shaft and the bottom of the cap is preferably placed at a distance equal to the cap depth above the normal water elevations for balanced appearance. It is suggested that this junction should be a minimum of 2 feet above the high water elevations for removal of temporary sheetings used for the cofferdams unless an alternate method for the removal is used or removal of piles is not required. Also, it is suggested that for good appearance the height and width ratio of the wall and hammerhead piers be proportioned and the shaft or wall be tapered.

**Pile Bent:**
On small stream crossings, a pile bent may be used for heights up to 15 feet. Since ice and debris are hazardous to a pile bent pier, its use is limited in northern states. When a pile bent is used, lateral buckling should be considered in design and bracing, if required, should be provided. Timber piles are not recommended for pile bent piers.

**Pier Footing:**
Generally, the bottom of pier footing should be below the frost line and to a minimum of 3 feet below the finished grade unless it is placed on solid rock. The pier footing of stream crossing should be placed a minimum of 6 feet below the lowest point of the stream bed unless it is founded on a rock. If scour problems exist, a greater depth may be required based on the estimated scour depth. It is desirable to have a minimum of 2 feet cover below the stream bed for footings.

It is suggested that the bottom of footings adjacent to railroad tracks to be 4 feet below the finished grade.

If the pier footings are located about 15 feet or more below the normal water elevation, a seal may be used for pier footings.

### 3-3.4 SELECTION OF FOUNDATION

The selection of the proper foundation is depending on the materials at the site, interpretation of the field data, the results of laboratory tests, cost, and engineering judgement. Where surface soil is in poor condition and a large amount of exca-

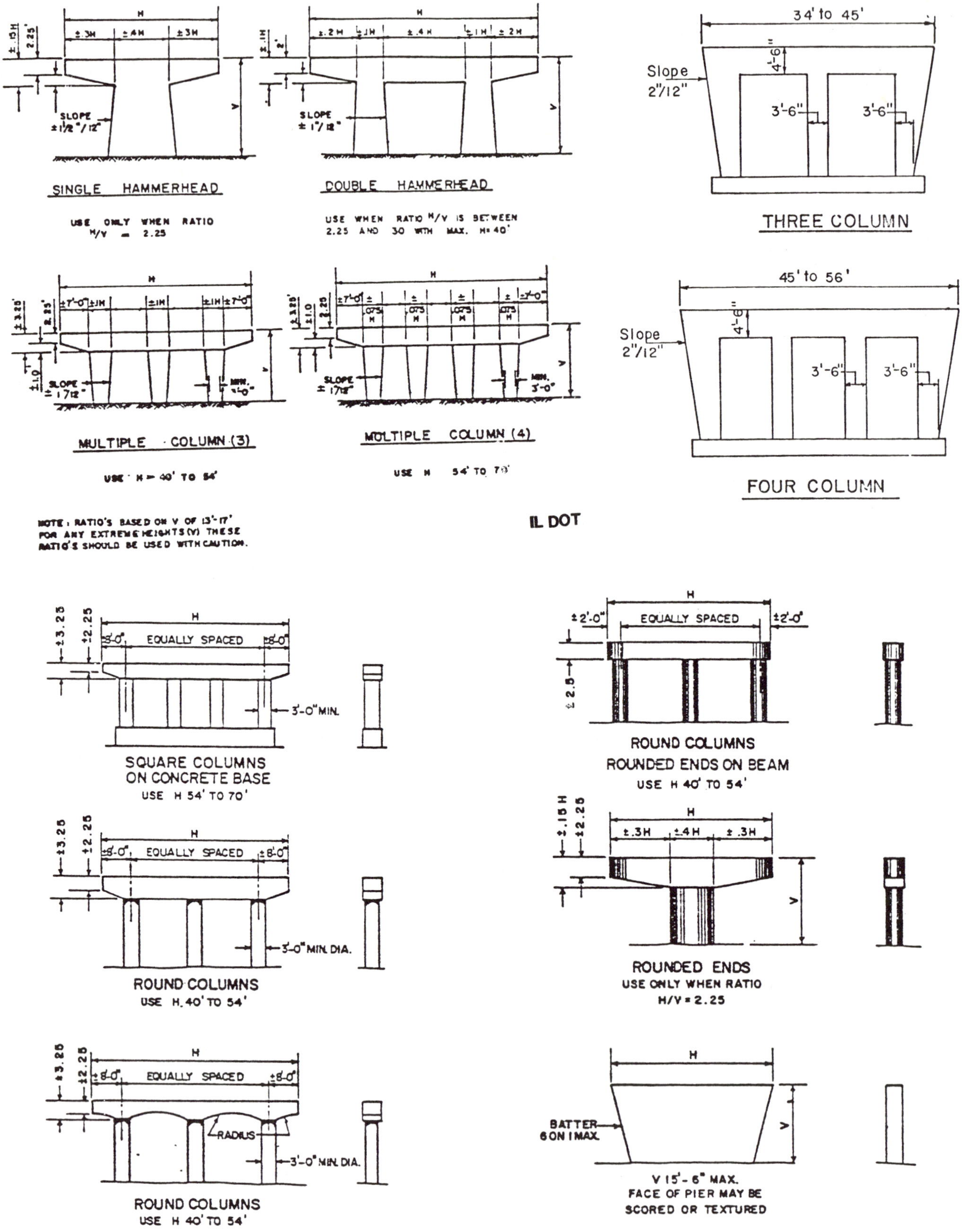


Fig. 3.10 Pier Types

## SOLID PIERS

Solid piers have been used primarily as river or stream piers but on occasion solid piers have been used on grade separations. Raised medians and cut sections have, in some cases, created low clearances at pier locations. Some types of solid piers which could be used are shown:

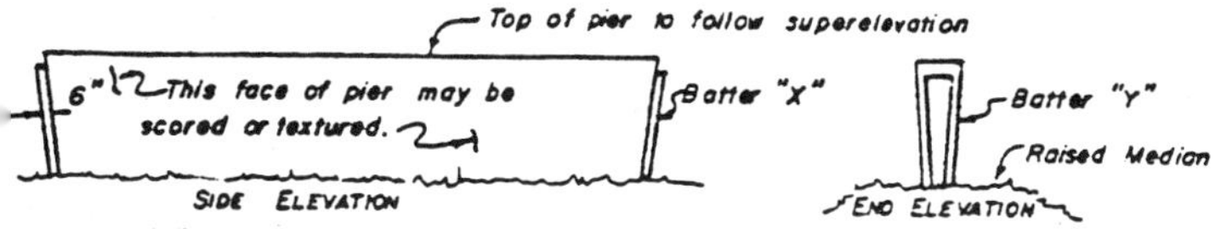


Batter "X" = 6 on 1 Max.
Batter "Y" = 12 on 1 Min.

The maximum height of this pier should be in the neighborhood of 9' above finished ground.

## TAPERED SIDE SOLID PIER

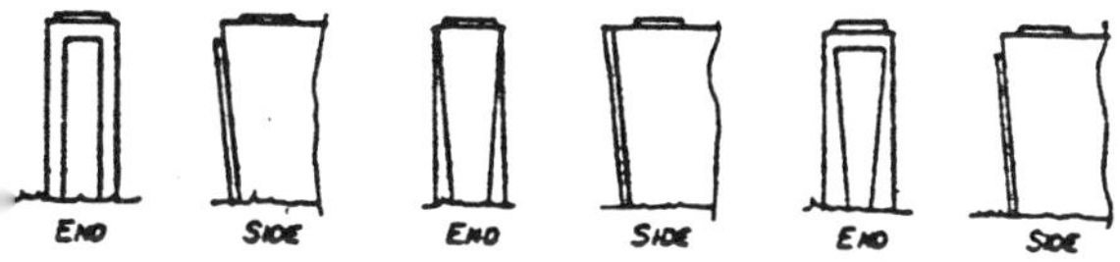


These piers may be as high as 15'-16' above finished ground if necessary.

## VERTICAL SIDE SOLID PIERS

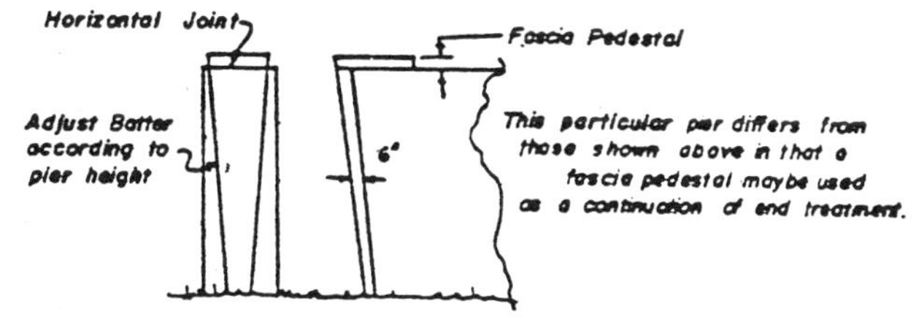


Variations of end treatments are only limited by the engineers imagination and good judgement.

## RIVER OR STREAM PIERS

In addition to the standard Bullnose Pier

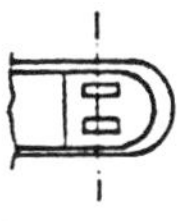

and Pointed End Pier

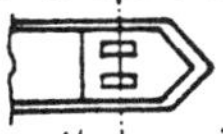

any of the piers indicated on the previous sheet can be used as waterway piers provided under clearance will permit a substantial leg above a plinth as shown here:

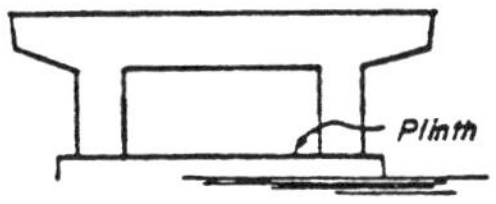


River piers should have thick stems if not completely solid from footing to bearings. Some suggestions are here:

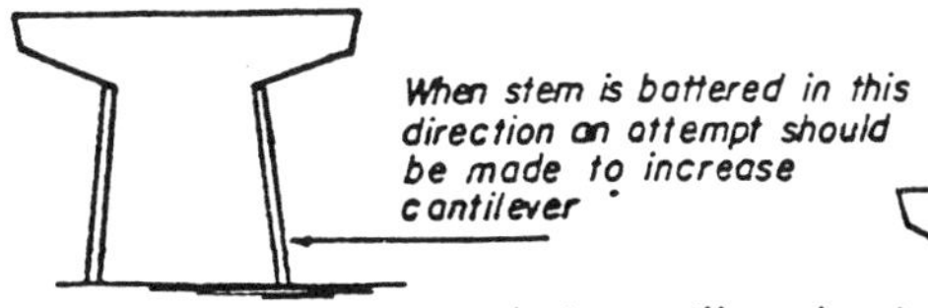


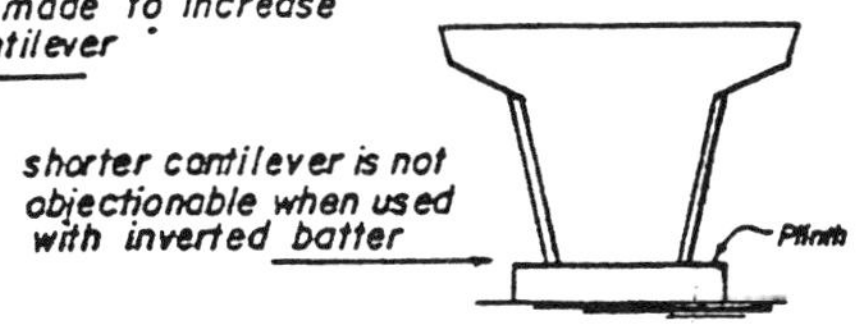


## VARIATIONS FOR TT PIERS

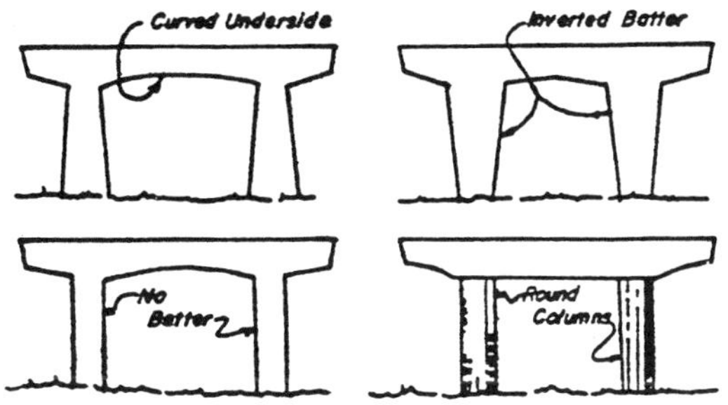


## HAMMERHEAD & TT PIERS

Note:
The shortest length of TT pier leg is approx. Cap. beam depth + 2' ±.

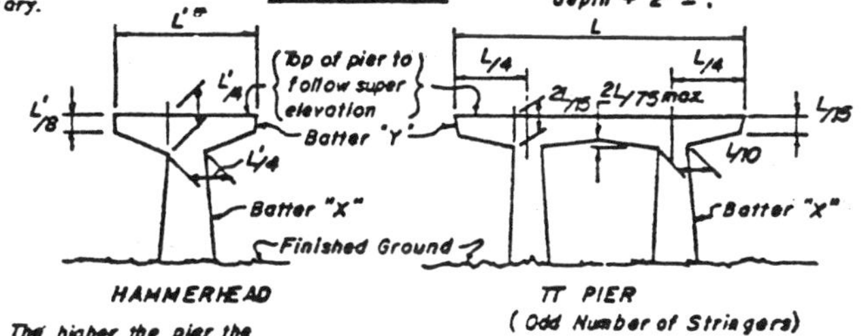


The higher the pier the longer L' may be and still look good.
Batter "X" - 12 on 1 to 60 on 1 depending upon height of leg.
Batter "Y" - 10 on 1 or nearest even dimension.

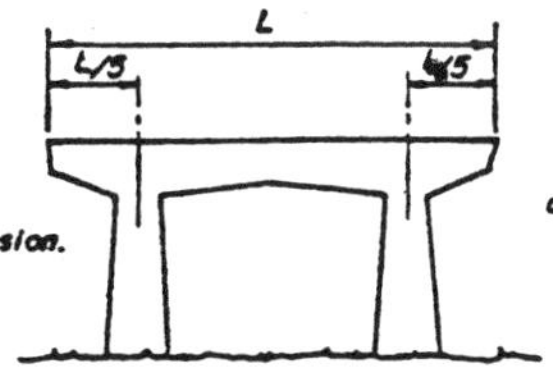


TT PIER
(Even Number of Stringers)

Proportions not indicated are similar to those above.

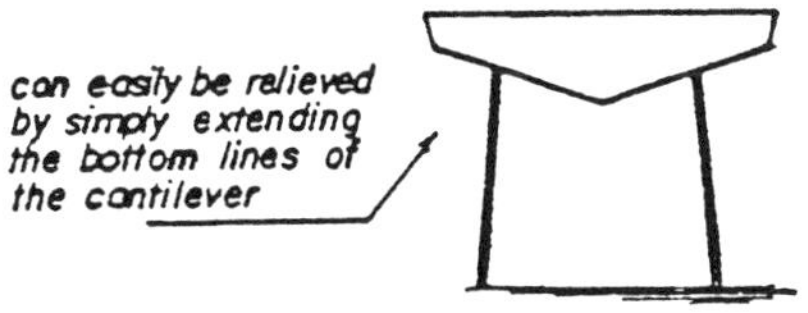

NY State DOT

Pier type - continued

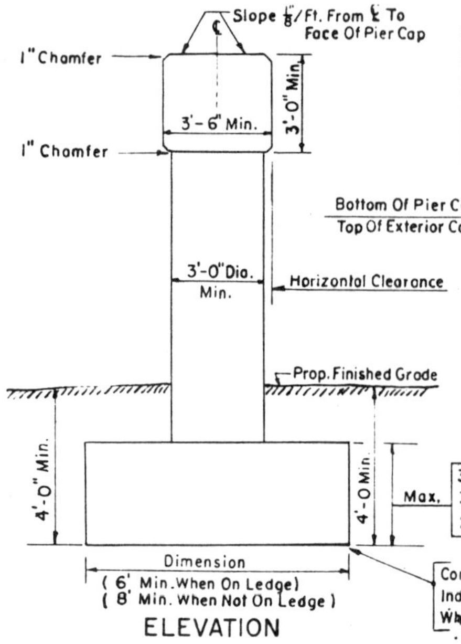


ELEVATION

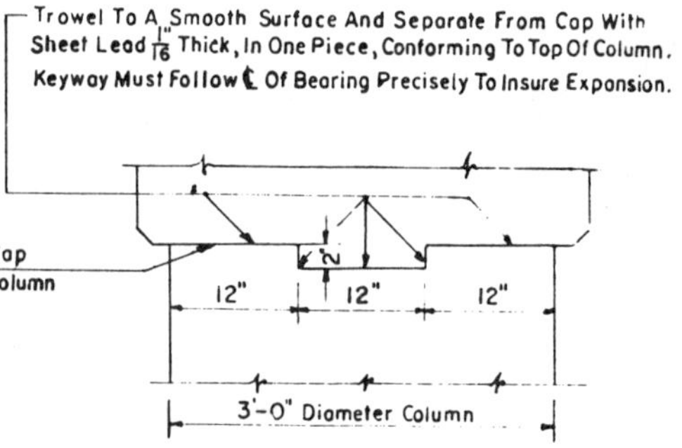


JOINT DETAIL AT EXTERIOR COLUMNS MORE THAN 80 FT. BETWEEN CENTERS

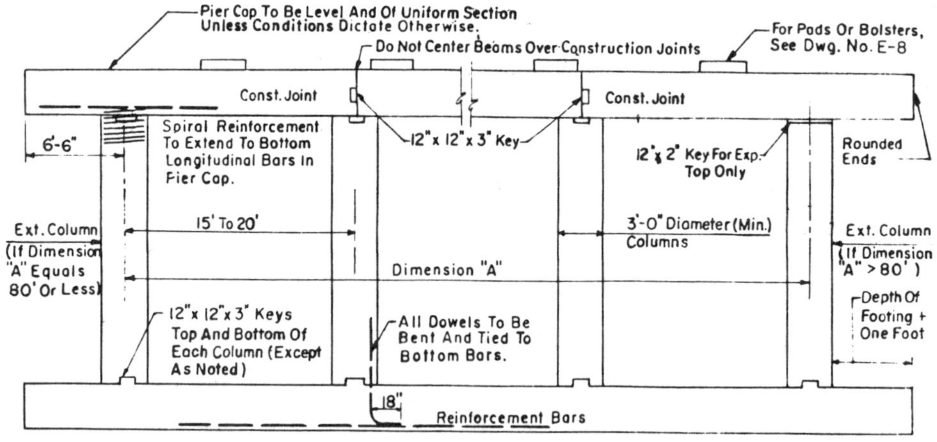


PIER SECTION

**Notes:**

If Dimension From ℄ To ℄ Of Exterior Columns Exceeds 80 Feet, Provide For Expansion Over Exterior Columns. See Detail Above.
All Keys To Be Slightly Tapered.
Columns To Be Designed In Accordance With Provisions Of A.A.S.H.O. Specifications, 1961 Edition And Addenda, Article 1.7.8 And Appendix C.

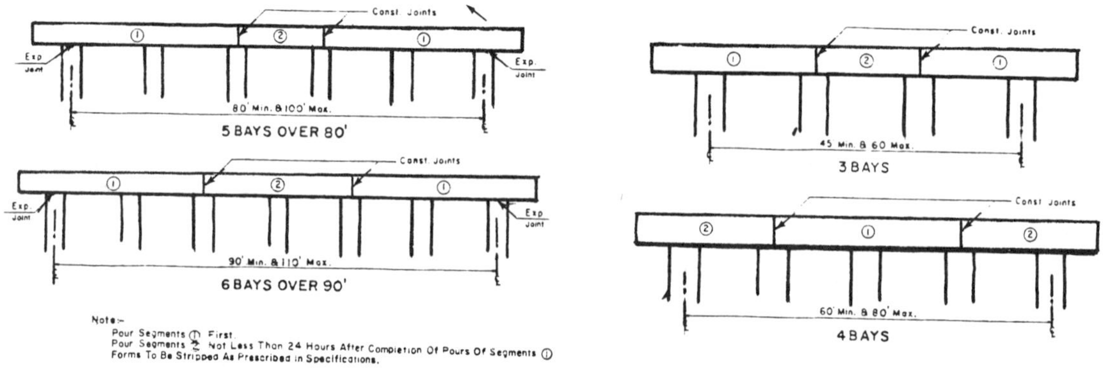


Note:–
Pour Segments ① First
Pour Segments ② Not Less Than 24 Hours After Completion Of Pours Of Segments ①
Forms To Be Stripped As Prescribed In Specifications.

POURING SEQUENCE OF PIER CAPS

Fig. 3.11 Pier Details

vation is needed to remove unsuitable materials for a spread footing, the foundation should be supported by piles or caissons. Also, when heavy scours are expected, piles should be used in place of a spread footing. It is recommended to use a spread footing where allowable soil bearing pressure is 2 tons per square feet or greater.

**Pile Foundation:**
The majority of abutments on new embankments are supported on piles to prevent their settlement. Most bridge approach embankments constructed of fill material usually undergo some settlement which may take several years. This may be due to changes in the embankment or the original foundation material under it. If piles are driven through the fill and well into the natural ground, the abutment usually does not settle with the embankment.

Generally, it is necessary to prebore non-displacement piles for any fill depth and displacement piles for fill depths over 10 feet to resist downdrag from a collapsing soil and to minimize uplift forces from expanding soils. Predrilling through the expansive soil may not be effective unless they are larger than the piles and non-cohesive material is used for filling around them. As an alternate, coating the pile surface with epoxy tar may be considered for bearing piles which penetrate thru expansive soil.

Battered piles may cause more problems than the vertical piles and they must be given special consideration. Fill embankments frequently shift laterally as well as vertically. A complete investigation of the foundation site and information on the fill material is a prerequisite for a successful pile design.

Piles in abutments are subject to lateral loads. The allowable lateral load on a pile is determined from an acceptable level of lateral displacement and not from the ultimate load that causes a stress failure in the pile. The allowable lateral load on a pile may be dependent more on the material that the pile is driven into than on the type of pile.

In the case of a pile bent type of substructure which is subject to large lateral forces, it is recommended that the piles be driven for a minimum penetration of 10 feet. In general, a minimum penetration of 10 feet in hard materials and one-third of the pile length or 20 feet in soft materials is suggested to sustain large bending forces.

The following are the some of the types of piles:

**Cast-in-place concrete pile** can be cased with steel shell (tapered, cylindrical and steel pipe) and the bottom of the casing is capped with a flat plate or is open. Open ended pipe piles are best suited for driving through the strata to bedrock.

**Precast concrete piles** may be of various cross-sectional shapes and can be prestressed. The precast prestressed concrete piles are gaining larger use because of moderate cost, easy driving, and durability. Previously the splice of precast piles was a problem but the use of the dowelled splice minimizes the difficulties.

**Steel H piles** are generally used where a hard layer of strata should be penetrated in order to rest on bedrock. The steel pile has a high capacity (9,000 psi on the tip) but is expensive.

**Timber piles** are inexpensive and easily obtainable but they have low capacity and susceptible to damages during driving.

**Spread Footing:**

Generally, spread footings are used where the original soil can sustain reasonable pressures without excessive settlement.

Spread footings can be used successfully on fill material with improved procedure and better control of embankment construction. It is important that construction be timed to permit the foundation material to consolidate before the spread footings are constructed. The advantage of spread footings is that the differential settlement between the approach fills and the abutments is minimized.

The use of spread footings is more favored for simple span bridges than for continuous spans because of their indifference for small amounts of settlement if they occur. But poor drainage for abutments on spread footings could cause problems.

**Continuous Footing:**
Continuous footings are used in pier frames of two or more columns when the use of isolated footings would result in a distance of less than 4'-6" between edges of adjacent footings. They are designed for the moments and shears produced by the frame action of the pier and the soil pressure under the footing. The soil pressure or pile load under the footing is assumed to be uniform. They are generally computed only from the vertical column loads and the soil and footing dead load. The moments at the base of the column are ignored for soil or pile loads.

To prevent unequal settlement, the continuous footing should be proportioned so that the soil pressures or pile loads are constant for Group 1 loading. The footing is kept relatively stiff between the columns to prevent upward deflections which cause high soil or pile loads under the columns.

**Seals:**
A seal is a mat of unreinforced concrete poured under water inside the sheet piling of a cofferdam. It is designed to withstand the hydrostatic pressure on its bottom when the water above it is removed. Dewatering of the cofferdam allows cutting of the piles, placement of reinforcing steel and pouring of the footing without the pressure of water. The seal also acts as a water stop to prevent water from entering through the underlying permeable soil.

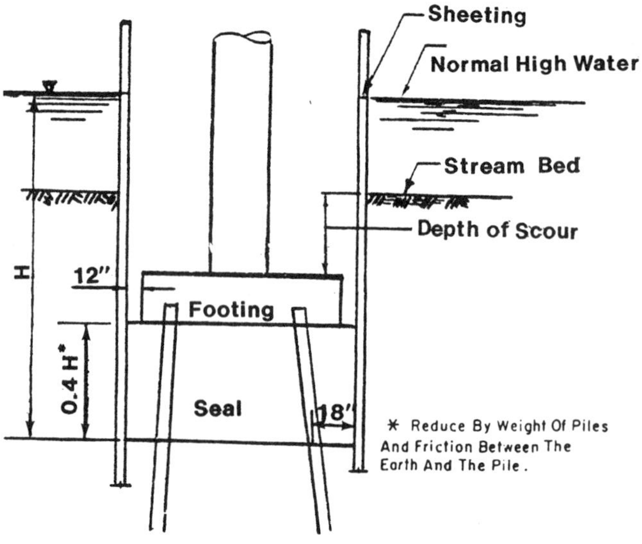


**Fig. 3.12 Cofferdam with Seal**

## 3-3.5 DECK DRAINAGE

Deck drainage system should not be used for a bridge of less than 500 feet and a catch basin should be provided at the approach pavement near the end of the bridge.

GUIDE FOR USE OF CATCH BASIN AT ENDS OF BRIDGE

| CASE | GRADE AT BRIDGE | RURAL PAVING (No curbs on approach pavement) | URBAN PAVIN (Curbs on appro pavement) |
|---|---|---|---|
| A | One end higher than other | Use Catch Basins at Low End Only | Use Catch Basins At Both Ends |
| B | Both ends lower than center | *Use Catch Basins At Both Ends | Use Catch Basins At Both Ends |
| C | Deck lower than ends | No Catch Basins Drains on Bridge Deck | Use Catch Basins At Both Ends Drains on Bridge Deck |
| D | Flat Grade | * Use Catch Basins At Both Ends Only When No Drains on Bridge Deck | Use Catch Basins At Both Ends |

*Check hydraulic runoff with Hydraulics Engineer to determine size and number of catch bas

**Fig. 3.13 Catch Basin Location** (MN DOT)

There are two reasons for this criteria, the principal reason is that the water draining off the bridge during construction leads to erosion and potential loss of the embankments and damage to the structure. Another reason is that several years after the construction of the bridge, seepage develops between the end of the bridge and the embankment if the inlet is not attached to the end. Also, the use of grate inlets attached to the ends of the structure simplifies the bridge design, the bridge construction and the bridge maintenance related to the drain.

However, deck drainage is required on some structures due to low points of sag vertical curve, flat grade, and superelevation cross over the point. It is suggested the drains be installed as close to the cross-over point as practical and as near to all joints practical to avoid gutter flow from passing over the joints. It is important to resolve the drain location in the preliminary plan phase because the drainage will influence the detailed structural design of the bridge.

The design of bridge deck drainage is generally based on the Rational Method and a storm frequency of 10 years with 5 minute durations. The average rainfall intensity (i) varies depend on the localities (average 6 inches per hour) and run-off coefficient (c) of 0.9 is used for concrete surface.

The maximum spread of the gutter width can be determined by the following conditions:

a. The maximum width of flow should be limited to the shoulder plus one third (1/3) of adjacent travel lane but the travel lane should have a minimum of 8 feet width clear of flow. This criteria varies from state to state. While one limits to the shoulder, other allows it to encroach one half (1/2) of the traffic lane.
b. The maximum flow depth is one-half (1/2) inch less than the curb height.

Free fall drain should be avoided from sections of the deck where the discharge would fall on the roadway, walkway, railroad, unprotected embankment and erodible area, and not be used in an urban area. The inlet should be placed directly above down spouts attached to the substructure. Mid span inlet locations should be avoided whenever possible because this requires a lengthy pipe and complex drainage system.

## 3-3.6 APPROACH SLAB

The reinforce concrete approach slabs are required to prevent the settlement of pavement adjacent to the bridges. In general, the minimum required length should be the height of abutment backwall plus 5 feet or the length of the U type wingwalls whichever is greater. The minimum required slab length is varies among states from 15 to 40 feet.

If the existing approach slab is in fair condition, this can be repaired by mud jacking, epoxy pressure grout and lift slab & grouting. If the slab is in poor condition, total replacement would be more economical. If the existing concrete pavement has no pressure relief joints adjacent to the bridges, new relief joint should be installed at the end of approach slab.

## 3-3.7 RAILROAD GRADE SEPARATION STRUCTURE

In designing railroad crossing structures, the following may be considered:

1. Structures carrying railroad tracks shall be designed according to the American Railway Engineering Association (AREA) Specifications and the modifications adopted by the railroad company involved.
2. The minimum horizontal clearance, approved by railroad, from the center line of the track to the face of abutment or pier shall be shown on the preliminary plans. If the track and the abutments or the piers are skewed relative to each other, minimum horizontal clearances to the structure should also be shown.
3. The elevation of the top of the rail at a point of minimum vertical clearance required by railroad and provided for each track shall be shown on the plans. If the track and the abutments or piers are skewed relative to each other, minimum vertical clearances to the structure should also be shown.
4. In order to provide for a drainage ditch parallel to the track, the elevation of the top of the footings adjacent to the track shall be at least 3'-6" below the elevation of the top of rail, unless rock is encountered.

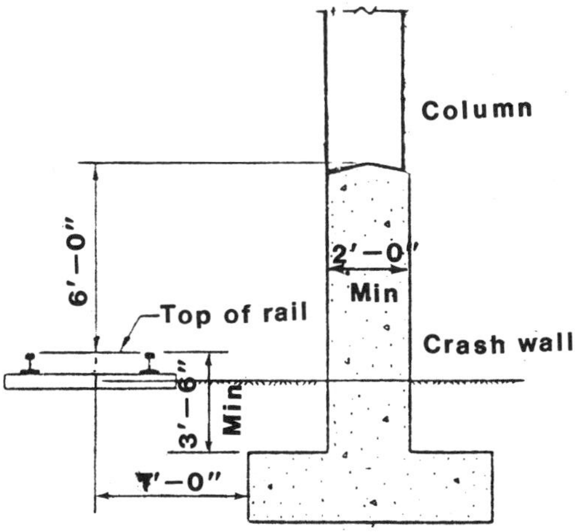


**Fig. 3.14 Pier adjacent to RR Tracks**

5. The edge of footing shall be at least 7'-0" from the centerline of the adjacent track.
6. Where piers are within 20 feet of the tracks, preference should be given to

solid wall piers or hammer head piers. If column piers are used within 20 feet of the tracks, columns should be protected by crash walls, at least 2'-0" thick and shall extend 6 feet above the top of rail and beyond the outside face of the outside columns. The crash wall should rest on the column footings and should be flush with the face of the columns adjacent to traffic, and should connect all columns in a pier bent. A minimum reinforcement consisting of horizontal bars of #6 at 12", each face and vertical loops of #4 at 12", is recommended. Only 50% of effects of shrinkage and temperature in the design of pier bents whose columns are founded on individual footings shall be considered.

7. Bridge scuppers shall not drain onto railroad tracks. Provisions shall be made to direct surface water from the bridge area into an adequate drainage facility along the railroad siding, in which case drainage approval by the Railroad Company is required prior to submitting final plans.
8. Safety provisions required during excavation in the close vicinity of railroad tracks and substructures shall be shown on the drawings.
9. Sheet piling used during excavation for protection of railroad tracks and substructures shall be designed according to AREA Specifications, and subject to approval by the railroad company.
10. Complete details of temporary track(s) or temporary bridge shall be shown on the design drawings.
11. For electrified railroads, special study should be made for relocation of catenary structures, line works, de-energizing during constructions, and safety. Even after a close consultation with the railroad company, obtaining the approval of plans from railroad company would take a long time. Therefore it is suggested to contact them in the stage of earlier development and use their standard drawings wherever applicable.

    If a railroad is electrified, it shall be so noted on the preliminary plans and protective barrier shall be provided on spans or on part of spans for structures over electified railroads as directed by the railroad company. Generally, the protective barrier shall extend at least 10 feet beyond the point at which any railroad wire passes under the bridge. However, in no case shall the end of the protective barrier be less than 10 feet from the wire measured in a horizontal plane and normal to the wire outside of the limit of the bridge, and less than 6 feet from the wire within the limit of the bridge.

    In the case of bridges cross over electrified railroad tracks, the details of catenary attachments and their locations, if attached or pertinent to the structure, shall be shown on the plans. Normally ground cables from structures to the ground are installed by the railroad crew. However, fastenings required to attach cables to the structure shall be shown on the drawings as requested by the railroad.
12. All open or expansion joints in parapets, dividers, sidewalks, and curbs within the limits of the barrier shall be covered or closed with joint materials. Details of such joints shall be shown on the plans.

## 3—4 PERMITS

Obtaining permits is a time consuming effort in relation to the overall project design schedule. Documents necessary for filing of permits should be prepared soon after the study report of the project have been approved and all matters relating to environmental assessments and/or impact statements have been resolved.

Some preliminary work on selection of the structure type should also be done at this time in order to preclude revisions and resubmission of the application for a permit.

### 1. U.S. Coast Guard Permit

If a navigable water or waterway limits of tidal influence is involved, the U.S. Coast Guard Permit is required for crossings, relocations or encroachment of the waterway. This permit application must describe and include a location map showing the greatest flood on record, the depth of channel, any proposed or existing obstruction, and the proposed vertical and horizontal clearance of the new structure. Environmental documents (EA or EIS) and a Water Quality Certificate must accompany the permit application.

**2. Corp of Engineer, Section 404 and 10 Permit**

The U.S. Army Corps of Engineers has responsibility to regulate the disposal of dredge or fill materials in all waters of the U.S. and/or adjacent of wetland by the Clean Water Act (formerly known as the Federal Water Pollution Control Act) Section 404 and to authorize certain structure or work in or affecting navigational waters of the United States pursuant to Section 10 of River and Harbor Act of 1899. The following are some of the activities requiring permits: artificial canals, breakwater, bulkheads, dams, dikes, weirs, discharging sand, stone, dolphins, dredging, filling riprap, road fills, piers and wharves, etc.

The Corps of Engineers permits will be either individual or general. An individual permit is the standard permit which is required to be processed through the public interest review procedures including public notice and receipt of comments. General permits are either regional or nationwide.

Regional permits may be issued for certain categories of structure or for work including discharges of dredge or fill material. If a regional permit has been issued, individual activities falling within those categories do not have to be further authorized unless the district engineer requires an individual review. Nationwide permits are issued for certain specific activities through out the nation. If certain conditions are met, the specific activities can be done without the need for individual or regional permit.

The interim rules issued on July 22, 1982 expanded their criteria for Section 404 and 10 Nationwide Permit authorization to include Categorical Exclusion. Basically, if a FHWA funded project is determined to be a Categorical Exclusion and the Army Corps of Engineers concurs with the determination, the project will not require an individual Section 404 or 10 Permit. Nation wide permits also allow bridge associated fills less than 200 cubic yards discharge below high-water and discharge of dredge or fill materials into water that do not exceed ten cubic yards as part of single and complete project provided no material is placed in wetland.

It is recommended to contact the district engineer's office of the Corps of Engineer and request their determination if an individual permit is required.

**3. State and Local Permit**

**a. Stream Encroachments Permits** are required for constructing new structures over, under or along existing streams, also when relocating or otherwise altering the existing water courses in a state. Channel relocation and major fill projects require EIS.

**b. Water Quality Certification** is required for construction of any facility which results in a discharge into a water body. It is a necessary prerequisite for Navigation permits, Wetland permit, and other state permits. When Federal navigation permit or the U.S. Army Corps of Engineers Permit is required, the necessary documents for the water quality permit should be prepared.

**c. Wetland Permits** are required for excavation, dredge, fill or construction which encroaches on the land designated as "Wetlands" under the National Wetland Inventory or state agencies (State Office of Coastal Zone Management, State Office of Natural Resource, etc.). Riparian and Corps of Engineers permits will be required.

After identifying the number, extent, and distribution of wetlands by consulting with the U.S. Fish and Wildlife Service, the U.S. Corp of Engineers, and State Agencies, the following general measure should be considered:

1. **Avoidance:**
   Avoiding project within the Wetlands, where the density of it and its related resources have a great impact on environment.
2. **Minimization:**
   Limiting the degree, extent or magnitude of impacts on the Wetlands by considering the following alternatives: Relocating the bridge beyond the Wetlands, improving existing bridges with

minimum disturbance of the Wetlands, and no action alternative.

3. **Rehabilitation**
Rectifying impacts or damage by repairing or restoring the Wetlands after construction.

**d. Riparian Permits** are required for any construction undertaking below the mean highwater line within any tidal or navigable waterway. Applicant must have legal document establishing right to use or occupy riparian land (grant or license).

**e. Coastal Area Facility Review Act (CAFRA) Permit** are required for major transportation facilities within designated "Coastal Zone".

## 4. Federal Aviation Administration

If a project is within two miles of an airport, an Airport Sketch Map should be prepared for FHWA and FAA approval.

**REFERENCE:**


1. AASHTO:
"Standard Specifications for Highway Bridges, 1977" and current interims.
"AASHTO Manual for Bridge Maintenance, 1976"
"Geometric Design Standards for Highway Other Than Freeway, 1969"
"Geometric Design Guide for Resurfacing, Restoration, and Rehabilitation (R-R-R) of Highways and Streets, 1977"
"A Guide for Highway Landscape and Environmental Design, 1970"
(AASHTO: 444N Capital St., Washington, D.C. 20001)
2. Calif. Dept. of Transportation: "Bridge Design Manual Vol. 1,2,& 3" "Project Development Procedure"
3. Conn. Dept. of Transportation: "Bridge Design Manual"
4. General Accounting Office:
"Excessive Truck Weight" CD-ED-79-94 Washington D.C.: GAO, July 1979.
5. Joseph Gibeily, George A. Harper, "Findings, Ideas, Conclusions and Recommendations Developed from Bridge Design Review-in-depth Program" Washington, D.C.: FHWA Bridge Division, March, 1973
6. Illinouis Dept. of Transportation:
"Bridge Manual"
7. Maine Dept. of Transportation:
"Bridge Design Manual"
8. Mass. Dept. of Public Works: "Bridge Manual"
9. Minesota Dept. of Transportation
"Bridge Design Manual",
"Bridge Details",
"Bridge Standard Plans Manual"
10. New Mexico Highway Department:
"Bridge Manual"
11. N.Y. State Dept. of Transportation
"Standard Details for Highway Bridges" (Accounting and Fiscal Service Bureau, NY State DOT, 1220 Washington Ave. State Campus, Albany, NY 12232)
12. Ohio Dept. of Transportation
"Ohio Design Regulations"
"Ohio Supplement to AASHTO"
(Bridge and Structure Design, 25 South Front St. P.O. Box 899, Columbus, OH 43216)
11. Penn. Dept. of Transportation
"Design Manual Part 4 - Structural Design" (Publication and Sales, P.O. Box 134 Middletown, PA 17057)
12. Underwood, Tom "Preliminary Design of Highway and Railroad Structures" Columbus Ohio, Columbus Engineering Ltd., 1966
13. Washington State Dept. of Transportation, "Bridge Design Manual", Bureau of Bridge Design, Highway Administration Bldg. Olympia, Washington, 98504
14. Wisconsin Dept. of Transportation, "Bridge Design Manual" Bridge Development Engineer, 4802 Sheboygan Ave. P.O. Box 7916, Madison, WI 53707
15. Branson, D.E. "Deformation of Concrete Structures" McGraw-Hill, NY, NY 1977
16. American Concrete Institute
"Manual of Concrete Practice - Bridge Analysis and Design" (American Concrete Institute, P.O. Box 19150, Redford Station, Detroit, MI 48219)
17. Florida Dept. of Transportation
"Manual for Bridge Maintenance Planning and Repair Methods"
18. New Jersey Department of Transportation
"Bridge Design Manual, 1983"
19. AREA "Manual of Railway Engineering" (American Railway Engineering Association, 2000 L St., N.W. Washington, D.C. 20036)
20. Transportation Research Board
"Bridge Engineering Volume 1 & 2" TRB Record 664 & 665 (TRB: 2101 Constitution Ave. N.W. Washington, D.C. 20418)

# 4 DECK RECONSTRUCTION

## 4—1 GENERAL

The deterioration of concrete bridge decks along the nation's highways is a major problem for most states. Bridge deck deterioration is generally caused by freezing and thawing cycles and the corrosion of the reinforcement steel due to the use of chloride containing deicing chemicals. With repeated application, deicing salt and moisture eventually reach the reinforcement steel and an electrical potential difference occurs in the steel, causing corrosion. Since the rusted steel occupies more space, pressure is exerted on the concrete by the corroded reinforcing steel and this eventually leads to spalls.

Bridge deck problems are getting worse every year becuase of the lack of available funds to provide necessary preventive maintenance. Most states use available funds first on deck replacement project and defer installing protective systems on the bridges requiring minor repairs. As a result, those bridges with minor defects continue to deteriorate and eventually require complete deck replacements at a much higher cost than that of a protective system. The U.S. General Accounting Office estimated that the potential benefit-cost ratio of repairing and protecting an existing bridge deck of minor defects versus a total deck replacement at a later date is 2.75 to 1.00.

Various methods and materials have been used to prevent corrosion in new decks and to repair deteriorated concrete decks. The following are alternate solutions for deck protection:

1. Methods to Keep Chlorides Out of Deck
   a. Surface protective sealer and coating
   b. Waterproofing membrane or Epoxy asphalt
   c. Low permeability concrete
      High density low slump conventional concrete
      Latex-modified concrete
      Polymer concrete
      Internally sealed concrete
      Polymer-impregnated concrete
2. Coated Reinforcement Steel
   a. Epoxy-coated reinforcing bar
   b. Galvanized reinforcing bar
3. Cathodic Protection
   a. Impressed current system
   b. Sacrificial system
4. Other Methods
   a. Electrochemical chloride removal
   b. Corrosion inhibiting admixture
   c. Deeper polymer impregnation
   d. Steel grid deck

The following table shows general approaches currently considered for deck rehabilitation:

**TABLE I DECK ALTERNATES**

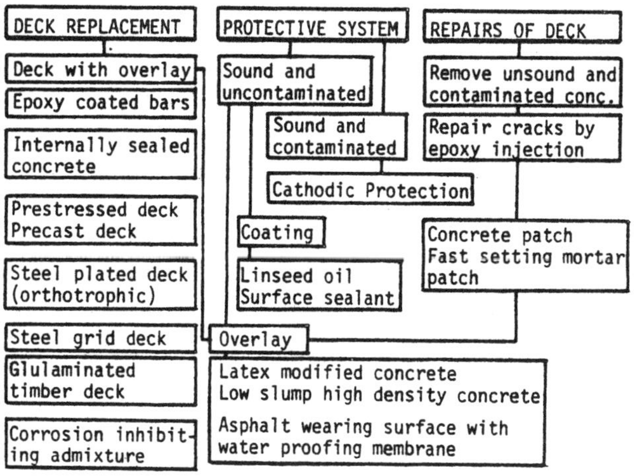


Among these approaches shown in the table, some have been successful in the field and others are still in the experimental research state. Each approach has its pros and cons and some methods are only applicable to new concrete decks while others are only applicable to the existing and uncontaminated deck or deteriorated and contaminated deck slab.

Early research indicated that the only way to be assured of the permanent repair of a deck was the removal of all concrete in potentially detrimental condition which is above the corrosion threshold of 2 pounds of chloride per cubic yard of concrete and has a electric potential of over .35 volts (CSE) and then preventing future de-icing salts from reaching the reinforcement steel. The removal of all chloride contaminated concrete bridge decks in the Nation may not be feasible due to high costs. Instead of removal of chlorides in concrete, the retardation of corrosion action in steel and neutralization of the chloride effects appear to be more cost effective approaches. Some of these techniques requires surface treatment with sealers, asphalt waterproofing membranes, low slump high density concrete or latex-modified concrete overlay, cathodic protection, internally sealed concrete or polymer impregnated concrete.

In 1983, FHWA revised the rules (23 CFR 650 Subpart F) of concrete bridge deck to allow states the flexibility to select a protective system and reconstruction method to protect a bridge deck from corrosion based on local conditions and experiences subject to approval by the FHWA. The reconstruction method chosen by the state will be considered acceptable and permanent. If the performance and cost effectiveness of a chosen protective system can not be adequately demonstrated to the FHWA, approval can be subjected to an experimental condition.

Pending development of these techniques, the Federal Highway Administration authorized an experience procedure to protect contaminated bridge decks by installing one of the waterproofing membrane or other protective overlay systems which are normally used for new concrete decks. The 1983 FHWA revision eliminated the strict provision requiring the removal of all contaminated bridge deck elements for permanent reconstruction, as well as the provisions for experimental reconstruction procedure. Therefore certain protective overlay system over contaminated concrete decks can be demonstrated to be effective and considered as permanent and acceptable methods. In general, the placement of waterproofing membrane over the chloride contaminated concrete deck with corroding reinforcing steel would not stop corrosion and would not reduce the rate of corrosion. But it would provide structural repairs and a smooth riding surface. It is better to remove all contaminated concrete before placing an overlay system.

## 4—2 CONCRETE DECK REPAIRS

The first step of deck repairs is to determine whether the areas showing deterioration need rehablitation or whether a total deck replacement is required. This decision normally should be made during the study phase by considering the location and importance of the structure, nature and extent of the deterioration, remaining service life of the deck, cost, traffic volume, duration of construction, future improvement, and available funds.

The second step is to decide how to repair the deck and these will fall into the following categories:

1. Preparation of the surface to be repaired
2. Patching spalls and cracks
3. Epoxy injection into delaminated areas
4. Overlaying with low slump concrete, latex modified concrete or polymer concrete
5. Sealing the deck with waterproofing membrane

### 4-2.1 PATCHING

The patch repair of a deteriorated concrete deck should be made prior to installation of a protective system and may be used as an interim repair until a permanent rehabilitation is made. If the deterioration is extensive, a total deck replacement is more economical than patching because the removal of deteriorated concrete and surface preparation will require a considerable amount of monies and time consuming hand labor while an entire deck removal could be made by a faster and economical mechanical method.

Repair material should be compatible with the existing concrete so that the new and old materials could react in the same manner to changes in temperature and loads. Most temporary repairs performed by the maintenance forces, where rapid restoration of riding surface is needed, · is a simple filling of spalls with asphalt because asphalt is relatively low in cost and easy to place. However, many asphalt patches do not last long and they actually accelerate deterioration of the surrounding concrete by retaining water under the patches. Even a concrete patch would last only a few years and more spalls will occur adjacent to the patch area. The patch should be used to restore the riding surface of the deck until a more permanent rehabilitation is undertaken. Where and when a more durable riding surface is required, permanent repairs should be considered.

To patch a spall in a bridge deck, the salt contaminated concrete which has greater than 2 pounds of chloride content per cubic yard of concrete, must be removed from the area around and beyond the immediate spalls. Without removal of the deteriorated concrete and cleaning of the corroded reinforcement steel, the new concrete patch will create a different corrosion environment cell. The repaired area will take the position of a strong cathode and a large potential difference will result in a current flow and accelerate deterioration of the existing concrete deck slab. Beside the removal of contaminated concrete, the concrete surface should be strong and sound. Also, the delaminated deck should be bonded before patching. It is recommended that an epoxy coating or cement grout be applied to the surface of the cleaned area before placing the patch mortar such as low slump concrete, epoxy resin mortar, polymer concrete or latex modified concrete.

**1. Preparing the Repair Area**

The success of a repair requires careful removal of contaminated concrete and a proper preparation of the existing concrete. All unsound, deteriorated and chloride contaminated concrete must be removed to provide durable repairs. Final determination of the deteriorated area to be removed shall be made as work progresses, first by removing the suspected area, then continuing around the perimeter, enlarging the area until a sound area is encountered by visual inspection or by other test methods.

Since the use of heavy equipment could cause fractures and possible damage to the surrounding concrete, the maximum pneumatic hammer weight should be limited to 30 pounds and the chipping hammer should have a chiesel point. For the removal of a shallow depth of concrete, scabbler scarifier, planer or rotahammer may be more effective than air chipping hammers. The top edge of the repair area should be sharp and undercut to the edge of sound concrete to eliminate feather edges and to provide a keyed patch. This can be achieved by slightly tilting the saw blade to undercut the edge of the repair area at a slight angle. One to 1 1/2 inches deep saw cutting around the perimeter of repair area is preferred but hand chipping would also give a satisfactory patch edge. Chipping should be directed from inside of the sawcut area toward the sawcut to prevent damaging cutface. If the remaining depth of the sound concrete is less than 50% of the original depth, removal of the full depth of concrete slab is recommended.

It is also recommended that the repair areas should be wetted after removal of a major portion of concrete to detect any cracks and breaks in the remaining concrete. A spray of moisture applied on the dry concrete surface will evaporate quickly, while the moisture in cracks would remain for a longer period.

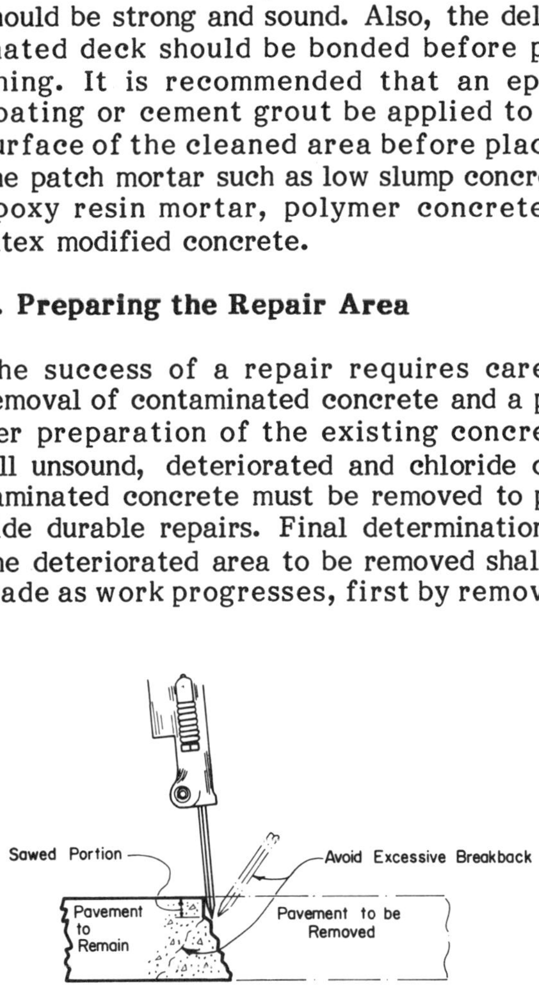


**Fig. 4.1 Concrete Removal**

**Fig. 4.2 Surface Preparation**

The cracks can be repaired by the following methods: 1) If the concrete is sound on both side of the crack, the crack should be opened to a width of 1" to 2" and minimum 1/2" (2" preferred) depth by chipping, then the area should be filled with grout after cleaning and priming or the crack can be repaired by epoxy pressure injection method. 2) If the concrete is deteriorated, the normal patching method of sawcutting, removal of unsound concrete and patching should be used for the entire area. After removal of deteriorated concrete, the repair area should be thoroughly cleaned with a high pressure water or vacuum cleaner. An air compressor equipped with an oil and moisture trap may be used for sand blasting and all existing concrete surface bonding to the new patch should be cleaned before applying the epoxy bonding compound or cement mortar.

## 2. Existing Reinforcement Steel

If the delamination or spalls extend to the top reinforcing bars or the concrete around the reinforcing bars are loose, concrete should be removed to a minimum of 1 inch below and around reinforcing bars to assure proper encasement and bonding. Also, loose concrete, rust and scale around bars should also be removed by sand blasting and where the reinforcing bars are corroded or weakened, the undamaged and clean reinforcing bars should be exposed to a sufficient length (min. 12 inches) from the sound concrete to provide an adequate overlap splice.

**Fig. 4.3 Existing Reinforcing Steel** to be remain for splice overlapping with new bars in deck widening.

The replacement reinforcement should have a sufficient lap splice with the existing bars by overlapping (generally 30 bar diameter for grade 40 steel bars) or a mechanical splice. A welding method could be used to conform to the Structural Welding Code (AWS D 1.4). However, welding to the existing bar technique is not preferred by some states due to additional cost of welding and the weldability of some of the old reinforcing steel. Many states use uncoated reinforcing bars in place of the existing bars to accomplish the same purpose. It is suggested to provide a minimum 2 inch clearance between end of new bars and existing concrete face to allow possible movement.

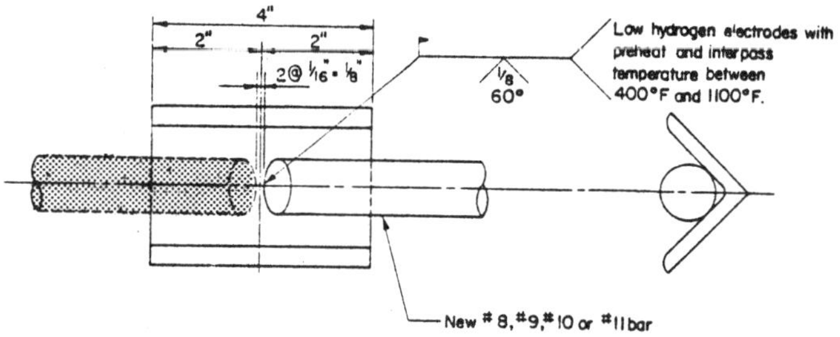


**Fig. 4.4 Welding Details for Existing Bars**

Uncoated reinforcement steel is generally used as replacement bars for repair areas. But for a complete replacement of a joint header or total section replacement, consideration should be given to place epoxy bars at the top and side of the header where concrete is exposed. Galvanized reinforcing bars are not recommended to be used as replacement steel with uncoated plain steel because the zinc on the new bars would become an anode and sacrificed to the existing uncoated bars by electrical potential differences.

## 3. Patching Material

There are wide range of patching materials available and can be categorized into the following:

- Portland cement
- Chemical-setting cement
- Thermosetting cement
- Calcium sulfate cement
- Bituminous material

For fast setting materials, there are many different products and their properties varies substantially from brand to brand so care should be exercised in selecting proper products for particular repairs.

**Portland Cement Concrete or Mortar:**
The commonly used patching material is Type I, II, or III portland cement with or without an admixture. The most widely used admixture or additive is an accelerator, retarder, air-entraining agent to control durability, scaling, set time and workability; but the accelerator admixture should not have more than 1 percent chloride ion by weight. Calcium chloride can be used as an accelerator but it should be added to ready-mix concrete at the site when temperature is above 70°F. Superplasticizer may be used to gain strength and to provide slump for workability (maximum 7" after adding). Higher slump and excessive plasticizer will cause segregation. If it is used, mix design should be changed to 2" slump and 8% air entrainment. Also, plasticizer should be only added at the site.

Portland cement Type III (High early strength) has been used widely for a long time for patch work because it is relatively inexpensive, simple to use and durable. But it has a high shrinkage, a tendency to crack in patch and a low rate of strength gain in cold weather therefore, caution should be exercised if high early strength cement is specified.

As a patch material, concrete with lower water-cement ratio is preferred. It is suggested to use a water-cement ratio of lower than 0.45 (5 gallons of water per bag of cement, generally one gallon of water is 8.34 pounds at 70°F and one sack of cement weighs 94 pounds) for moderate environments and less than 0.40 at severe environments.

**Low Slump High Density Concrete:**
This concrete developed by Iowa Department of Transportation could be used for patch work but workability could be a problem in the field. The water cement ratio should be about 0.32 and the slump of mix should not exceed 1 inch. Cement paster or grout is generally applied before placing the concrete and the repair area is frequently prewetted to improve bond. For low slump concrete, the surface should be completely dried before applying a bonding coat of epoxy bonding compound or latex modified concrete grout.

**High-Alumina Cement:**
This type of cement has been used in Europe and is recommended to use where corrosion is a problem. Fondu Calcium Aluminate, Lumnite, and other are available brands.

**Magnesia-Phosphate Concrete:**
This premixed cement generally come with dry magnesia component and liquid phosphate package. It appears to perform satisfactory for patch works. There are a few brands names such as Acmaset, Bostik 276, Fast-Crete 100, Horn 240, Set 45, FX 90 and others.

**Latex Modified Concrete:**
This is a mixture of concrete with latex emulsion admixture. It has not been used widely as portland cement mortar and epoxy mortars. However, it is widely used in conjunction with deck overlay systems.

**Epoxy Resin Mortar:**
Epoxy resin concrete has been widely used in the last several years and lately polymer concrete has been introduced. These materials have given satisfactory results for repairing spalls, popouts and partial depth repairs. But it may fail due to low modulus of elasticity. Also, preparing adequate specifications for needs and meeting requirements has been a problem because many different types of products are available for a specific purpose. Choosing the type to satisfy the job condition and requirement would be a difficult task for engineers. Since a wide variety of commercial epoxy patch materials are available (such as Cono /Crete, Colma Dur, FX 700, Epoxy, etc.), it is recommended that a technical representative of the manufacturer should be at the site during the entire operation from mixing to curing in order to provide guidance. Their service will outweigh the additional cost by resulting in satisfactory patches.

Epoxy resin mortar is generally used for small and shallow patch areas while portland cement mortar is used for larger areas. The cost of epoxy mortar is substantially higher than portland cement but the rapid hardening of the concrete would make it possible to reduce the required curing time, thus the bridge can be open to traffic earlier. This will save user's operational cost and reduce the motorists exposure to the hazard conditions during the construction period.

**Polymer Concrete:**
Polymer concrete is a premixed aggregate with methyl methacrylate (MMA) momomer that is later polymerized in place. It is fast curing and has great strength and durability. Most prepacked products come in two components, a monomer liquid and a fine powder, and course aggregate can be added to the mix. They should be mixed in the field just before use by hand or a rotating drum mixer. The main advantage of polymer concrete is the minimizing of inconvenience to the public and to reduce the traffic maintenance cost. Thus, it is suitable for use in deck repairs where the traffic is heavy and the working hours are limited to off-rush hours or night. There are a few brands such as Concresive 2020 Polymer, Crycon, FX 826, Duracryl, Silikal, and others.

**4. Superplasticizer (Super-water reducer)**

A superplasticizer is a synthetic product which reduces friction between the fine particles in the concrete mix by what is known as a distance keeping effect thus providing easy movement between particles. They should meet the requirements of ASTM C-494 Type F admixtures. Super-plasticizers have the ability to reduce the water content of the concrete mixture. Therefore it has the potential for producing high strength concrete and for providing workability.

Superplasticizers have a limited time of effect and must be added to the concrete mix just before placing the concrete (about 30 to 45 minutes). There are no apparent problems for concrete specimens produced in a laboratory regarding the concrete strength, workability, shrinkage, creep, freeze-thaw resistance, chloride penetration, and other physical properties but it has following serious drawbacks:

1. Rapid rate of slump loss.
2. Questionable effect on long term durability and freeze-thaw resistance
3. Need for high fines content in the mix to reduce bleeding and segregation.
4. Excessive amount of this additive promote segregation of the aggregates.

The main problem is that it is not a pure product but a by-product of the chemical industry. The impurities within the plasticizer depend on the method of manufacturing and may adversely effect the curing process of the concrete. These adverse effects should be determined in the laboratory before using in the field and adjustments should be made in the field based on slump, water-cement ratio, size of aggregates, etc.

This can be used for following bridge construction activities:

1. Improve the quality of patch concrete or overlay for reinforced concrete deck by using low water-cement ratio, thus increasing strength and reducing permeability.
2. Produce high strength precast members using lower temperatures and/or shorter production cycle. The rate of slump loss of concrete is low at lower temperatures and an increase of temperature will result in a substantial reduction of slump.
3. Allow easy placement of concrete in heavily reinforced members and inaccessible areas by creating flowing and self-leveling concrete, thus saving time and labor costs.

There are three major types of admixtures:
1) Naphthalene sulfonate formaldehyde condensates
2) Melamine sulfonate formaldehyde condensates
3) Modified ligno sulfanate.

The required dosage depends on the tricalcium aluminate content, alkali content, and fineness of the cement. Generally, the mix requires smaller size coarse aggregate (3/8" to 3/4"), fly ash or 5% more of the finer fraction of sand to prevent segregation. Also, it uses a higher cement contents (7 sacks or more per cubic yard) to furnish the necessary fine. The increased dosage will increase the slump, but beyond the certain limit, it will promote segregation and bleeding. After superplasticizer is added, a 2" to 3" conventional mix will become a 7" to 9" slump flowing mix.

Since concrete containing super water reducer has high rates of slump loss with time, it requires expeditious handling and is suitable for "on-site" production of dense concrete. The use of the admixture in the central mixing plant or in the truck haul is not recommended. The dosage of superplasticizer is based on the original slump of the concrete mix and the desired slump of the mix after adding superplasticizer. Normally the slump of base concrete is kept constant by delievering concrete with slightly below the desirable base slump and by adding water at the site to bring slump to the desirable level. Then the predetermined dosage is added for each batch. When a superplasticizer is used, the air loss is usually higher than normal concrete therefore additional air-entrainment agents are needed to compensate losses.

The following are some of the findings of FHWA research (FHWA/RD-80/132 "Evaluation of super-water reducers for highway applications"):

1. Concrete mixtures prepared with super-water reducers at low water-cement ratios (less than 0.35), exhibit a higher rate of loss of workability with time (slump loss) then corresponding control (admixture free) mixtures. The problem is especially severe in mixtures of moderate cement content produced in central-mix type operations. In some cases, the mixture may become unworkable within 20 minutes after addition of the admixture. The use of super-water reducers is not recommended for a central-mix operation when concrete is subsequently transported to the job site in a nonagitating vehicle.
2. Produce concrete by on-site volumetric batching and continuous mixing (concrete mobile).
3. The slump loss problems can be reduced by increasing the dosage of super-water reducer so as to obtain an initial slump between 5-6 inches, and by adding reducer at the job site immediately prior to placement. The concrete mixture should be redesigned so as to increase the paste content while maintaining the water-cement ratio but the dosage of reducer should be adjusted to obtain an initial slump between 5-6 inches.
4. In hot weather, a significant increase in rate of slump loss is to be expected.
5. The rate of loss of air from air entrained concrete containing reducer is greater than normal air entrained concrete. It is recommended entrained air contents of 7 to 8 percent which is 2 to 3 percent higher than those normally used.
6. Gain of strength with time, plastic shrinkage, elastic properties, chloride permeabilty, sulfate resistance, and resistance to D-cracking were all improved in concrete using super reducer. Little differences in creep, drying shrinkage, freeze-thaw resistance, or thermal expansion between control and super-water reduced concretes were seen. Resistance to abrasion was somewhat lower in the concretes containing reducers. Resistance to deicer-scaling was considerably lower.
7. For the Iowa type overlay concrete produced by using a continuous mixer and reducer, slump loss was moderate. Also, high strength were achieved, and the majority of mixture exhibited excellent air-void systems and good durability.

**5. Opening Patch to Traffic**

The methods of curing the patch concrete generally are covered by the standard specifications but special attention should be given at each site. There are several factors involved to determine the curing time for concrete to develop sufficient strength to resist traffic loads. Important factors are ambient temperature and the type of the patch material. Generally higher temperature reduce the curing time. The minimum time for opening traffic should be based on the actual cylinder compression test result. Minimum strength of concrete should be about 1,000 psi before opening for traffic.

## 4-2.2 EPOXY INJECTION

Where delamination exists, traffic impact may cause further damage and ultimately produce surface spalls. It is suggested that all delamination from the deck be removed after removal of the chloride contaminated concrete, or all fractures and cracks be sealed by injecting with epoxy grout and then patch the repair area.

Before using epoxy injection, it is necessary to determine the causes of the cracks and to correct the problems. There are two types of crack, one is moving and the other is stabilized. If the crack is moving, the problem should be solved by eliminating the cause. For example, if a crack occurred due to the overstress of a girder by loss of the section, the girder should be strengthened before injecting epoxy resin into the cracks. If the movement of the cracks can not be corrected then the cracks should be sealed or filled with expandable materials such as neoprene sealer or hot asphalt filler. Once the problem is corrected and the cracks are stabilized, the cracks can be effectively repaired by injection. If the structure has minor movement, the cracks can be repaired by using a flexible type of epoxy resin which has higher viscosity but requires longer curing time.

For wide cracks and voids in concrete structures, it is cost effective to use pressure grout with cement mortar of portland cement - fly ash mixture.

Cracks in the deck could be filled with premixed epoxy by gravity flow or injection by using a common cauking gun. The gravity method generally requires "V" cut of the cracks and blast cleaning of the cracks before filling the cracks. The gravity method does not always allow sufficient flow into the fractured plane or hollow area.

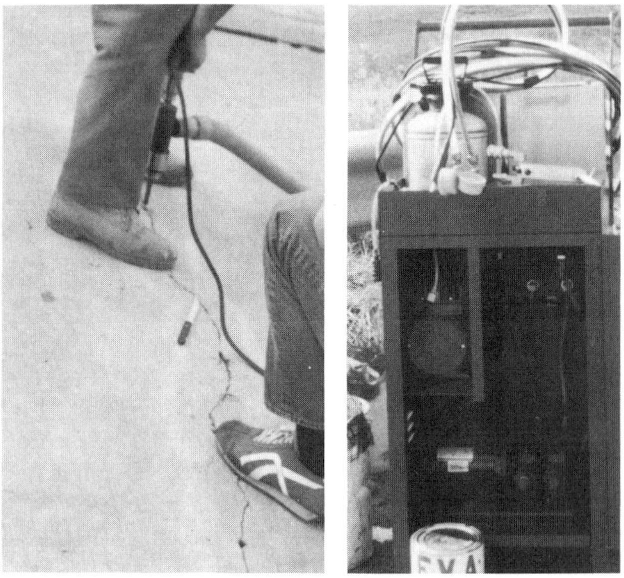

**Fig. 4.5 Epoxy Injection: a: Drilling holes with vaccum swivel at 4 to 10 spacing. b: Pumping system for epoxy injection.**

The pressure injection method, along with a pumping system, offers a sufficient injection pressure for full penetration into the cracks. The pumping system should be a positive displacement piston driven pump system because a gear driven pump may not meter the two component accurately under high pressure.

After removal of all loose and unsound concrete, the area to be injected should be cleaned by sand blasting. In some cases, grinding the surface is acceptable but sand blasting insures a proper bond line for sealing the cracks. To give the most effective penetration, the port spacing should be a range of 4 to 10 inches depending on the condition of the concrete and the depth to be injected. Holes should be drilled by hollow drill bits and a vacuum swivel to give the clean proper fit for the port. The injection port should be inserted into the drilled hole to about 1/2" above the bottom of hole. This allowance for a small reservoir below the bottom of the port will speed the resin flow. Overdrilling can cause loss of the crack and result in unsatisfactory injection. It is recommended not to set ports beyond cracks that run other than perpendicular to the surface and not to use a rotary hammer drill in drilling for ports.

If the cracks turn away from a perpendicular plane within 1/2 inch of the surface and under hydro-static conditions, or hairline cracks, chip out the crack so that its base can be recessed, then place the T type port. This exposes a larger area of the crack for admitting the epoxy resin. Before injection, the damaged and cracked area should be sealed so that injection can be made from one port to next with full penetration and without leaks or waste.

To test the system, it is recommended that water be injected through the entire crack area. This water injection serves several purposes. First, it effectively flushes the cracks of debris and dust which can interfere with the smooth flow of injection resin. Second, it points out trouble spots where pumping time from one port to another is quite long or where the pressures required may vary. The amount of material needed and time required to pump from one port to another is helpful information to avoid surprises when epoxy resin is actually used.

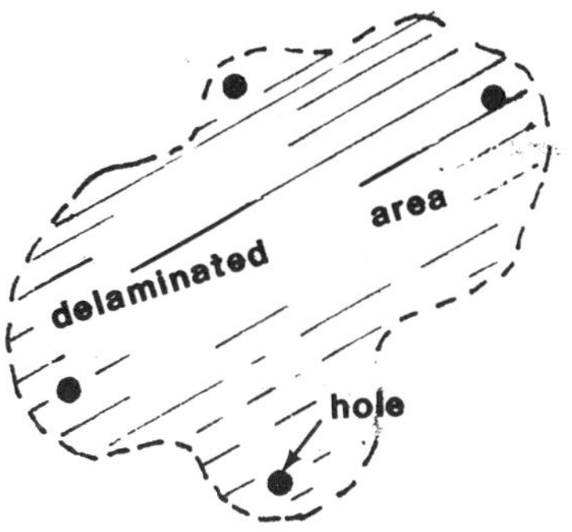


Locate the delaminated area then drill the holes at about 4 to 10" spacing

Do not set the port beyond the crack and leave about 1/2" at bottom. If crack turn away from perpendicular from surface, chip out crack then set the port

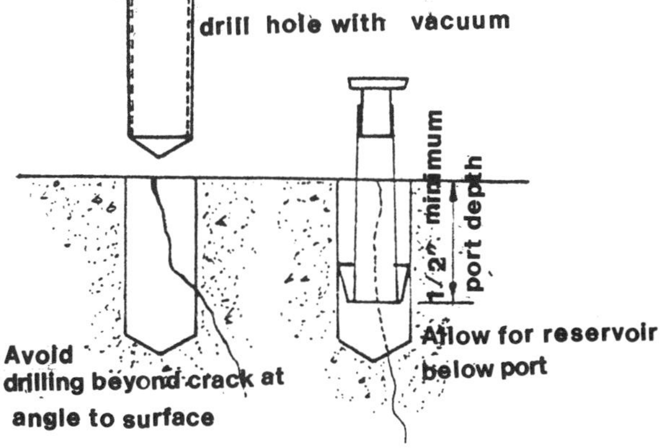

Chipping out crack before drilling holes

**Fig. 4.6 Epoxy Injection**

During the injection process, pressure should be adjusted according to the type of cracks. Excessive pressure can cause additional stresses in the cracks. On vertical cracks, injection should started at the lowest point and continue upward on the cracked area. While injecting on the lowest port, resin should flow to and out of the next higher port. When this flow is established, the lower port is capped and the injection is moved to the port showing resin flow. This procedure is continued until all ports have been injected and flow has been established between them. After injection the ports should be removed and the surface should be cleaned.

The State of Kansas uses the following procedure for injecting epoxy grout into a hollow spot in bridge decks:

1. Locate the hollow spots by sounding with a chain drag or hammer and mark with lumber crayon. Marking an outline of the hollow spot will help in placing drill holes.
2. With epoxy paste, seal visible spalls, popouts, or cracks within the area of the hollow planes that would allow epoxy to leak.
3. Determine the location of reinforcement steel with a pachometer.
4. Drill a hole near the center of the hollow spot (not on the steel). Keep the hole away from points of possible leakage such as spalls or cracks. Drill holes as perpendicular as possible to make it easier to get a good seal with an injection probe. Check the drill tip often to make sure that it does not become clogged. The hole need not be much deeper than 3 inches and usually less than 1.5 inches.
5. Inject epoxy resin into hollow spot at 0 to 50 pounds per square inch (psi). Check progress of epoxy with a hammer. If a hollow spot can not be completely filled by one hole, drill another in the remaining area to be filled and inject it. If epoxy is pushed from a previous hole, plug it with a rubber stopper. Keep injection pressure as low as possible. The pressure required will depend on the type of the epoxy used, the temperature, and the thickness of the spot. If the hollow plane is very thin, the pressure will

tend to rise faster and the pump speed should be turned back to prevent a pressure build up.

6. Scrape off excess epoxy and place waste in can. Sprinkle wet epoxy areas with a sand-cement mixture or a dust-sand mixture.

## 4—3 CONCRETE DECK PROTECTIVE SYSTEM

The protective system should be effective, durable and reasonably economical. If the system is waterproof and keeps chloride out of the deck or prevents corrosion of reinforcement, it is considered to be effective. If the system provides effective protection for 5 to 10 years under moderate service conditions, it may be considered durable.

The following are the currently recommended alternate protective systems: Low slump high density concrete, Latex-modified concrete, Waterproofing membrane system, and Protective sealer. A concrete overlay system requires a minimum of one-fourth inch scarification over the entire deck slab and each system requires removal of all delaminated and deteriorated concrete and patching to the scarifyng level or other proper level with concrete or mortar.

Membrane system and other overlay systems installed on the salt laden concrete deck slab have shown good performance of retarding active corrosion and sealing off corrosion attributing moistures and oxygens. However, it is believed that the removal of contaminated concrete will definitely give better results.

### 4-3.1 LOW SLUMP, HIGH DENSITY CONCRETE

The wide use of low slump concrete in Iowa has come to be known as "The Iowa Method of Bridge Deck Repairs and Overlay". The minimum overlay thickness should be 1.75 inches. But, when additional thickness can be achieved without reducing load carrying capacity or adversely affecting the roadway profile, two (2) inches of thickness may be desirable and more cost effective.

The following steps are recommended for the overlay:

1. The entire existing concrete deck shall be uniformly scarified and prepared to a depth of one-fourth (1/4) inch.
2. All deteriorated concrete shall be removed to sound concrete.
3. All exposed surfaces and reinforcing bars shall be sandblasted and cleaned before applying grout for bonding new concrete to old concrete. The grout may consist of equal parts by weight of portland cement and sand mixed with sufficient water to form a stiff slurry.
4. Concrete shall be placed continuously throughout the deck and patch repair shall be made monolithically with overlay course except for the large full depth repairs.
5. Curing of resurface area is very important. Immediately after final finishing, the surface shall be covered with a single layer of clean, wet burlap and cured at least 72 hours with the temperature above 55°F.

The concrete should be air entrained (6%) and have slump of 3/4 inch with tolerances of plus or minus 1/4 inch. It is recommended that high early strength portland cement (Type III) should not be used.

A research report indicated that the overall performance of a low slump concrete deck overlay has been satisfactory. The following are recommended:

1. All delaminated concrete in the deck should be identified and be removed prior to placing an overlay.
2. If the half-cell reading is -.35 volts or more, the concrete should be removed to the top layer of reinforcement steel bars.
3. The protection of concrete in the gutter line and curb should be made with a high quality concrete sealer.

Some states prefer the use of latex-modified concrete in place of a low slump high density concrete because of the relatively poor skid resistance and thicker depth of overlay of a low slump high density concrete.

## 4-3.2 LATEX MODIFIED CONCRETE

Latex modified concrete is a mixture of concrete with latex emulsion admixture (3.5 gallons per bag of cement or about 15% latex solid by weight of cement). Latex can be used in water and is compatible with portland cement. The particles which are dispersed in water are used in a concrete or mortar mix to improve the physical properties of the concrete such as bonding, tensile, compressive and flexural strength.

In 1976, the Federal Highway Administration approved the use of a 1.25 inch thick latex modified concrete overlay as an alternative protective system. An approved and widely used latex-modifier is a styrene butadiene type developed by the Dow Chemical Co. in 1957. Recently, several other companies have manufactured styrene butadiene latex modifier.

The preparation of overlay is similar to the low slump concrete with the following differences:

a. High slump (5" vs. 3/4")
b. Before placing modified concrete, the cleaned surface shall be wetted for at least one hour.
c. The surface shall be cured by wet burlap for first 24 hours following by additional minimum 72 hours (prefer 5 days) of air curing. Polyethylene film should be applied over burlap within one hour after placing burlap.
d. A 1.25 inch thick latex modified concrete will give equivalent chloride resistance of 1.75 inch thick low slump concrete. The advantage of a thin overlay is the reduction of dead loads of the bridge.

In preparing LMC overlay plans for existing bridges, the following should be considered:

1. Provision for transition at deck joints.
2. Necessity to adjust approach slab by LMC overlay or asphalt overlay.
3. Problems of raising existing joint sealers, armored steel deck joints, and scuppers.
4. Difficulties of providing transitions in gore area.
5. Necessity of additional scarification beyond 1/4 inch.
6. Repairs of deteriorated concrete deck slabs.

The latex modified concrete or mortar mix is placed over the wet surface and screeded to the proper elevation. In hot weather above 85°F, placement of concrete should be made in the early morning or at night. In cold weather, the air temperature should be above 45°F for minimum 8 hours after placement. During the curing period, if temperatures fall below 35°F, the work may be considered unsatisfactory. It is recommended not to place concrete in cold weather unless heating provisions are provided. Consolidation can be accomplished with a vibrating screed. Spud vibration should be used at the edges, gutter areas adjacent to joints and bulkheads and in all depressions. Finishing is then accomplished in the normal manner. Overfinishing of the latex mortar or latex concrete should be avoided.

In no case should cracks be covered with latex concrete. Such cracks should be cut out to a minimum depth of 2" (1/2" may be used sometimes) and a minimum width of 1", and filled with latex concrete prior to the resurfacing operation. All joints in the surface should be maintained during the placement of latex modified concrete.

Latex modified concrete offers good bond strength to a sound substrate deck, resistance to moisture and salt penetration, and wearing resistance. However, the most common problem is the development of moderate map cracks after placement and bonding failure between latex modified concrete and the existing concrete near the joints. This may be caused by improper application of either curing method, under high temperature, or shrinkage due to high slump.

Latex Modified Concrete Overlay defects are as follows:

a) Random Cracks: Most of these cracks appear shortly after placing, as surface cracks and may extend to the full depth after several winters. They may be caused by high air temperatures, high slump exceeding 5 inches, and high winds during placement due to differential shrinkages between the overlay and the concrete deck slab.

Scarifying Machine

Scarified and Sandblasted Bridge Deck Surface is covered with polyyethylene sheet

Self-propelled Finishing Machine

Spreading Latex Modified Concrete

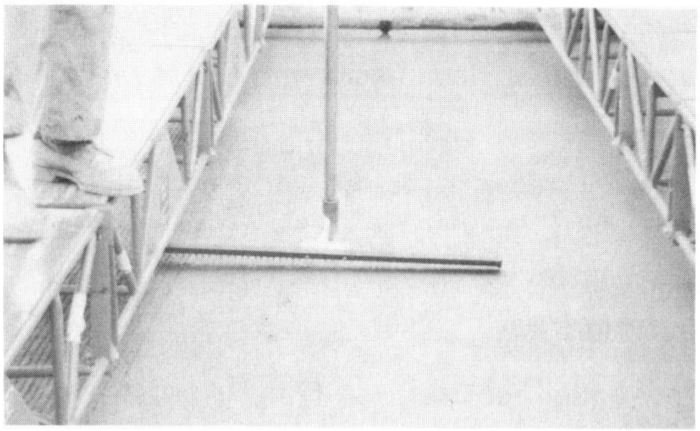

Applying Tine Surface Texture to the LMC

Finished Surface

**Fig. 4.7 Placement of Latex-Modified Concrete**

b) Transverse Cracks: These cracks are usually full depth cracks and occur over the piers on the continuous deck slab. They are reflective cracks which may extend through the deck slab and cause more damage than random cracks because the moisture could penetrate to the concrete deck slab and lead to spalls and stripping.
c) Debonding: Debonding is a hidden separation of overlay from the concrete deck slab.
d) Stripping: Stripping is an actual separation and failure between the overlay and the concrete deck.
e) Spalls: Failure of overlay and concrete slab exposing reinforcing steel.

The following are some of the findings from the 1979 Ohio Department of Transportation and FHWA Field Performance Analysis Report:

1. The increase in the number of the random or transverse cracks is rather slow in the first three years but become much higher in later years. The percentage of increase may become 100% after 7 years.
2. Relatively higher average daily traffic seems to increase slightly higher chloride content.
3. A continuous bridge deck tends to exhibit a higher degree of surface distress than a simply supported deck. In a continuous deck, the existing tension transverse cracks in the deck over the pier are likely to be intermittently widened by traffic induced vibration. These lead to more tensile strains in the overlay and form more transverse cracks.
4. Bridges that are open to traffic during latex overlay placements generally exhibit a higher degree of surface distress than those which are closed to traffic. Statistically, this increase is insignificant. Therefore it does not warrant closing the deck to traffic and an additional expense in detouring traffic. However, it is beneficial to restrict speed or place the overlay at night or at other times when traffic volume is low, particularly when the original surface conditions require deep scarifying of deck thus reducing its flexural rigidity.
5. A thicker overlay would result in the lower chloride content at a bond surface.
6. It is estimated that the useful life of a rehabilitated deck with an overlay should be over 10 years and a new construction should be in excess of 15 years.

### 4-3.3 WATERPROOFING MEMBRANE SYSTEM

The bituminous wearing surface with membrane has been used widely in Europe and the northeastern states, and its performance has been acceptable. The advantages of a membrane are its easy installation and its relatively low cost. The disadvantages are as follows:

1. Premature deterioration of the bituminous overlay in areas of high volume traffic and inadequate drainage.
2. Blistering, caused by expansion of trapped air and water vapor after placement.
3. Poor bonding at the protection layers in the area near the expansion joint.
4. Local instabilities due to bleeding and bubbling.
5. Replacement of the membrane whenever the surface is removed.

Most of the problems could be corrected by following proper construction procedures. This system could be a cost effective method for short term performance.

It is suggested that the minimum thickness of asphalt wearing surface over the membrane be a minimum of 2 1/2 to 3 inches and if placed in two courses, the lower course should be denser or more impermeable than the upper course in order to prevent trapping of water in the lower course.

Also, placement of a membrane should be made under warm, sunny and moderate wind conditions to provide a drier deck surface which promotes better primer penetration and adhesive bonding, easier handling of membrane and application, and more rapid curing of primers and adhesive thus preventing blistering.

In place of a membrane, a dense epoxy asphalt course or non-permeable modified asphalt mix (minimum 1/2 inch) can be used where night time work is required to minimize traffic interruptions. It replaces the waterproofing membrane system. It is easy to install, and capable of supporting traffic shortly after placement. Asphalt-rubber mixtures are used on asphalt pavements by applying a thin layer of hot mixture which contains the rubber made up of 1/4 of the mixture. This method is effective when used to level irregular surfaces, it is able to installed in night time without extensive surface preparation, and it provides speedy installation without troublesome blistering or bubbling. It also provides an good base for adhesion of the asphalt wearing surface.

On a resurfacing project for lower level of the George Washington Bridge, the Port Authority of NY & NJ developed an asphalt membrane using natural and asphalt in a ratio of 40 to 60%. The asphalt mix for the wearing course was 75% refined asphalt, 20% natural asphalt, and 5% liquid latex additive. As an experimental project, an asphalt-rubber mixture known as stress absorbing membrane was used on asphalt pavement by applying a thin layer of hot mixture. The rubber from discarded tires are added to make up about 1/4 of the mixture. Rosphalt, a modified rubberized waterproofing asphalt developed by Royston Laboratory had been used on Pittsburgh's Tenth Street Bridge. Also, Concresive, an epoxy asphalt patented by Adhesive Engineering Company can be used as an alternative to the waterproofing membrane.

### 4-3.4 POLYMER IMPREGNATED AND POLYMER CONCRETE

The impregnation of a concrete bridge deck surface with the methyl-methacrylate monomer system and in situ polymerization is a process to reduce chloride and freeze and thaw deterioration of the concrete by filling the pores in the concrete.

The polymer impregnation of bridge deck concrete which is salt contaminated but structurally sound has demonstrated to be an effective protective method to immobilize chloride ions in the concrete and to retard or eliminate additional chloride penetration, thus lessening its corrosive action. However, the costs are high and special safety precautions are required due to flammability and vapor pressure of the monomer mixture.

The FHWA sponsored several successful research projects for highway applications. However, this concrete is not currently practical due to high initial capital investments and a lengthy process. Further research is required for development of cost effective practical applications even though it has long expected service life and reduced maintenance costs.

The following four steps are suggested for the surface impregnation process:

a) Preparing of concrete surface to remove contaminations (oil, asphalt etc).
b) Drying the concrete to the depth which would permit the desired monomer penetration. In order to achieve this, which is normally top reinforcement steel level, the concrete should be heated until the temperature at the specific depth is 230°F by using a gas fired infrared heater and torch, and cooled off below 100°F before applying monomer.
c) Impregnation of the concrete with the liquid monomer to the desired depth by soaking. Field impregnation normally performed greater than atmospheric pressure by using a pressure chamber but a heat blanket system can be used in place of other drying methods utilizing enclosures.
d) Polymerization of the monomer in the concrete to form a protective barrier at 158 to 185°F by pumping hot water through the pressure impregnation chamber or other heating systems.

**Polymer Concrete Overlay:**

A polymer concrete overlay protective system developed by Brookhaven National Laboratory for FHWA (FHWA-TS-78-225 Polymer Concrete Overlays Interim Users Manual Method B) has been used as an experimental projects in several states. The overlay consists of an application of monomer to the deck surface followed by an application of the aggregate. The process is repeated until four layers have been placed. The overlay is relatively impermeable and skid resistant. Generally, the resin is sprayed over the deck and fine aggregate is covered over the resin. After polymerization, the excessive aggregate is removed and the process is repeated for other layers. The four layers produce a thickness of about .50 to .63 inch.

The overlay system consists of the following steps:

- Surface preparation.
- Mixing and application of monomer.
- Aggregate application and compaction.
- Polymerization of monomer and removal of excessive aggregate.

The main advantage of this system is that it can be placed without the removal of concrete and with minimal interruption to traffic. The disadvantage are handling problems, high cost, deterioration due to bond failure between the concrete and overlay and thermally induced cracks in the overlay.

As a research project, the Oklahoma Department of Transportation placed a 3/4" inch thick polymer concrete overlay on a sound bridge deck. The polymer concrete overlay was based on vinyl ester which has performed as well as other base materials like polyester styrene but is less sensitive to water. The promoter, dimethyl aniline was added to resin to improve the chemical reaction and to make polymerization occur at ambient temperature. The report " Polymer Concrete Overlay on SH51, Bridge Deck" conclude that the protection of a bridge deck with a polymer concrete overlay appears to be a rather complicated procedure. Besides constraints of weather, temperature, aggregate and resin availability, it is necessary to develop machinery which can accurately and efficiently mix and lay polymer concrete. It appears that polymer concrete is good for deck repairs and concrete patchs but not practical for deck overlays.

### 4-3.5 WAX BEAD CONCRETE (INTERNALLY SEALED CONCRETE)

The internally sealed concrete is also under a continuing research project which consists of adding wax particles to the concrete mix and a heat treatment after curing. Melting of wax beads is done by application of heat, which causes the wax beads to flow into the voids of the concrete and prevents the penetration of chloride.

Few states have successfully constructed internally sealed concrete deck overlays by utilizing gas fired infrared heaters. Considering the success of wax bead overlay for a bridge deck and its resistance to penetration of moisture and chloride, this method is a very promising practical application in the future. Improved internally sealed concrete which does not require heating is currently studied.

### 4-3.6 CATHODIC PROTECTION SYSTEM

The cathodic protection has been used to protect underground pipes and steel piles. This protection of reinforcing bars in the concrete is accomplished by sending a direct electric current to the anode. The current flows through the concrete to the corroded rebars so that the reinforcing bars(anodes) are prevented from discharging ions and making them cathodes which in turn stops the corrosion.

Following are two types of protective systems:

**A) Sacrificial System**
Corrosion occurs by galvanic action of the potential difference between metals. Zinc or Magnesium is used as an anode to be sacrificed. The advantage of the sacrificial system is that no outside power source is required and it is almost maintenance free. However, this system requires numerous anodes spaced closely together in order to function properly.

**B) Impressed System**
This system requires D.C. current from batteries or a D.C. rectifier to impose a current in the reverse direction of the corrosion current. There are two systems one is overlay system and other non-overlay system.

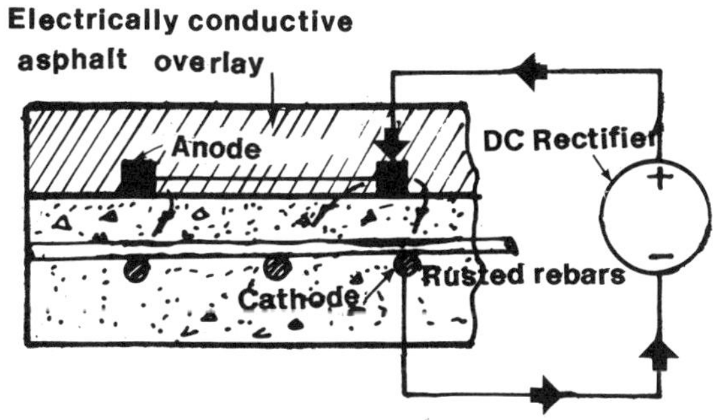


The imposed current prevents electrons from leaving rusted bars and stop all corrosion process

**Fig. 4.8 Overlay Cathodic Protection System**

On overlay system, the conductive asphalt wearing surface contains "coke breeze" as coarse aggregate and become electrically conductive. This system consists of placing coke asphalt concrete overlay over the evenly spaced platinum-clad wires and deck slab. The rectifier converts alternate current (AC) electricity to direct current (DC) and DC current is impressed through cables to the platinum wire anode. Then current is sent through and distributed by the conductive asphalt wearing surface.

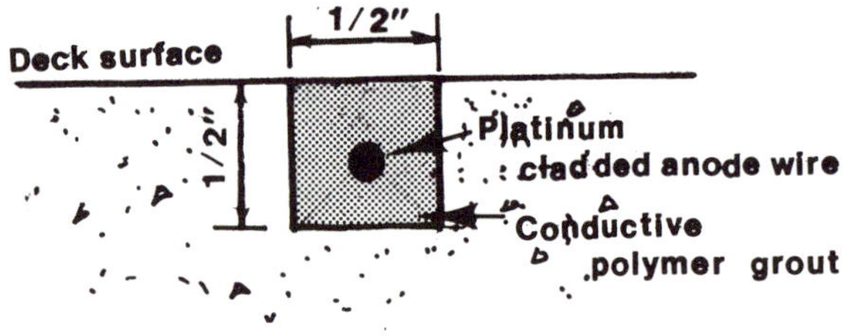


**Fig. 4.9 Non-overlay Cathodic System**

On the non-overlay system, the deck slab will have cut slots or strip which will hold platinum-clad anode and backfilled with conductive polymer grout. Generally main anode systems with platinum cladded copper wires are placed transversely at 20 to 25 foot intervals and secondary anode lines with carbon wires are placed at 2 feet intervals in the longitudinal direction.

The following should be considered in selecting a cathodic protection system: The existing deck should be evaluated to determine whether the concrete has adequate air contents to provide necessary freeze-thaw protection and to determine extent of delaminations and spalls. For non-air entrained concrete deck slabs, a protective sealer should be applied before installing a system. If the deck delamination is extensive, cathodic protection systems should not be considered. Minor delaminations and spalls should be patched with non-epoxy type portland cement mortar. Also, areas previously repaired by epoxy injection should be removed and be patched by normal concrete.

Cathodic protective systems for concrete deck slabs have been successfully used in several states as experimental projects and may require further improvements for the system to be economical.

### 4-3.7 CHLORIDE REMOVAL

The best way of protecting a bridge deck could be the removal of chloride from the concrete. In the laboratory, an attempt has been made to flush away salts by a pressurized the water supply, but it was found that this method was ineffective and impractical. Other methods involving the electrochemical removal of chloride by using an impressed voltage have been found to be effective but the permeability of the concrete was increased and the deck should be overlayed with an impermeable waterproofing system after the removal of chloride. Due to its initial high cost, the practical application of this system in the near future is limited.

### 4-3.8 PROTECTIVE COATINGS AND SEALANTS

Protective coatings have been used to protect a deck slab and top of the substructure bearing seats. The most common coatings are silicon, polymer, epoxies, linseed oil, and latex modified concrete. After two or three years of exposure, most of the coating shows some distress and no significant evidence of protecting the concrete. Apparently, the protective coatings retard the intrusion of chlorides, moisture and oxygen, and minimize corrosion actions for a few years. Linseed oil or mixtures containing linseed oil appears to be more effective than others for a deck against freeze-thaw induced scalings. However, it is not very effective against corrosion and water absorption.

For underside deck protection, epoxy and linseed oil coatings are widely used with 1 to 5 years of expected life. Copolymer acrylic mixed with styrene has been also successfully used with an expected life of 5 to 10 years. In the laboratory test for coating performance on a smooth surface of white concrete, the coatings based on acrylic resin generally perform better than other types. However, there is a wide difference in performance depending on the type of acrylic resin used as the major component and many of the specimens pitted or etched at the completion of the weathering test.

The following are some of the coatings rated as satisfactory:

a. Methyl methacrylate (MMA) and ethyl acrylate copolymer
b. Polyester-epoxy resin
c. Styrene-acrylic copolymer
d. Styrene-acrylic-silicone terpolymer
e. Alkylalkoxy silane
f. Urethane

Current research results indicated that epoxy, methyl methacrylate, and silane show excellent performance in reducing chrolide intrusion and water absorption and in providing a barrier to the salt water. The epoxy and methyl methacrylate require two coats whereas the silane requires only one coat. Generally a 5 day air drying period is required before applying the coat. For epoxy coating, the epoxy has low water vapor transmission characteristics, therefore, fresh concrete should be allowed a little longer period so that internal water can be evaporated before coating. Once the concrete is coated, the epoxy will allow sufficient water vapor transmission to handle vaporization of subsequently absorbed water.

Coating over the linseed oil treated concrete surface appears to be effective except for silane coating. Therefore it is not recommended to use silane over pretreated concrete by linseed oil. Silane and methyl methacrylate coatings provide added protection against corrosion for embedded reinforcing bars in concrete which has fine cracks.

Beside the use of penetrating sealant, the key considerations reducing chrolide intrusion into concrete are lower water-cement ratio of concrete, sufficient cover for reinforcing bars, lower chrolide contents of aggregates in the concrete and proper placement of the concrete.

**Penetrating Sealant**

Currently, the penetrating sealant offers good protection to concrete from freezing and thawing and corrosion by making the concrete surface water-repellent. The sealant functions by chemically reacting with concrete. When it contacts, it creates a very hydrophobic surface.

Commercially available silane materials are Chem-Trete Silane and Sil-Act from Advanced Chemical Technologies, Inc., and others. Other type of penetrant such as Hydrozo 600 appear to be performed satisfactory as the penetrant. Several states have used silane sealant for deck surface treatment as experimental deck protection projects.

Silane consists of about a 40% solution of Alkylalkoxy Silane in an ethanol or anhydrous Isopropanol solvent. The concrete deck surface should be cleaned by sandblast before application of the silane solution. Generally, it is recommended for application to the entire concrete deck, the exposed surface of sidewalks, the traffic side of parapets,and bridge seats. One state suggested the application of the silane to the bridge decks prior to placing the latex modified concrete to promote better bonding between the concrete deck and the latex modified concrete overlay.

It appears that the penetrated surface treatment is promising to be a cost effective method for protecting of the concrete deck slab. But for surfaces subject to abrasion, the life of protection will be limited. It will not prevent scaling damages on non-air entrained concrete.

### 4-3.9 CORROSION INHIBITING ADMIXTURE

Calcium nitrate called "Darex" is an anodic corrosion inhibiting mixture produced by W.R. Grace. It also functions as an acceleration admixture, increases compressive and flexural strength, and decrease shrinkage but has no effect on freeze-thaw durability.

In general, the calcium nitrate is a viable corrosion inhibitor for concrete. The Illinous Department of transportation uses this admixture for precast deck slabs.

The use of granuated slag blended with portland cement would reduce the corrosion potential by reduxing permeability, alkali aggregate reaction and resisting sulfate attack. Atlantic Cement Company, Inc. Introduced new product called New Chem in 1982. The manufacturer stated that when New Chem is blended with Type II or I portland cement in proportions from 40 to 60 percent, This mix will achieve the same mortar strength equal to the portland cement alone at three to seven days and ultimately reach greater strength.

## 4–4 DECK REPLACEMENT

### 4-4.1 CONCRETE DECK SLAB REPLACMENT

The deterioration of a concrete deck slab in new or replacement construction could substantially be reduced by producing a good quality concrete, strict quality control, good field construction practices, adoption of a lower water-cement ratio (.40 to .45) and increased concrete cover for reinforcement steel (minimum of 2 inches). The lower water-cement ratio in concrete and adequate cover have significant effects on preventing chloride to penetrate the concrete. The use of higher strength concrete will add to the initial cost but it will be cost effective in terms of life cycle cost by providing protection against corrosion. The deck thickness may be increased by the required cover and an allowance for placing tolerance rather than strength requirements. It is recommended that the minimum deck thickness should be 8 1/2" (9" preferred). The following methods have been used to prevent early deterioration of concrete decks:

**1. Epoxy Coated Reinforcing Bars:**

The epoxy coated bars are widely used in the top mat of new decks, in the barrier curbs, parapets, and the top of bridge seats under the joints to minimize corrosion damages. The Illinois Department of Transportation assessed the cost of coating the bottom reinforcement mat of deck slabs and found that the cost is less than 1 percent of the cost of a deck. Therefore they decided that it is prudent to epoxy coat all reinforcement bars in new deck slabs including all bars in parapets, curbs, sidewalks, and medians but excluding reinforcing bars in the solid concrete diaphragms used at the fixed and expansion piers and fixed abutments of precast prestressed concrete I-beam bridges.

**Fig. 4.11 Placement of Reinfocing Bars:** Top bars are epoxy coated and additional bars are placed at the acute corner of the deck. Shear connectors for composite action are extended between top and bottom mat of bars.

Epoxy coated bars are manufactured as follows: The reinforced bars are blast cleaned to a near white then heated by inducing coils to the temperature required for application of the epoxy powder (400°F to 450°F). The bars then pass through an electrostatic spray which applies the charged dry powder to the heated bars. These process provide about 7 mils of coating, thus protecting the bars from corrosion inducing elements of chloride, moisture, and oxygens. Care should be exercised during shipping and handling of the bars to minimize damages. If any damages occur, they should be repaired before placing the concrete.

According to past research, galvanized reinforcing bars appear to be ineffective in preventing corrosion and most states do not use them. However, the FHWA recently approved the use of galvanized bars on Federal aid projects as an experimental project. Stainless cladded reinforcing bars are used in experimental projects in some states.

**2. Latex-Modified Concrete Overlay and Low Slump High Density Concrete Overlay:**

These overlay systems have been used for the new and the existing concrete deck slabs to slow the penetration of chlorides into concrete, thus preventing a critical level of chloride build up from reaching the reinforcing steel.

**3. Waterproofing Membrane System with Asphalt Overlay:**

These systems are intended to seal the concrete against chloride intrusion. Most systems are sheets of roof felt and asphalt impregnated protection board. Waterproofing membrane retard corrosion activity in the reinforcing steel and significantly extend the service life of the deck. However, the system requires a total overlay replacement in less than 15 years, otherwise the cracks developed in the overlay penetrate through the membrane and permit local corrosion activities in the concrete deck slab.

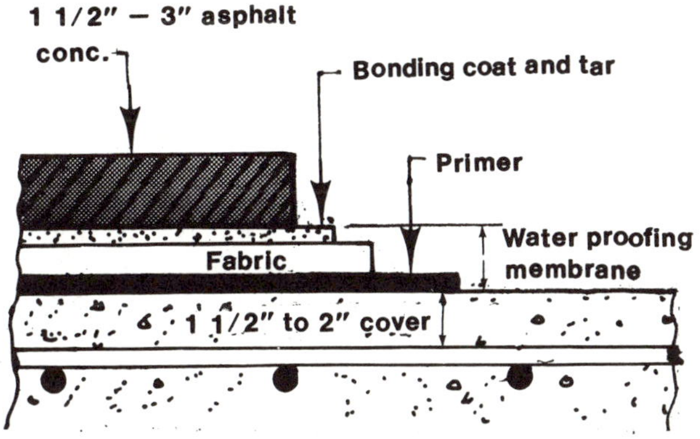


**Fig. 4.12 Waterproofing Membrane System**

Also, other problems associated with this system are as follows:

1. The deterioration of concrete could not be detected until an extensive distress has occurred.
2. Since chloride is accumulated on the membrane, when cracks occur, the salt water is trapped on the deck and it accelerates corrosion activities.
3. Asphalt is difficult to compact at the curb, joints and deck drains. These are the most critical areas which require waterproofing.
4. Improper bonding between the membrane and deck sometimes cause blistering.

Considering the above problems with the flexible overlay systems, most states use epoxy coated reinforcement steel in new construction.

**4. Double Protective System:**

Where the bridge has high traffic volumes and exposed to the heavy use of deicing chemicals, it is justified to use multiple protective systems. Generally, the deck has epoxy coated bars in the top mat and covered with another protective overlay system.

The Ontario Ministry of Transportation uses the epoxy coated bars with a conventional waterproofing system and their study concluded that epoxy coated bars with a waterproofing system is the least expensive alternative. The cost of this system included the total replacement cost of a waterproofing membrane and an asphalt overlay at every 15 years. The concrete overlay system has the disadvantage of high initial costs and the restriction of constructing LMC overlay in lower temperatures. However, a number of states use the double protective system of epoxy reinforcement steel bars and Latex-Modified Concrete overlay because of the maintenance free nature of concrete overlay.

## 4-4.2 OTHER FACTORS

In addition to a protective system for the replacement deck slab, the following factors should be studied:

**1. Thermal, Shrinkage and Creep Effects:**

It should be recognized that the existing concrete has different strength, modulus properties, and thermal shrinkage and creep effects than the concrete in the new deck slab, especially a slab which is built monolithically with main concrete members. The remaining main member can act as a restraint to the new deck's thermal expansion and shrinkage and form cracks in the new deck slab. Also the creep of the existing concrete could exert an additional stress to the deck slab. It is suggested that the depths of the replacement concrete slabs for monolithically constructed concrete bridges such as concrete arch and T beam bridges should have equal thicknesses as the existing concrete deck slabs so as to maintain the same rigidity and compatibility with the remaining main members.

As concrete shrinks, shrinkage forces are developed independent of the stress condition of the concrete. If it is restrained, shrinkage would cause cracks in the concrete and deflection of the member. In composite beam, differential shrinkage forces are caused by the restrained shrinkage of the concrete slab over the girders. This differential shrinkage can cause significant forces and stresses in the concrete deck and the girders. Generally the steel beam and concrete deck will have greater differential shrinkage effects than the concrete beam and concrete deck slab. For details of differential shrinkage forces, refer to "Deformation of Concrete Structure" by D.E. Branson, McGraw Hill, NY.

Also, The pouring sequence of concrete will influence the stresses. Bridge decks should be placed by first pouring the positive moment areas of all spans then the negative moment areas. For a three span bridge, the positive moment area of the center spans and one of the end span should be placed first, then the other positive moment area of end spans, before concrete is poured over the piers. For multi-span bridge, the deck should be placed from the middle of the bridge to the abutments by pouring in two adjacent spans.

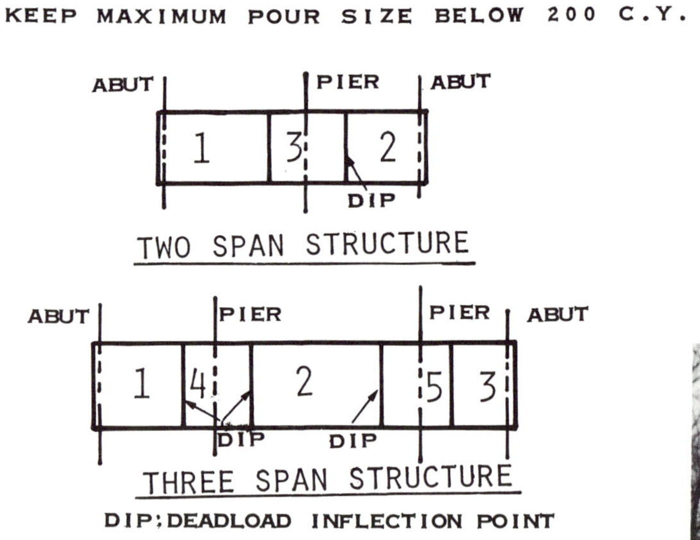


**Fig. 4.13 Deck Pouring Sequence for Continuous Bridge**

**2. Increasing Live Load Capacity:**

The capacity of a structure could be increased by reduction of the dead load. Therefore, different systems may be considered for a new deck.

The following systems may be used as alternatives:

- Steel Grid Deck
- Corrugated Metal Deck
- Orthotropic Steel Deck
- Laminated Timber Deck
- Light Weight Concrete Deck

**3. Maintenance of Traffic and Costs:**

Considering the inconvenience to the traveling public, their automobile operational costs, and safety during construction, the following system may be used:

- Permanent Stay-in-place Forms
- Precast Concrete Deck Slabs
- Precast Panels for Composite Concrete Decks

**4. Stage Construction of Deck:**

In order to minimize the public travelling cost on a detour route in some cases, the bridge should be open to traffic during the construction and proper provisions should be made in the design and details. For example, an open longitudinal deck joint with or without shear keys, construction joints at the deck slab etc. According to Texas Transportation Institute Research Report "Bridge Slab Concrete Placed Adjacent to Moving Live Load", traffic in lanes adjacent to the deck which is being widened or reconstructed, does cause deflections and vibrations in the fresh concrete deck, but no deterioration was found in the deck.

**Fig. 4.14 Stage Construction of Bridge**

## 4—5 SKID RESISTANCE

The use of the skid resistant design on the bridge deck reduces wet weather skidding accidents and saves lives and property damages. FHWA is promoting to establish a guideline for the design and use of the skid resistant pavement surface on the roadways and is recommending the use of "Guidelines for Skid Resistant Pavement Design" AASHTO 1976.

To provide adequate skid resistance for high speed traffic, two requirements must be satisfied. First, the pavement surface must provide adequate adhesion between the tire and the pavement under wet weather conditions. Second, the pavement must have sufficient surface texture and drainage potential to prevent a buildup of water pressure at the tire pavement interface. On a concrete riding surface, the drainage potential can be provided by texturing the concrete surface but the adhesion component depends on the wear and polish resistance characteristics of the aggregates. A proper finishing procedure for the concrete is required to provide an adequate and durable skid resistant surface texture on the bridge deck surface.

**Concrete Deck:**

The following are the FHWA guidelines for a skid resistant surface for a concrete bridge deck:

a. A burlap drag finish should not be used as the sole means of providing a surface texture on highway with design speeds of 40 miles per hour or greater.

b. A broom finish or an artificial grass drag finish, while it often offers a good initial surface texture, may not be durable unless a deep texture is produced by heavy bristles and high pressures. When a broom finish or a similar finish is specified for use on a project with a design speed of 40 miles per hour or greater, the specification requires a deep and durable texture and quality control.

c. Metal tine finish is recommended as being the most practical and dependable method of providing positive texture in concrete surfaces.

d. The use of deep surface texture on a bridge deck may warrant increased concrete cover over the top reinforcement steel beyond the 2.5 inches required for the minimum design cover and construction tolerance. The use of a heavy transverse texture may also increase the concentration of deicing salts along the curb line unless proper provisions are made to drain these areas. The last 12 inches of the deck next to the curb should be left untextured to facilitate drainage.

e. Special techniques may be required to produce a durable grooved finish on a latex-modified and low slump concrete deck overlay. The use of flexible tines on such a dense surface frequently fails to produce a groove depth sufficient for reasonable durability.

f. Regardless of the finishing or the texture method used, adequate durable skid resistant characteristic can not be attained unless the fine aggregate has the suitable wear and polish resistance. The crushed material will normally provide a higher skid resistance than the uncrushed gravel. Research conducted by the Portland Cement Association indicated that the siliceous particle content of fine aggregates should not be less than 25 percent.

**Timber Deck:**

A timber plank bridge deck gives poor resistance against skidding when it is wet or if the surface of it is bleeding of the preservative. The creosote or the heavy oil borne surface bleeding worsens the skid resistance. To increase the skid resistance and to provide a smooth riding surface, asphalt concrete overlays are used extensively on timber decks.

**Grid and Steel Plate:**

A bare steel deck or grid deck, particularly when wet, has poor skid resistance and exposes traffic to accident hazards. It is suggested that studs be welded on the steel plates or intersection of grids to minimize the hazards.

Placing Reinforcing Bars and Forms

Finishing Machine

Pouring Concrete

Thinning Concrete

Machine Finishing Operation

Broom Finishing

**Fig. 4.15 Bridge Deck Slab Construction**

# 4–6 DECK DESIGN

## 4-6.1 CONCRETE DECK SLAB DESIGN

**Monolithic Deck:**
The required thickness of a concrete deck slab varies with states. Generally the slab has a thickness of 8 to 9 inches including a monolithic wearing surface (Minimum 7 inches including 1/2 inch wearing surface). Normally, the top 0.5 to 1.5 inches is considered as a wearing surface and is ignored in the design. The cover for the top reinforcement steel is 2.5 inches with a 0.5 inch tolerance for construction and wearing surface and the bottom cover is 1 to 1.5 inches. For a new deck, it is recommended that epoxy coated reinforcement steel be used on the top layer of bars as the protective system.

**Concrete Deck with Overlay:**
The thickness of a concrete deck slab is generally 7 to 8 inches with 1.5 inches on the top cover and 1 inch on the bottom. Presently, the common type of protective overlay consists of one of the following systems:

1. Asphalt concrete wearing surface (minimum 2.5 inches thick) with waterproofing membrane.
2. Concrete wearing surface with 1.5 inches of latex-modified concrete or 2.0 inches of low slump concrete including 0.25 inch for tine surface finishing.

**Reinforcement:**
The concrete slab should be designed in accordance with AASHTO Bridge Specification Section 3 (Distribution of loads) using the Load Factor or the Working Stress Method.

**Reinforcement at Skewed Bridge:**
If the bridge is skewed 15 degrees or less (some states allow up to 30°) The transverse reinforcement steel may be placed parallel to the abutments and piers and the reinforcement steel shall be designed as if it is at a right angle to the stringers and multiply the required area by the secant of the skew angle squared ($As \times Sec^2\theta$). If the skew angle is over 15 degrees, The reinforcement steel should be placed at right angle to the girders but additional top reinforcement steel should be placed at the acute corner of deck slab to prevent cracks. Generally, #5 bars at 6" are used at the top layer of reinforcing bars.

**Reinforcement over Piers:**
Since the composite behavior may occur in the non-composite section of deck slab, the distributional reinforcement steel (1.3.2.E) of the transversly reinforced deck slab over a continuous stringer bridge should be supplemented in the negative moments areas with extra bars designed in accordance with AASHTO 1.7.48(D). In general, the bottom reinforcement steel designed in accordance with the AASHTO requirements may not require any additional steel but the top mat of the reinforcement steel should be upgraded to provide the increased steel requirements. The supplementary bars should be spaced uniformly within the effective composite action width and then placed symmetrically about the center line of pier bearings. The following are the guidelines for designing a slab on steel girders in the negative moment regions of continuous spans:

The minimum reinforcement steel including the longitudinal distribution reinforcement steel should equal or exceed 1 percent of the cross section area and two-thirds (2/3) of the steel should be placed in the top layer of slab within the effective width. The supplementary bar in the top mat shall be extended beyond the points of dead load contraflexure a distance of 40 diameters and as a rule of thumb, the total length of extra bar is about 40 percent of the adjacent longer span length.

The negative moment areas of continuous spans should be designed as a non-composite section, however, double studs spaced at 24 inches apart may be installed throughout the area in place of additional connectors as indicated in AASHTO article 1.7.48(E)(1)(c) except that studs shall not be installed in an area where the allowable fatigue stresses would exceed if they are installed.

## 4-6.2 PRECAST DECK

Since the first full depth precast prestressed deck slab construction in 1970 in Indiana, full depth precast modular deck systems have been used in the New York Thruway, the Pennsylvania Turnpike and other states. Recently the Massachusetts Turnpike used lightweight precast prestressed concrete deck panels along with longitudinal post tensioning. When the bridge can not be closed for traffic during reconstruction, stage construction should be used by allowing traffic to flow on one side of the structure while the adjacent side of the deck is under construction. This would cause the concrete setting and bonding problems at the junction of newly poured deck due to traffic vibrations of the adjacent section. However, precast deck systems have the advantage of avoiding the problems of vibrations from traffic, while cast-in-place construction will face these problems during concrete setting.

The precast slab offers greater structural efficiency by using higher concrete strength, eliminating the need of forms, reducing traffic interruption and potential increase of live load capacity.

But the disadvantage are as follows:

1. Precast slab can not be used for curved and skewed bridges.
2. Post tensioning is required across the joints.
3. The gap between precast deck slab and steel stringer should be filled by epoxy grout or non-shrink grout. The bonding grout is expensive.
4. The bonding between the precast unit requires epoxy bonding compound which requires careful field handling and is susceptible to fail due to temperature variations.
5. For composite action, positive mechanical tie-down connection between precast slabs and steel stringers need careful studies including welability of the existing steel.
6. The major disadvantage is the higher cost of the precast slab compare to a cast-in-place concrete deck (about 50%) apparently due to higher installation labor cost.
7. Construction time saving may not be realized as expected because the method of construction is new and unexpected field problems delay the completion of the project.

With proper planning, repetitive installation method and simple connection details, the construction time and cost can be reduced.

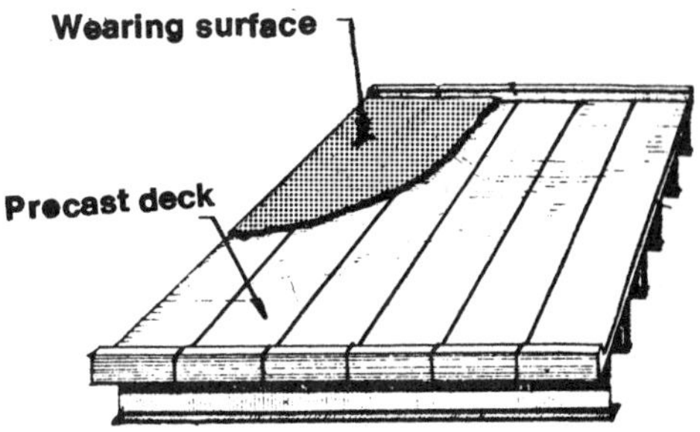


**Fig. 4.16 Precast Prestressed Concrete Deck**

A precast concrete deck may include the following:

1. A precast slab extends the full width of the bridge and attachs to the superstructure members with or without composite actions. The concept is to use precast and prestressed slabs which are placed on top of the girder, then tighten the tensioning or threaded rods, which run through ducts in the slab, to put together the new slab and the old slab. The slab is tied mechanically to the underlying girder or compositely connected to the girder. A tongue and grove joint by match casting or butt joining is sealed with epoxy bonding compound. Grouted joint filled with non-shrink grout is also used in some cases. Asphalt wearing surface with waterproofing membrane or Latex-Modified Concrete overlay is placed over the precast slab.
2. A precast panel spanning between stringers and cast-in-place concrete over the panel is another type of a precast deck system. The deck panel serves as a form for cast-in-place concrete deck and as bottom reinforcing for bridge deck. Sometimes, reinforcing shear connector bars are extended into the concrete to tie the girders to provide monolithic actions of the deck then the panel can be treated as simple slab spans for dead loads and continuous composite deck for live loads.

   The 1979 AASHTO Interim added Article 1.6.26 - Deck Panel to recognize the use of deck panel in the states. To provide continuity for live load without developing longitudinal cracks in the topping slab, it is recommend to provide positive bearing at the end of panel by grouting in place of fiber board, to extend the prestress strand beyond the panel and to

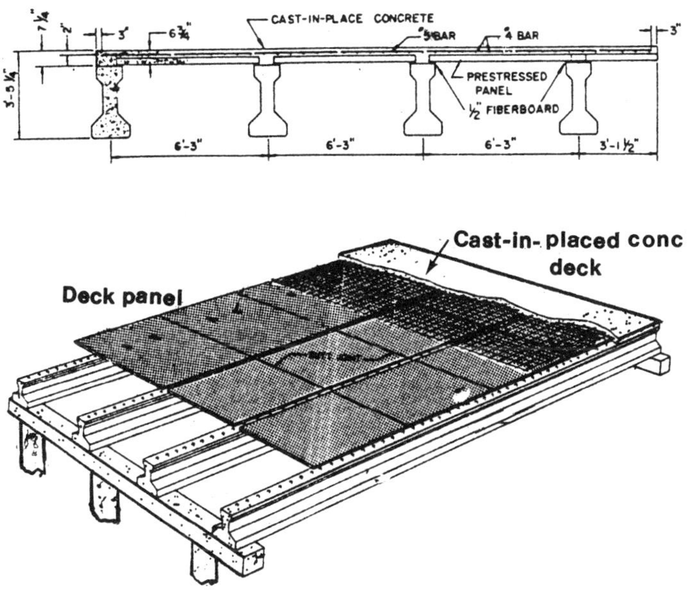


**Fig. 4.17 Precast Prestressed Deck Panel**

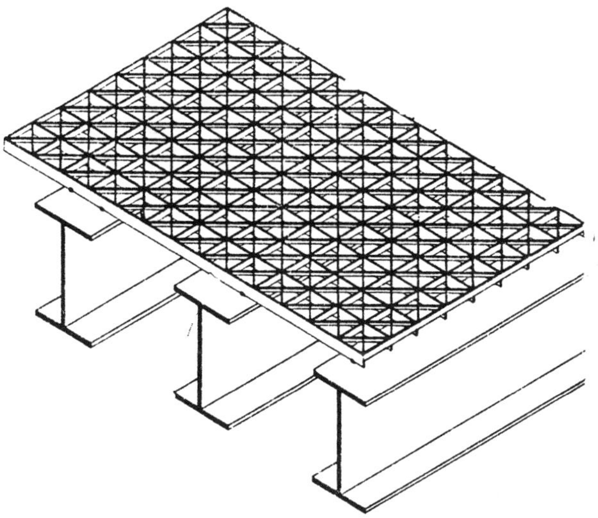

**Fig. 4.18 Steel Grid Deck System**

place #4 transverse reinforcement steel at a minimum 12 inch spacing in the top slab. The roughened surface of the panel may be adequate to develop composite actions between the panel and the topping slab.

## 4-6.3 STEEL GRID DECK

The steel grid deck has been used widely for movable bridges because of the light weight composition of the flooring and minimum maintenance. Steel grid floors may be left open or may be filled with concrete.

Old bridges which would otherwise have to be dismantled, abandoned or have their load-carrying capacity seriously impaired, will have many more years of usefulness added to them by replacement of their present floors with the lightweight steel grid flooring. The welded reinforcing provided by the floor panels adds rigidity and raises the carrying capacity of the bridge to modern traffic load demands.

Majority of open grid decks face severe maintenance problems and the following are common failures:

1. Failure of plug welds connecting primary bars and secondary bars which are caused by excessive vibration due to heavy truck and inadequate welds.
2. Failure of welds between the primary bars of the open grid and steel girders which are caused by inadequate field welds and lack of resistance to horizontal shear forces.
3. Cracks of cross bars and diagonal bars which resulted from inadequate shear resistance of secondary bars.
4. Failure of grid panels under truck wheel lines caused by inadequate resistance to additional braking or acceleration forces on steep grades or sharp horizontal curves.
5. Other failures are involved with fatigue stress and poor skid resistance.

In order to overcome the above failures, the following are suggested:

1. Fillet welds should be placed in place of plug welds to connect the main and secondary bars.
2. The differential stringer elevations would cause substantial stress on the main bars, therefore, shim plates should be used if the differential elevation is more than 1/2 inch.
3. The field welds between the stringer and the main bars should be designed to resist shear forces and be should made by a qualified welder.

The open grid also has the advantage of permitting snow, rain and debris to fall through the structure, thus minimizing maintenance. However, it becomes slippery when wet or ice-covered. The concrete filled grid offers better skid resistance.

The concrete filled grid floors are shop fabricated in large panels including form pans and all necessary longitudinal and transverse reinforcement. Panels are set on the bridge stringers and are welded in place. Concrete filled grid decks can be grouped into fully filled decks, partially filled decks (about half the depth of the main bars), and deck overlays (1/2" to 1" over the top main bars). The grid is filled with concrete and screeded flush to the top of the steel grid. A wearing surface of asphalt concrete can then be applied or traffic can ride on the armoured grid surface. Concrete filled grid deck eliminates skid problems and maintenance problems of broken bars but it would add the extra weights thus reducing the live load capacity, and requiring deck drainage provision for run off.

Recently, there have been several instances where expansion of a concrete-filled steel grid deck (deck growth) has caused extensive damage to the bearings, abutments, and decks. It is apparent that the corrosion of the transverse main supporting members with the resultant expansion in the longitudinal direction exerts sufficient force to elongate the lighter longitudinal members and crack the grid deck at the joints between panel where inadequate welds were made. This expansion causes failure of the abutment backwalls, bearings and concrete within the steel grid.

It is recommended that the concrete filled grid deck have a one-half (1/2) inch concrete cover above the grid and also either a 1 1/4" latex modified concrete or an asphalt wearing surface with waterproofing membrane system. Joints between panel should be weld spliced at all cross bars and the main bars to the stringer should be rigidly connected.

### 4-6.4 CORRUGATED METAL DECK

Corrugated steel flooring with asphalt cement has been used on local street bridges. This system can extend the service life of bridges by reducing the dead weight, however it may be necessary to install additional beams between the existing stringers to meet restricted deck span.

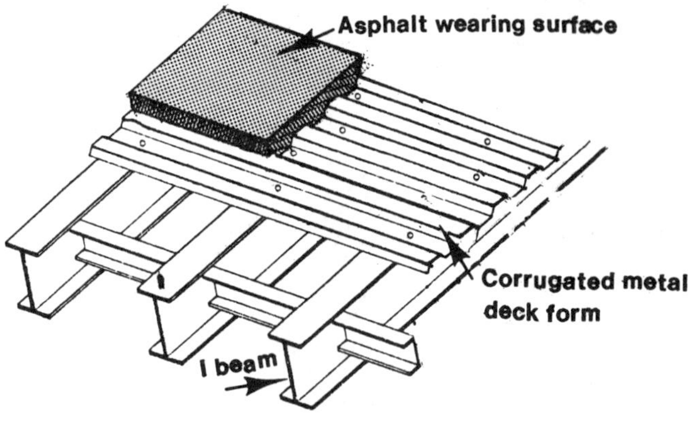


**Fig. 4.19 Corrugated Metal Deck**

Since the common problem of a corrugated metal deck is corrosion, the proper design of a drainage system for run off water should be a critical element.

### 4-6.5 TIMBER DECK

Common timber decks are plank floors or laminated wood floors:

**Plank Floors:**
This is a conventional type of floor and is widely used for timber deck.

**Glue-laminated Wood Floor:**
This system is relatively new and its performance has been satisfactory. It has a resistance to chemicals used in ice control but care should be exercised in connection and fastening details to avoid insect infestation, water leakage into laminated sheets, and shear failures.

A timber plank bridge deck gives poor resistance against skidding when it is wet. The surface bleeding of the creosote or heavy oil-borne preservative also reduces the skid resistance. To increase the skid resistance and to provide a smooth riding surface, asphalt concrete overlays are used extensively. However, there are the following problems: bonding of asphalt overlay to the deck, deterioration of the asphalt wearing surface such as cracks, breaks, spalls, etc. especially cracks over the joints between planks due to differential deck deflection.

**Fig. 4.20 Timber Deck**

It is recommended that the use of glulaminated decks in place of plank decks to eliminate the differential deflection between deck planks. To improve bonding to the treated timber and to neutralize the free preservatives, it is suggested that blotter-spread aggregate or dust be used to act as an absorbent and to increase surface roughness for better bonding. Also a minimum 1 inch of asphalt surface thickness is recommended.

## 4-6.6 ORTHOTROPIC STEEL DECK

The orthotropic steel system which is considered as an orthogonal anisotropic plate, has different flexural stiffeness of longitudinal rib in orthogonal directions or at right angles to the floor beam and the plate is anisotropic in two perpendicular directions of the floor beams and the ribs. The whole system became orthogonal-anisotropic, or, briefly, orthotropic.

The steel deck plate with its closely spaced ribs (open or closed) and transverse floor beam is considered as a homogeneous plate in flexure. It acts not only as the deck slab but also as the top flange of the girders The elimination of a concrete deck slab and reduction of the dead loads increases the load carrying capacity of the existing bridge. It is for this reason that engineers should consider the use of an orthotropic steel deck as a replacement for the existing floor slab.

### Closed Rib System

Most major orthotropic steel deck plate bridges use closed rib because it has a better wheel load distribution, uses less steel and provides smooth surfaces of the box rib. It is best suit for long spans (up to 25 feet). The disadvantages are complicated details and higher fabrication costs.

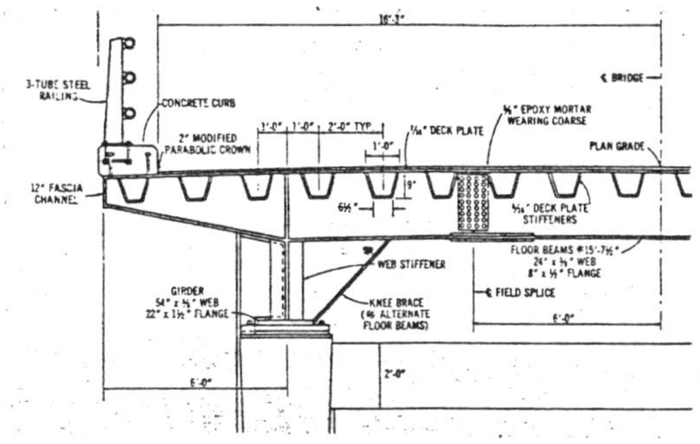

**Fig. 4.21 Orthotropic Deck**

### Open Rib System

Open rib systems generally use inverted T stiffener or bulb bars. The open rib system requires heavy steel but it can be considered as an alternate where the installation, accessibility and clearance are critical. A "T" stiffener may face a narrow space between the flanged rib while the flat stiffener does not have space problem but faces substantial weight increase problem. However the bulb type provide relatively light weight and accessibility.

**Fig. 4.22 Types of Rib for Orthotropic Deck**

### Wearing Surface

It is important to protect deck plates from corrosion and provide smooth riding surface. The basic types can be either the bituminous based thermoplastic or epoxy base thermosetting. Thermosetting can be heated until they become liquid. They set permanently on continued heating and can not be softened again by reheating while thermoplastics set when they are cooled and can be softened by reheating.

COAL TAR EPOXY:
Coal tar epoxy has been used since 1960 as seal or bond coats on steel deck along with inorganic zinc primer or red-lead epoxy primer. The final layer usually consists of a 2" asphalt wearing surface. Coal tar epoxy bond coat appears to be performed satisfactory on the Port Mann Bridge of Vancouver, B.C. since its installation in 1964.

EPOXY ASPHALT:
Epoxy asphalt is known as Concresive Asphalt, produced by Adhesive Engineering Company. It is reported to be performing well on several bridges. It maintains good bond to steel surfaces, adds stiffenes to the steel deck plate resisting for fatigue cracks and has high plastic characteristics, and high resistance to wear. Epoxy asphalt is mixture of epoxy resin, asphalt filler and aggregates.

The steel deck plate surface are sand blast cleaned to white metal and primed with a inorganic zinc primer. After spraying an epoxy tack coat, a layer of the epoxy asphalt mixture (3/4" to 5/8" thick) will be compacted and 1 1/4" to 2" thick bituminous concrete wearing surface will be placed.

POLYURETHANE TILE:
This is relatively new product manufactured from vulcanized fiberglass filled elastomeric compound using polyurethane.

WATERPROOFING MEMBRANE:
This consists of placing bituminous based fabric waterproofing membrane after sandblasting and primer coating, 3/4" modified non-permeable asphalt and 1 1/4" to 2" asphalt wearing surface.

The Port Authority of New York and New Jersey had selected and used orthotropic steel decks among other alternatives for replacement of the existing upper roadway concrete deck slab of the George Washington Bridge. A minimum section of deck was removed by saw-cutting and then replaced with prefabricated and prepaved deck panels. The deck panels consist of a 5/8 inch thick deck plate stiffened by WT 7x13 sections at 15 inches spacing and welded to the deck plate. All steel in the panel is A588 grade 50 steel and the asphalt concrete pavement is 1 1/2 inches thick. The main advantage of the orthotropic deck was a significant reduction of the stress in the main structure members of the bridge by reducing dead loads.

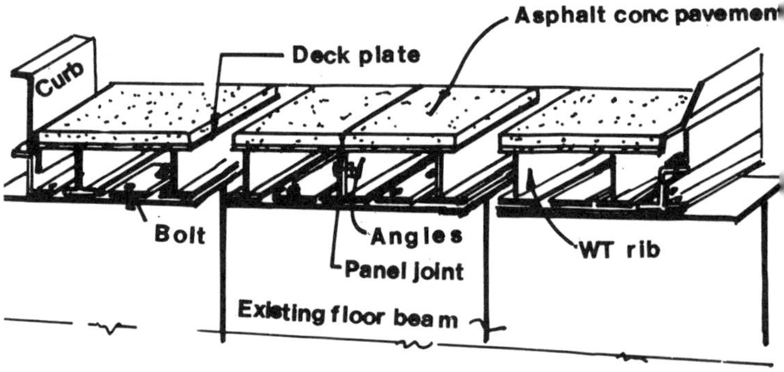


**4.23 Orthotropic Deck Replacement**

In Cornwell, Connecticut, an old, covered, wooden bridge superstructure was replaced with an orthotropic deck system. The system contains a 3/8 inch steel plate stiffened with closed ribs spaced at 2'-0". On the top of plate, 2"x10" wooden planks were bolted with thread welded studs.

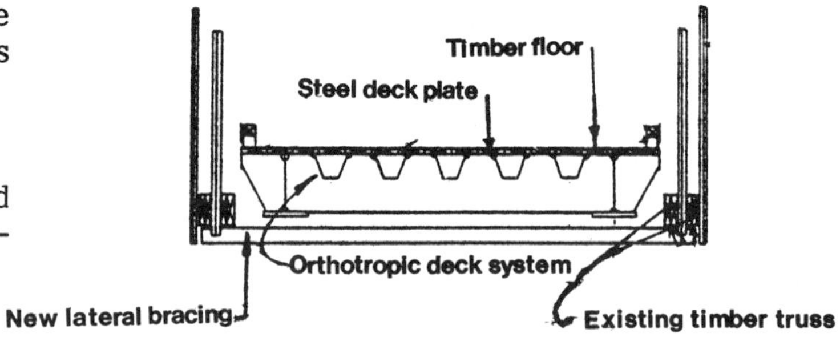


**Fig. 4.24 Orthotropic Deck Replacement**

When the deck replacement of the Golden Gate Bridge was studied, the steel deck stiffened by closed trapezoidal ribs with an epoxy asphalt wearing surface was selected over other alternates, (precast lightweight concrete deck and concrete filled steel grid deck). The steel deck provide the lightest weight thus reducing the stresses of the cable tower and the foundation. It also offers the longer life expectancy with low life cycle costs.

Fig. 4.25 Placement of Stay-In-Place Form

## REFERENCE


1. "Manual of Concrete Practice" ACI Detroit, Mich.
2. "Guide to Durable Concrete" ACI Committee 201, ACI Journal Dec. 1977. pp573-609
3. "Standard Practice for Concrete Highway Bridge Deck Construction" ACI Committee 345, Concrete International, Sep. 1981.
4. "Guide for Repairs of Concrete Superstructure" Report No. 546, ACI 1980.
5. Branson, D.E. "Deformation of Concrete Structure" McGraw Hill, NY, 1977.
6. Brown, B.C. " A Further Evaluation of Concrete Bridge Deck Surfacing in Iowa" Iowa DOT Special Report, 1979.
7. "Bridge Floor System" Reliance Catalog RSL-18B.
8. FHWA Publication:

- "Federal-Aid Highway Program Manual" Vol.6, Chapt.7, Sect.2, Subsection 7, 1976.
- "Cost effective Rigid Concrete Construction and Rehabilitation in Adverse Environment" 4K Annual Report (1982).
- "Extending the Service Life of Exisiting Bridge by Increasing Their Load Carrying Capacity" FHWA-RD-78-133, June 1978.
- "Evaluation of Super-Water Reducers for Highway Application" FHWA/RD-80/132, March, 1981.
- "Connections for Modular Precast Concrete Bridge Details" FHWA/RD-82/106, August, 1983
- "Evaluation of Portland Cement Concrete for Permanent Bridge Deck Repair" FHWA/RD-74/5, 1974.
- "Internally Sealed Concrete" FHWA/DP-49-2, Jan. 1982.
- "Latex Modified Concrete Bridge Deck Overlays: Field Performance Analysis" FHWA/OH-79/004, 1979.
- "Polymer Impregnation of New Concrete Bridge Deck Surfaces" FHWA Implementation Package 78-5, Jan. 1978.
- "Polymer Concrete Patching Manual" FHWA/IP-82-10, June 1982.
- "Polymer Concrete Overlays Interim User Manual" FHWA-TS-78-225 and TS-78-218.
- "Styrene-Butadiene Latex Modifiers for Bridge Deck Overlay Concrete" FHWA/RD-78-35, April, 1978.
- "Transverse Cracking of Asphalt Pavement" FHWA-TS-82-206, July 1982
- "Waterproofing Membranes for Bridge Deck Rehabilitation" FHWA/NY-77-59-1


- "Transverse Cracking of Asphalt Pavement" FHWA-TS-82-206, July 1982.

9. "Superplasticized Concrete for Rehabilitation of bridge Decks and Highway Pavements" US DOT/RSPA/DPB-50/81/31, Jan. 1981.
10. Furr, Howard L. "Bridge Slab Concrete Placed Adjacent to Moving Live Loads" Report FHWA/TX-81/11+266-IF, Texas Transportation Institute, 1981.
11. GangaRao, Hota V.S. "Feasibility Study of Steel Grid Decks for Bridge Floors" West Virginia University, July 1980.
12. "Gunite" Pressure Concrete Construction Co. Florence, AL 35630.
13. "Solving Corrosion Problems of Bridge Surfaces Could Save Billions" US General Accounting Office, Jan. 1979.
13. "Polymer Concrete Overlay on S4-51 Bridge Deck" Oklahoma DOT Research & Development Division, March, 1982.
14. Mont, R.M., Eugene Fasullo, Daniel M. Hahn, "Replacement of the Upper Deck of the George Washington Bridge" Annals of the NY Academy of Sciences, Vol.352, pp 143-156, Dec. 1980.
15. "Pre-Cast Slabs for Composite Bridge Decking" US Steel State of the Art Report, Nov. 1981.
16. Reilich, H.D "Golden Gate Bridge: Deck Investigation" ASCE Journal of Technical Council, Vol. 107, pp101-116, April 1981.
17. Seim, Charles "Studies Leading to the Golden Gate Bridge Deck Replacement Project" Proceeding of Bridge Maintenance and Rehabilitation Conference, pp727-744, 1980, W. VA College of Engineering, Morgantown, WV.
18. Stewart, C.F. "Consideration for Repairing Salt-Damaged Bridge Decks" ACI Journal Vol 72 pp685-698, 1975.
19. Timmer, Donald H. "A Study of Concrete filled Steel Grid Bridge Decks in Ohio" Proceeding of Bridge Maintenance and Rehabilitation Conference pp422-475, 1980, Morgantown, WV.
20. Transportation Research Board:

- Minkarah I. "Behavior and Repair of Deteriorated Reinforced Concrete Beams" Transportation Research Record 821 pp73-78.
- "Bridge Deck Repair" Research Results Digest 85, March 1976.
- "Bridges on Secondary Highways and Local Roads-Rehabilitation and Replacement" NCHRP Report 222 (1980).
- Biswas M. "Bridge Replacement with Precast Concrete Panel" TRB Special Report 148 pp136-148, 1974.
- Buth, Eugene "Use of Prestressed, Precast Concrete Panels in Highway Bridge Construction" Special Report 148 pp122-135, 1974.
- Nicholso, J.P. "New Approach Cathodic Protection of Bridge Deck and Concrete Structure" TR Record 762 pp13-17. 1980.
- Ellis, W "Cathodic Protection of Concrete Structures" Draft Report of NCHRP Project 12-19, Sep. 1980.
- Manning, D "Decision Criteria for Rehabilitation of Concrete Bridge Decks" TR Record 762 pp1-8, 1980.
- "Durability of Concrete Bridge Decks" NCHRP Syntheisis of Highway Practice 57 (1979).
- Manning, David G. "Effects of Traffic-Induced Vibrations on Bridge Deck Repairs" NCHRP Report 86 (Dec. 1981).
- Boyd, W.K. "Influence of Bridge Deck Repairs on Corrosion of Reinforcing Steel" Draft Report of NCHRP Project 12-16 (Jan. 1979).
- "Rehabilitation and Replacement of Bridges on Secondary Highways and Local Roads" NCHRP Report 243 (Dec. 1981).
- Kliethermess, John "Repair of Spalling Bridge Deck" Highway Research Record 400 pp83-91.
- Darter, Michael I. "Patching Continuously Reinforced Concrete Pavements" TR Record 800 pp12-17.
- Fromm, H.J. "Successful Application of Cathodic Protection to a Concrete Bridge Deck" TR Record 762
- "Use of Polymers in Highway Concrete" NCHRP Report 190.
- "Waterproof Membrane for Protection of Concrete Bridge Deck" NCHRP Report 165 (1976).
- "Rapid-Setting Materials for Patch ing of Concrete" NCHRP Practice 45.
- "Concrete Sealer for Protection of Bridge Structures" NCHRP Report 244, TRB, Dec., 1981
- "Evaluation of Methods of Replacement of Deteriorated Concrete in Structure" NCHRP Report 1. TRB

21. "Technical Data" E-Poxy Industry, Inc. Ravena, NY
22. Ryell, J, M. Owens, D.G. Manning "Durable Highway Structures - Ontario's Approach" Ontario Ministry of Transportation Dec. 1982.
23. Fondriest FF, M J Snydel "Paving Practice for Wearing Surface on Orthotropic Steel Deck" Jan. 1968 and "Addendum" Oct.,1971 American Iron and Steel Institute.
24. "Adhesive/Coating/Sealant Construction" Adhesive Engineering Co. San Carlos, CA
25. "Benjamin Franklin Bridge Design Engineering Report" Delaware River Port Authority, 1983

# 5 CONCRETE STRUCTURE

## 5–1 GENERAL DEFECTS

The deterioration of concrete structures is caused by many factors such as traffic collision damage, structural failures due to overstress and/or loss of section, freeze and thaw, settlement of foundation, chemical attack, failure of expansion joints, scours, silting, poor design details, construction deficiencies and corrosion of reinforcing steel.

**Cracks:** Usually any signs of distress is detected by the evidence of cracks. When a crack is seen, the location of it should be carefully observed and special efforts should be made to determine the cause of it so as to make proper remedial repairs. Cracks may be caused by shrinkage, overstress, restraint of bridge movement due to frozen bearings, settlement of soil, scour and chemical attack.

**Scaling:** Scaling is surface pitting due to the freeze and thaw cycles. It can spread over the entire surface. Air entrainment in concrete is proven to be effective for preventing scaling.

**Spalling:** Spalling is the breaking away of the surface concrete due to the expansive forces caused by corrosion of reinforcing steel. Spalling is aggravated by water and salt to seeping into cracks.

**Popouts:** Popouts are cavities in concrete caused by the expansion of porous aggregates or reactive ingredients. They usually appear shortly after placement, but may appear after a year depending upon the depth of embedment, volume instability of the aggregate and permeability.

**Honeycombing:** Honeycombing is the formation of small holes or voids in walls or columns due to a failure of the filling in between the spaces of coarse aggregate particles. Honeycombing is also caused by segregation of the concrete during construction.

**Traffic Collision Damage:** When major structural components of a bridge are severely damaged by vehicular collisions, there is a danger of collapse.

**Scour:** Channel blockade caused by accumulations of silt and debris may reduce the available waterway opening and cause severe scour damages to the structure. Consideration should be given to inspect the channel and the bridge over the waterway after a seasonal flood in order to detect any changes in the stream channel which indicates a tendency to scour and the build up of bars in the waterway opening.

**Settlement:** Foundation settlement can cause major structural failures, therefore, any foundation movement of a structure should be carefully monitored. Field surveys of the structure are helpful in determining the type and amount of settlement. Wide vertical cracks or the crushing of concrete members by other structural elements where contact is not required should be observed as possible signs of settlement.

**Chemical Attack:** Salt and deicing chemical agents are the greatest cause of concrete deterioration. Chemical agents disintegrate the concrete and separate the aggregate. It is recommended to use non-alkali aggregate and cement which has less than 0.6% alkali contents ($Na_2O$) to prevent deterioration.

**Design Defects:** Poor design details and practices can cause the deterioration of a concrete bridge and make maintenance of the structure difficult. The following are some details which result in the deterioration of a concrete structure:

1. Insufficient cover for reinforcement causes cracks and spalls.
2. Lack of consideration for creep, especially on skewed bridges.
3. Improper location of contraction and expansion joints.

4. Expansion joint details which cause water leakage.
5. Drainage details which permit clogging.

## 5–2 INVESTIGATION

The major problem of bridge maintenance results from the corrosion of top layers of reinforcement steel in the deck slab. However, Deterioration is not limited to the deck. Severe deterioration occurs in beams, columns, pier caps, bearing seats, piles and pier noses and the problems are very similar to those associated with bridge decks. The inspection methods are similar to the techniques used for a deck condition survey:

1. **Visual Inspection:** Most common method of detecting cracks, spalls, rust stains, discoloration and other deterioration.
2. **Sounding:** Using hammer to locate delaminated and disbonded area of concrete.
3. **Potential Measure:** Using Copper/ Copper sulfate half cells to measure electrical potentials. Potential difference of -.35 volts (CSE) or more indicates active corrosion.
4. **Chloride Analysis:** A quantitative measure of the chloride ion content in the concrete. The threshold chloride content needed to initiate corrosion is from 1.3 to 2.0 pound per cubic yard of concrete.
5. **Concrete Core:** Cores can determine the condition of the concrete and compressive strength, special gravity, percentage of void and absorption.
6. **Pachometer** measures insufficient cover of the reinforcing steel.

All of the above methods are useful for detecting and defining the deteriorated areas and they provide the data necessary for a complete evaluation of deterioration and repair.

**Fire and Collision Damage:** Fire and collision damage is different from normal defects and a careful visual examination of columns, slabs and beams gives valuable information for repairs and strengthening. The fire and collision damages may be classified into the following three (3) groups:

Minor
Moderate
Major

TABLE I FIRE DAMAGES

| Description | Minor | Moderate | Major |
|---|---|---|---|
| Smoke Deposit | Present. | Generally covered. | Completely covered and color changed. |
| Spalls | Minor. | Substantial on the concrete cover. | Extensive. |
| Cracks | None. | Several fine cracks. | Wide cracks. |
| Delamination | None. | Concrete cover sounds hollow when hit with a hammer. | Extensive beyond the concrete cover. |
| Number of main bars buckled. | None. | One or less. | Several. |
| Reinforcement bars exposed. | Little or None. | 50% of the main bars. | Almost all main bars. |
| Color change | None. | Pink. | Buff to gray. |
| Deflection | None. | None. | Substantial. |
| Required repairs | Cosmetic patch. | Removal of unsound conc. & reinf. bars and rebuilding damaged areas. | Replacement or strengthening. |

| TABLE II COLLISION DAMAGES | | | |
|---|---|---|---|
| Description | Minor | Moderate | Major |
| Spalls | Minor. | Substantial. | Extensive. |
| Cracks | None. | Few fine cracks. | Several wide cracks. |
| Delamination | None. | Concrete cover. | Beyond the conc. cover. |
| Number of reinf. bars cut | None. | One or less. | Several. |
| Required repairs | Cosmetic repairs | Removal of concrete and patch and/or epoxy injection. | Replacement or extensive strengthening. |

## 5–3 CONCRETE REPAIRS

### 5-3.1 GENERAL REPAIRS

The damaged concrete bridges are repaired by replacing the defective concrete with portland cement concrete, mortar or epoxy type mortars. Shotcrete, or so called "Gunite", has been used widely for repairs along with epoxy pressure injections and pressure grout. To achieve proper repairs, the cause of deterioration or distress must be removed or corrected. The repairs of concrete then proceed as follows:

**Fig. 5.1 Severely deteriorated concrete "T" beam. The only choice is total replacement of superstructure.**

The deteriorated area should be saw cut or jackhammered at the end of the repair area to provide a clean and sharp edge for the bonding between the new and the old concrete. The substrata for patch should be free of cracks and all cracks should be sealed before patching.

The reinforcing steel should be cleaned and if there is damage, it should be repaired or replaced with new reinforcing bars having a proper overlap with the existing bars.

When there is an active or a possible corrosion in the reinforcement, all contaminated concrete should be removed and the existing reinforcement bars or the new bars can be coated with epoxy bonding compound to improve bonding and to prevent galvanic corrosion in the future. However, the epoxy coating should not be used, if any future cathodic protection is planned, because it would shield the underlying steel from a protective current.

The choice of the repair method and the materials depends on the deterioration, availability of materials, cost, environment, adaptability of the proposed method of repair, aesthetics and nature of the required repairs (permanent or temporary). The following are some of the factors to be considered in selecting materials:

1. The new repair material must be completely bonded to the existing concrete and be able to withstand the weathering and the surrounding environment.
2. Repair materials must have low differential volume changes between the existing concrete and new repair materials. To minimize the initial shrinkage, the aggregate size should be maxi-

mum (low volume of cement paster) and water contents should be minimum (low water cement ratio) within the limits of acceptable mix design. A strict curing procedure should also be practiced.

3. The patch material should have low permeability so as to retard moisture penetration into the existing concrete.
4. The patching material should have the same matching color and texture as the existing concrete.

## 5-3.2 REPAIR MATERIALS

**Porland cement concrete or mortar** are most widely used for repairs. Admixture and additives such as an accelerator, a retarder, and air-entraining are used to improve workability, set-time and strength. Superplasticizer (high range water reducer) are recently used for good flowability, low water-cement ratio and high strength. Type III high early strength cement is used where rapid strength gain is required. For pressure grouting, cement-pozzolan slurry (mix 1:3 by volume) is recommended. Superplasticizer and non-shrink additive may be added to provide fludity and to offset shrinkage.

**Epoxy resin mortar** is popular and successful for small patch areas. This mortar with a curing agent has a wide range of set time and strength depending on manufacturers.

**Polymer concrete** is a two component methylmethacrylate monomer based mortar. The monomer penetrates the concrete and strengthens the concrete after polymerization.

**Latex modified concrete** is portland cement mixed with latex emulsion. This is widely used for deck protective systems.

**Dry pack** refers to a cement mortar with no slump. The mortar mix has a very low water-cement ratio. It forms a ball when you squeeze a small quantity of material in your hand, but leaves very little water.

**Fast-Setting Mortar:**

1. **Type III - High early strength cement:** This has been widely used but it has the disadvantages of high shrinkage and slow strength gain in cold weather.
2. **Chemical - Setting Cement:**
   **a) High-Alumina Cements (Mono-Calcium Aluminate)** are widely used in Europe and are used in several states. This type of concrete is expected to decrease in strength of about 50% in service therefore it is not recommended for use in structural concrete unless the strength obtained at 24 hours is about 200% of design strength. It is sulfate resistant and could be considered for use where steel corrosion is a problem.
   **b) Magnesia-Phosphate** It is supplied in two packages - dry magnesia and liquid phosphate. It produces a high strength, low permeability patch material. It is suitable for small and large patches.
3. **Thermosetting Material (Polymer Concrete)** Epoxy resin mortar has been used widely and successfully several years. However, the selection of a proper material from the many commercially available products could be a problem. Currently available types are epoxy resin, polyester resin and monomer.
4. **Calcium Sulfate Material** Patching materials containing calcium sulfate gain rapid strength and can be used at temperatures above freezing. But it has been found that some brands are not durable when exposed to moisture and freeze and thaw cycles.

In general, the fast-setting material is not cost effective for large repairs. However, the use of it is justified to save curing time where early completion of repairs is essential.

**Bituminous Material:**

1. **Bituminous Coatings** are used for water proofing and protecting concrete from weathering. Coal tars have been used extensively to retard water penetration. The disadvantage of coating is the periodic recoating and replacement of the water proofing membrane, if damaged.
2. **Hot Asphalt Sealer** is used to seal the cracks in an inexpensive way.

3. **Hot-Mixed Dense Grade Asphalt Cement** is used for large patch areas.
4. **Cold-Mixed Cutback Asphalt** is used for temporary patches.
5. **Guss Asphalt** This highly impervious asphalt is used for restoring surfaces in thin applications of 1/2" to 3/4".

**Coatings:**

1. **Linseed Oil** is used to prevent scalings and is usually used for a deck.
2. **Epoxy Waterproofing** Epoxy sealers are used at the bridge bearing seats to prevent deterioration from joint leakage.
3. **Penetrant Sealant** This consists of the application of penetrant solution to the

TABLE III CONCRETE REPAIR METHODS AND MATERIALS

| Defects | Repair Methods | Materials |
|---|---|---|
| Cracks (Active) | Caulking | Elastomeric sealer |
| | Pressure injection | Flexible epoxy resin mortar |
| | Jacketing | |
| | Strapping | Steel wire or rod |
| | Overlaying | Membrane or special mortar |
| | Strengthening | Steel plate, post tensioning, stitching |
| Cracks | Caulking | Cement grout or mortar, Fast-setting mortar |
| | Pressure injection | Epoxy grout or mortar |
| | Coating | Bituminous coating, tar |
| | Overlaying | Asphalt overlay with membrane |
| | Grinding & Overlay | Latex modified concrete, High density concrete |
| | Dry pack | Dry pack |
| | Shotcrete | Cement mortar, fast-setting mortar |
| | Patching | Cement mortar, Epoxy or Polymer concrete |
| | Jacketing | Steel rod |
| | Strengthening | Post tensioning |
| | Reconstruction | |
| Voids | Dry pack | Dry pack |
| | Patching | Portland cement grout, mortar, cement |
| | Resurfacing | Epoxy or Polymer concrete |
| | Shotcrete | Fast-setting mortar |
| | Preplaced aggregate | Coarse aggregate and grout |
| | Replacement | |
| Scaling | Overlaying | Portland cement conc., Latex-modified concrete |
| | | Asphalt cement, Epoxy or polymer concrete |
| | Grinding | |
| | Shotcrete | Fast-setting mortar, Cement mortar |
| | Coating | Bituminous, Linseed oil coat, Silane treatment |
| | Replacement | |
| Spalling | Patching | Concrete, Epoxy, Polymer, Latex, Asphalt |
| | Shotcrete | Cement mortar, Fast-setting mortar |
| | Overlay | Latex modified conc., Asphalt concrete, Conc. |
| | Coating | Bituminous, Linseed oil, Silane |
| | Replacement | |

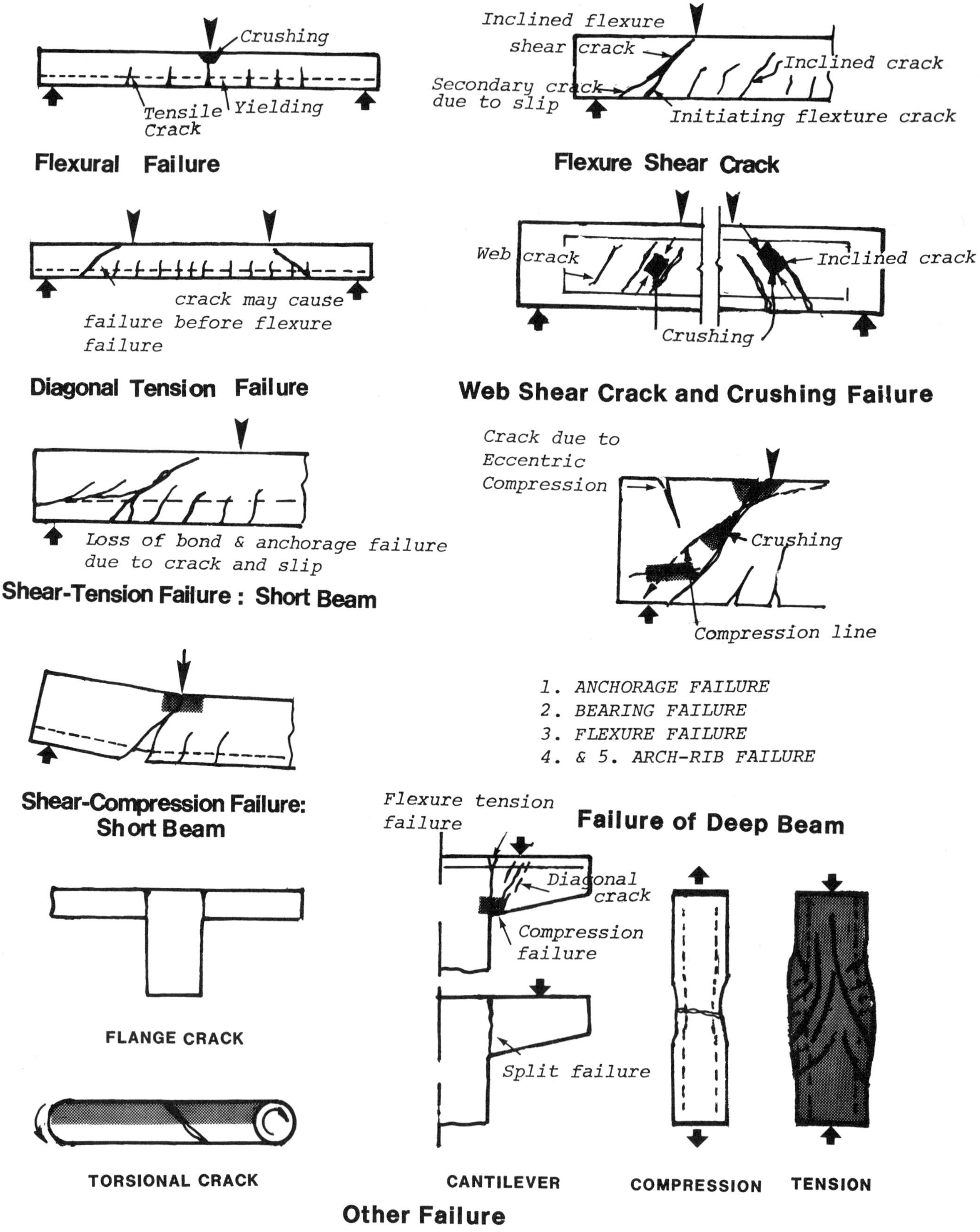


**Fig 5.** Typical Failures of Concrete Structure

concrete surface. This solution reacts with the concrete when it contacts, and creates a hydrophobic surface that prevent water penetration.

Several states have used silane bridge treatment on new bridges and its performance appear to be satisfactory. The advantage are low cost, easy application and minimal curing times. Before spraying the solution, the surface must be cleaned by sandblasting or another suitable way to be able to absorb the solution. Generally, the surface is treated at a minimum rate of 125 to 175 square feet per gallon.

### 5-3.3 REPAIR METHODS

Bridge repairs can be classified either as cosmetic repairs, rehabilitation in kind or strengthening. The following are the commonly used methods for repairs.

1. Dry pack method for deep and narrow cavities.
2. Preplaced aggregate method (Prepack method) for restoration of large areas such as walls, foundation and spillways.
3. Partial or full depth concrete replacement by casting or patching.
4. Shotcrete (Pneumatically applied mortar or Gunite).
5. Patching.
   a. Low slump high density concrete or mortar.
   b. Latex modified concrete or mortar.
   c. Epoxy mortar.
   d. Polymer concrete mortar.
6. Epoxy injection, caulking or pressure grouting.
7. Protective coating and plaster.
8. Strengthening.
   a. Jacketing.
   b. Stitching.
   c. Reinforcing.
   d. Post tensioning.

### 5-3.4 SURFACE PREPARATION AND REINFORCEMENT STEEL

**Surface Preparation:**

Since the patch will only be as strong as the surface to which it is bonded, the deteriorated concrete should be removed and the sound and uncontaminated concrete should be exposed to place repair materials.

The edge of patch area should be sharp cut 1" to 1 1/2" deep by a saw-cut or slightly undercut by chipping hammer. Removal of concrete should be done by hand tools or light weight air hammer of the 15 to 30 pound size with a spade shape bit (not a chisel-tipped bit) to avoid any formation of cracks during removal.

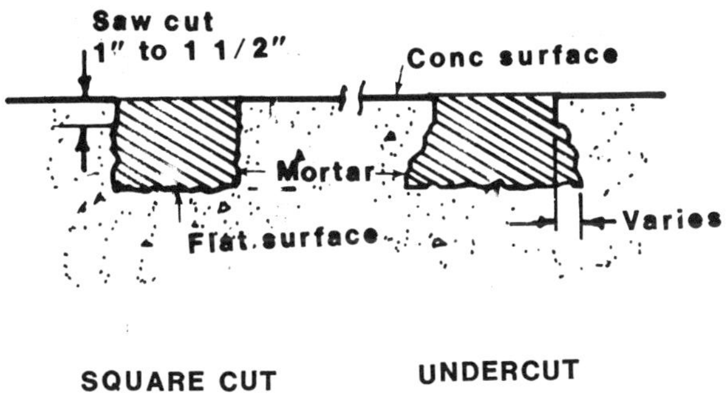


**Fig 5.3. Typical Cut for Patched Area**

**Crack Repairs:**

All deteriorated joints and cracks in concrete should be chipped out to a minimum 2 to 4 inches in width and 2 inches in depth. Where there are any movements in the structure, the cracks and joints which showed evidence of leakage can be sealed by 10 inch long cold rolled copper expansion plates or PVC waterstop which is preformed along the longitudinal centerline of sealer to make a modified V shape, 1 inch high and 1/2 inch open with round bend. The concrete should be chipped out enough to make space for installation of a water tight joint between the sealer material and concrete. A channel for water seepage and a drainage pipe at the base of joint or crack should be placed as shown below:

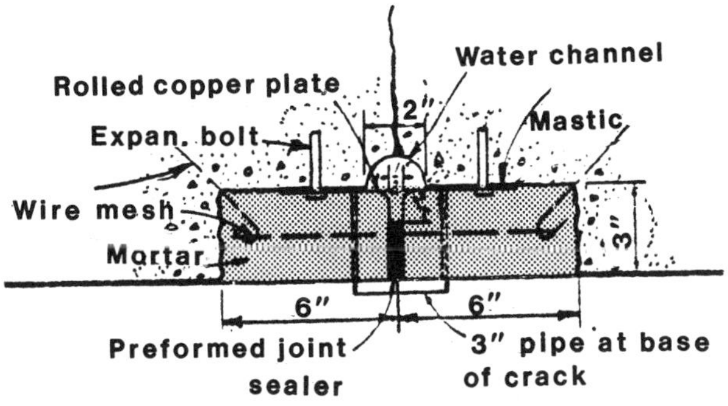


**Fig. 5.4 Repair of Active Crack**

Vertical cracks which is not leaking but subjects to movement only can be sealed with a flexible two component compound such as flexible epoxy or polysulphide joint sealing compound, or elastomeric joint sealer. Non-moving cracks are normally repaired by the epoxy injection method. On the detail shown above, the joint between the finished surface and V of the copper plate or PVC should be filled with a 1/2" thick premould asphalt sealer or preformed elastomeric joint sealer material and the patch should be reinforced by wire mesh or smaller size reinforcing bars.

**Dowel or Anchor Bars:**

When new concrete is joined to existing concrete by the dowel and grout method, the dowel should be deformed reinforcing bars or threaded bars and long enough to sustain design loads. The holes should be drilled in the existing concrete by methods that will not shatter or damage the concrete adjacent to the holes. Horizontal holes should be drilled downwards with slope of 12:1. The diameters of the drilled holes shall be 1/4" or 50% larger than the nominal diameter of the dowels.

The grout shall be a neat cement paster of portland cement and water. The water content should be not more than 4 gallons per sack of 94 pounds of cement. Retempering of grout should not be allowed. Before placing the dowels, the holes must be cleaned of dust and other foreign materials. The holes should be thoroughly saturated with water and then have all free water removed. The holes should be dried to saturated surface dry condition. Sufficient grout should be placed in the holes so that no voids remain after the dowels are inserted.

Grout should be cured for a period of at least 3 days or until the dowels are encased in concrete, whichever occurs first, by either the curing compound method or by keeping the surface of the grout continuously damp. Any dowels or grout which fail to bond or are damaged before the new concrete is placed should be replaced.

There are several ready made dowel and grout systems in the market, such as USM Corp's Parabond and Hilt HVA Adhesive Anchor, which include a sealed glass cartridge with quartz sand and hardening agent. After placing the cartridge inside the hole, the rod is driven into the hole breaking the glass cartridge and mixing the resin, sand and glass. Also, rock anchor with hollow hole in the rod can be used on certain case such as overhead installation where the filling the grout holes is difficult.

In place of dowel bars, expansion hooked anchor bolts may be used for patches is less than 9" thick but if the thickness of patch is more than 9 inches, the dowel should be used. Where the thickness of patch is less than 4 inches, 1/4" expansion bolts may be placed at an 18" spacing on vertical surfaces and an 12" spacing on overhead surface. Each 1/4" bolts should have sufficient contact in the sound concrete to sustain a pull strength of 150 pounds. The spacing of bolts should be based on supporting 3 times the total weight of suspended concrete and 2 times the weight of concrete on vertical surface. Each bolts should be set in sound concrete as specified by the manufacturer and be able to support loads without loosening.

The following table may be used as a guideline where the patch thickness is more than 4 inches:

**TABLE IV EXPANSION BOLTS**

| Patch Thickness | Bolt Size | Spacing of Bolts Overhead | Vertical | Top |
|---|---|---|---|---|
| 4 to 5" | 3/8" | 1'-6" | 1'-9" | 3'-0" |
| 6 to 7" | 3/8" | 1'-2" | 1'-6" | 3'-0" |
| 8 to 9" | 1/2" | 1'-6" | 1'-10" | 3'-0" |
| 10"-12" | 1/2" | 1'-3" | 1'-6" | 2'-0" |

**Reinforcement:**

If any reinforcing bars are exposed, the removal should be extended a minimum of 1" around the exposed bars. If the reinforcing bars show corrosion, the concrete removal should extend to minimum of 6" into sound concrete. The reinforcing bars should not be cut or damaged during removal. Upon completion of chipping, the repair area and reinforcing bars should be cleaned by sandblasting.

After the reinforcing bars are cleaned, they should be visually inspected to check for any loss of section or other damage. Where reinforcing steel is cut or the cross section of bars are reduced by more than 20%, additional new reinforcing bars should be placed and tied into the existing bars by welding, mechanical splices or overlap splices. Generally, the minimum splice lengths for structural bars are 30 diameters for grade 40 bars and about 45 diameters for grade 60 bars.

**Fig. 5.5 Concrete Removal**

Where the thickness patch is 4" or less, one layer of 2x2 No.12 or 3x3 No.10 gage electrically welded wire mesh should be attached to the dowel or expansion bolts spaced maximum 2'-0" at vertical surface and 3'-0" on top surface. If the patch thickness is more than 4", one layer of wire mesh should be used for each 4" thickness of patch or fraction on vertical and top surfaces. On overhead surfaces, one layer of mesh should be used for each 3" thickness of patch. All reinforcement steel or wire mesh should have a minimum 1 cover for normal patches and 3/4" for shotcrete. At corners, it is recommended to place double reinforcing wire mesh extending a minimum distance of 6" beyond the intersection of the two planes. The wire mesh slice should have a minimum of one lap of mesh wire together at a minimum of 1'-6" spacing.

Before placing patch materials, it is important to recheck the following:

1. Whether all unsound concrete has been removed.
2. Whether the edges of the patch area have neat cut possibly with undercut.
3. Whether the surface is rough but uniform and free from cracks, oils, dirt, and other foreign materials.
4. Whether the perimeter of the patch area is reasonably rectangular in shape or of good geometry.
5. Reinforcing bars or wire mesh are properly placed.

### 5-3.5 DRY PACK METHOD

The dry pack method is used in repairing areas such as the narrow slots, cone-bolts, she-bolts and grout insert holes which have a depth equal to, or greater than, the least surface dimension. Dry is usually a mix of one (1) part of cement to three (3) parts of washed plaster sand.

Since color is important, small amounts of white cement are added to produce matching concrete color as determined by trial. Enough water is added to produce a mortar which can be molded into a ball by a slight hand pressure and will not exude water but will leave the hand dry. The repair holes must be cleaned and dried before filling.

Repairs are done by prewetting the surface and brushing with a stiff mortar or bonding grout. The bonding grout mix is a one to one (1:1) mix by volume of cement and fine sand, mixed to a thick cream consistency. The dry pack mix should be promptly pacted into place before the bonding grout dries. The dry packed mix is compacted with a variety of wooden tools, the exact choice of which should be made by the user to best fit the job requirements. Wooden tools are preferred, as they do not polish the surface layer of the mix and consequently offer better bond between each successive layer of the dry pack.

### 5-3.6 PREPLACED AGGREGATE METHOD (PREPACKED METHOD)

This method is used most advantageously on large repair jobs, especially where placement of normal concrete would be most difficult. This process consists of the removal of deteriorated concrete and the forming a section to be repaired. Then, forms are filled with coarse aggregate and the grout is pumped into the voids between the aggregate. After the removal of deteriorated concrete using pneumatic hand hammers, the cracks are grouted and dowel bars are anchored into the sound concrete to fasten the temperature and shrinkage reinforcement bars (generally #4 ]12"). A form is placed around the perimeter of the repairs then a clean coarse-grade aggregate (minimum 5/8" size thick section and minimum 1/2" size thin section) is placed and compacted before pumping a specially designed grout into the aggregate by special insert fittings placed in the face of the forms.

The coarse aggregate may be of any suitable size depending on the thickness of the repairs. Normal size varies from 3/4" to 1 1/2" inches. The grout can consists of 2:2:3 volume ratio of type I portland cement, fly ash, sand and water (water-cement and fly ash ratio of 0.5).

The preplaced aggregate concrete offers low drying shrinkage because of its point to point contact of coarse aggregate and high bonding strength which are essential properties for all concrete repairs. Other features of the preplaced aggregate method are excellent resistance to freezing and thawing and greater impermeability at later stages in relation to conventional concretes. Preplaced aggregate method has been used with much success to repair dams, tunnels, retaining walls, bridge piers, reinforced concrete columns and beams in bridges.

## 5-3.7 PATCHING

The most common type of repairs consists of removal of unsound concrete and patching with portland cement concrete, epoxy mortar, latex modified concrete or other fast setting material.

**Portland Cement Concrete**

After surface preparation, the surface should be kept wet for a minimum of 1 hour before application of bonding mortar and no patch material should be placed on the dry surface. The bonding grout should be made of 1 part of cement and 1 part of sand with sufficient water to make a slurry mix. The bonding coat can be applied by brushes to completely fill all surface areas. If a positive bond is necessary, epoxy bonding compound can be used on the dry surface.

Portland cement mortar with a low water-cement ratio has been used due to its low cost, compatibility with existing concrete and easy application. For low slump concrete, a water cement ratio of about five (5) gallons of water per sack of cement (about .45 by weight) is recommended. Thorough consolidation of the patch must be emphasized, as this definitely aids in bonding of the new concrete to the old.

Retempering concrete can be used for hand patching. Retempered concrete should be placed and rammed in thin layer into the form to have good bonding between existing concrete and new patch mortar.

Particular care must be taken whenever a bonded patch extends through a joint in the existing structure. All original joints must be duplicated exactly in the bonded patch. Jointing in each project is an individual matter and must be properly designed to avoid future spalling.

Proper curing of the newly placed concrete is vitally important if the patched area is to withstand the rigors of adverse weather conditions and the impact of heavy truck traffic. The first 24 hours after concrete placement are especially important. During this period, be certain that the temperature surrounding the concrete does not change appreciably and that the concrete does not dry. It is recommended the patch area should be kept wet for minimum of 7 days after placement.

**Epoxy Mortar:**

Epoxy resin mortar has the advantages of good bonding and high early strength but it has a short pot life and high cost without any advantage of surface preparation over portland cement mortar. Polymer concrete mortar (MMA) has high strength, durability, and good bonding, but it is expensive and presents handling problems.

Epoxy mortar utilizes a group of plastic materials which is comprised of two liquid components of epoxy resin compounds. The basic ingredient of epoxy is the resin itself and a catalyst. Each individual component is stabilized for an indefinite period, however once mixed in a predetermined ratio, a chemical reaction begins which converts the liquid components into a tough plastic solid. Epoxy resin mortar provides excellent bonding between materials.

The repair area must meet the requirements of being clean and dry as does the aggregate to be used in the mix. In preparing the epoxy compound, it is necessary to follow the manufacturer's recommendations for best application techniques, procedures, temperature and allowable working life.

The two liquid components must be mixed thoroughly, but not excessively, as epoxy cures by an exothermic heat reaction. It is easy to accelerate the rapid hardening characteristics to such a state that the entire batch may have to be discarded.

The normal procedure is to apply a thin layer of an epoxy bond coat to the contact surface and as soon as the coat is tacky, the epoxy mortar (usually a 1:5 to 1:8 by weight of epoxy and total aggregate) is applied to the repaired area. The epoxy concrete is then screeded to a desired level and trowelled to a desired finish. Cracks or large surface areas can be repaired by this method.

Curing of the epoxy is most satisfactory at 70 degrees plus and minus 10 degrees (Fahrenheit). The bonding agent also serves well in place of the neat cement paster between the old and new concrete in overlay and full depth repairs. In cold weather, special attention should be given to the curing time and protective curing method against cold temperatures. Curing time may be extended beyond the manufacturer's recommended period with proper heating provisions.

### 5-3.8 SHOTCRETE (Pneumatically Applied Mortar or Gunite)

The shotcrete known as "Gunite" is a mixture of water, cement and aggregate and sprayed by pneumatic method to a large and shallow repair area. This method forms a good bond to the properly prepared existing concrete. But if the proper surface preparation is not made, a large amount of material bounces off the surface to be shotcreted. Usually, welded wire fabric mesh is placed as a patch reinforcement. The use of chicken wire to retain mortar should be prohibited in any circumstance.

There are dry mix fine aggregate and wet mix course aggregate processes. The dry mix process using fine aggregate was successfully used first and it is still widely used for small and thin patch projects. The effectiveness of the dry mix process depends on the moisture content of the fine aggregate. The dry aggregate results in an excessive rebound and poor quality when sprayed from the mix nozzle. On the other hand, the wet mix process has a high rebound due to the coarse aggregate and can not be used for thin layer repairs.

The shotcrete consists of higher cement and small size aggregate. It is consolidated by the impact of the high velocity. The shortcrete may have the problems of sand pockets and lamination. The dry mix shotcrete normally does not use air entrainment to withstand freeze-thawing cycle but the use of air entrained cement and admixture will increase durability. The wet mix shotcrete show modest benefit from air entrainment even if the air bubble is lost during application. Generally durability of the shotcrete are achieve by using dense shotcrete with a lower water-cement ratio. Shotcrete is very effective for repairs of vertical members, beams, wingwalls, piers and other substructure members.

After the surface preparation, a welded wire fabric is usually attached to the existing reinforcing bars or anchor bolts where reinforcing bars are not exposed. In most cases, forms are not required for shortcreting. The quality of work depends on the skillful handling of the nozzle for good bonding, minimized rebounding and proper cover of reinforcement steel. After application of the final coat, it should be kept wet for a minimum of 4 days.

### 5-3.9 PRESSURE INJECTION

The epoxy injection technique described in the previous chapter is a simple rebonding of concrete involving sealing the cracks by pumping epoxy through ports which are inserted at interval along the crack. This sealing would not prevent any further movement between crack interface and such movement would open new cracks in the epoxy.

For cracks that have water running from them, a fast setting hydraulic cement should first used to seal the cracks and set ports. The hydraulic cement seals the cracks and diverts all water flow through the injection ports. After the hydraulic cements is cured, the cracks and port should then be sealed with the epoxy bonding compound by overlapping the hydraulic cement about one inch on either side of the cracks. Although the hydraulic cement is fast setting, it does not have sufficient strength of epoxy grout that is necessary to prevent blow-outs while injecting. The injection process follows the previously described procedures in the chapter 4. Another method involves drilling holes crossing the crack plane at about 90 degrees, inserting reinforcing bars (#4 or #5) at full depth into the holes, filling the cracks and then bond the concrete back together with reinforcing steel.

### 5-3.10 PRESSURE GROUTING

In place of pressure injection, pressure grouting can be used for filling voids, cracks in concrete and stone masonry with grout of cement, fly ash, and water. The grout mix consists of 1 part of cement, 1/2 part of sand for stone masonry and neat cement grout for concrete structure. If required, fly ash or superplasticizer can be added.

Before grouting, all deteriorated material should be removed and the entire surface of the masonry and concrete should be carefully looked for points of leakage and any signs of voids. The insert port for grouting should be properly located that the grout will fill all voids and paths of leak. The cracks and joints should be cleaned and restored or repointed with hand pointing ot shortcrete mortar ( 1 part of cement and 3 parts of sand) If the cracks or joints are small, it may be repaired after completion of grouting to relieve any trapped air inside the cracks.

For concrete structures, 1 1/2" holes should be drilled to sufficient depth and spacing to fill the grout throughout these voids and cracks. For stone masonry, It is suggested to perform test drilling through the masonry and to the bottom of the footing to decide the thickness of stone. From these test drills, the depth of grout hole can be estimated for actual grouting.

Grout hole should be drilled at regular spacing such as at 10 ft horizontal spacing and 3 ft vertical spacing staggered to cover about 25 square feet of area per hole. Holes should be a minimum of 1 1/2" diameter and drilled to intercept internal void and joints to completely bond the structure. Holes should not be completely drilled through the section and these holes should be plugged before pressure grouting to avoid leakage. Holes should be drilled to intercept horizontal joints however, if can not intercept horizontal joints then vertical joint should be intercepted.

The hole below ground should be drilled diagonally downward to the natural ground so as to consolidate bottom course and fill the cavity under the footing. For arch rings, the hole should be drilled diagonally to intercept longitudinal joints which is parallel to the barrel. The holes should be staggered and cover about 12 square feet of area.

Generally, grouting should be started at the lowest hole and at the center of the structure. If grout comes out in adjacent holes at the same level, these holes should be plugged temporarily. Pressure grouting should be continued until grout appears in the next hole or the next higher hole. Then grouting the last hole where grout came out. The same procedure is continued until all holes are grouted. In grouting, care should be given to the adjacent area for possible leakage. If grout breaks out, the grouting operation should be moved to another location or suspended for 24 hours.'

### 5-3.11 OTHERS

**Caulking:**
Caulking is used for repairs of medium to small size stabilized cracks by filling

narrow fractures in the concrete with grout or dry pack mortar. For active cracks, elastomeric caulking material or compound may be used as a filler.

**Protective Coating:**
Most widely used coating materials are linseed oil, bituminous coating, silicones and epoxy waterproofing and silane treatment. All protective coatings retard the deterioration of concrete but recoating is required about every 5 years.

## 5—4 REINFORCING AND STRENGTHENING

### 5-4.1 REINFORCING

When loss of cover of reinforcing bars occur, there are no substantial loss of strength but when the loss of bond between concrete and reinforcing bars happened, the tensile stress is increased in the region of bond loss because it can not redistribute through bond stress. The concrete has a large stress concentration at the inside reentrance corner where flexural bond is provided between the steel and concrete. If this process continues, a large spall occurs and results in the loss of cover and flexure bond. Therefore, the strength of the beam will be reduced.

When large spalls and cracks occur at the bottom of the beam, the deterioration in the beam would reduce the capacity as well as the change in behavior by losing composite action of conventional reinforcing beam design. The large spalls at the bottom of the beam result in the catenary effects which is hanging wires between two supports. This catenary effects make the spalled concrete beam to act like a tied arch. Therefore, the end anchorage of a beam would influence the reducing the crack in the beam and carrying load for which the beam was originally designed. Its failure will be sudden ultimate type shear compression failure. In repairing these spalls, it is required to use bonding compound on the patched areas including the existing bars to provide good bonding between the patch material and the existing beam.

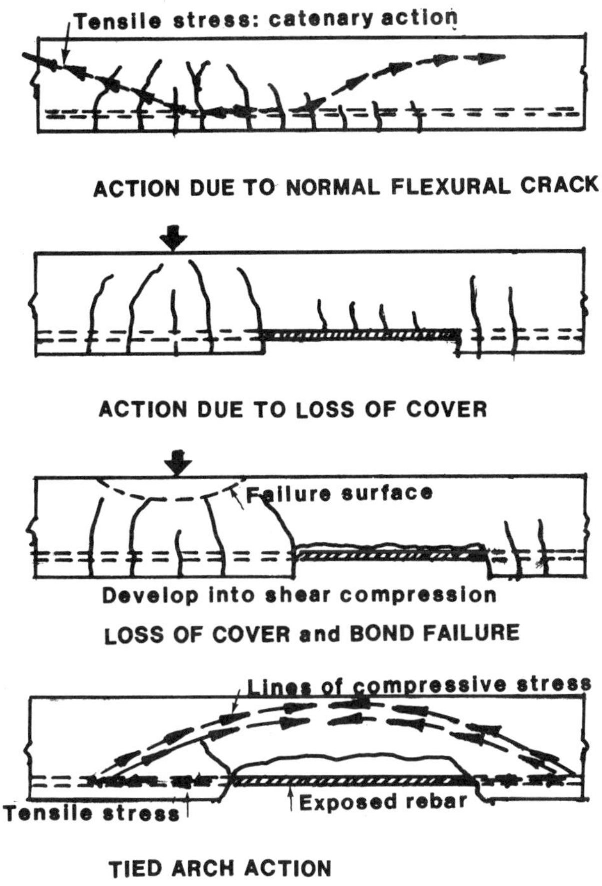


**Fig. 5.6 Behaviors of Cracked Beam**

Besides the reserved strength of the structure, most of our old structures were designed conservatively by using lower allowable stresses and higher safety factors. If analyzed, the load factor method gives a more realistic capacity of these structures, especially where dead load uses up most of the allowable stress. Before strengthening the existing structures which show no evidence of distress or deteriorations, structural analysis should be made by the load factor method to assess whether the structure could sustain the desirable live loads. If the investigation shows overstresses, then strengthening of the structure should be considered by adding a steel plate at the bottom flange or post tensioning the beam.

If a structure shows severe deterioration of concrete with exposed reinforcing steel, all unsound concrete should be removed and the loss of section of the reinforcement steel should be checked. Where the section loss is more than 20 %, new reinforcement steel should be added as required. In adding new reinforcement steel, the design should be based on the original design method, most likely working stress methods, so as to be compatible with the existing concrete. If the original design loadings are adequate, the replacement and/or supplementary reinforcing bars could be the same size and the same number of bars which can be overlap spliced, mechanically connected or welded to the existing reinforcing bars.

If the loss of section is less than 20%, the following reinforcing bars should be provided to prevent cracks:

1. On the top horizontal surfaces of patches, no reinforcement steel will be required.
2. The reinforcement on other surface of patches consists of reinforcing bars and/or welded steel wire fabric either 2"x2" of No. 12 gage wire or No. 10 gage 3"x3" wire. This fabric will cover the entire area of the patch and shall have a minimum of 1" cover from the completed surface of the patch.

Another important factor to be considered is adequate bonding between the new section and the old section. In order to prevent slippage at the interface as the new reinforcing steel pick up loads, the reinforcing steel and wire fabric should preferably be securely fastened to the existing reinforcing steel exposed by removing the disintegrated concrete. When there is no such exposed reinforcing bars or it is not practical to fasten the reinforcing steel and wire mesh to the existing steel, it should be fastened to dowels or expansion bolts which would be installed. The spacing of the dowels or bolts should not exceed 18 inches centers in both directions.

**Dowels** should be set at least 3 1/2 inches deep and the drilled hole shall be a minimum of 1/2 inch in diameter. Holes should be 1.5 times larger than diameter of the dowels. The expansion bolts must be driven tightly, but care should be used so as not to split the concrete. After installation, the bolts should be protected from being disturbed or loosened. Also, a shear key may be provided between the existing and the new concrete to prevent slippage.

**Severe Damage:** If severe structural damage has occurred due to overloads, collisions or fire, and the damaged structure shows large deflections, an adequate repair may cost more than replacing the damaged member. It is possible to replace the cast-in-place concrete member with a new prestressed concrete member having a shallow depth and thereby achieve savings in the construction time and cost, and increase the vertical clearance for the bridge.

Damaged beams or columns may be strengthened by jacketing and filling with concrete or by installing steel bandage strapping around the existing members after removing unsound concrete.

For severely damaged members, it may be presumed that the capacity of an existing member is only adequate for dead load and additional reinforcing steel should be provided for live loads and impacts. Moderate damages will require a structural analysis for evaluation.

**Shoring** is required for repairs to avoid transfer of construction loads to other members and to maintain uniform stress distribution between the existing and the new sections. Without shoring, the additional reinforcing bars will only pick up the extra loads but they would not release the residual stress of the existing reinforcing steel and the existing bars will be slightly more stressed before the new reinforcing bars are stressed.

If a beam is shored, the dead load stress in the old reinforcing steel will be relieved. After the new concrete is cured, the new and the old concrete section would be able to carry their shares of the loads and all reinforcing bars will be proportionally stressed without residual stress on particular bars.

**Reinforcing bars:** If the existing bars are weakened due to collision impacts or fire, replacement bars should be placed just far enough from the old bars so as to obtain a good bond. Also, stirrups should be replaced or added wherever required, especially near the end 1/3 of span length. New stirrups can be provided by anchoring with bolts or by welding the existing stirrup.

The end of new reinforcing bars should be adequately spliced to the existing bars by overlapping or mechanical coupler or by welding to the existing steel. Generally, a 30 diameter minimum overlap length is required for Grade 40 bars and a 40 to 45 diameter for Grade 60 bars. The minimum lap distance should be 12 inches. It is also beneficial to provide temperature and crack reinforcing bars in the new concrete. Wire fabrics are widely used for this purpose.

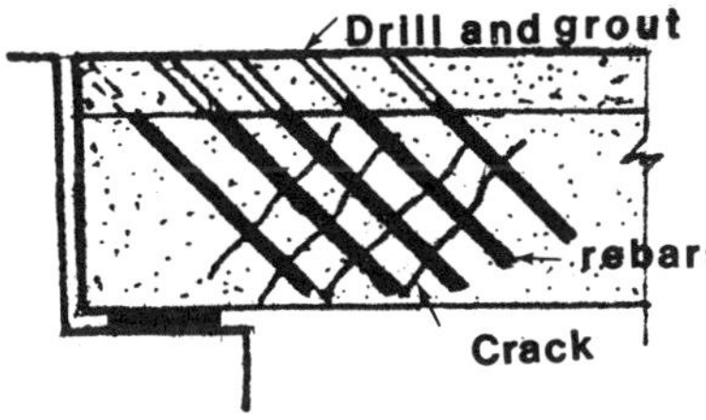


1. *Vacuum drill holes @ 6" apart and 45 degree to deck surface.*
2. *Insert a full length of reinforcing bars (min. 18") in the holes.*
3. *Fill the holes with epoxy grout under low pressure.*

**Fig. 5.7 Shear Crack Repair**

## 5-4.2 STRENGTHENING

Concrete structures can be externally strengthened by adding reinforcing plates, stitching reinforcing bars or post tensioning.

The cracks can be repaired by stitching. The stitching consists of drilling holes in the concrete in each side of the cracks and inserting U shape bars into the holes and filling them with non-shrink grout. It is important to investigate the cause of the cracks and correct the source of the problem before stitching, otherwise, the stitching will only cause other cracks to form somewhere else.

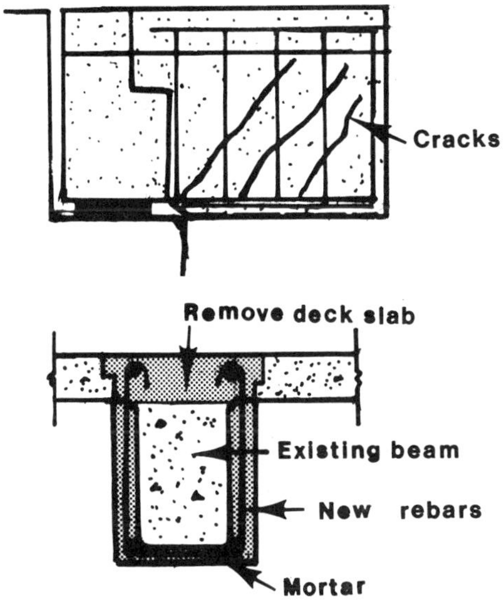


**Fig. 5.9 Jacketing Repair**

**Reinforcing Steel Plate:**

Steel plates can be added to the beam flange by anchoring with expansion bolts or with epoxy adhesive. The additional plate is transformed into an equivalent concrete section for live load or dead load. In practice, the new steel plate only be stressed under live load. Therefore it is required to find out the dead load stress in the existing section and the live load stress on the rebuilt section. Then the stresses should be added for final stress but if the plate is fully stressed, the original section may be overstressed unless the beam is jacked during installation of the plate.

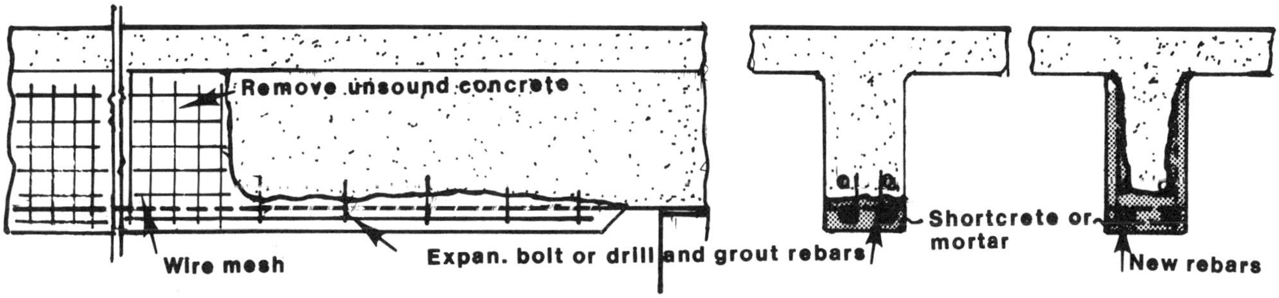


**Fig. 5.8 Repairs of Spalled Concrete**

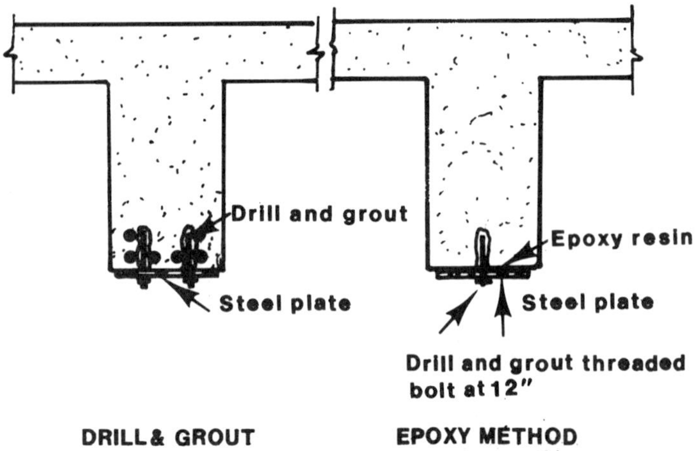


**Fig. 5.10 Strengthening by Adding a Plate**

In South Africa, applications of steel plate using epoxy-resin have been satisfactory in accordance with following procedure:

1. Holes are drilled for 5/8" threaded bolts at the center of the beam at about 12" spacing, then the bolts are inserted and the holes are grouted.
2. The concrete surface is scarified and a template is made for steel plate, then steel is drilled to match existing thread bolts.
3. The epoxy paster adhesive is mixed and spread on the plate in a wedge shape with apex along the center line of the plate. The concrete is primed with epoxy primer.
4. The steel plate is transferred into its final position, then bolts are tightened to secure the plate to concrete surface and squeeze out the extra epoxies.

Since the epoxy method of attachment is still in an experimental stage in the states, it is better to use the bolting method.

**Post Tensioning:**

The flexure cracks in concrete can be corrected by the post tensioning method. It closes the cracks by providing compression force to compensate for tension and adds a residual compression force. This method consists of anchorage of the tie rods or wires to the anchoring device, and the guide bracket angles attached to the beam. The rod wire are then tensioned by tightening of end nuts or by turning of turnbuckles at the rods against the anchoring device.

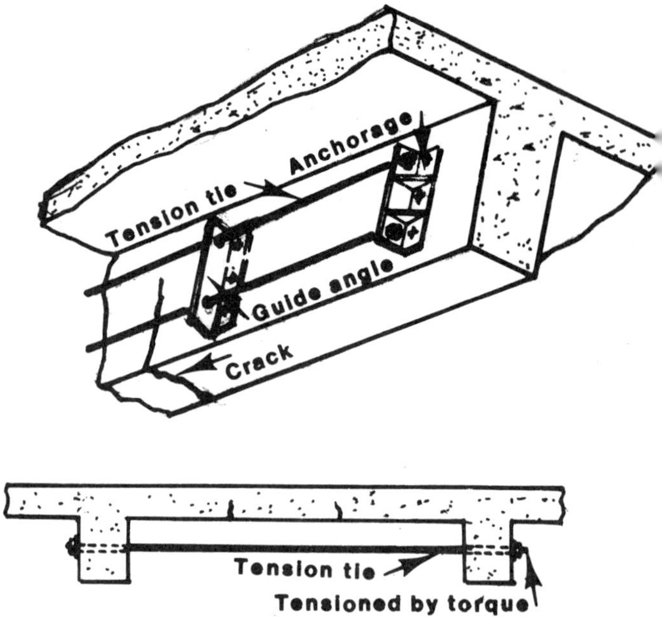


**Fig. 5.11 Post Tensioning**

**Jacketing:**

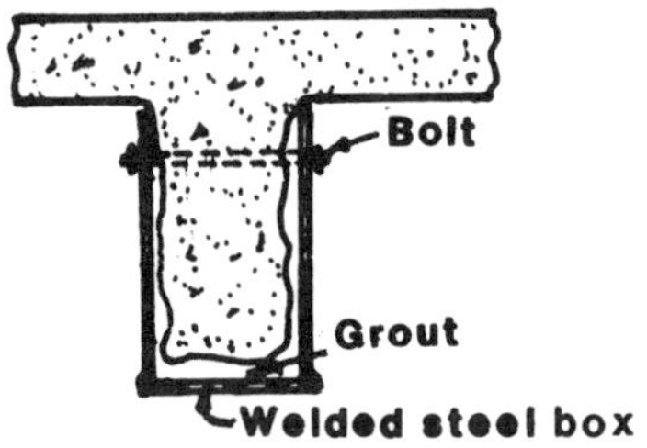


**Fig. 5.12 Jacketing by Metal Sleeve**

After removal of the unsound concrete, the beam can be jacketed with a steel form and filled with concrete for strengthening.

## 5–5 DESIGN DETAILS

Most of major structural members were designed in accordance with AASHTO or AREA specifications and are structurally adequate for design loadings. However, poor detailing of members usually causes deterioration and costly repairs such as insufficient concrete cover, poor joint details, improper joint sealers, inadequately detailed reinforcement steel, congested and complicated reinforcement details which can not be properly built, joint leakage, inadequate drainage, and improper bearings. These problems can be avoided if field condition, actual structural behaviors and construction operational restrictions are clearly thought out in advance. Unfortunately, most designers use the standard details without investigating their adaptability. Standard details will work in most cases, but sometimes they result in failure.

Other factors which should be considered are specifications, inspection and control during construction.

**Reinforcement Details:**

Epoxy coated bars are generally used in the top mat of concrete deck slab. Considering the deterioration of substructure members due to leaking deck joints, poorly located drainage pipes and splash from adjacent roadway surfaces, it is recommended that the epoxy reinforcement bars should be used in the following substructure elements:

1. directly exposed areas such as beneath the deck joints, abutment header and back wall, bearing seat, pier cap, and cap beams.
2. Directly exposed areas subjected to salt splash from roadway and marine environments such as pier columns, walls and abutment breast walls.

Due to limitations of fabrication, construction, manufacturing and transportation, the splicing of reinforcing bars is necessary to provide continuity of reinforcing steel. The knowledge of limitation of bar length would help to detail the bars. The usual stock length of No. 4 to No. 10 size reinforcing bar is 60 feet and the length of the short dimension of L-shaped bars should not be more than 8 feet in order that it does not extend beyond the maximum legal width of the truck. Also, considering the placement problems, the over-all length of the size No. 4 and No. 5 bars should be limited to 40 feet. When lapped splices are not structurally objectionable, about 30 feet maximum length of bars are preferred for No. 6 to No. 18 bars.

If a minor structure member is attached to a major member, such as a short cantilevered wingwall attached to the main U type abutment wingwall, adequate reinforcing steel must be provided in the minor members at the connection point to the main member because there is a probable presence of some movement and/or tensile stress due to the minor members. The additional reinforcing steel may be provided without design computation if the anticipated stress is small.

In deep concrete girders and pier caps, the cracks may develop at the side faces due to inadequate reinforcing bars. It is suggested to use the small size distribution bars at close spacing at each face of the beam in place of fewer larger bars. The maximum spacing is suggested to be smaller of 12" or one tenth of the depth of beams (D/10).

In bearing seats, the details of the location and spacing of the reinforcing steel should be shown to provide adequate clearance for pore-set anchor or for post drilled anchor bolt holes. Also, the location of anchor bolts and edge distance on masonry plates must be detailed so as to avoid uneven pressures on the edge of the seat.

**Fig. 5.13 Deteriorated Bearing Seat**

Theoretically, the top of the pier footing, the bottom of the toe and the top of heel of footing for an abutment and a cantilever retaining wall do not require any reinforcing steel, but the unreinforced section may develop a tensile stress due to unanticipated longitudinal or lateral movement. Therefore, it is advisable to provide some reinforcing steel (#4 or #5 at 18").

To control shrinkage cracks, it is necessary to provide small size bars (#3 or #4) at close spacing in the top of the deck slab and the other members. Since the concrete cover would influence the amount of salt intrusion and corrosion of reinforcing steel, a minimum 2 to 2 1/2 inches of cover is required for the concrete deck slab and a minimum of 1 1/2 to 2 inches cover is recommended for other locations.

**Fig. Rehabilitation and Widening of Existing Bridge** #1: Deteriorated existing beams. Because of the high water level at cofferdam, pumping was required during flood. #2: Unsound concrete was removed before patching. #3: The void beyond crack was filled with pressure grout. #4 The deteriorated pier wall was partially rebuilt by prepacked method and connected to the new pier wall.

**REFERENCE:**


1. "Manual of Concrete Practice: Part 1 Material and General Properties of Concrete; Part 4 Bridge, Substructure, Sanitary and Other Special Structures; Structural Property; Part 5 Masonry; Precast Concrete; Special Process." American Concrete Institute, Detroit, MI 48219
2. "AREA Manual for Railway Engineering: Chapter 8 Concrete Structure & Foundation, Part 13 Shotcrete; Part 14 Reparing and Solidifying Masonry Structure." American Railway Engineering Association, Washington, DC
3. Dussek, Ian "Strengthening of Bridge Beams and Similar Structure by Means of Epoxy-Resin Bonded External Reinforcement" Transp. Research Record 785 pp21-24, TRB
4. Shanafelt, G.O. "Damage Evaluation and Repair Methods for Prestressed Concrete Bridge Members" NCHRP Report 226 (1980), TRB
5. Clark, R; Paulay T "Reinforced Concrete Structures" John Wiley & Son Inc. 1975
6. Cordon, William "Chapter 48 Repairs of Concrete, Concrete Construction Handbook" McGraw Hill, NY
7. Johnson, Sidney M. "Deterioration, Maintenance, and Repairs of Structures" McGraw-Hill, NY 1965
8. "Cement and Concrete Technical Data" Kaiser Cement, Oakland, CA
9. "Concrete Information" Portland Cement Association, Skokie, IL

Application of Silane Sealant

# 6 PRESTRESSED CONCRETE STRUCTURE

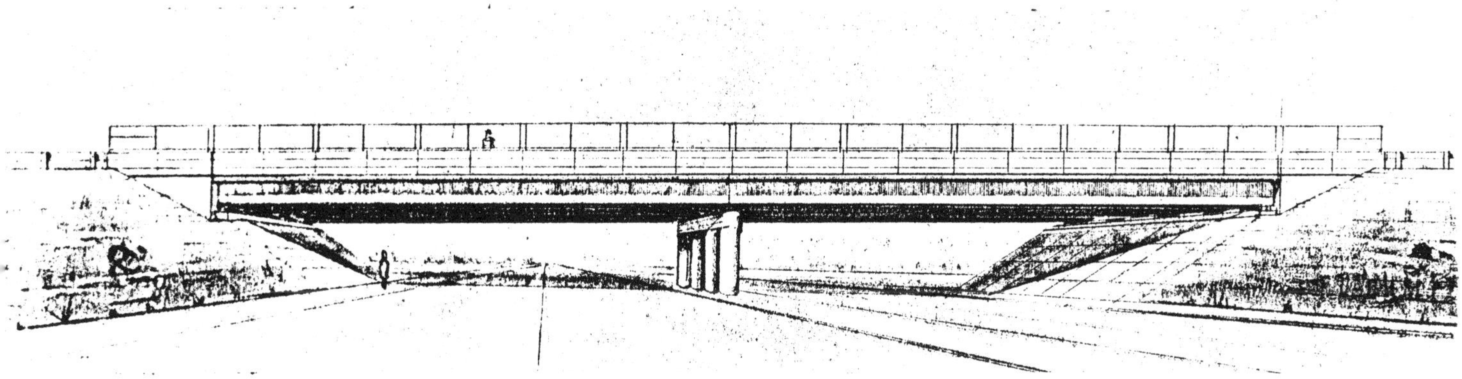

## 6—1 PRESTRESSED CONCRETE REPAIRS

### 6-1.1 GENERAL

The use of prestressed concrete bridges has become popular as a replacement for small to medium span length bridges due to cost, short construction time, simplicity and aesthetics. The older prestressed concrete bridges show some defects, apparently due to inadequate design details and defective construction. In the new prestressed concrete bridges, most of the earlier design problems have been overcome. However, distress resulting from defective construction still occurs.

One of the design problems is the estimation of prestress loss which would affect deflection, camber, and stress in service condition. The 1977 AASHTO has revised the provision for prestress loss to conform to the findings of updated research.

The most critical problem facing the transportation agencies today regarding prestressed concrete bridges is traffic collision damages. Collision damages may range from cracks and spalls on the lower flange of a girder to loss of a major portion of the girder sections with several strands cut. Most collision damage occurs at the exterior or the first interior girder.

Complete rehabilitation of a damaged prestressed concrete structure are almost impossible to make and impractical in terms of cost. Cosmetic repairs can be made by sealing cracks, painting a protective coating on exposed strands and patching the damaged concrete portions for minor and moderate damage. Acceptable repairs for severe damage prestressed concrete beams would require either jacking and external post tensioning or other methods to achieve a reversal of camber. Broken tendons cannot be fully replaced by conventional reinforcing methods but the lost prestress force can be restored by the use of exterior post tensioning at the face of the damaged girders.

### 6-1.2 DAMAGE EVALUATION

The determination of degree of damage of a prestressed girder is generally a matter of judgement based on detailed field inspections, monitoring by strain gage, crack gage and survey, and structural analysis of the damaged girder and compared with the original design and as-built shop drawings.

The field inspection should include the following:

1. Damage to prestress strands: location of damages, the number of severed or yielded strands and their locations at cross section of the damaged girder and comparison of original as-built plans and shop drawings.
2. Damages to concrete: location of damage, the size, patterns, and severity of

cracks and spalls, and determination of internal fractures by sounding, dye test or other methods.
3. Damages to adjacent structural members such as diaphragms, deck slab and parapets.

In assessing the damage, it may be classified into the following four groups: minor, moderate, severe, and critical damage.

**1. Minor Damage:**

Minor damage consists of damage to concrete portions of the beams only. These can be either slight to extensive surface spalls without exposing prestressing strands or fine cracks in concrete. Generally, minor damage does not require structural analysis. However, structural analysis is preferred if the spall damage is extensive.

**2. Moderate Damage:**

Moderate damages may consist of an extensive spall with exposed strands and fine to medium size cracks in the flange. Generally structural analysis is required for brief stress check and preloading stress calculation for repairs.

**3. Severe Damage:**

Severe damage may include following damages to the strands and concrete:

a. Substantial loss of flange section.
b. Loss of strands and severely deformed strands.
c. Cracks extending from flange to web.
d. Substantial loss of portions of web section.
e. Excessive horizontal and vertical misalignments of the girder.

It is apparent that some prestressing force will be lost in the damaged area due to reduced section, loss of prestressing strands or strand shortening. Structural analysis must be made based on the reduced section property of the remaining cross section of the damaged girder and number of effective strands to determine allowable capacity of the girder.

If the stress is slightly over the allowable capacity, then it may be considered as minor damage and the repair methods such as epoxy pressure injection and concrete patching may be adequate for the repairs. Another factor to be kept in mind is that whether the damaged girder is a fascia girder or a girder which is actually carrying live loads. AASHTO design specification requires full live load distribution to the fascia girder but, in analysis of the damaged girder, it is prudent to assume partial live load distribution by using simple beam actions between the fascia and interior girder or no truck live load distributions if the fascia girder only carries sidewalk area.

If the overstress is excessive, action should be taken to restrict live loads on the structure or use temporary supports until the repair is completed. If the remaining strands cannot meet the AASHTO required prestressing force, consideration should be given to restore the prestressing force by either jacking and external post tensioning or replacing the girder.

**4. Critical Damage:**

Critical damage includes the following defects where damage to the girder is considered as not repairable:

a. The loss of strands is extensive and the prestress force can not be restored at a reasonable cost.
b. Wider crack extend across the bottom flange and into the web.
c. Abrupt (kinked) lateral deflection of the prestressed girder and excessive vertical misalignment.
d. Any wide cracks indicate that the strands have exceeded their yield points and permanent deformation has occurred.

### 6-1.3 COUNTERMEASURE AND REPAIRS

In selecting a repair method, the following factors should be considered:

1. Load carrying capacity based on remaining section: Inventory rating of Service load capacity, Ultimate load capacity and Operating rating of Overload capacity. The load capacity rating

can be analyzed based on the AASHTO Manual for Maintenance Inspection of Bridge 1978 or current AASHTO design specifications.

It is desirable to upgrade a damaged girder to the service load capacity. However, if the bridge is to be widened or raised at a later date, the girder can be repaired at less than service load capacity without restoring prestress forces by using complex jacking and post tensioning methods. This simple repair will cost less and provide a minimum level of service for the public for a short period.

In many cases, a damaged prestressed girder provides higher ultimate load capacity than other type of structures which has a similar damage, therefore the selection of repair methods relies heavily on the engineering judgements and not on the calculations alone. It is also suggested that a fatigue stress check be made for the remaining strands.

2. Severity of damage: All loose concrete should be removed caused by vehicle impacts; the fractured area should be identified and sealed; and all compression cracks should be carefully examined for the durability of repairs.

3. Availability of material, equipment and personnel should be considered for particular repair schemes.

4. Cost comparison between alternative repairs should be made.

5. Public inconvenience and safety during repair should be considered. Since maintenance of the traffic is emerging as one of the most costly items for bridge repairs, special care should be exercised to select practical and cost effective traffic control schemes which will affect the choice of repair schemes.
6. Aesthetics: A pleasant looking bridge repair is desirable.
7. After considering the above factors, the following general repair procedure can be used:

- Determine whether preloading is required.
- Apply preload as required.
- Repair damaged concrete and splice strands or reinforcing bars if needed.
- After the concrete is fully cured and gained required strength, remove preloading.
- Post-tension girder if required.

## 6-1.4 MINOR DAMAGE

Minor damage requires superficial patches by the use of epoxy grout or shotcrete methods. Where necessary all damaged and delaminated concrete should be removed by hand tools to avoid the possibility of damaging strands. The exposed surface should be cleaned and applied with epoxy bonding compound before patching. All cracks should be sealed by the epoxy pressure injection method which is similar to the concrete repair. Pressure injection should be made from the bottom of flange area toward the top of cracks.

## 6-1.5 MODERATE DAMAGE

Moderate damages with extensive spalls are generally patched with epoxy grout or concrete mortar applicable to minor repairs. It is recommended that welded wire fabrics be attached to drilled dowels placed at about 2 feet spacing or other reinforcement steel in the patched area. Surface preparation is similar to the other concrete repairs.

If the strands or reinforcement steels are exposed, care should be exercised so as not to damage the steel during cleaning. The exposed strands may be coated with epoxy bonding compound or slurry cement grout before patching.

In order to restore the steel to its original capacity, pre-loading may be necessary before and during the repair operations. Pre-loading is an application of temporary vertical loads to restore partial or full prestress force of the damaged area and to reduce tension under live loads. Preload can be applied by the use of loaded trucks or vertical hydraulic jacking.

Collision damage- outside

Collision damage- inside

Preparation & installing port

Preloading

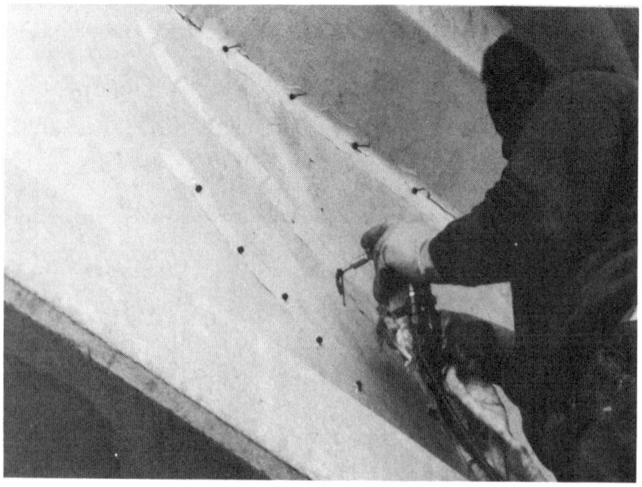

Pressure injection

Clean up

**Fig 6.2** Repairs by Epoxy Pressure Injection

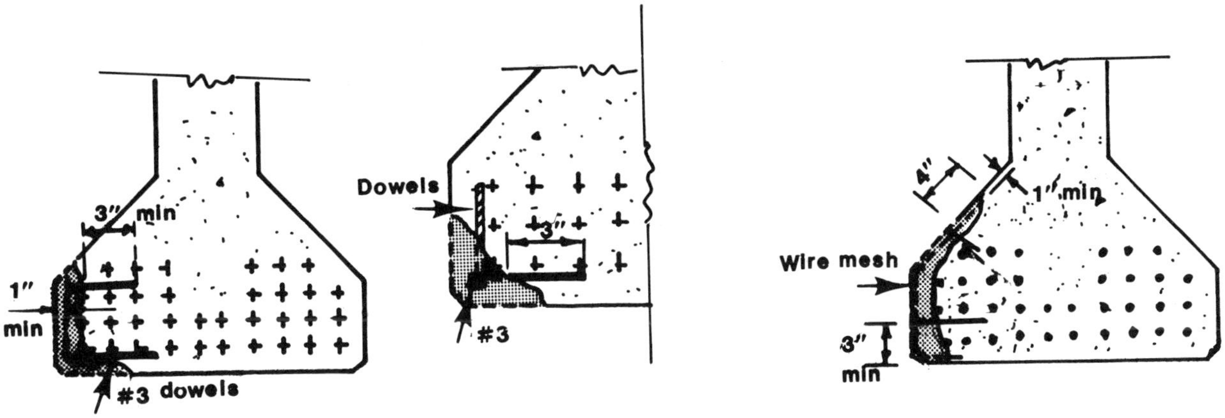


Unexposed strand

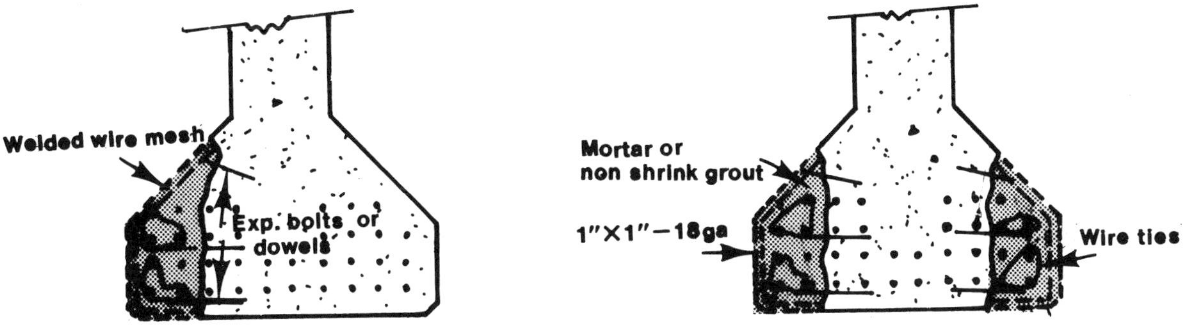


Exposed strands

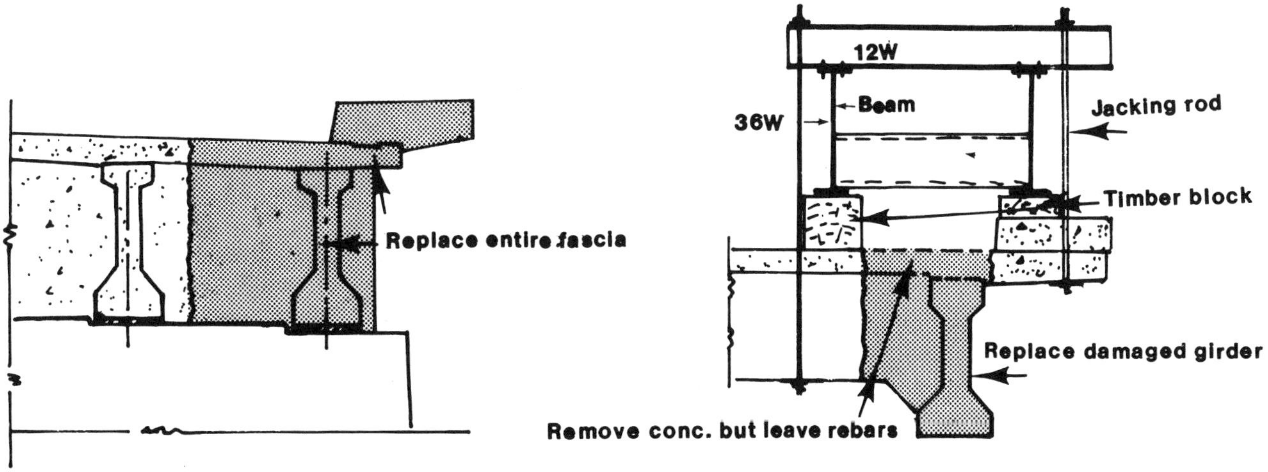


Replacement

**Fig 6.3** Repairs of Collision Damages

## 6-1.6 SEVERE DAMAGE

Repair of a severely damaged girder requires structural analysis of the damage based on the conditions of the damages and the best engineering assumptions and judgment. It is important to have a complete review of the calculations of damage because it would help selection of a cost effective repair method for the bridge to safely carry a load.

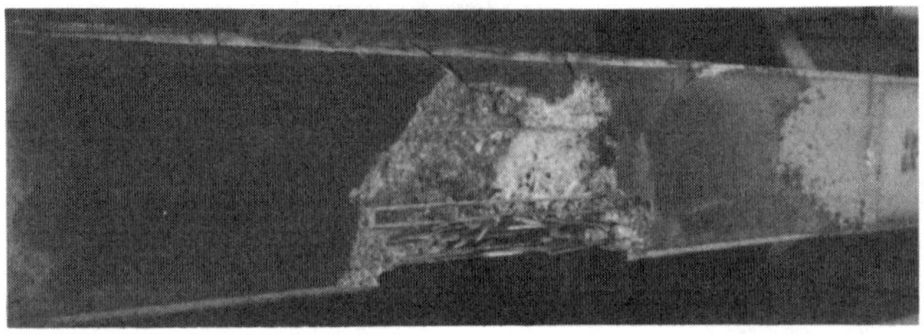

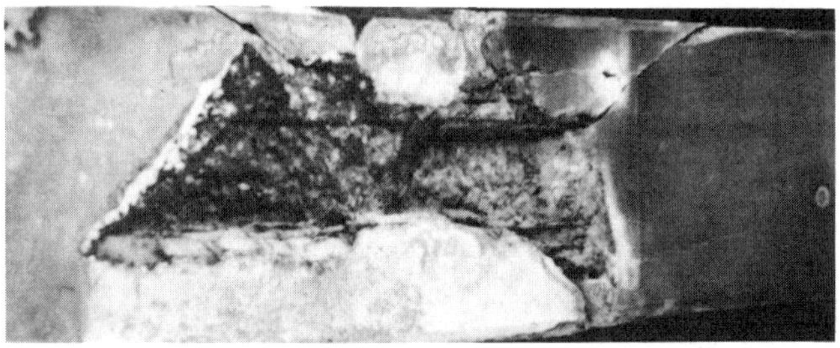

**Fig. 6.4 Severely damaged girder and cosmetic surface patch where total replacement or post-tensioning is required.**

The preloading should be used in making concrete repairs in order to restore full prestress force equal to its original capacity. The repair procedure for the use of patching concrete, epoxy pressure injection, shortcrete, and additional welded fabric with drilled anchors are similar to other methods of concrete repairs.

The following may be used as a general guideline for restoring the load carrying capacity of beams.

1) If the bridge was designed in accordance with the working stress method, the repairs should be designed to withstand the original design stress using the original design concepts.
2) The design should be made in accordance with the current AASHTO specifications. But the following modification may be used for repair design:

a. Maximum tensile stress after losses : 0.6f's
b. Maximum allowable stress for seven wire strands for LL+I : 10 ksi

If it is known in advance that prestress must be restored, determine the stress in the bottom fiber of the girder as originally designed due to dead loads, live loads and impact (DL+LL+I) prestress. (This will normally be about "0"psi).

Determine the additional preload which, when applied to the damaged girder in its existing condition, will result in the same stress. Take into account the reduced girder section, the effective composite section and any reduced prestress due to strand loss. Should the damage occur outside of the middle one-third of the span length, the shear stress with the preload applied should also be computed. Where strands are broken, consideration should be given to coupling and jacking them to restore their prestress comparable with illustrated samples.

In preparing a repair plan, a sketch should be drawn to show how the preload is to be applied and it should be specified that the damaged area be thoroughly prepared, coated with epoxy and repaired with grout equal in strength to the original concrete. It should be also specified that the preload is to remain in place until the grout has attained sufficient strength. The effect of this load is to restore lost prestress to the strands.

## 6-1.7 PRELOADING

Preloading is an application of a temporary vertical load during the repair of a damaged prestressed girder. Preloading can be used in many instances to restore the girder to its original condition without adding prestress. The section properties of the remaining cross-section of the girder should be computed. If the maximum allowable compressive stress in the reduced bottom flange is exceeded, enough preload during repair may be applied to bring the stress within allowable limits to restore partial or full prestress force to the repair area.

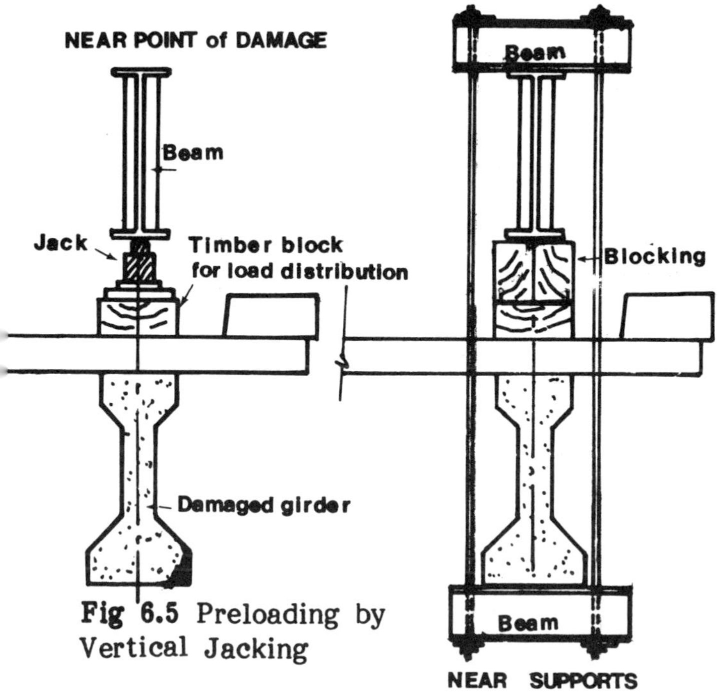


**Fig. 6.5 Preloading by Vertical Jacking**

Dead load weight of the curb and rail base is not included in the examples. For the actual bridge repair designs, any portion of the curb and rail base weight carried by the beam which is being repaired should be included in the calculations. Normally this weight acts on the composite section. If preload is provided by a single jack, consideration should be given to the temporary stresses induced in other bridge elements. Part of the jacking load will be transferred through the slab and diaphragms to the adjacent girders. If it is concluded that this transfer would cause excessive stress in the slab and diaphragms, more than one jack should be used, with a reasonable longitudinal spacing (about 14 feet).

## 6-1.8 RESTORATION OF PRESTRESS

In order to restore strength to the girders, the strand should be spliced or strengthened by external forces. The following are some of the splicing methods:

1) The Use of Two Post Tensioned Rod and Jacking Concrete Corbels located outside of the damaged areas: Construction procedures are as follow:

a. Apply preload to repair the damaged concrete;
b. After repaired concrete has gained strength, remove preload;
c. Construct jacking corbels and post-tension rod.

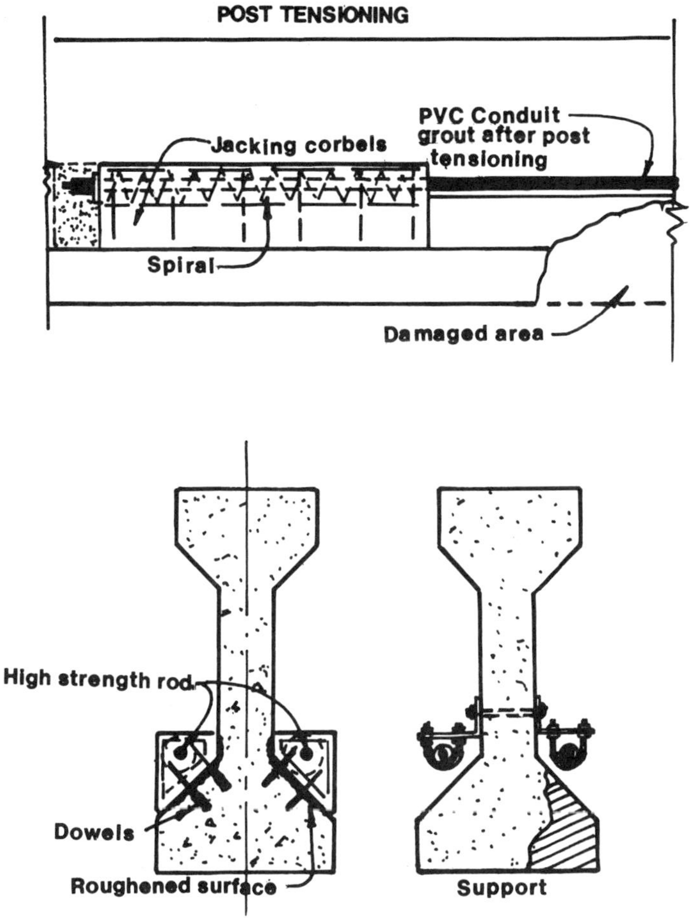


**Fig. 6.6 Restoration by Post-Tensioning**

2) Adding External Reinforced Concrete: The use of conventional reinforcement steel with concrete corbel throughout the girders.

a. After applying the preload, the damaged concrete should be repaired and the concrete corbel should be constructed with required reinforcement steel.
b. Upon the cure of the concrete, the preload is removed.

3) Installation of Metal Sleeve: This method does not restore the prestress loss except that a partial prestress may be gained by the preloadings. The pre-

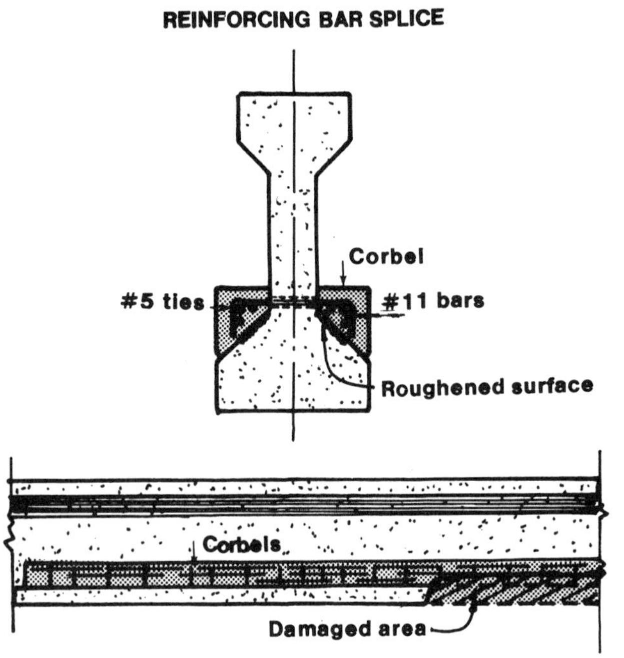


**Fig. 6.7 Restoration by External Reinforcing Concrete Corbels.**

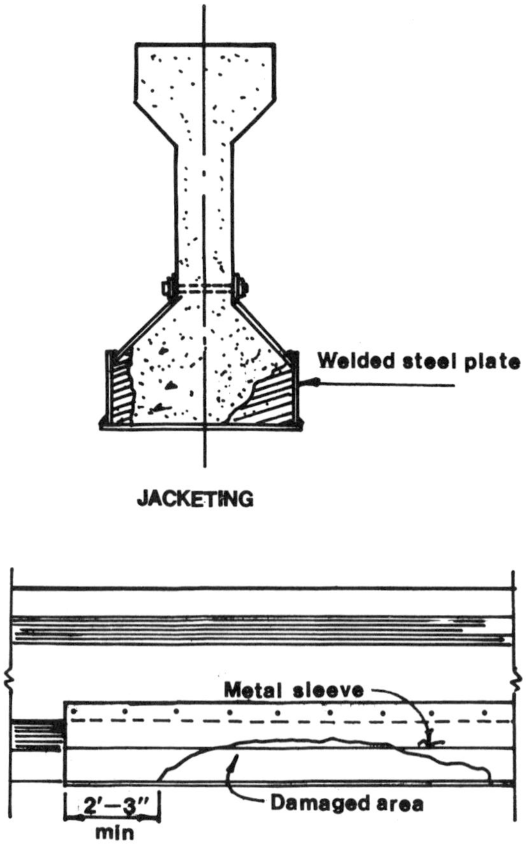


**Fig. 6.8 Restoration by Installing Steel Metal Sleeve.**

load should be applied prior to the repair of the damaged concrete and removed upon the completion of repairs. Then a metal sleeve should be installed around and beyond the damaged areas (2'-3" min.). The gap between the metal sleeve and girder is filled with epoxy grout by pressure injection.

4) Splicing Damaged Strand by the Heating Technique:

a. Cut the damaged strand then slip splicing strand chuck or coupler over the end of the broken strand.
b. Heat the exposed strand to expand and tighten the strand grip with wrench, then let strand cool to induce tension.
c. Apply preload.
d. Repair the damaged concrete.
e. Remove preload upon curing the damaged concrete.

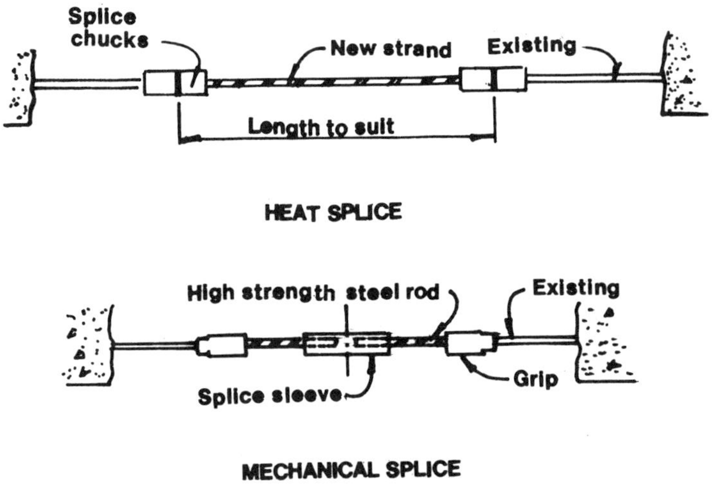


**Fig. 6.9 Restoration by Other Methods**

5) Splicing of Damaged Strand by Mechanical Torque:

a. Cut the damaged splice, assemble splice, locating splice sleeve and strand grip to allow seating of the grips and enough thread length in the splice.
b. Torque the splice to gain predetermined stress at the strand (about 22,000 lbs. per strand).
c. Apply preload.
d. Repair concrete and remove pre-loading upon curing.

### 6-1.9 GIRDER REPLACEMENT

The removal of a damaged girder could be done from below or above the structure. In general, the removal consists of cutting through the existing slab and diaphragms and lifting the damaged girders. However, the beam could be removed from the underside of the deck by saw cutting the slab along the girder without damaging the existing reinforcement, installing a temporary support, then sliding down the damaged girder from below. This method has the advantage of minimum disruption to the deck reinforcements but careful supervision is required during construction.

It is recommended that a sufficient length of existing reinforcement bars should be left in place for splicing with new reinforcement. Otherwise, a drilled dowel should be used for that purpose. Also, protective shielding and false work are required during removal operation for the safety of the public and workers. Generally, the protection details are designed by the contractor and are subject to the approval of the agency.

The exposed surface should be cleaned and coated with epoxy bonding compound before pouring a new slab and diaphragms.

In designing a replacement girder, the new girder should be designed to conform with the original design method and to provide strength equal to the original girder by using the same strand arrangements and type.

It is important that the camber of the new girder be matched with that in the old girders. Excess new girder camber can result in an inadequate slab thickness. Girder camber can be controlled by prestress, curing time or dimensional changes. It is suggested to consider differential shrinkage stresses in designing new composite deck slab on the existing girders.

Consideration should be given to pouring of the new slab and diaphragms simultaneously in order to avoid overloading of the existing girders in the structure. Extra bracing of the girder at the time of slab pouring is required.

Methods of construction should be selected to minimize inconvenience and dangers to the public while achieving satisfactory structural results. Grouts and concretes having high early strength should be considered. Plans must carefully spell out the materials and procedures in order to insure that this result is attained.

When pouring new concrete, vibration from live loads from the existing structure should be minimized until a minimum concrete strength is obtained, even though current research from the Texas Highway Research Board has indicated that the effect from live loads is not detrimental to the structural integrity.

## 6—2 PRESTRESSED CONCRETE BRIDGE

### 6-2.1 REPLACEMENT BRIDGE

Precast concrete has been used widely for replacing short to medium span bridges. The standard precast prestressed concrete sections developed by AASHTO and Prestressed Concrete Institute (PCI) are used as standard section in 24 States. Several States have developed their own standard of economical sections. Generally, precast sections are designed for composite action with cast-in-place deck slab but an integral deck slab can be achieved by the use of a slab, box, or T sections with wearing surface over the integral deck sections.

The advantages of using precast concrete bridges are minimum traffic interruption due to speed erections, relatively low construction cost, minimum maintenance, low span depth ratio which helps to increase vertical underclearance, and high quality control of concrete at the plant.

The disadvantages are the necessity for good connection details between members in order to fasten concrete members together and to prevent any future deteriorations at the joints, and lack of easy repair methods for traffic collision damages.

Fig. 6.10 Construction of Adjacent Prestressed Concrete Box Beams

Majority of the precast concrete bridge members are prestressed by pre-tensioning strands or post-tensioning tendons. Common precast bridge superstructures includes deck slab, box beam, I-girder, channel,and T-beam. For long span bridges, a precast segmental construction and stay-girder system can be used but the discussion in this chapter will be limited to medium span prestressed concrete bridges.

### 6-2.2 DESIGN CONSIDERATION

**General:**

1. If many bridges are to be replaced in certain portions of highways, they should be grouped under one project instead of breaking into separate bridge projects.
2. Bridge spans should be made equal as possible so as to minimize fabrication cost.
3. Skew angles should be limited to 30 degree or less.
4. Adequate vertical underclearance must be provided to prevent collision damages.
5. Integral deck system of adjacent boxes, T beams, and double T beams should be used to avoid falsework and speed of construction.
6. Requirements of transportation, availability of heavy equipment, and capacity of local precast concrete plants must be checked. Maximum span length for the AASHTO girders is limited to about 140 feet.
7. If composite design is utilized, the compatibility of the precast and cast-in-place concrete should be studied regarding the following:
   a) Shear transfer between precast and cast-in-place concrete.
   b) Overstress of precast concrete members during the cast-in-place concrete to attain its strength.
   c) Effects of differential material properties, temperature, shrinkage, creeps, etc.
8. Current FHWA Report "Optimized Section for Major Prestress Concrete Bridge Girder" concludes that the AASHTO standard bridge girders are not the most structurally efficient or cost effective

for spans of 80 to 140 feet. The most efficient sections are Bulb-T's and the next most cost effective section is the Washington States Standard with 5 inch thick web. The main problem of adopting new sections will be the necessity of large initial form retooling costs. Also the report concluded the following for girder bridges:

a) End diaphragms are sufficient, therefore intermediate diaphragms are not required. AASHTO recommends that one intermediate diaphragm at the mid span for spans of 40 to 80 feet and two at the third points of spans for spans in excess of 80 feet.
b) The increase of girder's concrete strength from 5,000 to 7,000 psi will increase the girder capacity by about 15%.
c) Bundling of strands at midspan to increase eccentricity of prestress does not result in cost saving.

9. Larger girder spacing will reduce initial construction cost however, it will cause flexibility of the deck slab and premature deck deterioration. It is recommended not to use the largest girder spacings. Generally girder spacing should be limited to about 9 feet.
10. The Prestressed Concrete Institute (PCI) Committee on Bridges recommends that reduction of the initial prestress forces by the elastic shortening loss is an acceptable practice for calculating initial concrete stresses. However, reduction due to strand relaxation is not recommended due to the uncertainty of elapsed time unless the time is controlled by specifications or other means.

    Lack of vents and drains in the voids of prestressed box beams could trap the moisture or leaked water inside the box and could seriously reduce the duarabilty by freeze and thaw cycles. Therefore it is recommended that void vents and drain provisions be included in the specifications.

**Concrete Deck and Box Beams:**

1. On an integral deck, the key joints between beams or deck slabs should be able to transmit all forces without causing any cracks or defects. Generally, the box beams or the slab beams are held together by tie rods with or without post-tensioning tie rods. A shear key is provided between beams and filled with cement grout. However, the grout bond between the beams is often broken due to failure of the key to transmit heavy truck loads to adjacent beams and the final result is that each beam acts independently. Subsequently, reflective cracks appear on the wearing surface over the longitudinal joints.

It is suggested that transverse tie rods be post tensioned and epoxy grout be used in the shear keys. Epoxy grout generally will have higher bond strength in tension than non-shrink grout. It is also recommended that shear key surface be sandblasted to provide better bond.

The transverse tie rod are required to provide sufficient horizontal restraint to keep the grouted shear keys closed but in reality they are subjected to transverse bending moments, lateral forces caused by vehicle changing lanes, distribution effects of vehicle loading, shrinkage and temperature changes. Shrinkage and temperature change result in the decrease of the width of the beams and the separation of the grout at the keys. To avoid this separation of

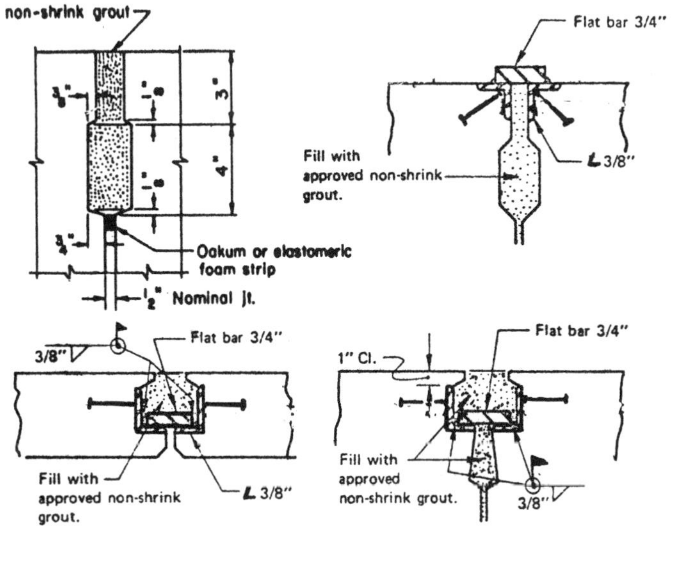


**Fig. 6.11 Typical Shear Keys**

# PCI - AASHTO STANDARD SECTION

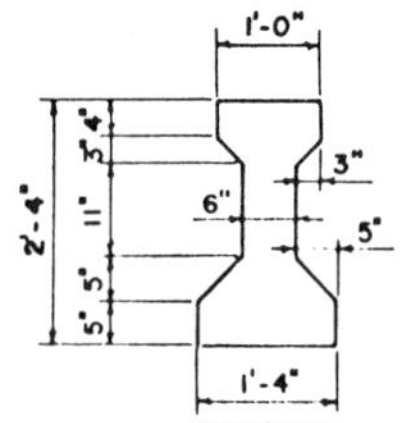


SECTION
**TYPE I BEAM**
30 FT. TO 45 FT. SPANS

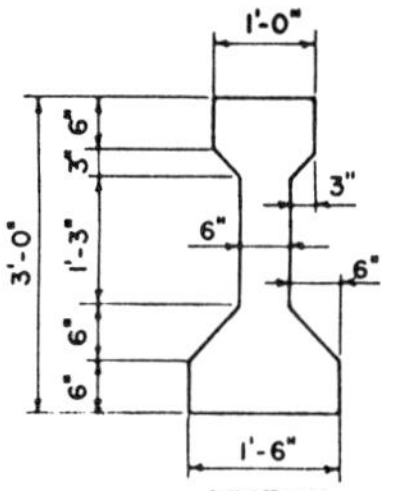


SECTION
**TYPE II BEAM**
40 FT. TO 60 FT. SPANS

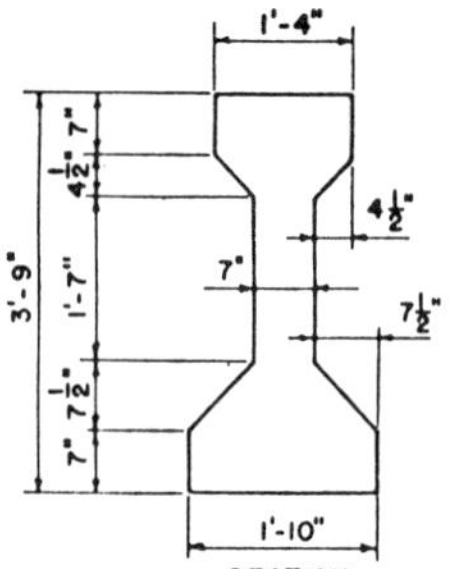


SECTION
**TYPE III BEAM**
55 FT. TO 80 FT. SPANS

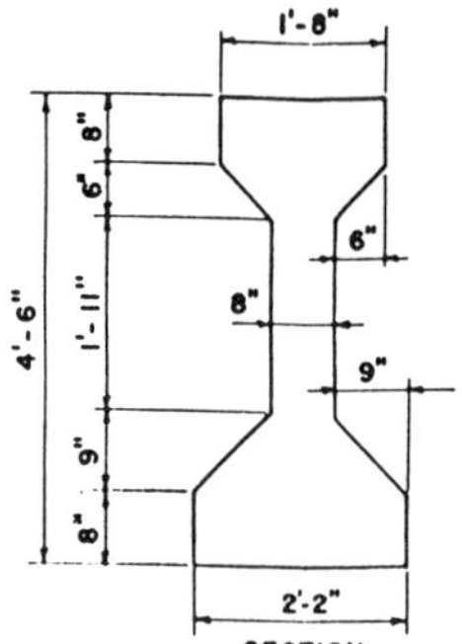


SECTION
**TYPE IV BEAM**
70 FT. TO 100 FT. SPANS

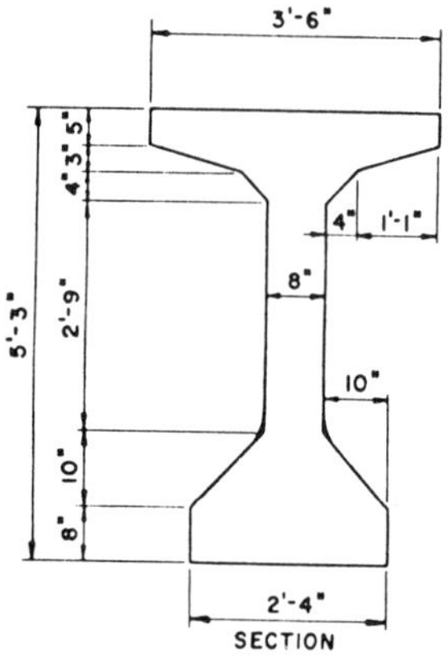


SECTION
**TYPE V BEAM**
90 FT. TO 120 FT. SPANS

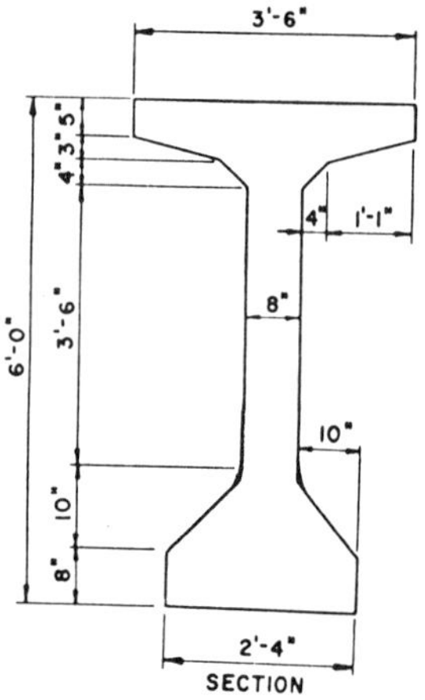


SECTION
**TYPE VI BEAM**
110 FT. TO 140 FT. SPANS

**BEAM PROPERTIES**

| BEAM TYPE | AREA Inches² | $\bar{y}$ Inches | Moment of Inertia Inches⁴ |
|---|---|---|---|
| I | 276 | 12.59 | 22,750 |
| II | 369 | 15.83 | 50,980 |
| III | 560 | 20.27 | 125,390 |
| IV | 789 | 24.73 | 260,730 |
| V | 1013 | 31.96 | 521,180 |
| VI | 1085 | 36.38 | 733,320 |

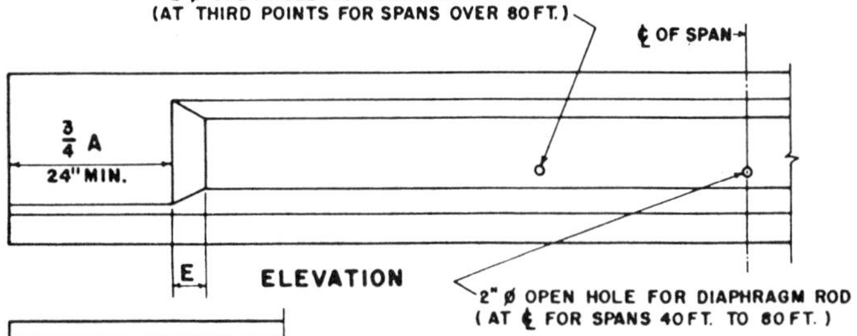


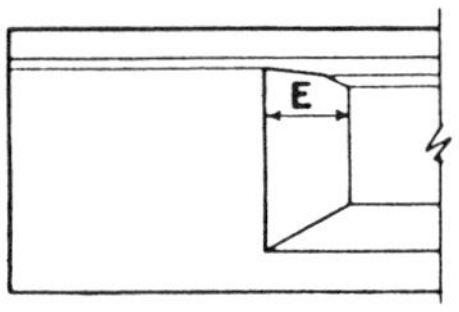


NOTE: 2" Ø OPEN HOLE TYPICAL FOR PRETENSIONE[D] AND POST-TENSIONED BRIDGE BEAMS.

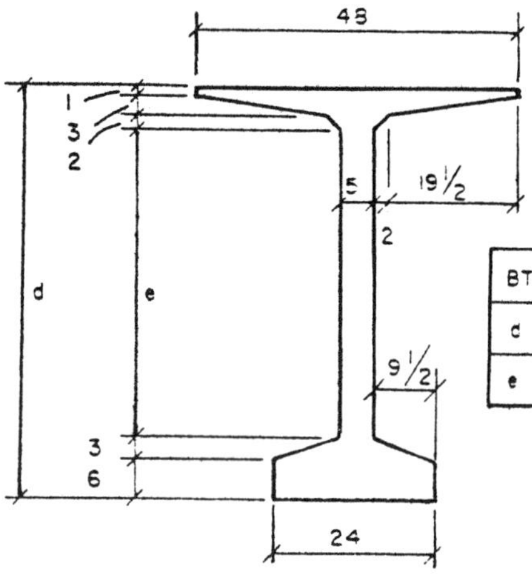


| BT | 48 | 60 | 72 |
|---|---|---|---|
| d | 48 | 60 | 72 |
| e | 33 | 45 | 57 |

**BULB T**

**WASHINGTON STATE**

**END BLOCK DIMENSIONS**

| BEAM TYPE | A | B | C | D | E |
|---|---|---|---|---|---|
| I | 2'-4" | 1'-9" | 1'-0" | 1'-4" | 6" |
| II | 3'-0" | 2'-3" | 1'-0" | 1'-6" | 6" |
| III | 3'-9" | 2'-11" | 1'-4" | 1'-10" | 9" |
| IV | 4'-6" | 3'-7" | 1'-8" | 2'-2" | 1'-0" |
| V | 5'-3" | 4'-8½" | 3'-6" | 2'-4" | 1'-8" |
| VI | 6'-0" | 5'-5½" | 3'-6" | 2'-4" | 1'-8" |

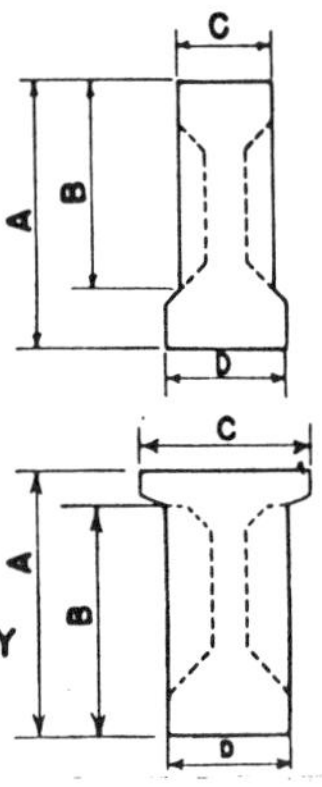


**END BLOCK DETAILS — FOR POST-TENSIONED BEAMS ONLY**

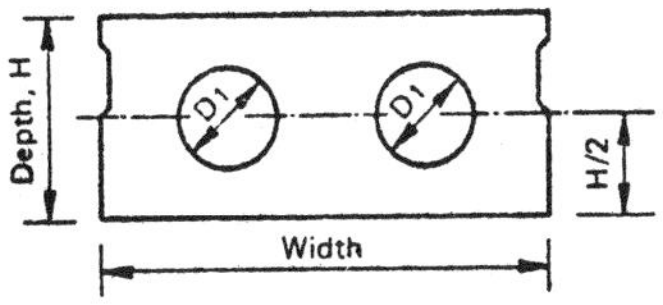


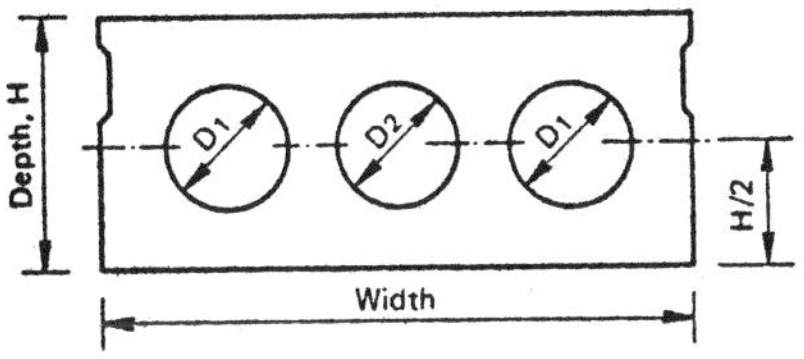


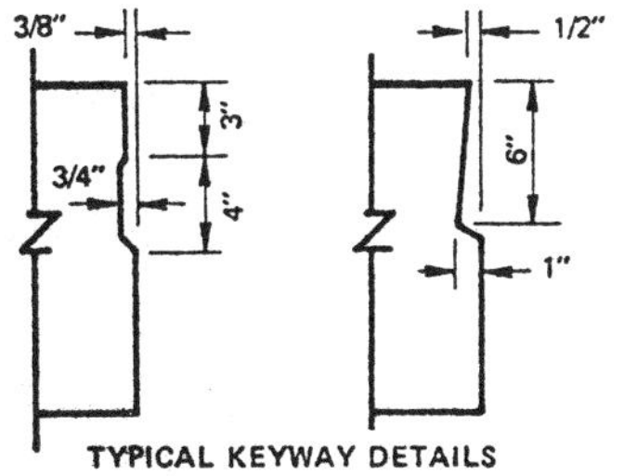


TYPICAL KEYWAY DETAILS

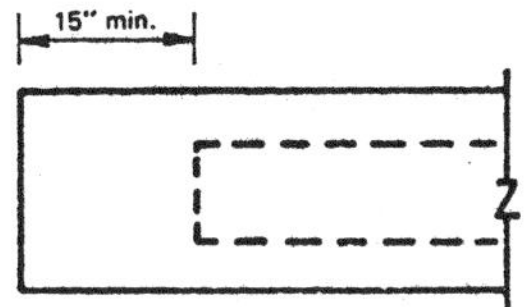


TYPICAL LONGITUDINAL SECTION

TYPICAL SECTION PROPERTIES

| Width ft | Depth in. | No. of Voids | Void Dia. in. D1 | Void Dia. in. D2 | Weight lb/ft | Net Area in.$^2$ | $I_x$ in.$^4$ | S in.$^3$ |
|---|---|---|---|---|---|---|---|---|
| 3 | 15 | 2 | 8 | – | 457 | 439 | 9,725 | 1296 |
| 3 | 18 | 2 | 10 | – | 511 | 491 | 16,514 | 1835 |
| 3 | 21 | 2 | 12 | – | 552 | 530 | 25,747 | 2452 |
| 4 | 15 | 3 | 8 | 8 | 593 | 569 | 12,897 | 1720 |
| 4 | 18 | 3 | 10 | 10 | 654 | 628 | 21,855 | 2428 |
| 4 | 21 | 3 | 12 | 10 | 733 | 703 | 34,517 | 3287 |

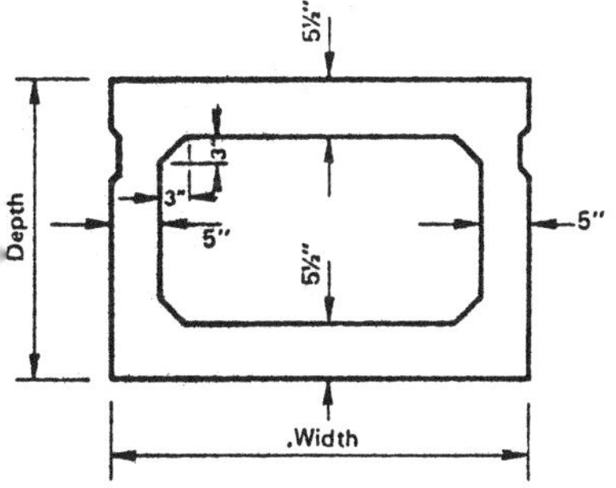


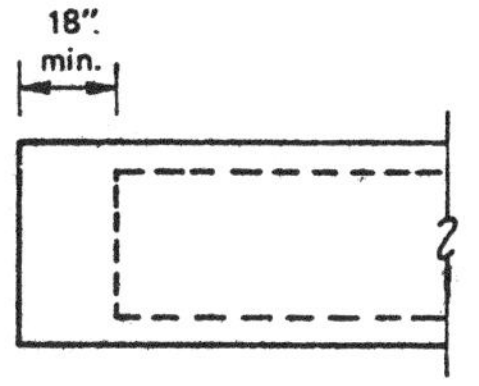


TYPICAL LONGITUDINAL SECTION

TYPICAL SECTION PROPERTIES

| Type | Width ft | Depth in. | Weight lb/ft | Net Area in.$^2$ | $I_x$ in.$^4$ | $y_b$ in. | $S_b$ in.$^3$ | $S_t$ in.$^3$ |
|---|---|---|---|---|---|---|---|---|
| B I-36 | 3 | 27 | 584 | 561 | 50,334 | 13.35 | 3770 | 3687 |
| B II-36 | 3 | 33 | 647 | 621 | 85,153 | 16.29 | 5227 | 5096 |
| B III-36 | 3 | 39 | 709 | 681 | 131,145 | 19.25 | 6813 | 6640 |
| B IV-36 | 3 | 42 | 740 | 711 | 158,644 | 20.73 | 7653 | 7459 |
| B I 48 | 4 | 27 | 722 | 693 | 65,941 | 13.37 | 4932 | 4838 |
| B II-48 | 4 | 33 | 784 | 753 | 110,499 | 16.33 | 6767 | 6629 |
| B III-48 | 4 | 39 | 847 | 813 | 168,367 | 19.29 | 8728 | 8542 |
| B IV-48 | 4 | 42 | 878 | 843 | 203,088 | 20.78 | 9773 | 9571 |

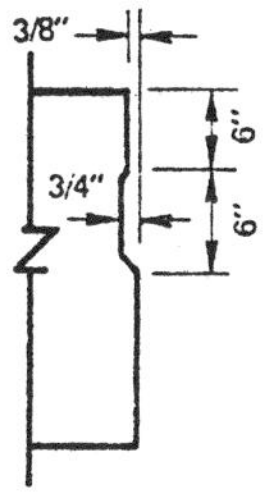


TYPICAL KEYWAY DETAIL

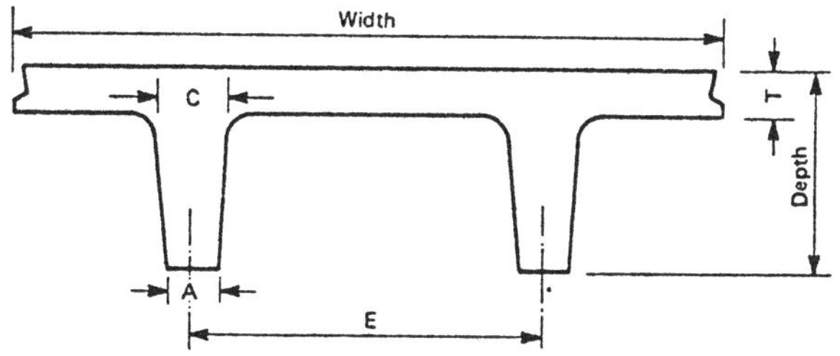


TYPICAL SECTION PROPERTIES

| SECTION | Width ft | Depth in. | Slab T in. | Stems A in. | Stems C in. | Stems E in. | Weight lb/ft | Area in.$^2$ | $I_x$ in.$^4$ | $y_b$ in. | $S_b$ in.$^3$ | $S_t$ in.$^3$ |
|---|---|---|---|---|---|---|---|---|---|---|---|---|
| "LIGHT" | 5 | 27 | 5 | 4.50 | 8 | 36 | 599 | 575 | 33,740 | 18.60 | 1812 | 4020 |
| | 6 | 23 | 5 | 4.50 | 6.50 | 36 | 582 | 558 | 21,366 | 16.61 | 1286 | 3345 |
| | 6 | 27 | 5 | 4.50 | 8.00 | 36 | 662 | 635 | 35,758 | 19.15 | 1866 | 4560 |
| | 8 | 27 | 5 | 3.75 | 5.75 | 48 | 718 | 689 | 32,888 | 20.64 | 1593 | 5171 |
| | 8 | 35 | 5 | 3.75 | 6.50 | 48 | 820 | 787 | 72,421 | 26.20 | 2764 | 8230 |
| "HEAVY" | 5 | 36 | 6 | 6 | 8 | 30 | 812 | 780 | 90,286 | 23.69 | 3811 | 7334 |
| | 6 | 36 | 5 | 6 | 9.75 | 48 | 876 | 840 | 90,164 | 23.30 | 3870 | 7706 |
| | 7 | 35 | 5 | 6 | 9.75 | 48 | 938 | 900 | 95,028 | 23.91 | 3974 | 8569 |
| | 8 | 35 | 5 | 6 | 9.75 | 48 | 1001 | 960 | 99,299 | 24.45 | 4061 | 9412 |
| | 6 | 27 | 5 | 7 | 9.75 | 48 | 761 | 731 | 45,084 | 18.09 | 2492 | 5060 |
| | 7 | 27 | 5 | 7 | 9.75 | 48 | 824 | 791 | 47,486 | 18.58 | 2556 | 5640 |
| | 8 | 27 | 5 | 7 | 9.75 | 48 | 886 | 851 | 49,566 | 19.00 | 2609 | 6196 |
| | 6 | 21 | 5 | 7.75 | 9.75 | 48 | 671 | 644 | 22,720 | 14.11 | 1610 | 3298 |
| | 7 | 21 | 5 | 7.75 | 9.75 | 48 | 733 | 704 | 23,903 | 14.48 | 1651 | 3666 |
| | 8 | 21 | 5 | 7.75 | 9.75 | 48 | 796 | 764 | 24,920 | 14.80 | 1684 | 4019 |

the beams, tie force are required to pull the beams together as they decrease their width. This force suppose to be resisted at the bearings if the bearings are propoerly designed. However, for several reasons, the bearings are not performed as intended and tie rods are required. The maximum force will be the friction between the beams and their supports. If the coefficient of friction is assumed about 1.0, The design tie force should be about half the weight of the superstructure.

Where a simple shear key is expected to have problems to transmit shear, other details may be used such as double tie rods, post tensioning ties and welded plated. In estimating transverse shear in multi-box beams, generally empirical methods (Design force are about equal to half the weight of the total beams or slabs) have been used but current research work based on the computer program of orthotropic plate theory give simplified method for determining shear forces. Refer to the paper "Transverse Shear in Multibeam Bridge" by Baidar Bakht, Leslie G. Jaeger and M. S. Cheung, ASCE Journal of Structural Engineering, Vol. 109, No.4, April 1983.

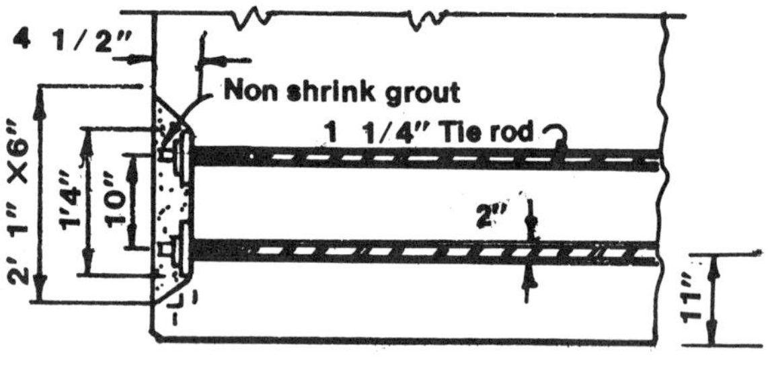


**Fig. 6.12 Double Transverse Tie Rods**

2. The use of concrete overlay (4 1/2 to 5") with epoxy coated reinforcement steel should be considered where the bridge is on skew angles or steep vertical grades. An asphalt wearing surface with waterproofing membrane is susceptible to tearing off the membrane and forming of reflective cracks at the joints. On steep grades, asphalt wearing surface may develop shoving or slippage cracks.

3. The bearings should be detailed to avoid the rocking or lateral movements of beams which can result in failure of the shear keys and development of reflective joint cracks.

4. For stage construction, the Illinois Department of Transportation recommends the use of shear key clamping devices. The stage construction keyway shall be the last to be grouted and in the stage traffic control plans for bridge, the wheel loads should be away from the two beams adjacent to the stage construction line.

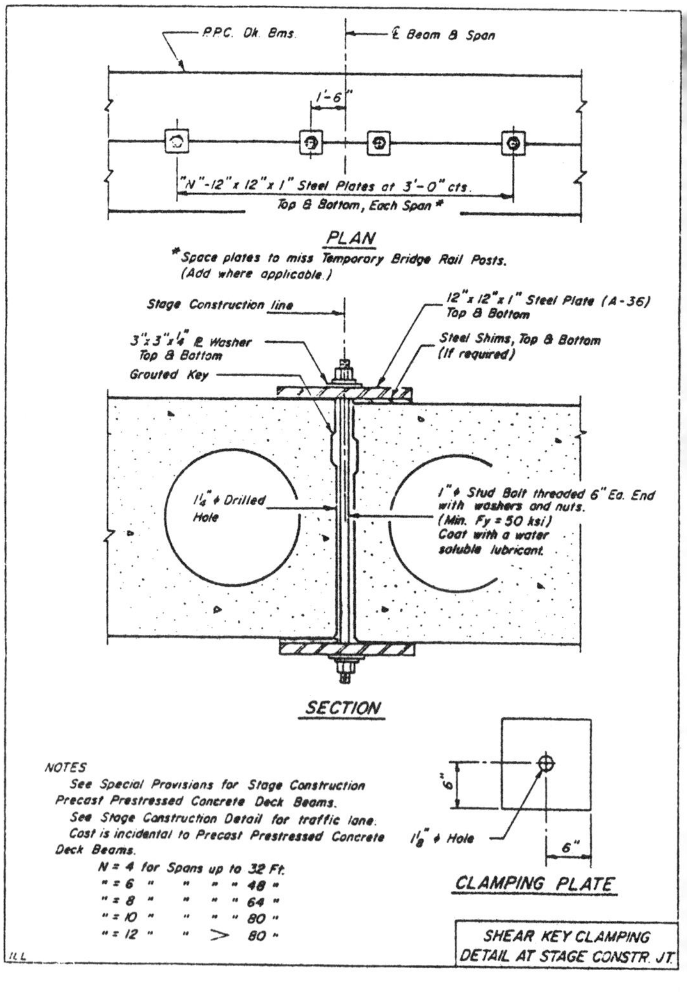


**Fig. 6.13 Shear Key Clamping Details**
(The llinois Department of Transportation)

The special provision covering the use of shear key clamp devices is as follows:

**Stage Construction Precast Prestressed Concrete Deck beams:**

When stage construction is specified, shear key clamping between stage I and stage II construction will be required as specified and shown on the plans.

After the deck beams for stage I construction are erected and grouted, the traffic shall be diverted to them. Temporary bituminous tapers shall be constructed by the contractor, as required, to permit the traffic to cross any expansion joint dams. The deck beams for stage II construction shall then be erected and all keyways grouted except the keyway along the stage construction line. After stage II construction is completed, including waterproofing and surfacing as shown on the plans, traffic shall be redirected onto stage II construction as shown.

With traffic on stage II construction and maintained as far as possible from the stage construction line, the shear key clamping devices may be installed.

The contractor shall furnish all material for the clamping devices, including sufficient 1/16 inch and 1/8 inch steel shims to adjust for differential elevations between the two deck beams.

The 1 1/4 inch hole for the clamping devices shall be drilled at the location shown on the plans. Care shall be taken to drill the holes perpendicular to the beams. The clamping devices shall be installed and pulled up tight so a full firm bearing is obtained between the clamping plates and deck beam concrete.

After the shear key clamping devices in a span are fully secured, the stage construction line key may be grouted. The clamping devices shall not be loosened or removed until the grout has fully cured as specified.

After the clamping devices are removed, the drilled holes and unfilled area of key adjacent shall be flushed out with water. The drilled hole and unfilled adjacent areas shall then be filled with grout to the bottom of the keyway.

## 6-2.3 MATERIALS

### Concrete:

As the technology of concrete advanced, higher strength concrete has been used for pre-cast concrete members. When high strength concrete (over 5000 psi) is used, investigation should be made regarding quality control, results of trial mix, buckling of thin sections, dimensional tolerances, camber and deflection, requirements of stirrups or tie bars at high compression zones, and details for anchorage zones.

In order to gain high strength and durability, admixture are used in concrete mix for precast members. A calcium nitrite corrosion inhibitor may be specified as an admixture to increase durability. Also a super plasticizers introduced by the concrete industry is a new method of casting precast members with increased workability and strength.

A current U.S. Dept of Transportation, University Research Report and FHWA Research Report stated that a well designed mix with the superplasticizer will have good flowability and sufficient cohesiveness. It would not cause bleeding, segregation or reduction in strength either during or after placing concrete, and would not decrease freeze-thaw or deicer scaling resistance. In addition, the super-plasticizer would increase the rate of the strength development at early ages.

With the addition of a superplasticizer, a workable, low water-cement ratio and a high strength concrete can be produced. However, concrete shows high rates of slump loss and air loss when super-plasticizers are used. The use of the admixture in a central mix is not recommended unless expeditious handling techniques are used. FHWA report recommends air entrainment of 7 to 8 percent.

### Prestressing Steel:

Prestressing steel should be free of defects and grease. In prestressing, light rusting has been considered as an advantage by providing better bonding with concrete. But severely rusted strands should not be used. The most common strand for new pretensioned girder is stress relieved Grade 270, 1/2" or 7/16" diameter. Grade 250 strand had been used for earlier girders but are not popular today.

A Blanket strand concept is used in some states. With the blanket method, all strands are straight and draped strands are eliminated. Some strands are debonded at the end of the girders to avoid sudden transfer of prestressing forces at the end blocks and these forces are spread over longer portion of girders. The blanket strand method can be done in one operations and no hold down device is required.

In post-tensioned concrete, corrosion protection is very important and all post tension tendons should be protected by pressure grouting on conduit, galvanizing, coal tar epoxy coatings and wrappings.

It is reported that hydrogen embrittleness of steel prestressing tendons has occurred in England and Europe. Hydrogen brittleness occurs along with normal corrosion in high strength steel at high stress level when hydrogen ion from atmosphere penetrates the steel and changes the molecular structure of the tendon. Most hydrogen brittle failures have been detected in the ungrouted or wrapped tendons of the post tension members, but the prestressing strand embedded in concrete could also be susceptible to failures and inspecting and finding of the brittle would be difficult. The designer should give a particular attention to the recent development of hydrogen brittleness problems in the post-tensioned members and the transverse tie bars on multi-bars.

## 6—3 PRELOADINGS AND REPAIRS

### 6-3.1 PRELOADING

The following samples are taken from National Cooperative Highway Research Program Report 226 "Damage Evaluation and Repair Methods for Prestressed Concrete Bridge Members".

**Case 1: Broken Concrete Only**

AASHTO Girder Type : IV
Span Length : 85'-0"
Girder Spacing : 7'-6"
f'c girder : 5,000 psi
Ec (slab)= 0.8 Ec (girder)
1/2" diameter 270 K strand with 22 kips of final tension.
Distribution Factor S/5.5=1.36
Impact 50/(L+125)=23.8%

Section Property:

Girder only:
A : 789 $in.^2$
I :260,700 $in.^4$
$S_{bot}$ : 10,540 $in.^3$

Girder and Slab:
A : 1,257 $in.^2$
I :573,000 $in.^4$
$S_{bot}$ : 15,570 $in.^3$

Damaged Girder:
A : 659 $in.^2$
I :184,000 $in.^4$
$S_{bot}$ : 7,610 $in.^3$

Damaged Girder and Slab:
A : 1,127 $in.^2$
I :402,000 $in.^4$
$S_{bot}$ : 11,230 $in.^3$

Dead Loads:
Girder: 822 lb/ft; Slab=650 lb/ft
Dead load Moment: $M_{(DL)}$=1,130 psi

Live Loads and Impacts:
M (LL+I)=1,058 psi for HS20

Stresses:
Stresses at the bottom of the girder:

Undamaged girder;
f(DL) =1,329 (12)/10,540=1,513 psi
f(LL+I) =1,058 (12)/15,570= 815 psi
f(total) =2,328 psi
$f_{bot}$=P/A + Pe/S - f(total):
=34 (22k)/789 + 748(18.73)/10,540 - f(t)
=948 + 1,330 - 2,328
= - 50 (tension)

Damaged girder and slab due to LL+I:
f=1,058 (12)/11,230=1,130 psi

Depending on the length of damaged area and number of exposed strands, stress redistribution upward through the roadway slab may or may not occur. If in doubt the following alternate calculations are recommended:

1. Assume that prestress plus dead load of girder and slab act on the damaged girder section alone.

```
f (prestress)=748/659 + 748(23.1)/7610
              =1135 + 2270    =3405 (comp)
f (dead load)=1329(12)/7610 =2096 (tens)
                     f(total)=1309 (comp)
```

2. Assume that prestress plus dead load of girder and slab act on the composite damaged girder and slab.

```
f(prestress)=748/1127 + 748(34.8)/11230
             = 664 +2320=2984 (comp)
f(dead load)=1329(12)/11230 =1420 (tens)
                     f(total)=1564 (comp)
Allowable   fc=0.4 f'c=.4(5000)=2000 psi
```

Therefore preloading is not needed in this case to reduce compressive stresses at the bottom of the damaged section.

Determine the amount of prestress that can be restored by preloading:
Damaged girder:

| | | |
|---|---|---|
| f(prestress+dead load) | : | 1,309 psi(c) |
| f(LL+I) | : | 1,130 psi(t) |
| Total | : | 179 psi(c) |

Since the stress at the bottom of the girder is compressive, the girder could be preloaded for full LL+I and original prestress restored at the bottom of the patched area.

49.8 kips preload per girder

```
P=4 M(LL+I)/span(L)
 =4(1058)/85
 =49.8 kips
```

If preloading is not used, the tensile stress in the repaired area at the bottom of the girder would be f=815 psi (tensile stress) due to LL+I. The normal allowable tension 6 f'c= 424 psi. The repaired area will be overstressed by 391 psi and likely have cracks. Therefore preloading is necessary and should be used on the girder before repairing damaged concrete. The preloading should remain on the girder until the repairs has gained required compressive strength.

**Case 2: Severed Strands**

Case 2 is similar to case 1 except that 4 strands are cut and remaining prestresses are less.

1. Determine if preloading is required: The prestress force at the bottom of the girder in the damaged area will be furnished by the 2-1" diameter post tensioned rods.

```
f (post-tension)=P/A + Pe/S
  =142/1257 + 142(19.8)/15570 =293 psi C
f (LL+I)= M/S= 1058(12)/15570 =815 psi T
                        Total =522 psi T
Assume allowable tension 6 f'c=424 psi T
                          Net = 98 psi
```

2. Determine if preloading plus the added post-tensioning can reduce the LL+I tensile stress to 50 psi of the original design.

Assume that prestress plus dead load of girder and slab act on the damaged girder section alone:

```
f(p) =30(22k)/659 + 660(22.63)/7610
     =1000 + 1960          =2,960 psi C
f(dead load)=1329(12)/7610=2,100 psi T
                   Total =  860 psi C
```

Assume that prestress plus dead load of girder and slab act on the composite damaged girder and slab:

```
f(p) =30(22)/1127 + 660(34.33)/11230
     = 586 + 2018          =2,604 psi C
f(dead load)=1329(12)/11230=1,420 psi T
                   Total  =1,184 psi C
```

Additional stresses are needed to reduce the tensile stress to 50 psi.

```
f(LL+I)          = 815 psi
f(post-tension)=(293)psi
f(50 psi)        =( 50)psi
           Total - 472 psi
```

```
M=fS= .472(15,570)/12=612 ft-kips
P=4M/L =4(612)/85     = 28.9 kips
       28.9 kips preload per girder
       85 ft. span
```

Stresses due to preload:
At bottom of damaged section,
f(Tension)=M/S= 612(12)/11,230=654 psi
f(Compression)= 860-654 =206 (with 28.9 kips preloads).

Review results:
Stresses at the bottom of repaired girder= f (LL+I) + f (post-tension) + f (preload)= 815(t) + 293(c) + 472(c)=50 psi (tension). Therefore post-tensioning plus preload restore the full loss of prestress due to 4 severed strands plus loss of section.

Procedure:
Place preload prior to replacing the lost concrete. Keep preload on the girder until the repair has gained required compressive strength. Remove preload and proceed with post-tensioning.

### 6-3.2 REPAIRS

**Case 1: Adding External Reinforcing**

Grade 40 reinforcing steel is used to restore the loss of strength of one severed 1/2" 270k strand. The use of preload is required to restore prestress. This is relatively inefficient since it requires six #11 bars to restore the strength of one strand for an AASHTO girder type III. It is believed to be economical because it does not required post-tensioning.

f'c (girder) : 5,000 psi
Ec (slab) : .8 Ec (girder)
Span length : 60'-0"
Girder spacing : 7'-6"

Section Properties:

Girder only
A : 559.5 $in.^2$
I : 125,400 $in.^4$
$S_{bot}$ : 6,190 $in.^3$

Girder and Slab
A : 1,027 $in.^2$
I : 326,600 $in.^4$
$S_{bot}$ : 9,895 $in.^3$

Girder, Slab, Corbels and 6-#11 bars
A : 1,190 $in.^2$
I : 379,000 $in.^4$
$S_{bot}$ : 12,450 $in.^3$

Moments:
Live load plus impacts:
M (LL+I)=697 ft.-kips for HS20
$f_{bot}$=697(12)/9,895= 845 psi

Preloads:
P=4M/L=4(697)/60=46.5 kips
46.5 kips

span=60 ft.

Stress at the bottom of girder when preload is released (Loads acts on composite section-girder, slab, corbel, and reinforcing steel)

f=697(12)/12,450=.672 ksi=672 psi

Prestressing:
Final prestressing force on 1/2" 270k:
Assume prestress loss of 45,000 psi.
(0.7(270)-45)(.153)= 22 kips per strand
$f_{bot}$=P/A + Pe/S
=22,000/559 + 22,000(17.8)/6190
=39 + 63 =102 psi
Assume corbel length = 40 ft.
corbel weight =122 lb./ft.
40 ft.

span= 60 ft. M=48.8 ft.-kips
Corbel weight acts on composite girder and slab: f=M/S=48.8(12000)/9895
= 60 psi

Prestressing gain:845-672-60 =113 psi
Prestress loss for one strand=102 psi
6-#11 bars will only replace one 1/2" 270k strand at service load.

Ultimate Strength:
a= depth of equivalent compression zone
a=AsFy/0.85(f'c)b; f'c=4000(slab) b=90"
As(reinf. steel)=6 x 1.56=9.36 sq. in.
AsFy = 9.36 x 40 =374 k

Assume 16 original strands with one broken. As (strand)=15(.153)=2.30 $in.^2$
AsFy =2.30(270).85=528 k
a=(374+528)/.85(4)(90)=3.0 (approx)

Assume center of gravity of 16 strands to be 5 inches above the bottom of the girder then c.g of 15 strands is 5.17 in. (-1.0 (2.5)/15) -5.0 =5.17.

Mu (6-#11) =374(36) /12=1,122 ft-kips
Mu (strand)=528(44.8)/12=1.971 ft-kips
Total =3,093 ft-kips
Girder :583 lb/ft.; Slab:650 lb/ft.
M (DL)=555 ft-k M (corbel)=48.8 ft-k
M (total dead load) = 604 ft-kips
M (live load + impacts)= 697 ft-kips
Required Mu=1.3 (DL + 5/3 LL+I)
=1.3(604 + 5/3 (697))
=2,300 < 3,093 ft-kips

Therefore ultimate strength is adequate.

Stagger cut-off points for #11 bundle bars, 40 bar diameter minimum (AASHTO Art. 1.5.5 governs). 40 bar diameters=56", use 4'-9" stagger at the ends.

Calculate number of #5 tie bars:

Avf=Vu/o fy u (AASHTO Art. 1.5.359
u=1.2; concrete intersurface to be roughened. As(one #11)=1.56 Fy=40 ksi

Avf= 1.56(40)/0.85(40)(1.2)=1.53 sq. in.
Area of One #5 tie=0.31 sq. in.
1.53/0.31= 5 ties are required.

Also, check shear along interface per AASHTO Art. 1.5.35E.

Allowable $v_u$=350; $V_u$=1.56(40)=62.4 k
62,400/9(57)=122 psi < 350 psi.

**Case 2 : Post-Tensioning**

Two post-tensioned one inch round ASTM A722 smooth Grade 160 rods are used to restore the loss of prestress of severed 1/2" 270k strands in an AASHTO beam type IV. This utilizes four foot long jacking corbels located outside the damaged area. The jacking corbels should be located in an area where holes can be drilled through the web without interference with harped or draped strands.

Between corbels the high strength rods are protected by 1-1/2 inch minimum inside diameter rigid plastic conduits which are pressure grouted after post-tensioning. No harmful reaction between PVC conduits and concrete has been found. Plastic tubes have been an accepted means of blanketing strands near the ends of prestressed girders for many years.

The defined yield point of A722 bars is at total strain of 0.7% and offsets of 0.2%. These values are compatible with ASTM A416 strand properties. The corbel length of 4 feet was determined by using a minimum tie spacing of 9 inches.

Span length : 85'-0"
Girder spacing : 7'-6"
f'c (girder) : 5,000 psi
Ec (slab) : 0.8 Ec (girder)

Section Properties:

Girder only
A = 789 $in^2$
I =260,700 $in^4$
$S_{bot}$ = 10,540 $in^3$

Girder and Slab
A = 1,257 $in^2$
I =573,000 $in^4$
$S_{bot}$ = 15,570 $in^3$

Moments:
Dead load: Girder=822 #/ft; Slab=650
M (DL)=1,329 ft-kips
Live loads plus impacts:
M (LL+I)=1,058 ft-kip

Stresses at the bottom of the girder:
f (LL+I)= 1058(12000)/15570=815 psi
f(prestress loss due to 4 severed strands)=PA + Pe/S
=4 (22k)/789 + 88(22.23)/10,540
=111 + 186 =297 psi (Prestress loss)
Adding 2-1" A722 Grade 160 rods:
Use P=0.6 f's =.6(160)=96 ksi

Assume conduit supports at 10 spacings and bending stress is about 5.5 ksi.

P of two rods=(96-5.5)(.785)2=142 k

f (post-tensioning)=P/A + Pe/S
=142/1,257 + 142(19.8)/15,570
= 113 + 180 = 293 (Gain)

Since the prestress loss due to 4 severed strand and the gain by post-tensioning is close, 2-1" rods will replace 4 severed 1/2" 270k strands. Strands are assumed to be in bottom layer.

Ultimate Strength:

a=AsFy/.85(f'c)b; f'c=4000, b=90"
2-1" rods:AsFy=1.57(160).85= 213 k
30 Strands:30(.153)(270).85=1,053 k

a=(213+1,053)/.85(4)(90)=4.0" (approx)

Assume the center of gravity of the original 34 strands to be 6" above the bottom of the girder and the center of gravity of 30 strands is 6.47".

Mu rods =213(41.5)/12 = 736 ft-k
Mu strands=1053(52)/12 =4,563 ft-k
Total=5,299 ft-k

Mu required=1.3(DL + 5/3 LL+I)
=1.3(1,329 + 5/3 (1,058))
=4020 ft-k < 5,299 ft-k

Therefore ultimate strength is adequate.

Calculate Corbel Reinforcing:

Area of bearing plate=41 sq. in.
Area of corbel=67 sq. in.

Pu=.785(160)(.95)=119 k (AASHTO 1.6.17)
Working load per corbel=142/2=71 kips

fc bearing=71/41=1,730 psi under plate.
fc allowable=3,000 psi (AASHTO 1.6.6)
fc on corbel=119/67=1,780 psi (Safe).

Calculate required ties:

Avf=Vu/o fy u (AASHTO Art. 1.5.35.)
=119/.85(40)(1.2)=2.9 sq. in.
Required Strength=2.9(40)=116 kips
5 #5 ties=5 x .31 x 40=62 k
5/8" expansion bolts,
use 80% of allowable.
10 x 7000 x .80 =56 k
Total each side =118 k > 116 k
Use grade 60 #5 tie bars for added strength across interface.

Also check shear along interface per AASHTO Art. 1.5.35.E.

Allowable $v_u$=350 psi
116,000/12(48)= 201 psi < 350 psi o.k.

## 6—4 SEGMENTAL CONCRETE BRIDGE

### 6-4.1 GENERAL

The construction of segmental concrete bridge has been increased during last 15 years. The segmental concrete bridge become a viable alternative to the steel bridge for the long span bridges. Standardized AASHTO type I girder were used extensively for span length of 80 to 140 feet in 1950's and early 1960's. For longer spans, these beams become separated and were placed in structures with cast-in-placed joints then post tensioned together in the field.

The first major structure using precast and segmental construction was the French Choisy-Le-Poi Bridge over the Seine River in Paris in 1962. The Livere Bridge in Quebec, Canada was constructed in 1967 with three spans of 130, 260 and 130 feet. It used precast segment with the balanced construction method. The first prestressed concrete segmental bridge in the United States was the JFK Memorial Causeway (100'-200'-100') in Corpus Christi, Texas which was completed in 1973 with using the balanced cantilever method.

Since 1973, about 30 segmental concrete bridges have been built and about 15 bridges under construction in 1983. Also about 80 bridges are in design or under studies. Because the Federal Highway Administration has required alternate design for major Federal-aid bridges to promote competition and to achieve cost reduction, segmental concrete bridge becomes cost saving alternative bridge for longer span lengths which have traditionally been constructed of steel. Traditionally designers in the states were concerned about the design methods and details but neglected the cost and construction methods. Alternate designs, especially segmental construction methods bring to the engineers attention the available methods of construction and their efficiencies.

## 6-4.2 CASTING OF SEGMENTAL BRIDGE

Precast segments are cast in a plant or near the job site then assembled in place. The cast-in-place segmental construction is cast in place in a series of steps. The adoption of cast-in-place or precast segment will depend on site condition, accessibility, available erection equipment, transportation, and completion time.

The precast segments have the following advantages and disadvantages:

Advantages:
1. Segment can be fabricated and cured while substructure is under construction thus saving construction time.
2. Better quality control will be achieved at the yard.
3. Sufficient curing time will be elapsed at the time of erection so shrinkage problem will be minimum.

Disadvantages:
1. It requires high degree of geometric control during casting and erection.
2. The use of epoxy joint material requires restriction of temperature and weather during erection although some bridges have been built without their use.
3. It lacks normal reinforcement steel bars cross the joints.
4. Size of segment is limited to about 50 feet.

The cast-in-place segmental construction has the following pros and cons:

Advantages:
1. Normal reinforcement steel bars across the joints.
2. Correction of girder grade can be easily made at the site.
3. There is flexibility of change of cross section and dimension to accommodate flare, skew and widening.

Disadvantages:
1. It requires lengthy construction time to cast and cure the concrete to reach the strength required for post tensioning.
2. It requires proper design to support false work.

In precast, the fabrication tolerance of the individual segments is very importance to achieve proper vertical and horizontal alignment of the bridge. Match casting is to cast a segment by casting each segment against the previously cast adjacent segment and this requires a perfect fit between the segment at the joint. There are the **Short-line System** and the **Long-line System** in precasting methods.

**Short-line System:**

All segments are cast in stationary by using a adjacent segment as the side form. After the casting, the adjacent segment is removed and recently complete one is used for casting the next segment. This permits the space for the short-line system to be relatively small and the entire process centralized. The major disadvantage is the geometric control of the vertical and horizontal alignment. To achieve proper alignment, all adjacent segments should be accurately positioned and a slight error in the process can be magnified at the end of the span.

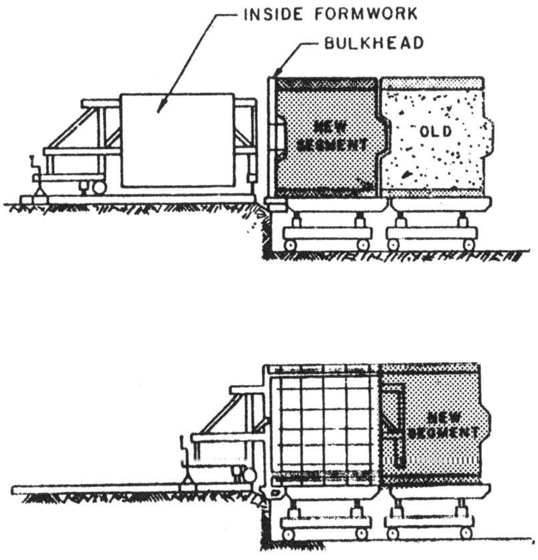


**Fig. 6.14 Shortline Casting Method**

Generally the designer furnish a theoretical casting curve which is a curve along which the perfect segment should be cast so that the final alignment will be achieved after all deformation. But in reality, each actual precast never fit the curve and each segment must be adjusted to compensate the casting error and attain desired geometry. Good computer program is required for this a function.

**Long-line System:**

All segments are cast in their relatively correct position using movable formwork on a preadjusted soffit. The soffit is adjusted to provide the proper vertical alignment of the segments when erected. It is easy to maintain control of alignment but it requires that formwork be substantially long, constructed on a firm foundation and that casting equipment be movable. These above requirements make the method relatively expensive.

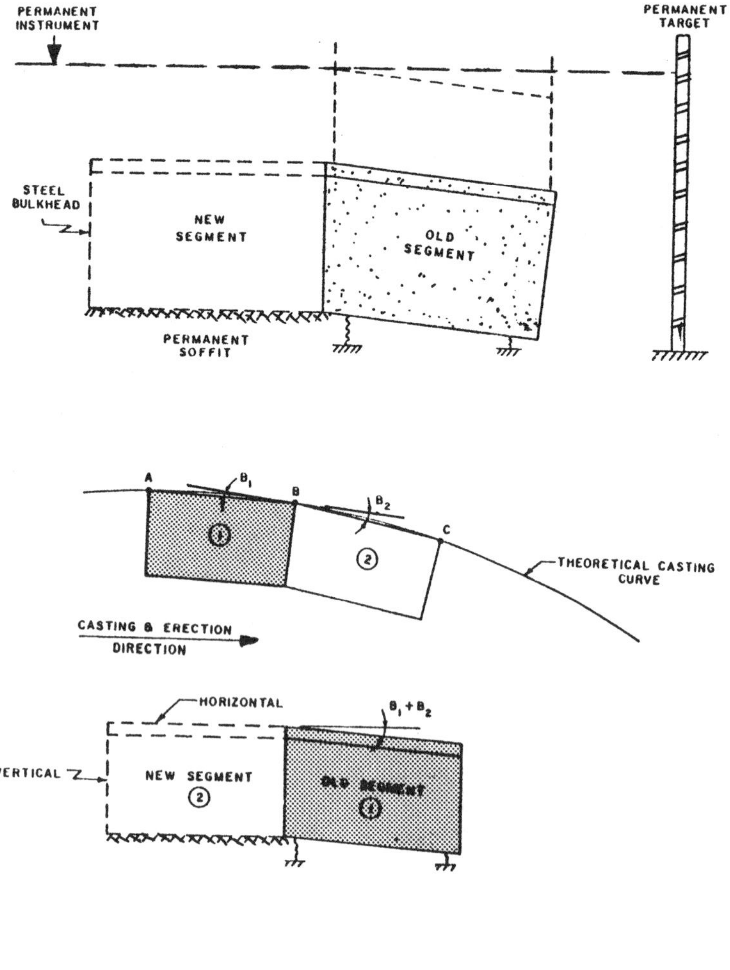


**Fig. 6.15 Alignment Control - Set Up**

**Cast-in-place Casting:**

This is alternate to the precasting system by casting the segment in their final position on the structure. It is an advantage for large and heavy segments by using form traveller. The superstructure is entirely dependent on the falsework for support until such time as the concrete is prestressed, so the temporary support system must be adequate to sustain all dead load and construction live loads before the post-tensioning of the tendons. The failure of the Riely Road Interchange Ramp Bridge, East Chicago, Indiana is blamed on 1) failure of a concrete pad for shoring towers, 2) lacked positive bracing in the longitudinal and transverse direction at the tops of shoring towers, and 3) the omission of specified wedges between stringers and crossbeams. It is recommended that the design criteria including maximum anticipated loads, allowable stresses and deflection of the falsework member should be well defined in the specifications and the details of falsework should be submitted for the engineer approvals.

## 6-4.3 ERECTION METHODS

Before designing a segmental concrete bridge, it is required to select the erection method. The erection method will establish the construction stresses based on the geometric configuration and the construction stresses are critical for design.

The major erection method of segmental concrete bridges as follows:
1. Balanced Cantilever
2. Span-by-Span Erection
3. Progressive Placement
4. Incremental Launching

The method selected for any bridge depends on the site condition and cost. Other factors influence the selection are length and height of bridge, geometry, weight and size of individual segment, location of casting yard, and local contractors expertise.

**Balanced Cantilever:**

This method is referred to free cantilever method and was the popular method used in the early stage of segmental bridge construction for long spans over navigational water. It eliminates falsework but requires balanced cantilever from a pier. This pier should be able to support the largest unbalanced overturning moments caused by unequal cantilevering as segments are attached to the cantilever end one at a time. This method generally requires a larger substructure than actually required for the finished structure. As the length of the cantilever increase, the moment at the pier increases and the post tensioning tendons are installed and stressed in increments as segments are attached. Cantilevering continues until two opposing half-spans meet at mid span where cast in place closure pours are made connecting adjacent cantilevers. Upon completion, the superstructure is post tensioned to induce final continuity with positive moment within the span.

**Span-by-Span Method:**

This requires the construction of an entire span on falsework by placing segments in position and post tensioning the span while on the false work. The ideal situation is to have the falsework capable of being moved for the construction of an adjacent span. This is the type of erection used on a long viaducts with relatively short spans such as the Long Key Bridge in Florida. It has advantages of mass production and permitting versatile adjustment in the field. It requires provision for adjusting the profile and glueing the segments together.

**Progressive Placement:**

This method is similar to the Span-by-Span method that construction starts at one end of the structure and progress continuously to the other end of the bridge. The precast segments are

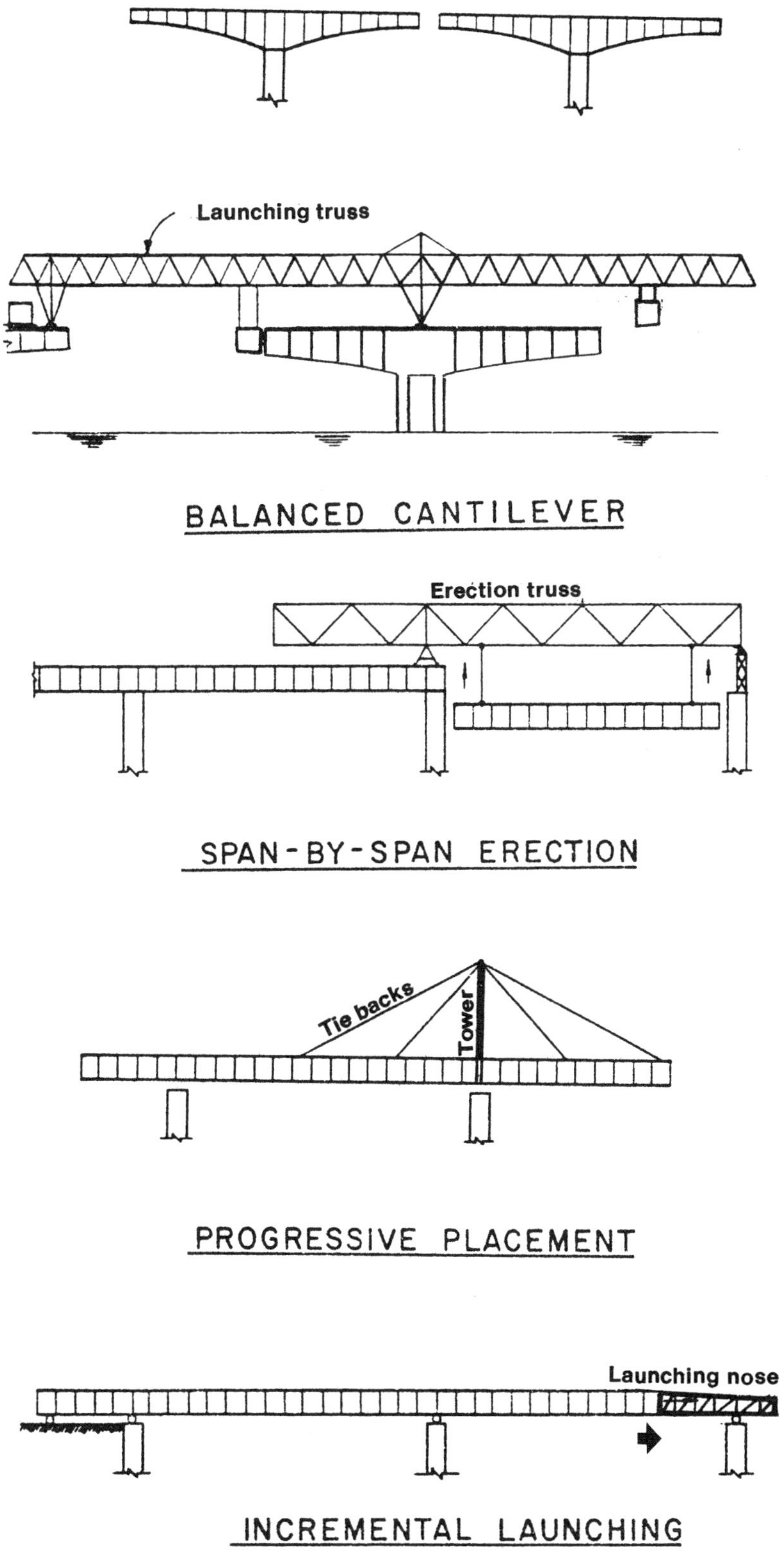


**Fig. 6.16 Erection Methods**

erected continuously in successive cantilever on the same side of the pier rather than by balanced cantilever on each side of the pier. This method is used for relatively short length of 100 to 160 feet spans where the balanced cantilever is not economical. On longer spans, this method usually requires temporary stays and movable tower to support the segment far from the starting pier.

**Incremental Launching:**

This is also called push-out construction and was used on the Rio Caroni Bridge in Venezuela in 1963. It is considered to be the latest type of segmental construction. This method has an on site casting bed located at one end of the structure. After a segment has reached sufficient strength, it is post-tensioned to the other completed segments and launched off the abutment. The succeeding segment is then cast against it. Normally a work cycle of one week is required to cast and launch a segment to attain sufficient concrete strength. Temporary sliding bearings are provided at each pier for the superstructure during construction. This method requires either a straight bridge or one of constant horizontal or vertical curvature.

### 6-4.4 DESIGN CONSIDERATION

For segmental concrete bridge, a box girder is generally used and the most economical types of bridges are as follows:

| Span (feet) | Type of Girder |
|---|---|
| 80-140 | Prestressed I Girder |
| 120-160 | Post-Tensioned Cast-In-Placed Concrete Box Girder |
| 150-300 | Precast Segmental Box (constant depth) |
| 200-350 | Precast Segmental Box (variable depth) |
| 300-600 | Cast-in-place Segmental |

In designing the segmental box bridge, the following basic factor should be determined based on the site condition:

1. Span length.
2. Span-depth ratio.
3. Steel alternate design. Whether both designs will use the same basic bridge layout or if they need changes such as skew angle, span length, etc.
4. Precast versus Cast-in-place.
5. Erection method should be shown on the plans but with provision for permitting alternate erection method.
6. Configuration of the box, Single cell, Multiple cell, Mutiple single cell joining with deck slab.
7. Constant depth versus Variable depth.

In 1978, the AASHTO interim include the design specification for precast segmental concrete box girder bridges and allows the use of the beam theory and elastic analysis in design. For girders of unusual proportions, the analysis should consider shear lag, which can be defined as "non-uniform distribution of bending stress over the cross section" This analysis is important on segments wider than 40 feet used on spans of 150 or less.

The following should be considered in design of segmental box:

1. The span-depth ratio should meet the following:

| Depth | Location | Ratio |
|---|---|---|
| Constant | All sections | 18 to 25 |
| Variable | Supports | 18 to 25 |
| Variable | Mid span | 40 to 50 |

These ratios should be balanced against other factors such as approaches, clearances and aesthetics. Ratios as high as 30 have been used in the past but less than 20 is preferred. Shallower section usually requires more tendons but have less available spaces in the cross section.

2. The width-to-depth ratio should be less than six (6). The larger the ratio, the more the structure will act as a slab. If the ratio exceeds 6, a multiple cell box should be considered.

3. Fillet and web dimension should be selected carefully to allow for stress trans-

fer around the box and adjacent room for tendon anchorage.

4. When required, the thickned bottom slab at the supports should be tapered or stepped down to the mid span thickness in the shortest length possible.

5. Access to the interior of each box or cell should be provided through manholes.

6. Where deicing chemical are used, all reinforcing bars in the top flanges including stirrups should be epoxy coated bars. In addition, a latex modified concrete overlay should be used. Transverse post-tensioning in the deck should be protected by non-metallic ducts and grouting.

7. In a coastal salt water environment, all reinforcing bars should be epoxy coated.

8. Provisions for future additional longitudinal post-tensioning should be considered. This may be in the form of external post-tensioning tendons inside boxes. It requires current construction of additional deviation blocks, saddles and anchorages. This additional provision for future additional post-tension is a safety factor for unforeseen losses, errors in design and construction. It is suggested to provide an additional 10% of design post tensioning.

9. Transverse design should consider the segment as a rigid box.

10. Transverse post tensioning of the top slab should be considered.

11. No tension should be permitted between segments under either erection or service loads.

12. Preformed shear keys should be provided between segments.

13. Torsional effects of unsysmmetrical loading should be considered.

The design of the substructure should be based on the normal loading condition with consideration for unbalanced cantilever moments due to segment weight and erection loads. The substructure design is controlled by these unbalanced overturning moments therefore the erection method should be selected before the design.

## 6-4.5 PROBLEMS OF SEGMENTAL CONCRETE BRIDGES

The following is a report titled "Prestressed Concrete Bridges in the United States as of September 30, 1982" prepared by Dr. Walter Podoly, Jr. of Bridge Division, Office of Engineering, Federal Highway Administration.

### Introduction

As with any type of new construction with any material, problems arise that require the attention of not only the designer, but contractors and subcontractors as well. No matter how good the design, if the structure is not properly constructed, there will be problems. Conversely, no matter how diligently the contractor, if the design details are poor, problems will result. Obviously, if the design and construction is poor, problems are compounded.

Problems are generally associated with quality control, poor design details, or lack of understanding as to how the structure will behave, either through ignorance or because a particular phenomenon is unknown to the current state-of-the art, or a combination of all these factors. The following documentation is of problems that have been encountered during the construction of segmental bridges in the United States.

**Vail Pass Bridges, Colorado**

Serious problems developed from the start, for the four precast segmental bridges, in the geometry control of segments during fabrication. Since the casting bed is fixed in a near level position and many segments will be erected in a different attitude in space during erection, there are many subtle geometric relationships that need to account for. If the state does not have the expertise to approve and accept responsibility for project geometry procedures and calculations, they should retain a consultant in responsible charge to perform this work as their agent.

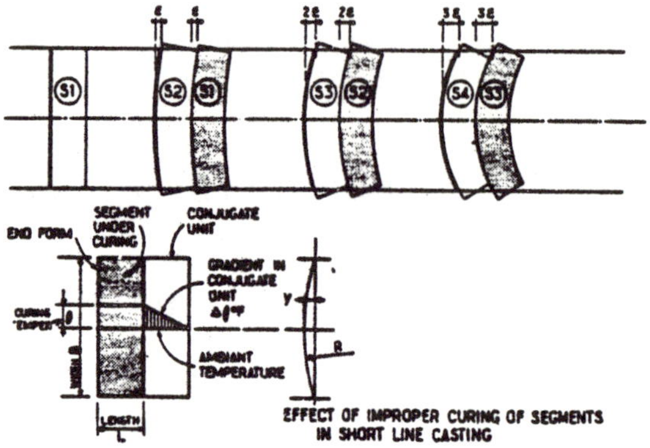


**Fig. 6.17 Effects of Curing Environments in Short Line Casting**

Another difficulty is that resulting from large temperature differences between the curing environment of the cast-against segment and the outside environment of the cast-against segment, Fig. 6.17. If a temperature gradient exists across the cast-against segment, its shape will change. The segment will return to its original shape when the temperature gradient falls back to zero. This will result in match cast surfaces that are not necessarily parallel the full length. This condition was noticed in the top slab joints, the joints were tight at the curbs and a small gap was noticed at the center. The problems become accumulative if this condition exists for subsequent castings. With the tremendous prestressing force, the segments developed longitudinal cracks as the segments are seated against each other. This also results in an unaccounted variable when evaluating the bridge camber.

For match-cast segments, it is important to enclose the formwork and the match forming segment is an isothermal enclosure. This will prevent longitudinal thermal gradients from being set up, which could lead to horizontal curvature of the segments and consequently lack of matching in the structure. This effect is particularly significant for segments with high width to length ratios (in excess of 6).

A problem occurred of longitudinal cracking in the top flange over the webs. The source of this problem was attributed to the monolithic barriers at the tip of the cantilevered top flange, Fig. 6.18. At the time the side forms are stripped the concrete is still relatively green and does not have sufficient strength to accommodate the bending stresses imposed by the relatively large concentrated force provided by the weight of the barrier. This is further aggravated in those segments near the pier where the area at the web is reduced by the voids for the longitudinal negative moment tendons.

**Fig. 6.18 Longitudinal Cracks in the Top Flange over the Web.**

It is preferable to have the barriers, either precast or cast-in-place, placed independently of the single cell box. This avoids the problem discussed above and further, allows to some extent, the opportunity to correct geometry errors in the superstructure.

The four cast-in-place segmental bridges, East and West Bound at Miller Creek and Black Gore Creek, have developed open joints at or near inflection points and have developed cracking behind the anchorages of the bottom flange continuity tendons. Diagonal cracking has developed in the bottom flange and propagated to and up the webs. The cause of this distress may be the result of any of the following causes or combinations thereof:

- Unanticipated magnitude of creep
- Unanticipated high level of thermal contraction.
- Thermal gradients

All bearings on the structure are pot bearings, all bearings at piers are fixed for longitudinal movement with the exception of short stiff piers in the Black Gore Creek Bridges. If a temperature drop in excess of what was anticipated in design had occurred, then the contraction movement of the superstructure would only occur to the extent of that allowed by the flexibility of the piers, at this point the pier become a fixed point and further contraction would produce a tension stress in the superstructure until a crack occurs or a joint opens up to relieve this stress. This is somewhat borne out by the fact that the expansion joints in the deck have failed by exceeding their design movement.

These structures have asphalt wearing surfaces which would mean that they store solar heat and could cause a larger thermal gradient than the structure could accept. Any or all of these conditions could have caused a tension in excess of the residual compression across the unreinforced joints causing the joint to open up and relieve the structure.

These structures were designed in the early 1970's, engineers in the U.S. did not start to become aware of thermal gradient stresses until approximately 1977.

**Kishwaukee River Bridge, Illinois**

This is a precast segmental structure with large shear key epoxy joints. As a result of poor quality control of the mixing of the two part component epoxy mix, the epoxy never set-up, resulting in a shear failure at the joints where the soft epoxy was located.

Two holes per web joint were drilled in critical shear stress areas to determine the condition of the epoxy. In the problem joints, up to fourteen 1 15/16" diameter pins were inserted in 2" holes through the joint in each web. These pins, 1" shorter than the web thickness, were glued in with epoxy.

**I-205 Columbia River Bridge, Oregon**

**Epoxy failure in precast joints:**
An epoxy failure was experienced in one joint, similar to that described above for the Kishwaukee Bridge, but for different reason. The epoxy was applied to the segment joint face during a driving rain and a great deal of surface moisture was present. The epoxy will accept some moisture, but the amount experienced during a driving rain is excessive and unacceptable. The epoxy in the joint failed to set properly. By the time the problem was discovered, disassembly was not practical.

To minimize reliance on the unsound epoxy, additional load transfer capacity was developed by coring 4" diameter holes in the joint, placing coil springs as a spiral reinforcement in the holes, and filling the holes with high strength mortar.

**Water Freezing in U ducts:**

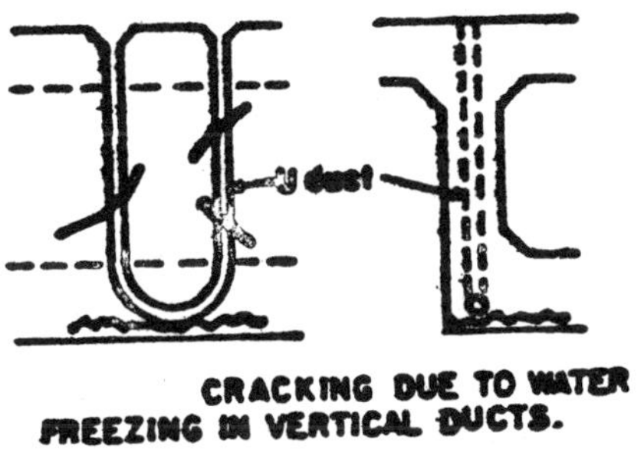


**Fig. 6.19 Cracks due to Water Freezing in "U" Ducts**

Shear at the pier is resisted by vertical prestressing installed in U ducts. Shortly after the main span construction had commenced, a period of heavy rain occurred followed by a sudden freeze. Because some of the ducts had not been adequately sealed at the deck level, water had entered the ducts and subsequently froze, resulting in multiple through web cracks in the web and laminar cracking in the bottom flange, Fig. 6.19.

Repairs to the through-web crack were carried out by epoxy injection. No problems were encountered in achieving full restoration. The 12" thick bottom flange had laminated horizontally approximately 4" above the soffit. The affected area was determined by hammer sound and corroborated by the drilling of cores. These were in plane laminations and not crossed by reinforcing steel. The problem here was a fracture mechanics problem of driving the crack by pressure injection further complicated by the fact that there was no cross reinforcement to arrest crack propagation. Stitch bolting was installed by drilling through the bottom slab. Epoxy injection was accomplished in stages, working inward from the edges of the affected area, and using low pressure so as to avoid hydraulic fracturing and crack propagation. Test cores and hammer testing indicated full homogeneity of the restored section.

Plastic caps taped on to the duct openings should be used to prevent entry of water. An alternative method would be to temporarily fill the ducts with anti-freeze until grouting can be accomplished.

**Slip of H-type dead end anchors:**
This is dead end anchorage used for a post-tensioned system whereby the anchorage is developed by bond over a portion of the tendon at the dead end, in addition, at the dead end, the individual wires in the strand are separated and bent to form a "bulb".

There were several instances where the H-type anchor failed during stressing operations, especially where there was an angle change in the tendon near the anchor. In some instances there was a measured elongation 35 to 50 percent greater than calculated and severe laminar and horizontal cracking occurred at these anchors. The apparent cause is in the manufacturing of the strand, i.e. a long lay reduces bond. Another reason appears to be the amount and type of lubricant used in drawing the wire. If stress relieving is performed by gas, it tends to burn off contaminants whereas if performed in electric furnaces it tends to bake the contaminants on as glaze.

Use of H-type anchors should not be recommended because of their sensitivity to strand manufacturing problems given above and furthermore because a portion of the strand has to be exposed to bond with the concrete in order to develop an anchorage stress transfer mechanism. There is no way to provide corrosion protection in this critical portion of the tendon.

**Misalignment of ducts:**
Although ducts are rigidly held in position at the joints by bulkheads during casting, the duct between joints may be deflected downward by the weight of the deposited concrete on top of them or the weight of workmen stepping on them. This then produces an angle break in the duct profile at each joint, Fig. 6.20. In addition to increased friction losses, a transverse localized pressure is created at the joints when the prestressing tendon is stressed. This will lead to spalling, bursting, and/or laminar cracks. A more rigid duct, or preferably a temporary mandrel in the duct should be used to maintain duct alignment.

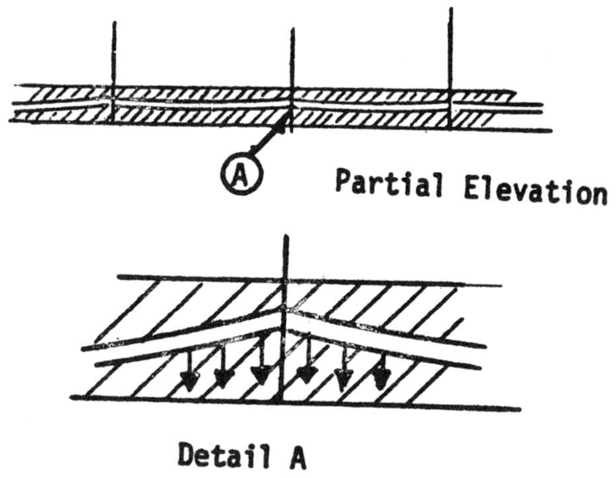


**Fig. 6.20 Misalignment of ducts leads to spalls and cracks**

**Laminar spalling:**
In one instance the upper portion of a deck slab over a pier segment laminated and spalled, affecting an area of approximately 16 by 32 feet. The lamination occurred at the underside of the longitudinal ducts. Investigation determined that there was an interference between the longitudinal and transverse duct profiles that had not been detected in the design. The longitudinal ducks were therefore deflected downward to clear the transverse ducts. The pressure produced by the upward component of the two stressing systems exceeded the tensile capacity of the remaining concrete area at that level.

The causes of other lamination spallings appear to be several and interacting. Because of the large number of ducts and their spacing in the deck area near the piers, the slab was effectively divided into an upper and lower portion, i.e., a "built-in" laminations. Although the design had the ducts spaced apart, in reality, in many instances they touched, especially where there was a horizontal curvature of the ducts. Because of this congestion, in many areas the area of the concrete on a horizontal plane near the mid-depth of the slab was less than half of the deck area.

In some instances where there was a horizontal curvature and there were closely adjoining ducts, stressing of the tendon on the outer side of the curve caused the duct to crush into the adjoining duct, distorting it vertically and causing spalling, Fig. 6.21.

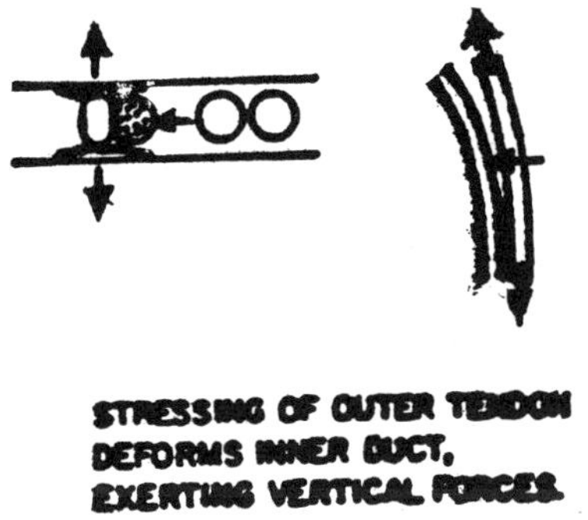


**Fig. 6.21 The duct crushed into the adjoining ducts**

Close spacing of longitudinal ducts combined with transverse ducts, longitudinal and transverse reinforcing steel in the top and bottom of the slab made it very difficult to effectively place the concrete. Concrete did not always work around and under the ducts and reinforcement. Concreting practices were not the best. Workmen often stood or walked in the fresh concrete. This pushed the ducts and reinforcement down and when the men had moved, it would spring back up, which may have caused a laminar separation in the concrete. Apparently vibration did not always extend deep enough to force concrete up between closely spaced ducts and knit the top and bottom layer together. Bleed water collected under flat ducks and in the V formed by two adjoining longitudinal ducts.

Much of the problems given above could be alleviated by selection of tendon size and duct spacing to maximize the concrete area at mid-depth of the slab. That is, increasing the area capable of transmitting tensile forces through the thickness of the slab. The use of larger size tendons (24 instead of 19 strands) and increase in size of strand (0.6" diameter instead of 0.5" diameter) would decrease the number of tendons and increase the mid-depth concrete area. Use of through slab stirrups or ties would help in the prevention of crack propagation. No. 3 vertical ties were installed on 24" centers between pairs of closely-spaced ducts. At least a 1" space should be left between adjacent ducts to assure adequate concreting. Heavier gauge ducts should be used to prevent local deformations of ducts. Angular changes and local bursting along with seams opening up are more prevalent with light gauge ducts. A thickness of 0.039" should be used instead of the widely used 0.023" thickness. Where curved tendons are used (either horizontally or vertically) the tendon on the inside of the curve should be stressed and grouted before the outside tendon is stressed.

**Blister failure:**
In span 26, a continuity tendon anchorage block failed where eight tendons were anchored between the webs, Fig. 6.22. The anchorage block (blister) simply moved forward about 11", taking the bottom slab with it. The concrete slab in front of the block crushed and the portion of slab adjacent to the block sheared on both sides of the block. The transverse reinforcement was bent into a U-form across the slab, and the segment separated from the rearward segment at the epoxy joint.

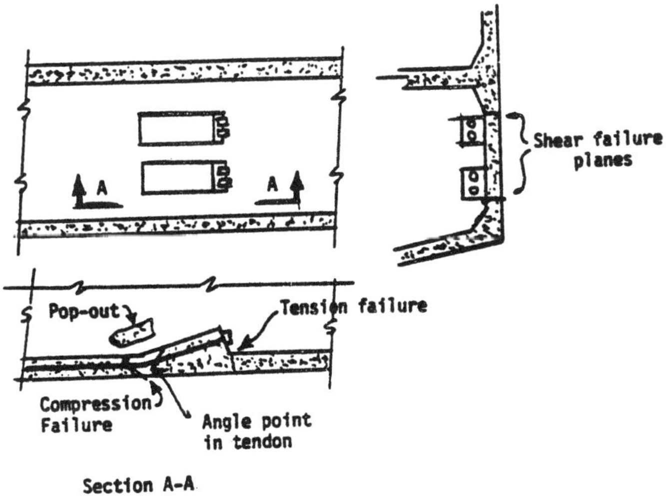


**Fig. 6.22 Blister Failure**

Repair of the span 26 segment includes a full width (between webs) beam equal to the depth of block, extending about 6 feet in back of anchorage location and about 14 feet in front of the anchorage location. This creates an anchorage buttress rather than an anchorage block or blister. The beam is tied into the bottom slab and into the webs with dowels and keys. Also, the tendons anchored in back of the failure were detensioned and entire stressing sequence repeated upon completion of the concrete repair.

Anchorage blisters located in the bottom of flange should be avoided. They should be located at the intersection of the flange and web. Where anchorage in the flanges cannot be avoided, then a full flange width buttress from web to web should be employed.

Whether the blisters are located in the slab or at the intersection of web and slab, certain precautions need to be addressed. The design assumes that tendon profile as it leaves the slab (or web) and enters the anchorage blister is a smooth curve, Fig. 6.23(a), with a uniform pressure along the curve. In reality, this is seldom accomplished in construction. The duct usually had an angular "kink" instead of a smooth curve transition, which produces a concentrated force at the front edge of the blister, Fig. 6.23(b). This then leads to a spalling or "pop-out" failure. The concrete area in the slab forward of the blister is then reduced, leading to a compression failure at that location, tension failure behind the blister and shearing along the longitudinal edge of the blister.

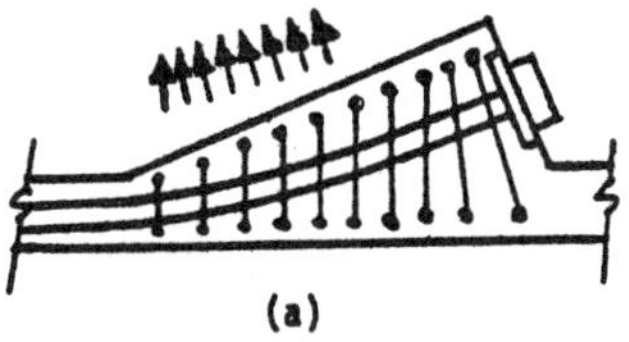


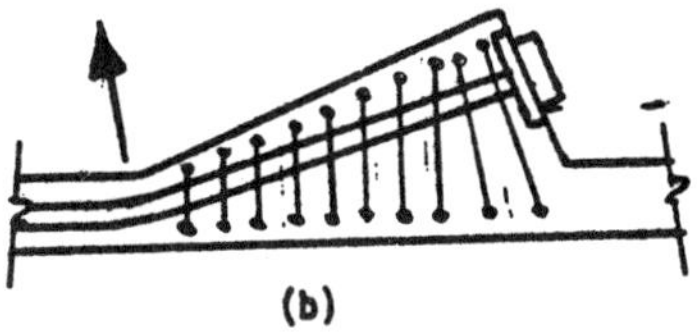


**Fig. 6.23 Blister Failure**

To avoid the "kink" in the duct, a heavier gauge duct or pre-bent tube should be used, tie confining reinforcement around the tendon should be extended well forward of the blister, adequate reinforcement extending from the blister into the slab behind it should be employed to cater to the tensile forces behind the blister and adequate transverse reinforcement should be used to cater to shear forces along side the blister.

## Long Key and Seven Mile Bridge, Florida

Spalling damage has occurred over some piers (spall area of approximately three square feet) in both of the Long Key and Seven Mile Bridges. This spalling has been attributed to tensile forces created by radial pressure in curved tendons. External tendons pass through the diaphragm along a curved profile and are anchored in the diaphragm, in Fig. 6.24. The curved profile then produces a radial pressure as illustrated in Fig. 6.25.

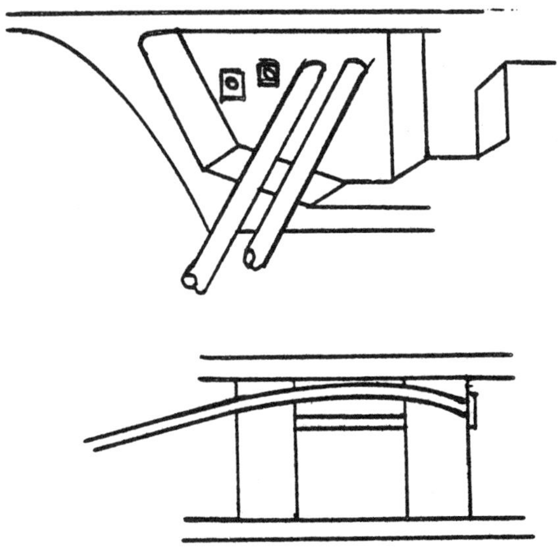

**Fig. 6.24**

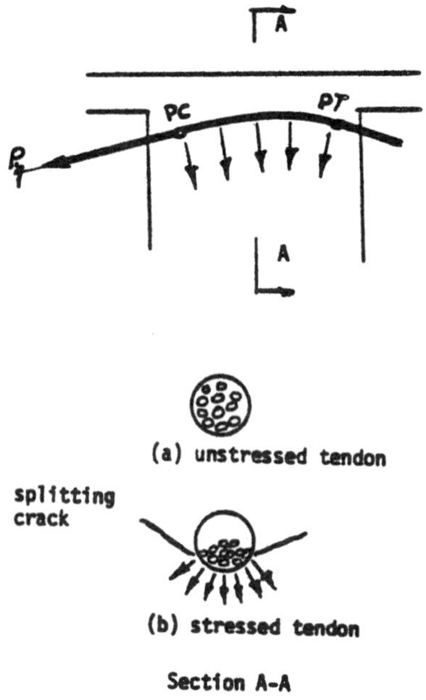


**Fig. 6.25**

In addition, as the tendon is stressed it becomes more compact towards the inside of the curvature which develops splitting cracks.

Also, because of strain compatibility a tensile field is set up above the tendons (similar to tension developed behind a blister block), Fig. 6.26. There is more concrete area in the vicinity of tension field A then in the vicinity of tension field B, to resist these tensile forces. This appears to be the reason why the spalling has occurred more towards the center line of the section. It appears that the splitting cracks may have propagated and in conjunction the tension field stresses developed into the spalling of the top surface. This is the first time that this phenomenon has been exhibited. However, recent research at the University of Texas at Austin on the "Design of Post-tensioned Girder Anchorage Zones" appears to some extent to verify this phenomenon. It appears that confinement reinforcement needs to be extended from the immediate anchorage zone to encompass the sharp curvature of the tendon.

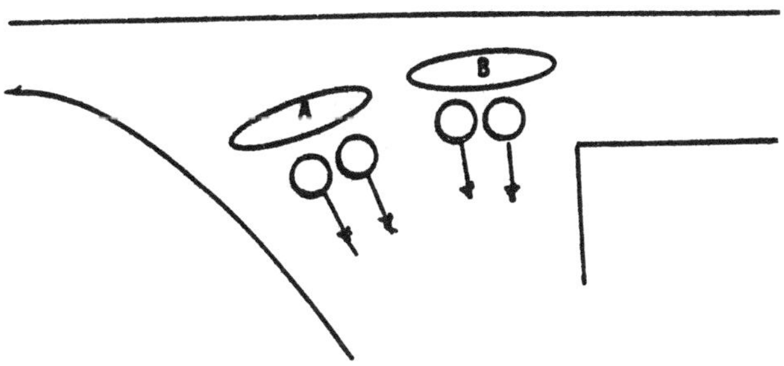


**Fig. 6.26**

**Zilwaukee Bridge, Michigan**

On August 27, 1982, while erecting the last precast segment by means of an erection truss in the span between piers 10 and 11 of the north bound structure, a mishap occurred. The cantilever south of pier 11 deflected down approximately 5 feet and the cantilever north of pier 11 deflected up approximately 3 feet. The cause of this accident is under investigation and no determination has been made yet, with regard to remedial action.

**REFERENCE**


1. Shanafelt, George O. "Damage Evaluation and Repair Method for Prestressed Concrete Members" NCHRP Report 226 , TRB Washington, DC, Nov. 1980
2. "Precast Concrete Element for Transportation Facilities" NCHRP Report 53, TRB 1978
3. "Precast Prestressed Concrete Short Span Bridge" Prestressed Concrete Institute, Chicago, IL
4. "Design Manual Part 4 Structure" Penn DOT
5. "Bridge Design Manual" Washington State DOT
6. "Optimized Section for Major Prestressed Concrete Bridge Girder" FHWA/RD-82/005, FHWA, Feb., 1982.
7. "Recommended Practice for Segmental Construction in Prestressed Concrete" PCI Journal Vol.20, No. 2, March-April 1975, pp22-41.
8. "Precast Segmental Box Girder Bridge Manual" Post-Tensioning Institute and PCI, 1978
9. Podoly, Walter Jr., "An overview of Precast prestressed Segmental Bridges" PCI Journal, Vol. 24, No. 1 Jan.-Feb. 1979, pp56-87.
10. Muller, Jean, "Construction of Long Key Bridge" PCI Journal, Vol. 25, No. 6, Nov.-Dec. 1980, pp97-111.
11. Barker, James, "Construction Techniques for Segmental Concrete Bridges" PCI Journal, Vol. 25, No.4, July-Aug. 1980, pp66-86.
12. FHWA Structural Engineering Series No.6 "Prestressed Concrete Segmental Bridge" FHWA Bridge Division, Aug. 1979.
13. Kulka,F, S J Thomas, T Y Lin, "Feasibility of Standard Sections for Segmental Prestressed Concrete Box Girder Bridges" FHWA/RD-82/024 FHWA, July 1982.
14. "Prestressed Concrete Segmental Bridges in the United States as of September 30, 1982" FHWA Bridge Division.
15. PCI Tchnical Bulletin, Number 1, 1983
16. Martin, L.D., A.E. Osborn "Connection for Modular Precast Concrete Bridge Decks" FHWA/RD-82/106 FHWA, August, 1983

# 7 STEEL STRUCTURE

## 7–1 CORROSION

### 7-1.1 GENERAL

Steel structures are subject to damage by corrosion and their vulnerability is increased where marine environments and industrial pollution exist. Where deicing salt is used in the winter and a bridge has leaking joints, the end of the beam become corroded. Bridges in shore area have special corrosion problems and the splash zone is one of the most critical area for the corrosion damage. The submerged and buried steel sections are more vulnerable to corrosion than the structural elements exposed only to the atmosphere.

The corrosion rate is dependent on many factors including saline water, carbon dioxide, hydrogen sulfide, and high temperatures in addition to the vital elements of moisture and oxygen.

Corrosion control in the atmospheric zone consists mainly of painting and galvanizing. Steel structures in the submerged and sea bed zones are also protected by painting and the cathodic method. Since corrosion in the tidal and splash zones can not be inhibited by the cathodic system, painting and sheathing are recommended at the splash zone. The repair and protection of steel structures against corrosion consists of proper surface preparation, strengthening corroded members, and applying primer and finishing coats of paint to a desirable dry film thickness.

Corrosion is an electro-chemical process of the flow of electricity and chemical changes of the metal. The metal ions (electrically charged particles) are removed from a metal surface and they chemically combine with other types of ions in the surrounding soil or water. When metal is installed in the ground or in an electrolyte, the separate particles at the metal surface may act as dissimilar electric poles. From the anodic points, galvanic current flows through the surrounding medium, such as soil or water, to the cathodic point. It will cause corrosion, pitting and eventual failure of metal by disintegration.

Underground corrosion generally falls into the following categories:

1. Galvanic corrosion
2. Spray or Electrolytic corrosion

**Galvanic corrosion** results from the differences in electrical potential along the structural steel when it is placed in soil which acts as an electrolyte. In the soil, the iron releases ions represented as $Fe^{++}$. Each iron ion is an atom of iron which has lost its electrons (negative electrical charges). As a result of losing the electrons, the ion become positively charged as represented by the "plus" sign. Each time an ion escapes, a negative charge (abandoned electrons) remains on the metal and is represented by a "minus" sign. If the conditions on the metal surface are uniform, a voltage (accumulation of electrons) is reached when ions no longer escape. At this stage, the corrosion stops.

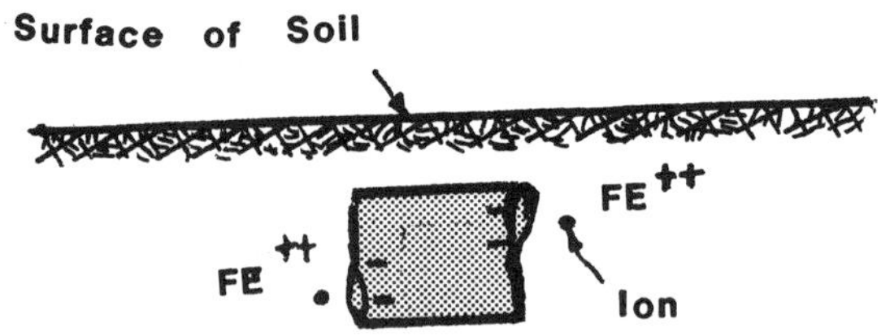


**Fig. 7.1 Formation of Ions in Soil**

**Electrolytic stray corrosion** occurs due to the direct current from outside neighboring sources such as an electrified railway and the cathodic protection systems. The direct current is introduced into the ground. The current entering the steel has no harmful effect but when it leaves the steel into the soil, which acts as an electrolyte, corrosion occurs at that point.

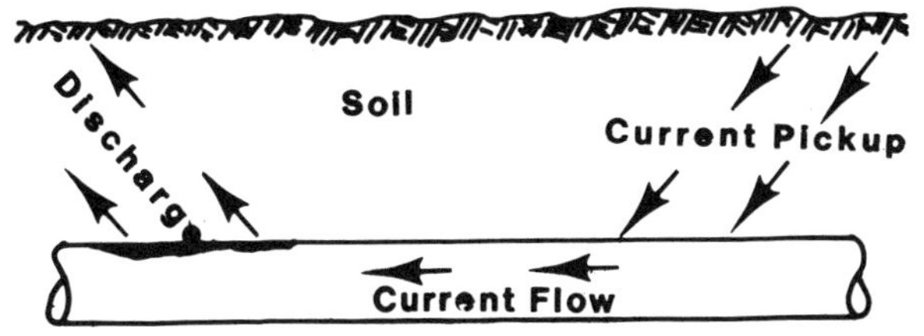


**Fig. 7.2 Corrosion due to Stray Current**

In water and in atmosphere, galvanic corrosion generally occurs due to differential environments, dissimilar metals or different external condition on the same metal.

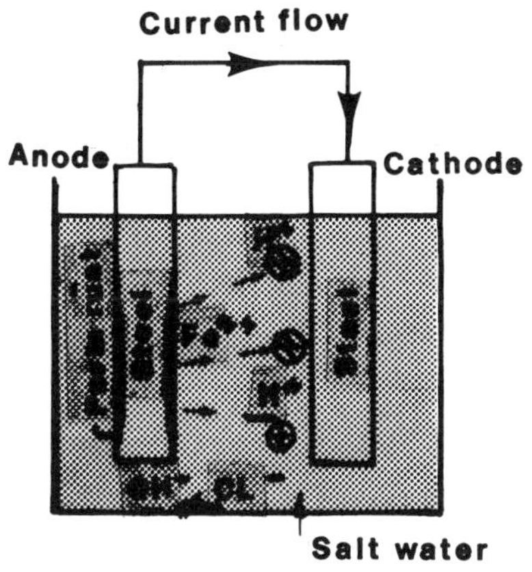


**Fig. 7.3 Corrosion Action**

In saltwater, the water molecule is broken into positively charged hydrogen ion ($H^+$) and negatively charged hydroxyl ions ($OH^-$) and the salt is split into a positively charged sodium ion ($Na^+$) and a negatively charged chloride ion ($CL^-$). As the chloride ion ($CL^-$) attracts a flow of electrons from the iron atoms at the surface of the steel, positively charged ferrous ions ($Fe^{++}$) are discharged into water and they combine with either negatively charged hydroxyl ($OH^-$) or chloride ($CL^-$) ions and form rust.

In air, oxygen and moisture are active causes of corrosion. The absorbed oxygen combined with the metal ion causes oxidation. With the presence of moisture, the hydrogen film exposes the metal to direct contact with water and permits the corrosion action deeper into the metal.

## 7-1.2 GALVANIC CORROSION

When two different metals are coupled together and placed in an electrolyte or soil and air, the dissimilarity in the metals cause one to be an anode and the other to be a cathode. The consequent current flow causes corrosion of one of the metals.

If iron and copper are connected together, a current flows to equalize the potential unbalance. As the iron ions move away from the surface of the iron, corrosion occurs as the current flow. The movement of positive ions toward the copper surface provides the copper with protection against corrosion while the iron is corroding with iron ions flowing from iron into the soil. The iron surface becomes the anode of the cell. The copper, which is receiving the corrosion protection is the cathode.

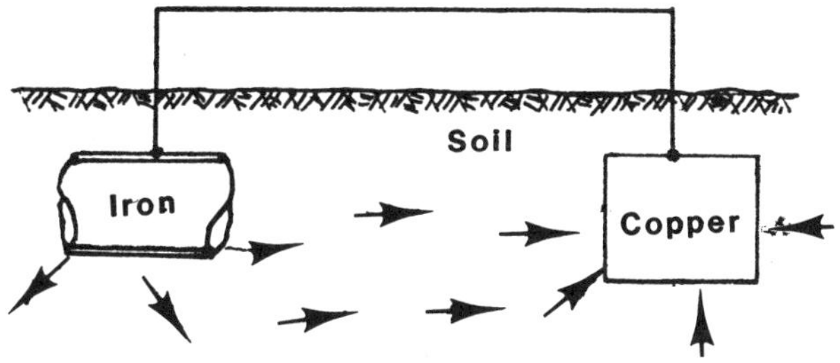


**Fig. 7.4 Galvanic Corrosion of Dissimilar Metals.** Direction of current is from anode to cathode. Copper becomes the cathode and protected from corrosion by current from anode.

The following metals or alloys will corrode and become the anodes, in decreasing order of the list:

- Magnesium
- Aluminum
- Duranium
- Zinc
- Iron
- Chromium
- Tin
- Lead
- Nickel
- Bronze
- Copper
- Silver

The following are methods for preventing galvanic corrosion:

a. Galvanize the surface of the metal.
b. Minimize surface contact between dissimilar metals.
c. Separate surface contact by using
   - Rubber
   - Neoprene
   - Stainless steel
   - Paint
d. Avoid the use of two different metals between which moisture could be trapped.

### 7-1.3 DIFFERENTIAL ENVIRONMENTAL CORROSION DUE TO CONCENTRATION CELL

The most typical corrosion is by local cell action, when the metal is surrounded by an electrolyte(sea water or mineral salts in soils). Even if the anode and cathode are the same material, the electrolytes may not be homogeneous because of the different conditions and the different concentration of oxygen and chemicals around the metal. In some moist soils, corrosion is caused by an electric current resulting from a chemical action between metal and soil.

The current flows through moist soil to the metal located in drier soil and from unaerated soil to aerated soil. For numerous reasons, considerably different environmental conditions may exist where steel is located and where corrosion occurs.

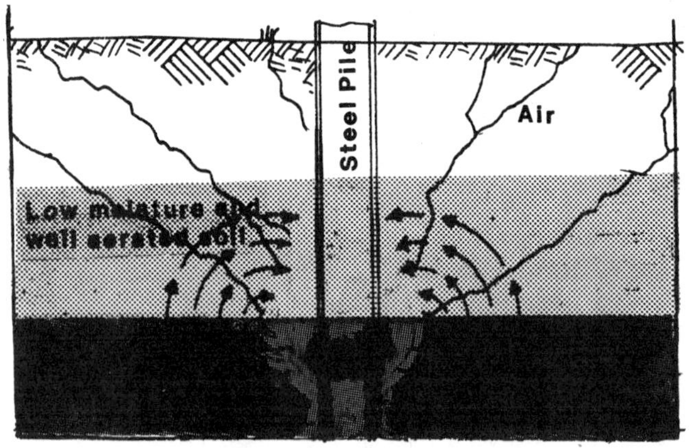


**Fig. 7.5 Corrosion Caused by Differential Aeration and Moisture of Soil**

### 7-1.4 STRESS CORROSION

Metals are susceptible to corrosion due to residual stresses or tensile forces which expose more of the grain.

### 7-1.5 FRETTING CORROSION

This type of corrosion generally occurs in the machinery of movable bridges on gears, on a shaft and metal fittings due to surface contact of two metals under vibration. The contact surface becomes pitted and forms a red colored deposit.

### 7-1.6 CHEMICAL CORROSION

Chemical corrosion may occur in industrial areas due to chloride (marine atmosphere), carbon dioxide, or sulfur oxide atmospheric containment which cause acid films and eventually corrosion.

### 7-1.7 STRAY CURRENT CORROSION

Stray current corrosion, also known as electrolysis occurs when direct current from outside source find their way into the buried steel structure member of pipe. Corrosion occurs wherever the flow of direct currency leaves the buried metal. Stray-current corrosion may be much more rapid than other types of corrosion but it is also more localized, occurring only in the vicinity of direct currency (DC) sources. Probable sources are pipeline-protection rectifiers, protected pipelines, DC electric facilities including high voltage DC (HVDC) transmission lines, DC traction systems including DC powered electrified railways and apparatus in mines.

## 7—2 CORROSION CONTROL AND REMEDIAL MEARSURES

### 7-2.1 GENERAL

Surveys show that about 40% of steel bridges have corrosion problems due to paint breakdown which allows corrosion of the base metal. If the bridge is neglected and not painted for a long period, the corrosion could cause a loss of section and result in a severe reduction of the load carrying capacity of the members.

The usual situation is that there are too little funds for the number of repairs.

Fig. 7.6 **Severely Corroded Girder** - Need Total Superstructure Replacement

Therefore, it is necessary to set up the following priorities for repairing the damage instituting corrosion control and preventing future problems.

## 1. Immediate Repairs

Severe loss of section of members could cause a failure of the structure and present immediate hazards to the travelling public. At this point there are three things that should be done:

1) Replace old sections or the structural members by adding new sections.
2) Remove the dead loads thereby increasing the live load capacity of the existing members.
3) Close the existing structure until rehabilitation or total replacement can be made.

Along with repairs, the unprotected and corroded structural steel surfaces should be blast cleaned and painted first with primer and then with finish coats of paint.

## 2. Priority Repairs

To prevent undesirable loss of section, costly damage and replacement, repair work should be scheduled to replace the existing members or reinforce the old members at a relatively small cost.

## 3. Routine Maintenance

Inspection of the condition of paint and repainting should be scheduled based on the locations and for best results at least cost.

## 4. Corrosion Control

To avoid future problems, improvements should be made in designs, specifications and materials for new and rehabilitation projects. If proper consideration is given in the choice of designs and materials during design and construction, many corrosion problems could be eliminated at little cost. The following are some of the factors to be considered.

### 1. Avoid the use of dissimilar metals

Avoid the use of dissimilar metals or burial of them in dissimilar soils to avoid galvanic action. Where dissimilar metals will be used, galvanize the steel or separate the contact surfaces of two different metals by stainless steel or organic coating materials.

### 2. Selection of suitable materials for a given environment.

In general, environmental characteristics may be grouped into following four categories according to the nature of the atmosphere:

- **Rural:** where no major harmful environment exists.
- **Marine:** where ocean spray, salt, and humidity affect the structural corrosion - coastal zone.
- **Industrial:** where high levels of carbon monoxide, sulfur oxide, nitrogen oxide exist due to burning of high sulfur coal, oil or other pollution resulting from industry.
- **Arid:** desert.

The above classification should be carefully examined and done according to individual locations instead of merely classifying by the geographic zones. For instance, a structure in a wet rural area would be more susceptible to corrosion than a dry industrial area. The PH of rain (high humidity) is influenced by the level of sulfur dioxide, the time of wetness, geographic location, and maintenance. Therefore, it should be carefully considered in the selection of materials.

Construction materials are usually selected for reasons other than corrosion resistance. The most common structural steels used for highway bridges are carbon steels which have no apparent corrosion resistance. However, low-alloyed weathering steel which normally does not require painting is used widely in some states.

In addition to corrosion, the following factors should be considered in the selection of steel members:

- Fatigue stress.
- Notch-toughness for low temperature.
- Weldability and through-thickness ductility to prevent lamellar tearing.
- Maintenance repairs.
- Life expectancy and life cycle cost.
- Environmental restrictions.

Where exotic and hostile environments are encountered, engineers currently prefer to use alloyed steel. In attempting to reduce detrimental corrosion, alloyed steel containing chromium, molybdenum, nickel, copper and other metals are used in place of the carbon steel normally used in structures.

The high chromium content retards corrosion and makes steel stainless. Most commonly used stainless steel is the chromium-nickel grade for its weldability and corrosion resistance.

The corrosion rates of steel in tropical areas are much higher than other marine environments due to their relatively high temperatures and humidity.

Low alloy steel, so called weathering steel, has been used effectively to resist atmospheric corrosion by the formation of protective rust oxidate films. However, current research indicates that their corrosion rate will rapidly increase in industrial and marine environments and the use of weathering (ASTM A 588) steel is not recommended in these locations unless it is protected by the suitable paint coat. Without a protective coating system, the use of weathering steel will be cost effective in rural environments. However, painting of such steel is strongly recommended in marine and polluted atmospheres and cost saving by using A 588 steel may not be realized in such areas. The useful life of the painted steel will be extended in absence of the rust film. It is recommended that weathering A588 steel should not be galvanized due to its silicon contents.

### 3. Care in Detailing

Details should be carefully examined to avoid water and soil penetration and not to create traps for water and debris. In corrosive environments, oversize sections may be used where the structural materials are placed underground and they are in the splash zone, or are in accessible for proper routine maintenance. For steel bridge members in corrosive environments, an additional 1/8" thickness of steel would extend an average life of the structure about 50% more.

### 4. Coating Systems

The use of a coating system: metallic coating (hot-dipping, clad steels or electroplating), organic coating (coal-tar or painting) and inorganic coating (cement coating) should be made to increase the corrosion resistance of the base metal

### 5. Cathodic Protection

Cathodic protection is generally a sacrificial system or an impressed system. Sacrificial system uses galvanic anodes which have a differential potential with respected to the steel. Generally, magnesium or zinc is used as anode. This anode is directly connected the steel structure to be

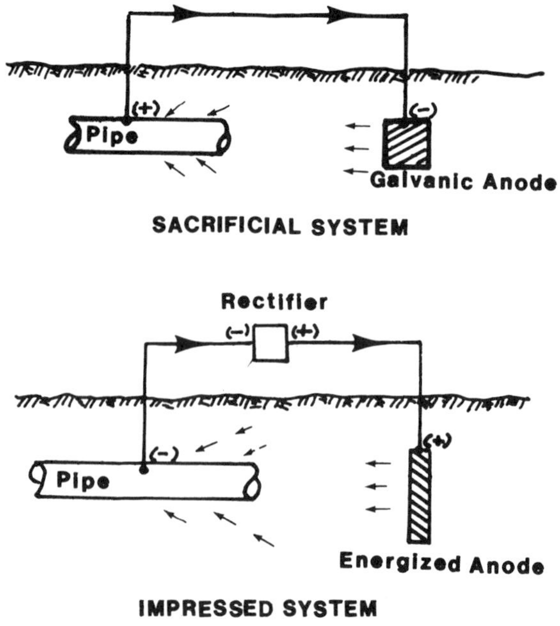


Fig. 7.7 Cathodic Protection System

protected. Impressed system is often called to as a rectifier system because the power is always furnished by a rectifier unit. This method uses anodes which are energized from an outside direct current (DC) power source. Anodes are placed in the soil or water to the positive terminal of rectifier.

Impressed system requires external power and can regulate voltage and current as required. But sacrificial galvanic system does not required external power and provides fixed voltage and lower current therefore it is used in lower resistivity electrolytes.

**6. Environmental Consideration**

In the selection of a protective coating system for bridge repairs, environmental considerations often require changes in the coating materials, paint application methods, and cleaning methods of the steel structures. These changes provide the public with the adequate protection from air and/or water pollution caused by air borne painting elements such as lead, silica, chromates, sandblasting abrasives, and paint dust caused during cleaning of the existing structure.

To solve the environmental problems, the industry is trying to produce paint materials which contain environmentally acceptable pigments, resins and solvents. Under the Performance of Alternate Coatings in the Environment (PACE) program, water borne epoxy coatings, high solid or waterborne version of chlorinated rubber, vinyl, urethane and acrylic coatings are being evaluated. Engineers should adopt safe procedures for sandblasting, cleaning, surface preparation and repainting of the existing structures. These procedures will help to avoid pollution of the surrounding environments while reducing the health hazards to workers. Also, consideration should be given to minimize the frequency of repainting by proper selection of coating materials or by the use of double protection system such as painting over the weathering steel.

## 7-2.2 PROTECTIVE METHODS

In general, protective methods may be divided into four groups:

- Metallic coating
- Organic coating
- Inorganic coating
- Cathodic protection system

In practice, one or combination of more than one of the above methods can be used as the most effective protections.

**1. Metallic Coating**

The most popular method of coating is hot dip galvanizing by brief immersion in a molten bath of metals such as zinc, aluminum, and lead. Other passive coating methods are electroplating, the plasma spray system (using plasma energy to apply powder alloy such as aluminum), and cementation of aluminum, nickel or stainless steel (Stainless clad steel).

**Galvanizing:** Galvanizing protects the steel by acting as a barrier in preventing oxygen from penetrating into the steel and as a sacrificial anode in case the steel is exposed. The base metal becomes a cathode and is protected from corrosion by the current from the anode. Zinc galvanizing has been used as a very effective method in reducing general corrosion problems. However, this option is available only for new structural members but the existing member should be protected by other methods.

Most ferrous materials are satisfactory for hot dip galvanizing but the selection of steel will affect the result of galvanizing. Normal carbon structural grade steel be galvanized with a satisfactory coating under most circumstances. The steel to be chemically suitable for galvanizing should contain carbon below 0.25%, phosphorous below 0.05%, or manganes below 1.35% and have the silicon contents less than 0.05%. Weathering steel has silicon content of 0.15 to 0.75% and is not suitable for galvanizing.

Bolts, nuts, washers and items of iron and steel hardware are galvanized in accordance with ASTM A 153 and steel members, assembles and shapes are galvanized after fabrication by the hot dip process in accordance with ASTM A 123.

The minimum desirable zinc coating thickness is about 5 mils for structural sections and about 2 mils for bolts and nuts. Since the zinc on the hardware has a shorter life, some engineers recommend an additional field galvanizing of the hardware after installation.

**2. Organic Coating**

The most common organic materials are asphalts, coal tar, paint and plastic wrap. There are following three principles to protect steel from corrosion:

1. Barrier by stopping ingress of water and oxygen.
2. Inhibition by providing source of inhibitive ions.
3. Galvanic by being electrically conductive.

Often more than one function is combined for improvement.

Organic coatings are used to stop corrosion by chemical inhibition and/or mechanical exclusion of the action of oxygen and moisture. Ordinary paints are a mixture of insoluble particles of pigments in organic vehicles and the pigments act as a barrier to ions.

In underground structures, bituminous mastic, cold applied tars, and hot asphalt enamel over primer are widely used after commercial blast cleaning as as impermeable barrier. For bridges, oil based lead and basic lead silicochromate primer with or without alkyd resins and zinc-rich primers are used widelyas inhibitive and galvanic primer. The high concentration of the zinc dust in zinc-rich primer provides sacrificial cathodic protection to the steel. Proper surface preparation is necessary to develop a proper bond between the paint and metal. It is recommended that the surface preparation and type of painting material should be selected according to the environment. The following table contains suggested painting systems to be used in various zones or environments:

**2.1 Painting:**

Since many terms of painting system are leading to confusion for the engineer, the following generally accepted definitions are included:

**Vehicle:** The liquid portion of paint, in which the pigment are able to disperse and to be distributed over the surface. It is composed of binder, thinner and additive. In general, the vehicle is used to indicate the oil or resin.

**Pigment:** A fine powdered solid which is basically insoluble when dispersed in the liquid vehicle to make paint. Pigment can be inorganic such as titanium or organic. Most paints include some white pigment because it is able to hide the underlying surfaces. The pigments provide the color, hardness, durability and corrosion resistance. A pigment is distinguished from a dye on the basis of solubility. Pigment is insoluble and dispersed in the material and dye is soluble in vehicle.

**Solvent:** Volatile liquids which are used to dissolve or disperse binder which evaporates during drying and thereby providing a consistency and character of the coating.

**Plasticizer:** A material included into a polymer to increase its flexibility or workability.

**Binder:** The film forming agent which binds the pigment particles to one another and to substrate.

**Resin:** Generally it applied to a wide variety of transparent and fusible product which may be natural or synthetic.

**Alkyd Resins:** The term "Alkyd" is an acronym for ALcohol and Acid, with the "c" changed to a "k" for sake of euphony. It is formed by the condensation of polyhydric alcohols with polybasic acids. They may be regarded as complex ester. The most common alcohol is glycerol and acid is phthalic anhydride. Generally resin is modified for practical purpose.

**Acrylic Resins:** A synthetic resin made from derivatives of acrylic acid. It is available as water based latex coating.

**Vinyl Resins:** It is generally refer to the resins resulting from polymerization of vinyl chloride, vinyl acetate and copolymers.

**Epoxy Resin:** One common type is resin made from epichlorhydrin and bisphenol A in presence of alkali.

**Basic Lead Silico Chromate:** Calcined basic lead chromate is used as a corrosion inhibitive pigment.

**Oleoresinous Vehicle:** A vehicle prepare by the addition of a resin to drying oil. These two components may or may not be further processed to obtain specific properties. Alkyd resins are sometime, but not generally included in this category.

**Primer:** First complete coat of paint applied to a surface. Primer are designed to provide adequate adhesion to new surface and are formulated to meet the special requirement of the surface, Primers contains special anticorrosive pigment such as red lead, zinc chromate, zinc powder.

Painting systems can be grouped into the following four categories:

I Oleoresinous system (oil & alkyds) on hand cleaned steel

II Oleoresinous system on commercial blasted steel

III Vinyls, polyurethanes, epoxies and synthetic polymers on near white metal

IV Zinc-rich primers with organic top coats on near white metal

## 2.2 Red Lead and Basic Lead Silico Chromate Primer:

Lead and chromate pigments in the conventional oil and alkyds vehicles (oleoresinous system) have performed satisfactorily to date. However, the Occupational Safety and Health Administration (OSHA) Regulations restrict the use of lead containing paint in work places. The OSHA rules apply only to coating in the shop, but the application of lead paints in the field will also be affected by the Environmental Protection Agency (EPA) and state and local regulations in their attempts to reduce or eliminate the amount of lead emitted in air or water during sand blasting and/or painting. The red lead primer should meet the requirement of Federal Specification TTP86 and basic lead silicochromate meet TTP615.

There are four different type of basic lead silicochromate primers which contains the following:

1. Linseed oil and alkyds
2. Base-iron oxide, and linseed oil-alkyds
3. Base-iron oxide and alkyd vanish primer
4. Base iron oxide-linseed oil and alkyds.

The basic lead silico chromate is widely accepted as rust inhibiting painting and its performance is equal to the red lead paint. The red lead primer has 100% toxic agent in its weight while the basic lead silico chromate has only 1/4 of total weight. Another advantage of basic lead paint is that it can be used as second and top coat but the red lead paint can be used only for primer. Also, basic lead paint requires inexpensive hand cleaning for surface preparation while zinc rich primer need expensive sand blasting type of surface treatment.

Some states are in the process of phasing out lead contained paints for use on structural steel, but none of the new alternates have been proven superior to lead or chromate paints in easy application, versatility and corrosion protection. and presently there is no cost effective alternate for lead paint.

## 2.3 Zinc-Rich Paint:

The main substitute to lead painting is the zinc-rich primer. Zinc is not poisonous nor detrimental to the environment and these coatings function as galvanic or sacrificial protection as opposed to the anodic inhibiting properties of lead pigments. There are two types of zinc-rich primers:

1. **Inorganic Zinc-rich:** These primers have minimum zinc dust contents of 75% by weight of total non-volatile contents. Separately packaged zinc dust is mixed at the time of application with an inorganic material like silicate, silicate ester, and phosphates
2. **Organic Zinc-rich:** These primers have minimum zinc contents of 80 to 95% by weight of non-volatile contents and zinc dust is mixed in an organic vehicle, typically phenoxy, chlorinated rubber, epoxies, urethane, vinyls or epoxy-polymide. This coating is more tolerant to the variation of surface preparation than inorganic system and has better compatibility with top coats.

The inorganic zinc-rich coatings are preferred to the organic zinc-rich paints and basic lead silico chromate paint in the severe exposure conditions and marine environment but the inorganic zinc-rich primers are less tolerant of improper surface preparations and applications. This system usually requires a near white blast cleaning. For new bridge in severe environment, inorganic zinc primer can be applied at the time of fabrication but field application of inorganic zinc primer on the existing structure is not recommended due to great efforts required to clean the existing surface before painting. Without proper surface cleaning, the coating will have earlier failures. The organic zinc-rich coatings are normally applied over a commercial blast cleaned surface and can be brush or spray applied and used for touch up.

A zinc-rich paint system has excellent corrosion resistance, however, these paints are highly variable in quality and other characteristics and require special preparation and application techniques when compared to highly tolerant conventional oleoresinous paint. This results in about 30 to 50% higher cost than lead or chromated paint.

The following lead free system are considered suitable for application of steel structure:

1. Organic or inorganic zinc-rich primer with or without alkyd top coat. This is used at normal atmosphere and mild exposure.
2. Organic zinc-rich primer with epoxy top coat (mostly two component epoxy polymide). This coating is generally used at tidal and splash zones.
3. Inorganic or organic zinc-rich primer with vinyl top coat. This coat is used at marine and industrial atmosphere and chemical exposures.
4. Vinyl primer (modified vinyl-alkyd) with or without vinyl top coats. This is recommended to use on bridges exposed to fresh water or at severe chemical exposures.
5. In marine environment and for underground structural members, organic coatings can be used. Splash zone epoxy compound (Koppers A-788) can be mixed and apply to dry surface of piles by hand to provide protection against corrosion of piles at splash zone. Coal tar epoxy coating has been applied on steel H and sheet piles, and the surface of concrete structure which is buried underground.

   A current reports conclude that in a marine environment, 7 mil zinc-rich system is more effective than 5 mil oleoresinous system and in an industrial environment, 7 mil zinc-rich system and 6 mil synthetic polymer (vinyl) show promising results.
6. Urethane Coating: A moisture curing urethane primer and aliphatic polyester urethane coating system or zinc-rich primer and epoxy intermediate coating followed by aliphatic urethane top coating system. The urethane coating system show advantages of good color retention; good chemical, salt spray and abrasion resistance; low temperature curing; and improved primer adhesion to marginally prepared steel surface. Where zinc-rich primer requires near white surface preparation (SSPC SP10), commercial blast cleaning (SSPC SP6) is adequate for urethane primer. Therefore the use of moisture curing urethane primer may be justified for repainting the bridges in moderate environment.

**For new bridges** in rural areas which will not subjected to a highly corrosive, industrial atmosphere and salt brine, weathering steel (A 588) is found to be an economical alternative in a mild environments, while painting weathering steel with a 7 mils zinc-rich system is recommended in marine environments. For a low carbon steel structure in a marine environment, application of 7 mils zinc-rich system or 6 mils synthetic system is recommended.

In severe environments, consideration should be given to paint over galvanized structural members to extend their lives for another 30 to 35 years. It is assumed that after the paint is gone, the unprotected zinc will last for 30 to 35 years. If the accessibility and other factors will pose problems for repainting of the structures, it would be prudent to provide double coating at a small amount of additional initial cost over the large repainting cost in future.

### 2.4 Surface Preparation for Repaint

Generally paints fail prematurely due to loss of adhesion or disruption of the film by corrosion. This failure can be minimized by proper surface preparation. The surface preparation and application costs are usually from 2 to10 times the paint cost therefore the extra cost of removing the last trace of contamination is not cost effective and the minimum degree of surface prepara-

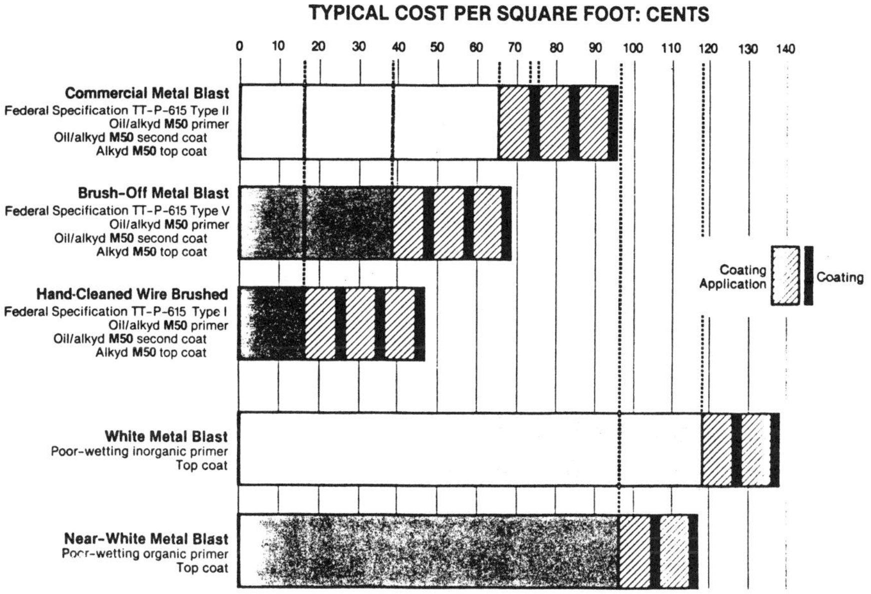


**Fig. 7.8 Total Painting Cost of Different System :** M50 is basic lead silico chromate anticorrosive pigment. This Graph is based on the 1983 cost and reproduced from the technical brochure of NL Industries, Inc. Hightstown, NJ. Special Attention should be given to the preparation cost. The cost of surface preparation prior to repainting is the key factor in determining the true cost of bridge maintenance painting.

tion should be used by considering the environment and the type of paint to be applied.

In bridge repairs, The surface should be cleaned by one or more of the following Steel Structures Painting Council (SSPC) Specifications:

I: Solvent cleaning (SSPC Specification No.1): Oil or grease shall be removed by this method.
II: Hand cleaning (SSPC Specification No.2): This cleaning shall remove all rust, foreign material, dirt, oil and all loose or peeling paint by the use of metal brushes or scrapers.
III: Power cleaning (SSPC Spec. No.3)
IV: Blast cleaning (SSPC Spec. No.6 Commercial blasting cleaning or N0. 10 Near white blast cleaning).

Heavy rusting and paint which shows cracks, peeling, brittleness, blistering, checking or general defects should be removed by Method II, Hand cleaning or Method III, Power cleaning, and before applying Method IV, Sand blasting cleaning. Any paint that can not be removed with a scraper may be kept in place and removal of paint should be carried back around the edge of the subject area until rust and the blistering underneath is removed.

There are four blast cleaning methods: **The white metal** grade is the most thorough method; the slightly less thorough **near white** is satisfactory for most paints; the next lower grade **commercial blast** can be used successfully at considerable saving in cost. **Brushoff blast** cleaning can be used in place of hand or power tool cleaning.

## 2.5 Painting System for Repairs:

The paint for the repaired area should be the same type as for the new structure. If the red lead and alkyd resin paint is no longer used for new repainting projects, complete cleaning of the existing structure to white blast cleaning is required before applying new zinc system coating. When the cleaning and repainting is limited to the portions of the existing structure, overlapping of the new zinc and old lead painting can not be avoided at the connection point of two paints and the use of inorganic zinc-rich coat is not recommended to apply over red lead primer or alkyd resin.

Often field top coating applied with inorganic zinc-rich primer shows excessive bubbling and blistering of top coats apparently due to inadequate surface cleaning, high temperature, poor application procedure and excessive thickness of zinc-rich primer.

It is reported that zinc silicate primer exhibits excellent performance in marine and atmospheric zones. In splash zones, zinc silicate and zinc epoxy primers are recommended over the oleoresinous primers, mixture of resin and oil vehicle-basic lead silico-chromate pigment and linseed oil-alkyd vehicle (1 to 1). Epoxy and vinyl based finished coats appear to perform well in the marine and splash zones. If epoxy finish coats are used in the splash zone, epoxy primer is recommended. In rural and light industrial environments, corrosion resistant silico-chromate oil-alkyds paint is suitable.

### 2.6 Painting A588 Steel:

Field observation of unpainted A588 steel bridge show corrosion problems which is associated with spilling off deicing salt on the steel. In addition to salt problems, galvanic corrosion and accumulation of debris are accelerating corrosion damages. It is generally accepted today that painting A588 steel will extend the service life of the bridge to 20 years or more before maintenance painting and the initial cost saving is not so substantial when life cycle cost is compared. Therefore it is recommended painting weathering steel (A588) in marine and industrial environment. In mild environment, the end ten(10) feet of all stringers at all abutments and piers, and all other structural steel within the ten feet area should be painted. At piers the area to be painted is ten feet in each direction from the centerline of the pier thus painting a total length of 20 feet at each pier. The following are suggested for A 588 steel:

1. A588 steel should not be galvanized due to its high silicone contents but it can be painted with one coating system rust-inhibitive shop primer, a two coating system of rust-inhibitive primer followed a top coat of synthetic resin paint or zinc rich primer with a vinyl or an epoxy top coat. Inorganic zinc primer with a vinyl top coat appears to be effective if properly applied. Some states use four coating system, shop, first field, second field, and final field coat. Immediately after blast cleaning the shop primer coat should be applied to the steel in the fabrication shop. The remaining paint to be applied in the field.
2. Before painting the surface, A588 steel should be blast cleaned to a near white or better. The cleaning should meet the requirements of the Steel Structural Painting Council (SSPC) surface preparation specification No. 6 (Commercial Blast) or No. 10 (Near White Blast) for severe exposure depending on the environment.
3. For structural bolt connections, the faying surface of unpainted A 588 steel are preferred to be blast cleaned and the bolts should be ASTM Type 3 instead of galvanized bolts because the zinc coat of bolts will be eroded rapidly when coupled with A-588 steel. When A-588 steel is painted, the faying surface may be painted with inorganic zinc primer.
4. If A-588 steel are selected because of its high strength, consider the use of ASTM A572, Grade 50 Which has the same strength as A588 if the thickness is 2 inches or less.

## 3. Inorganic Coating

On bridges, inorganic coatings are limited to portland cement applied by either spraying or casting. Generally, the cover is reinforced with wire mesh for shortcreting and cast-in-place construction. To be effective against corrosion, the cover should be more than 1 inch at atmospheric and 2 1/2 inches for subground, submerged or splash zones.

## 4. Cathodic Protection

Cathodic protection methods have been used extensively for underground pipe installation and offshore drilling. However, cathodic protection has a limited use on the bridge. There are two basic types of cathodic protection systems: the sacrificial anode and the impressed current methods. The fundamental concept of protection is that all galvanic corrosion on the structure will halt when all points on the surface of the

# TABLE I SUMMARY OF TYPICAL STEEL COATING RECOMMENDATION

| ZONE [a] | ENVIRONMENT | PREFERRED SYSTEM | ALTERNATES |
|---|---|---|---|
| 1A | Interior, normally dry (or temporary protection) Unusual in hwy. work, very mild (oil base paints would last 10 yr or more) | One coat of fast-drying shop paint (example: SSPC-Paint 13) over nominally hand-cleaned steel. Finish coat optional. (See SSPC-PS 7.01) | (1) Other one-coat primers (example: TT-P-636) (2) Rust proofing (SSPC-PS 8.01), or (3) More durable systems as per Zone 1B, or (4) Approved proprietary paint. |
| 1B | Exteriors, normally dry (Includes most highway areas where oil base paints now last 6 yr or more) | Apply 2 coats oil base primer (example: SSPC-Paint 14) over wire-brushed steel. 1-2 finish coats of long oil alkyd (SSPC-Paint 101 aluminum or SSPC-Paint 104 white, gray or green) 4.0 mils or more thickness (5.0 mils for 4 coats). (See SSPC-PS 1.01, 1.02, or 1.03) | (1) Blast clean (SSPC-SP 6) and use same paints or shorter oil alkyds. (2) Alternate primers (SSPC-Paint 2; TT-P-57, Type I; AASHO M72-57, Type I or II; or TT-P-615, Type V) or (3) Alternate intermediate TT-P-86, Type II or non-leafing aluminum, or (4) Equivalent state system, or (5) Same systems as Zone 2A or 2B, or (6) Proven proprietary system. |
| 2A | Frequently wet by fresh water Involves condensation, splash, spray, or frequent immersion. (Oil base paints now last 5 yr or less) | Near-white blast clean surface; 4 coats (4.5 mils) of vinyl system (example: SSPC-Paints 8 or 9) (See SSPC-PS 4.04 or 4.02) | (1) Pickle (SSPC-SP 8) instead of blast cleaning. (2) Alternate vinyls are VR 3 or approved proprietaries. (3) Epoxy system guide (SSPC-PS 13.00), coal tar epoxy (SSPC-PS 11.01), chlorinated rubber system, or approved proprietary system. |
| 2B | Frequently wet by salt water Involves condensation, splash, spray or frequent immersion. (Oil base paints now last 3 yr or less) | Near-white blast clean surface; apply zinc-rich primer (example: SSPC-PS 12.00 or MIL-P-23236 or California Highway Spec. 66-G-55) followed by approved wash primer and finish coat. (Example: SSPC-PT 3 plus SSPC-Vinyl Paint 8 or 9, 3+ mils) Assure satisfactory adhesion of finish coats. | (1) Use finish coat with same vehicle as zinc-rich primer (inorganic, epoxy, chlorinated rubber, vinyl, etc.) (2) Use vinyl paint system with wash coat and inhibitive primer (example: SSPC-PS 4.01 or 4.03) (3) Use as alternate finish coats or by themselves, coal tar epoxy (SSPC-PS 11.01), epoxy (guide SSPC-PS 13.00), or approved chlorinated rubber system, or other proven proprietary system. |
| 3 | Chemical exposures (Acidic, alkaline, oxidizing, solvents, etc.) | Same as for Zone 2B, but with chemically resistant finish coat system specially chosen to protect primer and base metal against specific chemical agent. (Zinc-rich unsatisfactory for very acid or very alkaline conditions.) Assure satisfactory adhesion of finish coats. | Same choices as for Zone 2B, but with special finish coats. (1) Coal tar epoxy (SSPC-PS 11.01) (at least 16 mils). (2) Straight vinyls for acid and alkali (SSPC-PS 4.01 or 4.03). (3) Epoxies for alkalies, salts, aliphatics, acid splash; not for strong solvents. (4) Neoprenes and other proven proprietary systems to resist specific conditions. |
| 4 | *Special Conditions* | | |
| | Painting galvanized steel | Solvent clean to remove oil and grease. Wire brush to remove any rust. Apply zinc dust-zinc oxide paint TT-P-641 (Type II for new steel, Type I for old, as per SSPC-PS 2.05 and 1.04). Somewhat better adhesion if surface is weathered before painting. | (1) Chemical pretreatment of new work by commercial hot phosphate or wash primer. (2) Zinc-rich primer (example: Guide SSPC-PS 12.00). (3) Prime with SSPC-Paint 5. (4) Prime with proven proprietary cement-base, polyvinyl acetate emulsion, or acrylic latex. |
| | Mildew | After surface preparation, wash mildewed surface with trisodium phosphate and dry. Add mildewcide to each coat of paint (example: 8-quinolinoleate). Vinyl, chlorinated rubber resins, and barium metaborate and zinc-rich pigmentations tend to resist mildew. | Alternate mildewcides and fungicides include copper naphthenate, chlorinated phenols, phenyl mercuric dodecylsueinate, proprietary agents. Add in amount recommended by the manufacturer. |
| | Temporary protection and rustproofing | See system for Zone 1A. Also see SSPC-PS 8.01, "Rust Preventive Compounds" (thick non-hardening films over minimum surface preparation) | Soft, heavy, or hard film compounds as per 52-MA-602, Type B, C, or D; or use proprietary rustproofing compounds. |
| | Painting welds | Before welding, do not paint within 2 in. of edges. Blast clean after welding. See SSPC-PA 1, Sections 3.5.2.4 and 3.5.2.5. | Chip and wire brush weld thoroughly. Wash with 5% phosphoric acid and rinse. See SSPC-SP 1, Section 3.1.6. |

[a] These are intended as specific exposure zones of the portion of the structure under consideration rather than geographic zones. Severity of exposure can change sharply over very short distances due to such factors as wind, spray, condensation, and use of de-icing chemicals.

# TABLE II MINIMUM SURFACE PREPARATION

| PAINT SYSTEMS | MINIMUM SURFACE PREPARATION |
|---|---|
| 1.00-1.06 Oil-base paint systems | Hand-tool cleaning (SSPC-SP 2) |
| 2.00-2.05 Alkyd paint systems for weather exposure | Blast cleaning (SSPC-SP 6) or pickling (SSPC-SP 8) |
| 3.00 Phenolic paint systems for fresh-water immersion | Blast cleaning (SSPC-SP 6) or pickling (SSPC-SP 8) |
| 4.00-4.05 Vinyl paint systems for chemical exposure | Blast cleaning (SSPC-SP 6 or 10) or pickling (SSPC-SP 8) |
| 6.00-6.03 Paint systems for vessels | Blast cleaning (SSPC-SP 6 or 10) or pickling (SSPC-SP 8) (most areas) |
| 7.00-7.01 One-coat shop paint for structural steel | Nominal cleaning |
| 8.00-8.01 Rust preventive compounds | Solvent cleaning (SSPC-SP 1) and/or nominal cleaning |
| 9.01 Asphalt mastic | Blast cleaning (SSPC-SP 6) |
| 10.00-10.02 Coal tar coatings | Blast cleaning (SSPC-SP 6) |
| 11.01 Coal tar epoxy | Blast cleaning (SSPC-SP 6 or 10) |
| 12.00 Zinc-rich systems | Blast cleaning (SSPC-SP 6 or 10) |
| 13.00 Epoxy systems | Blast cleaning (SSPC-SP 6) or pickling (SSPC-SP 8) |

structure have been polarized to equal potential.

**a. Sacrificial System:**
This galvanic type requires no external power supply but it uses magnesium, zinc, or aluminum as the anode and the metal becomes the cathode. For example, with zinc coated pipe lines or by placing anode zinc along the steel piles, the flow of current is induced by galvanic action.

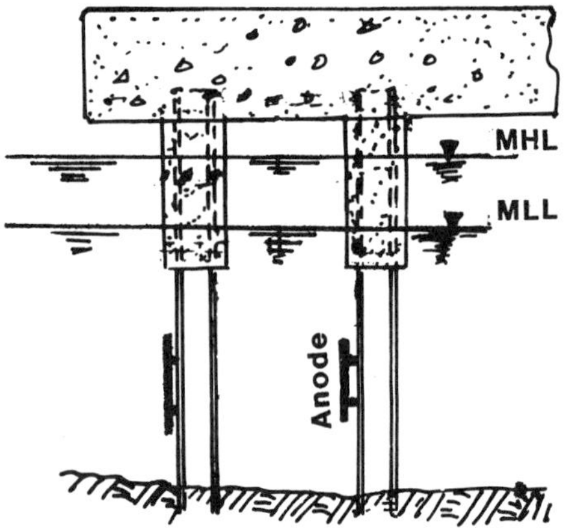


**Fig. 7.9a Sacrificial Galvanic Method:** Sacrificial anodes are zinc, aluminum, or magnesium. Zinc and aluminum are effective in salt water and magnesium in soil.

**b. Impressed Current System:**

In this applied potential type, an external direct current power source ( rectifier) is necessary to establish a potential of sufficient magnitude to offset and cancel out the corrosion action. Since the current flow does not depend on the potential of the anode and metal of the structure, selection of the anode is based on the capability of the conducting current and transmitting current to the electrolyte with minimum corrosion to itself. This type of system uses graphite, carbon, silicon or cast iron as the anode.

Since the current must pass through the soil or water from the anode to the structure to be protected, a cathodic protection system should be designed by considering such factors as conductivity, moisture content, soil and water characteristics, oxygen content, temperature, dissimilarity of metals, and environment. A cathodic system is used extensively to extend the life and reduce the maintenance of pipe lines, water tanks, and marine structures.

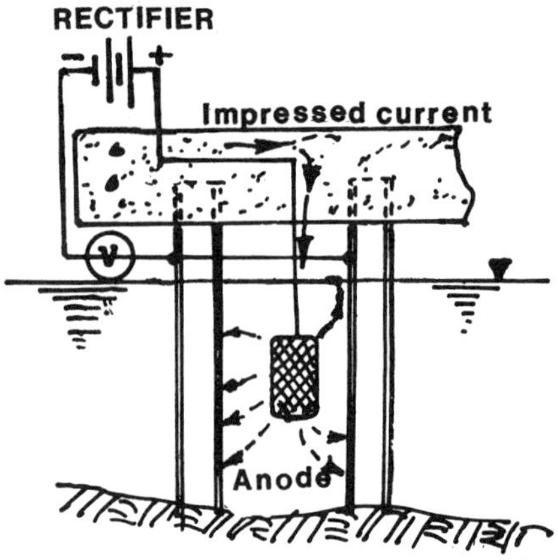


**Fig. 7.9b Impressed Current System:** Anode will be selected based on its capability to transmit the current with minimum corrosion. Energized anodes are graphite or high silicone cast iron. Most corrosion stops at -0.77 volts of half cell reading.

## 7—3 FRACTURE AND FATIGUE

Bridge cracks may be grouped as follows:

- Tension cracks due to overload
- Brittle fracture cracks
- Stress corrosion cracks
- Fatigue cracks

Fracture and fatigue cracks are far more critical in a welded structure than in a riveted structure. A riveted structure is built up of several components and a crack stops at interface between components. In welded bridges, if a crack starts in a weld due to an accidental impact, improper fabrication, or faulty details, the crack will progress until the member completely fails.

Critical Bridge Members: Any crack is considered undesirable but cracks are considered as being more critical than others. If the crack or failure occurs in the critical member, this can cause a sudden catastrophic failure without warning. The following nonredundant members are considered as critical bridge members:

1. Main suspension cables and anchorage of the suspension bridge.

2. A main eye bar of suspended member of truss: It is considered that eye bar systems are more critical than wire or rope suspension members.
3. Hanger bars and pins of suspended girders of two girder cantilever-suspended span system: Cantilever-suspended span system bridges are more critical than continuous multi-girder bridges.
4. Two through girders with floor beams and single or double box girder bridge: This type of bridge is more susceptible to total failure than multi-girder system bridge.
5. Tie girders of tied arch bridge.
6. Main tension chord of truss.
7. Steel pier cap of single or two column steel bent.
8. Counterweight hanger of bascule bridge.
9. Anchors for cable stayed bridge.

### 7-3.1 BRITTLE FRACTURE

Brittle fracture usually occurs without prior plastic deformation or other warning signs. Brittle fracture causes catastrophic failure in the structural steel and generally there is no time available to take corrective action.

The following are the major causes of brittle fracture:

1. Brittle fracture occur due to discontinuities such as a torch cut edge, a stress concentration, a mechanical gouge, welding defects, quench cracks, porosity, or lack of fusion. These discontinuities could grow to a critical size by fatigue or stress corrosion.

2. When metal has less ability to carry a load or deform plastically in the presence of a notch, this lack of metal toughness can cause brittle fracture. For example, silicon steel is susceptible to poor notch toughness.

3. When tensile stresses are required to initiate the fracture, brittle fracture may occur.

4. Other factors affecting brittle fracture are as follows:

- Temperature drop
- Dynamic loading
- Constraint
- Redundancy
- Stress corrosion
- Crack growth
- Plate thickness
- Fatigue crack growth
- Residual stress
- Flaws

In 1973, the American Association of State Highway and Transportation Officials (AASHTO) adopted material toughness (Charpy V-Notch Impact) requirement for main load carrying members subject to tensile stress and in 1978 issued the "Guide Specification for Fracture Critical Non-Redundant Steel Bridge Members" to assure responsibility of design, fabrication and non-destructive testing of Fracture Critical Members (FCMs).

Fracture Critical Non-Redundant bridge members are tension components with a single load path where a single fracture could lead to total failure. An example of such fracture critical non-redundant members are the flange and web of either girder of a two girder system, truss main chords, simple span floor beams carrying the entire reaction of unit, pins of a truss or two girder system bridge, a hanger plate, a steel cap of single or two column bent pier, tie girder of a tie arch bridge, counterweight hangers of bascule bridges and anchors for a cable-stayed bridge.

### 7-3.2 STRESS CORROSION CRACK

Tensile stress, normal or residual, is required to develop stress corrosion cracks. The tensile force tends to expose more metal at the grain boundaries, permitting corrosion to progress. As the corrosion continues to penetrate the surface, it will start pitting and will cause a small intergranular stress corrosion crack along the grain boundaries and eventually through the center of the grains. However, no significant cases of stress corrosion on bridge members are known except for a few cases with a prestressed strand and the machinery of movable bridges. Stress relieve heat treatment can be used to reduce residual stress and prevent stress corrosion.

## TABLE III FATIGUE CATEGORIES

CATEGORY A: *Category A is the least fracture critical condition and comprised of plain material, plates and rolled beams without attachment.*

CATEGORY B: *Category B includes built-up sections without attachments which are made by longitudinal welds in the direction of stress. The following are also Category B: Groove welded flange butt joints at curved radius and tapered transition; Bolted connection by high strength bolts.*

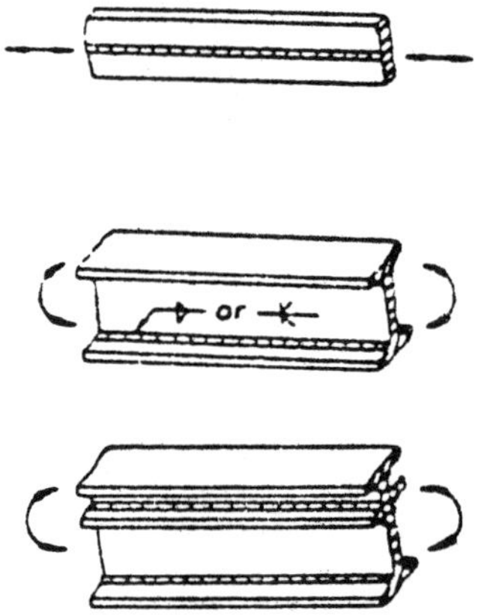


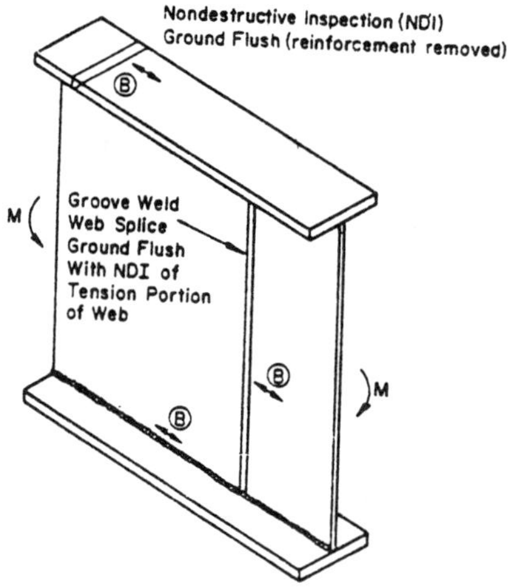


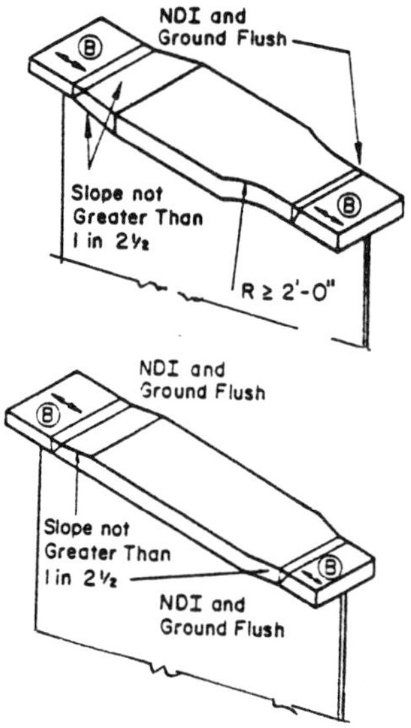


CATEGORY C; *Category C is consists of girder or beams with attachments such as toe of vertical stiffeners. If butt welded joint of flange is not ground smooth and is not tested by radiographic or ultrasonic method, It is classified as Category C.*

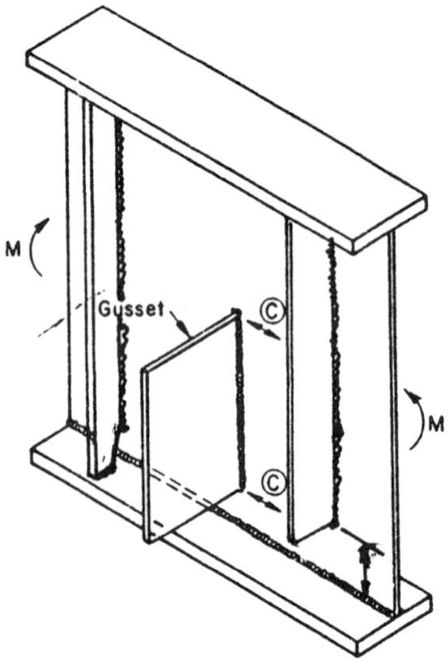


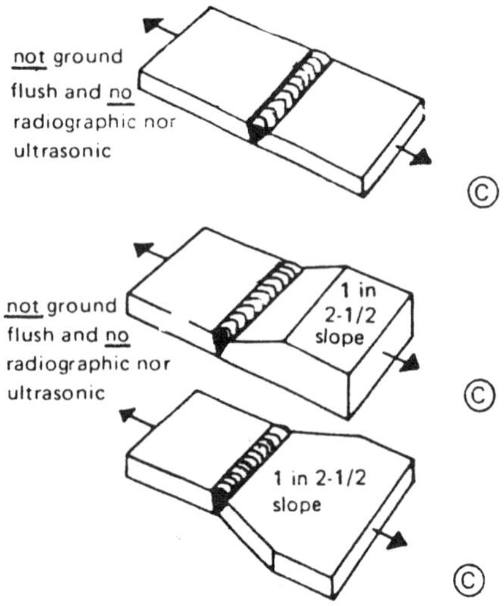

CATEGORY D: *Base metal adjacent to the welded attachment can be Category C to E depending on the length, and radius of termination. Short gusset plate less than 4 inches is classified Category D. The longer the attachment the more sensitive it is at the weld termination to crack.*

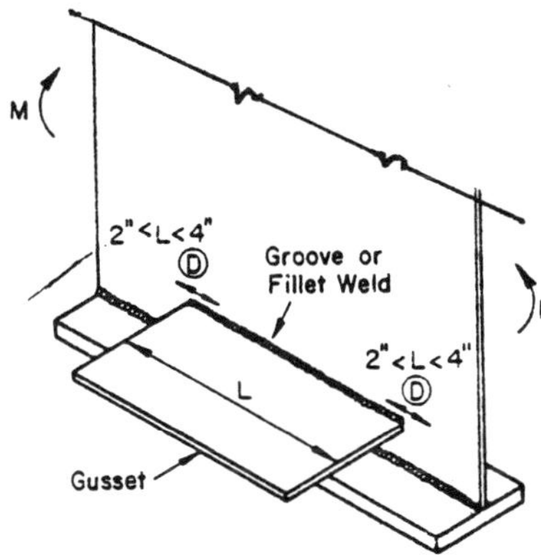


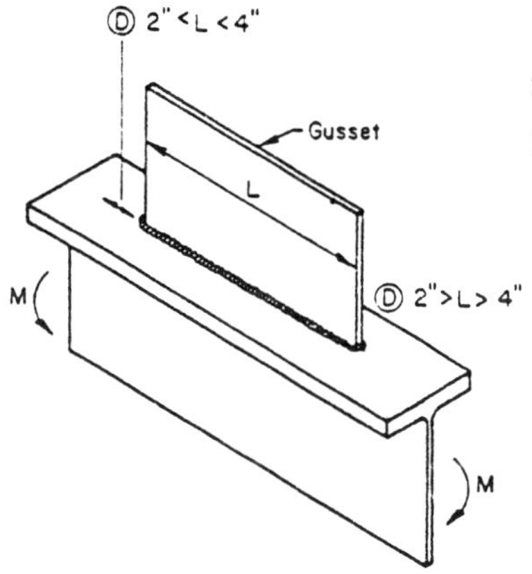


| | |
|---|---|
| L<2" | : C |
| 2"<L<4" | : D |
| L>4" | : E |

CATEGORY E: *Category E or E' is the transverse weld perpendicular to the main stress path such as toe weld of cover plated beams.*

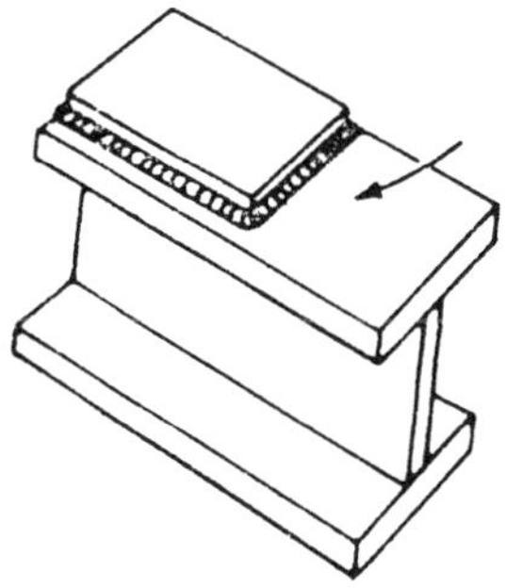

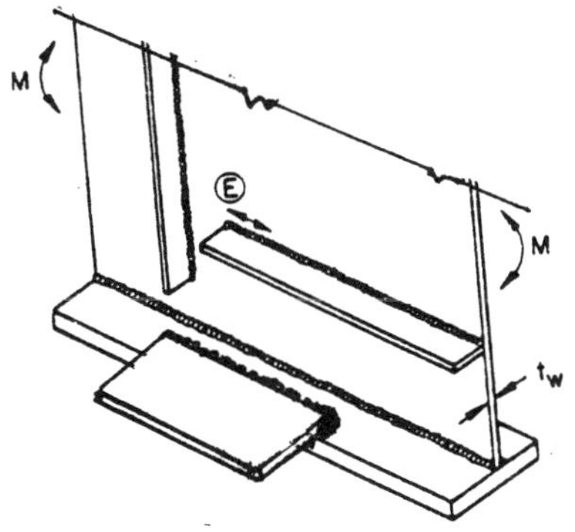


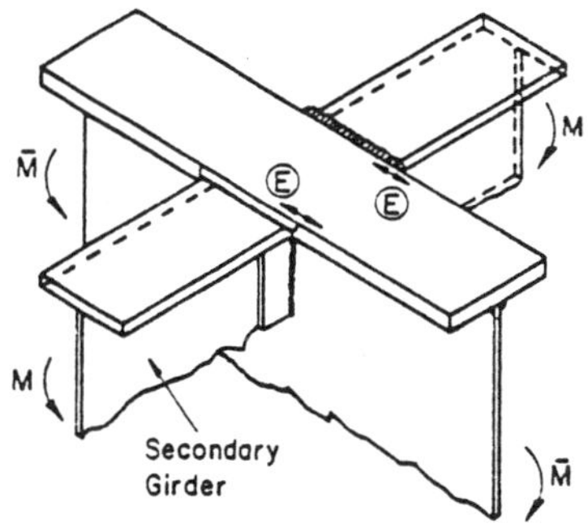


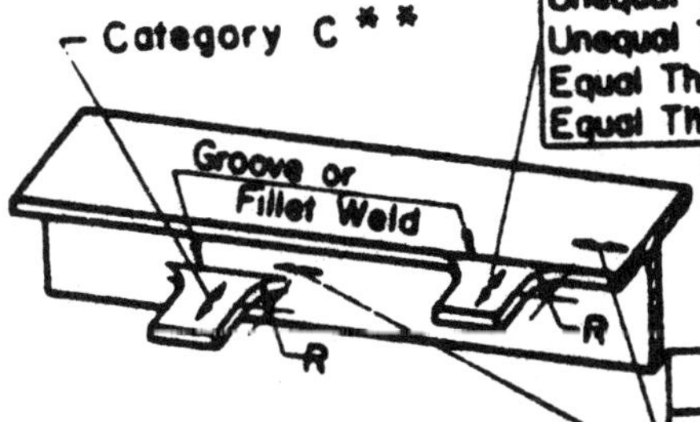


| WELD CONDITION* | CAT |
|---|---|
| Unequal Thickness - Reinf. in Place | E |
| Unequal Thickness - Reinf. Removed | D |
| Equal Thickness - Reinf in Place | C |
| Equal Thickness - Reinf. Removed | B |

*For transverse loading - check transition radius for possible lower category

| R** | CAT |
|---|---|
| R≥24"(.610m) | B |
| 24"(.610m)>R> 6"(.152m) | C |
| 6"(.152m)>R> 2"(.051m) | D |
| 2"(.051m)>R | E |

**Also applies to transverse loading

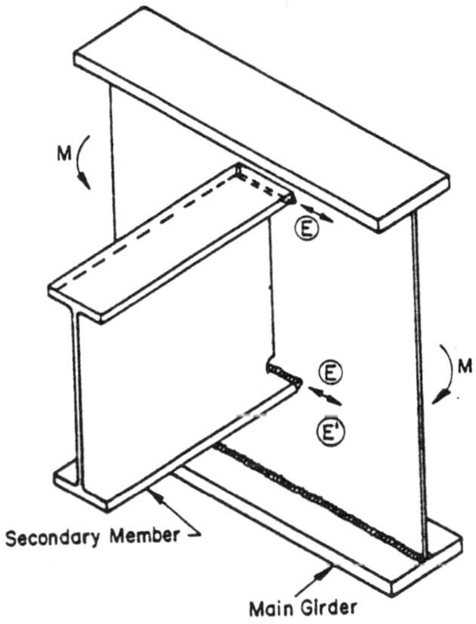

### 7-3.3 FATIGUE CRACKS

Steel may fracture at stress less than at its elastic limit if the stress is repeated a sufficient number of times between certain values of maximum and minimum stress. There have been numerous incidents of fatigue cracks in welded steel bridges under normal traffic conditions. The most notorious case of fatigue cracks are those that led to the Silver Bridge failure in 1967. A broken eye bar caused this catastrophic collapse which cost forty six lives.

The fatigue and fracture problems should be considered during all phases of project such as study phase, selection of structural type and materials, development of details, preparation of plans and specification, fabrication and erection of the bridge.

The minimum and maximum stresses affect the growth of fatigue cracks which occur mostly at the end of cover plates, stiffeners and gusset plates to beam webs or flanges. The following locations are also susceptible to fatigue damage:

1. A loose connection can cause unbalanced or excessive stresses.
2. Damaged, misaligned or bent member.
3. Areas of excessive vibration or torsion.
4. Corrosion weakened areas under tension.
5. The toe of fillet welds on a welded steel cover plate on a rolled beam.
6. Welded stiffeners without sufficient gap between the flange plate and stiffener or the web and stiffener.
7. Stringer and floor beam connection:
    - Throat of connection angle.
    - Damaged rivet head in the leg of connection angle attached to floor beam.
    - Coped top flange of stringers connected to the upper portion of floor beam web.
    - When end connection is rigid and it acts as fixed, the coped stringer web would subject to relative movement between web and flanges and result in a fatigue crack.
8. Knee braces of floor beams and top connection plates of floor beam which cross over the main girders.
9. Chord splice, eye bars, hanger and diagonal rods, and bottom chord bars of truss bridges.
10. Armored expansion joints and hanger and pin assembly of suspended girders.
11. Torn or cut out sections of web and flange.
12. Center and end gusset plate connections of lateral bracing.
13. Stress raiser such as rivet or bolt holes, craters in longitudinal fillet welds, intersections of longitudinal and vertical welds, sudden change of size in welded members without transition, severe porosity, discontinuity of back up bars, lack of fusion, notch, flame cut edge or mechanical indentations.
14. Open grating welded to the top flanges of stringers and floor beams.
15. Previous repairs which used indiscriminate welding or cutting.

Also, details in welding design have caused fatigue damages such as excessive welding, intersecting weldings, unnecessary member (stiffener perpendicular to applied loads, double diagonal members, etc.), neglect of strain compatibility, discontinued backing bars for welds, welding attachment in tension areas instead of bolting which is more forgiving fatigue problems, and other overlooking of the structural behaviors.

Structural member should have as few attachment or connections as possible and the details should be simple and buildable with enough room for welding. If suitable welding details can not be used, bolting should be considered.

### 7-3.4 TESTING METHODS

Non-destructive methods of fatigue crack detection are generally as follows:

- Ultrasonic Test
- Radiographic Test (X ray Test)
- Magnetic Particle Test
- Eddy Current Test
- Dye Penetration Test

**The Ultrasonic Test** is based on the wave properties of sound in a material to detect cracks. It is widely used in the field because it is portable and able to detect the location of defects in depth. However, it has the disadvantage of not giving permanent test records.

# TABLE IV

## Reference Guide To Major Methods For The Nondestructive Testing of Welds

| Inspection Method | Equipment Required | Enables Detection Of | Advantages | Limitations | Remarks |
|---|---|---|---|---|---|
| VISUAL | Magnifying glass<br><br>Weld-size gauge<br><br>Pocket rule<br><br>Straight edge<br><br>Workmanship standards | Surface flaws — cracks, porosity, unfilled craters, slag inclusions.<br><br>Warpage, underwelding, overwelding, poorly formed beads, misalignments, improper fitup. | Low cost<br><br>Can be applied while work is in process, permitting correction of faults<br><br>Gives indication of incorrect procedures. | Applicable to surface defects only.<br><br>Provides no permanent record. | Should always be the primary method of inspection, no matter what other techniques are required.<br><br>Is the only "productive" type of inspection.<br><br>Is the necessary function of everyone who in any way contributes to the making of the weld. |
| RADIOGRAPHIC | Commercial X-ray or gamma units, made especially for inspecting welds, castings, and forgings.<br><br>Film and processing facilities.<br><br>Fluoroscopic viewing equipment. | Interior macroscopic flaws — cracks, porosity, blow holes, nonmetallic inclusions, incomplete root penetration, undercutting, icicles, and burnthrough. | When the indications are recorded on film, gives a permanent record.<br><br>When viewed on a fluoroscopic screen, a low-cost method of internal inspection. | Requires skill in choosing angles of exposure, operating equipment, and interpreting indications.<br><br>Requires safety precautions.<br>Not generally suitable for fillet weld inspection. | X-ray inspection is required by many codes and specifications.<br><br>Useful in qualification of weldors and welding processes.<br><br>Because of cost, its use should be limited to those areas where other methods will not provide the assurance required. |
| MAGNETIC PARTICLE | Special commercial equipment.<br><br>Magnetic powders — dry or wet form; may be fluorescent for viewing under ultraviolet light. | Excellent for detecting surface discontinuities — especially surface cracks. | Simpler to use than radiographic inspection.<br><br>Permits controlled sensitivity.<br><br>Relatively low-cost method. | Applicable to ferromagnetic materials only.<br><br>Requires skill in interpretation of indications and recognition of irrelevant patterns.<br><br>Difficult to use on rough surfaces. | Elongated defects parallel to the magnetic field may not give pattern; for this reason the field should be applied from two directions at or near right angles to each other. |
| LIQUID PENETRANT | Commercial kits, containing fluorescent or dye penetrants and developers.<br><br>Application equipment for the developer.<br><br>A source of ultraviolet light — if fluorescent method is used. | Surface cracks not readily visible to the unaided eye.<br><br>Excellent for locating leaks in weldments. | Applicable to magnetic & nonmagnetic materials.<br><br>Easy to use.<br><br>Low cost. | Only surface defects are detectable.<br><br>Cannot be used effectively on hot assemblies. | In thin-walled vessels, will reveal leaks not ordinarily located by usual air tests.<br><br>Irrelevant surface conditions (smoke, slag) may give misleading indications. |
| ULTRASONIC | Special commercial equipment, either of the pulse-echo or transmission type.<br><br>Standard reference patterns for interpretation of RF or video patterns. | Surface and subsurface flaws including those too small to be detected by other methods.<br>Especially for detecting subsurface lamination-like defects. | Very sensitive.<br><br>Permits probing of joints inaccessible to radiography. | Requires high degree of skill in interpreting pulse-echo patterns.<br><br>Permanent record is not readily obtained. | Pulse-echo equipment is highly developed for weld inspection purposes.<br><br>The transmission-type equipment simplifies pattern interpretation where it is applicable. |

**The Radiographic Test** is based on the use of X ray radiation to take pictures of internal cracks including shape and size. This test does not show the depth of defects and is difficult to use in field location.

**Fig. 7.10 Radiographic Testing**

**The Magnetic Particle Test** is based on the use of magnetic fields. It is a nondestructive testing method for determining the existence and extent of possible defects in steel. Finely divided magnetic particles, applied to the magnetized parts, are attracted to and outline the pattern of any magnetic leakage fields caused by discontinuities and cracks. It is limited to flat surfaces or near surface defects and difficult to use in certain field locations.

**The Eddy Current Test** is based on electrical field in the material examined by inducing eddy current flow. Changes in the flow caused by variation in the test object are reflected into nearby coils for analysis by suitable instrumentation and techniques. This test is rarely used in the field.

**The Dye Penetration Test** is the most popular field test method for detecting surface cracks. After wetting the surface with the dye spray penetrant, the penetrant is wiped off from the surface, then a developer (normally a contrasting color) is applied to the surface. If there is a crack, the dye color appears on the developer.

Technique of Acoustic Emission is lately introduced for detecting delayed crack. This is used 72 hours after completion of welding and can be used in the field for locating and detecting cracks in existing beams that have vehicular collision damage.

## 7-3.5 WELDED COVER PLATE

The use of welded cover plates on rolled beams was popular among the states from 1940 to 1970 due to its simplicity and cost. However, welded cover plate failures are detected frequently as compared to other types of defects. The cover plate was generally used on the multi-stringer systems and its failure rarely resulted in sudden complete failures.

When an abrupt change in flange cross section occurs, a severe stress concentration is produced in the fillet weld at the end of the cover plate. Cracks may form at the toe of the weld and grow through the flange. If cracks occur throughout the weld, the beam will fail.

Cover plated beams induce a low fatigue strength at the toe of the welds of the cover plate terminations regardless of their shape and with or without end transverse weld because the stress concentration at the weld end is transversed to the applied stresses.

**Fig. 7.11 Crack at Flange and Web**

For these reasons, many states require that the bottom cover plates be extended beyond the allowable fatigue stress cut off points to the bearings (even if this means some waste of material and additional cost). Alternatives to lengthening cover plates are the following two improvements to the end of the cover plate:

1. Grinding the transverse weld to 1 to 3 taper ( current Maryland State Highway Practice).
2. Bolting the ends of plates instead of using transverse welds at the theoretical cut-off point.

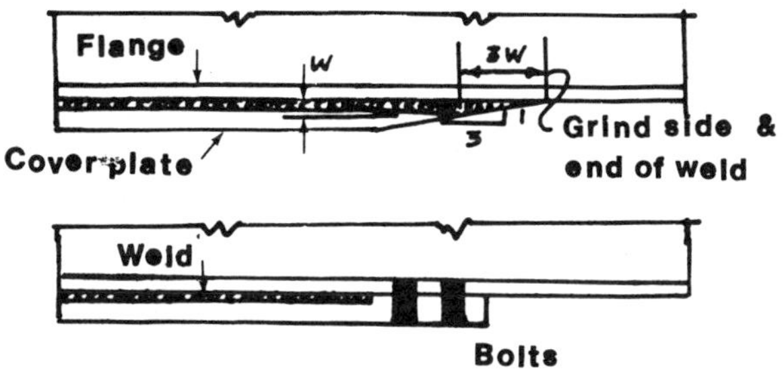


**Fig. 7.12 End Details of Cover Plate**

A 1978 Pennsylvania Department of Transportation Research Report on full scale cover plate beams ("Fatigue Resistance of Full Scale Cover Plate Beam" Penn DOT Project 72-3) suggested that cover plated steel beam bridges subjected to random variable stress range cycles will eventually crack if subjected to a sufficient number of stress cycles. Cover plated beams with flange thicknesses greater than 0.8 inch have less fatigue strength than that allowed by Category E. of the AASHTO bridge specifications. Those bridges with more than 1500 ADTT will experience fatigue cracking before their expected lives of 80 to 100 years. Bridges with low traffic volumes have less chances of fatigue distress. Therefore, it is suggested that existing cover plated beams which may require retrofitting in the future have it done based on their traffic volumes and their age.

When the steel bridge has AASHTO Category E or E' details, it is suggested that the fatigue stress level should be checked and remaining expected life should be computed based on the ADT and age.

## Retrofit:

In retrofitting of the welded cover plates if large fatigue cracks occur at the weld between the flange and web plate, the following repair sequence should be used:

1. Drill holes at the crack tips to arrest any further crack propagations.
2. Jack up the beams.
3. Bolt the web and flange with splice web and cover plates - similar to a normal beam splice.
4. Lower the jack after splicing.

If the cracks are limited to the weld toe, the following two methods may be used:

1. Peen the weld toe region for crack depths of less than 0.125 inch.
2. Apply a gas tungsten arc remelt pass for crack depths of less than 0.188 inch.

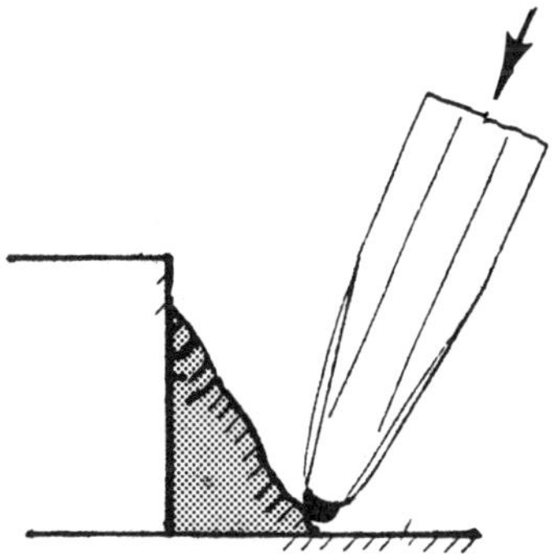

**Fig. 7.13 Peening the Toe of Weld**

## Repair Methods:

Some of the available methods to repair fatigue damage and improve or upgrade the structural details to prevent crack growth and increase fatigue strength are as follows:

- Replacement of the damaged members
- Strengthening
- Drilling hole
- Peening
- Gas tungsten remelt
- Replacing rivets with high strength bolts

**Replacement:** If the crack has propagated to a substantial length, the only practical solution is total replacement of the members.

**Strengthening:** After the crack is arrested by drilling a hole at the tip, the damaged member can be strengthened by bolting or welding new sections over the damaged members, adding new sections, or groove welding the cracks and grinding the surface smooth.

**Drilling Holes:** If the crack does not reduce the capacity of the member, drilling a hole at the tip of the crack may be adequate to terminate its growth. It is suggested to drill a hole about 1/2" beyond the identified tip of the crack so as to cover the undetected but propagated crack tip. The size of hole varies from 1/2" to 1" depending on the size of crack, location and stress level.

**Peening:** Air hammer peening of the weld toe introduces compressive residual stresses and the live load stress range is prevented from being effective in the compressive residual stress region. However, this method is less effective at high stress range level and high minimum stress. For deep crack, the crack may be sealed only at the top but leaving the crack active internally. This method may improve the fatigue strength from Category E to D.

**Gas Tungsten Arc Remelt:** This process removes the nonmetallic intrusions at the weld toe and reduces the magnitude of the stress concentration by smoothing the weld termination.

**Replacement of Rivets:** For riveted structures, the replacement of rivets in critical locations with high-strength bolts (A-325) can be effective if the fatigue damage is limited to the rivet head and rivet itself and the crack growth is limited to a length of 1 inch beyond the rivets.

The rivet head can be removed with a pneumatiac hammer and the remainder of rivet can be driven out with a back-out punch. After removing the rivets, high strength bolts are installed with hardened washers in the holes and the bolts are tightened to the required strength (minimum 70% of breaking strength). The clamping force is induced by tightening the bolts, creating a zone of compressive stress and this helps prevent crack initiation at previously critical locations.

__New Design:__

The New York State Department of Transportation has the following guideline for new design:

Generally cover plate shall be used only on simple span structure. Continuous span structures shall make use of fabricated plate girder.

Cover plates added to rolled beams shall be full length cover plates. Cover plates terminating 1'0" or less from the centerlines of bearing shall be considered full length.

Welded cover plates shall be limited to one on any flange. The maximum thickness of cover plate shall not be greater than 2 times the thickness of the flange to which it is attached. The thickness and width of a cover palte may be varies by butt welding parts of different thickness or width with transition conforming to the requirement of AASHTO Article 1.7.15. Such plate shall be assembled, welded and radiographed before attachment to the flange.

A cover plate may be wider or narrower than flange to which it is attached. However a cover plate shall not vary in width so that it is both wider and narrower than the flange to which it is attached. When a cover plate is narrower than the flange it shall be detailed for welding, "All around". When cover plate is wider than the flange it shall be detailed with transverse end welds.

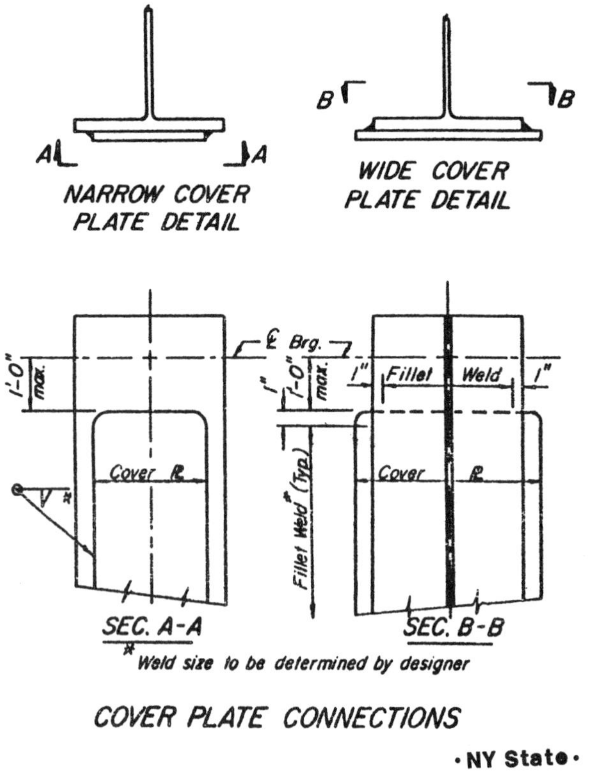


**Fig. 7.14 Cover Plate Connection**

The end welds shall not be returned around the beam flange but shall stop 1" from the flange edge. The longitudinal side welds shall stop 1" from each end of the cover plate.

All welds connecting cover plates to beam flange shall be continuous, except as noted above, and shall not be smaller than nominal size permitted by AASHTO Article 1.7.21 (B).

### 7-3.6 WELDED STIFFENER

Stiffener fatigue cracks occur at the end of the weld which fastens the stiffener to the web. These cracks are due to out of plane cyclic displacement and high bending stresses in the unstiffened web gap area between the bottom tension flange weld and the end of the stiffener. With all of these stresses concentrated in a short gap of the web, fatigue cracks develop under many thousands of cyclic movements.

Also, out of plane bending stresses during handling or shipping due to relative rotation between the web and the flange cause web cracks which start at the weld end. Generally, cracks start at the termination of a weld and propagate at a right angle to the principal stress. During fabrication, cracks may develop from large residual stress concentrations between the stiffener weld and the web to flange weld.

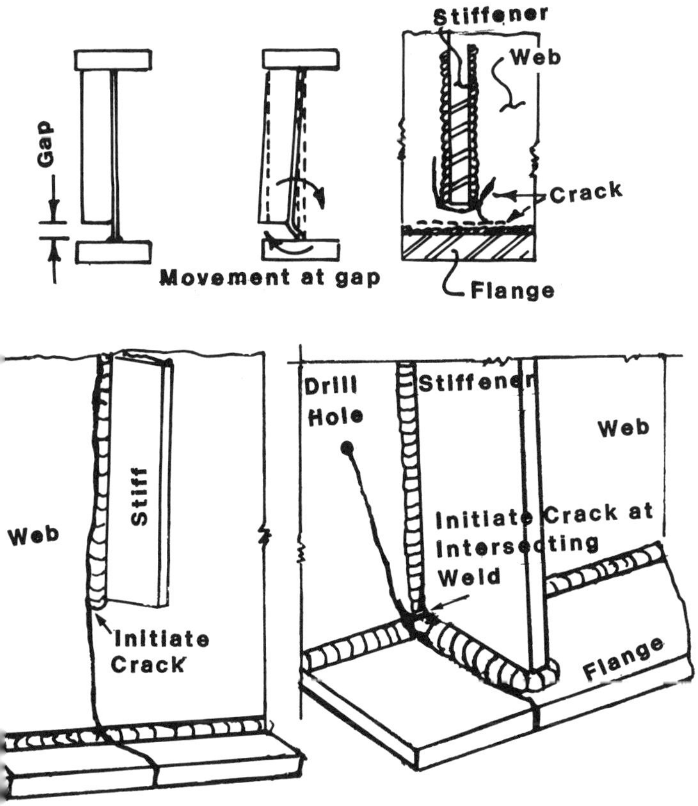


Fig. 7.15 Crack at Connection Weld

**Retrofit:**

For large fatigue cracks which extend into large areas of the web or flange, repairs should be made by the application of bolted splice or cover plate to the damaged web and flange plate after drilling holes at the tip of the cracks.

The cracks developed at the end of the stiffeners can be repaired by drilling and reaming 3/4 inch holes at the crack tips.

If there is a crack at the web to flange weld, the following procedure should be used:

1) Drill 3/4 inch holes at the crack tips
2) Ream the holes
3) Vee out the crack
4) Weld using 7018 electrodes
5) Grind smooth

However, if there are no visible cracks at any stiffener location, drill 3/4 inch holes in the web at each side of the end of the stiffener to prevent any future fatigue crack growth. Some designers use high strength bolts in the drilled hole to prevent the cracks from growing by inducing residual compression through the washer by bolting. However, the washers prevent the observation of any further crack development.

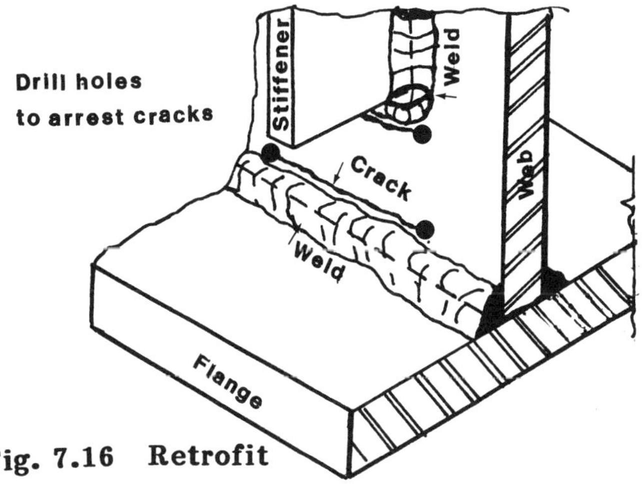


Fig. 7.16 Retrofit

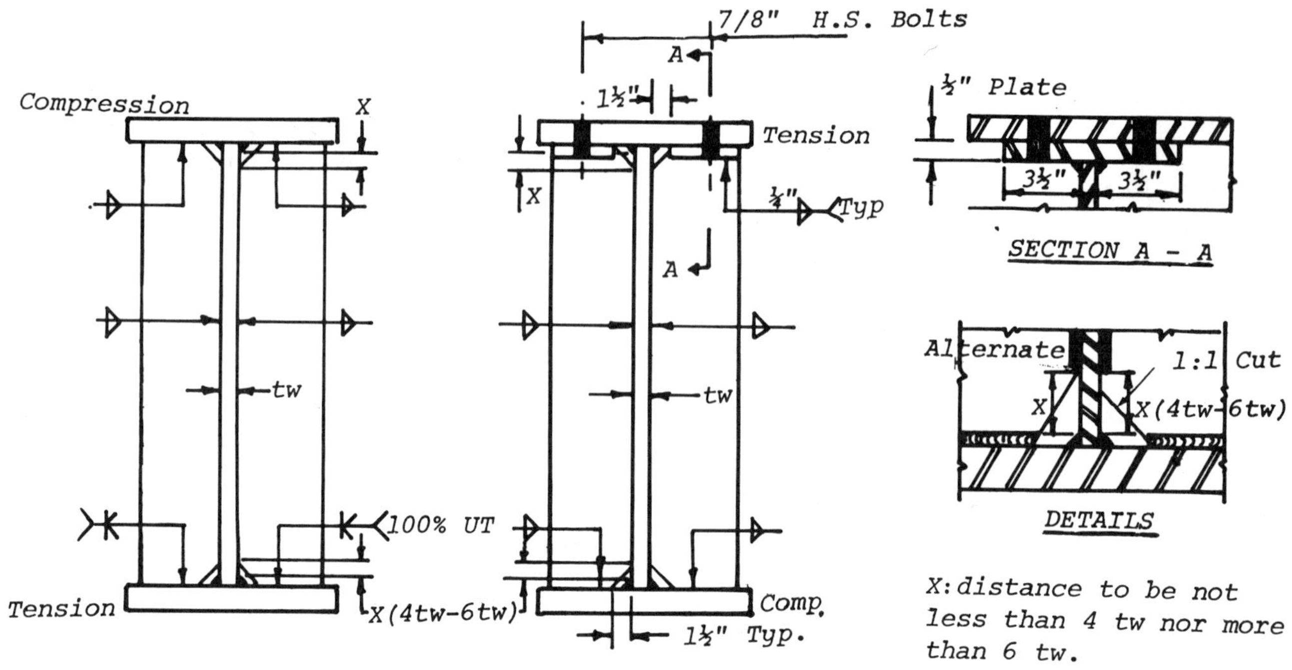


*All Abutments and Piers of Simple Span*

*Bearing stiffeners and Connection Plates over Pier (Negative Moment Zone : About 1/5 of span length either side of span)*

BEARING STIFFENER AND CONNECTION PLATE

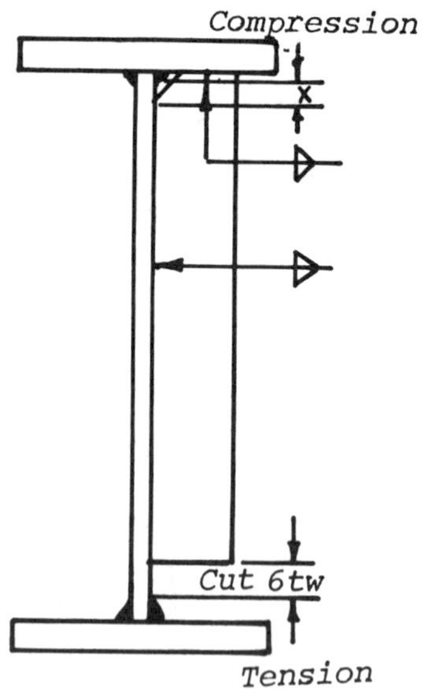


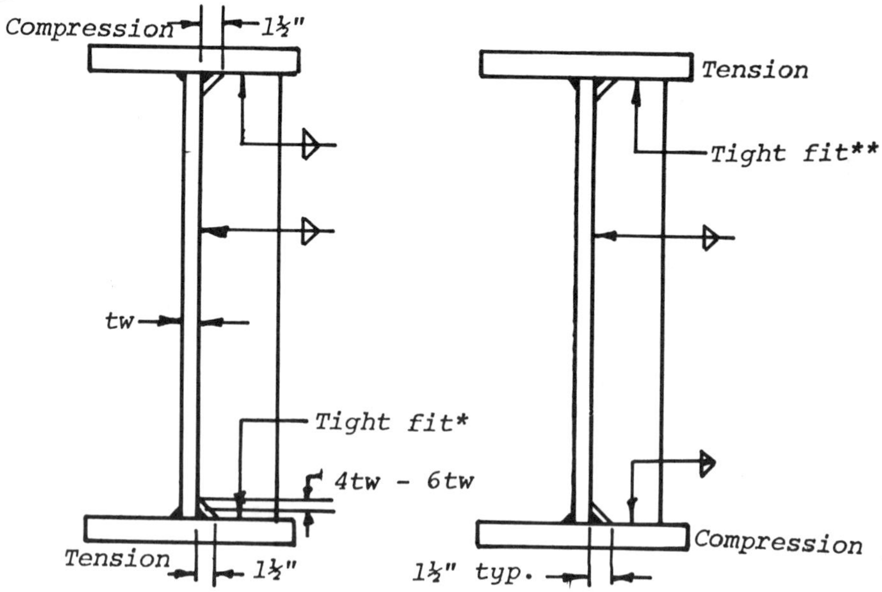


ALTERNATE DETAIL OF CONNECTION PLATE

INTERMEDIATE STIFFENER AND CONNECTION PLATE

* Connetion plate may be fillet welded to the flange near the support of simple span.

** At negative moment zone, connection plate can be bolted.

**STIFFENER and CONNECTION PLATE DETAILS**

**New Design:**

It is recommended that the gap should be a minimum of 4 and maximum of 6 times the thickness of the web to provide flexibility in the web plate. The connection plate should be distinguished from a stiffener in that connection plate of diaphragm or cross frame is subjected to secondary stresses and movements while the stiffener is not. The connection plates are recommended to be welded to the tension flange plate at least near the supports instead of just tight fit. Some states and FHWA's Standard details permit to weld the connection plates of diaphragm or cross bracing at both top and bottom flanges while majority of states still do not allow to weld the connection plate on the tension flange.

The Illinois Department of Transportation has the following guideline regarding connection plates: The connection plates for the cross frames near the support within a distance equal to twice the girder depth for all girders shall be welded for both flanges. This requirement shall apply to the cross frames in all other areas for skewed girders and for horizontally curved girders only. (In these cases, flange stress shall be investigated for fatigue under Category C.) For non-skewed girders, the connecting plates for the cross frames in all areas other than the areas near the supports should be welded to the compression flanges and to be undercut at the tension flanges. The distance between the near edge of the web-to-tension-flange weld and the end of the connecting plate shall be six times the web thickness.

If the stiffeners or connection plates are welded to the bottom high tension flange, the beam capacity will be reduced a Category C details and require to investigate stress range for fatigue stress. On the other hand, if connection plate was not welded to the flange, it will subject to potential out-of-plane deformations and cracks. Out-of-plane web crack is more severe on skewed and curved girder bridges.

It is suggested to weld connection plate (of cross frame or diaphragm) to tension flange near the abutments and the piers of simple span if the computed fatigue stress range is within AASHTO allowable stresss range for Category C details but to bolt at negative moment zone over the pier of continuous span. It is also recommended that bearing stiffener should be welded to the flange in place of mill to bear on all skewed welded girders and rolled beams with skew angle of 30® or more. The weld to the bottom flange should be groove weld and others (top flange and both side of stiffener) should be fillet weld.

Transverse stiffeners can be eliminated by increasing the thickness of web plate. On plate girders under 54 inches in depth, consider thickening the web to eliminate all intermediate stiffeners. In determining the cost effective structural steel element, the material saving alone should not be highly emphasized because simplification and repetition of details, reduction of fabrication operation and easy erection are other major cost factors. As rule of thumb, It is economical to increase the web thickness up to 5/8" without stiffeners. In cost comparison of using stiffener verse increased web thickness, the Illinois bridge manual suggests to assume that one foot of weld (one stiffener fillet welded both side) is equivalent to 40 pounds of structural steel.

## 7-3.7 FLOOR BEAM TO GIRDER OR STRINGER CONNECTION

### STRINGER AND FLOOR BEAM

A girder crack starts at the floor beam connection of sharp re-entrant and coped corners in the connection of the web then it propagates into the web.

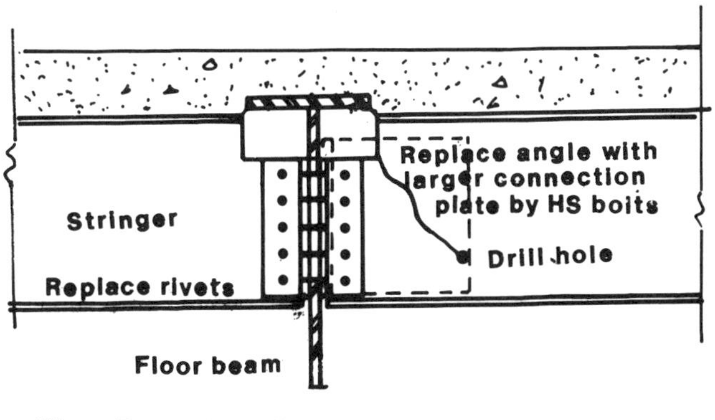


**Fig. 7.17 Crack at Coped Flanges**

Under normal conditions, these details will not act as simple pin connections and this unintended continuity of the stringer exists as a result of the end connection and restraint of the top flange from the continuous composite slab. It places the neutral axis near the top of the end connection and results in the bottom of the connection being in tension during part of the stress cycle. Hence, even though the stringers are assumed to be simply supported in the design, in fact they act as though they are continuous.

**Retrofit:**

For inadequate coping, if there are no cracks or fine cracks, the edge must be ground to a smooth radius so as to avoid stress concentrations.

If there is a crack extending into the web plate, a 3/4" (1/2" to 1" depending on size of the crack) hole at the tip of the crack should be drilled and the web be reinforced. The web can be reinforced by bolting a horizontal stiffener angle extending 4 times the angle leg size or a minimum of 12 inches beyond the coping or placing the wider connection plate covering beyond the coping. Also, a larger size web splicing plate (covering beyond cope) could be placed by drilling oversize holes in the portion above the tip of the cracks and the normal size holes below them followed by bolting with high strength bolts. If the crack is large, the size of hole should be determined by fracture mechanic method.

If the rivets attaching a connection angle to a floor beam web have broken heads, there is a continuity of the structure and a flexible connection is required to relieve the stresses. The following procedures may be used for it:

a) Replace the outside broken rivets with high strength bolts.
b) Remove inner broken rivets and leave the rivet holes empty. This would transmit shear force to the web and provide for a prying action. The existing rivet hole should not filled by weld metals because the shrinkage of weld will cause cracks.

**New Design:**

In the new design, the outstanding leg of the web angle in the tension region, (probably the upper one third) should have a wider gage and the connection plate should be thin for prying action.

**THROUGH GIRDERS AND FLOOR BEAMS**

In a through girder system, web cracks occur near the end of the girders when the end of the floor beam is rigidly connected to the connection plate, which is welded to the top flange and the web of the main girder but just a tight fit at the bottom flange. The end rotation of the floor beams is transmitted to the girder by the connection plate.

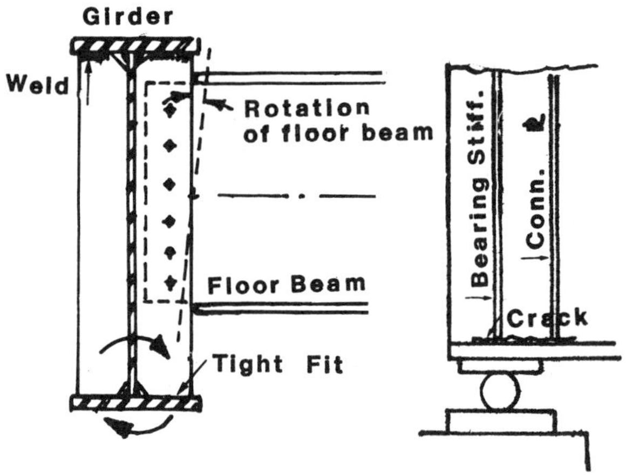


**Fig. 7.18 Crack due to Rotation of Floor Beam near the End of Girder**

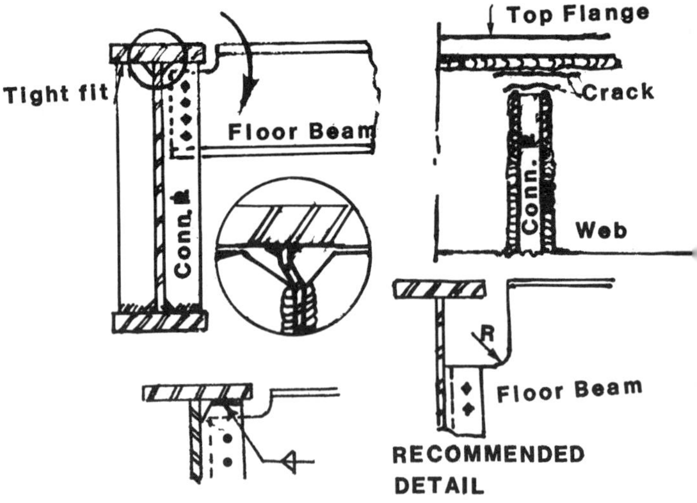


**Fig. 7.19 Cracks in Girder Web at Floor Beam Connection Plates in Negative Moment Area over the Pier**

As the floor beam deflects, the bottom flange of the floor beam pushes the main girder outward while the top flange pulls away from the web. This floor beam rotation creates an out of plane displacement and may cause cracks to form. The top flange of the girder is prevented from moving by the concrete slab and the bottom flange by the bearings.

This restraint causes a moment at the connection point of the web and the flanges of the girder. Since the bottom connection plate is only tight fitted, the connection plate can not sustain the stress and it is transmitted to the web. These stresses reach the yield strength whenever the floor beam is deflected by the live loads.

It is suggested that repairs be made by drilling holes in the web at the end of the cracks and welding the connection plate to the bottom flange plate.

**Retrofit Detail No. 1:**

1) Cut or drill 2" diameter hole in stiffener at about 10" from top flange (This crack development length due to floor beam deflection and rotation should be determined for each bridge).
2) Make horizontal cut through stiffener at the line of bottom of the drilled hole.
3) Make vertical cut thru stiffener near the web. Care should be exercised to cut far enough away from the web to avoid any under cut ot gouging of the web from cutting operation. Grind resulting surface to a finish of ANSI 2000 or smoother after removal of the portion of stiffener.
4) Grind any remaining vertical fillet weld and stiffener so that the resulting surface of web is ANSI 125 or better. If the 2" hole of Step 1) did not touch the edge of the web, the hole shall be grounded so that a smooth radius from the face of the web to horizontal top of stiffener results.
5) Grind top of stiffener to ANSI 250 finish.
6) Remove the paint from both side of the web by sand blasting cleaning, an area bounded by vertical lines about one foot either side of floor beam connecting location being retrofitted and a horizontal line at the top of the floor beam. Inspect area of web adjacent to retrofit area by visually and by use of dye penetrant to find the tip of the crack then drill a hole (1/2" to 1") depending on the size of the crack.

**Retrofit Detail No.2 :**

1) Remove the top two rows of bolt (4 bolts) from the floor beam to intermediate stiffener connection plate.
2) Field drill matching holes in each of the angle L8x8x1x0'-6" using the existing hole in the stiffener as template.
3) Bolt the 2 - L8x8x1 into position with high strength bolts.
4) Remove the 12"x20" portion of the deck floor. Transverse reinforcing bars should not disturbed. Outline the removal area should be saw cut on the top od the deck.
5) Drill 4 - 15/16" vertical holes thru the top flange and the legs of the angles. No slotted or oversize holes should be allowed.
6) Bolt the 2 - L8x8x1 angles to the top flange with 7/8" diameter high strength bolts.
7) Patch the deck slab with non-shrink mortar.
8) Follow Step 6) of Detail No.1 for drilling crack arrester holes.
9) Fill the holes which were drilled to inhibit the further growth of the crack with high strength bolts. Bolts may be 1/8" less than the holes. Due to the closeness of some holes, some holes may left open and oversize washers used on each side of web to cover two holes with one bolt. Any holes that extend into the flange to web fillet weld so that the nut can not be tighted shall not be filled with bolts. In installation of bolts, the turn of nut method will be adequate.

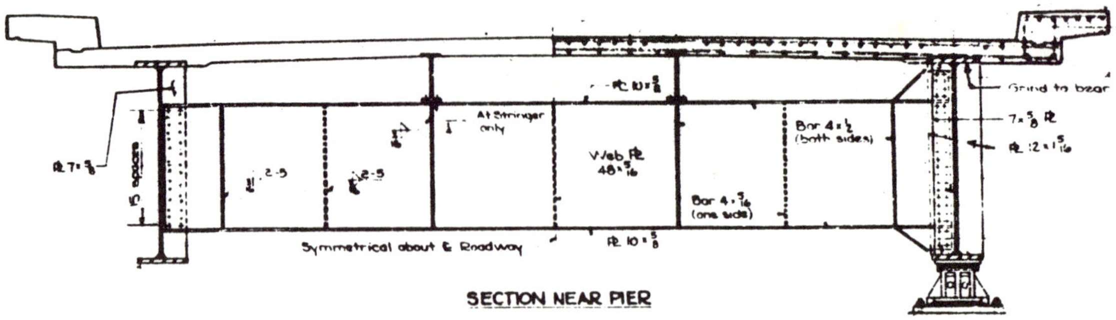


SECTION NEAR PIER

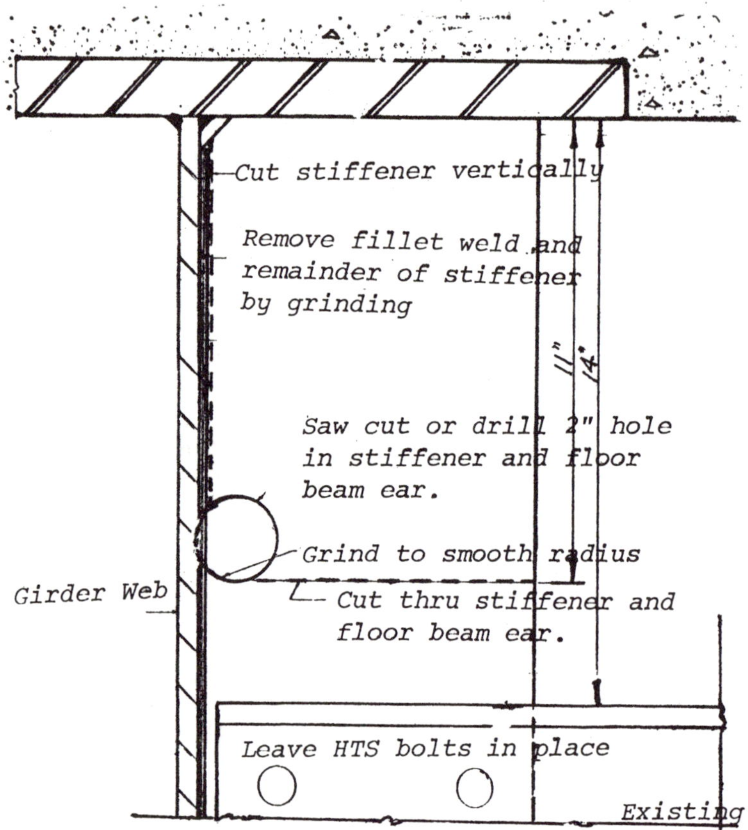


RETROFIT DETAIL No. 1

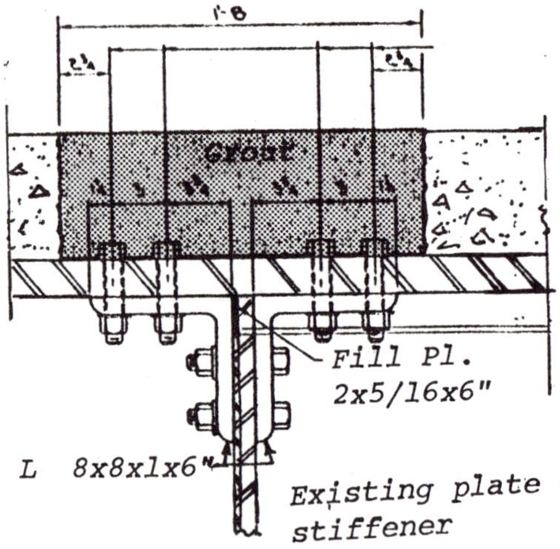


Section A-A

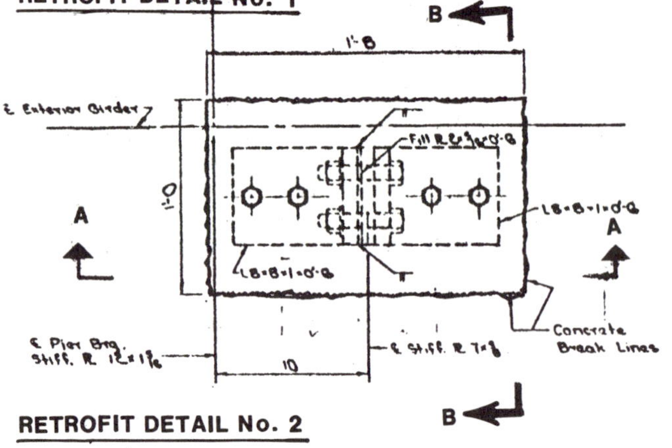


RETROFIT DETAIL No. 2

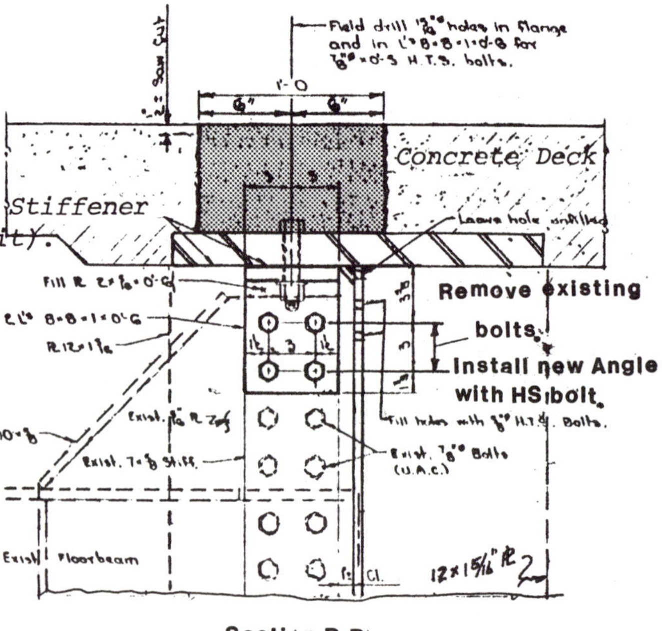


**VERTICAL STIFFENER RETROFIT**

**New Design:**

For design, the following details are recommended:

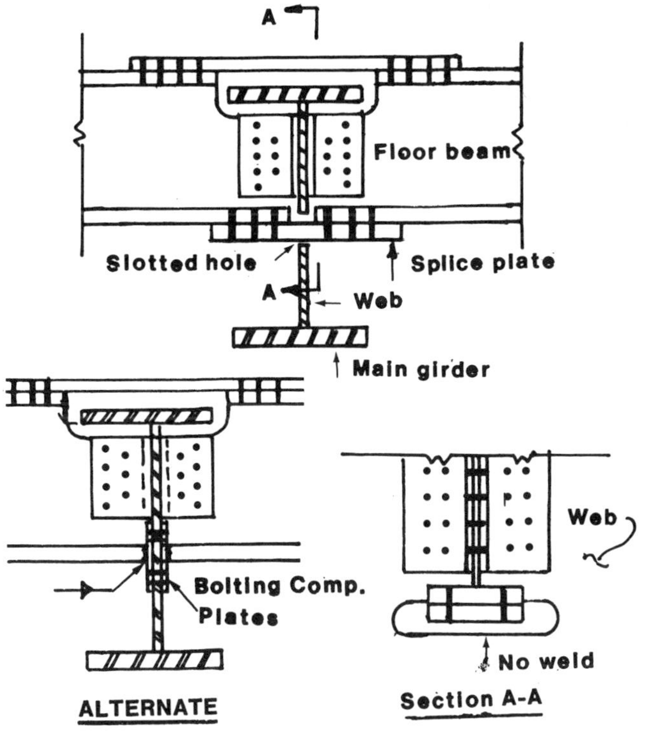


Fig. 7.21 Floor Beam Connection Detail

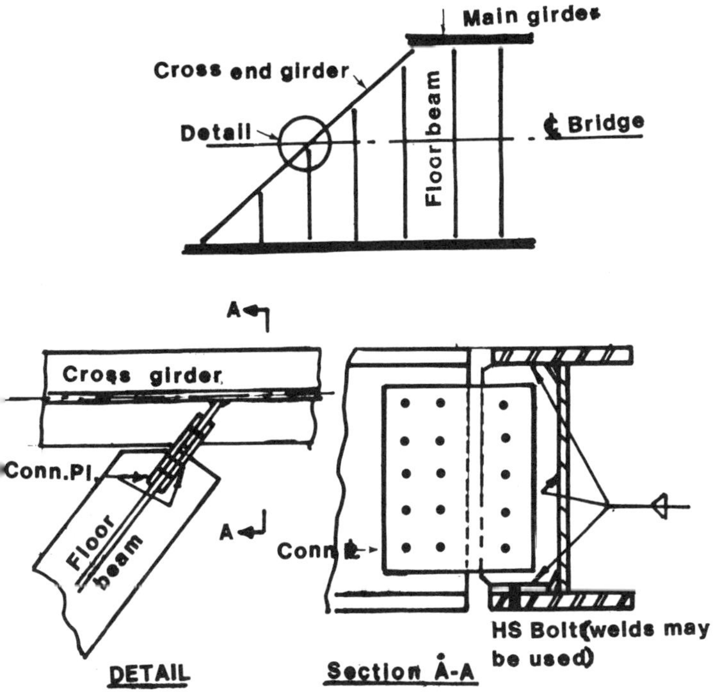


Fig. 7.22 Simple Connection at Skewed End of Floor Beam

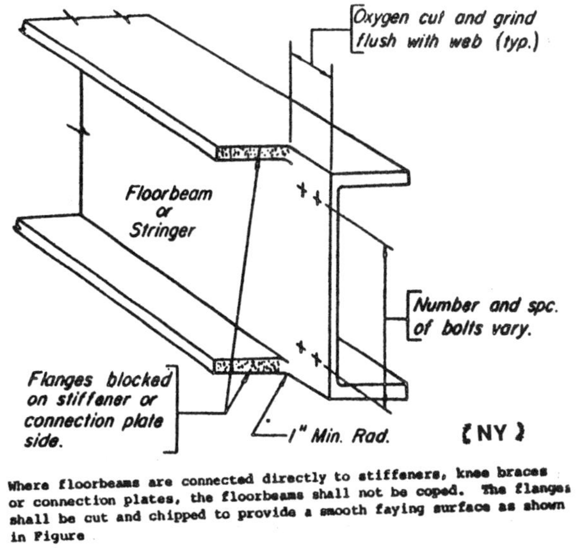


Fig. 7.23 End Connection

## GIRDER AND BOX GIRDER PIER CAP

To provide the continuity, often the longitudinal steel girder is interconnected to the cross box girder of steel pier bent. When flange plates are allowed to pass thru one or more box girder web, a slotted hole is provided to receive the lower flange but

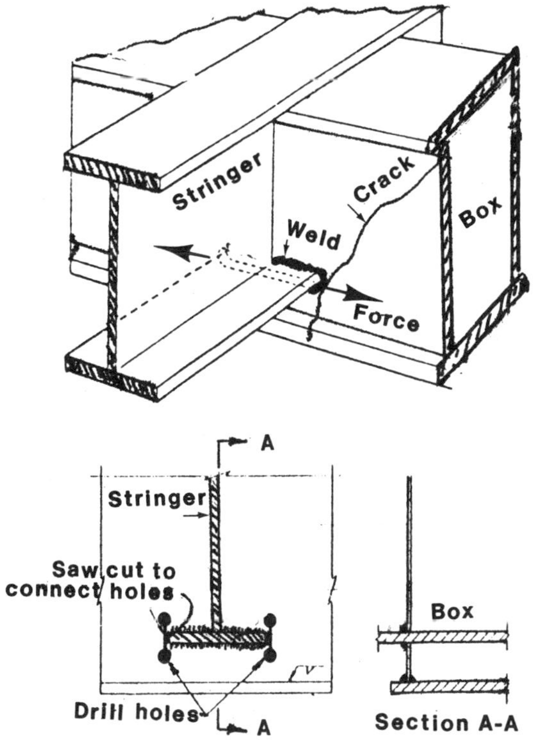


Fig. 7.24 Bottom Flange Pierce Web

the gap was generally welded to provide the air tightness of the box girder. If it is welded, The web of box girder will be stressed from bending and will elastically strain. As it strains, the weld will be pulled along the web plate and cause the crack at the intersection of flange tip and web.

**Retrofit:**

Drilling holes at the web directly above or below the flange tip then cutting between holes to form dumbbell shape to reduce the residual stress and minimize crack propagation.

**New Design:**

It is recommend that the rigid frame steel pier bent should be avoided if possible and the following should be considered as alternates:

1. Continuous steel multi-stringer superstructure supported on concrete column bent pier.
2. Simple span or continuous span steel girder supported on the steel cross girder supported on the concrete column pier. The simple span girder can be connected by simple connection to the web with or without seat bracket. For continuous span, the top flange may be connected by the tie plate but the bottom flange should pass through the web in a smooth cut slotted hole. Also, The girder can be seated on the top of the pier cap girder by the bearing with or without bolsters.

If the box section of steel rigid frame pier bent can not be avoid, The web should be cut to pass through the bottom flange of the longitudinal girders. The bottom flange can seated on the bolster type seat bracket which is bolted to the box and backed up internally in the box. Alternate detail is the use of wide bottom flange plate cut to a radius of 24" to provide flange connection with curved radius transition. This detail will reduce the Fatigue Category E' detail to Category B.

## 7-3.8 CANTILEVER BRACKET CONNECTION

The top of the bracket is connected by the tie plates to the top flange of the transverse floor beam by bolts or rivets to transmit the moment capacity of the bracket back to the floor beam.

The perpendicular intersection of the floor beams and the girder causes deformation in both members and lead to secondary bending moments. The floor beams have a lateral deflection under live loads while the main girder deflects vertically. This out of plane deformation causes the in-plane bending stresses.

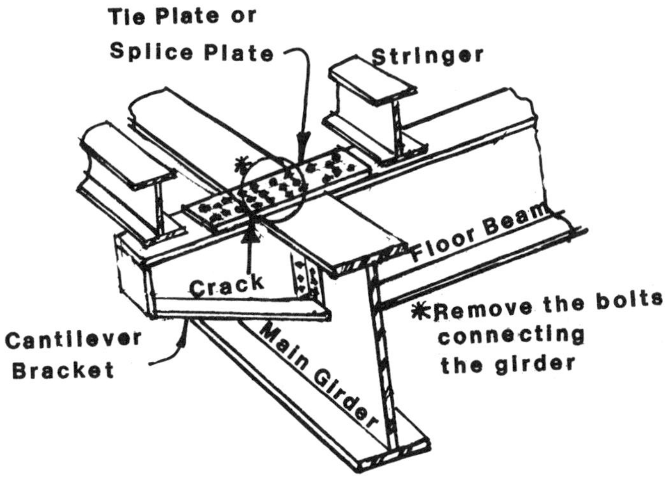


**Fig. 7.25 Cantilever Bracket Connection**

In other words, when the top flange of the main girder is shortening due to the cumulative effect of the compression stress, the shortening is not shared by the exterior stringer resting on the top flange of the floor beam bracket. As a result, a longitudinal displacement occurs in the end of the floor beam bracket relative to the part of the tie plate which is riveted or bolted to the longitudinal girder.

This relative displacement sets up in-plane bending stresses of a repetitive nature under live load, which are sufficient to fatigue the tie plate to a point of fracture.

This problem is not always evident in the situations where the stringers frame into the floor beam bracket and the flange of the bracket is in contact with the bottom of the concrete slab.

## Retrofit:

It is recommended that the bolts or rivets which connect the tie plate to the longitudinal girder be removed, and that the holes be protected against corrosion by the use of smaller bolts and suitable washers or plates.

## 7-3.9 LONGITUDINAL WELDS

### Web to Flange Welds:

The cracks that occur at the fillet weld connecting the compression flange to the web are caused by the out of plane movement due to original deflection during fabrication. This type of crack has a relatively low probability of occurrence in the field, but does happen in the shop. It originates with an internal weld discontinuity. Porosity (gas pocket) is a typical type of weld discontinuity. The crack propagates into both the web and the flange. The web crack will usually arrest itself. However, the crack may continue into the flange causing a failure.

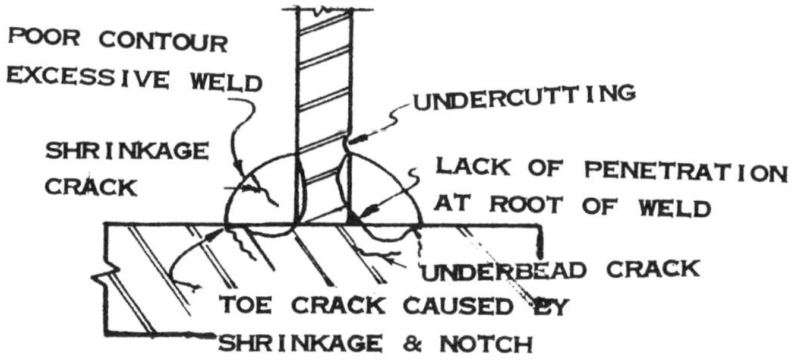


Bad Welds

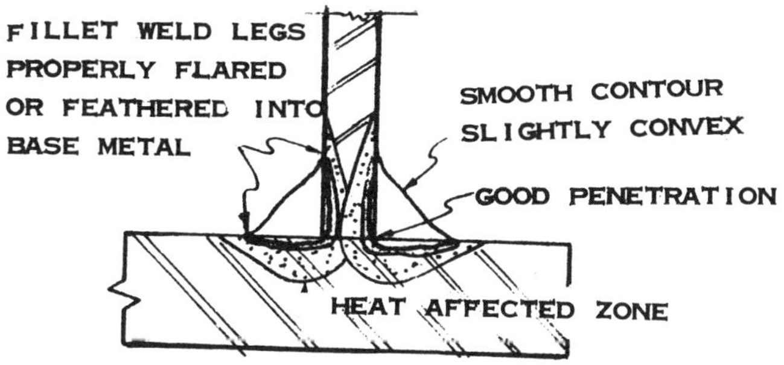


Good Welds

**Fig. 7.26 Fillet Welds: Web and Flange**

### Longitudinal Stiffener:

Longitudinal Stiffener is used to act as reinforcement for the web in compression zone. When longitudinal stiffeners are used only on the compression portion of the web, there are no problems but when additional longitudinal stiffeners are attached in the tension areas for aesthetic reasons or the longitudinal stiffener of continuous spans either continues to negative moment area or ends at the location near the contraflexure, there are possible fatigue problems.

Longitudinal welds which are parallel to the applied stresses have fatigue strength because discontinuities such as porosity, cold laps, or slag inclusions appear in these welds. However, the fatigue strength is governed by discontinuities that are perpendicular to the applied stresses, and not by the parallel ones. The longitudinal welds in the direction of stress is Category B while the end of longitudinal weld in a tension area is Category E.

For new design, the following should be considered:

1. Do not use additional longitudinal stiffeners for aesthetic purpose.
2. Place the longitudinal stiffener on one side of the web and place all transverse stiffeners on the opposite side of the web.
3. Terminate the welded stiffener in the compression zone and use a bolted connection through the stress reversal zone (Category B) or continue the stiffener through stress reversal zone but terminate weld with curved radius transition of stiffener plate then grind the weld termination (Category C).
4. In tension area and zone of stress reversal: Run the longitudinal stiffener continuously when it intersects with transverse stiffeners. Cut vertical stiffeners for tight fit and for longitudinal stiffener to pass through and cope for longitudinal welds. If tight fit is not possible because the longitudinal stiffener is wider than the vertical stiffener, the vertical stiffener can be cut and be fillet welded to the top and bottom of the longitudinal stiffener. If it is not possible, use curved radius termination.

In compression area: The stiffener may be discontinued by cutting short of vertical stiffeners.

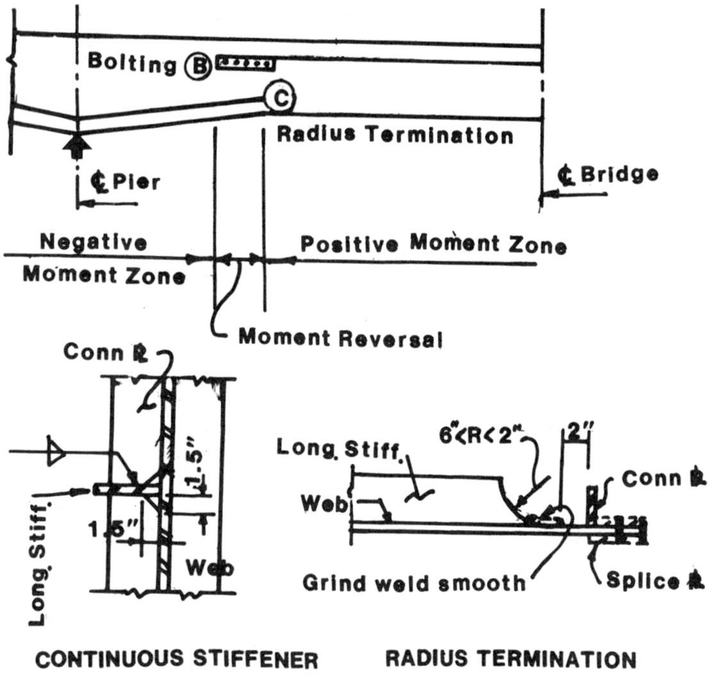


*In the zone of moment reversal, the longitudinal stiffener shall be made continuous thru connector plates. Radius Termination can be used at bolted field splice location and interference with stiffener*

**Fig. 7.27 Longitudinal Stiffener Details**

## Box Girder:

Larger box girders are sometimes made with a bevel groove weld, using an inside backing bar. The continuous backing bar is usually tack-welded in place so it fits tightly against the plates. As a practical shop procedure, the intermittent tack welds reduce the fatigue strength from Category B down to E which is fairly low. For small box girders, fillet weld can be used between the webs and flanges. It is suggested that wherever possible, groove weld should be made from both sides.

## Groove Welds

Since longitudinal stiffeners are made from length of plate, butt splice may be exist at several locations and the quality of groove weld may be poor. If any flaws exist in butt splice, it will propagate to the longitudinal fillet welds and eventually to the web. It is suggested to specify in the plans that all groove weld butt splices which intersect with other welds should be tested by non-destructive testing methods.

Some bridge girders used insert plates to change the cross section or patch the damaged section. Generally the insert plates were groove welded to the perimeter of cut area and longitudinal and transverse weld were intersected. If the quality of transverse weld which is perpendicular to the bending or stress of the section is poor, in tension area, it will cause fatigue crack. It is suggested to avoid sharp rectangular cut for insert plate and to test the weld at the insert plates.

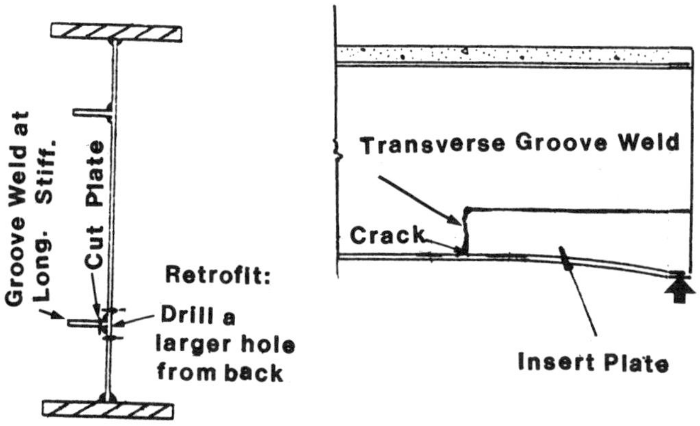


**Fig. 7.28 Transverse Groove Weld**

## 7-3.10 GUSSET PLATE CONNECTION

Rivet or bolt holes in gusset plates of the main structural members are points of stress concentrations. The shop fabrication of these holes creates discontinuities which are sources of crack. Cases of cracks in both gusset plates and the main structural members connected to them have been observed.

The crack in the metal usually occurs at the exterior lines of the rivets or bolts and at right angles to the main stress. On long span bridges, lateral bracing is often used. The lateral bracing vibrates when vehicles cross the structure and this vibration can create difficulties at the member ends. The movement of the member as it vibrates causes out-of-plane movement in the gusset plates attached to the web or the flange.

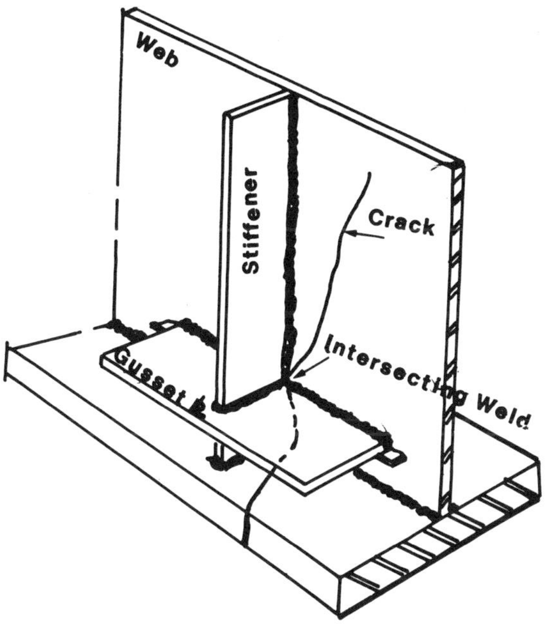


**Fig. 7.29 Gusset Plate Welded to Transverse Stiffener**

Cracks may originate where the gusset plates are attached to the transverse stiffener and have intersecting welds, and especially where backing bars are used for groove welds. At the fillet weld intersection, fatigue cracks may develop and these cracks can grow into the web. Also the notch created by an interrupted backing bar can cause cracks in the intersecting welds and these crack can propagate into the web and the flange plates.

## Retrofit:

a) **Gas Tungsten Arc Remelting:** If the crack in the weld is shallow(less than .125 inch deep) and has not propagated into the web, gas tungsten arc remelting can be used to increase the fatigue strength.

b) **Peening:** This method is applicable only for shallow and short cracks(.125 inch deep and 2 inches long)

c) **Drilling Holes:** Most cracks can be successfully treated by drilling 3/4" or 1" hole at the tip of the crack.

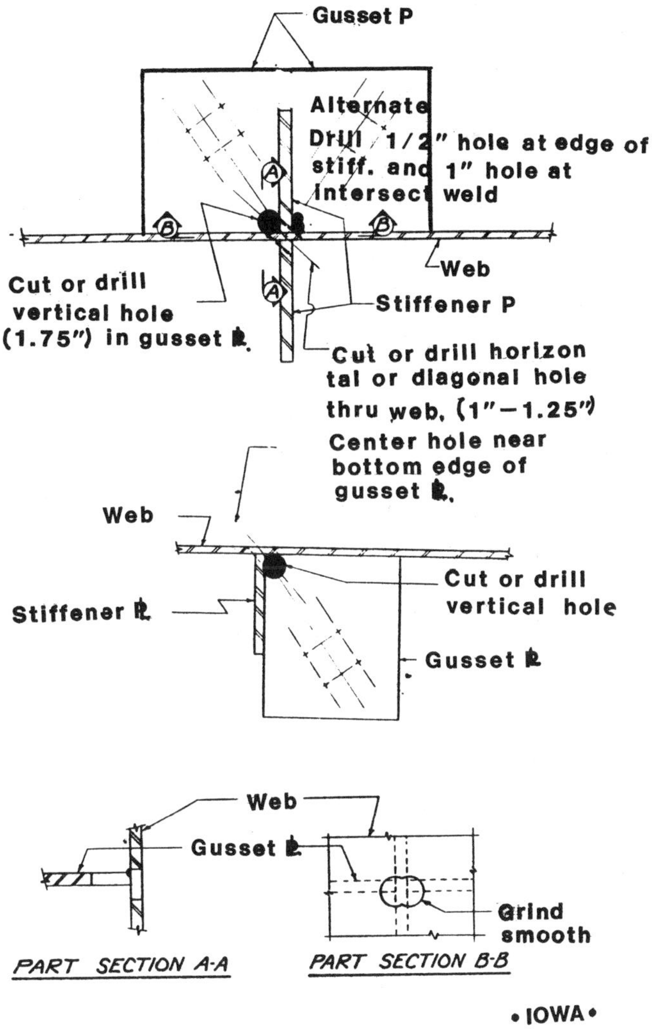


**Fig. 7.30 Gusset Plate Retrofit**

A vertical hole is drilled in the gusset plate, 1/2 inch diameter at the edge of the stiffener and 1 inch diameter hole, at the weld intersection of the gusset plate and web. Then the gusset plate and web surface is inspected for crack penetrations by using dye penetrant or other method. If there is no evidence of crack in the web, the holes are adequate. If a crack is detected in the web, diagonal holes are drilled across the web and/or a horizontal hole at the tip of the crack. After drilling, the holes should be reamed or ground smooth.

## New Design:

Gusset plates can be attached to flange edges using groove welds with an end radius (min. radius 3/4") transition or a gusset plate can be bolted to the top of the bottom flange or the web by using high strength bolts. This connection is classified as Category B while the welding connection varies from B to E depending on the transition radius and the welding condition. If a rectangular gusset plate is required, a small radius at the end of the groove weld should be ground smooth.

It is recommended to avoid welding an attachment to the flange surface.

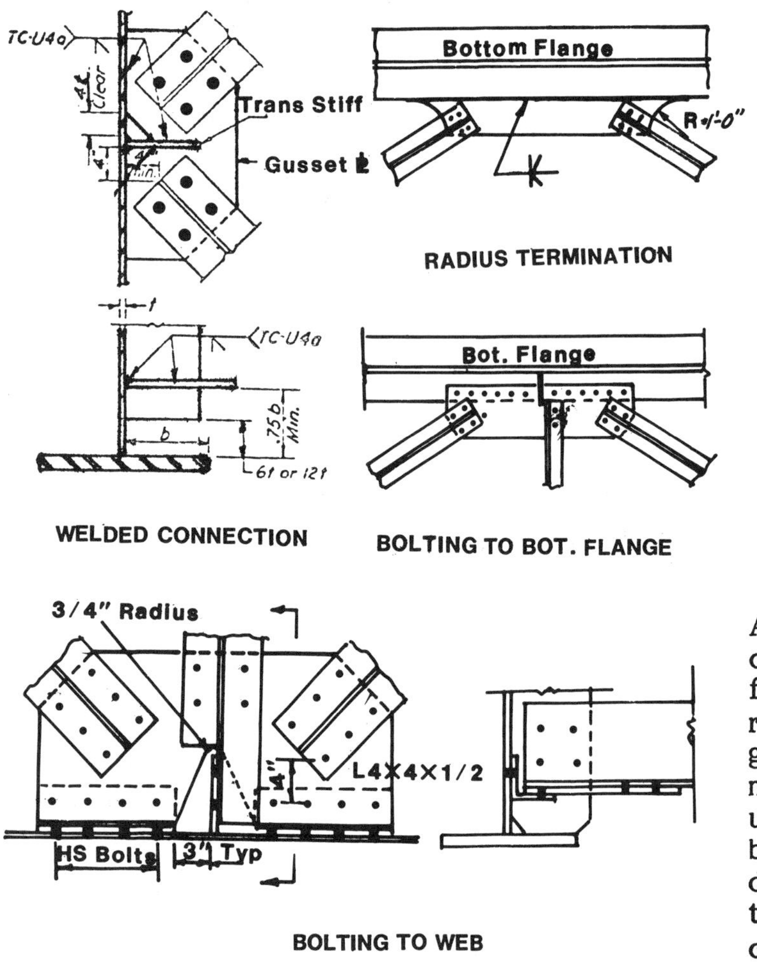


**Fig. 7.31 Gusset Plates Details**

1982 AASHTO Interim Article 1.7.17(B) drop the mandatory requirement of lateral bracing based on the recent study. The study found that the bottom lateral bracing has minimal effects on redistribution of wind loads. AASHTO requires to investigate the need for lateral bracing. Top flanges attached to concrete deck or other deck of comparable rigidity in manner which provide lateral restrain will not require top lateral bracing. Bottom lateral bracing may be placed at the discretion of the engineer after checking stresses in the bottom flanges due to wind loads in accordance with the Article 1.1.17(B).

The following are the New York State Department of Transportation criteria regarding lateral bracing:

> Lateral bracing will be required if the resulting lateral stress, when combined with stress due to all dead loads and the lateral stress due to curvature, if any, exceeds 125% of the allowable design stress.
>
> If lateral bracing is required, it shall be placed in the outside (fascia) bays between diaphragms or cross frames. A sufficient length of flange shall be braced so that the stress computation when repeated for the unbraced length shall result in a flange stress less than 125% of the allowable design stress. It may be desirable to provide lateral bracing in adjacent interior bays to limit the size of the bracing members. All required lateral bracing shall be placed in or near the plane of the flange being braced. Cross frames or diaphragms shall be placed in all bays.

Another recent study of "Load Distribution of Cracked Girder" concluded that the influence of bottom lateral bracing on load redistribution are significant for through girder system bridge which consists of two main girders, floor beams and stringers. The use of these secondary member (lateral bracing system) can develop effective redundancy for non-redundant structure like through girder bridge thus prevent immediate collapse when one of the main girder has a flange crack.

Considering this finding, It should be considered to install lateral bracing system on non-redundant structure as an extra safety measure even if the computation show that the bracing system is not required. On horizontally curved girder, the cross frame should be considered as an main member and lateral bracing system is preferred to be placed.

### 7-3.11 GEOMETRIC STRESS RISER AND REENTRANT CORNERS

Any time a structural member has a change in geometry, there is a change in stress. There are a large number of commonly used design and fabrication details that are stress risers. In most of the metal cracks observed in the field, geometric stress risers are present. A stress riser by itself will not cause harmful cracking, but in conjunction with other factors such as high stress and low material toughness, cracking does occur.

Some of the common sources of geometry changes that contribute to critical stress risers are bolt holes, sharp re-entrant corners, changes in section, and highly stressed areas.

**Retrofit:**

1. Grind ,corners smooth.
2. Drill holes at the tip of the cracks.
3. Reinforce the corner with a new plate.

### 7-3.12 LAMELLAR TEARING AND LAMINATION

**Lamellar Tearing:**

Lamellar tearing is the separation in grain structure of the base material. Failure usually occurs in thick plates in the thru thickness or Z-direction, if the X and Y axes are oriented in the planes of rolling and are associated with highly restrained welded joints with the plates generally over 1.5 inches thick which require large welds.

Lamellar tearing is the separation in the material caused by "thru-thickness" strains induced by weld metal shrinkage. At the highly restrained connection, localized strains due to weld metal shrinkage are higher than yield point strains while stress due to applied loads are small. When stress is carried into the plate in the short transverse direction (through thickness), any lack of cohesive strength in this direction will allow the plate to separate as if composed of two layers hold together by a weak adhesive. When stress is carried into the plate in the short transverse direction (through thickness), any lack of cohesive strength in this direction will allow the plate to be separated as if composed of two layers hold together by a weak adhesive.

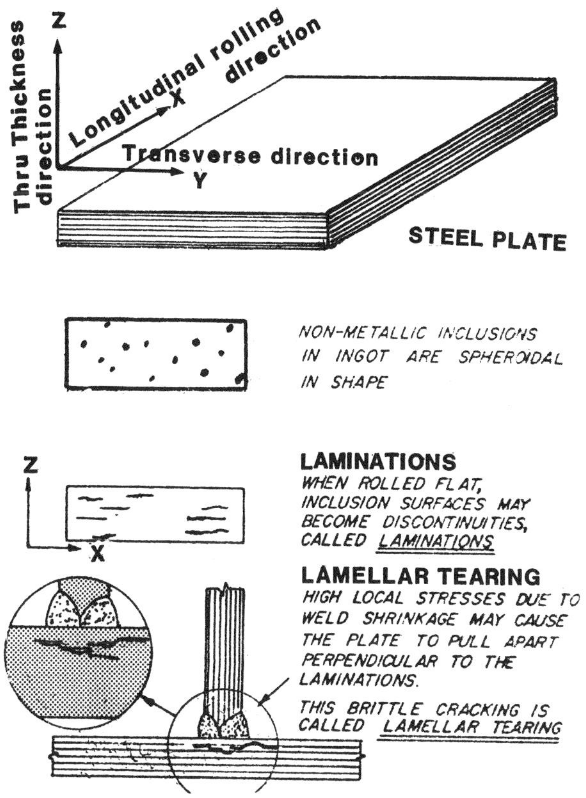


**Fig. 7.32 Lamellar Tearing**

These large localized strains occur as the solidified weld metal contracts When it cools. In the Z direction(thru-thickness) the steel deforms. Therefore, when the strains are over the yield point, the excessive force can not be redistributed to the less strained areas and a separation of the steel at grain boundaries can occur in the highly strained localized area. Generally lamellar tearing are detected during fabrications however this is not only material problems but also involving detail design and fabrication. For new structures special care should be exercised in detailing. The joint should be detailed to minimize the strain in the through direction. If the joint details can not be changed, specify the use of high ductility weld material in the direction of the strain.

Most instance of lamellar tearing have been associated with steel alloys having high sulfur content (0.02%). Sulfur level of 0.005 to 0.01 is desirable for optimum performance of steel. Since lamellar crack often occur on A588 steel, special attention should be given in detailing A588 steel.

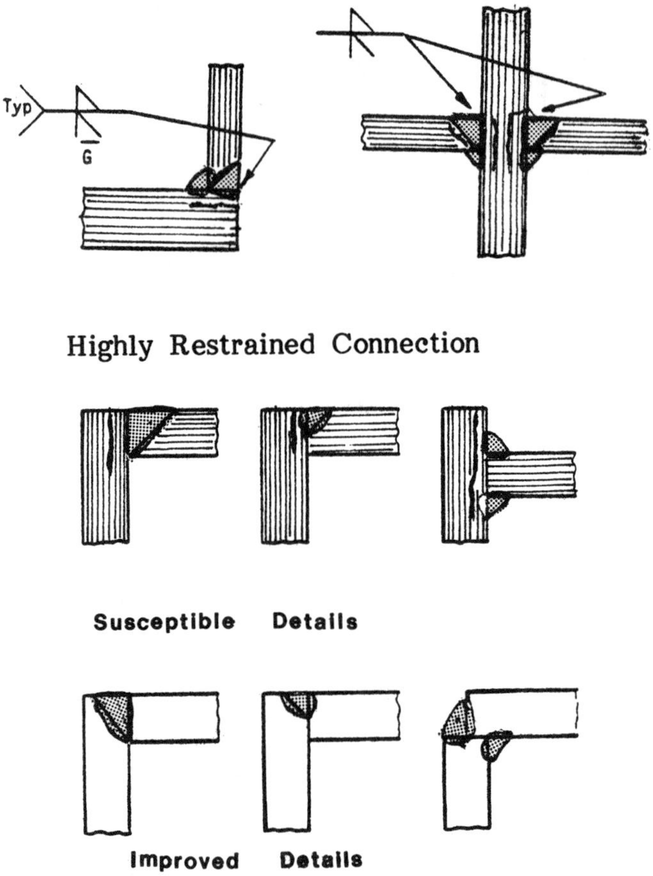


**Fig. 7.33 Susceptible and Improved Details**

**Hydrogen Induced Crack:**

This crack typically forms at the toe of weld and occurs during and after welding. Sometimes it is called toe crack, cold crack or delayed crack. Generally high strength steel is susceptible to this crack. It is caused by the diffusion of hydrogen from the weld metal into base metal heat affected zones. The hydrogen come from trapped moisture, hydrated oxide on plate, moisture in coating and fluxes, and from oil, paint or other contaminant. Other probable causes are unforeseen handling stresses after welding but before stress relieving and tack welds of inadequate size.

It can be prevented by the use of low hydrogen and controlled electrod, and preheating and postheating the plate to 200°F or more (400°) depending on the base metal.

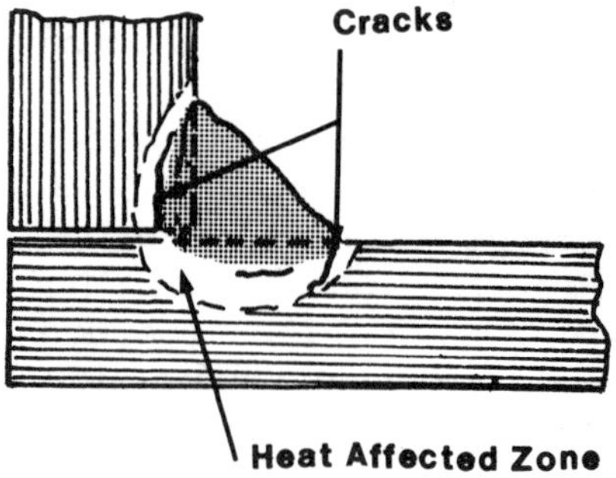


**Fig. 7.34 Hydrogen-Induced Crack**

**Laminations:**

Lamination is an extensive and continuous imperfection lying parallel to the plate surface resulting from shrinkage cavities, blisters or a large amount of non-metallic materials. Lamination has no resistance to through-thickness stresses.

Most of the laminations seldom become a major problem. However, laminations do exist undetected by routine inspection and there have been cases of structural failure due to laminations. Most laminations are the result of elongated imperfections or metal inclusions during the rolling process. They are usually aligned in the direction of principal stress, and this is why they seldom become critical.

Lamination may become critical if the steel is oriented so that the laminations are transverse to the direction of principal stress (direction of rolling).

**Fig. 7.35 Lamination of Bottom Flange of Concrete Encased Beam near the Bearing Seat.**

### 7-3.13 EYE BARS AND HANGER ASSEMBLY

The recent catastrophic bridge failures in U.S. were the Silver Bridge in 1967 and the Mianus River Bridge on Connecticut Turnpike in 1983.

The Silver bridge failure were cumulative fatigue damages of eye bars. There are no practical retrofit for eye bars therefore it is strongly recommend frequent inspection of eye bars to detect the signs of the fatigue crack. It is difficult to inspect visually the crack without proper crack detecting devices

**Fig. 7.36 Eye Bars of Truss**

The collapse of a 100 feet suspended span of cantilever system bridge on June 28, 1983 was apparently triggerred by a failure of a pin of the hanger bar. The one end of collapsed suspended span was connected to the cantilever girder by pin and hanger assemblies while other was supported on the cantilever girder by the bearings. The cause of the pin failure is not determined yet.

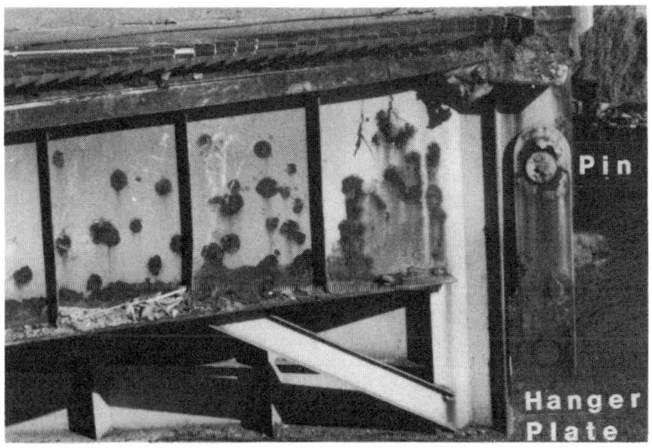


**Fig. 7.37 Failure of Hanger**

It appears that horizontal thrust and out of plane deformation of the hanger plate due to severe skew angle, and the rust of pin attributed the failure. As a retrofit, the existing hanger assemblys were replaced with larger hanger plates and a larger 9" threaded pin by enlarging the hole in the web. As alternates, new bearing and pier column can be built under the suspended girder or steel sling wire rope can be installed under the joint to avoid sudden fall down of the girder if assembly fails.

The cantilever span with suspended span of two girder system was used over multi-stringer continuous span in 1950's and 1960's due to easy installation of suspended spans and material savings. Today, this type of design is considered outdated because of non-redundancy and fatigue.

**Fig. 7.38 The Cantilevered Girder**

**Fig. 7.39 Types of Girders**

## 7-3.14 OTHER DEFECTS

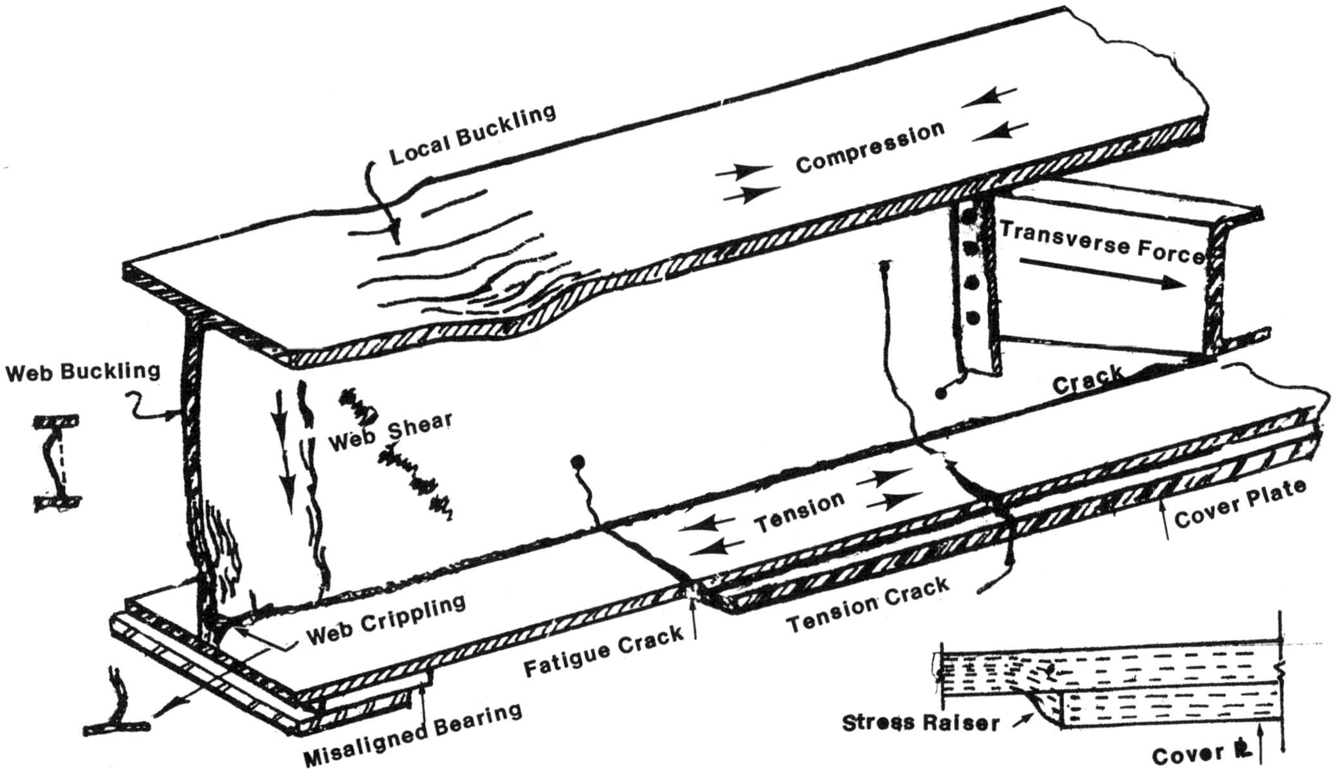


**Fig. 7.40 Common Defects of Steel Girder** Avoid sudden geometric changes to minimize stress riser. Perceive the stress as water flow in the pipe and make it flow smoothly without any obstruction. Also, avoid rigid connections and make them flexible.

Other types of defects which can develop are due to overstress, misalignment or impact. Buckling will develop in a compression member in the shape of a single bow or a double S due to overloading. The problem can be corrected by strengthening the member.

Web buckling can be caused by the eccentricity of the reaction due to misalignment between the centerline of the bearing stiffener and the bearing assembly or due to frozen bearings. Web buckling can be corrected by installation of bearing stiffeners and/or replacing the frozen bearings.

Minor collision damage can be straightened by heat treatment but severe damage may require total girder replacement.

**Fig. 7.41 Collision Damage**

## 7–4 FRACTURE CONTROL PLAN

There have been fatigue cracks and brittle fractures found in many bridges. It is necessary to examine the adequacy of the design, specification, fabrication and inspection of the steel structure in order to prevent catastrophic failures, expensive corrective repairs and to assure the safety and convenience to the public.

Fatigue and fractures are more serious problems in welded structures than in riveted structures mainly because of its component redundancy (built up with several members) and lower rigidity. The fracture problem is especially critical for structures containing non-redundant fracture critical members where a single fracture may cause a collapse of the bridge.

To prevent sudden collapse of bridge, the following three basic methods should be considered:

- Lower design stress level
- Tough Steel
- Avoid large flaws.

Besides the above three, extreme bridge geometry such as severe skew angle should be avoided. This would cause misalignment of structural member and create unbalanced forces.

Fracture control requires careful selection of materials, rigorous fracture-mechanics analysis, development of design details, evaluation of the existing standard details for quality control and quality assurance during fabrication and recognition of the maintenance related problems.

Fracture critical members are tension components whose failure would result in the collapse of the structure. Emphasis of fracture control is placed on the fracture critical non-redundant members. Careful attention should also be given to multiple-load-path redundant members for their premature failure due to fatigue crack, weld crack and brittle fractures.

In 1978, AASHTO adopted "Guide Specifications for Fracture Critical Non-Redundant Steel Bridge Members" to promote better understanding in application of the fracture control plan in the design and fabrication of non-redundant bridges. In 1979, the Federal Highway Administration (FHWA) published Structural Engineering Series No. 5 "A Proposed Fracture Control Plan for New Bridges with Fracture Critical Members" to discuss fracture control plans as a means to improve the specifications for design and construction and to address some controversial areas. The FHWA proposed fracture control plans appear to be more conservative than those by AASHTO fracture control plans and they should be used for references. The following are some of the FHWA fracture control plans concerning design:

1. The designer has the prime responsibility for determining which structural components fall in the fracture critical member (FCM) category.

2. Weld details on shop drawing and erection plans should be reviewed by a designer experienced in welding or by a welding engineer.

3. All groove welds in tension components such as groove weld butt splice in back up bars and any attachment welded to tension components of a FCM such as a gusset plate welded to a stiffener which is welded to the FCM should be nondestructively inspected.

4. The base metals, the carbon and low-alloy steel are commonly used in highway bridges. It is recommended that a special provision for FCMs do not preclude the use of quench and tempered steel to achieve the specific toughness level.

   The plate thickness should be limited to three (3) inches for yield strength up to 90 ksi, over 90 ksi yield strength, the thickness should be limited to two (2) inches. However, designers are encouraged to compensate for the 3 inch limitation by use of the quench and tempered steel.

   All plates used in FCMs shall meet the requirements of the Charpy V-Notch Impact Value instead of the AASHTO zone systems which requires that testing should be done at 70°F above the minimum service temperature.

5. In fracture critical members, plug and slot welds, intermittent fillet welds, partial penetration groove welds subject to tension normal to their longitudinal axis shall not be used. Holes or slots in a member or component to permit passage of another member or component through the member or component shall not be welded to seal or otherwise close the hole or slot (for example, bottom stringer flange going through the web of a box girder).

6. Fillet welds for FCMs shall be detailed as follows:
   a) Minimum required size of fillet weld shall be used. Large single pass fillet welds are prone to have low toughness when a fillet weld is transverse to the principal stress in FCMs. Low toughness may trigger a pop-in crack and a brittle crack.
   b) Fillet weld shall be used to develop shear only and shall not be used to transfer tension or compression normal to the axis of the weld.
   c) Sealing welds shall not be used in FCMs under any condition which will provide rigidity or restraint to the connected members.
7. Fatigue Categories D and E shall not be used

8. Details:
   The connection at the end of simple beams shall be designed with flexibility so as to avoid excessive secondary stresses due to bending. Details of the deck should be designed to avoid unintended participation in the chord or flange stresses.
   Lap joints shall not be used in FCMs.
   Corner and T joints subject to bending about an axis parallel to the joint shall have their welds arranged to avoid concentration of tensile stress at the root of any weld.

9. Welding Type:
   The following type of welds are prohibited:
   - Electroslag welding in main structural tension members
   - Butt joints not fully welded throughout their cross section
   - Groove welds made from one side only 1) without backing, 2) with backing other than steel. These prohibitions for groove welds made from one side only shall not apply to (1) secondary or non-stress carrying members, for shoes, etc., where such welds are not contiguous to an FCM, and (2) corner and T-joints parallel to the applied stress between components of built-up members designed primarily for axial stress.
   - Intermittent groove welds
   - Intermittent fillet welds
   - Bevel-grooves and J-grooves in butt joints for other than horizontal positions
   - Partial-penetration fillet welds and groove-weld butt-T and corner-joints perpendicular to the applied stress
   - For joints perpendicular to the applied stress, welds with backing bars left in place and backing fillet welds outside groove weld
   - Any temporary weld made in the steel mill, in fabrication, in transportation and/or in erection not shown on the drawings.

10. In attachment detail, the connection of the attachment, possibly, shall be a bolted. In place of the field welds, all field connections should be made by bolting.
11. Eccentricity of connection:
    Eccentricity between intersecting parts and members shall be avoided in so far as practicable. In designing the welded joints, adequate provision shall be made for bending stresses due to eccentricity, if any, in the disposition and section of the base metal parts and in the location and types of the welded joints.
    For members having symmetrical cross sections, the connection welds shall be arranged symmetrically about the axis of the member, or proper allowance shall be made for unsymmetrical distribution of stresses.
    For axially-stressed angle members, the center of gravity of the connecting welds shall lie between the line of the center of gravity of the angle's cross section and the center line of the connected leg. If the center of gravity of the connecting weld lies outside of this

zone, the total stresses, including those due to the eccentricity from the center of gravity of the angle, shall not exceed those permitted by this code.

12. Connections or splices:

Connections or splices of tension or compression members made by groove welds shall have complete joint penetration welds. If members subject to compression only, are spliced with full-milled bearing, the splice material and its welding shall be arranged, unless otherwise stipulated by the applicable general specifications to hold all parts in alignment and shall be proportioned to carry 50% of the computed stress in the member. Where such members are in full-milled bearing on base plates, there shall be sufficient welding to hold all parts securely in place.

When a member is made of two or more pieces, they shall be connected along their longitudinal joints by continuous welds to make the pieces act in unison.

13. Transition of Thickness or width at butt joints:

When butt joints are used to join material of different thickness or width, there shall be a smooth transition slope between the offset surfaces or edges. This slope shall not exceed that shown in AWS D1.1 (2 1/2 : 1 straight tapered transition for thickness and width; or 2 feet radius transition for width). The transition of thickness may be accomplished by sloping weld faces, chamfering the thicker part, or a combination of the two methods.

The transition of width shall be accomplished by sloping the edges of the wider part.

14. Girders and beams:

When connections or splices in beams or girders are made with groove welds, they shall have complete joint penetration welds.

Splices between sections of rolled beams or built-up girders shall preferably be made in a single transverse plane. Shop splices of webs and flanges in built-up girders, made before they are joined to each other, may be located either in a single transverse plane or multiple transverse planes, but the fatigue stress provisions of the general specification shall apply.

Girders (built-up I sections) shall preferably be made with one plate in each flange, i.e. without cover plates. The unsupported projection of the flange shall be no more than permitted by the applicable general specification. The thickness and the width of the flange may be varied by butt welding parts of different thickness or width with transitions conforming to the requirements.

15. Stiffeners:

If stiffeners are used on only one side of the web, they shall be welded to the compression flange.

Stiffener-to-web connections which is perpendicular to the applied stress in a pure tension member shall be a complete-penetration groove-welded T-joint.

Stiffener-to-web connections perpendicular to the applied stress in flexural members, in locations of tension stress the connections, shall be complete-penetration groove-welded T-joint for one-third (1/3) the depth of the girder but not less than two (2) feet. The remaining 2/3 of the connection may be an incomplete penetration fillet weld.

The groove-welded T-joint welding procedure for welds perpendicular to the applied stress shall be qualified by appropriate testing.

Diaphragm and hanger welds involving flange and/or web connections perpendicular to the applied stress shall be a complete-penetration groove-welded T-joint.

16. Cover Plate:

Cover plates shall be limited to one on any flange. The maximum thickness of cover plates on a flange shall be no greater than 1-1/2 times the thickness of the flange to which the cover plate is attached. The thickness and width of a cover plate may be varied by butt welding parts of different thickness or width with transitions conforming to the requirements. Such plates shall be assembled and welds ground flush before attaching to the flange. The width of the cover plate, with recognition of the dimensional tolerances allowed by ASTM Specification A6, shall allow suitable space for a fillet weld along each edge of the joint between the flange and the plate cover.

In FCMs, when partial-length cover plates are used, the ends of the cover plates shall be in sections where the stress is compressive. The ends of the cover plate shall be cut square, and a continuous fillet weld shall be deposited across the end and along both edges of the cover plate or the flange to connect the cover plate to the flange. The fillet welds shall not terminate at the corners of the cover plate, but shall be returned continuously full size around the corner for a length equal to twice the weld size where such return can be made in the same plane. Boxing shall be indicated on design and detail drawings.

Fillet welds connecting a cover plate to the flange in the region between terminal developments shall be continuous welds of sufficient size to transmit the incremental longitudinal shear between the cover plate and the flange. Fillet welds in each terminal development shall be of sufficient size to develop the cover plate's portion of the stress in the beam or the girder at the inner end of the terminal development length and in no case shall the welds be smaller than the minimum size permitted.

## 7–5 REHABILITATION

### 7-5.1 REHABILITATION PROCEDURE

The rehabilitation of structural members is required to restore the safe load carrying capacity of a bridge, to handle the designed traffic and to extend the useful life of the structure.

The rehabilitation can be accomplished by the following procedures:

1. Inspect the bridge for the nature & extent of damage and to determine the over-all physical condition of the structure.
2. Determine the capacity and ratings of the structure.
3. Evaluate the damage, serviceability and the life of the bridge.
4. Select a repair method.

**1. Inspection**

When a structure has any damage affecting its capacity, a detailed inspection of the structure should be made. The structural engineer assesses the damage, and makes recommendations as to the possible restriction in operation, reduction of load limits, and emergency repairs until the damage analysis is completed.

After the nature of the damage and the extent of the deterioration has been identified, the steel type and physical and chemical properties should be determined by examining the construction plans, specifications or by testing metal specimens obtained from the structure.

**2. Evaluation**

In evaluating the deterioration, the following should be considered:

1. Whether the damaged or deteriorated member is a redundant or non-redundant fracture critical member. Deterioration of the non-redundant fracture critical members should be given special attention in regard to reducing the level of service or closing the structure until repairs can be made.
2. Load-carrying capacity based on a working stress analysis can be improved by using a load-factor analysis which is believed to represent actual structural behavior or stresses. The decision toward bridge closing or reduced service will be affected by the methods of analysis used.
3. It should be checked whether the damaged members would affect adjacent members and cause actual stress redistribution or that can result in another member being overloaded.
4. A check should be made to determine the damaged members are in danger of buckling or collapse and are in need of additional support.

5. A check should be made as to whether the damage or the deterioration is localized or widespread. Actual field tests of through girder and floor beam system bridge might show that the actual loads are redistributed from the localized corrosion at the floor beam web to where the web is in good condition. Sometimes. the local distress could have serious impacts on the structure, therefore, caution should be exercised so as not to ignore localized defects.
6. Check whether a distress is located on the primary structural load path.

### 3. Selection of Rehabilitation Methods

The selection of repairs and rehabilitation methods is dependent on factors such as the extent of distress, the type of steel used, weldability, public safety, public inconvenience, availability of funds, remaining life of the structure, desirable live load capacity, fatigue stresses, and estimated cost.

In general, the five methods of rehabilitation are the following:

1. Strengthening of the critical members.
2. Adding supplementary members.
3. Reducing dead loads.
4. Modify the structural system.
5. Repair or replacement of the damaged members.

Current standard practices for strengthening steel bridges involve adding cover plates or new sections to the existing steel beams, or adding plates or structural shapes to the existing truss members. Welding is used extensively to attach the new sections to the existing members.

Sometimes, improper welding details and indiscriminate field welds cause fatigue failure, therefore, special care should be exercised in selecting proper welding details and in preparing specifications for the field welds. If the existing steel has a high carbon content (such as old A-7 steel), weldability should be carefully examined before selecting the weld details. Also, the thickness of existing members should be examined for possible lamellar tears due to welds. As an alternate to welding, bolting or riveting should be considered.

In place of adding cover plates, installing draped or straight post tensioned tendons on either side of the web of the beam can be used to reduce the existing tension stresses in the beam.

The capacity of the existing steel member could also be increased by adding a new steel beam between the existing beams or reinstalling the existing beams at closer spacings. Strengthening could be achieved by reducing excessive dead loads from the structure, for example, the removal of concrete encasements from the steel girders or replacement of a thicker asphalt wearing surface with one of minimum thickness.

The structural system can be modified to increase its capacity by installing shear connectors between the deck slab and steel beams to create composite action or by splice bolting the ends of the two simple beams to make it a continuous span or by replacing the deck floor system with different systems so that the dead loads are reduced.

Replacement of the deteriorated member can be an alternate solution to strengthening of the members.

## 7-5.2 REPAIR DESIGN CONSIDERATION

### 1. Stresses

The allowable unit stress for added material shall be in accordance with the design stress of the AASHTO or AREA Specifications.

During rehabilitation, the structure should be relieved of its dead loads. This relief can be obtained either by jacking of the structure or removing the dead loads from it. Without this relief, the existing members will continue to carry all of the dead load and its share of live load stresses while the added plate or section will only carry a portion of the live load stresses. By not carrying any of the dead load stresses, the new material will not reach the allowable stress without overstressing the existing members. It may be necessary to place a heavier new section to avoid an overstress on the existing members.

For example, the existing beam is W27x80. I=2875 in.$^4$ S=211.6 in.$^3$

| | | |
|---|---|---|
| Dead load Moments | : | 160 Kip-ft. |
| Live & Impact Moments | : | 215 Kip-ft. |
| Total Moments | : | 375 Kip-ft. |

f=(375x12)/211.6= 21.26 ksi
Allowable: 20.0 ksi

**Case 1: No Jacking**

All live loads are restricted during rehabilitation. Since no provision was made to relieve the residual dead load stress by jacking or shoring, original dead load stress will remain in the beam.

| | |
|---|---|
| Allowable stress | 20.00 ksi |
| Residual D.L. stress | 9.07 ksi |
| Max. all. L.L stress | 10.93 ksi |

For Trial size of Plate, assume the required bending moment is resisted by the flange section area then

M=Ahf A=M/(hf)

- M: Resistant moment
- A: Area of one flange
- h: Distant between flange (effective depth)
- f: Stress

A=215x129/(26x10.93)=8.74 sq.in.

Since the above approximate required area is based on the equal flange areas at the top and bottom, let's try section of 12"x1" (A=12 sq. in.) plate at bottom flange.

| | A | d | Ad | Ad$^2$ | I |
|---|---|---|---|---|---|
| W27x84 | 24.71 | | | | 2825 |
| 12"x1" | 12.00 | -13.85 | 166.2 | 2302 | 2302 |
| | 36.71 | | 166.2 | | 5127 |
| | | | | | (751) |
| | | | | $I_{n.a.}$ | 4376 |

d=166.2/36.71=4.52
-4.52x166.2=751

$d_{top}$ =13.35+4.52=17.87
$d_{bottom}$=13.35-4.52= 8.83

Residual D.L. Stress 9.07
Live Load Stress

$$f(top)=\frac{215x12x17.87}{4376}=10.53$$

$$f(bot)=\frac{215x12x\ 8.83}{4376}=5.20$$

$$\frac{5.2x7.83}{8.83}=4.61$$

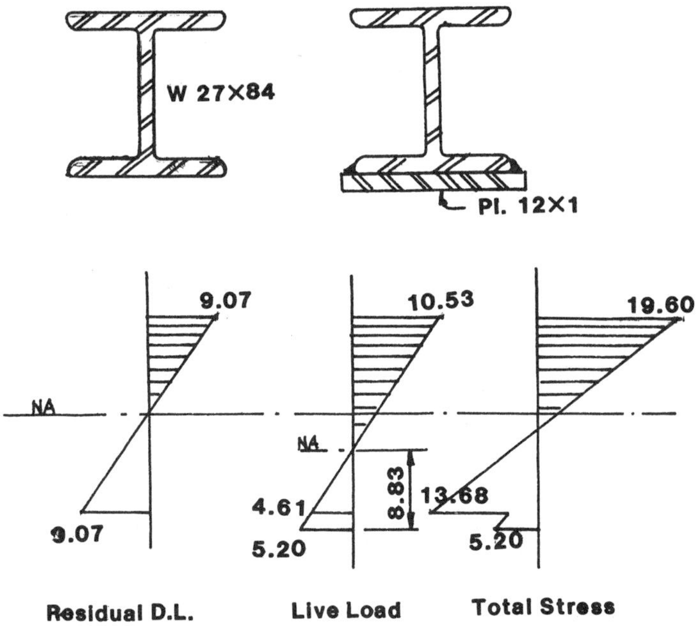


**Fig. 7.42 Stresses by Adding Cover Plate without Jacking the Beam**

**Case II: Jacking to relieve dead load stress**

Try 12"x1/2" plate

| | A | d | Ad | Ad$^2$ | I |
|---|---|---|---|---|---|
| W 27x84 | 24.71 | | | | 2825 |
| Plate | 6.00 | -13.6 | 81.6 | 1110 | 1110 |
| | 30.71 | | 81.6 | | 3935 |
| | | | -2.65x81.6 = | | (216) |
| | | | | | 3719 |

d=81.6/30.71=2.65

d(top)=13.35+2.65=16.00
d(bot)=13.35-2.65=10.70

f(top)=(375x12x16.0)/3719=19.36 ksi
f(bot)=(375x12x10.7)/3719=12.94 ksi

As shown by the above example, Case I (without jacking) requires a heavier section and the new cover plate has not been utilized to its maximum capacity. To avoid this condition, temporary shoring and jacking methods have been used to relieve the stress during the strengthening of the existing structure.

When temporary shoring with jacking or other methods, which introduce a proper dead load stress into the new materials and which can not be provided due to traffic conditions or other restrictions, the new material added to the existing structure member should be considered effective in carrying the live loads only and the stress of the old portion of the existing member should be investigated under the loads for which the member is strengthened as shown on the Case I.

For a structural member to be strengthened, a study should be made for an increase of dead load stress resulting from the temporary removal of cover plates, lacing bars, rivets, or other components. If the member is carrying a stress, then that member or an adjacent member may pop up unless it is relieved of stress and/or restraints. On the construction plans, the requirements of falsework or other jacking methods should be clearly shown if the design is made by assuming release of the dead load stress or other stresses during the strengthening operation.

Since the proportional ratio of load carrying capacity between the new material and the existing members are based on the assumption of equal deformations, it is important that the new and the existing material be compatible and meet the principle of consistent deformation.

If the existing structure was designed in accordance with the Ultimate Strength Design method, the section will be capable of sustaining fully a plastic state and allowance may be made for the redistribution of loads. The stage of loading is of no importance for analysis and all components participate to resist the total forces without requiring the removal of residual dead load stress by falsework and jacking methods. Under the ultimate strength design concept, the strengthening and existing sections could be considered as a single component for carrying the total load if the following requirements are satisfied:

1. New and existing material is able to undergo sufficient deformation without fracture.
2. Structural members should be able to deform without local and lateral buckling.

**2. Residual Stress**

The residual stress developed from the distortion and local buckling will be locked into the damaged members even after the member is straightened by heating or mechanical methods. It is suggested that buckling of a compression member be repaired by replacing or heat straightening the damaged member after it has been dismantled from the structure and then reinstalling it.

**3. Symmetry**

The new material or section should preferably be placed symmetrically to produce a balanced section and to eliminate or minimize the effects of eccentricity on the strengthened member. If this is not possible, the unbalanced section should be taken into consideration in determining the resultant stress or torsional forces on the member.

**4. Bolted and Welded Connections**

Generally, welds and rivets shall not be assumed to act together and if a weld is used on an existing riveted connection, the welds should be assumed to share the entire stress. If the bolted or riveted connection is strengthened by welding, the welds should be designed to carry the entire live load stress in the members and the bolts or rivets should be assumed to carry the existing unrelieved dead load.

For existing riveted connections, the new bolted reinforcement should be considered to carry both dead and live loads. For bolted connections, both new and old existing bolts should share the total stress.

For welded connections, it is preferred that a weld be used for strengthening the members instead of using new bolts. The existing weld should be assumed to carry the residual dead load stresses and the new weld, the entire live load.

If there are defective or damaged rivets, they should be replaced with bolts prior to welding and the bolts should be assumed to carry the residual dead load stresses. It is suggested that weld details be avoided wherever possible on tensile members and members subjected to fatigue stresses.

## 5. Structural Steel

The most commonly used structural steel in welded bridges are A36 steel (Carbon Structural Steel Fy=36 ksi) and A588 steel (High Strength Low Alloy Structural Steel). A572 (High Strength Low Alloy Columbium Vanadium Steel Grade 50 Fy=50 ksi) are used in place of A588 steel for steel thickness less than or equal to 2 inches. A514 (High Yield Strength, Quenched and Tempered Alloy Steel) may be considered for highly stressed area.

**A36**
This carbon steel is the most widely used for bridge construction. The carbon content is in range of 0.2-0.3% and manganes is added to about 1% to increase strength without increasing carbon level so that weldability will be maintained. If the manganes level exceed 1.5%, the toughness will be reduced.

**A514**
This readily hardened steel by using quench and temper procedure may be used where high strength is required. This steel contains chromium, molybdenum, and boron which has same effect as chromium and molybdenum on steel but less expensive. Molybdenum prevent temper embrittlement due to thermal effect of welding while chromium provide high strength.

A514 steel has high strength but it has low notch toughness, high applied stress level, and is susceptible to improper joint handling and welding procedure. For these reasons, A514 steel should be used with special care.

**A572**
This non-weathering steel is generally used as a substitute for A588 steel. In addition to about 1.3% manganes level, very small amount of columbium or vanadium is included to increase strength.

**A588**
A588 steel has been used widely due to corrosion resistance and weldability. It contains vanadium and nickel for strength. Zirconium is added to provide corrosion resistance. It is vulnerable to fatigue cracks at flame cut or tack weld. Also, it is susceptible to lamellar tearing.

## 7-5.3 STRENGTHENING

To increase the live load capacity, it is necessary to strengthen the deficient, deteriorated or inadequate members. Also, the capacity can be improved by removing portions of dead loads such as concrete encasement and asphalt wearing surface. They can be strengthened by adding new material to the existing members or by replacing the entire member or portions of it.

**Fig. 7.43 Increase Capacity by Removal of Concrete Encasement. The Surface Needs Near White Cleaning for Painting. Therefore this method may not provide great cost savings over the new stringers**

Generally, strengthening the structure under traffic is undertaken only to increase the strength of certain weak or deficient members to a desirable capacity rather than to strengthen it entirely. In some cases, a new structure may be more economical for a total structural strengthening. There are two different approaches to increase the load-carrying capacity of the existing steel structure. One is reinforcing the existing steel members by adding new ones. The other is prestressing the existing steel member.

## 1. Reinforcing:

The deficient structural members could be reinforced by bolting or welding of the plates, angles, beams, or other structural shapes to the web or the flange of beams or column components. This reinforcement increase the section modulus and thus result in increasing the capacity.

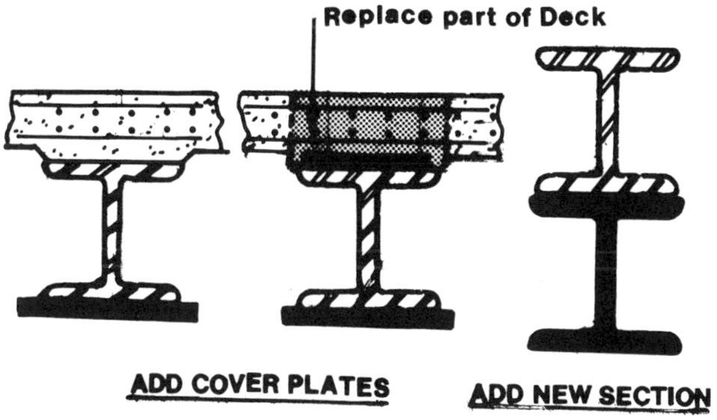


**Fig. 7.44 Methods of Reinforcing Beams**

The beam or girder section could be increased by adding cover plates to the flanges either by installing new beams over the existing, or by replacing the existing cover plates with new thicker ones.

In a riveted girder, if the exposed cover plates have excessive corrosion and the section is severely reduced, the cover plates should be replaced with new plates. If the corrosion damage extends into the second cover plate, a replacement with a new single cover plate should be considered. A single cover plate is recommended because, for a riveted plate girder, the repair would be laborious and costly for the removal of rivets while the cost of the additional material would be comparatively insignificant.

When a cover plate is added to a beam or plate girder without existing cover plates, the new plate should extend the full length of the beam instead of terminating at the theoretical cut-off points in order to avoid stress problems due to fatigue.

The cover plate may be connected by continuous fillet welds or by high strength bolts, but no intermittent welds should be used. If the cover plates are to be welded, consideration should be given to design the bottom flange cover plate wider than the existing flange and the top plate narrower than the existing flange so that overhead field welding can be avoided. The cover plates should have adequate thickness to prevent buckling, otherwise intermediate fasteners would be required during the field welding operations.

When the flange cover plates are replaced and the number of existing rivet holes is insufficient, the existing holes should be enlarged and larger size high strength bolts should be used to compensate for the lesser number of holes to make the connection.

When the rivets have broken heads, cracks, slippage or other defects, they should be replaced with high strength bolts. During construction, more defective rivets might be encountered than the specified repair limits shown on the plans. Situation of this type should be taken into account while preparing the plans by adding an extra item, "Replacement of Rivets"

It is desirable that new material be fabricated or subpunched in the shop and reamed in the field. Often, it is impractical to make totally accurate field measurements and drilling of the holes in the field may be permitted by drilling holes and inserting bolts immediately following each rivet removal. In no case should drilling of the holes be permitted after the total removal of rivets from a member or a connection.

Adding top cover plates to a beam requires removal of the concrete deck slab over the flange and the maintenance of traffic on the bridge. The removal of concrete should be made without damaging the existing reinforcing bars. The exposed reinforcing bars and concrete should be cleaned to provide proper bonding between the existing and new concrete. This type of rehabilitation would be costly and it is not recommended unless a total deck replacement is required in conjunction with strengthening of the beams.

If deck replacement is warranted, the installation of shear connectors on the top of flange should be studied in addition to the placing of the new cover plates on the top and the bottom flanges. Also, a non-composite girder could be strengthened by drilling circular holes in the deck slab over the top flange and installing shear connectors. The holes should be grouted with a non-shrink mortar or grout after installation of the connectors. The shear connectors provide for a composite action between the steel beam and the concrete deck slab thus increasing the live load-carrying capacity of the structure.

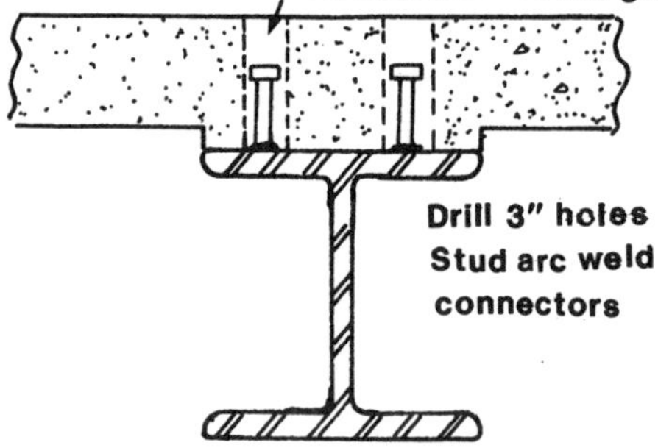


**Fig. 7.45 Increase Strength by Installing Shear Connector in Drilled Holes in Deck**

Where the removal and replacement of the deck is excessive, the flange can be strengthened by adding a full length longitudinal plate to the flange and/or to the angles just below them. This method could be also used to strengthen the bottom flange where it has a severe loss of section. In this case, the existing flange could be designed to carry no stresses and the new flange angles would carry the entire stresses.

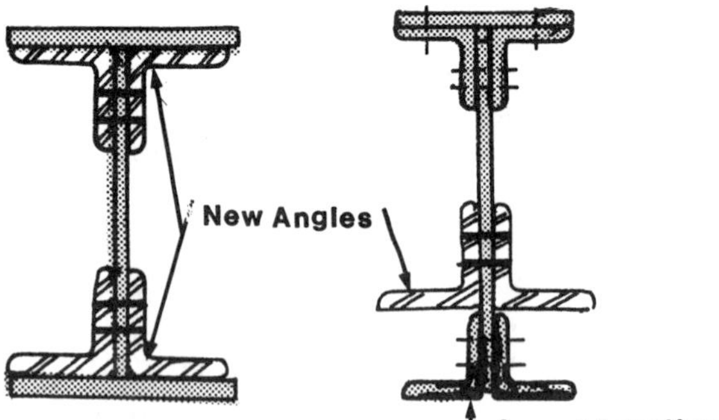


**Fig. 7.46 Adding Angles to the Flanges or Web to Increase the Capacity**

When vertical clearance is not a problem, a structural shape may be added to increase the section modulus by increasing the depth between the top and the bottom flanges. This method would produce inefficient use of material by making the two connecting flanges (bottom flange of existing beam and top of the new flange) structurally ineffective.

If there is a section loss in the web, it can be reinforced by bolting splice or side plates. Most of the problems occur in through-girder system where the floor beams are attached to the main plate girder.

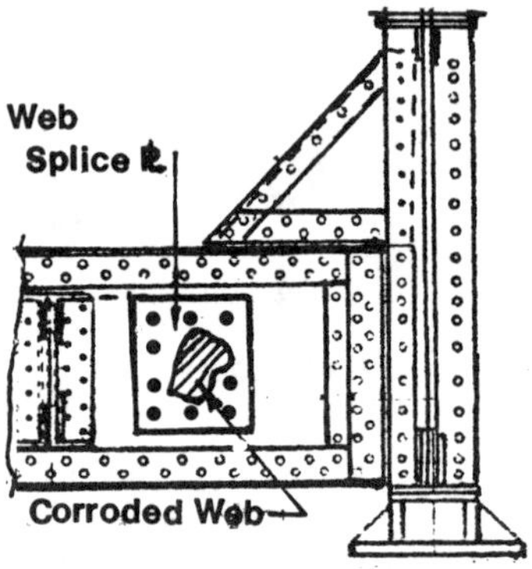


**Fig. 7.47 Repair of Floor Beam Web by bolting Splice Plate**

Generally, severe corrosion at the web is attributed to water leakage through the deck slab at the curb line. Since the floor beam web is the only primary load path for shear stress from the concrete deck slab to the main girder, priority should be given for its immediate repair.

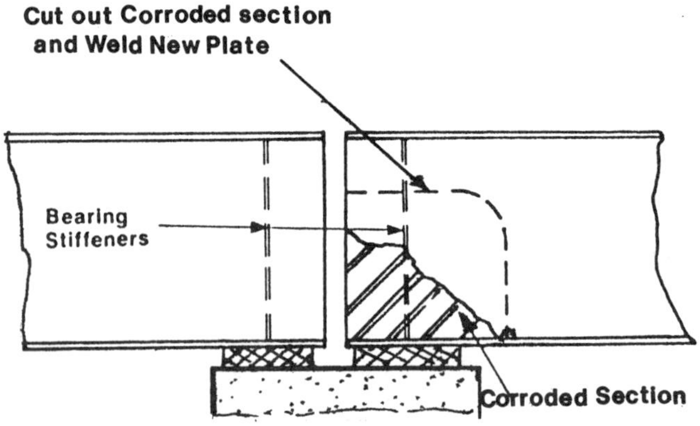


**Fig. 7.48 Repair of Corroded Beam Ends**

End corrosion of the steel beam occurs due to a joint leak and severe corrosion problems could be corrected by replacing portions of the beams. After all of the stringers are jacked up, the damaged end of the beam is cut and removed. Then the new section (WT section or pre-fabricated section) is welded into place by using full penetration welds. Upon completion of repairs, all stringers are lowered simultaneously to avoid unwanted damages during lowering.

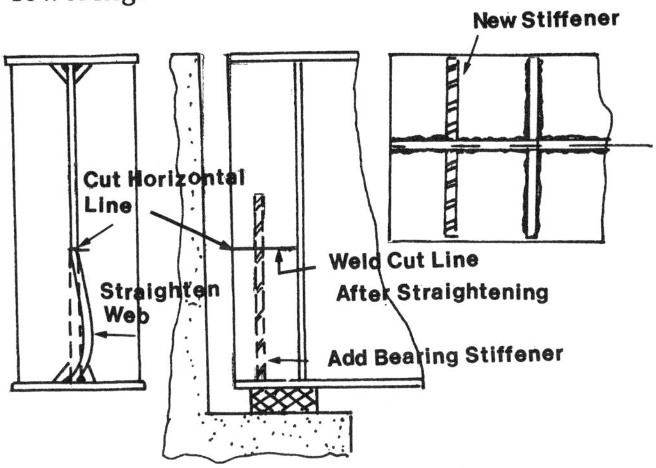


**Fig. 7.49 Repair and Strengthen the Web by Adding New Bearing Stiffener**

The lack of or the misalignment of bearing stiffeners, the misalignment of bearings, and the inadequate thickness of a stiffener plate can cause web crippling in the bearing area. The bearing stiffener could be reinforced by adding new angles or plates to the web. The end of the bearing stiffener should be a tight fit or welded to the flanges.

Before adding a stiffener plate, first the girder web is cut horizontally a few inches above the buckled web. Then clamping plates are placed on both sides and straightened by mechanical means (jacking). The damaged section of web is cut and replaced with new welded plate, followed by the addition of new bearing stiffener plates.

## 2. Details:

**Floor beam connection:** The end connections should be carefully examined for their fatigue susceptibility. They can be strengthened by using either longer connection angles, reaming existing holes and using larger high strength bolts or by welding plates to the bracket end of the beam to provide the additional support.

**Lateral bracing and connection:** Inadequate lateral bracing systems should be replaced with new structural members which are connected with high strength bolts. It is suggested that attaching lateral connection plates to the tension flange of short span stringers and floor beams be avoided wherever possible because of fatigue problem.

**Diaphragm:** Inadequate shallow diaphragms and cross frames should be replaced or strengthened to provide rigidity between the beams so that they will deflect together rather than individually. Special attention should be given to the end diaphragms because they are subjected to wheel loads transmitted directly from the deck slab.

## 3. Trusses:

The strengthening of trusses is more difficult and requires more analysis than that of girders or the floor beams. The connections are the most critical element for the strength of the span.

**Tension Members:**

Rolled or built-up sections may be reinforced by adding cover plates in the planes of the web plate, gusset plate and flanges by either bolting with high strength bolts or welding.

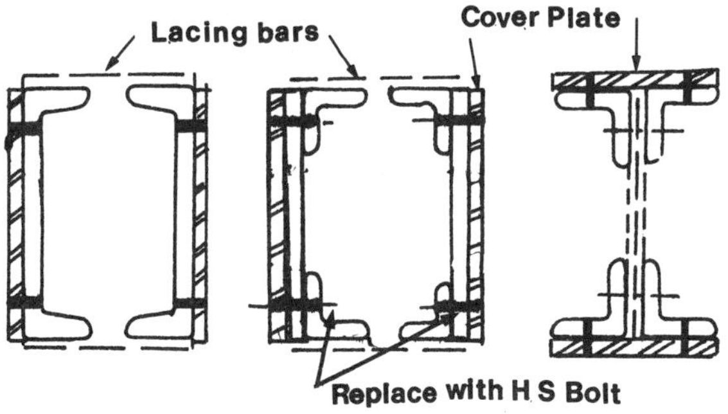


**Fig. 7.50 Reinforced by Cover Plates**

If the dead load stress in the original member is not relieved by jacking or falsework, the new cover plate should be considered effective in carrying its portion of the live load stress only. An analysis should be made to insure that the total stress in the existing tension member does not exceed the allowable stress when the locked in dead load stress of the original section adds its share of the live load stress. Also, care should be taken to protect against notches or other types of potential stress concentrations at the connection of cover plate, to the gusset plates or to the webplates.

Floor beam connections and hangers are frequently overstressed. In order to reduce the possibility of fatigue cracks in the highly stressed hangers, the sharp copes or the re-entrance cuts at floor beam connections should be modified or completely removed. High strength bolts should be used at the top connection of the floor beam to replace all of the existing rivets, so as to improve the load transfer to the web or gusset plates.

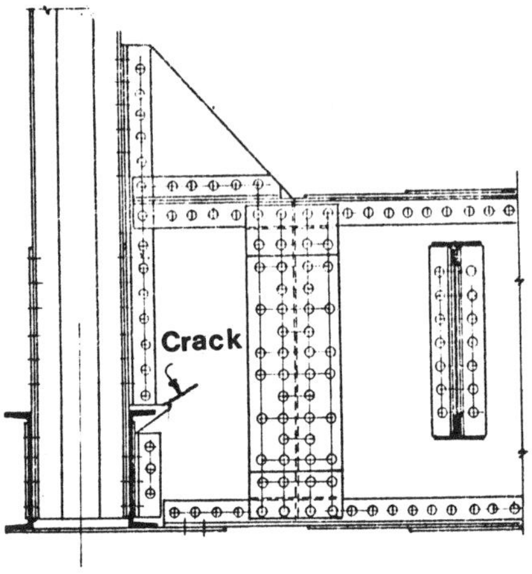


**Fig. 7.51 Crack at Reentrant Cut**

A tension member in a pin connected truss may be reinforced by adding either adjustable bars or cables. These may be loop bars or a single bar attached to a loop or a cable.

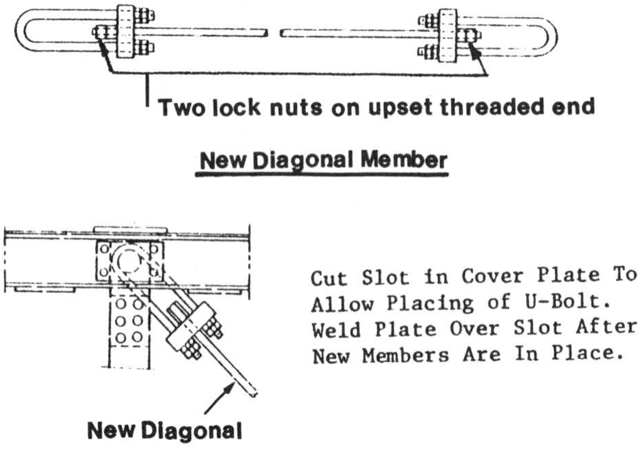


**Fig. 7.47 Reinforcing Pinned Tension Member by Adding or Replacing with U bars and Rod**

Post tensioning of truss tension members has been done effectively. Cables are strung along the truss member and attached to the end of it member or to the connection pin. Turnbuckles are provided to introduce tension. The compression stresses induced by post tensioning provides a reduction of tensile stresses in the member, permitting it to carry additional live loads.

**Compression Members:**

Reinforcing compression members require a thoughtful study and analysis. The chord members of most old bridges are not symmetrical in section and they most likely are eccentrically loaded. This problem may be solved by the addition of steel plates at the proper location. A small piece of steel plate placed at the critical point often will increase the capacity of the member substantially.

An additional steel plate could be added to the cover plate, usually between the existing lines of the rivets. This plate should be balanced by placing additional steel on the lower flanges. Also, a full length side plate can be added to either the web plate of the section or between the vertical legs of the upper and lower angles. This provides an adequate ways of transmitting stress to the connection or the adjacent members. If the existing cover plate is so much wider in proportion to its thickness that it has a limited resistance to buckling, then the condition could be corrected by adding a cover plate connected by bolts along the center line and through the angles.

Often, reinforcing a compression member will reduce its slenderness ratio and result in an increased allowable stress.

The most difficult problem in strengthening a compression member is introducing dead load stress to the new added material. Otherwise, the full value of the allowable stress can not be achieved in new material. If the maximum allowable stress is obtained in new material, the old member will be overstressed unless residual dead load stress is removed. So while the old material reaches a maximum stress, the new material will develop the stress for only its portion of the live loads.

There are a few methods to introduce a dead load stress to the new material. One commonly used in heavy reinforcing member is to calculate the shortening required to produce the desired stress under dead load, and then drill the holes in the old and new members in such a way that the new members will be shortened by the drifting or heating method before they are bolted.

Recent scale model test of truss rehabilitation utilize the concept of superimposing arches, hangers and additional floor beams to the existing trusses. It is assume that the truss would provide lateral support while the light weight arch will provide additional load carrying capacity. The additional floor beams and hanger may be necessary to strengthen the deteriorated and/or undercapacity floor beams and to provide equal distribution of the live loads. The

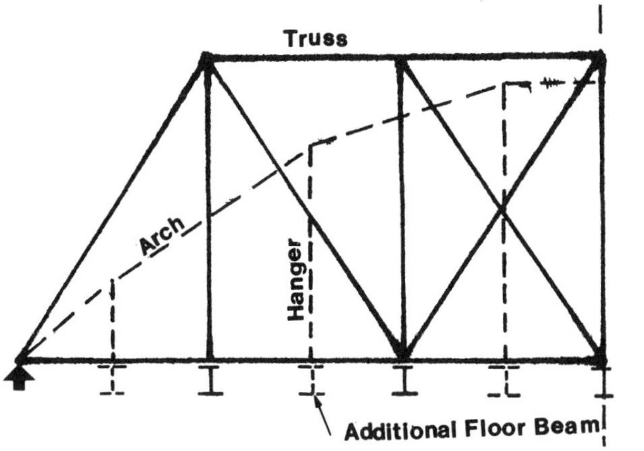


**Fig. 7.53 Arch-Hanger-Floor Beam System**

model appears to be satisfactory in regarding to easy constructions, minimum disruption of traffic, light weight and localized repairs. The design can be based on the tied arch or normal arch concept depending on the existing bridge condition to resist the thrust force. This retrofit would increase the capacity of the bridge but it would not improve the geometry of the narrow bridge.

## 4. Prestressing:

Prestressing is the application of a predetermined straight or eccentric force to a steel member so that any external loading will be counterbalanced by these forces.

There are three general ways of prestressing beams: The first is using end anchored high strength wires; the second, is to stress cover plates; and the third, is to cast a composite concrete slab to the deflected beam.

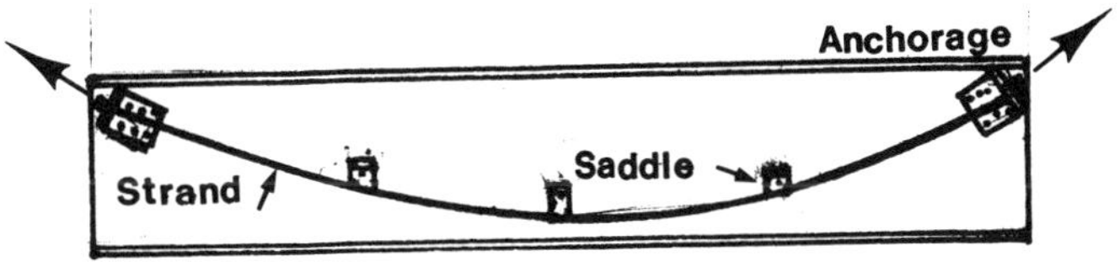


**Fig. 7.54 Prestressed Beam with Draped Strand**

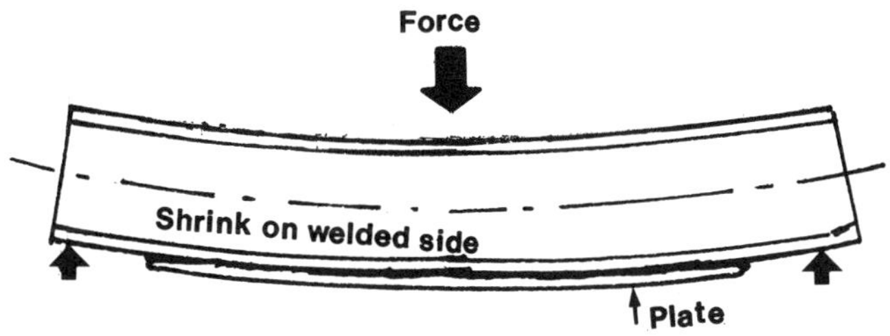


**Fig. 7.55 Prestressing by Deflecting a Beam and Attach Cover Plates**

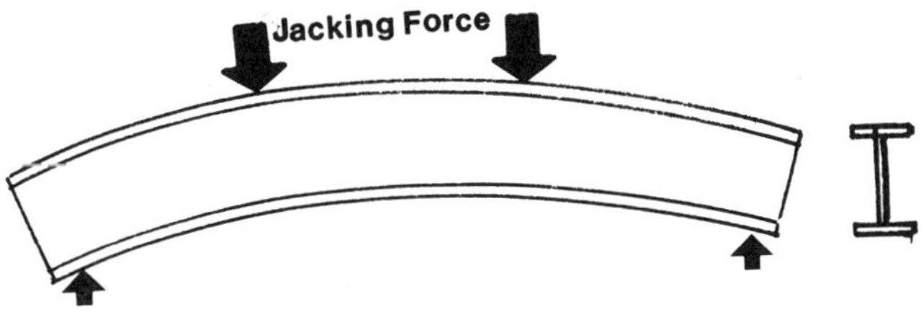


Jacking forces are applied to beam furnished by fabricator with camber.

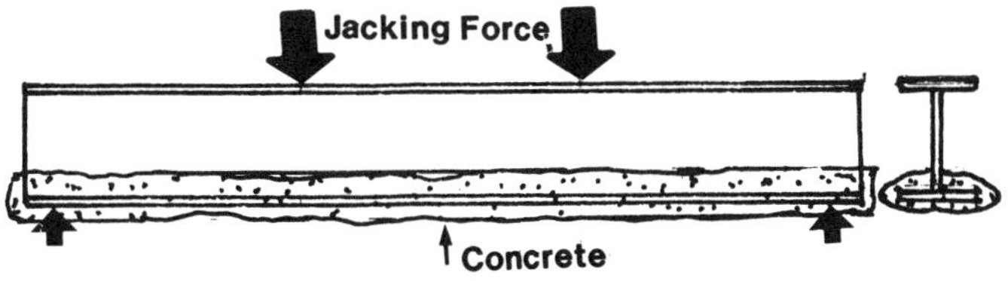


Concrete is placed while jacking forces are maintained.

**Fig. 7.56 Predeflected Beam Precompressed Concrete Composite with Steel Tension Flange**

This technique can be used where shallow depths are required. The beam is deflected in the direction of the loads. While it is deflected, the bottom flange is encased with concrete. The jacking force is released when the concrete gain the required strength to obtain precompressed concrete on the tension flange. The predeflected beam is erected in field then the web and top flange is encased monolithically with deck slab.

Since the second and third method are not suitable for bridge rehabilitation, we will discuss only the first method (prestressing by high strength steel). The prestressing method works much in the same way as in a concrete structure which is post-tensioned.

One of the common methods of prestressing is to place draped tendons below the centroid of the beam or below the beam with hold downs and anchor the tendons at each end of the beam. Other is simple straight post-tension tendons attached to the bolted anchor brackets which is located a few inches above the bottom flanges. The moments developed by the variable eccentric prestressing force will cancel out most of the moment produced by the external loads. Prestressing induces stress in the steel member which is similar but opposite in character to those produced by the dead load and live loads.

These procedures can be developed to carry live load or to reduce the dead load stress, thereby increasing the live load capacity. If it is designed to carry live load only, The tension cable should be drawn up snug under the dead load. The live loads then are carried jointly by the strand and flange. If the dead load reduction is required, the strand is stressed to the reverse the beam..

The stress distribution in a member due to prestressing consists of an axial stress and a bending stress caused by the eccentricity and it is shown by the following equation.

$$f = -H/A + He/S$$

H: Horizontal component of the prestressing force
e: Eccentricity of the force
S: Section modulus of the member

In a simply supported girder, the vertical components of the tension cable cause upward forces on the girder at the points of contact, and downward forces at the points of anchorage. The upward forces should be designed to counteract the downward dead loads. Direct or axial stress should be prevented from being transferred into the girder by providing a horizontal compression member to which the draped post-tensioning cable is anchored. The compression member is independent and free of the girder and in no way is connected to it during the post-tensioning of the steel, so that it can deform without transferring the axial force of the cable into the girder.

The draped cable produces moments and bending stresses that are similar but opposite in direction to those produced by the loads. The axial stress is eliminated by the use of a free compression member. The reduction or elimination of the dead load moments permits increased load carrying capacity of the structure without increasing the section modulus. In addition, it is possible to completely counteract the existing and any new dead loads to produce a member that is free of axial stress under the applied loads.

The post-tensioning cable may be held down by pins, brackets or other devices structurally connected to the beam. Supports for the compression member are made so that it can deform freely under the axial stress without transferring the stress into the existing beam, and at the same time the beam could provide lateral support to prevent buckling of the compression member.

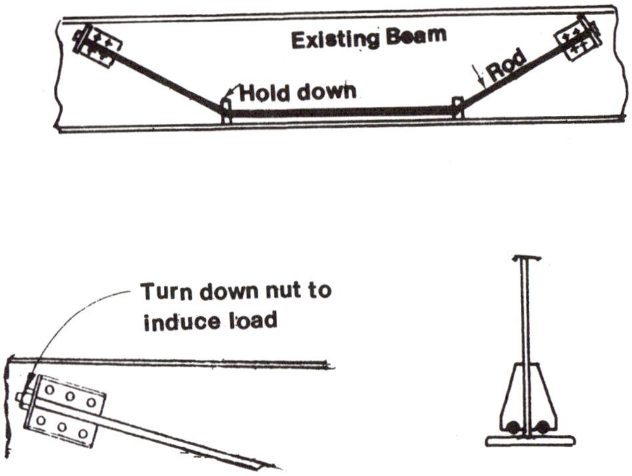


**Fig. 7.57 Prestressing Steel Beam**

The kingpost procedure is similar to the previous method. This procedure works by tensioning strands that are connected below the bottom flange with one or more posts. Treaded end connections are provided so that proper tension can be induced into the system.

Additional capacity can be obtained by changing the configuration of the truss and by adjusting the tension in the bottom chord. The installation should be monitored by controlling the number of turns of nuts at the anchors and by measuring the deflection induced in the existing member.

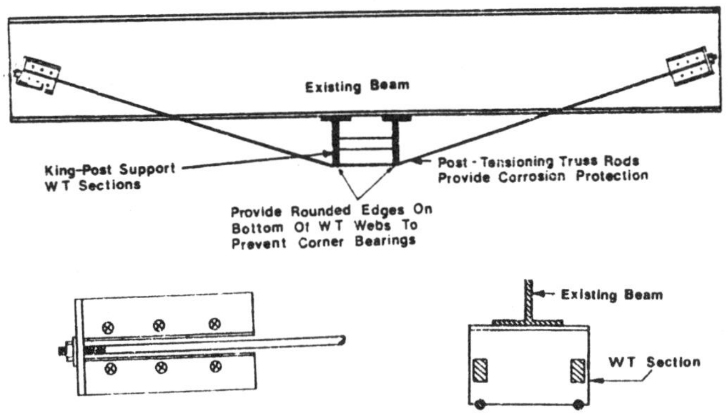


**Fig. 7.58 King Truss Type Prestressing**

The advantages of prestressing over the reinforcing method of increasing section modulus and higher load-carrying capacity of a steel member include the following:

- It may be more economical
- Normal traffic may be maintained staging and if detour is needed, the period will be short.
- Jacking of the beams to stress-free the members for the connection of new flange plates is eliminated
- In many cases, such as with existing riveted plate girders, increasing the section modulus is unfeasible and the technique described offers a feasible and economical solution

Disadvantages of prestressing include:

- Relaxation of the tendon steel can occur
- Tendons should be protected against corrosion
- Without relieving the axial compression force, the beam acts as a beam-column and the deflection may cause a significant change in stress distribution
- Possible cracking of the concrete deck

### 7-5.4 ADDING SUPPLEMENTARY MEMBER

**Girder:**

An inadequate structural system can be strengthened by placing additional stringers or floor beams between the existing members. Installation of a new stringer or a floor beam may require the removal or replacement of the deck slab if the deck is in poor condition. If the deck is in good condition, the new stringer can be installed from below the structure by jacking. Any void between the top flange of the beam and the underside of the deck can be filled by drilling holes through the deck and pressure grouting or by pressure grouting from below.

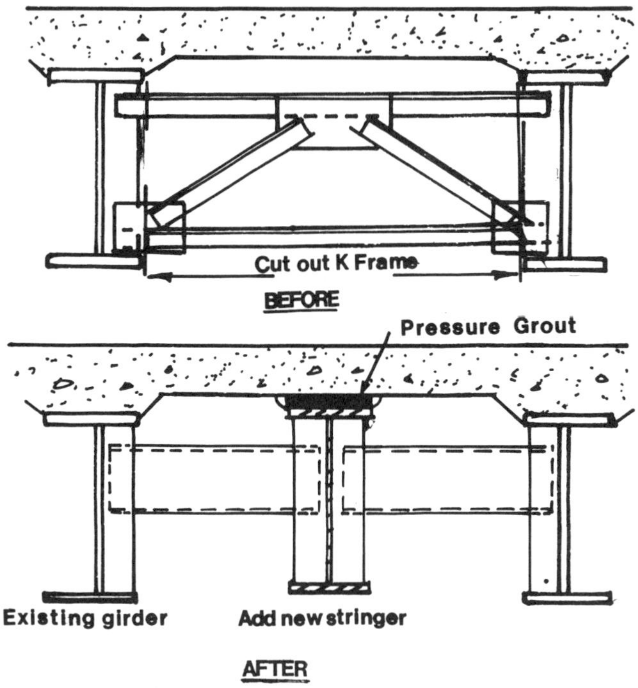


**Fig 7.59 Additional Stringer**

**Transverse Support for Stringer:**

Under certain favorable conditions where the vertical clearance and the geometric requirement allow, the load carrying capacity can be increased by shortening the effective span length of the bridge. Installation of auxiliary piers, bents or a transverse floor beam system with a main girder would convert a simple span bridge to a continuous span bridge and reduce the effective span length. This would introduce new and different stresses in the beams and a thorough analysis of the new system would have to be made to strengthen the deficient member to carry the revised loading before placing a new transverse support. Special attention should be given to the member over the new support.

This principle can also be used for increasing the capacity of a truss by installation of a bent or pier at the first or second interior panel points.

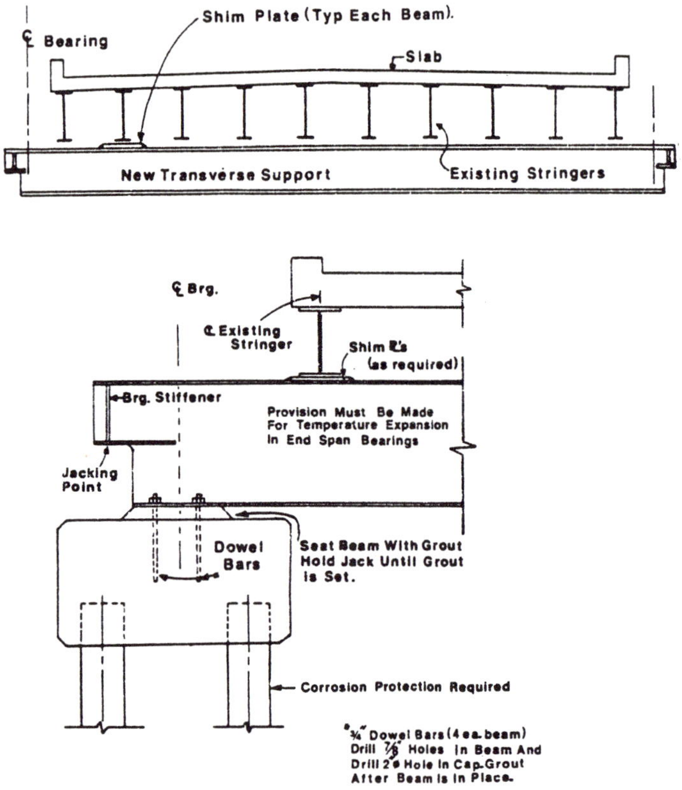


• NCHRP TRB •

**Fig. 7.60 New Transverse Beam**

**Truss:**

Deck type truss spans can be strengthened by adding a center truss. The center truss should not be too stiff but it should have equal deflection under the load as the old outer trusses. Otherwise excessive stresses will be induced in the center truss before the outer trusses have deflected enough to stress the members up to their safe limits of carrying capacity. If that equal deflection is not possible, heavy bracing will be required.

### 7-5.5 MODIFYING STRUCTURAL SYSTEM

The structural system of a bridge can be modified to improve its load carrying capacity and to avoid damages.

**Floor System:**

The existing deck system can be replaced with a lighter weight floor system. Since the dead load is reduced, the live load capacity can be increased. The following floor system should be considered for reducing the dead load:

- Open steel grid
- Concrete filled steel grid
- Laminated timber floor
- Orthotropic plate deck

**Composite Beam:**

An existing non-composite beam or a girder system can be modified to act compositely by installing shear connectors on the top flanges. If the existing deck slab shows severe deterioration, it should be replaced. Composite construction, with the concrete deck slab removed, can easily be accomplished by welding the shear connectors to the top flange of the existing beams and then placing the new concrete deck slab.

If the deck slab is in good condition, a large hole should be drilled through concrete deck slab over the beams and then the shear connector should be welded to the top flanges. After installation of the shear connectors, the hole should be filled with a non-shrink grout to provide composite action between the slab and the beams. It is recommended to have chemical analysis of existing steel for weldability before using composite design.

**Continuous Span:**

Simple span structures can be modified to continuous span bridges to gain additional capacity by splice bolting of the adjacent beams.

The portion of the deck over a pier should be removed, Two beams then can be spliced by high strength bolts to transmit negative moments and shears. New bearing stiffeners and new bearing assemblies should be

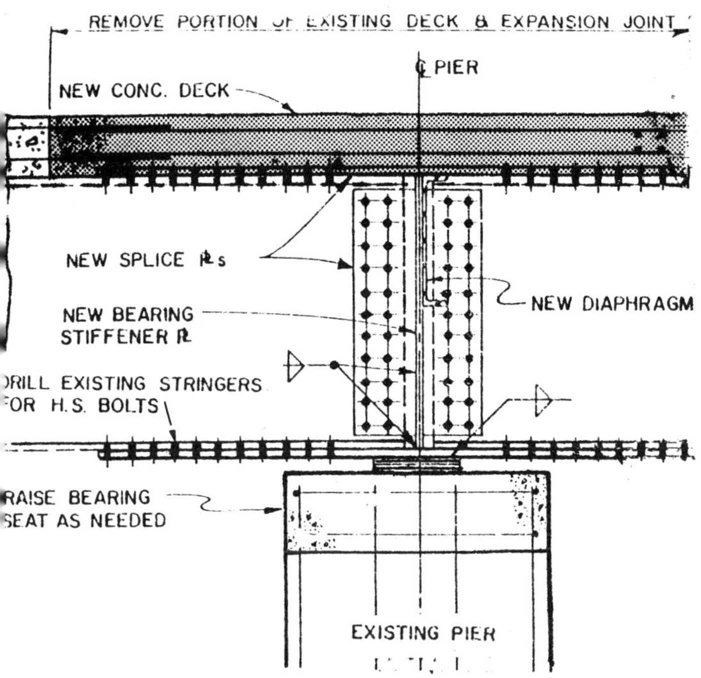


**Fig. 7.61 Converting Simple Span to Continuous Span**

provided. Additional deck reinforcing bars should be placed to compensate for the negative moments on the beams. After placing the reinforcement, new concrete should be placed in the deck slab.

## Replacement:

It is a common practice to remove the damaged section and replace it by either welding or bolting. The removal of the damaged portion can be cut by hand or by machine. The cut edge should be cleaned for welding with the new replacement part. To reduce fitting problems, it is suggested that the replacement part be cut after removal of the damaged portions. The following are some example of replacement.

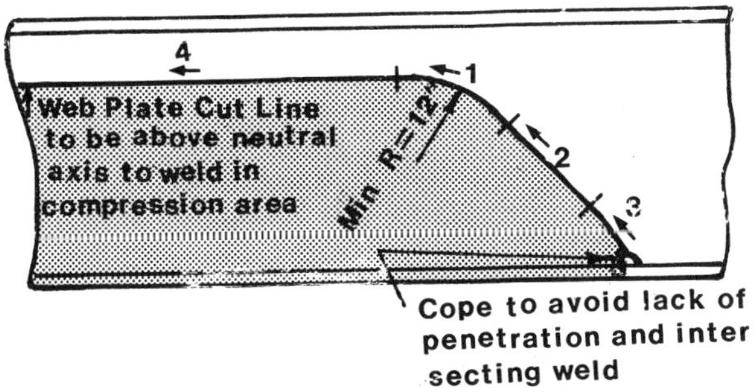


No. and Arrow show sequence and direction

**Fig. 7.62 Welding Patch Plate**

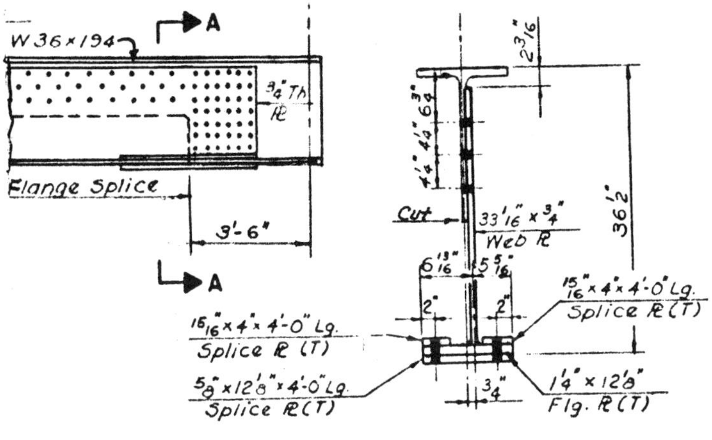


**Fig. 7.63 Bolting Patch Plate**

**Fig. 7.64 Removal of Non-Composite Fascia Girder while the deck is supported on the strong back beam by rods. New fascia beam is installed from underneath of the deck after removal of damaged girder.**

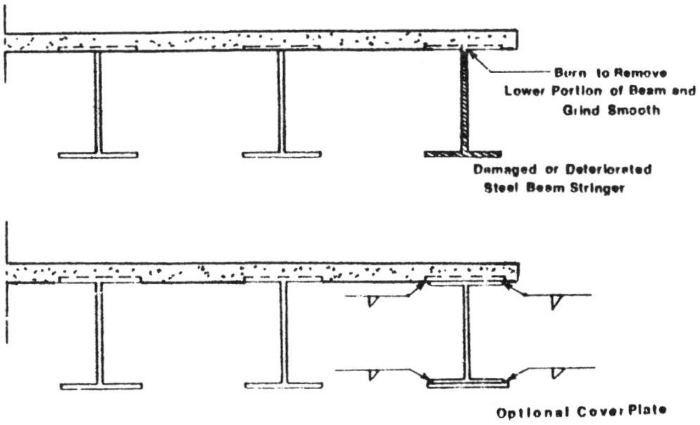


**Fig. 7.65 Replacing Steel Beam**

## Truss Member:

In order to increase the vertical clearance of a through girder, the portal frame system, sway frame and the knee bracket may be modified or replaced. The stress changes due to the modifications should be checked for the new configuration of the frame or the bracket.

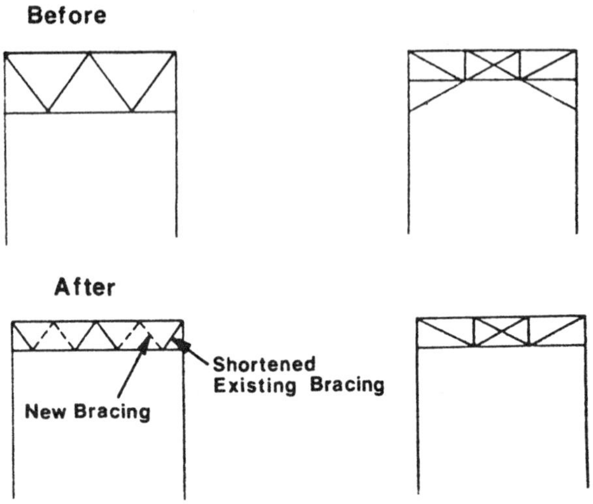


**Fig 7.66 Truss Portal Modification**

## 7-5.6 REPAIR OF COLLISION DAMAGED MEMBERS

The collision damage to structural member may results notches, misalignment, distortion of tension member which can result in high secondary stresses. Tearing or cut can cause fatigue failure and crack can lead into total failure. Collision damage in cold weather will have severe impacts on fracture because of lower toughness of steel in low temperature.

**Fig. 7.67 Severe Collision Damage**

Repairs to collision damage can be grouped as follows:

1) No action
2) Crack repairs
3) Straightening
4) Partial replacement
5) Total replacement.

**No Action:**

If the damage to the girder is minor and superficial such as a scratch or minor dent, no repair is required. However, the girder should be closely monitored for any further collision damage.

On the other hand, if the beam is severely damaged, emergency measures should be implemented to restore the beam's structural integrity and safety. However, if a total replacement of the damaged member is required for a permanent rehabilitation and an improvement in the vertical clearance, the no action alternative should be considered until a permanent improvement of the total structure can be implemented.

Emergency restoration can be achieved by installing a full span strong back support beam on the top of the deck near the damaged beam. The strong back beam is supported on the deck over the piers and deck slab is supported by the strong back beams through hangers.

**Crack Repairs:**

Cracks can be arrested by one of following method or combination:

1. Locally resisting crack propagation
2. Locally removing or decreasing forces of developing crack
3. Divert the crack direction where it can terminated in less harmful way.

If any cracks are detected in the steel components or welds, a hole can be drilled at the tip of the crack to stop further propagation. The crack may be removed by grinding, arc or flame gouging followed by welding and finally grinding it smooth. If the crack at the weld is fine, the peening method may be used.

Live loads of bridge can be removed or reduced by closing the bridge or partial opening of traffic lane away from the damaged girder.

## Straightening:

Bent and distorted members without tearing and wide cracks can be straightened by mechanical straightening or flame straightening.

### Mechanical Straightening:

This can be done with or without heat. Cold straightening is performed at normal temperature without applying heat to the damaged areas. Heat straightening is performed while the steel in the damaged area is heated to make it softer. In mechanical straightening sufficient straightening loads are applied by pulling or pushing to straighten the bent area by hydraulic cylinders or mechanical wrenches. Because of the adverse effects on stress and toughness of steel, cold straightening is not recommended.

### Flame Straightening:

Fundamentally, this method involves thermal expansion and shrinkage of the bent or kinked area of the beam. The same principle has been used in the fabrication of curved steel girders.

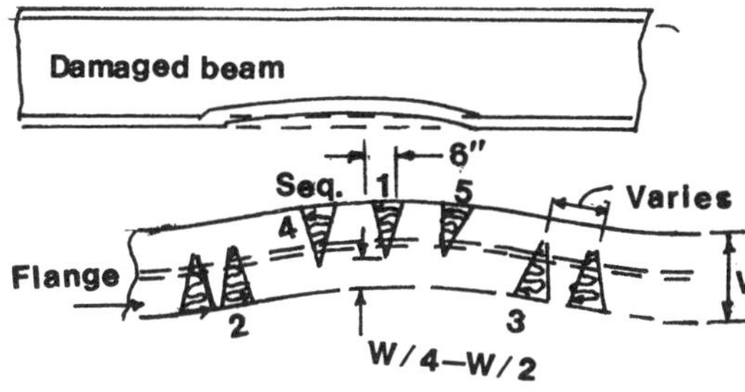


**Fig. 7.68 Flaming Straightening**

Bent areas can be straightened by applying heat to the wedge shape. The steel in the heated area is restricted by the cold steel around the wedge area. Due to this constraint, all of the thermal expansion happens in the direction of the thickness and when the heat is removed, the shrinkage will straighten the bend.

Since the flame straightening requires a great deal of field experience, it has not been used widely even though it appears to be an economical, easy and fast way of repairs.

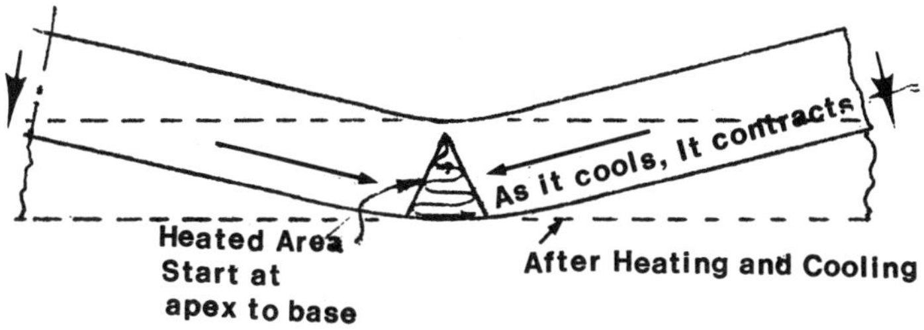


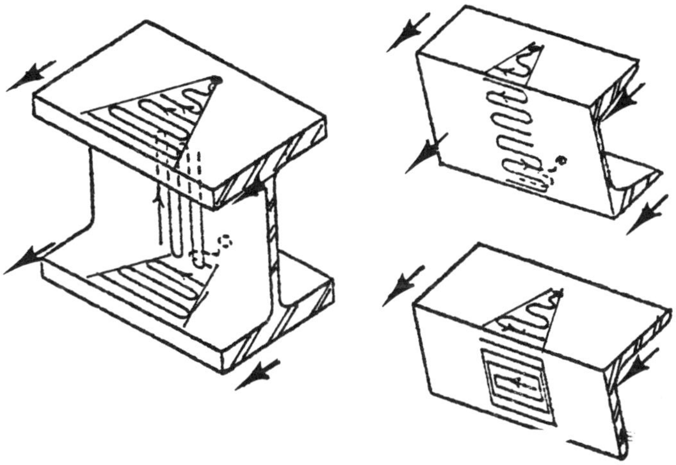

**Fig. 7.69 Bending Structural Shape by Heat**
Arrow show direction of bending.

## REFERENCE:


1. "Standard Specification for Highway Bridges" 1977 and Interim Specifications (1978 to 1982).
   "Guide Specification for Fracture Critical Non-Redundant Steel Bridge Members" Sept., 1978. AASHTO
2. "AREA Manual of Railway Engineering, Chapter 15 Steel Structure, Part 7 Existing Bridges" AREA
3. "Bridge Design Manual and Standard Drawings" of the following states: Ariz.; Calif.; Conn.; Colorado; Illinois; NY State; Minn.; Maryland; W. Va.; Maine; Mass.; New Mexico; Ohio; Penn; Washington State; Wisconsin.
4. "Extending the Service Life of Existing Bridges by Increasing Their Load Carrying Capacity" FHWA/RD-78-133, FHWA June 1978
5. "Design and Construction of Welded Bridge Members and Connection" FHWA Training Course Material, Sept. 1980

6. "Proposed Fracture Control Plan for New Bridges with Fracture Critical Member" FHWA Structural Engineering Series No.5 (3 vol.) June 1978
7. "Weld Detail" FHWA Notice N 5140.13 Dated April 24, 1978.
8. "Revised Connection Details for Cross Bracing in Steel Structure of FHWA Standard Plans for Highway Bridges" FHWA Memorandum Dated Sept. 23, 1981
9. "Development and Use of Prestressed Steel Flexural Member" ASCE Journal of Structural Division Vol. 94 No. ST9 (Sept. 1968) pp2033
10. Blodgett, Omer W. "Detailing to Achieve Practical Weld Fabrication" AISC Engineering Journal Vol 17, No. 4 (1980) pp106-119
11. Kandall, Charles "Increasing the Load-Carrying Capacity of Existing Steel Structure" ASCE Civil Engineering Oct. 1968. pp48-51
12. Johnson, Sidney M. "Deterioration, Maintenance, and Repair of Structure" McGraw-Hill, NY 1965
13. "Bridges on Secondary Highways and Local Roads: Rehabilitation and Replacement" NCHRP Report 222, TRB, May 1980
14. "Rehabilitation and Replacement of Bridges on secondary Highways and Local Roads" NCHRP Report 243, TRB, Dec. 1981
15. Mishler, H.W.; B.N. Leis "Evaluation of Repair Techniques for Damaged Steel Member" Draft copy of Phase I Final Report, May 1980
16. Heins, C.P.; H. Kato "Load Distribution of Cracked Girders" ASCE Journal of Structural Division Vol 108, No. ST8 (Aug. 1982) pp1909
17. Heins, C.P.; C.K. Hou; H. Kato "Lateral Bracing Requirements for Steel Composite Bridges" AISC Engineering Journal Vol 19, No.3 (1982) pp160
18. Heins, C.P. "Repairs and Strengthening of Old Steel Truss Bridges" ASCE NY, NY
19. Brungraber, Robert; Jai B. Kim; John M. Yadlosky "Truss Bridge Rehabilitation of An Arch Reinforcing Scheme" Dept. of Civil Engineering, Buckwell Univ. Lewisburg, PA
20. Johnson, R.N. "Assessment of Current Fabrication Technology for Welded Bridges" FHWA/RD-81/125 (March, 1982)
21. "Bridge Engineering Vol I & II" Transportation Research Record 664 & 665 TRB
22. "Steel Construction Manual" NY State DOT
23. Wade, Bill C. "Retrofit of Pin Connected Truss Bridge" Proceeding of Bridge maintenance and rehabilitation conference 1980, W. Va. Univ.
24. Keane, John D. "Protective Coating for Highway Structural Steel" NCHRP Report 74, TRB, 1969
25. Bruno, Joseph A.; John D. Keane "Evaluation of Low Solvent Maintenance Coating for Highway Structure Steel" FHWA/RD-81/019 FHWA Dec., 1981
26. Frondistou-Yannas, S "Coating and Corrosion Cost of Highway Structure" FHWA/RD-79/121, FHWA, 1980
27. "Manual of Underground Corrosion Control in Rural Electric System" REA Bulletin 161-23 (1977) US Dept. of Agriculture-REA
28. Uhlig, Herbert H. "Corrosion and Corrosion Control" John Wiley & Son, Inc.
29. "Handbook of Corrosion Protection for Steel Pile Structures in Marine Environments" American Iron and Steel Institute (AISI) Washington, DC 1981
30. Dolan, E.D. "Corrosion Control of Steel structure, Maintenance of Marine Structure" pp217-229 The Institute of Civil Engineer, London, England 1978
31. Bowman, Mark D; W.H. Munse "The Effect of Discontinuities on Fatigue Behavior of Transverse Butt Welds in Steel" Dept. of Civil Engr. Univ. of Ill. April, 1981
32. Fisher, John W. "Bridge Fatigue Guide Design and Detail" AISC, Chicago
33. Fisher, John W. "Inspection Steel Bridges for Fatigue Damage" PennDOT Research Project No. 72-3, March 1981
34. Slockbower, Robert; John W. Fisher "Fatigue Resistance of Full Scale Cover Plate Beam" PennDOT Proj. 72-3, 1978
35. Reemsnyder, Harold "Fatigue Life Extension of Riveted Connection" ASCE Journal of Structural Division Dec. 1975
36. Simon, Simon; Pedro Albrecht " Adding Fatigue Life to Cover Plate Ends" ASCE Journal of Structural Div. Vol 107 No. ST5, May, 1981
37. Fisher, John W. "Detection and Repair of Fatigue Damage in Welded Highway Bridge" NCHRP Report 206, TRB, 1979
38. Fisher, John W. "Fatigue Behavior of Full Scale Welded Bridge Attachment" NCHRP Proj. 12-15(3) Draft copy of Final Report, TRB
39. Sweeney, RAP "Some Example of Detection and Repair of Fatigue Damage in Railway Bridge Member" Transp. Research Record 676 pp8-14 , TRB
40. Sweeney, RAP "Importance of Redundancy in Bridge Fracture Control" TR Record 665, TRB
41. Hsiong, Wei "Repair of Popular Street Complex Bridge" TR Record 664 pp110-119, TRB

# 8 TIMBER STRUCTURE

## 8–1 DEFECTS

The most common defects to a timber structure are fire damage, insect attack, checking and splitting due to shrinkage, and decay. Protection against these problems is costly and sometimes incompatible. For example, the treatment against decay increases the vulnerability to fires.

Fire hazards exist when a bridge carries a gas main and a railroad bridge does not have fire protection against sparks or hot metal from brake shoes.

### 8-1.1 COMMON DEFECTS

**1. Decay**

Surface molds, stains and decay are caused by plants called fungi which require water, oxygen, and a favorable temperature of 50 to 90°(F) for growth. Timber which has a moisture content below 20% or that is soaked with water is safe from decay.

The fungi produce thousand of seed like structures called "spores" which are carried by the wind and grow on the timber through chemical cell enzymes. A sign of early decay is discoloration or a mushroom type growth. As the decay advances, the surface of the timber grows spongy, weak and highly absorbent and has a hollow sound upon hammering. The decay eventually destorys the timber.

**Prevention of Decay:**

1. Pressure treatment with preservatives, specially of cuts, holes, seasoning checks, and exposed surface areas.
2. Design for fast rain run-off, keeping wood dry and avoiding water traps.
3. Heavy treatment with coal-tar creosote or pentachlorophenol in heavy oil solvent (penta) can be used for deep penetration. For marine piles, the following treatments are used: Dual treatment with salt and creosote, Creosote and Waterborne salts.
4. For positive and improved uniform preservative penetrations, the incising method is recommended. It is a process of making a systematic pattern of punctures in the face of timber.

**2. Insect Damages**

Timber is damaged by beetles, carpenter ants and termites. Termites do not show any outward evidence of their presence until the damages is well advanced. Termites look like ants and live partly in the ground and must have a permanent water supply. Dark and poorly ventilated locations, where the wood is in moist soil, provide ideal conditions for termite attack.

**3. Marine Borers**

A major concern near salt water is the protection of the structure against marine borers. The main point of attack is usually between the high tide level and the mudline for growth. The most damage to timber are shipworms, toredos, and the crustaceans (limnoria).

**4. Other Defects**

Other defects are generally caused by fire, chemical attack, overloads and shrinkages.

**DEFINITION OF DEFECTS:**

**Decay** is the disintegration of wood substances by fungi.

**Check** is the lengthwise separation of wood that usually extends across the ring of annual growth and commonly results from stresses set up in during seasoning.

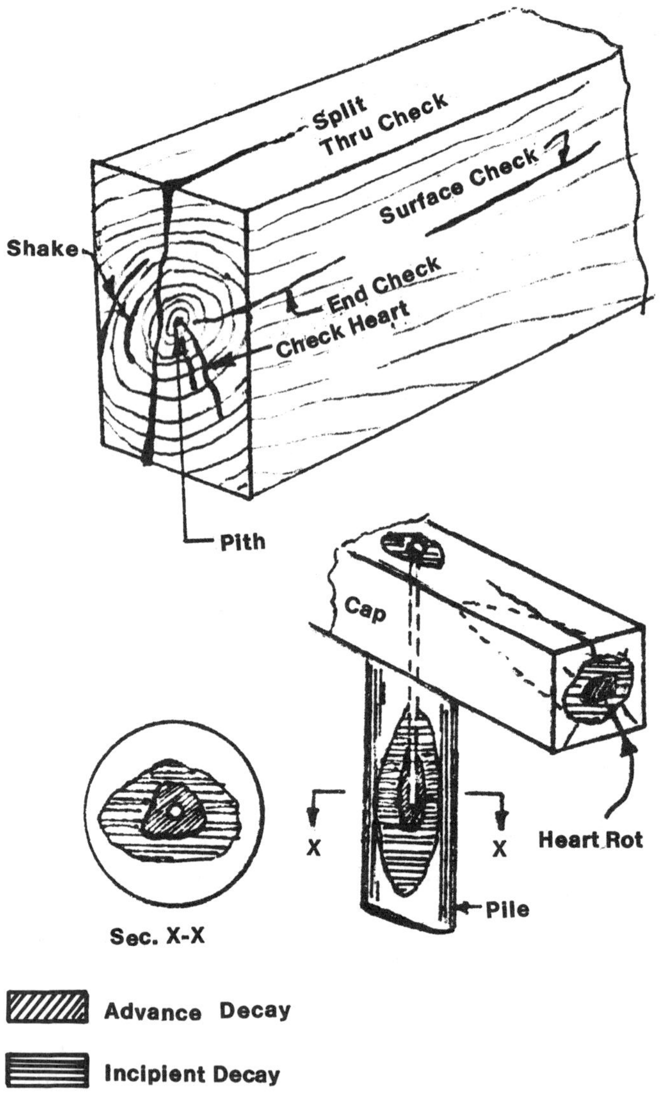


Fig. 8.1 Defects

**Split** is the lengthwise separation of the wood due to the tearing apart of the wood cells.

**Shake** is the separation along the grain, the greater part of which occurs between the rings of annual growth.

**Cross Break(Crack)** is the separation of wood cells across the grain. Such breaks may be due to internal strains resulting from unequal longitudinal shrinkage or to external forces.

**Holes** could extend partially or entirely through a piece and be from a knot, insects, marine borers, grub worms, and so on.

### 8-1.2 COMMON DAMAGES

**Joints in a Timber Stringer:**

Decay often starts along the contact surface of a field cut between stringers and spreads into the center of each member. A member may appear to be in good condition although it is hollow in the center. In later stages, crushing is visible along the contact surfaces. Accumulated debris between stringers over the caps and sills retains moisture and accelerates general decay in the stringers. When dirt or moisture comes in contact with the end of stringer, the decay often works into the cuts, rotting them hollow in a short time. All of these points should be examined with an increment borer.

**Laminated Deck with Poor Drainage:**

Plugged drains cause "spore" infected water to splash over the wearing surface and it eventually penetrates into the laminated timber deck.

**Piles and Posts at the Ground Line:**

When timber is neither always wet nor extremely dry for a long period of time, it will decay. A favorable place for such a condition is at the ground line. The earth should be dug out from around piles of posts to a depth of a foot or so and the timber should be probed thoroughly.

**Piles and Posts in Water:**

Fig. 8.2 Piles in the Water

Timber continuously submerged will not decay except in the splash zone. Even though it is in salt water it may be attacked by toredos or other marine borers. When any indication of the toredo action is discovered, the pile posts should be carefully checked for marine borer destruction. The infestation under these circumstances will occur at slightly above the point where the timber is continuously submerged. Toredos will not attack the outside of the pile below the mud-line.

**Shrinkage:**

Shrinkage in the laminated floors separates the lamination, breaks the bond between the stringers and the flooring, and permits movement which may break up the surfacing. Shrinkage of the stringers which results in unequal bearing and consequent working of the deck may also contribute to the destruction of the surfacing.

## 8–2 MAINTENANCE AND REPAIRS

Timber bridge maintenance could be grouped into following four categories based on the severity of the damages.

1. Replacement
2. Replacement of member
3. Remedial repairs
4. Preventative maintenance

### 8-2.1 REPLACEMENT

The timber bridges built in the late 1930's and 1940's are reaching their expected structure life with an accelerating rate of deterioration. Since many of them have been well maintained and have several more years of remaining useful life, age of the structure should not be the only criteria for replacement. The actual condition of the structure by a thorough inspection should considered before reaching a decision.

If the existing timber bridge shows severe deteriorations on the superstructure members, but minor defects in the substructure, only replacement of the superstructure may be required with reinforcing of the substructure members.

In a timber bridge, the bare plank bridge deck has poor skid resistance. When it is wet and the surface is bleeding the creosote, the bridge skid resistance decreases. In order to provide a smooth riding surface and improve skid resistance, asphalt overlay is used, but there are problems of bonding between the asphalt and the deck and the deterioration of asphalt especially cracks over the joints between planks. For replacement of the timber plank deck, it is suggested to use treated glu-laminated bridge materials to eliminate differential deck deflections and to reduce the dead loads.

### 8-2.2 REPLACEMENT OF MEMBER

Replacement of a member is required when its load carrying capacity is found to be reduced to 50 percent or more of the design capacity, otherwise the member could be strengthened by adding reinforcements or repairs. If the structure has several more years of remaining life, the replacement should comprise of pressure treated timbers and the adjacent member should be carefully inspected for defects. If a portion of the member shows defect, partial replacement of the member, preferrably two feet beyond the defected area, should be made. It is recommended that the area surrounding the defect and the newly exposed cut off area should receive field preservative treatment.

For a highly stressed short span tension the member, it is suggested to replace entire member instead of splicing or partial repair due to the difficulty of the designing splice. For a long span stringer or member, splicing or partial repair may be an economical alternative.

## 8-2.3 REMEDIAL REPAIRS

Remedial repairs required to extend the service life of the bridges, to increase or maintain the load-carrying the capacity of the structure and to improve bridge safety by adding an external reinforcement, performing remedial treatment of timber with internally applied liquid preservative, armoring, constructing concrete encasement, stitch bolting and clamping.

### Reinforcing:

Splice plate could be added to the damaged section by bolting or a steel cover plate can be added under the bottom.

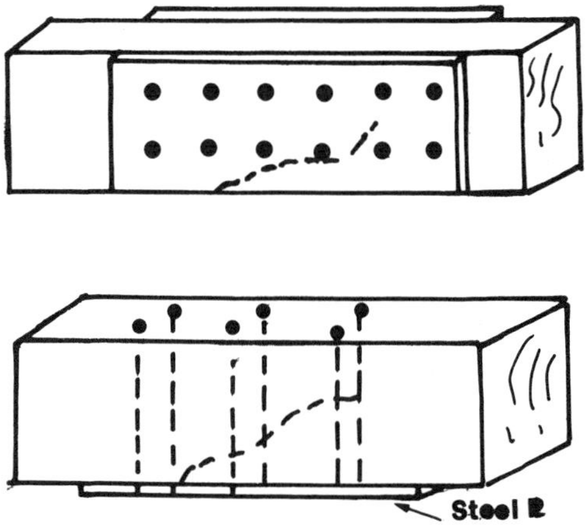


Fig. 8.3 Reinforcing

### Adding a Sister Member:

Adding a member of similar size to the existing stringers and piles have been commonly used to establish the capacity to replace the weakened members where actual removal of the old members and placement of the new members into the old positions is difficult and costly. If the old member or the damaged part is remaining in-place, a preservative treatment should be done to eliminate propagation of a decay infection or similar damage.

### Stitch Bolting:

It is made by drilling a hole through the member perpendicular to the split and placing a double treated bolt through the hole and clamping the cracks by turning the nuts over the clamping plate. Clamping is similar but the bolts are installed outside of the member by using a clamp plate.

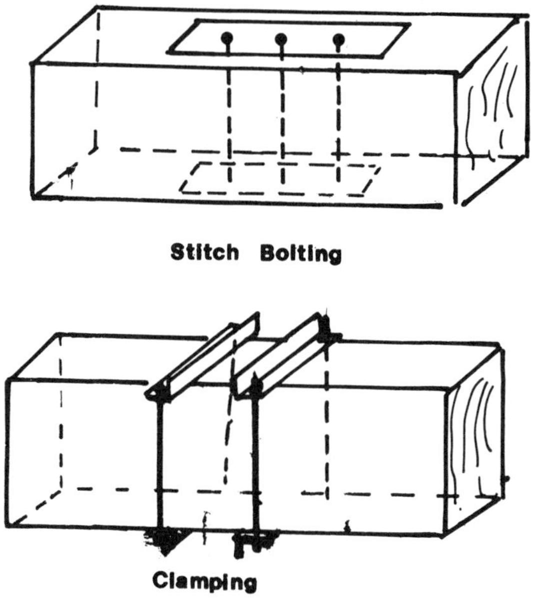


Fig.8.4 Clamping

**Tension Member Splice for Damaged Member Replacement.**

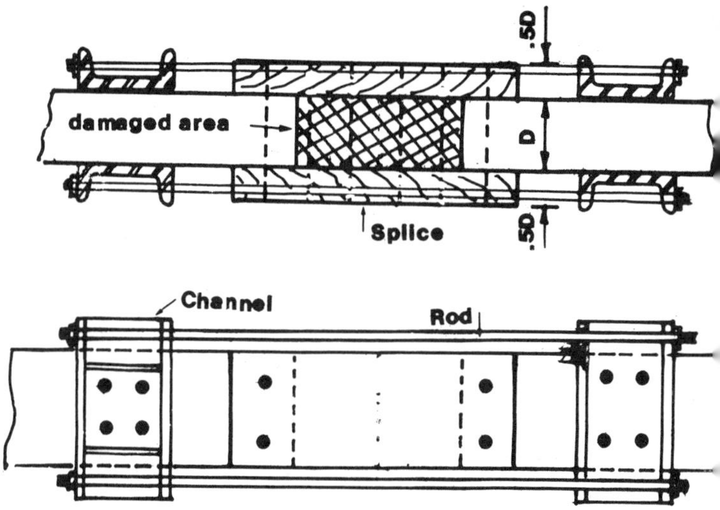


Fig. 8.5 Splice

The tension splice should have one contact surface on each side of the joint which could be tightened up. This splice detail should be designed to meet the following:

1) The flange of the channel should have the capacity to carry the bearing pressure, otherwise it must provided with a bearing stiffener.
2) The channel should be designed to withstand the bending force from the compression face of the timber to the tension rod.

**Protective Armors:**

The metallic or concrete armor jacket usually in sheeted form could be used as an effective protection against marine bore attack. The armor should not be punctured and should extend from the highwater elevation to far below the stream bed. Since this method is expensive, it is not widely used.

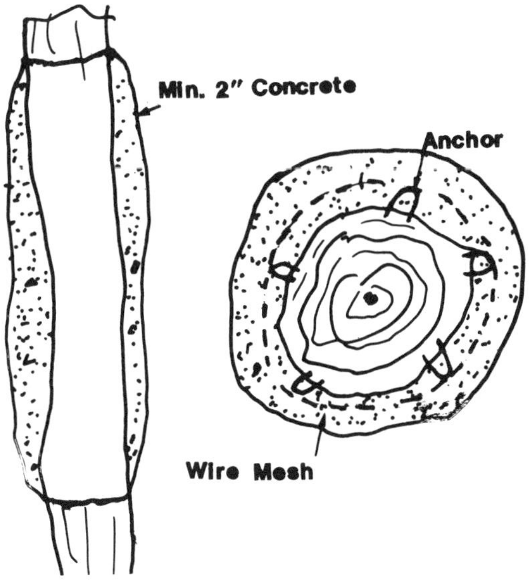


Fig. 8.6 Concrete Encasement

**Post-Tensioning:**

or longitudinally laminated wood deck, a transverse post-tesioning system will develop friction in the deck to promote load sharing of the laminated deck. It will close the deck tight and result in reducing the deflection and preventing passage of moisture.

After removal of asphalt wearing surface, a epoxy coated high strength bar is installed at top and below the deck at each post-stressing point along the bridge. The bar is enclosed in tight fitting PVC pipe for additional protection. The PVC pipe is connected steel sleeve of the steel anchorage plate. Post-tensioning bars can be made by using two jacks at each point to provide uniform contact pressure between steel and wood. After post-tensioning, the deck should be resurfaced.

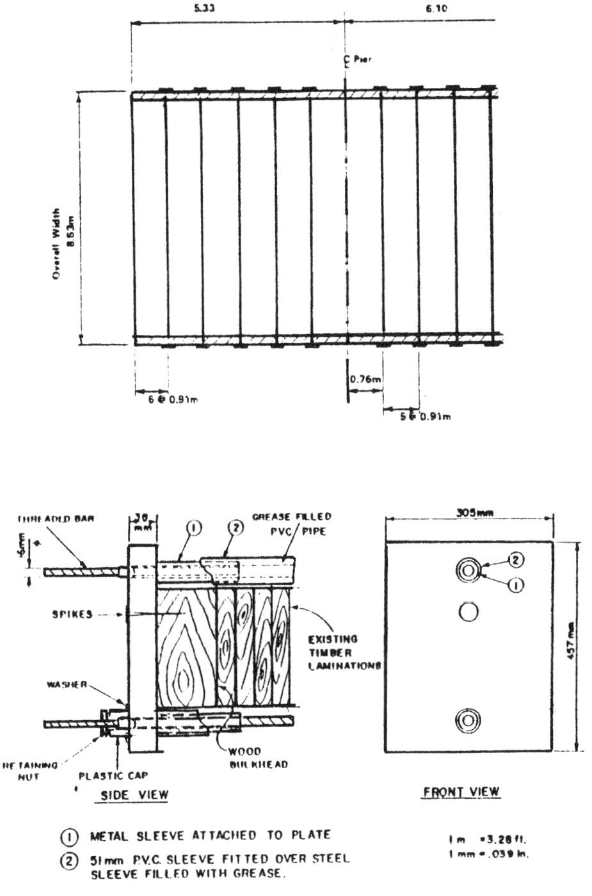


Ontario

Fig. 8.7 Post-Tensioning

**Composite Concrete Deck:**

The composite deck requires shear connectors and concrete or reinforced concrete wearing surface (minimum 3 1/2" thick concrete with minimum #3 bars spaced at 12"). The degree of composite action depends on the shape and size of connectors which is connected with the existing wood deck. Composite deck would add extra dead loads on the structure but it will improve the load distribution and overall load carrying capacity.

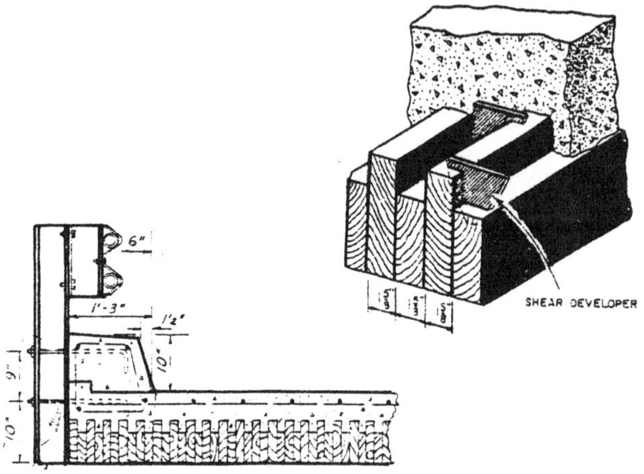


Fig. 8.8 Composite Deck

**Preservative:**

Remedial treatment for the decay in the timber with internally applied non volatile, liquid preservatives such as Vapam (sodium Methyl Dithiocarbamate) and Chloropicrin (Trichloro-nitro-methane) are effective. It is suggested that additional holes be drilled systematically through the decay and beyond the damaged areas so as to assure adequate spread of the preservative into the incipient decay areas.

In marine environment, soluble, insoluble salts, creosote oil and creosote coal tar solution injected preservative have been used to protect against marine borers. The piles in the marine environments can be treated as follows:

For areas where the toredo(shipworm) and pholad attacks are known and where the limnoria tripunotata attacks are not prevalent, creosote or creosote-coal tar solution treatment will probably provide an adequate protection.

For areas where toredo and limnoria tripunctata attacks are known and where pholad attack is infrequent, high retentions of chromated copper arsenate (CCA) treatment will provide an adequate protection.

For areas where limnoria tripunctata and pholad attack are known or expected, a dual treatment (creosote and CCA) provides a maximum protection.

## 8-2.4 PREVENTIVE MAINTENANCE

**Field Preservative Treatment:**

The field treatment of the existing highway bridge would extend the life of the structure. Treatment should be applied at regular intervals to the major structural component such as stringers, piles, caps and the exposed surface of the untreated part of the timber with collosion damage, delamination, check and split to protect them from infection. Also, the localized areas should be treated to prevent decay from excessive moisture accumulation in the untreated parts of the timber.

Liquid preservatives are applied with hand brush and sprayer and the heavier grease or emulsion may be applied with proper equipment to the existing member during rehabilitation.

**Moisture Control:**

Moisture control is required at location where the timber is subjected to frequent decay causing wetting. Proper and good bridge surface drainage is the key point for the maintenance practice. A plugged drain will cause the spore-infected water to leak through the deck and the joint between the stringers and the top of the caps or leak into the laminated deck. Also, improper location of the drainage can cause certain part of the timber structure to be vulnerable to infection.

The waterway must be cleared of debris, because the accumulation of debris against the piles will cause not only collision damage but also promote decay. Protection by metal or concrete jacketing is necessary.

## 8–3 NEW TIMBER BRIDGE

In designing new bridge the following should be considered:

**Wearing Surface:** Asphalt wearing surface is generally used for smooth ride. If asphalt wearing surface is used, the gluelaminated deck should be assumed to be in wet condition.

**Deck:** Gluelaminated panel is preferred to plank-nail system. The panel can be fastened to stringer by lag bolts, ring shank nails or steel clips attached to the stringers depending on the type of stringers. Sometimes horizontal steel dowels are used to transfer force between the panel. The holes for steel dowels for panel should be about 1/16" larger than dowels and the dowel should be galvanized.

**Stringer:** Gluelaminated stringer or steel rolled beam is used.

**Substructure:** Pile bent system or mudsill and post system can be used in place of masonry type substructure.

**Preservative Treatment:** Preservative Treatment should meet AASHTO Section 2.21. For douglas fir and larch, the use of incising is recommended.

### REFERENCE:


1. Clark, W, Wallace E. Eslyn "Decay in Wood Bridge" Forest Product Laboratory, Forest Service, U.S. Dept. of Agriculture.
2. McGee, Dennis "The Timber Bridge Inspection Program in Washington State" FHWA Olympia, Washington Jan. 1975
3. "Timber Structure Standard Plans" FHWA
4. AREA Manual, Chapter 7 Timber Structures, American Railway Engineering Association
5. "Standard Specification for Highway Bridges" 1977 and interims, AASHTO
6. "Glulam Wood Bridge System" Technical Manual, Weyerhaeuser, 1980
7. "Manual for Bridge Maintenance Planning and Repair Methods" Florida Dept. of Transportation, State Maintenance Office
8. "Standard Bridge Plan" West Virginia Dept. of Highways
9. Johnson, Sidney M. "Deterioration, Maintenance, and Repair of Structure" McGraw-Hill NY, NY
10. Ayers, James "Timber in Marine Structure" ASCE Journal of Structural Engineering Division, April 1967, pp 33
11. "Bridges on Secondary Highways and Local Roads Rehabilitation and Replacement" NCHRP Report 222, TRB, May 1980
12. "Rehabilitation and Replacement of Bridges on Secondary Highways and Local Roads" NCHRP Report 243, TRB, Dec. 1981
13. "Glulam Bridge Systems Plans & Details" The American Institute of Timber Construction, 1975
14. "Pole Inspection and Maintenance" REA Bulletin 161-4, Rural Electrification Administration, U.S. Dept. of Agriculture
15. Graham, Robert D., Guy G. Helsing "Wood Pole Maintenance Manual" Forest Research Laboratory, Oregon State University, Corvallis, Oregon 97331 Feb., 1979
16. Sprinkel, Michael M. "Glulam Timber Deck Bridge" FHWA, May 1979
17. "Transversely Post-Tensioning of Longitudinally Laminated Timber Bridge Decks" Ontario Ministry of Transportation and Communications.
18. Bruesch L. "Forest Service Timber Bridge Specification" ASCE Journal of Structural Division Vol. 108, No. ST 12 December 1982, pp 2737-2746.

# 9 BEARINGS

## 9—1 GENERAL

Bearings are designed to support the superstructure, and to carry all forces from it to the substructure. They provide free movement of the girders. Bearings should be cleared of debris and dust to prevent corrosive "freezing". The expansive movement of the bridge can be accommodated by flexible piers, rockers, sliding plates, rollers, elastomeric pads, and pot bearings.

### 9-1.1 FORCES

The following forces should be considered in designing and analyzing the bearings.

**1. Vertical Reaction**

Vertical reaction is assumed to act directly at the center of the bearing pins. It is generally consists of dead load and live loads.

**2. Transverse Force**

This force can be considered to act normal to the centerline of the bridge in a horizontal direction at the top of the bearing. It is consists of wind, earthquake and other horizontal forces, and should be resisted by keys, anchor bolts, and pintles. Sometimes, shear block can be used to resist transverse force, in this case the bearing itself does not need to resist the force. The transverse force will develop a moment within the bearing itself which is the product of the force times the height of the bearing. This moment can be significant for tall bearings and should be considered in the analysis.

**3. Longitudinal Force**

Longitudinal force acts parallel to the centerline of the bridge and includes traction force, thermal forces and forces due to concrete shrinkage. Generally longitudinal force will not be developed in an expansion bearing but a large longitudinal force may be developed due to friction and shear deformation in elastomeric and sliding bearings.

**4. Uplift Force**

On curved structure, the uplift force should be checked at the bearings. In earthquake zone, the bearing are generally designed for uplift forces due to earthquake in an amount equal to 10% of the vertical superstructure dead load reaction.

**5. Other Forces** such as torsional and combination of transverse and longitudinal forces should be included in the design when they are significant.

### 9-1.2 MOVEMENT

Allowance should be provided in the bridge for all anticipated movements. Usually these movement will be in the longitudinal direction. For extremely wide bridge and severely skewed bridge, transverse movement can be significant. The following should be considered in design.

**1. Temperature**

Expansion and contraction due to temperature changes will happen throughout the life of the structure. Proper expansion provisions are necessary to assure that the bridge will not be damaged by temperature movement. Where these movements are constrained due to poor design or construction, high force may affect other portions of the structure. It is important for setting bearings at proper temperature suitable at certain region. A normal temperature specified in the specification means that at that temperature, all plan dimensions are taken to be correct and the bearing can be set without adjustment.

As a rule of thumb, the expansion in cold regions is at rate of 1 1/4 inches per 100 foot spans. Care should be exercised that adequate provision is allowed for expansion movement.

**2. Shrinkage**

All concrete tends to shrink during curing unless retards are used. This shrinkage may need to be provided in designing the bearings. If the computed movements are significant, bearing for concrete bridge except elastomeric bearings should be set in the opposite direction to the anticipated shrinkage to be in the normal position after the shrinkage has taken full effect. Such adjustment should be shown on the plans.

**3. Creep**

In certain structures creep associated with applied loads should be taken into account in the bearing details. This is particularly true for post-tensioned bridges where the prestressing force will cause an immediate elastic shortening of the structure and associated long term creep effect. Long term creep effect should be considered for severely skewed through girder bridge. On unusual bridges, this effect could result from dead load sidesway forces. Bearings should be designed and set to account for this effect.

**4. Earth Pressure and Settlement**

Unexpected soil pressure behind the abutment can result in tilted bearings, closed expansion deck joints, and jamming joint openings. Where it is anticipated tilting and horizontal movement, bearings should be designed to accommodate such movement so that they can be adjusted or reset in position, if needed. Similar provision should be made where high soil settlement is anticipated. Consideration should be given to providing jacking pads to minimize the additional cost for future bearing adjustments. In any case, bearings should be designed to handle these movement without resulting any damages to the structure.

**5. Combination of Force and Movement**

Generally the bearing is designed in the normal position for dead load and live load. An analysis of bearing elements at high and low temperature position should be made. Forces in longitudinal direction can be assumed to act through the center of the pin and the moment applied to the base plate is the result of horizontal force times the height of the bearing. The transverse force can be assumed to act along the axis of the pin. The base plate of the bearing should be able to transmit all the horizontal forces to the concrete by the anchor bolts, shear lugs or other devices.

**6. Orientation of Bearings**

Bearings should be aligned to the actual direction of movement of the bridge. On curved and skewed bridges, care should be taken so that the details show the bearings to be set relatively to the actual axis of movement. On curved bridge, the axis generally corresponds with a chord between the end of the span.

## 9—2 BEARING TYPES

### 9-2.1 TYPES OF BEARING

Bridge bearings are generally either fixed or expansion type. The fixed bearings does not allow longitudinal movement but act as a pin or hinge to permit rotational movement. Also, the fixed bearings restrain the superstructure from longitudinally moving away from the bearing seats.

The expansion bearings provide for rotational movement of the girders due to the deflection and longitudinal movement for the expansion and contraction of the spans. If an expansion bearing develops a large resistance to the longitudinal movement due to corrosion or other causes, this frictional force opposes the expansion or contraction of the span, thus creating another force within the span which can lead to repair problems.

Both the expansion and fixed bearings transfer lateral forces such as wind and centrifugal loading from the superstructure to the substructure units. Both types of bearings are set parallel to the direction of the structural movement except when the bearings are placed under flared girders.

### 9-2.2 SLIDING PLATE TYPE

The sliding surface of the steel bearing plate usually freezes due to dirt and corrosion. This type of steel bearing plate was used for short spans. Another type is a bronze plate which is used between the sole and masonry plate. However, this also causes dirt and dust to get in between the plates and the sliding surface to become gall, increase friction and cause freezing.

Fig. 9.1 Critical Sliding Bearing

If the surface of the sliding plate is pitted, the plate should be removed, ground, or milled smooth and reinstalled or it should be replaced with a new plate.

### 9-2.3 ROCKER BEARINGS

Rocker bearings are used to allow the end of the bridge to move longitudinally along a horizontal line and are used as movable bearings for supporting heavy loads. Pintles are used to prevent walking of the rocker bearings from the base plate and to resist the transverse forces. The pin functions as an hinge to allow movements of the expansion bearing and rotation at fixed and expansion bearings.

Rocker type bearings should have uniform load distribution to the bearing seat. and the pins should carry uniform loadings throughout its length. Also the keeper plates or the side bars may be used to prevent lateral movements due to creep, skew effect and lateral forces.

Fig. 9.2 Tilted Pinned Rocker

The main problem with this type of bearing is that it allows small dirt particle to get in which will jam up the rollers and also will allow the moisture to be trapped, thus resulting in a freeze of the bearings.

### 9-2.4 ROLLER BEARINGS

Fig. 9.3 Frozen Roller Bearing

Roller bearings are similar to rocker bearings. They have rollers with a small radius. These bearings are used for moderate loads. The nest type (multiple roller in one bearing) of roller bearings are not recommended because of their susceptibility to corrosion and freezing, and difficulties of maintenance.

The roller nests collect dirt easily and readily become rusted. The large rolling friction causes horizontal pull on the abutment and increases the tendency for the abutment to tip over. The roller in such case should be replaced with rocker type bearings.

### 9-2.5 ELASTOMERIC BEARING

These bearings have few moving parts and they rely on the elastic property of neoprene in shear to allow them to move in any horizontal direction. Also they can accommodate longitudinal and as well as transverse rotations. All anticipated movements are provided by deformation of the elastomer within bearings. These are commonly used for prestressed concrete bridges, as well as on other types of bridges. For long bridges, the basic elastomeric bearing is also supplemented with sliding surfaces similar to those which are incorporated in other type of compound bearings.

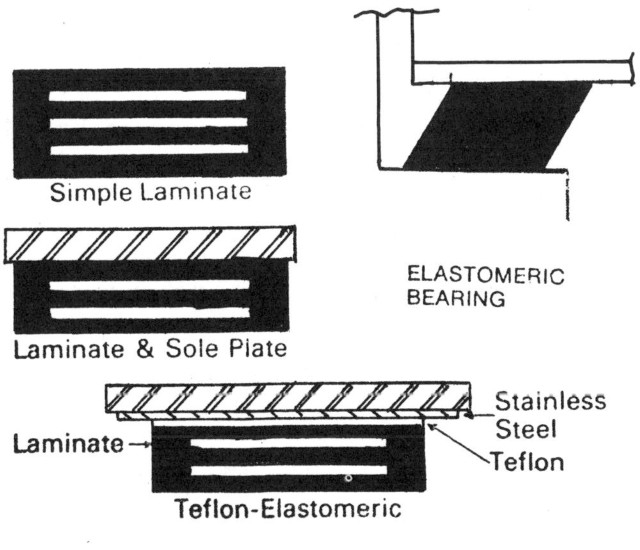


**Fig. 9.4 Elastomeric Bearings**

When these bearings are used in combination with a Teflon (TFE) sliding surface, they allow bridges to move by sliding the self lubricating tetra-fluro-ethylene (TFE) surface over the smooth, hard surface of stainless steel or other corrosion resistant metal.

Deterioration of the elastomeric material generally can be noticed by cracks, splits or tears, bulging and distortion. Initial evidence of the deterioration will be horizontal cracks at the junction between the rubber and steel laminate. Excessive rotation will show uneven thickness of the pad.

The deterioration of the elastomeric bearings are attributed to inadequate dimension, improper reinforcing and faulty fabrication. Because the elastomer stiffens at low temperature, the material also should be suitable for particular environment. For general use, both natural rubber and neoprene appear to be functioning satisfactory. Newly installed bearings seem to have less of these problems.

TFE (tetra-fluro-ethylene) has been used for sliding bearing in combination with elastomeric bearing. The advantages of using TFE are its non-sticking surface, low friction coefficiency, and corrosion resistance when used with stainless steel.

One research report concluded that the unfilled TFE surfaced bearing performed better than the graphite-impregnated self-lubricating bronze type sliding bearing. It is suggested that pure unfilled TFE material with 1 1/2 inch minimum backing material supported with a steel laminate over TFE material reinforced with glass fiber filler be used. For soft elastomeric material (Duro 50 to 60) a restraining device is recommended to restrict the horizontal shear when the anticipated movement exceeds 50% of the total thickness of neoprene.

Since many combination of TFE and backing materials are commercially available, and their performance is substantially different, the selection of a proper bearing will be difficult for bridge engineers.

## 9-2.6 POT BEARING

In 1957, Professor Fritz Leonhardt of Stuttgart University of Germany developed pot bearings and now they are widely used throughout the world. The pot bearing is used as an alternate to the heavy steel bearing. The use of this type of bearing is limited to the heavy girder or main girders of the through girder systems because of its high cost, even though it is efficient and compact compared to other bearings.

The pot bearing consists of a circular neoprene pad or a teflon coating which is enclosed by a steel pad to prevent it from bulging. Neoprene pad acts as a fluid on the bowl under high pressure, providing the required rotation with low friction coefficiency and point-contact effect of a rocker bearing at the center pressure point.

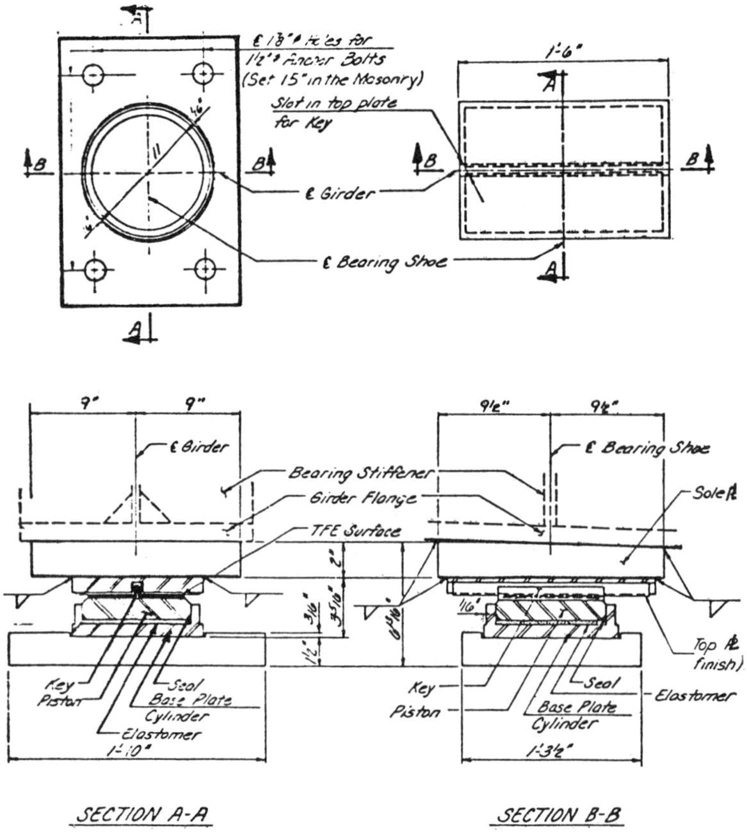


Fig. 7.5 Pot Bearing

Pot bearings are normally designed for an eccentricity of loading of 2 to 10% when fully rotated. Due to its uniform stress distribution, this bearing can use teflon coating which otherwise is subjected to creep under high edge stresses. Also, this uniform stress permits to increase the bearing stress of concrete bridge seat and results in a smaller masonry plate.

The following is the memorandum of June 2, 1982 from FHWA Region 10 concerning the pot bearings:

> With pot bearings gaining in use, it is becoming increasingly obvious that details of the bearing construction are a key feature for satisfactory performance. On several recent bearing installations, it has been found that elastomer has extruded out between the piston and cylinder. The cause of this is attributed to excessive clearance between the piston and cylinder and/or plastic deformation of a non-metallic sealing ring. Also, in some cases the extruded material contained fragments of the elastomer, which is believed to be the result of galling of the elastomer surface as the bearing moved through the piston-cylinder clearance.
>
> An evaluation of the bearing details and comparison of the problem bearings with similar ones that are performing satisfactorily has led to the conclusion that the extrusion problem can be overcome by limiting the piston-cylinder clearance to 0.03 inch and using a metallic (commonly brass) sealing ring. It is also believed that the galling problem can be solved by lubricating the top and bottom surface of the elastomer. One method of lubrication that seems to be successful is a polytetrafluoroethylene (TFE) sheet at the top and bottom of the elastomer.
>
> We recommend that these details be considered when specifying pot bearings.

## 9-2.7 SPHERICAL BEARING

Before development of pot bearings, spherical bearing had been used on the top or bottom of a column of steel bent piers to avoid the transmittal of bending moments into the columns and to provide movements. The spherical bearing is either a half sphere or a ball. It is susceptible to freeze and is difficult to erect. They can be used on curved and skewed bridges.

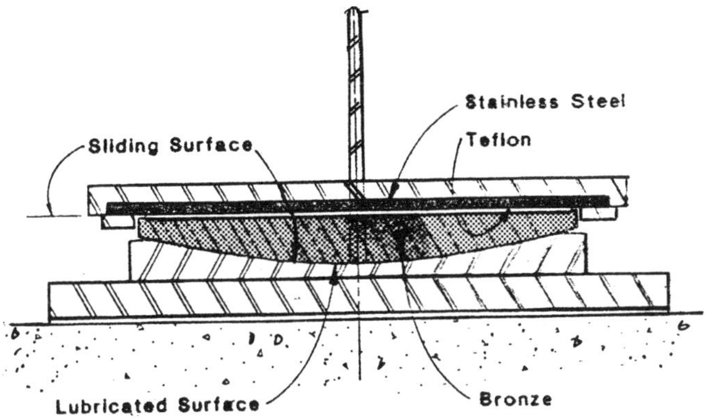


**Fig. 9.6 Spherical Bearing**

Modern spherical bearing usually has two sliding surfaces; a curved or spherical lower surface to provide for the transverse and longitudinal rotations and an upper flat surface to provide horizontal movement.

Regarding the sliding surface, spherical bearings were designed with two different types of surfaces. The spherical surfaces were lubricated bronze or stainless steel with TFE fabric, the upper flat surfaces were faced with stainless steel and teflon (TFE).

# 9–3 PROBLEMS AND REPAIRS

Bridge bearings are designed to transmit loads to the substructure and to permit a certain amount of movement. If bearings are not able to accommodate movements of the bridge, the excessive friction forces placed upon the bearings will cause damage to the girders or the bridge seat.

Jammed or closed expansion devices can cause movement of the abutment toward the end of the girders and this will reduce or eliminate the capacity of the bearing. Because of bearing and joint problems, many bridge designers tend to use flexible thin piers so that the entire bridge could take care of the movement without the joint and bearings.

## 9-3.1 RUSTING

1. Steel bearing plates are susceptible to corrosion-rusting. Most deterioration is caused by joint failure, high water elevation, environmental effects and moisture or corrosive acids. In most cases, the "frozen" joints are further evidenced by bending, buckling, or improper alignment of the members and cracks at the bridge seat.

**Fig. 9.7 Frozen Bearing**

2. Rusting or corrosion is usually caused by dirt and debris accumulation. Debris has the tendency to absorb and hold moisture and corrosive material close to the bearing surface and this hastens the corrosive action.

**Fig. 9.8 Cleaning Roller Bearing**

3. For expansion connections between girders where no moments are to be transferred across the joints, hanger bars are used for steel girder. There are many parts in the hanger and connection pin assembly which can be corroded and cause total failure of the non-redundant through girder bridge. Generally access to the pin hanger assembly is difficult for maintenance and inspection.

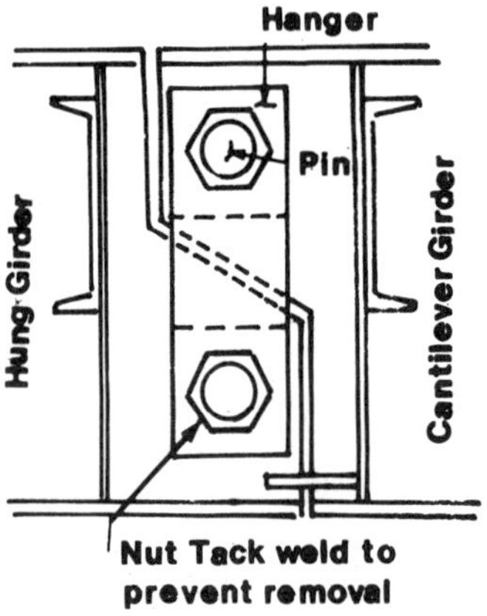


Fig. 9.9 Pin-Hanger Connection

**Repairs:**

After the bearings are cleaned, lubricant should be applied. If the bearings are jammed up critically, they should be repaired while the stringers are jacked up.

Fig. 9.10 Critical Bearing

## 9-3.2 ALIGNMENT

1. Each bearing is designed to perform a specific function at the bearing point. Movement of the bridge or span at the bearing point will either be rotational, longitudinal or both. Proper alignment of the bearing allows these movements with minimum resistance. If the bearings are not properly aligned for normal movements, stresses will occur at the support points, which in turn will be transmitted to other members of the bridge, causing damage.
2. The movement of the bridge span for a long period of time may cause parts of bearings to slip out of place. While transverse movement is normally restricted, constant vibration and repeated live loadings may cause bearing pads or bearing plates to move.
3. A severely skewed thru-girder or curved girder bridge can cause transverse movement of the bearings due to geometrics and concrete deck creep. Keeper plates and shear key may prevent this transverse movement problem.

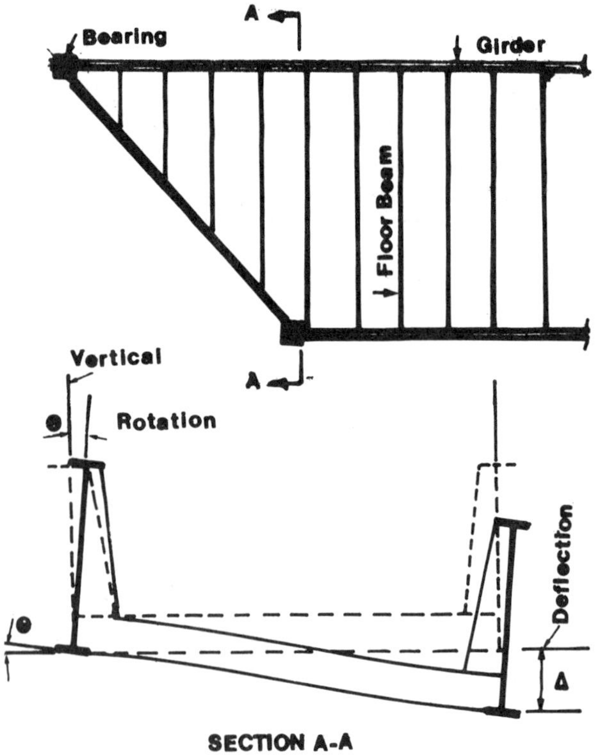


Fig. 9.11 Actual Structural Behavior on Skewed Bridge.

Another problem is lateral rotation along with normal longitudianl rotations. The combination of rotations cause abnormal rotation. The traditional bearings, rockers, rollers, sliding bearings and pinned bolsters are designed to accommodate a single member deformation and rotation of only the axis in the vertical plane and not designed for the entire structure as a unit or multiple directional rotations. The oversight of the actual bridge behaviors due to geometry in the heavy skewed or sharp curved bridges and the use of traditional bearings cause distress and failure of function and durability of the bearings. It is recommended to use spherical, pot, or elastomeric bearings to accommodate abnormal rotations and transverse movement. On curved girders, the expansion movement is rather parallel to a chord between the bearings than on a tangent to the curve therefore it is important to set the bearings on the chord line.

**Repairs:**

While the girder is supported by temporary supports, the masonry plate should be rearranged to make the bearing plumb after cutting off anchor bolts as necessary. Then the holes for new anchor bolts should be drilled and the swedge type of full length thread anchor bolts must be placed before inserting non-shrink grout into the holes.

For minor misalignment, a slotted or oversized hole in the sole plates or flange of beams may be drilled, and the alignment is adjusted between the sole plate and the bearings.

For severe tilt or critical bearing movements, instead of resetting, replacing of the damaged bearings with a more proper type should be considered such as pot or elastomeric bearings.

## 9-3.3 GIRDER BEARING

Girder bearings should be kept free of dirt, debris, and moisture. Bearings which are designed to be lubricated to prevent freezing should be greased regularly. However, the grease tends to attract dirt and dust which causes freezing. Bearings and supports under the deck joint and drainage should be spot cleaned and painted annually between the over-all structural steel paintings.

## 9-3.4 BEARING SURFACE

The full bearing surface should be in contact. If the surface contact is partial, damage can occur to the span, support and bearing. Where a girder span may have moved horizontally and the sole plate rests on only a portion of the masonry plate, the load transmitted to the abutment may exceed the bearing value of the masonry used in the abutment and may result in crushing of the support. Conversely, buckling in the web of the girder will occur if the load is supported by only a portion of the bearing surface. Where uplift exists, there may be gaps between the bearing and the superstructure. It is recommended that the bearings with hold-down devices be installed.

## 9-3.5 LOOSENESS IN BEARING

Looseness in a bearing can be detected by the noise made at the bearing when the bridge is subjected to live loads such as traffic or movements of the bearing parts.

Fig. 9.12 After jacking the girder, sliding plate is replaced

The following may be the causes for the looseness:

- Settlement or movement of the bearing support away from the portion of the bridge seat.
- Excessive rusting and corrosion which causes a material loss in the bearing itself.
- Excessive deflection or vibration in the bridge span members.
- Loose fasteners such as the rivets or bolts used to attach the bearing to either the bridge span or the bridge support.
- Worn bearing members such as the pins or pin shoes. Distress in elastomeric bearings will normally be noted by excessive and uneven bulging.

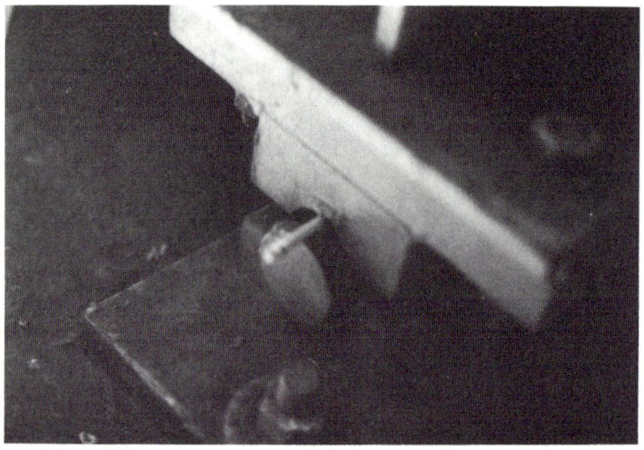

Fig. 9.13 Loose Bearing

## 9-3.6 BEARING SEAT

The bearing seat should be checked for any cracks or movement due to frozen bearings.

Fig. 9.14 Crack due to Frozen Bearing

Fig. 9.15 Deteriorated Bearing Seat

## 9-3.7 REPAIRS OF BEARINGS

If the bearing is severely tilted and out of alignment, usual maintenance can not provide a sufficient repair. It requires resetting of the bearing by jacking the girders and removing the masonry plate. The existing girder is raised and supported by jacks on suitable falsework, or a method which is capable of both raising and supporting the ends of all girders simultaneously. When a bearing is removed, care should be exercised to avoid any damages within the structure.

If tilts were the results of abutment movement, the abutment weep hole should be cleaned and provided with a drainage pipe behind the abutment backwalls and filled with backwall material. Also, relief joints should be placed adjacent to the abutment to prevent further movements.

## 9-3.8 GUIDELINE FOR NEW BEARING

1. The bridge should be designed with minimum expansion bearings possible by using flexible pier, longer continuous span or modifying abutment so as to limit the expansion movement at the joint.
2. The bearing may fail by one or another reason and need repair or replacement. Therefore it is recommended that provision for jacking should be included in

the construction plans so as to repair or replace them easily.

3. The bearing should be designed as an active moving element with minimum maintenance and easy cleaning.
4. On hinged joint of cantilever span, the expansion joints should have restrainer bolts across the joint so that the expansion movement will be equalized among the bearings instead of one or two bearings.

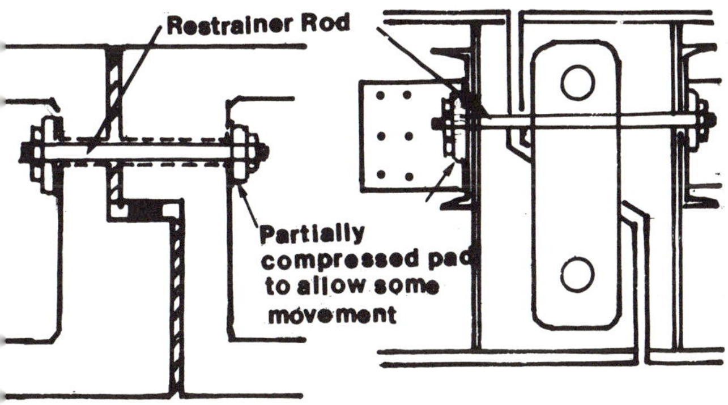


Fig. 9.16 Restrainer

5. In designing elastomeric bearings, special care should be exercised to select proper material. To accommodate the rotation due to beam deflection and grade slope and to avoid significant eccentricity on the bearings, tapered elastomers have been used but it is recommended to use tapered steel plate or curved bearing surface be used in place of tapered elastomers. Also sloped bearing seat may be used instead of tapered bearings.

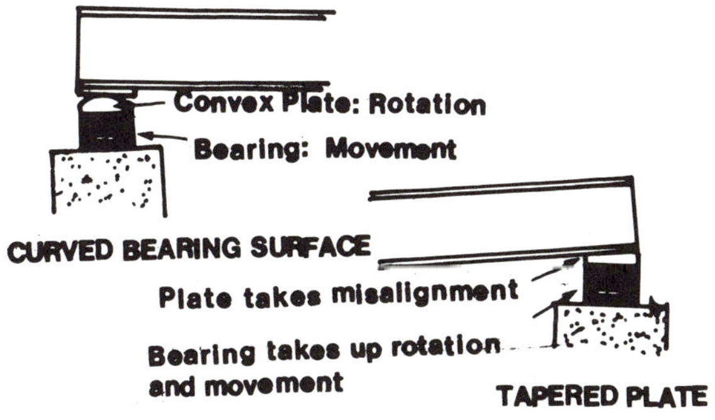


Fig. 9.17 Details

The National Cooperative Highway Research Program Report 248 "Elastomeric Bearings Design, Construction, and Materials" of Transportation Research Board suggested the revision of AASHTO Specifications. They are based on an improved understanding of bearing behavior and are intended to lead to bearing that have adequate and more uniform level of reliability. The major changes lies in the separate treatment of plain pads and reinforced bearings, revised provisions for allowable compressive stress (maximum 800 psi on plain bearing and 1,000 psi on steel reinforced bearing), deflection, anchorage, and the introduction of provisions for rotation and stress of the reinforcement.

6. The following type of bearing should not be used:

- Roller nest
- Sliding bearing with lead sheet in between.
- Bolster shoes pinned through girder web.
- Wheel type bearing with smaller axles.

**REFERENCE:**


1. Transportation Research Board "Bridge Bearings" NCHRP Practice Report 41, 1977
   "Elastomeric Bearings Design, Construction, and Materials" NCHRP Report 248, August, 1982, TRB
2. "Wabo-Fyfe" Watson Bowman Associates, Inc.
3. "Design Manual for Elastomeric Bearings" NY State DOT, March, 1979.
4. "Lubron TF" Allan P. James Co.
5. Con-Serv Inc. Technical Information.
6. "Bridge Manual" Washington State DOT
7. Jacobsen, Floyd K., "TFE Expansion Bearings For Highway Bridges" FHWA-IL-PR-71 Illinois Dept. of Transp. Physical Research Report No. 71, 1977.
8. "Felpro" Felpro Inc.

# 10 JOINTS AND DRAINAGE

## 10—1 JOINTS

### 10-1.1 EXPANSION JOINTS

The expansion joints should be sealed and be able to move freely. Debris, stone or other foreign materials will restrict the joint from moving freely and cause deterioration of the joint filler material which will result in water leaks into the substructure.

In the open joint, the accumulation of debris in the drainage trough causes an overflow and spillage on the bearings and bearing seats. In sealed joint, the sealers prevent water and foreign material from intruding into or leaking through the joint. The deterioration of the sealed expansion joint is caused by a loss of joint material and looseness of the joint. This loss may be caused by tearing of the sealer by tension forces, loss of bond between the sealers, failure of the sealer splice, extrusion of the sealer above deck surface and compression failure of the preformed elastomeric joint materials. The loss of joint material impairs the capability of the joint to accommodate movement and permits the deck drainage water to leak through the joints onto the bearings and the bridge seats.

Fig. 10.1 Accumulate Debris under Open Joint

If the joints have anchorage or armored steel angles, corrosion and traffic impact could cause a loosening of the steel joints and could lead to further deterioration such as cracks, spalls and disintegration of the concrete deck slab in the area adjacent to the joints.

The leakage of the joint, in turn, causes spalls on the bridge seats and "freezing" by corrosion of the bearings. Excessive wear and cracks show deterioration of the materials due to unusual functions of the joints or external forces by creep, severe temperature variation or unanticipated structural movements.

Also corrosion of the sliding plate and finger joints will cause "freezing" of the joints and cause stresses on the other parts of the structure. A bent, cracked or broken finger dam may be caused by insufficient clearance derived from the movements of the foundation and other structure, incorrect initial setting during construction, or inadequate provision for the movements in design.

Differential elevation at the joints will cause additional traffic impact forces on the joints and concrete deck slab and would lead to fracture in the deck slab.

The asphalt wearing surface over the joints may eventually cause cracks. The filling of the joints with a sealant may produce a bulge when they are compressed by the expansion of the slab.

Among the various type of sealer, the elastomeric joint sealer performs well in bridge joint and is moderate in price. Periodic maintenance cleaning and resealing of expansion joint is required for all type of joints and it is cost effective.

## 10-1.2 TYPES OF JOINTS AND PROBLEMS

Problems of Different Types of Joints are as follows:

### 1. Open Joint:

The major problem of the open joint is the debris accumulation on the substructure under the joint and subsequent concrete deterioration on the bearing areas and frozen bearing which will restrict its movements. The joint should be rebuilt with an elastomeric or a gland type strip seal joint with armors after removing portions of the concrete deck slab. Also, a study should be made to eliminate open joints by making a continuous deck slab.

Fig. 10.2 Deterioration of Joints

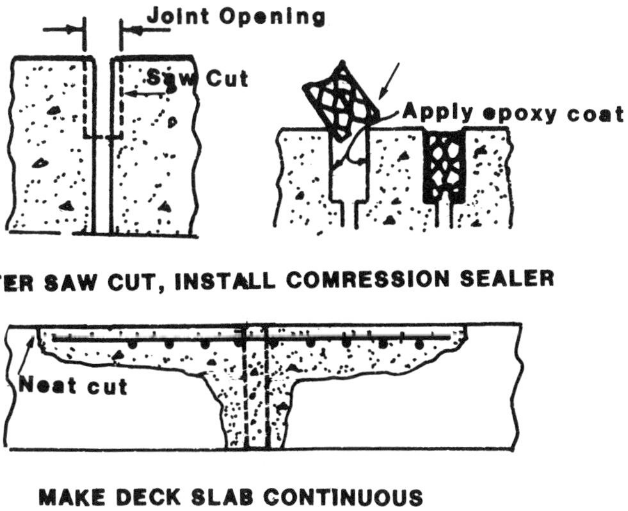


Fig. 10.3 Retrofit of Open Joint

### 2. Sliding Plate Joint:

In addition to the problem of debris accumulation, the plate is susceptible to freezing due to corrosion and it is easily deflected due to insufficient bearing area on the sliding side. Also, fatigue cracks at the connection between the sliding plate and the angle should be checked.

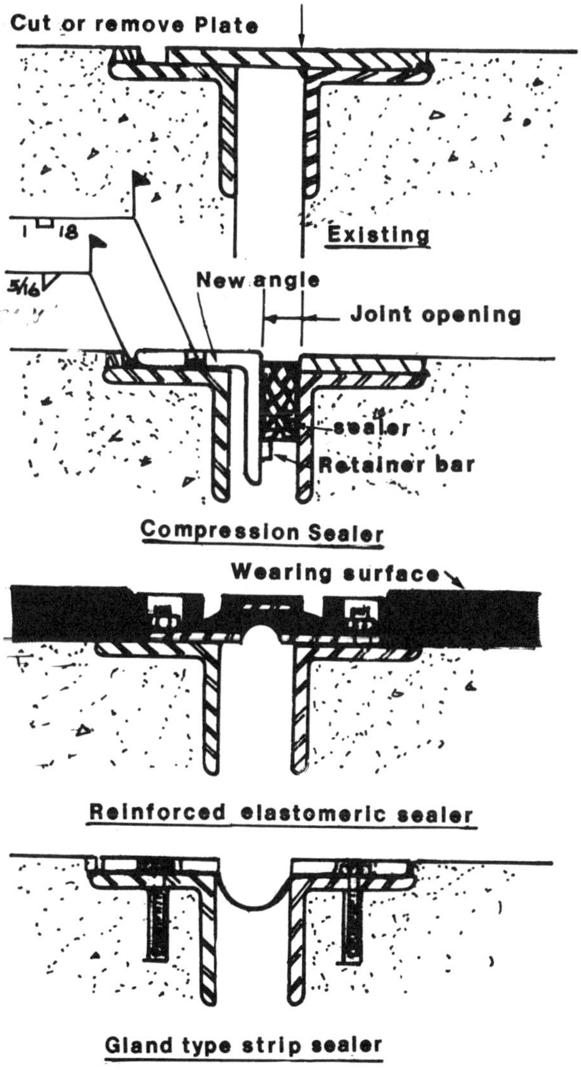


Fig 10.4 Retofit Sliding Plate Joint

### 3. Finger Type Joint:

The common problem with a finger joint is that the trough has a tendency to fill up with dirt and debris and requires frequent maintenance. Also, this type of joint requires a heavy assembly and tight controls during construction. If proper maintenance is neglected, it may lead to the spalling of the bearing seat and frozen bearing from the joint leaks.

The following methods have been used for resurfacing and rehabilitation:

1) Portions of the jammed finger are cut to provide free movement.
2) The anchor bolts are stud welded for installation of the steel reinforced elastomeric joints.
3) The elastomeric joint is continuously placed without any field splice.
4) A new wearing surface is placed.

Fig. 10.5 Finger Joint

To minimize the deterioration, the following is suggested for the design of a joint:

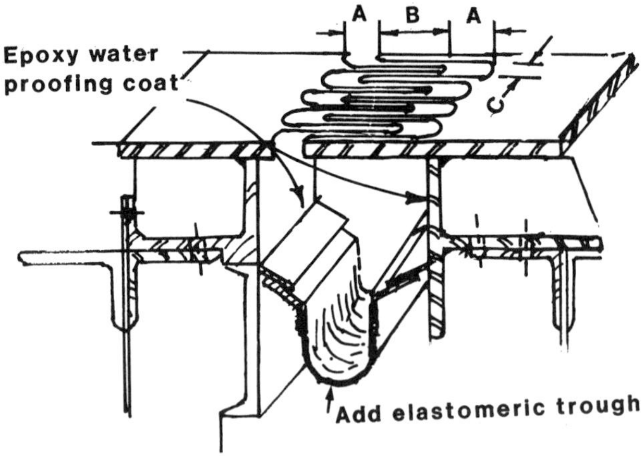


Fig. 10.6 Finger Joint

1) Minimum opening (at maximum design temperature 130 F.) should be 3/4" and at minimum design temperature (-30 F) the tooth overlap (B) should be minimum 1 1/2".
2) If "A" is over 8 inches, "C" should be less than 2" and if A is less than 8", "C" should be 3". This criteria is required for the safe operation of motor bike.
3) Slope of trough should be minimum of 1 inch per foot.

**4. Hot Poured Sealer Joint:**

The main problem of this type of joint is the necessity for frequent maintenance repairs. Also the procedures for cleaning the joint before pouring of the hot sealer and application of asphalt sealer have to be followed properly. Most of the failure occur due to the following:

- Hot poured sealer has either greater or lesser than the designed depth.
- It has irregular depth which causes adverse stresses in the sealer.
- It can also be overheated or underheated.
- Permitting traffic before the hot poured sealer has time to set (the sealer tracks over the bridge deck and dirt and stone gets in the sealer joint).
- Overflow of hot-poured sealer from the tops of the joints (the excess sealer disrupts the planned deck drainage and traps water adjacent to the joint ) can cause failure of the joint.

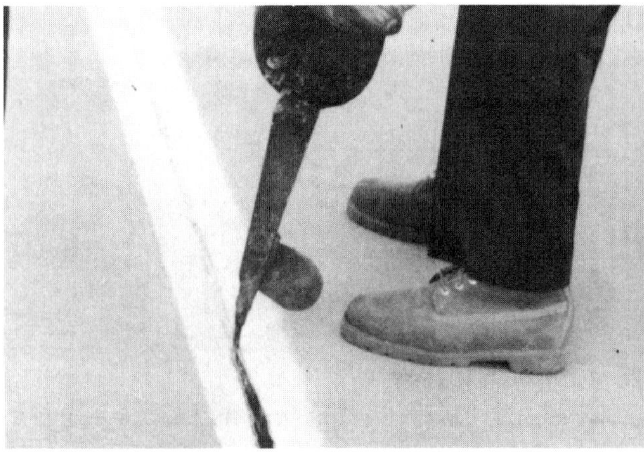

Fig. 10.7 Application of Hot Poured Sealer

The depth of the sealer (usually the upper 1" of the joint) shall be attained and maintained constantly. This depth is critical to the effectiveness of the sealed joint. A joint with the sealer constructed greater than the specified depth tends to fail by losing its bond with the concrete when the cold weather causes the joint to open.

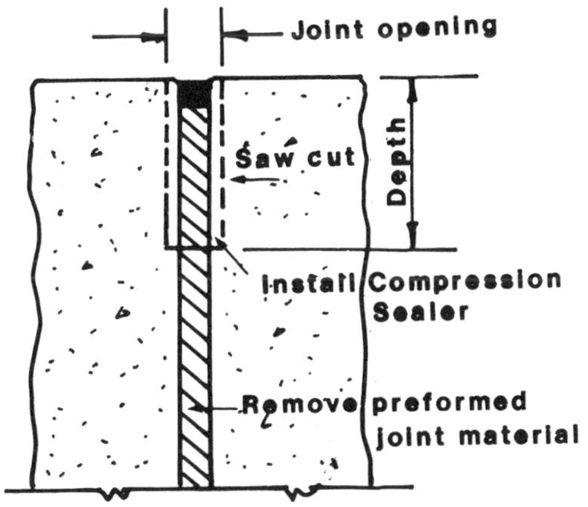


**Fig. 10.8 Retrofit**

**5. Closed Cell Elastomeric Sealer Joint:**

Elastomeric sealers are made of polychloroprene, commonly known as "Neoprene". The sealer manufactured from a vulcanized elastomeric compound using polychloroprene as the base polymer is reported to perform satisfactorily. The preformed sealer must be "functional" as described below:

1) The compression sealer should constantly exert adequate pressure against the joint sides.
2) Adequate width should be provided. If a joint is constructed too wide in warmer weather, the sealer may fail when the joint opens up more when it gets colder.

From the field investigations, most of the preformed elastomeric compression sealers appears to be performing properly in preventing leakages. There are some failures attributable to improper construction practices and impractical and inadequate design. The neoprene compression expansion joint seals is cost effective and functional joint sealer for small joint opening.

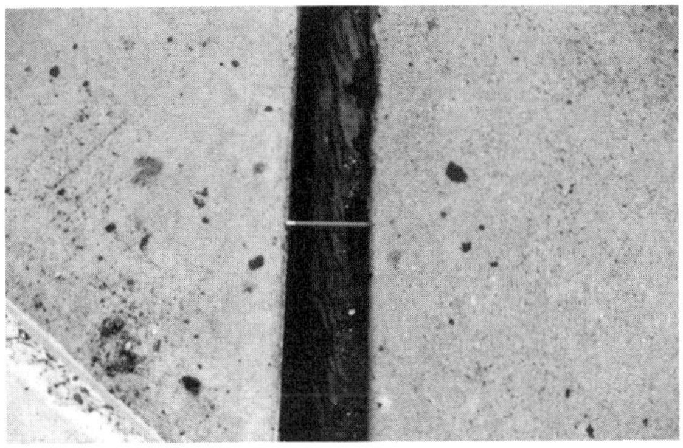

**Fig. 10.9 Failure of Sealer**

The leakage can be found usually at the curb line of the deck of a skewed bridge and splice sections. To improve the joint, the following are suggested:

1) The deck joint should be prefabricated and armored in the shop to ensure it to be water-tight to minimize field installation problems and to prevent exposure of the elastomeric component in direct contact with traffic.
2) The deck joint should be horizontally straight, out to out of the deck.
3) The field splice should not be permitted

Most neoprene joint sealers are acceptable for up to 2 inches of movement. If the movement is over 2 inches and up to 4 inches, or if the joints are on skewed bridges, a strip seal joint should be considered. The major drawback of a steel and neoprene gland type sealer is the high cost. Currently low density closed cell type form is used as an experimental joint and appears to be satisfactory for sealing longitudinal deck joint.

**6. Steel Reinforced Elastomeric Joint:**

New trend in bridge design toward longer spans with minimum deck joints and necessity of improved performance of joints have forced industries to develop various proprietary products.

This type of joint is used where the movement is substantial (over 4 inches). Most leakage occurs at the joint of the butted

**INSTALLING ARMORED JOINT**

**INSTALLING ROYSTON UNIDAM JOINT**

Fig 10.10 Installation of Joints

segment and the curb line at the gutter area of the deck slab, especially on skewed bridges.

Many of these proprietary deck joint systems have not performed satisfactorily. It is apparent that the problems of manufacturing joint assembly, improper installation and poor workmanship contribute to a leakage failure. Joints should be installed as one continuous unit instead of segmental installation to eliminate possible points of leakages.

**7. Integral Abutment:**

The main purpose for constructing integral abutment is to eliminate bridge expansion deck joints, thus reducing the construction and maintenance cost. The integral abutment concept is well received by FHWA. An abutment without joint could cause minor cracks in the approach slab. However, it apparently offsets the joint problems.

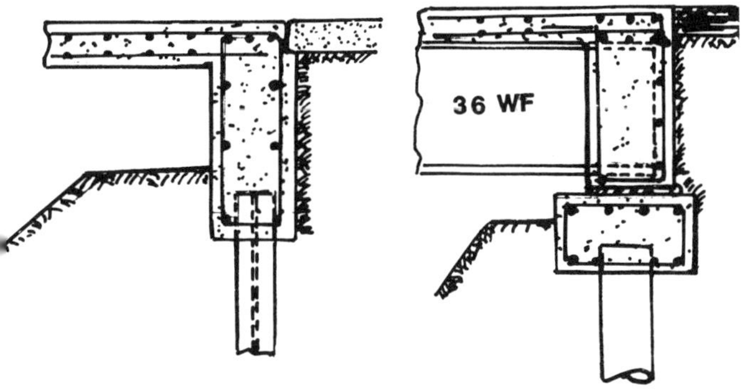


**Fig. 10.11 Integral Abutment**

When the total anticipated movement at the abutment is less than two (2) inches and the abutment is not restrained against movement, no joint is required and the superstructure beam, abutment and the reinforced pavement at the bridge end could be constructed integrally. An unrestrained abutment is one that is fred to rotate such as a stub abutment on row of piles or an abutment hinged at the footing. Continuous steel bridges with integral abutments have performed well for years up to 300 feet in length and with less than 30 degree skew angle.

## 10–2 DRAINAGE

The ponding and overflowing of water will cause damages to the structure and could cause serious traffic hazards during the cold months.

Short span bridges on the vertical curve normally are built without scuppers and water flows along downgrade in the gutter areas and sometimes without any catch basin near the end of the structure. But long span bridges require scuppers to prevent the flooding on the deck slab.

The clogged inlet and pipe is a major problem which causes the overflow to the deck slab. The clogged pipes retain water which eventually freeze and cause breaking of. the pipes. Once the pipes are broken, it becomes a drop drainage system and causes severe erosion underneath. Also, an improper drain outlet will result in the erosion of the fill and embankment or flooding of the roadway under the bridge.

Most of these problems could be eliminated through the following provisions in design:

1. The bridge should have a minimum of 2% cross slope and a minimum of 0.5% longitudinal grade to allow water to run quickly to the drains. Sag vertical curves on the bridge should be avoided.
2. For the short span bridges, a bridge drain may not be necessary, However, a catch basin should be provided at each end of the bridge regardless of the bridge profile. On a steep grade, normally a catch basin is placed at the lower end of the bridge. It is desirable to provide a catch basin near the upper end of the bridge to intercept as much run-off before it reaches the structure.

Fig. 10.12 Additional Catch Basin was added to avoid impounding of water on the deck slab.

3. An open drain free drop pipe through the deck should be a minimum of 4 inches in diameter, 6 feet apart. The inner surface drain hole should be smooth. Special attention should be given to extend the pipe far below the bottom flange of the beams to avoid any undesirable water splash to the structure. Also, free drop pipe should be properly located so the water will not be discharged on the roadway, sidewalk, embankment or any other portions of the structure. If downspouts are used splash block should be placed to minimize erosion.

4. If the inlet box is used, the location, spacing, and its capacity to collect the run-off should be analyzed and it should be designed to allow for bypass and to avoid spread of flow on travel lanes, which will cause a hazard to traffic. The grates are used to trap large debris should be hydraulically efficient and strong enough to carry highway traffic. In urban areas, the grates should be designed to prevent trapping the bicycle tires.

5. The pipe should be minimum of 6 inches in diameter (8 inches preferred) and have a minimum of an 18 inch radius in changing directions. A minimum 2% slope (8% preferred) shall be provided for longitudinal pipes between the scuppers or from scupper to the point of discharge. The cleanout plug and the elbow should be easily accessible and no elbow should be sharper than 45 degrees. The downspout should not be embedded in the pier stems, pier columns, abutments, wingwalls or retaining walls.

6. The most important part of a good drainage system is regular maintenance; removing debris and keeping pipes from clogging.

**REFERENCE:**


1. FHWA Technical Advisory:
   - "Integral, No-joint Structures and required Provision for Movement- T 5140.13" Jan., 28, 1980
   - "Expansion Devices for Bridge - T 5140.15" March 26, 1980
   - "Bridge Deck Joint Rehabilitation (Retrofit) - T 5140.16" March 26, 1980
2. FHWA Notice N-5140.11 and 12 "Development of Watertight Bridge Deck Joint Seals"
3. Kozlov, George "Preformed Elastomeric Joint Sealer for Bridges" FHWA/NJ-80/003 Nov., 1978
4. Azevedo, Vemon Wm., "Evaluation of Watertight Bridge Expansion Joints" Kenturky DOT Research Report UKTRP-81-12, July 1981
5. "Evaluation of Various Bridge Deck Joint Sealing System" Final Report Research Project 72F-128, Michgan Transportation Commission, July 1979
6. "Bridge Deck Joint Sealing System - Evaluation and Performance Specification" NCHRP Report 204, TRB, June 1979
7. "Design Manual Part 4 Structure Design" Penn DOT
8. "Drainage Manual" Manual No.5-294 Minnesota Dept. of Transportation
9. "Bridge Drainage System" NCHRP Practice 67, Dec., 1977

# 11 SUBSTRUCTURE

## 11–1 ABUTMENT

The abutments support the ends of the bridge and retain the earth behind them. The abutment may be a plain mass of concrete, reinforced concrete, or stone masonry. In addition to its own weight, the abutment should carry vertical and horizontal forces from the bridge ends and should withstand the lateral and/or hydraulic pressure from the soil behind the abutment.

### 11-1.1 CRACK IN ABUTMENT WALL

Most of the common cracks in the abutment wall are caused by the following:

**1. Non-Uniform Shrinkage Cracks**

Shrinkage in the abutment walls can be non-uniform because the front surface being exposed, dries out whereas the rear surface is kept moist by the adjacent fill. This shrinkage tends to bring tensile stresses in the breastwall. Most shrinkage cracks are fine cracks and have no significant effect on the structural integrity.

**2. Discrepancy between Design Assumptions and It's Actual Behavior**

In design, the breastwall is assumed to be a cantilever retaining wall and is reinforced accordingly, that is, the main reinforcement is placed near the rear face. However, some old abutment were constructed without vertical expansion joints between the breastwall and wingwall. Therefore, in reality the wall acts partly as a simple vertical cantilever and partly as a box retaining wall braced at the ends by the wingwalls acting as counterforts.

The walls have tensile stresses in the front face midway between the counterforts and therefore should be reinforced accordingly with horizontal bars near the front face. To accommodate the tensile force, most of the reinforcement bars are placed near the vertical face and the minimum number of horizontal bars or no horizontal bars are placed near the front face.

**3. Earth Pressure**

If the wingwalls are designed and built as cantilever retaining walls but are casted integrally with the breastwalls, the whole abutment acts as a unit and the earth pressure on the flare and the parallel wingwalls produces tension in the breastwalls.

Also, an additional hydraulic pressure may build up behind the walls due to an inadequate drainage system or weep holes causing overstresses to the wall reinforcement.

**4. Differential Settlement of Soil**

Damages by differential settlement can be accounted from an opening at the vertical joint between the wingwall and the abutment to a severe tipping and cracks of the wall. If the underlying soil consists of clay, a possible shear failure should be investigated for the stability of the soil beneath the structure before taking any remedial action.

### 11-1.2 CRACK REPAIRS

The selection of proper repairs requires a considerable effort to determine the cause of the crack, its's size, activity, magnitude, direction of it's future movements and possible strengthening requirements.

If the crack is active and moving, a sealer mortar which is flexible and non-brittle should be used. For non-moving crack repairs, epoxy injections, grouting, painting, and shotcrete methods may be used but these methods are not suitable for active cracks.

Most of the common crack repair methods for the non-moving cracks are epoxy pressure grout and injection, and conventional grouting by cleaning, chipping of the concrete along the cracks followed by sealing the crack with grout. In the epoxy injection method, occasional open ports or tubing inserts are placed for gun injection. The gun is placed against one of the inserts and a fast setting epoxy mortar is injected into the cracks using pressure to completely fill and seal the cracks and voids.

If the crack is active, it should be repaired by filling with a strip of premolded asphalt sealer, fast-cure polyurethane type sealant or polysulphide joint sealing compounds. If the deterioration of the wall is extensive, the loose and disintegrated area should be removed and new concrete should be poured.

A proper drainage system in the back fill such as weep holes and drainage pipe with porous backfill material quickly relieves the excess hydraulic pressure behind the backwall when the backfill is saturated. To correct the faulty drainage, all existing drains should be cleaned and new drains may be installed where required.

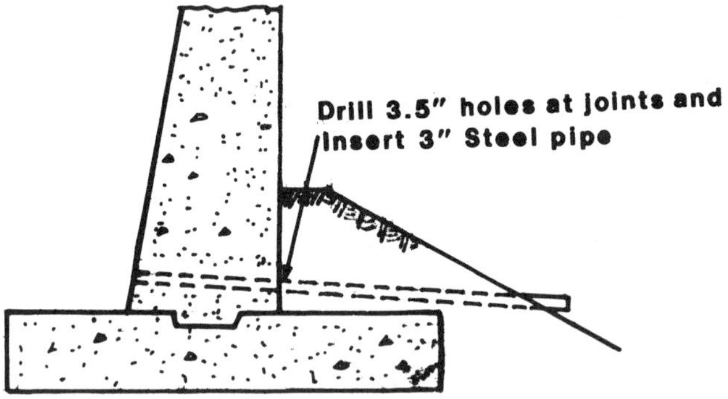


FIG. 11.1 Adding Weep Holes

A method of drilling 3 1/2" diameter holes in the abutment wall at or below the grade and driving a 3" diameter steel pipe into the opening has been used to provide weep holes.

## 11-1.3 BRIDGE SEAT

Bridge seats are built to transmit either vertical and/or horizontal loads with bending moments. They have expansion bearings, fixed bearings, or rigid corner connections to handle design loadings.

However, there are considerable problems concerning the freedom of movement in many types of expansion bearings. Corrosion of the metal and decomposition of other materials may ultimately interfere with the regular sliding action. The expansion bearings become "frozen" and the structure behaves as if the deck has fixed bearings. To avoid these undesirable effects, the bridge seats should be kept free of dirt and moisture.

Fig. 11.2 Spalled Bridge Seat

The most pronounced deterioration is spalls resulting from the joint leak. Therefore, water must be kept from leaking through the joints in the deck to prevent costly repairs. Conditions which give rise to constant seepage through porous concrete can be very destructive, especially in the areas where frequent climatic cycles of freezing and thawing occur.

The local damages may be prevented in the following ways:

1) Stop seepage through joints and cracks.
2) Use dense and nonabsorbent concrete.
3) Apply protective waterproofing seal coating the bridge seat.

Since the top of the bearing seats are generally subject to large amounts of localized pressure under the bearings, the shear cracks or concrete break-off arising from it may be prevented by placing additional reinforcement on the bearing seats.

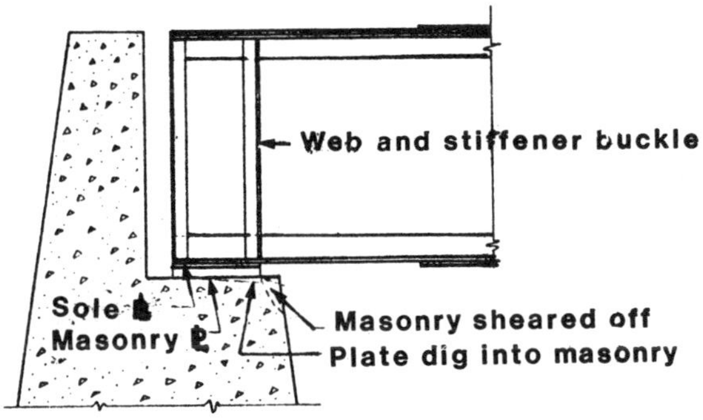


**Fig. 11.3 Crack at Bearing Seat**

Improper locations of the bearing stiffeners may bring problems too. A deflection of the girder pushes most of the load on the outer edge of the bridge seat causing cracks and spalls. To correct this, the bearing stiffener should be relocated or bearing seat should be partially rebuilt by extending the bridge seat.

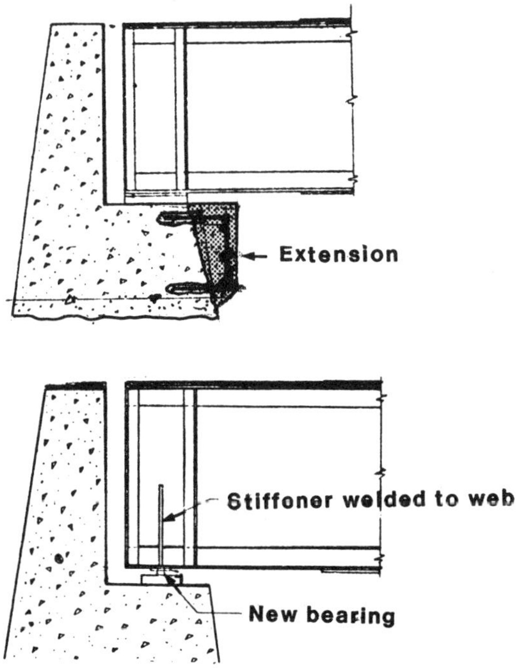


**Fig. 11.4 Retrofit**

Most bearing seat repairs require jacking of the stringers and a rebuilding of the beams.

## 11-1.4 SKEW EFFECTS

Numerous skewed bridges were built in order to be compatible with the approach roadways. Field observation show that a long diagonal skewed bridge deck has a tendency to lengthen during the service life of the bridge. A common problem is the lateral movement of the expansion bearings toward the acute corner of the deck. Cracks near the fixed bearings apparently resulting from the unusual loads caused by temperature variations and diagonal cracks at the acute corner of the deck slab are troublesome problems.

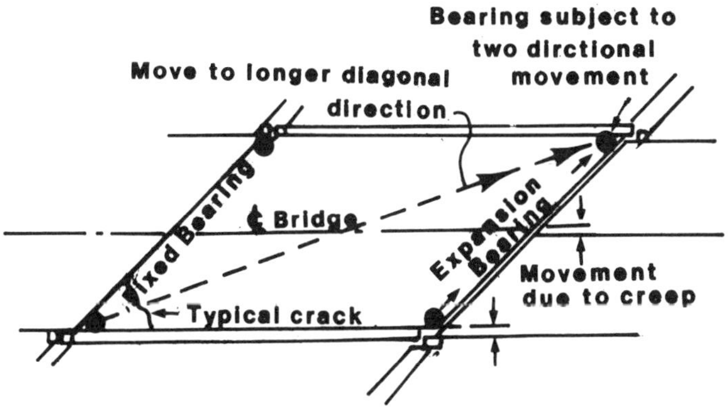


**Fig. 11.5 Movements of Skewed Bridge**

At the abutment, a lateral creep at the expansion bearing can cause damage to the wingwall. Cracks normally start at the edge of the bearing seat, where it is restrained from free movement.

In designing skewed bridges or curved bridges, problems of creeping effect should be considered so as not to permit the wingwalls to be overburdened by a considerable thrust. Damages due to creep may be prevented by omitting expansion bearings from the abutments with bridge span length of up to 45 feet. The fixed bearing will then be able to accommodate the thrust.

The use of expansion bearings in the long span bridges and the detrimental effects of the skew could be avoided by providing a raised concrete block on the bridge seats between girders. It counteracts creeping. Also, a transverse girder stop wall with bearings in the end diaphragm can be placed vertically against the girder stopwall to transmit the thrust while providing a longitudinal movement to the superstructure.

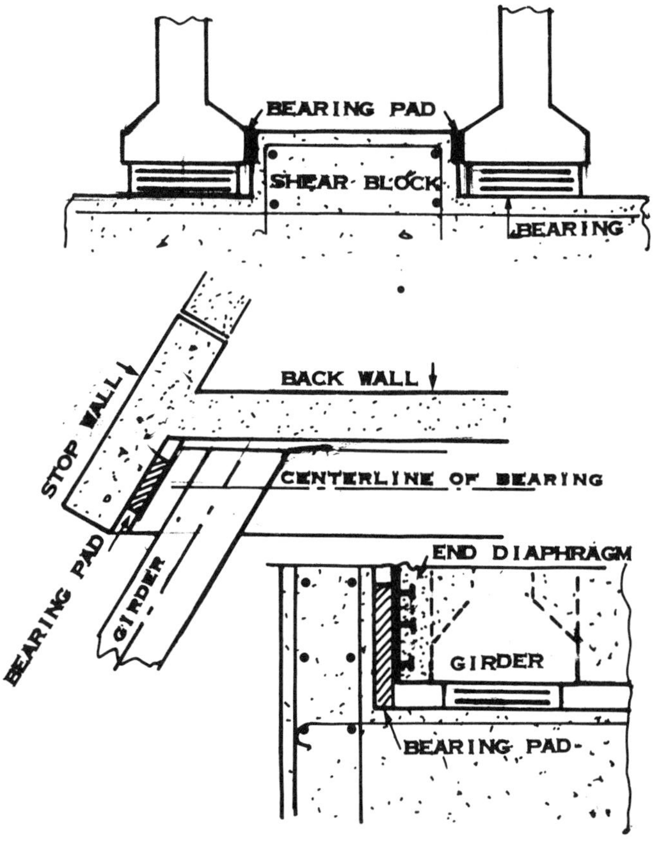


Fig. 11.6 Shear Block

If the expansion bearing is misaligned on a skewed bridge, it should be replaced with an elastomeric bearing with lateral restraint or a spherical type pot bearing. A reinforced concrete guide block may be provided adjacent to the center of the beam.

## 11-1.5 OTHER DEFECTS

Other abutment defects can be caused by the problems in the waterway and foundation outlined below:

1) Waterway
   a. Scour and erosion
   b. Change of stream direction
   c. Riprap failure

2) Foundation
   a. Bearing failure and settlement
   b. Overturning
   c. Sliding
   d. Pile deterioration

Waterway problems can be corrected by studying hydraulics and the following are some remedies for the defects:

1) Place the riprap around abutments and piers to protect against progressive erosion.
2) Install sheet-pile and cutoff walls to prevent erosion.
3) Place spur dikes or jetties for partial diversion of the stream to reduce scour depth with the bridge opening to protect the abutment.
4) Undermined areas should be filled with aggregate and cement by pressure injection or cast-in-place method.
5) Replace the bridge with a greater waterway opening.
6) Relocate the bridge.

Foundation stabilization can be achieved by the following methods:

1) Abutment tiebacks.
2) Soil anchor system.
3) Rebuilding the backwall drainage system.
4) Providing a relief joint for the approach pavement.
5) Soil stabilization by pressure grout.
6) Sheet piling in front of the abutment.

## 11-1.6 RAISING THE BRIDGE SEAT

In order to improve the restrictive vertical clearance, the bridge seat should be raised by reconstructing upper portions of the abutments or placing additional bridge seats on the existing one. The method or its construction are varied by its location, availability of the bridge closing, possible requirement of the deck replacing and other conditions of the structure. The possible occurrence of overstress in the bridge during the operation of jacking should be studied carefully. Jacking should be done uniformly across the bridge to avoid any unbalanced stress in the bridge. If the traffic cannot be detoured during repair as recommended, the bridge should be closed during the jacking operation. The stability of the abutments and wingwalls due to additional horizontal forces resulting from increased heights of the abutment also have to be considered.

The following are some of the techniques for increasing the vertical clearance:

**Jacking:**

The jacking method may vary depending on the structural member involved in jacking, maintenance of traffic, availability of equipment, etc,. The contractor should keep available alternative methods at hand. Jacking may require some modification of the structural members to accommodate jacking forces and also to ensure the structural integrity of the members.

A common method of jacking is constructing a temporary bent for supporting jacks and blocks. Jacking may be supported on the bearings between girders and in doing so the jacking forces could be transmitted to the girders by the reinforced end diaphragm members. The jacking bent may be supported in front of the face of the abutments by the shelve angles fastened into the breast walls.

Fig. 11.7 Jacking Seat

**Reconstruction:**

After jacking and blocking of the girders, the following procedure could be used:

1) Remove concrete pedestal, masonry plates and unsound concrete parts from the bridge seat.
2) Drill holes equal to the diameter of the dowel bars plus 1 inch minimum into the existing concrete and grout dowels for splicing reinforcement bars.
3) Clean the existing concrete surface, then apply epoxy bonding compound for new concrete.
4) Place the new concrete and install masonry plates after the concrete is set.
5) Lower the girders after the concrete is cured.

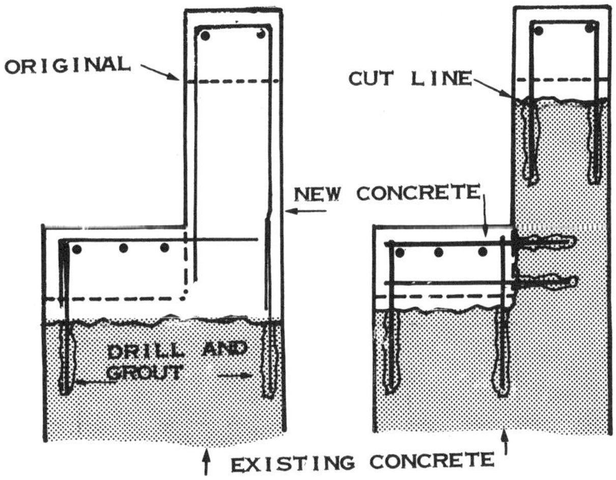


Fig. 11.8 Raising Concrete Seat

**Inserting:**

Prefabricate the new steel pedestal or bolster to support girders. After jacking, place 1/4 inch preformed bearing pad on the bearing area of the new pedestal and insert the prefabricated pedestal in position to receive the girder bearing. Weld or bolt the bracings between the prefabricated pedestals and lower the girders to the pedestal. Complete all welding or bolting after proper field adjustments of the bearings and pedestals before removing the jacks.

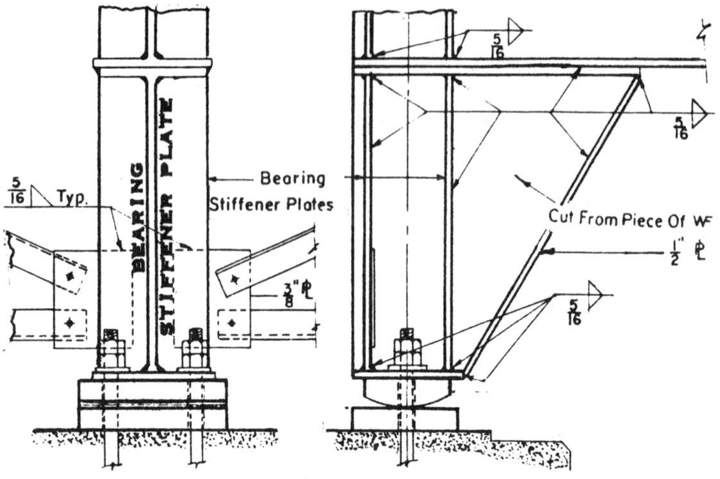


Fig. 11.9 Steel Bolster

11-1.7 TYPES OF ABUTMENT AND INTEGRAL ABUTMENT

**Type of Abutments**

There are several types of abutments based on their shape, function and construction.

**The full height abutment** is a cantilever abutment extending from the grade line to line of the road below and the road overhead.

**The stub abutment** is located within the top most portion of the end of an embankment or slope protection parallel to the embankment, so that it has a relatively small vertical height.

**The semi-stub abutment** is a cantilever abutment founded part way up the slope and it is intermediate in size between a full height abutment and a stub abutment.

**The cantilever or vaulted abutment** is completely closed by walls behind the abutment breastwall. Its inside vault can be filled with soil or it can be left unfilled, but spanned over by a reinforced concrete slab.

**The spill-through or open abutment** consists essentially of two or more columns or buttresses supporting a cap beam spanning the space between them. The approach embankment is retained only in part by the abutment since the embankment's sloped front and side portions extend with their normal slope to envelop the columns or buttresses.

**The gravity abutment** is a heavy abutment in which resists the horizontal earth pressure by its own dead weight. It is usually constructed with plain concrete or stone masonry.

**The cantilever abutment** is an abutment in which the stem or breastwall is fixed rigidly to the footing. The stem acting as a cantilever beam transmits the horizontal earth pressure to the footing, which maintains stability by virtue of the dead weight of the abutment and of the soil mass resting on the rear portion, or heel of the footing.

**The counterfort abutment** is an abutment which develops resistance to the bending moment (or horizontal force) in the stem using counterforts. This permits the breastwall to be designed as a horizontal beam or slab spanning between counterforts, rather than as a vertical cantilever slab.

**The pile bent abutment** is an abutment in which the bridge seat is supported by one or two rows of pile, and it's cap on spread footing on rock if soil is in stable condition. Since pile bent abutments have a tendency of leaning toward the superstructure, and its rotational movement can be avoided by increasing the pile embedment length of vertical and battered piles to a minimum of 1'-6" instead of 1'-0".

**Integral Abutment**

Several superstructures, abutments and approach slabs have been constructed without joints for short and medium length bridges in some states. FHWA recommends that bridges with their overall length of less than the following values be constructed continuous and be integral abutments if unrestrained:

Steel------------------- 300 feet
Cast-in-place----------- 500 feet
Pre-cast concrete------- 600 feet

This type of abutment allows rotation like a pile bent abutment on one row of piles or an abutment hinged at the footing.

For integral abutments, approach slabs are required to span the area behind the abutments to prevent traffic compaction material where the fill will be partially disturbed by the abutment's rotational movement. The approach slab should be anchored with reinforcement steel to the superstructure and it has a minimum approach slab length equal to the abutment depth plus 4 feet for soil bearing. Also it is necessary that the abutment be anchored to the superstructure.

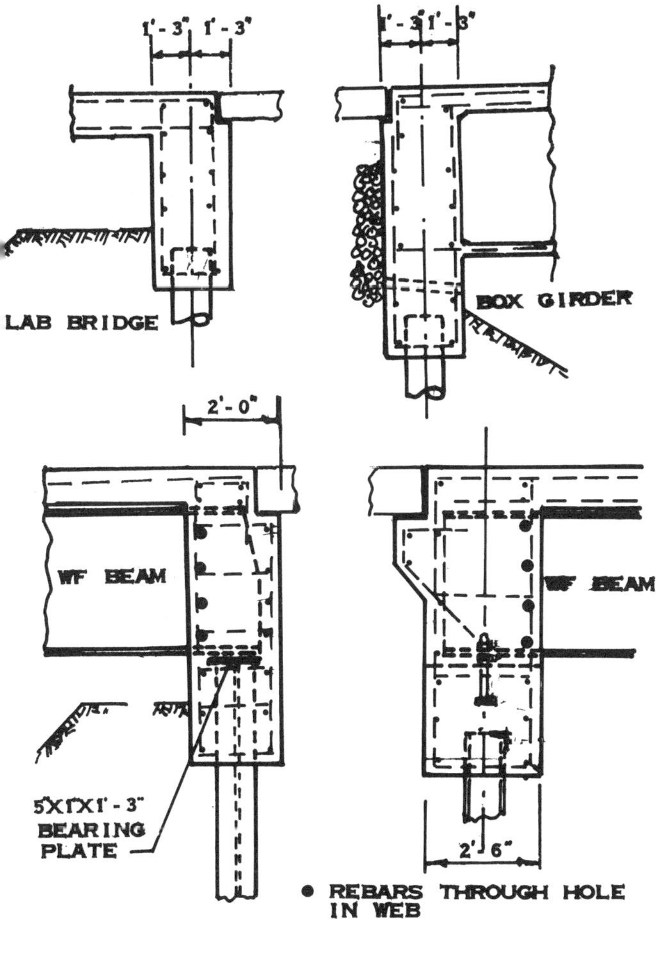


Fig. 11.10 Integral Abutments

For skewed integral abutment, The thermal forces would act along the centerline of the roadway not perpendicular to the centerline of the abutment while soil pressure acts against the abutment. Particular attention should be given to the rotational action caused by active soil pressure. The following should be considered for skewed integral abutment:

1. Some battered piles should be considered to resist the earth pressure.
2. Pile may be prebored for a distance of 10 to 20 feet below the pile cap and filling the void with gravel and sand to permit the thermal movements.
3. The pile may be placed along the centerline of the abutment so that main soil pressure will act mainly about the strong axis.
4. The backfill behind the abutment should be free drain granular materials so that thermal expansion will be accommodated and freezing will be avoided.

## 11—2 STONE MASONRY

Most defects for stone masonry include open joints and may be extended to the full depth of the stone. Some of the repairs are as follows:

**Restoration:**
The removal of all loose, unsound and disintegrated stone masonry to a minimum depth of 4" and the restoration thereof, together with all existing holes in the stone masonry with properly reinforced concrete securely anchored to the original structure.
**Repointing:**
The raking out of all faulty stone joints to a minimum depth of 1" and the repointing them with pressure injection of non-shrink grout. Also, the sealing of the open joints by pressure grouting.
**Reinforcing:**
The installation of vertical reinforcing steel dowels through the upper stone courses for stabilization of movement. The dowels to be securely grouted in holes drilled at 4'-0" centers around the perimeter from the top of the abutment downward a distance of approximately 4'-0" or into the sound stone course.

**Pressure Grouting:**

The drilling of 1 1/2" pressure grout holes properly located and of sufficient depth to intercept internal voids in the stone masonry. Diagonal downward holes, extending a minimum of 6'-0" below the waterline should be drilled to assure interior foundation consolidation.

The setting of grout inserts in the above holes and the testing, washing and lubricating of the void and grout passages insofar as practical. The pressure injection of the grout should be made to permeate the encountered internal void areas and consolidate the masonry into a homogeneous unit mass.

The following are some of the repair specifications:

**Raking of Stone Joints**

All designated exposed stone joints shall be chipped out to a minimum depth of 2" by means of pneumatic tools, followed by a thorough cleaning of all loose dust, and dirt, with high pressure air and water blast.

**Fig. 11.11 Repointing**

**Repointing of Stone Joints**

The above joints shall then be restored, cut flush with the stone face with pressure concrete mortar consisting of one part cement to one part sand, plus non-metallic, non-shrink additive and mortar shall be applied under pneumatic pressure, driving it well back into the joint.

**Chipping and Removal**

All loose, unsound and disintegrated stone or concrete material shall be removed from the areas to be repaired in a manner and to an extent as to expose a sound surface. Unless otherwise specified, chipping of the concrete is to be carried to a minimum depth of 1 1/2" and in general to a maximum depth of 8" and a minimum of 4" for stone masonry.

The boundaries of the areas to be repaired shall be cut at an approximate 45 degree angle to a depth of not less than one inch to avoid thin or feather edges in the finished surface.

All defective construction joints and cracks where necessary shall be chipped out to a minimum of four inches in width by two inches in depth. Shrinkage or fine cracks not impairing the structure shall not be chipped.

Where existing reinforcing rods are exposed at the surface, the concrete shall be cut back of the rods so that the reinforcing can be more thoroughly embedded in the concrete. Wherever this procedure is impractical, additional pressure concrete shall be applied over the reinforcing to obtain proper coverage. The chipping of all surfaces to be performed by means of pneumatic tools.

**Bonding**

All concrete chipped surfaces shall be rough, clean, sound concrete free of oil, dust, dirt and loose particles. The bonding surface shall be thoroughly wet prior to the application of epoxy bonding compound which shall be applied to all concrete surfaces to be restored.

**Replacement of Missing Stone**

All surfaces from which loose and disintegrated concrete and stone has been removed shall be rebuilt to their original lines and contours by the application of dense pressure concrete. The mortar shall consist of a mixture of one part masonry cement, (a mixture of 50% Portland cement and 50% masonry cement) and three parts sand by volume and shall be placed by means of the pressure concrete machine. All surfaces restored shall be brought to a true, even and uniform finish.

Sand shall be clean and sharp and evenly graded from fine to coarse with a "fineness modulus" between 2.0 and 3.25. Moisture content shall be as near to 4% by weight as practical. The sand and cement shall be thoroughly mixed before being placed in the cement gun. The mixture of sand and cement shall be screened to remove all particles 1/2" in size or over and shall be applied to the surface within one hour of time that the sand and cement are mixed.

**Anchorage and Reinforcement**

Over all areas to be restored, a 2"x2"x#12x#12 gauge galvanized electrically welded woven wire fabric shall be securely anchored to the original structure by means of hooked expansion bolts or drill and grouted anchor bolts. Top surface concrete over nine inches in depth shall be anchored with 3/4" dowels spaced 2'-0" on centers. No isolated area greater than two square feet shall have fewer than three anchors.

If the sound masonry has not been exposed at a depth of eight inches, long 1/2" round dowels shall be used for anchorage. In general, the dowels shall be securely grouted in holes drilled downward at about 30 degree angle to the depth of a minimum of 40 diameters of the bar and shall be spaced at 2'-0" centers both ways.

Where the thickness of replacement material is greater than 4" over any appreciable area, the reinforcing shall consist of one layer of 2"x2"x#12 gauge galvanized electrically welded woven wire fabric securely anchored in place by means of expansion anchors or drilled and grouted anchors. This layer of mesh to follow the contour of the chipped surface. In addition, 1/4" round pencil rods shall be spaced 2'-0" on centers both ways and set approximately 1 1/2" from the finished pressure concrete surfaces, to which a layer of 2"x2"x#12x#12 gauge galvanized electrically welded woven wire fabric shall be installed. This combination of reinforcing shall be in turn securely anchored to the expansion bolts or grouted anchor bolts.

All existing exposed reinforcing rods shall be thoroughly cleaned by sandblasting and/or wire brushing, and any exposed reinforcing steel that is badly corroded shall be removed and new reinforcing of equal size and type shall be installed. Splice bars shall lap at least 30 diameters, when wired, or the rods may be welded to the existing steel with sufficient lap to develop a full weld.

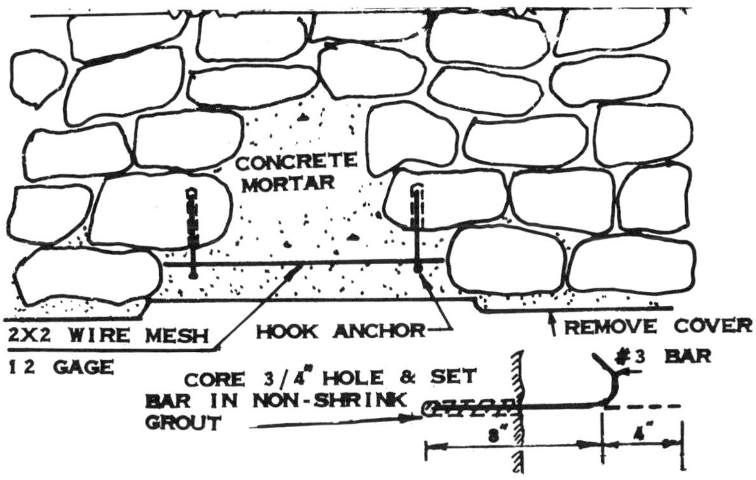


Fig. 11.12 Reinforcing ANCHOR DETAIL

Where splices of wire mesh are necessary, a lap of two meshes in each direction shall be required, wired together at intervals of not more than 18". At corners, an additional layer of mesh 12" wide shall be pre-formed symmetrically around its center line and securely wired to the mesh. Reinforcement shall be held away from the surface of the old concrete and in general shall be not less than one and one-half inch from the finished surface of the replacement.

**Placing Mortar**

In applying the pressure mortar, except as modified below, a pneumatic pressure of not less than 35 pounds per square inch shall be used in placing the material. Where lengths of hose exceed 100 feet, the pressure of the machine shall be increased to produce a nozzle velocity equivalent to 30 pounds of air pressure at the machine with 100 feet of hose.

The water used to hydrate the material at the nozzle shall be maintained at a pressure of at least 60 pounds per square inch. The stream of material forced from the nozzle shall be held from 2 to 4 feet from the surface and shall strike as nearly as possible at right angles to the surface being covered. Any previously placed pressure concrete shall be washed thoroughly with water and compressed air before new materials are applied. Any deposit of loose sand or other substance tending to impair the strength of the pressure concrete shall be removed. All mortar shall be applied by nozzle operators who are experienced in placing mortar and under the direction of the engineer thoroughly familiar with this class of work.

The mortar shall be applied with a sufficient number of coats to obtain the required thickness but in general not less than two coats shall be placed. The thickness of each coat shall be such that it will neither slough nor decrease the bond of or to the preceding coat. If more than two hours lapse between the placing of successive coats, the surface shall be thoroughly wetted prior to the placing.

Fig. 11.13 Pressure Grouting

At the end of each work day, or similar stopping periods, the pressure concrete shall be tapered to an approximate 45 degree edge and if succeeding coats are to be applied, the pressure concrete shall be thoroughly scored to provide a satisfactory bonding surface.

**Drilling for Pressure Grout**

Generally, at regular intervals, or at such other locations as may be necessary, over the exposed stone face, 1 1/2" holes shall be drilled into the stone masonry to necessary depth to intercept all internal voids so as to assure complete consolidation of all units of the structure with pressure grout. At grade, these holes shall be drilled diagonally downward so as to assure consolidation of the foundation or below the same where the structure is supported on timber mats and settlement is occurring.

**Setting Inserts and Cleaning Internal Voids**

In order to clean out the internal voids in the structure of dirt, sand, mud, clay or other deleterious material, the grout inserts shall be set in the above holes and the interior voids washed out with clean water and compressed air prior to applying the pressure grout.

**Pressure Grout**

Pressure grout consists in general of a mixture of one part of cement to one part of sand, with approved admixture. This mixture should be hydrated with sufficient water to produce a workable grout.

In general, the cement-sand grout mix shall be as stated above, however, where practical to introduce a mix containing a larger proportion of sand to the volume of cement, such as in more open voids, the leaner mixes shall be used. However, in any case, the amount of sand shall not exceed the proportion of two parts of it to one part Portland cement by volume. Where the voids are small, the sand shall be decreased in the mix as required to assure proper grout flow.

At grade or water line, the holes shall be drilled diagonally downward to assure consolidation of the foundation or below where the structure is supported on timber mats and settlement is occurring. The grout is to be forced into the clean open joints and the internal voids by means of pneumatic pressure or high pressure force pumping. Sufficient pressure to insure consolidation is to be used, but excessive pressure is to be avoided so that no stone movement occurs.

Grouting is to start at the lowest row of holes. Pumping is done at any given hole

## SUGGESTED SUBSTRUCTURE DETAILS FOR SUPERSTRUCTURE REPLACEMENT

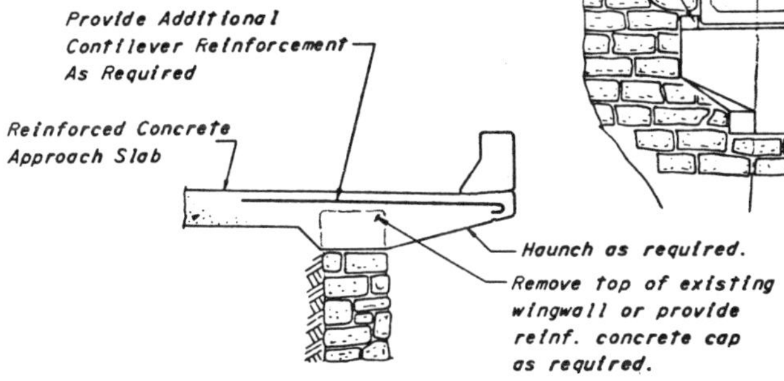


SECTION C-C

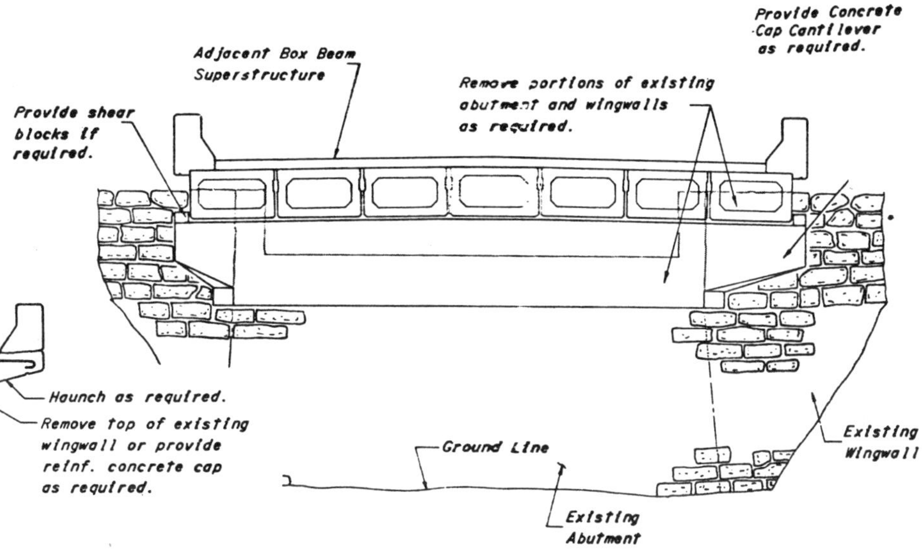


ABUTMENT ELEVATION

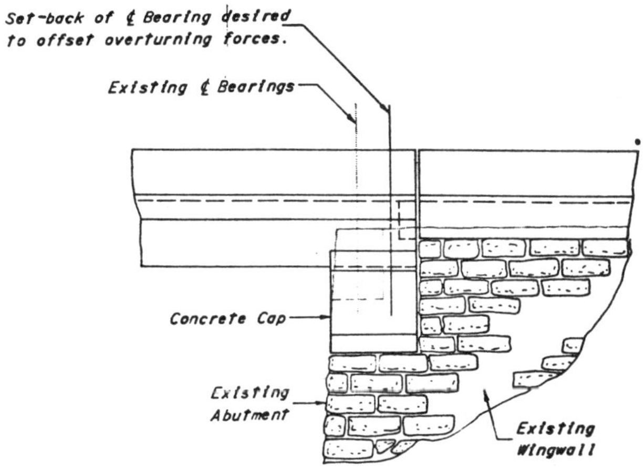


VIEW B-B

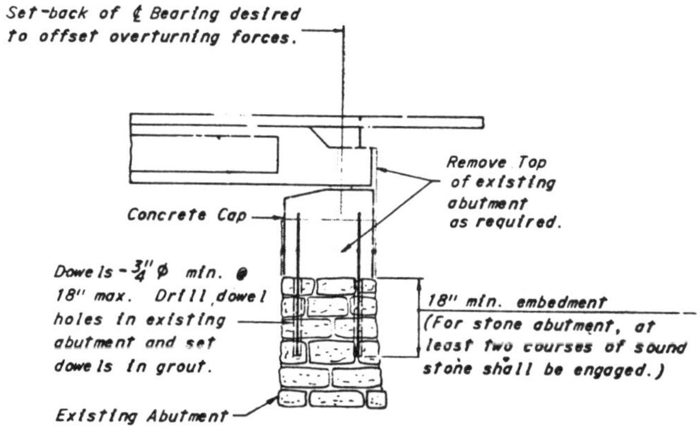


SECTION A-A

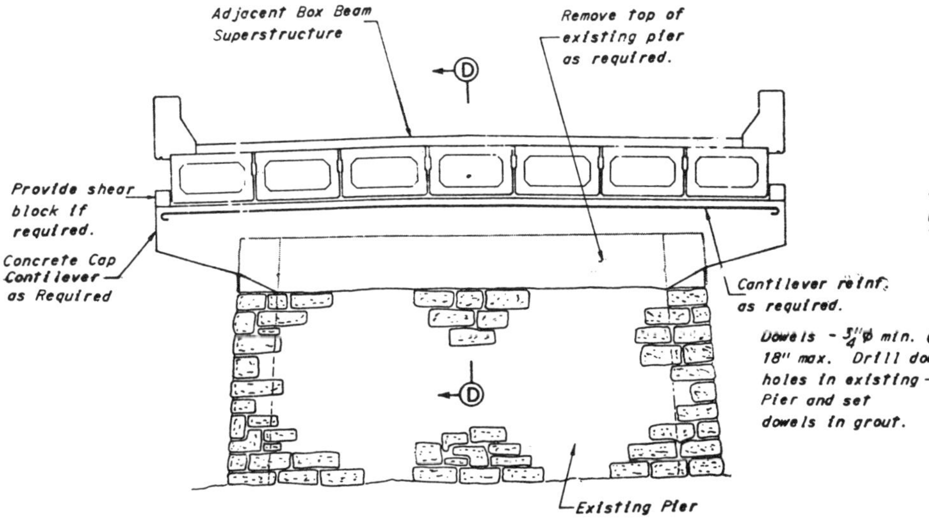


PIER ELEVATION

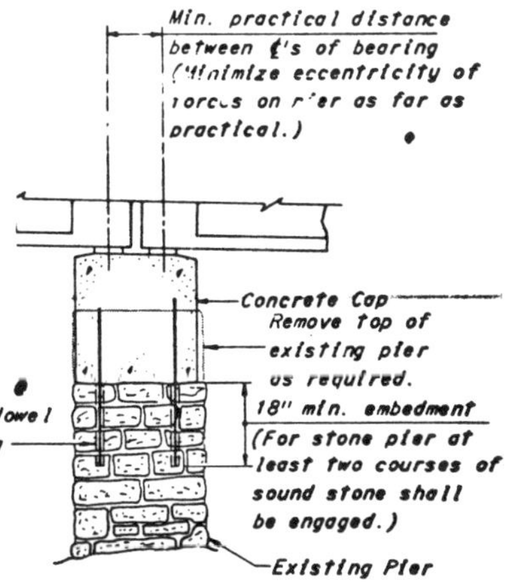


SECTION D-D

(PENN DOT)

until grout outflow occurs at some other hole. The original hole is plugged and the operation is repeated at all of the holes in the row. When all of the holes in the lowest row are grouted, the procedure is to be repeated in the next row above, and so on until the entire structure is completely filled.

During the grouting operation, care shall be given to observe that the surrounding ground, pavements and the stream bed does not break out of the grout and if it happens, the grout line should be moved to the other part of the structure until the grouting can be resumed to the original location the following day.

## 11–3 PIER

The pier which supports the bridge span, should be located as to minimize the constriction of traffic and/or the flow of water underneath the bridge. The pier may be of a plain mass of concrete, reinforced concrete, prestressed concrete, stone masonry, timber, steel, or any combination of these materials. Besides its own weight, the pier should be able to carry the vertical and horizontal forces transmitted by the ends of the two bridge spans and should be able to withstand impacts from collision, debris, ice, water flow and wave. The pier should also be able to resist the movement of the spans and the buoyancy of water.

If the pier is in water, it should be placed parallel to the flow of water and its foundation should be placed at a depth where scouring action of the current does not reach. The upstream end of the pier should be sloped or be in such a shape that it can minimize the resistance of flow patterns and avoid possible jamming by debris during a flood.

The cracks, collision damages, and scour can cause critical defects in the pier. The cracks can be resulted from frozen bearings and overstress, especially at the cantilever arm of the pier cap.

### 11-3.1 TYPE OF PIER

**Solid Pier:** This type of pier generally is used in streams where debris and fast current are present. A long span is desirable and it can be supported on spread footings or pile foundation.

**Column Bent or Open Pier:** This type is the most popular for highway structures and is usually used in place of a pile bent when spread footings are recommended. It can also be supported on spread footings or pile foundations.

**Cantilever and Hammerhead Pier:** This type is used where the geometry requires tight spaces such as ramp and city streets and to avoid skewed bents.

**Pile Bent:** It is widely used for concrete slab bridges over streams when debris problems are not so critical. It is the cheapest and usually used for slab spans, T beams and steel beam bridges.

**Integral Pier:** It can be used where the vertical clearance is critical. In integral pier cap design, the steel girders and the concrete pier cap are rigidly connected.

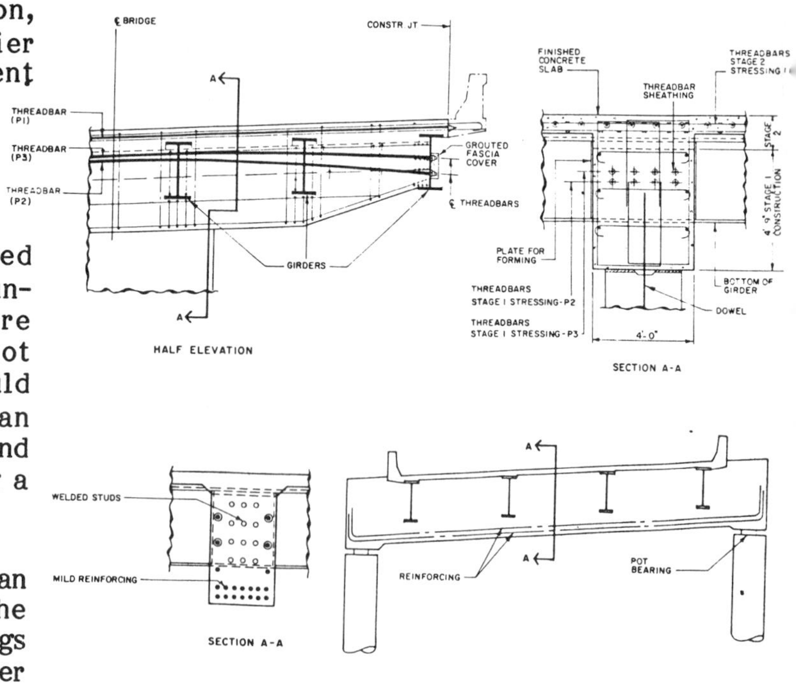


Fig. 11.13 Integral Pier

## 11-3.2 DEFECTS AND REPAIRS

Pier defects are most likely caused by the followings:

1. Movement (lateral and vertical)
2. Scour
3. Erosion
4. Deterioration of concrete, corrosion of steel, and decay of timber
5. Collision damage
6. Overstress due to heavy load
7. Floating ice in waterway can cause serious damages to the pier with its crushing force.

### Strengthening

1) A pile bent pier can be reinforced by adding supplementary piles and cap beams.
   a. Cut holes in the deck and drive piles.
   b. Place cap beams at the top of the piles.
   c. Place a shim plate under the existing cap while the beams are jacked up then lower the beams.
   d. Patch the deck slab.

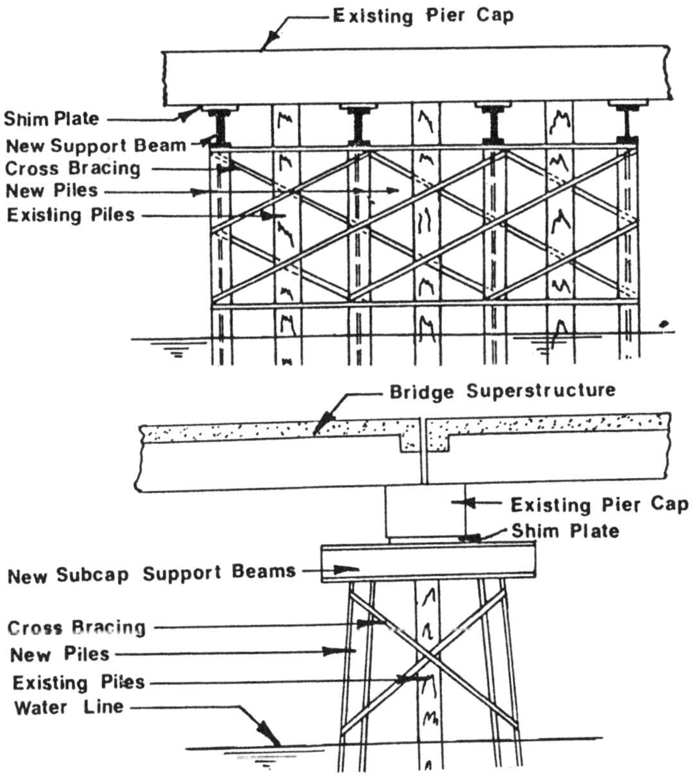


Fig. 11.15 Helper Pile Bent

2) Construct the pier columns outside the existing piers, then build new cap beams under the existing caps in order to accommodate additional strength.

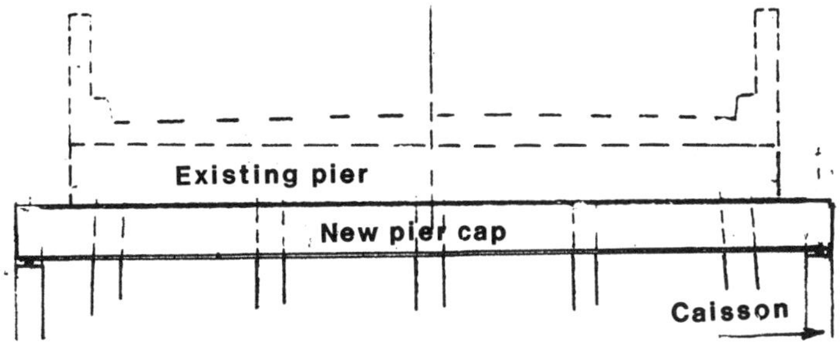


Fig. 11.16 New Helper Pier

### REFERENCE:

1. "Concrete Bridge Details" Portland Cement Association
2. FHWA Technical Advisory T 5140.13 "Integral, No-joint Structures and Required Provision for Movement" Jan. 28,1980
3. Greimann, L., A. Wolde-Tinsae, P. Yang "Skewed Bridge with Integral Abutments" Engineering Research Institute, Iowa State University, Dec. 1982
4. Jorgenson, James L. "Behavior of Abutment Piles in an Integral Abutment Bridge" Paper presented 1983 TRB Annual Meeting, Jan. 1983
5. "Underwater Inspection and Repair of Bridge Substructures" NCHRP Practice 88, TRB
6. "Bridge Manual" Wisconsin DOT
   "Bridge Manual" Conn. DOT
   "Bridge Manual" Illinois DOT
7. "Standard Details for Highway Bridges" NY State DOT
8. US Steel Bridge Report "Interstate Highway Interchanges, Knoxville, Tennessee" 1982

# 12 RETAINING WALL

## 12—1 GENERAL

The main function of a retaining wall is to laterally support a mass of earth behind it. It is subject to horizontal pressure from the back fill, live load, surcharge, seismic forces and unbalanced water pressure and these pressures have a tendency to overturn and slide the wall. Even though the wall is not subject to external vertical load, the foundation material should withstand the bearing pressure from its own weight and vertical pressure from the back fill.

When poor soil conditions exist near the bottom of the footings, a pile foundation has been used in place of a spread footing.

There are five main types of retaining wall:

- Gravity wall
- Cantilever wall
- Counterfort wall
- Crib wall
- Others: Gabion, Reinforced wall, Precast wall, Fabric reinforced wall, Tieback etc.

**1) Gravity Wall:**

This type of wall is made with very little tensile stress in any part of the wall. The mass can be reduced by including a small amount of reinforcing steel: the wall is then called "semi-gravity". It's cross-section is primarily trapezoidal, even though the back face may be stepped up to make placing concrete easy.

**2) Cantilever Wall:**

This type of wall is made of reinforced concrete to resist the active earth pressure and is subject to bending moments. It's cross-section is essentially an inverted T. The wall portion (or stem) performs structurally as a cantilever from the base slab. The concrete is reinforced along the back face where horizontal pressure causes tensile stress.

**3). Counterfort (Buttress) Wall:**

This type of wall is similar to a cantilever wall except that both the base slab and the wall are stiffened by the triangular vertical brackets. If the brackets are exposed at the back of the wall, it is called counterfort, and if they are exposed at the front, it is called buttress.

**4). Crib Wall:**

This is essentially a gravity type wall and is composed of interlocking precast or prefabricated members to form cells filled with soil. Its stability depends on the weight of the crib unit, its filling, and on the strength of the filling material. This type of wall is usually made of reinforced concrete, but both timber and fabricated metal crib can be used.

Fig. 12.1 Concrete Crib Wall

**5). Others:**

There are other types of retaining walls such as reinforced earth wall, gabion wall, interlocking precast retaining wall and synthetic fabric reinforced wall.

Several types of proprietary walls have been developed and used successfully in the States. They offer pleasing appearance, fast construction and economical. It appears that they are competitive as an alternate to the conventional concrete wall.

**Reinforced Earth® Structure:**

Reinforced Earth® method was developed in 1967 by French engineer Henry Vidal and has been used to construct retaining walls, bridge abutments, and other civil works.

Reinforced Earth is the composite material formed by the association of granular soil and linear metallic reinforcements, and has a similar behavior as reinforced concrete with steel bars. In a Reinforced Earth wall, the metal reinforcing strips are placed horizontally within the backfill and attached to the vertical precast concrete facing panels. The basic concept of Reinforced Earth lies in developing sufficient friction between thin reinforcement strips and earth backfill without causing internal shear. Design of the wall is based on the distribution of stresses in granular soil and the angles of internal friction. This friction can be developed without any slippage if the granular soil has a 15 to 25 degree internal friction angle.

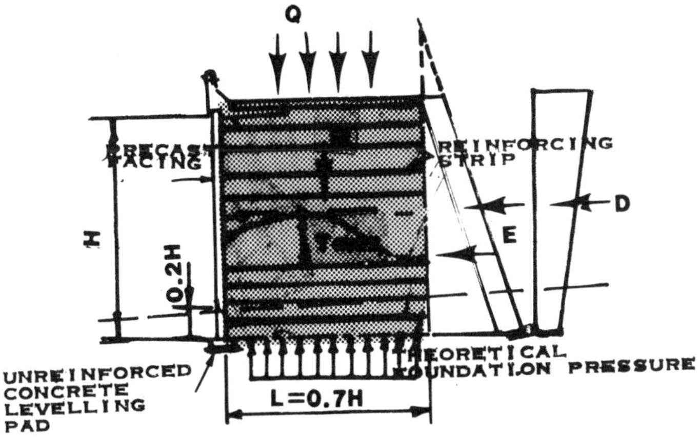


W WEIGHT OF REINFORCED EARTH
T MEASURED STRESS DISTRIBUTION IN REINFORCING STRIPS
Q SURCHARGE LOAD
D SEISMIC LOAD
E EARTH PRESSURE

Fig. 12-2 Reinforced Earth

Reinforced Earth system has three main component: 1. The facing precast panel; 2. Reinforcing strip of galvanized steel; 3. A granular backfill.

The panels are precasted with light reinforcing bars and 4,000 psi concrete. The galvanized reinforcing strips have transverse ribs and they are cut to the required length ( generally 70% of the wall height with 8.5 feet minimum ). The galvanized steel strips are bolted to protruding tiestrips, cast on the back side of the facing panels.

Reinforced Earth construction is a repetitive process. It begins with placing a lift of facing panels on an unreinforced concrete leveling pad. This first row of panels is braced for support. The backfill consists of gravel, sand or crushed stone and is spread and compacted. The contents of backfill materials are very important factors for wall's performance. Soil backfill is usually granular materials with not more than 15% of fine by weight passing a No.200 mesh sieve. The backfill material should not contain material which promote corrosion. A higher percentage of fines and moisture contents will reduce the strength in strip pull out capacity and could result in wall deformation by decreasing the interior stability. Fine grained residual soils are not generally used as backfill for Reinforced Earth walls. If residual soil is used, special quality control should be exercised during construction for highly variable fine contents of residual soil and their sensitivity to moisture. The coarse grained micaceous soil will face reduction of strength with increased moisture contents. A good drainage system may be required behind the wall to prevent saturation and development of hydraulic pressure. Generally the back fill width is about 70% of the wall height.

Next, the first layer of galvanized steel strips is laid perpendicular to the facing panels. They are bolted to the connections on the back of the panels. Another layer of backfill, about 2.5 feet deep, is spread and composite, keying the first strips into position and initiating the frictional bonding of earth and steel strips. Following rows of panels are installed into place and the entire procedure is repeated until the desired structural height is achieved.

The Reinforced Earth wall is used for reconstruction of landslides, embankments and as bridge abutments and ramps. It is an economical construction material with an attractive wall face but has a disadvantage of interference with the reinforced strips in case of future relocation.

For further information contact the Reinforced Earth Company, Rosslyn Center, 1700 North Moore St. Arlington, Va. 22209-1960.

Retained Earth® is a similar system to Reinforced Earth® except that it uses galvanized welded wire mesh in place of galvanized steel strips. For further information for Retained Earth contact the VSL Corp. 101 Albright Way, Los Gatos, CA 95030.

**Precast Concrete Modular Type Retaining Wall (DOUBLEWAL®)**

Doublewal® is recently developed gravity retaining wall system. This system consists of precast, interlocking, reinforced concrete modules of two face panels held rigidly and apart by connecting beams. Once in place without using fasteners, the units are backfilled with filler material then it functions as an economical gravity type retaining wall. The main advantages of this system are as follows: rapid construction; comparatively low cost; and reduced quantities of excavation and special backfill.

The foundation for the wall is graded level and compacted to the specifications depending on the soil condition. A concrete toe footing and rear concrete leveling pad are then cast-in-placed according to the profile line and grade. Then the first course of concrete modular units is set with a crane. All modular units above the first course interlock with lower courses. Vertical joints are staggered with each successive course. Front and rear bearing pads are used between horizontal courses, filter fabric is placed behind the vertical joints, and horizontal joint filler is installed at the front face. The joints at corners or angle points are closed.

The interior of each successive course of precast modular units is filled with granular fill material. Units 4 feet or less in height are filled on one layer and then thoroughly consolidated with a vibratory tamping device. The units which are more than 4 feet in height are filled in two equal layers and thoroughly consolidated after each layer is placed.

Backfill around the outside of the wall is placed and compacted in layers according to the specifications. When erecting a battered wall, placement of backfill behind the wall should closely follow erection of successive courses of units. At no time should the difference in elevation between the backfill and the top of the last erected course exceed six feet.

The Doublewal structure is designed to resist the overturning moment created by the soil pressures behind the wall based on Coulomb Equation assuming partial wall friction (2/3 θ).

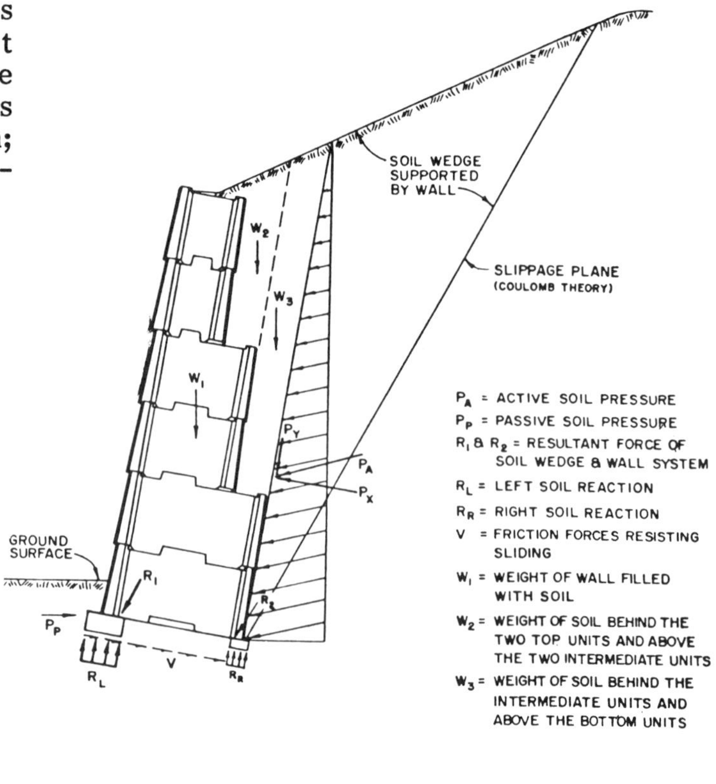


Fig. 12-3 Doublewal

A factor of safety for overturning should be at least 1.5. The Doublewal system functions somewhat differently than conventional retaining wall, as the wall starts to rotate, the vertical loads are translated into heel and toe, mainly toe to achieve a safety factor of 1.0 for overturning and frictions are developed between soils inside wall and wall as the wall tilts. A full scale field test of twenty-two feet high Doublewal confirmed the validity of design assumption based on pressure calculated from Coulomb Equation with wall friction.

The resistance against sliding is achieved by the soil friction at the base. When the horizontal force exceeds the allowable soil to soil friction, the excessive horizontal force is transmitted to the toe and heel. If there is excessive force at the toe and heel, consideration should be given to design the footing for the wall units.

The Doublewal Corporation provides complete construction plans and detail designs including external stability considerations for overturning moments and sliding. However, it is the engineer's responsibility to study and analyze other items such as allowable bearing pressure, foundation settlements, backfill characteristics, and performance requirements which are not analyzed by the Doublewal Corporation. It is recommend that the engineer should analyze and review the adequacy of the soil, footing and foundation requirements.

The Doublewal System® is patented by Doublewal Corporation, 59 East Main St. Plainville, Conn. 06062.

Fabric Reinforced Embankment

The concept of earth reinforcement is using synthetic fabric layers between the fills for earth strengthening. The fabric is covered with about 1 foot of fill and the outer edge of the fabric is folded back around the berm and lapped over to the next layer to form the vertical surface of the walls. Then the backfill is leveled with a dozer and compacted. The same procedure of unrolling a new fabric strip should be repeated. At the face of the wall the fabric retains soil from pushing out while in the internal fill, the fabric tensile strength is developed through friction with the soil layer. The wall face could be shortcreted after the completion of the wall.

Gabion:

Gabion has been used for embankment protection and also used for earth control for retaining wall. It can be subjected to various stresses and they must be able to settle, twist and conform to channel and foundation shifts and changes. At the same time, a gabion must have sufficient strength to contain the weight of the rocks with which they are filled and of additional filled gabions that may be placed upon them, as well as to resist the impact force of high velocity currents and soil pressure.

Fig. 12-4 Gabion Wall

A gabion wall is considered to be a mass gravity structure. The galvanized or polyvinyl chloride (PVC) coated wire mesh may be taken as an additional safety factor but should not be included in the design considerations. When water quality is in doubt (12<pH<6) or where high concentration of organic acid is suspected, PVC coated gabion should be used. The active soil pressure behind the wall can be calculated using the Coulomb Wedge theory and the mass of the wall is designed to balance the force exerted by that soil wedge. When designing a vertical face wall it should be battered at an angle of about 1:6 batter (6 degree) to keep the resultant force toward the back of the wall. The coefficient of friction between the base of a gabion wall and a cohesionless soil can be tan 0 (0: Angle of internal friction) for the soil and the angle of wall friction may be taken as 0.9. Where the retained material is mostly sand or silt, a filter cloth or granular filter is recommended to prevent any leaching of the soil. Determine the unit weight of the gabion by assuming the porosity to be 0.3. The specific gravity of the common material ranges between 2.2 (sandstone) to 3.0 (basalt). Along all exposed gabion faces the outer layer of stones should be hand placed to ensure proper alignment, and a neat compact appearance.

For clay slopes of cohesive soils, Counterforts systems are recommended. They should be used as headers and should be extended from the front of the wall to a point at least one gabion length beyond the critical slip circle of the bank. The counterforts may be spaced from 13 feet (vert soft clay) to 30 feet (stiff clay). A filter is required on the back of wall so that clay will not clog the free draining gabions.

Fig. 12.5 Gabion

Gabions shall be of a single unit construction. The front, base, back and lid shall be woven into a single unit. The ends and diaphragm(s) shall be factory connected to the base.

The galvanized wire mesh shall be fabricated to be non-raveling which is defined as the ability to resist pulling apart at any of the twists or connections forming the mesh when a single wire in a section of mesh is cut and the section of mesh then subject to the load. The gabion length shall be 1 1/2, 2, 3, or 4 times its horizontal width. The horizontal width shall not be less than 36 inches. However, all gabions furnished by the manufacturer shall be of uniform width. Where the gabion length exceeds 1 1/2 times its horizontal width, the gabion shall be equally divided into cells by diaphragm(s) of the same mesh and gauge as the gabion body.

The assembled gabion units shall be placed in the proper location. All adjoining empty gabions shall be placed along the perimeter of the gabion contact surfaces to obtain a monolithic structure.

The following method applies to 3 foot high gabions; once the gabion units are laced together, they shall be stretched to effective alignment. This operation shall be carried out after several empty gabion units have been positioned. The first gabion in the line shall be partially filled to provide the necessary anchorage. Any stretching shall be carried out using a means of at least one ton capacity.

While under tension, the gabion shall be carefully controlled against any possible unraveling. Whenever the gabion structures require more than one tier, the upper empty gabion tier under tension shall be laced to the top of the lower one.

Gabions shall be filled by earth handling equipment such as backhoe, gradall, crane and/or by hand. Care shall be taken when placing fill material to assure that the sheathing on coated gabions will no be broken or damaged. Gabions shall be filled in layers, 1 foot at a time. Two connecting wires shall be placed between each layer in all cells along all exposed faces of the gabion structure. All connecting wires shall be looped around 2 mesh openings and the wire terminals shall be securely twisted to prevent their loosening.

The cells in any row shall be filled in stages so that local deformation will be avoided. At no time shall any cell be filled to a depth exceeding 1 foot more than the adjoining cell. Along all exposed gabion faces, the outer layer of the stone shall be carefully placed and packed by hand to ensure proper alignment and a neat, compact, square appearance.

The last layer of stone shall be leveled with the top of the gabion to allow proper closing of the lid and provide an even surface for the next course. Gabions shall be well packed and full, without excessive bulging.

Fig. 12-6 Prefilled Gabion

Anchored Bulkheads:

Anchored bulkheads are formed of flexible sheeting restrained by tieback and by penetration of sheeting below dredge line. Since active pressures are redistributed on the wall by deflection, moving away from the position of maximum moment, the computed maximum moment can be reduced to allow for flexibility of sheeting. To limit differential water pressures, It is recommend to provide weep holes or drainage at a level above mean water.

For anchorage, a tie back may be carried to a buried deadman anchorage, to pile anchorage, parallel wall anchorage, or it may be drilled and grouted anchor. If a deadman must be positioned close to a wall, the anchorage resistance is decreased and an additional passive reaction is required for stability at the wall base. It is necessary to protect tie rods by wrapping, painting, galvanizing, or encasement to resist corrosion. Where backfill will have major settlement, to avoid overburden loading, tie rod should be enclosed in a rigid tube or supported.

The following precautions should be taken during construction.

1. Removal of soft material, or placement of fill in the passive zone should be made before the driving of sheet piles.
2. Deposit backfill by working away from the wall rather than toward it to avoid trapping soft material near sheeting.
3. Before anchorage is placed, sheeting is loaded as a cantilever wall, and safety during construction stages should be checked.
4. When granular backfill is scarce, a sand dike may be placed to form a plug across the potential failure of the active wedge. When such a dike rests on firm foundation soil, the lateral pressure on the bulkhead will be only the active pressure of the dike material.

**Cantilever Sheet Pile:**

A cantilever sheet pile derives support from the passive resistance below the dredge line to support the active pressure from the soil above the dredge line without an anchorage. This type of wall is suitable only for height up to 15 feet and can be used only in granular soil or stiff clays. For details, refer to U.S. Steel Steel Sheet Piling Design Manual.

**Internally Braced Flexible Walls:**

To restrain foundation or trench excavations, frequently laterally braced sheet piles are used. This restrains lateral movement of the soil and cause loads on the brace which exceed those expected from active earth pressure. Brace may be either long raking braces or short horizontal cross brace between trench walls.

**Cofferdams:**

Double wall or cellular cofferdams consists of a line of circular cells connected by smaller arcs, parallel semi-circular walls connected by straight diaphragms, or a succession of cloverleaf cells. The stability depends on ration of width to height, presence of an inboard berm, and type and drainage of cell fill materials. Ordinarily, a cell wall tilt 0.002 to 0.03 radian and active and passive pressure act on exterior face of the sheeting. A cell must be stable against sliding on its base, shear failure between sheeting and cell fill, shear failure on centerline of cell, and it must resist bursting pressures through interlocking tension. These factors are influenced by the type of foundation.

Clean coarse-grained, free draining soils are preferred for cell fill materials. They may be placed hydraulically or dumped through water without compaction or specific drainage.

**Tieback:**

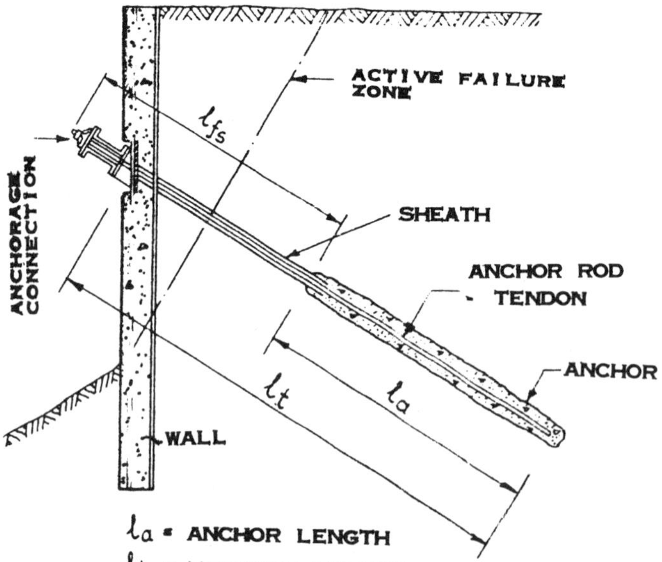


Fig. 12-7 Tieback

The tie back system as a permanent wall is a recent development. Sometimes, it is the effective way to provide horizontal resistance and to stabilize and control soil slide. It has been used to stabilize existing retaining walls, to resist uplift as tie down and as permanent wall near the stub abutment for road widening project. Tieback are used in cohesionless soil, rock but should not be used in fill material or soft cohesive soils.

Permanent tieback has advantages of elimination of foundation piles, large footings and backfilling but its main disadvantages are corrosion of the high strength tie rod (tendon) and long term creep at the anchor zone. As protective measure, grease tube or heat shrinking tube is generally provided at unbonded length of the tendon.

Tiebacks are suitable to use for particular soil conditions and the anchor system should demonstrate its capacity by proper inspection and testings during or after installation. Testing of anchor should have minimum 150% of design loads.

Since different construction methods are available depending on the fieled condition and availability of equipments and materials, it is recommended to use open and performance specifications by allowing the contractor to select the tendon system, corrosion protections, grouting, and method of drilling.

**Selection of Wall**

The following may be used as guideline for selecting wall:

**SELECTION OF WALL**
**Wall Height** - feet
(from bottom of footing to top of wall)

| **Wall Type** | Up to 10 | 10 to 25 | 25 to 45 | Over 45 |
|---|---|---|---|---|
| Reinforced Concrete Cantilever Wall | Yes | Yes | No | No |
| Counterfort Wall | No | No | Yes | Yes |
| Conc. Crib Wall | Yes | Yes | No | No |
| Gabion | Yes | Yes | No | No |
| Precast Wall | Yes | Yes | Yes | No |
| Reinforced Earth | No | Yes | Yes | Yes |

## 12–2 DEFECTS AND REPAIRS

**Overturning:**

If the total vertical loads are too small to overcome the overturning moment from horizontal earth pressure, the wall would have rotational movement. Proper drainage of the backfill is required to reduce the lateral pressure.

Repairing this defect includes drilling new weep holes in the wall, placing a slurry wall, installing a tie back and soil anchor system.

**Sliding:**

If the vertical loads do not develop sufficient friction forces to resist movement between the base and foundation soil, lateralmovementwilltakeplace.

Repairing this defect includes installing a pile system with soldier beams and a sheet pile or tie back system to remedy the movements.

**Settlement:**

If the shearing resistance of the foundation material is too low, rupture of the soil occurs. To improve this condition, concrete grouting and chemical grouting are recommended.

**Slope Failure:**

Slope failure of the embankment can occur when the soil lacks sufficient cohesion and/or internal friction, and the foundation is not placed at a sufficient depth below ground. To improve this condition, the back and soil anchor system has been used.

**Cracks due to Bending Stress:**

When insufficient reinforcement is placed at the tension area of the wall, tension cracks or total failure of the wall can occur. Remedies for this defect are strengthening the wall by installing sheet piles behind the walls or building new walls with the back system in front of the existing walls.

**Scour or Erosion:**

Countermeasures for the hydraulic problems are placing riprap, riveting, spur dike, pressure grouting, and installing sheet pile as in abutment repairs.

REFERENCE:


1. "Foundations and Earth Structures" Design Manual 7.2, Naval Facility Engineering Command, May 1982
2. Andersen, Paul "Substructure Analysis and Design" Second Edition, Ronald Press, NY 1956
3. Elias, Victor, Paul Swanson "Cautions of Reinforced Earth in Residual Soils" Paper presented 1983 TRB Annual Meeting
4. FHWA Memorandum dated Nov. 25,1981 "DOUBLEWAL - Interlocking Precast Retaining Wall System"
5. "Doublewal" Technical Manual Doublewal Corp.
6. "VSL Rock and Soil Stabilization System" Technical Manual VSL Corp.
7. "Maccaferri Gabions" Technical Manual Maccaferri Gabions, Inc.
8. "Galvanized Steel Strips Provide Long Life to Retaining Walls" The American Hot Dip Galvanizers Association Market News
9. Steward, J., R. Williamson, J. Mohney "Guidelines for Use of Fabrics in Construction and Maintenance of Low-Volume Roads" FHWA-TS-78-205 FHWA/FOREST SERVICE
10. "Recommendations for Prestressed Rock and Soil Anchors" Post-Tensioning Institute
11. "Tiebacks" FHWA/RD-82/046 & 047, July 1982 FHWA.
12. David E. Weatherby, James W. Sigourney "Tiebacks and their application" ASCE National Geotechnical Conference Jan 6-8 1981.
13. Ronald B. Reeves "Control of Landslides with Permanent Tiebacks"
14. Harry Schnabel, Jr. "Procedures for Testing Earth Tieback" ASCE National Structural Engineering Meeting April, 1976
15. Engineering Literature from Schnabel Foundation Company
16. "Rock and Soil Anchor Manual" Nicholson Construction Co.

# 13 FOUNDATION

## 13–1 FOUNDATION

### 13-1.1 GENERAL

All vertical loads and horizontal forces from the superstructure and substructure are ultimately supported by the underlying foundation soil. Most subsurface foundation materials are considerably varied and complex in their composition and their properties.

The engineer estimates the best soil profile showing the potential problem areas and the undesirable areas by selecting proper design methods and procedures with the soil boring information. It is difficult to determine the true subsurface soil profile with an economically limited number of borings and then implement this information into the design to avoid any foundation failures. Therefore, soil analysis and foundation design requires a great deal of experience and sound engineering judgement.

Foundations can be categorized by the supporting material, such as rock-bearing or soil-bearing, and can also be classified by the types of the substructure such as spread, continuous, mat pile, caisson, and grade beam.

### 13-1.2 SPREAD FOOTING

Spread footings are supported on soil or rock without piles. They are the most common type of footing used for the bridges where the soil is suitable. There are various types of footing such as isolated individual footings, combined footings, wall footings, cantilever footings, depending on the shape and the function of the footing. The simplest type of spread foundation is a continuous footing for a wall.

Where the superstructure loads are transmitted to the foundations through separate columns, the foundation is provided by an independent footing under each column. In situations when the column loads are significant or when the bearing capacity of the soil is poor, the size of individual footing become large. In this case, it is necessary to design the entire floor system as a foundation. This type of foundation is known as a raft foundation and used in building.

In designing spread foundations, it is necessary to ensure that the material under the foundations is suitable and there are no erosion problem. The soil beneath a spread footing should be able to support a minimum bearing pressure about 1.5 tons per square foot. An allowable bearing pressure can be increased to 3 tons per square foot if the compacted granular soil is used to support a spread footing. On stream crossings, a scour protection provision will be need. Soft clay can cause swelling damage if wet. It is suggested to remove potentially swelling unsuitable foundation soils such as silt, soft clay, organic soil, and replace with compacted fill of cohesionless or non-swelling material if economically feasible. If the above is not feasible, the following are suggested to minimize swelling effect.

1. Design footings with unit loads large enough to resist swelling pressure without exceeding bearing capacity to support the structure above ground.
2. Consider using spread footings or drilled under-reamed caissons below the zone.
   - Design the shafts for foundations with sufficient reinforcing to resist tensile forces applied to the shaft by friction or adhesion in the swelling materials.
   - At any depth, tensile forces exerted on a shaft should be equal to the circumferential area of the shaft times the difference between average swelling pressure above and below the point under consideration.
3. Schedule construction during or immediately after a prolonged rainy period to lessen potential volume change in the future where large seasonal changes in soil moisture are responsi ble for swelling.

Generally a spread footing on compacted fill is more economical than pile foundation. The movement of a spread footing may be slightly greater than pile foundations but the difference is insignificant in many cases and the overall performance of a spread footing is satisfactory. If settlement is expected, provisions can be made at the bridge seats for jacking so that the beams can be raised. For jacking purposes, a small concrete seat can be built under each girder or under the end diaphragm between girders. The diaphragm should be designed for lifting the adjacent girders. Where high settlement is anticipated, a simple span is usually used in place of a continuous spans to avoid stress increase by settlement with continuous span. The maximum allowable differential settlement is about one inch without causing significant distress.

### 13-1.3 PILE FOUNDATION

Piles are used where upper soil strata has soft or loose layers or has inadequate bearing capacity to sustain vertical and horizontal loads or uplift forces. Piles are also used where scour is likely to occur, where future excavation may be adjacent to the structure, and where expansive or collapsible soils extend for a considerable depth.

Piles can be classified as friction piles and end-bearing piles, according to the manner of transferring their loads into the underlying soil. A friction pile develops bearing capacity, either partly or wholly, by developing frictional forces between the embedded pile and the adjacent soil. An end-bearing pile transfers practically all of its load by bearing to the firm soil stratum on which the pile rests.

In addition to vertical foundation bearing piles, there are other types of plies. A batter pile makes an angle with the vertical plane and is designed to give lateral stability or to resist upward forces.

Piles are frequently used to overcome the inability of shallow footings to resist inclined, horizontal or uplift forces and overturning moments. Examples of this type of foundation are pile supported abutments and retaining walls. Piles are used to resist upward forces by negative friction around their shafts. A combination of vertical piles and battered piles is used to support the vertical load and horizontal forces.

Pile foundations are often required when scour around footings could cause undermining and erosion, even if sand or gravel layers exist to prevent structural failure due to scour. Piles are also used in urban areas, where future excavation is anticipated, to eliminate costly future underpinning.

In areas where expansive or organic soils extend to a considerable depth below the ground, pile foundations may be used to prevent vertical or lateral movements of the foundation. The function of piles in such situations is to transfer the foundation loads, including uplift from an expanding soil or downdrag from a collapsing soil, to a level where the soil is not affected by seasonal movements of moisture. A similar need for pile foundations can arise in areas of discontinuous permafrost to ensure transmission of loads to the part of the ground not affected by surface thaw.

In pile foundations, the soil stress is increased by overlapping bulbs of pressure, which is the volume of soil enclosed by contour lines. The pressure distribution is different between single piles and a group of piles. Pile foundations are normally constructed as groups of closely spaced piles. The spacing should be such that the capacity of group piles is equal to the sum of the capacities of the individual piles. However, the ultimate load of the group is less than the sum of individual pile capacity due to factors such as soil type, size and shape of the group, spacing and length of piles, and construction procedures. A lower pile pressure bulb will reduce settlement. Settlement of the group is more likely than that of a single pile because a group of piles produces higher pressure bulbs than a single pile.

All vertical loads and horizontal forces on piles should be transferred to the soil. The ultimate load is mostly determined by soil failure; punching shear under the point with or without frictional forces along the shaft. The ultimate load can be separated into a point load and a skin friction load. Usually, the ultimate skin friction (shear strength of bond between pile and soil) occurs first Therefore, the point load usually takes less than failure load.

When the water level drops in cohesive soil, soil around piles may move downward thus creating a negative friction force (drag force) along the pile. This problem also occurs where pile foundation is in or near an embankment constructed over a compressive soil deposit. To avoid dawn drag problem, the following should be considered:

1. Lengthen bridge span thereby avoid any compressive soil area or reduce the thickness of the approach fill to result in less settlement.
2. Use isolate type casing or bentonite slurry in predrilled holes.
3. Increase the number of piles to increase capacity of the pile group.
4. Use coating on the piles: Fly ash coating (3/8") or coal tar or bitumen coating (min. 1/4", preferred 3/8") is recommended. For steel pile, bitumen should meet the requirements of ASTM D 2521 (canal liner bitumen) with primer (AASHTO M 116). For Concrete piles, asphalt type bitumen meeting ASTM D 946 with primer (ASTM D 41) should be used.

Pile foundations generally cost more than a spread footings. It is estimated that the additional cost for pile foundation is about $80,000 (1982 price) for average bridge with two abutment. Besides the higher cost, the pile foundation does not guarantee to stop any movements and also can cause bump on the bridge approaches. Therefore the use of spread footing should be considered as a first choice of alternate for the foundation type.

## TYPES OF PILES:

Piles can be classified by the types of material or by method of placement. The choice of pile type is influenced by subsurface conditions, location and topography of the site, and structural and geometric characteristics of the structure to be supported.

Classification on the basis of material divides piles into timber, steel and concrete. These materials are sometimes combined to form composite piles. Timber and steel piles are usually prefabricated. Concrete piles can be precasted, or cast-in-place and can be either reinforced or prestressed. Cast-in-place is usually done under the protection of casting, which can be removed to be re-used or can be left in the ground as an integral part of the pile. Occasionally, in firm and stable soil, piles can be cast-in-placed without any casting shell.

Where conditions of excessively corrosive materials are expected, such as meadow mat, peat, unburned carbon, cinders or other chemically active materials which cannot be economically removed, additional protection of the piles may be necessary. Consideration shall also be given to additional protection in the event of nearby installations producing stray currents which can cause corrosion of the pile.

### Timber Pile:

Timber piles are used for temporary structures, water-front structures, bridges on stream crossings and for light loading conditions where the required pile length is not too great.

The material taper of these piles makes them ideal for short friction piles. Timber piles should be creosoted for protection from marine organisms. Untreated piles can be used to reduce cost in case the piles are submerged permanently in water. Untreated piles also can be used in conjunction with jacketing concrete encasement around piles above the mud-line using the shortcreting method.

### Pipe Pile:

Pipe piles are used where large vertical and lateral strength is required. Pipe pile also have the advantage of being able to check inside after each driving. Corrugated pipes and regular pipes filled with concrete are generally used with open or closed ends. On open end piles, driving shoes are used to protect the end of pipes and piles are driven into the rock or contact. Generally the pile is driven a minimum of 1 foot into the rock to provide a socket. The pipe is cleaned out then filled with concrete. Most pipe is driven with a closed end with flat plate or commercially available conical points. The conical end compresses soil aside, distributes the soil, and permits continuing movement of soil ahead to the direction of driving and surrounding soils.

**Cast-in-place Pile:**
Cast-in-place piles can be installed by driving pipe mandrel and withdrawing pipe as it is filled with concrete; and leaving pipe shell in-place or augering a hole for concrete.

**Steel H Pile:**
H piles are widely used to carry heavy loads. The use of cast steel end point protection will minimize the tip buckling and enable to drive critical ground by redistributing the load uniformly over the cross section of the piles.

**Precast Concrete Pile:**
Precast and prestressed piles have advantage of high load capacity. But it is difficult to splice and it is vulnerable to tip damage. Using an H-pile stub section to protect the tip of the pile enables these piles to penetrate into compact materials and drive on sloped rocks.

### 13-1.4 OTHER TYPES OF FOOTING

**Pedestal Footing:**
This type of footing is usually a mass of plain concrete and used to support relatively light loads and where foundation requires small soil-bearing areas.

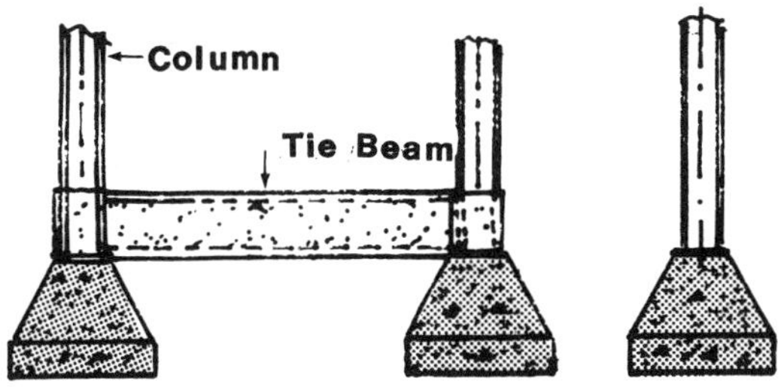


**Fig. 13.1 Pedestal Footing**

**Grillage Footing:**
This type of footing consists of two or more tiers or closely spaced structural steel beams resting on a concrete block, each tier being at right angles to the one below and made of steel encased in concrete.

**Underpinning:**
The construction of footings under the existing structure requires the temporary support of the existing structure and new foundation. Temporary support of the structure could be accomplished by the use of shores and/or needle beams. Upon completion of temporary support, new foundation could be built by excavation under the existing footings.

**Caisson:**
The caisson has enclosing shells to make load bearing foundation driving into desired elevation. Caissons may be open, pneumatic or floating. Cylindrical caissons are used widely because of their minimum perimeter and cost.

The principal difference between open and pneumatic caissons is that pneumatic caissons require compressed air due to their closed top. Open caissons are preferred over pneumatic where boulder or overburden obstacles are not present.

Also open caissons are often used in dry ground. A floating caisson is a fundamentally prefabricated box which could be towed to its final position and sunk to its final elevation.

## 13—2 FOUNDATION FAILURE

### 13-2.1 GENERAL

Foundation failure is mostly caused by unexpected movements resulting from soil subsidence, scours, water table changes, sliding and increasing lateral pressures. When these unexpected movements are small and uniform there shouldn't be any problem, but if there are large differential movements, foundation failure may occur.

Other deteriorations may occur due to corrosion of steel, insect attack to the timber piles, concrete spalls and cracks. Special attention should be given to splash zones of marine environment.

Foundation and bearing failure can be grouped as follow:

1. **Settlement**
   Unequal loads within elastic limits.
   Consolidation of organic soil or unstable soil.
   Loss of water content.
   Bearing failure.
   If the shearing resistance of the foundation material is too low, rupture of the soil will occur.

2. **Lateral Movement**
   Saturation and overflow of unconfined soil.
   Lower soil cohesion and friction resistance under heavy load.

3. **Sliding**
   Lack of friction between inclined layers of soil strata.
   Seepage after the soil is immersed in the water; the water level is quickly lowered but soil behind the structure is still fully saturated.
   Shear failure of soil under the structure and if the vertical loads do not develop sufficient friction forces to resist movement between the base and foundation soil, lateral movement will take place.
   Slope failure: slope failure of the embankment can occur when the soil lacks sufficient cohesion and/or internal friction, and the foundation is not placed at sufficient depth below ground.

4. **Erosion and Undermining**
   Scour may cause serious undermining and extensive settlement. Scour is caused by shear turbulence and high stream velocity.

5. **Chemical and insect attack** may cause deterioration of materials, and construction defects.

### 13-2.2 SETTLEMENT

Settlement is defined as vertical displacement of structures resulting from stress induced in the foundation soil by the imposed load of the structure. The displacement should be in a downward direction but upward displacement may occur in case of a structure resting on swelling soils. The amount of settlement which will occur on the bridge is important since the final grade of roadway will be affected.

**Uniform Settlement:**
A uniform settlement of all the foundations of a bridge will have little effect upon the structure.

**Differential Settlement:**
Differential settlement can cause serious damage to a bridge. Where differential settlement occurs between different substructure units, the magnitude of the damage depends on the bridge type and the span length. Damages due to differential settlement may vary from the opening of the vertical expansion joints between the wingwall and the abutment to severe tipping and cracking of walls or other members. If the underlying soil consists of clay, the possibility of a shear failure should be investigated.

Settlement considerations not only apply to individual substructure units, but to the stability of the soil beneath the entire structure. Simple span bridges can endure settlement without damage. A simply supported stringer bridge can be continuously jacked up to keep it at its original elevation although substructure settlement may occurs. However, pile bent or trestle bridges are very vulnerable since a large settlement or movement of a bent could cause the superstructure to fall off a narrow bridge seat and this leads to the loss of the bridge spans.

An indeterminate bridge is seriously affected by differential movements, since such movements over supports will deflect beams and redistribute the loads, possibly causing large overstresses. For example, a fixed ended arch can be severely damaged if a foundation rotates. Most continuous bridges have fewer joints than simple span bridges. These bridges are, however, more susceptible to the damages if subjected to excessive displacement over the support.

**Consolidation:**
Serious settlement can result in consolidation action in cohesive soils. Settlement of bridge foundations may be caused by changes in the ground water condition, loss of water content, the placement of additional embankments near the structure, or increasing embankment height.

Consolidation of soil can be defined as the removal of water from pore space, while compaction of soil is decrease of soil voids caused by removal of air.

### 13-2.3 SEEPAGE AND WATER TABLE VARIATION

The flow of water from a point of higher head (elevation or pressure) through the soil to a point of lower head is seepage. Seepage results in the lateral movement of retaining walls by:

- An increase in weight (and lateral pressure) of the backfill because of full or partial saturation.
- Reduced bearing capacity and shear strength due to lower friction resistance and cohesion of soil. Effective drainage provision behind the wall could be provided by drilling new weep holes at the walls.

As the amount of water is increased, cohesion will be decreased and bearing capacity reduced. Large cycle variations in the elevation of the water table in loose granular soil may lead to a compaction of the upper strata.

The effects of non-cyclical changes in the water table such as consolidation, slides, and seepage were previously discussed. Changes in the water table may also change the characteristics of the soil which supports the foundation, resulting in the lateral movement or the settlement of the foundation.

### 13-2.4 EXPANSIVE SOIL

The soil underneath the foundation is taking compressive loads, therefore, any failure will occur only by shearing due to a plane or zone of weakness in the soil. The shear strength of a cohesionless coarse-grained soil (sand and gravel) depends on the internal friction between the soil grains.

Granular soil is an excellent bed for spread footings. The bearing capacity of granular soil is high and settlement is low. Settlement also takes place almost immediately as soon as the load is applied for compaction. In general, the shear strength of a granular soil increases as the grain size of the particles increases. Under load, finer grains tend to roll, whereas larger ones interlock each other and smaller grains fill the voids between the larger ones. Such a soil is generally more compact and stable, and less permeable. It should be noted that the bearing capacity of granular soil depends also on the footing width, depth of surcharge, density of soil and position of the watertable.

Fine grained or cohesive soils with clay and silt are more difficult to evaluate, and are poor foundation materials. These type of soil may swell and slowly soften if the load is reduced or removed. Changes in the ground water condition can increase the water pressure within voids, and exposed fissures and cracks can admit water which causes a reduction in the shear strength. A cohesive soil thus, in general, is much less suitable foundation material than a granular soil. Shear strength of a cohesive soil is relatively low, and is greatly affected by water contents. Cohesive soils are usually plastic and compressible, and will deform, or creep, under constant load.

In view of the recent emphasis on environmental protection and preservation of nature, excavation of undesirable material may be difficult because of the problems of disposal and disruption of surrounding areas. If environmentally satisfying methods cannot be found, the engineer may be forced to examine alternative methods of design. Most of these alternative design methods involve problems during and after construction and with future maintenance.

With proper treatment, shearing strength of a soil can be increased, thus improving its capacity as a foundation material. Methods commonly used for increasing the shearing strength of soft clays and silts include overloading or a combination with installation of sand drains. Soft clays and silts have a low value of permeability and require a substantial time period for consolidation to take place.

With low embankments, where stability using in-situ undrained shear strength is not critical, only overloading the soil may be necessary to consolidate the soft clay and silt. In this case, the construction schedule must allow enough treatment time after the embankment is placed for complete consolidation to occur before any further construction takes place. However the in-situ undrained shear strength may not be sufficient for the stability of higher embankments, and the shear strength gain will have to be accelerated during the construction of embankment. In this event, sand drains are to be used with or without toe berms. Overload may be required for complete consolidation to occur within a reasonable time period. Also, to avoid the effect of expansive soil, drilled piers and grade berms are recommended.

## 13—3 MONITORING

### 13-3.1 GENERAL

Monitoring the behavior of the soil and the structure by using instruments is necessary for the following purposes:

- Diagnose the specific nature of failure and provide economical remedies.
- Provide adequacy of new construction methods and construction control.
- Provide safety.
- Verify satisfactory performance and design adequacy.

When a bridge or wall shows evidences of soil failure, remedial action should be taken to prevent further damage. These measures will depend on the magnitude and the location of any movement of the structure as well as other evidences of structural or soil failure. Without proper and accurate information, most engineers will use a conservative solution. When proper instruments are installed, the engineer can collect necessary data then can select an economical remedial measure. Also, the instruments can be used to monitoring the effectiveness of any repairs.

The following variables may be measured by using instruments:

| Variable | Instrument |
|---|---|
| Deformation: | Survey<br>Vertical pipe gage<br>Settlement platform<br>Portable gage<br>Heave gage |
| Horizontal movement : | Survey<br>Inclinometer<br>Crack gage |
| Tilt : | Tiltmeter |
| Groundwater & Pore pressure : | Observation well<br>Piezometer |
| Earth pressure : | Earth pressure cell |
| Load & Strain : | Calibered jack<br>Load cell<br>Strain gage |

### 13-3.2 INSTRUMENTATION

**Deformation:**
The movement of a soil can cause vertical settlement, horizontal deformation, and tilt. Normal survey methods can be used to monitor the magnitude of horizontal and vertical deformation of the structure and the ground surface. Other portable equipment and gages which can be used to monitor deformation are graduated scale, survey tape, wire, caliper, dial gage and crack gage. The most widely used instrument to measure settlement is vertical pipe settlement gage commonly known settlement platform.

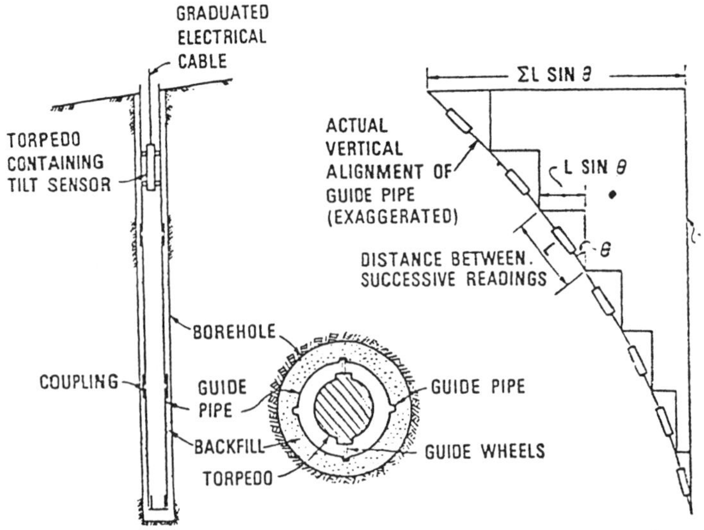


**Fig. 13.2 Inclinometer**

To measure horizontal movement of a soil, an inclinometer is used. A simple "Poor Man's" type inclinometer is inexpensive and consists of a thin wall PVC pipe containing a series of rigid rods of various length. After inserting this instrument in a vertical hole, as horizontal movement occurs, a rod will move until it can not move any more so that the series of rods will form the radius curvature or bend to show horizontal movement. A tiltmeter is used to monitor the tilt of bridge substructure and walls.

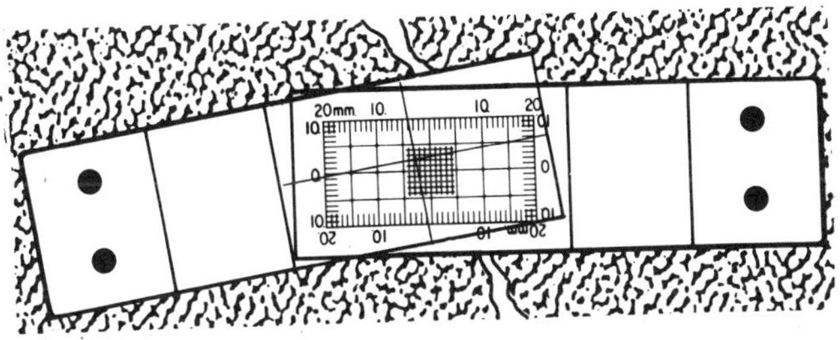

**Fig. 13.3 Crack Gage (Avongard Calibrated Crack Monitor)**

**Groundwater Level and Pore Pressure:**
The ground water level and the pore pressure indicate the flow of water through soil. They are used to determine the strength, pressure, stability and settlement of the soil. Monitoring deformation and pore pressure will give valuable information for planning and designing countermeasures for foundation failure, rock slope failure, embankment sliding and so on. The pore pressure can be monitored by an inexpensive observation well which is simply driven in place where monitoring is required. Open stan pipe and closed hydraulic type piezometer are two type of simple and inexpensive piezometers.

**Earth Pressure:**
To monitor the magnitude and distribution of the earth pressure on walls and bridge abutments, a pressure gage can be used.

**Loads and Strain:**
To establish the integrity of the structural system, it is necessary to have load cell and/or strain gage measurements for piles, tie back anchors, and rock anchors. Pile load tests are generally performed by using a calibrated jack along with load cells and strain gages.

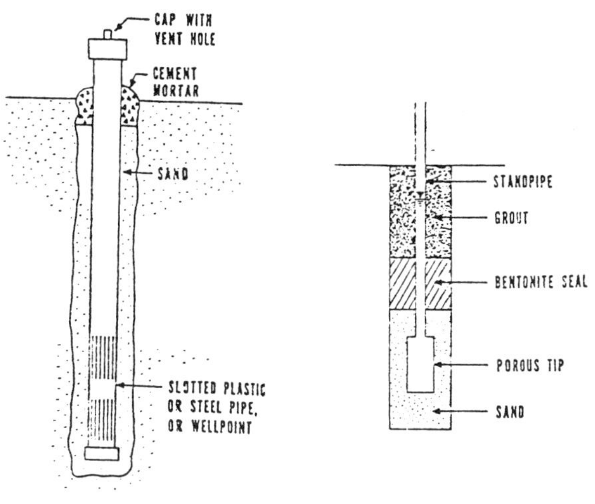


**Fig. 13.4 Well and Piezometer**

Load cell measurement are needed for tie backs and anchor rods to determine whether the required anchorage is attained. These measurements are also used to monitor any excessive or loss of tension forces of the rod due to anchorage slippage, improper installation, and soil movement. Strain gages will show any deformation of bridge piers, retaining walls, piles and bridge members.

**Implementation:**
The implementation of geotechnical instrumentation is usually undertaken by geotechnical engineers but the necessity of such instrumentation should be kept in mind during the project development. The following steps are suggested for any program of geotechnical instrumentation:

1. Define the condition and problems.
2. Define the purpose of the instrumentation.
3. Select the variable to be monitored.
4. Make prediction of behavior.
5. Select instruments.
6. Observe and record data.
7. Analyze and evaluate the data.
8. Develop countermeasure.

## 13–4 FOUNDATION REPAIRS

Foundation repairs vary considerably from project to project depending on the cause of failure, subsurface soil profiles, availability of materials, loading locations and economical factors. The following are a classification of foundation repairs:

- Upgrading capacity of existing foundation.
- Underpinning the existing foundation.
- Stabilizing the soils.
- Reconstruction of foundations.

The comparison of schemes is based on the accessibility of the area to be repaired, the type and quantities of materials, the length of time required for repairs, the determination of damages, appropriated structural analysis and the total cost.

## 13-4.1 STRENGTHENING

**Reinforcing:**

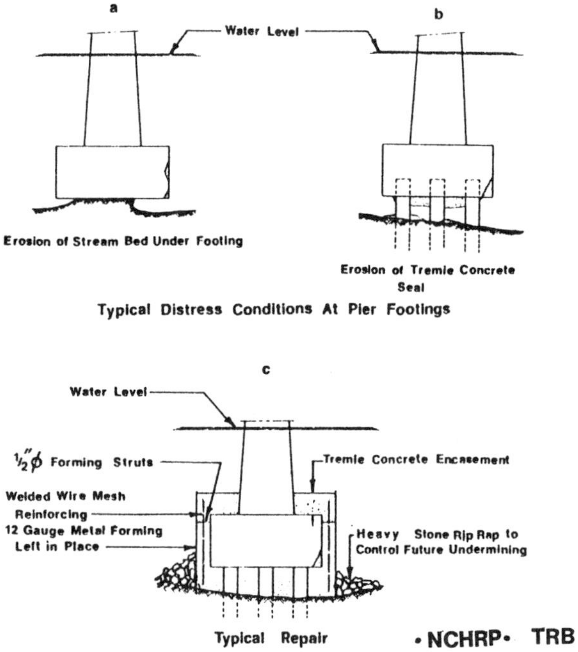


**Fig. 13.5 Reinforcing**

The existing foundation can be strengthened by widening and encasing existing footings with reinforced concrete and/or adding new piles around the perimeter of the footing. The new footing is then tied into the existing foundation.

**Drilling mini piles** behind existing wall: In this case the new piles relieve an unstable wall from the lateral soil thrust. It can be constructed to prevent or stop landslides.

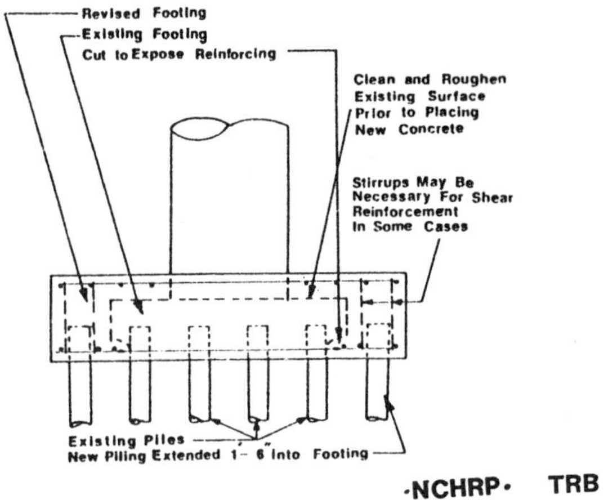


**Fig. 13.6 Increasing Capacity**

Drilling mini-piles to underpin an existing foundation: The mini-piles are rotary drilled through the structure to be underpinned. The load between the structure and the piles is transferred through concrete friction. It will increase the load bearing capacity of existing foundations and of new pile foundations.

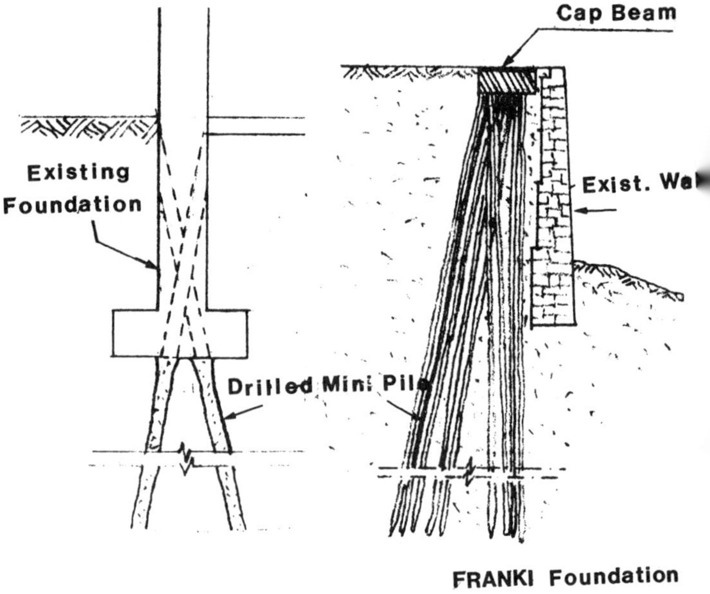


**Fig. 13.7 Drilling mini piles**

**Underpinning:**
Underpinning is used for constructing new footings under existing foundations. It can also be used for strengthening the existing foundation which can no longer safely sup-

port the existing loads. This condition causes soil settlements. Generally, it is used for constructing new foundations in small trench excavation underneath the existing footings. New piles are constructed near the existing footing and then loads are transferred by needle beams.

In the trench-pedestal method, the minimum size section of the trench is excavated in stages underneath the existing footings and a new pedestal is constructed in a small section by placing reinforcement and concrete in the excavation. Sometimes it is necessary to jack up the existing footing during excavation and construction to restore proper levels, and provide support. In all cases, a new pedestal footing should be built, section by section, to provide some support for the existing footings.

Sheeting and bracing for the excavation should be provided as necessary to prevent horizontal movement of the surrounding ground. Also, it may be required to provide for dewatering and avoid disturbance of bearing materials.

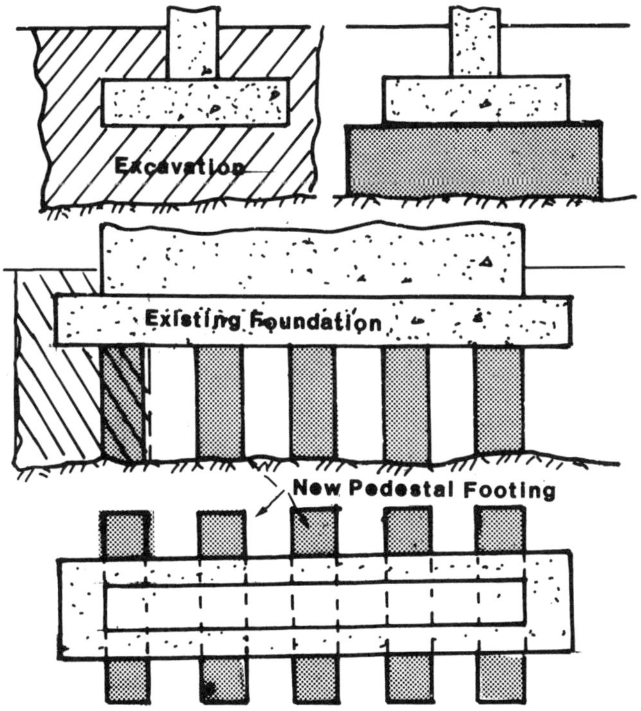


**Fig. 13.8 Underpinning**

In order not to damage adjacent footing areas due to potential cave in, a trench should be excavated by using the slurry wall method or some other protective system. The slurry wall method is applicable where the ground water table is high and soil is unstable for a narrow and deep trench. In slurry wall construction, as each bucket of soil is removed, bentonite slurry is poured into the excavated area to maintain hydraulic pressure from each side of the excavation. Finally, in the slurry filled excavation, the prefabricated reinforcement is placed and tremie concrete is pumped in and bentonite will be recovered.

**Needle Beam Method:**
This method requires the construction of a new foundation or driving piles adjacent to the existing footing. A horizontal supporting needle beam is then installed underneath the existing foundation. The needle-beams are supported on the piles, cribs, grillage, posts or new footings.

The final contact with structure is made by wedging between steel bearing plates or dry pack concrete. A careful examination of the structure and settlement should be made before and during underpinning in both the design phase and the actual construction. Maximum reduction of the existing load should be made before excavation. Also, stability and settlement of the structure and pressures on the braced excavation should be evaluated. For underpinning piles, the possible side frictional losses should be considered.

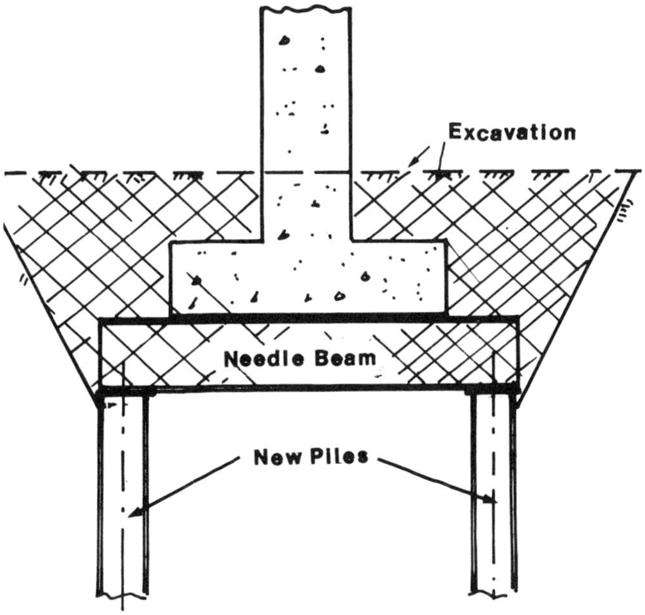


**Fig 13.8 Needle Beam**

## 13-4.2 STABILIZING SOIL

Soil stabilization has been used for increasing strength, decreasing permeability, reducing compressibility and increasing resistance to erosion. This can be generally achieved by drilling piles through weak strata of soils, replacing unsuitable soils with selected materials, and treating or injecting additives into soils to improve its properties.

**Slope and Wall:**
The following method is used for stabilizing soils for a wall:

1. Excavation of the top of the slope and flattening of the slope can be applied for deep cohesive soils.
2. Complete removal of unstable mass will be appropriate for a relatively small sliding movement. Rock or gravel fill at the toe of the slope is effective as this provides a shear key.
3. Pile walls or caissons will provide resistance to slide failure at the toe.
4. The use of the tie back system will prevent potential slide and settlements of the wall and slope by providing increased horizontal forces on the wall.

**Drainage Control:**
Since soil strength decreases with the increase of water in a soil, proper surface and subsurface drainage is needed. Some of these methods are as follows:

1. Place a drainage gutter at the top and behind the wall and place catch basins adjacent to the structure.
2. Surface treatment by seeding and paving. Place drainage facilities to collect and redirect surface water to prevent water penetration into unstable mass.
3. Seal joints and cracks to avoid seepage of water behind the wall.
4. Drill horizontal or sloping drain holes to drain out water trapped in soil.
5. Place deep drainage pipes or trenches to intercept ground water.
6. Place vertical sand drain wells to lower the water table and increase the bearing capacity by expulsion of water and consolidation of the soil.

**Seepage Control:**

1. A Sheet pile cutoff wall is suited especially for stratified soils with high horizontal and low vertical permeability or pervious hydraulic fill material. Interlocking steel sheet piling is used for deeper cutoffs. Steel sheeting must be carefully driven to maintain adequate interlocks between sheets. Steel H-pile soldier beams are used to minimize any deviation of sheeting in driving.
2. Grouting is applicable where the depth or the character of foundation materials make a sheet pile wall or cutoff trench impractical. Utilized extensively in major hydraulic structures, it may be used as a supplement below cutoff sheeting or trenches. A completely grouted cutoff is often difficult and costly to attain, requiring a pattern of holes staggered in rows with a carefully planned injection sequence and pressure control.
3. A slurry wall is suited for construction of an impervious cutoff trench below ground water table or for stabilizing the trench excavation when concrete foundation are placed with tremie method. In constructing slurry wall, A vertical sided trench is excavated by below ground water as a slurry with specific gravity generally between 1.2 and 1.8 is pumped back into the trench. Slurry may be formed by mixture of powdered bentonite with fine grained material removed from the excavation. For a permanent cutoff trench, well-graded backfill material is dropped through the slurry in the trench to form a dense mixture which is essentially incompressible after the trench is backfilled. The foundation wall is formed by placing concrete with tremie in the bottom of trench, thus displacing the slurry upward.

**Increasing Shear Strength:**
The following methods are used to increase the shear strength of soil:

1. Injection and grouting is commonly used for soils of moderate permeability. The common grouting methods are cement grouting,bitumen grouting, clay grouting and chemical injection.
2. Compaction is a common procedure used for embankments to squeeze out water by external forces, to remove and replace portions of unstable mass, and rearrange the soil particles by densification.

3. Soil and rock anchors are used for steep slopes of weathered rock, deteriorated retaining walls, tie back bulkhead systems, and tie down anchors to resist uplift force.
4. Freezing can be used for temporary stabilization during construction. In this process, in-situ porewater is converted into ice by refrigeration and the ice bond soil particles together to increase the strength and stiffness of the soil and make it impermeable. This process is slow and costly.
5. Drainage by the electro-osmosis method is applied for clays, silts and fine soils by passing direct electric current through saturated soil and stabilizing slopes of foundation excavations.
6. Riprap protection will prevent erosions and undermining of the footing.

**Grouting:**
Grouting will increase strength, decrease compressibility and permeability, and will seal large voids. This technique has been used for a variety of structures to change the character of the foundations to improve the deteriorated condition. The selection of the grouting method depends on foundation materials, durability of the grout, desirable improvements of foundation properties and costs. Studies should be made to determine overall stratification from test borings to locate cracks, cavities and loose strata, to determine seepage problems, to evaluate test performances of groutings using alternate methods and materials.

Cement grouting is effective if the voids are larger than the size of the grout mortars and also it is effective on coarse gravels or rocks. Cement grout is pumped into soil or rock under pressure to solidify and form a strong and durable foundation support. Disadvantages of cement grouting is the overflow of grout into surrounding areas. It is difficult to check whether the intended void is completely filled.

The design of the grout mix is fundamental to the success of a cement grouting program. A wide variety of properties can be given to cement based grout, depending on the additives and amount of water included in the mix. Fluid grouts can be made to flow long distances in rock cracks and crevices. In contrast, paste-like grouts with controlled set times can be used to carefully limit the grout flow range; special chemical additives have been developed to set up cement grouts almost instantaneously, even in flowing water conditions.

In addition to the water-cement ratio, additives such as sand, flyash, bentonite, accelerators, retarders, water-reducing agents, etc. must be considered. The flow properties of the grout under pressure must be balanced to the injection conditions. For example, water loss in grout under pressure may undesirably stop flow.

Placement of cement mortar grout should be carried out according to the plan. Split spacing drilling methods and vertical staging methods are often employed.

Compaction cement grouting techniques have been used to stabilize faulty support soils, to provide subsurface densification and improve inadequate support soils. Compaction cement grouting requires the injection of a stiff granular mortar grout into subsurface soils under high pressures to displace, densify and improve the adjacent soils.

The key element in any grouting system is the effective control and placement of the grout. In compaction cement grouting, this control is achieved by the use of a very stiff, granular grout mix with concrete slumps in the range of 1/2" to 2 1/2". The use of more fluid mixes, under the pressures required for effective compaction usually results in loss of control of the grout location, the splitting or fracturing of the ground, and migration of the grout away from the intended repair area. Injection of grout mortars with slumps of 3" to 5", although normally considered a stiff mix when pumping concrete, has been shown to easily disperse through consolidated soils and to run several feet away from the grout pipe tip, even under low pressures.

The selection of the mortar aggregate and water retaining agents such as cement, flyash, clay, etc. is extremely important to achieve a pumpable mix that still exhibits high internal shearing resistance to flow and can avoid sand blockages in the lines under high pumping pressures.

Compaction cement grouting techniques have been used for the following:

- Stabilize of poor backfill materials, retaining walls and under footings.
- Fill a large void under foundations.

- Treat and displace peak areas underlying structures.
- Eliminate settlements in weak strata.
- Replace and densify tunnel crown soils during soft ground tunneling operations to reduce surface settlement and eliminate the need for conventional underpinning.

Also, this technique could be applied to improve slope stability in conjunction with the reinforcing method.

The success of compaction cement grouting techniques depends on the correct understanding of the cause of the foundation distress or settlements and proper evaluation of the subsurface conditions as well as the competent application of the grouting procedure. In addition, a monitoring program should be carried out during grouting, to observe any potential surface heave and associated movements and to verify the effectiveness of the grouting operation by observation of slight surface heaves.

**Chemical Grouting:**
Chemical grouting has been used not only to rehabilitate a structure but as a construction method for cut-and-cover work and tunneling when conventional underground construction methods prove too costly or technically impractical. The injection of a chemical fluid into granular soils, rock, or concrete, which later gels and knits the soil particles together, chemical grouting has been used to solve soil stabilization problems on bridges. Chemical grouting can provide structural support of existing buildings, highways, railroad structures, utilities, stabilize excavations, and control groundwater. Generally it is used for ground water proofing and consolidations.

Chemical grouting is classified as following:

c: consolidation
w: waterproofing

1. Aqueous systems
   a. Silicate derivatives(c,w)
   b. Lingnosulfite derivatives(w)
   c. Acrylamide derivatives (w)
   d. Phenoplasts (c,w)

2. Non-aqueous systems
   Synthetic polyester, polyvinyl, polyepoxy (c,w)

3. Colloidal solutions in water
   a. Inorganic-bentonite (w)
   b. Organic-aliginate (w)

4. Emulsions
   a. Bituminous emulsions (w)
   b. Polyester, epoxy, vinyl (w)

**The Joosten's Method:**
Two fluids, sodium silicate and calcium chloride, are injected successively by introducing a solution of silicic acid followed by calcium chloride. Reaction between these two solutions is almost instantaneous and calcium silicate is precipitated in soil voids.

**Single Stage Sodium Silicate Grouting:**
Premixed sodium silicate with a setting agent such as sodium aluminate in water is injected into boreholes through pipe. Also available are a mixture of acrylamide and methylenebisacrylamide, epoxy and polyester resin grout.

The success of chemical grouting techniques depends on the correct evaluation of site problems, as well as competent application of grouting procedures. In addition, a concurrent monitoring program should be carried out during grouting, to observe any potential surface heave and to verify the strength and location of the grout underground. Where the soil has different properties, it may be necessary to change the type of grout during the grouting process by using a more penetrating type to compensate for reduction of permeability in the soil. Different types of grout should be used when one particular type of grout becomes ineffective and possibly cause rupture and heave. A fluid grout with greater penetràtion should be used as the operation progresses. It is important to select the proper grout for different types of the soil. Fluid grouts are excellent only in fine soils.

FHWA Research Report "Chemical Grouts for Soils" (FHWA-RD-77-51) have the following conclusions concerning grout suitability:

1. There is a series of grouts available and is suitable for a wide range of soils.
   a. Binghamian-type materials, with ce-

ment or clay in suspension, are applicable to coarse ground.

b. Somewhat viscous colloidal solutions (silica or lignochrome gels, organic or inorganic colloids) are suitable for soils of average particle size.

c. Pure non-colloidal solutions (organic monomers in aqueous solution) are suitable for very fine ground.

2. Grout users must know how to place the grout where it is needed (distribution systems all derive to some degree from the "sleeve tube" process).

3. When using a series of grouts at one site, one must know when to change grouts. The right moment comes when the more viscous grout is no longer effective, penetrating the ground only through force and rupturing it. One must monitor the heave benchmarks precisely to define the moment to change grouts.

Grouting not only concerns the selection of a grout, but involves the technology used for injection. Both the grout and the injection technology must be adapted to the terrain. Large test sites must often be carefully evaluated for the purpose of such matching.

The controls used at the worksite ensure the matching of grout and the injection technology to the site.

a. Concerning the grout, it should be verified that the specifications given by the manufacturer and by the operator are the same.

**Grain Size Ranges for Chemically Groutable Soils**

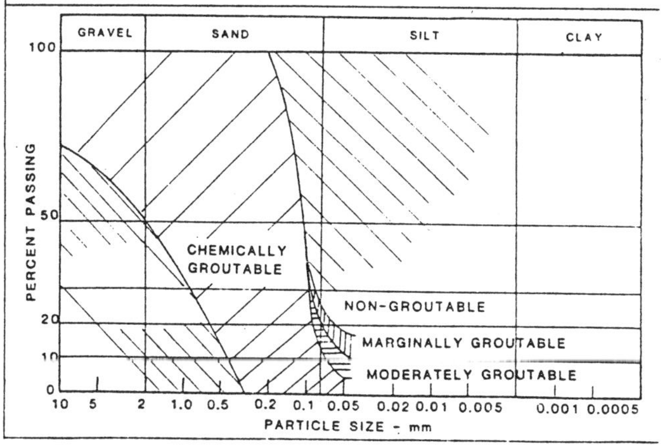


One rule-of-thumb suggests that soils with less than 12% fines (passing a No. 200 sieve are easily groutable, soils with 12 to 20% fines may be only moderately groutable, and soils with 20 to 25% fines are only marginally groutable. The accompanying figure shows typical ranges of chemically groutable soils.

**Fig. 13.10 Chemically Groutable Soil**
(Hayward Baker, Co.)

b. With regard to application, injection pressures, flow rates, ground movements, potential resurgences, etc., should be checked to ensure that the grout will penetrate the ground without ruptures.

**Soil and Rock Anchors:**

Rock and soil anchors represent a special application of the technique of prestressing in the field of rock and soil mechanics. The first applications of prestressed anchors began in the thirties, but they did not really become established until the sixties. The most common applications today are found in the stabilization of slopes, the anchoring of retaining walls and excavation faces and the anchoring of structures subjected to high uplift, overturning and concentrated forces, and the temporary support for sheeting and shoring of deep excavations.

Other applications of importance are those which have led to radical changes in methods of construction. If prestressed anchors had not been available, the raising and strengthening of dams would have been almost inconceivable, or at least would have required enormous quantities of materials. In either soil or rock, tiebacks can offer an effective design alternative to more expensive or less convenient techniques, such as internal struts, which often obstruct the work area.

Soil anchor systems have been used to stabilize existing retaining walls and abutments by providing increased horizontal forces on the wall. The anchor system has two categories: untensioned dowel system and post tensioned systems. The dowel system requires drilling holes to the required depth and inserting anchor bolts or rods filled with non-shrink grout into the holes. This will solidify and stabilize the soil. Post-tension anchors need high tensile strength rods or cables, anchors, bearing plates and nuts.

The length of an anchor is based on the required tension force, the types of soil involved and the interface strength between soil and grout. After the grout is set, the rod is tensioned by pulling on bolts with a hydraulic jack to the required strength. A major concern with the tension rod is corrosion. Tension rods should be galvanized or epoxy coated then fully grouted to protect against corrosion and to lock in post tension strength.

**Fig. 13.11 Tieback and Rock Anchor**

**Buttressed Wall:**
A wall and an abutment can be stabilized by constructing gabions or reinforced concrete block buttress at the toe of the failure slope or the wall. The strength of the buttress varies depend on the weight or the wall and the failure planes of soil. The rock filled gabion has advantages over a concrete buttress:

**Fig. 13.12 Buttress**

1) It is less expensive than a concrete structure.
2) A gabion wall does not require any drainage system. A concrete buttress requires weep holes to drain the water behind the wall.

**Grouting Scour:**

Obstruction to the flow of a river by bridge piers and abutments, and the water in the stream during high velocity flow and flood periods are factors causing undermining of bridge footings by scour if the foundations are not laid to a proper depth and have inadequate protection.

If scour occurs, the undermined area should be grouted with aggregate and pressure injected cement, cast-in-place cement, large size rock or cement bags. Cement bags come in different sizes and some have reinforcement steel wire mesh to fit the field conditions. First, divers place deflated bags at the undermined areas. Then, cement mortar is injected to fill the bags.

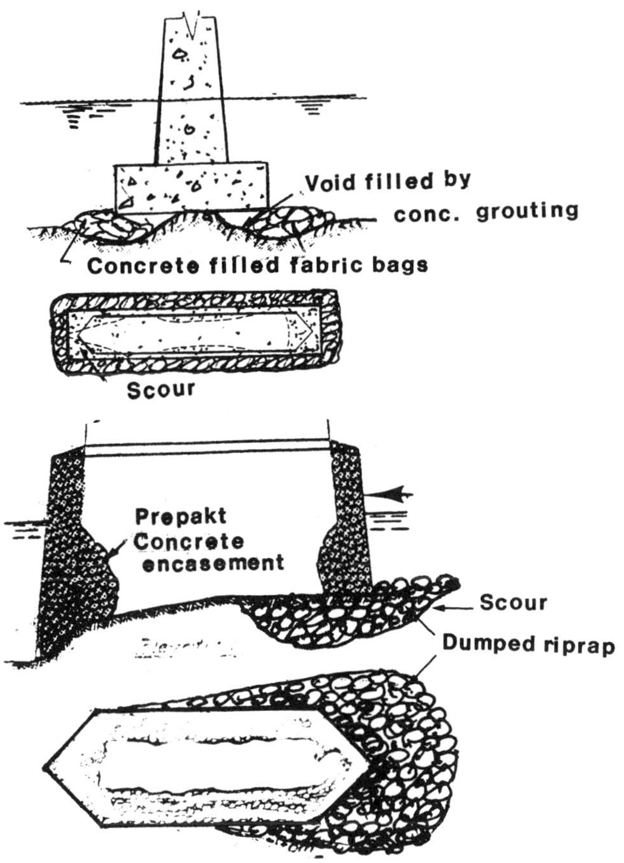


Fig 13.12 Scour Repair

## 13-4.3 PILE REPAIRS

The repair of piles requires field inspections and informations to determine the condition of the piles, interpretation of obtained data, the course of remedial action and evaluation of the repair technique.

Considerable damages can occur where piles are exposed to alternate wetting and drying cycles at splash and tidal zones and where steel piles are butt welded.

The thickness of steel piles should be measured from mean lower waterline to several feet below mud lines or to where the original pile thickness is intact. Since the loss of thickness is not directly proportional to the loss of capacity, the remaining capacity should be computed based on the new moments of inertia and areas of the existing piles. The location of damaged areas will influence the capacity and stability of the piles. Therefore, careful analysis should include such as accurately determining reduced cross sectional areas, moment of inertias and the length of each corroded section is essential to avoid costly unnecessary repairs. If the reduced cross section is adequate to carry the live loads to satisfy the minimum service level, it is preferred to apply a protective coating system to prevent further corrosion instead of replacing or strengthening weak members.

The pile repairs can be divided into repairs and protections. The repairs further classified as follows:

1. Jacketing.
2. Adding a supplementary section.
3. Replacing the section.

The protective systems are as follows:

1. Coating - Bitumen, Painting.
2. Barrier - Cladding, Jacketing, Plastic shell.
3. Sheathing - Monel®
4. Cathodic Protection System Sacrificial or Impressed

**PILE REPAIRS:**
Pile repairs include jacketing splicing, posting, underpinning and replacing:

**Jacketing:**
This system has been used extensively for concrete or steel piles to protect against corrosion and damage due to impact at,

Fig. 13.13 Jacketing Pier Foundation

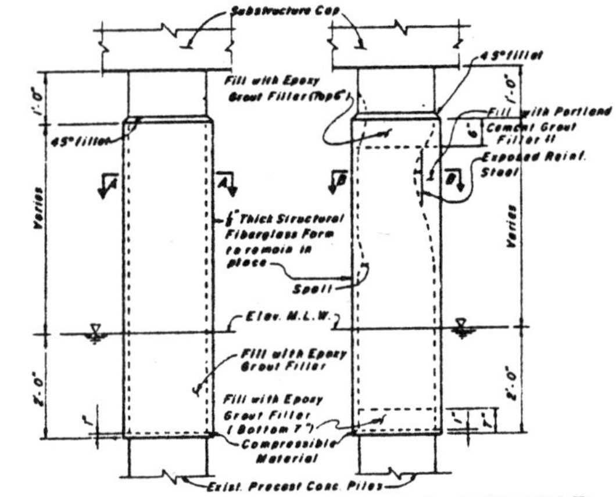


INTEGRAL PILE JACKET TYPE I INTEGRAL PILE JACKET TYPE II

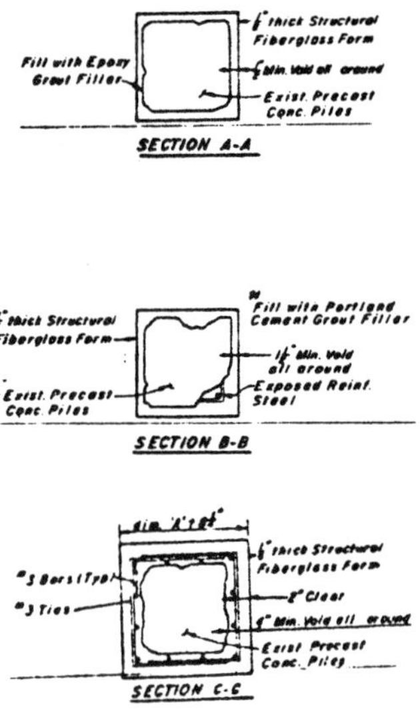


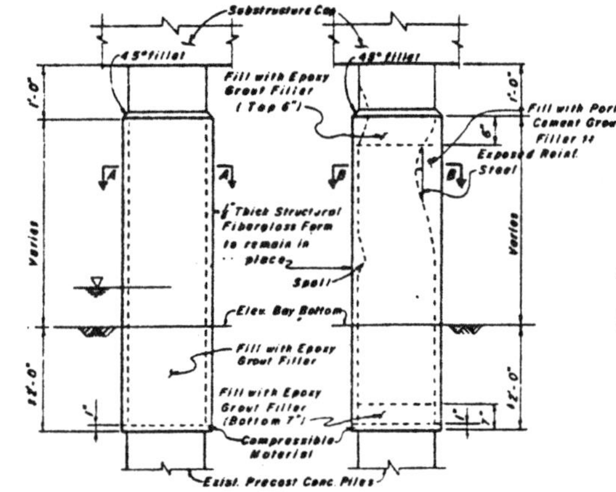


INTEGRAL PILE JACKET TYPE III INTEGRAL PILE JACKET TYPE IV

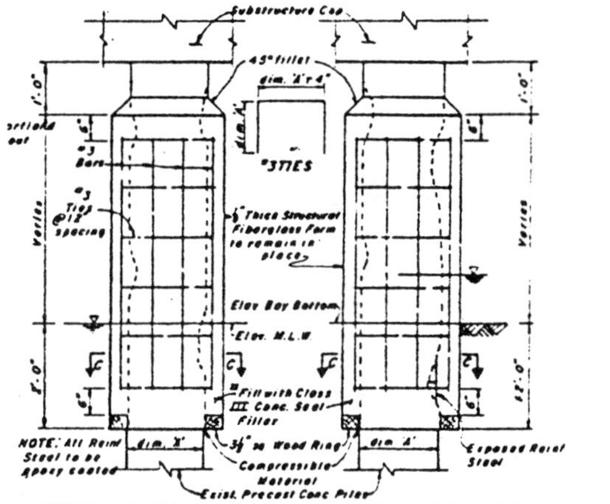


NOTE: An external bead shall be drawn against the wood ring seal during placement of Seal Concrete

INTEGRAL PILE JACKET TYPE V INTEGRAL PILE JACKET TYPE VI

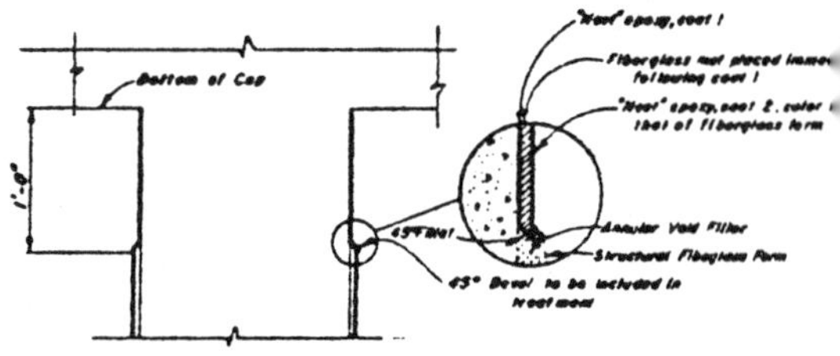


* Depth below bay bottom unless rock is encountered within the dimension shown. When rock is encountered the jacket shall extend to the required depth below the bay bottom or to the rock surface whichever is less.

DETAIL OF STANDARD TREATMENT OF TOP 12" OF PILING

**Florida DOT**

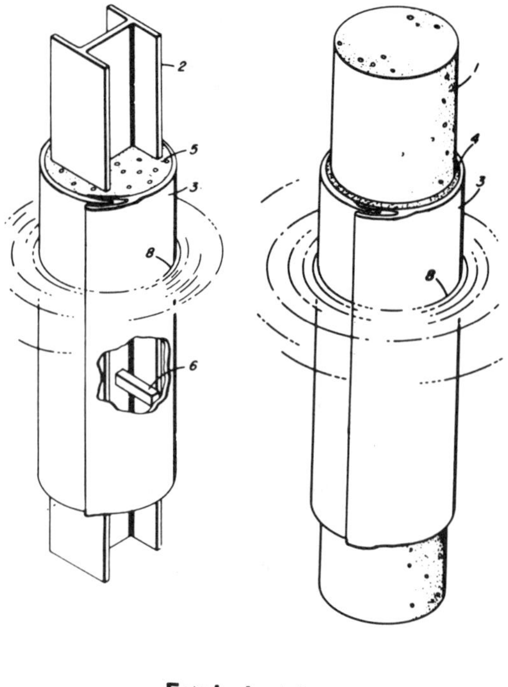

**Fox Industries**

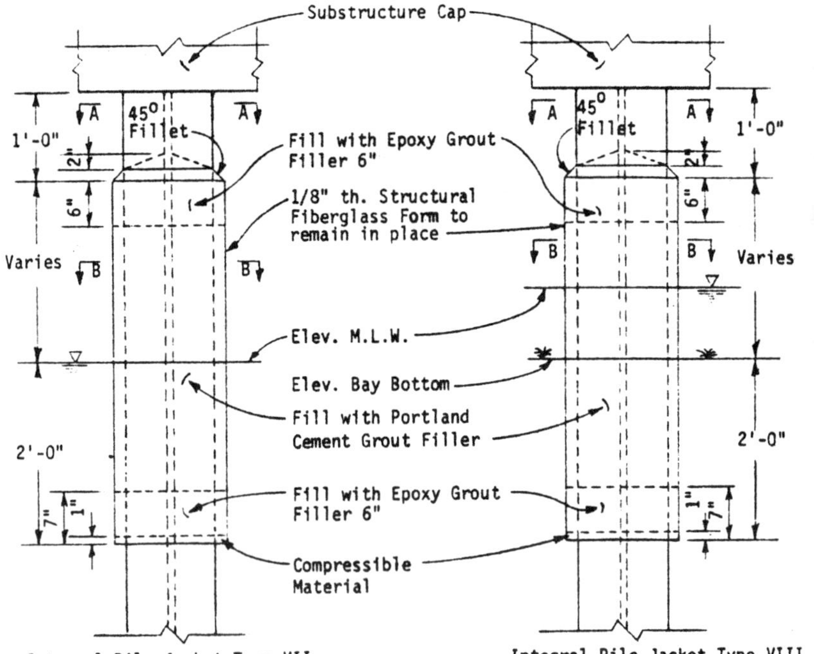


Integral Pile Jacket Type VII Integral Pile Jacket Type VIII

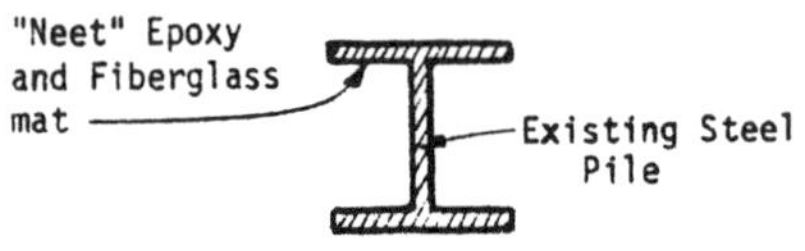


SECTION A-A

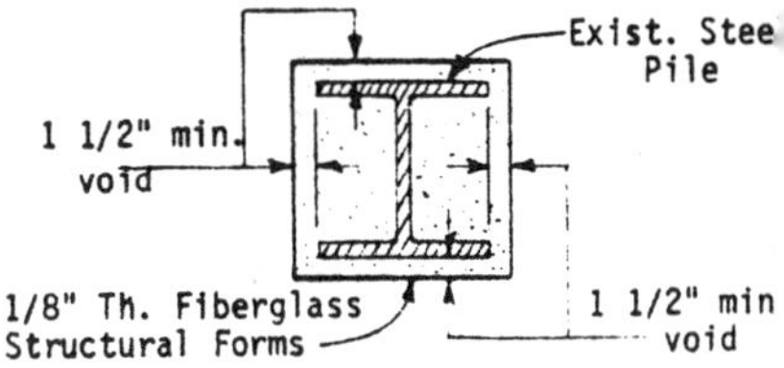


SECTION B-B

Pile Jacketing

above, and below water lines. It is also used on timber piles to protect from marine borer attack by jacketing the pile with concrete or by wrapping in metal sheathing.

Before jacketing, the deteriorated concrete and steel should be removed and the surface should be cleaned by means of sand blasting, water blasting or wire brushing. Steel wire fabric should then be installed around the existing piles. The grout mortar is filled after placing a jacket around it.

Commercially available products include the Symon Jacket® with Z-bead system of Symon Corp. which is composed of a fiberglass sleeve made of fiberglass reinforced polyester (FRP), the Fabriform® system of Intrusion-Prepakt, Inc. which is essentially nylon fabric forms holding concrete mortar while permitting the release of excess mixing water, the FX-70 system® of Fox industry and others.

**Underpinning:**
Underpinning requires an opening through the deck to drive new piles into the foundation soils. The new pile will be incorporated into the structure. However, if the superstructure is in poor condition, total superstructure replacement would eliminate this method.

**Posting**
Posting requires the removal of locally damaged portions of an individual pile and the splicing of a new section to the remaining portion of the old piles. The strengthening of piles does not protect piles and a protection system should be installed after strengthening the piles.

**Replacement**
Replacement of piles should be made to meet all current requirements of design and protections.

## PILE PROTECTION:

**Timber Piles:**
Deterioration likely occurs at the splash zone by marine borer attack and at the top of piles by fungus. The most common protection for timber piles is creosote treatment to retard deterioration. In areas where limnoria attack is expected, the piles are treated with either Chromated Copper Arsenate or Ammoniacal Copper Arsenate, followed by creosote treatment. For toredo or pholad attack areas, creosote or coal tar treatment is adequate with proper field treatment at the cut area.

**Concrete Piles:**
Deterioration occurs due to corrosion of reinforcing steel, freeze and thawing cycles, and wet and dry cycles at the splash zone. To repair concrete piles, it is suggested to remove unsound concrete before placing new concrete with a minimum of 3 inches cover.

**Steel Piles:**
Corrosion of steel piles can be prevented by coatings, claddings, concrete jacketing, sheatherings, and cathodic protection systems. The coatings and claddings provide a shield from moisture and oxygen thus preventing formation of differential aeration cells for rust. The cathodic protection system provide a sacrificial metal (anode) to reverse the flow of current and the corrosion will occur at an anode instead of the steel. This system is effective in the submerged areas but ineffective at the splash zone and in areas exposed to air.

Coal tar epoxy (Steel Structure Painting Specification SSPC No. 16) is frequently used as a protective coating for steel piles at the splash zone. Generally, epoxy is a two component system and requires two coatings but if it is shop painted, only one coat may be used. Hand applied epoxy type cladding materials are used in the splash zones and underwater. Other coating systems include metallic zinc, aluminum, and phenolic mastic.

Concrete jacketing is susceptible to crack on impact, therefore, this should not be used where excessive movement of the structure is expected by action of wave, impact, or debris. Jacketing is normally covered from a couple of feet below the mean low water line to several feet above high water lines or to full heights. It is recommended that the concrete cover should be 3 to 4 inches and the concrete mix should be aluminous cement, rich mixture, non-reactive selective aggregate and air-entrained.

The most widely used sheathing system is the metal sheathing known as "Monel 400"® a copper-nickel alloy manufactured by the International Nickel Company. This alloy shows excellent resistance to corrosion in a marine environment. However it is susceptible to damages by impact, scratching, cutting or wear out. Stainless steel can be used as sheathing materials.

Another type of sheathing material is an epoxy polymaide mastic which is used to wet the surface of splash zone by trowling two component system. Commercially available products are "Sikagard 694®" from Sika Corp., "A-788® Splash Zone Compound" from Koppers Co., and others. Flexible sheet plastic material such as vinyl, glass reinforced epoxy can be placed by clamps and premolded split shapes of plastic clamped in place and sealed with an epoxy polyamide adhesive.

**Cathodic Protective System**

Cathodic protection systems can be classified into impressed systems which require direct electric current and sacrificial systems using a sacrificial anode without an electric current.

The combination of cathodic protection and coating systems are sometimes recommended because cathodic protective systems will take care of the damaged coated areas under water and the coating will reduce the operational cost of the system. However, care should be exercised in selecting coating systems which are compatible with the cathodic protective system, to avoid blistering and loss of adhesion. Coal tar epoxy and epoxy offer excellent protection in conjunction with cathodic protection systems, but zinc based coating is not recommended to be used with impressive cathodic systems.

## REFERENCE:


1. "Subsurface Investigation for Design and Construction of foundation of Building" ASCE Journal of the Soil Mechanics and Foundation Division, May-August 1972
2. "American Wood Preserver's Association Standard" AWPA
3. "Soil Mechanics, Design Manual 7.1" and "Foundation and Earth Structures Design Manual 7.2" U.S. Dept. of Navy, Naval Facilities Engineering Command, May, 1982
4. "Handbook of Corrosion Protection for Steel Pile Structure in Marine Environments" American Iron and Steel Institute, 1981
5. "Maintenance of Maritime Structures" The Institute of Civil Engineers" London, England 1978
6. "Chemical Grouts for Soil" FHWA/RD-77-51, FHWA June 1977
   "Design and Control of Chemical Grouting" FHWA/RD-82/036 to 039, FHWA 1983
7. "Rock Slope" FHWA Training Course Text Material
8. "Geotechnical Instrument" FHWA Training Course Text Material
9. "Design of Pile Foundation" NCHRP Practice 42, TRB, 1977
10. "Underwater Inspection and Repair of Bridge Substructure" NCHRP Practice 88, TRB, 1982
11. "Geotechnical Instrumentation for Monitoring Field Performance" NCHRP Practice 89, TRB, April, 1982
12. Hunt, Hal W. "Current Practice in Design and Installation of Driven Piles" TRB Report 664 Bridge Engineering pp200
13. F. Jacob, "Lessons from Failures of Concrete Structures" ACI 1964
14. Goodman, Louis, R.H. Karol "Theory and Practice of Foundation Engineering" MacMillan Co. NY 1968
15. Heinz, Roney "Protection of Piles Woods, Concrete, Steel" Civil Engineering - ASCE Dec. 1975
16. Jumkis Alfred "Foundation Engineering" Intext Education Publisher, Scranton, PA 1971
17. Schnabel, H. "Procedures for Testing Earth Tiebacks" ASCE National Structural Engineering Meeting, 1974
18. Weatherby, David E., James W. Sigourney "Tieback and Their Application" ASCE National Geotechnical Conference, 1981
19. "Piling and Foundation News" Associated Pile & Fitting Corp, Clifton, NJ
20. "Franki Facts" Franki Foundation Inc. Boston, Mass.
21. "New Life for Deteriorated Piles" Intrusion-Prepakt Inc. Cleveland, Ohio
22. "Pile Restoration and Preservation System" Symons Corp. Des Plaines, IL
23. "Grouting Soil-Stabilization Water Control" Intrusion-Prepakt Inc. Cleveland Ohio
24. "Ground Modification" Hayward Baker Co. Odenton, MD
25. "Nicholson Rock and Soil Anchor Manual" Nicholson Anchorage Co. Bridgeville, PA
26. "Technical Information - Schnabel Foundation Co." Bethesda, MD
27. "Rock & Soil Stabilization System" VSL Corp, Los Gatos, CA

# 14 WATERWAY

## 14–1 GENERAL

The hydraulic problems of bridges are the results of the following three basic characteristics of the stream:

- Scours
- Natural and artificial degradation
- Aggradation

**Scour** is defined as the removal of stream bed material by stream or tidal currents. It may occur naturally as the result of channel constriction or changes in the flow pattern. The greatest scour usually occurs during floods and serious scour problems are encountered around piers and abutments.

**Degradation** is the gradual lowering of the stream bed material in the channel. Degradation may occur naturally with the stream as the means of erosion, or it may be caused due to stream straightening or the result of the stream flow, or the reduction of sediment loads. When the stream bed is lowered considerably due to degradation, the existing bridge foundation becomes unstable.

**Aggradation** is the deposition of sediment into the river channel. The problem caused by aggradation is the raising of the water surface. This will cause a reduction of the waterway opening and lead to possible scour.

**Waterway** opening is influenced by the geometry of the channel, type of stream (braided, straight and meandering) the amount of debris carried during a flood, and the adequacy of the waterway freeboard. Where large quantities of debris and ice are expected, sufficient freeboard is of the great importance to prevent structural damage during flood.

**Counter-measures** of hydraulic problems include replacement of the bridge, modification of the bridge, relocating the channel, installing an armor type revetment of channel, and constructing flow control features such as spur dikes and check dams. Before any determination is made regarding to in-kind replacement, relocation and repairs, the hydraulic adequacy of the existing bridge and stability of the stream should be carefully examined with cost, damages to property, highway functional requirements, backwater requirements, and environmental restriction by the state or National Flood Insurance Program of the Federal Emergency Management Agency (FEMA).

## 14–2 SCOUR

Scour is a natural phenomenon that takes place primarily in alluvial streams but it could happen in all stream beds including streams in coastal areas, lakes and reservoirs through wave and tide actions. Generally, scour is accelerated by the presence of flow constriction due to bridge abutments, piers, spur dikes, etc. The undermining of bridge foundations and approach roadways due to scour is a common problem. The cause of scour can be grouped as follows:

1. Increase in stream discharge, increased velocity and depth.
2. Temporary increase of discharge caused by flood, snow melt or dam gate opening.
3. Sharp bends and constrictions in the channel. A shift of channel will redirect the channel.
4. Increased sedimentation load.
5. Base level changed to lower level or lower level is raised by deposit.
6. Change in channel slopes which will affect discharge.
7. Obstructions in the path of flow such as piers, abutments, and debris.
8. Constriction of flow by riprap, other protective material at river bank or natural or man made changes.

Scour is classified either as local scour which is limited to localized small areas around the pier or abutment footings, or as general scour which happens around a bridge crossing over the entire channel.

**General Scour** is caused by the constriction of stream flows due to the reduction of waterway openings. This reduction increase flow velocity and shear stress, and will result in more bed materials and sediments being carried out of the area faster than it is transported into the area. As the stream bed is lowered, the input and output of sedimentation at the contracted section (bridge crossing) become equalized and the general scour will stop.

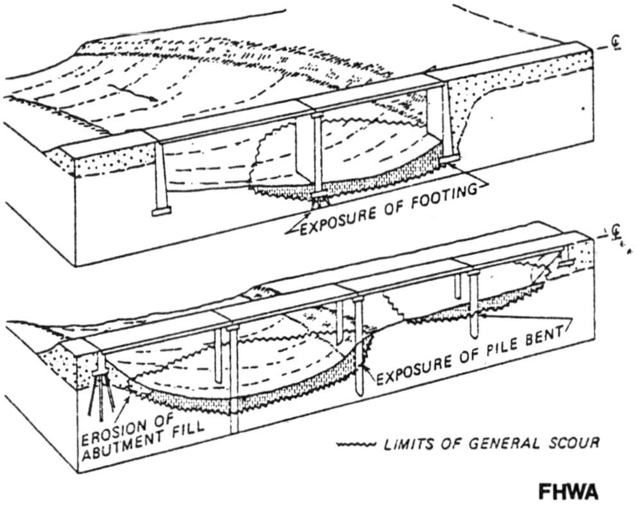


**Fig. 14.1 General Scour attributed to contaction of flow or to channel deepening at the outside of a bend.**

**Local Scour** is caused by the vortex of the stream from localized flow changes introduced by the presence of constrictions such as bridge piers, embankment ends and spur dikes. The vortex is a result of the accumulation of water at the upstream edge and subsequent acceleration of flow around the nose of the constrictions. The vortex action transports bed materials around the piers and abutments and causes undermining. The typical pattern is for the obstructed flow to dive down and pass by the obstruction as a roller in the scour. The shape of the obstruction will also influence the size of the scour.

Scour is associated with various type of stream behavior such as migration of the meander, shift of the deepest part of the channel, straightening of a bend, shift of a stream and change of the bends. The deepest scour will occur at a sharp bend where the channel is constricted.

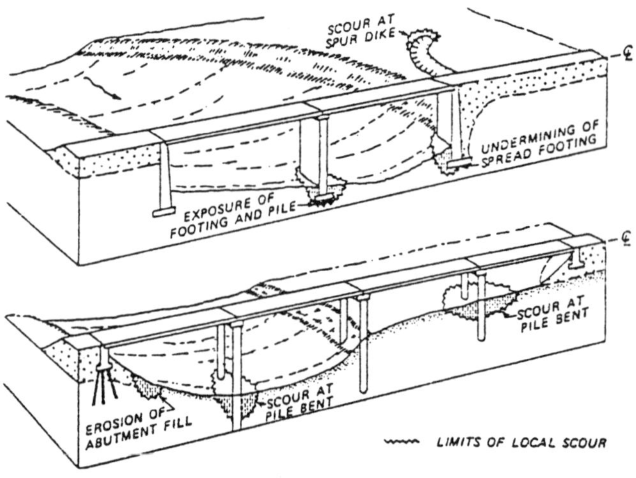


**Fig. 14.2 Local Scour attributed to the effects of obstructions to flow**

The lateral shift of the channel may affect the deep section of channel and cause braid or scour at the abutment.

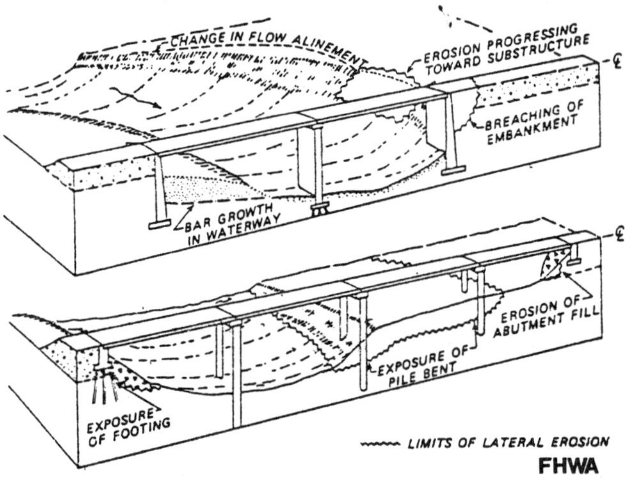


**Fig. 14.3 Lateral Stream Erosion attributed to erosion at a bend**

## 14—3 DEGRADATION AND AGGRADATION

Degradation and aggradation are long term changes in stream channel elevation due to the increase or decrease of the transportation of sedimentation. Although gradation takes place naturally, man made activities such as stream mining, construction of dams and structures, or channel alterations may cause severe changes. These changes can take place at the bridge crossings or upstream or down stream of the bridge site upsetting the natural equilibrium of the stream bed material. All alluvial rivers are a potential of grade changes by degradation or aggradation and it will gradually readjust to a normal stable condition.

**Degradation** is a result of bed materials being removed from the bed of a stream at a faster rate than deposited for a long extended period of time. This deficit of transportation of material accelerates the erosion of the stream bed and bank causing progressive down cutting of the bank. Most severe degradation occurs during floods and it is directly related to the availability of sedimentation and the rate of transporting them. Degradation can cause exposure of the foundations, piles, undermining of channel lining, lateral instability of the slopes and bank, and eventual bridge failure.

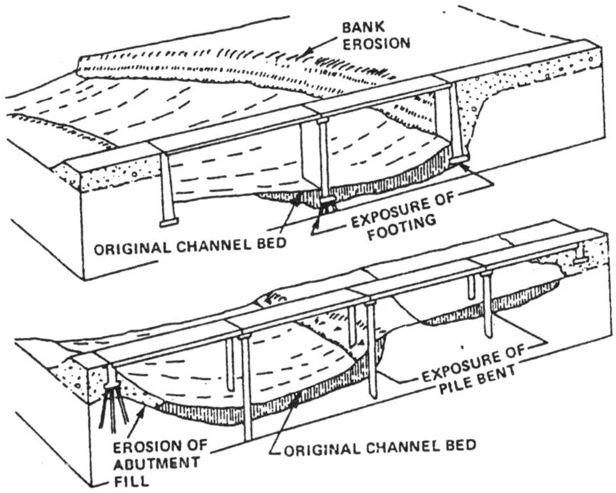


**Fig. 14.4 Degradation**

**Aggradation** occurs when more sedimentation is brought in than removed. This may be the result of an upstream dam or any construction activities at upstream and will result in the reduction of the bridge waterway opening thus forcing overflow on the roadway embankment during flood. The deposits will increase sinuosity and redirect the stream flow into a completely different course. Aggradation can also cause bank erosion and the formation of lateral bars in the channel which requires periodic channel cleaning.

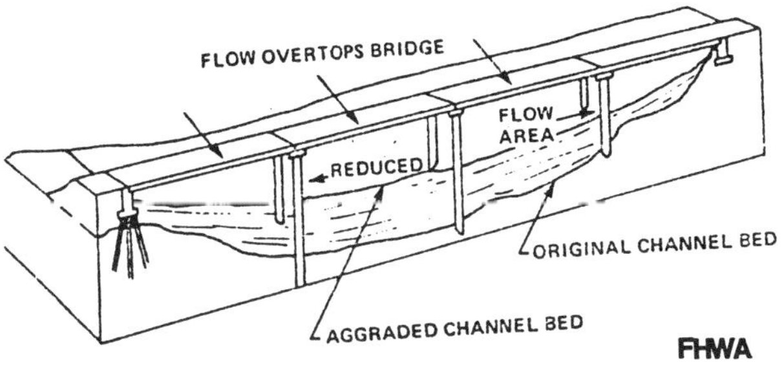


**Fig. 14.5 Aggradation**

## 14–4 COUNTERMEASURES

### 14-4.1 GENERAL

Countermeasures differ substantially from one state to another because each river system has a different hydraulic environment: stream type, climate and geology. However, general concepts of countermeasures can be used from one region to another. It is necessary to analyze the contributing hydraulic factors such as bridge length, foundation type, skew angle, pier shape and location, geomorphic factors of sinuosity, flood frequency, water level, channel and velocity before selecting appropriate countermeasures.

A general study includes following steps:
1. Obtain field data, background and historic information of the river.
2. Analyze the information and other impacting activities and perform an economic analysis as needed.
3. Draw conclusion and select countermeasures.

**Data Collection:**

Field data collection should be made on a permanent fixed basis such as a bridge deck and old data should be rechecked for any changes. Since the change of stream character at one section will affect others, field surveys should be extended to the upstream and downstream section of the channel.

To achieve the best scour measurements, it is suggested to collect extensive data at several different locations and different times, especially after flood. This must be done to prepare effective countermeasure to obtain realistic depth before the scour is filled back.

Aerial photographs and topographic maps should be used to obtain information regarding the location of the bridge, properties of the stream, and any commercial development in the surrounding area.

The required geomorphic and environmental information includes data and information concerning stream size, bank slope, bank lines, stream discharges, vegetation, wetland, fish and fish habitat, sedimentation load, flood plains, degree of sinuosity, braiding, flow habit, bed materials and man made activities (channel alteration, cleaning vegetation, gravel mining, dam, reservoir and land development).

The structural information should contain the location of the crossing, pier types and locations, types of structures and foundations, span length, constricted flow, the type and location of abutments and the waterway opening.

Hydraulic information should be gathered regarding water elevation, discharge, flood frequency, duration of flood, velocity of flow, channel geometry, slope, alignment, and opening.

**Analysis:**

After collecting the required information and performing an analysis of site characteristics, a hydrologic and hydraulic analysis should be performed to produce a backwater curve, evaluate scour conditions, establish foundation depths, and to determine waterway openings and structural requirements of flow control. If the bridge is located in the flood plain, economic and flood risk analysis is required to select the most economical alternative for the bridge crossing.

In selecting replacement or a relocation of the bridge, certain criteria should be used for hydraulic analysis.

1. Backwater should not increase substantial flood damage to properties upstream. Generally, the Federal Emergency Management Agency (FEMA) will approve a maximum one foot increase of elevation of the base 100 years flood. Some states, however, have stricter requirements..
2. The freeboard should allow free passage of debris or ice. (1 to 3 ft.)
3. The substructure configuration and spacing should be selected to minimize the obstruction to flow.
4. Basic design flood: Generally 50 years is accepted but in some locations the 100 year flood year is mandated by the State status. As an exception, lower flood years are accepted by allowing overtopping if this meets functional requirements and results in an economical design.
5. Foundation design criteria should be established to avoid scour or other failures.
6. Acceptable risks of damages to properties due to increased back water should be set.
7. Allowable velocities at stream crossing should be established. Higher velocities would increase the damage to downstream properties, the bridge itself and other features.
8. Highway profile and alignment: The vertical profile is generally selected based on topography, traffic requirements and geometrics. It is preferred to avoid a sag vertical profile at stream crossings to prevent jamming at the bridge fascia or overtopping the bridge during flood. If overtopping is unaviodable, overtopping can be minimized by using open type railings instead of solid parapets. It is desirable to have a crest vertical curve at stream crossings to provide passage of ice and debris, and permit overtopping at roadway embankments. The horizontal alignment should be chosen such way that the stream crossing sections are normal to flow directions without creating obstructions in the channel.

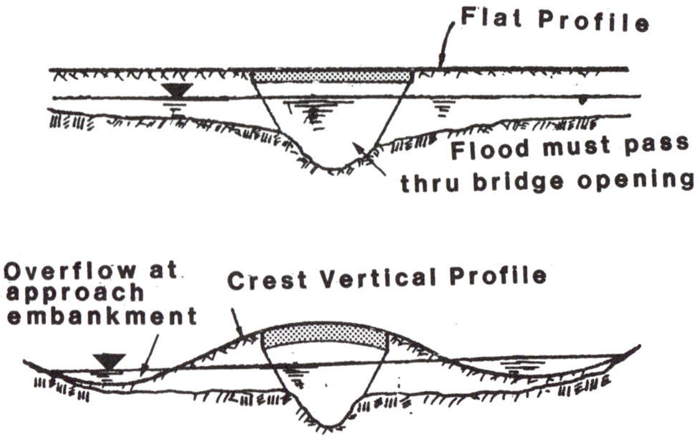


Fig. 14.6 Highway Profile

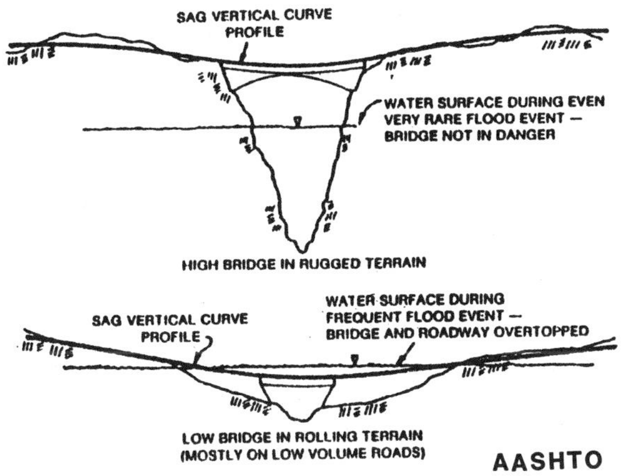


AASHTO

**Fig. 14.7 Sag Vertical Curve Profile**

9. In flood plains, tributary auxiliary openings may be constructed as relief openings in place of constructing viaduct crossing over entire flood plains.

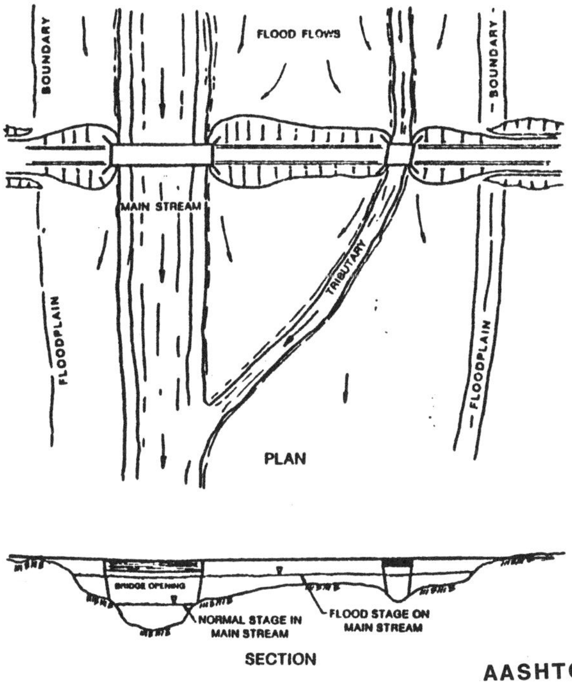


AASHTO

**Fig. 14.8 Auxiliary Bridge**

Analysis methods range from simple mathematical equation to rigorous analysis using computer programs for entire regions. The complete hydrologic study and risk analysis for an entire region would be expensive to apply at every bridge site. Simplified sedimentation and transportation relationships, scour analysis and general stream behavior based on a clear understanding of the difference of regional stream systems, hydraulic problems, and potential hazards at the bridge site may be adequate to draw conclusions of problems and make recommendation for countermeasures.

The economic analysis should include the cost of construction, the public cost of a traffic detour, annual maintenance cost, reconstruction and repair cost of the bridge and the surrounding properties and other cost such as litigation. The economic analysis of each alternative provides the comparison of cost and its engineering functions and enables the selection of the optimum design with minimum cost.

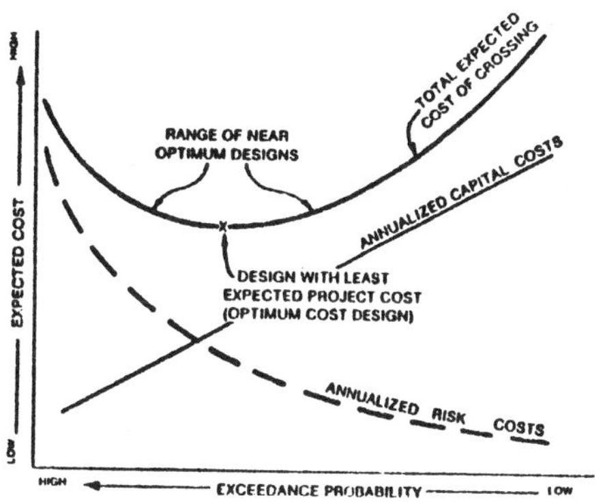


**Fig. 14.9 Optimum Design**

However, there are many other restrictions beyond the economics and engineering considerations. The selected final alternative may be strictly based on social, environmental or legal considerations rather than engineering and economic considerations.

**Countermeasure:**

The following is a summary of possible countermeasures:

**Modification of Bridge**

- Sheet pile protection
- Underpinning or Jacketing
- Pier protection
- Driving deeper pile and grouting undermined area
- Raising the bridge
- Replacing a solid parapet with an open parapet and railing for overtopping of stream during flood.

**Replacement of bridge**

- Larger waterway openings with longer spans and fewer piers
- Deeper pile foundation
- Raising the bridge
- Relocating the bridge.

**Roadway and Channel**

- Constructing overflow sections at roadway
- Constructing relief culverts
- Realignment of the channel
- Removal of debris from the channel.

**Armor**

**Flexible Revetment:**

- Riprap or Broken concrete
- Rock and wire mattress
- Gabion
- Precast concrete block
- Planted vegetation
- Car bodies.

**Rigid Revetment:**

- Concrete pavement
- Sacked concrete
- Precast concrete slope paving
- Asphalt concrete paving
- Grouted riprap
- Concrete filled fabric mat.

**Flow Control**

- Spur - Rock or earth embankments and sheet piles
- Retards- Timber piles, sheet piles, rail, or tetrahedron field
- Spur dike - Rock or earth embankments, sheet piling
- Drop structure - Check dam, cutoff wall.

## 14-4.2 SCOUR

The extent of countermeasures for scour is best guess work as to what may be required to correct a particular situation. In some cases, it may be possible to evaluate the scour damage and required contermeasures by visual and physical field inspection. In most cases, the countermeasure can be determined by taking depth soundings as a guide to evaluate the scour. When there is no involvement of gradation problems, the following countermeasures can be used:

1. **Underpinning** with replaced aggregate and pressure grouting, cast-in-place concrete or concrete filled fiber bags.

**Fig. 14.10 Underpinning**

2. **Placing riprap** around abutments and piers to guard against progressive erosion: Dumped riprap is widely and effectively used to fill scour holes and bank slopes on meandering streams and bends. It is proven to be effective when adequate quantities and proper sizes of riprap are used.

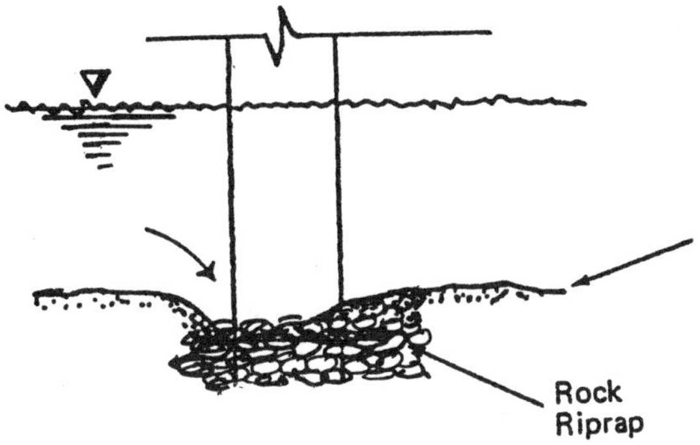


**Fig. 14.11 Dumped Riprap**

3. **Jacketing** the foundation by deepening and jacketing with concrete.

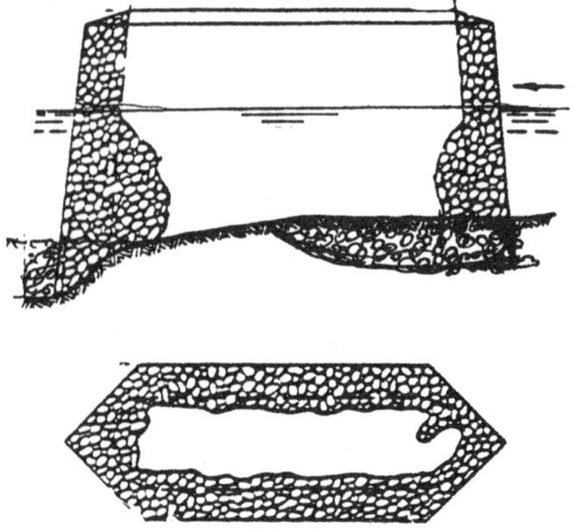

**Fig. 14.12 Concrete Jacketing**

**4. Sheet-pile** cofferdam enclosure and cut-off wall.

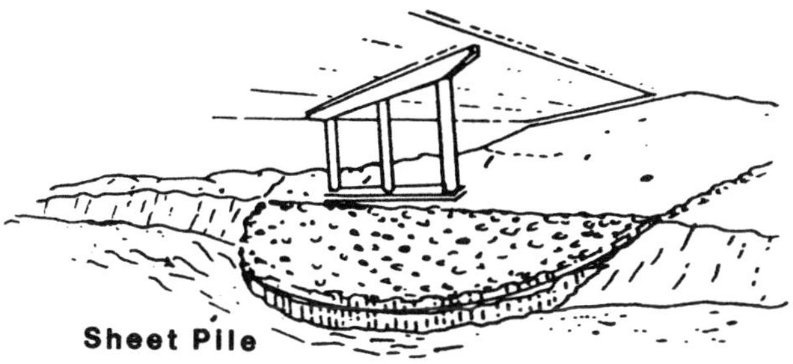


**Fig. 14.13 Cofferdam**

**5. Placing spur dikes or jetties** for partial diversion of the stream to reduce the scour depth within bridge openings and to protect abutments.

Spur dikes are used at bridge embankments to increase the effective waterway openings and improve flow conditions through the opening. Spur dikes reduce the scour depth at bridge sites. Therefore, the use of spur dikes provides the protection of existing bridge abutments and piers against scour.

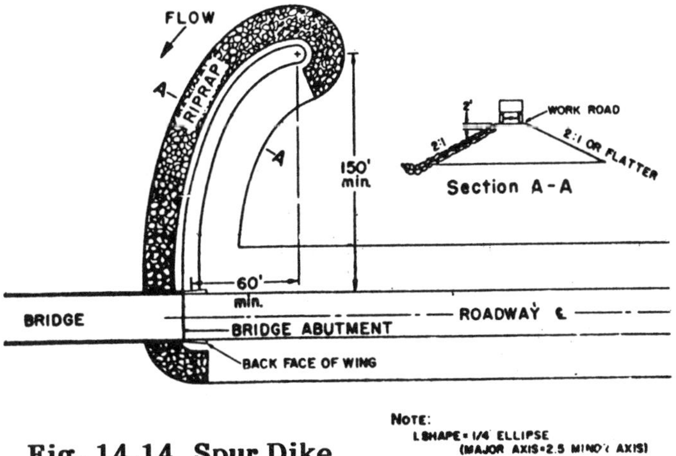


**Fig. 14.14 Spur Dike**

6. Replacing the bridge with greater waterway opening.

7. Relocating the bridge.

### 14-4.3 DEGRADATION

Countermeasures for degradation should be selected based on the level of hydraulic analysis and a clear understanding of the river system. The following are some effective methods used for small stream.

**1) Armor:**

Armor and revetments are used to protect the abutment, bank and embankment from erosion and include the following methods:

- Dumped riprap
- Sacked concrete
- Concrete slope pavement
- Gabion
- Grouted rock.

**Dumped Riprap:**

Dumped riprap is widely used as slope protection of abutments and banks. The rock should be sufficiently keyed into the stream bed below the anticipated scour depth and the banks should form a wide distance around the abutments to prevent bank slope failure due to lateral erosion induced by degradation. The success of riprap protection is depending on size of rock and interlocking capacity between rocks which are keyed into banks. The riprap keyed-in-place by tamping is found to be more stable than loose riprap. Thus, It is essential to make sure the dumped rock is properly keyed.

On mildly sloped embankments, properly keyed riprap will provide adequate protection of abutments against slumping or slope failure. On steeper slopes, concrete slope protection is suggested to be used except where internal slope failure is anticipated.

For general channel gradation, dumped riprap would likely cause failure even if the toe is well protected unless rock is used in conjunction with some type of spurs or retards which are supported on deeper piles.

**Gabion:**

Gabions which consists of rock and wire mattress, would be effective protection for slope stability of abutments on small channels. It is advantageous to use in regions of dry climate where slope is high and steep, and where stones of required size are ready available. In order to act as one unit, the gabion wire mattress should be well anchored from the top cf the bank to the entire length of the slope.

**Concrete Pavement:**

Concrete slope protection methods have been used with mixed results. The failures are generally attributed to the inadequate protection at the toe and end of the slope, and unstable foundations. It is desirable to provide heavy riprap or a cut-off wall at the toe to prevent undermining and to install drainage pipes or weep holes to relieve hydraulic pressures behind concrete pavement. The use of channel lining has been unsuccessful to solve degradation problems unless it is combined with the construction of a check dam.

**Sacked Concrete:**
Concrete grouted riprap and concrete filled fabric mats have been used as rigid revetment. Combinations of bulkheads and riprap are also used where bank slopes have steep cuts against mild fill slopes at abutments. The failure of bulkheads are likely caused by undermining or outflanking at the end so that the bulkhead should be driven into sufficient depth below the degradation line. Bulkhead are also used to control internal slope failure where the embankment slope is steep.

**2) Bridge Foundation**

The bridge foundation can be stabilized by the following:

- Underpinning
- Jacketing
- Increasing the depth of footing
- Armoring the foundation with riprap or concrete filled mats
- Additional piles.

For medium to small size bridges, the abutments should be located beyond the area where undermining is susceptible from bank slope failure. The minimum number of required piers should be placed in the channel, and all foundations should be placed well below degradation level.

Other factors to be considered for bridges are natural stream properties, behaviors, and possible effects of man-made works in the streams. The aerial photograph can be helpful to assess the lateral stability of a channel, the rate and direction of lateral erosions, stream behaviors and establishing the key bridge dimensions.

**3) Flow Control Structure**

Flow control structures are used to control the direction, velocity or depth of flow within or outside of the channel. The following are a few types of control structures:

- Spurs
- Retards
- Spur Dikes
- Check Dams
- Dike
- Jack Fields.

The important property of any flow control structure is its permeability. While an impermeable structure would deflect and redirect the stream, permeable structures would reduce the velocity of stream. The permeable structure provides openings through which water may pass through gaps between piles, wire mesh, etc.

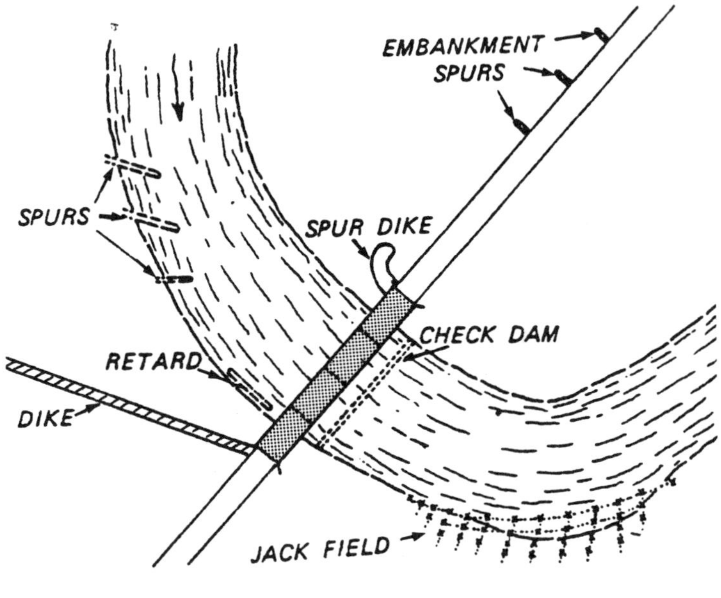


**Fig. 14.15 Placement of Flow Control Structure**

**Check Dam:**
Check dams have been a successful method of protecting the bridge foundations by maintaining original stream beds. When it is constructed just downstream of bridge crossing, this may cause downstream erosion. Therefore care should be exercized in designing check dam to minimize the downstream degradation. It can be built of sheet pile, rock riprap, gabions, concrete or concrete filled fabric mats.

**Spur:**
A spur is a linear structure projected from the bank into a channel to alter flow directions, induce deposition or reduce flow velocities along the bank. Spur structure types include the following: timber piles, sheet piles, steel piles, and earth or rock embankments.
For bank protection and control of flow at bends, spurs are more cost effective than riprap revetments. The permeable spurs generally provide adequate performance and are effective when placed at right angles to the banks or are inclined downstream. Impermeable spurs can cause flow disturbances and bank erosion. Spurs are suitable to use at shallow channels with slight upstream inclination.

While conventional impermeable riprap spurs are expensive and causes lateral stream erosion, permeable spurs offer low cost, flexibility, and easy maintenance.

**Retard:**
Retard is normally used at the toe of the bank to check velocity and to induce deposits which changes the trend of erosions and replace the lost materials or to maintain an existing flow alignment. These types include timber jacks, steel jacks, fence, tetrahedrons, piles and embankments.

Retards provide better flow lines along the bank line and can withstand the direct impingement of the flow on a bank. Retards have been effective for a small to medium width channel which have flow velocity not exceeding 1.5 to 1.8 m/sec and for channel with sand beds.

The required permeability is inversely proportion to the radius of curvature of the bend so that the sharper the bend, the greater impermeability of the retard is required to control lateral erosions.

Retard is best suitable in humid regions where natural revegetation would help bank slope stabilization but is susceptible to damage by ice and debris in cold regions.

**Spur Dike:**
The functions of spur dikes are to prevent erosion at abutments from flood flow concentrations at upstream face of embankments and redirect floodwater parallel to the desired channel alignment so that the full bridge waterway opening is utilized. The spur dike does not eliminate scour, but reduces the scour damage and moves the scour holes away from the abutment.

Most commonly used spur dikes are rivetted earth or rock embankments with quarter of an ellipse shape. It is recommended to use for the control of meanders which are migrating toward bridge or embankment.

### 14-4.4 AGGRADATION

Countermeasure for aggradation include the following:

- Channelization
- Bridge modification
  (Raising or Replacement)
- Continuing stream maintenance.

**Channelization**
This appears to be best solution to remedy aggradation problems but extensive channelization including dredging, clearing poor bed material have not been a complete success because it does not alter the natural instability of the river systems.

**Bridge Modification**
Increasing span length or raising super structure are very costly.

**Routine Maintenance**
Routine maintenance of the stream bed is the most cost effective where aggradation is caused by a temporary source or on small channel where problems are minor. As an alternate for continuing maintenance, controlled sand and gravel mining can be used from debris basin at upstream of the bridge, however, caution should be exercised and detailed analysis should be made not to alter natural balance of stream system.

## 14—5 PREVENTIVE MEASURES

There are several methods to minimize flood damages such as riprap protection, concrete slope protection, deep foundations, pile foundations and spur dikes.

In designing bridges over waterway, the estimate depth of scour is required because scour will occur unless the stream bed is inerodible rock. The following three design factors should be considered for preventive design:

1. The flood frequencies and magnitudes.
2. The flow pattern for each flood with geometry of given design.
3. The resulting scour: If a scour prediction can not be made explicitly, the foundation should be designed more conservatively to accommodate unexpected scour depth.

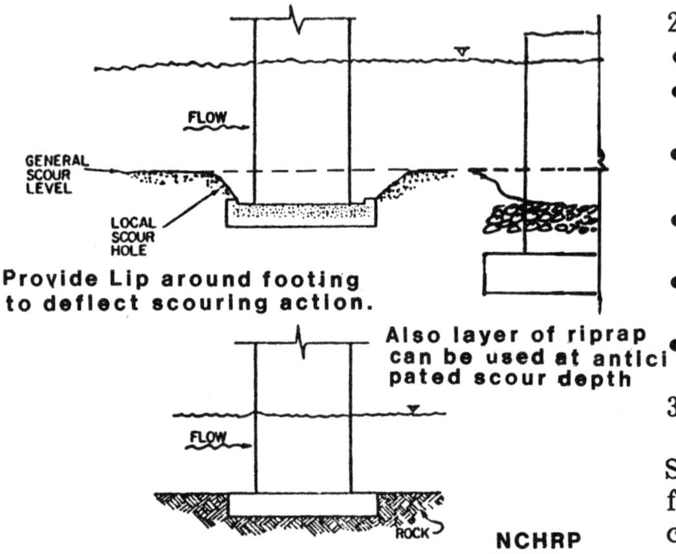


**Fig. 14.16 Spread Footing**

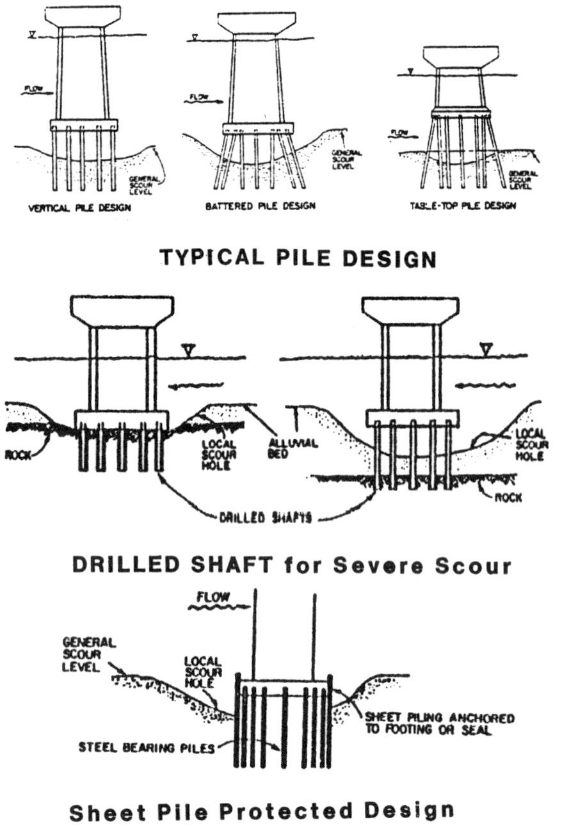


**Fig. 14.17 Pile Foundation**

The following preventative measures are suggested to minimize damages:

1. Abutment scour problem can be reduced by:
- Riprap protection
- Sheet piling
- Spur dike
- Deep foundation.

2. Pier scour can be minimized by
- Alignment with stream flow
- Place footing below estimated scour depth
- Provide riprap at expected scour depth
- Provide slope for scour holes so as not to disturb natural flow of water
- Use pile foundations or caissons to provide stability when scour occurs
- Sheet pile protection.

3. Erosion and silting prevention

Stream banks should be checked frequently for erosion. In order to minimize erosion, channel widening should be considered. This will permit discharge of the same volume of water with a lower velocity.

Accumulation of silt would reduce the freeboard between the water surface in a stream and the bottom of the bridge. Stream which are susceptible to significant silt deposits should be watched for the presence of silt and deposit material should be removed from the stream bed before serious damage occurs.

Gravel bars often form in a stream and divert the water from its normal channel and causes erosion of the bank and scour of bridge foundation. the removal of gravel bars would reduce the erosion and scour problems.

## 14–6 RISK ANALYSIS

### 14-6.1 GENERAL

Because of limited available funds from the Federal and states government for bridge repairs, maintenance and reconstruction, it is necessary to evaluate the cost of several alternatives and select an optimum economic scheme. In an economic analysis, the total cost should include the construction cost, preconstruction engineering cost, costs of traffic detour and maintenance, the user costs related to traffic interruption, reconstruction of flood damage, and potential backwater damages to surrounding properties.

An optimum design may require higher capital investment to reduce future repairs or flood damages. A risk analysis is require to evaluate the future backwater damages and their overall effects on the total life cycle cost. The risk analysis can be used to override the current criteria for a particular situation. For example, where the 100 year flood hydraulic design is mandated by the Flood Insurance Program or legislative requirements, and if the risk analysis assures that the 50 year design flood will furnish the best design with minimum cost and satisfy the social, environmental and community requirements, the agencies would allow the use of 50 year flood design.

### 14-6.2 RISK ANALYSIS

In accordance with Executive Order regarding floodplain management, all design selections for floodplain encroachment should include either an analysis or assessment of the inherent flood related risks to the highway facilities and surrounding properties.

Risk analysis includes the probable losses of the following categories:

- Direct damages to bridges and roadways.
- Traffic related losses.
- Losses due to additional backwater damage in the upstream.

It is assumed that all the losses can be determined and that an optimal design can be selected by minimizing the sum of expected losses and capital costs. The analysis involves the following:

1. Selecting several alternative designs.
2. Selecting several distinct discharges to adequately represent the flood frequency curve.
3. Repetitively perform hydraulic and economic analysis with each discharge for each alternative. These include the following:
   - Analysis of economic losses
   - Application of risk analysis
   - Determination of annual capital costs
   - Computing total the economic cost for each alternative
   - Perform sensitivity analysis to determine which variables are most influenctial in deciding least cost.
4. Finally, select the alternative that will provide the least total cost considering probable losses as well as capital costs.

Risk analysis furnishes a tool to quantify in dollars the flood damages to the bridge and the effects of the proposed bridge to properties and stream stability. The risk and construction cost should be measured on the same annual cost basis for easy comparison. A risk analysis is an useful tool to identify the items which imposes risks to the bridge and to be risked by the proposed bridges in term of construction cost. The least costly structure may turn out to be a short span across the river at valley level with occasional overtopping by a low frequency flood. Also, current high interest rates make lower initial construction cost with slightly higher risk the most attractive alternative. However caution should be exersized in selecting the alternative of the lowest cost which may not satisfy other conditions.

Risk could be reduced by spending extra money on the bridge. This analysis allows the engineer to assess the significance of flood damages and to revise the original design by adjusting embankment height, adding or deleting spans, placing channel protections, or relocating the bridge. The annual cost of improvements or additions should be computed and compared to the benefit or the risk that they are intended to prevent.

In the process of selecting alternatives, the result of the risk analysis is just one of main considerations. However, this provides a method of assessing the cost of constraints by comparing the cost of the encroachment to the cost of the alternative which has the least total cost.

### 14-6.3 DESIGN PROCEDURE

The following steps can be used for hydraulic design of bridges:

1. Define the scope of the project.

2. Select procedures for computing water-surface elevations, flow distributions, flood frequencies, and discharge and stage hydrographs. Collect field data.

3. Compute the natural water-surface profile, lateral flow distribution, and peak discharge frequency relation.

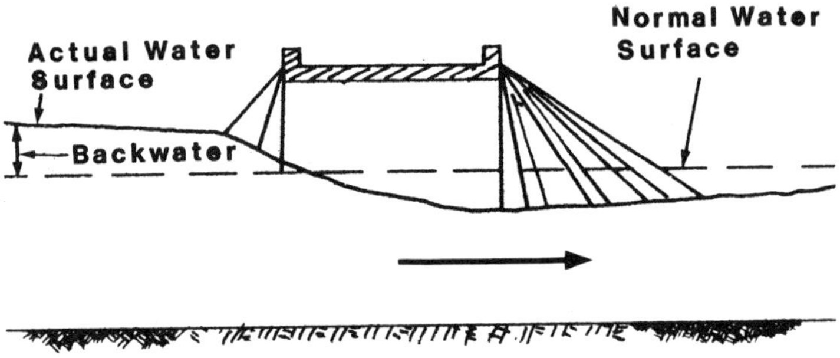


**Fig. 14.17 Water Surface Profile**

4. Define the combination of bridge span length, freeboard, embankment elevation, degree of longitudinal encroachment and countermeasures.

5. Determine the practical range of design alternatives including type of structures. Non-engineering constraints should be listed. In some cases, these restrictions limit the choice of alternatives.

6. By using a traditional discharge arrangement select a few alternatives.

7. Perform a hydraulic analysis for each alternative. Compute the expected water-surface profile, the design discharge, the velocity of flow through the waterway opening and the distribution of flow over the road.

8. Determine the cost of detouring traffic in case of an embankment overtopping or washout. Compute the overtopping time and depth of flow, and estimate the cost of interruption of normal traffic. The detour time is sum of the overtopping time and time to repair major damages. The interrupted traffic cost represents the difference between the cost of the normal route versus the detour route.

9. Evaluate the flood damage due to increase of backwater by computing the difference between the damages with the bridge in place and the natural condition without the bridge. The flood damages include only additional or incremental damages which is resulted from backwater related to the construction of a stream crossing. Assessing backwater damage requires considerable field data collection for hydraulic computations and defining specific flood plains. The data collection efforts should vary dependent on the complexity and importance of the individual encroachment.

10. Evaluate the potential damage of embankments and pavements and estimate the cost of repairs for erosion damage of the shoulders, fill material and loss of pavement surface.

11. Evaluate the potential damages to a bridge superstructure by the impact of large floating debris and ice. Also, scour damage around piers, abutments and spur dikes should be assessed. Estimated the potential cost of structural repairs.

12. Estimate capital cost. The capital cost includes the cost for the embankment pavement, bridge and protective measures and cost of easement. Engineering and right-of-way costs are excluded in analysis. It is important to determine the life span of the structure and applicable interest rate. Higher interest rate will tend to shift the emphasis on the lower construction cost while lower interest rate give favorable response to higher construction cost. A longer life span is insensitive to the annual cost but a shorter life span is very sensitive to the annual cost. The annual capital cost can be obtained by multiplying the capital cost by the capital recovery factor that is a function of the interest rate and life of the structure.

13. Perform risk analysis: The total loss is the sum of the incremental backwater damages, traffic related losses, embankment damages, and other structural losses. The annual risk is the total loss multiplied by the probability of occurrence for the given flood.

14. The total expected cost is a sum of the annual risk and annual construction cost. The minimum or least total expected cost alternatives are determined from plotting all total expected cost in relation to important bridge variables, bridge length and approach embankment height.

15. Plot total expected cost versus embankment elevation and plot design discharge and flood frequency interval versus embankment elevation.

16. Select a bridge length and embankment elevation at the point on the surface where total expected cost is minimum. Determine the corresponding design discharge and return interval. The optimum combination defines the lowest point on the lowest curve of total expected cost versus embankment elevation.

## 14–7 EROSION CONTROL

The soil erosion and sedimentation from construction activities will cause severe damage to the environment such as water pollution, restricting stream flow, damaging adjacent land, and disturbance of the natural ecology of the streams. Soil erosion requires additional expenses for repairs and delays the completion of the project. The erosion control measure should be considered from study phase and be incorporated into the plans and the specification. In assessing the degree of control need, the following should be considered: surrounding lands, proposed type of the structure, water supply, fish and wildlife, recreation facilities, soil type, climate and rainfall, vegetation on the bank, etc.

The following steps should be taken in preparing plans:

1. Identify and collect information about erosion sensitive and sedimentation problem zones around the construction site.
2. Determine soil erosion by using water-soil equation known as the Universal soil loss equation. If the soil loss is excessive, erosion control measures should be prepared to reduce the soil loss to the acceptable level.
3. One of the following should be used for erosion control measures:

**Mulch:** This is used to increase infiltration, decrease runoff, protect ground from erosion damages. Hay, straw and wood fiber are extensively used as mulch.
**Filter:** It is use to remove sediment from water. Widely used materials are berm, brush, staked straw bales, sand and gravel, and nylon cloth.
**Berm:** It is used to control and divert the runoff surface water.
**Drain Channel or Diversion Ditch** is used to stop the movement of sediment.
**Barrier** of straw bales or wire fence.
**Riprap Protection,** slope protection, walls, spur dike, jetties, check dams, and gabion.
**Vegetation** and seeding.

**REFERENCE:**


1. "Hydraulic Analysis for the Location and Design of Bridges" AASHTO, 1982
2. "Bank and Shore Protection in California Highway Practice" Calif. Dept. of Transportation, Nov., 1970
3. Gilje, Stephen "Stream Channel Degradation and Aggradation" FHWA Public Roads, Vol. 45, No.2, Sep., 1981 pp45–52
4. FHWA Hydraulic Publication
"Hydraulic of Bridge Waterway" Hydraulic Design Series No.1 (Revised Second Edition 1978) "Debris-Control Structures" Hydraulic Engineering Circular (HEC) No.9, March 1971 "Use of Riprap for Bank Protection" HEC No.11, June 1967 "The Design of Encroachments on Flood Plains Using Risk Analysis" HEC No. 17, October 1980
5. Brice, J.C., J.C. Blodgett "Countermeasures for Hydraulic Problems at Bridge; Vol. I Analysis and Assessment (FHWA/RD-78-162) and Vol. II Case Histories for Sites (FHWA/RD-78-163)" FHWA, September 1978
6. Brown, S.A., R.S. Mcquivey, T.N. Keefer "Stream Channel Degradation and Aggradation Analysis of Impacts to Highway Crossings" FHWA/RD-80-159, FHWA, March 1981
7. Hopkins, G.R., R.W. Vance, B. Kasraie "Scour around Bridge Pier" FHWA/RD-79-103, FHWA, Feb. 1980
8. Keepfer, T.N. "Stream Channel Degradation and Aggradation: Causes and Consequence to Highways" FHWA/RD-80-038, FHWA, June 1980
9. Schneider, V.R., K.V. Wilson "Hydraulic Design of Bridges with Risk Analysis" March 1980
10. "Scour at Bridge Waterway" NCHRP Practice 5, TRB, 1970
11. "Underwater Inspection and Repair of Bridge Substructure" NCHRP Practice 88, TRB, 1982
12. Israelsen "Erosion Control During Highway Construction Manual on Principles and Practices" NCHRP Report 221, TRB, 1980
13. "Rehabilitation and Replacement of Bridge on Secondary Highways and Local Roads" NCHRP Report 243, TRB, 1981
14. "Design Manual, Soil Mechanics, Foundation and Earth Structure" NAVDOCK DM-7, Dept. of Navy
15. "Bridge Design Manual" Wisconsin Dept. of Trnasportation
16. "Drainage Manual" Minnesota Dept. of Highway

# 15 CULVERT

## 15—1 GENELAL

A culvert should be able to handle the designed capacity of water flow through a channel and also carry traffic over the culvert. Therefore it should meet both hydraulic and structural requirements to function properly. Rigid (concrete) and flexible (corrugated metal) culverts are available depending on the magnitude and the distribution of loadings. For a small opening, a pipe is generally used while a concrete box, corrugated steel or arch culverts are used for larger waterway openings.

Corrugated metal, concrete, or vitrified clay are used in various ways to control the flow. Concrete or clay pipe use the bell and spigot type or the tongue and groove type. Most corrugated steel pipe is galvanized and coated with asphalt or bituminous material to protect the base steel from corrosion and wear out.

Reinforced concrete box culverts are commonly used for waterway crossings. Twin boxes are often built instead of a larger single box. Arch culverts are sometimes used for spans up to 20 feet and with waterway openings of up to 300 square foot.

The precast concrete culvert has following advantages over cast-in-placed concrete culvert:

- The cost is competitive because the casting become a repetitive and labor efficient therefore on-site labor cost is reduced.
- The construction time can be reduced substantially when compared with cast-in-placed culvert.
- If the standard section is adopted, further savings of material and labor cost will be realized.

The disadvantages are weather and temperature restriction of using epoxy bonding agent at joints, need of proper foundation preparation, and possible future joint leakage.

To recognize the benefit of precast culvert, 1980 AASHTO Interim added an article 1.15.7 for precast culvert and adopt precast culvert standard of ASTM C789 and C850 as AASHTO M259 and M273 for standard box culvert in the material specifications. These new AASHTO Standard Precast Box Culverts are designed according to load factor design method and 5,000 psi concrete strength therefore it is suggested to check required design criteria before using them.

A culvert has the advantages of providing an uninterrupted view of the highway. A culvert also provides easier and faster construction with relatively lower cost and increase of hydraulic capacity with water stage up to soffit. Disadvantages are joint separations; accumulation of silts on barrel; erosion of fill slope; scour at outlet; corrosion and abrasion problems.

For small stream crossings, culverts are considered a more cost effective replacement structures than bridges.

## 15—2 DEFECTS AND REPAIRS

### 15-2.1 STRUCTURE

**1. Reinforced Concrete Culvert:**

A box culvert is usually large enough to permit a man to walk or go through and repair defects without space problems. Most common defects such as cracks, spalls and excessive wear can be patched with non-shrink cement mortar. Deep wear of crumbled concrete is patched by pressure grouting or shotcreting.

Settlement of a section or a sagged invert occurs due to differential settlement of the soil, horizontal movements or built up water pressure caused by clogged weep holes. The settlement can be realigned by grout injection under the invert slab.

In a new design, settlement problems may be solved by adopting the following steps:

1. Construct the culvert with a camber. The camber will accommodate the settlement of the central portion of the embankment.
2. Remove all expansive and unsuitable material and replace them with crushed stone or compacted soil.
3. If removal of unsuitable material is not feasible, preload the area before placing the culvert or use a pile foundation.

Joint spalls and leakage between sections require removal of overburden and repacking of joint sealers.

A deteriorated culvert can be replaced with a slightly smaller sized corrugated metal culvert by inserting a new one into the existing openings. The void between the old and new can be filled with concrete grout and the capacity of the culvert can be increased or maintained by using improved inlet or outlet details.

For water that has PH value 4 or lower, a protective coating is desirable.

**2. Metal Culvert:**

The most serious defects of metal culverts are rusting and corrosion at the invert. Corrosion can be retarded by coating the invert with tar or bitumen or by paving the invert.

For a new design the following should be considered:

1. Determine the water PH value and flow velocity at peak design flow.
2. Where peak flow velocity is excessive, allowance should be made for wearing and tearing in selecting the thickness of metal. If velocity is more than 8 feet per second, consideration should be given to a paved invert.
3. Below are recommended types of culvert for various PH value and abrasion:

| PH Value | Type of Pipe or Culvert |
|---|---|
| 4 or less | Stainless steel; Vitrified clay; Reinforced concrete with vitrified clay liner. |
| 4 to 5 | Galvanized steel with full asphalt, epoxy coating or asbestos bonded bituminous coating.* |
| 5 to 8 | Galvanized steel without coating or Aluminum. Additional thickness of metal is desirable to allow corrosion. |
| 8 to 9 | Galvanized steel with full asphalt or epoxy coating. |
| Over 9 | Vitrified clay or Lined concrete culvert. |

* Asbestos bonded bituminous coating is factory applied protection to corrugated steel pipe or plate. The asbestos fibers are embedded in the molten zinc during galvanizing.

| Environment | Type and Protection |
|---|---|
| High sulfur: | Stainless steel<br>Vitrified Clay<br>Concrete with coal tar epoxy coating. |
| Salt: | Aluminum<br>Galvanized steel with asbestos bonded bituminous coating and paved invert. |
| Abrasion: | |
| • Modest | Bituminous coating. |
| • Substantial | Bituminous coating with paved invert. |
| • Severe | Welding several steel channel to the invert by planking. |

Among various existing coating systems for metal culverts, galvanized steel with asbestos bonded asphalt coating and paved invert appears to be the best system. Bituminous coating has adhesion problems but its durability can be improved through proper surface treatment and primer. Bituminous coating is satisfactory where abrasion is mild and the PH value is 5 or more. For PH values less than 5, asbestos bonded bituminous coating appears to perform better. Where the PH value is between 4 to 5, Concrete paving can be used for structural steel plated pipe. Since most deterioration of metal culverts occur on interior bottom half, protection may be necessary only on the bottom half of the culvert.

Besides abrasion and PH factor, the electrical resistivity of the water and soil adjacent to the pipe will influence the deteriorations. It is known that uncoated galvanized steel pipes perform well when the PH value is between 5 and 7 with resistivities above 10,000 Ohm-Cm and when the PH value is between 7 and 10 with minimum resistivity of 2,000 Ohm-Cm. Uncoated aluminum alloy pipe performs well when the PH value is 4 to 9 with minimum electrical resistivity of 500 Ohm-cm or greater. Aluminum has excellent corrosion resistance of all metals used for culverts.

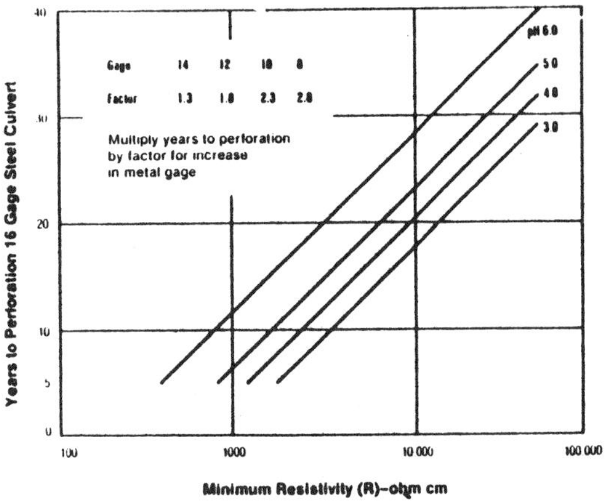


**Fig. 15.1 Estimating service life of galvanized steel culvert** (California)

4. It is recommended to increase the thickness of metal for corrosion protection. The rate of corrosion is fairly constant at the given locations and its increased life expectancy can be proportioned to the additional thickness of the metal. For galvanized steel culverts, the following rates may be used in absence of the exact information of the site:

| Type of Coating | Rate(In. per Yr.) |
|---|---|
| Uncoated | 0.0025 |
| Bituminous Coating | 0.0017 |
| Coated and paved | 0.0005 |

5. In addition to PH value, abrasion, and resistivity, sulfur and chlo-ride contents of the soil, degree of permeability, site preparation, beddding, compaction procedure should also be considered in selecting the culvert.

**3. Stone and Brick Arch Culverts:**
Common defects of stone arches are spalling, splitting and crumbling of the exposed stones of bricks due to weathering and abrasion caused by the stream at the waterline. The deteriorated surface can be jacketed by shotcrete and cracks be sealed by grout injection.

Another common defect is cracking of mortar joints between stones or bricks. The loosened stones or bricks could cause the collapse of the arch rings and lead to total arch failure. The joints should be repointed, loosened stone should be removed, and the voids should be grouted by concrete

**Fig. 15.2 Liner plate Strengthening: The gap between plate and intrados was filled with pressure grout and butresses were placed where the footing was unstabilized.**

mortar. Beside the grouting method, the arch can be strengthened by installing corrugate or steel plates under the arch barrel and the voids between the steel liner and the barrel is filled with mortar.

Movements of arch barrels can be stabilized by the installation of soil anchors.

An improper drainage system will saturate the earth fill and exert additional hydraulic pressure on the arch. The freezing and thawing of fill material above the arch accelerates the opening of mortar joints and the loosening of the stone and brick. This process will finally result in pushing out of the arch ring stones and total arch failure. It is recommended to install catch basins or drainage ficilities near the bridge to prevent any run-off go into the fill. Lateral drainage pipes or weep holes at arch barrel should be placed. Repointing joints, grouting loose stones and strengthening arch should be done at the same time for drainage improvement.

## 15-2.2 HYDRAULICS

Most culvert problems are local phenome-non. However, these problems may affected by long term changes in the streams or inadequate design or construction of the culverts. Change of slope, velocity, angle of flow, width and depth of the channel, headwater elevation, tailwater elevation and invert elevation will change the capacity and stability of the culvert. Hydraulic analysis may be required to select the best counter measures without upsetting the delicate balance between the hydraulic of stream flow and physical channel properties and geometrics. The countermeasures are as follows:

- Improved inlet
- Energy dissipator
  Transitional slab
- Channel protection at upstream and downstream
- Debris control devices

### 1. Scour at the inlet end

Scour is a result of turbulence when more water is collected at the inlet than can be rapidly discharged by the culvert. Possible causes are that the waterway opening is too small, poorly located, or the entrance is choked with silt, brush, boulders or other debris. Blocking of a culvert entrance is one of common causes of culvert failure. When sour occurs at the inlet end of the culvert, the cause should be determined before taking remedial actions. For limited local scours, it can be repaired by placing riprap in and around scour holes.

### 2. Scour at the outlet

Scour is caused by a rapid and uncontrolled discharge of a large volume of water into an outlet stream bed due to the steep slope of the culvert which increase the velocity of the flow greater than normal.

If the culvert is not properly aligned, it should be relocated to coincide with the flow at the stream bed. If the embankment and stream bed are not protected, a headwall and riprap or an apron slab should be installed respectively.

Clogging of the culvert by silt, branches, dead trees, and other debris can be caused by an insufficient grade. The grade can be increased to make it selfcleaning but a steeper slope is susceptible to outlet erosion. Also, clogging occurs when the culvert alignment is on a bend of the stream. An accumulation of silt and the formation of sand bars are apparent at the inner barrel of the culvert due to the slower velocity of the flow at the inner side of the bend and faster velocity at the outerside. The practical solution of clogging is a frequent removal of the debris and sedimentations after a heavy rain or flood.

The faster velocity causes scours and undermines of the wingwalls on the outerside of the bend. These local scours can be corrected by placing riprap protection or concrete apron slab with cut-off wall.

### 3. Gradation

Aggradation which is caused by the enclosure of dam at the down stream, increase sedimentation load from upstream and reduce the effective flow area. In return, the reduction of flow capacity will increase head-

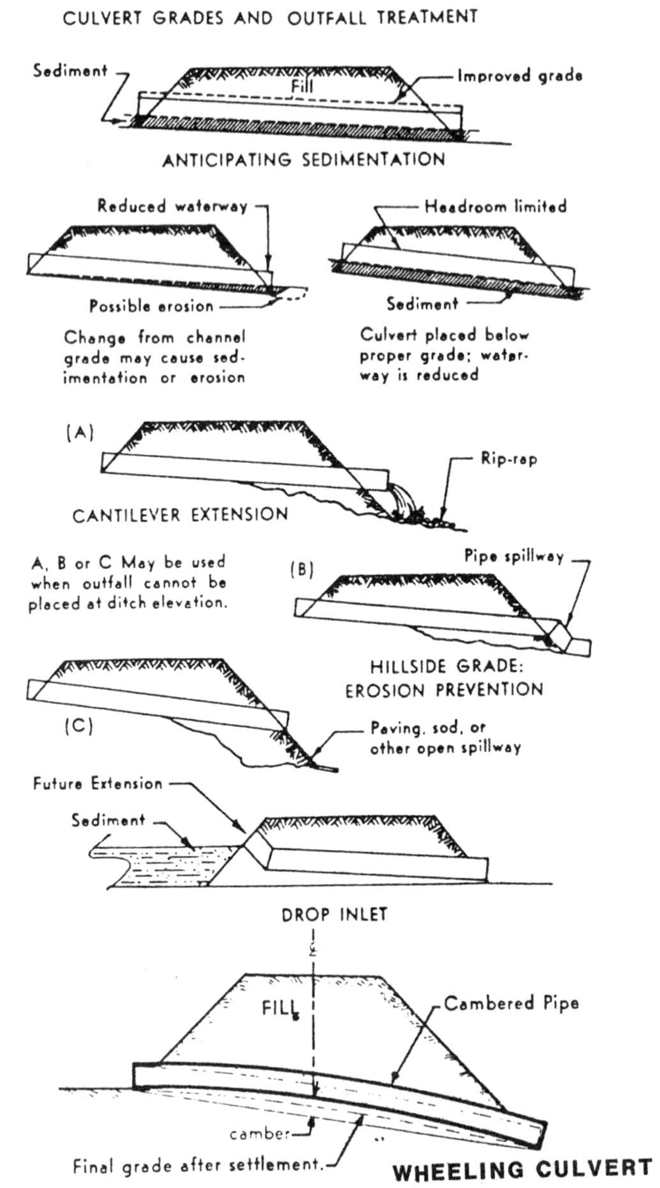


Fig. 15.3 Proper culvert grades

water and tailwater elevation and change the performance character of the culvert. The most critical results of partial clogging of the culvert due to aggradations are damming effects of embankment, increased frequency and magnitude of flood at upstream of culverts, overtopping embankment during flood, and potential failure of embankment.

Degradation will lower the tailwater level at the outlet and accelerate the local scour. Also, degradation can reduce the structural stability and undermine the culvert footing to total failure. The degradation at upstream of the culvert would result in local aggradation problems at the culvert if the culvert can not pass through all sediments and debris.

Countermeasure for gradation requires hydraulic studies but general corrective actions would be similar to the counter measures discussed in the previous chapter.

## 4. Hydraulic problems due to design

The culvert controls the stream environment by two aspects: first, vertical and horizontal control by grade, invert elevation and horizontal alignment, second, flow capacity. The lower culvert invert elevation would cause filling the culvert with sedimentation and reducing flow capacity. This

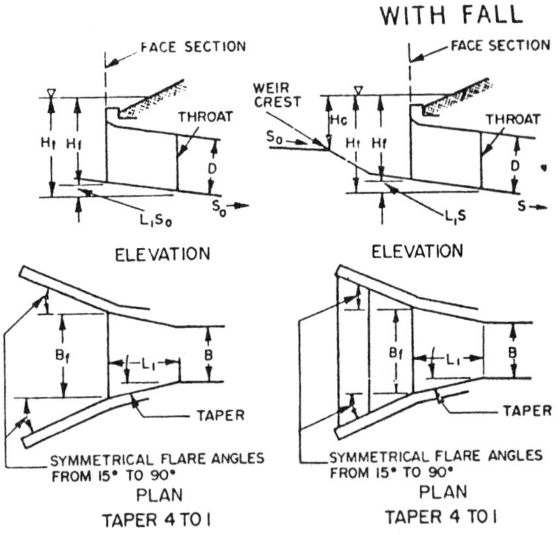


Fig. 15.4 Side tapered section

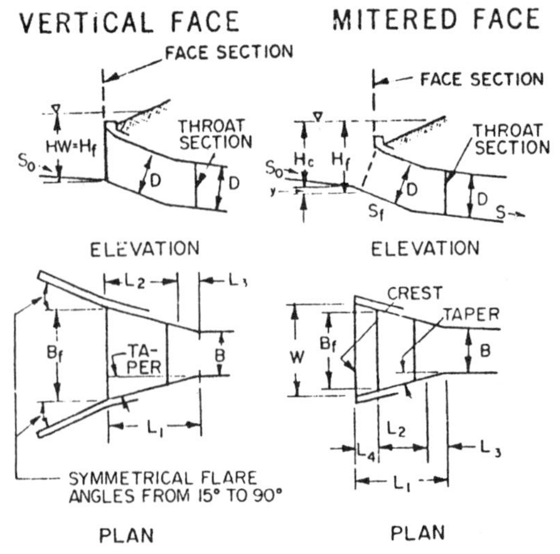


Fig. 15.5 Slope tapered section

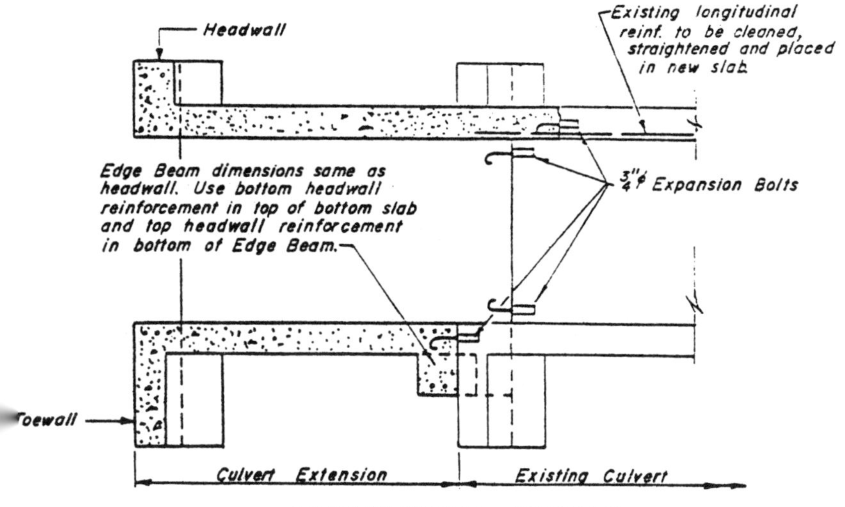


LONGITUDINAL SECTION

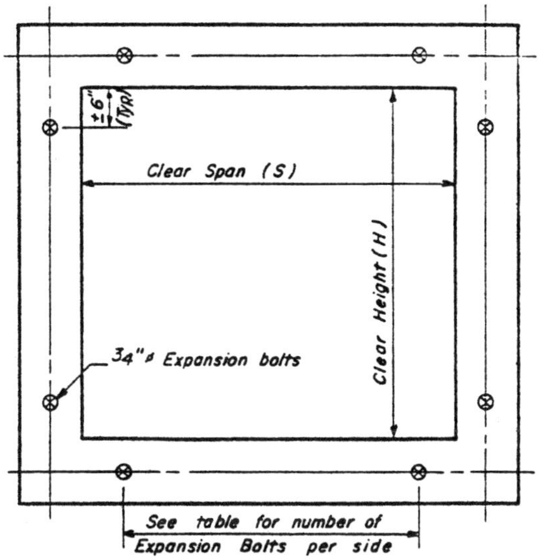


| H or S | No. Expansion Bolts Req'd per side | | | |
|---|---|---|---|---|
| | Extension ≤ 15' | | Extension > 15' | |
| | No.‡ | Spacing | No.‡ | Spacing |
| 2.0 | * | * | * | * |
| 2.5 | 2 | 18" | 2 | 18" |
| 3.0 | 2 | 24" | 2 | 24" |
| 4.0 | 3 | 18" | 3 | 18" |
| 5.0 | 4 | 16" | 3 | 24" |
| 6.0 | 5 | 15" | 4 | 20" |
| 7.0 | 5 | 18" | 4 | 24" |
| 8.0 | 6 | 17" | 5 | 21" |
| 9.0 | 6 | 19" | 5 | 24" |
| 10.0 | 7 | 18" | 6 | 21" |
| 11.0 | 8 | 17" | 6 | 24" |
| 12.0 | 8 | 19" | 7 | 22" |

* Note: Use minimum of 1 Expansion bolt at each corner.

‡ Note: Expansion bolts based on non-skewed culverts.

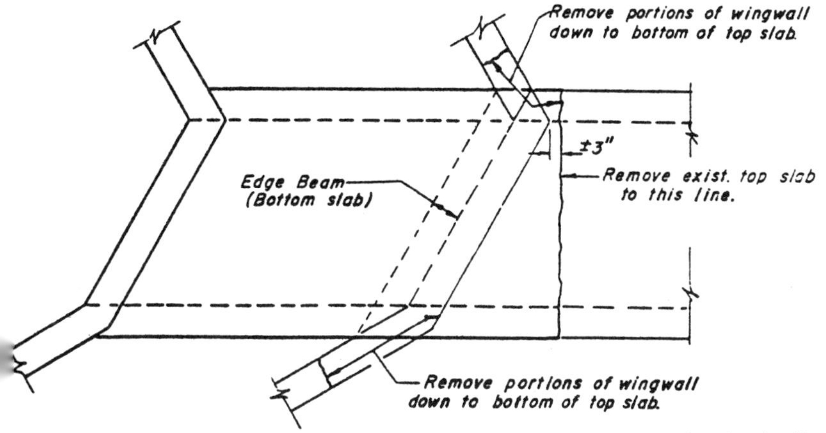


IL DOT

PLAN

For use when existing headwall is removed.

Example:

6' x 4' box culvert to be extended 18' at one end only.

From table find

6' side requires 4 - 3/4"ø Expansion bolts at 20" centers

4' side requires 3 - 3/4"ø Expansion bolts at 18" centers

Total Number required

(4 + 3) 2 = 14 - 3/4"ø Expansion Bolts.

Note: (to be placed on the plans) Expansion bolts shall consist of self drilling expansion shields and 3/4"ø hooked bolts. Hooked bolts shall extend a minimum of 9" into new concrete.

EXPANSION BOLTS REQUIRED FOR CULVERT EXTENSIONS

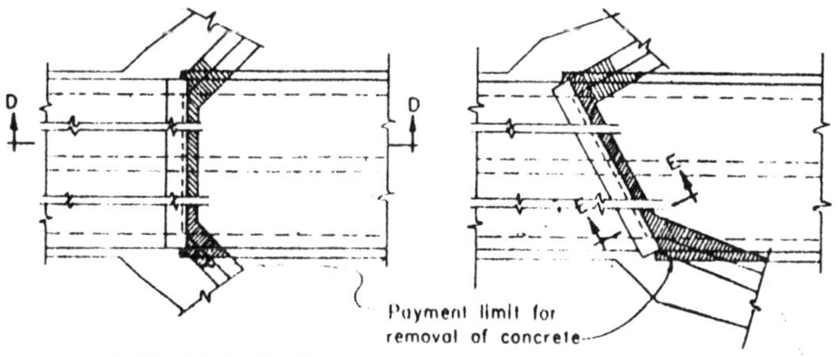


RIGHT ANGLE CULVERT

SKEWED CULVERT

PLAN

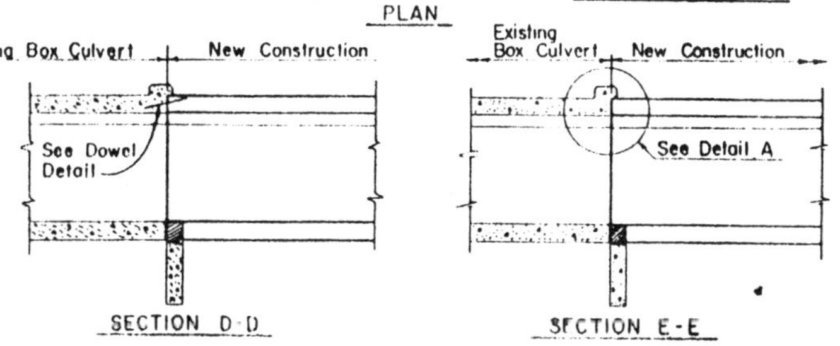


SECTION D-D

SECTION E-E

AR DOT

EXTENSION OF CULVERTS

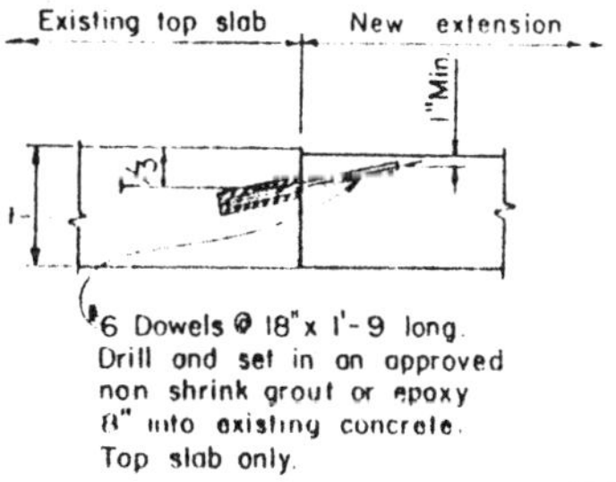


#6 Dowels @ 18" x 1'-9 long. Drill and set in an approved non shrink grout or epoxy 8" into existing concrete. Top slab only.

DOWEL DETAIL

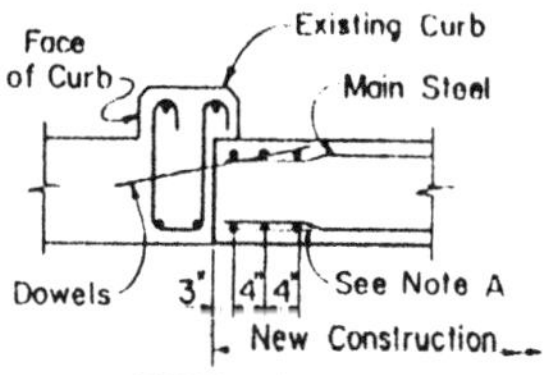


DETAIL A

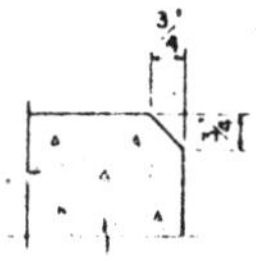


Chamfer all exposed corners thus unless otherwise noted. This note applicable to all culvert sheets pertaining to the project.

## CULVERT EXTENSION

would result in ponding actions at upstream. The higher invert elevation would cause a lesser degree of ponding actions but possible degradation at downstream of culvert.

An inadequate capacity causes ponding effects and induce aggradation at upstream or site and degradation at downstream. The capacity can be improved by reconstructing inlets with tapered and beveled shapes.

### 15-2.3 CULVERT EXTENSION

The widening of the roadway over the culvert will require an extension of the culvert. Generally the widening section should be the same as the original design but the current design specification should be used. The following is the design criteria used by the Illinois Department of Transportation:

**Design:**
Culvert extension shall be of the same design as the existing culvert; i.e., simple span or rigid frame. The culvert extension is to be connected to the existing barrel with 3/4 inch expansion bolts spaced at approximately 18 inch centers for extensions 15 feet or less and 24 inch center for extension greater than 15 feet.

**Headwall and Edge Beam Removal:**
If possible, the headwall of the existing culvert shall remain in place. If the existing culvert headwall is far enough removed from the shoulder line so that it will not require removal, then the end of the culvert extension against the existing barrel shall be constructed with edge beams supporting the top and bottom slabs. The cross sectional dimensions and reinforcement of the top slab edge beam against the existing culvert shall be identical to that used for the headwall of the culvert extension.

If the existing culvert headwall requires removal and if culvert is skewed, an edge beam shall be used to support the bottom slab of the culvert extension adjacent to the end at the existing culvert. The cross sectional dimension and reinforcement in the bottom slab edge beam shall be identical to that used for the headwall of the culvert extension. The top slab of the existing culvert shall be removed to a line perpendicular to the center lime of the existing culvert and approximately 3 inches behind the existing headwall. Special care should be taken when removing the concrete to retain longitudinal reinforcement and dowel rods.

If the existing culvert headwall requires removal and the culvert is straight, the headwall may be removed even with the top of the top slab or the headwall and top slab shall be removed to a line approximately 3 inches behind the headwall. An edge beam shall be used to support the bottom slab. The cross sectional dimensions of the bottom slab edge beam shall be identical to that used for the headwall of the culvert extension.

### REFERENCE:


1. "Guideline for the Hydraulic Design of Culverts" AASHTO Highway Drainage Guideline Series, 1975
2. "Handbook of Steel Drainage & Highway Construction Products" American Iron and Steel Institute
3. FHWA Hydraulic Engineering Circular:
   "Hydraulic Charts for the Selection of Highway Culverts" HEC No.5, Dec. 1965
   "Capacity Chart for the Hydraulic Design of Highway Culverts" HEC No. 10, Nov. 1972
   "Hydraulic Design of Improved Inlets for Culverts" HEC No.13, August 1972
   "Hydraulic Design of Energy Dissipators for Culverts and Channels" HEC No.14, Dec. 1975
4. "Culvert Design System" FHWA-TS-80-245, FHWA, Jan. 1980
5. "Improved Inlet for Culvert - Example Structural Plans" FHWA Technical Advisory T 5140.6, Jan. 6 1979
6. "Corrugated Metal Pipe Durability Guideline" FHWA Technical Advisory T 5040.12, October 22, 1979
7. Young, W.T. "Evaluation of Highway Culvert Coating Performance" FHWA/RD-80-059, FHWA, June 1980
8. Meacham, D.G., J.O. Hurd, W.W. Shisler "Ohio Culvert Durability Study" ODOT/L&D/82-1, Ohio DOT, Jan. 1982
9. "Maintenance Manual" Penn DOT
10. "Bridge Manual" Wisconsin DOT .
11. Market News - Winter 1982, The American Hop Dip Galvanized Association, 1982
12. "Culvert Manual" Illinois DOT
13. "Drainage Manual" Minnesota Dept. of Highway
14. "Planning Manual" California DOT

# 16 TRAFFIC SAFETY

## 16–1 GENERAL

Motor vehicle accidents are a major cause of death for people between 15 and 45 years of age. Each year, about 50,000 people are killed and 2,000,000 injuries occur on our Nation's highways. Highway safety improvement has been emphasized to reduce the number of highway deaths and injuries. This safety improvement will continue to focus on rehabilitating unsafe bridges and planning construction activities in order to reduce highway hazards.

There are numerous disciplines and agencies involved in influencing highway safety. The Federal Highway Administration makes decisions in regard to safety programs and availability of funds. An important part of planning is to determine network street layout and efficient and safe traffic operation. Design is the implementation of the planning process. The designer selects the alternate based on the criteria and operational requirements, but within the constraints imposed by the policy and the basic plans. Constructions carry out the intent of plannings and designs. Highway safety requires cooperative efforts at all level of activities including operation and maintenance.

The most frequently struck and unsafe fixed objects along highways are trees, traffic railings, utility poles, bridge piers and abutments, and traffic signs and supports. In this chapter, we will discuss the general design requirements related to bridges. The following are some areas to consider in design:

- Design consistency
- Horizontal & vertical alignment
- Skid resistance
- Drainage
- Lane width, curb, & median barriers
- Interchange geometry
- Gore area and impact attenuater
- Exit and entrance ramps
- Bridge railings
- Construction work zone safety
- Traffic control and signs for narrow bridges.

**Design Consistency:**

Geometric design should be consistent throughout a reasonable length of the highway. The uniform cross section of the roadway should carry over the bridges and be consistent throughout the highway. The same design speed should be used and a sufficient sight distance should be be provided throughout a reasonable stretch of the highway. A longer sight distance may be required at certain locations of potential hazards and at points where drivers process frequent informations from traffic signs and must make complex decisions such as areas which have closely spaced ramp, lane drops and interchanges.

**Horizontal Alignment & Approach Roadway:**

The safety problems associated with horizontal alignments are the number of lanes, average daily traffic (ADT), horizontal sight distance, the degree of curvature, frequencies of curve, speed, inadequate superelevation on curve, inadequate skid resistance, and pavement marking.

Abrupt curves should be avoided in horizontal alignment. When a curve is introduced, superelevation of the cross sections and the friction coefficient of the roadway surface are required to counteract centripetal forces and to provide comfort for the driver. When curves are hidden, isolated, and result in high accident rates, proper superelevation should be provided in accordance with AASHTO requirements. Increasing superelevation will reduce the run-off water depth at the curbs thus reducing the accident hazards caused by wet weather. If the existing conditions do not allow the use of a desirable superelevation rate, then a higher practical rate should be adopted. Also, the designer should make every effort to increase the friction coefficient to achieve the maximum skid resistance. The following factors influence the frictions:

1. A wet and icy road surface reduces the pavement frictions.
2. Aggregates in the paving mixture and the surface texture will influence the friction. Better anti-skid pavements will be obtained by the use of angular non-polishing aggregates in the mix. Bituminous surface treatment with non-polishing aggregates will improve skid resistance in asphalt concrete. For concrete pavement, brooming or tine finishing is generally used. The depth of texture should be deep enough to provide sufficient drainage in order to prevent forming hydroplane film. If the texture depth is between 0.01 to 0.50 mm, the surface texture is defined as microtexture which breaks through thin water films and provides direct tire-pavement contact. Macrotexture (0.5 to 10.00 mm) would provide direct tire-pavement contact and improve drainage functions and thus prevent skiddings. Macrotexture also offers additional increased friction. Therefore, it is recommended to make the use of macrotexture on bridge deck slabs.
3. Adequate surface drainage is needed to prevent hydroplaning and skidding. The pavement cross slope is a major factor in removing water from the pavement surface and the minimum cross slope should be 2%. Also, pavement grooving or tine finishing will provide a path for water to escape before forming hydroplane. Transverse grooving will rapidly drain away water but longitudinal grooving provide better traction.
4. Tire characteristics: A badly worn tire on fine textured wet pavement could lead to skidding.
5. Speed is inversely affected by friction. Friction will decrease as speed increases. Thus, a car traveling at a higher speed is more susceptible to hydroplane and skidding on wet pavements than a car traveling at a lower speed.

Problems of horizontal sight distance should be carefully studied in bridge rehabilitation. Solid barrier parapets or median barriers restrict the required horizontal sight distances, especially on diamond type interchange.

There are many old bridges which are built on right angles to the crossing instead of following the general roadway horizontal alignment. These right angle bridges are frequently accompanied by sharp curves at the bridge approaches and result in difficult and hazardous conditions. Most of these right angle bridges are truss or through girder type structures. It is often difficult to widen these bridges. The relocation of roadway and replacement of the bridge is usually preferred. The replacement and re-alignment will be costly and take several years to reach a higher priority list. If the bridge is located on a low volume traffic road and replacement is not imminent, the approach roadway should be properly warned and guarded by installing signs, signals, barriers, crash cushions and striping.

**Vertical Alignment:**

The factors to be considered in improving vertical alignment are the grade, the length of vertical curve and the desirable sight distance. For crest vertical curves, the sight distance is determined for drivers to see over the top of the crest to objects on the other side. The crest vertical curve is one of the main features of the roadway that limits sight distance and result in high accident frequencies. Most bridge rehabilitation projects would face steep grade and short length of the vertical curve to improve the vertical clearance under the bridge. At the same time it is required to make short transitions to the existing roadway and to minimize the impacts on existing facilities.

Often, desirable improvements are difficult to achieve without higher costs and major impacts on the surrounding areas. The engineer should identify all hazardous objects around the bridge and analyze any features which could contribute to accidents. If the causes of accidents were connected with grade, horizontal curve, stopping sight distance, and other geometrics, every effort should be made to improve hazard conditions, especially stop sight distance. If a physical improvement can not be made, consideration should be given to reduce the traffic speed at that particular road.

**Interchange Geometry:**

Since the bridge is an integral part of interchange design, the exit and entrance ramp on the bridge with a short length of tapered acceleration and deceleration lanes has become a hazard at the gore areas. Effective utilization of impact attenuaters are required as a countermeasures. For new bridges, ramp terminals should not be located near the bridge.

**Bridge Component:**

The narrow bridge width, curb, median, piers, abutments and parapet have been the most critical safety elements. Curbs are used to control drainage, to prevent vehicles from leaving the pavement and to delineate the edge of pavement. Higher curb height can cause a vaulting action when the car hits the curb. It is recommended that the curb height be limited to 6 inches.

An out of control car which penetrates through the median can cause violent head on collisions with opposite on-coming vehicles or other fixtures. To reduce the chances of these accidents, it is recommended to install solid concrete median barriers or guard rails in the median area of the bridge deck. The new trend of small cars which weigh less than 2000 pounds make it a necessity to develop new barrier systems in the future because small cars are vulnerable to roll over when they hit current concrete barriers in use.

As a cost effective measure, bridges have been designed for multiple short spans. Unfortunately, these bridges have caused safety hazards to the motorist. In many cases, the center and side piers did not have safe clear distance from travel lanes and adequate protection. Only one or two sections of guardrails were installed in front or in advance of the piers. These piers and abutments should be protected to prevent collisions and the nose of the piers should be shielded by impact attenuaters.

A bridge railing is provided for the safety of the public. The forgiving bridge rails should be able to stop a vehicle from penetrating and to redirect the vehicle.

**Work Zone Safety:**

The construction activities will interrupt the traffic patterns on the roads. The public should be adequately warned and guided in accordance with Part VI Traffic Controls for Street and Highway Construction and Maintenance Operation of the Manual of Uniform Traffic Control Device (MUTCD). Positive traffic control plans should be prepared to protect the travelling public as well as workers in work areas. This requires adequate training to all levels of workers and engineers on the projects and development of operational and acceptable traffic control plans involving local governments, police, contractors, and the public.

Most features for work zone safety are portable precast concrete barriers, steel beam guardrails, impact attenuaters, drums, and traffic signs.

To improve the safety at the work zone, New York State Department of Transportation developed a shorter and more portable 8 foot section concrete construction barrier curb. This 8 foot section performs well when compared with 20 foot section.

In general, the width of the travel lane at the work zone are preferred to be a minimum of 11'-0" but a 10'-0" lane can be used to increase the capacity through bottle neck sections of the temporary lanes at the work zone.

## 16—2 RAILING

Railings normally do not contribute to the structural strength of the bridge and are provided mainly for public safety. Railings are designed to prevent penetration and to safely redirect passenger cars impacting at certain angles and resonable speeds with minimal occupant injury potential.

The railings can be grouped into the following:

- Flexible Beam-Post System:
  This allows maximum dynamic deflection of up to one half the vehicle width.

This system consists of metal or wood rails with wood post or weak steel posts.

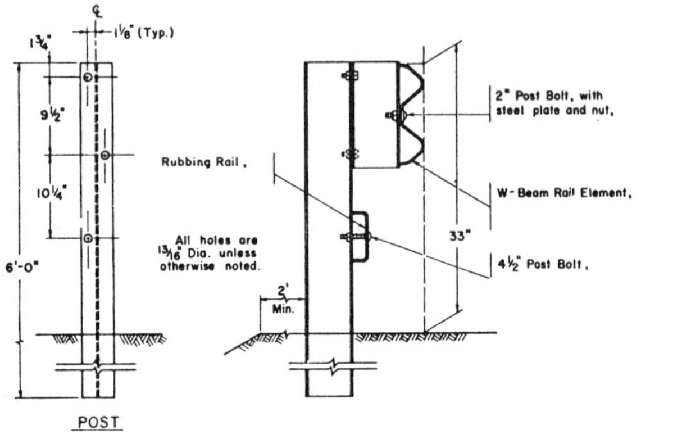


**Fig 16.1 Flexible Beam Post**

- Rigid Beam-Post System:
  This limits maximum deflection less than 6 inches and consists of tubular type trie rails with strong steel beam post.

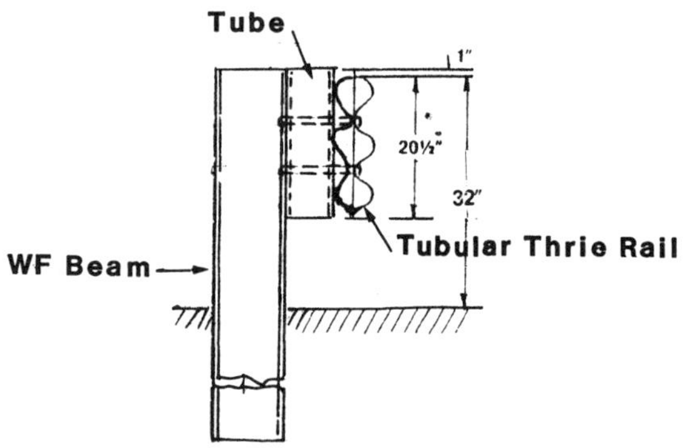


**Fig 16.2 Rigid Beam Post**

- Rigid Concrete Barrier or Parapet:

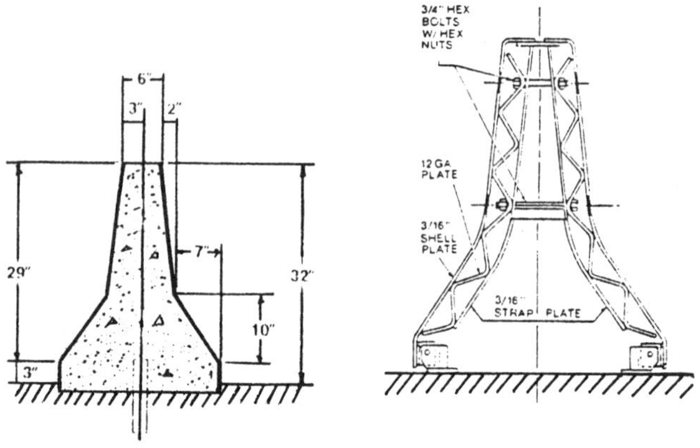


**Fig. 16.3 Concrete and Metal Barrier**

The bridge rail and parapet should be structurally and geometrically adequate to meet the current AASHTO static design requirement of a 10 kip force applied to the rail and post and/or to withstand the full scale crash test ( a 4000 to 4500 pound vehicle, traveling at a speed of 60 MPH speed at up to 25 degree impact angles).

Higher performance railing systems which can successfully redirect a 20,000 pound bus and a 32,000 pounds bus at 60 miles per hour speed and 15 degree enterance angle requires stronger railing (concrete or steel) and concrete deck to minimize crack damages due to impacts. Generally, a minimum of No.5 reinforcing bars spaced at 12" are required for concrete parapets of high type railings.

In addition to the structural strength, safety of the occupants of impacting vehicles and smooth redirection of the vehicles (without rollover) along the railing and parapets should be considered to evaluate the necessity of upgrading existing railings. Also, the vaulting and snagging problem should be taken into account.

The NCHRP Report 239 "Multiple Service Level Highway Bridge Railing Selection Procedures" categorized the railing into four (4) service levels for their performance criteria according to average daily traffics (ADT), and site condition.

The railing systems on primary and interstate highway can be categorized as normal service level (SL 2) and the secondary or local road can be classified as lower service level (SL 1). Service level 1 (SL 1) requires the containment of 4,500 pounds vehicle impacting at 60 mph and a 15 degree and SL 2 requires a 25 degree impact angle. There is no apparent relationship between the AASHTO 10 kips static load criteria and the crash test but it is believed that the AASHTO static design criteria meets the level 2 crash test performance requirements. Where primary and interstate systems have high traffic volume, heavy truck traffic, severe geometric or site conditions, Railing systems are classified as Level 3 or 4 (SL 3 or SL 4) and high performance railings should be used. Level 3 crash test requires a 20,000 lbs. school bus impacting at 60 mph and a 15 degree angles while Level 4 requires a 40,000 lbs commercial bus to be tested.

For service level 1, 32" high thrie beam guard railing with steel or wood post system appears to adequate along with NJ type con-

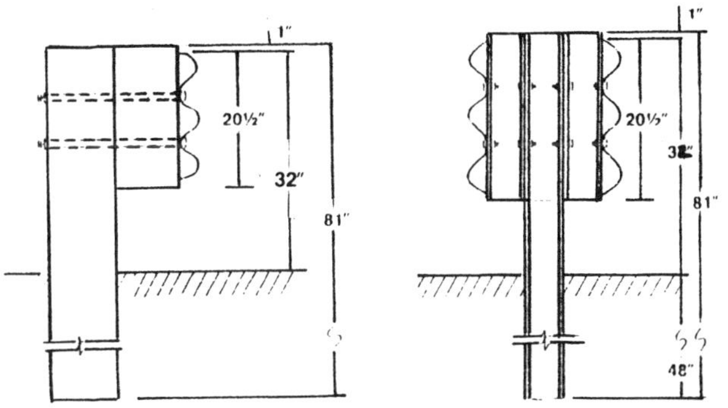


**Fig. 16.4 Thrie Beam Guard Rail**

crete barrier parapets. For concrete parapet of all service levels, the minimum vertical reinforcing bars of front face of barrier parapet is #5 bars at 12" with positive connections to the concrete deck slab.

The following are suggested to analyze the bridge railings:

**1. Geometry:** Minimum height of top of rail and spacing between posts: The height of rail varies from 27" to 34". The concrete safety type has 32" to 34" height. A barrier height of 32" is considered adequate but 34" is preferred. Most bridge guard rails have a 27" height but current research reports recommend guard rail heights of 27" for secondary or local roads and 34" to 38" for primary and interstate highways which have heavy traffic volumes, high percentages of truck traffic and severe geometric restrictions. The increase of height will reduce the possibility of vaulting but this would increase the chances of snagging the between posts. The trend of smaller cars would require a second look over the increasing height in regard to their performance.

**2. Structural strength** with respect to static loads and the crash strength test: Railing should be designed to the 10 kips static force or to withstand 60 MPH impacted by 4,000 pound vehicles of FHWA criteria or crash test requirements of NCHRP Report 237.

**3. Proper offset** and continuity of rails: The greater the offset distance from the travelway to the barrier, the greater the chance of regain control of the errant vehicle and the lesser the chance of collision damage.

**4. Proper material** selection and type of rails: Flexible type rail can fail in tension and often result in excessive deflections due to inadequate stiffness, splices, and anchorage. Since this system allows wheel snagging, space should be available behind the barrier to permit dynamic deflections of rails or a strong post system should be used for rigidity. In order to stiffen the rails, the use of double or thrie type railing should be considered.

**5. Continuity** with approach guardrail: A transition section is required to provide continuity between bridge barriers and approach guardrails. It is essential to increase the stiffness of the approach guard rail by reducing post spacing to half of the normal spacing (3'-1 1/2"). If there are deficiencies in one or more of these items, the deficient items should be investigated for potential modifications or upgradings. The upgrading may consist of replacing a rail section or providing continuity in the system.

The FHWA research report (FHWA RD-77-40) "Upgrading Safety Performance in Retrofitting Traffic Railing System" grouped existing bridge railings into four (4) categories according to profile geometry and features.

**CATEGORY I**

The railing is attached to the bridge deck or on low curb.

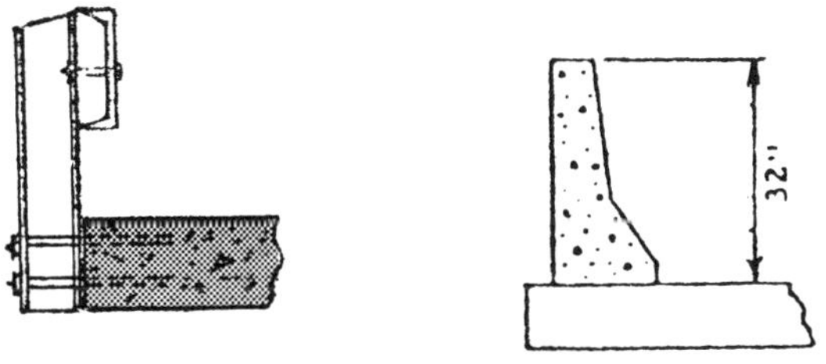


**Fig 16.5 Category I railing (no curb, parapets or walks)**

**CATEGORY II**
The railing is attached to the top of the parapet without curb.

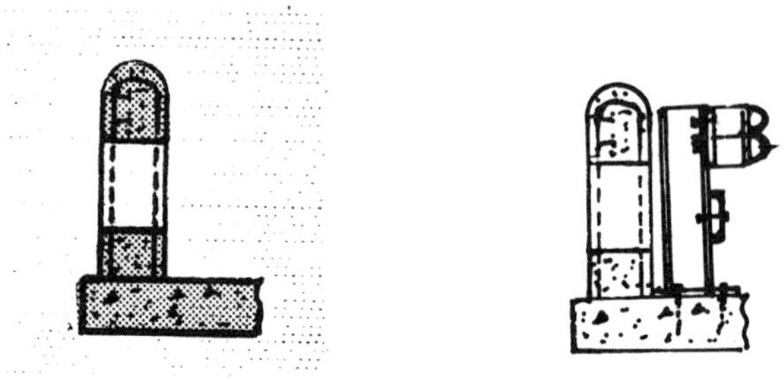

**Fig. 16.6 Category II Railings**

**CATEGORY III**
The railing is attached to the top of parapet or curb.

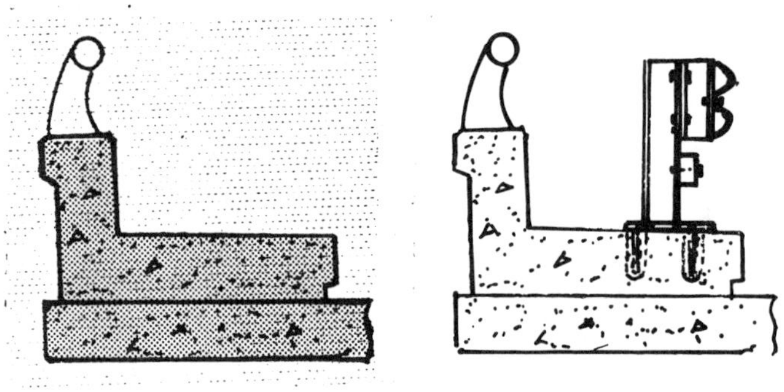

**Fig 16.7 Category III Railing**

**CATEGORY IV**
These bridge railings are characterized by safety shape concrete barriers with or without metal rails.

**Bridge Guard Rail**
The following are some bridge guard rail details used for new installation and retrofits.

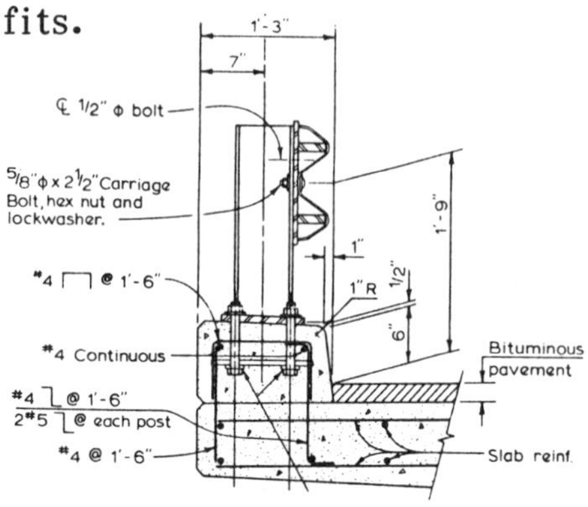


**Fig. 16.8 Bridge Guard Rail**

## 16–3 TRAFFIC BARRIER

### 16-3.1 GENERAL

The traffic barrier is required at any point where hazardous obstacles exist for vehicles which leave the roadway, where obstacles can not be removed or relocated, and where errant vehicles crossover into the opposing direction of traffic.

Roadside barriers are required for embankment and roadside obstacles. An embankment which is relatively flat (slope of 4:1 or flatter) and provides reasonable chances of recovery for an out-of-control vehicle does not require a barrier. Generally the following condition of embankments need barriers:

- Fills more than 6 feet in height with 2:1 slopes.
- Streams or waterways more than 2 feet deep.
- Shoulder drop-offs with slopes steeper than 1:1 and heights greater than 2 feet.

A clear unobstructed flat roadside is desirable to avoid hazard to the drivers of errant vehicles. Also, other elements of median such as cross slopes and obstructions and other irregularities will influence the safety on the median. The steep slopes do not provide reasonable recovery areas. Study indicates that depressed type median provide better chances for the encroaching vehicle to regain control and return to the roadway than mound type medians. It is common practice for highway engineers to use 30 or 40 feet clear recovery distance. The basic 30 feet distance may be appropriate for a cut slope but may not be sufficient for certain fill slopes.

It is necessary to provide barriers for the following conditions:

- Rough rock cuts
- Traffic sign supports
- Bridge piers
- Walls
- Utility poles
- Trees more than 6" in diameter

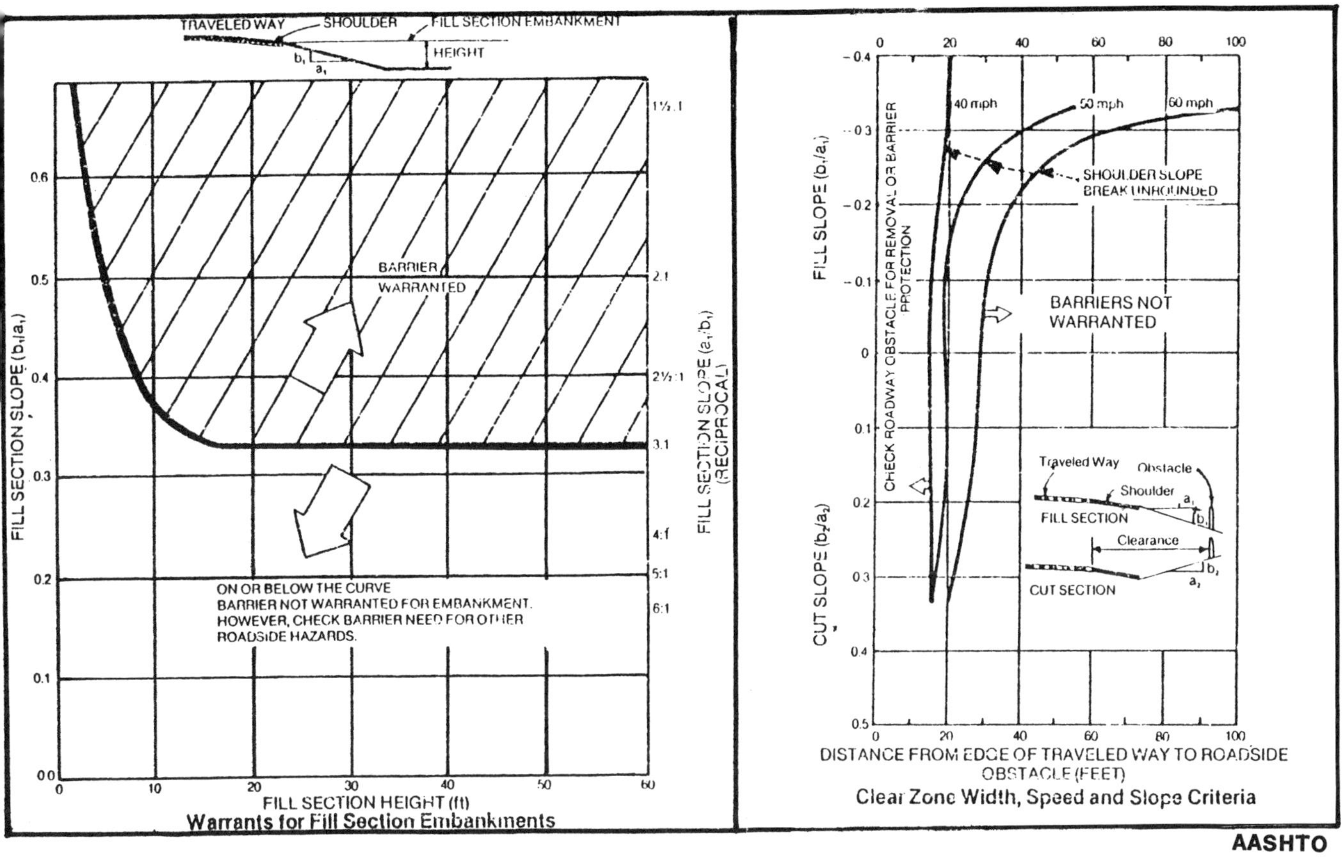


**Fig 16.9 Clear Zone Width, Speed and Slope Criteria**

The traffic barrier systems are classified into flexible, semi-rigid and rigid. The flexible barrier will deflect when struck and will require more space to deflect. Flexible barriers require more maintenance than other types of barriers. The semi-rigid barrier will deflect moderately ( max. 6") and will have moderate impact on the driver. The rigid concrete barrier will not deflect and practically no maintenance is required.

The deep rail "Thrie Beam" has been used widely to give more strength. Most rails have been blocked out from their posts about 6 inches in order to prevent the wheel to strike the post or pocket the cars.

In designing and upgrading traffic barriers, it is recommended the use of the 1977 AASHTO publication "Guide for Selecting, Locating, and Designing Traffic Barriers".

## 16-3.2 GUARD RAIL

**Guardrail:**
The guardrail is the most often struck highway element when vehicles lose control. The main object of the guardrail is to protect the driver from hitting fixed objects and to redirect the vehicle in order to minimize injuries to the driver.

In addition to being capable of safely redirecting an impacting vehicle, the approach guardrail must provide a transition to the bridge railing that will not cause snagging or pocketing of the impacting vehicle.

**Transition:**
The transition from the approach rail to the bridge railing requires that the approach rail be firmly attached to the bridge railing to develop barrier strength. It is also required that the approach rails be gradually stiffened as it comes closer to the bridge railing to provide flexural continuity in the connection. This is normally achieved by reducing the post spacings from 6'-3" to 3'-1 1/2".

The connection to the bridge end should be able to develop the tensile and flexural strength of approach guard rail. The connection should be flared or sloped so that vehicle will not snag at the connections and rails be blocked out to avoid snag.

The end of the curbs and safety walks should be gradually tapered out or shielded because the exposed end poses serious hazards to the motorist. The following are a few accepted systems:

1. Flare the concrete barrier end out of the clear zone at a proper angle or bury the end.
2. Use an impact attenuater.
3. Use a median barrier with a breakaway cable terminal.

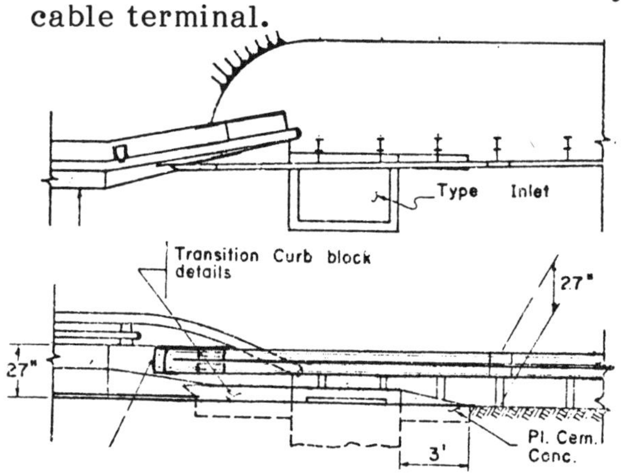


**Fig 16.10 Guard Rail Transition**

**Approach Rail Ends:**
The end of the approach guard rail to the bridge should be flared, buried, breakable, shielded or turndown or breakaway cable terminal (BCT) to avoid a spearing hazard to traffic. The end treatment should be crashworthy within the clear zone and should be able to develop the full tensile strength of the rail elements. The most desirable end treatment systems are the breakaway cable terminal (BCT) and the crash cushion.

If the site condition allows, the BCT should have 4 feet flared to avoid a spearing tendency during a direct hit. The posts should be breakaway when struck; therefore, the end posts normally are wood posts or steel posts with slip base. For the wood post system, the removal of the broken post from the concrete base creates maintenance problems. As an alternative to a concrete embedded wood post, the wood post may be inserted into a steel tube which is driven into soil or embedded into a concrete footing. For the crash cushion, there are several different systems, available for particular locations.

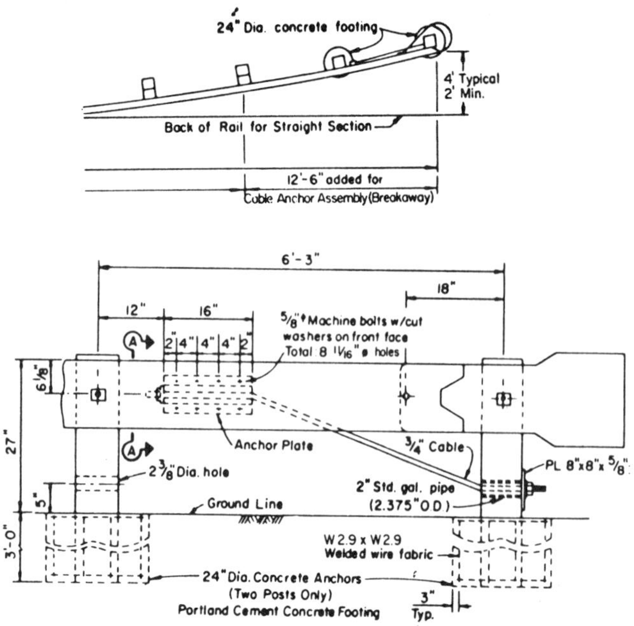


**Fig. 16.11 Breakaway Cable Terminal**

## 16–4 IMPACT ATTENUATER

### 16-4.1 GORE AREA ON BRIDGE

To reduce construction costs, many bridges were built with minimum shoulder widths, offsets and short deceleration lanes. This has caused very limited approach nose recovery areas at the gore area of the bridge exit ramp and was posed a hazard to traffic. Considerable thought and cost are required to make gore areas on bridges safe.

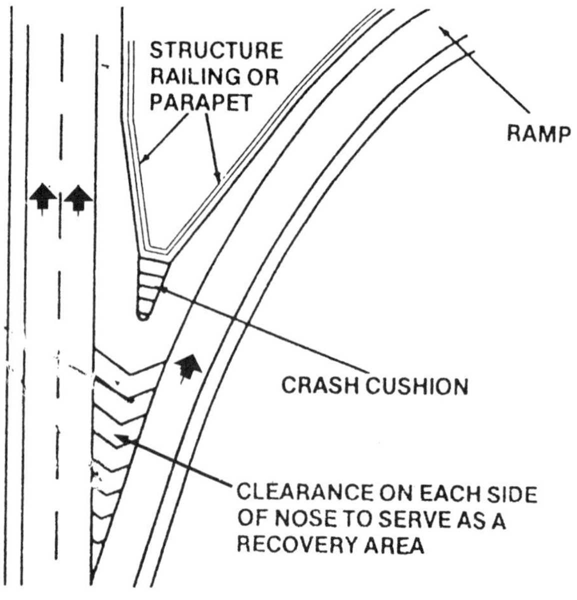


**Fig. 16.12 Gore Areas**

The illustration above shows a desirable type of gore on a bridge. The design for safety should include ample space for the installation of the crash cushion, pavement hazard markings, and full shoulder width clearance through the gore area. Impact attenuaters have been extensively used on the exit ramp gore areas but the crash cushion also is used as a roadside barrier for protecting rigid objects such as barrier terminals, the nose of bridge piers, and overhead sign supports.

## 16-4.2 IMPACT ATTENUATER

The crash cushion is based on the kinetic energy principle or momentum concepts. Upon impact by a vehicle, the kinetic type crash cushion absorbs the kinetic energy of a moving vehicle by the use of energy absorbing crushable material or hydraulic fluids in front of the fixed object. The momentum type inertia barrier transfers the momentum of the moving vehicle to the masses of material located in the path of the vehicles. Most of the inertial types use containers filled with sands and do not require rigid back up or support where kinetic type requires additional support.

**Steel Drum Crash Cushion:**
This kinetic system uses empty steel drums clustered together to provide slowly collapsing cushions when a vehicle hits. This type, known as the Texas Crash Pad, requires a back up structure and is no longer used in most states.

**Inertial Crush Cushion**
This momentum system consists of free standing plastic barrels containing sand. The barrels are filled with different levels of sand. The weight and amount of sand increases from nose to rear barrel to slow the vehicle to a safe stop by transferring momentum to the sand barrels. The inertial type cushion has no redirectional capabilities so that proper transition should be considered between the cushion to the fixed object. The original cost is low but it is not reusable, therefore this system can be cost effective where the probability of collision is low.

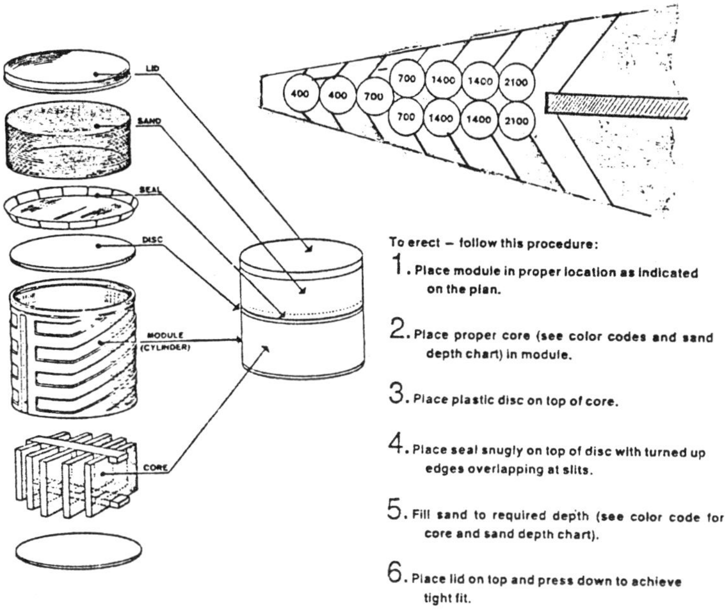


**Fig. 16.13 Inertial Type Cushion**

**Hydrocell Crush Cushion**
This system consists of small water filled plastic cylinders which, upon impact force out the water through orifices. The small plastic tubes are placed between plywood panels or solid diaphragms. This type of cushion is reusable, easy to maintain and can be cost effective where high accidents are expected. In cold regions, the use of anti-freezing solvent causes a pavement skidding problem after cushion is hit and water is spilled.

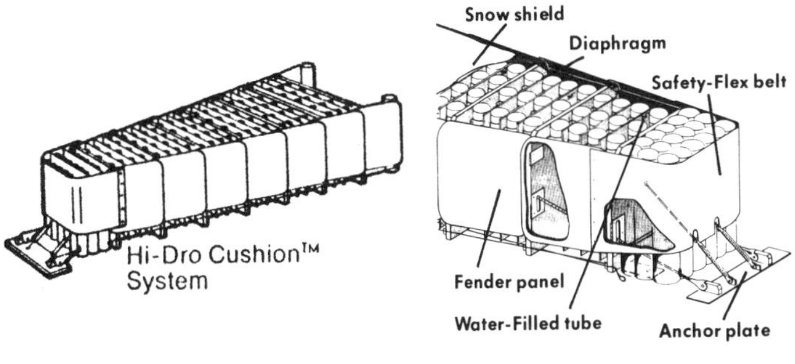


**Fig. 16.14 Hi-Dro Cushion System**

**Hi-Dri Cushion**
This system is identical to the Hi-Dro Cell system except that the water filled cells are replaced with lightweight concrete vermiculite concrete filled cartridges which are designed to collapse slowly when hit.

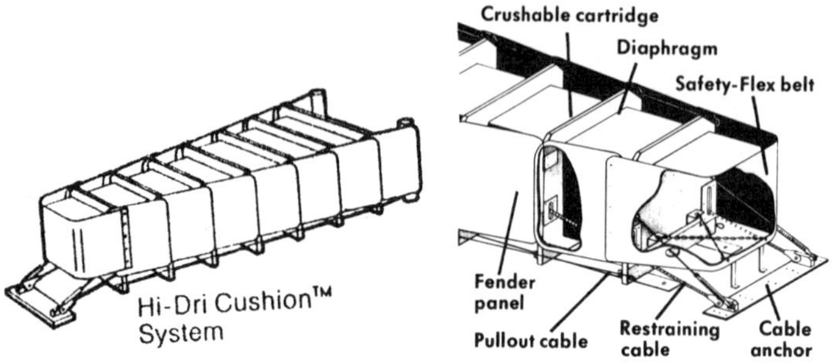


**Fig. 16.15 Hi-Dri Cushion System**

**G-R-E-A-T System**
This guardrail energy absorbing terminal system consists of triple corrugate guardrail fenders which enclose the collapsible light weight concrete cartridges or hex-foam cartridges of foam filled cardboard material. The GREAT system is costly but is effective, redirective, and highly renewable. This system appears to be very cost effective where a higher rate of accidents occur.

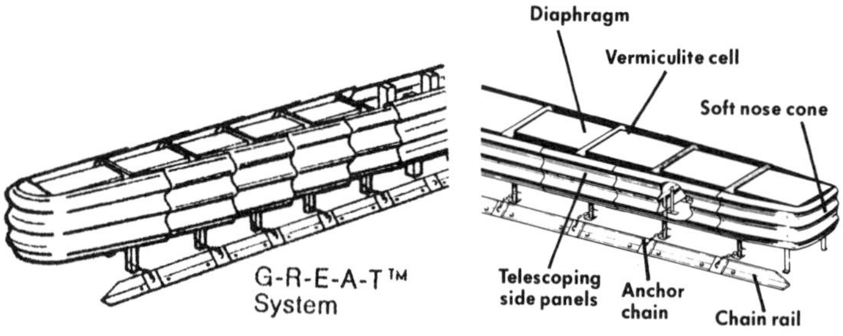


**Fig. 16.16 The GREAT System**

**Truck Mounted Attenuator (TMA)**
This movable attenuator can be used for protecting work crews and motorist in busy highway where repair works should be done within limited short time period. The attenuator consists of Hi-Dri Cells contained in a box, plus a crushable metal platform to support the unit, a steel back up, and the mounting hardware.

Commercially available inertial type systems are the Fitch Inertial Barrier System from FIBCO, Inc. and Energite System from the Energy Absoprtion System Inc. The Hi-Dro Cell, Hi-Dri Cell and GREAT Systems are available from the Energy Absorption System Inc.

## 16-5 NARROW BRIDGE

### 16-5.1 GENERAL

The functionally obsolete narrow bridges draw much attention. It is questionable whether they should receive high priority in bridge replacement. The choice of either rehabilitating structurally deficient and geometrically inadequate bridge or replacing such bridges with current design requirements requires flexibilities and judgements to meet the nation's need in terms of cost effectiveness. However, public safety should override the cost on certain cases.

The most critical narrow bridges are one lane bridges and two lane bridges where the curb to curb bridge width is narrower than the approach pavement width. A study showed that when bridge and approach roadway have the same width, the accident rate is 72 per 100 million vehicle while the accident rate of bridges which are 6 feet narrower than the approach roadway width is 120 accident per 100 million vehicles. The narrow bridge restricts the driver's manueverability, reduces the traffic capacity, increases fixed object collision potential and the accident rate.

When bridge width is less than 24 feet, the driver usually recognize the potential hazard of the bridge and an opposing vehicle. Therefore, if the bridge width is less than 24 feet, the bridge should be considered as a narrow bridge. Also, a driver perceive narrowness when the bridge width is less than the approach roadway width. If the bridge width is less than 80% of the approach roadway, this should be considered a narrow bridge. For example, a 24 foot approach roadway and a 20 foot wide bridge are considered as a narrow bridge.

A bridge less than 18 feet in width should be considered as a one-lane bridge because most drivers will place the left edge of their vehicle on the centerline of such a bridge. A bridge which is over 18 feet in width can be considered as two lane bridge.

When a vehicle approaches a narrow bridge, the driver tries to move over to the center of the bridge or tries to reduce his speed. This behavior can cause head on collisions or rear end collisions. Also, a slow down due to the restricted lane width or lateral clearance between the vehicle and bridge rail would reduce the overall highway capacity. The inadequate horizontal clearance and transition between approach guardrails and bridge rails would increase the frequency of accidents. Remedial treatment of approaches of narrow bridges actually reduce accident rates and increase the capacities.

## 16-5.2 REMEDIAL TREATMENT

Widening is the most desirable alternative where the ADT exceed 10,000 vehicles with more than 10% of truck. Widening can be achieved by eliminating sidewalks, widening or replacing existing bridges, or constructing a sister bridge so that the existing bridge would only have to handle one way traffic.

The change of approach roadway grade and alignment can reduce sight distance problems and thus improve the safety of the approach.

The following three general areas should be considered for safety improvements of narrow bridges until widening or replacement of the bridge can be undertaken.

- Bridge railing
- Approach guardrail
- Traffic control and warning systems.

**Bridge Railing:**
The rail should have strength, redirectional capacities, and sufficient height. The NJ type concrete barrier has been used as a countermeasure but if additional dead loads cause an overstress problem, a W beam with a strong post or thrie beam guardrail can be installed in front of the existing parapets or rails. Where a high curb poses potential vaulting situations, the rail should be blocked out. In certain locations where shoulders have not been provided on the bridge, the zero offset ( face of rail and curb face are vertically lined up ) rail cause frequent side collision damages of rail elements by trucks, therefore a minimum 6" offset is recommended. The higher rail heights would reduce the vaulting but will block the driver's view and can cause adverse behavior of the driver. It is suggested that the height of top of rail should be a minimum of 27 inches and 34 to 38 inches for heavy truck traffic routes. For retrofit for railing of narrow through truss, Thrie beam guard can be used.

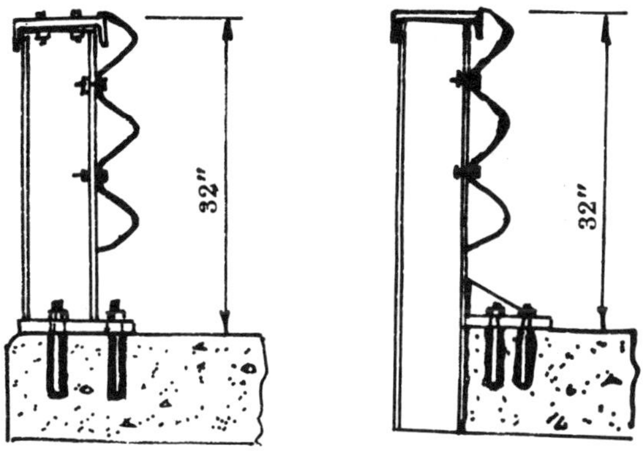


**Fig. 16.17 Bridge Railing**

**Approach Guardrail**
The approach guardrail should meet the same strength and redirectional requirements of roadway guardrails. The terminal ends should have a flared & embedded end, breakaway cable terminal (BCT) or be flared. The transition to the bridge rail should have a smooth transition and stiffness to provide compatibility with bridge rails.

**Traffic Control and Signs**
The installation of traffic control devices, delineators, signs, edge, and center stripes are essential to reduce the accidents and provide relatively safer roads until permanent improvements are made.

The placement of signs and traffic control devices should be made in accordance with 1978 FHWA "Manual on Uniform Traffic Control Devices" (MUTCD) and 1974 AASHTO "Highway Design and Operational Practices Related Highway Safety".

- Edge Lines:
  The lines are effective visual guides to show continuity and transition of the width from the approach roadway to the bridge on narrow bridges.
- Pavement Transition Marking:
  These markings are useful where large differences exist between the approach roadway width and the narrow bridge

width for positive identification of width reduction.

- Shoulder Transition:
  Gradual taper of the shoulder to the bridge is preferred rather than sudden drop off the shoulder at the bridge.
- Installation of Signs:
  Narrow bridge signs should be installed on all approaches of narrow bridges and on one lane bridges. Stop and yield signs or signalization are necessary to reduce accidents.
- Installation of Delineation:
  Delineation and warning devices should be placed for safer nighttime driving on narrow bridges.

Typical application of traffic control devices at narrow bridges are schematically shown below*

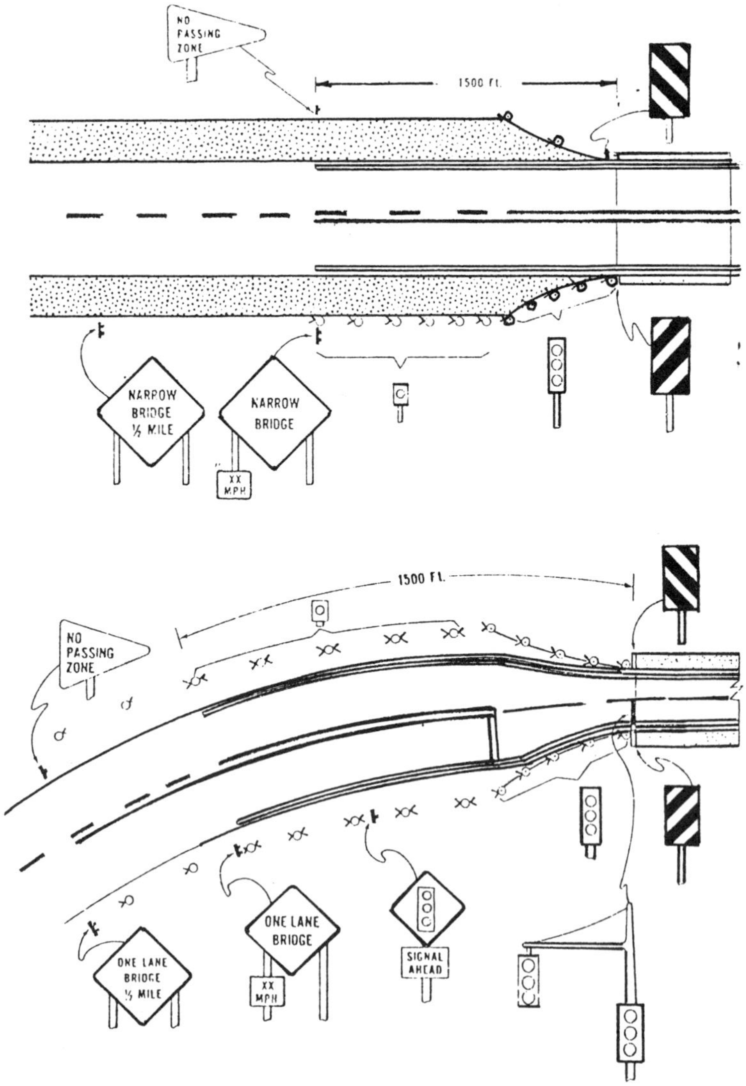


**Fig. 16.17 Narrow Bridge**

## 16–6 TRAFFIC CONTROL PLAN

### 16-6.1 WORK ZONE TRAFFIC CONTROL PLAN

Motorists, pedestrians and construction workers face dangers on or near the construction site and the bridge rehabilitation work zone. Disruption of traffic caused by construction activities would lead the driver confusion and could result in accident. It is necessary to provide adequate warnings and positive traffic guidance in the work zone to reduce traffic hazards.

The traffic control plan is a plan for handling traffic through a specific highway work zone and the preliminary traffic control plans should be developed during the studies and design phase by following steps:

**1. Collect Data**

a. Scope of bridge work:
   - Repairs
   - Replacement or widening
   - Limits of work
   - Schedule
   - Cost.

b. Traffic Data:
   - Traffic volume
   - Speed limits
   - Potential detour
   - Accident data.

c. Roadway Information:
   - Geometric
   - Right-of-way
   - Horizontal & vertical profiles.

**2. Determine Roadway Capacity**
Study the relationship between the reduction in roadway capacity and traffic demand and whether reduced roadway capacity through the work zone will cause extensive delay and congestion. The roadway capacity will be influenced by the type of work, the total length of the project, the working hours, the working days, and restriction to traffic.

**3. Identify Feasible Alternatives**
The alternatives can be categorized into the following eight groups:

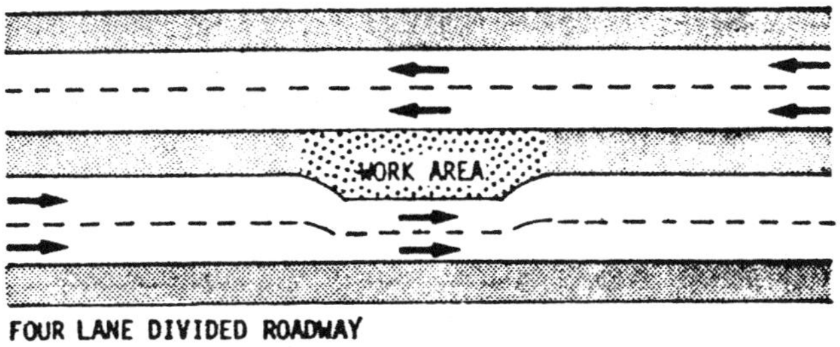

Figure 1. Lane Constriction Work Zone

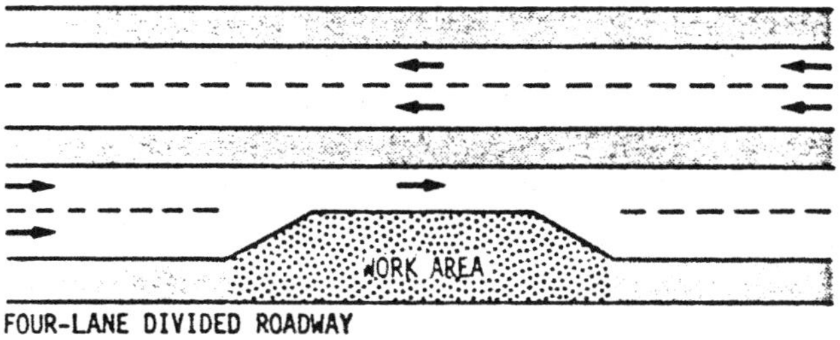

Figure 2. Lane Closure Work Zone

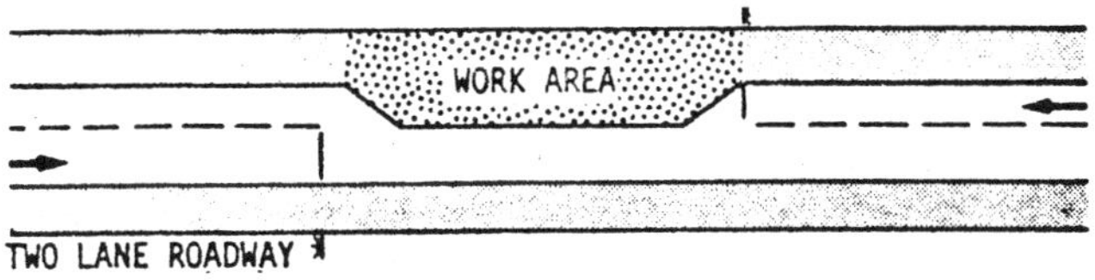

Figure 3. Shared Right-of-Way Work Zone

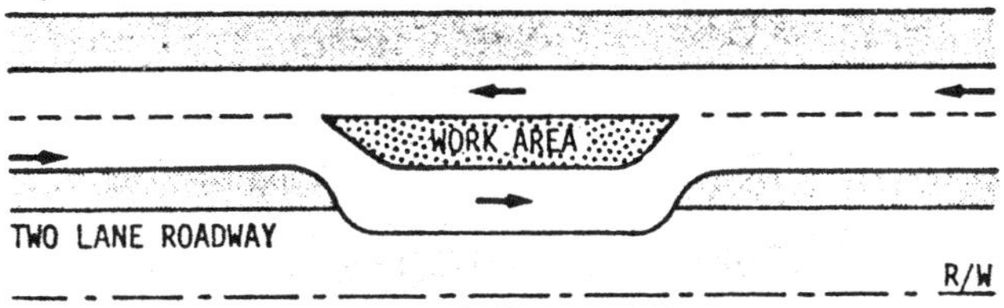

Figure 4. Temporary Bypass Work Zones

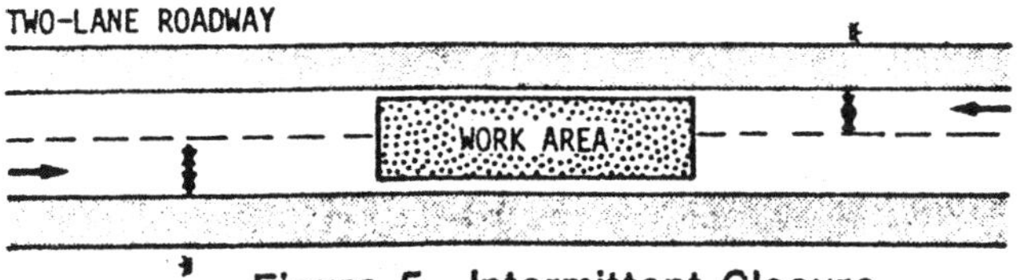

Figure 5. Intermittent Closure

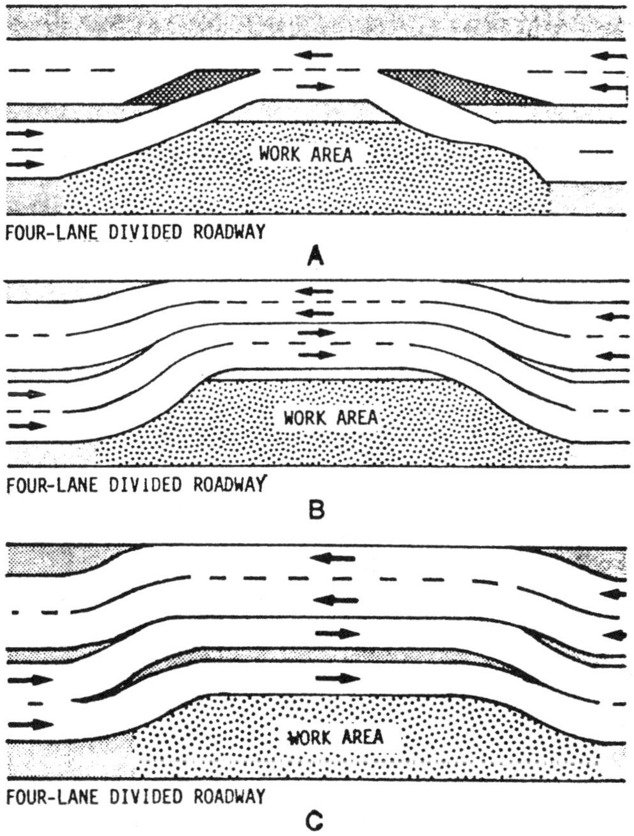

Figure 6. Crossover Work Zones

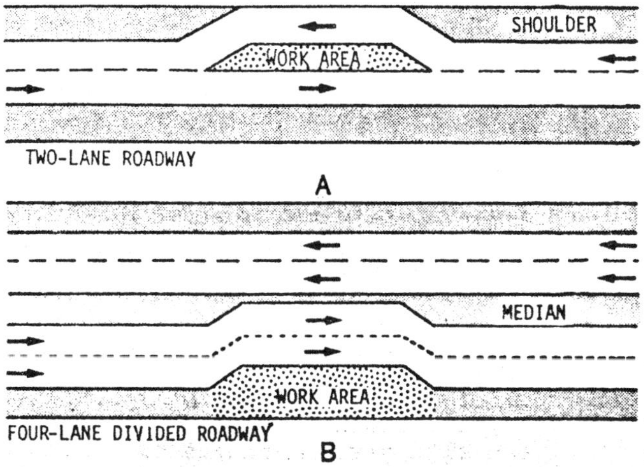

Figure 7. Shoulder or Median Use Work Zones

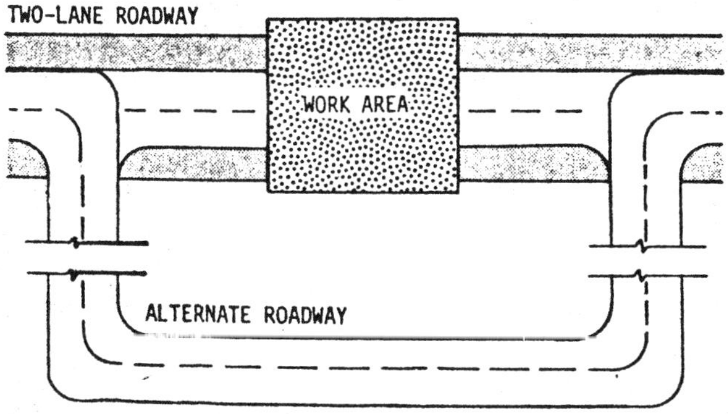

Figure 8. Detour Work Zones

- Lane constriction
- Lane closure
- Share right-of-way
- Intermitten closure
- Cross over
- Use of shoulder or median
- Detour
- Temporary structure.

Alternative is based on roadway type, shoulder width, traffic volume, lane closure requirements, roadway occupancy, restrictions imposed by the regulations and policies, and availability of right-of-way. Considering the above several factors, the feasible alternative can be reduced to only a few schemes.

**4. Analyze Volume-Capacity Relationship**
At this phase, the feasible alternative has been identified and a more detailed analysis for capacity constraints, que lengths and delays should be made. If an alternative does not have sufficient capacity to accommodate the traffic demand for a prolonged period, another scheme should be considered.

The roadway capacity at the work zone of various alternative should be compared regarding the traffic volume. When traffic volume exceeds capacity, an estimated que length should be calculated. An alternative may be disregarded depending on the maximum que length and duration of congestion or may be retained with improved measures.

**5. Analyze Capacity Improvement Methods**
If problems exist, delay and congestions may be reduced by the following techniques:

- Restrict work to be performed off-peak hour.
- Night time work
- Postpone the work to off season
- Work only on weekends & holidays
- Selective ramp closure
- Use reversible lane.
- Define Alternatives

After performing a volume-capacity analysis and capacity improvement methods, alternatives can be evaluated for their implementation and only practical alternatives should be selected for further development.

**7. Quantify Impacts**

Upon selection of the alternatives, traffic, social, economic, and environmental impacts should be evaluated. The impacts are categorized as follows:

| | |
|---|---|
| Cost: | Traffic control, Construction. |
| Traffic: | Delay, Stops, Fuel Consumption, Operation, Accidents. |
| Others: | Air Pollution, Business Loss. |

All factors should be considered in selecting the best alternative however the magnitude and impact of each factor to overall project should be influenced in selection.

**8. Modify Procedure**
After evaluating influencing factors, it may be apparent that certain factors have severe impacts on the alternative. At this time, it may be necessary to review the original bridge rehabilitation or replacement scheme to determine whether alternate bridge reconstruction scheme can be used to minimize the impact. If necessary, the following restrictions can be adopted in the traffic control plan: 1. Restricting working hours to off-peak hours only. 2. Performing work only during night time hours.

**9. Selection of Alternative**
After the analysis is completed, alternatives should be compared and ranked in terms of each impact category. Economic and trade-off analysis can be used in selecting the alternative. Economic and trade-off method includes cost-benefit analysis, cost effective evaluation and value engineering techniques.

## 16-6.2 TRAFFIC CONTROL FOR CONSTRUCTION AND OPERATION

These plans may varies from detailed traffic control plans for a specific project to a note on the plans referring standard drawings, or a section of the FHWA 1978 "Manual of Uniform Traffic Control Device" or FHWA 1980 "Standards and Guidelines, Work Zone Traffic Control" depending on the complexity of the project and traffic interference with construction activity.

In planning, designing, developing and excuting traffic control plans, the following should be considered:

The traffic control should be planned and scheduled to maximize safety and minimize the impact on traveling public.

Workmen, motorists and pedestrians should be protected by providing separate and clearly delineated area.

The traffic control plans should be closely coordinated with the local governments and other interested parties from planning stages.

The traffic control plans should at least meet the requirements of the MUTCD and state regulations.

Traffic control information should be installed sufficiently far in advance for drivers to react and respond, but not so much as to forget.

High type design should be used such as warning far in advance, large highly visible control devices, and gradual channelization for high speed conditions.

The traffic control should be designed based on the actual speeds , not posted speed. Sudden change of geometry or speed limits should be avoided. If speed reduction is required, reduction should be made in increments of 10 miles per hour.

For night operation, all devices should be reflectorized and illuminated.

Whenever two-way traffic should be maintained on one roadway of a normally divided highway, the separation of opposing traffic should be provided by either positive barriers (precast concrete barriers) or drums or cones throughout the length of two-way operations. For transition zones, positive barrier should be used.

Where terminal sections of concrete construction barriers are not tied to an existing bridges, the barrier shall be flared or fitted with impact attenuaters.

Since traffic control plans are an integral part of the projects and requires full cooperation of the contractor, provision are made to allow the contractor develop his own traffic plan if the proposed plans are as good as or better than those provided on the contract plans. The contractor designate a qualified person who is in charge and responsible for overall traffic controls and highway agency, also, should appoint a person who has responsibility and authority for assuring contractor's effective implementation of traffic control plans.

**REFERENCE:**


1. "Guide for Selecting, Locating, and Designing Traffic Barriers" AASHTO, 1977
2. "Highway Design and Operational Practices Related to Highway Safety" Second Edition, 1974, AASHTO
3. Manual on Uniform Traffic Control Device" 1978 FHWA
4. "Work Zone Traffic Control, Standard and Guideline" 1980 FHWA
5. "Planning and Scheduling Work Zone Traffic Control" FHWA-IP-81-6 FHWA, 1981
6. "Traffic Control for Street and Highway Construction and Maintenance Operations" FHWA, 1978
7. "Handbook of Highway Safety Design and Operating Practices" 1978 FHWA
8. "Safety Design and Operational Practices for Streets and Highways" 1980, Technology Sharing Report 80-228, FHWA
9. "Upgrading Safety Performance in Retrofitting Traffic Railing Systems" FHWA-RD-77-40, 1976,
10. "Location, selection, and Maintenance of Highway Traffic Barriers" NCHRP Report 118, 1971, TRB
11. "Effect of Curb Geometryand Location on Vehicle Behavior" NCHRP Report 150, 1974, TRB
12. "Multiple-Service-Level Highway Bridge Railing Selection Procedures" NCHRP Report 239, 1981, TRB
13. "Evaluation of Breakaway Cable Terminals for Guardrails" Research Results Digest 43, 1972, TRB
14. "A modified Foundation for Breakaway Cable Terminals" Research Results Digest 124, 1980, TRB
15. Ivey, Don L., Richard Robertson "Barrel/W-Section Barriers for Construction Zones" Transportation Research Record 796, TRB
16. Ivey, D.L., N.E. Walton "Safety at Narrow Bridge Sites" NCHRP Project 20-7 Final Report
17. Bronstad, M.E., L.R. Calcote, C.E. Kimball, Jr. "Development of retrofit Railing for Through Truss Bridges" Southwest Research Institute, 1983
18. Fortuniewicz, J.S., J.E. Bryden, R.G. Phillips "Crash Tests of Portable Concrete Median Barrier for Maintenance Zones" NY State DOT Research Report 102, 1982
19. "Implementing Highway Safety Improvements" ASCE, 1980, NY
20. Technical Information from Energy Absorption Systems, Inc. Chicago, IL
21. Design Manual, Fitch Inertial Barrier System Inc. Boston, MA

# 17 WIDENING

## 17–1 REVIEW OF STUDIES

During the study of the project, the conditions and informations of the existing structure has been investigated for structural soundness and economic benefits. The widening scheme is selected on the basis of its advantages over replacement or repair schemes. But a review of the following documents is necessary for a final design.

1. Bridge inspection and deck condition report. The portion of the existing bridge to be repaired and removed should be defined and shown on the construction plans based on the field inspection and other available reports.

2. "As-built" construction plans, special provisions and specifications of the existing bridge. The "As-built" plans will show the location and size of structural members and reinforcement steel, geometry, location of construction joints and details. Special provisions, specifications and standard AASHO or AASHTO and ASTM specifications, as indicated by the year on the plans, will show important information regarding materials, allowable stresses, construction methods and loading criteria used in constructing the existing bridge.

3. Original design calculation: Original computations will reveal any design assumptions or criteria used in the designs. The actual stress of the existing structure after widening should also be investigated because the actual stress caused by the widening will restrict the choice of method for widening.

4. Hydraulic and foundation reports: The existing waterway opening and hydraulic reports should be studied to see whether the prior studies and assumptions are still valid. Foundation reports should be reviewed if the same type of existing foundation can be used.

## 17–2 DESIGN CONSIDERATION

The widening of a section should improve the aesthetic value of the structure and make it compatible with the existing structure by using a material which has the same properties as the one in the original construction. Generally, the design concept of widening should be the same as in the original design method but the current specifications should be used for the design of a widened section.

The stage construction and the maintenance of traffic should be carefully planned in advance, before the design. The construction sequence and degree of interaction between the old and new structure should be studied in determining the load distribution and the stress for main member of the new and old portions of the structure, during and after the construction. If the old structure is overstressed by the working stress method, then a load factor analysis should be made to determine the acceptable range of overstress of the members. It would be prudent to accept the slightly overstressed structure by the working stress method rather than to replace it when the stress by load factor method is within the allowable limits.

When the existing shape and size of the substructure member are larger than the one proposed for widening, it is better to use the existing size and shape from the aesthetic view point.

If the existing structure has an asphalt wearing surface, consideration should be given to remove the existing wearing surface and replace it with asphalt overlay with a waterproofing membrane system. On the other hand, if the existing wearing surface is concrete, then latex modified concrete overlay should be considered. The proposed deck slab should have the same wearing surface as the original structure.

When larger cambers, deflections and heavy traffic are encountered on the existing structure during the concrete curing period, open longitudinal joints or longitudinal joints with a mechanical shear transfer system should be considered between the existing and the new section of the structure.

Contrary to previous belief that traffic induced vibrations will cause harmful effects on the widened bridge deck, current researches found that defects are caused by the excessive water in the concrete, insufficient cover over the reinforcing bars, and presence of voids around the reinforcing bars near the joint between the new and old concrete deck.

A research report from the Texas Transportation Institute "Bridge Slab Concrete Placed Adjacent to Moving Live Loads," concludes the following for bridges made of prestressed concrete and steel beams spanning up to approximately 100 feet:

1. No evidence of problems in the concrete, placed and cured while traffic was maintained, was found in bridges that have been in service for years after the new concrete was placed.
2. Traffic can be maintained during placement and curing without causing flexural problems in the fresh concrete.
3. Voids develop in the fresh concrete around dowels bent at a right angle in a horizontal plane upon emerging from the old concrete. No surface evidence was found that these voids cause problems in the performance of the deck. No such voids were found in the dowels that extended into the new concrete without bends.
4. Vibrations caused by normal bridge traffic have no detrimental effect on the concrete, the reinforcing steel, nor the interaction between the reinforcing steel and concrete.

   It recommends to extend all dowel bars about 24 bar diameters straight into the new deck area, to lap splice these dowels 20 bar diameters with the reinforcing bars that serve as transverse reinforcing for the new concrete, and to tie the top mat steel system to the dowels to prevent relative vertical movement between the dowels and the top mat bars.

Even though it is found that no adverse effects exist on widened deck from traffic induced vibration, designers should exercise every effort to avoid traffic on the deck by detouring traffic. Vibrations can be reduce by a maintaining smooth riding surface at the approach slab and the deck.

The following are suggested for widening:

1. The concrete for the widening section should have low water-cement ratio ( maximum 0.4 ).
2. The widening section should be connected to the existing structure by moment transfer methods of overlapping, welding, or mechanical splicing of the reinforcement steel.
3. If the capacity and camber of the fascia beam is different than the interior beams, it is preferred to replace it.
4. It is recommended to use a closure pour for smooth surfaces.

For widening narrow sections, consideration should be given to post tensioning the existing concrete cross beam to increase the moment capacity and eliminate additional substructure widening.

Where the vertical clearance is a problem, the prestressed concrete beam should be considered for the widening of the superstructure because the prestressed beam will provide a shallow depth.

In widening the narrow box girder and T beams, symmetry about the vertical section is desirable to prevent rotation and to maintain stability. If symmetry can not be achieved, lateral connection and false work should be provided. It is recommended that a minimum of two webs be provided for box girders, and the web of the T beam should be centered at the axis of symmetry of the deck, if only the web is necessary.

Extension of piers, abutments and wingwalls generally should be supported on the same foundation type as those of the existing structure. The new foundation should have sufficient bearing capacity to prevent possible differential settlement between the new and the old foundations. If necessary, foundation pedestals, foundation column and piles should be used to prevent settlement.

The following additional considerations should be given to the bridge site before proceeding to the detailed designs:

1. Traffic volumes and work area.
2. Overhead clearance for construction equipment.
3. Horizontal and vertical clearances.
4. Existing utilities and their locations.
5. Interference with existing piles.
6. Existing embankment.
7. Stream channel changes due to widening at pier.
8. Possible scour and drift due to widening.
9. Requirement of the lighting system under structure due to widening.

## 17–3 DESIGN

**Fig. 17.1 Widening bridge**

**Removal:**

The portion of the existing structure to be removed and the extent of removal are based on the limits of unsuitable materials and the structural adaptability of the fascia of the structure. The use of a saw cut at the edge of the removal areas to provide neat construction joints is recommended. The existing slab shall be saw cut to a depth of 1/2 to 3/4 inch depending on the concrete cover of the reinforcement steel. Care should be exercised not to damage the existing reinforcing steel. The rest of the slab shall be removed by chipping and breaking it off. The transverse bars used for dowels should be straightened, cleaned by a wire brush and given a coat of cement grout or epoxy bonding compound. The minimum length of the lapped splice of these bars should be based on the current AASHTO requirements.

The effects of the removal on the main members of the structure during, and after construction, should be investigated and this should be kept to a minimum.

**Closure Pour:**

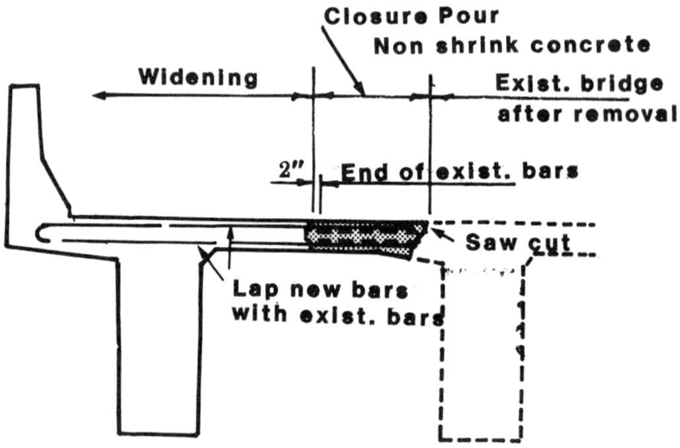


**Fig. 17.2 Closure pour**

In the reinforced concrete slab, box girders and T beams, once the supports for the form work have been removed, a closure pour should be placed which whould be wide enough to accommodate the splice bars. This closure pour should have minimum width for the overlap splice length of reinforcement steel and should be placed with non-shrink concrete to minimize the volume changes. If the reinforcement steel is extended through the section of the closure pour, the effects due to shrinkage, vibration and deflection should be considered.

**Shrinkage, Deflection and Vibration Effects:**

If the reinforcement steel is interconnected with the existing structure and the widening section, the connecting steel will reduce

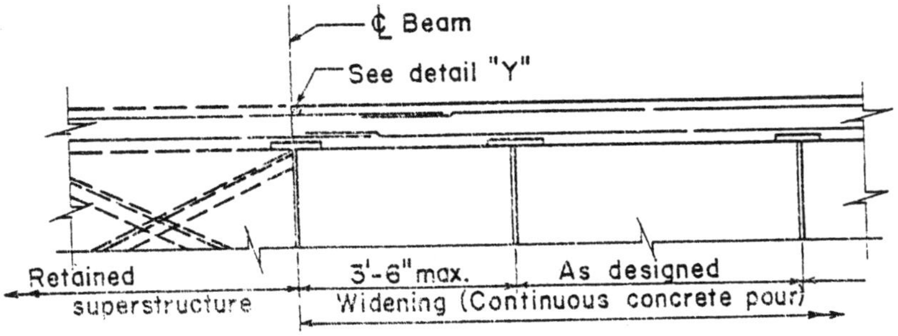


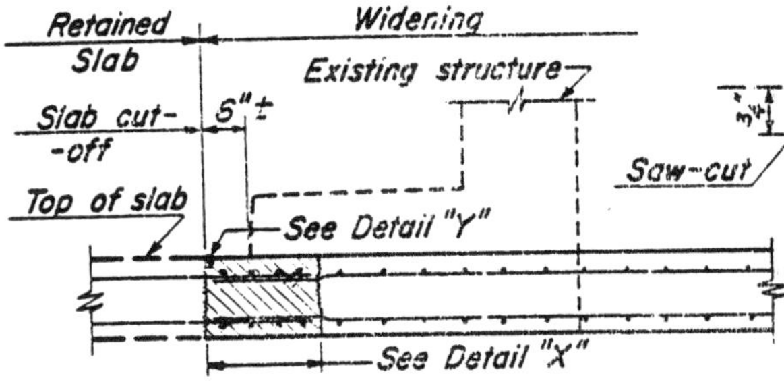


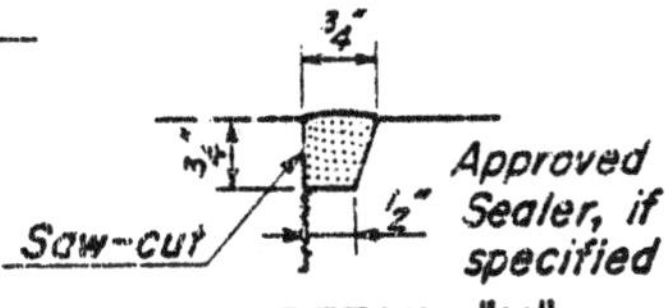


DETAIL "Y"

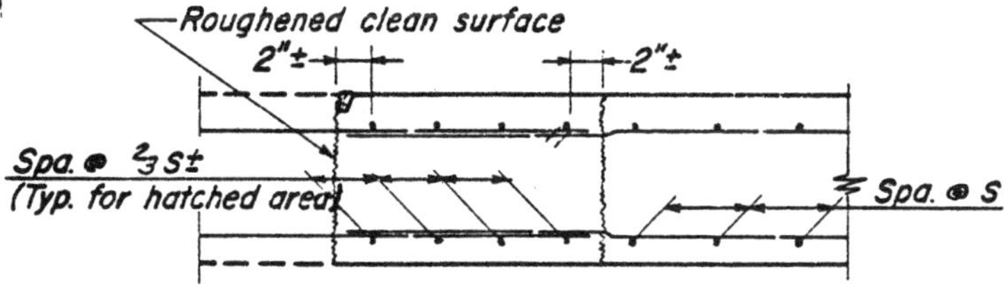


DETAIL "X"

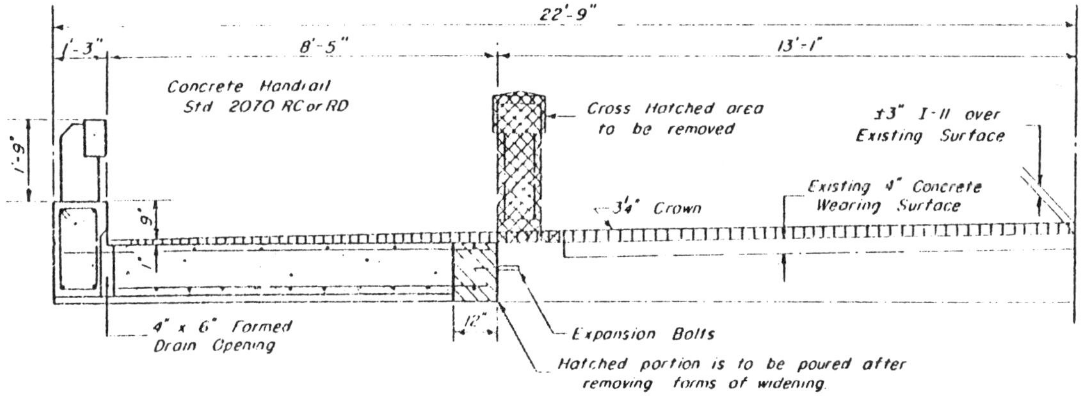


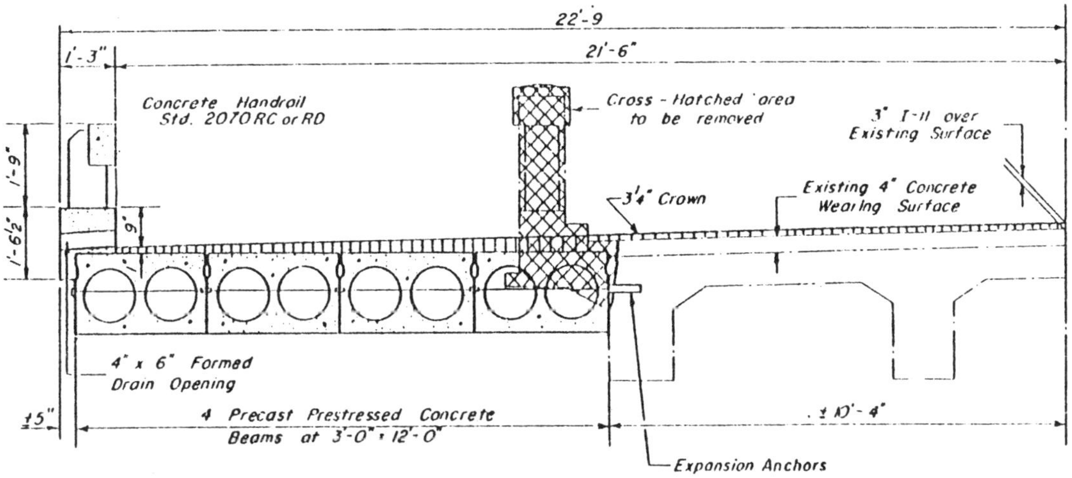


**WIDENING BRIDGE**

its capacity due to preloading. This is caused mainly by longitudinal shrinkage and deflection upon the removal of the false work. A wire rope or sleeved reinforcement steel can reduced the preloading conditions. It is believed that the preloading conditions do not have significant effects and most designers would provide a few additional bars to counteract these problems.

If the traffic is permitted on the existing structure during the pour or curing of the concrete, it is presumed that the new deck will be subject to vibrations due to the interaction between the existing and the new sections, and the debonding of reinforcing bars in the widening sections. But current research reports point out that the traffic in the adjacent lanes have no harmful effects on the widened section.

**Camber:**

The use of the following cambers are suggested:

Cast-in-placed Concrete Structures:
1/2 of ultimate camber of the existing bridge for all members except for full camber at outer edge.

Steel and prestressed girders:
full camber.

**Drilling and Dowel Installation:**

When drilling into heavily reinforced concrete, the use of a pachometer or chipping hammer should be specified to locate the main reinforcing steel. The hole should be at least 1/4 inch or 50% larger than the diameter of the dowels for epoxy grout and 1 inch larger for cement grout and filled with a non-shrink, non-metallic grout. Also, the drilled hole should be a minimum of 3" from the edge of the concrete and at least 1" from the reinforcing steel extending from the existing structure.

The minimum size of the dowels should be No. 4 reinforcing bars, It is preferred to have 2 feet minimum length, embedded at least 1 foot in the existing concrete and spaced about 1 foot apart. Sometimes, in place of dowels, expansion bolts, hooked anchor bolts or rock anchor are used, but care should be exercized in installing expansion anchor bolts because overdriving of the expansion bolts can cause splits or cracks in the concrete adjacent to the drilled holes.

If the existing concrete to be embedded has adequate strength to carry the loads to transfer, the following Dowel Bar Embedment Table for #4, #5 and #6 bars may be used: If the required embedment length can not be attained due to field conditions, The allowable stress in the reinforcing should be reduced by ratio of the actual embedment divided by the required embedment.

| Bar | Drilled Hole for | | Embedment Length | |
|---|---|---|---|---|
| | Cement | Epoxy | Gr. 40 | Gr. 60 |
| 4 | 1 1/4" | 3/4" | 6" | 7" |
| 5 | 1 1/2" | 1" | 7" | 8" |
| 6 | 1 3/4" | 1" | 8" | 9" |

The embedment length for #7 bars and larger should be at least equal to the minimum splice length required to transfer stress to the existing reinforcement. This length may be reduced where dowels are required to provide adequate development lengths rather than to transfer stress to existing reinforcement.

For main reinforcement bars, welding may be the preferred method. During welding process, the bars to be welded should be free from restraint at one end to avoid stress built up.

**REFERENCE**

1. "Bridge Slab Concrete Placed Adjacent to Moving Live Loads" Research Report 266-1F, Texas Transportation Institute, Jan. 1981
2. "Bridge Design Manual" Washington State Highway Commission.
3. "Bridge Manual" Illinois Department of Transportation.
4. "Bridge Manual" Wisconsin Department of Transportation.
5. "Design Manual Part IV" Pennsylvania Department of Transportation

# APPENDIX


**A: General** .......... 303

A.1 Sample Rehabilitation Report 303
A.2 FHWA Rules 316
A.3 Rehabilitation Procedure 322
A.4 Environmental Categories 324
A.5 Interest Table 326
A.6 Sufficiency Rating Program 330

**B: Deck** .......... 332

B. 1 Concrete Deck Slab 332
B. 2 Slab Overhang Design 340
B. 3 Deck Slab Pouring Sequence 342
B. 4 Stage Construction 343
B. 5 Stay-In-Placed Forms 346
B. 6 Precast Deck 348
B. 7 Precast Concrete Bridge Deck 350
B. 8 Timber Deck Details 354
B. 9 Steel Grid Deck Details 356
B.10 Bridge Flooring 358
B.11 Orthotropic Deck 360
B.12 Deck Repairs 362

**C: Concrete Structure** .......... 369

C.1 Reinforced Concrete Slab Bridge, Design and Details 369
C.2 Working Stress Design 386
C.3 Reinforcing Bar Table 388
C.4 Lap Splice Table 389
C.5 Welded Splice 392
C.6 Bar Placement Details 394
C.7 Torsional Reinforcement 396
C.8 Welded Wire Fabric 398

**D: Prestressed Concrete** .......... 401

D.1 Integral Precast Deck Slab 401
D.2 Prestressed Concrete Girder 409
D.3 Girder Selection Chart 418
D.4 Prestressed Concrete Bridge (Design and Details) 420
D.5 Prestressed Concrete Box Beam 459
D.6 Channel Beam, Design and Details 471
D.7 Plank Beam, Design and Details 474
D.8 Continuous Bridge Design 477
D.9 Defects and Countermeasure 484

**E: Steel Structure** .......... 486

E. 1 Typical Truss Bridge 486
E. 2 Methods of Upgrading Truss Bridge 487
E. 3 Roll-in Steel structure 490
E. 4 Addition of Pier to Increase Capacity 500
E. 5 Increase Vertical Clearance 503
E. 6 Adding New Stringer 504
E. 7 Steel Bridge Design Consideration 505
E. 8 Design Example 522
E. 9 Composite A36 Steel I-Beam Bridge Design Graph (Working Stress) 530
E.10 Simple Span Steel I-Beam Bridge 534
E.11 Two and Three Span Continuous I-Beam Bridge 536
E.12 FHWA Standard Details 538
E.13 Steel Bridge Details 542
E.14 Bolting 561
E.15 Welding Symbols 562
E.16 Types of Welds and Allowable Loads 565
E.17 Welding Terms 566
E.18 Effects of Alloy and Weldability 568
E.19 Common Welding Processes 570
E.20 Causes and Cure of Common Welding Troubles 572
E.21 Preheat Temperature 573

**F: Timber Structure** .......... 574

F.1 Types of Timber Bridge 574
F.2 Repairs and Replacement 576

**G: Deck Joints** .......... 582

G.1 Deck Joint Repairs 582
G.2 Preformed Joint Sealer 586
G.3 Strip Seal Joint 591
G.4 Finger Joint 596

**H: Deck Drainage** .......... 600

**I: Bearings** .......... 608

I.1 Resetting Bearing 608
I.2 Retrofit for Earth Quake 610
I.3 Rocker Bearing Design 612
I.4 Elastomeric Bearing Design 616
I.5 Bearing Details 626

## APPENDIX

**J: Substructure** .................... 634

J.1 Abutment 634
J.2 Substructure Retrofit 640
J.3 Armored Joints 644
J.4 Footing 646
J.5 Abutment Types and Details 647
J.6 Pier Design 651
J.7 Bridge Seat Elevation 666

**K: Wall** ............................ 667

K.1 Wall Design 667
K.2 Earth Pressure and Pressure Table 668
K.3 Bulkhead, Anchor and Cantilever Wall 672
K.4 Stabilization 674
K.5 Gabion and Reinforced Earth 675
K.6 Wall Details 676
K.7 Braced Wall 680
K.8 Tiebacks 681

**L: Foundation** ...................... 686

L.1 Stability 686
L.2 Settlement 687
L.3 Pile 690
L.4 Informations for Piles 696
L.5 Bearing Capacity 701
L.6 Cathodic Protection 702
L.7 Pile Repairs 704

**M: Culvert** ......................... 706

M.1 Culvert Hydraulic 706
M.2 Culvert Design 709
M.3 Culvert details 716

**N: Waterway** ........................ 722

N.1 Hydraulic 722
N.2 Flood Hazard Evaluation Guideline 725
N.3 Sample Hydraulic Report 729
N.4 Sample Bridge Site Report 731
N.5 Stream Property 738
N.6 Gabion and Bank Protection 739

**O: Traffic Safety** .................. 742

O.1 Bridge Railing Details 742
O.2 Guard Rail Details 745
O.3 Construction Barrier and Temporary Railing 749
O.4 Traffic Control Plan 751

**P: Widening** ........................ 753

P.1 Widening Schemes 753
P.2 Widening Details 755
P.3 Widening under the Bridge 760

**Q: Specification** ................... 763

Q.1 Repair Material 763
Q.2 Admixture 766
Q.3 Shotcrete 767
Q.4 Extension of Structure 769
Q.5 Welding Reinforcing Bars 770
Q.6 Painting Existing Steel 771
Q.7 Steel Structures 772
Fabrication 772
Bolted and welded structures 773
Repairs of defects 773
Heat curving 774
Straightening bent material 774
Bolted structures 775
Welded structures 775
Field Erection 777
Q.8 Structural Metal 778
Q.9 Structural Bolting 780
Construction methods 782
Q.10 Structural Welding 783
Filler metal 783
Preheat 784
Quality of weld 785
Q.11 Bank Protectioin 786
Q.11a Corrugate Metal Pipe Table 788
Q.12 Epoxy Bonding Compound 789
Q.13 Crack Repairs (Pressure Injection) 789
Q.14 Repairing Spalled Concrete 791
Q.15 Coring and Grouting 792
Q.16 Pier Repair Above and Below Waterline 793
Q.17 Repointing Stone Masonry 794
Q.18 Strengthening Structural Member 796
Q.19 Straightening Bent Steel Member 797
Q.20 Rivet Repalcement 798
Q.21 Cleaning Roller Bearing 798
Q.22 Bearing Replacement 799
Q.23 Bearing Repair 800
Q.24 Temporary Shield over Railroad 800
Q.25 Tiebacks 801

**R: Miscellaneous** .................... 804

R.1 Utility Support 804
R.2 Foundation Report 806
R.3 Soil Sample Identification 809
R.4 Erosion Control Practice 810
R.5 Seepage Control Methods 814

# APPENDIX A GENERAL

## A.1 SAMPLE REHABILITATION REPORT

In this sample report, Bridge Inspection and Deck Condition Survey Reports are included. For multi-bridge projects, it is recommended that these reports be prepared under separate covers and only a summary of the reports be included in the Rehabilitation Report.

The following format may be used in preparing the report:

- Table of Contents

- General and Site Location Map for the project

- Summary

1. Description of project area and structure
2. Condition of the structure
3. General consideration
4. Roadway and safety improvement
5. Structural rehabilitation and reconstruction scheme
6. Widening and/or vertical clearance improvement
7. Maintenance of traffic
8. Cost estimate for alternates and analysis
9. Recommendation

- Scope of work

- General:
  This section should include general considerations such as permit and environmental requirements, right-of-way, utilities, hydraulics, soils etc.

- Roadway and safety improvement:
  This section should identify the geometric design deficiencies such as horizontal and vertical alignment, cross slope, clearance, pavement conditions and recommended improvements. Also, studies should be made for alternate schemes of improvement including economic analysis. Safety improvement may include improvement of superelevation, elimination of obstacles such as removing and resetting guard rail, curb removal, installation of impact attenuator at gore areas, flatting slope, installation of beam guard rail attachment to the bridge parapets and guard rail terminal to current standard and its associated cost.

- Structural rehabilitation and reconstruction: Based on the inspection and deck condition survey, bridge should be grouped into rehabilitation or replacement categories. In rehabilitation, the bridges can be divided into deck replacement, deck overlay and deck reconstruction. After classification of bridge deteriorations for each bridge, further studies should be made for the necessity of widening or vertical clearance improvement in conjunction with roadway improvements. In some cases, even though the bridge is in good condition, total replacement of super structure would be warranted to achieve a better and safer roadway. All structural improvement schemes should be compatible with roadway requirements.

- Widening or vertical improvement:
  These schemes should be considered in conjunction with over-all requirements of the project. Vertical clearance should be based on the site condition by raising the bridges or lowering roadway underneath the structures. Studies should include traffic maintenance, stage construction, utility relocation, impacts on abutting properties and cost analysis.

- Maintenance of traffic during construction:
  Maintaining the flow of traffic, both safely and efficiently on main route and intersecting roads is most important in a reconstruction and rehabilition project. The bridge rehabilitation would be greatly affected by the selection of particular traffic schemes. It may require stage construction of bridge and roadway to maintain traffic on the mainline roadway at all times. In accordance with current safety requirement, the opposite traffic in

one roadway should be separated by a concrete construction barrier. For a multi-bridge project, a continuous diversion for entire length of the project is desirable for the stage construction.

- Cost estimate and analysis:
Upon completion of alternate studies, cost analysis should be made for each scheme. Selection procedure may require trade off between structure and roadway alternates. Sensitivity analysis may help in selecting the cost effective scheme. In this section, a rehabilitation scheme for each bridge can be included after general over-all project evaluation is completed.

- Conclusion and recommendation:
Conclusion and recommendation should be made for over-all roadway and safety improvement and structural rehabilitation or replacement for each bridge on basis of cost, engineering judgement and compliance with the current requirements as much possible.

## SAMPLE REHABILITATION REPORT

### SUMMARY

The New York Viaduct is a 907 feet long and carries Route 1 across the Raritan River and over Central Railroad in New York County. Major elements of the bridge are severely deteriorated and require major reconstruction or replacement.

In this study, three major schemes reconstructing the bridge with a new superstructure, relocating the bridge, and fill embankment with short span bridges are investigated. The cost comparison of the three major schemes as follows:

1. Total superstructure replacement: $ 6,000,000
2. Embankment fill: $ 8,000,000
3. New bridge on relocated alignment: $ 8,800,000

In addition to above three schemes, cost estimate had been made for rehabilitation because schemes 2 and 3 will require lengthy procedures and time to obtain required permits and approval of environmental documents while the bridge needs immediate repairs to provide minimum level of safe service for the public. Estimated rehabilitation cost is $ 1,000,000 and this additional cost should be considered as an extra cost to the schemes 2 & 3.

**Scheme 1 (Reconstruction):** Reconstruction would be less costly than a new bridge or embankment fill and minimum involvement with permit and environment requirements but will require stage construction to maintain traffic on the bridge.

**Scheme 2 (Embankment fill):** Fill will offer less construction cost, minimum interruption of traffic during construction and minimum future maintenance cost but it will disturb the existing wet land, encroach flood plains, and requires the preparation of Environmental Impact Statements and permits from various agencies.

**Scheme 3 (New Bridge):** Relocation of bridge has advantages of minimum traffic maintenance and longer expected life of the structure and its disadvantage is high initial construction cost and environmental involvement.

Considering cost, environmental involvements, encroachment of flood plains and advanced deterioration of the existing superstructure, We recommend the Scheme 1 (Reconstruction) including the following:

1. The entire superstructure should be replaced with a reinforced concrete deck on prestressed concrete boxes-simply supported for dead load and continuous for live loads.
2. The bridge deck should have 50 feet wide roadway flanked on south side by safety shape barrier parapets and north side with a 6 feet wide sidewalk and concrete parapet.
3. The abutments and the twin piers should be repaired and the piers at river crossing should be replaced with solid wall type piers. Also substructure should be extended to accommodate wider new superstructure.
4. All deteriorate concrete in the remaining substructure should be rehabilitated.

### SCOPE OF PROJECT

The scope of work involve the findings of the present condition of bridges, the bridge deck condition surveys, and studies of various schemes of rehabilitation or replacement of the bridges.

In evaluation of existing bridge capacities, it was assumed that original as-built construction plans show actual construction of the bridge and modifications. Adjustments are made to reflect field conditions.

In bridge deck survey inspection and testings, the following tests and procedures are used:

a. VISUAL INSPECTION:
The top of deck and the underside of deck was visually inspected for spalls, cracks, and signs of deterioration. The roadway surface was checked for potholes, patched areas and deteriorated asphalt.

b. CORES:
A total 20 random core samples were taken from the deck slab. Each of these cores was taken to determine the compressive strength of the concrete and to check for signs of corrosion of reinforcement steel and delamination in the concrete.

c. PACHOMETER TEST:
A total of 20 pachometer tests were taken to determine the existing concrete cover over the top reinforcement steel in the deck where chloride tests were performed.

d. CHLORIDE TEST:
Once the depth of the top reinforcement steel was determined by using the pachometer, concrete dust chloride samples were obtained by the use of drills. Each sample was taken at the level of the top reinforcement steel to determine the amount of chloride contents in the deck. After obtaining samples, the removed concrete was properly replaced with concrete.

e. HALF-CELL TEST (ELECTRICAL POTENTIAL TEST):
This test was made on a grid in the eastbound right shoulder lane of the bridge so as to minimize traffic interruptions. Test locations are located at about 2 feet, 6 feet and 8 feet from the curb and repeated at 10 feet intervals longitudinally throughout. Also, tests were made at the location where chloride samples were taken to compare the both test results.

In the Study, major repairs are considered as a temporary measure to provide a minimum level of services until construction of a new bridge or embankment. These schemes will require additional studies for Environmental Impact Statement, Wet land permit, Dredging and Fill permit and other permits. This would delay the starting date of actual construction and the existing bridge will deteriorated further into a critical condition.

Each scheme studied the requirements of maintaining traffic at all time during constructions and the possibility of a detour route. Also, permit requirements, environmental evaluations, right of way requirements and other factors have been analyze or considered in each scheme.

## BRIDGE DESCRIPTION AND HISTORY

The New York Viaduct carries Route 1 in the borough of Morris in New York County. It is comprised of 32 spans, 29 of which are cast-in-place reinforced concrete tee beams in continuous units of four or five spans. There are 2 spans of simply supported, concrete encased, riveted steel beams over spur track of the Central Railroad.

The structure is 907 feet long between the end of the slabs and 51'-9" wide, including a 40 feet roadway, an 8 feet sidewalk and a 9" south brush curb bounded by 1'-6" balustrades on each side.

The original deck was built with an 8.5 inches concrete slab and 2.5 inches asphalt wearing surface with .25 inch waterproofing membrane.

The substructure consists of reinforced concrete three column rigid frame piers, solid wall piers and plain concrete gravity abutment.

The original New York Viaduct was constructed in 1928 on Route 25, Section 10 in Morris Borough, New York County. In 1951, the 2.5 inches asphalt wearing surface was replaced by a 4" reinforced concrete wearing surface and in 1971, the sidewalk was reconstructed. The original Route 25 was renamed Route 7 in 1927 and it is now Route 1.

## BRIDGE INSPECTION AND DECK CONDITION SURVEY

### 1. BRIDGE INSPECTION

The inspection was conducted during August, 1980 in accordance with guidelines set forth in the 1978 AASHTO Manual for the Maintenance Inspection of Bridges.

The inspection was limited to the visual inspection of the major components of the bridge. Not included any coring of concrete or any underwater investigation. The purpose of the inspection was primarily to update the findings of the previous inspection, to determine the extent of further deterioration of structure since 1970 inspection and the immediate necessity of emergency repairs to insure the safety of the travelling public and gather information necessary to perform investigation and studies for rehabilitation or reconstruction.

### Observation:

Date: August 15, 1981 Inspected by: S.H. Park, PE
Temperature: 90 F Weather: Clear

### Deck

The top of the 4" concrete wearing surface exhibits numerous medium transverse cracks spaced about 5 to 10 feet apart and several large spalls and a few joint spalls. A few angles at the open deck joint are missing and several deck areas have been patrche[ with asphalt apparently to cover spalls and potholes.

The underside of the deck is in fair to poor condition. Most of the interior bays show only a few minor transverse cracks with efflorescence while the outside bays near the fascia beams show numerous cracks with efflorescence, exposed rebars and severe deterioration of concrete apparently attributed by water leak through the cracks at the gutter areas. Several areas in the deck had full depth repairs.

### Superstructure

Overall, the reinforced concrete tee beams are in poor condition and at few areas in critical condition. Several concrete beams show numerous fine map cracks with efflorescence and exposed rebars at the bottom of the beams. The end of the beams near the open joints had severely spalled and some of the bearing seat area have been reduced to almost half by severe spalling of the end of the beam and pier caps.

At the end of the riveted steel plate girder spans over the river and railroad, the concrete encasements are in poor condition with exposed corroded steel surfaces. It appears that open joints and runoff water leakage through the deck at the curb caused most of the spalls and deterioration.

### Substructure

In general, The elements of the substructure are in poor to fair conditions.

**Abutments:** Severe deterioration of bridge seats and erosion of the slope protection has occurred at both abutments.

**Bent Piers** Most bent piers over continuous span (without deck joints) appear to be in good condition with only a few minor scalings.

**Twin bent pier at open joint:** These are in poor condition. The bridge seats are severely deteriorated especially on exterior bearing seats and actual bearing area has been reduced to almost half at some locations. At Bent 17, the pier cap shows severe spalls and exposed tension reinforcement steel at the bottom. Bent pier 21 & 22 (at railroad) show deteriorated bearing seats, exposed rebars at pier caps and wide cracks with broken off concrete at the pier columns.

**River pier:** The reinforced concrete column bent piers supporting the river spans show severe spalls with loss of section on the bearing seats and pier caps. Upstream and downstream pier noses show heavy scaling.

There are many piers with wide cracks and large spalls. The main problem is the gradual disintegration of the concrete due to aging. The severe spalls of cap beams and columns and resulting exposure and corrosion of reinforcement are prominent at river piers due to joint leak.

### Waterway

During inspection the maximum water depth was 3 feet near the center of channel but the high water mark is about 2 feet above the existing ground adjacent the river. It appears that during a flood, a large portion near the river would be a flood plain. At the northwest corner of the south river pier, scour expose about one foot of the top of footing. Debris has accumulated at the south nose of the north river pier but the waterway opening appears to be adequate in all other respects.

### Approaches

An asphalt wearing surface has been used to correct for settlement of the concrete pavement but the surface shows depressions and pot holes. There are no catch basin near the bridge and runoff from surrounding areas pond on the north side of the approach. Beam guard rails are in good condition.

## 2. BRIDGE RATINGS

The bridge were rated in accordance with the AASHTO "Manual for Maintenance Inspection of Bridges, 1978", "Standard Specifications for Highway Bridge, 1977" and current interims. The allowable stresses used in this rating have been summarized in the following table:

| | Inventory | Operating |
|---|---|---|
| Structural Steel | 16.00 ksi | 22.50 ksi |
| Reinforcing Steel | 18.00 | 25.00 |
| Concrete | 0.88 | 1.30 |

The various parts of the bridge have been analyzed using the estimated dead load of the structure, as it now exists, and the dead load stress of each member has been calculated based on the properties of the actual member and reduced where necessary for the loss of section. The dead load stress were then deducted from the allowable stress to determined the capacity for live load plus impact. The live load capacity was compared to the stresses produced by AASHTO type legal loads 3, 3S2, and 3-3 as shown in the AASHTO Inspection Manual and HS 20 loads. The rating or allowable gross tonnage load for each truck type was then determined by proportion.

The structure was divided into five sections for ratings. The five section are as follows: 1). Spans 1 through 8, 2). Span 9 & 10, 3). Spans 11 to 20, 4). Spans Span 21 and 5). Span 22 to 31.

In computing the ratings of concrete encased plate girders, it is assumed that all live loads are carried by the riveted steel girders and no composite action will occur. This assumption would underestimate the actual capacity of the girders. There are no shear connector in the encased girders therefore it is difficult to determine the ratio of the contribution of the concrete to the strength of the girder by partial composite actions. However, additional ratings have been made by Load factor method to compensate partial composite actions and lower dead loads and live load ratio.

The structural analysis show that the reinforced concrete tee beams and the fascia riveted steel plate girder are critical members and their live load rating in present condition are as follows:

**Concrete T beams** (Working Stress)

| Truck type | | Inventory | Operating |
|---|---|---|---|
| HS 20 | (36 Tons) | 16 Tons | 30 Tons |
| Type 3 | (25 Tons) | 11 | 25 |
| Type 3S2 | (36 Tons) | 16 | 30 |
| Type 3-3 | (40 Tons) | 19 | 38 |

**Riveted Girder**

| | | |
|---|---|---|
| WORKING STRESS METHOD | | |
| HS 20 | 0 Tons | 58 Tons |
| Type 3 | 0 | 46 |
| Type 3S2 | 0 | 58 |
| Type 3-3 | 0 | 60 |
| LOAD FACTOR METHOD | | |
| HS 20 | 37 Tons | 62 Tons |
| Type 3 | 26 | 51 |
| Type 3S2 | 37 | 62 |
| Type 3-3 | 43 | 64 |

The reinforced concrete T beams in spans 1 through 8, 11 thru 20 and 22 to 31 have unsatisfactory ratings and a few beams were found to be slightly overstressed. Most of the concrete encased girders in span 21 (over railroad) were below the minimum inventory ratings while girders G1 and G7 were overstressed for dead load alone. All the girders of spans 9 and 10 were not meet minimum required inventory ratings but the operating ratings were satisfactory. These encased girders with a high ratio of dead load stress to live loads stresss, were reanalyzed by using the load factor design method.

Although these two members over railroad have a high ratio of dead load to live load stress, live load capacities increased by using load factor analysis. However, reusing these members would require removal of encasement, blast cleaning, application of protective coating which will reduce the cost effectiveness. Also, there is some question as to the actual existing condition of embedded steel surface which may have layers corrosion.

Beside the above factors, the age of the structures (54 Years), the remaining fatigue life of a steel girder, and lower ratings warrant the total superstructure replacement.

## 3. DECK CONDITION SURVEY

Deck condition survey may not be needed if the deck is severely deteriorated and the only logical choice is total deck replacement.

### Visual Inspection

The underside of the concrete deck slab shows fine cracks, scaling, and efflorescence, which is associated with leakage. Spalls occurs around many scuppers and open deck joints. Spalls at other localized areas and beneath open deck joint show rusted and corroded reinforced steel. Many areas have been reconstructed apparently due to larger size of pot holes and advanced stage of deterioration of concrete.

There is extensive deterioration of concrete overlay probably due to underlying concrete deterioration. The concrete near the joint is poor condition and many open deck joint is covered with asphalt and show missing protective joint angles.

### Delamination

No testing for delamination was made due to concrete overlay. However, the cores were taken to detect delamination and to determine the strength of concrete. Twenty cores were taken in the deck at random locations. Of the twenty cores only ten had sufficient height to be tested. The ten other cores were either broken or disintegrated. Delaminations were founded in the wall of recovered and broken cores.

### Corrosion Potential Test

The half cell test indicated that over 40% of the deck area is actively corroding with readings in excess of -0.35 volts to the copper-copper sulfate electrode.

### Chloride Sample

After removing concrete overlay, chloride samples at level of the top reinforcing bars were taken from the bridge deck at 20 locations. 11 were taken from the westbound lane and 9 from the east bound. The pachometer was use to determine the location of the top bars. The chloride contents was greater than 2 pounds per cubic yard in 11 of 20 samples. 7 others had 1.00 to 1.95 lbs/c.y and two others had less than 1.00 lbs./c.y.

### Cores

Of the 20 cores, only 10 had sufficient height for compression strength test. The compressive strength ranged from 3,605 to 7,080 lbs./sq. in. Although the compressive strength were satisfactory, these 10 cores were recovered from the locations where it had no apparent defects.

### CONCLUSION

The existing reinforced concrete slab is in poor condition and can be classified as Category 1 - Extensive Active Corrosion.

Spalls:

Total spalled area about 5000 sq. ft.
Percentile of spalled area 5000/36280=13.8%

Half-Cell:

This test show that 40% of the tested area is having active corrosion.

Chloride Sample:

This results summarized that 55% (11/20=.55) of the area of the bridge deck contains greater than 2 pounds of chloride per cubic yard at the level of the top reinforcing bars. The chloride results and electrical potential test results were compared and it was found that 9 out of 11 locations where samples having more than 2 pounds of chloride per cubic yard of concrete also had electrical potential readings in excess of -0.35 volts.

Summary of test results:

| | % of defected deck area | | % of non-duplicated deteriorated area |
|---|---|---|---|
| Spalls | 13.8 | | 13.8 |
| Half-cell | 40.0 | (100-13.8).40 = | 34.5 |
| | | sub total | 48.3 |
| Chloride | 55.0 | (100-48.3).55 = | 28.4 |
| | | total | 76.7 |

The total non-duplicated deteriorated area is about 77% and the deck is category 1. Therefore we recommend a complete deck replacement.

## 4. CONCLUSION AND RECOMMENDATION

The visual inspection and deck condition survey clearly indicated that the entire concrete superstructure (slab and beams) should be replaced and portions of a number of piers should be reconstructed. The major deterioration have been the result from the use of de-icing agents and joint leakage. The deck condition survey concluded that the deck should be categorized as Category 1 - Extensive Active Corrosion and the entire deck should be replaced. Although some of the continuous reinforced concrete beam themselves appear to be satisfactory, the deterioration of the deck slab which is monolithically constructed with the beams, would force the total replacement of beams in place of partial restoration of the beams at the end.

The concrete encased plate girder showed lower ratings and deterioration. In order to improve ratings, the concrete encasement should be removed then steel should be cleaned and painted.

However removal of concrete requires labor intensive hand chippings so the cost would be high. beside the high cost, the concrete encasement show cracks and evidences of corrosion in the riveted girder which may produce loss of section of the plate girder. Therefore, it is prudent to replace concrete encased girders in place of removal of encasement.

The twin bent which show severe deterioration are those located below open deck joints. At these locations, the piers support the end spans of continuous concrete T beams. Deterioration of concrete at pier cap and column is so extensive that complete reconstruction of pier above the footing or pedestals will be required. Most interior piers located at the interior spans of continuous beams are in good condition. The minor defects (cracks, spalls, scaling) can be effectively repaired by patching and sealing. After this minor restoration, these piers can be used to support a new superstructure.

We recommend the following:

1. Replace the entire superstructure.
2. Restore the abutments and all interior piers by patching and sealing.
3. Reconstruct all upper portion of twin bent piers and restore the pedestal at these locations.
4. Reconstruct all river piers.

## REHABILITATION REPORT

### 1. ALTERNATE SCHEMES

Considering the advance stage of deterioration of the superstructure, the following three (3) schemes are selected for studies along with temporary repairs:

I Total superstructure replacement: Steel rolled beam vs. Prestressed deck slab beam.

II New embankment and two new bridges spanning over river and railroad.

III New viaduct on new alignment: Composite steel rolled beam and plate girder vs. composite prestressed concrete girders.

IV Temporary repairs.

### 2. EVALUATION OF EXISTING STRUCTURE

**Superstructure**

Based on the inspection and rating, the entire superstructure will be removed. The following two schemes of superstructure arrangement were studied: Scheme I: The span length is double and every other interior pier is eliminated and Scheme II: Existing span length is kept and all existing pier will be used.

**Substructure**

The river piers show severe deterioration of cap and columns and these pier will be reconstructed above the footing level. In an analysis of pier capacity, the proposed superstructure is comprised of 5" concrete wearing surface and side walk over the prestressed concrete deck beams. The concrete deck beams are simply supported for dead loads and continuous for live loads. Under Scheme I (Longer span length and elimination of every other pier), the structural capacity of the intermediate pier cap will exceed by 30% of moment and 50% of shear capacity and the columns are highly overstressed. Therefore, Scheme I will be no longer considered in the studies. Under Scheme II, pier cap capacity is satisfactory but the column reinforcement would not meet 1% of the gross column area required by the AASHTO Article 1.5.11. All outer columns of intermediate piers have 0.70% steel area and inner column have 0.75%. This deficiency is not considered to be critical and this meet the requirement of 1% reinforcing steel of reduced effective area of the column under Article 1.5.11.(2). Therefore, we recommend to use the existing intermediate piers to support the new superstructure. Except for the ratio of steel area to gross column area, the column capacity is adequate to support the new superstructure.

The proposed footing pressure vary from 4 to 5 tons per square foot and appears to be excessive but we consider these pressures are satisfactory because the new superstructure will be lighter than existing superstructure and there are no evidence of soil bearing problems on existing footing.

A twin bent pier cap beam capacity for moments and shears are satisfactory. The column capacities are adequate except for minimum steel requirements.

## 3. GENERAL CONSIDERATION

### A. Deck Geometry

The existing approach roadway has two traffic lanes with two ten foot shoulders. Current traffic volume based on ADT requires widening of the roadway to four lanes but it would be costly and difficult due to commercial development along the route. Considering these factors, the proposed bridge deck will carry two traffic lanes with shoulders and a 6 foot side walk. Generally two 12 foot wide travel lanes and two 10 foot shoulders will be sufficient to handle traffic. Consideration had given that shoulder may be used as traffic lanes in future. If bridge carries 4 traffic lanes, median barrier will require to separate the directional traffic. Beside the provision for future traffics, the wider deck slab minimize the stage construction problems.

It is suggested that the typical section should be 2, 12 feet travel lane, two 12 feet shoulder and 2 feet wide painted median which will provide total 50 roadway width between parapet barrier to curb of sidewalks. This 50 feet roadway will reduce the traffic control problems and stage construction of the bridge. For the 50 feet roadway, 3 stage construction of bridge will be required and this will involves constructing the structure in stage with traffic being maintained on the existing structure and new superstructure and constructing side walk at the 3 stages. But for 44 feet roadway width, it will require 4 stage construction and this added stage will cost higher construction cost.

### B. New Alignment

Two alternate alignments for a new bridge are considered, one 100 feet north of the existing bridge and the other, south of the bridge. These two schemes were analyzed in detail using an aerial photography map. During the study, consideration was given to minimize the impact on established commercial development and wet land. The preferred alignment was the north alignment which would minimize the impact on wet land and required much less right-of-way cost. It is also compatible with the existing alignment with minor modifications but a disadvantage is that the the structure will be on curved alignment and requires superelevation. The alternate south alignment satisfies straight line alignment but requires more right-of-way cost and wet land excavations. Also, the interference with business establishment will be substantial. Based on the above analysis, it was recommended that the north alignment should be selected.

### C. Traffic Control Schemes

The maintenance and protection of traffic during construction is necessary and one of the most important considerations. The feasibility of a multiple stage construction and complete relocation were studied as a means of safe traffic control.

A two stage construction scheme will require that the first stage must be built with a sufficiently wide roadway to carry two traffic lanes and shift the center line of roadway to the north. Two stage scheme will be more costly than three stage scheme because wider new superstructure means a wider new substructure which will result in additional right-of-way and wet land involvement.

A three stage construction scheme was adopted for the superstructure replacement. With these stage arrangements, proposed roadway alignment will coincide or nearly so with the existing roadway alignment. The proposed and existing roadway centerline can be coincided or shift to the north of existing centerline up to 4 feet maximum.

In all stages, pedestrian traffic will not be maintained because the pedestrian traffic count shows that the volume of pedestrian traffic is negligible and additional cost of maintaining pedestrian traffic is not justified.

A new bridge along a relocated alignment will be best solution from the view point for traffic control during construction. Normal traffic will be maintained without interference with construction activities. Two acceptable alignments have been considered, north and south alignment. The north route is sufficiently relocated from the existing route so that traffic can use the full width of existing bridge for the entire length of bridge. But the south route will be located near the existing bridge and will require a partial removal of existing structure. It is recommended to select the north alignment after considering the traffic control scheme, right-of-way, wet land involvement and cost.

### D. Soil

#### 1. Soil Settlement

Subsurface conditions along the relocated alignment were such that the problem of settlement become apparent in design of roadway fill embankment. Almost the entire flood plain was underlain by deposit of organic silt averaging 5 feet in depth and ran up to 20 feet. A 2' to 5' thick layer of marsh vegetation, cinder and other fill covered at some locations.

The stable roadway fill embankment may be placed by the following method:

* Control of residual settlement by use of sand drain or wick drain and surcharge.
* Elimination of the settlement problem by removing unsuitable material where practical.
* Control of residual settlement by use of surcharge and limited height of fill

The use of sand drain was rejected because the sand drains itself being much less compressible than surrounding material, eventually developes into hard points in the area of soft compressive material and produces bumps in the roadway. The wick drain may be an alternate for consolidation but it was rejected due to lack of practical experience of installation in this region.

The removal of unsuitable material to the sand level was considered but disposal of unsuitable material was found to be unacceptable from environmental view points and this method is rejected.

Surcharge and fill appears to be a practical solution. It is decided that height of fill including permanent and temporary overload to be placed over wet land and should be limited to 15 feet so as not to destroy the natural cohesive strength of soil. Any such disturbance would greatly increase consolidation and settlement, both initial and residual.

This limitation in turn controlled the maximum height of embankment and, therefore, established the roadway profile and beginning of structure which will be supported on a pile foundation. To control differential settlement, a 10 feet height of surcharge was used throughout the fill area. The purpose was not to eliminate settlement in its entirety, but to reduce it to a reasonable amount and to even out any soft spots in the underlaying material prior to removing the surcharge and placing pavement.

**2. Foundation of Structure**

For the new structure, both pile foundations and spread footings were considered. The pile foundation will be supported on steel HP12 pile and spread footing will be placed on minimum 5 feet deep crushed stone after removal of unsuitable materials. The cost of the piles was found to be higher than the spread footings, but careful pile driving would minimize the environmental damages on wet lands. To satisfy the environmental restriction as much as feasible, it is suggested that pile foundations should be used on the wet land and spread footings on other areas.

**E. Hydraulic Consideration**

Based on data obtained from flood insurance studies of June 1978, U.S. Department of Housing and Urban Development, the design flow is 7890 cubic feet per second (100 years) and the 100 year flood elevation is 173.00 which was recorded high water elevation occurred during October 1913. The proposed superstructure will be designed to provide at least 1 foot of freeboard above 100 years flood elevation. It is desirable to design the lowest point of fill embankment to elevation 174.00 but due to soil surcharge requirement the lowest point of embankment is established to the elevation 173.0 without overtopping. With the bottom of superstructure at elevation 174.00 and no overtopping of embankment, the required bridge length is established to 250 feet and carries the 100 year flood.

The capacity analysis for the bridge is based on a typical channel cross-section taken through the bridge. This analysis requires that channel upstream and downstream from the bridge is of sufficient section to contain the estimated flows. However, the existing channel upstream is fairly shallow and most area is in a flood plains. The existing channel and downstream are similar.

In order to accurately determine the capacity through the proposed structure, a rigorous hydraulic analysis of upstream and down stream flood potential should be performed. The actual capacity of the proposed bridge opening is primarily a function of the difference in hydraulic head that would develop between points immediately upstream and downstream from the structure. In place of in-depth investigation of the entire region, the capacity is determined by the Manning equation with selected cross section obtained from the flood insurance studies.

If the fill scheme is selected, risk analysis should be performed to determine the waterway opening and impact of embankment fill in the flood plain. The risk analysis has not been made at this time because it require extensive back water studies and computer programs such as U.S. Army Corps of Engineer HEC-2 and U.S. Geological Survey Step-backwater program.

**F. Right-of-Way**

No significant additional amount of right-of-way will be needed if the bridge superstructure is replaced as Scheme I and elements of the existing substructure are used and widened by three construction stages. However relocation of bridge would require substantial increase of right-of-way cost.

The boundaries of property parcel in the vicinity of the structure have been mapped on the plans based on the municipal tax maps.

The following are estimated cost based on the areas necessary to be acquired from each parcel:

| | Type | Cost (Million) |
|---|---|---|
| Scheme I (Superstructure Replacement) | Construction easement | 0.1 |
| Scheme II (Embankment) North | Fee | 2.7 |
| | Easement | .2 |
| | Total | 2.9 |
| Scheme III (Relocation) North Route | Fee | 1.9 |
| | Easement | .2 |
| | Total | 2.1 |
| South Route | Fee | 2.1 |
| | Easement | .3 |
| | Total | 2.4 |

**G. Environmental Consideration**

The adverse effects of the proposed new bridge or fill embankment on the environment should be minimal and temporary in nature. At present there are several different species of vegetation flourishing in fresh water marshes along the river. During the construction of the bridge or fill, it is unavoidable that some damage to this wet land will occur in the immediately adjacent areas. Every reasonable effort was made in selecting alignment to minimize the damage to confine to the smallest area possible.

It is apparent that embankment fill would result in permanent damage to the wet lands and impact is much severe than new bridge construction. There are about 30 acres of fresh water marsh land down stream from the existing bridge to the river. There is also an appreciable area of marsh upstream from the bridge. Two and one-half (2 1/2) acres of the marsh on the north side of the bridge will be disturbed by the the project.

Embankment fill on new alignment will permanently disturb 2 1/2 acres of wet land while new bridge will permanently disturb 1 acre and the remaining 1 1/2 acres will be restored. Reconstruction would only require less than 1/4 acre disturbance of wet land. It is apparent that there will be no major adverse environmental impacts for new bridge but substantial impacts on embankment fill.

The actual method to be used for soil stabilization of the embankment and construction of foundation will be determined during the design phase of project. However, precaution should be taken to avoid damage to the remaining portion of marsh by proper selection of foundation and proper dumping of unsuitable material.

The relocation of schemes would commit a certain degree of detrimental impact to the environment and the reconstruction scheme would offer absolutely minimum taking of the wet land and it is anticipated that project will be classified as a Categorical Exclusion. New viaduct on relocated alignment would require Environmental Assessment (EA) and may be classified to be Found No Significant Impact (FONSI) but the embankment fill scheme will definitely require a detailed hydraulic risk analysis and Environmental Impact Statement (EIS). The preparation of an EIS would be a lengthy (estimated 5 years) and costly process while the existing bridge will further deteriorate.

**H. Permits**

This project will require extensive coordination with agencies of community, state and federal governments. The following agencies should be contacted for permits:

| | Scheme I Reconst. | Scheme II Fill | Scheme III New bridge |
|---|---|---|---|
| Corps of Engineer (404 permit) | May be | Yes | Yes |
| Coast Guard | (Not Applicable. Non navigational) | | |
| State Dept. of Environmental Protection | | | |
| Stream Encroachment | Yes | Yes | Yes |
| Wetlands Management | Yes | Yes | Yes |
| Water Quality | Yes | Yes | Yes |

**4. STUDIES OF ALTERNATE SCHEMES**

**A. Scheme I (Superstructure Replacement)**

Superstructure replacement will consist of replacement of superstructure and rehabilitation and reconstruction of substructure. Two studies are made for replacement one Non-composite continuous steel rolled beam bridge and other precast prestressed concrete deck beams which are simple in dead load and continuous on live loads.

The proposed roadway would be 50'-0" wide between curbs, flanked on the north side by a 6'-0" sidewalk and south side by barrier type parapets. The proposed steel viaduct would consist of 7 groups of continuous steel rolled beam structures, a two span continuous steel plate girder over the Raritan River and a simple span steel beam over the railroad. The grouping match the existing span arrangement of the concrete beams in the viaduct. The structure consists of the 9 inches thick reinforced concrete deck slab and a sidewalk over the slab supported on the 8 non-composite steel rolled beams or plate girder spaced at 7'-6" center to center, continuous or simple as describe above.

The proposed concrete alternate comprised of 7 groups of continuous prestressed concrete deck beams, a two span continuous prestressed box girders over the river and a prestressed box girder over the railroad. The prestressed concrete beams consists of 5" concrete wearing surface and sidewalk supported on sixteen (16) four feet wide prestressed deck beam or box girders. The prestressed bridge will be continuous for live loads and simply supported for dead loads.

The estimate construction cost of structural steel superstructure is $ 3,580,000 and prestressed structure $ 3,620,000.

Although the prestressed concrete structure is slightly higher in cost, prestressed concrete structure is recommended because it will reduce the construction time and require minimum maintenance.

## B. Scheme II (Fill Embankment)

Scheme II-1 is a two stage construction to allow the viaduct to be partially open to traffic. Scheme II-2 is a one stage construction under which new embankment will be built on relocated alignment at north of the bridge.

### Scheme II-1 (Existing location)

The river bridge will be a three span continuous steel bridge (80'-100'-80') supported on semi-stub abutment and solid wall type piers on a pile foundation. The bridge waterway opening will be adequate to carry 100 year flood.

The first embankment starts from station 124+20 between Bents #10 and #11 of the the easterly span from the river and ends at station 126+60 near Bent #21 located at the westside of the railroad. The railroad bridge will be 75 feet long and a simply supported steel bridge on full height abutments. The second embankment begins at station 127+50 near Bent#22 and stop at station 130+75, the east abutment.

The typical embankment section would be 80'-0" wide at the top with a 2:1 side slope. Original right-of-way is 75 feet wide and new embankment requires a minimum 200 feet width of right-of-way.

The next area studied is from the river to the railroad tracks and between the railroad and east abutment. The railroad tracks are still active and structures are required over railroad tracks and river. The remaining area would become a roadway placed upon compacted fill conform with the profile with a maximum 10 feet surcharge and maximum 15 feet fill height.

This scheme requires closing one-half of the bridge at each stage. The deck will be removed from one half, after which longitudinal sheet piling will be driven just off the centerline of the roadway. Transverse sheeting is to be driven at piers where new abutments will be built. This longitudinal and transverse sheeting allows to keep one half of bridge open at all times and to excavate unsuitable material for new foundation. Sheeting would be driven such that each pier footing would remain on one side of the centerline. The other side would be the first half of the viaduct to be demolished. New fill is then placed up to the 10 feet height as for surcharge in wet area and 15 feet at other locations for 90 days and the same procedure will be repeated. Roadways would then be built.

As an option fill may be placed in the lift with galvanized steel strip tiebacks placed at each lift as if reinforced earth. These reinforced strips would provide adhesion and resistance of creep under tension.

### Scheme II-2 (New Alignment)

This scheme is similar to scheme II-1, except for new alignment and one stage construction. The one stage construction would be simple, save construction time and eliminate on-bridge maintenance of traffic. Furthermore, the need for temporary sheeting is not required.

### Comparison of Fill Scheme

| | Scheme II-1 (Existing) | Scheme II-2 (Relocated) |
|---|---|---|
| Structure Cost | 2,300,000 | 2,300,000 |
| Roadway Cost | 3,700,000 | 2,800,000 |
| Right-of-way Cost | 1,200,000 | 2,900,000 |
| Total | 7,200,000 | 8,000,000 |

The total construction cost for scheme II-1 would be $7,200,000. This cost include land acquisition, four abutments, two piers, longitudinal and transverse sheet piling, and traffic maintenance for a two stage construction sequence.

The total cost of Scheme II-2 would be 8,000,000. A savings of $900,000 is made by eliminating the sheeting, maintenance of traffics on the bridge, temporary pavement, and reduction of overall construction time. But right-of-way cost is increased 1,700,000. and overall cost increase was $800,000.

The scheme II-2 would cost more to construct but it has advantage of minimizing traffic interruptions and user's operating cost. Therefore, the scheme II-2 is recommended.

**Scheme III (New Bridge)**

For a new bridge, the following span length and pier arrangement had studied:

* Total Bridge length: 875 feet
* River span length : 75-100-75feet
* Railroad span : 75 feet
* Other spans : 50 and 75 feet

The span length for river crossing and railroad crossing was selected by hydraulic and geometric requirement but selection of other span length will dependent on the cost.

The 50 feet spans requires less materials on stringers and this would reduce the superstructure cost but this arrangement would increase the substructure cost by adding more piers.

For the purpose of comparison, the bridge length was limited to 150 feet which would be comprised of three, 50 feet spans and 4 piers, or two, 75 feet spans and 3 piers. It is estimated that the 50 foot span consists of W36x150 rolled beams composite with a 9 inch concrete deck slab and a 75 foot span comprised with steel cover plated W36x230 rolled beams and compositely constructed 9 inch concrete deck slab.

The total increase of estimated cost for two, 75 feet spans over three, 50 feet span is $96,000. The total cost of a pier on spread footing and pile foundation is estimated $80,000 and $129,000 respectively.

The 50 foot span would be economical on spread footings and the 75 foot span has advantage on pile foundation. The longer 75 foot span would minimize environmental impacts on the wet land by reducing the number of piers to be built. Considering environmental impacts, the 75 foot span will be used on the wet land and the 50 foot spans on other areas. The 75 foot span will be used between the river and the railroad and the 50 foot spans between the railroad to the east abutment.

The superstructure alternate include the following:

| | |
|---|---|
| Scheme III-1 | Prestressed concrete I-girder, composite design |
| Scheme III-2 | Simple span steel beams, composite design |
| Scheme III-3 | Three span continuous steel beams |

Scheme III-1:

This scheme consists of eight (8) prestressed concrete I girders, 36" deep AASHTO Type II for 50 foot spans and 45" deep AASHTO Type III for 75' spans, supporting 9 inches thick compositely constructed reinforced concrete deck slab. The estimated cost of the variable superstructure components including river span and railroad span for this scheme is $3,600,000.

Scheme III-2:

This scheme consists of eight (8) composite W36x150 steel rolled beam supporting 9 inch reinforced concrete deck slab for span length of 50 feet.and W36x230 with 14"x1" cover plated beams for 75 foot spans. The cost estimate for superstructure is $3,850,000.

Scheme III-3:

This scheme uses a three span continuous steel beam wherever applicable. The superstructure consists of eight (8) W36x135 steel rolled beams for 50' spans and eight W36x230 beams for 75' spans. The estimated cost is $3,500,000.

Of the schemes studied, Scheme III-1 (Prestressed I Girder) is very close in term of cost and would offer minimum maintenance, therefore Scheme III-1 is recommended for the new structure.

**Scheme IV (Repairs)**

If the Scheme II (Fill) and Scheme III (New bridge) is selected, the lead time to meeting environmental and permit requirements would be lengthy, .probably another 3 to 5 years. Therefore, attention should be given to maintain the bridges for next 6 to 8 years until the new bridge or embankment is open for the public.

The following are the major temporary repairs necessary to provide minimum level of services:

1. The open joints should be sealed to prevent further run off water leakage onto pier caps and stringers.
2. The deteriorated deck slab should be repaired and resurfaced with asphalt wearing surface after placing waterproofing membrane.
3. The deteriorated end of concrete beams should be reconstructed at bent #17, 21 and 26.
4. All pier cap of river piers and twin bent piers should be restored to provide sufficient bearing areas.

The estimated cost of temporary repairs is $1,000,000.

## 5.COST COMPARISON

### Scheme I: Reconstruction

| | | |
|---|---|---|
| Superstructure Replacement | : | $ 3,600,000 |
| Substructure Repairs | : | $ 900,000 |
| Roadway & Traffic Control | : | $ 1,400,000 |
| Right-of-way | : | $ 100,000 |
| Total Cost: | | $ 6,000,000 |

### Scheme II: Embankment Fill

| | | |
|---|---|---|
| Superstructure | : | $ 1,500,000 |
| Substructure | : | $ 800,000 |
| Embankment Fill and Roadway | : | $ 2,800,000 |
| Right-of-way | : | $ 2,900,000 |
| Temporary Repairs | : | $ 1,000,000 |
| Total Cost: | | $ 9,000,000 |

### Scheme III: New Viaduct

| | | |
|---|---|---|
| Superstructure | : | $ 3,600,000 |
| Substructure | : | $ 2,300,000 |
| Roadway | : | $ 800,000 |
| Right-of-way | : | $ 2,100,000 |
| Temporary Repairs | : | $ 1,000,000 |
| Total Cost: | | $ 9,800,000 |

## 6. CONCLUSION AND RECOMMENDATION

### Scheme 1 (Reconstruction):

Reconstruction would be less costly than a new bridge or embankment fill and minimum involvement with permit and environment requirements but will require stage construction to maintain traffic on the bridge. Estimated construction cost is six million dollars ($ 6,000,000).

### Scheme 2 (Embankment fill):

Fill will offer less construction cost, minimum interruption of traffic during construction and minimum future maintenance cost but it will disturb the existing wet land, encroach flood plains, and requires the preparation of Environmental Impact Statements and permits from various agencies. Total estimated cost is $ 9,000,000.

### Scheme 3 (New Bridge):

Relocation of bridge has advantages of minimum traffic maintenance and longer expected life of the structure and its disadvantage is high initial construction cost ($ 9,800,000) and environmental involvement.

Considering cost, environmental involvements, encroachment of flood plains and advanced deterioration of the existing superstructure, We recommend the Scheme 1 (Reconstruction) including the following:

1. The entire superstructure should be replaced with a reinforced concrete deck on prestressed concrete boxes-simply supported for dead load and continuous for live loads.
2. The bridge deck should have 50 feet wide roadway flanked on south side by safety shape barrier parapets and north side with a 6 feet wide sidewalk and concrete parapet.
3. The abutments and the twin piers should be repaired and the piers at river crossing should be replaced with solid wall type piers. Also substructure should be extended to accommodate wider new superstructure.
4. All deteriorate concrete in the remaining substructure should be rehabilitated.

# A.2 FHWA RULES

## PART 650—BRIDGES, STRUCTURES, AND HYDRAULICS


**Subpart A—Location and Hydraulic Design of Encroachments on Flood Plains**

Sec.
650.101 Purpose.
650.103 Policy.
650.105 Definitions
650.107 Applicability.
650.109 Public involvement.
650.111 Location hydraulic studies.
650.113 Only practicable alternative finding.
650.115 Design standards.
650.117 Content of design studies.

**Subpart B—Erosion and Sediment Control on Highway Construction Projects**

650.201 Purpose.
650.203 Policy.
650.205 Definitions.
650.207 Plans, specifications, and estimates.
650.209 Construction.

**Subpart C—National Bridge Inspection Standards**

650.301 Application of standards.
650.303 Inspection procedures.
650.305 Frequency of inspections.
650.307 Qualifications of personnel.
650.309 Inspection report.
650.311 Inventory.

**Subpart D—Highway Bridge Replacement and Rehabilitation Program**

650.401 Purpose.
650.403 Definition of terms.
650.405 Eligible projects.
650.407 Application for bridge replacement or rehabilitation.
650.409 Evaluation of bridge inventory.
650.411 Procedures for bridge replacement and rehabilitation projects.
650.413 Funding.
650.415 Reports.

**Subpart F—Concrete Bridge Decks**

650.601 Purpose.
650.603 Definitions.
650.605 Construction.
650.607 Reconstruction.
650.609 Maintenance.
650.6'1 Additional stage of construction.
650.613 Protective system.

APPENDIX A—RECONSTRUCTION PROCEDURES


AUTHORITY: 23 U.S.C. 109(h), 116(d), 144, 315, 319(a), 319(b); 82 Stat. 829, sec. 26; sec. 6, 49 U.S.C. 1655; delegation of authority 49 CFR 1.48(b), unless otherwise noted.

### Subpart A—Location and Hydraulic Design of Encroachments on Flood Plains

AUTHORITY: 23 U.S.C. 109(a), 315; 23 CFR 1.32; 49 CFR 1.48(b); E.O. 11988—Floodplain Management, May 24, 1977 (42 FR 26951); Department of Transportation Order 5650.2, April 26, 1979 (44 FR 24678).

SOURCE: 44 FR 67580, Nov. 26, 1979, unless otherwise noted.

#### § 650.101 Purpose.

To prescribe Federal Highway Administration (FHWA) policies and procedures for the location and hydraulic design of highway encroachments on flood plains, including direct Federal highway projects administered by the FHWA.

#### § 650.103 Policy.

It is the policy of the FHWA:

(a) To encourage a broad and unified effort to prevent uneconomic, hazardous or incompatible use and development of the Nation's flood plains,

(b) To avoid longitudinal encroachments, where practicable,

(c) To avoid significant encroachments, where practicable,

(d) To minimize impacts of highway agency actions which adversely affect base flood plains,

(e) To restore and preserve the natural and beneficial flood-plain values that are adversely impacted by highway agency actions,

(f) To avoid support of incompatible flood-plain development,

(g) To be consistent with the intent of the Standards and Criteria of the National Flood Insurance Program, where appropriate, and

(h) To incorporate "A Unified National Program for Floodplain Management" of the Water Resources Council into FHWA procedures.

#### § 650.105 Definitions.

(a) "Action" shall mean any highway construction, reconstruction, rehabilitation, repair, or improvement undertaken with Federal or Federal-aid highway funds or FHWA approval.

(b) "Base flood" shall mean the flood or tide having a 1-percent chance of being exceeded in any given year.

(c) "Base flood plain" shall mean the area subject to flooding by the base flood.

(d) "Design Flood" shall mean the peak discharge, volume if appropriate, stage or wave crest elevation of the flood associated with the probability of exceedance selected for the design of a highway encroachment. By definition, the highway will not be inundated from the stage of the design flood.

(e) "Encroachment" shall mean an action within the limits of the base flood plain.

(f) "Floodproof" shall mean to design and construct individual buildings, facilities, and their sites to protect against structural failure, to keep water out or to reduce the effects of water entry.

(g) "Freeboard" shall mean the vertical clearance of the lowest structural member of the bridge superstructure above the water surface elevation of the overtopping flood.

(h) "Minimize" shall mean to reduce to the smallest practicable amount or degree.

(i) "Natural and beneficial flood-plain values" shall include but are not limited to fish, wildlife, plants, open space, natural beauty, scientific study, outdoor recreation, agriculture, aquaculture, forestry, natural moderation of floods, water quality maintenance and groundwater recharge.

(j) "Overtopping flood" shall mean the flood described by the probability of exceedance and water surface elevation at which flow occurs over the highway, over the watershed divide, or through structure(s) provided for emergency relief.

(k) "Practicable" shall mean capable of being done within reasonable natural, social, or economic constraints.

(l) "Preserve" shall mean to avoid modification to the functions of the natural flood-plain environment or to maintain it as closely as practicable in its natural state.

(m) "Regulatory floodway" shall mean the flood-plain area that is reserved in an open manner by Federal, State or local requirements, i.e., unconfined or unobstructed either horizontally or vertically, to provide for the discharge of the base flood so that the cumulative increase in water surface elevation is no more than a designated amount (not to exceed 1 foot as established by the Federal Emergency Management Agency (FEMA) for administering the National Flood Insurance Program).

(n) "Restore" shall mean to reestablish a setting or environment in which the functions of the natural and beneficial flood-plain values adversely impacted by the highway agency action can again operate.

(o) "Risk" shall mean the consequences associated with the probability of flooding attributable to an encroachment. It shall include the potential for property loss and hazard to life during the service life of the highway.

(p) "Risk analysis" shall mean an economic comparison of design alternatives using expected total costs (construction costs plus risk costs) to determine the alternative with the least total expected cost to the public. It shall include probable flood-related costs during the service life of the facility for highway operation, maintenance, and repair, for highway-aggravated flood damage to other property, and for additional or interrupted highway travel.

(q) "Significant encroachment" shall mean a highway encroachment and ny direct support of likely base flood-lain development that would involve ne or more of the following construc-ion-or flood-related impacts:

(1) A significant potential for inter-uption or termination of a transpor-ation facility which is needed for mergency vehicles or provides a com-nunity's only evacuation route.

(2) A significant risk, or

(3) A significant adverse impact on natural and beneficial flood-plain values.

(r) "Support base flood-plain devel-opment" shall mean to encourage, allow, serve, or otherwise facilitate ad-ditional base flood-plain development. Direct support results from an en-croachment, while indirect support re-sults from an action out of the base flood plain.

§ 650.107 Applicability.

(a) The provisions of this regulation shall apply to all encroachments and to all actions which affect base flood plains, except for repairs made with emergency funds (23 CFR Part 668) during or immediately following a dis-aster.

(b) The provisions of this regulation shall not apply to or alter approvals or authorizations which were given by FHWA pursuant to regulations or dir-ectives in effect before the effective date of this regulation.

§ 650.109 Public involvement.

Procedures which have been estab-lished to meet the public involvement requirements of 23 CFR Parts 771 and 795 or 790 shall be used to provide op-portunity for early public review and comment on alternatives which con-tain encroachments.

(a) Public notices issued in accord-ance with the above procedures shall make reference to significant en-croachments which are contained in alternatives under consideration.

(b) Public hearing presentations shall include identification of en-croachments.

§ 650.111 Location hydraulic studies.

(a) National Flood Insurance Pro-gram (NFIP) maps or information de-veloped by the highway agency, if NFIP maps are not available, shall be used to determine whether a highway location alternative will include an en-croachment.

(b) Location studies shall include evaluation and discussion of the prac-ticability of alternatives to any longi-tudinal encroachments.

(c) Location studies shall include dis-cussion of the following items, com-mensurate with the significance of the risk or environmental impact, for all alternatives containing encroachments and for those actions which would support base flood-plain development:

(1) The risks associated with imple-mentation of the action,

(2) The impacts on natural and beneficial flood-plain values,

(3) The support of probable incom-patible flood-plain development,

(4) The measures to minimize flood-plain impacts associated with the action, and

(5) The measures to restore and pre-serve the natural and beneficial flood-plain values impacted by the action.

(d) Location studies shall include evaluation and discussion of the prac-ticability of alternatives to any signifi-cant encroachments or any support of incompatible flood-plain development.

(e) The studies required by §§ 650.111 (c) and (d) shall be summa-rized in environmental review docu-ments prepared pursuant to 23 CFR 771.

(f) Local, State, and Federal water resources and flood-plain management agencies should be consulted to deter-mine if the proposed highway action is consistent with existing watershed and flood-plain management programs and to obtain current information on de-velopment and proposed actions in the affected watersheds.

650.113 Only practicable alternative find-ing.

(a) A proposed action which includes a significant encroachment shall not be approved unless the FHWA finds that the proposed significant en-croachment is the only practicable al-ternative. This finding shall be includ-ed in the final environmental docu-ment (final environmental impact statement or finding of no significant impact) and shall be supported by the following information:

(1) The reasons why the proposed action must be located in the flood plain,

(2) The alternatives considered and why they were not practicable, and

(3) A statement indicating whether the action conforms to applicable State or local flood-plain protection standards.

(b) A copy of the finding shall be made available to appropriate State and areawide clearinghouses following procedures established in accordance with 23 CFR Part 420, Subpart C.

§ 650.115 Design standards.

(a) The design selected for an en-croachment shall be supported by analyses of design alternatives with consideration given to capital costs and risks, and to other economic, engi-neering, social and environmental con-cerns.

(1) Consideration of capital costs and risks shall include, as appropriate, a risk analysis or assessment which in-cludes:

(i) The overtopping flood or the base flood, whichever is greater, or

(ii) The greatest flood which must flow through the highway drainage structure(s), where overtopping is not practicable. The greatest flood used in the analysis is subject to state-of-the-art capability to estimate the excee-dance probability.

(2) The design flood for encroach-ments by through lanes of Interstate highways shall not be less than the flood with a 2-percent chance of being exceeded in any given year. No mini-mum design flood is specified for In-terstate highway ramps and frontage roads or for other highways.

(3) Freeboard shall be provided, where practicable, to protect bridge structures from debris- and scour-re-lated failure.

(4) The effect of existing flood con-trol channels, levees, and reservoirs shall be considered in estimating the peak discharge and stage for all floods considered in the design.

(5) The design of encroachments shall be consistent with standards es-tablished by the FEMA, State, and local governmental agencies for the administration of the National Flood Insurance Program for:

(i) All direct Federal highway ac-tions, unless the standards are demon-strably inappropriate, and

(ii) Federal-aid highway actions where a regulatory floodway has been designated or where studies are under-way to establish a regulatory flood-way.

(b) Rest area buildings and related water supply and waste treatment facilities shall be located outside the base flood plain, where practicable. Rest area buildings which are located on the base flood plain shall be flood-proofed against damage from the base flood.

(c) Where highway fills are to be used as dams to permanently impound water more than 50 acre-feet ($6.17\times10^4$ cubic metres) in volume or 25 feet (7.6 metres) deep, the hydrologic, hydrau-lic, and structural design of the fill and appurtenant spillways shall have the approval of the State or Federal agency responsible for the safety of dams or like structures within the State, prior to authorization by the Division Administrator to advertise for bids for construction.

§ 650.117 Content of design studies.

(a) The detail of studies shall be commensurate with the risk associated with the encroachment and with other economic, engineering, social or envi-ronmental concerns.

(b) Studies by highway agencies shall contain:

(1) The hydrologic and hydraulic data and design computations,

(2) The analysis required by § 650.115(a), and

(3) For proposed direct Federal highway actions, the reasons, when applicable, why FEMA criteria (44 CFR 60.3, formerly 24 CFR 1910.3) are demonstrably inappropriate.

(c) For encroachment locations, project plans shall show:

(1) The magnitude, approximate probability of exceedance and, at appropriate locations, the water surface elevations associated with the overtopping flood or the flood of § 650.115(a)(1)(ii), and

(2) The magnitude and water surface elevation of the base flood, if larger than the overtopping flood.

## Subpart B—Erosion and Sediment Control on Highway Construction Projects

SOURCE: 39 FR 36332, Oct. 9, 1974, unless otherwise noted.

**§ 650.201 Purpose.**

The purpose of this subpart is to prescribe policies and procedures for the control of erosion, abatement of water pollution and prevention of damage by sediment deposition from Federal-aid highway projections and projects under the direct control of the Federal Highway Administration.

**§ 650.203 Policy.**

It is the policy of the Federal Highway Administration (FHWA) that Federal-aid highways and highways constructed under the direct supervision of FHWA shall be located, designed, constructed and operated according to standards that will minimize erosion and sediment damage to the highway and adjacent properties and abate pollution of surface and ground water resources.

**§ 650.205 Definitions.**

(a) Erosion control measures are installations used to inhibit dislodging of soil particles by water or wind.

(b) Sediment control measures are installations used to inhibit dislodging of remove settleable sediments from surface runoff.

(c) Permanent erosion and sediment control measures are installations which remain in place and in service on completion of the construction project.

(d) Temporary erosion or sediment control measures are installations used on an interim basis during construction.

(e) Pollutants are substances, including sediment, which cause deterioration of water quality when added to surface or ground waters in sufficient quantity.

**§ 650.207 Plans, specifications and estimates.**

(a) Emphasis shall be placed on erosion control in the preparation of plans, specifications and estimates.

(b) All reasonable steps shall be taken to insure that highway project designs for the control of erosion and sedimentation and the protection of water quality comply with applicable standards and regulations of other agencies.

**§ 650.209 Construction.**

(a) Permanent erosion and sediment control measures shall be installed at the earlist practicable time consistent with good construction practices.

(b) Temporary erosion and sediment control measures shall be coordinated with permanent measures to assure economical, effective and continuous control throughout the construction phase. Temporary measures shall not be constructed for expediency in lieu of permanent measures specified in the contract.

(c) Erosion and sediment control measures shall be adequately maintained to perform their intended function during construction of the project.

(d) Erosion and sediment control measures necessary because of contractor negligence, carelessness or failure to install contract measures as scheduled shall be installed at no cost to Federal-aid funds.

(e) Pollutants used during highway construction or operation and material from sediment traps shall not be stockpiled or disposed of in a manner which makes them susceptible to being washed into any watercourse by runoff or high water. No pollutants shall be deposited or disposed of in watercourses.

## Subpart C—National Bridge Inspection Standards

SOURCE: 36 FR 7851, Apr. 27, 1971, unless otherwise noted. Redesignated at 39 FR 10430, Mar. 20, 1974.

**§ 650.309 Inspection report.**

The findings and results of bridge inspections shall be recorded on standard forms. The data required to complete the forms and the functions which must be performed to compile the data are contained in section 3 of the AASHTO Manual.

[39 FR 29590, Aug. 16, 1974]

**§ 650.311 Inventory.**

(a) Each State shall prepare and maintain an inventory of all bridge structures subject to the Standards. Under these Standards, certain structure inventory and appraisal data must be collected and retained within the various departments of the State organization for collection by the Federal Highway Administration as needed. A tabulation of this data is contained in the structure inventory and appraisal sheet distributed by the Federal Highway Administration as part of the Recording and Coding Guide for the Structure Inventory and Appraisal of the Nation's Bridges (Coding Guide) in January of 1979. Reporting procedures have been developed by the Federal Highway Administration.

(b) All bridges subject to these Standards shall be inventoried by December 31, 1980, as required by section 124(a), and (c) of the Surface Transportation Assistance Act of 1978. Newly completed structures or any modification of existing structures which would alter previously recorded data on the inventory forms shall be entered in the State's records within 90 days.

## Subpart D—Highway Bridge Replacement and Rehabilitation Program

SOURCE: 44 FR 15665, Mar. 15, 1979, unless otherwise noted.

**§ 650.401 Purpose.**

The purpose of this regulation is to prescribe policies and outline procedures for administering the Highway Bridge Replacement and Rehabilitation Program in accordance with 23 U.S.C. 144.

**§ 650.403 Definition of terms.**

As used in this regulation:

(a) *Bridge.* A structure, including supports, erected over a depression or an obstruction, such as water, a highway, or a railway, having a track or passageway for carrying traffic or other moving loads, and having an opening measured along the center of the roadway of more than 20 feet between undercopings of abutments or spring lines of arches, or extreme ends of the openings for multiple boxes; it may include multiple pipes where the clear distance between openings is less than half of the smaller contiguous opening.

(b) *Sufficiency rating.* The numerical rating of a bridge based on its structural adequacy and safety, essentiality for public use, and its serviceability and functional obsolescence.

(c) *Rehabilitation.* The major work required to restore the structural integrity of a bridge as well as work necessary to correct major safety defects.

650.405 **Eligible projects.**

(a) *General.* Deficient highway bridges on all public roads may be eligible for replacement or rehabilitation.

(b) *Types of projects which are eligible.* The following types of work are eligible for participation in the Highway Bridge Replacement and Rehabilitation Program (HBRRP), hereinafter known as the bridge program.

(1) *Replacement.* Total replacement of a structurally deficient or functionally obsolete bridge with a new facility constructed in the same general traffic corridor. A nominal amount of approach work, sufficient to connect the new facility to the existing roadway or to return the gradeline to an attainable touchdown point in accordance with good design practice is also eligible. The replacement structure must meet the current geometric, construction and structural standards required for the types and volume of projected traffic on the facility over its design life.

(2) *Rehabilitation.* The project requirements necessary to perform the major work required to restore the structural integrity of a bridge as well as work necessary to correct major safety defects are eligible except as noted under ineligible work. Bridges to be rehabilitated both on or off the F-A System shall, as a minimum, conform with the provisions of 23 CFR Part 625, Design Standards for Federal-aid Highways, for the class of highway on which the bridge is a part.

(c) *Ineligible work.* Except as otherwise prescribed by the Administrator, the costs of long approach fills, causeways, connecting roadways, interchanges, ramps, and other extensive earth structures, when constructed beyond the attainable touchdown point, are not eligible under the bridge program.

§ 650.407 **Application for bridge replacement or rehabilitation.**

(a) Agencies participate in the bridge program by conducting bridge inspections and submitting Structure Inventory and Appraisal (SI&A) sheet inspection data. Federal and local governments supply SI&A sheet data to the State agency for review and processing. The State is responsible for submitting the six computer card format or tapes containing all public road SI&A sheet bridge information through the Division Administrator of the Federal Highway Administration (FHWA) for processing. These requirements are prescribed in 23 CFR 650.309 and 650.311, the National Bridge Inspection Standards.

(b) Inventory data may be submitted as available and shall be submitted at such additional times as the FHWA may request.

(c) Inventory data on bridges that have been strengthened or repaired to eliminate deficiencies, or those that have been replaced or rehabilitated using bridge replacement and/or other funds, must be revised in the inventory through data submission.

(d) The Secretary may, at the request of a State, inventory bridges, on and off the Federal-aid system, for historic significance.

[44 FR 15665, Mar. 15, 1979, as amended at 44 FR 72112, Dec. 13, 1979]

§ 650.409 **Evaluation of bridge inventory.**

(a) *Sufficiency rating of bridges.* Upon receipt and evaluation of the bridge inventory, a sufficiency rating will be assigned to each bridge by the Secretary in accordance with the approved AASHTO [1] sufficiency rating formula. The sufficiency rating will be used as a basis for establishing eligibility and priority for replacement or rehabilitation of bridges; in general the lower the rating, the higher the priority.

(b) *Selection of bridges for inclusion in State program.* After evaluation of the inventory and assignment of sufficiency ratings, the Secretary will provide the State with a selection list of bridges within the State that are eligible for the bridge program. From that list or from previously furnished selection lists, the State may select bridge projects.

§ 650.411 **Procedures for bridge replacement and rehabilitation projects.**

(a) Consideration shall be given to projects which will remove from service highway bridges most in danger of failure.

(b) *Submission and approval of projects.* (1) Bridge replacement or rehabilitation projects shall be submitted by the State to the Secretary in accordance with 23 CFR Part 630 Subpart A Federal-Aid Programs, Approval and Authorization.

(2) Funds apportioned to a State shall be made available throughout each State on a fair and equitable basis.

(c)(1) Each approved project will be designed, constructed, and inspected for acceptance in the same manner as other projects on the system on which the project is located. It shall be the responsibility of the State agency to properly maintain, or cause to be properly maintained, any project constructed under this bridge program. The State highway agency shall enter into a formal agreement for maintenance with appropriate local government officials in cases where an eligible project is located within and is under the legal authority of such a local government.

(2) Whenever a deficient bridge is replaced or its deficiency alleviated by a new bridge under the bridge program, the deficient bridge shall either be dismantled or demolished or its use limited to the type and volume of traffic the structure can safely service over its remaining life. For example, if the only deficiency of the existing structure is inadequate roadway width and the combination of the new and existing structure can be made to meet current standards for the volume of traffic the facility will carry over its design life, the existing bridge may remain in place and be incorporated into the system.

[44 FR 15665, Mar. 15, 1979, as amended at 44 FR 72112, Dec. 13, 1979]

§ 650.413 **Funding.**

(a) Funds authorized for carrying out the Highway Bridge Replacement and Rehabilitation Program are available for obligation at the beginning of the fiscal year for which authorized and remain available for expenditure for the same period as funds apportioned for projects on the Federal-aid primary system.

(b) The Federal share payable on account of any project carried out under 23 U.S.C. 144 shall be 80 percent of the eligible cost.

(c) Not less than 15 percent nor more than 35 percent of the apportioned funds shall be expended for projects located on public roads, other than those on a Federal-aid system. The Secretary after consultation with State and local officials may, with respect to a State, reduce the requirement for expenditure for bridges not on a Federal-aid system when he determines that such State has inadequate needs to justify such expenditure.

§ 650.415 **Reports.**

The Secretary must report annually to the Congress on projects approved and current inventories together with recommendations for further improvements.

## Subpart F—Concrete Bridge Decks

SOURCE: 41 FR 18297, May 3, 1976, unless otherwise noted.

§ 650.601 **Purpose.**

The purpose of this subpart is to prescribe policies and procedures for the construction, reconstruction, and additional stages of concrete bridge deck construction, with special emphasis on concrete bridge deck protective systems, within the existing controls on the expenditure of Federal-aid funds.

§ 650.603 Definitions.

(a) *Additional Stage of Construction*, as related to bridge decks, means a deferred or an additional stage of an initial (or new) construction project, or additional work not contemplated during the initial construction phase but accomplished as part of a subsequent or reconstruction project.

(b) *Construction* means the initial (or new) construction of any specific bridge deck.

(c) *Maintenance* means routine or incidental work necessary to keep a bridge deck functioning in a safe and efficient manner.

(d) *Protective System* means a system used to protect bridge decks from early deterioration due to reinforcing steel corrosion induced by highway deicing chemicals.

(e) *Reconstruction* means the restoration of the structural integrity of a bridge deck which may include partial restoration or complete removal and replacement of the existing deteriorated bridge deck. Partial restoration of an existing deck includes the partial removal and replacement of all deteriorated and/or inadequate components.

§ 650.605 Construction.

In view of bridge deck deterioration occurring in the early life of numerous bridges built in recent years, the following practices shall be implemented for all bridge decks to be constructed with Federal-aid funds:

(a) *Concrete Requirements.* (1) *Concrete Cover*—Design, specifications, construction controls and tolerances shall be established so that adequate concrete cover over the top deck rebars will be provided to assure durable, long-lasting bridge decks.

(2) *Water-Cement Ratio.* States shall adopt the lowest feasible maximum water-cement ratio for all concrete utilized in construction of bridge decks. Adequate mix design procedures shall be established to provide a method for determining water utilized in the actual construction and field procedures shall be employed to provide reasonable assurance that the specified maximum water-cement ratio is not exceeded.

(3) *Quality Concrete.* States shall adopt designs, specifications and procedures equal to or better than those specified in Standard Specifications for Highway Bridges.[1] Eleventh Edition, AASHO[2] 1973, as amended,[3] section 2.4.6, "Proportioning of Concrete," Class A(AE), Bridge Deck Concrete.

(4) *Consolidation.* With low water-cement ratio dense concretes, specifications shall include construction control procedures that insure consolidation of the in-place concrete.

(b) *Protective System.* Where bridge decks are likely to be subjected to potentially damaging applications of deicing salts, a protective system is required in accordance with the provisions of § 650.613.

[41 FR 18297, May 3, 1976; 42 FR 30835, June 17, 1977]

§ 650.607 Reconstruction.

In an effort to alleviate bridge deck deterioration and provide bridge decks with a reasonable service life, the following policies are established to assure that all Federal-aid bridge deck reconstruction will be as nearly permanent as judicious economics and current technology will permit.

(a) *Eligible Work.* Reconstruction work and procedures, necessary to assure acceptable performance of existing bridge decks, are set forth below and are eligible for Federal-aid participation from the appropriate funds:

(1) *Reconstruction Work.* Reconstruction shall include all concrete deck restoration work required to assure satisfactory performance of the concrete deck and of the adopted protective system if required. This may include items such as the removal of existing overlays, removal and replacement of all deteriorated reinforcing steel or the complete removal and replacement of an entire bridge deck when necessary. The installation of a protective system, when required, shall be considered an additional stage of construction;

(i) Reconstruction may also include removal and replacement of deteriorated concrete curbs, sidewalks, parapets, as well as rails, deck joints, bearings, or similar incidental items which are considered essential for proper functional restoration of the structure.

(ii) Consistent with 23 CFR 655, Subpart E, reconstruction may include safety improvements to be undertaken in conjunction with the above described work, where such improvements eliminate an established hazardous condition. Such safety improvements may include widening, elimination of hazardous walks and substandard railing systems, removal of hazardous fixed objects or installations of energy absorbing barrier systems, and any other features that are considered essential under current safety standards.

(2) *Reconstruction Procedures.* The following reconstruction procedures are acceptable for use on Federal-aid projects:

(i) *Acceptable Permanent Reconstruction.* (A) Permanent restoration requires a structurally sound bridge deck and the removal of all delaminated concrete, highly chloride contaminated concrete and also deteriorated concrete and rebars in areas of active corrosion. With proper restoration procedures and the installation of an approved protective system, this type of reconstruction is considered permanent and nonexperimental.

(B) Permanent restoration, however, will require a complete deck survey as to delaminations, corrosion potentials and chloride contents except where visual and delamination surveys indicate complete deck replacement as the obvious economical alternative.

(ii) *Experimental Cost Effective Reconstruction.* (A) This procedure is considered experimental, may be undertaken on concrete bridge decks found to be structurally sound, and requires restoration procedures sufficient to assure an estimated extended life for the deck of 10 to 15 years. This alternative allows salt contaminated concrete to be left in place, but the system of restoration must be considered effective based on previous successful experimental installations.

(B) Experimental considerations require that the first three installations and 10 percent of the remaining installations shall be identified as test structures and be completely surveyed prior to restoration.

(C) Immediately after restoration these test structures shall be resurveyed for corrosion potentials and, if membranes are used, resistivity readings shall be recorded. Periodically thereafter on all test structures, corrosion potential levels, resistivity readings where appropriate, and the presence of delaminations, where practical, shall be determined.

(b) *Miscellaneous.* (1) Acceptable restoration and protective systems are detailed in Appendix A.

(2) Reconstruction must be consistent with the applicable design and construction controls specified in § 650.605.

§ 650.609 Maintenance.

The following policies are applicable to maintenance of bridge decks as related to Federal-aid funding:

(a) Maintenance work is considered an obligation of the State and is not eligible for Federal-aid funding;

(b) Spot patching of a bridge deck is maintenance unless it is done as work incidental to reconstruction in accordance with § 650.607.

(c) The resurfacing of a bridge deck is maintenance when such work only restores the general condition of the riding surface and does not significantly increase the structural and/or durability characteristics of the original deck.

§ 650.611 Additional stage of construction.

The following policies have been established relative to additional stages of construction on existing or reconstructed bridge decks:

(a) *Eligible Work.* Where not used in the original construction, a moisture barrier, waterproofing, membrane, or other protective system, including a water tight deck joint system, is con-

[1] This publication establishes standards and specifications for Federal-aid projects pursuant to 23 CFR 625.3.

[2] American Association of State Highway and Transportation Officials, Suite 225, 444 North Capitol Street, NW., Washington, D.C. 20001

[3] AASHTO Interim Specifications Bridges, 1974.

sidered an additional stage of construction on an existing bridge where deicing chemicals are likely to be used. An asphaltic concrete protective wearing course, if required to maintain the integrity of a membrane, may likewise be considered an additional stage of construction.

(b) *Federal-Aid Participation.* A protective system, even though not included on the original construction plans, is eligible as an additional stage of construction for Federal-aid participation from the appropriate funds.

(c) *Procedures.* All additional stage construction procedures must be consistent with applicable design and construction controls herein.

§ 650.613 **Protective system.**

In view of the recognized need for protective systems on these bridges subjected to deicing chemicals, the following policies have been established:

(a) Contract plans and specifications for construction and/or reconstruction of bridge decks that are likely to be subjected to potentially damaging applications of deicing salts shall provide for a protective system that will effectively prevent chloride induced deterioration;

(b) Existing protective systems that prove effective based on field experience and/or laboratory research programs have been and will continue to be reported on in the National Experimental and Evaluation Program as conducted by FHWA Headquarters Office.

(c) Proprietary products, when used as a bridge deck protective system, must be used in accordance with Federal regulations on the use of proprietary products.

## 1983 REVISION

**§ 650.605 Construction, reconstruction and rehabilitation.**

The following policies are established for all bridge decks to be constructed, rehabilitated or reconstructed with Federal-aid funds.

(a) *Standard specifications.* States shall adopt design specifications and procedures equal to or better than those specified in Standard Specifications for Highway Bridges.[1]

(b) *Protective system.* When a bridge deck is likely to be subjected to potentially damaging applications of deicing salts or to be exposed to a seawater environment, a protective system is required.

(1) Approval of the type of protective system is made in accordance with the provisions of § 650.609.

(2) The installation of a protective system is eligible for Federal-aid funds from the appropriate category.

(3) The installation of a protective system in an unprotected bridge deck is eligible for Federal-aid funds from the appropriate category, even if rehabilitation of the existing bridge deck is not required.

(c) *Eligible work.* Reconstruction and rehabilitation procedures necessary to assure acceptable performance of existing structures are set forth below and are eligible for Federal-aid participation from the appropriate category. Reconstruction and rehabilitation work shall include all concrete deck restoration work required to assure satisfactory performance of the concrete deck. This may include items such as the removal of existing overlays, removal and placement of all deteriorated reinforcing steel, or the complete removal and replacement of the entire bridge deck if necessary.

(1) This work may also include removal and replacement of deteriorated concrete curbs, sidewalks, parapets, as well as rail, deck joints, bearings, or similar incidental items which are associated with proper functional restoration of the structure.

(2) Consistent with 23 CFR Part 924, safety improvements may be undertaken with the above described work, when such improvements eliminate an established hazardous condition. Such safety improvements may include widening, elimination of hazardous walks and substandard railing systems, removal of hazardous fixed objects or installation of energy absorbing barrier system, and any other features that are consistent with current safety standards.

**§ 650.607 Maintenance.**

The following policies are established for the use of Federal-aid funds as related to maintenance activities.

(a) Maintenance work is considered an obligation of the State and is not eligible for Federal-aid funding;

(b) Spot patching of a bridge deck is maintenance unless it is done as work incidental to the rehabilitation or reconstruction of the concrete bridge deck;

(c) The resurfacing of a bridge deck is considered maintenance when such work only restores the general condition of the riding surface and does not significantly increase the structural and/or durability characteristics of the original deck.

**§ 650.609 Protective system.**

In view of the recognized need for protective systems the following policies have been established for Federal-aid participation.

(a) The contract, plans and specifications (PS&E) for the construction, rehabilitation and/or reconstruction of all concrete bridge decks that are likely to be subjected to potentially damaging applications of deicing salts or seawater environme shall provide for a protective system that will effectively prevent chloride induced deterioration.

(b) Federal-aid approval. Approval of the protective system will be required as part of the approval of the Federal-aid construction, reconstruction, or rehabilitation project.

(1) Protective systems which, based on field experience and/or laboratory research, have not been demonstrated to be cost effective must be considered experimental. These systems are to be surveyed and evaluated on an experimental basis under the Federal Highway Administration (FHWA) experimental program.

(2) Proprietary products when used as a bridge deck protective system must be used in accordance with Federal regulations on the use of proprietary products.

*Acceptable reconstruction procedures for Federal-aid participation*[1]

| Category | Procedures | Acceptable permanent reconstruction | Experimental cost-effective reconstruction (estimated extended life 10 to 15 yr.) |
|---|---|---|---|
| Structurally inadequate. | .................................... | Complete deck replacement (unless restorable). | |
| Extensive deck deterioration. | Required restoration work. | Complete deck replacement ........................ | Removal of all deteriorated concrete. Follow the repair procedure approved for the protective system selected. |
| | Testing ............................ | Steps 1 through 5 as necessary (probably only steps 1 and 2). | Steps 1 and 2 only, except all steps on the first 3 plus 10 pct of the remaining decks. |
| | Suggested protective systems. | Epoxy coated rebars. Two course-Iowa system or latex modified concrete. Membrane W/ac overlay. Cathodic protection. | Two course-Iowa system or latex modified concrete. Membrane W/ac overlay. Cathodic protection. |
| Moderate deck deterioration. | .................................... | Same as for category 1 above.—or ............ Same as for category 3 below as determined by the State.. | Same as for category 1 above. |
| Light deck deterioration. | Required restoration work. | Removal and replacement of all areas of deterioration and chloride contaminated concrete as determined by corrosion potentials and/or chloride sampling (less than 5 pct of deck area).. | Same as for category 1 above.[2] |
| | Testing ............................ | Steps 1 through 5........... ............................ | Steps 1 and 2 only, except all steps on the first 3 plus 10 pct of the remaining decks. |
| Light deck deterioration. | Suggested protective systems. | Two course-Iowa system or latex modified concrete. Membrane W/ac overlay. Cathodic protection. | Two course-Iowa system or latex modified concrete. Membrane W/ac overlay. Cathodic protection. |

[1]Testing steps: 1 Visual, 2 Delamination, 3 Electrical potential, 4 Rebar Depth, 5 Chloride Content.
[2]For this category of condition permanent restoration is strongly recommended.

# A.3 REHABILITATION PROCEDURE

## From NY State DOT

The Bridge Inspection and Rehabilitation Design Program contains seven steps, as follows:

1. Project identification and initiation.
2. Bridge inspection.
3. Acquisition of supplemental data.
4. Load rating of the bridge.
5. Bridge Rehabilitation Project Report and preliminary plan preparation.
6. Rehabilitation/replacement decision.
7. Contract document preparation, if rehabilitation is the decision; or recycle as a replcement project if not.

The first five steps are common to projects for which the rehabilitate/replace decision must be made and those for which external contraints dictate the decision. If the decision to rehabilitate is made earlier, because of external constraints, step six, the rehabilitate/replace decision, is skipped and step seven, contract document preparation is started immediately after agreement is reached on the recommended scope of the work as defined by the Project Report and preliminary plan. If the decision was not made earlier, step six is included. Based on the decision in step six, step seven will consist of the preparation of rehabilitation plans and estimates or the development of the project as a bridge replacement project.

Details of each step are as follows:

1. Project Identification and Initiation. The selection of a bridge for the program is made by the Regional Office of the Department in which the bridge is located, based on detailed knowledge of the bridge including its age and condition, and bridge priorities within the Region.

   The selection process is facilitated by the information generated from the Department's biennial Bridge Inspection Program. This program consists of a statewide, standardized, visual inspection which results in the assignment of numerical ratings to the principal components of each bridge inspected. The inspector also provides an overall bridge condition rating and recommendation for each bridge. A listing of structurally deficient bridges is issued annually, showing bridges in priority order as determined by their condition rating.

   Regional Offices submit initiation requests for bridge rehabilitation and bridge replacement projects based on fund allocations, the structurally deficient bridge list, knowledge of the structure and the bridge needs of the Region. After review and acceptance in the Main Office of the Department, projects are placed in the Department's program and a designer, either from within the Department or a consulting engineer, is designated to perform the remaining activities.

2. Bridge Inspection. The bridge inspection used in this program is defined by the New York State Department of Transportation as a high quality, in-depth engineering study. The Department requires that the inspection, if done by a consulting engineering firm, be personally supervised by a Professional Engineer registered in the New York State with at least three years experience in bridge inspection or related bridge work. Inspections done by state staff are also supervised by experienced Professional Engineers.

   In order to ensure that there is uniformity in the quality and methods used for the inspection, the Department has written a specification for this purpose. This specification was developed from material in the AASHTO Manual for Maintenance Inspection of Bridges and the Federal Highway Administration's Bridge Inspector Training Manual and by gathering information and suggestions from the Department's Main Office and Regional Office personnel who have had experience in bridge inspection.

   Two points should be emphasized with regard to the specification for in-depth bridge inspection:

   1. Besides setting the minimum inspection requirements the specifications contains the following instructions for the engineer conducting the inspection. "Due to the varying nature of original designs, construction techniques, and materials, it is impossible to state every observation, measurement, and test that should be made to properly evaluate every bridge. The Engineer shall view each bridge separately and make observations, perform tests, and make measurements, etc., beyond the minimum criteria established in this specification as may be necessary to become reasonably certain as to the condition of the bridge and its ability to function properly.
   2. The specification is periodically revised based on the Department's assessment of the results obtained to that thime. This usually involves expanding or refining a particular inspection requirement.

For example, the original specification contained no provision for determining if the superstructure expansion joints were actually functioning or how much expansion was occurring at a particular location. After encountering problems in determining the sizes of joint seal assemblies needed on some steel bent structures it was decided to add a requirement for measuring the existing joint openings twice, with at least forty degree differential in temperature between the times of the measurements.

The in-depth inspection is literally a "hands on" inspection. All deteriorated and suspect areas of the bridge are closely examined and their condition recorded. Concrete is sounded and cored for laboratory analysis. Steel losses are measured. The structure is surveyed for settlement and alignment and clearance deficiencies. In addition, the approaches of the bridges are inspected. Occassionally these inspections reveal deterioration so extensive that the bridge cannot be rehabilitated. In these cases the in-depth inspection is terminated and the project is reprogrammed as a bridge replacement.

IN-DEPTH FIELD INSPECTION

The Designer will make an in-depth field inspection of the bridge. The Regional Bridge Maintenance Engineer will be informed of the inspection and the proposed rehabilitation and, at his option, he may either be present at the inspection or forward to the designer his comments. When the design is by Structures Subdivision or Consultant the Regional Structures Engineer will also be present at the field inspection. The Thorough inspection necessary may include several visits to the job site and the use of considerable inspection equipment. Arrangements must be made in advance with the Regional Maintenance Group for equipment needed to adequately perform the inspection. In the case of Structures Subdivision Inspection, arrangement will be made through the Regional Structures Engineer, who will use his judgement as to what equipment is needed. Equipment could include ladders, scaffolding, snooper, cherry picker, boat, etc.

The Designer will make a Field Inspection Report in which the following points must be addressed:

1. Purpose of inspection.
2. Scope of inspection.
3. Those participating.
4. Observations (will include color photographs in the original report and in all copies).
5. Conclusions.
6. Recommendations.

In the case of design by Consulting Engineer, an inspection specification may be included in the Consultant Agreement in place of the above format.

BRIDGE DECK EVALUATION REPORT

The Designer shall check to see if a Bridge Deck Evaluation Report has been prepared and is up-to-date. If not, the Regional Office, the Consulting Engineer, or the Local Government should be requested to prepare it, following the procedure in the current Bridge Deck Evaluation Procedure Manual. When required, substructure cores shall be requested through the Inventory and Inspection Unit (Structures).

3. Supplemental Data. In addition to the inspection data, additional information is required to provide a total overview of the bridge and its relationship to the highway it serves.

This information includes the highway design speed, traffic volumes and lane requirements, accident data on and in the vicinity of the bridge, a comparison of the existing geometrics to the current AASHTO Standards, cost estimates for upgrading the bridge to one or more feasible levels of geometric improvement, and an estimate for a complete replacement of the bridge to current geometric standards.

INFORMATION REQUIRED TO PROGRESS DESIGN

The following items of information are necessary to progress design. Some of the items may have already been secured under Steps of the Environmental Action Plan. If so, the appropriate report may be referred to as the source of the information.

1. Current traffic, projected traffic (20 years, or as specified) and design speed. If the design speed of the highway is not known, an evaluation shall be made as to whether the segment of highway which includes the bridge project has an equivalent design speed of (1) 50 mph or more or (2) less than 50 mph. When DHV's are furnished, both the 1-way DHV and the 2-way DHV shall be included.

2. A statement from the Regional Planning Group about proposed improvements to this section of the highway. Ordinarliy, the statement from the Planning Group will say that there are no plans in the foreseeable future. If there are plans, the Region must justify why the bridge is being rehabilitated at this time, rather than at the time of highway improvement.

3. A method for maintaining traffic: The Region will discuss the project with State and local police, and with local officials, including school boards and fire companies. Some of this discussion may have to take place during the preparationof the Preliminary Plan.

4. A stream bridge shall be inspected for hydraulic deficiencies: severe scouring, chronic flooding, etc. If none are found, a statement to that effect shall be made. If some are found or the rehabilitation work reduces the waterway area, a Hydraulic Report shall be made.

5. The bridge shall be inspected for foundation deficiencies: settlement, tilting, etc. If none are found, a statement to that effect shall be made. If some are found, Foundation Design Requirements must be obtained.

6. A statement on right-of-way: Any necessary right-of-way taking should be briefly described. Otherwise state that no taking is necessary.

7. The accident record at the bridge from the Regional Office.

8. A statement on any substandard features of the bridge and the immediately adjacent highway, and justification if they are to be retained. If no substandard features, so state. The Regional Office shall provide the Designer with information relative to substandard features on the immediately adjacent highway. The existing and proposed bridge roadway widths and highway sections shall be described.

   Relevant to the retention of substandard features is FHPM 6-2-1-1, paragraph 4e(2), "The determination to approve a project design that does not conform to the minimum (design) criteria is to be made only after due consideration is given to all project conditions such as maximum service and safety benefits for the dollar invested, compatibility with adjacent section of unimproved roadway, and the probable time before reconstruction of the section due to increased traffic demands or changed conditions".

4. Load Rating. Using the actual measurements obtained from the "hands on" inspection, a load rating anlysis of the existing bridge is made in accordance with the current AASHTO Manual for Maintenance Inspection of Bridges.

This analysis is done to the same level of accuracy and detail that would be used when a bridge is designed. Once computed, this rating goes into the bridge inventory files of the Department and is used to report inventory and operating ratings to the Federal Highway Administration. These ratings are also used to make determinations on overload permit requests, or to alert the maintaining agency to the need for emergency repairs or a posting of the bridge for a reduced load level.

5. Bridge Rehabilitation Project Report and Preliminary Plan. All of the information gathered in steps 1 through 4 is analyzed and combined into the Bridge Rehabilitation Project Report. This report contains complete recommendations regarding the details of the work to be done on the bridge and justifications for retaining any substandard conditions on the rehabilitated bridge or its approaches. A preliminary plan accompanies the report, showing the existing plan, profile and typical section; the proposed plan, profile, and typical section; and typical repair details. Cost estimates of the proposed rehabilitation work are also included.

After agreement is reached between the designer and the approving agency as to the most feasible approach to the restoration of a particular bridge, the preliminary plan is modified, if necessary, to reflect this approach and approval given to proceed with work on the final contract documents.

BRIDGE REHABILITATION PROJECT REPORT

The Designer shall prepare the Bridge Rehabilitation Project Report (BRPR) which will include the following when appropriate:

1. Current and projected traffic and design speed.
2. Statement from Planning Group.
3. Traffic Maintenance Plan and Cost.
4. Hydraulic Statement.
5. Foundation Statement.
6. Statement on right-of-way.
7. The accident record at the bridge.
8. Statement on substandard features.
9. Location map.
10. Field Inspection Report or In-Depth Inspection Report and Summary Report (Consulting Engineer).
11. Bridge Deck Evaluation Report.
12. Approach highway width.
13. Recommendations and conclusions, including the cost of minimum rehabilitation to existing geometrics, of rehabilitation to standard geometrics and of bridge replacement.

    If the Designer concludes that the bridge should be replaced instead of rehabilitated, he shall so recommend. In that case, he shall not proceed with further steps of this procedure, but shall forward the report and recommendation, through usual channels, to the Deputy Chief Engineer, Structures (DCES) for his review and confirmation.

    If the DCES concurs with the recommendation that the bridge be replaced rather than rehabilitated, he will recommend to the Regional Office that a scope change be submitted through appropriate channels and the project be rescheduled.

PRELIMINARY PLAN

If the original decision to rehabilitate is unchanged, the Designer will prepare the Preliminary Plan. This Preliminary Plan shall be prepared in accordance with the current Standard Details for Highway Bridges, Bridge Design Data Sheets and Guideline Drawings.

A Preliminary Plan shows basic concepts and major details, acquaints affected parties with the project and project components, serves as an instrument for initial approval by affected parties and serves as a basis for the development of final plans.

If traffic is to be maintained on the bridge, the location of the longitudinal deck joint, the type of traffic barrier, and the width and location of traffic lanes to be maintained during various phases of construction shall be shown.

6. Rehabilitation/Replacement Decision. If external constraints have not already dictated that the bridge must be rehabilitated, the decision to rehabilitate or replace is made at this time. The major influence on this decision is cost. A General criterion is that a rehabilitation project with a cost of more than half the cost of a new bridge should not be built, and instead the bridge should be replaced. This criterion is applied because:

    - The amount of work to be done inevitably grows during construction. Even with the best inspection and evaluation some items requireing repair are overlooked, some items cannot be found until the construction work uncovers them, deterioration continues during the period between the inspection and the start of construction work and the rate of deterioration is difficult to estimate and account for in the plans and estimated.

    - The bridge, after rehabilitation, is unlikely to have the same life expectancy as a new bridge built at the same time. The rehabilitated bridge is a hybrid of new and old material types and materials of different eras, and will contain details less desireable than those reflecting the current state of the art. Also, the inevitable dilemma of when to stop removing deteriorated concrete or steel results in compromises during construction which may affect useful life, or as an alternate, if removal is carried to a point where no question remains, the cost of the work will escalate dramatically.

    Other criteria applied when making the decision include the ability of the bridge to be readily rehabilitated. Some bridge types are relatively easy to rehabilitate and widen, such as stringer bridges, while others, such as through trusses and through girders, are not. Concrete bridges are difficult to rehabilitate unless the concrete is very sound; low strength or partially deteriorated concrete is not easily restored.

    The vertical and horizontal geometry, and load-carrying ability, of the bridge must also be considered. A bridge with poor geometrics and a bad accident record should not be retained, but instead should be replaced. One which must have a load restriction even after rehabilitation should be replaced, unless the load restriction will cause no economic disadvantage to the area served by the bridge.

    These are general criteria, and exceptions must be made on occasion. The decision is not easy in may cases, and is only made after much investigation and evaluation. Ultimately, it is often based on subjective judgements.

    Finally, many of these criteria can be applied before detailed inspection. Geometry, strengthening and ease of rehabilitation should be considered at step 1, and the final decision made at that time if possible. This results in a project which is completed more quickly with the resultant saving of sizeable amounts of engineering costs.

7. Contract Document Preparation. Following the rehabilitate/replace decision, plans are prepared for either a rehabilitation project or a replacement project. If a rehabilitation project, the inspection already performed is the basis of plans, specifications and estimates, the follow-up inspections during plan preparation as required to get detailed information. If a new bridge is to be built, the project is recycled in conformance with the agency's procedures for a bridge replacement project.

Because of the amount of work required for the inspection and other supporting activities, it is clear that much engineering time is lost in those cases where the replacement option is selected. For this reason it is very important to stop the process at any time it becomes clear that replacement becomes the favored option. It should also be understood that this type of inspection, because of its cost and limited time value, should not be used unless funds are available to immediately prepare plans and specifications for remedial work and let contracts promptly.

# A.4 ENVIRONMENTAL CATEGORIES

CLASSES OF ACTION

There are three classes of action which prescribe the level of documentation required. A Class I Action requires the preparation of a DEIS and a FEIS. The Class II Action constitutes a CE. Class III Actions require the preparation of an EA; the EA determines whether a FONSI or the preparation of a DEIS and a FEIS is required. The decision to proceed with a Class I Action or the preparation of a FONSI is dependent on the impacts identified and the comments received on the EA.

A. CLASS I ACTION **EIS**

## ENVIRONMENTAL IMPACT STATEMENT

A proposed action which will have significant environmental impacts will require the preparation of a DEIS and a FEIS. This includes actions which are likely to:

1. Have significant impacts on properties protected under Section 4(f) of the DOT Act or on properties protected under Section 106 of the Historic Preservation Act;

2. Be highly controversial on enrionmental grounds, i.e., opposed or unacceptable on environmental or legal grounds by a Federal, State or local agency;

3. Have a significant impact on natural, ecological or cultural resources, including such areas as threatened and endangered species, wetlands, floodplains, prime or unique farmland, ground water, natural resources, or fish and wildlife resources;

4. Have significant displacement impact or be controversial with respect to the availability of adequate relocation housing;

5. Cause substantial disruption of an established community, disrupt orderly, planned development, be inconsistent with plans or goals that have been adopted by affected communities, or adversely affect the economic vitality of an urban area;

6. Have a significant impact on noise levels of noise sensitive areas (schools, hospitals, residential areas);

7. Have a significant impact on air quality;

8. Have a significant impact on water quality or a surface or subsurface public water supply system.

Examples of these actions are:

- Any new controlled access freeway
- Any Highway project of four or more lanes on a new location

B. CLASS II ACTION

## CATEGORICAL EXCLUSION

An Action which does not individually or cumulatively have a significant effect on the human environment is called a Categorical Exclusion This action does not normally require additional environmental documentation. The FHWA national list of CEs is as follows:

1. Planning and technical studies which do not fund the construction of facilities or acquisition of capital equipment.

2. Grants for training and research programs which do not involve construction.

3. Approval of a unified planning work program and certification of a State or local planning process under FHPM 6-3-2-2.

4. Approval of Transportation Improvement Programs under FHPM 4-4-2 and statewide programs under FHPM 6-3-2-2.

5. Approval of project concepts under 23 CFR 476.

6. Engineering when undertaken to define the elements of a proposal or alternatives sufficiently so that environmental effects can be assessed.

7. Federal-aid highway system revisions under 23 USC 103, which establishes classes of highways on the Federal-aid highway system.

8. Approval of utility installations along or across a transportation facility.

9. Reconstruction or modification of an existing bridge structure on essentially the same alignment or location (e.g., widening less than a single travel lane, adding shoulders or safety lanes, walkways, bikeways, or pipelines) except bridges on or eligible for inclusion on the National Register of Historic Places or bridges providing access to barrier islands.

   Reconstruction of modification of an existing one lane bridge structure, presently services by a two-lane road and used for two-lane traffic, to a two-lane bridge on essentially the same alignment or location, except bridges on or eligible for inclusion on the National Register of Historic Places or bridges providing access to barrier islands.

10. Construction of bicycle and pedestrian lanes, paths and facilities.

11. Activity included in the State's "highway safety plan" under 23 USC 402.

12. Transfer of Federal lands pursuant to 23 USC 317 when the subsequent action is not an FHWA action.

13. Modernization of an existing highway by resurfacing, restoration, rehabilitation, widening less than a single lane wideth, adding shoulders, adding auxiliary lanes for localized purposes (e.g., weaving turning, climbing), and correcting substandard curves and intersections. This classification is not applicable when the proposed project requires acquisition of more than minor amounts of right-of-way or substantial changes in access control.

14. Highway safety or traffic operations improvement projects including the correction or improvement of high hazard locations; elimination of roadside obstacles; highway signing; pavement markings; traffic control devices; railroad warning devices; and lighting. This classification is not applicable when the proposed action requires acquisition of more than minor amounts of right-of-way or substantial changes in access control.

15. Alterations to existing buildings to provide for noise reduction and the installation of noise barriers.

16. Ridesharing activities and transportation corridor fringe parking facilities.

17. Landscaping.

18. Program administration and technical assistance activities by the applicant to administer Section 18 funds (rural public transportation program).

19. Installation of signs, small passenger and bus shelters, and traffic signs where no substantial land acquisition or traffic disruption will occur.

20. Acquisition of land in which the property will not be modified, the land use will not be changed, and displacements will not occur. For projects other than UMTA advance land loans, this categorical exclusion is limited to the acquisition of minor amounts of land. This is undertaken for the purpose of maintaining the current land use and preserving alternatives to be considered in the environmental process. Advance land acquisition shall not limit the evaluation of alternatives, including shifts in alignment for a construction project, which may be required in the NEPA process.

21. Track and rail bed maintenance and improvements when carried out within the existing right-of-way.

22. Rehabilitation or reconstruction of existing rail and bus buildings and ancilliary facilities where no additional lane is required and there is no substantial increase in the number of users.

23. Purchase and installation of operating or maintenance equipment to be located within the transit facility and with no significant physical impacts off the site.

24. Installation of signs, small passenger and bus shelters, and traffic signs where no substantial land acquisition of traffic disruption will occur.

25. Construction of new bus storage and maintenance facilities in areas used predominantly for industrial or transportation purposes where such construction is not inconsistent with existing zoning and located on or near a street with adequate capacity to handle anticipated bus and support vehicle traffic.

26. Acquisition of land in which the property will not be modified, the land use will not be changes, and displacements will not occur. For projects other than UMTA advance land loans, this categorical exclusion is limited to the acquisitions of minor amounts of land. This is undertaken for the purpose of maintaining the current land use and preserving alternatives to be considered in the environmental process. Advance land acquisition shall not limit the evaluation of alternatives, including shifts in alignment for a construction project, which may be required in the NEPA process.

27. Promulgation of regulations and directives for which a regulatory analysis is not required by Section 3 of Executive Order 12044.

28. Research activities as defined in 23 U.S.C. 307.

29. Emergency repairs under 23 U.S.C. 125 which do not substantially change the design and are commenced during or immediately after the occurrence of a natural disaster or catastrophic failure.

C. <u>CLASS III ACTIONS</u>

## ENVIRONMENTAL ASSESSMENT

An EA is required when the significance of the impact on the environment is not clearly established. All actions that are not Class I or II are Class III. Class III Actions require the preparation of an EA to determine that a Class I Action is required or that a Finding of No Significant Impact can be made. EA's are only prepared in those cases where it cannot be readily determined that a Class I Action or a Class II Action is appropriate.

# A.5 INTEREST TABLE

## DISCOUNT RATE

The concept of evaluating design alternatives by comparing life cycle cost is a valid approach if the present worth of all alternates is expressed in equivalent dollar value. In present worth analysis, a future cost can be estimated to an equivalent present worth by considering the interest rate and the period of the loan or life span of the structure. The interest rate is determined by average market rate of return, a percentage which reflects the value which society places on the use of funds. The interest rate is the theoretical return on the invested capital, but it also includes the inflation rate which is predicted over the period of the loan. A more accurate indicator of the average market rate of return of an investment is the **discount rate.**

The discount rate of any long term investment is defined as the interest rate minus the expected inflation rate. For example, an **12** percent market interest rate may be arrived at by considering **3** percent inflation rate and a **9** percent rate for opportunity cost of capital.

In order to neutralize any distortion created by inflation, the discount rate can be used in life cycle analysis to evaluate alternates by comparing the real cost of capital in term of constant dollar.

The selection of discount rate is a important factor in life cycle cost analysis. A higher rate presents a lower life cycle cost for those alternates whose major cost occurred in late stage of the economic life. Similarly, a lower discount rate means a much higher life cost for these same alternates. In other words, higher rate favors bridge repairs and replacement at later date while lower rate encourages major repairs or bridge replacement.

**7%** **Compound**

| | SINGLE PAYMENT | | EQUAL PAYMENT SERIES | | | |
|---|---|---|---|---|---|---|
| YEAR | COMPOUND AMOUNT FACTOR | PRESENT WORTH FACTOR | COMPOUND AMOUNT FACTOR | SINKING FUND FACTOR | PRESENT WORTH FACTOR | CAPITAL RECOVERY FACTOR |
| 1 | 1.070 | 0.9346 | 1.000 | 1.0000 | 0.9346 | 1.0700 |
| 2 | 1.145 | 0.8734 | 2.070 | 0.4831 | 1.8080 | 0.5531 |
| 3 | 1.225 | 0.8163 | 3.215 | 0.3111 | 2.6243 | 0.3811 |
| 4 | 1.311 | 0.7629 | 4.440 | 0.2252 | 3.3872 | 0.2952 |
| 5 | 1.403 | 0.7130 | 5.751 | 0.1739 | 4.1002 | 0.2439 |
| 6 | 1.501 | 0.6663 | 7.153 | 0.1398 | 4.7665 | 0.2098 |
| 7 | 1.606 | 0.6227 | 8.654 | 0.1156 | 5.3893 | 0.1856 |
| 8 | 1.718 | 0.5820 | 10.260 | 0.0975 | 5.9713 | 0.1675 |
| 9 | 1.838 | 0.5439 | 11.978 | 0.0835 | 6.5152 | 0.1535 |
| 10 | 1.967 | 0.5083 | 13.816 | 0.0724 | 7.0236 | 0.1424 |
| 11 | 2.105 | 0.4751 | 15.784 | 0.0634 | 7.4987 | 0.1334 |
| 12 | 2.252 | 0.4440 | 17.888 | 0.0559 | 7.9427 | 0.1259 |
| 13 | 2.410 | 0.4150 | 20.141 | 0.0497 | 8.3577 | 0.1197 |
| 14 | 2.579 | 0.3878 | 22.550 | 0.0443 | 8.7455 | 0.1143 |
| 15 | 2.759 | 0.3624 | 25.129 | 0.0398 | 9.1079 | 0.1098 |
| 16 | 2.952 | 0.3387 | 27.888 | 0.0359 | 9.4466 | 0.1059 |
| 17 | 3.159 | 0.3166 | 30.840 | 0.0324 | 9.7632 | 0.1024 |
| 18 | 3.380 | 0.2959 | 33.999 | 0.0294 | 10.0591 | 0.0994 |
| 19 | 3.617 | 0.2765 | 37.379 | 0.0268 | 10.3356 | 0.0968 |
| 20 | 3.870 | 0.2584 | 40.995 | 0.0244 | 10.5940 | 0.0944 |
| 21 | 4.141 | 0.2415 | 44.865 | 0.0223 | 10.8355 | 0.0923 |
| 22 | 4.430 | 0.2257 | 49.006 | 0.0204 | 11.0612 | 0.0904 |
| 23 | 4.741 | 0.2109 | 53.436 | 0.0187 | 11.2722 | 0.0887 |
| 24 | 5.072 | 0.1971 | 58.177 | 0.0172 | 11.4693 | 0.0872 |
| 25 | 5.427 | 0.1842 | 63.249 | 0.0158 | 11.6536 | 0.0858 |
| 26 | 5.807 | 0.1722 | 68.676 | 0.0146 | 11.8258 | 0.0846 |
| 27 | 6.214 | 0.1609 | 74.484 | 0.0134 | 11.9867 | 0.0834 |
| 28 | 6.649 | 0.1504 | 80.698 | 0.0124 | 12.1371 | 0.0824 |
| 29 | 7.114 | 0.1406 | 87.347 | 0.0114 | 12.2777 | 0.0814 |
| 30 | 7.612 | 0.1314 | 94.461 | 0.0106 | 12.4090 | 0.0806 |
| 31 | 8.145 | 0.1228 | 102.073 | 0.0098 | 12.5318 | 0.0798 |
| 32 | 8.715 | 0.1147 | 110.218 | 0.0091 | 12.6466 | 0.0791 |
| 33 | 9.325 | 0.1072 | 118.933 | 0.0084 | 12.7538 | 0.0784 |
| 34 | 9.978 | 0.1002 | 128.259 | 0.0078 | 12.8540 | 0.0778 |
| 35 | 10.677 | 0.0937 | 138.237 | 0.0072 | 12.9477 | 0.0772 |
| 36 | 11.424 | 0.0875 | 148.913 | 0.0067 | 13.0352 | 0.0767 |
| 37 | 12.224 | 0.0818 | 160.337 | 0.0062 | 13.1170 | 0.0762 |
| 38 | 13.079 | 0.0765 | 172.561 | 0.0058 | 13.1935 | 0.0758 |
| 39 | 13.995 | 0.0715 | 185.640 | 0.0054 | 13.2649 | 0.0754 |
| 40 | 14.974 | 0.0668 | 199.635 | 0.0050 | 13.3317 | 0.0750 |
| 41 | 16.023 | 0.0624 | 214.610 | 0.0047 | 13.3941 | 0.0747 |
| 42 | 17.144 | 0.0583 | 230.632 | 0.0043 | 13.4524 | 0.0743 |
| 43 | 18.344 | 0.0545 | 247.776 | 0.0040 | 13.5070 | 0.0740 |
| 44 | 19.628 | 0.0509 | 266.121 | 0.0038 | 13.5579 | 0.0738 |
| 45 | 21.002 | 0.0476 | 285.749 | 0.0035 | 13.6055 | 0.0735 |
| 46 | 22.473 | 0.0445 | 306.752 | 0.0033 | 13.6500 | 0.0733 |
| 47 | 24.046 | 0.0416 | 329.224 | 0.0030 | 13.6916 | 0.0730 |
| 48 | 25.729 | 0.0389 | 353.270 | 0.0028 | 13.7305 | 0.0728 |
| 49 | 27.530 | 0.0363 | 378.999 | 0.0026 | 13.7668 | 0.0726 |
| 50 | 29.457 | 0.0339 | 406.529 | 0.0025 | 13.8007 | 0.0725 |

# INTEREST TABLE

## 8% Compound

| YEAR | SINGLE PAYMENT: COMPOUND AMOUNT FACTOR | SINGLE PAYMENT: PRESENT WORTH FACTOR | EQUAL PAYMENT SERIES: COMPOUND AMOUNT FACTOR | EQUAL PAYMENT SERIES: SINKING FUND FACTOR | EQUAL PAYMENT SERIES: PRESENT WORTH FACTOR | EQUAL PAYMENT SERIES: CAPITAL RECOVERY FACTOR |
|---|---|---|---|---|---|---|
| 1 | 1.080 | 0.9259 | 1.000 | 1.0000 | 0.9259 | 1.0800 |
| 2 | 1.166 | 0.8573 | 2.080 | 0.4808 | 1.7833 | 0.5608 |
| 3 | 1.260 | 0.7938 | 3.246 | 0.3080 | 2.5771 | 0.3880 |
| 4 | 1.360 | 0.7350 | 4.506 | 0.2219 | 3.3121 | 0.3019 |
| 5 | 1.469 | 0.6806 | 5.867 | 0.1705 | 3.9927 | 0.2505 |
| 6 | 1.587 | 0.6302 | 7.336 | 0.1363 | 4.6229 | 0.2163 |
| 7 | 1.714 | 0.5835 | 8.923 | 0.1121 | 5.2064 | 0.1921 |
| 8 | 1.851 | 0.5403 | 10.637 | 0.0940 | 5.7466 | 0.1740 |
| 9 | 1.999 | 0.5002 | 12.488 | 0.0801 | 6.2469 | 0.1601 |
| 10 | 2.159 | 0.4632 | 14.487 | 0.0690 | 6.7101 | 0.1490 |
| 11 | 2.332 | 0.4289 | 16.645 | 0.0601 | 7.1390 | 0.1401 |
| 12 | 2.518 | 0.3971 | 18.977 | 0.0527 | 7.5361 | 0.1327 |
| 13 | 2.720 | 0.3677 | 21.495 | 0.0465 | 7.9038 | 0.1265 |
| 14 | 2.937 | 0.3405 | 24.215 | 0.0413 | 8.2442 | 0.1213 |
| 15 | 3.172 | 0.3152 | 27.152 | 0.0368 | 8.5595 | 0.1168 |
| 16 | 3.426 | 0.2919 | 30.324 | 0.0330 | 8.8514 | 0.1130 |
| 17 | 3.700 | 0.2703 | 33.750 | 0.0296 | 9.1216 | 0.1096 |
| 18 | 3.996 | 0.2502 | 37.450 | 0.0267 | 9.3719 | 0.1067 |
| 19 | 4.316 | 0.2317 | 41.446 | 0.0241 | 9.6036 | 0.1041 |
| 20 | 4.661 | 0.2145 | 45.762 | 0.0219 | 9.8181 | 0.1019 |
| 21 | 5.034 | 0.1987 | 50.423 | 0.0198 | 10.0168 | 0.0998 |
| 22 | 5.437 | 0.1839 | 55.457 | 0.0180 | 10.2007 | 0.0980 |
| 23 | 5.871 | 0.1703 | 60.893 | 0.0164 | 10.3711 | 0.0964 |
| 24 | 6.341 | 0.1577 | 66.765 | 0.0150 | 10.5288 | 0.0950 |
| 25 | 6.848 | 0.1460 | 73.106 | 0.0137 | 10.6748 | 0.0937 |
| 26 | 7.396 | 0.1352 | 79.954 | 0.0125 | 10.8100 | 0.0925 |
| 27 | 7.988 | 0.1252 | 87.351 | 0.0114 | 10.9352 | 0.0914 |
| 28 | 8.627 | 0.1159 | 95.339 | 0.0105 | 11.0511 | 0.0905 |
| 29 | 9.317 | 0.1073 | 103.966 | 0.0096 | 11.1584 | 0.0896 |
| 30 | 10.063 | 0.0994 | 113.283 | 0.0088 | 11.2578 | 0.0888 |
| 31 | 10.868 | 0.0920 | 123.346 | 0.0081 | 11.3498 | 0.0881 |
| 32 | 11.737 | 0.0852 | 134.214 | 0.0075 | 11.4350 | 0.0875 |
| 33 | 12.676 | 0.0789 | 145.951 | 0.0069 | 11.5139 | 0.0869 |
| 34 | 13.690 | 0.0730 | 158.627 | 0.0063 | 11.5869 | 0.0863 |
| 35 | 14.785 | 0.0676 | 172.317 | 0.0058 | 11.6546 | 0.0858 |
| 36 | 15.968 | 0.0626 | 187.102 | 0.0053 | 11.7172 | 0.0853 |
| 37 | 17.246 | 0.0580 | 203.070 | 0.0049 | 11.7752 | 0.0849 |
| 38 | 18.625 | 0.0537 | 220.316 | 0.0045 | 11.8289 | 0.0845 |
| 39 | 20.115 | 0.0497 | 238.941 | 0.0042 | 11.8786 | 0.0842 |
| 40 | 21.725 | 0.0460 | 259.057 | 0.0039 | 11.9246 | 0.0839 |
| 41 | 23.462 | 0.0426 | 280.781 | 0.0036 | 11.9672 | 0.0836 |
| 42 | 25.339 | 0.0395 | 304.244 | 0.0033 | 12.0067 | 0.0833 |
| 43 | 27.367 | 0.0365 | 329.583 | 0.0030 | 12.0432 | 0.0830 |
| 44 | 29.556 | 0.0338 | 356.950 | 0.0028 | 12.0771 | 0.0828 |
| 45 | 31.920 | 0.0313 | 386.506 | 0.0026 | 12.1084 | 0.0826 |
| 46 | 34.474 | 0.0290 | 418.426 | 0.0024 | 12.1374 | 0.0824 |
| 47 | 37.232 | 0.0269 | 452.900 | 0.0022 | 12.1643 | 0.0822 |
| 48 | 40.211 | 0.0249 | 490.132 | 0.0020 | 12.1891 | 0.0820 |
| 49 | 43.427 | 0.0230 | 530.343 | 0.0019 | 12.2122 | 0.0819 |
| 50 | 46.902 | 0.0213 | 573.770 | 0.0017 | 12.2335 | 0.0817 |

## 9% Compound

| YEAR | SINGLE PAYMENT: COMPOUND AMOUNT FACTOR | SINGLE PAYMENT: PRESENT WORTH FACTOR | EQUAL PAYMENT SERIES: COMPOUND AMOUNT FACTOR | EQUAL PAYMENT SERIES: SINKING FUND FACTOR | EQUAL PAYMENT SERIES: PRESENT WORTH FACTOR | EQUAL PAYMENT SERIES: CAPITAL RECOVERY FACTOR |
|---|---|---|---|---|---|---|
| 1 | 1.090 | 0.9174 | 1.000 | 1.0000 | 0.9174 | 1.0900 |
| 2 | 1.188 | 0.8417 | 2.090 | 0.4785 | 1.7591 | 0.5685 |
| 3 | 1.295 | 0.7722 | 3.278 | 0.3051 | 2.5313 | 0.3951 |
| 4 | 1.412 | 0.7084 | 4.573 | 0.2187 | 3.2397 | 0.3087 |
| 5 | 1.539 | 0.6499 | 5.985 | 0.1671 | 3.8897 | 0.2571 |
| 6 | 1.677 | 0.5963 | 7.523 | 0.1329 | 4.4859 | 0.2229 |
| 7 | 1.828 | 0.5470 | 9.200 | 0.1087 | 5.0330 | 0.1987 |
| 8 | 1.993 | 0.5019 | 11.028 | 0.0907 | 5.5348 | 0.1807 |
| 9 | 2.172 | 0.4604 | 13.021 | 0.0768 | 5.9952 | 0.1668 |
| 10 | 2.367 | 0.4224 | 15.193 | 0.0658 | 6.4177 | 0.1558 |
| 11 | 2.580 | 0.3875 | 17.560 | 0.0569 | 6.8052 | 0.1469 |
| 12 | 2.813 | 0.3555 | 20.141 | 0.0497 | 7.1607 | 0.1397 |
| 13 | 3.066 | 0.3262 | 22.953 | 0.0436 | 7.4869 | 0.1336 |
| 14 | 3.342 | 0.2992 | 26.019 | 0.0384 | 7.7862 | 0.1284 |
| 15 | 3.642 | 0.2745 | 29.361 | 0.0341 | 8.0607 | 0.1241 |
| 16 | 3.970 | 0.2519 | 33.003 | 0.0303 | 8.3126 | 0.1203 |
| 17 | 4.328 | 0.2311 | 36.974 | 0.0270 | 8.5436 | 0.1170 |
| 18 | 4.717 | 0.2120 | 41.301 | 0.0242 | 8.7556 | 0.1142 |
| 19 | 5.142 | 0.1945 | 46.018 | 0.0217 | 8.9501 | 0.1117 |
| 20 | 5.604 | 0.1784 | 51.160 | 0.0195 | 9.1285 | 0.1095 |
| 21 | 6.109 | 0.1637 | 56.765 | 0.0176 | 9.2922 | 0.1076 |
| 22 | 6.659 | 0.1502 | 62.873 | 0.0159 | 9.4424 | 0.1059 |
| 23 | 7.258 | 0.1378 | 69.532 | 0.0144 | 9.5802 | 0.1044 |
| 24 | 7.911 | 0.1264 | 76.790 | 0.0130 | 9.7066 | 0.1030 |
| 25 | 8.623 | 0.1160 | 84.701 | 0.0118 | 9.8226 | 0.1018 |
| 26 | 9.399 | 0.1064 | 93.324 | 0.0107 | 9.9290 | 0.1007 |
| 27 | 10.245 | 0.0976 | 102.723 | 0.0097 | 10.0266 | 0.0997 |
| 28 | 11.167 | 0.0895 | 112.968 | 0.0089 | 10.1161 | 0.0989 |
| 29 | 12.172 | 0.0822 | 124.135 | 0.0081 | 10.1983 | 0.0981 |
| 30 | 13.268 | 0.0754 | 136.308 | 0.0073 | 10.2737 | 0.0973 |
| 31 | 14.462 | 0.0691 | 149.575 | 0.0067 | 10.3428 | 0.0967 |
| 32 | 15.763 | 0.0634 | 164.037 | 0.0061 | 10.4062 | 0.0961 |
| 33 | 17.182 | 0.0582 | 179.800 | 0.0056 | 10.4644 | 0.0956 |
| 34 | 18.728 | 0.0534 | 196.982 | 0.0051 | 10.5178 | 0.0951 |
| 35 | 20.414 | 0.0490 | 215.711 | 0.0046 | 10.5668 | 0.0946 |
| 36 | 22.251 | 0.0449 | 236.125 | 0.0042 | 10.6118 | 0.0942 |
| 37 | 24.254 | 0.0412 | 258.376 | 0.0039 | 10.6530 | 0.0939 |
| 38 | 26.437 | 0.0378 | 282.630 | 0.0035 | 10.6908 | 0.0935 |
| 39 | 28.816 | 0.0347 | 309.066 | 0.0032 | 10.7255 | 0.0932 |
| 40 | 31.409 | 0.0318 | 337.882 | 0.0030 | 10.7574 | 0.0930 |
| 41 | 34.236 | 0.0292 | 369.292 | 0.0027 | 10.7866 | 0.0927 |
| 42 | 37.318 | 0.0268 | 403.528 | 0.0025 | 10.8134 | 0.0925 |
| 43 | 40.676 | 0.0246 | 440.846 | 0.0023 | 10.8380 | 0.0923 |
| 44 | 44.337 | 0.0226 | 481.522 | 0.0021 | 10.8605 | 0.0921 |
| 45 | 48.327 | 0.0207 | 525.859 | 0.0019 | 10.8812 | 0.0919 |
| 46 | 52.677 | 0.0190 | 574.186 | 0.0017 | 10.9002 | 0.0917 |
| 47 | 57.418 | 0.0174 | 626.863 | 0.0016 | 10.9176 | 0.0916 |
| 48 | 62.585 | 0.0160 | 684.280 | 0.0015 | 10.9336 | 0.0915 |
| 49 | 68.218 | 0.0147 | 746.866 | 0.0013 | 10.9482 | 0.0913 |
| 50 | 74.358 | 0.0134 | 815.084 | 0.0012 | 10.9617 | 0.0912 |

# INTEREST TABLE - Continued

## 10% Compound

| YEAR | SINGLE PAYMENT: COMPOUND AMOUNT FACTOR | SINGLE PAYMENT: PRESENT WORTH FACTOR | EQUAL PAYMENT SERIES: COMPOUND AMOUNT FACTOR | EQUAL PAYMENT SERIES: SINKING FUND FACTOR | EQUAL PAYMENT SERIES: PRESENT WORTH FACTOR | EQUAL PAYMENT SERIES: CAPITAL RECOVERY FACTOR |
|---|---|---|---|---|---|---|
| 1 | 1.100 | 0.9091 | 1.000 | 1.0000 | 0.9091 | 1.1000 |
| 2 | 1.210 | 0.8264 | 2.100 | 0.4762 | 1.7355 | 0.5762 |
| 3 | 1.331 | 0.7513 | 3.310 | 0.3021 | 2.4869 | 0.4021 |
| 4 | 1.464 | 0.6830 | 4.641 | 0.2155 | 3.1699 | 0.3155 |
| 5 | 1.611 | 0.6209 | 6.105 | 0.1638 | 3.7908 | 0.2638 |
| 6 | 1.772 | 0.5645 | 7.716 | 0.1296 | 4.3553 | 0.2296 |
| 7 | 1.949 | 0.5132 | 9.487 | 0.1054 | 4.8684 | 0.2054 |
| 8 | 2.144 | 0.4665 | 11.436 | 0.0874 | 5.3349 | 0.1874 |
| 9 | 2.358 | 0.4241 | 13.579 | 0.0736 | 5.7590 | 0.1736 |
| 10 | 2.594 | 0.3855 | 15.937 | 0.0627 | 6.1446 | 0.1627 |
| 11 | 2.853 | 0.3505 | 18.531 | 0.0540 | 6.4951 | 0.1540 |
| 12 | 3.138 | 0.3186 | 21.384 | 0.0468 | 6.8137 | 0.1468 |
| 13 | 3.452 | 0.2897 | 24.523 | 0.0408 | 7.1034 | 0.1408 |
| 14 | 3.797 | 0.2633 | 27.975 | 0.0357 | 7.3667 | 0.1357 |
| 15 | 4.177 | 0.2394 | 31.772 | 0.0315 | 7.6061 | 0.1315 |
| 16 | 4.595 | 0.2176 | 35.950 | 0.0278 | 7.8237 | 0.1278 |
| 17 | 5.054 | 0.1978 | 40.545 | 0.0247 | 8.0216 | 0.1247 |
| 18 | 5.560 | 0.1799 | 45.599 | 0.0219 | 8.2014 | 0.1219 |
| 19 | 6.116 | 0.1635 | 51.159 | 0.0195 | 8.3649 | 0.1195 |
| 20 | 6.727 | 0.1486 | 57.275 | 0.0175 | 8.5136 | 0.1175 |
| 21 | 7.400 | 0.1351 | 64.002 | 0.0156 | 8.6487 | 0.1156 |
| 22 | 8.140 | 0.1228 | 71.403 | 0.0140 | 8.7715 | 0.1140 |
| 23 | 8.954 | 0.1117 | 79.543 | 0.0126 | 8.8832 | 0.1126 |
| 24 | 9.850 | 0.1015 | 88.497 | 0.0113 | 8.9847 | 0.1113 |
| 25 | 10.835 | 0.0923 | 98.347 | 0.0102 | 9.0770 | 0.1102 |
| 26 | 11.918 | 0.0839 | 109.182 | 0.0092 | 9.1609 | 0.1092 |
| 27 | 13.110 | 0.0763 | 121.100 | 0.0083 | 9.2372 | 0.1083 |
| 28 | 14.421 | 0.0693 | 134.210 | 0.0075 | 9.3066 | 0.1075 |
| 29 | 15.863 | 0.0630 | 148.631 | 0.0067 | 9.3696 | 0.1067 |
| 30 | 17.449 | 0.0573 | 164.494 | 0.0061 | 9.4269 | 0.1061 |
| 31 | 19.194 | 0.0521 | 181.943 | 0.0055 | 9.4790 | 0.1055 |
| 32 | 21.114 | 0.0474 | 201.138 | 0.0050 | 9.5264 | 0.1050 |
| 33 | 23.225 | 0.0431 | 222.252 | 0.0045 | 9.5694 | 0.1045 |
| 34 | 25.548 | 0.0391 | 245.477 | 0.0041 | 9.6086 | 0.1041 |
| 35 | 28.102 | 0.0356 | 271.024 | 0.0037 | 9.6442 | 0.1037 |
| 36 | 30.913 | 0.0323 | 299.127 | 0.0033 | 9.6765 | 0.1033 |
| 37 | 34.004 | 0.0294 | 330.039 | 0.0030 | 9.7059 | 0.1030 |
| 38 | 37.404 | 0.0267 | 364.043 | 0.0027 | 9.7327 | 0.1027 |
| 39 | 41.145 | 0.0243 | 401.448 | 0.0025 | 9.7570 | 0.1025 |
| 40 | 45.259 | 0.0221 | 442.593 | 0.0023 | 9.7791 | 0.1023 |
| 41 | 49.785 | 0.0201 | 487.852 | 0.0020 | 9.7991 | 0.1020 |
| 42 | 54.764 | 0.0183 | 537.637 | 0.0019 | 9.8174 | 0.1019 |
| 43 | 60.240 | 0.0166 | 592.401 | 0.0017 | 9.8340 | 0.1017 |
| 44 | 66.264 | 0.0151 | 652.641 | 0.0015 | 9.8491 | 0.1015 |
| 45 | 72.890 | 0.0137 | 718.905 | 0.0014 | 9.8628 | 0.1014 |
| 46 | 80.180 | 0.0125 | 791.795 | 0.0013 | 9.8753 | 0.1013 |
| 47 | 88.197 | 0.0113 | 871.975 | 0.0011 | 9.8866 | 0.1011 |
| 48 | 97.017 | 0.0103 | 960.172 | 0.0010 | 9.8969 | 0.1010 |
| 49 | 106.719 | 0.0094 | 1057.190 | 0.0009 | 9.9063 | 0.1009 |
| 50 | 117.391 | 0.0085 | 1163.909 | 0.0009 | 9.9148 | 0.1009 |

## 11% Compound

| YEAR | SINGLE PAYMENT: COMPOUND AMOUNT FACTOR | SINGLE PAYMENT: PRESENT WORTH FACTOR | EQUAL PAYMENT SERIES: COMPOUND AMOUNT FACTOR | EQUAL PAYMENT SERIES: SINKING FUND FACTOR | EQUAL PAYMENT SERIES: PRESENT WORTH FACTOR | EQUAL PAYMENT SERIES: CAPITAL RECOVERY FACTOR |
|---|---|---|---|---|---|---|
| 1 | 1.110 | 0.9009 | 1.000 | 1.0000 | 0.9009 | 1.1100 |
| 2 | 1.232 | 0.8116 | 2.110 | 0.4739 | 1.7125 | 0.5839 |
| 3 | 1.368 | 0.7312 | 3.342 | 0.2992 | 2.4437 | 0.4092 |
| 4 | 1.518 | 0.6587 | 4.710 | 0.2123 | 3.1024 | 0.3223 |
| 5 | 1.685 | 0.5935 | 6.228 | 0.1606 | 3.6959 | 0.2706 |
| 6 | 1.870 | 0.5346 | 7.913 | 0.1264 | 4.2305 | 0.2364 |
| 7 | 2.076 | 0.4817 | 9.783 | 0.1022 | 4.7122 | 0.2122 |
| 8 | 2.305 | 0.4339 | 11.859 | 0.0843 | 5.1461 | 0.1943 |
| 9 | 2.558 | 0.3909 | 14.164 | 0.0706 | 5.5370 | 0.1806 |
| 10 | 2.839 | 0.3522 | 16.722 | 0.0598 | 5.8892 | 0.1698 |
| 11 | 3.152 | 0.3173 | 19.561 | 0.0511 | 6.2065 | 0.1611 |
| 12 | 3.498 | 0.2858 | 22.713 | 0.0440 | 6.4924 | 0.1540 |
| 13 | 3.883 | 0.2575 | 26.212 | 0.0382 | 6.7499 | 0.1482 |
| 14 | 4.310 | 0.2320 | 30.095 | 0.0332 | 6.9819 | 0.1432 |
| 15 | 4.785 | 0.2090 | 34.405 | 0.0291 | 7.1909 | 0.1391 |
| 16 | 5.311 | 0.1883 | 39.190 | 0.0255 | 7.3792 | 0.1355 |
| 17 | 5.895 | 0.1696 | 44.501 | 0.0225 | 7.5488 | 0.1325 |
| 18 | 6.544 | 0.1528 | 50.396 | 0.0198 | 7.7016 | 0.1298 |
| 19 | 7.263 | 0.1377 | 56.939 | 0.0176 | 7.8393 | 0.1276 |
| 20 | 8.062 | 0.1240 | 64.203 | 0.0156 | 7.9633 | 0.1256 |
| 21 | 8.949 | 0.1117 | 72.265 | 0.0138 | 8.0751 | 0.1238 |
| 22 | 9.934 | 0.1007 | 81.214 | 0.0123 | 8.1757 | 0.1223 |
| 23 | 11.026 | 0.0907 | 91.148 | 0.0110 | 8.2664 | 0.1210 |
| 24 | 12.239 | 0.0817 | 102.174 | 0.0098 | 8.3481 | 0.1198 |
| 25 | 13.585 | 0.0736 | 114.413 | 0.0087 | 8.4217 | 0.1187 |
| 26 | 15.080 | 0.0663 | 127.999 | 0.0078 | 8.4881 | 0.1178 |
| 27 | 16.739 | 0.0597 | 143.079 | 0.0070 | 8.5478 | 0.1170 |
| 28 | 18.580 | 0.0538 | 159.817 | 0.0063 | 8.6016 | 0.1163 |
| 29 | 20.624 | 0.0485 | 178.397 | 0.0056 | 8.6501 | 0.1156 |
| 30 | 22.892 | 0.0437 | 199.021 | 0.0050 | 8.6938 | 0.1150 |
| 31 | 25.410 | 0.0394 | 221.913 | 0.0045 | 8.7331 | 0.1145 |
| 32 | 28.206 | 0.0355 | 247.324 | 0.0040 | 8.7686 | 0.1140 |
| 33 | 31.308 | 0.0319 | 275.529 | 0.0036 | 8.8005 | 0.1136 |
| 34 | 34.752 | 0.0288 | 306.837 | 0.0033 | 8.8293 | 0.1133 |
| 35 | 38.575 | 0.0259 | 341.590 | 0.0029 | 8.8552 | 0.1129 |
| 36 | 42.818 | 0.0234 | 380.164 | 0.0026 | 8.8786 | 0.1126 |
| 37 | 47.528 | 0.0210 | 422.982 | 0.0024 | 8.8996 | 0.1124 |
| 38 | 52.756 | 0.0190 | 470.511 | 0.0021 | 8.9186 | 0.1121 |
| 39 | 58.559 | 0.0171 | 523.267 | 0.0019 | 8.9357 | 0.1119 |
| 40 | 65.001 | 0.0154 | 581.826 | 0.0017 | 8.9511 | 0.1117 |
| 41 | 72.151 | 0.0139 | 646.827 | 0.0015 | 8.9649 | 0.1115 |
| 42 | 80.088 | 0.0125 | 718.978 | 0.0014 | 8.9774 | 0.1114 |
| 43 | 88.897 | 0.0112 | 799.065 | 0.0013 | 8.9886 | 0.1113 |
| 44 | 98.676 | 0.0101 | 887.963 | 0.0011 | 8.9988 | 0.1111 |
| 45 | 109.530 | 0.0091 | 986.639 | 0.0010 | 9.0079 | 0.1110 |
| 46 | 121.579 | 0.0082 | 1096.169 | 0.0009 | 9.0161 | 0.1109 |
| 47 | 134.952 | 0.0074 | 1217.747 | 0.0008 | 9.0235 | 0.1108 |
| 48 | 149.797 | 0.0067 | 1352.700 | 0.0007 | 9.0302 | 0.1107 |
| 49 | 166.275 | 0.0060 | 1502.497 | 0.0007 | 9.0362 | 0.1107 |
| 50 | 184.565 | 0.0054 | 1668.771 | 0.0006 | 9.0417 | 0.1106 |

# INTEREST TABLE - Continued

## 12% Compound

| | SINGLE PAYMENT | | EQUAL PAYMENT SERIES | | | |
|---|---|---|---|---|---|---|
| YEAR | COMPOUND AMOUNT FACTOR | PRESENT WORTH FACTOR | COMPOUND AMOUNT FACTOR | SINKING FUND FACTOR | PRESENT WORTH FACTOR | CAPITAL RECOVERY FACTOR |
| 1 | 1.120 | 0.8929 | 1.000 | 1.0000 | 0.8929 | 1.1200 |
| 2 | 1.254 | 0.7972 | 2.120 | 0.4717 | 1.6901 | 0.5917 |
| 3 | 1.405 | 0.7118 | 3.374 | 0.2963 | 2.4018 | 0.4163 |
| 4 | 1.574 | 0.6355 | 4.779 | 0.2092 | 3.0373 | 0.3292 |
| 5 | 1.762 | 0.5674 | 6.353 | 0.1574 | 3.6048 | 0.2774 |
| 6 | 1.974 | 0.5066 | 8.115 | 0.1232 | 4.1114 | 0.2432 |
| 7 | 2.211 | 0.4523 | 10.089 | 0.0991 | 4.5638 | 0.2191 |
| 8 | 2.476 | 0.4039 | 12.300 | 0.0813 | 4.9676 | 0.2013 |
| 9 | 2.773 | 0.3606 | 14.776 | 0.0677 | 5.3282 | 0.1877 |
| 10 | 3.106 | 0.3220 | 17.549 | 0.0570 | 5.6502 | 0.1770 |
| 11 | 3.479 | 0.2875 | 20.655 | 0.0484 | 5.9377 | 0.1684 |
| 12 | 3.896 | 0.2567 | 24.133 | 0.0414 | 6.1944 | 0.1614 |
| 13 | 4.363 | 0.2292 | 28.029 | 0.0357 | 6.4235 | 0.1557 |
| 14 | 4.887 | 0.2046 | 32.393 | 0.0309 | 6.6282 | 0.1509 |
| 15 | 5.474 | 0.1827 | 37.280 | 0.0268 | 6.8109 | 0.1468 |
| 16 | 6.130 | 0.1631 | 42.753 | 0.0234 | 6.9740 | 0.1434 |
| 17 | 6.866 | 0.1456 | 48.884 | 0.0205 | 7.1196 | 0.1405 |
| 18 | 7.690 | 0.1300 | 55.750 | 0.0179 | 7.2497 | 0.1379 |
| 19 | 8.613 | 0.1161 | 63.440 | 0.0158 | 7.3658 | 0.1358 |
| 20 | 9.646 | 0.1037 | 72.052 | 0.0139 | 7.4694 | 0.1339 |
| 21 | 10.804 | 0.0926 | 81.699 | 0.0122 | 7.5620 | 0.1322 |
| 22 | 12.100 | 0.0826 | 92.503 | 0.0108 | 7.6446 | 0.1308 |
| 23 | 13.552 | 0.0738 | 104.603 | 0.0096 | 7.7184 | 0.1296 |
| 24 | 15.179 | 0.0659 | 118.155 | 0.0085 | 7.7843 | 0.1285 |
| 25 | 17.000 | 0.0588 | 133.334 | 0.0075 | 7.8431 | 0.1275 |
| 26 | 19.040 | 0.0525 | 150.334 | 0.0067 | 7.8957 | 0.1267 |
| 27 | 21.325 | 0.0469 | 169.374 | 0.0059 | 7.9426 | 0.1259 |
| 28 | 23.884 | 0.0419 | 190.699 | 0.0052 | 7.9844 | 0.1252 |
| 29 | 26.750 | 0.0374 | 214.583 | 0.0047 | 8.0218 | 0.1247 |
| 30 | 29.960 | 0.0334 | 241.333 | 0.0041 | 8.0552 | 0.1241 |
| 31 | 33.555 | 0.0298 | 271.293 | 0.0037 | 8.0850 | 0.1237 |
| 32 | 37.582 | 0.0266 | 304.848 | 0.0033 | 8.1116 | 0.1233 |
| 33 | 42.092 | 0.0238 | 342.429 | 0.0029 | 8.1354 | 0.1229 |
| 34 | 47.143 | 0.0212 | 384.521 | 0.0026 | 8.1566 | 0.1226 |
| 35 | 52.800 | 0.0189 | 431.663 | 0.0023 | 8.1755 | 0.1223 |
| 36 | 59.136 | 0.0169 | 484.463 | 0.0021 | 8.1924 | 0.1221 |
| 37 | 66.232 | 0.0151 | 543.599 | 0.0018 | 8.2075 | 0.1218 |
| 38 | 74.180 | 0.0135 | 609.831 | 0.0016 | 8.2210 | 0.1216 |
| 39 | 83.081 | 0.0120 | 684.010 | 0.0015 | 8.2330 | 0.1215 |
| 40 | 93.051 | 0.0107 | 767.091 | 0.0013 | 8.2438 | 0.1213 |
| 41 | 104.217 | 0.0096 | 860.142 | 0.0012 | 8.2534 | 0.1212 |
| 42 | 116.723 | 0.0086 | 964.359 | 0.0010 | 8.2619 | 0.1210 |
| 43 | 130.730 | 0.0076 | 1081.083 | 0.0009 | 8.2696 | 0.1209 |
| 44 | 146.418 | 0.0068 | 1211.813 | 0.0008 | 8.2764 | 0.1208 |
| 45 | 163.988 | 0.0061 | 1358.230 | 0.0007 | 8.2825 | 0.1207 |
| 46 | 183.666 | 0.0054 | 1522.218 | 0.0007 | 8.2880 | 0.1207 |
| 47 | 205.706 | 0.0049 | 1705.884 | 0.0006 | 8.2928 | 0.1206 |
| 48 | 230.391 | 0.0043 | 1911.590 | 0.0005 | 8.2972 | 0.1205 |
| 49 | 258.038 | 0.0039 | 2141.981 | 0.0005 | 8.3010 | 0.1205 |
| 50 | 289.002 | 0.0035 | 2400.018 | 0.0004 | 8.3045 | 0.1204 |

## 14% Compound

| | SINGLE PAYMENT | | EQUAL PAYMENT SERIES | | | |
|---|---|---|---|---|---|---|
| YEAR | COMPOUND AMOUNT FACTOR | PRESENT WORTH FACTOR | COMPOUND AMOUNT FACTOR | SINKING FUND FACTOR | PRESENT WORTH FACTOR | CAPITAL RECOVERY FACTOR |
| 1 | 1.140 | 0.8772 | 1.000 | 1.0000 | 0.8772 | 1.1400 |
| 2 | 1.300 | 0.7695 | 2.140 | 0.4673 | 1.6467 | 0.6073 |
| 3 | 1.482 | 0.6750 | 3.440 | 0.2907 | 2.3216 | 0.4307 |
| 4 | 1.689 | 0.5921 | 4.921 | 0.2032 | 2.9137 | 0.3432 |
| 5 | 1.925 | 0.5194 | 6.610 | 0.1513 | 3.4331 | 0.2913 |
| 6 | 2.195 | 0.4556 | 8.536 | 0.1172 | 3.8887 | 0.2572 |
| 7 | 2.502 | 0.3996 | 10.730 | 0.0932 | 4.2883 | 0.2332 |
| 8 | 2.853 | 0.3506 | 13.233 | 0.0756 | 4.6389 | 0.2156 |
| 9 | 3.252 | 0.3075 | 16.085 | 0.0622 | 4.9464 | 0.2022 |
| 10 | 3.707 | 0.2697 | 19.337 | 0.0517 | 5.2161 | 0.1917 |
| 11 | 4.226 | 0.2366 | 23.045 | 0.0434 | 5.4527 | 0.1834 |
| 12 | 4.818 | 0.2076 | 27.271 | 0.0367 | 5.6603 | 0.1767 |
| 13 | 5.492 | 0.1821 | 32.089 | 0.0312 | 5.8424 | 0.1712 |
| 14 | 6.261 | 0.1597 | 37.581 | 0.0266 | 6.0021 | 0.1666 |
| 15 | 7.138 | 0.1401 | 43.842 | 0.0228 | 6.1422 | 0.1628 |
| 16 | 8.137 | 0.1229 | 50.980 | 0.0196 | 6.2651 | 0.1596 |
| 17 | 9.276 | 0.1078 | 59.118 | 0.0169 | 6.3729 | 0.1569 |
| 18 | 10.575 | 0.0946 | 68.394 | 0.0146 | 6.4674 | 0.1546 |
| 19 | 12.056 | 0.0829 | 78.969 | 0.0127 | 6.5504 | 0.1527 |
| 20 | 13.743 | 0.0728 | 91.025 | 0.0110 | 6.6231 | 0.1510 |
| 21 | 15.668 | 0.0638 | 104.768 | 0.0095 | 6.6870 | 0.1495 |
| 22 | 17.861 | 0.0560 | 120.436 | 0.0083 | 6.7429 | 0.1483 |
| 23 | 20.362 | 0.0491 | 138.297 | 0.0072 | 6.7921 | 0.1472 |
| 24 | 23.212 | 0.0431 | 158.659 | 0.0063 | 6.8351 | 0.1463 |
| 25 | 26.462 | 0.0378 | 181.871 | 0.0055 | 6.8729 | 0.1455 |
| 26 | 30.167 | 0.0331 | 208.333 | 0.0048 | 6.9061 | 0.1448 |
| 27 | 34.390 | 0.0291 | 238.499 | 0.0042 | 6.9352 | 0.1442 |
| 28 | 39.204 | 0.0255 | 272.889 | 0.0037 | 6.9607 | 0.1437 |
| 29 | 44.693 | 0.0224 | 312.094 | 0.0032 | 6.9830 | 0.1432 |
| 30 | 50.950 | 0.0196 | 356.787 | 0.0028 | 7.0027 | 0.1428 |
| 31 | 58.083 | 0.0172 | 407.737 | 0.0025 | 7.0199 | 0.1425 |
| 32 | 66.215 | 0.0151 | 465.820 | 0.0021 | 7.0350 | 0.1421 |
| 33 | 75.485 | 0.0132 | 532.035 | 0.0019 | 7.0482 | 0.1419 |
| 34 | 86.053 | 0.0116 | 607.520 | 0.0016 | 7.0599 | 0.1416 |
| 35 | 98.100 | 0.0102 | 693.573 | 0.0014 | 7.0700 | 0.1414 |
| 36 | 111.834 | 0.0089 | 791.673 | 0.0013 | 7.0790 | 0.1413 |
| 37 | 127.491 | 0.0078 | 903.507 | 0.0011 | 7.0868 | 0.1411 |
| 38 | 145.340 | 0.0069 | 1030.998 | 0.0010 | 7.0937 | 0.1410 |
| 39 | 165.687 | 0.0060 | 1176.338 | 0.0009 | 7.0997 | 0.1409 |
| 40 | 188.884 | 0.0053 | 1342.025 | 0.0007 | 7.1050 | 0.1407 |
| 41 | 215.327 | 0.0046 | 1530.909 | 0.0007 | 7.1097 | 0.1407 |
| 42 | 245.473 | 0.0041 | 1746.236 | 0.0006 | 7.1138 | 0.1406 |
| 43 | 279.839 | 0.0036 | 1991.709 | 0.0005 | 7.1173 | 0.1405 |
| 44 | 319.017 | 0.0031 | 2271.548 | 0.0004 | 7.1205 | 0.1404 |
| 45 | 363.679 | 0.0027 | 2590.565 | 0.0004 | 7.1232 | 0.1404 |
| 46 | 414.594 | 0.0024 | 2954.244 | 0.0003 | 7.1256 | 0.1403 |
| 47 | 472.637 | 0.0021 | 3368.838 | 0.0003 | 7.1277 | 0.1403 |
| 48 | 538.807 | 0.0019 | 3841.475 | 0.0003 | 7.1296 | 0.1403 |
| 49 | 614.239 | 0.0016 | 4380.282 | 0.0002 | 7.1312 | 0.1402 |
| 50 | 700.233 | 0.0014 | 4994.521 | 0.0002 | 7.1327 | 0.1402 |

## A.6 SUFFICIENCY RATING PROGRAM

```
 CALCULATE-SUFFICIENCY-RATING.

      IF (W-ITEM-5A = '2') OR (W-ITEM-5A NOT < 'A' AND NOT > 'Z')
      OR NOT HIGHWAY-BRIDGE
           MOVE SPACES TO W-SUFF-RTNG
           GO TO CALCULATE-SUFFICIENCY-RATING-X.
      MOVE W-ITEM-12 TO H-ITEM-12.
      MOVE W-ITEM-19 TO H-ITEM-19.
      MOVE W-ITEM-28A TO H-ITEM-28A.
      MOVE W-ITEM-29 TO H-ITEM-29.
      MOVE W-ITEM-32 TO H-ITEM-32.
      MOVE W-ITEM-36 TO H-ITEM-36.
      MOVE W-ITEM-51 TO H-ITEM-51.
      MOVE W-ITEM-53 TO H-ITEM-53.
      MOVE W-ITEM-66 TO H-ITEM-66.
      MOVE ' ' TO W-ASTERISK.

* 1. STRUCTURAL ADEQUACY AND SAFETY

      IF (W-ITEM-59 = '0' OR '1' OR '2')
      OR (W-ITEM-60 = '0' OR '1' OR '2')
           MOVE 55 TO A
           MOVE 0 TO B C D E F G H
      ELSE
           MOVE 0 TO A
      IF (W-ITEM-59 = '3') OR (W-ITEM-60 = '3')
           MOVE 40 TO B
           MOVE 0 TO C D E F G H
      ELSE
           MOVE 0 TO B
      IF (W-ITEM-59 = '4') OR (W-ITEM-60 = '4')
           MOVE 25 TO C
           MOVE 0 TO D E F G H
      ELSE
           MOVE 0 TO C
      IF (W-ITEM-59 = '5') OR (W-ITEM-60 = '5')
           MOVE 10 TO D
           MOVE 0 TO E F G H
      ELSE
           MOVE 0 TO D
      IF (W-ITEM-59 = 'N' AND W-ITEM-60 = 'N')
           IF (W-ITEM-62 = '0' OR '1' OR '2')
                MOVE 55 TO E
                MOVE 0 TO F G H
           ELSE
                MOVE 0 TO E
           IF (W-ITEM-62 = '3')
                MOVE 40 TO F
                MOVE 0 TO G H
           ELSE

                MOVE 0 TO F
          IF (W-ITEM-62 = '4')
                MOVE 25 TO G
                MOVE 0 TO H
          ELSE
                MOVE 0 TO G
          IF (W-ITEM-62 = '5')
                MOVE 10 TO H
          ELSE
                MOVE 0 TO H
     ELSE
          MOVE 0 TO E F G H.
     REDUCTION FOR LOAD CAPACITY

     IF W-ITEM-66 NOT NUMERIC OR W-ITEM-66A = '0'
          MOVE '236' TO W-ITEM-66
          MOVE '*' TO W-ASTERISK.
     IF (W-ITEM-66A = '1')
          COMPUTE AIT ROUNDED = W-ITEM-66RB * 1.56
     ELSE
     IF (W-ITEM-66A = '2')
          COMPUTE AIT ROUNDED = W-ITEM-66RB * 1.00
     ELSE
     IF (W-ITEM-66A = '3')
          COMPUTE AIT ROUNDED = W-ITEM-66RB * 1.56
     ELSE
     IF (W-ITEM-66A = '4')
          COMPUTE AIT ROUNDED = W-ITEM-66RB * 1.01
     ELSE
     IF (W-ITEM-66A = '5')
          COMPUTE AIT ROUNDED = W-ITEM-66RB * 0.77
     ELSE
     IF (W-ITEM-66A = '6')
          COMPUTE AIT ROUNDED = W-ITEM-66RB * 0.67
     ELSE
     IF (W-ITEM-66A = '9')
          COMPUTE AIT ROUNDED = W-ITEM-66RB * 1.00
     ELSE
          MOVE 0 TO AIT.
     IF (36 - AIT) NOT > 0
          MOVE 0 TO I
     ELSE
          COMPUTE I ROUNDED = ((36 - AIT) ** 1.5) * 0.2778.
     IF I > 55
          MOVE 55 TO I
     ELSE
     IF I < 0
          MOVE 0 TO I.
     COMPUTE S1 ROUNDED = 55 - (A + B + C + D + E + F + G + H + I)
     IF S1 > 55
          MOVE 55 TO S1
     ELSE
     IF S1 < 0
          MOVE 0 TO S1.

* 2. SERVICEABILITY AND FUNCTIONAL OBSOLESCENCE

     RATING REDUCTIONS
     IF (W-ITEM-58 = '0' OR '1' OR '2' OR '3')
          MOVE 5 TO A
     ELSE
     IF (W-ITEM-58 = '4')
          MOVE 3 TO A
     ELSE
     IF (W-ITEM-58 = '5')
          MOVE 1 TO A
     ELSE
          MOVE 0 TO A.
     IF (W-ITEM-67 = '0' OR '1' OR '2' OR '3')
          MOVE 4 TO B
     ELSE
     IF (W-ITEM-67 = '4')
          MOVE 2 TO B
     ELSE
     IF (W-ITEM-67 = '5')
          MOVE 1 TO B
     ELSE
          MOVE 0 TO B.
     IF (W-ITEM-68 = '0' OR '1' OR '2' OR '3')
          MOVE 4 TO C
     ELSE
     IF (W-ITEM-68 = '4')
          MOVE 2 TO C
     ELSE
     IF (W-ITEM-68 = '5')
          MOVE 1 TO C
     ELSE
          MOVE 0 TO C.
     IF (W-ITEM-69 = '0' OR '1' OR '2' OR '3')
          MOVE 4 TO D
     ELSE
     IF (W-ITEM-69 = '4')
          MOVE 2 TO D
     ELSE
     IF (W-ITEM-69 = '5')
          MOVE 1 TO D
     ELSE
          MOVE 0 TO D.
     IF (W-ITEM-71 = '0' OR '1' OR '2' OR '3')
          MOVE 4 TO E
     ELSE
     IF (W-ITEM-71 = '4')
          MOVE 2 TO E
     ELSE
     IF (W-ITEM-71 = '5')
          MOVE 1 TO E
     ELSE
          MOVE 0 TO E.
     IF (W-ITEM-72 = '0' OR '1' OR '2' OR '3')
          MOVE 4 TO F
     ELSE
     IF (W-ITEM-72 = '4')
          MOVE 2 TO F
     ELSE

     IF (W-ITEM-72 = '5')
          MOVE 1 TO F
     ELSE
          MOVE 0 TO F.
     COMPUTE J = (A + B + C + D + E + F).
     IF J > 13
          MOVE 13 TO J
     ELSE
     IF J < 0
          MOVE 0 TO J.

     WIDTH OF ROADWAY INSUFFICIENCY

     IF W-ITEM-51 NOT NUMERIC OR W-ITEM-51R NOT > 000.0
          MOVE '*' TO W-ASTERISK
          MOVE '0360' TO W-ITEM-51.
     IF W-ITEM-28A NOT NUMERIC OR W-ITEM-28RA NOT > 00
          MOVE '*' TO W-ASTERISK
          COMPUTE W-ITEM-28RA ROUNDED = W-ITEM-51R / 12.
     IF W-ITEM-28RA NOT > 00
          MOVE '*' TO W-ASTERISK
          MOVE '01' TO W-ITEM-28A.
     IF W-ITEM-29 NOT NUMERIC
          MOVE '*' TO W-ASTERISK
          MOVE '000000' TO W-ITEM-29.
     COMPUTE X ROUNDED = W-ITEM-29R / W-ITEM-28RA.
     COMPUTE Y ROUNDED = W-ITEM-51R / W-ITEM-28RA.
     IF W-ITEM-32 NOT NUMERIC
          MOVE '*' TO W-ASTERISK
          MOVE W-ITEM-51R TO W-ITEM-32R.
     IF (W-ITEM-43B NOT = '19')
          IF ((W-ITEM-51R + 2.0) < W-ITEM-32R)
               MOVE 5 TO G
          ELSE
               MOVE 0 TO G
     ELSE
          MOVE 0 TO G.
     IF (W-ITEM-28A = '01')
          IF Y < 14
               MOVE 15 TO H
          ELSE
          IF Y NOT < 14 AND Y < 18
               COMPUTE H ROUNDED = 15 * ((18 - Y) / 4)
          ELSE
          IF Y NOT < 18
               MOVE 0 TO H
          ELSE
               MOVE 0 TO H
     ELSE
     IF (W-ITEM-28A = '02' AND Y NOT < 16)
     OR (W-ITEM-28A = '03' AND Y NOT < 15)
     OR (W-ITEM-28A = '04' AND Y NOT < 14)
     OR (W-ITEM-28A NOT < '05' AND Y NOT < 12)
          MOVE 0 TO H
     ELSE
     IF (Y < 9 AND X > 50)
          MOVE 15 TO H
```

```
ELSE
IF (Y < 9 AND X NOT > 50)
    MOVE 7.5 TO H
ELSE
IF (Y NOT < 9 AND X NOT > 50)
    MOVE 0 TO H
ELSE
IF (X > 50 AND X NOT > 125)
    IF Y < 10
        MOVE 15 TO H
    ELSE
    IF (Y NOT < 10 AND Y < 13)
        COMPUTE H ROUNDED = 15 * ((13 - Y) / 3)
    ELSE
    IF Y NOT < 13
        MOVE 0 TO H
    ELSE
        MOVE 0 TO H
ELSE
IF (X > 125 AND X NOT > 375)
    IF Y < 11
        MOVE 15 TO H
    ELSE
    IF (Y NOT < 11 AND Y < 14)
        COMPUTE H ROUNDED = 15 * ((14 - Y) / 3)
    ELSE
    IF Y NOT < 14
        MOVE 0 TO H
    ELSE
        MOVE 0 TO H
ELSE
IF (X > 375 AND X NOT > 1350)
    IF Y < 12
        MOVE 15 TO H
    ELSE
    IF (Y NOT < 12 AND Y < 16)
        COMPUTE H ROUNDED = 15 * ((16 - Y) / 4)
    ELSE
    IF Y NOT < 16
        MOVE 0 TO H
    ELSE
        MOVE 0 TO H
ELSE
IF X > 1350
    IF Y < 15
        MOVE 15 TO H
    ELSE
    IF (Y NOT < 15 AND Y < 16)
        COMPUTE H ROUNDED = 15 * (16 - Y)
    ELSE
    IF Y NOT < 16
        MOVE 0 TO H
    ELSE
        MOVE 0 TO H
ELSE
    MOVE 0 TO H.
IF (G + H) > 15


    MOVE 15 TO G
    MOVE 0 TO H
ELSE
IF (G + H) < 0
    MOVE 0 TO G H.

VERTICAL CLEARANCE INSUFFICIENCY

IF (W-ITEM-12A NOT NUMERIC OR W-ITEM-12B NOT ALPHABETIC)
AND (W-ITEM-12 NOT = '00000')
    MOVE '*' TO W-ASTERISK
    MOVE '0000 ' TO W-ITEM-12.
IF W-ITEM-53 NOT NUMERIC
    MOVE '*' TO W-ASTERISK
    MOVE '1600' TO W-ITEM-53.
IF W-ITEM-12A NOT = '0000'
    IF W-ITEM-53 NOT < '1600'
        MOVE 0 TO I
    ELSE
    IF W-ITEM-53 < '1600'
        MOVE 2 TO I
    ELSE
        MOVE 0 TO I
ELSE
IF W-ITEM-12A = '0000'
    IF W-ITEM-53 NOT < '1400'
        MOVE 0 TO I
    ELSE
    IF W-ITEM-53 < '1400'
        MOVE 2 TO I
    ELSE
        MOVE 0 TO I
ELSE
    MOVE 0 TO I.
COMPUTE S2 ROUNDED = 30 - (J + (G + H) + I).
IF S2 < 0
    MOVE 0 TO S2
ELSE
IF S2 > 30
    MOVE 30 TO S2.
```

```
* 3. ESSENTIALITY FOR PUBLIC USE

     COMPUTE K ROUNDED = (S1 + S2) / 85.
     IF K = 0
         MOVE 0.0001 TO K.
     IF W-ITEM-19 NOT NUMERIC
         MOVE '*' TO W-ASTERISK
         MOVE '00' TO W-ITEM-19.
     COMPUTE A ROUNDED =
         ((W-ITEM-29R * W-ITEM-19R) / (200000 * K)) * 15.
     IF A < 0
         MOVE 0 TO A
     ELSE
     IF A > 15
         MOVE 15 TO A.
     IF W-ITEM-12A NOT = '0000'


         MOVE 2 TO B
     ELSE
     IF W-ITEM-12A = '0000'
         MOVE 0 TO B
     ELSE
         MOVE 0 TO B.
     COMPUTE S3 ROUNDED = 15 - (A + B).
     IF S3 < 0
         MOVE 0 TO S3
     ELSE
     IF S3 > 15
         MOVE 15 TO S3.

* 4. SPECIAL REDUCTIONS

     IF (S1 + S2 + S3) < 50
         MOVE 0 TO S4
         GO TO FINAL-RATING.
     COMPUTE A ROUNDED =
         ((W-ITEM-19R) ** 4) * (5.205 * (10 ** -8)).
     IF A > 5
         MOVE 5 TO A
     ELSE
     IF A < 0
         MOVE 0 TO A.
     IF W-ITEM-43B = '10' OR '12' OR '13' OR '14' OR '15' OR '16'
     OR '17'
         MOVE 5 TO B
     ELSE
         MOVE 0 TO B.
     IF W-ITEM-36A NOT = '0' AND NOT = '1' AND NOT = 'N'
         MOVE '*' TO W-ASTERISK
         MOVE '1' TO W-ITEM-36A.
     IF W-ITEM-36B NOT = '0' AND NOT = '1' AND NOT = 'N'
         MOVE '*' TO W-ASTERISK
         MOVE '1' TO W-ITEM-36B.
     IF W-ITEM-36C NOT = '0' AND NOT = '1' AND NOT = 'N'
         MOVE '*' TO W-ASTERISK
         MOVE '1' TO W-ITEM-36C.
     IF W-ITEM-36D NOT = '0' AND NOT = '1' AND NOT = 'N'
         MOVE '*' TO W-ASTERISK
         MOVE '1' TO W-ITEM-36D.
     EXAMINE W-ITEM-36 TALLYING ALL ZEROS.
     IF TALLY = 2
         MOVE 1 TO C
     ELSE
     IF TALLY = 3
         MOVE 2 TO C
     ELSE
     IF TALLY = 4
         MOVE 3 TO C
     ELSE
         MOVE 0 TO C.
     COMPUTE S4 ROUNDED = A + B + C.


     COMPUTE RTNG ROUNDED = S1 + S2 + S3 - S4.
     IF RTNG < 0
         MOVE 0 TO RTNG
     ELSE
     IF RTNG > 100.0
         MOVE 100.0 TO RTNG.
     MOVE RTNG TO W-RTNG.
     MOVE W-SUFF-RTNG TO WORK-SUFF.

     IF W-ASTERISK = '*'
         MOVE H-ITEM-12 TO W-ITEM-12
         MOVE H-ITEM-19 TO W-ITEM-19
         MOVE H-ITEM-28A TO W-ITEM-28A
         MOVE H-ITEM-29 TO W-ITEM-29
         MOVE H-ITEM-32 TO W-ITEM-32
         MOVE H-ITEM-36 TO W-ITEM-36
         MOVE H-ITEM-51 TO W-ITEM-51
         MOVE H-ITEM-53 TO W-ITEM-53
         MOVE H-ITEM-66 TO W-ITEM-66.

*    THESE ITEMS MAY HAVE BEEN CHANGED DURING THE CALCULATION OF
*    THE SUFFICIENCY RATING.

 CALCULATE-SUFFICIENCY-RATING-X. EXIT.
```

## B.1 CONCRETE DECK SLAB

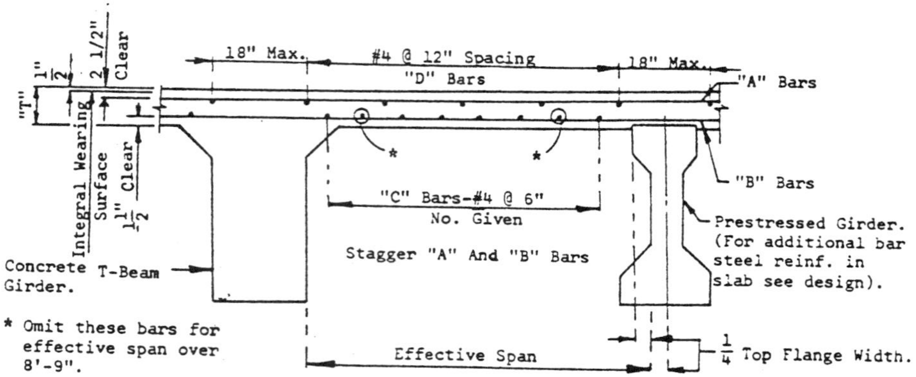


* Omit these bars for effective span over 8'-9".

SLAB ON CONCRETE GIRDERS

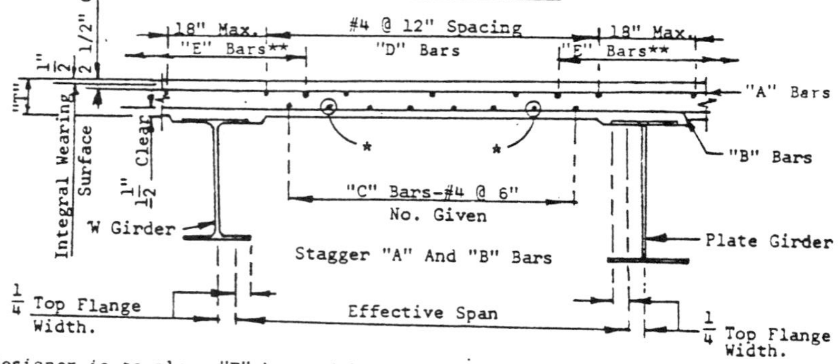


** Designer is to place "E" bars within the effective width and extend bars from Negative Moment regions of continuous spans at least 40 bar diameters beyond the last shear connectors into the Positive Moment region. If splicing is required employ Class B splice criteria.

SLAB ON STEEL GIRDERS

NOTES

Bottom transverse bars in slab shall be supported by continuous bar chairs on or adjacent to each girder and by individual bar chairs at 3'-0" centers approximately midway between girders. (If the center to center of girders exceeds 7'-6" individual bar chairs shall be placed at 3'-0" centers at approximately 1/3 points between girders.)

Top longitudinal bar steel shall be supported by continuous bar chairs at approximately 4'-0" centers.

For effective spans not listed in the tables, use lower values up to and including 1½" and higher values over 1½".

HS 20 and H 20 LOADING

| EFFECTIVE SPAN FT.-IN. | T=SLAB THICKNESS INCHES | "A" & "B" BARS SIZE NO. | "A" & "B" BARS SPACING INCHES | "C" BARS NO. OF BARS | "E" BARS* SIZE NO. | "E" BARS* NO. OF BARS |
|---|---|---|---|---|---|---|
| 4-6 | 7 | 5 | 7 | 6 | 5 | 6 |
| 4-9 | 7 | 5 | 7 | 6 | 6 | 6 |
| 5-0 | 7 | 5 | 6.5 | 6 | 6 | 6 |
| 5-3 | 7 | 5 | 6.5 | 6 | 6 | 6 |
| 5-6 | 7 | 5 | 6 | 6 | 6 | 6 |
| 5-9 | 7.5 | 5 | 6.5 | 8 | 6 | 6 |
| 6-0 | 7.5 | 5 | 6.5 | 8 | 6 | 6 |
| 6-3 | 7.5 | 5 | 6.5 | 8 | 6 | 8 |
| 6-6 | 7.5 | 5 | 6 | 8 | 6 | 8 |
| 6-9 | 7.5 | 5 | 6 | 10 | 6 | 8 |
| 7-0 | 7.5 | 6 | 8 | 10 | 6 | 8 |
| 7-3 | 8 | 6 | 9 | 10 | 6 | 10 |
| 7-6 | 8 | 6 | 8.5 | 10 | 6 | 10 |
| 7-9 | 8 | 6 | 8.5 | 12 | 6 | 10 |
| 8-0 | 8 | 6 | 8 | 12 | 6 | 10 |
| 8-3 | 8 | 6 | 8 | 12 | 6 | 10 |
| 8-6 | 8 | 6 | 7.5 | 12 | 6 | 10 |
| 8-9 | 8.5 | 6 | 8.5 | 14 | 6 | 10 |
| 9-0 | 8.5 | 6 | 8 | 12 | 6 | 12 |
| 9-3 | 8.5 | 6 | 8 | 12 | 6 | 12 |
| 9-6 | 8.5 | 6 | 8 | 12 | 7 | 10 |
| 9-9 | 8.5 | 6 | 7.5 | 14 | 7 | 10 |
| 10-0 | 8.5 | 6 | 7.5 | 14 | 7 | 10 |
| 10-3 | 9 | 6 | 8 | 14 | 7 | 12 |
| 10-6 | 9 | 6 | 8 | 14 | 7 | 12 |
| 10-9 | 9 | 6 | 7.5 | 16 | 7 | 12 |
| 11-0 | 9 | 6 | 7.5 | 16 | 7 | 12 |
| 11-3 | 9 | 6 | 7.5 | 16 | 7 | 12 |
| 11-6 | 9 | 6 | 7 | 16 | 7 | 12 |
| 11-9 | 9.5 | 6 | 7.5 | 18 | 7 | 14 |
| 12-0 | 9.5 | 6 | 7.5 | 18 | 7 | 14 |
| 12-3 | 9.5 | 6 | 7.5 | 18 | 7 | 14 |
| 12-6 | 9.5 | 6 | 7 | 18 | 7 | 14 |
| 12-9 | 9.5 | 6 | 7 | 20 | 7 | 14 |
| 13-0 | 9.5 | 6 | 7 | 20 | 7 | 14 |
| 13-3 | 10 | 6 | 7.5 | 20 | 7 | 16 |
| 13-6 | 10 | 6 | 7.5 | 20 | 7 | 16 |
| 13-9 | 10 | 6 | 7 | 22 | 7 | 16 |
| 14-0 | 10 | 6 | 7 | 22 | 7 | 16 |

Max. Allowable Design Stresses: $f_c'$ = 4000 psi, $f_y$ = 60 ksi, Top Steel 2-1/2" Clear, Bottom Steel 1-1/2" Clear, 20#/ft.$^2$ Future Wearing Surface.

* Use only for Slab on Steel Girders.

## LOAD FACTOR SLAB DESIGN

## CONCRETE SLAB REINFORCEMENT TABLE

*State of Wisconsin / Department of Transportation*

RECTANGULAR SECTION WITH TENSION REINFORCEMENT ONLY
STRENGTH METHOD STEEL YIELD STRENGTH = 60000. PSI CONCRETE COMPRESSIVE STRENGTH = 3500 PSI

$$R_u = \frac{M}{\phi\ bd^2}\ \text{psi} \qquad \rho = \frac{A_s}{bd}$$

| RU | RHO | RU | RHO | RU | RHO | RU | RHO | RU | RHO |
|---|---|---|---|---|---|---|---|---|---|
| 117.6 | 0.0020 | 306.4 | 0.0054 | 481.1 | 0.0088 | 641.9 | 0.0122 | 788.7 | 0.0156 |
| 123.3 | 0.0021 | 311.7 | 0.0055 | 486.1 | 0.0089 | 646.5 | 0.0123 | 792.9 | 0.0157 |
| 129.1 | 0.0022 | 317.0 | 0.0056 | 491.0 | 0.0090 | 651.0 | 0.0124 | 796.9 | 0.0158 |
| 134.8 | 0.0023 | 322.3 | 0.0057 | 495.9 | 0.0091 | 655.5 | 0.0125 | 801.0 | 0.0159 |
| 140.5 | 0.0024 | 327.6 | 0.0058 | 500.8 | 0.0092 | 659.9 | 0.0126 | 805.1 | 0.0160 |
| 146.2 | 0.0025 | 332.9 | 0.0059 | 505.7 | 0.0093 | 664.4 | 0.0127 | 809.2 | 0.0161 |
| 151.9 | 0.0026 | 338.2 | 0.0060 | 510.5 | 0.0094 | 668.9 | 0.0128 | 813.2 | 0.0162 |
| 157.6 | 0.0027 | 343.5 | 0.0061 | 515.4 | 0.0095 | 673.3 | 0.0129 | 817.2 | 0.0163 |
| 163.3 | 0.0028 | 348.7 | 0.0062 | 520.2 | 0.0096 | 677.7 | 0.0130 | 821.3 | 0.0164 |
| 168.9 | 0.0029 | 354.0 | 0.0063 | 525.1 | 0.0097 | 682.2 | 0.0131 | 825.3 | 0.0165 |
| 174.6 | 0.0030 | 359.2 | 0.0064 | 529.9 | 0.0098 | 686.6 | 0.0132 | 829.3 | 0.0166 |
| 180.2 | 0.0031 | 364.4 | 0.0065 | 534.7 | 0.0099 | 691.0 | 0.0133 | 833.2 | 0.0167 |
| 185.8 | 0.0032 | 369.6 | 0.0066 | 539.5 | 0.0100 | 695.3 | 0.0134 | 837.2 | 0.0168 |
| 191.4 | 0.0033 | 374.8 | 0.0067 | 544.3 | 0.0101 | 699.7 | 0.0135 | 841.2 | 0.0169 |
| 197.0 | 0.0034 | 380.0 | 0.0068 | 549.0 | 0.0102 | 704.1 | 0.0136 | 845.1 | 0.0170 |
| 202.6 | 0.0035 | 385.2 | 0.0069 | 553.8 | 0.0103 | 708.4 | 0.0137 | 849.1 | 0.0171 |
| 208.2 | 0.0036 | 390.3 | 0.0070 | 558.6 | 0.0104 | 712.8 | 0.0138 | 853.0 | 0.0172 |
| 213.7 | 0.0037 | 395.5 | 0.0071 | 563.3 | 0.0105 | 717.1 | 0.0139 | 856.9 | 0.0173 |
| 219.3 | 0.0038 | 400.6 | 0.0072 | 568.0 | 0.0106 | 721.4 | 0.0140 | 860.8 | 0.0174 |
| 224.8 | 0.0039 | 405.8 | 0.0073 | 572.7 | 0.0107 | 725.7 | 0.0141 | 864.7 | 0.0175 |
| 230.3 | 0.0040 | 410.9 | 0.0074 | 577.4 | 0.0108 | 730.0 | 0.0142 | 868.6 | 0.0176 |
| 235.8 | 0.0041 | 416.0 | 0.0075 | 582.1 | 0.0109 | 734.3 | 0.0143 | 872.4 | 0.0177 |
| 241.3 | 0.0042 | 421.0 | 0.0076 | 586.8 | 0.0110 | 738.5 | 0.0144 | 876.3 | 0.0178 |
| 246.8 | 0.0043 | 426.1 | 0.0077 | 591.4 | 0.0111 | 742.8 | 0.0145 | 880.1 | 0.0179 |
| 252.3 | 0.0044 | 431.2 | 0.0078 | 596.1 | 0.0112 | 747.0 | 0.0146 | 884.0 | 0.0180 |
| 257.7 | 0.0045 | 436.2 | 0.0079 | 600.7 | 0.0113 | 751.2 | 0.0147 | 887.8 | 0.0181 |
| 263.2 | 0.0046 | 441.3 | 0.0080 | 605.4 | 0.0114 | 755.5 | 0.0148 | 891.6 | 0.0182 |
| 268.6 | 0.0047 | 446.3 | 0.0081 | 610.0 | 0.0115 | 759.7 | 0.0149 | 895.4 | 0.0183 |
| 274.1 | 0.0048 | 451.3 | 0.0082 | 614.6 | 0.0116 | 763.9 | 0.0150 | 899.1 | 0.0184 |
| 279.5 | 0.0049 | 456.3 | 0.0083 | 619.2 | 0.0117 | 768.0 | 0.0151 | 902.9 | 0.0185 |
| 284.9 | 0.0050 | 461.3 | 0.0084 | 623.7 | 0.0118 | 772.2 | 0.0152 | 906.7 | 0.0186 |
| 290.3 | 0.0051 | 466.3 | 0.0085 | 628.3 | 0.0119 | 776.4 | 0.0153 | 910.4 | 0.0187 |
| 295.6 | 0.0052 | 471.2 | 0.0086 | 632.9 | 0.0120 | 780.5 | 0.0154 | | |
| 301.0 | 0.0053 | 476.2 | 0.0087 | 637.4 | 0.0121 | 784.6 | 0.0155 | | |

AASHTO 1.5.7 — $\rho$ MIN. (at 185.8 / 0.0032)

$\rho_{MAX} = .75\rho_b$ AASHTO 1.5.32(A) (at 910.4 / 0.0187)

RECTANGULAR SECTION WITH TENSION REINFORCEMENT ONLY
STRENGTH METHOD STEEL YIELD STRENGTH = 60000. PSI CONCRETE COMPRESSIVE STRENGTH = 4000 PSI

$$R_u = \frac{M_u}{\phi\ bd^2} \qquad \rho = \frac{A_s}{bd}$$

| RU | RHO | RU | RHO | RU | RHO | RU | RHO | RU | RHO |
|---|---|---|---|---|---|---|---|---|---|
| 117.9 | 0.0020 | 335.6 | 0.0059 | 537.1 | 0.0098 | 722.6 | 0.0137 | 892.0 | 0.0176 |
| 123.7 | 0.0021 | 340.9 | 0.0060 | 542.1 | 0.0099 | 727.2 | 0.0138 | 896.1 | 0.0177 |
| 129.4 | 0.0022 | 346.3 | 0.0061 | 547.1 | 0.0100 | 731.7 | 0.0139 | 900.2 | 0.0178 |
| 135.2 | 0.0023 | 351.6 | 0.0062 | 552.0 | 0.0101 | 736.2 | 0.0140 | 904.4 | 0.0179 |
| 141.0 | 0.0024 | 357.0 | 0.0063 | 556.9 | 0.0102 | 740.7 | 0.0141 | 908.5 | 0.0180 |
| 146.7 | 0.0025 | 362.3 | 0.0064 | 561.8 | 0.0103 | 745.2 | 0.0142 | 912.5 | 0.0181 |
| 152.4 | 0.0026 | 367.6 | 0.0065 | 566.7 | 0.0104 | 749.7 | 0.0143 | 916.6 | 0.0182 |
| 158.1 | 0.0027 | 372.9 | 0.0066 | 571.6 | 0.0105 | 754.2 | 0.0144 | 920.7 | 0.0183 |
| 163.8 | 0.0028 | 378.2 | 0.0067 | 576.5 | 0.0106 | 758.7 | 0.0145 | 924.8 | 0.0184 |
| 169.5 | 0.0029 | 383.5 | 0.0068 | 581.4 | 0.0107 | 763.1 | 0.0146 | 928.8 | 0.0185 |
| 175.2 | 0.0030 | 388.8 | 0.0069 | 586.2 | 0.0108 | 767.6 | 0.0147 | 932.8 | 0.0186 |
| 180.9 | 0.0031 | 394.1 | 0.0070 | 591.1 | 0.0109 | 772.0 | 0.0148 | 936.9 | 0.0187 |
| 186.6 | 0.0032 | 399.3 | 0.0071 | 595.9 | 0.0110 | 776.5 | 0.0149 | 940.9 | 0.0188 |
| 192.2 | 0.0033 | 404.6 | 0.0072 | 600.8 | 0.0111 | 780.9 | 0.0150 | 944.9 | 0.0189 |
| 197.9 | 0.0034 | 409.8 | 0.0073 | 605.6 | 0.0112 | 785.3 | 0.0151 | 948.9 | 0.0190 |
| 203.5 | 0.0035 | 415.0 | 0.0074 | 610.4 | 0.0113 | 789.7 | 0.0152 | 952.9 | 0.0191 |
| 209.1 | 0.0036 | 420.2 | 0.0075 | 615.2 | 0.0114 | 794.1 | 0.0153 | 956.8 | 0.0192 |
| 214.8 | 0.0037 | 425.4 | 0.0076 | 620.0 | 0.0115 | 798.4 | 0.0154 | 960.8 | 0.0193 |
| 220.4 | 0.0038 | 430.6 | 0.0077 | 624.8 | 0.0116 | 802.8 | 0.0155 | 964.7 | 0.0194 |
| 225.9 | 0.0039 | 435.8 | 0.0078 | 629.5 | 0.0117 | 807.2 | 0.0156 | 968.7 | 0.0195 |
| 231.5 | 0.0040 | 441.0 | 0.0079 | 634.3 | 0.0118 | 811.5 | 0.0157 | 972.6 | 0.0196 |
| 237.1 | 0.0041 | 446.1 | 0.0080 | 639.0 | 0.0119 | 815.8 | 0.0158 | 976.5 | 0.0197 |
| 242.7 | 0.0042 | 451.3 | 0.0081 | 643.8 | 0.0120 | 820.1 | 0.0159 | 980.4 | 0.0198 |
| 248.2 | 0.0043 | 456.4 | 0.0082 | 648.5 | 0.0121 | 824.5 | 0.0160 | 984.3 | 0.0199 |
| 253.7 | 0.0044 | 461.5 | 0.0083 | 653.2 | 0.0122 | 828.8 | 0.0161 | 988.2 | 0.0200 |
| 259.3 | 0.0045 | 466.6 | 0.0084 | 657.9 | 0.0123 | 833.1 | 0.0162 | 992.1 | 0.0201 |
| 264.8 | 0.0046 | 471.7 | 0.0085 | 662.6 | 0.0124 | 837.3 | 0.0163 | 996.0 | 0.0202 |
| 270.3 | 0.0047 | 476.8 | 0.0086 | 667.3 | 0.0125 | 841.6 | 0.0164 | 999.8 | 0.0203 |
| 275.8 | 0.0048 | 481.9 | 0.0087 | 671.9 | 0.0126 | 845.9 | 0.0165 | 1003.7 | 0.0204 |
| 281.3 | 0.0049 | 487.0 | 0.0088 | 676.6 | 0.0127 | 850.1 | 0.0166 | 1007.5 | 0.0205 |
| 286.8 | 0.0050 | 492.1 | 0.0089 | 681.3 | 0.0128 | 854.3 | 0.0167 | 1011.3 | 0.0206 |
| 292.2 | 0.0051 | 497.1 | 0.0090 | 685.9 | 0.0129 | 858.6 | 0.0168 | 1015.1 | 0.0207 |
| 297.7 | 0.0052 | 502.2 | 0.0091 | 690.5 | 0.0130 | 862.8 | 0.0169 | 1018.9 | 0.0208 |
| 303.1 | 0.0053 | 507.2 | 0.0092 | 695.1 | 0.0131 | 867.0 | 0.0170 | 1022.7 | 0.0209 |
| 308.6 | 0.0054 | 512.2 | 0.0093 | 699.7 | 0.0132 | 871.2 | 0.0171 | 1026.5 | 0.0210 |
| 314.0 | 0.0055 | 517.2 | 0.0094 | 704.3 | 0.0133 | 875.4 | 0.0172 | 1030.3 | 0.0211 |
| 319.4 | 0.0056 | 522.2 | 0.0095 | 708.9 | 0.0134 | 879.5 | 0.0173 | 1034.0 | 0.0212 |
| 324.8 | 0.0057 | 527.2 | 0.0096 | 713.5 | 0.0135 | 883.7 | 0.0174 | 1037.8 | 0.0213 |
| 330.2 | 0.0058 | 532.2 | 0.0097 | 718.1 | 0.0136 | 887.9 | 0.0175 | | |

AASHTO 1.5.7 — $\rho$ MIN. (at 192.2 / 0.0033)

$\rho_{MAX} = .75\rho_b$ AASHTO 1.5.32(A) (at 1037.8 / 0.0213)

(WIS DOT)

Sample problem: Using load factor design procedures determine the slab thickness and main reinforcement for a deck slab with an 8'-6" stringer spacing and an HS20-44 loading.

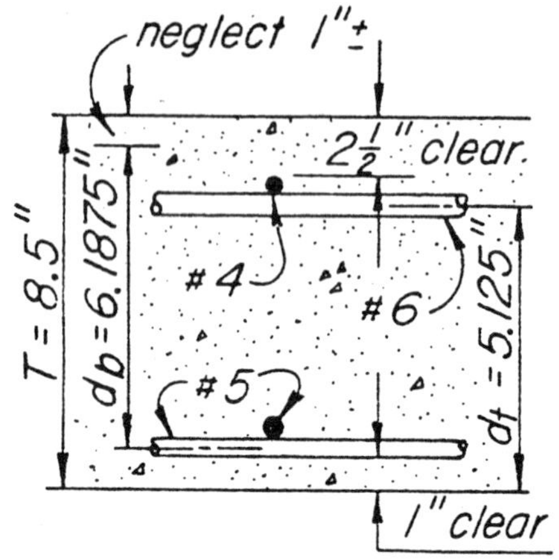


$S = 8'\text{-}6''$ minus $6'' = 8.0'$ (1.3.2(A))

$T_{min} = 17 + S/36 = 0.69'$, use $8\frac{1}{2}''$ (1.5.40)

$f'_c = 4{,}000$ p.s.i. (1.5.1(B))

$f_y = 60{,}000$ p.s.i.

$\phi = 0.9$

$Z = 130$ (top steel), 170 (bottom) (1.5.39)

$n = 8$ (1.5.23(D))

Dead load W:

Slab = (0.71)(.15) = 0.106

PWS + FWS = 60 p.s.f. = 0.060

} 0.166 kip/ft

Design Moments:

$DLM = (0.125)(W)(S^2) \times (0.8) = (.125)(.166)(8^2)(.8) = 1.06$ lk (1.3.2(C))

$LLM = (S+2)(16)(1.3)(0.8)/32 = (10)(16)(1.3)(.8)/32 = 5.20$ lk (1.3.2(C))

$M_u = 1.3[DL + 1.67(LL)] = 1.3[1.062 + 1.67(5.2)] = 12.67$ lk (1.2.22)

$M_r = LLM = 5.20$ lk (1.3.2(C))

$M_w$ = Service load moment = DLM + LLM = 6.26 lk (1.5.39)

## Top Reinforcement

$R = M_u/\phi bd^2$

$= (12.67)(1000)/(.9)(1)(5.125^2)$

$p = 536$ psi ∴

$= 0.00975$, from chart

$A_s = (.00975)(12)(5.125) = 0.60$ in²/ft.

~~Check fatigue stress (1.5.38)~~ *

~~$f_s = M_r/A_s jd$~~

~~$= (5.20)(12)/(0.60)(0.892)(5.125)$~~

~~$= 22.80$ ksi (no good)~~

~~Increase $A_s$~~

~~$A_s \cong (22.80)(0.60)/21.0$~~

~~$\cong 0.65$ in²/ft.~~

~~Try #6 bars at 8.11 in ($A_s = 0.65$)~~

Check steel spacing (1.5.39)

$f_s(\text{all.}) = z/(d_c A)^{1/3}$

$= 130/[(3.375^2)(2)(8.11)]^{1/3}$

$= 22.83$ or 36.0 max.

$f_s(\text{act.}) = M_W/A_s jd$

$= (6.26)(12)/(0.65)(0.89)(5.125)$

$= 25.30$ ksi (no good)

Reduce #6 bar spacing

$A_s \cong 25.30(0.65)/22.83 = \underline{\underline{0.72}}$ in²/ft.

∴ Use #6 bars @ 7¼" c/c

## Bottom Reinforcement

$R = (12.67)(1000)/(.9)(1)(6.19^2)$

$= 367$ psi ∴

$p = 0.00645$, from chart

$A_s = (.00645)(12)(6.19) = 0.48$ in²/ft.

~~$f_s = (5.20)(12)/(0.48)(.91)(6.19)$~~

~~$= 23.10$ ksi (no good)~~

~~Increase $A_s$~~

~~$A_s \cong 23.10(0.48)/21.0$~~

~~$\cong 0.53$ in²/ft ← (close)~~

~~Try #5 bars at 7¼" c/c ($A_s = 0.51$)~~

* In accordance with 1980 AASHTO, No fatigue check is required.

$f_s(\text{all.}) = 170/[(1.31^2)(2)(7.25)]^{1/3}$

$= 58.24$ or 36 ksi

$f_s(\text{act.}) = (6.26)(12)/(0.51)(0.91)(6.19)$

$= 26.15$ ksi (ok)

∴ Use #5 bars @ 7¼" c/c

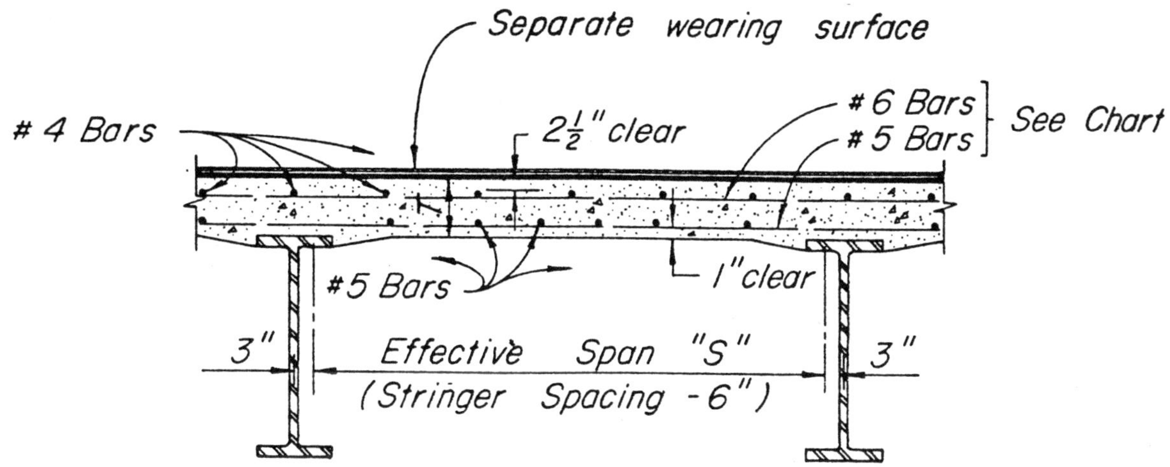


EXAMPLE: Stringer spacing of 8'-6"
S = 8'-6" minus 6" = 8.0
From chart: T = $8\frac{1}{2}$", $A_S$ (top) = 0.72, $A_S$ (bott.) = 0.53, (1.3.2 (A))

PRIMARY REINFORCEMENT: (See Design Regulations)
Use #6 bars (top) and # 5 bars (bott.), both at $7\frac{1}{4}$" c/c

DISTRIBUTIONAL REINFORCEMENT: (See Article 1.3.2 (E))
$A_S$ (top) = (0.33)(0.72) = 0.24 in²/ft
Use #4 bars at 11 equal spaces ($A_S$ = 0.26)
$A_S$ (bott.): $220/\sqrt{s}$ = 77.78 %, Use 67 % max.
= (0.67)(0.53) = 0.35 in²/ft in mid-half of span
= (0.50)(0.35) = 0.18 in²/ft in each outer quarter
Use 6 # 5 bars at 10" c/c in mid-half of span and 1 #5 bar in each outer quarter.

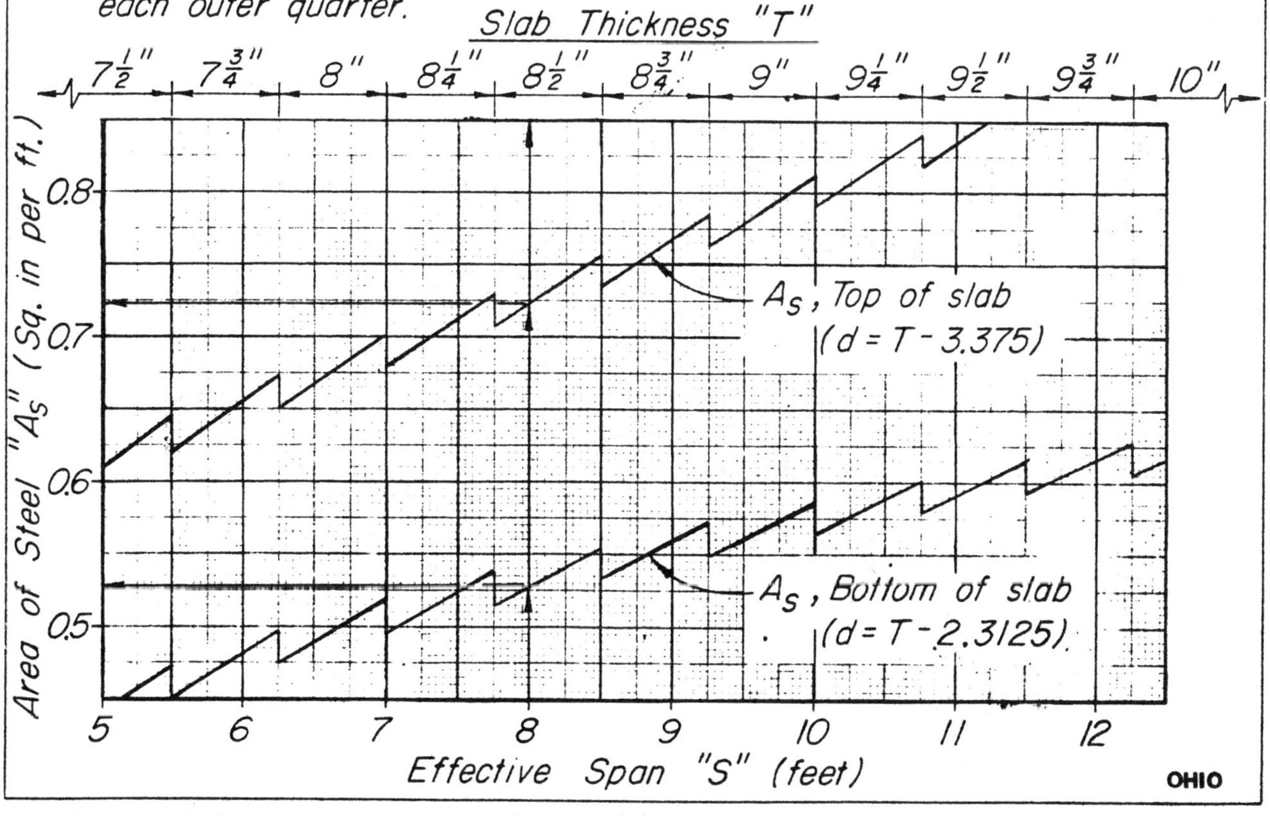


* By load factor procedures.

## STRENGTH DESIGN* OF CONCRETE FLEXURAL MEMBERS SINGLY REINFORCED

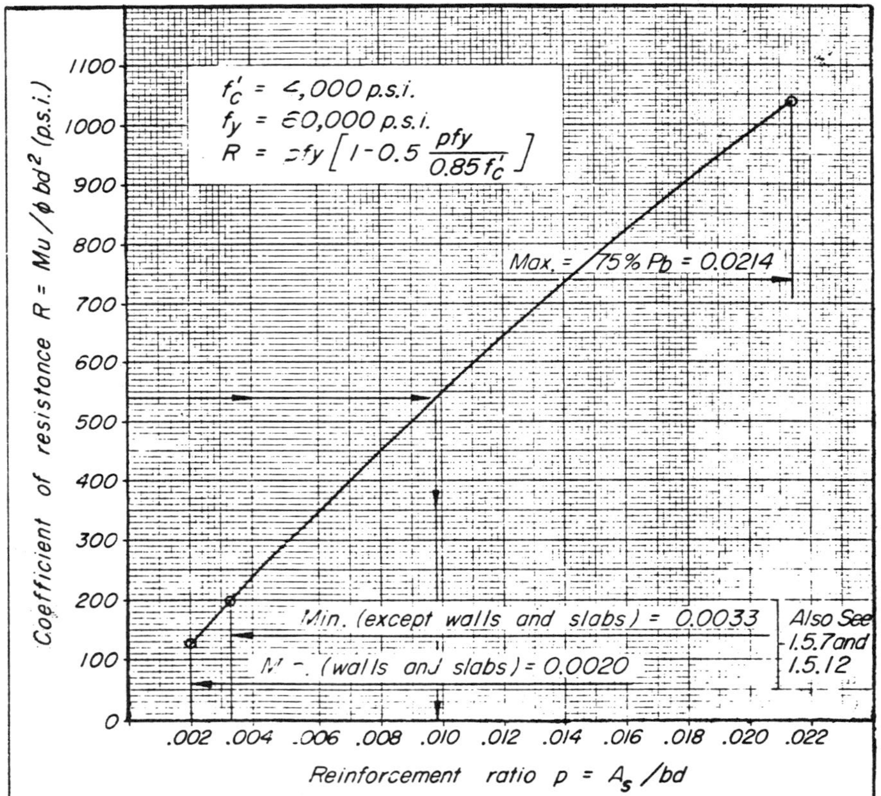


Sample problem (See sheet 3.2)

$M_u$ = 12.67 fk
$\phi$ = 0.9
$b$ = 12"
$d$ = 5.125

$R = M_u/\phi bd^2 = (12.67)(1000)/(0.9)(1)(5.125^2)$

$R$ = 536

$p$ = 0.00975, from chart

$\therefore A_s = pbd = (0.00975)(12)(5.125) = 0.60 \text{ in}^2/\text{ft}.$

* Or load factor design

## ANALYSIS OF CONCRETE FLEXURAL MEMBERS SINGLY REINFORCED $n = 8$

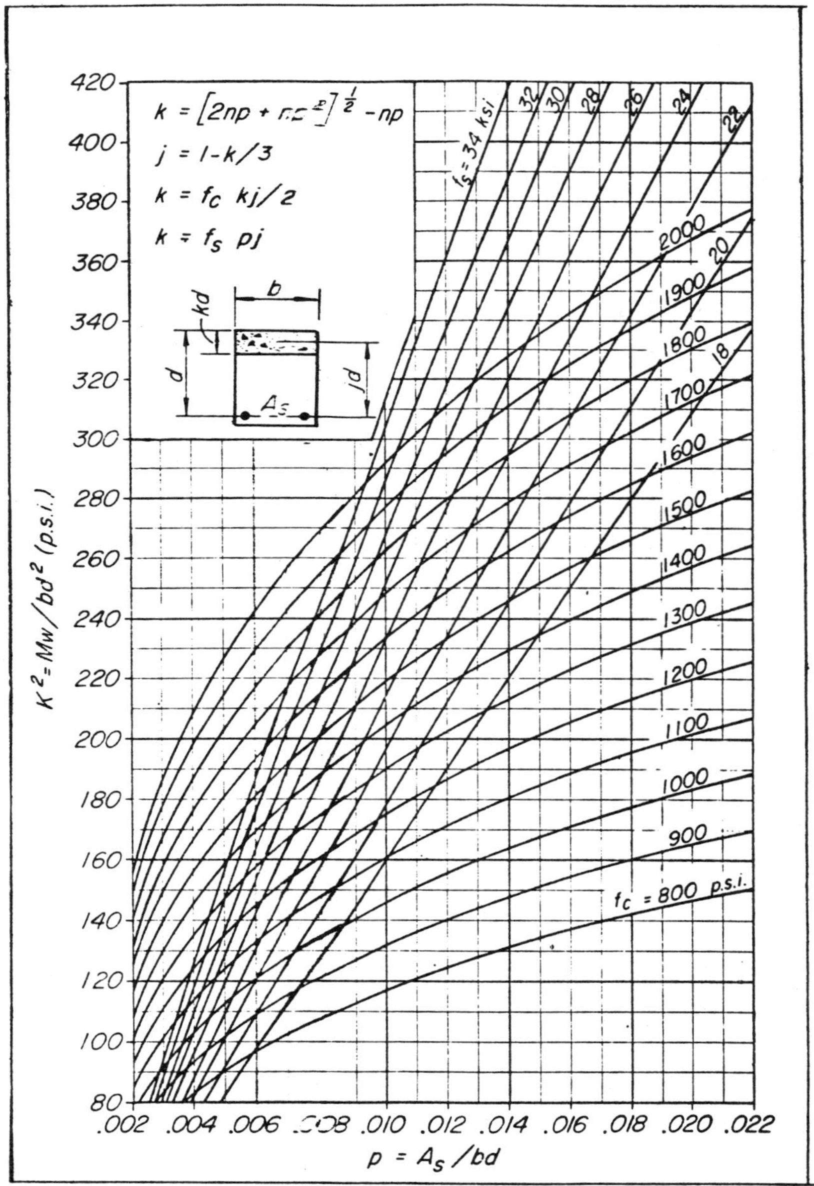


* Elastic Analysis

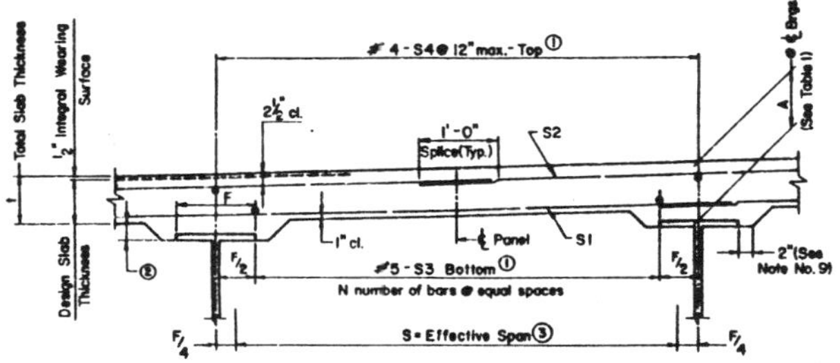

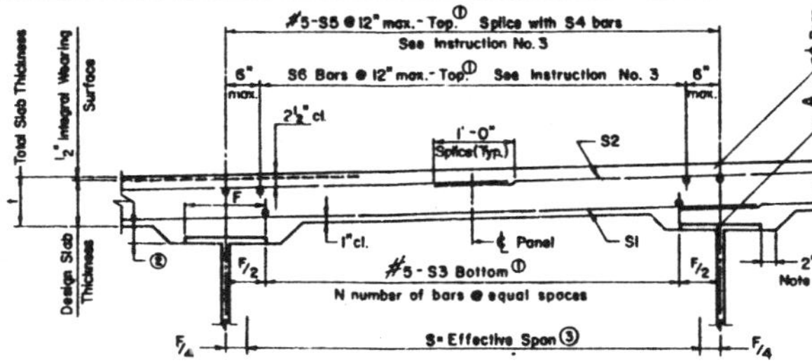

① Bars S3,S4,S5 and S6 are to be spaced symmetrically about the ℄ Panel.
② Depth of haunches will vary to compensate for irregularities in camber and type of surface layer.
③ Effective Span "S"
(a) Steel Stringers: as shown
(b) P/S I-Beams: For I-Beam 28/63 and AASHTO Type V & VI beams-
Distance between edges of Flanges (Clear Span) plus 10".
For all other I-Beams-clear span.
(c) P/S Spread Box Beams:
Distance between edges (clear span).

## TYPICAL SLAB PANEL 1

SIMPLE, CONTINUOUS NON-COMPOSITE and COMPOSITE (POSITIVE MOMENT REGION)

## TYPICAL SLAB PANEL 2

CONTINUOUS COMPOSITE AND NON-COMPOSITE (NEGATIVE MOMENT REGION)

S=Max. normal effective span permitted for given reinforcement bars S1 & S2 in table below.

| S | t | BARS S1 & S2 | BARS S3 N | BARS S3 MAX. SPACING | S6 |
|---|---|---|---|---|---|
| 3'-6" | 8" | #5 at 11" = 0.34 Sq. In. | 4 | 16" | #5 |
| 3'-9" | 8" | #5 at 10½" = 0.35 Sq. In. | 4 | 15½" | #5 |
| 4'-0" | 8" | #5 at 10" = 0.37 Sq. In. | 4 | 15" | #5 |
| 4'-3" | 8" | #5 at 9½" = 0.39 Sq. In. | 4 | 14½" | #5 |
| 4'-7" | 8" | #5 at 9" = 0.41 Sq. In. | 5 | 13½" | #5 |
| 5'-0" | 8" | #5 at 8½" = 0.44 Sq. In. | 5 | 13" | #5 |
| 5'-5" | 8" | #5 at 8" = 0.47 Sq. In. | 6 | 12" | #5 |
| 5'-10" | 8" | #5 at 7½" = 0.50 Sq. In. | 7 | 11" | #5 |
| 6'-3" | 8" | #5 at 7" = 0.53 Sq. In. | 8 | 10½" | #5 |
| 6'-9" | 8" | #5 at 6½" = 0.57 Sq. In. | 9 | 9½" | #5 |
| 7'-1" | 8" | #5 at 6" = 0.62 Sq. In. | 10 | 9" | #5 |
| | | | | | |
| 7'-5" | 8½" | #5 at 6½" = 0.57 Sq. In. | 10 | 9½" | #6 |
| 8'-0" | 8½" | #5 at 6" = 0.62 Sq. In. | 11 | 9" | #6 |
| 8'-5" | 8½" | #5 at 5½" = 0.68 Sq. In. | 13 | 8" | #6 |
| | | | | | |
| 8'-8" | 9" | #5 at 6" = 0.62 Sq. In. | 12 | 9" | #6 |
| 9'-3" | 9" | #5 at 5½" = 0.68 Sq. In. | 14 | 8" | #6 |
| | | | | | |
| | | | | | |

## INSTRUCTIONS

1. Place transverse reinforcement in deck slab parallel to ℄ Brgs. for skew angles 75° and more. For skew angles less than 75°, the bars shall be placed normal to ℄ of bridge.
2. For skew angles under 75°, a minium of 3-#5 bars @ 6" shall be placed in top and bottom of the deck slab parallel to abutment or pier joint over the end supports.
3. In continuous, composite and non-composite steel bridges, these bars (See Typical Slab Panel 2 Detail) shall be provided in the top of the deck slab over the piers, and shall be extended at least 3'-0" into the positive moment region beyond the dead load points of contraflexure in the adjacent spans.

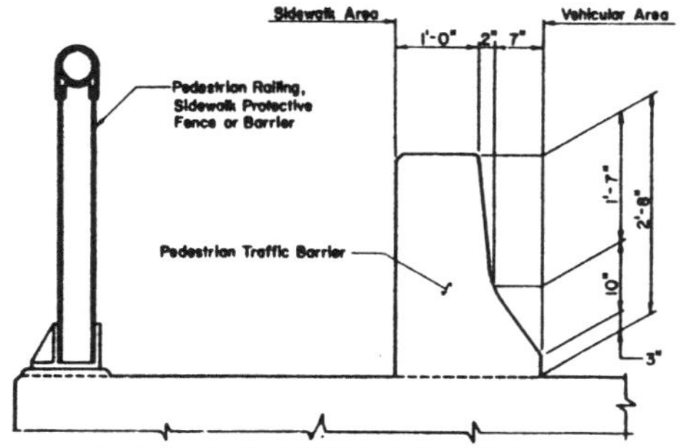

## ALTERNATE SIDEWALK DETAIL

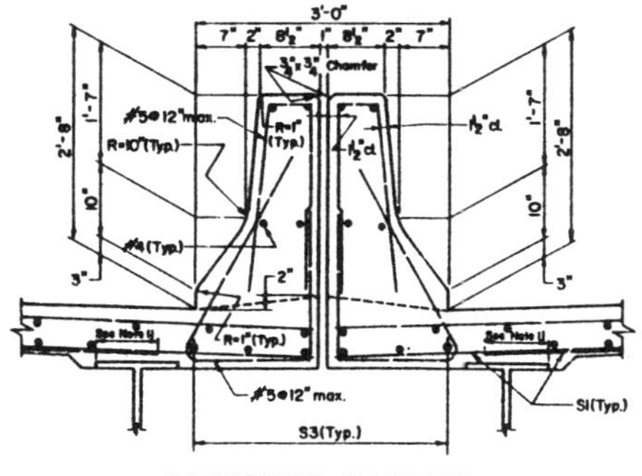

## CONCRETE BARRIER

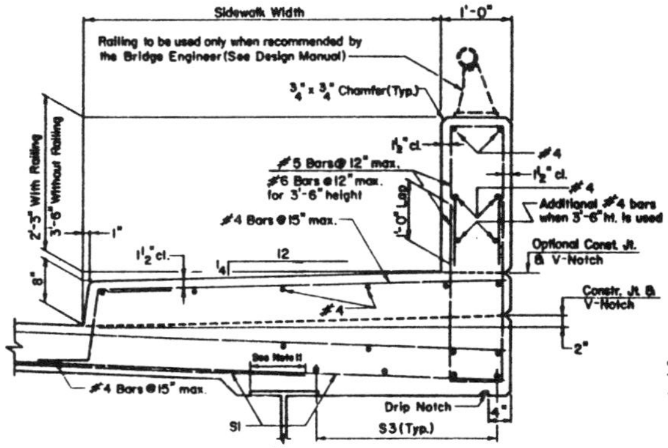

## SIDEWALK DETAIL

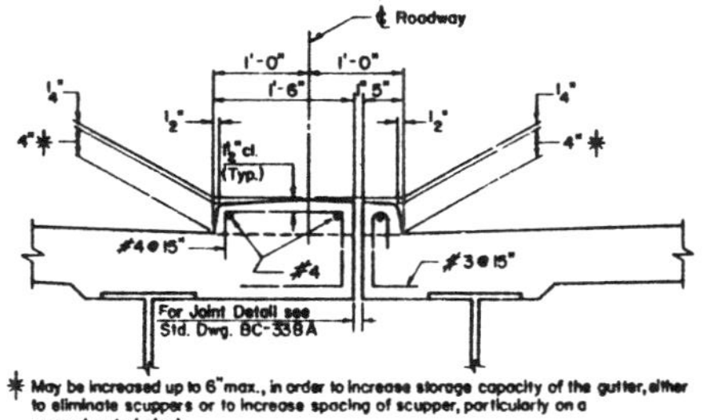

* May be increased up to 6" max., in order to increase storage capacity of the gutter, either to eliminate scuppers or to increase spacing of scupper, particularly on a superelevated deck.

## DIVISOR DETAIL

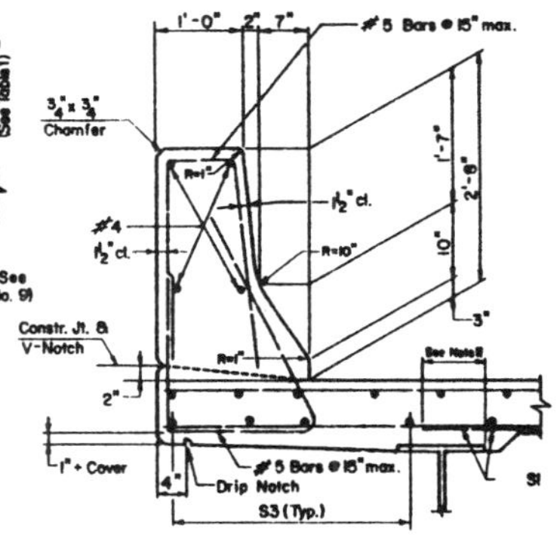

## SLOPED PARAPET DETAIL

## NOTES

1. Design Specifications: Design Division of 1977 AASHTO, "Standard Specifications for Highway Bridges" (Including 1981 Interim Specs.)
2. Live Load: HS20-44 or Alternate Military Loading.
3. Dead Load: Dead Load includes 30 lbs. per sq. ft. for future wearing surface on the deck slab and 15 lbs. per sq. ft. (wt. of forms plus wt. of concrete in valleys of forms) of deck area. 15 Lbs. per sq. ft. additional weight is not required when removable forms are used.
4. Design Stresses: Overstressing Allowed
   - $f_s$ = 24,000 lbs. per sq. in. — 15%
   - $f_c$ = 1,400 lbs. per sq. in. — 2.5%
   - n = 8
5. Dead Load Moment = $WS^2/10$ ft. kips/ft. width
   W = Dead Load Weight in kips/linear ft./ft. width
   S = Effective Span
6. (Live Load + Impact) Moment = $P(S+2)/32 \times 1.30 \times 0.80$ = 0.52(S+2) ft. kips/ft. width.
   Impact Factor = 1.30
   Continuity Factor = 0.80
   P = Wheel Load = 16 kips
7. Design is based on decks having 3 or more beams.
8. The effect of the haunch is not considered in the design.
9. Beam Haunch:
   Beam haunch details shown are to be used with removable deck forms. When permanent metal deck forms are used in placing the deck, beam haunch detail shall conform to Standard Drawing BC-332B.
10. When use of epoxy coated bars is approved, all bars shall be epoxy coated, and shall be so indicated on the Contract Drawings.
11. Extend alternate bottom bars 6" minimum over flange of fascia beam.
12. All bars shall be so detailed on the Contract Drawings.
13. All details shown also apply to P/S I-Beams and P/S Spread Box Beams. For effective Span "S", See ③.

| TABLE 1 | |
|---|---|
| PLAN CAMBER | A |
| UP TO 1½" | 1+½" |
| OVER 1½" TO 3" | 1+1" |
| OVER 3" | 1+1" |

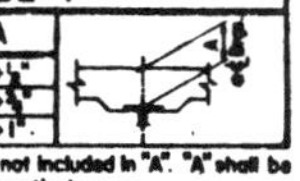

Variation in flange thickness is not included in "A". "A" shall be modified for a concave (sag) vertical curve.

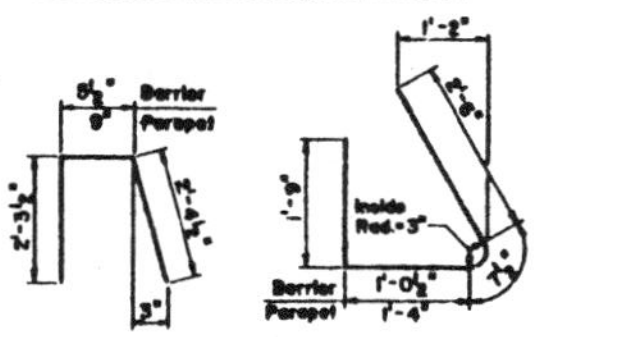

This Standard Drawing Supersedes Standard Drawing BD-101A, Dated Jan. 16, 1981

Commonwealth of Pennsylvania
DEPARTMENT OF TRANSPORTATION
BUREAU OF HIGHWAY DESIGN

STANDARD
# CONCRETE DECK SLAB
DESIGN & DETAILS
FOR BEAM BRIDGES

| Recommended Sept. 28, 1982 | Recommended Sept. 28, 1982 | Recommended Sept. 28, 1982 | Sheet 1 of 1 |
|---|---|---|---|
| B. F. Kotalik — Chief Bridge Engineer | W. Bowser — Dir., Bureau of Hwy. Design | [signature] — Chief Highway Engineer | BD-101B |

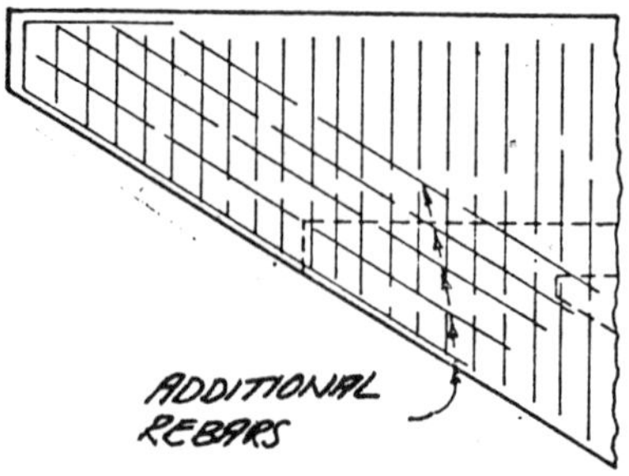


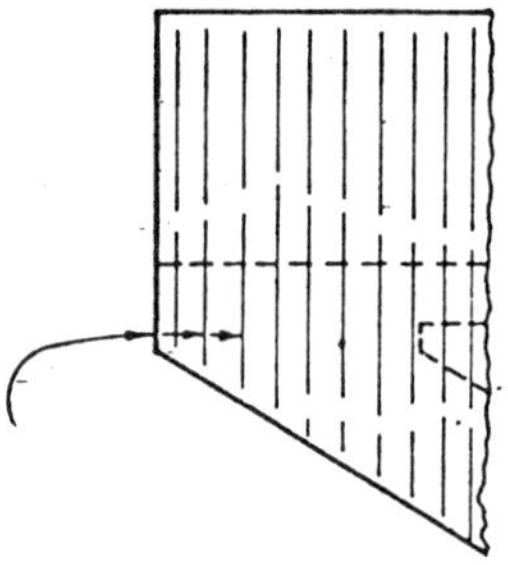

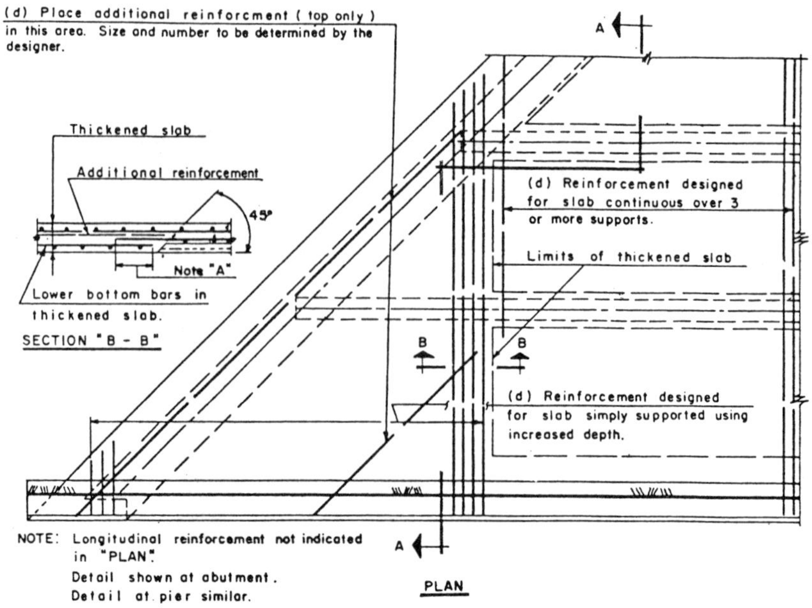


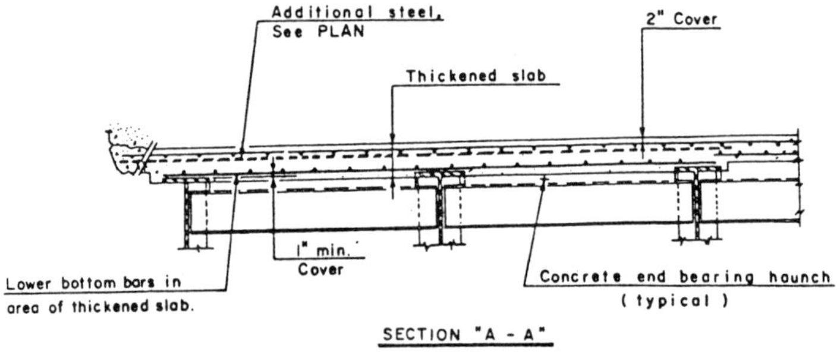


Note "A" Splice length shall be determined by the designer in accordance with the latest AASHTO design criteria.

ADDITIONAL SLAB REINFORCEMENT AT ACUTE CORNERS
( Skew angle greater than 20 degrees )

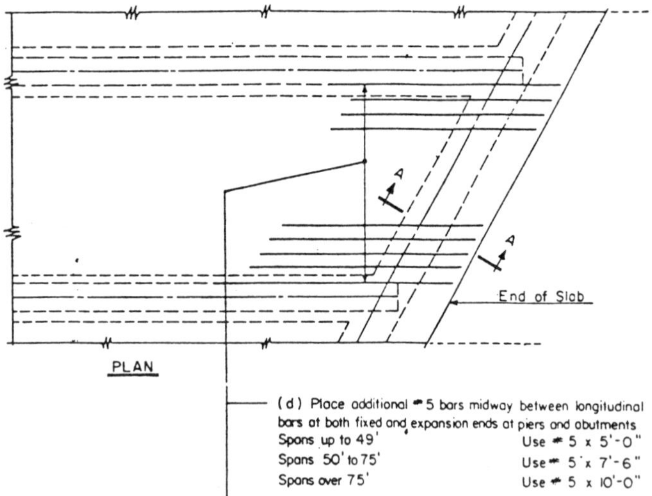


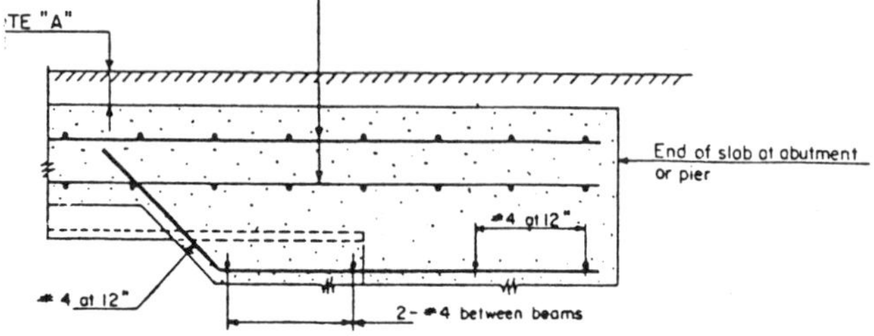


SECTION "A-A"
TYPICAL SECTION AT END OF SLAB

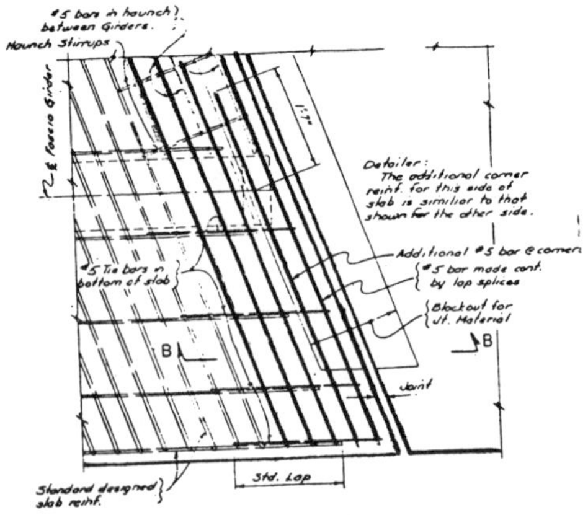


SLAB CORNER REINF. AT PIER
(SKEW 30° & UNDER)
Scale 1" = 1'-0"

DECK CORNER REINFORCEMENT

# REINFORCED CONCRETE WEARING SURFACE

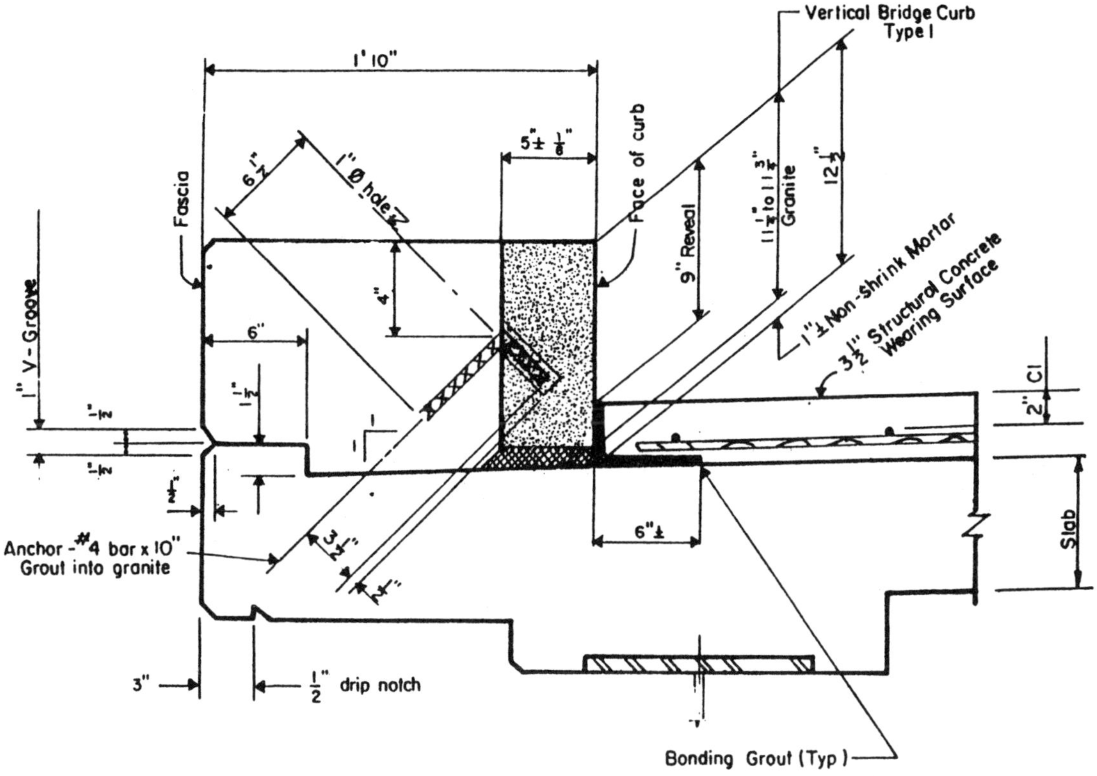


Curb Section

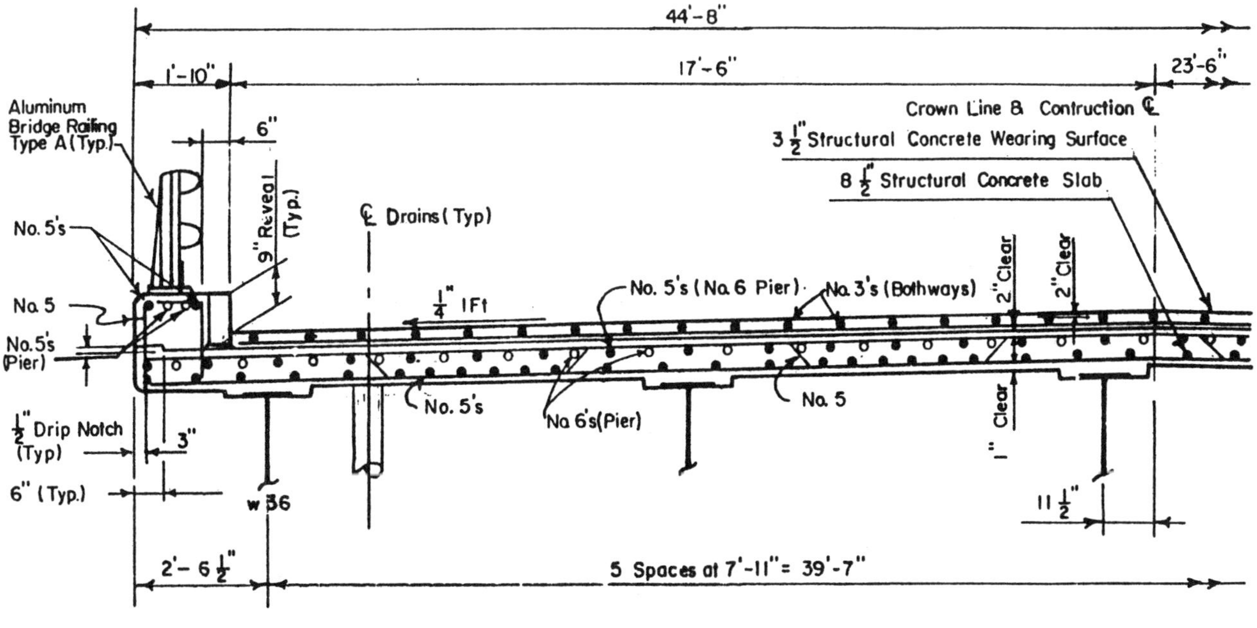


Maine

## B.2 SLAB OVERHANG DESIGN

Current design practice in the Bridge Section limits the slab overhang length to 3'-7 measured from the centerline of the exterior girder to the edge of a slab. The overhang length has been limited to prevent bending and rotation of the girder web during construction caused by the eccentric load carried by the forming brackets. Past construction experience indicates that a number of serious forming failures occurred when forming brackets were attached to steel or prestressed girder webs for support by using stud bolts or special inserts. Currently, forming brackets are not allowed to be supported by web attachment on any type of structure.

Transverse "A" bars are extended across the slab and employed to meet the overhang reinforcement requirements where practical. A minimum area of slab reinforcement, independent of overhang length, is required to carry the 10 kip transverse AASHTO concrete parapet live loading. At the parapet-slab interface, a 7'-0 longitudinal slab length is assumed for the live load distribution to the slab. For slab thickness "T", minimum area of slab reinforcement is as follows:

| T = Slab Thickness(IN) | 6.5 | 7.0 | 7.5 | 8.0 | 8.5 | 9.0 | 9.5 | 10.0 |
|---|---|---|---|---|---|---|---|---|
| Min. Area Steel ($IN^2$/FT) | .666 | .563 | .490 | .435 | .391 | .356 | .327 | .303 |

Total steel area required for varying overhang lengths and AASHTO live loading is given in Table        The slab thickness, t, equals the design slab thickness, T, plus the standard haunch thickness of 1-1/4 inches. The overhang length is redefined, to account for the variable top flange widths, as the distance from the edge of flange to the edge of parapet. The chart is designed on the basis of the controlling truck wheel or rail loading from the AASHTO specification. The two loading cases are as follows:

A. Truck Load

The design moment is computed for slab and parapet dead load plus a live loading of a truck wheel including impact. Reference is made to AASHTO Section 3 - Article 1.3.2(H)(1) for live load wheel distribution. The effective length of distribution equals $E_1 = 0.8\ X + 3.75$ in feet. X is the distance from point of load to the support in feet.

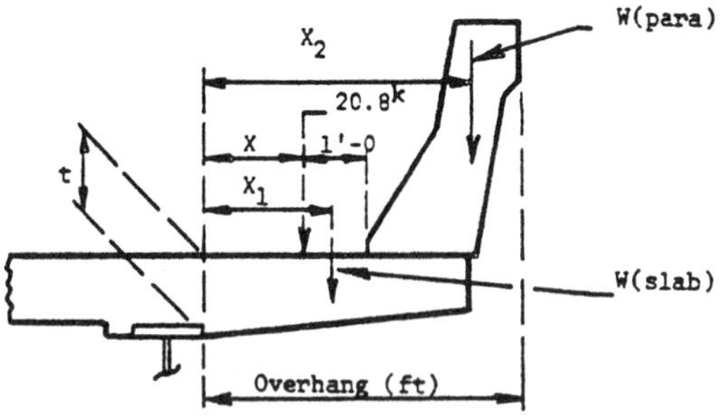


Moments are computed about the edge of the girder flange as follows:

$$M\ (slab) = W(slab)\ X_1$$

$$M\ (para) = W(para)\ X_2$$

$$M\ (LL) = \frac{20.8X}{E_1}$$

The design moment for truck live load equals the summation of the above moments multiplied by the strength design load factors.

| Over-hang* | t | | | | | | | |
|---|---|---|---|---|---|---|---|---|
| | 7-1/2" | 8" | 8-1/2" | 9" | 9-1/2" | 10" | 10-1/2" | 11" |
| 2'-6 | .665 | .594 | .540 | .496 | .460 | .430 | .404 | .382 |
| 2'-9 | .656 | .587 | .533 | .490 | .454 | .425 | .399 | .378 |
| 3'-0 | .649 | .581 | .528 | .485 | .450 | .421 | .396 | .374 |
| 3'-3 | .644 | .576 | .523 | .481 | .447 | .418 | .393 | .372 |
| 3'-6 | — | — | .521 | .479 | .445 | .416 | .391 | .370 |
| 3'-9 | -- | -- | -- | .639 | .495 | .458 | .426 | .399 |
| 4'-0 | -- | -- | -- | -- | .572 | .529 | .492 | .460 |
| 4'-3 | -- | -- | -- | -- | -- | .596 | .554 | .518 |
| 4'-6 | -- | -- | — | — | — | .661 | .613 | .573 |
| 4'-9 | -- | -- | -- | — | -- | .722 | .670 | .626 |
| 5'-0 | -- | -- | -- | -- | -- | .782 | .724 | .676 |

TOTAL STEEL AREA REQUIRED TOP OF SLAB ($IN^2$/FT)
$f_c$'=4000 psi, $f_y$=60000 psi, 2-1/2" CLEAR

**t = Slab thickness + 1 1/4" haunch thickness.**

* Overhang distance is as shown in sketch
Current Bridge Section design practice is to limit the Slab Overhang length to 3'-7 measured from the centerline of exterior girder to the edge of slab.

## B. Rail Load

The design moment is computed for slab and parapet dead loads plus a transverse rail live load of 10 kips placed at the top of parapet. Reference is made to AASHTO Section 3 - Article 1.3.2(H)(2) for live load rail distribution. If a parapet is employed the distribution length equals $E_2$ = 0.8X + 5.0 in feet. X is used as the distance from the center of gravity of the parapet to the point under investigation in feet.

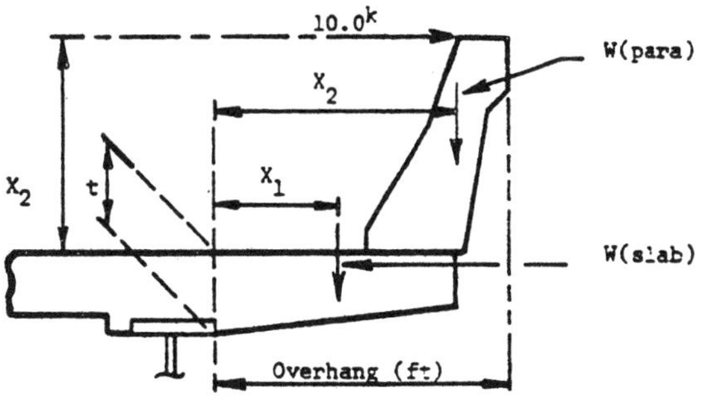


**Moments are computed about the edge of the girder flange:**

$$M(slab) = W(slab)\,X_1$$

$$M(para) = W(para)\,X_2$$

$$M(LL) = \frac{10.0(X_2 + 0.5\,t)}{E_2}$$

**The design moment for rail live load equals the summation of the above moments multiplied by the strength design load factors.**

**The procedures outlined can be used for values not on Table such as a haunch thickness greater than 1 1/4", overhang lengths greater than five feet plus half the top flange width, and median barrier requirements.**

# B.3 DECK SLAB POURING SEQUENCE

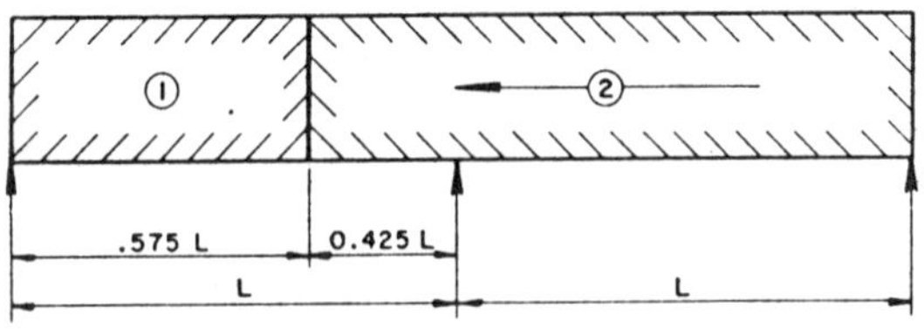


② — INDICATES POUR NUMBER AND DIRECTION OF POUR.

S = TOTAL NUMBER OF SPANS.

P = TOTAL NUMBER OF SUPPORTS.

L = LENGTH OF EXTERIOR SPAN.

$n$ = RATIO = $\frac{\text{INTERIOR SPAN}}{\text{EXTERIOR SPAN}}$

IDEAL POURS - 2 SPANS

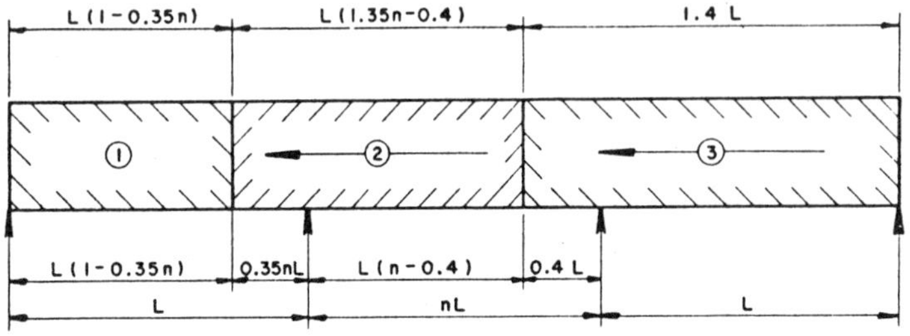


IDEAL POURS - 3 SPANS

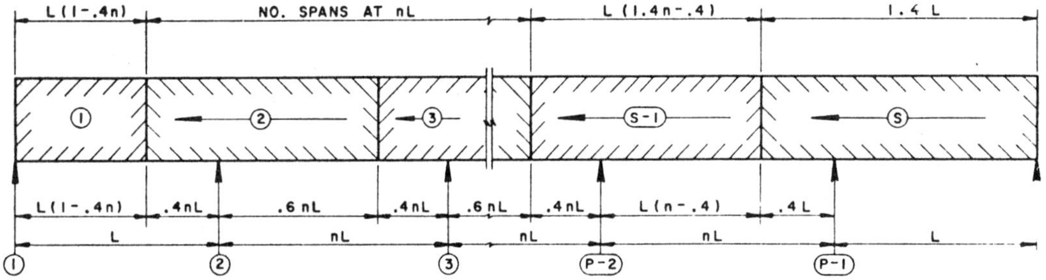


IDEAL POURS - ANY NUMBER OF SPANS

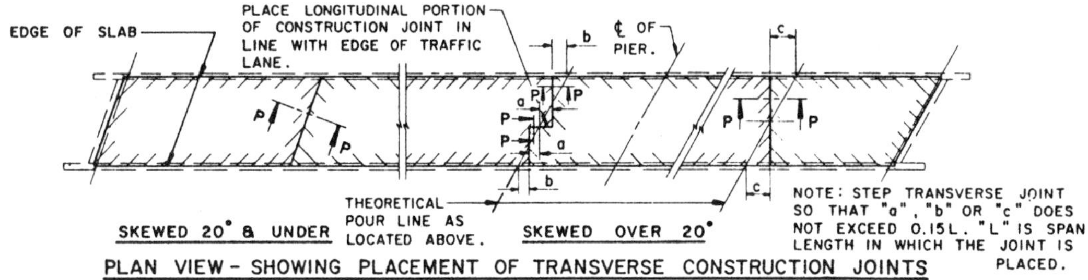


PLAN VIEW - SHOWING PLACEMENT OF TRANSVERSE CONSTRUCTION JOINTS

## NOTES

THE RATE OF POUR SHALL EXCEED 0.6 SPAN LENGTH PER HOUR. TRANSVERSE CONSTRUCTION JOINTS MAY BE OMITTED PROVIDED THE RATE OF POUR EXCEEDS 0.6 SPAN PER HOUR.

ON STRUCTURE LENGTHS OVER 300 FEET, TRANSVERSE CONSTRUCTION JOINTS, IF USED, SHALL BE PLACED AT THE 0.4 OF THE SPAN LENGTH BEYOND THE PIER IN THE DIRECTION OF POUR. FOR CONTINUOUS PRESTRESSED GIRDER BRIDGES, LOCATE THE OPTIONAL TRANSVERSE CONSTRUCTION JOINTS MIDWAY BETWEEN THE CUT-OFF POINTS FOR CONTINUITY REINFORCING STEEL OR AT THE 0.75 POINT OF THE SPAN, WHICHEVER IS CLOSEST TO THE PIER. AASHTO SPECIFICATION BAR DEVELOPMENT LENGTHS MUST BE MET BEYOND THE JOINTS.

OPTIONAL LONGITUDINAL CONSTRUCTION JOINTS, IF USED, SHALL BE APPROVED BY THE ENGINEER AND PREFERABLY BE PLACED UNDER THE MEDIAN BARRIER.

WHEN THE WIDTH OF A SUPERSTRUCTURE SLAB IS GREATER THAN 90 FEET, A LONGITUDINAL EXPANSION JOINT SHALL BE PROVIDED.

TWO OR MORE ALTERNATE POURS MAY BE PLACED ON THE SAME DAY.

DEADLOAD GIRDER DEFLECTIONS, BLOCKING, AND CAMBER SHOWN ON THE PLANS SHALL BE BASED ON NON-COMPOSITE ACTION. SPAN DEADLOAD DEFLECTIONS FROM POURS OCURRING AFTER CONCRETE WAS PLACED AT AN EARLIER DATE SHALL BE BASED ON COMPOSITE ACTION (BOTH POSITIVE AND NEGATIVE MOMENT ACTION) OF THE PREVIOUSLY PLACED CONCRETE.

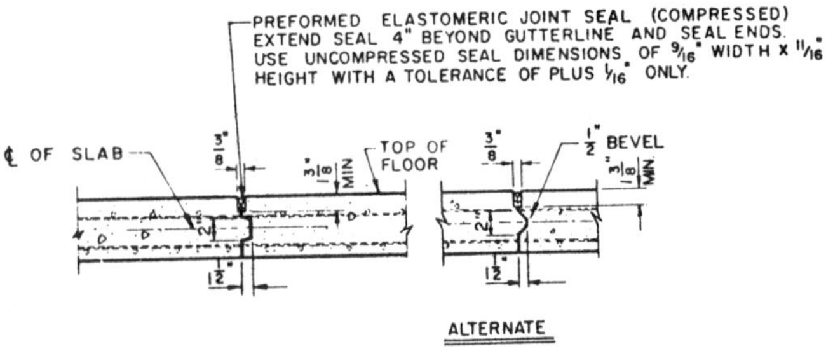


SECTION P

| SLAB POURING SEQUENCE |
| --- |
| STATE OF WISCONSIN<br>DEPARTMENT OF TRANSPORTATION<br>DIVISION OF HIGHWAYS<br>BRIDGE SECTION |

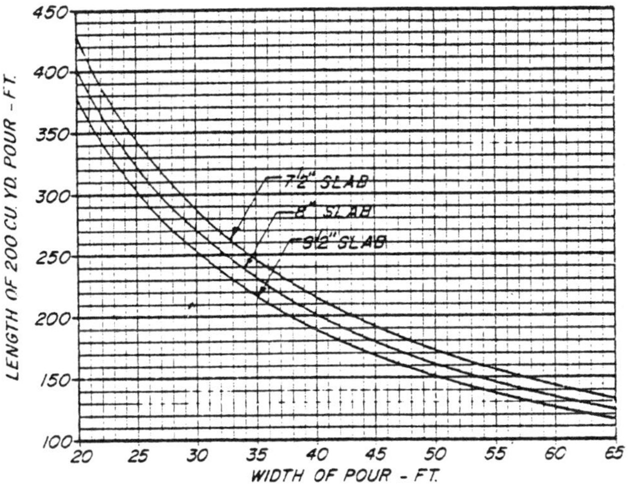


Transverse Construction Joints

If a deck pour is greater than 200 cu. yds. between the longitudinal joints or 300 cu. yds. when the longitudinal joints are omitted, an optional transverse construction joint shall be provided. The location of this transverse joint shall be near the point of dead load counterflexure with the day's pour terminating at the end of the positive moment area in the center span or spans.

On continuous bridges with the center span or spans exceeding 150 feet, the positive and negative moment areas shall be poured separately. The pouring sequence for these spans shall be determined on an individual basis.

**IL DOT**

# B.4 STAGE CONSTRUCTION

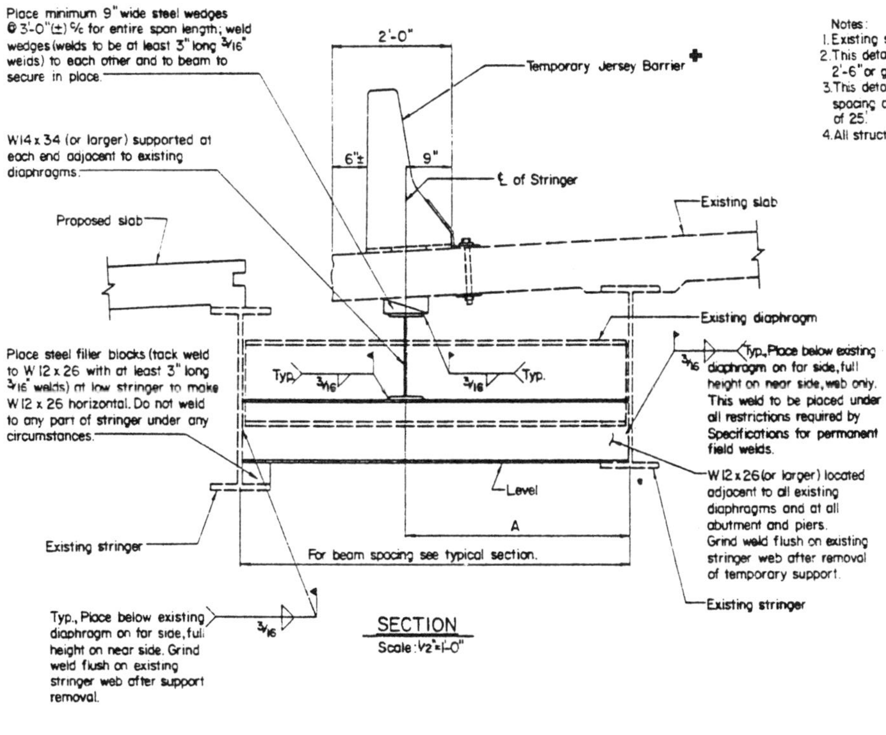


SECTION
Scale: ½"=1'-0"

Notes:
1. Existing structure shown in dashed lines.
2. This detail is only required where A dimension is 2'-6" or greater.
3. This detail can be used for maximum stringer spacing of 10' and maximum diaphragm spacing of 25'.
4. All structural steel to be ASTM A-36 or better.

**TEMPORARY SLAB UNDERPINNING**

**+For SINGLE FACE BARRIER**

METHODS OF ANCHORAGE CONNECTION TO CONCRETE DECKS

EXISTING BRIDGE DECK TO BE REMOVED.
Holes for anchor bolts in existing bridge deck shall be drilled. Use 1¼"Φ bolts with 5½" x 5½" x ¾" square washer under existing deck slab, as shown. Bolts shall be of sufficient length that when nut is tight, all the threads of the nut are engaged. Provide Type 'A' plain washer SAE N (narrow) for each 1¼"Φ bolt at connection plate.

EXISTING BRIDGE DECK TO REMAIN.
Holes for anchor bolts in existing bridge deck shall be cored. Use 1¼"Φ bolts with 5½" x 5½" x ¾" square washers under existing deck slab, as shown. Bolts shall be sufficient length that when nut is tight, all the threads of the nut are engaged. Provide Type 'A' plain washer SAE N (narrow) for each 1¼"Φ bolt at connection plate. The Contractor is alerted, that it is the intent, that as little damage as possible will be done to the reinforcing steel in the bridge deck. Therefore, the Contractor shall locate the reinforcing steel and space the bolts to miss the reinforcing steel, all as directed by the Engineer. Fill all cored holes with epoxy grout after barrier is removed. (See below for grout composition).

NEW BRIDGE DECK
1¼"Φ bolt to be placed in an epoxy coated open coil anchor insert (cast in slab) whose minimum working load tension strength is 16,000# and shear strength is 13,000# with a minimum length of 7½". Coil to be tapped for a 1¼" N.C. thread bolt. No insert shall be longer than slab depth, minus 1". Provide Type 'A' plain washer SAE N (narrow) for each 1¼"Φ bolt at connection plate. Fill all inserts with epoxy grout after barrier is removed. (See below for grout composition).

GROUT COMPOSITION

Any areas of bridge decks, to remain in place, damaged as a result of anchoring temporary concrete barriers (anchor holes, etc.) shall be repaired to the satisfaction of the Engineer using an epoxy grout. Epoxy grout shall consist of sand and epoxy, mixed by volume according to manufacturers recommendations. The epoxy grout shall be capable of developing a minimum compressive strength of 6,500 p.s.i. in 72 hours when tested in accordance with MSMT 501. Sand for epoxy grout shall conform to the Specifications Subsection 903.11.

## For DOUBLE FACE BARRIER

### METHODS OF ANCHORAGE CONNECTION TO CONCRETE DECKS

1. EXISTING BRIDGE DECK - TO BE REMOVED

Holes for anchor bolts in existing bridge deck shall be drilled. Use 3/4" ϕ bolts with 3½"x3½"x ½" square washer and 1¼" ϕ bolts with 6"x 6"x 3/4" square washers under existing deck slab, as indicated. Bolts shall be of sufficient length that when nut is tight all the threads of the nut are engaged. Provide standard washer for each 3/4" ϕ bolt and Type 'A' plain washer SAE N (narrow) for each 1¼" ϕ bolt at clip angles.

2. EXISTING BRIDGE DECK - TO REMAIN

Holes for anchor bolts in existing bridge deck shall be cored. Use 3/4" ϕ bolts with 3½"x 3½"x ½" square washers and 1¼" ϕ bolts with 6"x 6"x 3/4" square washers under existing deck slab, as indicated. Bolts shall be sufficient length that when nut is tight, all threads of the nut are engaged. Provide standard washer for each 3/4" ϕ bolt and Type 'A' plain washer SAE N(narrow) for each 1¼" ϕ bolt at clip angles. The Contractor is alerted, that it is the intent, that as little damage as possible will be done to the reinforcing steel. Therefore he shall locate the reinforcing steel and space the bolts to miss the reinforcing steel, as directed by the Engineer. Fill all cored holes with epoxy grout after barrier is removed. (See below for grout composition.)

3. NEW BRIDGE DECK

3/4" ϕ bolt to be placed in an epoxy coated open coil anchor insert whose minimum working load tension strength 6,000#, with a minimum length 4½". 1¼" ϕ bolt to be placed in an epoxy coated open coil anchor insert whose minimum working load tension strength is 17,000#, with a minimum length of 7½". Coil to be tapped for a 1¼" N.C. thread bolt. No insert shall be longer than slab depth, minus 1" inch. Provide standard washer for each 3/4" ϕ bolt and Type 'A' plain washer SAE N (narrow) for each 1¼" ϕ bolt at clip angles. Fill all inserts with epoxy grout after barrier is removed. (See below for grout composition).

GROUT COMPOSITION

Any areas of bridge decks, to remain in place, damaged as a result of anchoring temporary steel barriers (anchor holes, etc.) shall be repaired to the satisfaction of the Engineer using an epoxy grout. Epoxy grout shall consist of sand and epoxy, mixed by volume according to manufacturer's recommendations. The epoxy grout shall be capable of developing a minimum compressive strength of 6,500 p.s.i. in 72 hours when tested in accordance with MSMT 501. Sand for epoxy grout shall conform to the Specifications Subsection 903.11.

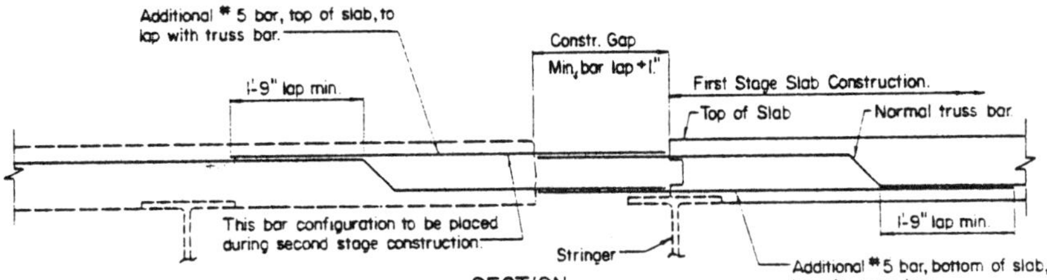


Note:
Straight bars, not shown, to be lapped in construction gap area.

SECTION
TYPE I SPLICE

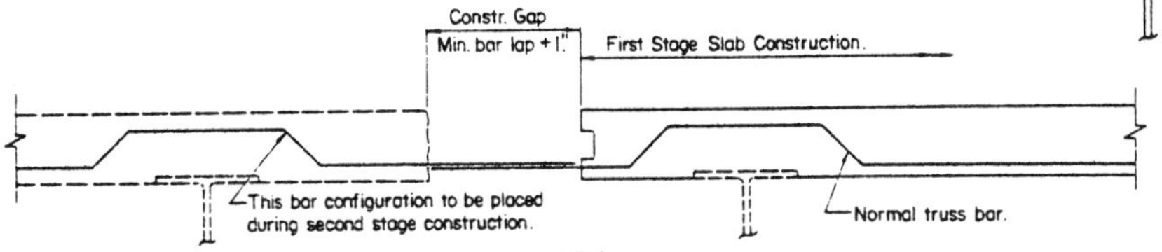


Note:
Straight bars, not shown, to be lapped in construction gap area.

SECTION
TYPE II SPLICE

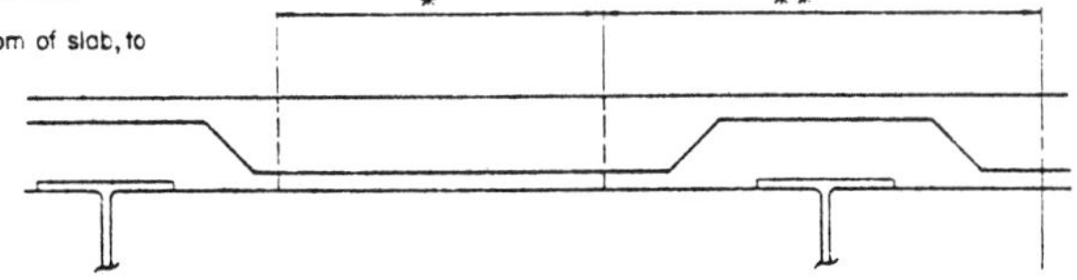


* If construction gap is located within these limits use Type II Splice.

** If construction gap is located within these limits use Type I Splice.

LAYOUT SECTION
Scale: ½"=1'-0"

### SPLICING DECK SLAB BARS

**MD DOT**

Note:
1. These additional bars are required in all areas where spirals or studs exist and top of spiral or stud does not extend into hatched area.
2. These bars to be epoxy coated.

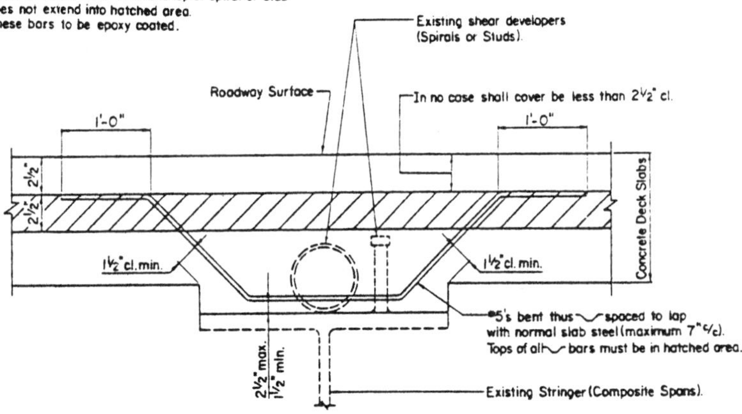


### ADDITIONAL REBARS AT EXISTING SHEAR CONNECTORS

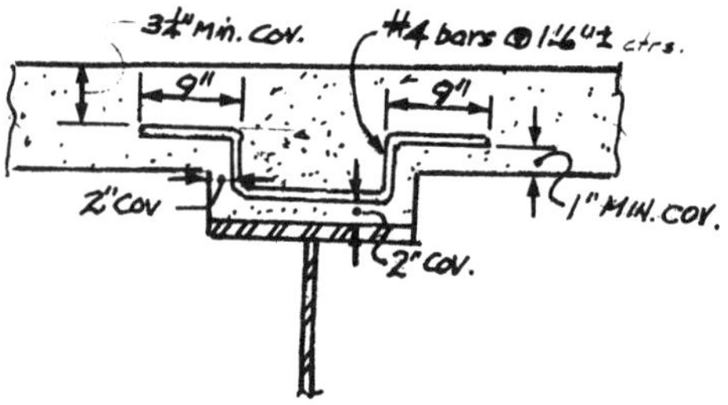


### ADDITIONAL REBARS AT DEEP HAUNCH

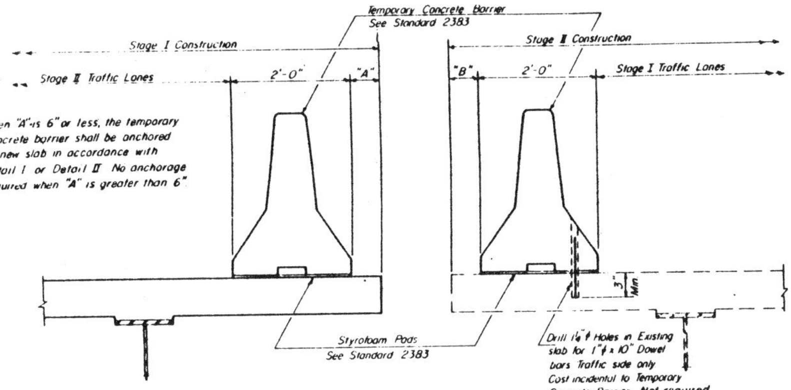


NEW SLAB

SECTIONS THRU SLAB

EXISTING SLAB

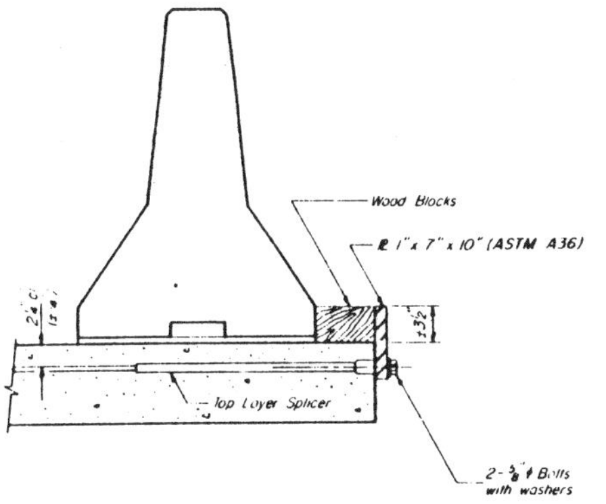


DETAIL I

The 1" x 7" x 10" Plate shall not be removed until Stage II Construction forms and reinforcement bars are in place

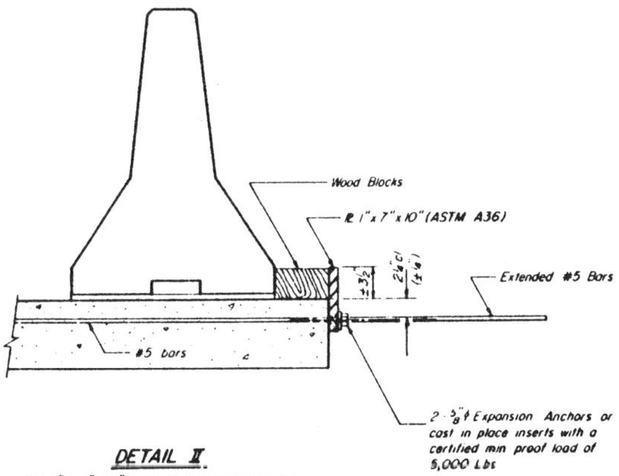


DETAIL II

The 1" x 7" x 10" Plate shall not be removed until Stage II Construction forms and all reinforcement bars are in place and the concrete is ready to be placed

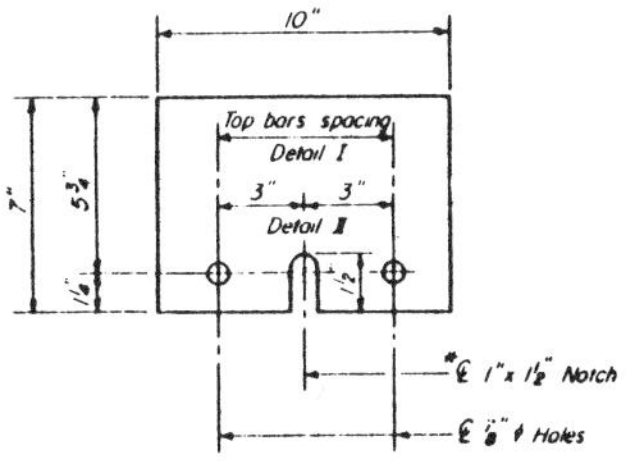


PL 1" x 7" x 10"

* Required only with Detail II

## NOTES

Detail I - With Bar Splicer or Couplers
Connect one (1) 1" x 7" x 10" steel PL to the top layer of couplers with 2-⅝" ϕ bolts screwed to coupler at approximate ℄ of each 10'-0" barrier panel

Detail II - With Extended Reinforcement Bars
Connect one (1) 1" x 7" x 10" steel PL to the concrete slab with 2-⅝" ϕ Expansion Anchors or cast in place inserts spaced between the top layer of reinforcement at approximate ℄ of each 10'-0" barrier panel

Cost of anchorage is incidental to Temporary Concrete Barrier.

| |
|---|
| For Design Only<br>"A"-Anchorage of Temporary Concrete Barrier to new slabs<br>a - When greater than 6" no need to anchor barrier<br>b - When 6" or less Temporary Concrete Barrier shall be anchored to new deck slab<br>"B"-Anchorage of Temporary Concrete Barrier to existing slabs<br>a - When greater than 1'-0" no need to anchor barrier<br>b - When 1'-0" or less Temporary Concrete Barrier shall be anchored to existing slab<br>Check the top reinforcement of slab for cantilever action (if applicable) due to dead loads and wheel loads only |

Design Notes will not appear in the contract plans

TEMPORARY CONCRETE BARRIER FOR STAGE CONSTRUCTION

STATE OF ILLINOIS
DEPARTMENT OF TRANSPORTATION

# STAGE CONSTRUCTION

# B.5 STAY IN PLACE FORMS

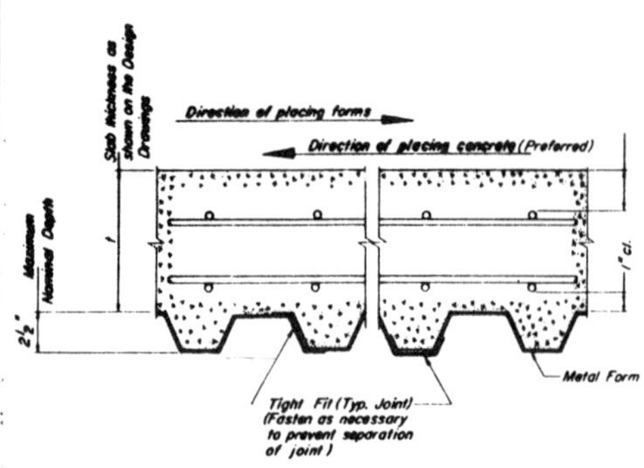


TYPICAL LONGITUDINAL SECTION

The configuration of metal form shall be such that the volume of concrete in the valleys will not exceed the equivalent of one inch of slab thickness.

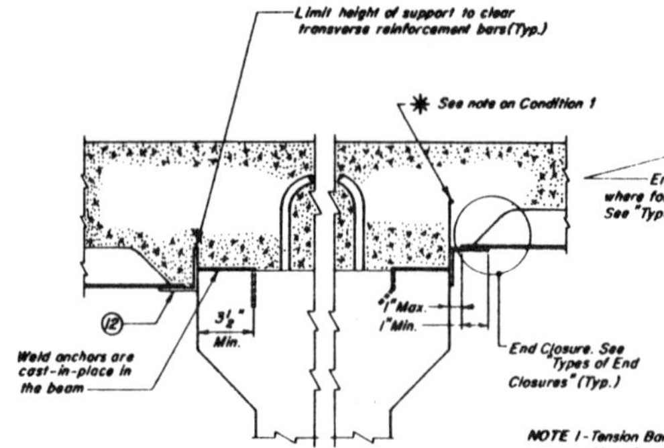


TYPICAL SUPPORT DETAIL

Detail shown for P/S I-Beam Bridge, detail similar for P/S Box Beam Bridge.

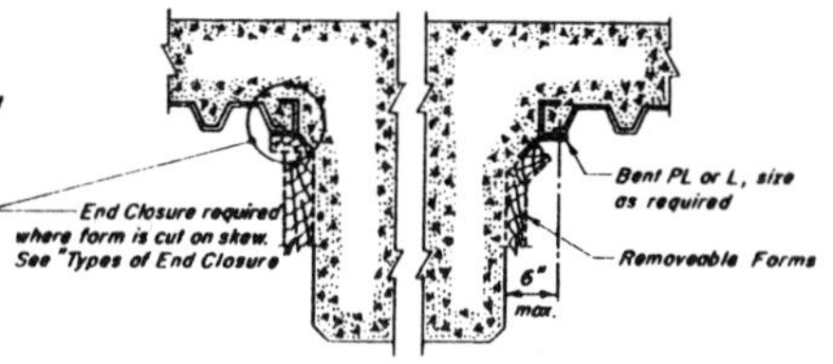


DETAIL AT CONCRETE DIAPHRAGM

NOTE 1 - Tension Bar, Angle or Channel, Size/Spacing as required. (Min. Bar 2" x 3/16" Max. Spacing 15", Angle or Channel Gage (12) Maximum. Spacing 24"). Welding Typical Each Side Unless Noted.

NOTE 2 - Haunch Angle or Channel, Size/Spacing as required. (Min. Gage (12), Max. Spacing 24"). Welding Typical Each Side Unless Noted.

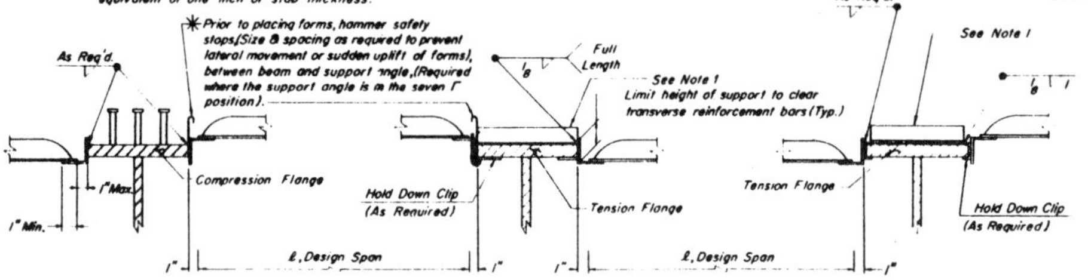


COMPOSITE DESIGN

NON-COMPOSITE DESIGN

CONDITION I

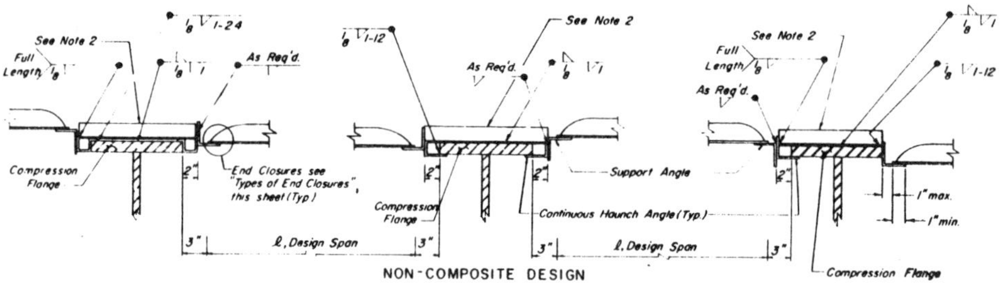


SUGGESTED SUPPORT DETAILS

FOR STEEL BEAM SYSTEMS, STEEL GIRDER SYSTEMS, GIRDER-FLOOR BEAM SYSTEMS AND GIRDER-FLOOR BEAM-STRINGER SYSTEMS

NON-COMPOSITE DESIGN

CONDITION 2

In non-composite compression flanges the haunch angle may be eliminated whenever the bottom of the metal deck form is at or below the bottom of the top flange.

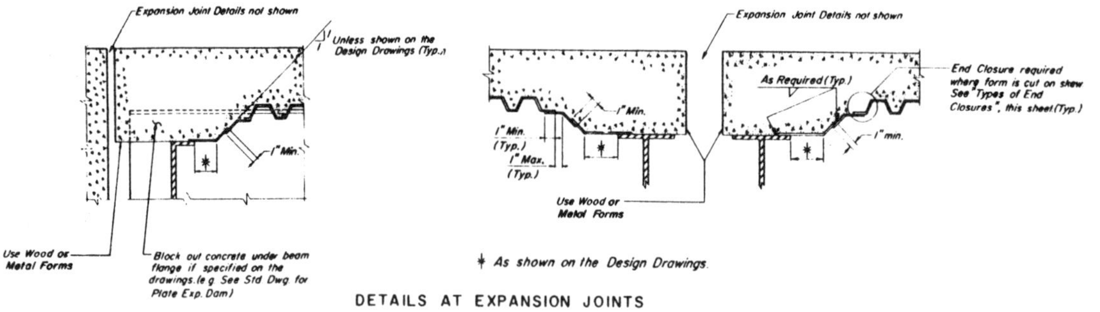


✱ As shown on the Design Drawings.

DETAILS AT EXPANSION JOINTS

*NOTES*

Materials and workmanship shall be in accordance with Specifications Forms 408 and 409.

This standard shall be used as a guide in the preparation of Shop Detail Drawings.

Gage and size of supporting elements and forms shall be shown on the shop drawings along with length, size and spacing of welds.

All form supports and their attachment shall be designed to carry dead load of deck slab (includes concrete in corrugations) plus 50 lbs per square foot for construction loads.

All forms shall be securely fastened to form supports and shall have a minimum bearing length of 1" at each end.

Form sheets shall be attached promptly to avoid hazards that can result from lateral movement or sudden uplift. Safety stops shall be provided where necessary.

Transverse deck construction joints shall be located at the bottom of a flute, and ¼"ϕ weep holes shall be field drilled at not less than 12" on center along the line of the joint.

When portions of the bridge deck are constructed adjacent to each other but are divided by an open or a preformed joint, metal forms in the cantilever area on either side of the joint may be used only with the permission of the Chief Bridge Engineer.

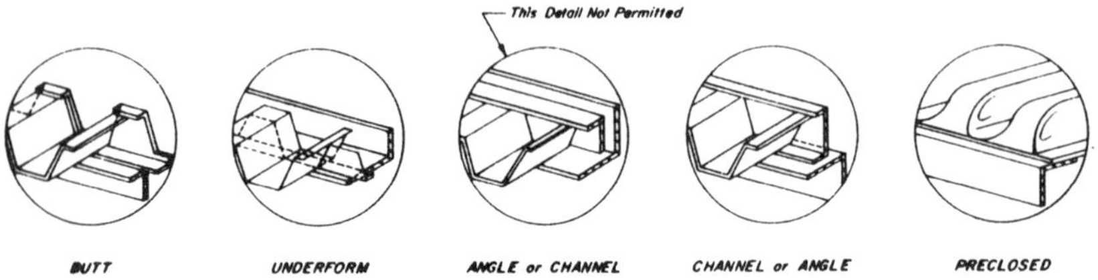


TYPES OF END CLOSURES

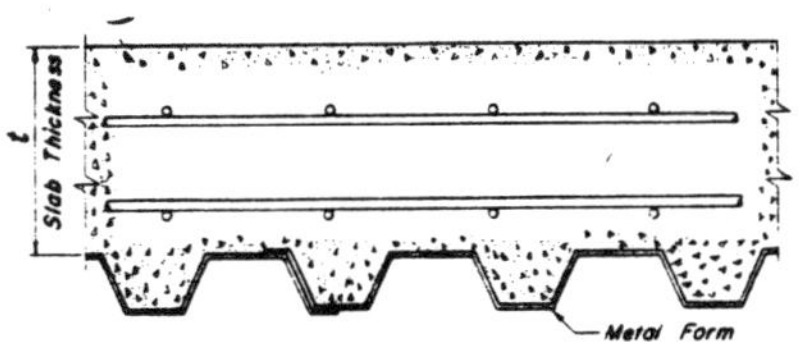


TYPICAL LONGITUDINAL SECTION OF DECK SLAB WITH METAL FORM

## DESIGN SPECIFICATIONS

*The metal forms shall be designed on the basis of dead load of the form, reinforcement and plastic concrete plus 50 pounds per square foot for construction loads. The unit working stress in the steel sheet shall be as follows:*

| Material Grade ASTM A446 | Yield (PSI) | Allowable Stress (PSI) |
|---|---|---|
| A | 33,000 | 23,900 |
| B | 37,000 | 26,800 |
| C | 40,000 | 29,000 |
| D | 50,000 | 36,000 |
| E | 80,000 | 36,000 |

*Deflection under the weight of the forms, plastic concrete and reinforcement shall not exceed 1/180 of the form span or 1/2 inch whichever is less, but in no case shall this loading be less than 120 PSF total. The permissible form camber shall be based on the actual dead load condition. Camber shall not be used to compensate for deflection in excess of the foregoing limits.*

## EXAMPLES

EXAMPLE # 1.

GIVEN: *Steel I-Beam Bridge, Composite Design*
*Beam Spacing = 7'-10" C/C*
*Beam Flange Width = 12"*
*Slab Thickness, t = 9" (2½" cl.)*

REQUIRED: *Section Modulus S, and Moment of Inertia I for Metal Form Section, Grade C Material.*

SOLUTION: *Design Span ℓ (from sheet 1 for Condition 1) =*
*(7'-10") – (12") – (2") = 6'-8"*
*From Table, for ℓ = 6'-8" and t = 9"*
*S = .3297 in.³/ft. and I = .4343 in.⁴/ft.*
*For Grade C Material, multiply S by 1.24*
*i.e. Sreqd. = 1.24 x .3297 = .4088 in.³/ft.*
*Ireqd. = .4343 in.⁴/ft.*

EXAMPLE # 2

GIVEN: *Steel I-Beam Bridge, Non-Composite Design*
*Beam Spacing = 7'-10" C/C*
*Beam Flange Width = 12"*
*Slab Thickness, t = 9" (2½" cl.)*

REQUIRED: *Section Modulus S, and Moment of Inertia I for Metal Form Section, Grade D Material, Support Detail as per Condition 2 as shown on sheet 1.*

SOLUTION: *Design Span ℓ (from sheet 1 for Condition 2) =*
*(7'-10") – (12") – (6") = 6'-4"*
*From Table, for ℓ = 6'-4" and t = 9"*
*S = .2976 in.³/ft. and I = .3724 in.⁴/ft.*
*For Grade D Material, multiply S by 1.00*
*i.e. Sreqd. = 1.00 x .2976 = .2976 in.³/ft.*
*Ireqd. = .3724 in.⁴/ft.*

## REQUIRED SECTION MODULUS AND MOMENT OF INERTIA OF FORMS

| S, Section Modulus In.³ Per Ft. * ★ | | | | | | | ⊗ ℓ, DESIGN SPAN | I, Moment of Inertia In.⁴ Per Ft. ★ | | | | |
|---|---|---|---|---|---|---|---|---|---|---|---|---|
| t, SLAB THICKNESS, INCHES | | | | | | | | t, SLAB THICKNESS, INCHES | | | | |
| 7 | 7½ | 8 | 8½ | 9 | 9½ | 10 | | ≤8 | 8½ | 9 | 9½ | 10 |
| .0574 | .0597 | .0621 | .0644 | .0667 | .0691 | .0714 | 3'-0" | .0371 | .0376 | .0395 | .0415 | .0434 |
| .0607 | .0632 | .0656 | .0681 | .0706 | .0730 | .0755 | 3'-1" | .0404 | .0410 | .0431 | .0452 | .0473 |
| .0640 | .0666 | .0693 | .0719 | .0745 | .0771 | .0797 | 3'-2" | .0438 | .0444 | .0467 | .0490 | .0512 |
| .0673 | .0701 | .0728 | .0756 | .0784 | .0811 | .0838 | 3'-3" | .0471 | .0478 | .0502 | .0527 | .0551 |
| .0709 | .0738 | .0767 | .0796 | .0825 | .0854 | .0883 | 3'-4" | .0510 | .0518 | .0544 | .0571 | .0597 |
| .0744 | .0775 | .0805 | .0836 | .0866 | .0897 | .0927 | 3'-5" | .0550 | .0558 | .0587 | .0615 | .0644 |
| .0780 | .0812 | .0844 | .0876 | .0908 | .0940 | .0972 | 3'-6" | .0589 | .0598 | .0629 | .0659 | .0690 |
| .0819 | .0853 | .0886 | .0919 | .0953 | .0986 | .1020 | 3'-7" | .0634 | .0644 | .0677 | .0710 | .0743 |
| .0858 | .0893 | .0928 | .0963 | .0998 | .1033 | .1068 | 3'-8" | .0679 | .0689 | .0725 | .0760 | .0795 |
| .0897 | .0934 | .0970 | .1006 | .1043 | .1080 | .1116 | 3'-9" | .0724 | .0735 | .0773 | .0810 | .0848 |
| .0938 | .0976 | .1014 | .1053 | .1091 | .1129 | .1167 | 3'-10" | .0776 | .0787 | .0828 | .0868 | .0908 |
| .0979 | .1019 | .1059 | .1099 | .1139 | .1179 | .1219 | 3'-11" | .0827 | .0840 | .0883 | .0926 | .0969 |
| .1020 | .1062 | .1103 | .1145 | .1187 | .1228 | .1270 | 4'-0" | .0879 | .0892 | .0938 | .0983 | .1029 |
| .1064 | .1108 | .1151 | .1194 | .1238 | .1282 | .1325 | 4'-1" | .0937 | .0951 | .1000 | .1048 | .1097 |
| .1108 | .1153 | .1198 | .1244 | .1289 | .1335 | .1380 | 4'-2" | .0996 | .1010 | .1062 | .1114 | .1166 |
| .1152 | .1199 | .1246 | .1293 | .1340 | .1387 | .1434 | 4'-3" | .1054 | .1069 | .1124 | .1179 | .1234 |
| .1198 | .1247 | .1296 | .1345 | .1394 | .1443 | .1492 | 4'-4" | .1120 | .1136 | .1194 | .1253 | .1311 |
| .1245 | .1296 | .1346 | .1397 | .1448 | .1498 | .1549 | 4'-5" | .1185 | .1202 | .1264 | .1326 | .1388 |
| .1291 | .1344 | .1396 | .1449 | .1502 | .1554 | .1607 | 4'-6" | .1251 | .1269 | .1334 | .1400 | .1465 |
| .1340 | .1395 | .1449 | .1504 | .1559 | .1613 | .1668 | 4'-7" | .1324 | .1343 | .1412 | .1482 | .1551 |
| .1390 | .1447 | .1503 | .1560 | .1616 | .1673 | .1730 | 4'-8" | .1398 | .1418 | .1491 | .1563 | .1636 |
| .1439 | .1498 | .1556 | .1615 | .1673 | .1732 | .1791 | 4'-9" | .1471 | .1492 | .1569 | .1645 | .1722 |
| .1491 | .1552 | .1612 | .1673 | .1733 | .1794 | .1855 | 4'-10" | .1553 | .1575 | .1656 | .1737 | .1818 |
| .1542 | .1605 | .1668 | .1731 | .1794 | .1857 | .1920 | 4'-11" | .1634 | .1658 | .1743 | .1828 | .1913 |
| .1594 | .1659 | .1724 | .1789 | .1854 | .1919 | .1984 | 5'-0" | .1716 | .1741 | .1830 | .1920 | .2009 |
| .1648 | .1716 | .1783 | .1851 | .1917 | .1985 | .2052 | 5'-1" | .1806 | .1833 | .1926 | .2021 | .2115 |
| .1702 | .1772 | .1842 | .1912 | .1981 | .2051 | .2120 | 5'-2" | .1897 | .1924 | .2023 | .2122 | .2220 |
| .1757 | .1829 | .1901 | .1972 | .2044 | .2116 | .2188 | 5'-3" | .1987 | .2016 | .2119 | .2223 | .2326 |
| .1814 | .1888 | .1963 | .2036 | .2111 | .2185 | .2259 | 5'-4" | .2086 | .2116 | .2225 | .2334 | .2442 |
| .1871 | .1947 | .2024 | .2100 | .2177 | .2254 | .2330 | 5'-5" | .2185 | .2217 | .2330 | .2444 | .2558 |
| .1928 | .2007 | .2086 | .2164 | .2243 | .2323 | .2401 | 5'-6" | .2284 | .2317 | .2436 | .2555 | .2674 |
| .1988 | .2072 | .2156 | .2240 | .2324 | .2408 | .2492 | 5'-7" | .2393 | .2427 | .2552 | .2677 | .2801 |
| .2048 | .2137 | .2226 | .2315 | .2404 | .2494 | .2583 | 5'-8" | .2501 | .2538 | .2668 | .2798 | .2929 |
| .2108 | .2202 | .2297 | .2391 | .2485 | .2580 | .2674 | 5'-9" | .2610 | .2648 | .2784 | .2920 | .3056 |
| .2170 | .2264 | .2359 | .2453 | .2547 | .2641 | .2735 | 5'-10" | .2729 | .2768 | .2911 | .3053 | .3195 |
| .2232 | .2326 | .2420 | .2514 | .2608 | .2703 | .2797 | 5'-11" | .2847 | .2889 | .3037 | .3185 | .3334 |
| .2294 | .2388 | .2482 | .2576 | .2670 | .2764 | .2858 | 6'-0" | .2966 | .3009 | .3164 | .3318 | .3473 |
| .2359 | .2456 | .2553 | .2649 | .2746 | .2842 | .2939 | 6'-1" | .3095 | .3140 | .3301 | .3462 | .3624 |
| .2425 | .2524 | .2623 | .2722 | .2821 | .2921 | .3020 | 6'-2" | .3223 | .3270 | .3439 | .3606 | .3774 |
| .2490 | .2592 | .2694 | .2795 | .2897 | .2999 | .3101 | 6'-3" | .3352 | .3401 | .3576 | .3750 | .3925 |
| .2558 | .2663 | .2767 | .2871 | .2976 | .3081 | .3185 | 6'-4" | .3491 | .3542 | .3724 | .3906 | .4088 |
| .2626 | .2733 | .2841 | .2948 | .3055 | .3163 | .3270 | 6'-5" | .3631 | .3684 | .3873 | .4062 | .4251 |
| .2694 | .2804 | .2914 | .3024 | .3134 | .3244 | .3354 | 6'-6" | .3770 | .3825 | .4021 | .4218 | .4414 |
| .2764 | .2877 | .2990 | .3103 | .3216 | .3329 | .3442 | 6'-7" | .3921 | .3978 | .4182 | .4386 | .4590 |
| .2834 | .2950 | .3066 | .3181 | .3297 | .3413 | .3529 | 6'-8" | .4071 | .4131 | .4343 | .4555 | .4767 |
| .2904 | .3023 | .3142 | .3260 | .3379 | .3499 | .3617 | 6'-9" | .4222 | .4284 | .4504 | .4723 | .4943 |
| .2977 | .3099 | .3221 | .3342 | .3464 | .3586 | .3708 | 6'-10" | .4384 | .4449 | .4677 | .4905 | .5133 |
| .3051 | .3176 | .3300 | .3425 | .3549 | .3674 | .3798 | 6'-11" | .4546 | .4613 | .4850 | .5086 | .5323 |
| .3124 | .3252 | .3379 | .3507 | .3634 | .3762 | .3889 | 7'-0" | .4709 | .4778 | .5023 | .5268 | .5513 |
| .3200 | .3331 | .3461 | .3592 | .3722 | .3853 | .3983 | 7'-1" | .4883 | .4955 | .5209 | .5463 | .5717 |
| .3275 | .3409 | .3543 | .3676 | .3810 | .3944 | .4078 | 7'-2" | .5058 | .5131 | .5395 | .5658 | .5922 |
| .3351 | .3488 | .3625 | .3761 | .3898 | .4035 | .4172 | 7'-3" | .5232 | .5308 | .5581 | .5853 | .6126 |
| .3429 | .3570 | .3710 | .3849 | .3989 | .4130 | .4270 | 7'-4" | .5419 | .5497 | .5780 | .6062 | .6344 |
| .3508 | .3651 | .3794 | .3938 | .4081 | .4224 | .4367 | 7'-5" | .5605 | .5687 | .5979 | .6270 | .6563 |
| .3586 | .3733 | .3879 | .4026 | .4172 | .4319 | .4465 | 7'-6" | .5792 | .5876 | .6178 | .6479 | .6781 |
| .3667 | .3817 | .3966 | .4117 | .4266 | .4416 | .4566 | 7'-7" | .6063 | .6151 | .6467 | .6782 | .7098 |
| .3747 | .3901 | .4053 | .4207 | .4360 | .4514 | .4666 | 7'-8" | .6333 | .6425 | .6755 | .7085 | .7415 |
| .3828 | .3985 | .4141 | .4298 | .4454 | .4611 | .4767 | 7'-9" | .6604 | .6700 | .7044 | .7388 | .7732 |
| .3912 | .4072 | .4232 | .4392 | .4551 | .4711 | .4871 | 7'-10" | .6902 | .7002 | .7362 | .7721 | .8081 |
| .3996 | .4159 | .4322 | .4486 | .4649 | .4813 | .4976 | 7'-11" | .7200 | .7305 | .7680 | .8055 | .8430 |
| .4080 | .4247 | .4413 | .4580 | .4746 | .4913 | .5080 | 8'-0" | .7498 | .7607 | .7998 | .8388 | .8779 |
| .4166 | .4336 | .4506 | .4677 | .4847 | .5017 | .5187 | 8'-1" | .7825 | .7939 | .8347 | .8754 | .9162 |
| .4253 | .4427 | .4600 | .4774 | .4948 | .5121 | .5295 | 8'-2" | .8153 | .8272 | .8697 | .9121 | .9546 |
| .4339 | .4516 | .4693 | .4871 | .5048 | .5225 | .5402 | 8'-3" | .8480 | .8604 | .9046 | .9487 | .9929 |
| .4428 | .4609 | .4790 | .4971 | .5151 | .5332 | .5513 | 8'-4" | .8839 | .8968 | .9428 | .9888 | 1.0348 |
| .4517 | .4702 | .4886 | .5071 | .5255 | .5440 | .5624 | 8'-5" | .9197 | .9331 | .9810 | 1.0289 | 1.0768 |
| .4606 | .4794 | .4982 | .5171 | .5359 | .5547 | .5735 | 8'-6" | .9556 | .9695 | 1.0193 | 1.0690 | 1.1188 |
| .4698 | .4890 | .5082 | .5274 | .5465 | .5657 | .5849 | 8'-7" | .9947 | 1.0092 | 1.0610 | 1.1128 | 1.1646 |
| .4789 | .4985 | .5180 | .5376 | .5572 | .5767 | .5963 | 8'-8" | 1.0339 | 1.0489 | 1.1028 | 1.1566 | 1.2105 |
| .4881 | .5080 | .5280 | .5479 | .5678 | .5878 | .6077 | 8'-9" | 1.0730 | 1.0886 | 1.1445 | 1.2004 | 1.2563 |
| .4975 | .5178 | .5381 | .5585 | .5788 | .5991 | .6194 | 8'-10" | 1.1157 | 1.1319 | 1.1900 | 1.2482 | 1.3063 |
| .5070 | .5277 | .5484 | .5691 | .5898 | .6105 | .6312 | 8'-11" | 1.1583 | 1.1752 | 1.2356 | 1.2959 | 1.3562 |
| .5164 | .5375 | .5586 | .5797 | .6007 | .6218 | .6429 | 9'-0" | 1.2010 | 1.2185 | 1.2811 | 1.3436 | 1.4062 |
| .5261 | .5476 | .5691 | .5906 | .6120 | .6335 | .6550 | 9'-1" | 1.2474 | 1.2656 | 1.3306 | 1.3956 | 1.4606 |
| .5358 | .5577 | .5796 | .6015 | .6233 | .6452 | .6671 | 9'-2" | 1.2938 | 1.3126 | 1.3801 | 1.4475 | 1.5149 |
| .5455 | .5678 | .5901 | .6124 | .6346 | .6569 | .6792 | 9'-3" | 1.3402 | 1.3597 | 1.4296 | 1.4994 | 1.5693 |
| .5555 | .5782 | .6009 | .6236 | .6462 | .6689 | .6916 | 9'-4" | 1.3905 | 1.4107 | 1.4832 | 1.5556 | 1.6281 |
| .5654 | .5885 | .6116 | .6347 | .6578 | .6809 | .7040 | 9'-5" | 1.4407 | 1.4617 | 1.5368 | 1.6118 | 1.6869 |
| .5754 | .5989 | .6224 | .6459 | .6694 | .6929 | .7164 | 9'-6" | 1.4910 | 1.5127 | 1.5904 | 1.6680 | 1.7457 |
| .5856 | .6095 | .6334 | .6574 | .6813 | .7052 | .7291 | 9'-7" | 1.5454 | 1.5679 | 1.6485 | 1.7289 | 1.8094 |
| .5958 | .6202 | .6445 | .6689 | .6932 | .7176 | .7419 | 9'-8" | 1.5999 | 1.6232 | 1.7065 | 1.7898 | 1.8732 |
| .6060 | .6308 | .6555 | .6803 | .7051 | .7298 | .7546 | 9'-9" | 1.6543 | 1.6784 | 1.7646 | 1.8507 | 1.9369 |
| .6165 | .6417 | .6669 | .6921 | .7173 | .7425 | .7677 | 9'-10" | 1.7131 | 1.7380 | 1.8273 | 1.9165 | 2.0057 |
| .6270 | .6526 | .6782 | .7039 | .7295 | .7551 | .7807 | 9'-11" | 1.7718 | 1.7977 | 1.8900 | 1.9822 | 2.0745 |
| .6375 | .6636 | .6896 | .7157 | .7417 | .7678 | .7938 | 10'-0" | 1.8306 | 1.8573 | 1.9526 | 2.0480 | 2.1433 |

* *Multiply values of S shown in table by:*
*1.00 for grades D and E*
*1.24 for grade C*
*1.34 for grade B*
*1.51 for grade A*

⊗ ℓ = *Design span measured parallel to the form flute, for steel structures as shown on sheet 1, and for (prestressed) concrete structures design span shall be taken as clear distance between (the flanges of) the supporting beams minus two inches.*

★ — *Physical design properties shall be computed in accordance with requirements of the American Iron and Steel Institute Specification for the Design of Cold Formed Steel Structural Members, latest published edition.*

Commonwealth of Pennsylvania
DEPARTMENT OF TRANSPORTATION
BUREAU OF HIGHWAY DESIGN

STANDARD
PERMANENT METAL DECK FORMS

## B.6 PRECAST DECK

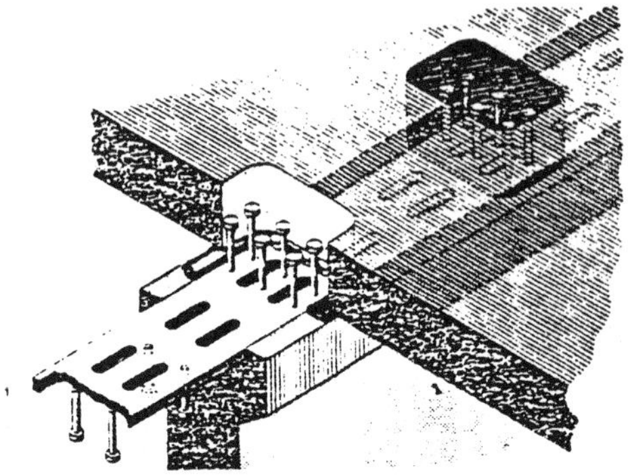

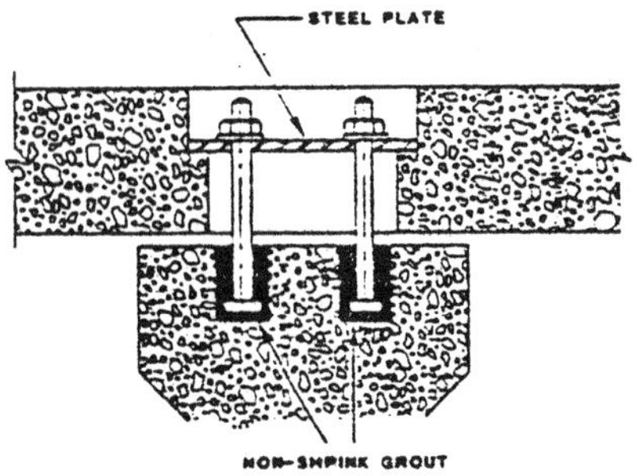


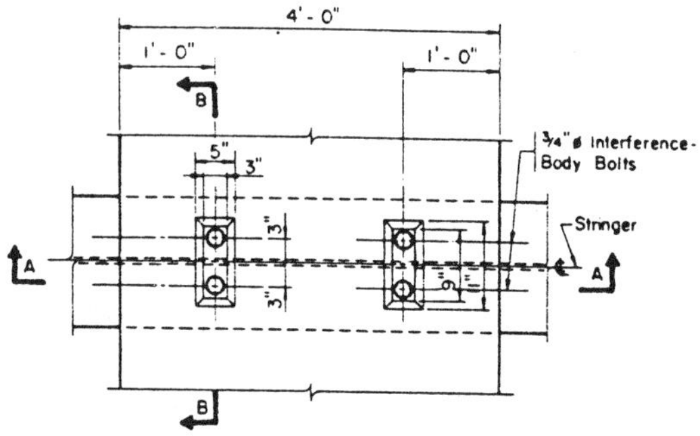


PLAN

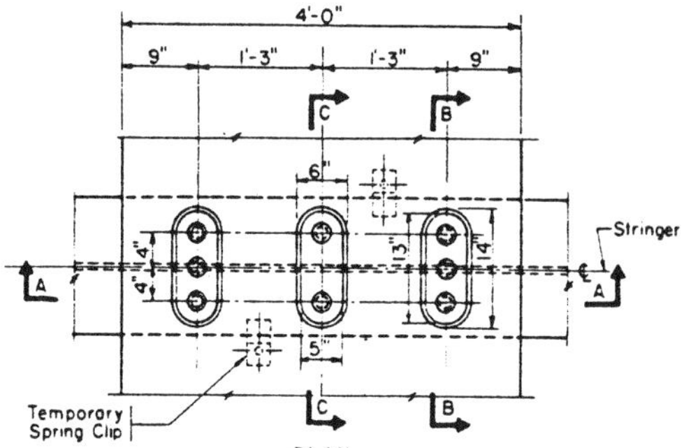


PLAN

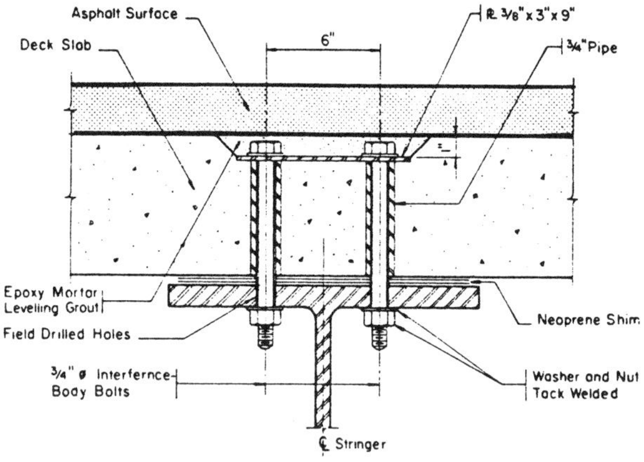


SECTION B-B

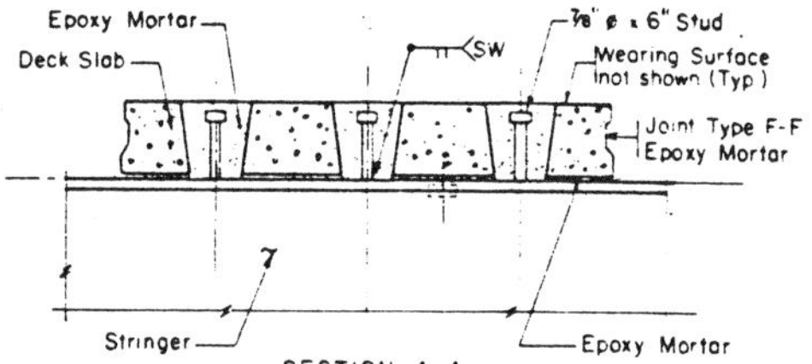


SECTION A-A

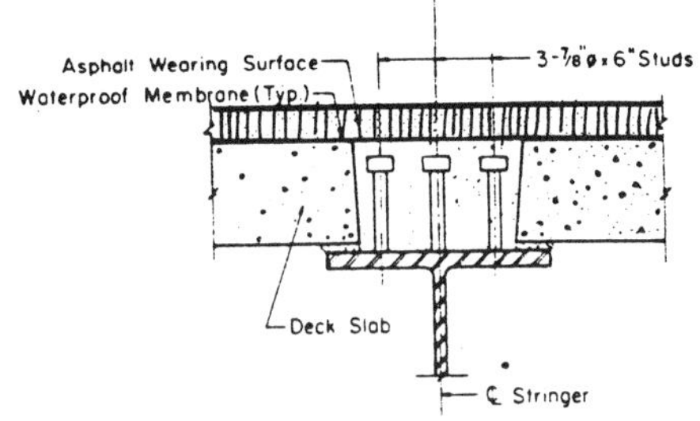


SECTION B-B

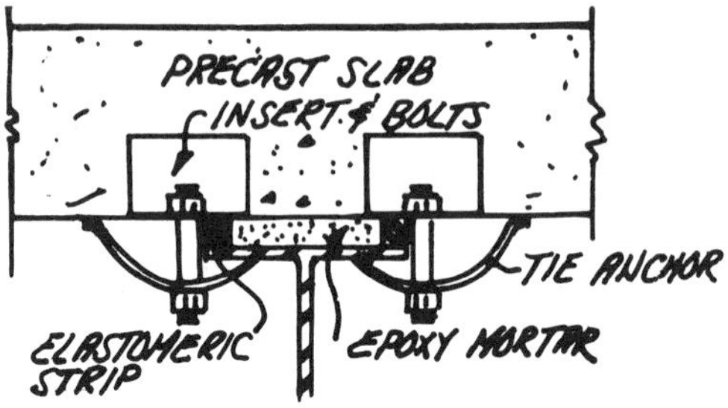


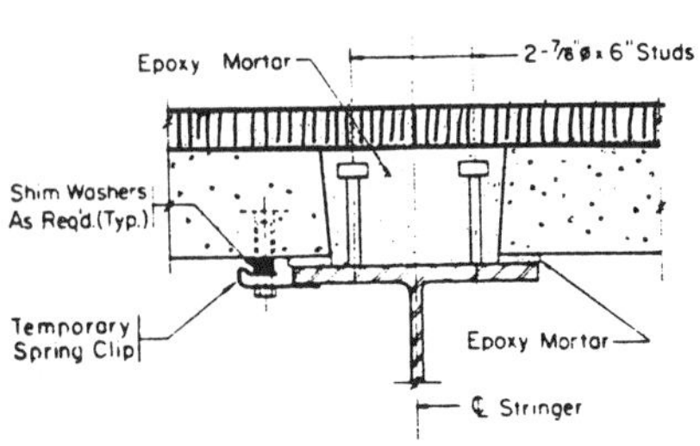


SECTION C-C

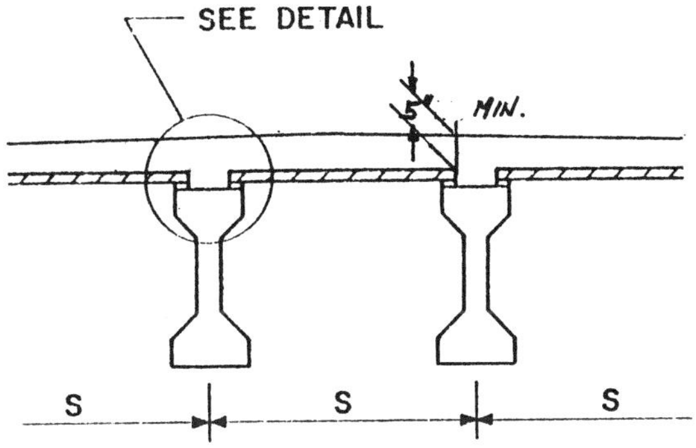


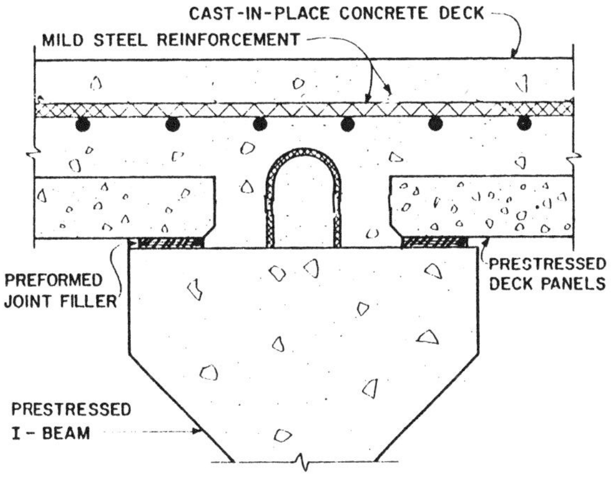


DETAIL

HS20 Loading    8'-0" Beam Spacing

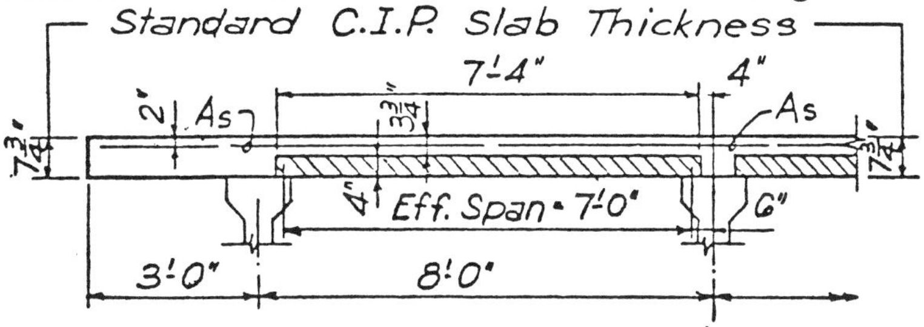


Try Standard Panels
4" Thick
$f'_{ci} = 4000$ p.s.i.
$f'_c = 5000$ p.s.i.
$\frac{3}{8}$"φ 270K Strands @ 6" c.c.
Cast-in-Place
$f'_c = 3000$ p.s.i.
$f_y = 60$ k.s.i.

$DLM = \frac{1}{8} \times .0969 \times 7^2 = 0.59^{K'}$

$LLM = \frac{.8(7+2)}{32} 16 \times 1.3 = \pm 4.68^{K'}$

Panels:

DL $S = \frac{12 \times 4^2}{6} = 32 \text{ in}^3$

DL $f_c = \frac{.59 \times 12000}{32} = -221$ p.s.i.

LL $S = \frac{12 \times 7.75^2}{6} = 120 \text{ in}^3$

LL $f_c = \frac{4.68 \times 12000}{120} = -468$ p.s.i.

Design Total $f_c = -689$ p.s.i

Prestress $= \frac{16{,}100 \times .8 \times 2}{4 \times 12} = +538$ p.s.i.

Final Panel $f_c = -151$ p.s.i.

Allow. $= 6\sqrt{5000} = -424$ O.K.

Cast-in-Place Slab:

Req'd d (Overhang) $= \sqrt{\frac{4.68}{.197}} = 4.87"$
$\underline{2.}$
$6.87"$

Provided 7.75" O.K.

$A_s = \frac{4.68 \times 12}{20 \times .875 \times 5.75} = .558 \square"$

#5 @ $6\frac{1}{2}$" c.c. O.K.

Check Ultimate Moment: +M

$f_{su}^* = f'_s\left(1 - \frac{.5 p^* f'_s}{f'_c}\right)$ AASHTO 1.6.9(C)

$= 270\left(1 - \frac{.5 \times .0854 \times 2 \times 270}{5.75 \times 12 \times 3}\right) = 240$ ksi

or

$f_{su}^* = \frac{L}{D} + \frac{2 f_{se}}{3}$ AASHTO 1.6.18

$= \frac{44}{.375} + \frac{2 \times .7 \times 270 \times .8}{3} = 218$ k.s.i. Use

$M_u = A_s^* f_{su}^* d\left(1 - \frac{.6 p^* f_{su}^*}{f'_c}\right)\phi$ AASHTO 1.6.9(A)

$= 2 \times .0854 \times 218 \times \frac{5.75}{12}\left(1 - \frac{.6 \times .0854 \times 2 \times 218}{5.75 \times 12 \times 3}\right).9$

$= 14.3^{K'}$

Check Ultimate Moment: −M

$M_u = A_s f_y d\left(1 - \frac{.6 \rho f_y}{f'_c}\right)\phi$ AASHTO 1.5.32(B)

$= .567 \times 60 \times \frac{5.75}{12}\left(1 - \frac{.6 \times .567 \times 60}{5.75 \times 12 \times 3}\right).9$

$= 13.2^{K'}$

Req'd $M_u = 1.3(DLM + \frac{5}{3} LLM)$
$= 1.3(.59 + 1.67 \times 4.68) = 10.9^{K'}$
O.K.

Standard Panels Adequate

● STAY-IN-PLACE FORMS

# B.7 PRECAST CONCRETE BRIDGE DECK

*Reproduced from FHWA/RD -82/106 CONNECTIONS FOR MODULAR PRECAST CONCRETE BRIDGE DECKS, Aug. 1982*

## 2.1 Description

In this type of construction thin, precast, prestressed concrete members are placed between the prestressed concrete or steel girders. A cast-in-place, reinforced concrete deck is then poured on top of these members (See Fig. 1). In most cases the precast members are designed to act integrally with the cast-in-place concrete and compositely with the girders. Numerous uses of these slabs have been reported in the literature[1,2]*, and the Prestressed Concrete Institute has published design and construction specifications[3]. Several variations of the slabs have been used. In most cases the top surface is intentionally roughened (to varying degrees). Sometimes metal ties, reinforcing bars, wire, etc. are projected from the surface of the precast slab. In other cases a machine made slab is used that has transverse (to direction of traffic) ribs as shown in Fig. 2a. In the machine made sections strand normally does not project from the end of the member. Some state specifications require a reinforcing bar to be placed between adjacent slabs when the strand does not project, as shown in Fig. 2b.

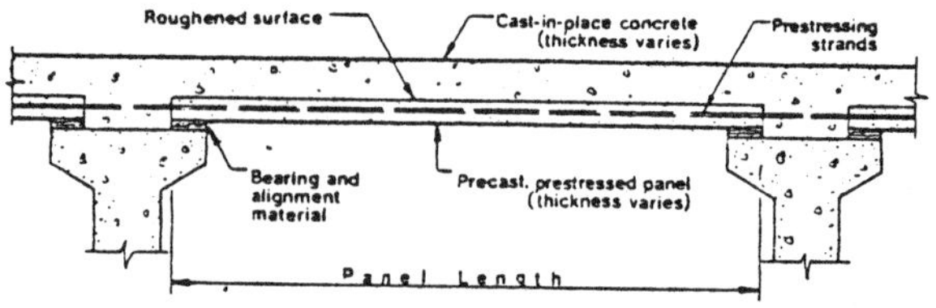


**Fig. 1.** **Typical construction using stay-in-place forms.**

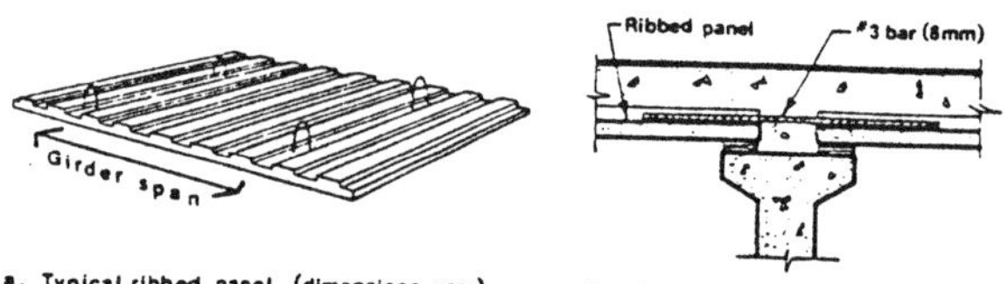


a. Typical ribbed panel (dimensions vary)  b. Construction using Continuity bars

**Fig. 2.** **Stay-in-place construction using machine-made panels.**

## 2.3 Evaluation and Conclusions

The stay-in-place form as a practical and economical means of construction of bridge decks appears to have proven itself, both in terms of longevity, considering the Beverly Road Bridge, and frequency of use. There has also been enough controlled research on this subject to verify that the system will perform adequately on high volume, high traffic bridges and will withstand significant overloads adequately.

The only problems that have been reported were minor cracking in the deck surface. In most cases, the cracks appeared to coincide with the joints between panels, or longitudinally along the girders at the ends of the panels. Since this cracking was sometimes noticed before the bridge was open to traffic, it seems likely that it is a result of shrinkage of the cast-in-place deck, rather than flexure. It was generally considered no more severe than normally found in decks formed conventionally.

The Beverly Road tests showed that a 2-1/2 in. precast slab was adequate. This would be more economical than the 3 to 4 in. slabs used elsewhere. Furthermore, the joints in the precast slabs act as contraction joints. By reducing the ratio of precast to cast-in-place slab thickness, it is likely that deck cracking would be reduced.

In addition to slab thickness, there are differences of opinion as to the value of projecting reinforcement from the face and projecting the strand from the ends of the slab. If these steel projections could be eliminated, it would open the way for more economical machine processes for manufacturing the slabs. Recent research by the Texas Department of Highways indicates equivalent performance with and without the steel projections. Various materials have been used for bearing of the deck unit to the girder, including mortar, fiberboard strips and neoprene, with no apparent difference in performance.

The Pennsylvania State University Research Program resulted in an "Implementation Statement" by the researchers for the Pennsylvania Department of Transportation[8], which details design and construction procedures. With modifications discussed above, this statement can serve as a model guide specification for the use of this construction method.

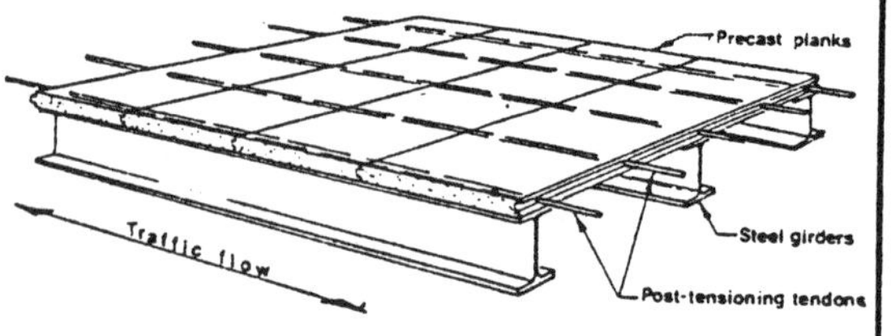


**Fig. 5.** **Schematic of Purdue test bridges.**

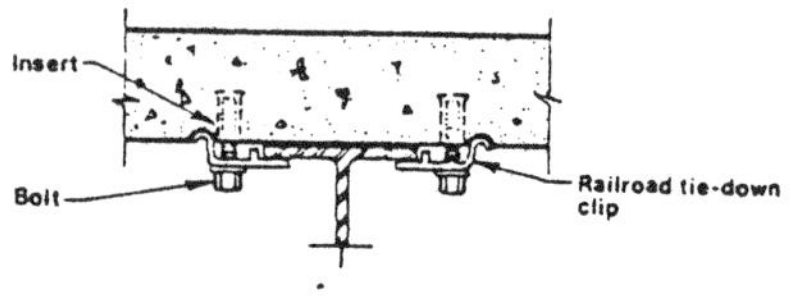


**Fig. 8.** **Tie-down connection used on Indiana experimental bridges.**

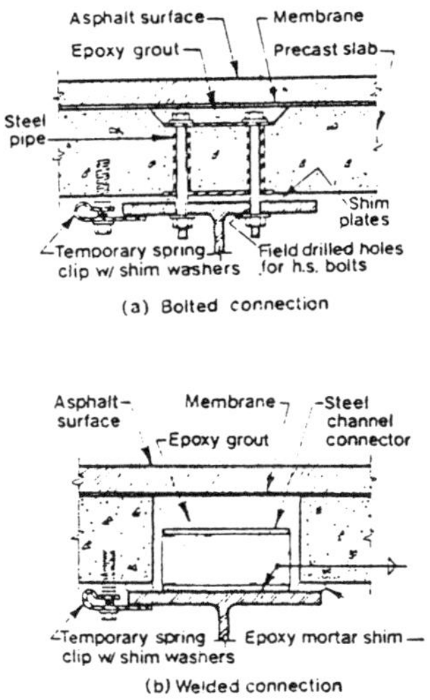


Fig. 13. Connections used on New York Thruway

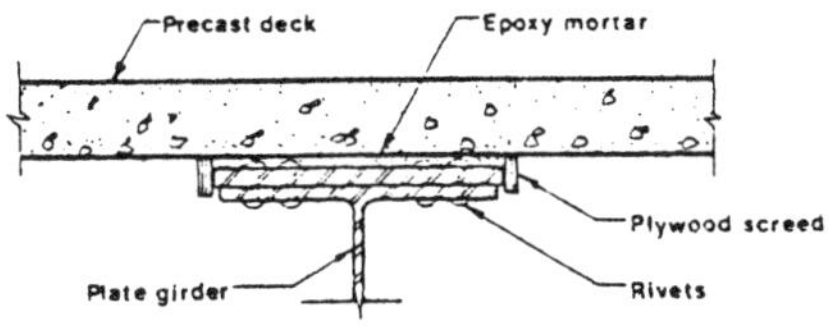


Fig. 23. Connection detail, Santa Fe Railroad bridges.

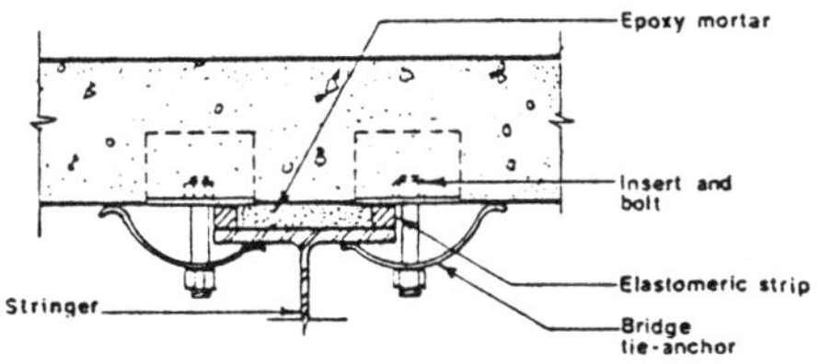


Fig. 26. Connection of slab to stringer, Clark's Summit bridge.

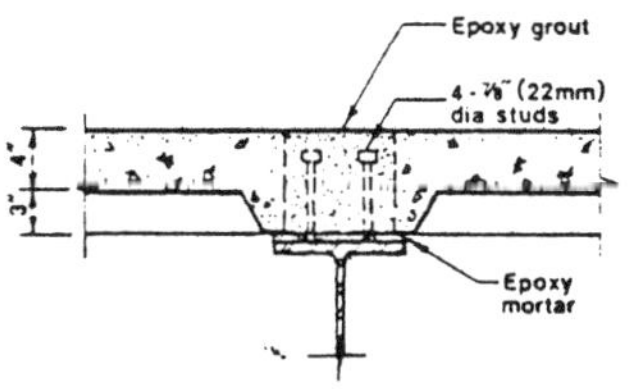


Fig. 27. Slab to beam connection. CERL tests.

## FULL DEPTH PRECAST CONCRETE DECKS ON GIRDERS

### 3.1 Description

In this type of construction, the entire bridge deck is of precast concrete. There is no additional field cast concrete that acts structurally, except that which may be used in the connections. There may be a riding surface of asphaltic concrete, or in some cases, high density portland cement concrete or other material applied for the purpose of combating deterioration.

The deck may or may not be designed to act compositely with the girder, depending on the connection of the deck to the girder (and the confidence of the designer).

### 3.2 Past and Present Usage

Most applications of full depth precast decks have been for replacement of deteriorated concrete decks on steel girders. These applications have been relatively few, and typically in situations where rapid reopening to traffic was important.

### 3.3 Summary and Evaluation of Connections

The type of construction described in this section involves two types of connections: 1) connection of the deck to the girders, and 2) the connection of adjacent panels.

#### 3.3.1 Connections of Deck Slabs to Girders

Of the several projects using precast concrete slabs on girders, some considered the deck and supporting beam to act compositely and some were designed non-compositely. The available research seems to indicate that it is entirely possible to develop composite action, and there seem to be very few reasons not to use it.

The New York Thruway projects are examples of assuring composite action. These projects used both an epoxy coating on the girders and a mechanical tie between the girders and the slabs through a block-out. In this case the design was based on the mechanical connector providing the full horizontal shear tie between the slabs and the girders. The epoxy mortar over the girders was primarily used as a leveling device, although it was recognized that some adhesion was present and that this would provide additional safety for assuring composite action.

The Oakland Bridge (Sect. 3.2.6) and the CERL tests (Sect.3.2.8) used only standard sand cement mortar between the slab and the beams. The purpose of this mortar was merely to provide vertical load bearing capacity. Leveling was done by other means. Composite behavior was established by the welded studs in grout pockets. Clark's Summit Bridge used epoxy mortar that was not designed compositely.

Conversely, the Santa Fe Railroad Bridges used epoxy mortar without any mechanical ties except for spring clips, which cannot be counted upon for significant shear resistance. Their field tests indicated they were getting satisfactory composite action with just the epoxy bonding , but the tests were limited. The Purdue-Indiana Highway Commission test bridges (Sect. 3.2.1) also seemed to verify that epoxy bonding will provide composite behavior. However, the load distribution characteristics of the slab may also be a variable parameter in their calculations. Thus it would seem that some additional testing is warranted.

In every case investigated it appeared that every connection tried worked satisfactorily for its intended purpose. In most cases, it would seem that the use of both epoxy mortar at the slab-beam interface and mechanical ties is unnecessarily conservative. However, more controlled testing of the mechanical ties would give additional assurance. Standard composite design calls for mechanical ties, usually headed steel studs, spaced uniformly along the steel girder. The type of connection studied here involves the mechanical ties being spaced somewhat further apart.

In many of the projects investigated, the epoxy mortar was used for leveling the slabs. The double-headed bolts used on the Oakland project (Sect. 3.2.6) would seem to be more controllable and faster. At Oakland, the space between the slab and the girder flange was formed from below and a flowable grout poured into the connector blockouts. This seems costly and possibly dangerous when working over traffic or on high bridges. Plywood strips, as used on the Santa Fe Railroad jobs, or neoprene strips, as used at Clark's Summit, seems preferable. Reference 15 suggests a neoprene tube or O-ring for this purpose. Its advantage is that it is more compressible and hence would allow more tolerance than plywood or solid neoprene. The disadvantage of any of these methods compared with forming from below is that it is one more operation that must be performed prior to placing the slabs.

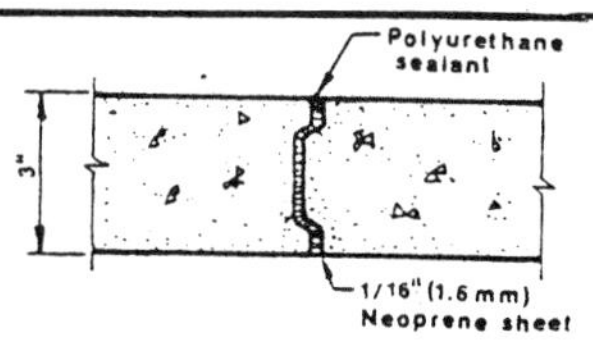


Fig. 7. Joint detail selected for prototype bridges.

3.3.2 Connections with Adjacent Slabs

This connection is intended to provide lateral load transfer across the joint and prevent moisture leakage. Several of the projects reported used standard grout keys between the slabs and either an epoxy mortar or a sand-cement grout in the joint. With the limited experience available it appears that either of these methods works satisfactorily. However, the sand-cement grout joint, which is usually used more extensively in the integral deck bridges described in Section 4, may not hold up in very high volume traffic. Grouted shear keys are discussed in Sect. 5.

The "dry" joint used on the Indiana bridges (Sect. 3.2.1) was not satisfactory. The joint configuration chosen was too sensitive to forming tolerances, causing stress concentration and resulting in concrete spalling. The simpler V-joint used on the Kingston, New York bridge (Sect. 3.2.3) may have worked better, but since it was covered with asphalt, it could not be determined if it did. The Indiana joint fell far short of preventing moisture leakage. It appears that dry joints (no cast-in-place concrete or grout) would only be applicable with some additional waterproof membrane and surfacing.

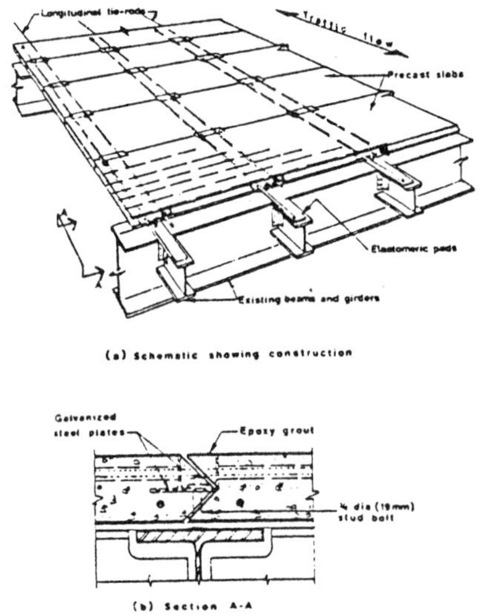


Fig. 17. Rondout Creek bridge deck Kingston, New Yor

3.4 Potential for Extended Use

The biggest problem with this type of construction seems to be its cost. The Erie County Bridge in New York State (Sect. 3.2.3) was estimated to cost about 50% more than a cast-in-place monolithic deck. Few other direct cost comparisons were made.

In most of the projects reported in Sect. 3.2 the precast deck units were chosen because of anticipated increase in speed of construction. Several users have reported disappointment in the time savings. The Indiana analysis, shown in Table 1 and discussed in Sect. 3.2.1 may show why.

Probably the primary reason that the cost and time saving did not live up to expectations is that in nearly every case, the method was considered "new". Standard methods of construction and design for this method have not been developed and contractors have not been able to gain repeat experience. The Oakland Interstate bridge, and the Santa Fe Railroad experience illustrate that on large or repetitive projects with proper planning, the anticipated time savings can be realized. Some cost and construction time could be saved with simplified connection details. However, to convince most design engineers that a "belt and suspenders" approach to design is not necessary would require some testing of the connection details.

Table 1

Time Spent for Various Operations - Bloomington Bridge

| Operation | Time Spent |
|---|---|
| Removal of old deck and the north and south approaches | 5 days |
| Sandblasting stringers, floor beams - applying epoxy coating | 7 days |
| Raising outside stringers, tack welding plate to stringers and floor beams | 3-1/2 days |
| Placing precast slabs and post tensioning | 2 days |
| Bolting slabs to stringers | 5 days |
| Joint preparation and applying joint sealant | 5 days |
| Epoxy mortar tuck pointing between the stringers and floor slabs | 3 days |
| Expansion joint installation | 2 days |
| Forming and pouring mudwalls | 4 days |
| Forming and pouring approaches - guardrail installation | 5 days |
| No work due to delivery delays | 4 days |
| No work due to weather conditions | 1 day |

3.5 Full-Depth Decks on Prestressed Concrete Girders

All of the applications of precast concrete full-depth decks as replacements have been on steel girder bridges. The method should also be applicable with precast, prestressed concrete girders, if a suitable connection capable of horizontal shear transfer between the slab and girder could be devised.

Section 1.6.14 of the 1977 AASHTO Bridge Specifications[24] prescribes the design requirements for horizontal shear transfer for precast concrete girders with cast-in-place decks. The transfer is basically dependent on bond between the two components at ultimate load. The girder shear reinforcement projects into the slab to assist in the shear transfer. In order to use the maximum allowable ultimate shear stress, 300 psi, the top surface of the precast girder must be "intentionally roughened".

With a precast deck slab placed on the girder, it would be necessary to have a grout bed between the slab and girder in order to provide a firm bearing for the slab. Thus there are two interfaces to be considered, one between the deck and the grout, and the other between the grout and the girder. The bond between the grout and the girder would undoubtedly be equivalent to the bond between a cast-in-place slab and a girder. The process of removal of an existing slab would leave the top surface of the girder at least as rough as an intentionally roughened top surface of a new girder. However, the bottom surface of a precast deck unit will normally be cast against smooth, oiled steel forms. Therefore, in order to get sufficient bond between the slab and grout, it may be necessary to "intentionally roughen" the part of the bottom surface of the slab that is in contact with the grout. This could be accomplished by placing a liner in the slab form that has castellations or some other rough surface.

The second problem is to provide mechanical ties through both interfaces that would be equivalent to the projected reinforcement required by the AASHTO Bridge Specifications. In order to place the precast slab on the existing girder, it would be necessary to burn off the projecting reinforcement, since it would be impossible to predetermine the location of these bars.

After considering several alternatives, a connection as shown in Fig. 30 was chosen as the most practical. The advantages of this connection are:

1. The coil rods shown are made of carbon steel bars meeting ASTM A-108, similar to the material used in steel studs. Properties of coil rods are shown in Table 2.
2. The nut threaded on the rod would provide anchorage in the slab. In the relatively small blocked out opening, the heads would not be as congested as reinforcing bar hooks.
3. The coarse threads of the coil rods would provide better anchorage in the epoxy grouted hole than a smooth bolt.
4. The use of epoxy grouts for anchor bolts has been shown to develop either the full strength of the grouted bolt or the shear strength of the surrounding concrete.

Laboratory tests of this connection would be required before use.

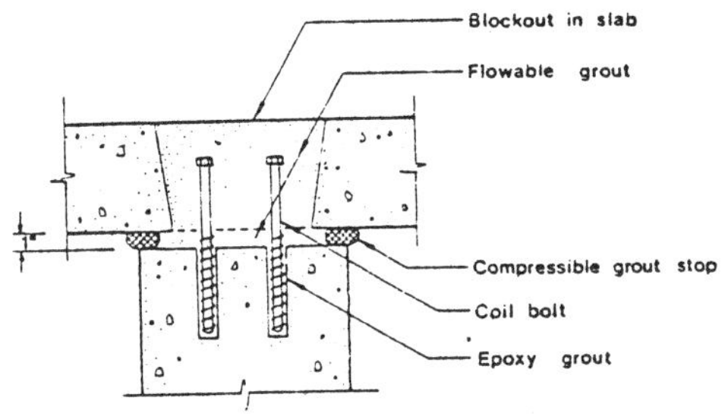


**Fig. 30.** Suggested connection detail - precast slab to precast girder.

In actual application, a typical construction procedure for a deck replacement would involve the following steps:

1. Remove the existing deteriorated deck in sections. This would require breaking out the slab over the girders with jack hammers. The width of the section would be equal to the girder spacing and the section could be sawed into convenient lengths for handling. This is similar to the procedure used on the Oakland bridge described in Section 3.2.6.
2. Burn off the projecting reinforcing bars with a cutting torch and remove loose and cracked material from the top of the girder.
3. Attach a compressible "grout-stopper" along the edges of the girder. This could be a tube as suggested in Reference 15, or other compressible filler material.
4. Place and level the new precast deck slabs in a preplanned sequence. Levelling may be done using a bolt threaded into a sleeve cast in the slab, as was done on the Oakland project. Depending on the method used to connect the precast slabs together, it may be necessary to perform some related operations at this time.
5. Drill holes into the top of the girder through the preformed holes in the deck slab.
6. Place epoxy grout in the holes and the dowels into the grout.
7. Pour high-strength flowable grout into the slab blockouts. The grout will flow under the slab between blockouts to provide a firm bearing on the girders. With blockouts at approximately two feet on centers, the grout flow is easily monitored.

Table 2: Root Diameter and Area of Coil Rods

| Nominal Diameter (in.) | Shank Area (sq. in) | Minimum Root Diameter (in.) | Root Area (sq. in.) |
|---|---|---|---|
| 1/2 | 0.20 | 0.411 | 0.133 |
| 3/4 | 0.44 | 0.617 | 0.299 |
| 1 | 0.79 | 0.830 | 0.541 |
| 1-1/4 | 1.23 | 1.075 | 0.908 |
| 1-1/2 | 1.77 | 1.340 | 1.410 |

Coil rods are made from "medium carbon steel", designated C-1045. Conforms to ASTM A-108. Estimated min. mechanical properties:

| | |
|---|---|
| Tensile strength: | 104,000 psi |
| Yield strength: | 91,000 psi |
| Elongation: | 15% in 2 in. |

# B.8 TIMBER DECK DETAILS

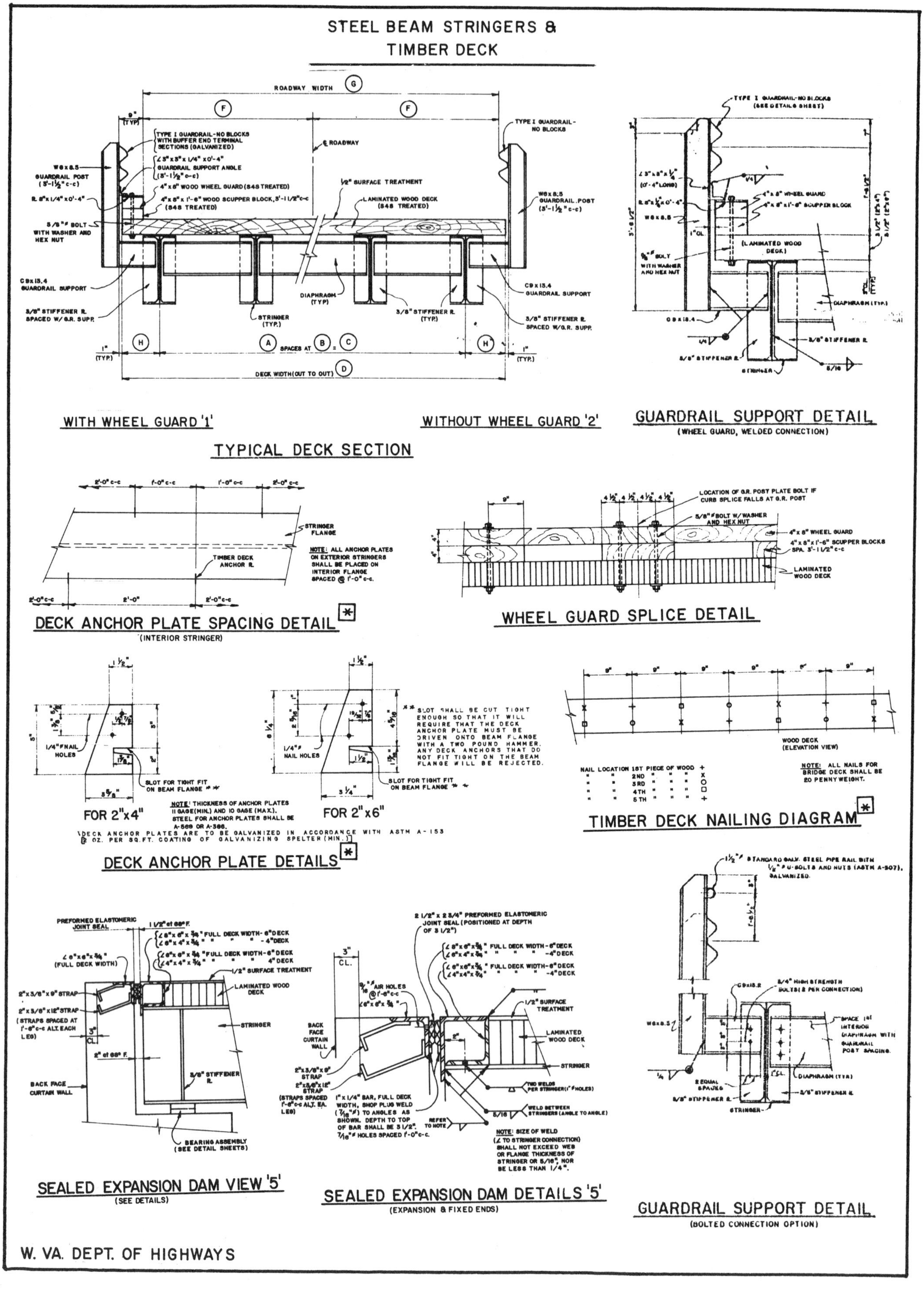

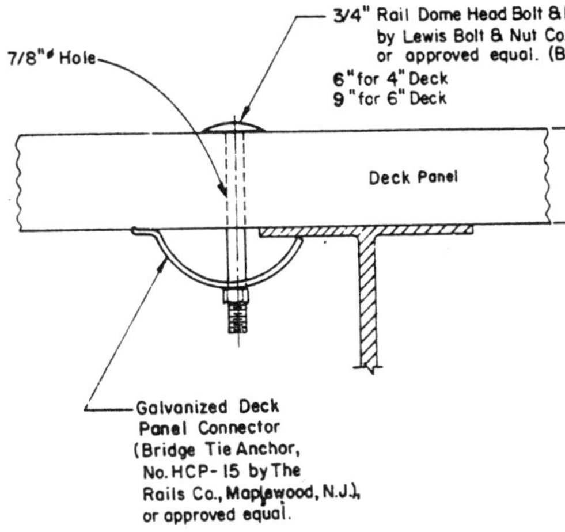


| DOWELED DECK PANEL CONNECTOR SCHEDULE | | |
|---|---|---|
| STRINGER SPACING | "N" | "S" |
| ≤ 1'-6" | 1 | 2 |
| > 1'-6" ≤ 3'-0" | 2 | 2 |
| > 3'-0" < 4'-6" | 2 | 1 |

Deck Panel to be attached to stringers with 2 Deck Panel Connectors to the exterior stringers and with (N) Connectors to each (S) interior stringer.(s)

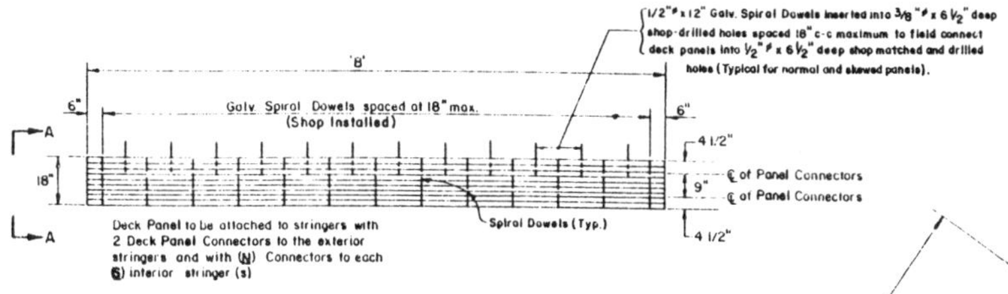


DOWELED DECK PANEL
NORMAL

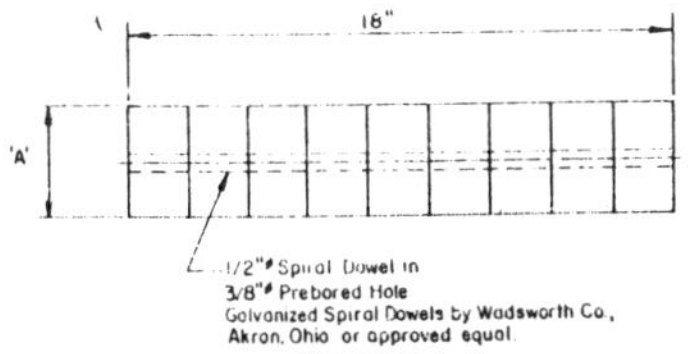


SECTION A-A

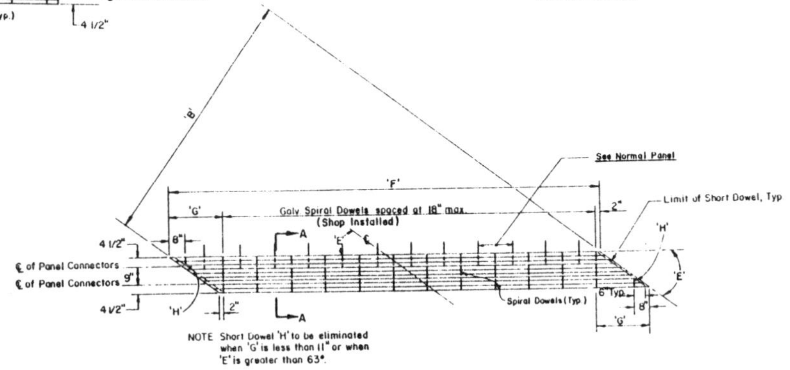


DOWELED DECK PANEL
SKEWED

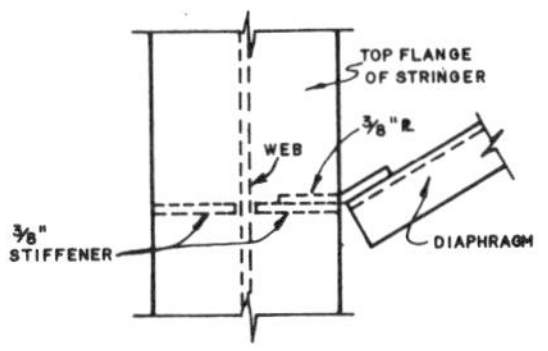


SKEW DETAIL

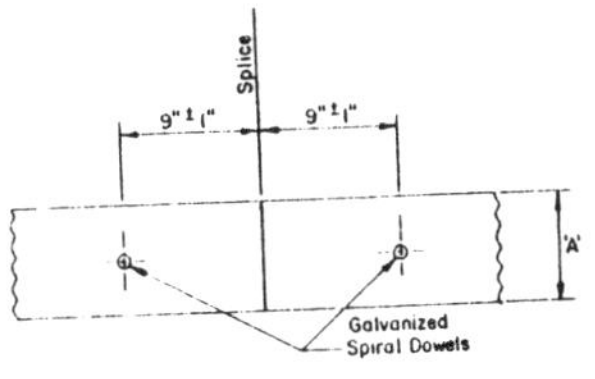


SHOP SPLICE DETAIL

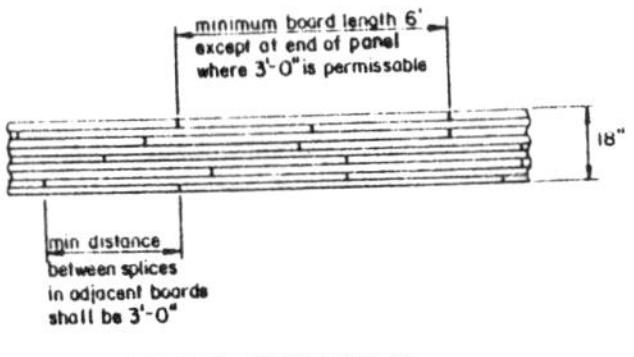


SPLICE SPACING & BOARD LENGTH OF DOWELED DECK PANEL

FIELD TREATMENT: All abrasions and cuts made in the field shall be carefully trimmed and given three brush coats of the original treatment solution; holes drilled in the field shall be poured full of the same preservative. Unused holes shall be poured full of preservative and plugged with tight-fitting treated plugs.

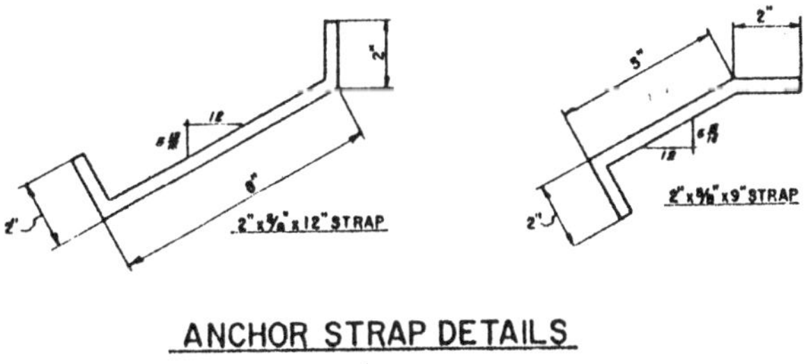


ANCHOR STRAP DETAILS

# B.9 STEEL GRID DECK DETAILS

## STEEL GRID DECK

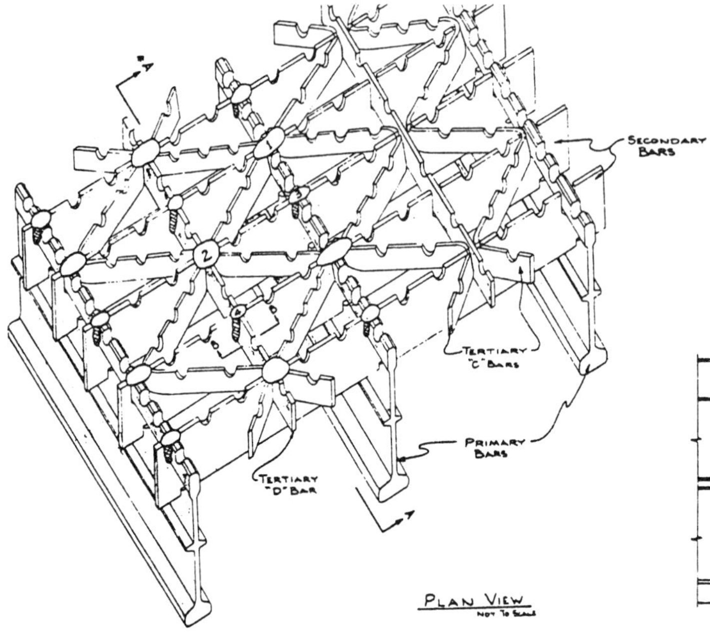


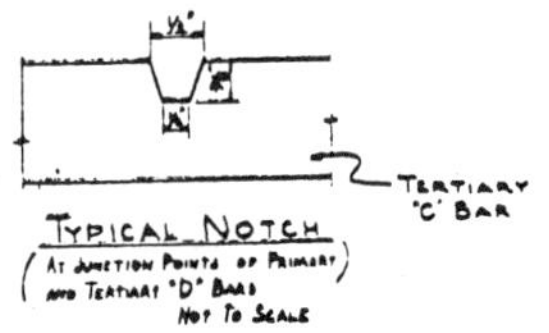


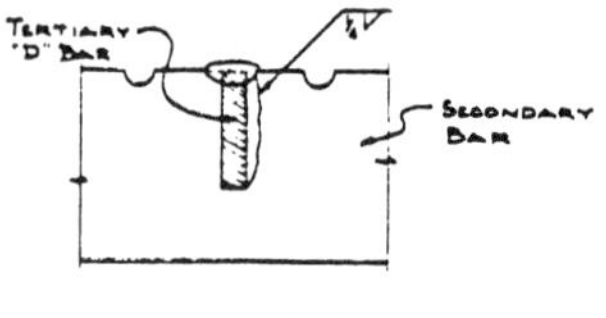


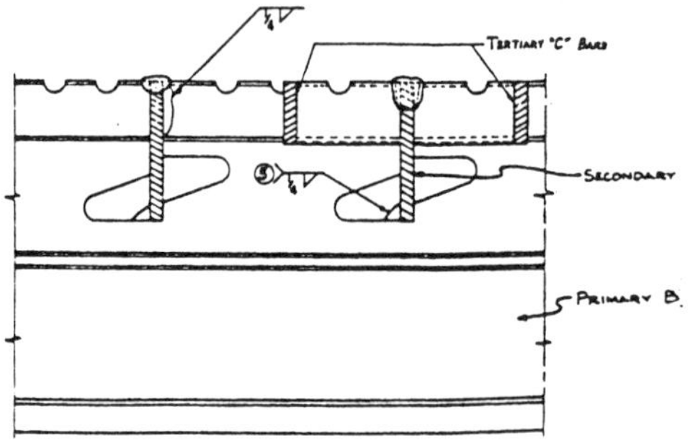


Modified 5" - 4 Way Steel Grid

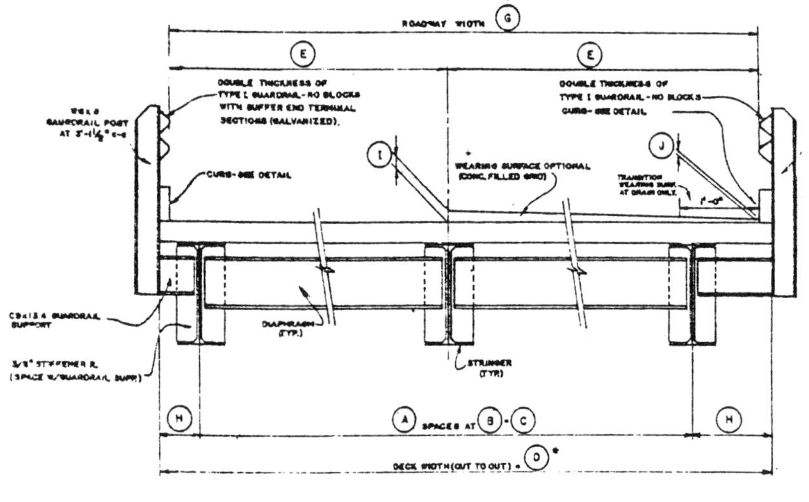


| CODE | DESCRIPTION |
|---|---|
| A | NUMBER OF SPACES BETWEEN STRINGERS |
| B | SPACING OF STRINGERS |
| C | DISTANCE CENTER-TO-CENTER BETWEEN OUTSIDE STRINGERS |
| D | DECK WIDTH (OUT-TO-OUT) |
| E | DISTANCE FROM FACE OF WHEEL GUARD TO ℄ ROADWAY |
| | |
| G | ROADWAY WIDTH |
| H | DISTANCE FROM ℄ EXTERIOR STRINGER TO EDGE OF DECK |
| I | THICKNESS OF WEARING COURSE AT ℄ ROADWAY |
| J | THICKNESS OF WEARING COURSE AT FACE OF CURB |

OPEN STEEL GRID (TYPICAL)

CONCRETE FILLED STEEL GRID (TYPICAL)

## TYPICAL DECK SECTION

NOTE: THE GRID DECK SHOULD BE CANTILEVERED SUFFICIENTLY TO PROVIDE ROOM FOR DRAINS BETWEEN THE FACE OF CURB AND THE EDGE OF THE OUTSIDE STRINGER.

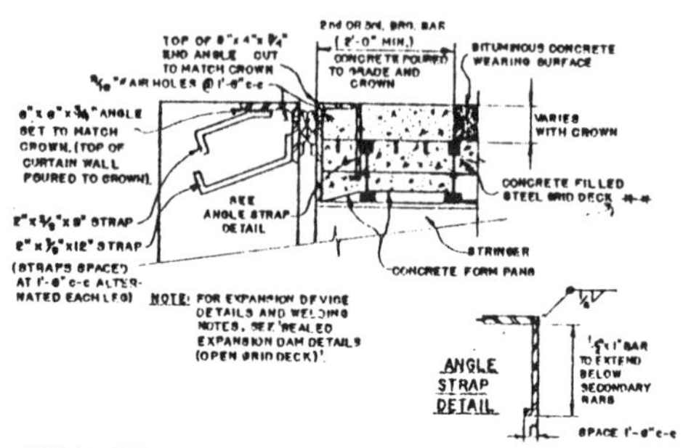


## SEALED EXPANSION DAM DETAILS

(CONCRETE FILLED GRID DECK)

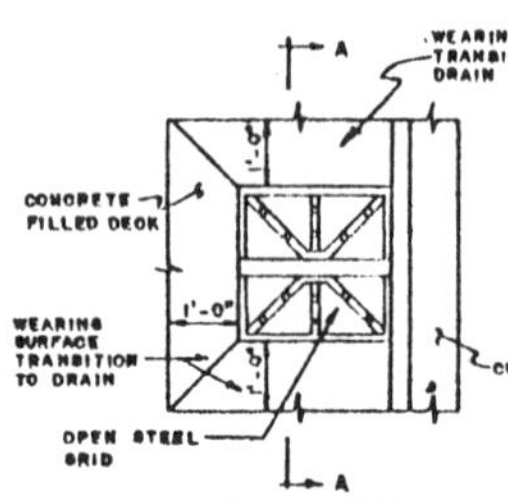


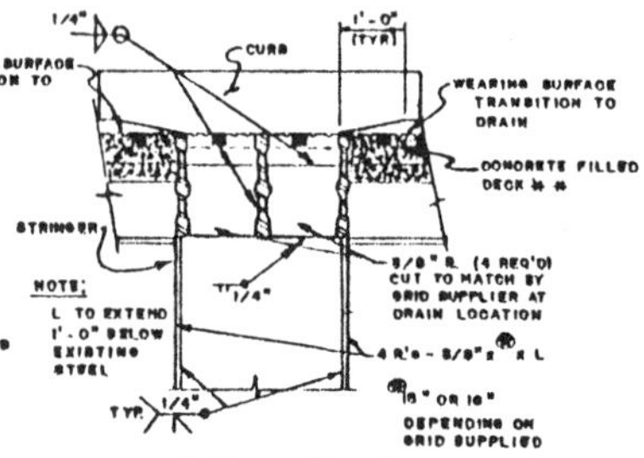


## DRAIN DETAIL

## Modified Steel Grid Flooring Detail

**Greulich**

Typ. Location 1: Notch Tertiary "C" Bars And Puddle Weld.

Typ. Location 2: Notch Tertiary "C" Bars And Puddle Weld.

Typ. Location 3: Vertical Fillet Weld As Shown On Section A-A Plus Puddle Weld On Top.

Typ. Location 4: Vertical Fillet Weld Full Depth Of Tertiary Bar Plus Puddle Weld As Shown On Section B-B.

Typ. Location 5: 1/4" Fillet Weld At Base Of Secondary Bar.

**Reliance**

Typ. Location 1: Puddle Weld, Plus Vertical Fillet Weld Full Depth Of Secondary Bar Slot.

Typ. Location 2: Vertical Fillet Weld Full Depth Of Tertiary Bar Plus Puddle Weld On Top.

Typ. Location 3: 1/4" Fillet Weld At Base Of Secondary Bar.

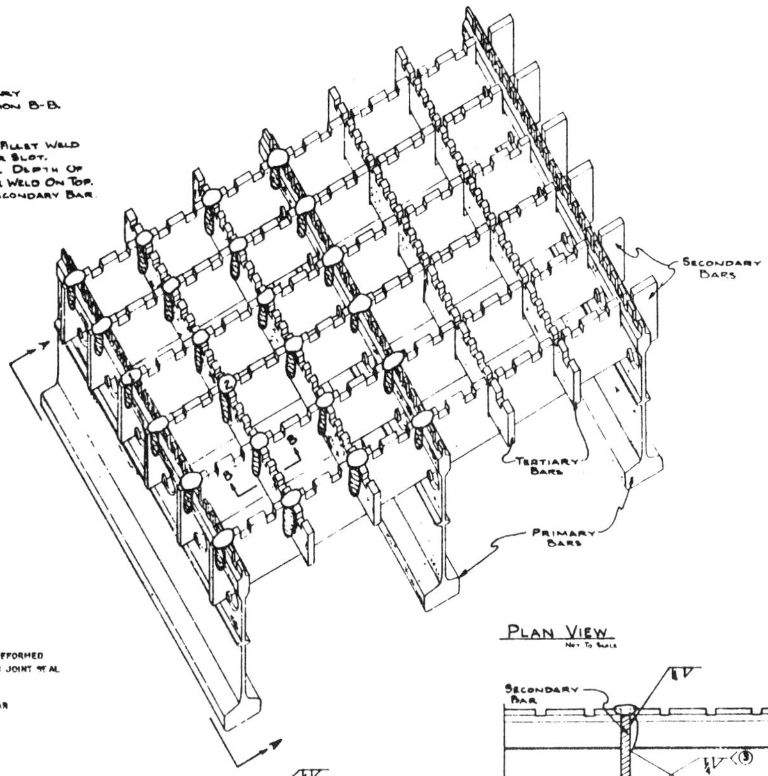


Plan View
Not To Scale

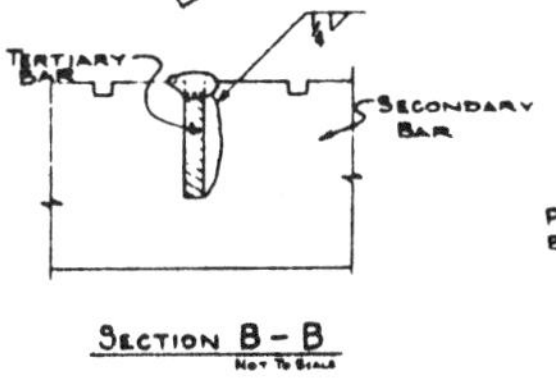


Section B-B
Not To Scale

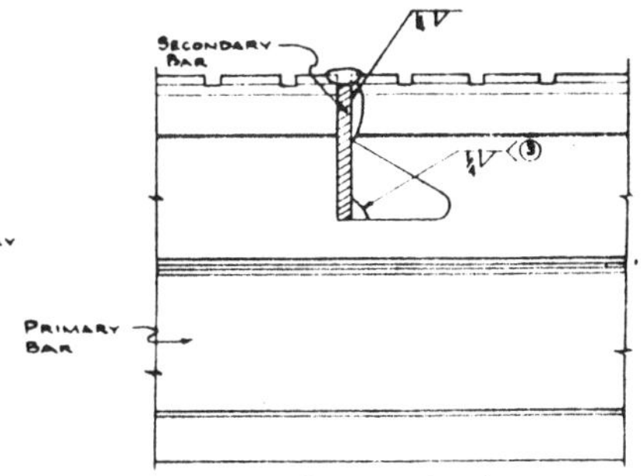


Section A-A
Not To Scale

Modified 5" Weldlock Type H

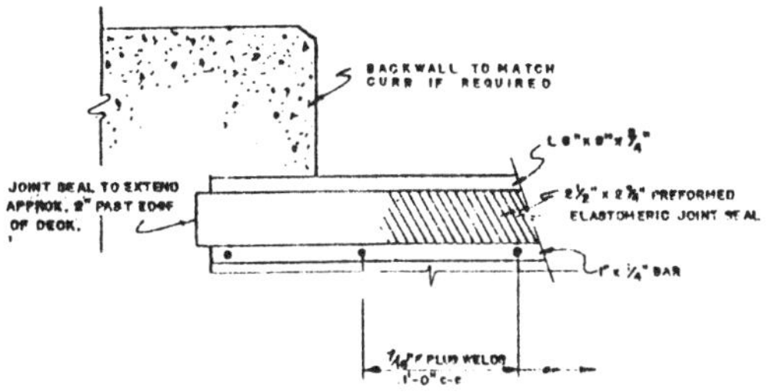


SEALED EXPANSION DAM AT ABUTMENT

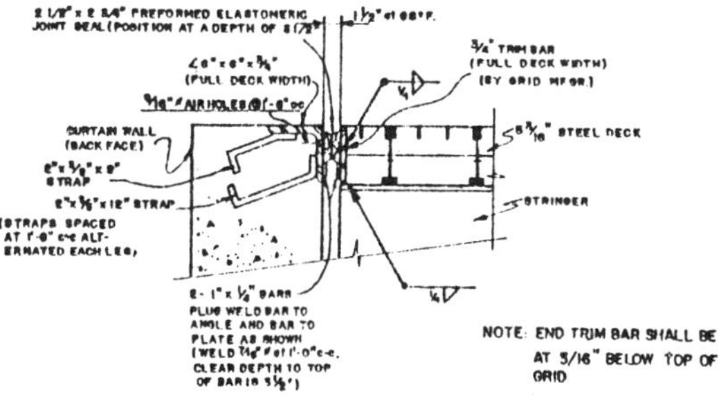


SEALED EXPANSION DAM DETAILS
(OPEN GRID DECK)

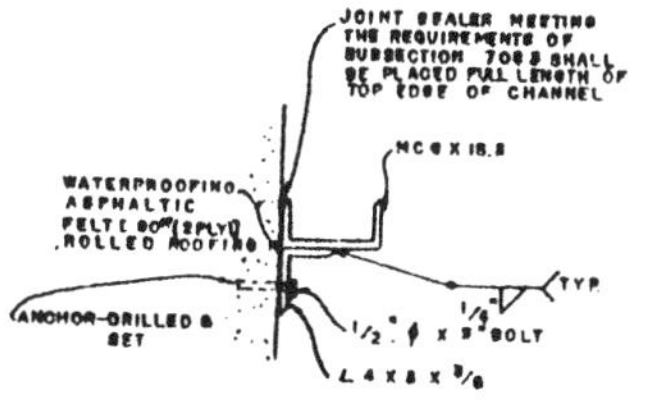


SECTION "D-D"

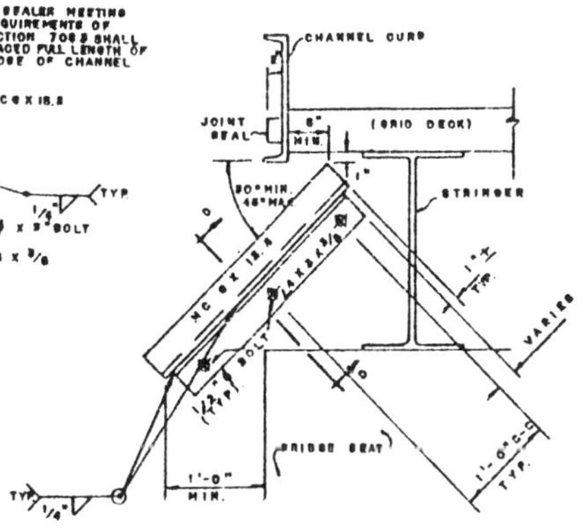


DRAIN TROUGH INSTALLATION

# Bridge Flooring roll-formed, pregalvanized

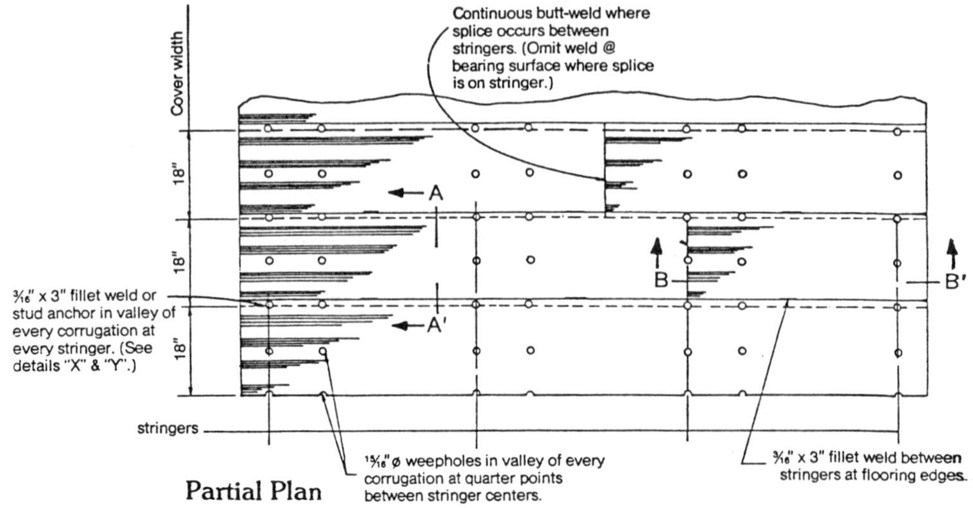


Partial Plan

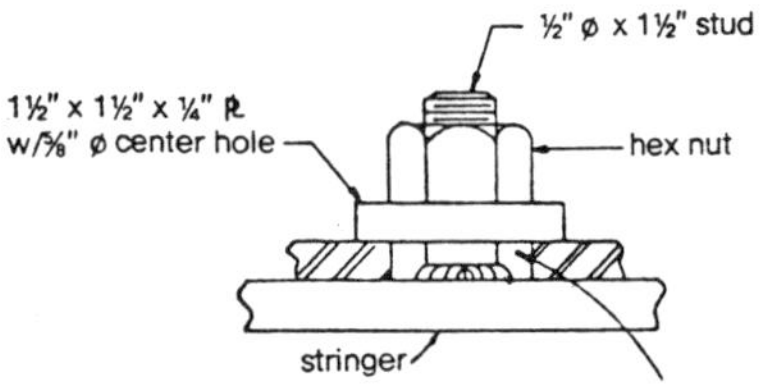


Detail "X"

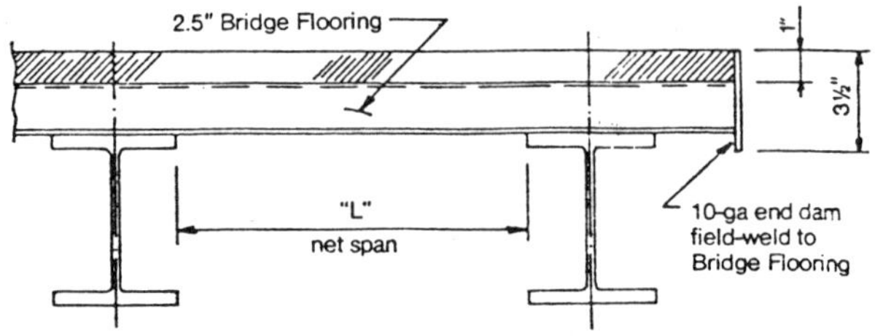


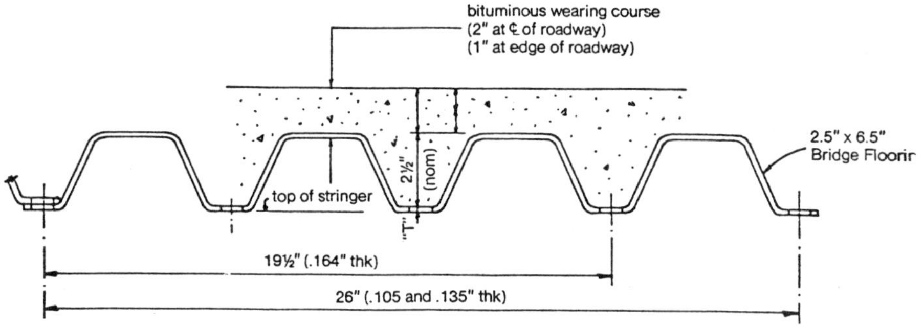


SECTION A-A'

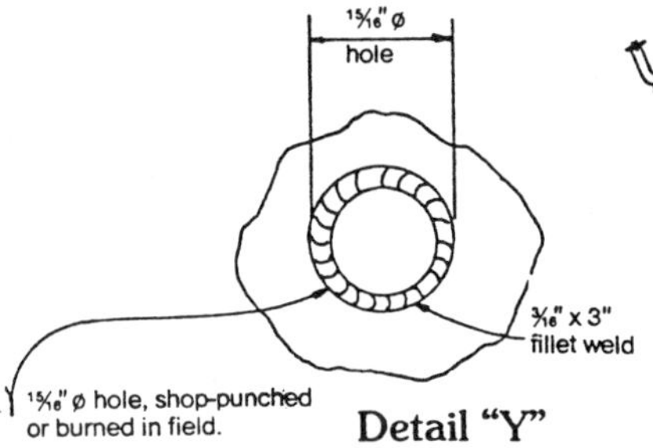


Detail "Y"

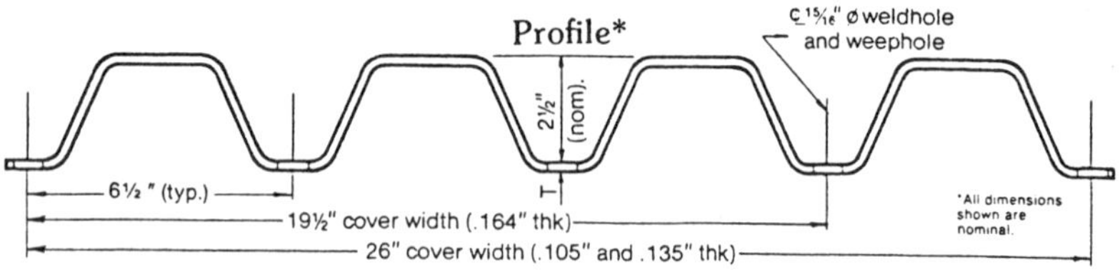


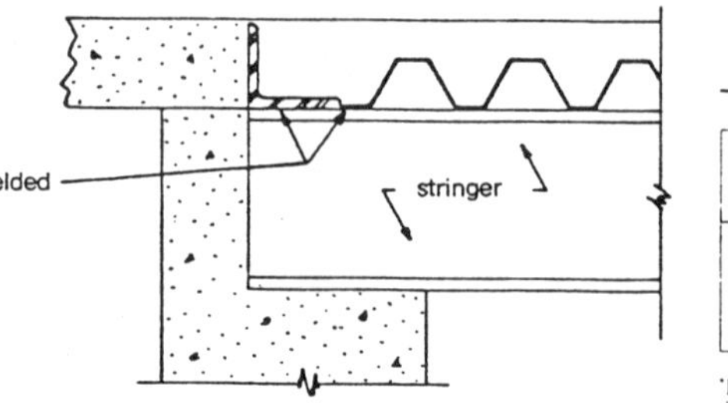


SECTION B-B'

Table 1. Engineering Data, 2.5-in. x 6.5-in. Bridge Flooring

| Nom Gage | Design Thickness "T" in. | Yield Strength min ksi | Weight approx. psf | Mom. of Inertia in.⁴/ft | Section Modulus in.³/ft | *Allowable Net Span "L" in. |
|---|---|---|---|---|---|---|
| 12 | .105 | 40 | 6.5 | 1.633 | 1.131 | 24 |
| 10 | .135 | 40 | 8.3 | 2.110 | 1.447 | 28 |
| 8 | .164 | 40 | 10.1 | 2.577 | 1.750 | 32 |

*For AASHTO Loadings HS20, H20, HS15, or H15, based on allowable stress of 24 ksi and an assumed wheel-load distribution of 20 by 20 in. (Net span "L" is clear span between stringer flanges.)

## Specifications

**Material:** pregalvanized steel sheet per ASTM A446, Grade C. Zinc coating per ASTM A525, Coating Designation G 210.
**Maximum panel length:** 30 ft (without splices)

**Weldholes and weepholes:** 15⁄16-in. diameter rounds or ¾- by 1¼-in. ovals.
**Bituminous fill:** an average of 1 in. of bituminous concrete is required to fill the valleys.

**Bethlehem Steel**

# ENGINEERING DETAILS AND DESIGN DATA

## 6" x 2" BRIDGE PLANK SECTION

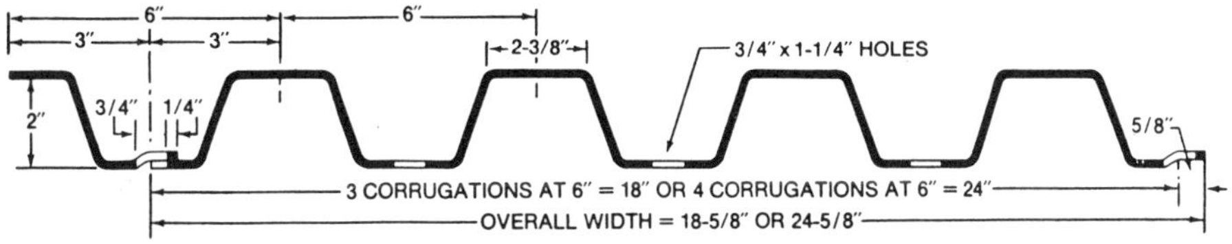


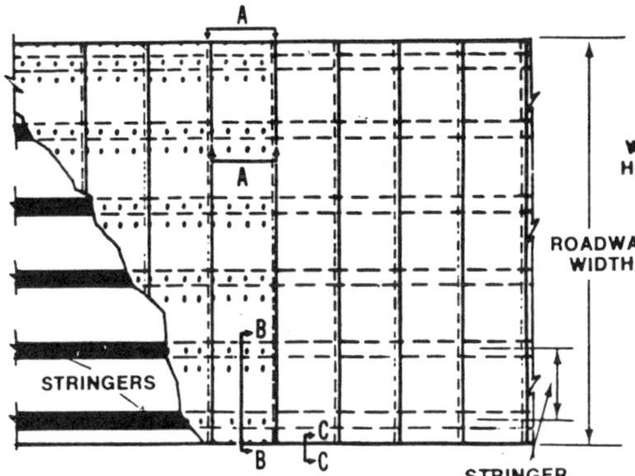


BRIDGE FLOOR

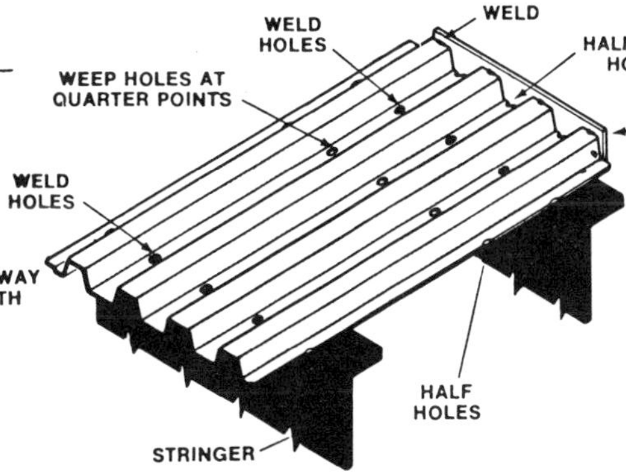


ISOMETRIC DETAIL A-A

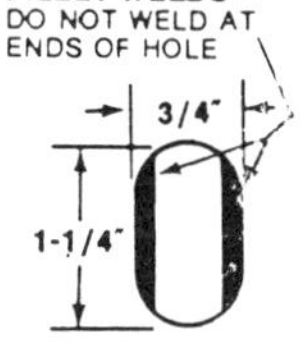


HOLES LOCATED OVER STRINGERS PROVIDE ROOM FOR TWO 1-INCH-LONG FILLET WELDS FOR ATTACHING PLANK.

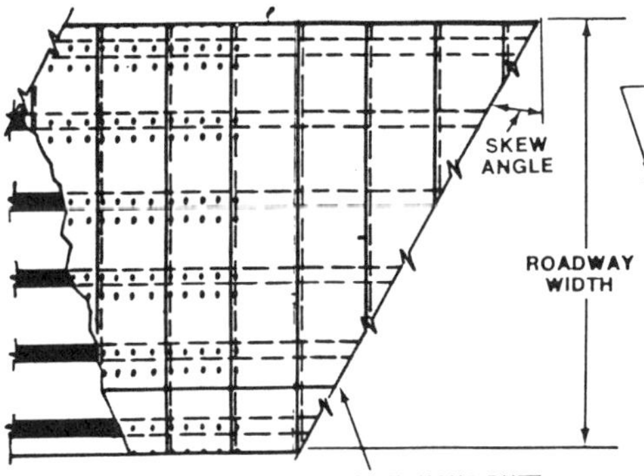


SKEW BRIDGE FLOOR

CONTINUOUS BUTT WELD SPLICES WHEN BETWEEN STRINGERS

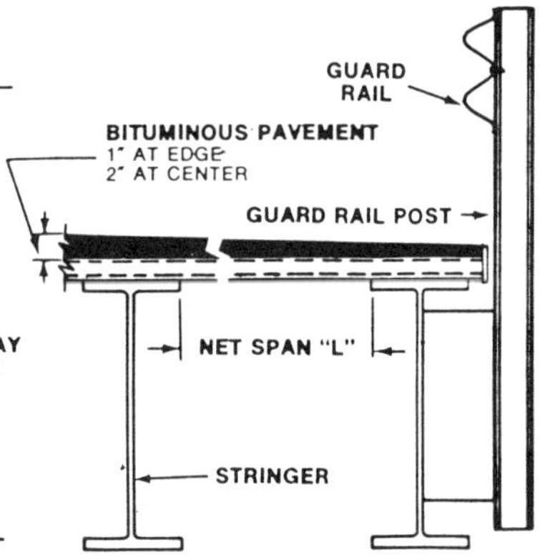


SECTION B-B

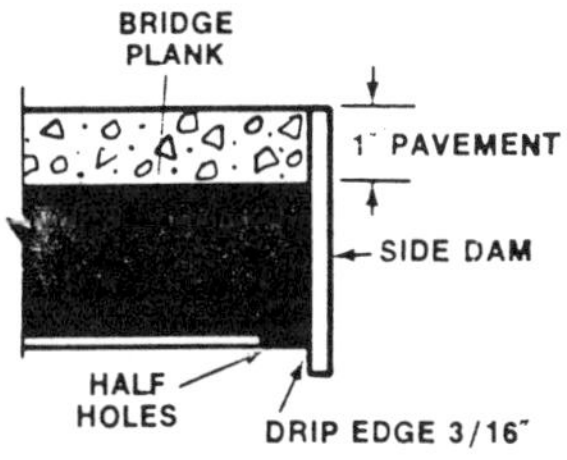


SECTION C-C

### TABLE 1. 6" X 2" CORRUGATION

| Spec. Thk. In. | (Gage) | Eff. Sec. Mod. In.$^3$/In. | Weight psf | Maximum allowable net span—Inches H10 | H15 | H20 |
|---|---|---|---|---|---|---|
| .179 | (7) | 0.162 | 10.7 | 39 | 30 | 30 |
| .134 | (10) | 0.124 | 8.0 | 32 | 25 | 25 |
| .109 | (12) | 0.098 | 6.2 | 28 | 22 | 22 |

NOTE: Allowable spans and effective Section Modulus are based on actual wheel load tests. Average weight and surfacing for 6" x 2" pank is 28.1 pounds per square foot, based on 2" depth over the corrugations at the center, tapering to 1" at the edge. For 9" x 3" plank, it is 34.0 psf.

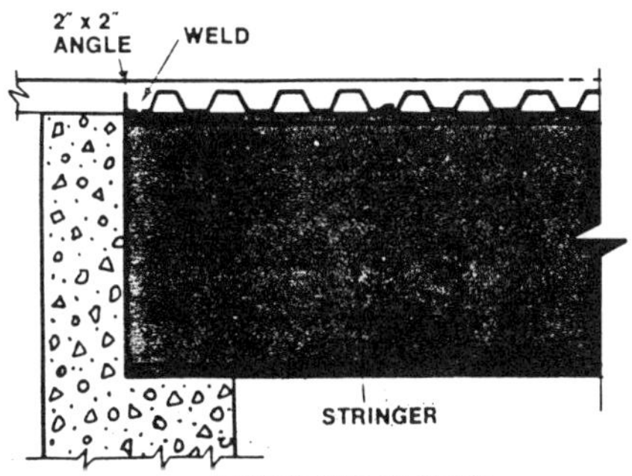


TYPICAL END OF FLOOR

Armco Inc.

# B.11 ORTHOTROPIC DECK

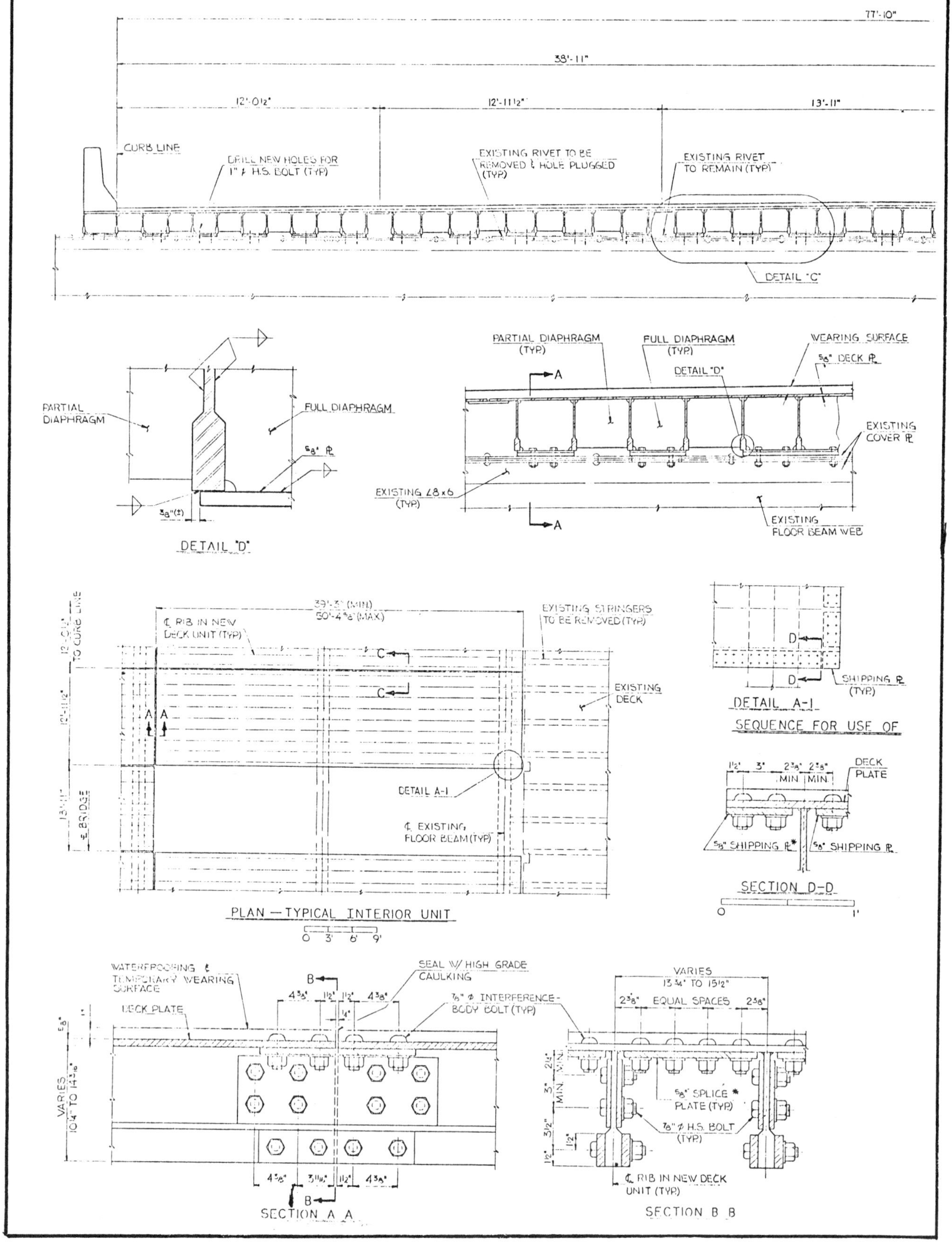

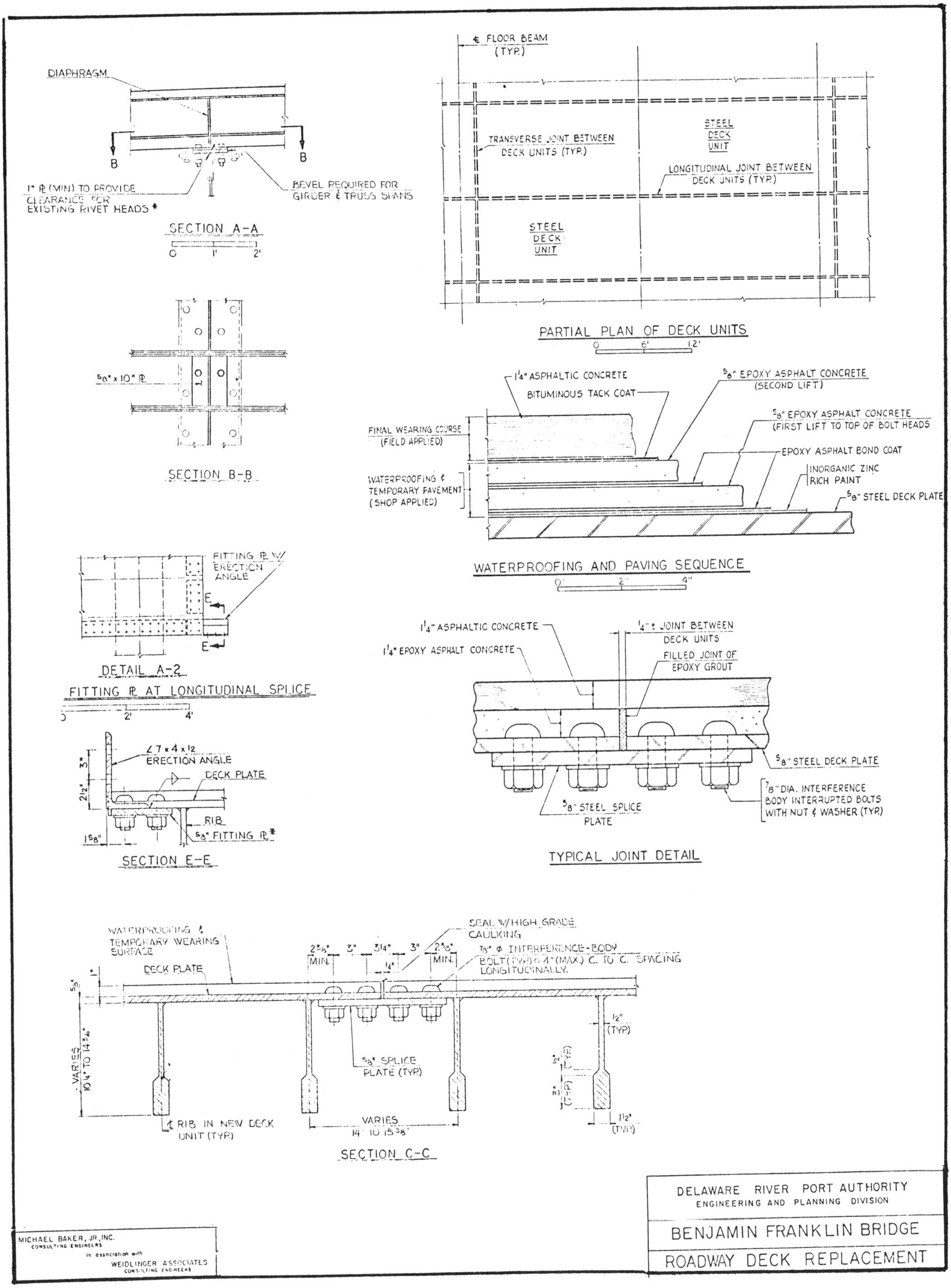
DIAPHRAGM
1" ℞ (MIN) TO PROVIDE CLEARANCE FOR EXISTING RIVET HEADS *
BEVEL REQUIRED FOR GIRDER & TRUSS SPANS
SECTION A-A
SECTION B-B
5/8" x 10" ℞
℄ FLOOR BEAM (TYP.)
TRANSVERSE JOINT BETWEEN DECK UNITS (TYP.)
LONGITUDINAL JOINT BETWEEN DECK UNITS (TYP.)
STEEL DECK UNIT
PARTIAL PLAN OF DECK UNITS
1 1/4" ASPHALTIC CONCRETE
BITUMINOUS TACK COAT
5/8" EPOXY ASPHALT CONCRETE (SECOND LIFT)
5/8" EPOXY ASPHALT CONCRETE (FIRST LIFT TO TOP OF BOLT HEADS
EPOXY ASPHALT BOND COAT
INORGANIC ZINC RICH PAINT
5/8" STEEL DECK PLATE
FINAL WEARING COURSE (FIELD APPLIED)
WATERPROOFING & TEMPORARY PAVEMENT (SHOP APPLIED)
WATERPROOFING AND PAVING SEQUENCE
FITTING ℞ W/ ERECTION ANGLE
DETAIL A-2
FITTING ℞ AT LONGITUDINAL SPLICE
∠7 x 4 x 1/2 ERECTION ANGLE
DECK PLATE
RIB
5/8" FITTING ℞ *
SECTION E-E
1 1/4" ASPHALTIC CONCRETE
1 1/4" EPOXY ASPHALT CONCRETE
1/4" ± JOINT BETWEEN DECK UNITS
FILLED JOINT OF EPOXY GROUT
5/8" STEEL DECK PLATE
7/8" DIA. INTERFERENCE BODY INTERRUPTED BOLTS WITH NUT & WASHER (TYP.)
5/8" STEEL SPLICE PLATE
TYPICAL JOINT DETAIL
WATERPROOFING & TEMPORARY WEARING SURFACE
DECK PLATE
SEAL W/ HIGH GRADE CAULKING
7/8" Φ INTERFERENCE-BODY BOLT (TYP.) @ 4" (MAX.) C. TO C. SPACING LONGITUDINALLY.
5/8" SPLICE PLATE (TYP.)
℄ RIB IN NEW DECK UNIT (TYP.)
VARIES 14" TO 15 3/8"
VARIES 10 1/4" TO 14 3/4"
SECTION C-C
DELAWARE RIVER PORT AUTHORITY
ENGINEERING AND PLANNING DIVISION
BENJAMIN FRANKLIN BRIDGE
ROADWAY DECK REPLACEMENT
MICHAEL BAKER, JR., INC. CONSULTING ENGINEERS
in association with
WEIDLINGER ASSOCIATES CONSULTING ENGINEERS

## B.12 DECK REPAIRS

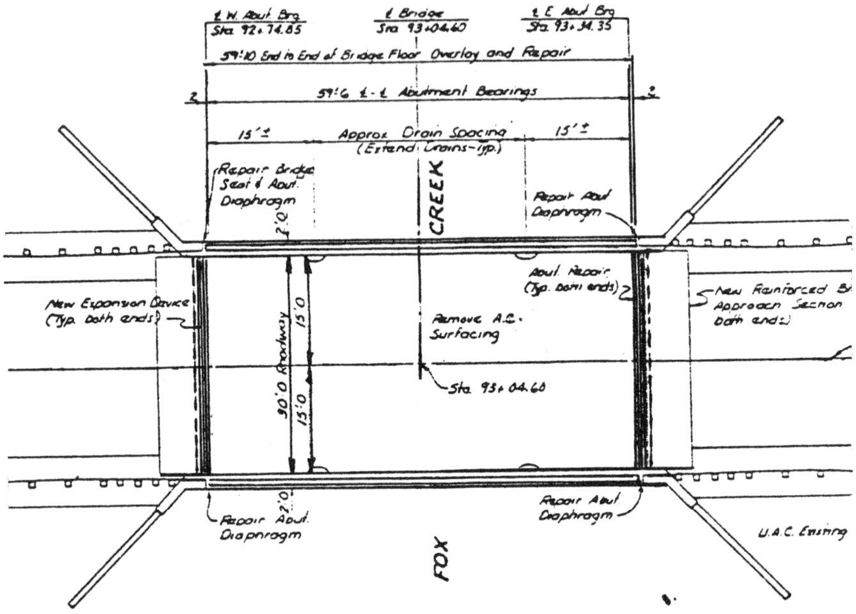


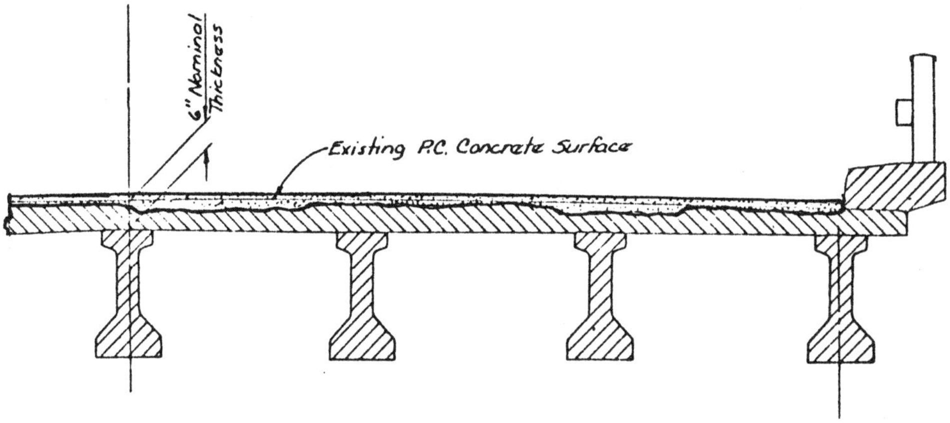


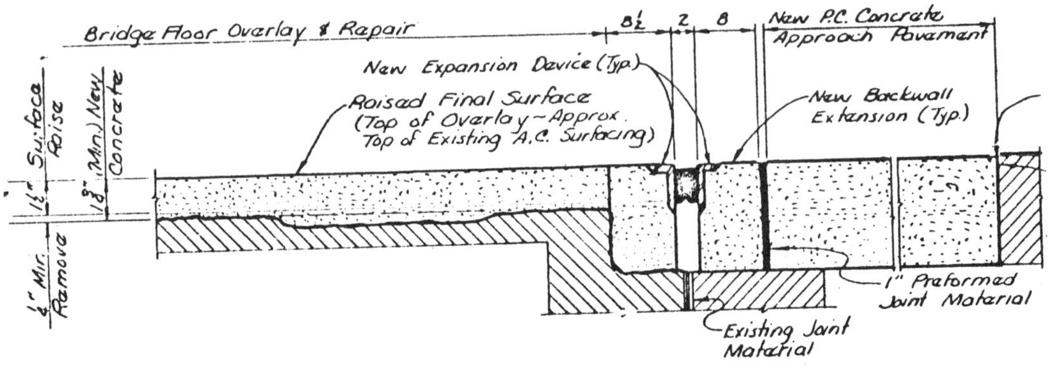

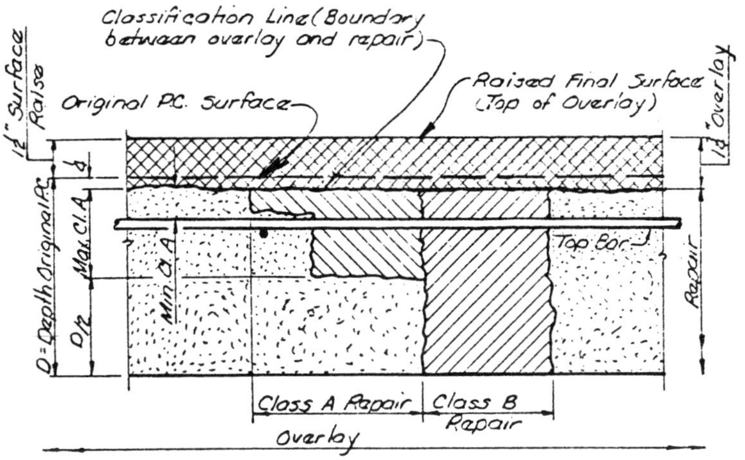


REPAIR AND OVERLAY DEFINITION

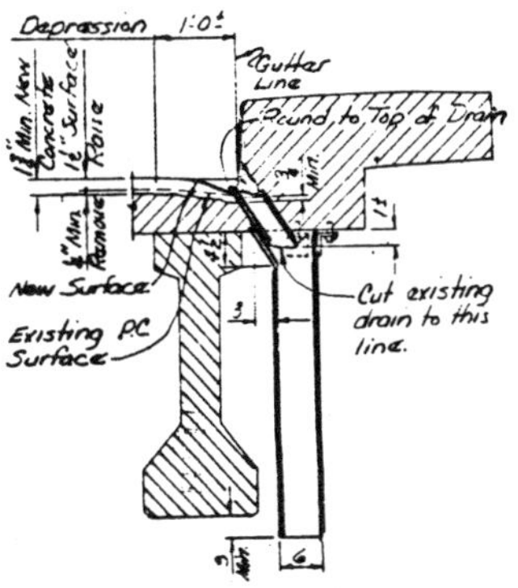


NOTE: Drain extension material shall be galvanized after welding. An electroplated coating is adequate for anchor assembly.

DRAIN EXTENSIONS
(4 Drain Extensions Wt. 99 Lbs each)
Note: Drain extensions are to be fabricated from ¼" plate (welded) TS10x6 x 0.250 may be substituted where applicable.

Construction shall be done in stages with a least one lane of traffic maintained at all times in accordance with "Traffic Control Plan" note.

Existing 1½ ± inches thickness of A.C. surfacing on the bridge floor shall be removed by a method acceptable to the engineer. The A.C. removed shall become the property of the contractor. Price bid for "Removal of Asphaltic Concrete Surfacing" shall include all costs of removal and disposal.

The quantity shown for "Class A Bridge Floor Repair" is based on the assumption that 50% of the deck will require Class A repair and is for bidding purposes only. Because of the existing A.C. surface on the bridge the actual areas of "Class A Bridge Floor Repair" to be removed and paid for will be determined at time of construction after the asphaltic concrete surfacing has been removed.

Upon completion of removal of the A.C. surfacing the engineer will survey the floor for areas of repair as they exist at time of construction by the following methods:

A. Areas of high corrosion potential will be determined by survey with an electrical corrosion detection device. (It has been determined that significant corrosion of reinforcing is present when the device registers 0.35 volt.) All areas with electrical readings of 0.35 volt or more shall receive Class A Bridge Floor Repair.

B. The engineer will conduct an audible and visual survey to determine spalled and hollow areas. Spalled and hollow areas, as they exist at the time of construction, shall be repaired by Class A methods.

The entire concrete surface of the floor, including areas over Class A repair shall have Bridge Floor Overlay. Surface raise, as shown on the plans, shall be considered a minimum. In order to limit the additional dead load, surface raise shall be restricted to a maximum of 1/2 inch more than shown on the plans. Profile shall be adjusted to the extent possible within these limits.

Present thickness of the P.C.C. floor is about 6 inches. The contractor will be required to exercise care in removing concrete in order to prevent unnecessary unbonding of reinforcing steel.

It is anticipated that the approach section surface, as an extension of the raised bridge floor surface, will closely approximate existing crown and profile. However, the contractor may be required to plane, or scarify, the adjacent A.C.C. approach Surface to meet the reconstructed approach. If A.C.C. patch or runout taper is required, this work will be performed by local Highway Division Maintenance forces. Approaches are to be adjusted before traffic is shifted to new construction. The contractor shall give adequate prior notice so that the A.C.C. may be placed during normal working hours and working days of Maintenance.

Before proceeding with bridge floor repair and bridge floor overlay, the contractor may complete all stages of other construction. Any construction shall have stage limits, temporary barrier rail and traffic control as detailed on these plans. If abutment repair, joint construction and approach reconstruction are completed and floor overlay is not a part of the same stage, the contractor shall provide a rideable surface by filling or wedging with A.C.C. Surfacing. The A.C. material shall be a commercial grade hot surfacing mix or a mix approved by the engineer. A.C. may be placed by hand methods and may be compacted by any approved method. All costs for additional traffic control, repositioning of barrier and A.C. Surfacing shall be borne by the contractor.

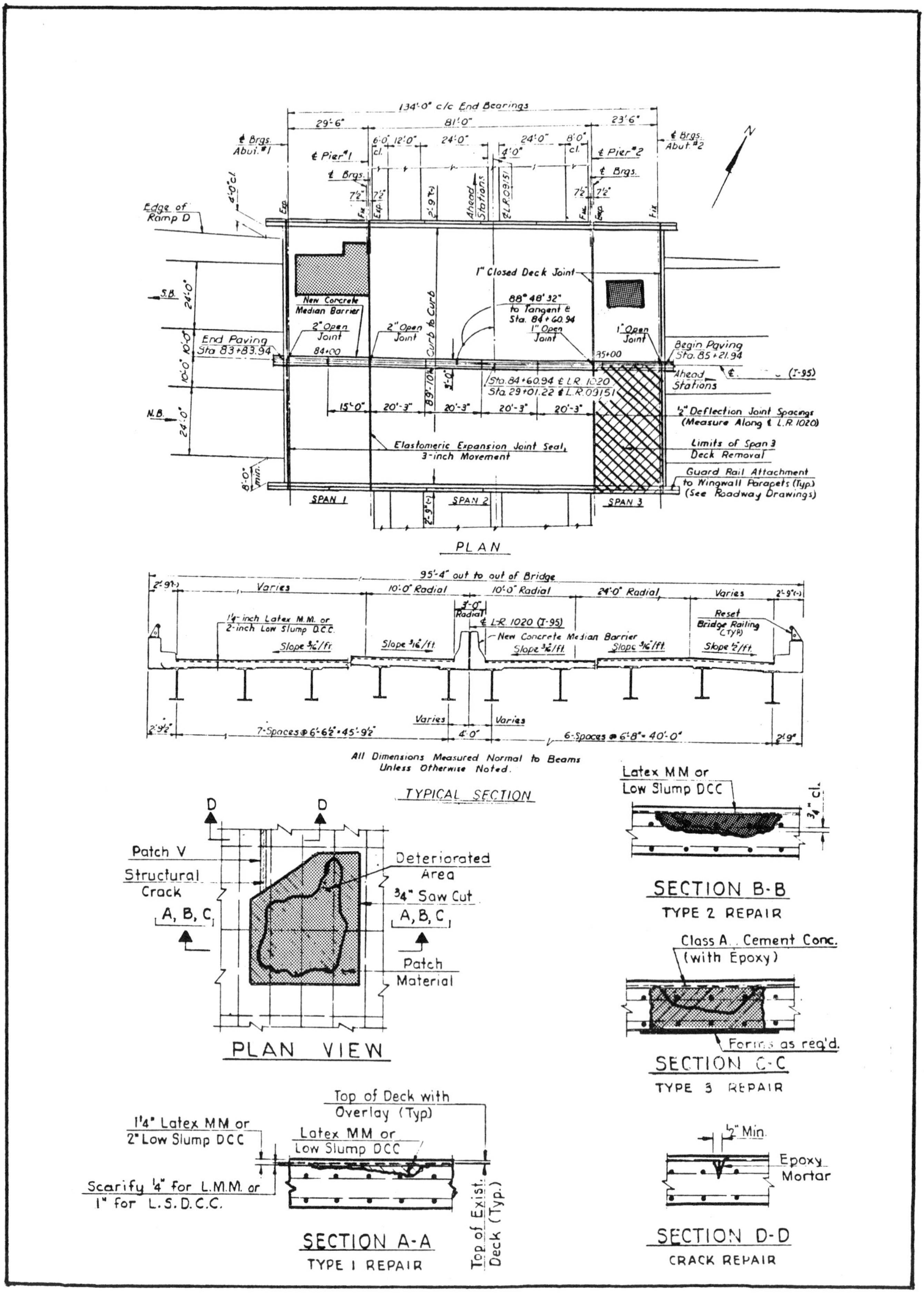
134'-0" c/c End Bearings
29'-6"
81'-0"
23'-6"
℄ Brgs. Abut. #1
℄ Pier #1
℄ Pier #2
℄ Brgs. Abut. #2
℄ Brgs.
Edge of Ramp D
1" Closed Deck Joint
New Concrete Median Barrier
2" Open Joint
1" Open Joint
Curb to Curb
88°48'32" to Tangent ℄ Sta. 84+60.94
End Paving Sta 83+83.94
84+00
85+00
Begin Paving Sta. 85+21.94
Ahead Stations
Sta. 84+60.94 ℄ L.R. 1020
Sta. 29+01.22 ℄ L.R. 09151
S.B.
N.B.
½" Deflection Joint Spacings (Measure Along ℄ L.R. 1020)
Elastomeric Expansion Joint Seal, 3-inch Movement
Limits of Span 3 Deck Removal
Guard Rail Attachment to Wingwall Parapets (Typ.) (See Roadway Drawings)
SPAN 1
SPAN 2
SPAN 3
PLAN
95'-4" out to out of Bridge
Varies
10'-0" Radial
24'-0" Radial
3'-0" Radial
1¼-inch Latex M.M. or 2-inch Low Slump D.C.C.
℄ L.R. 1020 (I-95)
New Concrete Median Barrier
Reset Bridge Railing (TYP)
7-Spaces @ 6'-6½" = 45'-9½"
6-Spaces @ 6'-8" = 40'-0"
All Dimensions Measured Normal to Beams Unless Otherwise Noted.
TYPICAL SECTION
Patch V
Structural Crack
Deteriorated Area
¾" Saw Cut
Patch Material
A, B, C
PLAN VIEW
Latex MM or Low Slump DCC
¾" cl.
SECTION B-B
TYPE 2 REPAIR
Class A Cement Conc. (with Epoxy)
Forms as req'd.
SECTION C-C
TYPE 3 REPAIR
Top of Deck with Overlay (Typ)
1¼" Latex MM or 2" Low Slump DCC
Latex MM or Low Slump DCC
Scarify ¼" for L.M.M. or 1" for L.S.D.C.C.
Top of Exist. Deck (Typ.)
SECTION A-A
TYPE 1 REPAIR
½" Min.
Epoxy Mortar
SECTION D-D
CRACK REPAIR

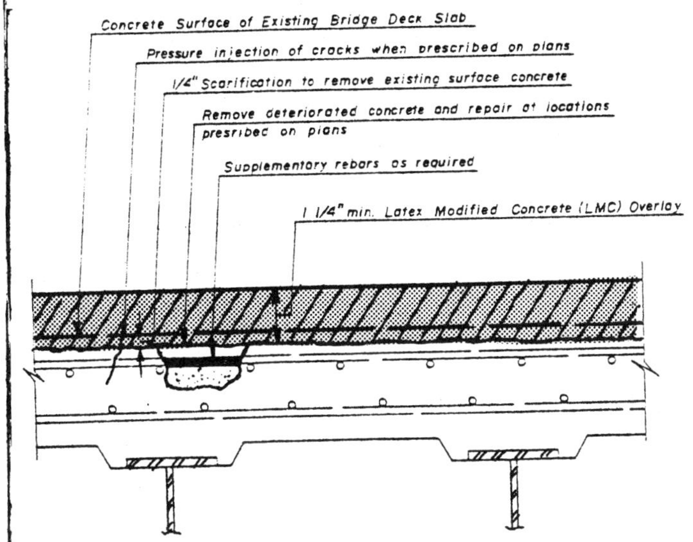


NOTE 1: "T" indicates nominal thickness of concrete deck slab.

NOTE 2: Sawcut (3/4" deep) for the entire periphery for all repair types.

NOTE 3: For Type A Repair, where the bond between existing concrete and reinforcement steel has been destroyed, or where more than half the diameter of the steel is exposed, 1 inch of concrete removal around the bar is required.

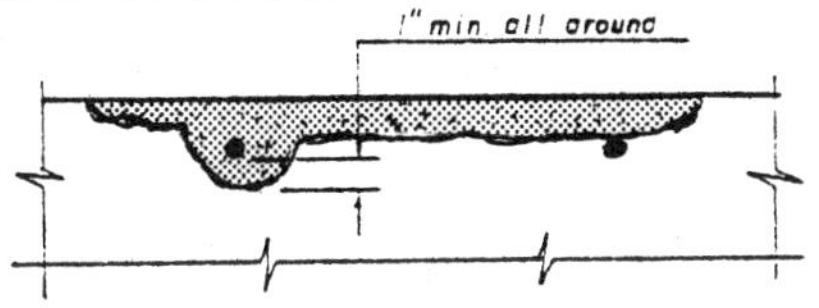


NOTE 4: New rebars shall be placed to supplement existing reinforcement when:

- Existing bar has lost 25% or more of the original cross-sectional area.
- Existing bar is broken.

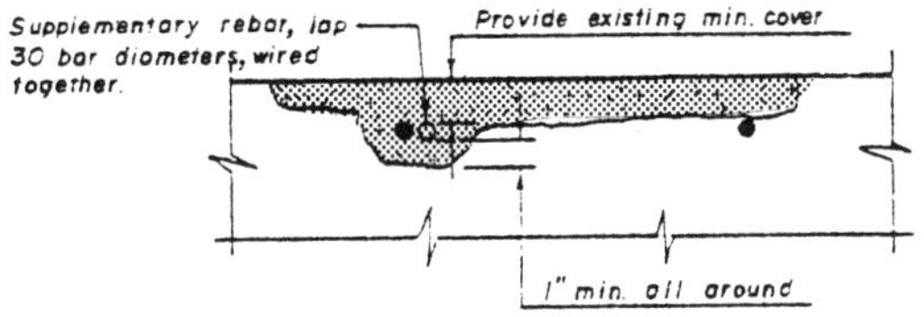


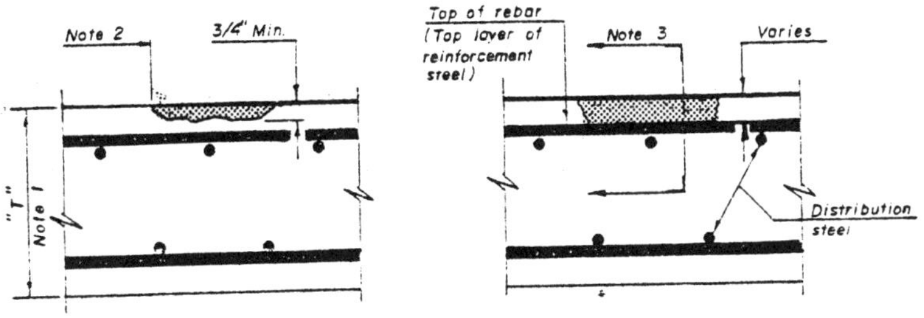


REPAIR TYPE A

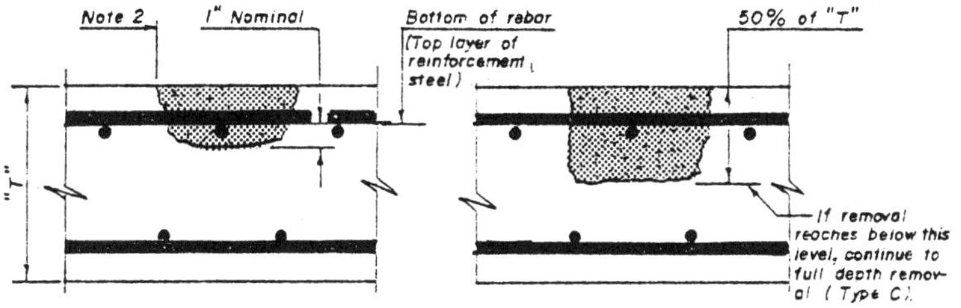


REPAIR TYPE B

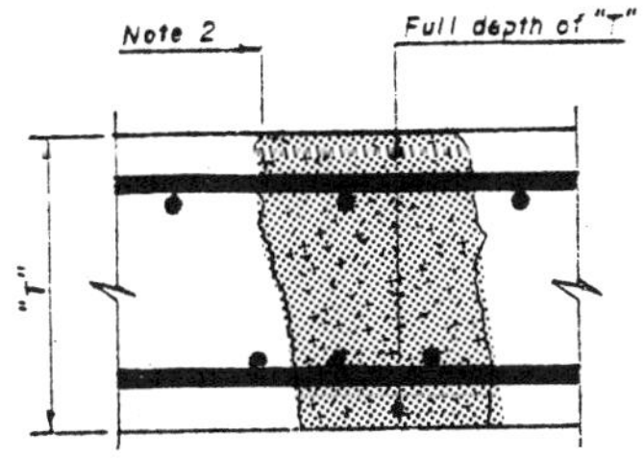


REPAIR TYPE C

CONSTRUCTION OF LMC OVERLAY
ON EXISTING DECK SLAB, BARE CONCRETE SURFACE

1. Deck Condition Survey (by DOT or consultant when directed).
2. Outline repair areas designated on Plans and as may be prescribed after Deck Condition Survey.
3. Pressure Injection of cracks (if scheduled) except in designated repair areas.
4. Sawcut repair areas.
5. Scarify to depth specified on Plans.
6. Determine if other areas beyond the initial sawcut peripheries should be prescribed for repairs.
7. New sawcuts for any additional repair limits designated.
8. Remove deteriorated concrete in designated areas.
9. Clean existing ReBars by sandblasting, waterblasting, or wire brushing.
10. Supplement existing reinforcement with new ReBars as required.
11. Fill repair areas with material specified.
12. Cure repair areas (time depends on repair material used).
13. Sandblast or waterblast entire surface to remove loosened chips of concrete, curing compound and laitance from repair areas, and all other residue. Chipping with hand tools or pneumatic scabblers may be required to remove oil intrusions. Wet area and cover with polyethylene.
14. Place LMC Overlay.

NJ

CONSTRUCTION OF BC OVERLAY
ON EXISTING DECK SLAB WITH BC OR AC SURFACE

1. Remove B.C. or A.C. Overlay.
2. Remove Waterproofing (if any).
3. Deck Condition Survey (by DOT or consultant when directed).
   (a) Outline repair areas.
   (b) Designate Repair Type (A, B, or C). Compare quantities with contract plan estimates.
   (c) Confirm that repair materials specified on Plans shall be used. (Construction change order required if change is warranted).
4. Pressure Injection of cracks (if scheduled) except in designated repair areas.
5. Sawcut repair areas.
6. Remove deteriorated concrete in repair areas.
7. Clean existing ReBars by sandblasting, waterblasting, or wirebrushing.
8. Supplement existing reinforcement with new ReBars as required.
9. Fill repair areas with material specified.
10. Cure repair areas (time depends on repair material used).
11. Sandblast or waterblast entire surface to remove loosened chips of concrete, curing compound and laitance from repair areas, and all other residue. Projections shall be removed by grinding or other approved method. Surface of concrete must be clean in accordance with membrane waterproofing manufacturer's recommendations.
12. Place membrane waterproofing in accordance with manufacturer's recommendations.
13. Place Bituminous Concrete Surface Overlay to thickness specified on Plans.
14. Prepare information for As-Built Plans showing:
   (a) Product name of membrane waterproofing placed.
   (b) Outline of repair areas with following information at each location:
      - Repair material used.
      - Repair Type (A, B, or C).
      - Nominal depth of repair.
   (c) Location and length of supplementary ReBars.
   (d) Outline of cracks sealed by pressure injection.

Commentary

An additional 10to 20% (depends on size of Overlay area) should be added to the estimated quantity (based on the 1 1/4 inch minimum plan thickness specified) if the Structural Design Engineer anticipates that additional thicknesses may be required at some locations to achieve proposed plan profile elevations and cross-slopes.

Where additional thickness of LMC Overlay is anticipated, the total Overlay thickness should be clearly prescribed on the Plans. In this case, the total Overlay thickness should be limited to 2 inches maximum. If the area of additional thickness is significant (compared to total overlay area) changes in the specifications for LMC mix design and curing time should be considered by special provisions in the Supplementary Specifications for the Project.

Situations which would produce on LMC Overlay thickness in excess of 2 inches require special solutions (for example, if substructure settlements have resulted in significant lowering of the original plan/as-built profile elevations, jacking of the superstructure at selected locations may be considered and scheduled as a contract pay item when warranted).

Specifications provide that when LMC is used as the repair material in Repair Types A and B, the volume of LMC used in the repair area is not included in the volume of LMC measured for payment, but is measured for payment under square feet of Repair of Concrete Deck of the Type which applies.

Scarification is always scheduled for rehabilitation projects using Latex Modified Concrete Overlay. The standard depth adopted for scarification is 1/4 inches. This depth shall be noted on the Plans. Where additional depth is required, the areas and depths should be clearly prescribed on the plans.

In the case of a newly constructed bridge deck which has not been opened to traffic (or subjected to traffic for a relatively short time) and an LMC Overlay is to be added as a "Protective System" scarification may not be necessary. Cleaning of the surface by sandblasting may suffice. Provisions should be included in the Supplementary Specifications accordingly.

Pressure Injection is scheduled when warranted, as determined by the recommendations of the Deck Condition Survey or visual observations on-the-site. The outline of the cracks to be sealed shall be shown on the plans.

Various other contract pay items are scheduled on a project to project basis when warranted. Special provisions shall be included in the Supplementary Specifications.

If new preformed elastomeric joint sealer and/or reinforced elastomeric expansion dams are proposed and approved for scheduling as a contract pay item, the ideal would be to provide a continous sealer across the full width of the deck slab. If this is impractical the Structural Design Engineer shall prescribe the location of splices on the Plans depending on construction traffic staging.

Latex modified concrete will not bond properly to oil, epoxy, other contaminantes, and rubber (neoprene). Use of reinforced elastomeric expansion dams exclusively as the "headers" is not recommended. Other details should be considered as part of the installation.

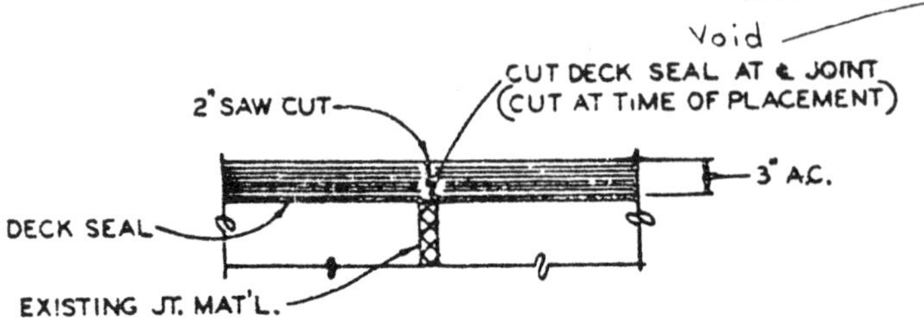


AC OVERLAY SAWCUT

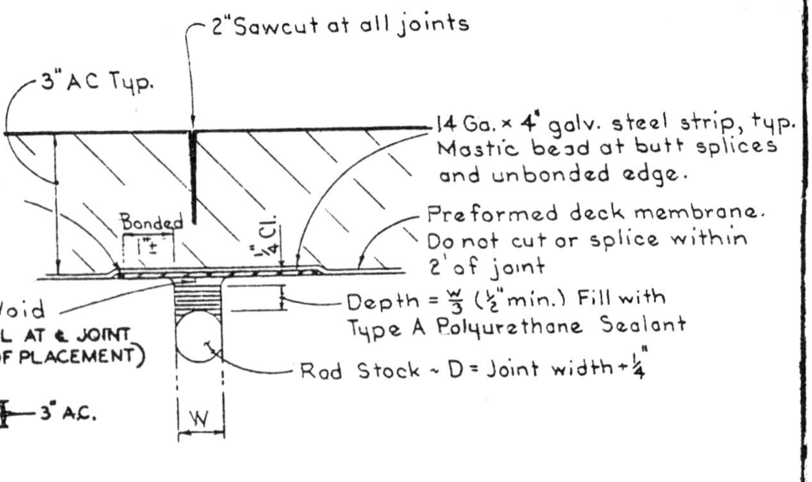


DECK JOINT SEALER

## EDGE TREATMENT OF MEMBRANE WATERPROOFING

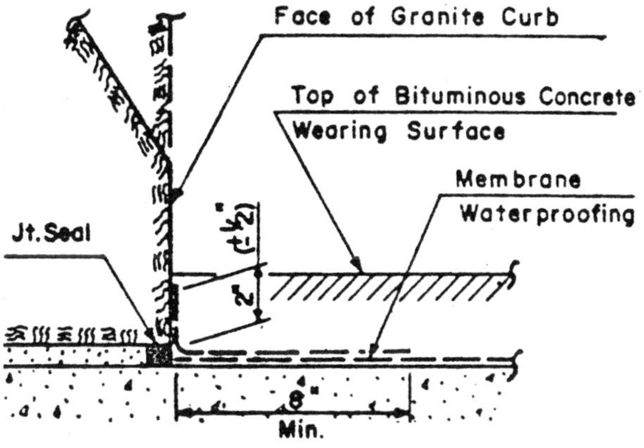


TREATMENT OF MEMBRANE WATERPROOFING AT GUTTER

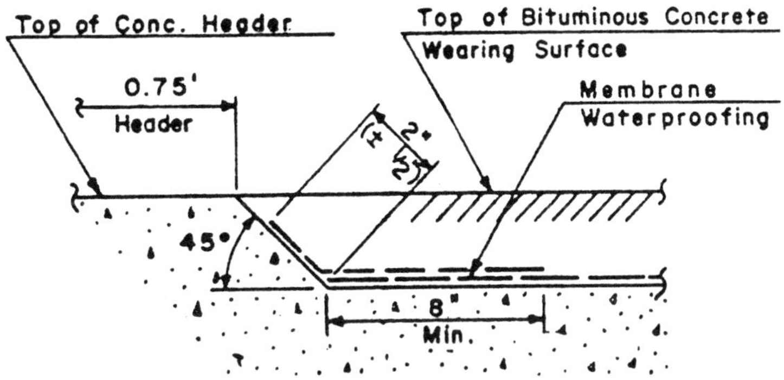


TREATMENT OF MEMBRANE WATERPROOFING AT HEADERS

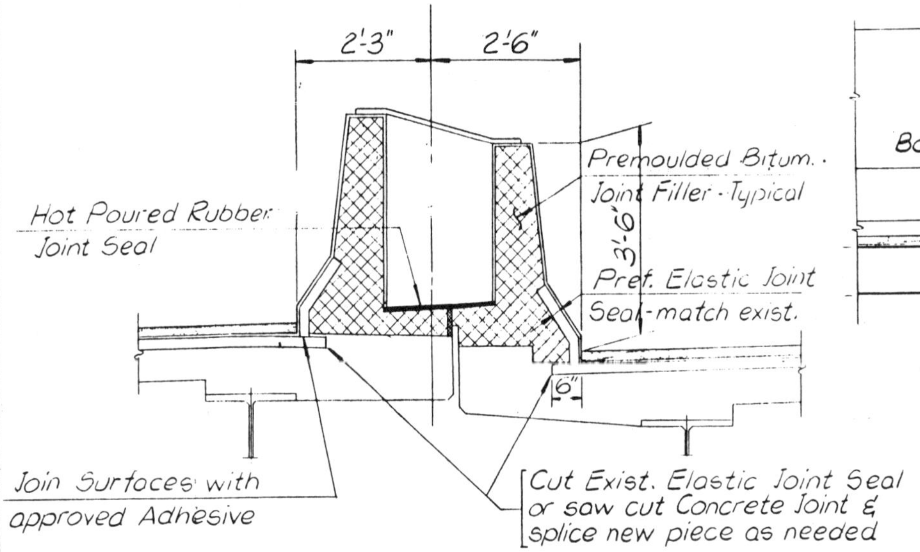


TYPICAL SECTION @ MEDIAN

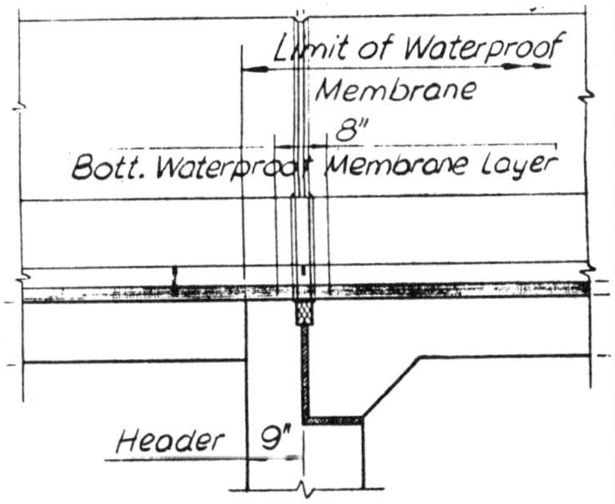


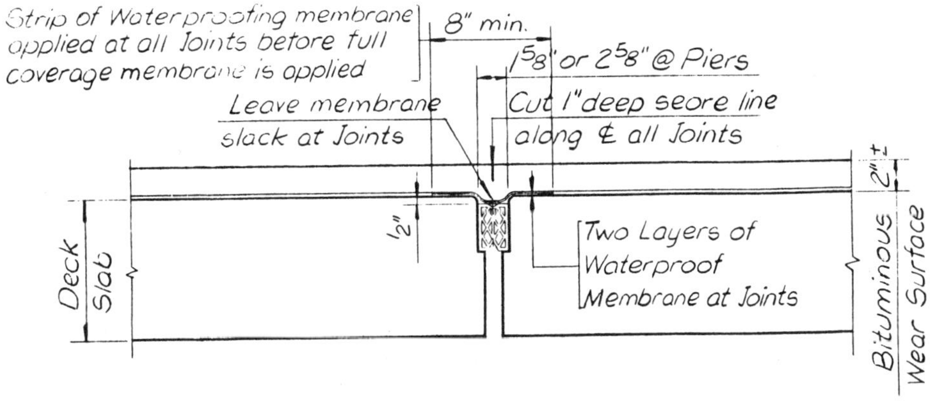

# DETAIL OF DRAIN PIPES THROUGH DECK SLAB

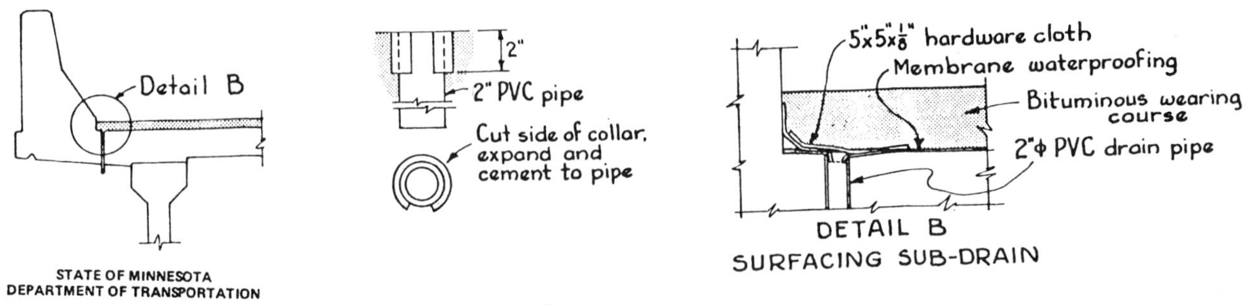


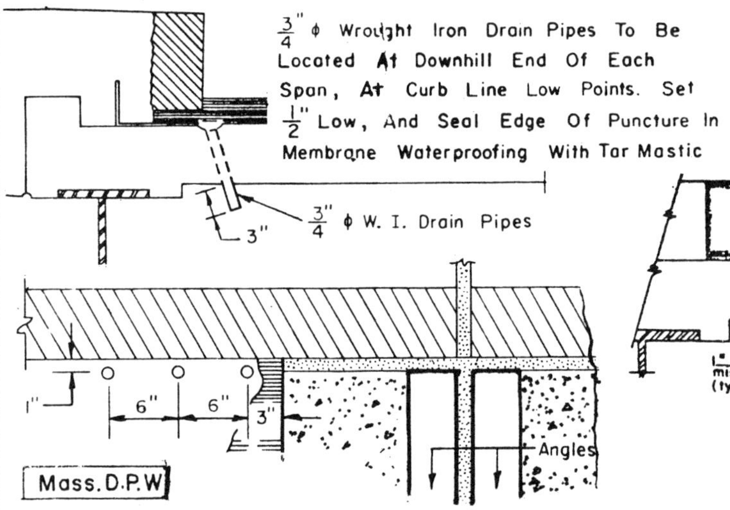


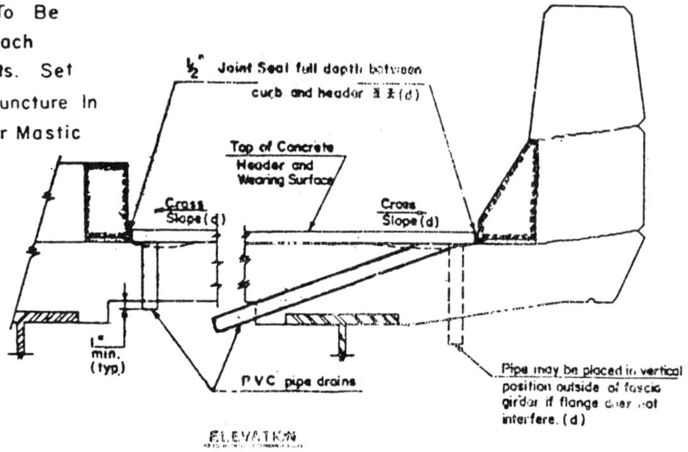


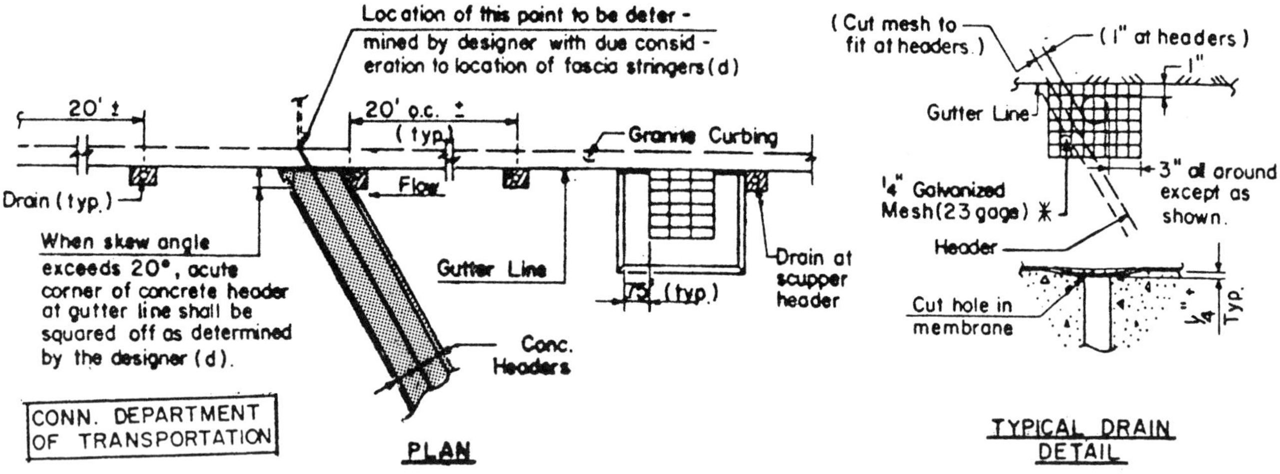


WEARING SURFACE INTERFACE DRAINAGE DETAILS

PLAN NOTES:

1 1/2" Polyvinyl Chloride Plastic Pipe Drains shall be placed thru slabs where shown on the plans.

* The cost of furnishing and installing 1/4" square Galvanized Mesh shall be included in the contract Bid price per ton for "Bituminous Concrete Wearing Surface."

DESIGN INFORMATION:

Pipe Drains to be located by designer as follows:

1 1/2" PVC Pipe Drains shall be located along the gutter on the low end of cross-slopes at approximately 20' o.c. starting at the low end of span. No drains shall be placed directly above travelway.

On structures over Railroads, the Pipe Drains, in general, shall not be located in spans over tracks. However, for long span structures, the pipes may be located in spans over tracks, in which case they shall not be located closer than 25' from the center line of outside tracks.

## C.1 REINFORCED CONCRETE SLAB BRIDGE - DESIGN AND DETAILS

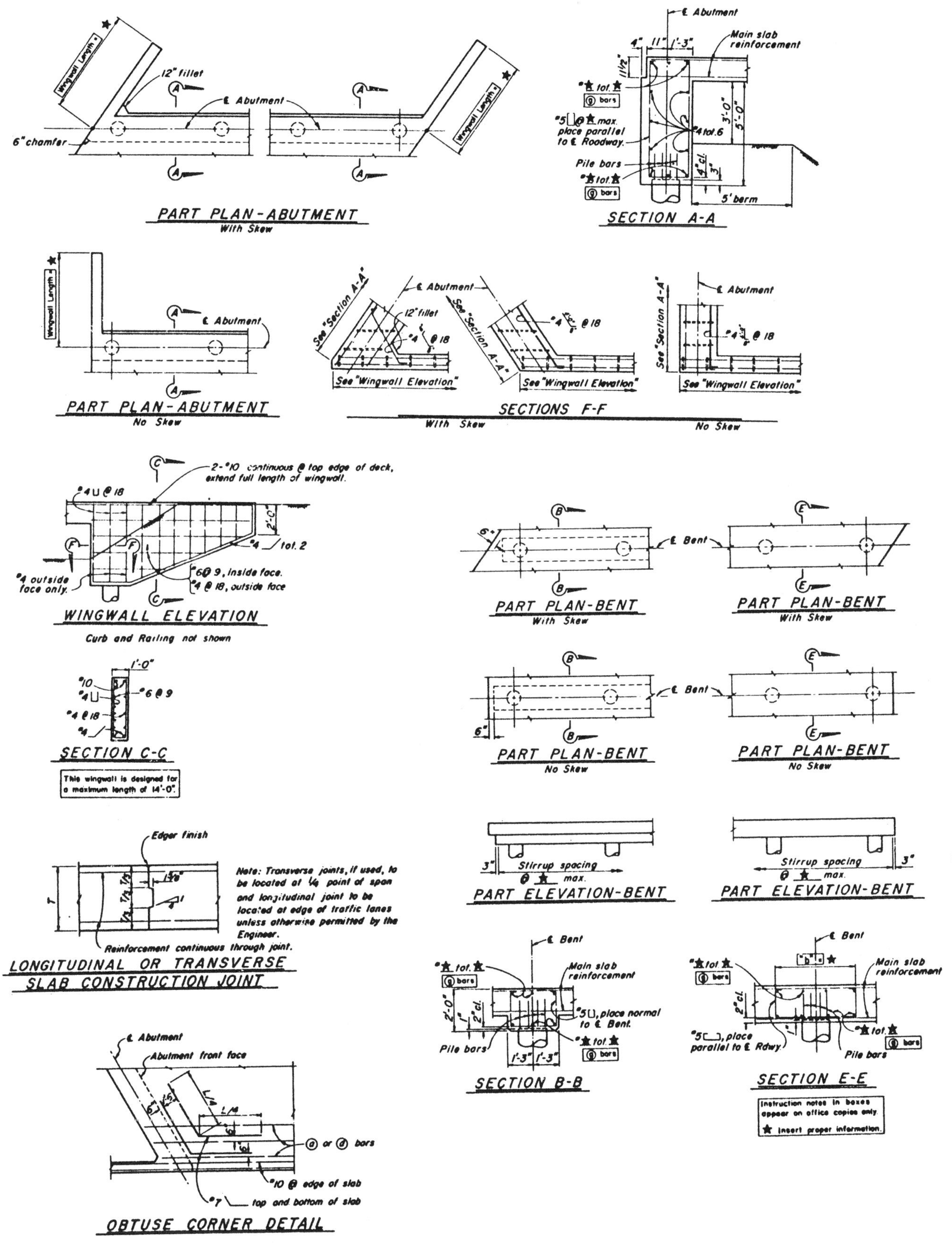


Calif

ABUTMENT AND BENT DETAILS

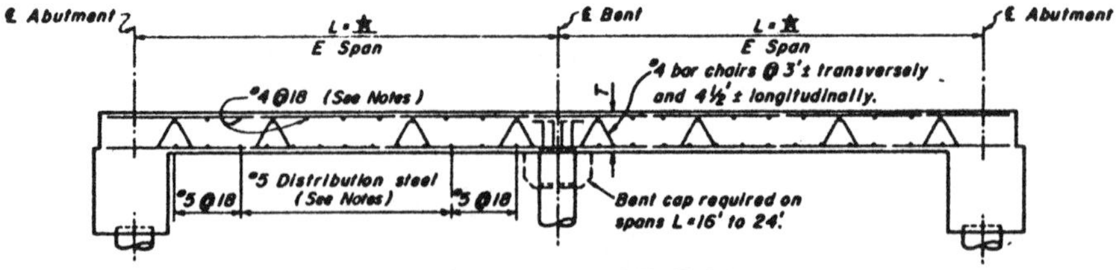


LONGITUDINAL SECTION

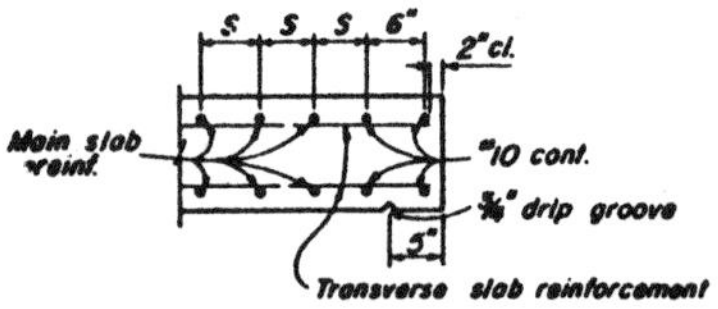


EDGE OF SLAB DETAIL

Curb and railing not shown

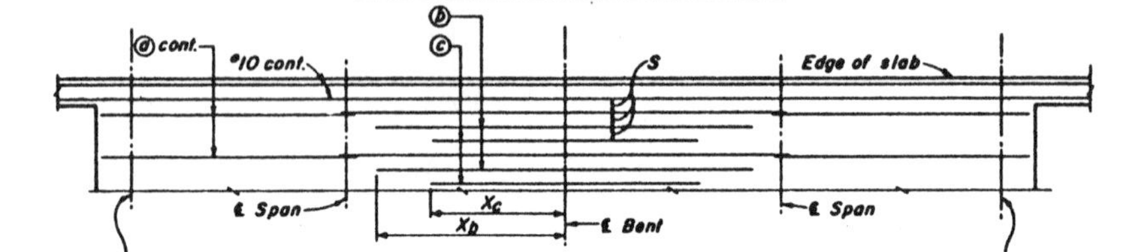


REINFORCEMENT-TOP OF SLAB

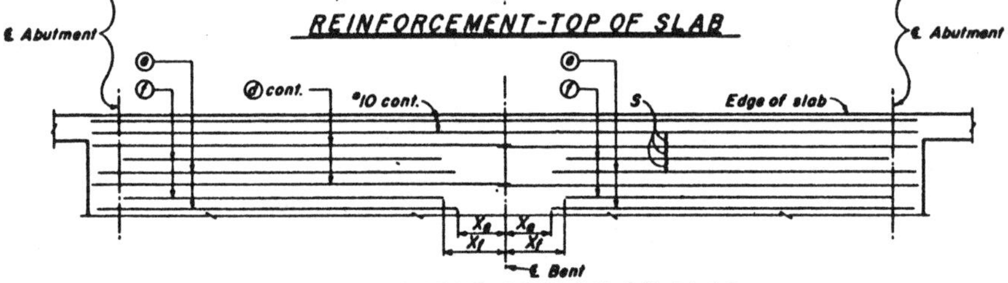


REINFORCEMENT-BOTTOM OF SLAB

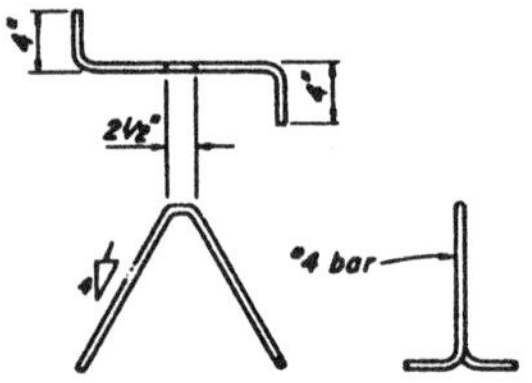


BAR CHAIR DETAIL

| L = Length of Span | | | 16' | 18' | 20' | 22' | 24' | 26' | 28' | 30' | 32' | 34' | 36' | 38' | 40' | 42' | 44' |
|---|---|---|---|---|---|---|---|---|---|---|---|---|---|---|---|---|---|
| REINFORCEMENT – Top of Slab | ⓐ bars | Size | #8 | #9 | #9 | #8 | #11 | #10 | #9 | #9 | #8 | #9 | #10 | #9 | #10 | #11 | #11 |
| | | Length | 34'-0" | 38'-0" | 42'-0" | 46'-0" | 50'-0" | 54'-0" | 61'-0" | 65'-0" | 69'-0" | 73'-0" | 77'-6" | 81'-0" | 86'-0" | 90'-0" | 94'-0" |
| | ⓑ bars | Size | #7 | #7 | #7 | #8 | #7 | #8 | #9 | #9 | #10 | #10 | #10 | #11 | #10 | #11 | #11 |
| | | Length | 12'-0" | 11'-6" | 12'-6" | 15'-0" | 11'-0" | 13'-0" | 15'-6" | 16'-6" | 19'-0" | 19'-0" | 18'-6" | 21'-6" | 20'-0" | 20'-0" | 21'-0" |
| | | $X_b$ | 6'-0" | 5'-9" | 6'-3" | 7'-6" | 5'-6" | 6'-6" | 7'-9" | 8'-3" | 9'-6" | 9'-6" | 9'-3" | 10'-9" | 10'-0" | 10'-0" | 10'-6" |
| | ⓒ bars | Size | #7 | #7 | #8 | #8 | #7 | #9 | #10 | #10 | #10 | #10 | #10 | #11 | #11 | #11 | #11 |
| | | Length | 6'-6" | 6'-6" | 7'-6" | 8'-0" | 7'-0" | 7'-6" | 9'-0" | 9'-6" | 10'-0" | 10'-0" | 10'-0" | 11'-0" | 13'-0" | 12'-0" | 12'-0" |
| | | $X_c$ | 3'-3" | 3'-3" | 3'-9" | 4'-0" | 3'-6" | 3'-9" | 4'-6" | 4'-9" | 5'-0" | 5'-0" | 5'-0" | 5'-6" | 6'-6" | 6'-0" | 6'-0" |
| REINFORCEMENT – Bottom of Slab | ⓓ bars | Size | #8 | #8 | #9 | #8 | #11 | #11 | #10 | #10 | #10 | #10 | #9 | #9 | #8 | #9 | #9 |
| | | Length | 34'-0" | 38'-0" | 42'-0" | 46'-0" | 50'-0" | 54'-0" | 58'-0" | 64'-0" | 68'-0" | 72'-0" | 76'-0" | 80'-0" | 84'-0" | 88'-0" | 92'-0" |
| | ⓔ bars | Size | #7 | #7 | #7 | #8 | #7 | #9 | #8 | #9 | #9 | #9 | #9 | #9 | #9 | #10 | #10 |
| | | Length | 14'-6" | 16'-6" | 17'-6" | 19'-6" | 19'-0" | 21'-0" | 23'-6" | 25'-6" | 26'-6" | 28'-0" | 30'-0" | 31'-6" | 33'-6" | 35'-0" | 36'-6" |
| | | $X_e$ | 2'-6" | 2'-6" | 3'-6" | 3'-6" | 6'-0" | 6'-0" | 5'-6" | 6'-0" | 6'-6" | 7'-0" | 7'-0" | 7'-6" | 7'-6" | 8'-0" | 8'-6" |
| | ⓕ bars | Size | #7 | #8 | #8 | #8 | #7 | #6 | #9 | #8 | #8 | #9 | #10 | #10 | #11 | #11 | #11 |
| | | Length | 10'-0" | 12'-0" | 13'-0" | 14'-0" | 11'-6" | 11'-0" | 17'-0" | 16'-0" | 17'-0" | 19'-0" | 22'-6" | 24'-6" | 27'-6" | 26'-0" | 27'-6" |
| | | $X_f$ | 4'-3" | 4'-6" | 5'-3" | 6'-6" | 8'-6" | 10'-0" | 8'-0" | 9'-9" | 10'-6" | 10'-9" | 10'-0" | 10'-3" | 10'-6" | 12'-0" | 12'-6" |
| Dist.-Steel Spacing | | | 11" | 11" | 11" | 10" | 10" | 10" | 10" | 9" | 9" | 9" | 9" | 9" | 9" | 9" | 9" |
| T = Thickness of slab | | | 10½" | 11½" | 11¾" | 12¾" | 13½" | 14½" | 15½" | 16½" | 17½" | 18½" | 19¾" | 20¾" | 22" | 23" | 24" |
| S = Bar Spacing | | | 6" | 6" | 6" | 5½" | 5½" | 6" | 6" | 5½" | 5½" | 5½" | 5½" | 5½" | 5" | 5½" | 5" |
| Clearance to main reinforcing | | | 1½" | 1½" | 1½" | 1½" | 2" | 2" | 2" | 2" | 2" | 2" | 2" | 2" | 2" | 2" | 2" |
| Quantities per ft. of slab width (For bridge without skew) | Conc. ft³ | | 14.9 | 17.6 | 20.5 | 24.5 | 28.1 | 32.7 | 37.5 | 42.7 | 48.2 | 54.0 | 60.9 | 67.5 | 75.0 | 82.4 | 90.0 |
| | Steel-lbs. | | 145 | 175 | 205 | 230 | 310 | 315 | 330 | 385 | 395 | 450 | 495 | 510 | 600 | 655 | 735 |
| Deflection at Midspan Δ | | | .00' | .00' | .00' | .02' | .02' | .03' | .03' | .04' | .04' | .05' | .05' | .06' | .07' | .07' | .08' |

NOTES :

Splices in top main bars to be 35 dia. and located near center of span.

Splices in bottom main bars to be 20 dia. and located near bent.

Spacing of #4 @ 18 is measured normal to bars.

Spacing of #5 bars in bottom of slab is measured along ℄ roadway. For spacing in center half of span see Table.

Place all transverse bars parallel to bent.

(Calif)

SLAB DETAILS-2 SPANS

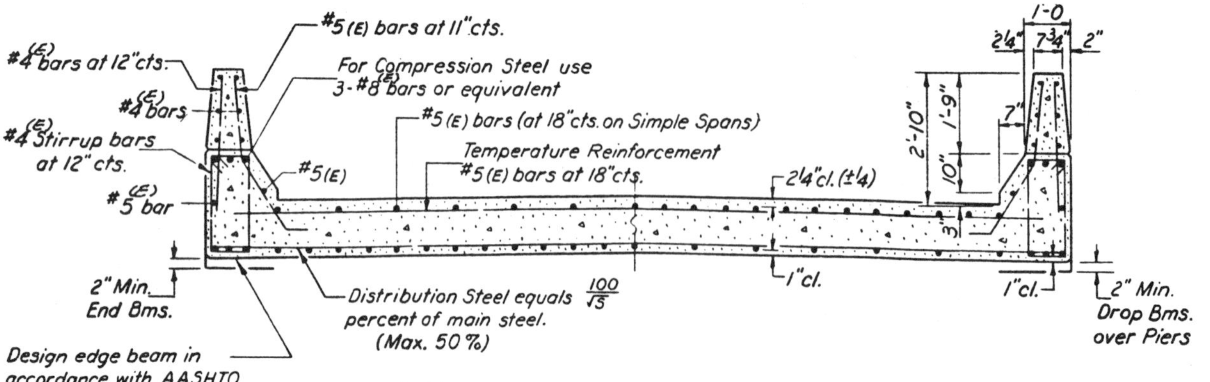


Design edge beam in accordance with AASHTO Art. 1.3.2(D), but use as a min. the same area of steel per ft. as in slab.

HALF SECTION

Simple Spans or Midspan of Continuous Spans

HALF SECTION

Near Pier of Continuous Spans

(ILL.)

* NOTE: Top reinforcement bars no less than #5 at 18"cts. any location.

Pt. of Theoretical cut-off & check for Fatigue Art. 1.5.38(B)

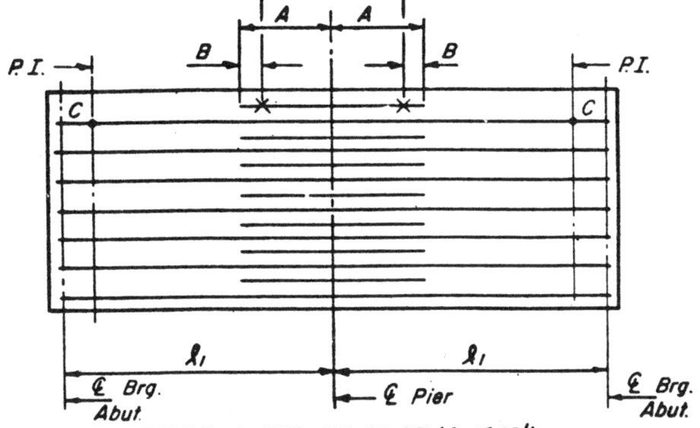


REINF. IN TOP OF SLAB ($\ell_1 \leq 18'$)

(Use same size bars, or two sizes where long bars are not less than $\frac{1}{3}$As. As = Total Neg. Moment steel area.)

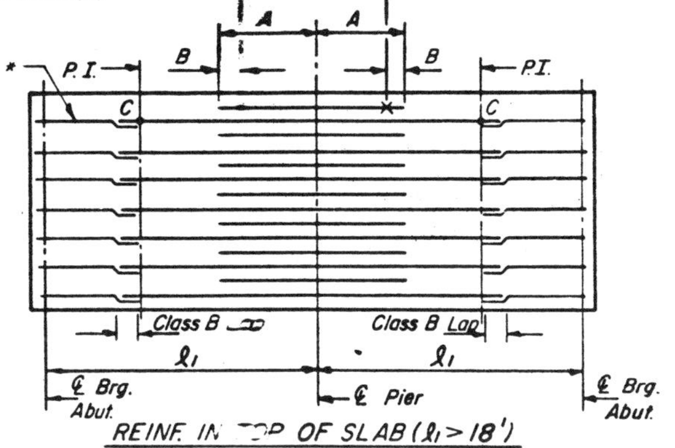


REINF. IN TOP OF SLAB ($\ell_1 > 18'$)

(Use same size bars, or two sizes where long bars are not less than $\frac{1}{3}$As. As = Total Neg. Moment steel area.)

REFERENCE TO A.A.S.H.T.O.:
- A - Article 1.5.14 & 1.5.13(c)(2)
- B - 1.5.13(A)(4)
- C - 1.5.13(C)(3)
- D - 1.5.13(B)(1) & (2)

NOTES:
- P.I. - The Point of Inflection as determined from the applied Moment Envelopes of

$$D + [-(L + I)] \text{ for Top Reinforcement}$$

$$D + [+(L + I)] \text{ for Bottom Reinforcement}$$

for either the W.S.D. or the L.F.D. methods. Max. Concrete Stresses shall be investigated at sections of stress reversal Art. 1.5.38

Pt. of Theoretical cut-off & check for Fatigue Art. 1.5.38

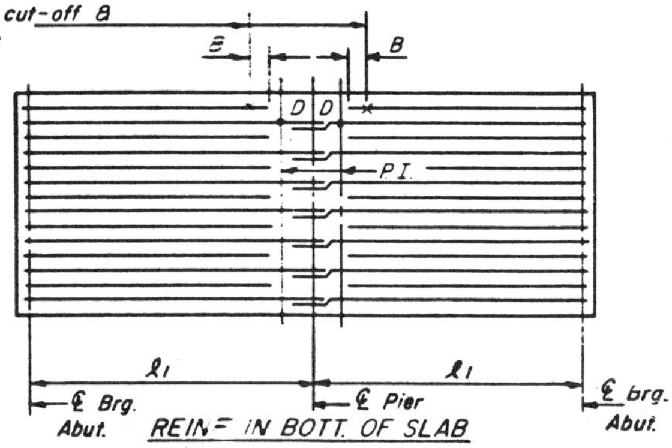


REINF. IN BOTT. OF SLAB

(Use same size bars)

## TWO SPAN SLAB BRIDGES

than #5 at 18"cts. any location.

by extending and lapping the long bars.

Pt. of Theoretical cut-off & check for Fatigue Art. 1.5.38(B)

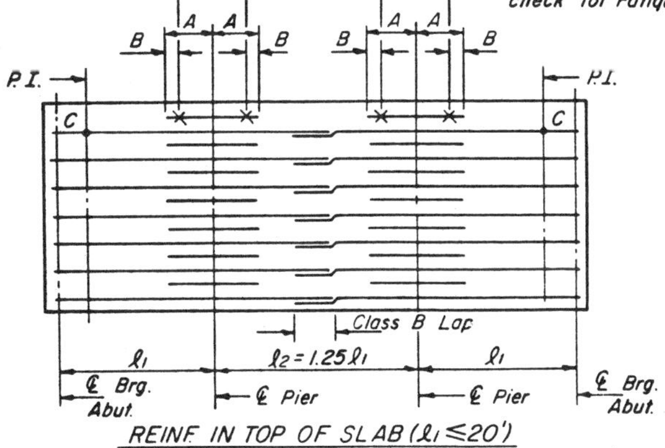


REINF. IN TOP OF SLAB ($\ell_1 \leq 20'$)

(Use same size bars, or two sizes where long bars are not less than $\frac{1}{3}$As. As = Total Neg. Moment steel area.)

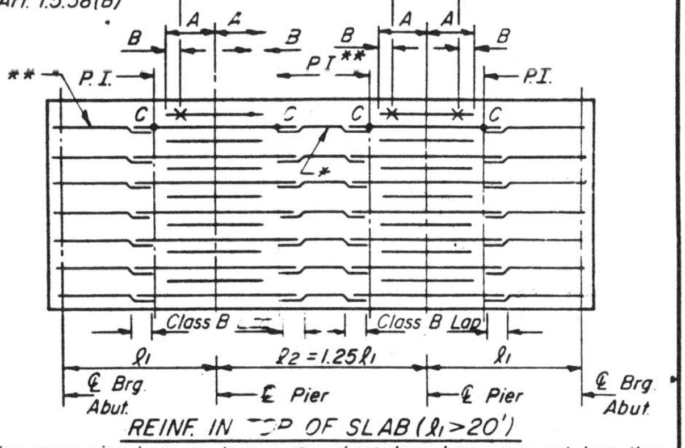


REINF. IN TOP OF SLAB ($\ell_1 > 20'$)

(Use same size bars, or two sizes where long bars are not less than $\frac{1}{3}$As. As = Total Neg. Moment steel area.)

REFERENCE TO A.A.S.H.T.O.:
- A - Article 1.5.14 & 1.5.13(c)(2)
- B - 1.5.13(A)(4)
- C - 1.5.13(C)(3)
- D - 1.5.13(B)(1) & (2)

NOTES:
- P.I. - The Point of Inflection as determined from the applied Moment Envelopes of

$$D + [-(L + I)] \text{ for Top Reinforcement}$$

$$D + [+(L + I)] \text{ for Bottom Reinforcement}$$

for either the W.S.D. or the L.F.D. methods. Max. Concrete Stresses shall be investigated at sections of stress reversal Art. 1.5.38

Pt. of Theoretical cut-off & check for Fatigue Art. 1.5.38

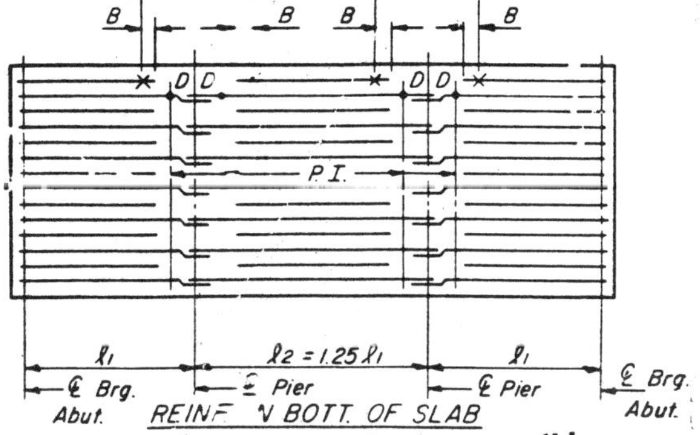


REINF. IN BOTT. OF SLAB

(Use same size bars)

## THREE SPAN SLAB BRIDGES

ILL

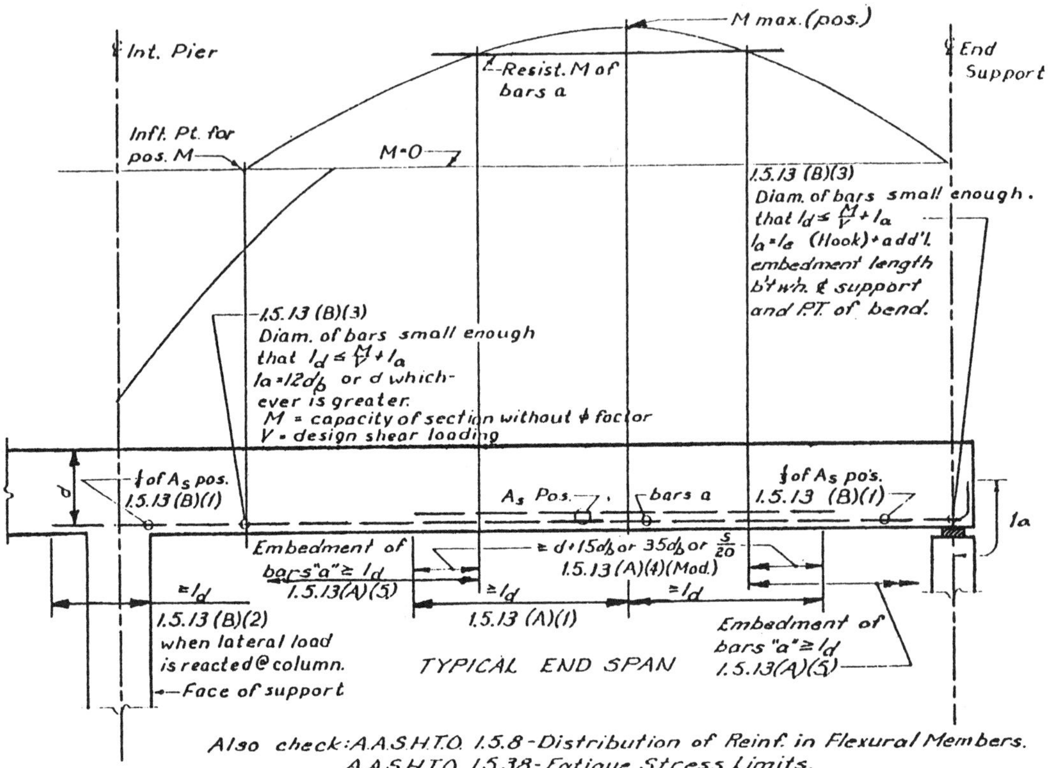


*Also check: A.A.S.H.T.O. 1.5.8-Distribution of Reinf. in Flexural Members.*
*A.A.S.H.T.O. 1.5.38-Fatigue Stress Limits.*
*A.A.S.H.T.O. 1.5.39-Distribution of Flexural Reinforcement.*

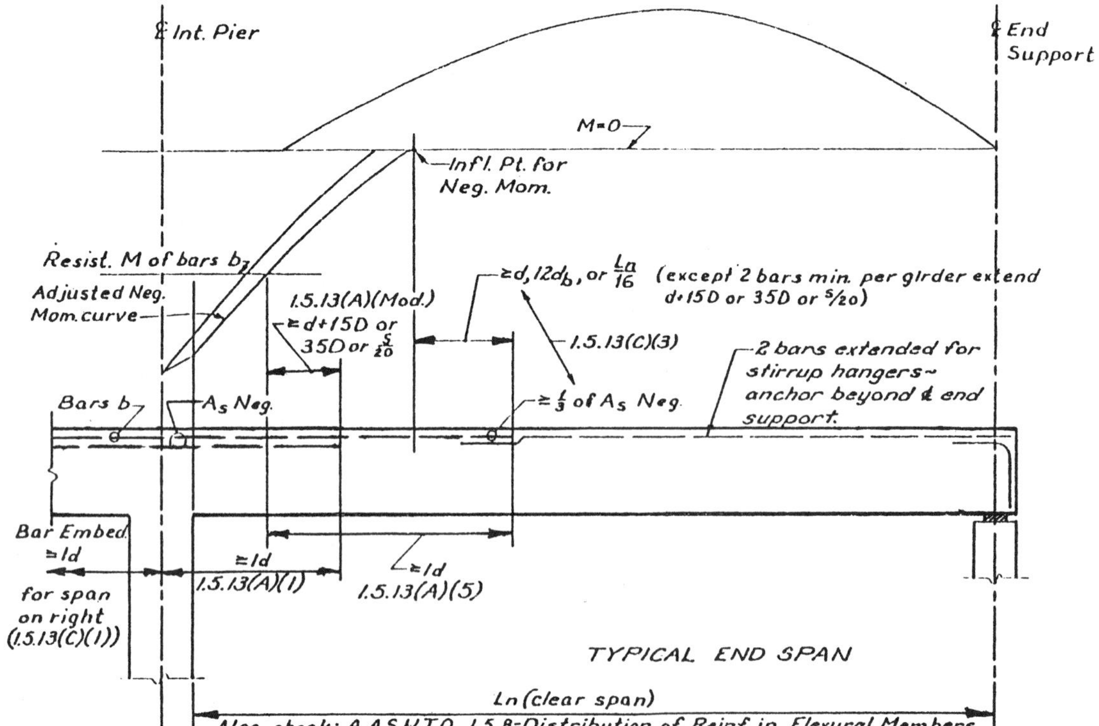


*Also check: A.A.S.H.T.O. 1.5.8-Distribution of Reinf. in Flexural Members.*
*A.A.S.H.T.O. 1.5.38-Fatigue Stress Limits.*
*A.A.S.H.T.O. 1.5.39-Distribution of Flexural Reinforcement.*

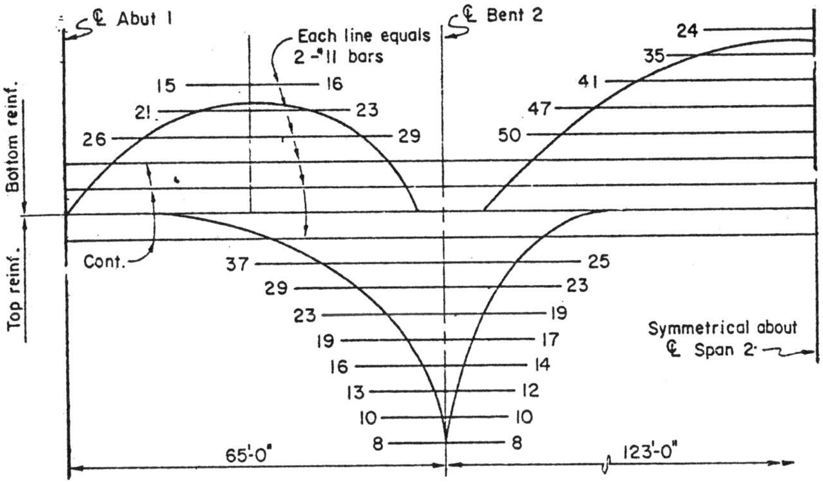


The main reinforcement should be placed so that the ends are staggered rather than being progressively shorter.

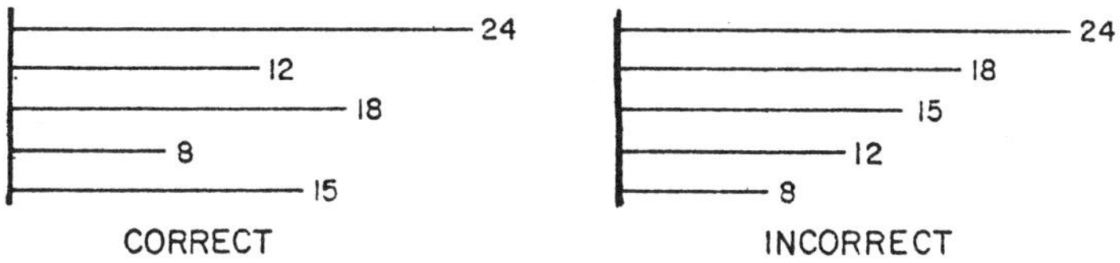


Bars should be symmetrical about the ℄ of bay and the shorter bars should generally be placed nearer the ℄ of bay.

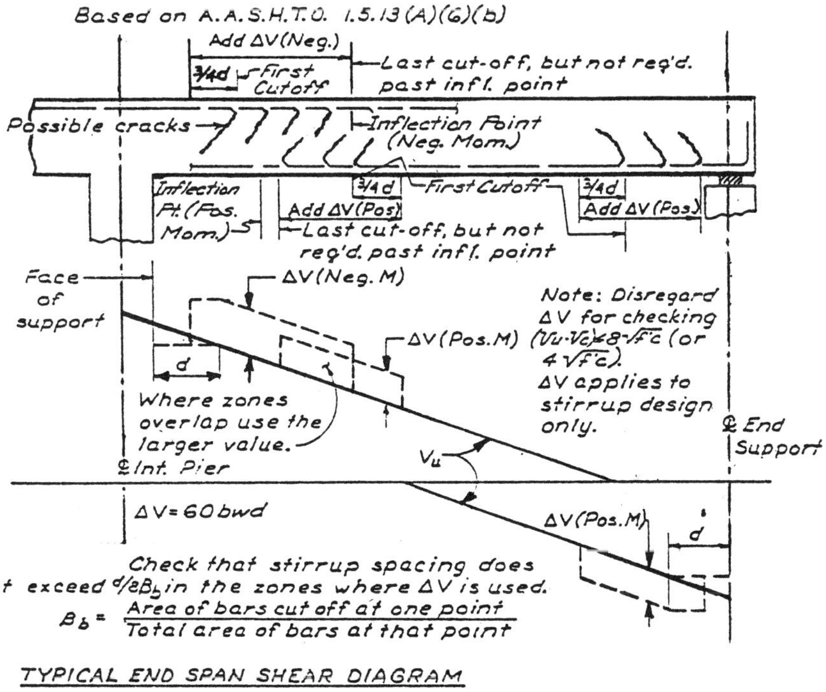


TYPICAL END SPAN SHEAR DIAGRAM

Where practical, A.A.S.H.T.O. 1.5.13(A)(6)(a) or (c) may be used instead of (b).

# DESIGN EXAMPLE CONCRETE SLAB BRIDGE

**From the Bridge Manual of Wisconsin Dept. of Transp.**

## DESIGN EXAMPLE

A continuous haunched slab structure is used for the design example. The same basic procedure is applicable to continuous flat slabs.

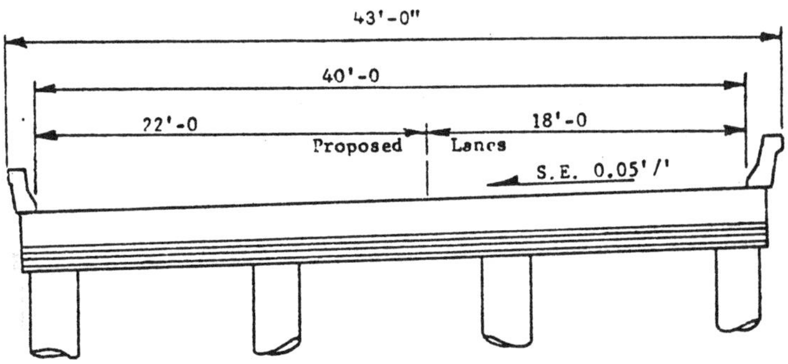


Span Lengths: 38'-0, 51'-0, 38'-0.
Live Load: HS 20.
Skew 6° 00' RHF.

**Dead Loads**

Refer to Figure , for a span length of 51 feet. The slab depth is estimated at 16 inches. The haunch depth (D) is approximately equal to $d_s$ divided by 0.6 where $d_s$ is the slab depth.

$$D = 16/0.6 = 27 \text{ in.}$$

The length of haunch is approximately 0.15 to $0.18L_2$ or say 8'-0 which equals $0.157\ L_2$.

**SLAB - HAUNCH TRANSITION**

The following procedure is used to compute the transition tangent length employing a 20'-0 radius

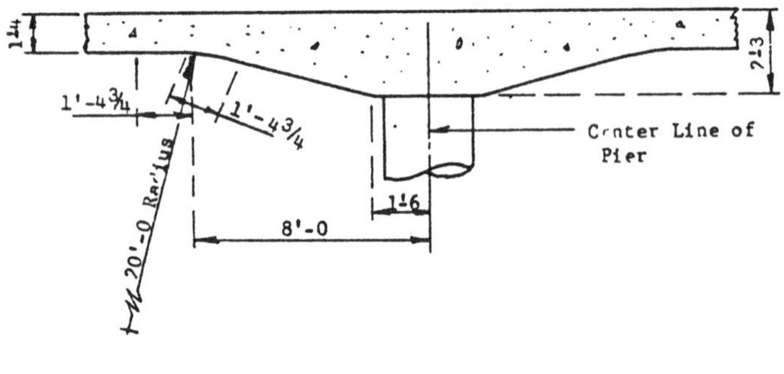


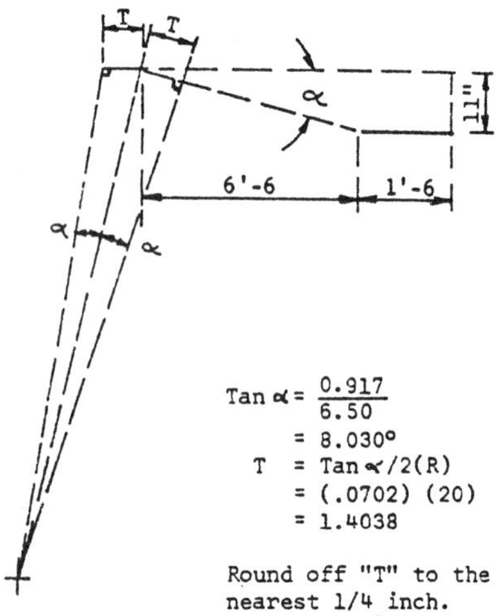


$$\text{Tan } \alpha = \frac{0.917}{6.50}$$
$= 8.030°$
$T = \text{Tan } \alpha/2 (R)$
$= (.0702)\ (20)$
$= 1.4038$

Round off "T" to the nearest 1/4 inch.
Use T = 1'-4¾

*State of Wisconsin/Department of Transportation*

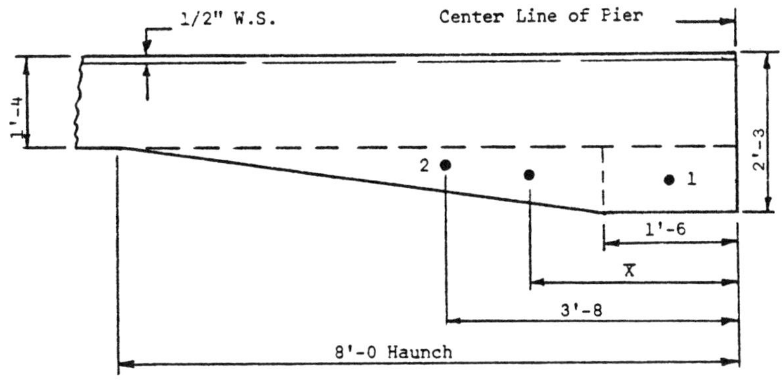


For hand computations determine partial haunch dead load. Determine value of $\bar{X}$ and distribute haunch weight uniformly over twice this distance. Haunch dead load is computed automatically by computer program.

Slab Dead Load = (16/12)(1)(150) = 200 #/ft

Rail, curb, and parapet dead load is distributed over the full width of slab for negative moment only.

$$\text{Parapet Dead Load} = \frac{(2)(338)}{42.17} = 16\ \#/\text{ft}$$

A post dead load of 20 $\#/ft^2$, for possible future wearing surface, plus the 1/2 inch wearing surface load (6 $\#/ft^2$) is input to the program.

Live Load Distribution

Wheel loads are distributed over a width, $E = 4.0 + 0.06(S) \leq 7'\text{-}0$ where S equals the effective span length in feet. Lane loading is distributed over a width of 2E.

The distribution factor, DF, is computed for a unit width of slab equal to one foot. The distribution factor is

$$DF = \frac{1}{E}$$

For spans 1 & 3:

$$E = 4.0 + 0.06(38) = 6.28'$$

$$DF = \frac{1}{6.28} = 0.159$$

For span 2:

$$E = 4.0 + 0.06(51) = 7.06' \quad \text{must be} \leq 7.0'$$

$$DF = \frac{1}{7.0} = 0.143$$

Service Load Moments (ft-k)

| Point | DLM | Curb DL | +(L+I) | -(L+I) | DLM+(L+I) | Curb DL+ DLM-(L+I) |
|---|---|---|---|---|---|---|
| 0.1 | 8.9 | .6 | 16.4 | -3.1 | 25.3 | 6.4 |
| 0.2 | 14.7 | 1.1 | 26.6 | -6.3 | 41.3 | 9.5 |
| 0.3 | 17.2 | 1.3 | 31.4 | -9.4 | 48.6 | 9.1 |
| 0.4 | 16.6 | 1.2 | 32.3 | -12.5 | 48.9 | 5.3 |
| 0.5 | 12.8 | .9 | 30.8 | -15.7 | 43.6 | -2.0 |
| 0.6 | 5.9 | .4 | 27.9 | -18.8 | 33.8 | -12.5 |
| 0.7 | -4.3 | -.3 | 21.1 | -22.0 | 16.8 | -26.5 |
| 0.8 | -17.6 | -1.3 | 11.7 | -25.1 | -5.9 | -44.0 |
| 0.9 | -34.3 | -2.5 | 4.6 | -28.2 | -29.7 | -65.0 |
| 1.0 | -55.4 | -3.9 | 8.0 | -35.5 | -46.8 | -94.8 |
| 1.1 | -28.1 | -2.0 | 4.7 | -22.1 | -23.4 | -52.2 |
| 1.2 | -8.0 | -.6 | 11.8 | -18.4 | 3.8 | -27.0 |
| 1.3 | 6.3 | .5 | 22.2 | -14.9 | 28.5 | -8.1 |
| 1.4 | 14.9 | 1.1 | 28.6 | -11.4 | 43.5 | 4.6 |
| 1.5 | 17.7 | 1.3 | 30.0 | -10.0 | 47.7 | 9.0 |

Longitudinal Slab Reinforcement

Positive moment reinforcement for span 1. At the 0.4 point of span 1, $(+M_u) = 1.3(16.6 + 5/3\ (32.3)) = 91.6$ ft-k

$d = 16 - (1.5 + 0.7 + .5) = 13.3$

Solve for steel area using Table 17.4:

$$R_u = \frac{Mu}{\phi bd^2} = \frac{(91.6)(12)(1000)}{(0.9)(12)(13.3)^2} = 575.4 \text{ psi}$$

$$\rho = .0106$$

$$A_s = (.0106)(12)(13.3) = 1.70 \text{ in}^2/\text{ft.}$$

Check fatigue by AASHTO formula for max. stress range of steel. At 0.4 point of span 1 the moment range is 44.8 ft-k.

Allowable $f_f = 21. - .33\ f_{min} + 8(.3) = 22.6$ ksi

where $.33\ f_{min} = .33 \left[\frac{(16.6 - 12.5) \times 12}{1.7 \times .9 \times 13.3}\right] \cong .8$ ksi

$$A_s\ (\text{min}) = \frac{M}{fs\ j\ d} = \frac{(44.8)(12)}{(22.6)(0.9)(13.3)} = 1.99 \text{ in}^2/\text{ft} \quad \underline{\text{controls}}$$

Fatigue check using max allowable concrete stress of 0.5 f'c @ 0.4 point of span 1.

$$f_c = \frac{2M}{k\ j\ bd^2} = \frac{(2)(44.8)(12)}{(0.30)(0.9)(12)(13.3)^2} = 1.88 \text{ ksi}$$

less than allowable stress of 2.0 ksi.

Crack control is governed by the equation

$$f_s = \frac{Z}{(d_cA)^{1/3}} \quad \text{not to exceed } 0.6\ f_y.$$

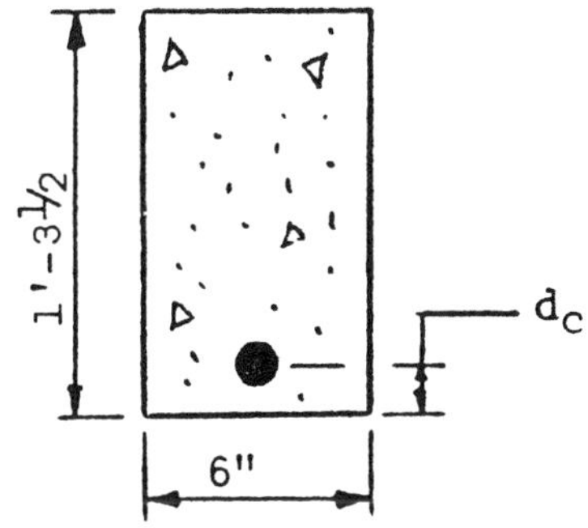


Z = 170 k/in for positive steel area

$A_s$ required is 1.99 in$^2$/ft

Try: 9's at 6" c-c
Spacing ($A_s$ = 2.00 in$^2$/ft)

$$d_c = 1.5 + (1.13)/2 = 2.065 \text{ in}$$

$$A = (2)(2.065)(6.) = 24.78 \text{ in}^2$$

$$f_s = \frac{170}{[(2.065)(24.78)]^{1/3}} = 46.4 \text{ ksi} > 36 \text{ ksi}$$

$$f_s\ (\text{act}) = \frac{M}{A_sjd} = \frac{(48.9)(12)}{(2.00)(0.9)(13.3)} = 24.5 \text{ ksi}$$

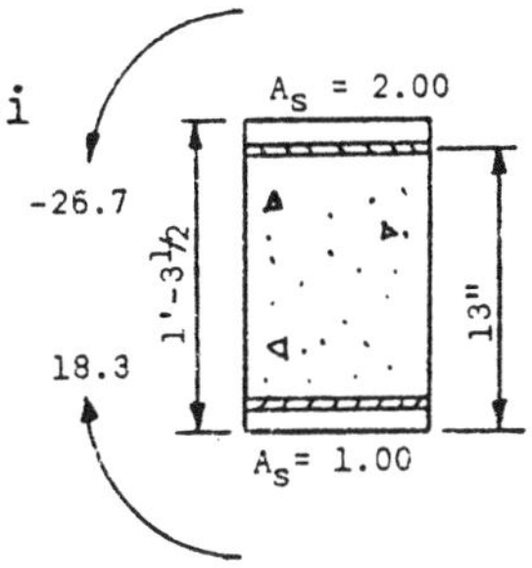


$f_s$ (act) is less than 36.0 ksi, o.k.

Use: #9's @ 6" c-c spacing in span 1.

Preliminary bar steel cutoff locations for positive moment are determined by one half the total steel area required for the ultimate moment. However, the service load moments must meet the fatigue requirements at the cutoff locations. Assume that approximately 85% of the allowable stress range (say 21,600 psi) is available for the positive moment.

Then (+M) = (0.85)(21.6)(0.9)(13.3)(1.)/12 = 18.3 ft-k which is at the 0.691 point of span 1. The corresponding (-M) = 26.7 ft-k. Checking the stress range

$$f_s = \frac{M}{A_sjd} = \frac{(18.3)(12)}{(1.00)(0.9)(13.3)} = 18.4 \text{ ksi.}$$

$$f_s' = \frac{(-M)}{A_sjd}\ \frac{k - (d'/d)}{1-k} = \frac{(26.7)(12)}{(2.00)(0.9)(13)}\ \frac{0.3-(2.135/12)}{1-0.3} = 2.3 \text{ ksi}$$

$$f_s + f_s' = 20.7 \text{ ksi} < 21.6 \text{ ksi o.k.}$$

Crack control check for #9's @ 12" c-c spacing.

$$f_s = \frac{170}{[(2.135)(55.5)]^{1/3}} = 34.6 \text{ ksi} > f_s\ (\text{act}) \text{ o.k.}$$

The first cutoff is at the 0.691 point of span 1 plus the bar extension. The remainder of the #9's are extended to the point of inflection (0.774 point) plus the bar extension length. Refer to Figure 18.4 showing typical longitudinal bar lengths and standard 18.1 for details.

Negative moment reinforcement for span 1.

At C/L Pier 1:

Max $(-M_u) = 1.3(55.4 + 3.9 + 5/3\ (35.5)) = 154.1$ ft-K, $d = 27-(2.5 + 0.5)$

$R_u = \dfrac{(154.1)(12)(1000)}{(0.9)(12)(24)^2} = 297.3$ psi  $\rho = .0052$  $= 24''$

$A_s = (0.0052)(12)(24) = 1.50\ in^2/ft$

Fatigue check for a moment range of 44.1 ft-k at centerline of pier 1.

$f_s$ (allow.) $= 21. - .33\ fmin + 8(.3) = 29.4$ ksi

where $.33\ fmin = .33 \dfrac{((55.4 - 3.9 + 8.6) \times 12)}{(1.5 \times .9 \times 24)} = +6.3$

$f_s = \dfrac{(44.1)(12)}{(1.5)(.9)(24)} = 16.3$ ksi $< 29.4$ o.k.

Check steel required for crack control. Try #9's @ 6" c-c spacing

$Z = 130$ k/in  $fs = \dfrac{130}{(3.065 \times 36.8)^{1/3}} = 26.9$ ksi

$As = \dfrac{94.8 \times 12}{26.9 \times .9 \times 24} = 1.96\ in^2/ft. < 2.00\ in^2/ft.$ o.k.

Use #9's @ 6"

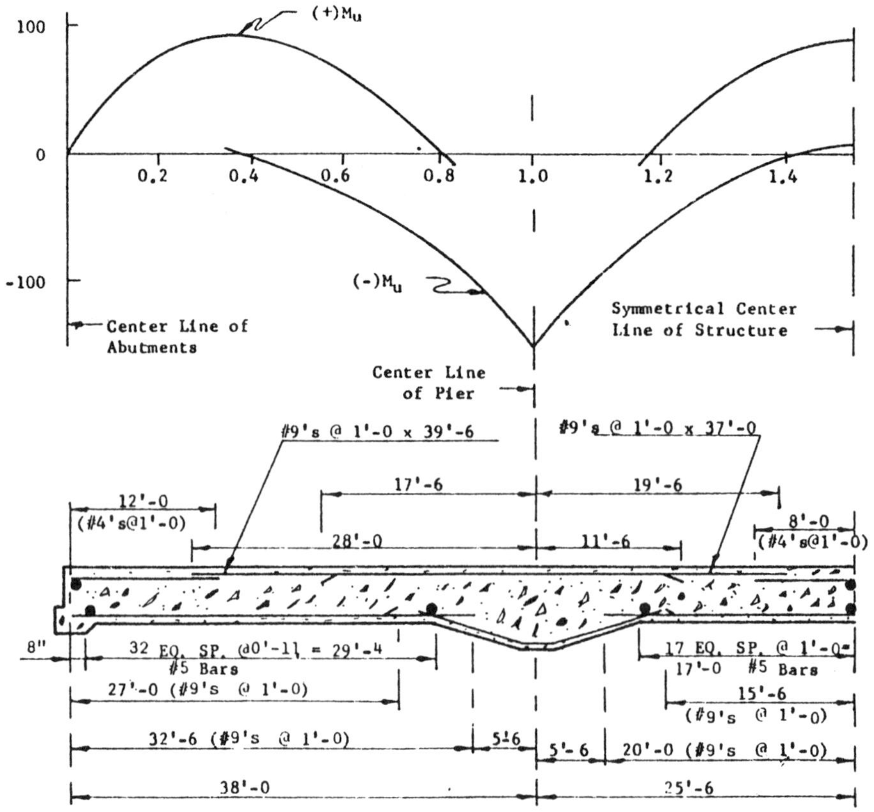

At the haunch-slab intercept:

Max $(-M_u)$ = 76.2 ft-k; d = 16-(2.5 + 0.5) = 13"

$$R_u = \frac{(76.2)(12)(1000)}{(0.9)(12)(13)^2} = 501.0 \text{ psi} \qquad \rho = .0091$$

$$A_s = (0.0091)(12)(13) = 1.42 \text{ in}^2\text{/ft}$$

Fatigue check for steel with a moment range of 37.4 ft-k at the haunch-slab intercept:

$$A_s = \frac{(37.4)(12)}{(22.9)(0.9)(13)} = 1.68 \text{ in}^2\text{/ft.} \quad \underline{\text{controls}}$$

Crack control check employing #9's @ 6" c-c spacing

($A_s$ = 2.0 in²/ft) at the haunch-slab intercept.
Use Z = 130 for negative steel area.

$$d_c = 2.5 + (1.13)/2 = 3.065 \text{ in.}$$

$$f_s = \frac{130}{[(3.065)(36.8)]^{1/3}} = 26.9 \text{ ksi}$$

$$A_s = \frac{(41.6)(12)}{(26.9)(0.9)(13)} = 1.59 \text{ in}^2\text{/ft.}$$

Preliminary bar steel cutoff locations for negative moment are determined by one half the steel area required for the ultimate moment. However, fatigue requirements must also be checked here. Assume 15% of allowable stress range for the negative moment and minimum stress in compression. Estimate the max negative moment at the bar cutoff to be (-M) = (0.85)(25.)(1.00)(0.9)(13)/12 = 20.7 ft-k.

The location corresponding to the (-M) is at the 0.659 point of span 1 and the (+M) = 23.8 ft-k.

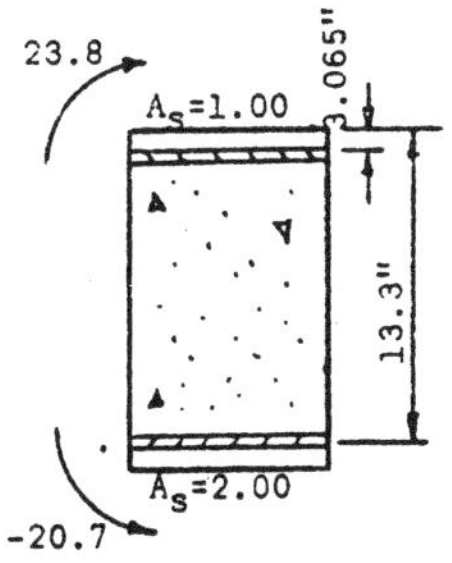


**The tensile part of the stress range in the top bars is computed as**

$$f_s = \frac{(-M)}{Asjd} = \frac{(20.7)(12)}{(1.0)(0.9)(13)} = 21.2 \text{ ksi.}$$

**The compressive part of the stress range in the top bars is computed as**

$$f_s = \frac{(+M)}{Asjd}\left(\frac{k-(d'/d)}{1-k}\right) = \frac{(23.8)(12)}{(2.00)(0.9)(13.3)}\left(\frac{0.3-(3.065/13.3)}{1.0-0.3}\right) = .8 \text{ ksi}$$

$f_s + f_s'$ = 21.2 + .8 = 22.0 ksi < 25.0 ksi o.k.

<u>Use: #9's @ 6" c-c spacing in span 1.</u>

Place the bar cutoff for one-half the negative reinforcement at the 0.65 point of span 1 less the bar extension.

The cutoff for the remainder of the reinforcement for fatigue consideration is located at the inflection point of the D + [-(L+I)] which is at the 0.466 point of span 1. The cutoff point at this location is not affected by fatigue considerations since most of the moment range is in compression. However, the actual cutoff point is governed by the location of the point of inflection based on strength design moment less the bar extension length.

Spans 1 & 3:

The dead load slab and safety parapet moments are placed on an edge beam width equal to the width of the parapet bearing on the slab (1'-1) plus one-half the depth of slab.

Edge Beam Width $= 1'\text{-}1 + \frac{1'\text{-}4}{2} = 1'\text{-}9$

DLM (slab) = (1.75 ft) (16.6 ft-k/ft) = 29.1 ft-k

DLM (para) $= \frac{\text{Parapet wt./ft}}{\text{Slab wt/ft}}$ (DLM(slab)/ft)

$= \frac{338}{200}$ (16.6) = 28.1 ft-k

DLM (w.s.) $= \frac{20}{200}$ (16.6) = 1.7 ft-k

DLM (total) = (29.1 + 28.1 + 1.7) = 58.9 ft-k

DL $M_u$ = 1.3 (58.9) = 76.6 ft-k

$$R_u = \frac{Mu}{\phi bd^2} = \frac{(76.6)(12)(1000)}{(0.9)(21)(13.3)} = 275 \text{ psi}$$

$\rho$ = 0.0048

$A_s$ = (0.0048)(12)(13.3) = 0.76 $in^2/ft$

The live load moment is placed on an edge beam having a width equal to the portion overlapped by the parapet (1'-1). The depth of the section is equal to the depth of the slab plus 1'-0 of the parapet.

b = 1'-1, d = 1'-4 + 1'-0 - 2" = 2'-2

$LLM_t$ = [LLMft-k/ft)(1/DF)] (4)/10

$LLM_t$ = [(32.3)(1/0.159)] 0.40 = 81.3 ft-k

$LLM_u$ = (1.3)(5/3)(81.3) = 176.1 ft-k

$$R_u = \frac{(176.1)(12)(1000)}{(0.9)(13)(26)^2} = 267.2 \text{ psi}$$ p = .0047

$$A_s = (0.0047)(13)(26) = 1.59 \text{ in}^2 \text{ or } 1.32 \text{ in}^2/\text{ft}.$$

$A_s$ (total) = 0.70 + 1.32 = 2.02 in$^2$/ft

Use existing #9 bars at 6 inch centers ($A_s$ = 2.00 in$^2$/ft)

Span 2:

In a similar manner to Span 1, the bar steel required for the edge beam in Span 2 is computed.

**Edge Beam Reinforcement**

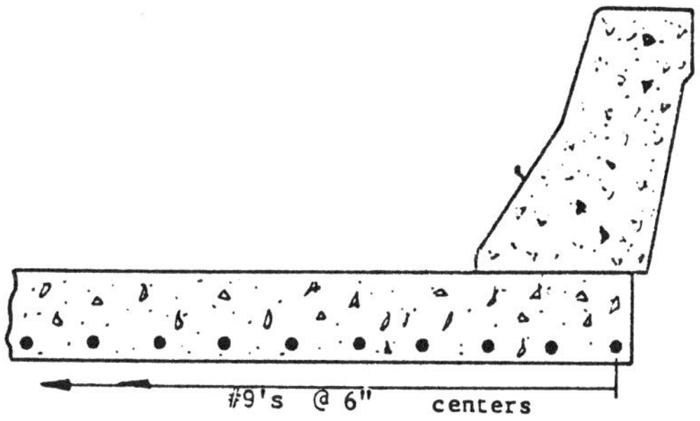


Note: Place deflection joints equidistant on either side point of maximum positive moment. The point of maximum positive moment in span 1 is at the 0.35 point, in span 2 is at the 0.50 point, and in span 3 is at the 0.65 point.

**Distribution Reinforcement**

The criteria for main reinforcement parallel to traffic is applied; the percentage of distribution reinforcement is based on a fraction of the positive moment steel.

**Spans 1 & 3:**

Percentage = $\frac{100}{\sqrt{S}}$ x 100% $\leq$ 50% MAX, where S is the span length in feet. Main positive reinforcement equals #9's @ 6" ctrs.

$$\text{Percentage} = \frac{100}{\sqrt{38}} \times 100\% = 16.2\% \leq 50\% \text{ MAX}$$

$$A_s = (0.162)(2.00) = \underline{0.33 \text{ in}^2/\text{ft}}$$

= #5's @ 0'-11 ctrs.

**Span 2:**

$$\text{Percentage} = \frac{100}{\sqrt{51}} \times 100\% = 14.0 \leq 50\% \text{ MAX.}$$

$$A_s = (0.140)(2.00) = \underline{0.28 \text{ in}^2/\text{ft}}$$

= 5's @ 1'-0 ctrs.

## Transverse Reinforcement at Piers

Transverse Haunch Length

Out-to-Out of Slab = 42'-8

| | |
|---|---|
| Length of haunch along skew | = 42'-8/cos 6° |
| | = 42.90 ft. |
| Less overhang from ℄ of columns | = 4.00 |
| | 38.90 ft |
| Center line to center line column spacing | = 38.90'/3 |
| | = 12'-11 1/2 along skew |

Dead Load Moment

From the computer analysis, WDL = 13.0 k/ft. at the piers. (DL shear span 1 = $6.5^k$ and DL shear span 2 = $6.5^k$.)

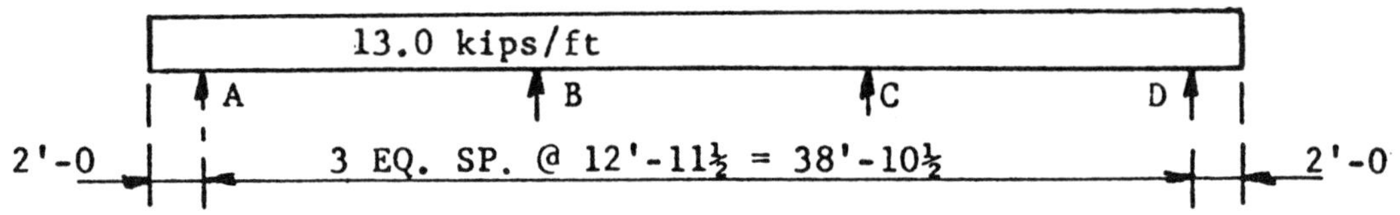


$M_A = 1/2\ WL^2 = (1/2)(13.0)(2.0)^2 = 26.0$ ft.-kips $= M_D$

Applying the three-moment equation for $M_B$ gives values of

$\frac{6A\bar{a}}{L} = \frac{6A\bar{b}}{L} = \frac{WL^3}{4}$. Other methods such as influence tables or moment distribution can also be used to obtain the dead load moments.

$$\frac{WL^3}{4} = \frac{(13.0)(12.96)^3}{4} = 7075 \text{ ft-k}$$

The three-moment equation is

$$M_A L_1 + 2M_B(L_1+L_2) + M_C L_2 + 6\,\frac{A1\bar{a}1}{L1} + 6\,\frac{A2\bar{b}2}{L2} = 0$$

If $M_A$ is known and due to symmetry $M_B = M_C$; the above equation reduces to one unknown, $M_B$, as follows:

$$(-26.0)(12.96) + (2)\,(M_B)(12.96 + 12.96) + (M_B)(12.96) +$$

$$7075. + 7075. = 0$$

$$M_B = 213. \text{ ft-kips} = M_C$$

The partial moment diagram for dead load is as follows:

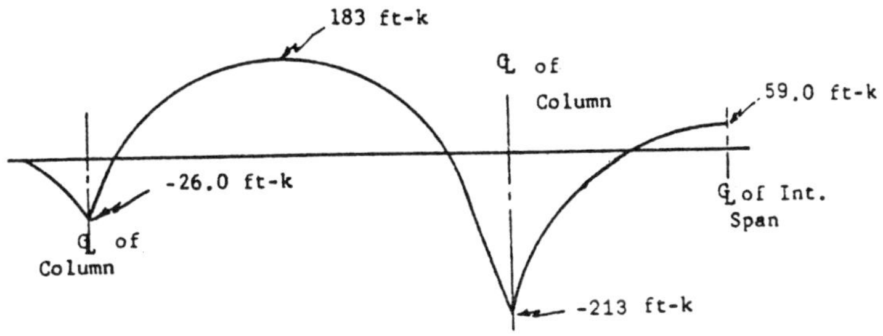


## Live Load Moments

Pier Lane = 61.2 k/lane
Pier Truck = 68.1 k/truck

Impact $= \dfrac{50}{13+125} = 36.2\%$; Use: 30% MAX.

Max. Wheel Load $= \dfrac{(68.1)}{2}(1.30)$

$= 44.3$ k/wheel

Using influence lines for a 3-span continuous beam; the following results are obtained.

+ LLM at 0.4 Point of Exterior Span

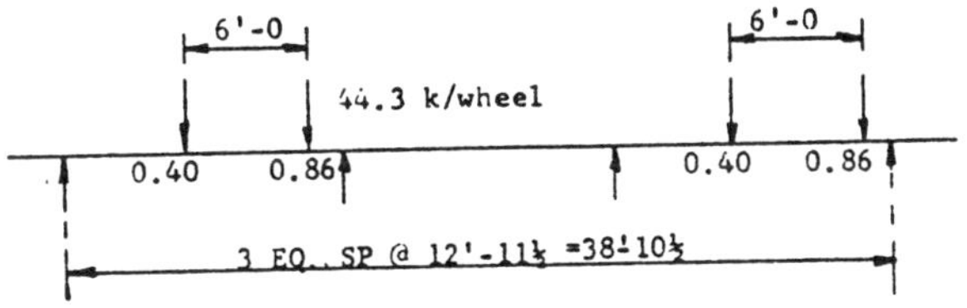


+ LLM @ 0.4 Pt. = (0.2042+0.0328+0.0102+0.0036)(44.3)(12.96)

= 144.0 ft-k (Max + LLM Ext. Span)

- LLM at Center Line of Support

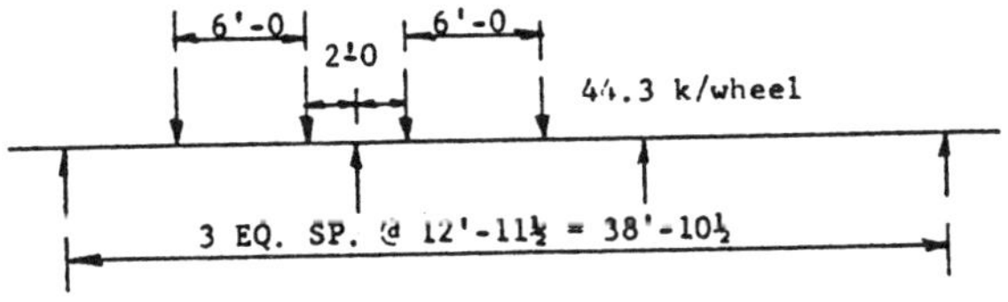


-LLM @ Support = (0.0862+0.0612+0.0515+0.0610)(44.3)(12.96)

= 149.2 ft-k (Max - LLM, Over Column)

## Positive Reinforcement at Pier Beam

$M_u$ = 1.3 (183.0 + 5/3(144.0)) = 548.6 ft-k

Design b = 1/2 column spacing

= 12.96/2 = 6.48'

Effective Depth = 27" - (1.5+0.6+0.6) = 24.3"

$$R_u = \frac{(548.6)(12)(1000)}{(0.9)(6.48)(12)(24.3)^2} = 159.3 \text{ psi}$$

$\rho$ = 0.0028

$A_s$(total) = (0.0028)(6.48)(12)(24.3) = 5.29 $in^2$

Try: 6 - #9's ($A_s$ = 6.00 $in^2$)

$$= \frac{6.0}{(36)(24.3)} = 0.0069 > \rho_{min}$$

Crack Control Check for #9's @ 6½" ctrs.

$d_c$ = 2.7

A = (2.7)(2)(6.5) = 35.1

$$f_s = \frac{170}{((2.7)(35.1))1/3} = 37.3 \text{ ksi} > 36.0 \text{ ksi}$$

$$f_s \text{ (act)} = \frac{(182+144)(12)}{(0.9)(6.0)(24.3)} = 29.8 \leq 36 \text{ ksi o.k.}$$

Use: 6 - #9's @ 6 1/2" c-c spacing

## Negative Reinforcement of Pier Beam

$M_u$ - 1.3 (213.0 + 5/3(149.2)) = 600.2 ft-k

Design b = 3'-0

Effective Depth = 27" - (2.5+1.2+0.5) = 22.8"

$$R_u = \frac{(600.2)(12)(1000)}{(0.9)(36)(22.8)^2} = 427.6 \text{ psi}$$

$\rho$ = 0.0077

$A_s$(total) = (0.0077)(3)(12)(22.8) = 6.32$in^2$

Check crack control for bundled bars:

Try: 14 - #7's @ 4 1/2" ctrs. ($A_s$ = 8.40 $in^2$)

$d_c$ = (2.50 + 1.13 + 0.44) = 4.07"

A = (2)(4.07)(4.5) = 36.6

$$f_s = \frac{(130)}{((4.07)(36.6))1/3} = 24.5 \text{ ksi}$$

$$f_s\text{(act)} = \frac{(213 + 149)(12)}{(0.9)(8.40)(22.8)} = 25.1 \text{ ksi o.k. within 3\%}$$

Use: 16 - #7's @ 4 1/2" ctrs.

## Shear Reinforcement of Pier Beam

Due to the geometry and loading on the pier cap, stirrups are generally not required or recommended where the column spacing does not exceed 16 feet. The vertical shear stress distribution is considered similar to that of drop panels in a building. A peripheral shear check at d/2 from the column face is required as follows:

Approximate shears at centerline of column.

DLV = (0.847)(13.0)(12.96) = $142.8^k$

LLV = (1.0 + 0.69 + 0.54 + 0.23)(44.3) = $109.0^k$

$V_u$ = 1.3 (142.8 + 5/3 (109.0)) = $421.8^k$

$$v_u = \frac{V_u}{\phi \pi D_d}$$

$$v_u = \frac{421{,}800}{(0.85)(\pi)(30 + 22.8)(22.8)}$$

= 131 psi < $4\sqrt{f'_c}$ O.K.

Stirrups are not required.

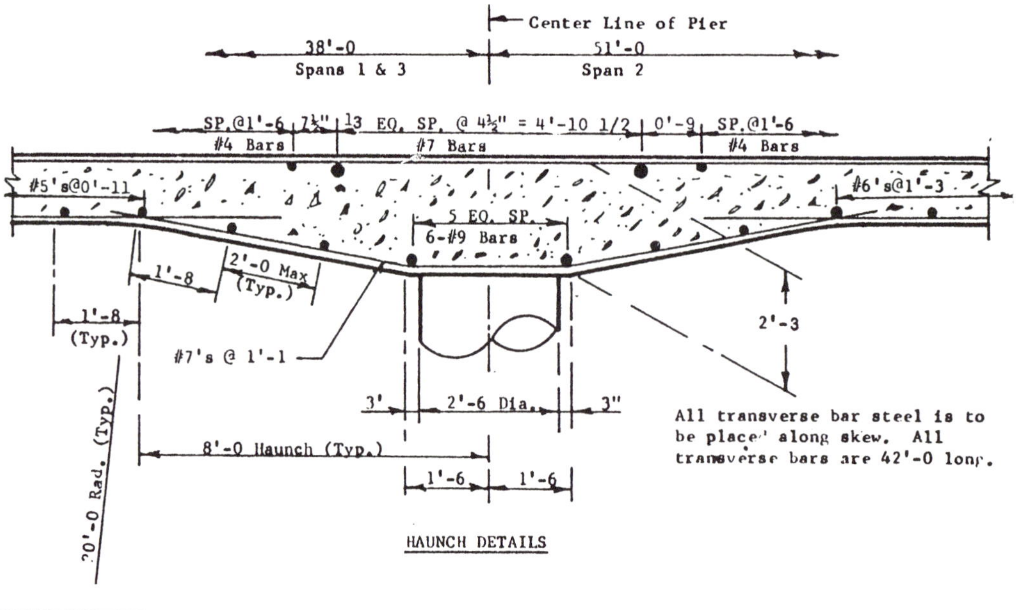


HAUNCH DETAILS

## Check For Uplift At Abutments

Maximum live load uplift at the abutments is obtained from the following influence line and loadings.

INFLUENCE LINE

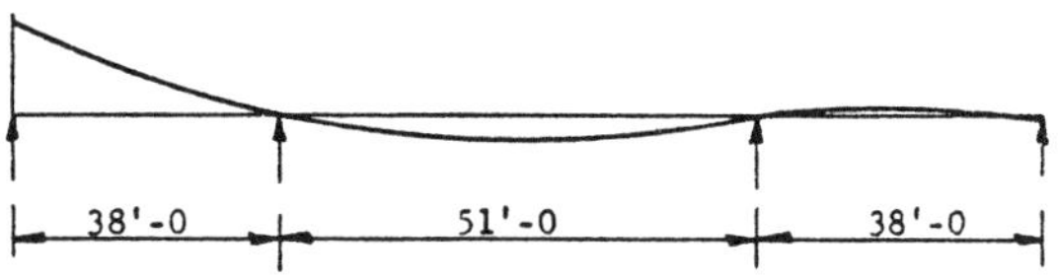


LANE LOADING

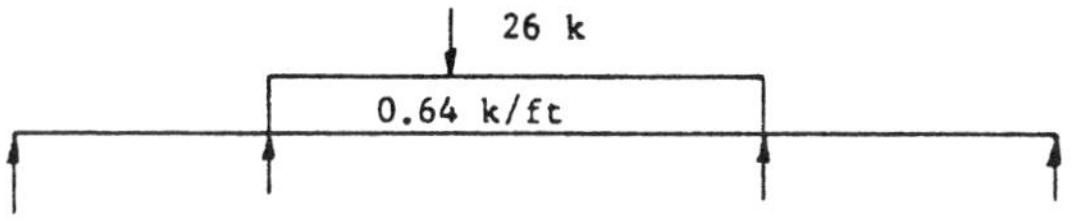


TRUCK LOADING

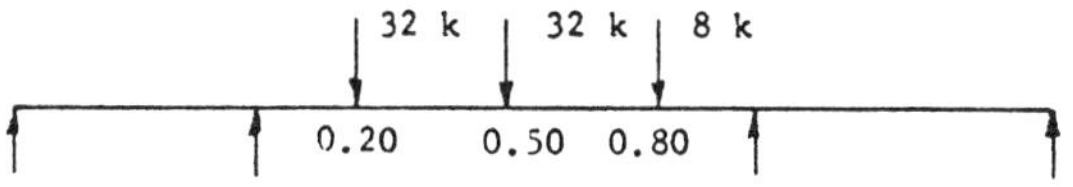


Lane Load = $(0.1107)(0.64)(38.0) + (0.1276)(26) = 5.9^{k}$

Truck Load = $(0.1035 + .1185)(32) - (0.0482)(8) = 7.5^{k}$ (Controls)

$$\text{Impact} = 1 + \frac{50}{51 + 125} = 1.284$$

With a Safety Factor of 2;

Total Uplift = $(7.5^{k})(1.284)(3 \text{ Lanes})(0.90)(2.0) = 52.0^{k}\uparrow$

Dead Load at Abutments = $(42.667 \text{ ft.})(2.8 \text{ k/ft.} + 0.3\text{k/ft.})$

$= 132.3^{k}\downarrow > 52.0^{k}\uparrow$

Since no uplift results at the abutments, the existing dowels at 1'-6 centers are adequate.

REFERENCES

1. Whitney, C. S., "Plastic Theory of Reinforced Concrete Design", ASCE Trans., 107, 1942, p. 251.

2. Anderson, A. J., Derthick, H. W. McDowell, R. C., and Seibert, J. A., "Load Factor Design", Standard Specifications for Highway Bridges Proposed Revision of Section 5 - Reinforced Concrete Design, 1973, pp. 6-1 to 6-25.

3. Notes on ACI 318-71 Building Code Requirements with Design Applications, Portland Cement Association, "Strength Requirements", 1972, pp. 3-1 to 3-8.

4. Singer, F. L., Strength of Materials, New York, N. Y., Harper and Row, Publishers, 1962.

5. Wang, C. K., and Salmon. G. G., Reinforced Concrete Design, Scranton, Penn., International Textbook Company, 1973.

6. Notes on Load Factor Design for Reinforced Concrete Bridge Structures with Design Applications, Portland Cement Association, 1974.

7. Building Code Requirements for Reinforced Concrete, (ACI Std. 318-71) American Concrete Institute, 1971.

8. Analysis and Design of Reinforced Concrete Bridge Structures, ACI Committee 443, American Concrete Institute, 1974.

# C.2 WORKING STRESS DESIGN

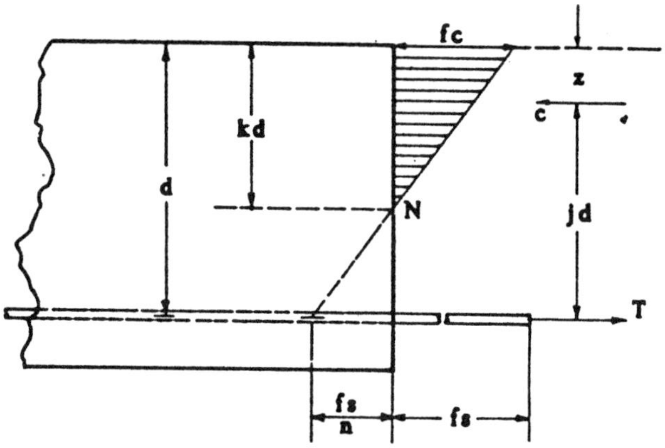


RECTANGULAR BEAM
WITHOUT COMPRESSIVE REINFORCEMENT

*Rectangular Beams.*

$f_s$ = tensile unit stress in longitudinal reinforcement.

$f_c$ = compressive unit stress in extreme fiber of concrete.

$E_s$ = modulus of elasticity of steel.

$E_c$ = modulus of elasticity of concrete.

$$n = \frac{E_s}{E_c}$$

M = bending moment, or moment of resistance in general.

$A_s$ = effective cross sectional area of tension reinforcement.

b = width of beam.

d = effective depth, or depth from compression surface of beam to center of tension reinforcement.

k = ratio of depth of neutral axis to effective depth, d.

j = ratio of lever arm of resisting couple to depth, d.

jd = d − z = arm of resisting couple.

p = ratio of effective area of tension reinforcement to effective area of concrete in beam $= \frac{A_s}{bd}$

z = depth from compression surface of beam to resultant of compressive stresses.

(1) Reinforced for tension only.
Position of neutral axis,

$$k = \sqrt{2pn + (pn)^2} - pn.$$

Arm of resisting couple,

$$j = 1 - \frac{k}{3}.$$

Compressive unit stress in extreme fiber of concrete,

$$f_c = \frac{2M}{jkbd^2} = \frac{2pf_s}{k}.$$

Tensile unit stress in longitudinal reinforcement,

$$f_s = \frac{M}{A_s jd} = \frac{M}{pjbd^2}$$

Steel ratio for balanced reinforcement,

$$p = \tfrac{1}{2}\,\frac{1}{\frac{f_s}{f_c}\left(\frac{f_s}{nf_c} + 1\right)}$$

Note: For approximate computations, the following assumptions may be made:

$$j = \tfrac{7}{8}$$
$$k = \tfrac{3}{8}$$

$$A_s = \frac{M}{\tfrac{7}{8} f_s d}$$

$$f_c = \frac{6M}{bd^2}$$

*(c) Shear, Bond and Web Reinforcement*

Diagonal tension and shear in reinforced concrete beams shall be calculated by the following formulas:

Shearing unit stress,

$$v = \frac{V}{bjd}.$$

Stress in vertical web reinforcement.

$$f'_v = \frac{V's}{A_v jd}.$$

## SERVICE LOAD~CONCRETE STRESSES & CONSTANTS

| | CLASS "AX" | CLASS "B" |
|---|---|---|
| $n$ (See $E_c$ below) | 8 | 10 |
| $f'_c$ | 4000 psi. | 3000 psi. |
| $f_c$ (Compression) | 1600 | 1200 |
| $f_c$ (Tension) Use only with special permission | 100 | 86 |
| $f_s$ (Grade 40) | 20,000 | 20,000 |
| $f_s$ (Grade 60) | 24,000 | 24,000 |
| $v_c$ (With web reinf.) | 313 | 271 |
| $v_c$ | 60 * | 52 * |
| Slabs & Footings (Peripheral Shear) | | |
| $v_c$ | 174 | 99 |
| $v_c$ (With web reinf.) | 190 | 164 |
| $K$ } Balanced rectangular sections | 272 | 197 |
| $k$ } Balanced rectangular sections | .390 | .375 |
| $j$ } Balanced rectangular sections | .870 | .875 |
| $p$ } Balanced rectangular sections | .0156 | .01125 |
| $E_c$ (for stress calc.) ($n$ as above) | 522,000 K/□" | 418,000 K/□" |
| $E_c$ (for short term defl. due to E. Q., etc.) ($n = 8$) | 522,000 | 522,000 |
| $E_c$ (for D.L. Camber of Slabs, T&S Moms., Settlement) ($n = 16$) | 261,000 | 261,000 |
| $E_c$ (for D. L. Camber, except slabs) ($n = 24$) | 174,000 | 174,000 |

Temp. Coeff. = .000006 %/° ~ 45° Drop to 35° Rise ~ All climates.
Shrinkage Coeff. = .0002% (Temp. rise & shrinkage cancel).

* For more detailed analysis $V_c = 0.9\,(f'_c)^{\frac{1}{2}} + 1100\,p_w\left(\frac{Vd}{M}\right)$
See 1974 A.A.S.H.T.O. Interim 1.5.29 (B)(2).

Stirrup spacing ; $S = \dfrac{A_s \times f_s \times jd}{V - V_c} = \dfrac{A_s \times 20 \times 7/8\, d}{V_s} = \dfrac{17.50\, A_s \times d}{V_s}$
(Kip & inch units)

$A_s$ = Total area of stirrup legs.
$V_s$ = Total shear taken by stirrups.
$V$ = Total shear on section.
$V_c$ = Total shear by conc. = $v_c \times bjd$

$d = \sqrt{M/bK}$ (Balanced rectangular section)

$f_c = \dfrac{2M}{kjbd^2}$ (Rectangular section)

$f_s = \dfrac{M}{A_s jd}$

$v = \dfrac{V}{bjd}$

Service Load Values of K & J For Rectangular Beams

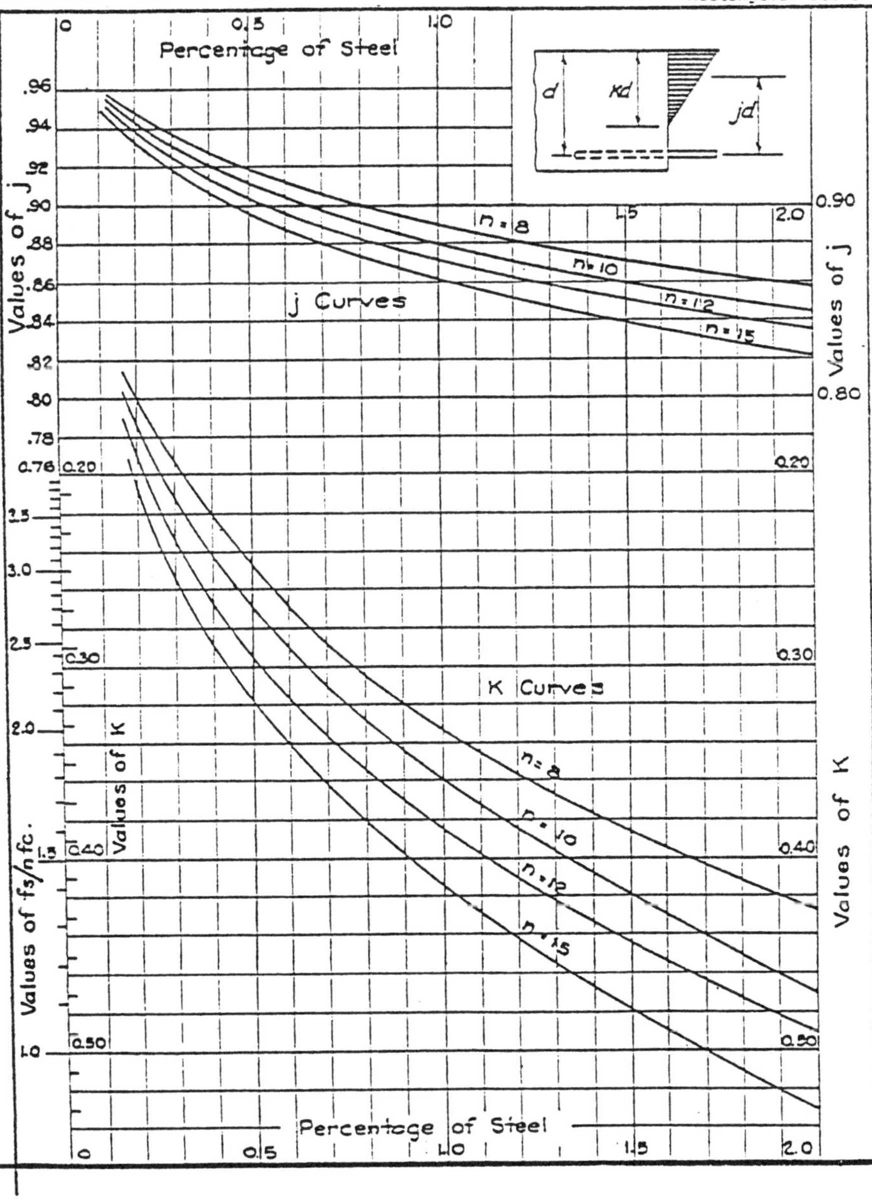

## AREAS AND PERIMETERS FOR VARIOUS BAR SIZES AND NUMBER OF BARS

**TOP NUMBERS ARE AREAS BOTTOM NUMBERS ARE PERIMETERS**

| No. \ Size | #3 | #4 | #5 | #6 | #7 | #8 | #9 | #10 | #11 | #14 | #18 | Size / No. |
|---|---|---|---|---|---|---|---|---|---|---|---|---|
| 1 | 0.11<br>1.18 | 0.20<br>1.57 | 0.31<br>1.96 | 0.44<br>2.36 | 0.60<br>2.75 | 0.79<br>3.14 | 1.00<br>3.54 | 1.27<br>3.99 | 1.56<br>4.43 | 2.25<br>5.32 | 4.00<br>7.09 | 1 |
| 2 | 0.22<br>2.36 | 0.40<br>3.14 | 0.62<br>3.93 | 0.88<br>4.71 | 1.20<br>5.50 | 1.58<br>6.28 | 2.00<br>7.09 | 2.54<br>7.98 | 3.12<br>8.86 | 4.50<br>10.63 | 8.00<br>14.18 | 2 |
| 3 | 0.33<br>3.53 | 0.60<br>4.71 | 0.93<br>5.89 | 1.32<br>7.07 | 1.80<br>8.25 | 2.37<br>9.43 | 3.00<br>10.63 | 3.81<br>11.97 | 4.68<br>13.29 | 6.75<br>15.95 | 12.00<br>21.26 | 3 |
| 4 | 0.44<br>4.71 | 0.80<br>6.28 | 1.24<br>7.85 | 1.76<br>9.42 | 2.40<br>11.00 | 3.16<br>12.57 | 4.00<br>14.18 | 5.08<br>15.96 | 6.24<br>17.72 | 9.00<br>21.26 | 16.00<br>28.35 | 4 |
| 5 | 0.55<br>5.89 | 1.00<br>7.86 | 1.55<br>9.82 | 2.20<br>11.78 | 3.00<br>13.75 | 3.95<br>15.71 | 5.00<br>17.72 | 6.35<br>19.95 | 7.80<br>22.15 | 11.25<br>26.58 | 20.00<br>35.44 | 5 |
| 6 | 0.66<br>7.07 | 1.20<br>9.43 | 1.86<br>11.78 | 2.64<br>14.14 | 3.60<br>16.49 | 4.74<br>18.85 | 6.00<br>21.26 | 7.62<br>23.94 | 9.36<br>26.58 | 13.50<br>31.90 | 24.00<br>42.53 | 6 |
| 7 | 0.77<br>8.25 | 1.40<br>11.00 | 2.17<br>13.74 | 3.08<br>16.49 | 4.20<br>19.24 | 5.53<br>21.99 | 7.00<br>24.81 | 8.89<br>27.93 | 10.92<br>31.01 | 15.75<br>37.21 | 28.00<br>49.62 | 7 |

| Spacing | #3 | #4 | #5 | #6 | #7 | #8 | #9 | #10 | #11 | #14 | #18 | Spacing |
|---|---|---|---|---|---|---|---|---|---|---|---|---|
| 5" | 0.26<br>2.8 | 0.48<br>3.8 | 0.74<br>4.7 | 1.06<br>5.7 | 1.44<br>6.6 | 1.90<br>7.5 | 2.40<br>8.5 | 3.05<br>9.6 | 3.74<br>10.6 | 5.40<br>12.8 | 9.60<br>17.0 | 5" |
| 5¼" | 0.25<br>2.7 | 0.46<br>3.6 | 0.71<br>4.5 | 1.01<br>5.4 | 1.37<br>6.3 | 1.81<br>7.2 | 2.29<br>8.1 | 2.90<br>9.1 | 3.57<br>10.1 | 5.14<br>12.2 | 9.14<br>16.2 | 5¼" |
| 5½" | 0.24<br>2.6 | 0.44<br>3.4 | 0.68<br>4.3 | 0.96<br>5.1 | 1.31<br>6.0 | 1.72<br>6.9 | 2.18<br>7.7 | 2.77<br>8.7 | 3.40<br>9.7 | 4.91<br>11.6 | 8.73<br>15.5 | 5½" |
| 5¾" | 0.23<br>2.5 | 0.42<br>3.3 | 0.65<br>4.1 | 0.92<br>4.9 | 1.25<br>5.7 | 1.65<br>6.6 | 2.09<br>7.4 | 2.65<br>8.3 | 3.26<br>9.2 | 4.69<br>11.1 | 8.34<br>14.8 | 5¾" |
| 6" | 0.22<br>2.4 | 0.40<br>3.1 | 0.62<br>3.9 | 0.88<br>4.7 | 1.20<br>5.5 | 1.58<br>6.3 | 2.00<br>7.1 | 2.54<br>8.0 | 3.12<br>8.9 | 4.50<br>10.6 | 8.00<br>14.2 | 6" |
| 6½" | 0.20<br>2.2 | 0.37<br>2.9 | 0.57<br>3.6 | 0.81<br>4.4 | 1.11<br>5.1 | 1.46<br>5.8 | 1.85<br>6.5 | 2.35<br>7.4 | 2.88<br>8.2 | 4.15<br>9.8 | 7.39<br>13.1 | 6½" |
| 7" | 0.19<br>2.0 | 0.34<br>2.7 | 0.53<br>3.4 | 0.75<br>4.0 | 1.03<br>4.7 | 1.35<br>5.4 | 1.71<br>6.1 | 2.18<br>6.8 | 2.67<br>7.6 | 3.86<br>9.1 | 6.86<br>12.2 | 7" |
| 7½" | 0.18<br>1.9 | 0.32<br>2.5 | 0.50<br>3.1 | 0.70<br>3.8 | 0.96<br>4.4 | 1.26<br>5.0 | 1.60<br>5.7 | 2.03<br>6.4 | 2.50<br>7.1 | 3.60<br>8.5 | 6.40<br>11.3 | 7½" |
| 8" | 0.17<br>1.8 | 0.30<br>2.4 | 0.47<br>2.9 | 0.66<br>3.5 | 0.90<br>4.1 | 1.19<br>4.7 | 1.50<br>5.3 | 1.91<br>6.0 | 2.34<br>6.7 | 3.38<br>8.0 | 6.00<br>10.6 | 8" |
| 8½" | 0.16<br>1.7 | 0.28<br>2.2 | 0.44<br>2.8 | 0.62<br>3.3 | 0.85<br>3.9 | 1.12<br>4.4 | 1.41<br>5.0 | 1.79<br>5.6 | 2.20<br>6.3 | 3.18<br>7.5 | 5.65<br>10.0 | 8½" |
| 9" | 0.15<br>1.6 | 0.27<br>2.1 | 0.41<br>2.6 | 0.59<br>3.1 | 0.80<br>3.7 | 1.05<br>4.2 | 1.33<br>4.7 | 1.69<br>5.3 | 2.08<br>5.9 | 3.00<br>7.1 | 5.34<br>9.5 | 9" |
| 9½" | 0.14<br>1.5 | 0.25<br>2.0 | 0.39<br>2.5 | 0.56<br>3.0 | 0.76<br>3.5 | 1.00<br>4.0 | 1.26<br>4.5 | 1.60<br>5.0 | 1.97<br>5.6 | 2.84<br>6.7 | 5.06<br>9.0 | 9½" |
| 10" | 0.13<br>1.4 | 0.24<br>1.9 | 0.37<br>2.4 | 0.53<br>2.8 | 0.72<br>3.3 | 0.95<br>3.8 | 1.20<br>4.3 | 1.52<br>4.8 | 1.87<br>5.3 | 2.70<br>6.4 | 4.80<br>8.5 | 10" |
| 10½" | 0.13<br>1.3 | 0.23<br>1.8 | 0.35<br>2.2 | 0.50<br>2.7 | 0.69<br>3.1 | 0.90<br>3.6 | 1.14<br>4.0 | 1.45<br>4.6 | 1.78<br>5.1 | 2.57<br>6.1 | 4.57<br>8.1 | 10½" |
| 11" | 0.12<br>1.3 | 0.22<br>1.7 | 0.34<br>2.2 | 0.48<br>2.6 | 0.65<br>3.0 | 0.86<br>3.4 | 1.09<br>3.9 | 1.39<br>4.4 | 1.70<br>4.8 | 2.45<br>5.8 | 4.36<br>7.7 | 11" |
| 11½" | | 0.21<br>1.6 | 0.32<br>2.0 | 0.46<br>2.5 | 0.63<br>2.9 | 0.82<br>3.3 | 1.04<br>3.7 | 1.33<br>4.2 | 1.63<br>4.6 | 2.35<br>5.6 | 4.17<br>7.4 | 11½" |
| 12" | | 0.20<br>1.6 | 0.31<br>2.0 | 0.44<br>2.4 | 0.60<br>2.8 | 0.79<br>3.1 | 1.00<br>3.5 | 1.27<br>4.0 | 1.56<br>4.4 | 2.25<br>5.3 | 4.00<br>7.1 | 12" |
| 13" | | 0.18<br>1.4 | 0.29<br>1.8 | 0.41<br>2.2 | 0.55<br>2.5 | 0.73<br>2.9 | 0.92<br>3.3 | 1.17<br>3.7 | 1.44<br>4.1 | 2.08<br>4.9 | 3.69<br>6.5 | 13" |
| 14" | | 0.17<br>1.3 | 0.27<br>1.7 | 0.38<br>2.0 | 0.51<br>2.4 | 0.68<br>2.7 | 0.86<br>3.0 | 1.09<br>3.4 | 1.34<br>3.8 | 1.93<br>4.6 | 3.43<br>6.1 | 14" |
| 15" | | 0.16<br>1.3 | 0.25<br>1.6 | 0.35<br>1.9 | 0.48<br>2.2 | 0.63<br>2.5 | 0.80<br>2.8 | 1.02<br>3.2 | 1.25<br>3.5 | 1.80<br>4.3 | 3.20<br>5.7 | 15" |
| 16" | | 0.15<br>1.2 | 0.23<br>1.5 | 0.33<br>1.8 | 0.45<br>2.1 | 0.59<br>2.4 | 0.75<br>2.7 | 0.95<br>3.0 | 1.17<br>3.3 | 1.69<br>4.0 | 3.00<br>5.3 | 16" |
| 17" | | 0.14<br>1.1 | 0.22<br>1.4 | 0.31<br>1.7 | 0.42<br>1.9 | 0.56<br>2.2 | 0.71<br>2.5 | 0.90<br>2.8 | 1.10<br>3.1 | 1.59<br>3.8 | 2.82<br>5.0 | 17" |
| 18" | | 0.13<br>1.1 | 0.21<br>1.3 | 0.29<br>1.6 | 0.40<br>1.8 | 0.53<br>2.1 | 0.67<br>2.4 | 0.85<br>2.7 | 1.04<br>2.9 | 1.50<br>3.6 | 2.67<br>4.7 | 18" |

## C.4 LAP SPLICE

TENSION LAP SPLICES fy= 60 ksi f'c= 4000 psi

| BAR SIZE | SPACING GREATER THAN 6" * | | | | | | SPACING LESS THAN 6" | | | | | |
|---|---|---|---|---|---|---|---|---|---|---|---|---|
| | Class A (1.0Ld) | | Class B (1.3 Ld) | | Class C (1.7 Ld) | | Class A (1.0 Ld) | | Class B (1.3 Ld) | | Class C (1.7 Ld) | |
| | Top | Other | Top | Other | Top | Other | Top | Other | Top | Other | Top | Other |
| 4 | 1-2 | 1-0 | 1-6 | 1-1 | 1-11 | 1-8 | 1-5 | 1-0 | 1-10 | 1-4 | 2-5 | 1-9 |
| 5 | 1-5 | 1-0 | 1-10 | 1-4 | 2-5 | 1-8 | 1-9 | 1-3 | 2-4 | 1-8 | 3-0 | 2-2 |
| 6 | 1-9 | 1-3 | 2-3 | 1-7 | 2-11 | 2-0 | 2-2 | 1-6 | 2-9 | 2-0 | 3-7 | 2-7 |
| 7 | 2-2 | 1-7 | 2-9 | 2-0 | 3-8 | 2-6 | 2-8 | 1-11 | 3-6 | 2-6 | 4-7 | 3-3 |
| 8 | 2-10 | 2-0 | 3-8 | 2-8 | 4-9 | 3-5 | 3-6 | 2-6 | 4-7 | 3-3 | 6-0 | 4-3 |
| 9 | 3-7 | 2-7 | 4-8 | 3-4 | 6-1 | 4-4 | 4-6 | 3-2 | 5-9 | 4-2 | 7-7 | 5-5 |
| 10 | 4-6 | 3-3 | 5-10 | 4-2 | 7-8 | 5-6 | 5-8 | 4-1 | 7-4 | 5-3 | 9-7 | 6-10 |
| 11 | 5-7 | 4-0 | 7-2 | 5-2 | 9-5 | 6-9 | 6-11 | 5-0 | 9-0 | 6-5 | 11-9 | 8-5 |
| | TENSION LAP SPLICE fy=60 ksi f'c=3500 psi | | | | | | | | | | | |
| 4 | 1-2 | 1-0 | 1-6 | 1-1 | 1-11 | 1-5 | 1-5 | 1-0 | 1-10 | 1-4 | 2-5 | 1-9 |
| 5 | 1-5 | 1-0 | 1-10 | 1-4 | 2-5 | 1-9 | 1-9 | 1-3 | 2-4 | 1-8 | 3-0 | 2-2 |
| 6 | 1-9 | 1-3 | 2-3 | 1-7 | 2-11 | 2-1 | 2-2 | 1-6 | 2-9 | 2-0 | 3-7 | 2-7 |
| 7 | 2-4 | 1-8 | 3-0 | 2-2 | 3-11 | 2-10 | 2-11 | 2-1 | 3-9 | 2-8 | 4-11 | 3-6 |
| 8 | 3-0 | 2-2 | 3-11 | 2-10 | 5-1 | 3-8 | 3-9 | 2-8 | 4-10 | 3-6 | 6-5 | 4-7 |
| 9 | 3-10 | 2-9 | 5-0 | 3-7 | 6-6 | 4-8 | 4-9 | 3-5 | 6-2 | 4-5 | 8-1 | 5-9 |
| 10 | 4-10 | 3-6 | 6-3 | 4-6 | 8-2 | 5-10 | 6-1 | 4-4 | 7-10 | 5-7 | 10-3 | 7-4 |
| 11 | 5-11 | 4-3 | 7-9 | 5-6 | 10-1 | 7-3 | 7-5 | 5-4 | 9-8 | 6-11 | 12-7 | 9-0 |
| | TENSION LAP SPLICES fy=60 ksi f'c=3000 psi | | | | | | | | | | | |
| 4 | 1-2 | 1-0 | 1-6 | 1-1 | 1-11 | 1-5 | 1-5 | 1-0 | 1-10 | 1-4 | 2-5 | 1-9 |
| 5 | 1-5 | 1-0 | 1-10 | 1-4 | 2-5 | 1-9 | 1-9 | 1-3 | 2-4 | 1-8 | 3-0 | 2-2 |
| 6 | 1-10 | 1-4 | 2-4 | 1-8 | 3-1 | 2-3 | 2-3 | 1-8 | 2-11 | 2-1 | 3-10 | 2-9 |
| 7 | 2-6 | 1-9 | 3-3 | 2-4 | 4-3 | 3-0 | 3-1 | 2-3 | 4-0 | 2-11 | 5-3 | 3-9 |
| 8 | 3-3 | 2-4 | 4-3 | 3-0 | 5-7 | 4-0 | 4-1 | 2-11 | 5-3 | 3-9 | 6-11 | 4-11 |
| 9 | 4-1 | 2-11 | 5-4 | 3-10 | 7-0 | 5-0 | 5-2 | 3-8 | 6-8 | 4-9 | 8-9 | 6-3 |
| 10 | 5-3 | 3-9 | 6-10 | 4-11 | 8-11 | 6-4 | 6-6 | 4-8 | 8-6 | 6-1 | 11-1 | 7-11 |
| 11 | 8-9 | 6-3 | 8-4 | 6-0 | 10-11 | 7-10 | 8-0 | 5-9 | 10-5 | 7-5 | 13-7 | 9-9 |

*Top bars are horizontal bars so placed that more than 12 inches of fresh concrete is cast in the member below the bar.

Class A: Low stressed bars - 75% or less spliced (Use 1.0 ℓd

Class B: Low stressed bars - more than 75% spliced.
High Stressed bars - 1/2 or less spliced.

Class C: High stressed bars - more than 50% spliced.

| TENSION LAP SPLICES fy= 40 ksi f'c=3000 psi | | | | | | | | | | | | |
|---|---|---|---|---|---|---|---|---|---|---|---|---|
| BAR SIZE | SPACING GREATER THAN 6" | | | | | | SPACING LESS THAN 6" | | | | | |
| | Class A | | Class B | | Class C | | Class A | | Class B | | Class C | |
| | Top | Other | Top | Other | Top | Other | Top | Other | Top | Other | Top | Other |
| 4 | 1-0 | 1-0 | 1-0 | 1-0 | 1-3 | 1-0 | 1-0 | 1-0 | 1-3 | 1-0 | 1-7 | 1-2 |
| 5 | 1-0 | 1-0 | 1-3 | 1-0 | 1-7 | 1-2 | 1-2 | 1-0 | 1-6 | 1-1 | 2-0 | 1-5 |
| 6 | 1-3 | 1-0 | 1-7 | 1-1 | 2-1 | 1-6 | 1-6 | 1-1 | 2-0 | 1-5 | 2-7 | 1-10 |
| 7 | 1-8 | 1-2 | 2-2 | 1-6 | 2-10 | 2-0 | 2-1 | 1-6 | 2-8 | 1-11 | 3-6 | 2-6 |
| 8 | 2-2 | 1-7 | 2-10 | 2-0 | 3-8 | 2-8 | 2-9 | 1-11 | 3-6 | 2-6 | 4-7 | 3-4 |
| 9 | 2-9 | 2-0 | 3-7 | 2-7 | 4-8 | 3-4 | 3-5 | 2-6 | 4-5 | 3-2 | 5-10 | 4-2 |
| 10 | 3-6 | 2-6 | 4-7 | 3-4 | 6-0 | 4-3 | 4-4 | 3-2 | 5-8 | 4-1 | 7-5 | 5-3 |
| 11 | 4-3 | 3-1 | 5-7 | 4-0 | 7-4 | 5-3 | 5-4 | 3-10 | 6-11 | 5-0 | 9-1 | 6-6 |

| BASIC TENSION EMBEDMENT (L) WITH STANDARD END HOOK | | | | |
|---|---|---|---|---|
| BAR SIZE | Fy=60ksi | | Fy=40 ksi | |
| | 3000 | 4000 | 3000 | 4000 |
| 4 | 11" | 10" | 7" | 6" |
| 5 | 14 | 12 | 9 | 8 |
| 6 | 16 | 14 | 11 | 10 |
| 7 | 19 | 17 | 13 | 11 |
| 8 | 22 | 19 | 15 | 13 |
| 9 | 25 | 21 | 17 | 14 |
| 10 | 28 | 24 | 19 | 16 |
| 11 | 31 | 27 | 21 | 18 |

| COMPRESSION LAP SPLICES | | |
|---|---|---|
| BAR SIZE | 40 (20D) | 60 (30D) |
| 4 | 10 | 1-3 |
| 5 | 1-1 | 1-7 |
| 6 | 1-3 | 1-11 |
| 7 | 1-6 | 2-3 |
| 8 | 1-8 | 2-6 |
| 9 | 1-11 | 2-10 |
| 10 | 2-1 | 3-2 |
| 11 | 2-4 | 3-6 |

| MINIMUM COMPRESSION DOWEL EMBEDMENT | | | | |
|---|---|---|---|---|
| BAR SIZE | Fy=40 ksi | | Fy=60 ksi | |
| | 3000 | 4000 | 3000 | 4000 |
| 4 | 8 | 8 | 11 | 10 |
| 5 | 9 | 8 | 1-21 | 1-0 |
| 6 | 11 | 10 | 1-4 | 1-2 |
| 7 | 1-1 | 11 | 1-7 | 1-5 |
| 8 | 1-3 | 1-1 | 1-10 | 1-7 |
| 9 | 1-5 | 1-2 | 2-1 | 1-9 |
| 10 | 1-7 | 1-4 | 2-4 | 2-0 |
| 11 | 1-9 | 1-6 | 2-10 | 2-3 |

*E=L x Factors*

- *Side cover more than 2½" & cover over hook more than 2"* ---------------------0.7
- *Hook enclosed within ties or stirrup* -------0.8

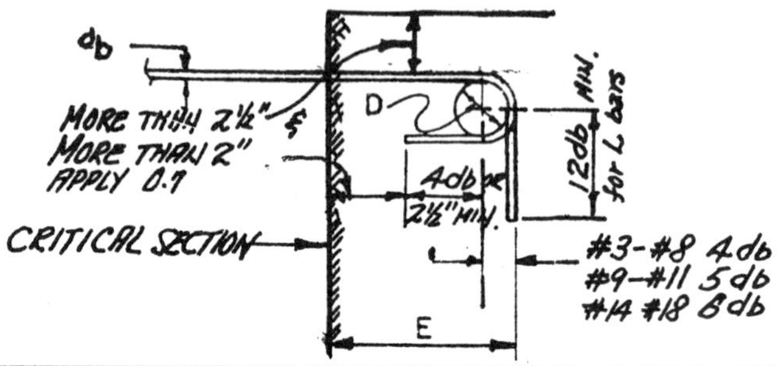

## REINF. BAR BEND RADIUS FOR HIGHLY STRESSED BARS

This table is for a 90° bend where the bar is stressed to its yield strength at the tangent to the bend.

| BAR SIZE | $f'_c$ = 4000 CONC. | | $f'_c$ = 3000 CONC. | |
|---|---|---|---|---|
| | $f_s$ = 40 K.S.I. | $f_s$ = 60 K.S.I. | $f_s$ = 40 K.S.I. | $f_s$ = 60 K.S.I. |
| | Radius | Radius | Radius | Radius |
| 6 | 8" | 11" | 10" | 1'-3" |
| 7 | 9" | 1'-1" | 1'-0" | 1'-5" |
| 8 | 11" | 1'-3" | 1'-2" | 1'-8" |
| 9 | 1'-0" | 1'-5" | 1'-3" | 1'-10" |
| 10 | 1'-1" | 1'-7" | 1'-5" | 2'-1" |
| 11 | 1'-3" | 1'-9" | 1'-7" | 2'-4" |
| 14 | 1'-6" | 2'-1" | 1'-11" | 2'-9" |
| 18 | 1'-11" | 2'-10" | 2'-6" | 3'-8" |

General equation for outside radius of bar

$$R = \frac{2 f_s A_s \sin\frac{\alpha}{2}(180°)}{f'_c K_1 \phi d \pi \alpha} + d$$

Outside radius for 90° bend

$$R = \frac{0.90 f_s A_s}{f'_c K_1 \phi d} + d$$

R = Outside radius (in.)
$f_s$ = Stress in bar (p.s.i.)
$A_s$ = Area of individual bar (sq. in.)
$f'_c$ = Ult. allow. conc. stress (p.s.i.)
$K_1$ = 0.85 = (For conc. up to 4000 p.s.i.)*
$\phi$ = 0.90 = (Capacity mod. factor)*
d = Nominal diam. of bar (in.)
$\alpha$ = Bend angle (Degrees)
T = Force developed by the bar = $f_s A_s$
* Per A.A.S.H.T.O. Interim Specifications-1974, Article 1.5.30(A)

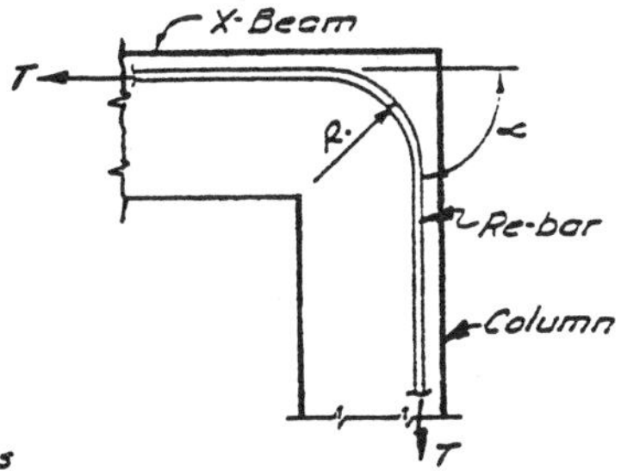


EXAMPLE

## ALLOWABLE LOADS FOR EXTERNALLY THREADED EXPANSION BOLT ANCHORS

Kwik-Bolt, Molly Parabolt or Equal.

| Bolt & Hole Diam. In. | Embedment (Min.) In. | Max. Bolt Length In. | Thread Length In. | Allowable Loads (Lb.) in 4000 psi Conc. | | | | Min. Edge Distance (in.) | | Min. Spacing (in) | |
|---|---|---|---|---|---|---|---|---|---|---|---|
| | | | | W.S.D. | | U.S.D. | | For 100% of Allow. Load | For 50% of Allow. Load | For 100% of Allow. Load | For 50% of Allow. Load |
| | | | | Tension | Shear | Tension | Shear | | | | |
| 1/2 | 2 3/4 | 7 | 1 1/4 | 2150 | 2150 | 4300 | 4300 | 3 | 3 | 6 | 3 |
| 5/8 | 3 1/2 | 7 | 1 1/2 | 3500 | 3500 | 7000 | 7000 | 4 | 3 | 8 | 4 |
| 3/4 | 4 1/4 | 10 | 1 1/2 | 5000 | 5000 | 10,000 | 10,000 | 5 | 3 | 10 | 5 |

NOTES:

1. For edge distances & spacings within the above limits, use linear interpolation to obtain the appropriate percentage of allowable loads.
2. Above data based on test data supplied by the manufacturers.
3. Bolt lengths longer than those shown under "Max. Bolt Length" are available on special order.
4. Do not use expansion bolts for shear-friction theory designs.
5. Kwik-Bolts or Parabolts are made of 100,000 p.s.i. yield steel & plated to resist corrosion.
6. Specify bolt size & embedment depth on Plans, e.g. "3/4" diam. expansion bolt, Kwik-Bolt, Parabolt or equal, with 4 1/4" embedment." (Nut & washer are included. Specify lockwasher when required.)
7. For bolts subject to combined tension & shear.
   $\frac{T}{T_{all'd}} + \left(\frac{V}{V_{all'd}}\right)^2 \leq 1.0$ where T = design tension, V = design shear
8. Concrete thickness shall be equal or greater than embedment + 2".

# C.5 WELDED SPLICE

| Joint Type |
| --- |
| Square End with Angle Backup |
| Single-Vee with Angle Backup |
| Single-Vee without Backup |
| Double-Vee |
| Double Bevel |
| Sleeve |

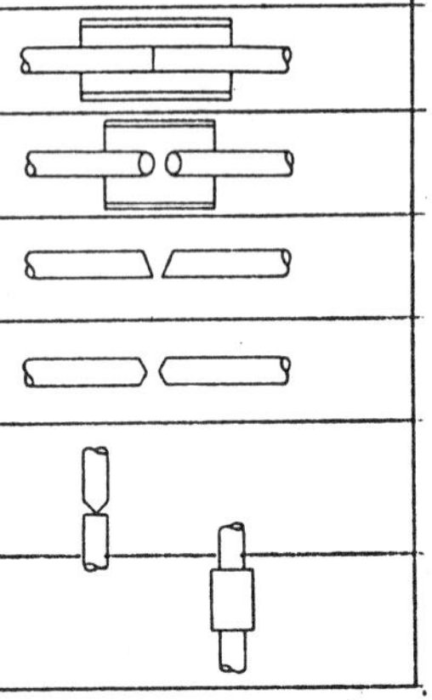

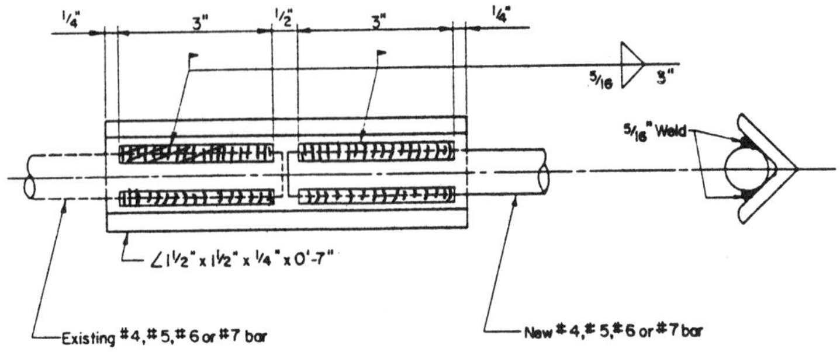


DETAIL OF WELD SPLICE FOR #4,#5,#6 OR #7 BARS

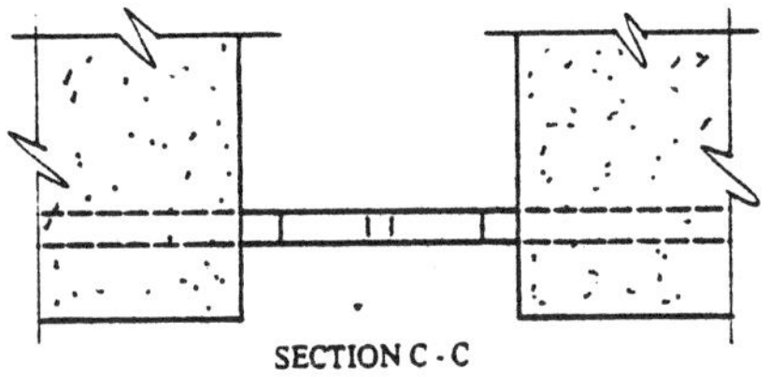

SECTION C - C

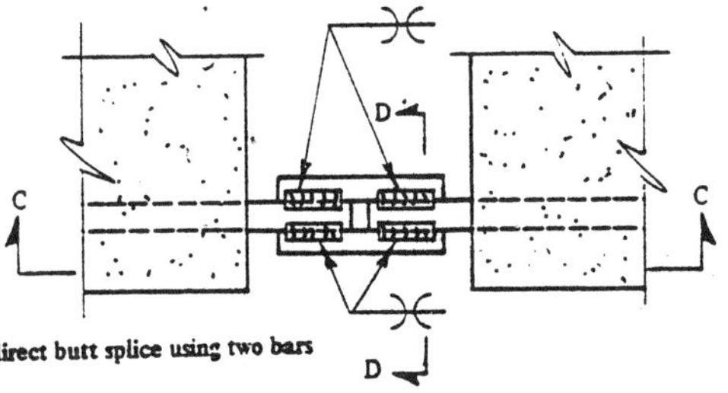


Indirect butt splice using two bars

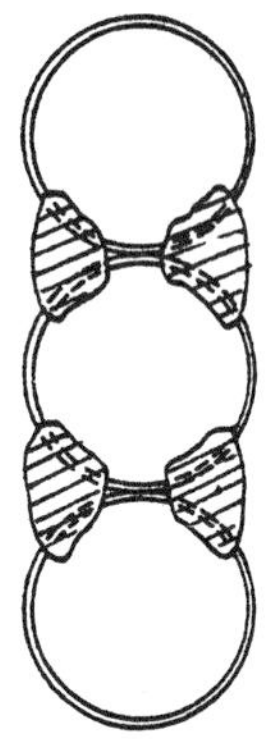

SECTION D - D
Double-flare-V-groove weld

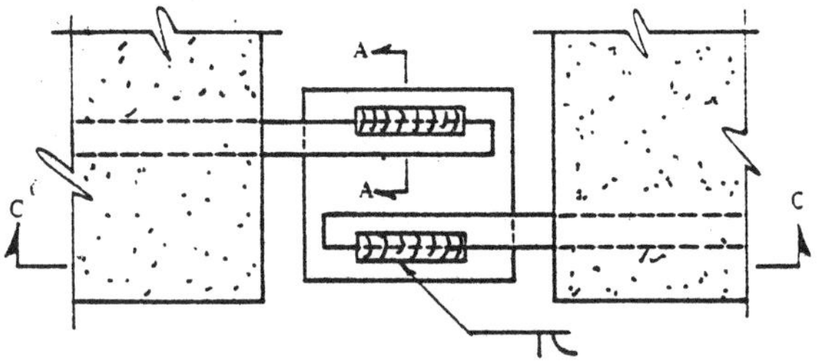


Single-flare-bevel-groove weld

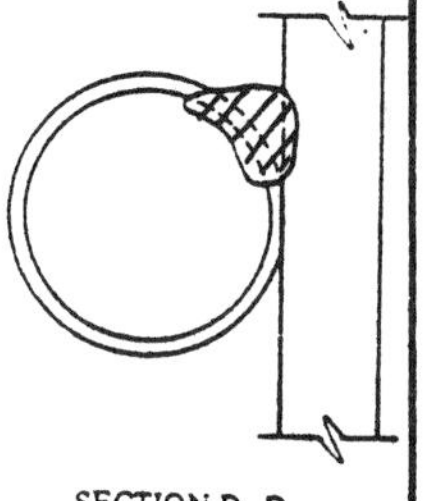

SECTION D–D

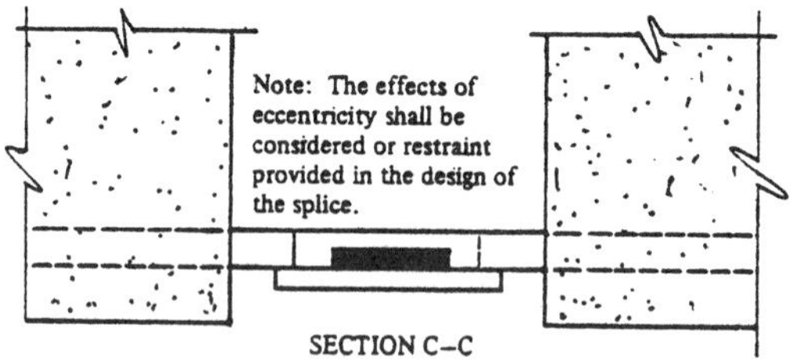


SECTION C–C

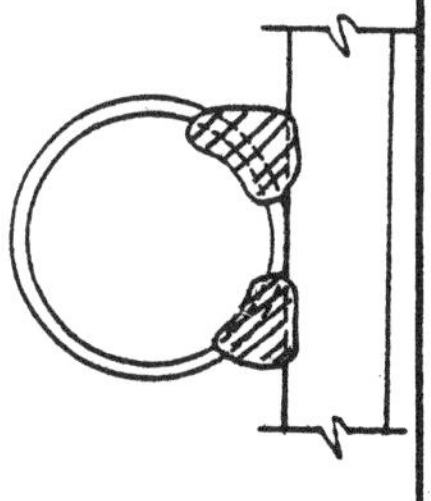


Double-flare-bevel-groove weld

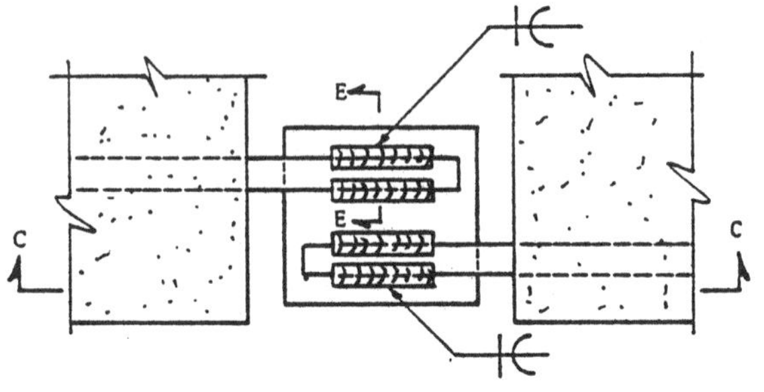

SECTION E–E

**INDIRECT SPLICES**

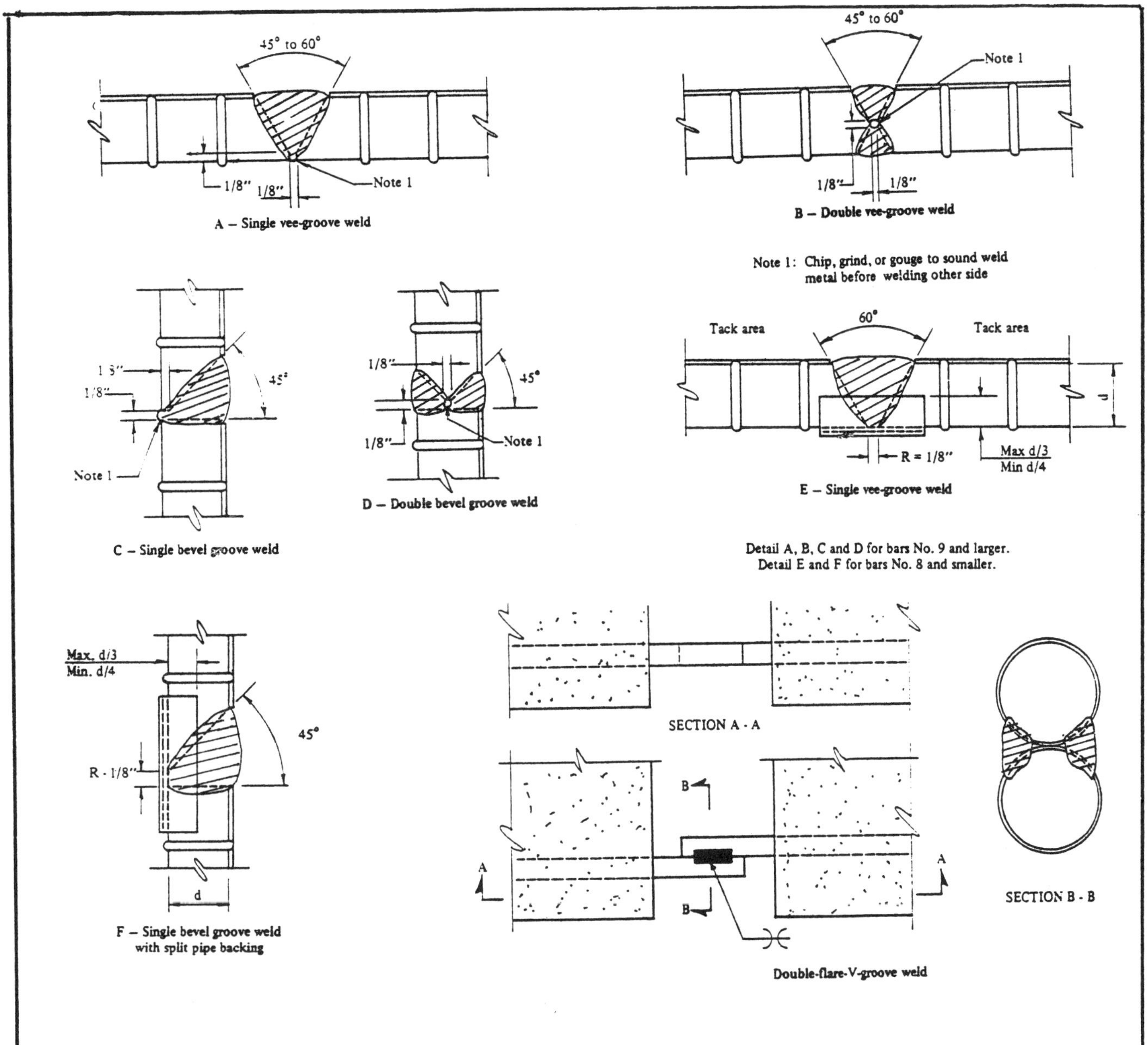


## DIRECT SPLICES

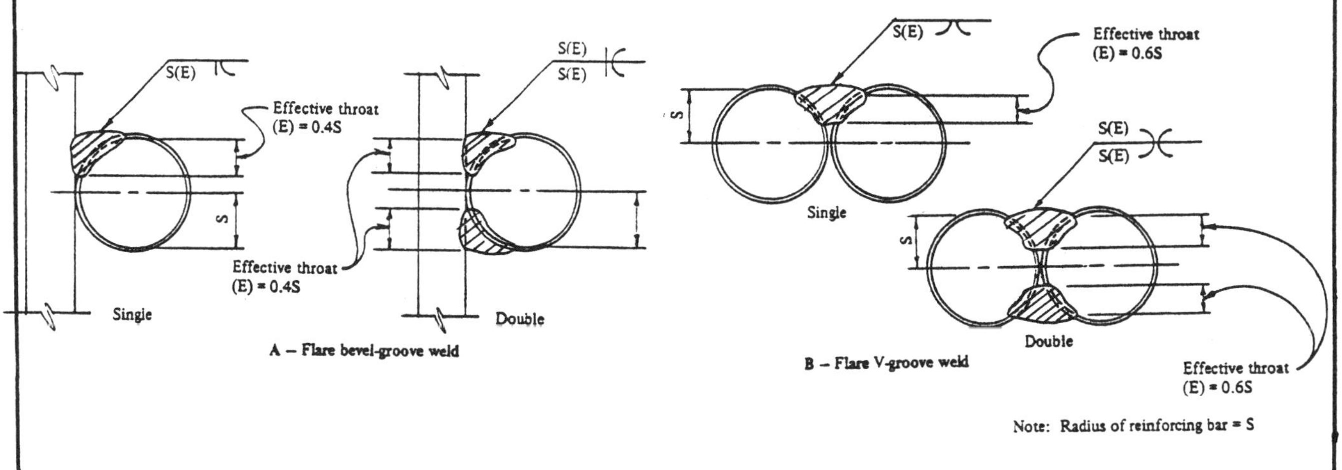


## EFFECTIVE THROATS for FLARE GROOVE WELDS

# C.6 BAR PLACEMENT DETAILS

## "Normal" right corners

Corners subjected to bending as in figure will crack radially in the corner outside of the main reinforcing steel. Smaller size reinforcing steel (similar to temperature reinforcing steel) shall be provided in the corner as shown to distribute the radial cracking.

For highly stressed reinforcing steel, the radius of bend may have to be increased to keep radial pressures against the concrete within safe limits.

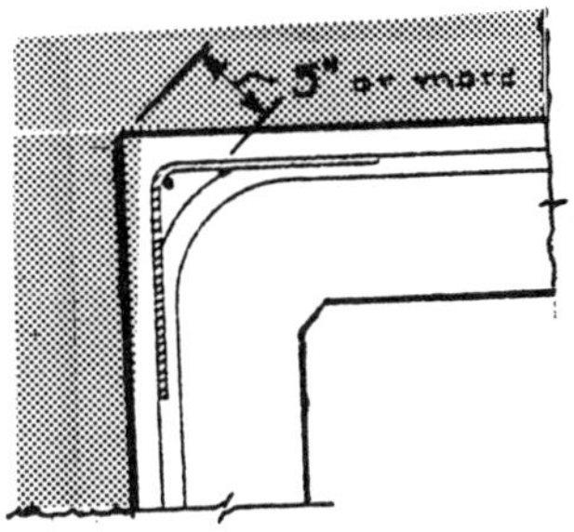


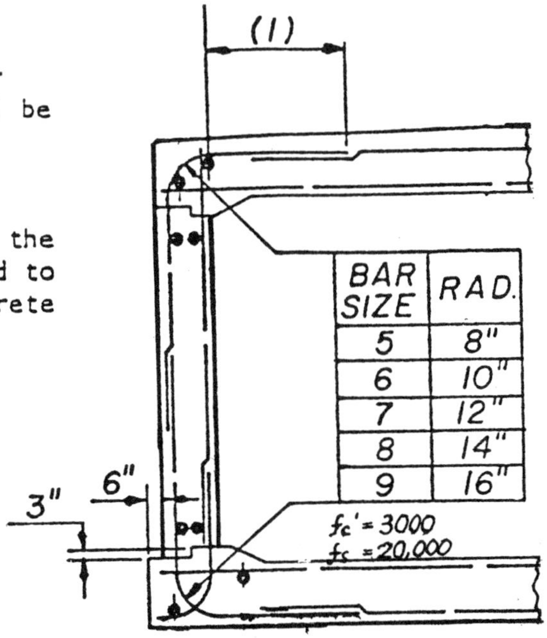


(1) *As required by Design, 40 Dia. Min.*

## Cantilever Retaining Wall

The footing near the corner, at the junction between the stem and the footing, with reinforcing as in figure will fully develop the resisting moment as long as the toe of the footing is long enough for anchorage, and stress at "A" (bottom) is not critical.

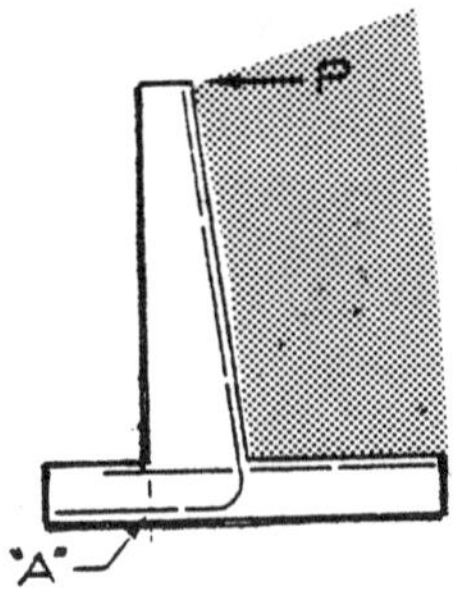


## T-Joint

The forces in figure form a tension crack at 45°. Reinforcement as shown is more than twice as effective in developing the strength of the corner than if the reinforcement were turned 180°.

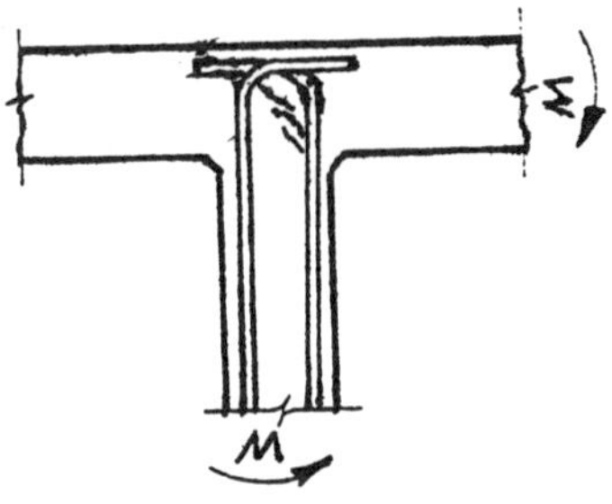

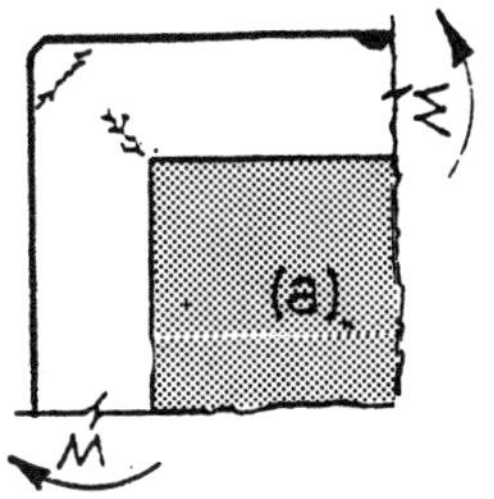


Right or obtuse angle corners

Corners subjected to bending as in figure (a) tend to crack at the reentrant corner and fail in tension across the corner. If not properly reinforced, the resisting corner moment is less than the applied moment.

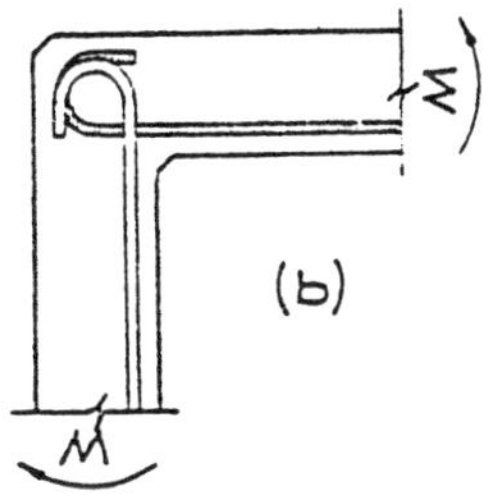


Reinforced as in figure (b), the section will develop 85% of the ultimate moment capacity of the wall. If the bends were rotated 180°, only 30% of the wall capacity would be developed.

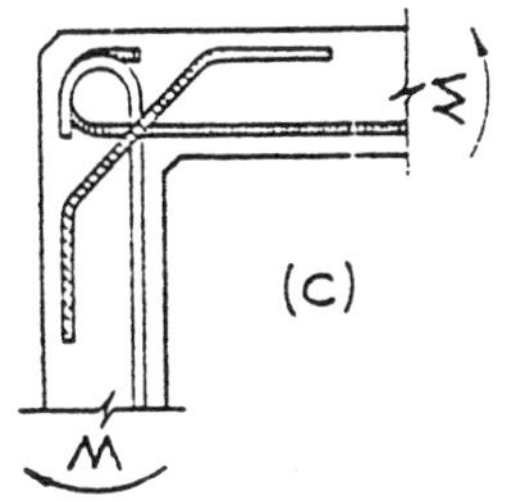


Adding diagonal reinforcing steel across the corner as in figure (c), approximately equal to 50% of the main reinforcing steel, will develop the corner strength to fully resist the applied moment. Extend the diagonal reinforcement past the corner each direction for anchorage (4 feet long for bars through #7 are adequate).

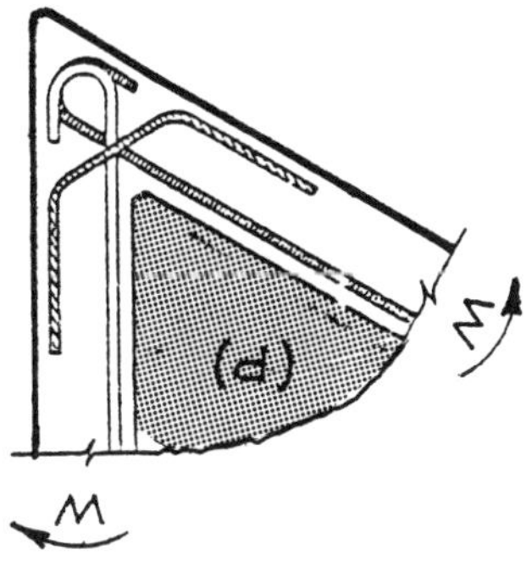


Acute angle corners

Like right angle corners, the section in figure (d) tends to crack at the reentrant corner. The section reinforced as shown in figure (d) will develop the corner strength to fully resist the applied moment.

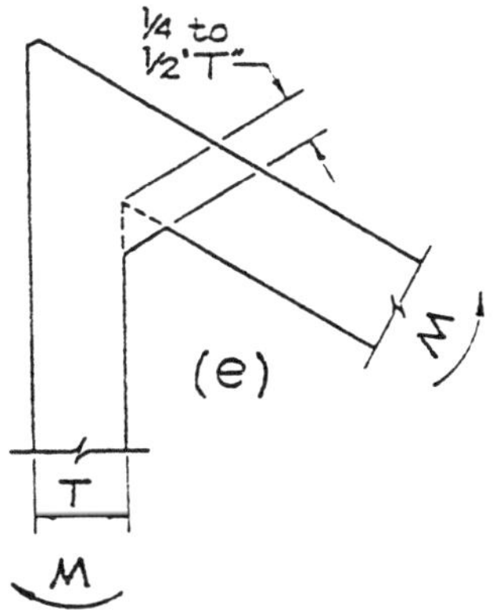


A fillet is not needed to develop the corner strength, but better resistance to cracking and better embedment of reinforcement can be accomplished with a fillet 1/4 to 1/2 of the wall or beam thickness. See figure (e). Use reinforcing steel across the corner approximately equal to 60% of the main reinforcing steel.

Since this bar arrangement will fully develop the resisting moment, a fillet in the corner is normally unnecessary.

For a more detailed discussion, refer to the June 1976, ASCE Journal of the Structural Division, "Reinforced Concrete Corners and Joints Subjected to Bending Moment" by Inguar H. E. Nilsson and Anders Losberg.

# C.7 TORSIONAL REINFORCEMENT Load Factor Design

## APPLICABILITY

On portions of structures where torsional stresses may be higher than a small nominal value, the following method of designing for torsion shall be used. This would include crossbeams of raised crossbeam prestressed bridges and other similar cases.

This criteria is based on the ACI code and for cases which are not covered herein that code should be consulted.

## DEFINITIONS

For purposes of this criteria, the expression vertical shear shall be taken to mean a shear caused by loads tending to produce bending in any direction.

Closed stirrups shall be taken to be stirrups completely enclosing the area of concrete or, as an alternative, a U stirrup with main deck steel forming the top tie.

## REFERENCES

ACI Standard 318-17-Building Code Requirements for Reinforced Concrete.

"How to Design for Torsion", Mattock, ACI Paper 18-18.

## PROCEDURE

1. Computation of Factored Torsional Shear Stress $V_{tu}$

The maximum torsional stress shall be taken as

$$V_{tu} = \frac{T_u}{\phi \, \Sigma \frac{(x^2 y)}{3}}$$

and occurs on the periphery at the middle of the longest side of a component rectangle.

($\Sigma \frac{x^2 y}{3}$ including all component rectangles of the section)

For a rectangular section with the long side vertical,

$$V_{tu} = \frac{3 T_u}{\phi \, x^2 y}$$

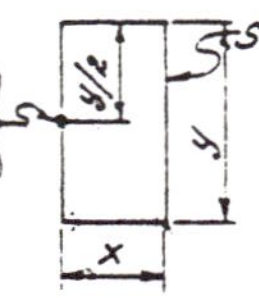


Eq (1)

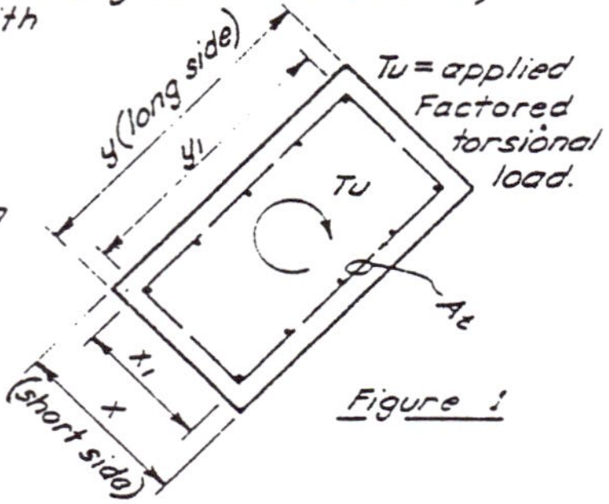


Figure 1

Since the shear stress due to vertical shear and the shear stress due to torsion add only on the vertical faces of the section, for the case where the rectangular section is placed with the long side horizontal the torsional stress should be multiplied by $\frac{x}{y}$ for purposes of combining with vertical shear stresses.

or $V_{tu} = \frac{3 T_u}{\phi \, x y^2}$ (1a)

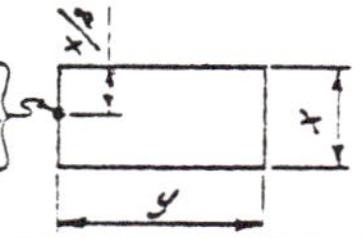


If the maximum torsional stress is less than $1.5 \phi \sqrt{f'_c}$ p.s.i. (or 80.6 p.s.i. for AX concrete) the effect of torsion may be ignored and the stirrups may be selected on the basis of vertical shear alone. (Note that $\phi = .85$ for all cases of shear).

2. Limitations on torsional stress

The nominal vertical shear stress shall be computed by

$$V_u = \frac{V_u}{b_w d} \quad (2)$$

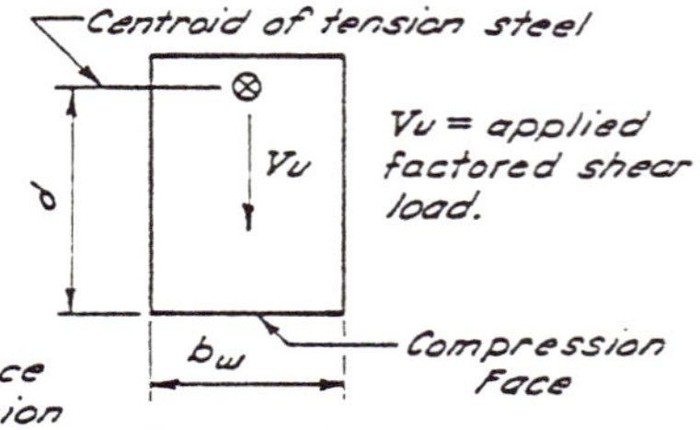


Figure 2

where $b_w$ is the web width or, for a rectangular section, the effective width of the section, and d is the distance from the extreme compression fiber to the centroid of the tension reinforcement. In order that the concrete section not be overloaded due to the combination of torsion and shearing stress, the total design torsion stress shall be limited to

$$V_{tu} \leqq \frac{12 \sqrt{f'c}}{\sqrt{1 + \left(1.2 \frac{V_u}{V_{tu}}\right)^2}} \quad (3)$$

or setting $\frac{V_{tu}}{V_u} = \gamma$; and using class AX concrete

$$V_{tu} \leqq 12 K \quad (3a)$$

The value of $K$ can be taken from Figure 3.

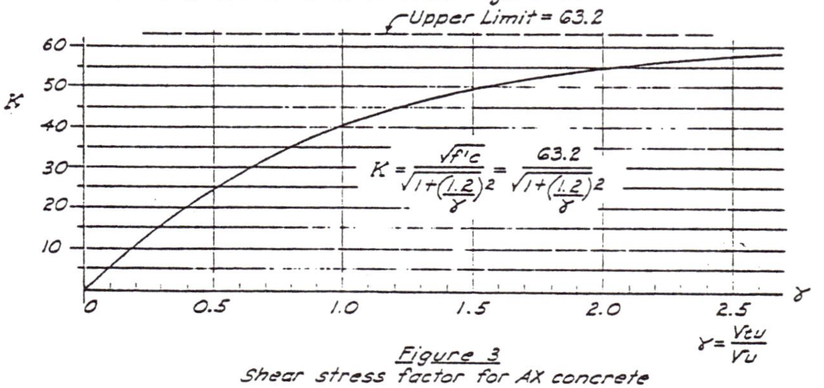


Figure 3
Shear stress factor for AX concrete

If $V_{tu}$ exceeds the allowable as calculated above, the dimensions of the concrete section must be changed to bring $V_{tu}$ within the allowable before proceeding with stirrup selection. (Note that $V_{tu}$ and $V_u$ are tied together in the equation so it is not necessary to look at $V_u$).

3. Proportioning Torsion to Concrete Section and Stirrups

(a) The nominal torsion stress carried by the concrete, $v_{tc}$, shall not exceed

$$v_{tc} = \frac{2.4 \sqrt{f'c}}{\sqrt{1+\left(\frac{1.2}{\gamma}\right)^2}} \quad \text{or } 2.4\ K \text{ for AX concrete.} \qquad (4)$$

(b) Likewise, at sections carrying torsion stress $v_{tu}$ exceeding $1.5\phi\sqrt{f'c}$, the nominal shear stress carried by the concrete, $v_c$, shall not exceed,

$$v_c = \frac{2 \sqrt{f'c}}{\sqrt{1+\left(\frac{\gamma}{1.2}\right)^2}} \quad \text{or } 2\left(\frac{1.2}{\gamma}K\right) = \frac{2.4K}{\gamma} \text{ for AX concrete.} \qquad (5)$$

(c) The portion of the applied torque carried by the concrete, $T_{uc}$, can then be taken to be

$$T_{uc} = T_u\left(\frac{v_{tc}}{v_{tu}}\right) \qquad (6)$$

Likewise the portion of the vertical shear carried by the concrete, $V_{uc}$, can be taken to be

$$V_{uc} = V_u\left(\frac{v_c}{v_u}\right) \qquad (6a)$$

4. Computation of Required Stirrup Area

The required area of closed stirrups for torsion shall be computed by

$$\frac{A_t}{s} = \frac{(T_u - T_{uc})}{a_t\, x_1 y_1 f_y} \qquad (7)$$

where $a_t = [0.66 + .33(y_1/x_1)]$ but not more than 1.5. Note that $A_t$ is here the area of the stirrup on one side of the section and the total stirrup area is $2A_t$

The required area of stirrups for vertical shear shall be computed by

$$\frac{A_{vs}}{s} = \frac{(V_u - V_{uc})}{d\, f_y} \qquad (8)$$

The total stirrup area and spacing must be ≧ the sum of the above values

$$\frac{A_v}{s} = 2\frac{A_t}{s} + \frac{A_{vs}}{s} \qquad (9)$$

Since only the stirrups near the outside face are effective for torsion, the stirrups at the face must provide a minimum of $\frac{A_t}{s}$. For designs involving large torsional stresses, if the spacing and size of face stirrups becomes a problem, interior stirrups may be used to partially fulfill torsion requirements at a reduced effective value. See Figure 4. If $A_t'$ is the outer stirrup area and $A_t''$ is the effective interior stirrup area then: $A_t'' = A_t'\left(\frac{x_2}{x_1}\right)^2$ When the inner stirrup is the same size & spacing as the outer stirrup.

The total stirrup area across the section shall be: $\frac{A_v}{s} = \frac{2A_t}{s[1+(\frac{x_2}{x_1})^2]} + \frac{A_{vs}}{2s}$

where $A_v$ is the area of 2 legs of the stirrups.

In this case special attention shall be given to showing the location of these bars clearly on the plans.

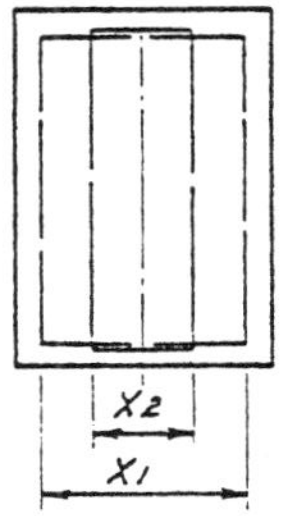


Figure 4

5. Minimum Stirrup Area

In order to control sudden cracking a minimum amount of stirrup reinforcement shall be provided as follows:

where $v_{tu} \leqq 1.5\sqrt{f'c}$ the minimum area in square inches shall be $\frac{A_v}{s} \geqq 50\frac{b_w}{f_y}$ (10a)

or where $v_{tu} > 1.5\sqrt{f'c}$ the minimum area of closed stirrups shall be $\frac{A_v}{s} = \frac{A_{vs} + 2A_t}{s} = 50\frac{b_w}{f_y}$ (10b)

6. Longitudinal Reinforcement

Longitudinal reinforcement in sections in which $v_{tu} > 1.5\phi\sqrt{f'c}$ shall be taken to be

$$A_L = \frac{2A_t}{s}(x_1 + y_1) \qquad (11)$$

but not less than #5 at 1'-0". This longitudinal reinforcement shall be distributed to each side of the section in accordance with it's proportion of the total perimeter. The tension steel should be increased by the amount of the longitudinal reinforcement calculated for the tension face.

7. Other Considerations

The spacing of closed stirrups shall not exceed $(x_1 + y_1)/4$ or 12 inches, whichever is smaller, in the area requiring closed stirrups. The spacing of longitudinal bars shall not exceed 12 inches and at least one bar shall be placed in each corner of the stirrups. The above spacing of closed stirrups shall extend from the face of the support to a point $(d + b_w)$ beyond the point theoretically required. Sections located less than a distance d from the face of the support may be designed for the same torsion $v_{tu}$ as that computed at a distance d from the support.

# C.8 WELDED WIRE FABRIC

## Nomenclature

### 2.2 Wire Size Designation

Both wire for welded wire fabric and wire for reinforcement are specified by cross-sectional area. Gage numbers were used exclusively for many years. The industry changed to a letter-number combination. The letters "W" and "D" are used in combination with a number. The letter "W" designates a smooth wire and the letter "D" denotes a deformed wire. The number following the letter gives the cross-sectional area in hundredths of a square inch. For instance, wire designation W4 would indicate a smooth wire with a cross-sectional area of 0.04 sq. in.; a D-10 wire would indicate a deformed wire with a cross-sectional area of 0.10 sq. in. This system has many advantages. Since the designer knows the cross-sectional area of a wire and the spacing, he can easily determine the cross-sectional area per lineal foot. For instance, a W6 wire on 4 inch centers would provide 3 wires per foot each with a cross-sectional area of 0.06 sq. in. or 0.18 sq. in. per lineal foot.

When describing metric wire or fabric, the letter "M" is added. MW describes metric smooth and MD metric deformed wire. The wire spacings in metric fabric are given in millimeters (mm) and the cross-sectional areas of the wires in millimeters squared (mm²). Thus, the sample style, using customary units 6 x 12 W16 x W8, would be expressed metrically as 152 x 305 MW103 x MW51.6. The cross-sectional area for fabric is expressed in mm² per meter.

Nominal cross-sectional area of a deformed wire is determined from the weight (mass) per foot of wire rather than the diameter.

### 2.3 Style

Spacings and sizes of wires, in welded wire fabric are identified by "style." A typical style designation is:
6 x 12—W16 x W8

This denotes a welded wire fabric in which:

| | |
|---|---|
| Spacing of longitudinal wires | = 6" (152mm) |
| Spacing of transverse wires | = 12" (305mm) |
| Size of longitudinal wires | = W16 (0.16 sq. in.) (103mm²) |
| Size of transverse wires | = W8 (0.08 sq. in.) (51.6mm²) |

A welded deformed wire fabric style would be noted in the same manner by substituting D-number wire sizes for the W-number wire sizes shown. Note that "style" gives spacings and sizes of wires only and does not show any other information such as width and length of sheet or roll.

Welded wire fabric with non-uniform wire spacings may be available upon request. In this case, special information is added to the style designation to describe the fabric.

It is very important to note that the terms "longitudinal" and "transverse" are related to the manufacturing process and have no reference to the position of the wires in a concrete structure. In the manufacture of welded wire fabric, longitudinal wires move continuously through the automatic welding machine. Transverse wires are individually welded at right angles as the fabric advances through the welder. In some fabric machines, the transverse wire is fed from a continuous coil; in others they are precut to length and hopper fed to the welding position.

### 3.3 Stock Items

Certain items of welded smooth wire fabric are carried in stock either at the producing mills or warehousing points. While practice varies somewhat with different manufacturers and localities, the items listed in Table 1 are usually available.

| Typical roll widths and lengths are: | | mm. m |
|---|---|---|
| U.S. Except West Coast | 60" x 150' | 1520 x 45 |
| U.S.—West Coast | 84" x 150' | 2134 x 45 |
| | 84" x 200' | 2134 x 60 |
| Canada | 60" x 200' | 1520 x 60 |
| | 72" x 200' | 1830 x 60 |

**Note:** Wire sizes other than those listed above including larger sizes may be produced provided the quantity required is sufficient to justify manufacture

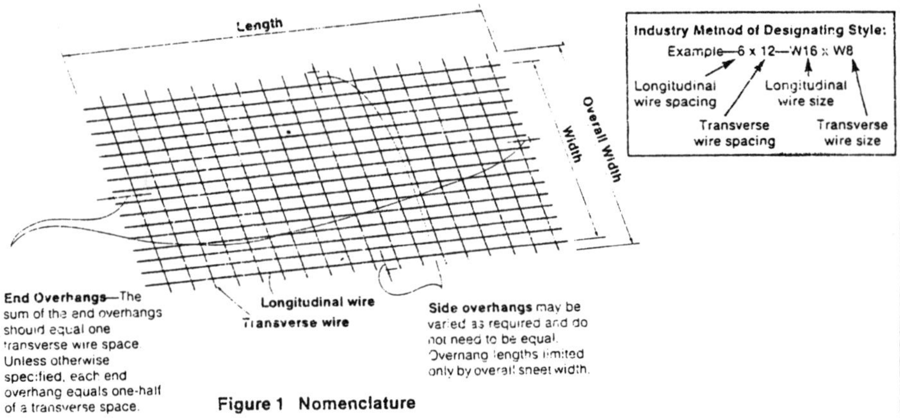


Figure 1 Nomenclature

TABLE 1 Common Stock Styles of Welded Wire Fabric

| STYLE DESIGNATION | | STEEL AREA sq. in. per ft. | | METRIC STYLE DESIGNATION |
|---|---|---|---|---|
| New Designation (By W-Number) | Old Designation (By Steel Wire Gage) | Longit. | Trans. | |
| ROLLS | | | | |
| 6 x 6—W1.4 x W1.4 | 6 x 6—10 x 10 | .028 | .028 | 152 x 152 MW9.1 x MW9.1 |
| 6 x 6—W2.0 x W2.0 | 6 x 6—8 x 8* | .040 | .040 | 152 x 152 MW13.3 x MW13.3 |
| 6 x 6—W2.9 x W2.9 | 6 x 6—6 x 6 | .058 | .058 | 152 x 152 MW18.7 x MW18.7 |
| 6 x 6—W4.0 x W4.0 | 6 x 6—4 x 4 | .080 | .080 | 152 x 152 MW25.8 x MW25.8 |
| 4 x 4—W1.4 x W1.4 | 4 x 4—10 x 10 | .042 | .042 | 102 x 102 MW9.1 x MW9.1 |
| 4 x 4—W2.0 x W2.0 | 4 x 4—8 x 8* | .060 | .060 | 102 x 102 MW13.3 x MW13.3 |
| 4 x 4—W2.9 x W2.9 | 4 x 4—6 x 6 | .087 | .087 | 102 x 102 MW18.7 x MW18.7 |
| 4 x 4—W4.0 x W4.0 | 4 x 4—4 x 4 | .120 | .120 | 102 x 102 MW25.8 x MW25.8 |
| SHEETS | | | | |
| 6 x 6—W2.9 x W2.9 | 6 x 6—6 x 6 | .058 | .058 | 152 x 152 MW18.7 x MW18.7 |
| 6 x 6—W4.0 x W4.0 | 6 x 6—4 x 4 | .080 | .080 | 152 x 152 MW25.8 x MW25.8 |
| 6 x 6—W5.5 x W5.5 | 6 x 6—2 x 2** | .110 | .110 | 152 x 152 MW34.9 x MW34.9 |
| 4 x 4—W4.0 x W4.0 | 4 x 4—4 x 4 | .120 | .120 | 102 x 102 MW25.8 x MW25.8 |

*Exact W-number size for 8 gage is W2.1.
**Exact W-number size for 2 gage is W5.4.

### 6.1 Sectional Areas of Welded Wire Fabric

| Wire Size Number Smooth | Deformed | Nominal Diameter, Inches | Nominal Weight Lbs./Lin. Ft. | $A_s$—Sq. In. Per Lin. Ft. Center to Center Spacing 2" | 3" | 4" | 6" | 8" | 10" | 12" |
|---|---|---|---|---|---|---|---|---|---|---|
| W20 | D20 | 0.505 | .680 | 1.20 | 80 | .60 | .40 | 30 | .24 | .20 |
| W18 | D18 | 0.479 | .612 | 1.08 | 72 | .54 | .36 | .27 | .216 | .18 |
| W16 | D16 | 0.451 | .544 | 96 | .64 | .48 | .32 | .24 | .192 | .16 |
| W14 | D14 | 0.422 | 476 | 84 | .56 | .42 | .28 | .21 | .168 | .14 |
| W12 | D12 | 0.391 | .408 | .72 | 48 | .36 | .24 | .18 | .144 | .12 |
| W11 | D11 | 0.374 | .374 | 66 | .44 | .33 | .22 | .165 | .132 | .11 |
| W10.5 | | 0.366 | .357 | .63 | .42 | .315 | .21 | .157 | .126 | .105 |
| W10 | D10 | 0.357 | .340 | .60 | .40 | .30 | .20 | .15 | 12 | .10 |
| W9.5 | | 0.348 | .323 | .57 | .38 | .285 | .19 | .142 | .114 | .095 |
| W9 | D9 | 0.338 | .306 | .54 | .36 | 27 | .18 | .135 | .108 | .09 |
| W8.5 | | 0.329 | .289 | .51 | .34 | .255 | .17 | .127 | .102 | .085 |
| W8 | D8 | 0.319 | .272 | .48 | .32 | 24 | .16 | .12 | .096 | .08 |
| W7.5 | | 0.309 | .255 | .45 | .30 | .225 | .15 | .112 | .09 | .075 |
| W7 | D7 | 0.299 | .238 | .42 | .28 | .21 | .14 | .105 | .084 | .07 |
| W6.5 | | 0.288 | .221 | .39 | .26 | .195 | .13 | .097 | .078 | .065 |
| W6 | D6 | 0.276 | .204 | .36 | .24 | .18 | .12 | .09 | .072 | .06 |
| W5.5 | | 0.265 | .187 | .33 | 22 | .165 | .11 | .082 | .066 | .055 |
| W5 | D5 | 0.252 | 170 | .30 | .20 | .15 | .10 | .075 | .06 | .05 |
| W4.5 | | 0.239 | .153 | .27 | .18 | .135 | .09 | .067 | .054 | .045 |
| W4 | D4 | 0.226 | .136 | .24 | .16 | .12 | .08 | .06 | .048 | .04 |
| W3.5 | | 0.211 | .119 | .21 | 14 | .105 | .07 | .052 | .042 | .035 |
| W3 | | 0.195 | .102 | .18 | .12 | .09 | .06 | .045 | .036 | .03 |
| W2.9 | | 0.192 | .098 | .174 | .116 | .087 | .058 | .043 | .035 | .029 |
| W2.5 | | 0.178 | .085 | .15 | .10 | .075 | .05 | .037 | 03 | .025 |
| W2 | | 0.160 | .068 | .12 | .08 | .06 | .04 | .03 | .024 | .02 |
| W1.4 | | 0.134 | .049 | .084 | .056 | .042 | .028 | .021 | .017 | .014 |

# DESIGN AIDS FOR STRUCTURAL WELDED WIRE FABRIC

TABLE 1

## Specifications Covering Welded Wire Fabric

| U.S. Specification | Canadian Standard | Title* |
|---|---|---|
| ASTM A 82 | CSA G 30.3 | Cold-Drawn Steel Wire for Concrete Reinforcement |
| ASTM A 185 | CSA G 30.5 | Welded Steel Wire Fabric for Concrete Reinforcement |
| ASTM A 496 | CSA G 30.14 | Deformed Steel Wire for Concrete Reinforcement |
| ASTM A 497 | CSA G 30.15 | Welded Deformed Steel Wire Fabric for Concrete Reinforcement |

*The titles of the ASTM Specifications and CSA Standards are identical.*

TABLE 2

## ASTM and CSA Minimum Properties of Steel Wires in Welded Wire Fabric

| Type of Fabric | Tensile Strength psi | Yield Strength psi | Weld Shear Strength |
|---|---|---|---|
| Welded Smooth Wire Fabric | 75,000 | 65,000 | 35,000 psi |
| Welded Deformed Wire Fabric | 80,000 | 70,000 | 20,000 psi |

*Yield strength is measured at 0.005 inch per extension of gauge length.*

### Tables 3 through 12. Equivalency Tables.

Styles of welded wire fabric with yield strengths, $f_y$, of 60,000, 65,000, 70,000, 72,500, 75,000 and 80,000 psi can be selected from these tables to replace reinforcement systems designed with Grade 40 and Grade 60 reinforcement except for the following:

1. When designs for slabs specify Grade 40 steel and only the minimum shrinkage and temperature reinforcement has been used, the fabric wire size must be selected from Table 14. This is necessary because the reinforcement to concrete ratio for Grade 40 steel is not the same as fabric with yield strengths ($f_y$) of 60,000 psi and larger. **(See Table 13)**

TABLE 3

## Equivalency Tables — Reinforcing Bars & High Strength Welded Wire Fabric

Reinforcing Bar 40,000 psi
Welded Wire Fabric 60,000 psi

| BARS | SPACE | #/CSF | 4 IN. | 6 IN. | 8 IN. | 10 IN. | 12 IN. | 16 IN | #/CSF |
|---|---|---|---|---|---|---|---|---|---|
| NO. 4 | 4" | 200 | W13.1 | W19.6 | | | | | 133 |
| NO. 4 | 5" | 160 | W10.5 | W15.7 | | | | | 107 |
| NO. 4 | 6" | 134 | W 8.7 | W13.1 | W17.5 | | | | 89 |
| NO. 4 | 7" | 114 | W 7.5 | W11.2 | W15.0 | W18.7 | | | 76 |
| NO. 4 | 8" | 100 | W 6.5 | W 9.8 | W13.1 | W16.4 | W19.6 | | 67 |
| NO. 4 | 9" | 89 | W 5.8 | W 8.7 | W11.6 | W14.5 | W17.5 | | 59 |
| NO. 4 | 10" | 80 | W 5.2 | W 7.9 | W10.5 | W13.1 | W15.7 | | 53 |
| NO. 4 | 11" | 73 | W 4.8 | W 7.1 | W 9.5 | W11.9 | W14.3 | D 19.0 | 49 |
| NO. 4 | 12" | 67 | W 4.4 | W 6.5 | W 8.7 | W10.9 | W13.1 | D 17.5 | 45 |
| NO. 4 | 13" | 62 | W 4.0 | W 6.0 | W 8.1 | W10.1 | W12.1 | D 16.1 | 41 |
| NO. 4 | 14" | 57 | W 3.7 | W 5.6 | W 7.5 | W 9.4 | W11.2 | D 15.0 | 38 |
| NO. 4 | 15" | 53 | W 3.5 | W 5.2 | W 7.0 | W 8.7 | W10.5 | D 14.0 | 36 |
| NO. 4 | 16" | 50 | W 3.3 | W 4.9 | W 6.5 | W 8.2 | W 9.8 | D 13.1 | 33 |
| NO. 4 | 17" | 47 | W 3.1 | W 4.6 | W 6.2 | W 7.7 | W 9.2 | D 12.3 | 31 |
| NO. 4 | 18" | 45 | W 2.9 | W 4.4 | W 5.8 | W 7.3 | W 8.7 | D 11.6 | 30 |

| BARS | SPACE | #/CSF | 4 IN. | 6 IN. | 8 IN. | 10 IN. | 12 IN. | 16 IN. | #/CSF |
|---|---|---|---|---|---|---|---|---|---|
| NO. 5 | 4" | 313 | | | | | | | 209 |
| NO. 5 | 5" | 250 | W16.4 | | | | | | 167 |
| NO. 5 | 6" | 200 | W13.6 | | | | | | 139 |
| NO. 5 | 7" | 179 | W11.7 | W17.5 | | | | | 119 |
| NO. 5 | 8" | 156 | W10.2 | W15,3 | | | | | 104 |
| NO. 5 | 9" | 139 | W 9.1 | W13.6 | W18.2 | | | | 93 |
| NO. 5 | 10" | 125 | W 8.2 | W12.3 | W16.4 | | | | 84 |
| NO. 5 | 11" | 114 | W 7.4 | W11.2 | W14.9 | W18.6 | | | 76 |
| NO. 5 | 12" | 104 | W 6.8 | W10.2 | W13.6 | W17.0 | | | 69 |
| NO. 5 | 13" | 96 | W 6.3 | W 9.4 | W12.6 | W15.7 | W18.9 | | 64 |
| NO. 5 | 14" | 89 | W 5.8 | W 8.8 | W11.7 | W14.6 | W17.5 | | 60 |
| NO. 5 | 15" | 83 | W 5.5 | W 8.2 | W10.9 | W13.6 | W16.4 | | 56 |
| NO. 5 | 16" | 78 | W 5.1 | W 7.7 | W10.2 | W12.8 | W15.3 | | 52 |
| NO. 5 | 17" | 74 | W 4.8 | W 7.2 | W 9.6 | W12.0 | W14.4 | D 19.2 | 49 |
| NO. 5 | 18" | 70 | W 4.5 | W 6.8 | W 9.1 | W11.4 | W13.6 | D 18.2 | 46 |

TABLE 4

## Equivalency Tables — Reinforcing Bars & High Strength Welded Wire Fabric

Reinforcing Bar 40,000 psi
Welded Wire Fabric 65,000 psi

| BARS | SPACE | #/CSF | 4 IN. | 6 IN. | 8 IN. | 10 IN. | 12 IN. | 16 IN | #/CSF |
|---|---|---|---|---|---|---|---|---|---|
| NO. 4 | 4" | 200 | W12.1 | W18.1 | | | | | 123 |
| NO. 4 | 5" | 160 | W 9.7 | W14.5 | W19.3 | | | | 99 |
| NO. 4 | 6" | 134 | W 8.1 | W12.1 | W16.1 | | | | 82 |
| NO. 4 | 7" | 114 | W 6.9 | W10.4 | W13.8 | W17.3 | | | 71 |
| NO. 4 | 8" | 100 | W 6.0 | W 9.1 | W12.1 | W15.1 | W18.1 | | 62 |
| NO. 4 | 9" | 89 | W 5.4 | W 8.1 | W10.7 | W13.4 | W16.1 | | 55 |
| NO. 4 | 10" | 80 | W 4.8 | W 7.2* | W 9.7 | W12.1 | W14.5 | D 19.3 | 49 |
| NO. 4 | 11" | 73 | W 4.4 | W 6.6 | W 8.8 | W11.0 | W13.2 | D 17.6 | 45 |
| NO. 4 | 12" | 67 | W 4.0 | W 6.0 | W 8.1 | W10.1 | W12.1 | D 16.1 | 41 |
| NO. 4 | 13" | 62 | W 3.7 | W 5.6 | W 7.4 | W 9.3 | W11.2 | D 14.9 | 38 |
| NO. 4 | 14" | 57 | W 3.5 | W 5.2 | W 6.9 | W 8.6 | W10.4 | D 13.8 | 35 |
| NO. 4 | 15" | 53 | W 3.2 | W 4.8 | W 6.4 | W 8.1 | W 9.7 | D 12.9 | 33 |
| NO. 4 | 16" | 50 | W 3.0 | W 4.5 | W 6.0 | W 7.6 | W 9.1 | D 12.1 | 31 |
| NO. 4 | 17" | 47 | W 2.8 | W 4.3 | W 5.7 | W 7.1 | W 8.5 | D 11.4 | 29 |
| NO. 4 | 18" | 45 | W 2.7 | W 4.0 | W 5.4 | W 6.7 | W 8.1 | D 10.7 | 27 |

| BARS | SPACE | #/CSF | 4 IN. | 6 IN. | 8 IN. | 10 IN. | 12 IN. | 16 IN. | #/CSF |
|---|---|---|---|---|---|---|---|---|---|
| NO. 5 | 4" | 313 | W18.9 | | | | | | 193 |
| NO. 5 | 5" | 250 | W15.1 | | | | | | 154 |
| NO. 5 | 6" | 209 | W12.6 | W18.9 | | | | | 129 |
| NO. 5 | 7" | 179 | W10.8 | W16.2 | | | | | 110 |
| NO. 5 | 8" | 156 | W 9.4 | W14.2 | W18.9 | | | | 96 |
| NO. 5 | 9" | 139 | W 8.4 | W12.6 | W16.8 | | | | 86 |
| NO. 5 | 10" | 125 | W 7.6 | W11.3 | W15.1 | W18.9 | | | 77 |
| NO. 5 | 11" | 114 | W 6.9 | W10.3 | W13.7 | W17.2 | | | 70 |
| NO. 5 | 12" | 104 | W 6.3 | W 9.4 | W12.6 | W15.7 | W18.9 | | 64 |
| NO. 5 | 13" | 96 | W 5.8 | W 8.7 | W11.6 | W14.5 | W17.4 | | 59 |
| NO. 5 | 14" | 89 | W 5.4 | W 8.1 | W10.8 | W13.5 | W16.2 | | 55 |
| NO. 5 | 15" | 83 | W 5.0 | W 7.6 | W10.1 | W12.6 | W15.1 | | 51 |
| NO. 5 | 16" | 78 | W 4.7 | W 7.1 | W 9.4 | W11.8 | W14.2 | D 18.9 | 48 |
| NO. 5 | 17" | 74 | W 4.4 | W 6.7 | W 8.9 | W11.1 | W13.3 | D 17.8 | 45 |
| NO. 5 | 18" | 70 | W 4.2 | W 6.3 | W 8.4 | W10.5 | W12.6 | D 16.8 | 43 |

*Note: Deformed wire can be used for all spacings up to and including 16 in.*

TABLE 7

## Equivalency Tables — Reinforcing Bars & High Strength Welded Wire Fabric

Reinforcing Bar 60,000 psi
Welded Wire Fabric 60,000 psi

| BARS | SPACE | #/CSF | 4 IN. | 6 IN. | 8 IN. | 10 IN. | 12 IN. | 16 IN. | #/CSF |
|---|---|---|---|---|---|---|---|---|---|
| NO. 4 | 4" | 200 | W19.6 | | | | | | 200 |
| NO. 4 | 5" | 160 | W15.7 | | | | | | 160 |
| NO. 4 | 6" | 134 | W13.1 | W19.6 | | | | | 134 |
| NO. 4 | 7" | 114 | W11.2 | W16.8 | | | | | 114 |
| NO. 4 | 8" | 100 | W 9.8 | W14.7 | W19.6 | | | | 100 |
| NO. 4 | 9" | 89 | W 8.7 | W13.1 | W17.5 | | | | 89 |
| NO. 4 | 10" | 80 | W 7.9 | W11.8 | W15.7 | W19.6 | | | 80 |
| NO. 4 | 11" | 73 | W 7.1 | W10.7 | W14.3 | W17.0 | | | 73 |
| NO. 4 | 12" | 67 | W 6.5 | W 9.8 | W13.1 | W16.4 | W19.6 | | 67 |
| NO. 4 | 13" | 62 | W 6.0 | W 9.1 | W12.1 | W15.1 | W18.1 | | 62 |
| NO. 4 | 14" | 57 | W 5.6 | W 8.4 | W11.2 | W14.0 | W16.8 | | 57 |
| NO. 4 | 15" | 53 | W 5.2 | W 7.9 | W10.5 | W13.1 | W15.7 | | 53 |
| NO. 4 | 16" | 50 | W 4.9 | W 7.4 | W 9.8 | W12.3 | W14.7 | D19.6 | 50 |
| NO. 4 | 17" | 47 | W 4.6 | W 6.9 | W 9.2 | W11.6 | W13.9 | D18.5 | 47 |
| NO. 4 | 18" | 45 | W 4.4 | W 6.5 | W 8.7 | W10.9 | W13.1 | D17.5 | 45 |

| BARS | SPACE | #/CSF | 4 IN. | 6 IN. | 8 IN. | 10 IN. | 12 IN. | 16 IN. | #/CSF |
|---|---|---|---|---|---|---|---|---|---|
| NO. 5 | 4" | 313 | | | | | | | 313 |
| NO. 5 | 5" | 250 | | | | | | | 250 |
| NO. 5 | 6" | 209 | | | | | | | 209 |
| NO. 5 | 7" | 179 | W17.5 | | | | | | 179 |
| NO. 5 | 8" | 156 | W15.3 | | | | | | 156 |
| NO. 5 | 9" | 139 | W13.6 | | | | | | 139 |
| NO. 5 | 10" | 125 | W12.3 | W18.4 | | | | | 125 |
| NO. 5 | 11" | 114 | W11.2 | W16.7 | | | | | 114 |
| NO. 5 | 12" | 104 | W10.2 | W15.3 | | | | | 104 |
| NO. 5 | 13" | 96 | W 9.4 | W14.2 | W18.9 | | | | 96 |
| NO. 5 | 14" | 89 | W 8.8 | W13.1 | W17.5 | | | | 89 |
| NO. 5 | 15" | 83 | W 8.2 | W12.3 | W16.4 | | | | 84 |
| NO. 5 | 16" | 78 | W 7.7 | W11.5 | W15.3 | W19.2 | | | 78 |
| NO. 5 | 17" | 74 | W 7.2 | W10.8 | W14.4 | W18.0 | | | 73 |
| NO. 5 | 18" | 70 | W 6.8 | W10.2 | W13.6 | W17.0 | | | 69 |

TABLE 9

## Equivalency Tables — Reinforcing Bars & High Strength Welded Wire Fabric

Reinforcing Bar 60,000 psi
Welded Wire Fabric 70,000 psi

| BARS | SPACE | #/CSF | 4 IN. | 6 IN. | 8 IN. | 10 IN | 12 IN. | 16 IN. | #/CSF |
|---|---|---|---|---|---|---|---|---|---|
| NO. 4 | 4" | 200 | W16.8 | | | | | | 171 |
| NO. 4 | 5" | 160 | W13.5 | | | | | | 138 |
| NO. 4 | 6" | 134 | W11.2 | W16.8 | | | | | 114 |
| NO. 4 | 7" | 114 | W 9.6 | W14.4 | W19.2 | | | | 98 |
| NO. 4 | 8" | 100 | W 8.4 | W12.6 | W16.8 | | | | 86 |
| NO. 4 | 9" | 89 | W 7.5 | W11.2 | W15.0 | W18.7 | | | 76 |
| NO. 4 | 10" | 80 | W 6.7 | W10.1 | W13.5 | W16.8 | | | 69 |
| NO. 4 | 11" | 73 | W 6.1 | W 9.2 | W12.2 | W15.3 | W18.4 | | 62 |
| NO. 4 | 12" | 67 | W 5.6 | W 8.4 | W11.2 | W14.0 | W16.8 | | 57 |
| NO. 4 | 13" | 62 | W 5.2 | W 7.8 | W10.4 | W12.9 | W15.5 | | 53 |
| NO. 4 | 14" | 57 | W 4.8 | W 7.2 | W 9.6 | W12.0 | W14.4 | W19.2 | 49 |
| NO. 4 | 15" | 53 | W 4.5 | W 6.7 | W 9.0 | W11.2 | W13.5 | W18.0 | 46 |
| NO. 4 | 16" | 50 | W 4.2 | W 6.3 | W 8.4 | W10.5 | W12.6 | W16.8 | 43 |
| NO. 4 | 17" | 47 | W 4.0 | W 5.9 | W 7.9 | W 9.9 | W11.9 | W15.8 | 40 |
| NO. 4 | 18" | 45 | W 3.7 | W 5.6 | W 7.5 | W 9.4 | W11.2 | W15.0 | 38 |

| BARS | SPACE | #/CSF | 4 IN. | 6 IN. | 8 IN. | 10 IN. | 12 IN. | 16 IN. | #/CSF |
|---|---|---|---|---|---|---|---|---|---|
| NO. 5 | 4" | 313 | | | | | | | 268 |
| NO. 5 | 5" | 250 | | | | | | | 215 |
| NO. 5 | 6" | 209 | W17.5 | | | | | | 179 |
| NO. 5 | 7" | 179 | W15.0 | | | | | | 153 |
| NO. 5 | 8" | 156 | W13.1 | W19.7 | | | | | 134 |
| NO. 5 | 9" | 139 | W11.7 | W17.5 | | | | | 119 |
| NO. 5 | 10' | 125 | W10.5 | W15.8 | | | | | 107 |
| NO. 5 | 11" | 114 | W 9.6 | W14.3 | W19.1 | | | | 98 |
| NO. 5 | 12" | 104 | W 8.8 | W13.1 | W17.5 | | | | 89 |
| NO 5 | 13" | 96 | W 8.1 | W12.1 | W16.2 | | | | 83 |
| NO. 5 | 14" | 89 | W 7.5 | W11.3 | W15.0 | W18.8 | | | 77 |
| NO. 5 | 15" | 83 | W 7.0 | W10.5 | W14.0 | W17.5 | | | 71 |
| NO. 5 | 16" | 78 | W 6.6 | W 9.9 | W13.1 | W16.4 | W19.7 | | 67 |
| NO. 5 | 17" | 74 | W 6.2 | W 9.3 | W12.4 | W15.5 | W18.6 | | 63 |
| NO. 5 | 18" | 70 | W 5.8 | W 8.8 | W11.7 | W14.6 | W17.5 | | 60 |

TABLE 16

## — Smooth Wire Fabric — Splice Length (1.5 $l_d$) —

Development Length ($l_d$) = Table Values ÷ 1.5

| $f'_c$ | Yield Strength ($f_y$) – (psi) | | | | | |
|---|---|---|---|---|---|---|
| (psi) | 60,000 | 65,000 | 70,000 | 72,500 | 75,000 | 80,000 |
| 3000 | 36.97 $A_s$ | 40.05 $A_s$ | 43.13 $A_s$ | 44.67 $A_s$ | 46.21 $A_s$ | 49.30 $A_s$* |
| 3500 | 34.23 $A_s$ | 37.08 $A_s$ | 39.93 $A_s$ | 41.36 $A_s$ | 42.79 $A_s$ | 45.64 $A_s$ |
| 4000 | 32.02 $A_s$ | 34.69 $A_s$ | 37.35 $A_s$ | 38.69 $A_s$ | 40.02 $A_s$ | 42.69 $A_s$ |
| 4500 | 30.19 $A_s$ | 32.70 $A_s$ | 35.22 $A_s$ | 36.48 $A_s$ | 37.73 $A_s$ | 40.25 $A_s$ |
| 5000 | 28.64 $A_s$ | 31.02 $A_s$ | 33.41 $A_s$ | 34.60 $A_s$ | 35.80 $A_s$ | 38.18 $A_s$ |

*$A_s$ = $A_s$ (sq. in./lin. ft.)

TABLE 18

## Deformed Wire Fabric — Splice Length (1.7 $l_d$)

— Development Length ($l_d$) = Table Values ÷ 1.7 —

| $f'_c$ (psi) | $S_w$ < 6" | $S_w$ ≥ 6" | $S_w$ < 6" | $S_w$ ≥ 6" | $S_w$ < 6" | $S_w$ ≥ 6" | $S_w$ < 6" | $S_w$ ≥ 6" | $S_w$ < 6" | $S_w$ ≥ 6" | $S_w$ < 6" | $S_w$ ≥ 6" |
|---|---|---|---|---|---|---|---|---|---|---|---|---|
| | Yield Strength ($f_y$) | | | | | | | | | | | |
| | 60,000 psi | | 65,000 psi | | 70,000 psi | | 72,500 psi | | 75,000 psi | | 80,000 psi | |
| | $\times\sqrt{A_sS_w}$ | | $\times\sqrt{A_sS_w}$ | | $\times\sqrt{A_sS_w}$ | | $\times\sqrt{A_sS_w}$ | | $\times\sqrt{A_sS_w}$ | | $\times\sqrt{A_sS_w}$ | |
| 3000 | 12.12 | 9.70 | 14.63 | 11.75 | 17.32 | 13.85 | 18.65 | 14.92 | 20.00 | 16.00 | 22.73 | 18.18 |
| 3500 | 11.22 | 8.98 | 13.60 | 10.88 | 16.03 | 12.83 | 17.27 | 13.82 | 18.52 | 14.82 | 21.04 | 16.84 |
| 4000 | 10.50 | 8.40 | 12.72 | 10.18 | 15.00 | 12.00 | 16.16 | 12.92 | 17.32 | 13.86 | 19.69 | 15.75 |
| 4500 | 9.90 | 7.92 | 11.99 | 9.59 | 14.14 | 11.31 | 15.23 | 12.19 | 16.33 | 13.07 | 18.56 | 14.85 |
| 5000 | 9.39 | 7.51 | 11.38 | 9.10 | 13.41 | 10.73 | 14.45 | 11.56 | 15.49 | 12.40 | 17.61 | 14.09 |

To determine minimum end and side wire overhangs subtract two in. from splice lengths and divide by two.

**From Structural Welded Wire Fabric Detailing Manual**

# APPENDIX D: PRESTRESSED CONCRETE

## D.1 INTEGRAL PRECAST DECK BRIDGE

4.1 Description

This category of precast concrete bridge decks includes those types in which the deck and the main supporting members are the same (such as a slab) or those in which the deck and the structural supporting unit (the beam or girder) are cast as a single unit. In some cases a cast-in-place composite deck is poured on top of these units. In others there is merely a wearing course or in some cases no additional treatment at all.

4.1.1 Types of Sections Used

A wide variety of sections have been used for integral deck bridges, these are shown in Fig. 31 and are listed below:

a) Solid and voided slabs - Solid slabs 10 to 14 in. (254 - 356 mm) deep and voided slabs 15 - 21 in. (380 - 530 mm) deep in widths from 3 to 8 ft (0.9 - 2.4 m) are shown in the Prestressed Concrete Institute publication "Short Span Bridges - Spans to One Hundred Feet"[38]. Solid slabs are normally used for spans of less than 35 ft (10.7 m), and voided slab sections can span up to 50 ft (15.2 m) with HS 20 loading. These slab bridges are widely used in many parts of the country but most extensively in South Florida where literally thousands of these bridges have been built by both governmental agencies and private developers in residential areas (Fig. 32).

b) Box beams - Standard box beam sections 3 ft (6.9 m) and 4 ft (1.2 m) wide, in depths of 27 to 42 in. (690 -1070 mm) sections are widely used in areas of the midwest and the southeast (Fig. 33). Texas has developed a similar standard with a five foot wide section. For the deeper sections, spans in excess of 100 ft (30.5 m) are possible with HS 20 loading. Box beams for railroad loadings have also been standardized by the American Railway Engineering Association.

c) Channels - Precast concrete channel sections are used in many parts of the country in depths to 35 in. (890 mm) and various widths from 40 to 66 in. (1020 to 1680 mm). Spans are usually from 20 to 60 ft (6.1 to 18.3 m) without a composite cast-in-place deck, and up to 70 ft (21.3 m) with a composite deck.

d) Double tees - Most double tee bridges (Fig. 34) have been on county, municipal or private roads, although a few state highway departments have used them. Many of these bridges use standard thin stemmed sections up to 36 in. (910 mm) deep, while others use a wide stemmed member developed especially for bridge use. Spans up to about 60 ft (18.3 m) and more with HS 20 loading are possible.

e) Single tees - The standard single tee developed for long spans in buildings has been used extensively for bridges (Fig. 35). Flanges are usually thickened for bridge use and may be cut back from the standard 8 ft (2.4 m) width to 4 or 6 ft (1.2 or 1.8 m), unless a cast-in-place topping is placed on top. Depths from 24 to 48 in. (610 to 1220 mm) are available in some parts of the country and these units can span up to 120 ft (36.6 m) in the four foot (1.2 m) width.

f) Bulb tees - A special bridge section called a bulb tee is used extensively in the Pacific Northwest and Alaska (Fig. 36). Sections with depths of 29 to 77 in. (740 to 1960 mm) have been used and spans of up to 180 ft (54.9 m) are possible in the very deep sections. Because of the high section modulus to weight ratio this section is gaining acceptance for a wide variety of bridge uses.

g) Multi-stemmed units - A relative newcomer to the short span bridge market is the multi-stemmed tee which is especially suitable for spans of 25 to 55 ft (7.6 to 16.8 m). Depths range from about 16 in. (400 mm) to about 24 in. (610 mm) and widths are typically at 4 ft (1.2 m). The most extensively used section has four stems as shown in Fig. 30, but other sections with 3 stems or in one case 5 stems (this section is 6 ft (1.4 m) wide) have also been used. Multi-stemmed units are used in the Pacific Northwest, Iowa, Minnesota and the Dakotas.

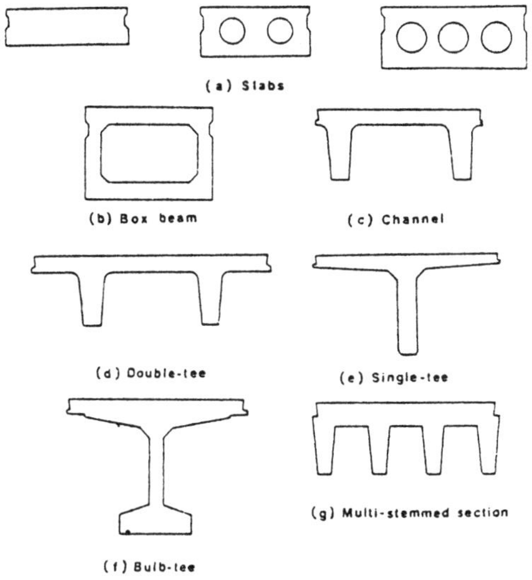


**Fig. 31.** Sections used for integral deck bridges.

*Reproduced from report FHWA/RD 82/106 "CONNECTIONS FOR MODULAR PRECAST CONCRETE BRIDGE DECKS" August, 1983*

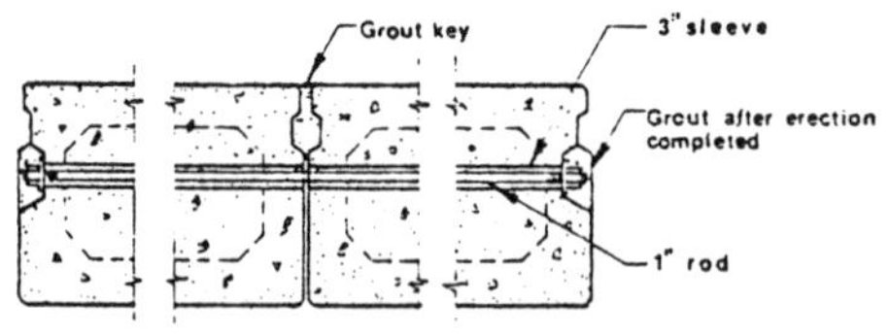


Fig. 37. Grout key with transverse tie rod.

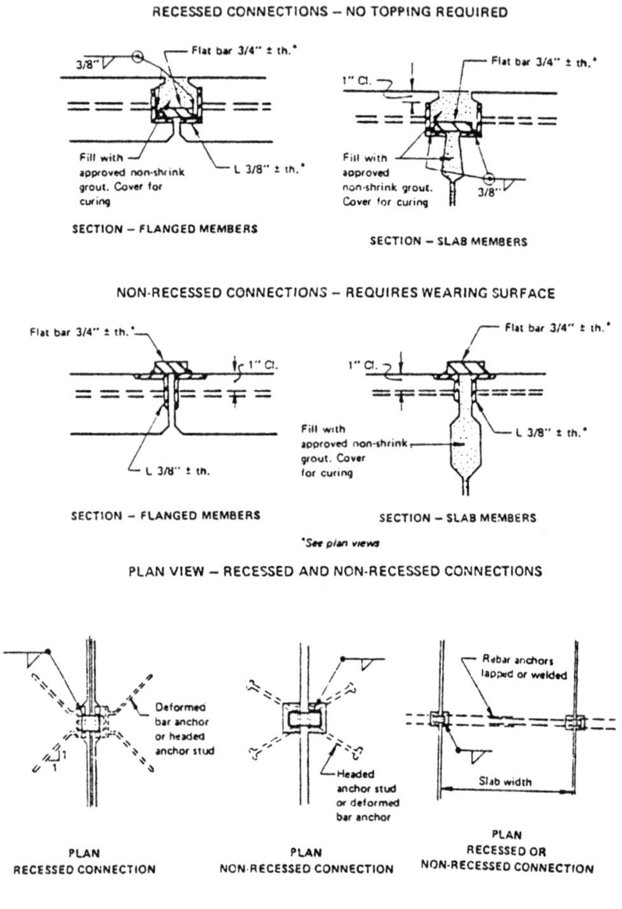


note: 1" = 25.4 mm

Fig. 39. Typical weld plate connections (From Ref. 38).

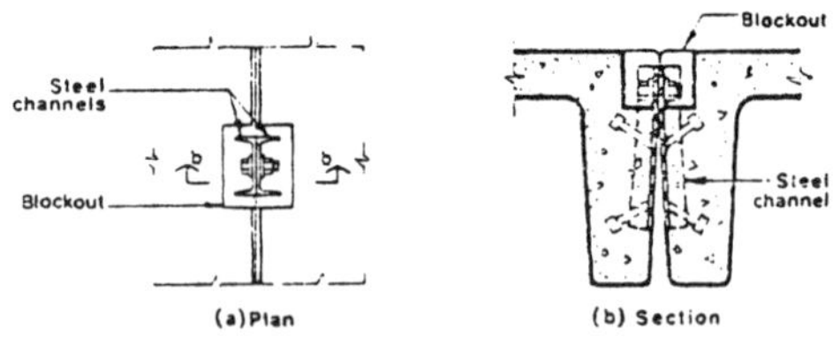


Fig. 40. Bolted connection used in Canada.

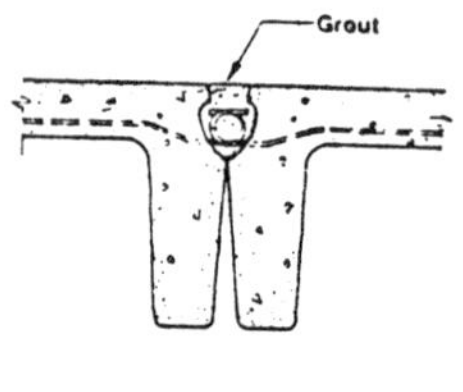


Fig. 41 Canadian hooked reinforcing bar connection

## 4.2 Types of Deck Connections

### 4.2.1 Sand Cement Grout Keys

The usual method of connecting adjacent integral deck units is with a key filled with a sand-cement grout (See Fig. 37). In some cases a non-shrink ingredient (aluminum powder) is added to the grout, or a commercial non-shrink grout mix is used. These joints typically have a mechanical tie between the units laterally to control transverse displacement. Various types of transverse ties are described below.

### 4.2.2 Transverse Bolts or Tie Rods

The most common method of tying box beams, slabs, and sometimes channels together transversely is with bolts or threaded rods as illustrated in Fig. 37. Very often these transverse bolts are placed in diaphragms, although in the case of the adjacent channel slabs used in the South, the bolts go directly between the stems of the units, as shown in Fig. 38.

### 4.2.3 Transverse Post-tensioning

After experiencing some cases of poor grout key performance when transverse bolts or tie rods were used, some states and some contractors doing private work have used transverse post-tensioning. The post-tensioning used consists of either strands or bars and is thought to give a more positive lateral tie. In the case of box girders these post-tensioning tendons are nearly always passed through sleeves in the diaphragms.

### 4.2.4 Weld Plates

In stemmed units the most common type of transverse connection is a weld plate placed in the flanges. These weld plates are spaced from 4 to 6 ft (1.2 to 1.8 m) apart along the length of the member. Size of the plate varies as does the connection detail into the flange of the member. Typical weld plate details which have been used are shown in Fig. 39.

### 4.2.5 Bolted connections

The bolted connection shown in Fig. 40 has been used in Canada. In this connection a steel channel is cast into the stem of the member and the channels are bolted together in the blocked-out portion of the flange. While this connection goes together easily, it has often been found to be too flexible to do a good job of tying the units together.

### 4.2.6 Hooked reinforcing bars

Another type of connection used in Canada is the one shown in Fig. 41. In this connection, a hooked reinforcing bar is projected from the flanges into the grout key. On longer spans this has not always kept the grout key from cracking.

### 4.2.7 Tongue & Groove Joints

The use of a match cast tongue and groove joint similar to that used on the Indiana experimental bridges (Sect. 3.2.1) has been suggested as a connection detail. Several such bridges using solid slabs are currently being cast in Florida. With this design a prestressing bed 24 ft (7.3 m) wide and 110 ft (33.5 m) long has been set up. Slabs are cast in full lane widths, and then the adjacent slabs are match cast alongside. The resulting longitudinal joint requires no grouting or special treatment. The transverse ties are tensioned as in the standard Florida DOT design (Sect. 4.4.3). There is no performance data available but based on a history of segmental box beam bridges it should be satisfactory.

## 4 Summary and Analysis of Integral Deck Bridge Connections

### 4.1 Longitudinal Connections Between Members

Integral deck precast, prestressed concrete bridges have over 25 years of generally excellent performance. The best connection is 4 in. (100 mm) or more of cast-in-place reinforced concrete. This, however, is more costly and adds significantly to the construction time. Also, when deicing salts are used, the same problems are encountered as when concrete decks are used on other types of supports.

The vast majority of the bridges which do not have a cast-in-place topping use grout keys to connect adjacent units together. There have been numerous cases of unsatisfactory performance of such shear keys, and a few reports of complete failure of the keys. Sect. 5 of this report presents a detailed discussion of grouted shear keys.

Alternatives to grout keys for longitudinal joints have usually proved impractical. Tongue-in-groove joints have been tried, but an exact fit between units is nearly impossible without match-casting. This is not practical for plant cast products, especially if they are pretensioned. Minor misfitting may be tolerated if an asphaltic concrete surface is used. The advantage is that construction could be accomplished in cold weather.

Another alternative would be to borrow a joint detail from typical parking structure construction (Fig. 51). This joint has yet to be tested under HS-20 loading, however, and the 3/4 in. (19 mm) of steel cover may lead to corrosion. A full scale bridge using this detail needs to be constructed and evaluated over a period of many years before a joint of this type can be safely recommended for integral deck bridges.

No direct relationship between the use of deicing salts and joint performance could be verified in this study — either on integral deck or precast deck-on-girder systems. Many of the applications were in parts of the country or on rural roads where salt is seldom used, and others were on relatively new projects. It is well established that deicing salts accelerate corrosion, and this may have negative implications for weld plate connections. Also, prestressing strand is more susceptible to corrosion damage than bars, so systems which use post-tensioning as part of the tee system should take this into consideration.

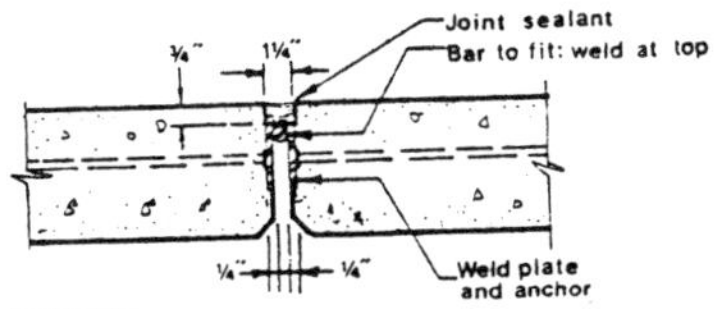


Fig. 51. Typical flange connection used in parking structures.

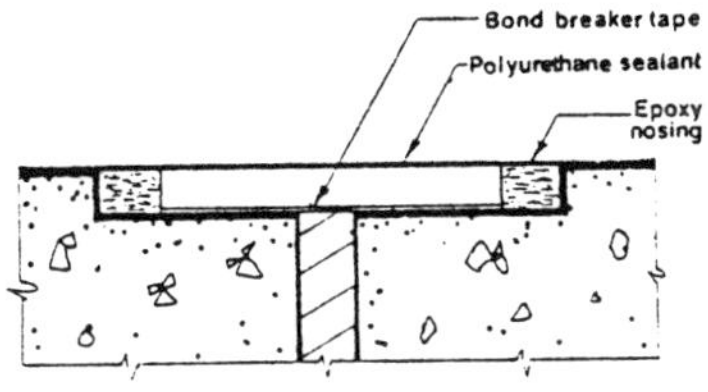


Fig. 52. Expansion joint used in Parking structures.

### 4.4.2 Transverse Deck Connections—Multi-Span Bridges

Multi-span bridges using integral deck units also have problems over fixed piers where the precast members abut. Under present practice, there is usually a precast or cast-in-place end diaphragm. Typical connections are shown in Fig. 53.

When a cast-in-place topping is used, the bridge is usually designed for live load continuity over the supports. Properly designed and constructed, these designs have exhibited few problems. With no cast-in-place topping, there are frequent problems over the fixed piers. Without continuity, rotation at the end supports of the members causes a constant "working" of the joint. To provide a watertight seal requires a "semi-expansion" device of some type. The problems caused by leakage of this joint are usually not severe, and often the most practical solution is to recognize the problem and detail the end bearings accordingly. Cracking in asphalt surfaces will occur and must be maintained.

A possible solution that should be investigated further is the type of expansion joint shown in Fig. 52. This type of joint has had good success in parking structures and other building applications.

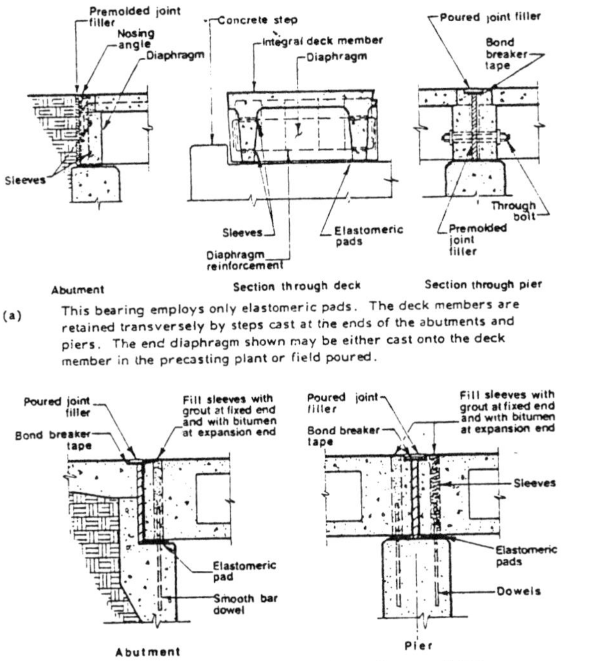


(a) This bearing employs only elastomeric pads. The deck members are retained transversely by steps cast at the ends of the abutments and piers. The end diaphragm shown may be either cast onto the deck member in the precasting plant or field poured.

(b) This bearing detail for slab or box units employs smooth bar dowels to retain the deck members transversely.

Fig. 53. Typical transverse deck connections.

### 4.4.3 Diaphragms

Typical diaphragms that have been used on integral deck bridges include: 1) field cast concrete (Fig. 53); 2) precast concrete attached in the field (Fig. 54); 3) precast concrete cast monolithically with the member in the plant (Fig. 55); and 4) steel bracing (Fig. 56).

Previous studies[26, 43, 44] have indicated that the value of diaphragms in concrete bridges is questionable. Many of the bridges investigated had no diaphragms between supports and some even eliminated them at the support. Significant cost savings can be realized by eliminating diaphragms and further study on this matter is encouraged.

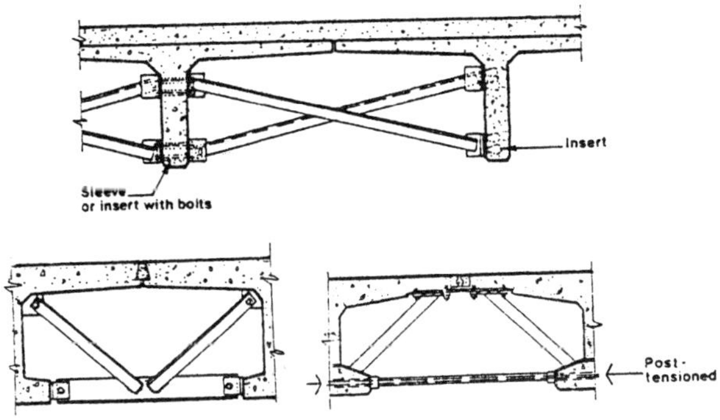


Fig. 56. Steel diaphragms used in integral deck bridges.

## 5. GROUTED SHEAR KEYS

The most frequent problems with connections of modular precast concrete bridge decks relate to grouted shear key connections between adjacent elements. Unsatisfactory performance in shear keys is most often evidenced by cracking in asphalt surfacing directly over the joint and moisture leakage. This apparently indicates that the grout has cracked or lost bond with the formed sides of the precast element. In extreme cases, a complete deterioration of the keyway material can occur. When this happens, it is probably caused by an initial crack or separation. The pounding from the traffic eventually breaks up the grout material.

In order to suggest methods of improving shear key performance, a review of grout key materials and functions is warranted.

### 5.1 Sand-Cement Grouts

Most grouts used in shear keys connecting bridge elements are a mixture of portland cement, sand, and water. Proportions are usually one part cement to 2.5 to 3 parts sand, with enough water to make the grout easily flowable and completely fill the keyway. When such grouts are used, the water-cement ratio is usually about 0.50. This is a relatively high value, resulting in low strength and high shrinkage. These mortars also exhibit a tendency for the solids to settle, leaving a layer of water on the top. Special ingredients or treatments can improve these characteristics, but add to the cost.

The Bureau of Reclamation(39) has shown that prolonged mixing of mortars can significantly reduce the settlement without affecting the workability. However, the mixing times required—2 hours or more—is impractical for routine field applications. Furthermore, the strength and shrinkage characteristics are apparently not changed significantly.

### 5.2 Non-Shrink Grouts

#### 5.2.1 Aluminum Powder

Aluminum powder added to the grout mix reacts chemically with alkaline constituents of the cement and generates hydrogen gas. Expansion of the mortar, which results from generation of the gas, causes the mortar to fit snugly in the space which confines it. Extremely small amounts of aluminum powder are required—50 to 60 millionths of the weight of cement used, about a teaspoonful per bag of cement. Reference 39 suggests that the performance of the aluminum powder mixes are not always predictable, and some brands do not react as expected. It recommends trial mixing with the particular cement and sands that are to be used.

#### 5.2.2 Proprietary Non-Shrink Grout Mixes

There are several proprietary products designed to expand sufficiently during initial hardening and curing to offset subsequent shrinkage of the grout. These mortars are prepared mixtures of cement, fine aggregate and an expansive ingredient. This ingredient may provide expansion by liberating gas, oxidation of metals, formation of gypsum or by using expansive cement(45). Assurance of suitability of these proprietary mixes should be determined by performance records or tests.

It should be kept in mind that the shear key constitutes a very low percentage of the total width of the bridge deck. Thus, while the unit shrinkage of sand-cement grout is much greater than that of the precast units, the total shrinkage of the deck members may actually contribute more to the separation at the joints than the shrinkage of the grout itself.

### 5.3 Epoxy Grouts

#### 5.3.1 Pre-Mixed Epoxy Mortar

Epoxy compounds are usually formulated in two parts, one containing the epoxy resin and one a hardener. Different formulas are used for different purposes such as bonding agent, pressure grouting, or with appropriate fillers, seal coats or wearing courses over concrete decks or floors.

Depending on the formulation and ratio of fillers, epoxy compounds have a wide range of properties. ASTM "Standard Specification for Epoxy-Resin-Base Bonding Systems for Concrete" ANSI/ASTM C881-78[27] divides epoxy systems into "Types", based on usage, "Grade" to define viscosity, and "Class" which determines setting time.

Since 1963 epoxy resins have been used to bond match cast precast segmental bridge members together. Based on research of epoxy shear keys between precast, prestressed box beams at the Association of American Railroad, the Chicago & Eastern Illinois Railroad used the concept on a 43 ft (13.1 m) span bridge in 1963[29]. In 1965 epoxy shear keys were used on a box beam highway bridge on I-487 on the border between New York and Connecticut[30]. Florida DOT uses epoxy grout for shear keys between solid and voided slabs in conjunction with lateral post-tensioning.

5.3.2 Pre-placed Aggregate with Epoxy Resin

Several railroads have been using epoxy keys for fifteen years or longer on box beam bridges. The Southern Pacific and Santa Fe Railroads place a well graded aggregate in the keyway, then pour in a low viscosity epoxy resin. Nominally tightened tie rods are used. Bridge engineers with Santa Fe and Southern Pacific report that their epoxy shear keys have been performing well. The Southern Pacific keys are 18 in. (460 mm) deep and the Santa Fe's are 12 in. (305 mm) deep. Both are 1-1/2 in. (38 mm) wide at the top and increase to 3 in. (76 mm) at mid-depth.

The railroads discovered that it is necessary to sand-blast the keyways to provide good adhesion between the precast concrete and the epoxy key. The precast concrete keyways have a form release agent residue that should be removed. The railroads require that the keyways be sand-blasted in the casting yard prior to shipment. This step is also recommended in the ACI Committee 503 documents (Refs. 28 and 32). Inadequate surface preparation was also cited as a frequent source of trouble in Ref. 35.

The railroads often need to replace bridge spans between passing trains, so construction speed is very important. The pre-placed aggregate method allows the work to be accomplished more quickly than the mixed mortar method. To meet schedules, it is also often necessary to work in the rain. Under such conditions, a moisture insensitive epoxy resin is needed. It is probably best to use moisture insensitive resins in all cases because it is not always possible to dry the keyways sufficiently prior to placement.

While both the epoxy mortar method and the pre-placed aggregate method appear to provide excellent results, the pre-placed aggregate method has several advantages over pre-mixed mortar:

1. When epoxy resin is mixed mechanically with the aggregate there is a tendency for the mix to foam up slightly. This action can result in a void structure which is vulnerable to freeze-thaw damage.
2. The pre-placed aggregate method allows the use of larger aggregate since pre-mixed mortars with large particles are difficult to mix and place. By using a graded aggregate it is possible to obtain a 70% aggregate ratio which makes it more economical.
3. The coefficient of thermal expansion comes closer to concrete when the aggregate-binder ratio is about 7 or greater. Mixtures in this range are not workable, but it is possible to achieve these high ratios with pre-placed aggregate. The particle-to-particle contact that results with the pre-placed aggregate method also contributes to a lower coefficient of thermal expansion, because the aggregate coefficient is closer to concrete. When there is a large difference in the coefficient of thermal expansion between concrete and the shear key mortar, high bond stresses are induced during large temperature changes.
4. The pre-placed method makes it easier to mix and place, since only the resin and hardener needs to be mixed. Problems with clean-up are simplified when only the resin is mixed.

There have been occasional problems with epoxies due to improper storage (temperature control) and mixing. These factors have been cited by some investigators as the cause of the difficulties in the bridge project reported in Ref. 34. Since the time this problem was discovered, the Illinois DOT requires that the A and B components be colored black and white. When mixed thoroughly the mixture has a uniform gray color. If it is not properly mixed, streaking is apparent.

## 5.4 Design Requirements of Transverse Ties

This study revealed no established procedure for the design of transverse tie systems. In general, standards appear to have been set on the basis of subjective evaluations, and modified when performance was unsatisfactory. Engineers sometimes apply different criteria to different systems. For example, welded ties were rejected because they were not sufficient to transfer vertical wheel loads, even though there was a full length grouted shear key between precast members. Obviously, no such requirement was imposed on tie rods or post-tensioning tendons. In this analysis, all vertical loads are transferred by the shear key. The transverse ties are only required to provide enough horizontal restraint to keep the grout keys closed.

The forces which tend to separate the precast members at the shear keys are difficult to identify, let alone quantify. The most likely causes are transverse bending moments, shrinkage and temperature change.

A qualitative analysis of transverse bending moments indicates that such moments tend to produce compression at the top. Thus, such moments would close rather than open the shear keys between members. Under cyclic loading, the "rebound" may cause some tension, and with enough cycles, the fatigue could cause some cracking at the keyway. This may be the justification for the common practice of placing transverse tie rods at mid depth of the members.

Shrinkage and negative temperature change cause the members to decrease in width and pull apart at the joints. In the absence of a positive tie force to pull the members together this would result in a crack at the grout key, or a separation of the grout from the precast member, depending on the bond strength of the grout. If there is a uniform loss of moisture and temperature change through the depth of the member, this decrease in width would be uniform throughout the depth.

To prevent the complete separation of the precast members, there must be a tie force to pull the members together as this decrease in width occurs. This force is resisted at the supports of the members. if the bearing is sufficiently flexible, e.g., with properly designed and performing elastomeric pads, the force will not be great. However, for a variety of reasons, these pads may not perform as intended and the forces can build up to the point that the only relief is through slip. Thus, the upper limit of the force would be the friction between the deck member and its support.

Coefficients of friction vary considerably, so the magnitude of this force is not precisely determinable. Several studies have indicated that an average value of the coefficient of friction can be assumed at about 1.0. Using this value, the total design tie force should be equal to approximately half the weight of the bridge.

This analysis should be recognized as idealized. Other factors will contribute to the behavior, such as the stiffness of the deck system, transverse connection details at piers and abutments, and the bearing conditions. Certainly, the magnitude of temperature changes and the shrinkage characteristics of the concrete will enter the picture.

### 5.4.1 Non-Prestressed Tie Rods and Bolts

The above analysis provides a clue as to why non-prestressed tie rods and bolts have often proven ineffective. With only nominal tightening of these rods, insufficient force is provided to overcome the friction at the supports. As an example, consider a typical slab bridge, 30 ft (9.1 m) long, 32 ft (9.8 m) wide and weighing 165 psf (7.90 kPa). The total weight of the bridge is then 30 x 32 x 165 = 158,400 lb (705 kN). If two 1 in. (25 mm) diameter rods are used, the tensile force required in each rod is 39,600 lb (176 kN). Using the empirical equation relating torque and axial tension:

$P = \frac{T}{kD}$ (P = axial tension, T = torque, k = a constant usually assumed = 0.2, and D = rod diameter), the torque required would be 39,600 x 0.2 x 1/12 = 680 ft-lb (922 N-m). Not only is this difficult to apply in the field, the 39,600 lb would yield a 1 in. rod made of A36 steel.

5.4.2 Lateral Post-Tensioning

In the above example, if post-tensioning is substituted for the tie rods, the analysis indicates a post-tensioning force on each tendon of about 40,000 lb (180 kN) should be applied. Experience in Florida (Sect. 4.3.3) indicates, however, that cracking may still be a problem with a force of this magnitude (the above example is typical of Florida slab bridges).

5.4.3 Weld Plate Connectors

Weld plate connections resist the forces described above by their strength in tension. These connection assemblies can be designed by methods shown in the ACI Design Handbook[40] and elsewhere. A connection similar to those shown in Fig. 39 would have a capacity of from 15 to 25 kips (67-110 kN), using load factor design methods. Thus, in the example of Sect. 5.4.2, and a weld plate assembly capacity of 20 kips, the number of connections required can be calculated by:

$$\frac{158{,}400/2 \times 1.3 \text{ (load factor)}}{20{,}000} = 5.15 \text{ say 6 connections}$$

This is the maximum number of connections required, and would only be necessary at the joints nearest the center of the bridge. Progressively fewer would be required nearer the outside edges, since less of the bridge needs to be "pulled back".

5.4.4 Epoxy Shear Keys

Epoxy compounds as described in Sect. 5.3 generally will have sufficient tensile and bond strength to overcome the frictional resistance at the supports, provided the shear key surfaces are properly prepared by sand blasting. Since epoxies are usually much stronger in tension than the concrete they are bonded to, the lateral tensile strength of an epoxy shear key is limited by the tensile strength of the plain concrete to which it is bonded. For concrete, this tensile strength can be assumed to be $4\sqrt{f'_c}$, or for the 5000 to 6000 psi (35-40 MPa) concrete normally used in precast, prestressed concrete members, about 300 (2.0 MPa) psi, using load factor design. In the example of Sect. 5.4.1, the minimum shear key depth to provide adequate lateral tensile resistance would be:

$$\frac{158{,}400/2 \times 1.3}{300 \times 30 \times 12} = 0.95 \text{ in. (24 mm)}$$

which is much less than usually provided.

5.5 Costs

There is a wide variation in the cost of the grout key materials in a typical shear key (Table 3), the volume of material will vary considerably, depending on the tightness of the slabs. If the slabs are spaced 1 in. (25 mm) apart at the bottom, the volume is almost double that with the slabs snug at the bottom. The comparison in Table 4 is based on a gap between slabs of 1/2 in. (13 mm). It is assumed that labor costs of placing the grout are the same for all materials.

The costs of various types of transverse tie systems will vary somewhat with the experience of the contractor, and the amount of work available at one time. For the typical slab bridge of Sect. 5.4.1, the costs are estimated in Table 4.

The various combinations of grout material and transverse ties are shown in Table 5.

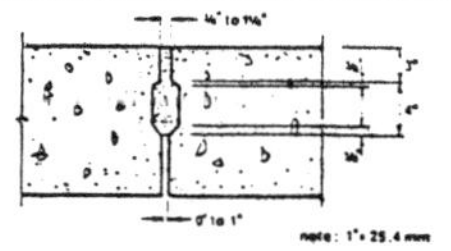

Table 3: Estimated Costs of Shear Keys, Dollars per foot*

| Type of Grout | Material | Labor | Joint Prep | Total |
|---|---|---|---|---|
| Sand-cement** | 0.10 | 1.17 | — | 1.27 |
| Proprietary Non-shrink | 1.84 | 1.17 | — | 3.01 |
| Epoxy Mortar | 3.48 | 1.17 | 0.60 | 5.25 |

* $1 per foot = $3.28 per meter

** Aluminum powder may be added at negligible cost

Table 4: Estimated Costs of Transverse Tie Systems

| Tie System | Cost | | | | |
|---|---|---|---|---|---|
| | Cost Per Unit | | | Number | $/ft |
| | Material | Labor | Total* | Req'd** | of Key*** |
| Weld plate assembly | 5.26 | 8.54 | 17.32 | 42 | 3.56 |
| Tie rods | 60.00 | 36.00 | 115.00 | 2 | 1.10 |
| Post-tensioned bars | 96.00 | 41.00 | 165.00 | 2 | 1.57 |

* Includes overhead and profit

** Per 30 ft span, 8 slabs wide

*** 210 lin. ft of grout key; $1/ft=$3.28 per meter

Table 5: Estimated Cost of Grout Keys Plus Transverse Tie, $ per foot

| Grout Material | Transverse Tie System | | | |
|---|---|---|---|---|
| | None | Weld Plate | Tie Rod | Post-tens. |
| None | — | 3.56 | 1.10 | 1.57 |
| Sand-Cement | 1.27 | 4.83 | 2.37 | 2.84 |
| Proprietary non-shrink | 3.01 | 6.57 | 4.11 | 4.58 |
| Epoxy Mortar | 5.25 | 8.81 | 6.35 | 6.82 |

## 5.6 Summary and Conclusions

The primary purpose of shear keys is to transfer vertical wheel loads to adjacent units and to prevent moisture leakage through the deck. The presence of a positive transverse tie force was shown in the tests at North Carolina State[22] to improve the load transfer when dry joints were used. The Lehigh University tests [23] on decks with grouted joints were less conclusive. Further study in this area is warranted.

This chapter presents the hypothesis that environmental conditions may cause separation of the deck units, leading to unsatisfactory performance. It is also suggested that if crack opening is prevented, the chance of grout key failure will be substantially reduced.

The analysis and investigations of existing structures leads tentatively to the conclusion that of those shear key construction methods which have been used, the following order of performance (from best to worst) could be expected:

1. Epoxy grout keys with post-tensioned ties (joints properly prepared)
2. Epoxy grout keys with non-prestressed ties (joints properly prepared)
   Non-shrink grout with weld plates
   Non-shrink grout with post-tensioned ties
3. Sand-cement grout with weld plates
   Sand-cement grout with post-tensioned ties
4. Non-shrink grout with non-prestressed ties
   Sand-cement grout with non-prestressed ties

This study did not uncover any place where epoxy grout with weld plates was used, but the ranking of that combination would probably be near the top.

At this point, the question should be asked: "What are the consequences of grout key failure?" It is apparent that if the members are designed assuming some load transfer to adjacent units (i.e., the member is designed for less than full wheel load), the member will be overstressed under the design wheel load. An analysis was made of typical span ranges and widths of some of the precast, prestressed integral deck members commonly used. Assuming the members are designed by the 1977 AASHTO Bridge Specifications, Section 1.3.1(D)[24], the indicated overstress resulting from a failed shear key is in the range of 50% for narrow slabs on short spans, and less for wider and longer members. Units 6 ft (1.8 m) and wider will usually not be allowed any distribution. This is well within normal load factors, so, while the joints should be quickly repaired, there would not appear to be a serious safety hazard.

There have been suggestions that the lateral load distribution equations for precast concrete members are too conservative, and should be changed. If this is done, the need for reliable connections between members is more crucial.

The damage that is caused by moisture leakage through the grout keys is relatively minor. If water becomes entrapped within the shear key, it sometimes freezes and causes localized spalling. Moisture leakage contributes to corrosion of some bearing devices, and streaking of the concrete may be unsightly. For an overpass, the dripping water is a nuisance to passing motorists.

These problems are not unusual in other types of construction. Thus, the responsible engineer will likely consider the risk acceptable, if the cost of a more reliable joint is significantly higher.

Many of the bridge problems on precast bridge decks are with the relatively thin asphalt surfaces often placed over them. While the joint pattern of the precast decks will often reflect through the asphalt, there was not conclusive evidence that the resulting cracks contributed to the deterioration of the asphalt surface, as long as the grout key maintained its integrity. Experience with the Indiana experimental bridges and the numerous untopped integral deck bridges in the Pacific Northwest and elsewhere indicates that if the leakage problem can be solved—or tolerated—the best solution to this problem may be to forego the asphalt surface. This is especially deserving of consideration on low-traffic bridges.

Additional research on this subject is warranted. Laboratory tests on the shear resistance of various grout materials are needed. Frictional resistance of bearing conditions should also be further studied. Finally, experimental full scale bridges using some of the suggestions in this report should be constructed to observe environmental effect.

## D.2 PRESTRESSED CONCRETE GIRDER

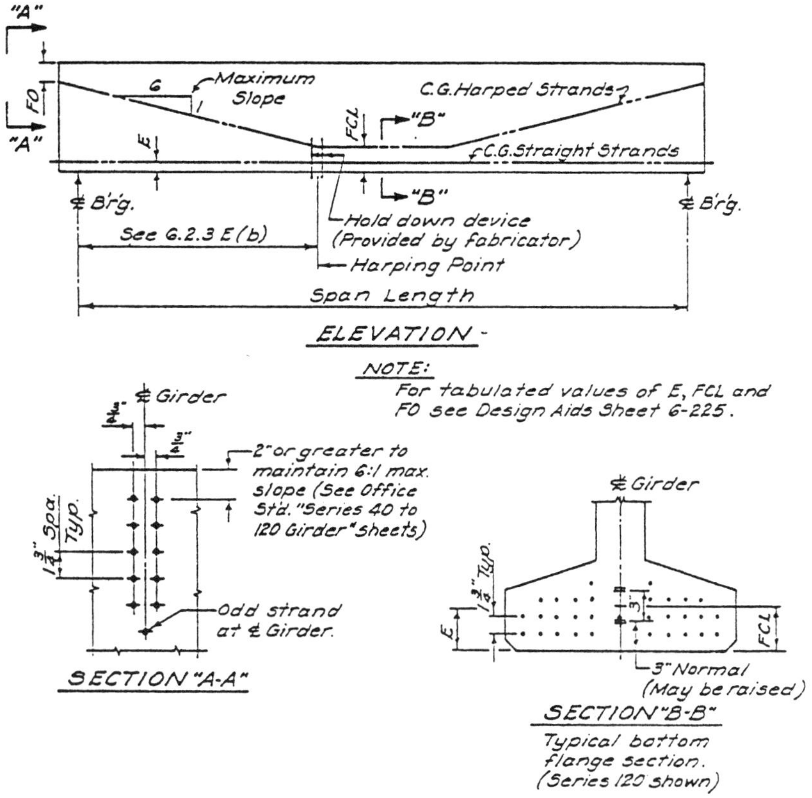


Note: FO may be increased in 1-inch increments to keep slope of harped strands below 6:1.
FCL may be increased in 1-inch increments in order to reduce tension at the top of the girder at harping point at time of strand release.

(a) Prestress

The prestress acting on the section shall be assumed to be NAs(.7f'-35) for final conditions where N = number of stressed strands passing through the section, As = the area of one strand and $f_s'$ is the ultimate strand strength in ksi. (270 for 270$^k$ strands).

Prestress at time of transfer of stress may be computed using elastic shortening plus two percent loss in the prestress steel due to relaxation. The steel stress change due to elastic shortening is assumed to be equal to the stress in the concrete adjacent to the strand times the ratio of elastic moduli (Es/Eci).

(b) Strand Patterns

Standard strand patterns are shown in figure 6.2.3C - 4 and Design Aid Sheet 6-225.

Design Procedure

(a) General

The procedure for design of a prestressed girder is generally a trial and error process. Design Aid Sheet 6-270 will give some guidance regarding spans and spacings. Published literature has suggested various direct means for establishing proper prestress force and position, but many of these means are quite complex. Usually final design will be arrived at by use of the W.S.H.D. "Prestressed Girder Design" computer program. This program uses trial and error methods to arrive at a solution which best satisfies the stress requirements.

(b) Stress Conditions

A number of conditions must be met by the girder and its prestress. Each condition is the result of the summation of stresses with each load acting on its appropriate section (Girder only, composite section n = 8, composite section n = 24, etc.). One or more of the conditions described below may govern design.

1. Compression, Bottom & Span

The compressive stress at bottom of the girder at centerline span with final loads (DL + LL + I + Prestress) shall not be less than zero. No tension is allowed. If a tension condition occurs, prestress must be added or the girder spacing must be reduced.

2. Compression, Top ℄ Span

The compressive stress at the top of the girder with final loads shall not exceed 0.4 f'c. If this value is exceeded additional prestress will help; however, this condition normally indicates that the girder capacity has been exceeded and girder spacing must be reduced. Some relief may be obtained by specifying higher strength girder concrete.

3. Compression Bottom, Harping Point at Release

The compressive stress at the bottom of the girder at the harping point shall not exceed 0.6 f'c at the time of transfer of prestress. (Girder D.L. + Prestress). An overstress condition normally indicates that the capacity of the girder has been exceeded. If the amount of prestress cannot be reduced the girder concrete strength must be increased or the girder spacing decreased. In some cases the harping point of the strands can be moved toward span centerline in order to take advantage of girder dead load moment.

4. Tension Top, Harping Point at Release

The tension stress in the top of the girder at the harping point shall not exceed $3\sqrt{f'_{ci}}$ at time of transfer of prestress. (Note: W.S.H.D. experience has shown that this value is satisfactory. The use of the lower AASHTO 200 psi limitation is being investigated.) This will normally be exceeded only on short, heavily loaded girders. The harped strands can be raised to reduce this tension and the total amount of prestress increased. Note that on series 60 girders, some harped strand arrangements may interfere with the diaphragm bar holes.

5. Tension Top, d/2 from End at Release

The tension stress in the top of the girder at a distance of 1/2 girder depth from the end of girder shall not exceed $3\sqrt{f'ci}$ at transfer of prestress. (See note in paragraph 4.) This location is assumed to be the point at which the strands are fully developed. For girders which are to be used in continuous bridges this value is limited to "zero" tension. In this way the girder concrete modulus of rupture will not be exceeded under the action of ultimate continuous live loads. The tension can be limited by increasing the number of harped strands.

Methods used to stress the strands may result in overstressing and breaking strands if the angle to the horizontal exceeds 6:1. For this reason the height of the harped strands at the end block is limited for short girders in the larger series.

6. Shear

Web shear at the end of the end block due to ultimate load shall not exceed $10\,\phi\sqrt{f'_c}$. This web shear computation shall be based on the web width and total depth from the top of the slab to the centroid of the total strands at centerline span.

E. Reinforcement

(a) Straight Strands

The position of straight strands in the flange has been standardized for each size of flange. Those strand positions and the girder flange size are summarized in Design Aid Sheet 6-225.

(b) Harped Strands

The harped strands are bundled at the 4/10 point of the girders for series 120 and 100 and at the 1/3 point at the girders for series 80, 60 and 40. The harped strands are bundled at the harping points. Bundles are limited to 12 strands each. 12 and fewer harped strands are placed in a single bundle with the centroid normally 3" above the bottom of the girder. Strands in excess of 12 are bundled in a second bundle with the centroid 6" above the bottom of the girder. At the girder ends the strands are splayed to a normal pattern such as that shown in figure 6.2.3 C-4. The centroid of strands at both the girder end and the harping point may be varied to suit girder stress requirements.

(c) Stirrups

Shear for computation of stirrup requirements is computed at the end of the girder end block taper. This is a minimum of 9'-6" from the end of the girder for series 120 girders. Ultimate shear is computed at this point based on 1.5 DL + 2.5 (LL+I). The portion of this shear which is carried by the concrete is $Vc = b_w\,(d - \frac{a}{2})\,(.180)$ in kips. in this equation d is the distance from the top of the slab to the centroid of the total strands at centerline span.

The stirrup spacing is then $s = \dfrac{2\,A_v f_y\,(d - \frac{a}{2})}{Vu - Vc - Vh}$

where $V_h$ is the vertical component of harped strand forces at the section. Standard Av required is that furnished by two number four bars. Maximum spacing is taken to be 1'-6". The computed s value or 1'-6" spacing (if smaller) is used throughout the portion of the girder between ends of end block taper. Grade 40 steel is specified.

(d) End Block Reinforcement

The Washington State Standard Prestressed Concrete girders are provided with a thickened end block section in order to provide for bursting stresses and to provide lateral bending resistance at reaction points. The reinforcement is detailed on the Office Standard Plans. Tie size and spacing is based on the requirements to resist bursting forces due to strand force development in this area.

F. Strand Development

(a) General

AASHTO specifies an embedment length for strand in order to develop full usable capacity. Near the ends of the girders the developed strands may cause overstress in the top fibers of the girder. For this reason, the conservative assumption is made that the strands will be fully developed at a point 1/2 of the girder depth from the end of the girder. This criteria is to be used only for the purpose of computing possible overstress conditions caused by the prestress force. Where the prestressing forces are too high, the strands may be raised or unbonded as discussed below.

(b) Unbonding Strands

Where it is necessary to prevent a strand from actively supplying prestress force near the end of a girder, it may be unbonded. This can be accomplished by taping a close fitting pvc tube to the stressed strand from the end of the girder to some point where the strand can be allowed to develop its load. Since this is not a common procedure, it should be carefully detailed on the plans. It is important when this method is used in construction that the taping of the tube be done in such a manner that grout cannot leak into the tube and provide an undesirable bond of the strand.

(c) Strand Development Outside of Girder

For girders made continuous for live load, extended bottom prestress strands are used to carry positive live load, creep and other moments from one span to another. Usually four strands per girder will provide an adequate resistance. Strands used for this purpose must be developed in the short distance between the two girder ends. This is normally accomplished by requiring the strands to be bent up around a reinforcement bar. The nominal development length is normally two feet. See Figure 6.2.3F(c)-1. For wide crossbeams the strands may be extended straight and a 2'-0" splice used. At back walls, which are connected to the superstructure, the extended strands may be used to withstand earthquake forces and, in this case, should be

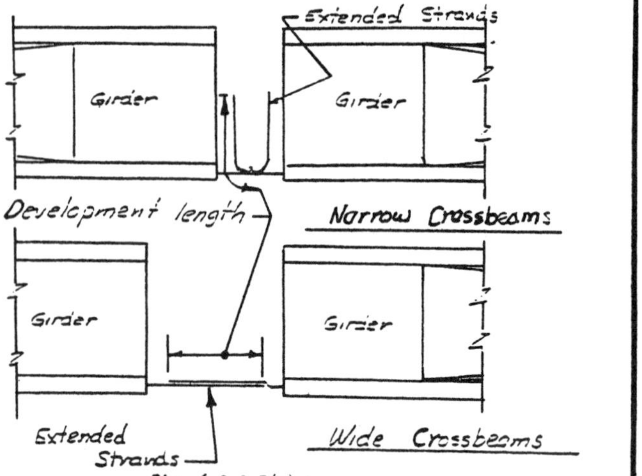


Fig. 6.2.3 F(c)-1

developed accordingly. Of course the number of strands to be extended can not exceed the number of straight strands available in the girder.

The appropriate strand stress available to resist ultimate load ($f_{su}^{*}$) at this section shall be no greater than $[(l'_d/D) - \frac{2}{3} f_{se}]$ (Say $\frac{25}{1/2} - \frac{2}{3}(35) = 25$ ksi.) where $l'_d$ is the developed length available.

G. Fabrication and Handling

(a) Shop Plans

Fabricators of prestressed girders are required to submit shop plans which show specific details for each girder that they construct. These shop plans are checked and approved by the project engineers office for conformance with the contract plans and specifications.

(b) Special Problems of Fabricators

1 Strand Tensioning

The method selected for strand tensioning may affect the design of the girders. The strand arrangements shown in the Office Standard Plans and included in the "Prestressed Girder Design" computer program are satisfactory for tensioning methods used by fabricators in this State. Harped strands are normally tensioned by pulling them as straight strands to a partial tension. The strands are then deflected vertically as necessary to give the required harping angle and strand stress. In order to avoid over-tensioning the harped strands by this procedure, the slope of the strands is limited to a maximum of 6:1. The straight strands are tensioned by straight jacking.

2. Hold Down Forces

Forces on the hold-down units are developed as the harped strands are raised. It is necessary that the hold-down device provided by the fabricator be capable of holding the vertical component of the harping forces. Normally a two unit hold-down is required. Standard commercial hold-down units have been pre-approved for use with particular strand groups.

3. Numbers of Strands

Since the prestressing beds used by the girder fabricators can carry several girders in a line, it is desirable that girders have the same number of strands where practical. This allows several girders to be set up and cast at one time and is a saving in terms of both time and strand material. Most fabricators in Washington State have the capability to pretension a maximum of 50 strands.

(c) Handling and Hauling

1. Pick Up Points

The office standard plans show pick-up points for the girders. These points are quite critical since the girder is in its most highly stressed condition just after strand release (girder tensioning). In some cases, fabricators have moved the pick-up points toward the center of the girder. A request to do this must be reviewed carefully since the moments in the girder due to its own weight are needed to keep the tension at the top of the girder within allowable construction ranges. A decrease in girder dead load moment near centerline span may cause failure of the girder. Similarly, the girders must never be stored by supporting at any point other than the centerline of bearing. The girders are also very sensitive to lateral loads and accordingly must be stored in a truly vertical position.

2. Girder Buckling

Long prestressed girders are very flexible and thus are highly susceptible to buckling failure. Such failures are very costly and involve a high danger to workmen. Buckling failures are sudden and catastrophic and must be guarded against. The office

(2) Girder Buckling (Cont.)

standard plans state that girders over certain given lengths must be laterally braced and that all girders must be handled carefully. It is the fabricators responsibility to provide adequate bracing and provide suitable handling facilities. On unusually long girders, however, additional consideration should be given to this matter by the designer. Published material on beam buckling should be consulted. See Technical File 3.6.4

(3) Hauling Loads

Girders are normally transported one to a truck over the State highways. The weight of these girders for hauling will be on the order of 50 tons for long series 120 girders and is not a problem. For special girders a hauling weight of up to 60 tons for girders over 100 feet long is satisfactory. For higher loads and shorter lengths of heavy girders the Bridge Construction Plans Engineer should be consulted. The decision to use those heavier sections should be based in part on his recommendation.

(d) Repair of Damaged Girders

This section pertains only to girders which have been damaged before becoming part of a final structure. Repair of damaged girders in existing bridges is covered in Section 6.3.6.

(1) Repairs to Girders Prior to Strand Release

When girders suffer defects during casting or damage prior to strand release, the repair procedures are well documented in reference 6.99.7. Normally, no designer action is required. In prescribing repairs for unusual situations not covered in reference 6.99.7, the designer must insure that the required strength and appearance of the girder can be maintained. Since stressing will occur after the repair is made normally no test loading is required, however such a test should be considered.

## PRESTRESSED GIRDER

Precast Girder Bridges

The precast prestressed girder bridge has proven to be an economical and rapid means of bridge construction in this state. This section is intended as a guide to the design of precast prestressed girder bridges. For design criteria not covered see Section 5.4, Box Girder Bridges.

6.3.1 Criteria for Girder Design

The following criteria is described for simple span bridges. Present practice in this Department is to use simple span girder designs in continuous prestress bridges. Effects of creep and shrinkage are not considered. This is a somewhat conservative procedure but it leads to great economy in engineering time. For continuous structures consisting of a large number of girders a more exact analysis should be used. Additional comments are added below concerning special problems in design of continuous bridges.

A. Support Conditions

The prestressed girders as erected are assumed to be supported on rigid permanent simple supports. These supports can be either bearing seats and elastomeric pads or in the case of continuous bridges, may be falsework supports. The design span length is then taken to be the distance center to center of bearings for simple spans. For continuous spans erected on falsework, the effective point of support for girder design is assumed to be the face of the crossbeam. For continuous spans on notched crossbeams the design span length is usually taken as the distance center to center of temporary bearings. See figure 6.2.3 C-3.

1. Flange Width

The effective width of slab on each side of the girder centerline which can be considered to act as a compressive flange shall not exceed any of the following:

One-fourth of the span length.

Six times the thickness of slab plus 1/4 of the girder flange width.

One-half the distance to the next girder.

The actual distance to the edge of slab.

For effective tension flange widths see AASHTO.

2. Flange Position

For purposes of calculating composite section properties the bottom of the slab shall be assumed to be directly on the top of the girder. This assumption may prove to be true at center of span when excess girder camber occurs.

For dimensioning the plans, an increased dimension from top of girder to top of slab is used at centerline of bearing. This is called the "A" dimension. This dimension accounts for effect of girder camber, vertical curve, slab cross slope, etc. See Design Aids Sheet 6-360 for method of computing.

3. Flange Thickness

For purposes of computing composite section properties the slab thickness shall be reduced by 1/2 inch to account for the wearing course. Where it is known that a bridge will have an asphalt overlay applied prior to traffic being allowed on the bridge the full slab thickness should be used as effective slab thickness. The effective slab thickness shall be reduced by the ratio Es/Eg. The effective modulus of composite section is then Eg. This ratio will be Es/Eg for short term load applications such as live load and Es/3Eg for long term loads such as asphalt overlay.

4. Section Dead Load

The slab dead load to be applied to the girder shall be based on full thickness. The full effective pad ("A"-t) weight shall be added to that load. This assumed pad weight is applied over the full length of the girder and is thus considered to compensate for extra diaphragm weight due to skew.

B. Composite Action

(a) General

The sequence of construction and loading is extremely important in the design of prestressed girders. The composite section has a much larger capacity than the basic girder section but it cannot, of course, take loads until the slab has obtained adquate strength. For assumptions used in computing composite section properties see figure 6.3.1 B(c)-1

(b) Load Application

The following sequence and method of applying loads is used in girder analysis:

1. Girder Dead Load

The dead load of the girder is applied to the girder section.

2. Slab Dead Load

The dead load of the slab and diaphragms is applied to the girder section.

3. Barrier & Overlay Deadload

Dead load of one traffic barrier is divided among three girders (provided the bridge has a minimum of six girders) and this uniform load is applied to the long term composite section. (Slab n = 24). Dead load of any overlay is applied to the same composite section.

4. Live Load

Live load plus impact is applied to the short term composite section. (Slab n = 8).

(c) Composite Section Properties

Deck slab thickness is specified as 7" by office practice but may be thicker if girder spacing dictates. This deck slab forms the top flange of the composite girder in prestressed girder bridge construction. The properties of this slab-girder composite section are affected by specification and by physical considerations. Figure 6.3.1 B(c)-1 shows some standard values to be used for design and detailing.

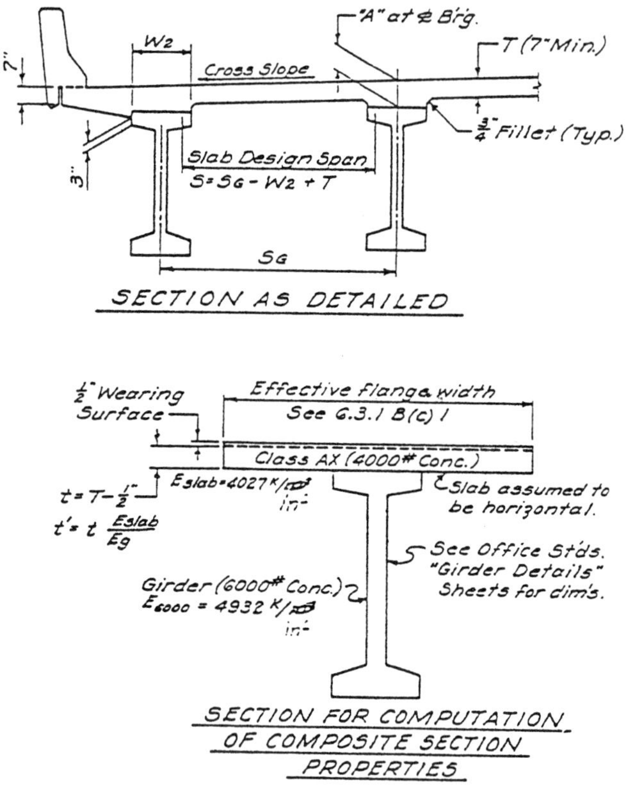


Figure 6.3.1 B(c)-1
Composite Prestressed Girder Section

(d) Shear Transfer

Transfer of shear forces in prestressed girder bridge design will be critical in three areas. The first has been previously discussed; the section through the web at the end of the end block taper. The other two critical areas for shear transfer are between slab and girder and at the end connection of the girder to the crossbeam for girders in continuous bridges. Shear in these areas will normally be resisted by reinforcement extending from the girder.

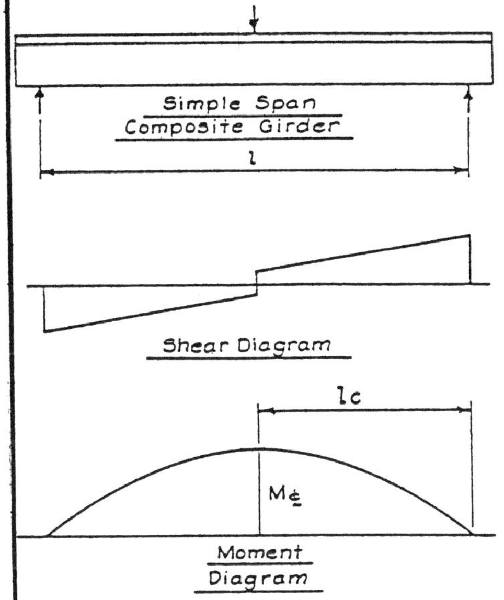


Figure 6.3.1 B(d) 1-1
Shear in Simple Girder

1. Shear Between Slab and Girder

This shear represents a rate of change of compression load in the flange of simple span girders or a rate of change of tension load in the flange near the piers of continuous girders. For a simple span girder, as represented by Figure 6.3.1.B(d)1-1, the top flange stress is the factored centerline moment divided by the section modulus of the composite girder at the centerline of the slab. The slab load is, of course, that stress times the area of the slab. The factored ℄ moment can be taken as total factored moment less 1.0 times dead load applied to girder.

The design composite section slab modulus is used for this calculation; however, full slab width should be used to compute force. As an alternate, a composite section can be calculated using the full slab width for a more accurate analysis, but this will usually not be necessary. Further explanation of this calculation and a solved example are available in reference 5.99.4, "PCA Notes on Load Factor Design."

This shear is resisted by the girder stirrups which extend up through the interface between the slab and the girder. This surface of the girder must be roughened. The force may be assumed to be carried uniformly over the entire girder top surface from centerline of bearing to centerline of span. All stirrups in this area can be assumed to be acting in accordance with the shear friction theory as described in Section 5.2.1 C.

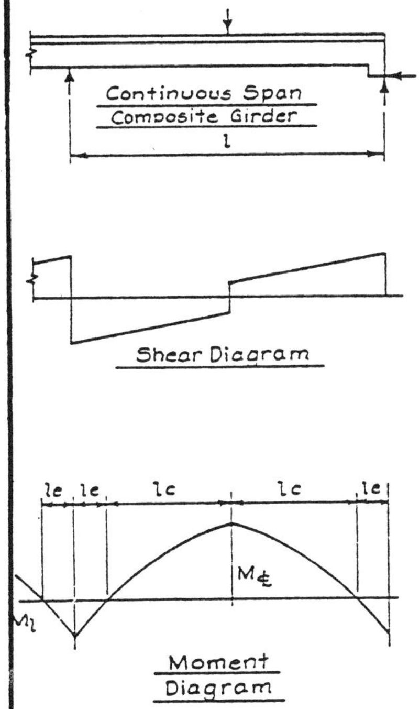


Figure 6.3.1 B(d)1-2
Shear in Continuous Girder

For continuous girders, the span, shear and moment relationships are shown in Figure 6.3.1 B(d)1-2. Similar methods are used to analyze slab to girder shear. For positive moment resistance, only those stirrups within length "lc" are considered effective in resisting the slab force due to M . Likewise, only those stirrups within one continuous length "le" are used to resist the negative moment slab force (tension) in that area.

For illustrative purposes, a single concentrated load has been shown. In actual practice, the point of factored maximum moment of the actual Moment diagram would be used.

Other flange shear problems are described in Section 5.4. These problems also need to be considered for prestressed girder bridges.

2. Shear at Girder End

A continuous prestressed girder will nearly always be required to carry end reaction shears at the

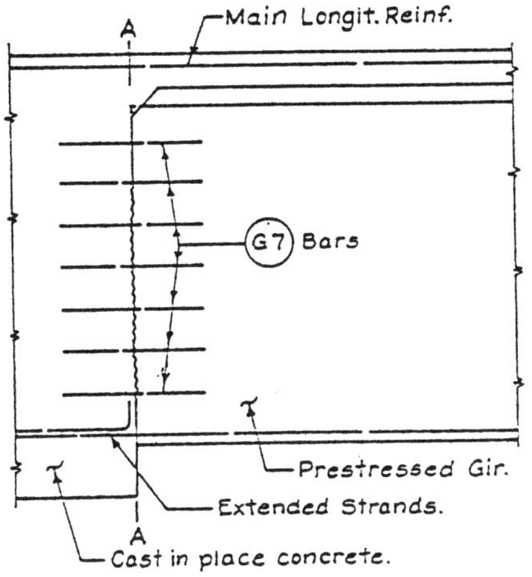


6.3.1B(d)2-1

End Connection for a
Continuous Prestressed Girder

surface of the end of the girder. An exception to this is girders with notched cross-beams where loads must be carried across the connections which act as hinges. See Section 5.4.10.

The usual end condition is similar to that shown in Figure 6.3.1 B(d)2-1. The shear which must be carried along the interface A-A is the actual factored dead load and live load shear acting on the section. The girder end is required by the plans to be roughened. The sawtoothed shear key shown on the office standard girder plans may be assumed to provide a friction factor of 1.0. Shear resistance must be developed using shear friction theory and assuming the "G7" bars and the extended strands to be actively participating. The main longitudinal girder reinforcement is already fully stressed by girder bending moments and thus cannot be considered for shear requirements. All bars, including the extended strands, must be properly anchored in order to be considered effective. This anchorage requirement must be clearly shown on the plans.

Note that similar requirements exist for connecting the end diaphragm at bridge ends where the diaphragm is cast on the girders. In this case, however, loads consist only of the factored diaphragm dead load, approach slab dead load and those wheel loads which can distribute to the interface.

C. Prestressed Girder Camber

(a) General

The amount of girder camber for prestressed girder bridges is normally established by use of the computer program "PSGDEF". This program computes the deflections due to prestress, girder dead load, slab dead load and live load. It is based on theory and on measured girder deflections.

(b) Calculation

Figure 6.3.1 C (b)-1 shows a typical pattern of girder deflection with time at centerline span. Portions of this characteristic curve are described below. The sub-paragraph numbers correspond with circled numbers on the curve.

1. Elastic Deflection Due to Prestress Force.

The prestress force produces moments in the girder tending to bow the girder upward. Resisting these moments, but overpowered by them, are girder section dead load moments. The result is a net upward deflection. In addition, a shortening of the girder occurs due to axial prestress loading.

2. Creep Deflection

The girder continues to deflect upward due to the effect of creep. This effect is computed using the equation stated in 6.1.2 C(b).

3. Slab Load Deflection

The load of the slab is applied to the girder section resulting in an elastic downward deflection. It is this deflection which is offset by the screed camber that is to be applied to the bridge deck during construction.

4. Final Camber

It might be expected that the above slab dead load deflection would be accompanied by a continuing downward deflection due to creep. Many measurements of actual structure deflections have shown, however, that once the slab is poured the girder tends to act as though it is locked in position. To obtain a smooth riding surface on the deck the deflection indicated on Figure 6.3.1 C(b)-1 as "Screed Camber" is added to the profile grade elevation of the deck screeds. The actual position of the girder at time of the slab pour has no effect on the screed camber.

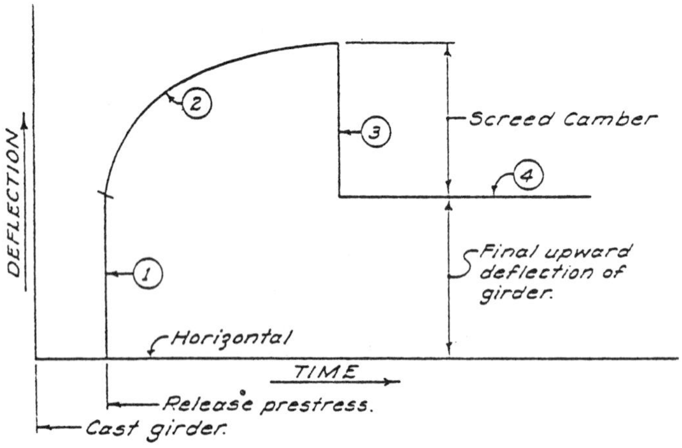


Figure 6.3.1 C (b)-1

Prestressed Girder Camber

Framing

A. Girder Selection and Spacing

Cost of the girders is a major portion of the cost of prestressed girder bridges. Much care is therefore warranted in the selection of girders and in optimizing their position within the structure. The following general guidelines should be considered.

(a) Girder Series Selection

All girders in a bridge will normally be of the same series. If vertical clearance is no problem, a larger girder series, utilizing fewer girder lines, may be a desirable solution. This must be balanced with considerations such as appearance. At the present time, the following relative girder series cost factors may be used as a guide for this decision:

| Series | Relative Cost Factors |
|---|---|
| 40 | 1.0 |
| 60 | 1.05 |
| 80 | 1.09 |
| 100 | 1.13 |
| 120 | 1.17 |

Note that the small marginal cost factors between series tends to make the larger series more economical.

The wider spacings expected when using larger series girders may result in extra reinforcement and concrete but less forming cost. These items must also be considered.

(b) Girder Concrete Strength

Higher girder concrete strengths should be specified where that strength can be effectively used to reduce the number of girder lines. See Section 6.1.2 A(a). When the bridge consists of a large number of spans, consideration should be given to using a more exact analysis than the usual design program in an attempt to reduce the number of girder lines. This analysis should take into account actual live load, creep and shrinkage stresses in the girders.

(c) Girder Spacing

Selection of a proper slab cantilever length is an important consideration in determining the most economical girder spacing. This matter is discussed in 6.3.2 B. Slab cantilever length should be made a maximum if a line of girders can be saved. The spacing of the interior girders must be considered at the same time. Once the positions of the exterior girders have been set, the positions and lengths of interior girders must be established. The following guidance should be considered.

1. Straight Spans

On straight constant width roadways, all girders should be parallel to bridge centerline and girder spacings should be equal.

2. Tapered Spans

On tapered roadways, the minimum number of girder lines should be determined as if all girder spaces were to be equally flared. As many girders as possible, within the limitations of girder capacity, should be placed parallel to bridge centerline. Slab thickness may have to be increased in some panels in order to accomplish this.

3. Curved Spans

On curved roadways, normally all girders will be parallel to each other. Proper positioning of the exterior girders is critical in this case, and is described in 6.3.2 B.

4. Geometrically Complex Spans

Spans which are combinations of taper and curves will require especially careful consideration in order to develop the most effective and economical girder arrangement. Where possible, girder lengths and numbers of straight and harped strands should be made the same for as many girders as possible in each span.

5. Number of Girders in a Span

Usually all spans will have the same number of girders. Where aesthetics of the underside of the bridge is not a factor and where a girder can be saved in a short side span, strong consideration should be given to using unequal numbers of girders. It should be noted that this will complicate crossbeam design by introducing torsion effects and that additional reinforcement will be required in the crossbeam.

B. Slab Cantilevers

The selection of the location of the exterior girders with respect to the curb line of a bridge, is a critical factor in the development of the framing plan. This location is established by setting the curb distance, which is that dimension from centerline of the exterior girder to the adjacent curb line. For straight bridges, the curb distance will normally be no less than 1'-6". Some considerations which affect this are noted below.

(a) Appearance

In the past, some prestressed girder bridges have been designed by placing the exterior girders directly under the curb (traffic barrier). This gives a very poor bridge appearance and is uneconomical. Normally, for best appearance, the largest slab overhang which is practical should be used.

(b) Economy

Fortunately, the condition tending toward best appearance is also that which will normally give maximum economy. Larger curb distances may mean that a line of girders can be eliminated, especially when combined with higher concrete strengths.

(c) Slab Strength

This is, of course, one of the governing conditions which limits the maximum practical curb distance. Design aid sheet 5-315 gives some guidance for cantilever design. It must be noted that for larger overhangs, the slab section between the exterior and the first interior girder may be critical and may require thickening. In some cases live load moments which produce transverse bending in the exterior girder should be considered.

(d) Drainage

Water from bridge drains on prestressed bridges is normally piped across the top of the girder and dropped inside of the exterior girder line unless all water can be drained to the bridge ends. A large slab cantilever length may severely affect this procedure and it must be considered when determining exterior girder location.

(e) Bridge Curvature

When straight prestressed girders are used to support curved roadways, the curb distance must vary. Normally the maximum slab overhang at the centerline of the long span will be made approximately equal to the overhang at the piers on the inside of the curve. At the point of minimum curb distance, however, the edge of the girder should be no closer than 6 inches from the slab edge. Where curvature is extreme and the difference between maximum and minimum curb distance becomes large, say 1'-6", other types of bridges should be considered. Straight girder bridges on highly curved alignments have a poor appearance and also tend to become structurally less efficient.

C. Diaphragm Requirements

(a) General

Diaphragms used with prestressed girder bridges serve two purposes. During the construction stage, the diaphragms provide girder stability for pouring the slab. During the life of the bridge the diaphragms act as load distributing elements, and are particularly advantageous for distribution of large overloads. Standard diaphragms and diaphragm spacings are given in the office standards for prestressed girder bridges. Diaphragms that fall within the limitations stated on the office standards need not be analyzed. Where large girder spacings are to be used or other unusual conditions exist, special diaphragm designs should be considered.

(b) Analysis

Where analysis is required, diaphragms shall be analyzed as transverse beam elements carrying both dead load and live load. Wheel loads for design shall be placed in positions so as to develop maximum positive moments, maximum negative moments, and maximum shears. Reinforcement shall be provided for these forces. This diaphragm reinforcement shall be so detailed that placement of girders is inhibited as little as possible, remembering the girders must be lowered vertically into position.

(c) Geometry

Diaphragms shall normally be oriented parallel to skew (as opposed to normal to girder centerlines). This procedure has the following advantages:

1. The build-up of higher stresses at the obtuse corners of a skewed span is minimized. This build-up has often been ignored in design.

2. Skewed diaphragms are connected at points of approximately equal girder deflections and thus tend to distribute load to the girders in a manner which more closely duplicates design assumptions.

On curved bridges diaphragms shall normally be placed on radial lines.

D. Skew Effects

Skew in prestressed girder bridges affects structural behavior and member analysis and complicates construction.

(a) Analysis

Normally, the effect of skew on girder analysis is ignored. The assumption that skew has little structural effect is valid for normal spans and normal skews. For short, wide spans and for extreme skews (say values over 50°) the effect of the skew on structural action should be investigated.

(b) Detailing

In order to minimize the labor in prestressed girder construction, and to avoid stress problems, the ends of girders for continuous spans with raised crossbeams shall normally be made square at the intermediate supports. This may not be possible for "notched crossbeam" designs. For these designs and for all simple beam ends, it may be necessary to skew the girder ends. When girder ends are skewed, the "θ" angle of the girder end should be rounded to the nearest five degrees. If this causes problems where the girder extends into the crossbeam, the angle can be specified to the nearest degree. See Standard Specifications 9-19.7(1) for girder tolerances.

E. Grade and Cross Slope Effects

The effect of large cross slopes is to increase the amount which must be provided in girder pad dimensions ("A" dimension) in order to insure that the structure can be built. See design aid sheet 6-360. This effect is especially pronounced if the bridge is on a horizontal or vertical curve. Care must be taken that deck drainage details reflect the cross slope effect. (See 6.3.2 B) Girder lengths may need to be modified to correct for added length along slope. It should be remembered that the girder is a rectangle in elevation. Thus the position of the girder top corner is affected by grade, girder camber, and tolerances. Details must account for this.

F. Curve Effect and Flare Effect

Curves and tapered roadways each tend to complicate the design of straight girders. The problem becomes that of determining what girder spacing should be used for dead load and live load design and whether or not a refined analysis taking into account actual load application is warranted. Normally, the girder spacing at centerline of span can be used for girder design, especially in view of the conservative assumptions made for the design of continuous girders.

6.3.3 Reinforcement

This section discusses reinforcement requirements for resistance of longitudinal memonts in precast girder bridges. This is limited to reinforcement in the top slab since capacity for resisting positive moment is provided by the prestressing of the girders.

A. Simple Spans

For simple span bridges the longitudinal slab reinforcement is not required to resist negative moments and therefore the reinforcement requirements are nominal. Figure 6.3.3 A-1

defines longitudinal reinforcement requirements for these slabs. The bottom longitudinal reinforcement is defined by the AASHTO requirements for "distribution reinforcement." The top longitudinal reinforcement is based on current office practice. The requirements of "Distribution of Flexural Reinforcement" do not apply to these bars.

B. Continuous Spans

(a) General

Longitudinal reinforcement of continuous spans is dominated by the main girder reinforcement bars. Where these bars are cut off, they are lapped by the nominal top longitudinal reinforcement described in 6.3.3 A. Typical arrangement of main girder reinforcement is shown in Figure 6.3.3 B(a)-1.

(b) Distribution of Flexural Reinforcement

The provision of AASHTO specifications dealing with this subject is provided to limit cracking. The value of "z" for the equation $F_s = z/(d_c A)^{1/3}$ shall be taken as 130 k/inch regardless of whether or not a deck seal is used. Figure 6.3.3 B(a)-1 shows the area to be used for computing "A". For unevenly spaced bars, this area can be computed as total flange area / $A_s$.

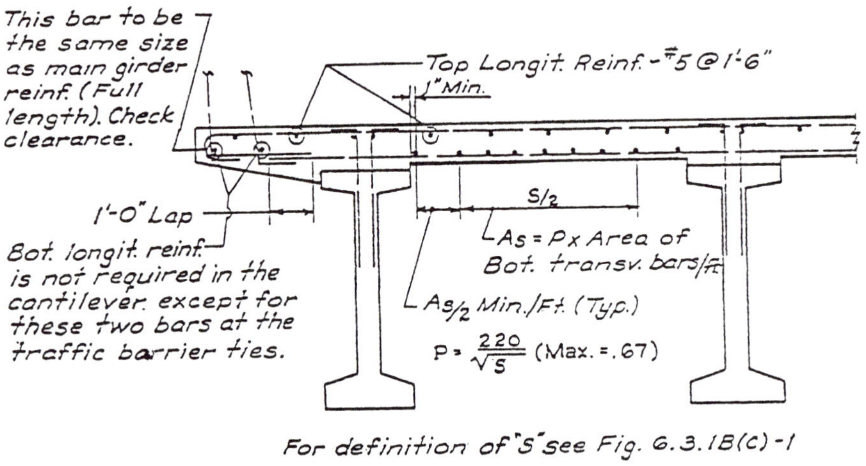


Figure 6.3.3A-1
Nominal Longitudinal Slab Reinforcement For Prestressed Girder Bridges

Figure 6.3.3 B(a)-1 shows typical arrangement of main girder reinforcement in the slab. Case I is considered to be the preferred method. Distribution steel shall be accounted for in the bottom longitudinal layer as follows:

1. Prestressed Girder Bridges with Girders Designed as Simple Spans

This is the case of bridges designed using the "Prestressed Girder Design" program. For these structures, it is not necessary to add to the steel area in the bottom longitudinal layer to account for distribution reinforcement. The bars in the bottom layer, however, shall provide an area not less than that required for distribution steel.

2. Other Prestressed Girder Bridges

On bridges where the effect of continuity is taken into account in reducing moments for girder design, additional longitudinal steel shall be provided as "distribution reinforcement." The sum of the areas in both layers of longitudinal bars shall be equal to the area required to resist negative moments plus the area required by the AASHTO specification for "distribution reinforcement." Equal areas shall be used in the top and bottom layers throughout the negative moment region. The total area of steel required in the bottom longitudinal layer shall not be less than that required for "distribution reinforcement." (For "Distribution reinforcement" see Figure 6.3.3 A-1)

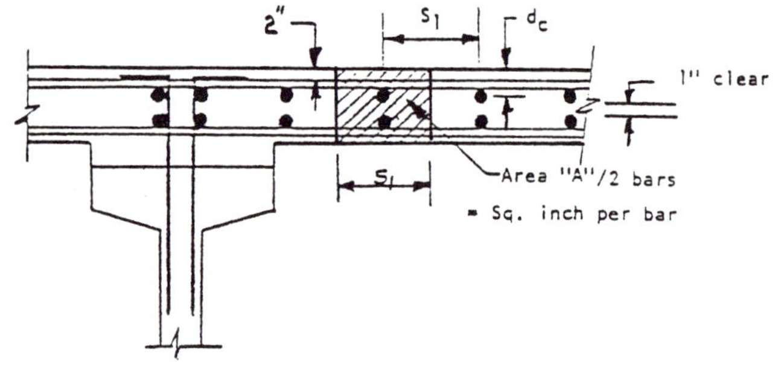


Case I
Girder Steel in Two Layers

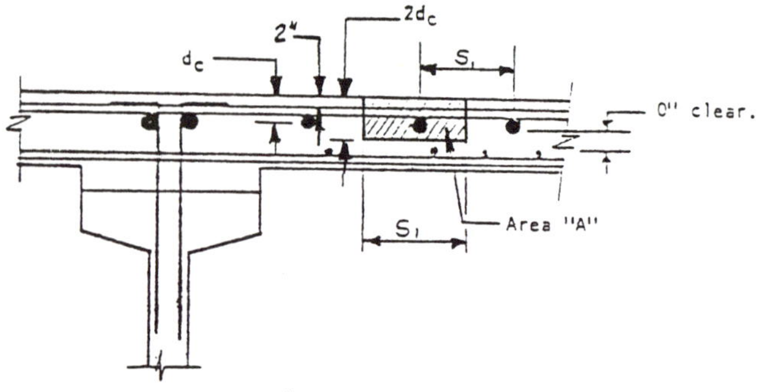


Case II
Girder Steel in Top Layer

Figure 6.3.1 B(a)-1
PLACEMENT OF REINFORCEMENT

When a two layer pattern as shown in case I requires an increase in slab thickness of more than one inch, case II should be used. In case II, when the zero vertical clearance between longitudinal bars is used, the bars should be dimensioned to insure a stagger between the bars in the top and bottom layers. The minimum clearance between top and bottom bars should then be 1", on the diagonal line between bars. Table 6.3.4 B(b)-1 shows required slab thickness for various bar arrangements.

| Transv. Bar / Longit. Bar | PLACEMENT CASE I | | | PLACEMENT CASE II | | |
|---|---|---|---|---|---|---|
| | #5 | #6 | #7 | #5 | #6 | #7 |
| #4 | 7* | - | - | - | - | - |
| 5 | 7 | 7 | 7¼ | - | - | - |
| 6 | 7 | 7¼ | 7½ | 7 | 7 | 7 |
| 7 | 7¼ | 7½ | 7¾ | 7 | 7 | 7 |
| 8 | 7½ | 8 | 8¼ | 7 | 7 | 7 |
| 9 | 8 | 8¼ | 8½ | 7 | 7 | 7 |
| 10 | 8¼ | - | - | 7 | 7 | 7 |
| 11 | - | - | - | 7 | 7 | 7¼ |
| 14 | - | - | - | 7 | 7¼ | 7½ |
| 18 | - | - | - | 7½ | 8 | 8¼ |

Slab Thickness (inches)

Table 6.3.4 B(b)-1

MINIMUM SLAB THICKNESS
FOR VARIOUS BAR SIZES

(c) Bar Patterns

Figure 6.3.3 B(c)-1 shows two typical top longitudinal reinforcing bar patterns. Care must be taken that bar lengths conform to the requirements of 5.1.2 F. Note that the reinforcement is distributed over a width equal to the girder spacing according to office practice, and does not conform to AASHTO Section 1.5.8(A).

The symmetrical bar pattern shown should normally not be used when required bar lengths exceed 60 feet. If the staggered bar pattern will not result in bar lengths within the limits specified in 5.1.2 F the method shown in Figure 6.3.3B(c)-2 may be used to provide an adequate splice. All bars shall be extended a length $l_d$ beyond the point where the bar is required.

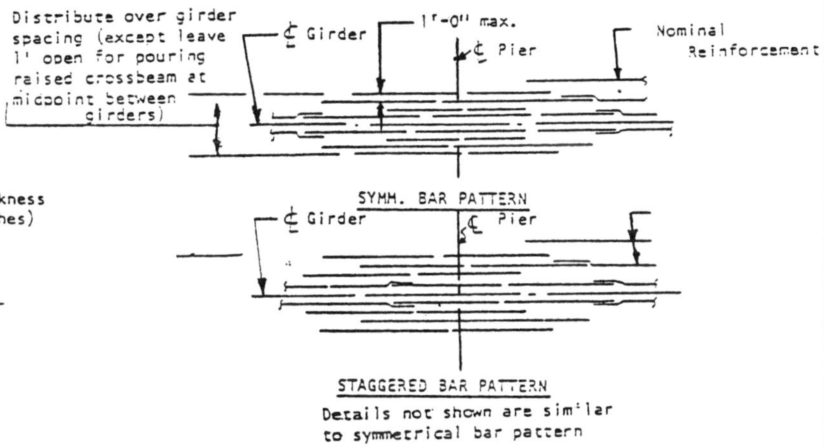


Figure 6.3.3 B(b)-1

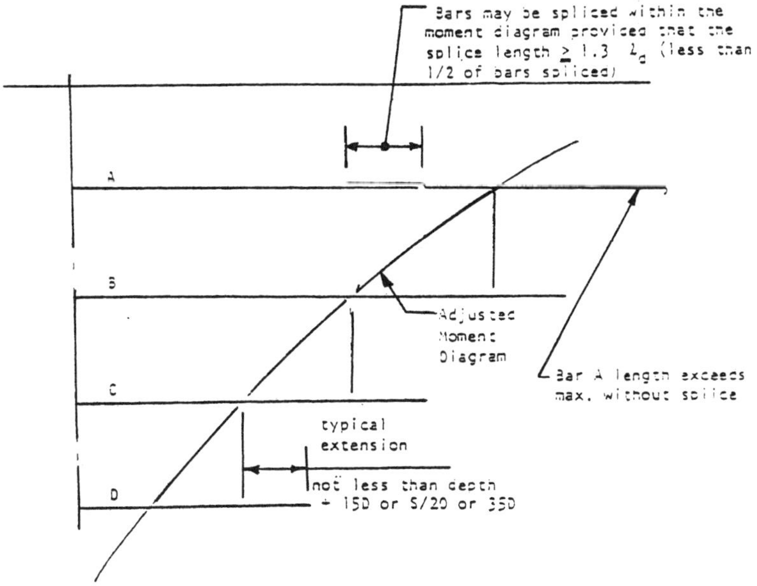


Figure 6.3.3 B(c)-2
BAR SPLICE WITHIN MOMENT ENVELOPE

In all bar patterns the reinforcement shall be well distributed between webs. Where this cannot be done without exceeding the 1'-0" maximum spacing requirement, the nominal longitudinal bars may be extended through to provide the 1'-0" maximum.

Normally no more than 20% of the main reinforced bars shall be cut off at one point. Where limiting this value to 20% leads to severe restrictions on the reinforcement pattern, an increase in this figure may be considered. Two main reinforcement bars shall be carried through the positive moment area as stirrup hangers.

# D.3 GIRDER SELECTION CHART

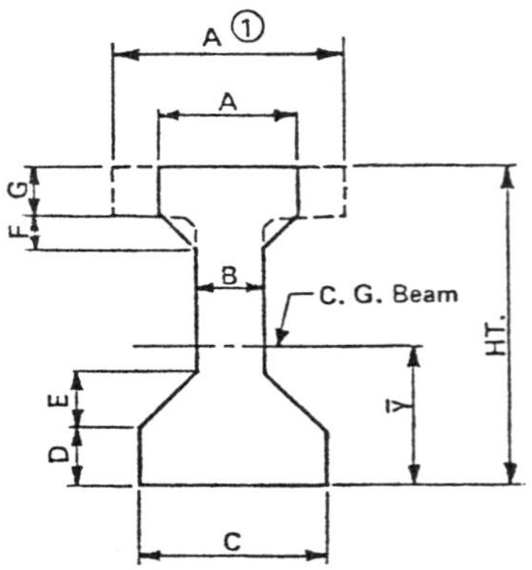


BEAM PROPERTIES

| AASHTO Type | HT. (In.) | AREA (In.$^2$) | W (lb./ft.) | $\bar{y}$ (In.) | I (In.$^4$) | $S_B$ (In.$^3$) |
|---|---|---|---|---|---|---|
| I | 28 | 276 | 288 | 12.59 | 22,750 | 1,807 |
| II | 36 | 369 | 384 | 15.83 | 50,980 | 3,221 |
| | 40 | 485 | 505 | 17.88 | 87,690 | 4,904 |
| III | 45 | 560 | 583 | 20.27 | 125,390 | 6,186 |
| IV | 54 | 789 | 822 | 24.73 | 260,730 | 10,543 |
| | 63 | 729 | 760 | 31.10 | 390,160 | 12,545 |
| | 72 | 783 | 815 | 35.50 | 545,220 | 15,358 |
| | 81 | 837 | 872 | 39.90 | 731,966 | 18,345 |

BEAM DIMENSIONS

| | HT. | A | B | C | D | E | F | G |
|---|---|---|---|---|---|---|---|---|
| | 28" | 1' 0" | 6" | 1' 4" | 5" | 5" | 3" | 4" |
| | 36" | 1' 0" | 6" | 1' 6" | 6" | 6" | 3" | 6" |
| | 40" | 1' 4" | 6" | 1' 10" | 6" | 8" | 5" | 6" |
| | 45" | 1' 4" | 7" | 1' 10" | 7" | 7-1/2" | 4-1/2" | 7" |
| | 54" | 1' 8" | 8" | 2' 2" | 8" | 9" | 6" | 8" |
| ① | 63 | 2' 6" | 6" | 2' 2" | 7-1/2" | 3-1/2" | 1-13/16"* | 6" |
| ① | 72 | 2' 6" | 6" | 2' 2" | 7-1/2" | 3-1/2" | 1-13/16"* | 6" |
| ① | 81 | 2' 6" | 6" | 2' 2" | 7-1/2" | 3-1/2" | 1-13/16"* | 6" |

* Small fillets ignored

DESIGN ASSUMPTIONS FOR GRAPH

1973 AASHTO "Standard Specifications for Highway Bridges"

Effective slab thickness minus 1/2". (where plan will provide for a slab waterproofing system and wearing course, use full slab depth for final design)

Steel strands are 1/2" dia. - 270 kips. (area = 0.153 sq. in.)

① Slab design as per MHD Concrete slab reinforcement (5-392.501)

Dead load applied to composite section ($DL_2$) includes 30 pounds per square foot, for wearing course and railing load. Generally, assume that 1/3 of type G railing load (160 lb/lin ft.) is applied to the first interior beam. One third of the composite section modulus is assumed in computing $DL_2$ stress. (Modify these assumptions as necessary for specific design.)

Initial stress calculated after elastic shortening loss.

Steel strands stressed to .70 ultimate tensile strength at initial force.

Transformed steel strand area used for LL composite.

$E'c$ = 4,400,000 for 5000 PSI concrete. $E'c$ = 4,800,000 for 6000 PSI concrete.

$E_c$ = 5,200,000 (concrete) $E_s$ = 28,000,000 (steel)

SH = 10,000 PSI (shrinkage) $CR_s$ = 20,000 - 0.125 (SH + ES + $CR_c$) (Steel relaxation)

$CR_c = \frac{E_s \, fcd}{E_c \times .435}$ (creep of concrete) $ES = \frac{E_s \, fcr}{E'c}$ (Elastic shortening)

PRESTRESSED CONCRETE BEAM.

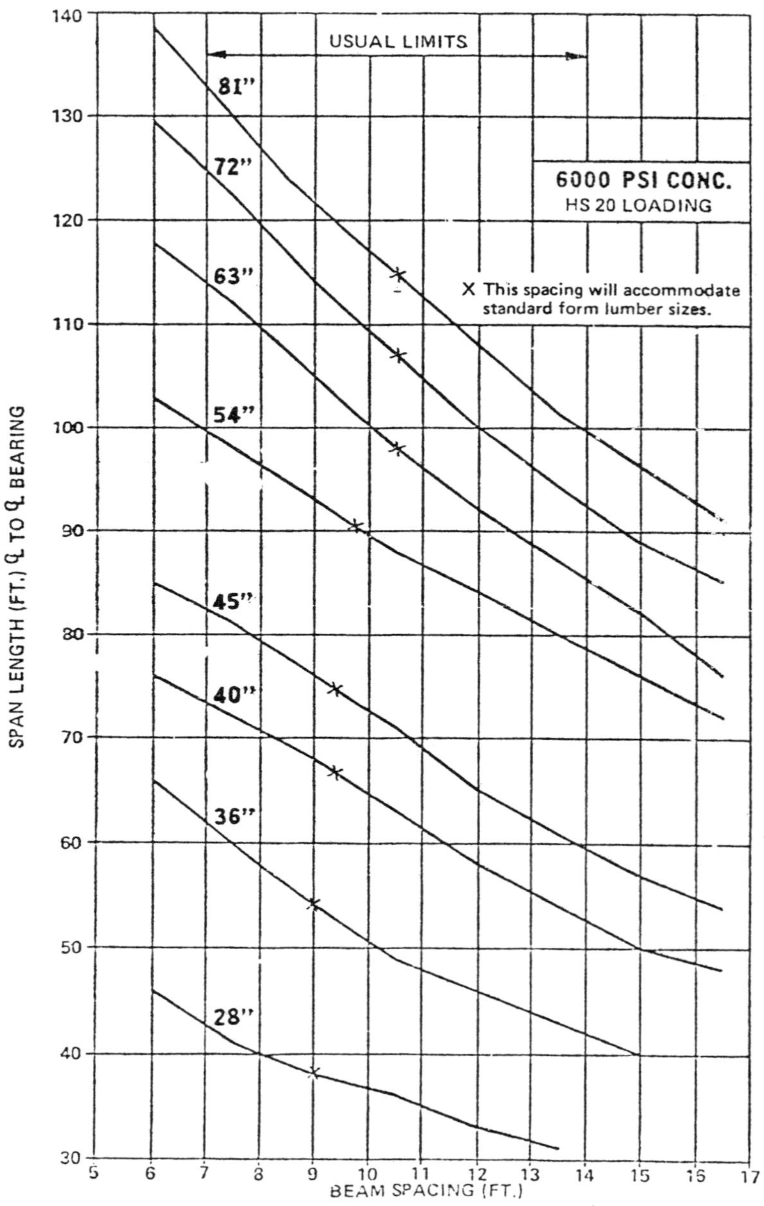


PRESTRESSED CONCRETE BEAM GRAPH

(Minn.)

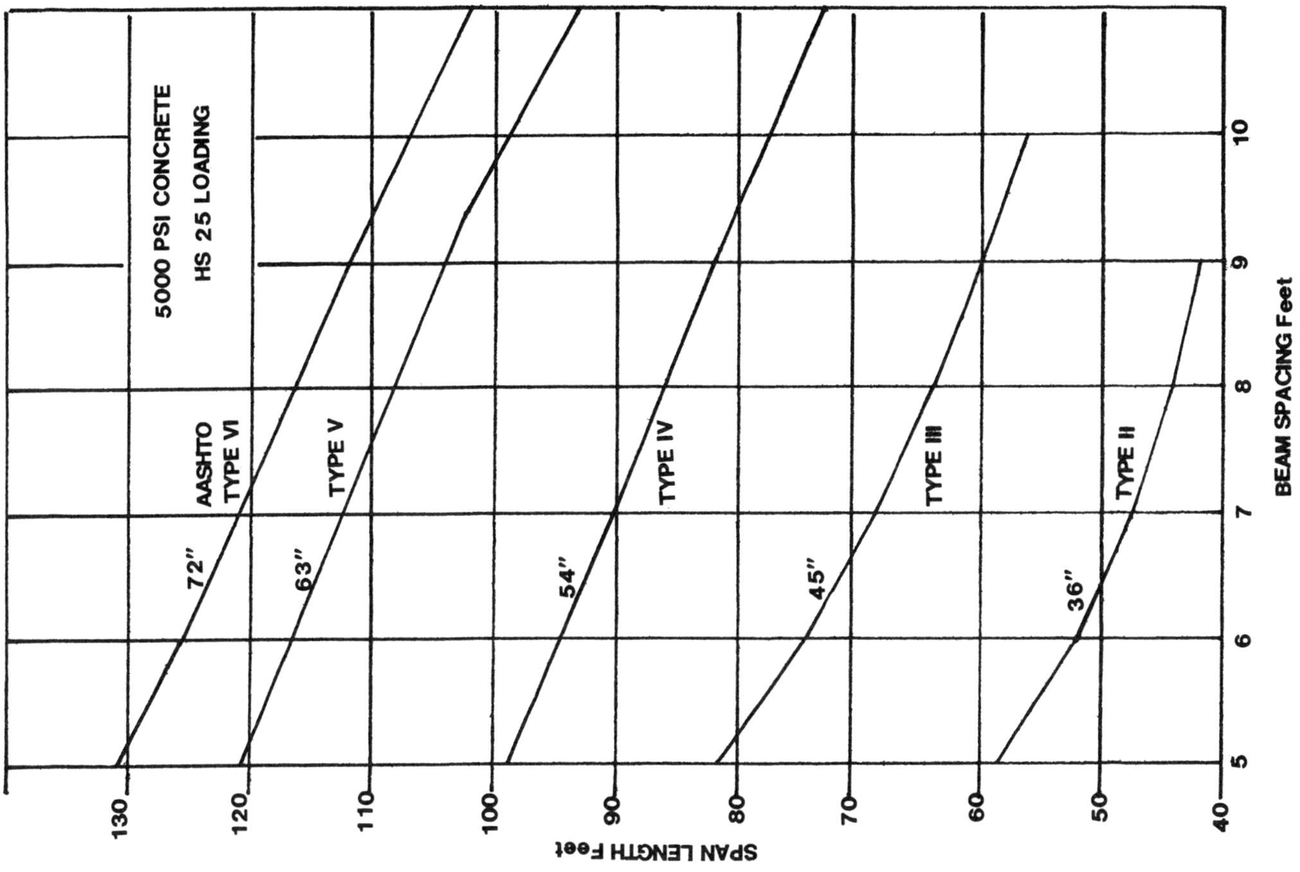


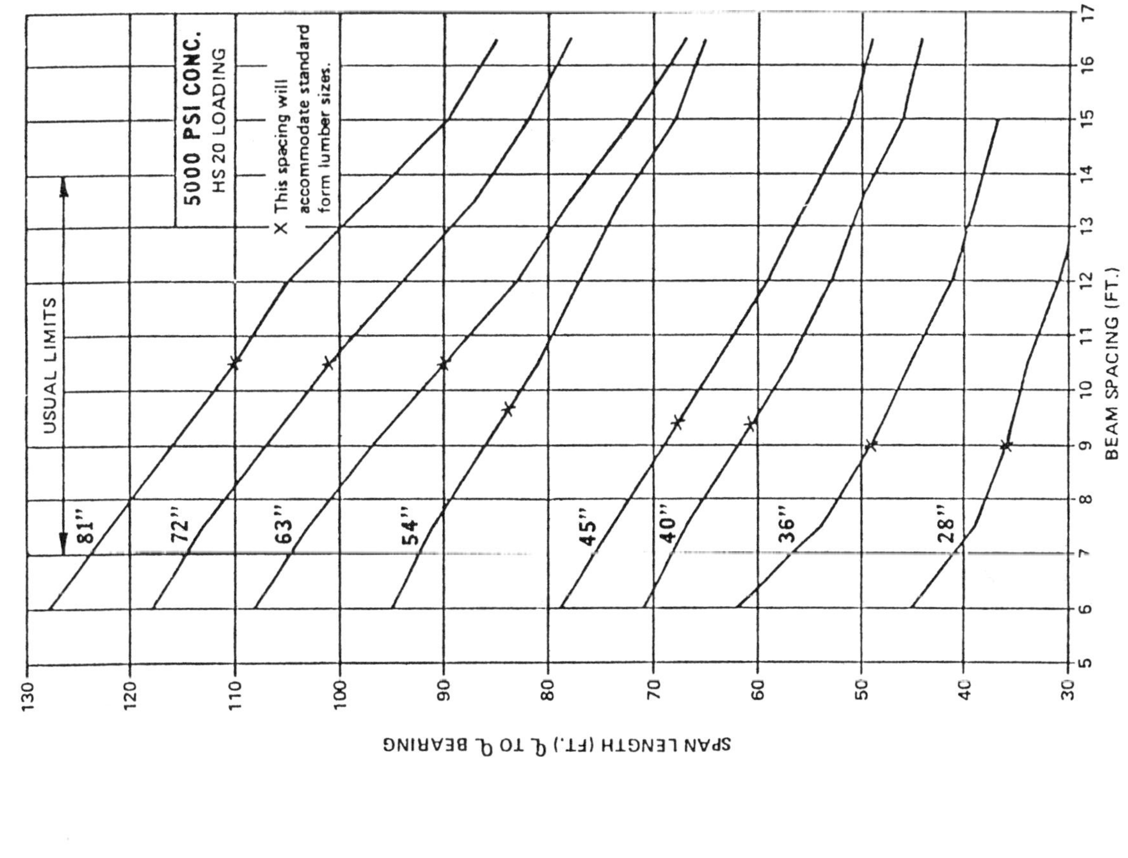


PRESTRESSED CONCRETE BEAM GRAPH
(INCLUDES WEARING COURSE)

Minn

# D.4 PRESTRESSED CONCRETE BRIDGE DESIGN AND DETAILS

**From Wisconsin DOT Bridge Manual**

## PRETENSIONED MEMBER DESIGN

This section outlines the aspects associated with the design of conventional pretensioned members.

### Allowable Stresses

The allowable stresses at different loading stages are defined in Section 6 of AASHTO. A 3% overstress in concrete is allowed if a line of girders can be eliminated. The basic ultimate stress for the different structural components are:

| | |
|---|---|
| 'I' Type Girder Concrete | 6000 psi |
| Slab and Box Type Girder Concrete | 5000 psi |
| Deck and Diaphragm Concrete | 4000 psi |
| Prestressed Reinforcement | 270 ksi |
| Grade 60 Reinforcement | 60 ksi yield |

The required concrete strength at time of prestress transfer is stated on the plans:

| | Min. | Max. |
|---|---|---|
| 'I' Type Girders | 4800 | 5200 psi |
| Slab and Box Type Girders | 4000 | 4500 psi |

### Loading Stages

The loads that a member is subjected to during its design life and those stages that generally influence a design are discussed below:

A. Prestress Transfer

This is the initial condition of prestress. This condition exists immediately following the release of the tendons (transfer of the tendon force to the concrete). The maximum strand stress at transfer before elastic shortening losses is $0.70\ f'_s$.

Because of the eccentricity of the prestress force an upward camber is produced in the member. This camber causes the member to span its full length thereby inducing a stress due to the dead load of the member itself.

This is a stage of temporary stress with the reduction in prestress due to elastic shortening of the member taking place and including dead load girder stress.

B. Losses

After losses the 'external' loading is the same as in (A), but the 'internal' stress (prestress) is further reduced by losses resulting from relaxation (creep) of the prestressing steel together with creep and shrinkage of the concrete. It is assumed that all losses occur prior to service loading.

Losses that are considered are:

Shrinkage - Loss from concrete shrinkage is 6,000 psi. This is based on an average ambient relative humidity in Wisconsin of 72%.

Elastic Shortening - Elastic shortening loss equals 7. times $f_{cir}$. For stress relieved strands $f_{cir}$ equals the concrete stress at the c.g. of the prestressing steel at the .5 point when the prestressing steel is stressed to .63 times its ultimate. For low relaxation strands $f_{cir}$ equals the concrete stress at the c.g. of the prestressing steel at the .5 point when the prestressing steel is stressed to .65 times its ultimate. ($f_{cir}$ is the stress due to the prestressing force and the dead load of beam and may be computed in lieu of using the previous values).

Creep of Concrete - Loss due to creep of concrete equals 12. times $f_{cir}$ - 7. times $f_{cds}$. $f_{cds}$ equals the concrete stress at the c.g. of the prestressing steel due to all dead loads except the dead load present at the time the prestressing force is applied. $f_{cds}$ is calculated at the .5 point of the span.

Relaxation of Prestressing Steel:

(1) For Stress Relieved Strand

$$CRS = 20{,}000. - .04\ ES - 0.2\ (SH + CRC)$$

where
CRS = loss due to creep of steel
SH = loss due to concrete shrinkage
ES = loss due to elastic shortening
CRC = loss due to creep of concrete

(2) For Low Relaxation Strand

$$CRS = 3{,}850 - .08ES - .04\ (SH + CRC)$$

Fabrication Losses - These losses are not considered by the designer but they affect the design criteria because they exist. Anchorage losses which occur during stressing and seating of the prestressed strands vary between 1 and 4 percent. Losses due to temperature change in the strands during cold weather prestressing are 6% for a 60° F. change. The construction specifications permit a 5% difference in the jack pressure and elongation measurement without any adjustment.

AASHTO specifications permit a temporary strand stress of 80 percent of ultimate steel stress with the stress at transfer to be 70 percent. The fabrication losses account for most or all of this temporary overstress. Therefore, the elastic shortening losses cannot be applied to this temporary condition because of those fabrication losses.

C. Service Load

Here the member is subjected to the same loads as stage (B) together with the effect of either of the following:

1. In the case of the 'I' girder type the dead load of the deck and diaphragm are always carried by the basic girder section on a simple span basis. This is because of the type of construction used, i.e., nonshored girders simply spanning from one substructure unit to another.

   The girders are constructed in this fashion for single as well as multi-span structures as shown.

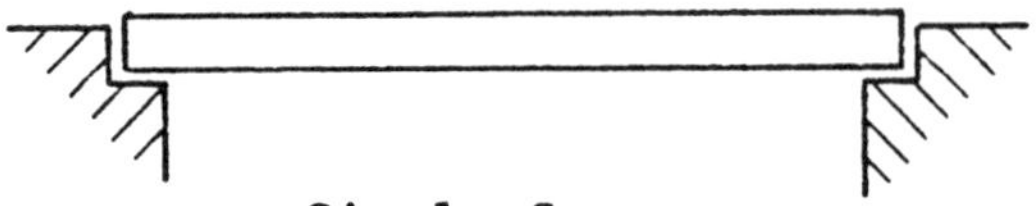

Single Span

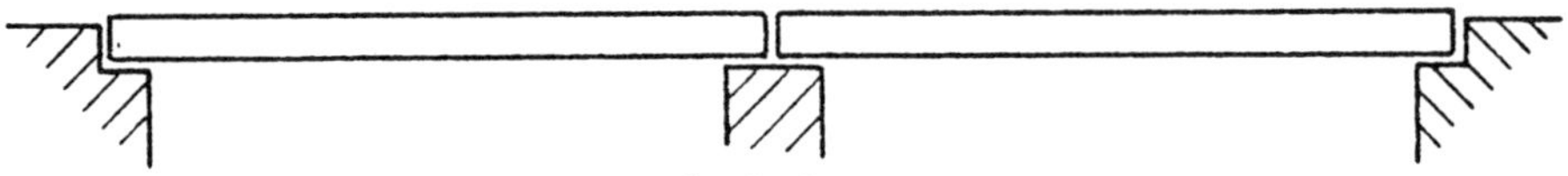

Multi-Span

On the other hand, the live load plus impact, together with any post-dead load (curb, parapet or median strip which is placed after the deck concrete has hardened) are carried by the composite section.

In the case of multi-span structures the longitudinal distribution for the live, impact and post-dead loads are based on a continuous span structure. This continuity is achieved by:

(a) Placing non-prestressed (conventional) reinforcement in the deck area over the interior supports.

(b) Casting concrete between and around the abutting ends of adjacent girders to form a diaphragm at the support.

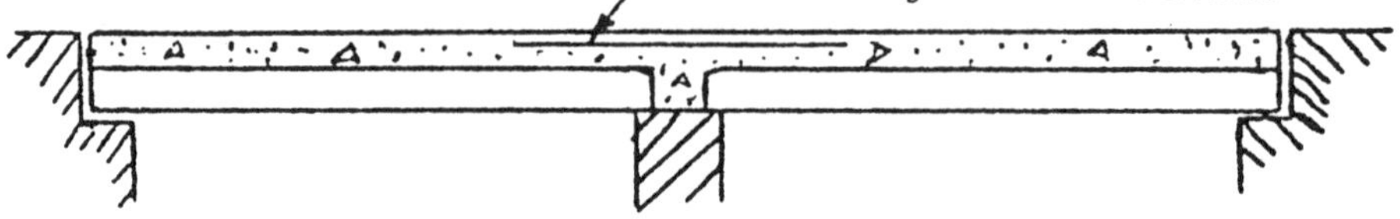


Deck slab and monolithic support diaphragm

Live load continuity is not considered for girders resting on expansion bearings where the diaphragm does not bear on the top of the pier unless a positive moment connection is made between the girders. If the span length ratio of two adjacent spans exceeds 1.5, design the girders as simple spans. In either case detail the stirrup spacing the same as for continuous spans and place bar steel over the supports equivalent to continuous span design. This value of 1.5 is not a structural limit as the positive moment reduces as this ratio increases. However, the reduction in strands for the reduced moment may not offset the increased girder cost in the shorter spans if the girders are not removed. For simple spans the girder spacing can increase in the end spans.

2. In the case of the slab and box type girders with a bituminous floor the dead load together with the live load plus impact are carried by the basic girder section.

   Whereas, when this type has a concrete floor the dead load of the floor is carried by the basic section and the live load, impact, and any post-dead loads are carried by the composite section (3" minimum composite floor).

   Note that the slab and box type girders are mostly used for single span structures. Therefore, both dead and live loads are always on a simple span basis.

   Occasionally, these girders are used on continuous spans with a concrete floor. Transverse cracking is likely to occur at piers with a shallow deck. To control this cracking, detail a compression seal joint at the piers or consider using a composite floor thickness of 5 inches.

D. Ultimate Load

At this stage the ultimate strength of the composite section is considered, except for the bituminous floor case where the basic section applies.

Since the member is designed on a service load basis it must be checked for its overload ability at the ultimate stress condition.

The need for both working and ultimate strength computations lies with the radical change in a member's behavior when cracks form. Prior to cracking the gross area of the member is effective. As a crack develops, all the tension in the concrete is picked up by the reinforcement. If the percentage of reinforcement is small, there is very little added capacity between cracking and failure.

(4) Design Procedure

The intent is to provide the engineer with a general outline of steps for the design of pretensioned members.

A. Transverse Member Spacing

A trial 'I' girder arrangement is made by using Table 19.1 as a guide. An ideal spacing results in equal strands for interior and exterior girders together with an optimum slab thickness.

The pretensioned slab or box is used as a 'multi-beam' system only, i.e., precast units placed side by side and locked together.

B. Live Load Distribution

This criteria is found in AASHTO Section 3 - "Distribution of Wheel Loads to Stringers, Longitudinal Beams and Floor Beams".

For exterior 'I' type girders the distribution factor (DF) for moment equals the reaction of a unit wheel load obtained by assuming the deck to act as a simple span between girders.

For interior 'I' type girders the DF = $\frac{S}{5.5}$ where (S) is the girder spacing.

C. Live Load and Impact

The impact formula is stated in AASHTO Section 2 - "Loads". It is applicable to all superstructure members.

The exterior 'multi-beam' is designed as a longitudinal 'edge beam' capable of resisting a live load moment LLM = 0.10 P x SPAN, where P is a wheel load (16 kips for H20 and HS20 loading).

D. Deck Design

The deck thickness and reinforcement is a function of the effective span and live load.

E. Dead Load

The dead load moments and shears are computed for simple spans. When post-dead loads are considered, these post-dead load moments are based on continuous spans for concrete deck slabs.

A post-dead load of 20 pounds per square foot is to be included in all designs which accounts for a possible future wearing surface.

F. Composite Section

A portion of the deck is taken as a composite top flange. For interior 'I'girders the effective width of this flange is taken as the least of the following:

1. One-fourth the span of the girder.

2. The center to center distance of girders.

3. Twelve times the effective deck thickness plus the width of the girder web.

The effective deck thickness equals the overall thickness minus the 1/2" wearing surface.

For exterior 'I' girders the effective flange width is taken as the full overhang from the C/L of girder plus the least of the following:

1. One-eighth the span of the girder. (Due to overhang meeting criteria for top flange on both sides of web.)

2. Six times the effective deck thickness plus 1/2 the girder web.

3. One-half the distance to the next girder.

The composite flange area for an interior 'multi-beam' is taken as the width of the member by the effective thickness of the floor. Minimum concrete overlay thickness is 3 inches.

The composite flange for the exterior member consists of the curb and the floor over that particular 'edge beam'.

Since the deck (or floor and curb) concrete has a lower strength than the girder concrete, it has a lower modulus of elasticity. Therefore, the effective flange width must be transformed by the modular ratio (N) which is assumed as 0.75. This value is based on tests which yield the modulus of elasticity (E) for various ultimate concrete strengths. The (E) for 4000 psi concrete is divided by the (E) for 6000 psi concrete.

The composite properties can be obtained from Table

G. Design Stress

The design stress is the bottom fiber stress due to dead load on the basic girder section together with live load, impact and post-dead load on the composite section.

The point of maximum stress is generally 0.5 of the span for both end and intermediate spans. But for longer spans (100 feet or so) the 0.4 point of the end span may control and is to be checked.

H. Prestress Force

With the design stress known, compute the required prestress needed to counteract all the design stress except that residual amount of tension which is allowed ($6.0\sqrt{f'_c}$).

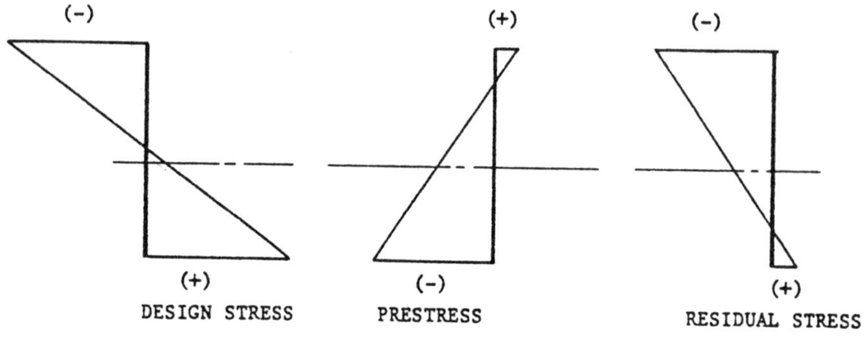


The required prestress ($f_p$) equals the design stress minus the allowable residual stress.

For slab type girders assume a strand pattern eccentricity (e) and apply this to Equation 19.2.6 to determine the final prestress force (P) and the approximate number of strands. Then a trial strand pattern is established using Standard 19.13 as a guide and a check is made on the assumed eccentricity.

Whereas for 'I' type girders, Equation 19.2.6 is solved for several predetermined patterns and is tabulated in Standards 19.8 thru 19.12.

Present practice is to detail all spans of equal length with the same number of strands, unless a span requires more than 3 additional strands. In this case the different strand arrangements are detailed along with a plan note stating: "The manufacturer may furnish all girders with the greater number of strands".

I. Service

Check top and bottom girder stresses along with stress at top of composite slab at the design point of the span. This confirms compliance with the allowable stress criteria for top of girder and top of slab since only the bottom stress is used to determine the prestress. Maximum allowable concrete compressive stresses after losses are: slab-1600. psi and girder-2400. psi.

J. Strand Drape

This consists of draping some of the strands in order to decrease stresses due to prestress at the ends of the 'I' girder where the stress due to loads are minimum.

The typical strand profile for this technique is shown below.

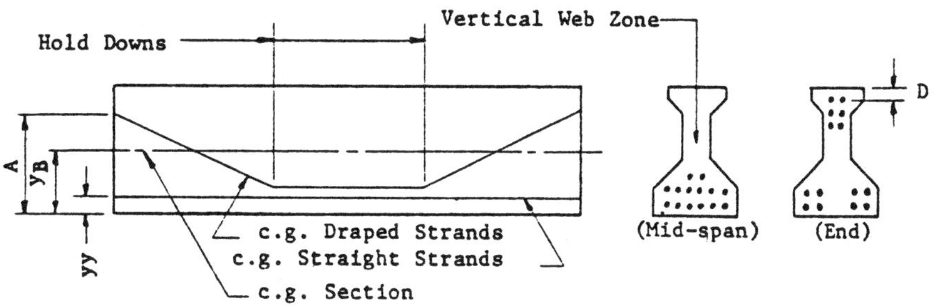


Note that all the strands that lie within the 'vertical web zone' of the mid-span arrangement are used in the draped group.

The required drape is determined by first computing the eccentricities, using Equation 19.2.6, that yield the following stresses at the girder ends:

1. Top fiber stress equal to zero.

2. Bottom fiber stress equal to 0.4 $f'_c$ after losses.

Using the smaller eccentricity sum moments, with respect to the number and distance of straight and draped strands, about the bottom of the girder and determine (A).

No. draped times (A) + No. straight times (yy) = total No. times ($Y_B$-e)

The drape must also meet the following conditions.

1. The minimum dimension (D), center of uppermost draped strand to top of girder at the ends, is 4 inches.

2. The maximum slope of the draped strands is 12%.

The holddown point of draped strands is taken as the 1/3 point. Actual field conditions may vary from this since fabricators use metal forms with pre-spaced holes to locate the holddown device with respect to the 1/3 point.

Note that certain tolerances for drape are set at the 1/4 point and are stated in Standards 19.1 thru 19.5.

An alternate to draping is that of raised strand patterns for 'I' type girders shown in Standard 19.8 thru 19.10. Present practice is to show a standard raised arrangement as an alternate to draping for short spans.

The strands in slab and box type girders are normally not draped, but instead are arranged to satisfy the stress requirements at mid-span and ends of girder.

K. Strength at Transfer

The required strength at transfer ($f'_{ci}$), which is stated on the plans, is computed based on the temporary compressive stress due to initial prestress and girder dead load.

$$f_c = f_{DL} + F_P \leq (0.6\ f_{ci})$$

The controlling bottom fiber stress occurs at either the hold-down or girder ends of draped designs and at girder ends of non-draped designs. The maximum allowable girder stress is 3120. psi.

L. After Losses

Check the bottom fiber stress at:

1. The above control point due to girder dead load and effective prestress.

2. The continuous end of the girder due to negative live load moment and effective prestress.

Note that the stress in 1. need not be checked at the simple end of draped designs since the drape criteria covers this.

Generally the stresses in 1. and 2. meet the 'design stress' criteria. See Section 'S' for explanation.

M. Non-prestressed Reinforcement

This consists of bar steel reinforcement used in the conventional manner. It is placed longitudinally along the top of the member to carry any tension which may develop after transfer of prestress.

The amount of reinforcement is that which is sufficient to resist the total tension force in the concrete based on the assumption of an uncracked section.

For draped designs the control is at the holddown point of the girder. Here initial prestress together with girder dead load stress is acting. This is where tension due to prestress is still maximum and compression due to girder dead load is decreasing.

For non-draped designs the control is at the end of the member where prestress tension exists but dead load stress does not.

Note that a minimum amount of reinforcement is specified in Standards 19.1 thru 19.6. This is intended to help prevent serious damage due to unforeseeable causes like improper handling or storing.

N. Ultimate Capacity

The required ultimate positive moment capacity is expressed as multiples of dead load (DL), live load (LL), and impact (I).

$$M_u = 1.3 (DL + 5/3 (LL + I))$$

These load factors are applicable for both simple and continuous span structures.

The equations for determining the ultimate flexural strength of a member are stated in AASHTO Specifications. Use a capacity reduction factor "Ø" equal to 0.90 for flexure.

O. Steel Percentage

The maximum and minimum steel percentage criteria is given in AASHTO-Section 6 and it applies to both girder types. But, this requirement need not be checked for 'I' girders since the preselected strand pattern is based on this criteria.

P. Horizontal Shear

This is the stress between the member and its composite deck at ultimate shear.

$$V_u = 1.3 (DL + 5/3 (LL + I))$$
$$\bar{V}_u = V_u/\phi$$

Use a capacity reduction factor "Ø" equal to 0.85 for shear.

The dead load in this case is only that which acts on the composite section.

The shear stress between the contact surfaces is

$$v = \frac{\bar{V}_u Q}{Ib} .$$

where:

Q = statical moment of the effective transferred area of the deck taken about the centroidal axis of the composite section.

I = moment of inertia of the composite section.

b = width of the contract area between the member and the composite deck.

Laboratory investigations 1, 2 indicate that the horizontal shear stresses permitted by AASHTO are very conservative. It is on the basis of these laboratory tests that the following criteria is based:

The allowable horizontal shear stress is 350 psi when the contact surfaces are clean and intentionally roughed. When the horizontal design shear is greater than 350 psi refer to AASHTO-Section 5 - article 1.5.35(D).

This criteria is applied at the support having maximum vertical shear. The stirrups in Standard 19.13 are considered adequate for slabs with composite concrete floors.

Q. Web Reinforcement

This consists of placing conventional reinforcement perpendicular to the axis of the 'I' type girder. The required area of web reinforcement is

$$A_v = \frac{(\bar{V}_u - V_c)S}{2f_{sy}jd} \quad \text{(or 100 b's/fsy min)}$$

where $V_c$ = 180 b'jd, (b') the web-thickness, (j) may be taken as 0.88 and (S) is the spacing of the reinforcement, which is limited to 3/4 of the girder depth. Use a capacity reduction factor "Ø" equal to 0.90 for shear.

The ultimate shear is taken as

$$V_u = 1.3\ (DL + 5/3\ (LL + I))$$

$$\bar{V}_u = V_u/\emptyset$$

Where (DL) is the dead load of the girder, deck and post-dead loads when applicable.

Note that the vertical component of the draped strands which helps to reduce the shear on the concrete section is neglected for conservatism.

For simple spans the shear is investigated in the middle half of the span (1/4 point thru 3/4 point). The amount of reinforcement required at the quarter points is used throughout the outer quarters of the span.

Whereas for continuous spans the shear is investigated over the full length of the intermediate spans and over the inner three-quarters of the end spans.

Reinforcement, in the form of vertical stirrups, is required at the extreme ends of the girder. The stirrups are designed to resist 4% of the total prestressing force at a unit stress of 20 ksi and are placed within d/4 of the girder end. This is a conservative design but is currently employed for crack control.

Slab used in a 'multi-beam' system are considered satisfactory for shear.

R. Continuity Reinforcement

The design of non-prestressed reinforcement for negative moment at the support is based on ultimate strength methods.

$$M_u = 1.3(DL + 5/3\ (LL + I))$$

$$V_u = 1.3(DL + 5/3\ (LL + I))$$

Use the appropriate "$\phi$" factors

AASHTO Specifications allow a capacity reduction factor equal to 1.0 for factory produced precast prestressed concrete members.

Here the dead load is only that which is placed after the monolithic support diaphragm and deck have set up.

The amount of reinforcement is based on Load Factor Design

The concrete between the abutting girder ends is usually of a much lesser strength than that of the girders. But, tests[4] have shown that due to lateral confinement of the diaphragm concrete, the girder itself fails in ultimate negative compression rather than the material between its ends. Therefore the ultimate compression stress ($f'_c$) of the girder concrete is used in place of that of the diaphragm concrete.

This assumption has only a slight effect on the computed amount of reinforcement, but it has a significant effect on keeping the compression force within the bottom flange.

The transverse spacing of the continuity reinforcement is usually taken as the whole or fractional spacing of the "D" bars Grade 60 bar steel is used for continuity reinforcement.

The length of continuity reinforcement is based on the following criteria:

1. When one-half the bars satisfy the moment diagram, terminate one-half the bars.

2. When the top fiber stress of the girder equals the modulus of rupture, terminate the remaining one-half of the bars. This point is to be at least 1/20 of the span length or 4 feet from point 1., whichever is greater.

Certain secondary features result when spans are made continuous.

That is, positive moments develop over piers due to creep[5] together with those produced by live plus impact loads in remote spans. The latter only exists for bridges with three or more spans.

These positive moments are somewhat counteracted by negative moments resulting from differential shrinkage[6] between the cast-in-place deck and precast girders along with those due to post-dead loads.

At the present time it is not considered necessary to provide a positive moment connection at the pier if the bearing is fixed and the diaphragm is full depth and anchored to the pier; this is based primarily on past experience.

S. Precompression

The precompression in the bottom flange, at the girder end due to prestress, is neglected in the ultimate moment computation if the following limitations are met:

1. The maximum precompression stress is less than 0.4 $f'_c$. This is always met because of the drape criteria.

2. The percentage of continuity reinforcement is less than 1.5%[4,5].

T. Camber and Deflection

The prestress camber and dead load deflection are used to establish vertical position of the deck forms with respect to the girder.

Fig. 19.3.6 typifies a girder with a draped strand profile.

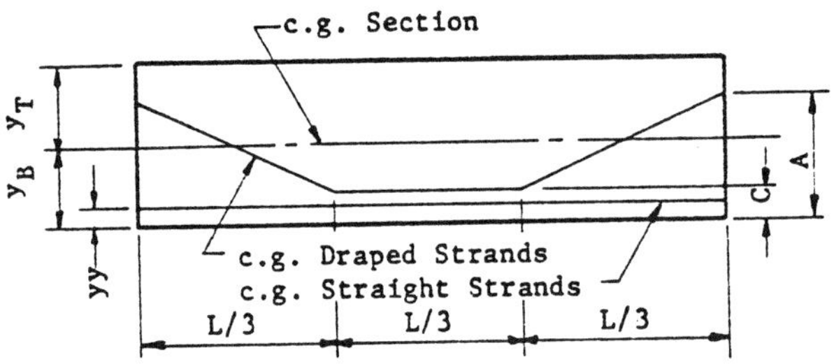


Fig. 19.3.6

1. Prestress Camber

The eccentric straight strands induce a constant moment of

$$M_1 = P_i^s\,(y_B - yy)$$

where ($P_i^s$) is the initial prestress force in the straight strands minus the elastic shortening loss. This moment produces an upward deflection at mid-span.

$$\Delta_S = \frac{M_1 L^2}{8EI}$$

The draped strands induce moments at the ends and within the span

$$M_2 = P_i^D \text{ (A-C) produces upward deflection}$$

$$M_3 = P_i^D \text{ (A-}y_B\text{) downward when (A) is greater than } (y_B)$$

where ($P_i^D$) is the initial prestress force in the draped strands minus the elastic shortening loss.
These moments produce a net upward deflection at mid-span.

$$\Delta D = \frac{L^2}{8EI}\left(\frac{23}{27} M_2 - M_3\right)$$ (See 2-200 of AISC Manual for formula)

The combined upward deflection due to prestress is

$$\Delta \text{ prestress} = \Delta_S + \Delta_D$$

$$= \frac{L^2}{8EI}\left(M_1 + \frac{23}{27} M_2 - M_3\right)$$

The downward deflection due to dead load is

$$\Delta = \frac{5WL^4}{384\ EI}$$

where (W) is the member weight per unit length, (I) is the moment of inertia and (E) the modulus of elasticity which is assumed at $3.5 \times 10^6$ psi for these deflections at this age.

Therefore, the anticipated prestress camber is the net prestress deflection minus the members dead load deflection.

2. Dead Load Deflection

The downward deflection due to the dead load of the deck and mid-span diaphragm is:

$$\Delta_{DL} = \frac{5WL^4}{384EI} + \frac{PL^3}{48EI}$$

where (W) is the deck weight per unit length and (P) the weight of the diaphragm, both based on that carried by a typical interior girder. The modulus of elasticity (E) is assumed at $5.5 \times 10^6$ psi for mature concrete and (I) is the moment of inertia of the basic member.

3. Residual Camber

This is the camber that remains after the prestress camber has been reduced by the dead load deflection.

U. Deck Forming

This consists of computing the relationship between the top of girder and bottom of deck necessary to achieve the desired vertical roadway alignment.

Present practice is to imbed the top of the girder in the deck a maximum of 1/2" at some point along the span. This usually occurs at mid-span but may occur at the ends due to certain grades or effects of adjacent spans. In Fig. 19.3.7 the two cases are shown.

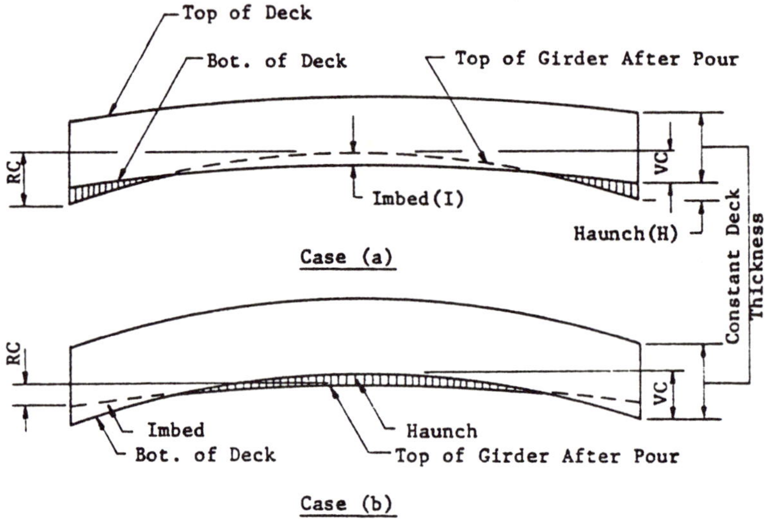


Fig. 19.3.7

In case (a) the vertical curve amount (VC) is <u>less</u> than the computed residual camber (RC) and therefore the imbedment is detailed at mid-span. Whereas in case (b), the (VC) is <u>greater</u> than (RC) which requires the imbedment to be detailed at the girder ends.

For <u>equal</u> span continuous structures, having <u>all</u> spans on the same vertical alignment, the deck forming is the same for each span. This is due to the constant change of slope of a vertical curve or tangent and the same (RC) per span.

The following equation is derived from Fig. 19.3.7.

$$\underset{\text{END}}{[(+H) \text{ or } (-I)]} = (RC)-(VC) + \underset{\text{MID-SPAN}}{[(+H) \text{ or } (-I)]}$$

where upward (RC) is positive and (VC) is positive for crest curves and negative for sag curves.

For unequal spans or when some spans are on the curve and others on the tangent a different approach is used. Generally the longer span or the one off the curve dictates the haunch or imbedment required at the common support. Therefore, it is necessary to pivot the girder about its mid-span in order to achieve an equal condition at the common support. This is done Mathematically by adding together the equation for each end (abutment and pier).

| LEFT END | | RIGHT END | | MID-SPAN |
|---|---|---|---|---|
| [(+H) or (-I)] | + | [(+H) or (-I)] | = 2 | [(RC)-(VC) + [(+H) or (-I)]] |

with the condition at one end known, due to the adjacent span the condition at the other end is computed.

V. Construction Joints

Locate the transverse construction joints in deck midway between the cut-off points of the continuity reinforcement or at the .75 point of the span, whichever is closest to the pier. Locate joint at least <u>one</u> foot from cut-off points.

This change keeps stresses in the slab steel from slab dead load at a minimum and makes deflections from slab dead load closer to the theoretical.

X. Strand Types

Currently stress relieved and low relaxation strand types are used in prestressed concrete 'I' girder design. Strand patterns and initial prestressing forces for both types of strands are given on the plans and deflection data is based only on stress relieved strands.

Y. Construction Dimensional Tolerances

Refer to the Standard Specifications for Road and Bridge Construction for the required dimensional tolerances.

## REFERENCES


1. Hanson, N. W., "Precast-Prestressed Concrete Bridges 2. Horizontal Shear Connections", Portland Cement Association Development Department, Bulletin D35.

2. Mattock, A. H. and Karr, P. H., "Precast-Prestressed Concrete Bridges 4. Shear Tests of Continuous Girders", Portland Cement Association Development Department, Bulletin D46.

3. Whitney, C. S., "Plastic Theory of Reinforced Concrete Design", ASCE Trans., 107, 1942, p. 251.

4. Karr, P. H., Kriz, L. B. and Hognestad, E., "Precast-Prestressed Concrete Bridges 1. Pilot Tests of Continuous Beams", Portland Cement Association Development Department, Bulletin D34.

5. Mattock, A. H. and Karr, P. H., "Precast-Prestressed Concrete Bridges 3. Further Tests of Continuous Girders", Portland Cement Association Development Department, Bulletin D43.

6. Freyermuth, Clifford L., "Design of Continuous Highway Bridges with Precast, Prestressed Concrete Girders (EB014.01E), Portland Cement Association, 1969.

## Design Examples

### A. 'I' Type Girder

This is an example of a three span continuous structure. The same basic procedure is applicable to single spans except live load continuity reinforcement computations are not required.

Trial Roadway Cross Section

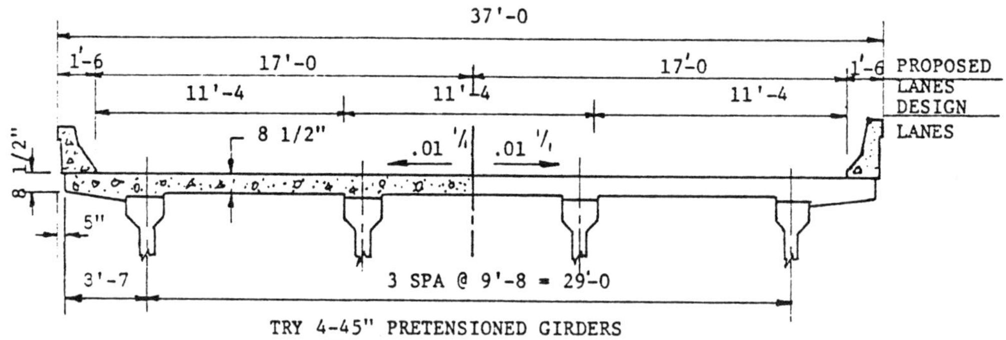


TRY 4-45" PRETENSIONED GIRDERS
H20-CONTINUOUS FOR LIVELOAD
* L.L. DESIGN SPANS: 74'-1 1/2", 74'-9, 74'-1 1/2
* D.L. " " : 73'-6", 73'-6, 73'-6 (c/c BRG)
OVERALL GIRDER LENGTH: 74'-6
1/2" Ø STRESS RELIEVED STRANDS

* It is OK to use the LL span lengths for dead load moment computations.

LL Distribution

Interior DF = 9.67/5.5 = 1.76 wheels/girder

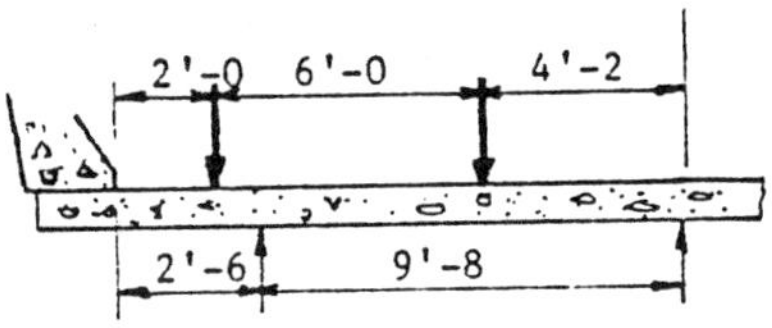


Exterior DF = (4.17 + 10.17)/9.67 = 1.48

LL + Impact

Values are from the "Continuous Beam Analysis"

| | Span 1 & 3 (POS) | PIER (NEG) | Span 2 (POS) |
|---|---|---|---|
| Int. LLM+I | = 8180 in-K | 8660 in-K | 6640 in-K |
| Ext. LLM+I | = 6880 | 7280 | 5580 |

Deck Design

Effective Span = (9'-8") - 8" = 9'-0"
Deck T = 8.5"
"D" Bars = #4's @ 12"

Dead Load

| | Interior | | Exterior | |
|---|---|---|---|---|
| Deck | = .708 x 9.67 x 150 | = 1025#/' | .708 x 8.42 x 150 | = 894#/' |
| Girder | = | 583# | | 583 |
| Diaphragm | = (9.67-.58) x 375 | = 3400# | 3400/2 | = 1700# |
| Deck DLM | = $1.025 \times (73.5)^2 \times 12/8$ | = 8310 in-K | .894 x 8100 | = 7245 in-K |
| Girder DLM | = .583 x 8100 | = 4730 | | = 4730 |
| Diaph. DLM | = 3.4 x 73.5 x 12/4 | = 750 | 750/2 | = 375 |
| | DLM | = 13790 in-K | | 12350 |

Post-Dead Load

Future Wearing Surface = 9.67 x 20 = 193#/' Int., 7.33 x 20 = 147#/' Ext.
Parapet = 338#/' (Assume Carried by Exterior Girder Only)
Span 1 & 3 PDLM; Int. = 950 in-K, Ext. = 2390 in-K
Span 2 PDLM: Int. = 338 in-K, Ext. = 850 in-K

Composite Section

Interior

$T_{eff}$ = 8-1/2 - 1/2 Wearing Surface = 8"  B = (12 x 8) + 7 = 103"

$I = 362814 \text{ in}^4$  $S_B = 10342 \text{ in}^3$  $S_T = 36574 \text{ in}^3$ (Values from Table 19.3)

$Y_T$ = 362814/36574 = 9.92"  $Y_{slab}$ = 9.92 + 8.50 - 0.50 = 17.92"

$S_{slab}$ = 362814/17.92 = 20230 $\text{in}^3$

Exterior

$T_{eff}$ = 8"  B = 43 + (6 x 8) + 3 = 94"

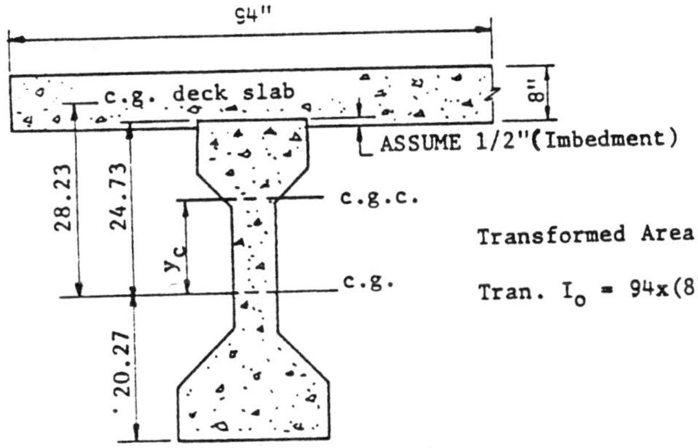


Transformed Area = 94x8x.75 = 564 $\text{in}^2$

Tran. $I_o$ = $94\text{x}(8)^3\text{x}.75/12$ = 3008 $\text{in}^4$

Sum Moments About c.g.

| Item | A | Y | Ay | $Ay^2$ | Io | $Io+Ay^2$ |
|---|---|---|---|---|---|---|
| Girder | 560 | 0 | 0 | 0 | 125390 | 125390 |
| Slab | 564 | 28.23 | 15920 | 449470 | 3008 | 452480 |
| Sum | 1124 | - | 15920 | - | - | 577870 |

$-A Y_c^2$ = -225540

$I_{c.g.c}$ = 352330 $\text{in}^4$

$Y_c$ = Ay/A - 15920/1124 = 14.16"

$Y_t$ = 24.73 - 14.16 = 10.57"  $S_t$ = 352330/10.57 = 33330 $\text{in}^3$

$Y_B$ = 20.27 + 14.16 = 34.43"  $S_B$ = 352330/34.43 = 10230 $\text{in}^3$

Design Stress (Bottom mid-span)

| | Spans 1 & 3 | Span 2 |
|---|---|---|
| Interior $f_{DL}$ | $= \frac{13790}{6186} = 2.229$ ksi | $\frac{13790}{6186} = 2.229$ ksi |
| $f_{LL+I}$ | $= \frac{8180}{10342} = 0.791$ | $\frac{6640}{10342} = 0.642$ |
| $f_{PDL}$ | $= \frac{950}{10342} = .092$; 3.112 | $\frac{338}{10342} = 0.034$; 2.905 |
| Minus Tension Allowed | = -.465; 2.647 ksi | = -.465; 2.440 ksi |
| Exterior $f_{DL}$ | $= \frac{12350}{6186} = 1.996$ ksi | $\frac{12350}{6186} = 1.996$ ksi |
| $f_{LL+I}$ | $= \frac{6880}{10230} = 0.670$ | $\frac{5580}{10230} = 0.543$ |
| $f_{PDLM}$ | $= \frac{2390}{10230} = 0.234$ | $\frac{850}{10230} = 0.083$ |
| | 2.900 | 2.622 |
| | -.465 ksi | -.465 |
| | 2.435 ksi | 2.157 ksi |

Prestress Force

| | 36" | 45" | 54" | 70" Girders |
|---|---|---|---|---|
| Assume total losses = | 45, | 50, | 55, | 57 ksi |

% Losses = 50/189 = 26%

Estimated initial prestress = Design Stress/(1-% Losses)
= 2.647/.74 = 3.577 ksi (Int. controls Span 1 & 3)

Try 28 Strands (See $f_B$ (Init.) on Standard 19.9)

**(Estimate the Elastic Shortening Concrete Loss as 7%).**

Initial Force = 0.63 x 270 x 0.153 x 28 = 729 Kips

$f_B$ = (-729/560) (1+ (-16.41 x -20.27/223.9)) = -3.235 ksi (Bot. Fiber)

$f_T$ = (-1.302) (1+ (-16.41 x 24.73/223.9)) = +1.058 (Top Fiber)

Initial Prestress + Beam Dead Load (Mid-Span, Elastic Shortening of Concrete Loss)

$f_B$ = -3.235 + 4730/6186 = -2.471 ksi

$f_T$ = +1.058 - 4730/5070 = +0.125 ksi

Elastic Shortening Loss

Stress at c.g. of Strands (5/10 pt)

$f_{cir}$ = -2.471 + ((2.471 + 0.125) ( 20.27 - 16.41)/45) = -2.248 ksi

ES = 7$f_{cir}$ = 7 x 2.248 = 15.7 ksi [100 (15.7/270) = 5.8%]

Creep of Concrete

Stresses due to Deck DLM and Diaphragm DLM (There is no Curb DLM.)

$f_B$ = 9060/6186 = 1.465 $f_T$ = -9060/5070 = -1.787

Stress at c.g. of Strands (5/10 pt)

$f_{cds}$ = 1.465 -((1.465 + 1.787) (20.27 - 16.41/45) = 1.186 ksi

CRC = 12 $f_{cir}$ - 7 $f_{cds}$ = 12 x 2.248 - 7 x 1.186 = 18.7 ksi

Shrinkage = SH = 6 ksi based on average relative humidity of 72%

Creep of Prestressing Steel

CRS = 20. - .4 ES - 0.2 (SH + CRC) = 20. - .4 (15.7) - 0.2 (6 + 18.7) = 8.8 ksi

Service Load (At 5/10 Point with all Losses)

Losses = 6. + 15.7 + 18.7 + 8.8 = 49.2 ksi

(189-49.2)/189 = 0.740

$f_B$ = 3.595 x 0.740 + 3.112 = -2.660 + 3.112 = +0.452 $<$ 0.465 OK

$f_T$ = (-810/560)(1 + (-16.41 x 24.73/223.9) x 0.740 - 13790/5070
- (8180 + 950)/36574 = .870 - 2.720 - 0.250 = -2.100 $<$ 2.4 ksi OK

$f_{slab}$ =-(8180 + 950) 0.75/20230 =-0.338 $<$ 1.4 ksi OK

Strand Drape

Find c.g. of Straight Strands by Summing Moments About Bottom of Girder

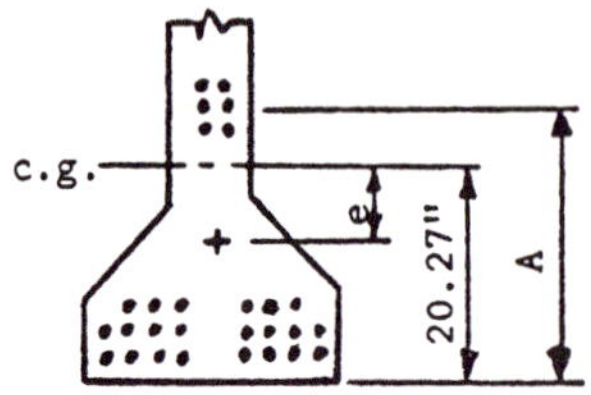

Top Fiber Stress Equal to Zero

$f_T = 0 = \frac{-P}{A}\ (1+(ey_T/r^2))$, $e = -r^2/y_t$ = -223.9/24.73 = -9.05"

Bottom Fiber Stress Equal to 2.4 ksi After Losses

$f_B$ = -2.4/0.740 = (-810/560) (1+(e(-20.27/223.9)))

e = (223.9/20.27) ((3.24 x 560/810) -1)) = 13.70 > -9.05  Use e = 9.05"

Sum Strand Moments About Bottom of Girder

6A + 8 x 2 + 8 x 4 + 6 x 6 = 28 (20.67 - 9.05)

A = 38.33"  Try 38"  (Round to nearest inch)

Check Clearance:  45-38-2 = 5 > 4" OK

Check Slope:  (38-4)100%/(74.5 x 12/3) = 11.4 < 12% OK

Determine Bottom End Stress Using Drape "A" = 38"  (Before Losses except ES Loss

6 x 38 + 84 = 28  (20.27 + e)   e = -9.13"

$f_B$ = (-729/560) (1+(-9.13 x -20.27/223.9)) = -2.378 ksi

Strength at Transfer ($f'_{ci}$)

Check at Hold Down (1/3 Point) Interior Girder After ES Loss
Assume Parabolic Distribution of Dead Load
1/3 Point Ordinate = 3.33 x 6.67/25 = 0.89
Girder $f_{DL}$ = 0.89 x 4730/6186 = 0.680 ksi

$f_B$ = -3.235 + 0.680 = 2.555

$f'_{ci}$ = 2.555/0.6 = 4.260 ksi < 4.8 ksi

Check at Girder End After ES Loss
$f_B$ = 2.378 ksi OK

Specify: (Plan Values)
28 Strands
Minimum Prestress Force at Initial Concrete Set is:
0.7 x 270 x 0.153 x 28 = 810 Kips
Maximum Allowable Temporary Prestress Force is:
0.8 x 270 x 0.153 x 28 = 926 Kips which allows for the inaccuracies and losses associated with stressing the strands.
Minimum Concrete Compressive Strength at time of Transfer is 4800 psi.

Non-prestress Reinforcement Required in Top Flange (Hold down point)

$f_T$ = + 1.058 - 0.89 x 4730/5070 = + 0.228 < $(7.5\sqrt{4800}$ = 0.520) OK

$f_B$ = - 2.555 ksi (See Str. at Transfer Comps.)

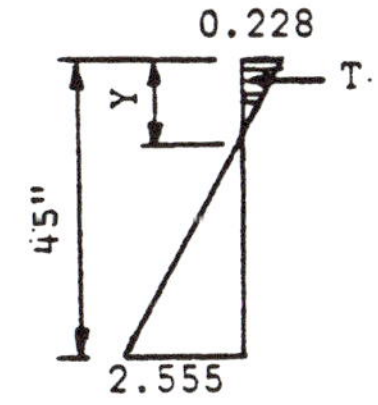


Similar Triangles

Y = (0.228 x 45)/(0.228 + 2.555) = 3.69"

T = 1/2(3.69) x 0.228 x 16 = 6.7 Kips

$A_S$ = 6.7/24 = 0.28 in$^2$

The Std. provides 2 - #5's = .62 in$^2$ OK

## Ultimate Capacity (Positive)

Interior Girder Span 1 & 3 Controls

$M_u$ = 1.3 (13790 + 950) + 2.17 x 8180 = 36,913 in-K

Assume Rectangular Section:

Cap. $M_u = \emptyset A_s f_{su} d (1-0.6pf_{su}/f_c')$ if $1.4\ dpf_{su}/f_c' < 8"$

| | |
|---|---|
| Girder depth | = 45.0" |
| Eff. deck | = 8.0 |
| Girder imbed | = -0.5 |
| c.g. strands | = -3.9 |
| d | = 48.6" |

$A_s$ = 28 x .1531 = 4.3 $in^2$ p = 4.3/(103 x 48.6) = .000859

$f_{su} = f_s'(1-.5pf_s'/f_c')$ = 270 (1 - .5 x .000859 x 270/4.0) = 262 ksi

1.4 x 48.6 (.000859) $\frac{262}{4.0}$ = 3.8 < 8"

Cap. $M_u$ = .9 x 4.3 x 262 x 48.6 (1 - 0.6 x .000859 x 262/4.0) = 47614 in-K OK

Cracking Moment (Minimum Steel Percentage)

Prestress After Losses: $f_B$ = - 2.660

$f_R = 7.5\sqrt{6000}$ = 0.580 ksi

Total stresses req'd to cause cracking = 0.580 + 2.660 = 3.240 ksi

$M_{CR}$ = f x $S_B$ = 3.240 x 10342 = 33508 in-K

Moment Ratio = 55299/33508 = 1.65 > 1.2 OK

## Horizontal Shear Between Slab and Girder (LLM and PDLM)

Int $V_u$ = 2.17 x 60.9 + 1.3 x 8.6 = 143.3 Kips $\bar{V}u = Vu/\emptyset$ = 143.3/.85 = 168.6 Ki

Y = $Y_{slab}$ - ($T_{eff}$/2) = 17.92 - 4. = 13.92"

Q = B x $T_{eff}$ x Y x Mod. Ratio

= 103 x 8 x 13.92 x .75 = 8602 $in^3$

v = (168.6 x 8602)/(352814 x 16) = 0.25 < 0.35 ksi OK

Min. Stirrup Area = (2 - #3 at 12") = 0.22 $in^2$/ft.

## Web Reinforcement

A check on the outer 1/4 point stirrups of Standard 19.2 is for demonstration purposes only.

Int. $V_u$ = 1.3 x 29.5 + 2.17 x 34.4 = 113.0 K $V_u/\emptyset$ = 126. K

d = 45.0 + 7.5 - 3.9 = 48.6" J = 0.88 jd = 42.7"

$V_c$ = 180 x 7 x 42.7 = 53.8 Kips

$A_v$ = Pair #4's = 0.4 $in^2$

S = (0.4 x 2 x 40 x 42.7))126.0 - 53.8) = 18.9 > 12" OK

## Continuity Reinforcement

Find Exterior Continuity Reinforcement

$M_u$ = 1.3 (3215 + 5/3 (7280) = 19953 in-K

d = 45.0 + 7.5 - 2.0 - 0.5 = 50.0 in.

$R_u = M/\emptyset bd^2$ = 19,953,000/(.9)(22)$(50)^2$ = 403 psi

p = 0.0072; $A_s$ = (0.0072)(22)(50) = 7.92 $in^2$

Check Fatigue Requirement (Assume $f_f$ = 23 Ksi)

$A_s$ = (7280)/(23)(0.9)(50) = 7.03 $in^2$

Find Interior Continuity Reinforcement

$M_u = 1.3(1284 + 5/3\ (8660)) = 20441$ in-K

$R_u = 20{,}441{,}000/.9(22)(50)^2 = 413$ psi

$p = 0.0074;\ A_s = (0.0074)(22)(50) = 8.14\ in^2$

Check Fatigue Requirement

$A_s = (8660)/(23)(0.9)(50) = 8.37\ in^2$ (Controls)

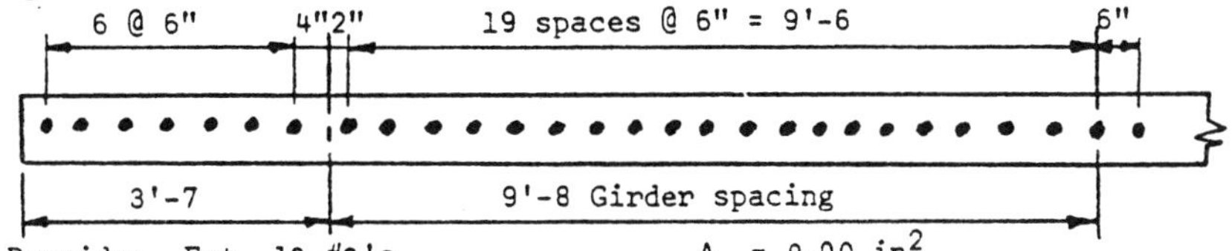


Provide: Ext. 18-#6's $A_s = 8.00\ in^2$

Int. 19-#6's $A_s = 8.40\ in^2$

Check Crack Control

Crack Control allowable stress is compared to fatigue allowable stress since both values are based on live load moment alone. Crack control when computed for an interior beam is solely a function of live loading. For all practical purposes, the future wearing surface acts to seal the cracks and is not expected to cause cracking in the same manner as live loading.

Allowable $f_s = Z/(d_c A_c)^{1/3}$ where Z = 130 and $d_c$ = 3".

At Pier: $A_c = 6 \times 6 = 36in^2$ $f_s = 130/(3 \times 36)^{1/3} = 27.2^{ksi}$ which is greater than the allowable $f_f = 23.0^{ksi}$ o.k.

At First Cutoff: $A_c = 6 \times 12 = 72in^2$ $f_s = 130/(3 \times 72)^{1/3} = 21.7^{ksi} < f_f$ requiring that the bar cut off be extended for crack control.

Ultimate Negative Moment Values (Span 1 & 3) see moment diagram

| Span point | 0.6 | 0.7 | 0.8 | 0.9 | C/L Pier |
|---|---|---|---|---|---|
| 2.17 x LLM (Int.) | -411 | -478 | -553 | -915 | -1565 |
| 1.30 x Post DLM | 0* | 0* | 0* | -63 | -138 |
| | -411 | -478 | -553 | -978 | -1703 |
| 2.17 x LLM (Ext.) | -344 | -403 | -464 | -770 | -1316 |
| 1.30 x Post DLM | +140 | +81 | -2 | -157 | -348 |
| | -204 | -322 | -466 | -927 | -1664 |

* Load Factor = 0.975 since moment is of opposite sign. Do not include positive moment of future wearing surface.

Negative Moment Cutoff Points (Span 1 & 3)
Moment capacity of one-half bar area (7 - #7's = 4.2 $in^2$)
$p = A_s/bd = 4.2/(22)(50) = .00382$ Ru = 220.4

$M = Ru/\emptyset bd^2 = 220.4 \times .9 \times 22 \times 50^2/12 = 909$ ft-K.
Terminate 1/2 the bars at about the .88 span point plus $L_d$.

(Top fiber = modulus of rupture = 580 psi = $7.5\sqrt{f_c'}$)
0.8 Span e = -14.40"
Prest. $f_t = (-810/560)\ (1-(14.40 \times 24.73)/223.9) \times 0.740 = .632$ ksi

Interior DL $f_t = 2.720 \times 16/25 = -1.740$

Prestress + DL $f_t = -1.740 + 0.632 = -1.108$

Negative Stress Req'd for cracking = 1.108 + 0.580 = 1.688 ksi
Negative Moment Req'd for cracking = 1.688 x 5070/12 = 713 ft-kips

0.9 Span e = -12.19"

Prest. $f_t = -1.446\ (1-(12.19 \times 24.73)/223.9) \times 0.740 = 0.371$ ksi

Int. DL $f_t = 2.720 \times 9/25 = -0.980$

$f_{CR} = 0.980 - .371 + .580 = 1.189$ ksi

$M_{CR} = 1.189 \times 5070/12 = 502$ ft-kips

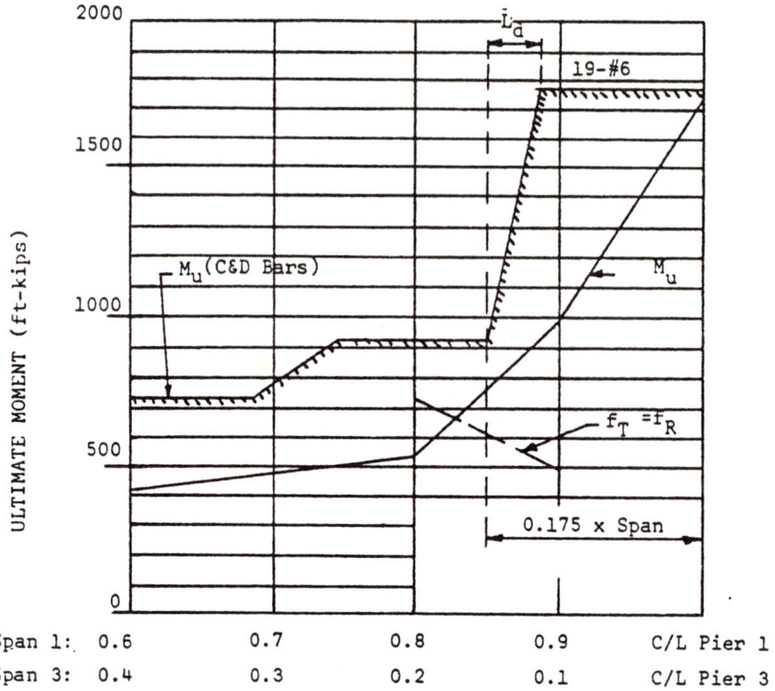


The top fiber stress equals the modulus of rupture at about the .83 point, which is 1/20 of the span length from the first cutoff point. Add $L_d$ to this point and terminate remaining one-half of bars.

Note that for demonstration purposes, only the interior girders of Span 1 & 3 have been considered.
Extension length (See Chapter 9, Table 9.4)

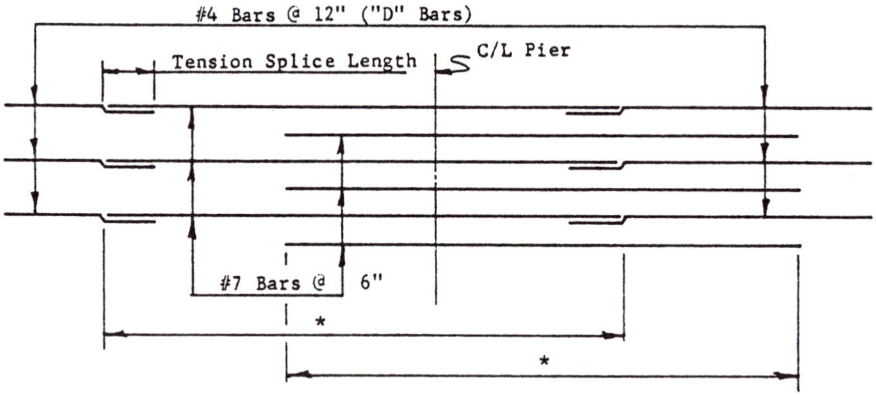


* Make lengths equal when practical.

Precompression

$p = A_s/bd = 8.4/22 \times 50.0 = .0076 < 015$ OK

Camber and Deflection

Prestress Camber (Based on 0.7 $f_s'$ minus ES loss)

Straight strands $P_s = 810 \times ((189 - 15.7)/189) \times 22 = 583$ Kips

Draped strands $P_D = 810 \times ((189 - 15.7)/189) \times 6/28 = 159$

$M_1 = 583\ (20.27 - 3.82) = 9590$ in-K

$M_2 = 159\ (38 - 4) = 5406$

$M_3 = 159\ (38 - 20.27) = 2819$

$L = 74.5'$ or $894"$ $E = 3.5 \times 10^3$ ksi $I = 125390.\ in^4$

Prestress $\Delta = (894)^2 \times (9590 + (23 \times 5406/27) - 2819)/(8 \times 3.5 \times 10^3 \times 125390.) = 2.59"$

DL $\Delta = 5 \times (583/12) \times (894)^4/(384 \times 3.5 \times 10^3 \times 125390.) = 0.92"$

Prestress Camber = 2.59 - 0.92 = 1.67 Say A = 1 5/8"

Dead Load Deflection

$\Delta_{DL} = 5 \times (1.025/12) \times (894)^4/(384 \times 5.5 \times 10^3 \times 125390) + 3.4 \times (894)^3$
$/(48 \times 5.5\ 10^3 \times 125390) = 1.03 + 0.07 = 1.10''$ Say B = 1 1/8"

Residual Camber = 1 5/8 - 1 1/8 = 1/2" Say C = 1/2"

Deck Forming

Span 1 is on a straight grade, therefore VC = 0.

Span 2:

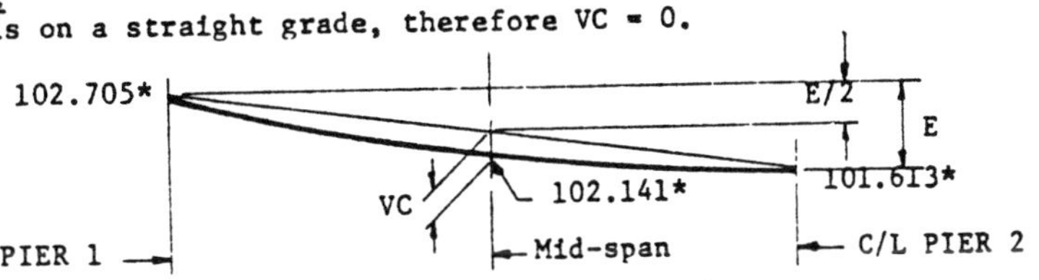


*Elevations along the profile grade line

E/2 = (102.705 - 101.613)/2 = 0.546
101.613
102.159 - 102.141 = 0.018' Say VC = 1/4"

Span 3: VC = 1/4"

Span 1: Try 1/2" imbed at mid-span
End Haunch: $H_{END}$ = 1/2 - 0 - 1/2 = 0"

Span 2: Try 1/2" imbed at mid-span
($H_{LT}$ = 1/4) + $H_{RT}$ = 2(1/2 + 1/4 - 1/2) = 1/2"

Right Haunch: $H_{RT}$ = 1/2 - 1/4 = 1/4"

Span 3: Try 1/2" imbed at mid-span
1/4 + $H_{RT}$ = 2(1/2 + 1/4 - 1/2) = 1/2"
Right Haunch: $H_{RT}$ = 1/4"

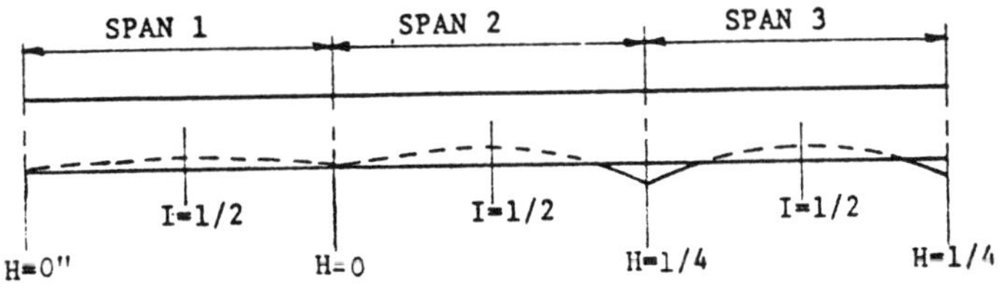


DECK FORMING DIAGRAM

## QUANTITY OF STEEL IN PRESTRESS GIRDERS

Prestress Steel Weight (lbs.)=0.525N(L+1.)

| *CONVENTIONAL REINFORCEMENT WEIGHT, POUNDS/GIRDER | | | |
|---|---|---|---|
| Size | Simple Span (No Continuity) | w/ Continuity at One End w/o Continuity at Other End | Continuity at Both Ends |
| 36 | 93.0 + 6.8 L | 78.0 + 11.7 L | 63.0 + 16.5 L |
| 45 | 131.0 + 7.7 L | 111.0 + 12.0 L | 91.0 + 16.4 L |
| 54 | 152.0 + 8.5 L | 128.0 + 13.5 L | 105.0 + 18.5 L |
| 70 | 360.0 +11.8 L | 323.0 + 13.2 L | 287.0 + 14.6 L |

where N = total number strands per girder

L = span length c/c/ bearings, feet

B. Slab Type Girder Example Problem

This is a 'multi-beam' structure with a 3" poured in place composite concrete floor.

Data from the 'preliminary structure' layout:

Span = 44'    Loading = HS 20    Roadway width = 30'
Use a Type 3 Slab

LL Distribution

Interior Load Fraction = $\frac{S}{D}$    $S = \frac{12N_L + 9}{N_G} = \frac{12(2) + 9}{11} = 3.0$

$D = 5 + \frac{NL}{10} + (3 - \frac{2NL}{7})(1 - \frac{C}{3})^2 = 5.2 + 2.43(1 - \frac{.68}{3})^2 = 6.64$

S/D = 0.452

LL + Impact

Impact = 50/(44 + 125) = 0.296

Int. LLM = 520.9/2 x 0.452 x 1.296 = 152.6 ft-K (AASHTO-App. A)

Ext. LLM = 0.1 x 16 x 1.3 x 44. = 91.5 ft-K

Dead Load

Interior

Floor = 0.27 x 3 x 1 x 0.15 = 0.121 K/ft (Includes 1/4" floor thickening but no future wearing surface is provided for)

Girder = 474/144 x 1 x 0.15 = 0.494
0.615

Int. DLM = 0.692 x $(44.)^2/8$ = 148.8 ft-K

Exterior

Ext. DLM = 148.8 ft-K

Ext. PDLM = 0.338 x $(44)^2/8$ = 81.8 ft-K

Composite Section

Transformed Area = 36 x 2.5 x 0.75 = 67.5 $in^2$

Transf. $I_o$ = 36 x $(2.5)^3$ x 0.75/12 = 35 $in^4$

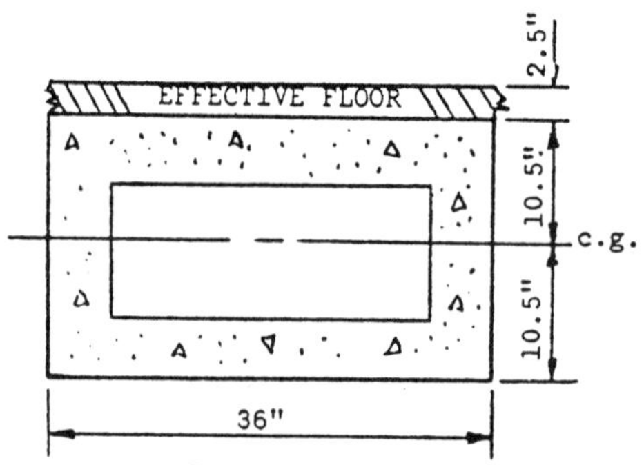

INTERIOR

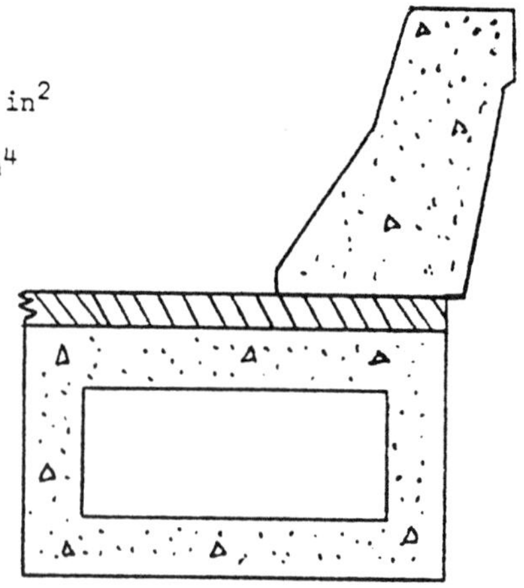
EXTERIOR

SUM MOMENTS ABOUT c.g.

| Item | A | Y | Ay | $Ay^2$ | Io | $Io+Ay^2$ |
|---|---|---|---|---|---|---|
| Girder | 474.0 | 0 | 0 | 0 | 24793 | 24793 |
| Floor | 67.5 | 11.75 | 793 | 9320 | 35 | 9355 |
| Sum | 541.5 | | 793 | | | 34148 |

$-Axy_c^2 = \underline{-1161}$

$y_c = Ay/A = 793/541.5 = 1.46"$ $\quad I_{c.g.c.} = 32987\ in^4$

$y_t = 10.50 - 1.46 = 9.04"$ $\quad S_t = 32987/9.04 = 3642\ in^3$

$y_B = 10.50 + 1.46 = 11.96$ $\quad S_B = 32987/11.96 = 2758$

$y_{FL} = 9.04 + 3. = 12.04$ $\quad S_{FL} = 32987/12.04 = 2740$

The exterior girder must be analyzed for the parapet DL on the composite section above and the live load on a composite section which includes the parapet. The Exterior girder should not have less capacity than the Interior girder. This example does not check the Exterior Girder.

Design Stress

Interior: $f_{DL} = 148.8 \times 12/2361 = 0.756$ ksi $\quad f_{DL\ GIR} = .607$ ksi

$f_{LL} = 152.6 \times 12/2758 = \underline{0.664}$
1.420

Minus tension allowed $\underline{-0.424}$
.996 ksi

Prestress Force

Assume e = -10.5 + 2.0 = -8.5", $r^2 = I/A = 52.3$, Assume losses at 25%
Final Req'd P = .996(474)/(1+(-8.5 x -10.5/52.3)) = 174 Kips
Max. force per strand = 28.93 x 0.75 = 21.7 Kips
N = 174/21.7 = 8.02 Try 8 strands P = 231 Kips

Estimate the Elastic Shortening Concrete Loss as 5%.

$f_B = (231/474)(1+(-8.5 \times -10.5/52.3)) = -1.319$ ksi x .95 = -1.253 ksi
$f_t = 0.487\ (1+(-8.5 \times 10.5/52.3)) = +0.344$ ksi x .95 = .327 ksi

Elastic Shortening Loss

Take Stress at 5/10 Pt. for ES Loss (Prestress + Girder DL, Assume ES Loss)
$f_B = -1.253 + 0.607 = -.646$
$f_t = +0.327 - 0.607 = -.280$

Stress at c.g. of strand (5/10 point)
$f_{cir} = -.646 + ((.646 - .280)(10.5 - 8.5)/21) = -.611$

$ES = 7f_{cir} = 7 \times .611 = \underline{4.3\ ksi}$ $\quad (189 - 4.3)/189 = \underline{.977}$

Creep of Concrete

Stress due to Floor DLM: $f_B = -f_T = 0.149$ ksi

$f_{cds} = .149 - ((.149 + .149)(10.5 - 8.5)/21) = .121$ ksi

$CRC = 12\ f_{cir} - 7\ f_{cds} = 12 \times .611 - 7 \times .121 = \underline{6.5\ ksi}$

Shrinkage = SH = 6 ksi based on average relative humidity of 72%

Creep of Prestressing Steel

CRS = 20. - .4ES - .2(SH + CRC) = 20. - .4 (4.3) - .2(6.+ 6.5) = 15.8 ksi

Service Load (5/10 Pt. with all losses)

Losses = 6. + 4.3 + 6.5 + 15.8 = 32.6
(189 - 44.8)/189 = 0.83

$f_B$ = -1.319 x .83 + 1.420 = .325 < 0.424 ksi OK

$f_t$ = .344 x .83 - .796 - (152.6 x 12/3649 = -1.012 < 2.4 ksi OK

Floor f = (-152.6 x 12/2740) x 0.75 = -501 < 1.4 ksi OK

Use 8 Strands

Strength at Transfer

Ends: $f_t$ = 0.344 x 0.77 = 0.336 ksi

$f_b$ = -1.319 x 0.77 = -1.289

Req'd $f'_{ci}$ = 1.289/0.6 = 2.148 ksi (Based on compression)

Req'd $f'_{ci}$ = $(336/7.5)^2$ = 2.007 (Based on tension)

Both are less than minimum, therefore specify $f'_{ci}$ = 4000 psi

Non-prestressed Reinforcement

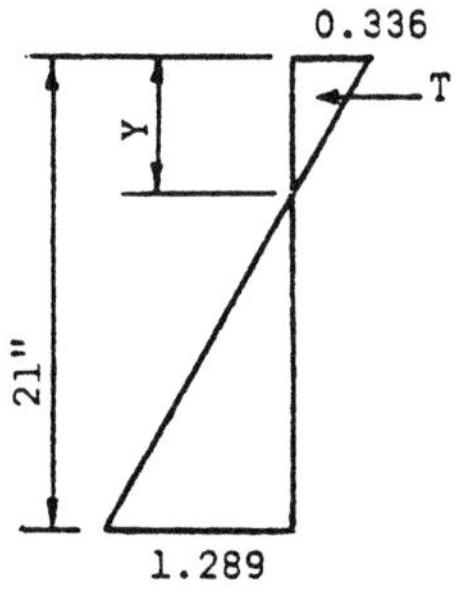


Y = 0.336 x 21/(1.289 + 0.336) = 4.34"
T = 0.336 x 0.5 x 36 x 4.34 = 26.3 Kips
$A_s$= 26.3/24 = 1.10
Use 4 - #5 Bars ($A_s$ = 1.24 sq. in.)

Ultimate Capacity

Int. $M_u$ = 1.3 (148.8 + 5/3 (152.6)) = 524.1 ft.-k

d = 21.0 + 2.5 - 2. = 21.5" p = (8 x 0.153)/(21.5 x 36) = 0.00158

$f^*_{su}$ = 270 (1. - 0.5(0.00158 x 270/5)) = 258 ksi

Cap $M_u$ = (.9 x 8 x 0.153 x 258 x 21.5 (1 - 0.6(0.00158 x 258/5)))/12 = 484.3

This section should be redesigned for 9 strands.

Cracking Moment (Minimum Steel Percentage)

Prestress After Losses: $f_B$ = -1.095

$f_R = 7.5\sqrt{5000}$ = 0.530 ksi

$f_{CR}$ = 0.530 + 1.095 = 1.625 ksi

$M_{CR}$ = 1.625 x 2758 = 4482 in-K

Ratio = (484.3 x 12)/4482 = 1.30 > 1.2 OK

Camber and Deflection

Overall length = 45' or 540" (Compute for 9 strands)
$M_1$ = 261 x 0.97 x 8.5" = 2152 in-K

Prestress $\Delta$ = $(540)^2$ x 2152/(8 x 3.5 x $10^3$ x 25747) = 0.87" ↑

Girder $\Delta_{DL}$ = 5 x (0.494/12) x $(540)^4$/(384 x 3.5 x $10^3$ x 25747) = 0.46" ↓

Floor $\Delta_{DL}$ = 5 x (0.121/12) x $(540)^4$/(384 x 5.5 x $10^3$ x 25747) = 0.08" ↓

Prestress Camber = 0.87 - 0.46 = 0.41" A = 3/8"
Dead Load Defl. = 0.08" B = 1/8"
Residual Camber = 3/8 - 1/8 C = 1/4"

Floor Forming

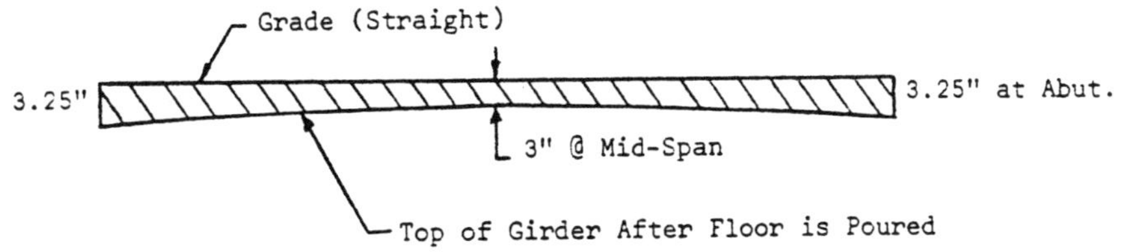


FLOOR THICKNESS DIAGRAM

(6) Field Adjustments of Prestress Force

When strands are tensioned in open or unheated areas during cold weather they are subject to loss due to change in temperature. This loss can be compensated for by noting the change in temperature of the strands between tensioning and initial set of the concrete. For purposes of uniformity the strand temperature at initial concrete set is taken as 80 degrees F.

Minor changes in temperature have negligible effects on the prestress force, therefore only at strand temperatures of 50 degrees F. and lower are increases in the tensioning force made.

Since "plan" prestress forces are based on 70% of the ultimate it is necessary to utilize the AASHTO allowable of temporarily overstressing up to 80% to provide for the losses associated with fabrication.

The following example outlines these losses and shows the elongation computations which are used in conjunction with jack pressure gages to control the tensioning of the strands.

Computation For Field Adjustment of Prestress Force

KNOWN:

22 - 1/2", 7 wire strands $A = 0.1531 \text{ in}^2$

$P_i$ = 629.2 Kips (Initial prestress force from plan)

$T_1$ = 40°F. (Air temperature at strand tensioning)

$T_2$ = 80°F. (Concrete temperature of initial set)

L = 300' = 3600" (Distance from anchorage to reference point)

$L_1$ = 240' = 2880" (Length of cast segment)

E = 29000 KSi (Modulus of elasticity, sample tested from each spool)

C = .0000065 (Coef. of thermal expansion, per degree F.)

COMPUTE:

Force per strand = P = 629.2/22 = 28.6 Kips

$$DL_1 = \frac{PL}{AE} = \frac{28.6 \times 3600}{0.1531 \times 29000} = 23.19"$$

Initial Load of 1.5 Kips to set the strands

$$DL_2 = \frac{1.5 \times 3600}{0.1531 \times 29000} = 1.22"$$

$DL_3$ = Slippage in Strand Anchors = 0.45" (Past Experience)

$DL_4$ = Movement in Anchoring Abutments = 0.25" (Same)

$DL_5 = C \times L_1 \times (T_2 - T_1) = 0.0000065 \times 2880 \times 40 = 0.75"$

$P_{Loss} = DL_5 \times A \times E/L = 0.749 \times 0.1531 \times 29000/36000 = 0.9$ Kips

Total Prestress Force = $P + P_{Loss} = 28.6 + 0.9 = 29.5$ Kips

Total Elongation = $DL_1 + DL_3 + DL_4 + DL_5 = 24.64"$

Elongation After Initial Load = 24.64 - 1.22 = 23.42"

(7) Prestress Girder Strand Patterns

The basic criteria used in establishing the 'standard strand patterns' is arranging the strands so that the c.g. of the patterns is as close to the bottom of the girder as possible.

Although this approach yields the most efficient capacity, it occasionally is modified to meet some of the stress conditions along the girder. These conditions are:

1) Maximum initial compressive stress in bottom fiber at transfer is $0.6\ f'_{ci}$.

2) Maximum initial tensile stress in top fiber at transfer is $7.5\sqrt{f'_{ci}}$.

   In order to compute these values the dead load stresses of the girder are combined with the prestressing stresses. Average span lengths are assumed for each strand pattern with the length increasing as the number of strands increase. Because of this assumption, certain strand patterns cause too high stresses when used on short girders with a wide spacing.

3) The number of draped strands, which lie in the center two columns of strands, must be adequate to fulfill either:

   A) Zero tension in the top fiber.
   B) $0.4\ f'_c$ compression in the bottom fiber.

   Also the amount of drape is controlled by the girder size and the 4" clearance from center of strand to top of girder.

4) The minimum and maximum amount of reinforcement together with strand clearance and cover requirements of AASHTO are also used.

TABLE 19.1 SPAN LENGTH VS. GIRDER SPACING FOR INTERIOR GIRDERS

| 36" Girder | | |
|---|---|---|
| Span Length | HS20 Multiple Span ▲ | H20 Multiple Span ▲ |
| 40' | 11'- 9"** | 13'- 3"** |
| 45 | 10 - 8 ** | 12 - 0 ** |
| 50 | 10 - 3 ** | 11 - 6 ** |
| 55 | 9 - 2 * | 10 - 8 * |
| 60 | 8 - 0 * | 9 - 2 * |
| 65 | 7 - 3 * | 8 - 0 * |

| 45" Girder | | |
|---|---|---|
| Span Length | HS20 Multiple Span ▲ | H20 Multiple Span ▲ |
| 65' | 12'- 2"* | 13'- 8"* |
| 70 | 10 - 10 * | 12 - 4 * |
| 75 | 10 - 0 * | 10 - 10 * |
| 80 | 9 - 0 * | 9 - 4 * |
| 85 | 7 - 10 | 8 - 2 |
| 90 | 6 - 4 | 7 - 2 |

| 54" Girder | | |
|---|---|---|
| Span Length | HS20 Multiple Span ▲ | H20 Multiple Span ▲ |
| 85' | 12'- 6"* | 13'- 3"* |
| 90 | 11 - 0 * | 11 - 6 * |
| 95 | 9 - 10 | 10 - 3 |
| 100 | 8 - 8 | 9 - 0 |
| 105 | 7 - 9 | 7 - 10 |

| 70" Girder | | |
|---|---|---|
| Span Length | HS20 Multiple Span ▲ | H20 Multiple Span ▲ |
| 105' | 11'- 5"* | 12'- 3"* |
| 110 | 10 - 7 * | 11 - 3 |
| 115 | 9 - 11 | 10 - 1 |
| 120 | 8 - 11 | 9 - 1 |
| 125 | 8 - 0 | 8 - 3 |
| 130 | 7 - 0 | 7 - 2 |

* Controlled by high initial compressive stress in bottom midspan region.
** Controlled by high initial tensile stress in top midspan region.
▲ Estimate the Girder Spacing for Single Spans as .86 to .90 times the Multiple Span Girder Spacing.

Notes:
1) This table includes 20 psf future wearing surface.
2) This table is based on using stress relieved strands.

# COMPOSITE SECTION PROPERTY

## Table 19.2
## 36" PRESTRESSED GIRDER

| T | B | INERT | SB | ST |
|---|---|---|---|---|
| 6.5 | 48.0 | 127026. | 5138. | 11266. |
| 6.5 | 54.0 | 132617. | 5227. | 12480. |
| 6.5 | 60.0 | 137723. | 5304. | 13724. |
| 6.5 | 66.0 | 142405. | 5373. | 15000. |
| 6.5 | 72.0 | 146715. | 5433. | 16308. |
| 6.5 | 78.0 | 150696. | 5487. | 17650. |
| 6.5 | 84.0 | 154386. | 5536. | 19028. |
| 7.0 | 48.0 | 132395. | 5247. | 12296. |
| 7.0 | 54.0 | 138207. | 5337. | 13680. |
| 7.0 | 60.0 | 143494. | 5415. | 15105. |
| 7.0 | 66.0 | 148328. | 5483. | 16574. |
| 7.0 | 72.0 | 152765. | 5544. | 18088. |
| 7.0 | 78.0 | 156853. | 5598. | 19650. |
| 7.0 | 84.0 | 160633. | 5647. | 21261. |
| 7.0 | 90.0 | 164139. | 5691. | 22924. |
| 7.5 | 48.0 | 137764. | 5355. | 13409. |
| 7.5 | 54.0 | 143785. | 5446. | 14984. |
| 7.5 | 60.0 | 149247. | 5524. | 16616. |
| 7.5 | 66.0 | 154224. | 5593. | 18307. |
| 7.5 | 72.0 | 158782. | 5654. | 20060. |
| 7.5 | 78.0 | 162972. | 5708. | 21879. |
| 7.5 | 84.0 | 166838. | 5757. | 23767. |
| 7.5 | 90.0 | 170419. | 5801. | 25728. |
| 7.5 | 96.0 | 173745. | 5842. | 27768. |
| 8.0 | 48.0 | 143140. | 5462. | 14615. |
| 8.0 | 54.0 | 149364. | 5553. | 16407. |
| 8.0 | 60.0 | 154992. | 5632. | 18275. |
| 8.0 | 66.0 | 160109. | 5701. | 20223. |
| 8.0 | 72.0 | 164783. | 5762. | 22255. |
| 8.0 | 78.0 | 169071. | 5817. | 24379. |
| 8.0 | 84.0 | 173022. | 5866. | 26599. |
| 8.0 | 90.0 | 176674. | 5911. | 28922. |
| 8.0 | 96.0 | 180064. | 5951. | 31356. |
| 8.0 | 102.0 | 183218. | 5988. | 33908. |
| 8.5 | 48.0 | 148533. | 5569. | 15926. |
| 8.5 | 54.0 | 154953. | 5660. | 17966. |
| 8.5 | 60.0 | 160743. | 5740. | 20105. |
| 8.5 | 66.0 | 165994. | 5809. | 22352. |
| 8.5 | 72.0 | 170782. | 5871. | 24715. |
| 8.5 | 78.0 | 175167. | 5926. | 27203. |
| 8.5 | 84.0 | 179200. | 5975. | 29825. |
| 8.5 | 90.0 | 182925. | 6020. | 32593. |
| 8.5 | 96.0 | 186378. | 6061. | 35518. |
| 8.5 | 102.0 | 189588. | 6098. | 38614. |
| 8.5 | 108.0 | 192583. | 6133. | 41897. |
| 9.0 | 48.0 | 153950. | 5674. | 17357. |
| 9.0 | 54.0 | 160561. | 5767. | 19679. |
| 9.0 | 60.0 | 166509. | 5847. | 22134. |
| 9.0 | 66.0 | 171892. | 5917. | 24733. |
| 9.0 | 72.0 | 176792. | 5979. | 27489. |
| 9.0 | 78.0 | 181273. | 6034. | 30417. |
| 9.0 | 84.0 | 185389. | 6084. | 33533. |
| 9.0 | 90.0 | 189187. | 6129. | 36854. |
| 9.0 | 96.0 | 192705. | 6170. | 40402. |
| 9.0 | 102.0 | 195973. | 6208. | 44200. |
| 9.0 | 108.0 | 199021. | 6243. | 48274. |
| 9.0 | 114.0 | 201872. | 6276. | 52656. |

## Table 19.3
## 45" PRESTRESSED GIRDER

| T | B | INERT | SB | ST |
|---|---|---|---|---|
| 6.5 | 49.0 | 252679. | 8870. | 15302. |
| 6.5 | 55.0 | 263253. | 9026. | 16627. |
| 6.5 | 61.0 | 273112. | 9164. | 17969. |
| 6.5 | 67.0 | 282327. | 9289. | 19329. |
| 6.5 | 73.0 | 290960. | 9402. | 20706. |
| 6.5 | 79.0 | 299065. | 9504. | 22100. |
| 6.5 | 85.0 | 306690. | 9597. | 23513. |
| 7.0 | 49.0 | 261987. | 9034. | 16373. |
| 7.0 | 55.0 | 273066. | 9193. | 17852. |
| 7.0 | 61.0 | 283362. | 9334. | 19354. |
| 7.0 | 67.0 | 292956. | 9460. | 20878. |
| 7.0 | 73.0 | 301917. | 9573. | 22426. |
| 7.0 | 79.0 | 310307. | 9676. | 23997. |
| 7.0 | 85.0 | 318180. | 9770. | 25593. |
| 7.0 | 91.0 | 325583. | 9855. | 27214. |
| 7.5 | 49.0 | 271289. | 9197. | 17501. |
| 7.5 | 55.0 | 282853. | 9358. | 19147. |
| 7.5 | 61.0 | 293565. | 9500. | 20821. |
| 7.5 | 67.0 | 303516. | 9628. | 22525. |
| 7.5 | 73.0 | 312786. | 9742. | 24260. |
| 7.5 | 79.0 | 321444. | 9845. | 26027. |
| 7.5 | 85.0 | 329549. | 9939. | 27826. |
| 7.5 | 91.0 | 337154. | 10025. | 29658. |
| 8.0 | 49.0 | 280596. | 9357. | 18692. |
| 8.0 | 55.0 | 292625. | 9520. | 20517. |
| 8.0 | 61.0 | 303734. | 9664. | 22379. |
| 8.0 | 67.0 | 314026. | 9793. | 24281. |
| 8.0 | 73.0 | 323589. | 9908. | 26222. |
| 8.0 | 79.0 | 332500. | 10012. | 28204. |
| 8.0 | 85.0 | 340824. | 10106. | 30229. |
| 8.0 | 91.0 | 348619. | 10192. | 32298. |
| 8.0 | 97.0 | 355934. | 10270. | 34413. |
| 8.0 | 103.0 | 362814. | 10342. | 36574. |
| 8.5 | 49.0 | 289916. | 9515. | 19950. |
| 8.5 | 55.0 | 302394. | 9681. | 21970. |
| 8.5 | 61.0 | 313884. | 9827. | 24037. |
| 8.5 | 67.0 | 324502. | 9956. | 26155. |
| 8.5 | 73.0 | 334344. | 10072. | 28324. |
| 8.5 | 79.0 | 343496. | 10176. | 30547. |
| 8.5 | 85.0 | 352028. | 10270. | 32825. |
| 8.5 | 91.0 | 360003. | 10356. | 35161. |
| 8.5 | 97.0 | 367475. | 10435. | 37557. |
| 8.5 | 103.0 | 374492. | 10507. | 40014. |
| 8.5 | 109.0 | 381095. | 10574. | 42535. |
| 9.0 | 49.0 | 299257. | 9673. | 21281. |
| 9.0 | 55.0 | 312168. | 9840. | 23513. |
| 9.0 | 61.0 | 324026. | 9987. | 25805. |
| 9.0 | 67.0 | 334957. | 10118. | 28162. |
| 9.0 | 73.0 | 345068. | 10234. | 30584. |
| 9.0 | 79.0 | 354450. | 10339. | 33075. |
| 9.0 | 85.0 | 363181. | 10433. | 35638. |
| 9.0 | 91.0 | 371329. | 10520. | 38276. |
| 9.0 | 97.0 | 378952. | 10598. | 40992. |
| 9.0 | 103.0 | 386101. | 10671. | 43789. |
| 9.0 | 109.0 | 392820. | 10738. | 46672. |
| 9.0 | 115.0 | 399148. | 10800. | 49644. |

## Table 19.4
## 54" Prestressed Girder

| T | B | INERT | SB | ST |
|---|---|---|---|---|
| 6.5 | 50.0 | 452516. | 14015. | 20841. |
| 6.5 | 56.0 | 469640. | 14248. | 22323. |
| 6.5 | 62.0 | 485853. | 14460. | 23815. |
| 6.5 | 68.0 | 501224. | 14654. | 25318. |
| 6.5 | 74.0 | 515818. | 14832. | 26833. |
| 6.5 | 80.0 | 529693. | 14996. | 28359. |
| 6.5 | 86.0 | 542902. | 15148. | 29896. |
| 7.0 | 50.0 | 466916. | 14241. | 22010. |
| 7.0 | 56.0 | 484976. | 14481. | 23646. |
| 7.0 | 62.0 | 502023. | 14698. | 25297. |
| 7.0 | 68.0 | 518140. | 14896. | 26962. |
| 7.0 | 74.0 | 533401. | 15078. | 28642. |
| 7.0 | 80.0 | 547874. | 15244. | 30337. |
| 7.0 | 86.0 | 561618. | 15397. | 32047. |
| 7.0 | 92.0 | 574688. | 15539. | 33772. |
| 7.5 | 50.0 | 481321. | 14465. | 23225. |
| 7.5 | 56.0 | 500285. | 14710. | 25025. |
| 7.5 | 62.0 | 518132. | 14933. | 26844. |
| 7.5 | 68.0 | 534960. | 15134. | 28681. |
| 7.5 | 74.0 | 550855. | 15318. | 30537. |
| 7.5 | 80.0 | 565893. | 15486. | 32413. |
| 7.5 | 86.0 | 580141. | 15641. | 34308. |
| 7.5 | 92.0 | 593661. | 15784. | 36223. |
| 7.5 | 98.0 | 606509. | 15917. | 38159. |
| 8.0 | 50.0 | 495742. | 14685. | 24490. |
| 8.0 | 56.0 | 515578. | 14937. | 26463. |
| 8.0 | 62.0 | 534196. | 15163. | 28460. |
| 8.0 | 68.0 | 551705. | 15368. | 30480. |
| 8.0 | 74.0 | 568203. | 15554. | 32525. |
| 8.0 | 80.0 | 583776. | 15724. | 34595. |
| 8.0 | 86.0 | 598501. | 15881. | 36689. |
| 8.0 | 92.0 | 612445. | 16025. | 38810. |
| 8.0 | 98.0 | 625670. | 16157. | 40956. |
| 8.0 | 104.0 | 638231. | 16281. | 43129. |
| 8.5 | 50.0 | 510186. | 14904. | 25808. |
| 8.5 | 56.0 | 530867. | 15161. | 27964. |
| 8.5 | 62.0 | 550227. | 15391. | 30150. |
| 8.5 | 68.0 | 568390. | 15598. | 32367. |
| 8.5 | 74.0 | 585465. | 15787. | 34613. |
| 8.5 | 80.0 | 601548. | 15959. | 36891. |
| 8.5 | 86.0 | 616724. | 16116. | 39201. |
| 8.5 | 92.0 | 631069. | 16261. | 41543. |
| 8.5 | 98.0 | 644650. | 16394. | 43919. |
| 8.5 | 104.0 | 657528. | 16518. | 46329. |
| 8.5 | 110.0 | 669756. | 16633. | 48773. |
| 9.0 | 50.0 | 524661. | 15121. | 27181. |
| 9.0 | 56.0 | 546162. | 15382. | 29533. |
| 9.0 | 62.0 | 566240. | 15616. | 31920. |
| 9.0 | 68.0 | 585032. | 15826. | 34346. |
| 9.0 | 74.0 | 602666. | 16016. | 36809. |
| 9.0 | 80.0 | 619230. | 16190. | 39312. |
| 9.0 | 86.0 | 634837. | 16348. | 41855. |
| 9.0 | 92.0 | 649562. | 16493. | 44439. |
| 9.0 | 98.0 | 663481. | 16627. | 47065. |
| 9.0 | 104.0 | 676658. | 16751. | 49734. |
| 9.0 | 110.0 | 689151. | 16866. | 52447. |
| 9.0 | 116.0 | 701017. | 16973. | 55205. |

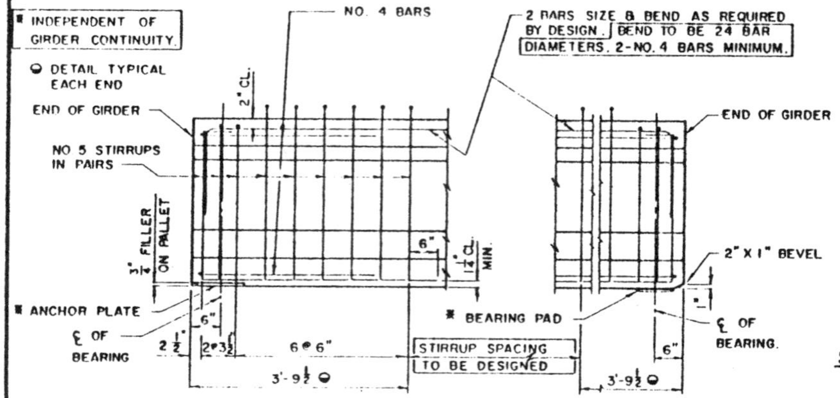


SUPPORT WITHOUT GIRDER CONTINUITY — SIDE VIEW OF GIRDER — SUPPORT WITH GIRDER CONTINUITY

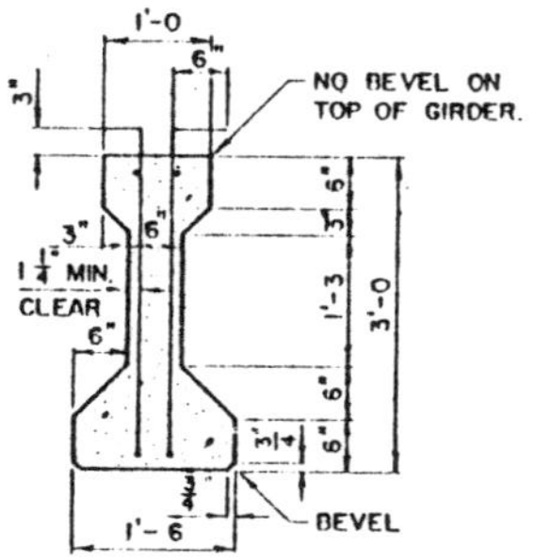


## SECTION THRU GIRDER

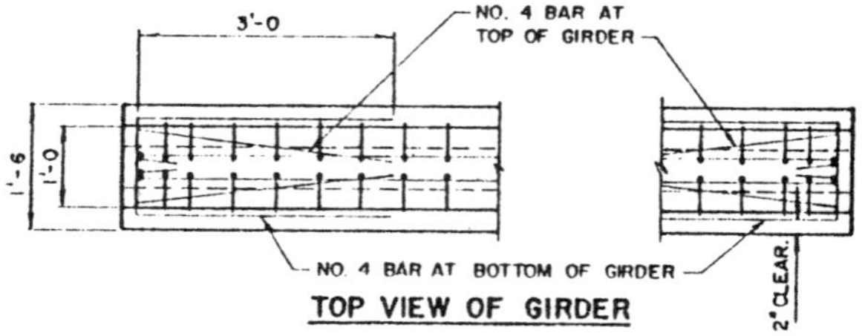


## TOP VIEW OF GIRDER

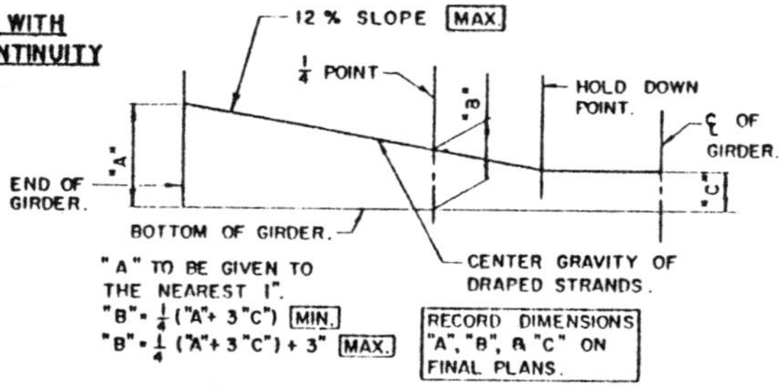


## LOCATION OF DRAPED STRANDS

## GENERAL NOTES

THESE NOTES APPLY TO ALL GIRDERS.

ALL STIRRUPS SHALL BE IN PAIRS.

THE GIRDERS SHALL BE PROVIDED WITH A SUITABLE LIFTING DEVICE FOR HANDLING AND ERECTING THE GIRDERS.

ALL GIRDERS TO BE CAST FULL LENGTH AS SHOWN.

STRANDS SHALL BE FLUSH WITH END OF GIRDER.

END OF STRANDS SHALL BE PAINTED WITH NON-STAINING GRAY NON-BITUMINOUS JOINT SEALER. APPLIES ONLY TO THOSE ENDS OF GIRDERS THAT ARE FINALLY EXPOSED.

TOPS OF GIRDERS TO BE ROUGH FLOATED AND BROOMED TRANSVERSELY FOR BONDING TO SLAB, EXCEPT THE OUTSIDE 2" OF GIRDER, WHICH SHALL BE TROWEL FINISHED.

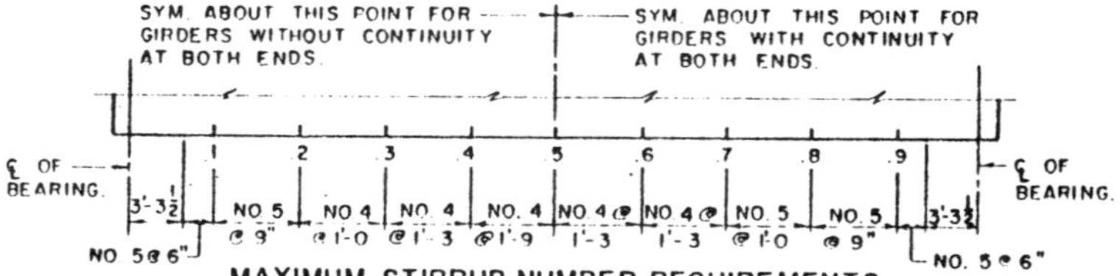


## MAXIMUM STIRRUP NUMBER REQUIREMENTS

VALUES SHOWN ARE FOR STIRRUPS IN PAIRS FOR A 40'-0 SPAN AND 11'-9 GIRDER SPACING. DESIGN STIRRUPS FOR ALL OTHER CASES. USE NO. 4 BARS AT 1'-9 AS MINIMUM STIRRUP AREA.

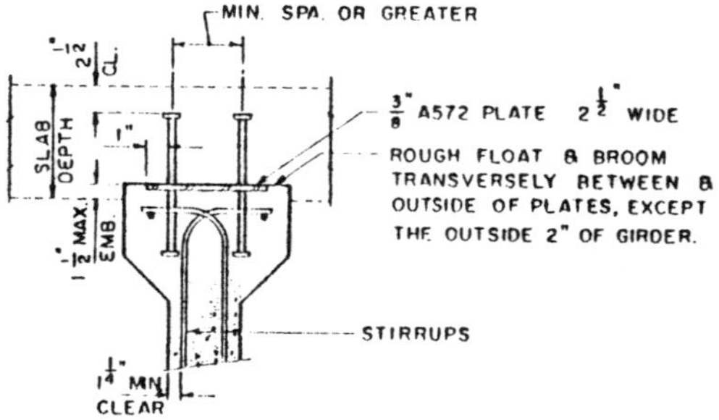


## SHEAR STUD ALTERNATE

ANCHOR SIZE AND SPACING TO BE DETERMINED BY DESIGN. DESIGN SHALL BE BASED ON SHEAR FRICTION CONCEPT USING THE EFFECTIVE COEFFICIENT OF FRICTION AND A SHEAR FORCE BASED ON THE YIELD STRESS OF STRANDS & CONTINUITY STEEL OR THE COMPRESSIVE STRENGTH OF THE CONC. SLAB, WHICHEVER IS OF LESS STRENGTH.

| 36" PRETENSIONED GIRDER DETAILS | |
|---|---|
| STATE OF WISCONSIN DEPARTMENT OF TRANSPORTATION DIVISION OF HIGHWAYS BRIDGE SECTION | |
| Approved. [signature] Chief Bridge Engineer | Date 1-'84 |

STANDARD 19.1

FOR GENERAL NOTES SEE STANDARD 19.1.

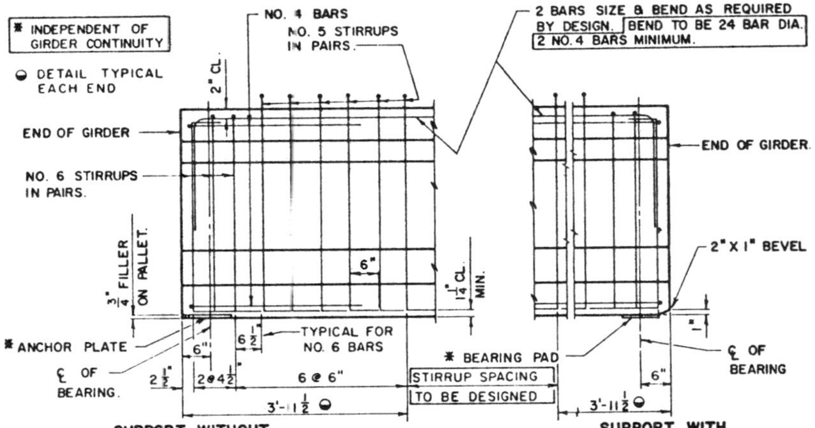


SUPPORT WITHOUT GIRDER CONTINUITY — SIDE VIEW OF GIRDER — SUPPORT WITH GIRDER CONTINUITY

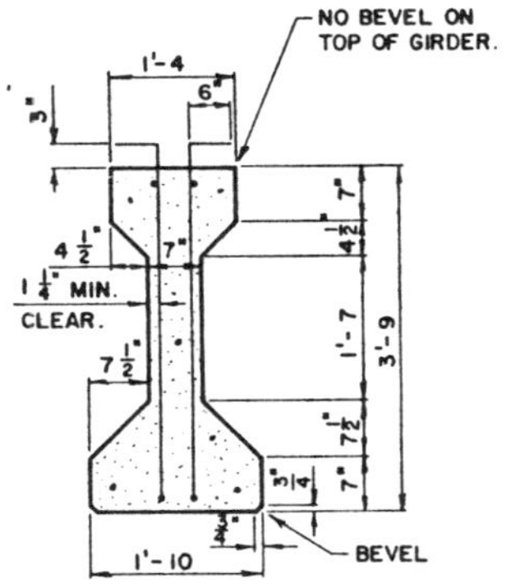


SECTION THRU GIRDER

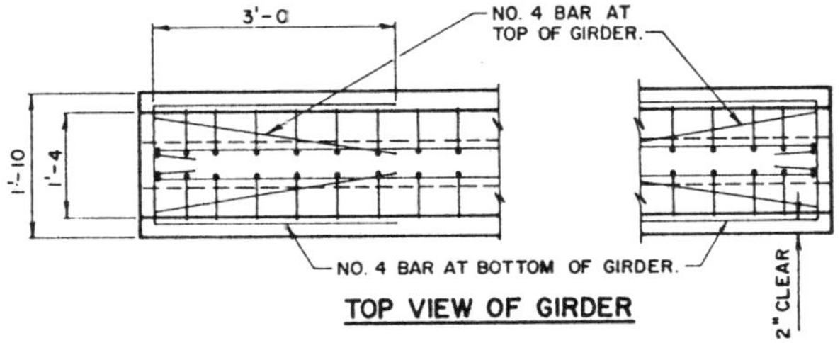


TOP VIEW OF GIRDER

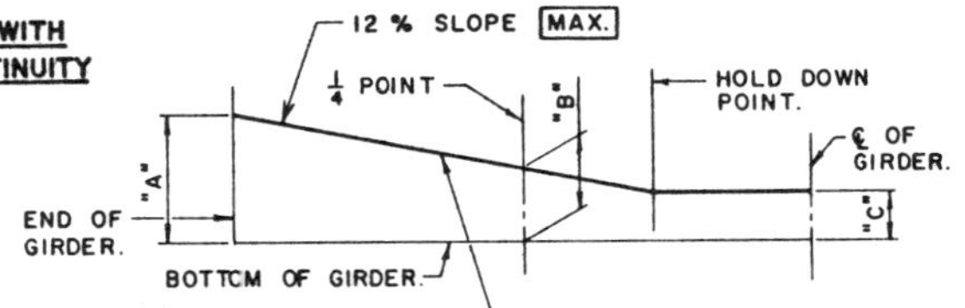


"A" TO BE GIVEN TO THE NEAREST 1".

$"B" = \frac{1}{4}("A" + 3"C")$ MIN.

$"B" = \frac{1}{4}("A" + 3"C") + 3"$ MAX.

RECORD DIMENSIONS "A", "B", & "C" ON FINAL PLANS.

LOCATION OF DRAPED STRANDS

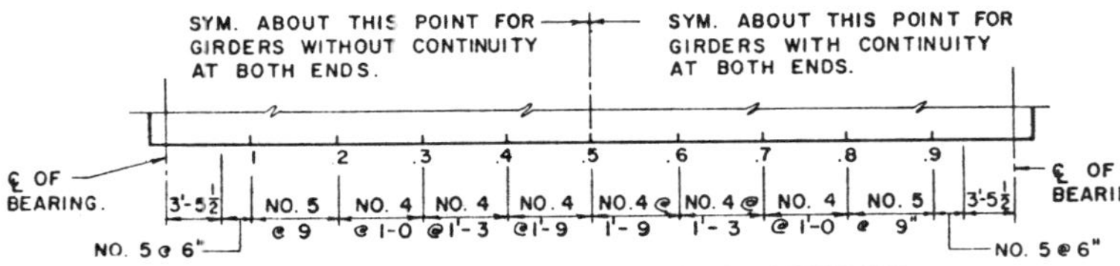


MAXIMUM STIRRUP NUMBER REQUIREMENTS

VALUES SHOWN ARE FOR STIRRUPS IN PAIRS FOR A 65'-0 SPAN AND 12'-0 GIRDER SPACING. DESIGN STIRRUPS FOR OTHER CASES, USE NO. 4 BARS AT 1'-9 AS MINIMUM STIRRUP AREA.

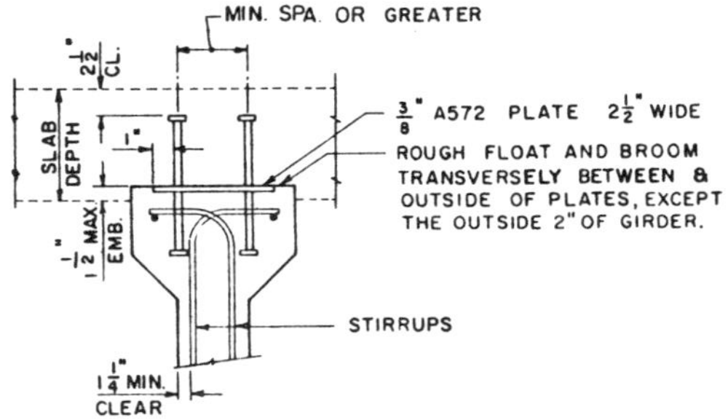


SHEAR STUD ALTERNATE

ANCHOR STUD AND SPACING TO BE DETERMINED BY DESIGN. DESIGN SHALL BE BASED ON SHEAR FRICTION CONCEPT USING THE EFFECTIVE COEFFICIENT OF FRICTION AND A SHEAR FORCE BASED ON THE YIELD STRESS OF STRANDS AND CONTINUITY STEEL OR THE COMPRESSIVE STRENGTH OF THE CONCRETE SLAB, WHICHEVER IS OF LESS STRENGTH.

**45" PRETENSIONED GIRDER DETAILS**

*STATE OF WISCONSIN*
*DEPARTMENT OF TRANSPORTATION*
*DIVISION OF HIGHWAYS*
*BRIDGE SECTION*

Approved: [signature] Chief Bridge Engineer — Date: 1-'84

STANDARD 19.2

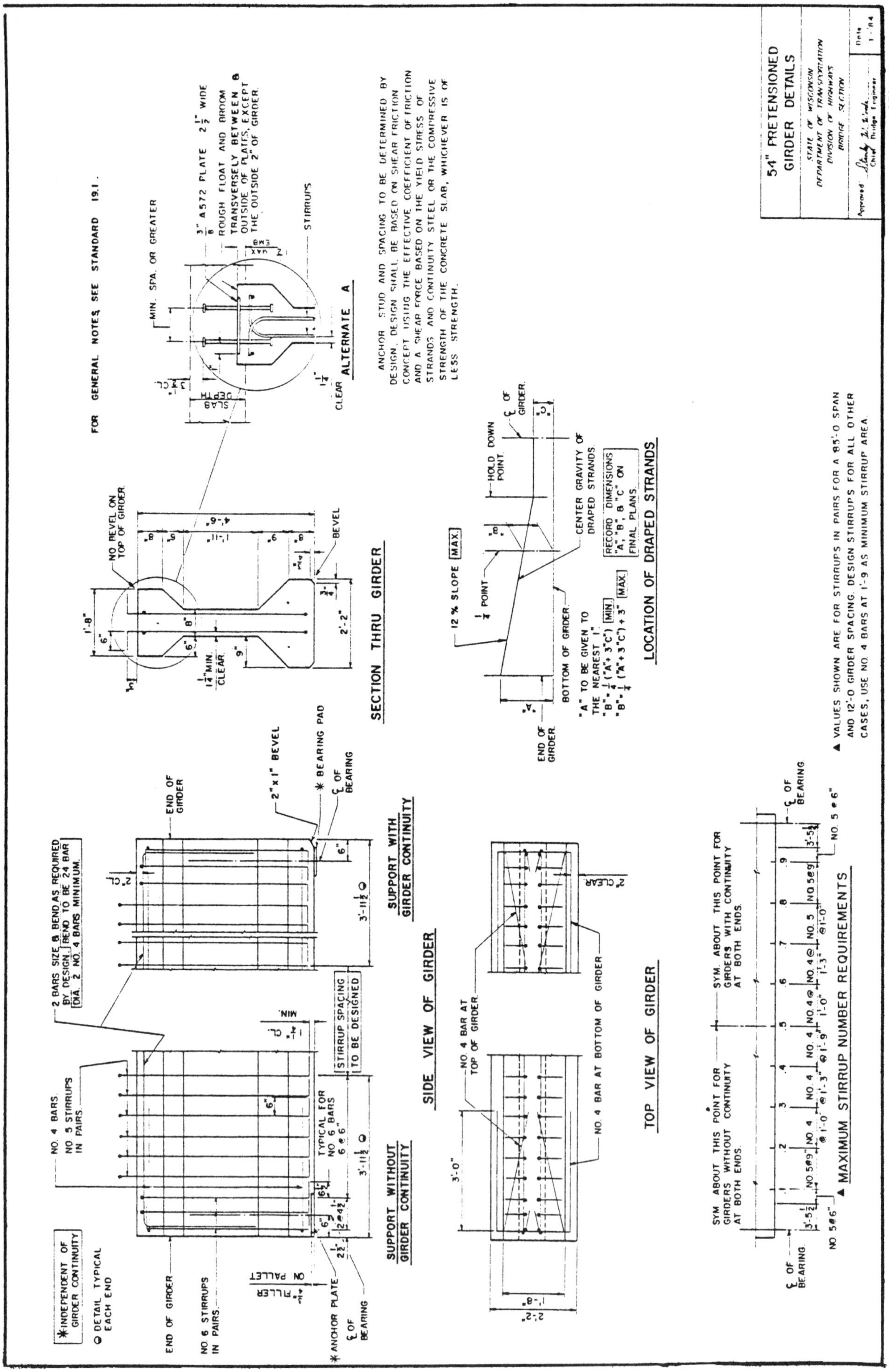
54" PRETENSIONED GIRDER DETAILS
STATE OF WISCONSIN
DEPARTMENT OF TRANSPORTATION
DIVISION OF HIGHWAYS
BRIDGE SECTION
STANDARD 19.3
FOR GENERAL NOTES SEE STANDARD 19.1.
SIDE VIEW OF GIRDER
SUPPORT WITHOUT GIRDER CONTINUITY
SUPPORT WITH GIRDER CONTINUITY
TOP VIEW OF GIRDER
SECTION THRU GIRDER
ALTERNATE A
LOCATION OF DRAPED STRANDS
MAXIMUM STIRRUP NUMBER REQUIREMENTS
ANCHOR STUD AND SPACING TO BE DETERMINED BY DESIGN. DESIGN SHALL BE BASED ON SHEAR FRICTION CONCEPT USING THE EFFECTIVE COEFFICIENT OF FRICTION AND A SHEAR FORCE BASED ON THE YIELD STRESS OF STRANDS AND CONTINUITY STEEL OR THE COMPRESSIVE STRENGTH OF THE CONCRETE SLAB, WHICHEVER IS OF LESS STRENGTH.
VALUES SHOWN ARE FOR STIRRUPS IN PAIRS FOR A 85'-0 SPAN AND 12'-0 GIRDER SPACING. DESIGN STIRRUPS FOR ALL OTHER CASES, USE NO. 4 BARS AT 1'-9 AS MINIMUM STIRRUP AREA.
"A" TO BE GIVEN TO THE NEAREST 1".
RECORD DIMENSIONS "A", "B", & "C" ON FINAL PLANS.
INDEPENDENT OF GIRDER CONTINUITY
DETAIL TYPICAL EACH END
STIRRUP SPACING TO BE DESIGNED
2 BARS SIZE & BEND AS REQUIRED BY DESIGN. BEND TO BE 24 BAR DIA. 2 NO. 4 BARS MINIMUM.

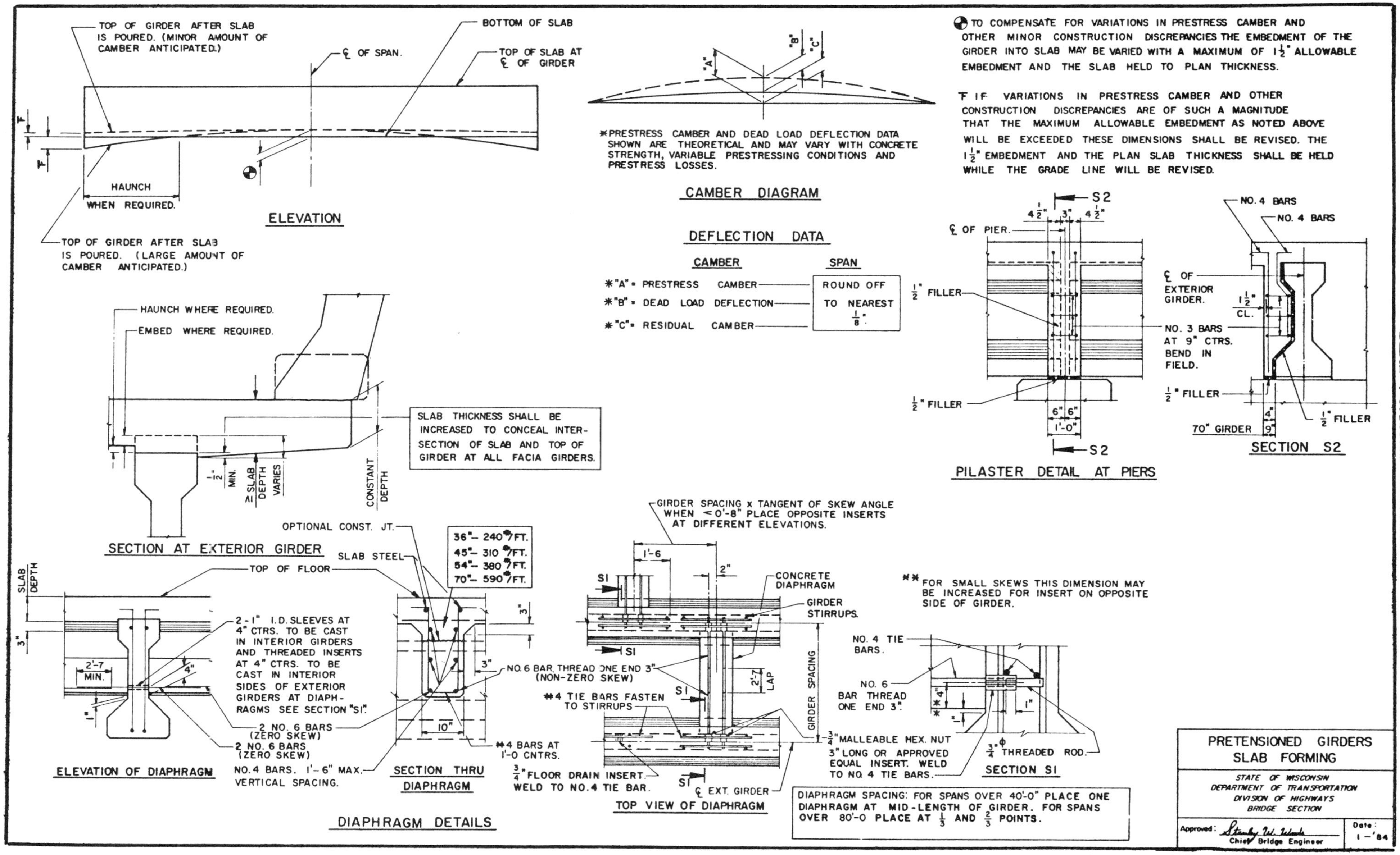


STANDARD 19.6

## 36" PRETENSIONED GIRDER DESIGN DATA

STATE OF WISCONSIN
DEPARTMENT OF TRANSPORTATION
DIVISION OF HIGHWAYS
BRIDGE SECTION

**36" BEAM**

$A = 369$ sq. in.
$r^2 = 138.15$
$y_T = 20.17$ in.
$y_B = 15.83$ in.

$I = 50,979 \text{ in.}^4$
$S_T = 2,527 \text{ in.}^3$
$S_B = 3,220 \text{ in.}^3$
WT. = 384 #/FT.

**PRE- TENSION**

$f'_s = 270,000$ psi.

$f_s = 0.7 \times 270,000 = 189,000$ p.s.i.

Pi per $\frac{1}{2}$"ϕ Strand = 0.1531 x 189,000 = 28.93 KIPS

$\frac{y_B}{r^2} = 0.114586$

(COMPRESSION IS NEGATIVE)

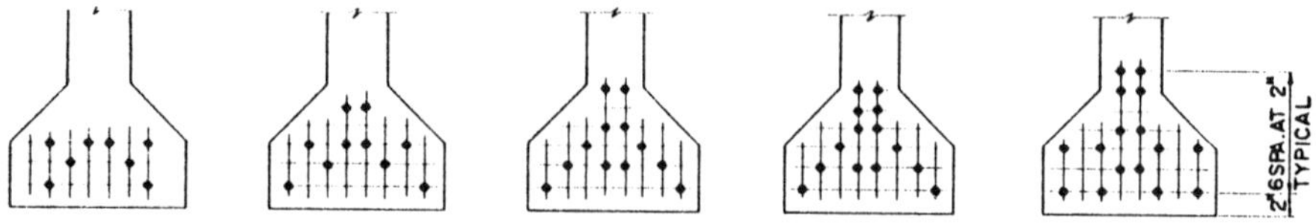

8 STRANDS | 10 STRANDS | 12 STRANDS | 14 STRANDS | 16 STRANDS

STANDARD ARRANGEMENTS TO RAISE C.G. TO AVOID DRAPING OF STRANDS

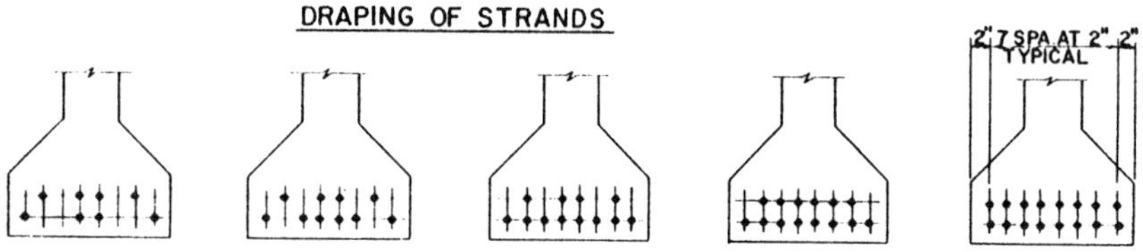

8 STRANDS | 10 STRANDS | 12 STRANDS | 14 STRANDS | 16 STRANDS

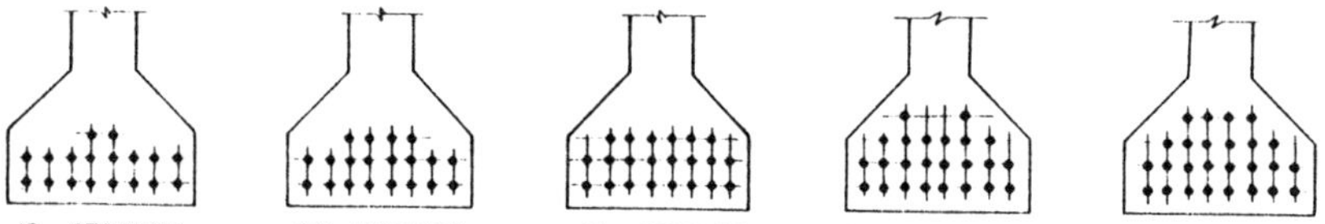

18 STRANDS | 20 STRANDS | 22 STRANDS | 24 STRANDS | 26 STRANDS

ARRANGEMENT AT ℄ SPAN — FOR BEAMS WITH DRAPED STRANDS

| N<br>No. Strands | (1)<br>$e_s$<br>(Inches) | (2)<br>$(1+\frac{e_s y_B}{r^2})$ | (3)<br>(A/(2))<br>(Sq. In.) | (4)<br>P(Init.) = $A_s f_s$<br>(Kips) | (5)<br>$f_B$ (Init) = $\frac{(4)}{(3)}$<br>(K/Sq. In.) |
|---|---|---|---|---|---|
| RAISED PATTERN | | | | | |
| 8 | 11.33 | 2.298 | 160.56 | 231.4 | -1441 |
| 10 | 10.23 | 2.1722 | 169.87 | 289.3 | -1.703 |
| 12 | 9.83 | 2.126 | 173.53 | 347.2 | -2.001 |
| 14 | 9.26 | 2.0609 | 179.04 | 405.0 | -2.262 |
| 16 | 9.08 | 2.0404 | 180.84 | 462.9 | -2.560 |
| STANDARD PATTERNS FOR DRAPED STRANDS | | | | | |
| 8 | 12.83 | 247014 | 149.38 | 231.4 | -1.549 |
| 10 | 13.03 | 24931 | 148.01 | 289.3 | -1.955 |
| 12 | 13.16 | 2.508 | 147.14 | 347.2 | -2.359 |
| 14 | 12.97 | 2.486 | 148.42 | 405.0 | -2.729 |
| 16 | 12.83 | 2.470 | 149.39 | 462.9 | -3.099 |
| 18 | 12.50 | 2.432 | 151.71 | 520.7 | -3.432 |
| 20 | 12.23 | 2.401 | 153.66 | 578.6 | -3.765 |
| 22 | 12.01 | 2.376 | 155.30 | 636.5 | -4.098 |
| 24 | 11.66 | 2.336 | 157.96 | 694.3 | -4.396 |
| 26 | 11.37 | 2.303 | 160.24 | 752.2 | -4.694 |

## 45" PRETENSIONED GIRDER DESIGN DATA

STATE OF WISCONSIN
DEPARTMENT OF TRANSPORTATION
DIVISION OF HIGHWAYS
BRIDGE SECTION

**45" BEAM**

$A = 560$ sq. in.
$r^2 = 223.91$
$y_T = 24.73$ in.
$y_B = 20.27$ in.
$I = 125,390 \text{ in}^4$
$S_T = 5,070 \text{ in}^3$
$S_B = 6,186 \text{ in}^3$
WT. = 583 #/FT.

**PRE - TENSION**

$f'_s = 270,000$ p.s.i.

$f_s = 0.7 \times 270,000 = 189,000$ p.s.i.

Pi per $\frac{1}{2}$"ϕ Strand = 0.1531 x 189,000 = 28.93 KIPS

$\frac{y_B}{r^2} = \frac{20.27}{223.91} = .09052744$

(COMPRESSION IS NEGATIVE)

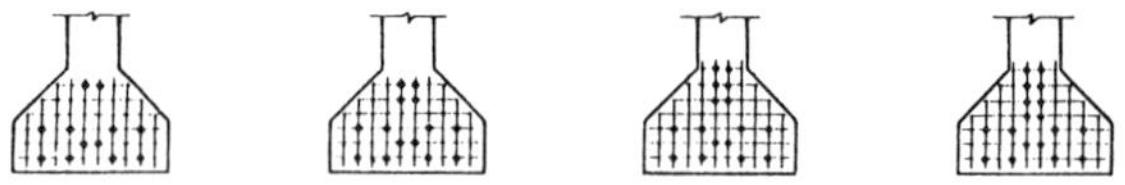

12 STRANDS | 14 STRANDS | 16 STRANDS | 18 STRANDS

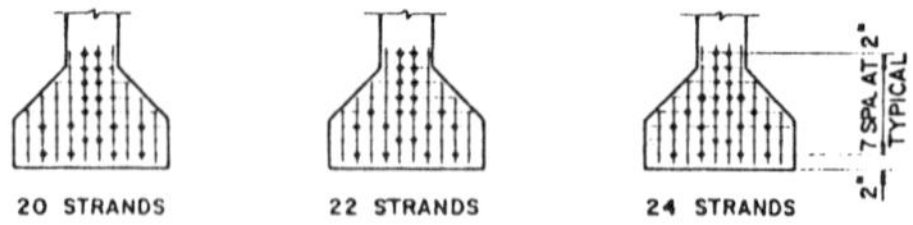

20 STRANDS | 22 STRANDS | 24 STRANDS

STANDARD ARRANGEMENTS TO RAISE CENTER OF GRAVITY TO AVOID DRAPING OF STRANDS

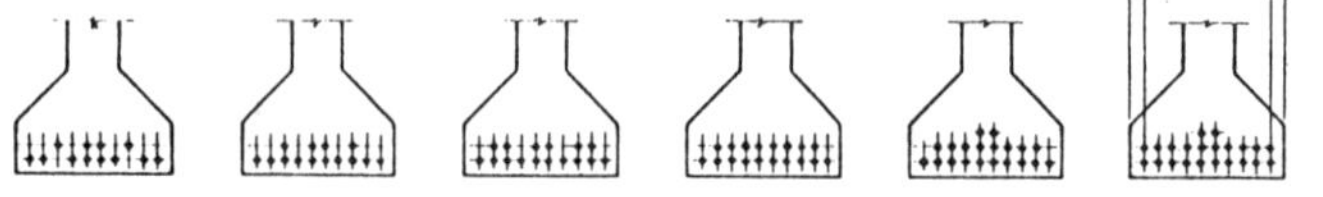

12 STRANDS | 14 STRANDS | 16 STRANDS | 18 STRANDS | 20 STRANDS | 22 STRANDS

24 STRANDS | 26 STRANDS | 28 STRANDS | 30 STRANDS | 32 STRANDS

ARRANGEMENT AT ℄ SPAN — FOR BEAMS WITH DRAPED STRANDS

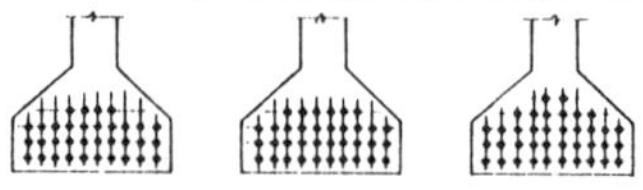

34 STRANDS | 36 STRANDS | 38 STRANDS

| N<br>No. Strands | (1)<br>$e_s$<br>(Inches) | (2)<br>$(1+\frac{e_s y_B}{r^2})$ | (3)<br>(A/(2))<br>(Sq. In.) | (4)<br>P(Init.) = $A_s f_s$<br>(Kips) | (5)<br>$f_B$(Init.) = $\frac{(4)}{(3)}$<br>(K/Sq. In.) |
|---|---|---|---|---|---|
| RAISED PATTERN | | | | | |
| 12 | 14.94 | 2.3525 | 238.05 | 347.2 | -1.458 |
| 14 | 14.27 | 2.2918 | 244.35 | 405.0 | -1.658 |
| 16 | 13.27 | 2.2013 | 254.40 | 462.9 | -1.820 |
| 18 | 13.15 | 2.1904 | 255.66 | 520.7 | -2.037 |
| 20 | 12.27 | 2.1108 | 265.31 | 578.6 | -2.181 |
| 22 | 12.27 | 2.1108 | 265.31 | 636.5 | -2.399 |
| 24 | 12.10 | 2.0954 | 267.25 | 694.3 | -2.598 |
| STANDARD PATTERNS FOR DRAPED STRANDS | | | | | |
| 12 | 17.60 | 2.5933 | 215.94 | 347.2 | -1.608 |
| 14 | 17.70 | 2.602 | 215.19 | 405.0 | -1.882 |
| 16 | 17.52 | 2.5860 | 216.55 | 462.9 | -2.138 |
| 18 | 17.38 | 2.573 | 217.61 | 520.7 | -2.393 |
| 20 | 17.07 | 2.5453 | 220.01 | 578.6 | -2.630 |
| 22 | 17.01 | 2.5399 | 220.48 | 636.5 | -2.887 |
| 24 | 16.77 | 2.5181 | 222.39 | 694.3 | -3.122 |
| 26 | 16.58 | 2.5009 | 223.92 | 752.2 | -3.359 |
| 28 | 16.41 | 2.486 | 225.30 | 810.0 | -3.595 |
| 30 | 16.13 | 2.460 | 227.62 | 867.9 | -3.813 |
| 32 | 16.02 | 2.450 | 228.55 | 925.8 | -4.051 |
| 34 | 15.80 | 2.430 | 230.42 | 983.6 | -4.269 |
| 36 | 15.60 | 2.412 | 232.15 | 1041.5 | -4.486 |
| 38 | 15.32 | 2.387 | 234.62 | 1099.3 | -4.686 |

## 54" PRETENSIONED GIRDER DESIGN DATA

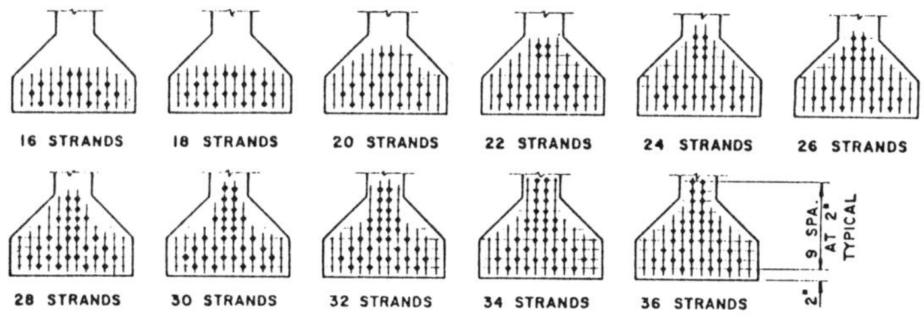


STANDARD ARRANGEMENTS TO RAISE CENTER OF GRAVITY TO AVOID DRAPING OF STRANDS

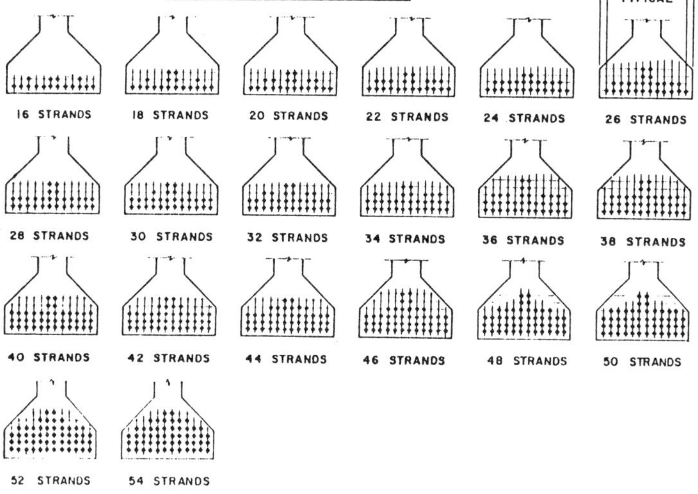


ARRANGEMENT AT ℄ SPAN - FOR GIRDERS WITH DRAPED STRANDS

### PRE-TENSION

**54" GIRDER**

A = 789 SQ. IN.

$r^2$ = 330.46

$y_T$ = 29.27 IN.

$y_B$ = 24.73 IN.

I = 260,730 $IN.^4$

$S_T$ = 8,908 $IN.^3$

$S_B$ = 10,543 $IN^3$

WT. = 822 #/FT.

$f'_s$ = 270,000 P.S.I.

$f_s$ = 0.7 X 270,000 = 189,000 P.S.I.

Pi PER $\frac{1}{2}$" STRAND = 0.1531 X 189,000 = 28.93 KIPS

$\frac{y_B}{r^2} = \frac{24.73}{330.46} = .07483598$

(COMPRESSION IS NEGATIVE)

| N No. Strands | (1) $e_s$ (Inches) | (2) $(1+\frac{e_s y_B}{r^2})$ | (3) (A/(2)) (Sq. In.) | (4) P(Init.) = $A_s f_s$ (Kips) | (5) $f_B$(Init.) = $\frac{(4)}{(3)}$ (K/Sq. in.) |
|---|---|---|---|---|---|
| RAISED PATTERN | | | | | |
| 16 | 20.23 | 2.513 | 313.85 | 462.88 | -1.474 |
| 18 | 19.84 | 2.484 | 317.54 | 520.74 | -1.639 |
| 20 | 19.13 | 2.431 | 324.48 | 578.60 | -1.783 |
| 22 | 18.37 | 2.374 | 332.25 | 636.46 | -1.915 |
| 24 | 17.55 | 2.313 | 341.06 | 694.32 | -2,035 |
| 26 | 17.18 | 2.285 | 345.19 | 752.18 | -2.179 |
| 28 | 17.02 | 2.273 | 347.01 | 810.04 | -2.334 |
| 30 | 16.33 | 2.222 | 355.07 | 867.90 | -2.444 |
| 32 | 16.23 | 2.214 | 356.27 | 925.76 | -2.598 |
| 34 | 15.54 | 2.162 | 364.78 | 983.62 | -2.696 |
| 36 | 15.50 | 2.159 | 365.28 | 1041.48 | -2.851 |
| STANDARD PATTERNS FOR DRAPED STRANDS | | | | | |
| 16 | 22.23 | 2.663 | 296.22 | 462.88 | -1.562 |
| 18 | 21.84 | 2.634 | 299.50 | 520.74 | -1.738 |
| 20 | 21.73 | 2.626 | 300.44 | 578.60 | -1.925 |
| 22 | 21.64 | 2.619 | 301.21 | 636.46 | -2.113 |
| 24 | 21.57 | 2.614 | 301.81 | 694.32 | -2.300 |
| 26 | 21.19 | 2.585 | 305.13 | 752.18 | -2.465 |
| 28 | 21.16 | 2.583 | 305.40 | 810.04 | -2.652 |
| 30 | 20.99 | 2.570 | 306.91 | 867.90 | -2.827 |
| 32 | 20.85 | 2.560 | 308.16 | 925.76 | -3.004 |
| 34 | 20.73 | 2.551 | 309.25 | 983.62 | -3.180 |
| 36 | 20.39 | 2.525 | 312.36 | 1041.48 | -3.334 |
| 38 | 20.31 | 2.519 | 313.11 | 1099.34 | -3.511 |
| 40 | 20.23 | 2.514 | 313.85 | 1157.20 | -3.687 |
| 42 | 20.06 | 2.501 | 315.45 | 1215.06 | -3.852 |
| 44 | 19.91 | 2.490 | 316.87 | 1272.92 | -4.017 |
| 46 | 19.60 | 2.467 | 319.85 | 1330.78 | -4.161 |
| 48 | 19.48 | 2.458 | 321.02 | 1388.6 | -4.326 |
| 50 | 19.37 | 2.450 | 322.10 | 1446.5 | -4.491 |
| 52 | 19.19 | 2.436 | 323.88 | 1504.4 | -4.645 |
| 54 | 19.03 | 2.424 | 325.48 | 1562.2 | -4.800 |

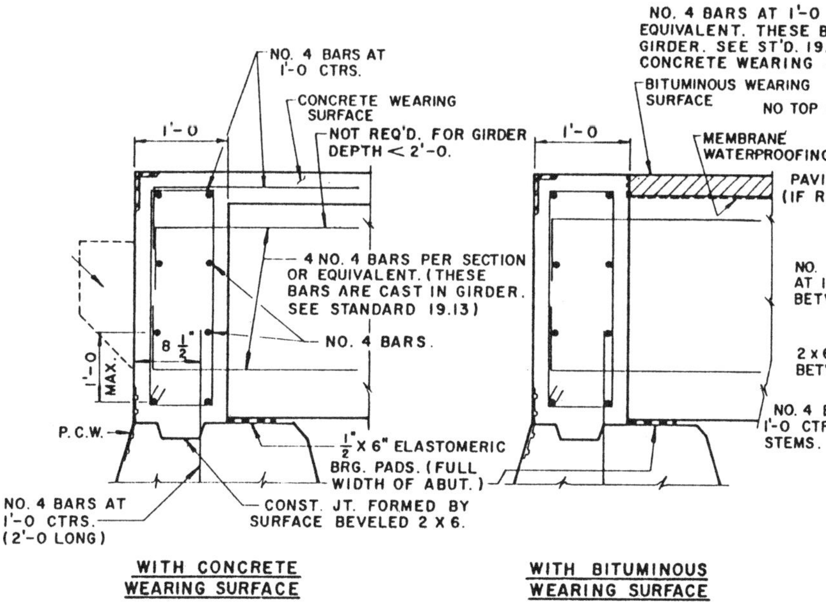


PRECAST BOX OR SLAB SECTION

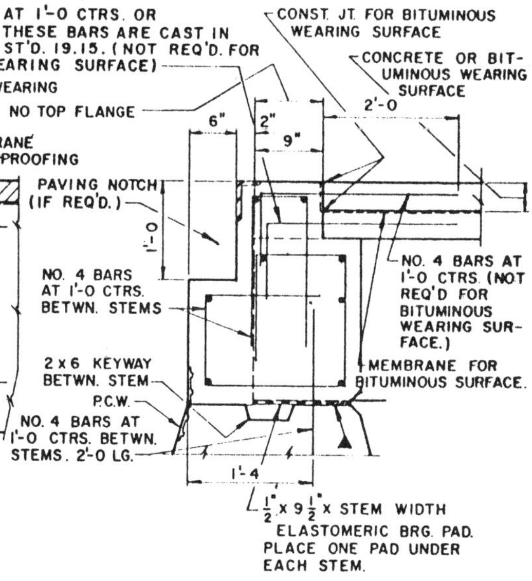


PRECAST DOUBLE TEE OR MULTI-STEM SECTION

▲ $\frac{3}{4}$ x 3 FILLER x LENGTH OF ABUT. PLACE ADDITIONAL FILLER UNDER FLANGES BETWN. BRG. PAD AND $\frac{3}{4}$ x 3 FILLER.

### NOTES

FOR SKEWED STRUCTURES CAST END OF PRECAST UNITS ALONG SKEW.

BRG. DETAILS FOR STEEL GDRS AND PRECAST UNITS ON A1 ABUTMENTS

*STATE OF WISCONSIN*
*DEPARTMENT OF TRANSPORTATION*
*DIVISION OF HIGHWAYS*
*BRIDGE SECTION*

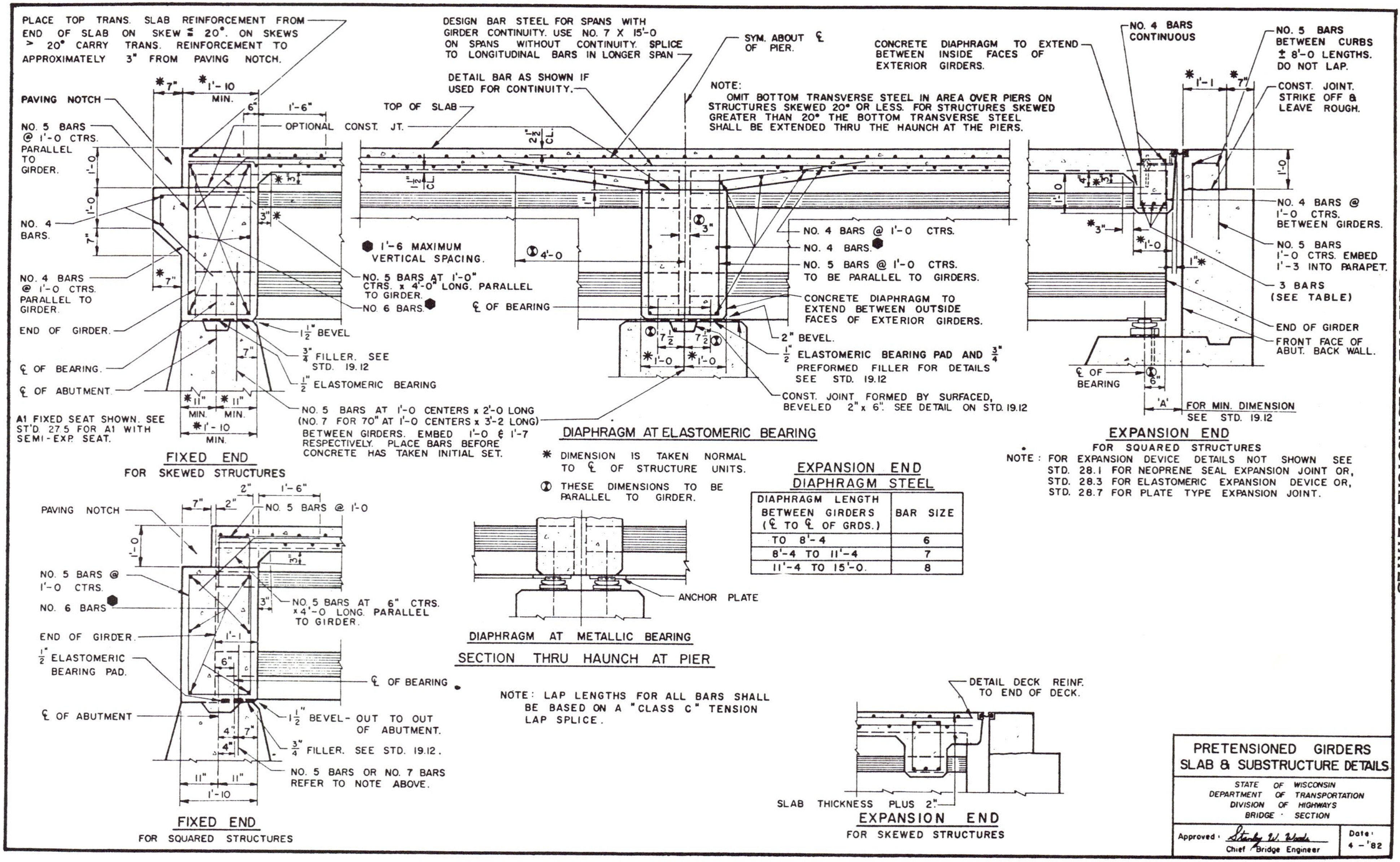


✱ DIMENSION IS TAKEN NORMAL TO ℄ OF STRUCTURE UNITS.

Ⓧ THESE DIMENSIONS TO BE PARALLEL TO GIRDER.

EXPANSION END DIAPHRAGM STEEL

| DIAPHRAGM LENGTH BETWEEN GIRDERS (℄ TO ℄ OF GRDS.) | BAR SIZE |
|---|---|
| TO 8'-4 | 6 |
| 8'-4 TO 11'-4 | 7 |
| 11'-4 TO 15'-0. | 8 |

NOTE: FOR EXPANSION DEVICE DETAILS NOT SHOWN SEE STD. 28.1 FOR NEOPRENE SEAL EXPANSION JOINT OR, STD. 28.3 FOR ELASTOMERIC EXPANSION DEVICE OR, STD. 28.7 FOR PLATE TYPE EXPANSION JOINT.

NOTE: LAP LENGTHS FOR ALL BARS SHALL BE BASED ON A "CLASS C" TENSION LAP SPLICE.

PRETENSIONED GIRDERS
SLAB & SUBSTRUCTURE DETAILS

STATE OF WISCONSIN
DEPARTMENT OF TRANSPORTATION
DIVISION OF HIGHWAYS
BRIDGE · SECTION

Approved: [signature] Chief Bridge Engineer — Date: 4 - '82

STANDARD 19.7

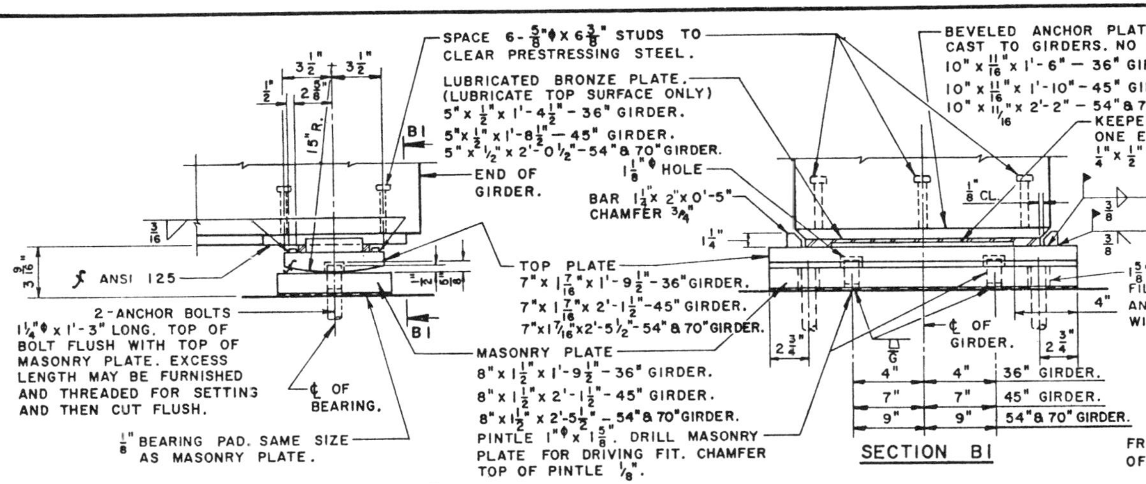


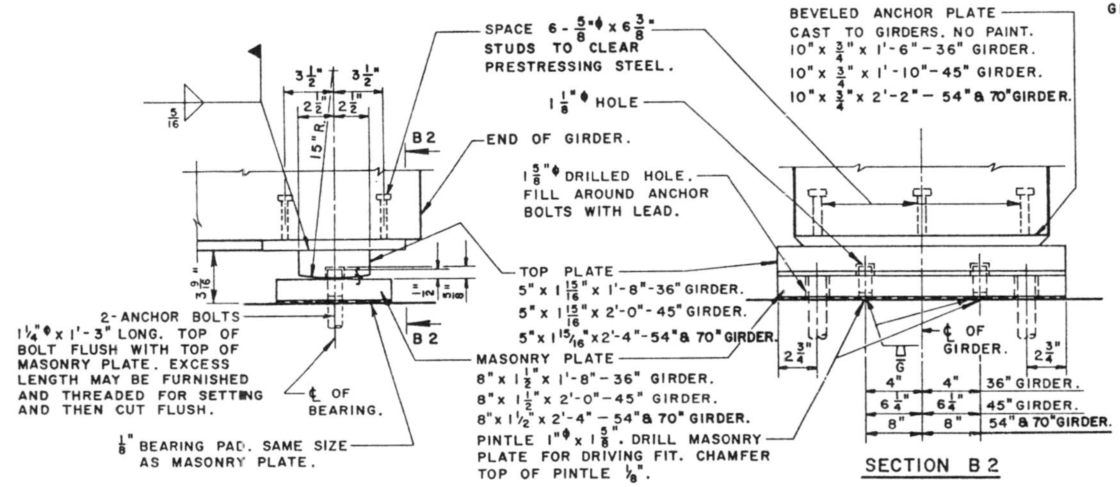


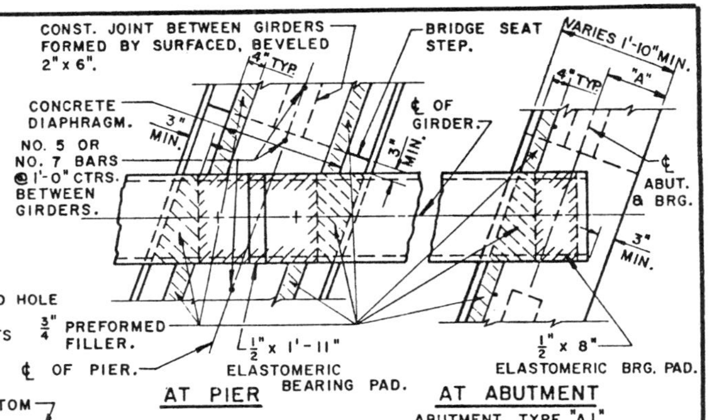


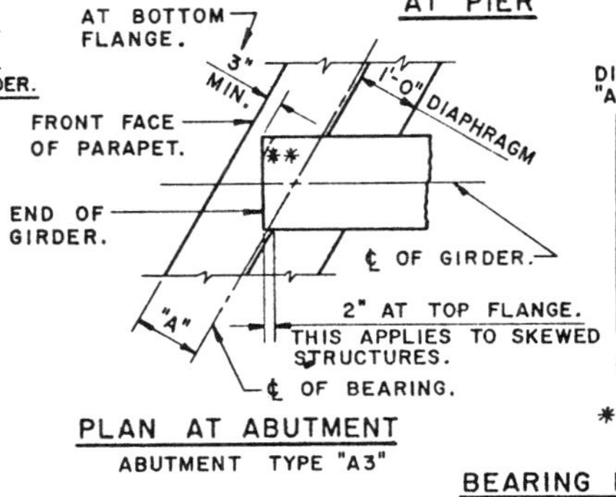


DIMENSION "A" APPLIES TO BOTH TYPE "A1" & "A3" ABUTMENT.

| SKEW ANGLE | 36" GIRDER MINIMUM "A" DISTANCE | MINIMUM "A" DISTANCE 45" GIR. | MINIMUM "A" DISTANCE 54" & 70" GIR. |
|---|---|---|---|
| 0°+-5° | * 11" | * 11" | 11" |
| 5°+-15° | 11" | 12" | 13" |
| 15°+-25° | 12" | 13" | 15" |
| OVER 25° | 14" | 15" | 17" |

* DIMENSION "A" TO BE 9" FOR TYPE "A3" AND "A4" ABUTMENT WITH 0° SKEW.

** FORM-OUT TOP FLANGE ON 70" GIRDER FOR CLEARANCE.

## BEARING NOTES

ALL BEARING MATERIAL EXCLUDING PINTLES AND ANCHOR BOLTS SHALL CONFORM TO A.S.T.M. SPECIFICATION TYPE A588 STEEL.

ALL BEARING MATERIAL EXCEPT BRONZE PLATES, BEARING PADS, AND ANCHOR PLATES SHALL BE PAID FOR AT THE UNIT PRICE BID FOR "HIGH STRENGTH STRUCTURAL STEEL".

ALL STRUCTURAL STEEL BEARING PLATES SHALL BE FLAT ROLLED STEEL PLATES WITH ALL SURFACES SMOOTH AND FREE FROM WARP AND ALL EDGES SMOOTH, STRAIGHT, AND VERTICAL.

ALL PLATE CUTS SHALL BE MACHINE OR MACHINE FLAME CUTS.

ALL SURFACES MARKED ƒ SHALL BE MACHINE ANSI 250 FINISHED, UNLESS SHOWN OTHERWISE.

**COMPUTATIONS FOR STRUCTURAL STEEL WEIGHTS HAVE INCLUDED THE WEIGHTS OF ANCHOR BOLTS, PINTLES, KEEPER BARS, AND ALL PLATES BASED ON THE WEIGHT OF THEIR FULL NOMINAL DIMENSION. NO WEIGHT HAS BEEN INCLUDED FOR WELDS.**

PINTLES & ANCHOR BOLTS SHALL CONFORM TO A.S.T.M. SPECIFICATION TYPES A449▲ & A36 RESPECTIVELY.

▲ OR MATERIAL OF EQUIVALENT YIELD STRENGTH AND ELONGATION.

| BEARING WEIGHTS | 36" GIRDER | 45" GIRDER | 54" & 70" GIRDER |
|---|---|---|---|
| WEIGHT OF ONE EXPANSION BEARING | 149.6 LBS. | 175.4 LBS. | 200.7 LBS. |
| WEIGHT OF ONE FIXED BEARING | 127.3 LBS. | 154.9 LBS. | 179.5 LBS. |
| WEIGHT OF ONE LUBRICATED BRONZE PLATE | 12.8 LBS. | 15.9 LBS. | 19.0 LBS. |

PRETENSIONED GIRDERS BEARING DETAILS

STATE OF WISCONSIN
DEPARTMENT OF TRANSPORTATION
DIVISION OF HIGHWAYS
BRIDGE SECTION

Approved: Stanley W. Woods, Chief Bridge Engineer — Date: 1-'84

STANDARD 19.12

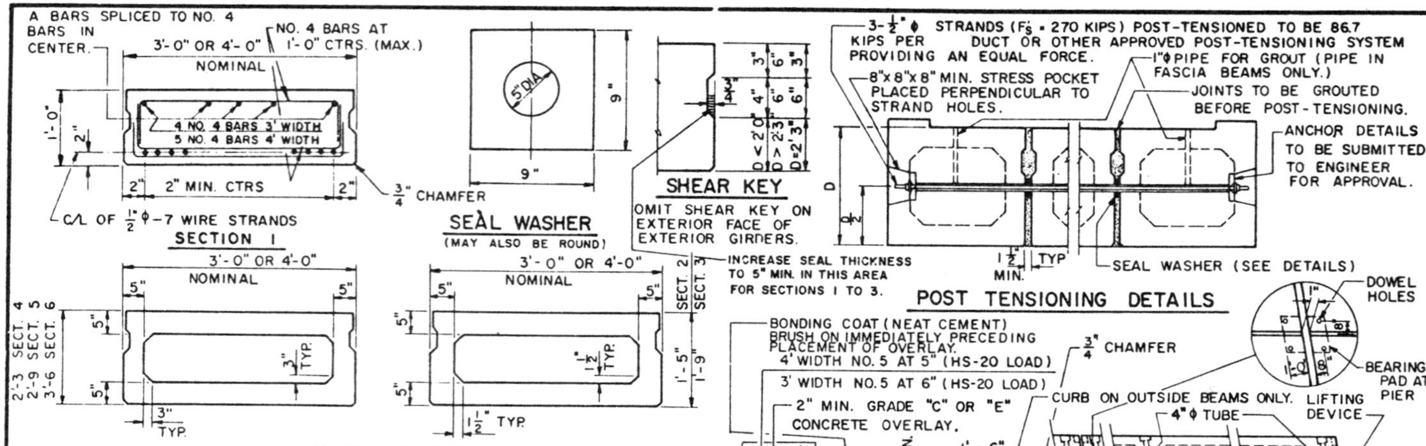


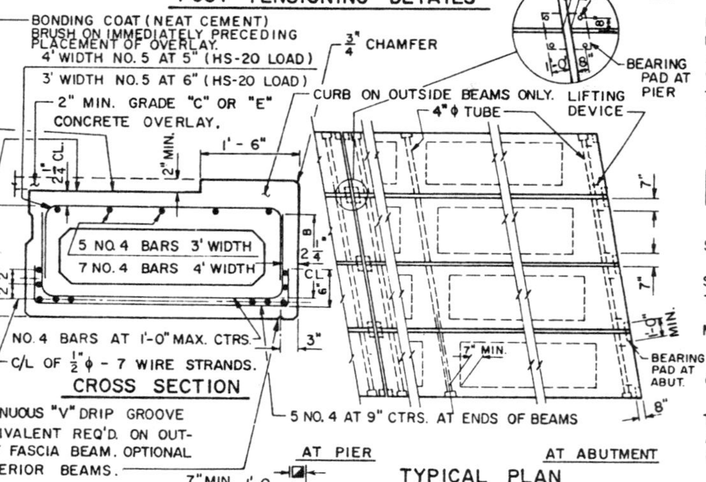


## SECTION PROPERTIES

| SECT NO. | DEPTH | AREA (IN)² 3 | AREA (IN)² 4 | MOMENT OF INERTIA (IN)⁴ 3 | MOMENT OF INERTIA (IN)⁴ 4 | SECTION MODULUS (IN)³ 3 | SECTION MODULUS (IN)³ 4 | MAX. RECOMMENDED SPAN IN FEET H-20 | MAX. RECOMMENDED SPAN IN FEET HS-20 | "A" | "B" |
|---|---|---|---|---|---|---|---|---|---|---|---|
| | | SECTION WIDTH (FT.) | | | | | | | | | |
| 1 | 1'-0" | 432 | 576 | 5,184 | 6,912 | 864 | 1,152 | 28 | 25 | 7 1/2" | 7 1/2" |
| 2 | 1'-5" | 434 | 554 | 14,043 | 18,613 | 1,652 | 2,190 | 43 | 35 | 9" | 1'-0 |
| 3 | 1'-9" | 474 | 594 | 24,793 | 32,723 | 2,361 | 3,116 | 54 | 50 | 1'-0" | 1'-4" |
| 4 | 2'-3" | 548 | 668 | 49,426 | 64,196 | 3,661 | 4,755 | 68 | 65 | 1'-3" | 1'-10" |
| 5 | 2'-9" | 608 | 728 | 83,443 | 107,213 | 5,057 | 6,498 | 82 | 75 | 1'-3" | 2'-4" |
| 6 | 3'-6" | 698 | 818 | 155,326 | 196,646 | 7,396 | 9,364 | 94 | 90 | 1'-3" | 3'-1" |

NO. 4 BARS OR EQUIVALENT. SEE STD. 27.5. THREADED INSERTS OR WELD PLATES MAY BE SUBSTITUTED PROVIDED THEIR PULLOUT CAPACITY IS APPROXIMATELY EQUAL TO THE YIELD STRENGTH OF THE NO. 4 BARS.

4" DIA. HOLE FOR POST-TENSIONING AT MIDSPAN FOR SPANS UP TO 50'; 2 HOLES 6 FEET FROM C/L (12' APART) FOR SPANS OVER 50' TO 62'. AT QUARTER POINTS FOR SPANS OVER 62' (3 HOLES).

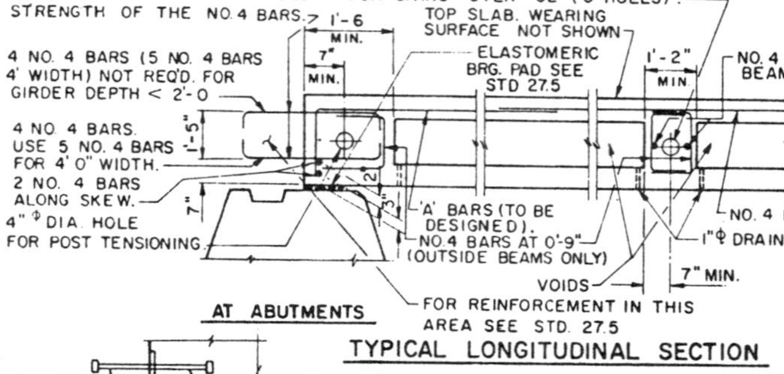


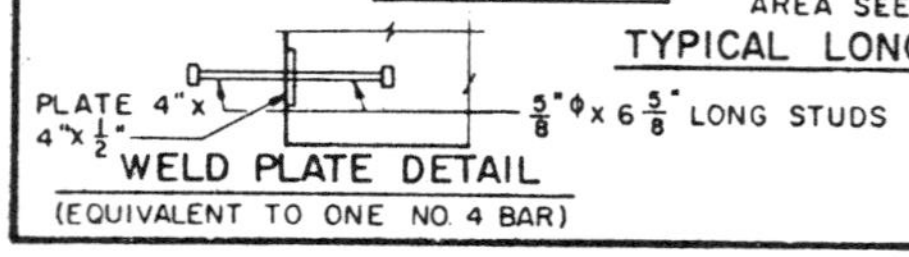


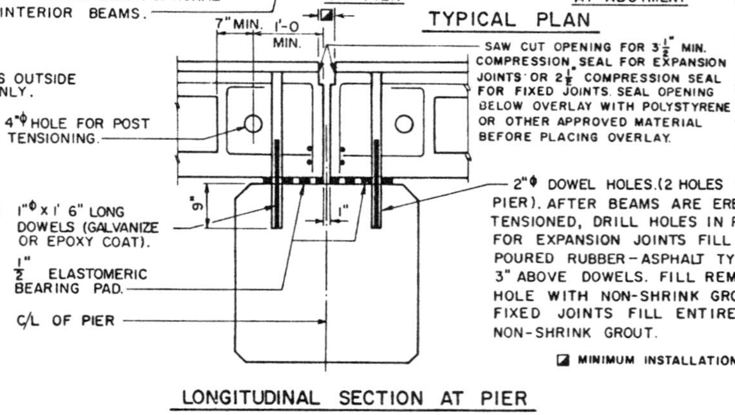


2"ϕ DOWEL HOLES. (2 HOLES PER BEAM AT PIER). AFTER BEAMS ARE ERECTED AND POST-TENSIONED, DRILL HOLES IN PIER FOR DOWELS. FOR EXPANSION JOINTS FILL WITH HOT-POURED RUBBER-ASPHALT TYPE FILLER TO 3" ABOVE DOWELS. FILL REMAINDER OF HOLE WITH NON-SHRINK GROUT. FOR FIXED JOINTS FILL ENTIRE HOLE WITH NON-SHRINK GROUT.

■ MINIMUM INSTALLATION WIDTH PLUS 1/4".

## GENERAL NOTES

4 WAY SLING MUST BE USED TO ENGAGE ALL 4 LIFTING DEVICES ON BOTH ENDS OF UNITS.

STRANDS SHALL BE FLUSH WITH END OF UNIT VOIDS SHALL BE VENTED AND DRAINED BY CASTING A 1"ϕ TUBE AT EACH END AND CENTER OF VOID SEGEMENT.

SLOPE BEAM SEATS TO MATCH ROADWAY CROWN.

POST TENSIONING OF THE TRANSVERSE TENDONS SHALL NOT BEGIN UNTIL THE GROUT BETWEEN THE PRECAST BEAMS HAS BEEN ALLOWED TO CURE FOR 48 HRS.

BAR STEEL REINF. SHALL BE GRADE 60. ($F_Y$ = 60 K.S.I.). PRESTRESSING STEEL ULTIMATE STRENGTH = 270 K.S.I. PRESTRESSED CONCRETE STRENGTH AT 28 DAYS = 5.0 K.S.I.

THE CEMENT AND FINE AGGREGATE FOR THE GROUT BETWEEN THE POST-TENSIONED BEAMS SHALL BE PROPORTIONED BY WEIGHT AS INDICATED IN THE FOLLOWING TABLE. THE CEMENT USED SHALL BE TYPE 1. WATER SHALL BE ADDED IN AMOUNTS AS NECESSARY TO OBTAIN APPROXIMATELY A 5' SLUMP OR TO A CONSISTENCY TO INSURE THAT THE VOIDS ARE COMPLETELY FILLED. THE GROUT SHALL BE RODDED TO INSURE THAT THE VOIDS ARE COMPLETELY FILLED.

| CEMENT PER CU. YD. | FINE AGGREGATE PER CU. YD. |
|---|---|
| 920 LBS. | 2350 LBS. SSD AT 2.65 SPECIFIC GRAVITY |

THE MAXIMUM ALLOWABLE SKEW ANGLE OF STRUCTURE SHALL BE 30°.

ABUTMENT BACKWALLS AND CONCRETE OVERLAY SHALL NOT BE POURED UNTIL AFTER THE POST-TENSIONING HAS BEEN COMPLETED.

SEAL WASHER SHALL BE SPONGE NEOPRENE GASKET 2 1/2 MIN., GRADE I MEETING A.S.T.M. C509 OR OTHER MATERIAL APPR. BY THE ENG. SEAL MAY BE LAMINATED.

STRESS POCKETS SHALL BE FILLED WITH NON-SHRINK GROUT AFTER POST-TENSIONING.●

TRANSITION BETWEEN CHANGING SLOPES OF POST-TENSIONING DUCTS SHALL BE PROVIDED BY EITHER A CIRCULAR OR PARABOLIC CURVE WITH A MINIMUM LENGTH OF 3'-0.

POST-TENSIONING DUCTS SHALL BE PRESSURE GROUTED FROM ONE GROUT PIPE UNTIL ALL ENTRAPPED AIR IS EXPELLED AND GROUT BEGINS TO FLOW FROM THE OPEN GROUT PIPE. THE OPEN GROUT PIPE SHALL BE CLOSED AFTER A PRESSURE OF 50 PSI IS MAINTAINED FOR 15 SEC. THE GROUT COMPOSITION SHALL BE 94 LBS. OF TYPE 1 CEMENT, 5 GALLONS OF WATER AND 1 LB. OF APPROVED PLASTICIZER OR A PRE-MIXED PACKAGED GROUT APPROVED BY THE ENGINEER.

● REFER TO SPECIAL PROVISION FOR NON-SHRINK GROUT SPECIFICATIONS

**PRETENSIONED SLAB AND BOX SECTIONS POST-TENSIONED TRANSVERSELY**

STATE OF WISCONSIN
DEPARTMENT OF TRANSPORTATION
DIVISION OF HIGHWAYS
BRIDGE SECTION

Approved: [signature] Chief Bridge Engineer | Date: 1-'84

STANDARD 19.14

# D.5 BOX BEAM DESIGN AND DETAILS

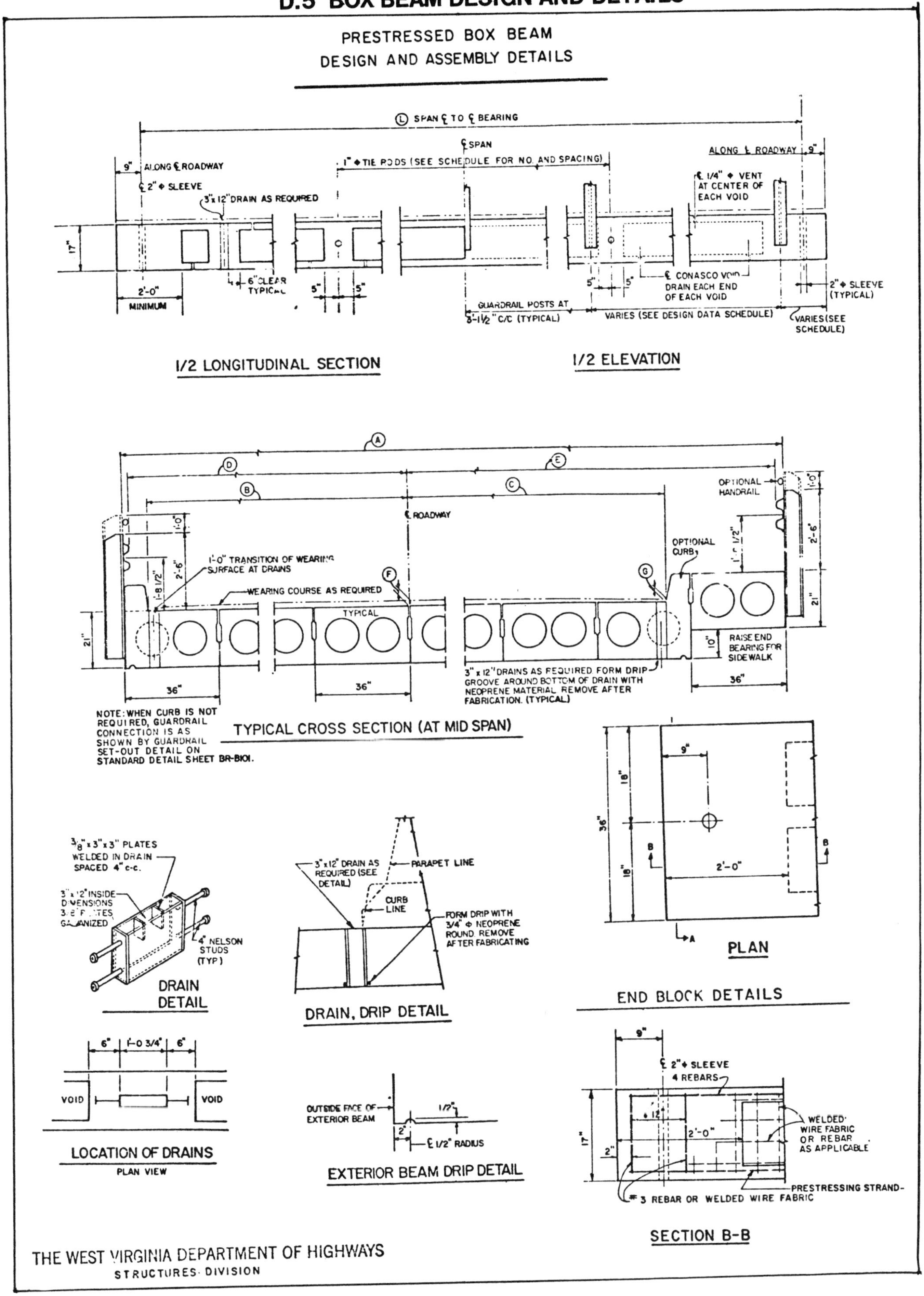

## TYPICAL REINFORCING DETAILS

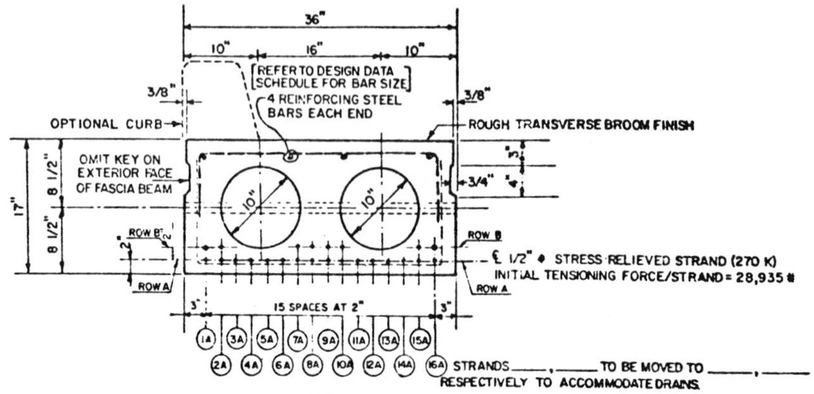


## SECTION

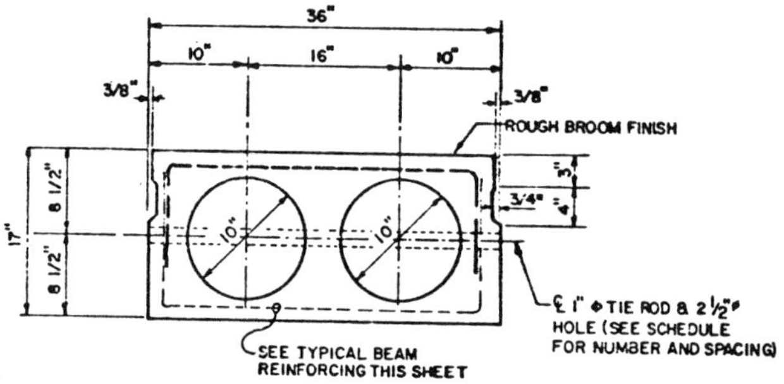


## TYPICAL SECTION

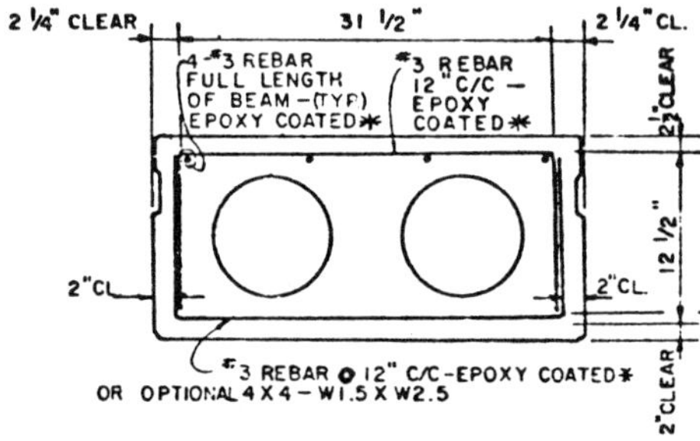


## TYPICAL BEAM REINFORCING

### DESIGN DATA FOR 17" HOLLOW CORED BEAM

| (L) SPAN ℄ TO ℄ BEARING | | 20' | 22' | 24' | 26' | 28' | 30' | 32' | 34' | 36' | 38' | 40' | | |
|---|---|---|---|---|---|---|---|---|---|---|---|---|---|---|
| OVERALL LENGTH ALONG ℄ ROADWAY | | 21'-6" | 23'-6" | 25'-6" | 27'-6" | 29'-6" | 31'-6" | 33'-6" | 35'-6" | 37'-6" | 39'-6" | 41'-6" | | |
| NO. OF 270K, 1/2" Φ STRAND | | 6 | 6 | 6 | 8 | 10 | 10 | 12 | 12 | 14 | 16 | 16 | | |
| STRAND POSITION | | 1A THRU 3A<br>14A THRU 16A | 1A THRU 3A<br>14A THRU 16A | 1A THRU 3A<br>14A THRU 16A | 1A THRU 4A<br>13A THRU 16A | 1A THRU 5A<br>12A THRU 16A | 1A THRU 5A<br>12A THRU 16A | 1A THRU 6A<br>11A THRU 16A | 1A THRU 6A<br>11A THRU 16A | 1A THRU 7A<br>10A THRU 16A | 1A THRU 16A | 1A THRU 16A, | | |
| INITIAL PRESTRESSING FORCE (KIPS/BEAM) | | 174 | 174 | 174 | 231 | 289 | 289 | 347 | 347 | 405 | 463 | 463 | | |
| FINAL PRESTRESSING FORCE (FOOT KIPS/BEAM) | | 132 | 132 | 132 | 176 | 220 | 220 | 265 | 265 | 309 | 353 | 353 | | |
| RESISTING MOMENT (FOOT KIPS/BEAM) | | 139 | 139 | 139 | 175 | 212 | 212 | 249 | 249 | 286 | 323 | 323 | | |
| BENDING MOMENT (FOOT KIPS/BEAM) | | 109 | 122 | 136 | 156 | 177 | 198 | 220 | 243 | 268 | 294 | 320 | | |
| END STEEL (EACH END) - EPOXY COATED ✱ | | 4-#4x9' → | | | 4-#4x10' | 4-#5x10' | 4-#5x10' | 4-#5x12' → | | | 4-#6x14' | 4-#6x14' | | |
| WEIGHT OF TYPICAL MEMBERS (TONS) INTERIOR BEAMS | | 6.5 | 7.0 | 7.7 | 8.3 | 8.9 | 9.5 | 10.1 | 10.7 | 11.3 | 11.9 | 12.5 | | |
| WEIGHT OF TYPICAL MEMBERS (TONS) EXTERIOR BEAMS | | 7.5 | 8.1 | 8.9 | 9.6 | 10.3 | 11.0 | 11.7 | 12.4 | 13.1 | 13.8 | 16.4 | | |
| NUMBER AND SPACING OF GUARDRAIL POSTS | NUMBER OF POSTS | 7 | 8 | 9 | 9 | 10 | 11 | 11 | 12 | 12 | 13 | 14 | | |
| | END OF BEAM TO ℄ OF FIRST POST | 1'-4-1/2" | 1'-0" | 1'-0" | 1'-3" | 1'-0" | 1'-0" | 1'-1-1/2" | 1'-0" | 1'-6¾" | 1'-0" | 1'-0" | | |
| | FIRST POST SPACING TO SECOND POST SPACING BOTH ENDS | 3'-1½" | 2'-11¼" | 2'-4½" | 3'-1½" | 2'-9¾" | 2'-3" | 3'-1½" | 2'-8¼" | 3'-1½" | 3'-1½" | 2'-6¾" | | |

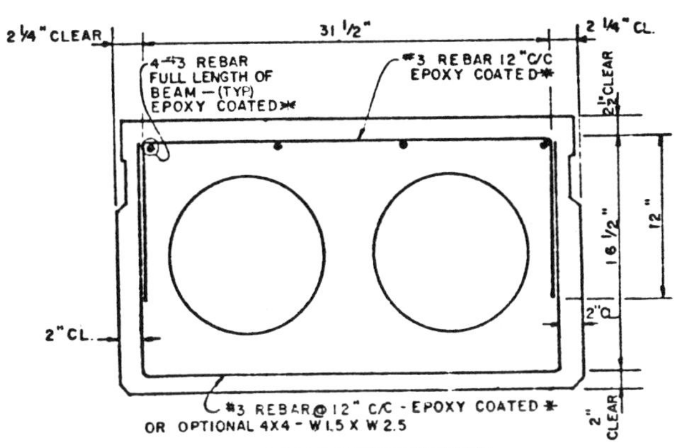


TYPICAL BEAM REINFORCING

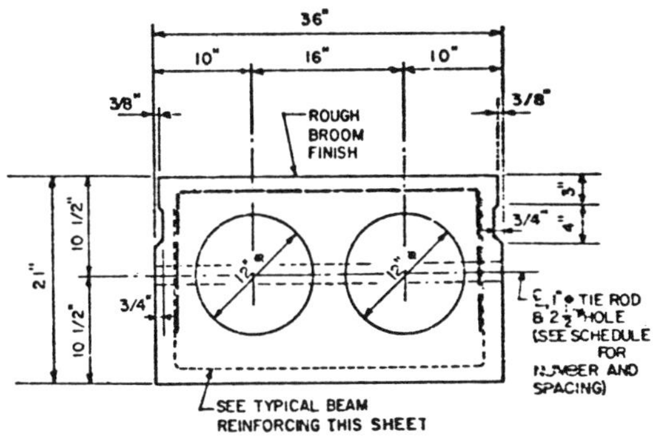


TYPICAL SECTION

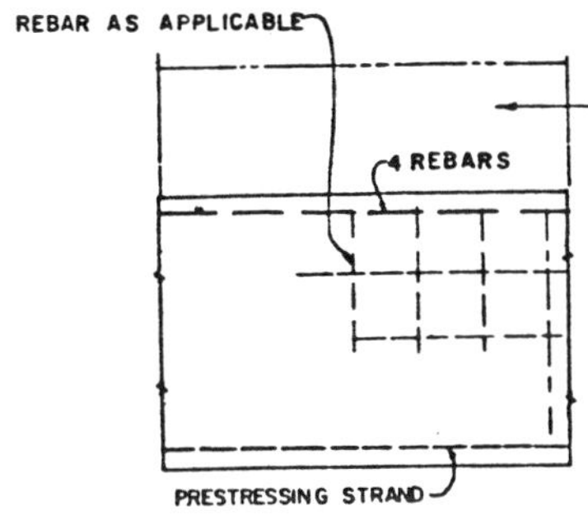


SIDE VIEW

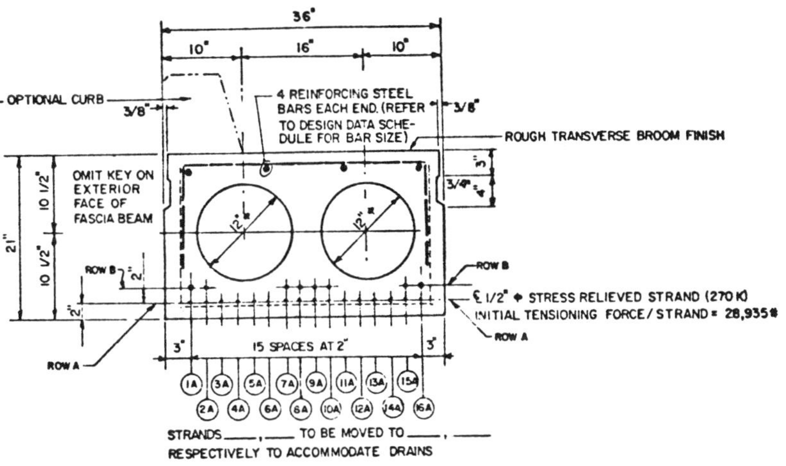


SECTION

TYPICAL REINFORCING DETAILS

DESIGN DATA FOR 21" HOLLOW CORED BEAM

| (L) SPAN ℄ TO ℄ BEARING | 30' | 32' | 34' | 36' | 38' | 40' | 42' | 44' | 46' | 48' | | | |
|---|---|---|---|---|---|---|---|---|---|---|---|---|---|
| OVERALL LENGTH ALONG ℄ ROADWAY | 31'-6" | 33'-6" | 35'-6" | 37'-6" | 39'-6" | 41'-6" | 43'-6" | 45'-6" | 47'-6" | 49'-6" | | | |
| NO. OF 270 K, 1/2" ø STRAND | 8 | 8 | 10 | 10 | 12 | 14 | 14 | 16 | 16 | 18 | | | |
| STRAND POSITION | 1A THRU 4A 13A THRU 16A | 1A THRU 4A 13A THRU 16A | 1A THRU 5A 12A THRU 16A | 1A THRU 5A 12A THRU 16A | 1A THRU 6A 11A THRU 16A | 1A THRU 7A 10A THRU 16A | 1A THRU 7A 10A THRU 16A | 1A THRU 16 A | 1A THRU 16A | 18, 1A THRU 16A, 16B | | | |
| INITIAL PRESTRESSING FORCE (KIPS/BEAM) | 231 | 231 | 289 | 289 | 347 | 405 | 405 | 463 | 463 | 521 | | | |
| FINAL PRESTRESSING FORCE (FOOT KIPS/BEAM) | 176 | 176 | 220 | 220 | 265 | 309 | 309 | 353 | 353 | 397 | | | |
| RESISTING MOMENT (FOOT KIPS/BEAM) | 235 | 235 | 283 | 283 | 331 | 379 | 379 | 427 | 427 | 475 | | | |
| BENDING MOMENT (FOOT KIPS/BEAM) | 207 | 230 | 254 | 281 | 308 | 336 | 364 | 393 | 422 | 453 | | | |
| END STEEL (EACH END) - EPOXY COATED * | 4-#4x10' | 4-#5x12' → | | | 4-#5x14' → | | | 4-#6x16' | 4-#6x16' | 4-#7x16' | | | |
| WEIGHT OF TYPICAL MEMBER (TONS) INTERIOR BEAM | 9.1 | 10.0 | 10.6 | 11.1 | 11.7 | 11.9 | 12.8 | 13.2 | 13.9 | 14.5 | | | |
| WEIGHT OF TYPICAL MEMBER (TONS) EXTERIOR BEAM | 10.6 | 11.7 | 12.3 | 12.9 | 13.6 | 13.9 | 14.6 | 15.3 | 16.1 | 16.8 | | | |
| NUMBER AND SPACING OF GUARDRAIL POSTS — NUMBER OF POSTS | 11 | 11 | 12 | 12 | 13 | 14 | 14 | 15 | 16 | 16 | | | |
| NUMBER AND SPACING OF GUARDRAIL POSTS — END OF BEAM TO ℄ OF FIRST POST | 1'-0" | 1'-1-1/2" | 1'-0" | 1'-6 3/4" | 1'-0" | 1'-0" | 1'-5 1/4" | 1'-0" | 1'-0" | 1'-3 3/4" | | | |
| NUMBER AND SPACING OF GUARDRAIL POSTS — FIRST POST SPACING TO SECOND POST SPACING BOTH ENDS | 2'-3" | 3'-1 1/2" | 2'-8 1/4" | 3'-1 1/2" | 3'-1 1/2" | 2'-6 3/4" | 3'-1 1/2" | 3'-0" | 2'-5 1/4" | 3'-1 1/2" | | | |

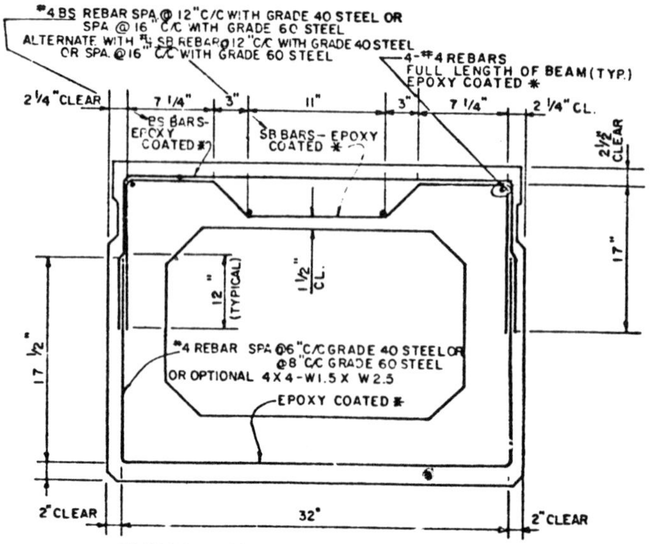


TYPICAL BEAM REINFORCING

## END BLOCK DETAILS

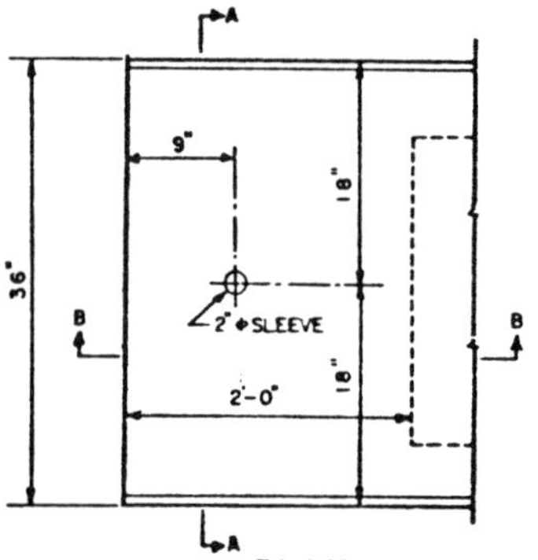


PLAN

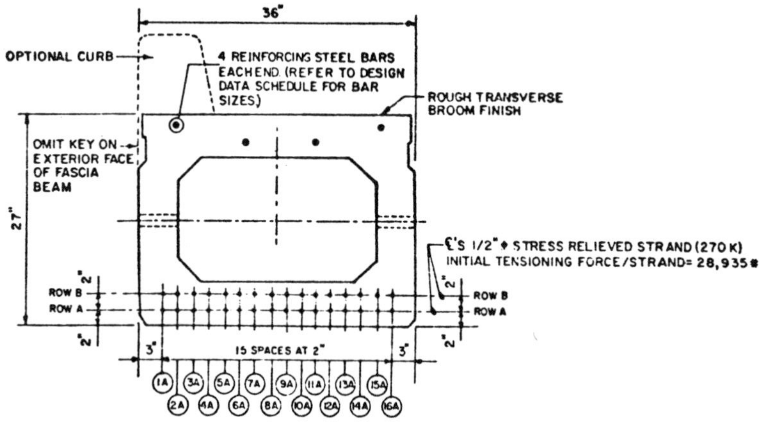


SECTION W/STRAND & END STEEL

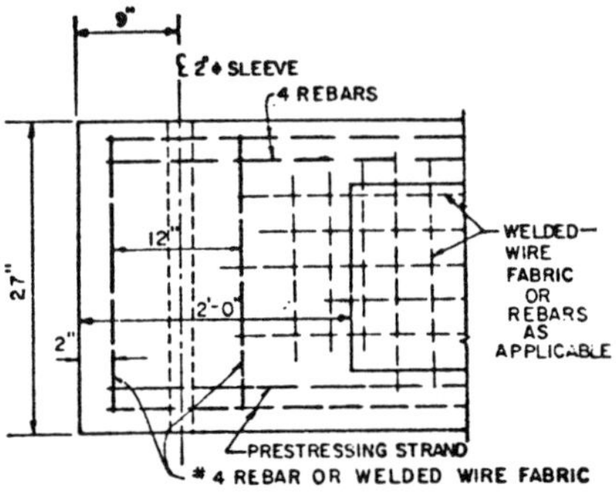


SECTION B-B

## DESIGN DATA FOR 27" DEPTH BOX BEAM

| | | | | | | | | | | | | | |
|---|---|---|---|---|---|---|---|---|---|---|---|---|---|
| (L) SPAN ℄ TO ℄ BEARING | | 40' | 42' | 44' | 46' | 48' | 50' | 52' | 54' | 56' | 58' | | |
| OVERALL LENGTH ALONG ℄ ROADWAY | | 41'-6" | 43'-6" | 45'-6" | 47'-6" | 49'-6" | 51'-6" | 53'-6" | 55'-6" | 57'-6" | 59'-6" | | |
| NUMBER OF 270K, 1/2" ⌀ STRANDS | | 10 | 10 | 12 | 12 | 14 | 14 | 16 | 16 | 18 | 18 | | |
| STRAND POSITION | | 1A THRU 5A<br>12A THRU 16A | 1A THRU 5A<br>12A THRU 16A | 1A THRU 6A<br>11A THRU 16A | 1A THRU 6A<br>11A THRU 16A | 1A THRU 7A<br>10A THRU 16A | 1A THRU 7A<br>10A THRU 16A | 1A THRU 16A | 1A THRU 16A | 1B,<br>1A THRU 16A,<br>16B | 1B,<br>1A THRU 16A,<br>16B | | |
| INITIAL PRESTRESSING FORCE (KIPS/BEAM) | | 289 | 289 | 347 | 347 | 405 | 405 | 463 | 463 | 521 | 521 | | |
| FINAL PRESTRESSING FORCE (KIPS/BEAM) | | 220 | 220 | 265 | 265 | 309 | 309 | 353 | 353 | 397 | 397 | | |
| RESISTING MOMENT (FT. KIPS/BEAM) | | 397 | 397 | 463 | 463 | 529 | 529 | 595 | 595 | 662 | 662 | | |
| BENDING MOMENT (FT. KIPS/BEAM) | | 349 | 378 | 408 | 439 | 470 | 502 | 535 | 568 | 602 | 637 | | |
| END STEEL (EACH END)-EPOXY COATED * | | 4-#5 x 8' → | | | | | 4-#5 x 10' | 4-#5 x 10' | 4-#6 x 12' → | | | | |
| WEIGHT OF TYPICAL MEMBER (TONS) INTERIOR BEAM | | 12.76 | 13.40 | 14.03 | 14.66 | 15.30 | 15.94 | 16.58 | 17.21 | 17.85 | 18.49 | | |
| WEIGHT OF TYPICAL MEMBER (TONS) EXTERIOR BEAM | | 14.71 | 15.44 | 16.17 | 16.70 | 17.60 | 18.37 | 19.11 | 19.84 | 20.58 | 21.32 | | |
| NUMBER AND SPACING OF GUARDRAIL POSTS | NUMBER OF POSTS | 14 | 14 | 15 | 16 | 16 | 17 | 18 | 18 | 19 | 19 | | |
| | END OF BEAM TO ℄ OF FIRST POST | 1'-0" | 1'-5¼" | 1'-0" | 1'-0" | 1'-3¾" | 1'-0" | 1'-0" | 1'-2¼" | 1'-0" | 1'-7-1/2" | | |
| | FIRST POST SPACING TO SECOND POST SPACING BOTH ENDS | 2'-6¾" | 3'-1½" | 3'-0" | 2'-5¼" | 3'-1½" | 2'-10½" | 2'-3¾" | 3'-1½" | 2'-9" | 3'-1½" | | |

# 33 INCH BOX BEAM

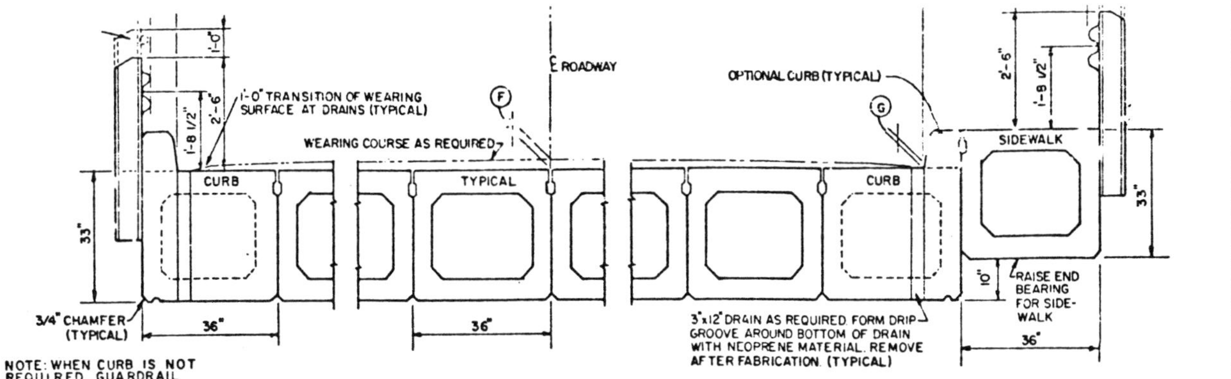


NOTE: WHEN CURB IS NOT REQUIRED, GUARDRAIL CONNECTION IS AS SHOWN BY GUARDRAIL SET-OUT DETAIL ON STANDARD DETAIL SHEET BR-B101.

TYPICAL CROSS-SECTION AT MID-SPAN

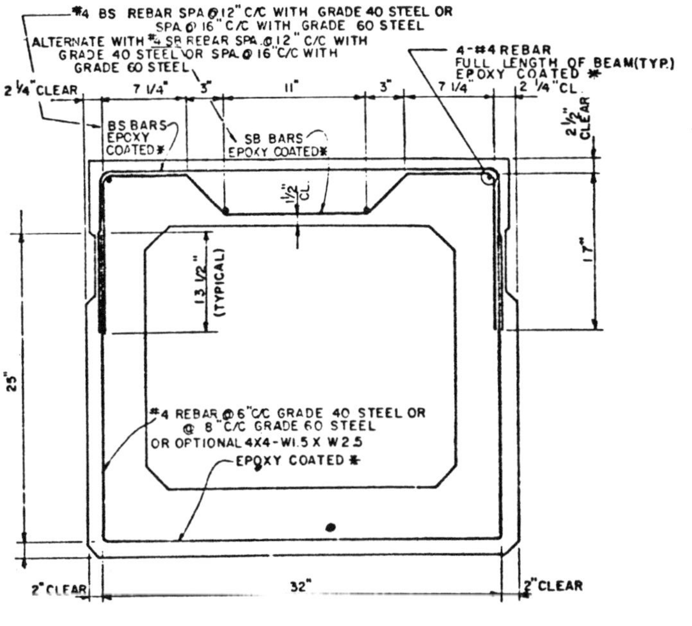


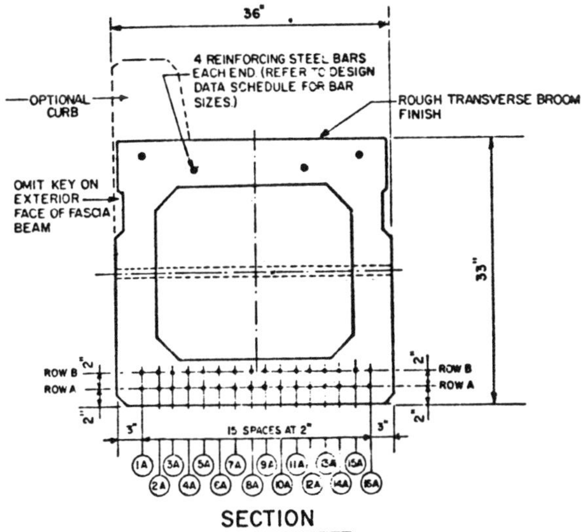


SECTION

**DESIGN DATA FOR 33" DEPTH BOX BEAM**

| (L) SPAN ℄ TO ℄ BEARING | | 50' | 52' | 54' | 56' | 58' | 60' | 62' | 64' | 66' | 68' | 70' | |
|---|---|---|---|---|---|---|---|---|---|---|---|---|---|
| OVERALL LENGTH ALONG ℄ ROADWAY | | 51'-6" | 53'-6" | 55'-6" | 57'-6" | 59'-6" | 61'-6" | 63'-6" | 65'-6" | 67'-6" | 69'-6" | 71'-6" | |
| NUMBER OF 270K, 1/2" Φ STRAND | | 12 | 12 | 12 | 14 | 14 | 16 | 16 | 18 | 18 | 20 | 20 | |
| STRAND POSITION | | 1A THRU 6A<br>11A THRU 16A | 1A THRU 6A<br>11A THRU 16A | 1A THRU 6A<br>11A THRU 16A | 1A THRU 7A<br>10A THRU 16A | 1A THRU 7A<br>10A THRU 16A | 1A THRU 16A | 1A THRU 16A | 1B,<br>1A THRU 16A,<br>16B | 1B<br>1A THRU 16A,<br>16B | 1B, 2B,<br>1A THRU 16A,<br>15B, 16B | 1B, 2B,<br>1A THRU 16A,<br>15B, 16B | |
| INITIAL PRESTRESSING FORCE (KIPS/BEAM) | | 347 | 347 | 347 | 405 | 405 | 463 | 463 | 521 | 521 | 579 | 579 | |
| FINAL PRESTRESSING FORCE (KIPS/BEAM) | | 265 | 265 | 265 | 309 | 309 | 353 | 353 | 397 | 397 | 441 | 441 | |
| RESISTING MOMENT (FOOT KIPS/BEAM) | | 592 | 592 | 592 | 676 | 676 | 759 | 759 | 843 | 843 | 926 | 926 | |
| BENDING MOMENT (FOOT KIPS/BEAM) | | 520 | 554 | 589 | 625 | 661 | 698 | 735 | 774 | 813 | 853 | 894 | |
| END STEEL (EACH END) - EPOXY COATED * | | 4-#5x10' | 4-#5x10' | 4-#6x12' | 4-#6x12' | 4-#6x12' | 4-#6x14' | 4-#6x14' | 4-#6x14' | 2-#7x15'<br>2-#6x15' | 2-#7x15'<br>2-#6x15' | 2-#7x15'<br>2-#6x15' | |
| WEIGHT OF TYPICAL MEMBER (TONS) INTERIOR BEAM | | 15.9 | 16.1 | 17.2 | 17.8 | 18.4 | 19.1 | 19.7 | 20.3 | 21.0 | 21.6 | 22.2 | |
| WEIGHT OF TYPICAL MEMBER (TUNS) EXTERIOR BEAM | | 18.3 | 18.6 | 19.8 | 20.5 | 21.2 | 22.0 | 22.7 | 23.4 | 24.2 | 24.9 | 23.6 | |
| NUMBER AND SPACING OF GUARDRAIL POSTS | NUMBER OF POSTS | 17 | 18 | 18 | 19 | 19 | 20 | 21 | 21 | 22 | 23 | 23 | |
| | END OF BEAM TO ℄ OF FIRST POST. | 1'-0" | 1'-0" | 1'-2¼" | 1'-0" | 1'-7-1/2" | 1'-0¾" | 1'-0" | 1'-6" | 1'-0" | 1'-0" | 1'-4-1/2" | |
| | FIRST POST SPACING TO SECOND POST SPACING BOTH ENDS | 2'-10½" | 2'-3¾" | 3'-1½" | 2'-9" | 3'-1½" | 3'-1½" | 2'-7½" | 3'-1½" | 3'-0¾" | 2'-6" | 3'-1½" | |

## DESIGN AND FABRICATION NOTES

ALL MEMBERS ARE DESIGNED FOR HS 20-44 LIVE LOADING IN ACCORDANCE WITH 1973 AASHTO STANDARD SPECIFICATIONS FOR HIGHWAY BRIDGES. THE DESIGN PROVIDES FOR A WEARING SURFACE OF 25 POUNDS PER SQUARE FOOT OF ROADWAY.

COMPRESSIVE STRENGTH OF CONCRETE AT 28 DAYS ($f'_c$) SHALL NOT BE LESS THAN 5,000 PSI.

COMPRESSIVE STRENGTH OF CONCRETE AT TIME OF PRESTRESSING ($f'_{ci}$) SHALL NOT BE LESS THAN 4,000 PSI.

PRESTRESSING STEEL SHALL BE SEVEN-WIRE UNCOATED STRESS-RELIEVED STRAND IN ACCORDANCE WITH 270K GRADE ASTM A416 WITH MINIMUM TENSILE STRENGTH ($f'_s$) OF 270,000 PSI.

ALLOWABLE TEMPORARY STRESSES: AT THE TIME OF TRANSFER THE UNIT STRESSES SHALL NOT EXCEED THE FOLLOWING: CONCRETE COMPRESSION = $0.60 f'_{ci}$; TENSION $= 3\sqrt{f'_{ci}}$; PRESTRESSING STEEL TENSION $0.7 f'_s$.

ALLOWABLE STRESSES AT DESIGN LOAD AFTER LOSSES OF PRESTRESS: THE UNIT STRESSES SHALL NOT EXCEED THE FOLLOWING: CONCRETE COMPRESSION = $0.40 f'_c$; TENSION $= 3\sqrt{f'_c}$ BUT NOT TO EXCEED 250 PSI, PRESTRESSING STEEL TENSION $= 0.60 f'_s$ OR $0.80 f_{dy}$, WHICHEVER IS SMALLER; REINFORCING STEEL ($f_s$) = 20,000 PSI.

ULTIMATE LOAD CAPACITY SHALL NOT BE LESS THAN $\frac{1.30}{\Phi}\left[D + 5/3(L+I)\right]$, $\Phi = 1$ FOR FACTORY PRODUCED PRECAST PRESTRESSED CONCRETE MEMBERS.

WHERE THE CALCULATED TENSILE STRESS IN CONCRETE EXCEEDS THE VALUES ALLOWED, AUXILIARY NON-PRESTRESSED REINFORCING SHALL BE PROVIDED TO RESIST THE TOTAL TENSION.

ALL ANCHOR BOLTS AND DOWELS (ASTM A-36), INSERTS, NUTS, WASHERS AND TIE BOLTS (ASTM A-36) AS WELL AS THE U-BOLTS AND NUTS (ASTM A-307) FOR THE PIPE RAIL AND SPECIAL GUARDRAIL PARTS SHALL BE GALVANIZED IN ACCORDANCE WITH ASTM A-153 (2 OZ. PER SQ. FT. COATING OF GALVANIZING SPELTER (MIN.)

GUARDRAIL POSTS, CONNECTION PLATES AND DRAIN PLATES (ASTM A-36) TO BE GALVANIZED IN ACCORDANCE WITH ASTM A-123.

THE GUARDRAIL WITH BUFFER END TERMINAL SECTIONS SHALL CONFORM TO THE REQUIREMENTS AS SET FORTH BY THE W. VA. DEPARTMENT OF HIGHWAYS STANDARD DETAIL SHEET G.R.1, CURRENT ISSUE.

TIE BOLTS SHALL CONFORM TO ASTM A-36. TIE BOLTS TO BE TIGHTENED SNUG TIGHT AFTER FINAL POSITIONING OF BOX BEAMS IN PLACE.

ALL WELDING TO BE ACCOMPLISHED PRIOR TO GALVANIZING.

ALL EXPOSED THREADED PORTIONS OF THE HANDRAIL SHALL BE COATED WITH LIQUID GALVANIZE "GALVICON" OR APPROVED EQUAL AFTER ASSEMBLY.

THE WELDED WIRE FABRIC SHALL CONFORM TO THE REQUIREMENTS OF ASTM A-185 AND IN ACCORDANCE WITH STANDARD SPECIFICATIONS AND PRACTICE ESTABLISHED BY WIRE REINFORCEMENT INSTITUTE, INC.

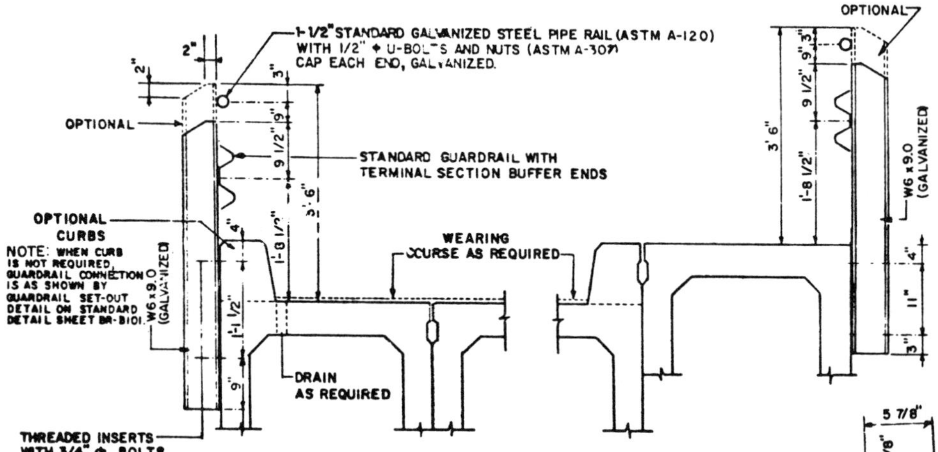


TYPICAL CROSS SECTION

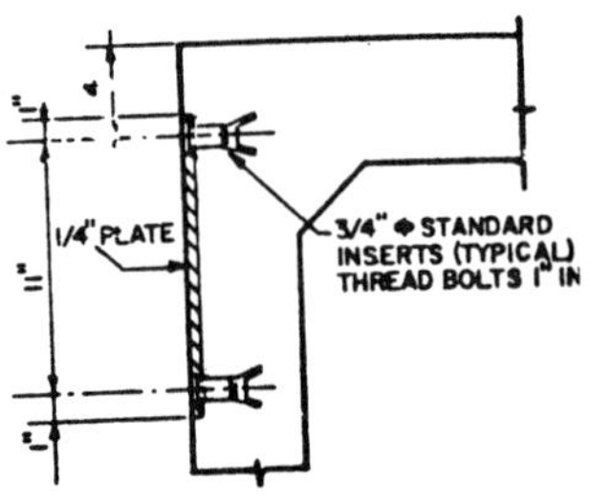


GUARDRAIL INSERT ASSEMBLY WITH SIDEWALK

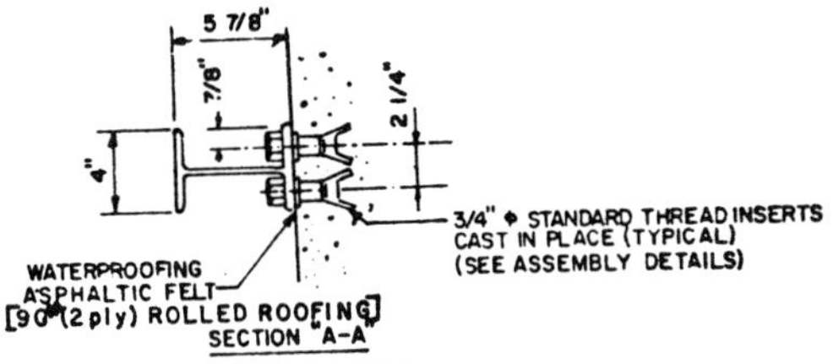


ANCHOR DETAIL

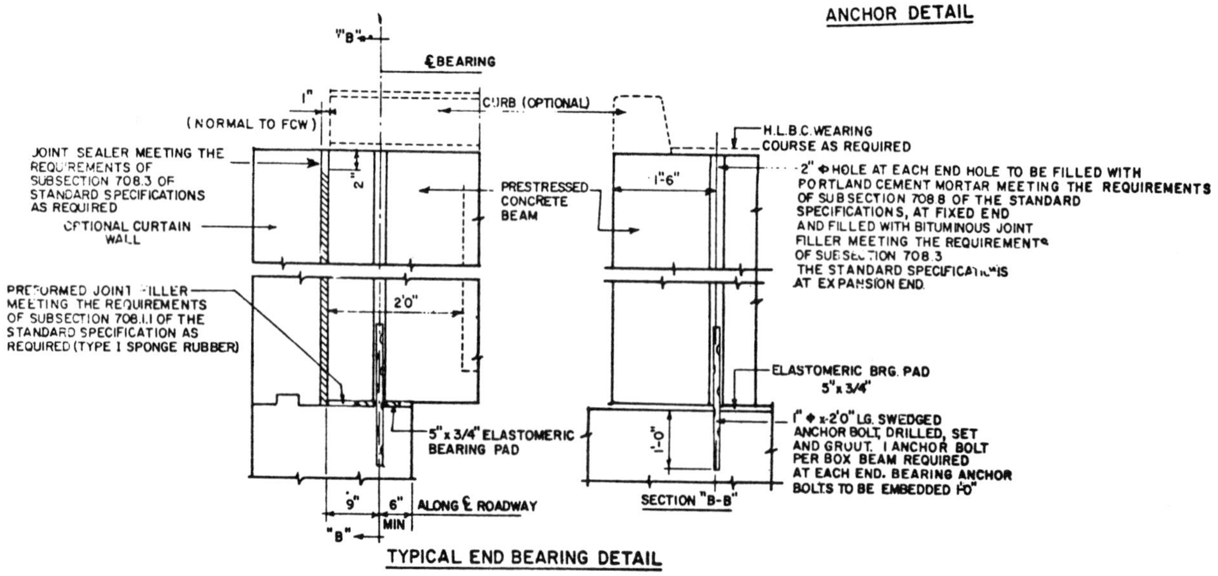


TYPICAL END BEARING DETAIL

(ALONG ℄ ROADWAY)

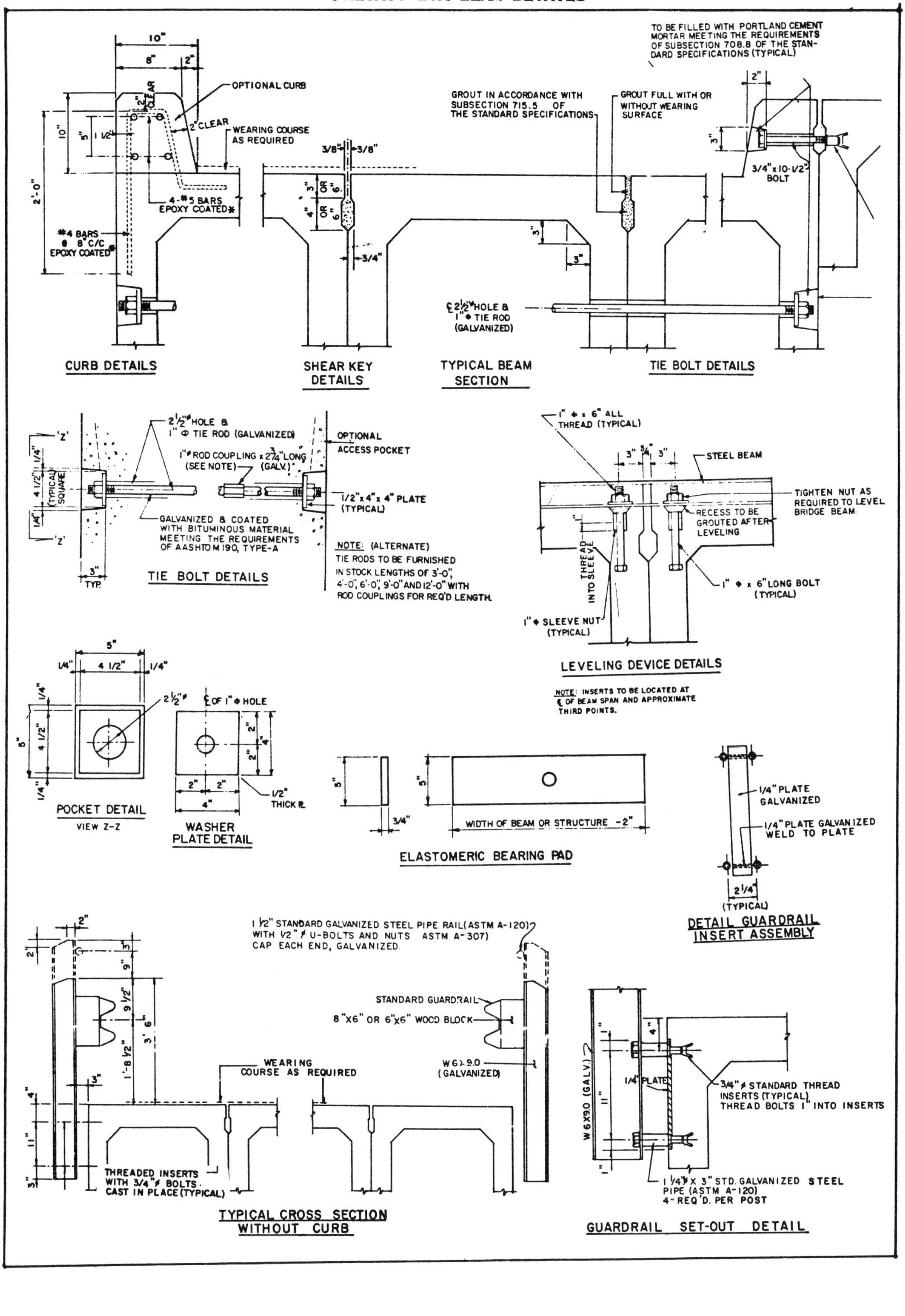
10"
8"
2"
2" CLEAR
OPTIONAL CURB
2" CLEAR
WEARING COURSE AS REQUIRED
10"
5"
1 1/2"
2'-0"
4-#5 BARS EPOXY COATED*
#4 BARS @ 8" C/C EPOXY COATED*
CURB DETAILS
3/8"
3/8"
3" OR 6"
4" OR 6"
3/4"
SHEAR KEY DETAILS
GROUT IN ACCORDANCE WITH SUBSECTION 715.5 OF THE STANDARD SPECIFICATIONS
GROUT FULL WITH OR WITHOUT WEARING SURFACE
3"
3"
℄ 2 1/4" HOLE & 1" ɸ TIE ROD (GALVANIZED)
TYPICAL BEAM SECTION
TO BE FILLED WITH PORTLAND CEMENT MORTAR MEETING THE REQUIREMENTS OF SUBSECTION 708.8 OF THE STANDARD SPECIFICATIONS (TYPICAL)
2"
3"
3/4" x 10-1/2" BOLT
TIE BOLT DETAILS
2 1/2" ɸ HOLE & 1" ɸ TIE ROD (GALVANIZED)
1" ɸ ROD COUPLING x 2 3/4" LONG (SEE NOTE) (GALV.)
OPTIONAL ACCESS POCKET
'Z'
1/4"
4 1/2" (TYPICAL) SQUARE
1/4"
'Z'
1/2" x 4" x 4" PLATE (TYPICAL)
GALVANIZED & COATED WITH BITUMINOUS MATERIAL MEETING THE REQUIREMENTS OF AASHTO M 190, TYPE-A
3" TYP.
TIE BOLT DETAILS
NOTE: (ALTERNATE)
TIE RODS TO BE FURNISHED IN STOCK LENGTHS OF 3'-0", 4'-0", 6'-0", 9'-0" AND 12'-0" WITH ROD COUPLINGS FOR REQ'D LENGTH.
1" ɸ x 6" ALL THREAD (TYPICAL)
3" 3/4" 3"
STEEL BEAM
TIGHTEN NUT AS REQUIRED TO LEVEL BRIDGE BEAM.
RECESS TO BE GROUTED AFTER LEVELING
THREAD INTO SLEEVE
1" ɸ x 6" LONG BOLT (TYPICAL)
1" ɸ SLEEVE NUT (TYPICAL)
LEVELING DEVICE DETAILS
NOTE: INSERTS TO BE LOCATED AT ℄ OF BEAM SPAN AND APPROXIMATE THIRD POINTS.
5"
1/4"
4 1/2"
1/4"
2 1/2" ɸ
POCKET DETAIL
VIEW Z-Z
℄ OF 1" ɸ HOLE
2"
2"
4"
2"
2"
4"
1/2" THICK PL.
WASHER PLATE DETAIL
5"
3/4"
5"
WIDTH OF BEAM OR STRUCTURE -2"
ELASTOMERIC BEARING PAD
1/4" PLATE GALVANIZED
1/4" PLATE GALVANIZED WELD TO PLATE
2 1/4" (TYPICAL)
DETAIL GUARDRAIL INSERT ASSEMBLY
2"
2"
3"
9"
9 1/2"
3'-6"
1'-8 1/2"
3"
4"
11"
3"
1 1/2" STANDARD GALVANIZED STEEL PIPE RAIL (ASTM A-120) WITH 1/2" ɸ U-BOLTS AND NUTS ASTM A-307) CAP EACH END, GALVANIZED.
STANDARD GUARDRAIL
8"X6" OR 6"X6" WOOD BLOCK
W6X9.0 (GALVANIZED)
WEARING COURSE AS REQUIRED
THREADED INSERTS WITH 3/4" ɸ BOLTS CAST IN PLACE (TYPICAL)
TYPICAL CROSS SECTION WITHOUT CURB
W6X9.0 (GALV.)
1"
4"
11"
1"
1/4" PLATE
3/4" ɸ STANDARD THREAD INSERTS (TYPICAL) THREAD BOLTS 1" INTO INSERTS
1 1/4" ɸ X 3" STD. GALVANIZED STEEL PIPE (ASTM A-120) 4-REQ'D. PER POST
GUARDRAIL SET-OUT DETAIL

# ADJACENT BOX BEAMS

## DESIGN CRITERIA AND NOTES

### DESIGN CRITERIA

Design Specifications – Design Division of 1969-AASHO Standard Specifications for Highway Bridges, as supplemented by the Department Design Manual, Part 4, Structures (including 7-72 revisions) and as noted herein.

1. Live Load
   - AASHO HS20-44 Standard Truck, Lane Loading, or Alternate Loading (2 axles of 24 kips each at 4'-0" c/c, 2 wheels of 12 kips each at 6'-0" c/c).
   - AASHO Impact.

2. Dead Loads
   - ½" integral wearing surface and 30 psf future wearing surface for composite box beams.
   - 6 psf additional slab dead load to compensate for thicker deck at the ends of bridge due to beam camber.
   - Additional beam dead load to provide for the possibility of undersized void forms. (¼" additional concrete assumed around perimeter of voids.)
   - Weight of standard parapet and railing.
   - Diaphragms – internal: one diaphragm from 45' to 90' and two diaphragms over 90', equally spaced.

3. Transverse Distribution of Loads for Beam Design

   Beams with composite slab or with bituminous surface:
   - Wheel load – Interior and exterior beams: S/6.0
   - Parapet and railing – 3 ft. composite boxes: considered as distributed equally to 5 beams.
     All other boxes: considered as distributed equally to 4 beams.

4. The designs of the precast prestressed concrete bridge members included in these Standards are based on the use of the following materials:

   Concrete: f'c = 5,250 psi, for f'ci = 4,500 psi
   f'c = 5,750 psi, for f'ci = 5,000 psi
   f'c = 6,250 psi, for f'ci = 5,500 psi
   Deck Slab (Class AA – Form 408/70)
   f'c = 3,500 psi  fc = 1,000 psi
   (3,750 psi design mix)

   Reinforcement Bars: (ASTM A615)
   fs = 20,000 psi

   Prestressing steel:

| Nom. Dia. (in.) | Grade | Nom. Steel Area of Strand (sq. in) | Ultimate Strength (lbs.) | Initial Force per Strand (lbs.) |
|---|---|---|---|---|
| 1/4 | 250 | 0.036 | 9,000 | 6,300 |
| 3/8 | 270 | 0.085 | 23,000 | 16,100 |
| 7/16 | | 0.115 | 31,000 | 21,700 |
| Properties from ASTM Designation A416-68 | | | | |

   The designs included in these Standards are based on 7/16" dia. strands except for spans less than 21', ¼" dia. strands are used and for spans less than 30', ⅜" dia. strands are used.

### DESIGN CRITERIA (CONT.)

5. Prestress Losses for Pretensioned Beams
   Total design loss – psi = 5,000 + 16 fcs + 0.08 fsi
   But not less than 20.0% of initial prestress force.

6. Allowable Stresses Used for Design of Pretensioned Simple Span Beams
   - A. Temporary Stresses Before All Design Losses
     - Tension in concrete:
       Straight strand design: $*6\sqrt{f'ci}$
       Draped strand design: 212 psi (Basic section @ ℄ bearing)
       $3\sqrt{f'ci}$ (@ drape point)
     - Compression in Concrete: 0.60 f'ci
       *At the shop drawing stage, the computed tensile stress for straight strand design shall be reduced to $3\sqrt{f'ci}$ through unbonding of strands at the end of beam.
     - Tension in prestressing steel: 0.70 f's
   - B. Stresses Under Design Loads After All Losses
     - Tension in concrete: $**3\sqrt{f'c}$
     - Compression in concrete: 0.40 f'c
     - Tension in prestressing steel: 0.60 f's or 0.80 fsy (whichever is less)

     ** precompressed tensile zone and other areas

7. Section Properties
   - The effect of camber irregularities and undersized voids is disregarded in computing section properties.
   - For composite section properties, modular ratio $n = E_{beam}/E_{slab}$ is determined in accordance with $E = w^{1.5} 33\sqrt{f'c}$ (Unit weight of the concrete for beam and deck slab are considered to be the same.)

8. Ultimate Load Factors
   - 1.5 (Dead Load) + 2.5 (Live Load Plus Impact)
   - For ultimate moment capacity, the modular ratio n = f'c beam/f'c slab

9. Live Load Plus Impact Deflection.
   - LL + Imp. Deflection < L/1000 except for smaller beams for which the design curves are extended beyond L/1000 limit.
   - Transverse Distribution same as in Item 3.

10. Horizontal Shear (Bond) for Composite Beams
    - Horizontal shear (bond) between slab and beam is calculated at the critical section 1.5d from the centerline of bearing, using 1.5 (superimposed dead loads such as loads from parapet, railing, future wearing surface, etc.) + 2.5 (Live load plus impact).
    - Horizontal shear stress does not exceed 225 psi.

11. Vertical Web Reinforcement
    - #4 Stirrups were used for design.
    - Shear and stirrup spacing were computed at the tenth points and quarter points of the span using distribution of loads as specified in Item 3, above. "Simple beam action" was not assumed at ends of beams since the stirrup spacing required at the quarter points was to be used throughout the end thirds of the beam.
    - Minimum steel requirements governed the designs included in these standards; hence, the stirrup spacings specified do not exceed three-fourths the depth of the beam or 160/total web thickness. For 12 inch deep solid beams, only the former of these two criteria was considered applicable, resulting in the 9" spacing specified.

# ADJACENT BOX BEAMS

## 3FT. AND 4FT. COMPOSITE

### SUPERSTRUCTURE DETAILS

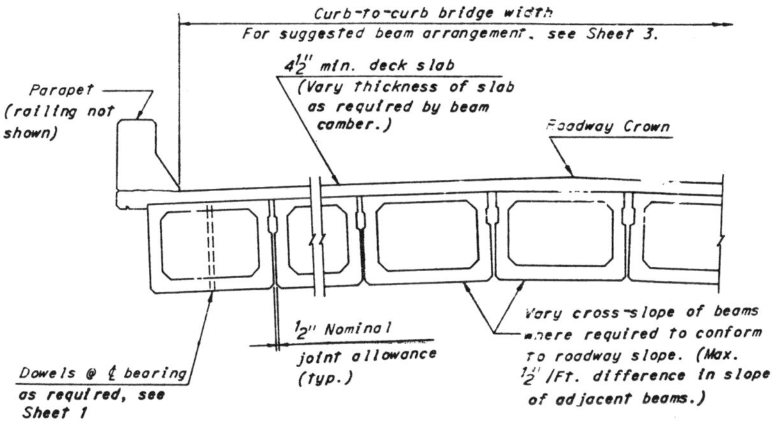


TYPICAL BRIDGE CROSS SECTION

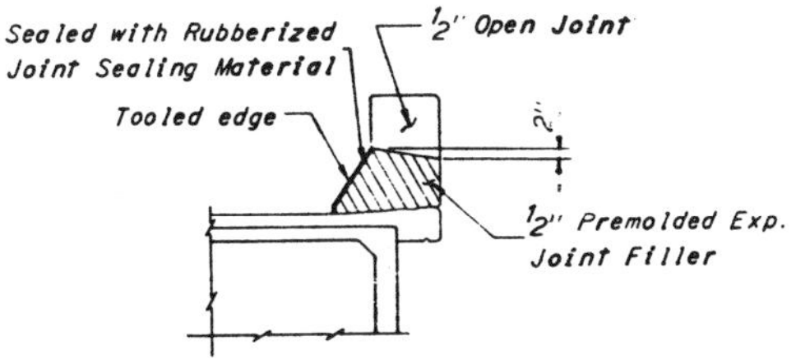


DEFLECTION JOINT DETAIL

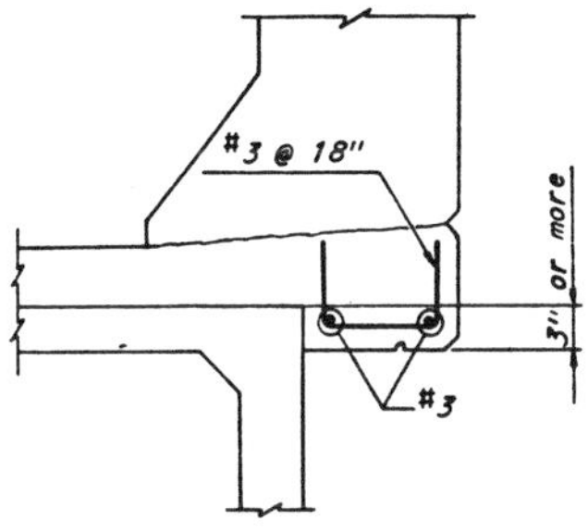


DETAIL F

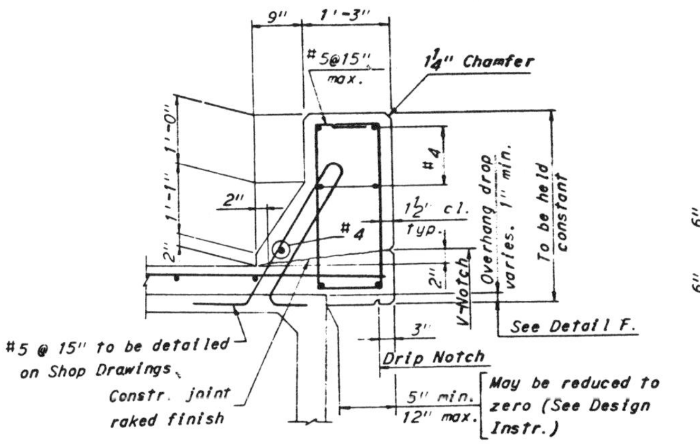


SECTION C-C

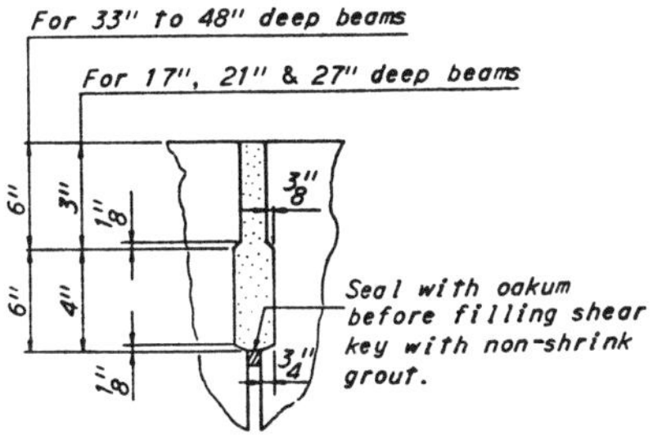


DETAIL OF SHEAR KEY

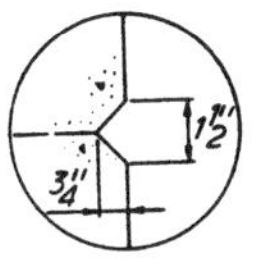


V-NOTCH DETAIL

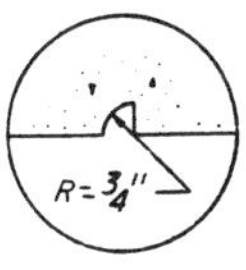


DRIP NOTCH DETAIL

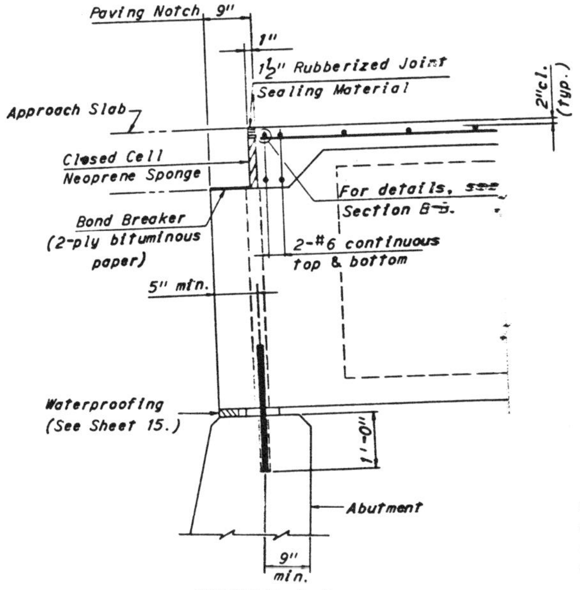


SECTION A-A

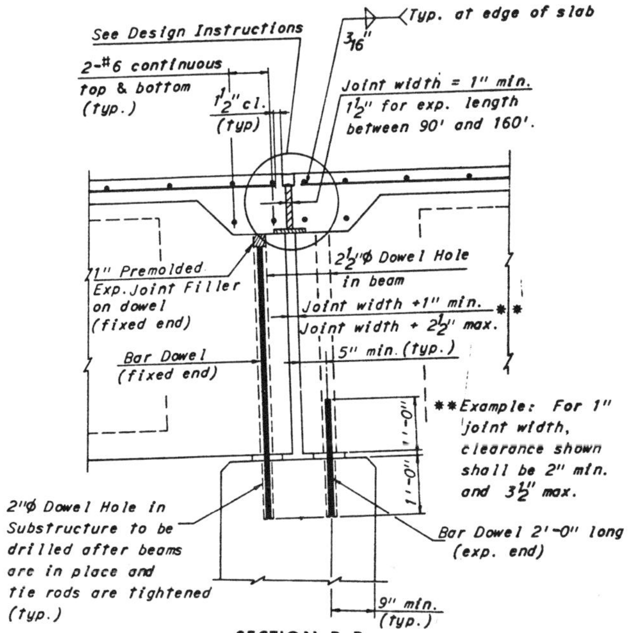


SECTION B-B

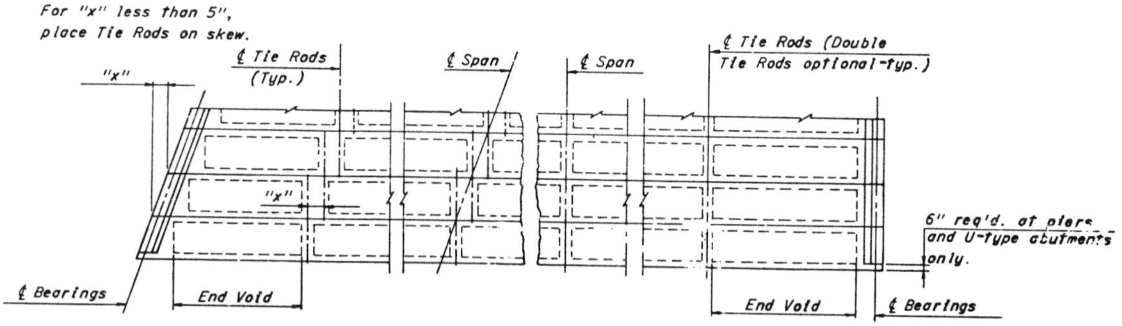


SKEW 80° OR LESS

SKEW 90° TO 80°

TYPICAL FRAMING PLAN

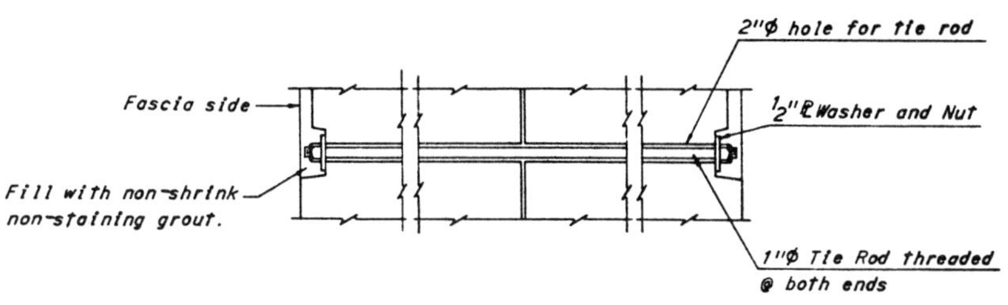


TIE ROD DETAIL (SKEW)

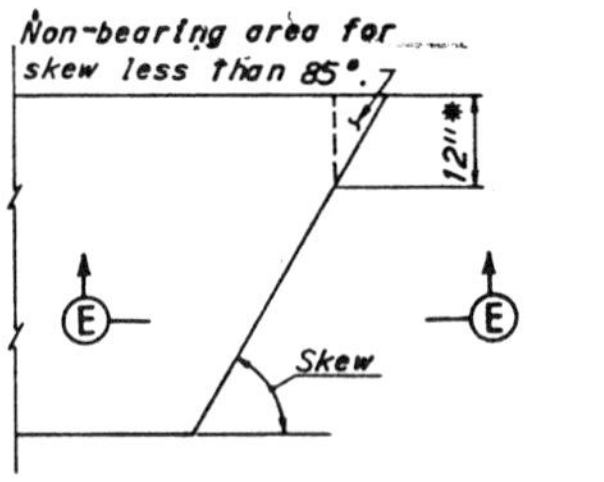


PLAN-END OF BEAM

*Modify if required to accommodate bearing pad arrangement for sharp skews.

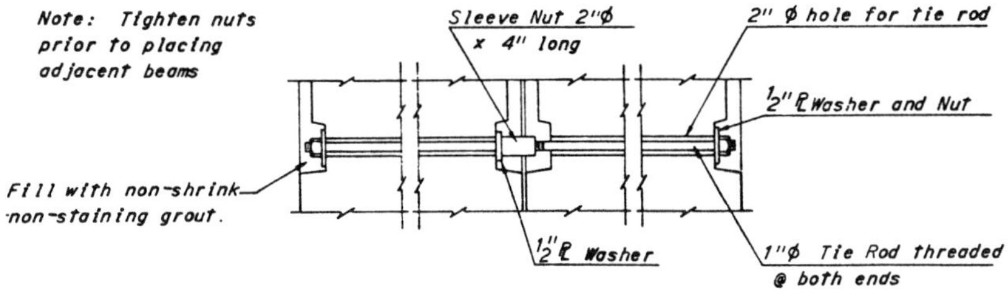


TIE ROD DETAIL (90°)
(Details similar for tie rods placed on skew)

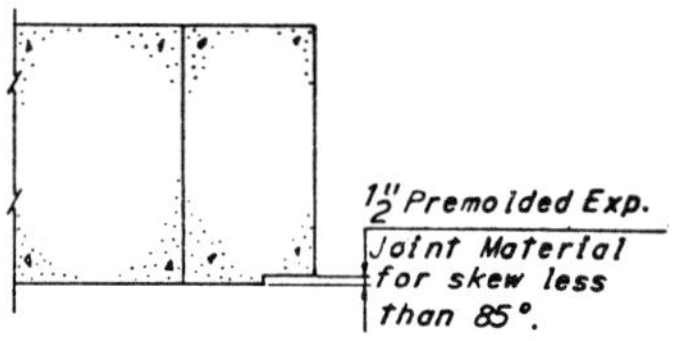


SECTION E-E

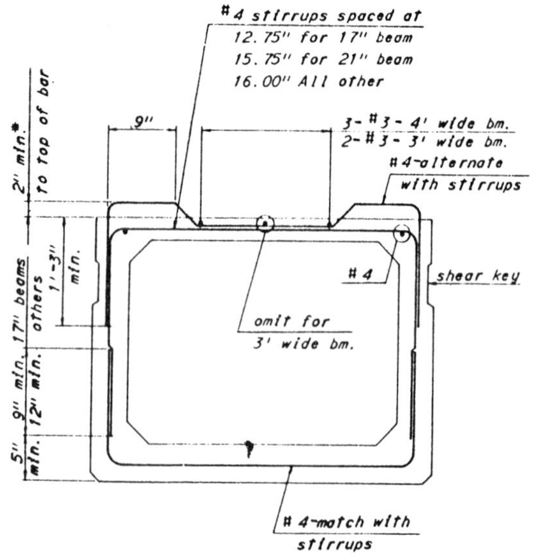


3' AND 4' ADJACENT BOX BEAM
COMPOSITE

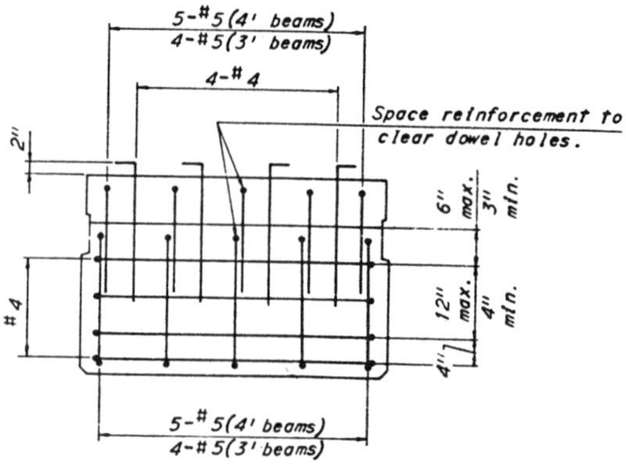


VIEW B-B

Increase reinforcement for sharp skew and highly concentrated prestressing force.

## ABUTMENT WATERPROOFING DETAILS

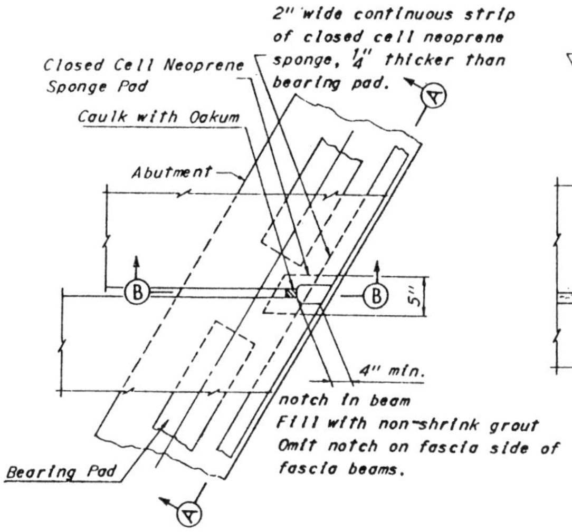


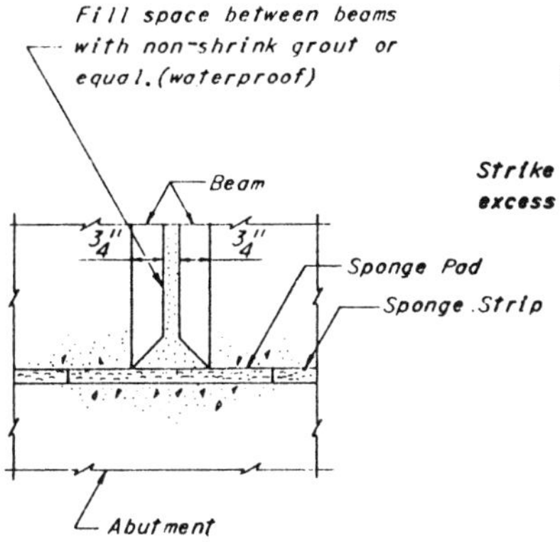


SECTION A-A

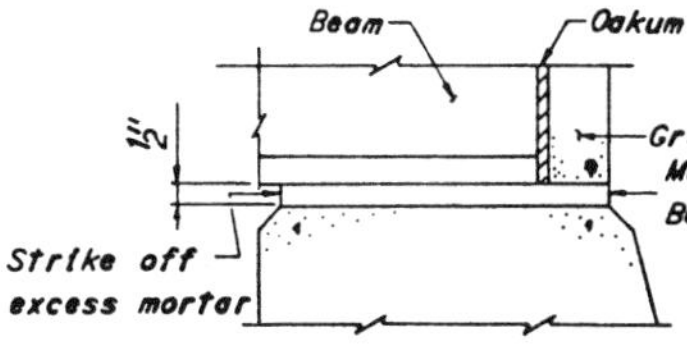


DETAIL E

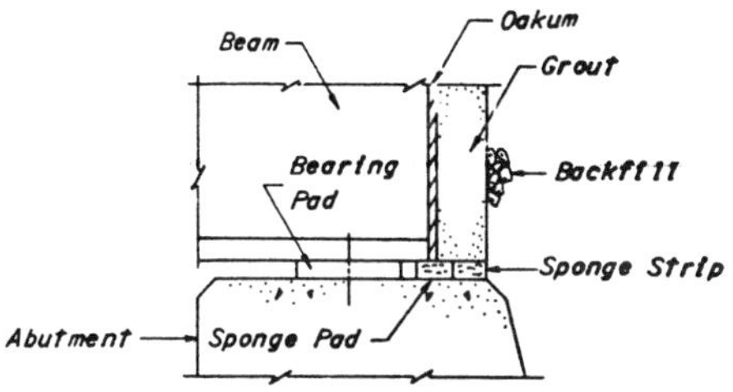


SECTION B-B

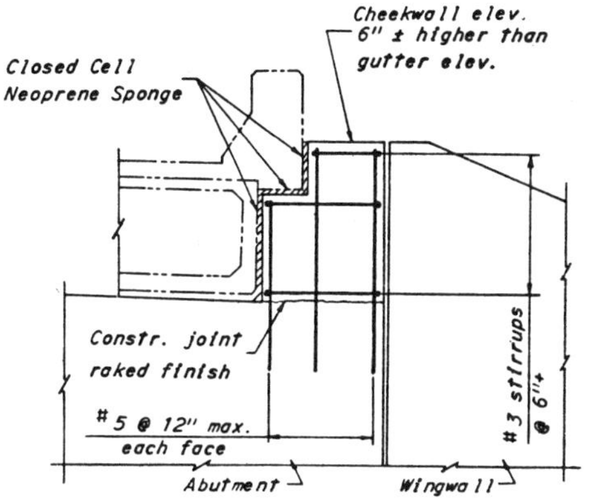


CHEEKWALL DETAIL

## NEOPRENE BEARING DETAILS

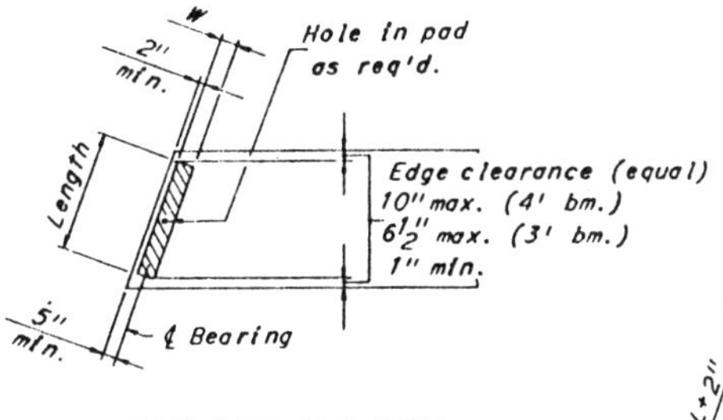


ONE PAD PER BOX

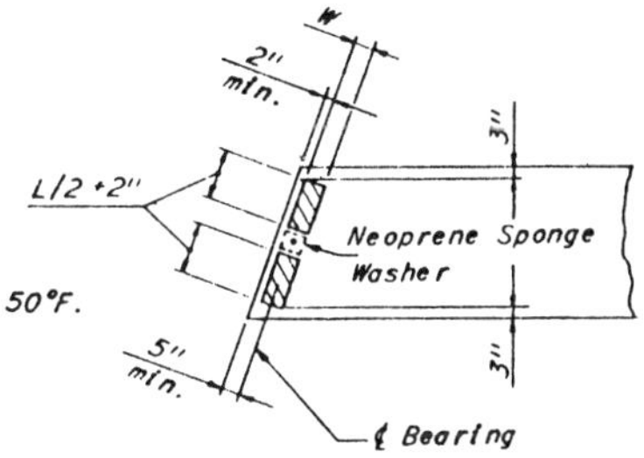


TWO PADS PER BOX

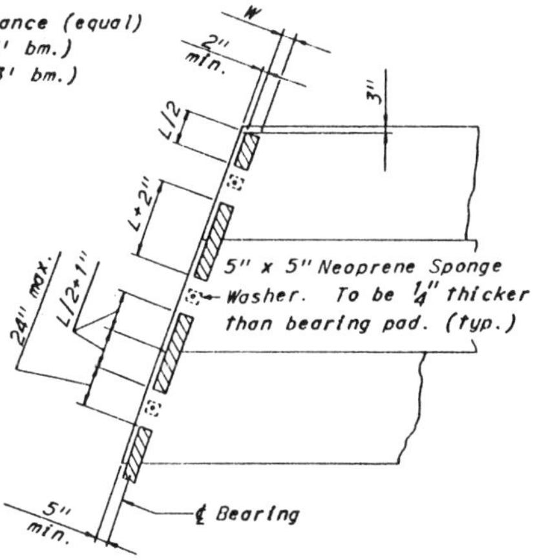


COMMON PAD

*Plain Neoprene Bearing Pads Design Criteria*

1. Maximum shear strain = 25%
2. Maximum change in temperature of concrete = 50°F.
3. Thermal coefficient of concrete = 0.000006.
4. Maximum D.L. + L.L. bearing pressure
   600 psi for cut pads (used for design)
   700 psi for moulded pads
5. Maximum compressive strain = 15%.
6. Minimum D. L. bearing pressure = 200 psi

*Instructions*

1. Neoprene to be 50 or 60 Durometer.
2. T = 0.0144 x Beam Length in Feet (1 1/4" max.)
3. Check for bearing hardness:

   $$\text{Compr. Stress (psi)} = \frac{D.L. + L.L.}{LW}$$

   $$\text{Shape Factor} = \frac{LW}{2(L + W))(T)}$$

   For compressive stress'strain curves and Shape Factor information, refer to manufacturers' recommendations.
4. Check Slippage (Expansion pads)

   Beam travel pad can absorb without

   $$\text{Slippage (in.)} = \frac{(D.L.\ in\ lbs.)}{5LW} T \times \frac{1.8}{K}$$

   K = 110 (50 Durometer)
   160 (60 Durometer)

   Beam travel anticipated (in.)
   = 0.00504 x Beam length in feet
   Temperature range = 70°

   Beam travel anticipated to be less than beam travel pad can absorb without slippage.

### DIMENSIONS (INCHES) OF 50 DUROMETER NEOPRENE BEARING PADS - EXPANSION

| BEAM WIDTH | SPAN RANGE FEET | T IN. | 17 W | 17 L | 21 W | 21 L | 27 W | 27 L | 33 W | 33 L | 36 W | 36 L | 39 W | 39 L | 42 W | 42 L | 45 W | 45 L | 48 W | 48 L |
|---|---|---|---|---|---|---|---|---|---|---|---|---|---|---|---|---|---|---|---|---|
| 4 FEET | 30-43 | 3/4 | 4 | 30* | 4 | 30* | | | | | | | | | | | | | | |
| | 44-49 | | 4 | 30* | 4 | 30 | 4 | 30 | | | | | | | | | | | | |
| | 50-52 | | | | 4 | 30 | 5 | 30 | | | | | | | | | | | | |
| | 53-60 | 1 | | | 5 | 30 | 5 | 32 | 5 | 34 | | | | | | | | | | |
| | 61-69 | | | | | | 5 | 34 | 5 | 34 | 5 | 34 | | | | | | | | |
| | 70-80 | 1 1/4 | | | | | 5 | 34 | 6 | 36 | 6 | 36 | 6 | 38 | 6 | 38 | | | | |
| | 81-87 | | | | | | | | 6 | 38 | 6 | 38 | 6 | 38 | 6 | 38 | 6 | 38 | 6 | 40 |
| 3 FEET | 30-43 | 3/4 | 4 | 24* | 4 | 24* | | | | | | | | | | | | | | |
| | 44-49 | | 4 | 24* | 4 | 24 | 4 | 24 | | | | | | | | | | | | |
| | 50-52 | | | | 4 | 24 | 5 | 24 | | | | | | | | | | | | |
| | 53-60 | 1 | | | 5 | 24 | 6 | 24 | 6 | 24 | | | | | | | | | | |
| | 61-69 | | | | | | 6 | 24 | 6 | 24 | 6 | 24 | | | | | | | | |
| | 70-80 | 1 1/4 | | | | | 7 | 26 | 7 | 26 | 7 | 26 | 7 | 26 | | | | | | |
| | 81-87 | | | | | | | | 7 | 26 | 7 | 26 | 7 | 26 | 7 | 26 | 7 | 26 | 7 | 26 |

(Beam depth columns: BEAM DEPTH - INCHES)

T = Thickness W = Width L = Length

*DL bearing pressure is less than 200 psi.

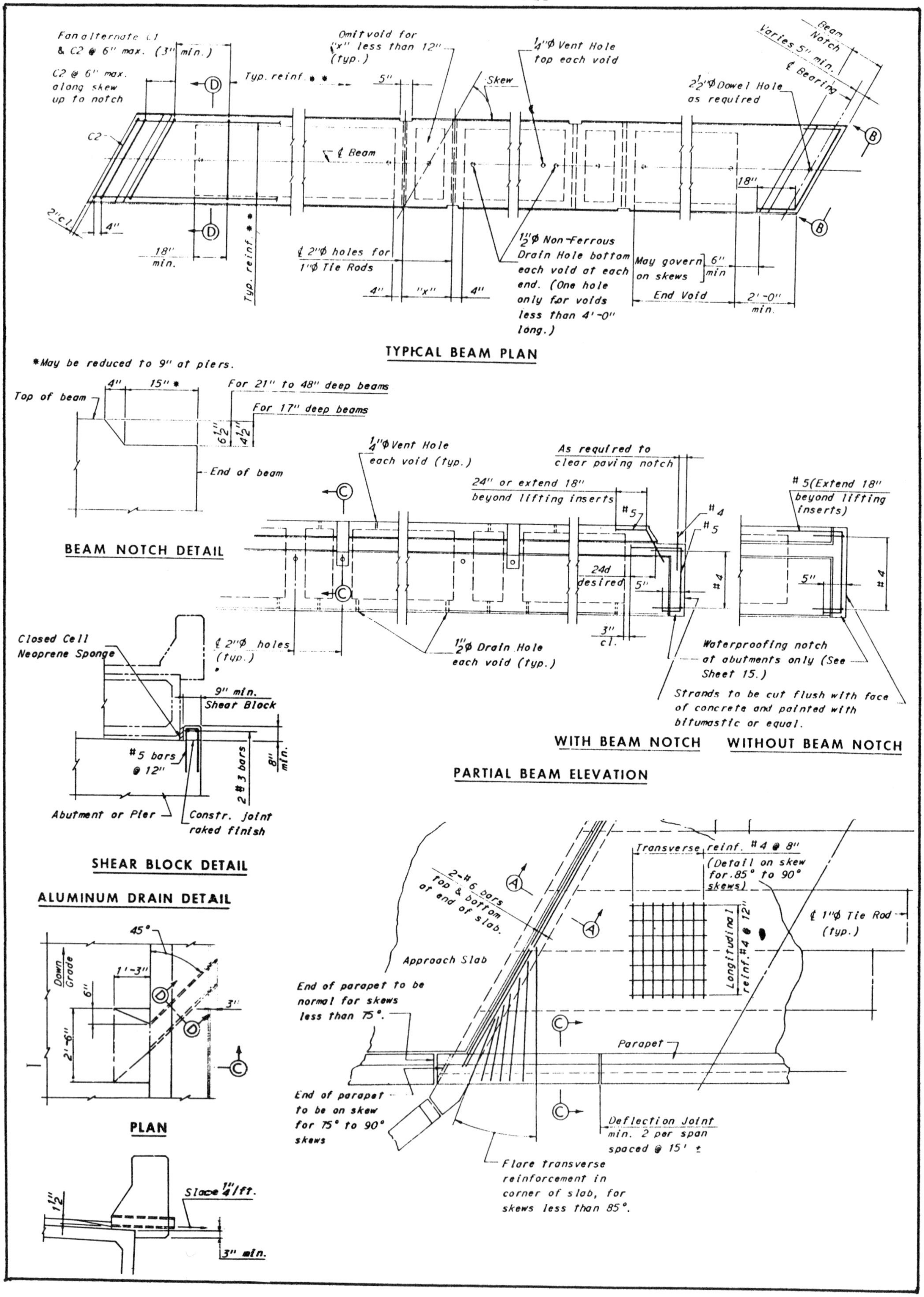
Fan alternate C1 & C2 @ 6" max. (3" min.)
C2 @ 6" max. along skew up to notch
C2
2" cl.
4"
18" min.
Typ. reinf. * *
5"
℄ Beam
Omit void for "x" less than 12" (typ.)
Skew
1/4"Ø Vent Hole top each void
Beam Notch
Varies 5" min.
℄ Bearing
2 1/2"Ø Dowel Hole as required
18"
℄ 2"Ø holes for 1"Ø Tie Rods
4"
"x"
1 1/2"Ø Non-Ferrous Drain Hole bottom each void at each end. (One hole only for voids less than 4'-0" long.)
May govern on skews
6" min
End Void
2'-0" min.
TYPICAL BEAM PLAN
*May be reduced to 9" at piers.
Top of beam
4"
15" *
For 21" to 48" deep beams
For 17" deep beams
6 1/2"
4 1/2"
End of beam
BEAM NOTCH DETAIL
1/4"Ø Vent Hole each void (typ.)
As required to clear paving notch
24" or extend 18" beyond lifting inserts
#5
#4
#5
#5 (Extend 18" beyond lifting inserts)
24d desired
5"
#4
3" cl.
℄ 2"Ø holes (typ.)
1 1/2"Ø Drain Hole each void (typ.)
Waterproofing notch at abutments only (See Sheet 15.)
Strands to be cut flush with face of concrete and painted with bitumastic or equal.
WITH BEAM NOTCH
WITHOUT BEAM NOTCH
PARTIAL BEAM ELEVATION
Closed Cell Neoprene Sponge
9" min. Shear Block
#5 bars @ 12"
2 #3 bars
8" min.
Abutment or Pier
Constr. joint raked finish
SHEAR BLOCK DETAIL
ALUMINUM DRAIN DETAIL
45°
Down Grade
1'-3"
6"
2'-6"
3"
PLAN
1 1/2"
Slope 1/4"/ft.
3" min.
2-#6 bars top & bottom at end of slab.
Approach Slab
Transverse reinf. #4 @ 8"
(Detail on skew for 85° to 90° skews)
Longitudinal reinf. #4 @ 12"
℄ 1"Ø Tie Rod (typ.)
End of parapet to be normal for skews less than 75°.
Parapet
End of parapet to be on skew for 75° to 90° skews
Deflection Joint min. 2 per span spaced @ 15' ±
Flare transverse reinforcement in corner of slab, for skews less than 85°.

# D.6 CHANNEL BEAM DESIGN AND DETAILS

## PRESTRESSED CHANNEL BEAMS
## DESIGN AND ASSEMBLY DETAILS

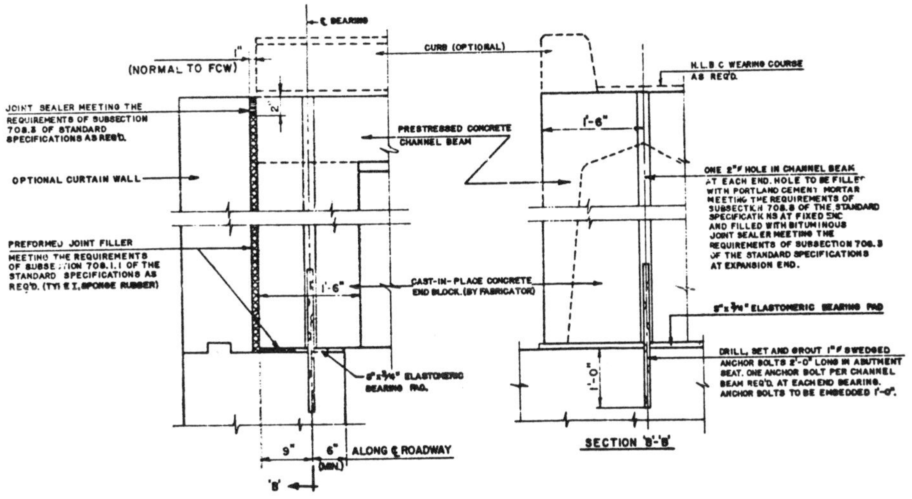


TYPICAL END BEARING DETAIL
(ALONG ℄ ROADWAY)

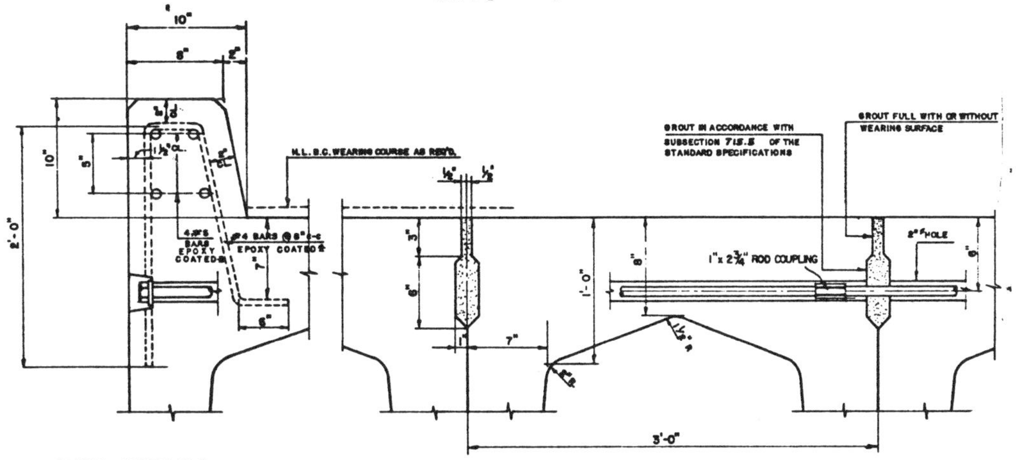


CURB DETAILS

SHEAR KEY DETAILS

TYPICAL BEAM SECTION

## DESIGN AND FABRICATION NOTES

REFER TO THE APPROPRIATE STANDARD PLAN SHEET FOR DESIGN STRESSES, SPECIFICATIONS OR NOTES.

ALTHOUGH THE PLANS ARE DETAILED FOR A PARTICULAR TYPE OF PRESTRESSED CONCRETE BEAM, ALTERNATE TYPES OR SHAPED PRESTRESSED CONCRETE BEAMS MAY BE FURNISHED WITH THE FOLLOWING STIPULATIONS:

(a) SUPPLIER MUST SUBMIT PROPOSED ALTERNATE WITH DESIGN COMPUTATIONS FOR REVIEW AND APPROVAL BY THE DEPARTMENT OF HIGHWAYS.
(b) CONTRACTOR MUST SUPPLY REVISED MODIFIED CONTRUCTION PLANS SHOWING ALL REVISIONS AND MODIFICATIONS AS REQUIRED BY THE USE OF THE ALTERNATE BEAM FOR REVIEW AND APPROVAL BY THE DEPARTMENT OF HIGHWAYS.
(c) COMPLETION DATE OF PROJECT WILL NOT BE EXTENDED DUE TO ANY DELAY ENCOUNTERED IN OBTAINING ALTERNATE BEAM AND REVISED MODIFIED PLAN APPROVAL BY THE DEPARTMENT OF HIGHWAYS.
(d) THE PROJECT CANNOT BE STARTED UNTIL THE REVISED MODIFIED PLANS ARE APPROVED BY THE DEPARTMENT OF HIGHWAYS.

ALL MEMBERS ARE DESIGNED FOR HS 20-44 LIVE LOADING IN ACCORDANCE WITH 1973 AASHTO STANDARD SPECIFICATIONS FOR HIGHWAY BRIDGES. THE DESIGN PROVIDES FOR A WEARING SURFACE OF 25 POUNDS PER SQUARE FOOT OF ROADWAY.

COMPRESSIVE STRENGTH OF CONCRETE AT 28 DAYS ($f'c$) SHALL NOT BE LESS THAN 5,000 PSI.

COMPRESSIVE STRENGTH OF CONCRETE AT TIME OF PRESTRESSING ($f'ci$) SHALL NOT BE LESS THAN 4,000 PSI.

PRESTRESSING STEEL SHALL BE SEVEN-WIRE UNCOATED STRESS-RELIEVED STRAND IN ACCORDANCE WITH 270K GRADE ASTM A416 WITH MINIMUM TENSILE STRENGTH($f's$) OF 270,000 PSI.

ALLOWABLE TEMPORARY STRESSES: AT THE TIME OF TRANSFER THE UNIT STRESSES SHALL NOT EXCEED THE FOLLOWING: CONCRETE COMPRESSION = $0.60f'ci$; TENSION $3\sqrt{f'ci}$ PRESTRESSING STEEL TENSION $0.7f's$.

ALLOWABLE STRESSES AT DESIGN LOAD AFTER LOSSES OF PRESTRESS: THE UNIT STRESSES SHALL NOT EXCEED THE FOLLOWING: CONCRETE COMPRESSION = $0.40f'c$; TENSION = $3\sqrt{f'c}$ BUT NOT TO EXCEED 250 PSI, PRESTRESSING STEEL TENSION = $0.60f's$ OR $0.80fsy$, WHICHEVER IS SMALLER; REINFORCING STEEL ($fs$) = 20,000 PSI.

ULTIMATE LOAD CAPACITY SHALL NOT BE LESS THAN: $\frac{1.30}{\phi}\left[D+\frac{5}{3}(L+I)\right]$, $\phi=1$ FOR FACTORY PRODUCED PRECAST PRESTRESSED CONCRETE MEMBERS.

WHERE THE CALCULATED TENSILE STRESS IN CONCRETE EXCEEDS THE VALUES ALLOWED, AUXILLIARY NON-PRESTRESSED REINFORCING SHALL BE PROVIDED TO RESIST THE TOTAL TENSION.

ALL ANCHOR BOLTS AND DOWELS (ASTM A-36), INSERTS, NUTS, WASHERS AND TIE BOLTS (ASTM A-36) AS WELL AS THE U-BOLTS AND NUTS (ASTM A-307) FOR THE PIPE RAIL AND SPECIAL GUARDRAIL PARTS SHALL BE GALVANIZED IN ACCORDANCE WITH ASTM A 153 (2 OZ. PER SQ. FT. COATING OF GALVANIZING SPELTER MIN.), EXCEPT THE 3/4"⌀ BOLTS AND INSERTS SPECIFIED IN CONJUNCTION WITH THE END BLOCKS WILL NOT REQUIRE GALVANIZING SINCE SAME ARE TOTALLY ENCASED IN CONCRETE.

GUARDRAIL POSTS, CONNECTION ℙ'S AND DRAIN ℙ'S (ASTM A-36) TO BE GALVANIZED IN ACCORDANCE WITH ASTM A-123.

THE GUARDRAIL WITH BUFFER END TERMINAL SECTIONS SHALL CONFORM TO THE REQUIREMENTS AS SET FORTH BY THE W. VA. DEPARTMENT OF HIGHWAYS STANDARD DETAIL. "GUARDRAIL ELEMENT", DATED OCT. 20, 1972.

TIE BOLTS SHALL CONFORM TO ASTM A-36. TIE BOLTS TO BE TIGHTENED SNUG TIGHT AFTER FINAL POSITIONING OF CHANNEL BEAMS IN PLACE.

ALL WELDING TO BE ACCOMPLISHED PRIOR TO GALVANIZING.

DEBONDING OF STRANDS SHALL BE ACCOMPLISHED BY TAPING OR TIGHT-FIT PLASTIC TUBING. LENGTHS OF DEBONDING SHALL BE MEASURED FROM ENDS OF CHANNEL BEAMS. LONGER LENGTHS OF DEBONDING SHALL BE APPLIED ON LOWERMOST STRANDS.

OPTIONAL CURB TO BE CAST MONOLITHIC WITH BEAM.

ALL EXPOSED THREADED PORTIONS OF THE HANDRAIL SHALL BE COATED WITH LIQUID GALVANIZE, "GALVICON", OR APPROVED EQUAL, AFTER ASSEMBLY.

WVA

# 24 INCH CHANNEL BEAM

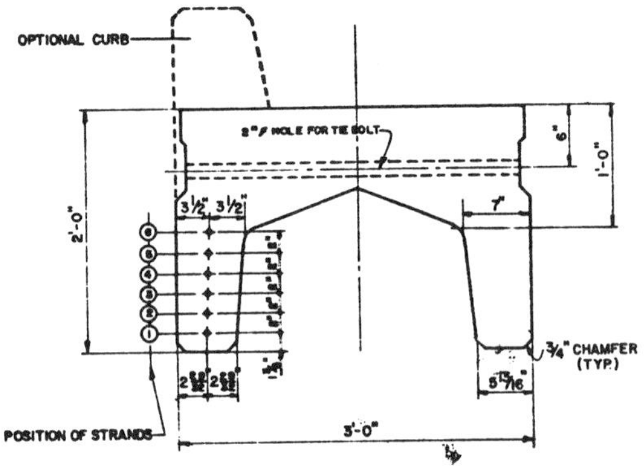


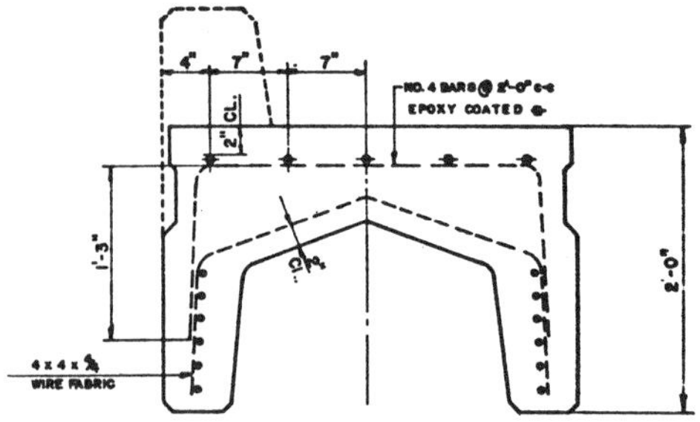


| DESIGN DATA FOR 24" DEPTH CHANNEL BEAM | | | | | | | | | | | | |
|---|---|---|---|---|---|---|---|---|---|---|---|---|
| SPAN Ⓛ C-C OF BEARING | | 20' | 22' | 24' | 26' | 28' | 30' | 32' | 34' | 36' | 38' | 40' |
| OVERALL LENGTH ALONG ℄ ROADWAY | | 21'-6" | 23'-6" | 25'-6" | 27'-6" | 29'-6" | 31'-6" | 33'-6" | 35'-6" | 37'-6" | 39'-6" | 41'-6" |
| NO. OF 270K ½"ϕ STRANDS | | 4 | | | 6 | | | 8 | | 10 | | 12 |
| STRAND POSITIONS | | ① AND ② | | | ① THRU ③ | | | ① THRU ④ | | ① THRU ⑤ | | ① THRU ⑥ |
| INITIAL PRESTRESSING FORCE (KIPS) | | 115 | | | 173 | | | 231 | | 289 | | 347 |
| FINAL PRESTRESSING FORCE (KIPS) | | 92 | | | 139 | | | 185 | | 231 | | 277 |
| RESISTING MOMENT (FT. KIPS) | | 141 | | | 187 | | | 226 | | 257 | | 280 |
| BENDING MOMENT (FT. KIPS) | | 102 | 116 | 138 | 145 | 161 | 177 | 196 | 212 | 230 | 248 | 268 |
| NO. OF #5 LONGITUDINAL BARS EPOXY COATED * | | 5 | 5 | 5 | 5 | 5 | 5 | 5 | 5 | 5 | 5 | 5 |
| NO. & SPACING OF TIE BOLTS | | 2@10' | 2@10' | 2@10' | 2@12' | 2@12' | 2@12' | 2@14' | 2@14 | 2@14' | 2@16' | 2@16' |
| NO. AND LENGTHS OF DEBONDING STRANDS | 2'-6" | | | | | | | | | | | |
| | 2'-0" | 2 | | | 2 | | | | | | | |
| | 1'-6" | | | | | 2 | | 2 | 2 | 2 | | 2 |
| | 1'-0" | | 2 | | | | 2 | 2 | | 2 | 4 | |
| WEIGHT OF TYPICAL MEMBER (TONS) | INTERIOR | 6.6 | 7.2 | 7.8 | 8.4 | 9.0 | 9.6 | 10.2 | 10.8 | 11.4 | 12.1 | 12.7 |
| | W/10" CURB | 7.6 | 8.3 | 9.0 | 9.7 | 10.4 | 11.1 | 11.8 | 12.5 | 13.2 | 13.9 | 14.6 |
| | W/BARRIER | 9.8 | 10.8 | 11.7 | 12.6 | 13.5 | 14.4 | 15.3 | 16.2 | 17.2 | 18.1 | 19.0 |
| NO. AND SPACINGS OF RAIL POSTS | NO. | 7 | 8 | 9 | 9 | 10 | 11 | 11 | 12 | 12 | 13 | 14 |
| | END OF BEAM TO ℄ 1ST POST | 1'-4½" | 1'-0" | 1'-0" | 1'-3" | 1'-0" | 1'-0" | 1'-1½" | 1'-0" | 1'-6¾" | 1'-0" | 1'-0" |
| | ℄ FIRST POST TO ℄ 2ND POST | 3'-1½" | 2'-11¼" | 2'-4½" | 3'-1½" | 2'-9¾" | 2'-3" | 3'-1½" | 2'-8¼" | 3'-1½" | 3'-1½" | 2'-6¾" |

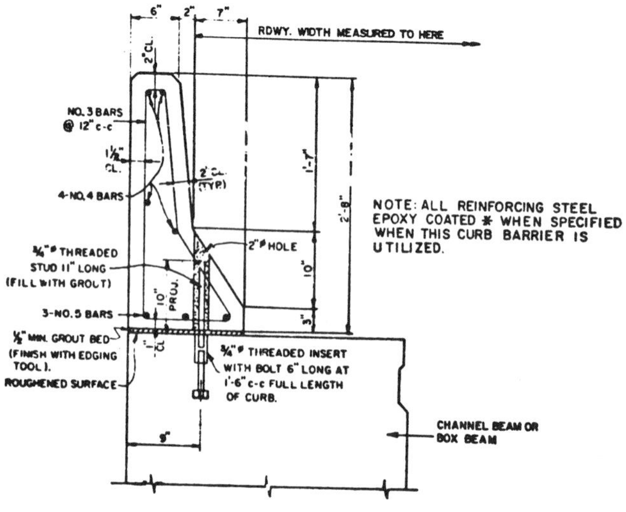


BARRIER CURB
TYPICAL SECTION

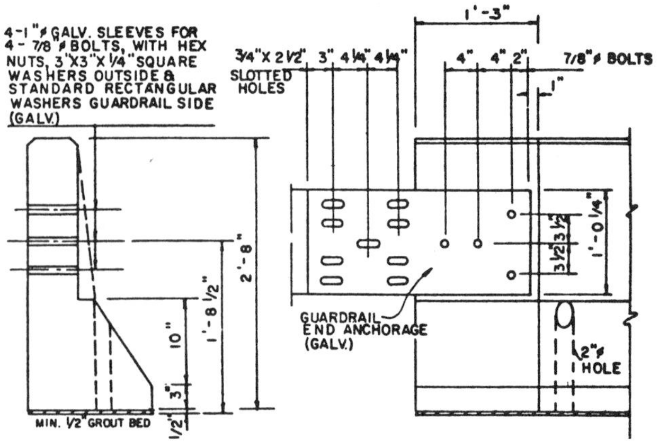


OPTION NO. 2
CURB BARRIER
GUARDRAIL ATTACHMENT

## 32 INCH CHANNEL BEAM

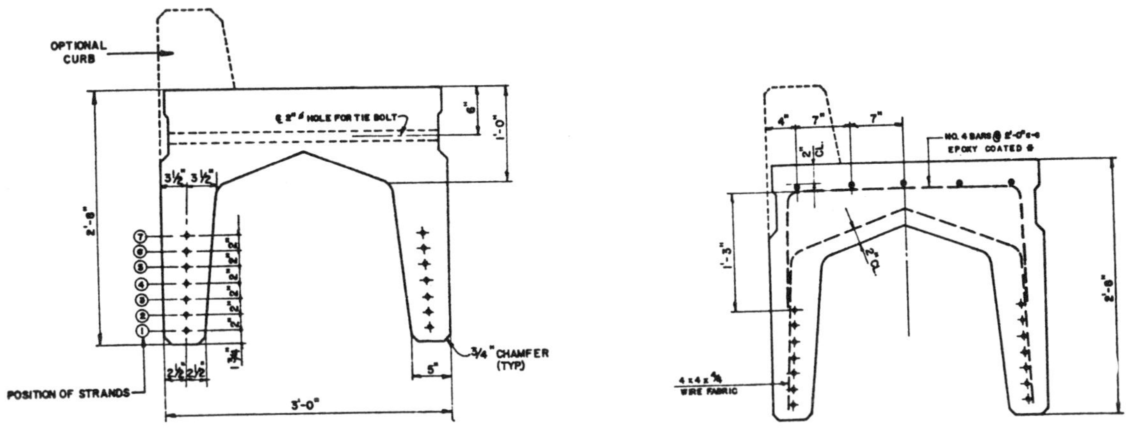


TYPICAL SECTION

| DESIGN DATA FOR 32" DEPTH CHANNEL BEAM | | | | | | | | | | | | | | |
|---|---|---|---|---|---|---|---|---|---|---|---|---|---|---|
| SPAN (L) C-C OF BEARING | | 30' | 32' | 34' | 36' | 38' | 40' | 42' | 44' | 46' | 48' | 50' | 52' | 54' |
| OVERALL LENGTH ALONG ℄ ROADWAY | | 31'-6" | 33'-6" | 35'-6" | 37'-6" | 39'-6" | 41'-6" | 43'-6" | 45'-6" | 47'-6" | 49'-6" | 51'-6" | 53'-6" | 55'-6" |
| NO. OF 270K 1/2" ⌀ STRANDS | | 6 | | | | | 8 | | | 10 | | 12 | | 14 |
| STRAND POSITIONS | | (1) THRU (3) | | | | | (1) THRU (4) | | | (1) THRU (5) | | (1) THRU (6) | | (1) THRU (7) |
| INITIAL PRESTRESSING FORCE (KIPS) | | 173 | | | | | 231 | | | 289 | | 347 | | 405 |
| FINAL PRESTRESSING FORCE (KIPS) | | 139 | | | | | 185 | | | 231 | | 277 | | 323 |
| RESISTING MOMENT (FT.-KIPS) | | 278 | | | | | 342 | | | 398 | | 446 | | 487 |
| BENDING MOMENT (FT.-KIPS) | | 188 | 205 | 225 | 244 | 264 | 286 | 307 | 330 | 354 | 378 | 404 | 430 | 459 |
| NO. OF #5 LONGITUDINAL BARS EPOXY COATED * | | 5 | 5 | 5 | 5 | 5 | 5 | 5 | 5 | 5 | 5 | 5 | 5 | 5 |
| NO. & SPACING OF TIE BOLTS | | 2 @ 12' | 2 @ 14' | 2 @ 14' | 2 @ 14' | 2 @ 16' | 2 @ 16' | 2 @ 16' | 2 @ 16' | 2 @ 16' | 2 @ 18' | 2 @ 18' | 2 @ 18' | 2 @ 20' |
| NO. AND LENGTHS OF DEBONDING STRANDS | 2'-6" | | | | | | | | | | | | | |
| | 2'-0" | | | | | | | | | 2 | | | 2 | |
| | 1'-6" | | | | | | 2 | | | | 2 | 2 | | 2 |
| | 1'-0" | 2 | 2 | 2 | | | | 2 | 2 | | | 2 | | 2 |
| WEIGHT OF TYPICAL MEMBER (TONS) | INTERIOR | 11.0 | 11.7 | 12.4 | 13.1 | 13.8 | 14.5 | 15.2 | 15.9 | 16.6 | 17.3 | 18.1 | 18.7 | 19.4 |
| | W/10" CURB | 12.5 | 13.3 | 14.1 | 14.9 | 15.7 | 16.5 | 17.3 | 18.1 | 18.9 | 19.7 | 20.4 | 21.2 | 22.0 |
| | W/BARRIER | 15.8 | 16.8 | 17.8 | 18.8 | 19.9 | 20.9 | 21.9 | 22.9 | 23.9 | 24.9 | 25.9 | 26.9 | 27.9 |
| NO. AND SPACINGS OF GUARDRAIL POSTS | NO. | 11 | 11 | 12 | 12 | 13 | 14 | 14 | 15 | 16. | 16 | 17 | 18 | 18 |
| | END OF BEAM TO ℄ 1ST POST | 1'-0" | 1'-1 1/2" | 1'-0" | 1'-6 3/4" | 1'-0" | 1'-0" | 1'-5 1/4" | 1'-0" | 1'-0" | 1'-3 3/4" | 1'-0" | 1'-0" | 1'-2 1/4" |
| | ℄ FIRST POST TO ℄ 2ND POST | 2'-3" | 3'-1 1/2" | 2'-8 1/4" | 3'-1 1/2" | 3'-1 1/2" | 2'-6 3/4" | 3'-1 1/2" | 3'-0" | 2'-5 1/4" | 3'-1 1/2" | 2'-10 1/2" | 2'-3 3/4" | 3'-1 1/2" |

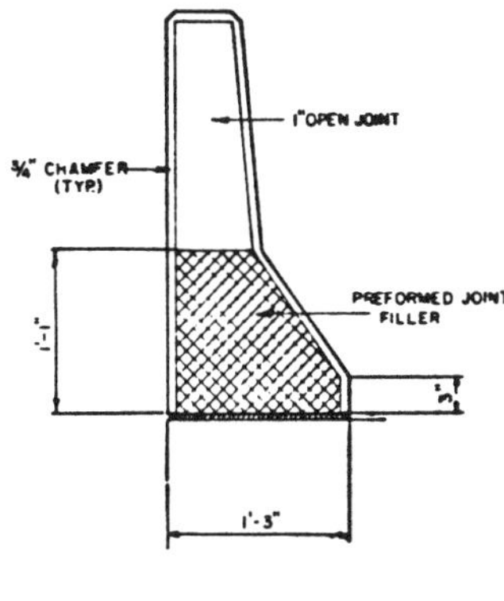


JOINT FILLER DETAIL

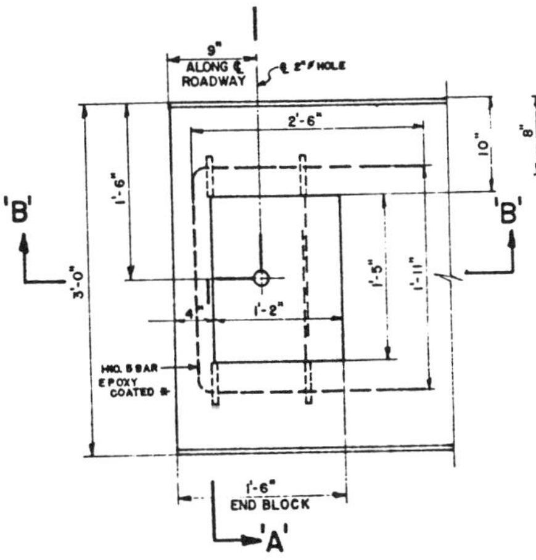


PLAN

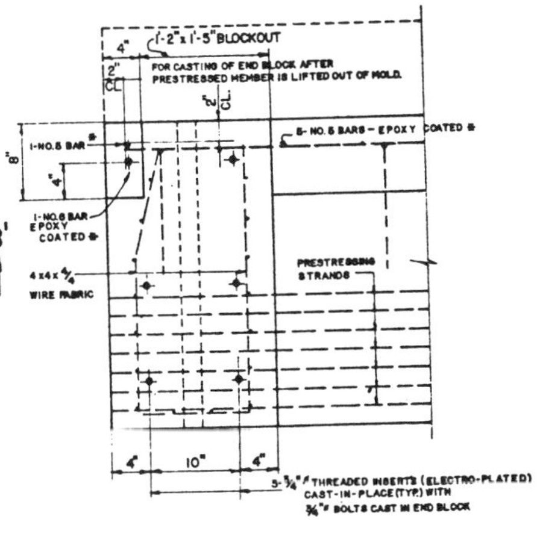


SECTION 'B'-'B'

END BLOCK DETAILS

## D.7 PLANK BEAM DESIGN AND DETAILS

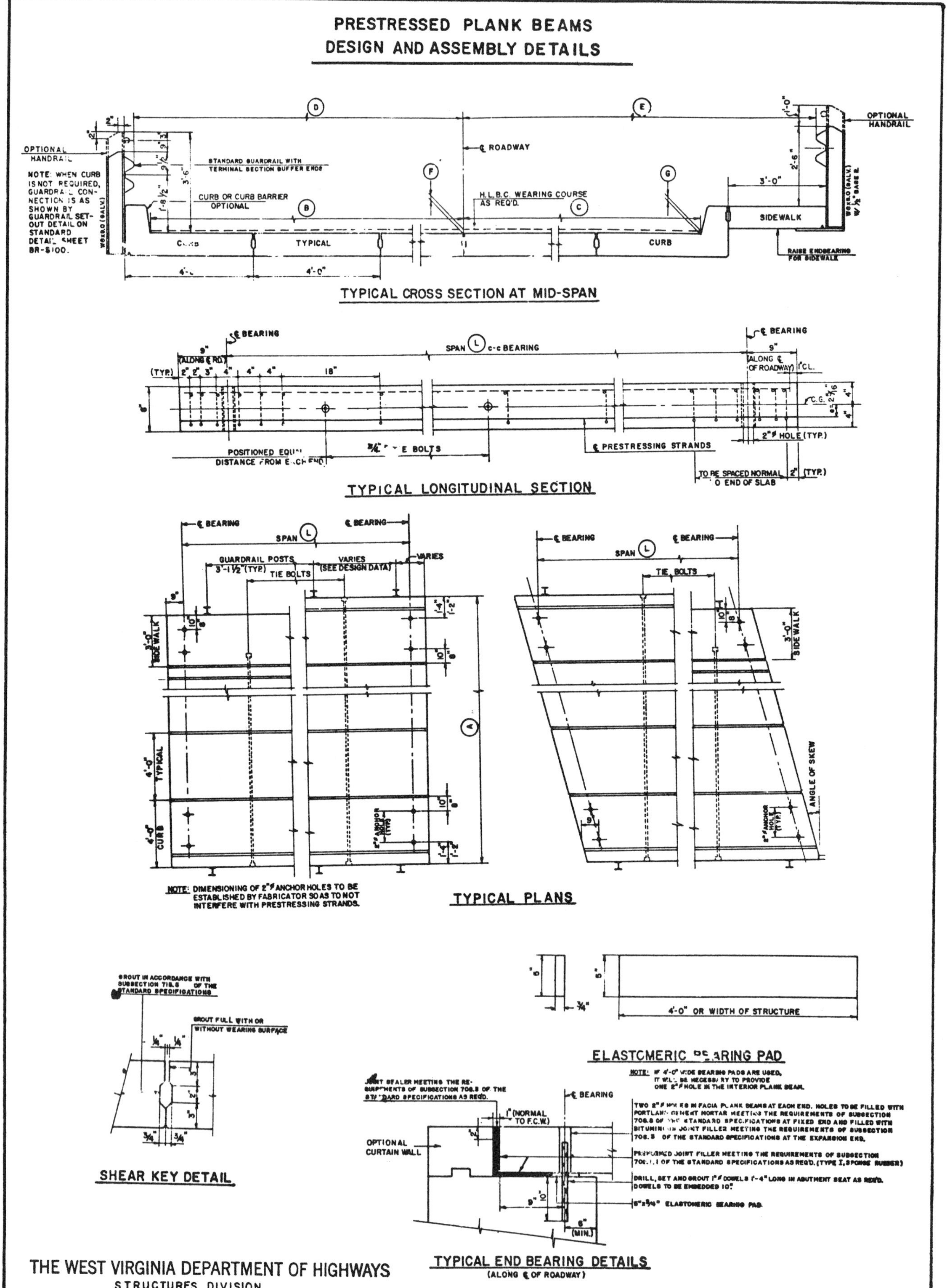


THE WEST VIRGINIA DEPARTMENT OF HIGHWAYS
STRUCTURES DIVISION

# PLANK SLAB BEAM

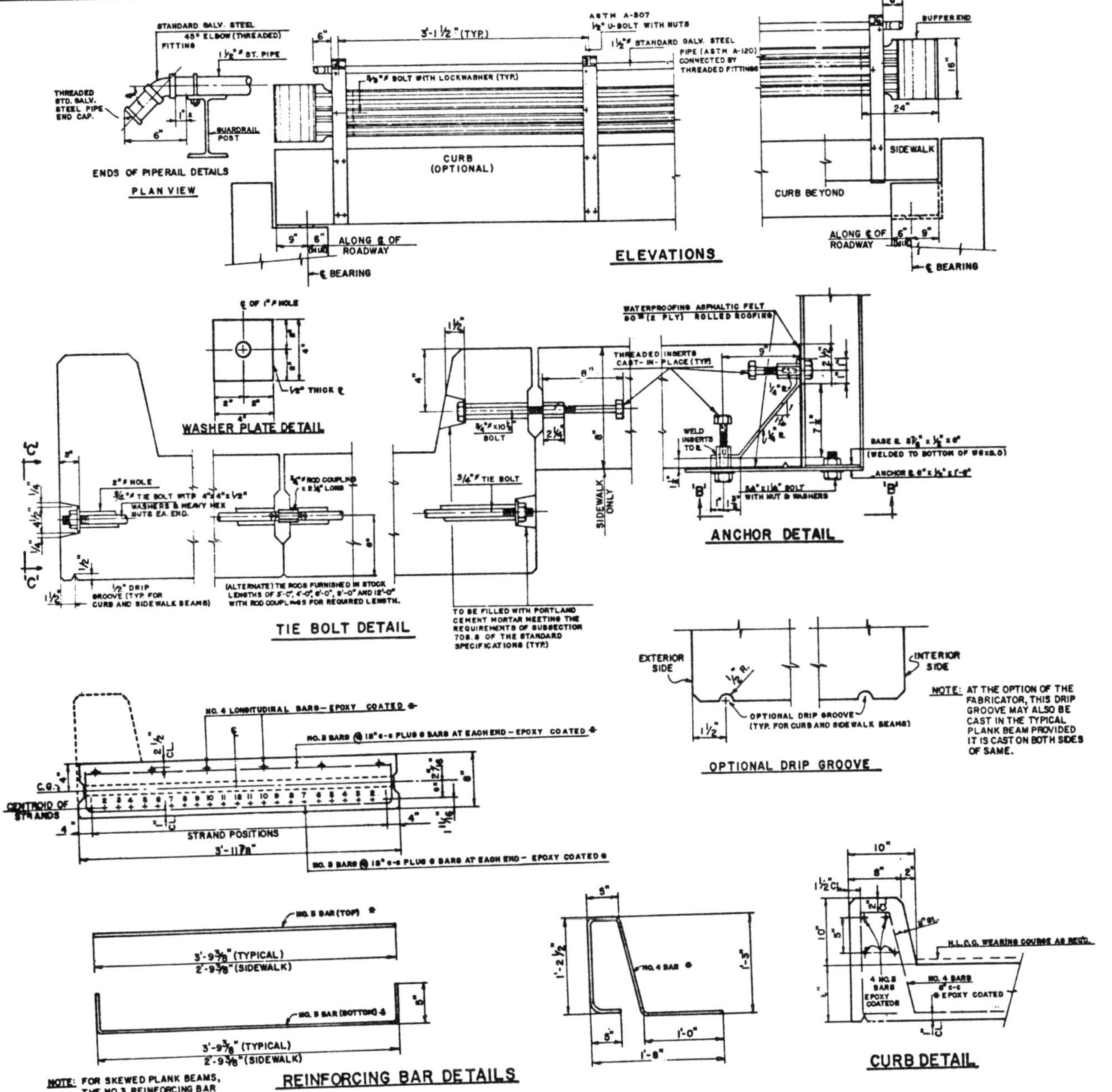


| DESIGN DATA FOR 8" x 4'-0" PLANK BEAM | | | | | | | | |
|---|---|---|---|---|---|---|---|---|
| SPAN (L) c-c OF BEARING | | 8'-0" | 9'-0" | 10'-0" | 11'-0" | 12'-0" | 13'-0" | 14'-0" |
| OVERALL LENGTH (ALONG ℄ OF ROADWAY) | | 9'-6" | 10'-6" | 11'-6" | 12'-6" | 13'-6" | 14'-6" | 15'-6" |
| NO. OF 270K 3/8"⌀ STRANDS (SEE SHOP DRAWINGS FOR STRAND POSITIONS) | | 9 | 10 | 12 | 13 | 15 | 17 | 19 |
| INITIAL PRESTRESSING FORCE (KIPS) | | 144 | 160 | 192 | 208 | 240 | 272 | 304 |
| FINAL PRESTRESSING FORCE (KIPS) | | 117 | 130 | 156 | 169 | 195 | 221 | 247 |
| BENDING MOMENT (FT-KIPS) | | 41.4 | 47.2 | 53.0 | 59.0 | 65.2 | 71.4 | 77.8 |
| TOTAL RESISTING MOMENT (FT-KIPS) | | 46.2 | 50.1 | 57.9 | 61.9 | 69.8 | 77.7 | 85.7 |
| NO. OF #4 LONGITUDINAL BARS EPOXY COATED * | | 4 | 4 | 6 | 7 | 7 | 8 | 9 |
| NO. AND SPACING OF TIE BOLTS | | 2 @ 4' | 2 @ 4' | 2 @ 4' | 2 @ 6' | 2 @ 6' | 2 @ 6' | 2 @ 6' |
| WEIGHT OF 4'-0" PLANK BEAMS (TONS) | INTERIOR | 1.90 | 2.10 | 2.30 | 2.50 | 2.70 | 2.90 | 3.10 |
| | 10" CURB | 2.34 | 2.59 | 2.84 | 3.08 | 3.33 | 3.58 | 3.82 |
| | CURB BARRIER | 3.35 | 3.70 | 4.06 | 4.41 | 4.76 | 5.11 | 5.47 |
| NUMBER OF GUARDRAIL POSTS REQUIRED | | 3 | 4 | 4 | 4 | 5 | 5 | 5 |
| END OF SLAB TO ℄ FIRST POST | | 1'-7 1/2" | 1'-0" | 1'-0 3/4" | 1'-6 3/4" | 1'-0" | 1'-0" | 1'-6" |
| ℄ OF FIRST POST TO ℄ SECOND POST | | 3'-1 1/2" | 2'-8 1/4" | 3'-1 1/2" | 3'-1 1/2" | 2'-7 1/2" | 3'-1 1/2" | 3'-1 1/2" |

* NOT REQUIRED WITH WELDED WIRE FABRIC REINFORCEMENT.

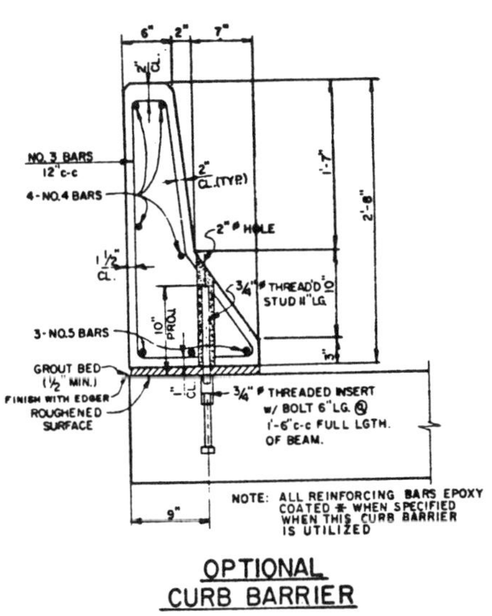


OPTIONAL CURB BARRIER

# PLANK SLAB BEAM

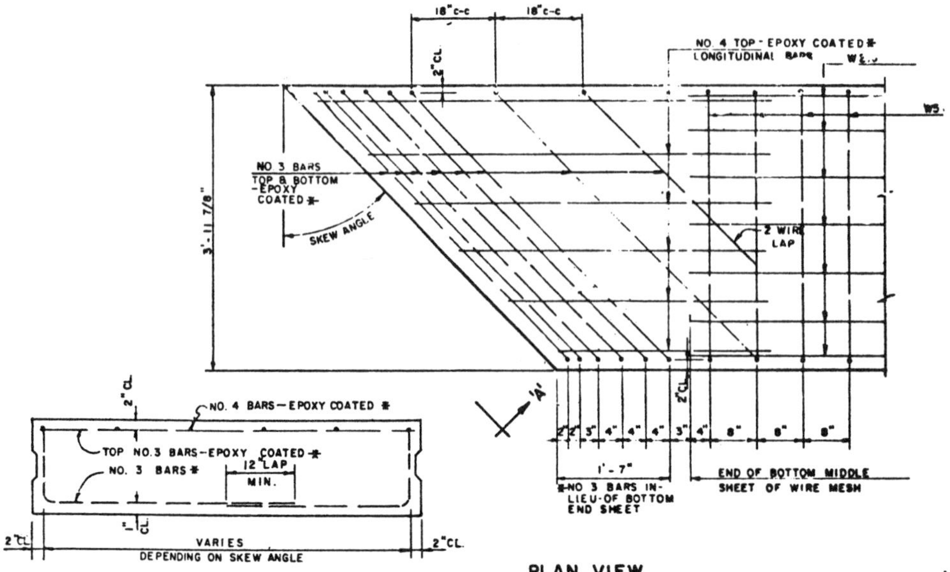


VIEW 'A'-'A'

PLAN VIEW

END SKEW DETAIL UTILIZING WELDED WIRE FABRIC

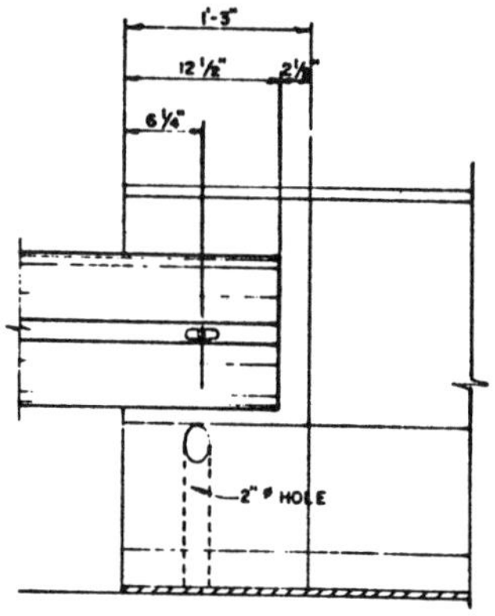


ALL REINFORCING BARS EPOXY COATED # WHEN SPECIFIED WHEN THIS CURB BARRIER IS UTILIZED

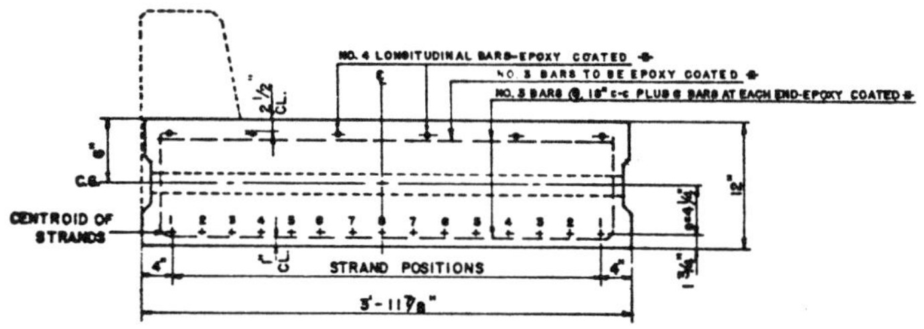


TYPICAL SECTION

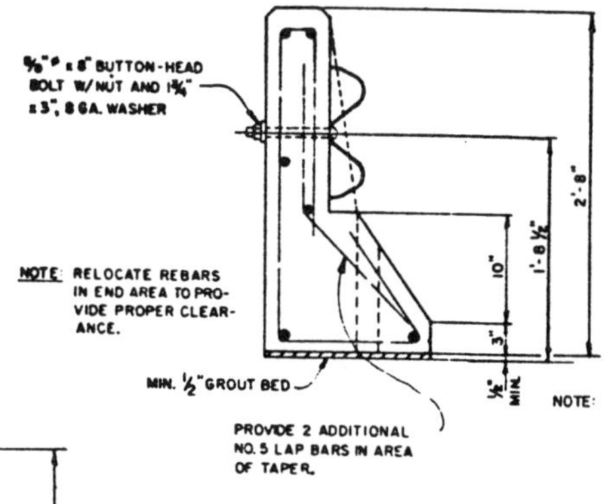


CURB BARRIER
GUARDRAIL ATTACHMENT

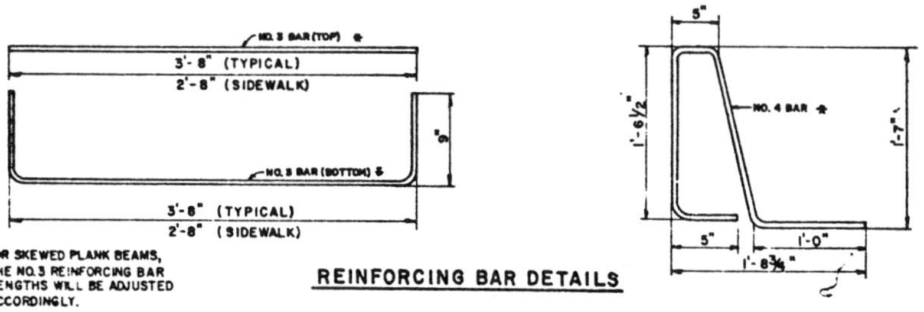


NOTE: FOR SKEWED PLANK BEAMS, THE NO. 3 REINFORCING BAR LENGTHS WILL BE ADJUSTED ACCORDINGLY.

REINFORCING BAR DETAILS

| DESIGN DATA FOR 12" x 4'-0" PLANK BEAM | | | | | | | | | | | | | |
|---|---|---|---|---|---|---|---|---|---|---|---|---|---|
| SPAN Ⓛ c. TO c. OF BEARING | | 15' | 16' | 17' | 18' | 19' | 20' | 21' | 22' | 23' | 24' | 25' | 26' |
| OVERALL LENGTH (ALONG ℄ OF ROADWAY) | | 16'-6" | 17'-6" | 18'-6" | 19'-6" | 20'-6" | 21'-6" | 22'-6" | 23'-6" | 24'-6" | 25'-6" | 26'-6" | 27'-6" |
| NO. OF 270K 1/2" Ø STRANDS (SEE SHOP DRAWINGS FOR STRAND POSIT.) | | 7 | 7 | 8 | 9 | 9 | 10 | 11 | 11 | 12 | 13 | 14 | 14 |
| INITIAL PRESTRESSING FORCE (KIPS) | | 202 | 202 | 231 | 260 | 260 | 289 | 318 | 318 | 347 | 376 | 405 | 405 |
| FINAL PRESTRESSING FORCE (KIPS) | | 162 | 162 | 185 | 208 | 208 | 231 | 254 | 254 | 278 | 300 | 324 | 324 |
| BENDING MOMENT (FT. KIPS) | | 89.9 | 97.3 | 104.9 | 112.7 | 120.5 | 128.6 | 136.7 | 145.3 | 153.8 | 162.6 | 171.7 | 180.8 |
| RESISTING MOMENT (FT. KIPS) | | 108.2 | 108.2 | 120.2 | 132.0 | 132.0 | 144.0 | 156.0 | 156.0 | 168.7 | 180.0 | 192.5 | 192.5 |
| ● NO. OF #4 LONGITUDINAL BARS EPOXY COATED # | | 6 | 6 | 6 | 6 | 6 | 6 | 6 | 6 | 6 | 6 | 6 | 6 |
| NO. & SPACING OF TIE BOLTS | | 2@8' | 2@8' | 2@8' | 2@10' | 2@10' | 2@10' | 2@12' | 2@12' | 2@12' | 2@12' | 2@12' | 2@12' |
| NO. & LENGTHS OF DEBONDING STRANDS | 2'-0" | | | | | | | | | | 2 | 4 | 2 |
| | 1'-6" | | | | | | | 2 | 2 | 2 | 2 | 2 | 4 |
| | 1'-0" | | | | 2 | 2 | 4 | 2 | 1 | 4 | 2 | 2 | 2 |
| WEIGHT OF 4'-0" PLANK BEAMS (TONS) | INTERIOR | 4.98 | 5.28 | 5.58 | 5.88 | 6.18 | 6.48 | 6.78 | 7.08 | 7.38 | 7.68 | 7.98 | 8.28 |
| | 10" CURB | 5.72 | 6.07 | 6.41 | 6.76 | 7.11 | 7.46 | 7.80 | 8.15 | 8.51 | 8.84 | 9.19 | 9.53 |
| | CURB BARRIER | 7.47 | 7.92 | 8.37 | 8.83 | 9.28 | 9.73 | 10.18 | 10.64 | 11.09 | 11.54 | 12.01 | 12.48 |
| NUMBER OF GUARDRAIL POSTS REQUIRED | | 6 | 6 | 6 | 7 | 7 | 7 | 8 | 8 | 8 | 9 | 9 | 9 |
| END OF SLAB TO ℄ FIRST POST | | 1'-0" | 1'-0" | 1'-5 1/4" | 1'-0" | 1'-0" | 1'-4 1/2" | 1'-0" | 1'-0" | 1'-3 3/4" | 1'-0" | 1'-0" | 1'-3" |
| ℄ OF FIRST POST TO ℄ SECOND POST | | 2'-6 3/4" | 3'-0 3/4" | 3'-1 1/2" | 2'-6" | 3'-0" | 3'-1 1/2" | 2'-5 1/4" | 2'-11 1/4" | 3'-1 1/2" | 2'-4 1/2" | 2'-10 1/2" | 3'-1 1/2" |

● NOT REQUIRED WITH WELDED WIRE FABRIC REINFORCEMENT.

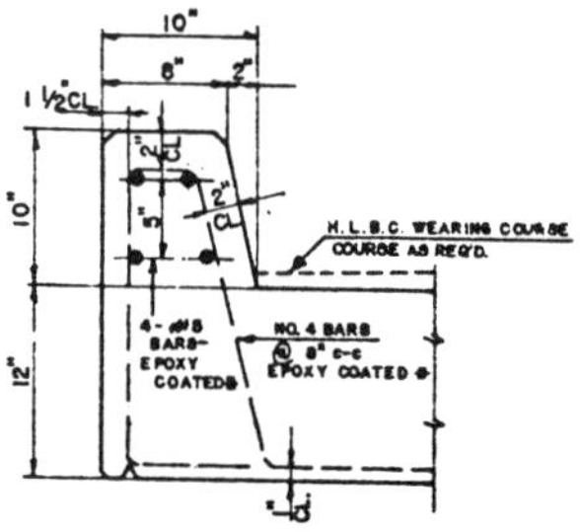


CURB DETAIL

WVA

# Design of Continuous Highway Bridges with Precast, Prestressed Concrete Girders

**To achieve continuity in a bridge using precast, prestressed single span units requires design considerations that also include the effects of creep and shrinkage. A complete design example illustrates these considerations following a general discussion of the principles.**

Continuous highway bridges with precast, prestressed girders have been built by a number of states. Examples of bridges of this type built by the states of Tennessee and California are presented in Figs. 1 and 2, respectively. The Big Sandy River Bridges in Tennessee were built in 1963-64. Performance has been excellent. The 639-ft. and 703-ft. continuous lengths of deck slab on these bridges are believed to be the longest built to date supported on precast, prestressed girders. The Los Penasquitos Bridge in California was a 1966 PCI award winner.

A primary reason for using continuity with precast, prestressed girders is the elimination of the maintenance costs associated with bridge deck joints and deck drainage onto the substructure. Continuity also improves the appearance and the riding qualities of this type of bridge. Due to the structural economy of continuous designs and the elimination of the deck joint details, some initial economic advantage may also be obtained.

The construction sequence for the type of bridge under consideration is shown in Fig. 3. Continuity is achieved for live load plus impact moments by use of non-prestressed reinforcement in the deck slab and in the diaphragms over piers. An extensive research program on this type of bridge was conducted by the Research and Development Laboratories of the Portland Cement Association during 1960-61. The results of this research, which provide most of the background for this publication, have been made available in a series of six PCA Development Department Bulletins[1-6]. While the above research and the applications of this procedure to date have been directed primarily towards bridges built with I-beams, the procedure applies equally well to bridges with box beams, tee beams, or other available shapes.

The first portion of this report contains a general description of the design features peculiar to this type of bridge and the background material necessary to make design calculations. The second portion consists of a design example.

Fig. 1. One of the Big Sandy River Bridges, Tennessee

## Part 1. Design Procedure

### PRESTRESSING REINFORCEMENT

Except for the determination of the governing live load plus impact moment, the design of the prestressing reinforcement for continuous bridges follows the well-known procedure used for simple-span bridges[7]. The use of continuity permits a reduction of 5 to 15 percent in the required prestress force when compared to simple-span designs. The larger reductions occur in shorter span bridges where the live load plus impact moments are a large portion of the total design moment. Illustrative calculations for determining the governing live load plus impact moments and the required prestressing reinforcement are presented in the design example.

For bridges made up of a series of equal spans, the governing live load plus impact moment occurs at about the 0.4 or 0.5 point of the end spans, with the moment at the mid-point of interior spans being considerably less. Therefore, in some cases it may pay to make separate designs for the prestressing reinforcement in the interior and exterior spans. Where the designer has freedom in choosing the span layout, another alternative is to shorten the exterior spans to obtain the same amount of prestressing reinforcement in interior and exterior spans.

The positive restraint moments over piers, as described in the following section, also increase the positive moments at mid-span sections. When a section at mid-span is subject to the maximum service load moment plus the limiting value of the positive restraint moment at that section, it is recommended that the stress at the bottom face of the girder be limited to about 80 percent of the 28-day modulus of rupture of the concrete. The modulus of rupture is usually taken as $7.5\sqrt{f'_c}$. This requirement will usually be satisfied by the prestressing reinforcement supplied for gravity loads calculated as described in the paragraphs above.

Fig. 2. Los Penasquitos Bridge, California

### POSITIVE MOMENTS AT PIERS

One of the unique features of the type of design under consideration is that positive moments develop over piers due to creep in the prestressed girders, as well as due to the effect of live loads in remote spans. These positive moments are partially counteracted by negative moments resulting from differential shrinkage between the cast-in-place deck slab and the precast girders. The positive live load plus impact moments over piers due to loads in remote spans are calculated by conventional procedures.

**The effect of creep.** The deformations and restraint moments induced in a two-span continuous beam by creep due to prestress force are illustrated in Fig. 4. Since creep is time dependent, with the more rapid deformation occurring during the early stages of loading, the amount of positive restraint moment induced by the prestress force depends on the time when the continuity connection is made, as well as on the inherent creep potential of the concrete mix, the exposure conditions, and the volume to surface ratio of the prestressed member.

The most accurate method of obtaining creep data is by extrapolation from a number of laboratory samples prepared in advance from the actual mix to be used in the girders. In instances where many beams are to be produced, it may be possible to take observations on the initial beams produced, and to develop the necessary creep data from those observations. In most design work, however, it will suffice to rely on available research data. Such research[8,9] indicates that the basic creep value for loading at 28 days can be predicted from the elastic modulus according to the curves in Fig. 5. For design purposes, the 20-year creep curve can be regarded as the ultimate creep.

The ultimate creep value for loading at 28 days, from Fig. 5, must be adjusted to account for the age when the girders are prestressed and for the volume/surface ratio of the girders. The effects of reinforcement on creep can usually be ignored for a prestressed beam. The variation of creep with age at loading is given by the curve in Fig. 6[8,10]. For loading when the concrete is one day old, for example, creep would be 1.8 times the value from the 20-year curve in Fig. 5. The variation of creep with the volume/surface ratio is shown in Fig. 7[8,11]. The volume/surface ratio of the AASHO-PCI I-beams, and the related volume/surface ratio creep cor-

1 2

rection factor, are listed in Table 1. The amount of creep which will take place after the continuity connection is made can be found from Fig. 8[8,10]. For example, for a connection made at 28 days, about 40 percent of the creep strains has taken place, leaving the remaining 60 percent to develop moments in the connection.

It has been shown[5] that the effects of creep under prestress and dead load can be evaluated by an elastic analysis assuming that the girder and slab were cast and prestressed as a monolithic continuous girder. The moments obtained from this analysis are multiplied by a factor $(1\text{-}e^{-\phi})$ to allow for the effect of creep. The factor $\phi$ is the ratio of creep to elastic strain ($\epsilon$ creep/$\epsilon$ elastic). This factor can be evaluated as follows:

1. For a Type VI AASHO-PCI I-beam with an elastic modulus of $4 \times 10^6$ psi at release of strands, the creep for loading at 28 days can be obtained from the 20-year loading curve in Fig. 5 as $0.34 \times 10^{-6}$ in./in./psi.
2. Assuming strand release two days after casting, the multiplication factor from Fig. 6 is about 1.7, and the adjusted creep becomes: $1.7 \times 0.34 \times 10^{-6} = 0.58 \times 10^{-6}$ in./in./psi.
3. The volume/surface ratio for a Type VI AASHO-PCI beam is 4.4 and the correction factor from Fig. 7 is 1.18. Therefore, the further adjusted creep becomes $1.18 \times 0.58 \times 10^{-6} = 0.67 \times 10^{-6}$ in./in./psi.
4. Assuming the moment connection is made at 28 days, Fig. 8 indicates that 40 percent of the ultimate creep will have occurred at the time the connection is made. The connection will have to be capable of developing moments that result from the remaining 60 percent, or, $0.60 \times 0.67 \times 10^{-6}$ in./in./psi $= 0.40 \times 10^{-6}$ in./in./psi.

$$\phi = \frac{\epsilon \text{ creep}}{\epsilon \text{ elastic}}$$

$$E = \frac{1}{\epsilon \text{ elastic}}$$

$$\phi = \epsilon \text{ creep} \times E$$

$$\phi = 0.40 \times 10^{-6} \times 4.0 \times 10^6 = 1.60$$

Thus, for a connection at 28 days after prestressing, the creep strains still to occur are 1.60 times the original elastic strain.

5. Having evaluated $\phi$, the value of $(1\text{-}e^{-\phi})$ can be taken directly from Fig. 9. This figure also contains a graph of the factor $(1\text{-}e^{-\phi})/\phi$ which is used to allow for the effect of creep in calculating the negative restraint moment due shrinkage.

Dead load creep partially counteracts prestress creep because dead load causes tension in the precompressed bottom fibers. The amount of creep bending that occurs depends on the residual compressive stress gradient under dead load. For this reason, more creep bending is likely to occur on shorter span bridges where a larger percentage of the precompression remains in the bottom of the beam under dead load.

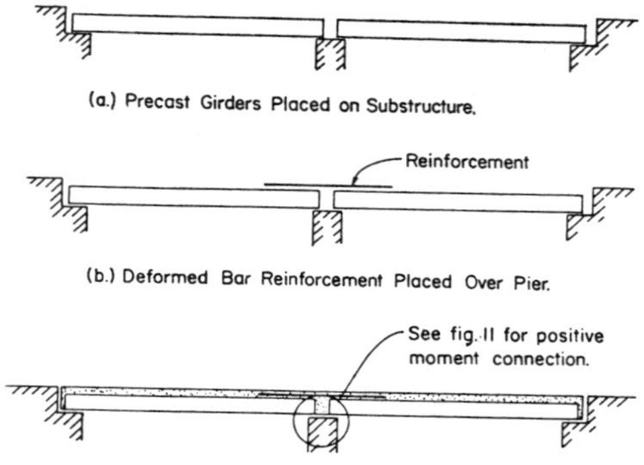


Fig. 3. Construction sequence for a two-span bridge with precast, prestressed girders made continuous for live loads

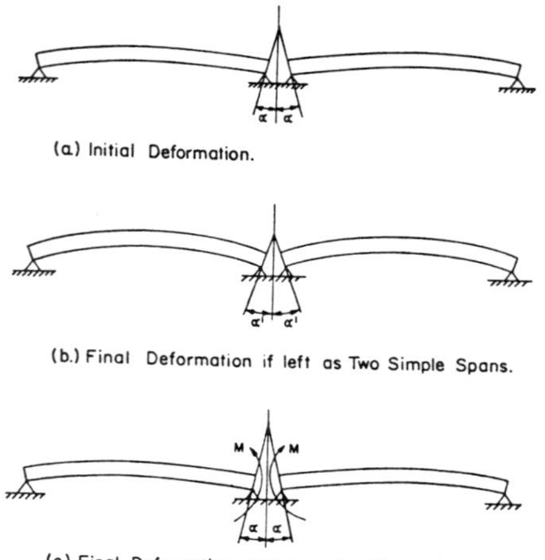


Fig. 4. Deformations and restraint moments in a two-span continuous beam caused by creep under prestress force

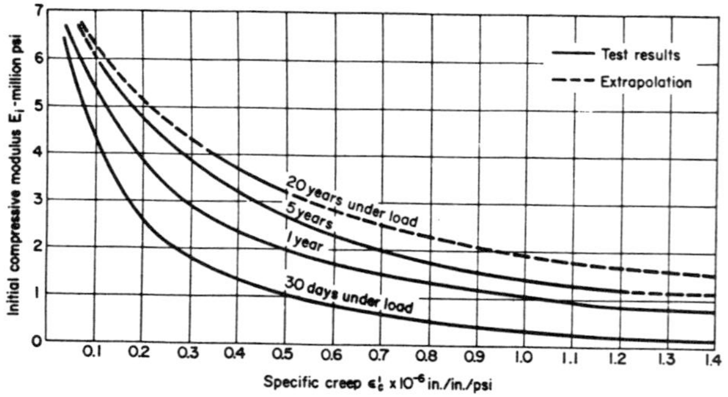


Fig. 5. Prediction of basic creep from elastic modulus

**The effect of shrinkage.** Shrinkage of the deck slab with respect to the girders causes negative restraint moments over piers that reduce the creep restraint moment. As shown in a PCA Development Department Bulletin[5], the restraint moments at the piers due to shrinkage are calculated as if the differential shrinkage moment were applied to the continuous composite girder along its entire length, assuming the girder to behave elastically. The differential shrinkage moment applied to the girder along its entire length is given by:

$$M_s = \epsilon_s E_b A_b \left(e_2' + \frac{t}{2}\right)$$

where $\epsilon_s$ = differential shrinkage strain (as described below)

$E_b$ = elastic modulus for the deck slab concrete

$A_b$ = cross-sectional area of deck slab

$\left(e_2' + \frac{t}{2}\right)$ = distance between the mid depth of the slab and centroid of composite section

$t$ = slab thickness

The most reliable means of estimating shrinkage values is from observations on specimens made from the mix to be used in the structure. With such observations, the final values of shrinkage can be projected by assuming the shrinkage-time curve can be represented by the following equations[11]:

$$\epsilon_s = \frac{\epsilon_{su} T}{N_s + T}$$

where $\epsilon_s$ = measured shrinkage strain at any time $T$ in days

$\epsilon_{su}$ = ultimate shrinkage strain at $T$ = infinity

$N_s = 26.0\, e^{0.36V/S}$

$V/S$ is the volume/surface ratio of the member under consideration (see Table 1 for $V/S$ ratios of the AASHO-PCI I-beams).

When test data are not available, the ultimate shrinkage under exposure at 50 percent relative humidity can be estimated[12,13] as $0.600 \times 10^{-3}$. This value should be corrected for very humid or very dry environments by multiplying by a humidity correction factor as presented in Fig. 10. The differential shrinkage strain between slab and girder for a given time lapse between castings can then be obtained by using the time-shrinkage relationship in Fig. 8. For example, with a 35-day lapse between casting of the girder and the slab, Fig. 8 indicates that about 0.42 of the girder shrinkage has taken place. Thus a slab exposed to 50 percent relative humidity would have $0.42 \times 0.600 \times 10^{-3}$ in./in. or $0.252 \times 10^{-3}$ in./in. more shrinkage remaining than the girder. The value of $0.252 \times 10^{-3}$ would be used for $\epsilon_s$ in the above equation for $M_s$. This value checks very closely the value of $0.245 \times 10^{-3}$ in./in. obtained in the PCA research[5] using 10 ft. long free shrinkage specimens of the slab and girder and a time lag between castings of 35 days.

After calculating the negative elastic moments at the piers due to shrinkage, they must be modified for

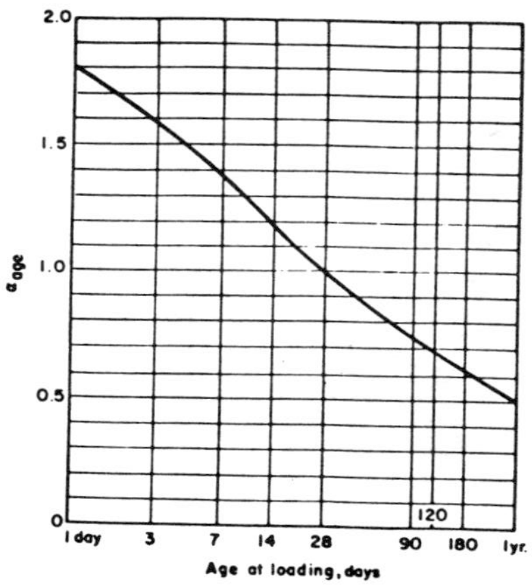


Fig. 6. Creep vs. age at loading

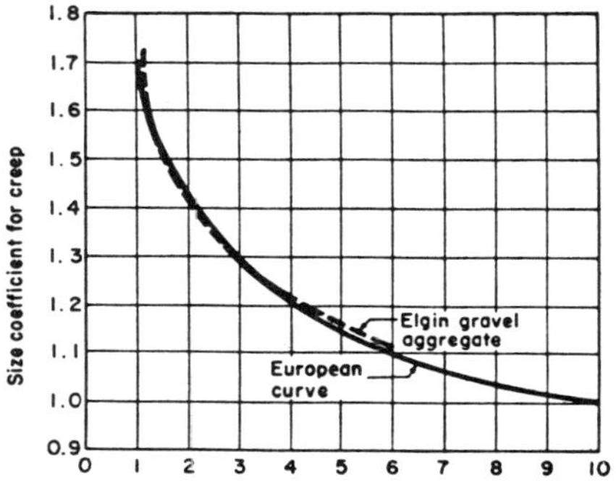


Fig. 7. Creep vs. volume-to-surface ratio

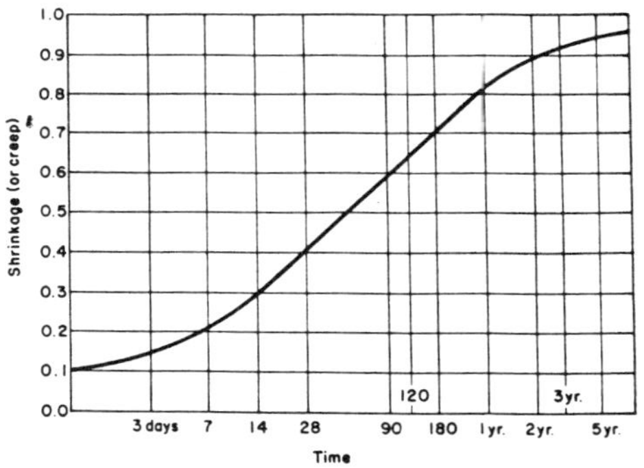


Fig. 8. Proportion of final shrinkage or creep vs. time

Table 1. Creep data for AASHO-PCI I-beams

| AASHO-PCI girder type | Volume/surface ratio in. | Creep volume/surface ratio correction factor (from Fig. 7) |
|---|---|---|
| I | 3.0 | 1.28 |
| II | 3.4 | 1.25 |
| III | 4.0 | 1.20 |
| IV | 4.7 | 1.16 |
| V | 4.4 | 1.18 |
| VI | 4.4 | 1.18 |

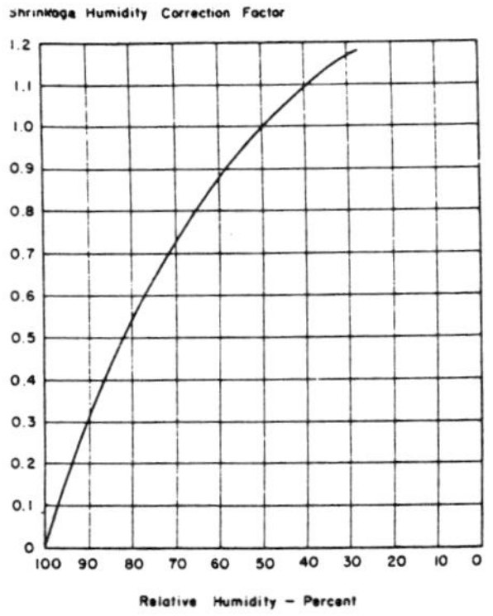


Fig. 10. Shrinkage humidity correction factor

the effect of creep by multiplying by the factor $(1-e^{-\phi})/\phi$. A graph of this factor with respect to $\phi$ is presented in Fig. 9.

**Summation of restraint moments.** The final positive restraint moment over piers is calculated as:

$$M_r = (Y_c - Y_{DL})(1 - e^{-\phi}) - Y_s\left(\frac{1-e^{-\phi}}{\phi}\right) + Y_{LL}$$

where $M_r$ = final restraint moment
$Y_c$ = restraint moment at a pier due to creep under prestress force
$Y_{DL}$ = restraint moment at a pier due to creep under dead load
$Y_s$ = restraint moment at a pier due to differential shrinkage between the slab and girder
$Y_{LL}$ = positive live load plus impact moment.

**Details of the positive moment connection.** In the PCA Development Department Bulletin on this subject[3], various connection details to resist the positive restraint moment were evaluated. The detail considered most practical is shown in Fig. 11.

In the tests run on this connection, the hooked bars were bent essentially at right angles and a relatively short horizontal projection was used to conserve on the width of the diaphragm. Although the connections developed substantial moments under static test loads, under fatigue testing (with a stress range of 20,000 psi) most of the connection bars failed after about 670,000 applications of the load. The bars fractured in a brittle manner at the knee of the hook. To avoid this type of failure, the live load plus impact stress range in the bars *at the point where the bend starts* should be limited to 50 percent of the fatigue strength[14]. A conservative value for this stress range is 12,000 psi[15]. Connections designed in accordance with the recommendations in the following paragraphs will usually conform to this stress range restriction.

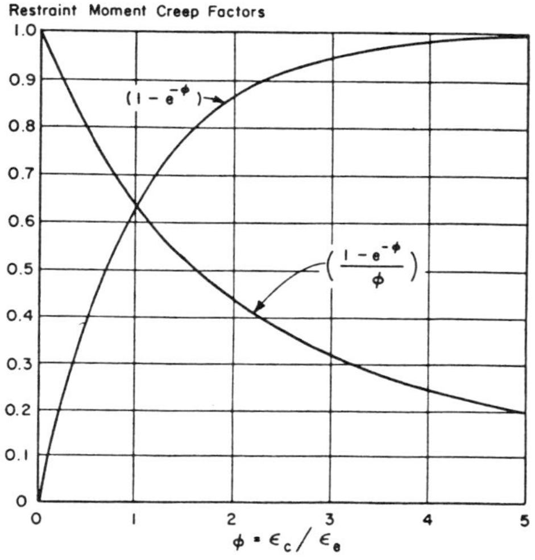


Fig. 9. Variations in restraint moment creep effect factors for both creep and shrinkage

The total length of embedment required for the hooked connection bars can be calculated considering bond to develop uniformly from the face of the beam around the bend to the end of the bar[16]. The unit bond stress should be $0.10\ f'_c$ (maximum 350 psi) as prescribed in the AASHO Specifications[19]. The distance from the inside face of the precast girder to the inside face of the hook should be at least 12 times the bar diameter. The minimum bend radius should conform to Table 2.

Table 2. Minimum bar bend radii

| Bar Size | Minimum Radii |
|---|---|
| #3, #4, #5 | 2½ bar diameters |
| #6, #7, #8 | 3 bar diameters |

As determined in the tests at the PCA Laboratories, the restraint moment over the piers does not affect the factor of safety of the structure. For this reason, it was suggested that the positive moment continuity reinforcement be designed to work at 75 percent of its yield point when subject to the limiting value of the restraint moment due to creep and differential shrinkage together with the maximum positive moment due to live load plus impact in remote spans. However, when general research data are used to estimate creep and shrinkage coefficients, it is recommended that the design be based on the lower stress of 0.6 times the yield strength. This lower stress will reduce the live load plus impact stress range, and it will also provide more assurance against the possibility of diaphragm cracking.

Due to the amount of calculation involved in the design of a positive moment connection, as described above, it is suggested that standard details be developed for the various prestressed sections commonly used. A practical consideration in designing the positive moment connection is to detail it to facilitate the removal of the end girder form after prestressing. This may require that the connection bars be cast in the beam straight and then field bent after removal of the end form. However, in some areas, end forms have been developed to accommodate shop bent bars. If field bending is required, No. 6 connection bars should probably be the maximum size used. Consultation with local prestressed concrete fabricators is recommended to develop practical and economical positive moment connection details for use in a particular area. The connection and bearing details for the Big Sandy River Bridges are shown in Fig. 12.

**NEGATIVE MOMENT REINFORCEMENT**

Since most of the dead load moments are carried by the beam acting as a simple span, the negative design moments over piers are those due to live load plus impact. In some designs, the dead load applied after continuity is achieved may also be considered in the negative design moment. The recommended design procedure for proportioning negative moment reinforcement is to use ultimate strength procedures with a load factor of 2.5 on live load and 1.5 on dead load. The effect of initial precompression due to prestress in the precast girders may be neglected

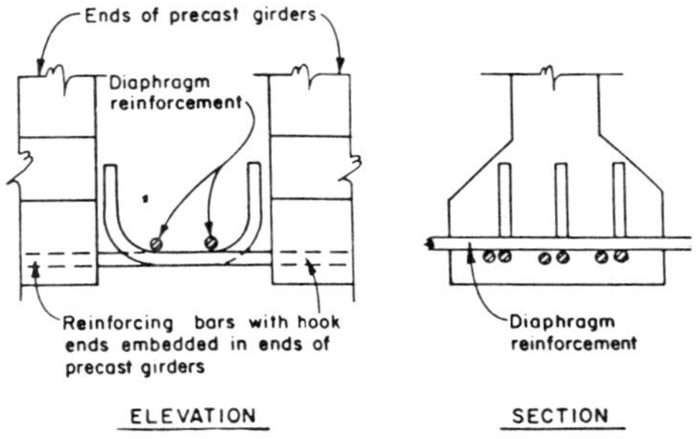


Fig. 11. Positive restraint moment connection details

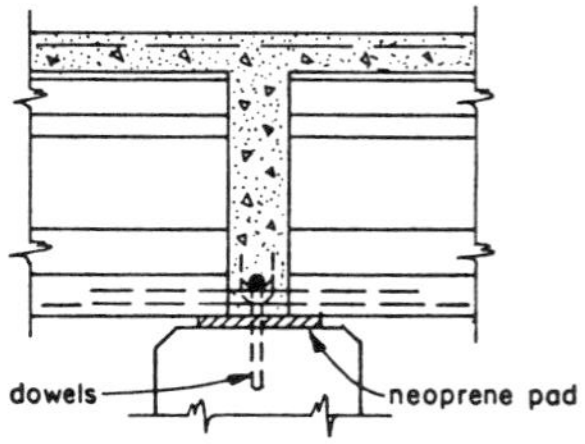


Fig. 12. Bearing and connection details at piers, Big Sandy River Bridges, Tennessee

in the negative moment computation of ultimate strength if the maximum precompression stress is less than 0.4 $f'_c$ and the continuity reinforcement is less than 1.5 percent[1]. A design aid to facilitate determination of the required steel percentage is presented in Fig. 13[17]. Use of this aid is illustrated in the design example.

It will usually be found that the depth of the compression block will be less than the thickness of the bottom flange of the precast girder. For this reason, the negative moment reinforcement required can be determined by assuming the beam to be a rectangular section with a width equal to the bottom flange width of the girder. Due to the lateral restraint of the diaphragm concrete, ultimate negative compression failure in the PCA tests[1] always occurred in the girders, even though the diaphragm concrete strength was about 2000 psi less than that of the girder concrete. For this reason, it is recommended that the negative moment reinforcement be designed using the compressive strength of the girder concrete.

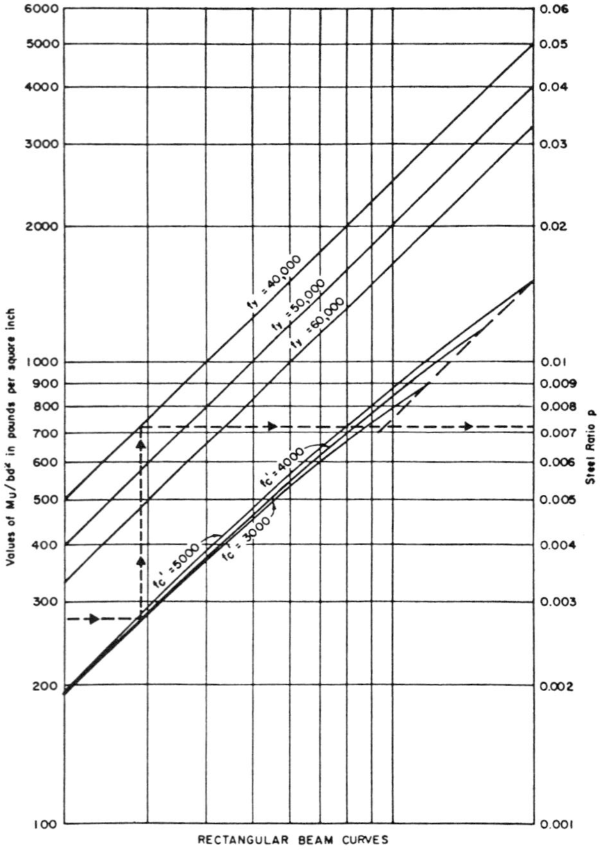


Fig. 13. Design aid for negative moment reinforcement

#### COMPRESSIVE STRESS IN GIRDERS AT PIERS

A point that concerns some designers is the high compression stresses that might theoretically occur due to addition of stresses due to prestress and negative live load bending over the supports. The PCA studies indicate that these stresses do not affect the ultimate strength for the following reasons:

1. Stress relief is obtained through inelastic strains.
2. Full compression does not occur at beam ends due to the strand transfer length.
3. Diaphragms strengthen the area by providing lateral restraint.

In addition, the effect of the negative live load plus impact moment is reduced by the positive creep restraint moment. However, when considered essential, the theoretical precompression can be reduced by raising the strand center of gravity and/or by blanketing some of the strands at the girder ends. Tests of girders with blanketed prestressing strands have been made at the PCA Laboratories[18].

#### SHEAR

The shear tests of continuous beams conducted at the PCA Laboratories[4] revealed that the present shear provisions in the American Association of State Highway Officials (AASHO) Specifications for Highway Bridges[19] provide a conservative estimate of the shear capacity of a continuous beam. However, on continuous beams the formulas must be applied over the full length of the beams rather than only over the middle half of the spans, as is usually the case for simply supported girders.

The horizontal shear stress between the cast-in-place slab and the precast girder was also investigated at the PCA Laboratories[4,5]. These tests found the horizontal shear stresses between slab and beam permitted by the AASHO Specifications to be very conservative.

## Part 2. Design Example

### DERIVATION OF GENERAL FORMULAS—FOUR EQUAL CONTINUOUS SPANS

The design example is for a bridge with four equal spans of 130 ft. Prior to the detailed calculations, development of general formulas for restraint moments due to shrinkage and creep is presented. Similar developments for bridges with any number of equal or unequal spans can be made by use of the same procedure or any other consistent procedure for the analysis of indeterminant structures. The conjugate beam theory is used to calculate the various fixed-end restraint moments. The final restraint moments are then obtained by moment distribution.

**Shrinkage restraint moments.** As previously discussed, differential shrinkage between the deck slab and the prestressed beam causes the following moment on the composite section:

$$M_s = \epsilon_s E_b A_b \left(e'_2 + \frac{t}{2}\right)$$

If the bridge were composed of 4 equal simple spans, this moment would cause bending as shown below:

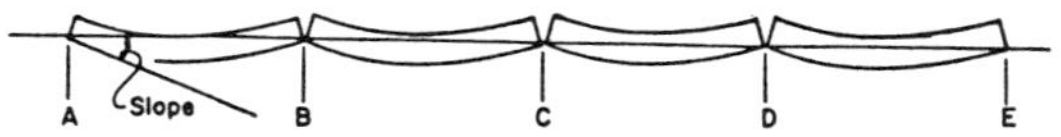


The slope at the ends of the simple-span beams may be obtained as the reaction of the $ML/EI$ diagram:

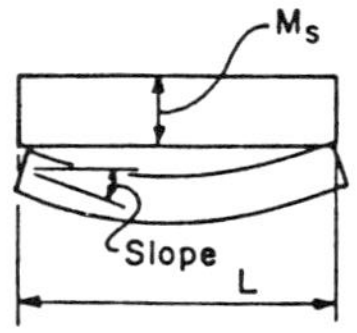


$$\text{Slope} = \frac{M_s L}{2EI}$$

Apply moments at B, C and D to return the beam ends to a horizontal position:

<u>End Span</u>

$$\text{Slope at B} = -\frac{M_B L}{2EI} \times \frac{2}{3}$$
$$= -\frac{M_B L}{3EI}$$

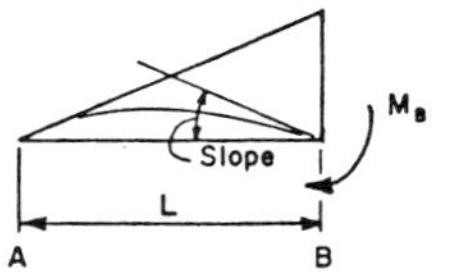

Set slope equal to slope due to shrinkage:

$$-\frac{M_B L}{3EI} = \frac{M_s L}{2EI}$$

$$M_B = -\frac{3}{2} M_s$$

**Interior Span**

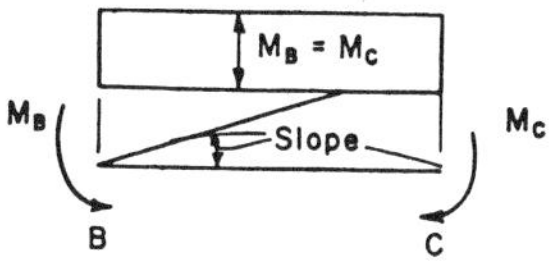


Slope at B and C $= -\dfrac{M_B L}{2EI}$

Set slope equal to slope due to shrinkage:

$$-\frac{M_B L}{2EI} = \frac{M_s L}{2EI}$$

$$M_B = -M_s = M_C$$

Since different moments are required to bring the various beams to a horizontal position, the moments must be distributed to obtain the final shrinkage restraint moments.

| | B 3/7 | B 4/7 | C 1/2 | C 1/2 | D 4/7 | D 3/7 |
|---|---|---|---|---|---|---|
| | $-1.5 M_s$ | $-M_s$ | $-M_s$ | $-M_s$ | $-M_s$ | $-1.5M_s$ |
| | $+.214 M_s$ | $-.286M_s$ | | | $-.286 M_s$ | $+.214M_s$ |
| | | | $+.143M_s$ | $+.143 M_s$ | | |
| A | $-1.286 M_s$ | | $-.857 M_s$ | | $-1.286 M_s$ | E |

Final shrinkage restraint moments (not reduced by the creep effect factor) are $-1.286\ M_s$ at B and D, and $-0.857\ M_s$ at C.

**Creep restraint moments.** The creep restraint moment is broken into two components, one due to dead load and the other due to prestress.

**Dead Load Creep Restraint Moment**

Slope $= \dfrac{M_D L}{3EI}$

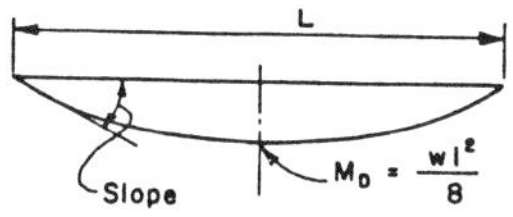


Apply moments to return beam ends to horizontal:

**End Span**

$$-\frac{M_B L}{3EI} = \frac{M_D L}{3EI}$$

$$M_B = -M_D$$

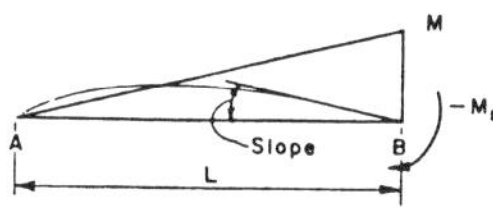


**Interior Span**

$$-\frac{M_B L}{2EI} = \frac{M_D L}{3EI}$$

$$M_B = 2/3\ M_D = M_C$$

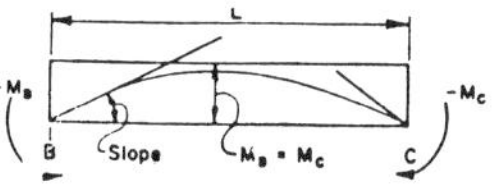


Distribute the unequal end moments due to creep under dead load:

| | D 3/7 | D 4/7 | C 1/2 | C 1/2 | D 4/7 | D 3/7 |
|---|---|---|---|---|---|---|
| | $-M_D$ | $-.667M_D$ | $-.667M_D$ | $-.667M_D$ | $-.667M_D$ | $-M_D$ |
| | $+.14M_D$ | $-.19M_D$ | | | $-.19M_D$ | $+.14M_D$ |
| | | | $+.095M_D$ | $+.095M_D$ | | |
| A | $-.86 M_D$ | | $-.572 M_D$ | | $-.86M_D$ | E |

Final dead load creep restraint moments (not reduced by the creep effect factor) are $-0.86\ M_D$ at B and D, and $-0.572\ M_D$ at C.

**Prestress Creep Restraint Moment**

The creep under the prestress force depends on the location of the various forces with respect to the center of gravity of the composite section. The tendons will be replaced in the calculations by the equivalent forces exerted on the member.

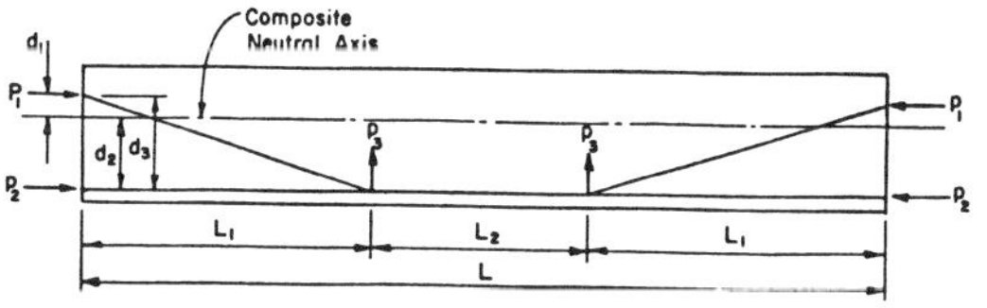


The moment diagram by parts for the above forces is as follows:

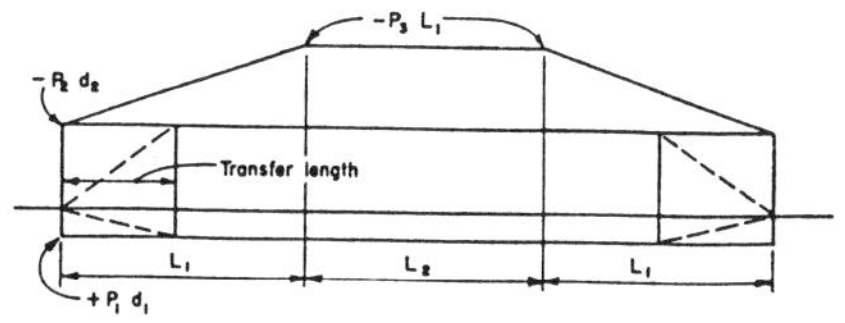


Moments might more realistically be considered to vary from zero to full value along the transfer length (dotted lines). This would reduce the creep under the prestress force. Also, any blanketed strand could be disregarded for the length of the blanketing. For convenience, the rectangular moment diagrams will be used in the following.

Slope due to prestress (abbreviated P/S)

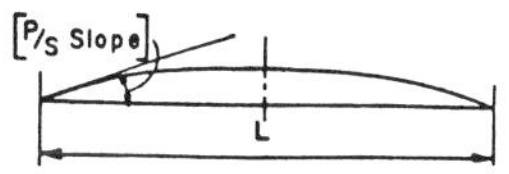


$$[\text{P/S Slope}] = \frac{P_1 d_1 L}{2EI} - \frac{P_2 d_2 L}{2EI} - \frac{P_3 L_1^2}{2EI} - \frac{P_3 L_1 L_2}{2EI}$$

Apply moments to bring beam ends at B, C and D to horizontal:

**End Span**

$$\frac{M_B L}{3EI} = -\ [\text{P/S Slope}]$$

$$M_B = -\frac{3EI}{L}[\text{P/S Slope}]$$

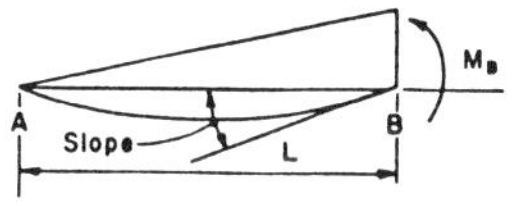


**Interior Span**

$$\frac{M_B L}{2EI} = -\ [\text{P/S Slope}]$$

$$M_B = M_C = -\frac{2EI}{L}[\text{P/S Slope}]$$

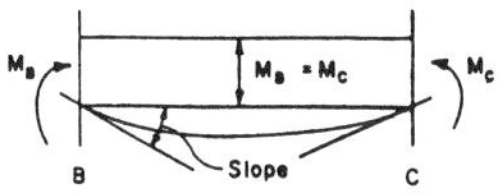


Distribute moments (for convenience distribute $-3$ and $-2$):

| | B 3/7 | B 4/7 | C 1/2 | C 1/2 | D 4/7 | D 3/7 |
|---|---|---|---|---|---|---|
| | -3 | -2 | -2 | -2 | -2 | -3 |
| | +.43 | -.57 | | | -.57 | +43 |
| | | | +.28 | +.28 | | |
| A | -2.57 | | -1.72 | | -2.57 | E |

Final creep restraint moments due to prestress (not reduced by creep effect factor) are $\dfrac{-2.57\ EI}{L}$ [P/S Slope] at B and D, and $\dfrac{-1.72\ EI}{L}$[P/S Slope] at C. Recognizing that the value of [P/S Slope] will be negative, the prestress restraint moment will then have a positive value.

Summing up the various restraint moments and multiplying by the appropriate factors, the final restraint moments due to creep and shrinkage can be calculated as:

$$\text{at B and D} = \left(\frac{-2.57\ EI}{L}[\text{P/S Slope}] - 0.86\ M_D\right)(1 - e^{-\phi})$$

$$-\ (1.286\ M_s)\left(\frac{1-e^{-\phi}}{\phi}\right)$$

$$\text{at C} = \left(\frac{-1.72\ EI}{L}[\text{P/S Slope}] - 0.572\ M_D\right)(1 - e^{-\phi})$$

$$-\ (0.857\ M_s)\left(\frac{1-e^{-\phi}}{\phi}\right)$$

For a given value of $\phi$, the values of $(1 - e^{-\phi})$ and $\left(\dfrac{1-e^{-\phi}}{\phi}\right)$ may be taken directly from Fig. 9.

### SAMPLE CALCULATIONS FOR A BRIDGE WITH FOUR EQUAL 130-FT. SPANS

This design example for the interior girder of a bridge with four equal 130-ft. spans was chosen largely to illustrate that this type of construction is applicable to the entire range of bridge spans for which precast, prestressed girders would usually be considered. More numerous applications will undoubtedly occur for shorter spans. In states which have laws prohibiting highway transportation of the longer and heavier precast girders, the girders may be cast in segments and connected at the bridge site. A comprehensive study of methods of connecting precast segments to achieve long spans has been published by the Prestressed Concrete Institute[20].

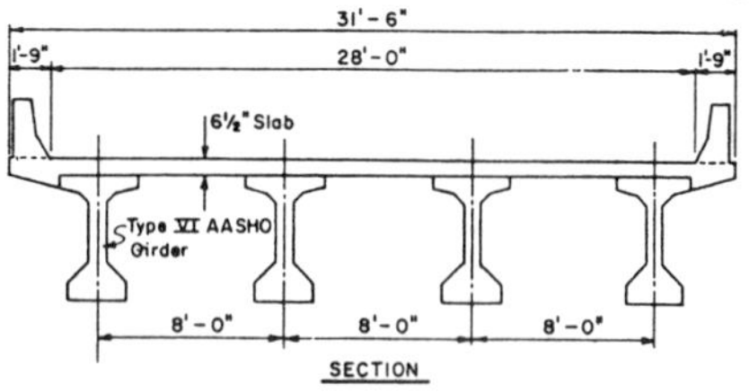


Design criteria:

| | Concrete Strength | E |
|---|---|---|
| Prestressed beam at release: | 4000 psi | $3.66 \times 10^6$ psi |
| Prestressed beam at 28 days: | 5000 psi | $4.05 \times 10^6$ psi |
| Deck slab and diaphragm: | 4500 psi | $3.88 \times 10^6$ psi |

Loading: AASHO HS20-44
Prestressing strand: 1/2" $\phi$, 270 k
Area = 0.153 in.$^2$
Final force (after losses) = 23.6 k/strand

Section properties:

| Prestressed beam | Composite girder (interior) |
|---|---|
| Area = 1085 in.$^2$ | Area = 1676 in.$^2$ |
| $\bar{y}$ = 36.38 in. | STC = 59,977 in.$^3$ |
| $I$ = 733,320 in.$^4$ | SBC = 26,266 in.$^3$ |
| ST = 20,587 in.$^3$ | STSC = 46,252 in.$^3$ |
| SB = 20,157 in.$^3$ | $I$ = 1,313,811 in.$^4$ |

**Design of prestressing reinforcement**

Impact = 50/255 = 19.6%

$$LL = \frac{8.0}{5.5 \times 2} = 0.727 \text{ lanes}$$

Moment at 0.5 Span AB:

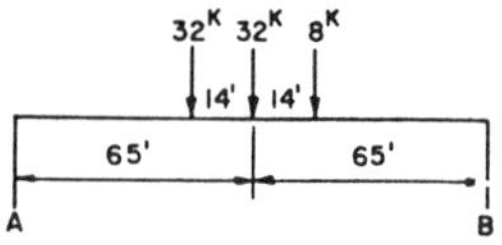


Using influence lines for 4 equal continuous spans:

| | |
|---|---|
| $0.1517 \times 32 \times 130 =$ | 631 ft. kips |
| $0.1998 \times 32 \times 130 =$ | 830 |
| $0.1449 \times 8 \times 130 =$ | 150 |
| Total for 1 lane = | 1611 ft. kips |

LL + I moment at
0.5 Span AB = $1.196 \times 0.727 \times 1611 = 1403$ ft. kips

Dead load moments at 0.5 Span AB:

| | |
|---|---|
| Beam = $1085/144 \times 0.150 \times 130^2/8$ = | 2390 ft. kips |
| Slab = $96 \times 6.5/144 \times 0.150 \times 130^2/8$ = | 1372 |
| Diaphragm = $3.30 \times 130/3$ = | 143 |
| | $M_D$ = 3905 ft. kips |

Stresses at 0.5 Span AB

| | Top | Bottom |
|---|---|---|
| Slab | 802 psi | 1425 psi |
| Beam | 1395 | 817 |
| Diaphragm | 83 | 86 |
| LL + I | 282 | 643 |
| | 2562 psi | 2971 psi |
| Allowable tension | 212 | 212 |
| | 2774 psi | 2759 psi |

Assume eccentricity of strand center of gravity from bottom of beam = 4.70 in. Eccentricity from center of gravity:
36.38 − 4.70 = 31.68 in.

Calculate force required to give a bottom flange stress of 2759 psi:

$2759 = F/1085 + F \times 31.68/20.157$
$F = 1{,}110{,}000$ lb.

Strands required = 1,110,000/23,600 = 47.1 strands
Use 48 − 1/2" $\phi$ 270 k strands

Check c.g. of strands compared to assumed value:

| | |
|---|---|
| 13 at 2 = | 26 |
| 13 at 4 = | 52 |
| 13 at 6 = | 78 |
| 9 at 8 = | 72 |
| 48 | 228 |

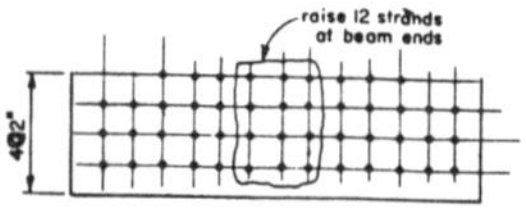


228/48 = 4.75 in.
(close enough to assumed value)

48 strands at 23.6 = 1132 kips
36 strands at 23.6 = 850 kips
12 strands at 23.6 = 283 kips

Check eccentricity at girder ends for no tension under working loads:

$0 = 1132/1085 - 1132e/20{,}587$
$e$ = 19 in. (17.38 in. from bottom)

Center of gravity of 12 harped strands must be about 55 in. from the bottom to provide this eccentricity. Raise center of gravity of 12 strands 60 in. at ends to reduce compressive stress in bottom fibers.

**Design for positive moments at piers**

**Prestress Creep Restraint Moments (not adjusted by creep effect factor)**

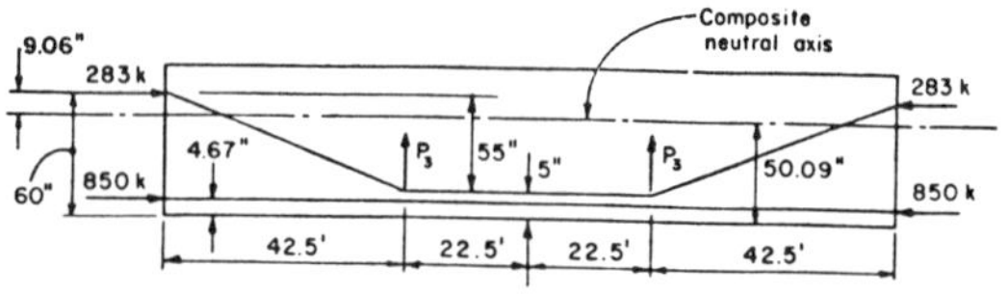


$P_3 = 55/42.5 \times 12 \times 283 = 30.5$ kips

$$[\text{P/S Slope}] = \frac{P_1 d_1 L}{2EI} - \frac{P_2 d_2 L}{2EI} - \frac{P_3 L_1^2}{2EI} - \frac{P_3 L_1 L_2}{2EI}$$

$$\frac{P_1 d_1 L}{2EI} = 283 \times 9.91 \times 130 \times 12/2 = +\ 2{,}190{,}000 \text{ in. kips}$$

$$-\frac{P_2 d_2 L}{2EI} = -850 \times 45.42 \times 130 \times 12/2 = -\ 30{,}100{,}000$$

$$-\frac{P_3 L_1^2}{2EI} = -30.5 \times 42.5^2 \times 144/2 = -\ 3{,}960{,}000$$

$$-\frac{P_3 L_1 L_2}{2EI} = -30.5 \times 42.5 \times 45 \times 144/2 = -\ 4{,}200{,}000$$

$$[\text{P/S Slope}] = -\frac{36{,}070{,}000}{EI} \text{ in. kips}$$

$$\text{at B} = \frac{-2.57\,EI}{130 \times 12}\left[-\frac{36{,}070{,}000}{EI}\right] = 59.4 \times 10^6 \text{ in. lb.} = 4960 \text{ ft. kips}$$

$$\text{at C} = \frac{-1.72\,EI}{130 \times 12}\left[-\frac{36{,}070{,}000}{EI}\right] = 39.8 \times 10^6 \text{ in. lb.} = 3320 \text{ ft. kips}$$

**Dead Load Creep Restraint Moments (not adjusted by creep effect factor)**

at B = −0.86 × 3905 = −3360 ft. kips

at C = −0.572 × 3905 = −2230 ft. kips

**Shrinkage Retraint Moments (not adjusted by creep effect factor)**

Assuming continuity connection made 28 days after prestress release, exposure at 50 percent relative humidity, and ultimate shrinkage = 0.0006 in./in.:

$\epsilon_s = 0.0006 \times 0.4 \times 1.0 = 0.00024$ in./in.
(factors from Figs. 8 and 10)

$$M_s = \epsilon_s \times E_b \times A_b \left(e_2' + \frac{t}{2}\right)$$

$M_s = 0.00024 \times 3.88 \times 10^6 \times 96 \times 6.5\,(21.91 + 3.25)$
$M_s = 14{,}600$ in. kips = 1215 ft. kips

The elastic shrinkage restraint moments are then:

at B = −1.286 × 1215 = −1565 ft. kips

at C = −0.857 × 1215 = −1040 ft. kips

**Live Load Positive Moments**

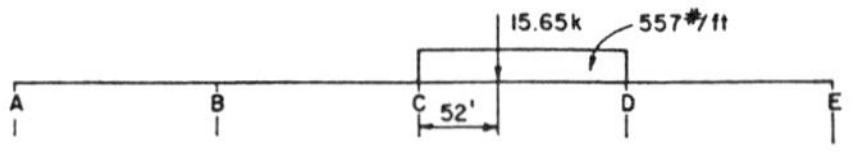


Using influence lines for 4 equal continuous spans, the positive LL + I moment at B is calculated to be = 169.5 ft. kips

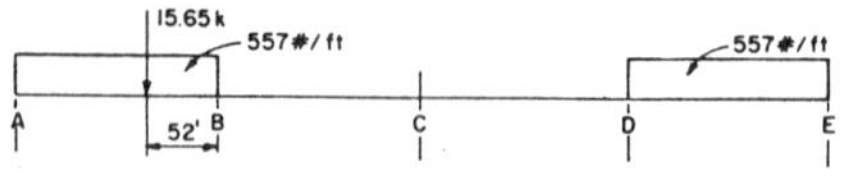


The LL + I moment for this loading at C = 392 ft. kips

**Evaluation of Creep Factors**

From Fig. 5, for an elastic modulus at strand release = $3.66 \times 10^6$ psi, specific creep at 20 years = $0.41 \times 10^{-6}$ in./in./psi.

For release of the prestress force at 2 days, the value of specific creep is adjusted by Fig. 6 to:

$1.66 \times 0.41 \times 10^{-6} = 0.68 \times 10^{-6}$ in./in./psi

From Fig. 7 and Table 1, the volume/surface ratio correction revises the specific creep to:

$1.18 \times 0.68 \times 10^{-6} = 0.802 \times 10^{-6}$ in./in./psi

From Fig. 8, at 28 days 0.4 of the creep has occurred, leaving 0.6 to occur or:

$0.6 \times 0.802 \times 10^{-6} = 0.481 \times 10^{-6}$ in./in./psi

The value of $\phi = \epsilon_c/\epsilon_e$ for the structure with a connection made at 28 days would then be:

$\phi = \epsilon_c/\epsilon_e = \epsilon_c \times E = 0.481 \times 10^{-6} \times 4.05 \times 10^6 = 1.95$

 

From Fig. 9, for $\phi = 1.95$:

$(1 - e^{-\phi}) = 0.88$

$$\left(\frac{1 - e^{-\phi}}{\phi}\right) = 0.45$$

### Summation of Positive Restraint Moments at Piers

Shrinkage plus creep:
at B = (4960 − 3360) × 0.88 − 1565 × 0.45 = 705 ft. kips
at C = (3320 − 2230) × 0.88 − 1040 × 0.45 = 491 ft. kips
Add LL + I:
at B = 705 + 169.5 = 874.5 ft. kips
at C = 491 + 392 = 883 ft. kips

### Design of the Positive Moment Connection

To design for the 883 ft. kips moment at C, use intermediate grade reinforcement with an allowable stress $= 0.6 \times f_y = 0.6 \times 40 = 24$ ksi. Obtain approximate area of steel required using an estimate of $jd = 68$ in.:

$A_s = M/f_s jd$
$A_s = 883{,}000 \times 12/24{,}000 \times 68 = 6.50$ sq. in.
Try 12 No. 7 bars, $A_s = 7.20$ sq. in.

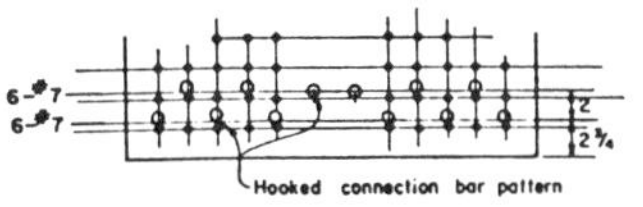


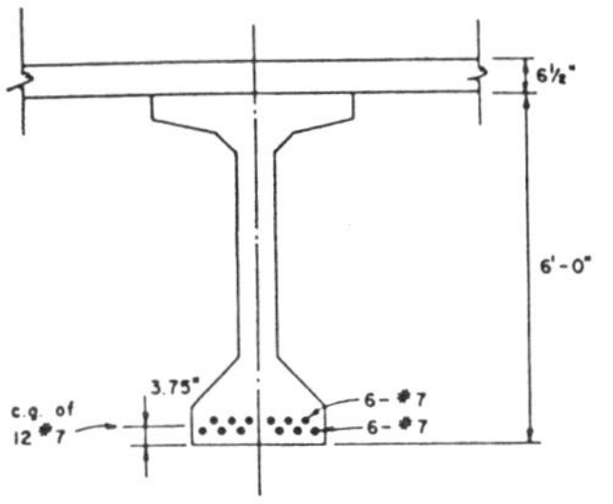


Assume $kd = 8.76$ in. (from trial computation)

| Section | A | y | Ay | Ay² | I₀ |
|---|---|---|---|---|---|
| Slab, 96 × 6.5 × 0.96 | 600 | − 3.25 | −1950 | 6,330 | 2530 |
| Top flange, 2.26 × 42 | 95 | − 7.63 | − 725 | 5,530 | — |
| Steel, 7.20 × 7.16 | 51.6 | −74.75 | −3860 | 289,000 | — |
| Totals | 746.6 | | 6535 | 300,860 | 2530 |

$$\frac{6535}{746.6} = 8.75 \text{ in. (ok)}$$

$$\begin{aligned} 300{,}860 + 2530 &= 303{,}390 \\ -746.6 \times 8.75^2 &= -57{,}200 \\ I &= 246{,}190 \text{ in.}^4 \end{aligned}$$

Distance from neutral axis to steel = 78.5 − 8.75 − 3.75 = 66.0 in.

$$\text{Steel stress} = \frac{883{,}000 \times 12 \times 66.0}{246{,}190} \times 7.16 = 20{,}300 \text{ psi}$$

Since the stress is less than 24,000 psi the design is satisfactory. However, in view of the low stress, consideration might be given to reducing the size or number of bars in the connection.

### Embedment Length and Bend Details

Allowable bond stress = 0.10 × 4500 = 450 psi, use 350 psi maximum. Perimeter of a No. 7 bar = 2.749 in.

Assume distance from the end face of the precast girder to the inside face of the hook = 12 × bar diameter = 12 × 7/8 = 10.5 in.

Minimum hook radius = 3 × 7/8 = 2⅝ in.

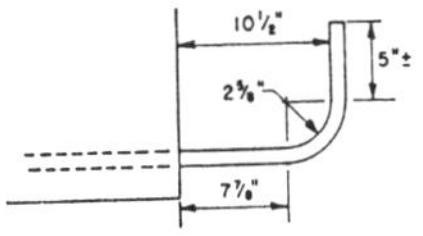


Required embedment length

$$= \frac{20{,}300 \times 0.60}{350 \times 2.749} = 12.2 \text{ in.}$$

1/4 bend = 4.80 in.

7.875 in. + 4.80 in. = 12.675 in.
12.2 − 12.7 = no extension required
Use 5 in. ± extension–gives 18 in. embedment

To ensure that bars extend into the part of the beam where the prestress force is effective, and to avoid terminating all connection bars at one point, extend 6 bars 3 ft. 0 in. into the beam, and extend the other 6 bars 4 ft. 6 in. into the beam.

**Design negative moment reinforcement over piers.** The live load plus impact negative moments over piers are calculated as 1499 ft. kips at B, and 1359 ft. kips at C. Design the negative moment reinforcement at B.

Ultimate negative moment = 2.5 × 1499 = 3740 ft. kips

$$\frac{M_u}{bd^2} = \frac{3{,}740{,}000 \times 12}{28 \times 75.25^2} = 283$$

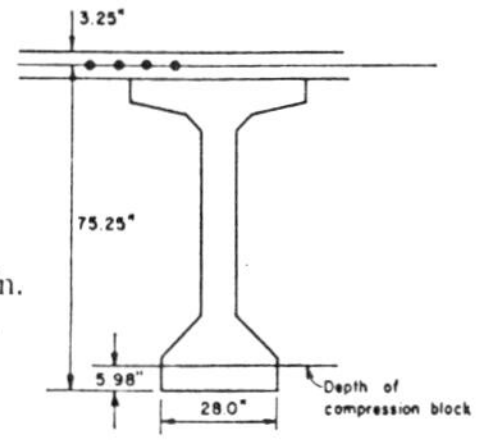


From Fig. 13 (the arrows in Fig. 13 apply to this design example):

$p = 0.0074$

$A_s = 0.0074 \times 28 \times 75.25 = 15.6$ sq. in.
Try 16 No. 9 bars, $A_s = 16.00$ sq. in.

Check ultimate moment provided:

$$a = \frac{A_s f_y}{0.85 f'_c b}$$

$$a = \frac{16 \times 40{,}000}{0.85 \times 4500 \times 28}$$

$a = 5.98$ in.

Note depth of compression block < bottom flange thickness.

$$M_u = A_s f_y \left(d - \frac{a}{2}\right) = 16.00 \times 40{,}000\,(75.25 - 2.99)$$

$M_u = 46{,}300{,}000$ in. lb. = 3850 ft. kips
$M_u$ provided > required 3740 ft. kips.

**Compressive stress in girders at piers.** From a trial calculation, assume $kd = 22.2$ in.

$$n = E_s/E_c = \frac{29 \times 10^6}{4.05 \times 10^6} = 7.16$$

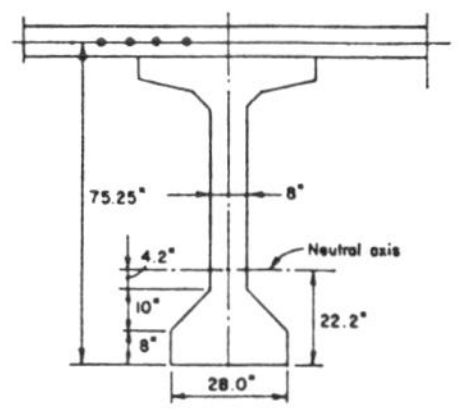


| Section | A | y | Ay | Ay² | I₀ |
|---|---|---|---|---|---|
| 16 × 7.16 (steel) | 114.2 | −75.25 | −8600 | 647,000 | — |
| 8 × 28 | 224 | − 4 | − 896 | 3,580 | 1195 |
| 10 × 18 | 180 | −11.63 | −2095 | 24,400 | 1221 |
| 8 × 4.2 | 33.6 | −20.1 | − 640 | 12,850 | — |
| Totals | 551.8 | | −12,231 | 687,830 | 2416 |

$$\text{N.A} = \frac{12{,}231}{551.8} = 22.2 \text{ in.}$$

$$\begin{aligned} 687{,}830 + 2416 &= 690{,}246 \\ -551.8 \times 22.2^2 &= -272{,}000 \\ I &= 418{,}246 \text{ in.}^4 \end{aligned}$$

$$f_c\,(\text{LL} + \text{I}) = \frac{1{,}499{,}000 \times 12 \times 22.2}{418{,}246} = 952 \text{ psi}$$

$$f_c\,(\text{prestress}) = \frac{1{,}132{,}000}{1085} + \frac{1{,}132{,}000 \times 18.88 \times 36.38}{733{,}320} = 2105 \text{ psi}$$

$$= 3057 \text{ psi}$$

This stress will not actually occur because of the tendon transfer length. The stress is also reduced by the positive moment induced by creep. The theoretical stress can be reduced to 2000 psi by blanketing 14 bottom strands for a short distance.

**Tensile stresses in girders at midspan.** In span AB, the positive creep restraint moment at midspan would be 0.5 × 705 = 352.5 ft. kips. This moment causes a tension of:

$$f_c = \frac{352{,}500 \times 12}{26{,}266} = 161 \text{ psi}$$

Adding the tension due to gravity loads of about 94 psi, the total tensile stress becomes 255 psi. The permissible value to avoid cracking would be $0.8 \times 7.5\sqrt{5000} = 426$ psi. Since the tensile stress is much less th. n the allowable, no additional prestressing reinforcement is required.

In span BC, the positive creep restraint moment at midspan is

$$\frac{491 + 705}{2} = 598 \text{ ft. kips.}$$ This oauses a tensile stress of 273 psi.

Sinoe there is no tensile stress in this span under gravity loads, this section is also satisfactory for tensile stresses due to creep.

# D.9 DEFECTS AND COUNTERMEASURES

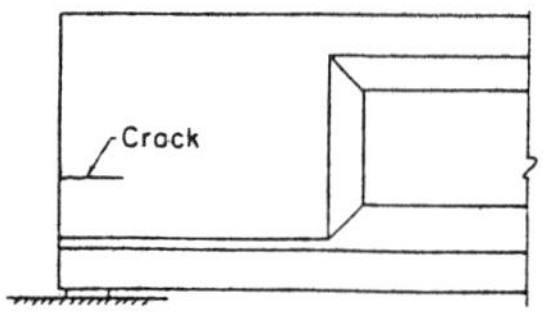


Horizontal Crack at Mid-Depth

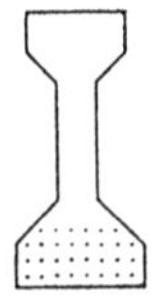

Pretensioned Beam (straight steel only)

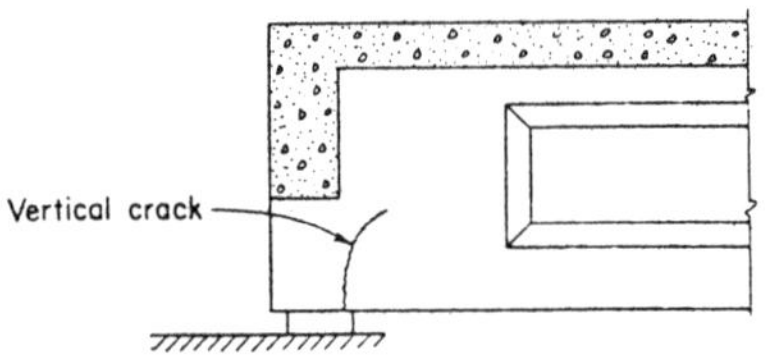


Vertical Crack in Vicinity of Bearing Plate

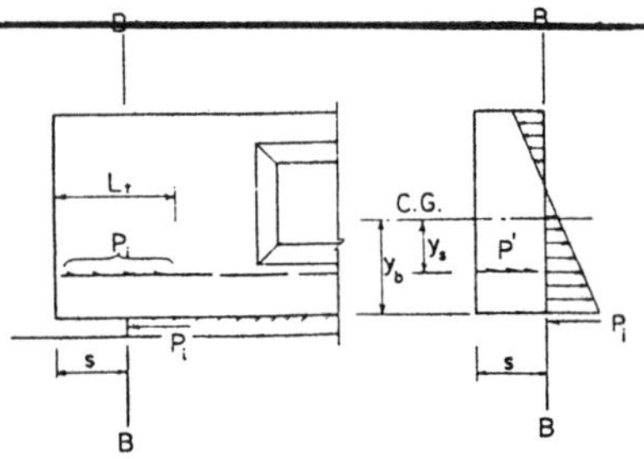


Stresses Resulting from Soffit Form Restraint

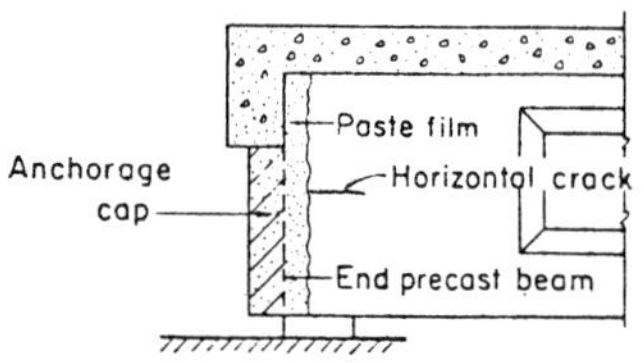


Posttensioned Beam with Horizontal Crack Not Reflected in Paste Film

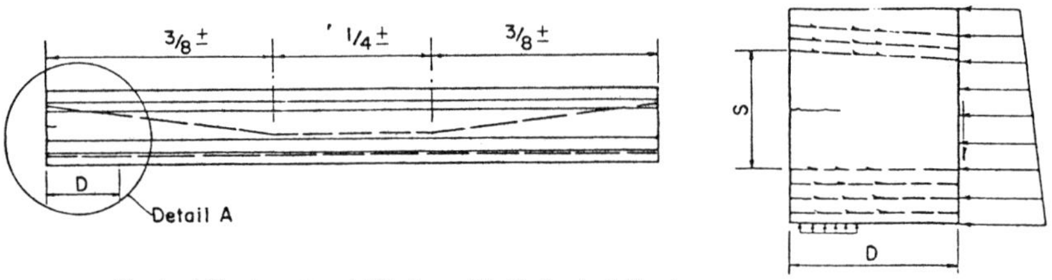


Typical Pretensioned Girder with Deflected Steel

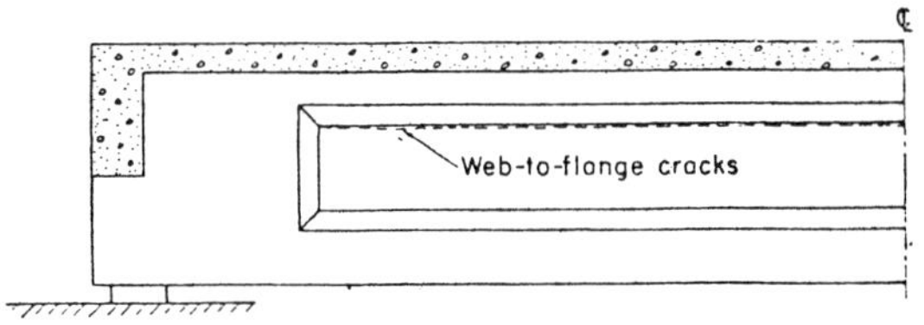


Typical Horizontal Cracks at Junction of Web and Top Flange

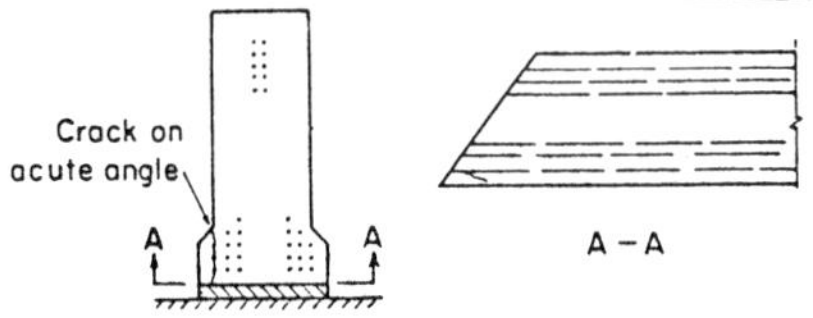


Spalling Tendency on Beam with Skewed End

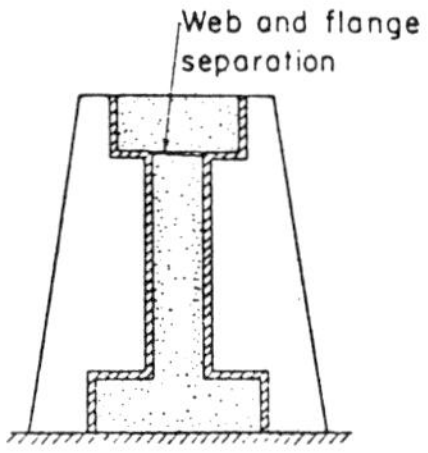


Effect of Web Settlement with Top Flange Restrained

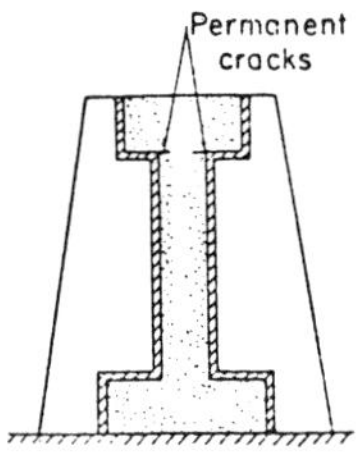


Cracks Resulting from Differential Settlement

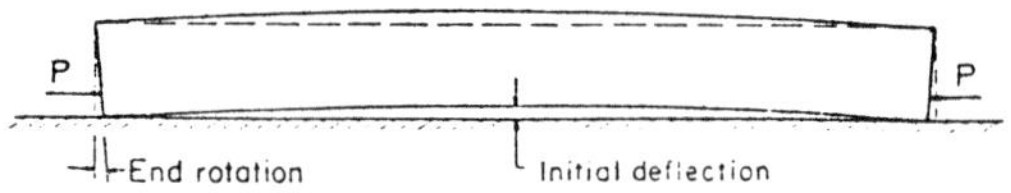


Initial Deflection of Prestressed Beam

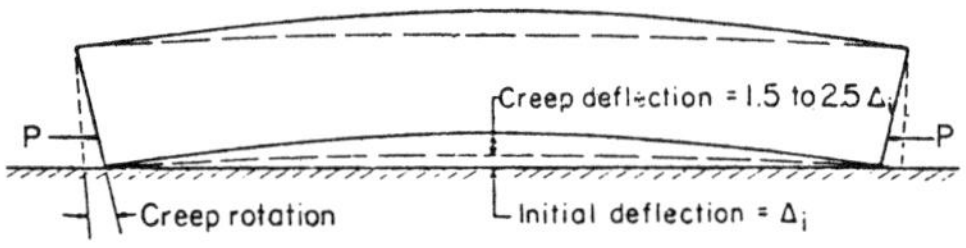


Initial Plus Creep Deflection of Prestressed Beam

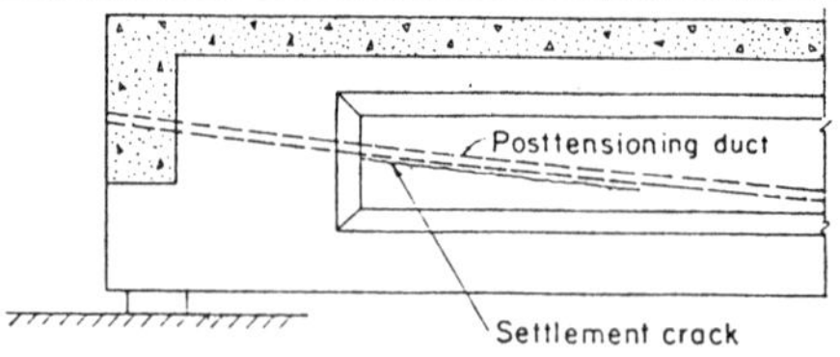


Settlement Crack Along Posttensioning Duct

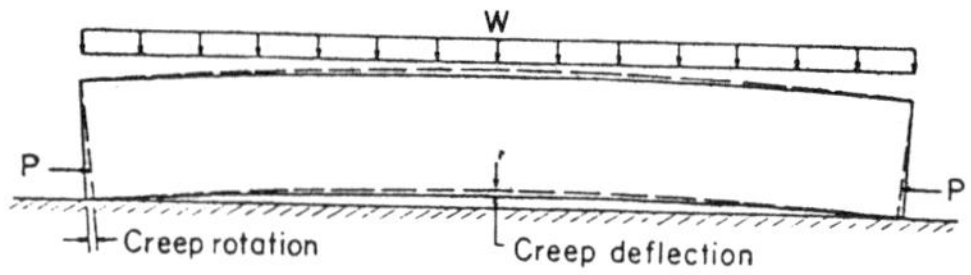


Creep Deflection Caused by Composite Concrete Deck Slabs

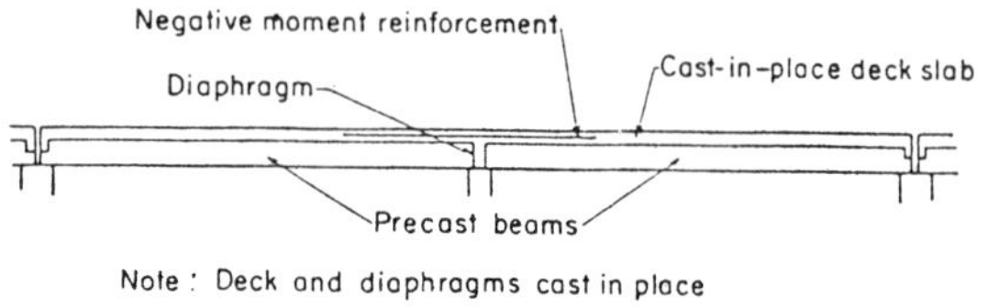


Construction for Live Load Continuity

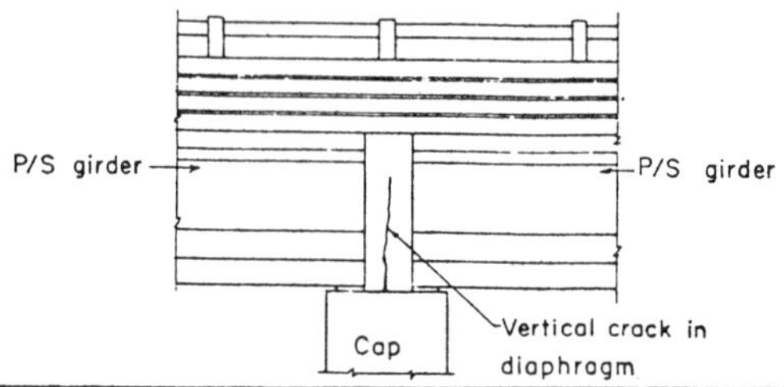


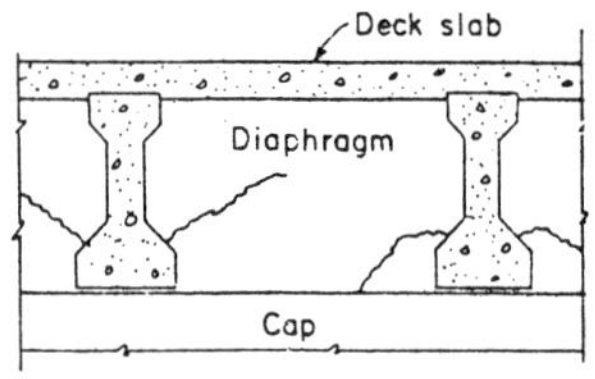

| Potential Defect | Suggestions for Elimination |
|---|---|
| Vertical cracking between protective end cap and main posttensioned beam | Beam end should be thoroughly cleaned, wetted and scrubbed with cement grout immediately prior to application of the driest cap mix practical. Curing of cap should begin as soon as possible. |
| Horizontal cracks at beam ends | Crack widths should be controlled by vertical reinforcing steel placed as close to the beam end as practicable. The amount of steel should be determined by a reliable design procedure. The minimum steel used should be sufficient to resist a vertical tensile force equal to about 4 per cent of the total initial prestress force. |
| Vertical cracks in bottom flange near beam end | Provisions should be made during manufacture of the beams to allow freedom for beam shortening upon application of the prestressing force. |
| Horizontal cracks along the junction of webs and flanges | Settlement and shrinkage should be reduced by using well-graded aggregates in a concrete having a low total water content and placed in three stages. Curing heat should be applied at a relatively slow rate, causing a rise in surrounding air temperature not to exceed 40° F. per hour. |
| Web cracks along post-tensioning ducts | Settlement and shrinkage should be reduced by using well-graded aggregates and a concrete having a low water content. Sufficient clearance should be provided for placing and compacting concrete. Grout mixtures and grouting procedures should be in accordance with the Prestressed Concrete Institute's Tentative Recommended Practices for Grouting Posttensioned Prestressed Concrete. Web reinforcement should be used. |
| Longitudinal cracks in bottom flanges of box beams | A minimum concrete cover of 1-1/2 in. should be provided under the prestressing steel. Transverse reinforcement should be placed in the bottom flange throughout the length of the beam. |
| Camber | Finishing concrete deck slabs on composite structures to the final desired elevation should give good results. Allowance for future increase in camber on non-composite structures should be made in the manufacture of the beam or in placing the wearing surface. |
| Cracks in diaphragms between abutting ends of precast beams in bridges built to act as continuous structures | Use horizontal ties designed to resist rotations of ends of abutting beams. |

**From "A Field Inspection of Prestressed Concrete Bridge" PCA**

## E.1 TYPICAL TRUSS BRIDGE

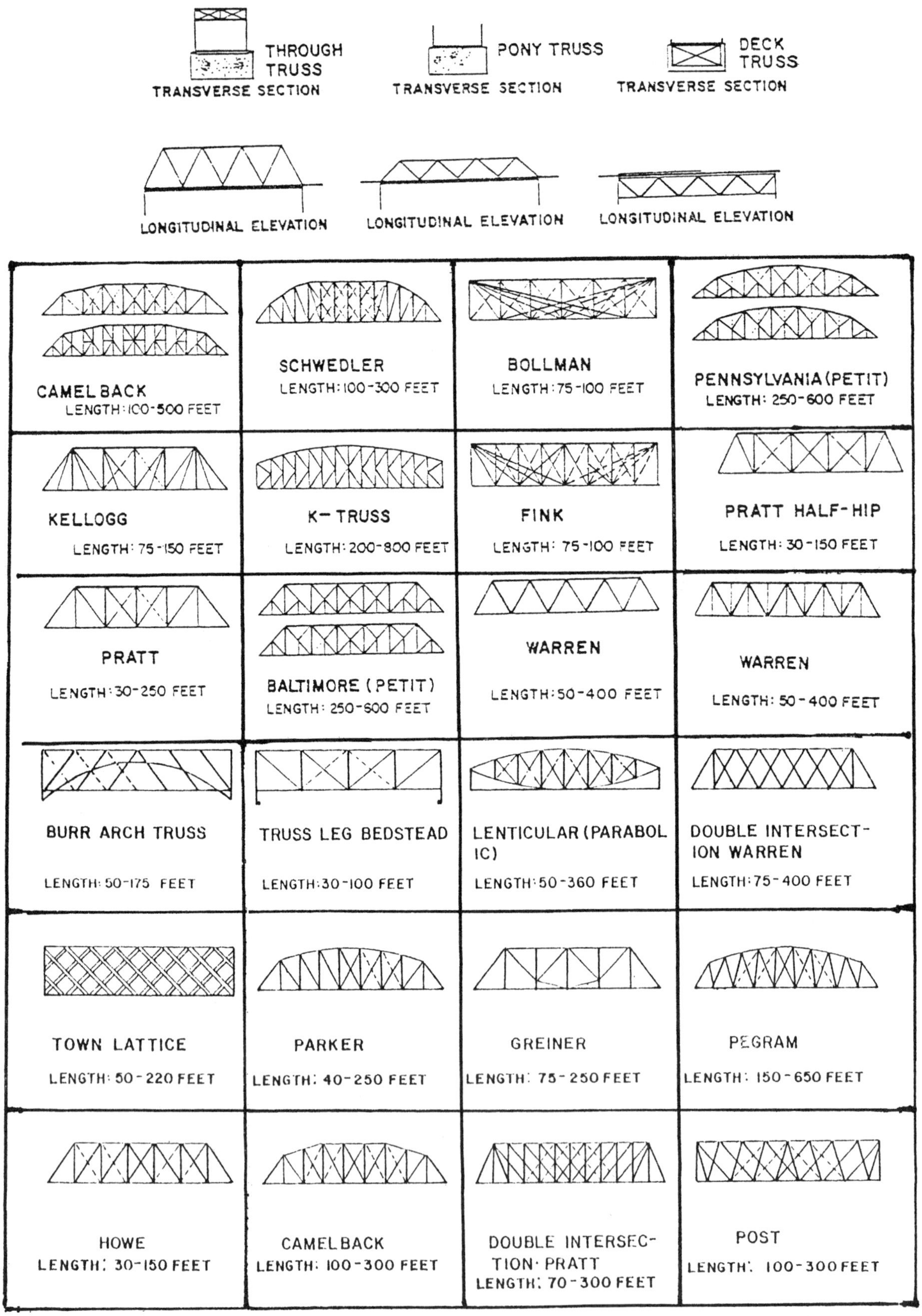

# E.2 METHOD OF UPGRADING TRUSS BRIDGE

## PRESTRESSING

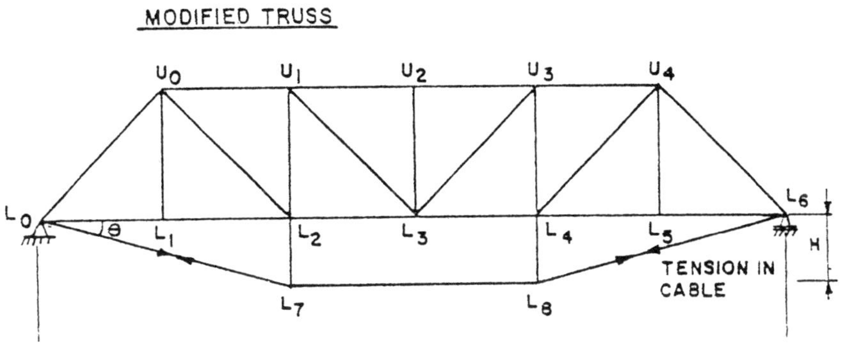

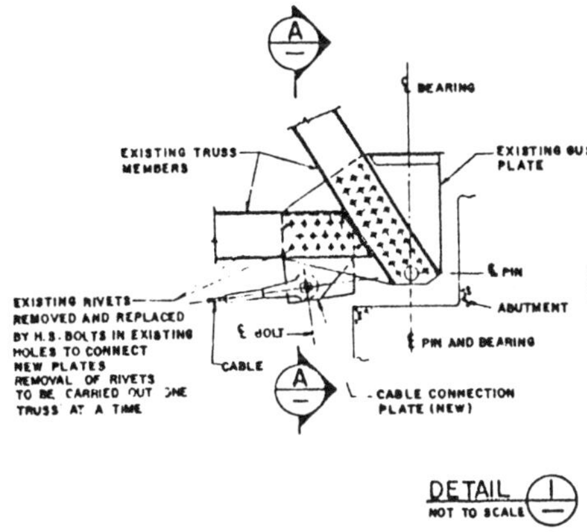


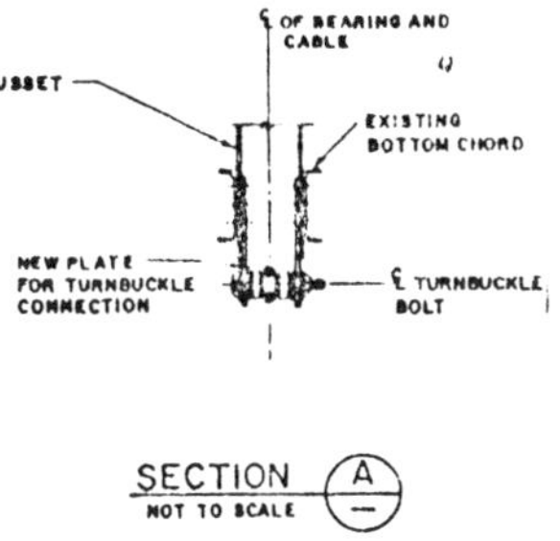


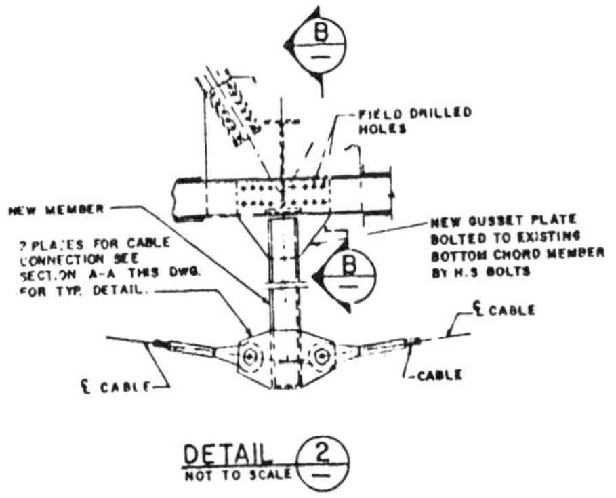


## STRENGTHENING COMPRESSION MEMBER

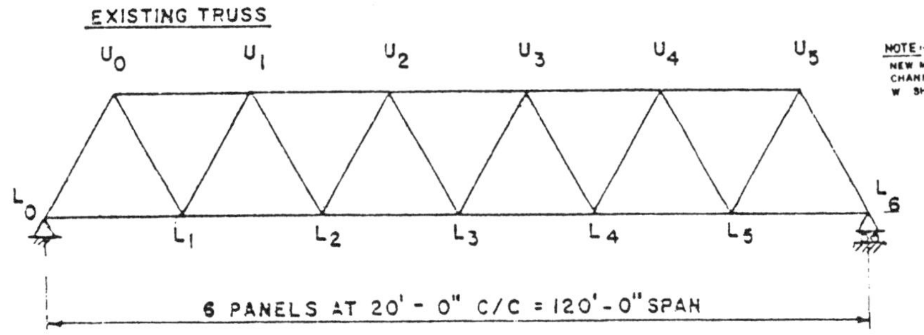


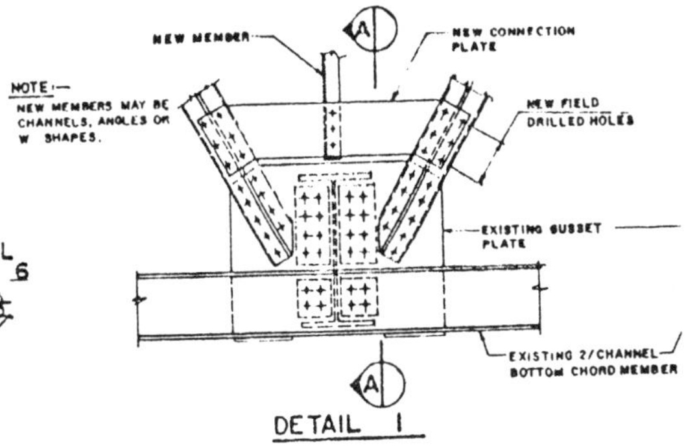


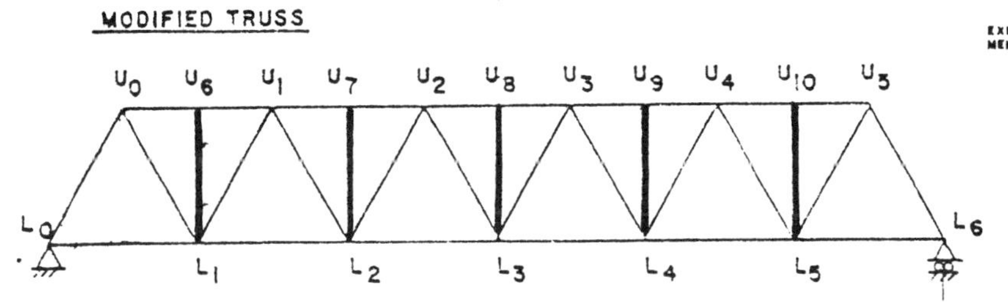


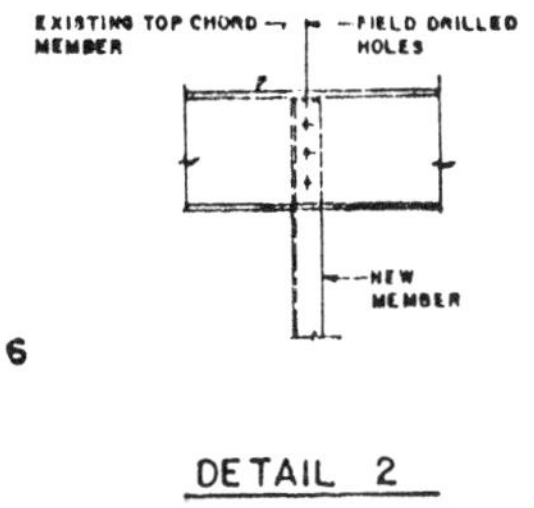


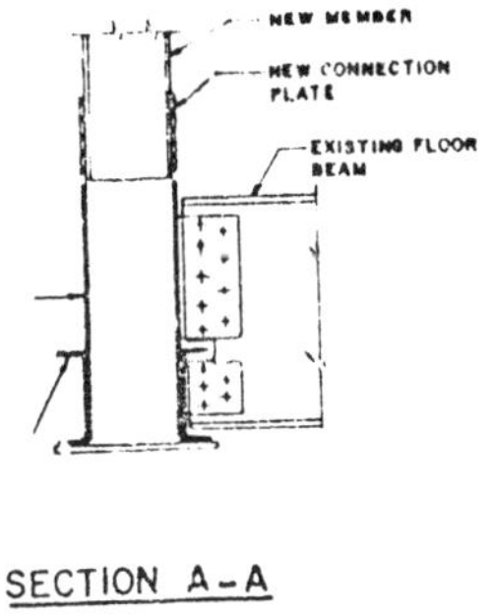

## ADDITIONAL SUPPORTS

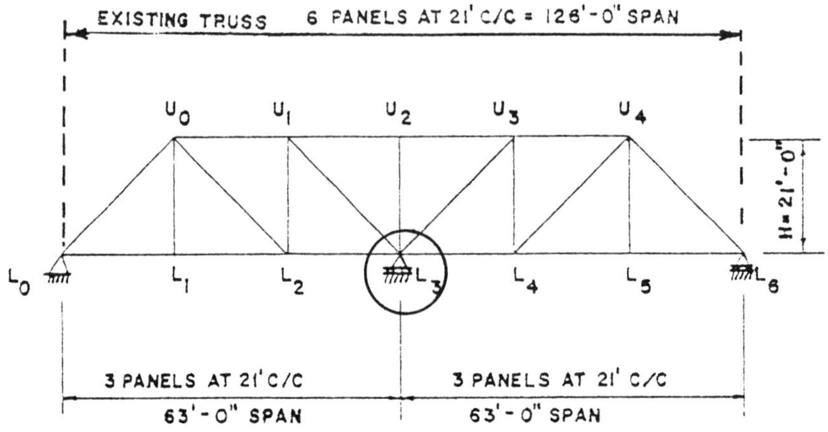


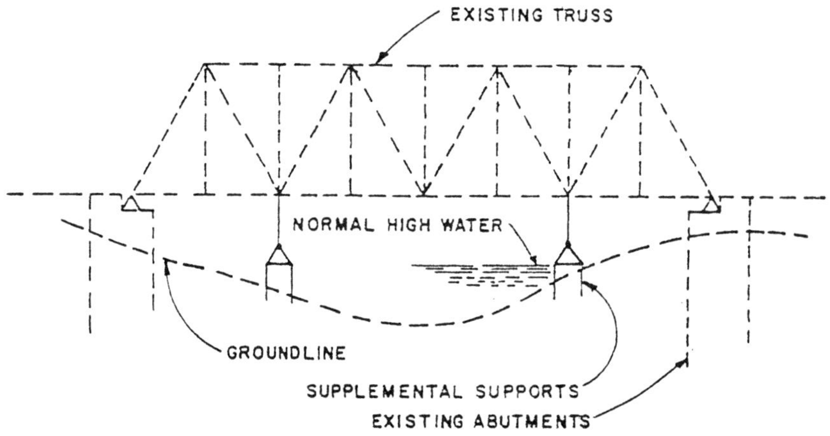


## SHIFTING SUPPORTS

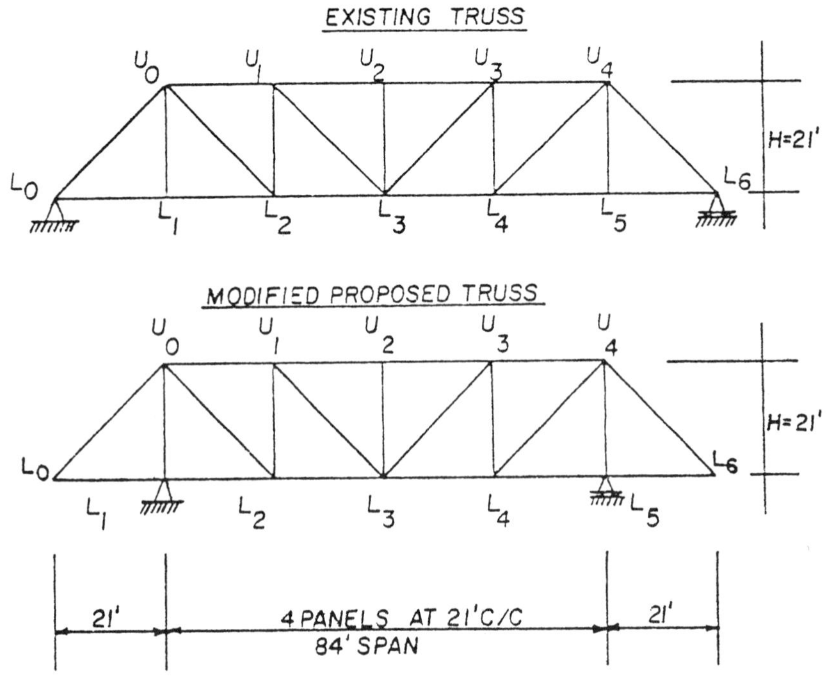

**SUBDIVISION**

METHOD OF SUBDIVISION (PRATT TRUSS)

EXISTING TRUSS

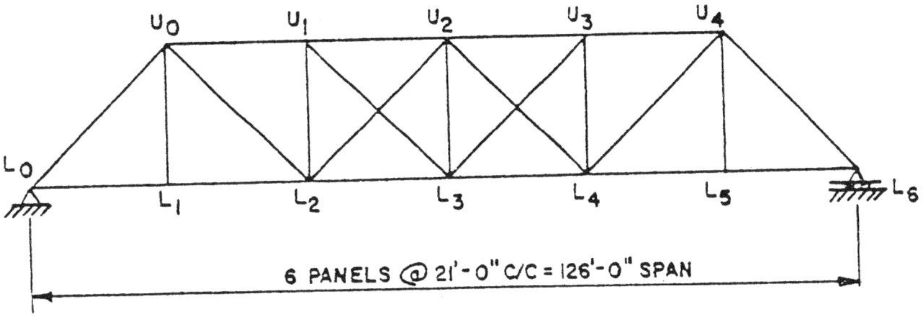


MODIFIED TRUSS

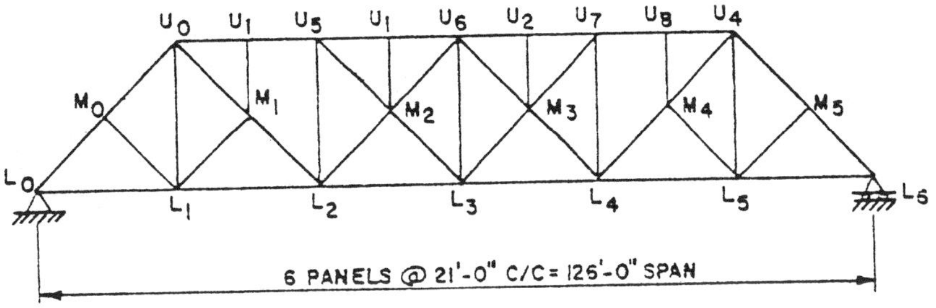


METHOD OF SUBDIVISION (WARREN TRUSS)

EXISTING TRUSS

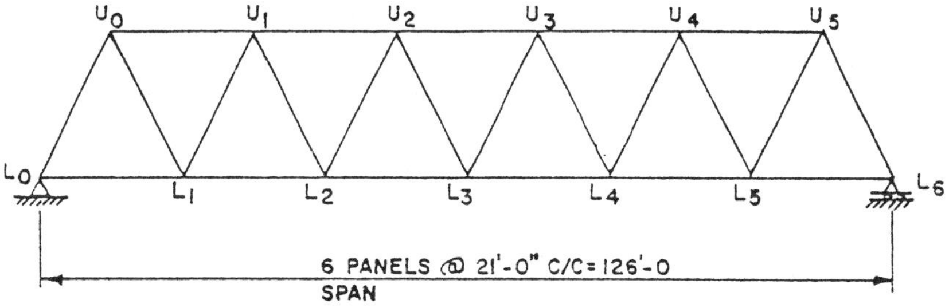


MODIFIED TRUSS

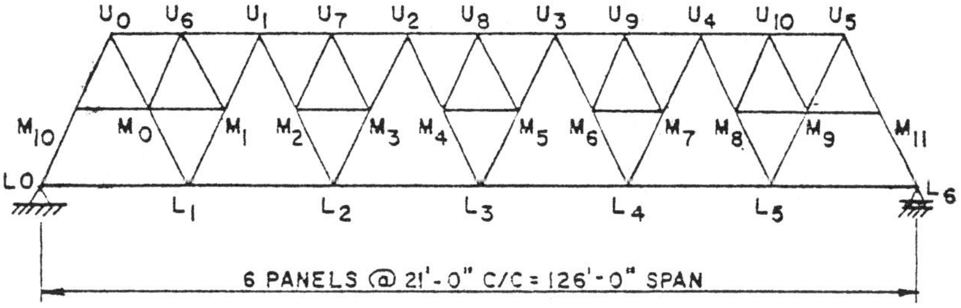

## E.3 ROLL-IN STEEL STRUCTURE

LOCATION: AMTRAK MAINLINE OVER WOOD AVE., ISELIN, NJ
OWNER: NJ DEPARTMENT OF TRANSPORTATION
CONTRACTOR: SCHIVONE CONSTRUCTION CO. SECAUCUS, NJ
FABRICATOR: AMERICAN BRIDGE DIVISION, U.S. STEEL CORP.
COMPLETION DATE; JUNE, 1981

AFTER CONSTRUCTING TEMPORARY STRUCTURES, EXCAVATIONS WERE MADE UNDER THE BRIDGES BY USING PILE, STRUT AND TIEBACK SYSTEM.

TEMPORARY TRACK BRIDGES FLANKING THE EXISTING ARCH BRIDGE

THE PREASSEMBLED BRIDGE ON THE ROLLED IN STRUCTURE WAS READY TO PLACED.

THE PROCESSES OF ROLLING OPERATION-CENTRAL CONTROL OF HYDRAULIC OPERATION AT BRIDGE DECK.

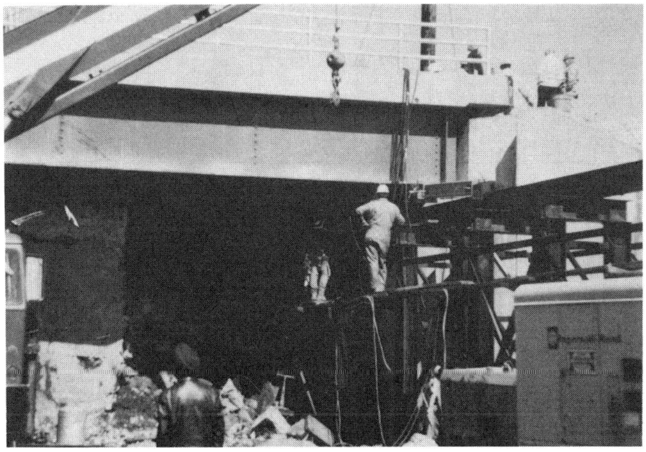

ROLLING IN OPERATION

| FED ROAD DIV NO | STATE | FEDERAL PROJECT NO | SHEET | TOTAL SHEETS |
|---|---|---|---|---|
| 1 | N.J. | M-7255(103)<br>M-7254(105)<br>MG-7253(102) | 111 | 131 |

492

| BILL OF MATERIALS | | |
|---|---|---|
| TEMPORARY STRUCTURE | | |
| DESCRIPTION | UNIT | ESTIMATED QUANTITY |
| Structural Steel (incl. tie rods) | LBS | 508,800 |
| Bent Piles (HP 14×89) | L.F. | 5,600 |
| Pile Casing (22"ϕ) | L.F. | 800 |
| Timber (including ties 10"×12") | M.B.F. | 68.6 |
| Temporary Track Supports for Installation of Pile Bent Cap Beams | | See table on Sheet B26 |

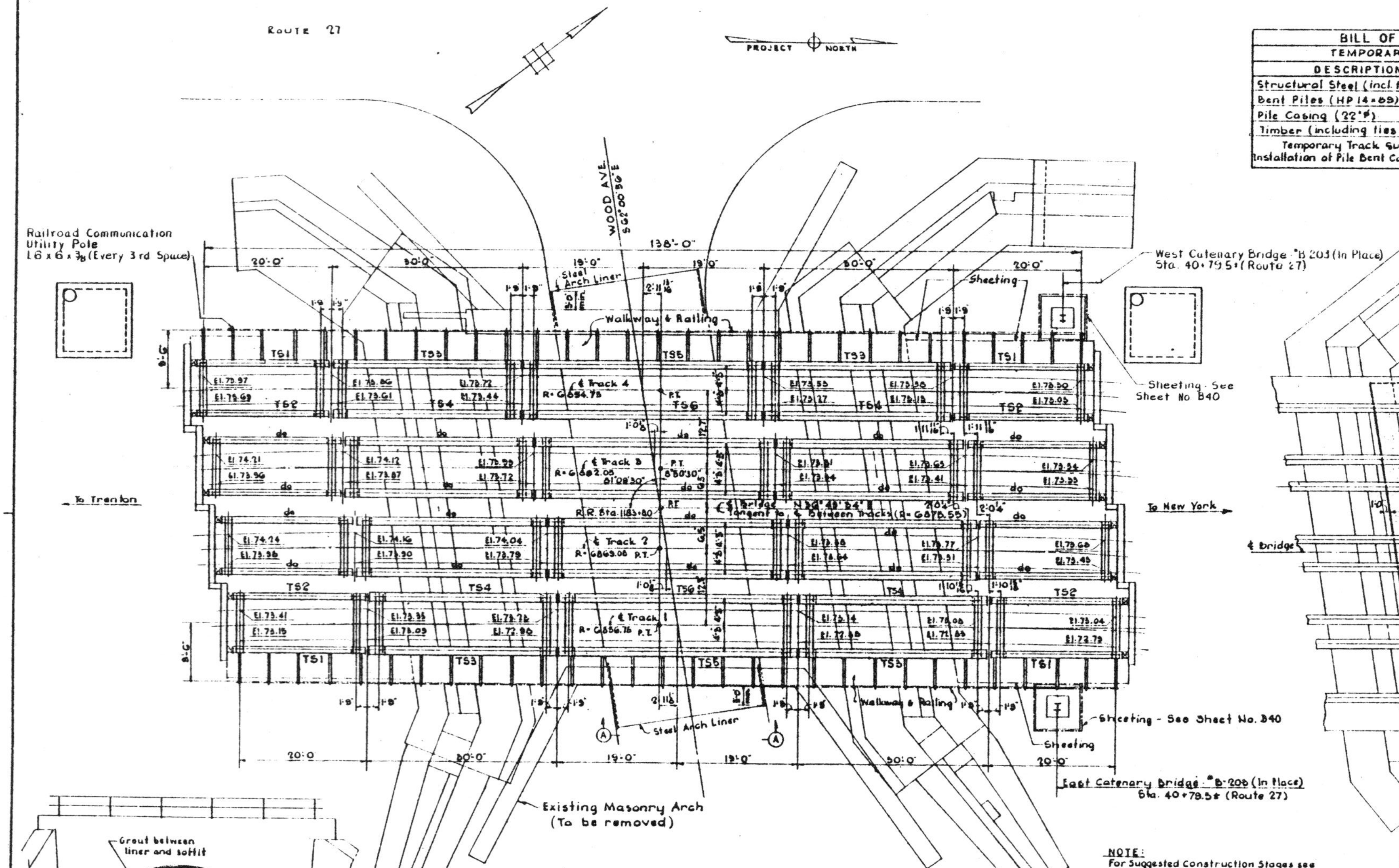


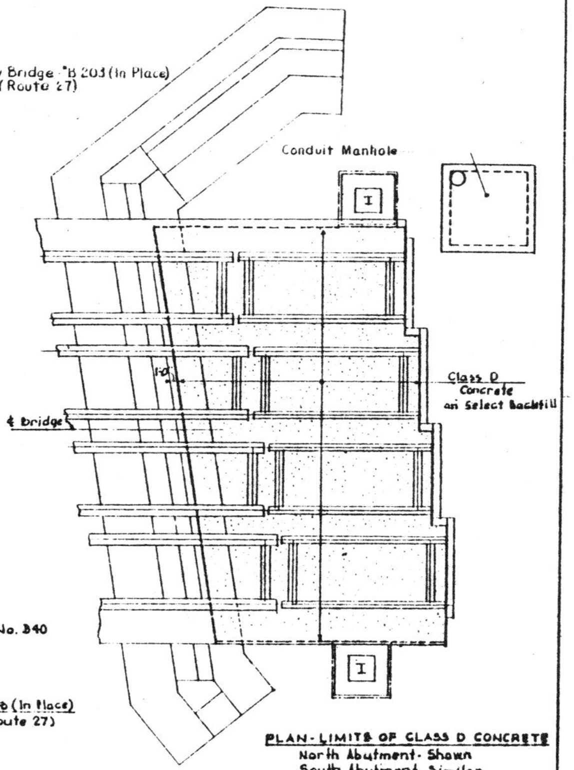


**PLAN - LIMITS OF CLASS D CONCRETE**
North Abutment - Shown
South Abutment - Similar

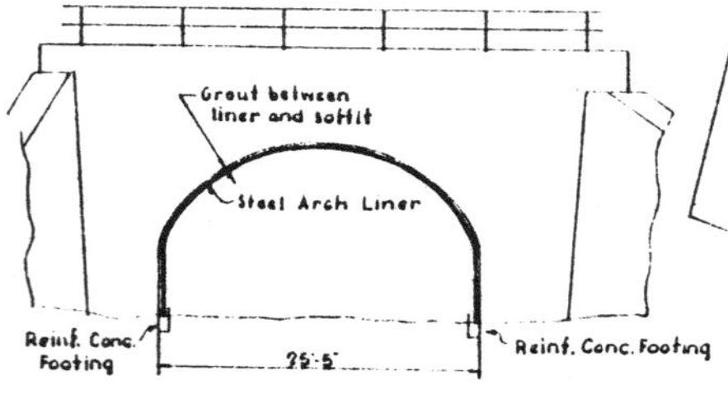


SECTION A-A

PLAN - TEMPORARY TRACK SUPPORT STRUCTURES

Note: Top of Rail Elevations & Geometry as per Date of Survey: Nov. 1971

NOTE:
For Suggested Construction Stages see Supplementary Specifications.

| | | | |
|---|---|---|---|
| REVISION | BY | CKD | DATE |

STATE OF NEW JERSEY
DEPARTMENT OF TRANSPORTATION
ROUTE 27 (1953) SECTION 6F
REPLACEMENT OF AMTRAK BRIDGE NO. 23.42
OVER WOOD AVENUE AT ROUTE 27
WOODBRIDGE AND EDISON TOWNSHIPS
MIDDLESEX COUNTY, ISELIN, N.J.

TEMPORARY STRUCTURE
GENERAL PLAN

| | | |
|---|---|---|
| SCALE: 1/8" = 1'-0" | DATE 7-14-78 | SHEET NO 111 / 131 B22 OF B42 |
| DWN BY ERS | CHECKED BY W.W.C. N.U.B. | |

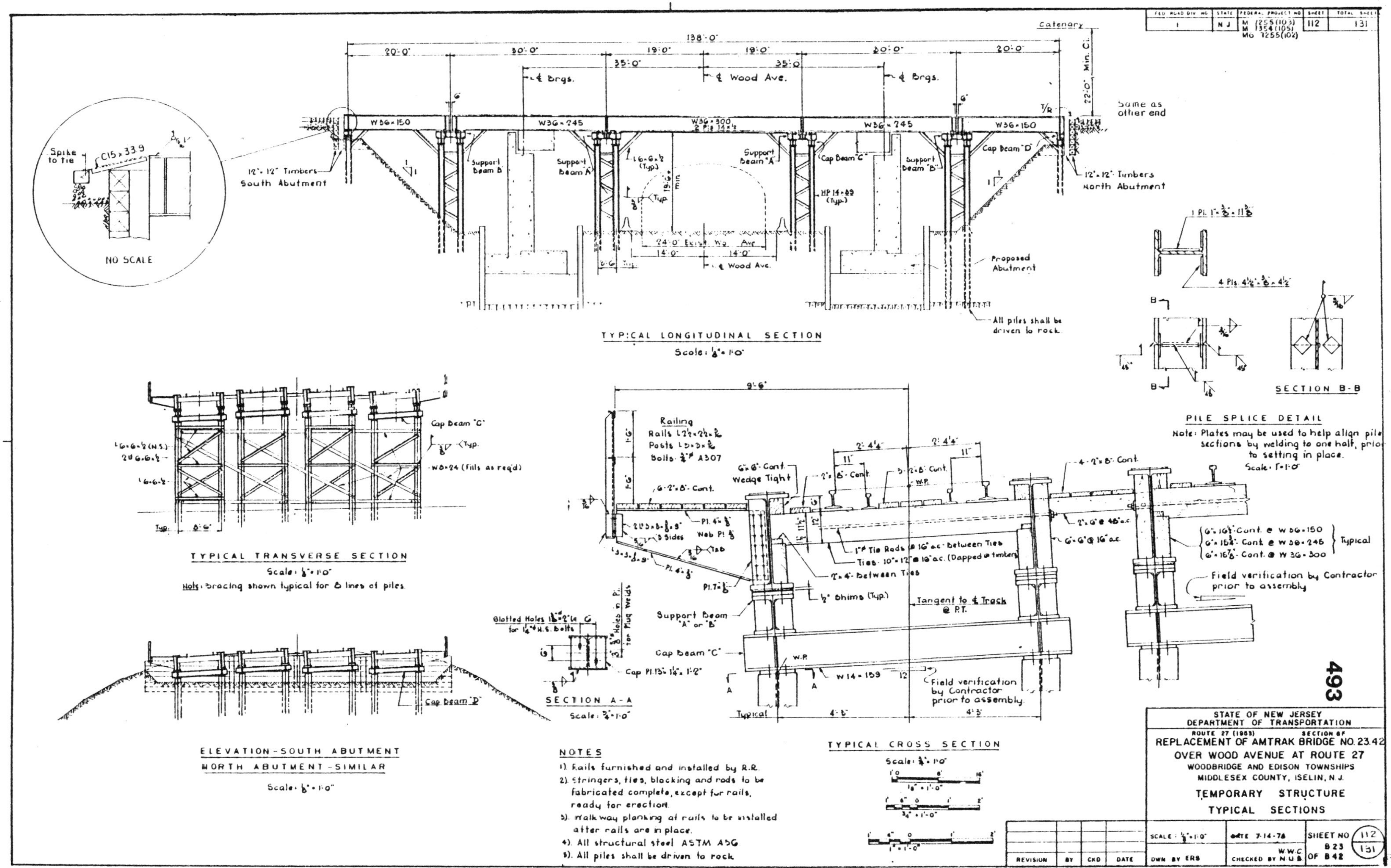

FED. ROAD DIV. NO. 1 | STATE N J | FEDERAL PROJECT NO. M 1255(103) M 1354(105) Mo 1255(102) | SHEET 112 | TOTAL SHEETS 131
Catenary
22'-0" Min. Cl.
138'-0"
20'-0"
30'-0"
19'-0"
19'-0"
30'-0"
20'-0"
35'-0"
35'-0"
℄ Brgs.
℄ Wood Ave.
℄ Brgs.
W36×150
W36×245
W36×300
2 Pls 14×½
W36×245
W36×150
T/R
Same as other end
Spike to tie
C15×33.9
12"×12" Timbers South Abutment
NO SCALE
Support Beam B
Support Beam A
L6×6×½ (Typ.)
19'-6" min.
Typ.
Support Beam A
Cap Beam C
HP 14×89 (Typ.)
Support Beam B
Cap Beam D
12"×12" Timbers North Abutment
24'-0" Exist. Wo. Ave
14'-0"
14'-0"
℄ Wood Ave.
6'-6" Typ.
Proposed Abutment
All piles shall be driven to rock.
TYPICAL LONGITUDINAL SECTION
Scale: ⅛" = 1'-0"
1 Pl. 1"×¾"×11⅜"
4 Pls. 4½"×¾"×4½"
SECTION B-B
PILE SPLICE DETAIL
Note: Plates may be used to help align pile sections by welding to one half, prior to setting in place.
Scale: 1" = 1'-0"
Cap Beam C
Typ.
L6×6×½ (H.S.)
2 L6×6×½
L6×6×½
W8×24 (Fills as req'd)
Typ.
8'-6"
TYPICAL TRANSVERSE SECTION
Scale: ⅛" = 1'-0"
Note: Bracing shown typical for 8 lines of piles.
9'-6"
Railing
Rails L2½×2½×¼
Posts L5×3×⅝
Bolts ¾"φ A307
6"×6" Cont. Wedge Tight
2'-4¼"
2'-4¼"
11"
11"
2"×8" Cont.
5-2"×8" Cont.
W.P.
4-2"×8" Cont.
6-2"×8" Cont.
Pl. 4×⅝
Web Pl. ⅝
3 Sides
Pl. 4×⅝
Pl. 7×⅝
1"φ Tie Rods @ 16" a.c. between Ties
Ties 10"×12" @ 16" a.c. (Dapped @ timbers)
2"×4" between Ties
½" Shims (Typ.)
2"×6" @ 48" a.c.
6"×6" @ 16" a.c.
6"×16¼" Cont. @ W36×150
6"×15¾" Cont. @ W36×245
6"×16⅞" Cont. @ W36×300
Typical
Field verification by Contractor prior to assembly
Tangent to ℄ Track @ P.T.
Support Beam A or B
Cap Beam C
W.P.
W14×159
Field verification by Contractor prior to assembly
Typical
4'-6"
4'-3"
TYPICAL CROSS SECTION
Scale: ¾" = 1'-0"
Slotted Holes 1⅜"×2" for 1¼"φ H.S. Bolts
Holes in Pl. for Plug Welds
Cap Pl. 13"×1¼"×1'-2"
SECTION A-A
Scale: ¾" = 1'-0"
Cap Beam D
ELEVATION-SOUTH ABUTMENT
NORTH ABUTMENT-SIMILAR
Scale: ⅛" = 1'-0"
NOTES
1). Rails furnished and installed by R.R.
2). Stringers, ties, blocking and rods to be fabricated complete, except for rails, ready for erection.
3). Walkway planking at rails to be installed after rails are in place.
4). All structural steel ASTM A36
5). All piles shall be driven to rock.
493
STATE OF NEW JERSEY
DEPARTMENT OF TRANSPORTATION
ROUTE 27 (1983) SECTION OF
REPLACEMENT OF AMTRAK BRIDGE NO. 23.42
OVER WOOD AVENUE AT ROUTE 27
WOODBRIDGE AND EDISON TOWNSHIPS
MIDDLESEX COUNTY, ISELIN, N.J.
TEMPORARY STRUCTURE
TYPICAL SECTIONS
SCALE: ⅛" = 1'-0"
DATE 7-14-78
SHEET NO. 112/131
B23 OF B42
DWN. BY ERB
CHECKED BY W.W.C. N.U.B.
REVISION | BY | CKD | DATE

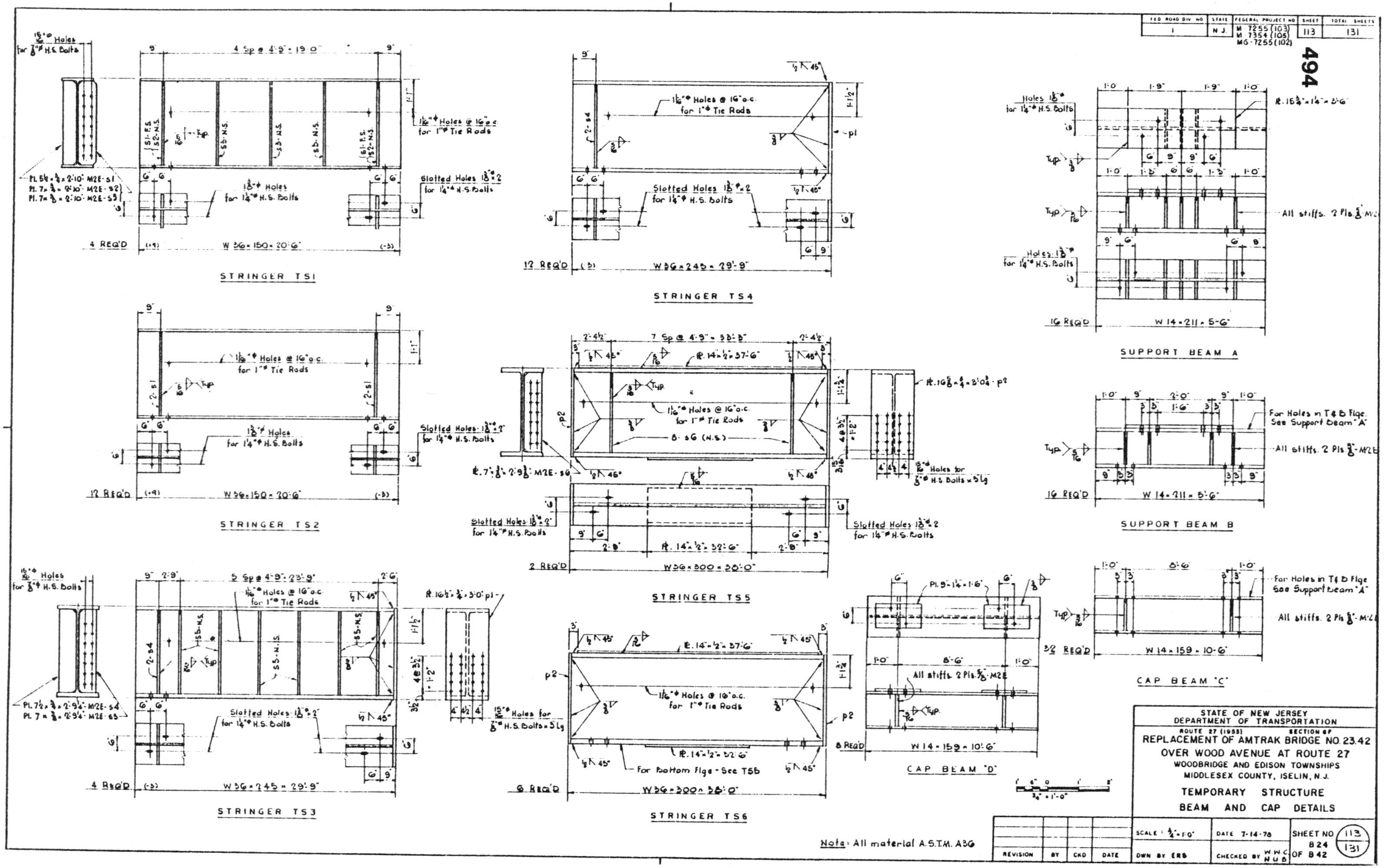

494
FED. ROAD DIV. NO. 1
STATE N.J.
FEDERAL PROJECT NO. M 7255 (103) M 7354 (105) MG-7255 (102)
SHEET 113
TOTAL SHEETS 131
STRINGER TS1
4 REQ'D
W 36 × 150 × 20'-6"
4 Sp @ 4'-9" = 19'-0"
STRINGER TS2
12 REQ'D
W 36 × 150 × 20'-6"
STRINGER TS3
4 REQ'D
W 36 × 245 × 29'-9"
5 Sp @ 4'-9" = 23'-9"
STRINGER TS4
12 REQ'D
W 36 × 245 × 29'-9"
STRINGER TS5
2 REQ'D
W 36 × 300 × 38'-0"
7 Sp @ 4'-9" = 33'-3"
STRINGER TS6
6 REQ'D
W 36 × 300 × 38'-0"
For Bottom Flge - See TS5
SUPPORT BEAM A
16 REQ'D
W 14 × 211 × 5'-6"
SUPPORT BEAM B
16 REQ'D
W 14 × 211 × 5'-6"
For Holes in T & B Flge. See Support Beam "A"
All stiffs. 2 Pls
CAP BEAM "C"
32 REQ'D
W 14 × 159 × 10'-6"
CAP BEAM "D"
8 REQ'D
W 14 × 159 × 10'-6"
1$\frac{1}{16}$"⌀ Holes @ 16" o.c. for 1"⌀ Tie Rods
Slotted Holes 1$\frac{3}{8}$"⌀ × 2 for 1¼"⌀ H.S. Bolts
Note: All material A.S.T.M. A36
STATE OF NEW JERSEY
DEPARTMENT OF TRANSPORTATION
ROUTE 27 (1953) SECTION 6F
REPLACEMENT OF AMTRAK BRIDGE NO. 23.42
OVER WOOD AVENUE AT ROUTE 27
WOODBRIDGE AND EDISON TOWNSHIPS
MIDDLESEX COUNTY, ISELIN, N.J.
TEMPORARY STRUCTURE
BEAM AND CAP DETAILS
SCALE: ¾" = 1'-0"
DATE 7-14-76
SHEET NO. 113/131
B24 OF B42
DWN BY ERB
CHECKED BY WWC NUB
REVISION BY CKD DATE

| FED. ROAD DIV. NO. | STATE | FEDERAL PROJECT NO. | SHEET | TOTAL SHEETS |
|---|---|---|---|---|
| 1 | N. J. | M-7255(103) M-7354(105) MG-7255(102) | 115 | 131 |



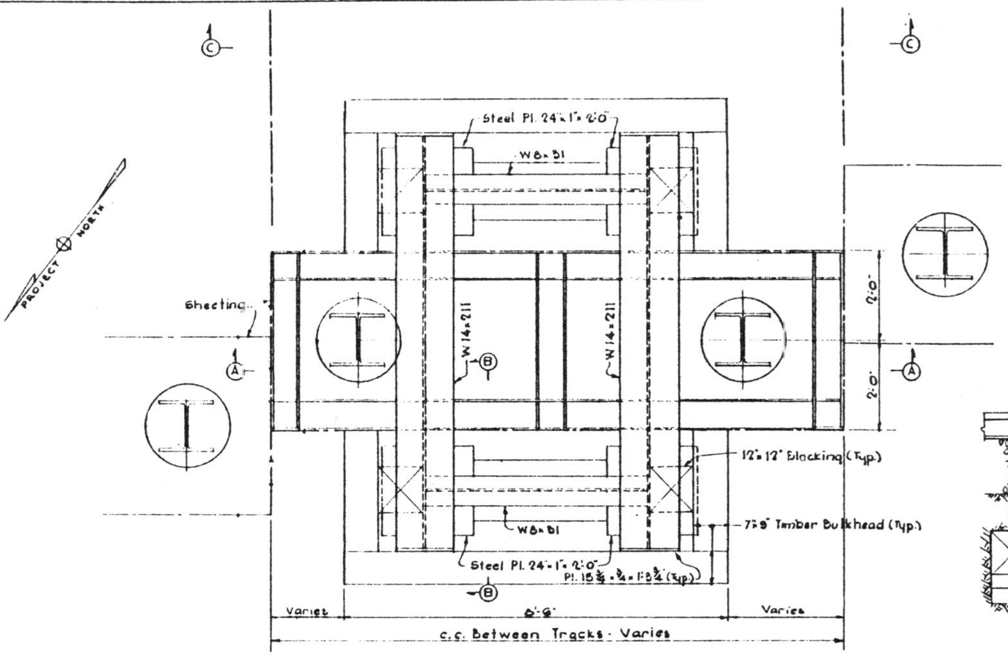


PLAN

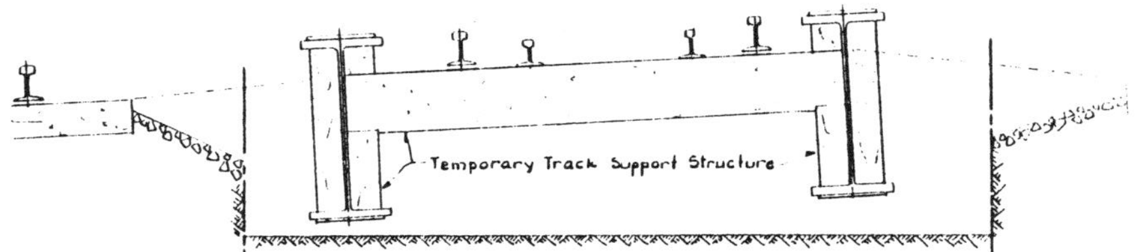


SECTION C-C

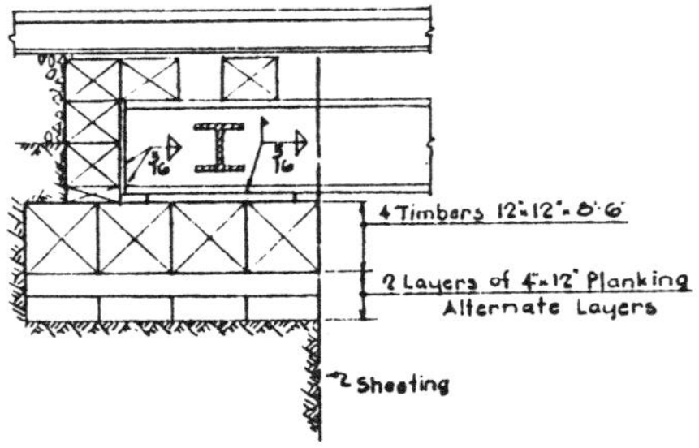


SECTION B-B

| BILL OF MATERIALS<br>TEMPORARY TRACK SUPPORTS FOR<br>INSTALLATION OF PILE BENT CAP BEAMS | | |
|---|---|---|
| DESCRIPTION | UNIT | ESTIMATED QUANTITY |
| Structural Steel * | LBS. | 138,200 |
| Temporary Sheet Piling | S.F. | 6,080 |
| Timber (Bulkhead) | M.B.F. | 41.8 |
| Ties (7"x9"x7'-0") | Unit | 200 |
| Ballast (Furnish, install) | C.Y. | 255 |

* Track Supports only.

NOTES

Sheeting to be PSA23.
Struts and wales to be W8x31.
Stringers, diaphragms and ties to be preassembled.
Structural Steel to be A.S.T.M. A36.

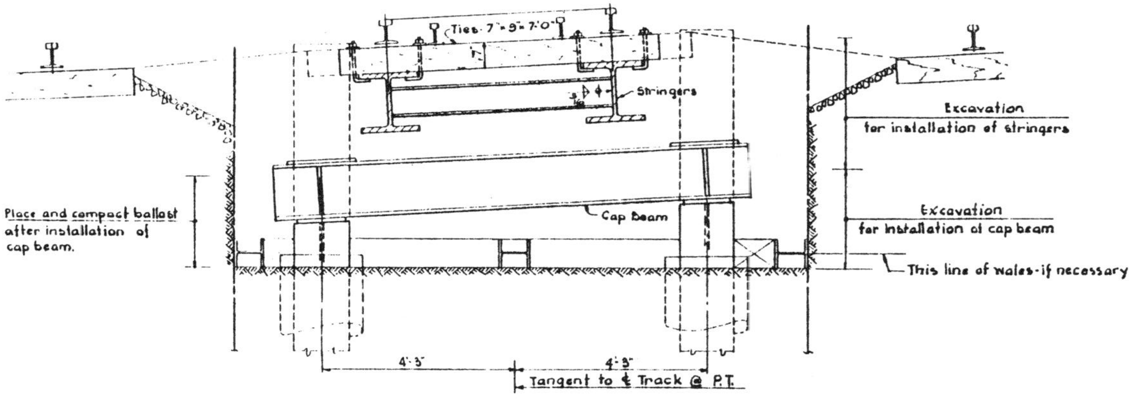


SECTION A-A

STATE OF NEW JERSEY
DEPARTMENT OF TRANSPORTATION
ROUTE 27 (1953) SECTION 6F
REPLACEMENT OF AMTRAK BRIDGE NO. 23.42
OVER WOOD AVENUE AT ROUTE 27
WOODBRIDGE AND EDISON TOWNSHIPS
MIDDLESEX COUNTY, ISELIN, N.J.
TEMPORARY STRUCTURE
PILE CAP INSTALLATION DETAILS

| REVISION | BY | CKD | DATE |
|---|---|---|---|
| | | | |

SCALE: ¾"=1'-0" | DATE 7-14-78 | SHEET NO. 115/131 B26 OF B42
DWN BY ERG | CHECKED BY WWC NUB

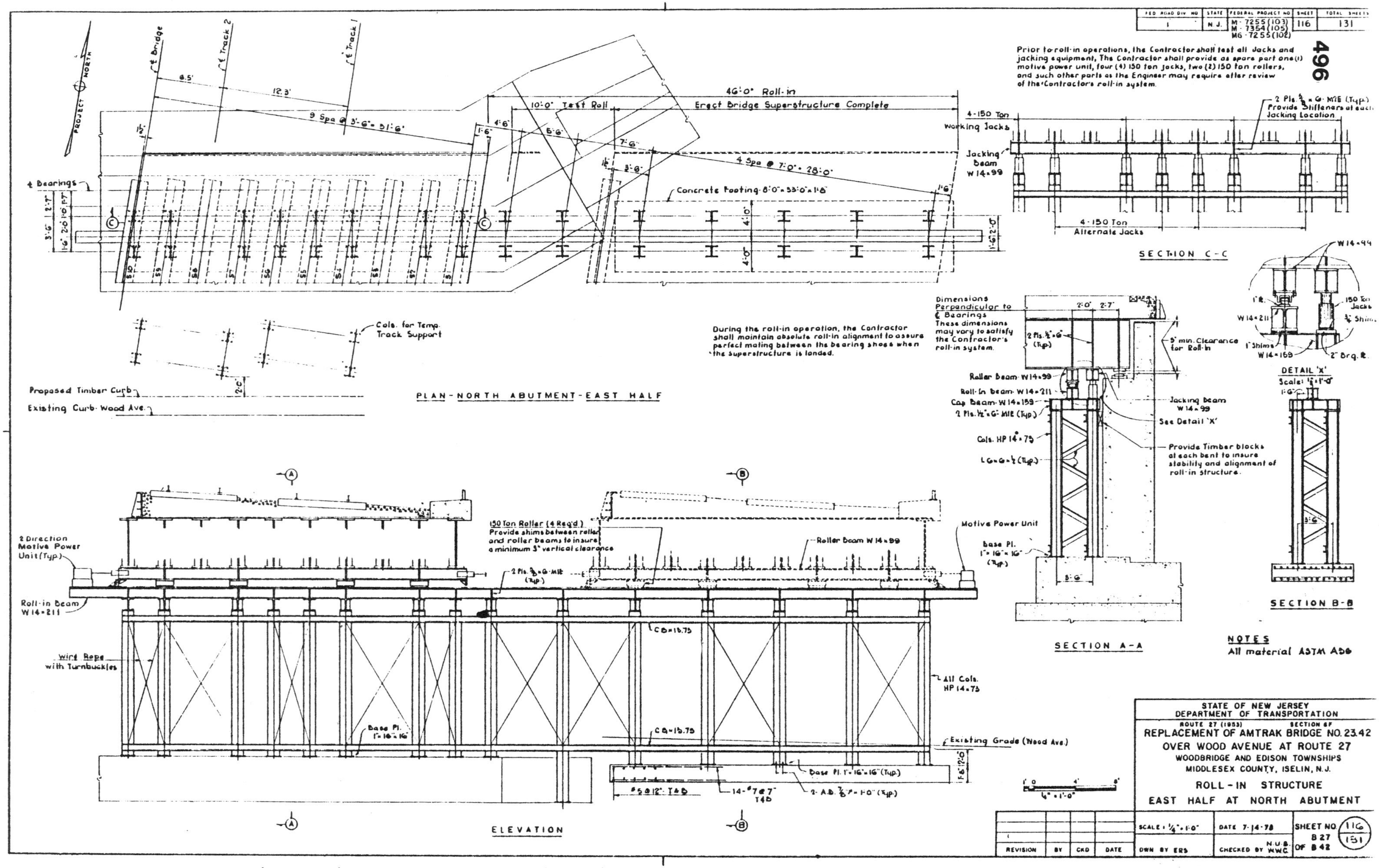
FED. ROAD DIV. NO. 1 | STATE N.J. | FEDERAL PROJECT NO. M-7255(103) M-7354(105) MG-7255(102) | SHEET 116 | TOTAL SHEETS 131
Prior to roll-in operations, the Contractor shall test all Jacks and jacking equipment. The Contractor shall provide as spare part one (1) motive power unit, four (4) 150 ton jacks, two (2) 150 ton rollers, and such other parts as the Engineer may require after review of the Contractor's roll-in system.
PROJECT NORTH
℄ Bridge
℄ Track 2
℄ Track 1
8.5'
12.3'
9 Spa @ 3'-6" = 31'-6"
46'-0" Roll-in
10'-0" Test Roll
Erect Bridge Superstructure Complete
4 Spa @ 7'-0" = 28'-0"
℄ Bearings
Concrete Footing 8'-0" x 33'-0" x 1'-6"
Cols. for Temp. Track Support
Proposed Timber Curb
Existing Curb Wood Ave.
PLAN - NORTH ABUTMENT - EAST HALF
During the roll-in operation, the Contractor shall maintain absolute roll-in alignment to assure perfect mating between the bearing shoes when the superstructure is landed.
4-150 Ton Working Jacks
Jacking Beam W14x99
2 Pls. 3/8" x 6" M1E (Typ.) Provide Stiffeners at each Jacking Location
4-150 Ton Alternate Jacks
SECTION C-C
W14x99
150 Ton Jacks
3/4" Shim
1" Pl.
W14x211
1" Shims
W14x159
2" Brg. Pl.
DETAIL 'X' Scale: 1/2" = 1'-0"
Dimensions Perpendicular to ℄ Bearings These dimensions may vary to satisfy the Contractor's roll-in system.
5" min. Clearance for Roll-in
Roller Beam W14x99
Roll-in beam W14x211
Cap Beam W14x159
2 Pls. 1/2" x 6" M1E (Typ.)
Jacking beam W14x99
See Detail 'X'
Cols. HP 14x73
Provide Timber blocks at each bent to insure stability and alignment of roll-in structure.
Motive Power Unit
Base Pl. 1" x 16" x 16" (Typ.)
SECTION A-A
SECTION B-B
NOTES
All material ASTM A36
2 Direction Motive Power Unit (Typ.)
Roll-in Beam W14x211
150 Ton Roller (4 Req'd.) Provide shims between roller and roller beams to insure a minimum 3" vertical clearance
2 Pls. 3/8" x 6" M1E (Typ.)
Roller Beam W14x99
C8x13.75
Wire Rope with Turnbuckles
All Cols. HP 14x73
Base Pl. 1" x 16" x 16"
Existing Grade (Wood Ave.)
Base Pl. 1" x 16" x 16" (Typ.)
#5 @ 12" T&B
14-#7 @ 7" T&B
2-A.B. 7/8" x 1'-0" (Typ.)
ELEVATION
STATE OF NEW JERSEY
DEPARTMENT OF TRANSPORTATION
ROUTE 27 (1953) SECTION 6F
REPLACEMENT OF AMTRAK BRIDGE NO. 23.42
OVER WOOD AVENUE AT ROUTE 27
WOODBRIDGE AND EDISON TOWNSHIPS
MIDDLESEX COUNTY, ISELIN, N.J.
ROLL-IN STRUCTURE
EAST HALF AT NORTH ABUTMENT
1/4" = 1'-0"
REVISION | BY | CKD | DATE
SCALE: 1/4" = 1'-0" | DATE 7-14-78 | SHEET NO. 116
DWN BY ERS | CHECKED BY N.U.B. W.W.C. | B27 OF B42

| FED. ROAD DIV. NO. | STATE | FEDERAL PROJECT NO. | SHEET | TOTAL SHEETS |
|---|---|---|---|---|
| 1 | N J | M-7266(103)<br>M-7354(106)<br>MG-7266(102) | 108 | 151 |

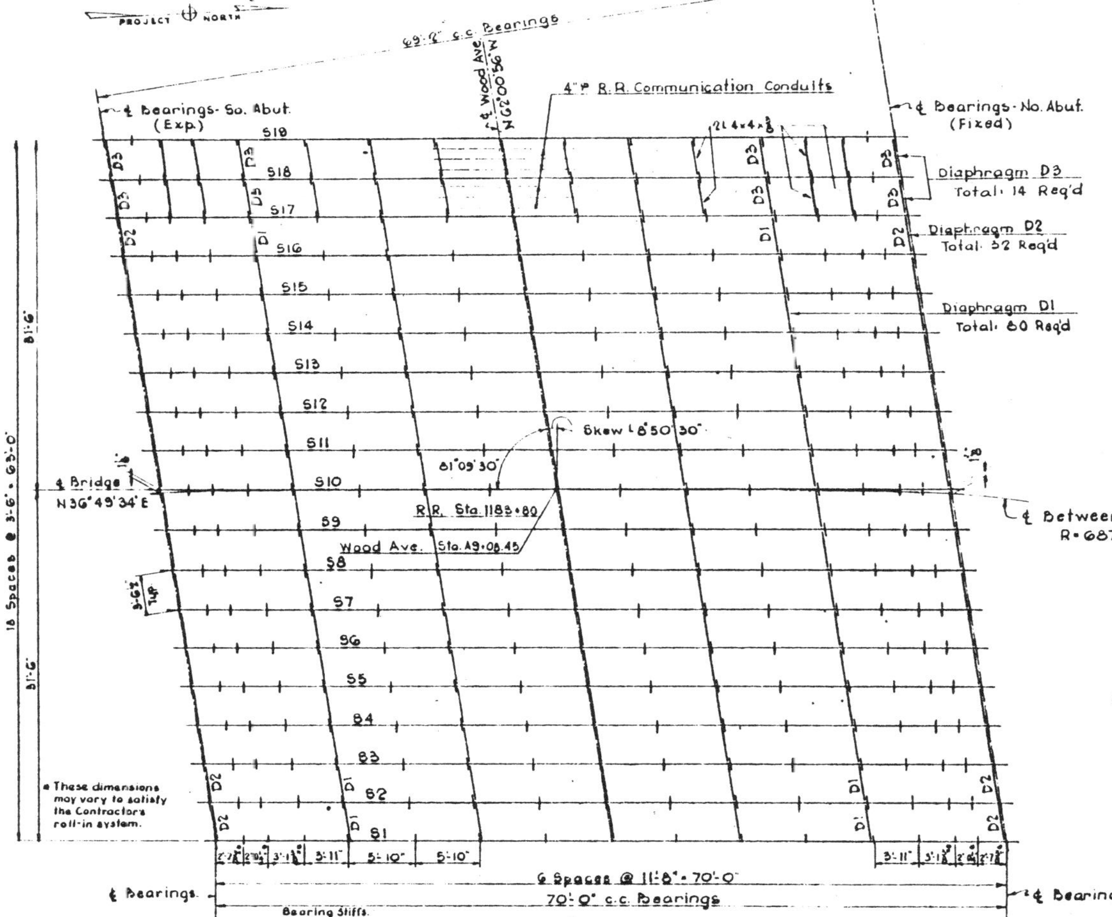


FRAMING PLAN

Scale: 1/16" = 1'-0"

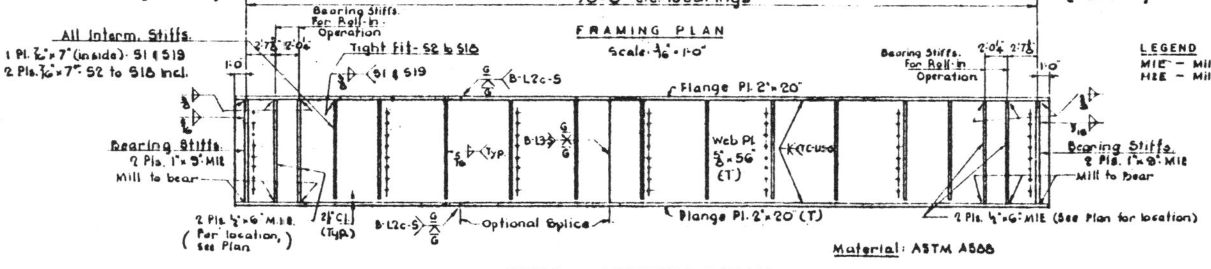


TYPICAL STRINGER DETAIL

Scale: 1/4" = 1'-0" Horiz. 1/2" = 1'-0" Vert.

Material: ASTM A588

LEGEND

M1E – Mill One End
M2E – Mill Two Ends

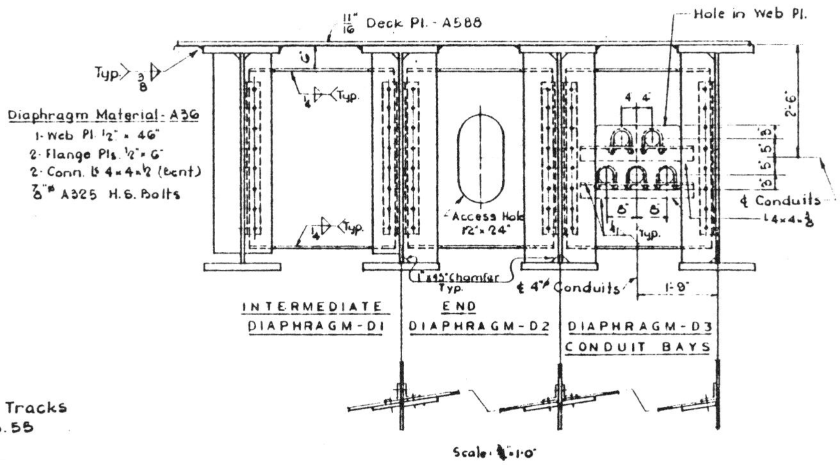


INTERMEDIATE DIAPHRAGM - D1

END DIAPHRAGM - D2

DIAPHRAGM - D3 CONDUIT BAYS

Scale: 3/4" = 1'-0"

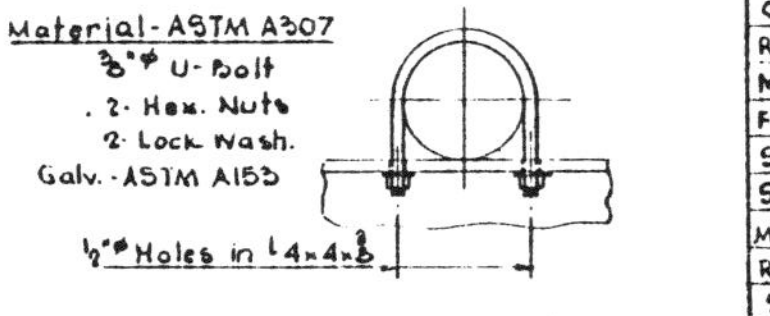


TYPICAL CONDUIT SUPPORT

Scale: 3" = 1'-0"

| SUPERSTRUCTURE | | QUANTITY | |
|---|---|---|---|
| ITEM | UNIT | ESTIMATED | AS BUILT |
| Class A Concrete in Structures | C.Y. | 54 | |
| Reinforcement Steel in Structures | LBS. | 3375 | |
| Membrane Waterproofing - Bridge Deck | S.Y. | 500 | |
| FA-BC-1 Sheet Asphalt Protection Course | TONS | 71 | |
| Structural Steel A36 | LBS. | 95,000 | |
| Structural Steel A588 | LBS. | 732,800 | |
| Metal Railing | L.F. | 154 | |
| Railroad Communication Conduits | L.F. | 960 | |
| Stone Ballast | C.Y. | 750 | |
| Ties | Unit | 188 | |

NOTES

All welding shall conform to the A.W.S. Standard Specifications for Bridges.

(T) = Charpy - Notch Test

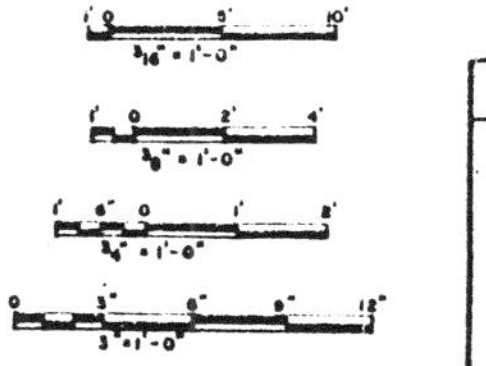

STATE OF NEW JERSEY
DEPARTMENT OF TRANSPORTATION

ROUTE 27 (1953) SECTION 8F

REPLACEMENT OF AMTRAK BRIDGE NO. 23.42
OVER WOOD AVENUE AT ROUTE 27
WOODBRIDGE AND EDISON TOWNSHIPS
MIDDLESEX COUNTY, ISELIN, N.J.

SUPERSTRUCTURE
FRAMING PLAN AND DETAILS

| REVISION | BY | CKD | DATE | SCALE: NOTED | DATE 7-14-78 | SHEET NO. 108 / 151 |
|---|---|---|---|---|---|---|
| | | | | DWN. BY ERB | CHECKED BY W.W.C. / N.U.B. | B 19 OF B 42 |

| FED. ROAD DIV. NO. | STATE | FEDERAL PROJECT NO. | SHEET | TOTAL SHEETS |
| --- | --- | --- | --- | --- |
| 1 | N. J. | M-7255(105) M-7254(106) MG-7255(102) | 109 | 131 |

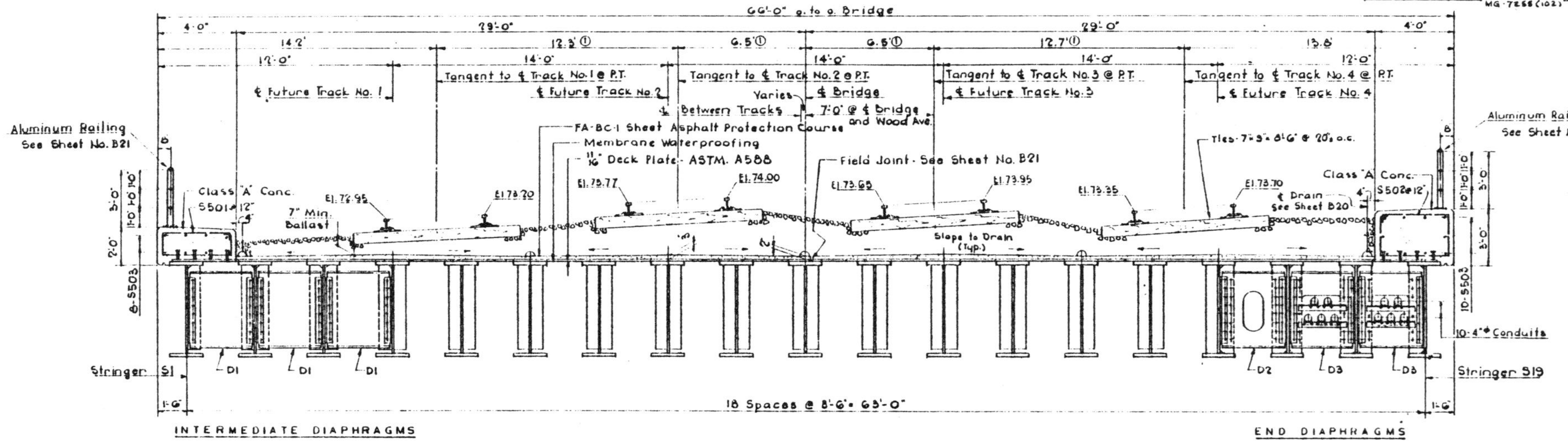


TYPICAL SECTION
LOOKING SOUTH
Scale: 3/8" = 1'-0"

| REINFORCEMENT STEEL | | | | | | |
| --- | --- | --- | --- | --- | --- | --- |
| MARK | SIZE | LENGTH | No. REQ'D | TYPE | DIMENSION A | DIMENSION B |
| S501 | 5 | 10'-10" | 77 | 1 | 1'-5" | 3'-9" |
| S502 | 5 | 12'-10" | 77 | 1 | 2'-5" | 3'-9" |
| S503 | 5 | 39'-3" | 36 | STR | | |

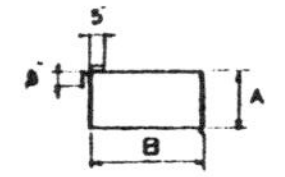

TYPE I

NOTES

Bearing anchor bolts shall not project above the sole plate.

After final assembly of bearings, nuts shall be drawn up finger tight, then backed off ⅛ turn and threads of bolts burred off face of nut with a pointed tool.

Survey rail elevations along centerline of Wood Avenue. Dimensions marked ① indicate existing track spacing. Survey dated Nov. 1971.

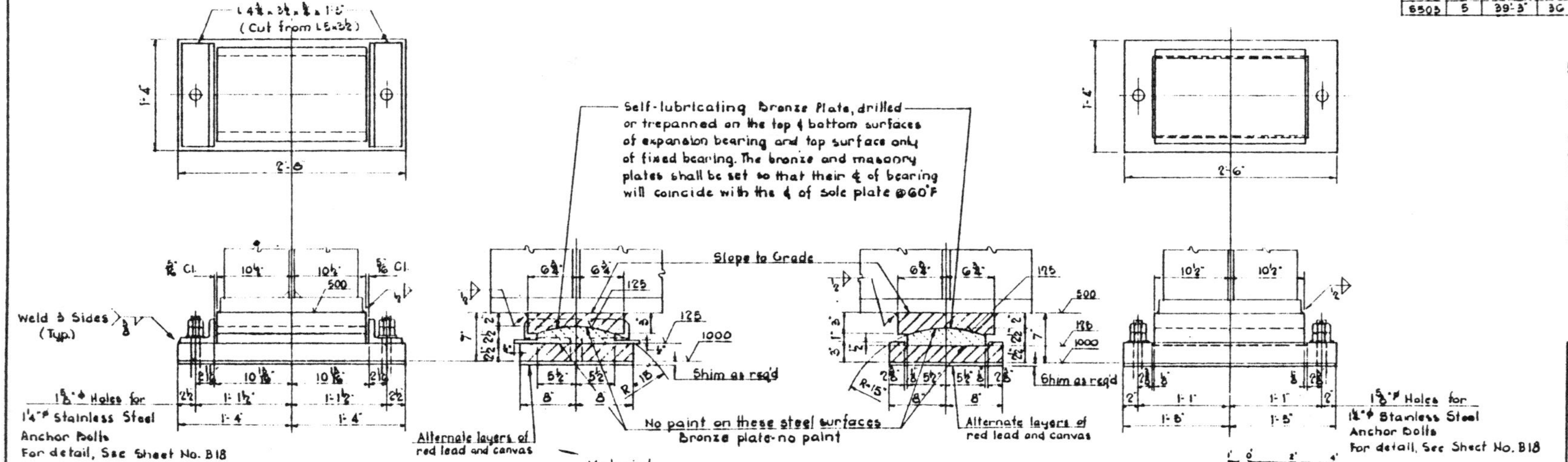


Material
Self-lubricating bronze plates - A.S.T.M. B22-911
Masonry and sole plates - ASTM A36

EXPANSION BEARING
Scale: 1½" = 1'-0"

FIXED BEARING
Scale: 1½" = 1'-0"

STATE OF NEW JERSEY
DEPARTMENT OF TRANSPORTATION
ROUTE 27 (1953) SECTION 6F
REPLACEMENT OF AMTRAK BRIDGE NO. 23.42
OVER WOOD AVENUE AT ROUTE 27
WOODBRIDGE AND EDISON TOWNSHIPS
MIDDLESEX COUNTY, ISELIN, N.J.

SUPERSTRUCTURE CROSS SECTION AND BEARING DETAILS

| SCALE: 3/8" = 1'-0" | DATE 7-14-78 | SHEET NO. 109 / 131 |
| --- | --- | --- |
| DWN BY ERS | CHECKED BY N.U.B. W.W.C. | B20 OF B42 |

| REVISION | BY | CKD | DATE |
| --- | --- | --- | --- |

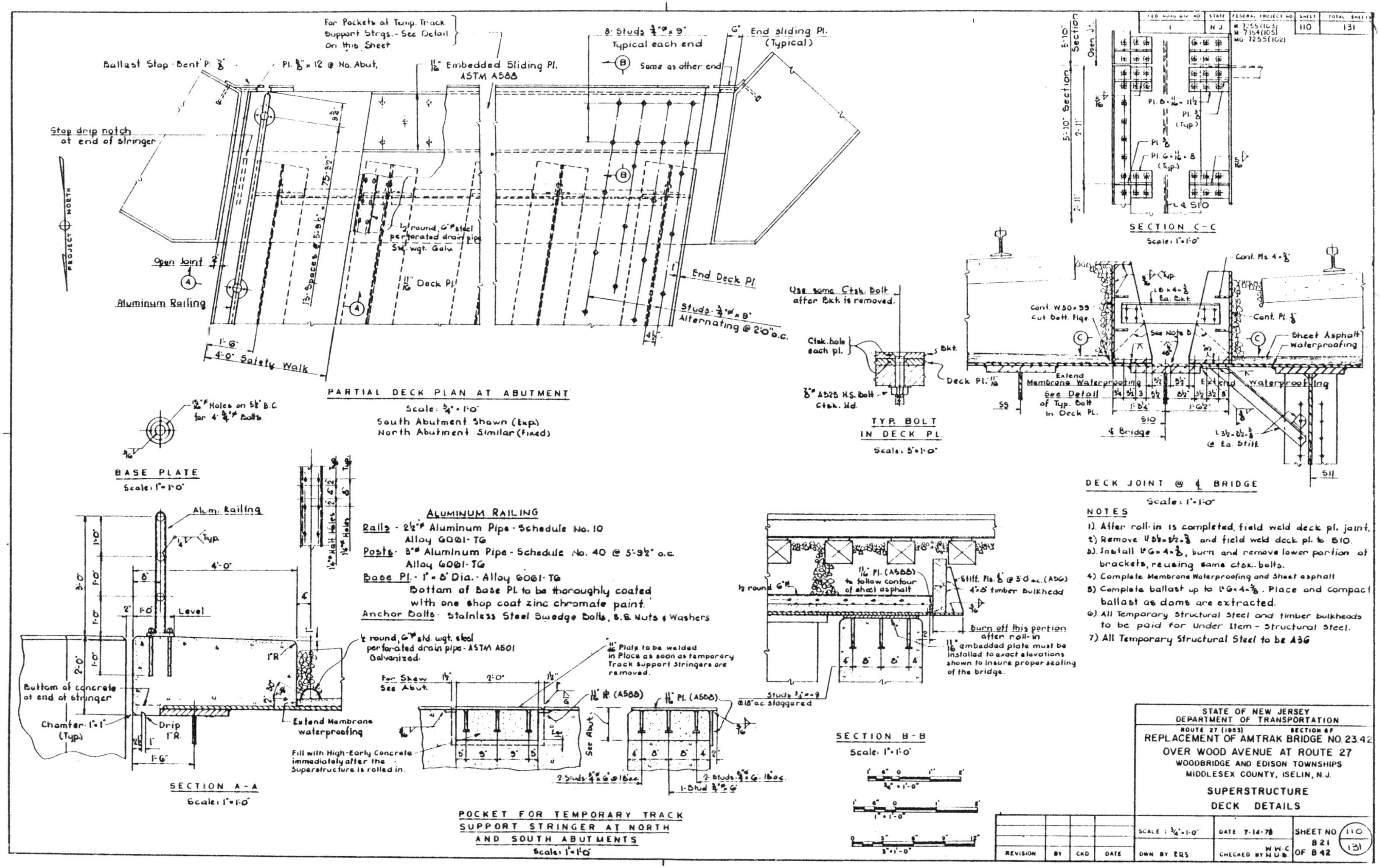
PARTIAL DECK PLAN AT ABUTMENT
Scale: 3/4" = 1'-0"
South Abutment Shown (Exp.)
North Abutment Similar (Fixed)
Ballast Stop - Bent Pl. 3/8"
Stop drip notch at end of stringer
Open Joint
Aluminum Railing
4'-0" Safety Walk
End Deck Pl.
End sliding Pl. (Typical)
Embedded Sliding Pl. ASTM A588
Deck Pl.
BASE PLATE
Scale: 1" = 1'-0"
SECTION A-A
Scale: 1" = 1'-0"
Extend Membrane waterproofing
Fill with High-Early Concrete immediately after the Superstructure is rolled in.
ALUMINUM RAILING
Rails - 2 1/2" Aluminum Pipe - Schedule No. 10 Alloy 6061-T6
Posts - 3" Aluminum Pipe - Schedule No. 40 @ 5'-9 1/2" o.c. Alloy 6061-T6
Base Pl. - 1" x 8" Dia. - Alloy 6061-T6 Bottom of Base Pl. to be thoroughly coated with one shop coat zinc chromate paint.
Anchor Bolts - Stainless Steel Swedge Bolts, S.S. Nuts & Washers
POCKET FOR TEMPORARY TRACK SUPPORT STRINGER AT NORTH AND SOUTH ABUTMENTS
Scale: 1" = 1'-0"
TYP. BOLT IN DECK PL.
Scale: 3" = 1'-0"
SECTION B-B
Scale: 1" = 1'-0"
SECTION C-C
Scale: 1" = 1'-0"
DECK JOINT @ ℄ BRIDGE
Scale: 1" = 1'-0"
NOTES
1) After roll-in is completed, field weld deck pl. joint.
2) Remove L 3 1/2 x 3 1/2 x 3/8 and field weld deck pl. to S10.
3) Install L 6 x 4 x 3/8, burn and remove lower portion of brackets, reusing same ctsk. bolts.
4) Complete Membrane Waterproofing and Sheet asphalt
5) Complete ballast up to L 6 x 4 x 3/8. Place and compact ballast as dams are extracted.
6) All Temporary Structural Steel and timber bulkheads to be paid for under Item - Structural Steel.
7) All Temporary Structural Steel to be A36
STATE OF NEW JERSEY
DEPARTMENT OF TRANSPORTATION
ROUTE 27 (1903) SECTION OF
REPLACEMENT OF AMTRAK BRIDGE NO. 23.42
OVER WOOD AVENUE AT ROUTE 27
WOODBRIDGE AND EDISON TOWNSHIPS
MIDDLESEX COUNTY, ISELIN, N.J.
SUPERSTRUCTURE
DECK DETAILS
SCALE: 3/4" = 1'-0"
DATE 7-14-78
SHEET NO. 110 / 131
B21 OF B42

## E.4 ADDITION OF PIER TO INCREASE CAPACITY

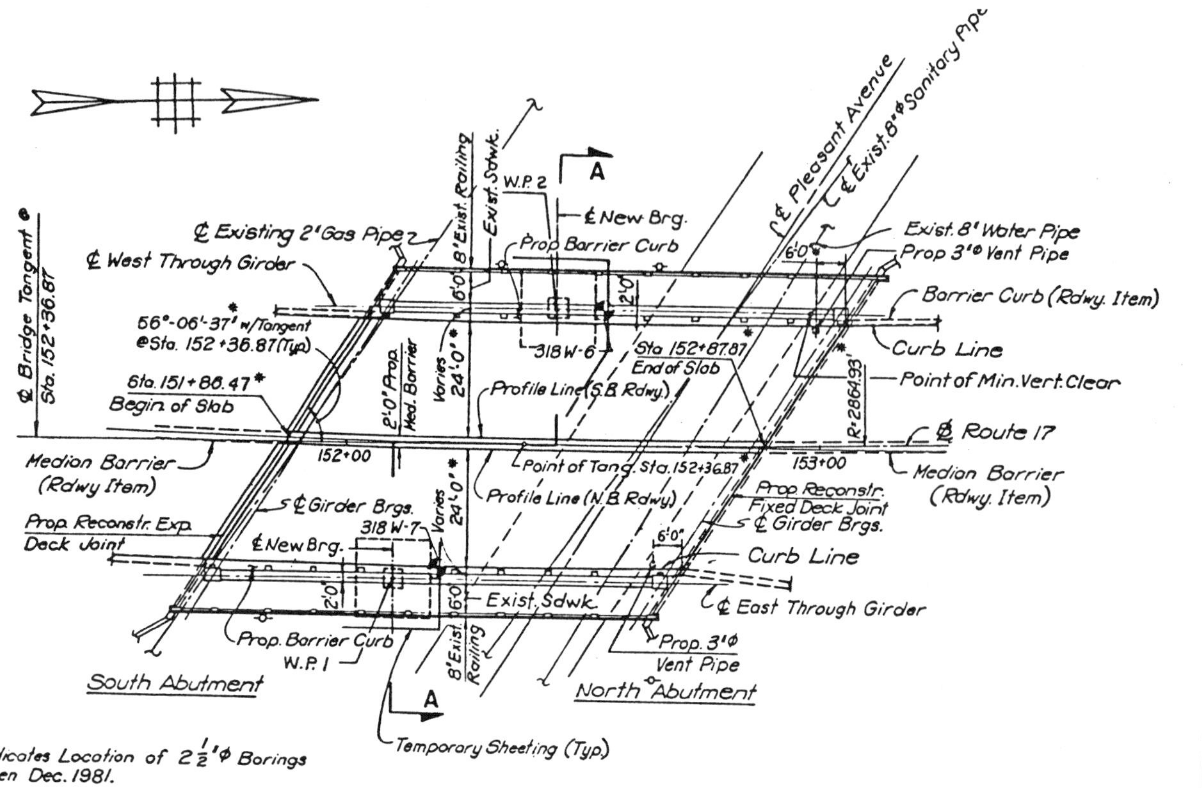


● Indicates Location of $2\frac{1}{2}$"Φ Borings Taken Dec. 1981.

-o- Indicates Location of Existing Utility Poles.

P L A N

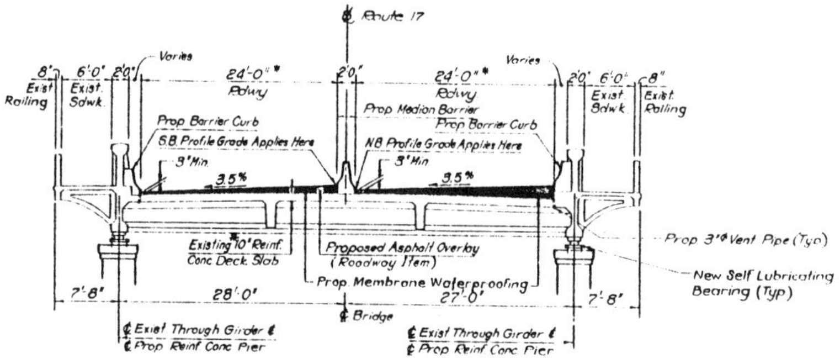


SECTION A-A

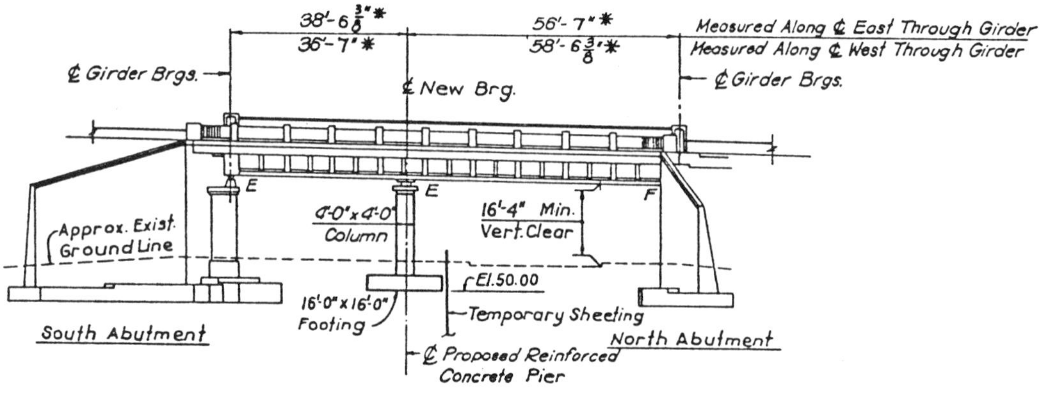


ELEVATION

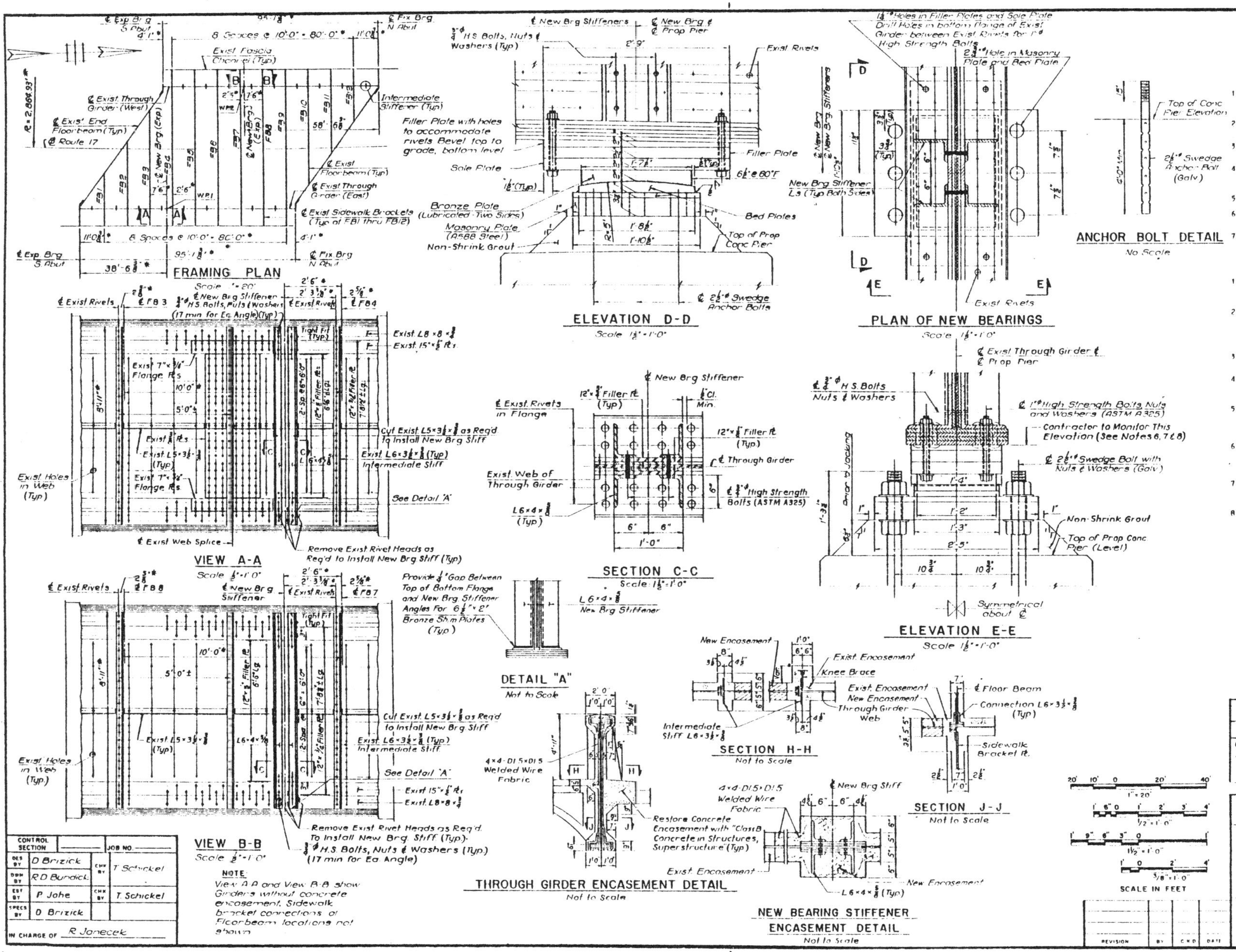


STRUCTURE NO. 0214-161

## BEARING NOTES:

1. MATERIAL FOR BEARINGS (EXCEPT FOR MASONRY PLATES AND SELF-LUBRICATING PLATES) SHALL BE STRUCTURAL STEEL CONFORMING TO THE REQUIREMENTS OF ASTM A36.
2. MASONRY PLATES SHALL BE STRUCTURAL STEEL CONFORMING TO THE REQUIREMENTS OF ASTM A588.
3. BRONZE PLATES SHALL CONFORM TO THE REQUIREMENTS OF ASTM B22, CAST BRONZE ALLOY NO. 911.
4. ANCHOR BOLTS, HEXAGON NUTS AND WASHERS SHALL CONFORM TO ASTM A36 AND SHALL BE HOT DIPPED GALVANIZED IN ACCORDANCE WITH ASTM A153.
5. FOR ANCHOR BOLT LOCATIONS SEE SHEET F4.
6. HIGH STRENGTH BOLTS, NUTS AND WASHERS SHALL CONFORM TO THE REQUIREMENTS OF ASTM A325.
7. BRONZE SELF LUBRICATING PLATES SHALL NOT BE PAINTED.

## CONSTRUCTION NOTES:

1. REMOVE THROUGH GIRDER ENCASEMENT AND PORTION OF CURBING AND SIDEWALK AS REQUIRED TO INSTALL NEW BEARING STIFFENERS AND NEW BEARINGS.
2. INSTALL NEW BEARING STIFFENERS WITH BRONZE SHIM PLATES AS SHOWN. LOCATION OF STIFFENERS IS APPROXIMATE AND IS BASED ON ORIGINAL PLANS OF THE BRIDGE. CONTRACTOR MUST VERIFY LOCATIONS OF NEW BEARING STIFFENERS BEFORE FABRICATION AND INSTALLATION BEGINS.
3. WHERE REQUIRED, FILLER PLATES SHALL BE FABRICATED WITH RECESSES OR HOLES TO ACCOMMODATE EXISTING RIVET HEADS.
4. AFTER NEW BEARING STIFFENERS ARE INSTALLED AND THE CENTER LINE OF BEARING IS ESTABLISHED, PROPOSED CONCRETE PIERS SHALL BE CONSTRUCTED (SEE SHEET F4.)
5. WHEN THE PIER CONCRETE HAS THOROUGHLY CURED, THE MASONRY AND BED PLATES SHALL BE PLACED IN POSITION AS SHOWN. THE BRONZE SELF-LUBRICATING PLATE, THE SOLE PLATE AND FILLER PLATES SHALL FOLLOW.
6. AFTER RECORDING THE ELEVATION OF THE BOTTOM OF STEEL GIRDER, AT THE CENTERLINE OF NEW BEARING, MAKE CONNECTION TO GIRDER AND JACK TO TRANSFER LOAD TO PIER (SEE SUPPLEMENTARY SPECIFICATIONS)
7. DURING THE ADDITION OF SUPERSTRUCTURE DEAD LOAD, THE ELEVATION OF THE BOTTOM OF GIRDER SHOULD BE MAINTAINED WITHIN 1/4 INCH. (SEE SUPPLEMENTARY SPECIFICATIONS).
8. UPON COMPLETION OF ALL BRIDGE WORK, FINAL ADJUSTMENTS SHOULD BE MADE TO THE BEARING AND NON-SHRINK GROUT AND EPOXY WATERPROOFING PLACED.

Based on Existing Dimensions
See General Note on Sheet F1.

ESTIMATED QUANTITIES

| ITEM | UNIT | ESTIMATED | AS BUILT |
|---|---|---|---|
| Class B Conc in Struct Superstruct | C.Y. | 6 | |
| Structural Steel (6,800 Lbs) | L.S. | L.S. | |

NEW JERSEY DEPARTMENT OF TRANSPORTATION
BUREAU OF STRUCTURAL DESIGN

ROUTE 17 OVER PLEASANT AVENUE

ROUTE 17 (1953) SEC 2J STA 152+36.87
MUNICIPALITY: TWP OF ROCHELLE PARK, BOROUGH OF PARAMUS COUNTY: BERGEN

STRUCTURAL STEEL DETAILS

EDWARDS AND KELCEY, INC.
LIVINGSTON, N.J.
CONSULTANT

Date MAR., 1983 Scale AS NOTED
Bridge Sheet No B39 of B42
Sheet No F5

| Fed. Road Div. No. | State | Federal Project No. | Sheet | Total Sheets |
|---|---|---|---|---|
| 1 | N.J. | F 55 (11) | 78 | 102 |

STRUCTURE NO. 0214-161

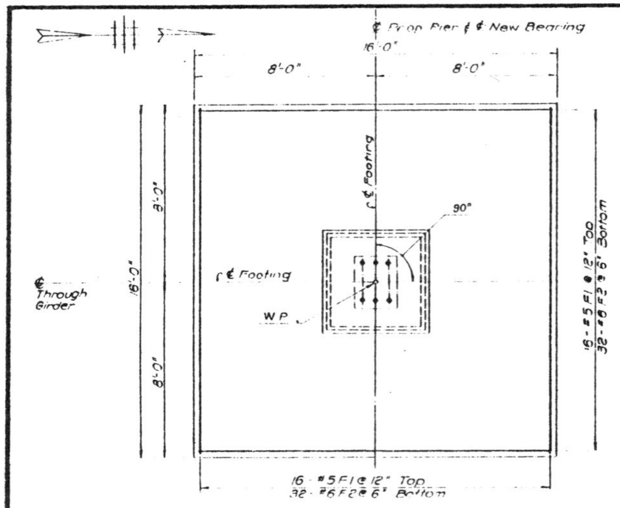


PLAN
Scale 3/8" = 1'-0"

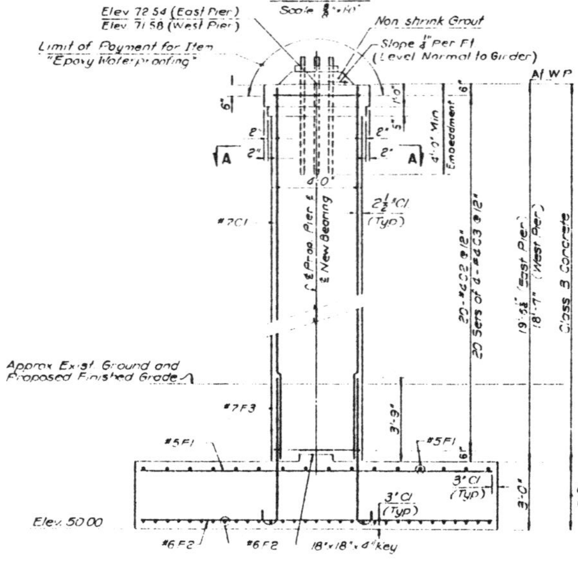


ELEVATION
Scale 3/8" = 1'-0"

| PIER | W.P. NO. | RT.17 STA. | OFFSET |
|---|---|---|---|
| East | 1 | 152+09.96* | 27.13' * |
| West | 2 | 152+44.78* | 27.99' * |

PIER TABLE

* Dimensions and Roadway Stationing Shown are Based on Existing Conditions, See General Note on Sheet F1.

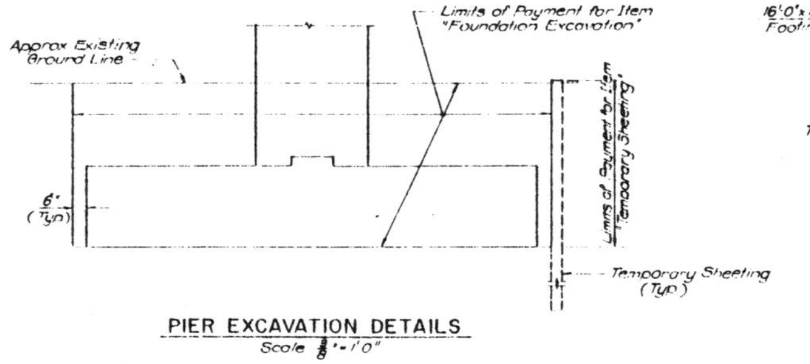


PIER EXCAVATION DETAILS
Scale 1/8" = 1'-0"

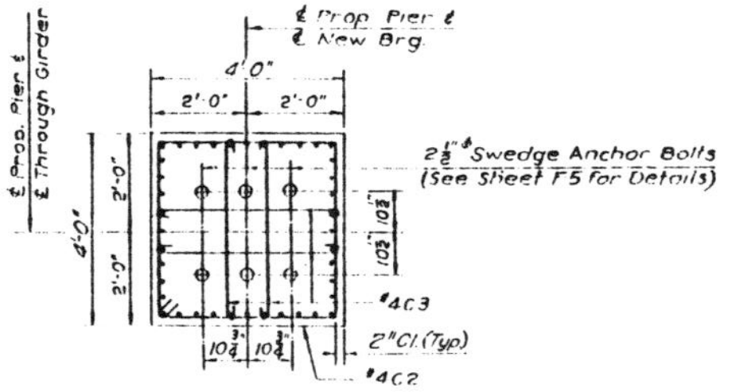


SECTION A-A
Scale 1/2" = 1'-0"

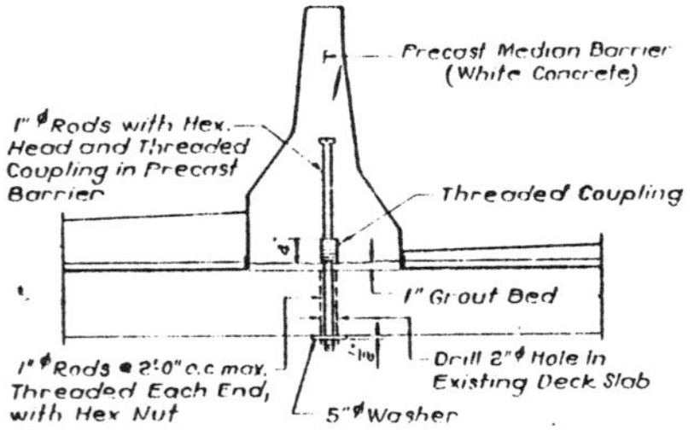


PRECAST BARRIER ALTERNATE

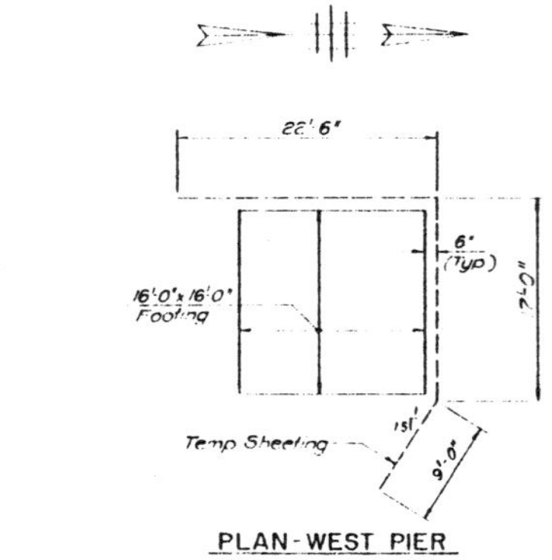


PLAN - WEST PIER
Not to Scale

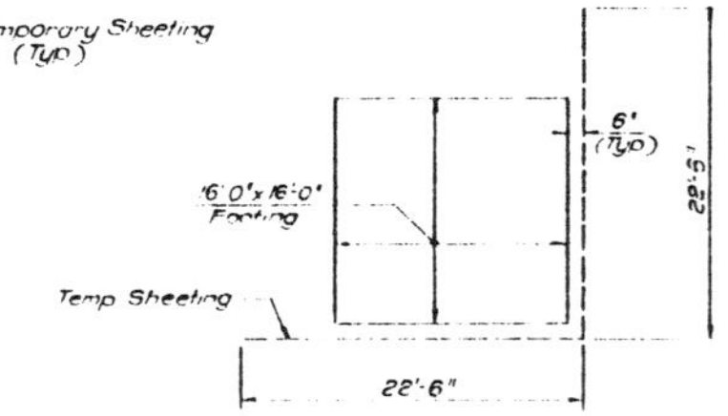


PLAN - EAST PIER
Not to Scale

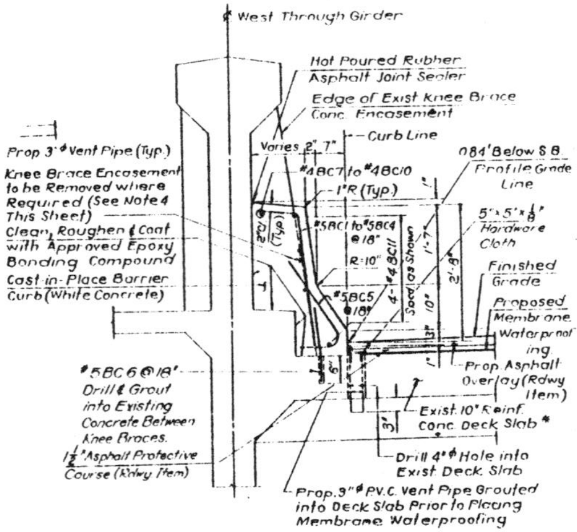


BAR SCHEDULE

| MARK | NO. | SIZE | LENGTH | TYPE | B | C | D | REMARKS |
|---|---|---|---|---|---|---|---|---|
| F1 | 64 | #5 | 15'-6" | Str. | | | | |
| F2 | 128 | #6 | 15'-6" | Str. | | | | |
| F3 | 80 | #7 | 7'-4" | I | | | | |
| C1 | 80 | #7 | 19'-8" | Str. | | | | |
| C2 | 40 | #4 | 15'-6" | II | | | | |
| C3 | 160 | #4 | 6'-9" | III | | | | |

TYPE I — TYPE II — TYPE III

NOTES:

1. All Reinforcement to have minimum cover of 2 inches, unless noted otherwise.
2. Chamfer all exposed corners 1 inch.
3. For Bearing Details see Sheet F5.

ESTIMATED QUANTITIES

| ITEM | UNIT | ESTIMATED | AS BUILT |
|---|---|---|---|
| Foundation Excavation | C.Y. | 129 | |
| Temporary Sheeting | S.F. | 560 | |
| Class B Concrete in Structures, Sub. | C.Y. | 23 | |
| Class C Concrete in Structures | C.Y. | 57 | |
| Reinf. Steel in Structures | Lbs. | 9360 | |
| Epoxy Waterproofing | S.F. | 63 | |

| CONTROL SECTION | | JOB NO. | |
|---|---|---|---|
| DES. BY | T. Schickel | CHK. BY | R. Little |
| DWN. BY | J. Budis | | |
| EST. BY | P. Johe | CHK. BY | T. Schickel |
| SPECS BY | D. Brzirk | | |
| IN CHARGE OF | R. Janecek | | |

| REVISION | BY | CHK'D | DATE |
|---|---|---|---|

NEW JERSEY DEPARTMENT OF TRANSPORTATION
BUREAU OF STRUCTURAL DESIGN

ROUTE 17 OVER PLEASANT AVENUE

ROUTE 17 (1953) TWP. OF ROCHELLE PARK, BOROUGH OF PARAMUS — SEC. 2J — STA. 152+36.87 — MUNICIPALITY — COUNTY: BERGEN

PIERS

EDWARDS AND KELCEY, INC.
LIVINGSTON, N.J.
CONSULTANT

Date MAR. 1983 — Scale AS NOTED

Bridge Sheet No. B38 of B42 — Sheet No. F4

# E.5 INCREASE VERTICAL CLERANCE BY JACKING

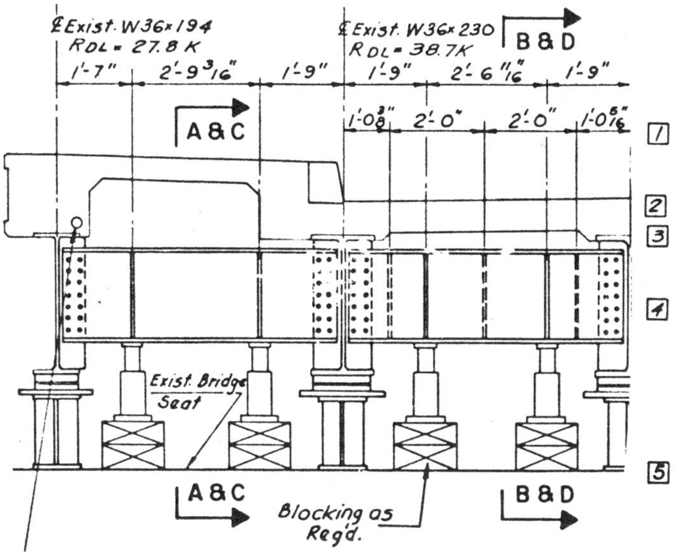


**JACK INSTALLATION**

## JACKING PROCEDURE:

Existing bridge shall be jacked to final grade in Step ⑥ of Stage III, as specified in Construction Sequence, Sh. B. 2 The procedure for this operation shall be as follows:

[1] Remove portions of existing deck slab, balustrade, concrete end diaphragms and approach slabs as shown hereon and on Sh. B 9.

[2] Install new W 24 × 68 end diaphragms.

[3] Install jacks. Jacks shall be hydraulic jacks with minimum rated capacity per jack of 50 Tons and minimum stroke of 6¼ inches. Total of ten jacks required at each end of individual span.

[4] Jack one span at a time, after first removing existing stringer bearing anchor bolt nuts. Raise span in equal lifts, cribbing stringers to bridge seats, directly beneath bearings, between lifts. Block jacks to bridge seats for each succeeding lift. Final cribbing for stringers, at completion of last lift, shall consist of welded steel bolsters, installed beneath masonry plates, as shown hereon.

[5] Install new anchor bolts as shown in Section E-E & reinstall anchor bolt nuts. Pour concrete pedestals on bridge seats, encasing steel bolsters, and construct new portions of existing abutment backwalls.

[6] Install diaphragms between existing bridge and new portion of superstructure as constructed in Stage I. Restore portions of deck slab removed in Step [1] and pour closure slab between halves of superstructure as specified in Step ⑦ of Construction Sequence, Sh. B. 2

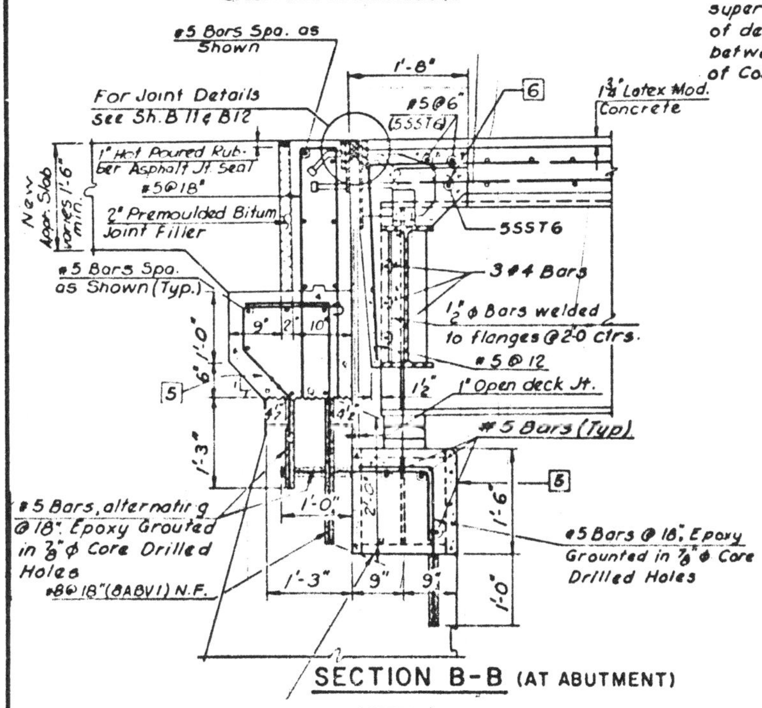


SECTION B-B (AT ABUTMENT)

**AFTER**

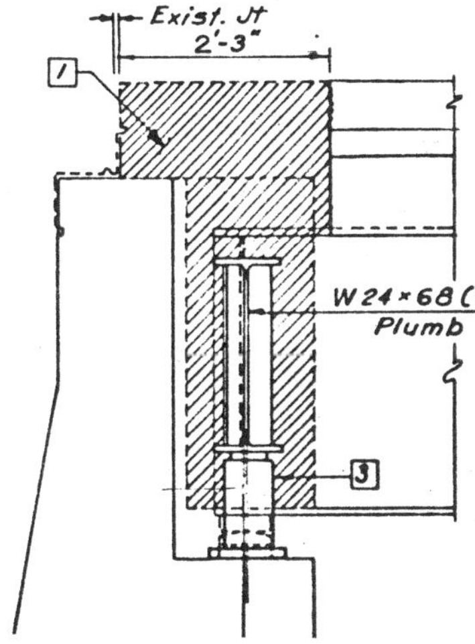


**BEFORE**

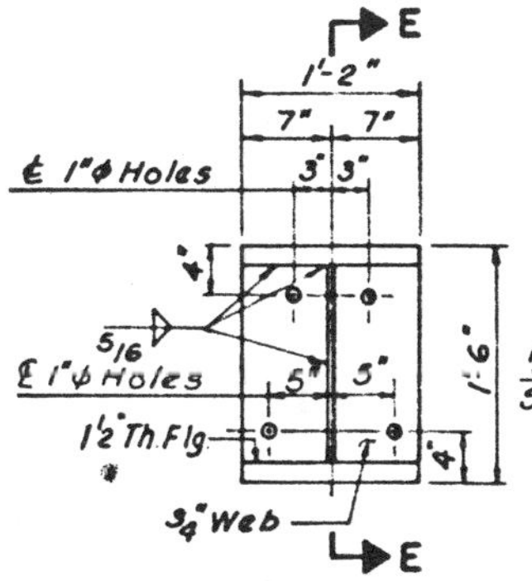


ELEVATION

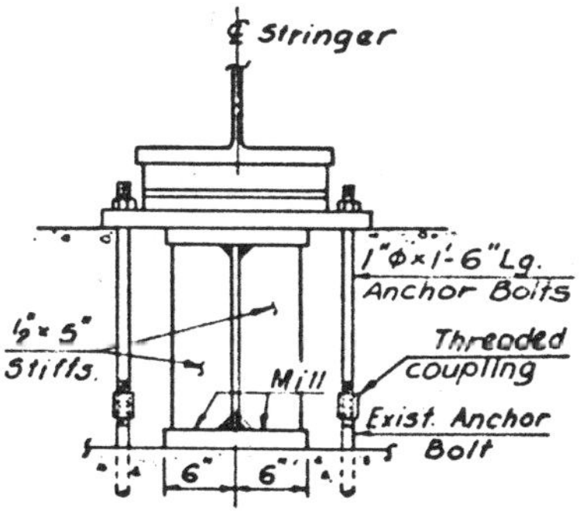


SECTION E-E

BOLSTER DETAILS

## NOTES:

1. Jacks at each end of span shall be synchronized to insure equal displacement of all pistons in order that ends of span be raised uniformly without racking steelwork.

2. Prior to core drilling existing concrete to install reinforcement dowels, the contractor shall locate and mark the existing reinforcement steel in order to avoid damaging same Spacing of dowels shall be adjusted as required to clear exist. reinforcement.

3. If a synchronous lift console is employed, in jacking operations, one additional jack will be required at each outside stringer resulting in one pair of jacks handling each stringer reaction.

# E.6 ADDING NEW STRINGERS

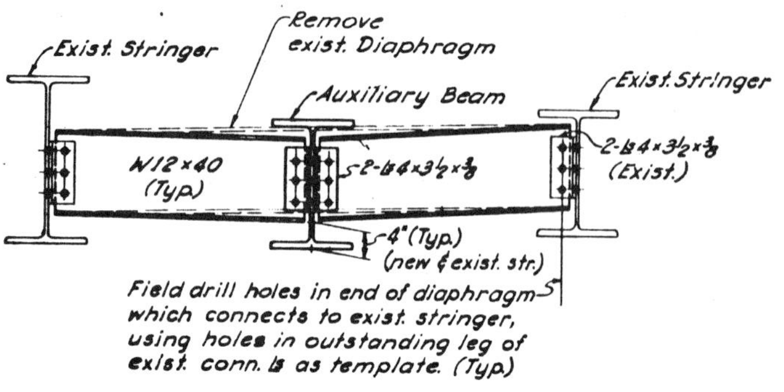


DIAPHRAGM DETAILS

Scale: ¾" = 1'-0"

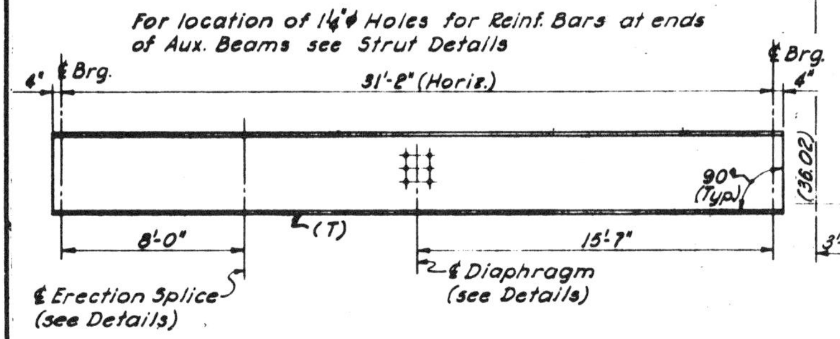


AUXILIARY BEAM ELEVATION

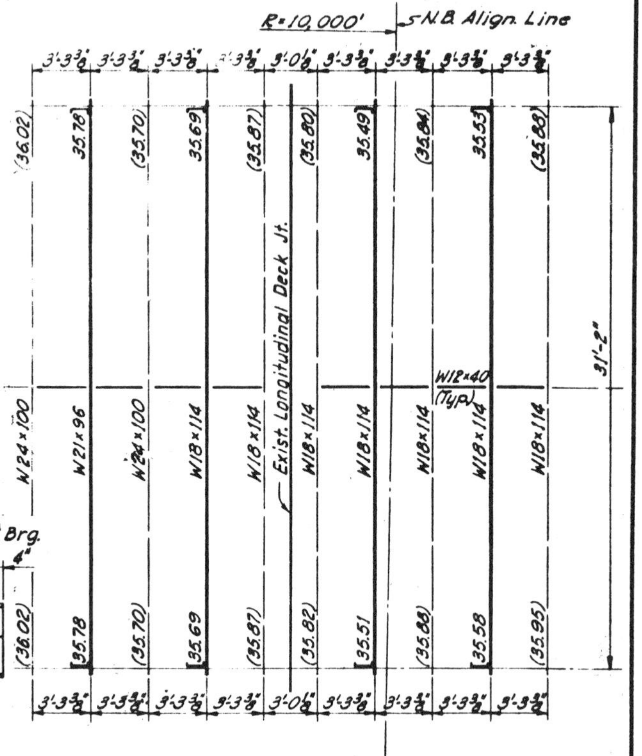


PARTIAL FRAMING PLAN

Scale: 3/16" = 1'-0"

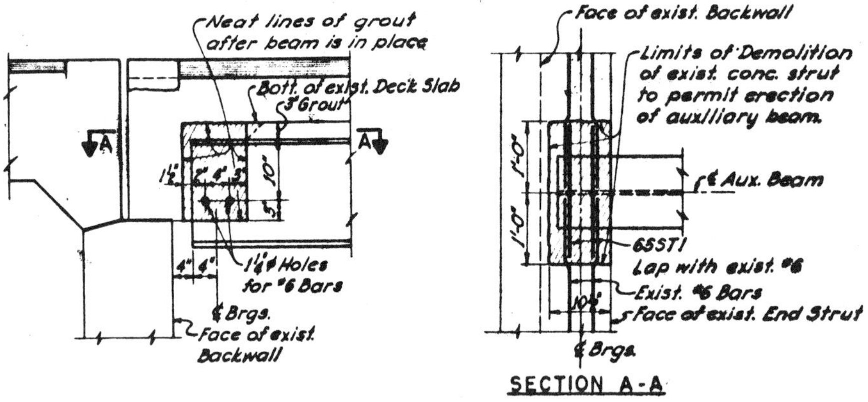


ALTERATIONS TO EXISTING STRUTS FOR ERECTION OF AUXILIARY BEAMS

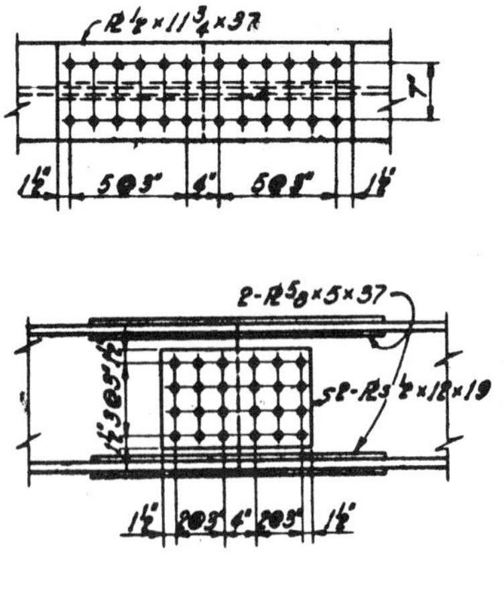


ERECTION SPLICE DETAILS

ERECTION PROCEDURE FOR AUXILIARY BEAMS

1. Demolish exist. conc. end struts in bays where auxiliary beams are to be installed (see Detail) and remove exist. intermediate diaphragms.
2. Erect beam by resting bottom flange of each section directly on bridge seat and supporting ends to be spliced on temporary shoring.
3. Connect splice plates, jack beam and install bearings, field welding sole plates to bottom flange and drill and grout anchor plates in place.
4. Install new intermediate diaphragms, restore conc. struts and grout top flange to underside of exist. deck slab.

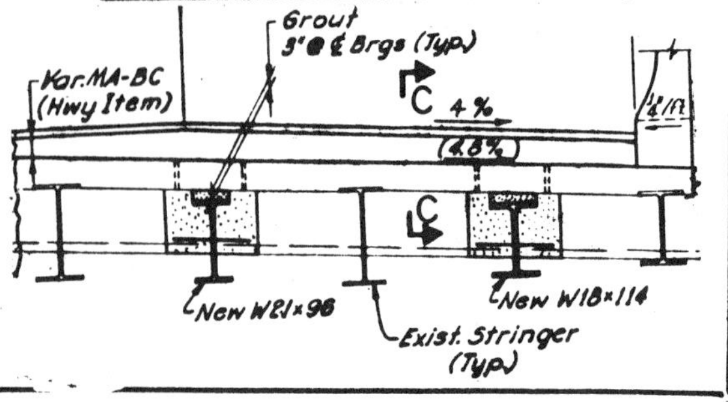

# E.7 STEEL BRIDGE DESIGN CONSIDERATIONS

## ACTUAL DEFLECTION

The distribution factor for computing live load deflection is not always the same as the momemt distribution factor because it is assumed that all the beams or girders act together and have an equal deflection. For example, a 40 foot width of bridge between curbs with 4 supporting girders carries a maximum of 3 trucks or lanes with a 90 percent reduction in load intensity. The wheel load distribution is as follows:

$$\frac{(3\ \text{Trucks})\ (2\ \text{Wheels/Truck})\ (90\%)}{4\ \text{Girders}} = 1.35\ \frac{\text{Wheels}}{\text{Girder}}$$

## UPLIFT AND POURING DIAGRAM

Permanent hold down devices are used to attach the superstructure to the substructure at the bearings when any combination of loading with a 100 percent increase in live load plus impact produces uplift. Also, permanent hold down devices are required on alternate girders that cross over streams with less than a 2 foot clearance for a 100 year flood where expansion bearings are used. These devices are required to prevent the girder from moving off the bearings during extreme flood conditions.

Uplift generally occurs under HS20 loading on continuous spans when the span ratio is greater than 1 to 1.75. It occurs for H20 loading when the span ratio is greater than 1 to 2. Under extreme span ratios the structure may be in uplift for dead load. When this occurs it is necessary to jack the girders upward at the bearings and insert shim plates to produce a downward dead load reaction. The use of simple spans or hinged continuous spans is als considered for this case.

On two span bridges of unequal span lengths, the slab is poured in the longer span first. Cracking of the concrete slab in the positive moment region has occurred on bridges with extreme span ratios when the opposite pouring sequence has been followed. When the span exceeds 120 feet, consider some method to control positive cracking such as limited pouring time, concrete retarder or other methods. On multiple span structures determine a pouring sequence that causes the least structure deflections and permits a reasonable construction sequence. Temporary hold down devices are placed at the ends of continuous girders where the slab pour ends if uplift does not occur from dead load and/or live load. The temporary hold down devices prevent uplift and unseating of the girders at the bearings during the pouring sequence.

Standard hold down devices having a capacity of 20 kips are attached symmetrically to alternate girders or to all the girders as required. Hold down devices are designed by considering line bearing acting on a pin. Refer to Standard 27.6 for permanent and temporary hold down details. To compute uplift, a shear influence line is first obtained from the continuous beam computer program. Next the wheel load distribution factor is determined in the same manner as for live load deflection. The number of loaded lanes is based on the width of the bridge between curbs. The live load plus impact is uniformly distributed to all the girders and is reduced in magnitude for multiple lane loadings. The truck or lane live load is increased 100 percent and applied to the shear influence line to produce maximum uplift. The allowance for future wearing surface should not be included in uplift computations when this additional dead load increases the end reaction.

From the Wisconsin Bridge Manual

·BRACING

All bracing systems must be attached to the main girder by bolted connections. On painted bridges connections within the system also must be bolted. On unpainted steel, these sites may be field welded.

## A. INTERMEDIATE DIAPHRAGMS

Diaphragms or cross frames are required at each support and throughout the span at 25'-0 maximum centers in all bays as specified by AASHTO. The spacing is adjusted to miss any splice material. The transverse bracing is placed parallel to the skew for angles up to and including 15 degrees and normal to the girders for skew angles greater than 15 degrees. When diaphragms are stepped slightly out of straight through alignment, the girder flanges will experience the greatest torsional stress. Larger steps in diaphragm spacing allow the torsional moment to distribute over a longer girder section. On curved girder structures, the diaphragms are placed straight through on radial lines to minimize the effects of torsion since the diaphragms or cross frames are analyzed as primary load carrying members.

Channel sections are used as diaphragms for rolled beams and plate girders with web depth up to and including 48 inches. Diaphragms are designed for a minimum of 1/3 and preferably 1/2 the member depth. Diaphragm details and dimensions are given on Standards 24.3 and 24.6. Diaphragms carry moment and tensile stresses caused by girder deflections. In the composite slab region, the steel section acts similar to the lower chord of a vierendeel truss and is in tension. A rigidly connected diaphragm resists bending due to girder deflection and tends to distribute the load. It is preferable to place diaphragms at the 0.4 point of the end spans on continuous spans and at the center of interior spans when this can be accomplished without an increase in total number. Also, if practical place diaphragms adjacent to a field splice between the splice and the pier. Bolted diaphragm connections may be used in place of welded diaphragm connections. ALL CROSS FRAMING IS ATTACHED TO THIS MAIN GIRDER VIA BOLTED GUSSET PLATES.

Cross framing is used for web depths over 48 inches. The bracing consists of two diagonal members connected at their intersection and one bottom chord member. The bottom chord is designed for a maximum KL/r of 140 for a secondary compression member. The diagonals are designed for a maximum KL/r of 280 for secondary tension members. The length of a minimum 1/4 inch fillet weld size is determined for each member based on a minimum of 75 percent of the member strength. The smallest angle used in cross framing is 3½" x 3 1/2" x 5/16".

On spans over 200 feet in length the stresses caused by wind load on part of the erected girders without the slab in place may control the size of the members. Construction loads are also considered in determining member size.

If welded connections are used, the members which make up the "X" portion of the bracing are attached with erection bolts which are left in place after welding the member ends. The "X" angles are field welded at this intersections.

On haunched steel girders, the cross framing follows the curvature of the girder. The lower chord is placed at 6 inches above the higher bottom flange of the adjacent girders. Refer to sketch below:

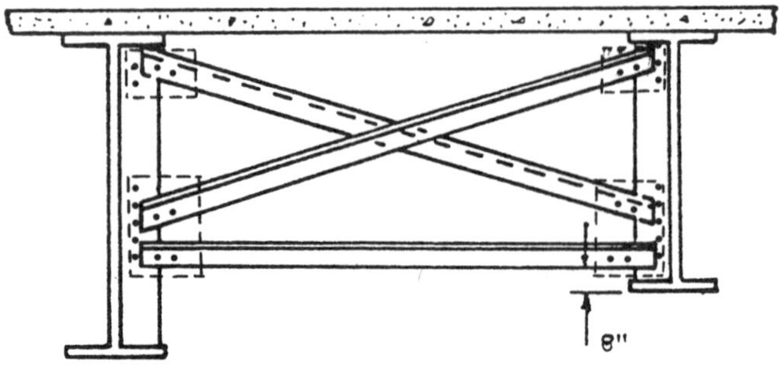

On girders where longitudinal stiffeners are used, the relative position of the stiffener to the cross frame is checked. When the longitudinal stiffener interferes with the cross frame, weld a gusset plate to the vertical stiffener and attach the cross frame to the gusset plates as shown below.

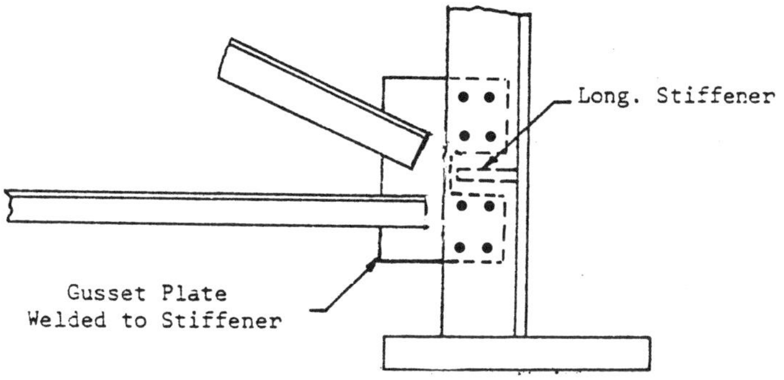


## B. END DIAPHRAGM

End diaphragms are placed horizontally along the abutment end of beams or girders and at other points of discontinuity in the structure. Channel sections are generally used for end diaphragms which are designed as simply supported edge beams. The live load moment plus impact is determined by placing one wheel load or two wheel loads 4 feet apart and correcting for the skew angle at the center line of the member. Generally, the dead load moment of the overlying slab and diaphragm is insignificant and as such is neglected. End diaphragm details and dimensions are given on Standard 24.4.

End diaphragms are either bolted or welded to the girders at points of discontinuity in the superstructure. The same connection detail is used throughout the structure. The connections are designed for shear only where joined at a web since very little moment is transferred without a flange connection. The connection is designed for the shear plus impact from the wheel live loads.

For web depth greater than 60 inches, a lower chord and two diagonals connected to the midpoint of the channel are used. The lower chords are designed as compression members using the same approach that is used in designing the cross framing. Refer to Standard 24.4 for details.

## C. LATERAL BRACING

Lateral bracing requirements for the bottom flanges are to be investigated as per AASHTO Specifications. Current Bridge Section practice is to eliminate the need for bracing by either increasing flange sizes or reducing the distance between cross frames. The controlling case for this stress is usually at a beam cutoff point. At cutoff points condition of maximum stress exists with the smallest flange size; here wind loads have the most effect. A case worth examining is the temporary stress that exists in top flanges during construction. These plates which are often only 12" wide can be heavily stressed by wind load. A temporary bracing system placed by the contractor may be in order.

On an adjacent span to one requiring lower lateral bracing, the bracing is extended one or two panel lengths into that span. The lower lateral system is placed in the exterior bays of the bridge and in at least 1/3 of the bays of the bridge. For spans under 200' the size of the lower lateral members is usually governed by the KL/r ratio. On longer spans the stresses caused by wind load during construction will generally govern the member size.

Lateral bracing consists of two diagonals connected at their intersection. The bracing is placed in the plan defined by the lower chords of the intermediate transverse cross bracing. The maximum KL/r ratio for a secondary tension member is 240. The length of a minimum 1/4 inch fillet weld size is determined for each diagonal based on 75 percent of the strength of the member. The smallest angle that is used for bracing is a 4" x 3" x 5/16". Since the effective length in one plane is half that in the other plane unsymmetrical angles are used.

The lateral bracing may consist of one diagonal member in place of two. For this case the maximum KL/r ratio is 140 for a secondary compression member. The member acts in both tension and compression as shown in the plan view below. Symmetrical angles are employed here since the effective length is equal in both directions.

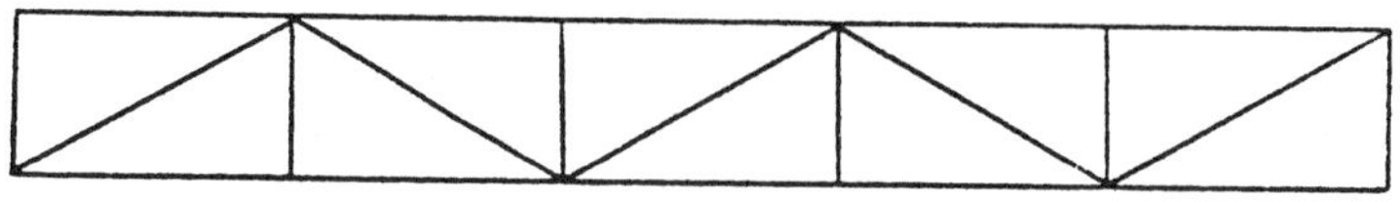

Lower Lateral System With Single Diagonal Member

## GIRDER SELECTION

The exterior girder section is always designed and detailed equal to or larger than the interior girder sections. Ratios of girder depth to length of span which is dependent on type of design are not to exceed values recommended by AASHTO specifications. The following criteria are used when determining the selection and sizes of girder sections.

### A. ROLLED BEAM

Rolled sections without cover plates are preferred. Cover plates are not recommended due to fatigue considerations and higher fabrication costs. The following guidelines are used for cover plate design and detailing.

1. Minimum size cover plates in positive and negative zones are 6 x 3/8 inches and 9 x 3/8 inches respectively. Cover plate widths are to be less than the flange width minus one inch. A larger width cover plate may be used if it is wider than the top flange only in very special cases. Maximum cover plate thickness is two times the thickness of the flange to which it is attached.

2. A partial length welded cover plate is extended beyond its theoretical end a terminal distance as specified in AASHTO.

3. The cutoff of cover plates is made square. If the flange is embedded in concrete, the end is not welded across. If the flange is exposed, the plate end is beveled and welded across the end with a slope of 1 in 2 1/2.

4. A shop welded butt splice is considered in place of adding cover plates. Also, a welded plate girder section is considered at piers where a rolled girder section requires heavy cover plates.

### B. PLATE GIRDER

1. Maximum change in flange plate thickness is 1 inch and perferably 3/4 inch or less. The thinner plate is not less than 1/2 the thickness of the thicker flange plate. Plate thicknesses are given in the following increments:

   1/16 inch through 9/16 inch (flanges)
   1/16 inch through 1 inch (webs)
   1/8 inch up to 2 inches
   1/4 inch above 2 inches.

2. Minimim plate size on the top flange of a composite section is variable depending on the depth of web, but in no case is less than 12 x 5/8 inches for web depths less than or equal to 66 inches. Flange plate widths and web plate depths are given in <u>even</u> inch increments. Changes in flange plate widths and web plate depths are to follow recommended standard transition distances and/or radii.

3. Flange plate sizes are detailed based on recommended maximum span lengths given in Tables 24.2 and 24.3 for haunched and parallel flanged girders. The most economical girder is generally the one having the least total weight, but is determined by comparing material costs and welding costs for added stiffener details. Plates over 60-65 feet are difficult to obtain and butt splices are detailed to limit flange plates to these lengths or less. It is better to detail more flange butt splices than required and leave the decision to utilize them up to the fabricator. All butt splices are made optional to the extent of available lengths and payment is based on the plate sizes shown on the plans.

4. Minimum web thickness is 3/8 inches for girder depths less than or equal to 60 inches.

   An economic analysis was made between using a thin web plate with both longitudinal and transverse stiffeners versus a thicker web plate with or without transverse stiffeners.

## WELDING

Welding design details shall conform to current requirements of AASHTO Standard Specifications for Welding of Structural Steel Highway Bridges. Refer to the AISC Engineering Journal (1973) for a commentary on highly restrained welded connections and lamellar tearing of materials.

Due to fatigue considerations, welding across the tension flange or welding to the edge of the tension flange is prohibited. Due to a number of failures in electroslag welds in 1976-77, this weld procedure is no longer permitted.

Weld details are not shown on the plans, but are indicated by using standard symbols as given on Chart 24.1. Weld sizes are based on the size required due to stress or the minimum size for plate thicknesses being connected. The strength of a fillet weld equals the fillet size times the allowable basic shear stress times (0.707) for equal leg weld sizes. The effective or net section is taken as the shortest distance from the root to the face of the weld.

The following formula is used to compute the weld size for connecting the flange of a plate girder to the web:

$$R = \frac{M'-M}{ad} = \frac{C'-C}{a}$$

where
- $R$ = Rate of change in flange stress, kips/inch
- $M'$ & $M$ = Moments at sections, inch-kips
- $C'$ & $C$ = Force in flange at the sections, kips
- $a$ = Distance between sections of $M'$ & $M$, inches
- $d$ = Effective web depth, inches

Knowing the rate of change in flange stress, the minimum size weld required to carry this force is chosen and compared to the minimum weld size required. Generally the minimum weld size governs due to the material thicknesses employed.

The following data is used to determine minimum fillet weld size based on the thickness of the thicker part joined.

| Fillet Weld Size | Thickness of Thicker Part Joined |
|---|---|
| 3/16 inch | to 1/2 inch |
| 1/4 inch | over 1/2 to 3/4 inch |
| 5/16 inch | over 3/4 to 1 1/2 inches |
| 3/8 inch | over 1 1/2 to 2 1/4 inches |
| 1/2 inch | over 2 1/4 to 6 inches |

The fillet weld size is not required to exceed the thickness of the thinner part joined. Refer to current AASHTO specifications for minimum effective fillet weld length and end return requirements.

Bevel welds are used to develop the full strength of a member. Specify the depth of the bevel as 1/8 inch less than plate thickness due to weld burn through of at least 1/8 inch. Since this type of weld develops full member strength, a strength check is not required.

CAMBER When straight girder sections between splice joints are erected, final girder elevations usually vary in height between the girder and roadway elevations due to dead load deflections and vertical curves. Since a constant slab thickness is detailed, a concrete haunch between the girder and slab is used to adjust these variations. If these variations exceed 3/4 inch, the girder is cambered to reduce the variation of thickness in the haunch. Straight line chords between splice points are sometimes used to create satisfactory camber.

Welded girders are cambered by cutting the web plates to a desired curvature. During fabrication all web plates are cut to size since rolled plates received from the mill are not straight. There is a problem in fabricating girders that have specified cambers between 0 and 3/4 inches. Therefore, cambers less than 3/4 inches are not detailed for plate girders.

Rolled sections are cambered by the application of heat in order that less camber than recommended by AISC specification may be used. For practical reasons cambers of less than 3/4 inches for rolled sections are not detailed. The concrete haunch is used to control the remaining thickness variations. A minimum haunch depth of 1 1/4 inches is set at the abutments and piers.

## REPETITIVE LOADING AND TOUGHNESS

Current AASHTO specifications for Structural Steel Design contain major revisions to previous fatigue design-detail guidelines and requires material impact testing for fracture toughness. These revisions and additions are the results of performance evaluations over the past decade on existing highways and bridges under the action of repetitive vehicle loading.

The direct application of fatigue specifications to main load carrying members has generally been apparent to most bridge designers. As a result main members have been designed with the appropriate details. However, fatigue considerations in the design of secondary members and connections has not always been so obvious. Many of these members interact with main members and receive more numerous cycles of load at a higher level of stress range than assumed. This accounts for most of the fatigue problems surfacing in recent years as cracking initiated by secondary members.

### FATIGUE STRENGTH

The main factors governing fatigue strength are the applied stress, the number of loading cycles, and the type of detail. The designer has the option of either limiting the stress range to acceptable levels or choosing details which limit the severity of the stress concentrations.

Details involving connections that experience fatigue crack growth from weld toes and weld ends where there is high stress concentration provide the lowest allowable stress range. This results for both fillet and groove welded details. Details which serve the intended function and provide the highest fatigue strength are recommended.

Generally, details involving failure from internal discontinuities such as porosity, slag inclusion, cold laps and other comparable conditions, will have a high allowable stress range. This is primarily due to the fact that geometrical stress concentrations at such discontinuities do not exist other than the effect of the discontinuity itself.

AASHTO specifications provide the designer with six basic design range categories for redundant and non-redundant load path structures. The stress range category is selected based on the highway type and the detail employed. The designer may wish to make reference to "Bridge Fatigue Guide Design and Details" by John W. Fisher.

## CHARPY V-NOTCH IMPACT REQUIREMENTS

Recognizing the need to prevent brittle fracture failures of main load carrying structural components, AASHTO Material Specifications adopted provisions for Charpy V-Notch impact testing in 1974. Impact testing offers an important measure of material quality, particularly its ductility. Brittleness is detected prior to placing the material in service to help prevent member service failures. Wisconsin Standard Specifications for Road and Bridge Construction require Charpy V-Notch tests on all girder flange and web plates, flange splice plates, hanger bars, links, rolled beams and flange cover plates.

## NON-REDUNDANT TYPE STRUCTURE

Previous AASHTO fatigue and fracture toughness provisions provided satisfactory fracture control for multi-girder structures when employed with good fabrication and inspection practices. However, concern existed that some additional factor of safety against the possibility of brittle fracture should be provided in the design of non-redundant type structures such as single box girders, two plate girder or truss systems where failure of a single element could cause collapse of the structure. A case in point was the collapse of the Point Pleasant Bridge over the Ohio River.

Primary factors controlling the susceptibility of non-redundant structures to brittle fracture are the material toughness, flaw size and stress level. One of the most effective methods of reducing brittle fracture is lowering the stress range imposed on the member. In 1977, AASHTO specifications provided an increased safety factor for non-redundant members by requiring a shift of one range of loading cycles for fatigue design with corresponding reduction of stress range for critical stress categories. The separate tables for redundant and non-redundant load path structures clearly indicates the need for special cases in the AASHTO specification. The restrictive ranges for certain categories will require the designer to investigate the use of details which do not fall in critical stress categories or induce brittle fracture.

# DESIGN APPROACH -STEPS IN DESIGN

(1) The following steps are taken in determining the Girder or Beam Spacing and the Slab Thickness:

The girder spacing (number of girders) for a structure is determined by considering the desirable girder depth and the span lengths.

The slab overhang on exterior girders is limited to 3'-7 measured from the girder centerline to the edge of slab. The overhang is limited to prevent rotation and bending of the web during construction caused by the forming brackets.

(2) The following steps are taken in estimating the Size of Beams and Girders:

Use design tables and similar structures previously designed in estimating the size of the members. Refer to Tables 24.2 and 24.3 for recommended girder depths for a given girder spacing and span length.

Use the same width of bottom flange throughout the bridge whenever possible. High bearing reactions at the piers of continuous girders may govern the width of the bottom flange.

Keep the flange plates on all girders the same size where possible to eliminate the problem of getting the wrong size flange on a given girder.

Make the thickness of A572 steel 2 inches or less to achieve maximum design economy. Consider the use of hybrid sections over the piers of continuous plate girders.

Estimate field splice locations at approximately the 7/10 point of continuous spans.

On haunched plate girders the length of parabolic haunch is approximately 1/4 of the span length. The haunch depth is 1 1/2 times the mid span depth.

(3) Determine the Dead Loads, Live Loads, and Live Load Distribution Factors for each member.

(4) Determine the Composite Section for each Member and estimate the Range of Composite Action for each Span.

On continuous spans estimate the end of composite action at a distance of approximately 2/10 the span length from the pier. On simple spans use composite action throughout the span.

(1) Composite Action

Composite design in steel pertains to structures composed of steel beams or girders with concrete slabs connected by shear connectors. The shear connectors prevent slip and vertical separation between the bottom of the slab and the top of the steel member. Unless temporary shoring is used the steel members deflect under the dead load of the wet concrete before the shear connectors become effective. Temporary shoring is not used in Wisconsin, therefore, composite action refers to the live load and portions of dead load placed after the concrete deck has hardened. The composite section in the positive moment area is the concrete in compression. In the negative moment area the composite section is the bar steel reinforcement in the slab. However, the effect of the composite section in the negative moment region is not used and shear connectors are not placed in this region. Composite sections are proportioned to have the neutral axis lie below the top flange of the steel member.

(2) Values of n for Composite Design

The effective composite concrete slab is converted to an equivalent steel area by dividng by n. For f'c = 4000 psi, use n = 8.

f'c = minimum ultimate compressive strength of the concrete slab at 28 days.

n = ratio of modulus of elasticity of steel to that of concrete.

The effect of creep is to be considered in the design of a composite member which has dead loads acting on the composite section. In such structures, stresses and horizontal shears produced by dead loads acting on the composite section are computed for a value of n as given above or for 3n, whichever gives the higher stresses and shears.

(3) Composite Section Properties

Two computer programs are available to compute composite section properties.

1. Composite plate girder properties.
2. Composite beam properties

The minimum effective slab thickness is equal to the nominal slab thickness minus 1/2 inch for wearing surface. The maximum effective slab width is defined by AASHTO specifications.

(4) Computation of Stresses

A. Composite stresses

$$f_b = \frac{\text{DLM (Slab \& Steel)}}{\text{S (steel)}} + \frac{\text{LLM (Traffic)}}{\text{S (composite)}} + \frac{\text{DLM (SDW, Curbs)}}{\text{S (composite)}} + \frac{\text{LLM (Pedestrian)}}{\text{S (composite)}}$$

B. Noncomposite stresses

$$f_b = \frac{\text{DLM (Slab \& Steel)}}{\text{S (steel)}} + \frac{\text{LLM (Traffic)}}{\text{S (steel)}} + \frac{\text{DLM (SDW, Curbs)}}{\text{S (steel)}} + \frac{\text{LLM (Pedestrian)}}{\text{S (steel)}}$$

where: $f_b$ is the computed steel bending stress (ksi)
DLM is the dead load moment (k-in)
LLM is the live load moment (k-in)
S is the elastic section modulus ($in^3$)

(5) Input the Preliminary Design into a Continuous Beam Analysis and Design Computer Program. Refer to Design Example at the end of this chapter.

(6) The following Steps are taken using the Output from the Computer Program:

Check the loads of the interior and exterior members to see if one or both members are to be designed.

Check the live load deflection of the initial design.

Establish the size of flange plates, web plates, rolled beams and cover plates. Check the bracing of the negative moment flange. The allowable stress may be reduced on the compression flange. Determine if a thinner stiffened plate girder web or a thicker unstiffened web is to be used. On deep plate girders the thinner stiffened web is more economical. Refer to Section 24.10. The use of a slightly heavier rolled section with an increased section modulus eliminates small cover plate requirements. Wide flange beams are used in preference to I-beams.

Establish butt splice and field splice locations and the length of cover plates. Where a change in steel section is needed, plate girder flanges are butt spliced rather than cover plated.

For a rolled beam consider a shop welded butt splice or adding cover plates to increase section size. Also, check the maximum rolling lengths of plates to see if additional butt splices are required.

Determine the range of composite action for each span. This is the point where shear connectors are omitted or where positive dead load and live load moment do not occur.

The stiffness of the composite section is used for determining live load moments and shears. Dead load moments and shears are based upon the stiffness of the noncomposite steel section.

(7) Input the Proposed Final Design into a Continuous Beam Analysis and Design Computer Program.

For plate girders include the change in section at butt splices but do not include the change in weight. The fabricator is given the option of extending the heavier plate and eliminating the butt splice at his expense. The fabricator has used this option on numerous occasions. Shim plates are provided at the bearing to allow for either option.

(8) In Continuous Spans it is necessary to re-input the Size of the Members into the Computer Program when the Final Design is appreciably different than the design originally input.

Changing the stiffness of members in the span or over a support effects the distribution of moments, shears, and reactions.

(9) Design the Bearings and Bearing Stiffeners.

The bearings are designed at an early stage in the design since they may govern the width of the bottom flange plates.

BEARING STIFFENERS

Bearing stiffeners are placed normal to the web of the girder except for skew angles of 15° or less. Here, they may be placed parallel to the skew at abutments and piers to support the end diaphragms or cross framing.

For structures on grades of three percent or more, the end of the girder section at joints is to be cut vertical. This is to eliminate the large extension and clearance problems at the abutments.

(1) Plate Girders

Bearing stiffeners are placed over the end bearings of welded plate girders and also over the intermediate bearings of continuous welded plate girders. The bearing stiffeners extend as near as practical to the outer edges of the flange plate. They consist of 2 or more plates placed on both sides of the web. They are ground to a tight fit at the top flange, welded to the web on both sides with a 1/4 inch minimum fillet weld, and attached to the bottom flange with full penetration groove welds.

(2) Rolled Beams

Bearing stiffeners are placed over the bearings of rolled beams when the unit shear in the web adjacent to the bearing exceeds 75 percent of the allowable shear stress in the web of the beam. The web of the steel beam carries the total external shear, neglecting the effect of the steel flanges and the concrete slab. The shear is assumed to be uniformly distributed across the gross area of the web.

(3) Design

Bearing stiffeners are designed as columns which transmit the entire end reaction to the bearings. For stiffeners consisting of 2 plates, the column section is assumed to consist of the 2 plates and a centrally located strip of web whose width is less than or equal to 18 times the thickness of the web. For stiffeners consisting of 4 or more plates, the column section is assumed to consist of the plates and a centrally located strip of web whose width is equal to that enclosed by the plates plus a width of not more than 18 times the web plate thickness. Use a spacing of 9 inches center to center of plates on bearing stiffeners consisting of 4 plates. No part of the web is considered to act in bearing on hybrid girders which have a ratio of minimum yield strength of the web to minimum yield strength of the tension flange less than 0.70. An effective length factor, K, of 0.75 is used.

Bearing stiffeners may be made of A36 steel unless higher strength or weathering steel is required. The allowable compression in the bearing stiffener is determined from the formula:

$$F_b = 16{,}980 - 0.53\ (KL/r)^2$$

where: L is the depth of the web, inches
K is the effective length factor
r is the radius of gyration, inches.

The thickness of bearing stiffener plates is not less than

$$t = \frac{b'}{12}\sqrt{\frac{F_y}{33{,}000}}$$

where: t is the thickness of one plate, inches
b' is the width of one stiffener plate extended as near practical to the outer edge of the flange, inches
Fy is the yield point strength of the bearing stiffener.

The bearing pressure on the edge of the bearing stiffener against the bottom flange of the member is checked. The allowable bearing on milled stiffeners and other parts of steel in contact is 0.80 Fy. In calculating the bearing pressure assume that the web transfers the entire reaction to the bearing stiffener plates. Deduct one inch from the plate width for each clipped corner required to clear the flange to web weld.

## (10) Check the Live Load Deflections and Uplift.

## (11) Design the Field Splices.

### FIELD SPLICES

(1) Computer Program

A computer program is used to design the field splices for wide flange beam and plate girders of A36, A572 and A588 steel. The computer outputs the following list of design information.

Net composite and noncomposite section moduli.

Actual stresses, range of stresses, and allowable fatigue range stresses.

Required number of vertical bolt lines for one side of the web splice and the number of bolts per line.

Required net area of top and bottom flange splice plates and the number of bolts for one side of the flange splice.

(2) Location of Field Splices

Field splices shall be placed at the following locations whenever it is practicable:

At or near a point of dead load contraflexure.

So that the maximum rolling length of the flange plates is not exceeded, thus eliminating one or more butt splices.

At a point where the fatigue in the net section of the base metal is held to a minimum.

To limit section length to 120' between splices unless special conditions govern.

(3) Splice Material

A36 steel is used for the field splice plates if the girder is designed using A36 material. In cases where both of the members being spliced are of higher strength steels such as A572 or A588, the splice material is the same as the members being spliced. Generally, 3/4 inch diameter high strength A325 bolted friction-type connectors are used unless the proportions of the structure warrant larger diameter bolts.

(4) Design

A. Web Splice

The method used to design the web splice is taken from "Design of Steel Structures" by Gaylord and Gaylord. It is based on the following equation:

$$P = \frac{(R)(n)}{t\sqrt{s^2 + v^2}}$$

where: P is the vertical spacing (pitch) of bolts equally spaced.
R is the allowable force on one bolt in double shear
n is the number of vertical bolt lines on one side of splice
t is the web thickness
s is the maximum bending stress in the web
v is the average shear stress in the web.

Two separate designs are output by the program for the web splice. One is referred to as the 75% criteria and the other as the average stress criteria. In the above equation s and v are set equal to the following values:

75% CRITERIA

s = 75% of .55 times the yield stress of the flange. If top and bottom flange have different yield stresses, the higher yield stress is used.
v = 75% of the input "Allowable Shear Stress". *

AVERAGE STRESS CRITERIA

s = The average of the maximum bending stress and .55 times the yield stress of the flange. Since there are four flanges, four "s" values are computed and used to compute four values of "p".
v = The average of the actual shear stress (computed from the input "Total Shear") and the input "Allowable Shear Stress". Two "v" values are computed (one for each side of the splice) and used with the corresponding four "s" values.

* For Plate Girders see AASHTO Figure 1.7.43D1 or Figure 1.7.43D2. For Rolled Beams see AASHTO Table 1.7.1A. Use 12 ksi maximum.

The design based on the average stress criteria is obtained by first selecting the required pitch on each side of the web. The computed pitches on one side of the web at the top and bottom flanges are compared and the smallest is the one that is saved. From the two pitches (one on each side of the web) remaining, the program uses the larger one for the final web splice design. The larger pitch is used because the program designs for the weaker of the two pieces being spliced.

B. Flange Splice

The net area of the flange splice is determined from the following equation:

$$A = \frac{F}{s},$$

where: A is the net area required for the flange splice plates. This is the total net area required for the splice plates on both sides of the flange. (Top and bottom of flange). The area of the splice plates on both sides of the flanges should be approximately the same.

F is the force in the flange obtained by multiplying the "Net Flange Area" by the following four stresses:

1. 75% of .55 times the yield stress of the flange.

2. The average of the maximum bending stress and .55 times the yield stress of the flange.

3. The range of stress in the flange allowed for truck loading. This is used to determine the net area of flange splice plates required to satisfy the allowable fatigue stress range.

4. The range of stress in the flange allowed for lane loading.

s is equal to .55 times the yield stress of the splice material for Number 1 and 2 above, and the allowable fatigue range for Number 3 and 4.

Using the four stresses mentioned above, four values of "A" will be computed for each flange. The maximum of these four values for each flange are then compared to the maximum of the four values of the adjacent flange and the minimum of the two maximums are output for the top flange and bottom flange. These two values are the "REQUIRED TOP NET AREA" and the "REQUIRED BOTTOM NET AREA."

The number of bolts required for the flange splices is calculated from the following equation:

$$N = \frac{F}{(\pi D^2/2)(F_v)}$$

where: N is the number of bolts required. ("N" is rounded up to the next highest even number).

F is the force in flange calculated by multiplying the "Net Flange Area" by the greater of the following two stresses.

1. Seventy five percent of .55 times the yield stress of the flange.

2. The average of the maximum bending stress and .55 times the yield stress of the flange.

D is the bolt diameter.

F is the allowable shear for a friction-type connection with a standard type hole.

The number of bolts required in the top flange splice plates is determined by selecting the minimum required for the two top flanges. The minimum number is selected because the program designs for the weaker of the two pieces being spliced. The number of bolts required in the bottom flange splice plates is determined by the same method used for the top flange splice plates.

(12) Determine the Shear Connector Spacing from Program Print Out.

Refer to Section for shear connector spacing. Determine the distance from the piers where shear connectors are omitted and field welding to the top flange for construction purposes is not permitted.

Refer to the Standard for shear connector details. Three shop or field welded 7/8 inch diameter studs, 4 inches long are placed on the top flange. The studs are equally spaced with a minimum clearance of 1 1/2 inches from the edge of the flange. On girders having thicker haunches where stud embedment is less than 2 inches into the slab, use longer studs to obtain the 2 inch minimum embedment.

Shear connectors are carried to the point nearest the support where positive moment for combined dead and positive live loading exists. This termination point is used; although theoretically, the shear connectors may be terminated when the steel section can carry the positive moment. The allowable stress in a tension flange is reduced by the fatigue criteria if shear connectors are attached. The top flange is the tension flange for negative moment which may control the flange plate size. Connectors which fall on the flange field splice plates are respaced near the ends of the splice plate. The maximum spacing of shear connectors is 2 feet. Begin connectors spacings 6 inches from centerline of abutments.

The shear connector spacing at the tenth points along the span can be determined by using the maximum value of composite moment of inertia in the span. Shear connector spacing is determined by the following formula based on fatigue and horizontal shear:

$$S = \frac{(P)(I)}{(V_r)(Q)}$$

where: S is the required shear connector spacing in inches
P is the horizontal shear capacity of shear connectors (taken as 23.8 kips for 3-4x7/8" studs)
I is the maximum composite moment of inertia in the span
$V_r$ is the range of shear due to live load and impact
The range in shear is the difference between the minimum and maximum live load shear evelopes
The computer adds any dead load shear carried by the composite member to the range of live load
Q is the statical moment of the transformed concrete area about the neutral axis of the composite section.

Shear connectors are placed to clear the bottom transverse reinforcement of the slab. In order to accomplish this, shear connectors are spaced in increments which are multiples of the bottom transverse bar spacing in the slab.

The number of shear connectors required for fatigue are checked to insure that adequate connectors are provided for ultimate strength. In most cases the connector spacing, three per set, using output based on fatigue requirements is more than adequate for ultimate strength design requirements. Additional shear connectors may be required for ultimate strength design on structures having spans in excess of 150 feet.

The number of shear connectors required for ultimate strength is checked between points of maximum positive moment and the adjacent end supports ($N_1$). The formulas given in the AASHTO specification are applied in the design example for shear connectors.

Additional shear connectors are required at the point of contraflexture when reinforcement steel embedded in the concrete is not used in computing section properties for negative moments. The additional connectors are required to fully develop the slab stresses. The number of additional connectors is computed from the following formula:

$$Nc = \frac{A_r^s\, fr}{Zr}$$

where: Nc is the number of additional connectors for each beam or girder at the point of contraflexure
$A_r^s$ is the total area of longitudinal slab reinforcement for each beam or girder over interior support
fr is the range of stress due to live load plus impact in the slab reinforcement over the support, in lieu of more accurate computations, use 10,000 psi.

Zr is the allowable range of horizontal shear on the shear connectors for a set of three studs, use 23,800 psi.

The additional connectors shall be placed adjacent to and centered on either side of the dead load point of contraflexure within a distance equal to one-third the effective slab width.

## (13) Design any Transverse and Longitudinal Stiffeners that are required.

### TRANSVERSE INTERMEDIATE STIFFENERS

Transverse intermediate stiffeners are plates which are welded to the web and compression flange of the member with a tight fit at the tension flange. Indicate on the plans the flange to which stiffeners are welded. The stiffeners are attached to the web with a continuous 1/4 inch fillet weld. The dead load moment diagram is used to define the compression flange.

If longitudinal stiffeners are required, the transverse stiffeners are placed on one side of the web of the interior member and the longitudinal stiffener on the opposite side of the web. Place intermediate stiffeners on one side of interior members when longitudinal stiffeners are not required. Transverse stiffeners are placed on the inside web face of exterior members. If longitudinal stiffeners are required, they are placed on the outside web face of exterior members as shown on Standard 24.2. The maximum projection of the flange beyond the stiffeners is three inches.

Transverse stiffeners can be eliminated by increasing the thickness of the web. On plate girders under 48 inches in depth, consider thickening the web to eliminate all transverse stiffeners. Within the constant depth portion of haunched plate girders over 48 inches deep, consider thickening the web according to the following ratios to eliminate the longitudinal stiffener and a portion of the transverse stiffeners within the span:

| | Less Trans. Stiff. | No Longit. Stiff. |
|---|---|---|
| A 36 compression flange 36,000 psi Y. P. | $t \geq \frac{D}{150}$ | $\frac{D}{165}$ |
| A572 & A588 compression flange 50,000 psi Y. P. | $t \geq \frac{D}{150}$ | $\frac{D}{140}$ |

For example, the difference in weight between a 64 x 3/8 inch web and 64 x 7/16 inch web is 13.6 pounds per foot. Using A36 steel the required thickness is:

$$\frac{64}{165} = 0.388, > 3/8", < 7/16"$$

$$\frac{64}{150} = 0.427, > 3/8", < 7/16"$$

Increasing to a 64 x 7/16 inch web eliminates the longitudinal stiffener throughout the range of span where the 64 inch depth is used. Furthermore, the 64 x 7/16 web eliminates a number of transverse stiffeners in the central portion of the span due to the increase in allowable shear stress.

In this case the 13.6 pounds per foot additional steel in the 7/16 inch web is more economical than welding the additional longitudinal and transverse stiffeners to the 3/8 inch web.

Current AASHTO specifications limit bending stress where the girder panel is subject to "tension field action" under the simultaneous action of shear and bending moment. If the full permissable shear stress is allowed, bending stress is limited to approximately 75 percent of the maximum allowable. If the shear stress is limited to 60 percent of the maximum allowable, full bending stress is allowed. This can be accomplished by decreasing the transverse stiffener spacing which in turn increases the allowable shear stress. The bending stress ($F_s$) is limited to (0.754 - 0.34 fv/Fv)Fy if the shear stress exceeds 60 percent.

The minimum size of transverse stiffeners is 5 x 5/16 inch. Minimum stiffener width, thickness and moment of inertia is specified by AASHTO.

Transverse stiffeners are placed on the inside face of all exterior girders where the slab overhang exceeds 2'-0 as shown on Standard 24.2. The stiffeners are to prevent web bending caused by construction of the deck slab where triangular overhang brackets are used to support the falsework.

If by special approval of the Bridge Engineer the slab overhang is allowed to exceed the recommended 3'-7 on exterior girders, the web and stiffeners should be analyzed to resist the additional bending during construction of the deck. Overhang construction brackets may overstress the stiffeners. It may also be necessary to provide longitudinal bracing between stiffeners to prevent localized web deformations which did occur on a structure having 5'-0 overhangs.

For stiffeners placed on one side of the web only the moment of inertia provided is:

$$I = \frac{th^3}{3}$$

where: I is the moment of inertia of one transverse stiffener about the face of the web plate, inches $^4$.
t is the thickness of one transverse stiffener, inches.
h is the width of one transverse stiffener, inches.

## 24.11 LONGITUDINAL STIFFENERS

Longitudinal stiffeners are plates which are placed on the web at 1/5 of the depth of the web from the compression flange. On the exterior members, the longitudinal stiffeners are placed on the outside face of the web as shown on Standard 24.2. If the longitudinal stiffener is required throughout the length of span on an interior member, the longitudinal stiffener is placed on one side of the web and the transverse stiffeners on the opposite side of the web. Longitudinal stiffeners are normally used in the haunch area of long spans and on a selected basis in the uniform depth section.

Where longitudinal stiffeners are used, place intermediate transverse stiffeners next to the web splice plates at a field splice. The purpose of these stiffeners is to prevent web buckling before the girder is erected and spliced.

AASHTO specifications give the web thicknesses where longitudinal stiffeners are required. Also, AASHTO specifies a minimim stiffener thickness and a minimum stiffener moment of inertia. The minimum size longitudinal stiffener used in Wisconsin is 5 x 3/8 inches.

The longitudinal stiffener cut-off point is determined from the following formula defined by AASHTO:

$$t_{(web)} \quad \frac{D\sqrt{f_b}}{23{,}000} \leq \frac{D}{170} \quad \text{whichever governs.}$$

where: $t_{(web)}$ is the thickness of the web, inches.
D is the depth of the web, inches.
$f_b$ is the calculated bending stress in the flange.

If computations indicate that the top or bottom flange carries a compressive stress, and the thickness of the web at that point is less than D/170, then a longitudinal stiffener is required at D/5 from that flange. It is possible to have an overlap of longitudinal stiffeners at D/5 from the top flange and D/5 from the bottom flange due to the variation between maximum positive and maximum negative moment.

(14) Design the Lateral Bracing and End Diaphragms.

(15) Determine the Dead Load Deflections, Blocking, Camber, Top of Steel and Top of Slab Elevations

(16) Analyze the Structure for Stresses caused during erection and construction for possible overstresses. Check the Lateral Bracing without the Deck Slab.

(17) Determine the Slab Pouring Sequence. Refer to Standard Determine the maximum amount of Concrete that can be poured in a day. Determine Deflections based on the Proposed Pouring Sequence.

CONSTRUCTION

When the deck slab is poured, the exterior girder tends to rotate between the diaphragms. This rotation causes the rail supporting the finishing machine to deflect downward and changes the roadway grade. To prevent this the contractor places bracing (wooden 4 x 4's, etc.) on the bottom flanges between the interior and exterior girders. This problem may result if the slab overhang is greater than the recommended 3'-7 on exterior girders.

Where built up box sections are used, full penetration welds provide a stronger joint than fillet welds and give a better looking joint.

The primary force on the member is tension or compression along the axis of the member. The secondary force is a torsion or racking force on the member cross section which produces a shearing force across the weld.

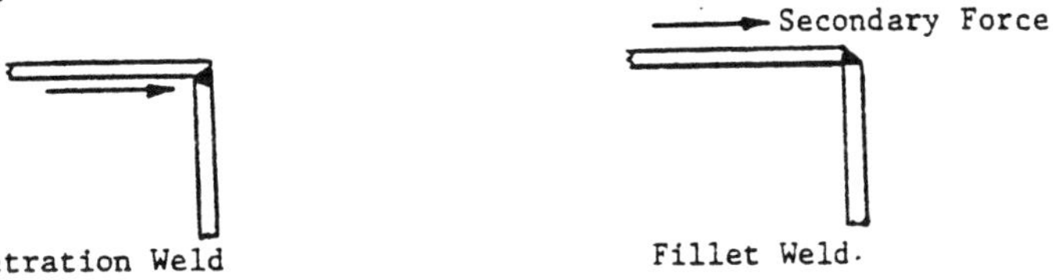


PAINTING

Unpainted weathering steel is used on bridges over streams and railroads. All highway grade separation structures require the use of painted steel as unpainted steel is subject to excessive weathering from salt spray distributed by traffic. On unpainted steel bridges the end six feet of the girder adjacent to an expansion joint and/or hinge is required to have two shop coats of paint. The second coat is to be a brown color similar to rusted steel. Do not paint the exterior girder faces for aesthetic reasons but do paint the hanger bar on the side next to web. If the girder web depth exceeds 6 feet, paint the girder end the distance of web depth. All unpainted steel is required to receive a blast cleaning equivalent to a No. 6 Commercial Blast as specified by the Steel Structures Painting Council.

FLOOR SYSTEMS

Floor systems may be used on long span bridges. It may be more economical to use two plate girders with floor beams and intermediate stringers than to use a multiple girder system if clearance is not a problem. However, the general Bridge Section policy is to use multiple girder systems for bridges with span lengths of up to 390 feet. Although there may be some cost savings with a two girder system, a multiple girder system is preferred for the ease of construction and because it is a redundant type bridge, i.e., failure of any particular member would not cause collapse of the bridge.

In a two girder system the main girders are designed equally to take the dead and live load unless the roadway cross section is unsymmetrical. The dead and live load carried by the intermediate stringers is transferred to the floor beams which transmit the load to the main girders. In designing the main girders it is an acceptable practice to assume the same load distribution along the stringers as along the girder and ignore the concentrated loads at the floorbeam connections.

The design criteria used for the girders with floor systems is the same as the criteria used for plate girders and rolled sections. Pay particular attention to the connection details to be sure they are adequate and to the lateral bracing requirements and connections. There are three types of connections between the floor beam and longitudinal girder. Refer to page 6-46 of the "Structural Engineering Handbook" by Gaylord and Gaylord for a discussion of Moment-Resistant Beam Connections.

BOX GIRDER

Initial considerations for the design of box girders indicated a material savings in the structure due to a more liberal live load distribution factor and to lesser requirements for transverse bracing between girders. Box girders present a smooth, uncluttered appearance under the bridge deck due to the lack of transverse bracing and to their closed section.

In the design of box girders, the concrete slab was designed as a portion of the top flange and also as the support between the two girder webs. During construction the box girder required additional bracing to support the web until the slab was in place. This was either temporary bracing removed after the slab was cured or permanent bracing left in place. In either case, it was an additional cost item. In addition, temporary bracing was required between the girders to transfer wind loads until the deck slab was poured.

Current experience shows that box girders may require more material than conventional plate girders. They require more effort to fabricate and erect than plate girders. On longer spans additional bracing between girders is required to transfer lateral loads.

# E.8 DESIGN EXAMPLE

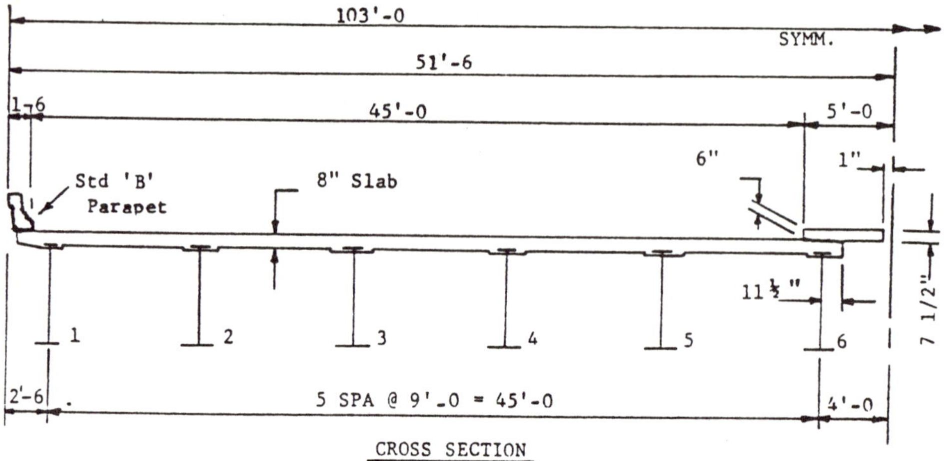


CROSS SECTION

HS20 loading, 2 span continuous 148'-6, 138'-6.

8" slab, (9-0) - (1/2)(14" top FLG) = 8'-5 eff. span.

Make all girders same.

Full curb load to exterior girder.

Median girder checked for HS20LL on median @ 1'-0" from centerline open joint @ 150% of allowable stress.

Structure is designed for a future wearing surface of 20 pounds per square foot.

Allowable Design Stesses:

Concrete Masonry Slab------------$f'_c$ = 4000 psi
All Other-------$f'_c$ = 3500 psi
Bar Steel Reinforcement, Grade 60------$f_y$ = 60,000 psi
Structural Carbon Steel (A 36)---------$f_s$ = 20,000 psi
Stuctural Low Alloy Steel (A 572)
to and including 2" thick--------------$f_s$ = 27,000 psi

Median Design - Slab

Place HS20 LL wheel @ 1'-0 from center line of open joint.

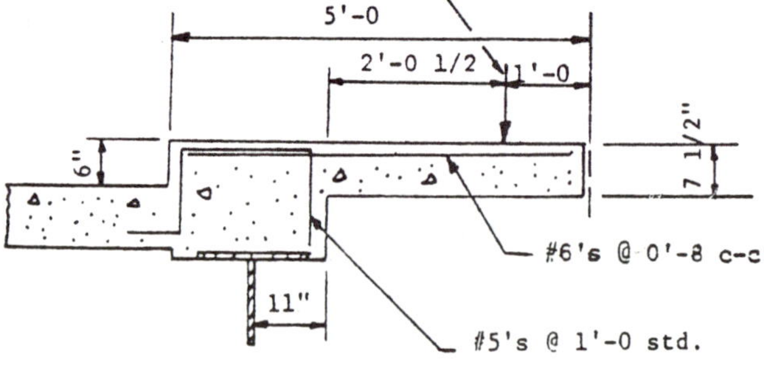


Wheel distribution (E) = (0.8)(2.04) + 3.75 = 5.38'

$$LLM = \frac{(16)(1.30)}{5.38} \times 2.04 = 7.99^{\prime\text{-k}/\prime}$$

Assume 7 1/2" depth of median section:

$$DLM = (1/2)(3.00)^2(0.625)(0.15) = 0.42^{'-k/'}$$

$$Mu = 1.3[(0.42) + 1.67(7.99)] = 17.9^{'-k/'}$$

$$R_u = \frac{Mu}{\phi b d^2} \quad \text{where } d = 7\ 1/2 - (2\ 1/2\ \text{Cl.} + 3/8) = 4.62''$$

$$R_u = \frac{(17.9)(12000)}{(0.9)(12)(4.62)^2} = 932 \text{ psi}$$

Ru/150% = 932/150% = 621 psi

$$\rho = 0.0116 \quad A_s = \rho\, bd$$

$$A_s = (0.0116)(12)(4.62) = 0.64 \text{ in}^2/\text{ft.}$$

Use: #6's @ 0'-8 c-c

COMPUTER INPUT

Interior Girders:

$$\text{LL Distr.} = \frac{9.0}{5.5} = 1.636 \text{ wheels/girder}$$

$$LLR = 1 + \frac{(3.0) + (5.0)}{9.0} = 1.888 \text{ wheels/girder}$$

Typical interior girder spacing with wheel loadings for LLR

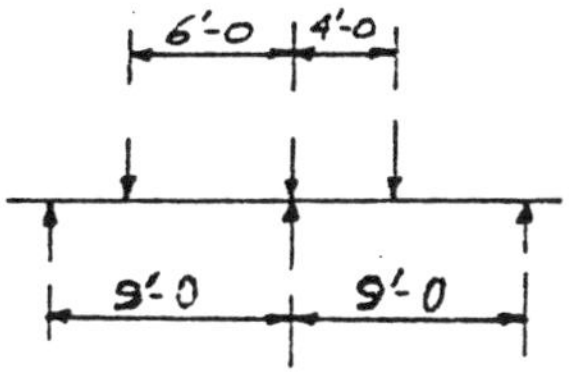


Concrete Dead Load:

| | | |
|---|---|---|
| Slab | = (9.0)(0.667)(150) | = 900 |
| Haunch | = (1.833)(0.104)(150) | = 29 |
| -Top Flg. | = -(1.25)(0.083)(150) | = -16 |
| Connections | = 30# /lin. ft. | = 30 |
| | | 943#/' |

Curb Dead Load = (9.0)(20) = 180#/' due to possible placement of future wearing surface. Curb and median dead loads are to be placed on the composite section.

Exterior Curb Girders:

$$\text{LL Distr.} = \frac{9.0}{4 + (0.25)(9.0)} = 1.440 \text{ wheels/girder}$$

Location of wheel loadings for LLR

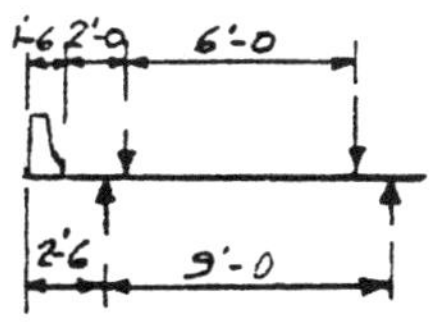


$$LLR = \frac{(2.0) + (8.0)}{9.0} = 1.111 \text{ wheels/girder}$$

Concrete Dead Load:

| | | |
|---|---|---|
| Slab | = (6.58)(0.667)(150) | = 659 |
| Haunch | = (2.37)(0.104)(150) | = 37 |
| -Top Flg. | = (1.25)(0.083)(150) | = -16 |
| Connections | = 20#/lin. ft. | = 20 |
| | | 700#/' |

Curb Dead Load = 338#/' + (5.5)(20) = 448#/'

Median Girders:

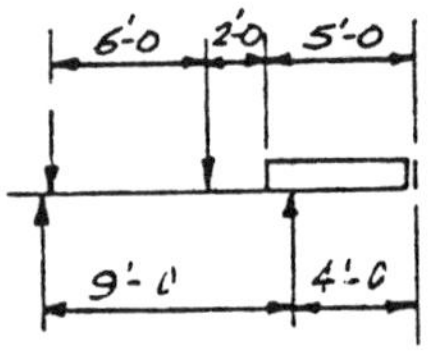


@ 100% — LL Distr. = $\frac{9.0}{4 + (0.25)(9.0)}$ = 1.440 wheels/girder

Location of wheel loadings for LLR

LLR = $\frac{(0.) + (6.0)}{9.0}$ = 0.667 wheels/girder

The designer may wish to consider the posibility of a wheel on top of the median section.

@ 150% — Location of wheel loadings on median

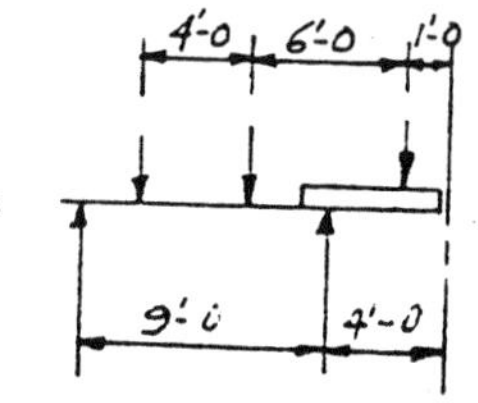


LL Distr = LLR = 1 + $\frac{(2.0) + (6.0)}{9.0}$ = 1.888 wheels/girder

Concrete Dead Load:

| | | |
|---|---|---|
| Slab | = (5.45)(0.667)(150) | = 545 |
| Haunch | = (1.83)(0.104)(150) | = 29 |
| -Top Flg. | = -(1.25)(0.083)(150) | = -16 |
| Connections | = 20#/lin. ft. | = 20 |
| | | 578#/' |

Median Dead Load = (1.92)(0.50)(150) + (3.04)(0.625)(150) = 429#/'

Note: Girder section dead load is computed and accounted for by the Continuous Beam Analysis program based on trial input sections.

COMPUTER OUTPUT

Final Plate Sizes

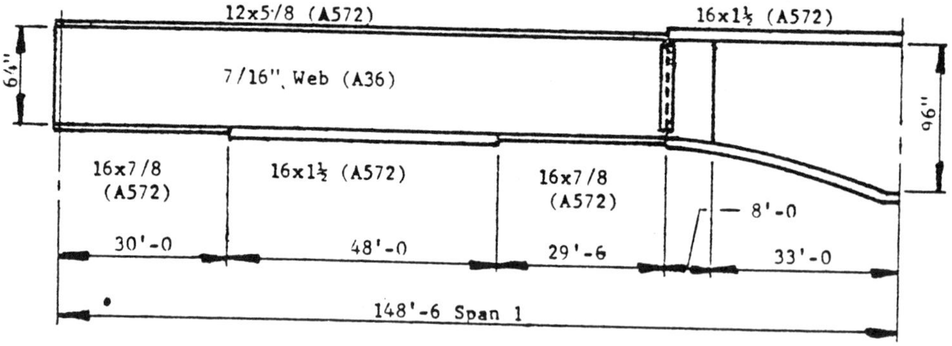


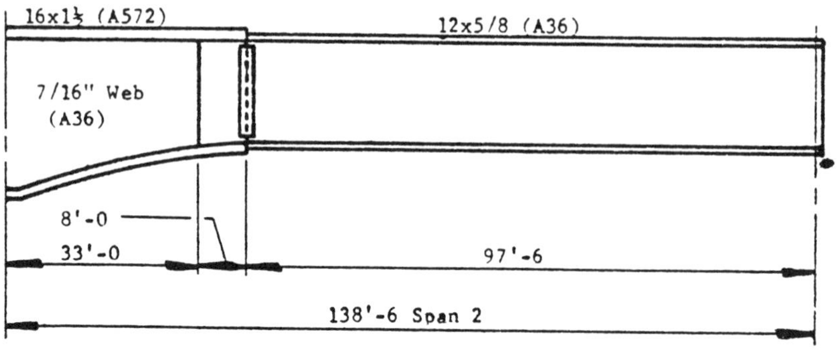

Live Load Deflections

From the computer output for exterior and interior girders, span 1, the live load deflections are 2.11" and 2.26" respectively. Refer to Article 24.4(3) for deflection considerations. The wheel load distribution per girder is computed as follows:

$$\frac{(3\text{ trucks})(2\text{ wheels/truck})(0.90)}{6\text{ girders}} = 0.90\text{ wheel/girder}$$

Actual deflection = (0.90)(2.11) / 1.44 = 1.32" Ext. girder

Actual deflection = (0.90)(2.26) / 1.636 = 1.24" Int. girder

Checking with the allowable deflection of L/1000 = (148.5)(12)/1000 = 1.78", the actual deflection is well within specification.

Bearing Stiffener Design - Pier 1

| | |
|---|---|
| Dead load reaction (slab and steel) | = 221 |
| Live load reaction (brg. truck or brg. lane) | = 141 |
| Dead load reaction curb | = 33 |
| Interior girder reaction controls | $395^k$ |

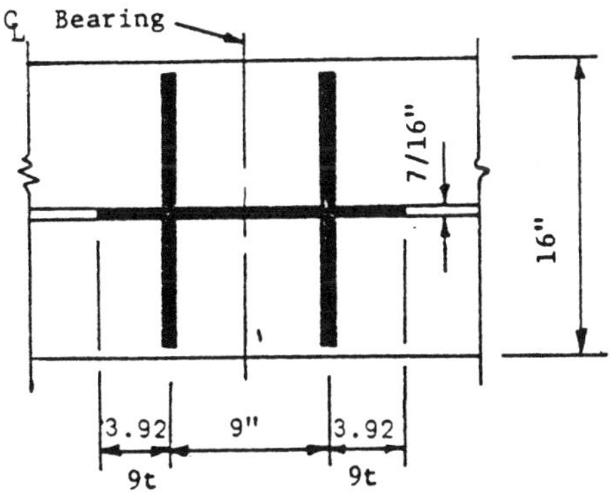


Cross section of girder at pier

| | | |
|---|---|---|
| Top flg. | 16 x 1-1/2 | (A572) |
| Web | 96 x 7/16 | (A36) |
| Bottom flg. | 16 x 1-1/2 | (A572) |

Referring to above sketch, try 4 PLs-7 1/2 x 3/4" (A36).

$$t_{min} \geq \frac{7.5}{12}\sqrt{\frac{36}{33}} = 0.653" < 3/4" \text{ o.k.}$$

Ratio of yield points of web to flange equals 36/50 = 0.72 > 0.70 min AASHTO required for web bearing on the tension flange.

Stiffener column area is as follows:

$$A = (4)(7\ 1/2)\ (3/4) + (16.84)(7/16) = 29.9\text{ in}^2$$

$$f_{(actual)} = \frac{395{,}000}{29.9} = 13{,}210\text{ psi}$$

$$f_{(allow)} = 16{,}980 - 0.53\ (KL/r)^2 \quad \text{where}$$

$K = 0.75$
$L = 96"$
$r = \sqrt{I/A}$
$I = bh^3/12 = (1.5)(15.4375)^3/12 = 460\text{ in}^3$
$A = 29.9$
$r = \sqrt{460/29.9} = 3.92$
$f_{(allow)} = 16{,}980 - 0.53\ [(0.75)(96/3.92)]^2$
$= 16{,}801\text{ psi} < 13{,}210\text{ psi}$ o.k.

Stiffener bearing area deducting for 1 1/4" clipped corners is as follows:

$$A = (4)(7\ 1/2 - 1\ 1/4)(3/4) = 18.75\text{ in}^2$$

$$f_{(brg)} = \frac{395{,}000}{18.75} = 21{,}067\text{ psi}$$

$$f_{(allow)} = 0.8\ Fy = 28{,}800\text{ psi} > f_{(brg)} \quad \text{o.k.}$$

Use: 4 PLs-7 1/2 x 3/4) (A36)

Uplift

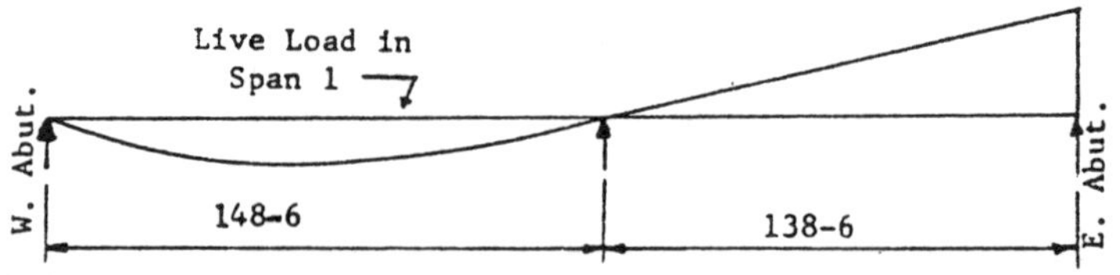


Check uplift at east abutment, by employing influence line coefficients for shear.

Lane Loading = (9.3)(.64) + (0.097)(26) = $8.55^k\uparrow$
Truck Loading = (0.097)(32) + (0.094)(8) = $3.85^k\uparrow$

Lane loading governs the uplift at $8.55^k$.

$$\text{Impact} = \frac{50}{148.5 + 125}(100) = 18.3\%$$

The total uplift force is as follows:

= (loading)(impact)(no. of lanes)(lane reduction)(100% increase)
= (8.55)(1.183)(3)(0.90)(2) = $54.6^k\uparrow$

The dead load reaction is as follows:

= Exterior girders + Interior girders
= (2)(43.4 + 22.6) + 4 (51.8) = $339.2^k\downarrow$

Uplift does not occur at the abutments, permanent hold down devices are not required.

Field Splice Design

| | | |
|---|---|---|
| Top flg. | 12 x 5/8 (A572) | spliced to 16 x 1 1/2 (A572) |
| Web | 64 x 7/16 (A36) | spliced to 64 x 7/16 (A36) |
| Bottom flg. | 16 x 7/8 (A572) | spliced to 16 x 1 1/2 (A572) |

All field splice plates are A36 steel and A325 - 3/4 inch diameter bolts are used throughout having an allowable double shear value of 16 kips per square inch.

Top Flange Splice

Computer output: A = 5.64 $in^2$ and 10 bolts/side (2 bolts/line)

(12)(3/8) - (2)(3/4 + 1/16)(3/8) = 3.89
(2)(5)(7/16) - (2)(3/4 + 1/16)(7/16) = 3.67 } = 7.56 > 5.64 $in^2$
(1 1/4 + (4)(2 1/2 1/2) + 1 1/2) 2 = 2'-1 1/2" Lg

Use: 1 PL - 12 x 3/8 x 2' - 1 1/2 Lg
2 PL - 5 x 7/16 x 2' - 1 1/2 Lg
20 - 3/4" H.S. Bolts

Bottom Flange Splice

Computer output: A = 9.78 $in^2$ and 18 bolts/side (4 bolts/line)

(16)(7/16) - (4)(3/4 + 1/16)(7/16) = 5.58
(2)(7)(1/2) - (4)(3/4 + 1/16)(1/2) = 5.37 } = 10.95 > 9.78 $in^2$
(1 1/4 + (4)(2 1/2 + 1 1/2) x 2 = 2'-1 1/2 Lg

Use: 1 PL - 16 x 7/16 x 2'-1 1/2 Lg
2 PL - 7 x 1/2 x 2'-1 1/2 Lg
36 - 3/4" H.S. Bolts

Web Splice

Computer output: 2 vert. lines per side with 21 bolts each.

(1 1/4 + (1)(2 1/2 + 1 1/2) 2 = 10 1/2" wide

Use: 2 PL - 10 1/2" x 3/8 x 5'-2 Lg
4 vert. lines, 21 bolts per line pitched at 2 7/8"
84-3/4" H.S. Bolts

## Transverse Intermediate Stiffener Design

Transverse intermediate stiffener spacing is limited to one half the web depth at the abutments and the maximum spacing is limited to 1.5 times the web depth. Current AASHTO specifications require transverse stiffeners if the web thickness is less than D/150.

Reference is made to Section 24.10 for discussion of current computer design. Spacings are given from computer output at 10th points as follows:

| Tenth Point | Span 1 | Span 2 |
|---|---|---|
| 0.0 | 96" | 62.0" |
| 0.1 | 96" | 66.2" |
| 0.2 | 96" | 81.2" |
| 0.3 | 0 | 96" |
| 0.4 | 0 | 96" |
| 0.5 | 0 | 0 |
| 0.6 | 96" | 0 |
| 0.7 | 96" | 0 |
| 0.8 | 80.5" | 0 |
| 0.9 | 64.5" | 96" |
| 1.0 | 60.0" | 96" |

## Stiffener Size Criteria

Minimum Width $\geq (1/4)(16) = 4"$
$\geq (2)(64/30) = 4.27"$
$\geq (2) + (96/30) = 5.20"$

Minimum Thickness $\geq (1/16)(5) = 5/16"$
$\geq (1/16)(5\ 1/2) = 11/32"$ say: 3/8"

Try: 5 x 5/16 up to the 0.75 span point of span 1
Try: 6 x 3/8 over the 0.75 span point of span 1

The required moment of inertia is checked to the 0.8 point of span 1 where D is approximately 65 inches.

$$I\ (\text{req'd}) = \frac{(d_o)(t_w)^3(J)}{10.92} \quad \text{where}$$

$$J = 25\left(\frac{D}{d}\right)^2 - 20 \geq 5.0$$

$$= 25\left(\frac{65}{80.5}\right)^2 - 20 < 5.0 \quad \therefore \text{Use: } 5.0$$

$$I\ (\text{req'd}) = \frac{(96)(0.4375)^3(5)}{10.92} = 3.68\ \text{in}^4$$

$$I\ (\text{prov'd}) = \frac{(t)(h)^3}{3} = \frac{(0.3125)(5)^3}{3}$$

$$= 13.0\ \text{in}^4 \quad \text{o.k.}$$

The required moment of inertia is checked from the 0.8 point of span 1 to the pier where

D (max) = 96"

$$J = 25\left(\frac{96}{60}\right)^2 - 20 = 44.0$$

$$I\ (\text{req'd}) = \frac{(60)(0.4375)^3(44.0)}{10.92} = 20.2\ \text{in}^4$$

$$I\ (\text{prov'd}) = \frac{(0.375)(6)^3}{3} = 27.0\ \text{in}^4 \quad \text{o.k.}$$

Use: 5 x 5/16" stiffeners up to the 0.75 point of span 1.

Use: 6 x 3/8 stiffeners from the 0.75 point of span 1 to the pier.

## Lateral Bracing Check

Determine the need for Lateral Diagonal Bracings.

Depth of Section = 2.67' + .67' + 5.50' = 8.84 $ft^2/ft$

Wind Force - 50#/$ft^2$ x 8.84 $ft^2/ft$ = 442#/ft Apply ½ of this (221 #/ft) to lower Lateral System. For the sake of simplicity in this computation assume for all points considered:

$f_D = 18$ ksi $f_L = 9$ ksi Check first cutoff point on Span 1 at 30' from the support. Try for a spacing of 25' using AASHTO Spec.

$R = (.2272xL-11)\ S_d{-2/3}$

$R = (.2272 \times 148.5^d - 11)\ 25^{-2/3} = 22.7/8.5 = 2.7$

$F_{cb} = 72M_{cb}/tb^2 = 72x11/.75x16^2 = 4.1$ ksi

$M_{cb} = .08xWxS_d^2 = .08 \times .221 \times 25^2 = 11$ Ft - kips

$F_w = R\ F_{cb} = 4.1x2.7 = 11.1$ ksi

Group II D + W = 18 + 11.1 = 29.1 ksi

Group III D + L + .3W = 18 + 9 + .3 x 11.1 = 30.3 ksi

$f_b$ (allowable) = 1.25 x 27 ksi = 33.75 ksi $>$ 30.3 ksi

Lower lateral bracing is not required in the final design condition. Temporary construction loadings may warrant bracing in top flanges for at least during placement of the superstructure. A computation is made below:

$M_{cb} = .08 \times WxS_d^2 = 11$ Ft - kips

$F_{cb} = 72M_{cb}/tb^2 = 72x11/.62x12^2 = 7.2$ ksi

$F_w = RF_{cb} = 2.7 \times 7.2 = 19.5$ ksi

$F_d$ Due to Steel Alone = 3 ksi

Group II Temporary = 3 ksi + 19.5 ksp = 22.5 ksi

Comp. Flg. Flexure $F_b = 27{,}000 - 14.4\left(\frac{25x12^2}{12}\right) = 19{,}000$ psi

Allowable $F_T = 1.25 \times 1.25 \times 19{,}000 = 28{,}000$ psi $>$ 22,500 psi - No temporary bracing is required in this case.

Current Bridge Section practice on this matter is discussed more fully in the Manual Section on Lower Lateral Bracing.

Longitudinal Stiffener Design

Longitudinal stiffeners are not required by AASHTO specifications if the actual web thickness is greater than D/170 as well as greater than $D\sqrt{f_b} / 23{,}000$.

$$D/170 = 64/170 = 0.38'' < 7/16''$$
$$D\sqrt{f_b} / 23{,}000 = (64) \times \sqrt{27{,}000}/23{,}000 = 0.46 > 7/16''$$

Consider using one size of stiffener (5 x 3/8") for the full girder length. A check is made at the pier for the haunched section.

$$t(\text{req'd}) = b'\sqrt{f_b} / 2250 = 5\sqrt{27{,}000}/2250 = 0.37'' \quad \text{o.k.}$$

$$I\ (\text{req'd}) = (D)(t_w)^3(2.4(\frac{d_o}{D})^2 - 0.13)$$

$$= (96)(0.4375)^3(2.4(\frac{60}{96})^2 - 0.13) = 6.59\ \text{in}^4$$

$$I\ (\text{prov'd}) = \frac{(t)(h)^3}{3} = \frac{(0.375)(5)^3}{3} = 15.6\ \text{in}^4 \quad \text{o.k.}$$

USE: 5 x 3/8" longitudinal stiffeners

Locate longitudinal stiffeners at a distance of D/5 from the compression flanges, note the overlapping region between (DLM + LLM) and (DLM - LLM) requiring stiffeners at a distance D/5 from both flanges.

Slab On Steel Girder Design

Effective Span = 9'-0 - (1/2)(1'-2) = 8'-5
Slab Thickness (T) = 8"

"A" and "B" Bars: #6's @ 7 1/2" c-c
"C" Bars: #4's (12 + 2 = 14 total)
"D" Bars: #4's (9 total)
"E" Bars: #6's (10 total)

Note in addition to "D" bars, "E" bars are required in the negative moment regions of continuous spans. The minimum reinforcement including the longitudinal distribution reinforcement must equal or exceed 1 percent of the cross-section area. Two-thirds of this required reinforcement is to be placed in the top layer of the slab within the effective width.

The area required in the negative moment regions is generally supplied by replacing the "D" and "E" bars by one larger bar size. Note #7 bar selection and spacing as shown on the following sketch. The "E" bars are extended from the negative moment regions of continuous spans at least 40 bar diameters beyond the additional shear connectors into the positive moment regions.

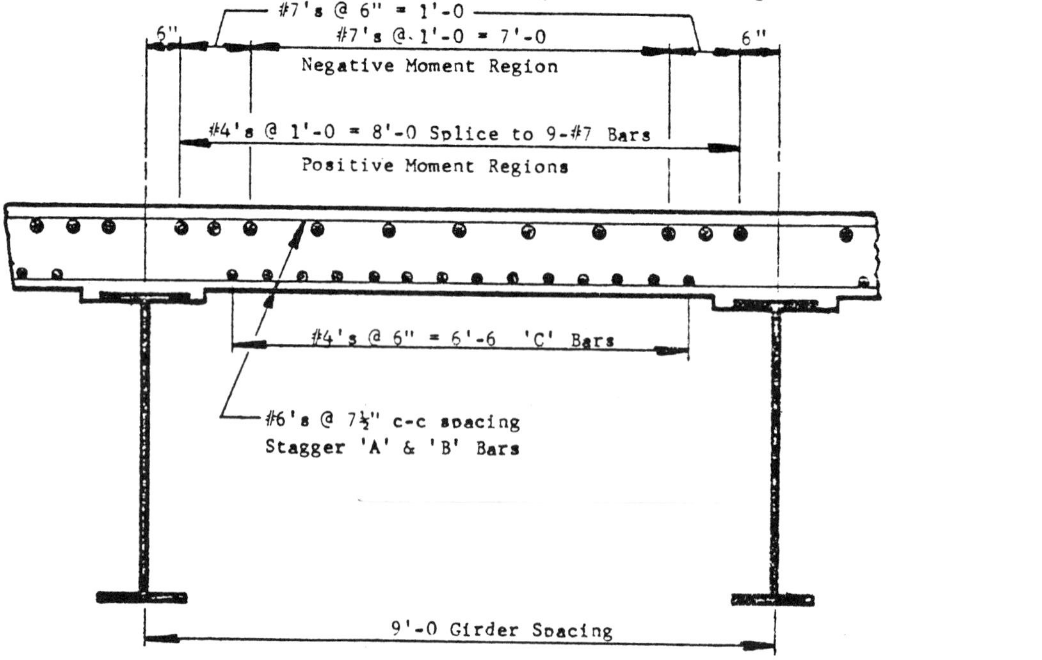


SECTION THRU SLAB

Shear Connector Design

Shear connectors are welded in sets of three to the top flange in either the shop or field. The shear capacity of 3-4 x 7/8"$\phi$connectors is 23.8 kips. The maximum shear connector spacing by AASHTO specifications is 2 feet and they are terminated nearest to the continuous support where positive moment for combined dead load and positive live loading exist.

The following shear connector spacings at tenth points are given on the computer output for Span 1:

The shear connector termination point in Span 1 is located between the 7th and 8th tenth points as follows:

| Tenth Point | Output Spacing | Actual Spacing |
|---|---|---|
| 0.0 | 22.1" | 21" |
| 0.1 | 25.9" | 24" |
| 0.2 | 28.7" | 24" |
| 0.3 | 29.4" | 24" |
| 0.4 | 29.2" | 24" |
| 0.5 | 28.9" | 24" |
| 0.6 | 28.5" | 24" |
| 0.7 | 26.6" | 24" |
| 0.8 | 23.1" | 21" |

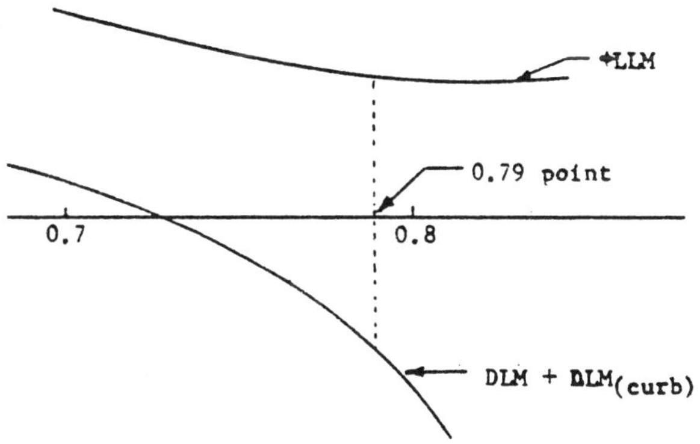


By scale, the ordinates are equal to the 0.79 point of Span 1 which is approximately 31 feet from the pier. For detailing purposes, the connectors are terminated at an even spacing beyond the 0.79 point.

The following formulas are used to determine the shear connector requirements for ultimate strength. The number of connectors ($N_1$) required from the points of maximum positive moment and adjacent end supports are as follows:

$$N_1 = \frac{P}{\phi Su}$$ where P is the smaller

of $P_1$ or $P_2$, $\phi$ = 0.85, and Su is the ultimate strength capacity of the shear connector.

$P_1 = A_s Fy$, As = 70 in$^2$

$P_1$ = (70)(50,000) = 3,500,000#

$P_2 = 0.85\ f'_c\ b\ c$

= (0.85)(4000)(90)(7.5) = 2,295,000#

$P = P_2$ = 2,295,000#

$Su = 0.4\ (d)^2\ \sqrt{f'c\ Ec}$ for studs

$= (0.4)(7/8)^2\ \sqrt{(4000)(3,834,000)}$ = 37,926#

$$N_1 = \frac{2,295,000}{(0.85)(37,926)} = 71.2 \text{ studs}$$

The actual number based on fatigue and horizontal shear from the 0.0 to 0.4 point of span 1 is

from 0.0 to 0.1 @ 21" spacing 9 sets
0.1 to 0.4 @ 24" spacing 22 sets
Total provided = (31)(3 per set) = 93 studs o.k.

Additional shear connectors to develop slab stress at the point of contra-flexure are required by AASHTO specifications when reinforcement steel embedded in the concrete is not used in computing section properties for negative moments. The additional sets are computed from the following formula:

$Nc = A_r^s\ (fr/Zr)$

$A_r^s$ = 9.35 in$^2$

Nc = (9.35)(10,000) / 23,800 = 3.93 sets

Use four extra sets of three studs, place two sets on either side of the field splice.

# E.9 COMPOSITE A36 STEEL I-BEAM BRIDGES

## DESIGN GRAPHS

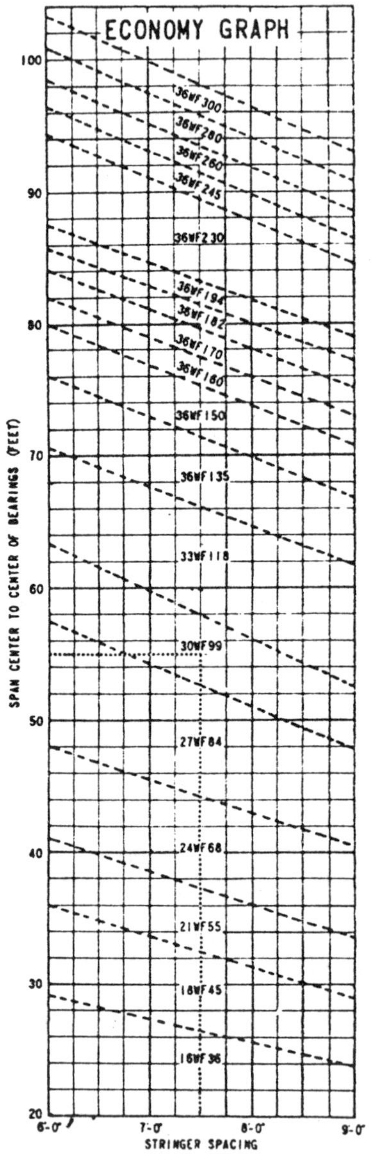


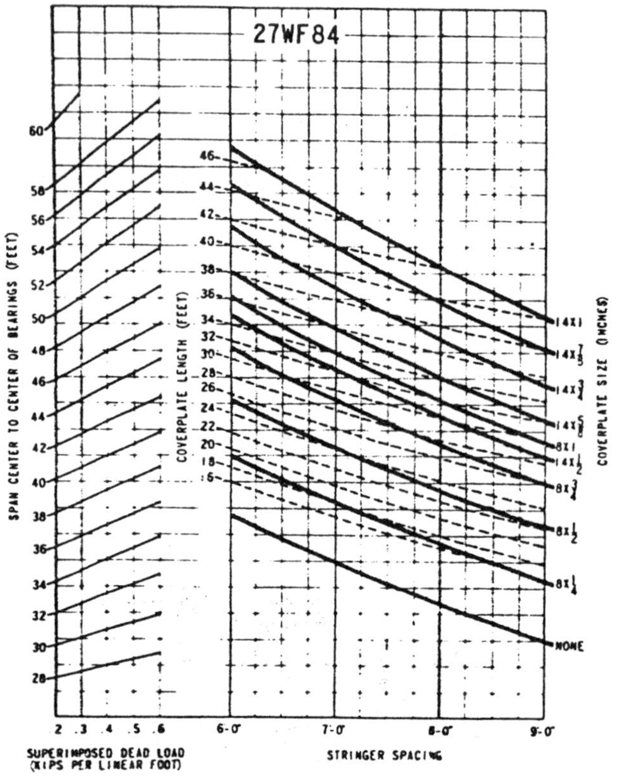


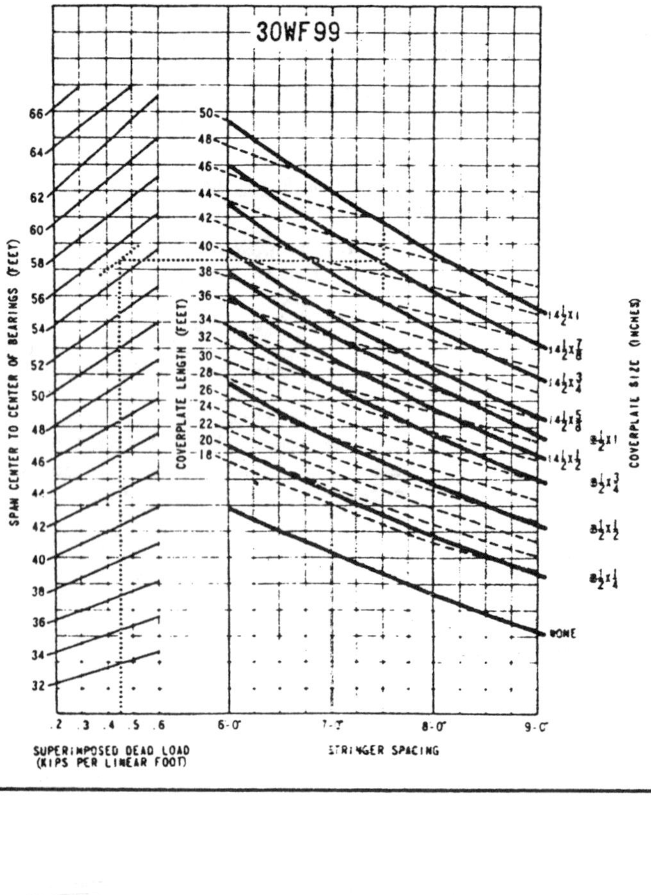


1. THE FOLLOWING INFORMATION IS KNOWN:
   A. SPAN CENTER TO CENTER OF BEARINGS=55 FEET.
   B. STRINGER SPACING=7'-6".
   C. SUPERIMPOSED DEAD LOAD=0.45 KIPS PER LINEAR FOOT OF BEAM.
2. REFERRING TO THE ECONOMY GRAPH, IT IS FOUND THAT THE 30WF99 GRAPH WILL PROBABLY YIELD THE MOST ECONOMICAL DESIGN.
3. ENTER THE 30WF99 GRAPH AT 0.45 KIPS PER LINEAR FOOT OF BEAM AND MOVE VERTICALLY TO THE 55-FOOT SPAN. MOVING FROM THIS POINT HORIZONTALLY TO THE 7'-6" STRINGER SPACING, IT IS SEEN THAT THE COVERPLATE IS $14\frac{1}{2}$" x $\frac{7}{8}$" x 43'-0" LONG. NOTE THAT IF THE SPAN HAD BEEN 48 FEET THE DESIGN WOULD HAVE BEEN IN THE SHADED UNECONOMICAL PORTION OF THE GRAPH, INDICATING THAT A LIGHTER BEAM AND COVERPLATE COULD HAVE BEEN USED.

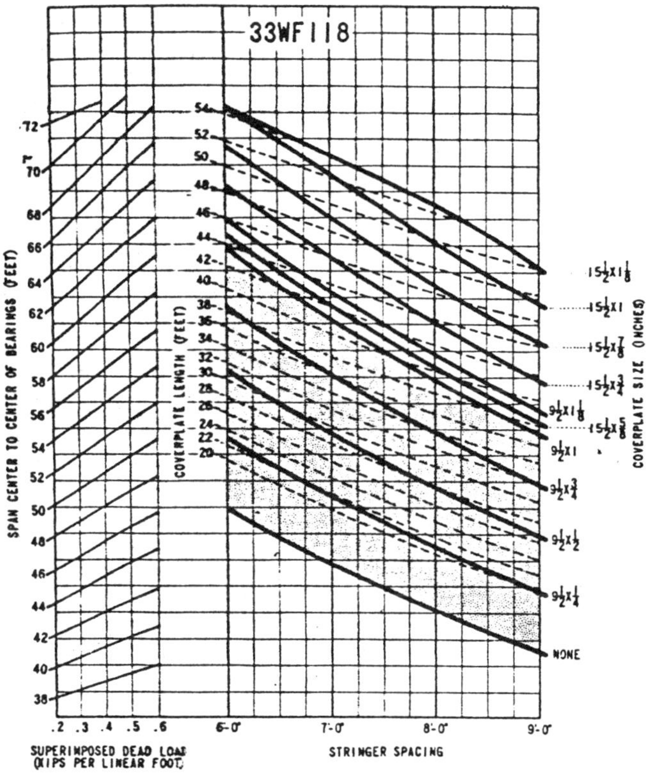


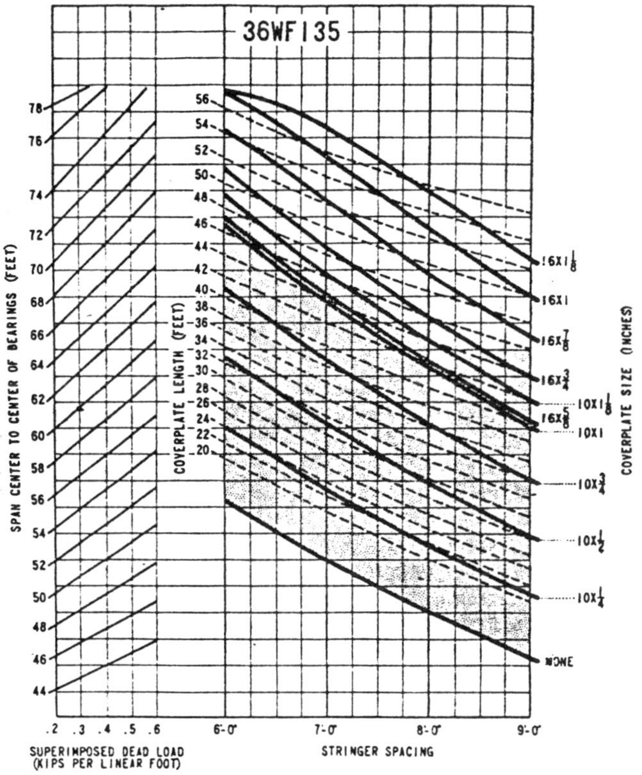


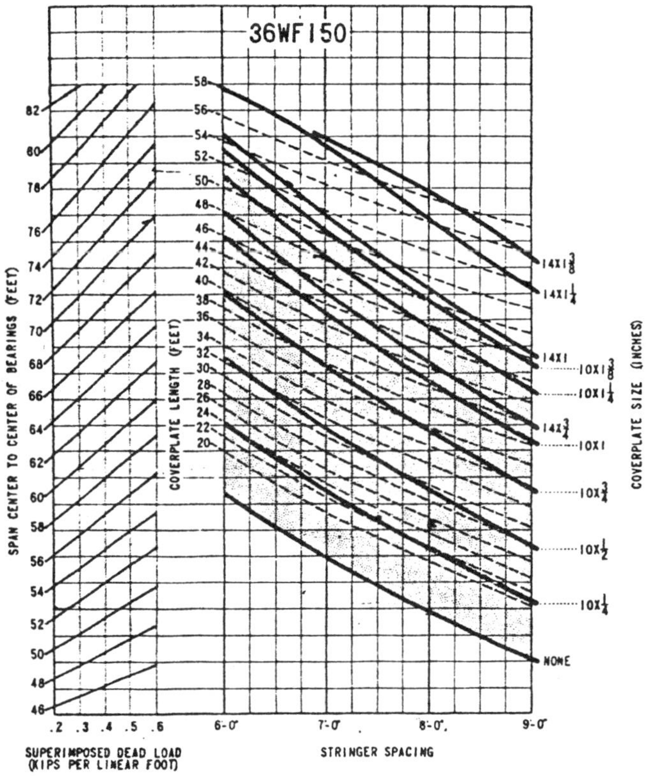


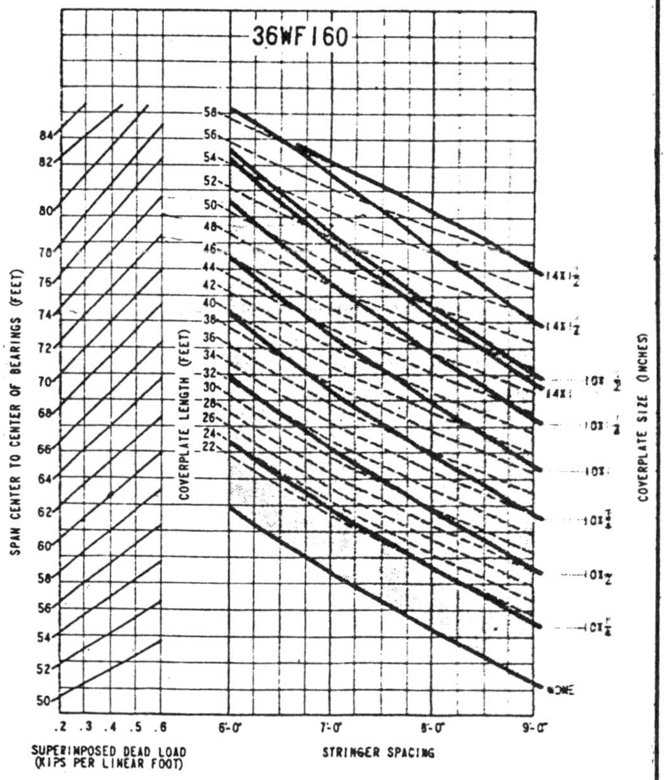


- LIVE LOAD: HS 20-44.
- SUPERIMPOSED DEAD LOAD INCLUDES: FUTURE WEARING SURFACE ON THE DECK SLAB, WEIGHT OF RAILINGS, PARAPETS, CURBS OR SIDEWALKS AND ANY OTHER DEAD LOAD APPLIED AFTER DECK SLAB HAS BEEN CURED, DISTRIBUTED EQUALLY TO ALL STRINGERS IN KIPS PER LINEAL FOOT.
- DESIGN SPECIFICATIONS: DESIGN DIVISION OF 1959 A.A.S.H.O., "STANDARD SPECIFICATIONS FOR HIGHWAY BRIDGES."

NOTE: SHADED AREAS INDICATE THAT LIGHTER BEAM AND COVERPLATE COMBINATIONS ARE AVAILABLE ELSEWHERE IN THE CHARTS.

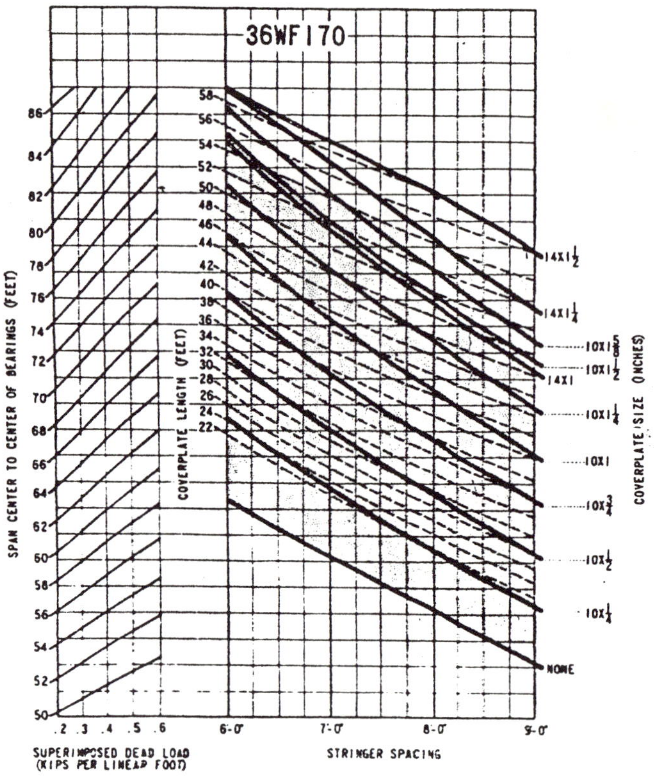
36WF170
SPAN CENTER TO CENTER OF BEARINGS (FEET)
COVERPLATE LENGTH (FEET)
COVERPLATE SIZE (INCHES)
14X1½
14X1¼
10X1⅝
10X1½
14X1
10X1¼
10X1
10X¾
10X½
10X¼
NONE
SUPERIMPOSED DEAD LOAD (KIPS PER LINEAR FOOT)
STRINGER SPACING

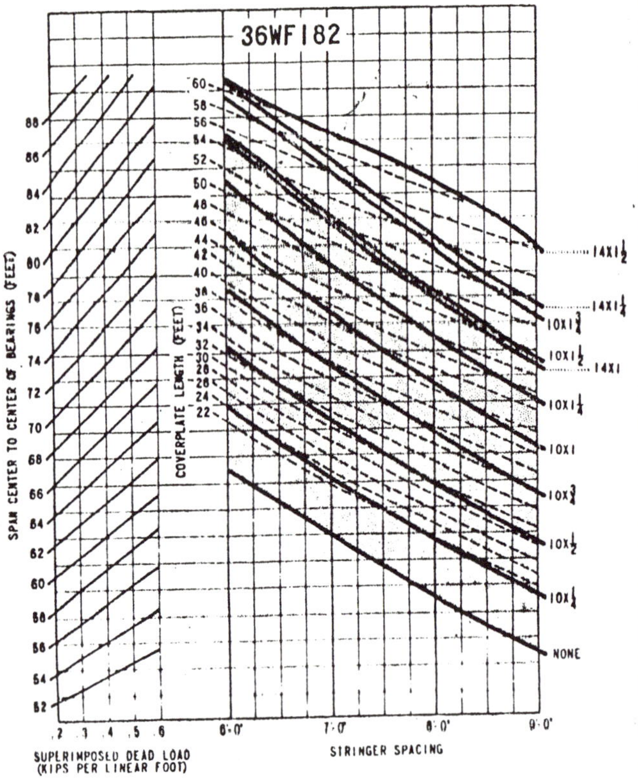
36WF182
SPAN CENTER TO CENTER OF BEARINGS (FEET)
COVERPLATE LENGTH (FEET)
14X1½
14X1¼
10X1¾
10X1½
14X1
10X1¼
10X1
10X¾
10X½
10X¼
NONE
SUPERIMPOSED DEAD LOAD (KIPS PER LINEAR FOOT)
STRINGER SPACING

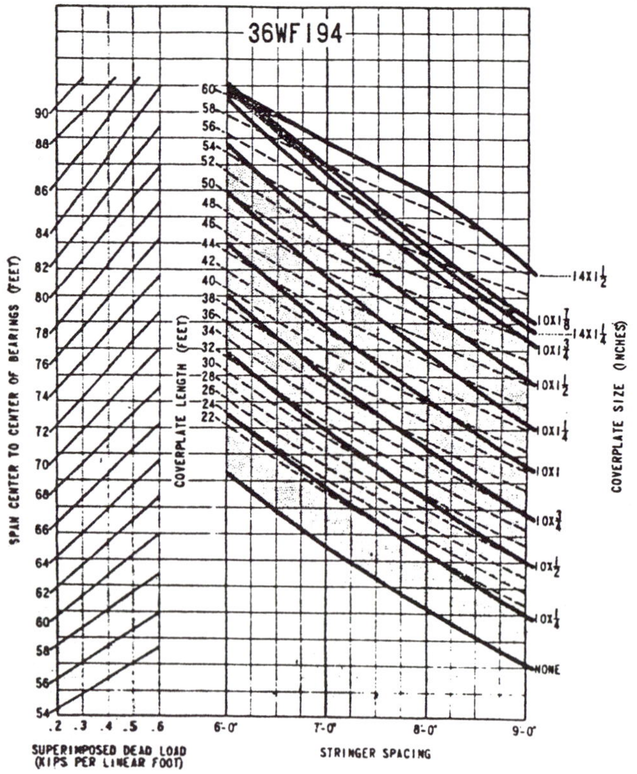
36WF194
SPAN CENTER TO CENTER OF BEARINGS (FEET)
COVERPLATE LENGTH (FEET)
COVERPLATE SIZE (INCHES)
14X1½
10X1⅞
14X1¼
10X1¾
10X1½
10X1¼
10X1
10X¾
10X½
10X¼
NONE
SUPERIMPOSED DEAD LOAD (KIPS PER LINEAR FOOT)
STRINGER SPACING

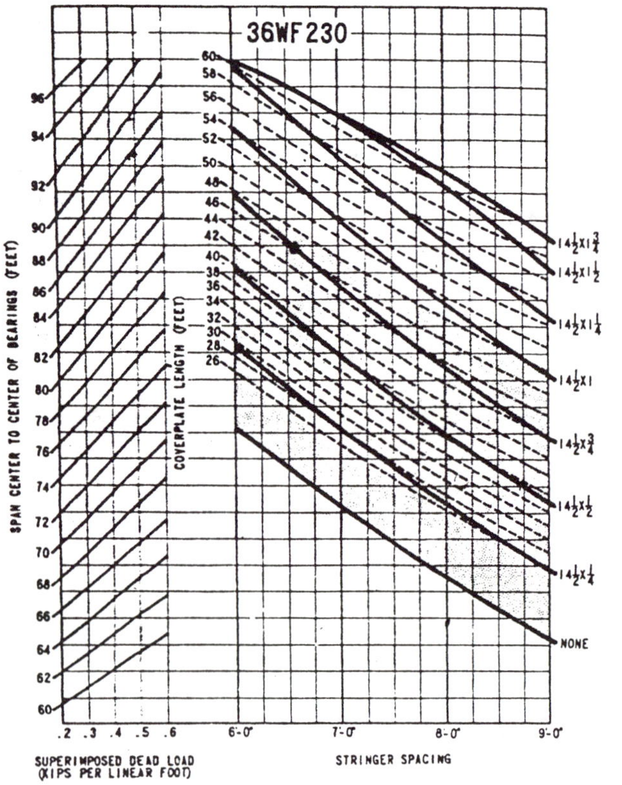
36WF230
SPAN CENTER TO CENTER OF BEARINGS (FEET)
COVERPLATE LENGTH (FEET)
14½X1¾
14½X1½
14½X1¼
14½X1
14½X¾
14½X½
14½X¼
NONE
SUPERIMPOSED DEAD LOAD (KIPS PER LINEAR FOOT)
STRINGER SPACING

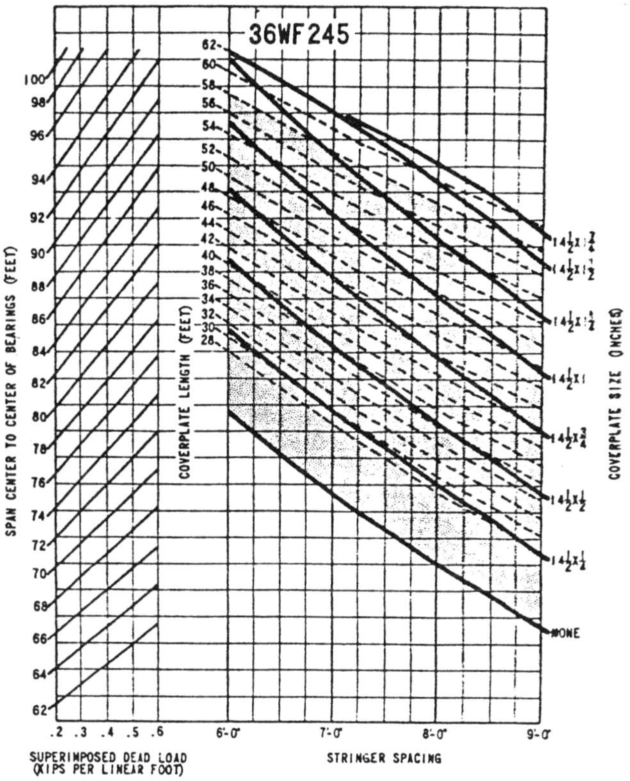


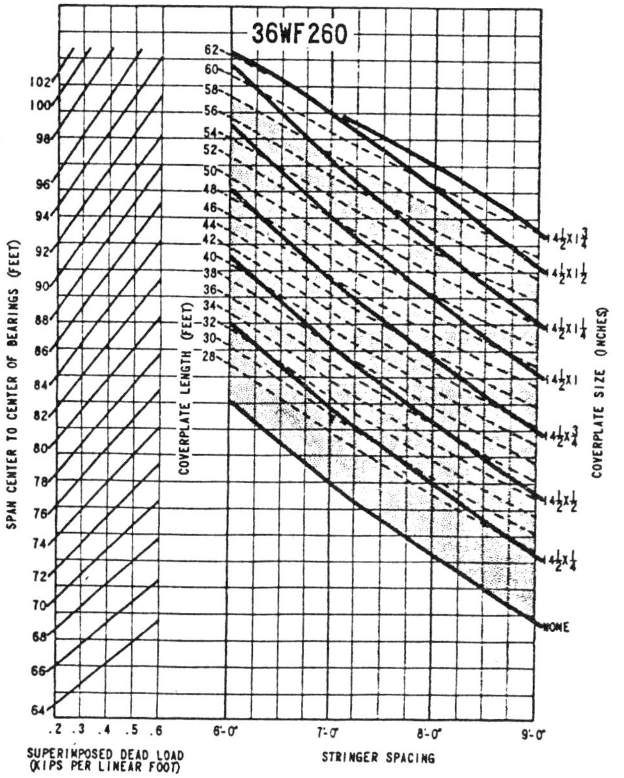


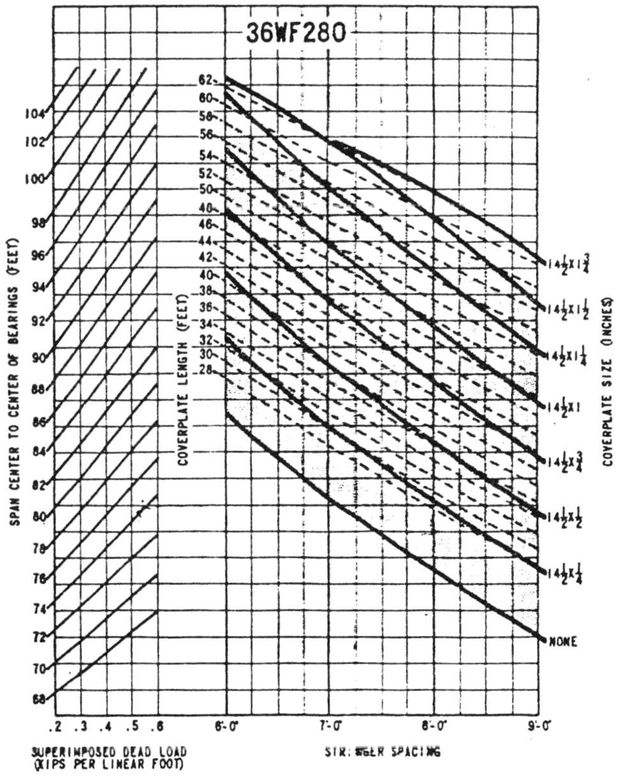


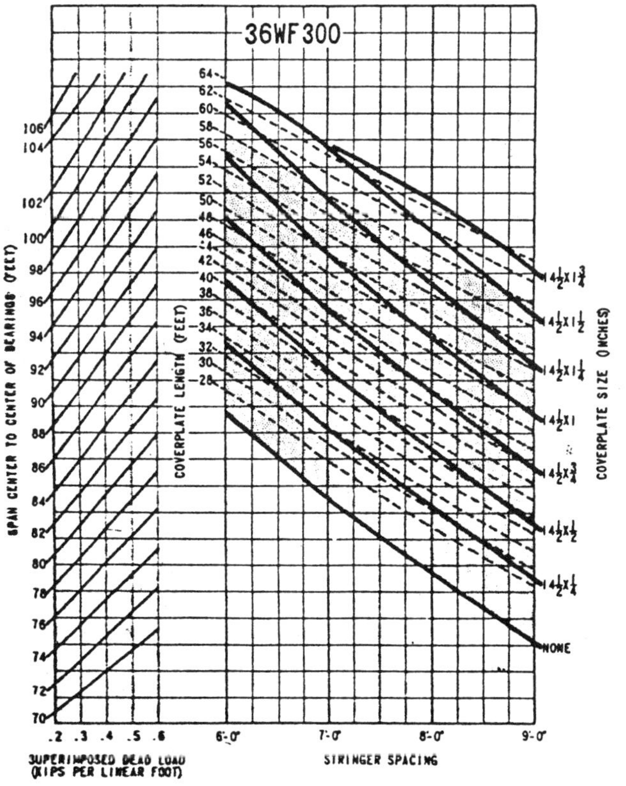


Commonwealth of Pennsylvania
DEPARTMENT OF TRANSPORTATION

# E.10 SIMPLE SPAN STEEL I-BEAM (A572) BRIDGE

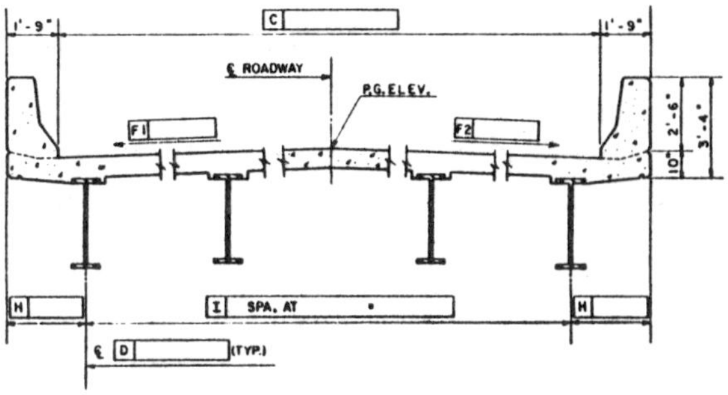


TYPICAL SECTION
LOOKING AHEAD

## GENERAL NOTES:

- MATERIALS AND WORKMANSHIP SHALL BE IN ACCORDANCE WITH PENNSYLVANIA DEPARTMENT OF TRANSPORTATION SPECIFICATION FORMS 408/76 AND 409/73 AND THE CONTRACT SPECIAL PROVISIONS.
- STRUCTURAL STEEL SHALL CONFORM TO ASTM A572 DESIGNATION EXCEPT WHERE NOTED OTHERWISE.
- DESIGN SPECIFICATIONS:
  DESIGN DIVISION OF 1977 AASHTO "STANDARD SPECIFICATIONS FOR HIGHWAY BRIDGES" (INCLUDING INTERIMS THROUGH 1981), AND AS SUPPLEMENTED BY THE PENNA. DEPT. OF TRANSPORTATION DESIGN MANUAL, PART 4, STRUCTURES (INCLUDING SEPT. 1976 REVISION).
- LIVE LOADS:
  HS20-44 LOADING AND ALTERNATE LOADING (2 AXLES OF 24 KIPS EACH AT 4'-0" C/C, 2 WHEELS OF 12 KIPS EACH AT 6'-0" C/C).
- DEAD LOADS:
  INCLUDES 30 LBS. PER SQ. FT. FOR FUTURE WEARING SURFACE ON THE DECK SLAB.
  INCLUDES 15 LBS. PER SQ. FT. OF DECK AREA FORMED USING PERMANENT METAL DECK FORMS.
- 2" CONCRETE COVER SHALL BE PROVIDED ON REINFORCEMENT BARS UNLESS OTHERWISE NOTED.
- CLASS AA CEMENT CONCRETE SHALL BE USED IN DECK SLAB, CURBS, PARAPETS AND CONCRETE DIAPHRAGMS.
- CLASS A CEMENT CONCRETE SHALL BE USED IN ABUTMENTS, WINGWALLS AND FOOTINGS.
- EXPOSED CONCRETE EDGES SHALL BE CHAMFERED 1" x 1" EXCEPT AS NOTED.
- BEARING AREAS SHALL BE PREPARED AS SPECIFIED IN SECTION 1054.04(a) OF FORM 409
- ALL REINFORCEMENT STEEL SHALL MEET THE REQUIREMENTS OF AASHTO DESIGNATION M-31, GRADE 60.
- MINIMUM LAP AND EMBEDMENT LENGTH SHALL BE 30 DIAMETERS UNLESS OTHERWISE NOTED.
- ALL BOLTS ARE 7/8" ⌀ DIAMETER, HIGH STRENGTH, EXCEPT AS NOTED.
- ALL WELDING SHALL CONFORM TO THE REQUIREMENTS OF THE AASHTO "STANDARD SPECIFICATIONS FOR WELDING OF STRUCTURAL STEEL HIGHWAY BRIDGES", 1981 PLUS INTERIMS.
- ALL DIMENSIONS ARE HORIZONTAL.
- SUPERSTRUCTURE DIMENSIONS ARE BASED ON A NOMINAL TEMPERATURE OF 68° F.
- THE FOOTINGS FOR ABUTMENTS MAY BE ORDERED BY THE ENGINEER TO BE AT ANY ELEVATION OR ANY DIMENSIONS NECESSARY TO PROVIDE A PROPER FOUNDATION.
- THE CONTRACTOR SHALL NOTIFY ALL INVOLVED UTILITY COMPANIES PRIOR TO STARTING WORK. UTILITIES WILL NOT BE DISTURBED OR ENDANGERED DURING CONSTRUCTION OPERATIONS.
- SHOP DRAWINGS SHALL CONFORM TO SPECIFICATION FORM 409/73, AND SHALL BE SUBMITTED TO AND APPROVED BY THE ENGINEER PRIOR TO BEAM FABRICATION.
- THE DECK DRAINS SHOWN ON THESE DRAWINGS [ ] REQUIRED.
- SUPERSTRUCTURE REINFORCING BARS [ ] BE EPOXY COATED.
- PRECAST PARAPETS AS DETAILED ON STANDARD DRAWING BC-309 MAY BE USED IN LIEU OF CAST-IN-PLACE PARAPETS SHOWN ON THESE DRAWINGS.

### TABLE 5
### BEAM SECTIONS (A572 STEEL)

| ROADWAY WIDTH-C-(FEET) | | SPAN LENGTH-A-(FEET) 30 | 35 | 40 | 45 | 50 | 55 | 60 | 65 | 70 | 75 | 80 | 85 | 90 |
|---|---|---|---|---|---|---|---|---|---|---|---|---|---|---|
| NON-COMPOSITE DESIGN | 24 | W27x84 | W27x94 | W30x108 | W33x130 | W36x135 | W36x160 | W36x194 | W36x230 | W36x260 | W36x280 | X | X | X |
| | 28 | | W27x84 | W30x99 | W33x118 | W33x130 | W36x150 | W36x160 | W36x182 | W36x210 | W36x245 | X | X | X |
| | 32 | | W27x94 | W30x108 | | W36x135 | | W36x170 | W36x210 | W36x230 | | X | X | X |
| | 36 | | | | | | | | | | W36x260 | X | X | X |
| | 40 | | | | W33x130 | | W36x160 | W36x182 | | | | X | X | X |
| | 44 | | | | W33x118 | | W36x150 | W36x170 | W36x194 | | W36x230 | X | X | X |
| | 48 | | | | | | | W36x182 | W36x210 | | W36x260 | X | X | X |
| COMPOSITE DESIGN | 24 | X | X | X | X | W30x116 | W33x130 | W36x150 | W36x160 | W36x182 | W36x210 | W36x230 | W36x260 | W36x280 |
| | 28 | X | X | X | X | x108 | W30x124 | x135 | x150 | x170 | W36x194 | W36x210 | x230 | x260 |
| | 32 | X | X | X | X | x116 | W33x130 | x150 | x160 | x182 | W33x221 | W36x230 | x260 | x280 |
| | 36 | X | X | X | X | x116 | x118 | x135 | x150 | x170 | W36x194 | W33x221 | x230 | x260 |
| | 40 | X | X | X | X | x124 | x130 | x150 | x170 | x194 | W36x210 | W36x230 | x260 | x280 |
| | 44 | X | X | X | X | x116 | x130 | x135 | x160 | x170 | W36x194 | | x245 | x280 |
| | 48 | X | X | X | X | x124 | x130 | x150 | x160 | x182 | W36x210 | | x260 | x280 |

### TABLE 6
### SPACING FOR FRAMING & SLAB REINFORCEMENT

| ROADWAY WIDTH (FEET) | DECK WIDTH (FEET) | SLAB OVERHANG (FEET) | BEAM SPACING-NORMAL NUMBER | SPACING (FEET) | TOTAL (FEET) | SLAB TOP BAR SPA. | SLAB BOT. BAR SPA. |
|---|---|---|---|---|---|---|---|
| C | G | H | I1 | I2 | I3 | J1 | J2 |
| 24 | 27'-6" | 3'-3" | 3 SPA. | 7'-0" | 21'-0" | 7 EQ. SPA. | 8 EQ. SPA. |
| 28 | 31'-6" | 2'-9" | 4 | 6'-6" | 26'-0" | 7 | 8 |
| 32 | 35'-6" | 3'-3" | 4 | 7'-3" | 29'-0" | 8 | 9 |
| 36 | 39'-6" | 2'-10 1/2" | 5 | 6'-9" | 33'-9" | 7 | 8 |
| 40 | 43'-6" | 3'-0" | 5 | 7'-6" | 37'-6" | 8 | 9 |
| 44 | 47'-6" | 2'-9" | 6 | 7'-0" | 42'-0" | 7 | 8 |
| 48 | 51'-6" | 3'-3" | 6 | 7'-6" | 45'-0" | 8 | 9 |

### TABLE 7
### STEEL BEAM DIMENSIONS

| BEAM SIZE | BEAM DEPTH (INCHES) | FLANGE WIDTH (INCHES) | BEAM SIZE | BEAM DEPTH (INCHES) | FLANGE WIDTH (INCHES) |
|---|---|---|---|---|---|
| D | K1 | K2 | D | K1 | K2 |
| W27x84 | 26.71 | 9.960 | W36x150 | 35.85 | 11.975 |
| W27x94 | 26.92 | 9.990 | x160 | 36.01 | 12.000 |
| W30x99 | 29.65 | 10.450 | x170 | 36.17 | 12.030 |
| W30x108 | 29.83 | 10.475 | x182 | 36.33 | 12.075 |
| W30x116 | 30.01 | 10.495 | x194 | 36.49 | 12.115 |
| W30x124 | 30.17 | 10.515 | x210 | 36.69 | 12.180 |
| W33x118 | 32.86 | 11.480 | x230 | 35.90 | 16.470 |
| W33x130 | 33.09 | 11.510 | x245 | 36.08 | 16.510 |
| W33x221 | 33.93 | 15.805 | x260 | 36.26 | 16.550 |
| W36x135 | 35.55 | 11.950 | x280 | 36.52 | 16.595 |

### TABLE 9
### BEARING COMPONENT SIZES
### (FIXED & EXPANSION)

| BEAM SECTION | NEOPRENE PAD LENGTH (IN.) | SOLE PLATE THICKNESS (IN.) |
|---|---|---|
| D | M1 | M2 |
| W27x84 | 15 | 1 1/4 |
| x94 | | |
| W30x99 | 16 | |
| x108 | | |
| x116 | 17 | 1 1/2 |
| W33x118 | 18 | |
| W30x124 | 17 | 1 1/4 |
| W33x130 | 18 | 1 1/2 |
| W36x135 | | |
| x150 | | |
| x160 | 20 | |
| x170 | | |
| x182 | 21 | |
| x194 | | |
| x210 | 22 | |
| W33x221 | 24 | 1 3/4 |
| W36x230 | 25 | |
| x245 | | |
| x260 | | |
| x280 | | |

### TABLE 8
### CAMBER ORDINATES

NON-COMPOSITE DESIGN

| SPAN FEET | BEAM SIZE | CAMBER VALUES (INCHES) STEEL D.L. | CONC. D.L. | TOTAL |
|---|---|---|---|---|
| A | D | L1 | L2 | L3 |
| 30 | W27x84 | 0.023 | 0.234 | 0.257 |
| 35 | x84 | 0.040 | 0.392 | 0.432 |
| | x94 | 0.039 | 0.375 | 0.414 |
| 40 | W30x99 | 0.056 | 0.480 | 0.536 |
| | x108 | 0.054 | 0.471 | 0.525 |
| 45 | W33x118 | 0.071 | 0.565 | 0.636 |
| | x130 | 0.068 | 0.500 | 0.568 |
| 50 | x130 | 0.101 | 0.699 | 0.800 |
| | W36x135 | 0.091 | 0.662 | 0.753 |
| 55 | x150 | 0.130 | 0.816 | 0.946 |
| | x160 | 0.127 | 0.762 | 0.889 |
| 60 | x160 | 0.178 | 1.002 | 1.180 |
| | x170 | 0.176 | 1.005 | 1.181 |
| | x182 | 0.175 | 0.916 | 1.091 |
| | x194 | 0.172 | 0.867 | 1.039 |
| 65 | x182 | 0.237 | 1.201 | 1.438 |
| | x194 | 0.236 | 1.102 | 1.338 |
| | x210 | 0.234 | 1.083 | 1.317 |
| | x230 | 0.224 | 0.957 | 1.181 |
| 70 | x210 | 0.311 | 1.415 | 1.726 |
| | x230 | 0.301 | 1.319 | 1.620 |
| | x260 | 0.292 | 1.139 | 1.431 |
| 75 | x230 | 0.394 | 1.599 | 1.993 |
| | x245 | 0.390 | 1.609 | 1.999 |
| | x260 | 0.385 | 1.465 | 1.850 |
| | x280 | 0.378 | 1.374 | 1.752 |

COMPOSITE DESIGN

| SPAN FEET | BEAM SIZE | CAMBER VALUES (INCHES) STEEL D.L. | CONC. D.L. | TOTAL |
|---|---|---|---|---|
| A | D | L1 | L2 | L3 |
| 50 | W30x108 | 0.128 | 0.911 | 1.039 |
| | x116 | 0.126 | 0.907 | 1.033 |
| | x124 | 0.123 | 0.833 | 0.956 |
| 55 | W33x118 | 0.159 | 1.030 | 1.189 |
| | W30x124 | 0.182 | 1.116 | 1.298 |
| | W33x130 | 0.154 | 0.994 | 1.148 |
| 60 | W36x135 | 0.191 | 1.124 | 1.315 |
| | x150 | 0.183 | 1.050 | 1.233 |
| 65 | x150 | 0.248 | 1.320 | 1.568 |
| | x160 | 0.246 | 1.327 | 1.573 |
| | x170 | 0.242 | 1.236 | 1.478 |
| 70 | x170 | 0.322 | 1.561 | 1.883 |
| | x182 | 0.320 | 1.545 | 1.865 |
| | x194 | 0.318 | 1.447 | 1.765 |
| 75 | x194 | 0.415 | 1.794 | 2.209 |
| | x210 | 0.412 | 1.776 | 2.188 |
| | W33x221 | 0.445 | 1.822 | 2.267 |
| 80 | W36x210 | 0.539 | 2.073 | 2.612 |
| | W33x221 | 0.585 | 2.163 | 2.748 |
| | W36x230 | 0.521 | 2.025 | 2.546 |
| 85 | x230 | 0.658 | 2.361 | 3.019 |
| | x245 | 0.651 | 2.237 | 2.888 |
| | x260 | 0.644 | 2.248 | 2.892 |
| 90 | x260 | 0.803 | 2.586 | 3.389 |
| | x280 | 0.793 | 2.594 | 3.387 |

NOTE: TURN MILL CAMBER UP FOR TOTAL CAMBER VALUES LESS THAN 3/4".

### TABLE 10
### SHEAR CONNECTOR SPACING (W1) (1)

| ROADWAY WIDTH | SPAN LENGTH - FEET 50 | 55 | 60 | 65 | 70 | 75 | 80 | 85 | 90 |
|---|---|---|---|---|---|---|---|---|---|
| 24' | 11" | 12" | 13" | 13" | 13 1/2" | 14" | 14 1/2" | 14 1/2" | 15" |
| 28' | 11 1/2" | 11 1/2" | 13 1/2" | 14" | 14" | 14 1/2" | 15" | 15 1/2" | 16" |
| 32' | 10 1/2" | 11 1/2" | 12 1/2" | 13" | 13" | 13" | 14" | 14 1/2" | 14 1/2" |
| 36' | 11 1/2" | 12" | 13" | 13 1/2" | 14" | 14" | 14" | 15 | 15 1/2" |
| 40' | 10 1/2" | 11 1/2" | 12 1/2" | 12 1/2" | 13" | 13" | 13 1/2" | 14 | 14 1/2" |
| 44' | 11" | 12" | 12 1/2" | 13" | 13 1/2" | 14" | 14 1/2" | 14 1/2" | 15 1/2" |
| 48' | 10 1/2" | 11 1/2" | 12 1/2" | 12 1/2" | 13" | 13" | 13 1/2" | 14" | 14 1/2" |

(1) USE ONLY WITH COMPOSITE DESIGN SECTIONS OF TABLE 5.
WHEN NON-COMPOSITE DESIGN SECTIONS ARE SELECTED, SHEAR CONNECTORS ARE NOT REQUIRED.

Commonwealth of Pennsylvania
DEPARTMENT OF TRANSPORTATION
BUREAU OF HIGHWAY DESIGN

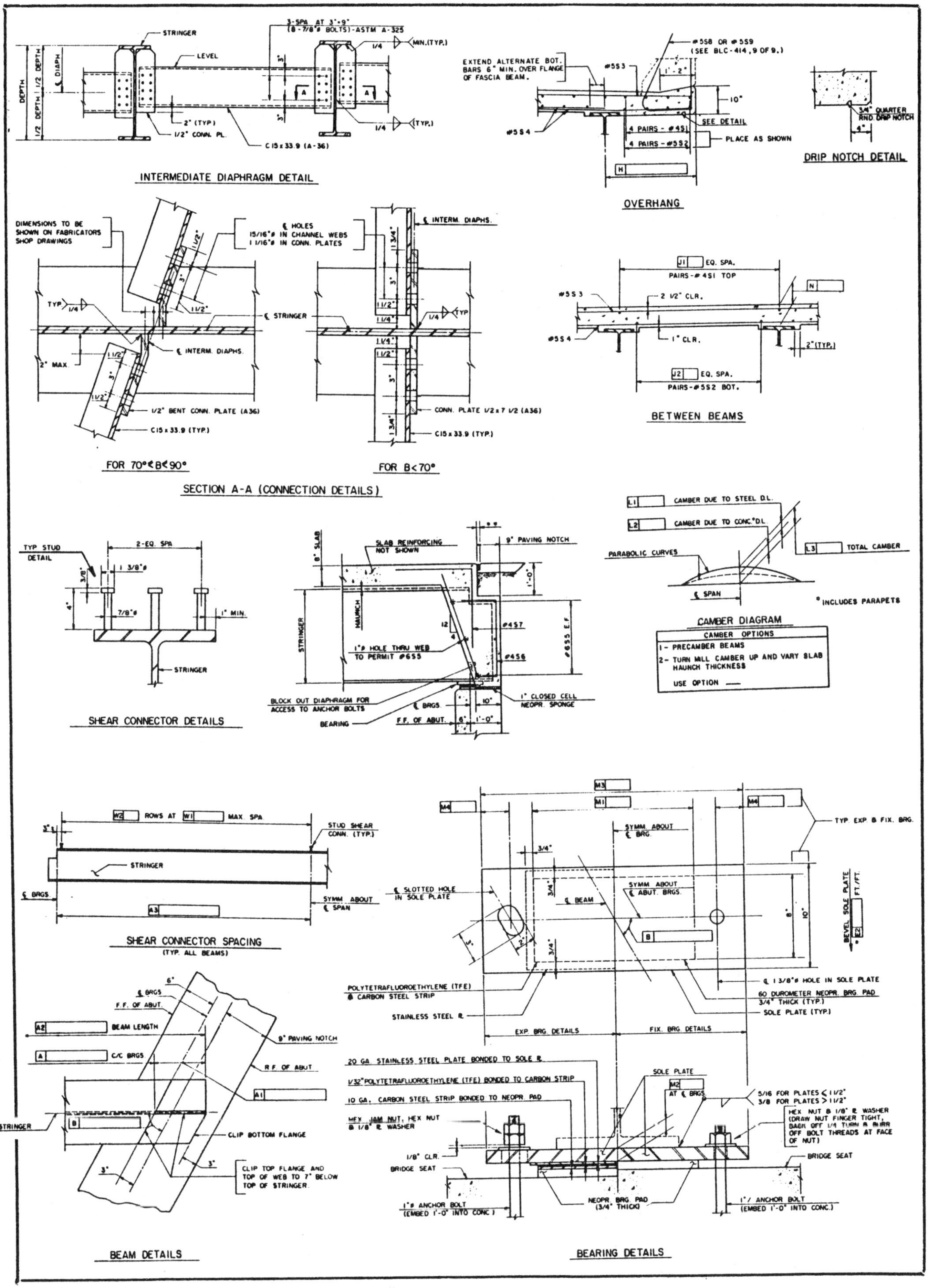

INTERMEDIATE DIAPHRAGM DETAIL
STRINGER
LEVEL
DEPTH
1/2 DEPTH
℄ DIAPH
3-SPA. AT 3"=9"
(8 - 7/8"ø BOLTS)-ASTM A-325
1/4 MIN.(TYP.)
2" (TYP.)
1/2" CONN. PL.
C15 x 33.9 (A-36)
1/4 (TYP.)
EXTEND ALTERNATE BOT. BARS 6" MIN. OVER FLANGE OF FASCIA BEAM.
#5S8 OR #5S9 (SEE BLC-414, 9 OF 9.)
#5S3
1'-2"
10"
#5S4
SEE DETAIL
4 PAIRS - #4S1
4 PAIRS - #5S2
PLACE AS SHOWN
H
OVERHANG
3/4" QUARTER RND. DRIP NOTCH
4"
DRIP NOTCH DETAIL
DIMENSIONS TO BE SHOWN ON FABRICATORS SHOP DRAWINGS
℄ HOLES 15/16"ø IN CHANNEL WEBS 1 1/16"ø IN CONN. PLATES
℄ INTERM. DIAPHS.
℄ STRINGER
2" MAX.
1/2" BENT CONN. PLATE (A36)
C15 x 33.9 (TYP.)
CONN. PLATE 1/2 x 7 1/2 (A36)
FOR 70°≤B≤90°
FOR B<70°
SECTION A-A (CONNECTION DETAILS)
J1 EQ. SPA.
PAIRS-#4S1 TOP
N
#5S3
2 1/2" CLR.
#5S4
1" CLR.
2"(TYP.)
J2 EQ. SPA.
PAIRS-#5S2 BOT.
BETWEEN BEAMS
L1 CAMBER DUE TO STEEL D.L.
L2 CAMBER DUE TO CONC.*D.L.
L3 TOTAL CAMBER
PARABOLIC CURVES
℄ SPAN
*INCLUDES PARAPETS
CAMBER DIAGRAM
CAMBER OPTIONS
1- PRECAMBER BEAMS
2- TURN MILL CAMBER UP AND VARY SLAB HAUNCH THICKNESS
USE OPTION ——
TYP STUD DETAIL
2-EQ. SPA
1 3/8"ø
7/8"ø
1" MIN.
STRINGER
SHEAR CONNECTOR DETAILS
8" SLAB
SLAB REINFORCING NOT SHOWN
9" PAVING NOTCH
1'-0"
HAUNCH
#4S7
#6S5 E.F.
1"ø HOLE THRU WEB TO PERMIT #6S5
#4S6
BLOCK OUT DIAPHRAGM FOR ACCESS TO ANCHOR BOLTS
℄ BRGS.
BEARING
F.F. OF ABUT.
1" CLOSED CELL NEOPR. SPONGE
W2 ROWS AT W1 MAX. SPA
STUD SHEAR CONN. (TYP.)
STRINGER
℄ BRGS.
A3
SYMM. ABOUT ℄ SPAN
SHEAR CONNECTOR SPACING
(TYP. ALL BEAMS)
℄ BRGS
F.F. OF ABUT.
A2 BEAM LENGTH
A C/C BRGS
9" PAVING NOTCH
R.F. OF ABUT
A1
℄ STRINGER
B
CLIP BOTTOM FLANGE
CLIP TOP FLANGE AND TOP OF WEB TO 7" BELOW TOP OF STRINGER
BEAM DETAILS
M3
M1
M4
SYMM. ABOUT ℄ BRG.
TYP. EXP. & FIX. BRG.
℄ SLOTTED HOLE IN SOLE PLATE
℄ BEAM
SYMM. ABOUT ℄ ABUT. BRGS.
BEVEL SOLE PLATE FT./FT.
E2
POLYTETRAFLUOROETHYLENE (TFE) & CARBON STEEL STRIP
STAINLESS STEEL ℞
℄ 1 3/8"ø HOLE IN SOLE PLATE
60 DUROMETER NEOPR. BRG. PAD 3/4" THICK (TYP.)
SOLE PLATE (TYP.)
EXP. BRG. DETAILS
FIX. BRG. DETAILS
20 GA. STAINLESS STEEL PLATE BONDED TO SOLE ℞
1/32" POLYTETRAFLUOROETHYLENE (TFE) BONDED TO CARBON STRIP
10 GA. CARBON STEEL STRIP BONDED TO NEOPR. PAD
HEX. JAM NUT, HEX NUT & 1/8" ℞ WASHER
SOLE PLATE
M2
AT ℄ BRGS.
5/16 FOR PLATES ≤ 1 1/2"
3/8 FOR PLATES > 1 1/2"
HEX NUT & 1/8" ℞ WASHER (DRAW NUT FINGER TIGHT, BACK OFF 1/4 TURN & BURR OFF BOLT THREADS AT FACE OF NUT)
1/8" CLR.
BRIDGE SEAT
NEOPR. BRG. PAD (3/4" THICK)
1"ø ANCHOR BOLT (EMBED 1'-0" INTO CONC.)
BEARING DETAILS

# E.11 TWO AND THREE SPAN CONTINUOUS I-BEAM BRIDGE

TWO SPAN CONTINUOUS STEEL BEAMS - A588 (SPEC. 3309)
COMPOSITE WITH COVER PLATES OVER PIER - LOAD FACTOR DESIGN ($\phi$ = 0.9)
HS 20 44 LOADING - 40'-0" ROADWAY - 100,000 FATIGUE CYCLES

| Span Lengths (Total) (Ft.) | Beam Size Req'd | Cover Plates at Pier | Area "A" (Ft. In.) | Dimensions (Ft.-In.) W | Dimensions (Ft.-In.) X | Shear Conn. Spa. (In.) | Reactions Abut/Pier (Kips) DL Min | DL Max | LL † | ℄ Pier to Field Splice | Dead Load Deflections (Ft.) 1.1 | 1.2 | 1.3 | 1.4 | 1.5 | 1.6 | 1.7 | 1.8 | 1.9 | Wt. Steel % Defl. | Camber Diagram Ordinates (In.) 1.4 | Lt. F.S. | Rt. F.S. |
|---|---|---|---|---|---|---|---|---|---|---|---|---|---|---|---|---|---|---|---|---|---|---|---|
| 30 - 30 (60) | W24x55 | 5 x 3/8 x 7'-0" | 8'-3" | 23'-6" | 13'-0" | 9" | 12<br>51 | 18<br>58 | 41<br>51 | 8'-0" | .007 | .012 | .016 | .018 | .017 | .014 | .010 | .005 | .002 | 5 | – | – | – |
| 35 - 35 (70) | W24x61 | 6x1/2 x 8'-0" | 9'-6" | 27'-3" | 15'-6" | 9" | 14<br>60 | 20<br>68 | 44<br>54 | 10'-0" | .010 | .020 | .026 | .028 | .027 | .022 | .016 | .008 | .002 | 5 | – | – | – |
| 40 - 40 (80) | W24x76 | 8x3/8 x 9'-0" | 10'-9" | 30'-6" | 19'-0" | 12" | 17<br>69 | 24<br>78 | 46<br>56 | 11'-0" | .014 | .025 | .033 | .036 | .035 | .029 | .021 | .011 | .003 | 6 | – | – | – |
| 45 - 45 (90) | W24x84 | 8 x 5/8 x 12'-0" | 12'-9" | 34'-6" | 21'-0" | 12" | 19<br>79 | 26<br>89 | 48<br>57 | 13'-0" | .019 | .035 | .045 | .049 | .047 | .039 | .027 | .014 | .004 | 7 | – | – | – |
| 50 - 50 (100) | W27x94 | 8 x 5/8 x 13'-0" | 14'-0" | 38'-3" | 23'-6" | 13" | 21<br>88 | 29<br>99 | 49<br>58 | 14'-0" | .021 | .039 | .051 | .055 | .053 | .044 | .030 | .016 | .004 | 8 | – | – | – |
| 55 - 55 (110) | W30x99 | 9½ x 5/8 x 14'-0" | 15'-6" | 41'-0" | 28'-0" | 14" | 24<br>98 | 32<br>110 | 50<br>59 | 16'-0" | .025 | .046 | .060 | .066 | .062 | .051 | .035 | .018 | .005 | 8 | 1-7/8" | 1-3/8" | 1/2" |
| 60 - 60 (120) | W30x116 | 9½x3/4 x 15'-0" | 16'-9" | 44'-9" | 30'-6" | 13" | 26<br>108 | 35<br>121 | 51<br>63 | 17'-0" | .029 | .054 | .070 | .077 | .073 | .060 | .042 | .022 | .006 | 9 | 2" | 1-7-7/16" | 9/16" |
| 65 - 65 (130) | W33x130 | 10 x 3/4 x 16'-0" | 18'-0" | 49'-1" | 31'-10" | 13" | 29<br>119 | 39<br>132 | 52<br>66 | 18'-0" | .030 | .056 | .073 | .080 | .076 | .063 | .044 | .023 | .007 | 10 | 2-1/16" | 1-7/16" | 5/8" |
| 70 - 70 (140) | W36x135 | 10 x 7/8 x 17'-0" | 19'-9" | 51'-10" | 36'-4" | 16" | 32<br>128 | 42<br>143 | 53<br>70 | 20'-0" | .035 | .063 | .083 | .090 | .086 | .071 | .049 | .026 | .008 | 11 | 2-3/16" | 1-1/2" | 11/16" |
| 75 - 75 (150) | W36x150 | 11 x 1 x 18'-0" | 21'-3" | 55'-10" | 38'-4" | 16" | 35<br>139 | 45<br>155 | 54<br>73 | 21'-0" | .040 | .073 | .095 | .103 | .098 | .081 | .056 | .029 | .008 | 12 | 2-3/8" | 1-9/16" | 3/4" |
| 80 - 80 (160) | W36x170 | 11 x 1-1/8 x 19'-0" | 22'-6" | 59'-10" | 40'-4" | 16" | 38<br>151 | 48<br>168 | 54<br>77 | 23'-0" | .045 | .082 | .106 | .116 | .110 | .091 | .063 | .033 | .010 | 13 | 2-5/8" | 1-3/4" | 7/8" |
| 85 - 85 (170) | W36x194 | 11 x 1-1/4 x 20'-0" | 24'-0" | 62'-6" | 45'-0" | 15" | 40<br>162 | 52<br>180 | 55<br>80 | 24'-0" | .050 | .091 | .119 | .129 | .123 | .101 | .070 | .036 | .011 | 15 | 2-13/16" | 1-13/16" | 15/16" |

† Does not include impact

All shear connectors shall be 7/8" $\phi$ x 5" studs at constant spacing.

Dead Load deflections include the weight of slab, wearing course and railing. Beam deflections are the indicated % of the deflections shown.

Camber diagram ordinates are given for span over 50 ft. without any vertical curve. (Modify values for vertical curve). Ordinates are based on 1" residual camber at point of maximum D.L. deflection.

The cover plates are centered on centerline of pier.

All bearing stiffeners shall be 5/8" in thickness. Their width shall be 7" except on the outside of the facia beams where they shall end 1/2" ± inside flange.

Area "A" is the portion of top beam flange which is in tension due to dead load. F.S. means Field Splice.

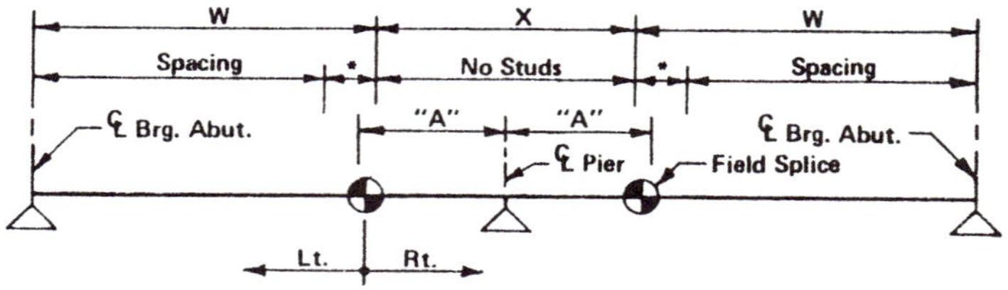


4-7/8" $\phi$ studs/row (12" Flgs.)
3-7/8" $\phi$ studs/row (9" to 11-1/2" Flgs.)
2-7/8" $\phi$ studs/row (7" Flgs.)

*3 spaces @ 10" = 2'-6" (12" Flgs.)
4 spaces @ 7-1/2" = 2'-6" (9" to 11-1/2" Flgs.)
6 spaces @ 5" = 2'-6" (7" Flgs.)

⊕ Field Splice, use as required. Camber ordinates are given at 5' either side of the ℄ of splice.

## 9'-6" BEAM SPACING

THREE SPAN CONTINUOUS STEEL BEAMS - A588 - COMPOSITE WITH
COVER PLATES OVER PIERS - LOAD FACTOR DESIGN METHOD ($\phi$ = 0.9)
HS 20-44 LOADING 40'-0" ROADWAY 100,000 FATIGUE CYCLES

| Span Lengths (Total) (Ft.) | Beam Size Req'd. | Cover Plates at Piers | Area "A" (Ft.-In.) "$A_1$" | "$A_2$" | Dimensions (Ft.-In.) W | X | Y | Shear Conn. Spa. (In.) | Reactions Abut/Pier (Kips) DL Min. | DL Max | LL † | ℄ Pier to F.S. | Dead Load Deflections (Ft.) 1.1 | 1.2 | 1.3 | 1.4 | 1.5 | 1.6 | 1.7 | 1.8 | 1.9 | 2.1 | 2.2 | 2.3 | 2.4 | 2.5 | Wt. Steel % Defl. | Camber Diagram Ordinates (In.) 1.4 | Lt. F.S. | Rt. F.S. | 2.5 |
|---|---|---|---|---|---|---|---|---|---|---|---|---|---|---|---|---|---|---|---|---|---|---|---|---|---|---|---|---|---|---|---|
| 28-35-28 (91) | W24x55 | 5 x 3/8 x 7'-0" | 7'-9" | 8'-0" | 22'-9" | 11'-3" | 11'-6" | 9" | 14<br>45 | 17<br>55 | 40<br>52 | 7'-0" | .005 | .009 | .012 | .013 | .013 | .011 | .008 | .004 | .001 | .002 | .007 | .012 | .015 | .017 | 5 | – | – | – | – |
| 32-40-32 (104) | W24x61 | 5 x 1/2 x 8'-0" | 9'-0" | 9'-6" | 25'-0" | 14'-9" | 12'-3" | 9" | 15<br>52 | 19<br>64 | 43<br>54 | 9'-0" | .007 | .014 | .018 | .019 | .018 | .015 | .010 | .005 | .002 | .003 | .009 | .015 | .020 | .022 | 5 | – | – | – | – |
| 36-45-36 (117) | W24x68 | 8 x 3/8 x 9'-0" | 9'-9" | 10'-3" | 28'-6" | 16'-8" | 13'-4" | 13" | 17<br>59 | 22<br>72 | 44<br>55 | 10'-0" | .010 | .019 | .025 | .027 | .026 | .022 | .015 | .008 | .002 | .004 | .013 | .023 | .030 | .033 | 6 | – | – | – | – |
| 40-50-40 (130) | W24x76 | 8 x 1/2 x 10'-0" | 11'-0" | 11'-6" | 30'-8" | 19'-11" | 14'-5" | 13" | 19<br>65 | 24<br>80 | 46<br>57 | 11'-0" | .014 | .025 | .033 | .036 | .034 | .029 | .020 | .010 | .003 | .005 | .016 | .029 | .039 | .042 | 6 | – | – | – | – |
| 44-55-44 (143) | W27x84 | 9 x 1/2 x 11'-0" | 12'-6" | 13'-0" | 33'-11" | 22'-1" | 15'-6" | 13" | 21<br>73 | 27<br>89 | 47<br>58 | 12'-0" | .015 | .028 | .036 | .039 | .037 | .031 | .021 | .011 | .003 | .005 | .017 | .031 | .041 | .045 | 6 | 1-1/2" | 5/16" | 1-1/8" | 1-9/16" |
| 48-60-48 (156) | W27x94 | 9 x 5/8 x 12'-0" | 13'-6" | 14'-0" | 36'-6" | 25'-0" | 16'-6" | 12" | 23<br>80 | 29<br>97 | 49<br>59 | 13'-0" | .018 | .034 | .044 | .048 | .045 | .038 | .026 | .013 | .004 | .007 | .022 | .038 | .050 | .055 | 7 | 1-5/8" | 7/16" | 1-3/16" | 1-11/16" |
| 52-65-52 (169) | W30x99 | 9-½ x 5/8 x 14'-0" | 14'-6" | 15'-3" | 39'-4" | 26'-5" | 18'-9" | 13" | 25<br>88 | 31<br>107 | 50<br>60 | 15'-0" | .021 | .038 | .049 | .054 | .051 | .042 | .029 | .015 | .004 | .008 | .024 | .042 | .055 | .060 | 8 | 1-11/16" | 1/2" | 1-1/4" | 1-3/4" |
| 56-70-56 (182) | W30x116 | 9 ½ x 3/4 x 17'-0" | 15'-9" | 16'-9" | 42'-6" | 29'-0" | 19'-6" | 12" | 27<br>96 | 34<br>116 | 50<br>63 | 16'-0" | .023 | .042 | .054 | .059 | .056 | .047 | .033 | .017 | .005 | .008 | .025 | .045 | .060 | .065 | 9 | 1-3/4" | 9/16" | 1-1/4" | 1-7/8" |
| 60-75-60 (195) | W33x118 | 10 x 3/4 x 16'-0" | 16'-9" | 17'-9" | 44'-9" | 31'-10" | 20'-11" | 13" | 30<br>103 | 37<br>125 | 51<br>66 | 17'-0" | .025 | .046 | .060 | .065 | .062 | .051 | .035 | .018 | .005 | .009 | 028 | .050 | .066 | .072 | 9 | 1-7/8" | 9/16" | 1-5/16" | 1-15/16" |
| 64-80-64 (208) | W33x130 | 10-½x7/8 x 16'-0" | 18'-0" | 19'-0" | 47'-6" | 35'-0" | 21'-6" | 12" | 32<br>110 | 39<br>133 | 52<br>69 | 18'-0" | .028 | .052 | .068 | .074 | .070 | .058 | .040 | .020 | .006 | .010 | .032 | 057 | .075 | .082 | 10 | 2" | 5/8" | 1-3/8" | 2-1/16" |
| 68-85-68 (221) | W36x150 | 10 x 7/8 x 17'-0" | 19'-3" | 20'-3" | 51'-3" | 35'-6" | 23'-9" | 15" | 34<br>119 | 42<br>144 | 53<br>72 | 20'-0" | .027 | .050 | .065 | .071 | .067 | .055 | .038 | .019 | .005 | .009 | .030 | .053 | .071 | .077 | 11 | 1-15/16" | 11/16" | 1-3/8" | 2" |
| 72-90-72 (234) | W36x160 | 11 x 1 x 18'-0" | 20'-3" | 21'-3" | 53'-9" | 38'-3" | 25'-0" | 15" | 36<br>127 | 45<br>153 | 53<br>75 | 21'-0" | .032 | .059 | .076 | .083 | .079 | .065 | .044 | .023 | .006 | .011 | 036 | .063 | .084 | .092 | 12 | 2-1/8" | 3/4" | 1-1/2" | 2-1/4" |
| 76-95-76 (247) | W36x182 | 11x1-1/4 x 20'-0" | 21'-6" | 22'-9" | 56'-3" | 41'-0" | 26'-3" | 15" | 38<br>137 | 47<br>164 | 54<br>78 | 22'-0" | .034 | .063 | .082 | .089 | .084 | .069 | .047 | .024 | .006 | .011 | .036 | .065 | .087 | .095 | 13 | 2-3/16" | 3/4" | 1-1/2" | 2-5/16" |

†No Impact

All shear connectors to be 7/8" $\phi$ x 5" studs at constant spacing.

Dead Load deflections include the weight of slab, wearing course (30 p.s.f.) and railing. Beam deflections are the indicated % of the deflections shown.

Camber diagram ordinates are given for spans over 50 ft. without any vertical curve. (Modify values for vertical curve). Ordinates are based on 1" residual camber at point of maximum D.L. deflection.

The cover plates are centered on ℄ Pier.

**All bearing stiffeners shall be 5/8" in thickness. Their width shall be 7" at all locations except on the outside of the facia beams where they shall end approximately 1/2" inside the edge of the flange.**

Area "A" is the portion of top beam flange which is in tension due to dead load. F.S. means Field Splice.

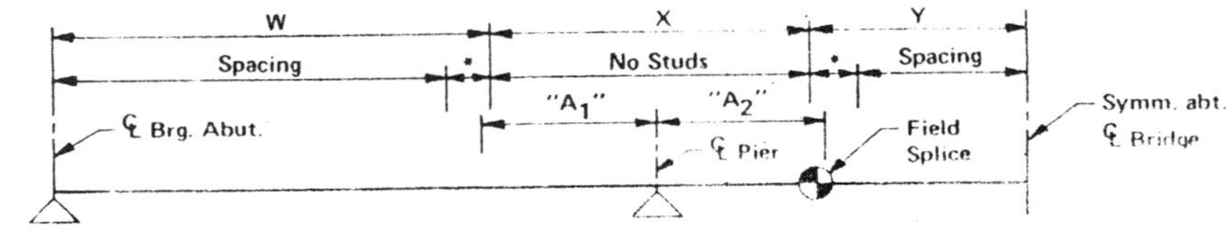


4-7/8" $\phi$ studs/row (12" Flgs.)
3-7/8" $\phi$ studs/row (9" to 11-1/2" Flgs.)
2-7/8" $\phi$ studs/row (7" Flgs.)

*3 spaces @ 10" = 2'-6" (12" Flanges)
4 spaces @ 7-1/2" = 2'-6" (9" to 11-1/2" Flanges)
6 spaces @ 5" = 2'-6" (7" Flanges)

## E.12 FHWA STANDARD DETAILS

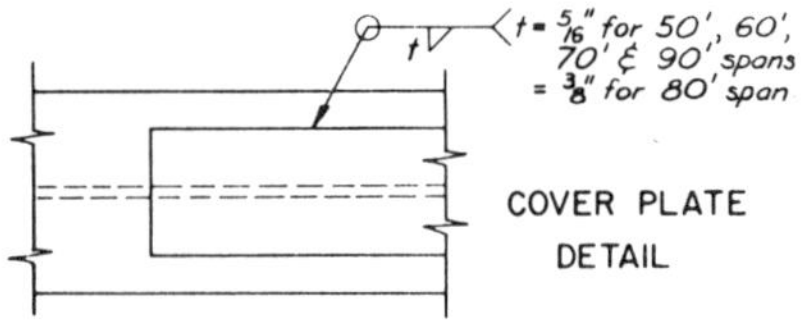


INTERSECTING WELDS: Welds used for attachments should be prevented from intersecting with main or secondary member weldments. Corners should be coped or welds stopped short from the point of intersection by a distance of 1½".

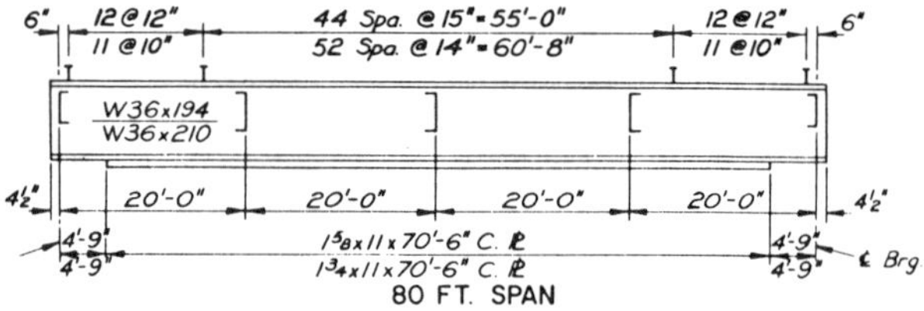


COVER PLATES: Cover plates should be extended to within about 5 feet of supports. It has been determined that a tapered end transition has no significant beneficial effect; therefore, its use is no longer advocated.

TRANSVERSE STIFFENERS:
Single or paired stiffeners may be used. For paired stiffeners, use same details as shown for a single stiffener. When longitudinal stiffeners are required, all transverse stiffeners should be placed on one side of web, place longitudinal stiffener on opposite side.

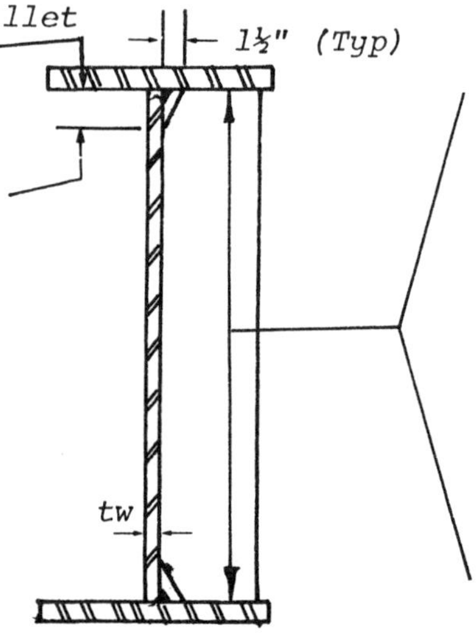


| *Web Thickness* | *x (a)* |
|---|---|
| *W33; W36* | *1½"* |
| *1/2"* | *2½"* |
| *5/8"* | *3 "* |
| *3/4"* | *3½"* |

*Fillet weld (both sides) to both flanges when used as a crossframe or diaphragm connection plate; otherwise, use a tight fit to both flanges. As an Option, when the stiffener is not used as a connection Plate, the stiffener may be cut short from the tension flange by a distance equal to the vertical leg of the cope.*

FHWA COMMENTS;

Bolting of gusset plate to the tensionflange, while being a better fatigue detail, will not eliminate the controlling details; i.e., the connection plate to web fillet weld. Since the difference between the stress range at the inside face of the tension flange and the stress range at the end of the connection plate to web fillet weld is insignificant in term of additional fatigue life (for those plate girder bridges in our standards) the cost of bolting will not be offset by an incerease fatigue life.

Fillet welding the connection plate to the tension flange should be more economical detail for our standards than a bolted connection. The bolted connection is, of course, a good alternate detail and for many classes of bridges, may be the preferred detail.

**BEARING STIFFENER:** At piers and at abutments, fillet weld both sides of the stiffener to the top flange, use full penetration groove welds to the bottom flange. Use same cope dimensions as used for transverse stiffener.

**FHWA COMMENTS;**
Our reasons for welding the bearing stiffeners to the top flanges at piers in continuous design are;

a. Comments in regarding to welding connection plate apply here also.

b. The cross frames over the piers transmit wind reactions to the bearings; therefore, out-of-plane deformations are likely. Thus, our main reason for welding the bearing stiffeners to the top flanges is to make this cross section less prone to out-of- plane deformations.

**LONGITUDINAL STIFFENERS:**
Preferably, the stiffener should be placed on one side of the web, place all transverse stiffeners on the opposite side. In zones of stress reversal, where possible, longitudinal stiffeners should be continuous plates and/or transverse stiffeners. If not possible, a radiused termination should be used.

**LATERAL BRACING:** Use of WT for members and connect to web with WT or T-shape cut from W shape. Bolted connections are shown on the new details; however, as an option, bracing member may be fillet welded to the T-gussets.

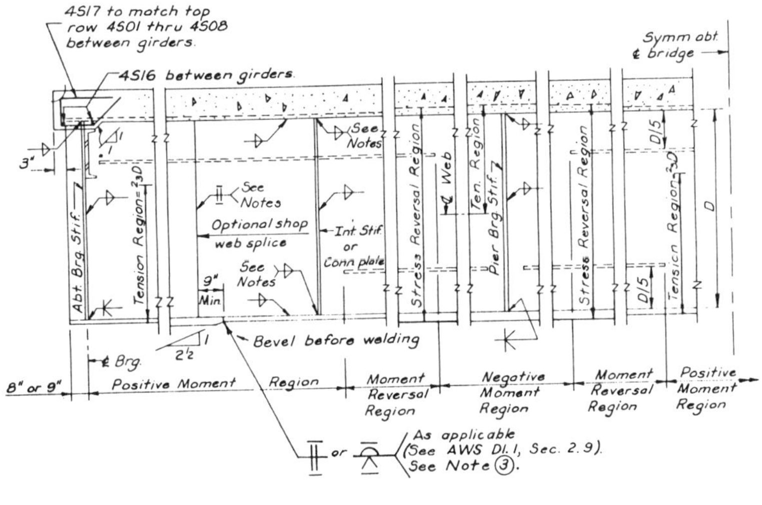


TYPICAL GIRDER DETAILS

| GIRDER WELD TABLE<br>Flanges to Webs; Stiffeners & Connection Plates to Webs & Flanges | |
|---|---|
| Thickness of Thicker Part Joined | Minimum Fillet Weld Size ** |
| ½" or less | 3/16" |
| Over ½" to ¾" | ¼" |
| Over ¾" to 1½" | 5/16" |
| Over 1½" | ⅜" |
| ** Weld size need not exceed thickness of thinner part joined. | |

NOTES:

① Flange and web plates over piers, between field splices, shall be ASTM A572 Grade 50 steel. All other steel shall be ASTM A36.

② For transverse stiffeners, fillet weld (both sides) to both flanges when used as a cross frame connection plate; otherwise, use a tight fit to both flanges. As an option, when the stiffener is not used as a connector plate, the stiffener may be cut short from the tension flange a distance of 2½ inches. When not cut short, the cope dimensions shown on Sht. 403, DETAILS A and B shall be used. #

③ Butt joints of flanges, web, and longitudinal stiffeners subject to tension or reversal of stress shall, as a minimum, be ground flush by grinding parallel to the direction of principle stress in accordance with section 3.6.3 of the AASHTO Welding Specifications.

④ Web, flange, girder splice plates, and longitudinal stiffeners subject to tension or reversal of stress shall meet supplemental Charpy V-Notch impact properties for specified temperature zones as described in AASHTO 1.7.2 (c).

⑤ See Appendixes for bearing details.

# Where transverse stiffeners are not welded to the flange, special care shall be exercised during shipping and handling of the girder to prevent relative rotation between the web and flange.

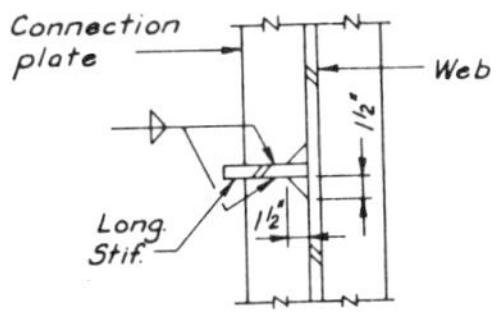


CONTINUOUS STIFFENER

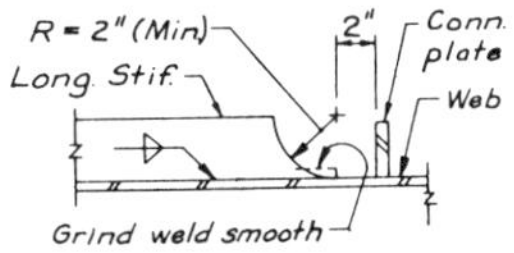


RADIUSED TERMINATION

In zones of moment reversal the longitudinal stiffener shall be made continuous through connector plates, unless the stiffener interferes with cross frame gusset plates or lateral bracing T-gussets. In these cases the radiused termination shall be used. Use the radiused termination at bolted field splice locations.

LONGITUDINAL STIFFENER DETAILS

Bottom flange plates and the lower two thirds of web plates which are butt welded to obtain required lengths shall, as a minimum, be ground flush by grinding parallel to the direction of principle stress in accordance with section 3.6.3. of the AASHTO Welding Spec.

Web and bottom flange plates (including girder splice plates) shall meet supplemental Charpy V-Notch impact properties for specified temperature zones as described in AASHTO 1.7.2 (c).

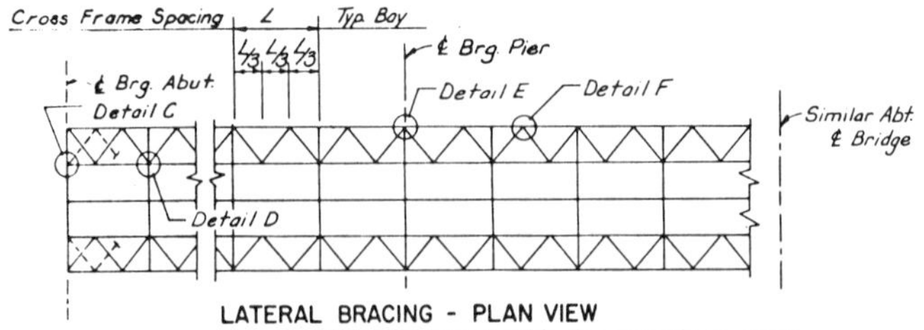


LATERAL BRACING - PLAN VIEW

Note: See Sheet 402 for cross frame spacing.

NOTES:

- All bolts shall be 78" Ø A325, and contact surfaces shall be Class A.
- For bolted girder splice use fill plates as needed.
- Lateral bracing required for 470' thru 620' girders only.
- Dashed line shown on LATERAL BRACING - PLAN VIEW indicates end diagonal on 520' and 570' girders.
- Lateral bracing T-gusset shall be cut to required depth from W 27 x 94.
- For bridges with lateral bracing, use T-gusset connection at piers on all interior girders.
- Position transverse stiffener to provide 6" clearance between stiffener and T-gusset when stiffener interfers with DETAIL F connection.
- Where appropriate, DETAIL G-OPTIONAL may be used to eliminate the need for coping the T-gusset shown in DETAILS A & D.
- As an option, bracing members may be welded to gusset plates and T-gussets by balanced fillet welds.
- See Sheet 404 for fillet weld size.

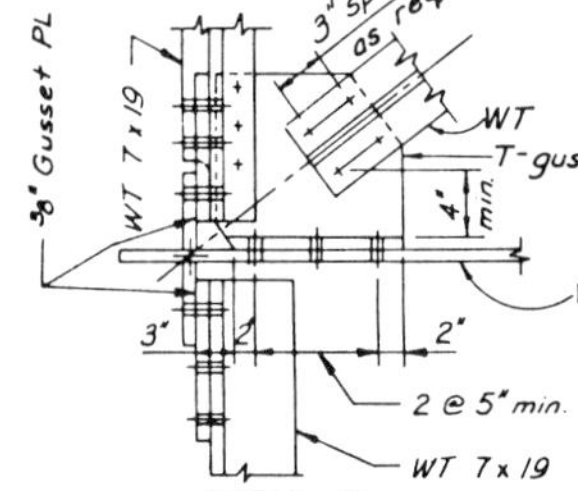


DETAIL C

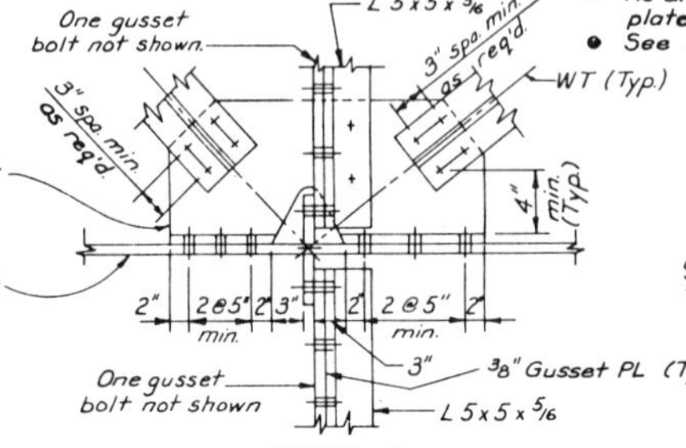


DETAIL D

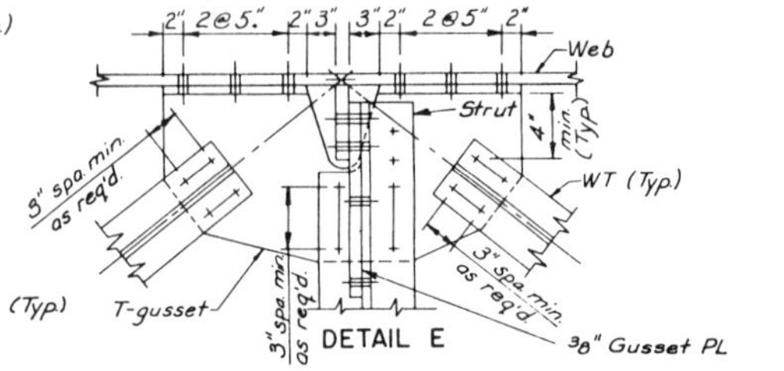


DETAIL E

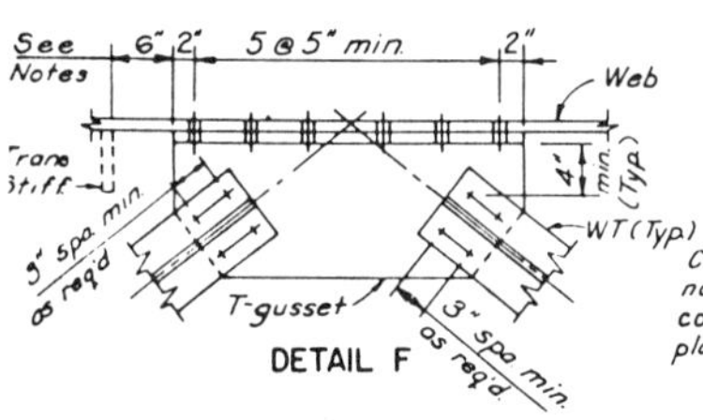


DETAIL F

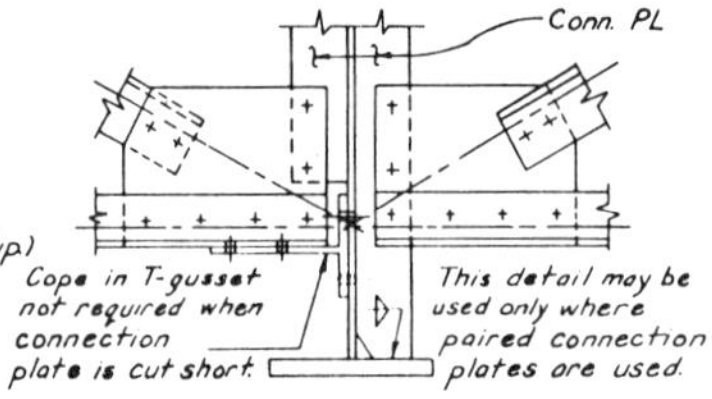


DETAIL G - OPTIONAL

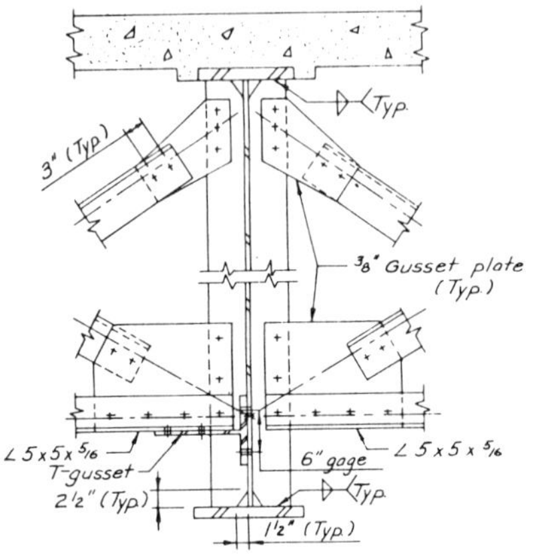


DETAIL A

(See alternate DETAIL G-OPTIONAL)

**BRACING MEMBER SIZES**

| SPANS | CROSS FRAMES | | | | | LATERAL BRACING | |
|---|---|---|---|---|---|---|---|
| | DIAGONALS | STRUT | | | | DIAGONALS | |
| | | ABUTMENT | INTERMED. | PIER | | | |
| | | | | | FLANGE BOLTS | | FLANGE BOLTS |
| 100-120-100 | L 5x5x 5/16 | L 5x5x 5/16 | L 5x5x 5/16 | L 5x5x 5/16 | N/A | Not Req'd. | N/A |
| 115-140-115 | " | " | " | " | " | " | " |
| 130-160-130 | " | " | " | " | " | " | " |
| 145-180-145 | " | WT 7 x 19 | " | WT 8 x 33.5 | 6 | WT 8 x 22.5 | 4 |
| 160-200-160 | L 6x6x 3/8 | " | " | WT 10.5 x 36.5 | 8 | WT 8 x 25 | 4 |
| 175-220-175 | " | " | " | WT 9 x 43 | 8 | WT 8 x 28.5 | 6 |
| 190-240-190 | " | " | " | WT 10.5 x 50.5 | 10 | WT 8 x 33.5 | 6 |

## FHWA COMMENTS

In regard to the lateral bracing Optional New Detail, we are requiring the full depth connector plate opposite the cut-short connector plate to insure that web is stiffened against out-of-plane deformations. For this reason, the full depth connector plate is welded to the tension flange.

For bridges like those in FHWA's Standard Plans, we suggest that the gusset plates bolted to the flanges be used as an alternative means of attachment rather than the primary means.

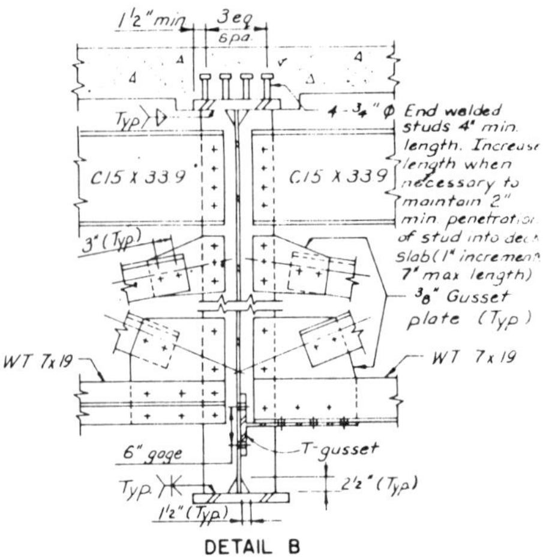


DETAIL B

**CROSS FRAMES:** Use angles for diagonals and struts, except that when lateral bracing is used, use WT shape for strut at abutment and pier bearings.

Bolted connections are shown in the new details; however, as an option, bracing members may be fillet welded to the gusset plates and T-gussets.

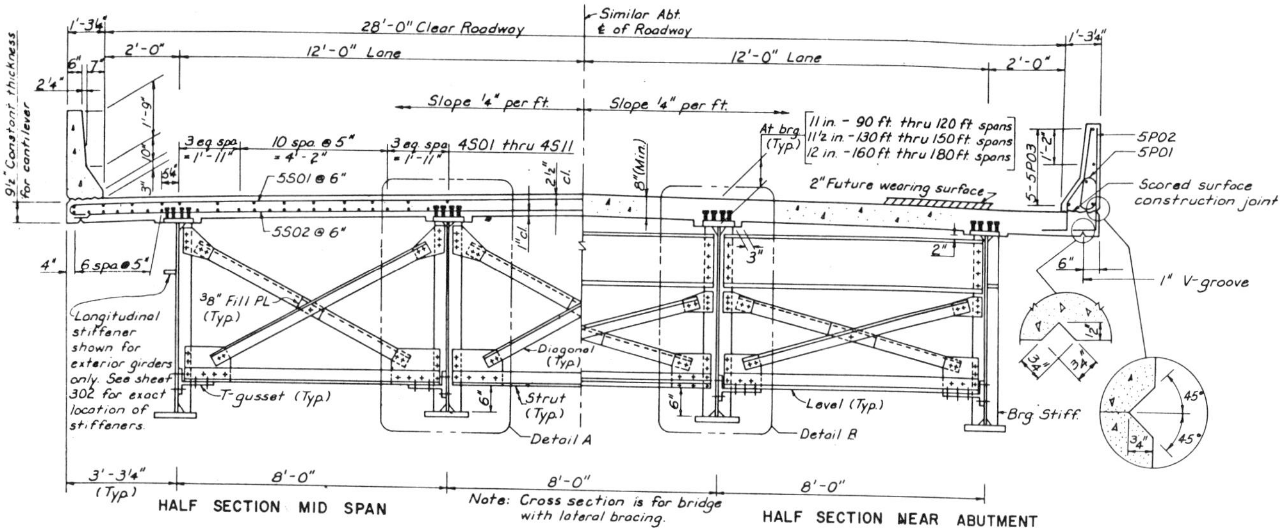


TYPICAL CROSS SECTION

**DIAPHGRAM CONNECTION PLATES;** Provide sufficient gap as shown. The bolted alternate may be also used.

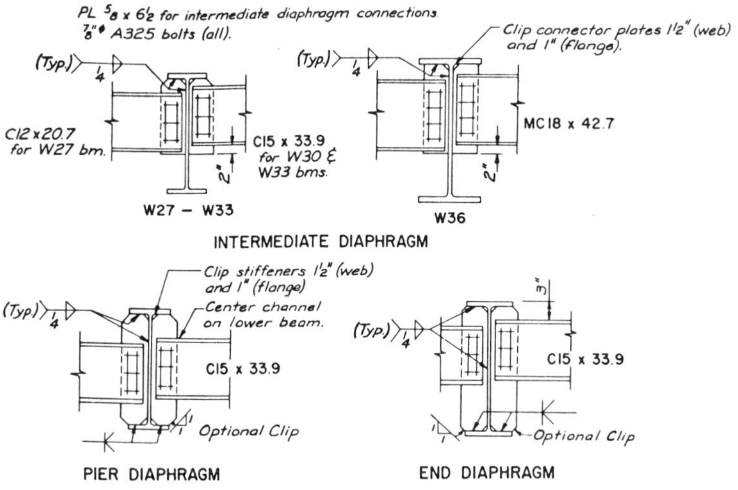


DIAPHRAGM DETAILS

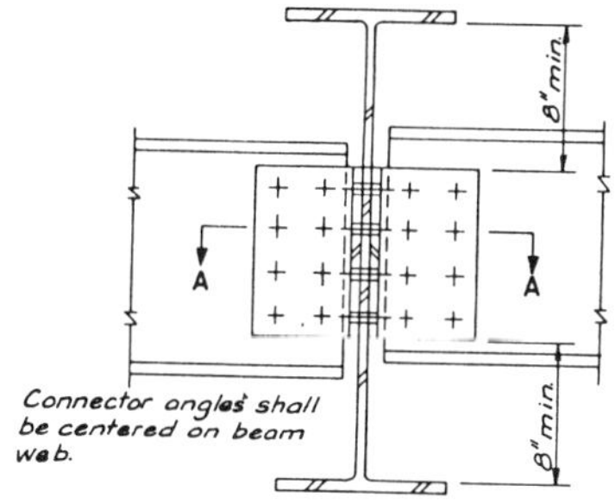


BOLTED CONNECTOR ANGLES

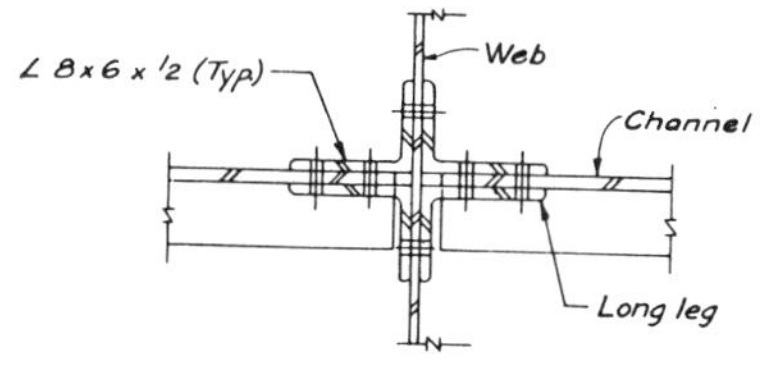


SECTION A–A

ALTERNATE INTERMEDIATE DIAPHRAGM CONNECTIONS

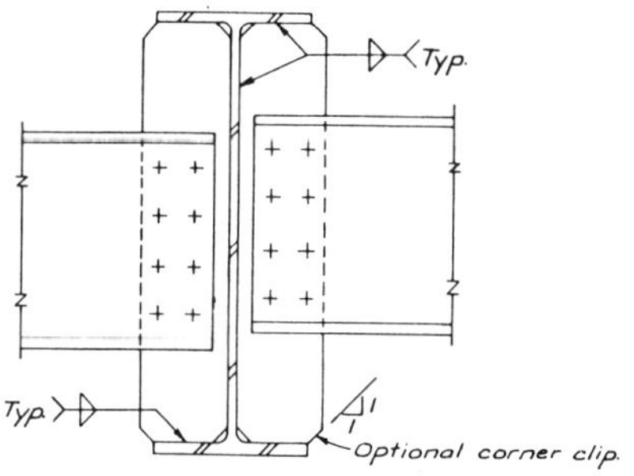


FULL DEPTH CONNECTOR PLATE

## FHWA COMMENTS

As a final comments on the details used, We point out that although out-of-plane deformations are more severe on skewed and curved bridges, the use of a straight girder bridge does not necessarily preclude fatigue problems resulting from these out-of-plane deformations, even when the AASHTO gap criteria is satisfied. However, the prudence of welding connection plates to tension flanges may vary from bridge to bridge.

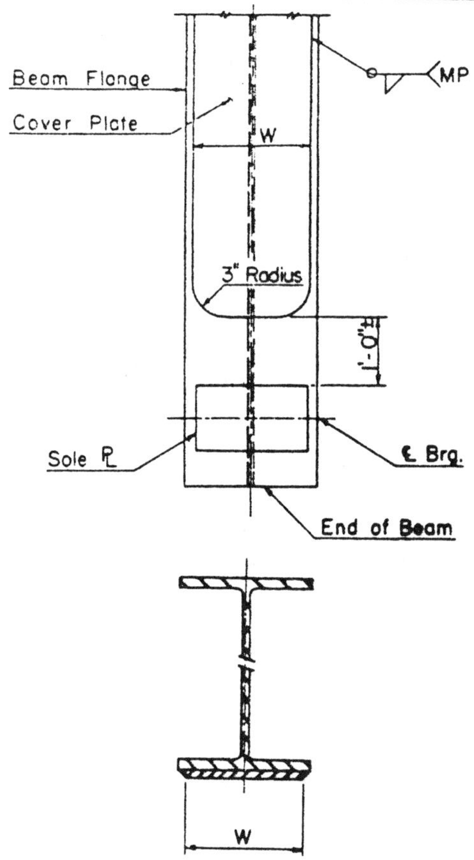


W (Max.) = Flange width - 1¼"
to next smaller inch.

COVER PLATE NARROWER
THAN FLANGE

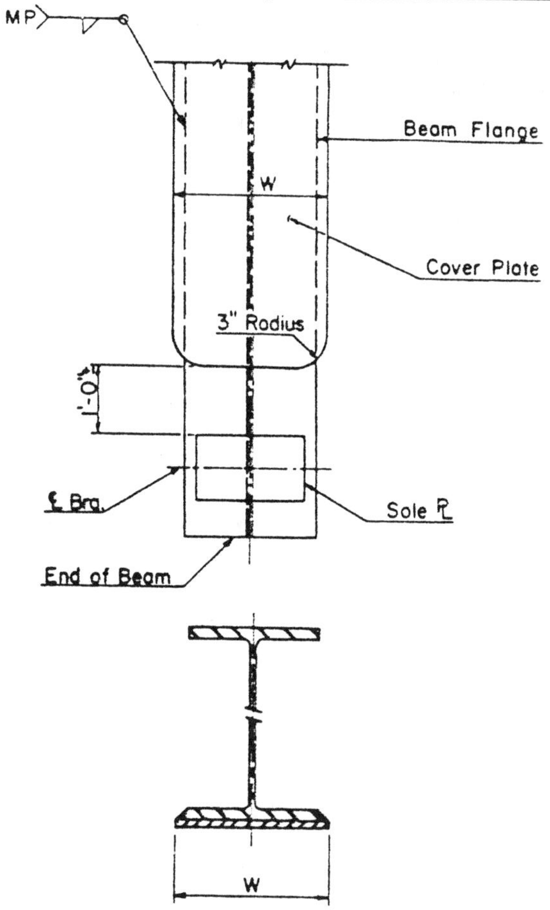


W (Min.) = Flange width + 1½"
to next larger inch.

COVER PLATE WIDER
THAN FLANGE

CONN.

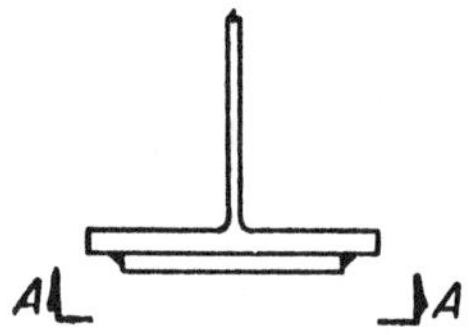


NARROW COVER
PLATE DETAIL

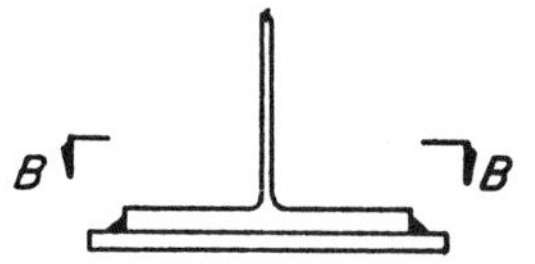


WIDE COVER
PLATE DETAIL

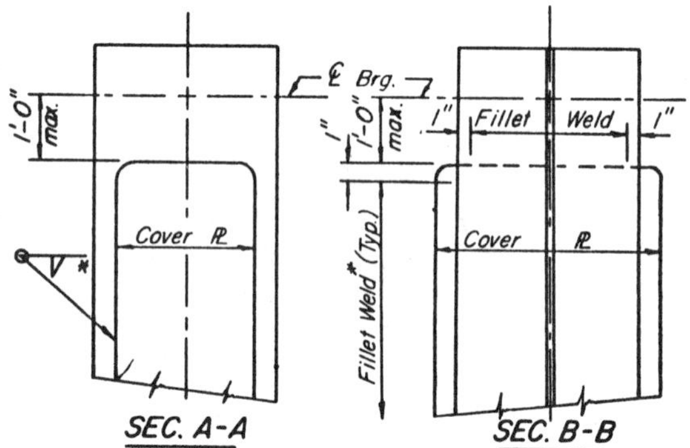


SEC. A-A SEC. B-B

*Weld size to be determined by designer

COVER PLATE CONNECTIONS

(NY STATE)

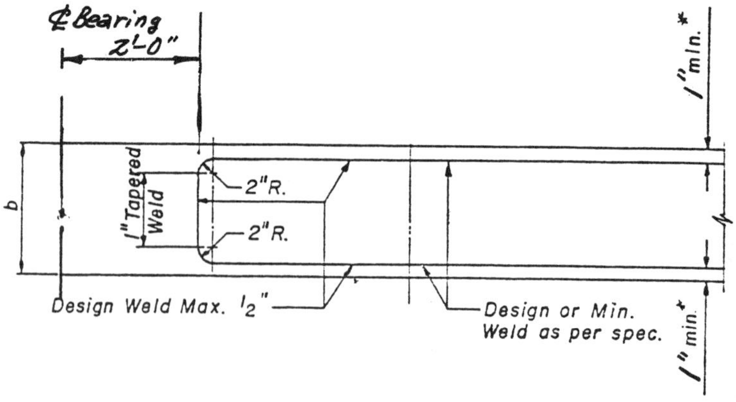


* Designer to select plate width to
provide minimum distance shown.

NEW JERSEY

ROLLED BEAM COVER
PLATE DETAILS

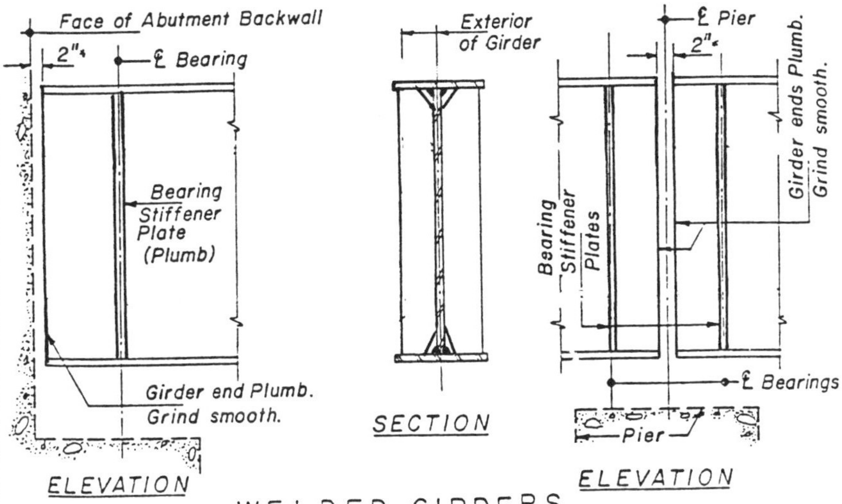


WELDED GIRDERS

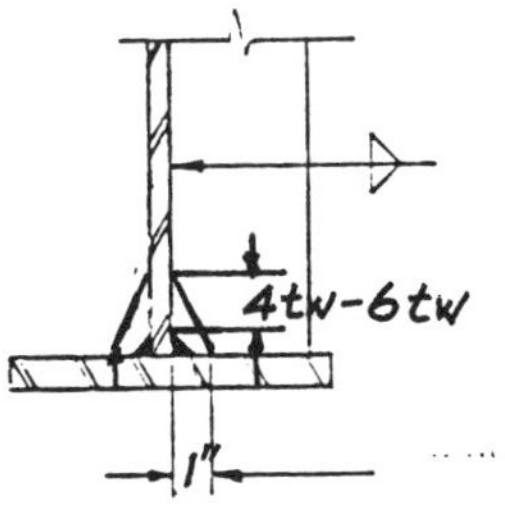


COPING DETAIL

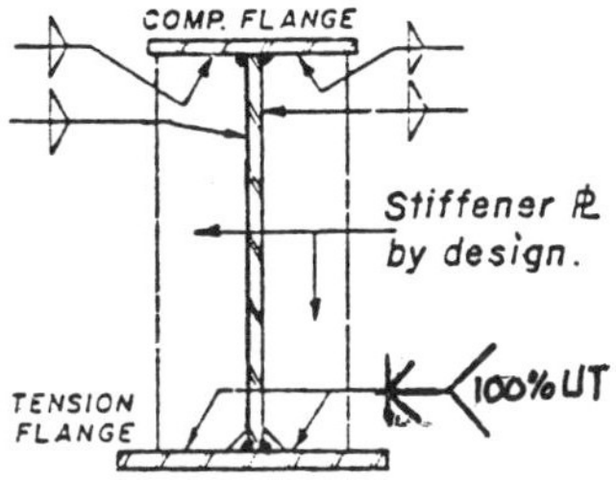


END STIFFENERS

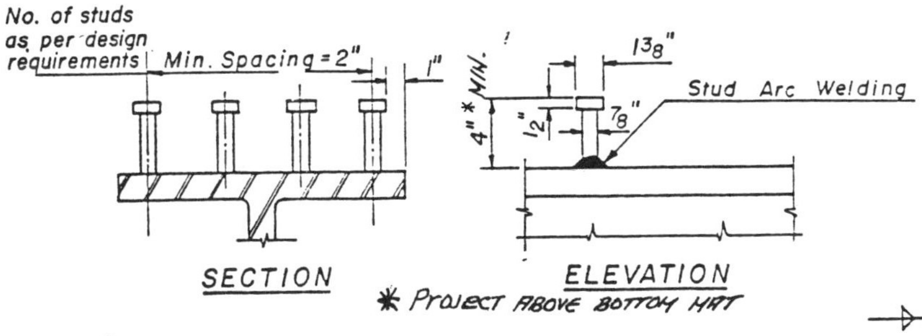


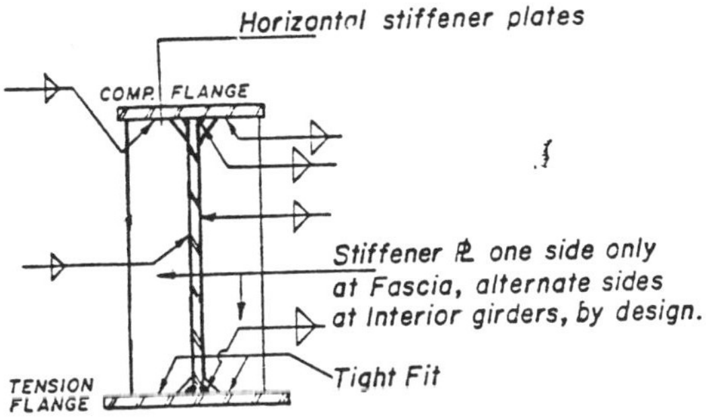


INTERMEDIATE STIFFENERS

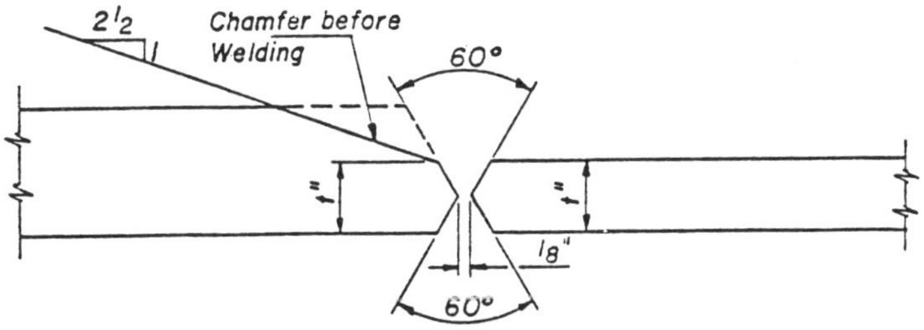


FLANGE TRANSITION WELD

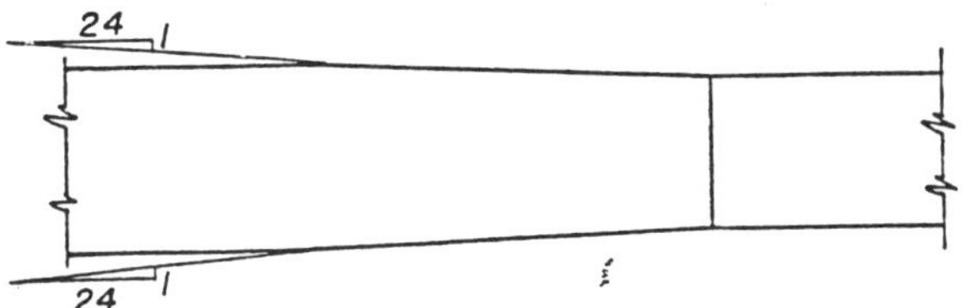


FLANGE TRANSITION

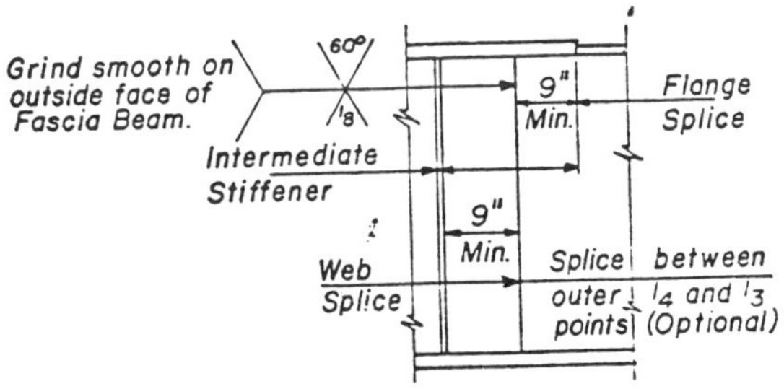


SHOP WEB SPLICE

NOTES:

* 1. Maximum thickness of Flange Plate = 4"

2. Do not change width of Top Flange in Simply Supported Girders.

* 3. Only change width of Bottom Flange to keep thickness of Flange at 4" maximum.

4. Minimum length of any Flange ℘s = 10 Ft.

5. Minimum change of thickness of Flange ℘ 1/2".

6. Maximum change of thickness of Flange ℘ not to exceed one half the thickness of the thicker plate plus 1/8

* 7. FHWA Notice N 5040.23 dated February 16, 1977 recommends that the use of main member tension material in excess of 2 inch thickness is to be avoided wherever possible by modification in the structural components of the individual structural member.

WELDED STRINGER DETAILS

# STIFFENER DETAILS

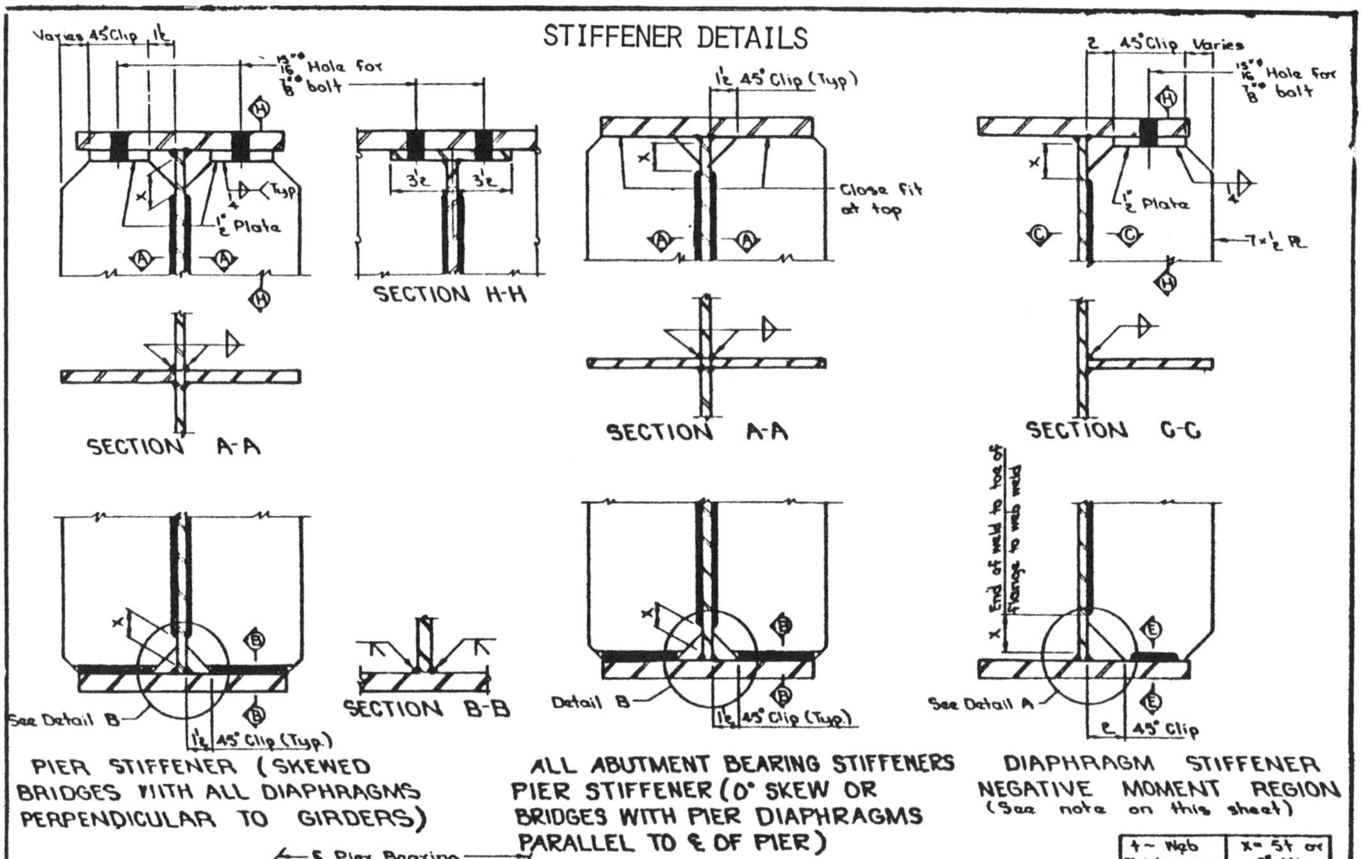


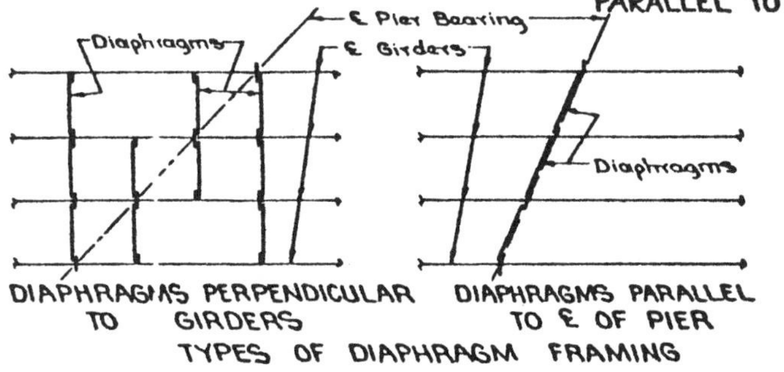


NOTE:
The positive moment region is the center 3/5 of all spans between piers and 4/5 of the end span from the abutment bearing. The negative moment region is 1/5 of the span length either side of the piers.

| t ~ Web Thickness | x = 5t or 2" Min. |
|---|---|
| 3/8 | 2 |
| 7/16 | 2 3/16 |
| 1/2 | 2 1/2 |
| 9/16 | 2 13/16 |
| 5/8 | 3 1/8 |
| 11/16 | 3 7/16 |
| 3/4 | 3 3/4 |

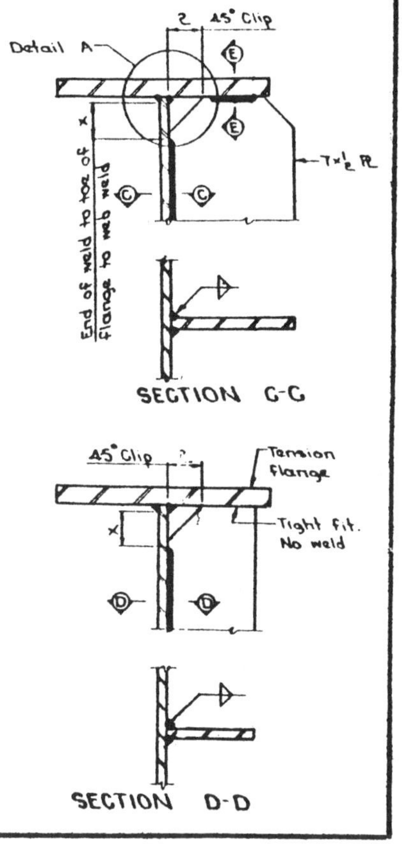


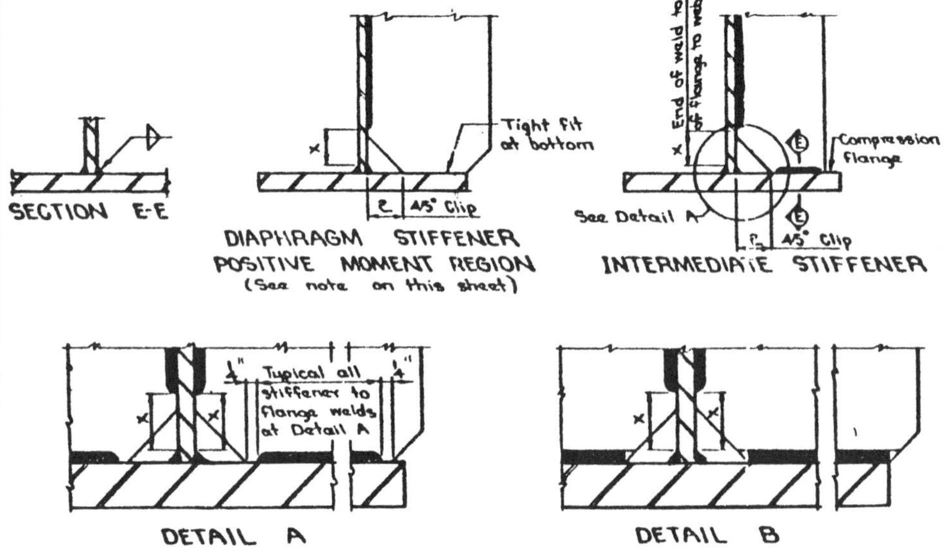


Iowa

# SHEAR LOCK AT LONGITUDINAL OPEN JOINTS IN DECK SLAB (Typical)

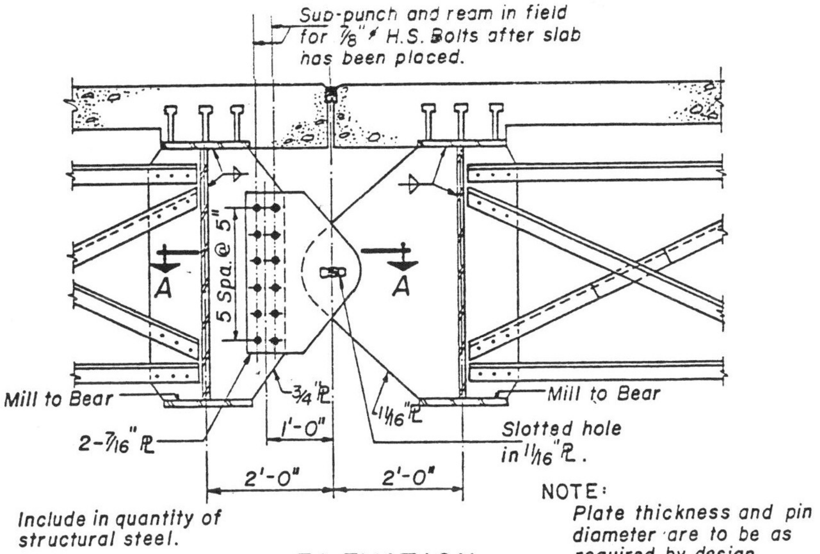


ELEVATION

Scale 1/2" = 1'-0"

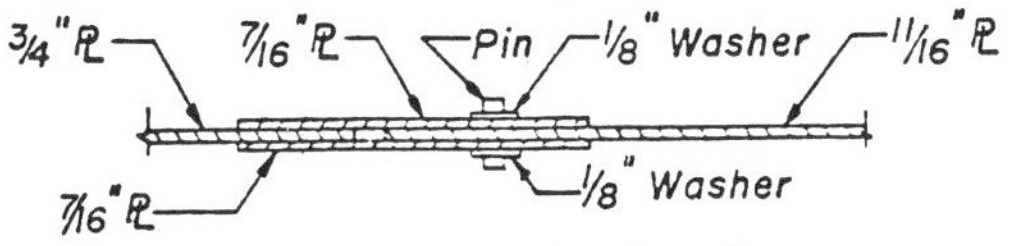


SECTION A - A

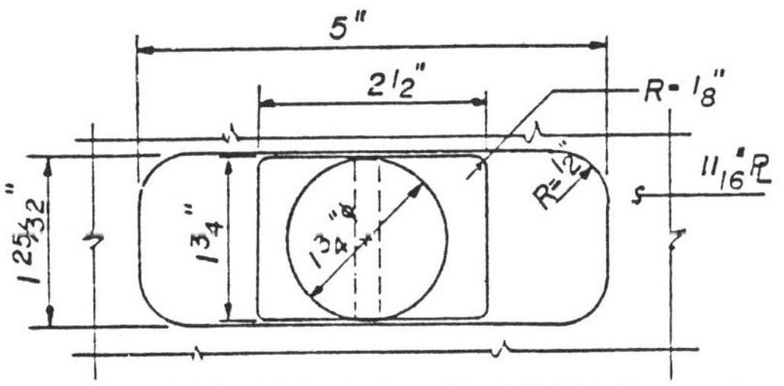


SLOTTED HOLE DETAIL (NJ)

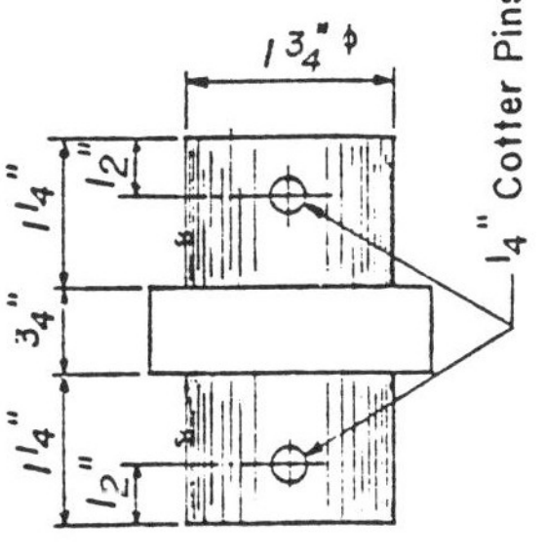


DETAIL OF PIN

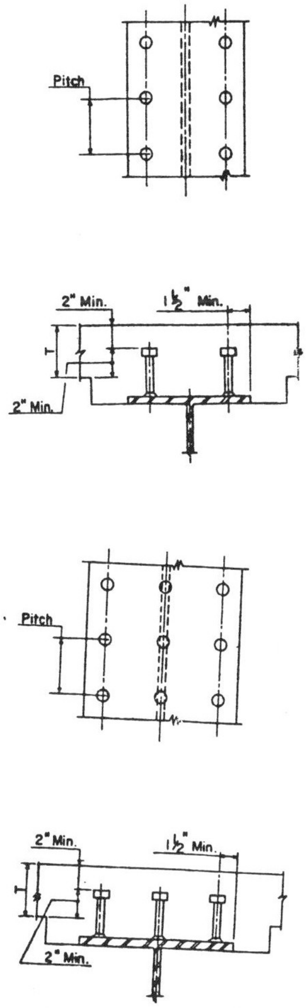


(CONN)

SHEAR CONNECTORS - (WELDED STUD)

DESIGN INFORMATION: Preference shall be given to the use of 7/8"⌀ studs.
Lengths of studs shall be in the increments of 1"
Maximum length of studs = 8"
If the stud length required is in excess of 8", the required length may be obtained by stacking. Stacking of studs is provided for in the specifications.
T = Slab thickness.

Design or Detailing by substitution will not be accepted. Each type shall be separately designed in accordance with current AASHTO Specifications.

| Material Thickness of Thicker Part Joined (Inches) | Minimum Size of Fillet Weld (Inches) |
|---|---|
| To 1/2 inclusive | 3/16 |
| Over 1/2 to 3/4 | 1/4 |
| Over 3/4 to 1 1/2 | 5/16 |
| Over 1 1/2 to 2 1/4 | 3/8 |
| Over 2 1/4 to 6 | 1/2 |

MINIMUM SIZE OF FILLET WELDS

NOTE:

The size of fillet welds specified above represents the minimum size of fillet welds that should be used on various thicknesses of steel when requirements of calculated stress or other design considerations do not require larger fillet welds.

The minimum size seal weld shall be 3/16" fillet weld.

(CONN)

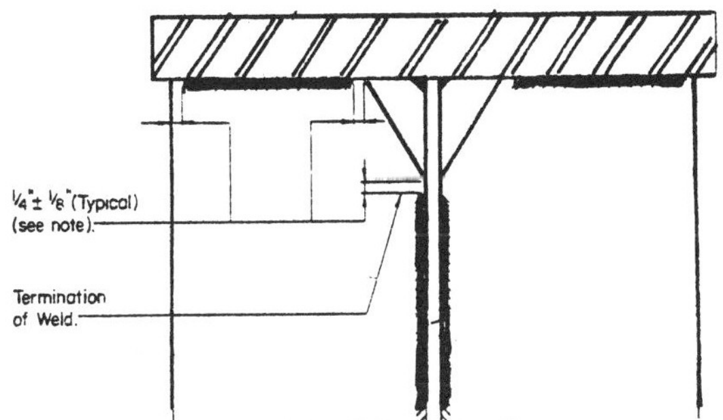


Weld Termination

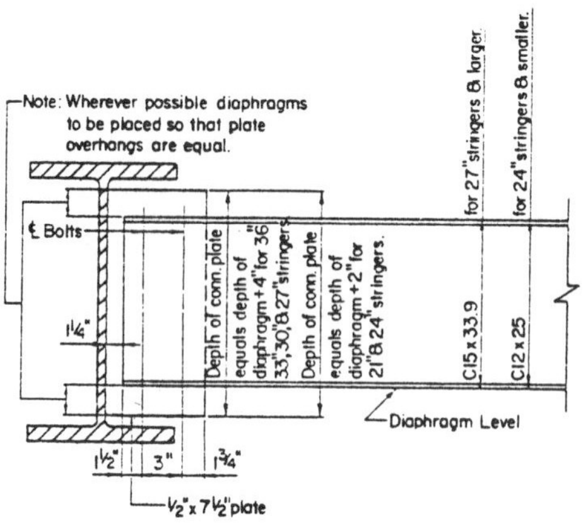


Note: Dimensions shown are for 90° connections.

VIEW 'A'

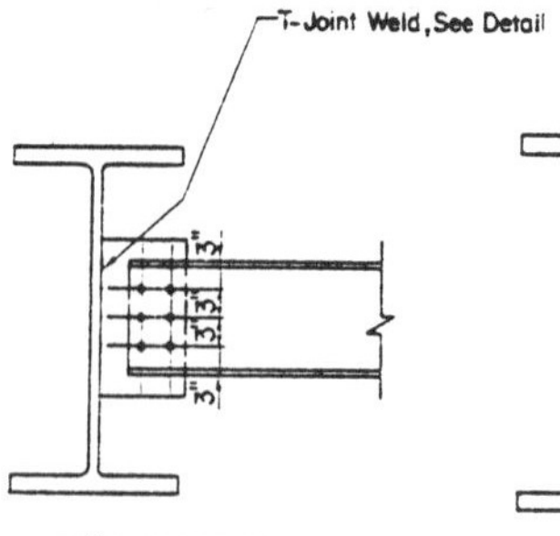


24" & SMALLER STRINGERS

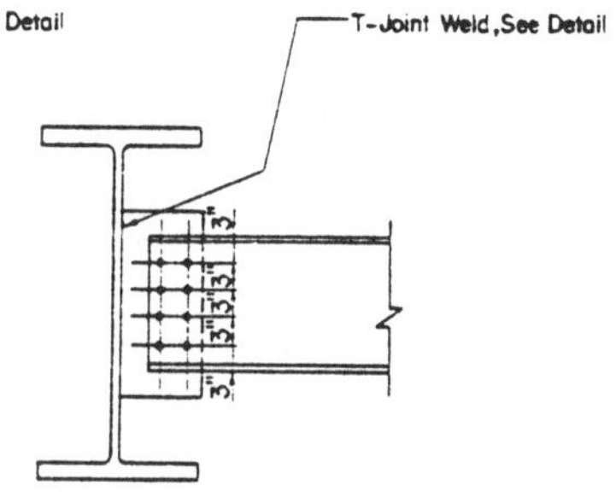


27" & LARGER STRINGERS

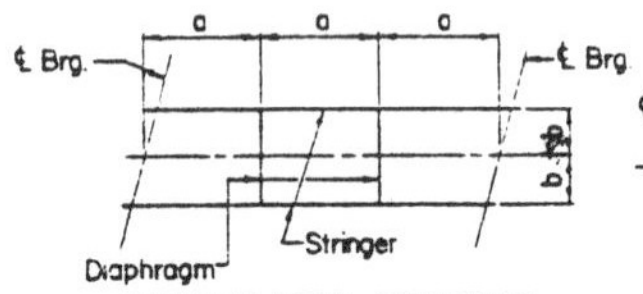


DIAPHRAGM SPACING 70° OR LESS SKEW

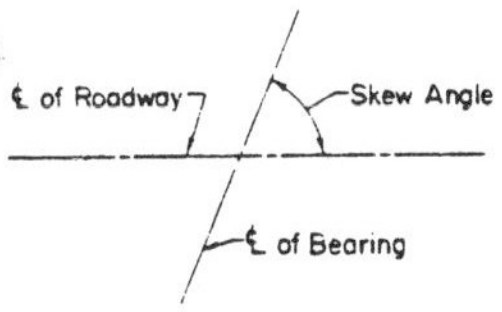


SKEW ANGLE

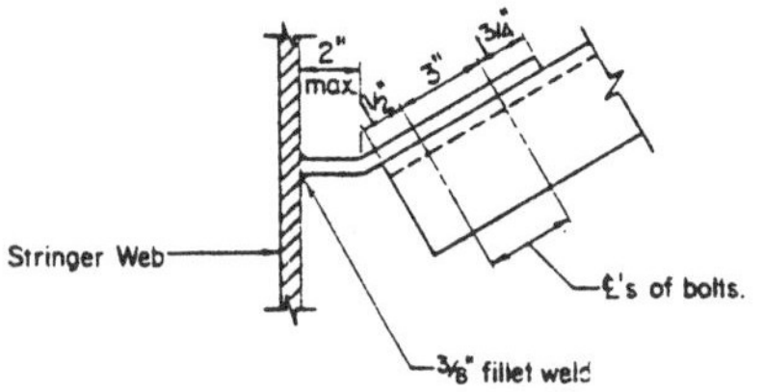


T-JOINT WELD DETAIL FOR SKEW ANGLE OVER 70° TO 90°

Note:

1. Where the angle between the center line of roadway and center line of bearing is 70° or less place diaphragms at 90° to the stringers. Diaphragms shall be spaced as shown in detail this sheet and as noted below.
2. Where aforementioned angle is greater than 70°, the diaphragms shall be parallel to the center line of bearing of the stringers.
. Space intermediate diaphragms at 20'± to 25'± ℅; i.e. for spans.

Up to 25' ℅ bearings - No Intermediate Diaphragm.
From 25' to 50' ℅ bearings - One Intermediate Diaphragm.
From 50' to 75' ℅ bearings - Two Intermediate Diaphragms, etc.
(See Framing Plan.)

4. All diaphragms are to be completely connected to stringers before deck slab is poured.

All bolts to be ⅞" ¢ A.S.T.M. A-325.
All bolts holes in diaphragm to be 15/16" ¢.
Bolt holes in connector plates may be 1 1/16" ¢.
Bolt spacing applies regardless of skew.

MD

*Note: Hardened washers shall be required over all 15/16" ¢ holes.*

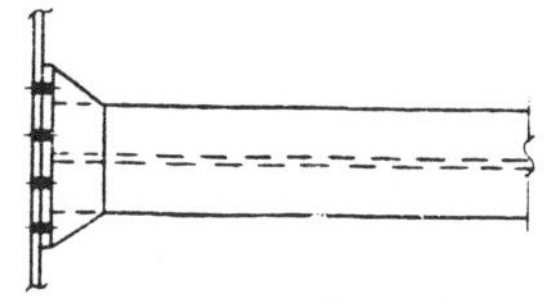

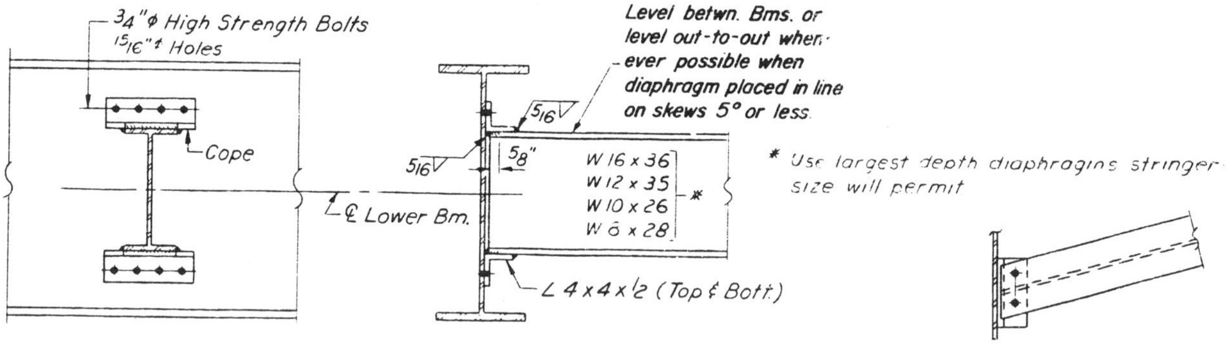


*INTERIOR DIAPHRAGM*

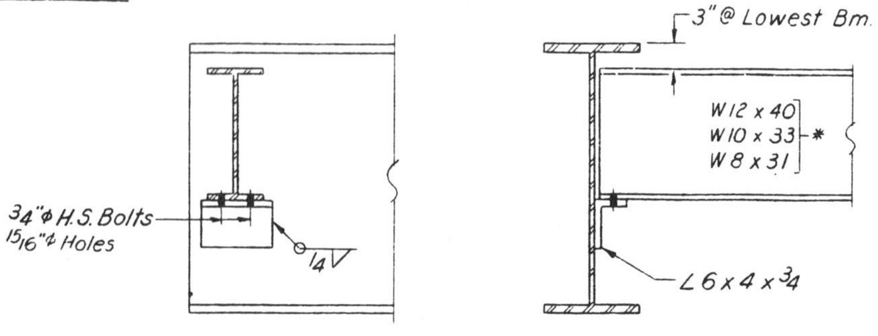


*END DIAPHRAGM*



ILL.

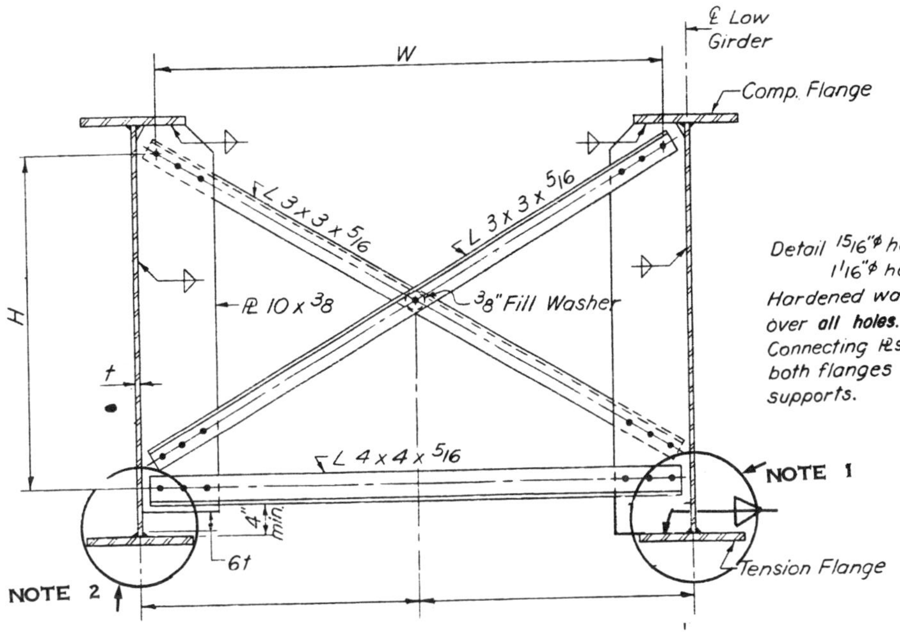


Detail 15/16"ϕ holes for all 3/4"ϕ bolts.
1 1/16"ϕ holes for all 7/8"ϕ bolts.
Hardened washers shall be required over all holes.
Connecting ℞s should be welded to both flanges for cross frames near supports.

*TYPICAL INTERIOR CROSS FRAME*
*Welded Girders greater than 48" in depth.*

**NOTE 1**
CONNECTION PLATE FOR SKEWED AND CURVED GIRDERS AND NEAR SUPPORTS FOR NON SKEWED GIRDER.

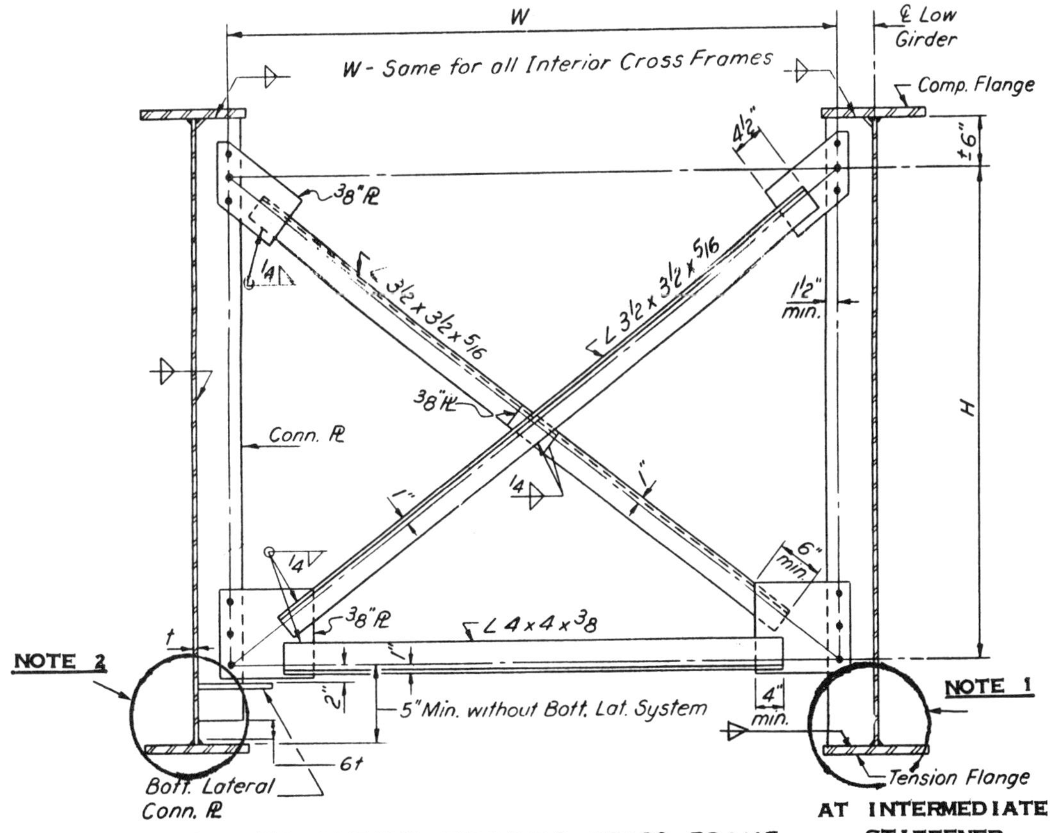


AT CONNECTION PLATE *TYPICAL INTERIOR CROSS FRAME* AT INTERMEDIATE STIFFENER
*Alternate Detail for Welded Girders greater than 48" in depth.*

**NOTE 2**
CONNECTION PLATE FOR NON SKEWED GIRDER.

ILL

*INTERIOR CROSS FRAME AWAY FROM SUPPORT*

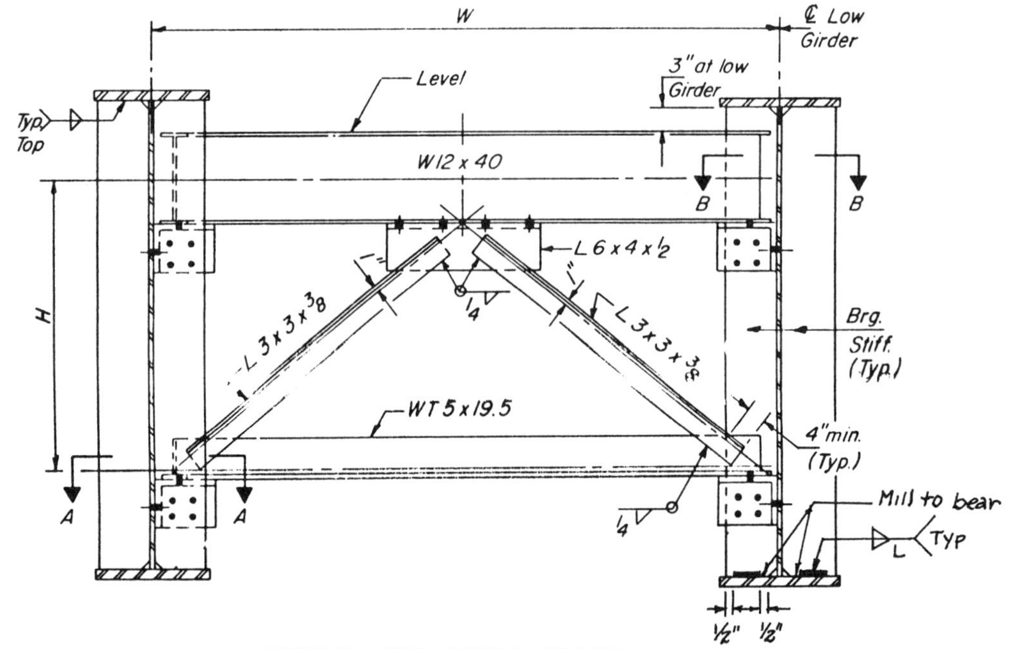


TYPICAL END CROSS FRAME
Welded Girders greater than 48" in depth

**FOR SKEW OVER 45°**

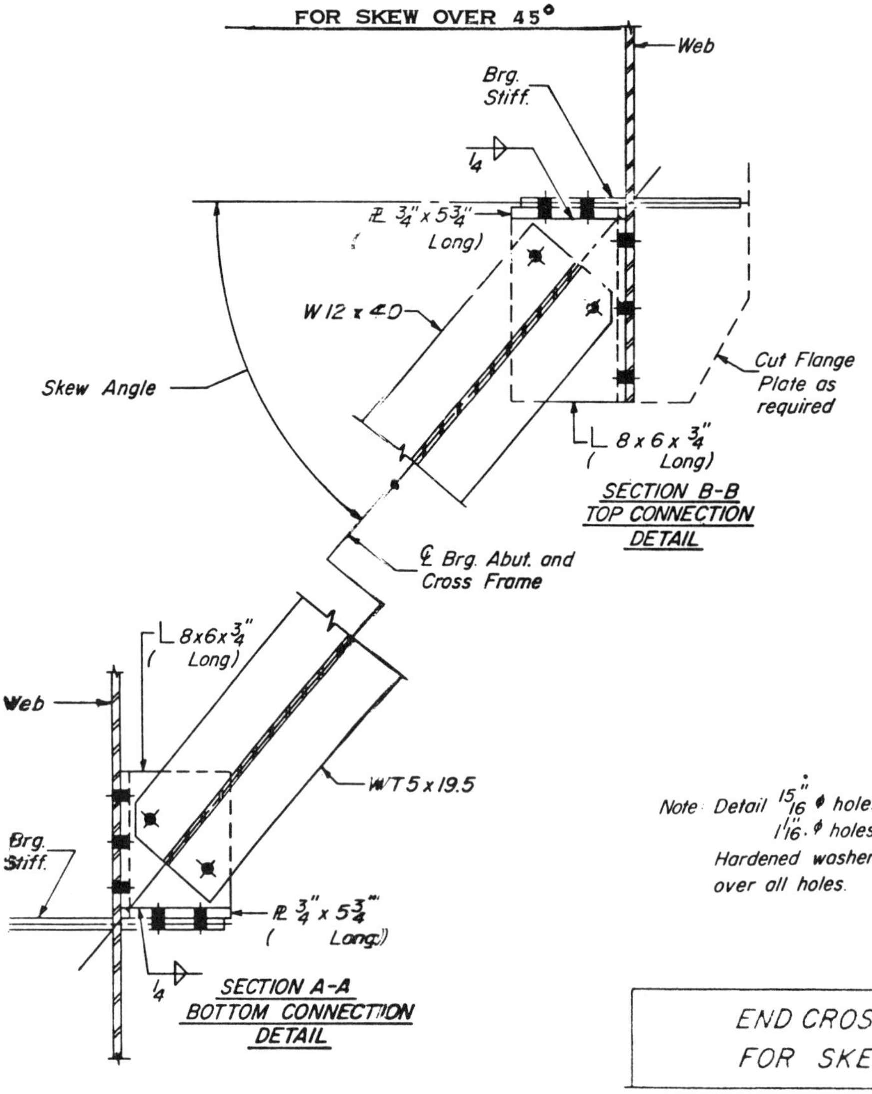


Note: Detail $\frac{15}{16}$" ϕ holes for all $\frac{3}{4}$" ϕ bolts.
$1\frac{1}{16}$" ϕ holes for all $\frac{7}{8}$" ϕ bolts.
Hardened washers shall be required over all holes.

END CROSS FRAME
FOR SKEW ≥ 45°

ILL

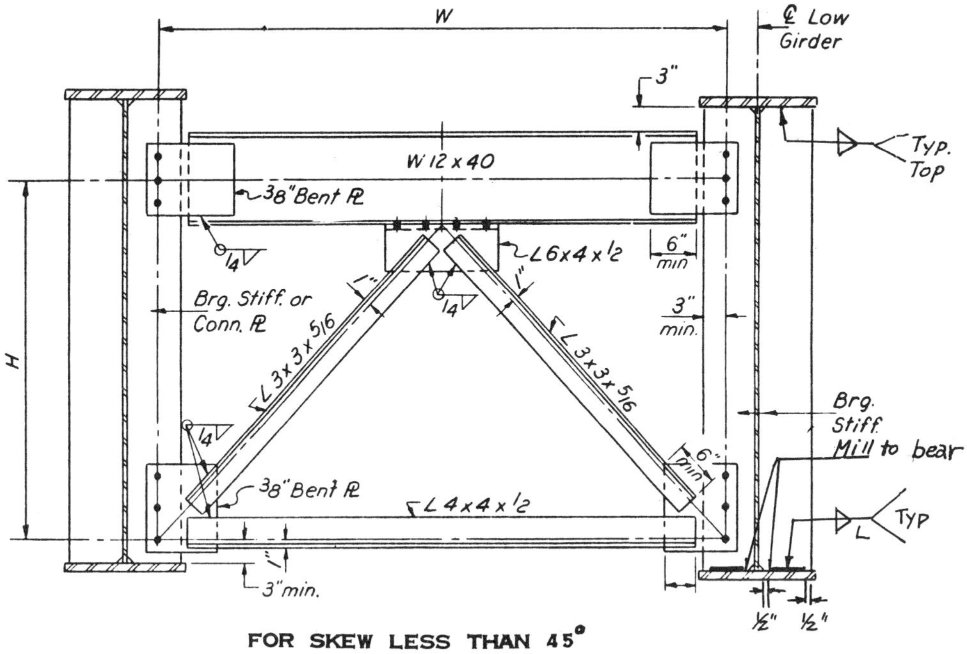


FOR SKEW LESS THAN 45°

TYPICAL END CROSS FRAME

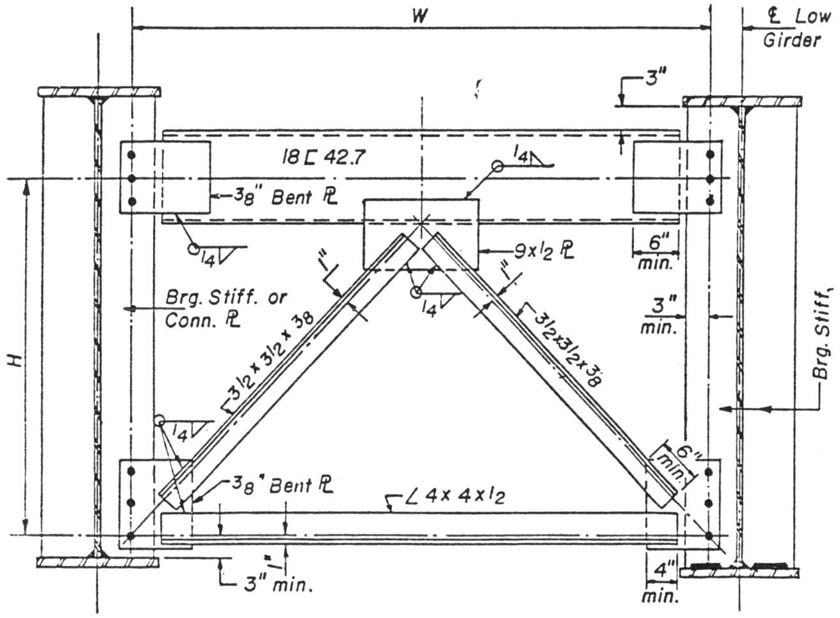


TYPICAL END CROSS FRAME

NOTE:
Detail 1 1/16" ϕ holes in gusset PLs for 7/8" ϕ bolts. Hardened washers shall be required over holes in gusset PLs.

ILL

BEARING STIFFENERS SHALL BE WELDED FOR ALL SKEWED AND CURVED GIRDERS WHEN THESE STIFFENERS ARE USED AS CONNECTION PLATES.

END CROSS FRAME

## LATERAL BRACING

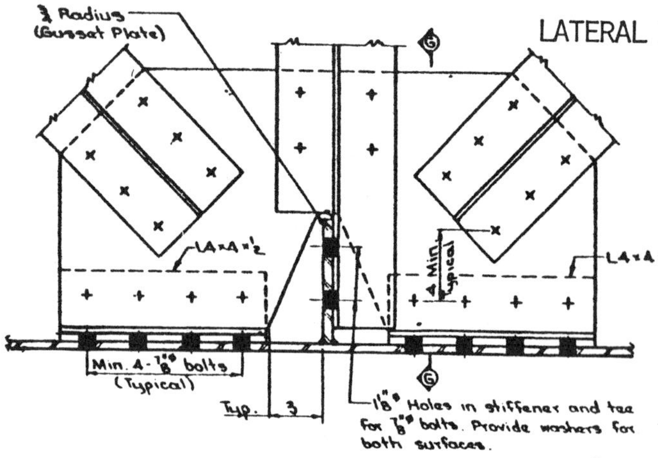


Iowa GUSSET PLATE TO WEB DETAIL

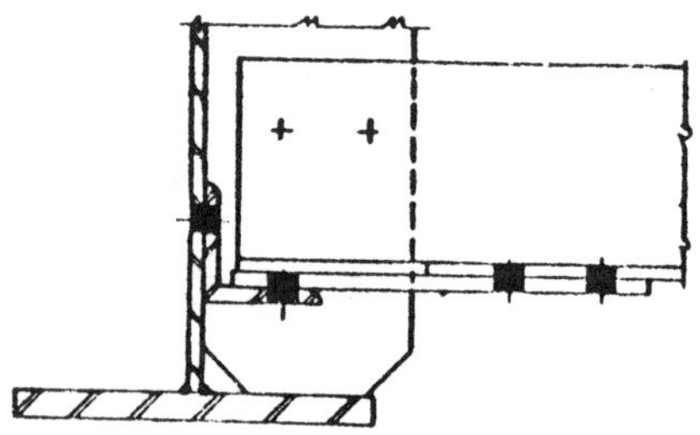

SECTION G-G (IOWA)

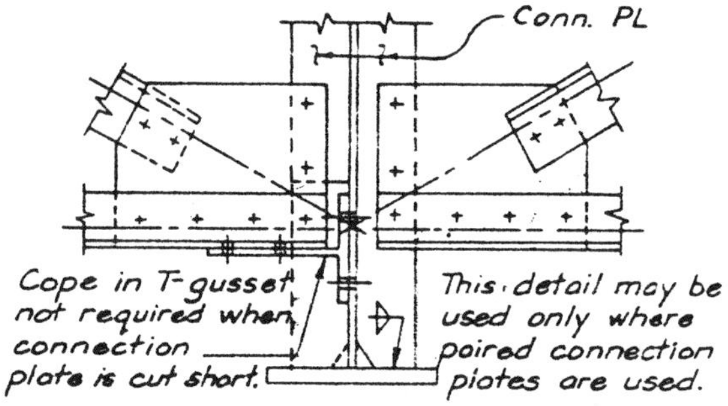


DETAIL F- OPTIONAL

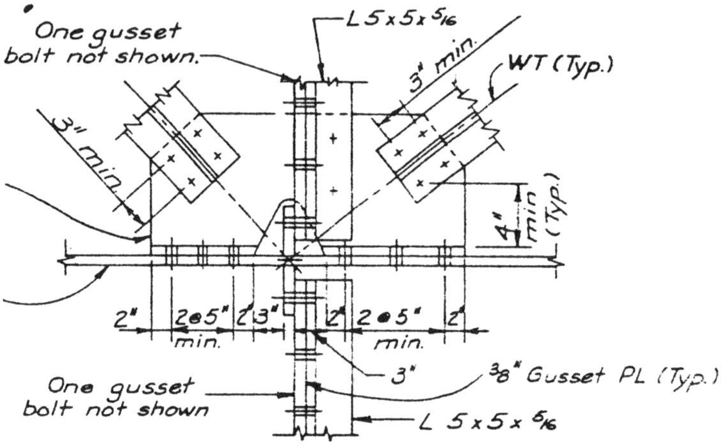


DETAIL D

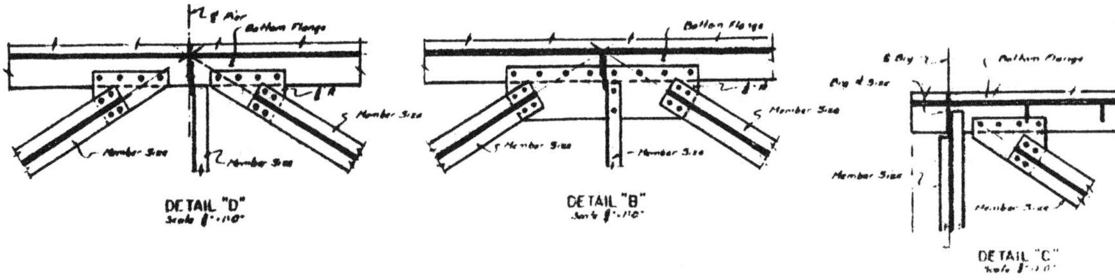


BOLTING GUSSET PLATE TO BOTTOM FLANGE NY

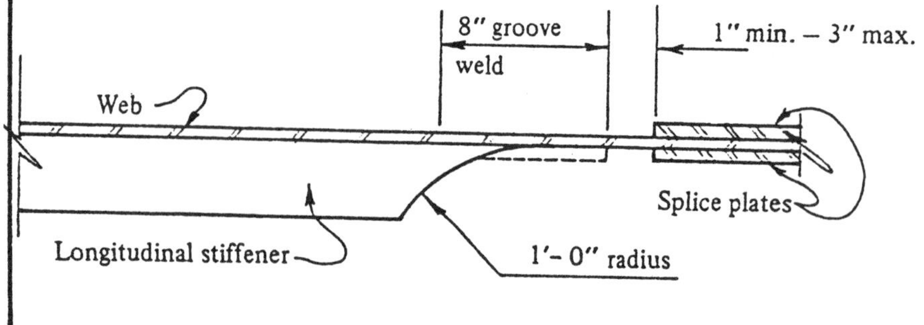


LONGITUDINAL STIFFENER AT BOLTED SPLICE

(NY STATE)

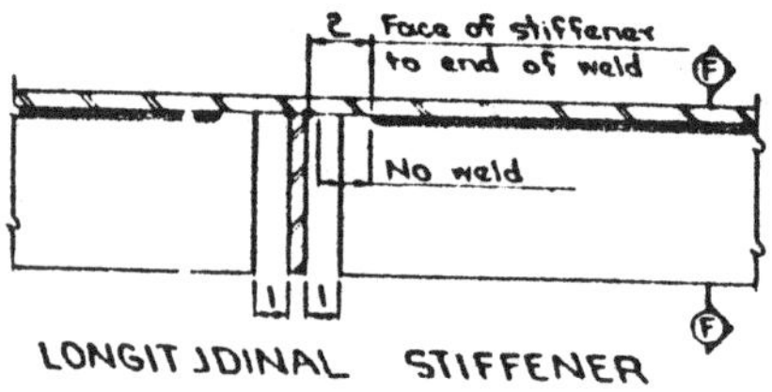


LONGITUDINAL STIFFENER

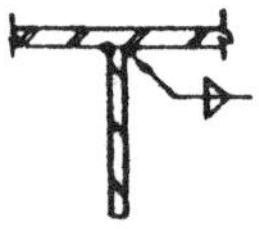

SECTION F-F (IOWA)

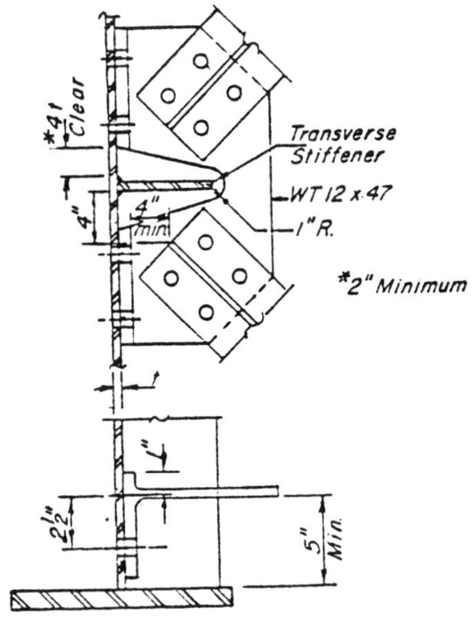


WITH TRANSVERSE STIFFENER and CROSS FRAME CONN. PLATE

ILL

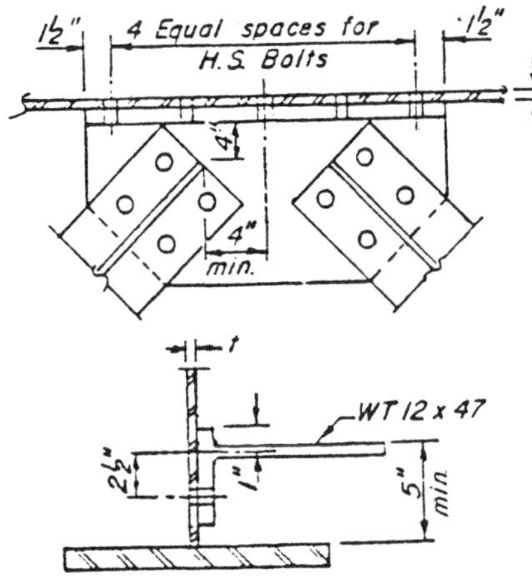


WITHOUT CROSS FRAME CONN. PLATE

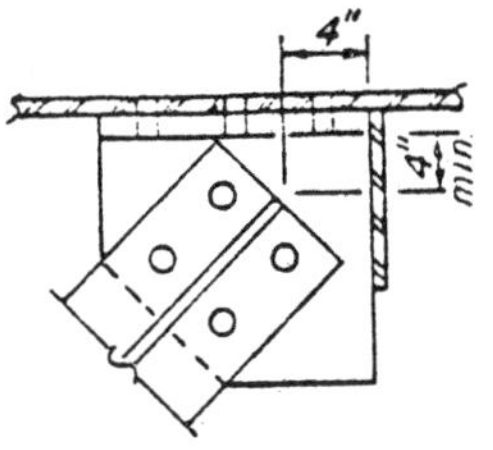


END CONNECTION

(ILL DOT)

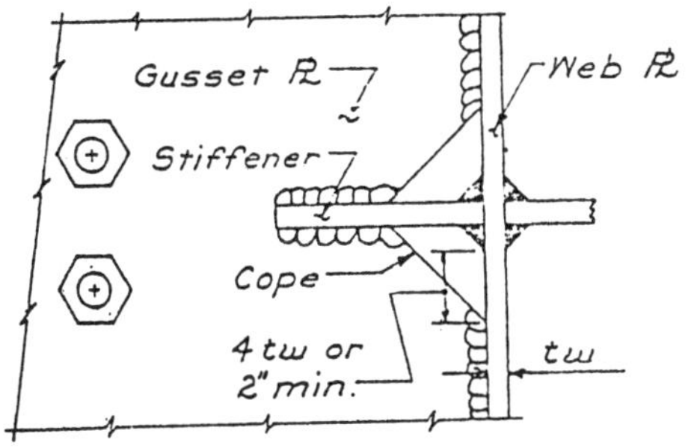


Detail $\frac{13}{16}$"ϕ holes for all $\frac{3}{4}$"ϕ bolts
$1\frac{1}{16}$"ϕ holes for all $\frac{7}{8}$"ϕ bolts.

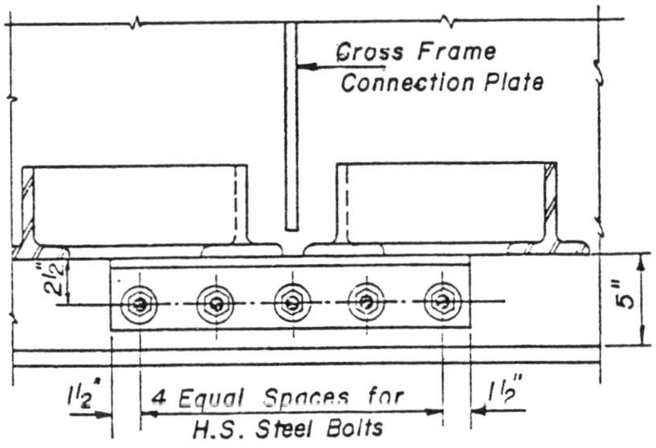


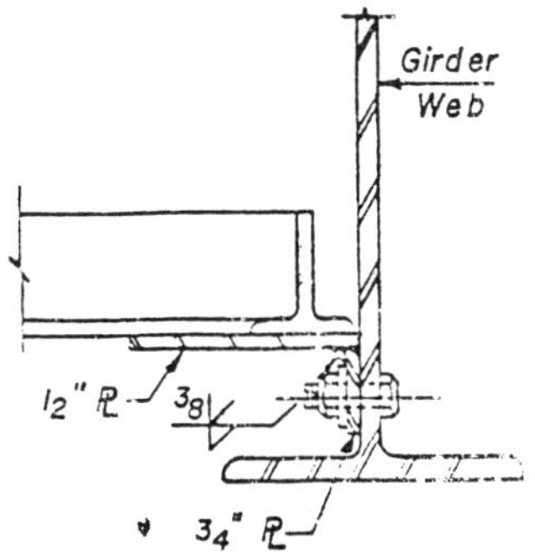


BOTTOM LATERAL BRACING CONNECTION

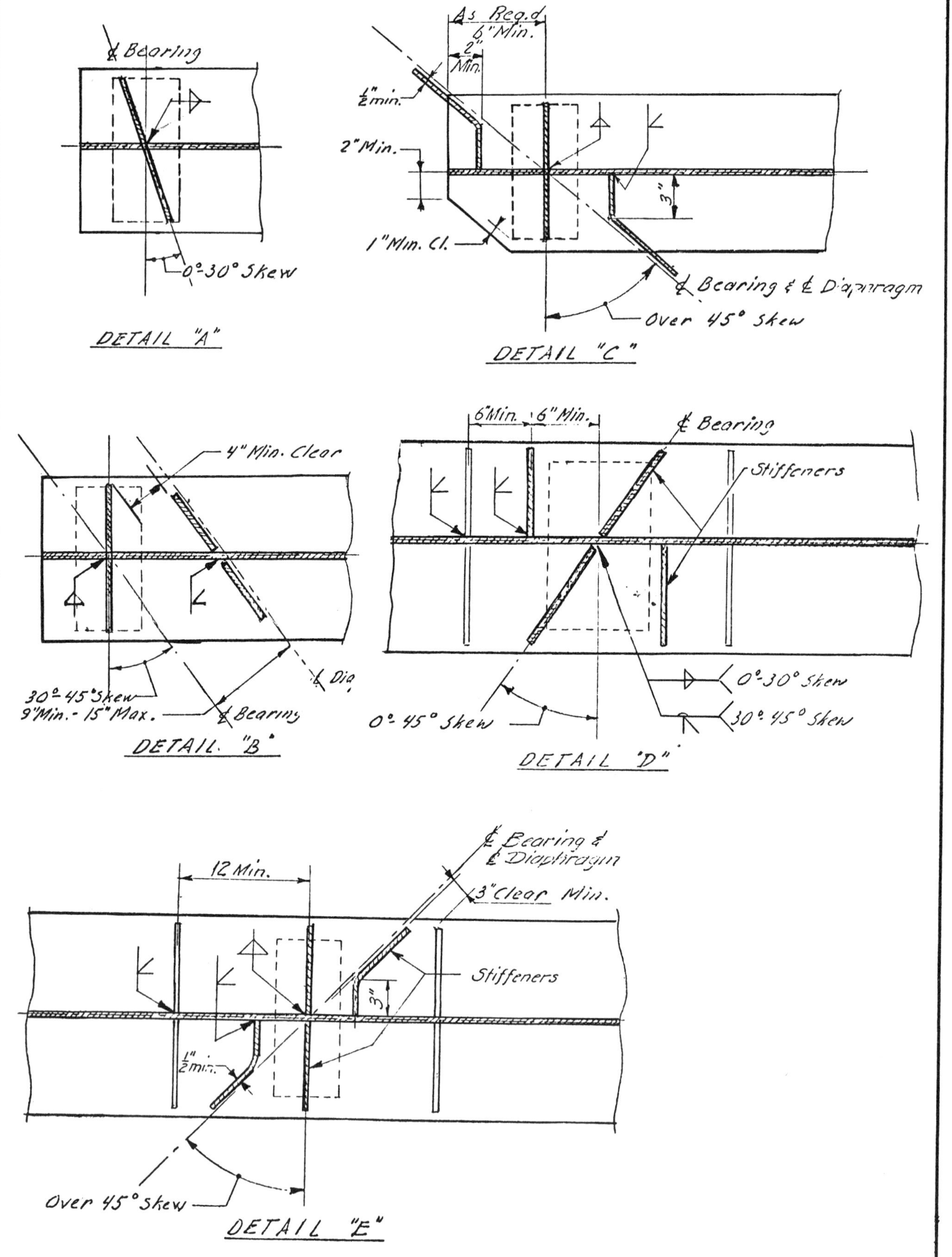


CONNECTION PLATE LOCATION

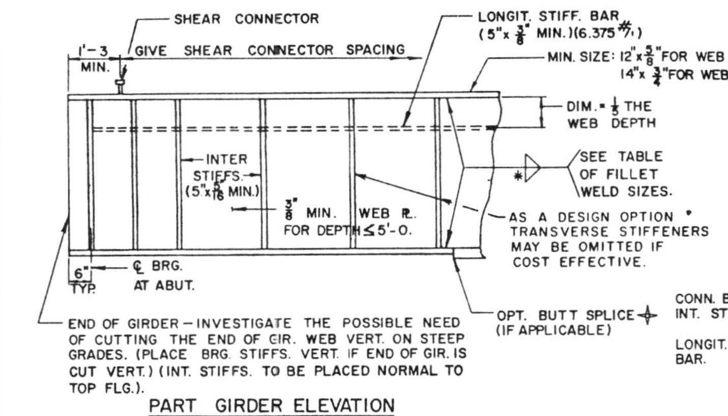


PART GIRDER ELEVATION

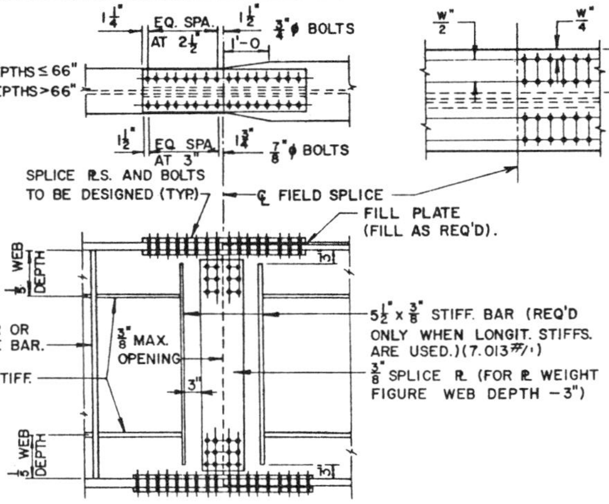


FIELD SPLICE DETAILS

SEE STANDARD 24.7 FOR KINKED GIRDER DETAILS.

## NOTES

OPTIONAL WELDED SHOP SPLICES MAY BE USED FOR ALL FLANGE AND WEB PLATES OVER 60'-0 LONG. IF USED, THE LOCATION OF THE SPLICE SHALL BE SHOWN ON SHOP DRAWINGS AND WILL BE SUBJECT TO THE APPROVAL OF THE ENGINEER.

⌖ OPTIONAL FLANGE BUTT SPLICE. A FLANGE PLATE OF THE LARGER SIZE MAY BE FURNISHED FULL LENGTH, BUT PAY WEIGHT SHALL BE BASED ON SECTIONS AS DETAILED.

(REMINDER – BASE BEARING SEAT ELEVATIONS AT ABUTMENT ON THICKER FLANGE AND DETAIL SHIM PLATES TO ACCOMMODATE THINNER FLANGE.)

AT EXTERIOR GIRDERS PLACE INTERMEDIATE STIFFENERS ON INTERIOR FACE OF GIRDER.

AT INTERIOR GIRDERS PLACE INTERMEDIATE TRANSVERSE STIFFENERS ON ONE SIDE OF GIRDER AND LONGITUDINAL STIFFENER ON THE OPPOSITE SIDE OF GIRDER. KEEP INTERMEDIATE STIFFENERS ON ONE SIDE WHEN LONGITUDINAL STIFFENERS ARE NOT REQUIRED.

AT EXTERIOR GIRDERS PLACE INTERMEDIATE STIFFENERS ALONG ENTIRE LENGTH OF GIRDER AT A MAX. SPACING EQUAL TO THE DEPTH OF WEB. THIS REQUIREMENT IS NECESSARY TO SUPPORT THE FALSEWORK FOR THE SLAB OVERHANG AND MAY BE DISREGARDED IF THE SLAB OVERHANG, MEASURED FROM ℄ WEB, IS 1'-6 OR LESS. PLACE LONGITUDINAL STIFFENERS ON THE OUTSIDE FACE OF EXTERIOR GIRDERS. RUN CONTINUOUS THE LENGTH OF THE GIRDER.

*TABLE OF FILLET WELD SIZES

| MATERIAL THICKNESS OF THICKER PART JOINED. | MIN. SIZE‡ OF FILLET WELD |
|---|---|
| TO 1/2" INCLUSIVE | 3/16" |
| OVER 1/2" TO 3/4" | 1/4" |
| OVER 3/4" | 5/16" |
| | |

‡ EXCEPT THAT THE WELD SIZE SHALL NOT EXCEED THE THICKNESS OF THE THINNER PART JOINED.

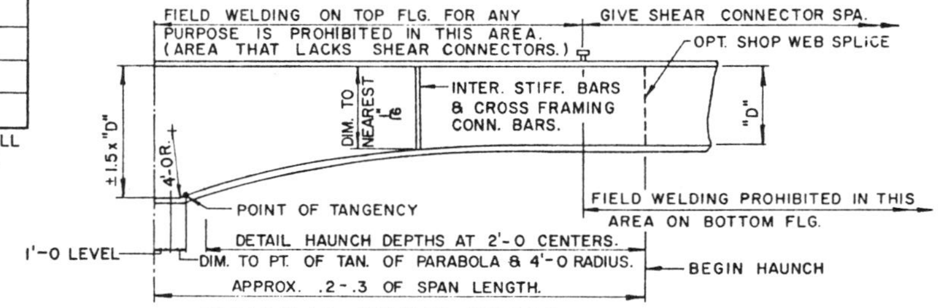


PARABOLIC HAUNCH DETAILS

| WEB | DISTANCE X |
|---|---|
| 3/8" | 2" |
| 7/16" | 2 1/4" |
| 1/2" | 2 1/2" |
| 9/16" | 2 3/4" |
| 5/8" | 3" |
| 11/16" | 3 1/4" |
| 3/4" | 3 1/2" |

TABLE "E"

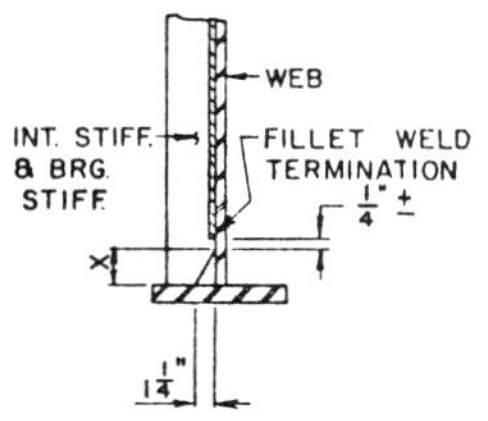


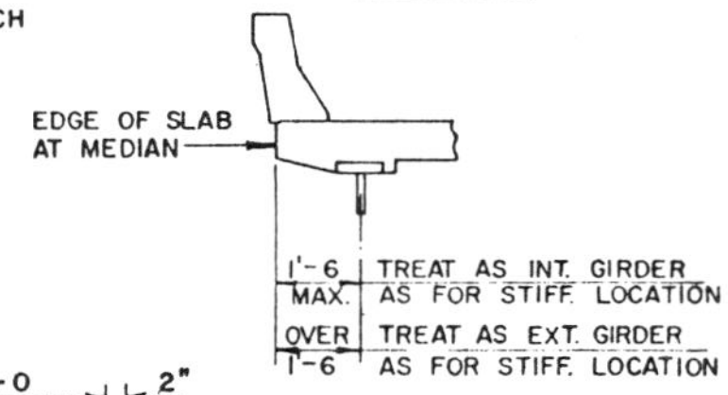


MEDIAN NOTE

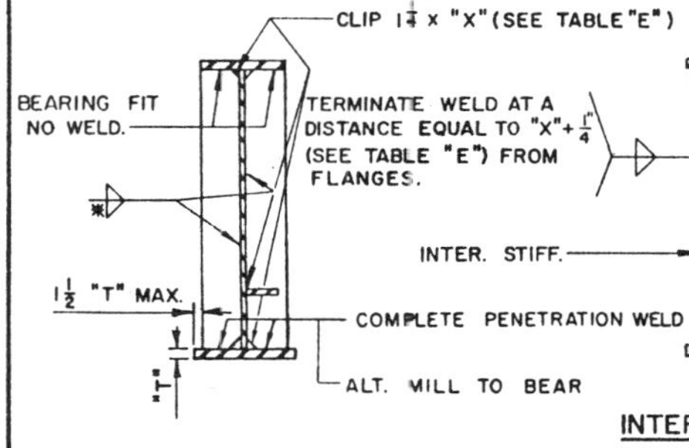


BRG. STIFF. DETAILS
TYP. AT ABUT. & PIER

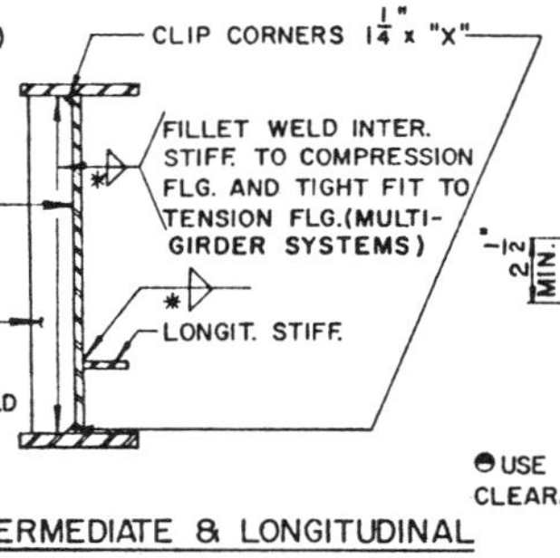


INTERMEDIATE & LONGITUDINAL
STIFF. DETAILS
(ALL GIRDERS)

NOTE: USE THREE SHOP OR FIELD WELDED 7/8" DIA. x 4" LG.● STUDS EQUALLY SPACED WITH A MIN. OF 1 1/2" CL. FROM THE FLG. EDGE. STUDS SHALL NOT BE PLACED OVER FIELD SPLICE PLATES. (2.422 # PER SET OF 3, 4" LONG).

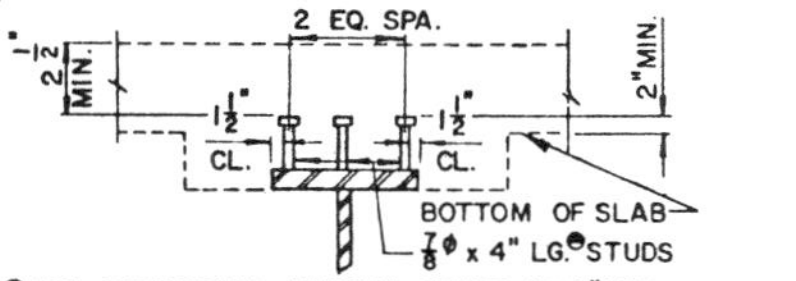


● USE DIFFERENT LENGTH STUDS IF 2" MIN. CLEARANCE OR EXTENSION CRITERIA IS VIOLATED.

SHEAR CONN.
DETAILS

FOR TWO GIRDER SYSTEMS FILLET WELD DIAPHRAGM OR CROSS FRAME CONN. BAR TO BOTH FLANGES.

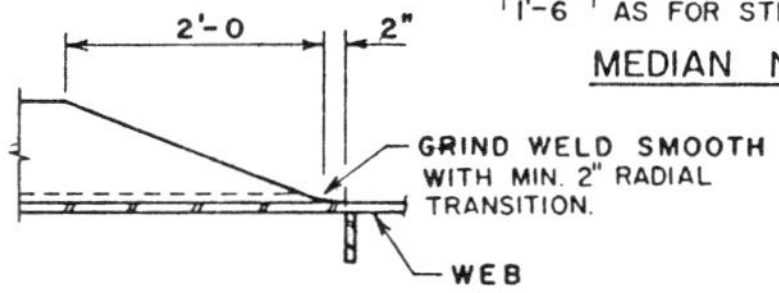


LONGIT. STIFF.
TERMINATION

PLATE GIRDER DETAILS

STATE OF WISCONSIN
DEPARTMENT OF TRANSPORTATION
DIVISION OF HIGHWAYS
BRIDGE SECTION

Approved: Stanley W. Woods, Chief Bridge Engineer | Date: 1-'84

STANDARD 24.2

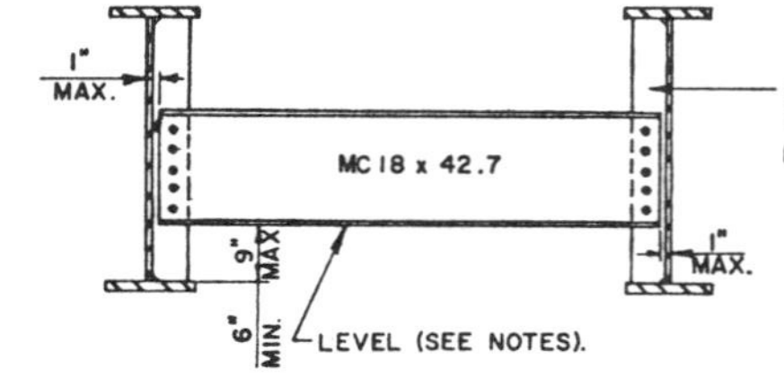


WEB PLATE ≤ 48"

TYP. IN SPAN & AT PIER

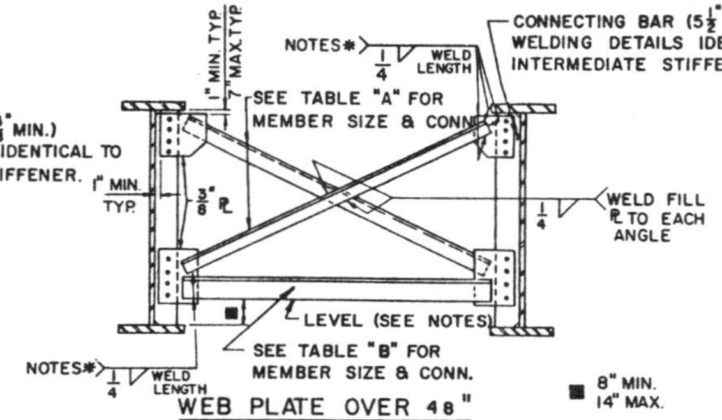


WEB PLATE OVER 48"

TYP. IN SPAN & AT PIER

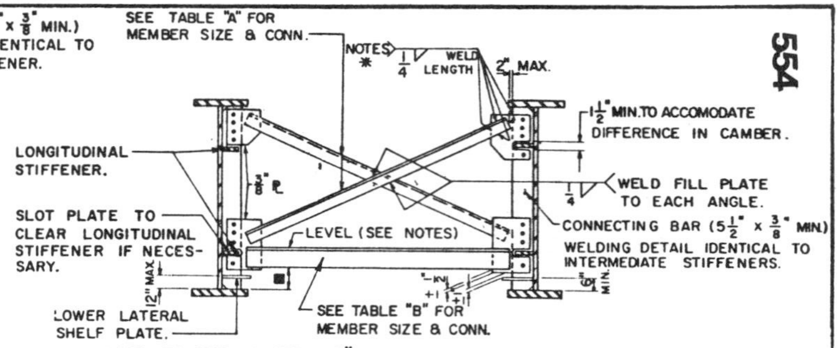


WEB PLATE OVER 48" WITH LONGITUDINAL STIFFENERS

TYP. IN SPAN & AT PIER

TABLE "A"

| SIZE | MAX. LENGTH OF MEMBER | WELD LENGTH | NO. OF ¾"ϕ BOLTS | WEIGHT PER FT. |
|---|---|---|---|---|
| L 3½ x 3½ x $\frac{5}{16}$ | 21'-6 | 9" | 4 | 7.2# |
| L 4 x 4 x $\frac{5}{16}$ | 25'-0 | 11" | 4 | 8.2# |
| L5 x 5 x $\frac{5}{16}$ | 31'-0 | 14" | 5 | 10.3# |

TABLE "B"

| SIZE | MAX. LENGTH OF MEMBER | WELD SIZE | WELD LENGTH | NO. OF ¾"ϕ BOLTS | WEIGHT PER FT. |
|---|---|---|---|---|---|
| L 5 x 5 x $\frac{5}{16}$ | 11'-6 | ¼" | 11" | 4 | 10.3# |
| L 6 x 6 x ⅜ | 13'-6 | $\frac{5}{16}$" | 13" | 6 | 14.9# |
| ⅜" T SECTION SEE DETAIL "A" | 17'-6 | $\frac{5}{16}$" | 14" | 7 | 16.575# |
| ⅜" T SECTION SEE DETAIL "B" | 22'-0 | ⅜" | 13" | 7 | 18.49# |

## NOTES

ALL BOLTED CONNECTIONS SHALL BE FRICTION TYPE USING ¾ ϕ HIGH STRENGTH BOLTS (AST.M. A 325) WITH DOUBLE WASHERS.

FOR SPANS OVER 200', THE CROSS FRAMES AT THE PIERS SHALL BE DESIGNED TO RESIST THE LATERAL LOADS THAT ARE TRANSFERRED TO THE PIERS.

DIAPHRAGMS OR LOWER CROSS FRAME MEMBERS ARE SLOPED WHEN DIFFERENCE IN ADJACENT BOTTOM FLANGE ELEVATIONS EXCEEDS 6". HOLD 8" FROM TOP OF ADJACENT FLANGES TO BOTTOM OF DIAPHRAGMS OR LOWER CROSS FRAME WHEN THESE MEMBERS ARE SLOPED.

DIAPHRAGMS OR LOWER CROSS FRAME MEMBERS THAT ARE LEVEL SHALL BE PLACED 8" ABOVE THE TOP OF THE HIGHER BOTTOM FLANGE OF ADJACENT GIRDERS.

HOLES IN CROSS FRAME PLATES TO BE OVERSIZED @ $\frac{15}{16}$" ϕ DIA.

✱ WELD LENGTH — TERMINATE ALL WELDS ¼" SHORT OF END PIECE. FOR UNPAINTED STEEL AT STREAM CROSSINGS THIS MAY BE A FIELD WELD. ALL AREAS SUBJECT TO FIELD WELDS PROTECTED BY WELDABLE RUSTPROOF COATING. PLACE PRIOR TO WELDING.

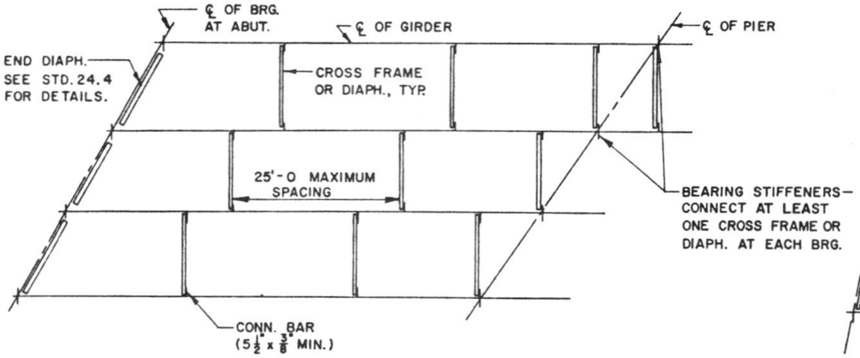


FRAMING PLAN FOR SKEW > 15°

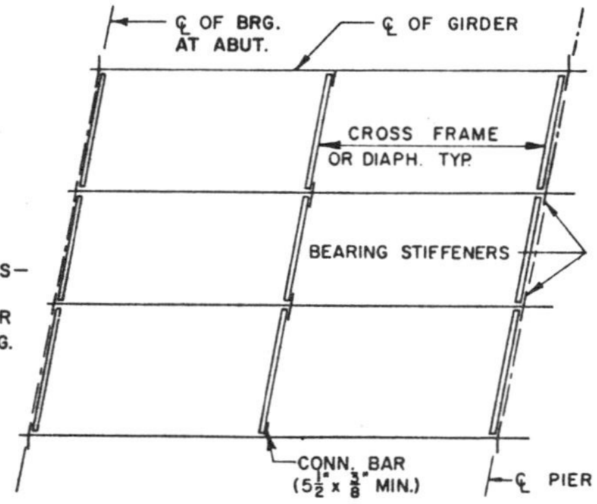


FRAMING PLAN FOR SKEW ≤ 15°

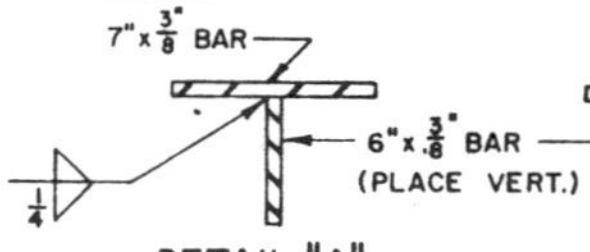


DETAIL "A"

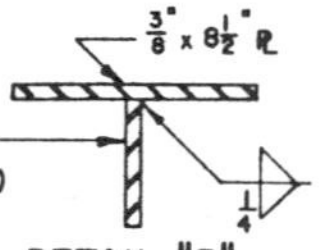


DETAIL "B"

| PLATE GIRDER DIAPHRAGMS AND CROSS FRAMES | |
|---|---|
| STATE OF WISCONSIN DEPARTMENT OF TRANSPORTATION DIVISION OF HIGHWAYS BRIDGE SECTION | |
| Approved: Stanley W. Woods, Chief Bridge Engineer | Date 6-'83 |

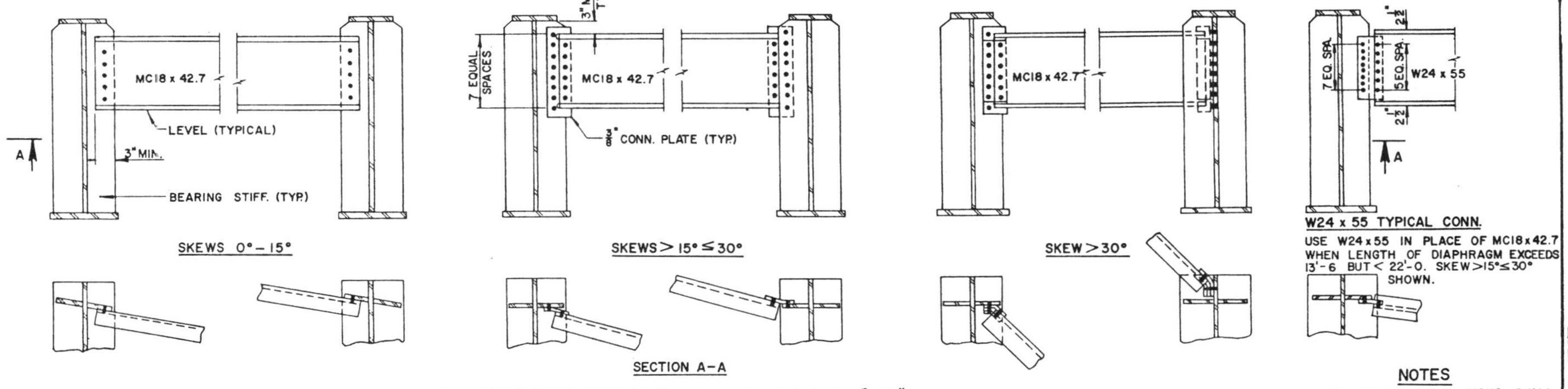


**W24 x 55 TYPICAL CONN.**

USE W24 x 55 IN PLACE OF MC18 x 42.7 WHEN LENGTH OF DIAPHRAGM EXCEEDS 13'-6 BUT < 22'-0. SKEW > 15° ≤ 30° SHOWN.

## END DIAPHRAGM CONNECTIONS – WEB DEPTHS ≤ 48"

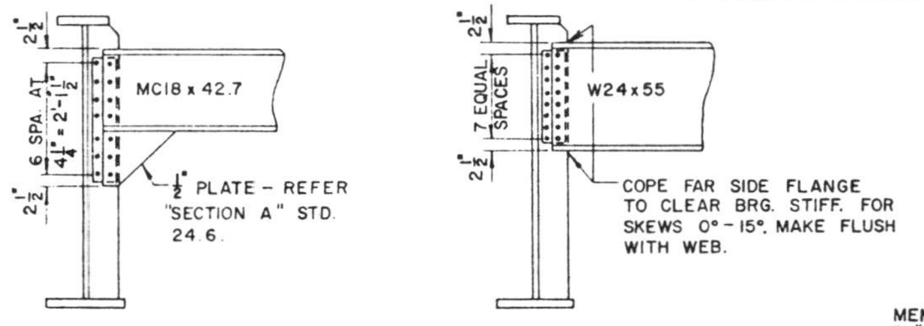


TYPICAL CONNECTION FOR MC18 x 42.7 AND W24 x 55. USE MC18 x 42.7 WHEN DIAPHRAGM LENGTH IS ≤ 13'-6. USE W24 x 55 FOR LENGTHS > 13'-6 < 22'-0. SKEW > 15° ≤ 30° SHOWN.

## END DIAPHRAGM CONNECTIONS – WEB DEPTHS > 48" ≤ 60"

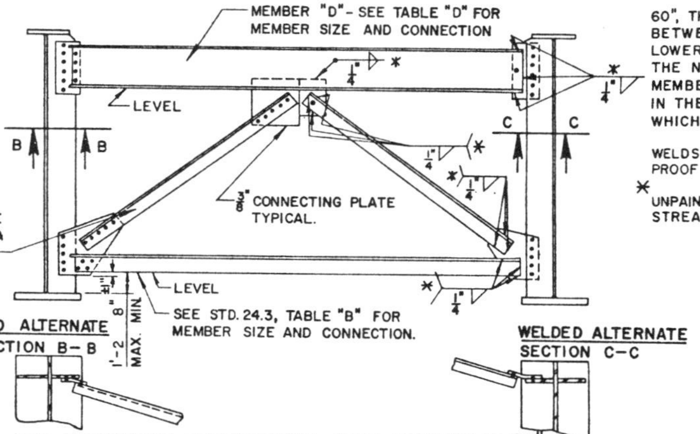


## END DIAPHRAGM CONNECTIONS – WEB DEPTHS > 60"

SKEWS > 15° ≤ 30° SHOWN

### NOTES

ALL BOLTED CONNECTIONS SHALL BE FRICTION TYPE MADE WITH 3/4" ϕ HIGH STRENGTH BOLTS. (A.S.T.M. A325)

FOR WEB DEPTHS GREATER THAN 60", THE NUMBER OF BOLTS REQUIRED BETWEEN BEARING STIFFENERS AND LOWER CONNECTING PLATES EQUALS THE NUMBER OF BOLTS REQUIRED IN MEMBER "C" OR THE NUMBER REQUIRED IN THE LOWER HORIZONTAL MEMBER, WHICH EVER IS GREATER.

ALL AREAS SUBJECT TO FIELD WELDS PROTECTED BY WELDABLE RUST PROOF COATING PRIOR TO WELDING.

* FIELD WELDS ALLOWED ONLY ON UNPAINTED STEEL AT DESIGNATED STREAM CROSSINGS

### TABLE "D"

| MEMBER "C" MAXIMUM LENGTH | WEB DEPTH 5'-0 – 6'-6: MEMBER "C" SIZE | 5'-0 – 6'-6: NO. OF 3/4" ϕ BOLTS | 5'-0 – 6'-6: LENGTH OF 1/4" WELD | 6'-6 – 7'-6: MEMBER "C" SIZE | 6'-6 – 7'-6: NO. OF 3/4" ϕ BOLTS | 6'-6 – 7'-6: LENGTH OF 1/4" WELD | 7'-6 – 8'-9: MEMBER "C" SIZE | 7'-6 – 8'-9: NO. OF 3/4" ϕ BOLTS | 7'-6 – 8'-9: LENGTH OF 1/4" WELD | MEMBER "D" SIZE | MEMBER "D" CONN. NO. OF 3/4" ϕ BOLTS: CONN. PLATE TO BRG. STIFF. | MEMBER "D" CONN. NO. OF 3/4" ϕ BOLTS: MEMBER "D" |
|---|---|---|---|---|---|---|---|---|---|---|---|---|
| 11'-6 | 4x4x5/16 | 5 | 13 | 4x4x5/16 | 5 | 12 | 4x4x5/16 | 5 | 11 | C12 x 20.7 | 6 @ 2 1/2" | 4 @ 2 1/2" |
| 13'-6 | 5x5x5/16 | 6 | 17 | 5x5x5/16 | 6 | 16 | 5x5x5/16 | 6 | 15 | C12 x 20.7 | 6 @ 2 1/2" | 4 @ 2 1/2" |
| 17'-6 | 6x6x5/16 | 8 | 20 | 5x5x5/16 | 7 | 18 | 5x5x5/16 | 6 | 16 | C15 x 33.9 | 7 @ 2 1/2" | 5 @ 2 1/2" |
| 22'-0 | 6x6x3/8 | 9 | 23 | 6x6x5/16 | 8 | 21 | 6x6x5/16 | 7 | 19 | MC18 x 42.7 | 7 @ 2 1/2" | 6 @ 2 1/2" |

NOTE: ALL MEMBER "C" SIZES REPRESENT ANGLES.

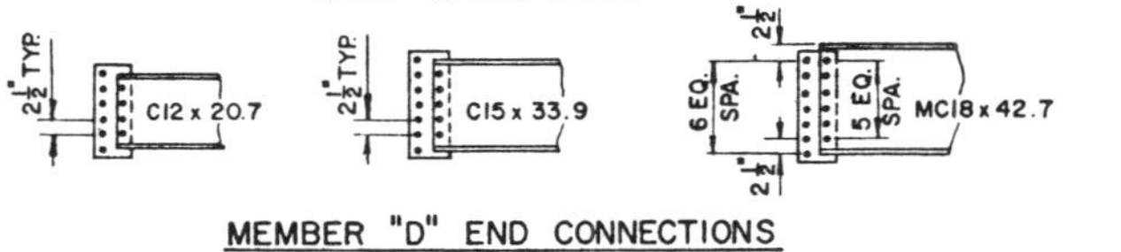


## MEMBER "D" END CONNECTIONS

OMIT END CONNECTING PLATE FOR SKEWS 0° - 15°

| END DIAPHRAMS | |
|---|---|
| STATE OF WISCONSIN DEPARTMENT OF TRANSPORTATION DIVISION OF HIGHWAYS BRIDGE SECTION | |
| Approved: [signature] Chief Bridge Engineer | Date: 1 - 84 |

STANDARD 24.4

| STATE PROJECT NUMBER | SHEET NO |
| --- | --- |
| | |

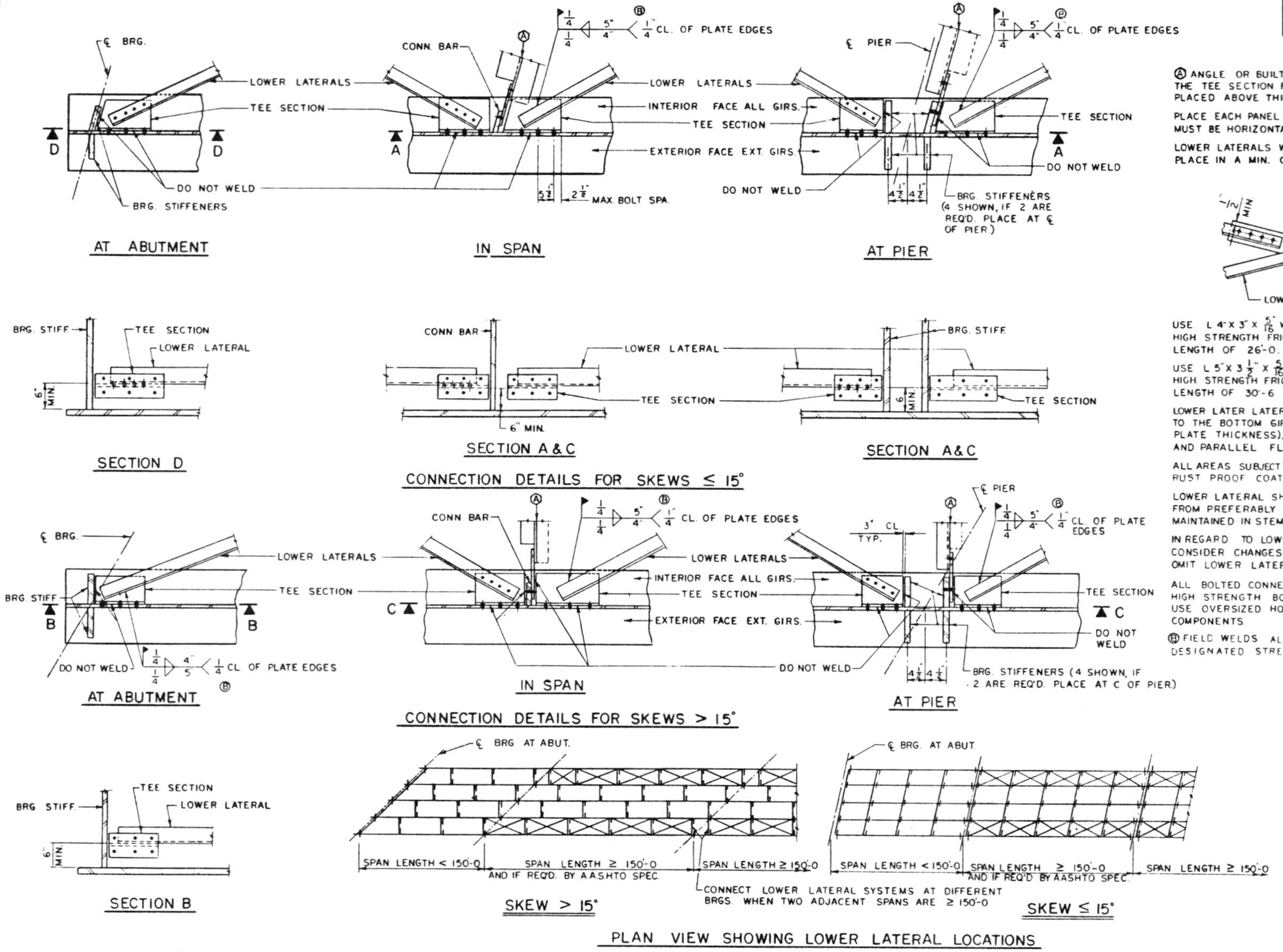


## NOTES

Ⓐ ANGLE OR BUILT-UP SECTION FROM TABLE B, STANDARD 24.3. THE TEE SECTION FOR THE LOWER LATERAL CONNECTION MAY BE PLACED ABOVE THIS MEMBER WHEN THIS MEMBER IS SLOPED.

PLACE EACH PANEL OF LOWER LATERALS IN ONE PLANE. PANELS MUST BE HORIZONTAL IN THE TRANSVERSE DIRECTION ONLY.

LOWER LATERALS WHEN REQ'D. IN SPANS OF 150'-0 OR GREATER. PLACE IN A MIN. OF $\frac{1}{3}$ OF THE SPAN BAYS.

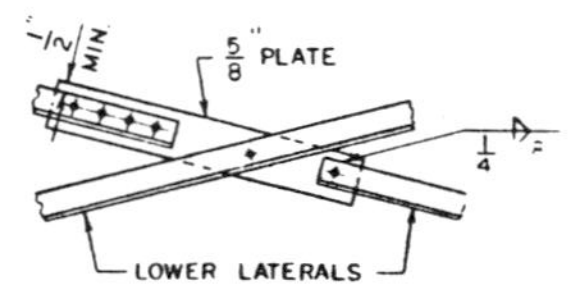


USE L 4"x 3" x $\frac{5}{16}$" WITH $10\frac{1}{2}$" LENGTH OF $\frac{1}{4}$" WELD AND 4 – $\frac{3}{4}$"Ø HIGH STRENGTH FRICTION TYPE BOLTS FOR MAX. LOWER LATERAL LENGTH OF 26'-0. (PLACE 4" LEG VERT) (WT. = 7.2 LBS./FT.)

USE L 5"x $3\frac{1}{2}$" x $\frac{5}{16}$" WITH $12\frac{1}{2}$" LENGTH OF $\frac{1}{4}$" WELD AND 4 – $\frac{3}{4}$"Ø HIGH STRENGTH FRICTION TYPE BOLTS FOR MAX. LOWER LATERAL LENGTH OF 30'-6 (PLACE 5" LEG VERT.)(WT. = 8.7 LBS./FT.)

LOWER LATER LATERAL CONNECTION STEMS SHALL BE AS CLOSE TO THE BOTTOM GIRDER FLANGE (BUT NO CLOSER THAN 6" PLUS PLATE THICKNESS), AS FEASIBLE. THIS APPLIES TO BOTH HAUNCHED AND PARALLEL FLANGE GIRDERS.

ALL AREAS SUBJECT TO FIELD WELDS PROTECTED BY WELDABLE RUST PROOF COATINGS PRIOR TO WELDING.

LOWER LATERAL SHELFS TO BE PROVIDED BY TEE SECTION CUT FROM PREFERABLY S 18 X 55 SECTIONS. $\frac{3}{8}$" + THICKNESS TO BE MAINTAINED IN STEM.

IN REGARD TO LOWER LATERALS THE DESIGNER IS ADVISED TO CONSIDER CHANGES OF PLATE SIZES OR DIAPHRAGM SPACING TO OMIT LOWER LATERAL BRACING.

ALL BOLTED CONNECTIONS SHALL BE FRICTION TYPE USING $\frac{3}{4}$"Ø HIGH STRENGTH BOLTS (ASTM A325) WITH DOUBLE WASHERS. USE OVERSIZED HOLES $\frac{15}{16}$"Ø IN LOWER LATERAL BRACING COMPONENTS.

Ⓑ FIELD WELDS ALLOWED ONLY ON UNPAINTED STEEL AT DESIGNATED STREAM CROSSINGS

**LOWER LATERAL DETAILS**

*STATE OF WISCONSIN*
*DEPARTMENT OF TRANSPORTATION*
*DIVISION OF HIGHWAYS*
*BRIDGE SECTION*

Approved: Stanley W. Woods, Chief Bridge Engineer — Date 1-84

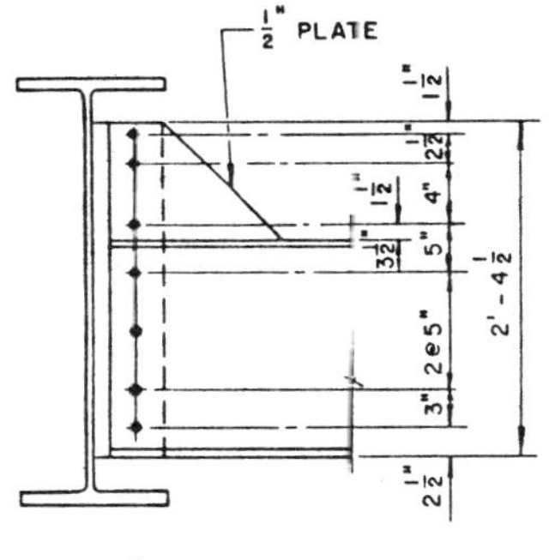


36" W. GIRDER

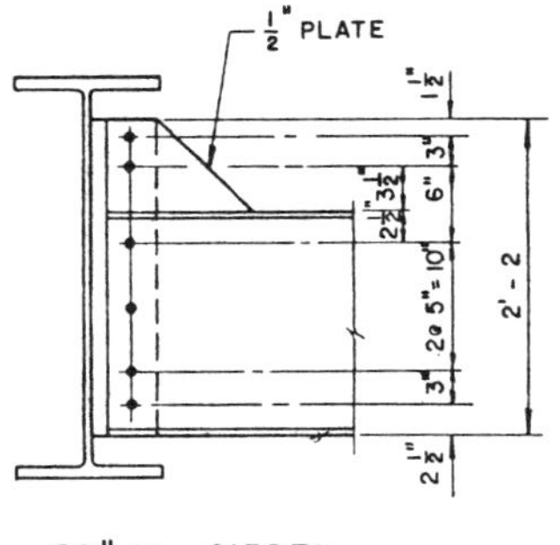


33" W. GIRDER

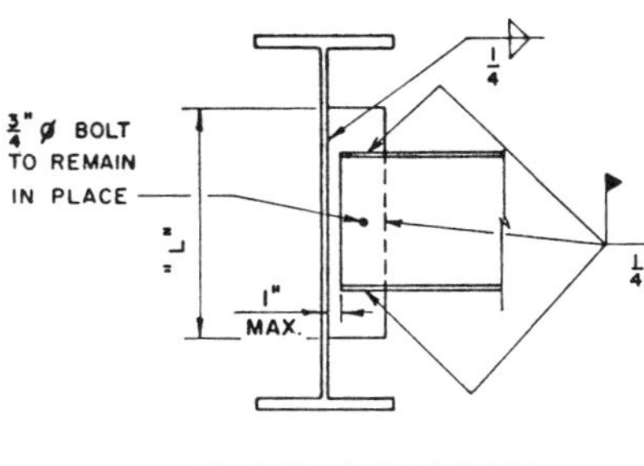


WELDED CONNECTION

TYPICAL ALL GIRDERS
SEE ALSO NOTES

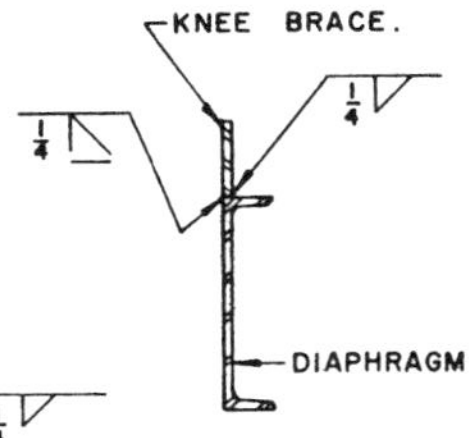


SECTION A

TYPICAL ALL
BOLTED CONNECTIONS

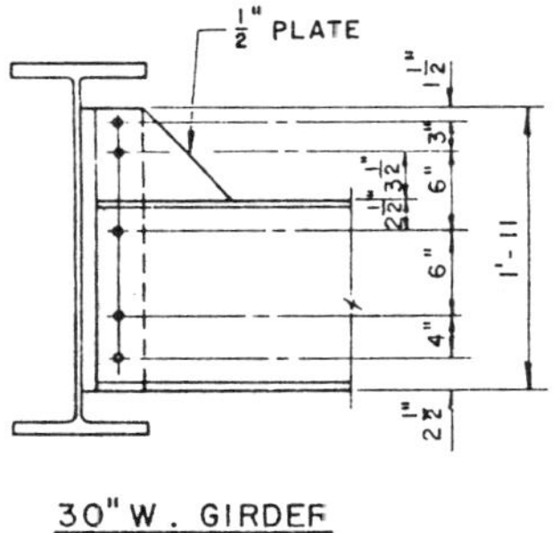


30" W. GIRDEF

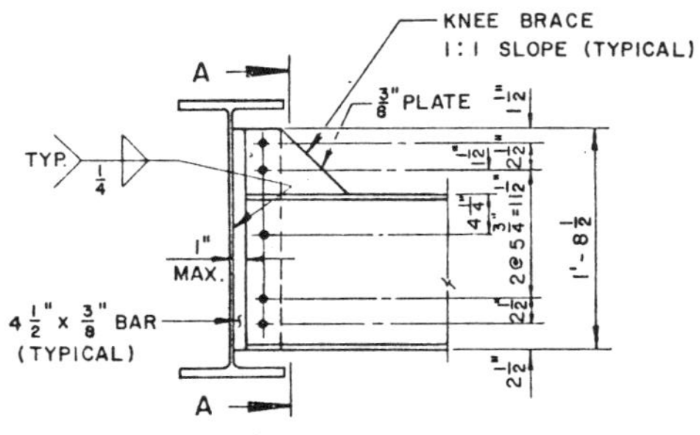


27" W. GIRDER

## INTERMEDIATE DIAPHRAGM SIZES

| ALL INTERMEDIATE CONNECTIONS | | WELDED INTERMEDIATE CONNECTION ONLY | |
|---|---|---|---|
| GIRDER DEPTH | INTERMEDIATE DIAPHRAGMS | CONNECTING BARS | "L" |
| 36" | MC18 x 42.7 | $5\frac{1}{2}$" x $\frac{1}{2}$" | 2'-0 |
| 33" | MC18 x 42.7 | $5\frac{1}{2}$" x $\frac{1}{2}$" | 2'-0 |
| 30" | C15 x 33.9 | 5" x $\frac{1}{2}$" | 1'-9 |
| 27" | C15 x 33.9 | 5" x $\frac{1}{2}$" | 1'-9 |
| 24" | C12 x 20.7 | $4\frac{1}{4}$" x $\frac{1}{2}$" | 1'-6 |
| 21" | C10 x 15.3 | 4" x $\frac{1}{2}$" | 1'-4 |
| 18" | C8 x 11.5 | 4" x $\frac{1}{2}$" | 1'-1 |

## NOTES

PLACE WELDED DIAPHRAGM CONNECTION SYMMETRICALLY ABOUT X-X AXIS OF LOWER GIRDER.

DIAPHRAGMS SHALL BE HORIZONTAL EXCEPT WHEN THE DIFFERENCE IN ADJACENT GIRDER ELEVATIONS IS OF A MAGNITUDE THAT NECESSITATES SLOPING THE DIAPHRAGMS.

WHEN DIAPHRAGMS ARE SLOPED, PLACE CONNECTING BARS SYMMETRICAL ABOUT X-X AXIS OF GIRDERS.

USE A $4\frac{1}{2}$" x $\frac{1}{2}$" CONNECTING BAR FOR ALL BOLTED CONNECTIONS.

ALL BOLTED CONNECTIONS SHALL BE MADE WITH $\frac{3}{4}$" HIGH STRENGTH BOLTS. (A.S.T.M. A325)

THE WELDED CONNECTION MAY BE USED ONLY IN CASES OF UNPAINTED STEEL AT STREAM CROSSINGS WHERE THE SURFACES TO BE FIELD WELDED ARE PROTECTED WITH A COATING OF WELDABLE RUST PROOFING.

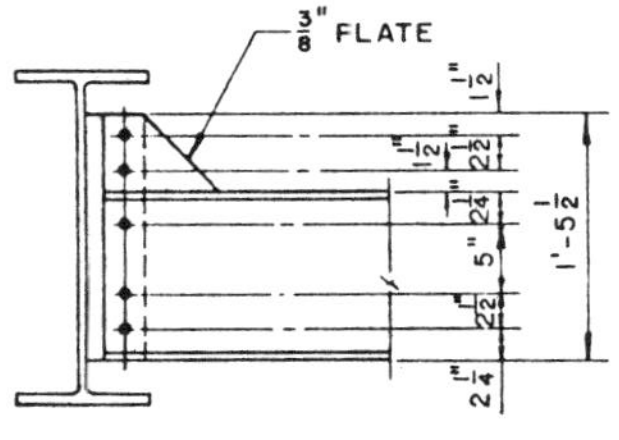


24" W. GIRDER

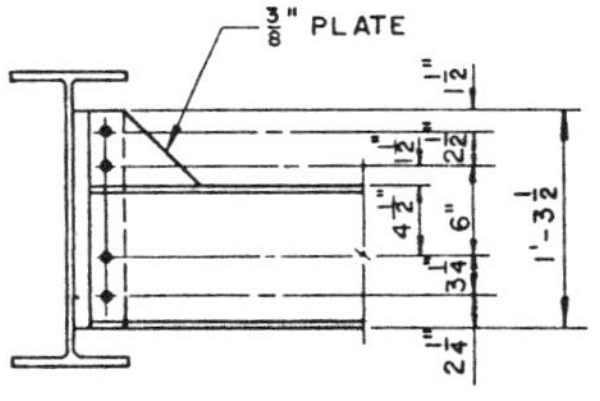


21" W. GIRDER

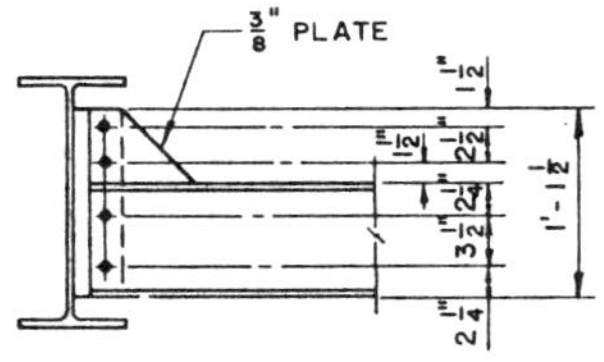


18" W. GIRDER

ROLLED GIRDER DIAPHRAGMS

STATE OF WISCONSIN
DEPARTMENT OF TRANSPORTATION
DIVISION OF HIGHWAYS
BRIDGE SECTION

Approved: [signature] Chief Bridge Engineer | Date: 1-'84

STANDARD 24.6

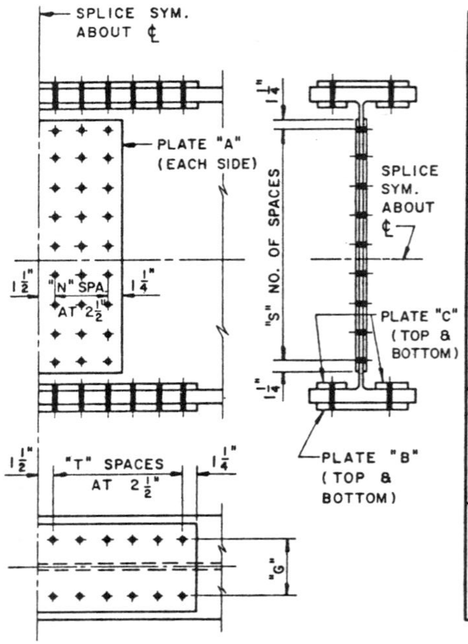


## GIRDERS OF A36 STEEL

| SECTION | | WEB SPLICE | | | FLANGE SPLICE | | | | MOMENT CAPACITY OF SPLICE FT. KIPS. | WEIGHT OF SPLICE LBS. |
|---|---|---|---|---|---|---|---|---|---|---|
| | | "N" | "S" | PLATE "A" | "T" | "G" | PLATE "B" | PLATE "C" | | |
| W 24 | 68 | 1 | 6 | 10 1/2 x 3/8 | 3 | 5 | 8 x 3/8 x 1-8 1/2 | 3 x 3/8 x 1-8 1/2 | 180 | 122 |
| | 76 | 1 | 6 | 10 1/2 x 3/8 | 3 | 5 | 8 x 3/8 x 1-8 1/2 | 3 x 1/2 x 1-8 1/2 | 205 | 131 |
| | 84 | 1 | 7 | 10 1/2 x 3/8 | 4 | 5 1/2 | 9 x 3/8 x 2-1 1/2 | 3 1/2 x 7/16 x 2-1 1/2 | 225 | 158 |
| W 27 | 84 | 1 | 8 | 10 1/2 x 3/8 | 4 | 5 1/2 | 9 x 3/8 x 2-1 1/2 | 3 1/2 x 3/8 x 2-1 1/2 | 245 | 159 |
| | 94 | 2 | 5 | 15 1/2 x 3/8 | 4 | 5 1/2 | 9 x 3/8 x 2-1 1/2 | 3 1/2 x 9/16 x 2-1 1/2 | 285 | 203 |
| W 30 | 99 | 2 | 7 | 15 1/2 x 3/8 | 4 | 5 1/2 | 9 x 3/8 x 2-1 1/2 | 3 1/2 x 1/2 x 2-1 1/2 | 320 | 209 |
| | 108 | 2 | 7 | 15 1/2 x 3/8 | 5 | 5 1/2 | 9 x 7/16 x 2-6 1/2 | 3 1/2 x 1/2 x 2-6 1/2 | 349 | 241 |
| | 116 | 2 | 7 | 15 1/2 x 3/8 | 5 | 6 | 9 1/2 x 7/16 x 2-6 1/2 | 3 1/2 x 5/8 x 2-6 1/2 | 385 | 259 |
| | 124 | 2 | 7 | 15 1/2 x 3/8 | 5 | 6 1/2 | 10 x 7/16 x 2-6 1/2 | 3 1/2 x 3/4 x 2-6 1/2 | 425 | 279 |
| | 132 | 2 | 8 | 15 1/2 x 3/8 | 6 | 6 1/2 | 10 x 1/2 x 2-11 1/2 | 3 1/2 x 3/4 x 2-11 1/2 | 450 | 323 |
| W 33 | 118 | 2 | 8 | 15 1/2 x 3/8 | 5 | 6 | 10 x 7/16 x 2-6 1/2 | 4 x 1/2 x 2-6 1/2 | 440 | 269 |
| | 130 | 2 | 8 | 15 1/2 x 3/8 | 5 | 6 | 10 x 7/16 x 2-6 1/2 | 4 x 5/8 x 2-6 1/2 | 480 | 286 |
| | 141 | 2 | 8 | 15 1/2 x 3/8 | 6 | 7 | 11 x 7/16 x 2-11 1/2 | 4 x 5/8 x 2-11 1/2 | 515 | 324 |
| | 152 | 2 | 9 | 15 1/2 x 3/8 | 7 | 7 | 11 x 1/2 x 3-4 1/2 | 4 x 3/4 x 3-4 1/2 | 580 | 393 |
| W 36 | 135 | 2 | 9 | 15 1/2 x 3/8 | 6 | 7 | 11 x 7/16 x 2-11 1/2 | 4 x 1/2 x 2-11 1/2 | 530 | 314 |
| | 150 | 2 | 10 | 15 1/2 x 3/8 | 7 | 7 | 11 x 1/2 x 3-4 1/2 | 4 x 5/8 x 3-4 1/2 | 600 | 381 |
| | 160 | 2 | 10 | 15 1/2 x 3/8 | 7 | 7 | 11 x 1/2 x 3-4 1/2 | 4 x 3/4 x 3-4 1/2 | 645 | 404 |
| | 170 | 2 | 11 | 15 1/2 x 3/8 | 8 | 7 | 11 x 9/16 x 3-9 1/2 | 4 x 3/4 x 3-9 1/2 | 690 | 458 |

## GIRDERS OF A572/A588 STEEL

| SECTION | | WEB SPLICE | | | FLANGE SPLICE | | | | MOMENT CAPACITY OF SPLICE FT. KIPS. | WEIGHT OF SPLICE LBS. |
|---|---|---|---|---|---|---|---|---|---|---|
| | | "N" | "S" | PLATE "A" | "T" | "G" | PLATE "B" | PLATE "C" | | |
| W 24 | 68 | 2 | 6 | 15 1/2 x 3/8 | 4 | 5 | 8 1/2 x 7/16 x 2-1 1/2 | 3 1/2 x 1/2 x 2-1 1/2 | 270 | 193 |
| | 76 | 2 | 6 | 15 1/2 x 3/8 | 4 | 5 | 8 1/2 x 7/16 x 2-1 1/2 | 3 1/2 x 9/16 x 2-1 1/2 | 270 | 199 |
| | 84 | 2 | 6 | 15 1/2 x 3/8 | 5 | 5 | 8 1/2 x 1/2 x 2-6 1/2 | 3 1/2 x 1/2 x 2-6 1/2 | 280 | 225 |
| W 27 | 84 | 2 | 7 | 15 1/2 x 3/8 | 5 | 5 | 8 1/2 x 1/2 x 2-6 1/2 | 3 1/2 x 9/16 x 2-6 1/2 | 330 | 245 |
| | 94 | 2 | 8 | 15 1/2 x 3/8 | 6 | 5 1/2 | 9 x 9/16 x 2-11 1/2 | 3 1/2 x 5/8 x 2-11 1/2 | 380 | 297 |
| W 30 | 99 | 2 | 9 | 15 1/2 x 7/16 | 6 | 6 | 9 1/2 x 1/2 x 2-11 1/2 | 3 1/2 x 5/8 x 2-11 1/2 | 430 | 316 |
| | 108 | 3 | 7 | 20 1/2 x 7/16 | 6 | 6 1/2 | 10 x 1/2 x 2-11 1/2 | 3 1/2 x 3/4 x 2-11 1/2 | 480 | 371 |
| | 116 | 3 | 7 | 20 1/2 x 7/16 | 7 | 6 1/2 | 10 x 9/16 x 3-4 1/2 | 3 1/2 x 3/4 x 3-4 1/2 | 510 | 417 |
| | 124 | 3 | 7 | 20 1/2 x 7/16 | 7 | 6 | 10 x 9/16 x 3-4 1/2 | 4 x 3/4 x 3-4 1/2 | 550 | 433 |
| | 132 | 3 | 8 | 20 1/2 x 7/16 | 8 | 6 | 10 x 5/8 x 3-9 1/2 | 4 x 3/4 x 3-9 1/2 | 580 | 488 |
| W 33 | 118 | 3 | 8 | 20 1/2 x 7/16 | 7 | 7 | 11 x 1/2 x 3-4 1/2 | 4 x 5/8 x 3-4 1/2 | 570 | 426 |
| | 130 | 3 | 8 | 20 1/2 x 7/16 | 8 | 7 | 11 x 5/8 x 3-9 1/2 | 4 x 5/8 x 3-9 1/2 | 640 | 494 |
| | 141 | 3 | 9 | 20 1/2 x 7/16 | 9 | 6 1/2 | 11 x 5/8 x 4-2 1/2 | 4 1/2 x 3/4 x 4-2 1/2 | 715 | 583 |
| | 152 | 3 | 9 | 20 1/2 x 7/16 | 10 | 6 1/2 | 11 x 5/8 x 4-7 1/2 | 4 1/2 x 7/8 x 4-7 1/2 | 765 | 659 |
| W 36 | 135 | 3 | 10 | 20 1/2 x 7/16 | 8 | 6 1/2 | 11 x 9/16 x 3-9 1/2 | 4 1/2 x 5/8 x 3-9 1/2 | 715 | 509 |
| | 150 | 3 | 10 | 20 1/2 x 7/16 | 9 | 6 1/2 | 11 x 5/8 x 4-2 1/2 | 4 1/2 x 3/4 x 4-2 1/2 | 805 | 598 |
| | 160 | 3 | 11 | 20 1/2 x 1/2 | 10 | 6 1/2 | 11 x 5/8 x 4-7 1/2 | 4 1/2 x 7/8 x 4-7 1/2 | 855 | 676 |
| | 170 | 3 | 11 | 20 1/2 x 9/16 | 11 | 6 1/2 | 11 x 3/4 x 5-0 1/2 | 4 1/2 x 7/8 x 5-0 1/2 | 930 | 768 |

## DESIGN CRITERIA

THE SPLICE IS DESIGNED WITH A CAPACITY OF 80% OF THE STRENGTH OF THE NET SECTION OF THE SMALLER MEMBER BEING JOINED CHECK MOMENT CAPACITY OF SPLICE WITH EXISTING CONDITIONS.

THE NET SECTION IS TAKEN THRU THE GIRDER WHERE ONLY THE FLANGE CONNECTORS OCCUR. THE SHEAR CAPACITY IS BASED ON THE "T" DEPTH AS GIVEN IN THE A.I.S.C. HANDBOOK.

ALL SPLICE DESIGNS ARE BASED ON 3/4"ϕ HIGH STRENGTH BOLTS IN A CONNECTION.

ALL SPLICE PLATES ARE MADE OF STEEL CONFORMING TO A.S.T.M. SPECIFICATIONS FOR A36.

THE ALLOWABLE STRESSES USED IN THE DESIGN ASSUMING NO ALTERNATING STRESSES (A.A.S.H.O. LOADING COND. NO. 1) ARE:

| | A36 | A572 / A588 |
|---|---|---|
| BENDING: | 20,000 p.s.i. | 27,000 p.s.i. |
| SHEAR: | 12,000 p.s.i. | 17,000 p.s.i. |

THESE STRESSES ARE APPLIED TO ALL SECTIONS REGARDLESS OF THICKNESSES.

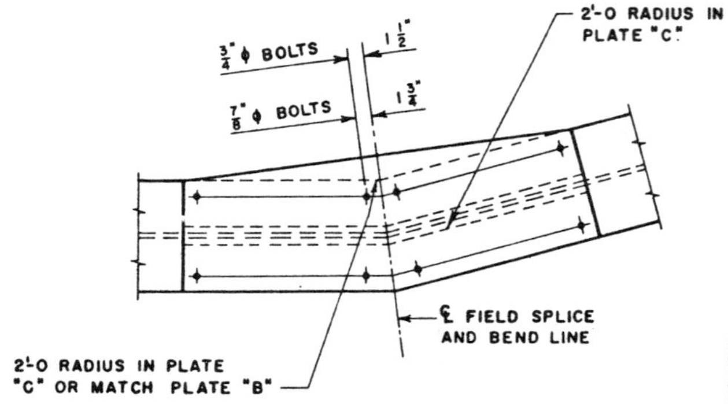


SPLICE DETAIL FOR KINKED GIRDERS

ROLLED GIRDER SPLICE DETAILS

STATE OF WISCONSIN
DEPARTMENT OF TRANSPORTATION
DIVISION OF HIGHWAYS
BRIDGE SECTION

Approved: Stanley W. Woods, Chief Bridge Engineer — Date: 4-82

STANDARD 24·7

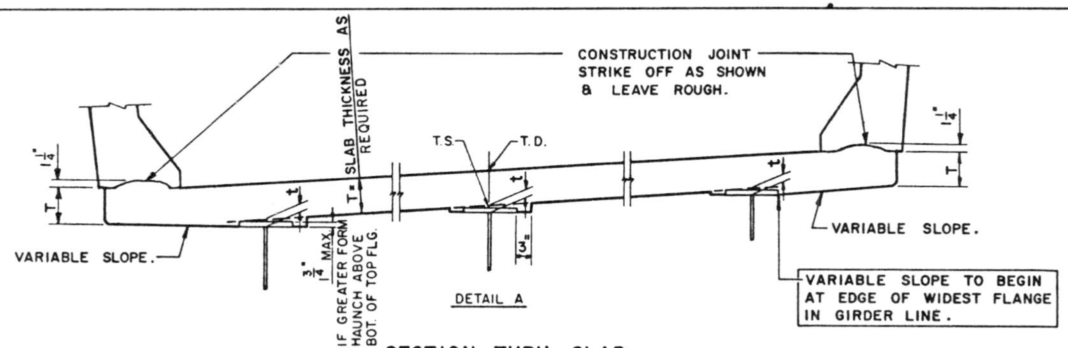


## SECTION THRU SLAB

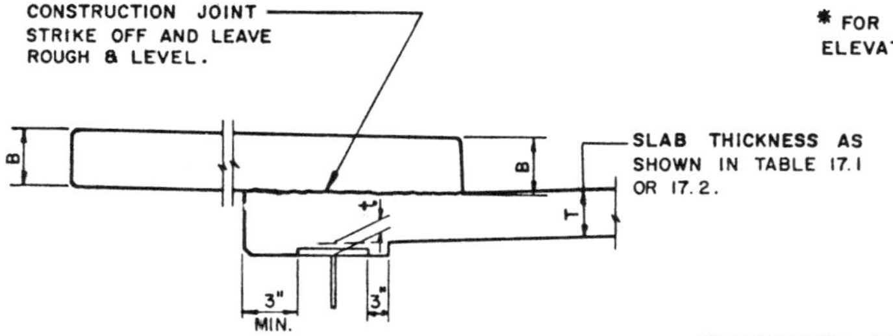


## TREATMENT OF EXTERIOR GIRDER AT SIDEWALK OVERHANG

* FOR SPANS OVER 100' LONG INCLUDE ELEVATIONS AT $\frac{1}{8}$ POINTS ALSO.

## ELEVATIONS AT TOP OF DECK (T.D.) & TOP OF STEEL (T.S.)

| | | W. ABUT. | 1/4 SPAN | 1/2 SPAN | 3/4 SPAN | ℄ PIER | ℄ SPLICE | | | ℄ ABUT. |
|---|---|---|---|---|---|---|---|---|---|---|
| GIRDER 1 | T.D. | 861.17 | 861.13 | 861.08 | 861.04 | 860.99 | 860.95 | | | 860.69 |
| | T.S. | 860.48 | | | | 860.35 | 860.35 | | | 860.00 |
| GIRDER 2 | T.D. | 860.62 | 860.58 | 860.53 | 860.49 | 860.45 | 860.40 | | | 860.16 |
| | T.S. | 859.93 | | | | 859.80 | 859.80 | | | 859.59 |
| GIRDER X | T.D. | | | | | | | | | |
| | T.S. | | | | | | | | | |

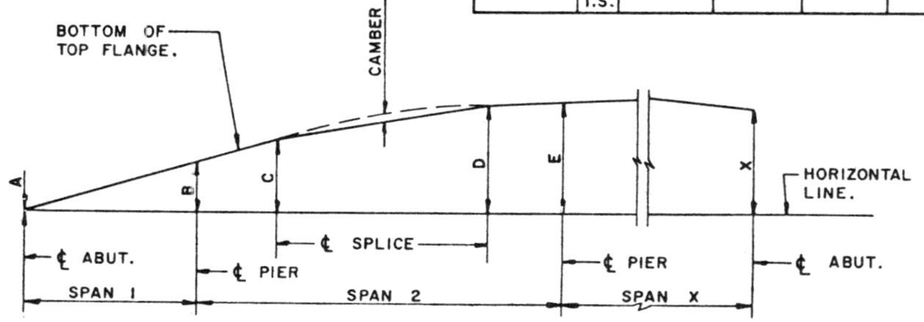


## BLOCKING DIAGRAM

## NOTES

t = HAUNCH DEPTH AT CENTERLINE OF GIRDER.

HAUNCH DEPTHS WILL NORMALLY BE MADE ONE-TENTH OF A FOOT AT ABUTMENTS, AT HINGES, AND FIELD SPLICES, AFTER THE DEADLOAD DEFLECTION AT THESE LOCATIONS HAS OCCURRED. WHEN THIS PROCEDURE CREATES A HAUNCH DEPTH OF LESS THAN ONE-TENTH OF A FOOT AT SOME POINT ALONG THE GIRDER ANOTHER APPROACH SHOULD BE USED.

HAUNCH DEPTH VARIATIONS WILL NOT BE SHOWN ON THE PLANS.

THE FOLLOWING NOTE SHOULD BE RECORDED ON THE PLANS TO CLARIFY THE PROCEDURE USED TO DETERMINE THESE DEPTHS:

[TO DETERMINE "t": AFTER ALL STRUCTURAL STEEL HAS BEEN ERECTED, ELEVATIONS OF THE TOP FLANGES, TOP OF SPLICE PLATES, OR TOP OF COVER PLATES, WHICHEVER APPLIES, SHALL BE TAKEN AT CENTERLINE OF BEARINGS, CENTERLINE OF FIELD SPLICES, AND AT QUARTER POINTS * OF EACH SPAN WHICH ARE MORE THAN SIX FEET FROM A FIELD SPLICE.]

THE PROCESS IS SHOWN BELOW AND IN "DETAIL A."

TOP OF DECK ELEV. AT FINAL GRADE.
− TOP OF STEEL ELEV. AFTER PLACEMENT.
+ THICKNESS OF TOP FLANGE & TOP SPLICE PLATE.

+ DEFLECTION; DOWNWARD DEFLECTION IS ADDED, UPWARD DEFLECTION IS SUBTRACTED.
− SLAB THICKNESS ('T' ON "DETAIL A")

= "t" VALUE FOR SETTING HAUNCH

[THESE ELEVATIONS ARE TO TOP OF STEEL (SPLICE AND COVER PLATE THICKNESS, IF APPLICABLE, ARE ACCOUNTED FOR) AND THEY ARE FOR THE MATERIAL AS ERECTED. THE ELEVATION OF THE TOP OF STEEL AT THE FIELD SPLICE POINTS SHALL BE CHECKED, AND CORRECTED, IF POSSIBLE, AFTER ERECTION AND BEFORE PERMANENTLY WELDING OR BOLTING THE DIAPHRAGMS IN PLACE.]

BLOCKING & SLAB HAUNCH DETAILS

STATE OF WISCONSIN
DEPARTMENT OF TRANSPORTATION
DIVISION OF HIGHWAYS
BRIDGE SECTION

Approved: Stanley W. Woods, Chief Bridge Engineer — Date: 12-'83

STANDARD 24·9

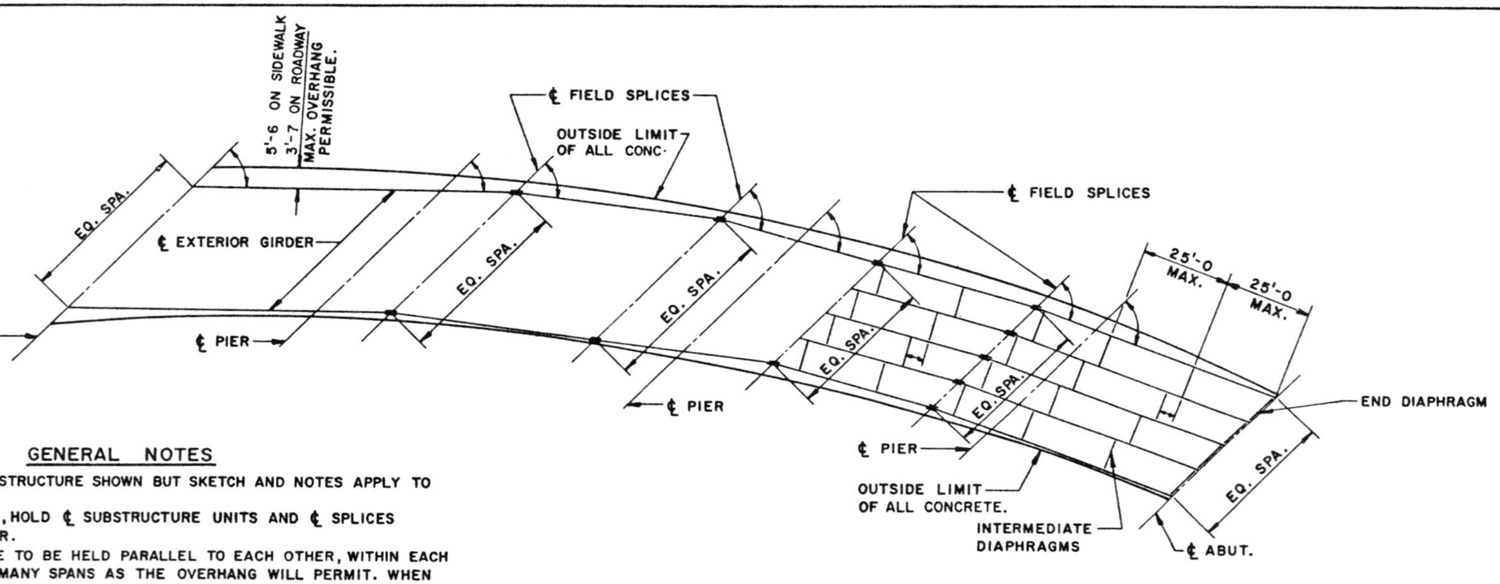


## GENERAL NOTES

FOUR SPAN STRUCTURE SHOWN BUT SKETCH AND NOTES APPLY TO ANY NUMBER OF SPANS.

IF POSSIBLE, HOLD ℄ SUBSTRUCTURE UNITS AND ℄ SPLICES PARALLEL TO EACH OTHER.

GIRDERS ARE TO BE HELD PARALLEL TO EACH OTHER, WITHIN EACH GIRDER LENGTH, FOR AS MANY SPANS AS THE OVERHANG WILL PERMIT. WHEN OVERHANG IS EXCEEDED, THE GIRDER SPACING SHALL BE CHANGED IN ONE GIRDER LENGTH, AFTER WHICH THE GIRDERS SHALL AGAIN BE HELD PARALLEL TO EACH OTHER FOR AS LONG AS THE OVERHANG PERMITS.

FOR HORIZONTAL CURVES EQUAL TO OR GREATER THAN 7° THE GIRDERS SHALL BE FABRICATED ALONG THE CURVE.

NUMBER AND SIZE OF GIRDERS AND LOCATION OF FIELD SPLICES TO BE DETERMINED BY DESIGN.

| GIRDER LAYOUT ON CURVE | |
|---|---|
| STATE OF WISCONSIN<br>DEPARTMENT OF TRANSPORTION<br>DIVISION OF HIGHWAYS<br>BRIDGE SECTION | |
| Approved: Stanley W. Woods<br>Chief Bridge Engineer | Date:<br>4 - '82 |

STANDARD 24·10

# E.14 BOLTING

## HIGH STRENGTH BOLTS

| Allow. Shear = 13,500 psi – Friction Type (16,000)<br>= 20,000 psi – Bearing Type (27,000)<br>Allow. Bearing = 40,000 psi | Diameter of Bolt | | |
|---|---|---|---|
| | 3/4" | 7/8" | 1" |
| Friction Type Connection<br>Value in Single Shear<br>Value in Double Shear | <br>5.96 (7.06)<br>11.93(14.13) | <br>8.12 (9.62)<br>16.24(19.24) | <br>10.60(12.56)<br>21.21(25.13) |
| Bearing Type Connection<br>Value in Single Shear<br>Value in Double Shear | <br>8.83(11.92)<br>17.66(23.85) | <br>12.02(16.23)<br>24.04(32.47) | <br>15.70(21.20)<br>31.40(42.41) |

| | DIAMETER OF BOLT OR RIVET | | |
|---|---|---|---|
| | 3/4" | 7/8" | 1" |
| Diameter of Hole (Design) | 0.875 | 1.00 | 1.125 |
| Minimum Pitch | 2-1/2 | 3 | 3-1/2 |
| Minimum Width of Angle | 2-1/2 | 3 | 3-1/2 |
| Minimum Edge Dist. Rolled Edge | 1-1/8 | 1-1/4 | 1-1/2 |
| Sheared Edge | 1-1/4 | 1-1/2 | 1-3/4 |
| Weight of Two Rivet Heads – lbs. | 0.32 | 0.48 | 0.70 |
| Weight of bolt nut and washer–lbs. (Excl. shank within grip). | 0.58 | 0.86 | 1.22 |

## USUAL GAGES FOR ANGLES, INCHES

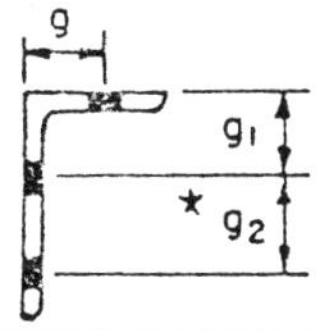


| Leg | 8 | 7 | 6 | 5 | 4 | 3 1/2 | 3 | 2 1/2 | 2 | 1 3/4 | 1 1/2 | 1 3/8 | 1 1/4 | 1 |
|---|---|---|---|---|---|---|---|---|---|---|---|---|---|---|
| g | 4 1/2 | 4 | 3 1/2 | 3 | 2 1/2 | 2 | 1 3/4 | 1 3/8 | 1 1/8 | 1 | 7/8 | 7/8 | 3/4 | 5/8 |
| $g_1$ | 3 | 2 1/2 | 2 1/4 | 2 | | | | | | | | | | |
| $g_2$ | 3 | 3 | 2 1/2 | 1 3/4 | | | | | | | | | | |

*When<3", check AISC for minimum pitch to maintain 3 dia.

FHWA Reasearch Report "An Experimental Studies of Bolted Shear Connection" (FHWA/RD/81/148) recommended coating the faying surface of bolted joints when steel itself should be painted. Because coating the faying surface, or other forms of surface treatment such as blast cleaning, reduces the scatter and provide more uniform contacts in the slip performance of a bolted joints relative to the uncontrolled condition of a millscale surface. Also, this will reduce the fabrication cost by eliminating the masking of faying surfaces. The coating will reduce early coating failures by eliminating the unprotected edges at the periphery of a joints.

# STANDARD WELDING SYMBOLS



**Basic Welding Symbols and Their Location Significance**

| Location Significance | Fillet | Plug or Slot | Spot or Projection | Seam | Back or Backing | Surfacing | Scarf for Brazed Joint | Flange / Edge |
|---|---|---|---|---|---|---|---|---|
| Arrow Side | | | | | Groove weld symbol | | | |
| Other Side | | | | | Groove weld symbol | Not used | | |
| Both Sides | | Not used | Not used | Not used | Not used | Not used | | Not used |
| No Arrow Side or Other Side Significance | Not used | Not used | | | Not used | Not used | Not used | Not used |

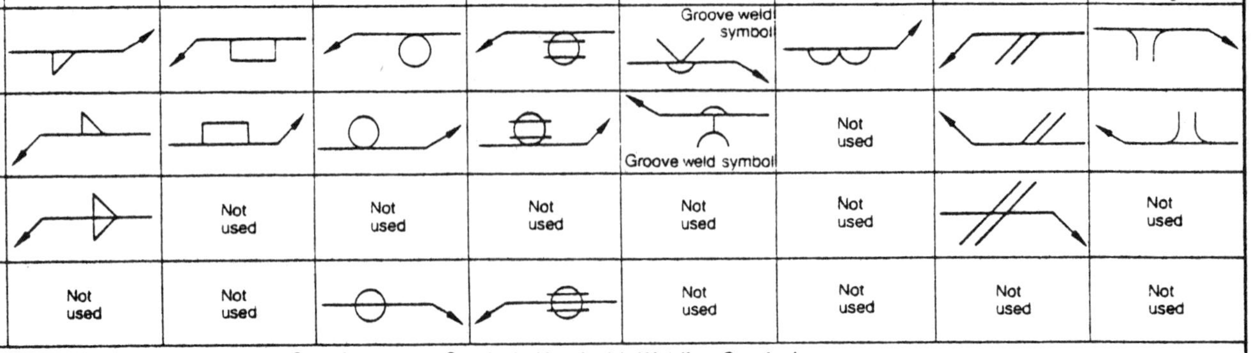

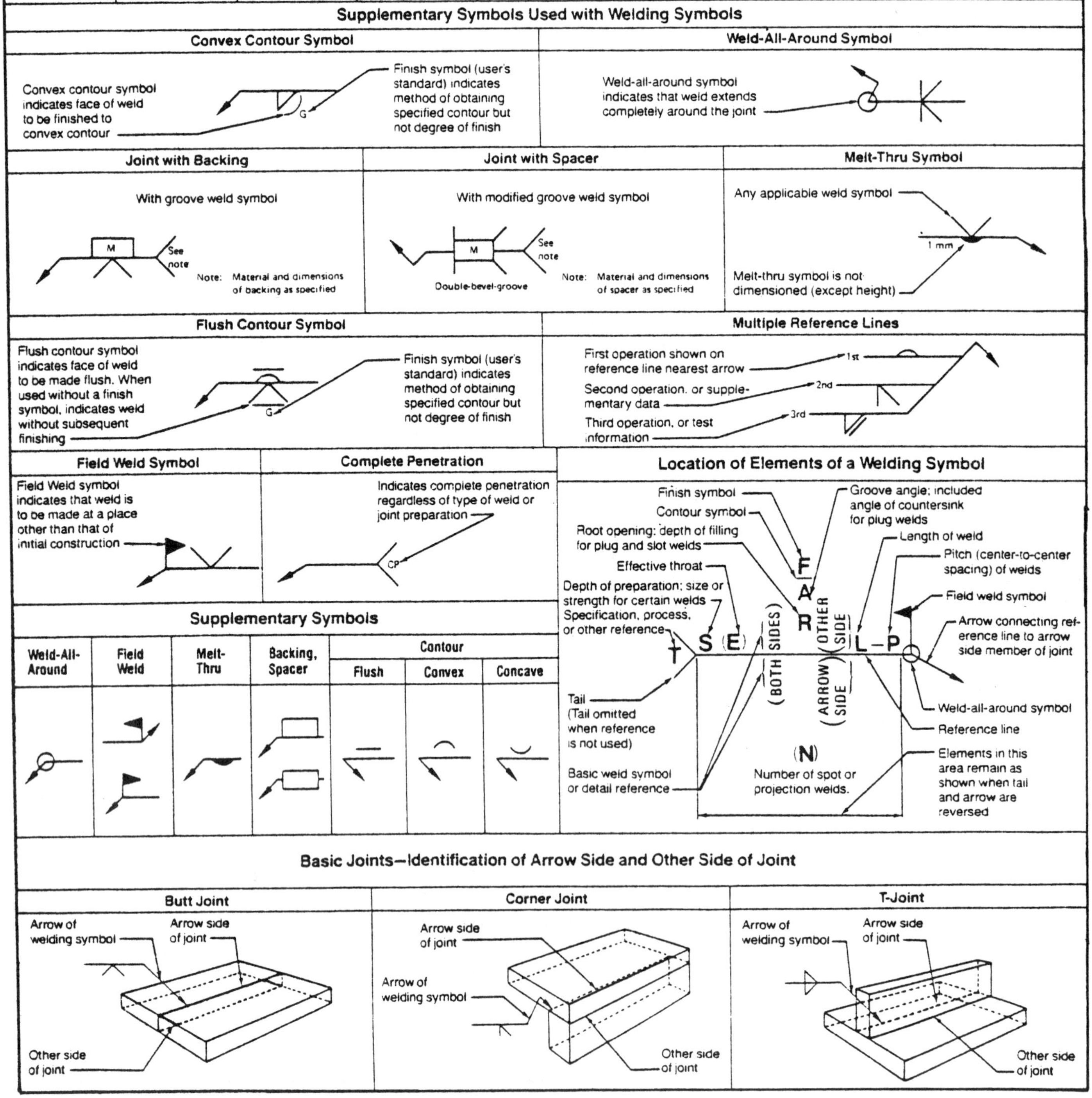

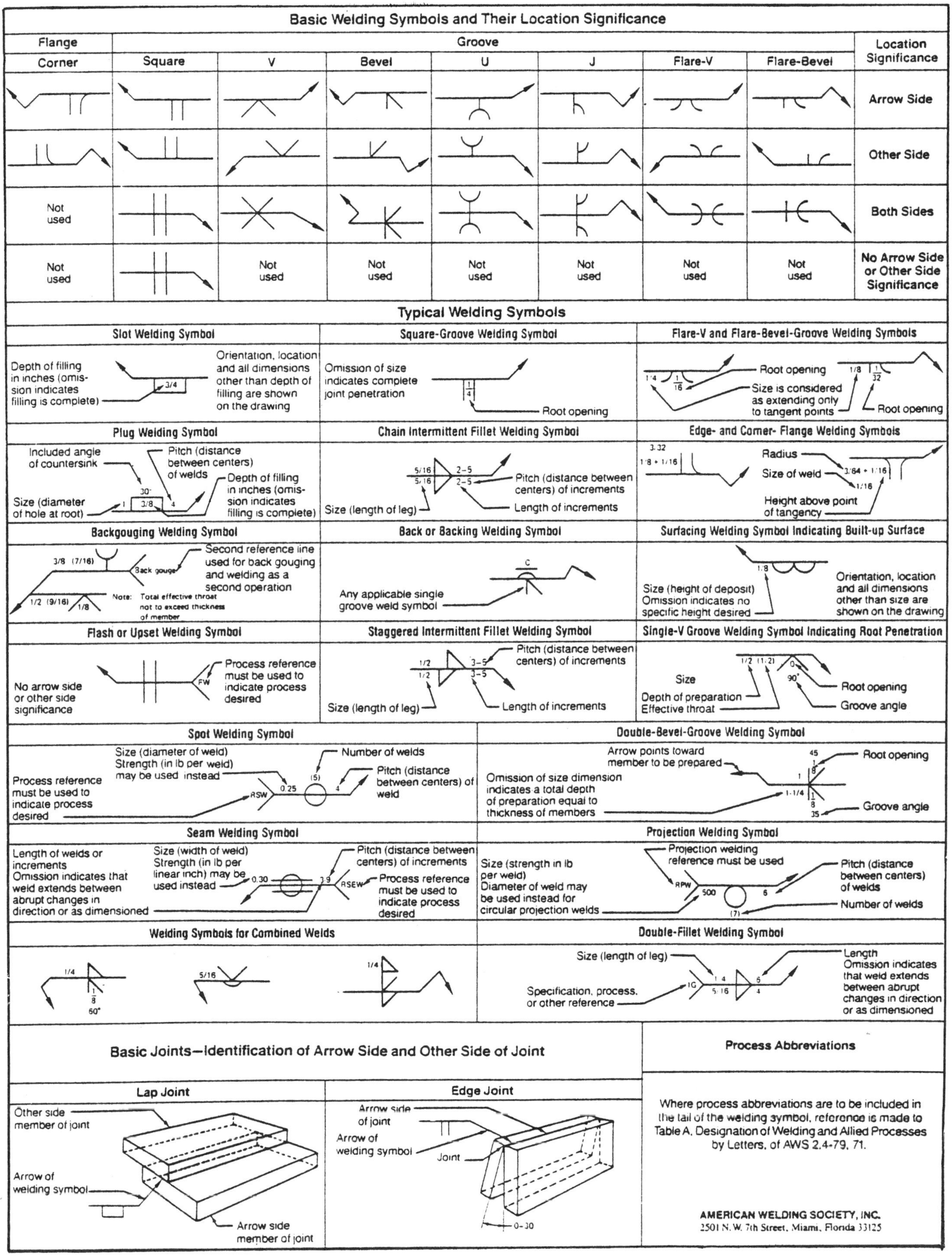
Basic Welding Symbols and Their Location Significance
Flange
Groove
Corner
Square
V
Bevel
U
J
Flare-V
Flare-Bevel
Location Significance
Arrow Side
Other Side
Not used
Both Sides
Not used
Not used
Not used
Not used
Not used
Not used
Not used
No Arrow Side or Other Side Significance
Typical Welding Symbols
Slot Welding Symbol
Depth of filling in inches (omission indicates filling is complete)
3/4
Orientation, location and all dimensions other than depth of filling are shown on the drawing
Square-Groove Welding Symbol
Omission of size indicates complete joint penetration
1/4
Root opening
Flare-V and Flare-Bevel-Groove Welding Symbols
1/4
1/16
Root opening
Size is considered as extending only to tangent points
1/8
1/32
Root opening
Plug Welding Symbol
Included angle of countersink
Pitch (distance between centers) of welds
30°
Depth of filling in inches (omission indicates filling is complete)
Size (diameter of hole at root)
1
3/8
4
Chain Intermittent Fillet Welding Symbol
5/16
5/16
2-5
2-5
Pitch (distance between centers) of increments
Size (length of leg)
Length of increments
Edge- and Corner- Flange Welding Symbols
3/32
1/8 + 1/16
Radius
Size of weld
3/64 + 1/16
1/16
Height above point of tangency
Backgouging Welding Symbol
3/8 (7/16)
Back gouge
Second reference line used for back gouging and welding as a second operation
1/2 (9/16)
1/8
Note: Total effective throat not to exceed thickness of member
Back or Backing Welding Symbol
C
Any applicable single groove weld symbol
Surfacing Welding Symbol Indicating Built-up Surface
1/8
Size (height of deposit) Omission indicates no specific height desired
Orientation, location and all dimensions other than size are shown on the drawing
Flash or Upset Welding Symbol
No arrow side or other side significance
FW
Process reference must be used to indicate process desired
Staggered Intermittent Fillet Welding Symbol
1/2
1/2
3-5
3-5
Pitch (distance between centers) of increments
Size (length of leg)
Length of increments
Single-V Groove Welding Symbol Indicating Root Penetration
Size
1/2 (1/2)
0
90°
Root opening
Depth of preparation
Effective throat
Groove angle
Spot Welding Symbol
Size (diameter of weld) Strength (in lb per weld) may be used instead
Number of welds
Process reference must be used to indicate process desired
RSW
0.25
(5)
4
Pitch (distance between centers) of weld
Double-Bevel-Groove Welding Symbol
Arrow points toward member to be prepared
45
1/8
1
1-1/4
1/8
35
Root opening
Omission of size dimension indicates a total depth of preparation equal to thickness of members
Groove angle
Seam Welding Symbol
Length of welds or increments Omission indicates that weld extends between abrupt changes in direction or as dimensioned
Size (width of weld) Strength (in lb per linear inch) may be used instead
0.30
3-9
RSEW
Pitch (distance between centers) of increments
Process reference must be used to indicate process desired
Projection Welding Symbol
Size (strength in lb per weld) Diameter of weld may be used instead for circular projection welds
Projection welding reference must be used
RPW
500
6
(7)
Pitch (distance between centers) of welds
Number of welds
Welding Symbols for Combined Welds
1/4
1/8
60°
5/16
1/4
Double-Fillet Welding Symbol
Size (length of leg)
Specification, process, or other reference
1G
1/4
5/16
6
4
Length Omission indicates that weld extends between abrupt changes in direction or as dimensioned
Basic Joints—Identification of Arrow Side and Other Side of Joint
Lap Joint
Other side member of joint
Arrow of welding symbol
Arrow side member of joint
Edge Joint
Arrow side of joint
Arrow of welding symbol
Joint
0-30
Process Abbreviations
Where process abbreviations are to be included in the tail of the welding symbol, reference is made to Table A, Designation of Welding and Allied Processes by Letters, of AWS 2.4-79, 71.
AMERICAN WELDING SOCIETY, INC.
2501 N.W. 7th Street, Miami, Florida 33125

# Supplementary welding symbols

## LOCATION OF INFORMATION ON WELDING SYMBOLS

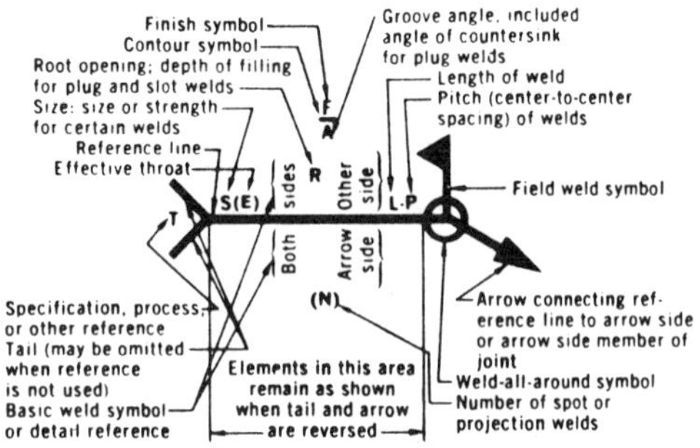


**FIELD WELD SYMBOL**

**WELD-ALL-AROUND SYMBOL**

**CONVEX-CONTOUR SYMBOL**
Convex-Contour Symbol indicates face of weld to be finished to convex contour.

**FLUSH-CONTOUR SYMBOL**
Flush-Contour Symbol indicates face of weld to be made flush. When used without a finish symbol, indicates weld to be welded flush without subsequent finishing.

**CONCAVE-CONTOUR SYMBOL**
Concave-Contour Symbol indicates face of weld to be finished to concave contour.

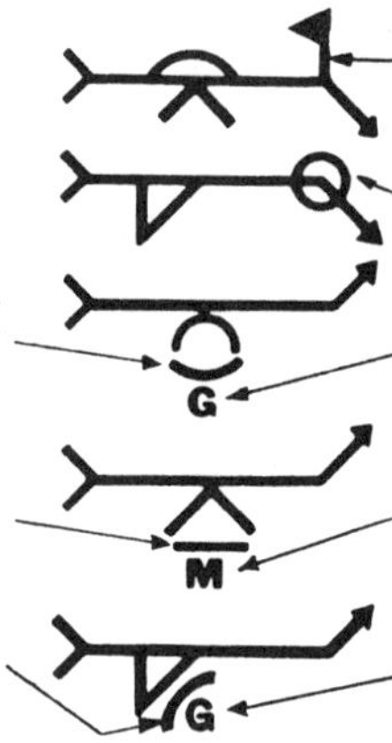


Field Weld Symbol indicates that weld is to be made at a place other than that of initial construction.

Weld-All-Around Symbol indicates that weld extends completely around the joint.

Finish Symbol (user's std.) indicates method of obtaining specified contour but not degree of finish.

Finish symbol (user's std.) indicates method of obtaining specified contour but not degree of finish.

Finish symbol (user's std.) indicates method of obtaining specified contour but not degree of finish.

## typical welding symbols

**UNEQUAL DOUBLE FILLET WELDING SYMBOL**

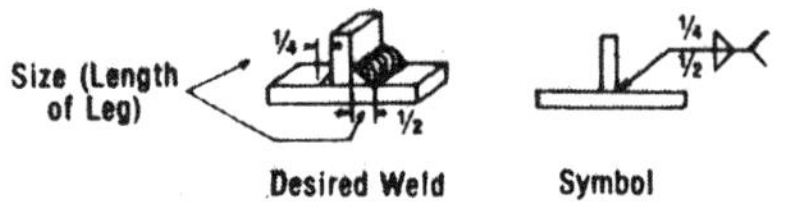


**CHAIN INTERMITTENT—FILLET WELDING SYMBOL**

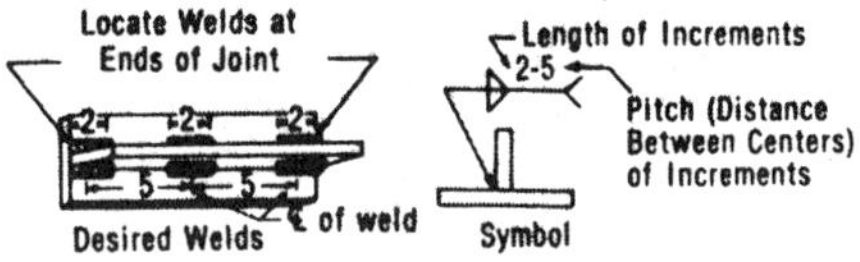


**STAGGERED INTERMITTENT—FILLET WELDING SYMBOL**

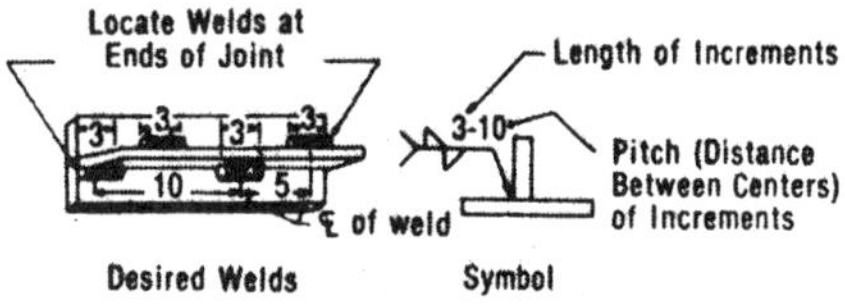


**SINGLE V-GROOVE WELDING SYMBOL**

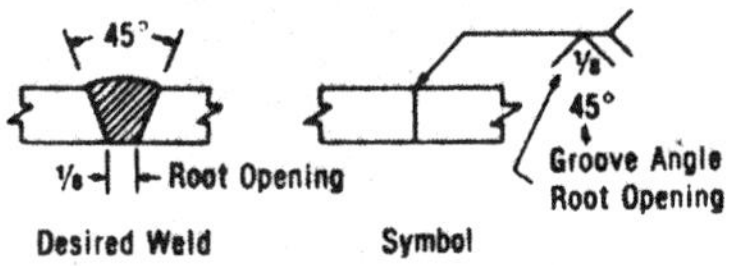


**SINGLE V-GROOVE WELDING SYMBOL INDICATING ROOT PENETRATION**

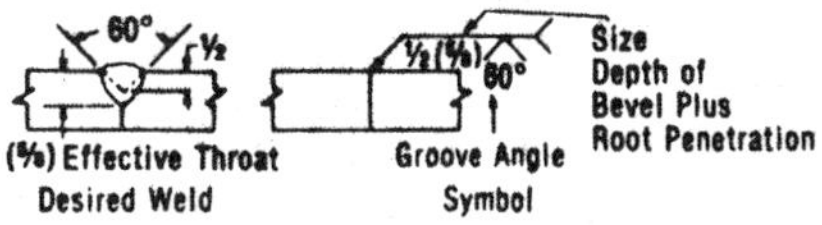


**USE OF BEAD WELD SYMBOL TO INDICATE SINGLE-PASS BACK WELD**

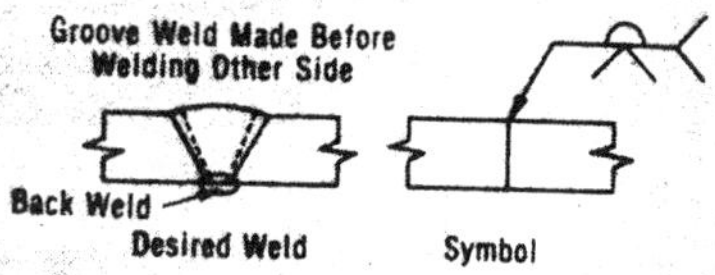


**DOUBLE V-GROOVE WELD INDICATING A SPACER IS TO BE USED**

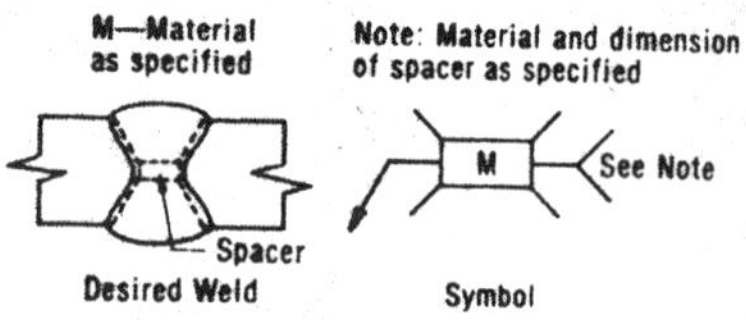


**SINGLE V-GROOVE WELD INDICATING A BACKING STRIP IS TO BE USED**

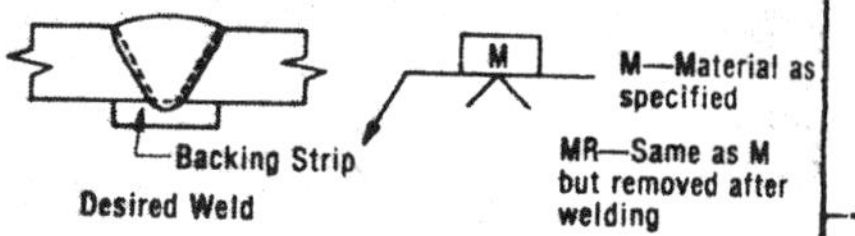


**DOUBLE-BEVEL GROOVE WELDING SYMBOL**

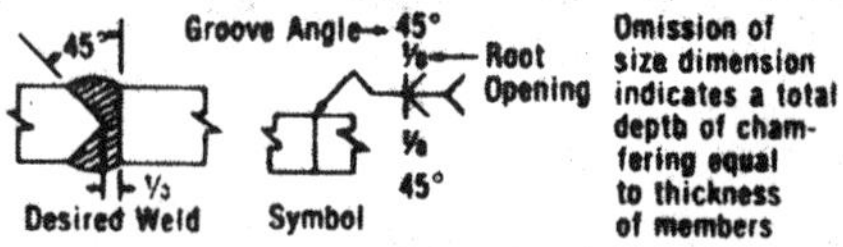


**PLUG WELDING SYMBOL**

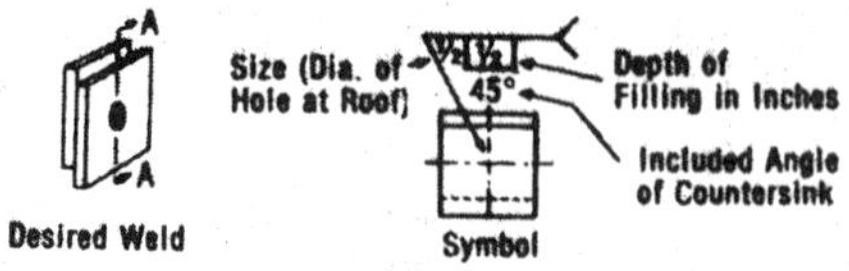


**SLOT WELDING SYMBOL**

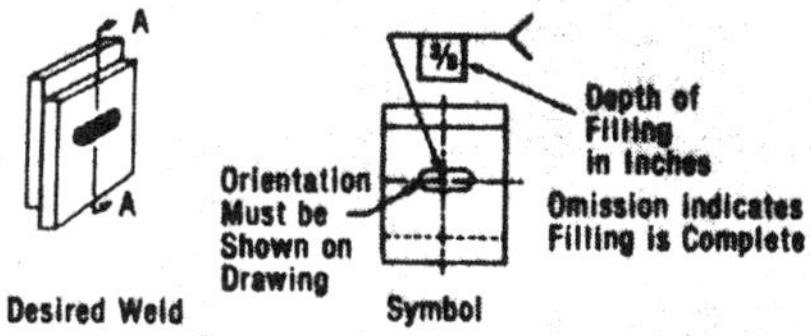


**SPOT WELDING SYMBOL**

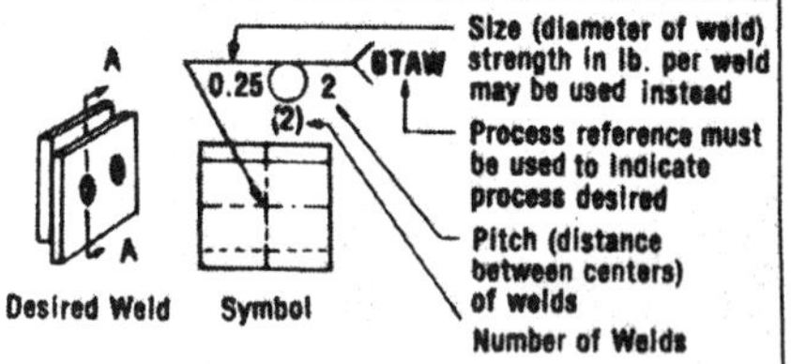


**SIZE OF SURFACE BUILT UP BY WELDING**

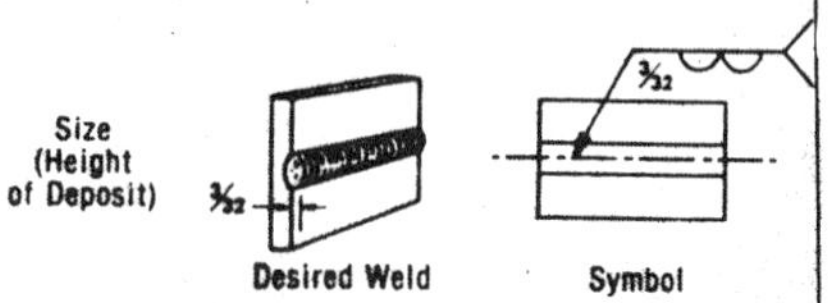


**SYMBOLS WTH MULTIPLE REFERENCE LINES**

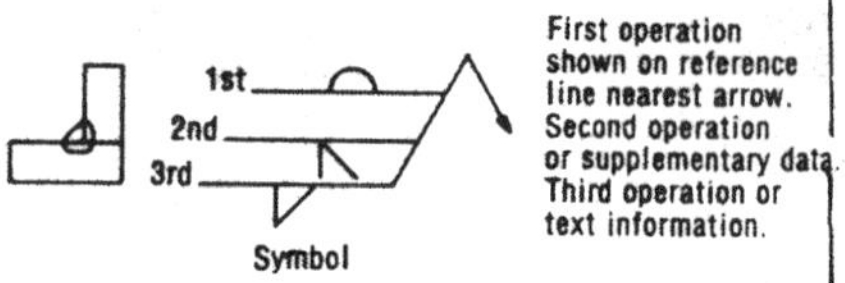


**EDGE FLANGE WELD SYMBOL**

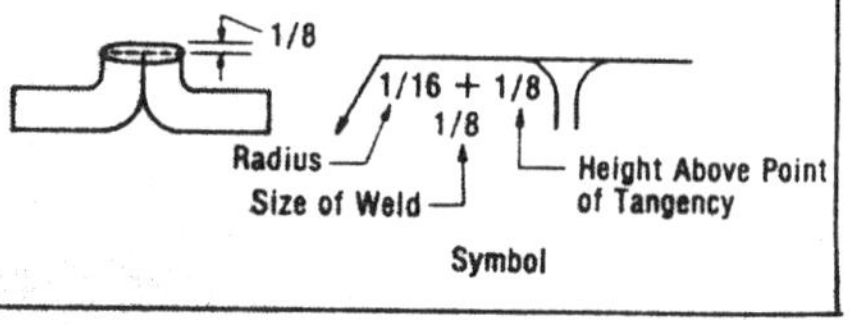

# E.16 TYPES OF WELDS AND ALLOWABLE LOADS

TYPES OF JOINTS

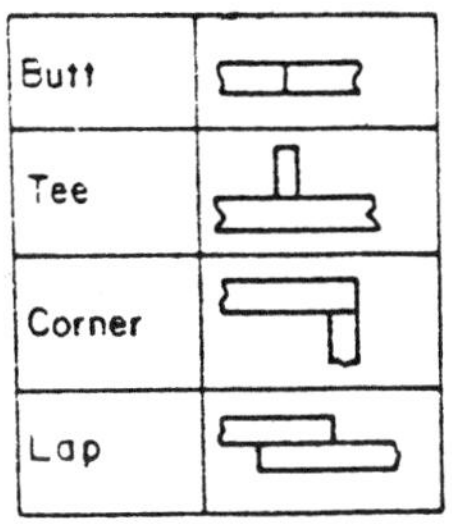


TYPES OF SINGLE WELDS

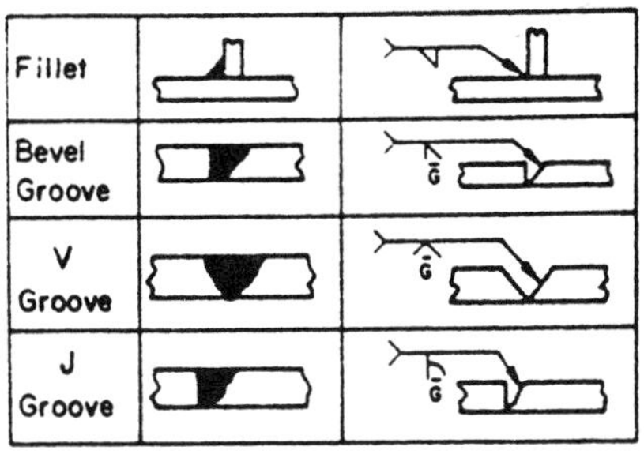


TYPES OF DOUBLE WELDS

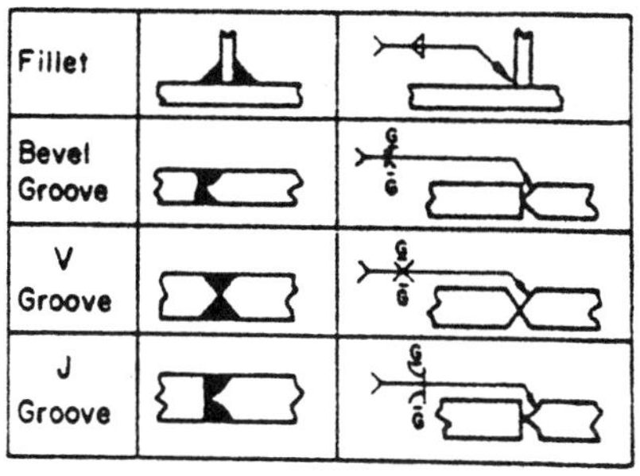


### Table 1 – Allowable Load for Various Sizes of Fillet Welds

| | Strength Level of Weld Metal (EXX) | | | | | | |
|---|---|---|---|---|---|---|---|
| | *60 | *70 | 80 | *90 | 100 | *110 | 120 |
| **Allowable Shear Stress on Throat ksi (1000 psi) of Fillet Weld or Partial Penetration Groove Weld** | | | | | | | |
| $\tau$ = | 18.0 | 21.0 | 24.0 | 27.0 | 30.0 | 33.0 | 36.0 |
| **Allowable Unit Force on Fillet Weld kips/linear in.** | | | | | | | |
| f = | 12.73$\omega$ | 14.85$\omega$ | 16.97$\omega$ | 19.09$\omega$ | 21.21$\omega$ | 23.33$\omega$ | 25.45$\omega$ |
| **Leg Size $\omega$, in.** | **Allowable Unit Force for Various Sizes of Fillet Welds kip/linear in.** | | | | | | |
| 1 | 12.73 | 14.85 | 16.97 | 19.09 | 21.21 | 23.33 | 25.45 |
| 7/8 | 11.14 | 12.99 | 14.85 | 16.70 | 18.57 | 20.41 | 22.27 |
| 3/4 | 9.55 | 11.14 | 12.73 | 14.32 | 15.92 | 17.50 | 19.09 |
| 5/8 | 7.96 | 9.28 | 10.61 | 11.93 | 13.27 | 14.58 | 15.91 |
| 1/2 | 6.37 | 7.42 | 8.48 | 9.54 | 10.61 | 11.67 | 12.73 |
| 7/16 | 5.57 | 6.50 | 7.42 | 8.35 | 9.28 | 10.21 | 11.14 |
| 3/8 | 4.77 | 5.57 | 6.36 | 7.16 | 7.95 | 8.75 | 9.54 |
| 5/16 | 3.98 | 4.64 | 5.30 | 5.97 | 6.63 | 7.29 | 7.95 |
| 1/4 | 3.18 | 3.71 | 4.24 | 4.77 | 5.30 | 5.83 | 6.36 |
| 3/16 | 2.39 | 2.78 | 3.18 | 3.58 | 3.98 | 4.38 | 4.77 |
| 1/8 | 1.59 | 1.86 | 2.12 | 2.39 | 2.65 | 2.92 | 3.18 |
| 1/16 | .795 | .930 | 1.06 | 1.19 | 1.33 | 1.46 | 1.59 |

*Fillet welds actually tested by the joint AISC-AWS Task Committee.

### Table 2 – Minimum Fillet Weld Size ($\omega$) or Minimum Throat of Partial Penetration Groove Weld (t)

| Material Thickness of Thicker Part Joined (in.) | | in. |
|---|---|---|
| ** to 1/4 incl. | | 1/8 |
| over 1/4 | to 1/2 | 3/16 |
| over 1/2 | to 3/4 | 1/4 |
| * over 3/4 | to 1-1/2 | 5/16 |
| over 1-1/2 | to 2-1/4 | 3/8 |
| over 2-1/4 | to 6 | 1/2 |
| over 6 | | 5/8 |

Not to exceed the thickness of the thinner part.

* For minimum fillet weld size, Table does not go above 5/16" fillet weld for over 3/4" material.

** Minimum size for bridge application does not go below 3/16".



## DETAILS FOR FILLET WELDS

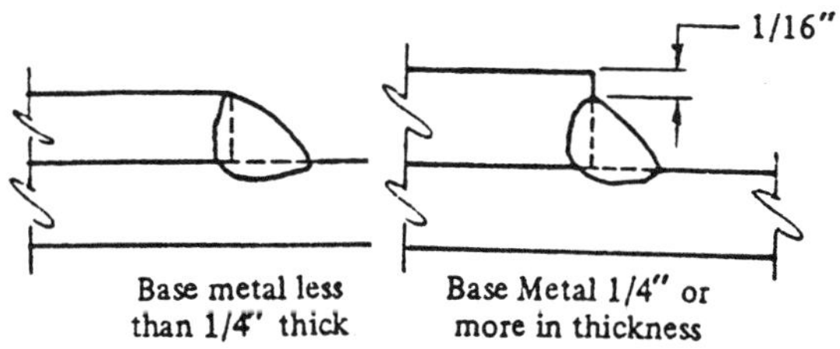


A B

Maximum Size of Fillet Weld Along Edges

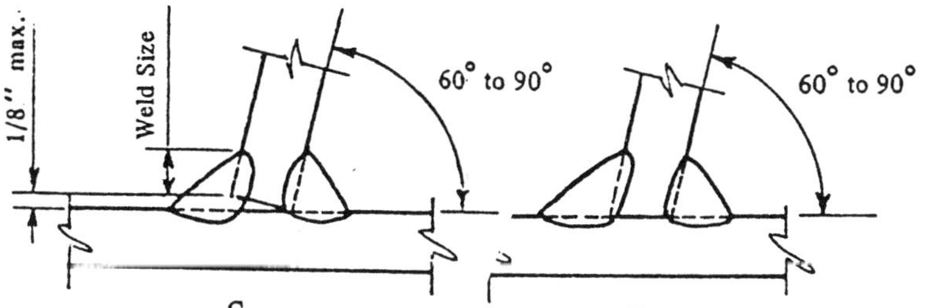


C D

Skewed Tee Joints

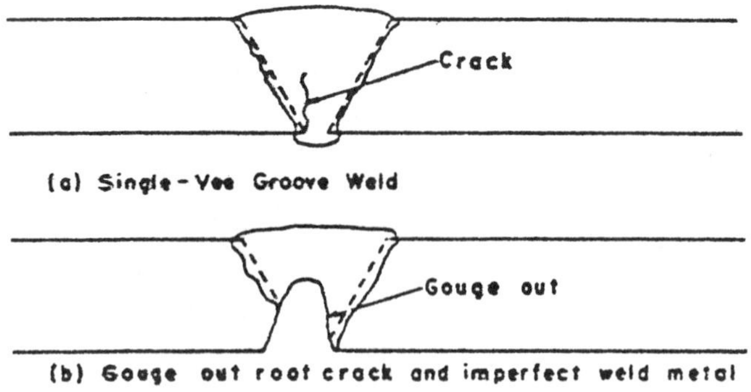


BUTT WELD ROOT CRACK

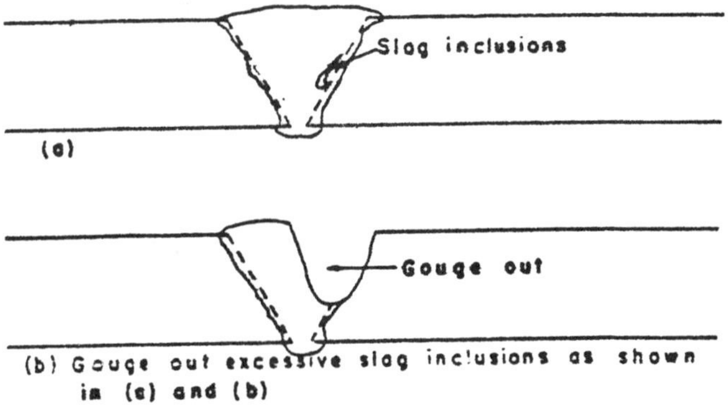


EXCESSIVE SLAG INCLUSIONS

PROPER USE OF BACK GOUGING

## E.17 WELDING TERMS

# Definitions of welding terms

**AC OR ALTERNATING CURRENT**—Is that kind of electricity which reverses its direction periodically. For 60 cycle current, the current goes in one direction and then in the other direction 60 times in the same second, so that the current changes its direction 120 times in one second.

**ARC BLOW**—The deflection of an electric arc from its normal path because of magnetic forces.

**ARC LENGTH**—The distance from the end of the electrode to the point where the arc makes contact with work surface.

**ARC VOLTAGE**—The voltage across the welding arc.

**AS-WELDED**—The condition of weld metal welded joints, and weldments after welding but prior to any subsequent thermal, mechanical, or chemical treatments.

**AUTOMATIC WELDING**—Welding with equipment which performs the welding operation without adjustment of the controls by a welding operator. The equipment may or may not perform the loading and unloading of the work.

**BACKING**—A material (base metal, weld metal, carbon, or granular material) placed at the root of a weld joint for the purpose of supporting molten weld metal.

**BACKSTEP SEQUENCE**—A longitudinal sequence in which the weld bead increments are deposited in the direction opposite to the progress of welding the joint.

**BARE ELECTRODE**—A filler metal electrode consisting of a single metal or alloy that has been produced into a wire, strip, or bar form and that has had no coating or covering applied to it other than that which was incidental to its manufacture or preservation.

**BASE METAL (material)**—The metal (material) to be welded, brazed, soldered, or cut.

**BUTT JOINT**—A joint between two members aligned approximately in the same plane.

**CONCAVITY**—The maximum distance from the face of a concave fillet weld perpendicular to a line joining the toes.

**CONVEXITY**—The maximum distance from the face of a convex fillet weld perpendicular to a line joining the toes.

**COVERED ELECTRODE**—A composite filler metal electrode consisting of a core of a bare electrode or metal cored electrode to which a covering sufficient to provide a slag layer on the weld metal has been applied. The covering may contain materials providing such functions as shielding from the atmosphere, deoxidation, and arc stabilization and can serve as a source of metallic additions to the weld.

**CRATER**—In arc welding, a depression at the termination of a weld bead or in the molten weld pool.

**DEPTH OF FUSION**—The distance that fusion extends into the base metal or previous pass from the surface melted during welding.

**DC OR DIRECT CURRENT**—Electric current which flows only in one direction. In welding, an arc welding process wherein the power supply at the arc is direct current.

**FACE OF WELD**—The exposed surface of a weld on the side from which welding was done.

**FILLET WELD**—A weld of approximately triangular cross section joining two surfaces approximately at right angles to each other in a lap joint, T-joint or corner joint.

**FLAT POSITION**—The welding position used to weld from the upper side of the joint; the face of the weld is approximately horizontal.

**FLUX**—Material used to prevent, dissolve, or facilitate removal of oxides and other undesirable surface substances.

**FLUX CORED ARC WELDING**—(FCAW) An arc welding process which produces coalescence of metals by heating them with an arc between a continuous filler metal (consumable) electrode and the work. Shielding is provided by a flux contained within the tubular electrode. Additional shielding may or may not be obtained from an externally supplied gas or gas mixture.

**GAS METAL ARC WELDING**—(GMAW) An arc welding process which produces coalescence of metals by heating them with an arc between a continuous filler metal (consumable) electrode and the work. Shielding is obtained entirely from an externally supplied gas or gas mixture. Some methods of this process are called MIG or $CO_2$ welding (nonpreferred terms).

**GAS TUNGSTEN ARC WELDING**—(GTAW) An arc welding process which produces coalescence of metals by heating them with an arc between a tungsten (non-consumable) electrode and the work. Shielding is obtained from a gas or gas mixture. Pressure may or may not be used and filler metal may or may not be used. (This process has sometimes been called TIG welding, a nonpreferred term.)

**GROOVE WELD**—A weld made in the groove between two members to be joined.

**HEAT-AFFECTED ZONE**—That portion of the base metal which has not been melted, but whose mechanical properties or microstructure have been altered by the heat of welding, brazing, soldering, or cutting.

**HORIZONTAL POSITION**—

- **Fillet Weld**—The position in which welding is performed on the upper side of an approximately horizontal surface and against an approximately vertical surface.
- **Groove Weld**—The position of welding in which the axis of the weld lies in an approximately horizontal plane and the face of the weld lies in an approximately vertical plane.

**JOINT PENETRATION**—The minimum depth a groove or flange weld extends from its face into a joint, exclusive of reinforcement. Joint penetration may include root penetration.

**LAP JOINT**—A joint between two overlapping members.

**LEG OF A FILLET WELD**—The distance from the root of the joint to the toe of the fillet weld.

**MACHINE WELDING**—Welding with equipment which performs the welding operation under the constant observation and control of a welding operator. The equipment may or may not perform the loading and unloading of the work.

**MANUAL WELDING**—A welding operation performed and controlled completely by hand.

**MELTING RATE**—The weight or length of electrode melted in a unit of time.

**MOLTEN WELD POOL**—The liquid state of a weld prior to solidification as weld metal.

**OPEN-CIRCUIT VOLTAGE**—The voltage between the output terminals of the welding machine when no current is flowing in the welding circuit.

**OVERHEAD POSITION**—The position in which welding is performed from the underside of the joint.

**OVERLAP**—The protrusion of weld metal beyond the toe, face, or root of the weld; in resistance seam welding, the area in the preceding weld remelted by the succeeding weld.

**PASS**—A single progression of a welding or surfacing operation along a joint, weld deposit, or substrate. The result of a pass is a weld bead, layer, or spray deposit.

**PEENING**—The mechanical working of metals using impact blows.

**POROSITY**—Cavity type discontinuities formed by gas entrapment during solidification.

**POSTHEATING**—The application of heat to an assembly after a welding, brazing, soldering, thermal spraying or cutting operation.

**PREHEATING**—The application of heat to the base metal immediately before welding, brazing, soldering, thermal spraying, or cutting.

**RADIOGRAPHY**—The use of radiant energy in the form of X-rays or gamma rays for the non-destructive examination of metals.

**REVERSE POLARITY**—The arrangement of direct current arc welding leads with the work as the negative pole and the electrode as the positive pole of the welding arc. A synonym for direct current electrode positive.

**ROOT OF WELD**—The points, as shown in cross section, at which the back of the weld intersects the base metal surfaces.

**ROOT OPENING**—The separation between the members to be joined at the root of the joint.

**ROOT PENETRATION**—The depth that a weld extends into the root of a joint measured on the centerline of the root cross section.

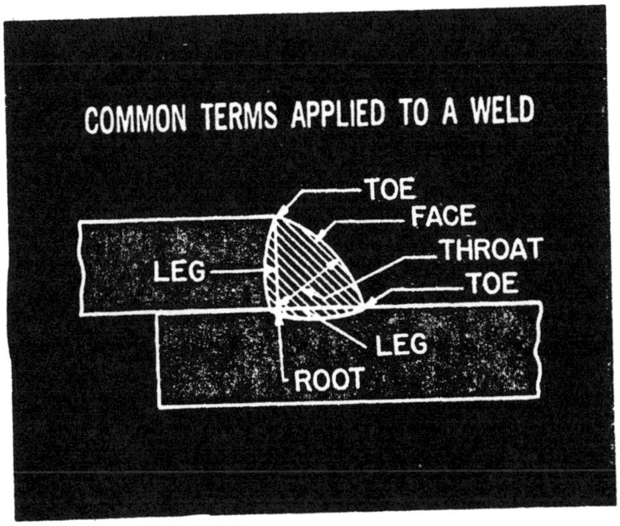


**SEMIAUTOMATIC ARC WELDING**—Arc welding with equipment which controls only the filler metal feed. The advance of the welding is manually controlled.

**SHIELDED METAL ARC WELDING**—(SMAW) An arc welding process which produces coalescence of metals by heating them with an arc between a covered metal electrode and the work. Shielding is obtained from decomposition of the electrode covering. Pressure is not used and filler metal is obtained from the electrode.

**SIZE OF WELD**—

**Groove Weld**—The joint penetration (depth of bevel plus the root penetration when specified). The size of a groove weld and its effective throat are one and the same.

**Fillet Weld**—For equal leg fillet welds, the leg lengths of the largest isosceles right triangle which can be inscribed within the fillet weld cross section.

For unequal leg fillet welds, the leg lengths of the largest right triangle which can be inscribed within the fillet weld cross section.

**SLAG INCLUSION**—Non metallic solid material entrapped in weld metal or between weld metal and base metal.

**SPATTER**—In arc and gas welding, the metal particles expelled during welding which do not form a part of the weld.

**STRAIGHT POLARITY**—The arrangement of direct current arc welding leads in which the work is the positive pole and the electrode is the negative pole of the welding arc. A synonym for direct current electrode negative.

**STRESS RELIEF HEAT TREATMENT**—Uniform heating of a structure or a portion thereof to a sufficient temperature to relieve the major portion of the residual stresses, followed by uniform cooling.

**STRINGER BEAD**—A type of weld bead made without appreciable weaving motion.

**TACK WELD**—A weld made to hold parts of a weldment in proper alignment until the final welds are made.

**THROAT OF A FILLET WELD**—

**Theoretical Throat**—The distance from the beginning of the root of the joint perpendicular to the hypotenuse of the largest right triangle that can be inscribed within the fillet weld cross section.

**Actual Throat**—The shortest distance from the root of a fillet weld to its face.

**Effective Throat**—The minimum distance minus any reinforcement from the root of a weld to its face.

**TOE OF WELD**—The junction between the face of a weld and the base metal.

**TUNGSTEN ELECTRODE**—A non-filler metal electrode used in arc welding or cutting, made principally of tungsten.

**ULTIMATE TENSILE STRENGTH**—The maximum tensile stress which will cause a material to break (usually expressed in pounds per square inch).

**UNDERBEAD CRACK**—A crack in the heat-affected zone generally not extending to the surface of the base metal.

**UNDERCUT**—A groove melted into the base metal adjacent to the toe or root of a weld and left unfilled by weld metal.

**VERTICAL POSITION**—The position of welding in which the axis of the weld is approximately vertical.

**WEAVE BEAD**—A type of weld bead made with transverse oscillation.

**WELD**—A localized coalescence of metals or non-metals produced either by heating the materials to suitable temperature, with or without the application of pressure, or by the application of pressure alone, and with or without the use of filler material.

**WELD METAL**—That portion of a weld which has been melted during welding.

**WELDING PROCEDURE**—The detailed methods and practices, including all joint welding procedures, involved in the production of a weldment.

**WELDING ROD**—A form of filler metal used for welding or brazing which does not conduct the electrical current.

**WELDMENT**—An assembly whose component parts are joined by welding.

## EFFECTS OF ALLOY

Steel is a combination of iron, carbon, and other alloying chemical components. Some of the common alloys included in the steel are impurities in the original iron ore, from which steel is made. Other alloys are intentionally added to the iron-carbon mixture to improve various characteristics of the steel. Common alloys found in typical bridge steels include carbon, aluminum, chromium, cobalt, manganese, molybdenum, nickel, phosphorus, silicon, sulfur, and vanadium.

Pure metals are only required in certain specialized applications. Generally, alloys are added to steel in order to increase the strength of the steel but usually at the expense of reducig the ductility of the steel. When selecting a type of steel for a specific purpose, it is important to consider such factors as its alloy contents, its strength, ductility, notch toughness, durability and cost.

### Carbon

Carbon is the principal hardening agent of carbon steel and high strength low alloy steels. Alloy steels with carbon contents up to .20% increase the strength of the steel by the carbonization process. Steels with carbon contents from 0.20% to 1% are considered as heat treated steel and are heat treated by quenching and drawing to increase their strength. Carbon has adverse effects on the ductility and weldability of steel.

### Aluminum

Aluminum is used as an alloying metal in order to function as a deoxidizer in the steel mixing process, thus purifying the steel and retarding grain growth during the cooling process. Aluminum also contributes to the surface hardness of the steel.

### Chromium

Chromium is used as an alloying elements to increase the strength and the hardness of steel. In high percentages (12 -30%), chromium improves the corrosion resistance of steel therefore it is used as one of the basic ingredients of stainless steel. Chromium also increases the strength of steel at high temperature. Chromium steels show little or no loss of strength at temperatures up to 900 F.

### Cobalt

Cobalt is used as an alloying element in steel when high strength or increased hardness at high temperatures are required. However, cobalt is not always necessary as an alloy in bridge steels.

### Manganese

Manganese is the most indispensible alloying element in steel. Up to about 0.80%, it is generally intended for the sole purpose of combining with sulfur or phosphorus to offset embrittlement and "hot shortness". In addition to reducing brittleness of steel, it increases the toughness and the strength of steel. Manganese is used in percentage of 0.3% to 1.9% by weight in bridge steel.

### Molybdenum

Molybdenum is used to replace other strategic metals. It increase the hardness and the endurance limits of steel. It decrease the tendency toward high temperature creep or slow stretching steel under stress at high temperature.

### Nickel

Nickel is used as a steel alloy to increase strength and toughness at low temperature. It also increases the corrosion resistance and the strength without decreasing the toughness of steel. For 0 to 4% of nickel, it is estimated that tensile strength is increased 6,000 psi for each 1% of nickel.

### Phosphorus

It is impurity in steel, In high percentages, it decreases the toughness of steel and in low percentage, it improves the machinability of high and low carbon steels. Phosphorus increases the corrosion resistance in steel when it is used in percentage between 0.10%-0.15%. Usually, the percentage of phosphorus should be kept below 0.04% in bridge steels. It has little adverse impacts on the weldability when steel has less than 0.30% contents. Most low carbon steels have only 0.01 to 0.02 phosphorus, but as much as 0.12% phosphorus is found in low carbon, low alloy high tensile steels. It strengthens these steels, but the same amount of phosphorus in high carbon steel would make them brittle. Low welding currents and fast travel speed are recommended for welding the the high phosphorus, low carbon steels in order to reduce the possibility of iron phosphide segregation.

**Silicon**

It increases the fluidity of the molted steel bath by controlling the oxygen content of the steel, thus simplifying the rolling process of the steel. Silicon is used in high percentage to create certain magnetic characteristics for steel used in specific electrical magnetic applications. It is also used in iron-carbon steel to increase the corrosion resistance of the steel. It also contributes to the strength of low-alloy steel.

**Vanadium**

It is used to control the grain size of the steel as it tends to promote the growth of a fine grain structure during the heat treating process. Because of this, vanadium often eliminates the adverse effects of overheating. It tends to improve hardenability and causes secondary hardness, yet resists tempering.

## WELDABILITY

Most new steels can be welded with good results. The weldability of a metal means to relative ease of producing a satisfactory, crackless, sound joint.

If you select steel within the "Normal Range"as listed in the table below prepared by the Lincoln Electrical Company, these steel can be welded at fast welding speeds with minimum cracking tendencies.

PREFERRED ANALYSIS

Of Carbon Steel for Good Weldability

| Element | Normal Range, % | Steels Exceeding Any of Following Percentages Probably Require Extra Care in the Selection and Proving of a Procedure |
|---|---|---|
| Carbon | .06 - .25 | .35 |
| Manganese | .35 - .80 | 1.40 |
| Silicon | .10 max. | .30 |
| Sulphur | .035 max. | .050 |
| Phosphorus | .030 max. | .040 |

However, when welding the thicker plates in even these low alloy carbon steels, the increased rigidity and restraint and the drastic quench effects requires the use of the special precautions and proper procedure. Also, thick plates usually have higher carbon content.

Today, the use of higher strength low alloy steels and heat treated high strength steels have been increased. These steels contains alloying ingredients in addition to those shown in Table for carbon steel. as a result, special precaution are required in welding these steels.

Steels with high sulphur and phosphorus contents are not recommended for production welding. If they must be welded, use low hydrogen electrodes, small electrode size and low current for minimum admixture, and slow travel speed to keep the puddle molten allowing gas bubbles time to boil out.

Frequently advance planning and qualified welding procedures are required to assure crackfree welds when joining thicker plates or old high carbon steel (A7) or the alloy steels. These procedure usually requires the following:

1. Proper bead shape and joint configuration
2. Minimized penetration to prevent dilution of the weld metal with alloy element in the plate.
3. Preheating, controlled interpass temperature and some times even controlled heat input from the welding procedure to retard the cooling rate and reduce shrinkage stresses.

For optimum economy and quality, under either favorable or adverse conditions, the welding procedure should be based on the steel's actual chemistry rather than the maximum alloy contents allowed by the specification. This is because a mill's average production normally runs considerably under the maximum limits.

Generally a Mill Test Report of the steel should be available for the inspectors. This report gives the specific analysis of any given heat of steel. For existing bridges, a chemical analysis should be made from samples taken out from the existing steels. Once this information is available, a welding procedure can be set to produce crackfree welds at the minimum cost.

## SMAW

**Shielded Metal Arc Welding,** also called Stick welding and Manual Metal welding is an electric arc welding process which fuses together the parts to be welded by heating them with an arc between a covered consumable metal electrode and the work. Shielding is obtained from decomposition of the electrode covering. The process is normally manually applied and is capable of welding thin and thick steels and some non-ferrous metals in all positions. The process requires a relatively high degree of welder skill.

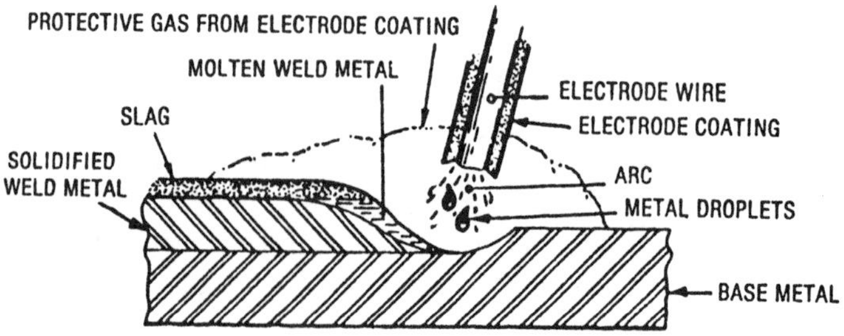


## GMAW

**Gas Metal Arc Welding,** also known as MIG welding, $CO_2$ welding, Micro Wire Welding, short arc welding, dip transfer welding, wire welding etc., is an electric arc welding process which fuses together the parts to be welded by heating them with an arc between a solid, continuous, consumable electrode and the work. Shielding is obtained from an externally supplied gas or gas mixture. The process is normally applied semi-automatically; however, the process may be operated automatically and can be machine operated. The process can be used to weld thin and fairly thick steels and some nonferrous metals in all positions. A relatively low degree of welding skill is required for the process.

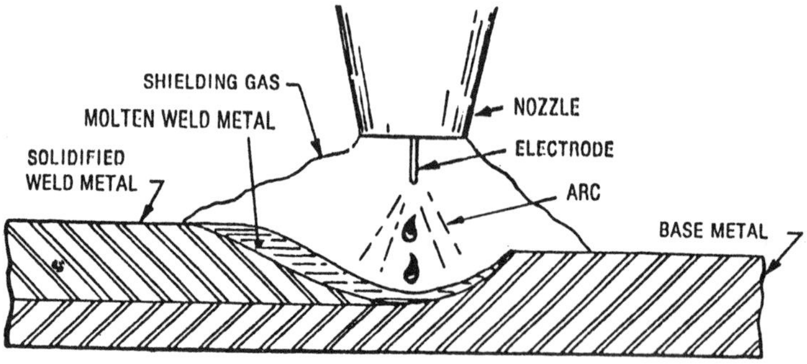


## FCAW

**Flux Cored Arc Welding,** also known as Dual-Shielded, Innershield, Self Shield, FabCO, etc., is an electric arc welding process which fuses together the parts to be welded by heating them with an arc between a continuous flux filled electrode wire and the work. Shielding is obtained through decomposition of the flux within the tubular wire. Additional shielding may or may not be obtained from an externally supplied gas or gas mixture. The process is normally applied semi-automatically, but can be applied automatically or by machine. It is commonly used to weld medium to thick steels using large diameter electrodes in the flat and horizontal position and small electrode diameters in all positions. The process is used to a lesser degree for welding stainless steel and for overlay work. The skill level required for FCAW is similar to GMAW.

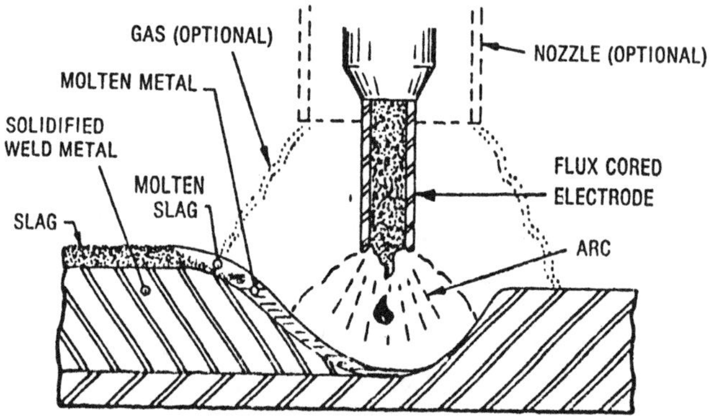


## SAW

**Submerged Arc Welding,** also known as Union Melt, Hidden Arc, Welding Under Powder, etc., is an arc welding process which fuses together the parts to be welded by heating them with an arc or arcs between a bare electrode or electrodes and the work. The arc is shielded by a blanket of granular flux on the work. The process is normally applied by machine or automatically, but is used on a limited basis semi-automatically. It is used to weld medium to thick steel in the flat and horizontal position only. Manual welding skill is not required; however, a technical understanding of the equipment and welding procedures is necessary to operate the process.

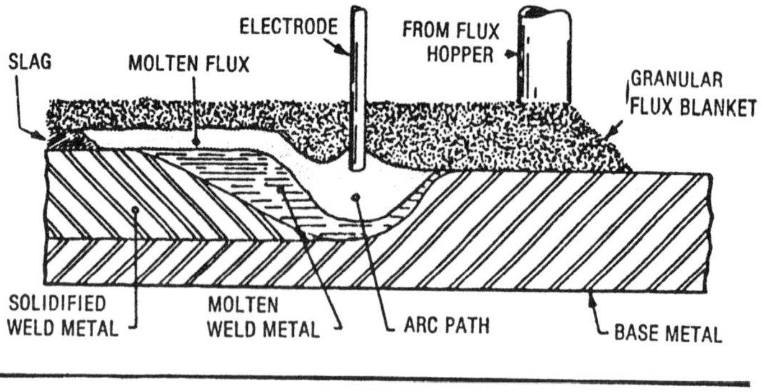


## GTAW

**Gas Tungsten Arc Welding,** also known as TIG Welding, Heliarc Welding, Heli-Welding, Argon-Arc Welding and Tungsten Arc Welding, is an electric arc welding process which fuses together the parts to be welded by heating them with an arc between a nonconsumable tungsten electrode and the work. Filler may or may not be used. Shielding is obtained from an inert gas or an inert gas mixture. The process is normally applied manually and is capable of welding steels and nonferrous metals in all positions. The process is commonly used on thin metals and for the root and hot pass on tubing and pipe. Requires a relatively high degree of welder skill.

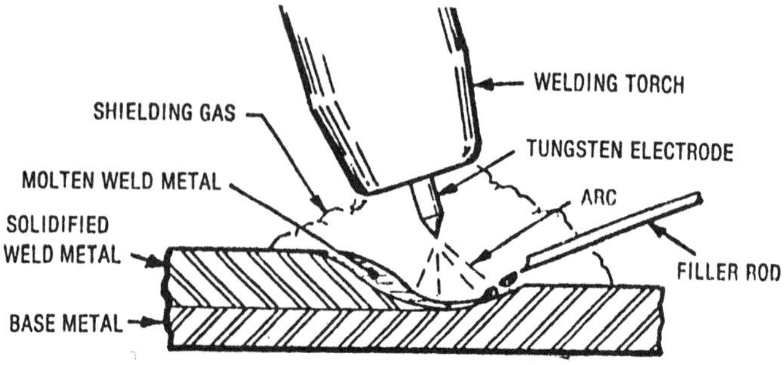

## PAW

**Plasma Arc Welding,** sometimes referred to as Needle Arc and Micro Plasma, is an electric arc welding process which fuses together the parts to be welded by heating them with a constricted arc between a tungsten electrode and the work (transferred arc) or the electrode and the constricting nozzle (non-transferred arc). Shielding is obtained from the hot ionized gas issuing from the orifice. An auxilliary inert shielding gas or mixture of inert gases may supplement the system. The process is commonly applied manually, but may be automatic to increase welding speeds. It can be used to weld almost all metals and can be all position at lower currents. Normally used on thinner metals, the process requires a slightly lesser degree of welder skill than Gas Tungsten Arc Welding, but a greater knowledge of equipment set-up.

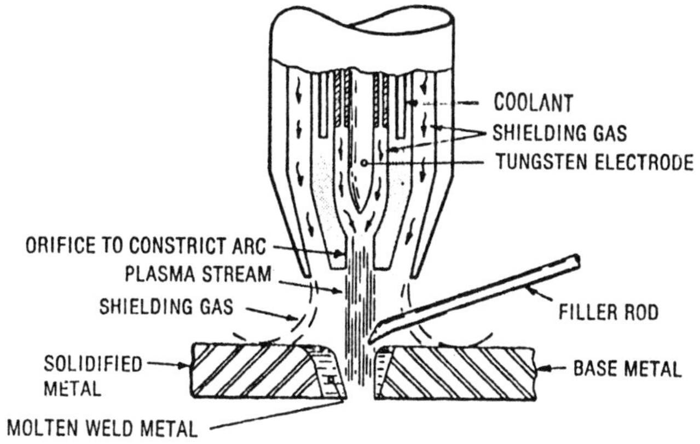


## CAW

**Carbon Arc Welding** is an electric arc welding process which fuses together the parts to be welded by heating with an arc between a carbon electrode and the work. No shielding is used. A variation of the process uses two carbon electrodes with an arc between them. The process is normally applied manually and is capable of welding thin metals. The process is also commonly used for brazing. It requires a relatively high degree of welding skill. This process has limited industrial popularity.

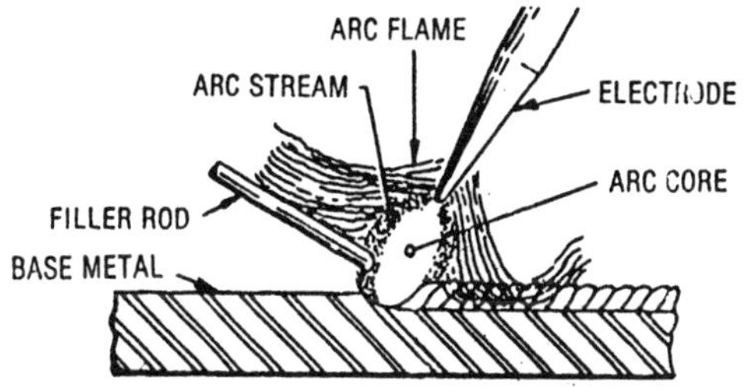


## ESW

**Electroslag Welding,** also known as Porta-Slag or Slag Welding, is a welding process which fuses together the parts to be welded with molten slag which melts the filler metal and the surfaces of the work to be welded. The molten weld pool is shielded by a slag covering which moves along the joint as welding progresses. The process is not an arc welding process, except that an arc is used to start the process. After stabilization the molten slag provides the necessary heat for welding. The process is always applied automatically. It is a limited application process used only for making vertical welds on medium to heavy thickness of mild steel. Manual welding skill is not required, but a technical knowledge of the process is required to operate the equipment.

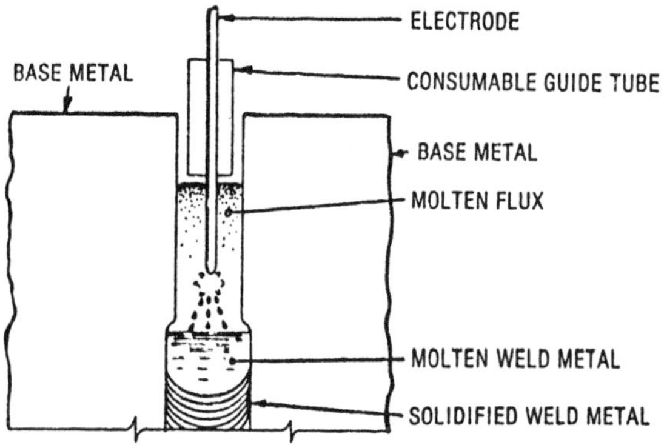


## EGW

**Electrogas Welding,** also known as Verti-Matic and Automatic Vertical Aircomatic, is an arc welding process which fuses together the parts to be welded by heating them with an arc between a continuous filler metal (consumable) electrode and the work. Molding shoes are used to confine the molten weld metal for vertical position welding. The electrode may be either flux cored or solid. Shielding may or may not be obtained from an externally supplied gas or gas mixture. The process is always applied automatically. It is a limited application process used only for making vertical welds on medium to heavy thickness mild steel. Manual welding skills are not required, but like electroslag welding, a technical knowledge of the process is required to operate the equipment.

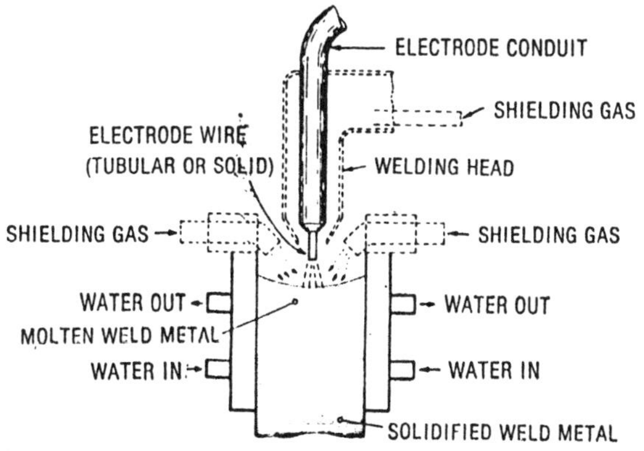


## SW

**Stud Welding,** also known as Stud Arc Welding, is a special purpose arc welding process used to attach studs to base metal. Partial shielding is obtained by a ceramic ferrule surrounding the stud. It is a machine welding process, using a specialized gun that holds the stud and makes the weld. The process is normally used on steels in the flat and horizontal position. A low degree of welding skill is required for stud welding operation.

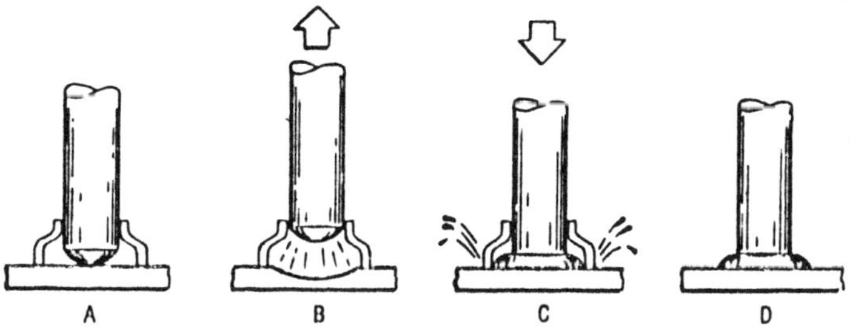

## porous welds

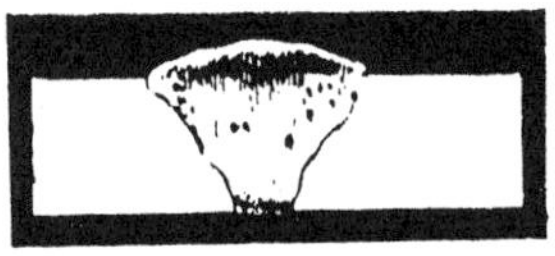

**WHY**

1. Excessively long or short arc length
2. Welding current too high
3. Insufficient or damp shielding gas
4. Too fast travel speed
5. Base metal surface covered with oil, grease, moisture, rust, mill scale, etc.
6. Wet, unclean or damaged electrode

**WHAT TO DO**

1. Maintain proper arc length
2. Use proper welding current
3. Increase Gas Flowrate & check gas purity
4. Reduce travel speed
5. Properly clean base metal prior to welding
6. Properly maintain and store electrode

## cracked welds

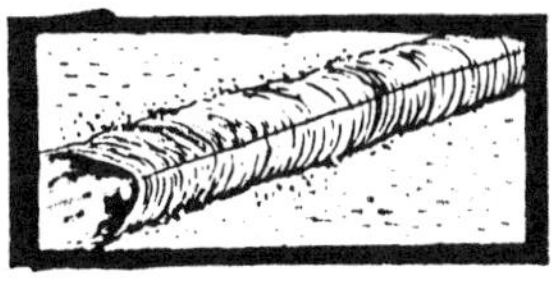

**WHY**

1. Insufficient weld size
2. Excessive joint restraint
3. Poor joint design and/or preparation
4. Filler metal does not match base metal
5. Rapid cooling rate
6. Base metal surface covered with oil, grease, moisture, rust, dirt or mill scale

**WHAT TO DO**

1. Adjust weld size to part thickness
2. Reduce joint restraint through proper design
3. Select the proper joint design
4. Use more ductile filler
5. Reduce cooling rate through preheat
6. Properly clean base metal prior to welding

## undercutting

**WHY**

1. Faulty electrode manipulation
2. Welding current too high
3. Too long an arc length
4. Too fast travel speed
5. Arc blow

**WHAT TO DO**

1. Pause at each side of the weld bead when using a weaving technique
2. Use proper electrode angles
3. Use proper welding current for electrode size and welding position
4. Reduce arc length
5. Reduce travel speed
6. Reduce effects of arc blow

## distortion

**WHY**

1. Improper tack welding and/or faulty joint preparation
2. Improper bead sequence
3. Improper set-up and fixturing
4. Excessive weld size

**WHAT TO DO**

1. Tack weld parts with allowance for distortion
2. Use proper bead sequencing
3. Tack or clamp parts securely
4. Make welds to specified size

## spatter

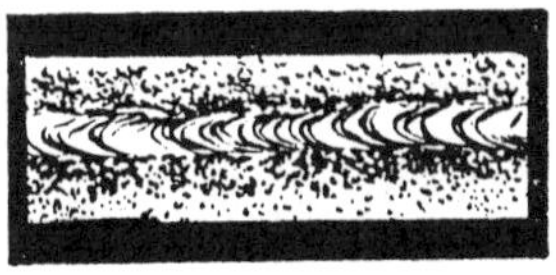

**WHY**

1. Arc blow
2. Welding current too high
3. Too long an arc length
4. Wet, unclean or damaged electrode

**WHAT TO DO**

1. Attempt to reduce the effect of arc blow
2. Reduce welding current
3. Reduce arc length
4. Properly maintain and store electrodes

## lack of fusion

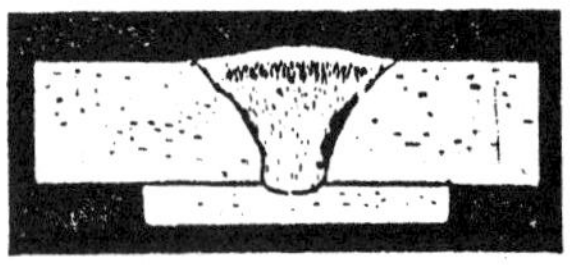

**WHY**

1. Improper travel speed
2. Welding current too low
3. Faulty joint preparation
4. Too large an electrode diameter
5. Magnetic arc blow
6. Wrong electrode angle

**WHAT TO DO**

1. Reduce travel speed
2. Increase welding current
3. Weld design should allow electrode accessibility to all surfaces within the joint
4. Reduce electrode diameter
5. Reduce effects of magnetic arc blow
6. Use proper electrode angles

## overlapping

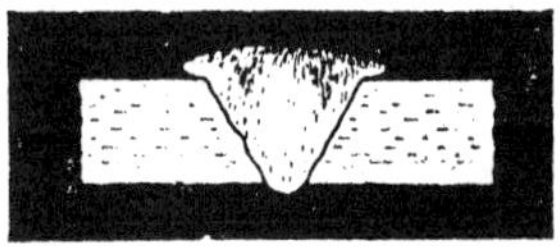

**WHY**

1. Too slow travel speed
2. Incorrect electrode angle
3. Too large an electrode

**WHAT TO DO**

1. Increase travel speed
2. Use proper electrode angles
3. Use a smaller electrode size

## poor penetration

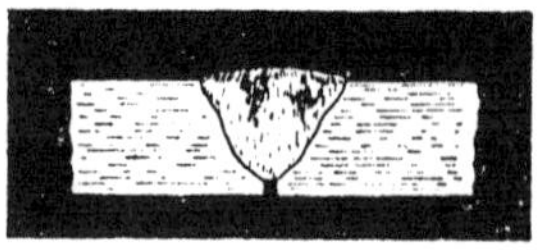

**WHY**

1. Travel speed too fast
2. Welding current too low
3. Poor joint design and/or preparation
4. Electrode diameter too large
5. Wrong type of electrode
6. Excessively long arc length

**WHAT TO DO**

1. Decrease travel speed
2. Increase welding current
3. Increase root opening or decrease root face
4. Use smaller electrode
5. Use electrode w/deeper penetration characteristics
6. Reduce arc length

## magnetic arc blow

**WHY**

1. Unbalanced magnetic field during welding
2. Excessive magnetism in parts or fixture

**WHAT TO DO**

1. Use alternating current
2. Reduce welding current and arc length
3. Change the location of the work connection on the workpiece

## inclusion

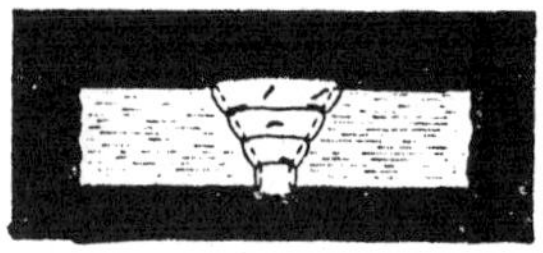

**WHY**

1. Incomplete slag removal between passes
2. Erratic travel speed
3. Too wide a weaving motion
4. Too large an electrode
5. Letting slag run ahead of arc
6. Tungsten spitting or sticking

**WHAT TO DO**

1. Completely remove slag between passes
2. Use a uniform travel speed
3. Reduce width of weaving technique
4. Use a smaller electrode size for better access to joint
5. Increase travel speed or change electrode angle or reduce arc length
6. Properly prepare tungsten and use proper current

Hobart

REPRODUCED FROM THE HOBART POCKET WELDING GUIDE AT THE COURTESY OF OF HOBART BROTHERS COMPANY.

# E.21 PREHEAT TEMPERATURES

## PREHEAT AND INTERPASS TEMPERATURES

Preheat and interpass temperatures shall be based on the carbon equivalent of the reinforcing steel bars and/or the splice material, whichever number is higher (See Table 742). The carbon equivalent number shall be calculated using the chemical composition as shown on the mill test report which represents the material to be welded. The following carbon equivalent formula shall be used:

$$C.E. = \%C + \frac{\%Mn}{6} + \frac{\%Cu}{40} + \frac{\%Ni}{20} + \frac{\%Cr}{10} - \frac{\%Mo}{50} - \frac{\%V}{10}$$

If mill test reports are not available, chemical analysis may be made on bars representative of the bars to be welded. When the mill test report does not establish the value for Molybdenum (Mo) and Vanadium (V), the carbon equivalent shall be determined by using Carbon (C), Manganese (Mn), Copper (Cu), Nickel (Ni), and Chromium (Cr). If the chemical composition of the bars is not known, or not obtained, the carbon equivalent shall be assumed to be above 0.75 for that material.

TABLE 742 – MINIMUM PREHEAT AND INTERPASS TEMPERATURES

| Carbon equivalent range, % | Size of reinforcing bar | Degrees F |
|---|---|---|
| 0.40 max. | up to 11 inclusive<br>14 and 18 | 50<br>50 |
| 0.41-0.45 inclusive | up to 11 inclusive<br>14 and 18 | 50<br>100 |
| 0.46-0.55 inclusive | up to 6 inclusive<br>7 to 11 inclusive<br>14 and 18 | 50<br>50<br>200 |
| 0.56-0.65 inclusive | up to 6 inclusive<br>7 to 11 inclusive<br>14 and 18 | 100<br>200<br>300 |
| 0.66-0.75 | up to 6 inclusive<br>7 to 18 inclusive | 300<br>400 |
| above 0.75 | up to 18 inclusive | 500 |

Notes:

1) When reinforcing steel is to be welded to main structural material, the preheat requirements of the structural material shall also be considered (see Table 708). The minimum preheat requirement to apply in this situation shall be the higher requirement of the two tables.

TABLE 708 – MINIMUM PREHEAT AND INTERPASS TEMPERATURE

| Thickness of Thickest Part at Point of Welding (inches) | ASTM A36, A53, A 252, A441, A500, A501, A572 | ASTM A588, A242 |
|---|---|---|
| To 3/4, inclusive | 50°F | 100°F |
| Over 3/4 to 1 1/2, inclusive | 70°F | 200°F |
| Over 1 1/2 to 2 1/2, inclusive | 150°F | 300°F |
| Over 2 1/2 | 225°F | 350°F |

2) Welding shall not be done when the ambient temperature is lower than 0° F. When the base metal is below the temperature listed for the welding process being used and the size and carbon equivalent range of the bar being welded, it shall be preheated in such a manner that the cross section of the bar for not less than 6 in. on each side of the joint shall be at or above the specified minimum temperature. Preheat and interpass temperatures must be sufficient to prevent crack formation.

3) After welding is complete, bars shall be allowed to cool naturally to ambient temperature. Accelerated cooling is prohibited.

4) Where it is impractical to obtain chemical analysis, the carbon equivalent shall be assumed to be above 0.75.

## F.1 TYPES OF TIMBER BRIDGE

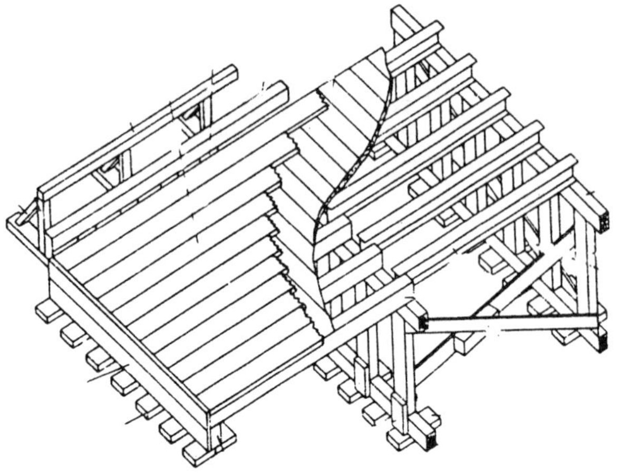

TIMBER TRESTLE

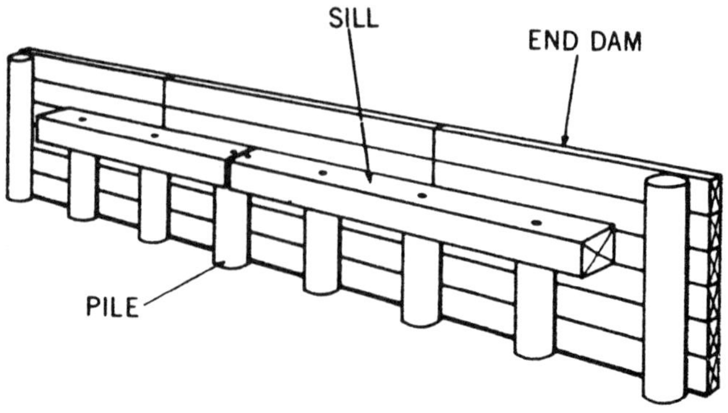


TIMBER PILE ABUTMENT

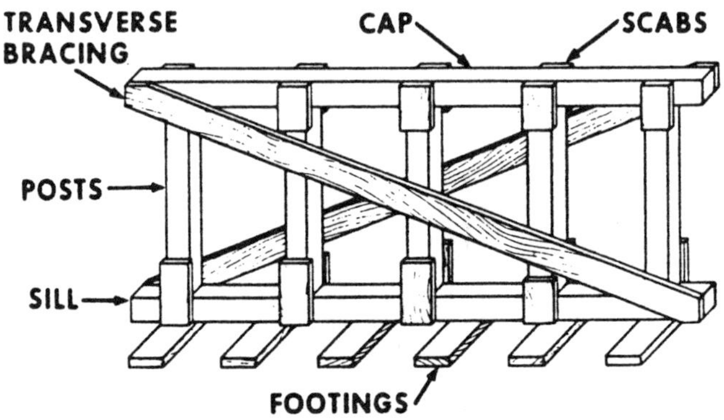


PILE BENT

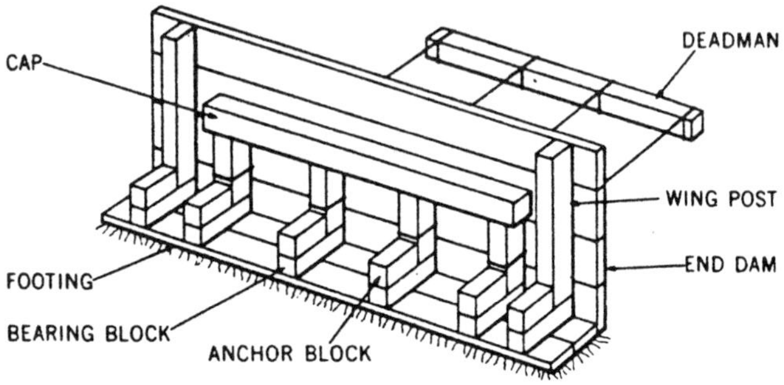


TIMBER BENT ABUTMENT

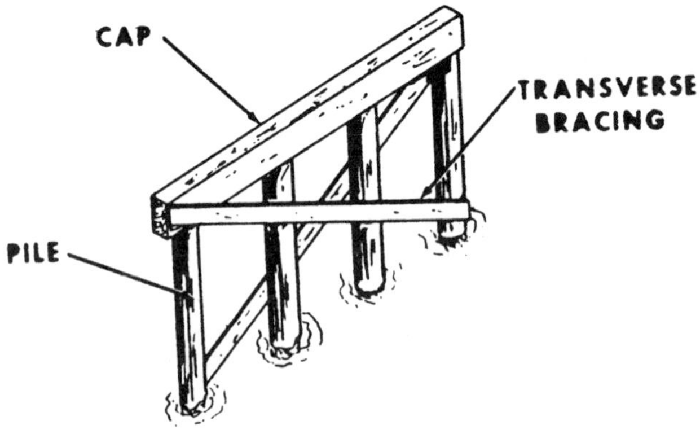


TIMBER TRESTLE BENT

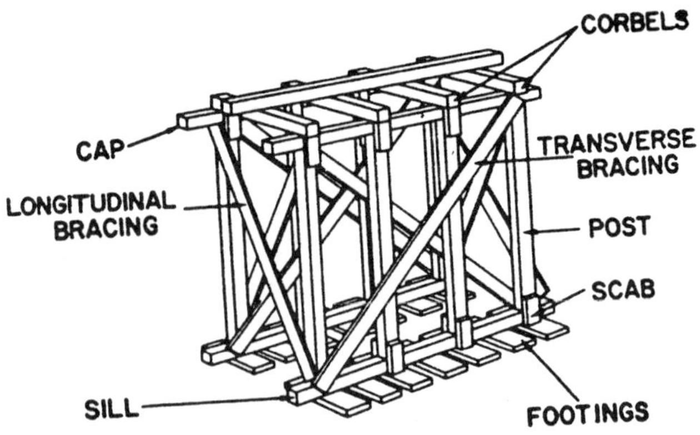


TIMBER TRESTLE PIER

ARMY FIELD MANUAL

## LONGITUDINAL LAMINATED DECK

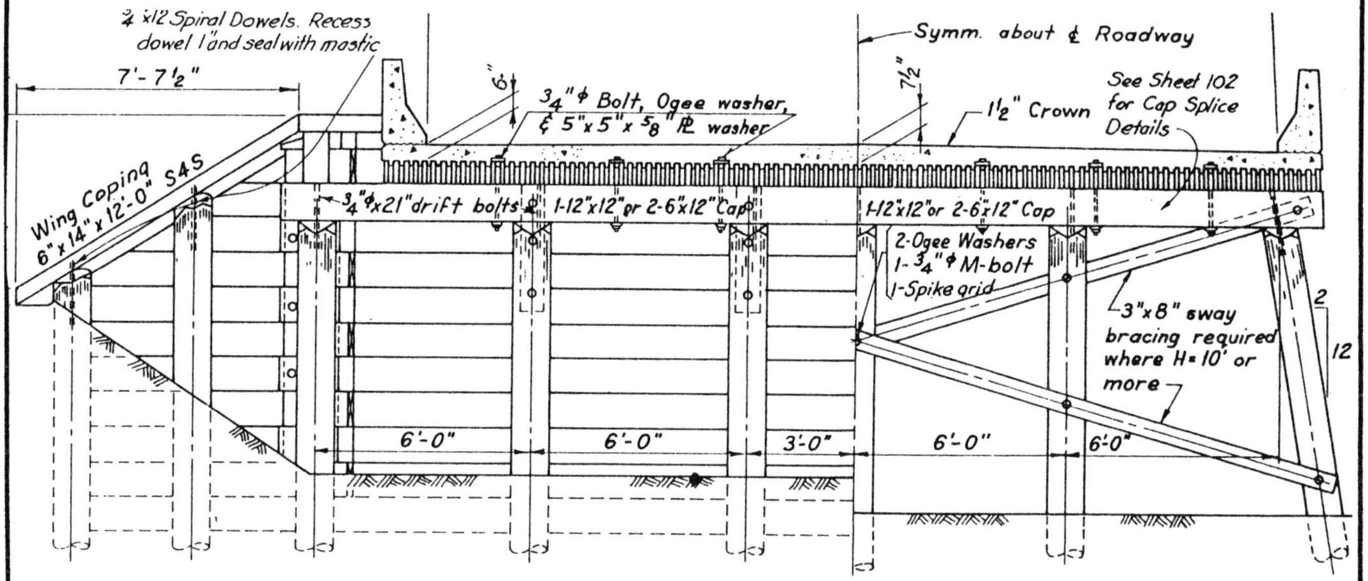


HALF TRANSVERSE SECTION

NEAR ABUTMENT

HALF TRANSVERSE SECTION

NEAR PIER

## TIMBER STRINGERS AND LAMINATED DECK

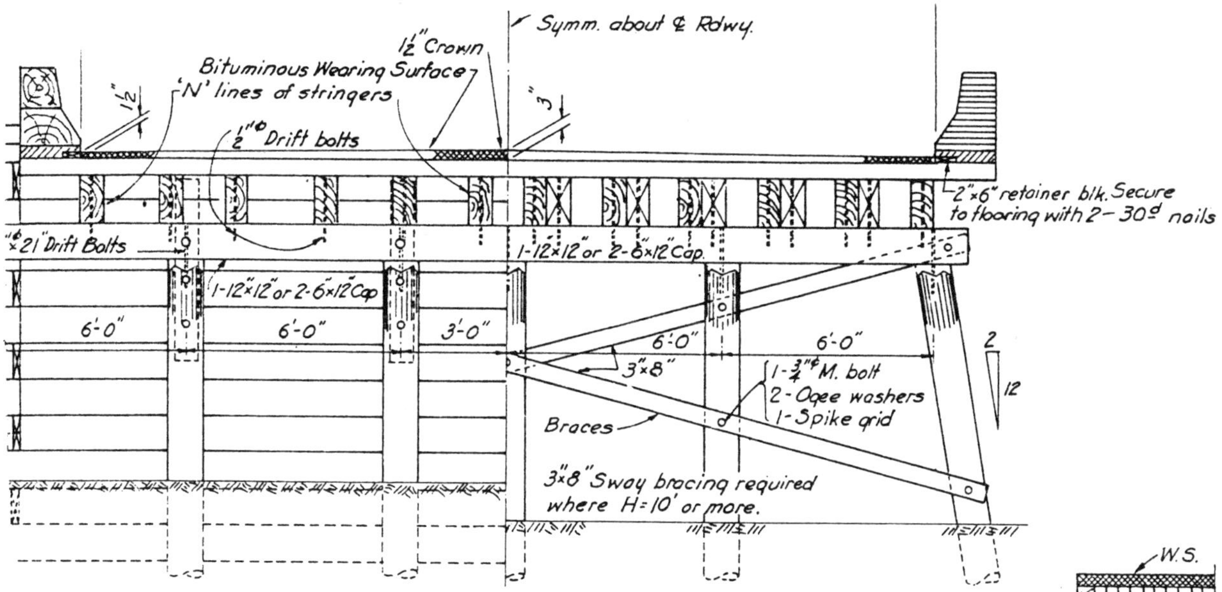


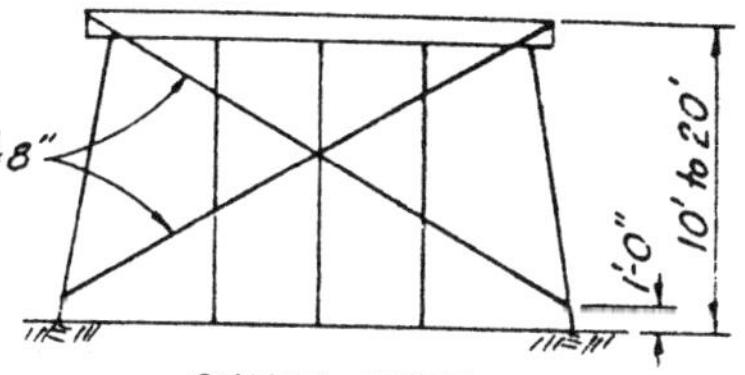


SINGLE STORY

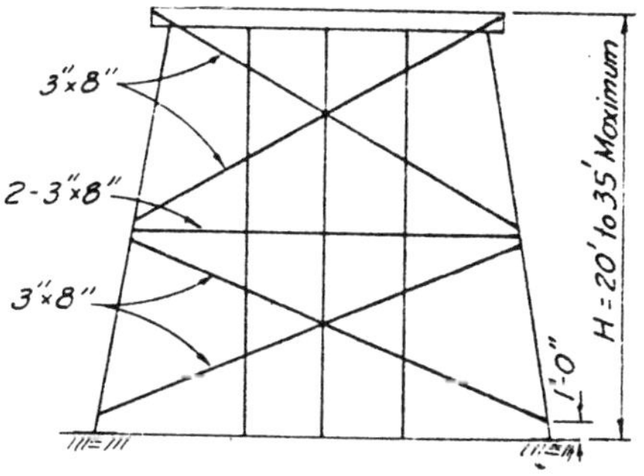


DOUBLE STORY

INTERIOR BENT SWAY BRACING

*Maximum H for H20 Loading = 30'*

*Maximum H for H15 Loading = 35'*

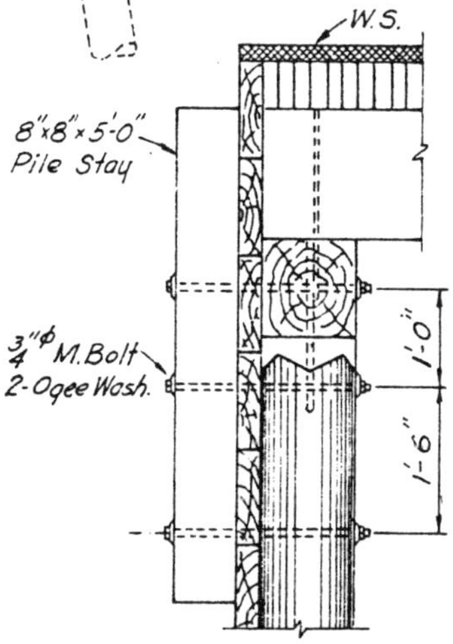


SECTION THRU ABUTMENT

## TIMBER STRINGER REPLACEMENT

DESCRIPTION: Replace or add timber stringer: Method A - Under deck; Method B - Above deck

APPLICATION: Use this method to replace a deteriorated stringer or add an additional stringer adjacent to a stringer whose function has been impaired.

CONSTRUCTION METHOD:

Method A: Under deck

1. Locate work platform or work barge as needed.
2. Place two jacks on each cap in adjoining bays next to stringer to be replaced.
3. Use steel plates under deck to jack against.
4. Remove traffic from jacking area if possible.
5. Jack up until stringer clears cap 1/4 to 1/2 inch.
6. Cut wedge out of one end of stringer and bevel corners on other end. If new stringer is warped, always place camber up to provide bearing on all decking timbers.

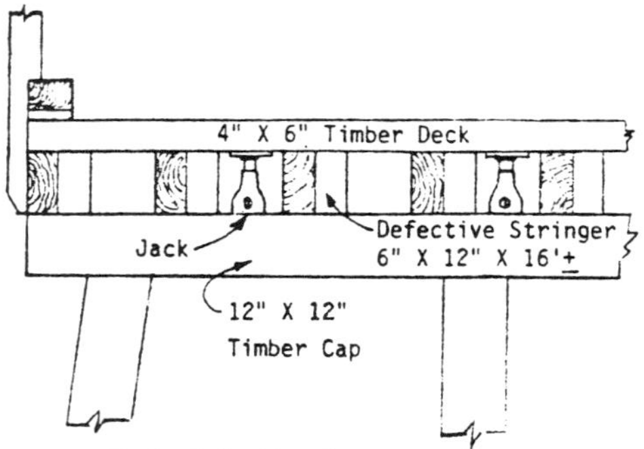

Typical Jacking Arrangement

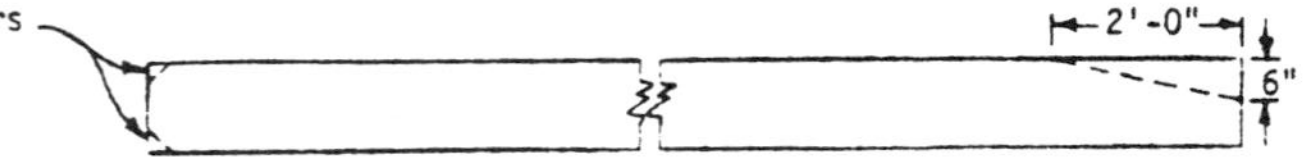

New Stringer

7. Place wedged end of stringer on cap far enough to allow opposite end to be placed on other cap.

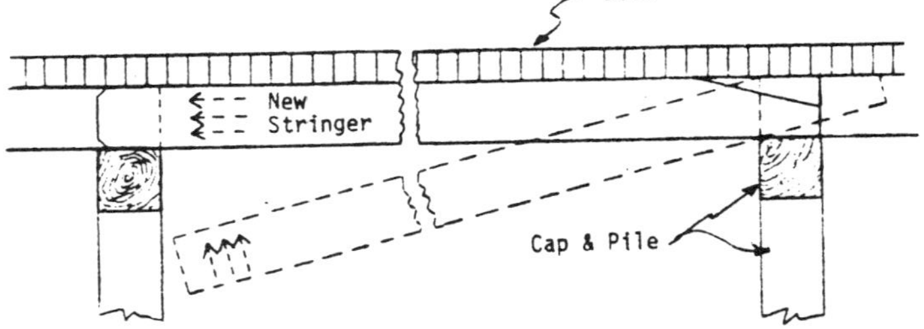

Erection Scheme

8. Place a come-along cable or chain around cap at beveled end of stringer and attach to wedged end of stringer.
9. Pull stringer with come-along to final position.
10. Remove jacks and come-alongs.
11. Place cut wedge at wedged end of stringer and nail decking to stringer.

Method B: Above Deck

1. Cut decking down center of stringer on each side of stringer to be replaced. Remove decking to allow one end of stringer to be placed on cap enough for the opposite end to clear cap when raised.
2. Raise stringer through pre-cut hole in decking high enough to clear cap.
3. Jack stringer to final position.
4. Replace decking, nailing to new stringer and adjacent stringers.

## REPAIR OF CRACKED STRINGER (WYOMING)

Steel plates and bolts are used to prevent further deterioration of cracked timber stringers.

Calculate the load that would be necessary to jack at a location near the cracked position in the stringer to eliminate the dead load moment. Calculate the corresponding deflection this load would produce. The stringer to be repaired should be raised exactly that amount.

Remove the asphalt overlay in the area where the holes are to be drilled.

Drill holes for draw-up bolts. See Figure

Place the retaining and support plates as shown in Section AA and tighten the draw-up bolts.

Replace the wearing surface.

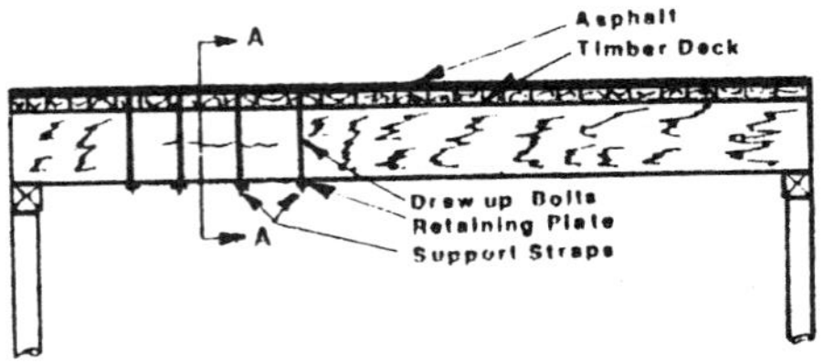


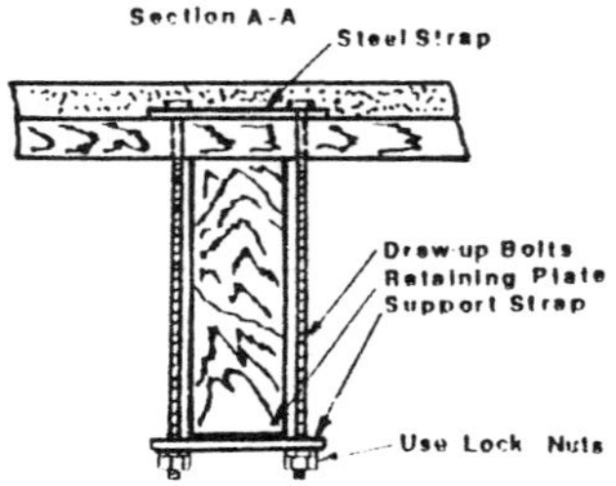

# TIMBER HELPER BENT

DESCRIPTION: Driving timber piles through deck parallel and adjacent to existing pile bent and placing timber cap on top of piles.

APPLICATION: This repair is used when existing piling have lost bearing and settlement occurs or when bent piling have deteriorated to the point that their load carrying capacity is in question.

CONSTRUCTION METHOD:

1. If deck is timber overlaid with asphalt or concrete, locate centerline of stringers closest to each pile location for the helper bent. If deck is reinforced concrete, locate piles so that deck beams will not interfere with driving.

2. Cut holes in only one lane of traffic at a time.

3. Cut through overlay and timber deck along centerline of stringer. Remove sufficient number of timber decking to permit pile to go through hole. For reinforced concrete deck, remove sufficient amount of concrete in a square pattern to permit pile to go through hole. Cut reinforcing steel at center of hole and bend back out of way of pile.

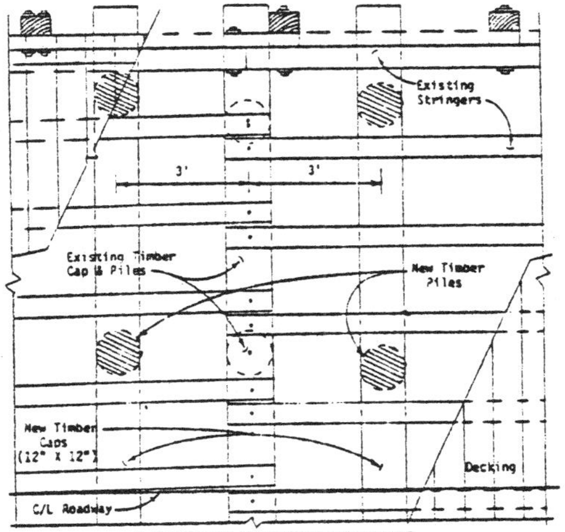

PLAN VIEW FOR INTERMEDIATE HELPER BENT

NOTE: If timber deck is not overlaid, cut timber decking at centerline of stringer near centerline of bridge and remove decking across travel lane.

4. Set piling and drive to required bearing.

5. Cut off piling approximately 1/4" above bottom of existing cap. If existing cap has settled, allowance must be made for grade differential.

6. Place cover plates over deck holes, open lane to traffic and move to adjacent, repeating operation.

7. After all piles have been driven and cut off, jacket up superstructure approximately 1/2" using existing pile bent.

8. Place timber cap over both rows of piling. For end bents, only one row of piles and cap is required.

9. Lower superstructure onto new caps and strap cap to piling. Shimming may be required to obtain bearing between superstructure and timber cap.

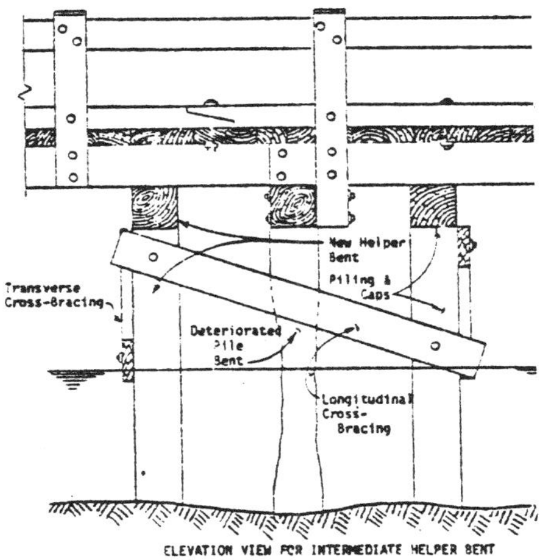

ELEVATION VIEW FOR INTERMEDIATE HELPER BENT

10. Remove deck plates and reconstruct deck. If deck is reinforced concrete, splice cut bars. Replace deck in one lane at a time. Concrete deck repair may be reopened to traffic if protected by steel plates.

11. Erect crossbracing on new pile bent. For intermediate bents, crossbracing between the two new bents is also required.

# HELPER BENT-TEMPORARY

CONSTRUCTION METHOD:

1. Set piles and drive to specified bearing.

2. Cut off piles to allow for beams, neoprene pads and wedge plates.

3. Connect 12" X 12" timber cap to plates with 7/8" Ø X 1'-9" headless drive spikes.

4. Connect timber brace to piles.

5. Set steel beams into position. (Beams may be W21 X 55 or W21 X 62.)

6. Jack beams off cap to obtain temporary bearing against superstructure.

7. Set neoprene pads into position. Cut Type IV in half for two 9" X 9" pads.

8. Place and drive wedge plates as necessary to obtain bearing.

9. Remove jacks and drive lag bolts.

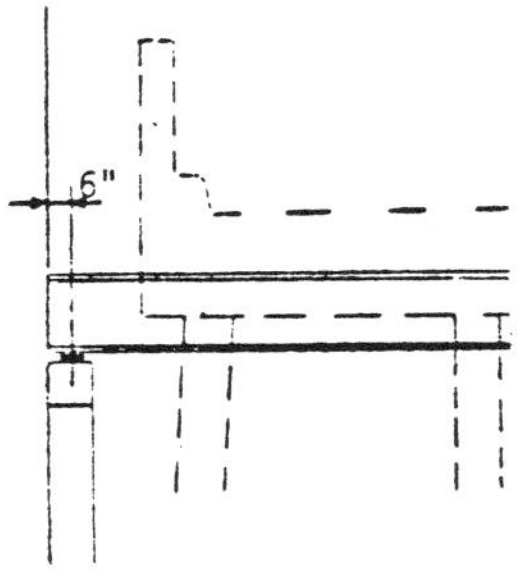


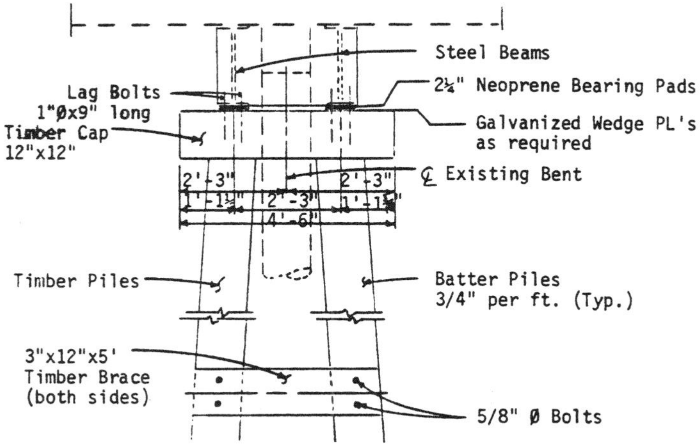

# TIMBER CAP REPLACEMENT

DESCRIPTION: Remove load from existing timber cap, remove cap and replace with new timber cap

APPLICATION: Use to replace deteriorated timber cap at an intermediate pile bent or end bent.

CONSTRUCTION METHOD:

General: Two different methods may be employed for removing load from existing cap. These are: Jacking from cribbing and Jacking up bent piles. Details for both of these methods are shown.

CRIBBING

1. Place cribbing for end bent caps in front of bent, leaving room for removal and replacement of cap. For intermediate bents, place cribbing on both sides of bent. Place jacks on cribbing. For timber spans, use false caps to jack against on each side of the bent. On spans with steel beams, jack against the bottom flange of the beams on each side of the bent.
2. Hang new cap along side of existing cap. For end bents, leave room to remove old cap.
3. Jack up span 1" to 2".
4. Cut dowels connecting piles and cap with burning torch or hacksaw.
5. Slide old cap off piling and move new cap into position.
6. Lower jacks.
7. Connect cap to piles with steel straps and spikes.
8. Remove cribbing.

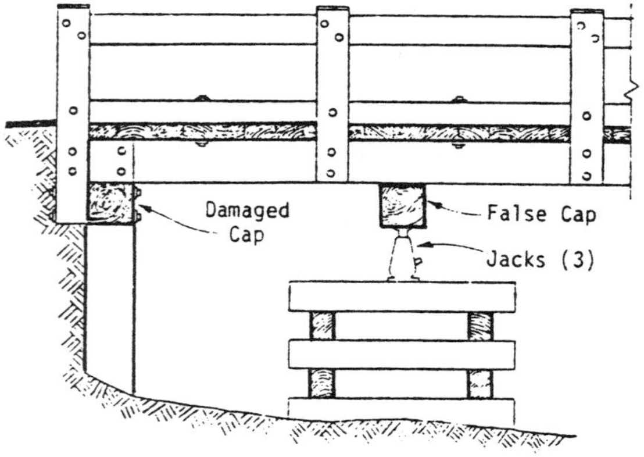


Cribbing Arrangement for End Bent Cap Removal

JACKING FROM BENT PILE

NOTE: This method is used only on intermediate bents for timber structures or steel beam spans where beams are continuous over the intermediate bent.

1. Place two U-clamps and blocking on each pile in the bent (4). Hang 6" X 12" timber above blocking and place jacks.

Steps 2 through 7 are the same as specified for cribbing method.

8. Remove jacks and U-clamp rigging.

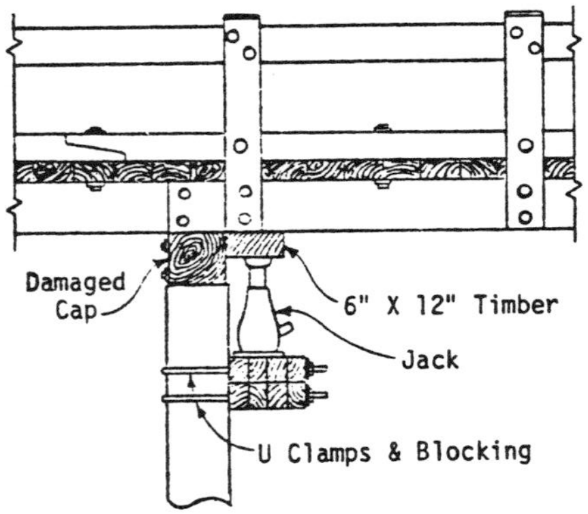


Rigging is required on each pile

# STRENGTHENING PIER CAP

**(NCHRP)**

Additional timbers are bolted to an existing bent or pier to provide required capacity.

CONSTRUCTION PROCEDURE:

1. Construct scaffolding as required around existing bent.
2. Notch existing piles or columns in order to accommodate new timber cap members.
3. Place new members snug against existing cap and stringers; temporarily clamp in place. See Figure 57 and Section AA.
3. Drill 13/16" ϕ holes for 3/4" ϕ bolts.
4. Place bolts, tighten, and remove clamps.
5. Remove scaffolding, if used.

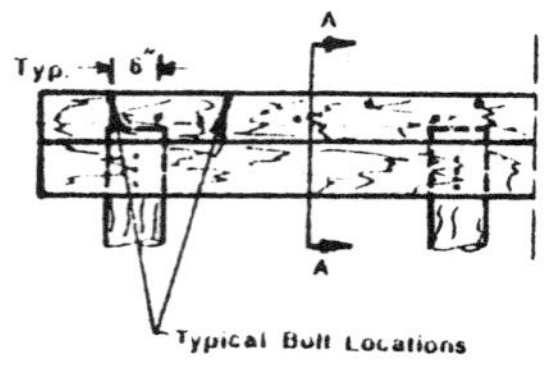


Partial Front Elevation Of Timber Bent

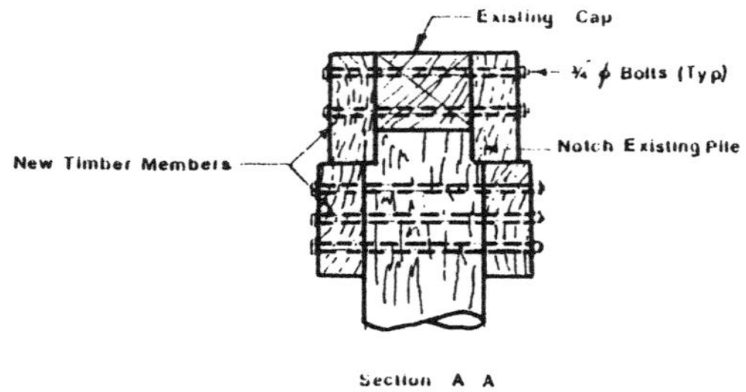


Section A A

## TIMBER CAP SCAB

DESCRIPTION: Attach 6" X 12" treated timber scabs on each side of an existing timber cap using split ring connectors and bolts.

APPLICATION: Apply to caps to extend bearing area where bottoms of stringers or laminated decking has deteriorated over existing cap. Existing cap must be sound timber. Generally, this is used as a temporary repair where bridge is scheduled to be replaced in a few years.

MATERIALS: 2 - 6" X 12" treated timbers ------ M.S. #955
3/4" X 30" Galvanized bolts ------- M.S. #470-11
O.G. galvanized washers ----------- M.S. #470-11
4" Ø split ring timber connectors

CONSTRUCTION METHOD:

1. Locate work platform or work barge as needed.
2. Use template to layout Ø ring locations on cap and timber scabs.
3. Cut Ø rings and drill bolt holes.
4. Insert Ø rings in cap.
5. Raise scabs into position and clamp to cap or use bolts to temporarily hold in position.
6. Place O.G. washers and nuts on one end of bolts, then insert bolts in predrilled holes.
7. Place O.G. washers on opposite end of bolts and tighten to final fit.

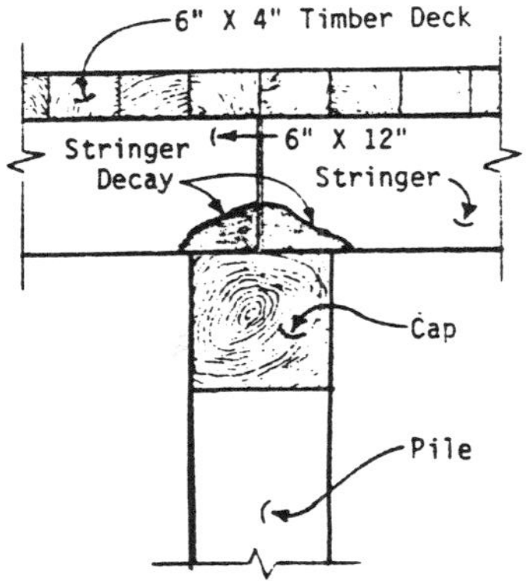


Typical condition where timber cap scabs should be used.

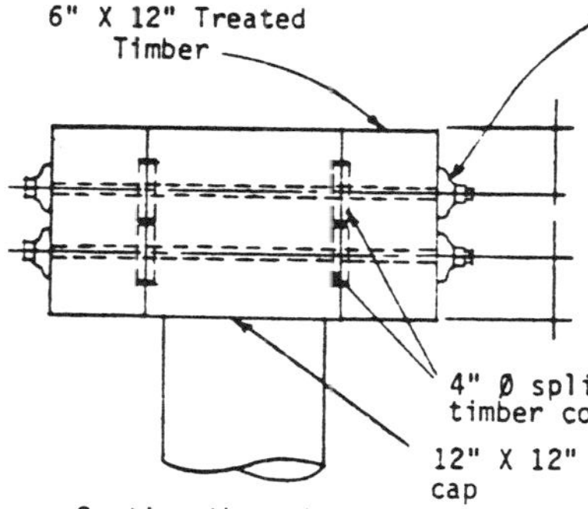


Section through cap

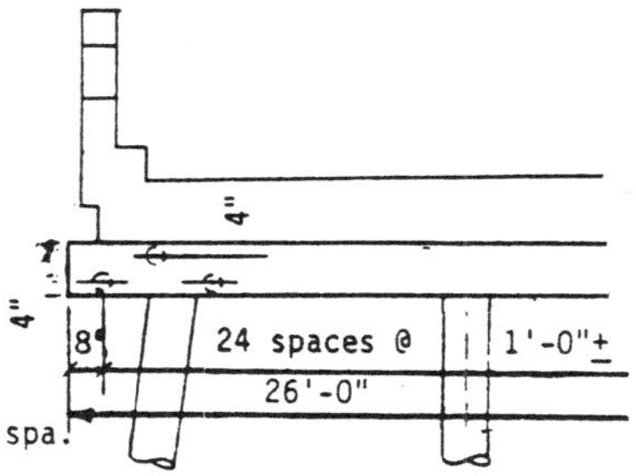


## SHIMMING TIMBER PILES

DESCRIPTION: Placement of a timber wedge between cap and pile head.

APPLICATION: Use to restore bearing that has been lost between the pile cap and pile due to settlement. Also may be used when top few inches of pile is decayed.

CONSTRUCTION METHOD:

1. Place cribbing or strut adjacent to pile to be shimmed. If several piles in one bent need shimming, cribbing should be placed to allow shimming all piles at one time.
2. Set jacks on cribbing (strut) and raise approximately 1/2" higher than desired elevation.

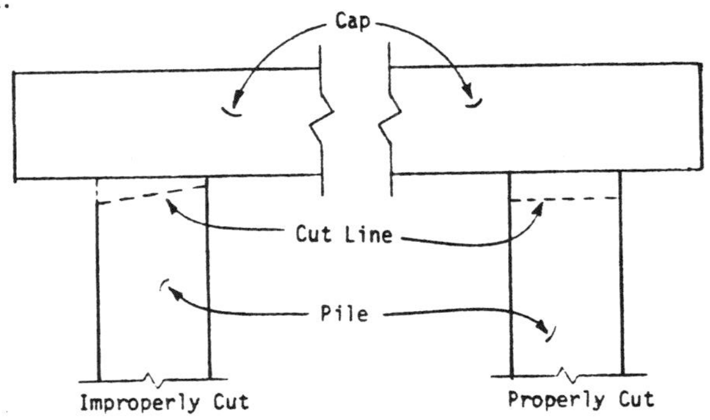


DECAYED PILE TOP

3. Cut shim 1/4" less than opening between cap and pile head, then place into position.
4. Lower jacks and toenail shim to pile.
5. Remove cribbing.

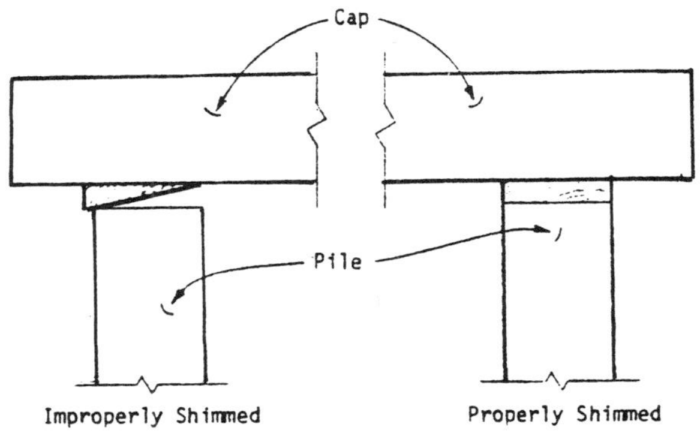


PILE SETTLEMENT

# TIMBER PILE SWAY BRACING

DESCRIPTION: Erecting timber sway bracing on existing or cutting and replacing sections of existing sway bracing.

APPLICATION: When scour affects the lateral stability of a pile bent where sway bracing was not originally specified, erect new sway bracing. Also, where original sway bracing has deteriorated, replace in whole or part as necessary.

TRAFFIC CONTROL AND SAFETY: Refer to Manual on Uniform Traffic Control & Safe Practices.

MATERIALS: Treated Timber (3" X 10") -------------- M.S. #955
Galvanized Bolts, sq. hd. & nut (3/4") - M.S. #470-11
C.I. Washers, O.G. --------------------- M.S. #470-11

CONSTRUCTION METHOD:

Pile Bents without Sway Bracing: When sway bracing is deemed necessary by the engineer, measurements should be taken and timber cut to dimension before treatment.

1. Temporarily attach sway bracing to piling in its final position with galvanized nails. Nails should be driven so as not to interfere with bolt installation.
2. Locate bolt holes and drill through bracing and pile. Use same diameter drill bit as bolt diameter.
3. Treat all holes with hot oil preservative before installing bolts.
4. Place bolts, washers and nuts, then tighten.

NOTE: Some bents may require horizontal bracing at bottom of sway bracing. Use same bolt hole on exterior piles for this bracing as for sway brace members.

Sway bracing needing repair

Repair of Existing Sway Bracing:

1. Locate end of deteriorated or damaged section of sway brace. Cut off brace at the next pile.
2. Measure length of new bracing required to re-connect piles and cut new sections.
3. Re-bolt through existing holes in pile where possible. Drill new holes in piling when new sway brace must be realigned. Treat both new bolt holes and old ones to be reused with hot oil preservative.
4. Treat all timber cuts, both old and new sway bracing, with hot oil preservatives followed by a coating of hot tar.

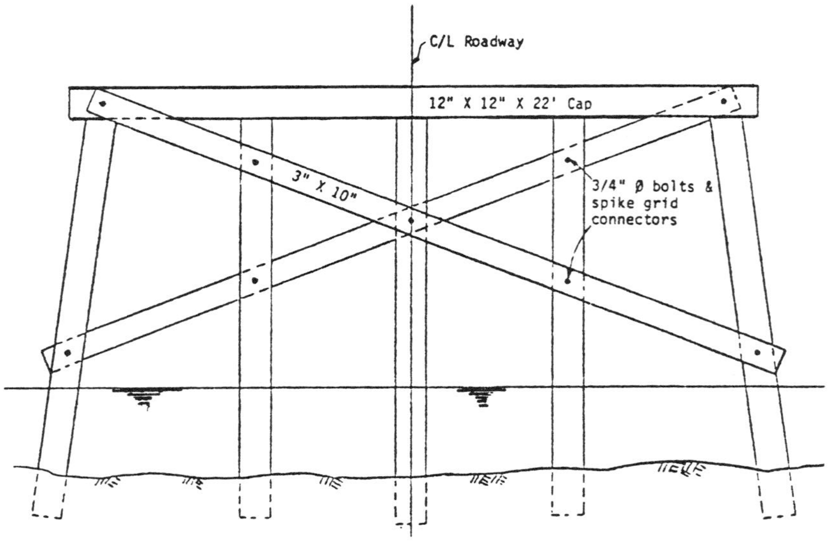


TYPICAL SWAY BRACING

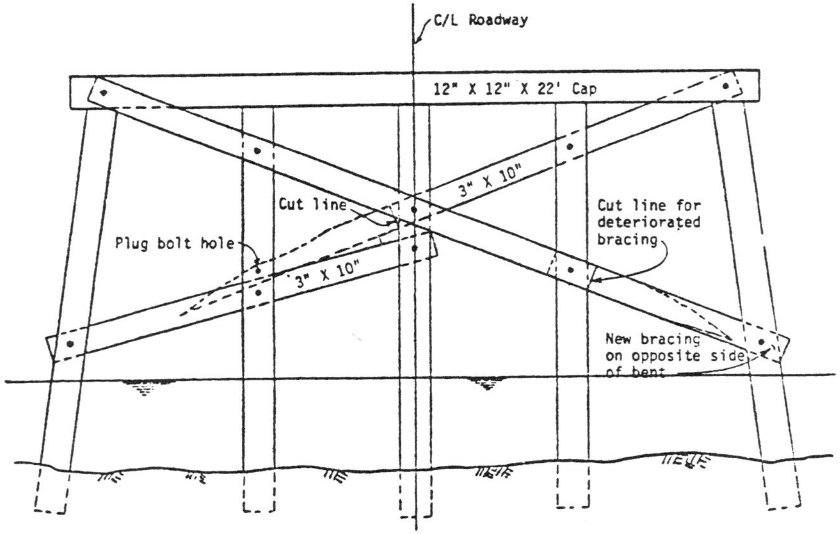


TYPICAL REPAIRS TO SWAY BRACING

# TIMBER PILE REPLACEMENT

DESCRIPTION: Drive treated timber pile through deck and fasten in place under cap.

APPLICATION: Use this method to replace a deteriorated timber pile where conditions for splicing are not practical.

CONSTRUCTION METHOD:

1. If deck is overlaid with asphalt or concrete, locate centerline of stringers closest to pile replacement.

2. Cut through overlay and timber deck along centerline of stringer. Remove sufficient number of timber decking pieces to permit pile to go through hole adjacent to cap.

NOTE: If deck is not overlaid, cut timber decking at centerline of stringer near centerline of bridge and remove decking across travel lane.

3. Remove cross bracing from between the piles and on the side of the bent where new pile will be driven.

4. Set pile at slight batter so when driven and pulled under cap, it will be plumb. Drive to specified bearing.

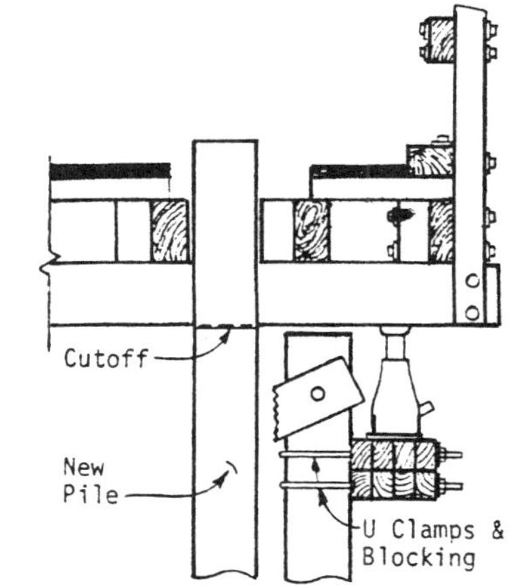


Method of Jacking Up Cap

5. Install U clamps and blocking around pile to be replaced. Place jack on blocking and jack cap up approximately 1/2".

6. Cut pile 1/4" below cap and pull pile into position under cap.

   Positioning pile can be accomplished by using come-alongs to pull against adjacent bents. Place copper sheeting on pile head.

7. Lower jack and strap in place.

8. Reconstruct cross bracing on bent piles.

# TIMBER PILE SPLICE

DESCRIPTION: Partial replacement of deteriorated pile with new section of treated timber pile.

APPLICATION: Used for repairing timber piles that are deteriorated at the ground line or above. No more than 50% of the piles in a bent should be repaired with this method.

MATERIALS:

| | | |
|---|---|---|
| Treated timber pile | M.S. #955-1 | Bolts (2 @ 5/8" X 3/4") |
| Concrete (Class II) | M.S. #345-2 | Steel plate straps (2 @ 1/4" X 2" X 18") |
| Reinforcing Steel | M.S. #931-1 | Jacket form |
| Copper sheeting and lg. hd. nails | M.S. #470-5 | Cribbing |

CONSTRUCTION METHOD:

1. Remove soil around pile to depth well below permanent moisture.
2. Construct cribbing or place struts for supporting jacks.
3. Place jacks and jack up cap 1/2" to 1". (See sketch.)
4. Cut steel dowel connecting cap and pile with torch or hacksaw.
5. Cut deteriorated pile off below permanent moisture line. Make cut at right angle to centerline of pile. (See sketch.)
6. Cut new pile section 1/4" longer than removed section.

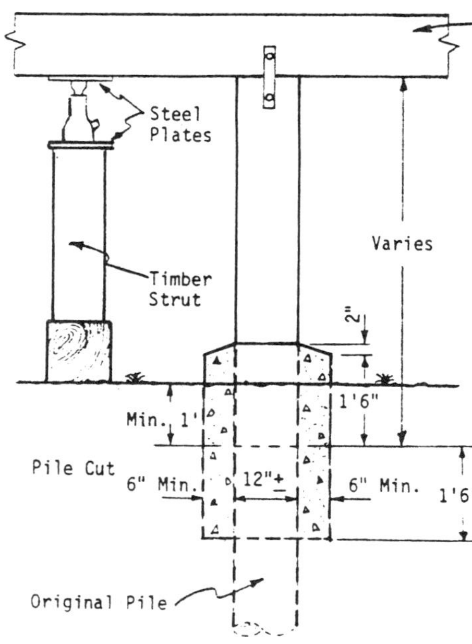


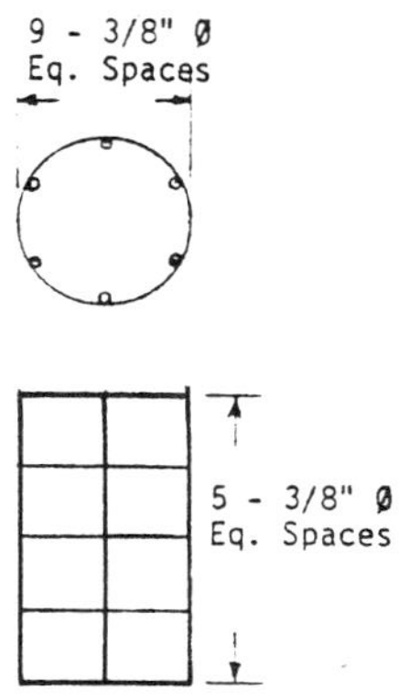


Reinforcing Detail

Place new pile section on pile stump and under cap at same location as original pile. Make sure there is even contact at bearing points. (Top of stump and bottom of cap.) Place copper sheeting on pile head. Lower jacks.

Place form and reinforcing steel for concrete jackets.

Pour concrete (Class II) a minimum of 6" cover around pile. (Top of jacket should be sloped slightly to allow water to run off.

Drill holes and attach steel plate to pile and cap as shown in details.

## G.1 DECK JOINT REPAIRS

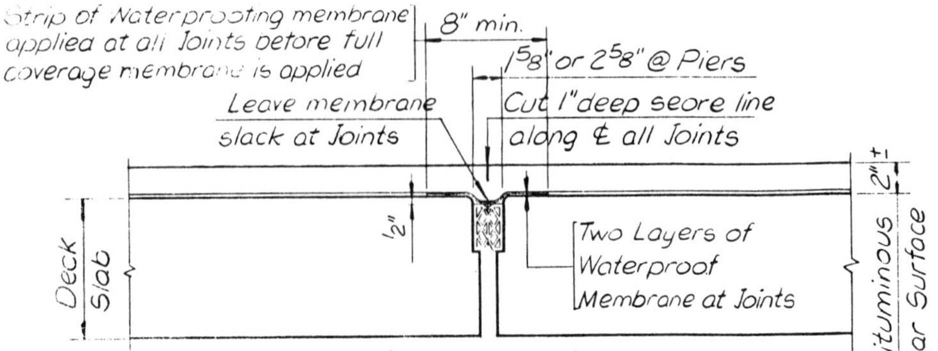


(Typical @ Piers)

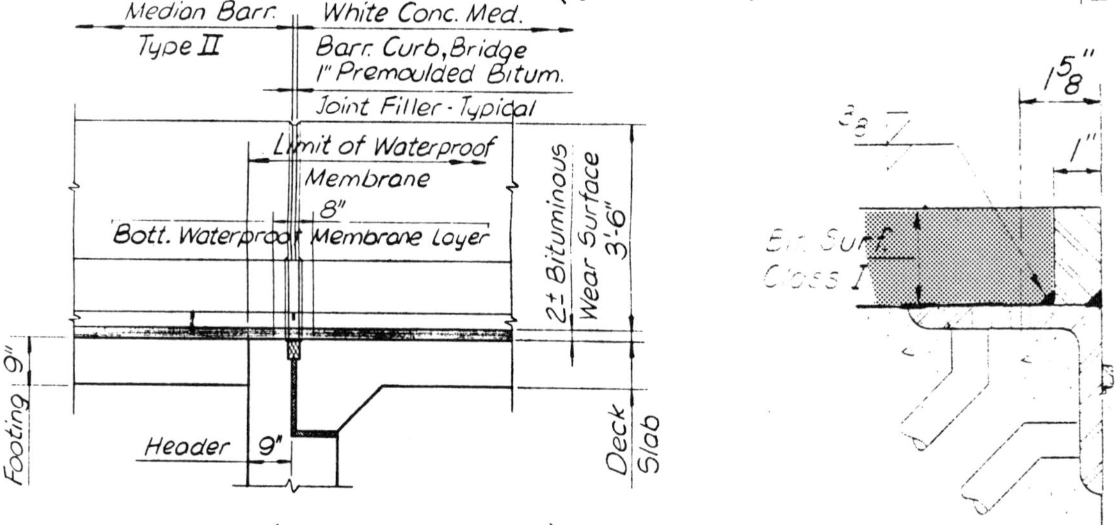


(Typical @ Abutments)

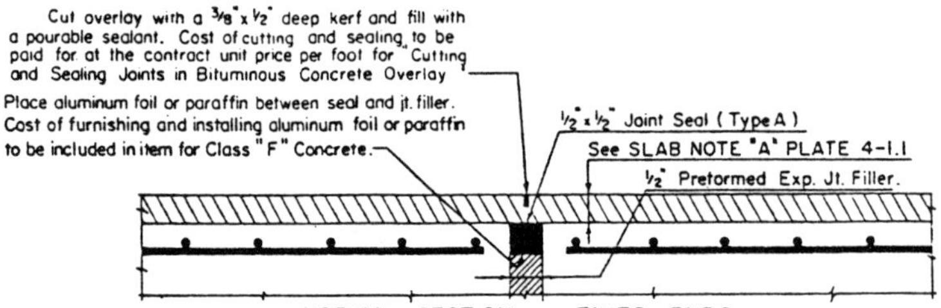


NORMAL SECTION — FIXED ENDS

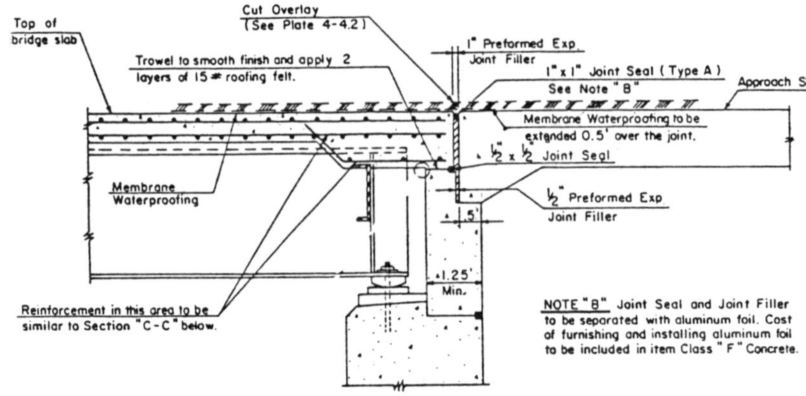


NORMAL SECTION AT FIXED ENDS

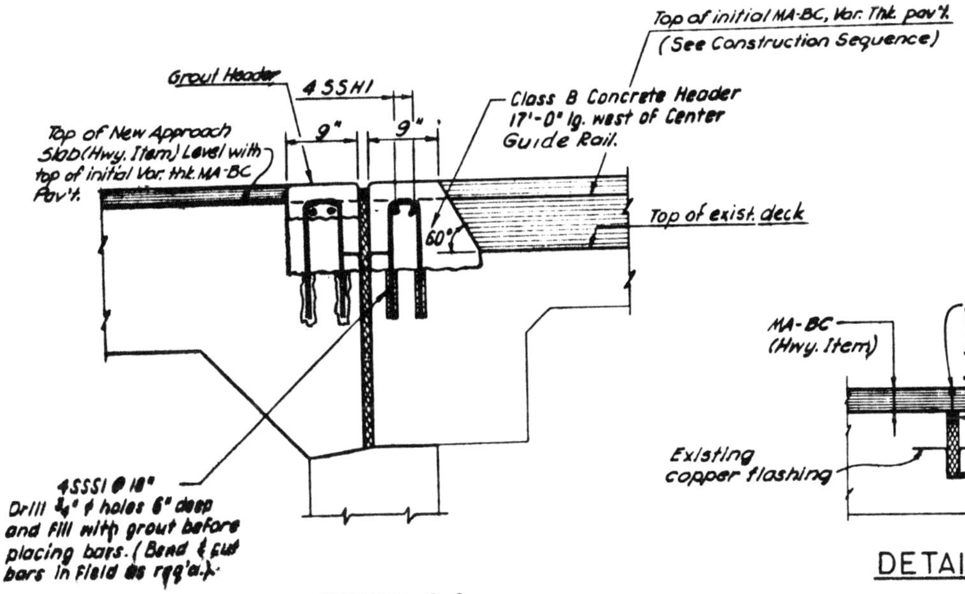


SECTION C-C

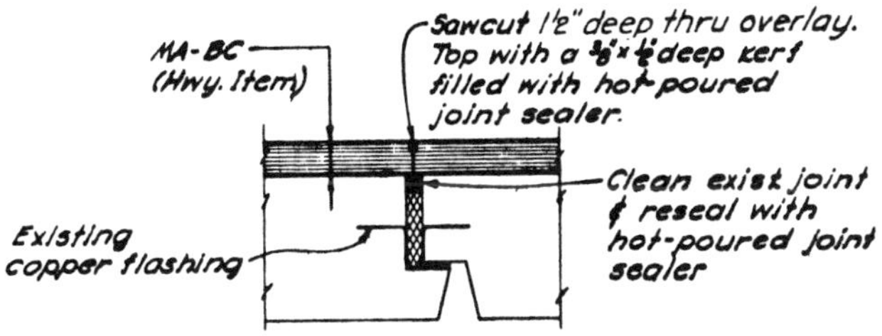


DETAIL B

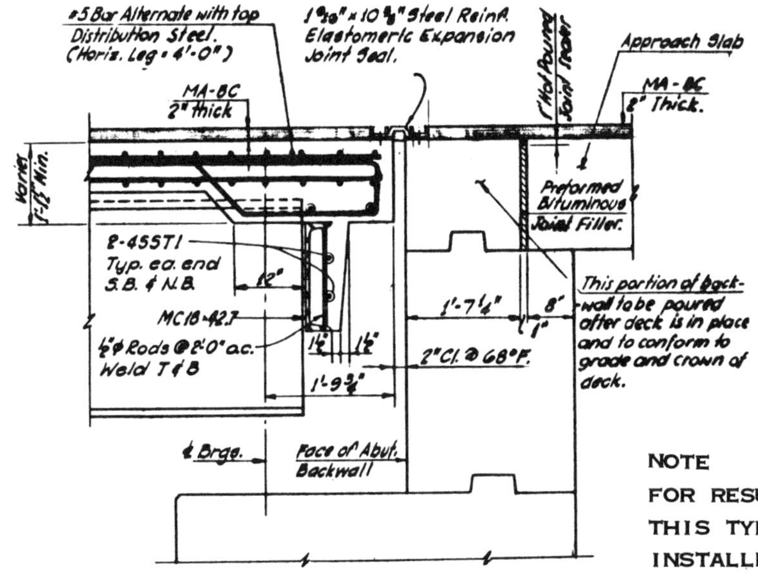


NOTE
FOR RESURFACING PROJECT,
THIS TYPE JOINT MAY BE
INSTALLED IN PLACE OF
REBUILDING JOINT.

SECTION

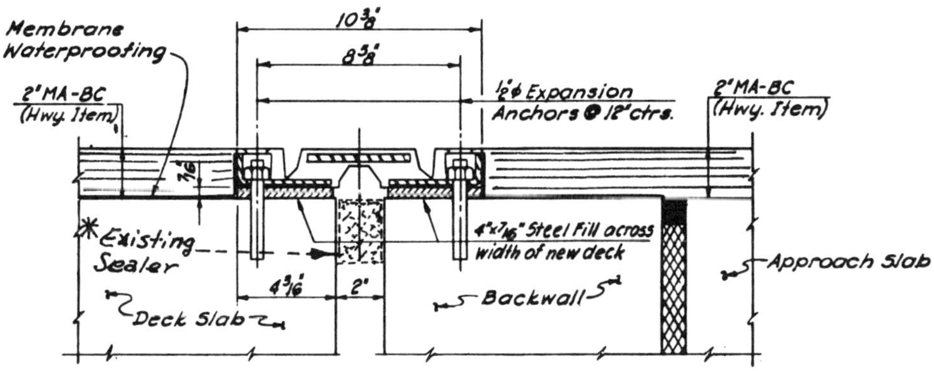

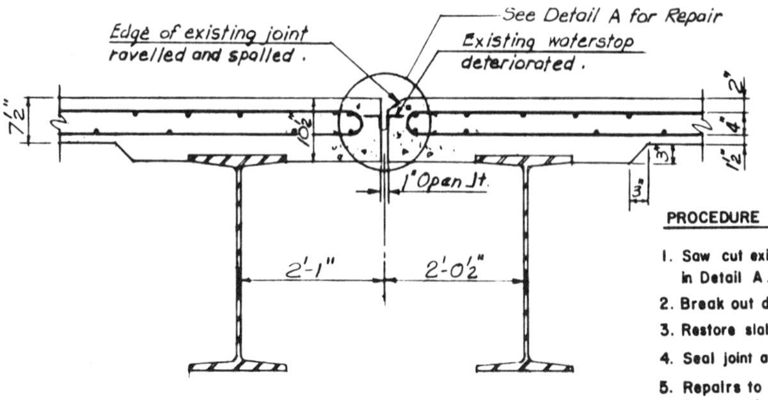


SECTION B-B
Scale: ¾" = 1'-0"

**PROCEDURE FOR REPAIRING LONGITUDINAL JOINT**

1. Saw cut existing slab 1 1/2" deep on either side of joint as shown in Detail A. Reduce depth of saw out if reinforcing steel is encountered.
2. Break out deteriorated concrete and install new 16 oz. copper waterstop.
3. Restore slab with Epoxy Mortar.
4. Seal joint as shown in Detail A.
5. Repairs to longitudinal joint may be performed during step 10 of Stage III Construction Operations (see Hwy. Stage Construction Plans, Sh. No. 144) utilizing single lane closing procedures and hours set for such closings as directed by the Engineer. This will necessitate paving the two inside lanes of the widened 4 lane northbound roadway first so that two lanes of traffic may be accomodated therein during the aforesaid single lane closing period. Length of joint repaired within prescribed lane closing period shall shall be limited to such as will insure that patch has gained sufficient strength to withstand load upon resumption of traffic, successive lane closings being used as required to complete entire length of joint.

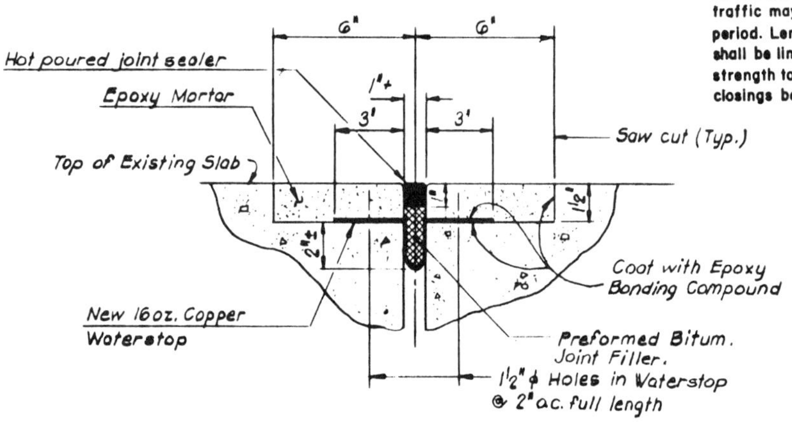


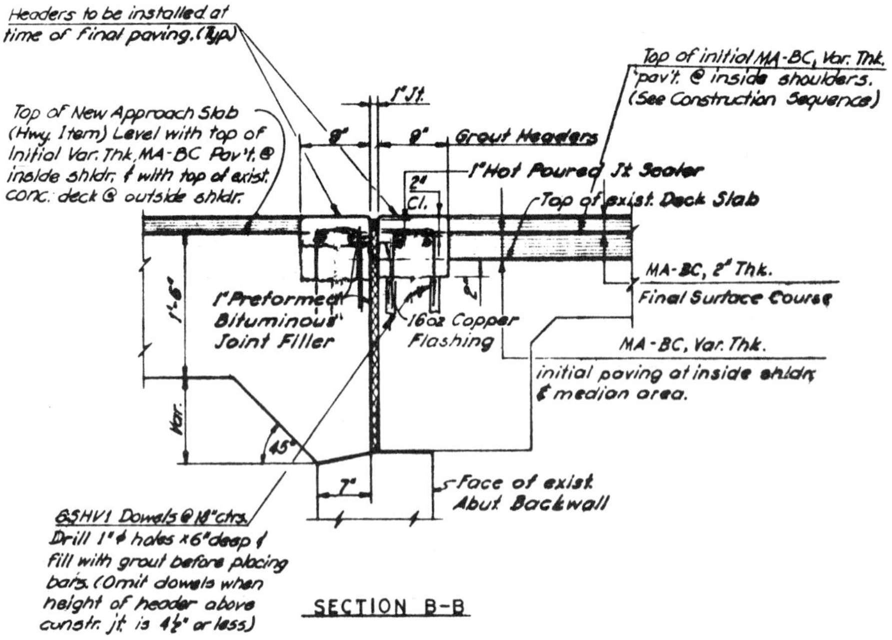


SECTION B-B

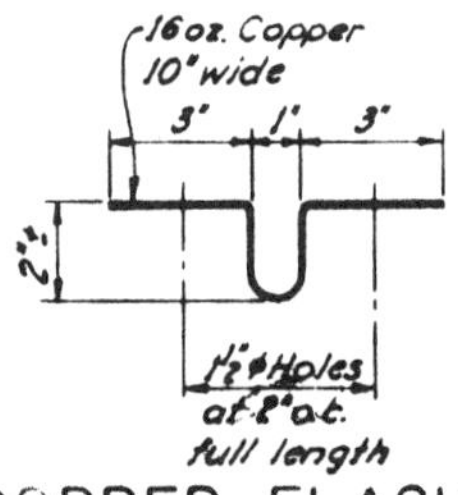


COPPER FLASHING DETAILS

DECK JOINT DETAILS

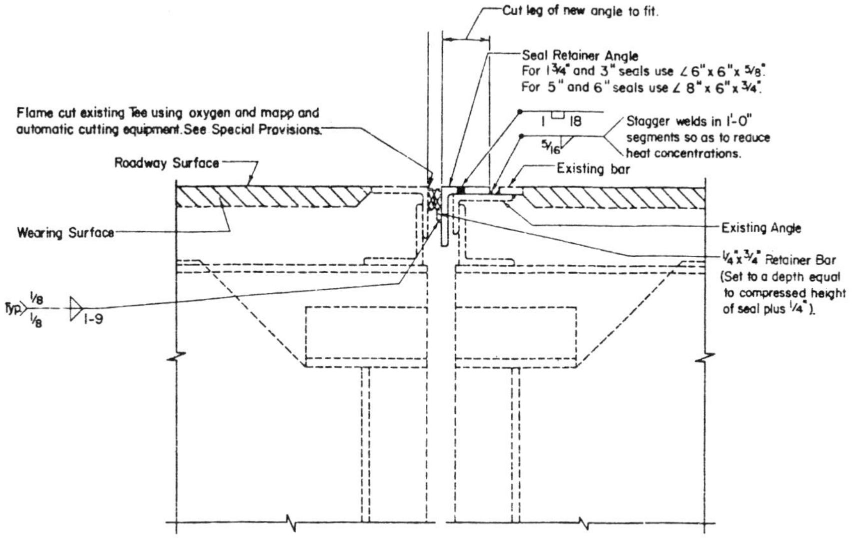


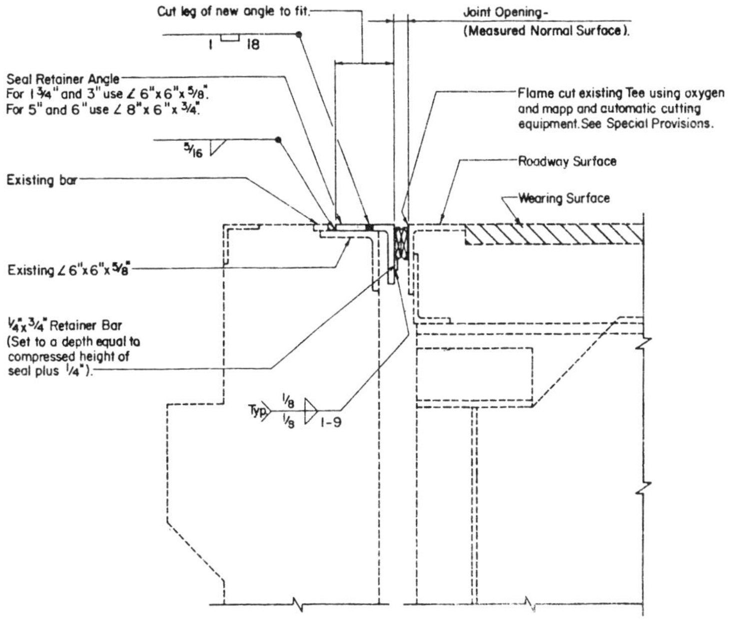


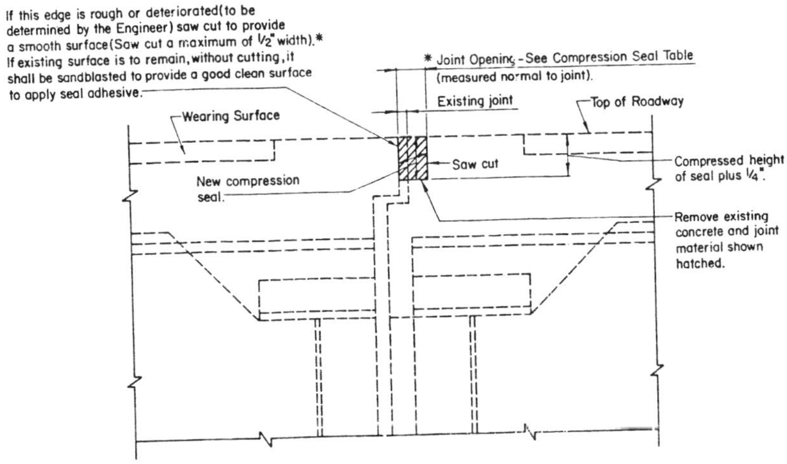


* Prior to ordering joint material each joint shall be evaluated to determine width of saw cutting required. If at 60°F or below the joint opening is 2" or less (measured perpendicular to ℄ of joint) the 3" seal may be used. For openings greater than above, contact Division of Bridge Development. Contractor will be paid for the larger seal, if required, based on the differential in cost of material between the two seals being compared.

SECTION

Scale: 1"=1'-0"

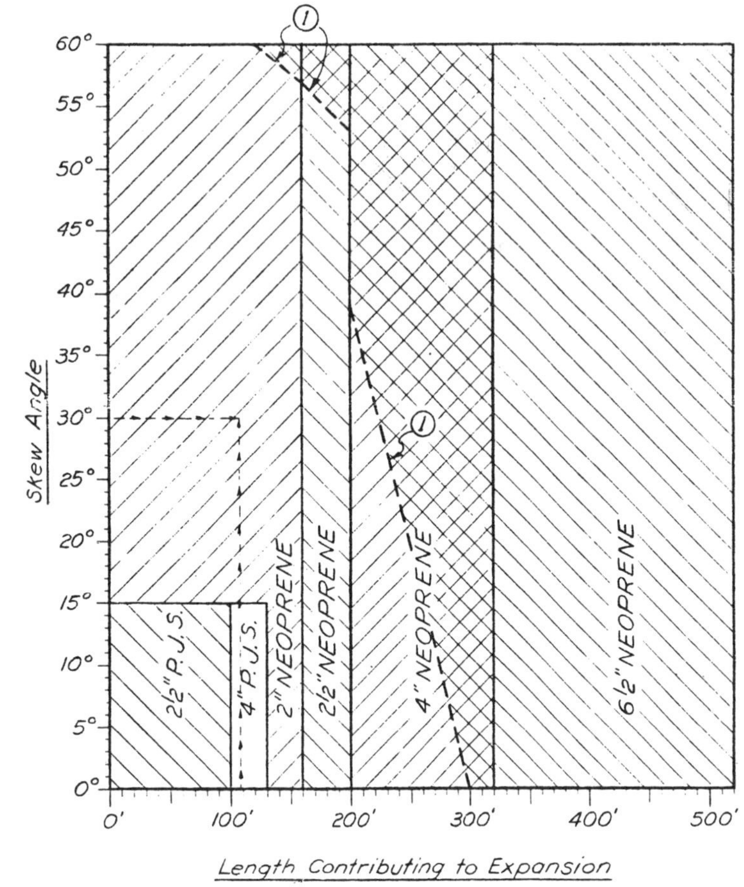


Example: Length of bridge contributing to expansion = 107 ft.
Skew angle of joint = 30°; USE **2" Neoprene Joint.**

(1) The Fel-Span and Wabo Elastodam Neoprene Jts. shall not be used in cross hatched area above dotted line.

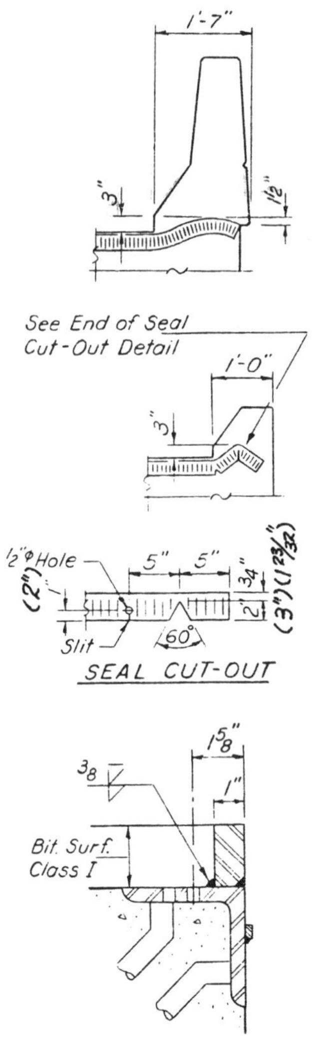


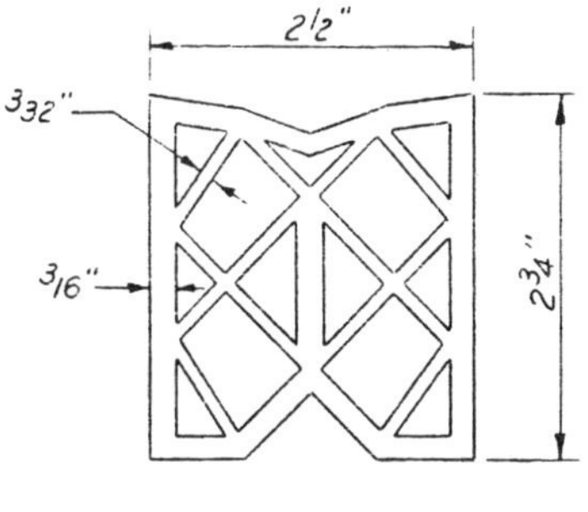


PREFORMED JOINT SEAL (2½")

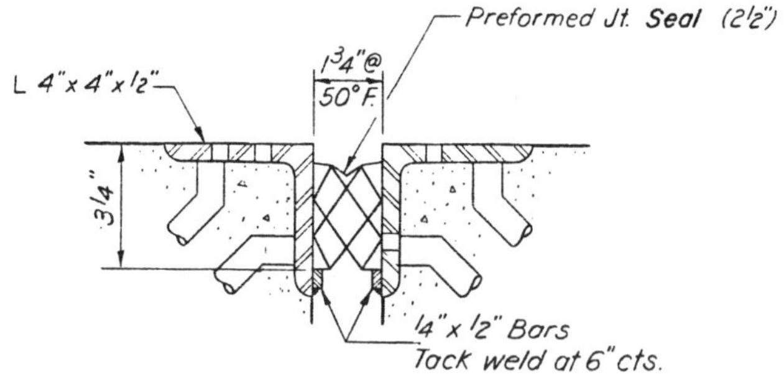


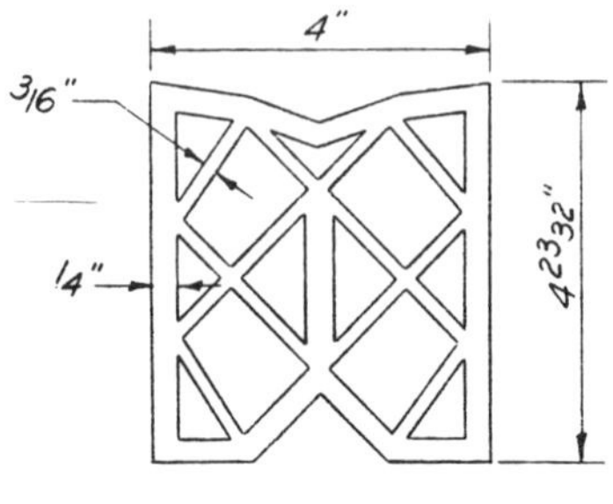


PREFORMED JOINT SEAL (4")

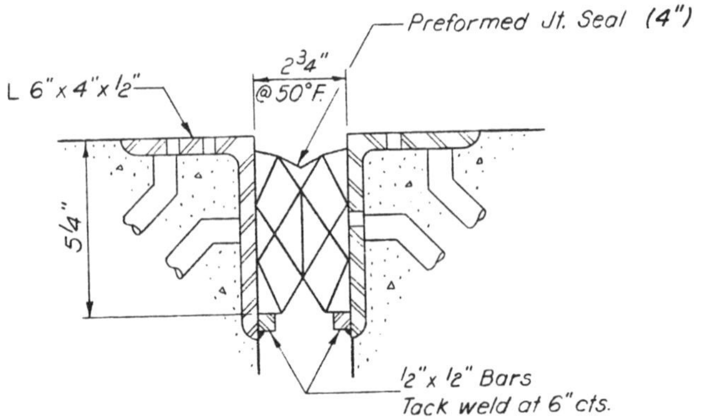


ILL

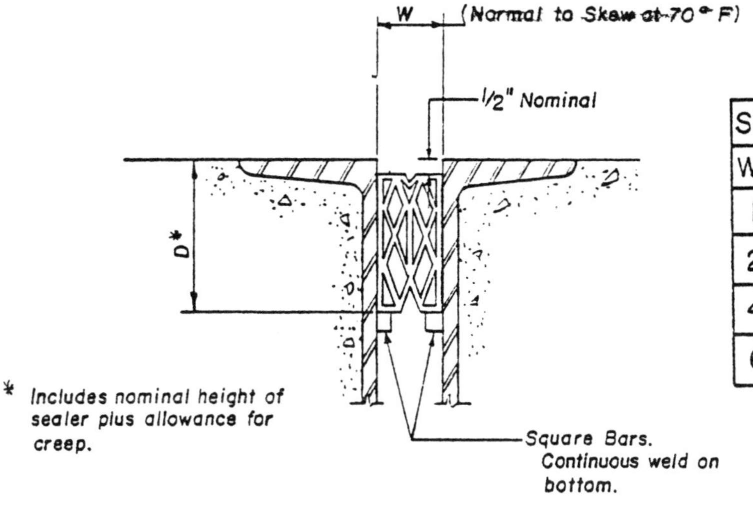


| Sealer Size | | W | D | Square Bar Size |
|---|---|---|---|---|
| Width | Height | Inches | Inches | Inches |
| 1¾ | 2 | 1 | 2¾ | ¼ |
| 2½ | 2¾ | 1⅝ | 3½ | 5/16 |
| 4 | 4¾ | 2⅝ | 5⅝ | 9/16 |
| 6 | 5¾ | 4 | 7⅜ | ⅝ |

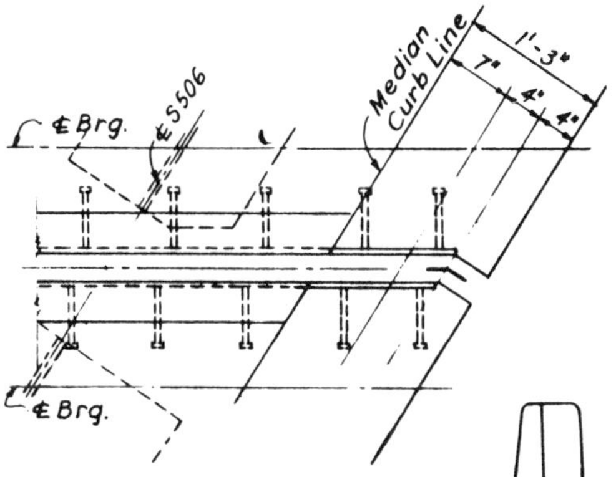


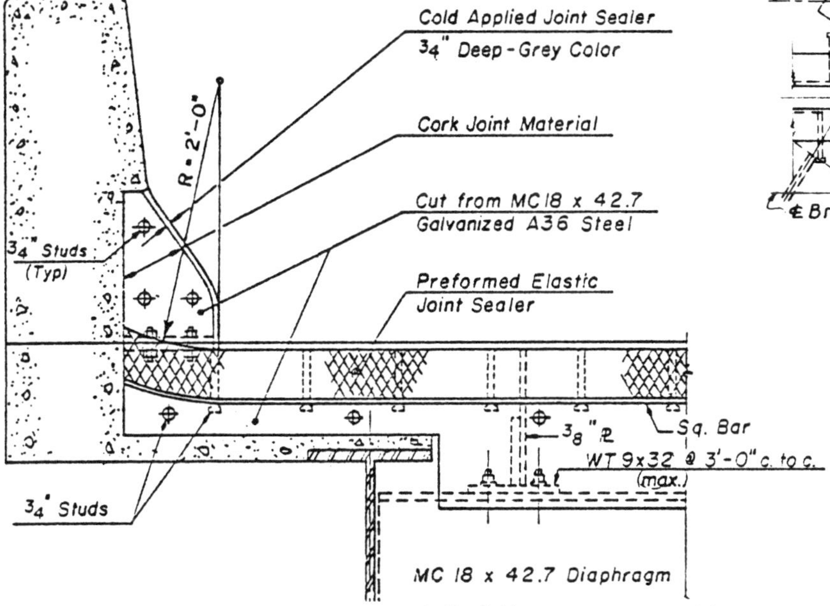


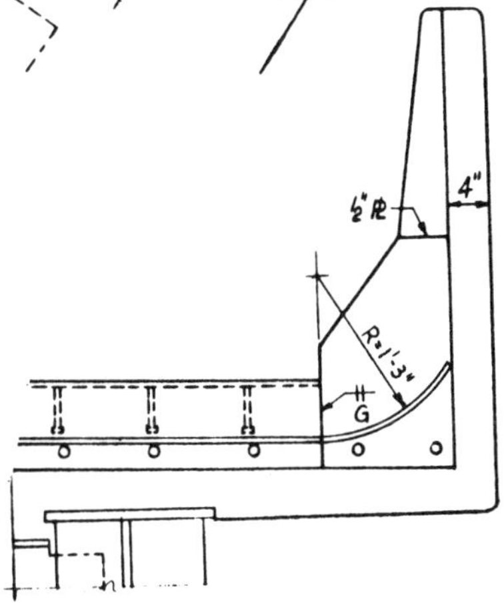


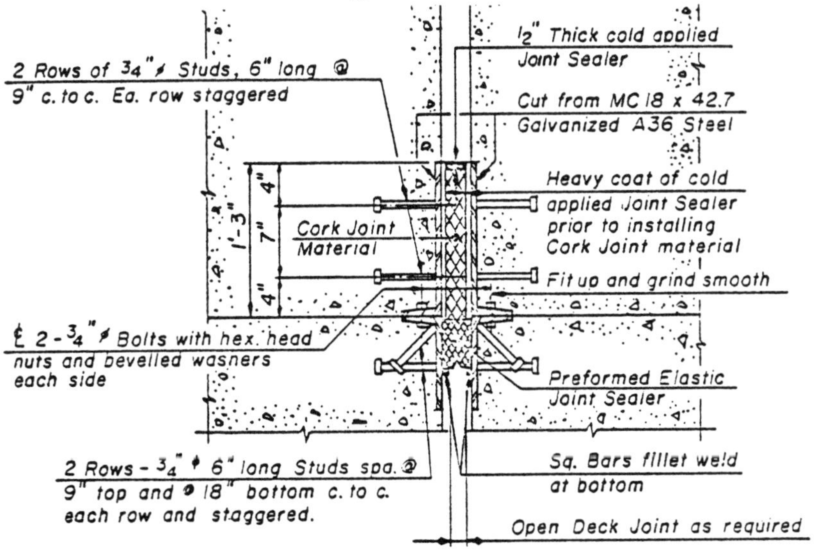


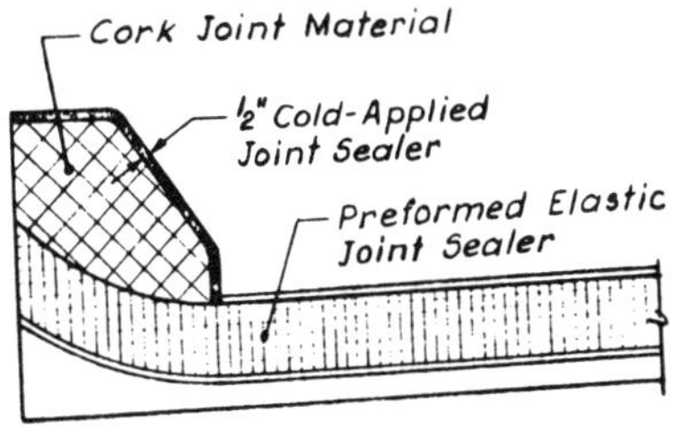


DETAIL

SECTION THRU BARRIER CURB AT ABUTMENT AND PIER

ARMORED JOINT DETAIL

TABLE 28.1.A

PREFORMED NEOPRENE COMPRESSION SEALS
ACME HIGHWAY PRODUCTS CORP.

| | CAT. NO. | SEAL SIZE WIDTH | SEAL SIZE HEIGHT | MIN. JOINT WIDTH | MAX. JOINT WIDTH | *MIN. INSTALL. WIDTH | MIN. JOINT DEPTH | MAXIMUM MOVEMENT | MAX. EXPAN. LENGTH** STEEL | MAX. EXPAN. LENGTH** CONC. *** |
|---|---|---|---|---|---|---|---|---|---|---|
| ◆ | J-125 | 1 1/4" | 1 5/16" | 0.570" | 1.060" | 0.74" | 2.000" | 0.49" | 42' | 85' |
| ◆ | J-162 | 1 5/8 | 1 5/8 | 0.72 | 1.38 | 0.97 | 2.25 | 0.66 | 56 | 114 |
| ◆ | J-175 | 1 3/4 | 1 25/32 | 0.81 | 1.48 | 1.02 | 2.50 | 0.67 | 57 | 116 |
| ▲ | J-200 | 2 | 2 1/32 | 0.92 | 1.70 | 1.19 | 2.875 | 0.78 | 66 | 135 |
| ▲ | J-225 | 2 1/4 | 2 5/8 | 1.06 | 1.91 | 1.36 | 3.125 | 0.85 | 72 | 147 |
| ▲ | J-250 | 2 1/2 | 2 9/16 | 1.13 | 2.13 | 1.49 | 3.250 | 1.00 | 85 | 173 |
| | J-300 | 3 | 3 1/16 | 1.13 | 2.55 | 1.74 | 3.825 | 1.42 | 121 | 247 |
| | J-350 | 3 1/2 | 3 9/16 | 1.37 | 2.97 | 2.01 | 4.25 | 1.61 | 138 | 280 |
| | J-400 | 4 | 4 1/32 | 1.50 | 3.40 | 2.35 | 5.25 | 1.90 | 162 | 330 |

TABLE 28.1.B

PREFORMED NEOPRENE COMPRESSION SEALS
D. S. BROWN CO.

| | CAT. NO. | SEAL SIZE WIDTH | SEAL SIZE HEIGHT | MIN. JOINT WIDTH | MAX. JOINT WIDTH | *MIN. INSTALL. WIDTH | MIN. JOINT DEPTH | MAXIMUM MOVEMENT | MAX. EXPAN. LENGTH** STEEL | MAX. EXPAN. LENGTH** CONC. *** |
|---|---|---|---|---|---|---|---|---|---|---|
| ◆ | H1250 | 1 1/4" | 1 5/16" | 0.500" | 1.000" | 0.750" | 2.000" | 0.580" | 49' | 99' |
| ◆ | H-1625 | 1 5/8 | 1 1/2 | 0.695 | 1.375 | 0.875 | 2.250 | 0.805 | 68 | 139 |
| ◆ | H-1750 | 1 3/4 | 1 3/4 | 0.735 | 1.484 | 1.250 | 2.500 | 0.900 | 76 | 156 |
| ▲ | H-2000 | 2 | 1 3/4 | 0.890 | 1.700 | 1.313 | 2.750 | 0.950 | 81 | 164 |
| ▲ | H-2250 | 2 1/4 | 2 1/8 | 1.000 | 1.900 | 1.500 | 3.000 | 1.050 | 89 | 182 |
| ▲ | H-2503 | 2 1/2 | 2 5/8 | 1.065 | 2.125 | 1.625 | 3.500 | 1.275 | 109 | 221 |
| | H-3000 | 3 | 3 1/4 | 1.200 | 2.550 | 1.750 | 4.25 | 1.550 | 132 | 269 |
| | H-3500 | 3 1/2 | 3 3/4 | 1.325 | 2.975 | 2.000 | 4.75 | 1.785 | 152 | 310 |
| | H-4000 | 4 | 4 1/4 | 1.600 | 3.400 | 2.500 | 5.25 | 2.030 | 173 | 352 |

◆ Limited to joints not subject to regular vehicle traffic.
*This is the minimum practical limit as suggested by the seal manufacturer.
**These values apply for skews up to and including 20°, presently these seals are not used for skews in excess of this amount.

▲ These seals are recommended only for fixed joints.

*** For Prestressed Girders, use same expansion lengths as Steel Girders.

TABLE 28.1.C

PREFORMED NEOPRENE COMPRESSION SEALS
WATSON-BOWMAN

| | CAT. NO. | SEAL SIZE WIDTH | SEAL SIZE HEIGHT | MIN. JOINT WIDTH | *MAX. JOINT WIDTH | *MIN. INSTALL. WIDTH | MIN. JOINT DEPTH | MAXIMUM MOVEMENT | MAX. EXPAN. LENGTH** STEEL | MAX. EXPAN. LENGTH** CONC. *** |
|---|---|---|---|---|---|---|---|---|---|---|
| ◆ | WA-175 | 1 3/4" | 2" | 0.687" | 1.487" | 1.00" | 2.50" | 0.80" | 68' | 139' |
| ▲ | WA-200 | 2 | 2 | 0.875 | 1.700 | 1.25 | 2.75 | 0.825 | 71 | 143 |
| ▲ | WA-250 | 2 1/2 | 2 3/4 | 1.000 | 2.125 | 1.50 | 3.50 | 1.125 | 96 | 195 |
| | WA-300 | 3 | 3 3/8 | 1.125 | 2.550 | 1.75 | 4.125 | 1.425 | 122 | 247 |
| | WA-350 | 3 1/2 | 3 1/2 | 1.375 | 2.975 | 2.25 | 4.50 | 1.600 | 137 | 278 |
| | WA-400 | 4 | 4 3/8 | 1.625 | 3.40 | 2.5 | 5.250 | 1.775 | 152 | 308 |

Commentary on Tables 28.1A Thru C

The maximum movements vary between manufacturers for a given seal size. The reason seems to lie with what each manufacturer considers the "min. width" to which his seal can be compressed. Whereas the maximum width (85%) is agreed upon by all three manufacturers for all their seals. We are told that the minimum values are based on laboratory and field tests which are a function of size and web configuration. Therefore the tables reflect what is published by each manufacturer.

The exception to this is the value of "min. joint depth." Here the manufacturer has tried to cover all cases of recess depth and lower web roll. From the discussions with the manufacturers representatives and physical inspection of seal samples it is concluded that for seals 2 1/2" and under a value of 3/8" for recess plus 3/8" for web roll is adequate. For seals 2 1/2" and over a value of 3/8" recess plus 5/8" web roll is sufficient. However, current Bridge Section practice is to use a 1/4" recess for all compression seals. In service evaluation of compression seals indicate that a 1/4" recess is adequate.

Skew Criteria

Until more knowledge is gained on sizing with respect to angular deflection, general seal usage is limited to structures with skews less than or equal to 20°. Exceptional cases with larger skews require approval of the Bridge Engineer. Limited observations indicate that seals under compression, which are subject to lateral movement and shear due to skew, function normally up to an "angular deflection" of about 10°. The angular deflection is the internal deformation resulting from the shearing force. For the popular seal sizes (2 to 4") this angular deflection limit is reached by seals subjected to their full potential thermal movement and skewed at about 20°. To accommodate skews in excess of 20° it is necessary to provide wider seals, for given expansion length, based on an angular deflection criteria.

In all cases the skew angle is taken into account when determining the thermal expansion length. The "maximum expansion length" as shown in Table 28.1 is the length normal to the joint.

Standard 28.1 shows the "straight through" treatment of the joint at the gutter line for skewed as well as square structures. This detail applies to sidewalks and brush curbs as well as safety parapets.

Seal Size

The width of the seal to be used in a given joint opening is computed by adding the anticipated change in joint width to the as-constructed joint width (at some temperature) plus a small width safety factor. In essence this is what is shown in Table 28.1.

The width limits to which the seal is expected to function are the controlling criteria in the size selection.

When the temperature is maximum the joint width is miminum and care is taken not to have the seal compressed to a width less than about 35% of the preformed width, which is the state where no air space exists and the webbing tends to rupture or extrude from the joint.

When the temperature is minimum the joint is at maximum width which should not be greater than about 85% of the preformed width or lack of sealing pressure results.

These width limits, as shown in Table 28.1 and applied in the examples, are based on laboratory tests and field experiences and vary slightly with each manufacturer.

The first three seals by both manufacturers, as shown in Table 28.1 are considered light-duty due to their somewhat thin internal web thickness. For this reason they are limited to special non-vehicle structure cases.

For best results oversize the seal by a minimum of 1/2" to compensate for unexpected movements from abutments, incorrect opening settings, neoprene relaxation, etc. Seals smaller than 3" are generally recommended only for fixed joints.

Special Details
Where a longitudinal and transverse seal intersect, the seal subjected to the greater movement shall be made continuous and the other shall abut it. The seal manufacturers shall be consulted so the most effective up-to-date cemented or vulcanized joint can be detailed based on their field experience.

When special situations require the seal to conform to certain shapes, the data in Figure 28.2 can be used as a guide. Bending is facilitated by removing the bottom half of the seal. This detail is limited to seals with D less than 3".

For horizontal directional changes in excess of 45 degrees a 12 inch radius is recommended.

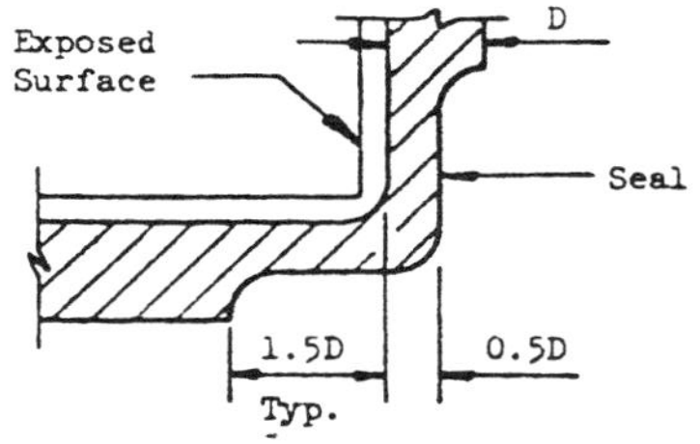


Maintenance
Manual removal of incompressible material which tends to collect within the joint opening is desirable but not completely necessary. In most cases the tire forces the material against the elastic neoprene seal which rebounds causing the material to bounce up and out.

It is essential to the operation of the seal that <u>no form of hot or cold joint filler be placed over the top of the seal</u>. This includes resurfacing mats or overlays. The reasons are:

Hot fillers either melt the seal or seriously effect the elastomeric properties.

The filler acts as a constant media of transmitting undue vertical tire forces to the seal.

Size Selection
The basic procedure using Table 28.1 is:

1. Locate the size which corresponds to the expansion length and type of main supporting member (steel or concrete). Proceed to the next size larger seal as recommended in 28.2(4). Note details are given on standards 28.1 and 28.2.
2. Determine the difference between the "maximum joint width" and the "minimum installation width".
3. Compute the change in temperature that corresponds to this change in length.
4. Add this to the minimum temperature (-30° F for prestressed/steel, +5° for concrete) to arrive at the maximum practical installation temperature.
5. Determine the difference between the minimum temperature and the assumed erection temperature (70° F).
6. Compute the change in length corresponding to this temperature change.
7. Subtract this value from the "maximum joint width" to arrive at the erection opening. Complete the temperature table values from 30 to 90° F for erection opening settings.

EXAMPLE

GIVEN: 96' Concrete structure with a fixed joint

(1) Select the minimum size seals recommended as follows:

Acme --------------- J-200
Brown -------------- H-2000
Watson-Bowman ------ WA-200

(2) Standard practice is to set the joint erection opening at or near the minimum installation widths which are 1 1/4", 1 3/16", and 1 1/4" respectively. These values correspond to the size of the formed opening at the fixed joint.

Compression Seal Selection Examples

Example #1

Given: 112' Prestressed/Steel Girder Expansion length having a $20^{\circ}$ skew angle.

(1) Calculate total movement perpendicular to joint.

$M_{\perp} = \Delta T(\alpha)(L)(\text{Cos skew angle})$

$= (150)(6.5\text{x}10^{-6})(112')(12''/')(0.940) = 1.24''$

(2) Calculate total movement parallel to joint.

$M_{\parallel} = \Delta T(\alpha)(L))(\text{Sin skew angle})$

$= (150)(6.5\text{x}10^{-6})(112')(12''/')(0.342) = 0.45''$

(3) Select a D.S. Brown Seal based on one size larger for perpendicular movement and based on allowing 15% of seal width for parallel movement to the joint which takes into account the racking movement on the seal due to the skew of the structure.

Refer to Table 28.1.B, Select a H-3500 and check requirements for 15% times 3.5" width equals 0.52" which is greater than the required 0.45". The H-3500 seal will be used for the following example calculations.

(4) Maximum installation temperature range.

$T = (2.975''-1.875'')/(112')(12''/')(6.5\text{x}10^{-6}) = 125^{\circ}F$

Practical installation temperature range $= 125^{\circ}F - 30^{\circ}F = 95^{\circ}F$

(5) Compute joint erection setting at $70^{\circ}F$.

$T = 70^{\circ}\ F - (-30^{\circ}F) = 100^{\circ}\ F$

Corresp. $L = (100)(112')(12''/')(6.5\text{x}10^{-6}) = 0.88''$

Erection Opening @ $70^{\circ}F = 2.95-0.88 = 2.095''$ <u>Say: 2 1/8"</u>

(6) A further computation is made at the maximum temperature of $120^{\circ}F$ for checking the minimum joint width.

Corresp. $L = (150)(112')(12''/')(6.5\text{x}10^{-6}) = 1.32''$

Opening @ $120^{\circ}\ F = 2.975-1.32 = 1.655'' > 1.50''$; therefore, adequate room remains at maximum temperature of $120^{\circ}F$.

(7) Temperature table value for erection opening @ $30^{\circ}F$

Corresp. $L = (60)(112')(12''/')(6.5\text{x}10^{-6}) = 0.53''$

Erection Setting @ $30^{\circ}F = 2.975-0.53 = 2.445''$ <u>Say: 2 7/16"</u>

Computations for the Acme (J350) and Watson-Bowman (WA-350) alternates are similar to above procedures for size selection and erection opening setting. A temperature table of values for erection opening settings from $30^{\circ}$ F to $90^{\circ}$ F is required per Standard 28.1.

**Wisconsin**

TABLE 5
EXPERIMENTALLY DETERMINED PERPENDICULAR MOVEMENT CAPABILITIES (IN INCHES) OF EVALUATED SYSTEMS VS. ANGLE OF CROSSING

| Joint System | Angle of Crossing | | | | | | |
|---|---|---|---|---|---|---|---|
| | 90° | 80° | 70° | 60° | 50° | 40° | 30° |
| Onflex 25 | 1.7 | 1.7 | 1.7 | 1.7 | 1.7 | 1.7 | 1.7 |
| Onflex 20 | 1.8 | 1.8 | 1.8 | 1.8 | 1.8 | 1.8 | 1.6 |
| Pro-Span 2-in. system (low-profile) | 2.0 | 2.0 | 2.0 | 2.0 | 1.7 | 1.5 | 1.4 |
| Acme 3-in. Strip Seal (AS 300) | 3.0 | 3.0 | 3.0 | 2.7 | 1.9 | 1.9 | 1.2 |
| Acme 3-in. Trojan (TR 300) | 3.0 | 3.0 | 3.0 | 3.0 | 2.7 | 2.0 | 1.4 |
| Watson Bowman 3-in. system (S 300) | 3.0 | 3.0 | 3.0 | 3.0 | 2.9 | 2.2 | 1.4 |
| Acme 4-in. Trojan (TR 400) | 3.2 | 3.2 | 3.0 | 2.5 | 1.5 | 1.0 | 0.6 |
| Onflex 40 | 3.8 | 3.8 | 3.8 | 3.8 | 3.6 | 2.8 | 2.0 |
| Acme 4-in. Strip Seal (AS 400) | 4.0 | 4.0 | 4.0 | 2.0 | 1.7 | 1.3 | 1.0 |
| Watson Bowman 4-in. system (S 400) | 4.0 | 4.0 | 4.0 | 3.8 | 2.8 | 1.8 | 1.3 |
| Pro-Span 4-in. system (low profile) | 4.0 | 4.0 | 4.0 | 3.8 | 3.2 | 2.8 | 2.2 |
| Onflex 45 | 4.1 | 4.1 | 4.0 | 3.7 | 3.6 | 2.8 | 2.0 |

# STRIP SEAL APPLICATION

Example: Strip Seal Application
Given: Prestressed concrete girder, 278' expansion length with a 33 degree skew.

(A) From Table 28.2, under Prestressed I, select strip seal models AS-300A, S-300(A-2) and SS-300 based on an expansion length of 280'. However, the skew angle is in excess of 30 degrees, due to stress from racking the next higher capacity (4 inch) model is recommended. Refer to Section 28.3(4).

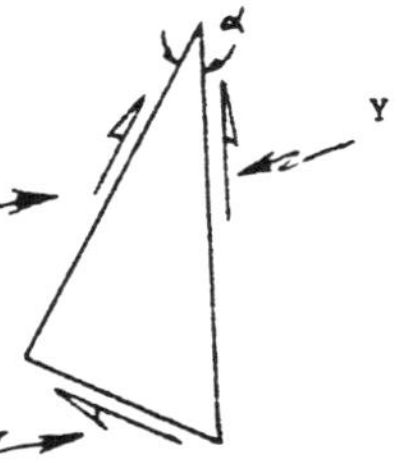


$Y = (278')(12''/')(6.0 \times 10^{-6})(120°F - (-30°F))$
$= 3.00''$

$\perp$ comp $= Y \cos\alpha = 3.00 \cos 33° = 2.52''$
$\parallel$ comp $= Y \sin\alpha = 3.00 \sin 33° = 1.63''$

In this case parallel racking as a percentage of joint capacity has been reduced from 54% to 41% of the 4 inch model capacity. Hence, strip seal models AS-400A, S-400(A-2) and SS-400 will be used.

(B) Refer to Table 28.2 for joint opening at 45°F which is 2 3/4".

(C) Computations for a Setting Data Table for joint forming for a temperature range of 30 to 90°F in 10 degree increments is as follows:
As an example the joint opening for a 4 inch strip seal at 90°F equals 2 3/4" - (6.0 x $10^{-6}$/°F)(90°F - 45°F)(278')(12"/')(Cos 33°) equals 2.75" - 0.76" = 1.99". SAY: 2"

Setting Data Table

| $O_F$* | Joint Opening - Inches AS-400(A)/S-400(A-2)/SS-400 |
|---|---|
| 90 | 2 |
| 80 | 2 3/16 |
| 70 | 2 3/8 |
| 60 | 2 9/16 |
| 50 | 2 3/4 |
| 40 | 2 15/16 |
| 30 | 3 1/8 |

* Shaded underside deck temperature.

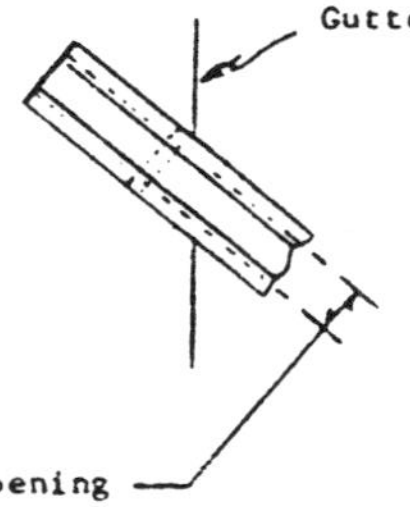


TABLE 28.2

| Acma/Wabo/Brn * Model No. | Travel (inches) | Max. Expansion Lengths (feet) Conc | Prest I | Steel | Jt. Opening** @ 45°F (inches) |
|---|---|---|---|---|---|
| AS-200A/S-200(A-2)/ SS-200 Δ | ± 1 | 350 | 190 | 170 | 2 1/8 |
| AS-300A/S-300(A-2)/ SS-300 | ± 1 1/2 | 520 | 280 | 260 | 2 3/8 |
| AS-400A/S-400(A-2)/ SS-400 | ± 2 | 690 | 370 | 340 | 2 3/4 |
| AS-500A/S-500(A-2)/ SS-500 | Recommended for expansion movement requirements ranging from 3 to 4 inches on skews greater than 30 degrees. Use the same design criteria as with the 400 models. | | | | |
| Acma/Wabo Low Profile | | | | | |
| AS2LP/S-200E Δ | ± 1 | 350 | 190 | 170 | 1 7/8 |
| AS3LP/S-300E Δ | ± 1 1/2 | 520 | 280 | 260 | 2 1/4 |
| AS4LP/S-400E Δ | ± 2 | 690 | 370 | 340 | 2 5/8 |

* The Acma, Brn and Wabo are manufactured by Acme, D. S. Brown and Watson-Bowman companies respectively.

** The joint opening at 45°F is given at mean shaded under side deck temperature normal to the joint for zero degree skew of structure.

Δ These models are not recommended for new construction.

STRIP SEAL

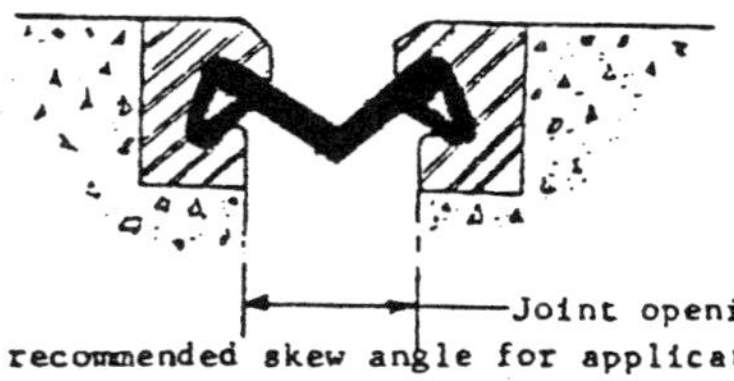


The maximum recommended skew angle for application is 45 degrees.

| FHWA Region No. | STATE | | SHEET NO. | TOTAL SHEETS |
|---|---|---|---|---|
| 6 | NEW MEXICO | | | |

# ALLOWABLE MOVEMENT LENGTHS & INITIAL GAP SETTINGS ("A" DIMENSIONS) FOR ACCEPTABLE JOINT SEALS

## SEALS

### FEL-PRO "PRO SPAN"

| SKEW Δ | Movement Lengths: Concrete Bridges | Movement Lengths: Steel Bridges | Model № | "A" at 68° F: Concrete Bridges | "A" at 68° F: Steel Bridges |
|---|---|---|---|---|---|
| 0° THRU 15° | 10-174 | 10-142 | 0-2" | 1 1/16" | 1" |
| | 174-261 | 142-213 | 0-3" | 1 1/2" | 1 3/8" |
| | 261-326 | 213-266 | 0-4" | 1 13/16" | 1 5/8" |
| 15+° TO 30° | 10-164 | 10-136 | 0-2" | 15/16" | 7/8" |
| | 164-246 | 136-204 | 0-3" | 1 5/16" | 1 3/16" |
| | 246-307 | 204-255 | 0-4" | 1 9/16" | 1 3/8" |
| 30+° TO 45° | 10-142 | 10-116 | 0-2" | 3/4" | 11/16" |
| | 142-213 | 116-174 | 0-3" | 1" | 7/8" |
| | 213-266 | 174-217 | 0-4" | 1 3/16" | 1 1/16" |

### WATSON BOWMAN "WABO MAURER" OR "ALU-STRIP"

| SKEW Δ | Movement Lengths: Concrete Bridges | Movement Lengths: Steel Bridges | Model № | "A" at 68° F: Concrete Bridges | "A" at 68° F: Steel Bridges |
|---|---|---|---|---|---|
| 0° THRU 15° | 10-174 | 10-142 | S200 | 1 1/16" | 1" |
| | 174-261 | 142-213 | S300 | 1 1/2" | 1 3/8" |
| | 261-326 | 213-266 | S400 | 1 13/16" | 1 5/8" |
| 15+° TO 30° | 10-164 | 10-136 | S200 | 15/16" | 7/8" |
| | 164-246 | 136-204 | S300 | 1 5/16" | 1 3/16" |
| | 246-307 | 204-255 | S400 | 1 9/16" | 1 3/8" |
| 30+° TO 45° | 10-142 | 10-116 | S200 | 3/4" | 11/16" |
| | 142-213 | 116-174 | S300 | 1" | 7/8" |
| | 213-266 | 174-217 | S400 | 1 3/16" | 1 1/16" |

### D.S. BROWN "DELASTIFLEX"

| SKEW Δ | Movement Lengths: Concrete Bridges | Movement Lengths: Steel Bridges | Model № ✱ | "A" at 68° F: Concrete Bridges | "A" at 68° F: Steel Bridges |
|---|---|---|---|---|---|
| 0° THRU 15° | 10-87 | 10-71 | CP100 | 15/16" | 7/8" |
| | 87-174 | 71-142 | CP200 | 1 5/16" | 1 1/4" |
| | 174-261 | 142-213 | CP300 | 1 3/4" | 1 5/8" |
| | 261-326 | 213-266 | CP400 | 1 13/16" | 1 5/8" |
| 15+° TO 30° | 10-82 | 10-68 | CP100 | 7/8" | 13/16" |
| | 82-164 | 68-136 | CP200 | 1 3/16" | 1 1/8" |
| | 164-246 | 136-204 | CP300 | 1 9/16" | 1 7/16" |
| | 246-307 | 204-255 | CP400 | 1 9/16" | 1 3/8" |
| 30+° TO 45° | 10-71 | 10-58 | CP100 | 3/4" | 11/16" |
| | 71-142 | 58-116 | CP200 | 1" | 15/16" |
| | 142-213 | 116-174 | CP300 | 1 1/4" | 1 1/8" |
| | 213-266 | 174-217 | CP400 | 1 3/16" | 1 1/16" |

✱ Series LM seal is also acceptable.

### OLD NORTH "ONFLEX"

| SKEW Δ | Movement Lengths: Concrete Bridges | Movement Lengths: Steel Bridges | Model № | "A" at 68° F: Concrete Bridges | "A" at 68° F: Steel Bridges |
|---|---|---|---|---|---|
| 0° THRU 15° | 10-217 | 10-177 | 25 | 1 5/16" | 1 3/16" |
| | 217-326 | 177-266 | 45 | 1 13/16" | 1 5/8" |
| 15+° TO 30° | 10-205 | 10-170 | 25 | 1 1/8" | 1" |
| | 205-307 | 170-255 | 45 | 1 9/16" | 1 3/8" |
| 30+° TO 45° | 10-177 | 10-145 | 25 | 7/8" | 13/16" |
| | 177-266 | 145-217 | 45 | 1 3/16" | 1 1/16" |

### ACME "ACMA-STRIP"

| SKEW Δ | Movement Lengths: Concrete Bridges | Movement Lengths: Steel Bridges | Model № ⊛ | "A" at 68° F: Concrete Bridges | "A" at 68° F: Steel Bridges |
|---|---|---|---|---|---|
| 0° THRU 15° | 10-174 | 10-142 | AS200 | 1 5/16" | 1 1/4" |
| | 174-261 | 142-213 | AS300 | 1 3/4" | 1 5/8" |
| | 261-326 | 213-266 | AS400 | 1 15/16" | 1 13/16" |
| 15+° TO 30° | 10-164 | 10-136 | AS200 | 1 3/16" | 1 1/8" |
| | 164-246 | 136-204 | AS300 | 1 9/16" | 1 7/16" |
| | 246-307 | 204-255 | AS400 | 1 11/16" | 1 9/16" |
| 30+° TO 45° | 10-142 | 10-116 | AS200 | 1" | 15/16" |
| | 142-213 | 116-174 | AS300 | 1 1/4" | 1 1/8" |
| | 213-266 | 174-217 | AS400 | 1 3/8" | 1 1/4" |

⊛ Series TR seal is also acceptable. Δ

### GENERAL TIRE "GEN-STRIP" Δ

| SKEW Δ | Movement Lengths: Concrete Bridges | Movement Lengths: Steel Bridges | Model № | "A" at 68° F: Concrete Bridges | "A" at 68° F: Steel Bridges |
|---|---|---|---|---|---|
| 0° THRU 15° | 10-217 | 10-177 | CCL | 1 15/16" | 1 13/16" |
| 15+° TO 30° | 10-205 | 10-170 | CCL | 1 3/4" | 1 5/8" |
| 30+° TO 45° | 10-177 | 10-145 | CCL | 1 1/2" | 1 7/16" |

## G E N E R A L   N O T E S

1. Seals of equal performance characteristics to those shown may be substituted if the Engineer's written approval is obtained at least two weeks prior to the day of the bid opening.
2. Joint seal sizes are based on an installation temperature of 68° F. and joint movements for temperatures ranging from -10° F. to 120° F. for concrete bridges and -30° F. to 120° F. for steel bridges. For installation temperatures below 68° F. increase dimension "A" by the amount shown below. For installation temperature above 68° F. decrease dimension "A" by the amount shown below.
   For concrete bridges:
   Change in "A" = 0.000072 x L x cos.ø x ΔT
   For steel bridges:
   Change in "A" = 0.000078 x L x cos.ø x ΔT
   Where:
   Change in "A" is in inches.
   L = movement length in feet.
   ø = skew angle.
   ΔT = change in temperature from 68° F. in degrees F.

   Example:
   For T = 50° F., L = 100 ft., ø = 28° concrete bridge, ΔT = 68 - 50 = 18° F.
   Change in "A" = 0.000072 x 100 x 0.8829 x 18 = 0.1144 inches; say 1/8 inches.
3. Joints shall be installed so that dimension "X" on sheet 2 of 2 will not exceed 4 inches when the joint is at its maximum opening.
4. Δ For Movement Length (L) See Superstructure Details.

| IDENT. NO. | DESCRIPTION | DATE | BY |
|---|---|---|---|
| 5 | CHANGE SERIAL NUMBER | 8-7-79 | E.R.D. |
| 4 | CHANGE SKEW ANGLE | 8-7-79 | E.R.D. |
| 3 | ADDED TRADE MARK SEAL | 8-7-79 | E.R.D. |
| 2 | ADDED GENERAL NOTE 4 | 8-7-79 | E.R.D. |
| 1 | ADDED "GEN-STRIP" TABLE | 8-7-79 | E.R.D. |

REVISIONS (OR CHANGE NOTICES)

NEW MEXICO
STATE HIGHWAY DEPARTMENT

ELASTOMERIC EXPANSION
JOINT SEALS FOR BRIDGES

DESIGNED BY J.A.H. — APPROVED, RECOMMENDED — BRIDGE ENGINEER — DATE 2-12-72
DRAWN BY A.J.S.
CHECKED BY E.R.D. — APPROVED — DIRECTOR ROAD DEVELOPMENT DIVISION — DATE

Δ SERIAL BJE-001-05 — SHEET 1 OF 2

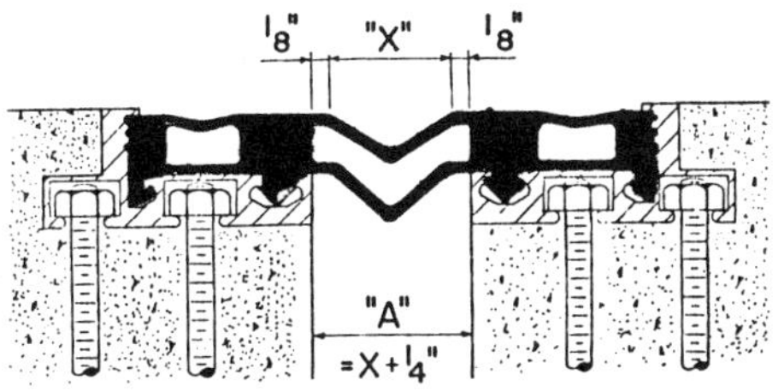


DELASTIFLEX CP MODEL EXPANSION JOINT

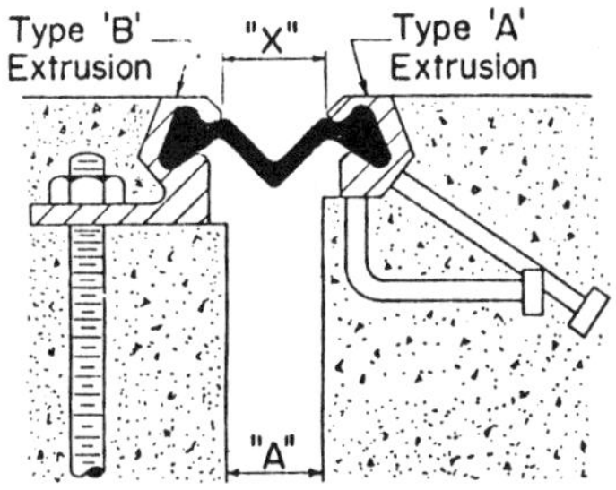


WABO MAURER EXPANSION JOINT

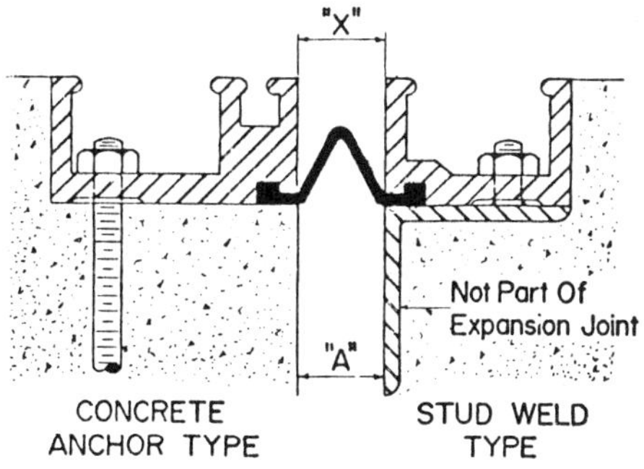


PRO-SPAN EXPANSION JOINT

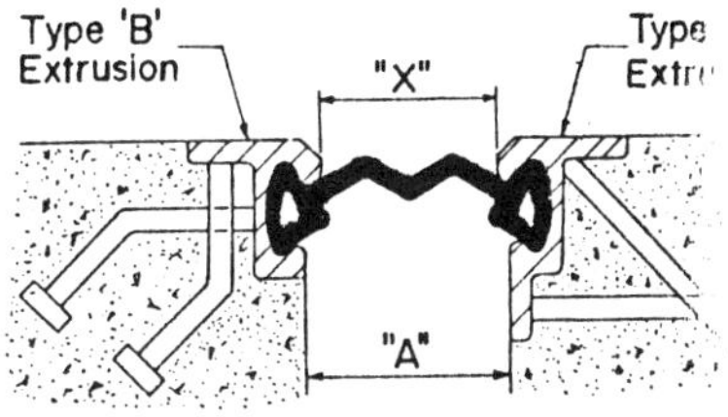


ACMA-STRIP EXPANSION JOINT

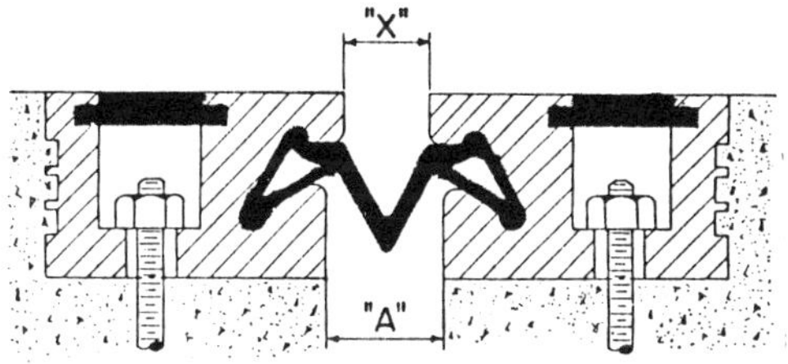


WABO ALU-STRIP EXPANSION JOINT

MOVEMENT (TOTAL) < 4"

GLAND TYPE ALTERNATES

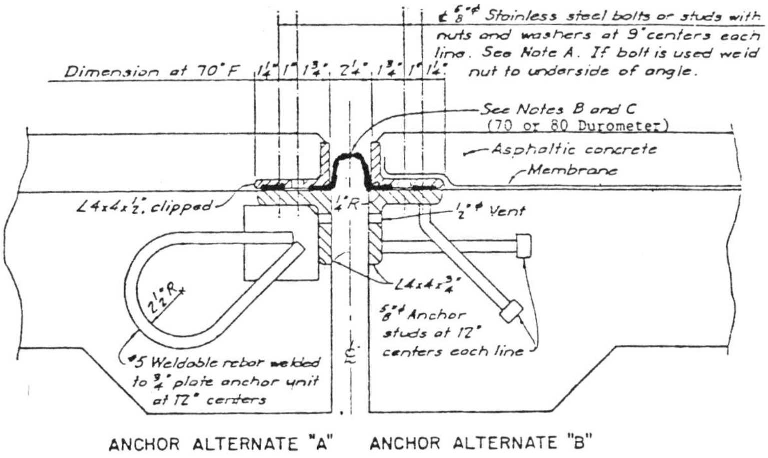


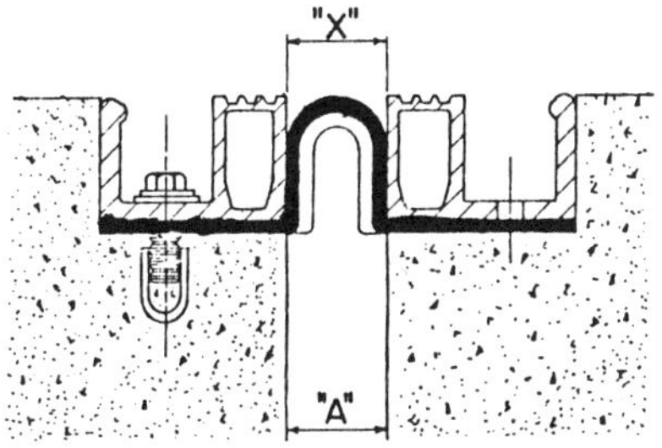


ONFLEX EXPANSION JOINT

Note A: Bolts or studs, nuts and washers shall be Type 304 stainless steel. Use 60 foot-pounds torque (5/8" ).

Note B: 1/8" thick continuous synthetic fabric reinforced sheet elastomeric gland. Constant width - 12".

Note C: Special shaped glands may be advantageous for severe skews.

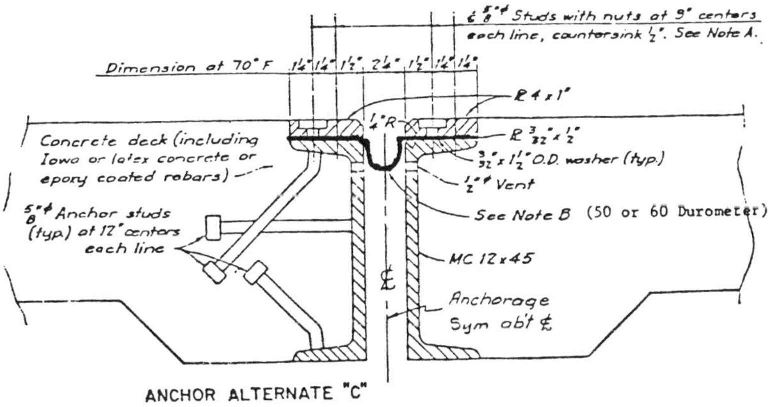


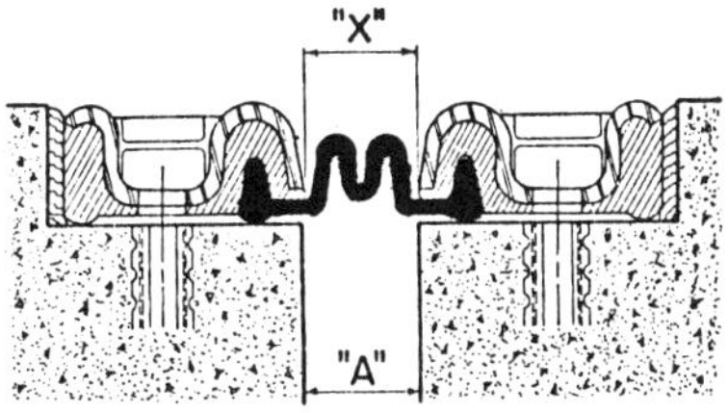


GEN-STRIP ⚠ EXPANSION JOINT

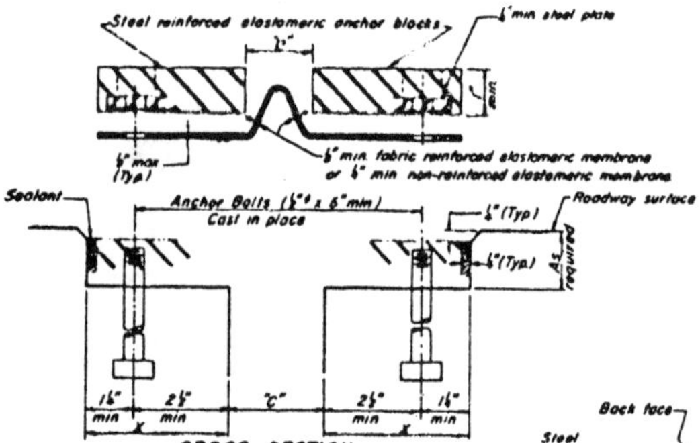


## CROSS SECTION

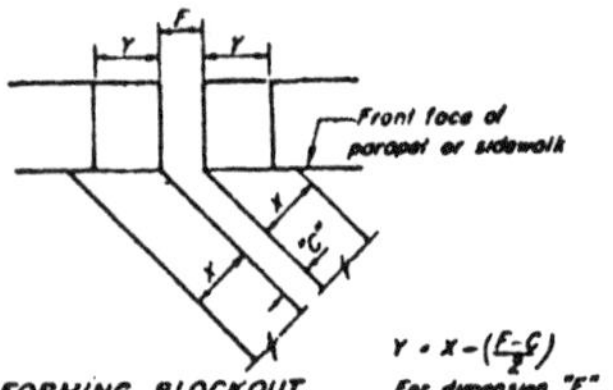


## FORMING BLOCKOUT SKETCH

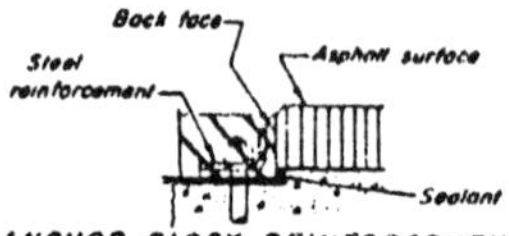


## ANCHOR BLOCK REINFORCEMENT with ASPHALT SURFACE

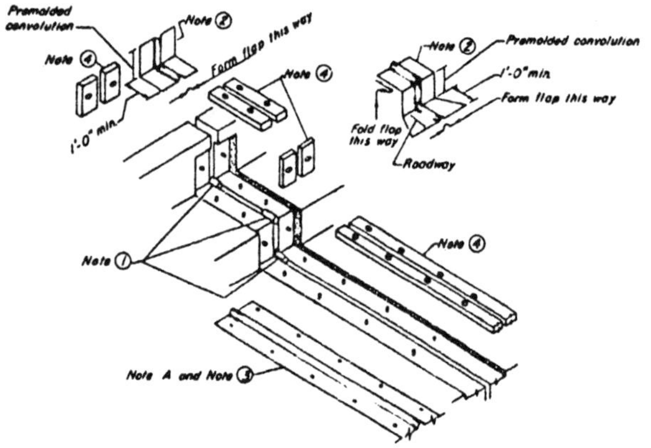


## AT SIDEWALK or MEDIAN

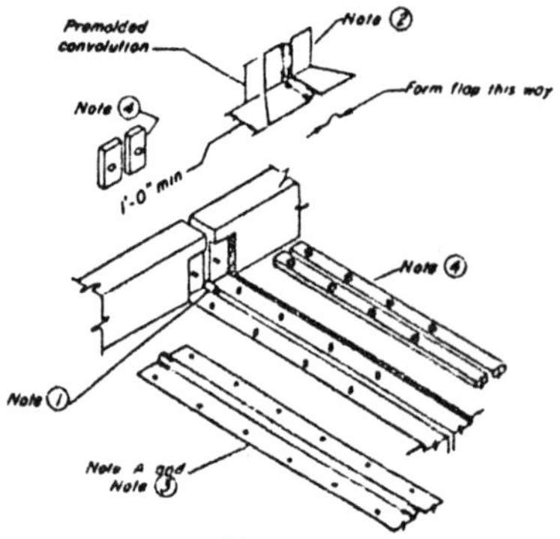


## AT WALL

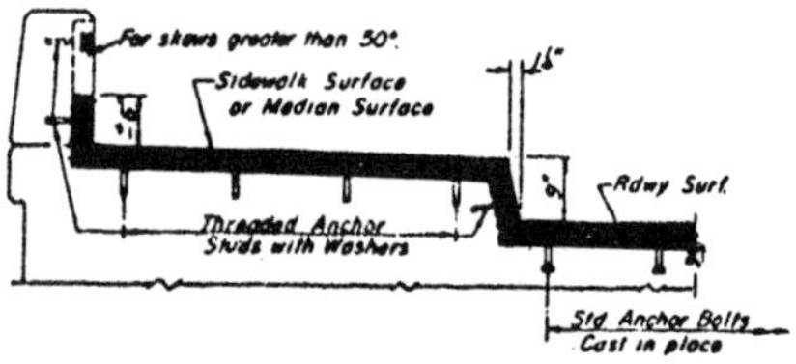


## AT SIDEWALK or MEDIAN

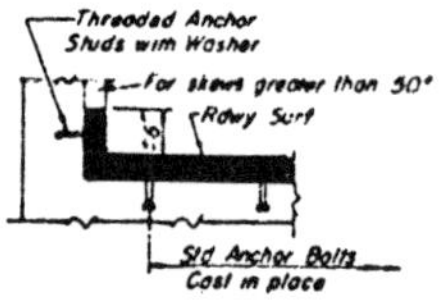


## AT WALL

## TYPICAL END TREATMENTS

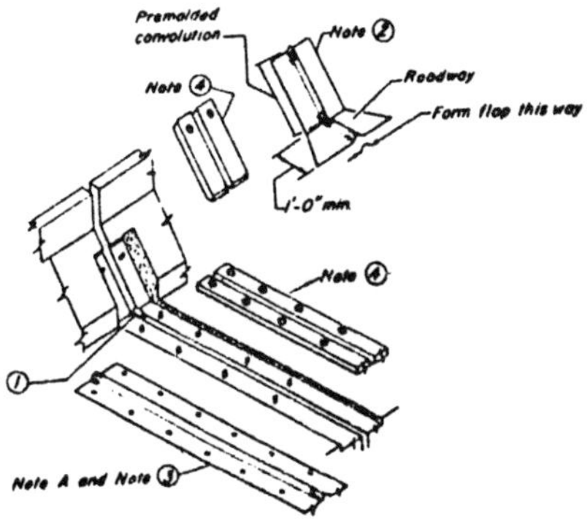


## AT PARAPET

## INSTALLATION NOTES

Use anchor blocks and continuous seal as anchor bolt location templates.

① Install sponge mandrels into positions shown to form flap convolution.

② Install parapet or sidewalk piece (trim roadway flap to fit before applying epoxy).

③ Install continuous seal in roadway.

④ Install anchor blocks as indicated.

NOTE A — Maximum spacing of anchor bolts shall be 12" centers

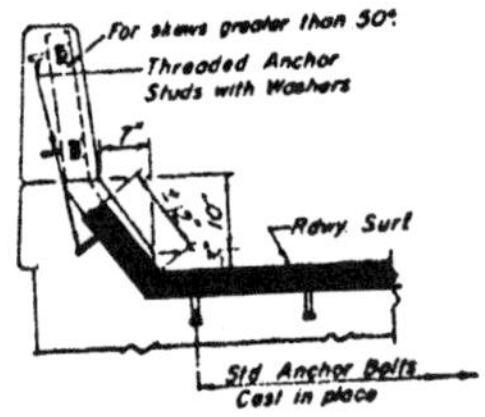


## AT PARAPET

## SKEW LIMITATIONS

The details of the anchor blocks and the elastomeric membrane in the parapet, as shown, are fu- up to 50° skews.

For skews greater than 50°, the anchor blocks and the elastomeric membrane, installed in accordance with dimension "D", might require modifications to insure a minimum clearance of 1½" from centerline of anchor studs to edge of parapet opening. The anchor blocks and the elastomeric membrane shall also be installed to the top of the parapet with the anchor studs spaced at ±12" cts.

## GENERAL NOTES

Continuous Seal Neoprene Expansion Joint shall consist of molded anchor blocks of elastomer and steel, field assembled over continuous lengths of elastomeric membrane. See Special Provisions.

The elastomeric membrane shall be premolded with a single or a double upward convolution that will have a "memory" to return to its molded position upon joint closure.

The steel reinforcement must extend up the back face of anchor blocks when asphalt surfaces are used but is optional in concrete blockout.

The convolution length shall be such that the extended length will not be greater than the manufactured length when the joint is fully expanded in its design range and will not protrude above the anchor blocks when the joint is fully compressed.

Joint openings shall be adjusted in accordance with Article 503.07(c) of the Standard Specifications when the deck is poured at an ambient temperature other than 50° F.

The parapet and sidewalk flaps may be furnished factory vulcanized to the roadway membrane provided the centerline of the convolution is maintained and the process and method meet the approval of the Engineer.

STATE OF ILLINOIS
DEPARTMENT OF TRANSPORTATION

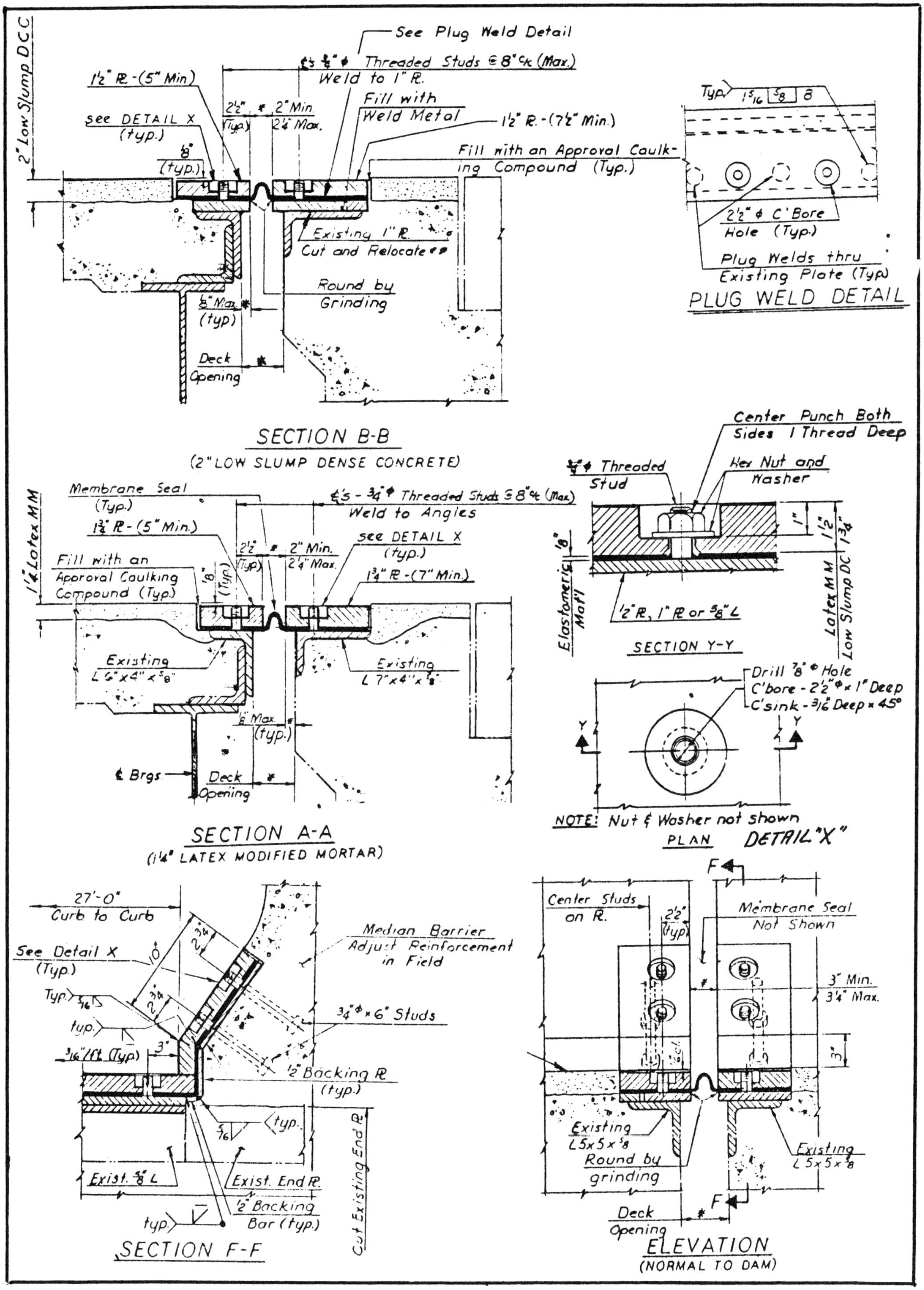

2" Low Slump DCC
See Plug Weld Detail
℄ 5 ¾"ϕ Threaded Studs @ 8" c/c (Max.)
Weld to 1" ℞.
1½" ℞.-(5" Min)
See DETAIL X (typ.)
2½" (Typ.)
2" Min. 2¼" Max.
Fill with Weld Metal
1½" ℞.-(7½" Min.)
⅛" (typ.)
Fill with an Approval Caulking Compound (Typ.)
Existing 1" ℞. Cut and Relocate
Round by Grinding
⅛" Max. (typ)
Deck Opening
Typ. 1 5/16 ⅝ 8
2½" ϕ C'Bore Hole (Typ.)
Plug Welds thru Existing Plate (Typ.)
PLUG WELD DETAIL
SECTION B-B
(2" LOW SLUMP DENSE CONCRETE)
1¼" Latex MM
Membrane Seal (Typ.)
℄ 5 - ¾"ϕ Threaded Studs @ 8" c/c (Max)
Weld to Angles
1¾" ℞.-(5" Min.)
2½" (Typ.)
2" Min. 2¼" Max.
See DETAIL X (typ.)
1¾" ℞.-(7" Min.)
Fill with an Approval Caulking Compound (Typ.)
⅛" (Typ.)
Existing L 6"x4"x⅝"
Existing L 7"x4"x⅝"
⅛" Max. (typ.)
℄ Brgs.
Deck Opening
SECTION A-A
(1¼" LATEX MODIFIED MORTAR)
¾"ϕ Threaded Stud
Center Punch Both Sides 1 Thread Deep
Hex Nut and Washer
⅛" Elastomeric Mat'l
1" 1½" 1¾"
½" ℞, 1" ℞ or ⅝" L
Latex M.M. Low Slump DC
SECTION Y-Y
Drill ⅞"ϕ Hole
C'bore - 2½"ϕ x 1" Deep
C'sink - 3/16 Deep x 45°
Y Y
NOTE: Nut & Washer not shown
PLAN
DETAIL "X"
27'-0" Curb to Curb
See Detail X (Typ.)
Median Barrier Adjust Reinforcement in Field
2¾ 10 2¾
Typ. 5/16
typ.
3/16"/Ft. (Typ.)
3"
¾"ϕ x 6" Studs
½" Backing ℞ (typ.)
5/16 typ.
Exist. ⅝" L
Exist. End ℞.
½" Backing Bar (typ.)
typ.
Cut Existing End ℞
SECTION F-F
F F
Center Studs on ℞.
2½" (typ)
Membrane Seal Not Shown
3" Min. 3¼" Max.
3"
Existing L5x5x⅝
Round by grinding
Existing L 5x5x⅝
Deck Opening
ELEVATION
(NORMAL TO DAM)

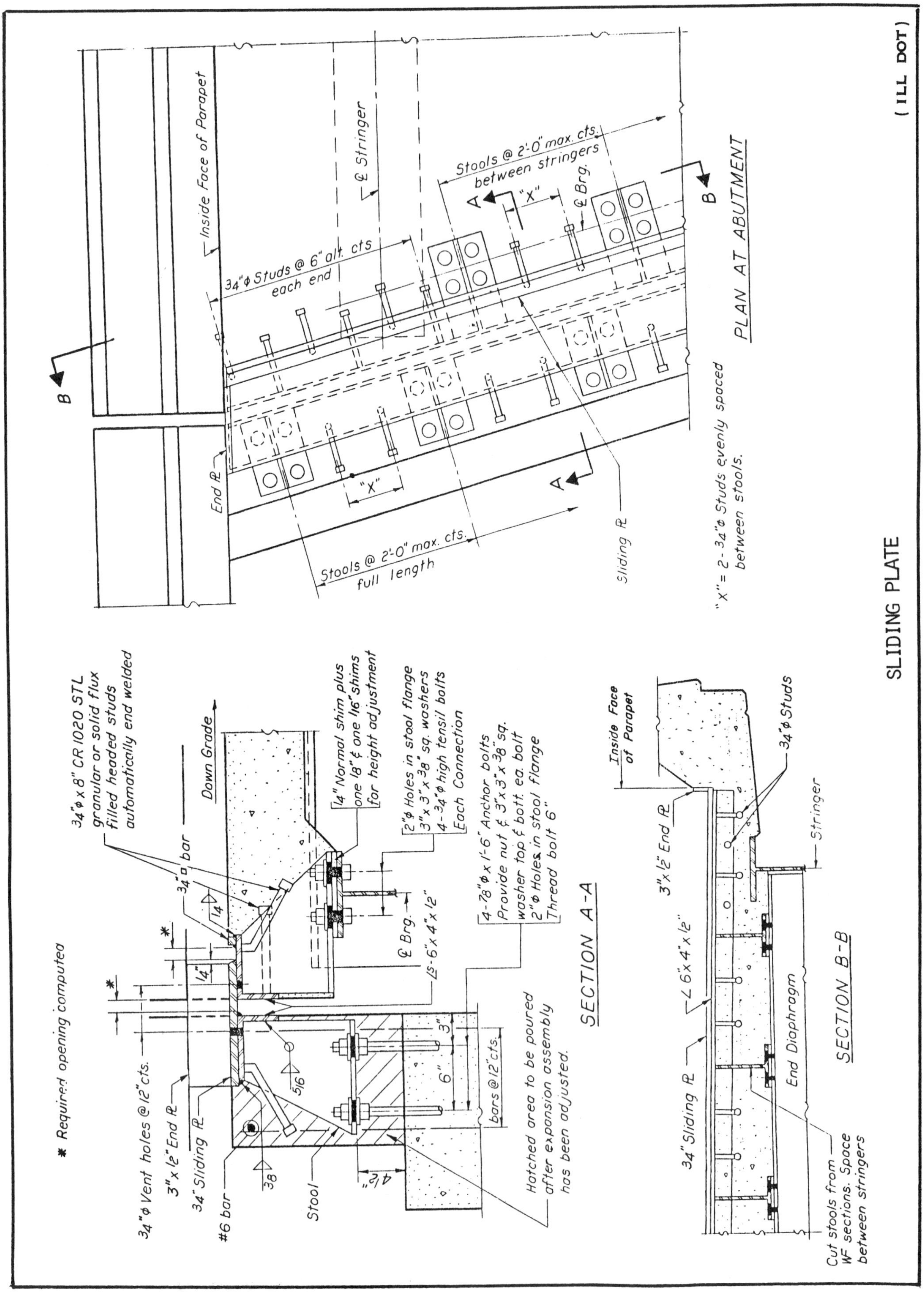


SLIDING PLATE

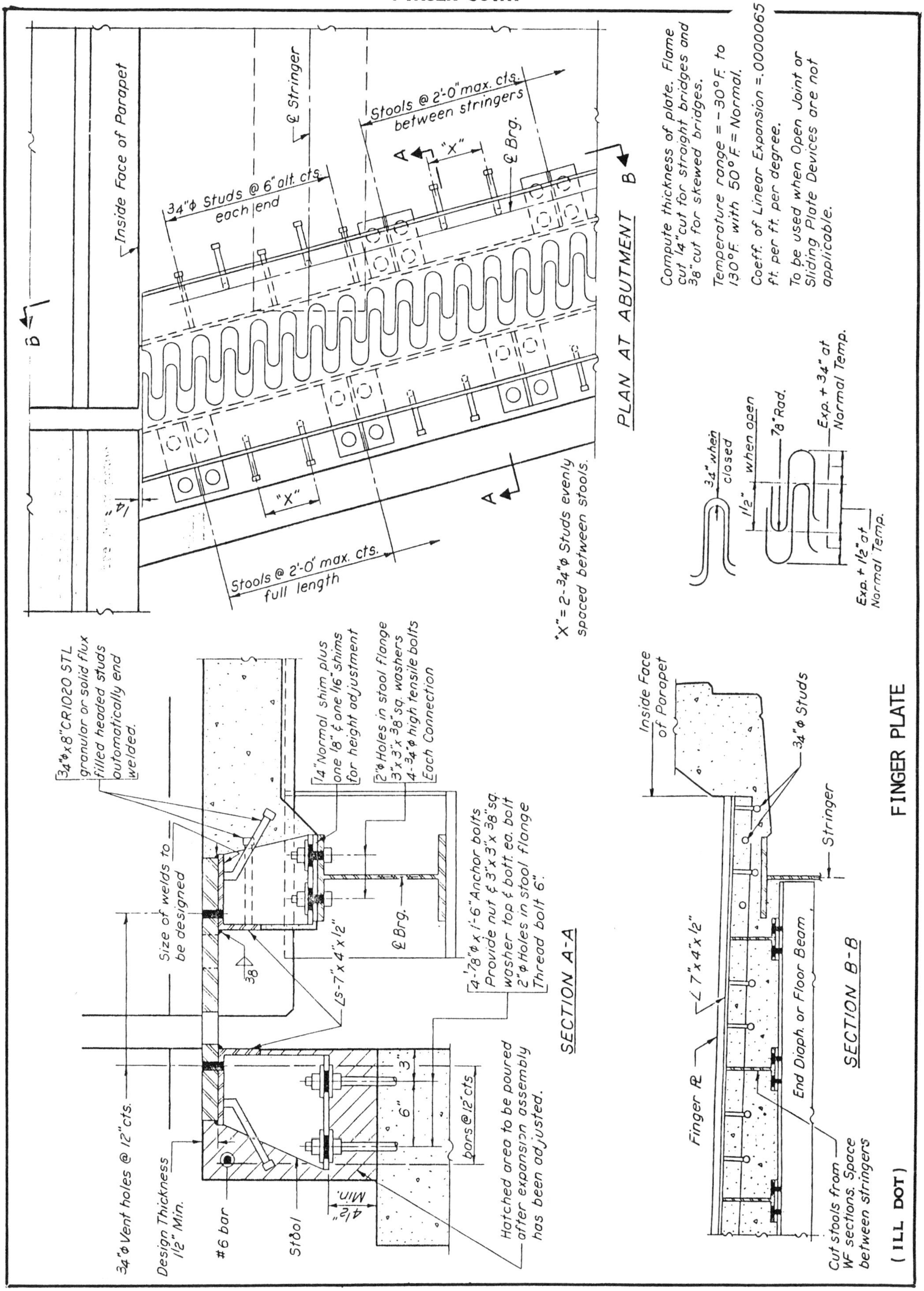
Inside Face of Parapet
℄ Stringer
Stools @ 2'-0" max. cts. between stringers
"X"
℄ Brg.
3/4"φ Studs @ 6" alt. cts. each end
A
B
1/4"
"X"
Stools @ 2'-0" max. cts. full length
PLAN AT ABUTMENT
"X" = 2-3/4"φ Studs evenly spaced between stools.
Compute thickness of plate. Flame cut 1/4" cut for straight bridges and 3/8" cut for skewed bridges.
Temperature range = -30°F. to 130°F. with 50°F. = Normal.
Coeff. of Linear Expansion = .0000065 ft. per ft. per degree.
To be used when Open Joint or Sliding Plate Devices are not applicable.
3/4" when closed
1 1/2" when open
7/8" Rad.
Exp. + 3/4" at Normal Temp.
Exp. + 1 1/2" at Normal Temp.
3/4"φ x 8" CR1020 STL granular or solid flux filled headed studs automatically end welded.
1/4" Normal shim plus one 1/8" & one 1/16" shims for height adjustment
2"φ Holes in stool flange 3"x3"x3/8" sq. washers 4-3/4"φ high tensile bolts Each Connection
Size of welds to be designed
3/8
LS-7"x4"x1/2"
℄ Brg.
4-7/8"φ x 1'-6" Anchor bolts Provide nut & 3"x3"x3/8" sq. washer top & bott. ea. bolt 2"φ Holes in stool flange Thread bolt 6".
SECTION A-A
3"
6"
bars @ 12" cts.
3/4"φ Vent holes @ 12" cts.
Design Thickness 1/2" Min.
#6 bar
Stool
4 1/2" Min.
Hatched area to be poured after expansion assembly has been adjusted.
Inside Face of Parapet
3/4"φ Studs
Stringer
∠ 7"x4"x1/2"
Finger ℙ
End Diaph. or Floor Beam
SECTION B-B
Cut stools from WF sections. Space between stringers
FINGER PLATE
(ILL DOT)

## MOVEMENT 4" to 16"

| Movement | 4" | 8" | 12" | 16" |
|---|---|---|---|---|
| Dim. A | 1½" | 2" | 2½" | 3" |
| Dim. B | 4½" | 5½" | 6½" | 6½" |
| Dim. C | 3" | 5" | 7" | 9" |
| Dim. D | 4" | 6" | 8" | 10" |

Note A: Use minimum preheat and interpass temperature of 250° F.

Note B: Apply weld in 3 passes using the "Annealing Bead" technique.

Note C: Hand pack concrete into this area.

Note D: Position end diaphragm to provide support and anchorage for WT's.

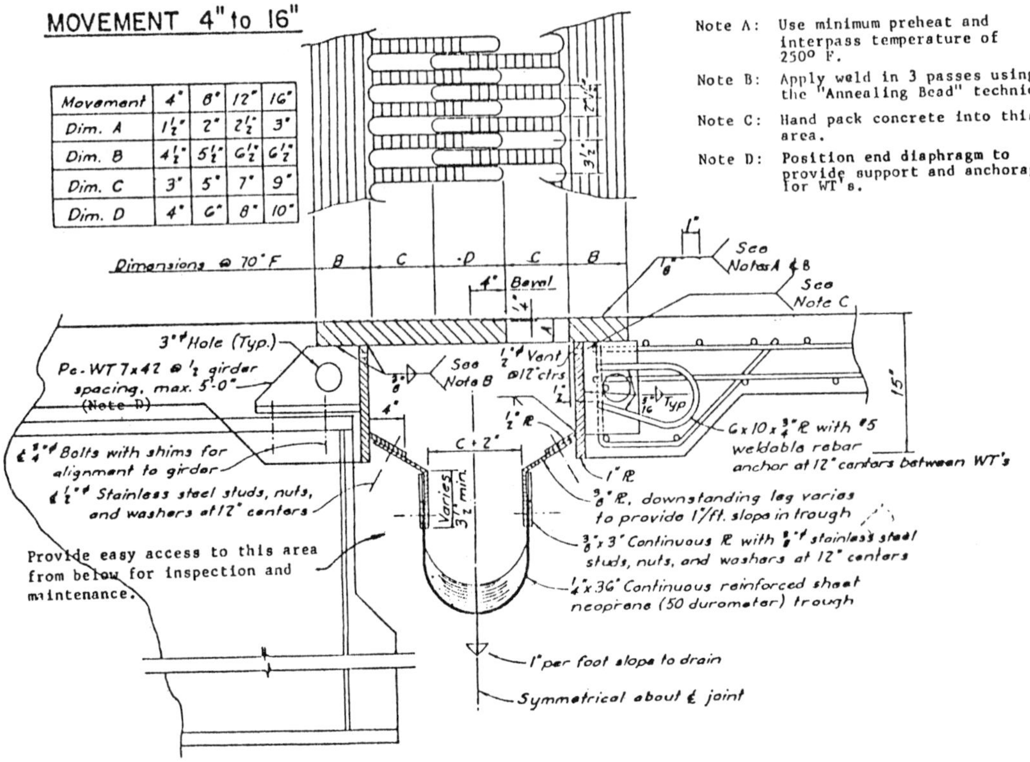


## MOVEMENT 16" to 24"

| Movement | 16" | 20" | 24" |
|---|---|---|---|
| Dim. A | 2½" | 2½" | 2½" |
| Dim. B | 6½" | 6½" | 6½" |
| Dim. C | 9" | 11" | 13" |
| Dim. D | 10" | 12" | 14" |

Note A: Use minimum preheat and interpass temperature of 250° F.

Note B: Apply weld in 3 passes using the "Annealing Bead" technique.

Note C: Hand pack concrete into this area.

Note D: Position end diaphragm to provide support and anchorage for WT's.

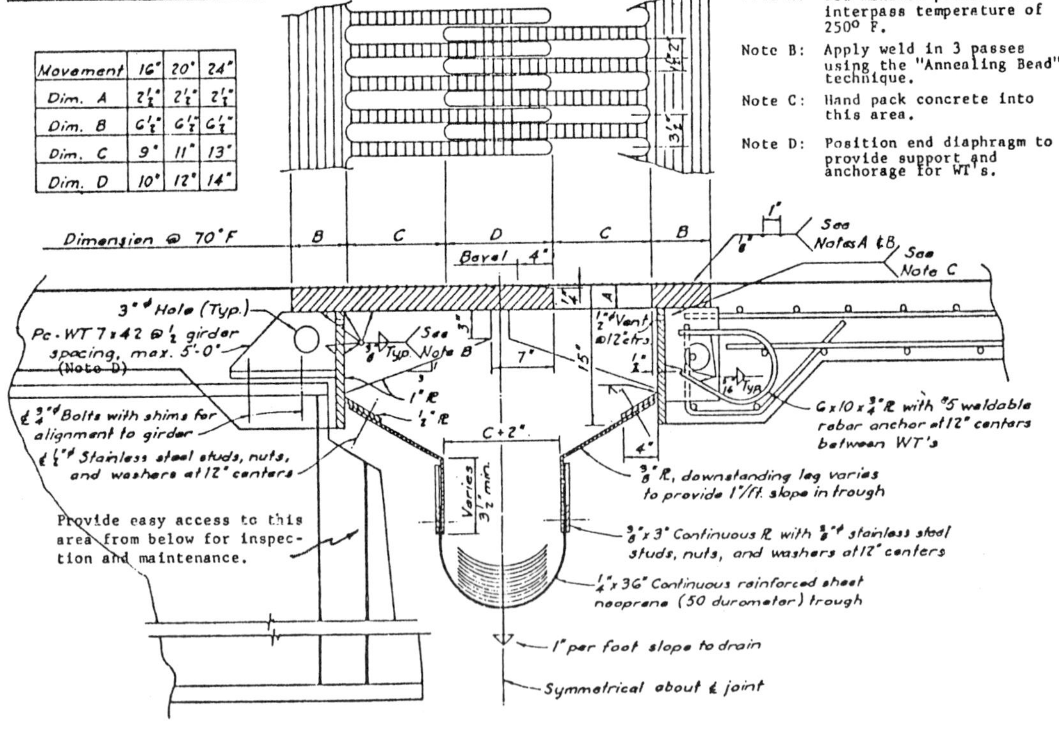

# MOVEMENT (TOTAL) < 4"

## SLIDING PLATE ALTERNATES

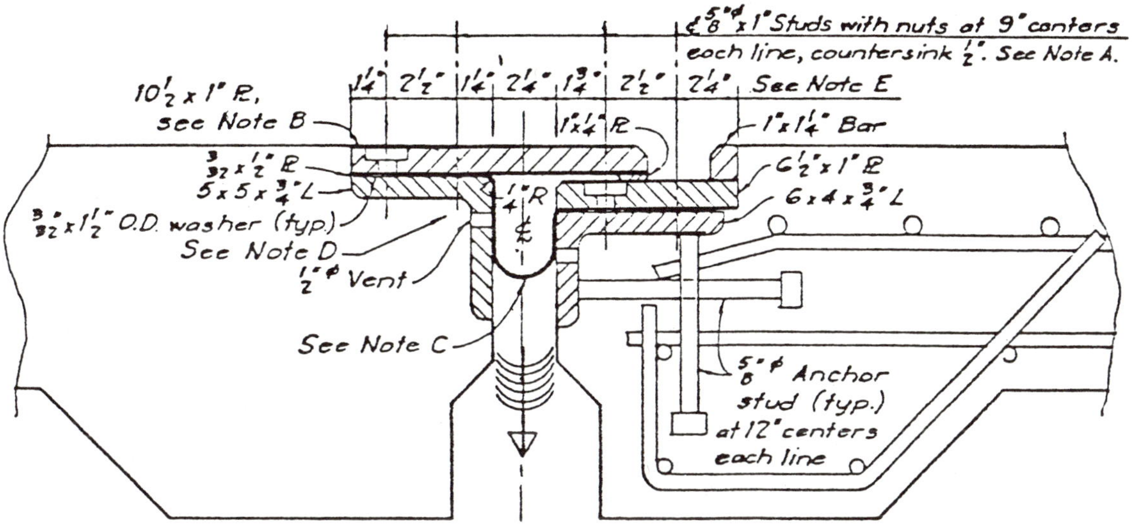


Note A: Studs and nuts shall be Type 304 stainless steel with 60 foot-pounds of torque (5/8" ).

Note B: Construct sliding plate side of joint last.

Note C: 1/8 " continuous synthetic fabric reinforced elastomeric trough (50 or 60 durometer). Vary width to slope 1/2 "/foot to drain.

Note D: Hand pack concrete into these areas.

Note E: Dimensions at 70° F.

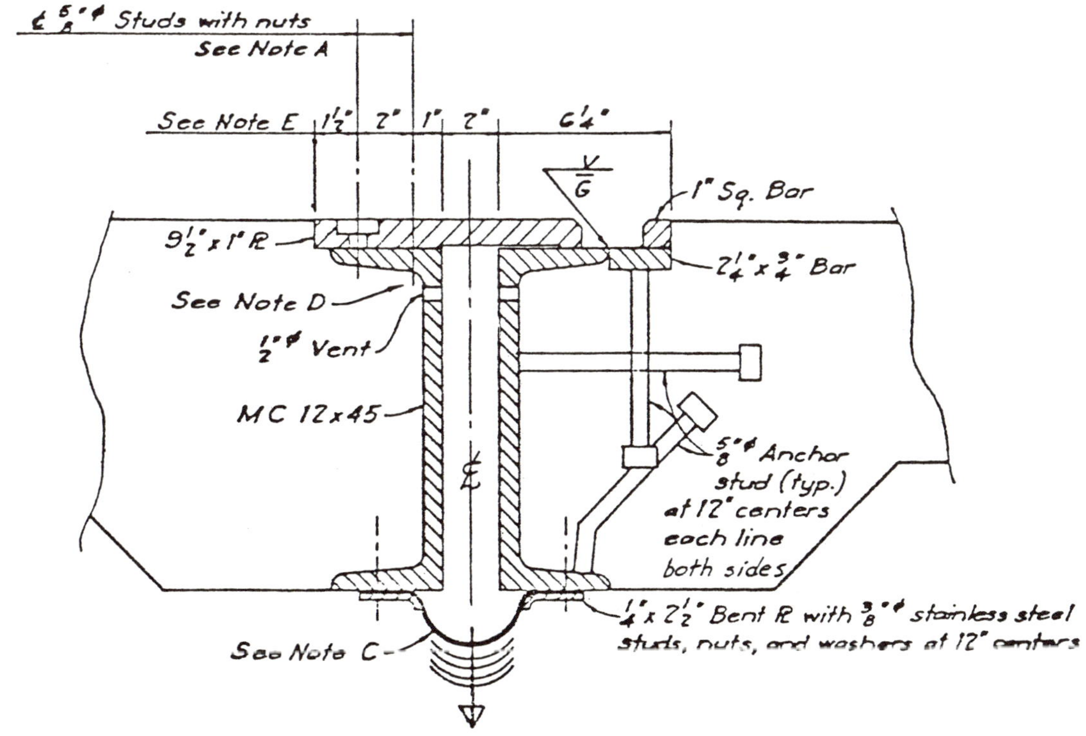

## PROCEDURE FOR DETERMINING SCUPPER SPACING

The procedure for spacing of scuppers shall be based on the following criteria:

"The width of flow in the gutter for a 25 year frequency - 5 minute duration shall not exceed the width of shoulder." (For narrow shoulders, a portion of the adjacent traffic lane, up to a maximum of one-fifth of the lane, may be included in the width of flow, if directed by the Department).

(3) Notations

$Q_R$ = maximum runoff on bridge deck, cfs.

$Q_G$ = maximum gutter storage capacity, cfs.

$Q_S$ = scupper capacity, cfs.

S1 = distance of scupper along the gutter from the point where flow begins to accumulate, and to have the gutter completely full assuming no flow of water at the beginning or starting point, ft.

S2 = spacing of successive scuppers necessary to keep the water at the edge of the roadway, including all water collected on the area covered by the spacing plus overflow past the previous scupper, ft.

n = roughness coefficient of deck surface = .013

(4) Values of $Q_R$, $Q_G$, and $Q_S$

(a) Value of $Q_R$

$Q_R = CIA$

where C = runoff factor = .95 for bridge decks

I = rate of rain fall, inches per hour for a given storm

A = drainage area, acres = W x L / 43,560

in which W and L are the width and length of deck to be drained by the scuppers, ft.

$Q_R$ for a 25 year 5 minute storm (I = 7.50 inches per hour)

$$= \frac{.95 \times 7.50 \times W \times L}{43,560} = \frac{W \times L}{6,114} = \frac{W \times L}{K}$$

i.e. $K = \frac{43,560}{C\,I} = 6,114$ for a 25 year 5 minute storm.

(b) Value of $Q_G$

The maximum gutter storage capacity shall be determined assuming the full shoulder as the width of gutter.

Given the roadway grade (slope of channel) G and the shoulder cross slope, determine $Q = Q_G$ from CHART I

(c) Value of $Q_S$

Smaller of the values

$$Q_{s1} = \frac{a \times b\,(y + y')}{\sqrt{(y + d)}} \qquad Q_{s2} = Q_y - Q_{y'}$$

where a = length of grate between walls, ft.

b = width of grate between walls, ft.

d = depth of grate bar, ft.

y = depth of flow at the curb, ft.

y' = depth of flow at the edge of the grate away from the curb, ft.

$Q_y$ = rate of flow at depth y, cfs.

$Q_{y'}$ = rate of flow at depth y', cfs.

(5) Values of S1 and S2

(a) If $Q_G \geq Q_R$, Scuppers are not required.

(b) If $Q_G < Q_R$, Scuppers are required.

$$S1 = \frac{K \times Q_G}{W} \quad \text{and} \quad S2 = \frac{K \times Q_S}{W}$$

(6) Spacing of scuppers on high side of tooth dam shall be such as to limit the width of flow over the dam to 2'-6" but not to exceed the width of shoulder. (See Example 2).

(7) Spacing of scuppers in a sag (See Example 3).

(a) Scuppers shall be required in a sag vertical curve, one at low point and one on each side of this point, where the grade elevation is approximately 0.20 foot higher than that at the low point unless required by design.

(b) Capacity of scuppers in a sag can be calculated from the following formula:

$$Q_{is} = 3.0\ P\ y^{1.5}$$

where $Q_{is}$ = scupper capacity in a sag, cfs.

P = perimeter of grate opening, in feet, disregarding bars and neglecting the side against the curb.

y = depth of flow at the curb, feet

Capacity of scuppers at other locations shall be computed from the formula given in Paragraph (4)(c).

Grade G at the scupper shall be determined from the elevations of two points, one at the scupper and the other 50 feet behind the scupper in relation to flow direction.

(8) Values of a, b, d, P, z/n

| Scupper type | a feet | b feet | d feet | P feet |
|---|---|---|---|---|
| A | 1.312 | 1.646 | .125 | 4.604 |
| B | 1.312 | 2.396 | .125 | 6.104 |
| C | 1.312 | 3.146 | .125 | 7.604 |

| Gutter slope Inch/foot | 3/8 | 7/16 | 1/2 | 9/16 | 5/8 | 11/16 | 3/4 | 13/16 |
|---|---|---|---|---|---|---|---|---|
| Values of z/n | 2460 | 2110 | 1846 | 1641 | 1477 | 1344 | 1231 | 1136 |

| Gutter slope Inch/foot | 7/8 | 15/16 | 1 |
|---|---|---|---|
| Values of z/n | 1055 | 985 | 923 |

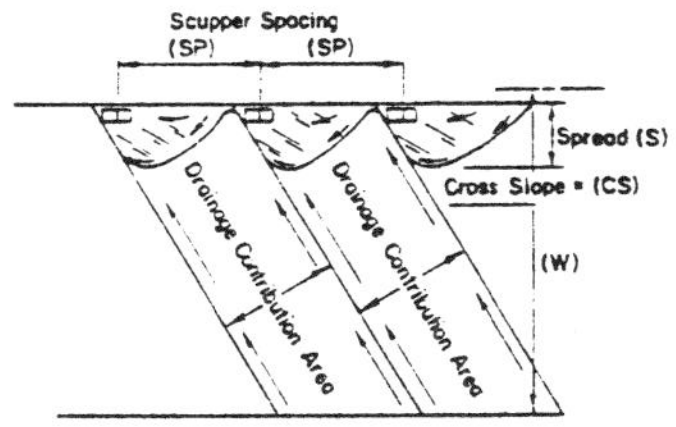


PART PLAN BRIDGE DRAINAGE

PLAN-STANDARD SCUPPER GRATING

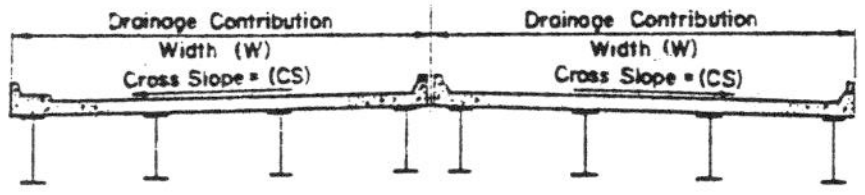


TYPICAL CROSS SECTION

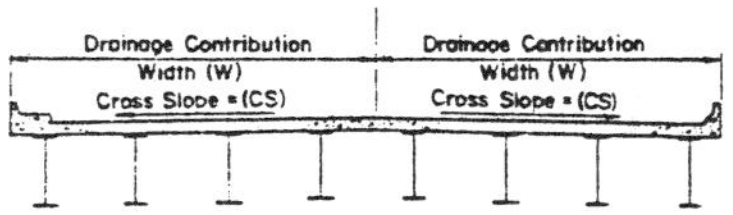


TYPICAL CROSS SECTION

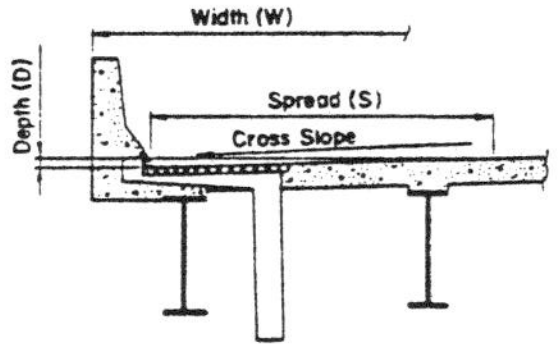


SECTION AT SCUPPER

TYPICAL DECK DRAINAGE TERMS

## DESIGN EXAMPLE

### EXAMPLE 1

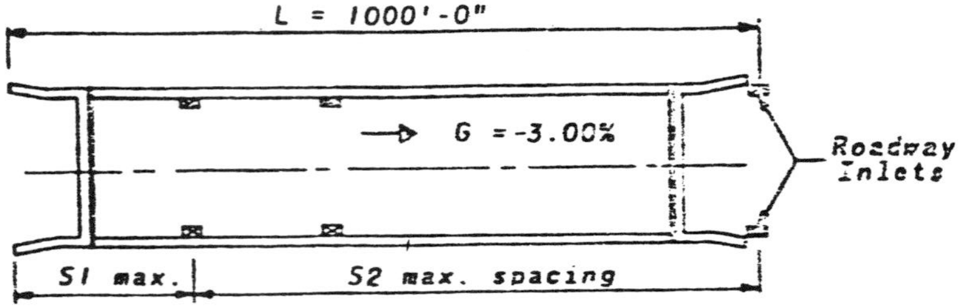


PLAN

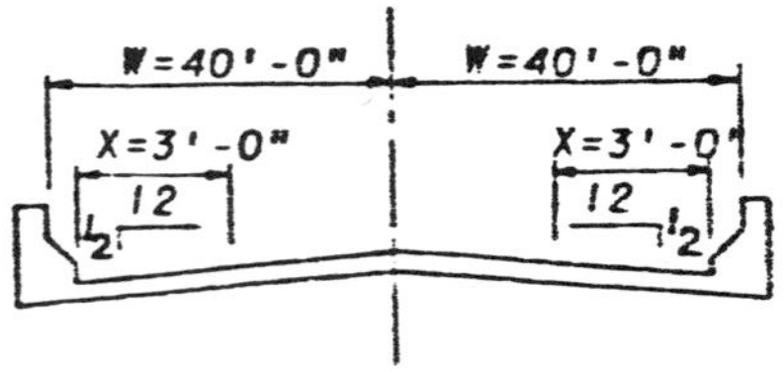


CROSS SECTION

| W | L | $Q_R$ | X | $y$ | $Z/n$ | G | $Q_G$ |
|---|---|---|---|---|---|---|---|
| 40 | 1000 | 6.54 | 3.0 | .125 | 1846 | .03 | .70 |

Since $Q_G < Q_R$, Scuppers are required.

| Type | a | b | d |
|---|---|---|---|
| A-Scupper | 1.312 | 1.646 | .125 |

| $y$ | $y'$ | Qy | Qy' | $Q_{s2}$ | $Q_{s1}$ | Qs |
|---|---|---|---|---|---|---|
| .125 | .056 | .70 | .08 | .62 | .78 | .62 |

$S1 = \frac{KQ_G}{W} = 107$ ft. $\quad S2 = \frac{KQ_S}{W} = 95$ ft.

### EXAMPLE 2

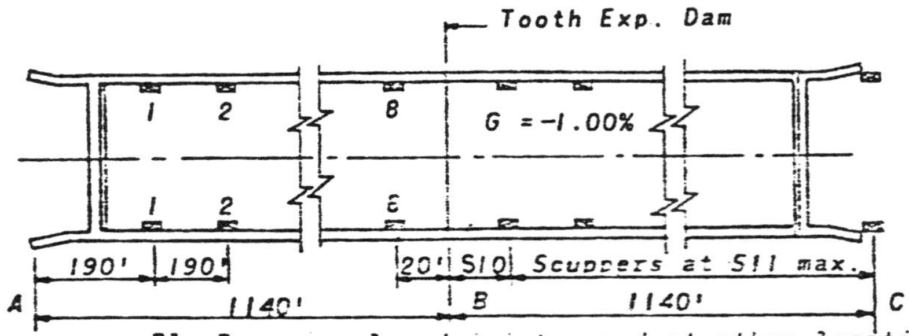


Pl. Dams or closed joints used at other locations.

PLAN

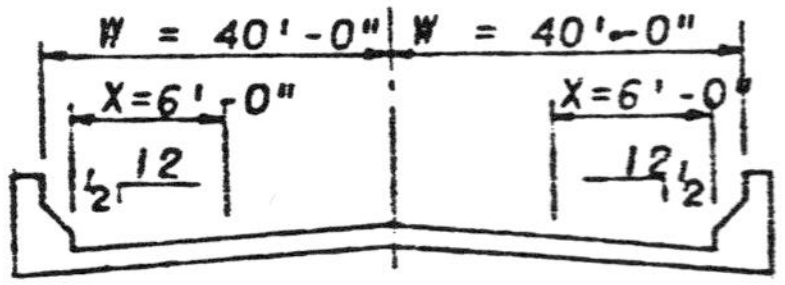


CROSS SECTION

Portion A-B (Trial-Error Solution)

| W | L | $Q_R$ | X@B | $y$ | $Z/n$ | G | $Q_G$ |
|---|---|---|---|---|---|---|---|
| 40 | 1140 | 7.46 | 2.50 | .104 | 1846 | .01 | .25 |

Since $Q_G < Q_R$, Scuppers are required.

| Type | a | b | d |
|---|---|---|---|
| A-Scupper | 1.312 | 1.646 | .125 |

Calculations for Qs

| Qy=Qt | $y$ | $y_1$ | $Q_{y1}$ | $Q_{s2}$ | $Q_{s1}$ | Qs |
|---|---|---|---|---|---|---|
| 1.24 | .193 | .123 | .32 | .92 | 1.21 | .92 |
| 1.56 | .205 | .135 | .50 | 1.06 | 1.28 | 1.06 |
| 1.74 | .215 | .145 | .60 | 1.14 | 1.34 | 1.14 |
| 1.65 | .212 | .142 | .58 | 1.07 | 1.16 | 1.07 |
| 1.43 | .200 | .130 | .50 | .93 | 1.25 | .93 |
| 1.25 | .193 | .123 | .32 | .93 | 1.22 | .93 |
| .88 | .170 | .100 | .23 | .65 | 1.08 | .65 |
| .62 | .143 | .073 | .10 | .52 | .90 | .52 |

## EXAMPLE 3

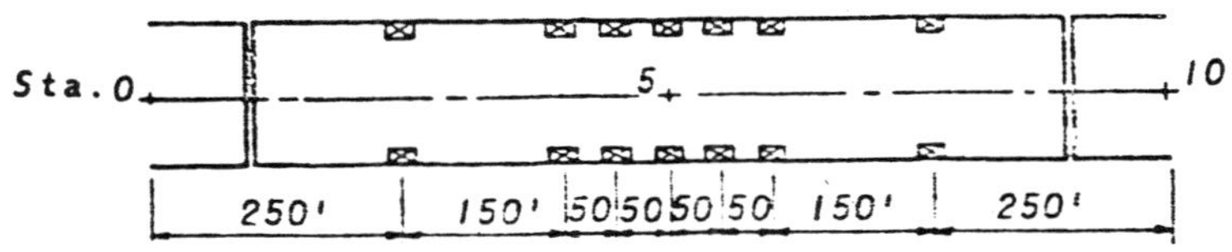


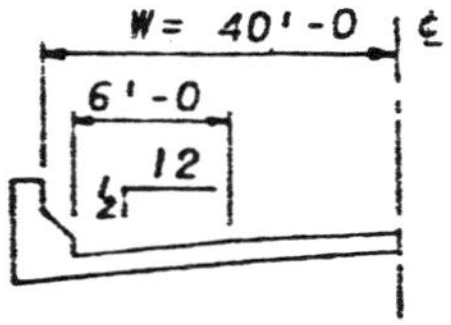


$y_{max.} = \frac{6 \times \frac{1}{2}}{12}$

$= .25$ ft.

P.V.I.sta. 5+00
El.= 1.00
V.C.= 1,000 ft.

| Type | a | b | d | P |
|---|---|---|---|---|
| A-Scupper | 1.312 | 1.646 | .125 | 4.604 |

Trial-Error Solution

| Calculations for Qs | | | | | | | |
|---|---|---|---|---|---|---|---|
| Qy=Qt | G | *y* | *y'* | Qy' | Qs2 | Qs1 | Qs |
| 1.64 | 1.66% | .20 | .13 | .53 | 1.11 | 1.26 | 1.11 |
| 1.51 | .76% | .217 | .147 | .50 | 1.01 | 1.35 | 1.01 |
| .828 | .10% | .19 | .12 | .25 | .578 | 1.19 | .578 |

| Sta. | Elev. | G | Scupper Spacing | Qc | $\Delta Q=\frac{WS}{K}$ | Qt | Qs | Qc |
|---|---|---|---|---|---|---|---|---|
| 2+00 | 5.20 | | | | | | | — |
| 2+50 | 4.37 | 1.66% | 250 | — | 1.64 | 1.64 | 1.11 | .53 |
| 3+50 | 3.18 | | | | | | | |
| 4+00 | 2.80 | .76% | 150 | .53 | .98 | 1.51 | 1.01 | .50 |
| 4+50 | 2.55 | .50% | 50 | .50 | .328 | .828 | .578 | .25 |
| 5+00 | 2.50 | .10% | 50 | .25 | .328 | .578 | .72* | 0 |

For Qy=Qt=.578 & G=.10%, *y* from fig.1 =.22

$\therefore$ $Qis = 3.0Py^{1.5} = 3.0 \times 4.604 \times .22^{1.5} = 1.44$

* ½ of Qis =.72 Since greater than Qt at sta.5+00, Scupper spacings O.K.

| Scupper No. | Assumed Spacings | Qc | $\Delta Q=\frac{WS}{K}$ | Qt=Qc+ΔQ | Qs | Qc=Qt-Qs |
|---|---|---|---|---|---|---|
| A | — | — | — | — | — | — |
| 1 | 190 | - | 1.24 | 1.24 | .92 | .32 |
| 2 | 190 | .32 | 1.24 | 1.56 | 1.05 | .50 |
| 3 | 190 | .50 | 1.24 | 1.74 | 1.14 | .60 |
| 4 | 160 | .60 | 1.05 | 1.65 | 1.07 | .58 |
| 5 | 130 | .58 | .85 | 1.43 | .93 | .50 |
| 6 | 115 | .50 | .75 | 1.25 | .93 | .32 |
| 7 | 85 | .32 | .56 | .88 | .65 | .23 |
| 8 | 60 | .23 | .39 | .62 | .52 | .10 |
| B T-Dam | 20 | .10 | .13 | .23 | — | .23 * |

* Since it is <.25, Assumed Spacing O.K.

Portion B-C

| Qc@B | Portion B-C | | | Total |
|---|---|---|---|---|
| | W | L | $Q_R$ | $Q_R$ |
| .23 | 40 | 1140 | 7.46 | 7.69 |

| X | *y* | Z/n | G | $Q_G$ |
|---|---|---|---|---|
| 6.00 | .25 | 1846 | .01 | 2.50 |

Since $Q_G < Q_R$ total, Scuppers are required.

| Type | a | b | d |
|---|---|---|---|
| A-Scupper | 1.312 | 1.646 | .125 |

| *y* | *y'* | Qy | Qy' | Qs2 | Qs1 | Qs |
|---|---|---|---|---|---|---|
| .25 | .18 | 2.50 | 1.00 | 1.50 | 1.52 | 1.50 |

$S10 = \frac{K(Q_G - Qc@B)}{W} = 347'$ $S11 = \frac{KQs}{W} = 230'$

*EXAMPLE 4*

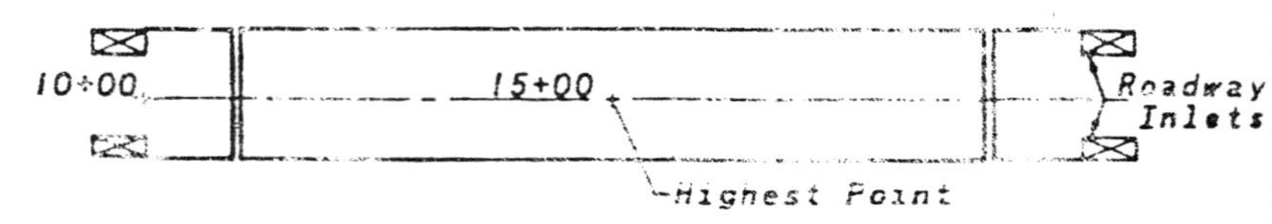


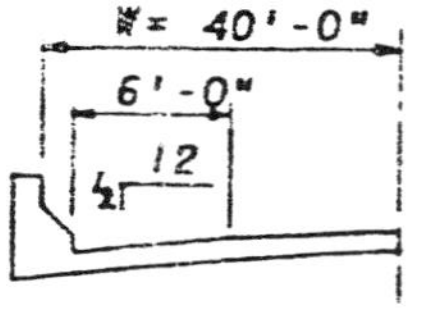


$y_{max} = \frac{6 \times 4}{12}$

$= .25$ ft.

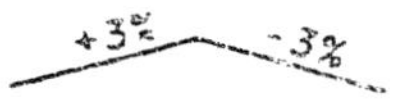


P.V.I. sta. = 15+00
El. = 40.00
V.C. = 1,000 ft.

| Sta. | Elev. | G | Total $Q_R$ | $Q_c$ |
|---|---|---|---|---|
| 15+00 | 32.50 | | | |
| 14+50 | 32.45 | .10% | .326 | .75 |
| 14+00 | 32.20 | .50% | .652 | 1.70 |
| 13+50 | 31.82 | .76% | .978 | 2.25 |
| 13+00 | 31.30 | 1.04% | 1.304 | 2.50 |
| 12+50 | 30.63 | 1.34% | 1.630 | 2.90 |
| 12+00 | 29.80 | 1.66% | 1.956 | 3.10 |
| 11+50 | 28.82 | 1.96% | 2.282 | 3.30 |
| 11+00 | 27.70 | 2.24% | 2.608 | 3.40 |
| 10+50 | 26.42 | 2.56% | 2.934 | 3.70 |
| 10+00 | 25.00 | 2.84% | 3.260 | 4.00 |

*Since $Q_c > Q_R$, Scuppers are not required.*

For design a storm frequency of 10 years with a duration of 5 minutes is used. This gives an average rainfall intensity (i) of approximately 6 inches per hour. A run-off coefficient (C) of 0.9 is used for concrete surfaces.

The Rational Method converts rainfall intensity for a given design frequency storm to run-off by the following equation:

$Q = C\,i\,A$ where:

- Q = peak rate of run-off in cfs.
- C = run-off coefficient for surface type.
- i = rainfall intensity in inches/hour.
- A = drainage area in acres and
- $A = \frac{(L)(W)}{43560}$ where
- L = floor drain spacing in feet.
- W = contributing structure width in feet.

The Manning equation modified for triangular flow is used to compute Q and Q bypass for the given gutter section. The modified equation is:

$Q = 0.56\ (Z/n)(S_o)^{1/2}(d)^{8/3}$ where:

- Q = discharge in cfs.
- Z = reciprocal of cross slope.
- n = Manning's coefficient of roughness, use n = 0.014 for concrete.
- $S_o$ = longitudinal slope in feet/foot.
- d = depth of flow at the deepest point (gutter line) in feet.

| CROSS SLOPE, $S_c$ | | | $1/S_c$ | VALUES OF Z/n | | | | |
|---|---|---|---|---|---|---|---|---|
| in/ft | in/ft | ft/ft | Z | Z/0.012 | Z/0.013 | Z/0.014 | Z/0.015 | Z/0.016 |
| 1/8 | .1250 | .0104 | 96.00 | 8,000 | 7,385 | 6,857 | 6,410 | 6,000 |
| | .1320 | .0110 | 90.90 | 7.575 | 6,992 | 6,493 | 6,060 | 5,681 |
| | .1440 | .0120 | 83.33 | 6,944 | 6,410 | 5,952 | 5,555 | 5.208 |
| 5/32 | .1563 | .0130 | 76.80 | 6,400 | 5,908 | 5,486 | 5,128 | 4.800 |
| | .1680 | .0140 | 71.43 | 5,952 | 5,495 | 5,102 | 4,762 | 4,464 |
| | .1800 | .0150 | 66.67 | 5,556 | 5,128 | 4,762 | 4,445 | 4,167 |
| 3/16 | .1875 | .0156 | 64.00 | 5.333 | 4,923 | 4,571 | 4,273 | 4.000 |
| | .1920 | .0160 | 62.50 | 5,208 | 4,808 | 4,464 | 4,167 | 3,906 |
| | .2000 | .0167 | 59.88 | 4,990 | 4,606 | 4,277 | 3,992 | 3.742 |
| | .2040 | .0170 | 58.82 | 4,902 | 4,525 | 4,201 | 3,921 | 3,676 |
| | .2160 | .0180 | 55.56 | 4,630 | 4,274 | 3,969 | 3,704 | 3.472 |
| 7/32 | .2188 | .0182 | 54.86 | 4,572 | 4,220 | 3,919 | 3,663 | 3.429 |
| | .2280 | .0190 | 52.63 | 4,386 | 4,046 | 3,759 | 3,509 | 3.289 |
| | .2409 | .0200 | 50.00 | 4,167 | 3,846 | 3,571 | 3.333 | 3,125 |
| 1/4 | .2500 | .0208 | 48.00 | 4.000 | 3.692 | 3,429 | 3.205 | 3,000 |
| 9/32 | .2813 | .0234 | 42.67 | 3.556 | 3,282 | 3.048 | 2,849 | 2,667 |
| 19/64 | .2969 | .0247 | 40.42 | 3.368 | 3,109 | 2,887 | 2,695 | 2,526 |
| | .3000 | .0250 | 40.00 | 3,333 | 3,077 | 2,857 | 2,667 | 2,500 |
| 5/16 | .3125 | .0260 | 38.40 | 3.200 | 2,954 | 2,743 | 2,564 | 2,400 |
| 21/64 | .3281 | .0273 | 36.57 | 3,047 | 2,813 | 2,612 | 2,438 | 2,285 |
| 11/32 | .3438 | .0286 | 34.94 | 2,912 | 2,688 | 2,496 | 2,331 | 2,184 |
| | .3600 | .0300 | 33.33 | 2.777 | 2,564 | 2,381 | 2.222 | 2.083 |
| 3/8 | .3750 | .0313 | 32.00 | 2,667 | 2,462 | 2,286 | 2,137 | 2,000 |
| | .4000 | .0333 | 30.00 | 2,500 | 2,308 | 2,143 | 2,000 | 1,875 |
| 13/32 | .4063 | .0339 | 29.54 | 2,462 | 2,272 | 2,110 | 1,967 | 1,846 |
| | .4200 | .0350 | 28.57 | 2.381 | 2,198 | 2,041 | 1.905 | 1,786 |
| 7/16 | .4375 | .0364 | 27.43 | 2,286 | 2,110 | 1.959 | 1,831 | 1,714 |
| 15/32 | .4688 | .0391 | 25.60 | 2,133 | 1,969 | 1,829 | 1,704 | 1,600 |
| | .4800 | .0400 | 25.00 | 2,083 | 1.923 | 1,786 | 1,667 | 1,562 |
| 1/2 | .5000 | .0417 | 24.00 | 2,000 | 1,846 | 1,714 | 1,603 | 1,500 |
| 17/32 | .5313 | .0443 | 22.59 | 1,882 | 1,738 | 1,614 | 1,505 | 1.412 |
| | .5400 | .0450 | 22.22 | 1,852 | 1,709 | 1.587 | 1.481 | 1,389 |
| 9/16 | .5625 | .0469 | 21.33 | 1.777 | 1,641 | 1,524 | 1.421 | 1,333 |
| 19/32 | .5938 | .0495 | 20.21 | 1,684 | 1,555 | 1,444 | 1,347 | 1,263 |
| | .6000 | .0500 | 20.00 | 1,667 | 1,538 | 1,429 | 1,333 | 1,250 |
| 5/8 | .6250 | .0521 | 19.20 | 1,600 | 1,477 | 1,371 | 1,282 | 1,200 |
| 21/32 | .6563 | .0547 | 18.29 | 1,524 | 1,407 | 1,306 | 1,219 | 1,143 |
| | .6600 | .0550 | 18.18 | 1,515 | 1,398 | 1,299 | 1,212 | 1,136 |
| 11/16 | .6875 | .0573 | 17.47 | 1,456 | 1,344 | 1,248 | 1,163 | 1,092 |
| | .7000 | .0583 | 17.15 | 1,429 | 1,319 | 1,225 | 1,143 | 1.072 |
| 23/32 | .7188 | .0599 | 16.70 | 1,392 | 1,285 | 1,193 | 1,112 | 1,044 |
| | .7200 | .0600 | 16.67 | 1,389 | 1,282 | 1,191 | 1,111 | 1,042 |
| 3/4 | .7500 | .0625 | 16.00 | 1,333 | 1,231 | 1,143 | 1,066 | 1,000 |
| | .7800 | .0650 | 15.38 | 1,282 | 1,183 | 1,099 | 1,025 | 961 |
| 25/32 | .7813 | .0651 | 15.36 | 1,280 | 1,182 | 1,097 | 1,024 | 960 |
| | .8000 | .0666 | 15.00 | 1,250 | 1,154 | 1,071 | 1,000 | 938 |
| 13/16 | .8125 | .0677 | 14.77 | 1,231 | 1,136 | 1,055 | 985 | 923 |
| | .8400 | .0700 | 14.28 | 1,190 | 1,098 | 1,020 | 952 | 892 |

# NOMOGRAPH FOR FLOW IN TRIANGULAR CHANNELS

**Equation**

$Q = 0.56\left(\frac{Z}{N}\right) S^{1/2} Y^{8/3}$

N is roughness coefficient in Manning formula

Z is reciprocal of cross slope.

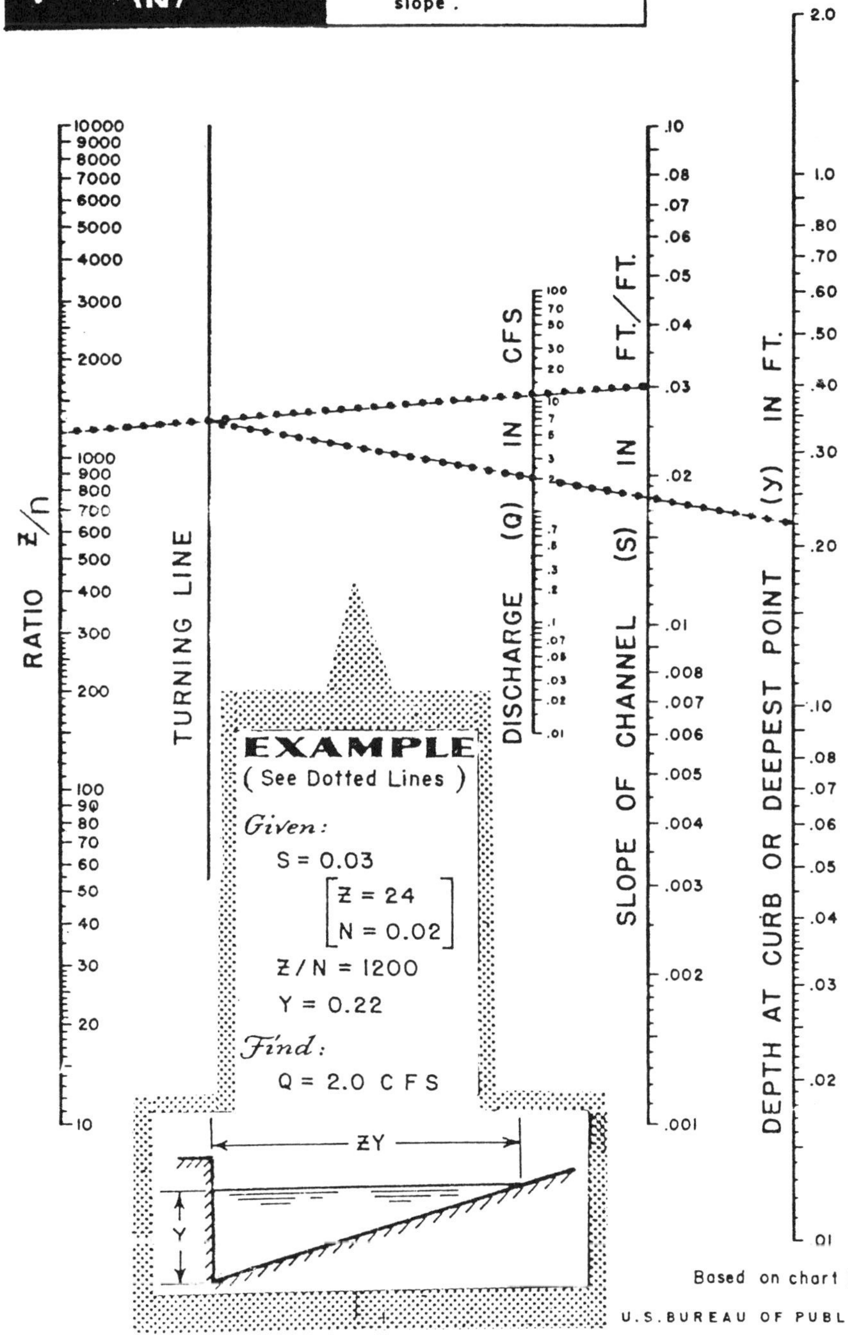


**EXAMPLE**
( See Dotted Lines )

*Given:*
S = 0.03
Z = 24
N = 0.02
Z / N = 1200
Y = 0.22

*Find:*
Q = 2.0 C F S

Based on chart by
U.S. BUREAU OF PUBLIC ROADS

## INSTRUCTIONS

1. Connect Z/N ratio with slope (S) and connect discharge (Q) with depth (Y). These two lines must intersect at TURNING LINE for complete solution.

2. For shallow v-shaped channel as shown use Nomograph with $Z = \frac{T}{Y}$

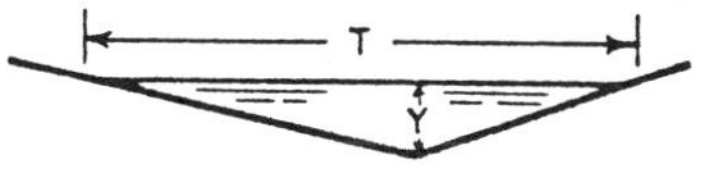


3. To determine discharge $Q_X$ in portion of channel having width X : determine depth Y for total discharge in entire section A. Then use Nomograph to determine $Q_B$ in section B for depth $Y' = Y - \left(\frac{X}{Z}\right)$

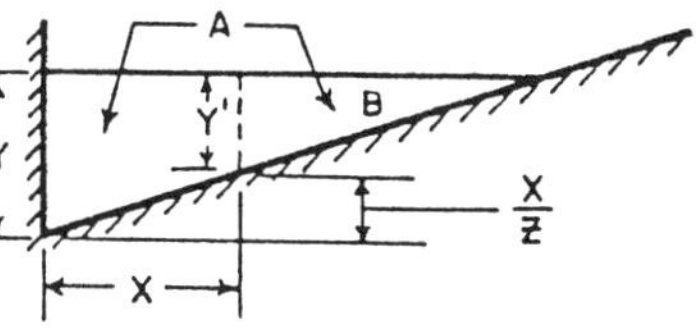


4. To determine approx. discharge in composite section :- Follow instruction 3 to obtain discharge in section A at assumed depth Y ; obtain $Q_B$ for slope ratio $Z_B$ and depth Y', then $Q_T = Q_A + Q_B$

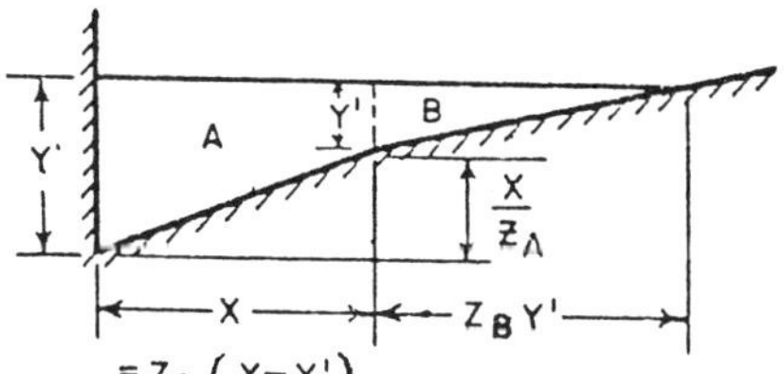


CHART I

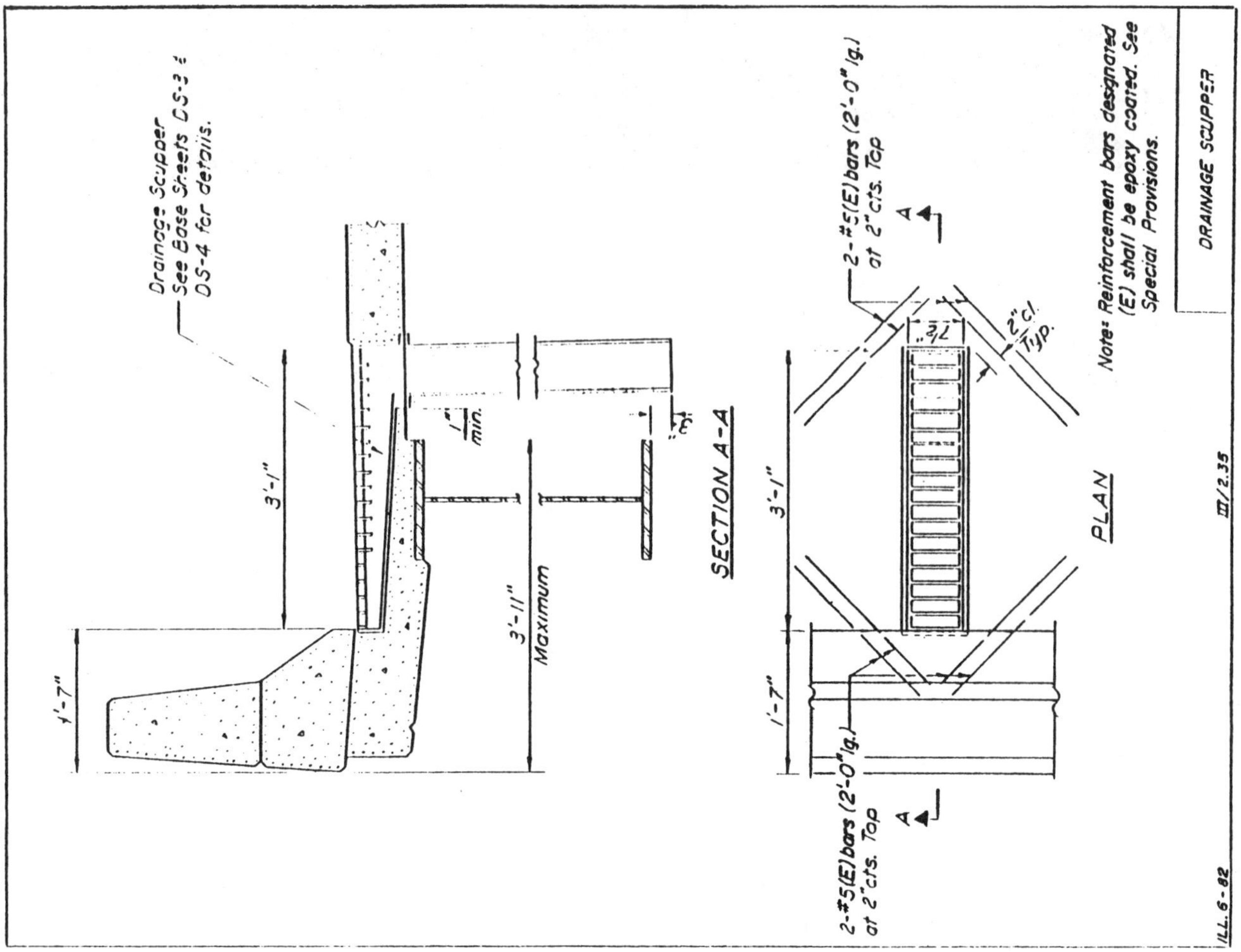
Drainage Scupper
See Base Sheets DS-3 & DS-4 for details.
1'-7"
3'-1"
3'-11"
Maximum
1" min.
SECTION A-A
2-#5(E) bars (2'-0" lg.) at 2" cts. Top
2-#5(E) bars (2'-0" lg.) at 2" cts. Top
A
A
2" cl. Typ.
1'-7"
3'-1"
PLAN
Note: Reinforcement bars designated (E) shall be epoxy coated. See Special Provisions.
DRAINAGE SCUPPER
ILL. 6-82
III/2.35

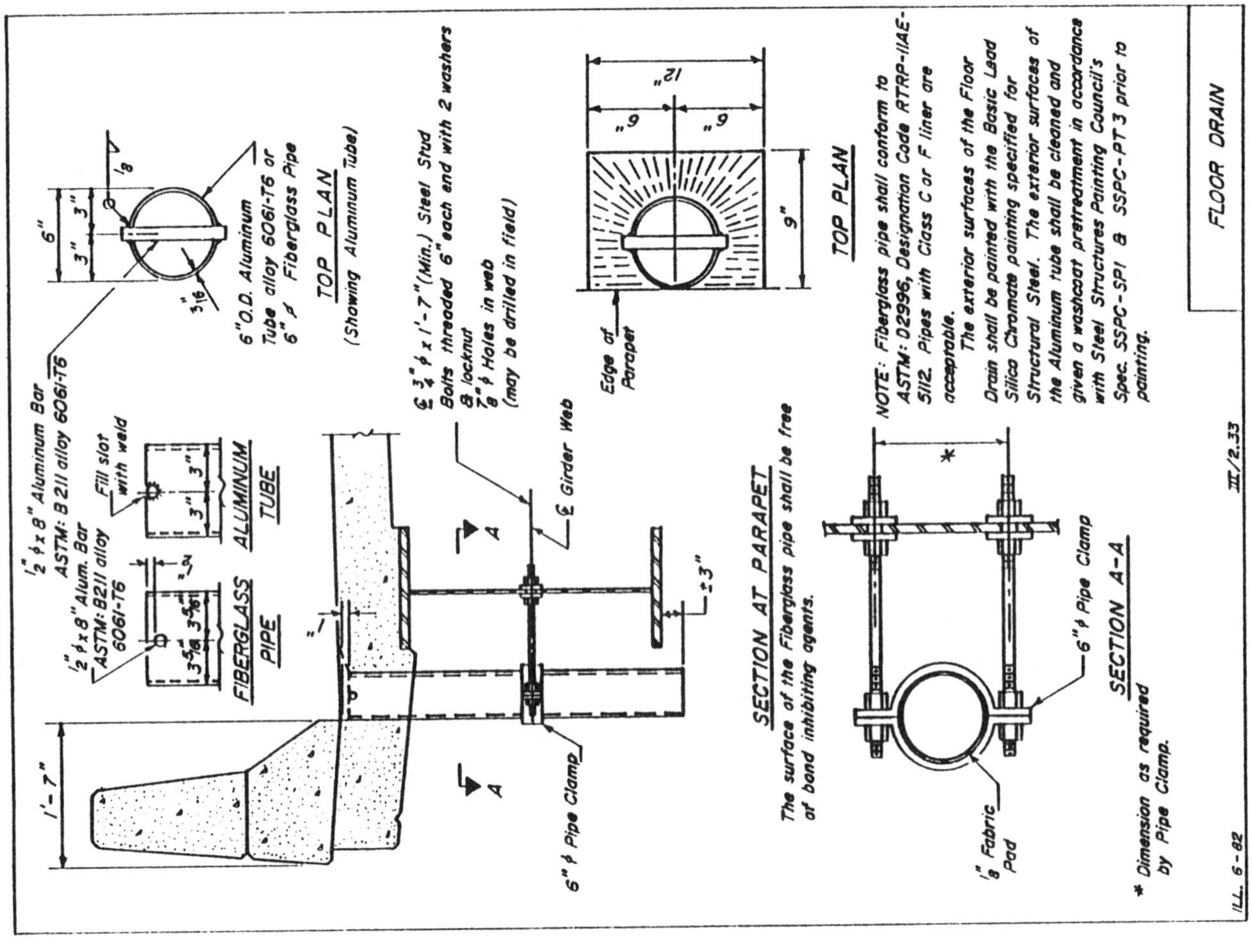
1'-7"
6" ϕ Pipe Clamp
A
A
1"
± 3"
℄ Girder Web
SECTION AT PARAPET
The surface of the Fiberglass pipe shall be free of bond inhibiting agents.
1/2" ϕ x 8" Alum. Bar ASTM: B211 alloy 6061-T6
1/2" ϕ x 8" Aluminum Bar ASTM: B211 alloy 6061-T6
Fill slot with weld
FIBERGLASS PIPE
ALUMINUM TUBE
6"
3"
3"
1/8
3/16
6" O.D. Aluminum Tube alloy 6061-T6 or 6" ϕ Fiberglass Pipe
TOP PLAN
(Showing Aluminum Tube)
℄ 3/4" ϕ x 1'-7" (Min.) Steel Stud Bolts threaded 6" each end with 2 washers & locknut
7/8" ϕ Holes in web (may be drilled in field)
Edge of Parapet
12"
6"
6"
9"
TOP PLAN
1/8" Fabric Pad
6" ϕ Pipe Clamp
SECTION A-A
* Dimension as required by Pipe Clamp.
NOTE: Fiberglass pipe shall conform to ASTM: D2996, Designation Code RTRP-11AE-5112. Pipes with Class C or F liner are acceptable.
The exterior surfaces of the Floor Drain shall be painted with the Basic Lead Silico Chromate painting specified for Structural Steel. The exterior surfaces of the Aluminum tube shall be cleaned and given a washcoat pretreatment in accordance with Steel Structures Painting Council's Spec. SSPC-SP1 & SSPC-PT3 prior to painting.
FLOOR DRAIN
ILL. 6-82
III/2.33

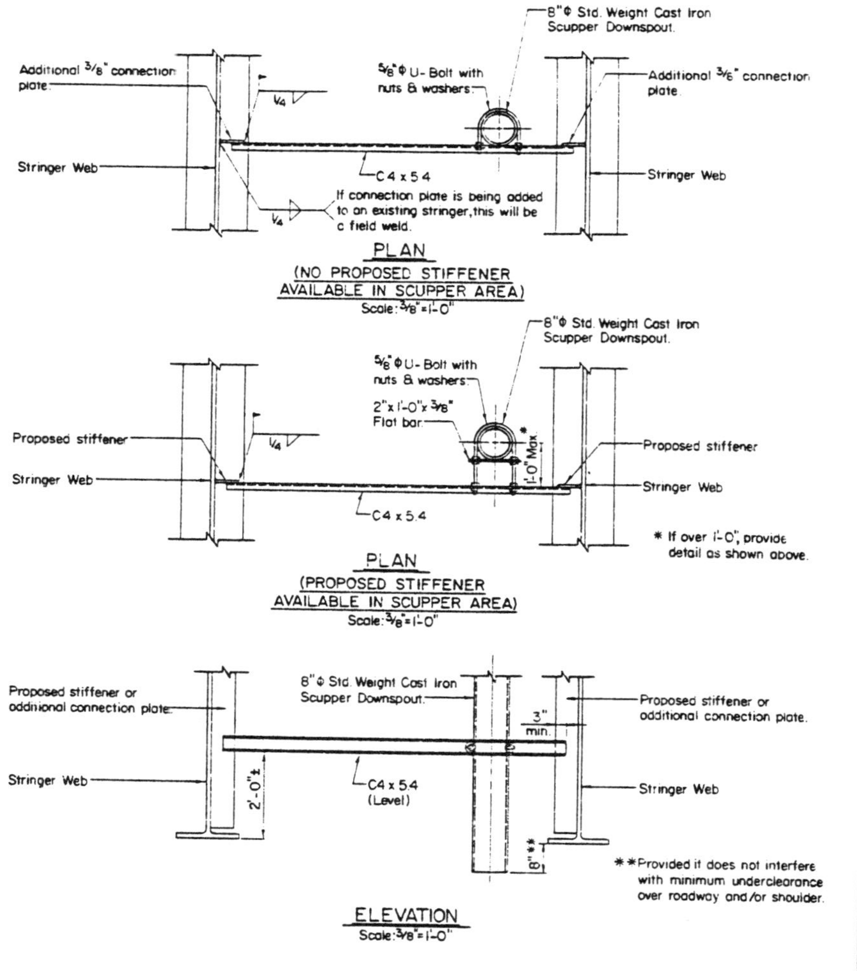


Notes:
1. U bolts, nuts, washers, bars, channels, etc. shall receive the same number of coats and same color field paint as required for structural steel. If weathering steel is utilized for any elements it need not be painted.
2. Exposed portions of cast iron downspout to be coated with coal tar epoxy conforming to Specification SSPC-16.

| APPROVAL | |
|---|---|
| E.S. Freedman ASST. CHIEF ENGR. BRIDGE DEVEL. | |
| DATE 5/2/83 | |
| REVISIONS | |
| SHA | FHWA |
| 5-25-83 | |
| 9-8-83 | |
| FHWA APPROVAL | |
| DATE | |

**STATE OF MARYLAND**
**DEPARTMENT OF TRANSPORTATION**
**STATE HIGHWAY ADMINISTRATION**
DIVISION OF BRIDGE DEVELOPMENT
SCUPPER DOWNSPOUT SUPPORT BRACKET FOR LONG CAST IRON OUTLET PIPE

STANDARD NO. BR-SS(0.08)-83-146 SHEET 1 OF 1

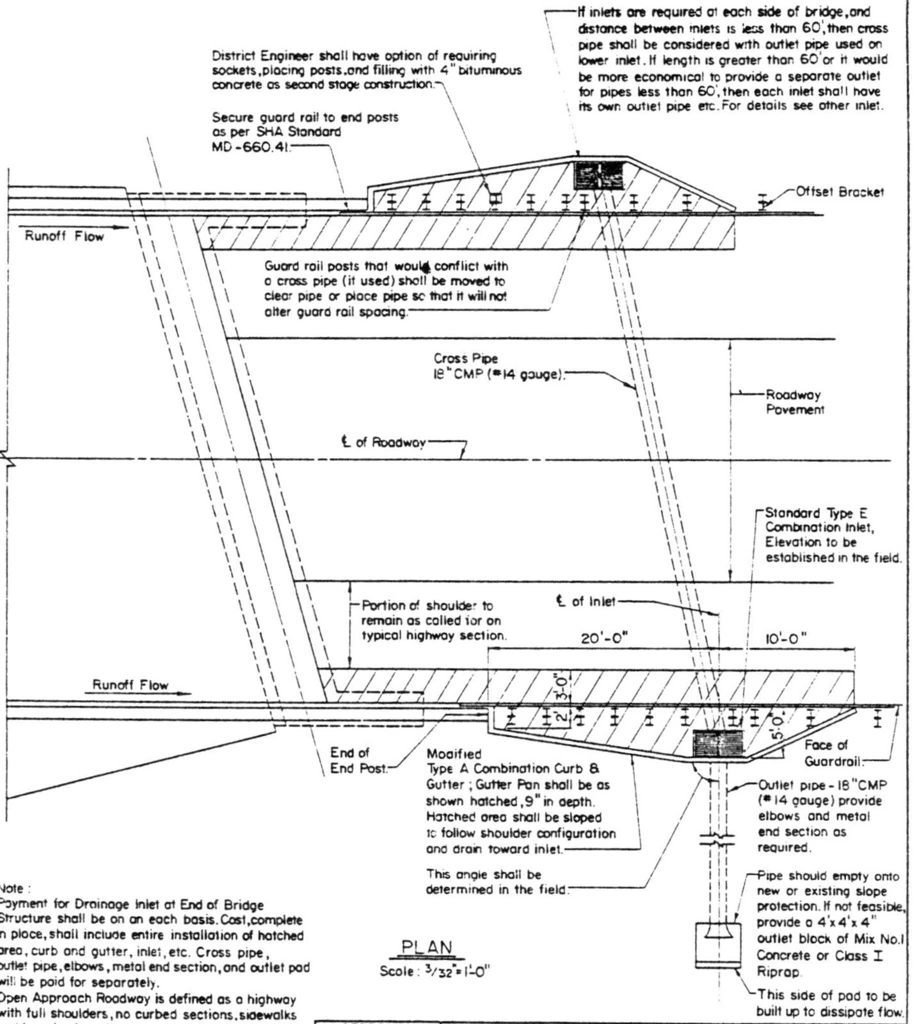


Note:
Payment for Drainage Inlet at End of Bridge Structure shall be on an each basis. Cost, complete in place, shall include entire installation of hatched area, curb and gutter, inlet, etc. Cross pipe, outlet pipe, elbows, metal end section, and outlet pad will be paid for separately.
2. Open Approach Roadway is defined as a highway with full shoulders, no curbed sections, sidewalks and/or raised medians.
3. See note on General Plan and Elevation as to which ends of structure will require inlets.
4. Guard rail posts shall be driven prior to placing concrete gutter pan except if option above is utilized.
5. Details shown on Standards MD 660.21, MD 660.22 & MD 660.23 shall be modified when used in conjunction with standard.

| APPROVAL | |
|---|---|
| E.S. Freedman ASST. CHIEF ENGR. BRIDGE DEVEL. | |
| DATE 12/23/80 | |
| REVISIONS | |
| SHA | FHWA |
| 4-20-81 | 5-6-81 |
| 11-13-81 | 12-18-81 |
| 3-7-83 | |
| 9-8-83 | |
| FHWA APPROVAL | |
| DATE 5-6-81 | |

**STATE OF MARYLAND**
**DEPARTMENT OF TRANSPORTATION**
**STATE HIGHWAY ADMINISTRATION**
DIVISION OF BRIDGE DEVELOPMENT
DRAINAGE INLET AT END OF BRIDGE STRUCTURE (OPEN APPROACH ROADWAY)

STANDARD NO. M(0.03)-80-123 SHEET 1 OF 1

# I.1 RESETTING BEARING

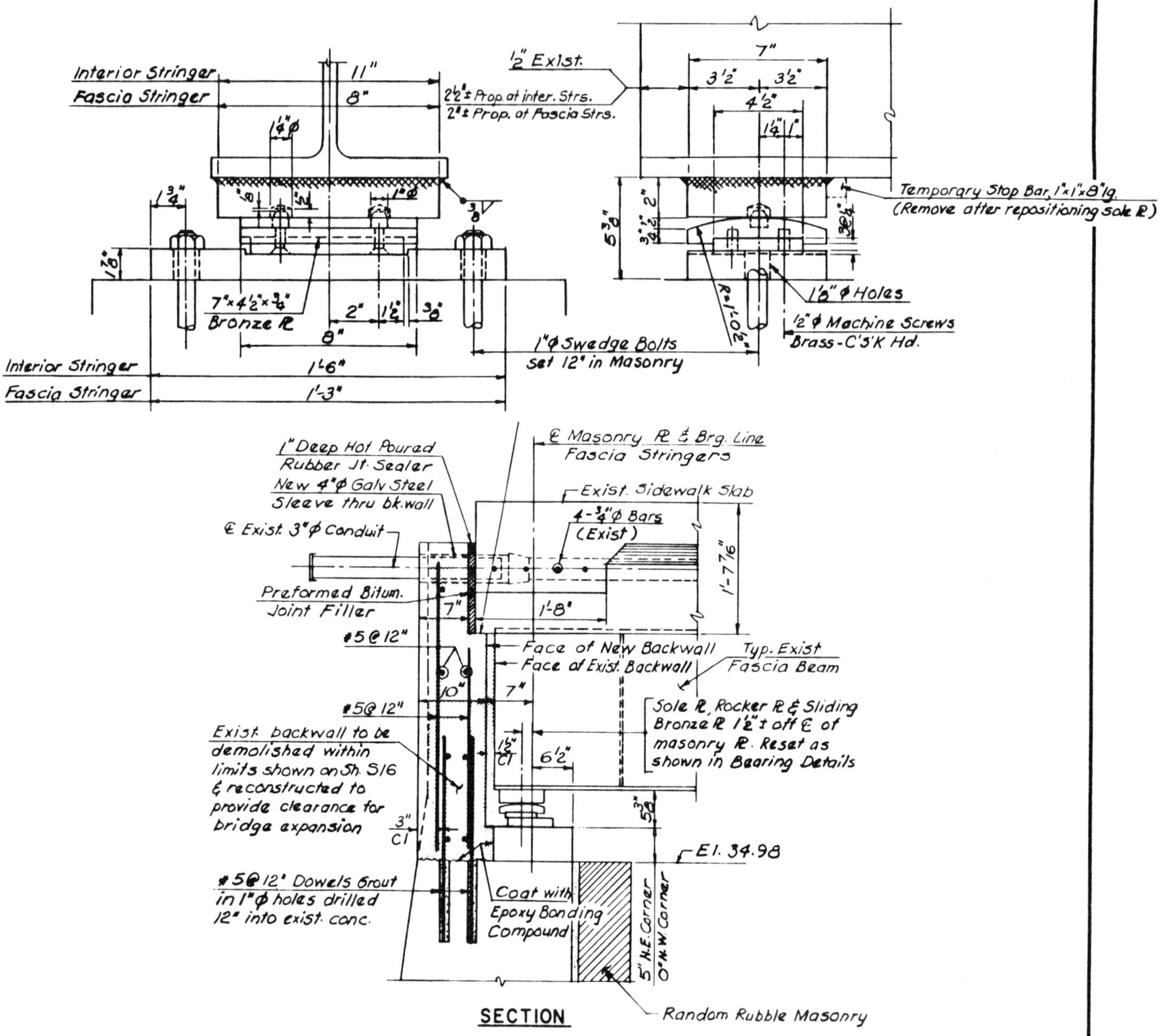


## EXISTING EXPANSION BEARINGS TO BE RESET

Scale: 3"=1'-0"

### PROCEDURE FOR RESETTING EXIST. EXPANSION BEARINGS

1. Complete one bearing at a time. (Total of 26 existing beam bearings to be reset.)
2. Remove welds securing sole plate to stringer flange using the Air-Arc Method.
3. Install temporary stop to bottom flange of stringer for new position of sole plate
4. Jack stringer while simultaneously forcing sole plate to new position.
5. Remove Jacks & temporary stop and reweld sole R to stringer flange.
6. Exist. bearings to be reset prior to installing steelwork required by repairs to exist. longitudinal Joint.
7. Contractor to submit details of his proposed method of Jacking to the Engineer for Approval.
8. All exist. bearings, both fixed and expansion, to be cleaned and painted upon completion of jacking *

Maximum stringer reaction = 64.0ᴷ (D.L.+L.L.).

* No separate payment will be made for cleaning & painting exist. bearings, but all costs there & shall be included in the lump sum price for the Item "Resetting Exist. Expansion Bearings."

# DETAILS FOR RESETTING EXISTING EXPANSION ROCKERS

*(Applies to existing stringers which are not to be replaced)*

**PROCEDURE:**

1. Determine amount by which each rocker is out of plumb at 68° (dimension A) and check dimensions marked by asterisks in Section A-A.
2. Fabricate new masonry PLs with pintles offset from ℄ of anchor bolts so rockers will be plumb at 68°F when new PLs are installed.
3. Jack entire span simultaneously at South Abutment enough to remove existing rockers and masonry PLs. Take precautions so rockers do not fall off bridge seat.
4. Install new masonry PLs, reinstall rockers and lower span.

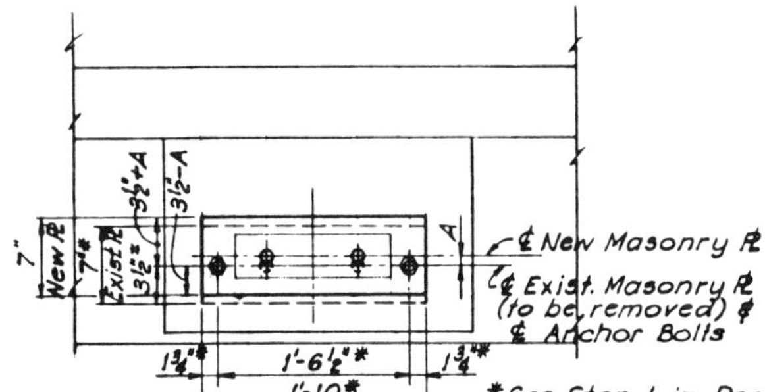


SECTION A-A
Scale: 1"=1'-0"

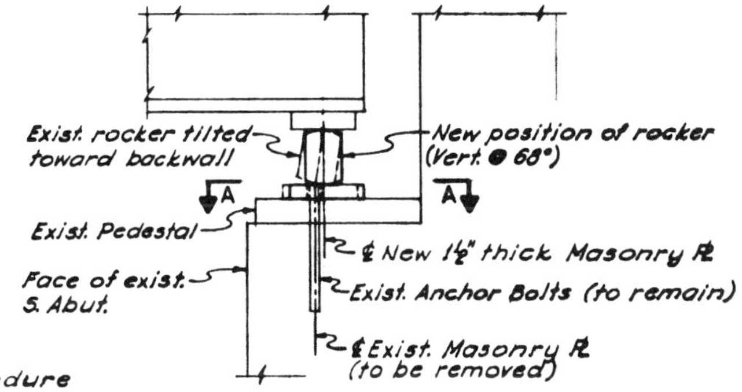


ELEVATION

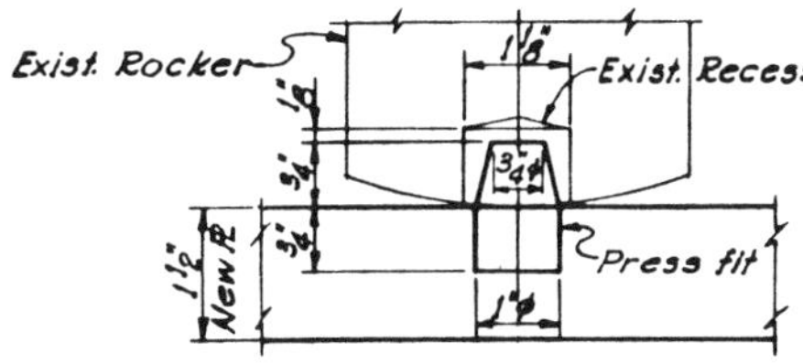


PINTLE DETAIL
Half Size

**NOTES:**

1. No traffic to be allowed on span while it is being raised or lowered by jacks.
2. Immediately after jacking, a ramp of asphalt shall be placed against the raised edge of deck to provide a smooth transition on and off the bridge. Remove asphalt immediately after span is lowered.
3. All jacks to be controlled by a common manifold so as to prevent racking of steelwork or cracking of deck slab.
4. Contractor shall submit detailed jacking procedures to the Engineer for approval before starting any work.
5. Existing rockers shall be reset before proceeding with any other items of work on the superstructure.

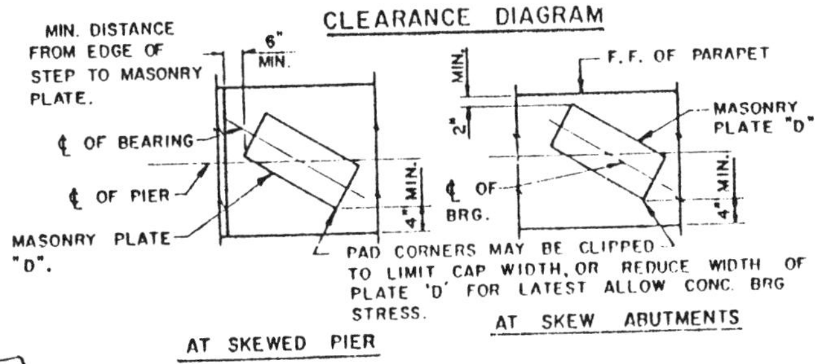


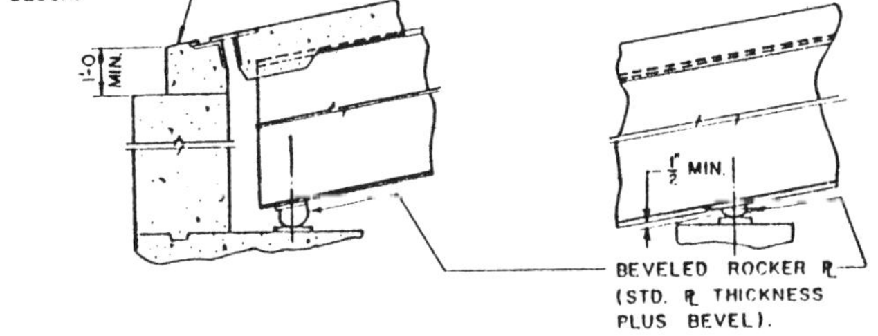


BEVELED ROCKERS OVER 3% GRADE

# I.2 RETROFIT FOR EARTH QUAKE

**From Earthquake Resistant Bridge Bearing FHWA RD 82 165**

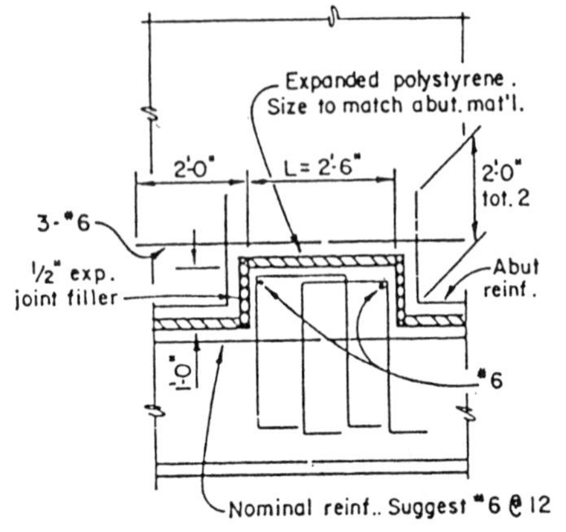


a. INTERNAL KEY

Note: The use of internal key should be avoided when possible because of the difficulty in repairing them when damaged.

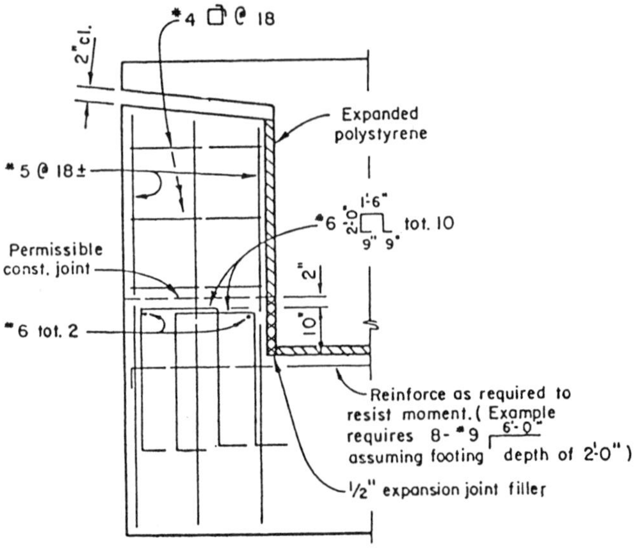


b. EXTERNAL KEY

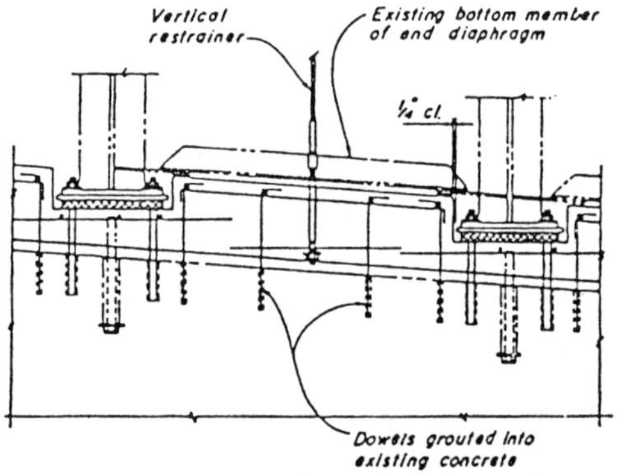


*EXPANSION ROCKER BEARING*
*PART ELEVATION*

RETROFIT DETAILS FOR TRANSVERSE RESTRAINT USING A REINFORCED CONCRETE SHEAR KEY

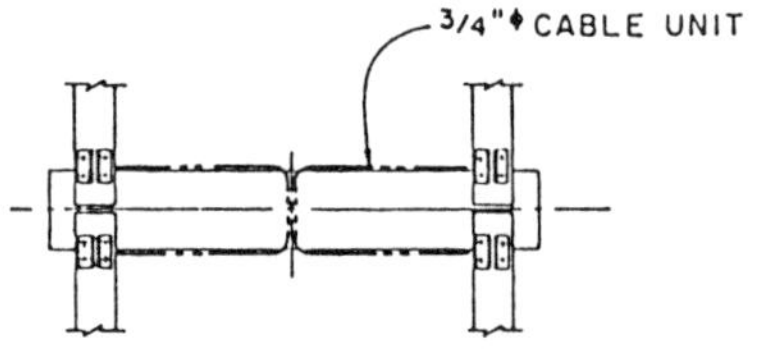


PART PLAN

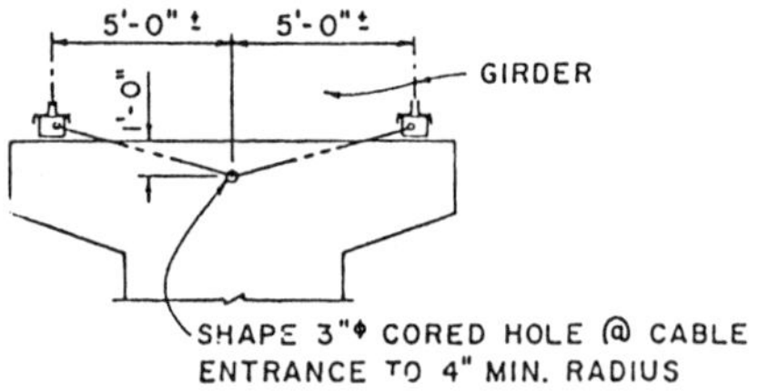


ELEVATION

RETROFIT DETAILS FOR TRANSVERSE RESTRAINT USING STEEL CABLE UNITS

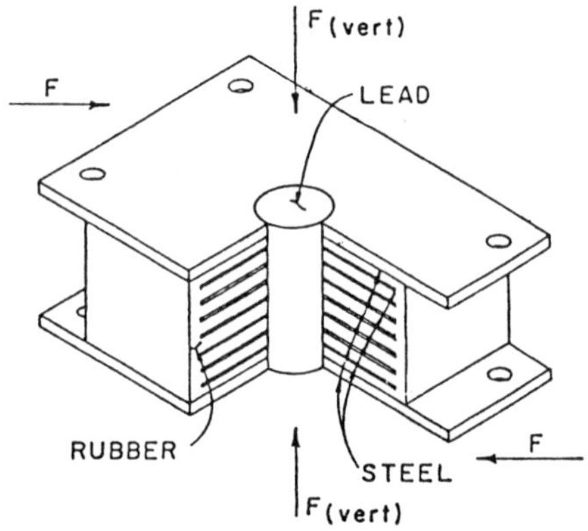


LEAD-RUBBER ENERGY DISSIPATING BEARING DEVICE

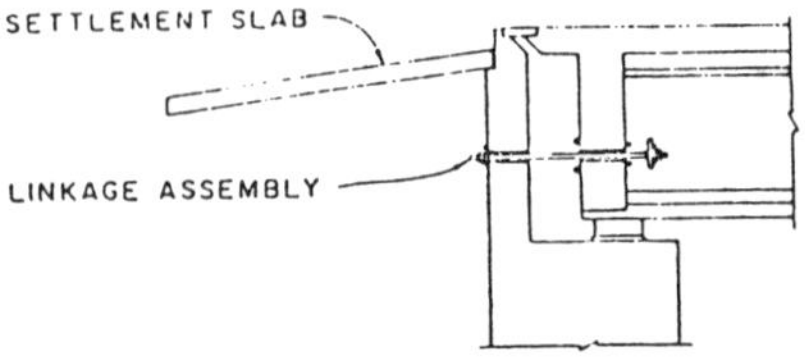


a. ABUTMENT BACKWALL CONNECTION AND SETTLEMENT SLAB

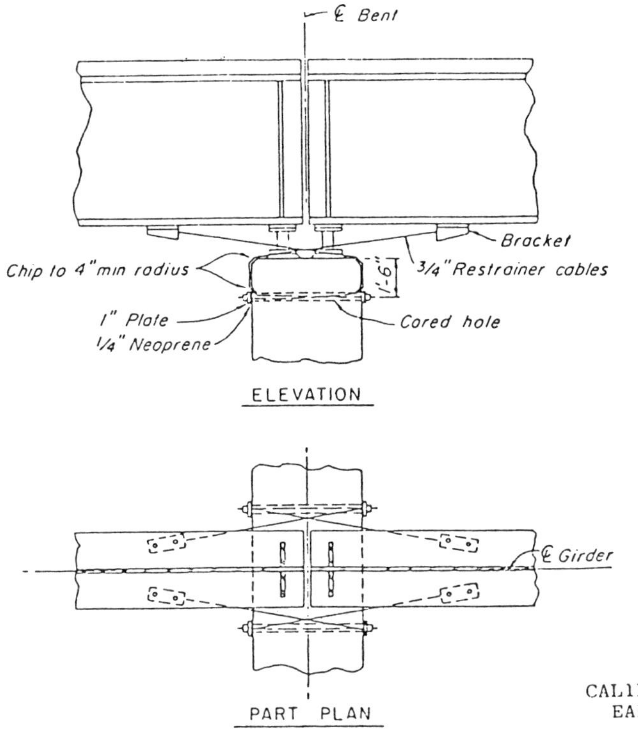


Restrainer Unit Details for Retrofitting Simply Supported I-Girder Spans Tied to Common Cap Member

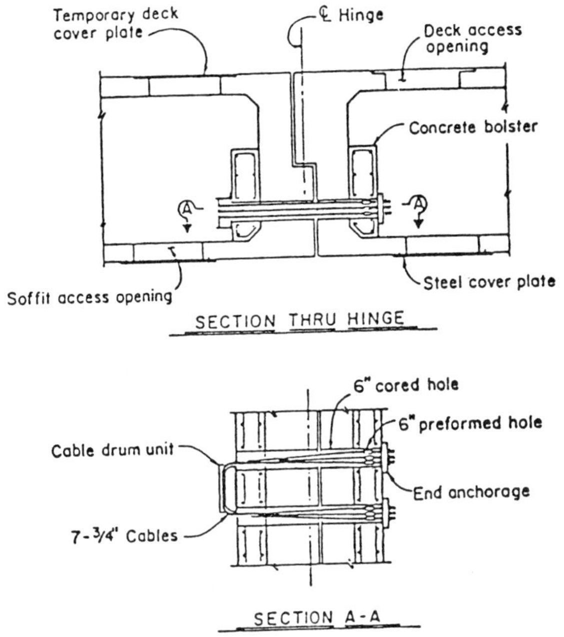


CALIFORNIA DEPARTMENT OF TRANSPORTATION DESIGN DETAILS FOR EARTHQUAKE RESTRAINER UNITS FOR COMPLETED CONSTRUCTION

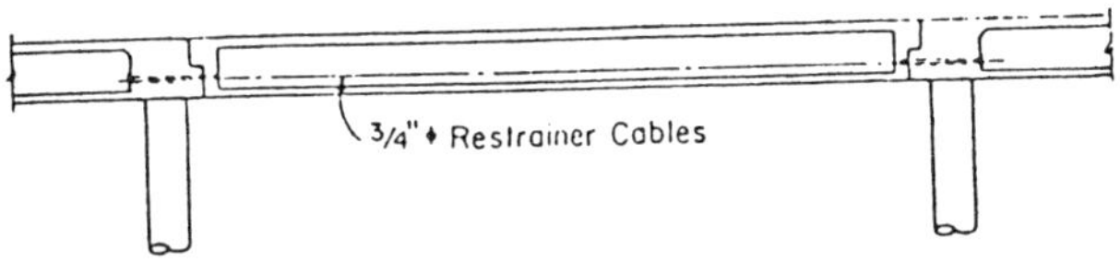


Hinge Retrofit Details for Bridge Drop-In Span Units.

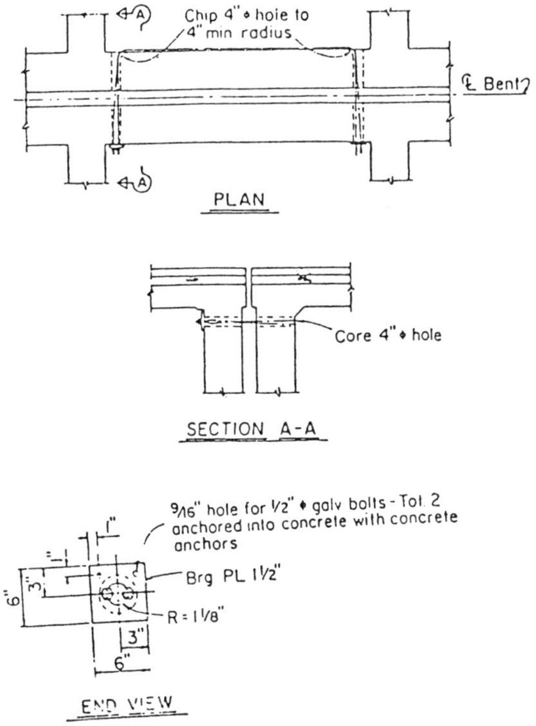


Hinge Retrofit Details for Concrete T-Beam Bridge

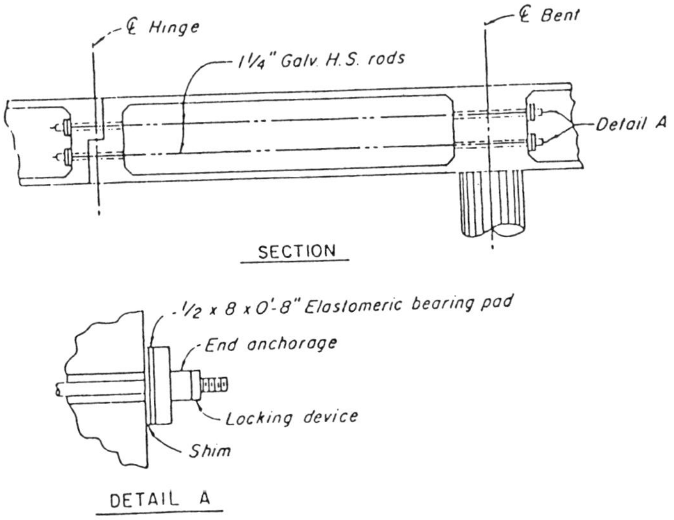


Bridge Hinge Details for Retrofitting with High Strength Steel Rods

## DESIGN EXAMPLE

A design example for the rocker bearing employing A588 steel is given to show how the dimensions in Standard 27.3 were arrived at and to assist the designer if a bearing capacity greater than 1500 kips is required.

(1) Sole Plate Design

Given the following values: N = 1200 kips, G = 2'-3, K = 2'-6, and R = 1'-9.

A uniform load distribution over a width equal to 9" is used due to the paired stiffener spacing; the sole plate uniform loading is equal to

$$Pu = \frac{N}{(G)(9")} \quad \text{where}$$

N equals the bearing reaction and G equals the length of the sole plate.

$$Pu = \frac{1200^k}{(27")(9")} = 4.94 \text{ k/in}^2$$

An assumed web to sole plate contact width of 1/2 inch is used and the approximation that the load spreads through the sole plate at an angle of 45 degrees as shown below. The uniform loading within the dashed lines is transferred from the sole plate to the web in direct axial stress.

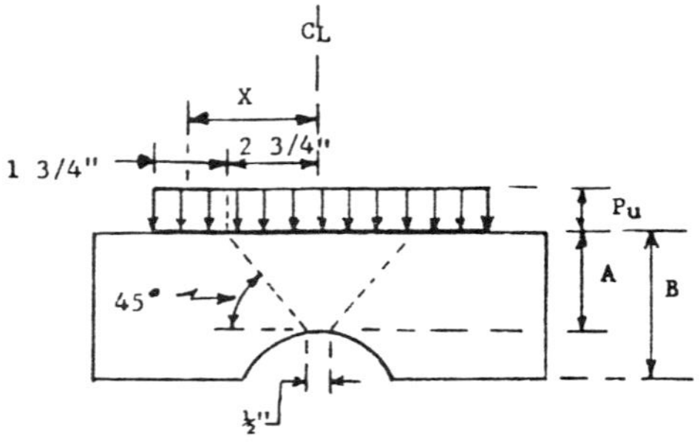


Sole Plate Loading

The value of "A" is assumed as 2 1/2 inches and for geometric reasons the "B" dimension is 4 inches. The "A" dimension is used to determine the width of the load transferred in direct stress and checked in bending from the moment of the loading transferred in bending about the centerline of the sole plate. Using a 1 inch strip, the moment, M, is as follows:

$$M = (1.75")(Pu)(X) \text{ where X equals}$$

$$(5.5 + 1.75)/2 = 3.625"$$

$$M = (1.75")(4.94 \text{ k/in})(3.625")$$

$$M = 31.34 \text{ in-k.}$$

The required sole plate thickness, "A", using an overstress of 50 percent is

$$A = \sqrt{\frac{6M}{1.5\ F_s}}$$

$$A = \sqrt{\frac{(6)(31.34 \text{ in-k})}{(1.5)(27.0 \text{ ksi})}}$$

$$A = 2.16".$$

USE: A = 2 1/2"

(A) Contact Stresses

Contact stresses are included for informational purposes and are not part of the design procedure.

The contact width, b, and the maximum contact stress, $S_c$, are computed from "Formulas For Stress and Strain" by R. J. Roarke, pages 318 through 329. Assuming the modulus of elasticity, $E_1 = E_2$, and Poisson's ratio $V_1 = V_2 = 0.30$, then the contact width equals

$$b = 2.15 \sqrt{\frac{P}{E}\ \frac{D_1 D_2}{D_1 - D_2}} \quad \text{where}$$

p equals the load per linear inch, $D_1$ equals the diameter of the sole plate, and $D_2$ equals the diameter of the web plate.

$$p = \frac{N}{G} = \frac{1200^k}{27"}$$

$$p = 44.44 \text{ k/in}$$

$$\text{If } E = 29{,}000 \text{ ksi},$$

$$D_1 = 4.406" \;\&\; D_2 = 4.375";$$

$$b = 2.15 \sqrt{\frac{44.44}{29000}\ \frac{(4.406)(4.375)}{(4.406-4.375)}}$$

$$b = 2.10" > 1/2" \text{ assumed for sole plate comp.}$$

$$S_c(max) = 0.591 \sqrt{pE\ \frac{D_1 - D_2}{D_1 D_2}} \quad \text{from Roarke;}$$

$$S_c(max) = 0.591 \sqrt{\frac{(44.44)(29000)(4.406-4.375)}{(4.406)(4.375)}}$$

$$S_c(max) = 26.91 \text{ ksi.}$$

The actual contact stress equals

$$S_c(act) = \frac{N}{(G)(b)}$$

$$S_c(act) = \frac{1200\ ^k}{(27")(2.10")} = 21.16 \text{ ksi}^* \text{ which}$$

for comparison is less than $S_c(max)$.

Web And Stiffener Design

Given the following information: Maximum rocker movement, AB = 5", C = 4", D = 1'-6, and E = 4".

Refer to the following sketch for bearing movement and resultant forces.

* AASHTO Specifications allow up to 0.40 $F_y$ bearing stress on rockers and pins subject to rotation.

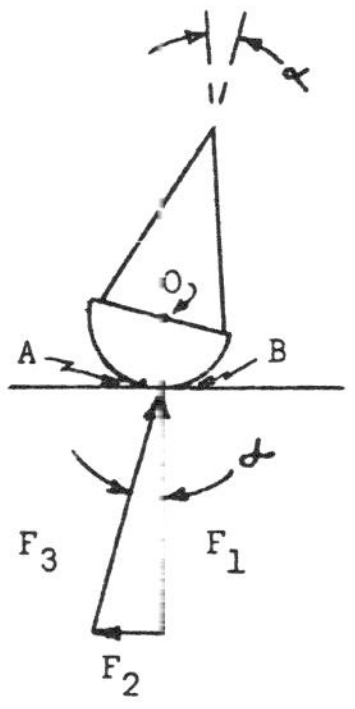


$$\alpha = \left(\frac{AB}{R}\right)\left(\frac{180}{\pi}\right)$$

$$\alpha = \frac{5''}{21''}(57.3^{o}) = 13.64^{o}$$

$$F_1 = 1200^{k}$$

$$F_2 = (1200)(\sin 13.64^{o})$$

$$F_2 = 283.0^{k}$$

$$F_3 = (1200)(\cos 13.64^{o})$$

$$F_3 = 1166.2^{k}$$

The summation of moments about point 0, neglecting the variable friction component, equals

$$(+)\ M_o = (5.0'')(1166.2^{k}) - (4.0'')(283.0^{k})$$

$$M_o = 5831.0 - 1132.0 = 4699.0^{in-k}.$$

The available section modulus at the web-rocker plate intercept, assuming 2 inch thick stiffeners, is as follows:

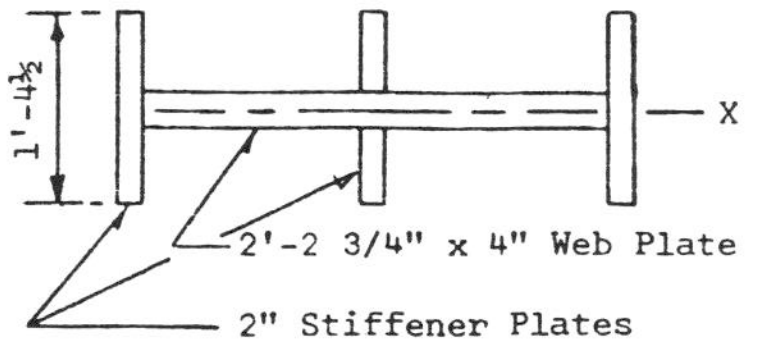


$$I_x = (26.75)(4)^3/(12) + (2)(2)(16.5)^3/(12) + (2)(16.5^3-4^3)/12$$

$$I_x = 142.67 + 1497.38 + 738.02 = 2378.07\ in^4$$

$$S_x = \frac{2378.07}{8.25} = 288.2\ in^3$$

$$Area = (26.75)(4) + (2)(2)(16.5) + (2)(2)(6.25) = 198.0\ in^2$$

The combined axial and bending stress, $f_s$, equals

$$f_s = \frac{F_3}{A} \pm \frac{M_o}{S_x}.$$

$$f_s = \frac{1166.2}{198.0} \pm \frac{4699.0}{288.2}$$

$$= 5.9\ ^{ksi} \pm 16.3\ ^{ksi}$$

$f_s$ (max) = $22.2^{ksi}$ (C) Less than $1.25F_s$

$f_s$ (min) = $10.4^{ksi}$ (T) Less than $1.25F_s$

USE: 4" Web and 2" Stiffener Thickness

(3) Weld design at web-rocker intercept

The fillet weld size will be determined using E-70 electrodes and an overstress of 25 percent. The overstress is allowing for weld metal at the web-rocker intercept using the same reasoning as discussed in Section 27.3(2) of this chapter for the steel overstress. The allowable stress for the weld is 14,700 psi as recommended by AASHTO, Section 7, Article 1.7.2.

The approximate fillet weld length, $L_w$, equals

$$L_w = (2)(26.75) + (2)(16.5) + (2)(12.5) + (6)(2) + (4)(6.25)$$

$$L_w = 148.5^{in}.$$

The shear component of stress is

$$f_v = \frac{F2}{L_w} = \frac{283.0^{k}}{148.5''}$$

$$f_v = 1.91\ ^{k/in}.$$

The bending moment component for stress is computed from the combined stresses;

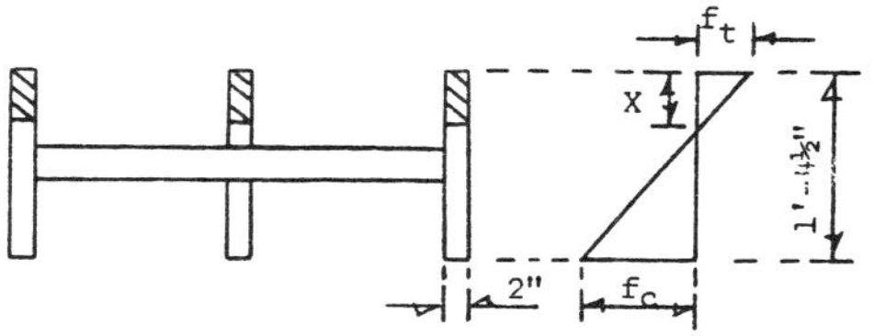


By similar stress triangles the following equation is

$$\frac{f_t}{X} = \frac{f_c}{16.5 - X}$$

Where $f_t = 10.4^{ksi}$ and $f_c = 22.2^{ksi}$; solving for X,

$$10.4(16.5 - X) = X(22.2),$$

$$171.6 - 10.4X = 22.2X,$$

$$X = 5.26''.$$

The weld metal must resist the following force

$$F = (3)(2")(10.4^{ksi})(5.26")/(2) = 164.1^{k}$$

and the resulting moment of

$$M = (164.1^{k})(5.26")(2/3) = 575.4^{in\text{-}k}.$$

The moment of inertia of the weld lines is computed as follows:

$$I = (6)(5.26")^3/(3) + (3)(2")(5.26")^2$$

$$I = 291.1 + 166.0$$

$$I = 457.1^{in^3}$$

The bending moment component, $f_b$, equals

$$f_b = \frac{(575.4^{in\text{-}k})(5.26")}{457.1^{in^3}}$$

$$f_b = 6.62^{k/in}.$$

The resultant stress, $f_r$, is equal to

$$f_r = \sqrt{(f_v)^2 + (f_b)^2} = \sqrt{(1.91)^2 + (6.62)^2}$$

$$f_r = 6.89^{k/in}.$$

The weld size, $W_s$, is obtained from the allowable weld metal stress as follows:

$$(0.707)(14.7^{ksi})(1.25)(W_s) = 6.89^{k/in}$$

$$W_s = 0.53^{in}.$$

Refer to Std. 27.3

USE: 3/4" Fillet Weld Size

(4) Rocker Plate Design

The required rocker plate length, F, is determined from the equation

$$F = \frac{N}{P} + 2 \text{ (Pintle dia.) where}$$

P is the allowable line bearing and N is the bearing reaction. The allowable line bearing is dependent on the allowable yield stress, $F_y$, and the diameter, d, of the rocker plate. Refer to AASHTO, Section 7, Article 1.7.4 for the following equations:

If $d \leq 25"$, $p = \frac{F_y - 13000}{20000}(600\ d)$ and

if $25" > d \leq 125"$, $p = \frac{F_y - 13000}{20000}(3000\sqrt{d})$.

The allowable line bearing value, p, is computed assuming d = 42 inches, and for A588 steel, $F_y$ = 50 psi.

$$p = \frac{50000 - 13000}{20000}(3000\sqrt{42})$$

$$p = 35.97\ k/in$$

Referring to the previous equation for the required rocker plate length, F, equals

$$F = \frac{1200^{k}}{35.97\ k/in} + 2(2\ 1/2")$$

$$F = 33.4 + 5.0 = 38.4"; \text{ however,}$$

$$F \geq G + 6" = 2'\text{-}3 + 6" = 2'\text{-}9.$$

USE: 3'-3 Line Bearing

The rocker plate thickness is computed by assuming the reaction, N/2, acts at an eccentricity of one-fourth the rocker width, D/4, from its centerline. The bending moment is equal to (N/2)(D/4) and the section modulus is equal to $(F)(E)^2/(6)$ where E is the required rocker plate thickness.

If $S = \frac{M}{F_s}$ by substituting

$$\frac{(F)(E)^2}{6} = \frac{(N/2)(D/4)}{F_s} \quad \text{then,}$$

$$(E)^2 = \frac{6(N)(D)}{8(F)(F_s)},$$

$$E = \sqrt{\frac{(3)(1200)(18)}{(4)(39)(27)}} = 3.92"$$

USE: 4" Rocker Plate Thickness

(5) Masonry Plate Design

The masonry plate dimensions are computed neglecting the small effect of the area loss due to pintle and anchor bolt holes. The masonry plate length, L, is designed on the basis of an assumed width, K, and a uniform pressure distribution of load between the plate and concrete. Using an allowable concrete masonry bearing stress of 800 psi, the required plate length is

$$L = \frac{N}{(0.8)(K)} = \frac{1200^{k}}{(0.8)(30")} = 50.0".$$

USE: 4'-2 Masonry Plate Length

The masonry plate thickness, M, in the longitudinal direction of the structure, is based on a pressure diagram equal to 2P average triangular at the centerline of the masonry plate. The triangular pressure diagram is used as a better approximation of the load distribution under the rocker plate due to possible yielding of the masonry plate edges. The pressure diagram is given as follows.

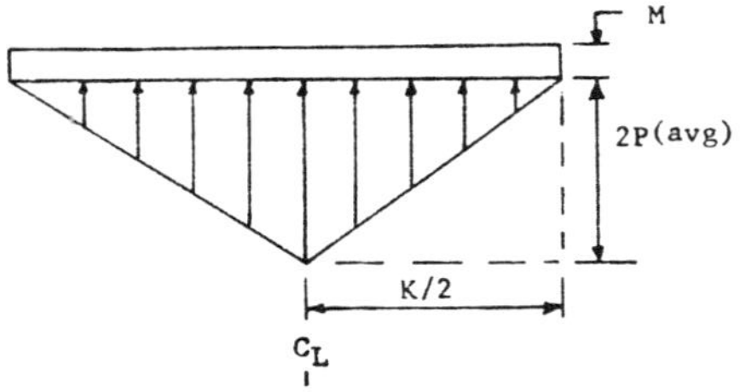


The maximum effective length, $L_r$, for masonry plate thickness design is equal to

$$L_r = F + 3M$$

where M is approximated for the first trial computation. $L_r$ must also be equal to or less than L, the actual masonry plate length.

A trial value of M = 4", gives

$$L_r = 3'\text{-}3 + 3(4") = 4'\text{-}3 > L \quad \text{N.G.}$$

Therefore, $L_r = L = 4'\text{-}2.$

The value of $P(avg) = \frac{N}{KL_r}$ and from the pressure diagram the bending moment, $M_b$, equals

$$M_b = (1/2)(2P(avg))(K/2)(K/6)(L_r)$$

$$M_b = (1/12)(P(avg))(K^2)(L_r)$$

For a rectangular shape, the section modulus, S, equals

$$S = (M^2)(L_r)/(6).$$

Substituting into the flexural equation

$$S = \frac{M_b}{F_s},$$

$$\frac{(M^2)(L_r)}{6} = \frac{(1/12)(P(avg))(K^2)(L_r)}{F_s},$$

substituting $N/KL_r$ for P(avg) and dividing out common terms, the plate thickness is

$$M = \sqrt{\frac{NK}{2F_sL_r}}$$

$$M = \sqrt{\frac{(1200^k)(30")}{(2)(27.\text{ksi})(50")}} = 3.65".$$

USE: 4" Masonry Plate Thickness

The end projection beyond the rocker plate is also checked for bending using a uniform pressure distribution equal to

$$P(avg) = \frac{N}{KL} = \frac{1200^k}{(30")(50")} = 0.80 \text{ ksi}.$$

Refer to the following diagram for end projection.

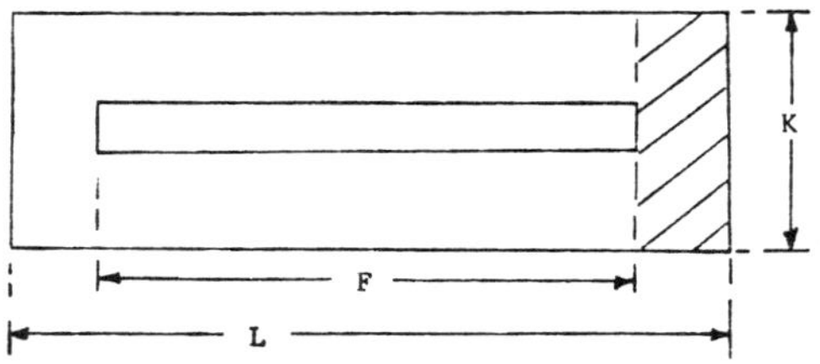


End Projection = (L-F)/2

= (50-39)/2

= 5.5" = $L_e$

The bending moment, $M_b = (1/2)(P(avg))(K)(L_e)^2$ and the section modulus equals

$$S = \frac{KM^2}{6}.$$ Solving for M in the flexural equation gives

$$M = \sqrt{\frac{(3.0)(P(avg))(L_e)^2}{F_s}} \qquad M = \sqrt{\frac{(3.0)(0.80 \text{ ksi})(5.5")^2}{27 \text{ ksi}}} = 1.64".$$

The previous masonry plate thickness requirement of 4 inches governs.

(6) Pintle Design

Assume a reaction of 1,400 kips, an average depth of 18 feet, and a span length of 360 feet.

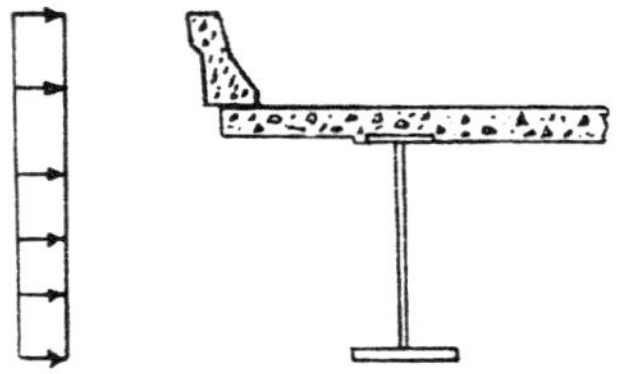

Transverse Wind Loading

From AASHTO Specifications Article 1.2.14(B), transverse wind loading is given as 50 pounds per square foot for girder type structures. The total transverse wind force is equal to (.050)(18)(360) = 324 kips. Assume one half of the total force is transferred to the remaining girders and that a full bearing type pintle connection is used with an allowable shear of 20 ksi.

The required pintle area = 162 kips/20 ksi = 8.1 square inches. Therefore, 2 - 2 1/2 inch diameter pintles per bearing are adequate.

Note that this procedure is on the conservative side since the dead load friction is not considered in resisting part of the wind force. However, if dead load friction is considered, an assumption is required for wind uplift on the superstructure which tends to lower the dead load reaction.

# I.4 ELASTOMERIC BEARING DESIGN

V. DESIGN PROCEDURE

A. FLOW CHART

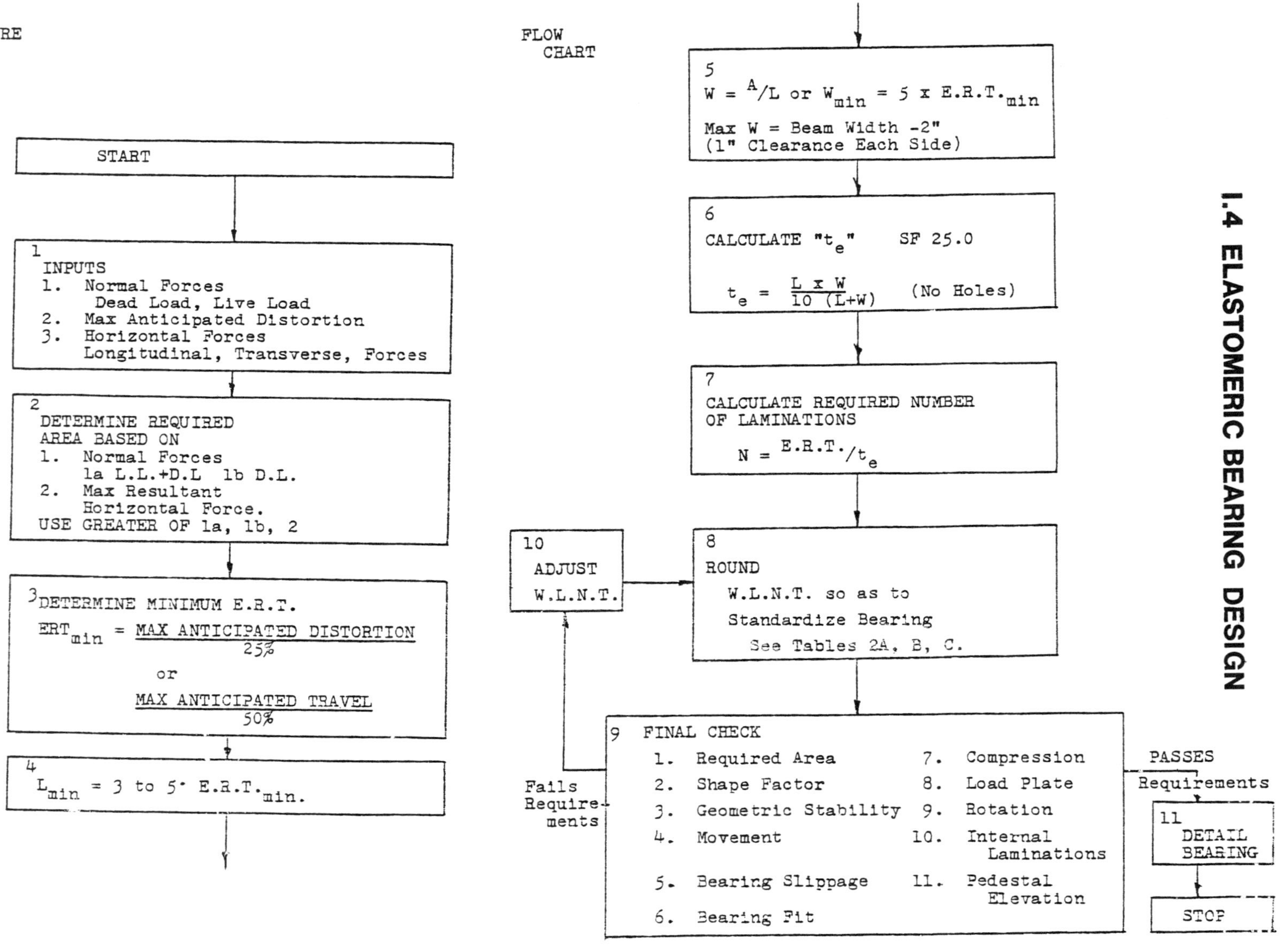


FROM DESIGN MANUAL FOR ELASTOMERIC BEARING
MARCH 1979, NEW YORK STATE DEPT. OF TRANS.

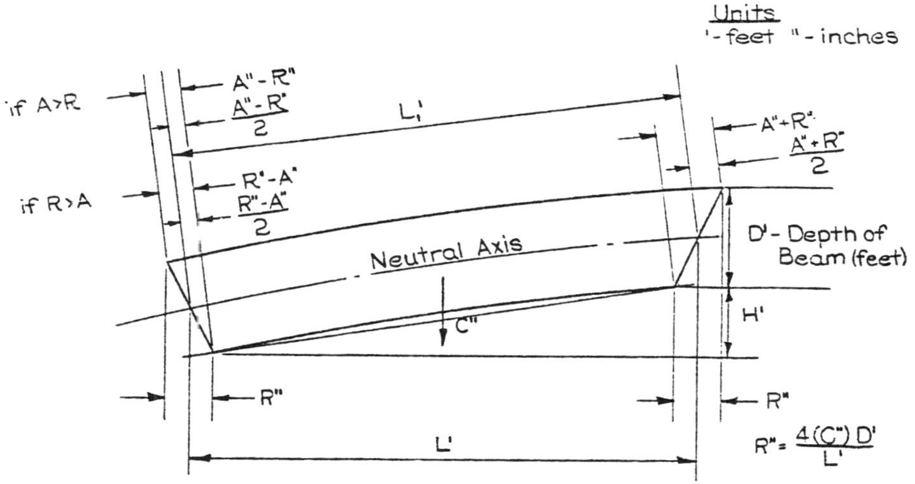


GIRDER BEFORE DEFLECTIONS HAVE TAKEN PLACE

C" = Deflection Due to Permanent Loads on the Beam i.e. S.D.L. D.L.

Note: Ends of Girder Vertical

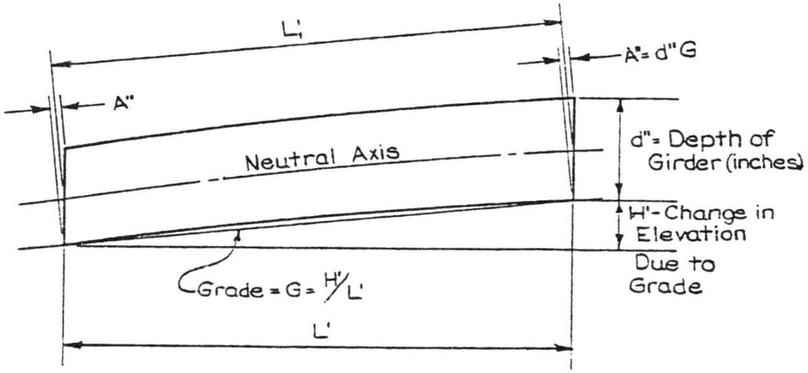


GIRDER AFTER DEFLECTIONS HAVE TAKEN PLACE

FIG. 3 MOVEMENT DUE TO DEFLECTIONS

B. FORCES — BRIDGE SEATS LEVEL (Forces at Top of Bearing)

1. Weight of Structure

Situation A - Bridge Seats Level Tapered Load Plate

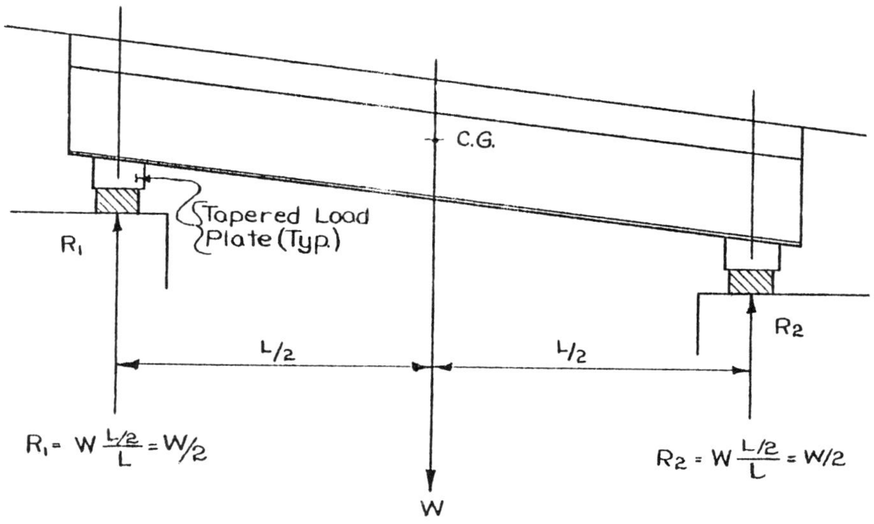


$R_1 = W\frac{L/2}{L} = W/2$

$R_2 = W\frac{L/2}{L} = W/2$

Horizontal Component of Weight = Zero

FIG. 4 FORCES — BRIDGE SEATS LEVEL

B. FORCES - BRIDGE SEATS SLOPED
(Forces at Top of Bearing)

1. Weight of Structure

Situation B.- Bridge Seats Parallel to Grade of Road.
No Tapered Load Plate
No Mechanical Fixation

Note:
It is contrary to present Office Policy to slope bridge seats. By sloping the bridge seats (Situation B) the bearings are subjected to a constant, non-fluctuating gravity shearforce.

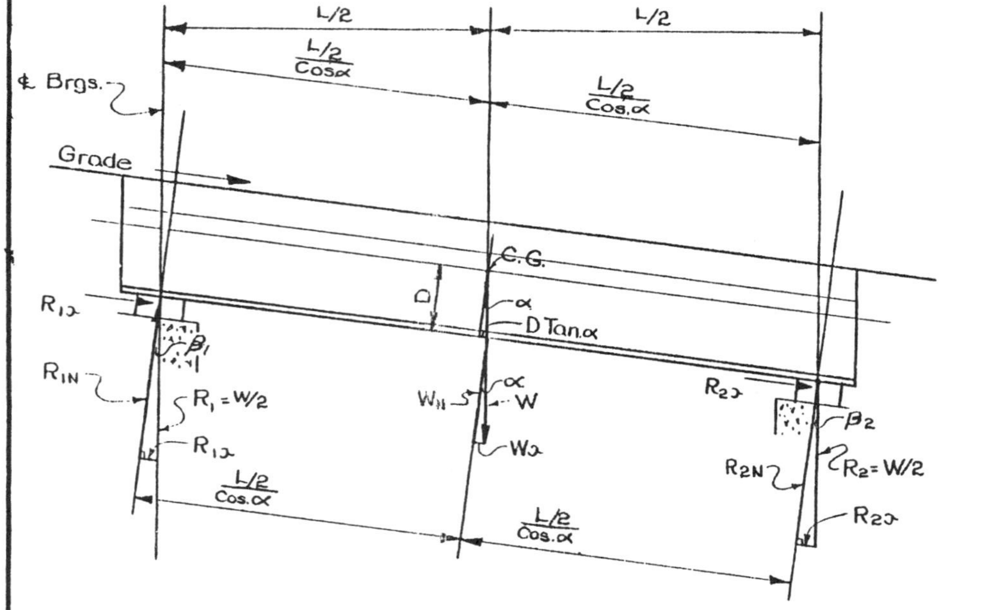


$R_{1\nu} = R_1 \sin \beta_1$ $R_{2\nu} = R_2 \sin \beta_2$

Notes:
1. - $\beta_1 = \beta_2 = \alpha$
2. $W_\nu = R_{1\nu} + R_{2\nu}$ $R_{1\nu} = R_{2\nu}$ ∴ $R_\nu = W_\nu/2$
$R_{IN} = R_{2N}$

FIG. 5. FORCES BRIDGE SEATS SLOPED

2. TRACTIVE FORCE

Ref: 1.2.13 LONGITUDINAL FORCES
N.Y.S. Standard Specification for Highway Bridges

2a TRACTIVE FORCE - No positive point of fixation

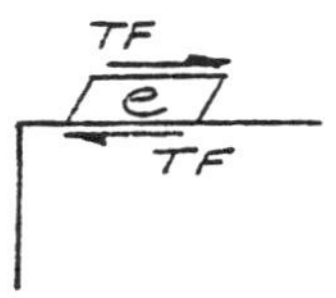


$$TF = 5\% \ \frac{LLR \cdot R}{N}$$

TF = Tractive force per bearing

LLR = Live load in all lanes carrying traffic headed in the same direction.

R = Load reduction factor

N = Number of Bearings Resisting Tractive Force

REF.
Standard Specification for Highway Bridges
1.2.9 Reduction in Load Intensity
1.2.13 Longitudinal Force

2b TRACTIVE FORCE - Positive point of fixation

If span is fixed at one end, the entire tractive force is transferred to the abutment by these fixed bearings.

$$TF^* = 5\% \ \frac{LLR \ R}{F}$$

TF* = TRACTIVE force per Fixed bearing
F = Number of FIXED Bearings

3. WIND FORCES

REF. 1.2.14 WIND LOADS
N.Y.S. Standard Specification for Highway Bridges

4. EARTHQUAKE FORCES

REF. 1.2.20 EARTHQUAKE STRESSES
N.Y.S. Standard Specification for Highway Bridges

5. CENTRIFUGAL FORCES

REF. 1.2.21 CENTRIFUGAL FORCES
N.Y.S. Standard Specification for Highway Bridges

6. Beam Rotation for live load Deflection

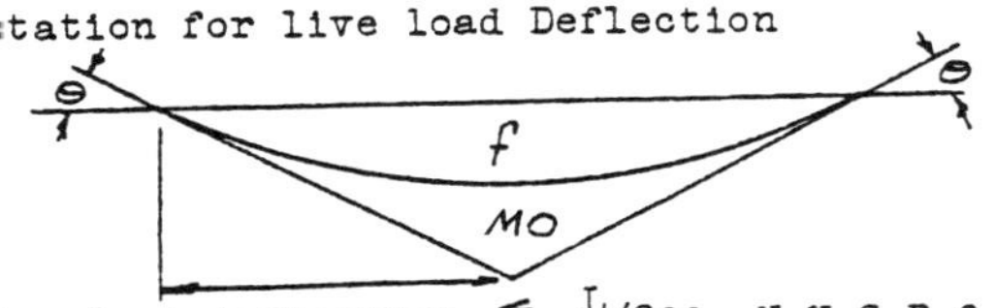


f = live load deflection $\leq L/800$ N.Y.S.D.O.T. Spec. 1.7.13

f = MC (By definition of a Parabola)

ANGLE OF ROTATION θ (radian)

$$\theta = \frac{2f}{L/2} = \frac{4f}{L} \text{ radians}$$

Compression of Bearing due to live load deflection

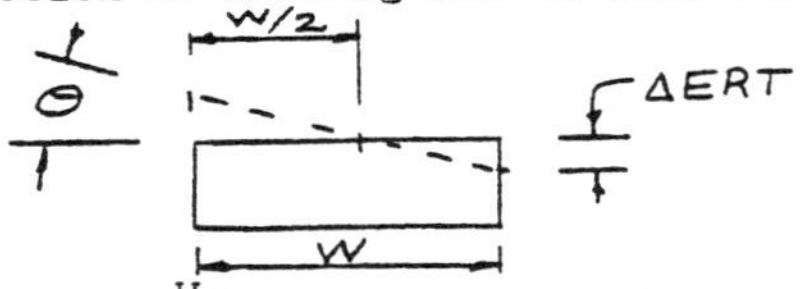


$\Delta E.R.T. = \theta \, W/2$   E.R.T. = Effective Rubber Thickness

% Compression of E.R.T. due to live load deflection $= \frac{\theta \, W/2}{ERT}$

7. Shear Force - per bearing.

$$G = \frac{(t)\ (\Delta P_s)}{ab\ \ (\Delta d)} \text{ psi},$$

where

| | | |
|---|---|---|
| G | = shear modulus, (Reference Fig. 1) | psi |
| t | = sample thickness, | inches |
| a | = sample width, | inches |
| b | = sample length, | inches |
| $\Delta P_s$ | = shear load over part of curve, | pounds |
| $\Delta d$ | = shear deflection over same part of curve, | inches |

III MATERIAL

A. GENERAL REQUIREMENTS*

1. ELASTOMER

a. Compression strain of each bearing shall not exceed 5% of the effective rubber thickness at 500 psi, nor 8% of the effective rubber thickness at 800 psi. For bearings designed for stresses higher than 800 psi, compression strain shall not exceed 8% at the higher stress.

b. Hardness (Shore A Durometer) . . . . . . 50.0 ± 5

c. The bearing shall be conditioned at -20 ± 1°F for 96 ± 1 hours and subject to a compressive force of 500 psi and a shear force such that the bearing exhibits a shear strain of 25% of the effective rubber thickness. The shearing force required to produce the 25% shear shall not exceed 50 psi when measured 15 minutes after the ultimate shear strain is achieved.

2. Steel sheets and Bonded Load Bearing Plates.

a. The internal steel plates shall be rolled mild steel. t ≥ 3/16" A.S.T.M. A36 or A282
t < 3/16" A.S.T.M. A242 or A569

b. External load bearing plates shall conform to the requirements of A.S.T.M. A-36.

3. Laminated Elastomeric Bearings shall be composed of multiple laminates of elastomeric material separated by steel plates.

*Excerpts from Material Specification

a. Internal Steel Plates
$t_s$ minimum = 0.075" (14 gage)

There shall be a minimum of 1/8" edge seal, integral with the bearing over all internal plates. No rubber cover will be required over internal details which will not be exposed to moisture after erection, specifically vertical holes covered by bearing seats or flanges.

b. External Load Plates
Steel - 3/4" minimum thickness - Steel or Concrete Superstructures
Grout Pad - 2" minimum thickness - Concrete Superstructures

Bearings shall have tapered load plates when nonparallel load surfaces would otherwise produce a compressive deflection in excess of 0.06 ERT under dead load.

## IV ANALYSIS

### A MAXIMUM ANTICIPATED DISTORTION

1. Thermal - The change in length of a member as a result of the change in temperature.

Air Temperature range - $30^{\circ}F$ to $120^{\circ}F$

Erection Temperature range $40^{\circ}F$ to $90^{\circ}F$

| | Rise | Fall | |
|---|---|---|---|
| Steel Beams | $80^{\circ}$ | 120 | |
| Concrete Beams | 35** | 45** | **A.A.S.H.O. |

Coefficients of Linear Expansion

| | | |
|---|---|---|
| Concrete | 0.000 006 | in/in per °F |
| Steel | 0.000 0065 | in/in per °F |

2. Concrete Shrinkage - The change in length of concrete beams as a result of the curing of the concrete.

Assume the following percentages of the total changes in length has already taken place. (Shrinkage)

| | | % REMAINING |
|---|---|---|
| Precast members | 50% | 50% |
| Cast in Place | 0% | 100% |

Shrinkage Coefficient = 0.0002 ft/ft

3. Elastic Shortening - The change in length of concrete beams as a result of prestressing.

Assume the following percentage of the total change in length has already taken place. (Elastic Shortening Concrete)

| | | % REMAINING |
|---|---|---|
| Precast Prestressed | 50% | 50% |
| Posttensioned | 0% | 100% |

$\Delta$* Approx. = $\frac{PL}{AE}$

*ECCENTRICITY NEGLECTED

$\Delta$ = Change in Length (in)
P = Total prestressing force (#)
L = Total length (in)
A = Area a Cross Section (in
E = Modulus of Elasticity (#/in

MAD = Maximum Anticipated Distortion, the maximum movement in one direction.

MAT = Maximum Anticipated Travel. The total maximum movement in both directions.

From Portland Cement Association

$$E = 1,800,000 + 500f$$

f = Compressive Strength of the Concrete (#/in²)

| f (#/in²) | E (#/in²) |
|---|---|
| 3,000 | 3,300,000 |
| 4,000 | 3,800,000 |
| 5,000 | 4,300,000 |
| 6,000 | 4,800,000 |

4. Beam Deflection - The change in length of members as a result of deflection.

If each end is free to displace the movement R" due to deflection will be 2(C")D'/L' at each end.

If one end is Fixed in place (Displacement is prevented) the movement R" due to the release of the camber will be 4(C")D'/L' at the free end.

R" = Horizontal movement (in)

C" = Deflection due to permanent loads on the beam i.e. S.D.L D.L. (in)

D' = Depth of member (ft)

L' = Length of member (ft)

SF = SHAPE FACTOR

SF = For each laminate the shape factor is the net area of the loaded face divided by the net area free to bulge. Holes, slots and other variations shall be considered.

For a rectangular bearing with no holes or slots:

$$SF = \frac{\text{Net Area}}{t_e\ (\text{Net Perimeter})} = \frac{L_n W_n}{2t_e \ (L_n+W_n)}$$

At no time should the shape factor be less than 5.0 nor more than 12.0. Laminates with shape factors as close to 5.0 as practical are most desirable, since this reduces the cost.

ERT = Effective Rubber Thickness - Summation of the $t_e$'s with $5.0 < SF < 12.0$.

T = Total thickness of elastomer and plates.

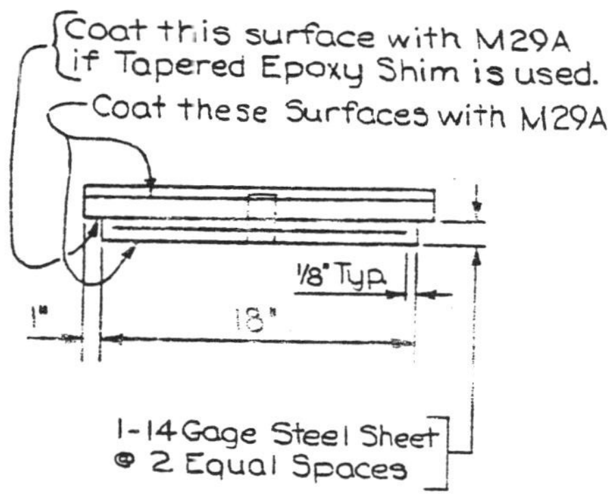


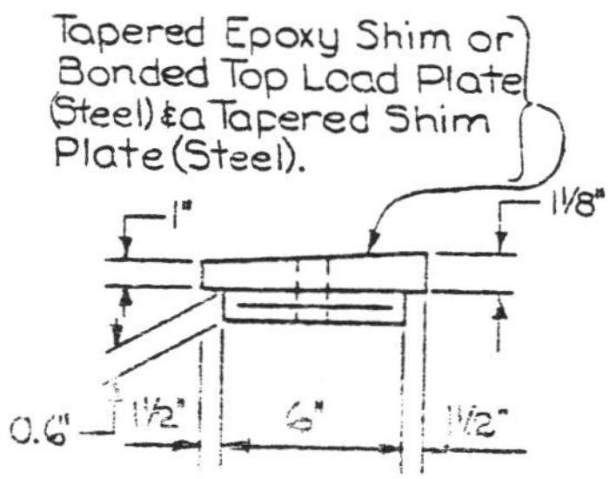


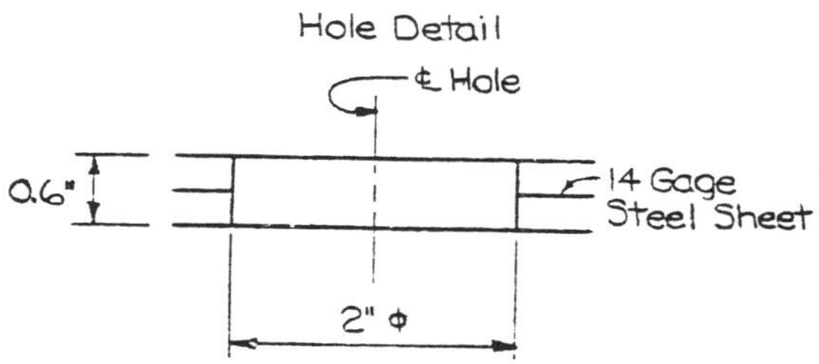


VI DETAILS:

SUGGESTED DETAILS

1. The size (width, length or diameter) of the elastomeric bearing shall be such that both surfaces are in complete contact with the bearing area.

2. There shall be a minimum of 1/8" edge seal, integral with the bearing, over all internal plates. No rubber cover will be required over internal details which will not be exposed to moisture after erection.

3. There shall be a minimum of 1/4 inch elastomeric bedding surface on both the top and bottom of the bearing.

4. When a bonded external load plate is required, there shall be an effective rubber laminate between the bonded load plate and the internal steel sheet.

5. Continuous sealing ribs on both the top and bottom of the bearing will not be required. When used, the ribs shall be 3/16" wide and 1/8" deep with a tolerance on each dimension of -0, +1/16". A sealing rib shall not be present on a surface where a bonded load plate is required.

GEOMETRIC PROPORTIONS

a. To insure stability, the following proportions shall be observed:

Bearing length to height ratios, as referred to in this specification, shall be defined as the plan dimension (L) in the direction of shear divided by the effective rubber thickness (ERT).

Plain Bearings - minimum L = 5ERT
minimum W = 5ERT
D = 6ERT

Laminated Bearings - minimum L = 3ERT
minimum W = 3ERT
D = 4ERT

b. To insure proper translation qualities, the following proportions shall be observed:

Plain and Laminated Bearings -
Maximum Anticipated Distortion (MAD) = 0.25 ERT
Maximum Anticipated Travel (MAT) = 0.50 ERT

c. To insure proper compression qualities, the following proportions shall be observed:

The average unit pressure on elastomeric bearings shall not exceed 800 psi under a combination of dead load plus live load, not including impact. The average unit pressure due to dead load only shall not exceed 500 psi.

Compressive strain of each bearing shall not exceed 5% of the effective rubber thickness at 500 psi, nor 8% of the effective rubber thickness at 800 psi.

d. To insure a uniform distribution of the load across the bearing, the size (width, length or diameter) of the elastomeric bearing shall be such that both surfaces shall come in complete contract with the bearing areas.

e. To insure permanent positioning, the bearing shall be secured against horizontal crawling preferably by positive attachment to the top and bottom surfaces when:

The dead load plus live load uplift (i.e. minimum pressure) reduces the average pressure to less than 200 psi.

The maximum shear force applied to, or developed by the bearing exceeds the frictional force developed by the minimum bearing pressure.

## Sample Problem No. 1

### Max. Anticipated Distortion

<u>Given:</u>

Concrete Box Beam - Type B II 36

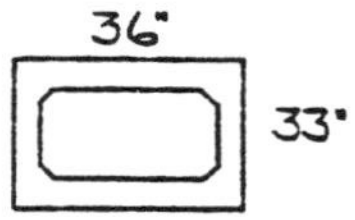


D. L. = 31.9K  Span 74'-0" ℄ to ℄ Brgs.

L. L. = 22.9K  Grade 1.2%

54.8K  Live Load Deflection = 0.49" ↓

Deflection due to Superimposed Dead Loads and all other Dead Loads on Beam. = 0.51"

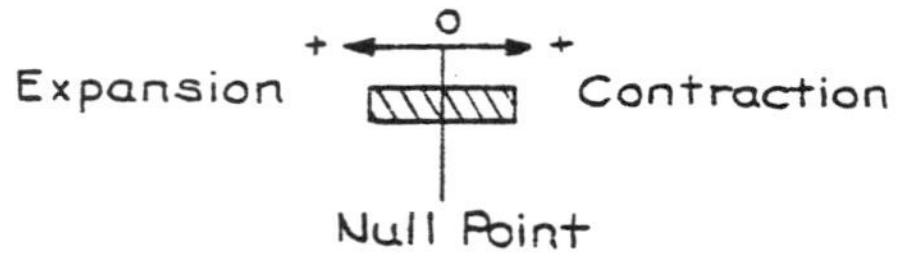


Temperature

Temperature Rise 35°F, Fall 45°F } A.A.S.H.O.

Expansion = $\alpha \cdot \Delta$ T · L = 6.0+10$^{-6}$ · 74 · 12 · 35 = 0.186"

Contraction = $\alpha \cdot \Delta$T · L = 6.0+10$^{-6}$ · 74 · 12 · 45 = 0.240"

*Shrinkage

Expansion (Contraction Only) ⟶ − 0.088"

Contraction = % left to go · Coef. · Length = 0.50 · 2.0×10$^{-4}$ · 74 · 12 = 0.088"

*Movement Due to Prestressing (Inelastic Shortening)

Expansion (Contraction Only) ⟶ − 0.123"

Contraction = (% left to go) × $\frac{P/A}{E}$ × L

$$P/A = \frac{\text{* Strands} \cdot \text{Load}}{\text{Area}} = \frac{24 \cdot 28{,}910^{\#}}{620.5} = 1{,}120^{\#}/in^2$$ use ~ 1,200#/in.² Average Concrete Section Stress

$$= 0.50 \cdot \frac{1.200}{4.3 \times 10^{6}} \cdot 74 \cdot 12 = \longrightarrow 0.123"$$

Use P = 28,910 − Release Cable Force
E = 4.3 · 10$^{+6}$ #/in.² Modulus of Elasticity ($f'c$ = 5000psi)

* For a more exact analysis see; Tentative Recommendations for Estimating Prestress Losses. PCI Committee on Prestress Losses.

### Max. Anticipated Distortion

Movement Due to Deflection

C = 0.51 - Beam Deflections Due to all Dead Loads.

Expansion = $\frac{4(C)D}{L} = \frac{4 \cdot 0.51" \cdot 28'}{74'}$ = +0.077"
Δ @ one end only

Contraction (Expansion Only) — −0.077"

Summation

<u>Short Term</u> (Temp & Deflection)

Expansion = 0.186 + 0.077 = +0.263"
Contraction = 0.240 − 0.077 = +0.163"
} 0.426"

<u>Long Term</u> (Temp., Shrinkage, Inelastic Shortening & Deflection)

Expansion = +0.186 −0.088 −0.123 + 0.077 = +0.052"
Contraction = +0.240 + 0.088 + 0.123 − 0.077 = +0.374"
} 0.426"

Maximum Anticipated Distortion (M.A.D.) = 0.374"

Maximum Anticipated Travel (M.A.T.) = 0.426"

## Sample Problem II

### Bearing Design
Expansion

**Step I**

1. Normal Force: D.L. = 31.9 K/brg., L.L. = 22.9 K/brg. max. } Given
2. Max. Anticipated Distortion (M.A.D.) = 0.374"; Max. Anticipated Travel (M.A.T.) = 0.426" } Problem #1
3. Horizontal, Longitudinal, Transverse Forces = Zero. Bridge Pinned at one end.

**Step II**

1. $A_1 = 31{,}900^{\#}/500 = 64 \text{ in.}^2$
   $A_2 = 54{,}800^{\#}/800 = 69 \text{ in.}^2$ } Min. Area = 69.0 in.²

**Step III**

$$\text{E.R.T. min.} = \frac{\text{Max. Anticipated Distortion}}{25\%} \text{ or } \frac{\text{Max. Travel}}{50\%}$$

$$= 4(0.374) = 1.496$$
$$= 2(0.426) = 0.852"$$
} Use E.R.T. ~ 1.50 ± Min.

**Step IV**

L min. = 3 to 5 × E.R.T. = 3(1.50) = 4.50; 5(1.50) = 7.50 Use 6"

**Step V**

W min. = A Max. / 6 = 69.0/6 = 11.5 in (Min.)

Beam Width = 36" − 4" (chamfer) = 32" max.

Use 18" ~ 2/3 Available Width = 2/3 (32) = 21"

**Step VI**

Calculate $t_e$ S.F. = 5.0 Start

$$\text{S.F.} = \frac{L_n \cdot W_n}{t_e \cdot 2(L_n + W_n)}$$

$$5.0 = \frac{5.75 \cdot 17.75}{t_e \cdot 2(5.75 + 17.75)} \qquad t_e = \frac{102.06}{10(23.5)} = 0.434"$$

**Step VII**

Calculate Required Number of Laminations

N = 1.500/0.434 = 3.4 Use 5 Laminations

**Step VIII** Standardize Bearing

T = 2.0" W = 18.0" L = 6.0" No Holes. Assume 14 Gage Sheets for lamination

$$t_{e\,final} = \frac{2.0 - 4 \cdot 0.075}{5} = 0.34"$$

$$\text{S.F.}_{final} = \frac{5.75 \cdot 17.75}{0.34 \cdot 2 \cdot 23.5} = 6.4^{\text{OK}} \quad 5.0 \le \text{S.F.} \le 12.0$$

E.R.T. = 5 × 0.34 = 1.70 in.

Max. Movement = 0.25 · 1.70 = 0.425" > 0.374

**Step IX** Final Check

1. Required Area = 69.0 Provided = 102.0 in.² OK
2. Shape Factor = 6.4 5.0 ≤ S.F. ≤ 12.0 OK
3. Geometric Stability L/E.R.T. = 6/1.70 = 3.53 3 to 5 Acceptable OK
4. Max. Allowable Distortion = 25% E.R.T. = 0.25 · 1.70 = 0.425"
   0.425 > 0.37 OK
5. Check Bearing Slippage (Capacity vs. Actual Force)
   Coefficient of Friction $\mu$ ~ 0.20 (bear concrete)
   Approx. Frictional Interface Capacity = 0.2(min. load) = .2 · 31.9 = 6,380#
   Approx. Max. Bearing Shear Resistance – Ref. Fig. 1
   $G_{A\mu}$ @ −20° F = 125#/in.² 50 Durometer Neoprene.
   $$p = \frac{G \cdot L_n \cdot W_n \cdot (\text{M.A.D.})}{\text{E} \cdot \text{R} \cdot \text{T}} = \frac{125 \cdot 5.75 \cdot 17.75 \cdot 0.37}{1.70} = 2{,}777 < 6{,}380^{\#} \text{ OK}$$
6. Check Bearing Fit
   6" × 18" × 2" —— OK on fit by inspection
   Bearing fits under beam and fits on pedestal.
7. Bearing Compression
   $$\text{D.L. Compression} = \frac{31{,}900^{\#} \text{(Min. LOAD)}}{5.75 \cdot 17.75} = 313^{\#}/\text{in.}^2 < 500^{\#}/\text{in.}^2 \text{ OK}$$; for minimum pressure less than 200 psi, secure bearing from horizontal crawling by positive attachments.
   Secure bearing with M29A even though min. pressure is over 200p
   Ref. Fig. 2 for S.F. = 6.4 & p = 313#/in.² Comp. = 3% < 5% Allowable.
   $$\text{D.L.} + \text{L.L. Compression} = \frac{54{,}800^{\#}}{5.75 \cdot 17.75} = 537^{\#}/\text{in.}^2 < 800^{\#}/\text{in.}^2 \text{ OK}$$
   Ref. Fig. 2 for S.F. = 6.4 & p = 537#/in.² Comp. = 4% < 8% Allowable OK

8. Tapered Load Bearing Plate - External
Grade + D.L. ≤ 0.06 E.R.T.
Grade = 6.0"·0.012 = 0.072"   % E.R.T. = $\frac{0.072}{1.70}$ = 4.24 %

4.2% + 3.0% = 7.2% > 6% - use tapered load plate.
Since Bearing Load Plate is not subject to bending and is in direct compression, use min. tapered load plate.
9·0.012 = 0.108" ~ 1/8"
use 9" by 1'-8" with a tapered shim tapering from 1" to 1 1/8"

9. Bearing Rotation - Loading Situation Below

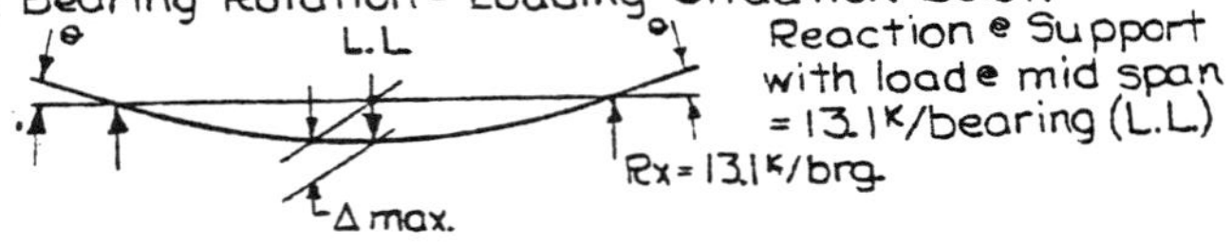


max. deflection = $\frac{L}{800}$ = $\frac{12.74}{800}$ = 1.11"† - max allowable deflection

† If actual deflection is known use that value.

θ radians = $\frac{2\cdot\Delta}{L/2}$ = $\frac{2\cdot 1.11}{12\cdot 74/2}$ = 5.0 × $10^{-3}$ radians

Δ E.R.T. = θ · $\frac{W}{2}$ = 4.96×$10^{-3}$ · $\frac{6}{2}$ = 14.88×$10^{-3}$

% Compression of E.R.T. due to rotation = 14.88×$10^{-3}$/1.70 = 0.88

Normal Pressure = $\frac{13.1(L.L.) + 31.9(D.L.)}{5.75 \cdot 17.75}$ = 440#/in.$^2$

% Compression E.R.T. due to Normal Pressure Ref. Fig. 2
for S.F. = 6.4 & p = 440#/in.$^2$ Comp. = 4%.
0.88 + 4.0 = 4.9% < 8.0% ok (8% Max. Compression of Elastomer)

10. Internal Laminations

$t_s = t_e \cdot \frac{\text{Max. Elastomer Pressure}}{\text{Tensile Strength Laminations}}$
te min. = 0.075" 14 gage

$t_s$ = 0.34 · $\frac{800}{20{,}000}$ = 0.0136" required thickness

use 14 gage = 0.075" min. sheet thickness allowable

Min. Recommended Internal Plate Thicknes

| Brg. Area in.$^2$ | Min. Interna Plate |
|---|---|
| 0-100 | 14ga.(0.07 |
| 100-200 | 3/32(0.09 |
| 200-350 | 1/8(0.125 |
| 350-600 | 3/16(0.18 |
| 600 + | 1/4(0.25 |

11. Pedestal Elevations
Comp. = 4.0% E.R.T.† + 25% [3.0% E.R.T.]* = 0.0806"

Bearing Height at ℄ Brg. = 2.0 + $\frac{1 + 1\frac{1}{8}}{2}$ = 3.0625"

Bearing Height under Service Conditions
= 3.0625 - 0.0806 = 2.9819" ~ 0.248'

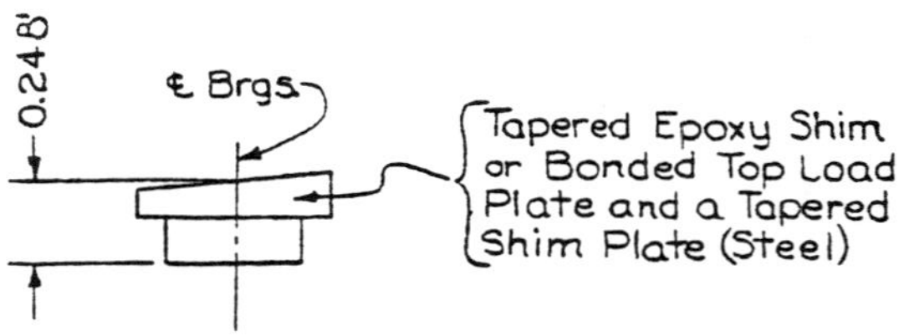


* Creep is generally taken as 25% - 100% D.L. Comp.

† Dead Load Compression

Step XI - Detail Bearing

Recommended Shim Material:
Concrete Beams - Epoxy Grout Pads
Steel Beams - Steel Plate Arrangement

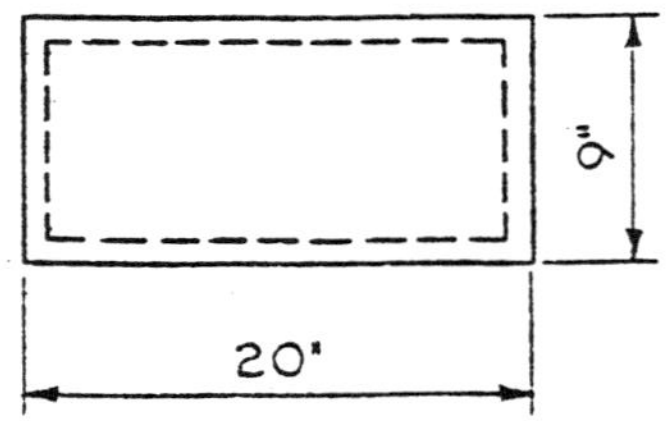


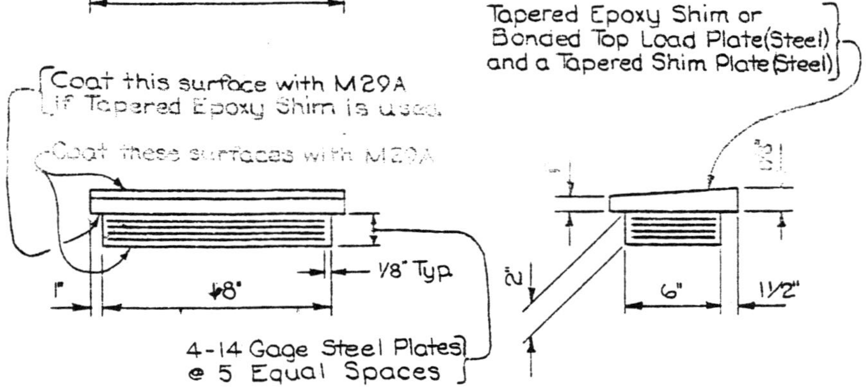


Materials Conforming to N.Y.S. Specification M6EA

## Bearing Design
### Fixed

Ref. Bearing Design - Expansion

1. Mechanical Fixation
   Horizontal (Tractive) Force Per Bearing

   $$\frac{5\% \cdot \text{Wt} \cdot \text{No. Lanes} \cdot \text{Reduction}}{\text{No. Bearings}} = \frac{5\% \cdot 72{,}000 \cdot 3 \cdot 0.9}{12} = 812^{\#}/\text{Brg.}$$

   Ref.: Section 2B Tractive Forces

   Max. Horizontal Force from Expansion Bearings
   Max. = 2,777#/Bearing
   Ref.: Step IX #5 Expansion Bearing, Sample Problem II.

   Max. Force = 2,777 + 812 = 3,589# = 3,600#/ Bearing
   If applicable include wind, centrifugal ... forces

   Dowels - one #8 Rod - Full Embedment - Grout Shut
   F shear resisting = $f_s \cdot A_s = 14^{k}/in^2 \cdot 1 \cdot 0.785 \cdot 1^2 = 11{,}000^{\#}$
   11,000# > 3,600# ok

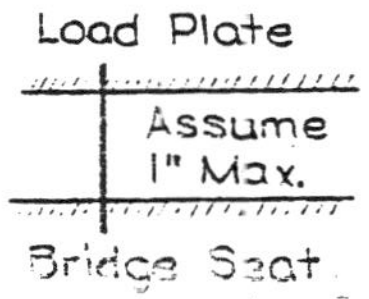


Assume ends of pin are fixed

$$\text{Moment} = \frac{PL}{2} = \frac{3{,}600 \times 1''}{2} = 1800 \text{ in.}^{\#}$$

$$\text{Section Modulus} = \frac{\pi D^3}{32} = \frac{3.14 \times 1^3}{32} = 0.098 \text{ in}^3$$

$$\text{Stress } M/S = \frac{1{,}800}{0.098} = 18{,}350^{\#}/\text{in.}^2$$

18,350#/in.² < 29,000#/in.² ok

2. Thickness of Elastomer
   % Comp. Normal Pressure ~ 4% ~ Step IX #9 Expansion Bearing

   Max. Compression = 8%
   % Compression Rotation (Remaining) = 8% - 4% = 4%

   $$\Delta ERT = \theta \cdot \frac{W}{2} = 14.88 \times 10^{-3} \text{ - Step IX \#9 Expansion Bearing}$$

   $$\% \text{ Comp. Rotation} = \frac{\Delta ERT}{ERT} = \frac{\theta \cdot W/2}{ERT \min.}$$

   $$4\% = \frac{14.88 \times 10^{-3}}{ERT \min.}$$

   ERT min. = 0.372    Try 2 Laminates @ 0.3 each

   $$te = \frac{6.0 - 0.075}{2} \sim 0.30$$

$$SF = \frac{5.75 \times 17.75 - 0.785 \times 2^2}{0.3[2(5.75 + 17.75) + 3.14 \times 2]} = \frac{98.92}{0.3 \times 53.28} = 6.2 \quad 5.0 \le SF \le 12^{ok}$$

2" ϕ Hole in Bearings for Fixation Dowel.

Check % Comp.

1.) $$\text{Max. Load Pressure} = \frac{54{,}800^{\#}}{98.92} = 555^{\#}/\text{in.}^2 \sim 5\%^{*} < 8\% \text{ ok}$$

2.) $$\text{Load @ Center Pressure} = \frac{31{,}900 + 13{,}100^{\#}}{98.92} = 455^{\#}/\text{in}^2 \sim 4.5\%^{*}$$

$$\% \text{ Comp. Due to Rotation} = \frac{14.88 \times 10^{-3}}{0.6} = 2.5\%^{*}$$

(4.5% + 2.5%) = 7.0% < 8.0%

* Ref. Fig. 2

Use Same Load Plate as Expansion Bearing.

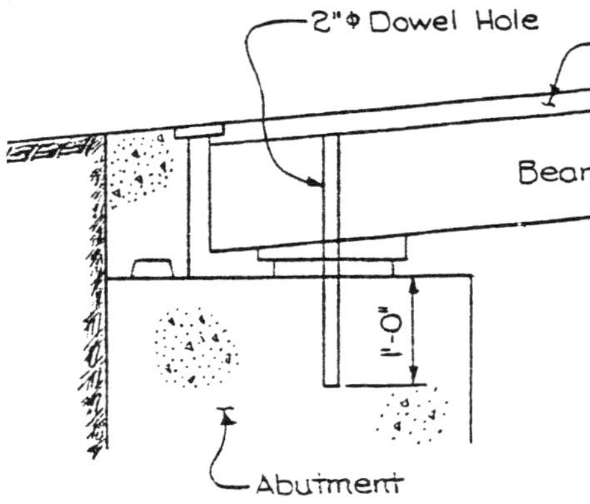


Fill dowel hole with sufficient grout such that, when dowel is inserted, the grout will completely fill the hole.

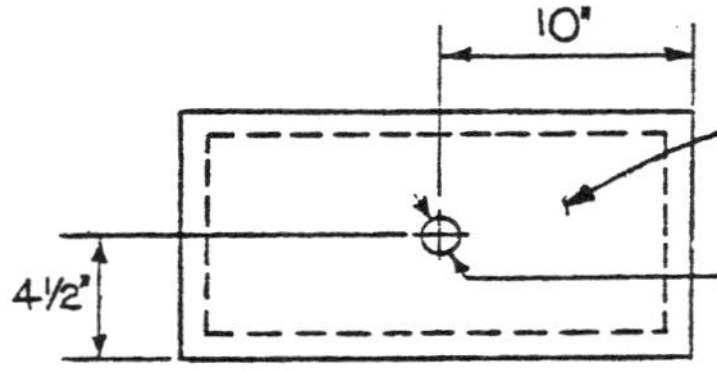

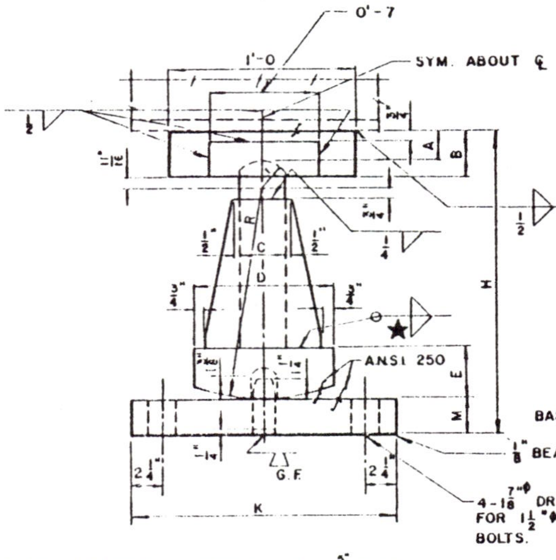


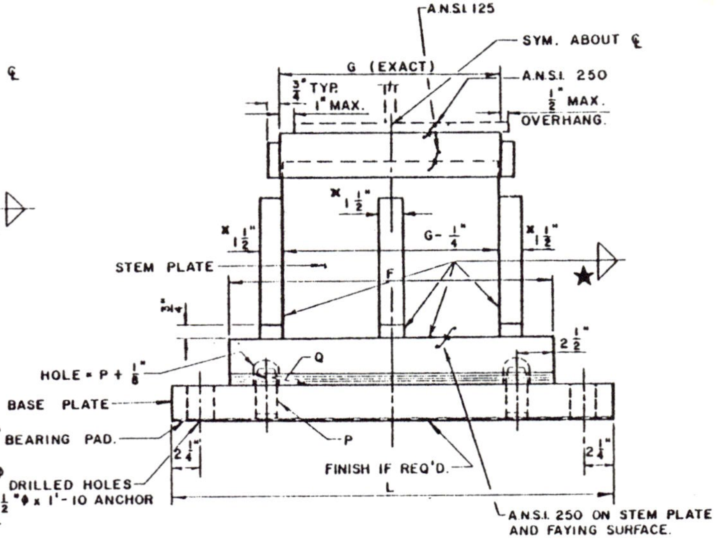


ROCKER

★ 400 K < REACTION ≤ 1000 K, USE 5/8" WELD.
1000 K < REACTION ≤ 1500 K, USE 3/4" WELD.

* FOR REACTION GREATER THAN 1000 KIPS USE 2" STIFFENERS.

● OR MATERIAL OF EQUIVALENT YIELD STRENGTH AND ELONGATION.

## GENERAL NOTES

ALL STRUCTURAL STEEL BEARING PLATES SHALL BE FLAT ROLLED STEEL PLATES WITH ALL SURFACES SMOOTH AND FREE FROM WARP AND ALL EDGES SMOOTH, STRAIGHT AND VERTICAL.

ALL PLATE CUTS SHALL BE MACHINE OR MACHINE FLAME CUTS. ON WELDED BEARINGS, FINAL MACHINING CAN BE PERFORMED BEFORE WELDING IS COMPLETED.

ALL MATERIALS IN TYPE "B" BEARINGS SHALL BE PAID FOR AT THE UNIT PRICE BID FOR "HIGH STRENGTH STRUCTURAL STEEL."

ALL MATERIALS FOR BEARINGS INCLUDING SHIMS BUT EXCLUDING PINTLES, ANCHOR BOLTS, NUTS AND WASHERS SHALL CONFORM TO A.S.T.M. SPECIFICATION TYPE A588 STEEL.

PINTLES SHALL CONFORM TO A.S.T.M. SPECIFICATION TYPE ●A449 STEEL. PINTLES SHALL BE MACHINED TO A DRIVING FIT.

ALL ANCHOR BOLTS, NUTS, AND WASHERS SHALL CONFORM TO A.S.T.M. SPECIFICATION TYPE A36 STEEL. PROJECT ANCHOR BOLTS "M" PLATE THICKNESS + 2 1/4" ABOVE TOP OF CONCRETE MASONRY. CHAMFER ANCHOR BOLTS PRIOR TO THREADING.

RADIAL SURFACES ON ROCKER SHALL BE MACHINE FINISHED AFTER WELDING.

ALL SURFACES MARKED "ƒ" SHALL BE MACHINE FINISHED BY AN AUTOMATIC PROCESS.

ANCHOR BOLT EDGE DISTANCE ALONG "L" MAY BE INCREASED FROM MINIMUM SHOWN WHEN A COMMON GRID DETAIL IS DESIRED FOR SEVERAL BEARINGS.

## ROCKER SETTING DATA

| TEMPERATURE TIME OF SETTING - °F | PIER (+) | PIER (VERTICAL) | PIER | PIER (-) |
|---|---|---|---|---|
| 120 | | | | |
| 100 | | | | |
| 80 | | | | |
| 60 | | | | |
| 40 | | | | |
| 20 | | | | |
| 0 | | | | |
| -20 | | | | |

ROCKER BEARING SHALL BE SET VERTICAL AT 45°F.

ROCKER BEARING SHALL BE USED WITH A MINIMUM FRICTION VALUE OF 2% AND A MAXIMUM FRICTION VALUE OF 4%.

MAXIMUM MOVEMENT FROM 45°F = (D-1)/2 BUT ACTUAL MOVEMENT NOT TO EXCEED R/3

## TABLE OF DIMENSIONS

| REACTION (KIPS) | A | B | C | D | E | G VALUES: G = 1'-7 F | G = 1'-7 L | G = 1'-9 F | G = 1'-9 L | G = 1'-11 F | G = 1'-11 L | G = 2'-1 F | G = 2'-1 L | G = 2'-3 F | G = 2'-3 L | G = 2'-5 F | G = 2'-5 L |
|---|---|---|---|---|---|---|---|---|---|---|---|---|---|---|---|---|---|
| 400-500 | $1\frac{15}{16}$ | $2\frac{15}{16}$ | 3 | 1-2 | $2\frac{7}{8}$ | 2-0 | 2-11 | 2-2 | 2-11 | 2-4 | 3-0 | 2-6 | 3-2 | — | — | — | — |
| 500-600 | $1\frac{15}{16}$ | $2\frac{15}{16}$ | 3 | 1-2 | $2\frac{7}{8}$ | 2-1 | 3-4 | 2-2 | 3-4 | 2-4 | 3-4 | 2-6 | 3-4 | — | — | — | — |
| 600-700 | $1\frac{15}{16}$ | $2\frac{15}{16}$ | 3 | 1-2 | $2\frac{7}{8}$ | — | — | 2-3 | 3-8 | 2-4 | 3-8 | 2-6 | 3-8 | 2-8 | 3-8 | — | — |
| 700-800 | $2\frac{3}{16}$ | $3\frac{7}{16}$ | $3\frac{1}{2}$ | 1-4 | $3\frac{3}{8}$ | — | — | — | — | 2-6 | 3-10 | 2-6 | 3-10 | 2-8 | 3-10 | 2-10 | 3-10 |
| 800-900 | $2\frac{3}{16}$ | $3\frac{7}{16}$ | $3\frac{1}{2}$ | 1-4 | $3\frac{3}{8}$ | — | — | — | — | 2-7 | 3-11 | 2-7 | 3-11 | 2-8 | 3-11 | 2-10 | 3-11 |
| 900-1000 | $2\frac{3}{16}$ | $3\frac{7}{16}$ | $3\frac{1}{2}$ | 1-4 | $3\frac{3}{8}$ | — | — | — | — | 2-11 | 4-0 | 2-11 | 4-0 | 2-11 | 4-0 | 2-11 | 4-0 |
| 1000-1100 | $2\frac{7}{16}$ | $3\frac{15}{16}$ | 4 | 1-6 | $3\frac{7}{8}$ | — | — | — | — | — | — | 3-1 | 4-1 | 3-1 | 4-1 | 3-1 | 4-1 |
| 1100-1200 | $2\frac{7}{16}$ | $3\frac{15}{16}$ | 4 | 1-6 | $3\frac{7}{8}$ | — | — | — | — | — | — | 3-3 | 4-2 | 3-3 | 4-2 | 3-3 | 4-2 |
| 1200-1300 | $2\frac{7}{16}$ | $3\frac{15}{16}$ | 4 | 1-6 | $3\frac{7}{8}$ | — | — | — | — | — | — | — | — | 3-5 | 4-4 | 3-5 | 4-4 |
| 1300-1400 | $2\frac{7}{16}$ | $3\frac{15}{16}$ | 4 | 1-6 | $3\frac{7}{8}$ | — | — | — | — | — | — | — | — | 3-7 | 4-7 | 3-7 | 4-7 |
| 1400-1500 | $2\frac{7}{16}$ | $3\frac{15}{16}$ | 4 | 1-6 | $3\frac{7}{8}$ | — | — | — | — | — | — | — | — | 3-9 | 4-9 | 3-9 | 4-9 |

| REACTION (KIPS) | H | K | M | R | r STEM | r PLATE | PINTLE P | PINTLE Q | WEIGHT OF ANCHOR BOLTS PER BEARING |
|---|---|---|---|---|---|---|---|---|---|
| 400-500 | $1\text{-}7\frac{1}{2}$ | 1-6 | $2\frac{7}{8}$ | 1-1 | $1\frac{11}{16}$ | $1\frac{45}{64}$ | 2ø | $3\frac{1}{2}$ | 50 |
| 500-600 | $1\text{-}8\frac{1}{2}$ | 1-7 | $2\frac{7}{8}$ | 1-2 | $1\frac{11}{16}$ | $1\frac{45}{64}$ | 2ø | $3\frac{1}{2}$ | 50 |
| 600-700 | $1\text{-}9\frac{1}{2}$ | 1-8 | $2\frac{7}{8}$ | 1-3 | $1\frac{11}{16}$ | $1\frac{45}{64}$ | 2ø | $3\frac{1}{2}$ | 50 |
| 700-800 | $1\text{-}11\frac{1}{2}$ | 1-10 | $3\frac{3}{8}$ | 1-4 | $1\frac{15}{16}$ | $1\frac{61}{64}$ | 2ø | $3\frac{1}{2}$ | 50 |
| 800-900 | $2\text{-}0\frac{1}{2}$ | 2-0 | $3\frac{3}{8}$ | 1-5 | $1\frac{15}{16}$ | $1\frac{61}{64}$ | 2ø | $3\frac{1}{2}$ | 50 |
| 900-1000 | $2\text{-}1\frac{1}{2}$ | 2-2 | $3\frac{3}{8}$ | 1-6 | $1\frac{15}{16}$ | $1\frac{61}{64}$ | 2ø | $3\frac{1}{2}$ | 50 |
| 1000-1100 | $2\text{-}3\frac{1}{2}$ | 2-4 | $3\frac{7}{8}$ | 1-7 | $2\frac{3}{16}$ | $2\frac{13}{64}$ | $2\frac{1}{2}$ø | $3\frac{3}{4}$ | 50 |
| 1100-1200 | $2\text{-}4\frac{1}{2}$ | 2-6 | $3\frac{7}{8}$ | 1-8 | $2\frac{3}{16}$ | $2\frac{13}{64}$ | $2\frac{1}{2}$ø | $3\frac{3}{4}$ | 50 |
| 1200-1300 | $2\text{-}5\frac{1}{2}$ | 2-7 | $3\frac{7}{8}$ | 1-9 | $2\frac{3}{16}$ | $2\frac{13}{64}$ | $2\frac{1}{2}$ø | $3\frac{3}{4}$ | 50 |
| 1300-1400 | $2\text{-}6\frac{1}{2}$ | 2-8 | $3\frac{7}{8}$ | 1-10 | $2\frac{3}{16}$ | $2\frac{13}{64}$ | $2\frac{1}{2}$ø | $3\frac{3}{4}$ | 50 |
| 1400-1500 | $2\text{-}7\frac{1}{2}$ | 2-9 | $3\frac{7}{8}$ | 1-11 | $2\frac{3}{16}$ | $2\frac{13}{64}$ | $2\frac{1}{2}$ø | $3\frac{3}{4}$ | 50 |

| REACTION (KIPS) | WEIGHT PER TOP PLATE, DIMENSION "G": 1'-7 | 1'-9 | 1'-11 | 2'-1 | 2'-3 | 2'-5 | WEIGHT PER ROCKER, PINTLES & BASE PLATE ▲, DIMENSION "G": 1'-7 | 1'-9 | 1'-11 | 2'-1 | 2'-3 | 2'-5 |
|---|---|---|---|---|---|---|---|---|---|---|---|---|
| 400-500 | 201 | 221 | 241 | 262 | — | — | 1109 | 1154 | 1213 | 1287 | — | — |
| 500-600 | 201 | 221 | 241 | 262 | — | — | 1256 | 1289 | 1335 | 1381 | — | — |
| 600-700 | — | 221 | 241 | 262 | 282 | — | — | 1429 | 1464 | 1512 | 1559 | — |
| 700-800 | — | — | 282 | 306 | 330 | 354 | — | — | 1941 | 1970 | 2030 | 2090 |
| 800-900 | — | — | 282 | 306 | 330 | 354 | — | — | 2104 | 2134 | 2180 | 2243 |
| 900-1000 | — | — | 282 | 306 | 330 | 354 | — | — | 2317 | 2350 | 2382 | 2415 |
| 1000-1100 | — | — | — | 350 | 377 | 404 | — | — | — | 3036 | 3075 | 3113 |
| 1100-1200 | — | — | — | 350 | 377 | 404 | — | — | — | 3263 | 3304 | 3345 |
| 1200-1300 | — | — | — | — | 377 | 404 | — | — | — | — | 3515 | 3555 |
| 1300-1400 | — | — | — | — | 377 | 404 | — | — | — | — | 3767 | 3813 |
| 1400-1500 | — | — | — | — | 377 | 404 | — | — | — | — | 3989 | 4037 |

▲ WEIGHT INCLUDES ANCHOR BOLTS

ROCKER BEARING
TYPE "B" – STEEL GIRDERS

STATE OF WISCONSIN
DEPARTMENT OF TRANSPORTATION
DIVISION OF HIGHWAYS
BRIDGE SECTION

Approved: [signature] Chief Bridge Engineer
Date: 1-'84

STANDARD 27.3

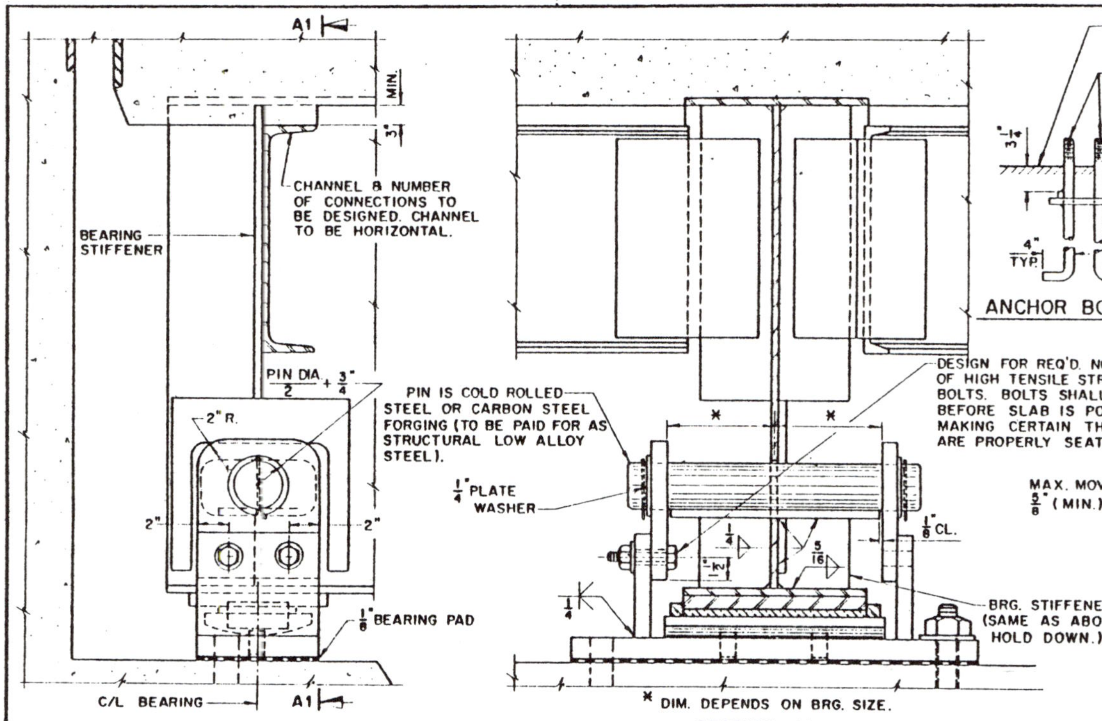


ELEVATION

SECTION A1

PERMANENT HOLD DOWN DEVICE

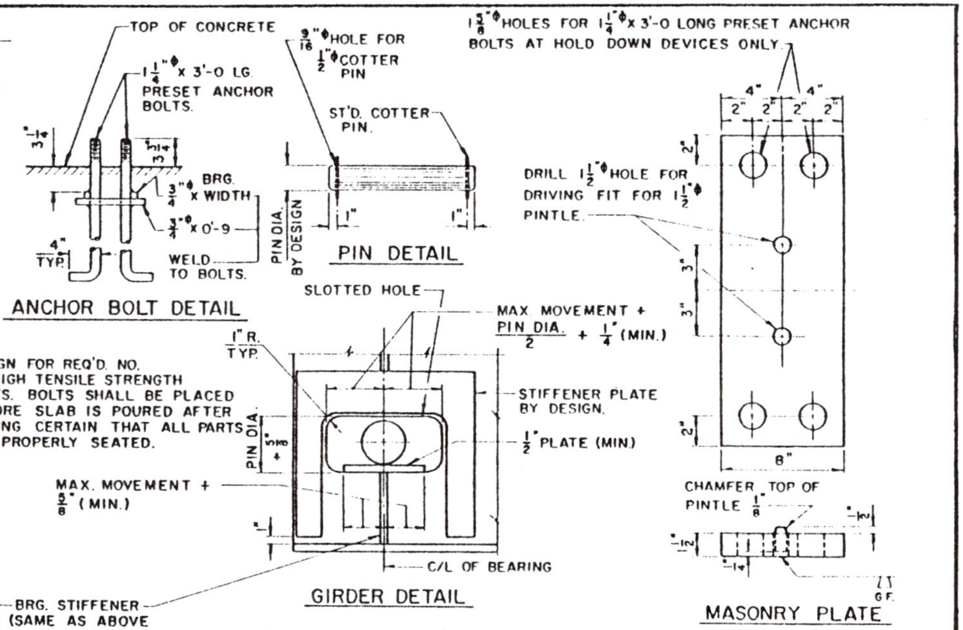


GENERAL NOTES (PERMANENT HOLD DOWN DEVICE)

ALL STRUCTURAL STEEL PLATES SHALL BE FLAT ROLLED STEEL PLATES WITH ALL SURFACES SMOOTH AND FREE FROM WARP AND ALL EDGES SMOOTH, STRAIGHT AND VERTICAL.

ALL PLATE CUTS SHALL BE MACHINE OR MACHINE FLAME CUTS.

ANCHOR BOLTS SHALL BE THREADED 3". PROVIDE ONE STANDARD WROUGHT WASHER AND ONE HEX. NUT PER BOLT. CHAMFER TOP OF ANCHOR BOLTS PRIOR TO THREADING.

ALL MATERIAL EXCEPT ANCHOR BOLTS, NUTS, WASHERS AND PINS SHALL BE MADE OF A588 STEEL.

ALL MATERIAL IN BEARINGS, EXCEPTING BRONZE PLATES AND BEARING PADS SHALL BE PAID FOR AT THE UNIT PRICE BID FOR "HIGH STRENGTH STRUCTURAL STEEL."

WHEN REQUIRED, HOLD DOWN DEVICES SHALL BE PLACED SYMMETRICALLY ABOUT LONGIT C/L OF FRAMING PLAN. MAXIMUM SPACING OF HOLD DOWNS SHALL BE AT ALTERNATE GIRDERS. HOLD DOWN DEVICE TO BE DESIGNED FOR MIN. UPLIFT CAPACITY OF 20 KIPS.

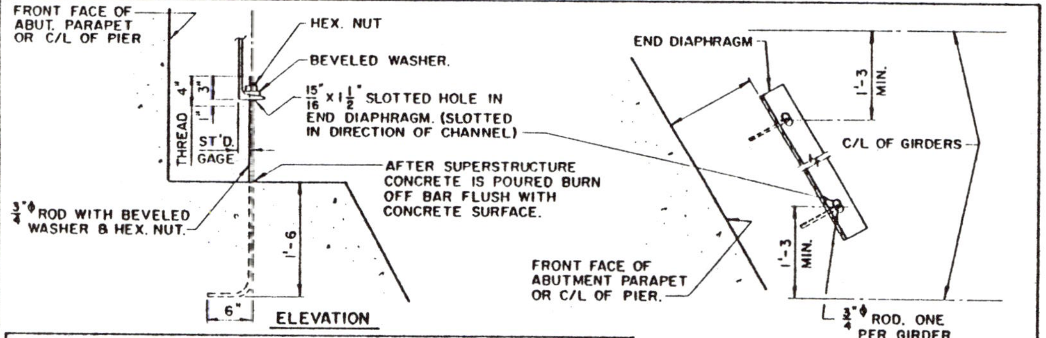


ELEVATION

SECTION

(TO BE PAID FOR AS STRUCTURAL CARBON STEEL)

TEMPORARY HOLD DOWN DEVICES SHALL BE PLACED AT THAT END OF ALL CONTINUOUS STEEL GIRDER UNITS WHERE THE SLAB POUR TERMINATES, EXCEPT WHERE PERMANENT HOLD DOWN DEVICES ARE PLACED AT THIS LOCATION.

TEMPORARY HOLD DOWN DEVICE

HOLD DOWN DEVICES

STATE OF WISCONSIN
DEPARTMENT OF TRANSPORTATION
DIVISION OF HIGHWAYS
BRIDGE SECTION

Approved: Stanley W. Woods, Chief Bridge Engineer | Date: 1-'84

STANDARD 27.6

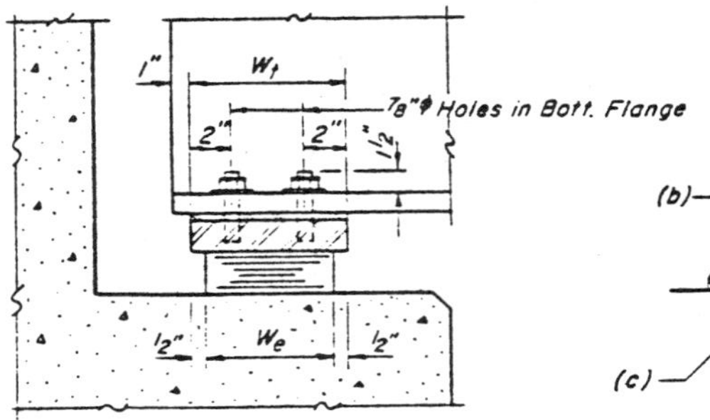


*May be 1½" for smaller W shapes.

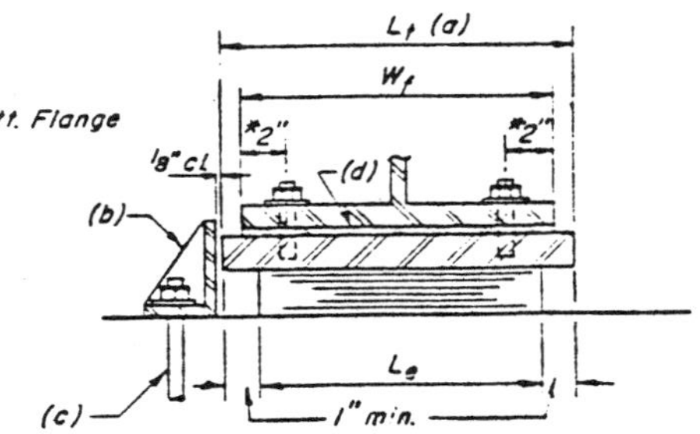


Note:

- a $L_f = W_f$ or $L_e + 2"$ whichever is greater.
- b Side retainer. (Place at inside face of exterior girders only)
- c 1"φ x 12" anchor bolt.
- d Shim plate if required.
- e 1½" Minimum thickness.

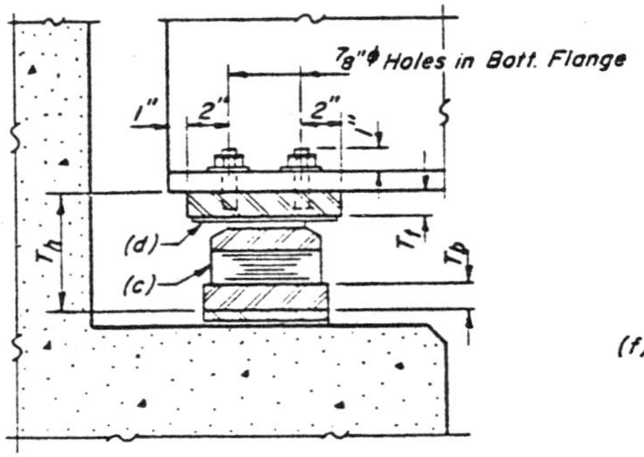


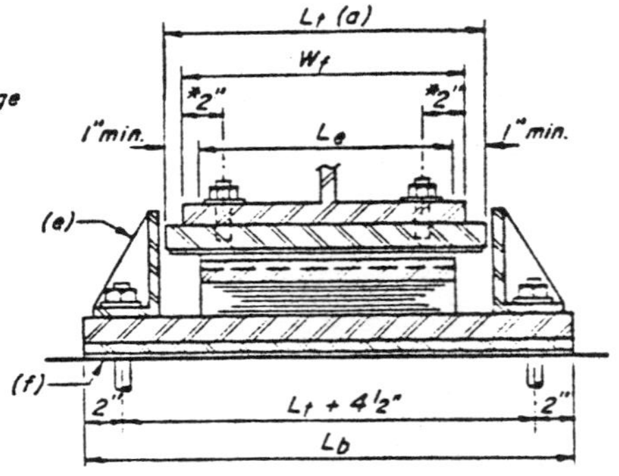


*May be 1½" for smaller W shapes.

Note:

- a $L_f = W_f$ or $L_e + 2"$ whichever is greater.
- b $W_f = W_e - 2A + 2E + 2"$, E = Expansion from 50° normal temperature.
- c TFE - Elastomeric bearing.
- d 1/16" Stainless steel sliding plate, A240, Type 304, 2B finish.
- e Side retainer.
- f Fill plate for crown and 1/8" lead plate.

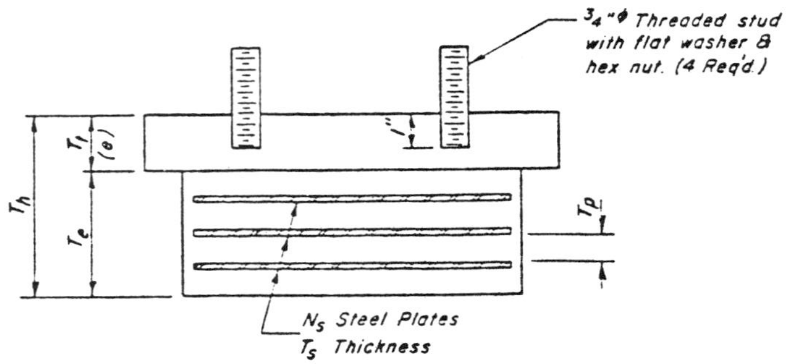


TYPE I

ELASTOMERIC

EXPANSION BEARING

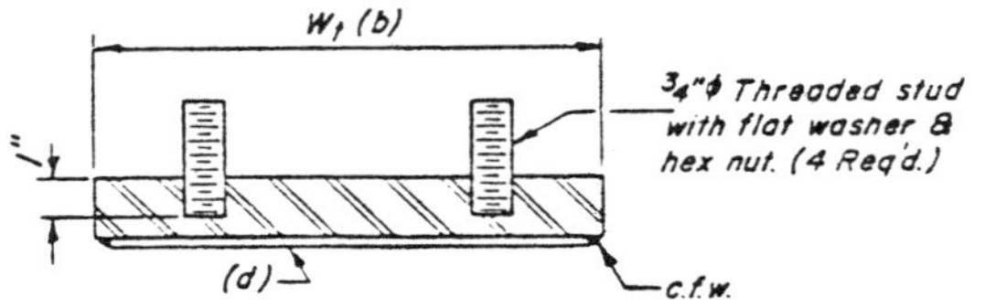


TOP BEARING ASSEMBLY

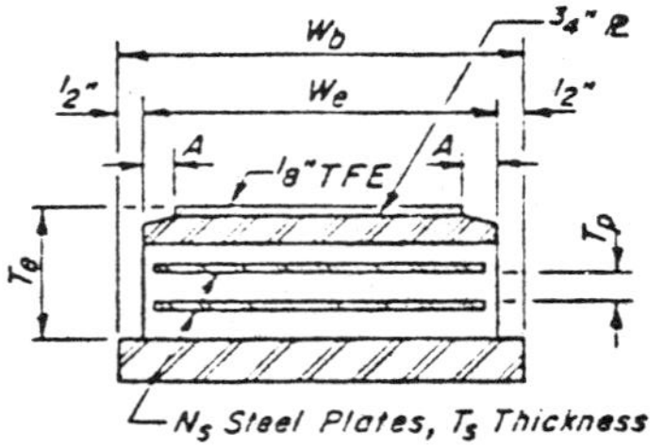


BOTTOM BEARING ASSEMBLY

TYPE II

TFE ELASTOMERIC

EXPANSION BEARING

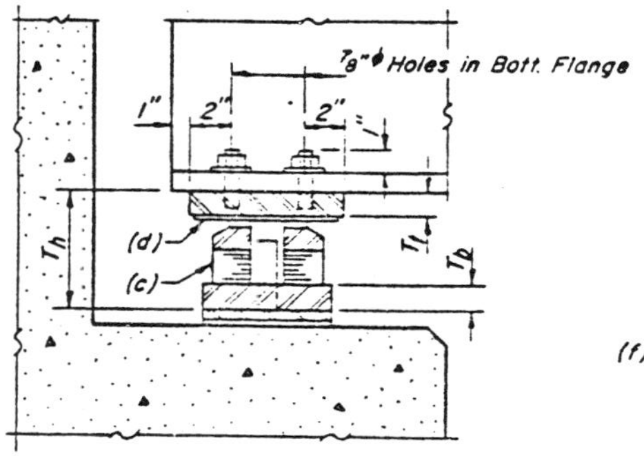


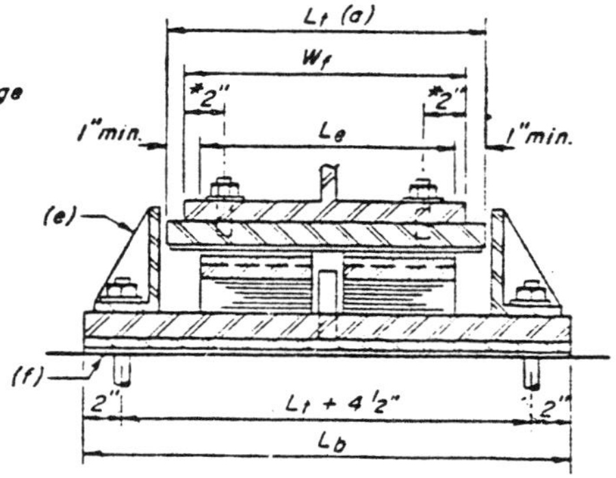


*May be 1 1/2" for smaller W shapes

Note:

- a $L_t = W_t$ or $L_e + 2"$ whichever is greater.
- b $W_t = W_e - 2A + 2E + 2"$, E = Expansion from 50° normal temperature.
- c TFE-Elastomeric bearing.
- d 1/16" Stainless steel sliding plate, A240, Type 304, 2B finish.
- e Side retainer.
- f Fill plate for crown and 1/8" lead plate.
- g Shear restrictor pin, AISI 4340, quenched and tempered.

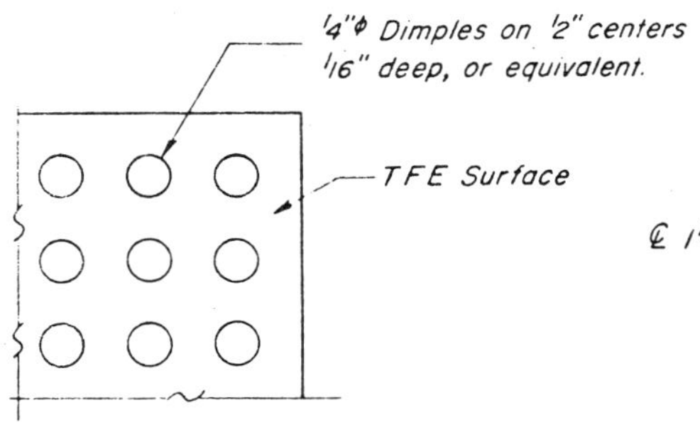


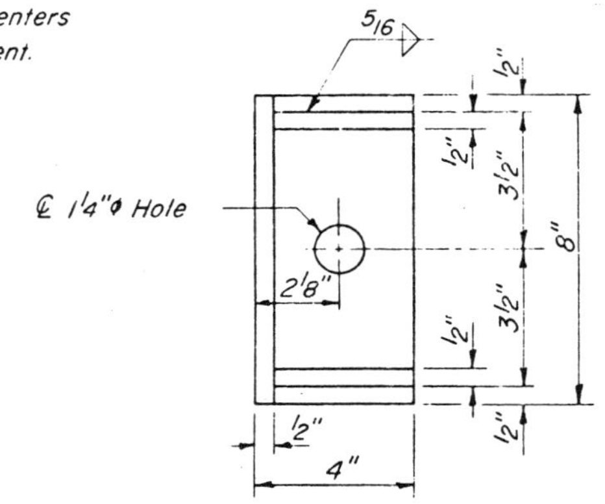


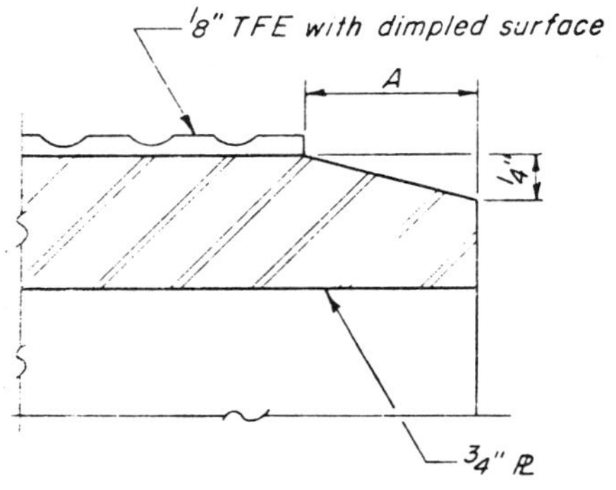


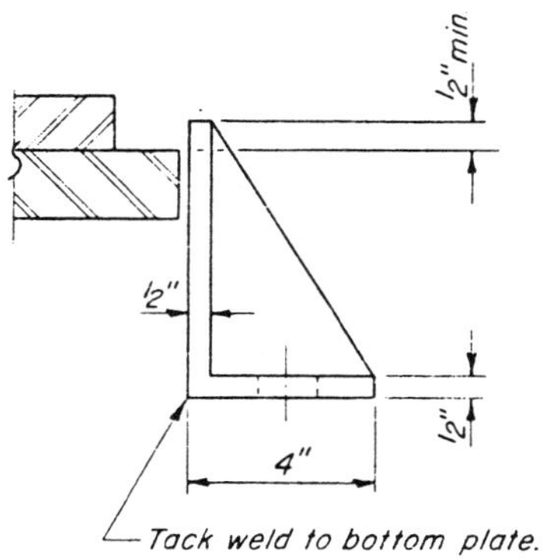


SIDE RETAINER

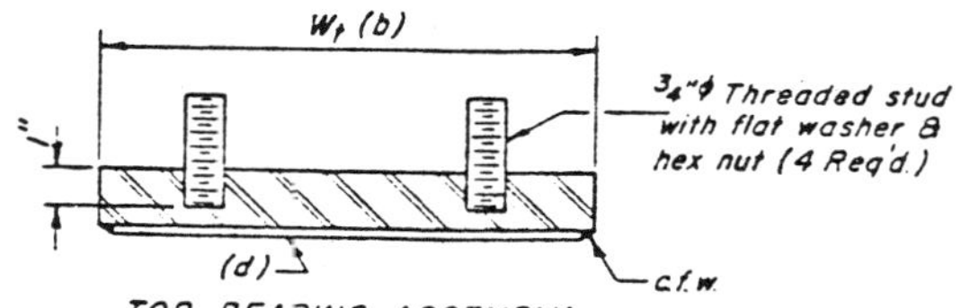


TOP BEARING ASSEMBLY

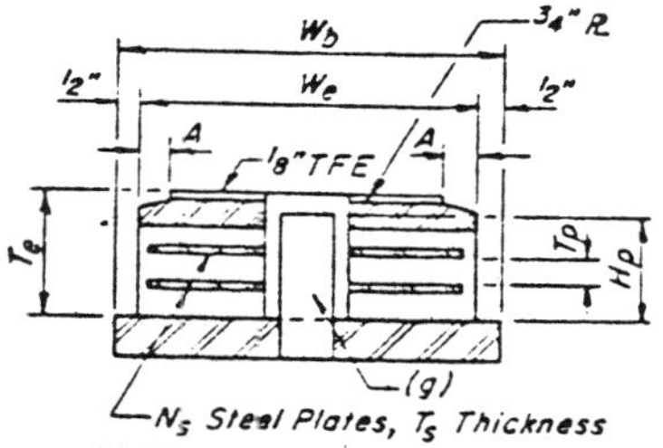


BOTTOM BEARING ASSEMBLY

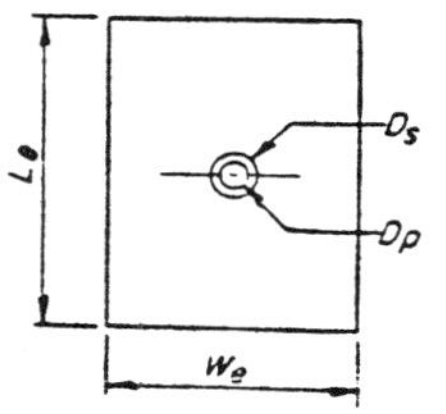


SLOT DETAIL

| $W_e$ | 6" | 7" | 9" | 10" | 11" | 12" |
|---|---|---|---|---|---|---|
| A | 1" | 1" | 1 1/2" | 1 1/2" | 1 1/2" | 1 1/2" |

Note: The 1/8" TFE sheet shall be bonded directly to the top steel plate with a two-component, medium viscosity epoxy resin, conforming to the requirements of the Federal Specification MMM-A-134, Type I. The bond agent shall be applied on the full area of the contact surfaces.

TYPE III
TFE ELASTOMERIC
EXPANSION BEARING

MISCELLANEOUS DETAILS
FOR ELASTOMERIC
EXPANSION BEARINGS

ILL DOT

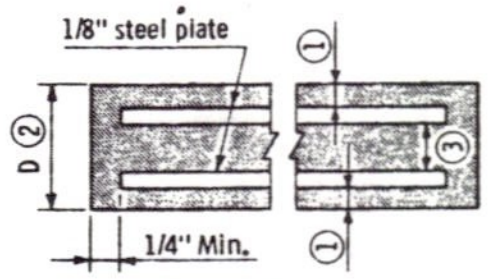


SECTION THRU PAD

① 1/8" for 1/2" and 3/4" laminate. 3/16" for 3/8" laminate.

② Total thickness including the steel plates.

③ For maximum laminate thickness, see Elastomeric Pad Table

## STEEL LAMINATE PAD DETAIL

(1" pad with 3/8" laminate shown)

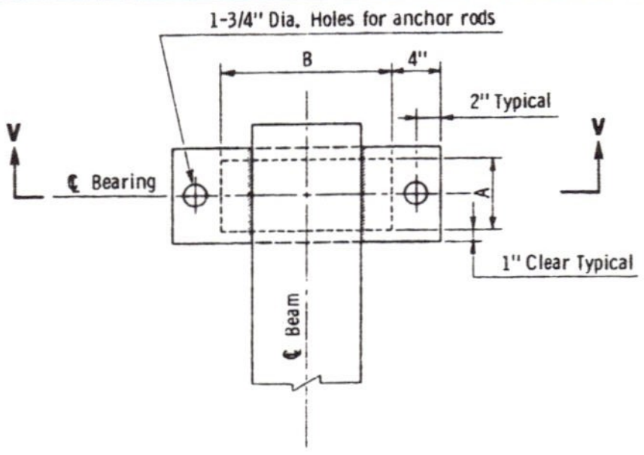


PLAN

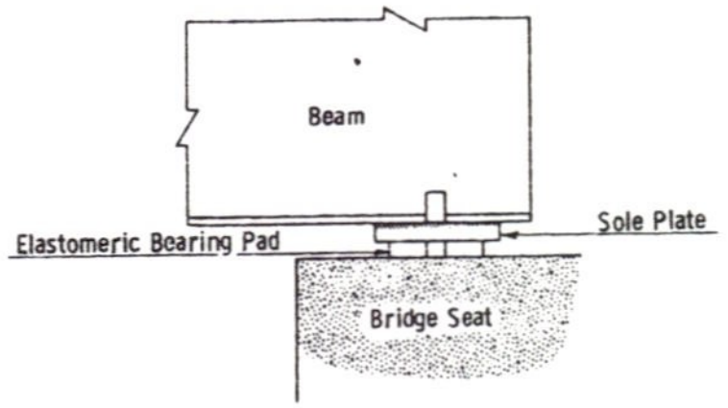


SIDE ELEVATION

(At beam end shown)

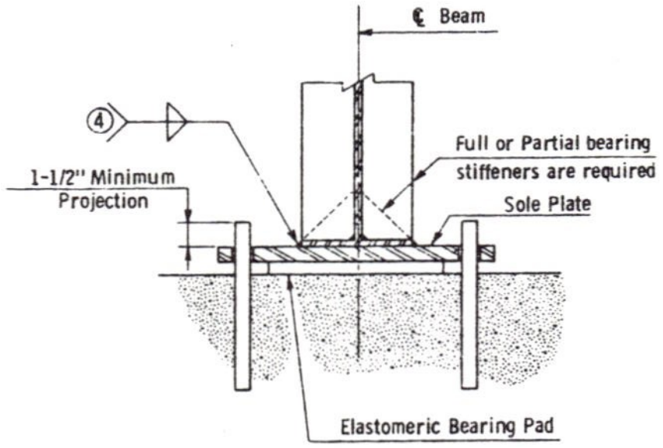


SECTION V-V

④ 5/16" minimum fillet weld for 3/4" up to and including 1-1/2" thick sole plates. 3/8" minimum fillet weld for over 1-1/2" to 2-1/4" thick sole plates.

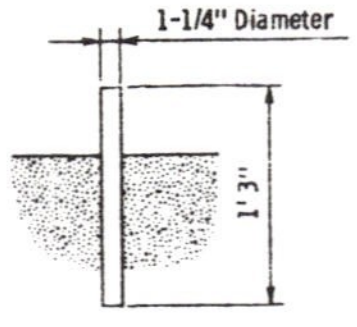


ANCHOR ROD DETAIL

### ELASTOMERIC PAD TABLE

| Pad Size A | Pad Size B | D | DL + LL Min. (Kips) | DL Max. (Kips) | DL + LL Max. (Kips) | Shape Factor | Sole Plate Size For Beam Flanges 9" to 10-1/2" | Sole Plate Size For Beam Flanges 11-1/2" to 12" | Sole Plate Size For Beam Flanges 15" to 16" | Max. Lam. Thk. | Assy. Type |
|---|---|---|---|---|---|---|---|---|---|---|---|
| 6" | 14" | 1" | 17 | 42 | 67 | 6 | 8" x 22" x 1-1/4" | 8" x 22" x 1" | 8" x 22" x 1" | 3/8" | |
| 6" | 16" | 1" | 19 | 48 | 77 | 6 | 8" x 24" x 1-1/2" | 8" x 24" x 1-1/4" | 8" x 24" x 1" | 3/8" | |
| 8" | 14" | 1" | 22 | 56 | 90 | 7 | 10" x 22" x 1-1/4" | 10" x 22" x 1" | 10" x 22" x 1" | 3/8" | |
| 8" | 16" | 1" | 26 | 64 | 102 | 7 | 10" x 24" x 1-1/2" | 10" x 24" x 1-1/4" | 10" x 24" x 1" | 3/8" | |
| 10" | 14" | 1" | 28 | 70 | 112 | 8 | 12" x 22" x 1-1/4" | 12" x 22" x 1". | 12" x 22" x 1" | 3/8" | |
| 10" | 16" | 1" | 32 | 80 | 128 | 6 | 12" x 24" x 1-1/2" | 12" x 24" x 1-1/4" | 12" x 24" x 1" | 1/2" | |
| 12" | 14" | 1" | 34 | 84 | 134 | 9 | 14" x 22" x 1-1/4" | 14" x 22" x 1" | 14" x 22" x 1" | 3/8" | |
| 12" | 16" | 1" | 38 | 96 | 154 | 7 | 14" x 24" x 1-1/2" | 14" x 24" x 1-1/4" | 14" x 24" x 1" | 1/2" | |
| 14" | 16" | 1" | 45 | 112 | 179 | 7 | 16" x 24" x 1-1/2" | 16" x 24" x 1-1/4" | 16" x 24" x 1" | 1/2" | |
| 16" | 20" | 1" | 64 | 160 | 256 | 9 | 18" x 28" x 2" | 18" x 28" x 1-1/2" | 18" x 28" x 1-1/4" | 1/2" | |
| 18" | 20" | 1" | 72 | 180 | 288 | 9 | 20" x 28" x 2" | 20" x 28" x 1-1/2" | 20" x 28" x 1-1/4" | 1/2" | |
| 20" | 20" | 2" | 80 | 200 | 320 | 7 | 22" x 28" x 2" | 22" x 28" x 1-1/2" | 22" x 28" x 1-1/4" | 3/4" | |
| 22" | 20" | 2" | 88 | 220 | 352 | 7 | 24" x 28" x 2" | 24" x 28" x 1-1/2" | 24" x 28" x 1-1/4" | 3/4" | |
| 24" | 20" | 2" | 96 | 240 | 384 | 7 | 26" x 28" x 2" | 26" x 28" x 1-1/2" | 26" x 28" x 1-1/4" | 3/4" | |

NOTES:

For elastomeric materials and pad construction, see Spec. 3741.

Anchor rods shall comply with Spec 3306.

Galvanize anchor rods as per Spec 3394.

Payment for bearing assembly shall include elastomeric bearing pad and anchor rods.

Sole Plate is to be of the same material Spec as the steel beam and is to be included in the weight of structural steel. Shop weld sole plate to the steel beam.

All bearing pads are to bear evenly on the bearing seats.

APPROVED: April 26, 1978

Developed by: ENGINEERING STANDARDS BRIDGES AND STRUCTURES

Issued by: ENGINEERING STANDARDS

STATE OF MINNESOTA DEPARTMENT OF TRANSPORTATION

**ELASTOMERIC BEARING ASSEMBLY STEEL BEAMS (FIXED)**

DETAIL NO. **B354**

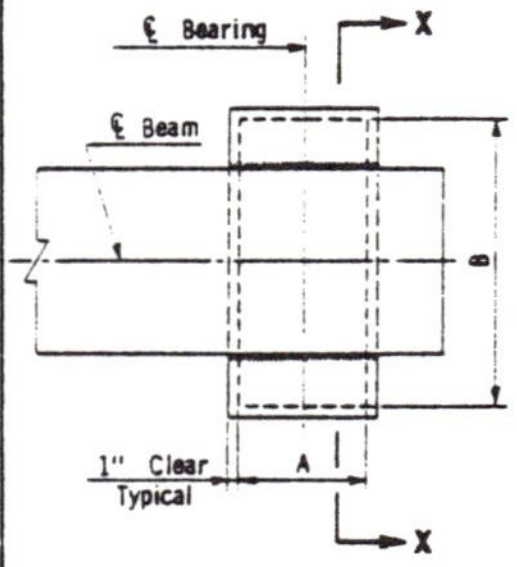


PLAN

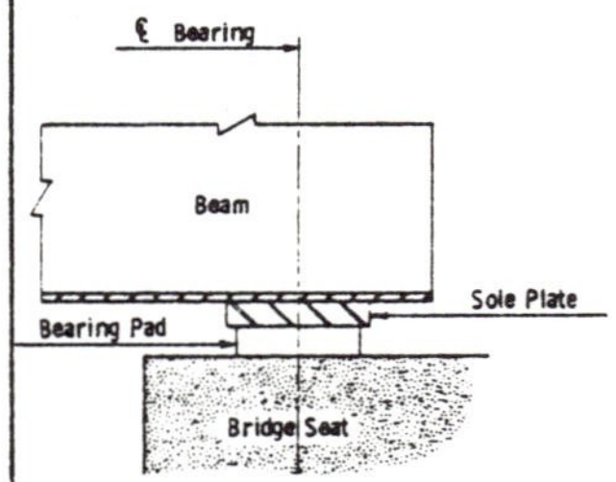


SIDE ELEVATION

(At beam end shown)

### ELASTOMERIC PAD TABLE

| Bearing Pad Size A | Bearing Pad Size B | D | DL + LL Min. (Kips) | DL Max. (Kips) | DL + LL Max. (Kips) | Shape Factor | Sole Plate Size For Beam Flanges 9" to 10-1/2" | Sole Plate Size For Beam Flanges 11-1/2" to 12" | Sole Plate Size For Beam Flanges 15" to 16" | Lam. Thk. | Pad Type |
|---|---|---|---|---|---|---|---|---|---|---|---|
| 6" | 14" | | 17 | 42 | 67 | 6 | 8" x 16" x 1-1/4" | 8" x 16" x 1" | 8" x 16" x 1" | 3/8" | |
| 6" | 16" | | 19 | 48 | 77 | 6 | 8" x 18" x 1-1/2" | 8" x 18" x 1-1/4" | 8" x 18" x 1" | | |
| 8" | 14" | | 22 | 56 | 90 | 7 | 10" x 16" x 1-1/4" | 10" x 16" x 1" | 10" x 16" x 1" | | |
| 8" | 16" | | 26 | 64 | 102 | 7 | 10" x 18" x 1-1/2" | 10" x 18" x 1-1/4" | 10" x 18" x 1" | | |
| 10" | 14" | | 28 | 70 | 112 | 8 | 12" x 16" x 1-1/4" | 12" x 16" x 1" | 12" x 16" x 1" | | |
| 12" | 14" | | 34 | 84 | 134 | 9 | 14" x 16" x 1-1/4" | 14" x 16" x 1" | 14" x 16" x 1" | ▼ | |
| 10" | 16" | | 32 | 80 | 128 | 6 | 12" x 18" x 1-1/2" | 12" x 18" x 1-1/4" | 12" x 18" x 1" | 1/2" | |
| 12" | 16" | | 38 | 96 | 154 | 7 | 14" x 18" x 1-1/2" | 14" x 18" x 1-1/4" | 14" x 18" x 1" | | |
| 14" | 16" | | 45 | 112 | 179 | 7 | 16" x 18" x 1-1/2" | 16" x 18" x 1-1/4" | 16" x 18" x 1" | | |
| 16" | 20" | | 64 | 160 | 256 | 9 | 18" x 22" x 2" | 18" x 22" x 1-1/2" | 18" x 22" x 1-1/4" | | |
| 18" | 20" | | 72 | 180 | 288 | 9 | 20" x 22" x 2" | 20" x 22" x 1-1/2" | 20" x 22" x 1-1/4" | ▼ | |
| 20" | 20" | | 80 | 200 | 320 | 7 | 22" x 22" x 2" | 22" x 22" x 1-1/2" | 22" x 22" x 1-1/4" | 3/4" | |
| 22" | 20" | | 88 | 220 | 352 | 7 | 24" x 22" x 2" | 24" x 22" x 1-1/2" | 24" x 22" x 1-1/4" | | |
| 24" | 20" | | 96 | 240 | 384 | 7 | 26" x 22" x 2" | 26" x 22" x 1-1/2" | 26" x 22" x 1-1/4" | ▼ | |

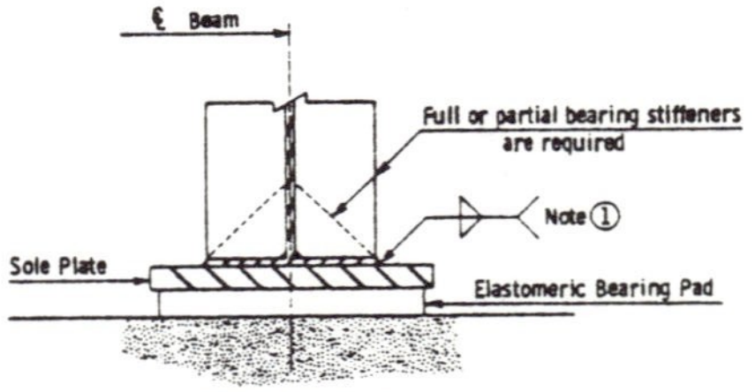


SECTION X-X

① 5/16" minimum fillet weld for 3/4" up to and including 1-1/2" thick sole plates. 3/8" minimum fillet weld for over 1-1/2" to 2-1/4" thick sole plates.

NOTES: All bearing pads are to bear evenly on the bearing seats.

For elastomeric material and pad construction, see Spec. 3741.

The Sole Plate will be the same material spec. as the steel beam and will be included in the weight of structural steel. Shop weld the sole plate to the steel beam.

Minimum A dimension shall be 3(D - 1/8 (No. of steel plates))
Minimum B dimension shall be 2(D - 1/8 (No. of steel plates))

### MOVEMENT AND LAMINATE TABLE

| MAXIMUM MOVEMENT 3/8" Lam. | MAXIMUM MOVEMENT 1/2" Lam. | MAXIMUM MOVEMENT 3/4" Lam. | PAD THICKNESS D | MAXIMUM NO. OF LAMINATES 3/8" Lam. | MAXIMUM NO. OF LAMINATES 1/2" Lam. | MAXIMUM NO. OF LAMINATES 3/4" Lam. |
|---|---|---|---|---|---|---|
| 3/16" | 1/4" | — | 1" | 1 | 1 | — |
| 3/8" | 7/16" | — | 1-1/2" | 2 | 2 | — |
| 9/16" | 5/8" | 11/16" | 2" | 3 | 3 | 2 |
| 3/4" | 13/16" | 7/8" | 2-1/2" | 4 | 4 | 3 |
| 15/16" | 1" | 1-1/8" | 3" | 5 | 5 | 3 |
| 1-1/8" | 1-1/4" | 1-5/16" | 3-1/2" | 6 | 5 | 4 |
| 1-5/16" | 1-7/16" | 1-1/2" | 4" | 7 | 6 | 5 |
| 1-1/2" | 1-5/8" | 1-3/4" | 4-1/2" | 8 | 7 | 5 |
| 1-11/16" | 1-13/16" | 1-15/16" | 5" | 9 | 8 | 6 |
| 1-7/8" | 2" | 2-3/16" | 5-1/2" | 10 | 9 | 6 |
| 2-1/16" | 2-1/4" | 2-3/8" | 6" | 11 | 9 | 7 |
| 2-1/4" | 2-7/16" | 2-5/8" | 6-1/2" | 12 | 10 | 7 |
| 2-7/16" | 2-5/8" | 2-13/16" | 7" | 13 | 11 | 8 |
| 2-5/8" | 2-13/16" | 3" | 7-1/2" | 14 | 12 | 9 |
| 2-13/16" | 3" | 3-1/4" | 8" | 15 | 13 | 9 |

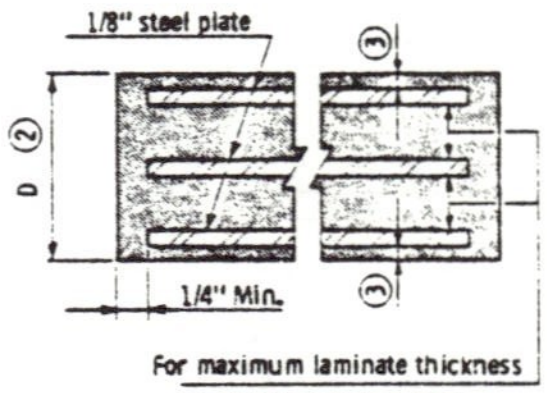


For maximum laminate thickness See Elastomeric Pad Table

SECTION THRU PAD

② Total thickness including the steel plates

③ 1/8" for 1/2" and 3/4" laminates 3/16" for 3/8" laminates

## STEEL LAMINATE PAD DETAIL

(1-1/2" Pad With 1/2" Laminates Shown)

APPROVED: April 26, 1978

Developed by: ENGINEERING STANDARDS BRIDGES AND STRUCTURES

Issued by: ENGINEERING STANDARDS

STATE OF MINNESOTA DEPARTMENT OF TRANSPORTATION

**ELASTOMERIC BEARING PAD STEEL BEAMS (EXPANSION)**

DETAIL NO. **B355**

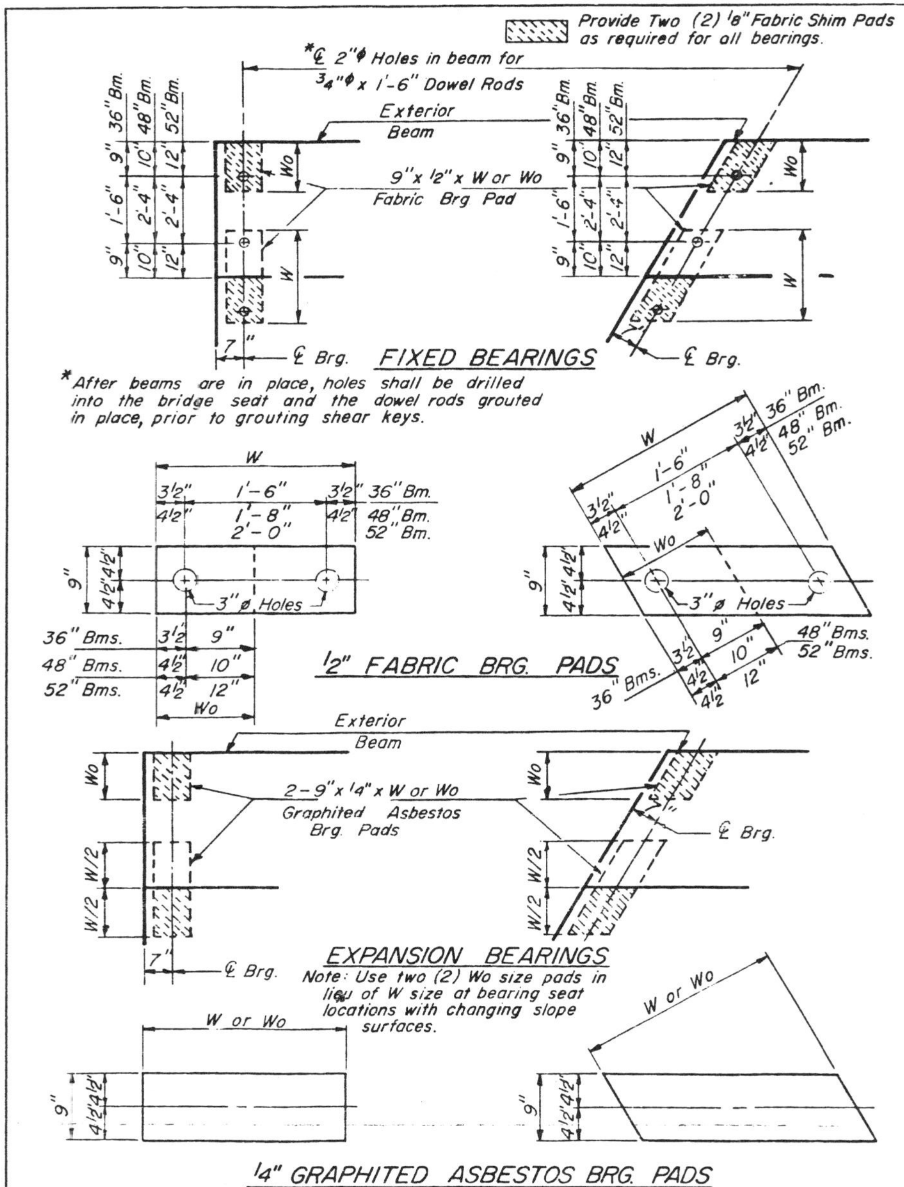


*After beams are in place, holes shall be drilled into the bridge seat and the dowel rods grouted in place, prior to grouting shear keys.

Note: Use two (2) Wo size pads in lieu of W size at bearing seat locations with changing slope surfaces.

W = 2'-1" & Wo = 1'-0½" for 36" Bms.
W = 2'-5" & Wo = 1'-2½" for 48" Bms.
W = 2'-9" & Wo = 1'-4½" for 52" Bms.

II/3.21

ILL. 3-81

BEARING PADS

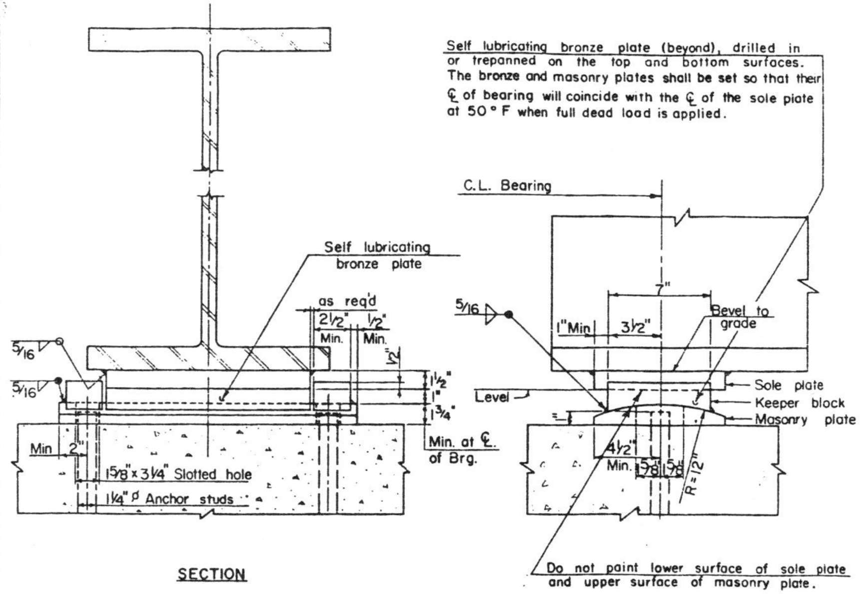


NOTES:

Anchor studs shall not project above masonry plate.

Caulk holes with lead wool.

All steel in bearings shall conform to A S T M A 588.

Self lubricating bronze plates shall be paid for as structural steel.

DESIGN INFORMATION:

For purpose of design, use coefficient of friction = .10 and an allowable bearing pressure of 2000 p.s.i. over the entire area of self lubricating bronze plate.

The bearing area shall be proportionately designed to the rest of the installation.

The 7 inch standard width of bronze plate may be modified under special conditions.

The length of bronze plate shall be detailed to the nearest inch.

A minimum projection of 3 in. will be required from edge of masonry plate to edge of concrete pad.

THIS TYPE OF BEARING MAY BE USED FOR SPANS BETWEEN 50' and 125'.

| | |
|---|---|
| SELF LUBRICATING BRONZE PLATE EXPANSION BEARING DETAIL | CONN. DEPARTMENT OF TRANSPORTATION |
| REVISED: 11-66, 5-67, 12-69, 5-71, 5-72, 4-73, 6-75, 1-76, 1-79 | PLATE: 4-5.1 |

# Illinois Department of Transportation
## Memorandum

**To:** All Bridge Designers 82.5

**From:** Carl E. Thunman, Jr.

**Subject:** WOVEN TEFLON BUSHINGS

**Date:** October 13, 1982

We have experienced in the past the corrosion and the freezing of the links used in connecting steel girders of structures with suspended spans. Other structural components incurred damage as a result of the freezing of the links. To alleviate the problems of corrosion and freezing, we have investigated the possibility of using available teflon material that could sustain the design loading. As a result, it shall be the practice of this office, to provide self lubricating woven teflon bushings in conjunction with the pins for this type of connection as shown on the attached sheet.

The following available types of bushings could sustain a bearing stress of 20,000 psi which is equivalent to the bearing stress allowed by the AASHTO specifications for pins subject to rotation when M 222 or M 223 (GR.50) structural steel is used.

1. Duralon Bearings. A self lubricating woven teflon bearing backed by a filament wound epoxy matrix. Produced by Rexnord Bearing Division.

2. Fiber Glide. A self lubricating metal backed woven teflon bearing. Produced by LSI Transport Dynamics Divisions.

This new practice shall not be retroactive to design plans already completed or under progress but shall be incorporated in designs for which TSL plans have not been completed.

SYK:dp

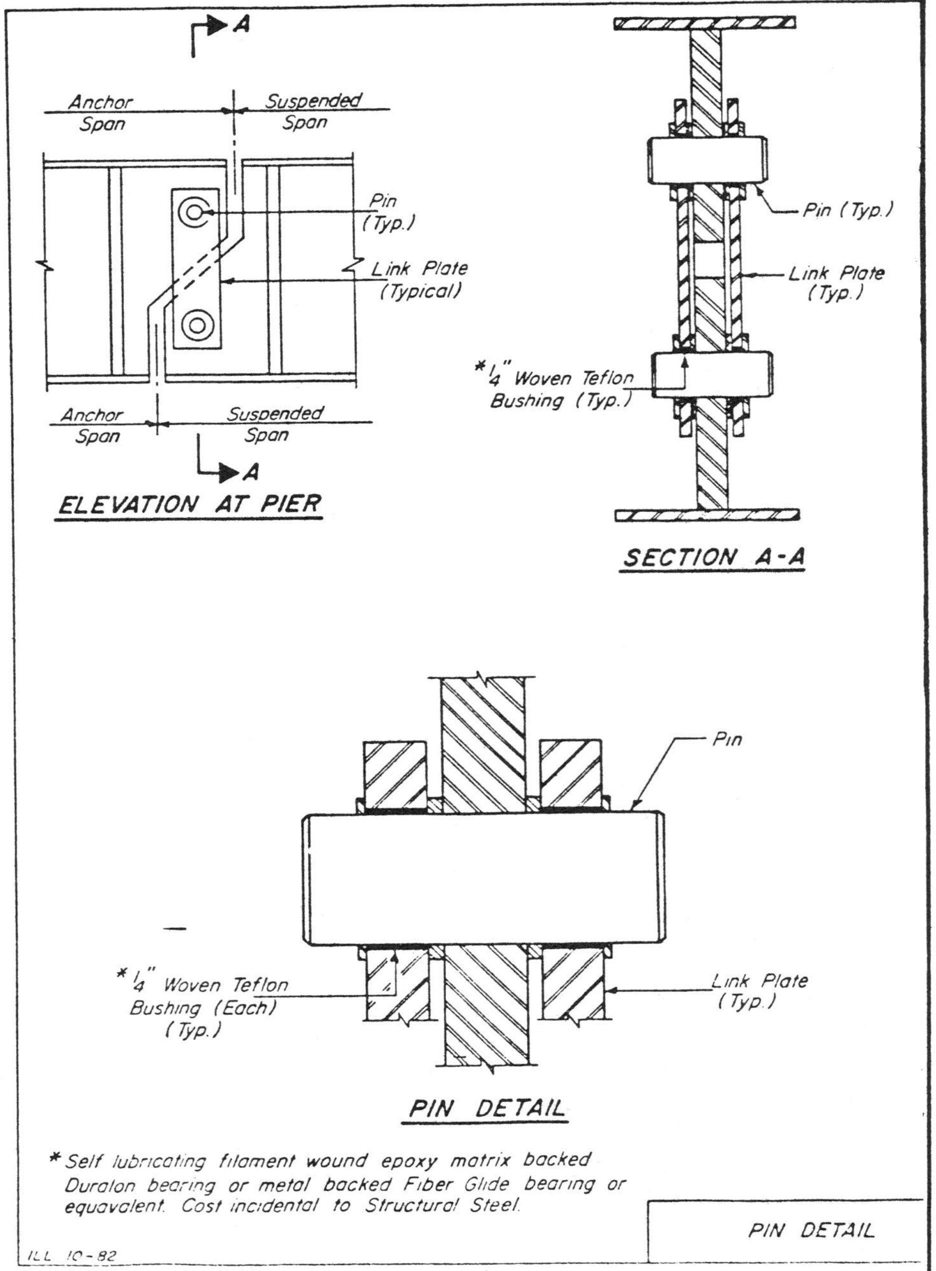

## BEARING PLATE DESIGN

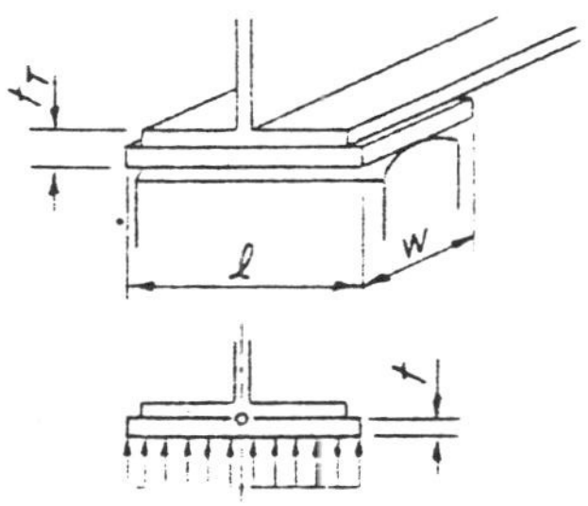


### TOP BEARING PLATES

R = Reaction in kips including impact for top plate only.
C = .194 for $f_s$ = 20 ksi
C = .167 for $f_s$ = 27 ksi

$$t_T^2 = \frac{6R\ell}{8wf_s}$$

$$t_T = C\sqrt{\frac{R\ell}{w}}$$

t = $t_T$ − flange thickness
Min. t = 1"

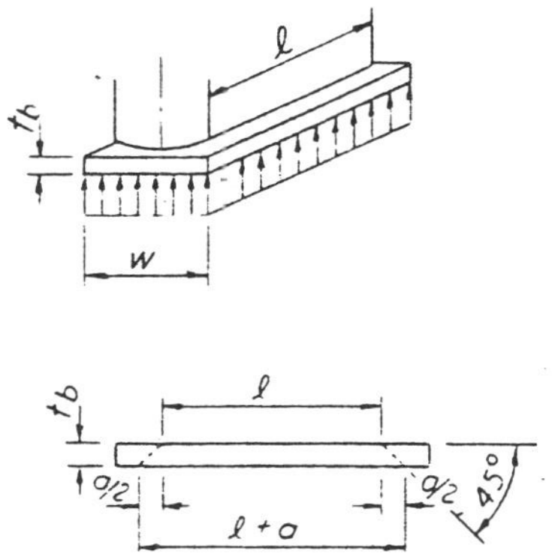


### BOTTOM BEARING PLATES

$$w = \frac{R}{f_{\substack{MASONRY\\ALLOW}}(\ell + a)} \qquad \text{Minimum } w = 9''$$

$$t_b^2 = \frac{6Rw}{8(\ell + a)f_s}$$

$$t_b = C\sqrt{\frac{Rw}{(\ell + a)}}$$

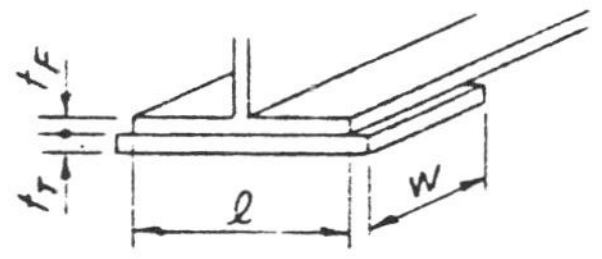


### TOP PLATE WELD DESIGN

* (No Bearing Stiffeners)

$S_w$ = Stress per inch of weld

$$S_w = \frac{3Rt_Ft_T\ell}{4(t_F + t_T)^3(\ell + w)}$$

* With bearing stiffeners use ½" weld.

## BOLSTER DETAILS

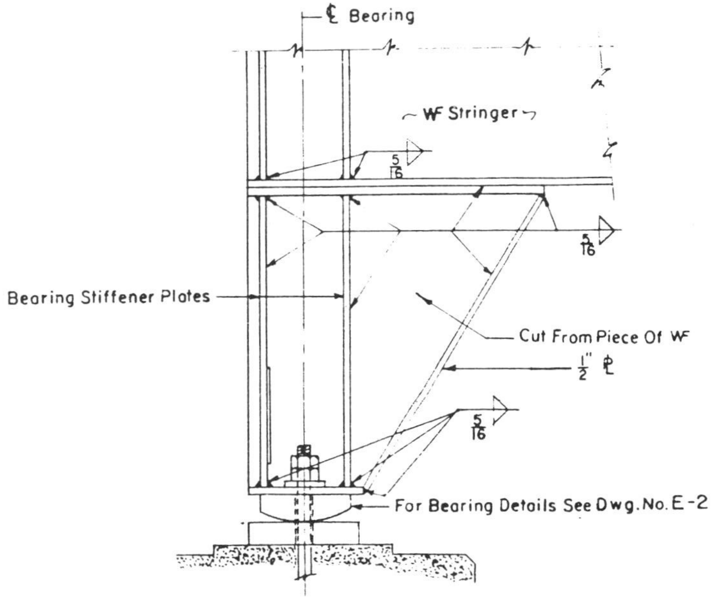


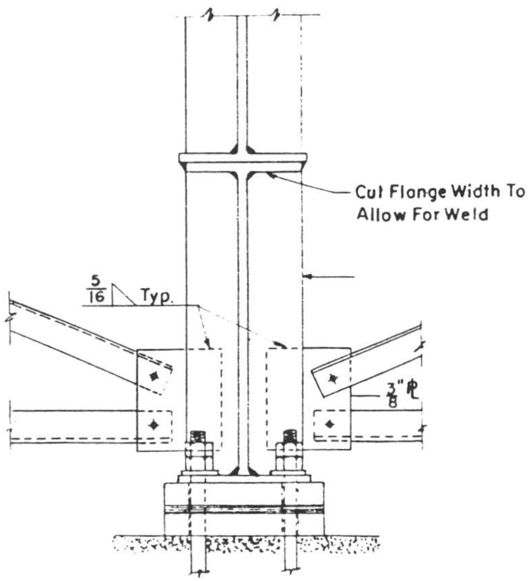




(MASS DPW)

## J.1 ABUTMENT

### BREASTWALLS



Three types of breastwall are commonly used: namely, the gravity wall of plain concrete, the cantilever wall of reinforced concrete, and the type of wall that acts as a vertical beam supported

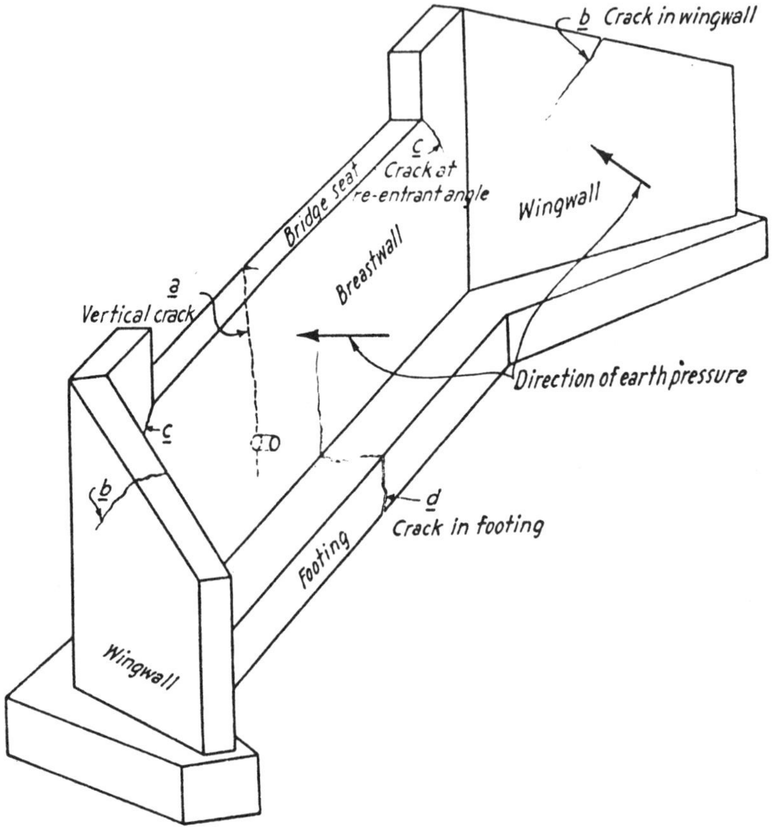


FIG. 3 Typical cracks that may develop in the common abutment type when the conventional analytical methods take little or no account of the actual behavior of the abutment.

**Reinforcing Footing to Prevent Cracks**

horizontally by the deck and by the foundation. There are also several modified types such as the semi-gravity and the buttressed retaining wall.

**Causes of Tension in Breastwalls**

The cantilever retaining wall, probably the type most commonly used for breastwalls, sometimes develops cracks similar to *a* in Fig. 3. Such cracks may be prevented as will be seen after a brief discussion of their contributing causes, among which are: (a) non-uniform shrinkage, (b) a discrepancy between design assumptions and actual behavior, and (c) the earth pressure on wingwalls. The relative effects of these phenomena are uncertain, but all of them act to create tension in the front face of the breastwall.

Shrinkage in abutment walls may be non-uniform because the front surface, being exposed, dries out in relation to the rear surface that is kept moist by the adjacent fill. This shrinkage tends to set up tensile stresses in the breastwall.

**Design Discrepancies**

In studying the discrepancy between design assumption and actual behavior, consider an abutment of the type shown diagrammatically in Fig. 3. The breastwall is assumed in the design to be a cantilever retaining wall and is reinforced accordingly; that is, the reinforcement is placed near the rear face. In reality, the wall acts partly as a simple vertical cantilever, and partly as a box retaining wall braced at the ends by the wingwalls acting as counterforts. Wall slabs of this type have tensile stresses in the front face midway between the counterforts and should be reinforced accordingly with horizontal bars near the front face. Unfortunately, all the reinforcement in ordinary breastwalls is placed near the rear face; but the horizontal bars would be more effective if placed near the front face.

The wingwalls in Fig. 3 are also designed and built as cantilever retaining walls but are cast integrally with the breastwall, with two results: the whole abutment acts as a unit, and the earth pressure on flared and parallel wingwalls produces tension in the breastwall.

**Bars in Front Face**

According to this discussion, an abutment of the type shown in Fig. 3 can be improved by adding horizontal bars near the front face. Bars *a* in Fig. 4 provide for both horizontal beam action and shrinkage, and are recommended for use with the regular reinforcement provided for cantilever action. In addition, it is sometimes advisable to place vertical bars near the front face. These bars will take the tensile stresses developed in case the breastwall acts as a vertical beam supported horizontally by the deck and by the foundation, a loading condition that will be discussed in the section on *Abutment Movements.*

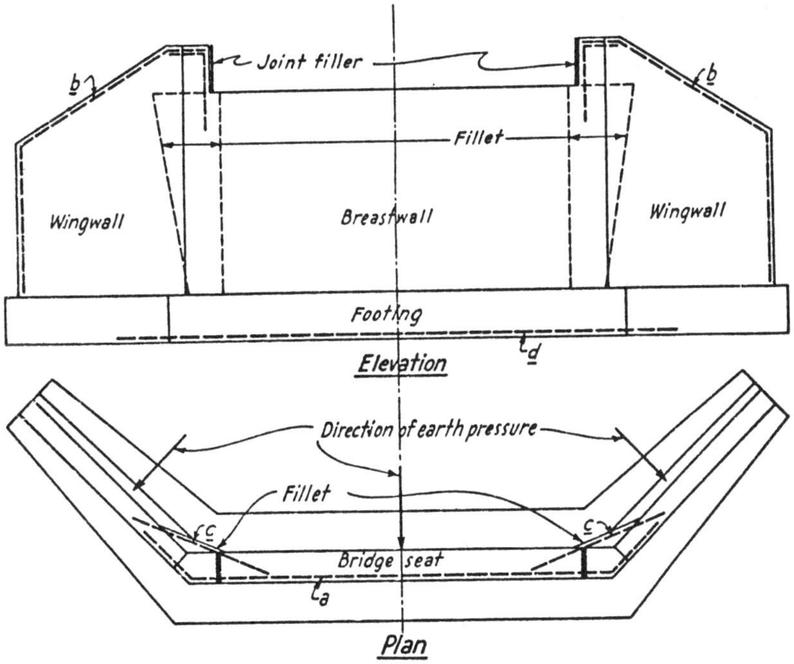


FIG. 4 Common abutment type, reinforced and strengthened. The ordinary abutment reinforcement is not shown.

## WINGWALLS

The wingwalls are usually of the same type as the breastwall but differ from it in that they are topped by a simple coping without provision for support of any superstructures. They will be referred to as parallel, flared, curved, or straight wingwalls.

Cracks that may develop in wingwalls and at the junction of wingwall and breastwall are designated as *b* and *c* in Fig. 6*a* and in

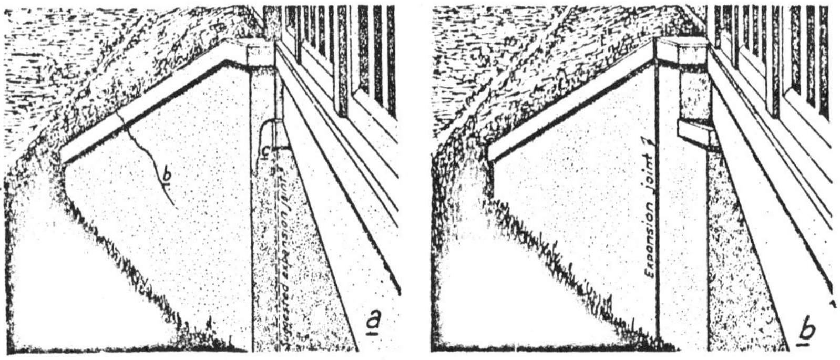


FIG. 6 Cracks *b* and *c* in abutment with wingwalls as illustrated in left-hand sketch may be avoided if the wingwalls are separated from the breastwall by expansion joints. Two types of joint layout are indicated.

Fig. 3 (page 8). What is the cause of these cracks and how may they be prevented?

**Stresses along Coping**

An abutment of the type shown in Fig. 6*a* may behave partly as a box retaining wall in which the wingwalls act as counterforts. Tensile stresses are developed accordingly along the coping of the wingwall, but adequate reinforcement is seldom provided. The development of cracks of type *b* may therefore be a result of the discrepancy between assumption and reality in abutment action. Additional tensile strains are created since flared or parallel wingwalls have a tendency to pull away from the breastwall. Because this tendency is rarely considered in the design, cracks of type *c* may develop as sketched in Figs. 3 and 6*a*.

Fig. 7 illustrates cracking that may occur in wingwalls built in direct extension of the breastwall. Such cracks are sometimes observed where the abutting surfaces in the vertical joint between the wingwall and the outer girder have not been kept apart with a soft filler. In such cases, relative movement of the abutment with respect to the deck—tilting of the abutment or lateral creep of the deck—may be harmful.

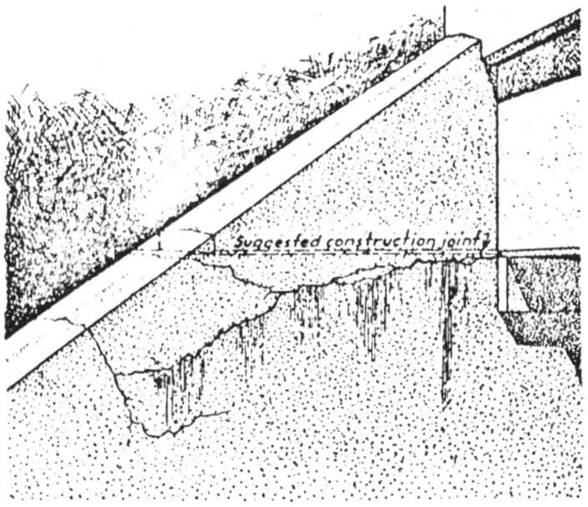


FIG. 7 The cracks in this type of abutment may be avoided by use of a construction joint as indicated.

Two conclusions may be drawn from the observations made: (a) tensile stresses of considerable magnitude may be set up in the concrete by phenomena frequently originating outside the structure, and (b) no structural analysis is available by which the stresses may be determined.

**Improved Wingwall Design**

Accordingly, there appear to be only two methods whereby improvements may be made: (a) suitable reinforcement may be provided—by judgment or empirical rules—in the conventional type of abutments, or (b) improved abutment layouts may be adopted in which the critical strains are eliminated.

Using the first method, wingwall cracks of type *b* in Fig. 3 may be eliminated by adding sufficient reinforcement as indicated by bars marked *b* in Fig. 4.

Cracks like *c* at the end of the bridge seat shown in Fig. 3 can

**Additional Reinforcing in Conventional Wingwalls**

also be avoided; the methods indicated in Fig. 4 have been used with good results for relatively shallow abutments with heights up to about 15 feet. The deck and wingwall are separated by a soft joint filler at least ¾ inch thick. The junction between wingwall and breastwall is strengthened by means of a concrete fillet which is reinforced with bars marked *c*. It is advisable to use several *c*-bars in the plane just below the bridge seat. The proportions of the filler and the amount of reinforcement must be based upon judgment Such methods of reinforcing the conventional abutment often have been found effective.

**Vertical Expansion Joints**

An example of an improved design adopted to eliminate critical strains is illustrated in Fig. 6*a*, in which the double dotted line indicates a vertical expansion joint. The tensile strains responsible for cracks are then eliminated and the wingwall acts as a simple retaining wall and not simultaneously as a counterfort. It is evident that a lateral pressure of the deck against the top of the wingwall will merely widen the space in the expansion joint, and critical tensile strains are eliminated where crack *c* might otherwise develop. A similar effect is obtained in the construction sketched in Fig. 6*b*.

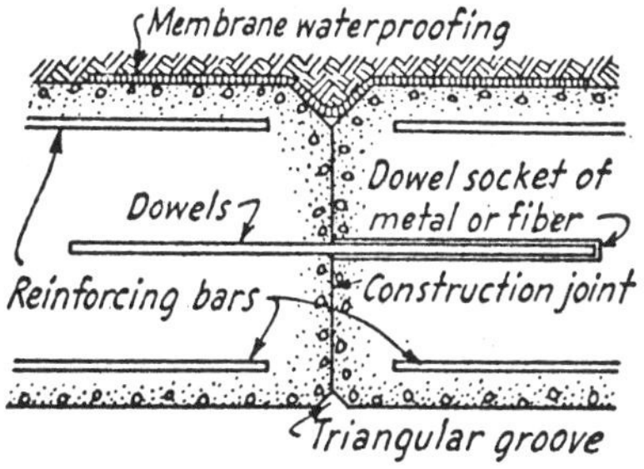


FIG. 23 Horizontal section of vertical construction joint sometimes used in abutments. The dowels are bonded to the concrete on one side only.

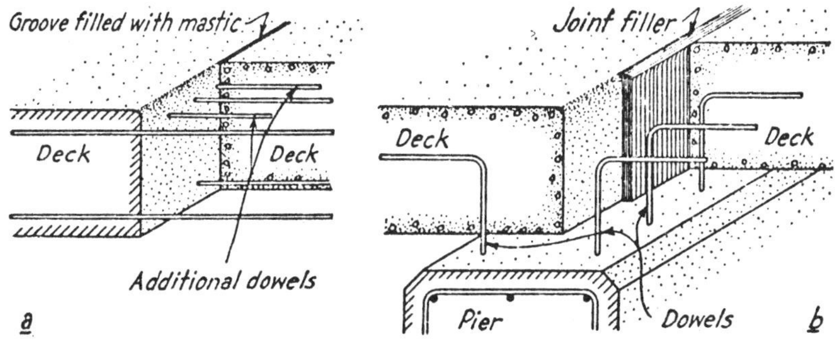


FIG. 27 Joints suitable for concrete decks. The joint in type *a* may be keyed and is used where the reinforcement is continuous, while type *b* is used where the reinforcement is not continuous.

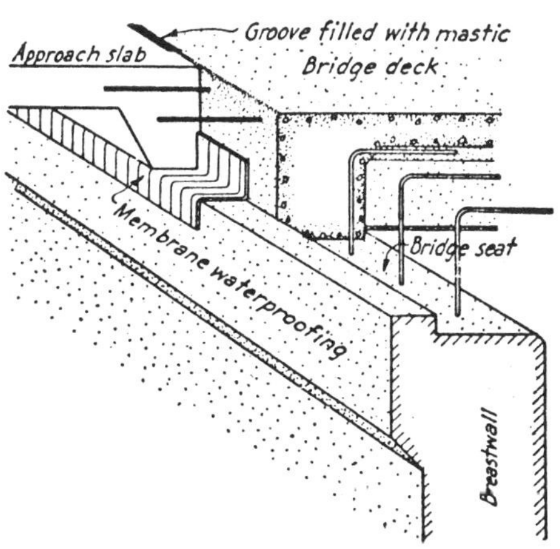


FIG. 30 Isometric view illustrating bearing arrangement at the fixed end of a concrete deck. The horizontal dowels across the vertical joint may be bonded to the deck concrete only.

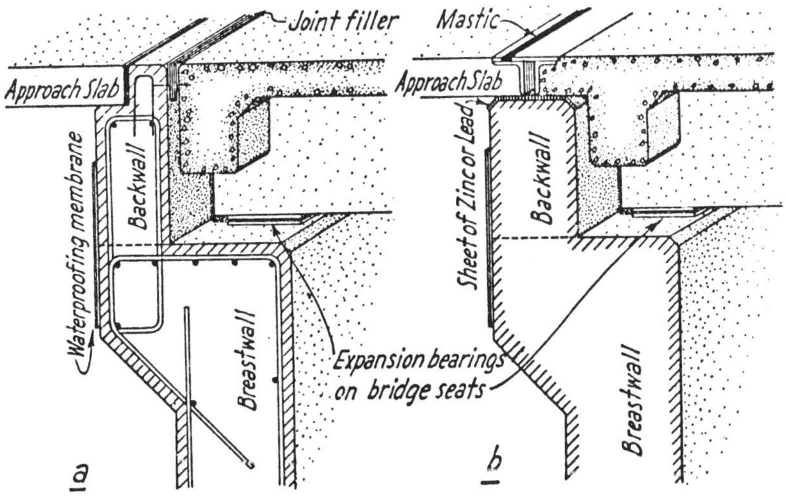


FIG. 31 Two arrangements of joints at the expansion end of concrete decks. The backwall is extended to the roadway surface in type *a*, while it stops at the bottom of the slabs shown in type *b*.

# CREEP IN SKEW BRIDGES

Many bridges are built with an oblique angle between the direction of the roadway and the direction of the piers or abutments. These so-called skew bridges have some characteristic features that are generally given little attention in detailing and construction.

**Observed Lengthening of Longer Diagonals**

Numerous observations revealed that the longer diagonal of skew bridge decks had a tendency to lengthen during a long period of service. The most distinct evidence was usually observed at the expansion bearing, where the deck had moved laterally upon the bridge seat in the direction toward the acute angle of the deck as indicated in Fig. 46.

Lateral creep has been observed to amount to as much as 4 inches. Even much smaller movements may cause damage. In single-span bridges, for example, the pressure accompanying the

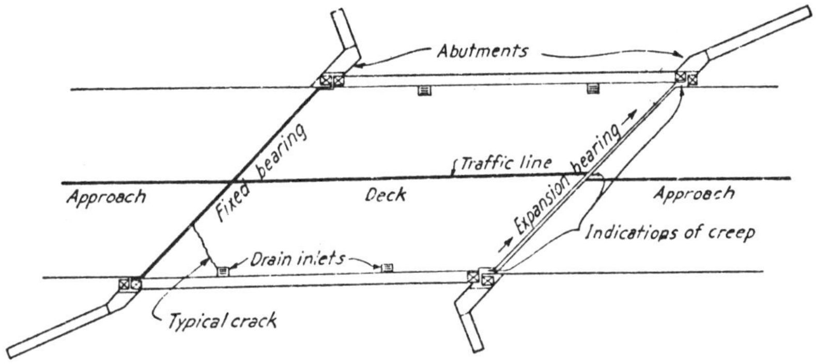


FIG. 46 Plan view of skew bridge illustrating how the long diagonal tends to elongate. The free end then creeps laterally on the expansion bearing as indicated.

movement may be great enough to break a large piece out of the wingwall and displace it several inches. Fig. 47 is a sketch of possible damage to an abutment through failure to prevent creep in a skew bridge. Note that the damage is greatest at the bottom of the abutment where it is restrained from moving.

When a skew slab as shown in Fig. 46 is loaded and deflects, the deck deformation will tend to raise the acute corner of the deck. Cracks of the type sketched in Fig. 46 may then develop. It is therefore advisable to provide suitable reinforcing bars in the top of the slab at its acute angle and preferably parallel to the long diagonal of the slab.

**Evidence of Creep**

It may be significant that the crack at the end of the bridge seat in Fig. 47 resembles very closely cracks $c$ in Figs. 3 and 6$a$. The crack in Fig. 6$a$ and the damage in Fig. 47 may represent different stages in the development of damage done by creep. Creep may possibly be another reason for cracks like $c$ in Fig. 3.

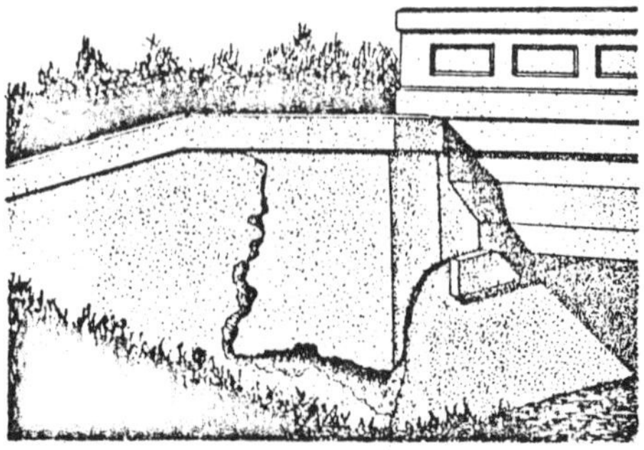

FIG. 47 Abutment damaged by creep at expansion bearing in skew bridge. It is evident that creep is accompanied by great forces.

**Suggested Practice in Bearing Layout**

In view of the discussion in this section, as well as in preceding sections, the following rules appear to warrant considerable attention:

1. In single-span layouts, use two fixed bearings on the abutments when the span length does not exceed 45 feet.
2. In two-span layouts, use fixed bearings on the abutments and expansion bearings on the pier.

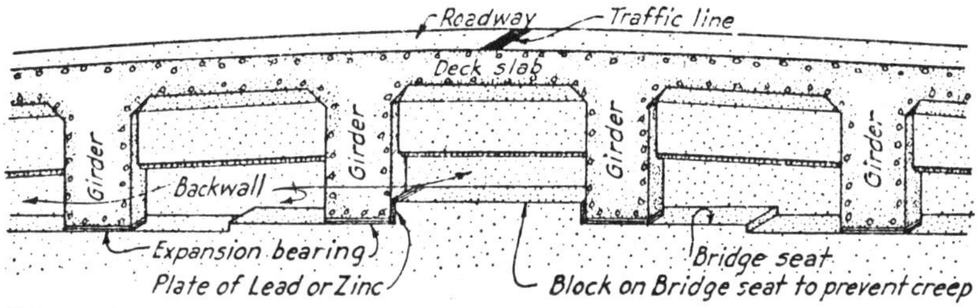


FIG. 51 Vertical section through deck shows expansion bearing in which creep is counteracted by raising the central portion of the bridge seat.

3. In three-span layouts, use fixed bearings on the abutments, two fixed bearings on one pier, and two expansion bearings on the other pier.
4. In four-span layouts, use fixed bearings on the abutments and on the center pier; use expansion bearings on the other two piers.

All expansion bearings in skew bridges or in bridges on curves should have the block between girders indicated in Fig. 51.

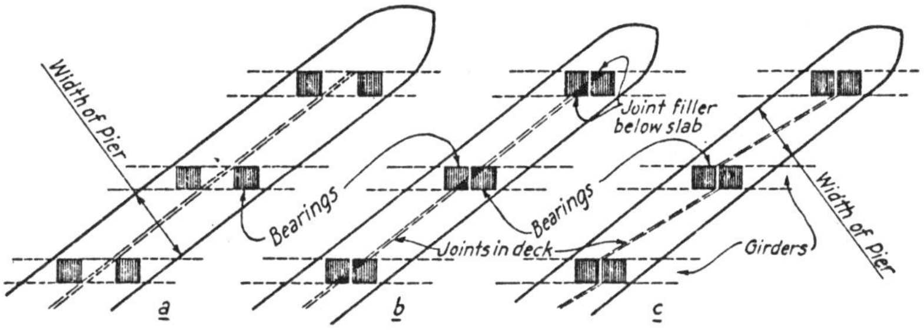


FIG. 52 Three arrangements of girders bearing on piers in a skew bridge. The pier width is often determined by the position of the bearings as indicated in type *a*. The width may be reduced considerably by use of the layouts in types *b* and *c*.

Valuable space may be wasted on piers supporting skew bridges when the bearings are arranged as indicated at *a* in Fig. 52. This arrangement is often adopted for the purpose of obtaining straight deck joints. A narrower pier seat—and a more economical pier and bridge construction—may be obtained by placing the bearings much closer together as indicated at *b* and *c* in Fig. 52. It has been suggested that the deck joint be kept straight by means of a construction as indicated at *b* in Figs. 52 and 53. Some bridge designers use a saw-toothed joint as indicated at *c* in these two illustrations.

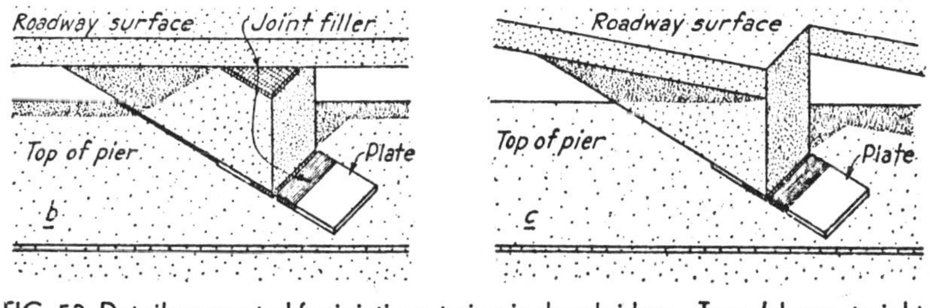


FIG. 53 Details suggested for jointing at piers in skew bridges. Type *b* has a straight deck joint, while the joint in the deck is saw-toothed in type *c*.

## CONCRETE CAP EXTENSION

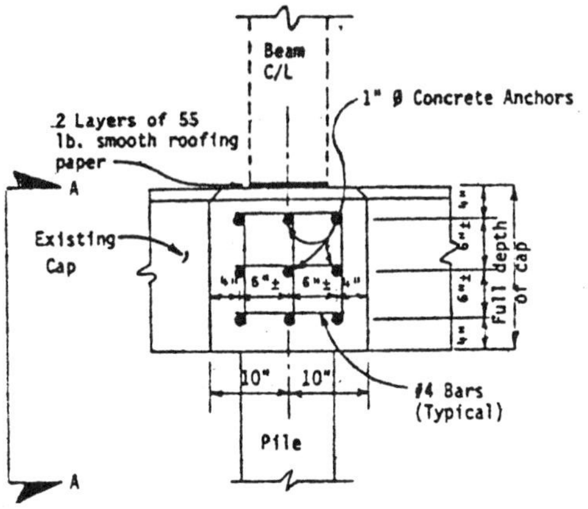


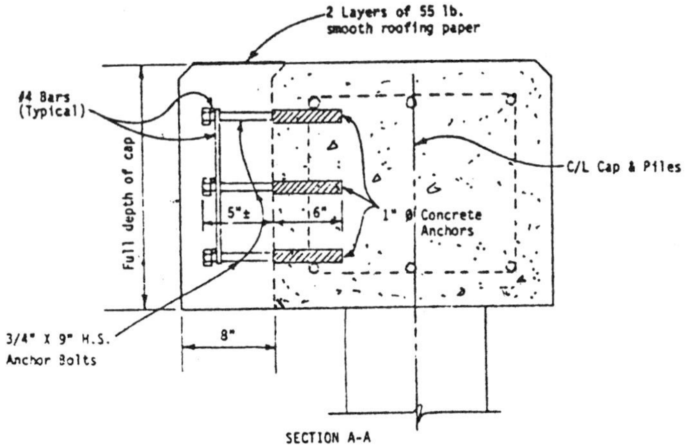


SECTION A-A

1. Locate and drill holes 6" deep for concrete anchors. If steel is encountered, shift hole to clear.

2. Thoroughly clean holes of dust, etc., prior to placing concrete anchors.

3. Set anchors and place anchor bolts.

4. Place reinforcing steel.

5. Form cap extension for placing concrete.

6. Place roofing paper against bottom of beam prior to placing concrete.

7. Remove forms after 3 days.

## CRUTCH BENT

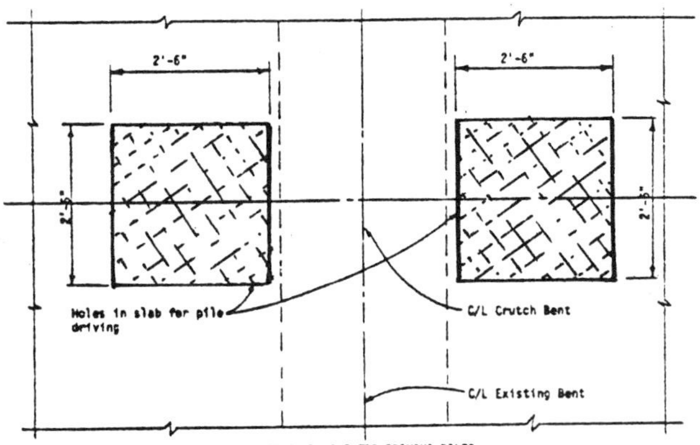


DETAIL OF SLAB HOLE FOR DRIVING PILES

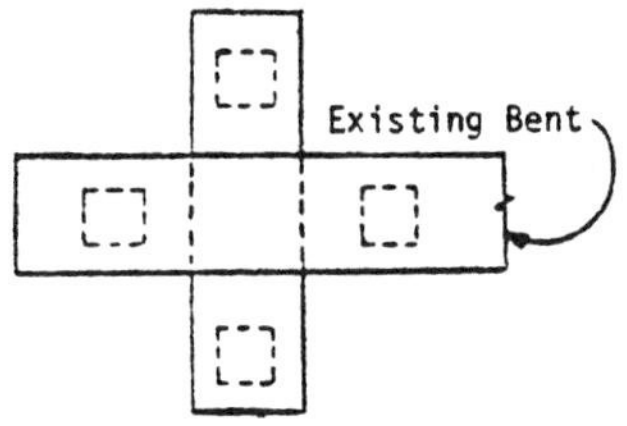


PLAN VIEW

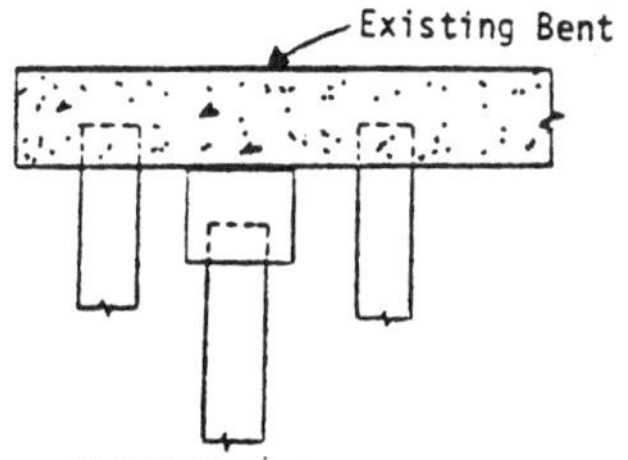


ELEVATION VIEW

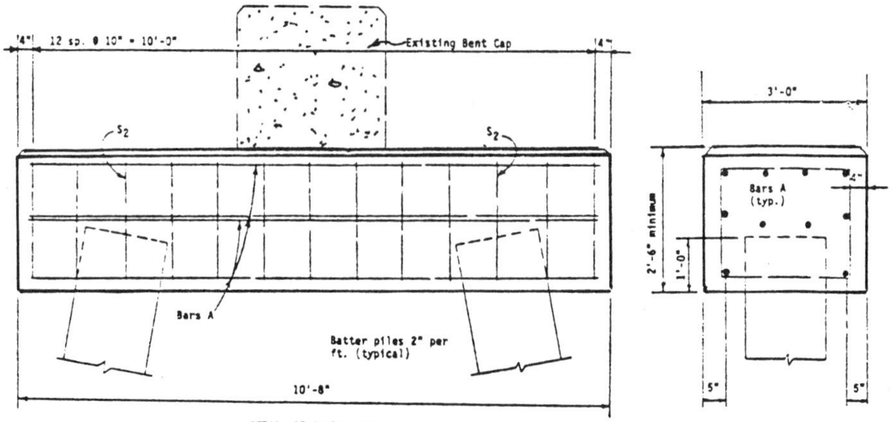


DETAIL OF CRUTCH CAP

**CRUTCH BENT**

CONSTRUCTION METHOD:

1. Layout location on deck for crutch bent piles.
2. Cut 2'-6" square hole in deck at each pile location. Cut conflicting reinforcing steel at center of hole and bend out of the way to allow pile to pass.
3. Set piles and drive to specified bearing.
4. Cut off piles 1'-6" below bottom of existing cap.
5. Bend deck reinforcing steel back into position and cover holes in deck temporarily by placing a 3/8" steel plate over hole. (See detail drawing.)
6. Set bottom cap form around pile.
7. Place reinforcing for cap.
8. Set and form tie end and side cap forms.
9. Remove temporary deck hole covers and place tremie through holes for pouring concrete.
10. Place Class IV concrete in cap form in layers. Vibrate each layer by letting vibrator penetrate into top of preceeding layer.
11. Trowel finish top of cap and apply curing compound.
12. Remove side and end cap form after 48 hours. Remove bottom cap forms after 5 days.

Deck Repair

13. Splice cut reinforcing bars in deck with #6 rebars 2'-4" long.
14. Form bottom of deck holes and place Class III concrete. Broom finish and cure concrete.

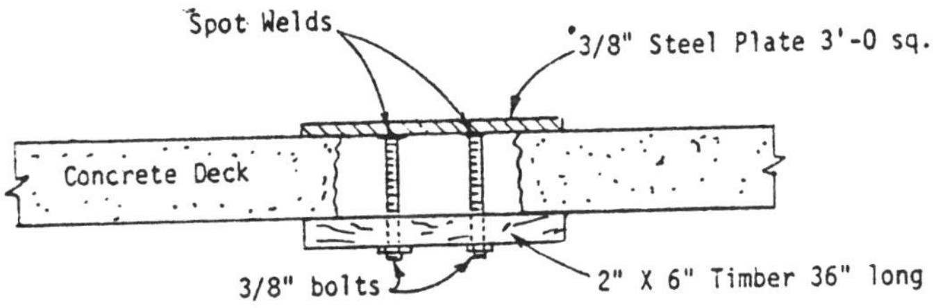


DECK PLATE DETAIL

## BEAM SADDLE

CONSTRUCTION METHOD:

1. Check plans for width of beam and cap or obtain field measurements.
2. Request Engineer to design saddle.
3. Fabricate and paint saddle members.
4. Prepare top of cap on both sides of beam and bottom of beams for good bearing contact between saddle and concrete.
5. Cut neoprene bearing pads to cover concrete cap areas in contact with the saddle on both sides of the beam. Also prepare pads for areas between bottom of beams and saddle.
6. Set saddle members at right angle to cap on each side of beam on top of neoprene pad.
7. Drop one bolt through holes in each end of set members. Hang sections under beam, placing bearing pads before tightening bolts.
8. Install remaining bolts and tighten until 100% bearing is obtained under beams or beam is raised to desired elevation.

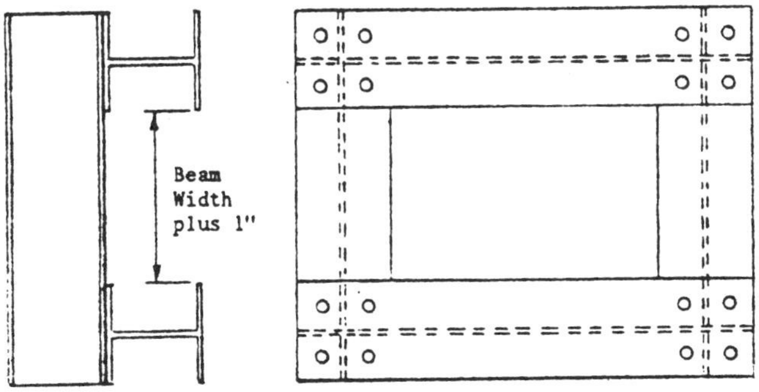


BEAM SADDLE

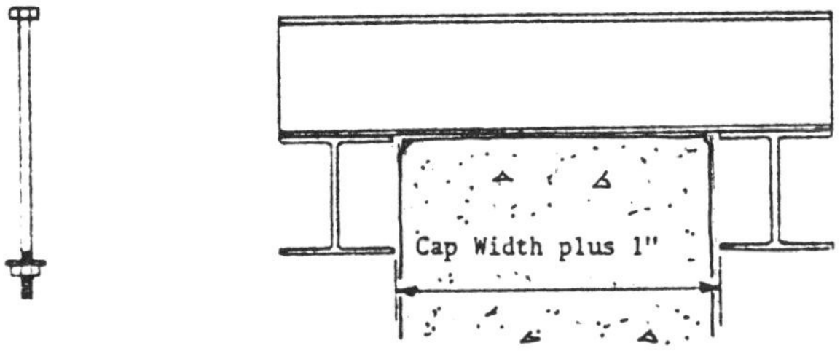

# PIER SADDLE ALTERNATES

## SADDLE ALTERNATE:A

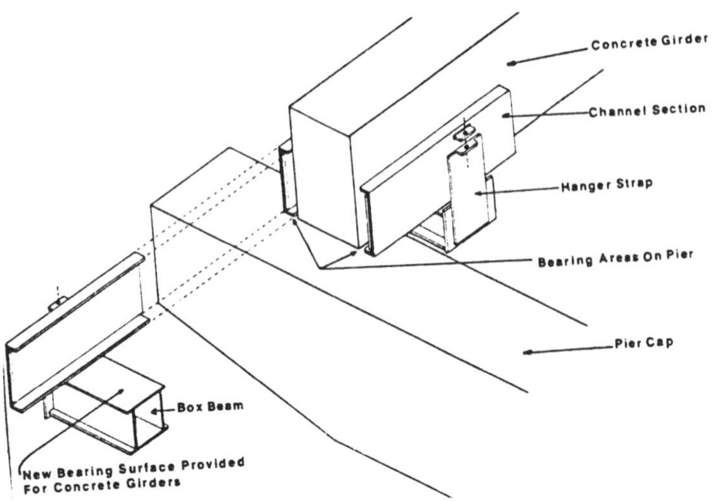


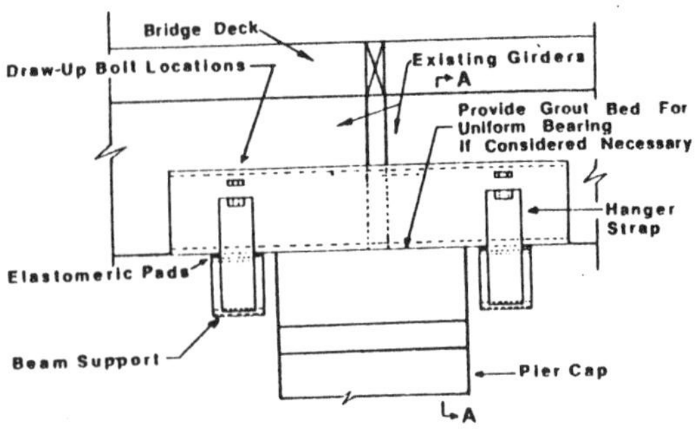


Figure 4a. Side Elevation

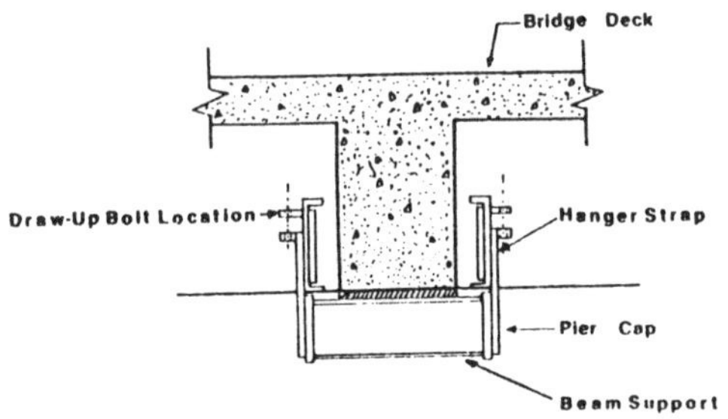


## SADDLE ALTERNATE:B

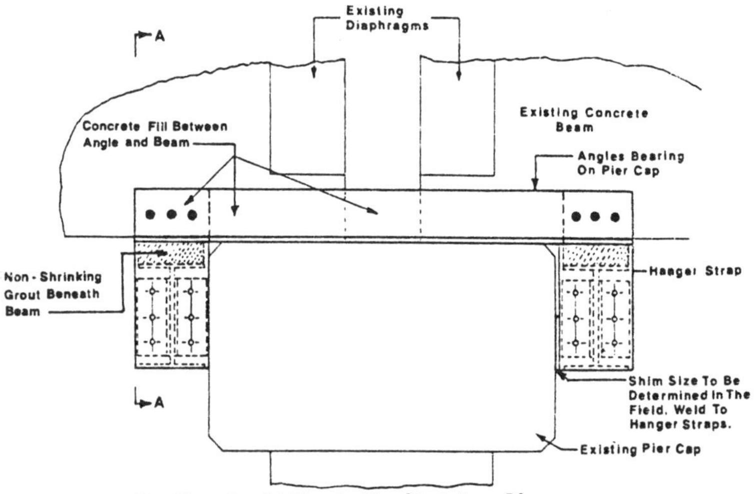


Side Elevation Of Pier Saddle [Technique B]

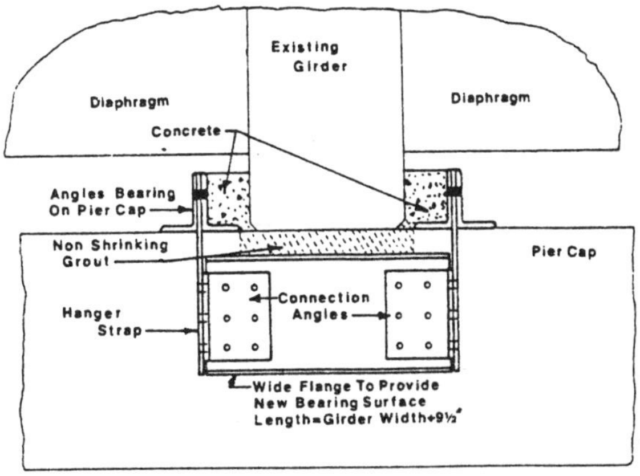


NCHRP TRB

## SADDLE ALTERNATE:C

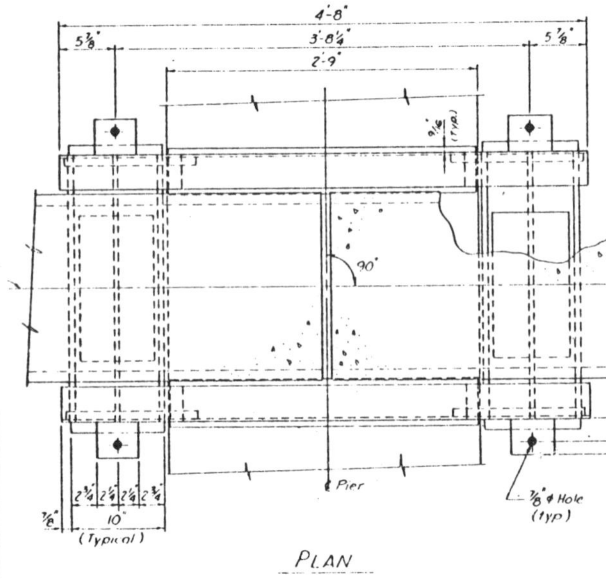


PLAN

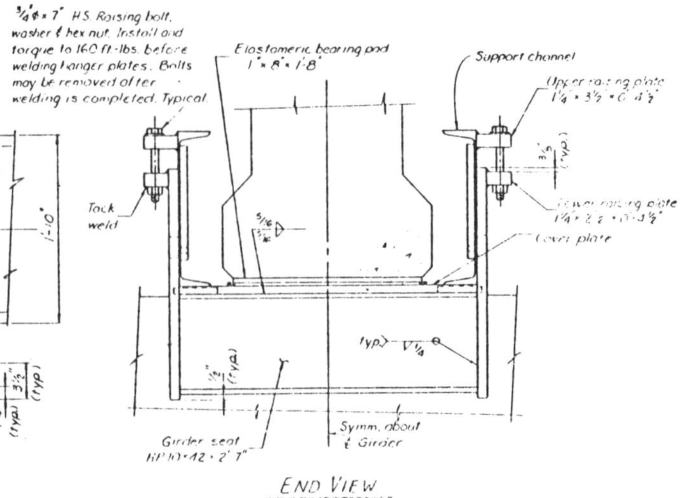


END VIEW

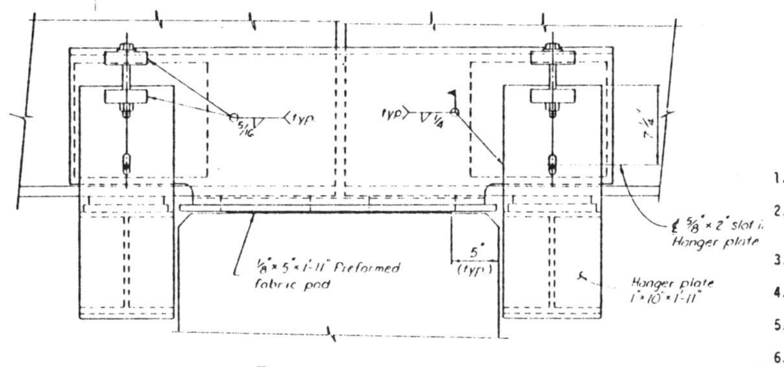


ELEVATION

**SUGGESTED CONSTRUCTION PROCEDURE**

1. LOCATE AND MARK CENTERLINE OF GIRDER ON THE PIER CAP.
2. MARK LOCATION OF BOLT HOLES ON PIER CAP, SEE ELEVATION VIEW AND SECTION THROUGH PIER CAP.
3. DRILL HOLES IN PIER CAP, HOLES MAY BE MOVED ONE INCH TO AVOID REINFORCING STEEL.
4. CUT HOLES IN PLYWOOD FORM TO MATCH THE HOLES IN THE PIER CAP.
5. REMOVE LOOSE CONCRETE AND CLEAN THE REINFORCING STEEL.
6. BOLT THE PLYWOOD FORM IN PLACE AND RESTORE THE PIER CAP TO ITS ORIGINAL SECTION USING AN APPROVED GROUT. (DURACAL, ETC.)
7. ALLOW CURING TIME AS RECOMMENDED BY THE MANUFACTURER BEFORE REMOVING THE FORM.
8. USING THE PLYWOOD FORM FOR A PATTERN, CUT THE BOLT HOLES IN THE FRONT PLATE.
9. ATTACH THE FRONT PLATE AND TIGHTEN THE BOLTS (100 LB-FT, TORQUE).
10. RAISE THE SUPPORT ASSEMBLY UNTIL THE ELASTOMERIC PAD IS COMPRESSED. (24K = JACKING FORCE)
11. FIELD WELD THE SUPPORT ASSEMBLY TO THE FRONT PLATE.
12. PAINT ALL STEEL WITH AN APPROVED PAINT.

## BEAM BRACKET

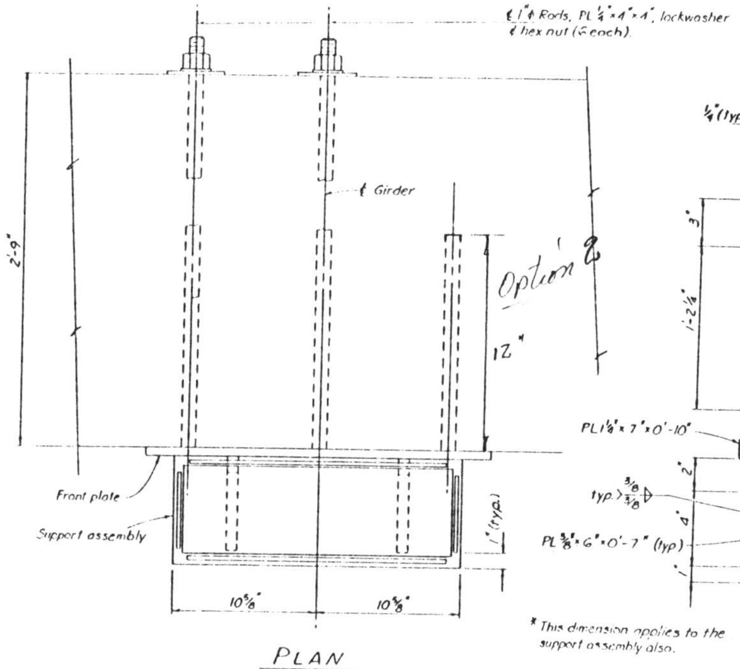


* This dimension applies to the support assembly also.

PLAN

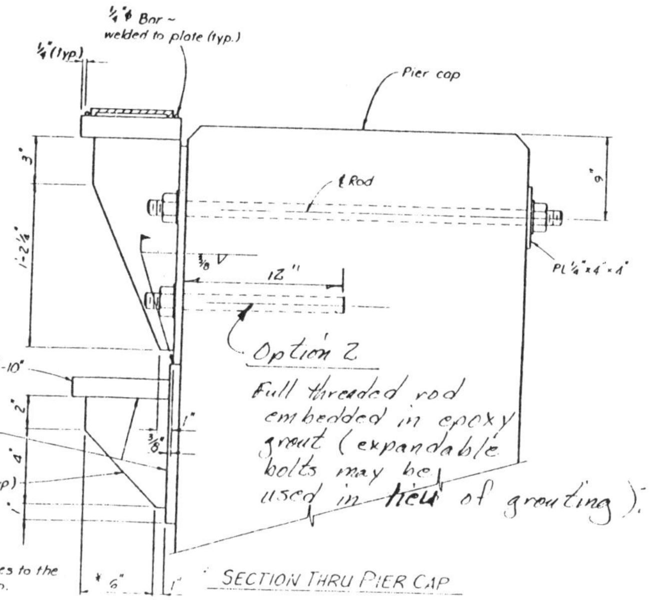


SECTION THRU PIER CAP

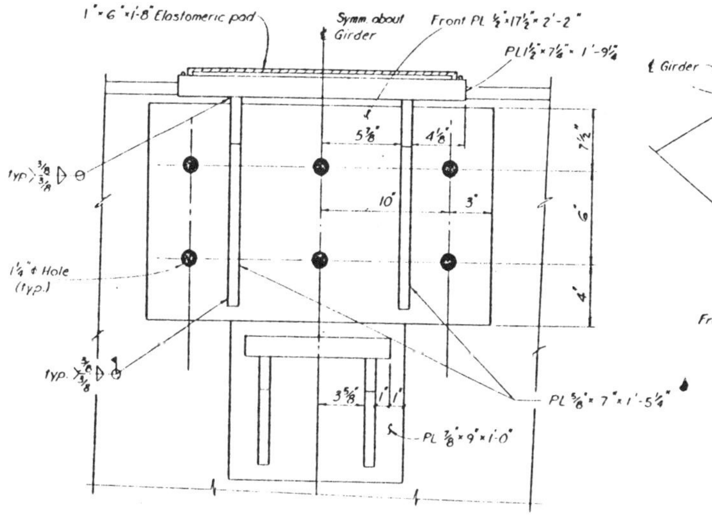


ELEVATION

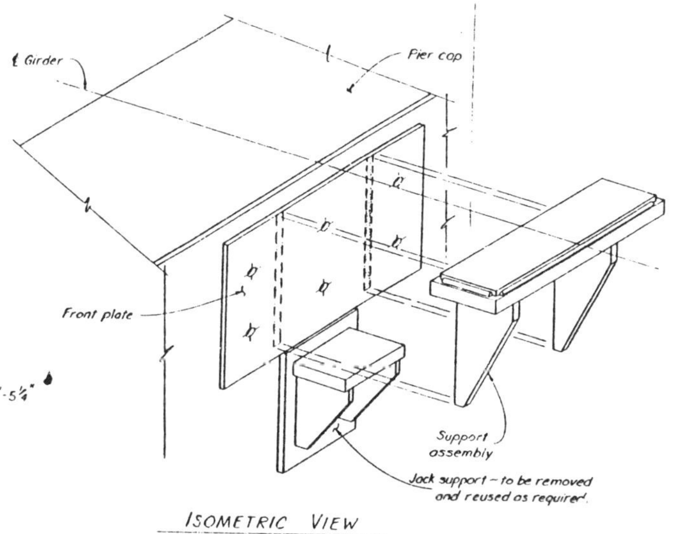


ISOMETRIC VIEW

Colorado

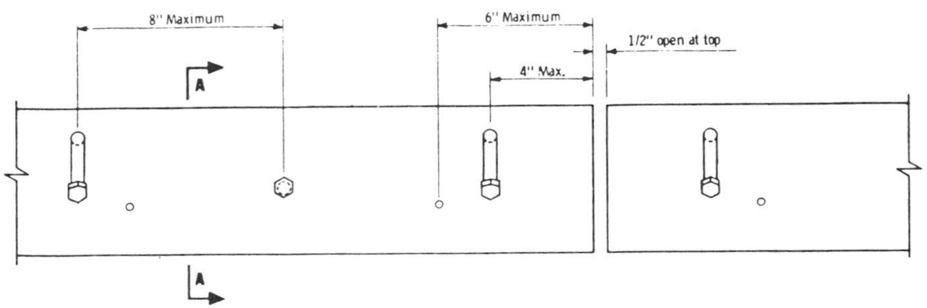


**ELEVATION**
(Concrete not shown)

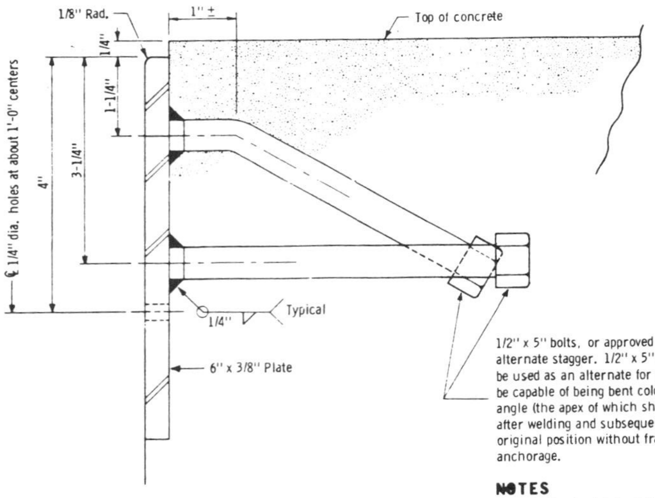


1/2'' x 5'' bolts, or approved equal, at 8'' centers with alternate stagger. 1/2'' x 5'' stud welded fastners may be used as an alternate for bolts. The anchorage shall be capable of being bent cold by hammering thru a 30° angle (the apex of which shall be at the fusion point) after welding and subsequently straightened to its original position without fracture of the weld or anchorage.

**SECTION A-A**

**NOTES**

Plate shall extend full width of roadway between gutter lines with a 1/2'' open joint at each break in crown profile. Maximum length 22 feet.

Materials: Structural steel per spec. 3306. Galvanize after fabrication per spec. 3394

Set plate to proper grade and crown. Plates shall be straightened to a tolerance of 1/16'' in 10 feet after galvanizing.

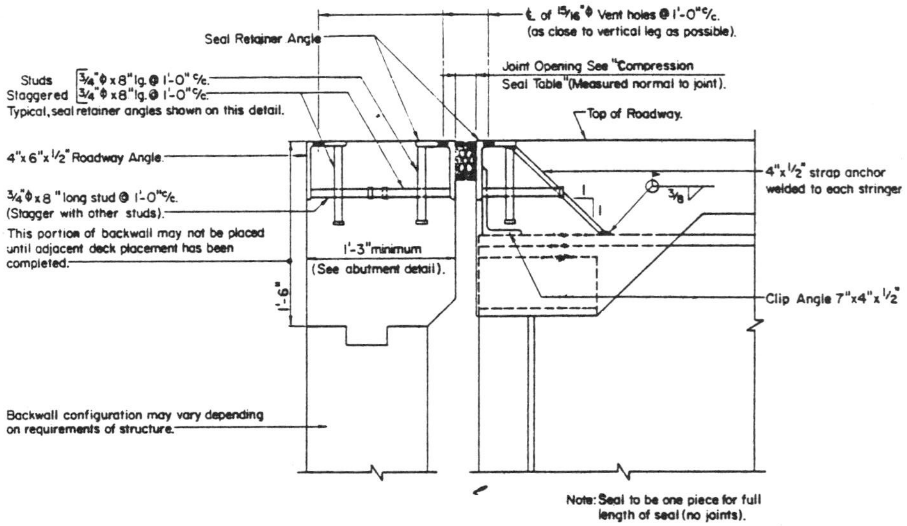


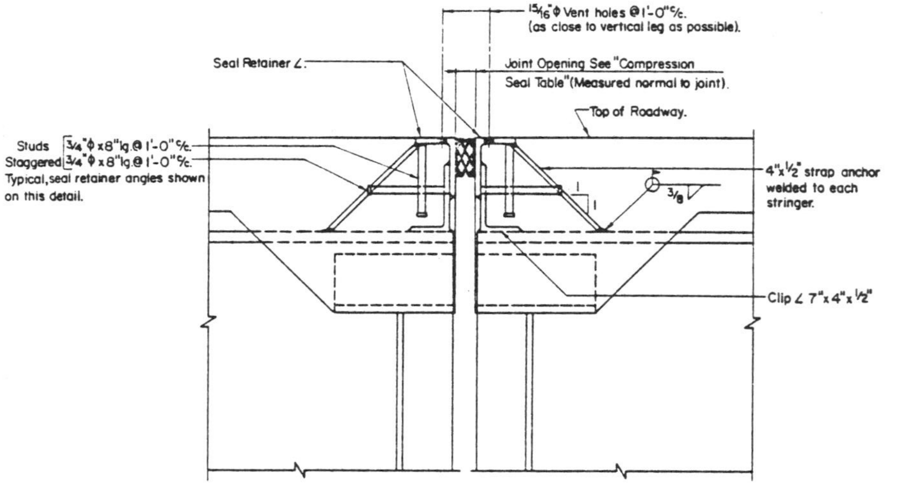


STATE OF MINNESOTA
DEPARTMENT OF TRANSPORTATION

**PROTECTION PLATE**
**FOR END OF SLAB**

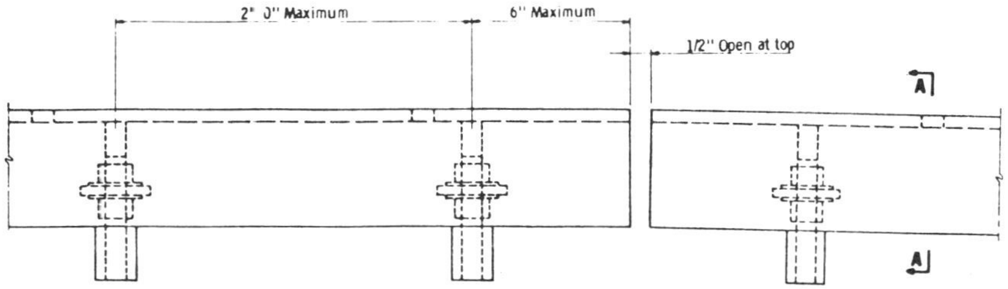


**ELEVATION**
(Concrete not shown)

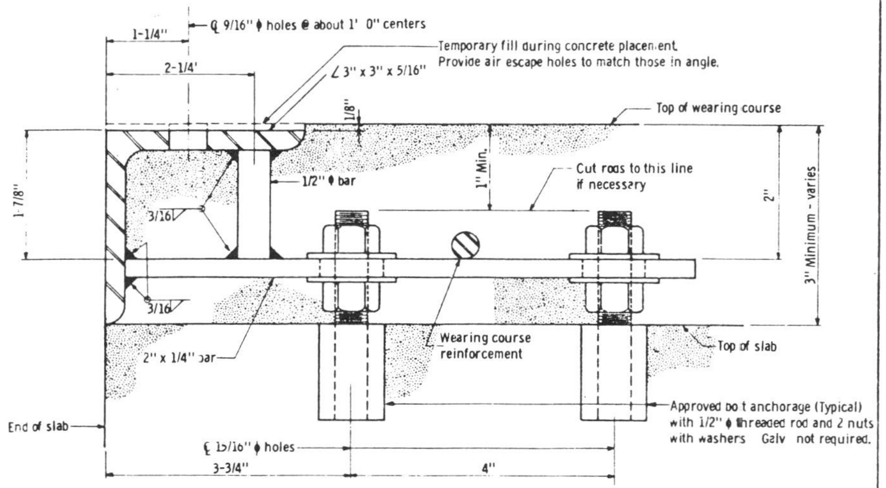


**SECTION A - A**

**NOTES:**

Angles shall extend full width of roadway between curbs with a 1/2" open joint at each break in crown profile. Maximum length 22 ft.

Angles shall be straightened to a tolerance of 1/16 inch in 10 feet after galvanizing.

Material: Structural steel per M. H. D. 3306. Galvanize after fabrication per M. H. D. 3394.

Set angle to proper grade and crown.

All material on this sheet except anchorage assembly will be paid for as structural steel M. H. D. 3306. Anchorage assembly will be included in price bid for other items.

See Special Provisions for bolt anchorages.

STATE OF MINNESOTA
DEPARTMENT OF HIGHWAYS
**PROTECTION ANGLE**
**FOR WEARING COURSE**

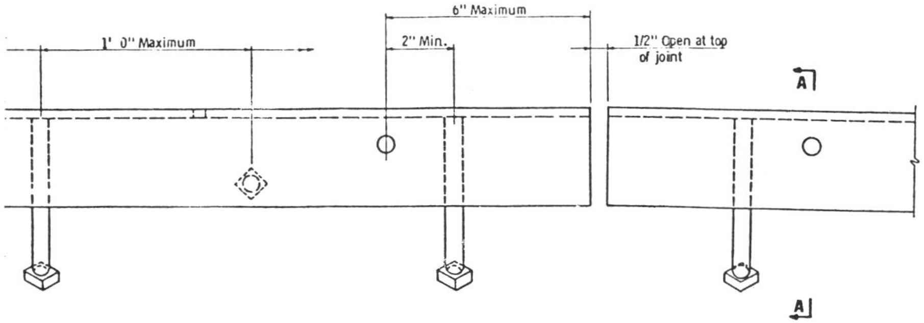


**ELEVATION**
(Concrete not shown)

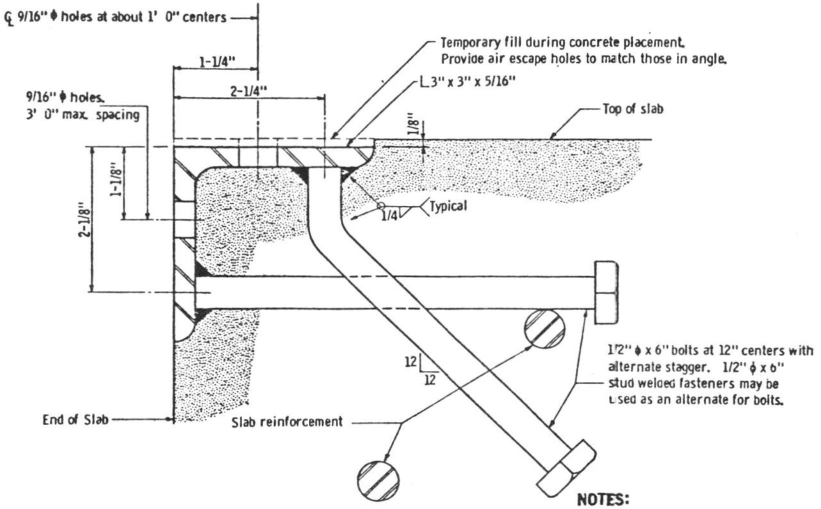


**SECTION A - A**

**NOTES:**

Angles shall extend full width of roadway between curbs with a 1/2" open joint at each break in crown profile. Maximum length 22 feet.

Angles shall be straightened to a tolerance of 1/16" in 10 feet after galvanizing.

Material: Structural steel per M. H. D. 3306. Galvanize after fabrication per M. H. D. 3394.

Set angle to proper grade and crown.

All material will be paid for as structural steel M. H. D. 3306.

STATE OF MINNESOTA
DEPARTMENT OF HIGHWAYS
**PROTECTION ANGLE**
**FOR END OF SLAB**

# J.4 FOOTING

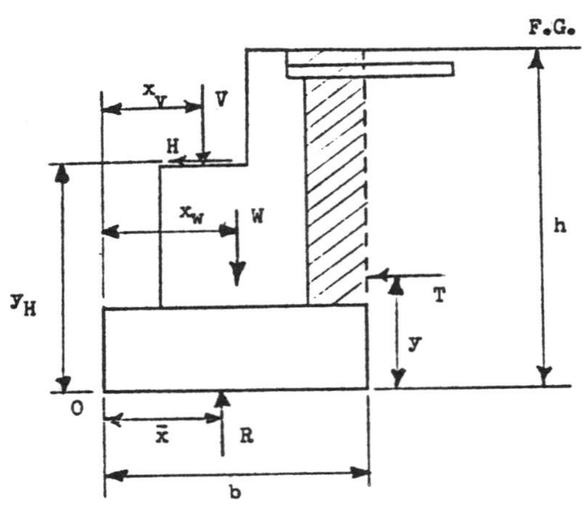


Concrete = 150 #/c.f.
Soil Weight = 120 pcf.

W = Weight of abutment plus the weight of the soil on the heel and toe (if any).

$h_1$ = Height of equivalent soil surcharge when no approach slab is used.

R = Resultant of all vertical loads.

H & V = Superstructure Forces

T = Horizontal Earth Thrust.

**Maine**

Piles or Drilled Shaft

The majority of abutments built are supported on piles. The chief reason for this is to prevent abutment settlement. Most bridge approach embankments are constructed of fill material. This fill material usually experiences some settlement which may take place for several years. This settlement may be the result of changes in the embankment or the original foundation under the embankment. By driving piles through the fill embankment and well into original ground the abutments usually do not settle with the embankment.

A settling embankment may become supported by the abutment piles by friction between piles and fill material. The added load to the piles may cause the piles to be displaced along with the fill. To prevent this from occurring, holes are drilled through the fill at the pile locations. These prebored holes are at least as large as the piles. Predrilling through expansive soils may not be effective unless the hole is larger than the pile and non-cohesive material is used for filling around the pile.

Generally, it is not necessary to prebore non-displacement piles for any fill depths or displacement piles for fill depths under 10 feet. However, for some problem soils this may not apply. Battered piles may cause more of a problem than vertical piles and are given special consideration. Fill embankments frequently shift laterally as well as vertically. A complete foundation site investigation and information on fill material is a prerequisite for successful pile design.

Piles in abutments are subjected to lateral loads. The allowable lateral load on a pile is usually determined from an acceptable level of lateral displacement and not the ultimate load that causes a stress failure in the pile. The allowable lateral load on a pile may be more dependent on the material that the pile is driven into than on the type of pile. See Chapter 11 for a more thorough discussion on piles and allowable pile loads.

Spread Footings

Generally, abutments on spread footings are used only in cut sections where the original soil can sustain reasonable pressures without excessive settlement. The allowable soil pressure is determined by the Materials Section.

With improved procedure and better control of embankment construction spread footings can be used successfully on fill material. It is important that construction be timed to permit the foundation material to consolidate before the spread footings are constructed. An advantage of spread footings is that the differential settlement between approach fills and abutments is minimized.

The use of spread footings is given more consideration for simple span bridges than for continuous spans. For continuous spans under special conditions, the superstructure is designed to accept small amounts of settlement, if they occur. Drainage for abutments on spread footings can be very critical. In general, pile footings are preferred by the Bridge Section.

Pile Encased

Pile encased abutments offer the advantage of economics and ease of construction. Economy results since pile abutments allow using shorter end span lengths. However, if cofferdams are required, recent letting cost data indicates that pile encased abutments do not have an economic advantage over sill abutments.

Pile encased abutment wall height is limited to a maximum of ten feet since more wall height along with the superstructure will increase soil pressure resulting in uneconomical pile design due to their size or spacing requirements. Reinforcement in the abutment body was designed based on two feet of live load surcharge and soil pressure on the back wall. Pile encased abutments are limited to a maximum skew of 15 and 30 degrees with girder superstructures and slab spans respectively in order to limit damage due to thermal expansion or contraction of the superstructure. Also, wing skew angles are limited to 45 degree to prevent cracking between the abutment body and wings.

Special Designs

Many different styles and variations of standard abutment types can be designed. The reasons for their special designs may be based on aesthetics, unique soil, or structural problems.

The Drummondville bridge abutment in Quebec, Canada, is designed for aesthetics. It consists of a hollow triangular base and a projecting rectangular top section. The reinforced roadway slab over the abutment is supported by the front abutment wall and rests on fill inside the abutment.

Another special abutment is the cellular or hollow type that is designed to lessen both vertical and horizontal earth loads by eliminating the approach embankment. This structure may be considered a part of the bridge but is actually a transition section between bridge and normal embankment.

Another method is to cantilever the end spans which removes all bridge loads from the abutment and transfers them to the first pier. Adequate support for the structure is of more concern with this type of design than approach rideability.

Tie rods and deadman can be utilized to provide resistance against horizontal earth loads.

If the abutment is to be used as a future pier, it is important that the wings and backwall be designed for easy removal. Construction joints are separated by felt or other material. Bar steel is not extended through the joints. Bolts with threaded inserts are used to carry tension stresses across joints.

**WisDOT**

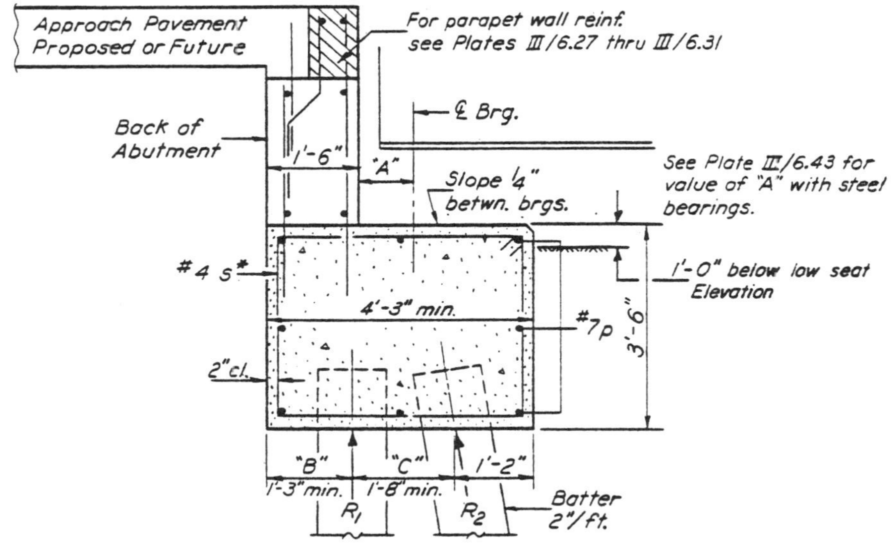


*** Space "s" bars to satisfy minimum shear requirements.**

PILE SPACING Use n piles front row and n + 1 piles back row at 8'-0" max. alternate centers. Values of "B" & "C" shall be selected so that the pile loads between the back row and the front row are as equal as possible.

PILE CAPACITY The specified pile capacity for all abutment piles shall be the maximum pile load determined by the following procedure:

The total load carried by each row of piles shall be assumed to be the static reactions, $R_1$ and $R_2$ acting on the bottom of the cap.

Therefore: $\text{Back Row Pile Load} = \dfrac{R_1 \max.}{n+1}$

$\text{Front Row Pile Load} = \dfrac{R_2 \max.}{n}$

The following loads shall be applied to produce maximum values of $R_1$ or $R_2$.

A. Dead Load of Pile Cap.
B. Dead Load of Superstructure.
C. Dead Load of Approach Slab (If Approach Slab is to be provided).
D. Earth Pressure on Back of Pile Cap. Assume 40 psf. Equivalent Fluid Pressure (Use earth pressure for all combinations of loading).
E. Live Load. The live load may be placed either on the bridge span or the approach span.
F. Live Load Surcharge. A live surcharge of 2 ft. shall be assumed in determining the earth pressure when no approach slab is provided.
G. Thermal force as specified on sheet III/4.2 as applicable.

PILE BENT ABUTMENT

ILL. 6-82 III/6.3

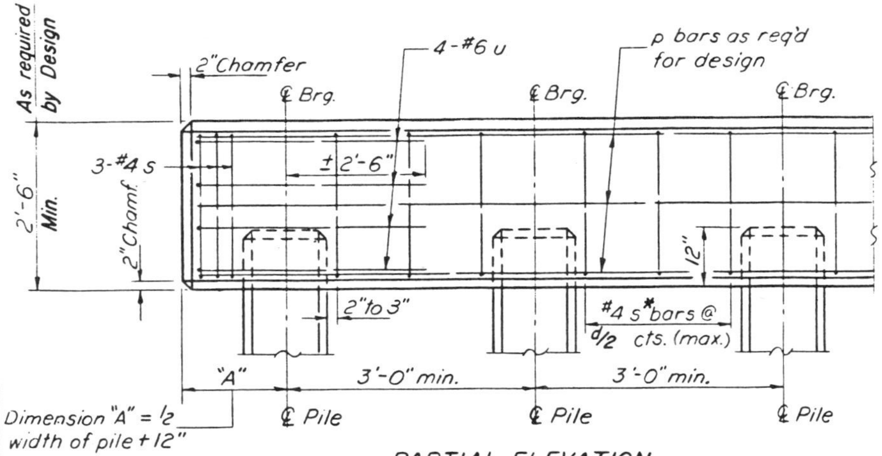


PARTIAL ELEVATION

Note: In all cases give dimensions from ℄ to ℄ of Pile Bents along ℄ of Roadway.

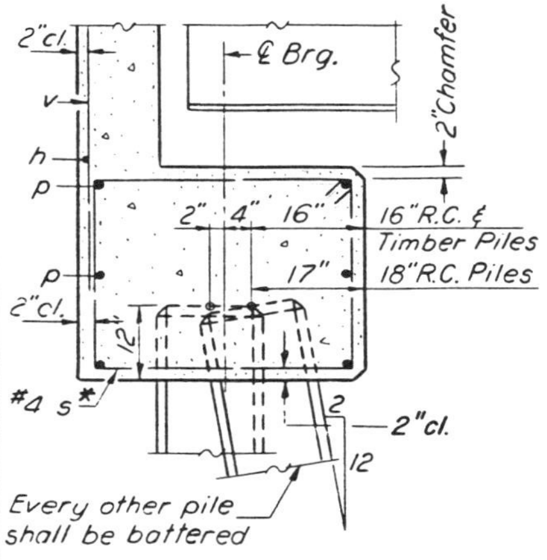


SECTION THRU END BENT

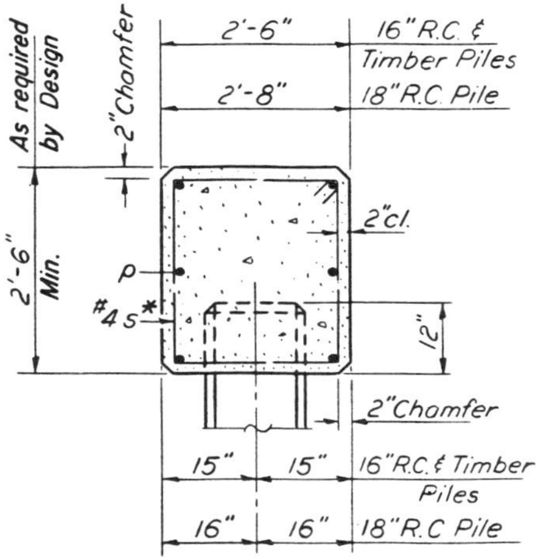


SECTION THRU INTERIOR BENT

*** Space "s" bars to satisfy minimum shear requirements.**
d = effective depth

REINFORCED CONCRETE CAPS FOR PILE BENTS

ILL. 8-80 III/7.15

# PILE ENCASED ABUTMENT WITH FIXED SEAT

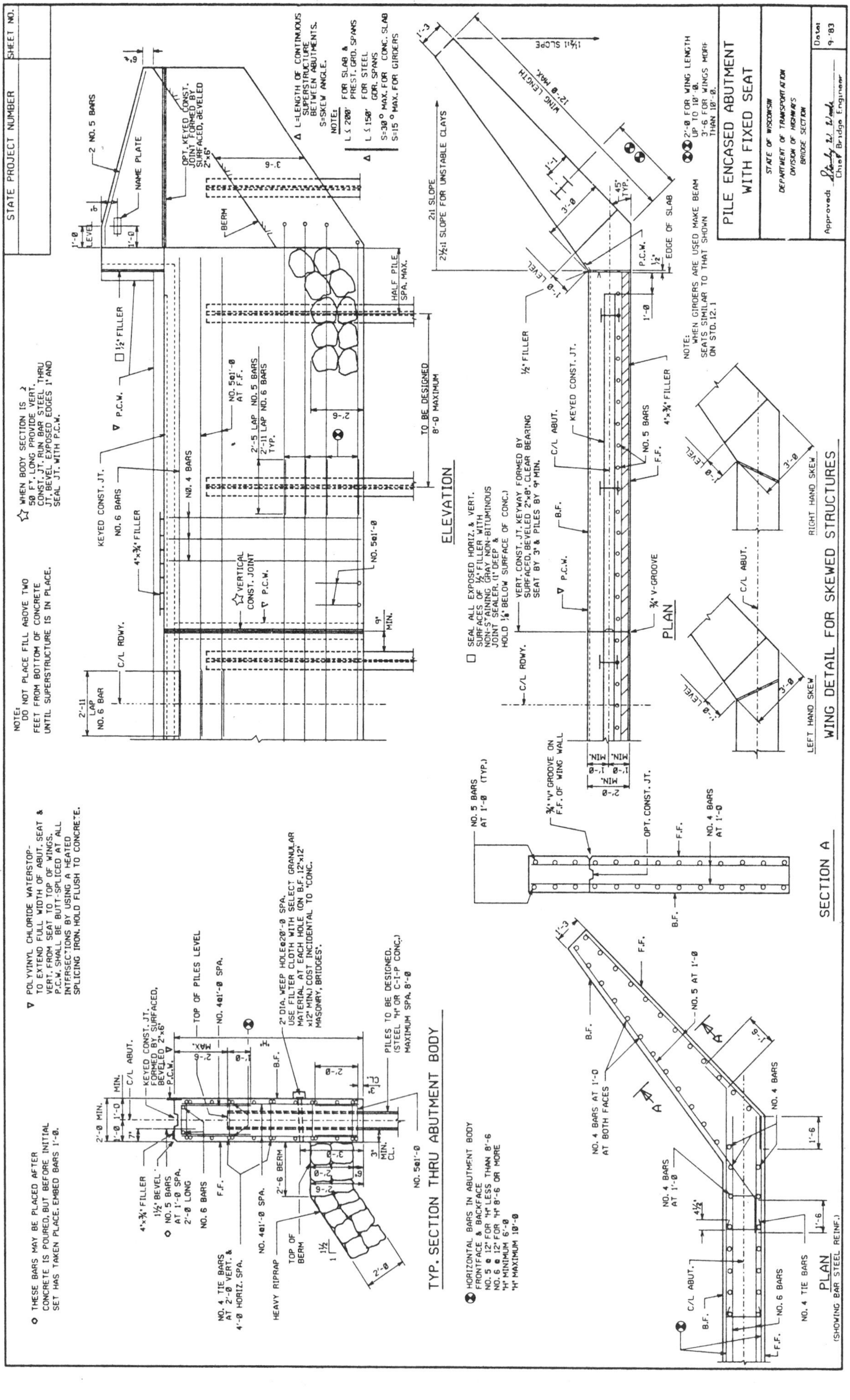

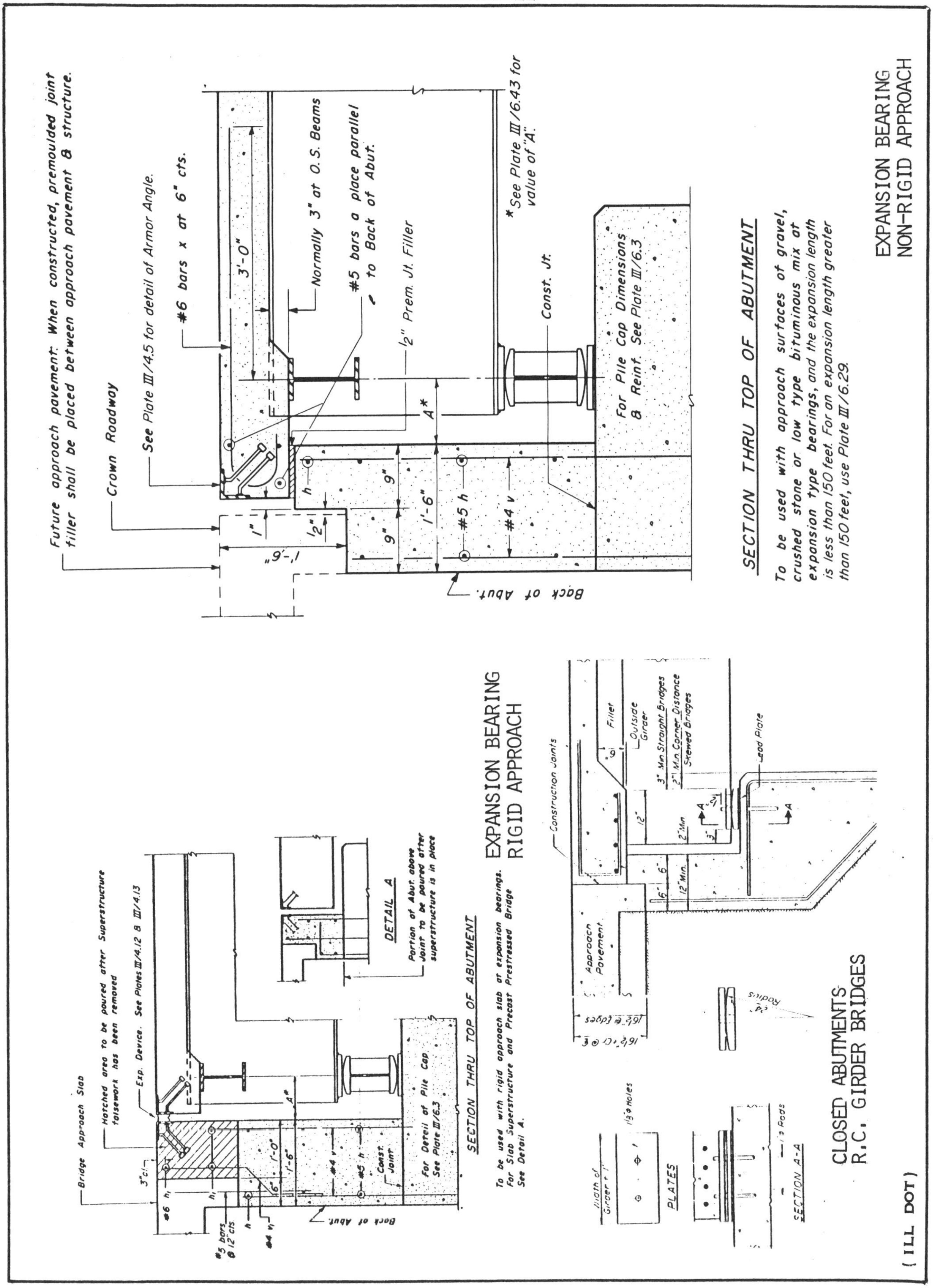

Future approach pavement: When constructed, premoulded joint filler shall be placed between approach pavement & structure.
Crown Roadway
See Plate III/4.5 for detail of Armor Angle.
#6 bars x at 6" cts.
3'-0"
Normally 3" at O.S. Beams
#5 bars a place parallel to Back of Abut.
1/2" Prem. Jt. Filler
A*
*See Plate III/6.43 for value of "A".
Const. Jt.
For Pile Cap Dimensions & Reinf. See Plate III/6.3
1"
1/2"
1'-6"
9"
9"
1'-6"
h
#5 h
#4 v
Back of Abut.
SECTION THRU TOP OF ABUTMENT
To be used with approach surfaces of gravel, crushed stone or low type bituminous mix at expansion type bearings, and the expansion length is less than 150 feet. For an expansion length greater than 150 feet, use Plate III/6.29.
EXPANSION BEARING
NON-RIGID APPROACH
Bridge Approach Slab
Hatched area to be poured after Superstructure falsework has been removed
Exp. Device. See Plates III/4.12 & III/4.13
3"cl.
#6 h
#5 bars @ 12" cts
#4 v
h
1'-0"
1'-6"
6"
#4 v
#5 h
Const. Joint
For Detail of Pile Cap See Plate III/6.3
Back of Abut.
A*
DETAIL A
Portion of Abut. above Joint to be poured after superstructure is in place
SECTION THRU TOP OF ABUTMENT
To be used with rigid approach slab at expansion bearings. For Slab Superstructure and Precast Prestressed Bridge See Detail A.
EXPANSION BEARING
RIGID APPROACH
Construction Joints
Fillet
9"
Outside Girder
3" Min Straight Bridges
2" Min Corner Distance Skewed Bridges
Lead Plate
12"
2" Min
3"
6"
6"
12" Min.
Approach Pavement
16 1/2" @ Edges
16 1/2" C.L. @ E
Radius
Width of Girder + 1"
1 3/8" Holes
PLATES
1 1/4" Rods
SECTION A-A
CLOSED ABUTMENTS
R.C. GIRDER BRIDGES
(ILL DOT)

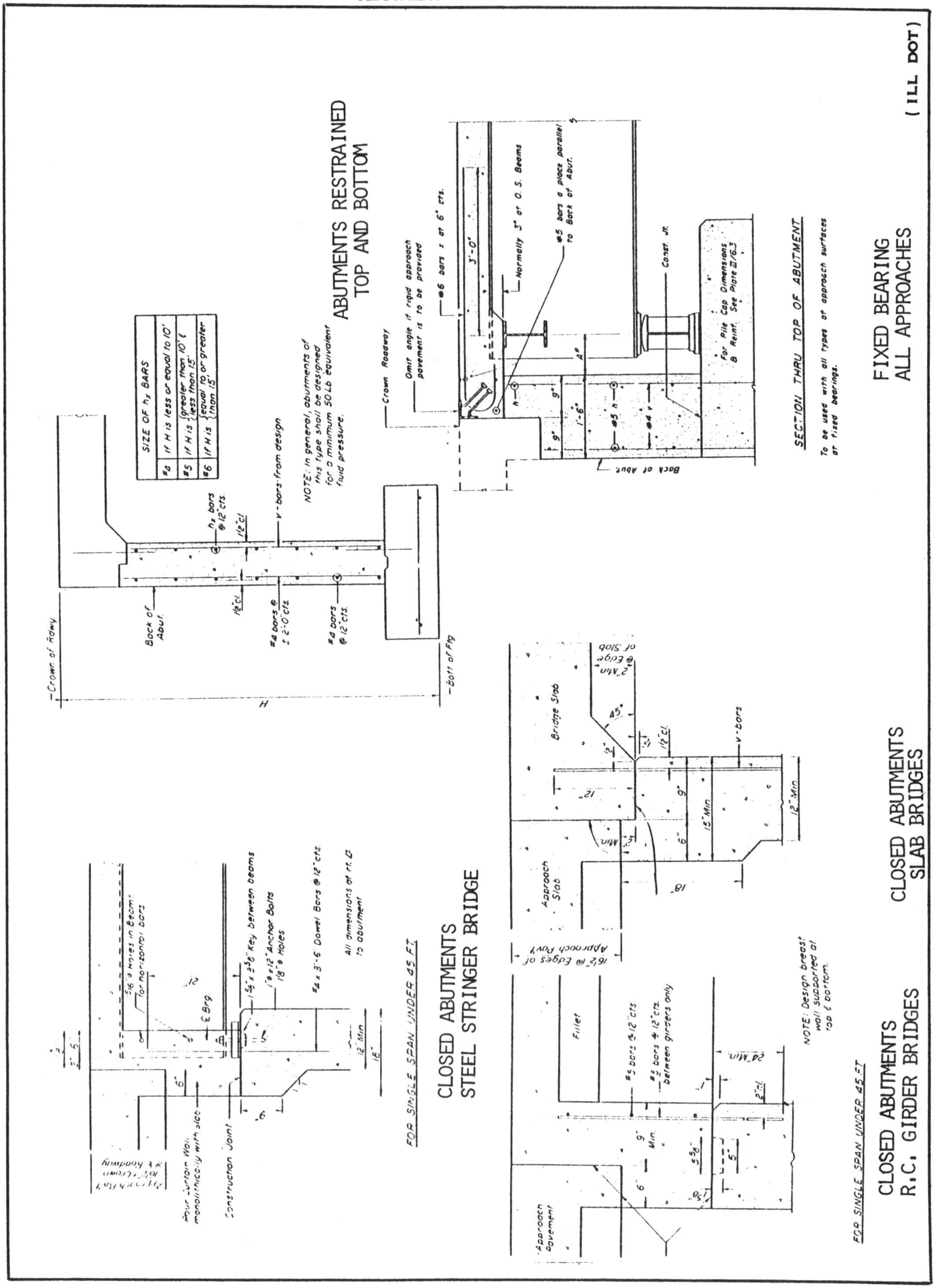
ABUTMENTS RESTRAINED TOP AND BOTTOM
SIZE OF $h_x$ BARS
#4 If H is less or equal to 10'
#5 If H is greater than 10' & less than 15'
#6 If H is equal to or greater than 15'
NOTE: In general, abutments of this type shall be designed for a minimum 50 Lb equivalent fluid pressure.
SECTION THRU TOP OF ABUTMENT
To be used with all types of approach surfaces or fixed bearings.
FIXED BEARING ALL APPROACHES
CLOSED ABUTMENTS STEEL STRINGER BRIDGE
FOR SINGLE SPAN UNDER 45 FT.
CLOSED ABUTMENTS SLAB BRIDGES
CLOSED ABUTMENTS R.C. GIRDER BRIDGES
FOR SINGLE SPAN UNDER 45 FT.
NOTE: Design breast wall supported at top & bottom.
(ILL DOT)

# J.6 PIER DESIGN

From Wisconsin Bridge Manual

## 13.0 GENERAL

A pier cap with round columns supported on individual footings is used mostly on grade separation structures. With wide girder spacings the cap may be omitted and the superstructure girders are set on top of the columns for both economy and appearance. This is also an advantage for future widening of the bridge. For long spans a more substantial pier may be needed so rectangular columns may be used.

A solid shaft type of pier is preferred for most stream crossings because of ice forces; debris and floating ice may collect between columns. Solid shaft piers are useful on structures with deep haunches or on stream crossings where a hammerhead pier cap is too close to the water. They may also be used on railroad structures where crashwalls are required, particularly high crashwalls. The hammerhead pier type with a single shaft is usually more economical than the solid shaft type because it requires a smaller cofferdam. If the hammerhead type is used, the junction between the shaft and bottom of the cap is to be about a distance equal to the cap depth above normal water for aesthetics. This junction must also be a minimum of 2 feet above high water to allow removal of the sheet piling unless an alternate method is developed. For aesthetics the solid shaft and hammerhead piers are proportioned to get a pleasing height to width ratio and the shaft is usually tapered.

For structures over trout streams, a pile bent type of pier causes a minimum of damage to spawning beds during construction. Ice and debris are a hazard to this type structure and its use is limited. In general, this type pier is not to be used on the state system. Where it is used, design the piles for the required loads and consider lateral buckling. Provide bracing if necessary. Timber piles are not allowed for use on this type pier.

The bottoms of pier footings are to be a minimum of 4 feet below finished groundline unless they are founded on solid rock. For pier footings used on stream crossings, the bottom of the footing is to be a minimum of 6 feet below the lowest point of the streambed unless founded on solid rock. Scour conditions may require a greater depth. Pier footings are to have a minimum cover of 2 feet below the streambed.

The bottoms of pier footings adjacent to railroad tracks are to be 4 feet below finished ground line.

A. Location

Piers are to be located to provide navigational clearance requirements and to give a minimum interference to flood flow. In general, place the piers parallel with the direction of stream current at flood stage. Make adequate provision for drift and ice by increasing span length and vertical clearances, by selecting proper pier types and by using debris deflectors. Special precautions against scour are required in unstable stream beds.

In the case of railroad and highway separation structures the spacing and location of supports (piers and abutments) is usually controlled by the clearances required for the roadways or tracks below the structure as well as the necessity to avoid existing improvements such as storm drains and sewers where it is not feasible to remove them. Requirements for vertical and horizontal clearances are specified in chapter 3.

## 13.1 LOADS ON PIERS

The following loads are considered by the Bridge Section in the design of piers.

(1) Live Loads

The live load is either the AASHTO truck or lane loading that produces the maximum vertical reaction at the pier. Impact and distribution are not included. The concentrated load for shear is included for lane load.

The truck and lane loadings are assumed to occupy a width of 10 feet. The lane load is uniformly distributed over the 10 feet. The truck load is applied as two equal concentrated loads at 6 foot spacing in the center of the 10 foot width.

The governing load is placed in "design traffic lanes" having a width $W = W_c/N$, where W = width of design traffic lane, $W_c$ = roadway width between curbs exclusive of median strip, and N = number of design traffic lanes. (See AASHTO) The governing load is placed anywhere within the individual design traffic lanes to produce maximum stresses. When a pier is loaded with three or more "design traffic lanes" the live load is reduced by a probability factor.

If fewer lane loads are used than what the roadway width can accommodate, keep the loads within their "design traffic lanes". "N" in the formula $W = W_c/N$ is the number of lane loads being applied.

For girder type superstructures the truck or lane load is transmitted to the pier through the girders. Simple beam distribution is used. The skew of the structure is not considered when calculating these girder reactions. Use the actual girder spacing and lane widths to determine the girder reactions applied at their actual locations on the pier.

For slab type superstructures the truck or lane load is assumed to be transmitted directly to the pier without any transverse distribution. This same assumption is used even if the pier cap is not integral with the superstructure. A truck load is applied as concentrated loads and a lane load as a uniform load. The skew of the structure is considered when applying these loads to the cap. The lane width is divided by the cosine of the skew angle and the load is distributed over the new lane width along the pier centerline. Impact is included in the load if there is no pier cap.

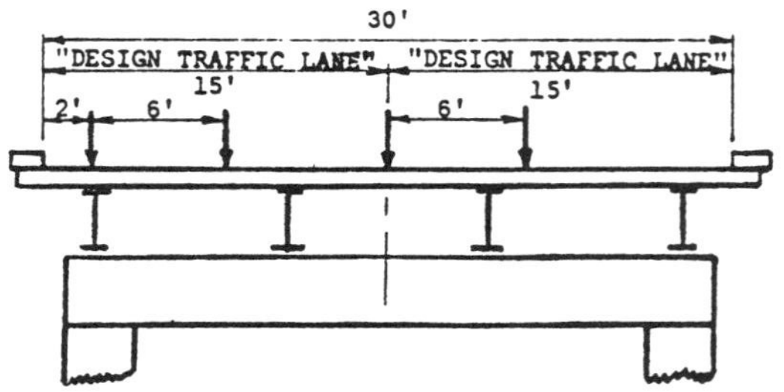


Cross section of superstructure at pier showing reactions of two truck loads.

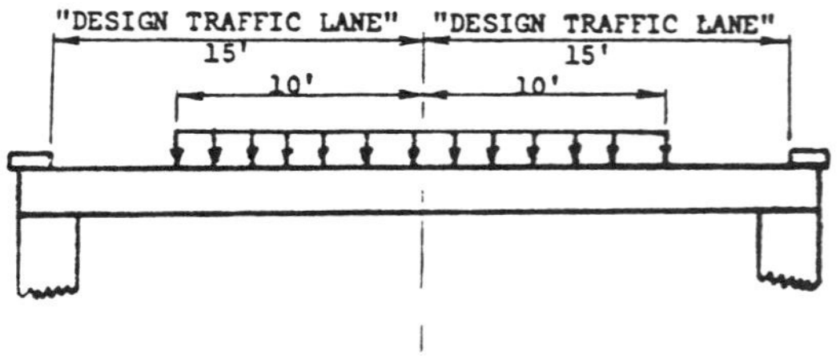


Cross section of superstructure at pier showing uniform load from two lane loads.

(2) Longitudinal Force

AASHTO requires that provision be made for a longitudinal force equal to 5 percent of the lane live load with concentrated load for moment and no impact. This force acts in all lanes carrying traffic in the same direction and acts 6 feet above the floor. It is meant to simulate the forces caused by vehicles braking or accelerating.

When this force is applied to skewed piers, the transverse component is applied 6 feet above the roadway and the longitudinal component is applied at the bearings. It is not possible to transfer the moment of the longitudinal component acting above the bearing to the footing on typical bridge structures.

(3) Wind Loads

The basic wind load is for 100 mph wind speeds assumed to act horizontally from any direction. Wind loads are divided into three types; (1) wind on live load, (2) wind on superstructure, and (3) wind on substructure.

AASHTO specifications are used for the usual girder and slab bridges having maximum span lengths of 125 feet:

A. Wind Load on Live Load

100 pounds per linear foot, transverse
40 pounds per linear foot, longitudinal
Both forces are applied simultaneously.

B. Wind Load on Structure

50 pounds per square foot, transverse
12 pounds per square foot, longitudinal
Both forces are applied simultaneously.

These forces are more conservative than the forces given for more precise wind directions but are used for convenience since the difference is small.

C. Wind Load on Substructure

The wind force applied to substructure units is specified as 40 pounds per square foot both longitudinally and transversely.

An overturning force as specified is also applied.

(4) Thermal Forces

The change in length of the superstructure and pier cap due to temperature changes cause deflections in the pier columns. These deflections cause "thermal forces" in the columns.

In determining the thermal forces applied to each substructure unit the entire bridge superstructure is considered. There is a neutral point on the superstructure which does not move due to temperature changes. This point is determined by examination or by trial and error. Using trial and error a neutral point is assumed and all horizontal forces are computed. The point is shifted until all the horizontal forces are in equilibrium. The horizontal forces computed are the forces produced at each substructure unit by deflections due to temperature.

The thermal force on an expansion pier is the minimum of:

1. Dead load reaction times maximum bearing coefficient of friction.
2. Force caused by deflection of pier for given temperature change. The friction coefficients of bearings are given in Chapter 27.0 Bearings.

The thermal force on a single fixed pier in a bridge is the resultant of the unbalanced forces acting on the substructure units. Maximum friction coefficients are assumed for expansion bearings on one side of the bridge and minimum coefficients on the other side to produce the greatest unbalanced force on the fixed pier.

The thermal changes in length of the superstructure are assumed as being along the longitudinal axis of the superstructure regardless of the substructure skew angle. This assumption is more valid for steel structures than concrete structures.

The force on a column due to a thermal change in length of the superstructure is:

$$F = \frac{3EI\ a\ TL}{h^3 \times 144}$$

where:
E = Modulus of Elasticity of column, Ksi
I = Moment of Inertia of column, $In.^4$
a = Coefficient of Thermal expansion of superstructure
T = Temperature change of superstructure, Degrees
L = Expansion length of superstructure, Feet
h = Column height, Feet
F = Force per column, Kips

Using figure 13.1, the force on a concrete column is:

$$P = \text{Factor} * \frac{L}{100}$$

where:
Factor = Value taken from the curves
L = Expansion length of superstructure, Feet
P = Column Force, Kips

Moment of Inertia for concrete columns ($\pi D^4/64$):

30" Diameter - I = 39,761. $in.^4$ = 1.9175 $ft.^4$
36" Diameter - I = 82,448. $in.^4$ = 3.9761 $ft.^4$
42" Diameter - I = 152,745. $in.^4$ = 7.3662 $ft.^4$

The following values apply to Wisconsin bridges for computing temperature forces. Do not confuse temperature change with temperature range used for expansion joints.

| | Reinforced Concrete | Steel |
|---|---|---|
| Temperature Change | 45°F | 90°F |
| Coefficient of Thermal Expansion | .000006 | .0000065 |

Using figure 13.1, the force for other section sizes and/or temperatures is:

$$F = \frac{P}{100} \times \frac{I}{1.92} \times \frac{T}{90}$$

where:
P = Force for 30" ϕ column from curve.
I = Moment of inertia of new section, $Ft.^4$

Temperature forces on bridges with two or more fixed piers are based on the movement of the superstructure along its centerline. These forces are assumed to act normal and parallel to the longitudinal axis of the pier as resolved through the skew angle. The lateral restraint offered by the superstructure is usually ignored.

(5) Force of Stream Current, Floating Ice and Drift

The force of flowing water on piers is given by AASHTO as $P = KV^2$

where:
P = Pressure, pounds per square foot

V = Water velocity, Ft. per second

K = Shape constant; 2/3 for circular piers, 1-3/8 for square ends, and 1/2 for angle ends of 30° or less.

Use the water depth and velocity at flood stage with the force acting at one-half the water depth. Normally this force does not govern the pier design.

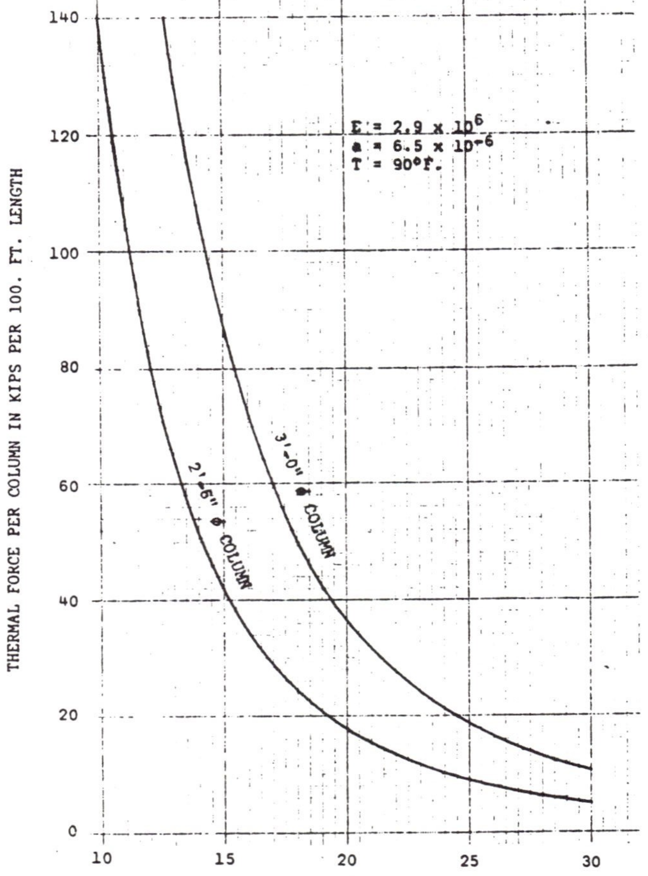


FIGURE 13.1 THERMAL FORCE FOR STEEL SUPERSTRUCTURES

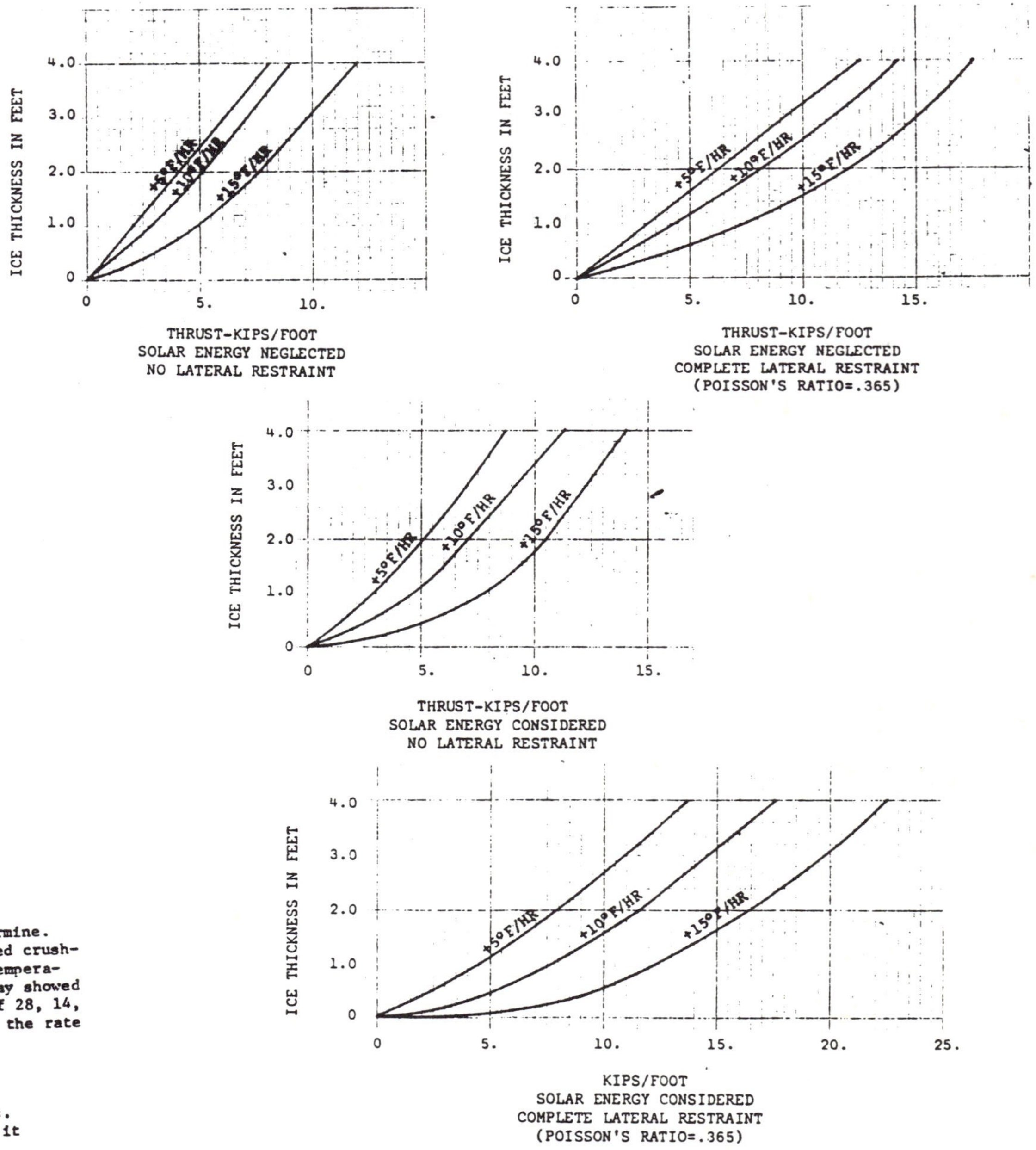


FIGURE 13.2 FORCES EXERTED BY EXPANDING ICE

The exact force of floating ice on a pier is difficult to determine. AASHTO sets this force at 400 psi which is the generally accepted crushing strength of ice. This is based on the ice thickness and temperature at the time of breakup. Tests at the St. Lawrence Waterway showed crushing strengths of 300, 693, and 811 psi for temperatures of 28, 14, and 2°F respectively. The crushing strength of ice depends on the rate of load application and temperature.

The force of floating ice on piers is caused by two conditions. First is the force of floating ice colliding with the pier as it floats downstream. By Newton's third law of motion, Force = M $(V_1-V_2)$

where: $M$ = Mass of ice, Weight in lbs. divided by 32.2

$V_1$ = Velocity of ice before impact, Ft. per second

$V_2$ = Velocity of ice after impact, Ft. per second.

Second is the force that develops when ice jams against the pier. The ice acts as a dam and transfers hydrostatic pressure to the pier, as well as the force due to current drag under the ice and wind action against the ice.

Investigate each site for existing conditions. If no data is available, use the following data as minimum design criteria:

1. Ice pressure is 400 psi.

2. Minimum ice thickness is 6 inches.

3. Height on pier where force acts is midway between high and low water elevations.

4. Pier width is the projection of the pier perpendicular to stream flow.

(6) Force Exerted By Expanding Ice Sheet

Expansion of an ice sheet, as the result of a temperature rise after a cold wave, can develop considerable force against abutting structures. This force can result if it is restrained between two adjacent bridge piers or between a bluff type shore and bridge pier. Its direction is therefore transverse to the direction of stream flow.

Force from ice sheets depends upon thickness, maximum rate of air-temperature rise, extent of restraint of ice, and extent of exposure to solar radiation. Figure 13.2 taken from an article by Row (ASCE TRANS, VOL 112, 1947) shows thrusts for different thicknesses of ice under various conditions.

It is not necessary to design all bridge piers for expanding ice. If one side of a pier is exposed to sunlight and the other side is in the shade along with the shore in the vicinity of the pier, consider the development of pressure from expanding ice. If the central part of the ice is exposed to the sun's radiation, consider the effect of solar energy which causes the ice to expand.

Forces from floating ice and expanding ice do not act on a pier at the same time. Consider each force separately when applying these design loads.

(7) Buoyancy

The footings of piers under water are designed for the buoyant effect of the water. Full hydrostatic pressure based on the water depth measured from the bottom of the footing is assumed to act on the bottom of the footing. The upward buoyant force equals the volume of concrete below the water surface times the unit weight of water. (62.4 lbs. per cu. ft.) The effect of buoyancy on column design is usually ignored. Use high water elevation when analyzing the pier for over-turning. Use low water elevation to determine the maximum vertical load on the footing.

The submerged weight of the soil above the footing is used for calculating the vertical load on the footing. Typical values in lbs. per cu. ft. are as follows:

| | Sand | Sand & Gravel | Silty Clay | Clay | Silt |
|---|---|---|---|---|---|
| MIN. (Loose) | 50. | 60. | 40. | 30. | 25. |
| MAX. (Dense) | 85. | 95. | 85. | 70. | 70. |

(8) Centrifugal Force

Centrifugal force is included in the design of piers for structures on horizontal curves. The centrifugal force is the percentage $0.00117\ S^2D = 6.68\ S^2/R$ of each axle load, excluding impact, of one standard truck in each traffic lane, where S = design speed, mph, D = the degree of curve, R = radius of curve, ft.

The centrifugal force is assumed to act radially and horizontally 6 feet above the roadway surface. The point 6 feet above the roadway surface is measured from the centerline of roadway. The design speed may be determined from the Wisconsin "Highway Design Manual", Chapter 11.

It is not necessary to consider the effect of superelevation when centifugal force is used. The point of application of centrifugal force considers this.

## 13.2 OFFICE PRACTICE FOR LOAD APPLICATIONS

(1) Loading Combinations

Piers are designed for the loading combinations specified by AASHTO except:

1. I and F forces are not included.

2. Group VII is not used.

(2) Expansion Piers

Transverse forces applied to expansion piers include one-half the adjacent span lengths. Longitudinal forces except for temperature include one-half the adjacent span lengths provided these forces do not exceed the maximum friction force (dead load times maximum friction coefficient of bearings). The maximum friction force is used for the temperature force as it is assumed the superstructure moves due to temperature changes.

(3) Fixed Piers

Transverse forces applied to fixed piers include one-half the adjacent span lengths. Longitudinal forces except for temperature include the entire bridge length. The magnitude of these forces is reduced by the amount of force taken by expansion piers and abutments. If there are two or more fixed piers, proportion the total bridge length between them. When temperature forces are considered, it is assumed the expansion bearings do not take any longitudinal forces. In this case the fixed pier carries longitudinal forces for the bridge length plus the unbalanced friction force in the expansion bearings. Do not design the fixed piers for longitudinal forces that are less than the forces based on one-half the adjacent span lengths.

(4) Point of Application of Forces

Longitudinal forces are transmitted from the superstructure to the pier through the bearings. The effect of load application above the roadway has a small effect in axial loads applied to the adjacent piers and does not affect the horizontal loads. Transverse forces are applied at the points specified by AASHTO as these forces are effective in increasing transverse moments.

Transverse and longitudinal superstructure forces are converted to transverse and longitudinal forces of the pier by using the sine and cosine functions of the skew angle of the pier.

Wind loads are applied to both sides of the structure to get the maximum effect. Use the precise wind loading specified by AASHTO instead of the approximate wind loading (50 pounds transverse and 12 pounds longitudinal) if a significant cost saving results.

Temperature changes in the superstructure cause it to expand and contract along its longitudinal axis. These length changes induce forces in the substructure units which are applied normal and parallel to the pier centerline as resolved by the sine and cosine of the skew angle. This assumption is applied to most structures.

Theoretically the superstructure expands along its centerline and the substructure unit must follow. This means the forces transmitted to the substructure are based on the stiffnesses of the pier in both the transverse and longitudinal directions. Where there are two or more fixed piers, there is movement in the piers due to temperature changes. For skewed piers these forces become excessive as the stiffness along the pier centerline is large. Except in unusual cases, this approach is ignored and the forces are resolved by the sine-cosine functions.

## 13.3 PIER FRAME ANALYSIS

The dimensions of the pier frame are assumed for the first trial for the loads to be carried by the pier. The results of the first trial are used to design the individual components of the pier; cap, columns, and footings. If the original size assumptions are incorrect, the pier dimensions are adjusted and the pier analyzed again until a satisfactory design is made.

Pier frames in general use round columns. Pier caps may be designed to any length. However, if the length between the outer columns of the pier cap exceeds 65 feet, special design considerations may be necessary due to changes in length caused by temperature.

The minimum thickness of spread footings is 2 feet and 2'-6" for pile footings. Isolated footings are used under individual columns unless the clearance between footings is less than 4'-6"; then a continuous footing is used. The degree of restraint of the footings is assumed as 100% for footings on rock (less if footing is in uplift), 70% for footings on piles, and 50% for footings on soil.

The column spacing on pier frames is to be 12 feet minimum and 25 feet maximum. Column height is determined by the roadway elevation and the location of the footing criteria. Columns are assumed rigidly connected to the pier cap where used. A minimum of three columns is required on all piers as a safety precaution against vehicle impact causing column failure.

Generally the pier cap is omitted on continuous slab span structures (a cap is provided on stream crossings) and the slab bears on the column ends. In this case the slab concrete adjacent to the columns is used as a cap and designed accordingly.

**If the girder spacing in a structure exceeds 12 feet, consider placing a column under each girder and eliminating the pier cap. The diaphram connections of the girders are used to transfer the lateral loads between tops of the columns. If this type of structure is used in water, do not eliminate the concrete pier cap.**

The pier is analyzed as a frame bent by any of the available analysis procedures. A computer program in the bridge section uses the Moment Distribution routine. Sidesway of the frame due to loading is taken into account.

The gross concrete areas of the components are used to compute their moments of inertia for analysis purposes. The effect of the steel reinforcing is neglected.

Vertical loads are applied to the pier through the girders or the slab of a concrete bridge. The vertical loads are varied to produce the maximum moments and shears at various positions throughout the structure in combination with the horizontal forces. The effect of length changes in the cap due to temperature is also considered in computing maximum moments and shears. All these forces produce several loading conditions on the structure which must be separated to get the maximum effect at each point in the structure.

The maximum moments, shears, and axial loads from the analysis routines are used to design the individual components of the pier.

## 13.4 PIER CAP DESIGN

The following steps are recommended for the design of a pier cap.

(1) Draw the maximum moment and shear curves using the results of the analysis routine.

(2) Use the assumed cap cross section to determine the bar steel required for positive and negative moments based on load factor design. Bars are placed straight in the cap. Determine bar cutoff points.

If the pier cap is cantilevered over exterior columns, the top negative bar steel may be bent down at the ends to develop end anchorage.

The minimum cap dimension to be used is 3 feet deep by 2'-6" wide except a 2'-6" deep section may be used for caps under slab structures. If a larger cap is needed, use 6 inch increments to increase the size. The cap width is the same as the column width.

If a pier cap is used on continuous slab structures, the moment and shear forces are proportioned between the transverse slab section and the cap by the ratio of their moments of inertia. The effective slab width assumed for the transverse beam is the minimum of 1/2 the center to center column spacing or 8 feet.

$$\text{Cap Moment} = (\text{Total Moment}) \times \frac{(\text{I of Cap})}{(\text{I of Cap}) + (\text{I of Slab})}$$

If a pier cap is not used on slab structures, use the assumed slab section as the cap of the pier. Check the peripheral shear in the cap. One way shear is not checked because if stirrups are required using the assumed slab width, the width is increased so stirrups are not required since experience shows this is not a problem area.

The pier cap is to extend a minimum of 3 inches beyond the edge of slab for ease of forming the superstructure. If the cap is rounded, measure from the tangent line on the cap.

The pier cap is to extend a minimum of 9 inches beyond the centerline of steel girders and 14 inches of prestressed girders. The minimum dimension is measured on the side of the cap to the tangent line.

(3) Determine the temperature steel required at the center of the cap on the sides. Number 5 bars are required for a 3' deep cap. Consider using number 4 bars at 18 inches for deeper caps.

(4) Determine the need for stirrups to resist shearing forces and the size and spacing required. Do not place stirrups closer than 4 inch centers. Usually only double stirrups are used, but triple stirrups may be used to increase the spacing. If these do not work, increase the size of the cap. Stirrups are generally not placed over the columns. The first stirrup is placed one-half the spacing distance from the edge of the column into the span.

The cap to column connection is made by extending the bar steel in the column straight into the cap far enough to develop the bond strength.

Stirrup details and bar details at the end of the cap are shown in Standard 13.1.

A. Design of Inverted T-Beams

An inverted T-beam is used as a bent cap beam in order to reduce the clearance requirements beneath the bent cap and to enhance the appearance of the bridge superstructure.

The application of loads to the lower portion of the beam creates tensile forces not ordinarily encountered in T-beams. Reinforcement of the flange of the T presents special problems of shear, flexure and bar anchorage. Recommendations are made for the design of these beams based on a report by the University of Texas published in July, 1971.

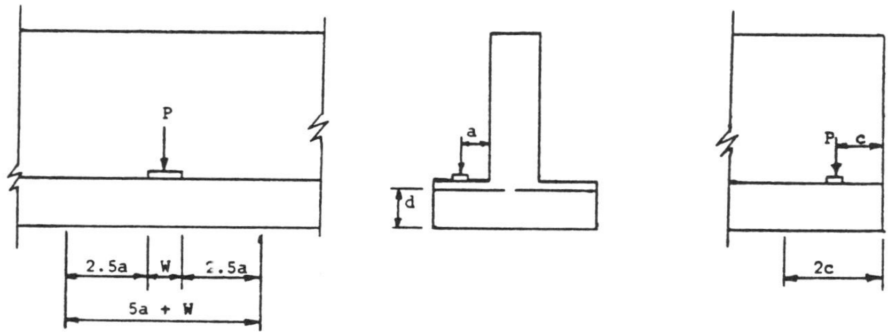


The above sketches show the width of bracket that is effective in flexure to resist the concentrated load. For the interior portion of the beam:

Effective Width = 5a + W

For loads near the end of the girder:

Effective Width = 2c provided c is less than 2.5a

For flexural computations for the bracket reinforcement, the effective distance jd between the centroid of compression and the centroid of tension is taken as 0.8d.

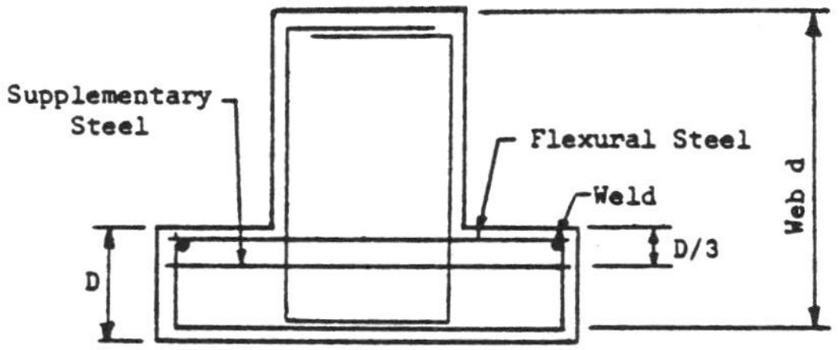


The flexural steel has to be hooked to develop the tensile strength of the bar. In addition, Supplementary Steel with an area not less than half the Flexural Steel is placed as shown in the sketch. Nominal flexural steel reinforcing is to be provided between points of concentrated loads.

The depth of bracket d required for shear friction is to be not less than:

$$d_{min} = \frac{1.5\ Vu}{af'c}$$

where: Vu = ultimate load applied to bearing plate, kips
f'c = concrete design compressive strength, ksi
a = shown in sketch

The effective width for shear friction capacity is 4a + w. Within this effective width provide an area of reinforcement (A) that extends into the flange in order to maintain the shear friction force.

$$A = \frac{Vu}{1.2\ fy}$$ fy = yield strength of bar steel, ksi.

At every concentrated load applied to the flange of the inverted T-beam, provide stirrups that act as hangers that have a capacity greater than the applied ultimate load. Place the stirrups within a web depth d centered about the load. The hangers must be closed across the bottom to develop the stirrup bar forces.

The design of the beam to resist flexure along its horizontal axis is the same as any rectangular concrete beam. Stirrups along this axis are designed to resist all ultimate shears not resisted by the concrete.

## 13.5 COLUMN DESIGN

Use an accepted analysis procedure to determine the axial load and longitudinal and transverse moments acting on the column. These forces are found at the top and bottom of the column. Apply the designated load factors for each design group condition; assume $B_d$ is equal to 1. Choose the larger loading cases for designing the column.

Columns that are part of a pier frame have transverse moments induced by frame action from vertical loads, wind loads on the superstructure and substructure, wind loads on live load, thermal forces, and centrifugal forces. Longitudinal moments are produced by the above forces plus the longitudinal live load force. These forces are resolved through the skew angle of the pier to act transversely and longitudinally to the pier frame. Longitudinal forces are divided equally among the columns.

Wisconsin uses tied columns designed by the strength method following AASHTO specifications. The minimum column size used is 30 inches in diameter. The minimum bar steel area is by specifications and the maximum steel area permitted is shown in Table 13.1. Bar details are shown in Table 13.2.

The computed column moments are magnified for slenderness effects by the moment multiplier ($\delta$) for the following conditions:

(1) Transverse Moments.

There are two conditions specified for checking the slenderness ratio. These conditions are combined into one since the frame is braced somewhat by the superstructure and the moment ratio is probably around .90. $\delta$ = 1.0 is assumed if $kl_u/r$ is less than 24.

$$\text{Otherwise } \delta = \frac{C_m}{1. - (P_u/\phi P_c)} \geq 1.0$$

where: $C_m = 1.0$

$P_c = \pi^2 EI/(kl_u)^2$

$k = 1.0$ (Top of column is braced)

$$EI = \frac{E_c I_g/2.5}{1. + B_d}$$

$B_d = 0.1$

$E_c = 57\sqrt{3500}$

$\phi = .7$

$I_g = \pi d^4/(64 \times 144)$ (Round column)

$l_u$ = Column height

(2) Longitudinal Moments.

$\delta = 1.0$ if $kl_u/r$ is less than 22.

$$\text{Otherwise } \delta = \frac{C_m}{1. - (P_u/\phi P_c} \geq 1.0$$

where $k = 2$ (Top of column not braced)

$k = 1$ (Top of column braced, such as slab bridges or fixed bearings)

$l_u$ = Column height + cap depth

Other values same as transverse moments.

The computed moments are multiplied by the magnification factors and then the column is designed for the computed loads following the specifications. The design strength under combined flexure and axial load is based on stress and strain compatibility. A computer program is recommended for solution due to the complexity of the problem.

Large single column pier shafts such as "Hammerhead Piers" are designed as a wall. In this case the minimum steel allowed is 0.1 percent of gross concrete area.

On large river crossings it may be necessary to protect the piers from damage. The noses may be protected against damage by using steel with anchors embedded in the concrete. One state used weathering steel.

Pier columns adjacent to railroad tracks may be protected with a concrete crash wall. If they are less than 20 feet from the tracks, the Soo Line and Chicago, Milwaukee, St. Paul and Pacific Railroads require crash walls. The distance is not determined for other railroads. The crash wall thickness varies from 1'-0" to the diameter of the columns depending on the railroad. The height of the wall above the tracks also varies with the railroad.

The column to cap connection is designed as a rigid joint considering axial and bending stresses. Column steel is run through the joint into the cap to develop the compressive stresses or the tensile steel stresses at the joint.

In general the column to footing connection is designed as a rigid joint. However, in cases of high transverse moments, it may be desirable to pin the connection in the transverse direction. If this is done, the bar steel is placed at the neutral axis of the column in the transverse direction as it passes through the joint. In all cases the column to footing connection is designed as a rigid joint in the longitudinal direction. The bar steel from the column is usually terminated at the top of the footing. Dowel bars are used to transfer the steel stress from the footing to the column.

(1) Tapered Columns of Concrete and Timber

Design these columns using the existing column formulas, taking the cross sectional area at the small end, but d, the dimension used in L/d, is taken as follows:

1. For round columns or rectangular columns tapered in both directions, use $d = d_B$
2. For rectangular columns tapered in the plane of bending only, use $d = (d_A)^{.2} \times (d_B)^{.8}$
3. For rectangular columns tapered perpendicular to the plane of bending, use $d = (d_A)^{.7} \times (d_B)^{.3}$

Where: $d_A$ = dimension at the small end

$d_B$ = dimension at the large end

## 13.6 FOOTING DESIGN

(1) Isolated Spread Footings

Spread footings are designed according to AASHTO specifications. The footing is proportioned so when it is loaded with the Group Loadings, the basic allowable soil stress with the percent increase is not exceeded. The following criteria is used to design footings.

A. Minimum depth of spread footings is 24 inches. Depth is generally determined from shear strength requirements. Shear reinforcement is not used.

B. A maximum of 25% of the area of the footing is allowed in uplift. When part of a footing is in uplift its section properties for analysis are based only on the portion of the footing in compression. When determining the percent of a footing in uplift, apply the actual load without any reductions for overload. For Groups 2-9 the basic allowable soil stress is increased but the allowable percent of area in uplift is 25. No uplift is allowed for Group 1 loading.

C. Soil weight on footings is based only on the soil directly above the footing.

D. The minimum depth for frost protection from top of ground to bottom of footing is four feet.

E. Spread footings on seals are designed by either method as follows:

a. The footing is proportioned so the pressure between the bottom of the footing and the top of the seal does not exceed the allowable soil stress and not more than 25% of the footing area is in uplift.

b. The seal is proportioned so that pressure at the bottom of the seal does not exceed the allowable soil stress and the area in uplift between the footing and seal does not exceed 25%.

F. The reinforcing steel in spread footings is determined from the load factor requirements for bending as stated in AASHTO. The moment used is determined from the volume of the pressure diagram under the footing which acts outside of the section being considered. The weight of the footing and soil above the footing is used to reduce the bending moment.

G. The negative moment which results from 25 percent or less of the footing area being in uplift is ignored. No negative reinforcing steel is used in spread footings.

H. Shear strength is determined by two methods.

a. Two-way action

The volume of the pressure diagram on the area of the footing outside lines at a distance d/2 from the face of column. The shear stress is determined by applying the forces from the volume of the pressure diagram over the length $2(L + d + W + d)$ for rectangular columns and $3.14(2R + d)$ for round columns where R is the radius of the column and d the effective footing depth.

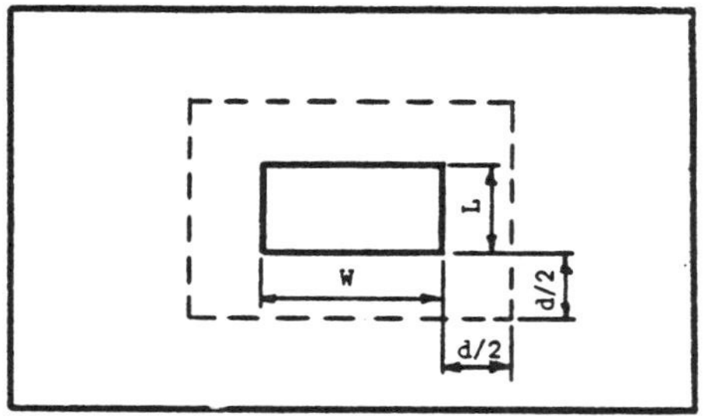


b. One-way Action

**The volume of the pressure diagram on the area enclosed by the footing edges and a line at distance "d" from the face of the column.** The shear stress is determined **by applying the forces from the volume of the pressure diagram over the entire footing width or length.**

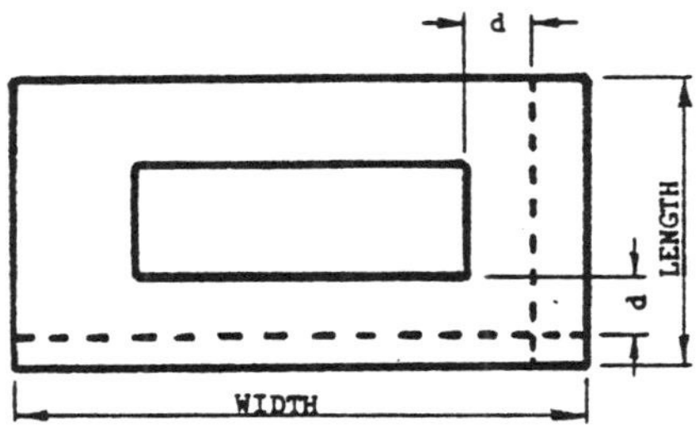


**The weight of the footing and soil above the areas is used to reduce the shear force.**

I. **The bottom layer of reinforcing steel is placed 3 inches clear from the bottom of the footing.**

J. **If adjacent edges of isolated footings are closer than 4'-6", use a continuous footing.**

(2) **Isolated Pile Footings**

**Pile footings are designed according to** AASHTO specifications. The **footing is proportioned so when it is loaded with the Group** Loadings **the basic allowable pile stress, with the percent increase, is not exceeded. The following criteria is used to design the footings.**

A. Minimum depth of pile footing is 24 inches. The minimum pile embedment is 6 inches.

B. Pile footings in uplift are usually designed by method (a) stated below. However, method (b) may be used if the designer feels there is a substantial reduction in cost.

a. Over one-half of the piles in the footings must be in compression. The section properties used by analysis are based only on the piles in compression. When analyzing the footing, apply the actual load and use the allowable stress increases as stated in AASHTO. All piles must be in compression for Group 1 loading.

b. Piles may be designed for upward forces provided an anchorage device, sufficient to transfer the load, is provided at the top of the pile. Provide reinforcing steel to resist the tension stresses at the top of footing.

C. Same as Spread Footing.

D. Same as Spread Footing.

E. The minimum number of piles per footing is four.

F. Pile footings on seals are analyzed above the seal. The only effect of the seal is to reduce the allowable pile stress above the seal by the portion of the seal weight carried by each pile.

If no seal is required but a cofferdam is, design the piles to use the minimum required batter. This reduces the size of the cofferdam necessary to clear the battered piles.

G. The reinforcing steel in pile footings is determined from the requirements for bending as stated in AASHTO. The moment and shear used is determined from the force of the piles which act outside of the section being considered. The weight of the footing and soil above the footing is used to reduce the bending moment and shear force.

H. Shear strength is determined by two methods.

a. Two-way Action

The summation of the pile forces is on the area outside lines at a distance d/2 from the face of the column ("d" = the effective footing depth). The shear stress is determined by applying the force over the length 2 (L + d + W + d) for rectangular columns and 3.14(2R + d) for round columns.

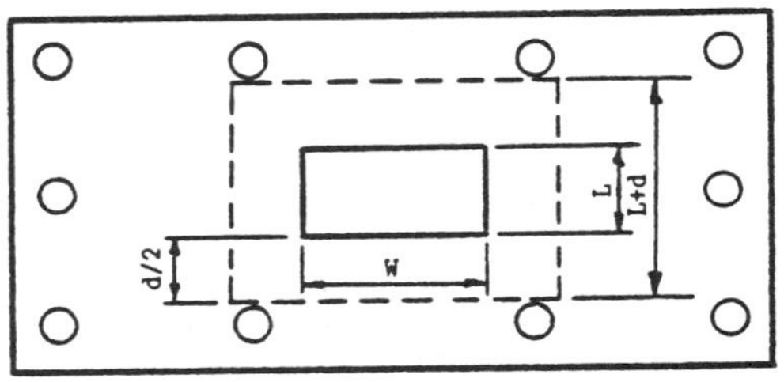


If the center of a pile falls on a line, 1/2 of the pile force is assumed to act on each side of the line.

b. One-way Action

The summation of the pile forces is within the area enclosed by the footing edges and a line at distance "d" from the face of the column. The shear stress is determined by applying the force over the entire footing width or length.

**The weight of the footing and soil above the areas is used to reduce the diagonal tension force.**

**I. The bottom layer of reinforcing steel is placed directly on top of the piles.**

**(3) Continuous Footings**

**Continuous footings are used in pier frames of two or more columns when the use of isolated footings would result in a distance of less than 4'-6" between edges of adjacent footings. They are designed for the moments and shears produced by the frame action of the pier and the soil pressure under the footing.**

**The soil pressure or pile load under the footing is assumed to be uniform. The soil pressures or pile loads are generally computed only from the vertical column loads and soil and footing dead load. The moments at the base of the column are ignored for soil or pile loads.**

**To prevent unequal settlement, proportion the continuous footing so the soil pressures or pile loads are constant for Group 1 loading. The footing is kept relatively stiff between columns to prevent upward footing deflections which cause high soil or pile loads under the columns.**

**Refer to the book "Foundations of Structures" by C. W. Dunham (p.117) for an example of a continuous footing design. Apply the AASHTO specifications.**

**(4) Seals**

**A seal is the mat of unreinforced concrete poured under water inside the sheet piling of a cofferdam. It is designed to withstand the hydrostatic pressure on its bottom when the water above it is removed. Dewatering the cofferdam allows cutting of piles, placement of reinforcing steel and pouring of the footing in an air environment.**

**Seals are used for piers founded approximately 15'-0" or more below normal water level. At some locations it may be desirable to place a seal at lesser depths if the boring log indicates that the material at the bottom elevation and below would permit water to percolate up under the sheeting, making dewatering a cofferdam difficult or impossible. An example is a footing on bedrock where a seal should be used if the water depth exceeds 10 feet. The rock makes it difficult to seal the bottom of the sheeting without a seal.**

The hydrostatic pressure under the seal is resisted by the weight of the seal, the friction between the seal perimeter and walls of the cofferdam, and friction between seal and piles for pile footings. The friction values to use for seal design are:

| | |
|---|---|
| Bond on Piles | - 10 pounds per square inch |
| Bond on Sheet Piling | - 2 pounds per square inch on (seal depth minus 2 feet) x perimeter. |

To compute the capacity of piles in uplift see Chapter 11 on piling.

A rule of thumb for seal thickness is 0.40 H for spread footings and 0.25 H for pile footings where H is the depth of water from bottom of seal to top of normal water. The minimum seal size is 1'-6" larger than the footing size on all sides.

Example: Determine seal thickness for a 9' x 12' footing with 12 - 12" diameter piles. Uplift capacity of one pile = 10. kips. Water depth to top of seal is 16'.

Assume 12' x 15' x 3'-0" seal

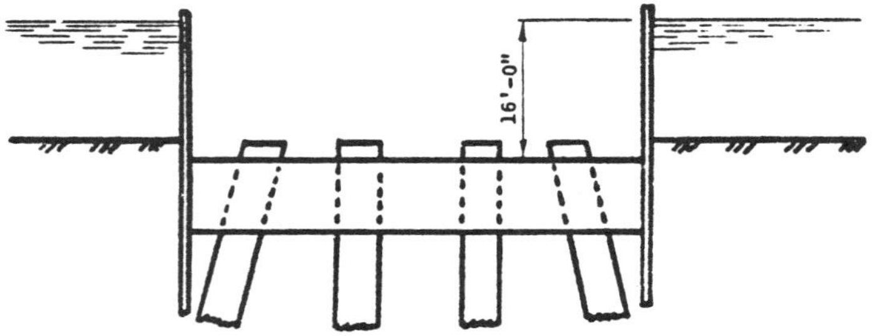

Uplift force of water - 12 x 15 x 19 x .0624 = 214 kips up

Weight of seal course - 12 x 15 x 3 x .15 = 81 kips down

Friction of sheet piling
2 x (12+15) x 1 x 144 x .002 = 16 kips down

Friction on 12 inch diameter pile
$\pi$ x 12 x 36 x .010 = 13.6 kips

Maximum uplift/pile = 10 kips

Total Available force from piles = 12 x 10 = 120 kips down

Summation of downward forces = 217 kips

217 > 214 OK

USE 3'-0" THICK SEAL

On a site with 60-75 feet of peat with 15-25 feet of weak clay below this, the Washington Bridge designers used precast seals. Precast seals have not been tried in Wisconsin.

(5) Isolated Spread Footing Sample Problem

Problem: Design a spread footing for a group 2 loading condition. Axial load is 400 kips, transverse moment is 500 ft. kips, and longitudinal moment is 800 ft. kips. Allowable soil pressure equals 6 kips/sq. ft. Column size is 2' x 2'. fc' = 3500 psi, fy = 60000 psi. A 25% soil pressure overstress is allowed for group 2 loading.

A. Determine Footing Size

Estimate a footing size.
Assume DL of footing and earth = 50 kips
From P/A = 6.0 x 1.25 = 7.5 ksf
(Group 2 Allowable soil pressure)
A = 450./7.5 = 60 sq. ft.

Because of the high moments assume the soil stress in one corner of the footing is near zero. Therefore the average stress on the bottom of the footing is half of 7.5 and the approximate required area is 120 sq. ft.

Try a 12' x 12.5' x 2' footing:

Concrete DL = 12 x 12.5 x 2 x .150 = 45.0 Kips
Earth DL = (150-4) x 2 x .120 = 35.0 Kips
Total DL = 80.0 Kips

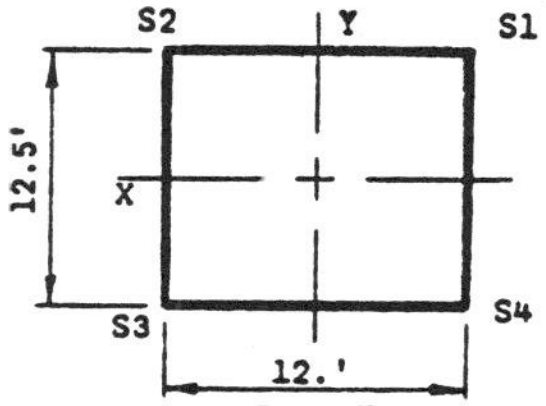


A = 12. x 12.5 = 150 sq. ft.
$S = 1/6\ bh^2$
$S_T = 12.5/6 \times 12^2 = 300.$
$S_L = 12/6 \times 12.5^2 = 312.$

$S1 = \frac{P}{A} + \frac{MT}{S} + \frac{ML}{S} = \frac{480}{150} + \frac{500}{300} + \frac{800}{312} = 7.43$ Ksf

$S2 = \frac{P}{A} - \frac{MT}{S} + \frac{ML}{S} = 4.09$ Ksf

$S3 = P/_A - \frac{MT}{S} - \frac{ML}{S} = -1.03$ Ksf

$S4 = P/_A + \frac{MT}{S} - \frac{ML}{S} = 2.31$ Ksf

Since S3 is in tension, reanalyze the footing using only the portion of the footing in compression.

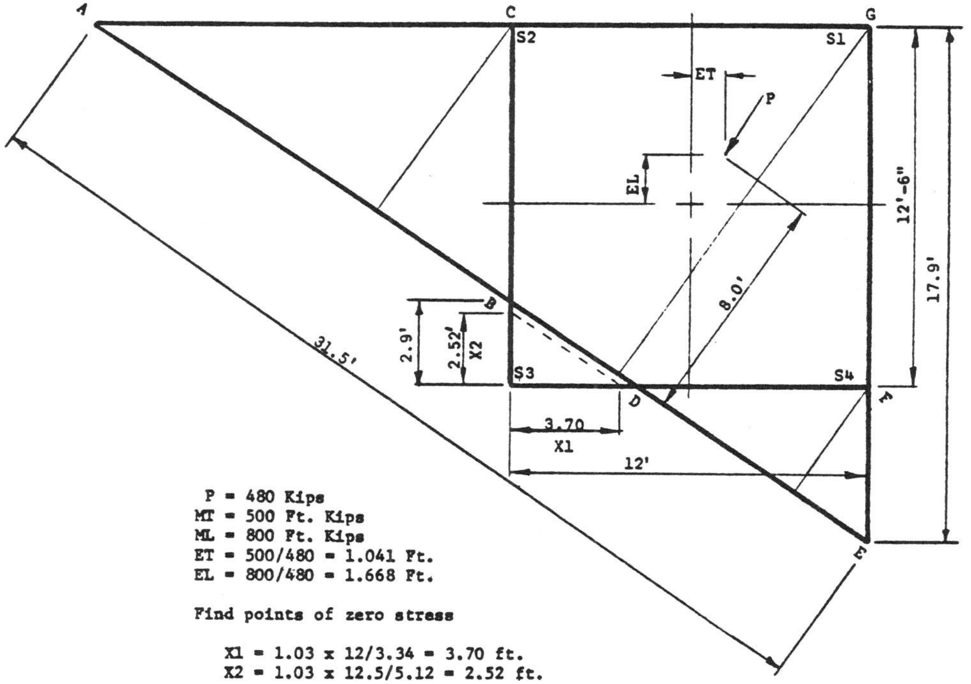


P = 480 Kips
MT = 500 Ft. Kips
ML = 800 Ft. Kips
ET = 500/480 = 1.041 Ft.
EL = 800/480 = 1.668 Ft.

Find points of zero stress

X1 = 1.03 x 12/3.34 = 3.70 ft.
X2 = 1.03 x 12.5/5.12 = 2.52 ft.

The actual line of zero stress is parallel to the line passing through the points of zero stress computed but closer to the point of maximum stress.

Select a trial zero axis 3 inches closer to the point of maximum stress. The footing is drawn to scale and the triangles ABC and DEF are drawn. The moment of inertia of BCGFD is computed about axis AE using $I = bh^3/12$ for the large triangle minus the I's of the two smaller triangles. Distances b and h are determined by scaling.

$I = 31.5 \times 14.75^3/12 - 17 \times 8^3/12 - 9.5 \times 4.4^3/12 = 7618\ ft.^4$

Moment about the line of zero stress is P x 8' = 480 x 8 = 3840 ft. kips

S1 = M/S = 3840/(7618./14.75) = 7.43 Ksf.

S2 = 14/26 x 7.43 = 4.00

S4 = 5.3/17.9 x 7.43 = 2.20

Check the volume of the pressure diagram to see if it is equal to P. If the discrepancy is large, assume a new location of the line of zero stress and repeat the procedure.

Volume of AEG = 1/3 Ah = 1/3 x 1/2 x 31.5 x 14.75 x 7.43
= 574. Kips

Volume of two small triangles:

ABC = 1/3 x 1/2 x 17 x 8 x 4.00 = 90.6 Kips
DEF = 1/3 x 1/2 x 9.5 x 4.4 x 2.2 = 15.3 Kips

Total Volume = 574 - 91 - 15 = 468 Kips

Since 468 is less than 480, the assumed position of the line of zero stress is too close to the point of maximum stress. However, the stresses obtained are close enough.

If the corner stresses exceed the allowable value or the pressure volume is less than the axial load, it may be necessary to increase the footing size to get a satisfactory design.

B. Check Shear Strength

Both one-way action and two-way action are checked. For group 2 loading the load factor is 1.3 for all loads. S1 and S2 therefore increase by a factor of 1.3.

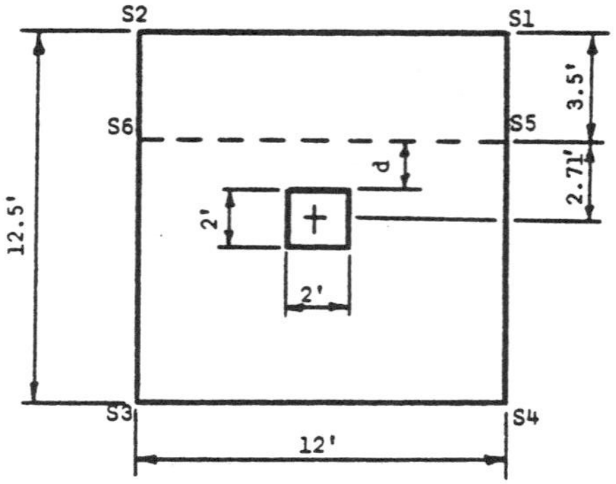


S1 = 1.3(7.43) = 9.66 kips/sq. ft.

S2 = 1.3(4.00) = 5.20

d = 24 - 3.5 = 20.5" = 1.71'

By proportioning:

S5 = 9.66(17.9 - 3.5)/17.9 = 7.77

S6 = 5.20(9.6 - 3.5)/9.6 = 3.30

Compute the volume of the soil pressure diagram that has the four corner stresses of S1, S2, S6 and S5.

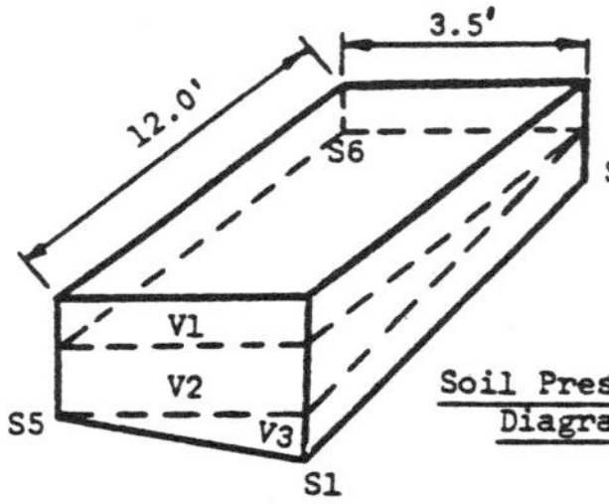


Soil Pressure Diagram

V1 = S6(12)(3.5)= 138.6

V2 = (S5 - S6) (12)(3.5)(.5) = 93.9

V3 = (S1 - S5 + S2 - S6)(12)
(3.5)/4 = 39.8

The dead load of the footing and earth must be deducted since they are included in S1, S2, S5 and S6.

Total Shear $V_u$ = V1 + V2 + V3 - 3.5(12)(.300 + .240) = 272.3 -22.6 = 249.7 kips

Actual Shear Stress $v_u = V_u/\phi b_w d$

$v_u$ = 249.7 x 1000/(.85 x 144 x 20.5) = 99 psi

Allowable Shear Stress = $2\sqrt{fc'} = 2\sqrt{3500}$ = 118 psi

Two - way action or peripheral shear must also be checked although it seldom governs. It is only necessary to consider the vertical reaction from the column.

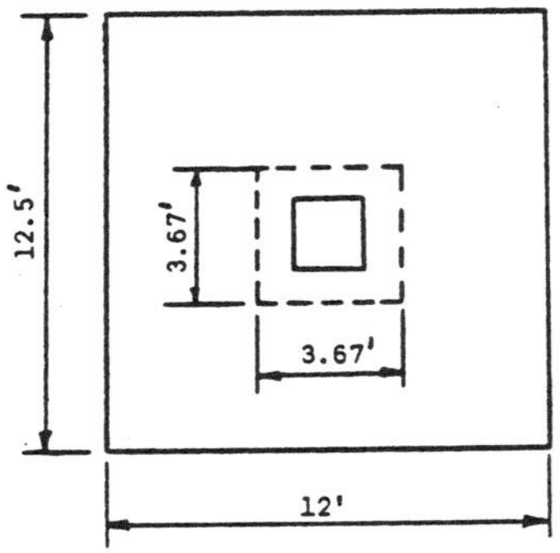


P = 1.3(400) = 520. kips

$V_u$ = 520(12 x 12.5 - 3.67$^2$)/(12 x 12.5)

Total Shear $V_u$ = 520(136.5)/150 = 473.2

$v_u = Vu/\phi b_o d$

$v_u$ = 473.2/(.85 x 4 x 12 x 3.67 x 20)

Actual Shear Stress $v_u$ = 473.2 x 1000/2994.7 = 158. psi

Allowable Shear Stress = $4\sqrt{fc'}$ = 236. psi

Use 2'-0 for footing depth

C. Design Longitudinal Steel

Determine moment at face of column

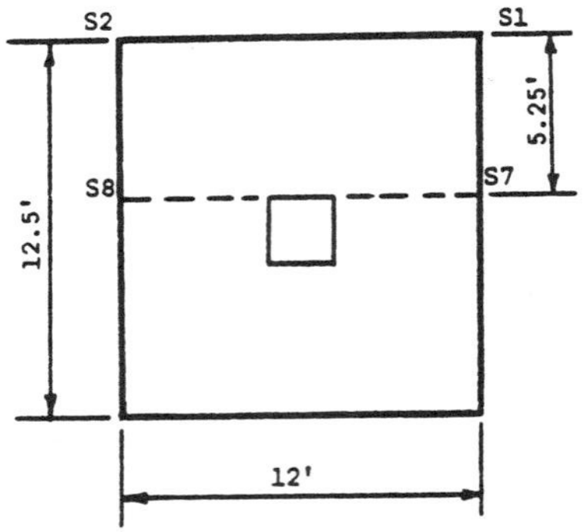


S1 = 9.66 kips/sq. ft.

S2 = 5.20

By proportioning:

S7 = 9.66(17.9 - 5.25)/17.9 = 6.83

S8 = 5.20(9.6 - 5.25)/9.6 = 2.36

Compute volume of soil pressure diagram of S1, S2, S8 and S7.

R1 = S8(5.25)(12.00) = 148.7

R2 = (S7 - S8)(12)(5.25)/2 = 140.8

R3 = (S1 - S7 + S2 - S8)(12)(5.25)/4 = 89.3

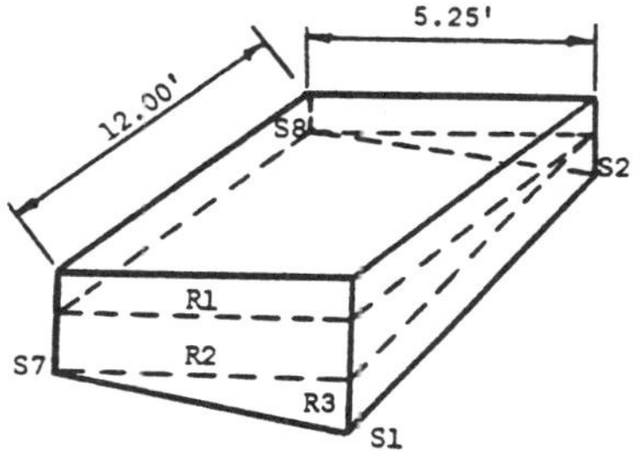


Soil Pressure Diagram

Compute moment of pressure diagram about line S7 to S8.

M = (R1 + R2)5.25/2 + R3(.667)(5.25)

M = 759.9 + 312.7 = 1072.6 ft. kips

Deduct moment due to footing and soil weight.

W = 12(5.25)(.3 + .24) = 34 kips

Total moment:
$M_u$ = 1072.6 - 34(1.3)(5.25)/2 = 956.6 ft. kips

Effective Footing Width:
From AASHTO article 1.4.6(F) the band width = 24 + 2(21) + .5(144 - 66) = 105 inches

Compute steel area required.
$R_u = M_u/\phi bd^2$
$R_u$ = 12(956.6)/(.9 x 105 x 20.5²) = .289

From design tables for fy = 60000. psi and fc' = 3500 psi
RHO = .0051
Area of steel = .0051(105)(20.5) = 10.98 sq. in.

The required area and perimeter of steel per foot in the band width = 10.98/8.75 = 1.25 sq. in. per foot.

Use #9 @ 9" within band and #9 @ 1'-6 outside of band

The transverse steel is designed in a similar manner. The shear strength in the transverse direction does not govern.

(6) **Isolated Pile Footing Sample Problem**

**Problem: Design a pile footing for a group 2 loading condition. Axial load is 400 kips, transverse moment is 500 ft. kips and longitudinal moment is 800 ft. kips.** Allowable single pile capacity = 110 kips (100% friction pile). Column size is 2' x 2'. Assume piles offer no resistance to uplift forces. A 25% pile overstress is allowed for group 2 loading.

A. Determine Footing Size and Piles Required

**The number of piles is first estimated. Assume DL of footing and earth = 50. kips and average pile force = 60 kips. From P/A = 450/60 = 8 piles.**

**Try a 6'-6" x 12'-0" x 3' footing**

**Concrete DL** = 12 x 6.5 x 3 x .15 = 35.1
**Earth DL** = (78-4) x 1 x .12 = 8.9
**Total DL** = 44.0

**Section Properties**

$S_T = 8 \times 2^2/2 = 16 \text{ ft.}^3$

$S_L = (4 \times 4.75^2 + 4 \times 2.25^2)/4.75 = 23.3 \text{ ft.}^3$

N = 8

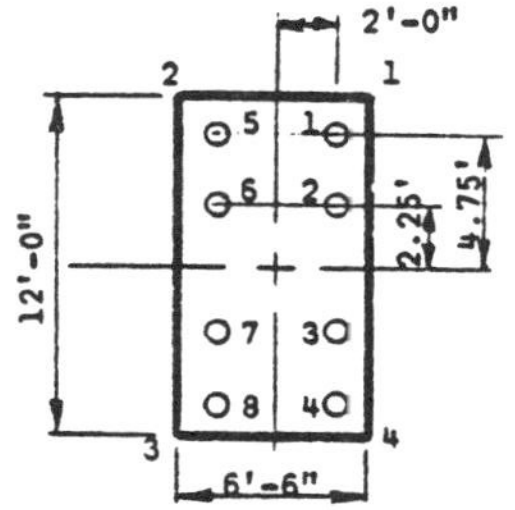


**Pile loads are:**

$$FL = \frac{P}{N} + \frac{M_T}{S_T} + \frac{M_L}{S_L}$$

**F1 = 444./8. + 500/16 + 800/23.3 = 121.2 Kips**
**F8 = 55.5 - 31.3 - 34.4 = -10.2 Kips**
**F5 = 55.5 - 31.3 + 34.4 = 58.6 Kips**
**F4 = 55.5 + 31.3 - 34.4 = 52.4 Kips**

**Since F 8 is negative, reanalyze the footing using only the piles in compression. Determine the line of zero stress based on the previously computed pile loads:**

**Distance from Pile 8:** $Y = \frac{10.2}{68.8} \times 9.5 = 1.41$ **feet**

$X = \frac{10.2}{62.6} \times 4.0 = .65$ **feet**

**Determine eccentricities of axial load:**

$e_x = \frac{500}{444} = 1.13$ **feet**

$e_y = \frac{800}{444} = 1.80$ **feet**

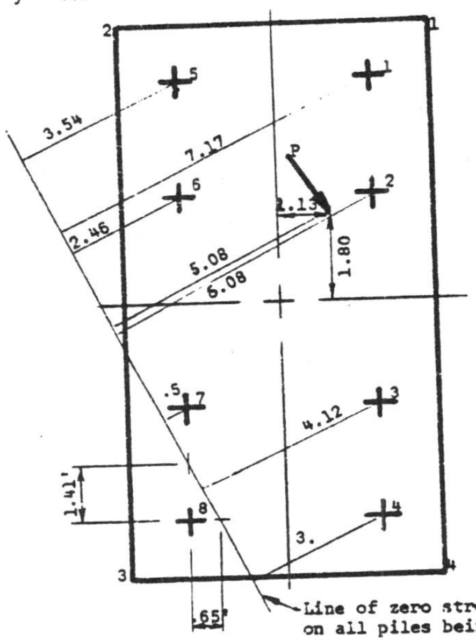


**Compute new section properties and pile loads:**

**About line of zero stress,** $I = \Sigma d^2$

$I = 3.54^2 + 7.17^2 + 2.46^2 + 6.08^2 + .5^2 + 4.12^2 + 3^2$
$= 133.21 \text{ ft.}^4$

$S = \frac{I}{d} = 133.21/7.17 = 18.60 \text{ ft.}^3$

**Moment = 444 x 5.08 = 2255. Ft. Kips**

* **F1 = M/S = 2255/18.6 = 121 Kips**
**F2 = 121 x 6.08/7.17 = 102.5**
**F3 = 121 x 4.12/7.17 = 69.5**
**F4 = 121 x 3/7.17 = 50.6**
**F5 = 121 x 3.54/7.17 = 59.6**
**F6 = 121 x 2.46/7.17 = 41.5**
**F7 = 121 x 5/7.17 = 8.4**

**Summation of pile loads = 453 Kips**

**Since 453 is greater than 444, the actual position of the line of zero stress is closer to the point of maximum stress. However, the pile forces obtained are well within the acceptable limits of accuracy.**

**The distances used in this method of analysis are scaled. The answer obtained is identical to the one obtained using the section properties of all piles. It is not necessary to perform an exact analysis when a small percent of the area between the outside piles is in uplift.**

* Allowable pile stress for Group 2 Loading:

110 Kips x 1.25 = 137.5 Kips.

B. Shear Strength

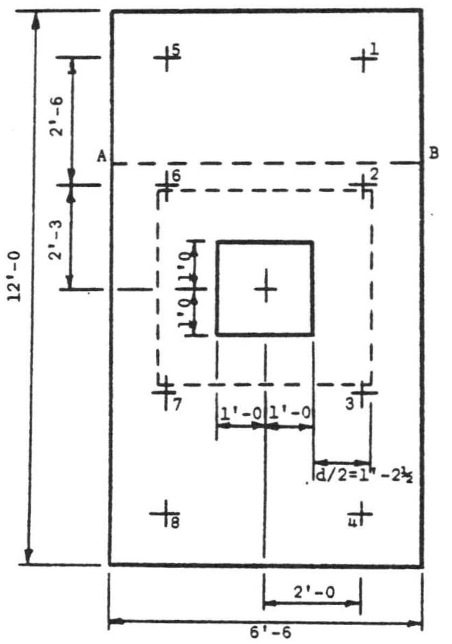


Because piles 1, 2, 3 and 4 fall inside a line at distance d from the face of the column a check of shear strength in the transverse direction is not required.

For shear in the longitudinal direction one-way action is checked.

Shear at AB = pile forces minus footing and soil loads.

For 3' footing, d = 36 - 7 = 29"

$V_u = 1.3(F1 + F5)$
$\quad -(6.5 \times 2.58 \times 3 \times .15)1.3$
$\quad -(6.5 \times 2.58 \times 1 \times .12)1.3$

$V_u = 235 - 9.8 - 2.6 = 222.6$

Actual Shear Stress:

$v_u = V_u/\phi b_w d$

$v_u = 222.6/(.85 \times 78 \times 29) = .116 \text{ ksi}$

Allowable Shear Stress = $2\sqrt{f_c'}$ = $2\sqrt{3500}$ = 118 psi

Two way action or peripheral shear:

$P = 1.3(400) = 520 = V_u$

$v_u = V_u/\phi b_o d = 520/(.85 \times 4 \times 4.42 \times 12 \times 29)$

$v_u = .099 \text{ ksi}$

C. Design Longitudinal Steel

Moment at face of column

$M_u = [(F1 + F5) \times 3.75 + (F2 + F6) \times 1.25]1.3$
$\quad - (5 \times 6.5 (3.08 \times .15 + 1 \times .12) \times 2.5)1.3 = 1055.6 \text{ ft Kips}$

Determine steel area required:

$$R_u = \frac{M_u}{\phi b d^2} = \frac{1055.6}{.9(6.5) \times 29^2} = .214$$

From tables p = .0037
Area of steel = .0037 x 78 x 29 = 8.37 in.$^2$

Use 11 #8 bars.

D. Design Transverse Steel

Because the piles are less than 2/3 of the effective footing depth away from the face of the column, bending is not considered in the transverse direction. Place a nominal amount of reinforcing steel in the footing, like #5 at 12".

E. Pile Group Action

This example assumes that a reduction for Pile Group Action must be computed for a full friction pile. Refer to Chapter 11 of this manual, or the AASHTO specifications for a definition of the Converse - LaBarre equation.

$E = 1 - (\phi(n - 1)m + (m - 1)n)/90mn$ (AASHTO 1.4.4(G))

$n = 4,\ m = 2,\ d = 1,\ s = 4$

$E = 1 - [14(3)(2) + 1(4)]/(90 \times 4 \times 2) = .88$

Allowable Pile Load = 110 x 1.25 x .88 = 121 kips (Drive pile to 110 kips)

## 13.7 QUANTITIES

**Consider the "Upper Limits for Excavation" for piers at such a time when the quantity is a minimum. This is either the existing ground line or the finished graded section. Indicate in the general notes which value is used.**

**Granular backfill is not used at piers except for special conditions.**

**Compute the concrete quantities for the footings, columns, and cap and show them for each one on the final plans.**

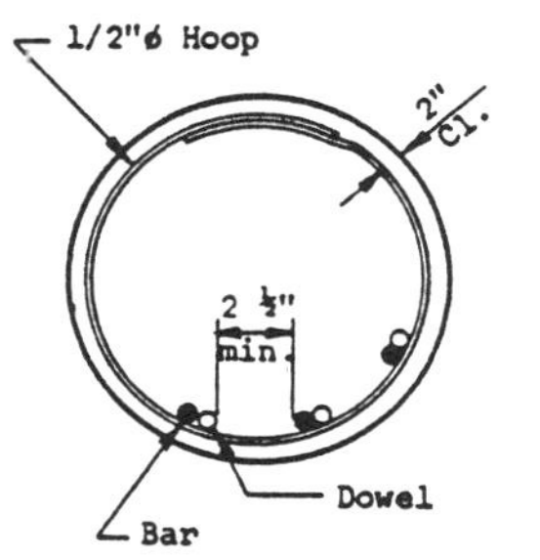

CASE I

Circumferentially lapped splices

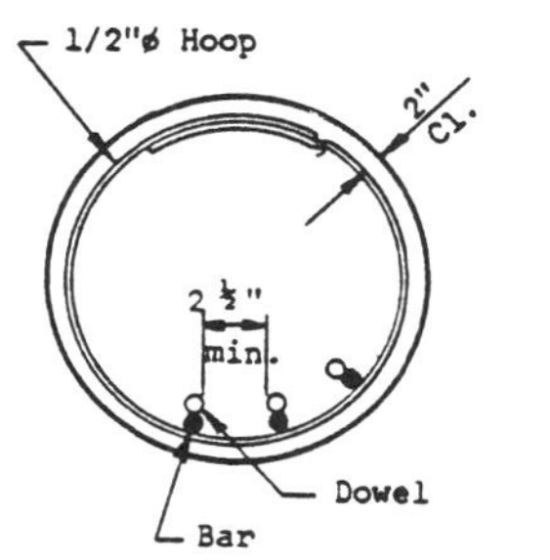

CASE II

Radially lapped splices

| COL. DIA. | | BAR SIZE NO. (Case I) | | | | | | BAR SIZE NO. (Case II) | | | | | |
|---|---|---|---|---|---|---|---|---|---|---|---|---|---|
| | | 7 | 8 | 9 | 10 | 11 | 14S | 7 | 8 | 9 | 10 | 11 | 14S |
| 30" | BARS PER COL. | 17 | 16 | 15 | 14 | 13 | 12 | 20 | 19 | 18 | 17 | 16 | 14 |
| 36" | | 22 | 20 | 19 | 18 | 17 | 15 | 26 | 25 | 23 | 22 | 21 | 19 |
| 42" | | 26 | 25 | 23 | 22 | 20 | 18 | 31 | 30 | 29 | 27 | 26 | 23 |

TABLE 13.1 MAXIMUM NUMBER OF BARS PERMITTED IN ROUND COLUMNS

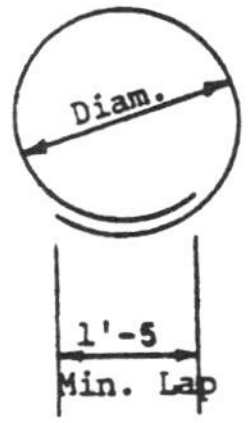


| COLUMN DIAMETER | BAR SIZE | DIAM. | LENGTH |
|---|---|---|---|
| 2'-6 | 4 | 2'-2 | 8'-2 |
| 3'-0 | 4 | 2'-8 | 9'-5 |
| 3'-6 | 4 | 3'-2 | 11'-8 |

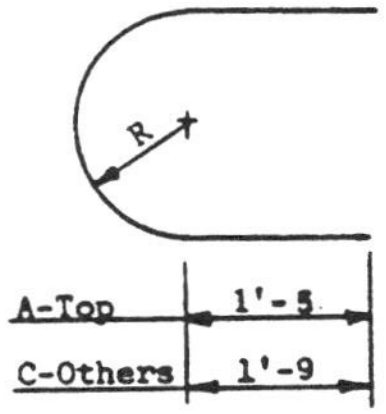


| CAP WIDTH | BAR SIZE | RADIUS R | A-TOP LENGTH | C-OTHERS LENGTH |
|---|---|---|---|---|
| 2'-6 | 5 | 1'-1 | 6'-3 | 6'-11 |
| 3'-0 | 5 | 1'-4 | 7'-1 | 7'-9 |
| 3'-6 | 5 | 1'-7 | 7'-10 | 8'-6 |

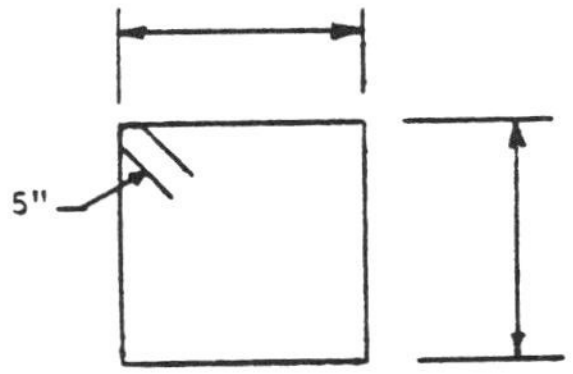


Stirrups - Determine bar size and dimensions. Refer to example.

TABLE 13.2 BENDING DETAILS - PIER BARS

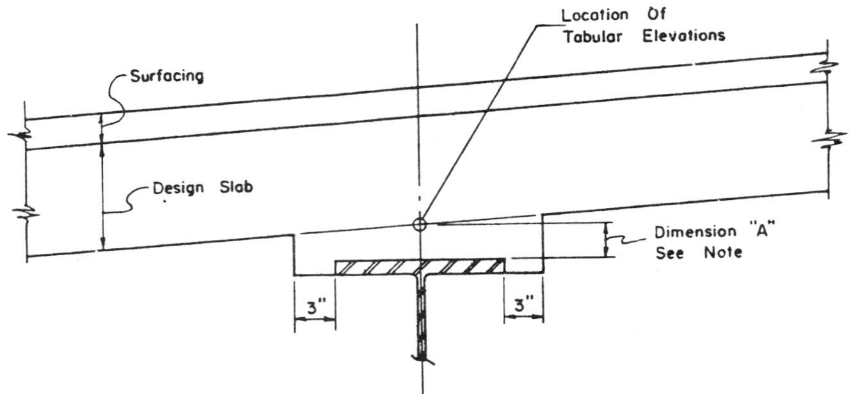


HAUNCH DETAIL

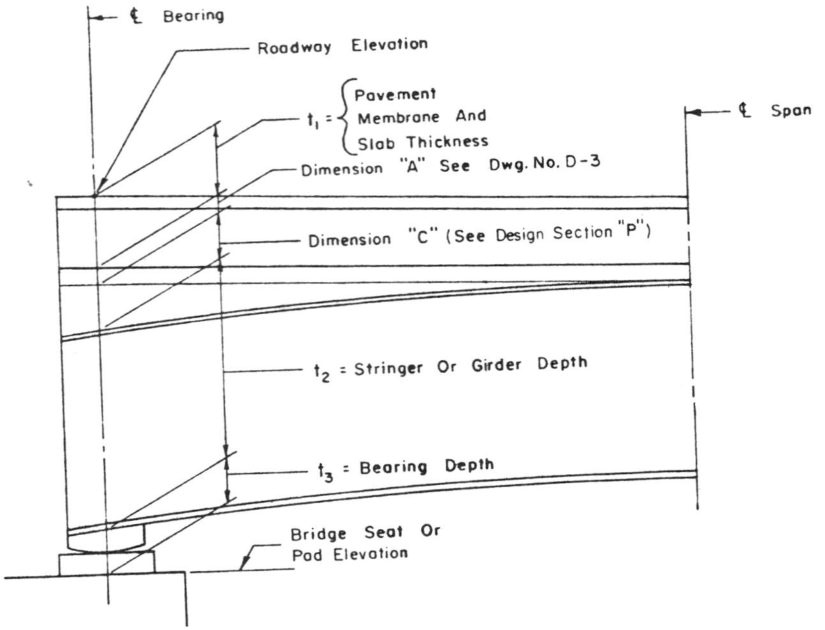


BRIDGE SEAT ELEV. = RDWY. ELEV. − $(t_1 + A + C + t_2 + t_3)$

BRIDGE SEAT ELEVATION DETERMINATION DIAGRAM

- At The Point Of Maximum Camber Dimension "A" Shall Be 1" For Spans Up To 50' And 1-1/2" For Spans Over 50'. Use Dimension "A" For The Computation Of Bridge Seat Elevations But Do Not Show On The Plans. Dimension "A" Shall Be Considered As 0" When Calculating The Physical Properties Of Composite Beams, However, The Weight Of The Concrete Shall Be Included In The Design Calculations.

P.

Camber. (Minimum Theoretical) For All Rolled Beams And Welded Girders.

The Minimum Theoretical Camber Shall Be Posted On The Construction Plans By Either Camber Diagram Or A Table Of Cambers. The Cambers Shall Be Specified By The Ordinate At The Mid-Length Of The Segment To Be Curved. Ordinates At Other Points Of Reverse Or Compound Curves Are Not Acceptable.

In The Calculation Of Minimum Theoretical Camber Do Not Include Or Show Tolerances Indicated In The "Standard Specifications".

The Minimum Theoretical Camber Shall Consist Of The Sum Of The Following Values:

A = Total 100% Dead Load Deflection

B = Middle Ordinate Of The Vertical Curve

C = Value From Following Table

VALUE OF C

| GRADE CONDITION | SINGLE SIMPLE SPAN | MULTIPLE SIMPLE SPANS & CONTINUOUS SPANS |
|---|---|---|
| VERTICAL CURVE | 1/16" PER 10' | 0 |
| STRAIGHT GRADE | 1/8" PER 10' | 1/16" PER 10' |

Camber Shall Not Be Computed For Prestressed I-Beams Or Prestressed Slabs.

Note: This Sheet Is To Be Issued As A Part Of The Written Material Of The Bridge Manual At A Later Date.

Mass

# K.1 WALL DESIGN

## DESIGN CRITERIA OF WALL

### RIGID WALL

| TYPE OF WALL | LOAD DIAGRAM |
|---|---|
| GRAVITY | 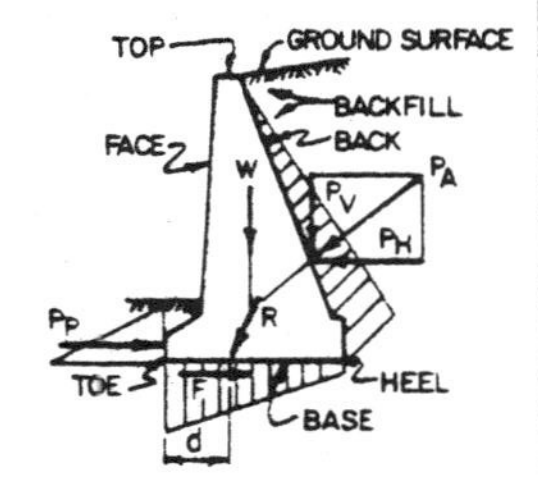  |
| SEMIGRAVITY | 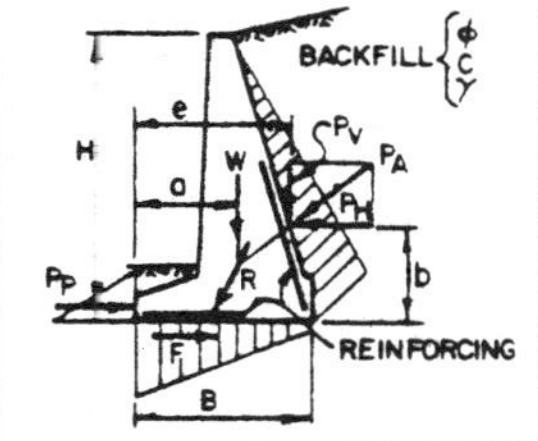  |
| CANTILEVER | 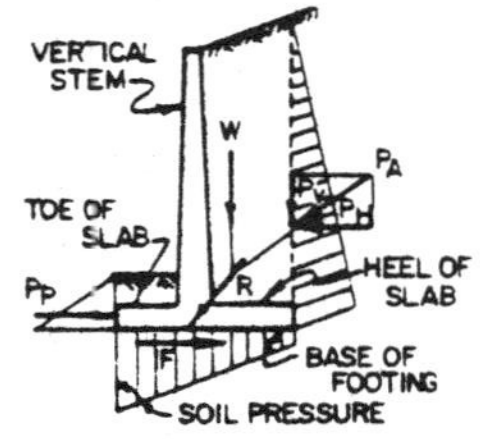  |
| COUNTERFORT | 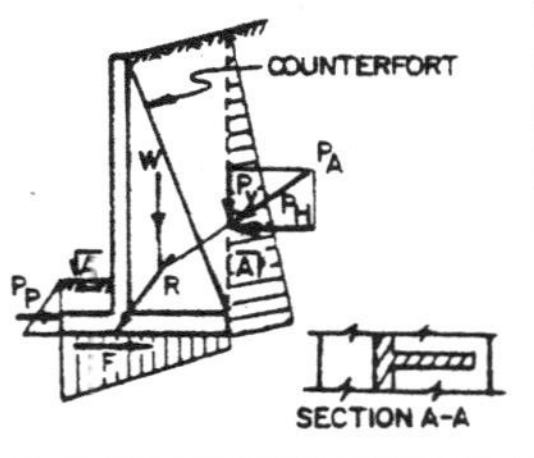  |

**DESIGN FACTORS**

LOCATION OF RESULTANT

MOMENTS ABOUT TOE:

$$d = \frac{Wa + P_V e - P_H b}{W + P_V}$$

ASSUMING $P_P = 0$

OVERTURNING

MOMENTS ABOUT TOE:

$$F_S = \frac{Wa}{P_H b - P_V e} \geq 1.5$$

IGNORE OVERTURNING IF R IS WITHIN MIDDLE THIRD (SOIL), MIDDLE HALF (ROCK).

CHECK R AT DIFFERENT HORIZONTAL PLANES FOR GRAVITY WALLS.

RESISTANCE AGAINST SLIDING

$$F_S = \frac{(W+P_V)\tan\delta + C_a B}{P_H} \geq 1.5$$

$$F_S = \frac{(W+P_V)\tan\delta + C_a B + P_P}{P_H} \geq 2.0$$

$$F = (W+P_V)\tan\delta + C_a B$$

FOR COEFFICIENTS OF FRICTION BETWEEN BASE AND SOIL SEE TABLE-1

$C_a$ = ADHESION BETWEEN SOIL AND BASE

$\tan\delta$ = FRICTION FACTOR BETWEEN SOIL AND BASE

W = INCLUDES WEIGHT OF WALL AND SOIL IN FRONT FOR GRAVITY AND SEMIGRAVITY WALLS. INCLUDES WEIGHT OF WALL AND SOIL ABOVE FOOTING, FOR CANTILEVER AND COUNTERFORT WALLS.

CONTACT PRESSURE ON FOUNDATION

FOR ALLOWABLE BEARING PRESSURE FOR INCLINED LOAD ON STRIP FOUNDATION, SEE CHAPTER 4

FOR ANALYSIS OF PILE LOADS BENEATH STRIP FOUNDATION, SEE CHAPTER 7.

OVERALL STABILITY

FOR ANALYSIS OF OVERALL STABILITY, SEE DM-7.1, CHAPTER 7.

### CRIB AND METAL BIN

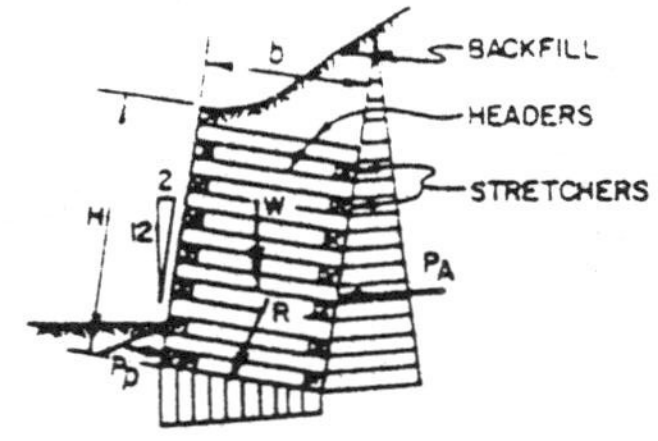


TYPICAL SECTION

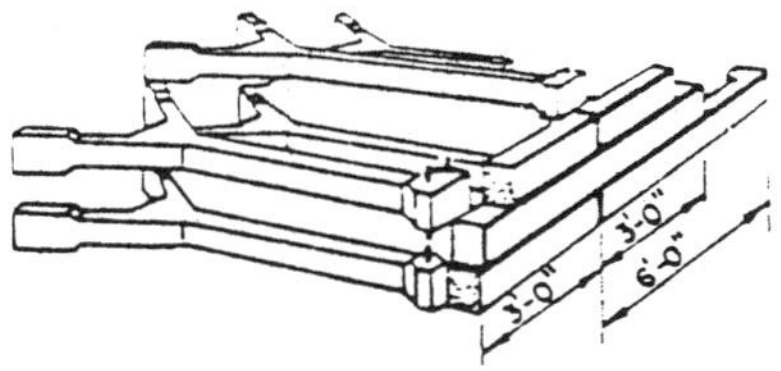


FISH TAIL TYPE ASSEMBLY

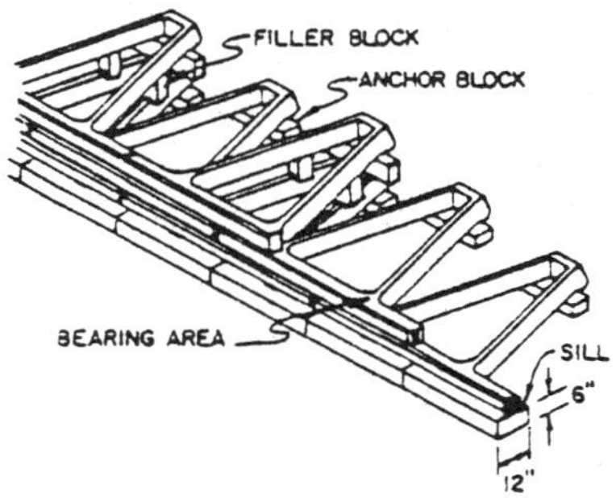


CLOSE FACE ASSEMBLY

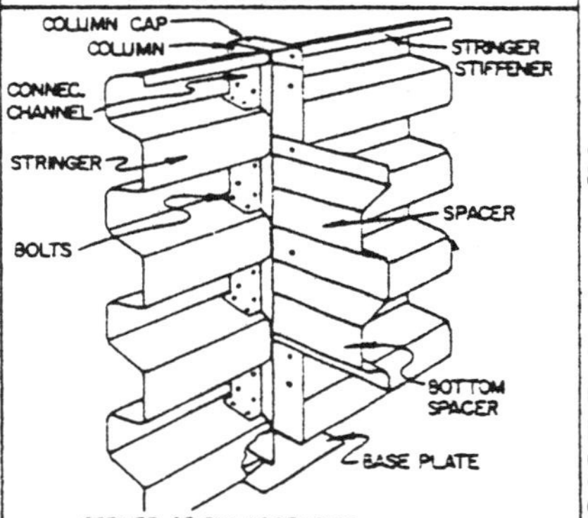


CORNER OF BIN ASSEMBLY

TYPES - COMMON TYPES OF CRIBS SHOWN ON ACCOMPANYING DIAGRAMS.

CRIBBING MATERIALS - TIMBER, CONCRETE, AND METAL.

FILL - CRUSHED STONE, OTHER COARSE GRANULAR MATERIAL, INCLUDING ROCK LESS THAN 2 INCHES IN SIZE.

DESIGN - DESIGN CRITERIA FOR GRAVITY WALLS APPLY. WALL SECTION RESISTING OVERTURNING IS TAKEN AS A RECTANGLE OF DIMENSION (H x b). WEIGHT OF CRIB IS EQUAL TO THAT OF MATERIAL WITHIN (H x b), INCLUDING WEIGHT OF CRIB MEMBERS. LOW WALLS (4 FT. HIGH AND UNDER) MAY BE MADE WITH A PLUMB FACE. HIGHER WALLS ARE BATTERED ON THE FACE AT LEAST 2 INCHES PER FOOT. FOR HIGH WALLS (12 FT. HIGH AND OVER) THE BATTER IS INCREASED OR SUPPLEMENTAL CRIBS ADDED AT THE BACK. SUCH WALLS ARE VERY SENSITIVE TO TRANSVERSE DIFFERENTIAL SETTLEMENTS. WALLS WITH CONVEX BACK ARE MORE DESIRABLE FOR GREATER HEIGHT. IN OPEN FACE CRIBS, THE SPACE BETWEEN STRETCHERS SHOULD NOT EXCEED 8 INCHES SO AS TO PROPERLY RETAIN THE FILL. EXPANSION JOINTS FOR CONCRETE AND METAL CRIBBING ARE SPACED NO MORE THAN 90 FEET.

FILLING - THE WALL SHOULD NOT BE LAID UP HIGHER THAN 3 FEET ABOVE THE LEVEL OF THE FILL WITHIN THE CRIB.

BIN TYPE RETAINING WALL - COMPOSED OF METAL BINS OR CELLS JOINED TO SPECIAL COLUMNAR UNITS AT THE CORNERS. THE DESIGN REQUIREMENTS ARE THE SAME AS FOR CRIB WALLS EXCEPT THAT SUITABLE DRAINAGE BEHIND THE WALLS IS NEEDED. INTERNAL STRESSES ARE INVESTIGATED IN ACCORDANCE WITH CRITERIA FOR CELLULAR WALLS.

(NAVFAC DM-7 SOIL MECHANIC AND FOUNDATION FROM U.S. NAVY)

# EARTH PRESSURE TABLE

## TABLE III

EARTH PRESSURES AND MOMENTS FROM RANKINE'S FORMULA WITHOUT SURCHARGE

Based on an equivalent fluid pressure of 40 pounds per cubic foot.

H = Height of filling in feet

p = Unit pressure at depth H, in lbs. per sq. ft.

$p = 40H$

P = Total pressure per linear ft. = $20H^2$ lbs.

M = Moment per linear ft. = $\frac{20}{3}H^3$ ft. lbs.

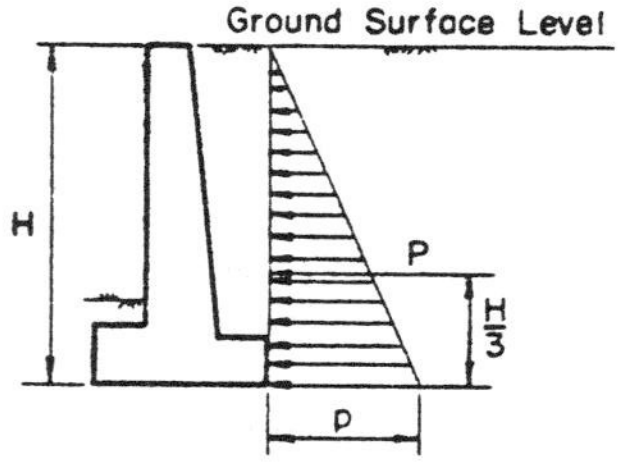


| H | p | P | M | H | p | P | M |
|---|---|---|---|---|---|---|---|
| 4 | 160 | 320 | 427 | 19 | 760 | 7,220 | 45,727 |
| 5 | 200 | 500 | 833 | 20 | 800 | 8,000 | 53,333 |
| 6 | 240 | 720 | 1,440 | 21 | 840 | 8,820 | 61,740 |
| 7 | 280 | 980 | 2,287 | 22 | 880 | 9,680 | 70,987 |
| 8 | 320 | 1,280 | 3,413 | 23 | 920 | 10,580 | 81,113 |
| 9 | 360 | 1,620 | 4,860 | 24 | 960 | 11,520 | 92,160 |
| 10 | 400 | 2,000 | 6,667 | 25 | 1,000 | 12,500 | 104,167 |
| 11 | 440 | 2,420 | 8,873 | 26 | 1,040 | 13,520 | 117,173 |
| 12 | 480 | 2,880 | 11,520 | 27 | 1,080 | 14,580 | 131,220 |
| 13 | 520 | 3,380 | 14,647 | 28 | 1,120 | 15,680 | 146,347 |
| 14 | 560 | 3,920 | 18,293 | 29 | 1,160 | 16,820 | 162,593 |
| 15 | 600 | 4,500 | 22,500 | 30 | 1,200 | 18,000 | 180,000 |
| 16 | 640 | 5,120 | 27,307 | 31 | 1,240 | 19,220 | 198,607 |
| 17 | 680 | 5,780 | 32,753 | 32 | 1,280 | 20,480 | 218,453 |
| 18 | 720 | 6,480 | 38,880 | 33 | 1,320 | 21,780 | 239,580 |

## TABLE IV

EARTH PRESSURES AND MOMENTS FROM RANKINE'S FORMULA WITH SURCHARGE

H = Height of filling in feet

h = Height of surcharge in feet

p = Unit pressure at depth (H+h) = 40(H+h) lbs. per sq. ft.

p' = Unit pressure at depth h = 40h lbs. per sq. ft.

P = Total pressure per linear ft. = 20H(2h+H) pounds.

M = Moment per linear foot = $\frac{20}{3}H^2(3h+H)$

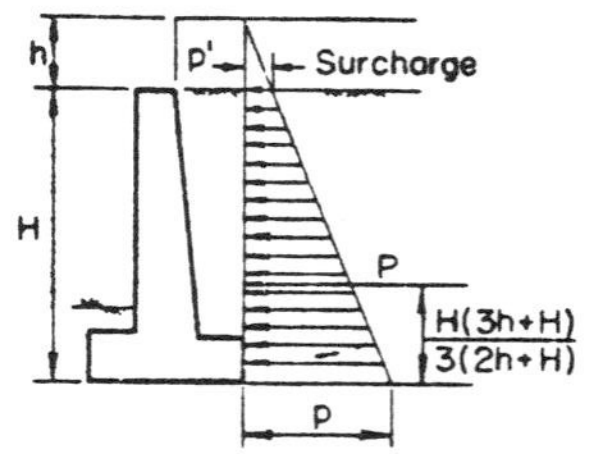


| H | h = 1 ft. | | h = 2 ft. | | h = 3 ft. | | h = 4 ft. | | h = 6 ft. | |
|---|---|---|---|---|---|---|---|---|---|---|
| | P | M | P | M | P | M | P | M | P | M |
| 4 | 480 | 747 | 640 | 1,067 | 800 | 1,387 | 960 | 1,707 | 1,280 | 2,347 |
| 5 | 700 | 1,333 | 900 | 1,833 | 1 100 | 2,333 | 1,300 | 2,833 | 1,700 | 3,833 |
| 6 | 960 | 2,160 | 1,200 | 2,880 | 1,440 | 3,600 | 1,680 | 4,320 | 2,160 | 5,760 |
| 7 | 1,260 | 3,267 | 1,540 | 4,247 | 1,820 | 5,227 | 2,100 | 6,207 | 2,660 | 8,167 |
| 8 | 1,600 | 4,693 | 1,920 | 5,973 | 2,240 | 7,253 | 2,560 | 8,533 | 3,200 | 11,093 |
| 9 | 1,980 | 6,480 | 2,340 | 8,100 | 2,700 | 9,720 | 3,060 | 11,340 | 3,780 | 14,580 |
| 10 | 2,400 | 8,667 | 2,800 | 10,667 | 3,200 | 12,667 | 3,600 | 14,667 | 4,400 | 18,667 |
| 11 | 2,860 | 11,293 | 3,300 | 13,713 | 3,740 | 16,133 | 4,180 | 18,553 | 5,060 | 23,393 |
| 12 | 3,360 | 14,400 | 3,840 | 17,280 | 4,320 | 20,160 | 4,800 | 23,040 | 5,760 | 28,800 |
| 13 | 3,900 | 18,027 | 4,420 | 21,407 | 4,940 | 24,787 | 5,460 | 28,167 | 6,500 | 34,927 |
| 14 | 4,480 | 22,213 | 5,040 | 26,133 | 5,600 | 30,053 | 6,160 | 33,973 | 7,280 | 41,813 |
| 15 | 5,100 | 27,000 | 5,700 | 31,500 | 6,300 | 36,000 | 6,900 | 40,500 | 8,100 | 49,500 |
| 16 | 5,760 | 32,427 | 6,400 | 37,547 | 7,040 | 42,667 | 7,680 | 47,787 | 8,960 | 58,027 |
| 17 | 6,460 | 38,533 | 7,140 | 44,313 | 7,820 | 50,093 | 8,500 | 55,873 | 9,860 | 67,433 |
| 18 | 7,200 | 45,360 | 7,920 | 51,840 | 8,640 | 58,320 | 9,360 | 64,800 | 10,800 | 77,760 |
| 19 | 7,980 | 52,947 | 8,740 | 60,167 | 9,500 | 67,387 | 10,260 | 74,607 | 11,780 | 89,047 |
| 20 | 8,800 | 61,333 | 9,600 | 69,333 | 10,400 | 77,333 | 11,200 | 85,333 | 12,800 | 101,333 |
| 21 | 9,660 | 70,560 | 10,500 | 79,380 | 11,340 | 88,200 | 12,180 | 97,020 | 13,860 | 114,660 |
| 22 | 10,560 | 80,667 | 11,440 | 90,347 | 12,320 | 100,027 | 13,200 | 109,707 | 14,960 | 129,067 |
| 23 | 11,500 | 91,693 | 12,420 | 102,273 | 13,340 | 112,853 | 14,260 | 123,433 | 16,100 | 144,593 |
| 24 | 12,480 | 103,680 | 13,440 | 115,200 | 14,400 | 126,720 | 15,360 | 138,240 | 17,280 | 161,280 |
| 25 | 13,500 | 116,667 | 14,500 | 129,167 | 15,500 | 141,667 | 16,500 | 154,167 | 18,500 | 179,167 |
| 26 | 14,560 | 130,693 | 15,600 | 144,213 | 16,640 | 157,733 | 17,680 | 171,253 | 19,760 | 198,293 |
| 27 | 15,660 | 145,800 | 16,740 | 160,380 | 17,820 | 174,960 | 18,900 | 189,540 | 21,060 | 218,700 |
| 28 | 16,800 | 162,027 | 17,920 | 177,707 | 19,040 | 193,387 | 20,160 | 209,067 | 22,400 | 240,427 |
| 29 | 17,980 | 179,413 | 19,140 | 196,233 | 20,300 | 213,053 | 21,460 | 229,873 | 23,780 | 263,513 |
| 30 | 19,200 | 198,000 | 20,400 | 216,000 | 21,600 | 234,000 | 22,800 | 252,000 | 25,200 | 288,000 |
| 31 | 20,460 | 217,827 | 21,700 | 237,047 | 22,940 | 256,267 | 24,180 | 275,487 | 26,660 | 313,927 |
| 32 | 21,760 | 238,933 | 23,040 | 259,413 | 24,320 | 279,893 | 25,600 | 300,373 | 28,160 | 341,333 |
| 33 | 23,100 | 261,360 | 24,420 | 283,140 | 25,740 | 304,920 | 27,060 | 326,700 | 29,700 | 370,260 |

## TABLE VI

### FORMULAS FOR EARTH PRESSURE FOR MEMBERS LOCATED WITHIN A SLOPING EMBANKMENT

Referring to Table V, the earth pressures may be computed by the diagrams, table and equations on this page.

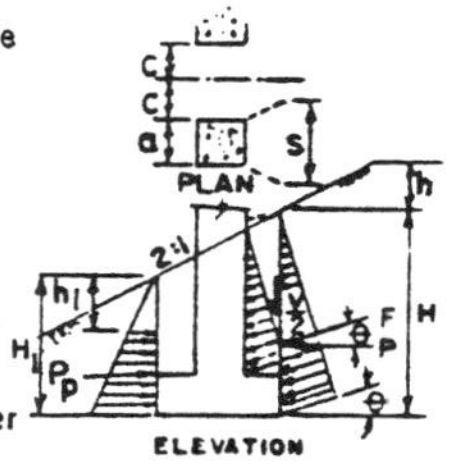


SHOWING PROJECTED PRESSURE DIAGRAM ON WALL OR COLUMN FOR COMPUTING MOMENTS.

ACTIVE PRESSURE: The active pressure shown in the diagram and described on Table V may be assumed to act against an adjusted member of width "S", to allow for the pressure due to embankment directly behind and adjacent to a wall or column of width "a". The extension of footing beyond the member should be ignored. The total active pressure $P_T$ on a member, may therefore be based on the following equation:

| TYPE OF MEMBER | | VALUES OF S |
|---|---|---|
| single column or wall | | a + H |
| interior columns | $c \geq H$ | a + H |
| | $c < H$ | $a + H - (H-c)^2/H$ |
| exterior columns | $c \geq H$ | a + H |
| | $c < H$ | $a + H - (H-c)^2/2H$ |

$P_T = SP = Sk\cos\theta H^2$, lbs. where,

$k\cos\theta$ = coeff. from Table V

S = adjusted member width, see adjacent table.

PASSIVE PRESSURE: For structures on piles or yielding soils, and allowing for the sloping embankment surface, the total passive pressure resistance $P_p$ (see diagram above) may be based on the equation:

$$P_p = \left[0.45 + 0.03(\phi - 26.5^\circ)\right] w \left[H_1^2 - h_1^2\right], \text{ lb. per lin. ft. where,}$$

$\phi$ = angle of internal friction, deg.

w = unit soil weight, lb. per cu. ft.

COMMENTS: For locating resultant forces, for designing structures on bedrock, and for computing bending moments on walls or columns, disregard passive pressure. Where used, the passive pressure may be assumed to act against the actual width of the member and/or footing being considered.

All dimensional properties are in ft. units.

## TABLE V

### FORMULAS FOR EARTH PRESSURE ON WALLS WITH A SLOPING SURCHARGE OF A DEFINITE HEIGHT

Earth fill with a sloping surcharge exerts a pressure which is inclined as indicated by the force F in the diagram. For computing pressures and overturning forces on the wall, the horizontal component and one-half the vertical component, P and V/2, of the force F shall be used. They shall be considered to be applied directly above the heel of the footing as shown in the diagram and as noted below.

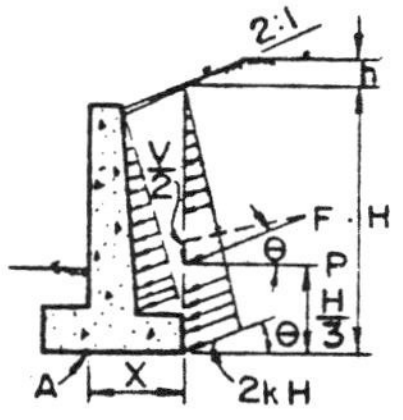


These forces may be computed as follows:

$F = kH^2$, lb. per lin. ft. of wall

$P = F\cos\theta = k\cos\theta H^2$, lb. per lin. ft. of wall

$V/2 = 0.5\,F\sin\theta = 0.5\,k\sin\theta H^2$, lb. per lin. ft. of wall

The overturning moment about any point "A" at bottom of footing is given by the following:

Overturning moment due to $P = P\frac{H}{3} = KH^3$, ft. lb. per lin. ft.

Restraining moment due to $V/2 = \frac{VX}{2} = 0.5\,k\sin\theta\,H^2X$, ft. lb. per lin. ft.

For computing bending moments on the stem, portions of the total pressure, as computed above, may be projected directly against the back of the stem, as indicated in the diagram.

For design of the heel, the vertical component V shall be used, but this force shall be considered as uniformly distributed over the rear projection of the footing. The computed value of V/2, located as shown, shall be used for all other design computations.

| h/H | k | θ | kcosθ | ksinθ | K |
|---|---|---|---|---|---|
| 0.00 | 18.50 | 10.49 | 18.19 | 3.37 | 6.06 |
| 0.10 | 21.00 | 16.00 | 20.19 | 5.79 | 6.73 |
| 0.20 | 23.30 | 19.69 | 21.94 | 7.85 | 7.31 |
| 0.30 | 25.20 | 21.99 | 23.37 | 9.44 | 7.79 |
| 0.40 | 26.80 | 23.60 | 24.56 | 10.73 | 8.19 |
| 0.50 | 28.00 | 24.90 | 25.40 | 11.79 | 8.47 |
| 0.75 | 30.40 | 26.45 | 27.22 | 13.54 | 9.07 |
| 1.00 | 31.80 | 26.55 | 28.44 | 14.21 | 9.48 |
| Inf. | 32.00 | 26.55 | 28.62 | 14.30 | 9.54 |

Coefficients based on the following:

Unit soil weight = 120 lb. per cu. ft.

Angle of internal friction, $\phi = 30^\circ$

Dimensions H, h and X in ft.

# COMPUTATION OF WALL PRESSURE

1. CONDITIONS. The pressure on retaining walls, bulkheads, or buried anchorages is a function of the relative movement between the structure and the surrounding soil.

a. Active State. Active earth pressure occurs when the wall moves away from the soil and the soil mass stretches horizontally sufficient to mobilize its shear strength fully, and a condition of plastic equilibrium is reached. (See Figure 1 from Reference 1, Excavations and Retaining Structures, by the Canadian Geotechnical Society.) The ratio of the horizontal component or active pressure to the vertical stress caused by the weight of soil is the active pressure coefficient ($K_a$). The active pressure coefficient as defined above applies only to cohesionless soils.

b. Passive State. Passive earth pressure occurs when a soil mass is compressed horizontally, mobilizing its shear resistance fully (see Figure 1). The ratio of the horizontal component of passive pressure to the vertical stress caused by the weight of the soil is the passive pressure coefficient ($K_p$). The passive coefficient, as defined here, applies only to cohesionless soil. A soil mass that is neither stretched nor compressed is said to be in an at-rest state. The ratio of lateral stress to vertical stress is called the at-rest coefficient ($K_o$).

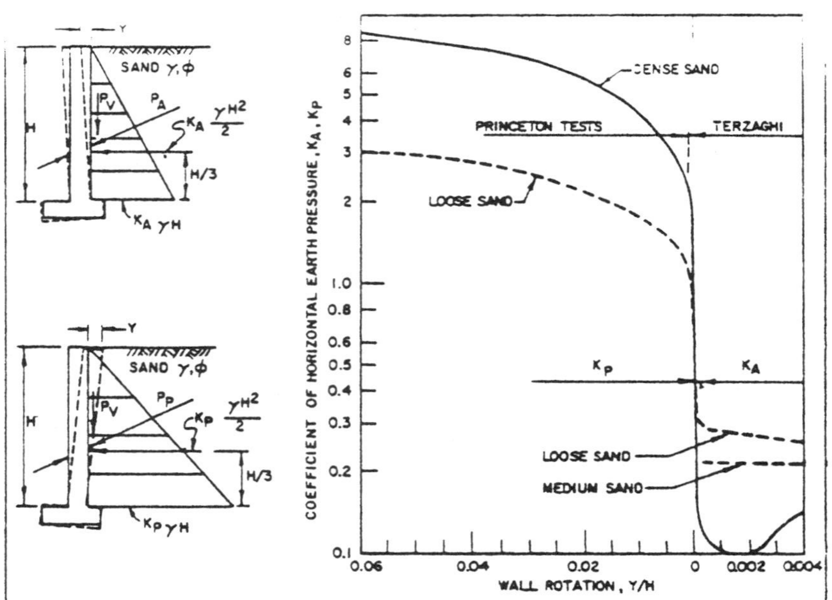


MAGNITUDES OF WALL ROTATION TO REACH FAILURE

| SOIL TYPE AND CONDITION | ROTATION Y/H* | |
|---|---|---|
| | ACTIVE | PASSIVE |
| DENSE COHESIONLESS | .0005 | .002 |
| LOOSE COHESIONLESS | .002 | .006 |
| STIFF COHESIVE | .01 | .02 |
| SOFT COHESIVE | .02 | .04 |

* Y = HORIZONTAL DISPLACEMENT
H = HEIGHT OF THE WALL

FIGURE 1
Effect of Wall Movement on Wall Pressures

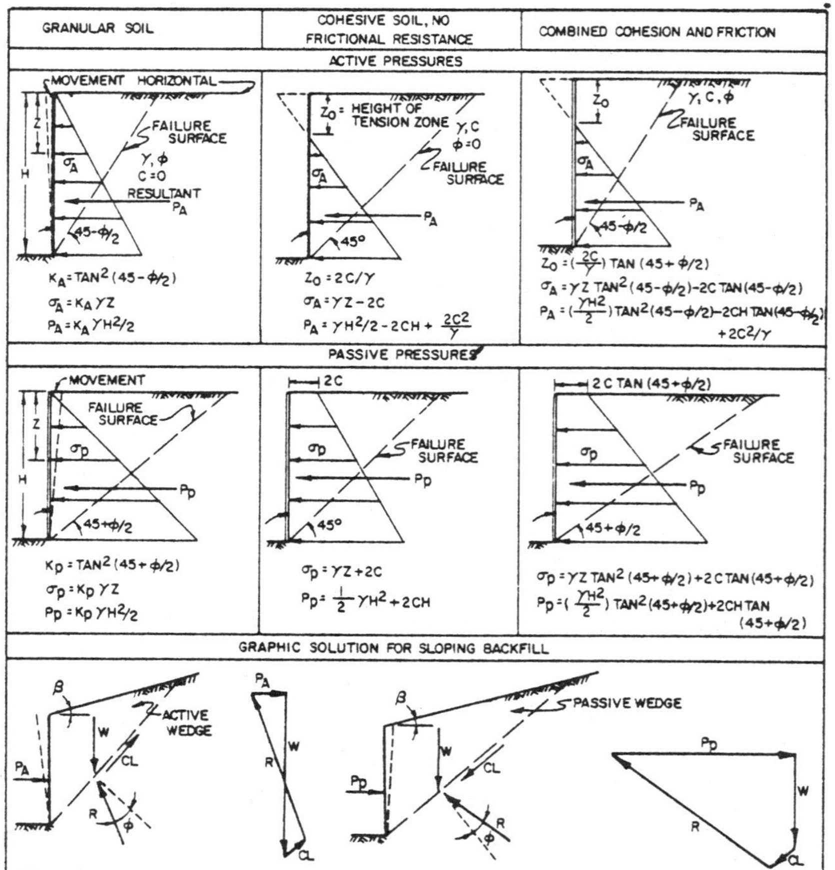


FOR COHESIONLESS SOILS WITH SLOPING BACKFILL, VALUES OF $K_A$ AND $K_P$, AND POSITIONS OF FAILURE ARE GIVEN IN FIGURE 4.
FOR SOIL WITH C AND φ, THE POSITION OF THE FAILURE SURFACE IS DETERMINED BY ANALYZING TRIAL WEDGES TO OBTAIN MAXIMUM VALUE OF $P_A$ AND MINIMUM VALUE OF $P_P$.

THE CASES SHOWN INVOLVE THE FOLLOWING ASSUMPTIONS.
1. MATERIALS ARE HOMOGENEOUS.
2. SUFFICIENT MOVEMENT HAS OCCURRED SO SHEAR STRENGTH ON RUPTURE SURFACE IS COMPLETELY MOBILIZED.
3. WALL IS VERTICAL. NO SHEAR FORCES ARE PRESENT ON BACK OF WALL. RESULTANT FORCES ARE HORIZONTAL.
UNDER THESE CONDITIONS, RESULTANT PRESSURES ARE ACTIVE AND PASSIVE VALUES, AND RUPTURE SURFACE IS A STRAIGHT PLANE THROUGH HEEL OF WALL. EFFECTS OF SURCHARGE AND GROUNDWATER PRESSURES ARE NOT INCLUDED.

FIGURE 2
Computation of Simple Active and Passive Pressures

## SHEAR KEY

ASSUME W = 120 pcf
ASSUME ANGLE OF INTERNAL FRICTION = 35° (Sand)

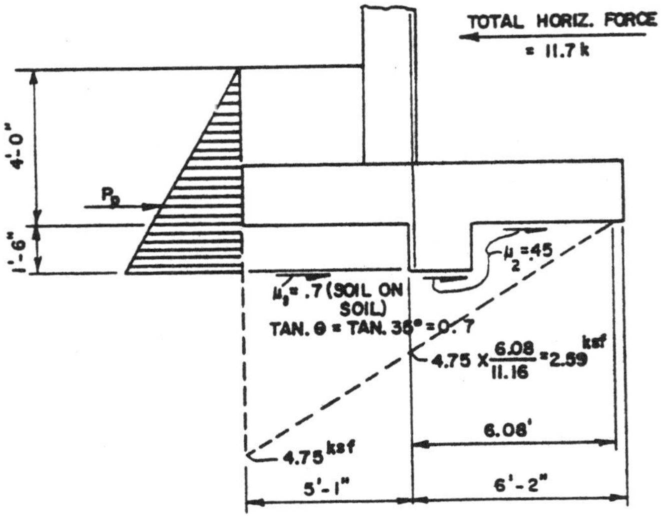


TOTAL FRICTIONAL FORCE $F = \mu_1 R_1 + \mu_2 R_2$

$= 0.7\,(1/2)\,(4.75 + 2.59)\,5.08 + 0.45 \times 1/2 \times 2.59 \times 6.08$

$= 13.03 + 3.54 = \underline{\underline{16.57^k}}$

$P_p = \frac{1}{2} C_p\, W\, (h_2^2 - h_1^2) = \frac{.300}{2}\,(5.5^2 - 0.0^2) = \underline{\underline{4.54^k}}$ (with $C_p$ = 2.5, $W$ = .120)

$F.S. = \frac{16.57 + 4.54}{11.7} = \underline{\underline{1.80}}$ OK > 1.5

**FIGURE 14.1 - FACTOR OF SAFETY AGAINST SLIDING USING SHEAR KEY**

May 1977 | *State of Wisconsin / Department of Transportation*

(WISCONSIN DOT)

## FOUNDATION PEDESTALS

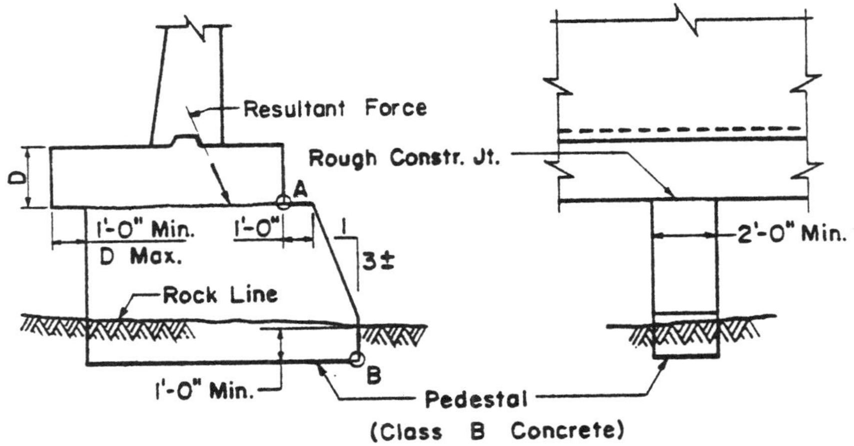


(1) Dimensions of the footing and pedestals shall be kept such as to satisfy the following requirements:

   (a) Factor of safety against overturning (about A for footing and about B for pedestals) shall be minimum 1.75.

   (b) The ratio of vertical loads to the horizontal loads shall be minimum 2.

   (c) Lateral earth pressures on pedestals may be disregarded.

(2) The clear distance between pedestals shall be kept approximately 10 ft. and preferably more if conditions are suitable.

(PENN DOT)

# DESIGN CRITERIA FOR ANCHORED BULKHEAD

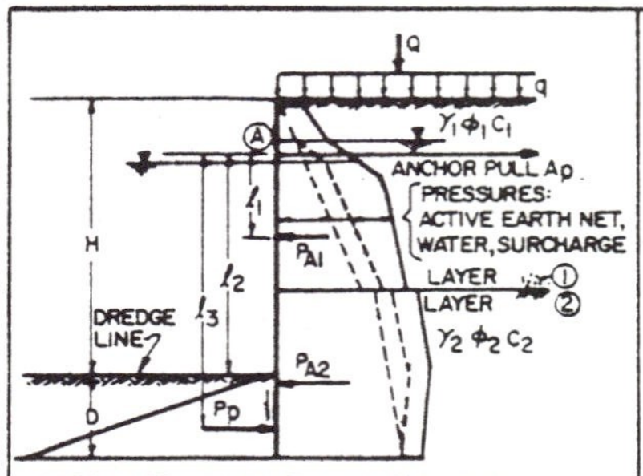


FREE EARTH SUPPORT - GENERAL CASE

1. COMPUTE PRESSURES BY METHODS OF FIGURES 2 TO 7 PASSIVE PRESSURES FOR CLEAN COARSE GRAIN SOILS INCLUDE WALL FRICTION (δ). TABLE I. FOR ACTIVE OR PASSIVE PRESSURES IN ALL OTHER SOIL TYPES, IGNORE WALL FRICTION.
2. DEPTH OF PENETRATION REQUIRED: TAKE MOMENTS ABOUT POINT Ⓐ AND SOLVE FOR D: $P_{A1}\ell_1 + P_{A2}\ell_2 = \frac{P_p}{F_S}\ell_3$
   $F_S$ = 2 TO 3 FOR COARSE GRAINED SOILS
   $F_S$ = 1.5 TO 2 FOR FINE GRAINED SOILS
3. ANCHOR PULL: $A_p = [P_{A1} + P_{A2} - P_p/F_S]\, d$, d = ANCHOR SPACING
4. MAXIMUM BENDING MOMENT ($M_{MAX}$) IN SHEETING COMPUTED BY THE FREE EARTH SUPPORT METHOD AND APPLYING $P_{A1}, P_{A2}, P_p/F_S$ AND $A_p$. FOR SHEETING IN SAND APPLY MOMENT REDUCTION FOR FLEXIBILITY OF FIGURE 19.
5. INCRESE PENETRATION COMPUTED (D) BY 20% TO ALLOW FOR DREDGING, SCOUR, ETC.

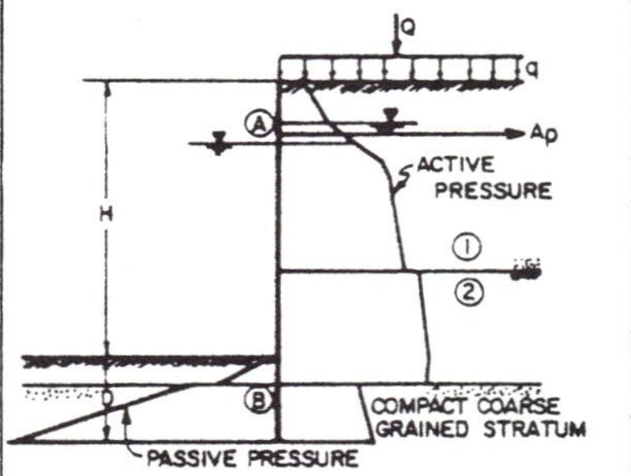


PENETRATION IN COMPACT COARSE GRAINED STRATUM

DESIGN STEPS 1, 2, AND 3 SAME AS ABOVE EARTH SUPPORT.
4. COMPUTE MAXIMUM BENDING MOMENT ($M_{MAX.}$) IN SHEETING BY FREE EARTH SUPPORT METHOD APPLYING $P_A$, $P_p/F_S$ AND $A_p$.
5. COMPUTE $\rho$ ACCORDING TO FIGURE 19. IF $\rho \geq 20$, $M_{DESIGN}$ IS COMPUTED FOR THE SPAN ⒶⒷ ASSUMING SIMPLE SUPPORT AT POINT Ⓑ
   IF $\rho < 20$ OBTAIN MOMENT REDUCTION FOR FLEXIBILITY FROM FIGURE 19.
6. INCREASE PENETRATION COMPUTED (D) BY 20% TO ALLOW FOR DREDGING, SCOUR, ETC.

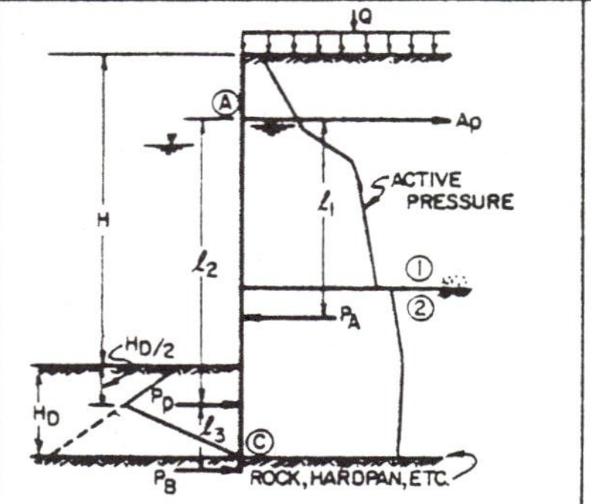


PENETRATION TO TOP OF HARD UNYIELDING STRATUM

1. COMPUTE PRESSURES AS ABOVE. EXCEPT THAT PASSIVE PRESSURE DECREASES TO ZERO AT TOP OF HARD STRATUM.
2. PENETRATION IN HARD STRATUM: TAKE MOMENTS ABOUT POINT Ⓐ AND SOLVE FOR $P_B$:
   $P_A\ell_1 - \frac{P_p}{F_S}\ell_2 = \frac{P_B}{F_S}(\ell_2 + \ell_3)$
   ESTIMATE IF REACTION $P_B$ CAN BE PROVIDED BY SHALLOW PENETRATION IN HARD STRATUM.
3. ANCHOR PULL: $A_p = \left[P_A - \frac{P_p}{F_S} - \frac{P_B}{F_S}\right] d$
4. MAXIMUM BENDING MOMENT IN SHEETING COMPUTED BY APPLYING $P_A$, $P_p$ AND $A_p$ TO SPAN Ⓐ-Ⓒ ASSUMING SIMPLE SUPPORT AT Ⓒ. NO REDUCTION FOR FLEXIBILITY.

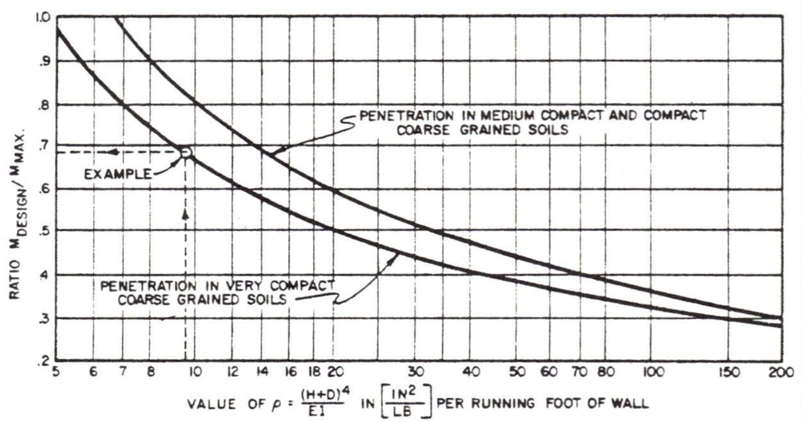


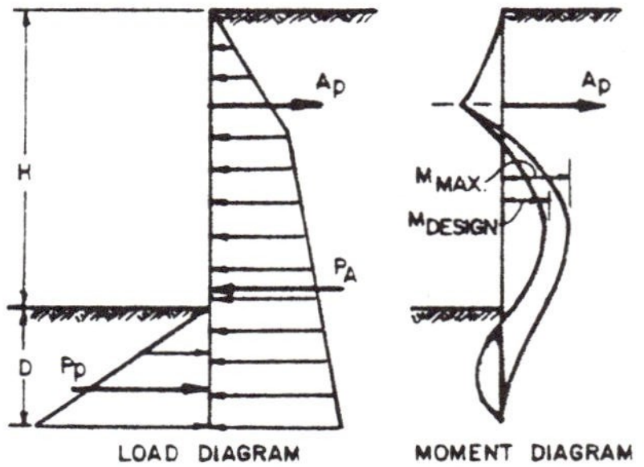


EXAMPLE: PENETRATION IN VERY COMPACT SAND

$M_{MAX}$ = 950,000 IN.LB/FT.

H = 33 FT, D = 15 FT

$f_s$ = 25,000 PSI, E = 30,000,000 PSI

TRY ZP 32, I = 385.7 IN.$^4$, S = 38.3 IN.$^3$

$$\rho = \frac{(33+15)^4 \times 12^4}{30{,}000{,}000 \times 385.7} = 9.5\ \frac{IN^2}{LB}$$

$\frac{M_{DESIGN}}{M_{MAX}}$ = 0.68, $M_{DESIGN}$ = 645,000 IN.LB/FT

$$f_s = \frac{M}{S} = \frac{645{,}000}{38.3} = 16{,}800 \text{ PSI}$$

16,800 < 25,000 PSI

TRY A SMALLER SECTION.

LEGEND

$M_{MAX}$ = MAXIMUM POSITIVE MOMENT IN SHEETING COMPUTED BY METHODS OF FIGURE 18.

$M_{DESIGN}$ = MAXIMUM POSITIVE MOMENT FOR DESIGN OF SHEETING.

$\rho$ = FLEXIBILITY NUMBER = $\frac{(H+D)^4}{EI}$, E = SHEETING MODULUS OF ELASTICITY, PSI
I = SHEETING MOMENT OF INERTIA, IN.$^4$ PER RUNNING FOOT OF WALL.

NOTES

1. $M_{DESIGN}$ IS OBTAINED BY SUCCESSIVE TRIALS OF SHEETING SIZE UNTIL MAX. BENDING STRESS IN SHEETING EQUALS ALLOWABLE BENDING STRESS.
2. NO REDUCTION IN $M_{MAX.}$ IS PERMITTED FOR PENETRATION IN FINE GRAINED SOILS OR LOOSE OR VERY LOOSE COARSE GRAINED SOILS.
3. FLEXIBILITY NUMBER IS COMPUTED ON THE BASIS OF LUBRICATED INTERLOCKS.

Reduction in Bending Moments in Anchored Bulkhead from Wall Flexibility

(NAVFAC DM-7)

# DESIGN CRITERIA FOR DEADMAN ANCHORAGE

## EFFECT OF ANCHOR LOCATION RELATIVE TO THE WALL

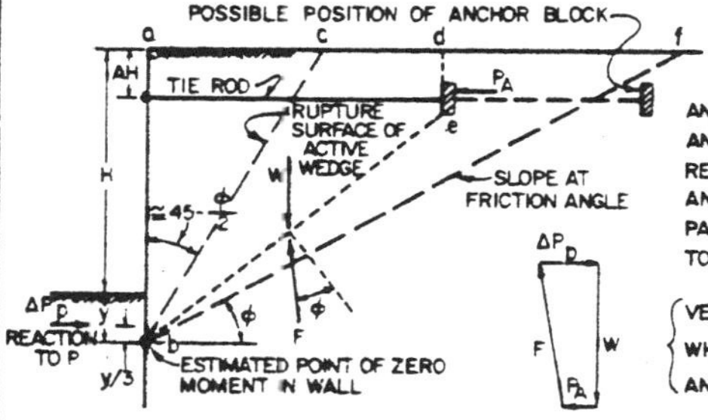


ANCHOR BLOCK LEFT OF bc PROVIDES NO RESISTANCE.
ANCHOR BLOCK RIGHT OF bf PROVIDES FULL RESISTANCE WITH NO LOAD TRANSFERRED TO WALL.
ANCHOR BLOCK BETWEEN bc AND bf PROVIDES PARTIAL RESISTANCE AND TRANSFERS LOAD $\Delta P_p$ TO BASE OF WALL.

VECTOR DIAGRAM FOR FREE BODY abed
WHERE $P_A$ = ACTIVE FORCE ON BACK OF de AT ANCHOR BLOCK.

## CONTINUOUS ANCHOR WALL LOCATED BETWEEN RUPTURE SURFACE AND SLOPE AT FRICTION ANGLE

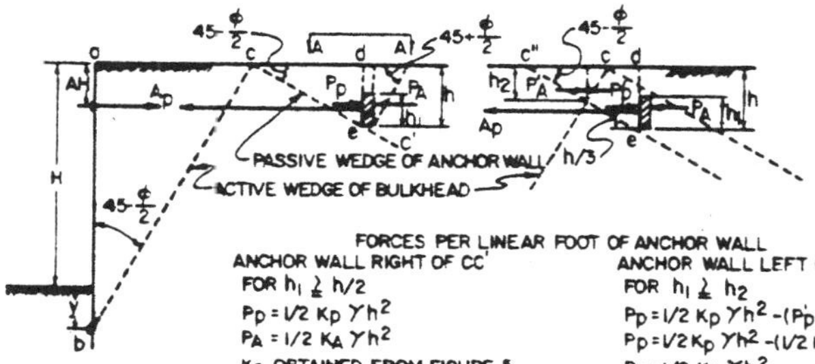


FORCES PER LINEAR FOOT OF ANCHOR WALL

ANCHOR WALL RIGHT OF CC'
FOR $h_1 \geq h/2$
$P_P = 1/2\ K_P \gamma h^2$
$P_A = 1/2\ K_A \gamma h^2$
$K_P$ OBTAINED FROM FIGURE 5
USING - $\delta/\phi = 0.5$

ANCHOR WALL LEFT OF CC'
FOR $h_1 \geq h_2$
$P_P = 1/2\ K_P \gamma h^2 - (P_P' - P_A')$
$P_P = 1/2 K_P \gamma h^2 - (1/2 K_P \gamma h_2^2 - 1/2 K_A \gamma h_2^2)$
$P_A = 1/2\ K_A \gamma h^2$
$K_A$ IS OBTAINED FROM FIGURE 3

## EFFECT OF DEPTH AND SPACING OF ANCHOR BLOCKS

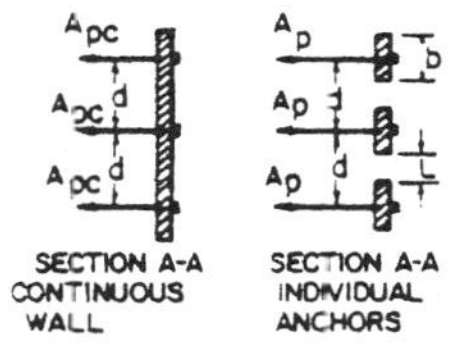


ANCHOR RESISTANCE FOR $h_1 \geq \frac{h}{2}$

1. CONTINUOUS WALL:
   ULTIMATE $A_{pc}/d = P_p - P_A$ WHERE $A_{pc}/d$ IS ANCHOR RESISTANCE AND $P_p$, $P_A$ TAKEN PER LINEAL FOOT OF WALL.
2. INDIVIDUAL ANCHORS:
   IF $d > b+h$, ULTIMATE $A_p = b(P_p - P_A) + 2P_o \tan\phi$, WHERE $P_o$ = RESULTANT FORCE OF SOIL AT REST ON VERTICAL AREA cde OR c"de.
   IF $d = h+b$, $A_p/d$ IS 70% OF $A_{pc}/d$ FOR CONTINUOUS WALL.
   L FOR THIS CONDITION IS L' AND L' = h.
   IF $d < h+b$, $A_p/d = A_{pc}/d - \frac{L}{L'}(.3\ A_{pc}/d)$, L' = h.

ANCHOR RESISTANCE FOR $h_1 < \frac{h}{2}$

ULTIMATE $A_p/d$ OR $A_{pc}/d$ EQUALS BEARING CAPACITY OF STRIP FOOTING OF WIDTH $h_1$ AND SURCHARGE LOAD $\gamma(h - \frac{h_1}{2})$, SEE FIGURE 1, CHAPTER 4
USE FRICTION ANGLE $\phi'$ : WHERE $\tan\phi' = 0.6 \tan\phi$.

## GENERAL REQUIREMENTS:

1. ALLOWABLE VALUE OF $A_p$ AND $A_{pc}$ = ULTIMATE VALUE/2, FACTOR OF SAFETY OF 2 AGAINST FAILURE.
2. VALUES OF $K_A$ AND $K_P$ ARE FOR COHESIONLESS MATERIALS. IF BACKFILL HAS BOTH $\phi$ AND c STRENGTHS, COMPUTE ACTIVE AND PASSIVE FORCES ACCORDING TO FIGURES 7 AND 9 FINE GRAINED SOILS OF MEDIUM TO HIGH PLASTICITY SHOULD NOT BE USED AT THE ANCHORAGE.
3. SOILS WITHIN PASSIVE WEDGE OF ANCHORAGE SHALL BE COMPACTED TO NO LESS THAN 90% OF MAX. UNIT WEIGHT (ASTM D698 TEST).
4. TIE ROD IS DESIGNED FOR ALLOWABLE $A_p$ OR $A_{pc}$. TIE ROD CONNECTIONS TO WALL AND ANCHORAGE ARE DESIGNED FOR 1.2 (ALLOWABLE $A_p$ OR $A_{pc}$).
5. TIE ROD CONNECTION TO ANCHORAGE IS MADE AT THE LOCATION OF THE RESULTANT EARTH PRESSURES ACTING ON THE VERTICAL FACE OF THE ANCHORAGE.

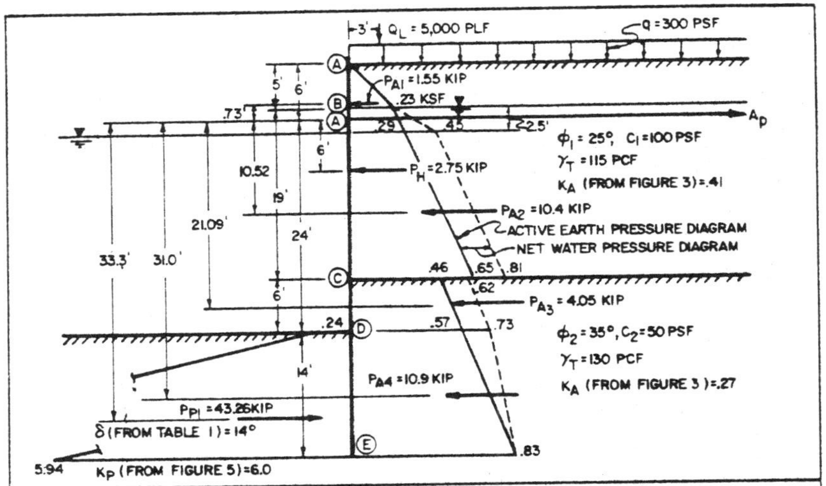


### ACTIVE EARTH PRESSURE (SEE FIGURE 2) INCLUDING UNIFORM SURCHARGE q

$\sigma_H = \gamma Z K_A - 2C\sqrt{K_A}$

Ⓐ, $\sigma_H = .30 \times .41 - 2 \times .10\sqrt{.41} = 0$
Ⓑ, $\sigma_H = (.30 + 5 \times .115).41 - 2 \times .10\sqrt{.41} = .23$ KSF
Ⓒ, $\sigma_H = (.30 + 5 \times .115 + 19 \times .053).41 - 2 \times .10\sqrt{.41} = .65$ KSF
Ⓒ, $\sigma_H = (.30 + 5 \times .115 + 19 \times .053).27 - 2 \times .05\sqrt{.27} = .46$ KSF
Ⓓ, $\sigma_H = .46 + 6 \times .068 \times .27 = .57$ KSF
Ⓔ, $\sigma_H = .46 + (6 + 14) \times .068 \times .27 = .83$ KSF

### PRESSURE OF LINE LOAD SURCHARGE

(SEE FIGURE 11)

$m = \frac{X}{H} = \frac{3}{30} = 0.1$

$P_H = 0.55\ Q_L = 0.55 \times 5 = 2.75$ KIP

LOCATION OF RESULTANT:
R = .60 H = .60 × 30 = 18'

### NET WATER PRESSURE

$\sigma_W = \gamma_W Z = .0625 \times 2.5 = .16$ KSF

### PASSIVE PRESSURE

$\sigma_H = \gamma Z\ K_P + 2C\sqrt{K_P}$

Ⓓ, $\sigma_H = 0 + 2 \times .05\sqrt{6.0} = .24$ KSF
Ⓔ, $\sigma_H = .068 \times 14 \times 6.0 + 2 \times .05\sqrt{6.0} = 5.94$ KSF

### SAFETY FACTOR AGAINST TOE FAILURE:

TAKE MOMENTS ABOUT Ⓐ

$F_S = \frac{\sum \text{MOMENTS OF PASSIVE FORCES}}{\sum \text{MOMENTS OF ACTIVE FORCES}}$

$= \frac{43.26 \times 33.3}{1.55 \times 0.73 + 2.75 \times 6 + 10.4 \times 10.52 + 4.05 \times 21.09 + 10.9 \times 31.0}$

$= 2.61 > 2.5$

### ANCHOR PULL

$A_p = \Sigma P_A - \Sigma P_p / F_s$

$= 1.55 + 2.75 + 10.4 + 4.05 + 10.9 - \frac{43.26}{2.5} = 12.37$ KIPS

### MAXIMUM BENDING MOMENT IN SHEETING

POINT OF ZERO SHEAR:

$12.37 - 1.55 - 2.75 - .45X - .022 \times \frac{X^2}{2} = 0$

X ≅ 13.6' BELOW OUTSIDE WATER LEVEL

$M_{MAX} = 1.55 \times 15.7 + 12.37 \times 15.1 - 2.75 \times 9.1 - .45 \times \frac{13.6^2}{2} - .022 \times \frac{13.6^2}{2} \times 4.52 = 86.9$ FT-KIPS

MOMENT REDUCTION:

ASSUME: $f_s$ = 27,000 PSI, E = 30,000,000 PSI

TRY ZP 32, I = 385.7 IN$^4$, S = 38.3 IN$^3$

$\rho$ = (FROM FIGURE 19) : $\frac{(H+D)^4}{EI}$

$\rho = \frac{(30 \times 12 + 14 \times 12)^4}{30{,}000{,}000 \times 385.7} = 6.7\ \frac{IN^2}{LB}$

$\frac{M_{DESIGN}}{M_{MAX}} = .83$; $M_{DESIGN} = .83 \times 86.9 = 72.1$ FT-KIPS

$f_s = \frac{M}{S} = \frac{72.1 \times 1000 \times 12}{38.3} = 22{,}600$ PSI < 27,000

TRY A SMALLER SECTION

## ANALYSIS OF CANTILEVER WALL

NOTE: WATER LEVELS CAN BE DIFFERENT ON OPPOSITE SIDES DUE TO PUMPING, TIDAL FLUCTUATIONS AND OTHER REASONS.

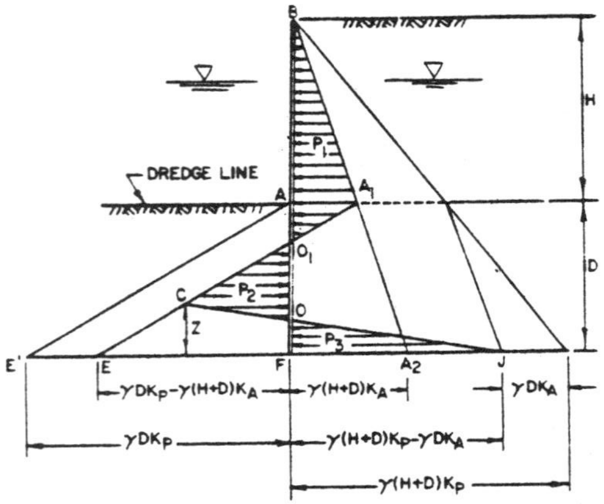


1. Assume a trial depth of penetration, D. This may be estimated from the following approximate correlation.

| Standard Penetration Resistance, N Blows/foot | Depth of Penetration* |
|---|---|
| 0 - 4 | 2.0H |
| 5 - 10 | 1.5H |
| 11 - 30 | 1.25H |
| 31 - 50 | 1.0H |
| +50 | 0.75H |

* H = height of piling above dredge line

2. Determine the active and passive lateral pressure using appropriate coefficients of lateral earth pressure. If the Coulomb method is used, it should be used conservatively for the passive pressure.
3. Satisfy the requirements of static equilibrium: the sum of the forces in the horizontal direction must be zero and the sum of the moments about any point must be zero. The sum of the horizontal forces may be written in terms of pressure areas:

$$\Delta(\overleftarrow{EA_1A_2}) - \Delta(\overleftarrow{FBA_2}) - \Delta(\overleftarrow{ECJ}) = 0$$

Solve the above equation for the distance, Z. For a uniform granular soil,

$$Z = \frac{K_P\,D^2 - K_A\,(H+D)^2}{(K_P - K_A)\,(H+2D)}$$

4. Take moments about point F. If sum of moments is other than zero, readjust D and repeat calculations until sum of moments around F is zero.
5. Compute maximum moment at point of zero shear.
6. Increase D by 20% - 40% to result in approximate factor of safety of 1.5 to 2.

## METHODS OF STABILIZING EXCAVATION SLOPE

| Scheme | Applicable Methods | Comments |
|---|---|---|
| 1. Changing Geometry<br>EXCAVATION<br>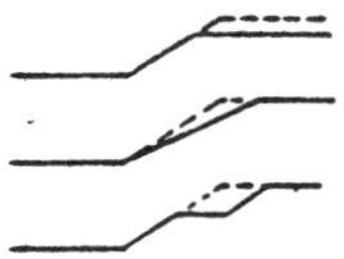 | 1. Reduce slope height by excavation at top of slope.<br>2. Flatten the slope angle.<br>3. Excavate a bench in upper part of slope. | 1. Area has to be accessible to construction equipment. Disposal site needed for excavated soil. Drainage sometimes incorporated in this method. |
| 2. Earth Berm Fill<br> | 1. Compacted earth or rock berm placed at and beyond the toe. Drainage may be provided behind berm. | 1. Sufficient width and thickness of berm required so failure will not occur below or through berm. |
| 3. Retaining Structures<br>RETAINING STRUCTURES<br>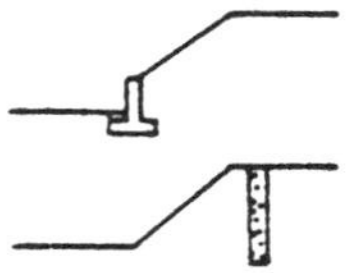 | 1. Retaining wall - crib or cantilever type.<br>2. Drilled, cast-in-place vertical piles, founded well below bottom of slide plane. Generally 18 to 36 inches in diameter and 4- to 8-foot spacing. Larger diameter piles at closer spacing may be required in some cases to mitigate failures of cuts in highly fissured clays. | 1. Usually expensive. Cantilever walls might have to be tied back.<br>2. Spacing should be such that soil can arch between piles. Grade beam can be used to tie piles together. Very large diameter (6 feet ±) piles have been used for deep slides. |
|  | 3. Drilled, cast-in-place vertical piles tied back with battered piles or a deadman. Piles founded well below slide plane. Generally, 12 to 30 inches in diameter and at 4- to 8-foot spacing. | 3. Space close enough so soil will arch between piles. Piles can be tied together with grade beam. |

(NAVFAC DM-7)

| | | |
|---|---|---|
|  | 4. Earth and rock anchors and rock bolts. | 4. Can be used for high slopes, and in very restricted areas. Conservative design should be used, especially for permanent support. Use may be essential for slopes in rocks where joints dip toward excavation, and such joints daylight in the slope. |
| 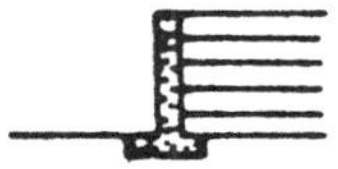 | 5. Reinforced earth. | 5. Usually expensive. |

## GABION

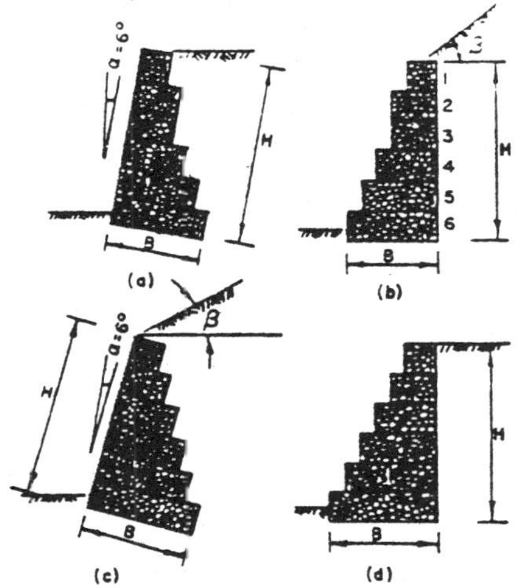


Gabion Retaining Wall

Types - Common Gabion walls shown on accompanying diagrams are:

a) Battered face wall with horizontal backfill.
b) Stepped face wall with sloped backfill.
c) Battered face wall with sloped backfill.
d) Stepped face wall with horizontal backfill.

The choice of either battered or stepped faces rests with designer; stepped face recommended if wall is more than 10 feet high.

Gabion Fill - Hard, durable, clean stone 4 to 8 inches in size or other approved size.

**Design:** Design criteria for gravity walls apply. Wall section resisting overturning and sliding. To increase wall stability, recommended to tilt the wall at an angle of 6° (i.e. 1:10).

The angle of friction between the base of gabion wall and granular soil may be assumed 0.9 times the angle of internal friction of soil.

For retaining clay slopes, a system of gabion counterforts is recommended.

Compute active soil pressure behind the wall using Coulomb Wedge theory and design mass of the wall to balance the force exerted by that soil wedge. (Higher than active pressures may be used depending on compaction conditions and limitations on deformations.)

Maximum pressure at the base of gabion wall must be less than the anticipated bearing capacity of the soil under the wall.

When water quality is in doubt (pH below 6 or greater than 12) or where high concentration of organic acids may be present, use of PVC (polyvinylchloride) coated gabions is recommended.

## REINFORCED EARTH

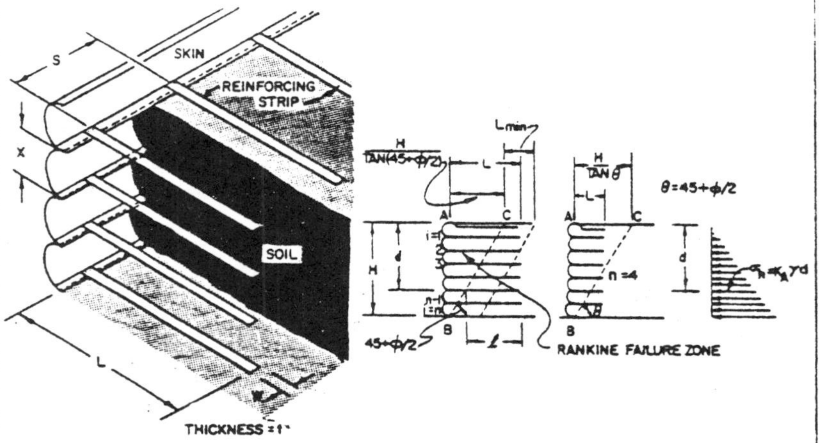


Safety against breaking of reinforced strips.

$$F_S = \frac{f_s\,W\,t}{K_A\,\gamma\,H\,S\,X}$$

S = Horizontal spacing between strips   X = Vertical Spacing between strips

$f_s$ = allowable stress of reinforced strips.

Typically W = 3". A high factor of safety, $F_s$ = 3.2, is used even though allowable metal stress is utilized in computing strip thickness. This is done to account for unknowns such as durability and corrosion.

SAFETY AGAINST PULLOUT $F_s = \frac{2\,L_{min}\,W\,\tan\delta}{K_A \cdot S \cdot X}$

$L_{min}$ is measured beyond zone of Rankine failure. The upper strips may not have enough length to fulfill this requirement, but as long as the average length of all the strips satisfies this conditon the wall is considered satisfactory.

d = depth beneath top of wall

t = thickness of strip

$\gamma$ = unit weight of backfill

B = width of wall

$K_A$ = coefficient of each active pressure (higher than active value may be used depending on compaction conditions and limitations on deformations).

$\delta$ = angle of friction between reinforcing strip and the backfill material

$\ell$ = effective length of tie beyond potential sliding surface

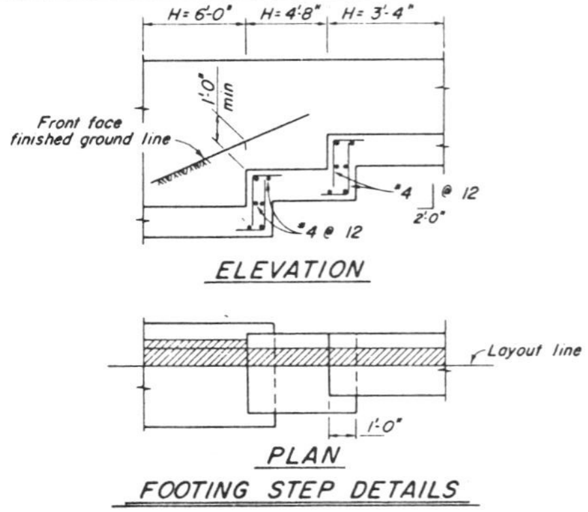


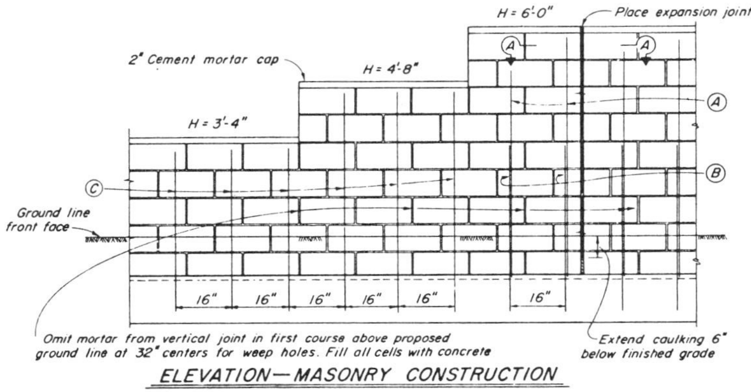


**DESIGN DATA:**

Masonry: $f_m$ = 500 psi  $f'_m$ = 1500 psi  $f_s$ = 24,000 psi  n = 20

Reinf Conc: $f_c$ = 1300 psi  $f'_c$ = 3250 psi  $f_s$ = 24,000 psi  n = 10

Earth = 120 pcf.

2' Level Surcharge:
Equivalent fluid pressure = 36 pcf for determination of toe pressure.
27 pcf for determination of heel pressure

2:1 Unlimited Surcharge: Earth pressure determined from Rankine's formula $\phi$ = 33°-42'.

Minimum allowable soil bearing capacity of foundation material = 1 TSF.

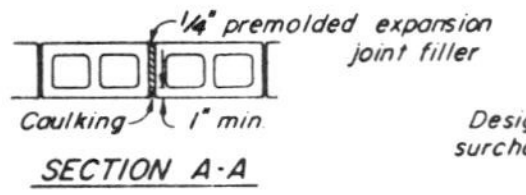


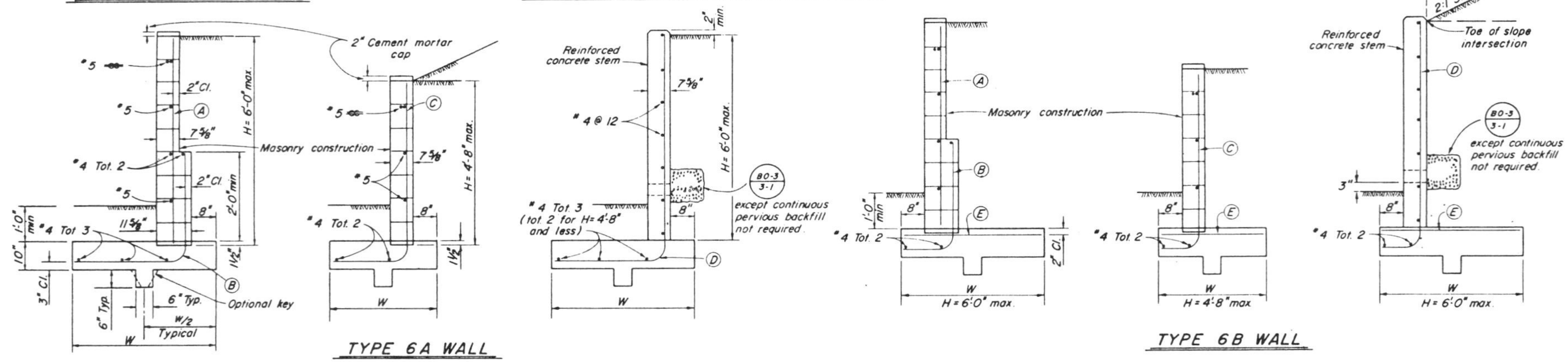


| Type | Design H | 3'-4" | 4'-0" | 4'-8" | 5'-4" | 6'-0" |
|---|---|---|---|---|---|---|
| 6A | W | 3'-2" | 3'-6" | 3'-10" | 4'-2" | 4'-6" |
| 6A | Ⓐ | | | | #4@16 | #4@16 |
| 6A | Ⓑ | | | | #4@16 | #5@16 |
| 6A | Ⓒ | #4@16 | #4@16 | #5@16 | | |
| 6A | Ⓓ | #4@18 | #4@18 | #4@14 | #4@10 | #5@12 |
| Footing conc. Cu. ft./L.F. | | 2.9 | 3.2 | 3.4 | 3.7 | 4.0 |
| Reinf. Lbs./L.F. Masonry / Reinf. Conc. Stem | | 8.5 / 6.6 | 8.9 / 7.7 | 11.6 / 9.8 | 12.8 / 13.7 | 15.0 / 16.7 |

NOTES

1. For details not shown at "6B", see "6A"
2. Area of wall for payment is based upon horizontal length and height from top of footing to top of wall.
3. Type 6 Retaining Wall shall be limited to use for walls of Design H of 6 feet or less.
4. Where traffic is adjacent to the top of wall, guard rails should be set back from the top front face of wall at least 3' for masonry walls or 2' for reinforced concrete walls.
5. Unless otherwise stipulated, the contractor will have the option of constructing the Type 6 Walls of either masonry or reinforced concrete.

| Type | Design H | 3'-4" | 4'-0" | 4'-8" | 5'-4" | 6'-0" |
|---|---|---|---|---|---|---|
| 6B | W | 2'-8" | 3'-0" | 3'-4" | 3'-8" | 4'-0" |
| 6B | Ⓐ | | | | #4@16 | #4@16 |
| 6B | Ⓑ | | | | #4@16 | #5@16 |
| 6B | Ⓒ | #4@16 | #4@16 | #5@16 | | |
| 6B | Ⓓ | #4@16 | #4@16 | #5@16 | #5@16 | #4@8 |
| 6B | Ⓔ | #4@16 | #4@16 | #4@16 | #4@16 | #5@16 |
| Footing conc. Cu. Ft./L.F. | | 2.5 | 2.8 | 3.0 | 3.3 | 3.6 |
| Reinf. Lbs./L.F. Masonry / Reinf. Conc. Stem | | 9.1 / 6.6 | 9.6 / 7.7 | 11.8 / 9.8 | 12.9 / 13.7 | 15.4 / 16.7 |

For reinforced concrete stem joints required see B0-3 3-3 & B0-3 3-4

JULY 1, 1980

STANDARD DRAWING
FILE No. XS-3-80
DRAWING DATE 1/80

State of CALIFORNIA
DEPARTMENT OF TRANSPORTATION

STRUCTURES-DESIGN

RETAINING WALL TYPE 6 - 6'-0" max.

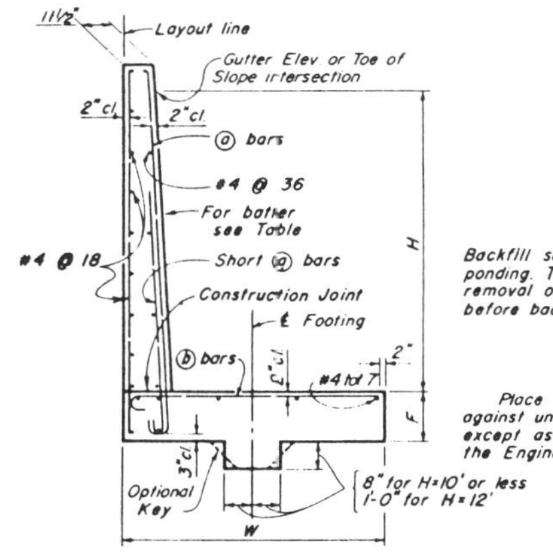


## SPREAD FOOTING SECTION

Backfill sufficiently to prevent ponding. To be done after removal of wall forms and before backfilling behind wall.

Place concrete in toe against undisturbed material except as permitted by the Engineer.

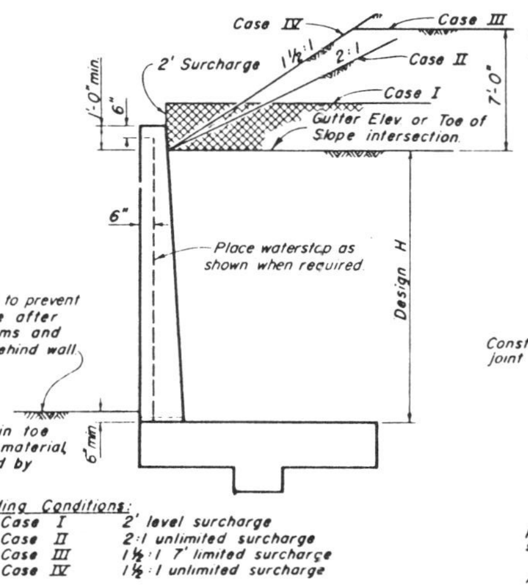


Loading Conditions:

- Case I — 2' level surcharge
- Case II — 2:1 unlimited surcharge
- Case III — 1½:1 7' limited surcharge
- Case IV — 1½:1 unlimited surcharge

## DESIGN

For drainage notes and other details, see (B3-8)

## MAX. PILE SPACING FOR 45 TON PILES

| Design H | Front Row 1:3 Batter | Back Row Vertical |
|---|---|---|
| 4 | 18'-0" | 18'-0" |
| 6 | 12'-0" | 18'-0" |
| 8 | 9'-0" | 18'-0" |
| 10 | 6'-0" | 12'-0" |
| 12 | 4'-0" | 8'-0" |

For actual spacing, see Wall Layout.

Pile layout does not apply to Case IV conditions.

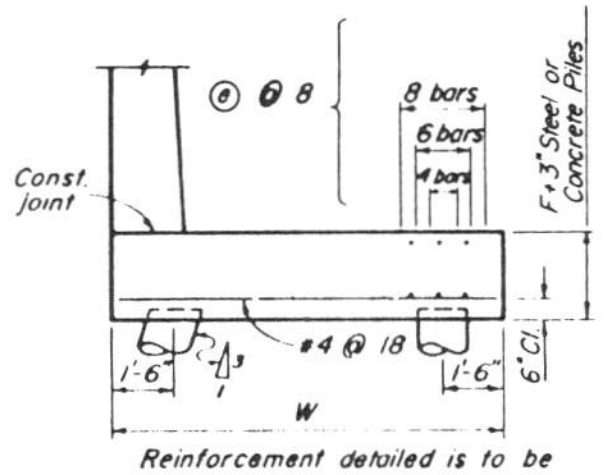


Reinforcement detailed is to be placed in addition to that shown for spread footings. ② ◉ For Design H=4' use W=5'-0". All others from table.

## 45T PILE FOOTING SECTION

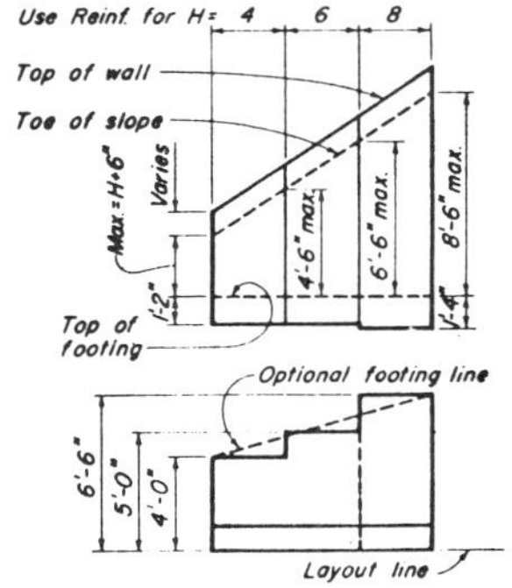


## TYPICAL LAYOUT EXAMPLE

For joints required, see (BO-3 / 3-3) & (BO-3 / 3-4)

## TABLE OF REINFORCING STEEL DIMENSIONS AND DATA

| Design H ft. | ◉ 4' | 6' | 8' | 10' | 12' |
|---|---|---|---|---|---|
| W | 4'-0" | 5'-0" | 6'-6" | 8'-0" | 9'-6" |
| F Spread Ftg. | 1'-2" | 1'-2" | 1'-4" | 1'-6" | 1'-10" |
| Batter | None | None | None | 3/8 : 12 | 3/4 : 12 |
| ⓐ bars | #4 @ 18 | #5 @ 18 | #5 @ 17 | #6 @ 17 | #6 @ 14 |
| Short ⓐ bars | None | None | #5 @ 17 | #6 @ 17 | #6 @ 14 |
| ⓑ bars | #4 @ 18 | #5 @ 18 | #5 @ 8½ | #6 @ 8½ | #6 @ 7 |
| Total ⓔ bars | 6 - #7 | 6 - #7 | 8 - #7 | 6 - #7 | 4 - #7 |
| Toe Pressure Case I k/sf | 1.6 | 2.2 | 2.5 | 3.0 | 3.5 |
| Toe Pressure Case II k/sf | 1.5 | 2.1 | 2.7 | 3.4 | 4.1 |
| Toe Pressure Case III k/sf | 1.6 | 2.3 | 2.9 | 3.8 | 4.4 |
| Toe Pressure Case IV k/sf | 2.0 | 3.2 | 4.2 | 5.3 | 6.5 |
| Spread Ftg. Steel | 16 | 22 | 35 | 55 | 73 |
| Spread Ftg. Conc. | 9.4 | 12.5 | 17.2 | 24.4 | 36.1 |
| Pile Footing Steel | 31 | 36 | 54 | 70 | 85 |
| Pile Footing Conc. | 10.9 | 12.9 | 17.9 | 25.5 | 36.5 |

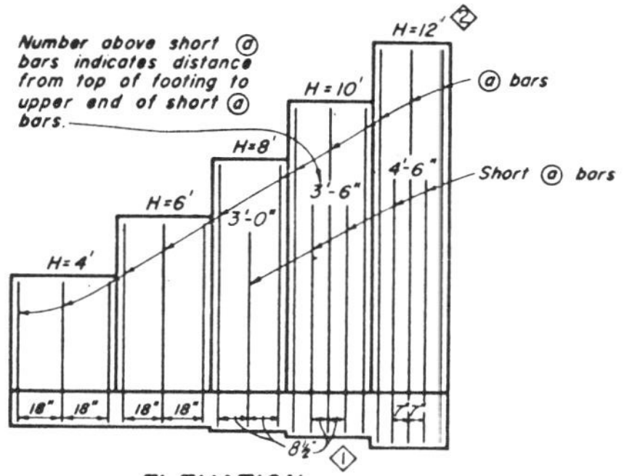


## ELEVATION

## NOTES

Design Conditions:
Design H may be exceeded by 6" before going to the next size.
③ Special footing design is required where foundation material is incapable of supporting toe pressure loads listed in table.

Design Data:
$f_c$ = 1300 psi  $f'_c$ = 3250 psi  $f_s$ = 24,000 psi  n = 10
earth = 120 pcf  Case I - Wall design for equivalent fluid pressure = 27 and 36 pcf. Case II, III, IV - Wall design is based on Rankine's formula with $\phi$ = 33°-42'.

Quantities:
Quantities do not include the wall portion above "Gutter Elevation" and are for design purposes only.

R. C. Cassano, Chief, Office of Structures Design — July 1, 1980

① Rev Value
② Rev Note
③ Deleted Note

STANDARD DRAWING — File No. XS-3-49 — Drawing Date 1/80

State of CALIFORNIA DEPARTMENT OF TRANSPORTATION — STRUCTURES-DESIGN

RETAINING WALL TYPE 5

| DIST | COUNTY | ROUTE | POST MILES–TOTAL PROJECT | SHEET No. | TOTAL SHEETS |
|---|---|---|---|---|---|
| | | | | | |

CHIEF, OFFICE OF STRUCTURES DESIGN
REGISTERED CIVIL ENGINEER NO 9013
DATE APPROVED July 1, 1980

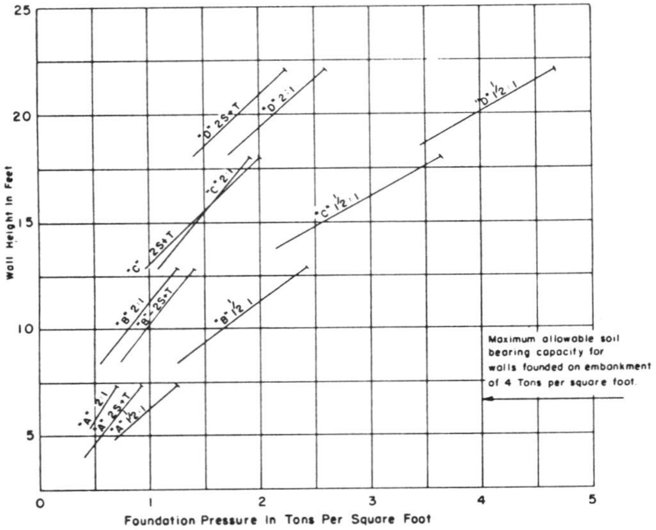


1:6 BATTERED WALL

## DESIGN NOTES

1. Wall base in embankment: A minimum depth of 5 feet of embankment at 95% relative soil compaction is required below the base of all walls in order to constitute an embankment condition. When the foundation pressure is between 2.5 & 4.0 Tons/S.F. embankment below the wall shall consist of "Structure Backfill" material as set forth in Section 19-3.06 of the Standard Specifications. The limits of relative compaction (95%) shall be as set forth in Section 19-5.03 of the Standard Specifications.

2. Wall base in original ground: Allowable soil pressure at toe of wall shall be determined by foundation site investigation. Walls that are to retain cut slopes shall be designed for lateral and toe pressures determined from site investigation data. Overall stability of slope with wall in place must be analyzed. If original ground slopes away from toe of wall, reduction in allowable bearing capacity due to slope must be considered. Walls shall not be founded in orig. gd. having an allowable bearing capacity of less than 1.5 Tons/S.F. Consideration should be given to removal and replacement of unsuitable material with "Structure Backfill" material as set forth in Section 19-3.06 of the Standard Specifications. The limits of relative compaction (95%) shall be set forth in Section 19-5.03 of the Standard Specifications.

3. Weight of soil = 120 pcf. For 2 foot level surcharge with traffic loading, an equivalent fluid pressure of 36 pcf. was used. Earth pressures for 2:1 slope and 1½:1 slope determined from Rankine's formula with $\phi = 33° 42'$.

## LEGEND FOR GRAPHS

A, B, C, D = Wall type
2S+T = 2 foot level surcharge with traffic loading
2:1 = 2:1 Slope above wall
1½:1 = 1½:1 Slope above wall
(2:1 and 1½:1:) 15' Max. difference in elevation from toe of wall to top of slope

Solid lines indicate normal range of wall use.
Upper end of line indicates maximum wall height for a given wall type and loading.

## DESIGN EXAMPLES

**EXAMPLE NO. 1**

Given: Wall Height 14'
2:1 embankment slope to be retained
Base in embankment (5' depth min. See Note 1)

Select: 1:6 Batter "C" Wall gives 1.25 T.S.F. foundation pressure. Vertical "C" Wall gives 2.1 T.S.F. foundation pressure. (See chart for required numbers and sizes of timber members.)

**EXAMPLE NO. 2**

Given: Wall Height 19'-4"
1½:1 embankment to be retained. Base is in original ground
Foundation sight investigation determines the allowable soil bearing capacity at 3 T.S.F.

Select: 1:6 batter "D" Wall, gives 3.75 T.S.F. foundation pressure. Vertical wall not permitted. Since foundation pressure is greater than allowable bearing capacity of native material, replace original material with "Structure Backfill" to increase base bearing capacity. (See Note 2.)

**EXAMPLE NO. 3**

Given: Wall Height 22'
2:1 embankment to be retained. Base is in embankment. (See Note 1.)

Select: 1:6 batter "D" Wall, gives 2.6 T.S.F. foundation pressure. Vertical wall not permitted. Foundation pressure is greater than 2.5 T.S.F. Base material below wall shall consist of "Structure Backfill" (See Note 1.)

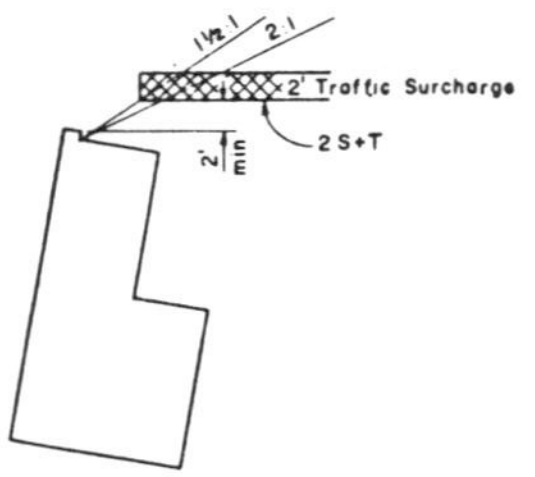


DESIGN SURCHARGES

STATE OF CALIFORNIA
BUSINESS AND TRANSPORTATION AGENCY
DEPARTMENT OF TRANSPORTATION

DESIGN DATA FOR
TIMBER CRIB WALL
TYPES A, B, C, D — C9-B

| DISTRICT | COUNTY | ROUTE | Post Miles | Total Project | Sheet No. | Total Sheets |
|---|---|---|---|---|---|---|
| | | | | | | |

CHIEF, OFFICE OF STRUCTURES DESIGN
REGISTERED CIVIL ENGINEER NO. 9013
APPROVED July 1, 1980

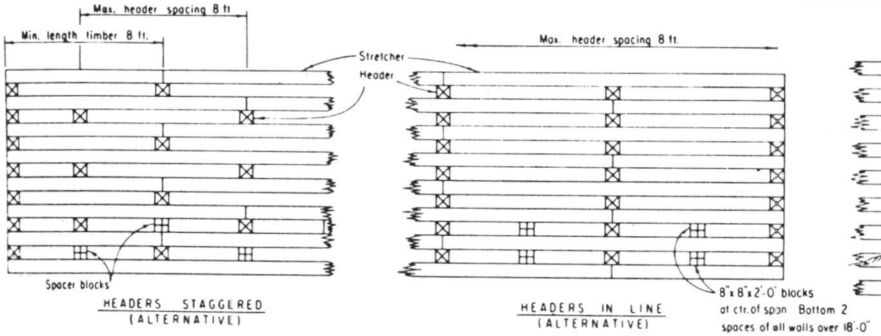


HEADERS STAGGERED (ALTERNATIVE)

HEADERS IN LINE (ALTERNATIVE)

OPEN FACE CRIB ELEVATION

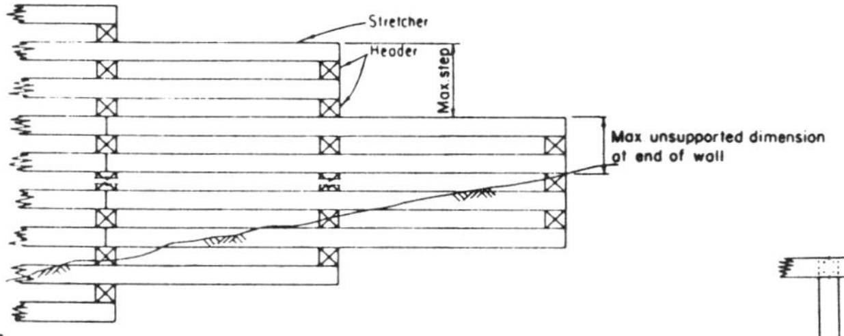


WALL LAYOUT DETAILS

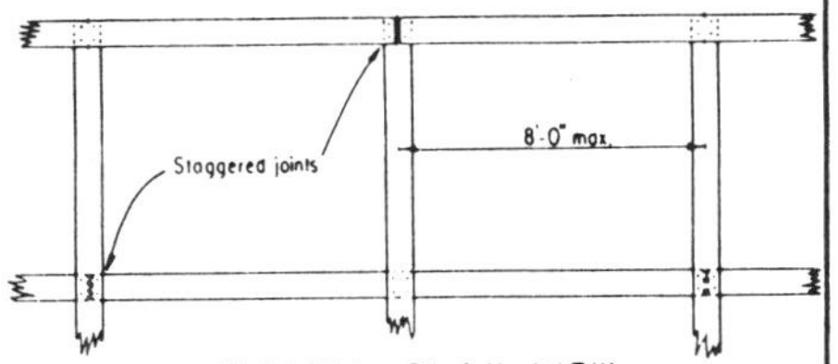


TYPICAL PLAN VIEW

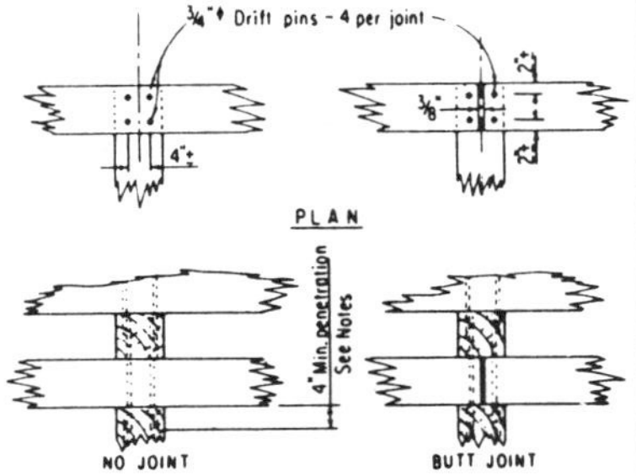


TYPICAL END CONNECTIONS

| TIMBER PER 8 FT CRIB SECTION | | | | | |
|---|---|---|---|---|---|
| HEIGHT OF WALL | HEADERS | | | | STRETCHERS |
| | 6'-0" | 8'-0" | 10'-0" | 12'-0" | |
| 4'-8" | 3 | | | | 8 |
| 6'-0" | 4 | | | | 10 |
| 7'-4" | 5 | | | | 12 |
| 8'-8" | 4 | 2 | | | 14 |
| 10'-0" | 5 | 2 | | | 16 |
| 11'-4" | 5 | 3 | | | 18 |
| 12'-8" | 5 | 4 | | | 20 |
| 14'-0" | 5 | 3 | 2 | | 22 |
| 15'-4" | 5 | 4 | 2 | | 24 |
| 16'-8" | 5 | 4 | 3 | | 26 |
| 18'-0" | 5 | 4 | 4 | | 28 |
| 19'-4" | 5 | 4 | 3 | 2 | 32 |
| 20'-8" | 5 | 4 | 4 | 2 | 38 |
| 22'-0" | 5 | 4 | 4 | 3 | 43 |

For Design Data see Plan Sheet C9-B

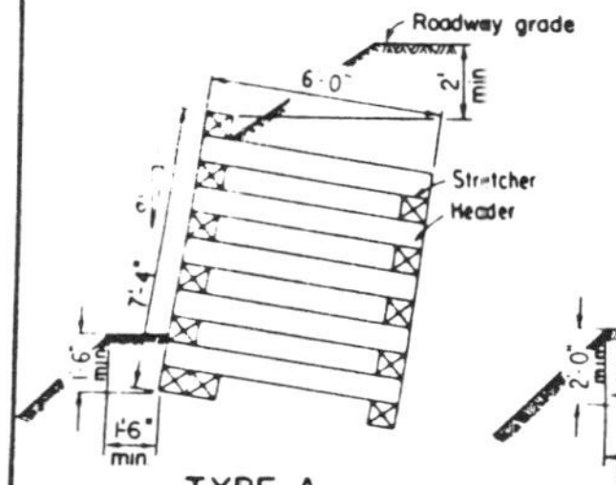


TYPE A

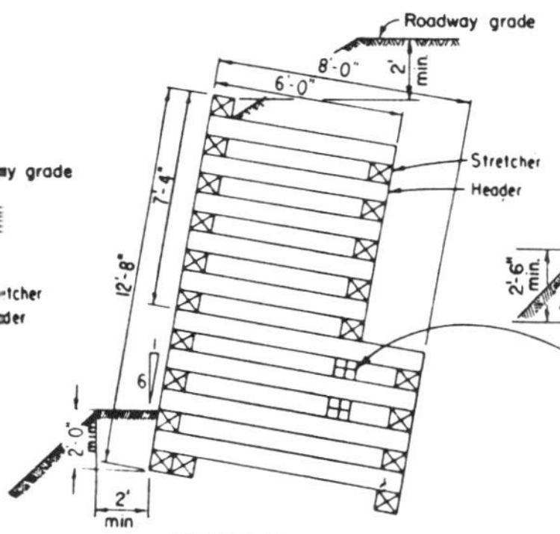


TYPE B

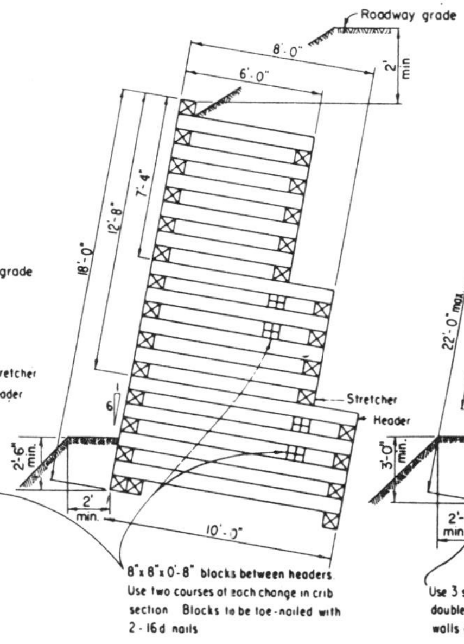


TYPE C

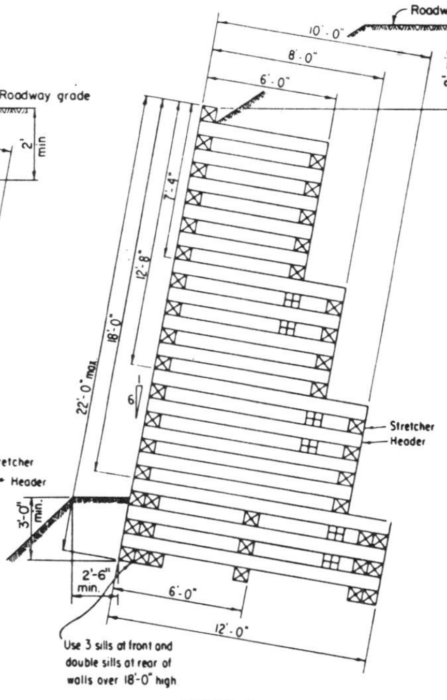


TYPE D

NOTES:

1. All timber to be 8"x 8" rough, No. 1 Structural Douglas Fir-Larch, treated after fabrication. All required cuts or bored holes to be swabbed with 2 coats of the same type of preservative used for treating the timber.
2. Drift pins and holes to be 3/4"φ.
3. Drift pins to be of sufficient length to penetrate thru 2 members and 4" min. into the third member.
4. All stretchers to be laid horizontal.
5. Vertical walls are not permitted.
6. Maximum allowable wall height is 22'-0".
7. All walls 18'-0" high or under shall have double sills at front and single sill at rear. When wall exceeds 7'-4", 12'-8" or 18'-0", use a minimum of two courses of next longer headers.

STATE OF CALIFORNIA
BUSINESS AND TRANSPORTATION AGENCY
DEPARTMENT OF TRANSPORTATION

CONSTRUCTION DETAILS FOR
TIMBER CRIB WALL - TYPES A, B, C, & D

C9-A

# K.7 BRACED WALL

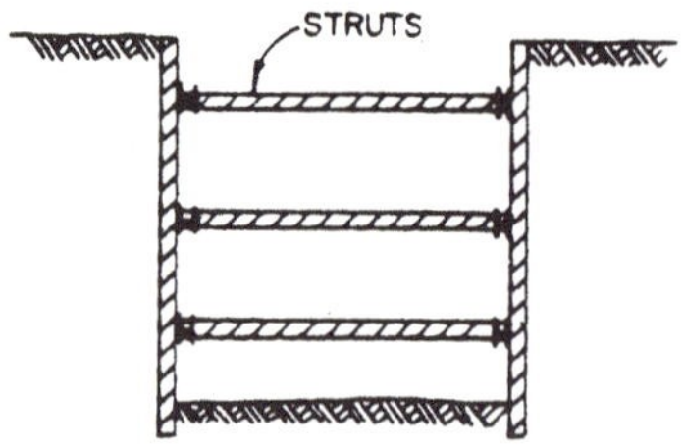


B. CROSS-LOT BRACED WALL

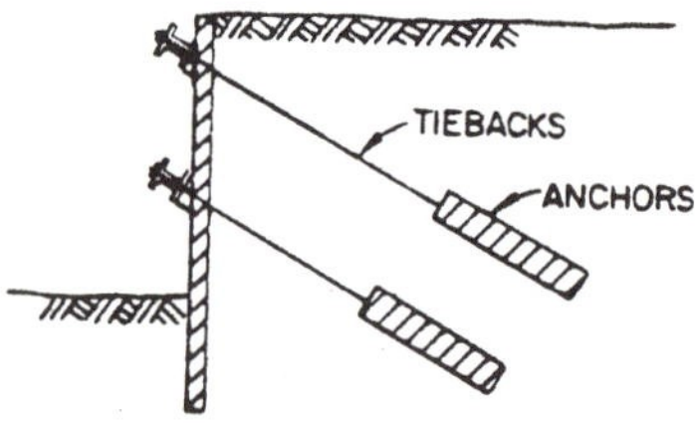


D. ANCHOR OR TIEBACK WALL

| Design Factor | Comments |
|---|---|
| 1. Water Loads | Often greater than earth load on impervious wall. Recommended piezometers during construction to monitor water levels. Should consider possible lower water pressures as a result of seepage through or under wall. Dewatering can be used to reduce water loads. Seepage under wall reduces passive resistance. |
| 2. Stability | Consider possible instability in any berm or exposed slope. Sliding potential beneath the wall or behind tiebacks should be evaluated. Deep seated bearing failure under weight of supported soil to be checked in weak soils. Stability should consider weight of surcharge or the weight of other facilities in close proximity to excavation. |
| 3. Piping | Loss of ground caused by high groundwater table and silty and fine sand soils. Difficulties occur due to flow beneath wall, through bad joints in walls, or through unsealed sheetpile handling holes. Dewatering may be required. |
| 4. Movements | Movements can be minimized through use of stiff wall supported by preloaded tieback or braced system. |
| 5. Dewatering - recharge | Dewatering reduces loads on wall systems and minimizes possible loss of ground due to piping. May cause settlements and will then need to recharge outside of support system. |
| 6. Surcharge | Construction materials usually stored near wall systems. Allowance should always be made for surcharge. |
| 7. Prestressing of tie backs or struts | Useful to remove slack from system and minimize soil movements. |
| 8. Construction Sequence | The amount of wall movement is dependent on the depth of excavation. The amount of load on the tie backs is dependent on the amount of wall movement which occurs before they are installed. Movements of wall should be checked at every major construction stage. Upper struts should be installed early. |
| 9. Temperature | Struts subject to load fluctuation due to temperature loads; may be important for long struts. |
| 10. Frost Penetration | In very cold climates, frost penetration can cause significant loading on wall system. Design of upper portion of system should be conservative. Anchors may have to be heated. Freezing temperatures also can cause blockage of flow and thus unexpected buildup of water pressure. |
| 11. Earthquakes | Seismic loads may be induced during earthquake. See DM-7.3, Chapter 1. |
| 12. Factors of Safety | See table below. |

| Item | Suggested Minimum Design Factor of Safety for Overall Stability: Permanent | Temporary |
|---|---|---|
| • Earth Berms | 2.0 | 1.5 |
| • Cut Slopes | 1.5 | 1.3 |
| • Bottom heave above foundation level | 1.5 | 1.5 |
| • General stability | 1.5 | 1.3 |
| • Bottom heave at foundation level | 2.0 | 1.5 |

Note: These values are suggested guidelines only. Design safety factor depends on project requirements.

# K.8 TIEBACKS

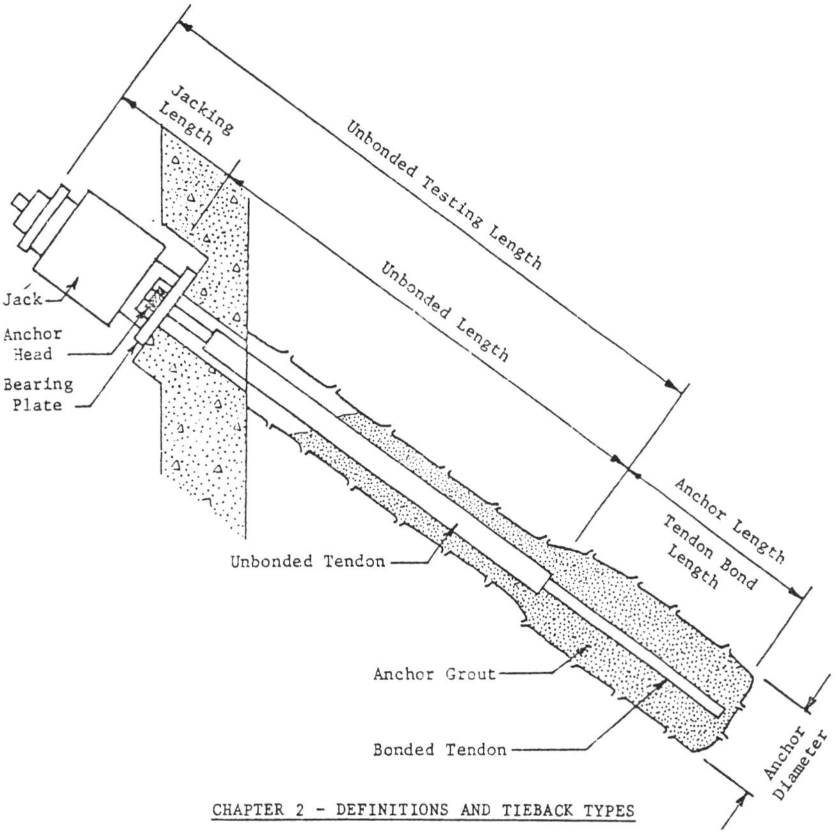


CHAPTER 2 - DEFINITIONS AND TIEBACK TYPES

A tieback system is a structural system which uses an anchor in the ground to secure a tendon which applies a force to a structure. Figures 1 and 2 show typical tiebacks and their components. Vertical or near vertical tiebacks are called tiedowns. Tiebacks are also referred to as ground anchors.

Tiebacks are used for temporary or permanent applications. A temporary tieback is used during the construction of a project; its service life is usually less than two years. A permanent tieback is required for the life of a permanent structure.

The tendon is made up of prestressing steel with sheathing, and an anchorage. The anchor transmits the tensile force in the prestressing steel to the ground. Cement grout, or polyester resin, or mechanical anchors are used to anchor the steel in the ground. The anchorage is made up of an anchor head or nut, and a bearing plate.

The anchor head or nut is attached to the prestressing steel, and transfers the tieback force to a bearing plate which evenly distributes the force to the structure. Anchor heads can be restressable or nonrestressable. A restressable anchor head is one where the tieback force can be measured or increased any time during the life of the structure. The load can not be adjusted when a nonrestressable anchor head is used. A coupling can be used to transmit the anchor force from one length of prestressing steel to another.

The anchor length is the designed length of the tieback where the tieback force is transmitted to the ground. The tendon bond length is the length of the tendon which is bonded to the anchor grout. Normally the tendon bond length is equal to the anchor length. The unbonded length of the tendon is the length which is free to elongate elastically. The jacking length is that portion which is required for testing and stressing of the tieback. The unbonded testing length is the sum of the unbonded length and the jacking length. A sheath or bond breaker is installed over the unbonded length to prevent the prestressing steel from bonding to surrounding grout. The anchor diameter is the design diameter of the anchor.

Anchor grout is used to transmit the tieback force to the ground. The anchor grout is also called the primary grout. Secondary grout is injected into the drill hole after stressing to provide corrosion protection for unsheathed tendons.

Tiebacks carry various loads during their lifetimes. The design load is the maximum anticipated load that will be applied to the tieback. The test load is the maximum load applied during testing. The lock-off load or transfer load is the load transferred to the tieback upon completion of stressing. The alignment load is a nominal load maintained on a tieback during testing to keep the testing equipment in position. The lift-off load is the load required to lift the anchor head or nut from the bearing plate. The residual load is the load carried by the tieback at any time. The load transfer rate is the tieback capacity per unit length of anchor.

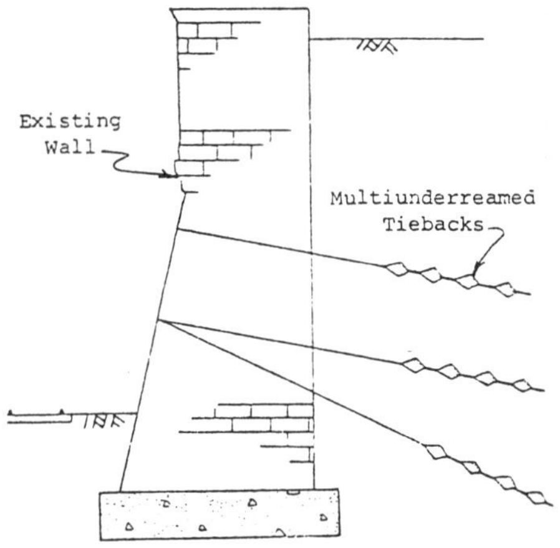


The basic types of tiebacks are: pressure-injected; low-pressure-grouted, straight-shafted; single-underreamed; multiunderreamed; and postgrouted. These tiebacks are shown in Figure

Pressure-injected tiebacks are used in sandy or gravelly soils. Grout pressures in excess of 150 psi (1034 kPa) are used to achieve high load transfer rates.

Low-pressure-grouted, straight-shafted tiebacks are installed in rock, cohesive soils, and sandy or gravelly soils. They can be made using a variety of drilling and grouting techniques. The grout pressure is less than 150 psi (1034 kPa).

Single-underreamed tiebacks are installed primarily in the United States using large uncased drill holes in cohesive soils. Sand-cement grout or concrete is used in grouting the tieback and the grout or concrete is not placed under pressure.

Multiunderreamed tiebacks are used in stiff cohesive soils and weak rocks. The spacing of the underreams is selected in order to induce a shear failure along the cylinder determined by the tips of the underreams.

Postgrouted tiebacks are primarily used in cohesive soils. In granular soils and rock, postgrouting is used to increase the rate of load transfer.

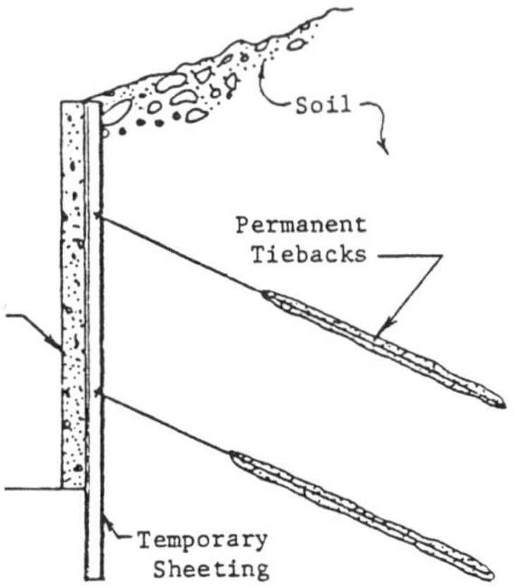


b) Permanent tiedback wall.

A tieback is a structural element which uses a grouted anchor in the ground to secure a tendon which applies a force to a structure. Figure 1 shows the components of a tieback. The anchor length is the length of the tieback which is bonded to the soil, and it is where the tieback force is transmitted to the ground. Each tieback has an unbonded length between the anchor and the structure. There the tieback tendon is not bonded to the soil, and it is free to elongate elastically. Force is applied to the tieback by post-tensioning.

Tiebacks are relatively new construction elements. They have been developed, in a large part, by speciality contractors who design and build temporary excavation support systems. Each contractor has evolved his own method of performing the work, and many of the techniques are proprietary.

Permanent tiebacks have been used to support structures in Europe since the mid-1960's, and since the early 1970's in the United States. They can be protected from corrosive attack, and they are tested to evaluate both their short-term performance and their long-term load holding capacity.

## B. DESIGN

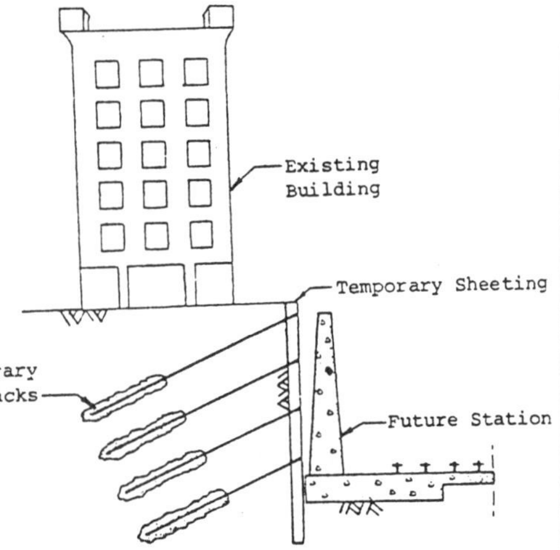


An initial evaluation must be made to determine if tiebacks can be used at a particular site, and whether or not they will be able to develop the necessary capacity without excessive movement or loss of load. The capacity of soil tiebacks are estimated using empirical relationships developed for the particular tieback type. Generally all rock materials can be considered as suitable ground in which to found anchors.

Permanent tiebacks are routinely installed in noncohesive soils with a standard penetration resistance greater than ten blows per foot. They should not be anchored in fill. Past experience, including testing and monitoring of many actual installations, indicates that permanent tiebacks installed in sandy soils will have satisfactory long-term performance.

Permanent tiebacks are not routinely installed in soft to medium cohesive soils because their long-term load holding capacity is questionable. These soils can often be avoided by installing the tiebacks at a steeper angle and to a depth where better soil or rock may be found. Soil strength, Atterberg limits, natural water content, and experience in similar soils will provide the best indication of the long-term performance of a permanent tieback installed in a cohesive soil.

Tiebacks installed in soils with a high organic content, in normally consolidated clays, and in cohesive soils with an unconfined compressive strength less than 1.0 ton/ft.$^2$ (96 kPa) and remolded strengths less than 0.5 ton/ft.$^2$ (48 kPa) may be creep susceptible. Tiebacks installed in soils that exceed these strengths, and have a consistency index ($I_c$) [1] greater than 0.8 have not experienced significant loss of load or movement with time. The consistency index is given by the relationship:

$$I_c = \frac{W_L - W}{W_L - W_P}$$

$W_L$ = Liquid limit
$W$ = Natural water content
$W_P$ = Plastic limit

In order to establish load holding characteristics and thus establish confidence in long-term performance, a tieback test program is recommended if permanent tiebacks are to be anchored in a cohesive soil, or in sandy soil with a standard penetration resistance less than ten blows per foot.

The most economical tieback installation will be obtained if the design specifications permit the contractor to select the tieback type, the construction method, and the tieback capacity. The designer should specify the minimum unbonded length, the minimum total tieback length, and the loading diagram for determining the tieback loads. In lieu of specifying a loading diagram, a unit tieback capacity for each tieback level could be specified. Finally, each production tieback should be tested to verify that the anchor will carry the design load.

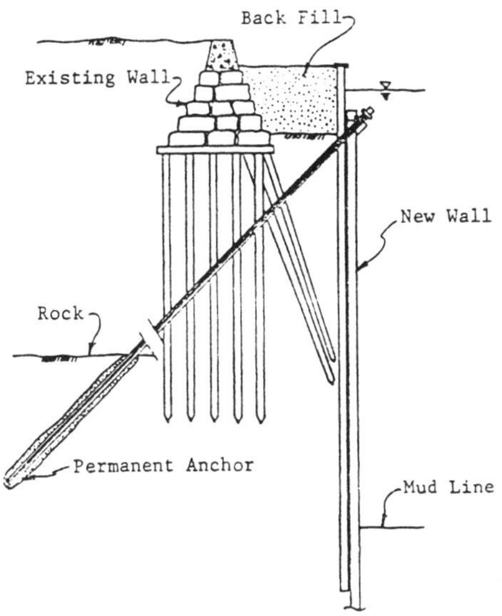


## C. CORROSION PROTECTION

Permanent tiebacks have been installed routinely since the mid-1960's in Europe, and since the early 1970's in the United States. They are performing well in a variety of environments. Most tiebacks use cement grout for protection over their anchor length. Portier [2], and Herbst [3] reported that there is no evidence of a corrosion failure where the tieback tendon was encased in grout. Corrosion failures have occurred along the unbonded length of unprotected tendons, with most of them located within 6.65 feet (2 m) of the anchor head. A significant number of the tieback corrosion failures were in tendons fabricated using quenched and tempered prestressing steels. These steels do not meet ASTM specifications for prestressing wires, strands, or bars.

Most permanent tiebacks can be protected by portland cement grout along the anchor length, and a grease-filled tube or heat shrinkage sleeve over the unbonded length, Figure 2. Grout protected tiebacks should be electrically insulated from the structures they support, and the tendon should have a minimum of 0.5 inches (12.7 mm) of grout protection.

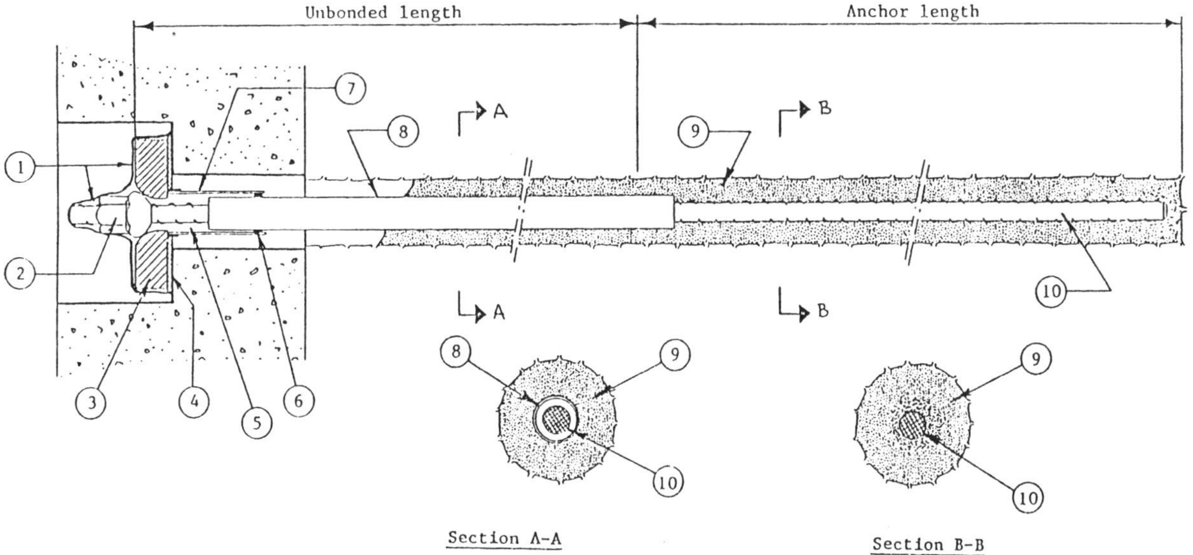


Legend:

1) Insulating cover of preformed plastic, heat shrinkable cover, or moldable tape.
2) Nut
3) Bearing plate
4) Bearing plate insulation
5) Anticorrosion grease
6) Seal
7) PVC trumpet
8) Grease-filled PVC or polyethylene sheath
9) Anchor grout
10) Tendon

Figure 2. Insulated simple corrosion protected tieback.

If the soil surrounding the anchor length has a pH less than 4.5, or a resistivity less than 2,000 ohm-cm, or if sulfides are present, then a local corrosion system could develop on the tendon. Figure 4 shows a local corrosion system. When the aggressive environments are encountered, then the tendon should be completely encapsulated in a plastic or steel tube. Figure 5 shows an encapsulated tieback. The encapsulation will interrupt any long-line and stray current corrosion system, and prevent the local corrosion system from developing.

Figures 2 and 5 show two ways to provide corrosion protection for the anchorage and the tendon below the bearing plate. Care must be taken to insure that this area is well protected since most known corrosion failures have occurred near the anchor head. The corrosion protection under the anchorage should be designed to accommodate small movements.

The American Water Works Association (AWWA) [4] describes how the pH, resistivity, and sulfide content can be measured in the field. The soluble sulfate content of the soil is determined in the laboratory. If the soluble sulfate content exceeds 2,000 mg/kg, then ASTM Type V cement should be used. When the pH of the soil is less than 4.5 or when nearby buried concrete structures are suffering from chemical attack, portland cements should not be used for the anchor grout.

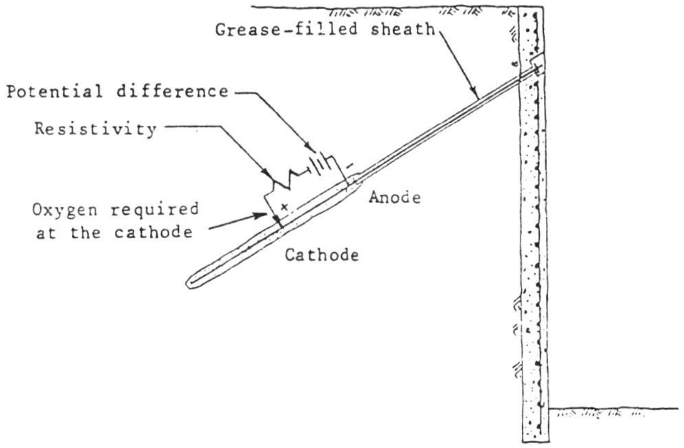


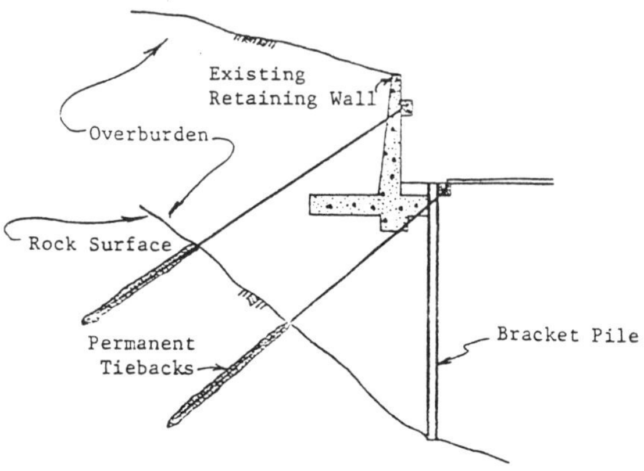


Figure 4. Local corrosion system that could affect a simple corrosion protected tieback.

## D. SPECIFICATION

The specification should establish a quality level without eliminating suitable proprietary tieback systems or methods. The designer may require the prequalification of the tieback contractor. The prequalification can be based on experience, or a list of acceptable contractors could be included in the specifications.

An alternative type of prequalification warrants evaluation. This method would require the submission and approval of the tieback system, and the corrosion protection prior to bid. The submission must be detailed enough to enable the designer to determine if his design is satisfied. This form of prequalification would also allow the contractor to know if his proprietary techniques would be acceptable, and the owner would be able to take advantage of any cost savings. Preparation and review of the submittal would not require a great deal of time, and this contracting procedure would encourage alternate tieback types, and continued tieback development.

## E. TESTING

Every tieback should be tested to verify that it will carry the design load without excessive movement. Tiebacks are one of the few structural systems where every member can normally be tested before placing them into service. Three types of tests are recommended; performance, proof, and creep tests.

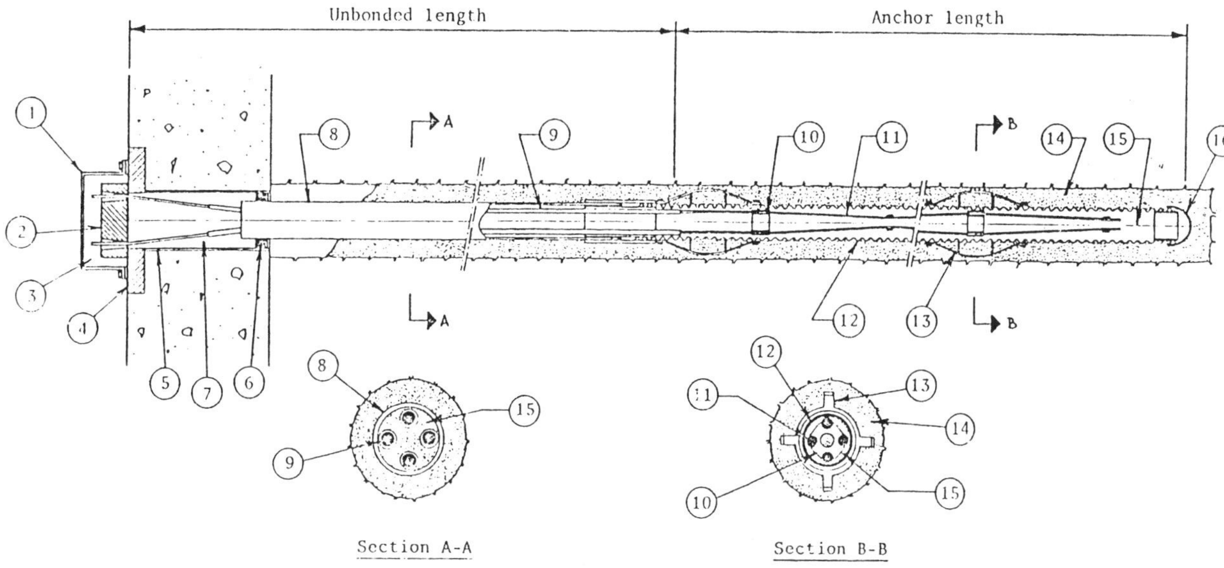


Legend:

1. Anchorage cover
2. Anchor head and wedges
3. Anticorrosion grease or grout
4. Bearing plate
5. Trumpet
6. Seal
7. Anticorrosion grease or grout
8. PVC or polyethylene tube
9. Individually greased & sheathed strands
10. Spacer
11. Strand tendon
12. Corrugated polyethylene or PVC
13. Centralizer
14. Anchor grout
15. Grout or polyester resin
16. End cap

Figure 5. Encapsulated strand tieback.

A hydraulic jack and pump are used to apply the load. The entire tieback tendon should be simultaneously loaded during testing. The movement of the tieback is measured with a dial gauge or a vernier scale supported on a reference which is independent of the tiedback structure. Movement cannot be accurately monitored by measuring the jack ram travel.

The first few tiebacks and a selected percentage of the remaining tiebacks should be performance tested. The performance test is used to establish the load-deformation behavior for the tiebacks at a particular site. It is also used to separate and identify the causes of tieback movement, and to check that the unbonded length has been established. The movement patterns developed during the performance test are used to interpret the results of a simpler proof test.

Each production tieback which is not performance tested should be proof tested. A proof test is a simple test which is used to measure the total movement of the tieback. Proof testing is done by measuring the load applied to the tieback and its movement during incremental loading

Creep tests are performed on tiebacks installed in cohesive soils. They are normally made on the initial two performance tested tiebacks. During a creep test, each increment of load is held and the elongations are recorded and plotted.

Tieback tests are used to identify the load deformation behavior of each tieback, and provide data that will enable the engineer to make a decision as to their adequacy. The total movement curve is helpful in quickly identifying any unusual behavior. However, the primary purpose of the test is to verify that the tieback will carry the load without excessive movement. The tieback behavior during the load hold or the creep test provides the best indication of the load carrying ability of the tieback.

## F. CONCLUSION

Permanent tiebacks can be effective tools to support a variety of different structures. They can be installed in rock and sandy soils without concern about their long-term performance. Permanent tiebacks can also be made in cohesive soils. However, a careful testing program, and a proven tieback system should be used in cohesive soils. The tieback tendon can easily be protected from corrosion. Each tieback should be tested to verify that it will carry the design load for the service life of the structure.

# L.1 STABILITY

## EMBANKMENT STABILITY

### APPROACH EMBANKMENT STABILITY

Approach embankment stability must be assured prior to consideration of other foundation related items. Embankment foundation problems involve the support of the embankment by natural soil. Problems with embankments and structures occasionally occur which could be prevented by initial recognition of the problem and appropriate design. Stability problems most often occur where the embankment is to be built over soft weak soils such as low strength clays, silts, or peats. Once the soil profile, soil strengths, and depth of water table have been determined by field explorations and field and lab testing, the stability of the approach embankment can be analyzed and factor of safety estimated.

There are three major types of stability problems that should be considered in the design of approach embankments over weak foundation soils. These are:

1. Circular Arc Failure.

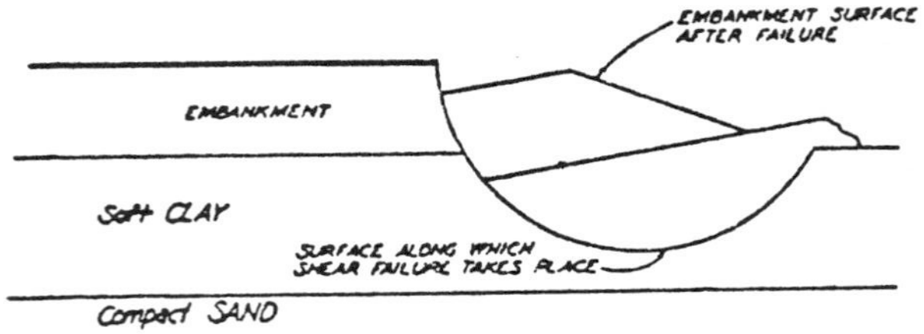


2. Sliding Block Failure.

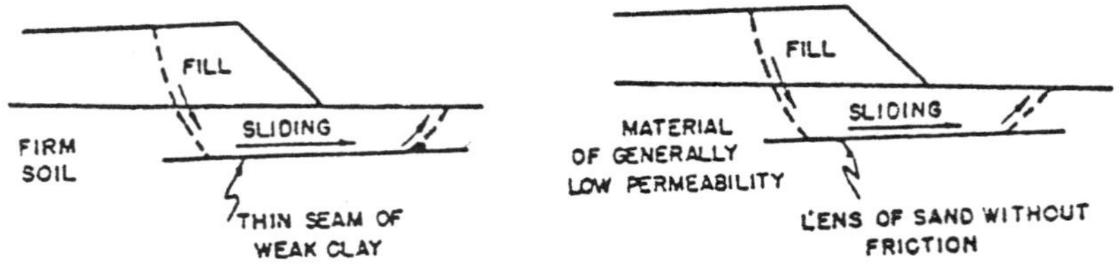


3. Lateral Squeeze of Foundation Soil.

The following are recommendations on how to recognize, analyze, and solve each of these three problems.

Note: The stability problems outlined above are "external" stability problems. "Internal" embankment stability problems involve selection of embankment materials and placement requirements. Internal stability may be "ordered" in project specifications by specifying minimum gradation and compaction requirements.

### Effects Of Water On Slope Stability

#### Importance Of Water

Next to gravity, water is the most important factor in slope stability.

#### Effect Of Water On Frictional Soils

In cohesionless soils, water does not affect the angle of internal friction ($\phi$). The effect of water on cohesionless soils below the water table is to decrease the intergranular (effective) pressure between soil grains (due to buoyancy), and this decreases the frictional shearing resistance.

#### Effect Of Water On Clays

An increase in absorbed moisture is a major factor in the decrease in strength of cohesive soils. Water is absorbed by clay minerals and high water contents decrease cohesion of all clayey soils.

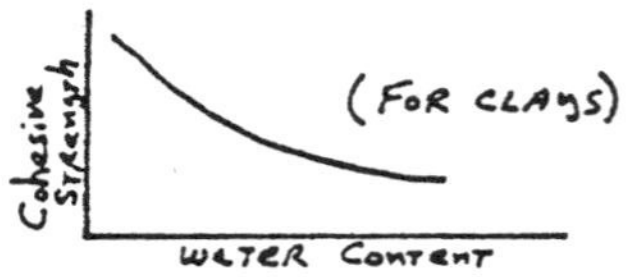


#### Fills On Clays

Excess pore pressures are created when fills are placed on clay or silt. As the pore pressure dissipates, consolidation occurs, and the clay or silt strength increases. This is the reason the factor of safety increases with time.

#### Cuts In Clay

As a cut is made in clay the effective stress is reduced. This will allow the clay to expand and absorb water, which will lead to a decrease in the clay strength with time. This is the reason the factor of safety of a clay cut slope decreases with time. Cut slopes in clay should be designed using effective strength parameters and the effective stress which will exist after the cut is made.

#### Slaking - Shales, Claystones, Siltstones, Etc.

Sudden moisture increase in a dry soil can produce a pore pressure increase in trapped pore air accompanied by local soil expansion and strength decrease. The "slaking" or sudden disintegration of hard shales, claystones, and siltstones result from this mechanism. If placed as rock fill, water percolating through the fill causes these materials to disintegrate to a clay soil, which often leads to settlement and/or shear failure of the fill.

### How To Estimate Amount Of Horizontal Abutment Movement

The amount of horizontal movement the abutment may undergo toward the fill can also be estimated in design. The following table contains case history information for nine structures where measurements of abutment movements have been made:

| Foundation | Fill Settlement (in.) | Abutment Settlement (in.) | Abutment Tilting (in.) | Ratio of Abutment Tilting to Fill Settlement |
|---|---|---|---|---|
| Steel H-piles | 16 | Unknown | 3 | 0.19 |
| Steel H-piles | 30 | 0 | 3 | 0.10 |
| Soil bridge | 24 | 24 | 4 | 0.17 |
| Cast-in-place piles | 12 | 3.5 | 2.5 | 0.19 |
| Soil bridge | 12 | 12 | 3 | 0.25 |
| Steel H-piles | 48 | 0 | 2 | 0.06 |
| Steel H-piles | 30 | 0 | 10 | 0.33 |
| Steel H-Piles | 5 | 0.4 | 0.5 to 1.5 | 0.1 to 0.3 |
| Timber Piles | 36 | 36 | 12 | 0.33 |

This data provides a basis for estimating horizontal abutment movement for similar problems, providing a reasonable estimate of the post-construction fill settlement is made, using data from consolidation tests on high quality undisturbed shelby tube samples. Note that the data for the structures listed in the previous summary showed horizontal abutment movement to range from 6 to 33 percent of the vertical fill settlement, with the average being 21 percent.

Therefore, if the fill load exceeds the 3C limit, then the horizontal abutment movement that may occur can reasonably be estimated as 25 percent of the vertical fill settlement, i.e.,

Horizontal Abutment Movement = 0.25 X Fill Settlement

#### Design Solutions To Prevent Abutment Tilting

**The best way to handle the abutment tilting problem is to prevent it by getting the fill settlement out before the abutment piling are driven.**

**If the construction time schedule or other factors do not permit the settlement to be removed before the piling can be driven, then the problems resulting from abutment tilting can be provided for by the following design provisions:**

1. **Use sliding plate expansion shoes large enough to accomodate the anticipated horizontal movement.**
2. **Make provisions to fill in the bridge deck expansion joint over the abutment by inserting either metal plate fillers or larger neoprene joint fillers.**
3. **Design piles for downdrag forces due to settlement.**
4. **Use steel H-piles for the abutment piling since steel H-piles are capable of taking large tensile stresses without failing.**
5. **Use backward battered piles at the abutment and particularly the wingwalls.**

Movements should also be monitored so that predicted movement can be compared to actual.

## EMBANKMENT SETTLEMENT

### APPROACH EMBANKMENT SETTLEMENT

Approach embankment settlement is the most prevalent foundation problem in highway construction. Unlike stability problems, the results are seldom catastrophic but the cost of perpetual maintenance of continuing settlement are immense. The difficulty in preventing these problems is not as much a lack of technical expertise as it is a lack of communication between personnel involved in the roadway design and those involved in the structure design.

#### The Problem

The design of a roadway embankment can utilize a wide range of soil materials and permit substantial amounts of settlement without affecting the performance of the highway. Roadway designers necessarily permit such materials to reduce project costs by utilizing cheap locally available soils. Structures are necessarily designed for little or no settlement to maintain specified highway clearances and to insure integrity of structural members. The approach embankment must affect a transition between roadway and structure while providing adequate structural foundation support. In most agencies the responsibility for approach embankment design is not defined which results in roadway criteria being used across the structure. This is wrong; the approach embankment requires special materials and placement criteria to prevent internal consolidation and to moderate external consolidation.

#### The Solution

1. Eliminate settlement within the approach embankment.

   A well constructed soil embankment, using quality control with regard to material and compaction, will not consolidate. Standard specifications and construction drawings should be prepared for the approach embankment area (normally designated as 50 feet behind the wingwall). The structural designer should have the responsibility for selecting the appropriate approach embankment cross section depending on selection of structure foundation type. A typical suggested approach embankment cross section is shown on figure 9 for spread footing and pile foundations.

   Special attention must be given to the interface area between the structure and the approach embankment as this is where the famous "bump at the end of the bridge" occurs. The reason for the bump are twofold; poor compaction of embankment material near the structure and migration of fine soil into drainage material. Poor densification is caused by restricted access of standard compaction equipment. Proper densification can be achieved by optimizing the soil gradation in this area to permit maximum density with minimum effort. Figure 10 shows a suggested procedure and page 141 contains a material gradation specification to prevent the problem.

2. Estimate settlement of the approach embankment caused by consolidation of the subsoil. Many and varied procedures exist for computation of embankment settlement. Two methods will be presented herein; one each for cohesionless and cohesive soils. However, certain steps are common to either method, namely pressure distribution.

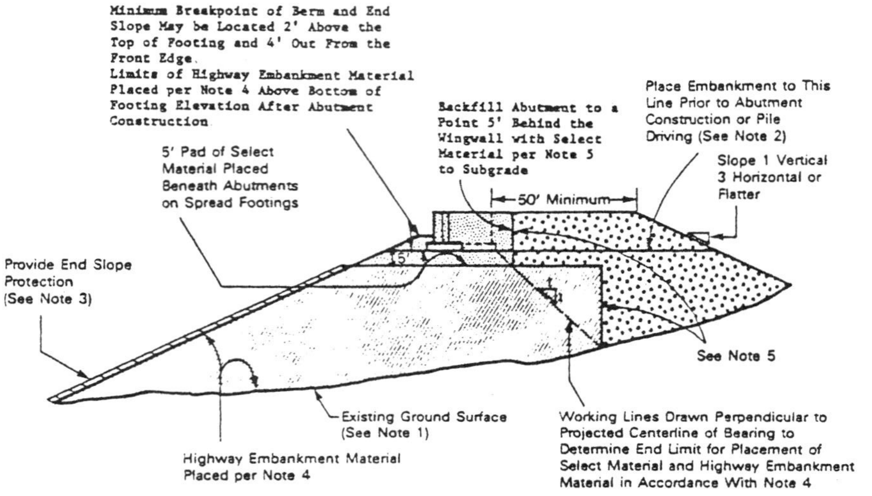


Note 1: Topsoil shall be stripped beneath abutments and piers less than 20' in height from a rectangular or trapezoidal area abounded by lines 15 feet outside the abutment and wingwall footings, or to the toe of slope, whichever is less. The depth or stripping shall be determined by the Soils Engineer and displayed on the highway cross sections by the Design Engineer.

Note 2: At some sites, fill is to be placed to the subgrade of the roadway and allowed to stand, in order to consolidate underlying material, before piles are driven.

Note 3: Slope protection treatment shall be as specified by the Bridge Engineer.

Note 4: Highway embankment material placed within these limits shall have a maximum dimension of 6 inches and shall be compacted to 95% of maximum density. Quantity to be included in highway estimate.

Note 5: Highway embankment material and select material shall be placed simultaneously of the vertical payment line.

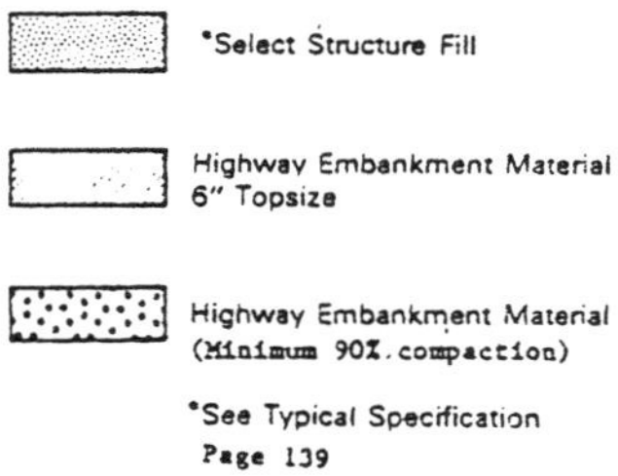


SUGGESTED APPROACH EMBANKMENT DETAILS

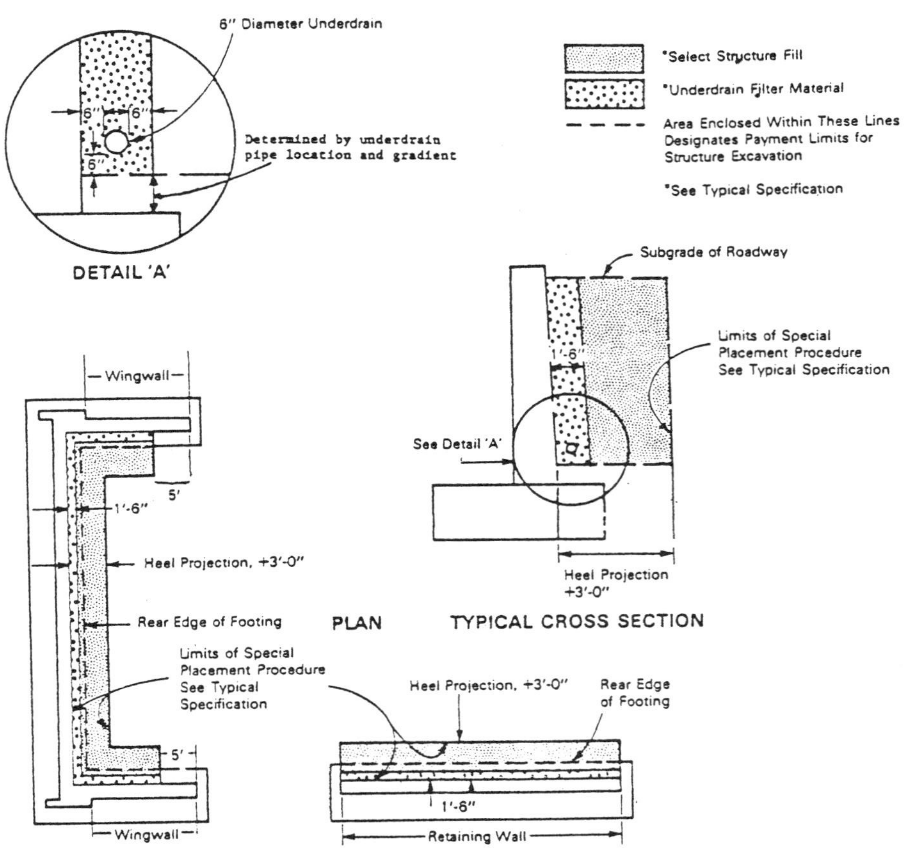


*FIGURE* 10 ***STRUCTURE BACKFILL PLACEMENT LIMITS FOR POROUS DRAINAGE AGGREGATE***

# DESIGN SOLUTIONS

Design Solutions - Settlement Problems

Reducing Settlement Amount

Several of the methods for solving embankment stability problems can also be used to reduce the amount of settlement. These include:

1. Reduce grade line.
2. Excavate and replace soft soil.
3. Lightweight fill.

Reducing Settlement Time

Often the major design consideration when faced with a settlement problem is the time it will take for the settlement to occur. Low permeability clays and silt-clays can take a long time to consolidate (water squeezed out). The settlement time is generally what will get the chief engineer "excited," since this can affect construction schedules, increase project costs due to inflation, etc.

The two most common methods used to accelerate settlement and reduce settlement time are:

1. Surcharge treatment.
2. Sand or Wick Drain treatment of subsoil.

Surcharge Treatment

The embankment is built up 1 to 10 feet above final grade elevation and allowed to remain for a predetermined waiting period (typically 3 - 12 months). The length of waiting period can be estimated using consolidation test data. The actual settlement occurring during embankment construction is then monitored with geotechnical instrumentation. When the settlement with surcharge equals the settlement originally estimated for the embankment the surcharge may be removed, as illustrated below:

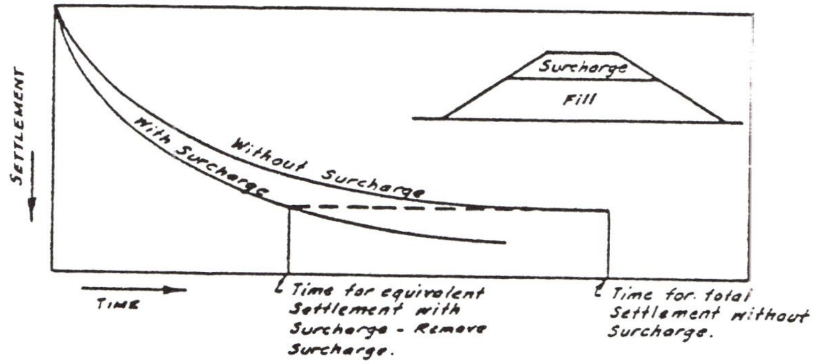


Note that the stability of a surcharged embankment must be checked to insure that an adequate safety factor exists to permit placement of the surcharge load.

Sand or Wick Drains

Some highly plastic clays of extremely low permeability can take many years for settlement to be completed. Surcharging alone may not be effective in reducing settlement time sufficiently. In such cases, sand or wick drains can be used to accelerate the settlement. Surcharging should always be considered first, since sand and wick drains are much more expensive. The reason such drains accelerate the settlement is that they shorten the drainage path the water must travel to escape from the impervious soil, as illustrated below.

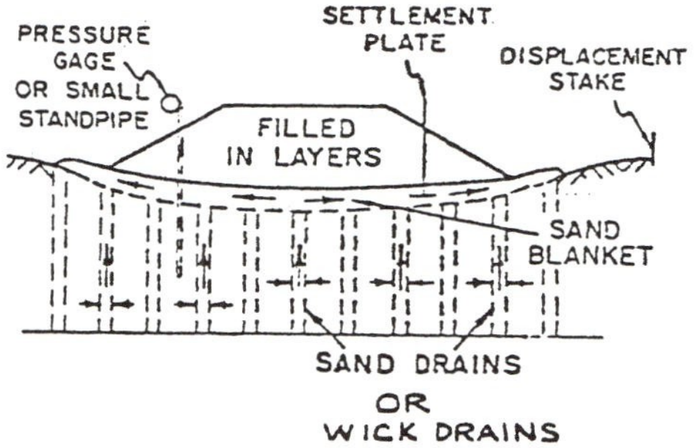


Recall that the settlement time is proportional to the square of the length of the drainage path, thus if the drainage path length can be cut in half, the time is reduced by a factor of four. The drains and sand blanket must have high permeability to allow water squeezed out of the subsoil (due to the fill pressure) to go up the drains and out through the blanket. Sand drains are typically installed by jetting or augering 12 to 18 inch diameter holes to the bottom of the compressible soil and backfilling with a high permeability graded sand. Typical sand drain spacings are 8 to 15 feet center to center.

Wick drains are small prefabricated drains consisting of a plastic core which is wrapped by a piece of filter fabric. Wick drains are approximately 4 inches wide and about 1/4 inch thick. Two examples are the "Alidrain" and "Mebra drain" (both proprietary). Wick drains are installed by pushing or vibrating a mandrel into the ground with the wick drain inside. When the bottom of the compressible soil is reached, the mandrel is withdrawn and the wick drain left in the ground. To minimize smear of the clay, it is recommended that the cross-sectional area of the mandrel be limited to a maximum of about 12 square inches. Since the wick drains have a much smaller surface area than sand drains, more are required. About twice as many wick drains will be required versus sand drains. Use of wick drains in the United States began about 5 years ago and is quickly gaining acceptance. Wick drains are now used almost exclusively in Japan. Wick drain projects will typically be 20-30 percent less costly than if sand drains were used. This is primarily due to much faster speed of installation of wick drains versus sand drains.

(Refer to the workshop design problem solution following for an example of a sand drain and wick drain design and cost comparison.)

Practical Aspects of Embankment Settlement

Few engineers realize the influence of embankment placement on the subsoils. The total weight of an embankment has an impact on the type of foundation treatment that may be selected. For instance a relatively low height embankment of 10' may be effectively surcharged because the additional surcharge weight could be 30 to 40 percent of the proposed embankment. However, when the embankment height exceeds 50' the influence of a 5 or 10 foot trapezoid of soil on top of this heavy 50 foot mass is small and probably not cost-effective. Conversely as the embankment height (and, therefore, weight) increase, the use of a spread footing abutment becomes more a attractive. A 30' high, 50' long approach embankment weighs about 15,000 tons compared to the insignificant weight of a total abutment loading which may equal 1,000 tons. Besides weight, the width of an embankment has an effect on total settlement. Wider embankments cause a pressure increase deeper into the subsoil. As might be expected, wide embankments will cause more settlement and take longer for consolidation to occur.

# PILE DESIGN

(1) Bearing Capacity of Pile Foundations

A pile foundation transfers load into the underlying strata by the following methods: skin friction, point bearing, or a combination of skin friction plus point bearing. Realistically, there is usually friction on any pile and there is generally some point bearing at the pile tip making distinct classifications a bit of a misnomer. However, a pile that receives the majority of its support by friction or adhesion from the soil along its shaft is referred to as a friction pile, whereas a pile that receives the majority of its support from soil near its tip is an end bearing pile.

The load applied to a single pile is carried jointly by the soil beneath the tip of the pile and by the soil around the shaft and the maximum load that the pile can support (the pile capacity) is:

$$Q = Qp + Qs$$

where

Q = pile capacity.
Qp = point resistance.
Qs = shaft resistance.

This is also shown in the following Figure 11.1.

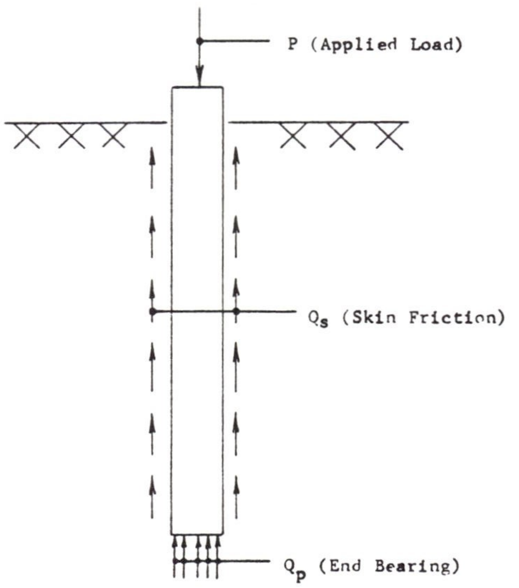


FIGURE 11.1

A. Shaft Resistance

Total shaft resistance can be calculated from the equation:

$$Q_s = \sum_{L=1}^{L=n} A_c L f_s$$

where

$A_c$ = circumferential area of pile per given length of pile.
L = length of pile.
$f_s$ = skin friction per given length of pile.

The skin friction developed in a given soil is approximated by the soils shear strength.

$$f_s = c + kP_o \tan \phi$$

where

c = cohesion of soil.
k = coefficient of lateral earth pressure (Use 1.0 for CIP, 1.2 for timber, 1.1 for concrete, and .8 for H-piles).

$P_o$ = effective vertical pressure at midheight of given length of pile.
$\phi$ = angle of internal friction.

In clay soils $\phi$ and $\tan \phi = 0$; thus the skin friction is equal to the cohesive force.

Average values of friction for various soils are given in Tables 11.1 and 11.2. The values given are average ranges and are intended to give orders of magnitude only. Other conditions such as layering sequences, drilling information, ground water, thixotropy, clay sensitivity, etc., must be evaluated by experienced soil engineers and analyzed by principles of soil mechanics.

Theoretically, skin friction values increase with depth. Actually, in non-cohesive material skin friction values reach a constant final value at a depth of 15 to 25 pile diameters in loose sands and 25 to 35 pile diameters in firm sands.

COHESIVE MATERIAL

| Soil Type | qu (1) tons/sq. ft. | Skin Friction (S.F. = 2) (PSF) |
|---|---|---|
| Very soft clay | 0 - .25 | --------- |
| Soft clay | .25 - .50 | 100 - 300 |
| Medium clay | .50 - 1.0 | 250 - 450 |
| Stiff clay | 1.0 - 2.0 | 550 - 950 |
| Very stiff clay | 2.0 - 4.0 | 1000 - 1500 |
| Hard clay | 4.0 | 1500 - 2000 |
| Silt | | 50 - 250 |
| Silty clay | | 200 - 400 |
| Sandy clay | | 200 - 400 |
| Sandy silt | | 300 - 500 |
| Dense silty clay | | 450 - 750 |

TABLE 11.1

(1) Unconfined Compression Strength
(2) Standard Penetration Value

GRANULAR MATERIAL

| Soil Type | N(2) | Skin Friction (S.F. = 2) (PSF) |
|---|---|---|
| Very loose sand and silt or clay | 0 - 6 | 25 - 75 |
| Medium sand and silt or clay | 6 - 30 | 200 - 300 |
| Dense sand and silt or clay | 30 - 50 | 300 - 400 |
| Very dense sand and silt or clay | over 50 | 400 - 500 |
| Very loose sand | 0 - 4 | 350 - 850 |
| Loose sand | 4 - 10 | 350 - 850 |
| Firm sand | 10 - 30 | 350 - 850 |
| Dense sand | 30 - 50 | 350 - 850 |
| Very dense sand | over 50 | 350 - 850 |
| Sand and gravel | | 500 - 1500 |
| Gravel | | 750 - 1750 |

TABLE 11.2

In computing shaft resistance, the method of installation must be considered as well as the soil type. The method of installation significantly effects the degree of soil disturbance, the lateral stress acting on the pile, the friction angle and the area of contact. Shafts of prebored piles do not always fully contact the soil therefore the effective contact area is less than the shaft area. Driving a pile in granular material densifies the soil and increases the friction angle. Driving also displaces the soil laterally and increases the horizontal stress acting on the pile. Disturbance of clay soils from driving can break down soil structure and increase pore pressures which greatly decreases soil strength. However, some or all of the strength recovers following reconsolidation of the excess pore pressure. Since piles in a foundation are not subjected to their full load until after the structure is completed, use the initial strength values for design purposes. Type and shape of a pile also affects the amount of skin friction developed. Refer to Section 11.2.

Negative skin friction, in which soil pulls down the pile instead of supporting the load, can occur when settlement of the soil through which the piling are driven takes place. Natural consolidation may provide this effect which is accelerated by embankment loads and sedimentation. When this condition is found to be present, the working pile loads are reduced to compensate for it, or consolidation is allowed to occur before driving piling. Other alternatives are to preauger the piling or drive through a permanent pipe sleeve.

Uplift forces may also be present, both permanently and intermittently, on a pile system. Such forces may occur from hydrostatic uplift on cofferdam seals, ice uplift resulting from ice grip on piles and rising water, wind uplift due to pressures against high structures, frost uplift, etc. In the absence of pulling test data, friction values on the order of magnitude as given in Tables 11.1 and 11.2 can be assigned. Values from this table are average ranges only and are to be used judiciously.

Generally, the type of pile with the largest perimeter is the most efficient in resisting uplift forces. Factors of safety to be used depend upon the expected frequency and duration of the uplift forces. AASHTO specifications allow only a maximum of 40 percent of the downward friction capacity in just the intermittent uplift condition. Group action is also considered, refer to Section 11.1 (1)D.

B. Point Resistance

The end bearing capacity of a pile is estimated from modifications to bearing capacity formulas developed for shallow footings. These formulas take the form of:

$$Q_p = A_b\ (cN_c + \gamma h N_q + \frac{\gamma B}{2} N_\gamma)$$

where $A_b$ = cross-sectional area of pile tip.
$c$ = cohesion of soil.
$N_c$, $N_q$, $N_\gamma$ = bearing capacity factors (functions of $\phi$) determined from tests.
$h$ = overburden pressure.
$\gamma$ = unit weight of soil.
$B$ = least lateral dimension of pile.

For clays $Q_p \doteq A_b(8c)$.

An end bearing pile surrounded by soil is not a structural member like a column. Both experience and theory demonstrate that there is no danger of an end bearing pile buckling on account of inadequate lateral support if it is surrounded by even the very softest soils. Therefore, pile stresses can exceed allowable column stresses.

C. Lateral Load Capacity

Recommended values of lateral resistance for battered or vertical piles used by the Bridge Section in lieu of the following detailed analysis are:

| Soil Conditions | Timber | 10-3/4" CIP | 12" Precast/ 12" CIP | Steel-H |
|---|---|---|---|---|
| Poor | $4^K$ | $7^K$ | $8^K$ | $7^K$ |
| Average | 5 | 9 | 11 | 10 |
| Good | 5 | 9 | 11 | 15 |

Structures supported by single piles or pile groups are frequently subjected to lateral forces from lateral earth pressures, live load forces, wave action and wind forces. Piles subjected to lateral forces must be designed to meet allowable stress and deflection criteria to prevent impairment or premature failure of the foundation or superstructure. To solve the soil-structure interaction problem, the designer determines the following:

1. Characteristics of the pile including:

   a. Stiffness of the pile.

   b. Rotation restrictions imposed on the pile top by the cap.

   c. Maximum bending moment imposed on the pile and the distribution of the bending moment along the pile length.

   d. Probable points of fixity on the pile.

2. Stiffness of the soil.

3. Allowable deflection of the pile permitted by the superstructure.

D. Pile Group Action

Friction piles in group action do not carry as much load per pile as an individual pile acting alone due to overlapping zones of pressure around each pile. The amount of reduction per pile depends upon the size and shape of the pile group, as well as the size, spacing, and length of the piles. No reduction due to grouping occurs with end bearing piles. For combined end bearing and friction piles, only the load carrying capacity of the frictional portion is reduced. Experience in Wisconsin indicates that in most thrixotropic clays where piles are driven to a hammer bearing as determined by dynamic formulas, reduction for pile group action is not required. Since reductions are not usually made, bridge computer programs do not include an efficiency formula for pile group reduction.

In special cases, where piles are driven in groups into plastic materials the designer may want to consider a reduction for pile group action. The Converse-Labarre efficiency formula is suggested in the AASHTO Specifications. This method is derived under the assumption that the area of a pile available for developing shear is reduced by the influence of adjacent piles in the same row and by the closest pile of the adjacent row. However, it disregards pile length, battered piles capped by a seal, and distances of embedment in friction strata. When applying this formula, consider battered piles by increasing the value "S" to "S" plus the amount of batter at the third point from the top of pile. The reduced pile capacity is compared to Group I axial loading for adequacy.

(2) Pile Driving Hammers

Pile drivers are generally powered by compressed air, steam pressure, or diesel units. The diesel hammer, a self-contained unit, is the most popular due to its compactness and adoption in most construction codes. Also, the need for auxiliary power is eliminated and the operation cost is nominal. Vibratory and sonic type hammers are employed in special cases where speed of installation and/or noise from impact is prohibited. The vibrators convert instantly from a pile driver to a pile extractor by merely tensioning the lift line. However, the cost of this equipment is high and the oscillators require frequent, costly maintenance.

The hammer is raised and allowed to fall either by gravity or with the assistance of power. If the fall is due to gravity alone, the hammer is referred to as single-acting. The single-acting hammer is suitable for all types of soils but is most effective in penetrating heavy clays. The major disadvantage is the slow rate of driving due to the relatively slow rate of blows from 50 to 70 per minute. Wisconsin construction specifications call for a minimum hammer weight of from 3000 to 5000 pounds depending on the required bearing value of the pile being driven. In order to avoid injury to the pile, the fall of the gravity hammer is usually limited to 10 feet with an absolute maximum of 15 feet.

If power is added to the downward falling hammer, the hammer is referred to as double-acting. This type of hammer works best in sandy soil but also performs well in clay. Double-acting hammers deliver from 100 to 250 blows per minute which increases the rate of driving considerably over the single-acting hammers. Wisconsin construction specifications call for a rated minimum energy of 15 percent of the required bearing of the pile or a minimum of 7200 ft.-lb. per blow on all piles. A rapid succession of blows at a high velocity can be extremely inefficient as the hammer bounces on heavy piles.

Differential-acting hammers overcome the deficiencies found with both single- and double-acting hammers by incorporating higher frequency of blows and more efficient transfer of energy. The steam cycle, which is different from that of any other hammer, makes the lifting area under the piston independent of the downward thrusting area above the piston. Sufficient force can be applied for lifting and accelerating these parts without affecting the dead weight needed to resist the reaction of the downward acceleration force. The maximum delivered energy per blow is the total weight of the hammer plus the weight of the downward steam force times the length of the stroke. Differential-acting hammers are becoming increasingly more popular due to their economy, efficiency, and reliability.

Generally the selection of the pile hammer is dependent on the following:

1. The hammer weight and rated energy are selected on the basis of supplying the maximum driving force without damaging the piles.

2. The hammer types dictated by the construction specification for the given pile type.

3. The hammer types available to the contractor.

4. Special situations such as sites adjacent to existing buildings that require consideration from the driving impact or noise aspect.

Most specifications require the heads of all piling to be protected by caps during driving. The pile cap serves to protect the pile as well as modulate the blows from the hammer which aids in eliminating high inefficient hammer forces. When penetration per blow is used as driving criteria, constant cap-block material characteristics are required. The cap-block characteristics are also assumed constant for all empirical formula computations to determine the rate of penetration equivalent to a particular dynamic resistance.

The material which was used for cap-blocks for many years is wood. However, the cost of wood plus the change in its characteristics as it wears down and ages must be considered. The above factors lead contractors to look for a permanent cap-block. The micarta-aluminum combination has a spring constant comparable to wood and a higher coefficient of restitution leading to high efficiency of energy transfer from hammer to pile.

(3) Pile Driving Formulas

Formulas used to estimate the bearing capacity of piles are of four general types: empirical, static, dynamic and wave equation.

Empirical formulas are based upon tests under limited conditions and are not suggested for general use.

Static formulas are based on soil stresses and try to equate skin friction and end bearing to the load-bearing capacity of the piles.

Dynamic pile driving formulas are based upon the theory that ultimate carrying capacity is equal to the ultimate driving resistance. These formulas are derived starting with the relation:

Energy input = energy used + energy lost

The energy used equals the driving resistance times the pile movement. Thus by knowing the energy input and estimating energy losses driving resistance can be calculated from pile movement observations. Numerous dynamic formulas have been proposed. They range from the simpler *Engineering News Formula*, to the more complex *Hiley Formula*. The State of Wisconsin, Standard Specifications call for use of a modified Engineering News Formula that takes the form:

$$P = \frac{2WH}{S+0.5} \quad \text{for gravity hammers}$$

and

$$P = \frac{2E}{S+0.2} \quad \text{for air, steam or diesel hammers.}$$

Because of the difficulty of evaluating the many energy losses involved with pile driving, these dynamic formulas can only approximate pile driving resistance, thus giving approximate results when used as a safe means of determining pile length and bearing requirements. Despite the obvious limitations, the dynamic pile formulas take into account the best information available and do have considerable utility to the engineer in securing reasonably safe and uniform results over the entire job.

The wave equation is based upon the theory of longitudinal wave transmission. This theory, proposed by Saint-Venant a century ago, did not receive widespread use until the advent of computers due to its complexity. The wave equation can predict impact stresses in a pile during driving and estimate static soil resistance at the time of driving by solving a series of simultaneous equations. The advantages of this method are adaption to any shape pile, the frictional and viscous resistance of the soil, and the point resistance of the pile. The effect of the hammer and cushion block can be included in the computations.

WISDOT presently utilizes a Pile Driving Analyzer in an advisory capacity for evaluating if sufficient pile penetration has been achieved to develop the design capacity, if pile damage may have occurred, or if a driving system is performing satisfactorily.

The Analyzer provides a method of dynamic pile testing both for pile design and construction control. Testing is accomplished by attaching reusable strain transducers and accelerometers directly on the pile. Piles can be tested while being driven, or during restrike. The instrumentation mounted on the pile allows the measurement of force and acceleration signals for each hammer blow. This data is transmitted to a small field computer for processing and recording on a magnetic data tape for a permanent record. Data is also fed into an oscilloscope to observe the dynamic response of the pile to driving. Calculations made by the computer, based upon one-dimensional wave mechanics, provide an immediate printout of maximum stresses in the pile, energy transmitted to the pile, and a prediction of the ultimate pile bearing capacity for each hammer impact. Monitoring of the force and velocity wave traces with the oscilloscope during driving also makes it possible to detect if structural pile damage has occurred. Polaroid photographs of selected force and velocity wave traces are also made to provide additional testing documentation. The Pile Analyzer can be used on all types of driven piles with any impact type pile driving hammer.

(4) Pile Investigation

Test piles are employed on the job site for two purposes:

1. For test driving, to determine the length of timber pile required prior to placing purchasing orders.

2. For load testing, to verify actual pile capacity versus computed design capacity.

A. Driving of Test Piles

If timber piles are called for on Wisconsin Bridge plans, test piles are required to determine the length of piles to be ordered. Specify lengths sufficiently longer than the estimated test pile lengths (about 50 percent longer) to assure getting the bearing capacity. Test piles are located so they can be used in the finished structure. If test piling cannot be incorporated in the substructure unit, the test piling is removed at least 2 feet below the stream bed or finished ground line.

The Bridge Section does not require special test piles for spliceable types of piles. Previous experience indicates that contractors order total plan quantities for cast-in-place or steel H-piling in 60 foot lengths. The contractor uses one of the driven structure piles as a test pile at each designated location.

Test piling is driven near a location of a soil boring where the soil characteristics are known and representative of the most unfavorable conditions at the site. The test pile must be exactly the same type and dimension as the piles to be used in the construction and are installed by the same equipment and manner of driving. A penetration record is kept for every foot of penetration for the entire length of pile. This record may be used as a guide for future driving on this project. Any pile encountering a smaller resistance is considered as having a smaller bearing capacity than the test pile.

Do not design piles for thixotropic loading capacities unless there is no alternative. It is more costly to drive piling when there is a waiting period on each pile.

If a rest period is allowed before application of loads on test piles in clays or silts, it is usually a few days but in no case less than twenty four hours (48 - 72 hours recommended) in order to permit the disturbed soil to regain most of its strength. The Construction Section recommends twenty four hours for economic reasons. If the pile bearing capacity is low (less than 20 tons) call for a gravity hammer in the Special Provisions to drive the pile, as a diesel hammer will not work.

B. Load Testing of Piles

A pile load test is usually conducted to furnish information to the soil engineer to develop design criteria or obtain test data to substantiate design loads. A load test is the only reliable method of determining the bearing capacity of a single pile but it is expensive and can be quite time consuming. The decision to embark on an advance test program is based upon the scope of the project and the complexities of the foundation conditions. Such test programs often result in substantial savings in foundation costs which can more than offset the test program cost.

Pile load testing usually involves the application of a direct axial load to a single vertical pole. However, load testing can involve uplift or axial tension tests, lateral tests applied horizontally, group tests or a combination of these applied to battered piles. Most static test loads are applied with hydraulic jacks reacting against either a stable loaded platform or a test frame anchored to reaction piles.

The basic information to be developed from the pile load test is usually the deflection of the pile head under the test load. Movement of the head is caused by elastic deformation of the piles and the soil. Soil deformation may cause undue settlement and must be guarded against and its value is the significant one to be obtained from load tests rather than the total downward movement of the pile head.

Most routine tests are carried to one, one and one-half or twice the design load for a single pile. Testing techniques most often utilized are the cycle loading and constant rate of penetration method. Cycle loading generally involves loading and unloading in increments of 25% of the proposed load, and is used to establish design loadings. The constant rate of penetration involves applying a load of sufficient magnitude to maintain a constant pile penetration rate into the ground. This method is more applicable to friction type piles. In the State of Wisconsin, the AASHTO criteria is used to determine the safe allowable load capacity of the piles based on the test pile loading as follows:

The safe allowable load is 50 percent of that load which after a continuous application of 48 hours produces a permanent settlement not greater than 1/4 inch measured at the top of the pile. This maximum settlement may not be increased by continuous application of test load for 60 hours or longer.

It is impractical to test every pile on a project, thus test results can be applied to other piles, providing that the following conditions exist:

1. The other piles are of the same type, material, and size as the test piles.

2. Subsoil conditions are comparable to those at the test pile locations.

3. **Installation methods and equipment used are the same as or comparable to those used for the test piles.**

4. **Piles are driven to the same penetration depth or resistance or both as the test piles to compensate for variations in the vertical position and density of the bearing strata.**

## 11.3 PILE EXAMPLE PROBLEM

Given: Soils Site Investigation Report, estimate various types of pile lengths and recommend a pile type for a 55 ton bearing capacity.

Subsurface Conditions: The soils encountered from surface elevations 620 to 600 are predominantly firm silts with some clay. Below elevation 600, soils are predominantly medium dense sand with some silt. Numerous gravel layers and boulders were also encountered.

Bearing Capacity: The following table can be used to estimate pile lengths.

| ELEVATION (ft) | PILE SKIN FRICTION (psf) (s.f.=2) | END BEARING, % (Displ. Piles) | (H-Piles) |
|---|---|---|---|
| 620-600 | 400 | 20-25 | 5-10 |
| 600-560 | 700 | 35-45 | 15-25 |

Estimated Pile Lengths:

1. Cast-in-place Concrete - 10 3/4" Diameter

Perimeter = $(\pi)(10.75)/12 = 2.81$ ft

Capacity $Q = Q_S + Q_p$, assume pile penetrates below elevation 600.

$$110 = (0.4)(20)(2.81) + (0.7)(L_1)(2.81) + (110)(40\%)$$
$$= 22.5 + 1.97L_1 + 44.0 \qquad L_1 = 22.1 \text{ ft}$$

$L = L_1 + 20 = \underline{42.1 \text{ ft}}$ **Use: 45 ft**

2. **Prestressed Precast Concrete - 12" x 12"**

$$110 = (0.4)(20)(4.0) + (0.7)(L_1)(4.0) + (110)(40\%)$$
$$= 32.0 + 2.80L_1 + 44.0 \qquad L_1 = 12.2 \text{ ft}$$

$L = L_1 + 20 = \underline{32.2 \text{ ft}}$ **Use: 35 ft**

3. **Steel H-Pile - 10HP42**

**Using projected H-pile perimeter = (2)(10.08 + 9.72)/12 = 3.3 ft**

$$110 = (0.4)(20)(3.3) + (0.7)(L_1)(3.3) + (110)(20\%)$$
$$= 26.4 + 2.31L_1 + 22.0 \qquad L_1 = 26.7 \text{ ft}$$

$L = L_1 + 20 = \underline{46.7 \text{ ft}}$ **Use: 50 ft**

**Recommendation: Initially, displacement piles appear to be an economical selection. However, difficulty in driving, due to the presence of boulders, would very likely result. Therefore, steel H-piles are recommended at this site.**

## 11.4 PILE DRIVING REPORTS

A study was made of the pile driving reports received for the years 1969 and 1970. The actual field lengths were compared to the estimated plan lengths and the following observations were made.

1. There were very few cases where the field lengths exceeded the plan lengths. In most cases, the field lengths were either equal to plan lengths or less than the plan lengths. The latter two cases were about equally divided.

2. There was good agreement between plan and field lengths for non-displacement type piles that were driven to bearing on rock.

3. Where the field lengths exceeded the plan lengths, the piles were usually driven in clay. The preliminary report recommended a 48-72 hour wait period for the soil to consolidate if bearing values were not achieved. It was not clear if this was done, but the preliminary report did point to possible problems.

4. Where the pile lengths in the field were less than the plan lengths, the conditions were:

   a. End bearing capacity of displacement piles was not used to determine plan lengths in sand.
   b. Piles were driven in granular material with large boulders which reduced pile penetration.
   c. Very conservative values were used for the plan lengths based on the soil reports and other data. In several cases, piles were driven on widening jobs where the old pile driving data was available. In one case, the soils report showed rock at 35', but the plan length was given as 40'. As noted in (a.), the designers were conservative by not computing end bearing capacities in determining plan lengths.

5. As a general rule, it appears that plan pile lengths were computed to be longer than required if questionable soil values were given. This was true in cases where erratic pile lengths were predicted in the soils report.

6. The overall sample was fairly reliable based on plan lengths versus actual lengths. Almost all of the discrepancies were explainable and could probably have been predicted.

7. Computations for plan lengths were based on:

   a. Bearing - the distance to bedrock plus a distance for penetration into the rock was used.
   b. Displacement - Recommended skin friction was used to determine length. End bearing values were not used. It is not known if the designers checked their estimated lengths against the boring log to see what material they would stop in or not. Obviously, pile tips in weak material after penetrating hard would be in error.

## 11.5 SHEET PILING

Sheet piling consists of interlocking steel, precast concrete or wood pile sections driven side by side to form a continuous unit. Individual pile sections usually vary from 12 to 21 inches in width allowing for flexibility and ease of installation. The most common use of sheet piling is for temporary construction of cofferdams, retaining walls or trench shoring. The structural function of sheet piles is to resist lateral pressures due to earth and/or water.

Steel sheet piles are by far the most extensively used in temporary construction due to versatility and high salvage value. They are also very adaptable to permanent structures such as bulkheads, seawalls and wharves if properly protected from salt water. Sheet pile shapes are generally Z, arched or straight web. The Z and the medium to high arched sections have high section moduli and can be used for substantial cantilever lengths or relatively high lateral pressures. The shallow arched and straight web sections have high interlocking strength and are employed for cellular cofferdams. The Z-section has a bowl-and-ball interlock and the arched and straight web sections have a thumb-and-finger interlock capable of swinging 10 degrees. The thumb-and-finger interlock provides high tensile strength and considerable contact surface to prevent water passage. Steel sheet piling sections are usually 3/8 inch or 1/2 inch thick. Refer to Table 11.5 for typical sections.

### Driving of Sheet Piling

All sheets in a section are set, if practical, before any are driven. There is a tendency for sheet piles to lean in the direction of driving producing a net "gain" over their nominal width. Most of this "gain" can be eliminated if the piles are driven a few feet at a time, say from 5 feet to one third of their length before any single pile is driven to its full length. If some sheet piles strike an obstruction, move to the next pile that can be driven; then return to the piles that resisted driving. With interlock guides on both sides and perhaps a heavier hammer it may be possible to drive the sheet to the desired depth.

Sheet piles are installed by driving with gravity, steam, air or diesel powered hammers, or by vibration, jacking or jetting depending on the material and pile type. A vibratory or double acting hammer of moderate size is best for driving sheet piles. For final driving of long heavy piles a single acting hammer may be more effective. A rapid succession of blows is generally more effective when driving in sand and gravel; slower, heavier blows are better for penetrating clay. For efficiency and impact distribution, where possible, two sheets are driven together. If sheets adjacent to those being driven tend to move down below required depth, they are stopped by welding or bolting to the guide wales. It is generally better to fill in with a short length at the top rather than trying to pull the sheet back up to plan location.

Sheet piles are pulled with air or steam powered extractors or inverted double acting hammers rigged for this use. Vibratory hammers are also effective in pulling. Usually a 25 ton lift on the extractor is adequate; occasionally a derrick with multiple-reeved blocks is needed. If piles are difficult to pull, slight driving is effective in breaking them loose. Pulled piles are to be handled carefully as they may be used again; perhaps several times.

## LOAD CARRING CAPACITY OF SINGLE PILE IN GRANULAR SOIL

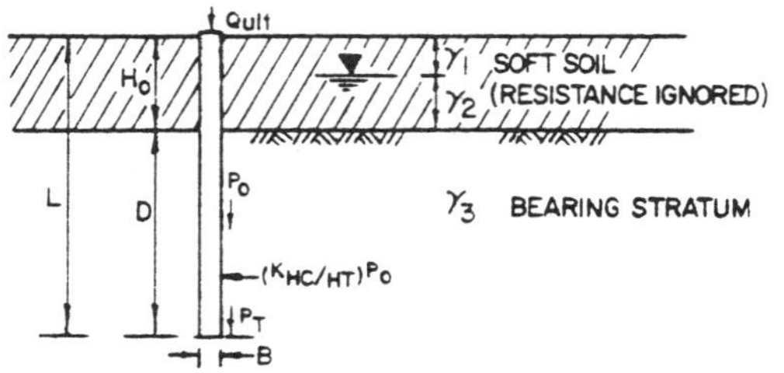


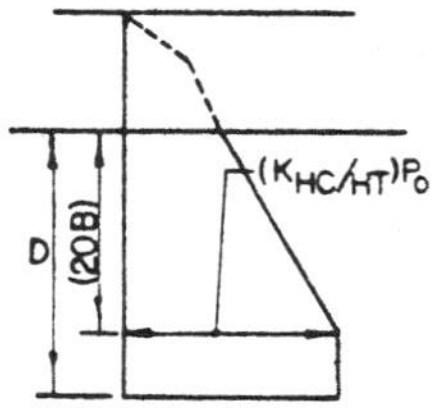


PRESSURE DIAGRAM
$K_{HC}P_0$ AND $K_{HT}P_0$

(A) ULTIMATE LOAD CAPACITY IN COMPRESSION

$$Q_{ult} = P_T N_q A_T + \sum_{H=H_0}^{H=H_0+D} (K_{HC})(P_0)(TAN\ \delta)(S)$$

WHERE $Q_{ult}$ = ULTIMATE LOAD CAPACITY IN COMPRESSION
$P_T$ = EFFECTIVE VERTICAL STRESS AT PILE TIP (SEE NOTE I)
$N_q$ = BEARING CAPACITY FACTOR (SEE TABLE, FIGURE I CONTINUED)
$A_T$ = AREA OF PILE TIP
$K_{HC}$ = RATIO OF HORIZONTAL TO VERTICAL EFFECTIVE STRESS ON SIDE OF ELEMENT WHEN ELEMENT IS IN COMPRESSION.
$P_0$ = EFFECTIVE VERTICAL STRESS OVER LENGTH OF EMBEDMENT, D (SEE NOTE I)
$\delta$ = FRICTION ANGLE BETWEEN PILE AND SOIL (SEE TABLE, FIGURE I CONTINUED)
S = SURFACE AREA OF PILE PER UNIT LENGTH

FOR CALCULATING $Q_{all}$, USE $F_S$ OF 2 FOR TEMPORARY LOADS, 3 FOR PERMANENT LOADS. (SEE NOTE 2)

(B) ULTIMATE LOAD CAPACITY IN TENSION

$$T_{ult} = \sum_{H=H_0}^{H=H_0+D} (K_{HT})(P_0)(TAN\ \delta)(S)(H)$$

WHERE: $T_{ult}$ = ULTIMATE LOAD CAPACITY IN TENSION, PULLOUT
$K_{HT}$ = RATIO OF HORIZONTAL TO VERTICAL EFFECTIVE STRESS ON SIDE OF ELEMENT WHEN ELEMENT IS IN TENSION

FOR CALCULATING $T_{all}$, USE $F_S = 3$ ON $T_{ult}$ PLUS THE WEIGHT OF THE PILE ($W_P$), THUS $T_{all} = \frac{T_{ult}}{3} + W_P$ (SEE NOTE 2)

NOTE-1: EXPERIMENTAL AND FIELD EVIDENCE INDICATE THAT BEARING PRESSURE AND SKIN FRICTION INCREASE WITH VERTICAL EFFECTIVE STRESS $P_0$ UP TO A LIMITING DEPTH OF EMBEDMENT, DEPENDING ON THE RELATIVE DENSITY OF THE GRANULAR SOIL AND POSITION OF THE WATER TABLE. BEYOND THIS LIMITING DEPTH (10B ± TO 40B ±) THERE IS VERY LITTLE INCREASE IN END BEARING, AND INCREASE IN SIDE FRICTION IS DIRECTLY PROPORTIONAL TO THE SURFACE AREA OF THE PILE. THEREFORE, IF D IS GREATER THAN 20 B, LIMIT $P_0$ AT THE PILE TIP TO THAT VALUE CORRESPONDING TO D = 20B.

NOTE-2: IF BUILDING LOADS AND SUBSURFACE CONDITION ARE WELL DOCUMENTED IN THE OPINION OF THE ENGINEER, A LESSER FACTOR OF SAFETY CAN BE USED BUT NOT LESS THAN 2.0 PROVIDED PILE CAPACITY IS VERIFIED BY LOAD TEST AND SETTLEMENTS ARE ACCEPTABLE.

FRICTION ANGLE - $\delta$

| PILE TYPE | $\delta$ |
|---|---|
| STEEL | 20° |
| CONCRETE | 3/4 $\phi$ |
| TIMBER | 3/4 $\phi$ |

(NAVFAC DM-7)

BEARING CAPACITY FACTORS - $N_q$

| $\phi$* (DEGREES) | 26 | 28 | 30 | 31 | 32 | 33 | 34 | 35 | 36 | 37 | 38 | 39 | 40 |
|---|---|---|---|---|---|---|---|---|---|---|---|---|---|
| $N_q$ (DRIVEN PILE DISPLACEMENT) | 10 | 15 | 21 | 24 | 29 | 35 | 42 | 50 | 62 | 77 | 86 | 120 | 145 |
| $N_q$ ** (DRILLED PIERS) | 5 | 8 | 10 | 12 | 14 | 17 | 21 | 25 | 30 | 38 | 43 | 60 | 72 |

EARTH PRESSURE COEFFICIENTS $K_{HC}$ AND $K_{HT}$

| PILE TYPE | $K_{HC}$ | $K_{HT}$ |
|---|---|---|
| DRIVEN SINGLE H-PILE | 0.5 - 1.0 | 0.3 - 0.5 |
| DRIVEN SINGLE DISPLACEMENT PILE | 1.0 - 1.5 | 0.6 - 1.0 |
| DRIVEN SINGLE DISPLACEMENT TAPERED PILE | 1.5 - 2.0 | 1.0 - 1.3 |
| DRIVEN JETTED PILE | 0.4 - 0.9 | 0.3 - 0.6 |
| DRILLED PILE (LESS THAN 24" DIAMETER) | 0.7 | 0.4 |

* LIMIT $\phi$ TO 28° IF JETTING IS USED

** (A) IN CASE A BAILER OR GRAB BUCKET IS USED BELOW GROUNDWATER TABLE, CALCULATE END BEARING BASED ON $\phi$ NOT EXCEEDING 28°.
(B) FOR PIERS GREATER THAN 24-INCH DIAMETER, SETTLEMENT RATHER THAN BEARING CAPACITY USUALLY CONTROLS THE DESIGN. FOR ESTIMATING SETTLEMENT, TAKE 50% OF THE SETTLEMENT FOR AN EQUIVALENT FOOTING RESTING ON THE SURFACE OF COMPARABLE GRANULAR SOILS.

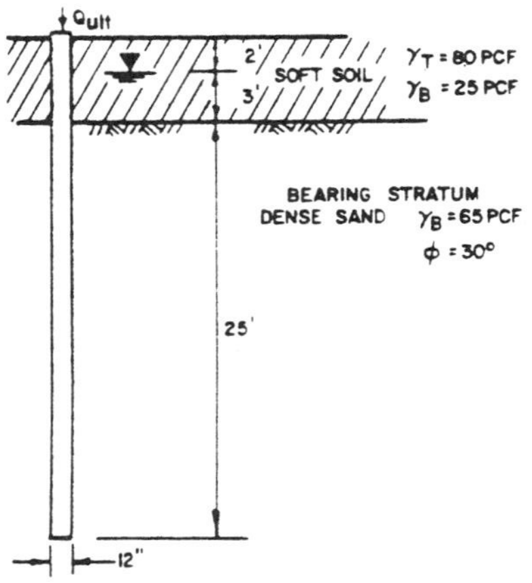


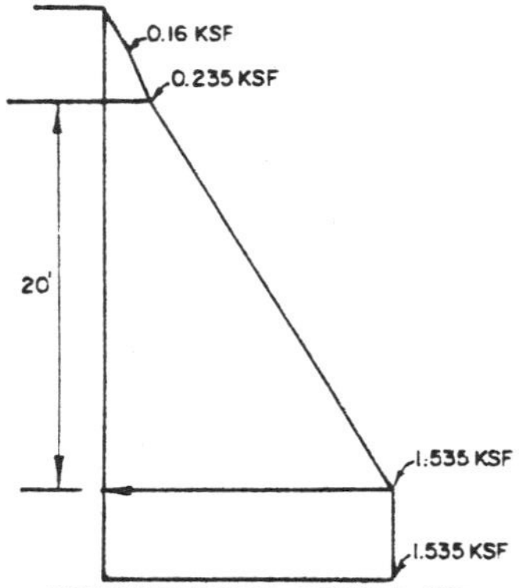


EFFECTIVE VERTICAL STRESS, $P_0$, FOR PILE DESIGN

FOR A 12" DIAMETER CLOSED END, DRIVEN PIPE PILE, CONCRETE FILLED, FIND $Q_{all}$ AND $T_{all}$ FOR A 30 FOOT LONG PILE.

$P_o$ MAX OCCURS AT 20B, OR 20' INTO BEARING STRATUM.

$\phi = 30°$, $N_q = 21$

$K_{HC} = 1.5$, $\delta = 20°$

$K_{HT} = 1.0$

$A_T = \pi \times 0.5^2 = 0.78$ SF

CIRCUM. AREA / lf $= 1 \times \pi = 3.14$ SF/lf

$$Q_{ult} = 1.535 \times 21 \times 0.78 + \left[ (1.5 \times (\frac{0.235+1.535}{2}) \times \tan 20 \times 20 \times 3.14) + (1.5 \times 1.535 \times \tan 20 \times 5 \times 3.14) \right]$$

$$= 25.14 + [30.34 + 13.16]$$

$$= 68.64 \text{ K}$$

FOR $F_S = 3$, $Q_{all} = \frac{68.64}{3} = 22.9$K

$$T_{ult} = 1.0 \times (\frac{0.235+1.535}{2}) \times \tan 20 \times 20 \times 3.14 + 1.0 \times 1.535 \times \tan 20 \times 5 \times 3.14$$

$$= 20.23 + 8.77$$

$$= 29.0\text{K}$$

$W_P = .117$ K/lf $\times 30' = 3.5$K

$T_{all} = \frac{29.0}{3} + 3.5 = 13.2$K

## ULTIMATE LOAD CAPACITY OF SINGLE PILE IN COHESIVE SOIL

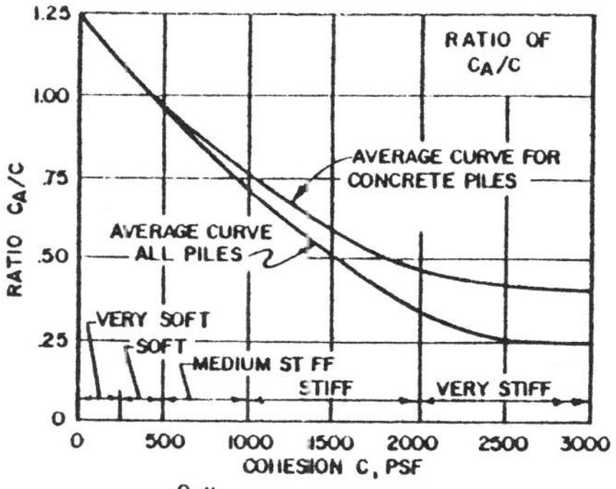


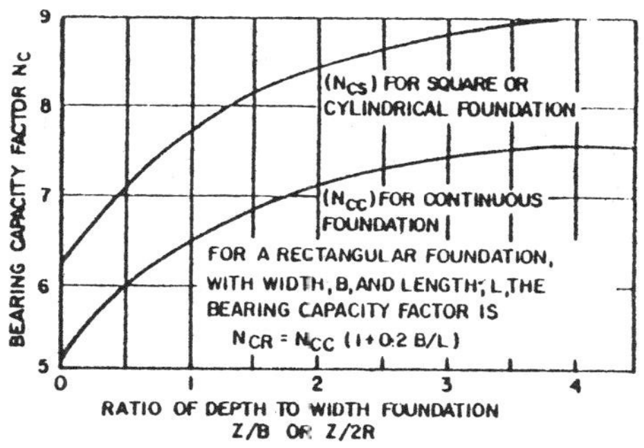


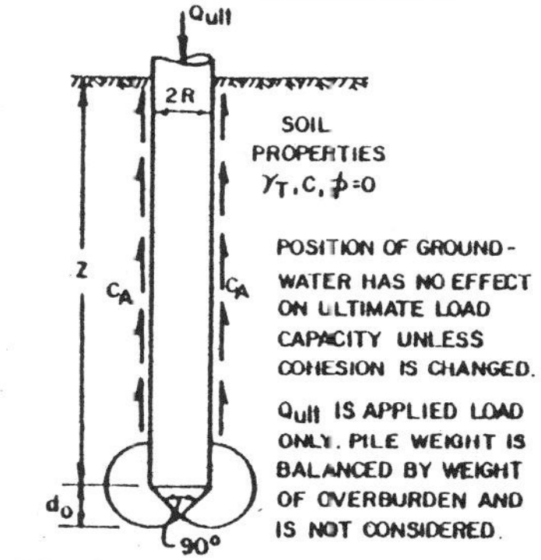


ULTIMATE LOAD CAPACITY IN COMPRESSION

$Q_{ult} = C \, (N_{CS}) \, \pi R^2 + C_A \, 2\pi RZ$
$(N_{CC})$

RECOMMENDED VALUES OF ADHESION

| PILE TYPE | CONSISTENCY OF SOIL | COHESION, C PSF | ADHESION, $C_A$ PSF |
|---|---|---|---|
| TIMBER AND CONCRETE | VERY SOFT | 0 - 250 | 0 - 250 |
| | SOFT | 250 - 500 | 250 - 480 |
| | MED. STIFF | 500 - 1000 | 480 - 750 |
| | STIFF | 1000 - 2000 | 750 - 950 |
| | VERY STIFF | 2000 - 4000 | 950 - 1300 |
| STEEL | VERY SOFT | 0 - 250 | 0 - 250 |
| | SOFT | 250 - 500 | 250 - 460 |
| | MED. STIFF | 500 - 1000 | 460 - 700 |
| | STIFF | 1000 - 2000 | 700 - 720 |
| | VERY STIFF | 2000 - 4000 | 720 - 750 |

ULTIMATE LOAD CAPACITY IN TENSION

$T_{ult} = C_A \, 2\pi RZ$

$T_{ult}$ UNDER SUSTAINED LOAD MAY BE LIMITED BY OTHER FACTORS, SEE TEXT.

<u>Pile Capacities</u>

## PILE CAPACITIES

The maximum pile capacities shall be as follows:

| | |
|---|---|
| Timber Piles (Untreated or Creosoted) | |
| 12" butt diameter | 24 tons |
| Concrete Piles (Equal Alternates) | |
| 14" Precast Concrete | |
| 14" Precast Prestressed Concrete | |
| 12" Metal Shell | 45 tons |
| Concrete Piles | |
| 14" Metal Shell | 55 tons |
| 16" Precast Concrete | 55 tons |
| Steel Point Bearing Piles | 9000 p.s.i. |

All piles under a foundation unit shall be driven to the same capacity. The required design capacity per pile shall be shown on the plans. The maximum pile capacities stated above shall be called for only if required by design calculations.

<u>Pile Spacing</u>

The minimum pile spacing is 3'-0" and the maximum pile spacing is limited to 3.5 times the effective footing thickness plus 1'-0" but not to exceed 8'-0".

| Capacity ∮ | Pile Material | Spacing of ● Friction Piles | Spacing of Point Bearing Piles | Edge Distance | Pile Size |
|---|---|---|---|---|---|
| 40T | Timber | 3'-3" | 3'-3" | 1'-6" | Spec. |
| 55T | Concrete<br>Steel | 4'-0" | 3'-0" | 1'-6" | ≠<br>HP 12 x 53 * |
| 70T | Concrete<br>Steel | 4'-6"<br>4'-6" | 3'-3"<br>3'-3" | 1'-5"<br>1'-6" | ╪<br>HP 12 x 53 |

Footnotes to Table 9.6.1 B.

* 10 BP 42 may be used if the pile is point bearing for this capacity.

≠ 12" Ø Min. for Concrete Filled Casing.
13" Ø Min. for Precast or Precast Prestressed.
14" Ø Min. Butt for Tapered.
10" or 12" Square Precast Prestressed.
12" Ø Min. for Hollow Prestressed Spun Piles

╪ 14" Ø Min. for Concrete Filled Casing.
16" Ø Min. for Precast or Precast Prestressed.

∮ Capacity shown is rated Basic (working) load value and shall include the effect of any downdrag forces.

● The Converse-Labare Formula (1973-AASHO 1.4.4G) need not be applied to pile values shown here.

## TYPE OF PILE DRIVING HAMMER

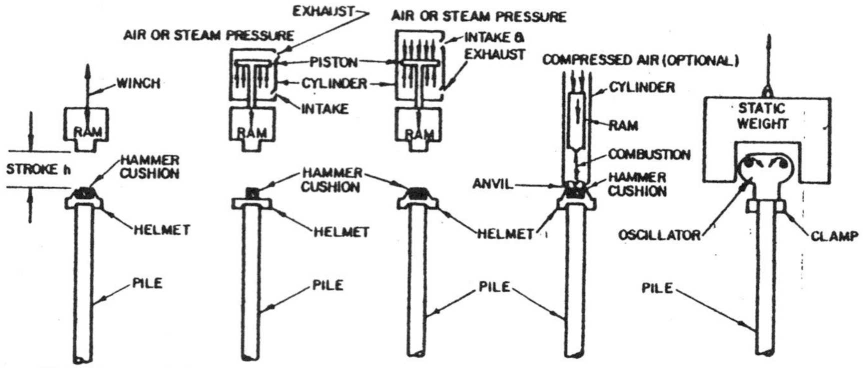


(A) DROP HAMMER (B) SINGLE-ACTING HAMMERS (C) DIFFERENTIAL AND DOUBLE-ACTING HAMMERS (D) DIESEL HAMMERS (E) VIBRATORY DRIVER

(1) Drop Hammer. Generally, it is only appropriate on small, relatively inaccessible jobs due to their slow rate of blows.

(2) Single Action Steam or Air Hammers. Blow rate is higher than drop hammer with maximum speeds generally ranging from about 35 to 60 blows per minute. Single acting hammers have an advantage over double acting hammers when driving piles in firm cohesive soils since the slower rate allows the soil and pile to relax before striking the next blow; thereby giving greater penetration per blow. In driving batter piles, single acting hammers can lose considerable energy due to the shortening fall and increases in friction.

(3) Double Acting Steam or Air Hammers. They provide a blow rate nearly double that of the single acting hammers and lose less energy driving batter piles. They are generally best suited for driving piles in granular soils or in soft clays. The energy per blow delivered by a double-acting hammer decreases rapidly as its speed of operation drops below the rated speed.

(4) Diesel Hammers. They have a relatively low fuel consumption, operate without auxiliary equipment, and can operate at low temperatures and are more efficient for driving batter piles. Maximum blow rates are about 35 to 60 blows per minute for single acting and about 80 to 100 blows per minute for double acting. Diesel hammers operate best in medium to hard ground; in soft ground the resistance and resulting compression may be too low to ignite the fuel.

(5) Vibratory Hammers. They are best suited to wet soils and low displacement piles but occasionally have been used successfully in cohesive soils and with high displacement piles. They can also be effective in extracting piles. When conditions are suitable, vibratory hammers have several advantages over impact hammers including lower driving vibrations, reduced noise, greater speed of penetration and virtually complete elimination of pile damage. However, there is the possibility that the pile may not be efficiently advanced, obstructions generally can not be penetrated, and there is no generally accepted method of determining ultimate pile capacity based on the rate of penetration.

## SUPPLEMENTARY PROCEDURE FOR PILE DRIVING PILE DRIVERS

| Method | Equipment and procedure utilized | Applicability |
|---|---|---|
| Means of reducing driving resistance above bearing stratum: | | |
| Temporary casing | Open end pipe casing driven and cleaned out. May be pulled later. | a. To drive through minor obstructions.<br>b. To minimize displacement.<br>c. To prevent caving or squeezing of holes.<br>d. To permit concreting of pile before excavation to subgrade of foundation. |
| Precoring | By continuous flight auger or churn drill, a hole is formed into which the pile is lowered. Pile is then driven to bearing below the cored hole. | a. To drive through thick stratum of stiff to hard clay.<br>b. To avoid displacement and heave of surrounding soil.<br>c. To avoid injury to timber and thin shell pipes.<br>d. To eliminate driving resistance in strata unsuitable for bearing. |
| Spudding | Heavy structural sections or closed end pipes are alternately raised and dropped to form a hole into which pile is lowered. Pile is then driven to bearing below the spudded hole. | a. To drive past individual obstruction<br>b. To drive through strata of fill with large boulders or rock fragments. |
| Jetting | Water, air, or mixture of both forced through pipe at high pressures and velocity, jets are sometimes built into piles. | a. Used to facilitate penetration, should not be permitted in fine grained, poorly draining soils where frictional support may be permanently destroyed. Piles should be driven to final embedment after jetting. |
| Means of increasing driving resistance in bearing stratum: | | |
| Upside down piles | Tapered piles, specifically timber, driven with large butt downward. | a. For end bearing timber piles, where it is necessary to minimize penetration into bearing stratum.<br>b. To avoid driving through to incompressible but unsuitable bearing material. |
| Lagging | Short timber or steel sections connected by bolting or welding to timber or steel pipes. | a. To increase frictional resistance along sides of pile.<br>b. To increase end bearing resistance when mounted near tip. |
| Means of overcoming obstructions: | | |
| Shoes and reinforced tips. | Metal reinforcing, such as bands and shoes for all types of piles. | a. To provide protection against damage of tip.<br>b. To provide additional cutting power. |
| Explosives | Drill and blast ahead of pile tip | a. To remove obstructions to open end piles under very severe conditions. |
| Preexcavation | Hand or machine excavation | a. Used for removal of obstruction close to ground surface. |
| Special equipment for advancing piles: | | |
| Jacking | Hydraulic or mechanical screw jacks are used to advance pile. Pile is built up in short, convenient lengths. | a. To be used instead of pile hammer where access is difficult.<br>b. To eliminate vibrations. |
| Vibration | High amplitude vibrators | a. Advantageous for driving in waterlogged sands and gravel.<br>b. Advantageous for driving sheetpiling. |
| Follower | Temporary filler section between hammer and pile top, preferably of same material as pile. | a. To drive pile top to elevation below reach of hammer or below water. |

# PILE DRIVERS

1. IMPACT PILE HAMMER

| Rated Energy** Kip - ft. | Make of Hammer* | Model No. | Types* | Blows per min | Stroke at Rated Energy | Weight Striking Parts Kips | Total Weight Kips |
|---|---|---|---|---|---|---|---|
| 180.0 | Vulcan | 060 | S-A | 62 | 36 | 60.0 | 121.0 |
| 130.0 | MKT | S-40 | S-A | 55 | 39 | 40.0 | 96.0 |
| 120.0 | Vulcan | 040 | S-A | 60 | 36 | 40.0 | 87.5 |
| 113.5 | S-Vulcan | 400C | Diff. | 100 | 16.5 | 40.0 | 83.0 |
| 97.5 | MKT | S-30 | S-A | 60 | 39 | 30.0 | 86.0 |
| 79.6 | Kobe | K42 | Dies. | 52 | 98 | 9.2 | 22.0 |
| 60.0 | Vulcan | 020 | S-A | 60 | 36 | 20.0 | 39.0 |
| 60.0 | MKT | S20 | S-A | 60 | 36 | 20.0 | 38.6 |
| 56.5 | Kobe | K32 | Dies. | 52 | 98 | 7.0 | 15.4 |
| 50.2 | S-Vulcan | 200C | Diff. | 98 | 15.5 | 20.0 | 39.0 |
| 48.7 | Vulcan | 016 | S-A | 60 | 36 | 16.2 | 30.2 |
| 48.7 | Raymond | 0000 | S-A | 46 | 39 | 15.0 | 23.0 |
| 44.5 | Kobe | K22 | Dies. | 52 | 98 | 4.8 | 10.6 |
| 42.0 | Vulcan | 014 | S-A | 60 | 36 | 14.0 | 27.5 |
| 40.6 | Raymond | 000 | S-A | 50 | 39 | 12.5 | 21.0 |
| 39.8 | Delmag | D-22 | Dies. | 52 | N/A | 4.8 | 10.0 |
| 37.5 | MKT | S14 | S-A | 60 | 32 | 14.0 | 31.6 |
| 36.0 | S-Vulcan | 140C | Diff. | 103 | 15.5 | 14.0 | 27.9 |
| 32.5 | MKT | S10 | S-A | 55 | 39 | 10.0 | 22.2 |
| 32.5 | Vulcan | 010 | S-A | 50 | 39 | 10.0 | 18.7 |
| 32.5 | Raymond | 00 | S-A | 50 | 39 | 10.0 | 18.5 |
| 32.0 | MKT | DE-40 | Dies. | 48 | 96 | 4.0 | 11.2 |
| 30.2 | Vulcan | OR | S-A | 50 | 39 | 9.3 | 16.7 |
| 26.3 | Link-Belt | 520 | Dies. | 82 | 43.2 | 5.0 | 12.5 |
| 26.0 | MKT | C-8 | D-A | 81 | 20 | 8.0 | 18.7 |
| 26.0 | Vulcan | 08 | S-A | 50 | 39 | 8.0 | 16.7 |
| 26.0 | MKT | S8 | S-A | 55 | 39 | 8.0 | 18.1 |
| 24.4 | S-Vulcan | 80C | Diff. | 111 | 16.2 | 8.0 | 17.8 |
| 24.4 | Vulcan | 8M | Diff. | 111 | N/A | 8.0 | 18.4 |
| 24.3 | Vulcan | 0 | S-A | 50 | 39 | 7.5 | 16.2 |
| 24.0 | MKT | C-826 | D-A | 90 | 18 | 8.0 | 17.7 |
| 22.6 | Delmag | D-12 | Dies. | 51 | N/A | 2.7 | 5.4 |
| 22.4 | MKT | DE-30 | Dies. | 48 | 96 | 2.8 | 9.0 |
| 24.4 | Kobe | K13 | Dies. | 52 | 98 | 2.8 | 6.4 |
| 19.8 | Union | K13 | D-A | 110 | 24 | 3.0 | 14.5 |
| 19.8 | MKT | 11B3 | D-A | 95 | 19 | 5.0 | 14.5 |
| 19.5 | Vulcan | 06 | S-A | 60 | 36 | 6.5 | 11.2 |
| 19.2 | S-Vulcan | 65C | Diff. | 117 | 15.5 | 6.5 | 14.8 |
| 18.2 | Link-Belt | 440 | Dies. | 88 | 36.9 | 4.0 | 10.3 |
| 16.2 | MKT | S5 | S-A | 60 | 39 | 5.0 | 12.3 |
| 16.0 | MKT | DE-20 | Dies. | 48 | 96 | 2.0 | 6.3 |
| 16.0 | MKT | C5 | Comp. | 110 | 18 | 5.0 | 11.8 |
| 15.1 | S-Vulcan | 50C | Diff. | 120 | 15.5 | 5.0 | 11.7 |
| 15.1 | Vulcan | 5M | Diff. | 120 | 15.5 | 5.0 | 12.9 |
| 15.0 | Vulcan | 1 | S-A | 60 | 36 | 5.0 | 10.1 |
| 15.0 | Link-Belt | 312 | Dies. | 100 | 30.9 | 3.8 | 10.3 |
| 13.1 | MKT | 10B3 | D-A | 105 | 19 | 3.0 | 10.6 |
| 12.7 | Union | 1 | D-A | 125 | 21 | 1.6 | 10.0 |
| 9.0 | Delmag | D5 | Dies. | 51 | N/A | 1.1 | 2.4 |
| 9.0 | MKT | C-3 | D-A | 130 | 16 | 3.0 | 8.5 |
| 9.0 | MKT | S3 | S-A | 65 | 36 | 3.0 | 8.8 |
| 8.8 | MKT | DE-10 | Dies. | 48 | 96 | 11.0 | 3.5 |
| 8.7 | MKT | 9B3 | D-A | 145 | 17 | 1.6 | 7.0 |
| 8.2 | Union | 1.5A | D-A | 135 | 18 | 1.5 | 9.2 |
| 8.1 | Link-Belt | 180 | Dies. | 92 | 37.6 | 1.7 | 4.5 |
| 7.2 | Vulcan | 2 | S-A | 70 | 29.7 | 3.0 | 7.1 |
| 7.2 | S-Vulcan | 30C | Diff. | 133 | 12.5 | 3.0 | 7.0 |
| 7.2 | Vulcan | 3M | Diff. | 133 | N/A | 3.0 | 8.4 |
| 6.5 | Link-Belt | 105 | Dies. | 94 | 35.2 | 1.4 | 3.8 |
| 4.9 | Vulcan | DGH900 | Diff. | 238 | 10 | .9 | 5.0 |
| 3.6 | Union | 3 | D-A | 160 | 14 | .7 | 4.7 |
| 3.6 | MKT | 7 | D-A | 225 | 9.5 | .8 | 5.0 |
| .4 | Union | 6 | D-A | 340 | 7 | .1 | .9 |
| .4 | Vulcan | DGH100A | Diff. | 303 | 6 | .1 | .8 |
| .4 | MKT | 3 | D-A | 400 | 5.7 | .06 | .7 |
| .3 | Union | 7A | D-A | 400 | 6 | .08 | .5 |

* Codes

MKT - McKiernan-Terry
S-Vulcan - Super-Vulcan
S-A - Single-Acting
D-A - Double-Acting
Diff. - Differential
Dies. - Diesel
Comp. - Compound

** In calculations of pile capacities by dynamic formula, effective energy delivered by hammer should be used. Hammer energy is affected by pressures used to operate the hammer, stroke rate, etc. Double-acting, differential, and diesel hammers may operate at less than rated energies; double-acting hammers deliver significantly less than rated energy when operated at less than rated speed. Consult manufacturers.

2. VIBRATORY DRIVERS

| Make | Model | Total Weight Kips | Available HP | Frequency Range cps | Force Kips***, Frequency cps |
|---|---|---|---|---|---|
| Foster (France) | 2-17 | 6.2 | 34 | 18-21 | |
| | 2-35 | 9.1 | 70 | 14-19 | 62/19 |
| | 2-50 | 11.2 | 100 | 11-17 | 101/17 |
| Menck (Germany) | MVB22-30 | 4.8 | 50 | | 48/ |
| | MVB65-30 | 2.0 | 7.5 | | 14/ |
| | MVB44-30 | 8.6 | 100 | | 97/ |
| Muller (Germany) | MS-26 | 9.6 | 72 | | |
| | MS-26D | 16.1 | 145 | | |
| (Russia) | BT-5 | 2.9 | 37 | 42 | 48/42 |
| | VPP-2 | 4.9 | 34 | 25 | 49/25 |
| | 100 | 4.0 | 37 | 13 | 44/13 |
| | VP | 11.0 | 80 | 6.7 | 35/7 |
| | VP-4 | 25.9 | 208 | | 198/ |
| Bodine (USA) | B | 22 | 1000 | 0-150 | 63/100 - 175/100 |
| Uraga (Japan) | VHD-1 | 8.4 | 40 | 16-20 | 43/20 |
| | VHD-2 | 11.9 | 80 | 16-20 | 86/20 |
| | VHD-3 | 15.4 | 120 | 16-20 | 129/20 |

*** Forces given are present maximums. These can usually be raised or lowered by changing weights in the oscillator.

## VALUES FOR DESIGN AND DRIVING OF STEEL BEAM PILES

IN TONS PER PILE

| SECTION NUMBER | WEIGHT PER FOOT | fs ALLOW. = 6,000 Lbs/Sq. In. | | | | | fs ALLOW. = 9,000 Lbs/Sq. In. | | | | |
|---|---|---|---|---|---|---|---|---|---|---|---|
| | | DESIGN | DRIVING | | | | DESIGN | DRIVING | | | |
| | | ALLOW. PER PILE | TEST PILES | | BEARING PILES | | ALLOW. PER PILE | TEST PILES | | BEARING PILES | |
| | | | A | B | C | D | | A | B | C | D |
| | Lbs | 100% | 180 % | 225 % | 150 % | 185 % | 100 % | 180 % | 225 % | 150 % | 185 % |
| HP14 * | 117 | 103 | 185 | 230 | 155 | 190 | 155 | 280 | 350 | 230 | 285 |
| | 102 | 90 | 160 | 205 | 135 | 165 | 135 | 245 | 305 | 205 | 250 |
| | 89 | 79 | 140 | 175 | 120 | 145 | 118 | 210 | 265 | 175 | 220 |
| | 73 | 64 | 115 | 145 | 95 | 120 | 97 | 175 | 215 | 145 | 180 |
| HP12 * | 74 | 65 | 120 | 145 | 100 | 120 | 98 | 175 | 220 | 145 | 180 |
| | 53 | 47 | 85 | 105 | 70 | 85 | 70 | 125 | 160 | 105 | 130 |
| HP10 * | 57 | 50 | 90 | 115 | 75 | 95 | 75 | 135 | 170 | 115 | 140 |
| | 42 | 37 | 65 | 85 | 55 | 70 | 56 | 100 | 125 | 85 | 105 |
| HP8 * | 36 | 32 | 55 | 70 | 50 | 60 | 48 | 80 | 105 | 70 | 90 |

(PENN DOT)

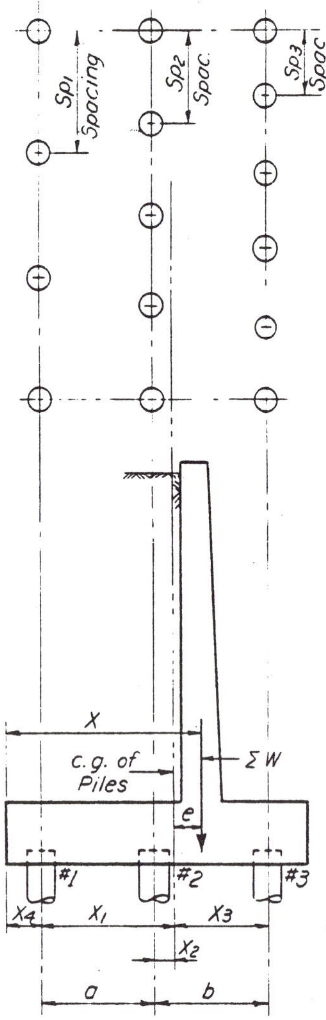


To locate c.g. of Piles $= X_1 = \dfrac{\left(\dfrac{a}{Sp_2} + \dfrac{a+b}{Sp_3}\right)}{A}$

The eccentricity of the resultant $= X - (X_1 + X_4) = e$

The moment of the resultant $= e \Sigma W$

$$S = \frac{P}{A} \pm \frac{MC}{I}$$

In this case:

$S$ = Pile Load

$P = \Sigma W$

$A$ = Area = portion of piles per ft.:

$$= \frac{1}{Sp_1} + \frac{1}{Sp_2} + \frac{1}{Sp_3}$$

$C$ = distance of pile from c.g. of Piles:

$= X_1, X_2$ or $X_3$

$I$ = Moment of inertia of Piles $= \Sigma X^2$

$$= \frac{X_1^2}{Sp_1} + \frac{X_2^2}{Sp_2} + \frac{X_3^2}{Sp_3}$$

Load on Pile #1 $= S_1 = \dfrac{\Sigma W}{A} - \dfrac{e \Sigma W}{I} \cdot X_1$

$$* \; S_2 = \frac{\Sigma W}{A} - \frac{e \Sigma W}{I} \cdot X_2$$

$$S_3 = \frac{\Sigma W}{A} + \frac{e \Sigma W}{I} \cdot X_3$$

Determine magnitude and position of resultant in usual manner.

(ILL DOT)

*Equation is for c.g. as shown. If c.g. falls between Piles #1 & #2 equation becomes:~

$$S_2 = \frac{\Sigma W}{A} + \frac{e \Sigma W}{I} \cdot X_2$$

METHOD OF COMPUTING PILE LOADINGS

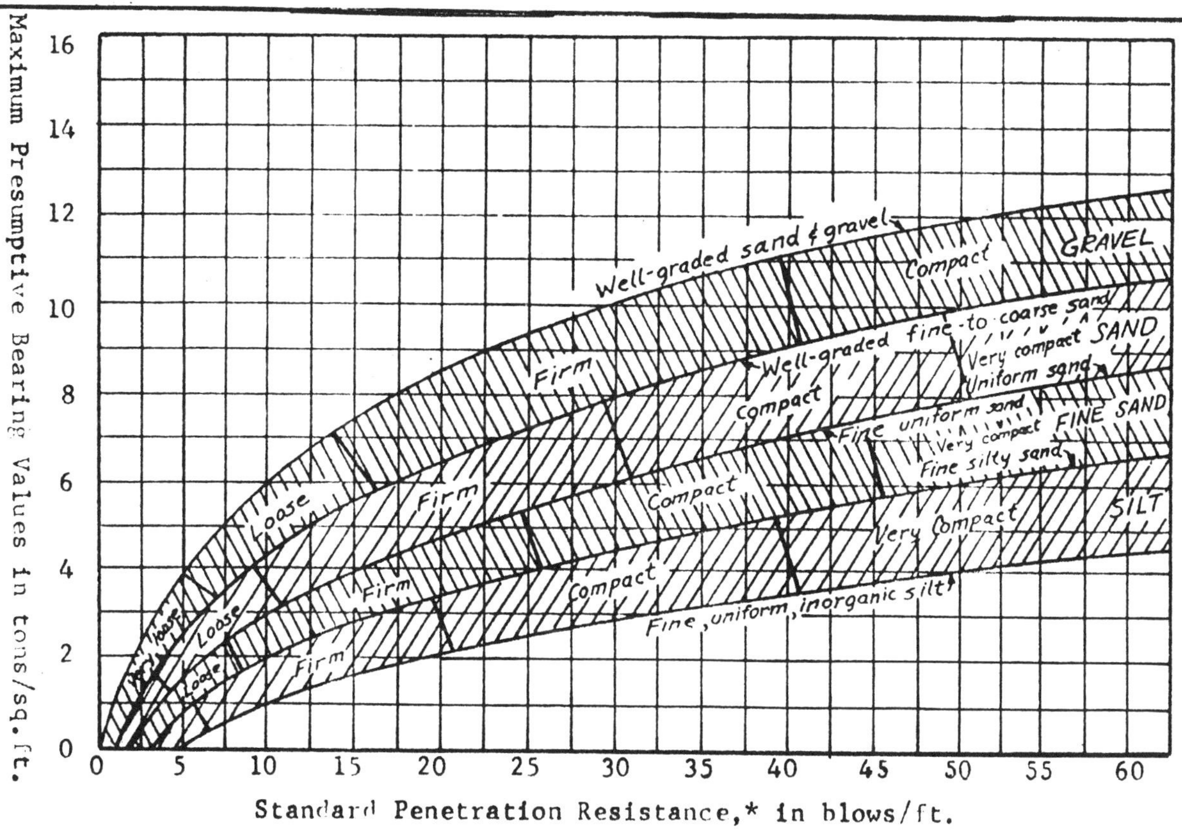


*Number of blows of 140-lb. pin-guided weight falling 30 in. per blow required to drive a split-barrel sample spoon with a 2-in. outside diameter 12 in.

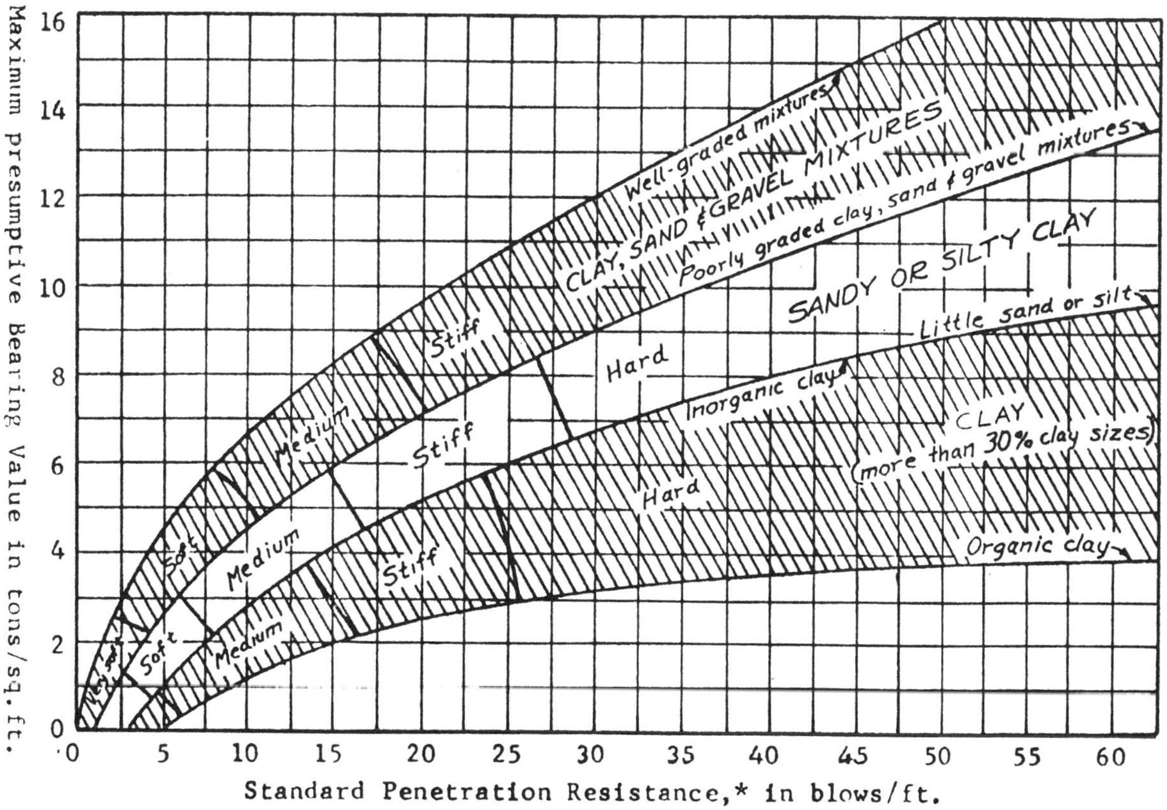


*Number of blows of 140-lb. pin-guided weight falling 30 in. per blow required to drive a split-barrel sample spoon with a 2-in. outside diameter 12 in.

(WIS DOT) **BEARING CAPACITY VERSUS STANDARD PENETRATION RESISTANCE**

| STATE OF FLORIDA DEPARTMENT OF TRANSPORTATION | ITEM NO. 844 | Page 1 of 3 |
|---|---|---|
| BRIDGE REPAIR MANUAL | UNIT: Each Pile | |

ITEM NAME: Cathodic Protection (Zinc Anodes - Small)

DESCRIPTION: Attachment of small zinc anodes to steel H piles or reinforced concrete piles below water by use of a clamping device.

APPLICATION: To abate corrosion of steel H piles or reinforced concrete piles located in salt or brackish water when less than 8 lineal feet of pile is exposed. Applicable for all sizes of steel and reinforced concrete piles.

TRAFFIC CONTROL AND SAFETY: Refer to Manual on Uniform Traffic Control & Safe Practices.

MATERIALS: Zinc Anodes - 4, 7 or 10 pound - M.S. #964

CONSTRUCTION METHOD:

Steel H Pile

1. Clean area on steel large enough to accommodate the clamp and hardware.
2. Place two anodes on opposite flanges 12" below low water line and then alternate one anode on each flange every 18 inches to the mud line. If the pile has concrete jackets in the splash zone, place the first two anodes 6" below the jacket in the water. (See Drawings on Pages 2 & 3.)

Reinforced Concrete Pile

1. Clean area on exposed reinforcing bar large enough to accommodate the clamp and hardware.
2. Place one anode on exposed reinforcing bar for each six lineal feet of pile in contact with brackish or salt water.

NOTE: The normal application consists of using 7-pound anodes. The use of 4 or 10 pound anodes requires additional site testing. Less brackish water or fast moving sea water will affect the two-year life expectancy requiring a change in the size of the anode. Contact the Bridge Maintenance and Operations Engineer in Orlando for coordinating design changes.

| STATE OF FLORIDA DEPARTMENT OF TRANSPORTATION | ITEM NO. 844 | Page 2 of 3 |
|---|---|---|
| BRIDGE REPAIR MANUAL | UNIT: Each Pile | |

ITEM NAME: Cathodic Protection (Zinc Anodes - Small)

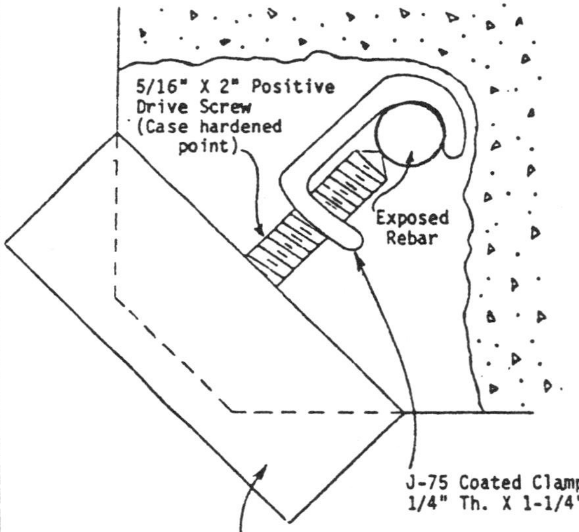


PLAN VIEW

Zinc Anode Assembly & Rebar Attachment

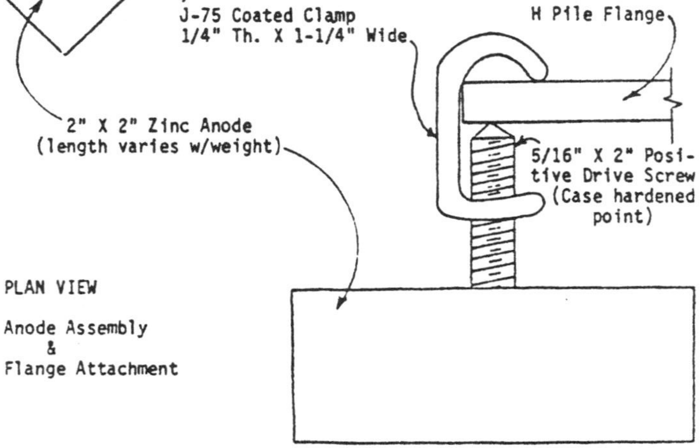


PLAN VIEW

Zinc Anode Assembly & H Pile Flange Attachment

STATE OF FLORIDA DEPARTMENT OF TRANSPORTATION | ITEM NO. 844 | Page 3 of 3

BRIDGE REPAIR MANUAL | UNIT: Each Pile

ITEM NAME: Cathodic Protection (Zinc Anodes - Small)

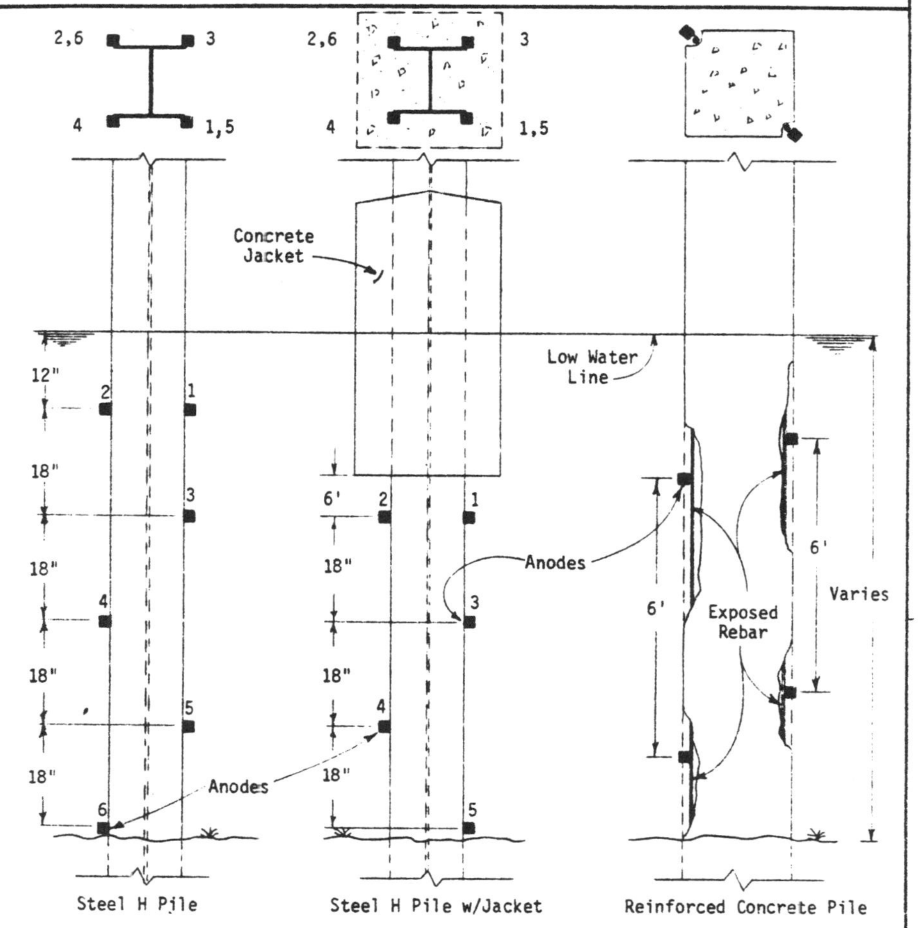


STATE OF FLORIDA DEPARTMENT OF TRANSPORTATION | MATERIAL NO. 964 | Page 1 of 1

BRIDGE REPAIR MANUAL | MATERIAL SPECIFICATION SECTION

MATERIAL NAME: Aluminum and Zinc Anodes

All Zinc Anodes (4, 7, 10 and 48 lb.)

Zinc anodes shall be High-Amp, zinc-anode type and shall conform to specifications MIL-A-18001 and ASTM B418-67, Type I. Composition will be within the limits listed below:

| | |
|---|---|
| Aluminum | 0.1 to 0.3 percent |
| Cadmium | 0.025 to 0.06 percent |
| Iron | 0.005 percent maximum |
| Zinc | Hi-Grade - Balance |

NOTE: The above zinc known as High-Amp zinc should not be confused with Hi-Purity zinc.

Zinc Anodes (4, 7 and 10 lb.)

ASARCO Hi-Amp anodes complete with epoxy coated J-75 clamp. Anodes contain threaded cone-pointed, hardened set screw, extending 2" beyond anode surface.

Weight ____________

(Specify 4, 7 or 10 lb. size.)

Zinc Anodes (48 lb. size)

ASARCO Hi-Amp zinc anodes, Series TZ-50-W, complete with clamp assemblies. (These anodes are comparable to ASARCO Aluminum, Type MK-IV-C-43, Hi-Amp 21 lb. anodes.)

Aluminum Anodes

ASARCO Aluminum anodes, Type MK-IV-C-43, Hi-Amp, 21 pounds each with #3 clamps and mounting hardware.

Anodes to be "High-Amp Output" containing aluminum, zinc, indium and cadmium.

(These anodes are comparable to the 48 lb. ASARCO Hi-Amp zinc anodes, Series TZ-50-W.)

| STATE OF FLORIDA DEPARTMENT OF TRANSPORTATION BRIDGE REPAIR MANUAL | ITEM NO. 840 | Page 2 of 3 |
|---|---|---|
| | UNIT: Linear Foot | |

ITEM NAME: Integral Pile Jacket - Type VII & VIII

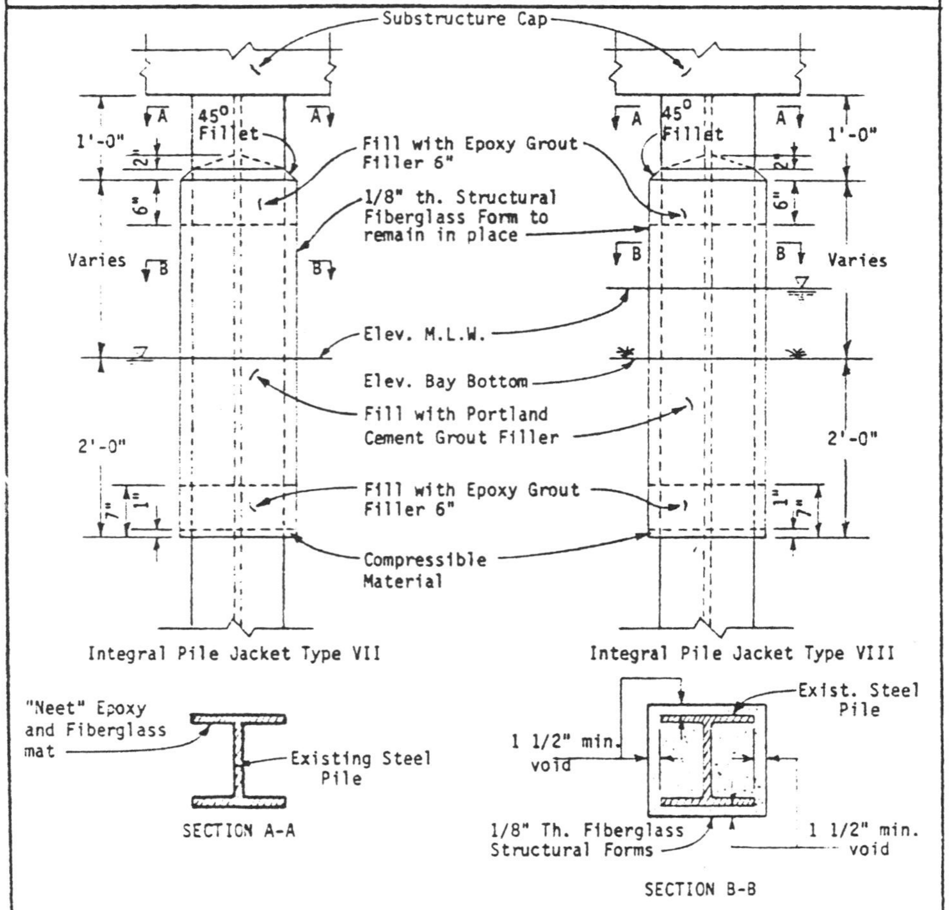


| STATE OF FLORIDA DEPARTMENT OF TRANSPORTATION BRIDGE REPAIR MANUAL | MATERIAL NO. 457-1 | Page 1 of 2 |
|---|---|---|
| | MATERIAL SPECIFICATION SECTION | |

MATERIAL NAME: Fiberglass Forms

The forms shall be composed of a durable inert corrosion resistant material with an interlocking joint along one side which will permit the form to be assembled and sealed in place around the pile. The forms shall be fabricated from fiberglass and polyester resins. The form dimensions shown in the plans are minimum dimensions permitted. The minimum thickness of the forms shall be 1/8 inch unless otherwise shown on the plans. Upon opening to place around a pile, the form shall be capable of returning to its original shape without assistance or damage. The inside face of the jacket shall have no bond inhibiting agents in contact with the cementitious or epoxy grouts. The forms shall be provided with bonded-on, non-corrosive standoffs, which will maintain the forms in the required positions. The inside surface of the forms shall be sandblasted or scored with an abrasive material to provide a texture equal to a sandblasted or scored surface. The forms shall be equipped with a compressible sealing strip at the bottom which will effectively seal the annular space between the pile and form.

The material furnished must meet the following physical property requirements:

(a) Water Absorption (ASTM D570)....................................1% Max.
(b) Ultimate Tensile Strength (ASTM-D638)*.....................9,000 psi Min.
(c) Flexural Strength (ASTM-D638)*............................16,000 psi Min.
(d) Flexural Modulus of Elasticity (ASTM-D790)...............700,000 psi Min.
(e) IZOD Impact (ASTM-D256)....................15 lb/inch minimum (unnotched)
(f) Barcol Hardness (ASTM-D2583)....................................45 Min.
(g) Color.........................................Similar to Federal Color Standard No. 595 - Table VIII - Shade No. 36622 - The color shall be integral in the form material.
(h) Accelerated Weathering......The fabricated form material shall be subjected to a 1 500 hour exposure test in a Twin-Carbon-ARC-Weather-Ometer (ASTM G-23, Type D) at an operating temperature of 145°F. Said test to be made at twenty minute cycles consisting of seventeen minutes of light and three minutes of water spray plus light. At the end of the exposure test the exposed samples shall not show any chipping, flaking or peeling. The test panels shall be prepared from the materials meeting the physical property requirements above and they shall be in accordance with the manufacturer's recommendations.

*On original specimen whose flat surfaces are not machined to disturb the fiberglass.

PILE JACKETING

| STATE OF FLORIDA DEPARTMENT OF TRANSPORTATION<br>BRIDGE REPAIR MANUAL | ITEM NO. 837 | Page 1 of 2 |
|---|---|---|
| | UNIT: Each | |

ITEM NAME: Steel H Pile Repair

DESCRIPTION: Reinforcing deteriorated steel H pile by bolting steel channel to weakened section.

APPLICATION: This repair method is used to increase the structural strength of a deteriorated section of H pile. This method of repair is applicable when pile replacement is not practical. Considered short term repair unless pile is jacketed.

TRAFFIC CONTROL AND SAFETY: Refer to Manual on Uniform Traffic Control & Safe Practices.

MATERIALS: 3/4" Ø High strength bolts/nuts & washers - M.S. #962-2
Steel channel (pile size 2")--------------- M.S. #962
Self-curing inorganic zinc-rich coating --- APL
(if above water)

CONSTRUCTION METHOD:

1. Clean damaged pile.
2. Locate extreme limits of deteriorated section. Channel section shall be 18" longer than the distance between these limits.
3. Thoroughly clean area to which channel is to be bolted.
4. Clamp channel section in place against pile.
5. Locate and drill holes through channel and pile for high strength bolts.
6. Place bolts and secure.
7. Remove clamps.
8. If pile repair is above water, coat in accordance with Item No. 817. If underwater, protect with zinc anodes.

NOTE: For permanent repair, steel piles should be encased with a concrete jacket.

Damaged pile requiring reinforcement

| STATE OF FLORIDA DEPARTMENT OF TRANSPORTATION<br>BRIDGE REPAIR MANUAL | ITEM NO. 837 | Page 2 of 2 |
|---|---|---|
| | UNIT: Each | |

ITEM NAME: Steel H Pile Repair

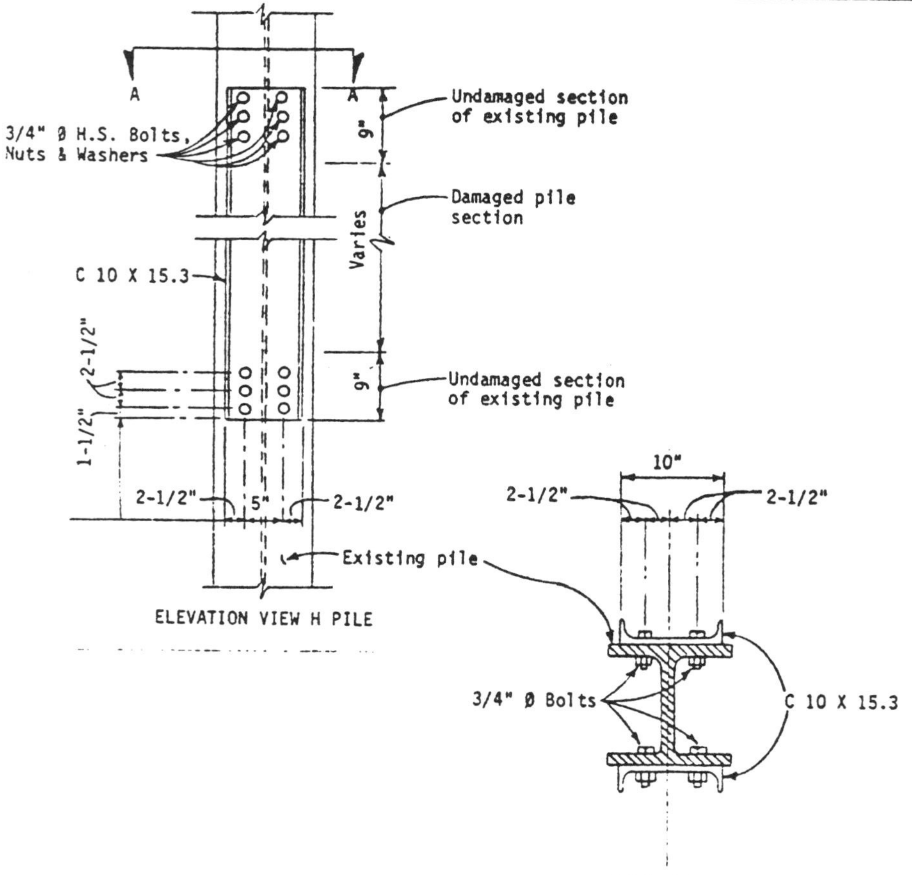


SECTION A-A

## M.1 CULVERT HYDRAULIC

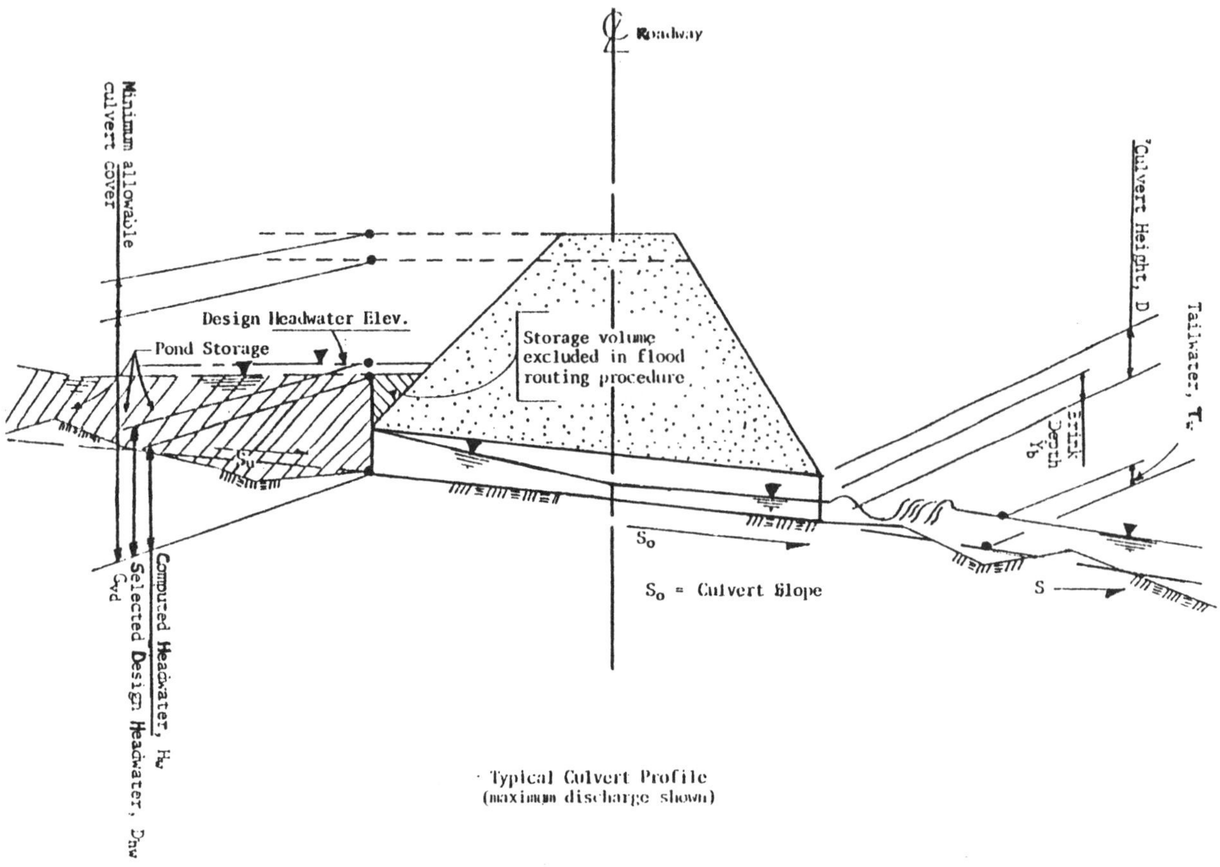


Typical Culvert Profile
(maximum discharge shown)

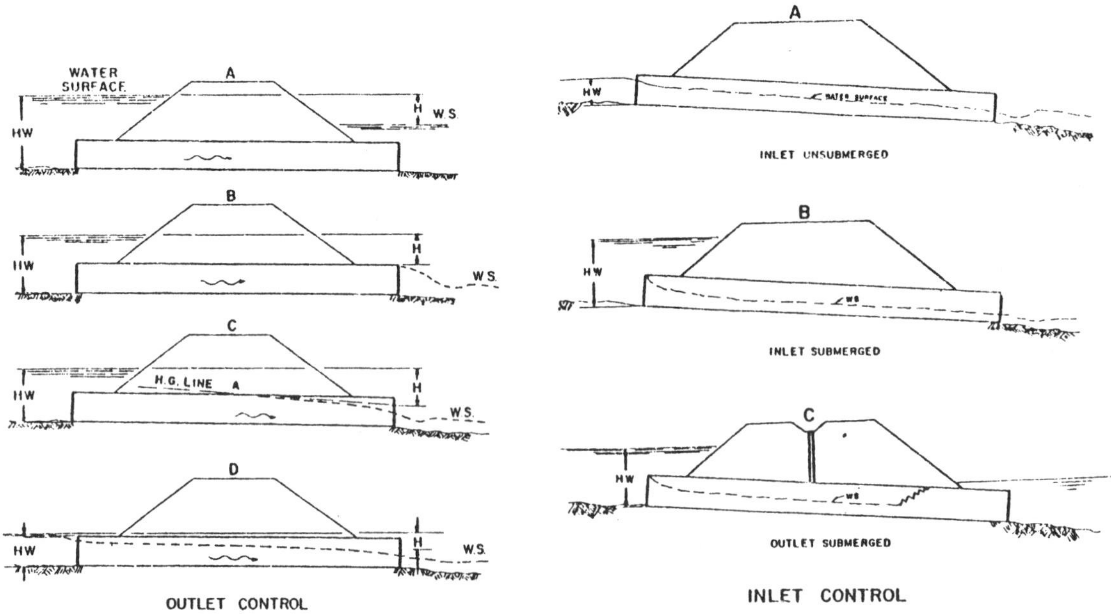

# NOMOGRAPH FOR SOLUTION OF MANNING EQUATION

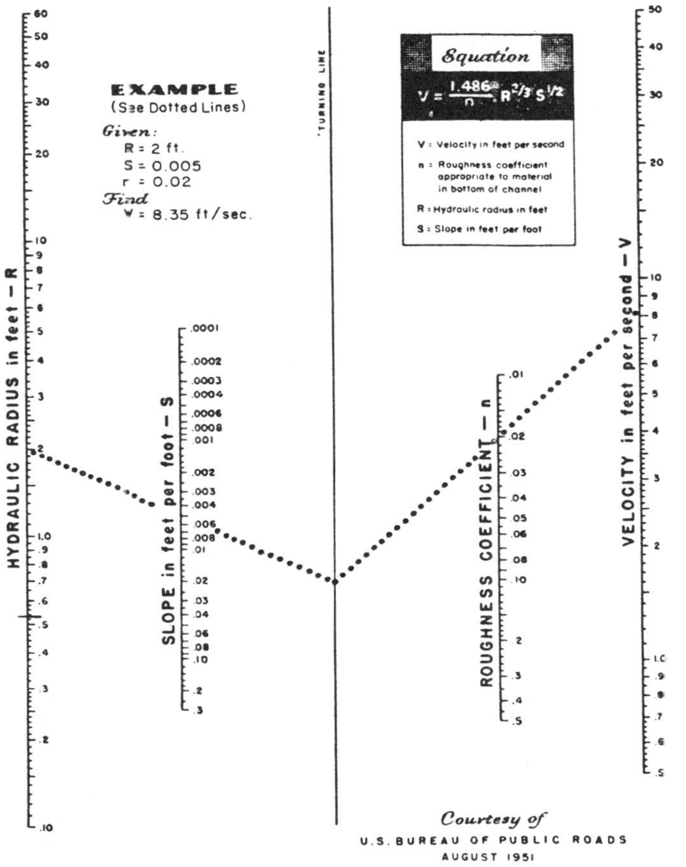


**Recommended Permissible Velocities for Unlined Channels**

| Type of Material in Excavation Section | Permissible Velocity (Feet per Second) | |
|---|---|---|
| | Intermittent Flow | Sustained Flow |
| Fine Sand (Noncolloidal) | 2.5 | 2.5 |
| Sandy Loam (Noncolloidal) | 2.5 | 2.5 |
| Silt Loam (Noncolloidal) | 3.0 | 3.0 |
| Fine Loam | 3.5 | 3.5 |
| Volcanic Ash | 4.0 | 3.5 |
| Fine Gravel | 5.0 | 4.0 |
| Stiff Clay (Colloidal) | 6.0 | 4.5 |
| Graded Material (Noncolloidal) | | |
| Loam to Gravel | 6.5 | 5.0 |
| Silt to Gravel | 7.0 | 5.5 |
| Gravel | 7.5 | 6.0 |
| Coarse Gravel | 8.0 | 6.5 |
| Gravel to Cobbles (Under 6 Inches) | 9.0 | 7.0 |
| Gravel and Cobbles (Over 8 Inches) | 10.0 | 8.0 |

**Standards for Channel Linings**

| Mean Velocity Feet Per Second | Thickness of Lining (Inches) | | Minimum Reinforcement |
|---|---|---|---|
| | Sides | Bottom | |
| Asphaltic Pavement | | | |
| Less Than 8 | 2 | 2-3 | None |
| 8-10 | 3 | 3-4 | None |
| Portland Cement Concrete (Poured or Pneumatically Applied) | | | |
| Less Than 10 | 2-3½ | 2-4 | 6″ x 6″ 10 Ga. Wire Mesh |
| 10 to 16 | 4-5 | 4-6 | #3 Bars at 15″ centers both ways |
| 16 or More | 6 or More | 7 or More | #3 Bars at 12″ centers both ways |

**7-833.9 Channel Linings**

(1) *Purpose and Type.* The main purposes of channel linings are:

(a) To prevent erosion damage,

(b) Increase velocity to prevent excessive sedimentation, and

(c) Increase capacity.

Two types of linings are commonly used: asphaltic pavement and portland cement concrete. Asphaltic pavement is used mainly for temporary construction. It cannot withstand hydrostatic pressure and is not recommended for steep side slopes. Portland cement concrete is used for permanent construction under all conditions of service. Seal coats are not recommended for a permanent installation.

(2) *Standards for Thickness of Lining.* Table 7-833.9 gives approximate channel lining thicknesses for both asphaltic pavement and portland cement concrete. The given side thicknesses normally apply to slopes flatter than the angle of repose of the soil and thus do not allow for earth pressure. In favorable soils and in the absence of hydrostatic pressure, side linings are often placed on a 1:1 slope. Under unusually favorable conditions, the slope may be as steep as 3/4:1. In situations where neither hydrostatic pressure nor wear are factors, the floor thickness should be the same as the sides. Under unfavorable conditions, where it is necessary to use a slope equal to or steeper than the angle of repose of the soil, the side lining either may be thickened or designed as a sloping retaining wall.

Portland cement concrete linings may be poured or placed pneumatically.

(3) *Notes on Design*

(a) Inflow From the Sides—Where the channel intercepts a surface flow from the sides, the following measures are essential:

- Carry the lining to an elevation slightly below ground level;
- To prevent undermining, place a cutoff lip or wall at the top of the lining; and
- When pipes discharge into the channel, cut them flush with the channel lining.

(b) Drainage—Whenever hydrostatic pressure is anticipated, both weep holes and a drainage system behind the side lining are required.

(4) *Slope Paving.* The object of this type of channel paving is bank protection, usually on one side of the channel. To guard against undercutting the ends, the slope paving must be turned into the bank a sufficient depth to provide good anchorage, and the bottom of the footing must be below the scour line.

## TABLE 1
## MANNING'S n FOR CULVERTS, CHANNELS AND NATURAL STREAMS

| | Manning's n Range | Manning's n Recommended Design Value |
|---|---|---|
| 1. Closed Conduits | | |
| a. Concrete pipe or pipe-arch | 0.011 - 0.013 | 0.012 |
| b. Concrete friction pipe | | 0.024 |
| c. Corrugated metal pipe or pipe arch (Small corrugation - 2.5" x 0.5") | | |
| (1) plain or fully coated | | 0.024 |
| (2) paved invert - full flow | | |
| 25% corcumference paved | | 0.021 |
| 50% circumference paved | | 0.018 |
| d. Structural plate pipe or pipe arch | 0.030 - 0.033 | 0.030 - 0.033 |
| e. Monolithic concrete (box culvert) | 0.012 - 0.014 | 0.012 |
| f. Vitrified clay pipe | 0.012 - 0.014 | 0.012 |
| 2. Open Channels - Lined | | |
| a. Concrete - smooth forms or trowelled | 0.012 - 0.014 | 0.012 |
| b. Asphalt | 0.013 - 0.016 | |
| (1) Smooth | | 0.013 |
| (2) Rough | | 0.016 |
| 3. Open Channels - Excavated (Straight alignment & natural lining) | | |
| a. Earth - uniform section | | |
| (1) clean - new to weathered | 0.016 - 0.020 | |
| (2) with short grass, few weeds | 0.022 - 0.027 | |
| (3) in gravelly soil, clean | 0.022 - 0.025 | |
| b. Earth - fairly uniform section | | |
| (1) no vegetation | 0.022 - 0.025 | |
| (2) grass, some weeds | 0.025 - 0.030 | |
| (3) dense weeds or plants in deep channel | 0.030 - 0.035 | |
| (4) sides clean, gravel bottom | 0.025 - 0.030 | |
| (5) sides clean, cobble bottom | 0.030 - 0.040 | |
| c. Dragline excavated or dredged | | |
| (1) no vegetation | 0.028 - 0.033 | |
| (2) light brush on banks | 0.035 - 0.050 | |
| d. Channels not maintained, weeds & brush uncut | | |
| (1) dense weeds, high as flow depth | 0.08 - 0.12 | |
| (2) clean bottom brush on sides | 0.05 - 0.08 | |
| (3) same, highest stage of flow | 0.07 - 0.11 | |
| (4) dense brush, high stage | 0.10 - 0.14 | |
| 4. Natural Stream Channels | | |
| a. Minor streams (surface width at flood stage < 100 feet) | | |
| (1) Fairly regular section | | |
| (a) some grass and weeds, little or no brush | 0.030 - 0.035 | |
| (b) dense growth of weeds, depth of flow materially greater than weed height | 0.035 - 0.05 | |
| (c) some weeds, light brush on banks | 0.035 - 0.05 | |
| (d) some weeds, heavy brush on banks | 0.05 - 0.07 | |
| (e) some weeds, dense willows on banks | 0.06 - 0.08 | |
| (f) for trees within channel with branches submerged at high stage increase all above values by | 0.01 - 0.02 | |
| (2) Irregular section, with pools, slight channel meander | | |
| (a) Channels (a) to (e) above, increase all values about | 0.01 - 0.02 | |
| (3) Mountain streams, no vegetation in channel, banks usually steep, trees and brush along banks submerged at high stage. | | |
| (a) bottom: gravel, cobbles, and few boulders | 0.04 - 0.05 | |
| (b) bottom: cobbles with larger boulders | 0.05 - 0.07 | |
| b. Flood plains (adjacent to natural streams) | | |
| (1) Pasture, no brush | | |
| (a) short grass | 0.30 - 0.035 | |
| (b) high grass | 0.035 - 0.05 | |
| (2) Cultivated areas | | |
| (a) no crop | 0.03 - 0.04 | |
| (b) mature row crops | 0.035 - 0.045 | |
| (c) mature field crops | 0.04 - 0.05 | |
| (3) Heavy weeds, scattered brush | 0.05 - 0.07 | |
| (4) Light brush and trees | | |
| (a) winter | 0.05 - 0.06 | |
| (b) summer | 0.06 - 0.08 | |
| (5) Medium to dense brush | | |
| (a) winter | 0.07 - 0.11 | |
| (b) summer | 0.10 - 0.16 | |
| (6) Dense willows, summer, not bent over by current | 0.15 - 0.20 | |
| (7) Cleared land with tree stumps, 100 - 150 per acre | | |
| (a) no sprouts | 0.04 - 0.05 | |
| (b) with heavy growth of sprouts | 0.06 - 0.08 | |
| (8) Heavy stand of timber, a few down trees, little undergrowth | | |
| (a) flood depth below branches | 0.10 - 0.12 | |
| (b) flood depth reaches branches | 0.12 - 0.16 | |

c. Major streams (surface width at flood stage 100 feet) - Roughness coefficient is usually less than for minor streams of similar description on account of less effective resistance offered by irregular banks or vegetation on banks. Values of "n" may be somewhat reduced. The value of "n" for larger streams of most regular sections, with no boulders or brush, may be in the range of 0.028 to 0.033.

## TABLE 2
## Entrance Loss Coefficients

Coefficient $k_e$ to apply to velocity head $\frac{V^2}{2g}$ to determine loss of head at entrance to a structure, such as a culvert or conduit, operating full or partly full with control at the outlet.

$$\text{Entrance head loss } H_e = k_e \frac{V^2}{2g}$$

| Type of Structure and Design of Entrance | Coefficient $k_e$ |
|---|---|
| **Pipe, Concrete** | |
| Projecting from fill, socket end | 0.2 |
| Projecting from fill, sq. cut end | 0.5 |
| Headwall or headwall and wingwalls | |
| Socket end of pipe | 0.2 |
| Square edge | 0.5 |
| Rounded (radius = 1/12 D) | 0.2 |
| Mitered to conform to fill slope | 0.7 |
| * End-Section conforming to fill slope (Apron) | 0.5 |
| **Pipe, or Pipe-Arch, Corrugated Metal** | |
| Projecting from fill (no headwall) | 0.9 |
| Headwall or headwall and wingwalls | |
| Square-edge | 0.5 |
| Mitered to conform to fill slope | 0.7 |
| * End-section conforming to fill slope (Apron) | 0.5 |
| **Box, Reinforced Concrete** | |
| Headwall parallel to embankment (no wingwalls) | |
| Square-edged on 3 edges | 0.5 |
| Rounded on 3 edges to radius of 1/12 barrel dimension | 0.2 |
| Wingwalls at 30° to 75° to barrel | |
| Square-edge at crown | 0.4 |
| Crown edge rounded to radius of 1/12 barrel dimension | 0.2 |
| Wingwalls at 10° to 30° to barrel | |
| Square-edge at crown | 0.5 |
| Wingwalls parallel (extension of sides) | |
| Square-edged at crown | 0.7 |

*Note: "End Section conforming to fill slope", made of either metal or concrete, are the sections commonly available from manufacturers. From limited hydraulic tests they are equivalent in operation to a headwall in both inlet and outlet control.

MnDOT

# CULVERT DESIGN

## 36.1 GENERAL

Box culverts are reinforced concrete closed rigid frames which must support vertical earth and truck loads and lateral earth pressure. They may be either single or multi-cells. The most common usage is to carry water under roadways but they are frequently used for pedestrian or cattle underpasses. The minimum size for pedestrian underpasses is 8' high by 5' wide. Minimum size for cattle is 6' high by 5' wide.

Typical sections for the most frequently used box culverts are shown below.

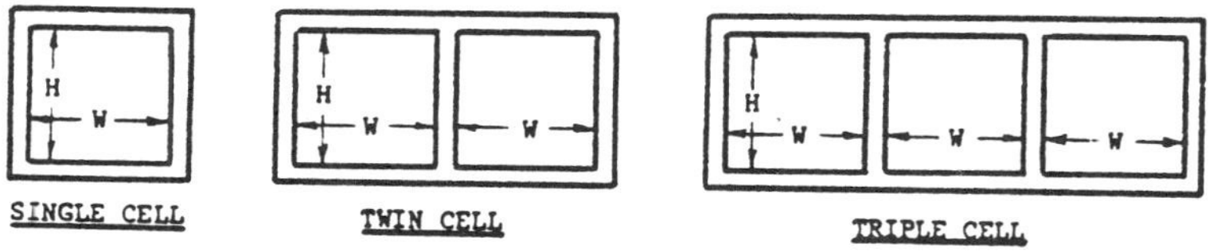


Fig. 36.1

Hydraulic and other requirements at the site determines the required height and area of the box.

Hydraulic design of box culverts is described in Chapter 8. Once the required height and area is determined the selection of a single or multi-cell box is determined entirely from economics. Barrel lengths are computed to the nearest 6 inches. For multi-cell culverts the cell widths are kept equal.

## 36.2 DEAD LOADS AND EARTH PRESSURE

The weight of soil above buried structures is taken as 120 pounds per cubic foot. For the lateral pressure from the soil an equivalent fluid pressure of 30 pounds per cubic foot is used.

When computing the maximum positive moment in the top and bottom exterior span, use an equivalent fluid pressure of 15 pounds per cubic foot. Also for multi-cell culverts use 15 pounds per cubic foot when computing the maximum negative moment over the interior wall in the end cells.

Earth pressures or loads on culverts may be computed as the weight of earth directly above the structure. A load factor of 1.3 is used for vertical earth pressure and dead loads and 1.69 (1.3 x 1.3) is used for lateral earth pressure.

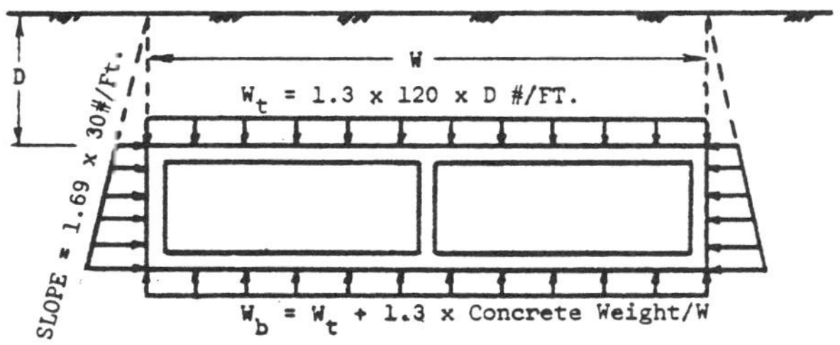


Fig. 36.2

Figure 36.2 shows the factored earth load pressures acting on a box culvert. The earth pressure from the dead load of the concrete is distributed equally over the bottom of the box. When designing the bottom slab of a culvert do not forget that the weight of the concrete in the bottom slab acts in an opposite direction than the bottom soil pressure and thus reduces the design moments and shears. For box culvert designs use:

| | |
|---|---|
| Soil Unit Weight | = 120#/cu. ft. |
| Concrete Unit Weight | = 150#/cu. ft. |
| Equivalent Fluid Pressure | = 30# or 15#/cu. ft. |

## 36.3 LIVE LOADS

Live load consists of the standard AASHTO trucks. All culverts are designed for HS20 modified loading. When the depth of fill over the box is less than 2 feet the wheel load is distributed as in slabs with concentrated loads. When the depth of fill is 2 feet or more the concentrated wheel loads are uniformly distributed over a square, the sides of which are equal to 1.75 times the depth of fill. When areas from several concentrations overlap, the total load is considered as uniformly distributed over the area defined by the outside limits of the individual areas.

For distributed loads two trucks are placed side by side. For single cell boxes the effect of live load may be neglected when the depth of fill is more than 8 feet and exceeds the span length. For multiple cells it may be neglected when the depth of fill exceeds the distance between faces of end supports.

Assume that the truck passes over the box in a direction perpendicular to the centerline regardless of the culvert skew angle. The load factor for live load equals 1.3(5/3)(L + I).

(1) Concentrated Live Loads

Distribute wheel load according to the formula E=4+.06S, where

E=width of slab in feet over which a wheel is distributed
S=clear span of cell

(2) Distributed Live Loads

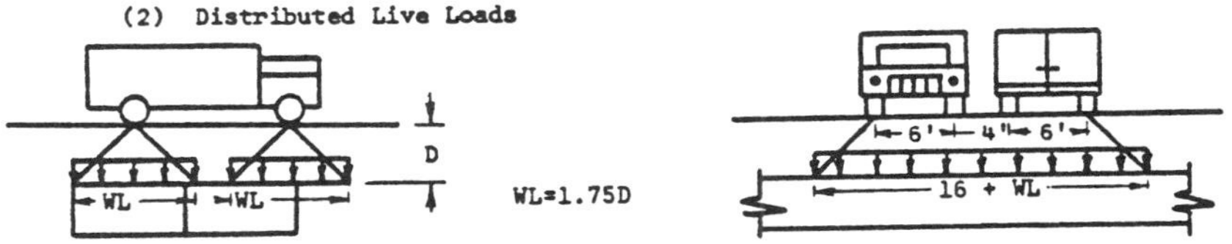


(3) Live Load Soil Pressures

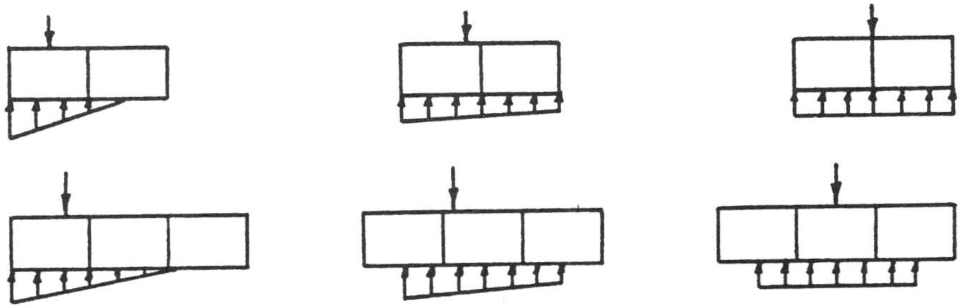

Soil pressures on the bottom of the box from live load are given a triangular or trapezoidal distribution. The resultant of the soil pressure diagram must keep the system in equilibrium.

For single and twin cell culverts, if the live load falls within the middle 1/3 of the box, distribute the soil pressure over the entire width of the box. For triple cell box culverts, if the live load falls within the middle 1/3, distribute the soil pressure between the mid-points of span 1 and 3.

(4) Impact decreases as the depth of fill increases. The following impact factors are listed in AASHTO and are used for culverts.

| Depth of Fill | Impact |
|---|---|
| 0'-0" to 1'-0" inc. | 30% |
| 1'-1" to 2'-0" inc. | 20% |
| 2'-1" to 2'-11" inc. | 10% |
| 3'-0" or more | 0% |

(5) The relative positions of the maximum ordinates of influence lines for various maximum moments are practically constant regardless of the height to width ratios of a culvert or the relative stiffnesses of the slabs and walls. It is therefore possible to assume an axle position of a truck which gives a particular maximum moment for any culvert analyzed. These axle positions are summarized below.

| MAXIMUM MOMENT LOCATION | OUTSIDE SPAN AXLE LOCATIONS | | |
|---|---|---|---|
| | 1 - Cell | 2 - Cells | 3 - Cells |
| Top Outside Corner - Negative | .3 | .35 | .4 |
| Top Inside - Negative | - | .65 | .6 |
| Bottom Outside Corner - Negative | .3 | .3 | .3 |
| Bottom Inside - Negative | - | .5 | .4 |
| Top Positive - Outside Span | .5 | .45 | .5 |
| Bottom Positive - Outside Span | .5 | .45 | .35 |

When span length permits, consideration must also be given to multiple axles on the culvert.

## 36.4 ANALYSIS ASSUMPTIONS

1. Sidesway of the box is not considered because of the lateral support of the soil.
2. Haunches at the center walls of multi-cell box culverts are not considered for analysis.
3. Centroid of rebar is 2 1/2" from surface except bottom steel in bottom slab which is 3 1/2" from surface.
4. The centerlines of the walls and top and bottom slabs are used for computing section properties and dimensions for analysis.
5. For skews of 20 degrees or less, culverts are analyzed as if the reinforcing steel is perpendicular to the centerline of box, even though the steel is actually placed along the skew. The only change is to the horizontal bar lengths which are increased in length by dividing by the cosine of the skew angle. For skews of over 20 degrees the reinforcing steel is placed perpendicular to the centerline of box.
6. Water pressure in culvert barrels is ignored.

## 36.5 DESIGN INFORMATION

Even though AASHTO specifications require a stronger box for 2 feet of fill than 4 feet, design box culverts for the actual loading condition producing the strongest box. Do not anticipate future placing or removal of fill which may require a stronger box.

The minimum thickness of the top and bottom slab is 6½". Wall thickness minimums are:

For clear height of less than 6 feet - 8 inches.
For clear height of at least 6 feet and less than 10 feet - 9 inches.
For clear height of 10 feet and over - 10 inches.
All slab and wall thicknesses are rounded to the next largest ½".

All bar steel is detailed as being 2 inches clear except the bottom steel in the bottom slab. This steel is detailed as being 3 inches clear.

Top and bottom slab thicknesses are determined by shear and moment requirements. When calculating shear, live load is not considered. Slabs designed for bending moment based on a wheel distribution of E = 4 + 0.065 are considered satisfactory in bond and shear.

A haunch is provided only when the slab depth required at the interior wall is more than an inch greater than that required for the remainder of the span. Minimum haunch depth is 3 inches and minimum length is 6 inches. Haunch depth is increased in 3 inch increments and haunch length is increased in 6 inch increments. The haunch length is not to exceed 7 times the depth.

Fatigue requirements in bar steel are satisfied by limiting the truck service load stress by AASHTO using working stress analysis. In addition to the truck live load, an equivalent fluid pressure of 15 or 30 pounds/sq. foot is added to the outside of the box. Vertical earth pressure and concrete dead load are not considered in the fatigue analysis. The fatigue stress limit for concrete usually does not govern.

Reinforced concrete design is based on the strength design method and AASHTO specifications. Material strengths are:

Concrete - $f_c'$ = 3500 psi.
Bar Steel - $f_y$ = 60,000 psi.

The slab thickness required is determined by moment or shear, whichever governs. Thickness based on moment is determined from using 37.5% of the reinforcement ratio producing balanced conditions.

The shear in the top and bottom slabs is assumed to occur at a distance "d" from the face of the walls. "d" equals the distance from the centroid of the reinforcing steel to the face of the concrete in compression. When a haunch is used, shear must also be checked at the end of the haunch.

For multi-cell culverts make interior and exterior walls of equal thickness.

For culverts under high fills use a separate design for the ends if the reduced section may be used for at least two panel pours per end of culvert. Maximum length of panel pour is 40 feet.

Barrel lengths are based on the roadway sections and wing lengths are based on a 2½:1 slope of fill from the top of box to apron.

Dimensions on drawings are given to the nearest ½ inch only.

## 36.6 DETAILING OF REINFORCING STEEL

To calculate the required bar steel area and cut off points a maximum positive and negative moment envelope is computed. Bar cut off points may also be determined by an approximate method based on the points of dead load contraflexture. It is assumed that the required bar lengths in the top slab are longer than those in the bottom slab. Therefore, cut off points are computed for the top slab and are also used in the bottom slab.

### (1) Corner Steel

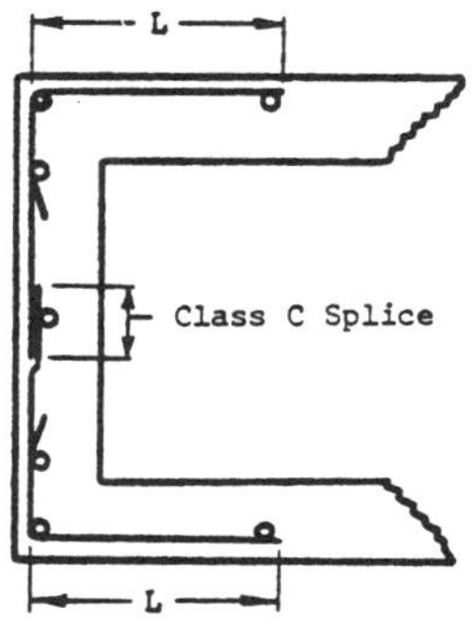


The area of steel required is the maximum computed from using the top and bottom corner moments and the thickness of the slab or wall, whichever controls. Identical bars are then used in the top and bottom. Top and bottom negative steel is cut in the walls and detailed in two lengths when a savings of over one foot in a single bar length can be obtained. Corner steel is always lapped at the center of the wall. If two bar lengths are used, only alternate bars are lapped.

Distance "L" is computed from the maximum negative moment envelope for the top slab.

An approximate method for computing "L" is to extend the bar past the point of dead load contraflexture a distance equal to 15D + EXT, where EXT=.06 x EL - .006 x D x EL but not less than .25 feet.

D = effective slab depth in inches
EL= effective span length in feet

### (2) Positive Moment Slab Steel

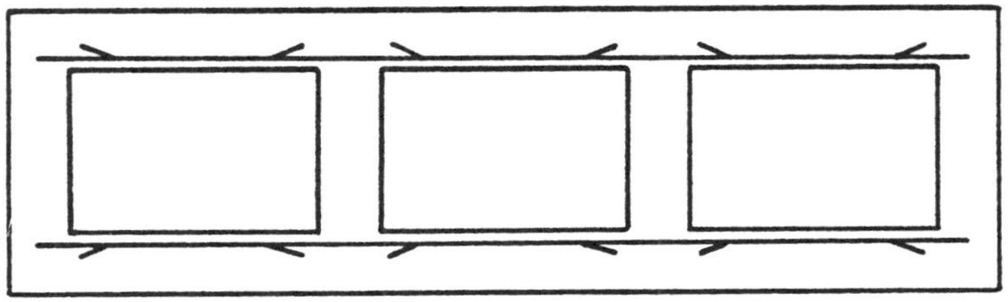

The area of steel required is determined by the maximum positive moments in each span. Top and bottom steel may be of different size and spacing. For single cell culverts detail two bar lengths if 2 1/2 feet or more of bar steel can be saved. For multi-cell culverts the criteria is a 3 foot savings for bar size 6 and smaller and a 2.25 foot savings for bar size 7 and larger. The above criteria may be altered at the discretion of the designer.

When two bar lengths are detailed in multi-cell culverts, run every other positive bar across the entire width of box. If this requires a length longer than 30 feet for a number 4 bar or 40 feet for a larger size bar, lap them over an interior wall.

The cut-off points of alternate bars are determined from the maximum positive moment envelope for the top slab. These same points are used in the bottom slab.

The following approximate method may also be used to compute cut off points.

At the outside wall of the exterior span extend the bar past the point of dead load contraflexture a distance equal to 15D+EXT, where EXT=.04 x EL - .004 x EL x D but not less than .25 feet.

At the inside wall of the exterior span extend the bar past the point of dead load contraflexture a distance equal to 15D+EXT where EXT=.08 x EL - .008 x EL x D but not less than .25 feet.

For an interior span extend the bar steel past the points of dead load contraflexture a distance equal to 15D+EXT where EXT=.08 x EL -.008 x EL x D but not less than .25 feet.

### (3) Negative Moment Steel over Interior Walls

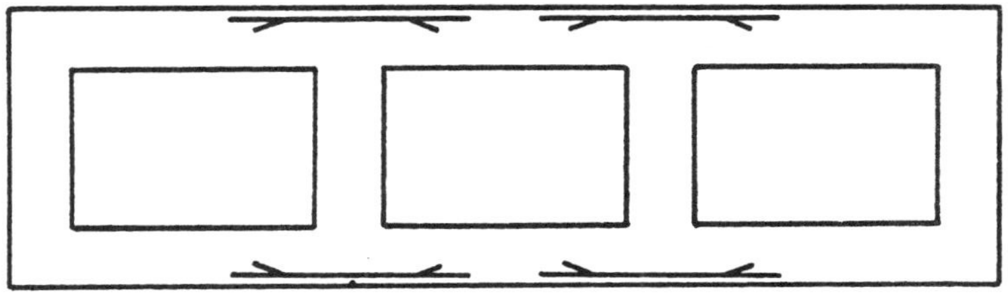

The area of steel is determined from the moment at the C/L of the interior wall and the effective depth of slab at the face of the interior wall. If the slab is haunched the steel area required at the end of the haunch must also be checked. Top and bottom steel may be of different size and spacing. If a savings in bar length of more than 2'-0 can be realized, two bar lengths are detailed.

Cut off points are determined from the maximum negative moment envelope of the top slab. The same bar lengths are then used in the bottom slab. The minimum length of any bar is 2 times the development length. For culverts of 3 or more cells, if the clear distance between negative bars of adjacent spans is one foot or less, make the bar continuous across the interior spans.

(4) Exterior Wall Positive Moment Steel

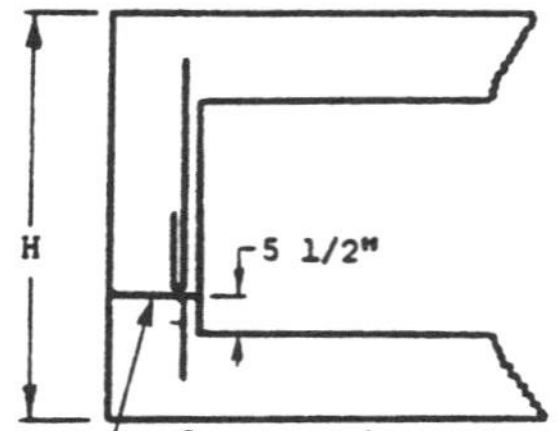


The area of steel is determined by the maximum positive moment in the wall but a minimum of #4 bars at 24" is supplied. The maximum positive moment may be assumed to occur at the .45 point from the bottom. The bars are extended into the top and bottom slab a distance of at least ten bar diameters. A construction joint, 5 1/2" above the bottom slab, is always used so a dowel bar must be detailed.

(5) Interior Wall Moment Steel

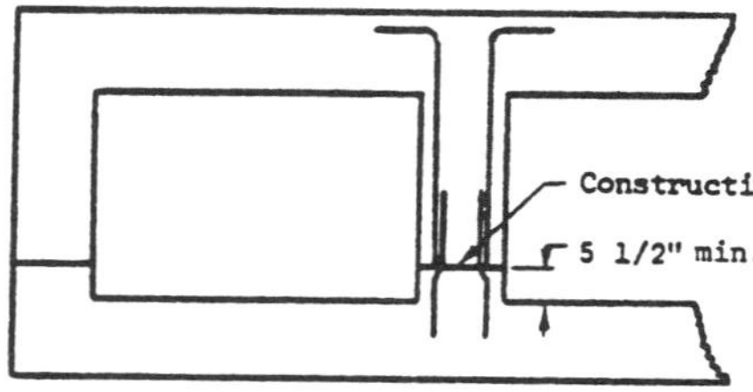


The area of steel is determined from the maximum moment at the top of the wall and the effective wall thickness. A minimum of #4 bars at 24" is supplied. Identical steel is provided at both faces of the wall. A one foot 90 degree bend is provided in the top slab with the horizontal portion being just below the negative moment steel. A construction joint, 5 1/2 inches above the bottom slab, is always used so a dowel bar must be detailed. If the haunch depth is 6 inches or greater the construction joint is placed a distance above the bottom slab equal to the haunch depth.

(6) Distribution Reinforcement

No distribution steel is required when the depth of fill over the slab exceeds two feet. The amount is a percentage of the required positive moment reinforcing steel in the top slab as given by the following formula :

$$\text{Percentage} = \frac{100}{\sqrt{S}}\text{, Maximum 50\%}$$

where S = the effective span length, in feet.

A minimum of #4 bars at 18 inches is provided for temperature reinforcement when the depth of fill exceeds 2 feet.

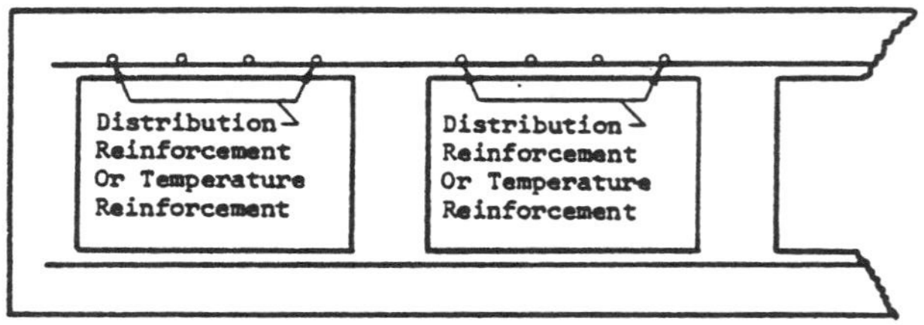


(7) Temperature Reinforcement

Temperature reinforcement is required at the top surface of the bottom slab, the inside face of the exterior walls, both faces of interior walls, and in the top of the top slab when the depth of fill over the box is one foot or less. The only exception is for single cell culverts where the temperature steel in the top of the top slab is not used, unless the top slab is an intergal part of wearing surface. This is because the positive dead load moment is not decreased due to live load in adjacent spans as is the case for multi-cell box culverts.

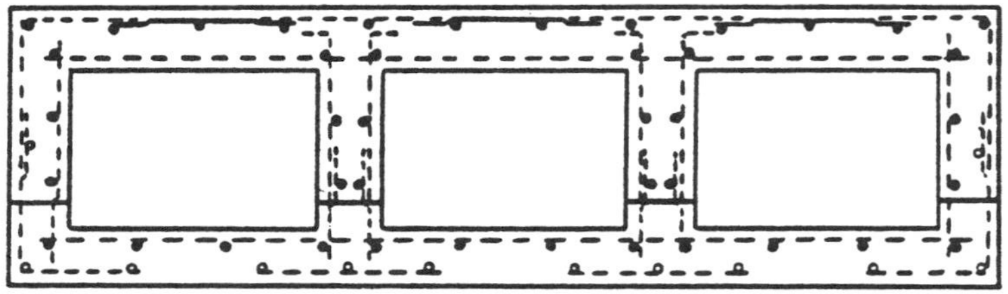

Temperature steel is always #4 bars at a maximum spacing of 18 inches. The longitudinal bars shown as solid circles represent the required temperature steel. The bars shown as hollow circles are #4 bars provided primarily to support the main reinforcement. The transverse bar in the top of the top slab, shown as a solid line, is also required when top slab temperature steel is used. It is a #4 bar with a spacing not to exceed 21 inches and preferably at a spacing which is a multiple of the negative moment steel over the wall and the corner steel. When the top slab is an intergal part of the wearing surface use a minimum of number 4 bars at 1'-0" centers in both directions in the top of the top slab.

36.7 BOX CULVERT APRONS

Five types of box culvert aprons are used. They are referred to as Type A, B, C, D and E. The angle that the wings makes with the direction of stream flow is the main difference between the five types. The allowable headwater and other hydraulic requirements are what usually determines the type of apron required. Physical characteristics at the site may also dictate a certain type. For hydraulic design of different apron types see charts 8.3.1 and 8.3.2 in chapter 8.

(1) Type A

Type A, because of its poor hydraulic properties, is generally not used except for cattle or pedestrian underpasses.

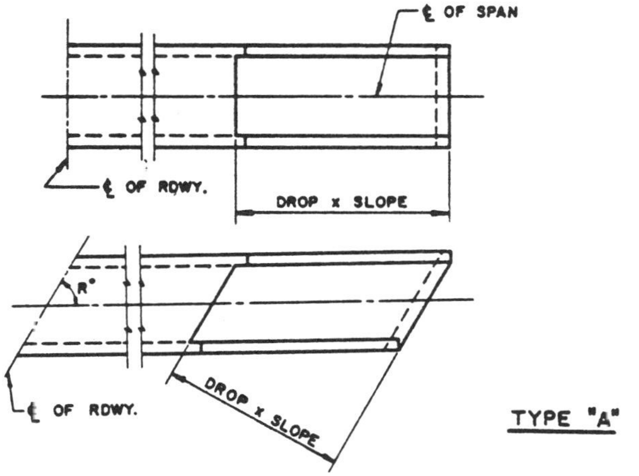


(2) Type B, C, D

Type C & D are of equal efficiency but type C is used most frequently. Type D is used for inlets when the water is entering the culvert at a very abrupt angle. Type B is used for outlets.

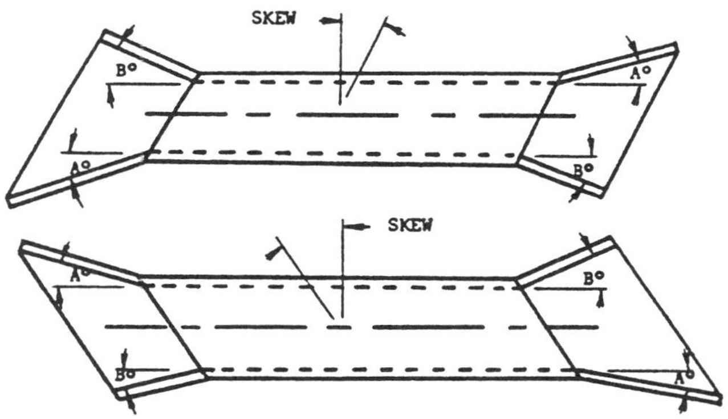


| SKEW | | WING TYPE B | | WING TYPE C | | WING TYPE D | |
|---|---|---|---|---|---|---|---|
| Greater Than | To Incl. | Angle A | Angle B | Angle A | Angle B | Angle A | Angle B |
| 0 | 7.5 | 15° | 15° | 30° | 30° | 45° | 45° |
| 7.5 | 15.0 | 15° | 15° | 25° | 30° | 40° | 45° |
| 15.0 | 22.5 | 10° | 15° | 25° | 30° | 35° | 45° |
| 22.5 | 37.5 | 10° | 15° | 20° | 30° | 30° | 45° |
| 37.5 | 45.0 | 10° | 15° | 15° | 30° | 25° | 45° |
| 45.0 | 52.5 | 5° | 15° | 15° | 30° | 20° | 45° |
| 52.5 | 67.5 | 5° | 15° | 10° | 30° | 15° | 45° |
| 67.5 | 75.0 | 5° | 15° | 5° | 30° | 10° | 45° |
| 75.0 | 82.5 | 0° | 15° | 5° | 30° | 5° | 45° |
| 82.5 | 90.0 | 0° | 15° | 0° | 30° | 0° | 45° |

(3) Type E

Type E is used primarily in urban areas where a sidewalk runs over the culvert and it is necessary to have a parapet and railing along the sidewalk. For type E the wingwalls run parallel to the roadway just like the abutment wingwalls of most bridges. It is also used where R/W is a problem and the aprons would extend beyond the R/W for other types.

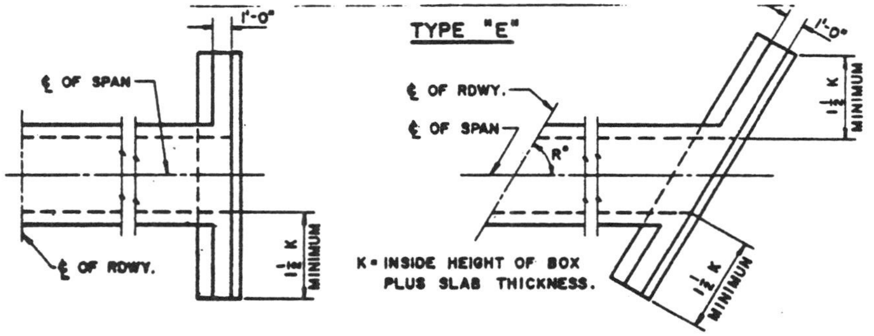


(4) Wing Wall Design

Culvert wing walls are designed for an equivalent fluid pressure of 33 pounds per foot and a 1 foot surcharge. For wing heights of seven feet or less determine the area of steel required by using the maximum wall height and use the same bar size and spacing along the entire wing wall length. The minimum amount of steel used is number 4 bar at 1'-0" spacing. Wingwall thickness is made equal to the barrel wall thickness.

For walls over seven feet, detail different bar sizes and/or spacing. Do not detail less than six bars in a set of bars of the same size and spacing.

36.8 BOX CULVERT CAMBER

Camber of culverts is a design compensation for anticipated settlement of foundation soil beneath the culvert. The culvert strength has little effect on this settlement and is ignored. Responsibility for the recommendation and calculation of camber belongs to the district soils section. Severe settlement problems with accompanying large camber are to be checked with the Central Soils Office.

Calculation of settlements is based on Terzaghi's theory of consolidation. Using Terzaghi's theory it is possible to reasonably estimate the probable settlement, (or decrease in volume), caused by the roadway embankment of a compressible layer of soil beneath a culvert. If several different compressible layers exist under the box the settlement of each layer is computed separately. The total settlement is the summation of the settlements of each layer.

The compressible layers of material may not be identical along the entire length of the box. It may be necessary to calculate the settlement at more than one point along the length of the box.

(1) Computation of Settlement

Settlement of a compressible layer is computed from the following basic consolidation equation:

$$S = H \log_{10} \left[\frac{P_o + \Delta P}{P_o}\right] \frac{Cc}{1+e_o} \quad \text{where}$$

S = settlement in feet

H = thickness of compressible layer in feet

$P_o$ = initial stress at the midpoint of the compressible layer. This includes buoyancy of soils below water table, Lbs./sq.ft.

$\Delta p$ = increase in stress due to the fill construction, Lbs./sq.ft.

$e_o$ = initial void ratio at pressure $p_o$

$C_c$ = compression index which is a measure of the compressibility of a soil. It is the slope of the straight-line part of the stress-void ratio curve from a consolidation test.

A further description of the above equation and consolidation tests can be found in most textbooks on soil mechanics.

For preliminary investigations $C_c$ can be determined from the following approximate formulas:

Non-organic soils: $C_c = .007\ (LL-10)$

Organic soils: $C_c = .007$ (MOISTURE CONTENT)

where LL = liquid limit expressed as whole number and moisture content is expressed as a whole number, although both are percents by weight.

If the in-place moisture content approaches the plastic limit the computed $C_c$ is decreased by 75%. If the in-place moisture content is near the liquid limit use the computed value. If the in-place moisture content is twice the liquid limit the computed $C_c$ is increased by 75%. For intermediate moisture contents the percent change to the computed $C_c$ is determined from a straight line interpolation between the corrections mentioned above.

If settlements computed by using the approximate value of $C_c$ exceed 1.5 feet, a consolidation test is performed. As in-place moisture content approaches twice the liquid limit, settlement is caused by a shear failure and the consolidation equation is no longer applicable.

The consolidation equation is applied to only compressible silts and clays. Sands are of a lower compressibility and no culvert camber is required until the fill exceeds 25 feet. When the fill exceeds 25 feet for sand, a camber of .01 feet per foot of fill is used.

(2) Configuration of Camber

The following guides are to be followed when detailing camber.

(a) It is unnecessary to provide gradual camber. "Brokenback" camber is closer to the actual settlement which occurs.

(b) Settlement is almost constant from shoulder point to shoulder point. It then reduces to the ends of the culvert at the edge of the fill.

(c) The ends of the culvert tend to come up if side slopes are steeper than 2 1/2 to 1. With 2 to 1 side slopes camber is increased 10% to compensate for this rise.

(3) Numerical Example of Settlement Computation

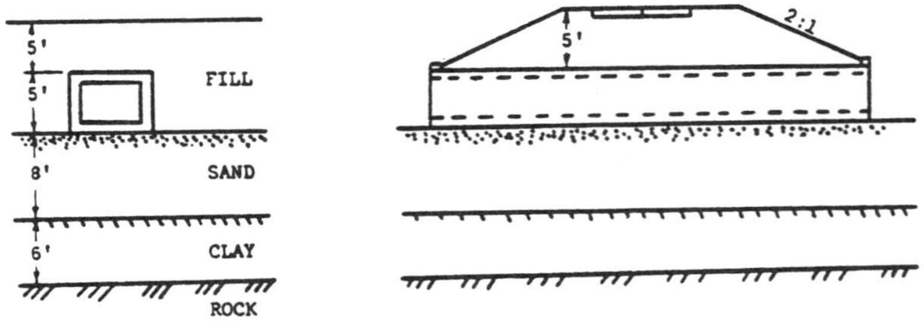


A box culvert rests on original ground consisting of 8' of sand and 6' of clay over bedrock. Estimate the settlement of the culvert if 10 foot of fill is placed on the original ground after the culvert is constructed. The in-place moisture content and liquid limit equal 40%. The initial void ratio equals .98. The unit weight of the clay is 105 lb/cu.ft. and that of the fill and sand is 110 lb/cu.ft. There is no water table.

$P_o = 8 \times 110 + 3 \times 105 = 1195.$ lb/cu.ft.

$\Delta p = 10 \times 110 = 1100.$ lb./cu.ft.

$$S = H \log_{10} \left[\frac{P_o + \Delta p}{P_o}\right] \frac{C_c}{1+e_o}$$

$C_c = .007\ (30) = .21$ (approximate value)

$$S = 6 \log_{10} \left[\frac{1195 + 1100}{1195}\right] \frac{.21}{1.98}$$

$$S = 6 \log_{10} 1.92\ (.106) = 6 \times .283 \times .106 = .18 \text{ ft.}$$

## 36.9 BOX CULVERT STRUCTURAL EXCAVATION AND GRANULAR BACKFILL

All excavations for culverts and aprons, unless on bedrock or fill, are undercut a depth of 6 inches. The upper limit of excavation is the existing ground line.

All spaces excavated and not occupied by the new structure are backfilled with granular backfill to the elevation and section existing prior to excavation within the length of the box. The backfill is placed to help eliminate settling problems on culverts. No granular backfill is placed behind the wings or around the cutoff walls. Granular backfill is placed in the undercut area under the apron. Usually 6" of granular backfill is placed under all boxes for construction purposes, which is covered by specification.

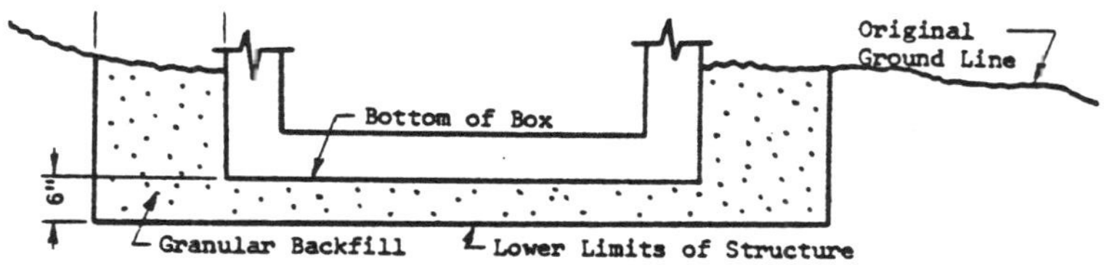


No granular backfill is placed under the box for culverts built on fills. The purpose of the backfill is to provide a solid base to pour the bottom slab. It is assumed that fill material provides this base without the addition of granular backfill.

## 36.10 BOX CULVERT HEADERS

For skews of 20 degrees and less the main reinforcing steel is parallel to the end of the barrel. A header is not required for structural purposes but is used to prevent the fill material from spilling into the apron. A 1'-0" wide by 6" high (above the top of top slab) header with nominal steel is therefore used for skews of 20 degrees and less on the top slab. No header is used on the bottom slab.

For skews over 20 degrees the main reinforcing is not parallel to the end of the barrel. The positive reinforcing steel terminates in the header and thus the header must support, in addition to its own dead load, an additional load from the dead load of the slab and fill above it. A portion of the live load may also have to be supported by the header.

The calculation of the actual load that a header must support becomes a highly indeterminate problem. For this reason a rational approach is used to determine the amount of reinforcement required in the headers. The design moment strength of the header must be equal to or greater than 1.3 times the header dead load moment (based on simple span) plus 1.3 times a live load moment from a 16 kip load assuming .5 fixity at ends.

To prevent a traffic hazard culvert headers are designed not to protrude above the ground line. For this reason the height of the header above the top of the top slab is allowed to be only 6 inches. The width of the header is standardized at 1'-6".

The header in the following sketch gives the design moment strengths listed using fy = 60,000 psi and d = 8.5 inches.

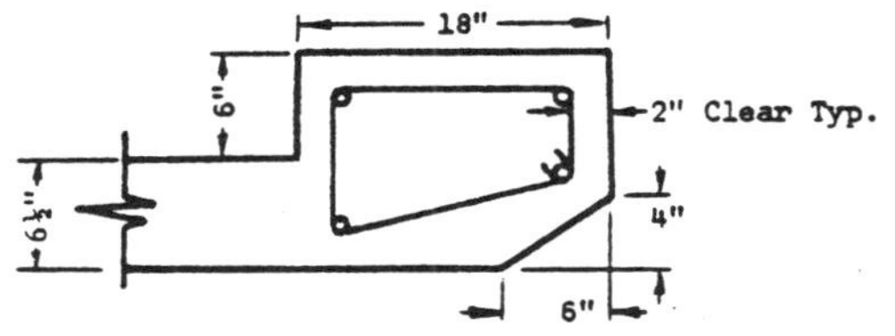


| BAR SIZE | ULTIMATE MOMENT (Ft.-Kips) |
|---|---|
| #6 | 31.7 |
| #7 | 42.1 |
| #8 | 54.0 |
| #9 | 66.5 |
| #10 | 80.9 |

The following size bars are recommended for the listed header lengths where "HEADER LENGTH" equals the distance between C/L of walls in one cell measured along the skew.

| HEADER LENGTH | BAR SIZE * |
|---|---|
| To 10'-0" | #7 |
| Over 10'-0" to 13'-0" | #8 |
| Over 13'-0" to 16'-0" | #9 |
| Over 16'-0" to 20'-0" | #10 |

* Use the bar size listed in each header and place 2 bars on top and 2 on bottom. Use a header on both the top and bottom slab. See standard 36.3 for details.

## 36.11 WEEPHOLES

Investigate the need for weep holes for culverts in cohesive soils. These holes are to relieve the hydrostatic pressure on the sides of the culverts. Where used, place the weep holes one foot above normal water elevation but a minimum of one foot above the lower sidewall construction joint. Do not use weepholes if they are closer than one foot to the top slab.

## 36.12 CUTOFF WALLS

Where dewatering the cutoff wall in sandy terrain is a problem, the concrete may be poured in the water and this can be noted on the plans.

## 36.13 NAMEPLATE

Designate a location on the wingwall for placement of the nameplate.

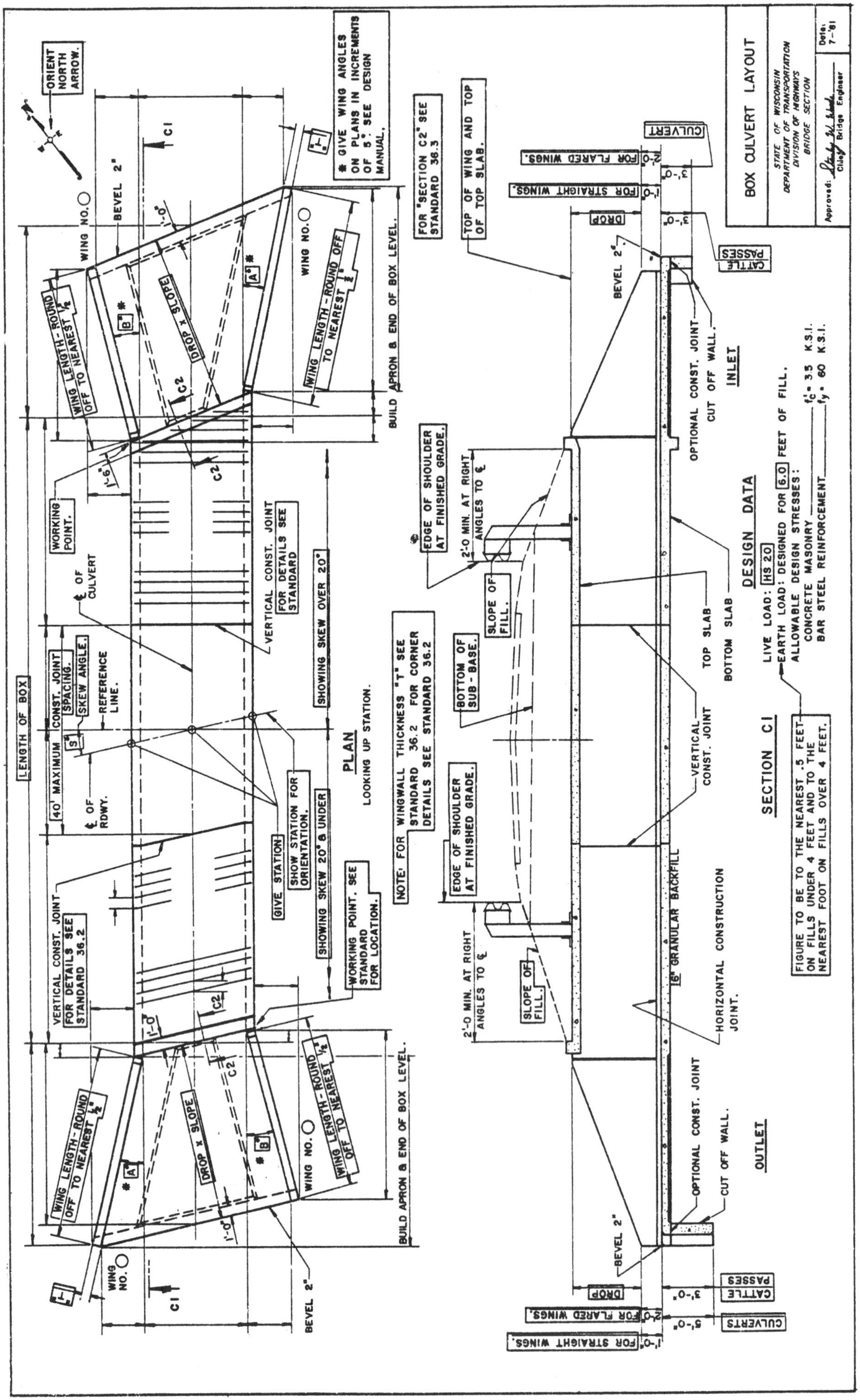
BOX CULVERT LAYOUT
STATE OF WISCONSIN
DEPARTMENT OF TRANSPORTATION
DIVISION OF HIGHWAYS
BRIDGE SECTION
Approved: Chief Bridge Engineer
Date: 7-81
ORIENT NORTH ARROW.
* GIVE WING ANGLES ON PLANS IN INCREMENTS OF 5°. SEE DESIGN MANUAL.
FOR "SECTION C2" SEE STANDARD 36.3
TOP OF WING AND TOP OF TOP SLAB.
LENGTH OF BOX
40' MAXIMUM CONST. JOINT SPACING.
SKEW ANGLE.
REFERENCE LINE.
℄ OF RDWY.
℄ OF CULVERT
WORKING POINT.
WING NO.
BEVEL 2"
WING LENGTH - ROUND OFF TO NEAREST ½
DROP x SLOPE
VERTICAL CONST. JOINT FOR DETAILS SEE STANDARD
VERTICAL CONST. JOINT FOR DETAILS SEE STANDARD 36.2
SHOWING SKEW OVER 20°
SHOWING SKEW 20° & UNDER
GIVE STATION.
SHOW STATION FOR ORIENTATION.
WORKING POINT. SEE STANDARD FOR LOCATION.
BUILD APRON & END OF BOX LEVEL.
PLAN
LOOKING UP STATION.
NOTE: FOR WINGWALL THICKNESS "T" SEE STANDARD 36.2 FOR CORNER DETAILS SEE STANDARD 36.2
EDGE OF SHOULDER AT FINISHED GRADE.
2'-0 MIN. AT RIGHT ANGLES TO ℄
SLOPE OF FILL.
BOTTOM OF SUB-BASE.
TOP SLAB
BOTTOM SLAB
VERTICAL CONST. JOINT
16" GRANULAR BACKFILL
HORIZONTAL CONSTRUCTION JOINT.
OPTIONAL CONST. JOINT
CUT OFF WALL.
INLET
OUTLET
DROP
CATTLE PASSES
CULVERT
CULVERTS
1'-0 FOR STRAIGHT WINGS.
2'-0 FOR FLARED WINGS.
3'-0"
5'-0"
SECTION C1
DESIGN DATA
LIVE LOAD: HS 20
EARTH LOAD: DESIGNED FOR 6.0 FEET OF FILL.
ALLOWABLE DESIGN STRESSES:
CONCRETE MASONRY f'c = 3.5 K.S.I.
BAR STEEL REINFORCEMENT fy = 60 K.S.I.
FIGURE TO BE TO THE NEAREST .5 FEET ON FILLS UNDER 4 FEET AND TO THE NEAREST FOOT ON FILLS OVER 4 FEET.

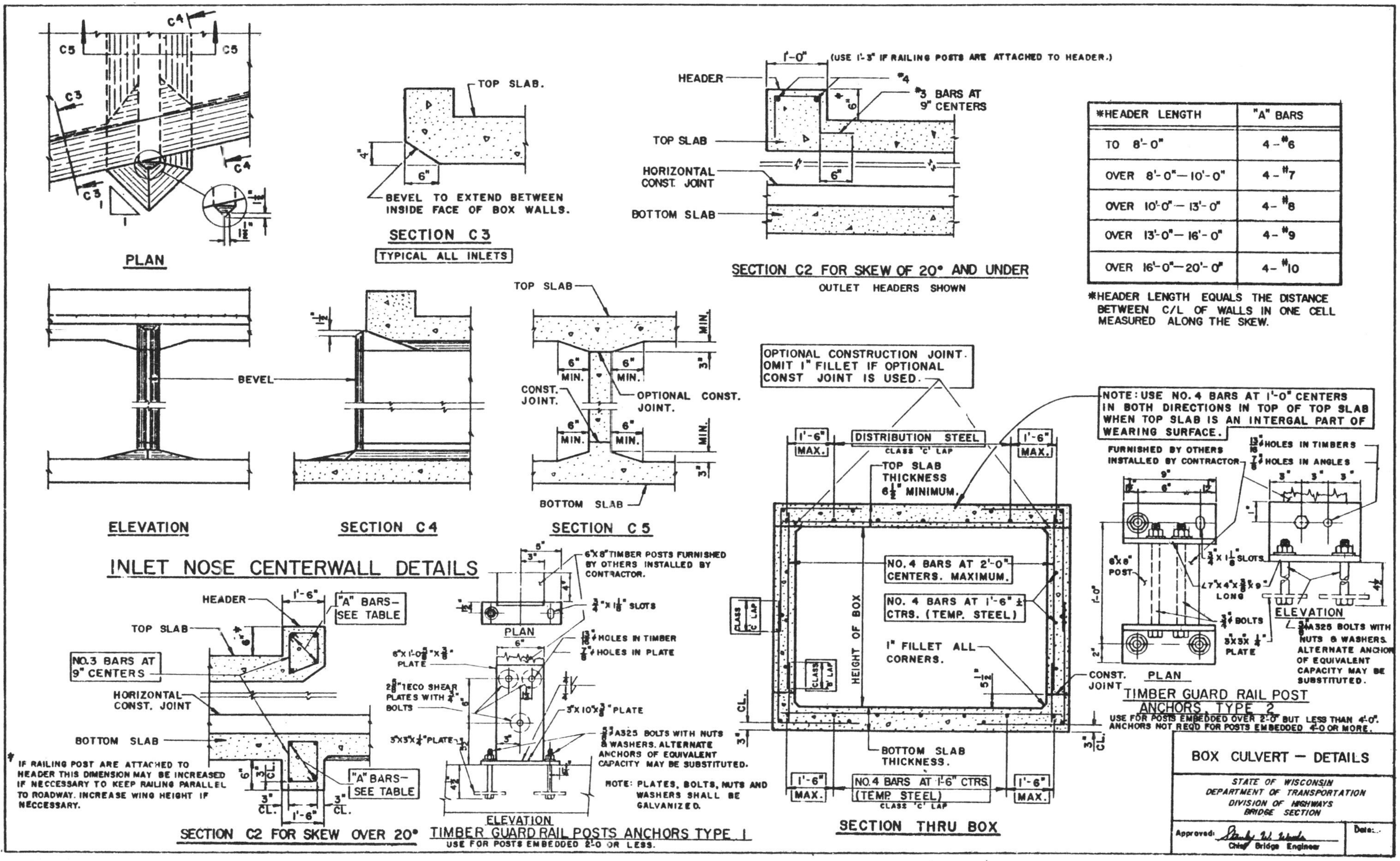


| *HEADER LENGTH | "A" BARS |
|---|---|
| TO 8'-0" | 4-#6 |
| OVER 8'-0"—10'-0" | 4-#7 |
| OVER 10'-0"—13'-0" | 4-#8 |
| OVER 13'-0"—16'-0" | 4-#9 |
| OVER 16'-0"—20'-0" | 4-#10 |

*HEADER LENGTH EQUALS THE DISTANCE BETWEEN C/L OF WALLS IN ONE CELL MEASURED ALONG THE SKEW.

| DESIGN EARTH COVER (FT.) | "M" MIN. (IN.) | CIRCUMFERENTIAL REINFORCEMENT AREAS ($IN^2$/FT.) | | | |
|---|---|---|---|---|---|
| | | $A_s1$ | $A_s2$ | $A_s3$ | $A_s4$ |
| 6'x4'x7" (SxRxT) | | | | | |
| 2 | 28 | 0.27 | 0.34 | 0.27 | 0.17 |
| 3 | 25 | 0.19 | 0.24 | 0.23 | " |
| 4 | 24 | 0.19 | 0.21 | 0.21 | " |
| 5 | 23 | 0.19 | 0.21 | 0.22 | " |
| 6 | 23 | 0.19 | 0.22 | 0.23 | " |
| 8 | 23 | 0.21 | 0.25 | 0.26 | " |
| 10 | 22 | 0.22 | 0.29 | 0.30 | " |
| 12 | 22 | 0.25 | 0.33 | 0.33 | " |
| 14 | 22 | 0.28 | 0.37 | 0.38 | " |
| 16 | 22 | 0.31 | 0.41 | 0.42 | " |
| 7'x6'x8" (SxRxT) | | | | | |
| 2 | 38 | 0.25 | 0.41 | 0.33 | 0.19 |
| 3 | 31 | 0.19 | 0.29 | 0.28 | " |
| 4 | 29 | 0.21 | 0.27 | 0.28 | " |
| 5 | 28 | 0.21 | 0.27 | 0.29 | " |
| 6 | 27 | 0.21 | 0.29 | 0.30 | " |
| 8 | 26 | 0.21 | 0.32 | 0.33 | " |
| 10 | 26 | 0.23 | 0.37 | 0.38 | " |
| 12 | 26 | 0.25 | 0.42 | 0.43 | " |
| 14 | 25 | 0.28 | 0.47 | 0.49 | " |
| 16 | 25 | 0.32 | 0.52 | 0.54 | " |
| 8'x6'x8" (SxRxT) | | | | | |
| 2 | 36 | 0.32 | 0.48 | 0.37 | 0.19 |
| 3 | 31 | 0.25 | 0.36 | 0.32 | " |
| 4 | 30 | 0.24 | 0.33 | 0.33 | " |
| 5 | 29 | 0.26 | 0.34 | 0.36 | " |
| 6 | 28 | 0.26 | 0.36 | 0.37 | " |
| 8 | 27 | 0.29 | 0.40 | 0.42 | " |
| 10 | 27 | 0.33 | 0.47 | 0.49 | " |
| 12 | 27 | 0.37 | 0.54 | 0.55 | " |
| 10'x6'x10" (SxRxT) | | | | | |
| 2 | 41 | 0.37 | 0.48 | 0.36 | 0.24 |
| 3 | 37 | 0.32 | 0.38 | 0.33 | " |
| 4 | 34 | 0.30 | 0.35 | 0.33 | " |
| 5 | 33 | 0.33 | 0.37 | 0.37 | " |
| 6 | 33 | 0.36 | 0.39 | 0.41 | " |
| 8 | 33 | 0.40 | 0.44 | 0.46 | " |
| 10 | 32 | 0.45 | 0.50 | 0.52 | " |
| 12 | 32 | 0.51 | 0.58 | 0.60 | " |
| 10'x8'x10" (SxRxT) | | | | | |
| 2 | 45 | 0.32 | 0.54 | 0.41 | 0.24 |
| 3 | 40 | 0.28 | 0.43 | 0.38 | " |
| 4 | 37 | 0.28 | 0.41 | 0.40 | " |
| 5 | 36 | 0.30 | 0.43 | 0.44 | " |
| 6 | 35 | 0.31 | 0.45 | 0.48 | " |
| 8 | 34 | 0.35 | 0.50 | 0.53 | " |
| 10 | 33 | 0.39 | 0.59 | 0.62 | " |
| 12 | 33 | 0.45 | 0.66 | 0.69 | " |
| 10'x10'x10" (SxRxT) | | | | | |
| 2 | 75 | 0.30 | 0.59 | 0.47 | 0.24 |
| 3 | 50 | 0.24 | 0.47 | 0.43 | " |
| 4 | 45 | 0.24 | 0.44 | 0.44 | " |
| 5 | 42 | 0.27 | 0.46 | 0.48 | " |
| 6 | 40 | 0.29 | 0.49 | 0.53 | " |
| 8 | 38 | 0.32 | 0.54 | 0.58 | " |
| 10 | 37 | 0.36 | 0.63 | 0.67 | " |
| 12 | 36 | 0.40 | 0.71 | 0.76 | " |

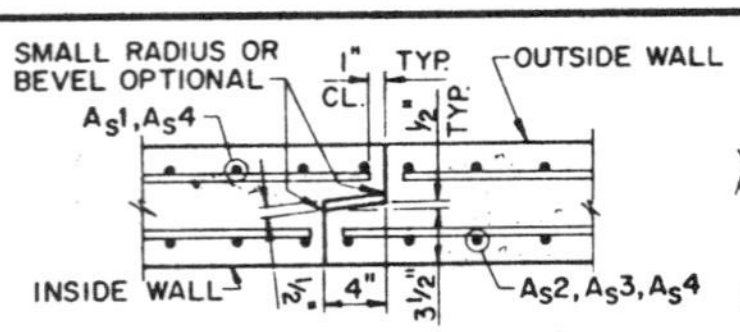


TONGUE AND GROOVE JOINT DETAIL

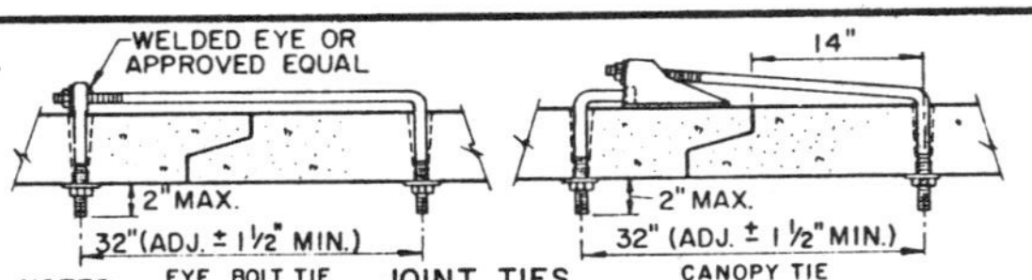


JOINT TIES

NOTES: EITHER EYE BOLT TIES OR CANOPY TIES MAY BE USED. THREADS MAY BE CUT OR ROLLED. TIE NUTS SHALL BE TIGHTENED AS DIRECTED BY THE ENGINEER. (2 TIES REQ'D. PER JOINT) (TIES TO BE GALVANIZED)

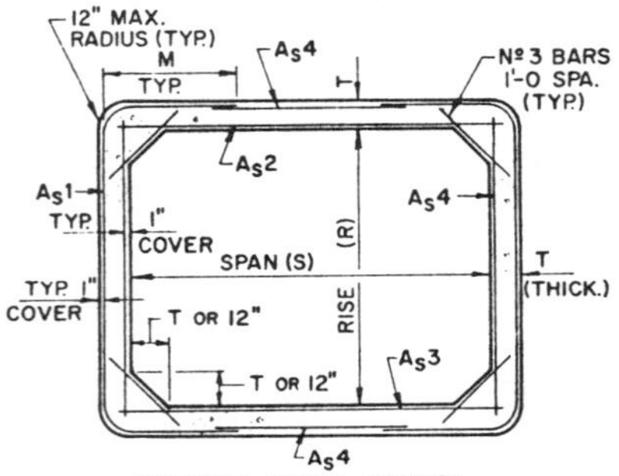


SECTION THRU BARREL

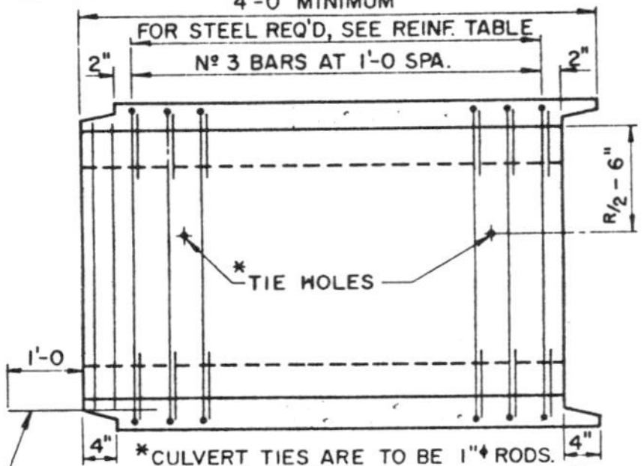


FOR EXTERNAL PANELS (ADJACENT TO APRONS), PLACE Nº 5 BARS 2'-0 LONG AT 1'-0 SPA. IN BOTTOM SLAB. OMIT TONGUE OR GROOVE ADJACENT TO APRON FOR 0° SKEW.

LONGITUDINAL SECTION

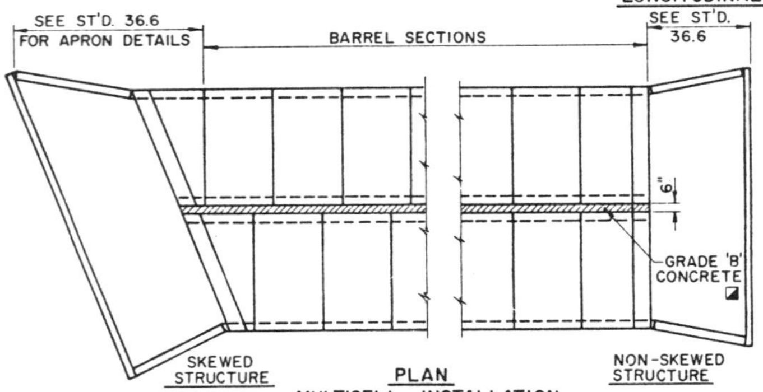


PLAN
MULTICELL INSTALLATION

◪ COST OF GRADE 'B' CONCRETE SHALL BE INCIDENTAL TO THE BID ITEM PRECAST CONCRETE BOX CULVERT.

BOX CULVERT DATA

| DIMENSIONS | | | EARTH COVER (FT.) | | | | | | | | |
|---|---|---|---|---|---|---|---|---|---|---|---|
| S (FT.) | R (FT.) | T (IN.) | | | | | | | | | |
| REINFORCEMENT | | | AREA | LENGTH | M | AREA | LENGTH | M | AREA | LENGTH | M |
| $A_s1$ | | | | | | | | | | | |
| $A_s2$ | | | | | | | | | | | |
| $A_s3$ | | | | | | | | | | | |
| $A_s4$ | | | | | | | | | | | |
| TOTAL BARREL LENGTH | | | | | | | | | | | |

## BARREL NOTES

DETAILS OF FABRICATION, CONSTRUCTION, MATERIALS & WORKMANSHIP NOT SHOWN OR STATED ON THIS DRAWING SHALL BE IN ACCORDANCE WITH A.A.S.H.T.O. M259 AND THE PERTINENT PARTS OF THE ST'D. SPECIFICATIONS & APPLICABLE SPECIAL PROVISIONS.

THE MINMUM EARTH COVER UNDER TRAFFIC AREAS SHALL BE 2'-0.

THE COVER OF CONCRETE OVER THE REINF. SHALL BE 1 INCH WITH AN ALLOWABLE VARIATION OF - 3/8 INCH TO + 1/2 INCH.

THE SPACING CTR. TO CTR. OF THE CIRCUMFERENTIAL WIRES SHALL NOT BE LESS THAN 2 INCHES NOR MORE THAN 4 INCHES. THE SPACING CTR. TO CTR. OF THE LONGIT. WIRES SHALL NOT BE MORE THAN 8 INCHES.

NOT MORE THAN FOUR (4) HOLES MAY BE CAST, DRILLED, OR OTHERWISE NEATLY MADE IN THE SHELL OF EACH PIECE OF BOX SECTION FOR HANDLING. THE HOLES SHALL BE TAPERED UNLESS DRILLED. HOLES SHALL BE FILLED WITH PORTLAND CEMENT MORTAR EXCEPT TAPERED HOLES MAY BE FILLED WITH CONCRETE PLUGS SECURED WITH PORTLAND CEMENT MORTAR OR OTHER APPROVED ADHESIVE.

THE JOINT ON THE BOTTOM OF THE CULVERT & THE BOTTOM 1 FOOT OF THE JOINTS ON THE SIDE OF THE CULVERT SHALL BE SEALED WITH A PREFORMED MASTIC⊕ A 2'-0 STRIP OF **GEOTEXTILE FABRIC** SHALL BE PLACED OVER THE JOINTS ON THE TOP AND ON THE SIDES OF THE CULVERT. (FILTER CLOTH NOT REQ'D OVER INSIDE WALL JTS. OF MULTI-CELL INSTALLATION)

WHEN TWO OR MORE BARRELS ARE UTILIZED IN PARALLEL FOR MULTICELL INSTALLATIONS THE CLEAR SPACING BETWN BARRELS SHALL BE 6 INCHES AND THE SPACE BETWN ADJACENT BARRELS FROM TOP OF BEDDING TO TOP OF TOP SLAB SHALL BE FILLED WITH GRADE 'B' CONCRETE. ◪

## DESIGN INFORMATION

STEEL AREAS ARE BASE ON AN HS20 LIVE LOAD, UNIT WEIGHT OF SOIL OF 120 LB/FT.³, EQUIVALENT FLUID PRESSURE OF 30 LB/FT., AND THE STRENGTH DESIGN METHOD.

LOAD FACTOR – CONC. DEAD LOAD ........ 1.5
" " – LIVE LOAD ........ 2.2
" " – VERT. EARTH LOAD ........ 1.3
" " – LATERAL EARTH LOAD ........ 1.0 TO 2.0

MATERIAL PROPERTIES
WELDED WIRE FABRIC ........ $F_y$ = 65,000 P.S.I.
CONCRETE ........ $F_c'$ = 5,000 P.S.I.

⊕ PREFORMED MASTIC MUST CONFORM TO A.A.S.H.T.O. MATERIALS SPEC. M198, **TYPE B.**

PRECAST CONCRETE BOX CULVERT BARREL DETAILS

*STATE OF WISCONSIN*
*DEPARTMENT OF TRANSPORTATION*
*DIVISION OF HIGHWAYS*
*BRIDGE SECTION*

Approved: Stanley W. Woods
Chief Bridge Engineer

Date: 1-'84

| | S(FT.) | R(FT.) | T(IN.) | SKEW | ANGLE A | ANGLE B | WL 1 | WL 2 | AL | AH | WH | BE 1 | BE 2 |
|---|---|---|---|---|---|---|---|---|---|---|---|---|---|
| INLET | | | | | | | | | | | | | |
| OUTLET | | | | | | | | | | | | | |

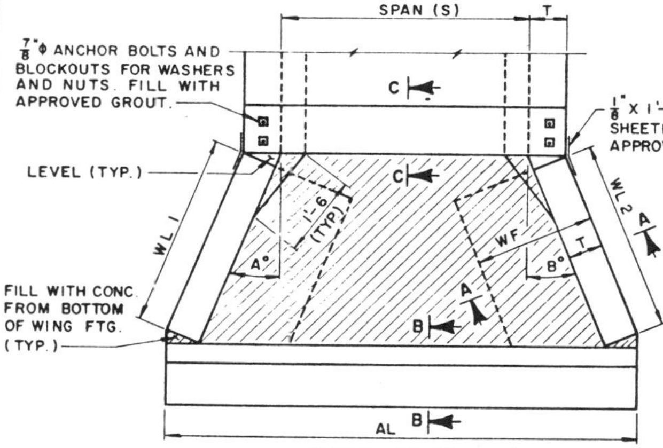


APRON PLAN
(NON-SKEWED STRUCTURE)

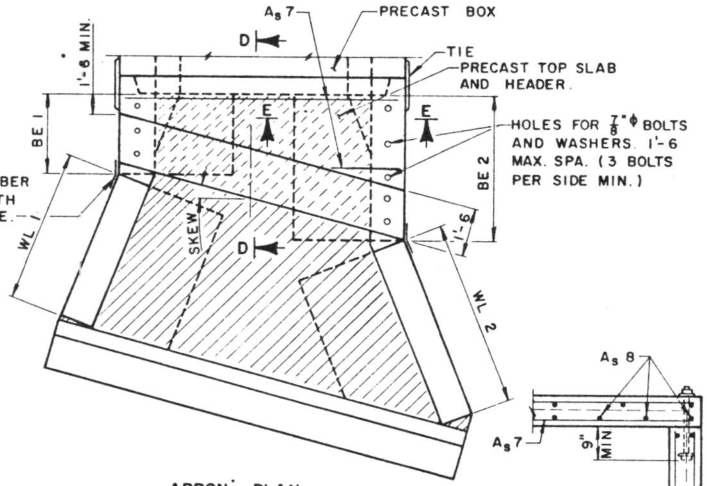


APRON PLAN
(SKEWED STRUCTURE)

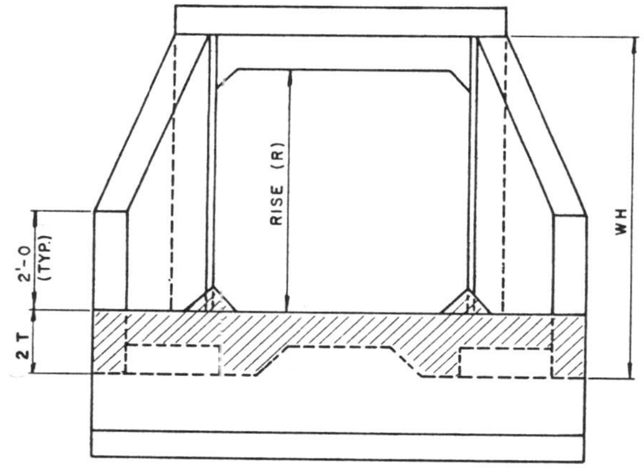


END VIEW

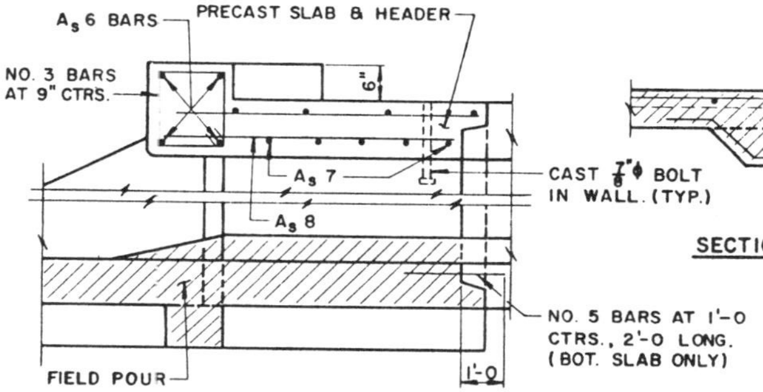


SECTION D

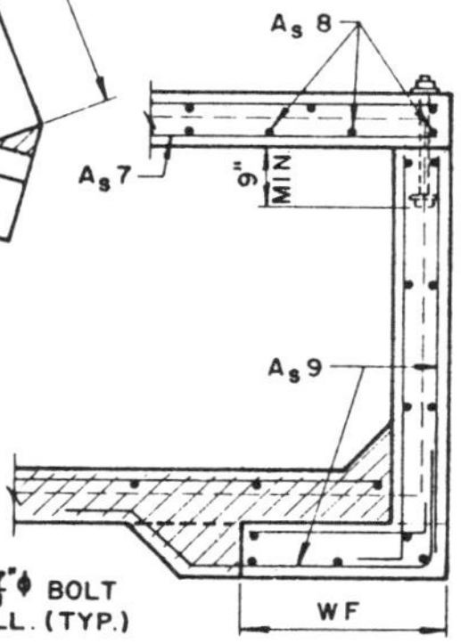


SECTION E

## GENERAL NOTES

CONCRETE COVER ON ALL REINFORCEMENT IN THE PRECAST ELEMENTS SHALL BE $1\frac{1}{2}$" UNLESS SHOWN OR NOTED OTHERWISE.

STEEL REINFORCEMENT MAY BE EITHER GRADE 60 DEFORMED BARS ($F_Y$ = 60,000 P.S.I.) OR WELDED DEFORMED-WIRE FABRIC OF EQUIVALENT STRENGTH ($F_Y$ = 65,000 P.S.I.)

THE ULTIMATE COMPRESSIVE STRENGTH OF THE FIELD POURED CONCRETE SHALL BE 3,500 P.S.I.

ALTERNATE DETAILS OF EQUAL STRENGTH AND HYDRAULIC CAPACITY TO THE DETAILS SHOWN ON THIS SHEET **MAY BE SUBMITTED** TO THE ENGINEER **FOR HIS APPROVAL.**

THE ULTIMATE COMPRESSIVE STRENGTH OF THE CONCRETE IN THE PRECAST ELEMENTS SHALL BE 4,000 P.S.I.

THE PRECAST ELEMENTS SHALL BE PROVIDED WITH SUITABLE LIFTING DEVICES FOR HANDLING AND PLACEMENT OF THE ELEMENTS.

VERTICAL CONSTRUCTION JOINTS THRU THE WALLS AND FOOTING WILL BE ALLOWED ONLY WITH THE APPROVAL OF THE ENGINEER. DETAILS MUST BE SHOWN ON THE SHOP DRAWINGS FOR APPROVAL.

THE AREA OF REINFORCING STEEL NOT IDENTIFIED IN SECTIONS EQUALS 0.17 IN²/FT.

THE MAXIMUM BAR SIZE OF GRADE 60 DEFORMED BARS, OTHER THAN THE $A_s$6 BARS, SHALL BE NO. 5.

THE $\frac{7}{8}$"ϕ ANCHOR BOLTS SHALL CONFORM TO THE REQUIREMENTS OF A.S.T.M. A575.

ALL EXPOSED CORNERS SHALL BE BEVELED $\frac{3}{4}$" ON THE SIDES OR TOOL EDGED WITH A $\frac{1}{2}$" MINIMUM RADIUS EDGER.

PRECAST CUT OFF WALLS MAY BE FIELD SPLICED BY EXTENDING THE REINFORCING STEEL FROM BOTH SEGMENTS TO BE SPLICED 1'-6 INTO THE SPLICE ZONE, LAPPING THE STEEL ±1'-6, AND FIELD POURING A SECTION OF CUT OFF WALL 1'-6 LONG.

PRECAST ELEMENTS MAY BE POURED IN PLACE AT THE OPTION OF THE CONTRACTOR.

| SPAN (S) | $A_s$6 BARS SKEW 0°-15° | 16°-30° | 31°-45° | $A_s$7 IN²/FT. | $A_s$8 IN²/FT. SKEW 0°-15° | 16°-30° | 31°-45° |
|---|---|---|---|---|---|---|---|
| 6'-0 | 4 NO. 6 | 4 NO. 6 | 4 NO. 7 | .35 | .17 | .17 | .17 |
| 7'-0 | 4 NO. 6 | 4 NO. 7 | 4 NO. 7 | .35 | .17 | .17 | .19 |
| 8'-0 | 4 NO. 6 | 4 NO. 7 | 4 NO. 8 | .42 | .17 | .19 | .24 |
| 10'-0 | 4 NO. 7 | 4 NO. 8 | 4 NO. 8 | .48 | .19 | .24 | .24 |

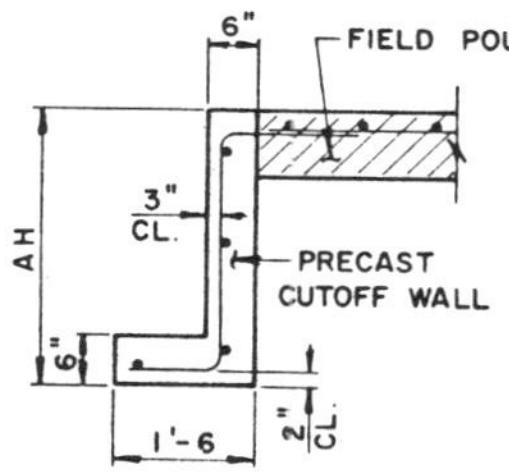


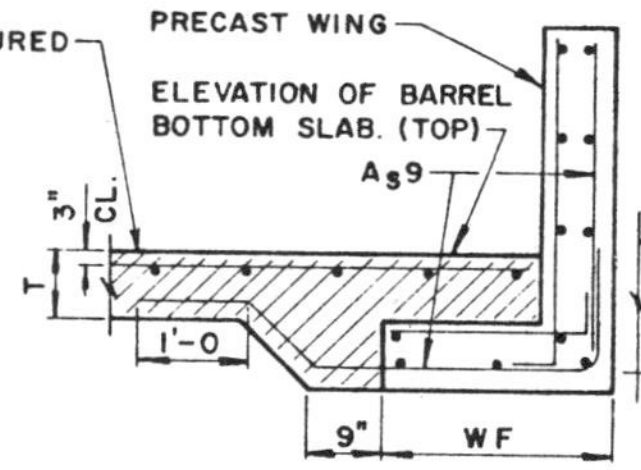


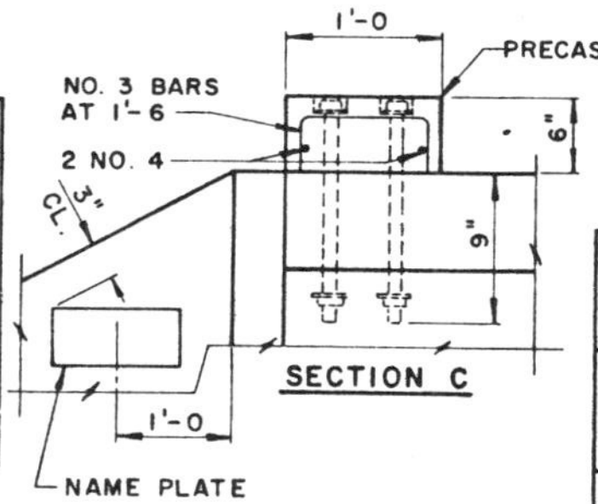


SECTION C

| RISE (R) | $A_s$9 IN²/FT. | WF |
|---|---|---|
| 4'-0 | .19 | 2'-6 |
| 6'-0 | .24 | 3'-6 |
| 8'-0 | .31 | 4'-0 |
| 10'-0 | .34 | 4'-9 |

**PRECAST WINGS, HEADERS AND CUT OFF WALLS FOR PRECAST CONCRETE BOX CULVERT**

STATE OF WISCONSIN
DEPARTMENT OF TRANSPORTATION
DIVISION OF HIGHWAYS
BRIDGE SECTION

Approved: Stanley W. Woods, Chief Bridge Engineer
Date: 1-'84

| DISTRICT | COUNTY | ROUTE | P. M. Miles - Total Project | SHEET NO. | TOTAL SHEETS |
|---|---|---|---|---|---|
| | | | | | |

CHIEF, OFFICE OF STRUCTURES DESIGN
REGISTERED CIVIL ENGINEER NO. 9013

DATE JULY 1, 1980

| SPAN | 4 | | | | | |
|---|---|---|---|---|---|---|
| HEIGHT | 2 | | 3 | | 4 | |
| Maximum Earth Cover ' | 10 | 20 | 10 | 20 | 10 | 20 |
| Conc. Roof T1 " | 7 | 7 | 7 | 7 | 7 | 7 |
| Conc. Walls T2 " | 6 | 6 | 6 | 6 | 6 | 6½ |
| Conc. Invert T3 " | 6 | 6 | 6 | 6 | 6 | 6½ |
| Reinf. Spacing " | 7 | 7 | 7 | 7 | 7 | 7 |
| Reinf. "a" Size Bar # | 6 | 6 | 6 | 6 | 6 | 6 |
| Reinf. "e" Size Bar # | 4 | 4 | 4 | 5 | 5 | 6 |
| Quan. Concrete CF/LF | 7.4 | 7.4 | 8.4 | 8.4 | 9.4 | 10.1 |
| Quan. Reinforcement lbs/LF | 48 | 48 | 50 | 55 | 58 | 67 |

| SPAN | 5 | | | | | | | |
|---|---|---|---|---|---|---|---|---|
| HEIGHT | 2 | | 3 | | 4 | | 5 | |
| Maximum Earth Cover ' | 10 | 20 | 10 | 20 | 10 | 20 | 10 | 20 |
| Conc. Roof T1 " | 7½ | 7½ | 7½ | 7½ | 7½ | 7½ | 7½ | 7½ |
| Conc. Walls T2 " | 6 | 6 | 6 | 6 | 6 | 7 | 6½ | 7½ |
| Conc. Invert T3 " | 6½ | 7 | 6½ | 7 | 6½ | 7 | 6½ | 7 |
| Reinf. Spacing " | 8½ | 6 | 8½ | 5½ | 8½ | 5½ | 8½ | 5½ |
| Reinf. "a" Size Bar # | 7 | 6 | 7 | 6 | 7 | 6 | 7 | 6 |
| Reinf. "e" Size Bar # | 4 | 4 | 4 | 4 | 5 | 5 | 6 | 6 |
| Quan. Concrete CF/LF | 8.9 | 9.2 | 10.0 | 10.2 | 11.0 | 12.0 | 12.5 | 13.7 |
| Quan. Reinforcement lbs/LF | 58 | 60 | 58 | 68 | 67 | 81 | 82 | 105 |

| SPAN | 6 | | | | | | | |
|---|---|---|---|---|---|---|---|---|
| HEIGHT | 3 | | 4 | | 5 | | 6 | |
| Maximum Earth Cover ' | 10 | 20 | 10 | 20 | 10 | 20 | 10 | 20 |
| Conc. Roof T1 " | 8 | 8 | 8 | 8 | 8 | 8 | 8 | 8 |
| Conc. Walls T2 " | 6 | 6 | 6 | 7 | 6½ | 7½ | 7 | 8½ |
| Conc. Invert T3 " | 7 | 8 | 7 | 8 | 7 | 8 | 7 | 8 |
| Reinf. Spacing " | 7½ | 5 | 7½ | 5 | 7½ | 5 | 7½ | 4½ |
| Reinf. "a" Size Bar # | 7 | 6 | 7 | 6 | 7 | 6 | 7 | 6 |
| Reinf. "e" Size Bar # | 4 | 4 | 5 | 5 | 6 | 6 | 7 | 6 |
| Quan. Concrete CF/LF | 11.7 | 12.3 | 12.7 | 14.2 | 14.2 | 15.9 | 15.9 | 18.3 |
| Quan. Reinforcement lbs/LF | 70 | 81 | 82 | 96 | 97 | 120 | 124 | 148 |

| SPAN | 7 | | | | | | | | | |
|---|---|---|---|---|---|---|---|---|---|---|
| HEIGHT | 3 | | 4 | | 5 | | 6 | | 7 | |
| Maximum Earth Cover ' | 10 | 20 | 10 | 20 | 10 | 20 | 10 | 20 | 10 | 20 |
| Conc. Roof T1 " | 8 | 9 | 8 | 9 | 8 | 9 | 8½ | 9 | 8½ | 9 |
| Conc. Walls T2 " | 6 | 6 | 6 | 7 | 6½ | 7½ | 7 | 8½ | 8 | 9½ |
| Conc. Invert T3 " | 8 | 9 | 8 | 9 | 8 | 9 | 8 | 9 | 8 | 9 |
| Reinf. Spacing " | 6½ | 5 | 6½ | 4½ | 6½ | 5½ | 6½ | 5½ | 6½ | 5½ |
| Reinf. "a" Size Bar # | 7 | 6 | 7 | 6 | 7 | 7 | 7 | 7 | 7 | 7 |
| Reinf. "e" Size Bar # | 4 | 4 | 5 | 5 | 6 | 6 | 6 | 7 | 7 | 7 |
| Quan. Concrete CF/LF | 13.7 | 14.9 | 14.6 | 16.8 | 16.1 | 18.5 | 18.1 | 21.0 | 20.7 | 23.8 |
| Quan. Reinforcement lbs/LF | 94 | 94 | 105 | 118 | 121 | 147 | 130 | 177 | 160 | 192 |

| SPAN | 8 | | | | | | | | | |
|---|---|---|---|---|---|---|---|---|---|---|
| HEIGHT | 4 | | 5 | | 6 | | 7 | | 8 | |
| Maximum Earth Cover ' | 10 | 20 | 10 | 20 | 10 | 20 | 10 | 20 | 10 | 20 |
| Conc. Roof T1 " | 8½ | 9½ | 8½ | 9½ | 8½ | 10 | 8½ | 10 | 8½ | 10 |
| Conc. Walls T2 " | 6 | 7 | 6½ | 7½ | 7 | 8½ | 8 | 9½ | 8½ | 10½ |
| Conc. Invert T3 " | 8 | 9½ | 8 | 10 | 8 | 10 | 8 | 10 | 8 | 10 |
| Reinf. Spacing " | 6 | 5 | 6 | 5 | 6 | 4½ | 6 | 5 | 6 | 4½ |
| Reinf. "a" Size Bar # | 7 | 7 | 7 | 7 | 7 | 7 | 7 | 7 | 7 | 7 |
| Reinf. "e" Size Bar # | 5 | 5 | 5 | 6 | 6 | 6 | 6 | 7 | 7 | 7 |
| Quan. Concrete CF/LF | 16.4 | 19.0 | 17.9 | 21.1 | 19.5 | 24.1 | 22.1 | 26.9 | 24.1 | 30.1 |
| Quan. Reinforcement lbs/LF | 123 | 151 | 131 | 171 | 137 | 198 | 160 | 215 | 192 | 252 |

| "d" bars, for earth covers of 2' and less | | | | | | | | |
|---|---|---|---|---|---|---|---|---|
| Span | 4' | 5' | 6' | 7' | 8' | 10' | 12' | 14' |
| Number | 5 | 6 | 7 | 8 | 9 | 10 | 12 | 16 |

| SPAN | 10 | | | | | | | | | | | |
|---|---|---|---|---|---|---|---|---|---|---|---|---|
| HEIGHT | 5 | | 6 | | 7 | | 8 | | 9 | | 10 | |
| Maximum Earth Cover ' | 10 | 20 | 10 | 20 | 10 | 20 | 10 | 20 | 10 | 20 | 10 | 20 |
| Conc. Roof T1 " | 8 | 10½ | 8 | 10½ | 8 | 10½ | 8 | 10½ | 8 | 10½ | 8½ | 11 |
| Conc. Walls T2 " | 8 | 8 | 8 | 9 | 8 | 9 | 8 | 10 | 8½ | 11½ | 9½ | 12½ |
| Conc. Invert T3 " | 8 | 11 | 8 | 11 | 8 | 11 | 8 | 11 | 8 | 11 | 8½ | 11 |
| Reinforcement Spacing " | 13 | 12 | 12 | 12 | 11 | 11 | 10 | 11 | 11 | 10 | 11 | 10 |
| "a" ⑥ Size Bar # | 6 | 7 | 6 | 7 | 6 | 7 | 6 | 7 | 6 | 6 | 6 | 6 |
| "b" Size Bar # | 6 | 7 | 6 | 7 | 6 | 7 | 6 | 7 | 6 | 6 | 7 | 7 |
| "b" Dimension "B" ' " | 2-11 | 2-11 | 2-11 | 3-0 | 2-11 | 3-9 | 2-11 | 3-9 | 2-11 | 3-9 | 2-11 | 3-9 |
| "b" Dimension "Bw" ' " | 4-10 | 5-0 | 4-10 | 5-0 | 4-10 | 5-0 | 4-10 | 5-0 | 4-10 | 5-0 | 4-7 | 4-7 |
| "c" Size Bar # | 6 | 7 | 6 | 7 | 6 | 7 | 6 | 7 | 7 | 7 | 7 | 7 |
| "c" Dimension "C" ' " | 3-4 | 3-4 | 7-4 | 4-8 | 7-7 | 4-8 | 8-2 | 4-8 | 7-7 | 4-8 | 7-7 | 4-8 |
| "e" ⑤ Size Bar # | 4 | 4 | 4 | 4 | 4 | 4 | 4 | 4 | 4 | 4 | 5 | 5 |
| Quan. Concrete CF/LF | 21.6 | 26.8 | 23.0 | 29.5 | 24.3 | 31.0 | 25.6 | 34.1 | 27.8 | 37.7 | 32.1 | 42.3 |
| Quan. Reinforcement lbs/LF | 191 | 267 | 233 | 285 | 260 | 325 | 300 | 339 | 314 | 327 | 360 | 373 |

| SPAN | 12 | | | | | | | | | | | | | |
|---|---|---|---|---|---|---|---|---|---|---|---|---|---|---|
| HEIGHT | 6 | | 7 | | 8 | | 9 | | 10 | | 11 | | 12 | |
| Maximum Earth Cover ' | 10 | 20 | 10 | 20 | 10 | 20 | 10 | 20 | 10 | 20 | 10 | 20 | 10 | 20 |
| Conc. Roof T1 " | 8½ | 12 | 8½ | 12 | 8½ | 12 | 8½ | 12 | 8½ | 12 | 9½ | 12 | 9½ | 12 |
| Conc. Walls T2 " | 8 | 10 | 8½ | 10½ | 8½ | 10½ | 8½ | 11½ | 9 | 12½ | 10½ | 13½ | 11½ | 15 |
| Conc. Invert T3 " | 8½ | 12½ | 9 | 12½ | 9 | 12½ | 9 | 12½ | 9 | 12½ | 9½ | 13 | 10 | 13 |
| Reinforcement Spacing " | 9 | 10 | 9 | 10 | 9 | 10 | 8 | 10 | 8 | 10 | 10 | 9 | 9 | 8 |
| "a" ⑥ Size Bar # | 6 | 7 | 6 | 7 | 6 | 7 | 6 | 7 | 6 | 7 | 6 | 6 | 6 | 6 |
| "b" Size Bar # | 6 | 7 | 6 | 7 | 6 | 7 | 6 | 7 | 6 | 7 | 7 | 7 | 6 | 7 |
| "b" Dimension "B" ' " | 3-1 | 3-2 | 3-9 | 3-9 | 3-9 | 3-9 | 3-7 | 3-9 | 3-7 | 4-4 | 3-7 | 4-4 | 4-4 | 4-4 |
| "b" Dimension "Bw" ' " | 6-0 | 5-10 | 6-2 | 6-7 | 6-2 | 6-7 | 4-11 | 4-11 | 4-11 | 4-11 | 4-11 | 4-11 | 4-4 | 4-4 |
| "c" Size Bar # | 6 | 7 | 6 | 7 | 6 | 7 | 6 | 7 | 6 | 7 | 7 | 7 | 7 | 7 |
| "c" Dimension "C" ' " | 4-4 | 3-7 | 8-1 | 5-7 | 8-1 | 5-7 | 8-1 | 5-7 | 8-1 | 5-7 | 8-1 | 5-7 | 8-1 | 5-7 |
| "e" ⑤ Size Bar # | 4 | 4 | 4 | 4 | 4 | 4 | 4 | 4 | 4 | 4 | 5 | 5 | 5 | 5 |
| Quan. Concrete CF/LF | 26.8 | 37.7 | 29.3 | 40.1 | 30.3 | 41.9 | 32.2 | 45.4 | 34.5 | 49.8 | 40.7 | 54.2 | 45.4 | 59.9 |
| Quan. Reinforcement lbs/LF | 295 | 362 | 353 | 409 | 365 | 420 | 402 | 413 | 415 | 424 | 440 | 471 | 468 | 534 |

| SPAN | 14 | | | | | | | | | | | | | | | |
|---|---|---|---|---|---|---|---|---|---|---|---|---|---|---|---|---|
| HEIGHT | 7 | | 8 | | 9 | | 10 | | 11 | | 12 | | 13 | | 14 | |
| Maximum Earth Cover ' | 10 | 20 | 10 | 20 | 10 | 20 | 10 | 20 | 10 | 20 | 10 | 20 | 10 | 20 | 10 | 20 |
| Conc. Roof T1 " | 9½ | 14 | 9½ | 14 | 9½ | 14 | 9½ | 14 | 10 | 14 | 10 | 14 | 10½ | 14 | 11 | 14 |
| Conc. Walls T2 " | 9½ | 11 | 9½ | 11½ | 9½ | 11½ | 10 | 12½ | 10½ | 14 | 11½ | 15 | 12½ | 16½ | 13½ | 17½ |
| Conc. Invert T3 " | 10 | 14½ | 10 | 14½ | 10 | 14½ | 10 | 14½ | 10½ | 14½ | 10½ | 15 | 11 | 15 | 11½ | 15 |
| Reinforcement Spacing " | 10 | 9 | 10 | 9 | 10 | 8 | 10 | 8 | 10 | 8 | 8 | 10 | 10 | 10 | 10 | 9 |
| "a" ⑥ Size Bar # | 7 | 7 | 7 | 7 | 7 | 7 | 7 | 7 | 7 | 7 | 6 | 7 | 6 | 7 | 6 | 7 |
| "b" Size Bar # | 6 | 7 | 7 | 7 | 7 | 7 | 7 | 7 | 7 | 7 | 6 | 8 | 7 | 8 | 7 | 8 |
| "b" Dimension "B" ' " | 2-9 | 4-2 | 3-4 | 4-4 | 3-4 | 4-6 | 3-4 | 4-6 | 3-6 | 4-6 | 3-6 | 4-6 | 3-6 | 4-6 | 4-9 | 6-0 |
| "b" Dimension "Bw" ' " | 6-5 | 7-0 | 7-0 | 8-4 | 7-0 | 8-4 | 6-0 | 6-0 | 6-0 | 6-0 | 6-0 | 6-0 | 5-0 | 6-0 | 4-6 | 6-0 |
| "c" Size Bar # | 7 | 7 | 7 | 7 | 7 | 7 | 7 | 7 | 7 | 7 | 7 | 8 | 8 | 8 | 8 | 8 |
| "c" Dimension "C" ' " | 4-0 | 4-2 | 6-9 | 6-9 | 5-10 | 5-10 | 6-0 | 6-4 | 8-11 | 6-4 | 8-11 | 6-4 | 8-11 | 6-4 | 8-11 | 6-4 |
| "e" ⑤ Size Bar # | 4 | 4 | 4 | 4 | 4 | 4 | 4 | 4 | 4 | 4 | 4 | 5 | 6 | 6 | 6 | 6 |
| Quan. Concrete CF/LF | 36.2 | 51.0 | 37.8 | 53.1 | 39.5 | 54.9 | 41.9 | 58.7 | 45.9 | 64.1 | 49.9 | 69.5 | 55.6 | 75.8 | 61.6 | 81.3 |
| Quan. Reinforcement lbs/LF | 374 | 471 | 449 | 522 | 446 | 587 | 449 | 580 | 492 | 596 | 537 | 605 | 564 | 634 | 604 | 759 |

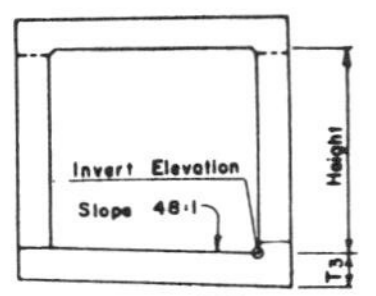


SLOPED INVERT

For exposed top, provide #4 @ 18 each way (2' lap with "c" bars) and adjust quantities

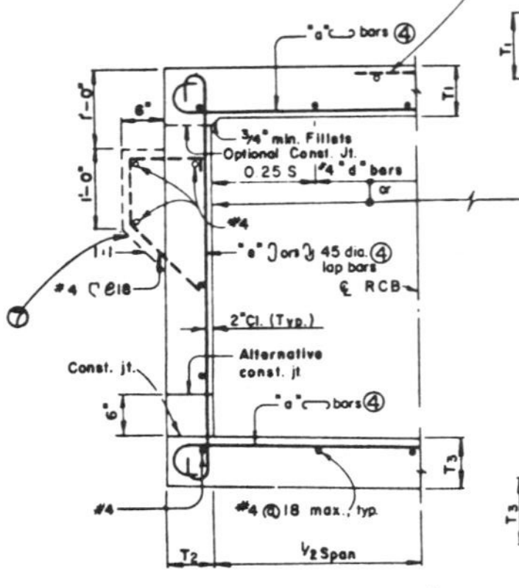


TYPICAL SECTION-SPANS 4' THRU 8'

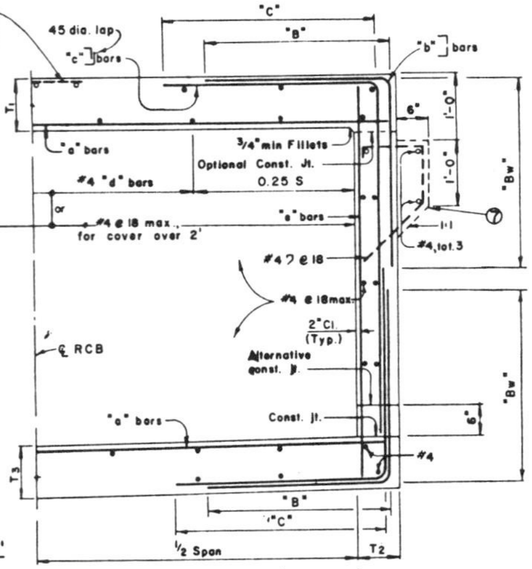


TYPICAL SECTION-SPANS 10' THRU 14'

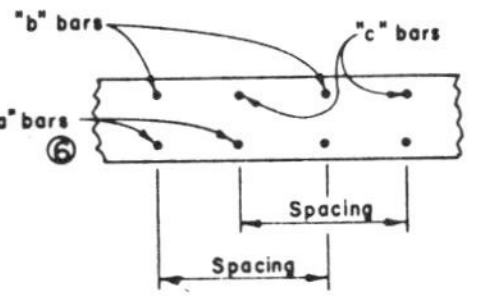


ROOF SECTION
SPANS 10' THRU 14'
Invert similar

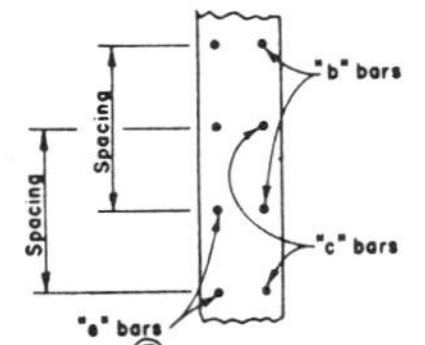


WALL SECTION
SPANS 10' THRU 14'

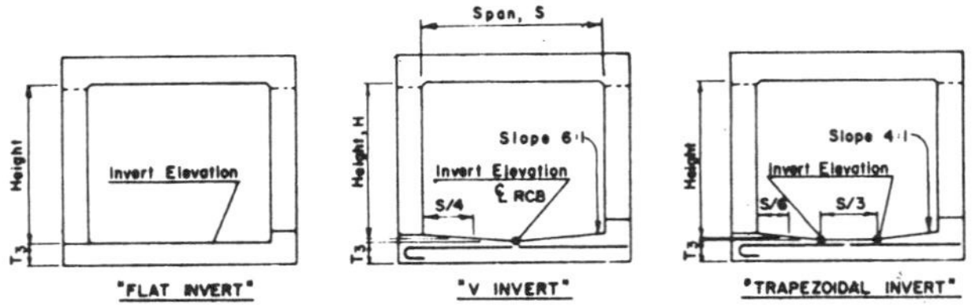


"FLAT INVERT" "V INVERT" "TRAPEZOIDAL INVERT"

ALTERNATIVE INVERTS

GENERAL NOTES

① For boxes with Span or Height less than any of those shown in table, use next greater size box concrete dimensions and reinforcement. Make necessary changes in bar lengths and quantities.

② Quantities are approximate and for design purposes only.

③ For boxes with Span or Height or cover greater than those shown in tables, a special design is required.

④ It is permissable to eliminate the 180° hooks on every other bar.

⑤ "e" bars are at half spacing.

⑥ "a" bars are at half spacing.

⑦ Provide paving notch when top is exposed and when pavement is P.C.C., and adjust quantities.

STATE OF CALIFORNIA
BUSINESS AND TRANSPORTATION AGENCY
DEPARTMENT OF TRANSPORTATION

SINGLE BOX CULVERT

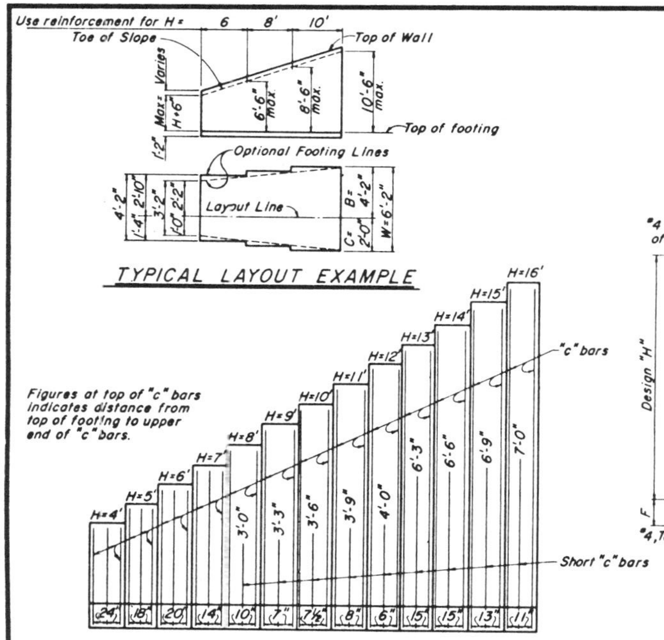


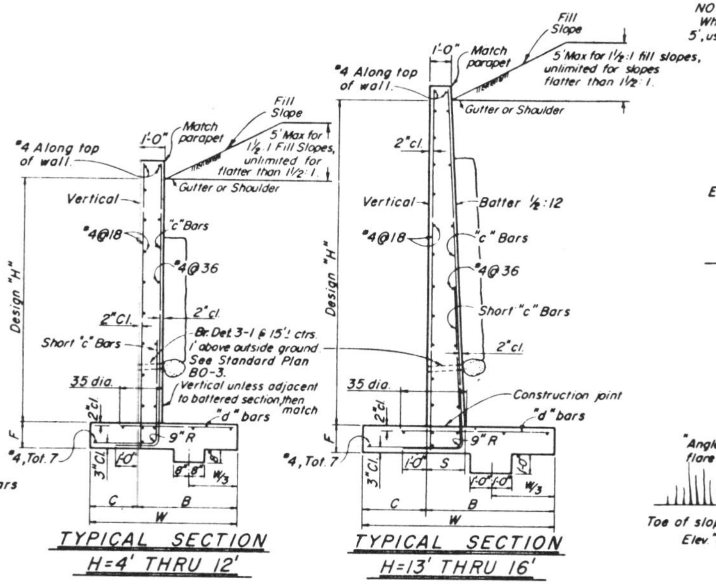


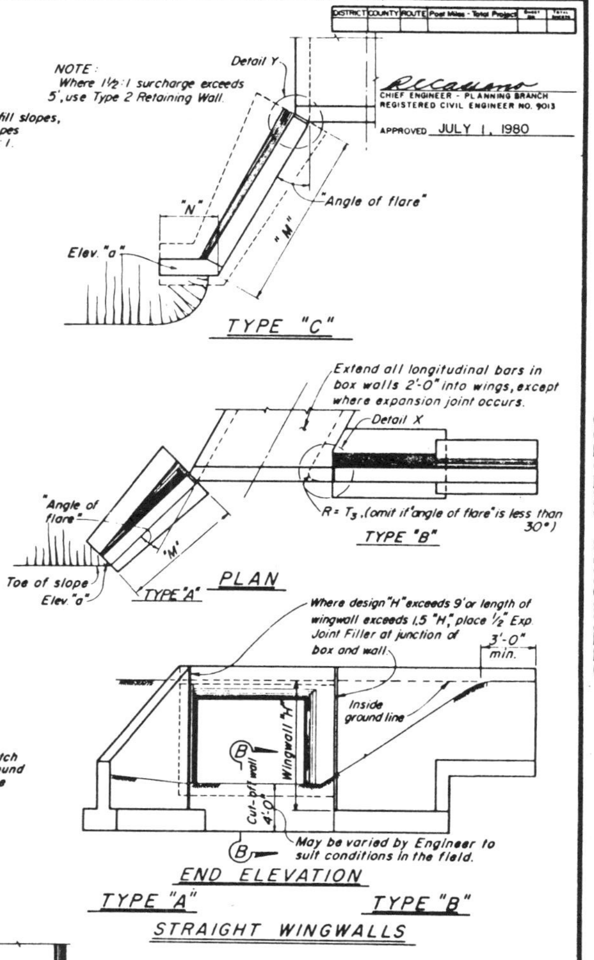


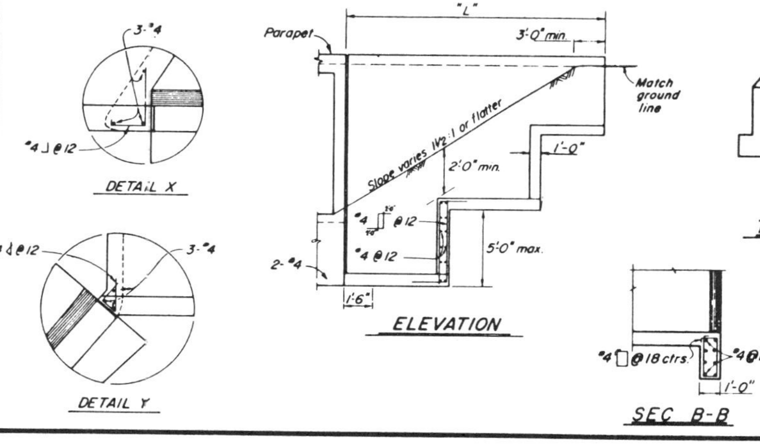


**REINFORCED CONCRETE WINGWALLS**

| "H" | 4' | 5' | 6' | 7' | 8' | 9' | 10' | 11' | 12' | 13' | 14' | 15' | 16' |
|---|---|---|---|---|---|---|---|---|---|---|---|---|---|
| W | 3'-2" | 3'-8" | 4'-2" | 4'-8" | 5'-2" | 5'-8" | 6'-2" | 6'-8" | 7'-2" | 7'-8" | 8'-2" | 8'-8" | 9'-2" |
| C | 1'-0" | 1'-2" | 1'-4" | 1'-6" | 1'-8" | 1'-10" | 2'-0" | 2'-2" | 2'-4" | 2'-6" | 2'-8" | 2'-10" | 3'-0" |
| B | 2'-2" | 2'-6" | 2'-10" | 3'-2" | 3'-6" | 3'-10" | 4'-2" | 4'-6" | 4'-10" | 5'-2" | 5'-6" | 5'-10" | 6'-2" |
| F | 1'-2" | | | | | | | | | | | | |
| Batter | None | | | | | | | | | ½:12 | | | |
| S | 1'-0" | 1'-0" | 1'-0" | 1'-0" | 1'-0" | 1'-0" | 1'-0" | 1'-0" | 1'-0" | 1'-6½" | 1'-7" | 1'-7½" | 1'-8" |
| "c" Bars | #4@24 | #4@18 | #5@20 | #5@14 | #5@10 | #5@7 | #6@7½ | #7@8 | #7@6 | #9@15 | #10@15 | #10@13 | #10@11 |
| "d" Bars | #4@24 | #4@18 | #5@20 | #5@14 | #5@10 | #6@14 | #7@15 | #8@16 | #7@12 | #8@15 | #9@15 | #9@13 | #9@11 |
| Conc. CY/L.F. | 0.32 | 0.38 | 0.44 | 0.49 | 0.55 | 0.61 | 0.67 | 0.73 | 0.79 | 1.02 | 1.10 | 1.18 | 1.26 |
| Reinf. lbs/L.F. | 13 | 16 | 19 | 25 | 30 | 37 | 49 | 62 | 76 | 73 | 90 | 104 | 125 |

Quantities do not include that portion above the design "H" limit.

Design Notes:

Unit Stresses: $f_s$ = 24,000 p.s.i, $f_c$ = 1,300 p.s.i, n = 10

Maximum Toe Pressure = 1½ Tons/sq.ft.

Elevations, length and angle of flare of wings may be varied by the Engineer to suit conditions encountered in the field. Walls designed for 2' liveload surcharge, 1½:1 sloping surcharge not to exceed 5' in elevation plus 2' liveload surcharge, or unlimited 2:1 surcharge.

Dimensions "H", "L", "M" "N", Elev. "a" and "Angle of flare" (as apply) are shown on the plans.

Wall height may be exceeded by 6" before going to next greater "H"

Eliminate cut-off wall if adjacent channel is paved and skew is 20° maximum.

| DISTRICT | COUNTY | ROUTE | Post Miles - Total Project | SHEET No. | TOTAL SHEETS |
|---|---|---|---|---|---|
| | | | | | |

R. E. Cassano
CHIEF ENGINEER - PLANNING BRANCH
REGISTERED CIVIL ENGINEER NO. 9013

APPROVED JULY 1, 1980

STATE OF CALIFORNIA
BUSINESS AND TRANSPORTATION AGENCY
DEPARTMENT OF TRANSPORTATION

## BOX CULVERT WINGWALLS
## TYPES "A", "B" & "C"

D84

## HYDRAULICS OF BRIDGE WATERWAY

8.4 HYDRAULICS OF BRIDGE WATERWAYS

With the information furnished by the District, as listed on the Structure Survey Report for Stream Crossings, the designer is able to proceed with the design of an adequate bridge size.

The main tool for selecting the bridge size is the "Hydraulics of Bridge and Culvert Waterways" program developed by FHWA and added to by the Wisconsin DNR. The outline of procedures to be used are discussed in this chapter.

(1) Procedures

A. Compute Discharge. See Section 8.2 for procedures.

The Corps of Engineers, Soil Conservation Service, State Department of Natural Resources and other agencies have made studies of some streams for which information is available.

B. Determine Slope. Survey crew going 1500 feet upstream and downstream or any distances needed to get representative slope of reach.

Water surface elevation from flood of record on upstream and downstream structure or any other rare flood.

From U.S.G.S. topographic maps.

From studies made by other agencies, the designer can get the water surface profile.

In some cases the streambed slope may not agree with the slope of the water surface profile at the 100 year flood. The slope may be less for downstream control and more for stream shortening due to channel changes or the flood waters not following the meandering streambed.

C. Select Section. The section must be along a line that is perpendicular to the stream flow at flood stage. Whenever a skewed crossing is studied, the skewed bridge opening must be projected onto the perpendicular cross section in its reduced size.

Make the section representative of the reach. Make use of aerial photos, topographic maps, or field investigations as an aid in this determination.

Eliminate slack water portions of the flood plain or portions not contributing to the downstream movement of water.

Plot the elevation view of the Section on cross-section paper for ease in dividing off sub-channels and using the program.

D. Assign "n" Values to Section. Mannings "n" values are assigned to the sub-channels. This is done by on site inspection, pictures taken at the section and use of aerial photos defining extent of each "n" value. Sample "n" values are published in the following books as an aid in picking values:

Open Channel Flow Hydraulics - Chow
FHWA Hydraulic Manuals
Roughness Characteristics of Natural Channels - U.S.G.S.

E. Determine Highwater Elevation. Superimpose the existing bridge on the cross-section. By use of the HOBCW computer program determine the $Q_{100}$ highwater value caused by the existing bridge. Check to see how it compares to the historic highwater. The designer should be aware that the historic highwater may not be a 100 year frequency storm.

Set allowable highwater elevation taking into consideration damage potential, past history, existing highwater, reliability of hydraulic information and criteria listed in 8.1(3).

F. Determine Bridge Size. Assume a bridge length and superimpose the bridge on the cross-section. By use of the HOBCW computer program determine the $Q_{100}$ highwater value. Compare to allowable for proper length.

Check the velocity through the opening and evaluate the scour damage probability. If roadway overflow is present, check to see those velocities are within tolerable limits.

A freeboard of 2 feet above the 100 year flood frequency highwater is the desirable minimum freeboard for all bridge type structures. On ~~secondary roads with~~ road overflow sections, economics may dictate a lesser freeboard, but precaution should be taken to evaluate the kind and amount of debris. Also, the gradeline may be controlled by near-by constraints such as railroad tracks, driveways, and businesses.

(2) Bridge Backwater Program

"Hydraulics of Bridge and Culvert Waterways" program (HOBCW) is used for the hydraulic analysis of waterways at bridge and culvert sites for determining backwater and velocities caused by structures.

The program was originally written by the FHWA based primarily on the principles discussed in "Hydraulics of Bridge Waterways", Hydraulic Series No. 1, by the FHWA, and revised by Roger D. Jutting of the Iowa Natural Resources Council. The Wisconsin Department of Natural Resources, using the 1973 version of Series No. 1 added the pressure flow and road overflow capability and three FHWA culvert review programs. One for box culverts, pipes, and plate arches. The program applies strictly to sub-critical flow through the bridge constriction. Flow is sub-critical when the Froude Number is less than one.

$$\text{Froude No.} = VN2/Vc$$

$$Vc = \sqrt{g \times AN2/T}$$

$$VN2 = \text{total } Q/AN2$$

Vc = critical velocity
g = acceleration of gravity
T = water surface width in opening
AN2 = area in opening
VN2 = velocity in opening

If the average velocity in the main channel is equal to or greater than the critical velocity, this program cannot be used to analyze bridge backwater. For super-critical flow see page 57 of "Hydraulics of Bridge Waterways" by FHWA. If the main channel flow is sub-critical, lengthen the bridge and re-run the program.

(4) Overflow Sections[1]

It is sometimes necessary to compute flow over highway embankments in combination with flow through bridge openings. The discharge equation and coefficients for flow over a highway embankment are given in this section.

The geometry and flow pattern for a highway embankment are illustrated in Figure 8.4.4. Under free flow conditions critical depths occur near the crown line. The head is referred to the elevation of the crown, and the length, L, in direction of flow is the distance between the points of the upstream and downstream embankment faces (edge of shoulder). The height of the embankment has no influence on the discharge coefficient.

The discharge equation is: $Q = CbH^{3/2}$

$Q$ = discharge
$C$ = coefficient of discharge
$b$ = length of flow section along the road normal to the direction of flow
$H$ = total head = $h + V_1^2/2g$

To measure b assume water surface at crown line to be 5/6 of maximum value of H. C is obtained by computing h/L and using Chart 8.4.1 or Chart 8.4.2, and curve for paved or gravel road.

The degree of submergence of a highway embankment is defined by ratio ht/H. The effect of submergence on the discharge coefficient (C) is expressed by the factor Kt as shown in Chart 8.4.3. The factor Kt is multiplied by the discharge coefficient (C) for free-flow conditions to obtain the discharge coefficient for submerged conditions. If the degree of submergence is greater than 0.9, the computed discharge may not be reliable. However, the portion of the total flow which passes over the road as compared to that which goes through the bridge is usually small, and thus a greater error can be tolerated in this computation.

To solve with use of a computer use DNR program "Hydraulics of Bridge and Culvert Waterways". (Bridge Program J5200104, RTP 44-2)

[1]Reference: Techniques of Water-Resources Investigations of the U.S. Geological Survey, Book 3, Chapter A5, page 26

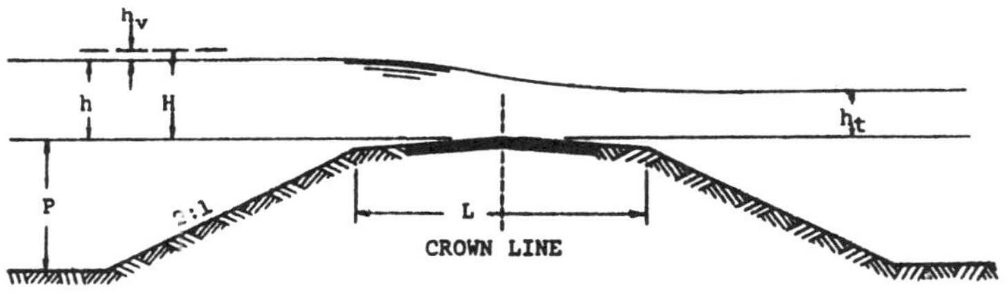


FIGURE 8.4.4 - DEFINITION SKETCH OF FLOW OVER A HIGHWAY EMBANKMENT

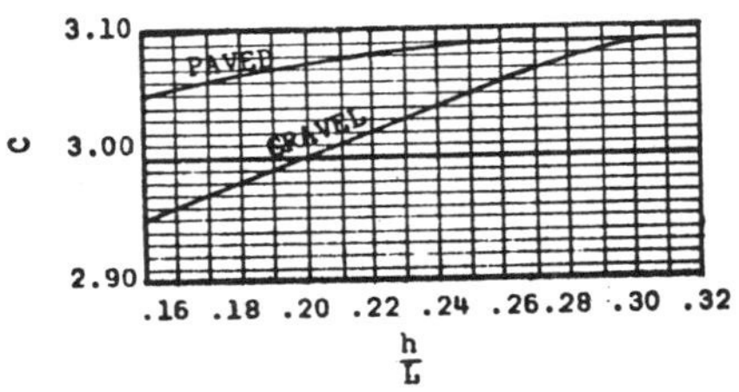


CHART 8.4.1 - DISCHARGE COEFFICIENTS FOR HIGHWAY EMBANKMENTS FOR h/L RATIOS > 0.15

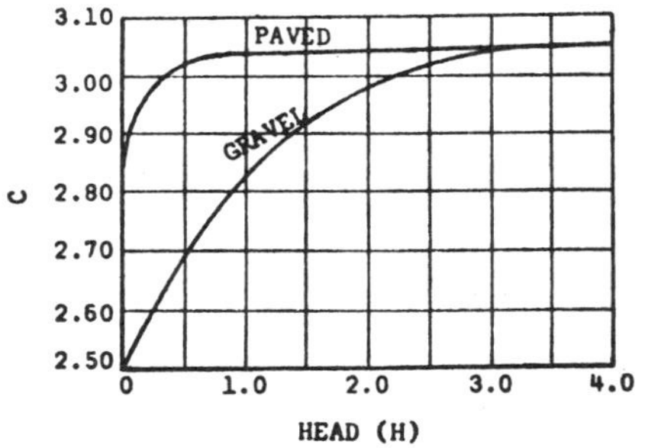


CHART 8.4.2 - DISCHARGE COEFFICIENTS FOR HIGHWAY EMBANKMENTS FOR h/L RATIOS < 0.15

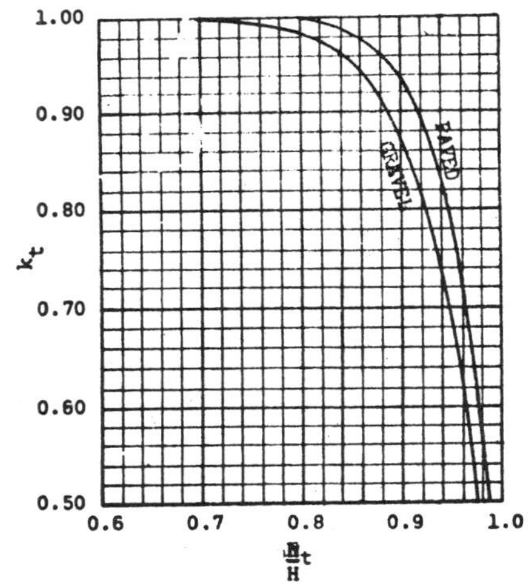


CHART 8.4.3 · DEFINITION OF ADJUSTMENT FACTOR, $k_t$, FOR SUBMERGED HIGHWAY EMBANKMENTS

(5) Local Scour[2]

The scour referred to here is the local scour around a pier and is measured from the general level of the bed at the bridge crossing. Excessive scour may endanger the structure, and all lowering of the streambed is considered additive in determining the requirements for a bridge foundation. If the bridge proper spans a natural or man-made contraction of the flooded stream, general scour may occur in addition to the local scour. General degradation may also occur due to stream straightening or an upstream dam.

The design criteria presented herein is subject to two restrictions; these being that the criteria are not applied to supercritical flow conditions ($F = V/\sqrt{gy} > 1$), and that there must exist a general movement of the bed material as bed load. Thus, since there may not be general movement of the flood plain, the procedure is not applicable to the design of relief bridges. For fine bed materials which might be carried in suspension, the design values should be used with caution or not at all, depending on the fineness.

Under certain circumstances of unsteady flow, scour depths in excess of equilibrium depth can occur. Because of this, it is recommended that the predicted depth of scour be increased by at least fifty per cent if the river is subject to flash floods.

The design curve presented in Chart 8.4.4 was drawn somewhat conservatively and represents the predicted local depth of scour for a rectangular pier, the sides of which are aligned with the current. If the pier is skewed to the current, the scour predicted from Chart 8.4.4 must be multiplied by a factor greater than unity. Values of this factor $K_{aL}$ are presented in Chart 8.4.5 as a function of the angle of attack and the length-width ratio of the pier at streambed.

For pier shapes other than rectangular, Table 8.4.1 presents multiplying factors for relative scour. These factors should be applied to the depth of scour found in Chart 8.4.4 but must not be used if the multiplying factors of Chart 8.4.5 are employed.

[2]Reference: Iowa Highway Research Board Bulletin No. 4, Scour around Bridge Piers and Abutments by Laursen & Toch; page 41

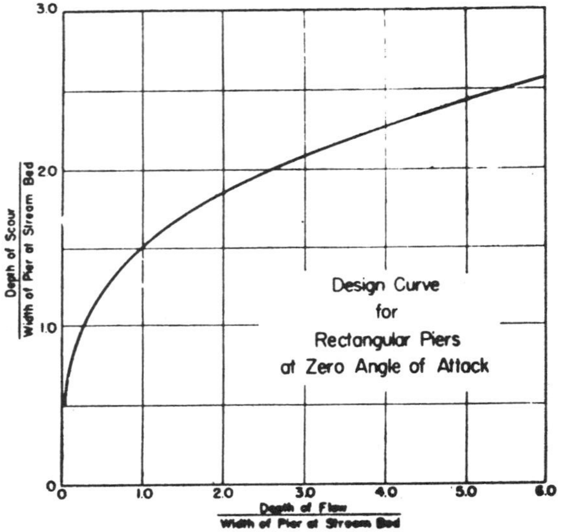


Basic design curve for depth of scour.

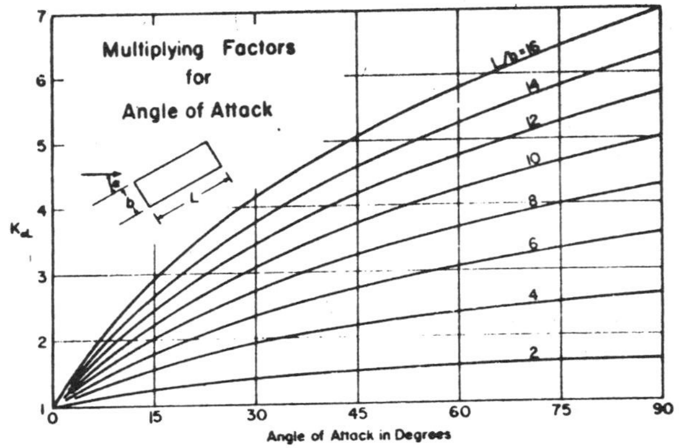


Chart 8.4.5 - Design factors for piers not aligned with flow

Table 8.4.1 - Shape coefficients $K_s$ for nose forms

(To be used *only* for piers aligned with flow)

| Nose form | Length-width ratio | $K_s$ |
|---|---|---|
| Rectangular | | 1.00 |
| Semicircular | | 0.90 |
| Elliptic | 2:1 | 0.80 |
| | 3:1 | 0.75 |
| Lenticular | 2:1 | 0.80 |
| | 3:1 | 0.70 |

(6) Spur Dikes

Flood waters, upstream of a long embankment across a flood plain, tend to rejoin the main stream by flowing against and parallel to the embankment. This water than enters the bridge constriction as a concentrated jet normal to the flow in the main channel, and the scour resulting from the junction of the two jets may be critical. Where a condition of excessive scour may develop, it may be remedied by constructing a spur dike curving upstream and tangent to the abutment face as shown in Figure 8.4.5. A spur dike serves to reduce the velocity of flow along the embankment, to direct the flow, and to alleviate the scour underneath the bridge.

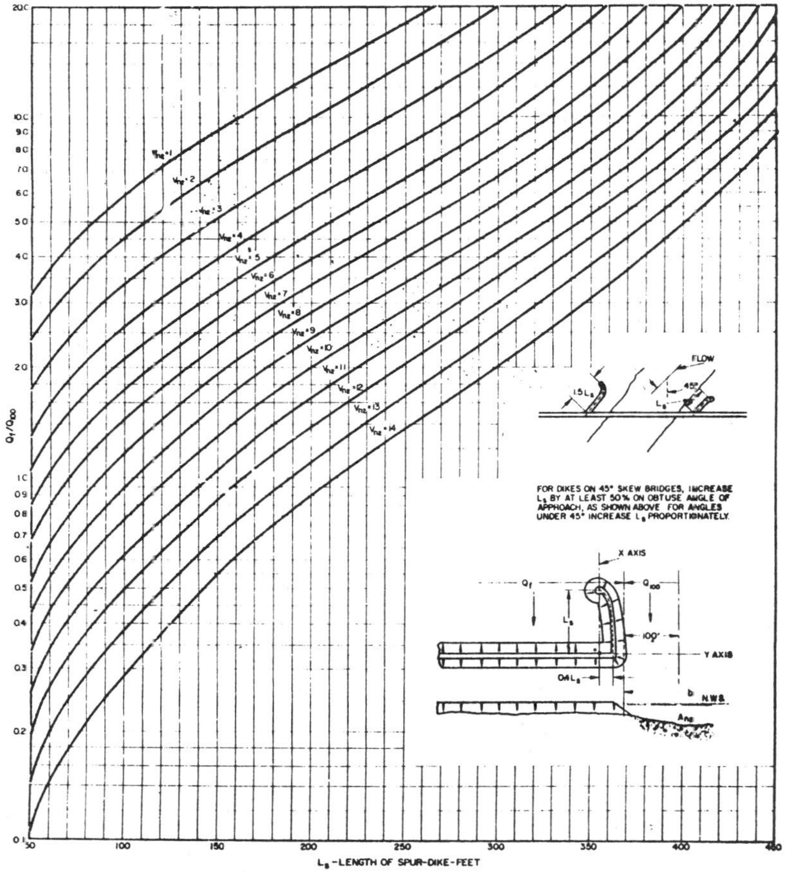


CHART 8.4.6 - CHART FOR DETERMINING LENGTH OF SPUR DIKES

The parameters are a spur dike discharge ratio, Qf/Q100, relating the flow over the left or the right flood plain to a specific portion of the flow under the bridge, a representative velocity adjacent to the abutment of the bridge, and the length of spur dike needed. The discharge ratio is shown as the ordinate, the length of dike as the abscissa, and the family of curves are for different values of the velocity, $V_{n2}$. (Average velocity through bridge opening in f.p.s.)

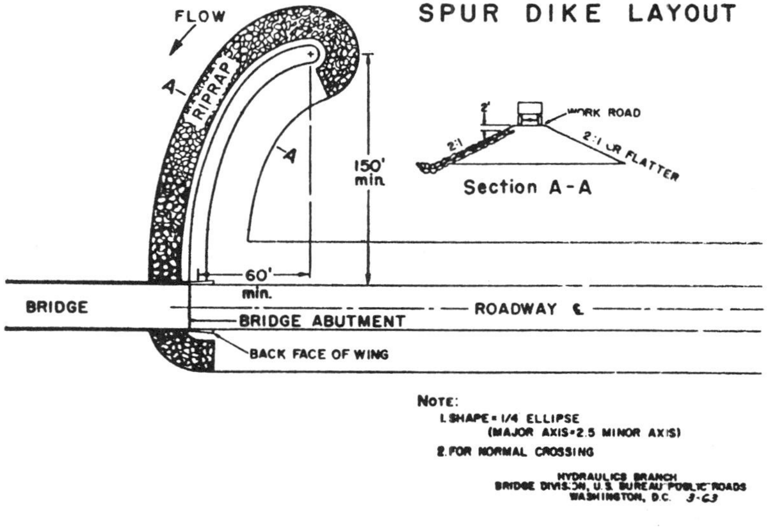


Figure 8.4.5 - Recommended layout for Spur Dike

## FLOOD HAZARD EVALUATION GUIDELINE

The following extracts from "Flood Hazard Evaluation Guidelines for Federal Executive Agencies," United States Water Resources Council, May, 1972, are applicable to the location, design and construction of Federal and Federal-aid highways. The WRC publication has been distributed by various sources. Additional copies are available from FHWA on request.

"(1) Determine first, ..., whether there is any need to evaluate the flood hazard at the site or structure location being considered."

"(3) Use the following to identify and evaluate the flood hazard:"

"(3A) The 100-year flood as the basic flood:"

"(3B) The flood hazard zone, defined as the area inundated by the basic flood:"

"(3C 2) In the Case Approach, a procedure to assure that any encroachment on the flood plain will permit conveyance of the basic flood without increasing flood heights or velocities to an extent which would cause significant upstream or downstream damage to existing or reasonably anticipated future development:"

"(3D) Floods greater or less than the basic flood as appropriate."

"(4) Determine whether there are existing laws or statutes of the Federal Government, rules or regulations of other Federal agencies, or laws, statutes, ordinances, etc., of State or local governments that provide standards for regulation of the floodplain under study. In cases where those standards are either more stringent than those based on these guidelines, or are applicable to situations or conditions not covered by these guidelines, they should be considered for the evaluation of flood hazard in that area."

"(5) Decide on the conditions under which an evaluation must be made to determine the impacts of including or excluding the use of site in a floodplain. Such evaluation must demonstrate clearly that the use of the site is to the advantage of society as well as to the user of such site."

"(6) Select the floods to be used in a flood hazard evaluation to fit conditions of the area being investigated."

"(10) Determine the effects of proposed highway construction in the floodplain and its vicinity, and of proposed upstream or local flood prevention or control measures, if any, on the elevations of the evaluation floods."

"(12) Adopt the policy of discouraging the construction of those roads, utilities, and other public facilities (except those crossing streams) within the most hazardous portions of the floodplain that aggravate flooding and encourage undesirable developments in that zone."

"(14) Delineate, or ensure the delineation of, on Federally owned properties, the elevation of the 100-year flood, and the elevations and dates of occurrence of floods of record whose magnitudes should be known by the public."

"(15) Encourage State and local agencies to keep a permanent record of information on each floodplain evaluated, the flood hazard evaluation procedure and decisions, the flood prevention or control measures proposed for upstream or local construction, the flood elevation delineations..."

# CHECKLIST FOR DRAINAGE STUDIES AND REPORT

Checklist for Drainage Studies and Reports

(This is a checklist of items to be considered for inclusion in hydraulic studies and reports. Numbers in parenthesis indicate references cited at the end of the checklist).

1. Preliminary drainage surveys (1)

a. Investigate potential problems

(1) Flood hazard - land use and development in flood plain, flood hazard studies(2)

(2) Channel stability - bank stability, bends and meanders, aggradation, degradation, necessity for channel change.

(3) Effects on the environment, fish and other wildlife, domestic water supplies, recreational resources, etc.

(4) Debris and ice

(5) Skew of crossing

b. Coordination with other agencies

(1) Permits required

(2) Existing and proposed water resource projects (PPM 50-4.2 (Volume 6, Chapter 1, Section 1, Subsection 4, of the Federal-Aid Highway Program Manual)

(3) Possibility of cooperative projects

2. Design

a. Site data

(1) Vicinity map

(a) Purpose - to show proposed highway alinement and reach of river, bends and stream meanders, general flow directions

(b) Type

1. USGS quadrangle sheet or map of equal detail

2. Aerial photo

(2) Site map:

(a) Purpose - for use in estimating flood flow distribution; to locate cross sections of stream; to show location of proposed encroachment and structure(s), alinement of piers, skew of crossing, stream controls, existing encroachments, existing highway structures.

(b) Type

1. Specially prepared map showing 1- or 2-foot contours, vegetation and manmade improvements

2. In some cases, cross sections normal to floodflow are acceptable in lieu of map. A minimum of three sections are desirable including cross sections upstream, at crossing, and downstream

(3) Existing structures (including relief or overflow structures):

(a) Locate existing structures with respect to proposed crossing or encroachment

(b) Describe each fully, giving-

1. Type, including span lengths and number of spans, bent design, pier orientation, culvert size, number of cells

2. Foundation type (spread footing, piling) founding depth

3. Scour history at abutments and bents, culvert outlets; head cutting; stream aggradation, degradation

4. Cross section beneath structures, noting clearance to superstructure and skew with direction of current during extreme floods

5. Flood history, highwater marks (dates and elevation), nature of flooding (including overtopping), damages and sources of information

6. Damage from abrasion, corrosion, wingwall failure, culvert end failure

(4) Locate and determine elevations of highwater marks along stream, giving dates of occurrence

(a) Describe or list critical flood elevations of interest in evaluating possible damage (record datum used)

(b) Refer to flood hazard studies for area[2]

(5) Comment on drift, ice, nature of streambed, bank stability, bends, meanders, vegetative cover, land use

(6) Photographs showing existing structures, past floods, main channel and flood plain to help in evaluating a location and documenting conditions existing prior to construction

(7) List factors affecting water stages

(a) Highwater from other streams

(b) Reservoirs - existing or proposed and approximate date of construction

(c) Flood control projects ( give status )

(d) Tides

(e) Other controls

b. Hydrologic analysis (site inspection should be made by engineer making hydrologic and hydraulic analysis)(3)(7)

(1) List available flood records

(2) Determine drainage area above proposed construction[3]

(3) Evaluate potential for changes in watershed characteristics which would change magnitude of flood peaks; e.g., urbanization, channelization

(4) Plot flood-frequency curve

(5) Plot stage-discharge-frequency curve

(6) Determine distribution of flow and velocities for several discharges or stages in natural channel for existing conditions.

c. Hydraulic analysis

(1) Design of bridge waterways[5]

(a) Determine permissible upstream water surface elevations

(b) Compute backwater for various trial bridge lengths and discharges

(c) Select design flood and waterway design (Paragraph 4.c.)

(d) Provide for conveyance of 100-year flood

(e) Estimate scour depth at piers and abutments[6]

(f) Design riprap for bank protection and scour attenuation devices, if required.

(g) Investigate need for spur dikes

(h) Show final layout in plan and profile

1. Show design discharge, elevations, and frequency

2. Show discharge and elevations of 100-year flood

(i) Comment on:

1. Selection of design flood

2. Conveyance of 100-year flood

3. Channel change (if provided)

4. Effects of construction

5. Need for stream controls to protect highway

(2) Design of culverts

(a) Determine allowable headwater elevation

(b) Compute and plot performance curves for trial culvert sizes[4]

(c) Evaluate abrasion and corrosion potential

(d) Select design flood and culvert design (Paragraph 4.c.)

(e) Provide for conveyance of 100-year flood

(f) Evaluate need for debris control

(g) Evaluate need for outlet protection

(h) Investigate need for protection against buoyancy and/or failure by separation at joints[9]

(i) Show final layout in plan and profile

1. Show design discharge, elevations, and frequency

2. Show discharge and elevations of 100-year flood

(j) Comment on:

1. Selection of design flood

2. Conveyance of 100-year flood

3. Channel change (if provided)

4. Effects on stream stability

5. Provision for fish passage

(3) Design of longitudinal encroachments

(a) Determine the effect of proposed encroachment on water-surface profile using various roadway design alternatives

(b) Select design flood and roadway design (Paragraph 4.c.)

(c) Provide for conveyance of 100-year flood

(d) Evaluate effects on scour and deposition in channel

(e) Design embankment, bank and channel protection needed

(f) Show final layout in plan and profile

1. Show design discharge, elevations, and frequency

2. Show discharge and elevations of 100-year flood

(g) Comment on:

1. Selection of design flood

2. Conveyance of 100-year flood

3. Channel change (if provided)

4. Effects on stream stability

5. Effects on stream biology

# FHWA Flood Plain Management Decisionmaking Process

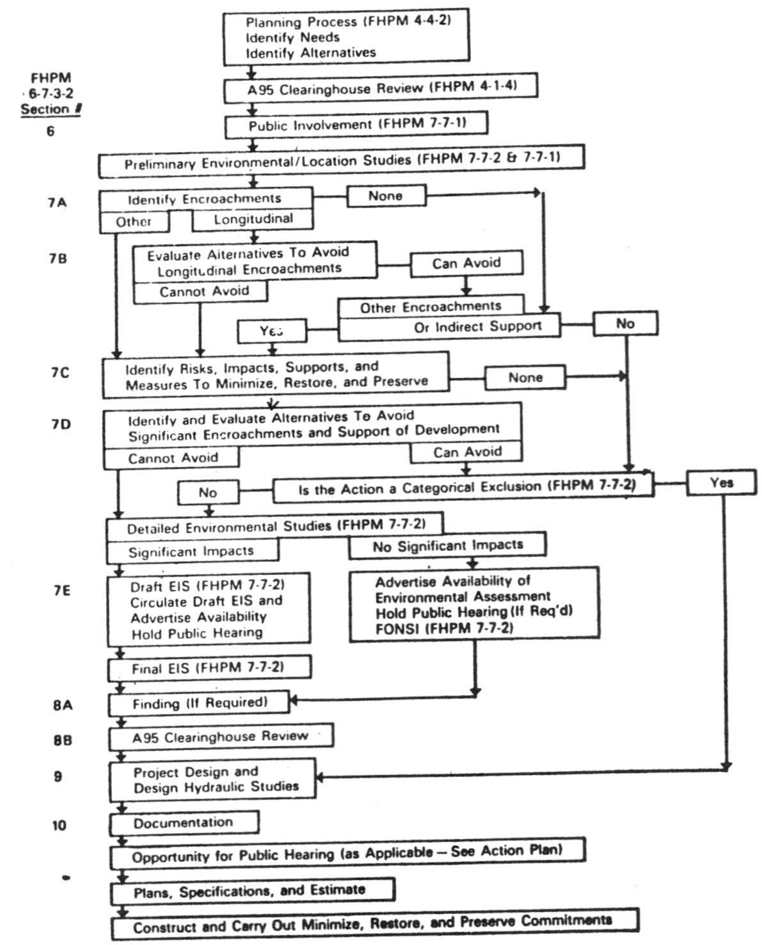

SAMPLE HYDRAULIC REPORT

For
PLATTE ROAD BRIDGE & APPROACHES
PLATTE RIVER
STRUCTURE B-22-68
Town of Harrison
Grant County
State Project I.D. 5629-0-00

INTRODUCTION

The Town of Harrison, with the approval of the Grant County Highway Commission, proposes to replace a narrow and structurally deteriorated bridge carrying Platte Road over the Platte River. The project is located in Section 4, T 3 N, R 2 W.

This report shows the hydraulic effects of the proposed structure on the flow characteristics of the Platte River at the structure site. Included in this report is information concerning the magnitude and frequency of floods, river stage to discharge relationships, backwater and highwater elevation computations, and additional hydraulic related information.

DATA AVAILABLE

Hydraulic information in this report is based upon accepted procedures as outlined in the following publications:

1. "Estimating Magnitude and Frequency of Floods in Wisconsin" by Duane H. Conger.
2. "Hydraulics of Bridge Waterways", Hydraulic Design Series No. 1 - Federal Highway Administration.
3. "A Uniform Technique for Determining Flood Flow Frequencies" - Water Resources Council.
4. "Wisconsin Bridge Manual" by the Wisconsin Department of Transportation - Bridge Section.
5. "Drainage-Area Data for Wisconsin Streams", by the United States Department of the Interior, Geological Survey.

All river and flood-plain characteristics are based upon conditions as observed on July 17, 1980. All elevations are referred to mean sea level as established by the United States Geological Survey.

SITE DESCRIPTION

The site description of the proposed bridge replacement is illustrated in the following enclosures to this report:

1. Preliminary Plan and Profile Sheets
2. Contour Map
3. Site Location Exhibits
4. Photographs

The size of the Platte River drainage basin at the proposed site is approximately 130 square miles. Both the upstream and downstream reaches of the river are well defined with all normal flow being confined within a definite floodway. The adjacent upstream and downstream reaches are lined with heavy grasses, light brush, and trees.

Based upon field measurements, the slope of the channel is approximately 0.00096 feet per foot.

The upstream low steel truss carries Ellenboro Road over the Platte River. The crossing is approximately 4.5 miles upstream from the project site. The upstream structure is a single span with a 67.5 foot clear opening and supported on full retaining wall abutments. A roadway overflow section does exist. The overtopping flood will have an exceedance probability of 33% annually.

The downstream structure is also a single span low steel truss providing a clear opening of 78.0 feet and supported on full retaining wall abutments. This structure is located approximately 0.5 miles downstream from the project site. A roadway overflow section also exists at this structure.

MAGNITUDE AND FREQUENCY OF FLOODS

A flood frequency - discharge analysis of the Platte River drainage basin at the proposed site was made to estimate the design discharge for the base flood. For Conger's flood frequency equations the following parameters were used:

1. Drainage area = 130 square miles
2. Main channel slope = 11.5 feet per mile
3. Lake and marsh area = 0.0% + 1.0 = 1.0%
4. Areal factor = 1.80
5. Annual snowfall = 39" - 20 = 19"

The relationship between flood-frequency and discharge assuming Conger's Equations as shown in Table 1, of the enclosed Hydraulic Report Exhibits. (Sample of Table 1 as follows):

TABLE 1

Flood Frequency Vs. Discharge
(Conger's Method)

| Recurrence Interval (Years) | Exceedance Probability | Discharge (Cubic Feet Per Second) |
|---|---|---|
| 2 | 50% | 4590 |
| 5 | 20% | 8080 |
| 10 | 10% | 10780 |
| 25 | 4% | 14870 |
| 50 | 2% | 22360 |
| 100 | 1% | 28460 |

A log-Pearson Type III analysis was made on the Platte River near Rockville (5-4140) drainage basin as a basis of comparison between Conger's Equation and those of known gaging records. Comparing the sites, assuming all factors to be similar except drainage area size, the relationship between flood frequency and discharge is established as shown in Table 2 in the Hydraulic Report Exhibits.

The Flood Frequency vs. Discharge correlation shown in Table 2 is the direct result of relating the log-Pearson III analysis of the gage near Rockville to the project site by the ratio of their areas. A weighted skew value of the gaging station records was used for the log-Pearson Type III analysis.

The relationship in Table 2 was plotted with the relationship given by Conger's Equations for the site. The following relationship shown in Table 3 in the Exhibits is based upon the above analysis and assumed for the project site.

STAGE - DISCHARGE RELATIONSHIP

Based upon the natural upstream cross-section of the channel and immediate overbanks, as shown in Exhibit "A", the relationship between stage elevation and discharge can be established as shown in Table 4. (Exhibits)

Interpolating the curve for a discharge of 21550 cfs, (base flood discharge) the maximum stage elevation under natural conditions is approximately elevation 694.68.

ANALYSIS OF HYDRAULIC CHARACTERISTICS

The hydraulic characteristics of the channel for both the existing and proposed condition were calculated to determine the backwater and highwater elevation. Based upon the natural upstream cross-section shown in Exhibit "A" and a design discharge of 21550 cfs, the main channel carries only 40% of the flow while the remaining discharge is carried by the overbanks.

EXISTING STRUCTURE AND APPROACHES

The existing 3-span structure is a combination single span high steel truss with single span steel deck girder approach spans. The abutments are vertical face semi-retaining wall and the piers are a combination solid shaft with steel caissons. According to reports by local residents, the historic high water elevation has been above the bridge deck elevation of 697$\pm$. It has also been noted that roadway overflow occurs on an annual basis.

Assuming a low roadway elevation of 689.0$\pm$, the frequency of roadway overflow is approximately 1.6 years for an equivalent discharge of 3950 cfs.

At the base flood discharge of 21550 cfs, a combination bridge low flow and roadway overflow condition exists which results in less than 0.1 feet of backwater and a highwater elevation of 694.8$\pm$. Flow through the structure opening accounts for 4990 cfs with the remaining 16560 cfs consisting of roadway overflow. The waterway opening is approximately 1440 square feet resulting in a velocity thru the opening of 3.5 feet per second.

DISCUSSION OF STRUCTURE SIZING

The potential for flooding of the Platte River at this site is extremely severe and therefore has considerable impact on the structure type and length proposed. Historic records furnished by local residents indicates the Platte River overflows the approach roadways on an annual basis. In 1950, a low frequency flood occurred and destroyed several bridges downstream from the project site. It has also been noted that considerable debris is carried by the Platte River at flood stage.

Adjacent land within the Platte River floodplain is primarily agricultural cropland and has little susceptibility to damage by flood. There are no buildings or residents within the floodplain which would be affected by an increase in the highwater elevation of the regional flood. Dependent on the season in which a major flood occurs, only minor damage and loss of cropland is anticipated. An increase in the highwater elevation is appropriate at this site to reduce the overall structure length to convey the design flood frequency.

This portion of the Platte River is not classified as trout water by the Wisconsin Department of Natural Resources. Water quality is somewhat substandard due to surface runoff from the adjacent farmland. Placement of fill material within the river banks is not considered to be an environmental concern; however, is not considered to be desirable.

The existing structure provides adequate navigational clearance above the normal river stage elevation. Passage of floating debris is not considered to be a major concern because of the roadway overflow provided both north and south of the structure. There is no evidence of structural damage to the existing bridge due to floating debris or flood related instances.

Several alternative alignments and structure types were considered and evaluated in terms of cost and potential for damages. Because of the extremely wide floodplain and severity of flooding it was not considered cost effective to raise the profile gradeline to eliminate roadway overflow. In addition to construction cost, raising the profile gradeline will also require major relocation to provide adequate horizontal alignment.

Replacement of the structure at the existing location was not considered feasible because of poor horizontal alignment with the approach roadways. This alternative was not considered to be in the best interest of the traveling public.

The recommended alternative consists of minor relocation downstream to improve the horizontal alignment and place the structure on a tangent section. This alternative will require the acquisition of additional right-of-way consisting of a narrow strip of agricultural cropland.

Because of the highwater elevation, feasible structure types were extremely limited without a major revision to the profile gradeline. The use of a steel or prestressed girder structure was restricted because of inundated flow. A multiple span reinforced concrete slab structure was recommended to reduce the potential for damages and allow for a lower profile gradeline to minimize increases in the highwater elevation.

Several alternative profile gradelines were evaluated to allow for passage of the flood waters. The first alternative resulted in no additional backwater but consisted of three back-to-back vertical curves providing minimum stopping sight distance. The major drawback of this alternative involved no improvement of the annual roadway overflow frequency (100% exceedance probability).

The second alternative improved the roadway overflow frequency to 3 years (33% exceedance probability) but still utilized three back-to-back vertical curves again providing minimum stopping sight distance.

The third or recommended alternative is the use of a constant grade to provide a roadway overflow frequency of 3+ years. (33% exceedance probability) This alternative extends the overall project length beyond Federal Aid participation but is endorsed by both the Town of Harrison and the Grant County Highway Department. A considerable amount of fill material is required south of the river to meet the proposed profile gradeline. Because of anticipated roadway overflow at a 33% exceedance probability, heavy riprap will be required along the roadway fill slopes.

The overall structure length considered ranged from 88.8-feet to 120.8-feet. The minimum structure length placed the toe of the abutment fill slope at the bottom of the natural river bank. This geometry resulted in the minimum length without placement of fill material into the Platte River. The maximum velocity through the waterway opening is 9.2 feet per second at a discharge less than the base flood. This high velocity was a prime consideration for not reducing the waterway opening by placement of additional fill material into the Platte River.

Increasing the structure length beyond the minimum did not result in a proportional increase in waterway opening or a reduction in backwater effects. Raising the profile gradeline to eliminate the combination of pressure flow and roadway overflow was also considered; however, because of excessive backwater and additional construction cost, this alternative was not considered feasible.

## PROPOSED STRUCTURE AND APPROACHES

The proposed structure is a three span reinforced concrete haunched slab with span lengths of 27.0, 37.0, and 27.0 feet having an overall length of approximately 93 feet. The structure provides for a 24.0 foot clear roadway width and is placed on 5° right hand forward skew to better fit the river geometry.

It is proposed to utilize concrete encased pile bent piers to reduce the construction cost by elimination of the need for additional piling and a pier footing.

The abutments will be standard sill type supported on steel bearing piles. Heavy riprap will be placed along the abutment fill slopes and river banks to mitigate the affects of erosion.

For the base flood, a combination pressure flow and roadway overflow condition exists resulting in a backwater 1.1 feet above the existing condition and a highwater elevation of 695.9±. Flow through the waterway opening of 970 square feet consists of 7140 cfs, while roadway overflow accounts for the remaining 14410 cfs. The velocity through the waterway opening is 7.3 feet per second for the base flood but 9.3 fps maximum at a lower stage elevation.

The proposed approach roadways provide for an overtopping flood with a 33% chance of being exceeded annually, which is an improvement to the existing condition. This frequency appears to be adequate based on the level of service provided by this facility.

## SUMMARY

The hydraulic data for the proposed structure and approaches can be summarized as follows:

**Base Flood**

| | |
|---|---|
| $Q_{100}$ | 21550 cfs |
| Drainage Area | 130 square miles |
| $Highwater_{100}$ Elevation | 695.9± ft. |
| Velocity (Thru Bridge) | 7.3 fps |
| Velocity (Overflow) | 5.6 fps |
| Waterway Area (Thru Bridge) | 970 square feet |
| Waterway Area (Overflow) | 2575 square feet |

**Design Roadway Frequency**

| | |
|---|---|
| Overflow Frequency | 3 years |
| $Q_3$ | 6350 cfs |
| $Highwater_3$ Elevation | 691.7± ft. |

# SAMPLE BRIDGE SITE REPORT

**ARKPORT RIVER AT EXAMPLESVILLE, PARAGON**
**FLOODFLOW CHARACTERISTICS AT PROPOSED BRIDGE SITE**
**State Highway 17**
**Whitehall - Pitt Section**
**Lancaster County**

## INTRODUCTION

The Paragon State Department of Transportation plans to construct a bridge over the Arkport River at Exampleville, Lancaster County. To aid in the design of the highway grade, structure, and waterway opening, an analysis of floodflow characteristics was requested by Mr. H. J. Skowron, Deputy Chief Engineer. This report provides information on magnitude and frequency of floods, relation of stage to discharge, flow distribution, backwater and scour.

Data and computations supporting the information given in this report are available for inspection at the Department of Transportation's regional office in London, Paragon.

## DATA AVAILABLE

Hydrologic information appearing in this report is based primarily on flood data collected at gaging stations and compiled by the U.S. Geological Survey in annual reports of surface water records. Additional data includes:

1. Plan drawings for the site and proposed structure prepared by the Department of Transportation.
2. Investigations and field work by the U.S. Geological Survey and Department of Transportation.
3. U.S. Geological Survey topographic maps for determination of drainage area at the site.

(WIS DOT)

Computations are based on channel and flood-plain conditions as observed in April 1970. All elevations mentioned refer to mean sea level (MSL), datum of 1929, which is used by the Department of Transportation.

SITE DESCRIPTION

Figure 1 is a vicinity map which shows the proposed highway alinement, reach of river, floodflow pattern, channel cross sections, location of proposed bridge, and alinement of piers at the proposed construction site. At the site the size of the Arkport River's drainage area is about 751 square miles. Flood regulation

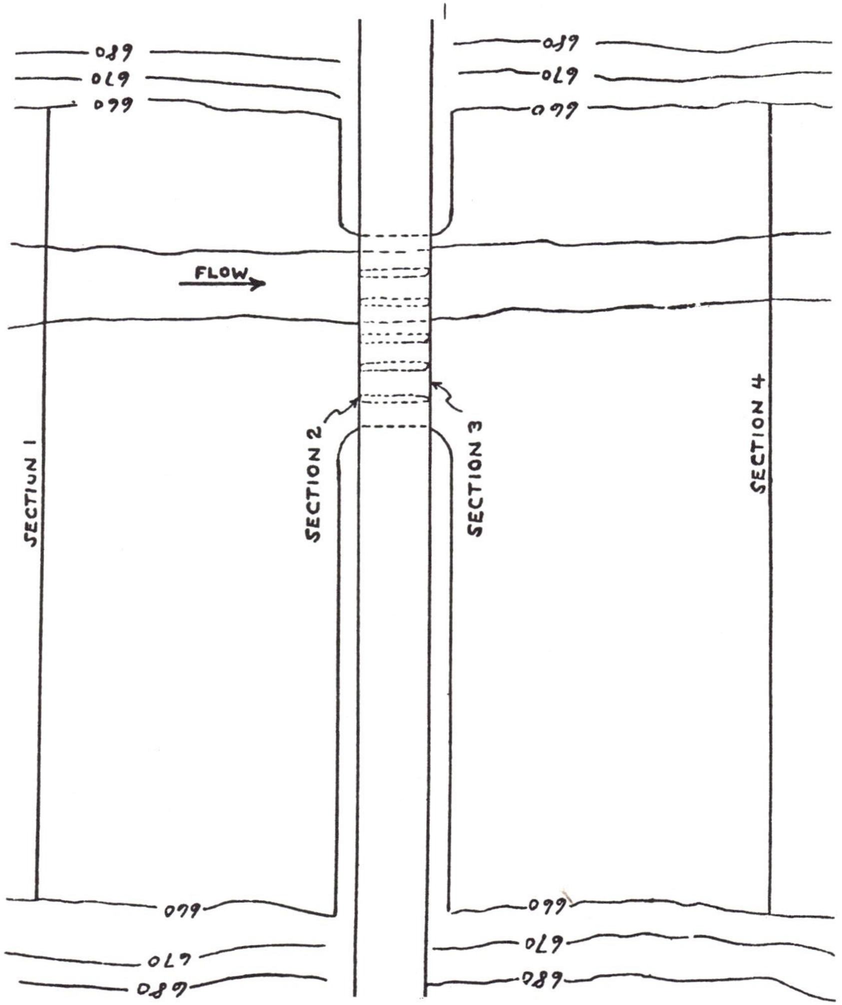


is not a consideration. For several hundred feet upstream and downstream from the site, the main channel is well defined, straight, and has steep banks. At flow approximating the design discharge, the slope of the water-surface profile has been determined to be 0.00024 foot per foot. The stream bed and flood plain consists of sand and loam overlying limestone and appears to be susceptible to scour. A wide flood plain having a large hydraulic capacity extends landward from the right bank.

A bridge on U.S. Highway 15 spans the Arkport River about 2,500 feet downstream from the site. Under conditions approximating or exceeding the 50-year flood, this structure would influence the water-surface elevation in the reach for some distance upstream. Analysis of 1964 high-water marks left by a flood having an estimated discharge of 48,000 cfs and a recurrence interval of about 11 years seems to indicate that some ponding existed in the 2,500-foot reach at that time. Futhermore, local residents reported that U.S. Highway 15 was temporarily closed during the 1964 flood when water from the flood plain, on the right bank of the river, crested the roadway.

MAGNITUDE AND FREQUENCY OF FLOODS

Flood Discharge

Records of flood discharge are obtained at the Geological Survey stream gaging station on the Arkport River at West Cameron, Paragon. These records, through 1965, are summarized in Geological Survey Water-Supply Papers 1302 and 1722. For the years subsequent to 1965, the records are available in the annual report, Surface Water Records of Paragon.

The largest flood at the West Cameron gaging station, within the period 1935-1969, occurred on May 6, 1944. Peak discharges for this flood, at the bridge site, is estimated at 92,000 cfs. On the basis of records for the gaging station at West Cameron, outstanding floods also occurred in 1950 and 1956. At the proposed construction site, peak discharges for these floods were about 83,000 cfs and 86,000 cfs respectively.

Historically the maximum discharge known at the West Cameron gaging station was about 96,000 cfs and occurred during the flood of July 30, 31, 1932.

Flood Frequency

Data on flood probabilities for a drainage area of 751 square miles were derived from a regional analysis of flood frequency relations in the State of Paragon (Smith, 1965). The computed relationship between frequency and discharge, is shown in Table 1 and Figure 2.

Table 1.--Frequency of Flood Discharge, Arkport River, drainage area 751 sq. mi.

| Recurrence interval (years) | Discharge (cubic feet per second) |
|---|---|
| 1.1 | 10,800 |
| 2.33 | 21,000 |
| 5.0 | 32,200 |
| 10 | 45,500 |
| 25 | 70,000 |
| 50 | 98,000 |
| 100 | 135,000 |

As applied to flood events, recurrence interval is the average interval of time within which a given flood will be exceeded once.

STAGE-DISCHARGE RELATION

Under existing channel conditions, the 50-year peak discharge of 98,300 cfs would reach an elevation 653.5 ft. above MSL at the approach section. This is shown on a curve relating stage to discharge (figure 3), which was computed on the basis of the hydraulic properties of the river and other considerations. The natural water-surface profile for the 50-year flood is such that an elevation of 653.2 ft. above MSL will occur at the downstream side of the proposed structure. A stage-discharge relation for this downstream (tailwater) section is shown in figure 4.

FLOW DISTRIBUTION IN NATURAL CHANNEL

A cross section, referred to as the approach section, about 1,300 ft. upstream from the proposed bridge site was selected to study the distribution of flow in the natural approach channel.

Figure 5 shows the valley cross section and the manner in 5 percent increments (4,915 cfs) of the 50-year flood (98,300 cfs) are distributed across it. Of the 98,300 cfs, about 41,000 cfs (41.7%) would be carried by the overflow section with the remaining 57,300 cfs (58.3%) being carried by the main channel.

ANALYSIS OF HYDRAULIC CHARACTERISTICS

To aid in design of highway grade, length of structure, and waterway opening, hydraulic characteristics of the channel and flood plain at the approach section as well as channel characteristics at a constricted section have been analyzed for 50-year flood conditions.

Natural Channel

A graphical summary of characteristics of the approach section can be found in figure 4 which is a set of curves relating elevation to discharge, area, and velocity in the main channel and its principal components. By following the dashed line as indicated, from a point (98,300 cfs, 653.5 ft.) on the stage-discharge relation for the composite channel, the hydraulic characteristics of the channel and/or flood plain at the approach section can be determined for 50-year flood conditions. The results are summarized as follows:

Table 2.--Hydraulic Characteristics of Natural Channel - before construction.

| Channel | Water-surface elevation (ft. above MSL) | Discharge (cfs) | Approach Section area (sq. ft.) | Mean velocity (fps) |
|---|---|---|---|---|
| Main | 653.5 | 57,300 | 10,200 | 5.61 |
| Overflow | do. | 41,000 | 17,000 | 2.41 |
| Composite | do. | 98,300 | 27,200 | 3.61 |

Constricted Channel

To illustrate the consequences of an imposed flood plain encroachment and channel constriction, it has been assumed that all of the flow will be restricted to the main channel. Hydraulic characteristics of such a constricted section, for four different bridge lengths, have been analyzed in a manner similar to that for the approach section. A graphical summary of these results can be found in figure 4 which is a set of design curves relating water-surface elevation at the downstream side of the constricted section to discharge, waterway area, velocity and to the amount of backwater created by constriction. The maximum water-surface elevations, upstream from the constriction, due to backwater is given also for each bridge length. A solution line for 50-year flood conditions is shown as a dashed line in figure 4. The results are summarized as follows:

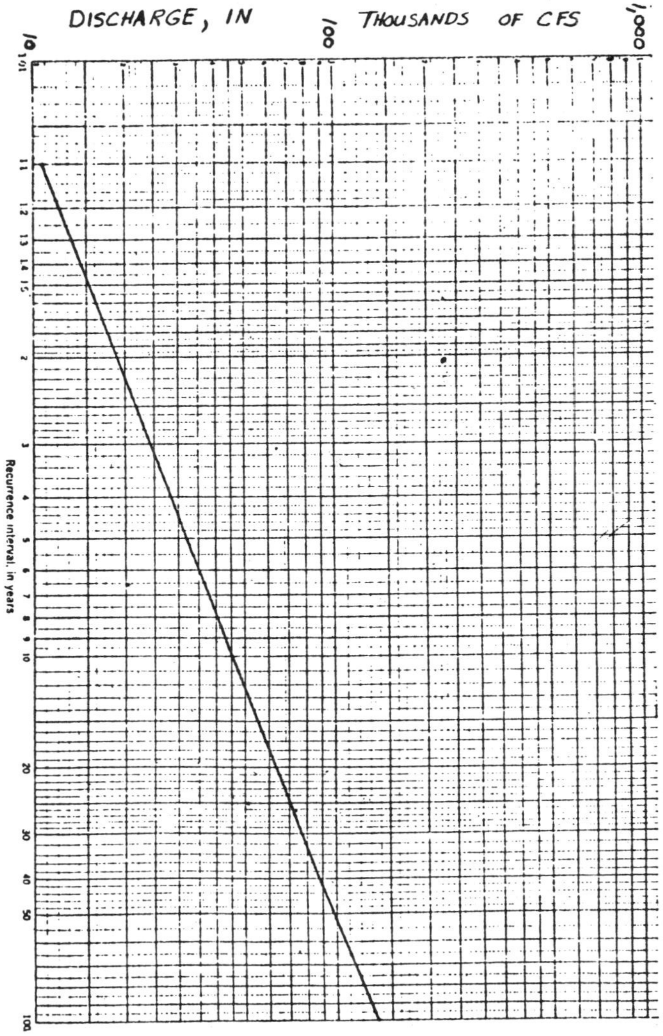


FIG. 2.-- FLOOD-FREQUENCY RELATIONSHIP, ARKPORT RIVER AT EXAMPLESVILLE, PARAGON

Table 3.--Hydraulic Characteristics of Bridge Openings

| | |
|---|---|
| Tailwater elevation | 653.2 ft. above mean level |
| Discharge | 98,300 cfs |
| Waterway area, in sq. ft. | |
| 600 ft. bridge | 10,800 |
| 900 do. | 14,100 |
| 1,100 do. | 16,100 |
| 1,300 do. | 18,000 |
| Velocity, in fps | |
| 600 ft. bridge | 9.01 |
| 900 do. | 6.97 |
| 1,100 do. | 6.10 |
| 1,300 do. | 5.46 |
| Backwater, in ft. | |
| 600 ft. bridge | 1.79 |
| 900 do. | 0.85 |
| 1,100 do. | 0.50 |
| 1,300 do. | 0.29 |
| Maximum upstream water-surface elevation | |
| 600 ft. bridge | 655.2 |
| 900 do. | 654.3 |
| 1,100 do. | 653.9 |
| 1,300 do. | 653.7 |

Figures 3 and 4 can be used to determine channel characteristics for floods of any particular recurrence interval given in figure 2. By projecting a ray horizontally from any point on the stage-discharge relation parallel to the "solution line" shown for the 50-year flood, corresponding hydraulic characteristics may be determined.

SELECTION OF BRIDGE LENGTH

By using backwater as a parameter, the family of curves shown in figure 6 can aid the designer in selecting the length of structure required to span the channel at the design discharge. This figure permits several possible lengths of structure to be considered and provides a ready cost estimate for each. Using this approach many possible options can be considered so that the one best satisfying the design criteria can be selected. The 900-foot structure would cause approximately 0.85 ft. of backwater when carrying 98,300 cfs.

Figure 7 relates the water-surface elevation at the upstream side of the embankment to the discharge, for various bridge lengths. This enables the designer to determine the backwater associated with a given span which, in turn, facilitates his selection of a bridge length that will not cause the upstream water-surface elevation to exceed its permissible limit. The 900-foot structure would cause the water-surface elevation to rise to approximately 654.3 ft. above mean sea level about 1,210 feet upstream. This is considered acceptable under design conditions.

DESIGN OF SPUR DIKE

Due to the crossing's extreme eccentricity, a spur dike may be needed on the flood-plain end of the bridge. Bridge length is 900 feet; design discharge is 98,300 cfs at a water-surface elevation of 653.5 ft. above mean sea level; and the abutment is of the spill-through type with 2:1 side slopes.

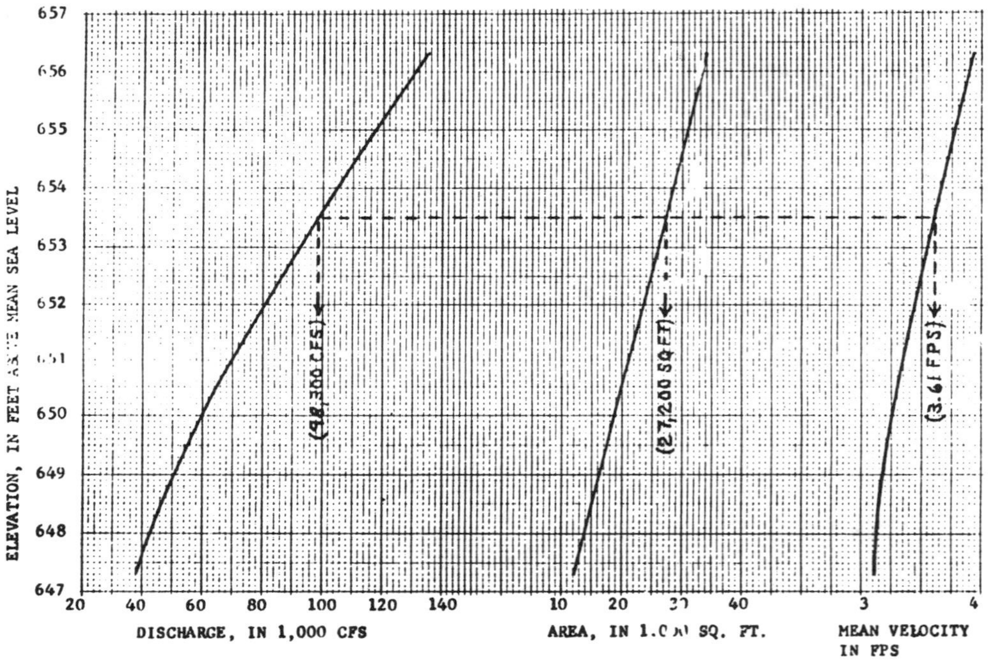


FIGURE 3 -Hydraulic Characteristics at Approach Section Before Construction.

$Q_f$ = 30% of 98,300 cfs from figure 5 - flow distribution diagram

$Q_{100}$ = 5% of 98,300 cfs

$V_{n2}$ = 6.97 fps (figure 4 - 900 ft. bridge)

$$\frac{Q_f}{Q_{100}} = \frac{30}{5} = 6$$

From figure 30, HDS #1, $L_s$ = 285 ft. (Length of required spur dike)

Compute coordinates of spur dikes - see figure 8 for layout

$$\frac{X^2}{X_s^2} + \frac{Y^2}{(0.4L_s)^2} = 1 \qquad \text{where } L_s = 285$$

$$0.4L_s = 114$$

$$L_s^2 = X^2 + 6.25Y^2$$

$$X = (L_s^2 - 6.25Y^2)^{1/2}$$

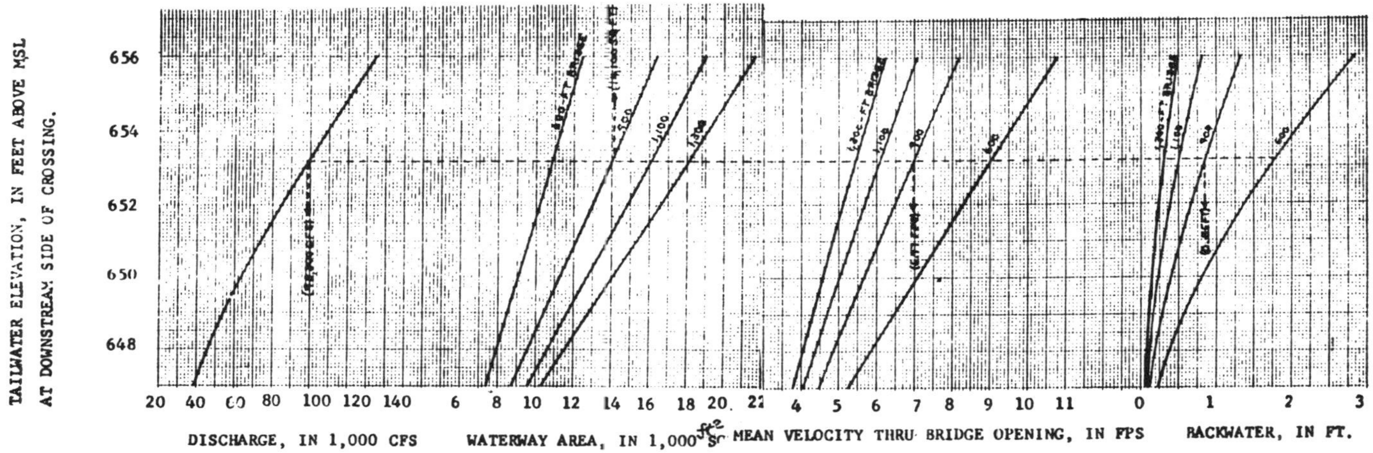


FIGURE 4 - Hydraulic Characteristics of Bridge Openings.

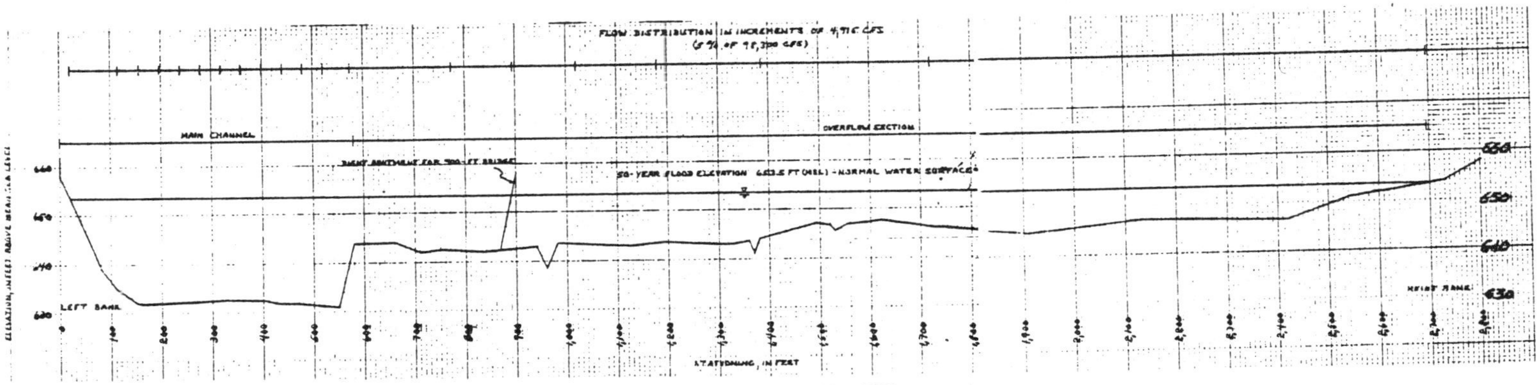


APPROACH SECTION LOCATED ABOUT 1,300 FEET UPSTREAM FROM PROPOSED CONST. SITE

FIGURE 5 - Valley Cross Section and Flow Distribution Diagram, Arkport River at Examplesville, Paragon.

Table 4.--Coordinates of spur dike

| Y | X |
|---|---|
| 0 | 285.00 |
| 20 | 280.58 |
| 40 | 266.88 |
| 60 | 242.33 |
| 80 | 203.04 |
| 100 | 136.84 |
| 114 | 0.00 |

SCOUR

The problem of scour around the bridge piers and abutments was carefully investigated because the field survey and soils investigation indicates that the stream bed and flood plain, consisting of sand and loam overlying limestone, appear to be scour prone. Using the Laursen method the depth of scour at the piers was estimated to be about 12 feet in the main channel and about 5 feet in the flood plain. Figure 9 relates water-surface elevation at the bridge to the estimated depth of scour.

SUMMARY

(1) Estimated flood discharge of an expected frequency of occurrence of 50 years is 98,300 cfs and would reach an elevation of 653.2 ft. above MSL at the proposed bridge site.

(2) In the natural channel, 41.7% of a flood of 98,300 cfs would be carried by the flood plain; the remaining 58.3% by the main channel.

(3) If a 98,300 cfs flood was restricted to the main channel and the flow was spanned, mean velocity would be increased, backwater would be created, and water-surface elevations in the vicinity of the proposed span would be raised as follows:

| Bridge Length (ft.) | Velocity (fps) | Backwater (ft.) | Water-surface elevation at section 1 and along upstream face of embankment (ft. above MSL) |
|---|---|---|---|
| 600 | 9.01 | 1.79 | 655.2 |
| 900 | 6.96 | 0.85 | 654.3 |
| 1,100 | 6.10 | 0.50 | 653.9 |
| 1,300 | 5.46 | 0.29 | 653.7 |

(4) The depth of scour at the main-channel piers was estimated at about 12 feet for 98,300 cfs. For the piers located on the flood plain, the scour depth was estimated at about 5 feet.

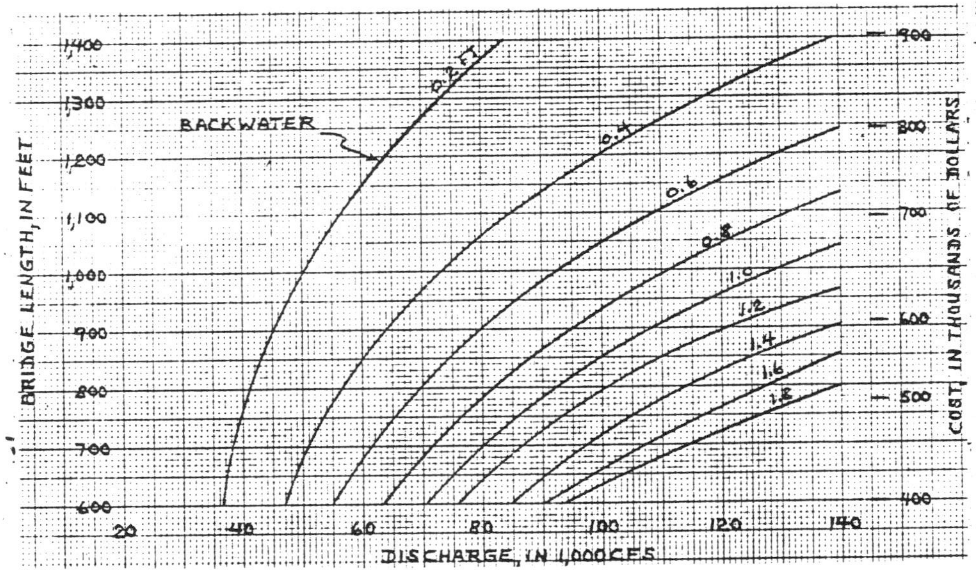


FIGURE 6 - Design Curves- Selection of Bridge Length and Cost Analysis.

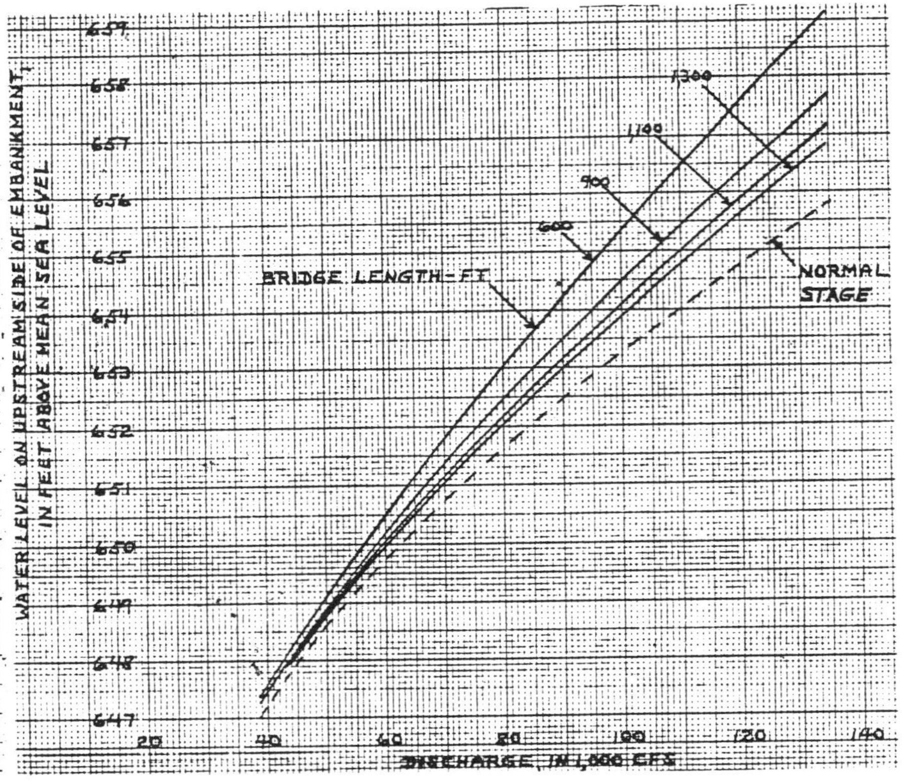


FIGURE 7 - Design Curves- Water Surface Elevation on Upstream Side of Embankment.

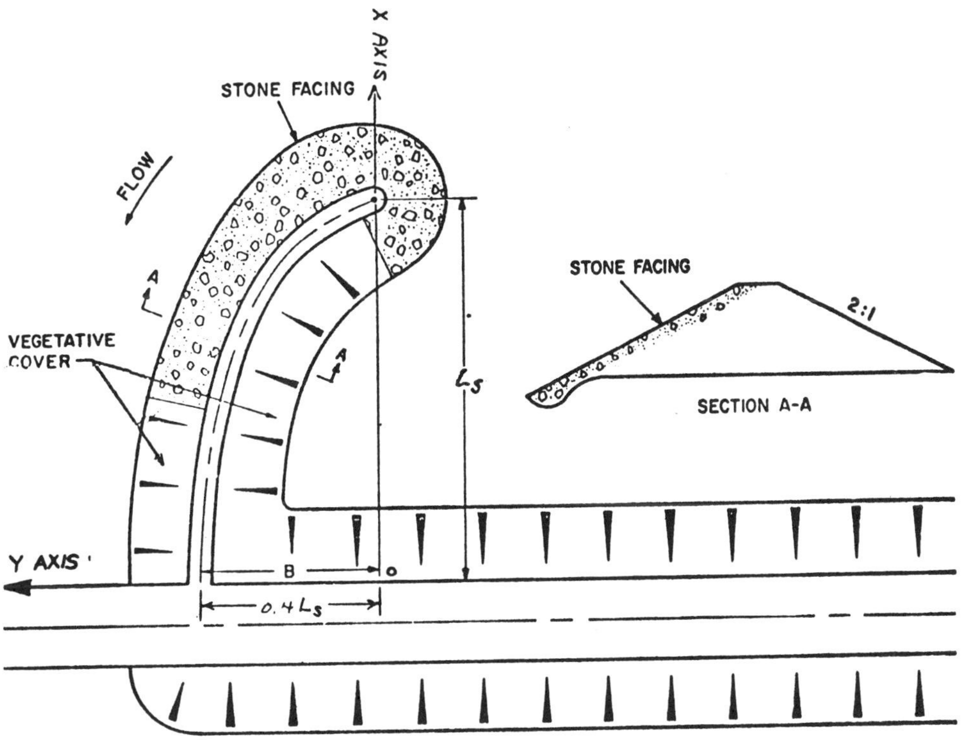


FIGURE 8- Suggested Plan and Cross Section of Spur Dike at Bridge Abutment. The Upstream End of Spur Dike is Protected by Riprap.

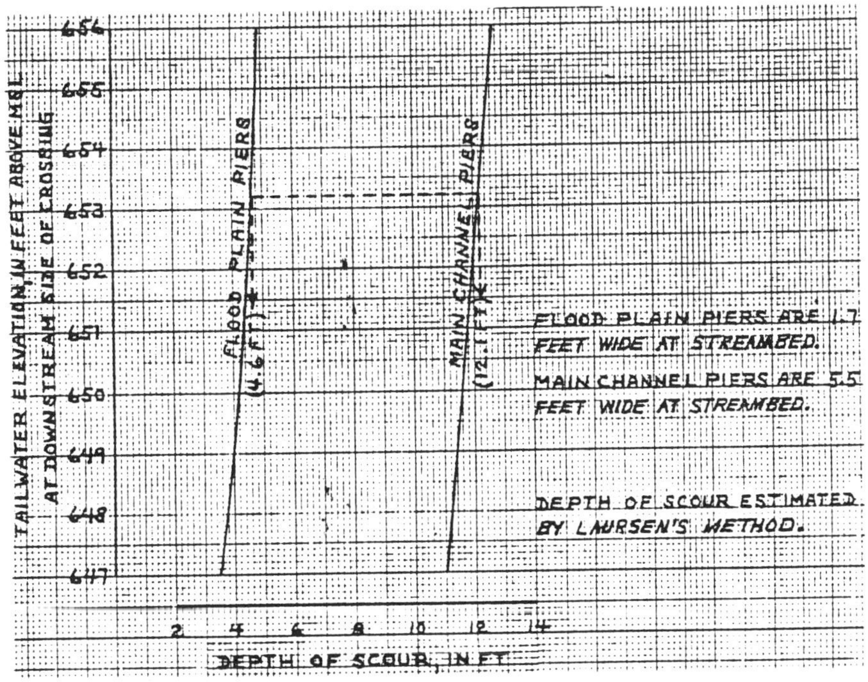


FIGURE 9 - Scour Curves for Design of Bridge Piers.

(WIS DOT)

# N.5 STREAM PROPERTY

| | |
|---|---|
| CHANNEL WIDTH | SMALL (30 M WIDE); MEDIUM (30-150 M); WIDE (150 M) |
| FLOW HABIT | EPHEMERAL (INTERMITTENT); PERENNIAL BUT FLASHY; PERENNIAL |
| CHANNEL BOUNDARIES |  ALLUVIAL; SEMI-ALLUVIAL; NON-ALLUVIAL |
| BED MATERIAL | SILT-CLAY; SILT; SAND; GRAVEL; COBBLE OR BOULDER |
| VALLEY; OR OTHER SETTING | 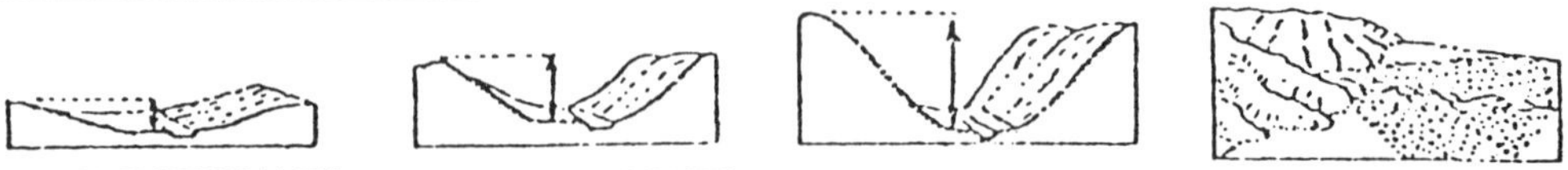 LOW RELIEF VALLEY (< 100 FT. OR 30 M DEEP); MODERATE RELIEF (100-1000 FT. OR 30-300 M); HIGH RELIEF (> 1000 FT. OR 300 M); NO VALLEY; ALLUVIAL FAN |
| FLOOD PLAIN | 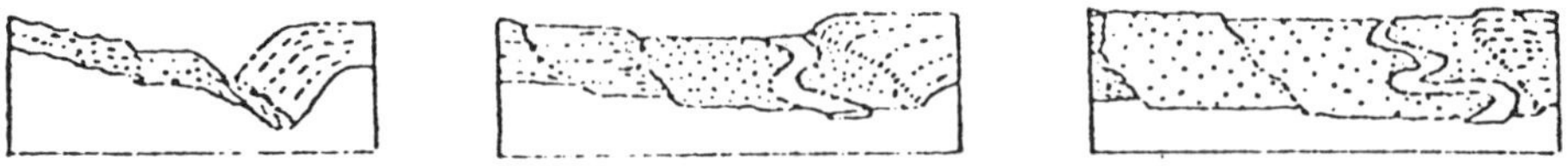 LITTLE OR NONE (< 2x CHANNEL WIDTH); NARROW (2-10x CHANNEL WIDTH); WIDE (> 10x CHANNEL WIDTH) |
| DEGREE OF SINUOSITY | 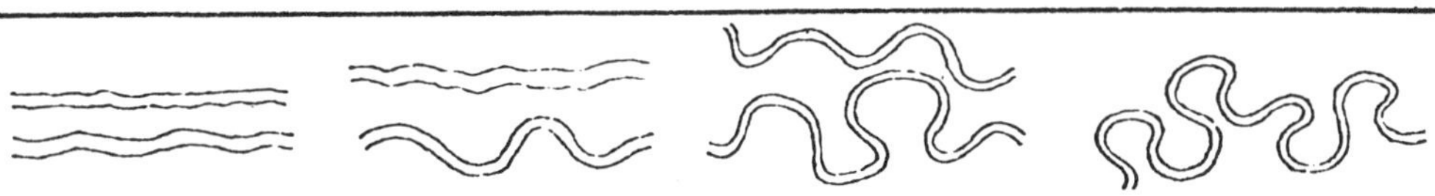 STRAIGHT (SINUOSITY 1-1.05); SINUOUS (1.06-1.25); MEANDERING (1.26-2.0); HIGHLY MEANDERING (> 2) |
| DEGREE OF BRAIDING |  NOT BRAIDED (< 5 PERCENT); LOCALLY BRAIDED (5-35 PERCENT); GENERALLY BRAIDED ( > 35 PERCENT) |
| DEGREE OF ANABRANCHING |  NOT ANABRANCHED (< 5 PERCENT); LOCALLY ANABRANCHED (5-35 PERCENT); GENERALLY ANABRANCHED (> 35 PERCENT) |
| VARIABILITY OF WIDTH AND DEVELOPMENT OF BARS |  EQUIWIDTH; WIDER AT BENDS; RANDOM VARIATION<br>NARROW POINT BARS; WIDE POINT BARS; IRREGULAR POINT AND LATERAL BARS |
| APPARENT INCISION |  NOT INCISED; PROBABLY INCISED |
| CUT BANKS | RARE; LOCAL; GENERAL |
| BANK MATERIAL | COHERENT<br>RESISTANT BEDROCK<br>NON-RESISTANT BEDROCK<br>ALLUVIUM<br>NON-COHERENT<br>SILT; SAND<br>GRAVEL;<br>COBBLE BOULDER |
| TREE COVER ON BANKS | 50 PERCENT OF BANKLINE; 50-90 PERCENT; > 90 PERCENT |

STREAM PROPERTIES FOR CLASSIFICATION AND STABILITY ASSESSMENT
FROM FHWA/RD-78/162 COUNTERMEASURES FOR HYDRAULIC PROBLEMS AT BRIDGES

# N. 6 GABION AND BANK PROTECTION

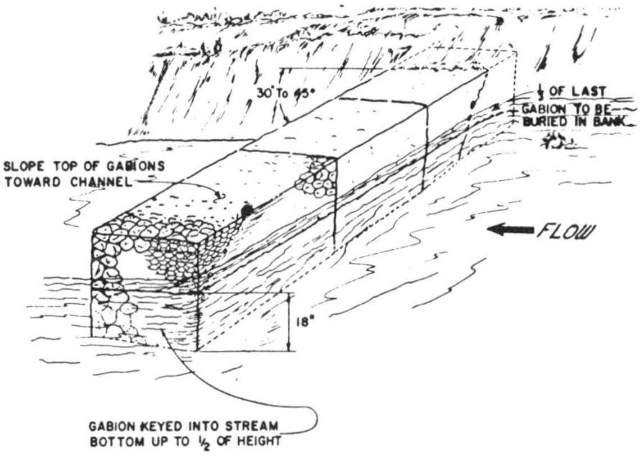


## CHANNEL DEFLECTOR

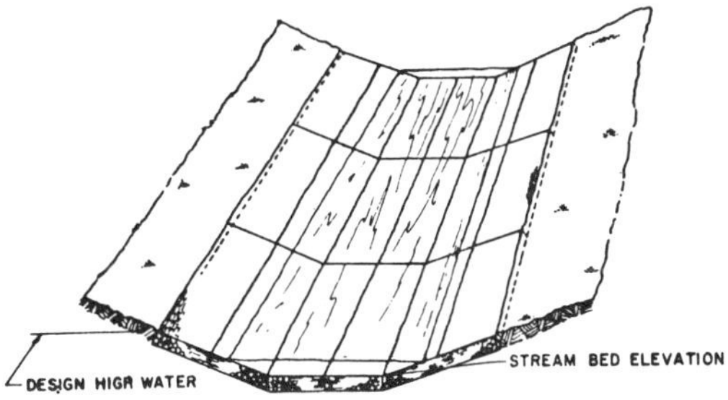


## CHANNEL LINING

- Type A Gabions shall consist of wire baskets filled by hand placement of coarse aggregate at least along the exposed faces for a uniform appearance.
- Type B Gabions shall consists of wire baskets filled completely with small power equipment or by hand.
- Corrosion Resistant Type A and B Gabions shall be the same as Type A and B Gabions except that the basket wire shall be sheathed in poly-vinyl chloride plastic.

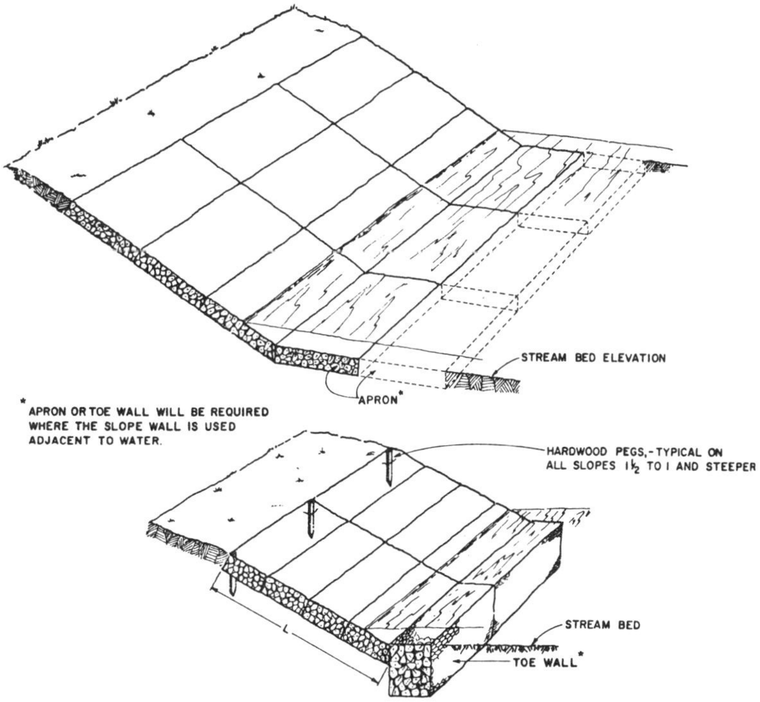


## SLOPE WALLS

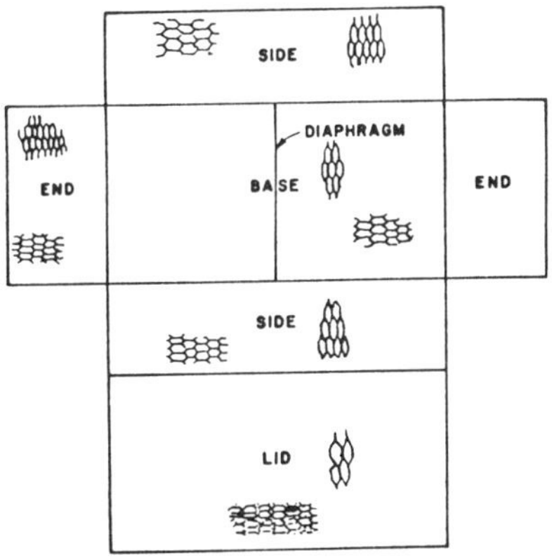


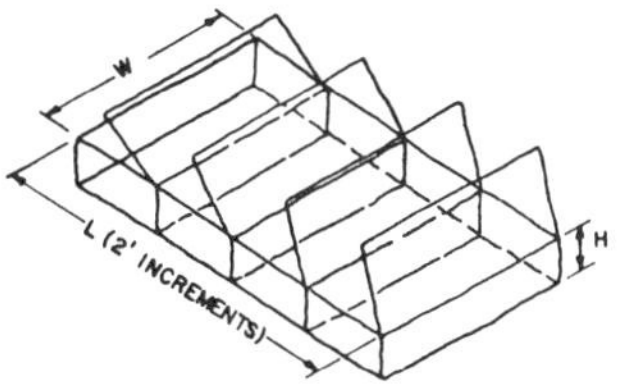


## MATTRESS TYPE

| GABIONS | | |
|---|---|---|
| W | H | L |
| 3'-3" | 3'-3" | 6'-6" |
| 3'-3" | 3'-3" | 9'-9" |
| 3'-3" | 3'-3" | 13'-1" |
| 3'-3" | 1'-8" | 6'-6" |
| 3'-3" | 1'-8" | 9'-9" |
| 3'-3" | 1'-8" | 13'-1" |
| 3'-3" | 1'-0" | 6'-6" |
| 3'-3" | 1'-0" | 9'-9" |
| 3'-3" | 1'-0" | 13'-1" |
| GABION (MATTRESS) | | |
| 6'-6" | 6" | 8'-0" |
| 6'-6" | 6" | 10'-0" |
| 6'-6" | 6" | 12'-0" |
| 6'-6" | 10" * | 8'-0" |
| 6'-6" | 10" * | 10'-0" |
| 6'-6" | 10" * | 12'-0" |
| | | |

* 9" For Corrosion Resistant

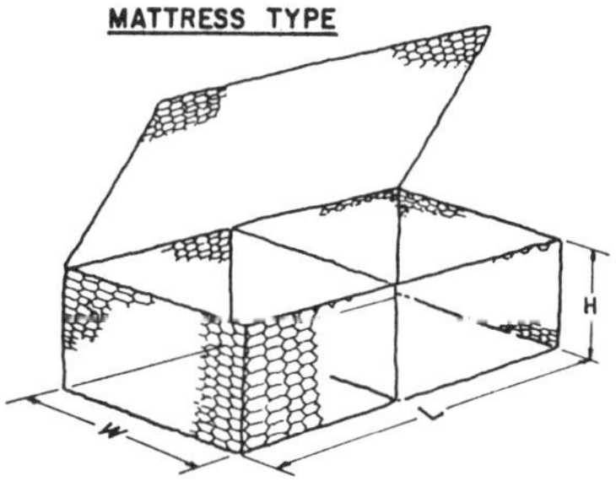


## WIRE MESH BASKETS

PENNDOT

# BANK PROTECTION

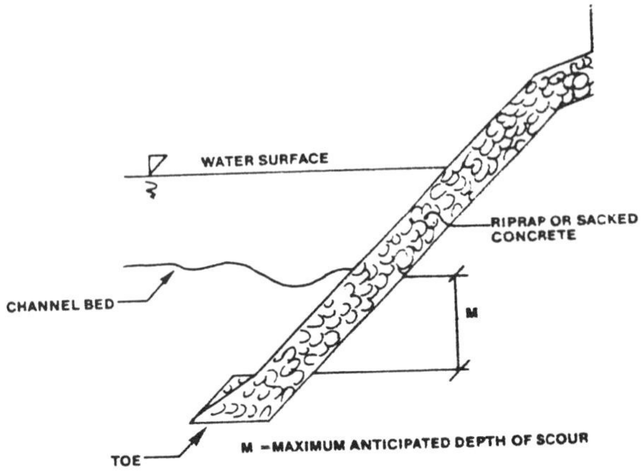


CONCRETE GROUTED RIPRAP

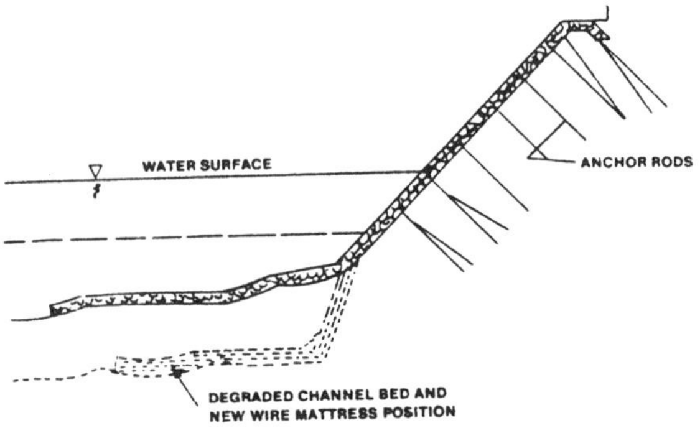


ROCK-AND-WIRE MATTRESS

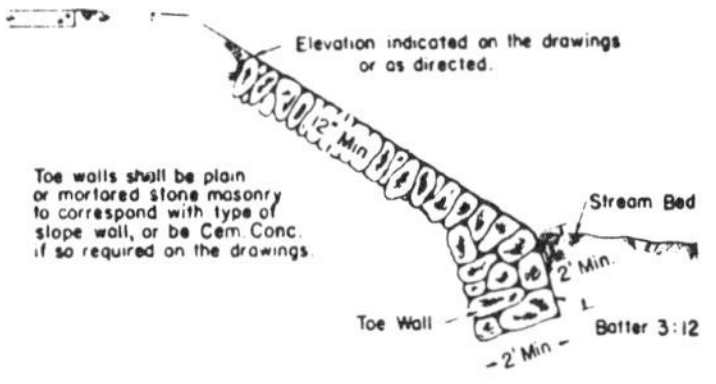


PLAIN AND MORTARED STONE SLOPE WALL

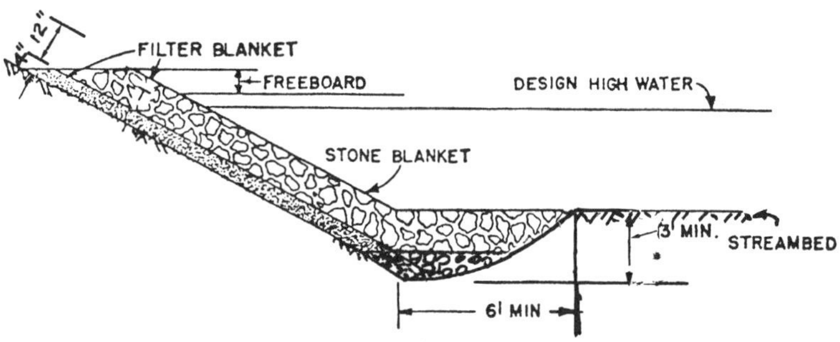


B – STONE BLANKET AND TOE DETAIL

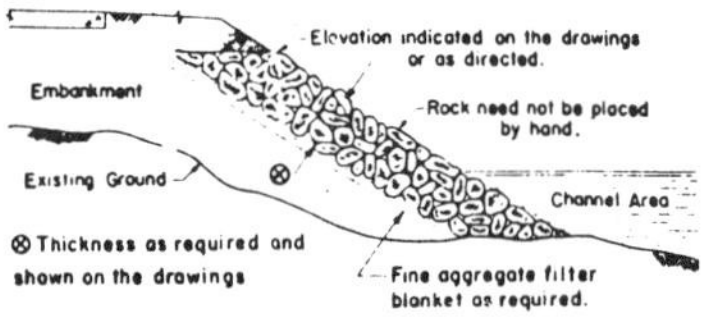


ROCK EMBANKMENT FOR SLOPE PROTECTION

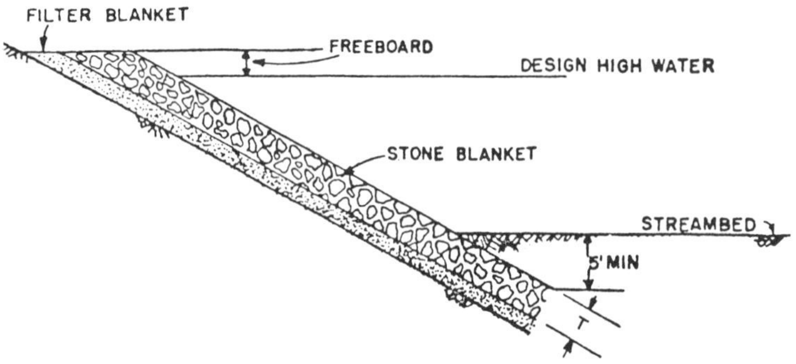


A – STONE BLANKET AND TOE TRENCH DETAIL

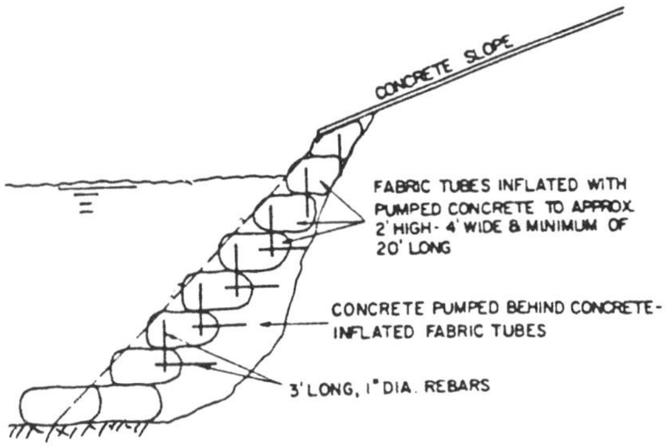

# RAIL BANK PROTECTION

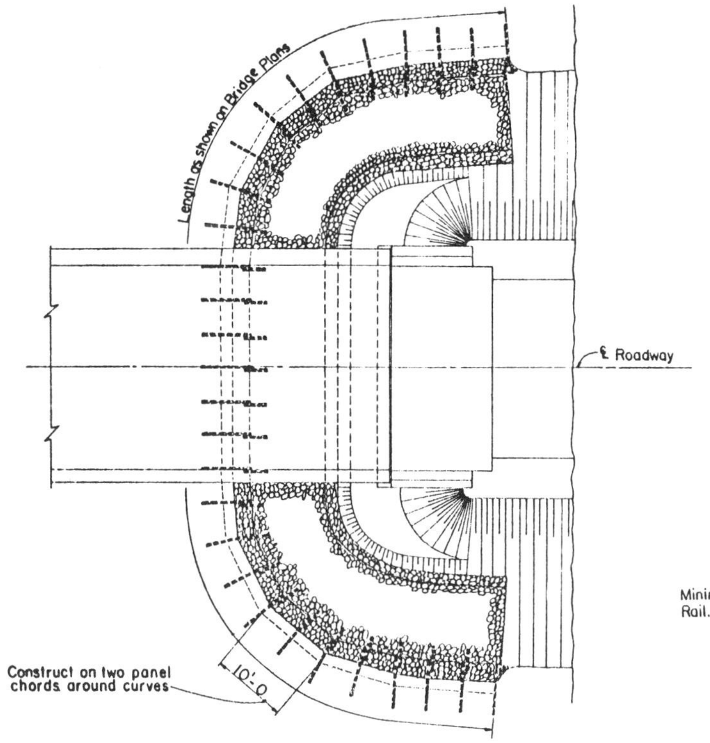


PLAN OF BANK PROTECTION AT ABUTMENT
Scale: 1" = 10'

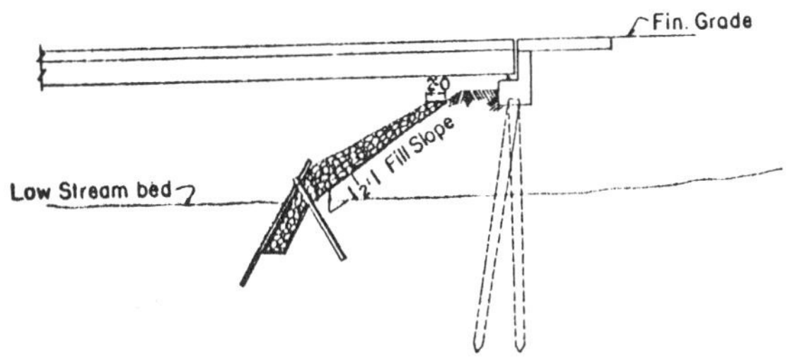


SECTION ON ℄ ROADWAY

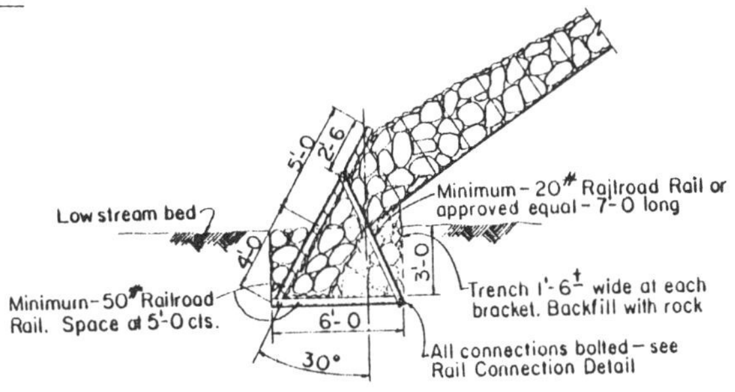


DETAIL - B
Scale ¼" = 1'-0

Use this Detail ONLY when called for on Bridge Plans

NOTE:
See Detail - A for dimensions and other details not shown

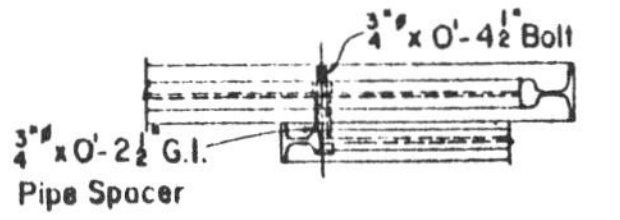


RAIL CONNECTION DETAIL
Scale 1½" = 1'-0
Burn holes through rails in field and bolt together as shown

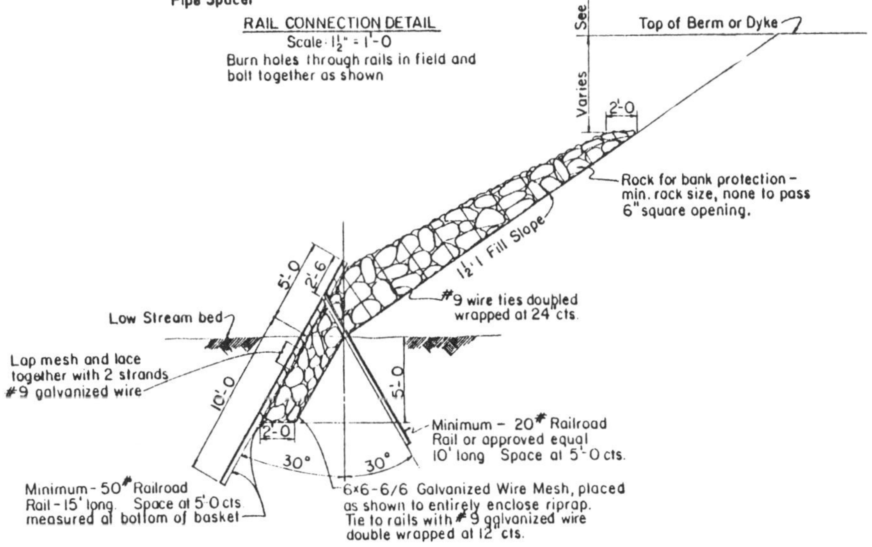


DETAIL - A

(ARIZONA)

## O.1 BRIDGE RAILING DETAILS

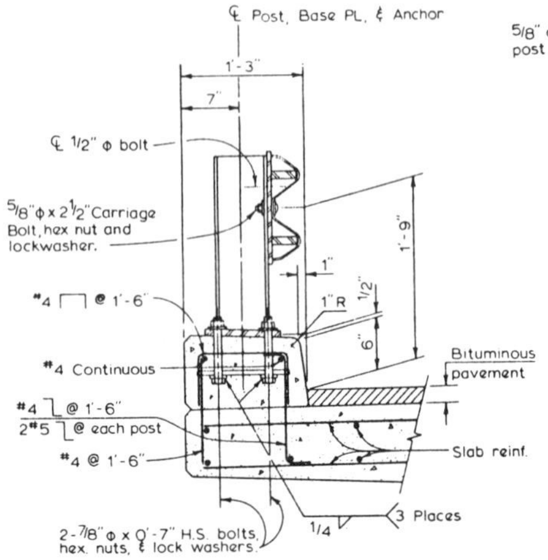


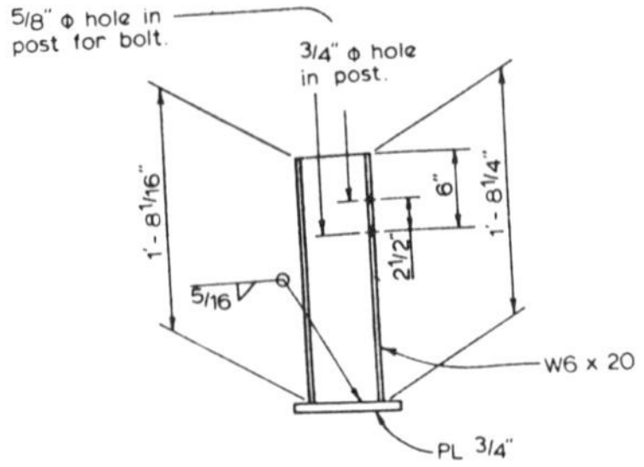


ELEVATION
POST DETAIL

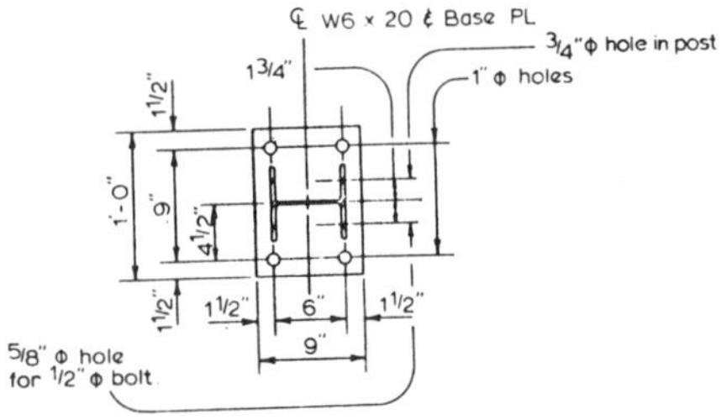


PLAN

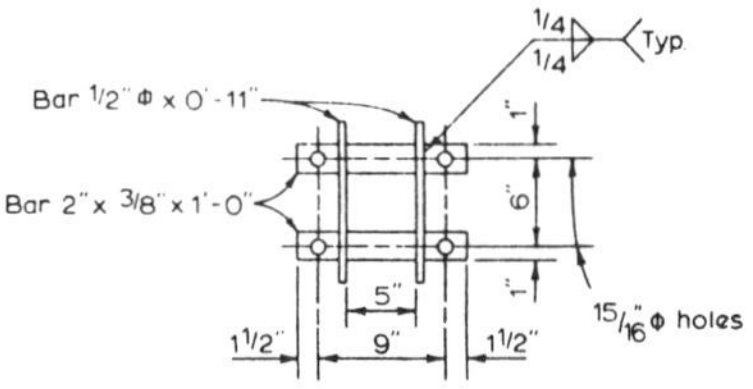


ANCHOR DETAIL

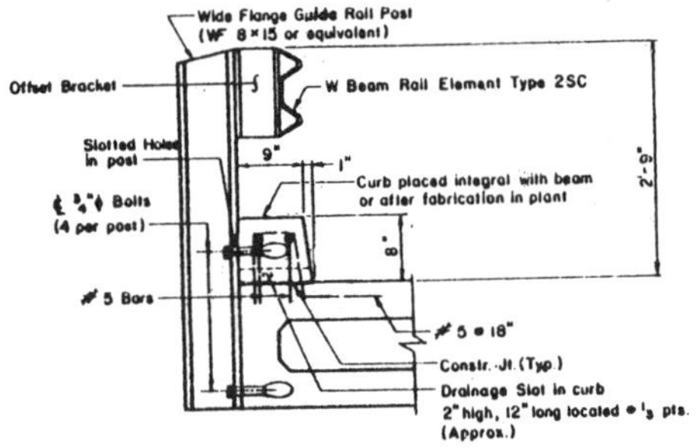


GUIDE RAIL & POST DETAIL
TYP. FOR 17" & DEEPER BEAM SIZES

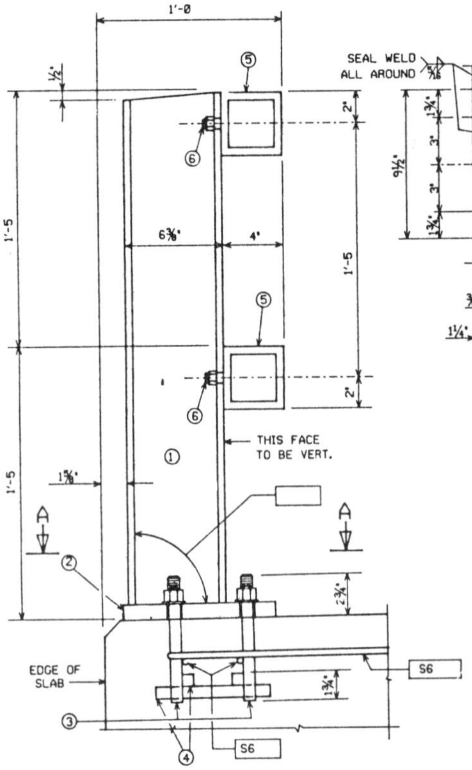


SECTION THRU RAILING

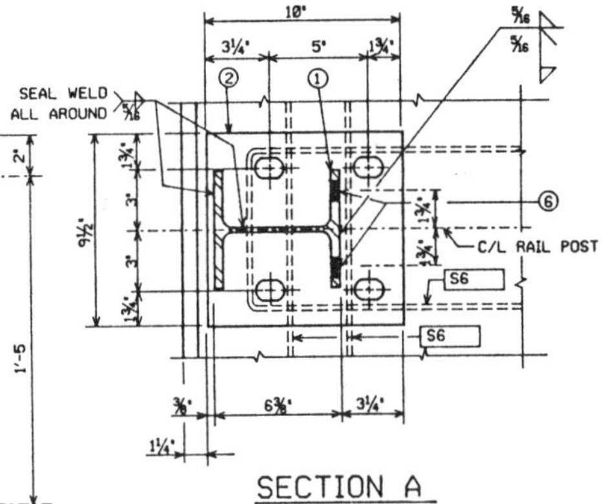


SECTION A

### LEGEND

① W6x25 WITH 1¼"DIA. HOLES ON EACH SIDE OF POST FLANGE. FOR STUD NO. 6. CUT BOTTOM OF POST TO MATCH CROSS SLOPE OF ROADWAY. PLACE POST VERTICAL. PLACE POSTS NORMAL TO GRADE LINE

② PLATE 1"x9½"x0'-10", WITH 1⅙"x1½"SLOTTED HOLES FOR ANCHOR BARS NO. 3. WELD TO NO. 1 AS SHOWN.

③ A449 OR MATERIAL OF EQUIVALENT YIELD STRENGTH AND ELONGATION ANCHOR BAR ⅞" DIA. (MIN. YIELD OF 92 K.S.I. AND ELONGATION OF 14%) WITH A325 NUT AND WASHER. 4 REQ'D. PER POST. THREAD 3" AND PLACE NORMAL TO PLATE NO. 2.

④ BAR ¾" SQ.x0'-8" LONG. WELD TO ANCHOR BAR NO. 3.

⑤ TS 4x4x.25 STRUCTURAL TUBING, CONFORMING TO A.S.T.M. DESIGNATION A36. ATTACH TO NO. 1 WITH STUDS NO. 6.

⑥ ⅝" DIA.x1½"LG. SHOP WELDED STUDS, WITH HEX. NUT AND 2 WASHERS. 4 PER POST REQ'D. (2 REQ'D. AT EACH LOCATION.)

⑦ PLATE ¾"x1'-0x1'-6. WELD TO END RAIL POST AS SHOWN IN DETAIL. REQUIRED AT BEAM GUARD ATTACHMENTS ONLY.

⑧ 1"DIA. HOLES IN PLATE NO. 7 FOR ⅞" DIA. A325 BOLTS W/HEX NUTS AND WASHERS.

⑨ SQUARE SLEEVE FABRICATED FROM ¼" PLATE. PROVIDE "SLIDING FIT" WITH A MINIMUM OUT TO OUT DIMENSION OF 3 13/32".

⑩ TS 3x3x.25x1'-10 LONG. PROVIDE ½ DIA. SURFACE WELDS ON ALL SIDES AS SHOWN. GRIND WELDS TO FIT FREE INTO I.D. OF NO. 5. PROVIDE ⅜" DIA.x½" WELDING STUDS ON TOP AND BOTTOM SURFACES AT CENTERLINE.

# THRIE BEAM RAILING RETROFIT

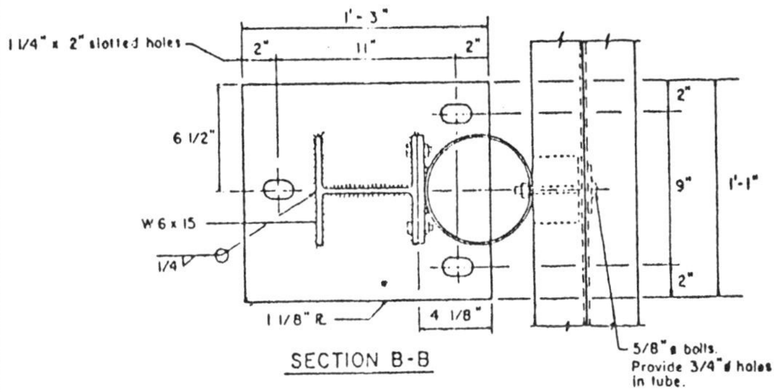


SECTION B-B

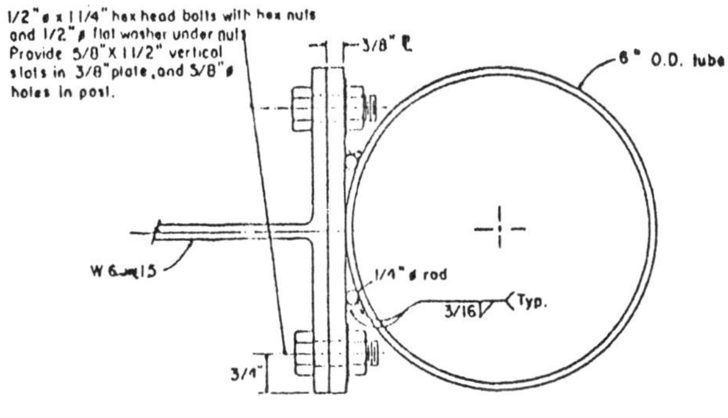


DETAIL OF ATTACHING TUBE TO POST

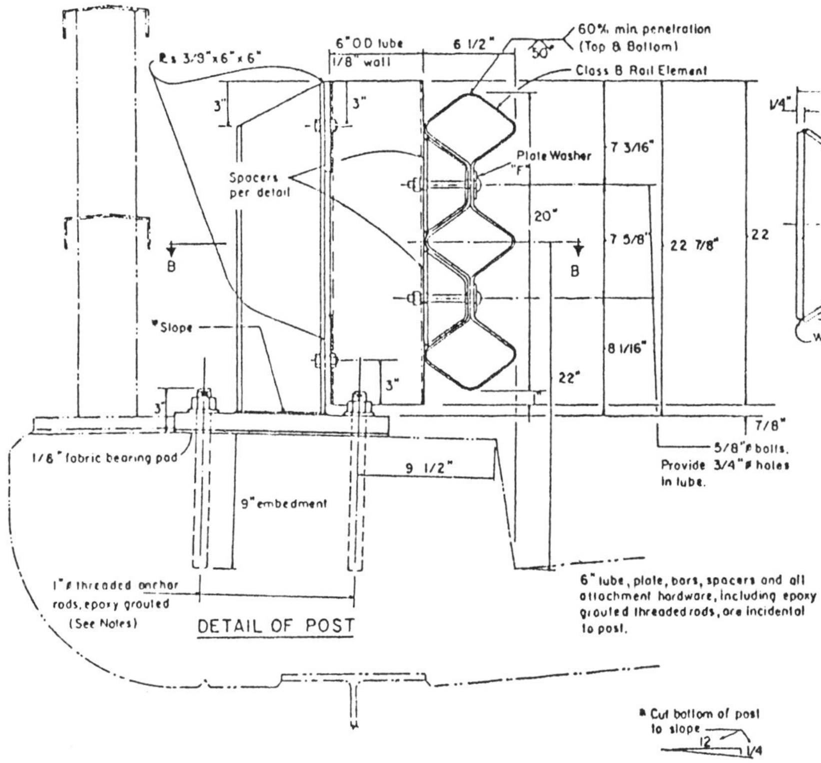


DETAIL OF POST

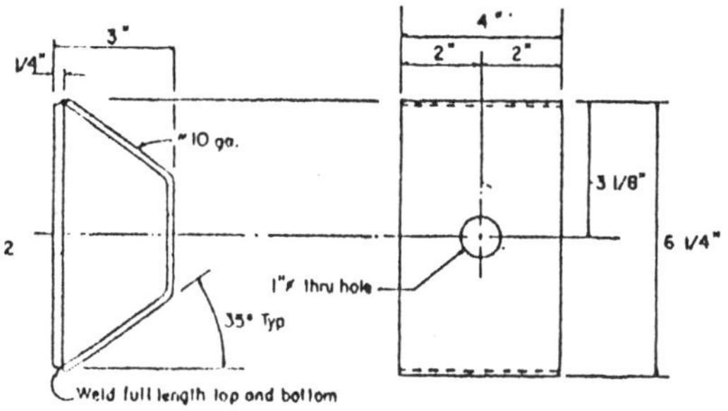


DETAIL OF SPACER

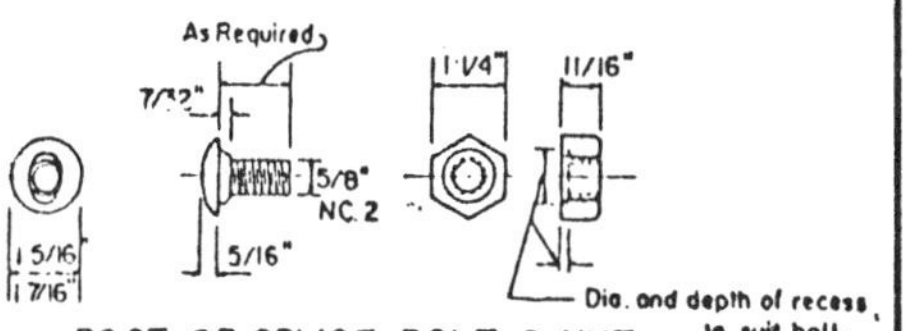


POST OR SPLICE BOLT & NUT

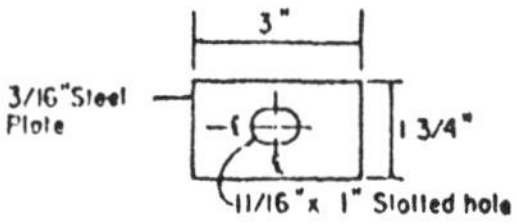


PLATE WASHER "F"

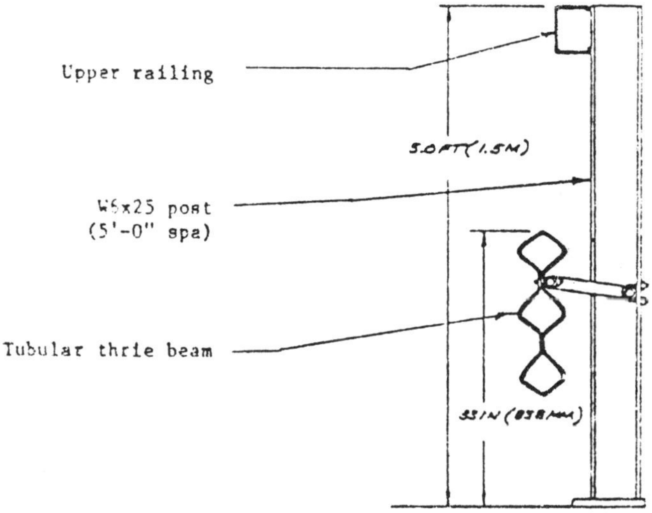


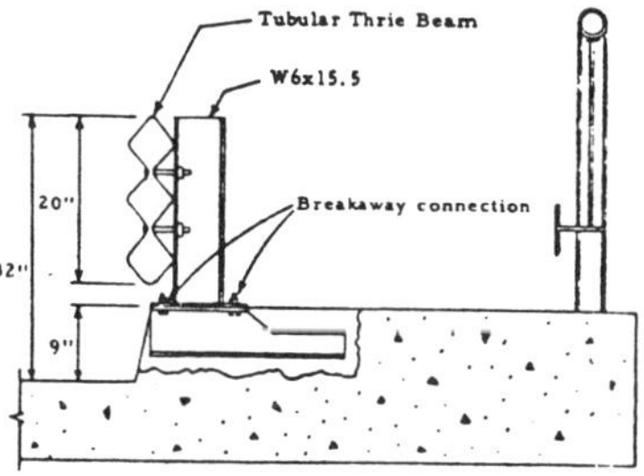

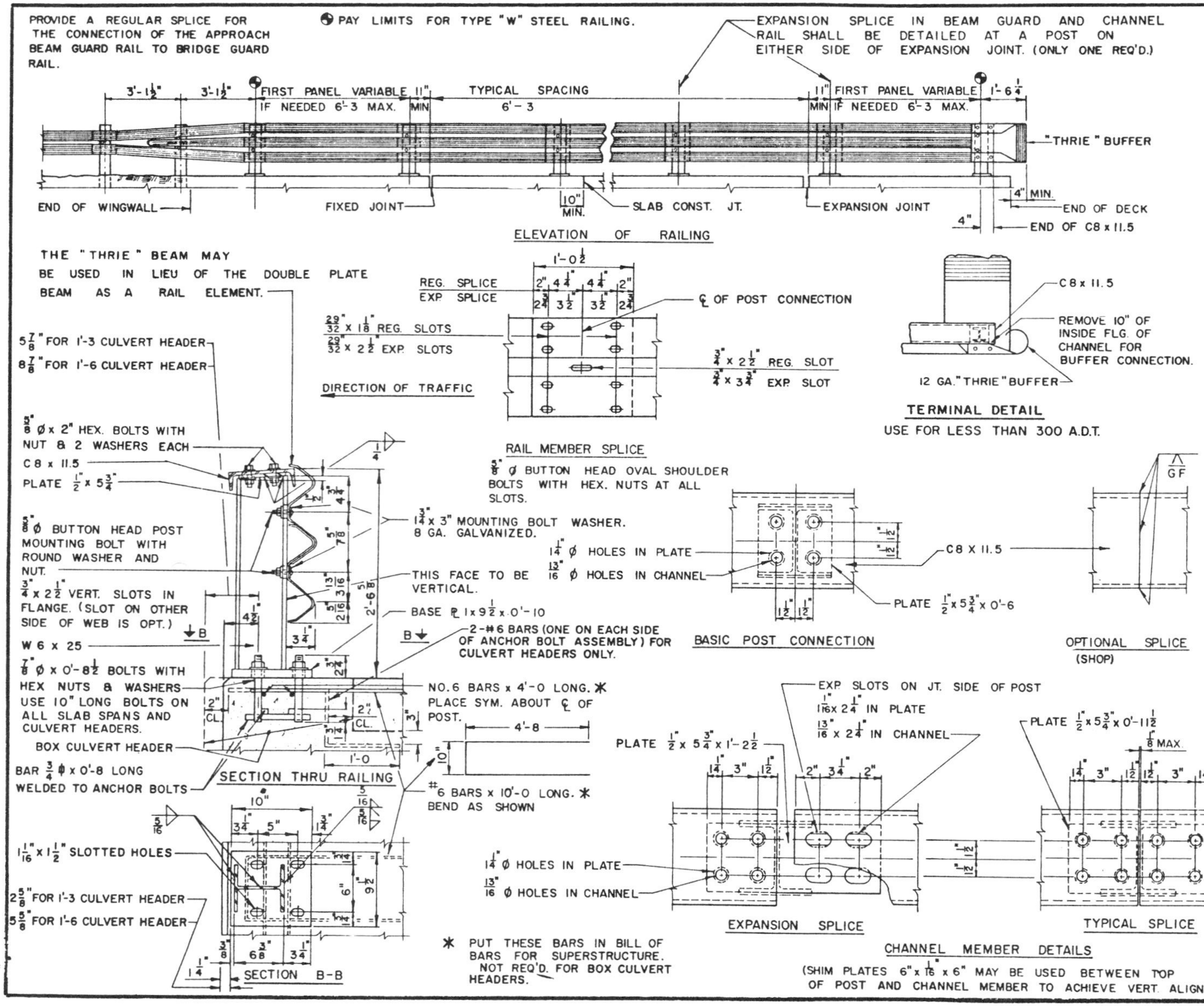


## GENERAL NOTES

1. BID ITEM SHALL BE STEEL RAILING TYPE "W" WHICH INCLUDES ALL ITEMS SHOWN

2. POST BASE PLATES SHALL BE FLAT WITH ALL SURFACES SMOOTH AND FREE FROM WARP AND ALL EDGES SMOOTH, STRAIGHT AND VERTICAL. ALL PLATE CUTS SHALL BE MACHINE OR MACHINE FLAME CUTS.

3. MAIN POST, BASE PLATES AND SHIM SHALL BE MADE FROM MATERIALS CONFORMING TO A.S.T.M. A36. CUT BOTTOM OF POST TO MATCH CROSS SLOPE.

4. PAY LIMITS FOR ▭ "GUARD RAIL" SHALL BE FROM ℄ TO ℄ OF END POSTS.

5. RAIL MEMBERS SHALL BE "W" OR "THRIE". BEAMS CONFORMING TO A.A.S.H.T.O. M180 CLASS A TYPE 2.

6. ANCHOR BOLTS SHALL BE ⅞ Ø NOMINAL CONFORMING TO A.S.T.M. A-449● WITH 3" THREAD.

7. CHANNEL MEMBER SHALL BE ATTACHED CONTINUOUSLY TO A MINIMUM OF FOUR POSTS AND A MAXIMUM OF EIGHT (EXCEPT AT ABUTMENTS), AND SHALL CONFORM TO A.S.T.M. DESIGNATION A36.

8. PLACE ANCHOR BOLTS NORMAL TO BASE PLATE.

9. AT EXPANSION SLOTS IN RAIL AND CHANNEL MEMBERS, TIGHTEN BOLTS, BACK OFF ONE HALF TURN AND BURR THREADS.

10. ALL MEMBERS, INCLUDING UPPER 4" OF ANCHOR BOLTS, SHALL BE GALVANIZED AFTER FABRICATION.

11. RAIL MEMBERS SHALL BE LAPPED IN THE DIRECTION OF TRAFFIC AND THE UPPER RAIL SHALL LAP THE LOWER RAIL.

12. FILL EXPOSED OPENING BETWEEN SHIMS AND POST ANCHOR BOLT HOLES WITH NON-STAINING GRAY NON-BITUMINOUS JOINT SEALER.

13. CUT BOTTOM OF POST TO MATCH CROSS SLOPE OF ROADWAY. PLACE POST NORMAL TO GRADE LINE.

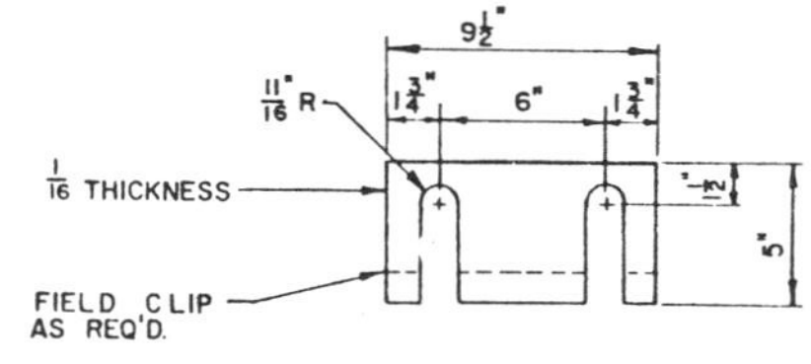


POST SHIM DETAIL
(4 PER POST)

● OR MATERIAL OF EQUIVALENT YIELD STRENGTH AND ELONGATION

STEEL RAILING
TYPE "W"

STATE OF WISCONSIN
DEPARTMENT OF TRANSPORTATION
DIVISION OF HIGHWAYS
BRIDGE SECTION

Approved: Stanley W. Woods, Chief Bridge Engineer
Date 6-83

# O.2 GUARD RAIL DETAILS



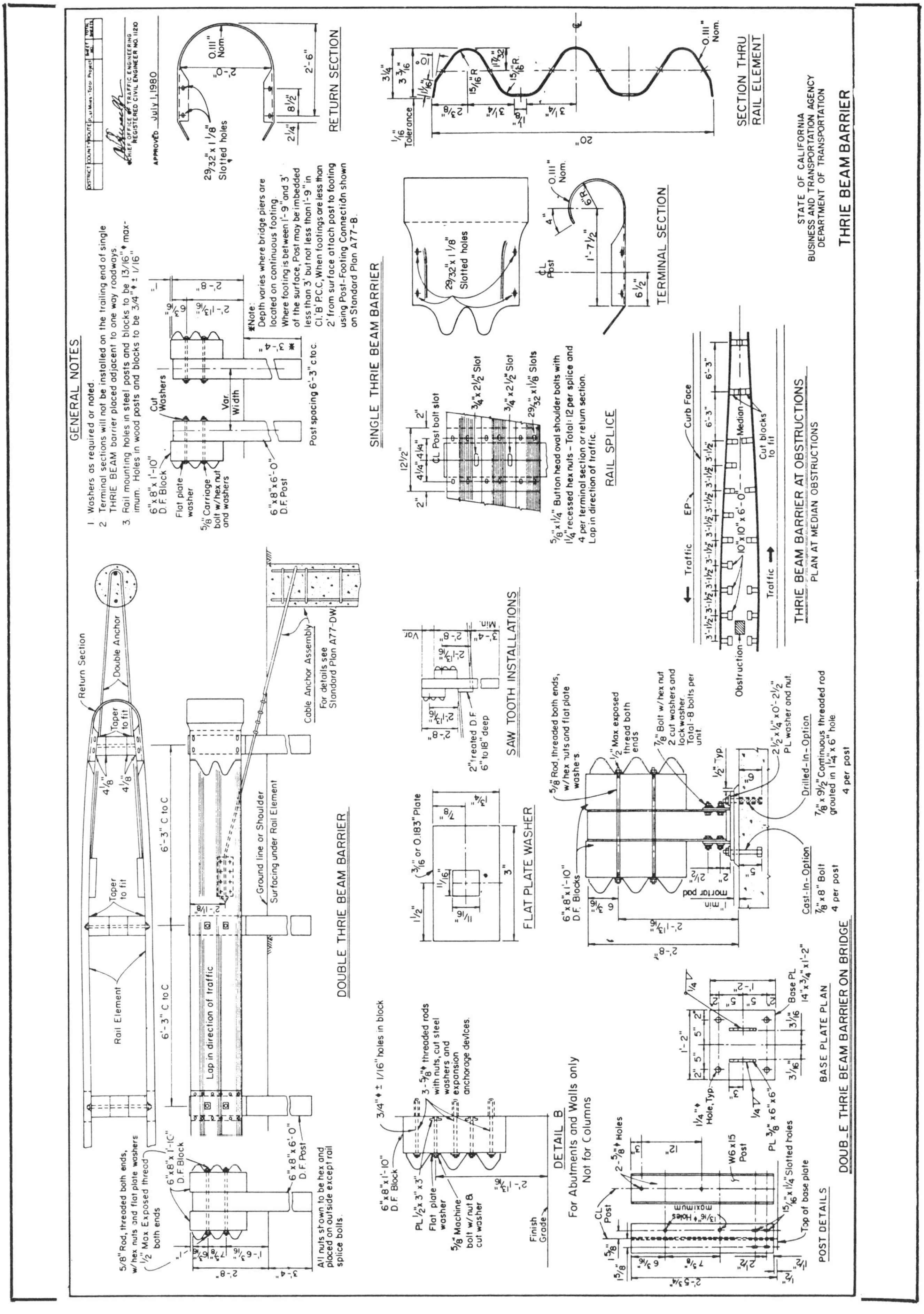

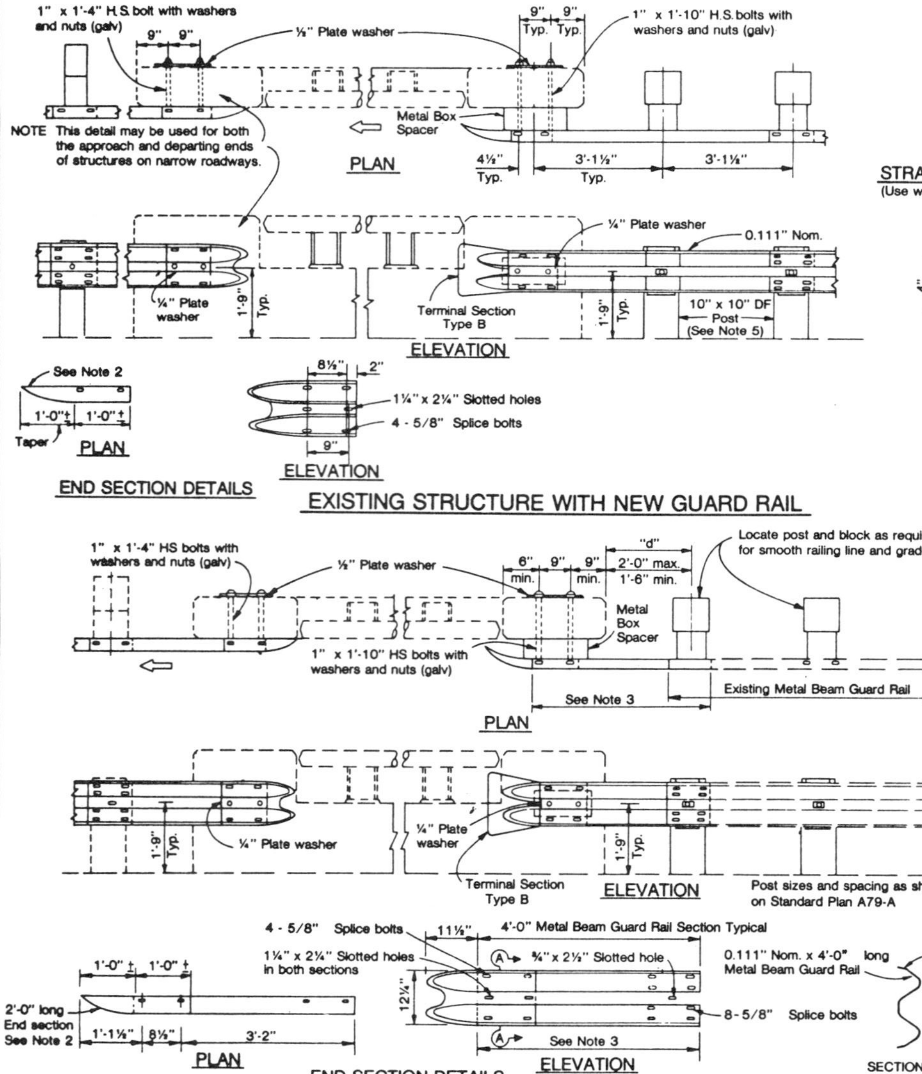


END SECTION DETAILS

EXISTING STRUCTURE WITH NEW GUARD RAIL

END SECTION DETAILS

EXISTING STRUCTURE AND EXISTING GUARD RAIL

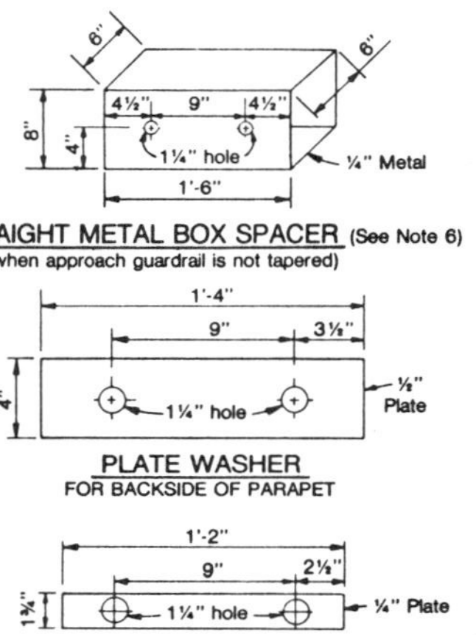


STRAIGHT METAL BOX SPACER (See Note 6)
(Use when approach guardrail is not tapered)

PLATE WASHER
FOR BACKSIDE OF PARAPET

PLATE WASHER
FOR GUARD RAIL

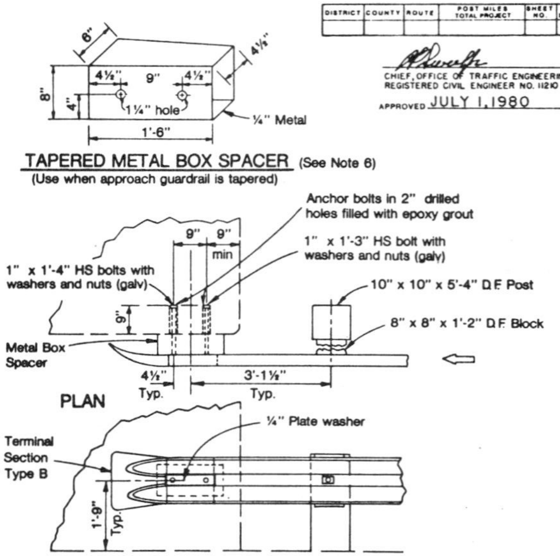


TAPERED METAL BOX SPACER (See Note 6)
(Use when approach guardrail is tapered)

GUARD RAIL ANCHORAGE TO WALL OR ABUTMENT FACE

| DISTRICT | COUNTY | ROUTE | POST MILES TOTAL PROJECT | SHEET NO. | TOTAL SHEETS |
|---|---|---|---|---|---|
| | | | | | |

CHIEF, OFFICE OF TRAFFIC ENGINEERING
REGISTERED CIVIL ENGINEER NO. 11210
APPROVED JULY 1, 1980

## GENERAL NOTES

1. These connection details apply to bridge rails, abutments and retaining walls. Do not connect guardrail to bridge columns.
2. End sections may be cut from Type B terminal section or fabricated.
3. The 4'-0" end section makes provision for a variable dimension "d". If the limits are exceeded at existing installations the Metal Beam Guard Rail end section must be modified to fit.
4. Direction of traffic indicated by ⇨
5. For post size and spacing See Type 1 Flare on Standard Plan A 79-A.
6. When metal box spacer is installed, place 1¼" x 5" and 1¼" x 4" pipe spacers on 1" bolts passing through interior of tapered box and 1¼" x 5¼" pipe spacers on 1" bolts passing through interior of straight box.
7. 6" x 8" D.F. Posts and Blocks shown. W6x8.5 or W6x9 Steel Posts and Blocks may be specified where applicable. 10"x10" D.F. Posts and 8"x8" D.F. Blocks shown. W6x15 Steel Posts and T.S. 6"x6"x1'-2" (0.1875" thick) Blocks may be specified where applicable. (See Standard Plan A77-CS).
8. Terminal Sections will not be installed on trailing end of guard rail placed adjacent to one way roadways. When terminal section is not installed, use backup plate. (See Standard Plan A77-CS or A77-CW).

STATE OF CALIFORNIA
BUSINESS AND TRANSPORTATION AGENCY
DEPARTMENT OF TRANSPORTATION

# GUARD RAIL CONNECTIONS TO BRIDGE RAILS, RETAINING WALLS AND ABUTMENTS

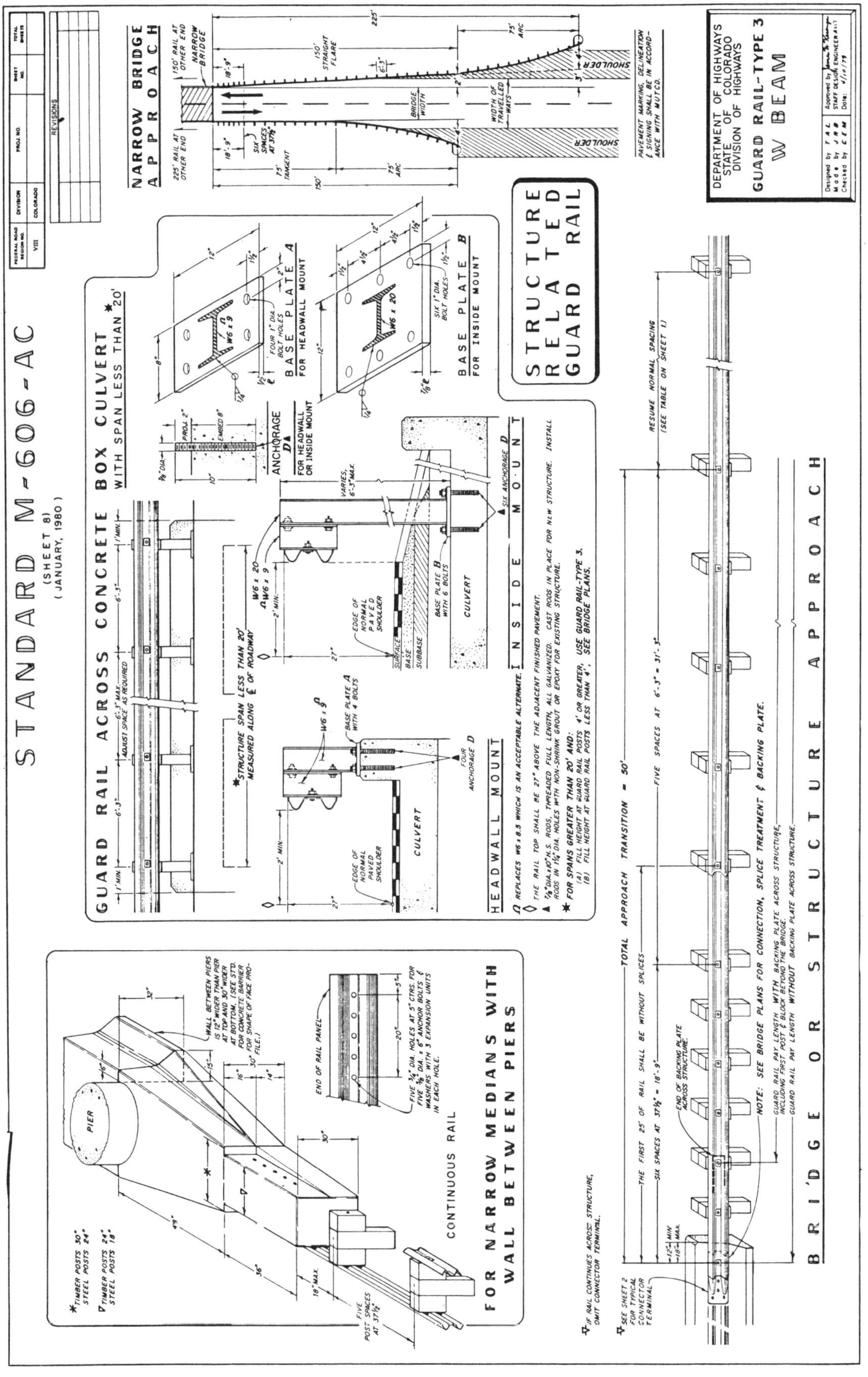
STANDARD M-606-AC
(SHEET 8)
(JANUARY, 1980)
FEDERAL ROAD REGION NO. VIII
DIVISION COLORADO
PROJ. NO.
SHEET NO.
TOTAL SHEETS
REVISIONS
*TIMBER POSTS 30" STEEL POSTS 24"
∇TIMBER POSTS 24" STEEL POSTS 18"
PIER
WALL BETWEEN PIERS IS 12" WIDER THAN PIER AT TOP AND 30" WIDER AT BOTTOM. (SEE STD. FOR CONCRETE BARRIER FOR SHAPE OF FACE PROFILE.)
FIVE POST SPACES AT 37½"
END OF RAIL PANEL
FIVE ¾" DIA. HOLES AT 5" CTRS. FOR FIVE ⅝" DIA. x 6" ANCHOR BOLTS & WASHERS WITH 3 EXPANSION UNITS IN EACH HOLE.
CONTINUOUS RAIL
FOR NARROW MEDIANS WITH WALL BETWEEN PIERS
GUARD RAIL ACROSS CONCRETE BOX CULVERT WITH SPAN LESS THAN 20'*
ADJUST SPACE AS REQUIRED
*STRUCTURE SPAN LESS THAN 20' MEASURED ALONG ℄ OF ROADWAY
EDGE OF NORMAL PAVED SHOULDER
CULVERT
BASE PLATE A WITH 4 BOLTS
FOUR ANCHORAGE D
HEADWALL MOUNT
SURFACE
BASE
SUBBASE
BASE PLATE B WITH 6 BOLTS
SIX ANCHORAGE D
INSIDE MOUNT
ANCHORAGE D▲
FOR HEADWALL OR INSIDE MOUNT
FOUR 1" DIA. BOLT HOLES
BASE PLATE A
FOR HEADWALL MOUNT
SIX 1" DIA. BOLT HOLES
BASE PLATE B
FOR INSIDE MOUNT
Ω REPLACES W6 x 8.5 WHICH IS AN ACCEPTABLE ALTERNATE.
◊ THE RAIL TOP SHALL BE 27" ABOVE THE ADJACENT FINISHED PAVEMENT.
▲ ⅞"DIA. x 10" H.S. RODS, THREADED FULL LENGTH, ALL GALVANIZED. CAST RODS IN PLACE FOR NEW STRUCTURE. INSTALL RODS IN 1¼" DIA. HOLES WITH NON-SHRINK GROUT OR EPOXY FOR EXISTING STRUCTURE.
* FOR SPANS GREATER THAN 20' AND:
(A) FILL HEIGHT AT GUARD RAIL POSTS 4' OR GREATER, USE GUARD RAIL-TYPE 3.
(B) FILL HEIGHT AT GUARD RAIL POSTS LESS THAN 4', SEE BRIDGE PLANS.
STRUCTURE RELATED GUARD RAIL
NARROW BRIDGE APPROACH
225' RAIL AT OTHER END
150' RAIL AT OTHER END
NARROW BRIDGE
SIX SPACES AT 37½"
75' TANGENT
150' STRAIGHT FLARE
75' ARC
BRIDGE WIDTH
WIDTH OF TRAVELLED WAYS
SHOULDER
PAVEMENT MARKING, DELINEATION & SIGNING SHALL BE IN ACCORDANCE WITH MUTCD.
IF RAIL CONTINUES ACROSS STRUCTURE, OMIT CONNECTOR TERMINAL.
SEE SHEET 2 FOR TYPICAL CONNECTOR TERMINAL
TOTAL APPROACH TRANSITION = 50'
THE FIRST 25' OF RAIL SHALL BE WITHOUT SPLICES
SIX SPACES AT 37½" = 18'-9"
FIVE SPACES AT 6'-3" = 31'-3"
RESUME NORMAL SPACING (SEE TABLE ON SHEET 1)
END OF BACKING PLATE ACROSS STRUCTURE.
NOTE: SEE BRIDGE PLANS FOR CONNECTION, SPLICE TREATMENT & BACKING PLATE.
GUARD RAIL PAY LENGTH WITH BACKING PLATE ACROSS STRUCTURE, INCLUDING FIRST POST & BLOCK BEYOND THE BRIDGE.
GUARD RAIL PAY LENGTH WITHOUT BACKING PLATE ACROSS STRUCTURE.
BRIDGE OR STRUCTURE APPROACH
DEPARTMENT OF HIGHWAYS
STATE OF COLORADO
DIVISION OF HIGHWAYS
GUARD RAIL-TYPE 3
W BEAM
Designed by T A L
Made by J R B
Checked by C E M
Approved by STAFF DESIGN ENGINEER
Date: 4/10/79

## GUARD RAIL END ANCHORAGE

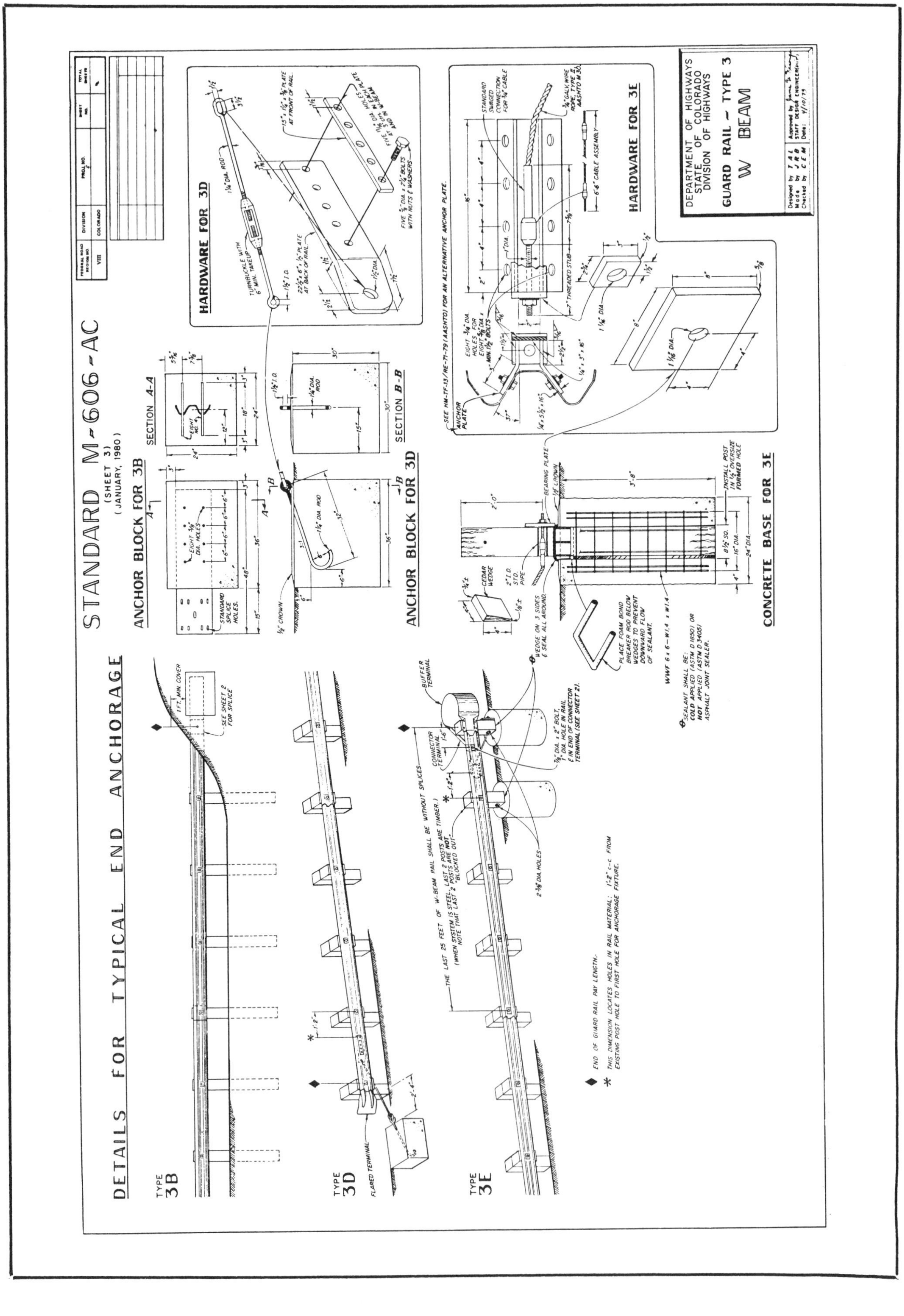

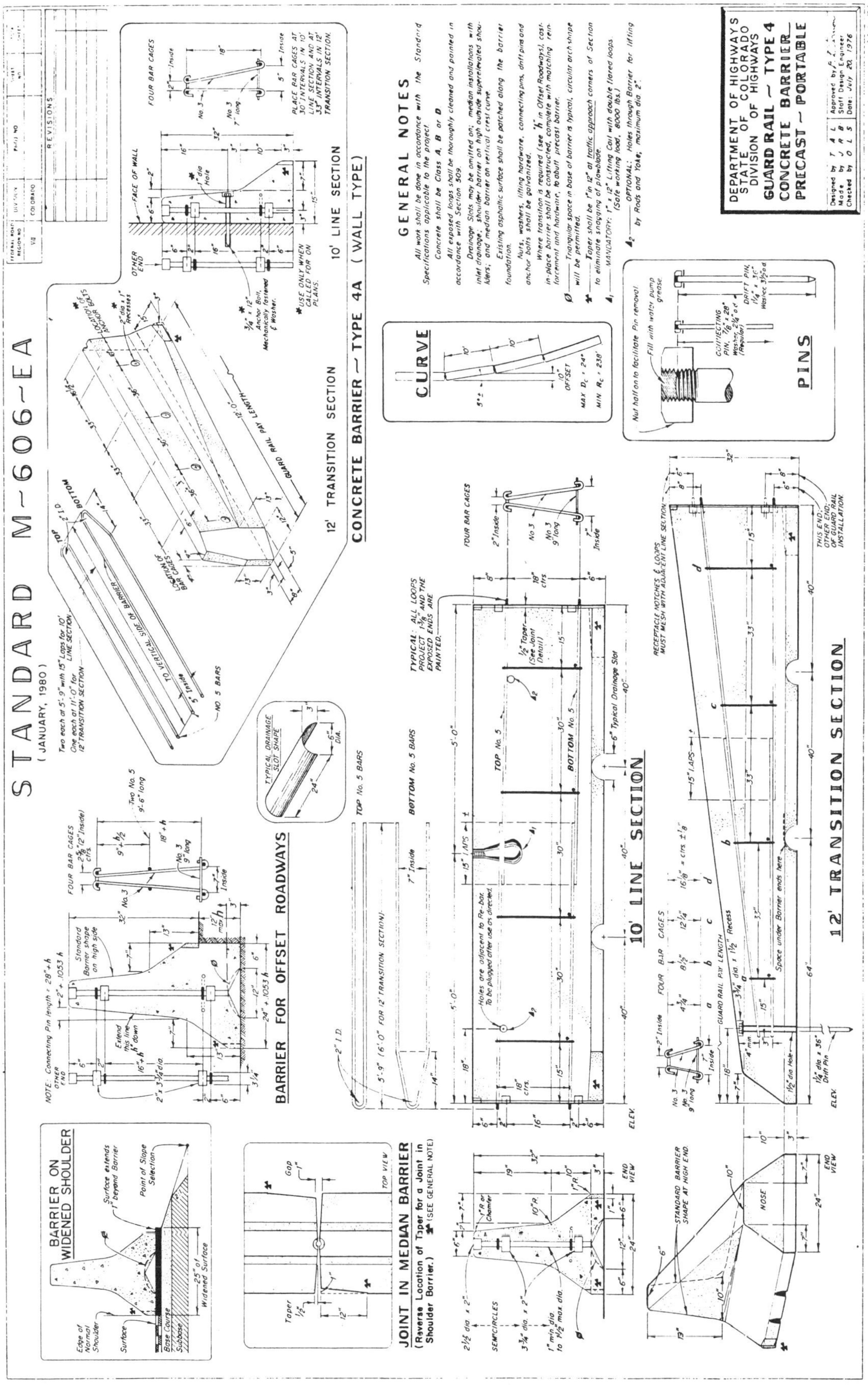

STANDARD M-606-EA
(JANUARY, 1980)
12' TRANSITION SECTION
CONCRETE BARRIER — TYPE 4A (WALL TYPE)
10' LINE SECTION
GENERAL NOTES
CURVE
PINS
BARRIER FOR OFFSET ROADWAYS
10' LINE SECTION
12' TRANSITION SECTION
BARRIER ON WIDENED SHOULDER
JOINT IN MEDIAN BARRIER
DEPARTMENT OF HIGHWAYS
STATE OF COLORADO
DIVISION OF HIGHWAYS
GUARD RAIL — TYPE 4
CONCRETE BARRIER
PRECAST — PORTABLE

# Illinois Department of Transportation

## Memorandum

**To:** All Bridge Designers 79.2

**From:** Carl E. Thunman, Jr.

**Subject:** (1) Minimum Bridge Roadway Width - Stage Construction
(2) Temporary Bridge Rail

**Date:** February 9, 1979

Effective immediately on all future bridge contracts the following policies shall be applicable.

(1) Minimum Bridge Roadway Width during Stage Construction

The minimum horizontal clearance required for a single lane roadway during any stage of a stage construction bridge project shall be:

(a) 14'-0" where a wide load detour is not provided

(b) 10'-0" where a wide load detour is provided

(2) Temporary Bridge Rails

All temporary bridge rails required for stage bridge construction shall have sufficient strength to carry the loading specified in Article 1.1.8 - Railings of the AASHTO Standard Specification for Highway Bridges.

Temporary bridge rails shall be provided in accordance with the attached applicable details unless alternate designs or proposals are submitted for analysis and approval to the Engineer of Bridge and Traffic Structures. A temporary concrete median barrier (MB 5) will be considered as an acceptable alternate design when it can be safely supported by the bridge deck.

RAK/OJR/lsp

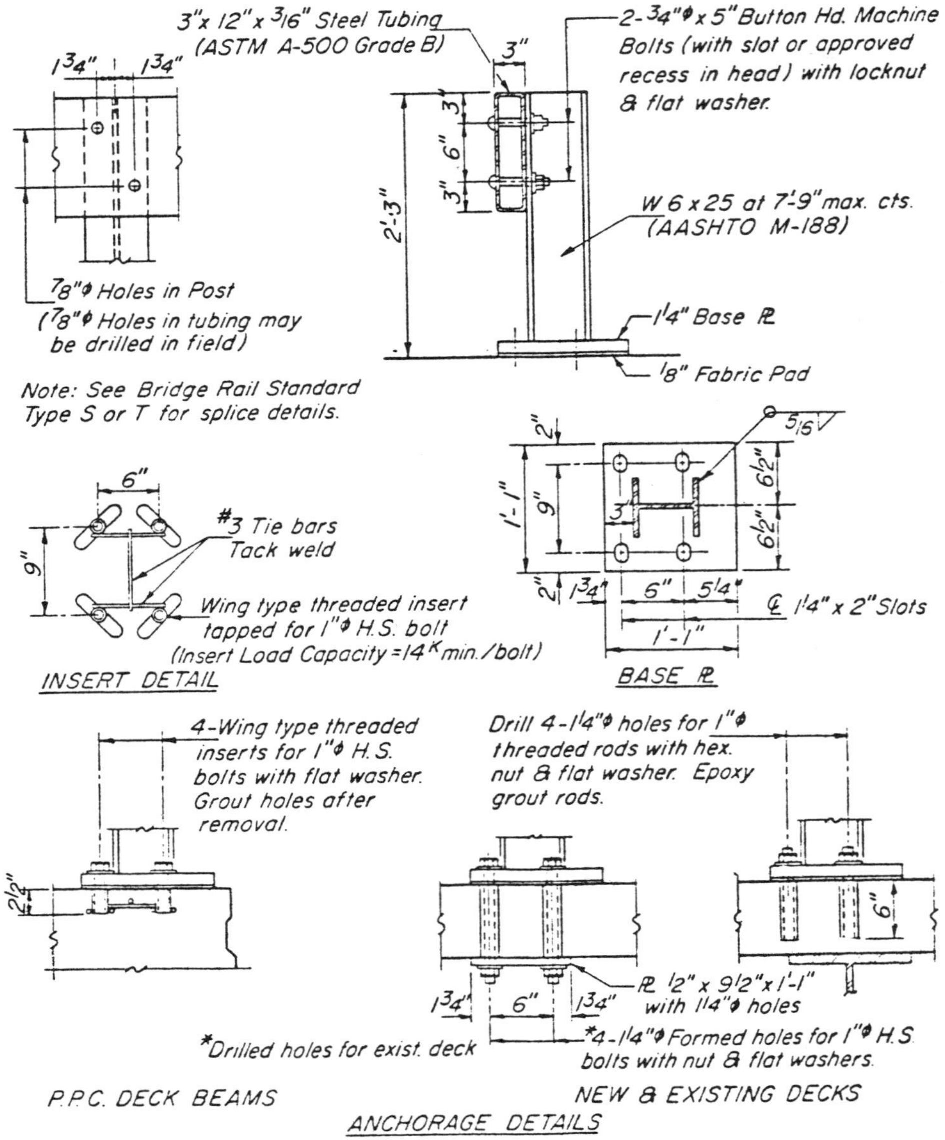


TEMPORARY BRIDGE RAIL DETAILS

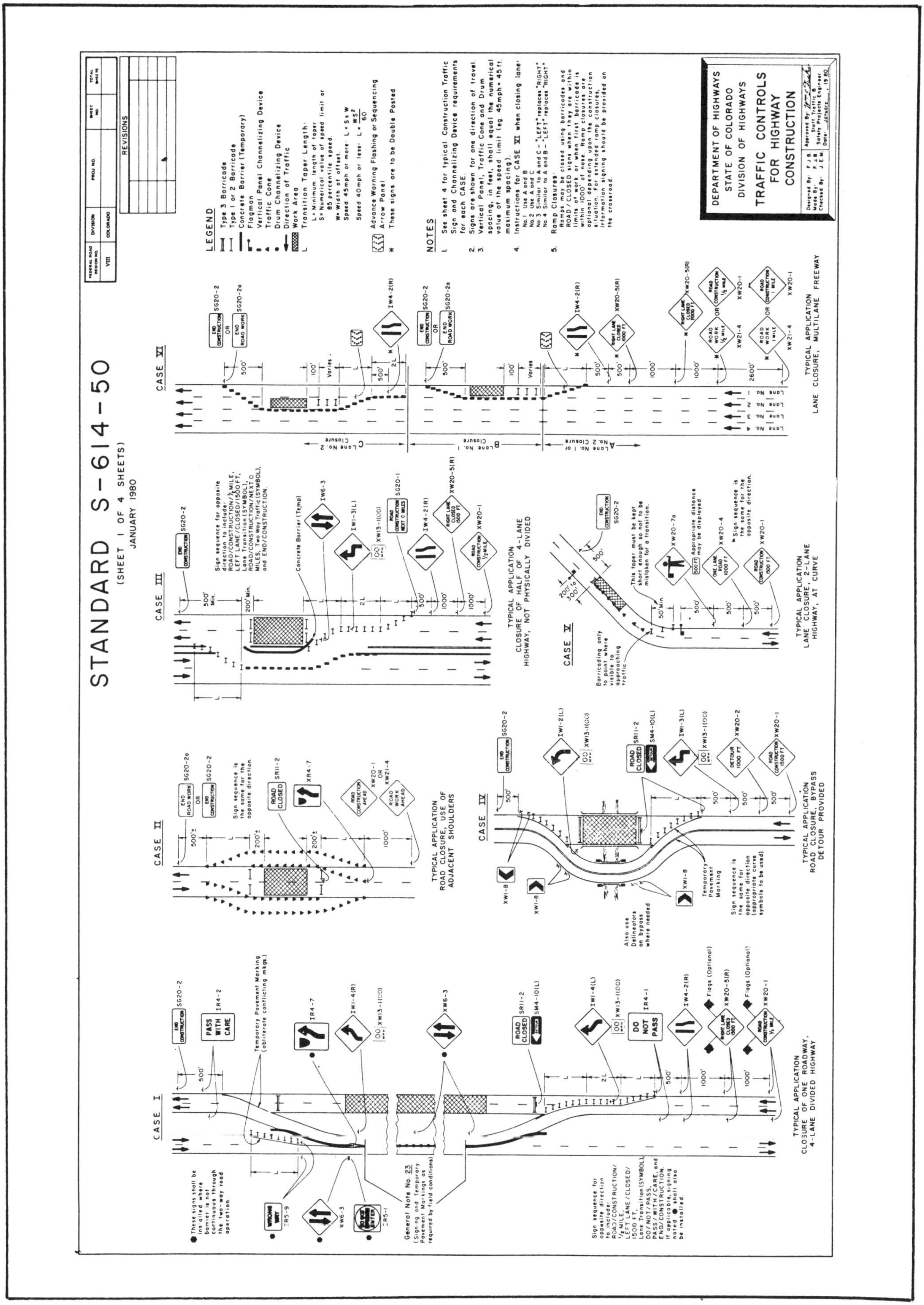
STANDARD S-614-50
(SHEET 1 OF 4 SHEETS)
JANUARY 1980
FEDERAL ROAD REGION NO. VIII
DIVISION COLORADO
PROJ. NO.
SHEET NO.
TOTAL SHEETS
REVISIONS
CASE I
CASE II
CASE III
CASE IV
CASE V
CASE VI
These signs shall be installed where barrier is not continuous through the two-way road operation.
General Note No. 23 (Signing and Temporary Pavement Markings as required by field conditions)
Temporary Pavement Marking (obliterate conflicting mkgs.)
Sign sequence for opposite direction to include: ROAD/CONSTRUCTION/½ MILE, LEFT LANE/CLOSED/1500 FT., Lane Transition (SYMBOL), DO/NOT/PASS, PASS/WITH/CARE, and END/CONSTRUCTION. If applicable, signing noted ● shall also be installed.
TYPICAL APPLICATION CLOSURE OF ONE ROADWAY, 4-LANE DIVIDED HIGHWAY
Sign sequence is the same for the opposite direction.
TYPICAL APPLICATION ROAD CLOSURE, USE OF ADJACENT SHOULDERS
Also use Delineators on bypass where needed
Temporary Pavement Marking
Sign sequence is the same for opposite direction (appropriate curve symbols to be used).
TYPICAL APPLICATION ROAD CLOSURE, BYPASS DETOUR PROVIDED
Sign sequence for opposite direction to include: ROAD/CONSTRUCTION/½ MILE, LEFT LANE/CLOSED/1500 FT, Lane Transition (SYMBOL), ROAD/CONSTRUCTION/NEXT 0 MILES, Two Way Traffic (SYMBOL), and END/CONSTRUCTION.
Concrete Barrier (Temp)
TYPICAL APPLICATION CLOSURE OF HALF OF 4-LANE HIGHWAY, NOT PHYSICALLY DIVIDED
Barricading only to point where visible to approaching traffic
This taper must be kept short enough so not to be mistaken for a transition.
Appropriate distance may be displayed
Sign sequence is the same for the opposite direction.
TYPICAL APPLICATION LANE CLOSURE, 2-LANE HIGHWAY, AT CURVE
TYPICAL APPLICATION LANE CLOSURE, MULTILANE FREEWAY
LEGEND
Type 3 Barricade
Type 1 or 2 Barricade
Concrete Barrier (Temporary)
Flagman
Vertical Panel Channelizing Device
Traffic Cone
Drum Channelizing Device
Direction of Traffic
Work Area
L Transition Taper Length:
L = Minimum length of taper
S = Numerical value of speed limit or 85 percentile speed.
W = Width of offset.
Speed 45mph or more: L = S x W
Speed 40mph or less: L = WS²/60
Advance Warning Flashing or Sequencing Arrow Panel
* These signs are to be Double Posted
NOTES
1. See sheet 4 for typical Construction Traffic Sign and Channelizing Device requirements for each CASE.
2. Signs are shown for one direction of travel.
3. Vertical Panel, Traffic Cone and Drum spacing, in feet, shall equal the numerical value of the speed limit (eg. 45mph = 45 ft. maximum spacing).
4. Instructions for CASE VI when closing lane:
No. 1 Use A and B
No. 2 Use A and C
No. 3 Similar to A and C - "LEFT" replaces "RIGHT"
No. 4 Similar to A and B - "LEFT" replaces "RIGHT"
5. Ramp Closures:
Ramps may be closed using barricades and ROAD/CLOSED signs when they are within limits of work or when first barricade is within 1000' of nose. Ramp closures are optional depending upon the construction situation. For extended ramp closures, information signing should be provided on the crossroad.
DEPARTMENT OF HIGHWAYS
STATE OF COLORADO
DIVISION OF HIGHWAYS
TRAFFIC CONTROLS FOR HIGHWAY CONSTRUCTION
Designed By: F.J.B.
Made By: F.J.B.
Checked By: J.E.M.
Staff Traffic & Safety Projects Engineer
Date: January, 1980

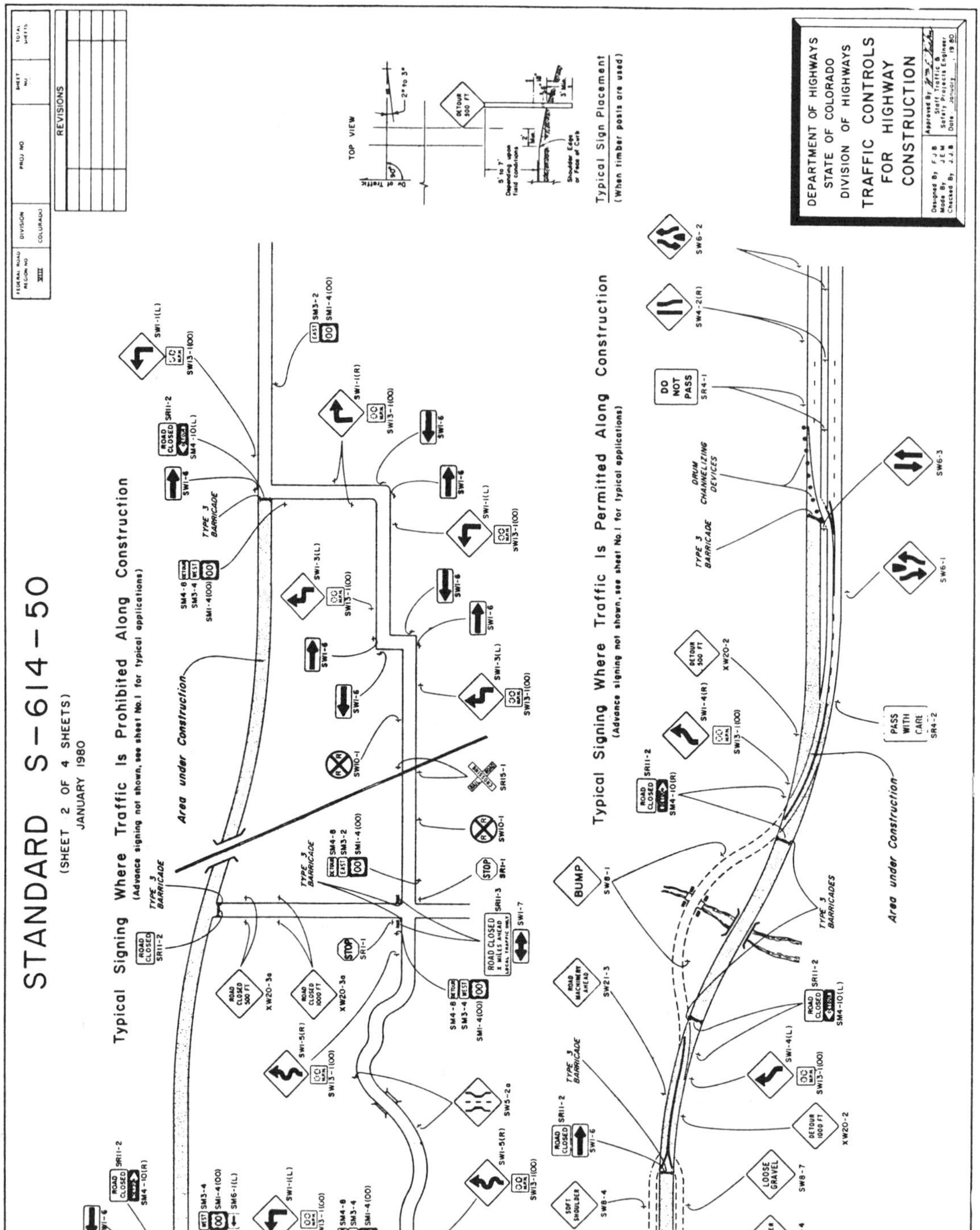
STANDARD S-614-50
(SHEET 2 OF 4 SHEETS)
JANUARY 1980
Typical Signing Where Traffic Is Prohibited Along Construction
(Advance signing not shown, see sheet No. 1 for typical applications)
Typical Signing Where Traffic Is Permitted Along Construction
(Advance signing not shown, see sheet No. 1 for typical applications)
Typical Sign Placement
(When timber posts are used)
TOP VIEW
Area under Construction
TYPE 3 BARRICADE
TYPE 3 BARRICADES
DRUM CHANNELIZING DEVICES
ROAD CLOSED
DETOUR 500 FT
DETOUR 1000 FT
ROAD CLOSED 500 FT
ROAD CLOSED 1000 FT
ROAD CLOSED 8 MILES AHEAD LOCAL TRAFFIC ONLY
DO NOT PASS
PASS WITH CARE
BUMP
LOOSE GRAVEL
SOFT SHOULDER
ROAD MACHINERY AHEAD
STOP
DEPARTMENT OF HIGHWAYS
STATE OF COLORADO
DIVISION OF HIGHWAYS
TRAFFIC CONTROLS
FOR HIGHWAY
CONSTRUCTION
REVISIONS
COLORADO

## P.1 WIDENING SCHEMES

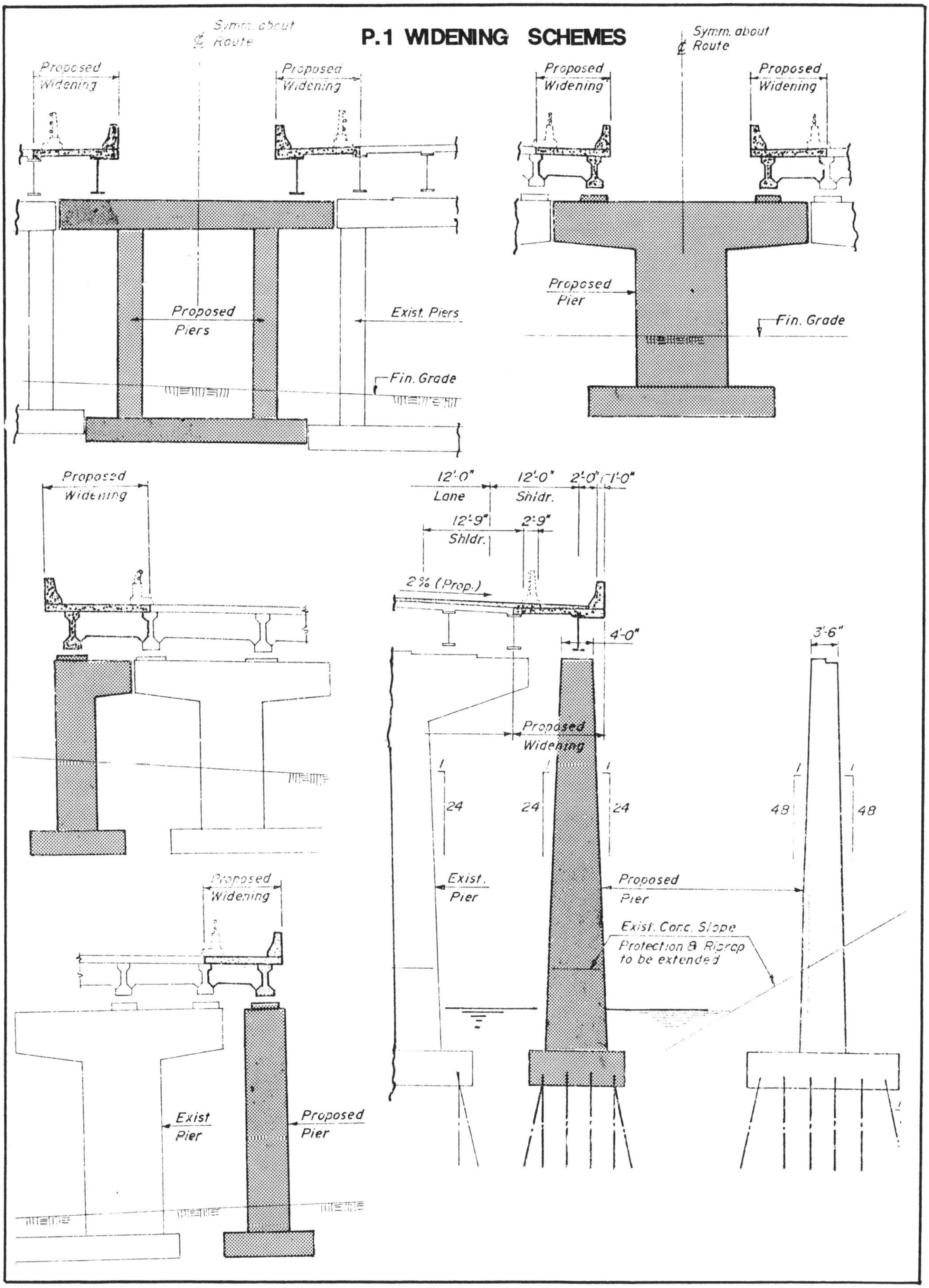

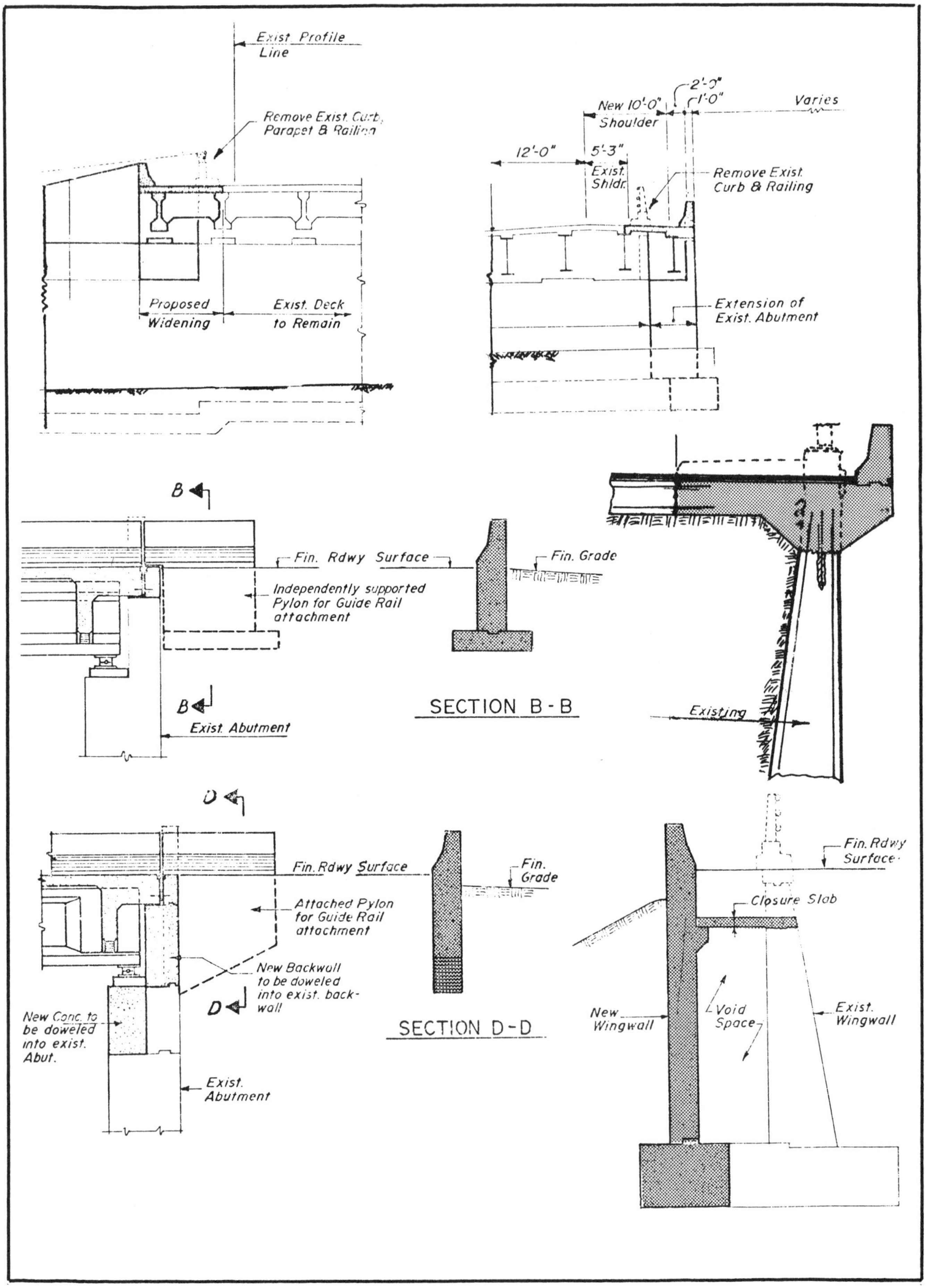
Exist. Profile Line
Remove Exist. Curb, Parapet & Railing
Proposed Widening
Exist. Deck to Remain
New 10'-0" Shoulder
2'-0"
1'-0"
Varies
12'-0"
5'-3"
Exist. Shldr.
Remove Exist. Curb & Railing
Extension of Exist. Abutment
B
Fin. Rdwy Surface
Fin. Grade
Independently supported Pylon for Guide Rail attachment
B
Exist. Abutment
SECTION B-B
Existing
D
Fin. Rdwy Surface
Fin. Grade
Attached Pylon for Guide Rail attachment
New Backwall to be doweled into exist. backwall
D
New Conc. to be doweled into exist. Abut.
Exist. Abutment
SECTION D-D
Fin. Rdwy Surface
Closure Slab
New Wingwall
Void Space
Exist. Wingwall

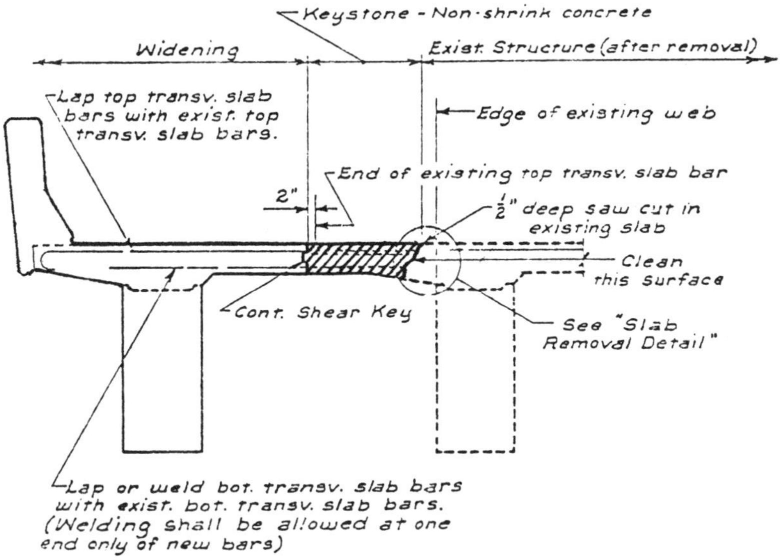


T-BEAM ~ SECTION IN SPAN

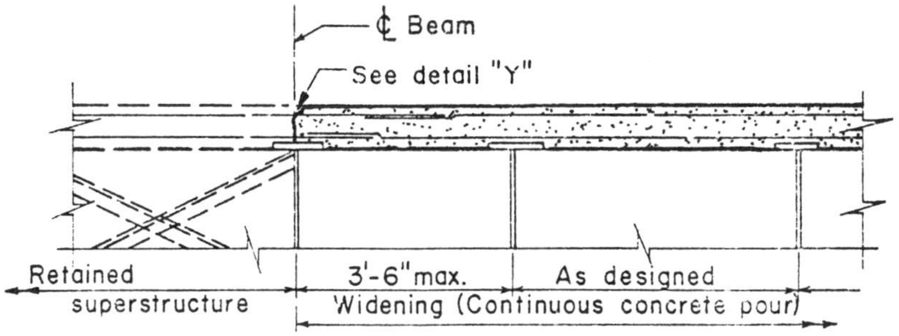


III. I-BEAM BRIDGES

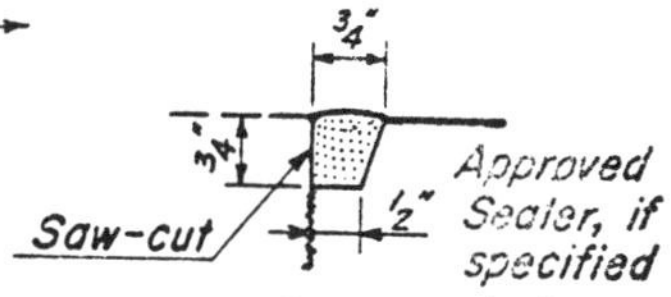


DETAIL "Y"

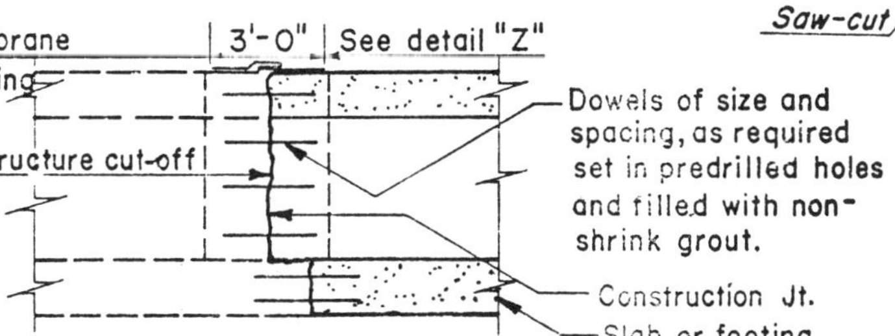


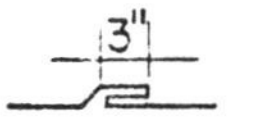


DETAIL "Z"

(A) INTERIOR PANEL

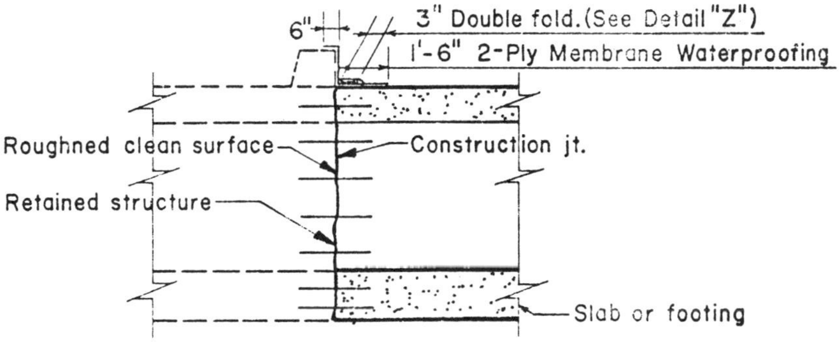


(B) END PANEL IF SUITABLE

IV. R. C. CULVERTS UNDER FILL

Because the transverse slab reinforcing is only distribution steel, it is not necessary to remove any portion of the existing slab to bare the existing transverse reinforcing for splicing purposes. The transverse slab reinforcing for the widening may be doweled directly into the existing structure without meeting the normal splice requirement.

For the moment connection details, see "Flat-Slab - Section Through X-Beam,"

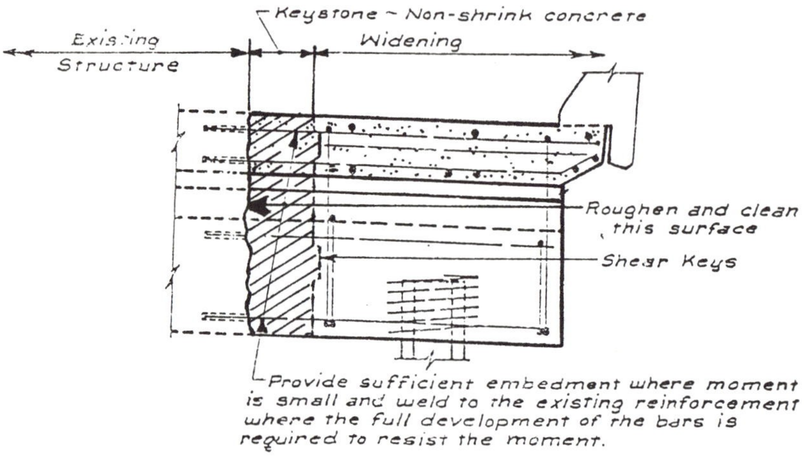


FLAT SLAB ~ SECTION THROUGH X-BEAM

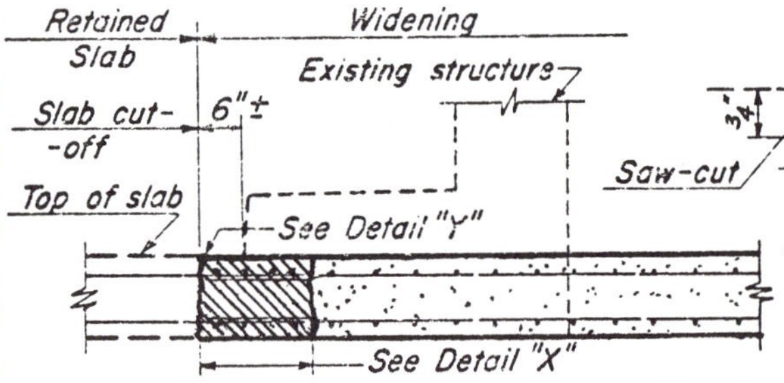


(A) OUTER PORTION UNSUITABLE FOR WIDENING

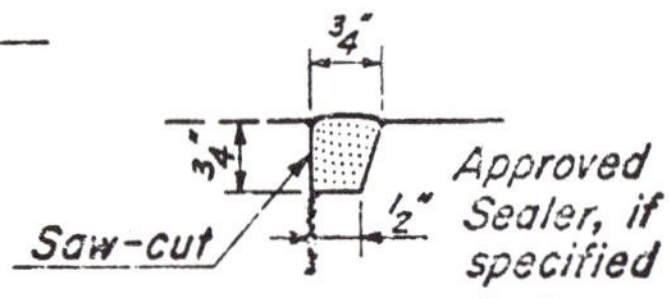


DETAIL "Y"

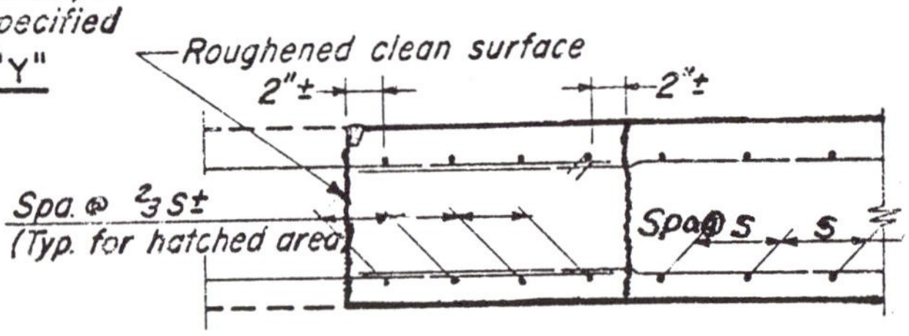


DETAIL "X"

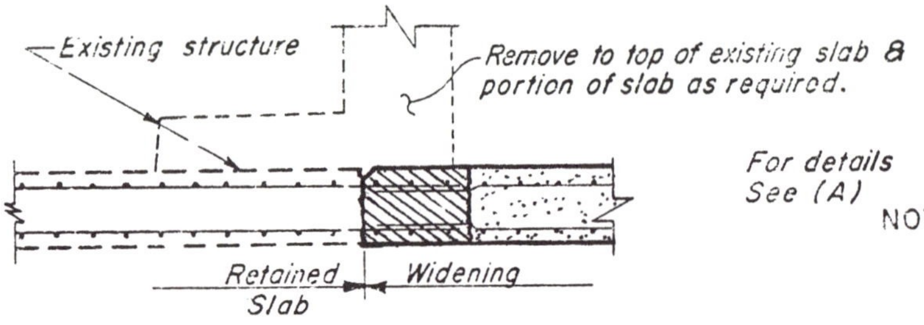


NOTE: Hatched areas to be cast after supports under widening are removed.

(B) OUTER PORTION SUITABLE FOR WIDENING
(For bituminous wearing surface only)

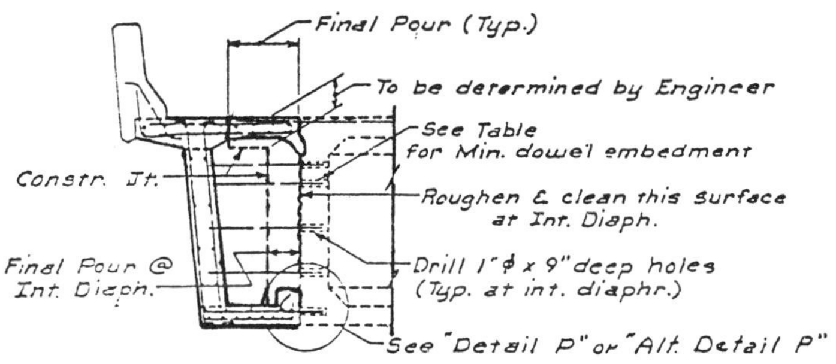


BOX GIRDER
SECTION IN SPAN
ALTERNATE I

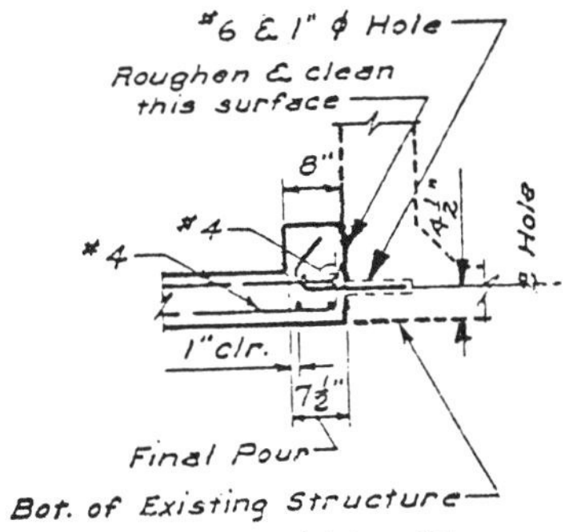


DETAIL P

Note: This alternate applies to narrow widenings where shear in the exterior web is not critical. This is typical for short to medium spans or where the existing slab overhang is considerably less than half the web spacing.

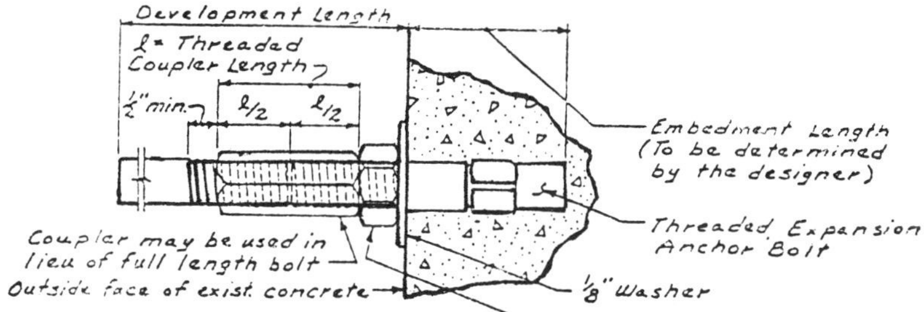


ALT. DETAIL P

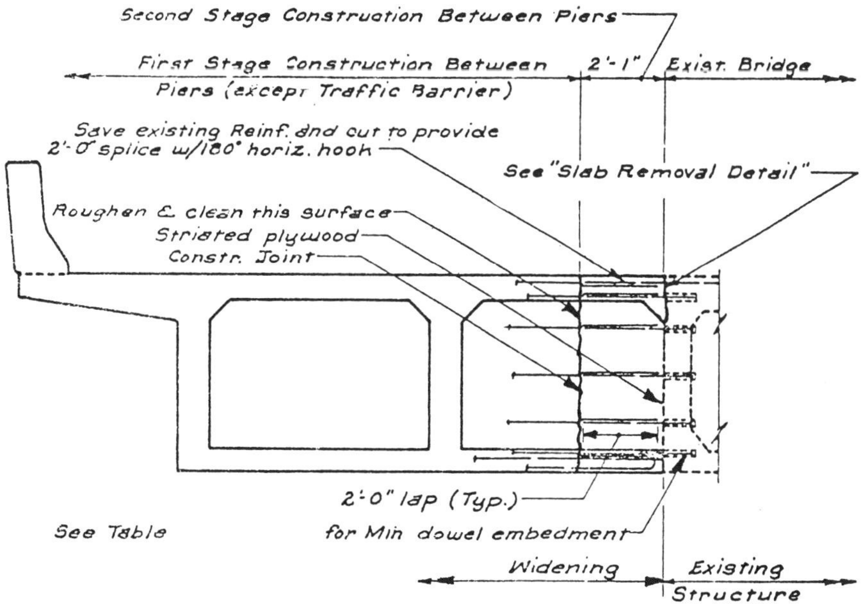


Note: This method is useful when the width of the widening is such that two or more rows of reinforcing is required at the top of the cross beam(s).

BOX GIRDER – SECTION NEAR ℄ SPAN
ALTERNATE II

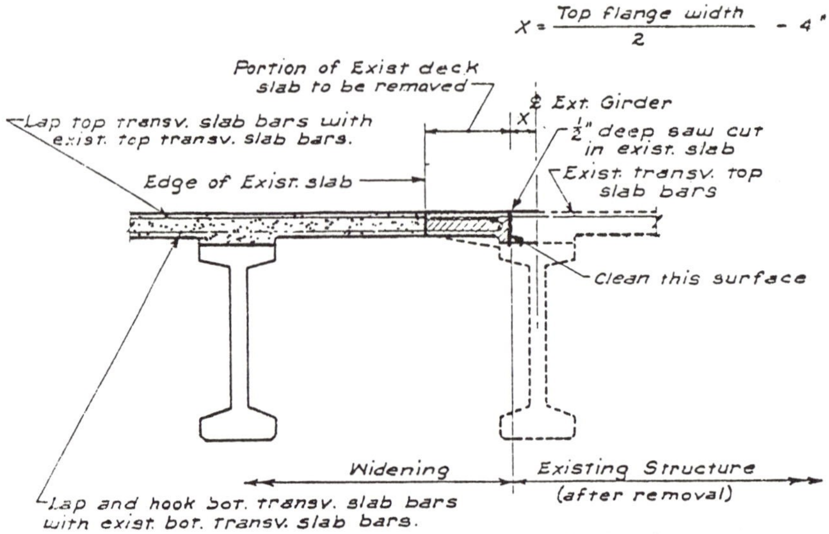


Note: A closure pour is not req'd. for this type of widening.

PRESTRESSED GIRDER ~ SECTION IN SPAN

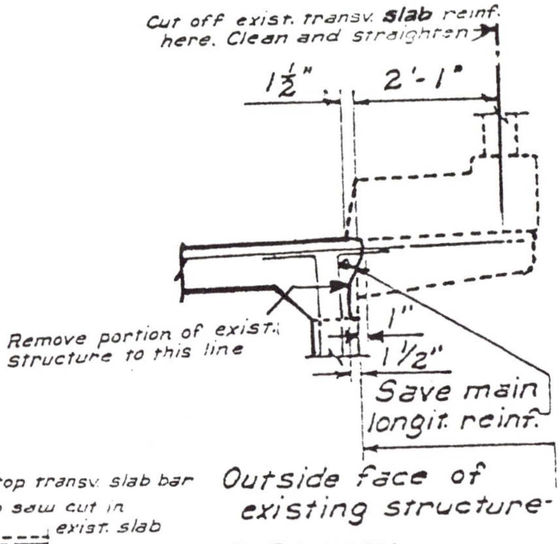


SLAB REMOVAL DETAIL

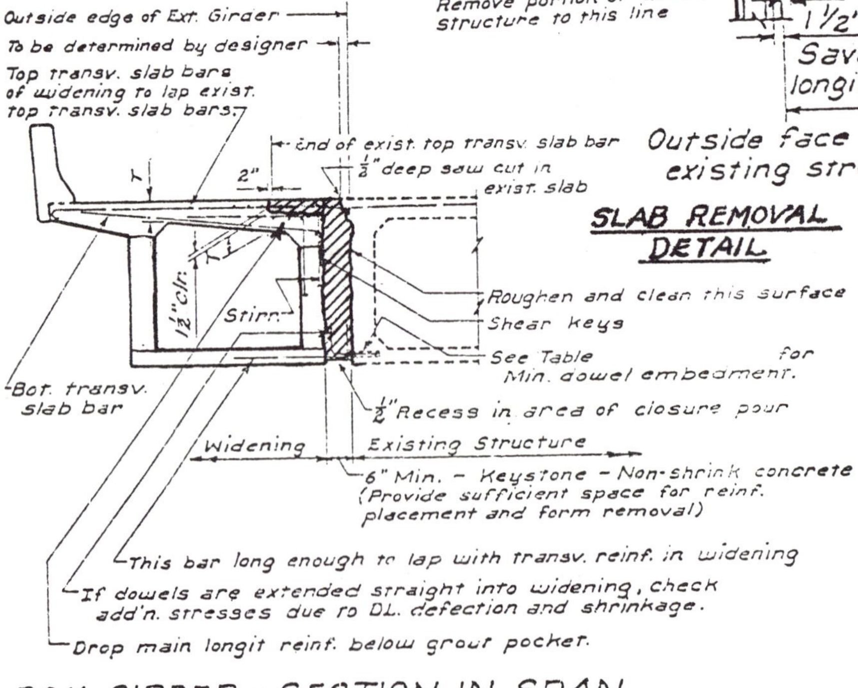


BOX GIRDER ~ SECTION IN SPAN

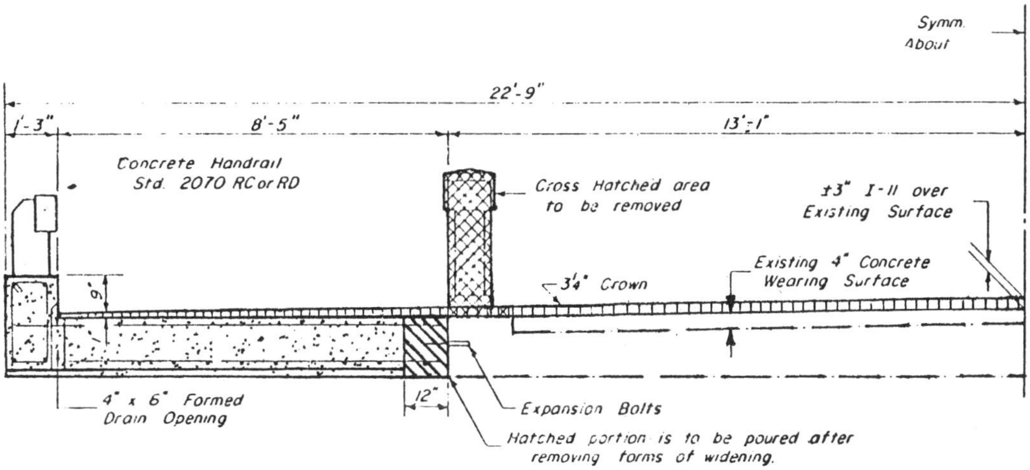


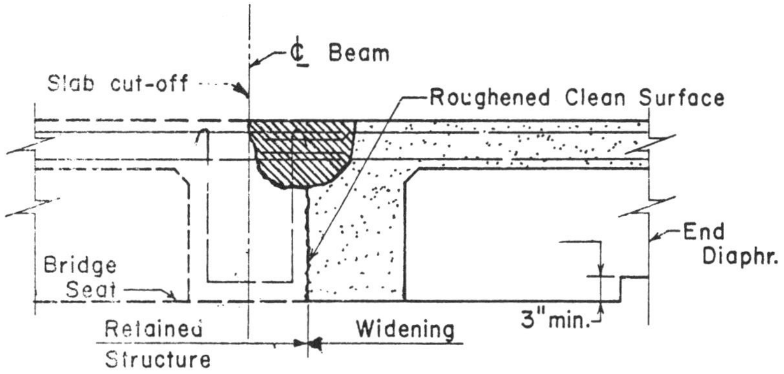


(A) OUTER PORTION UNSUITABLE FOR WIDENING

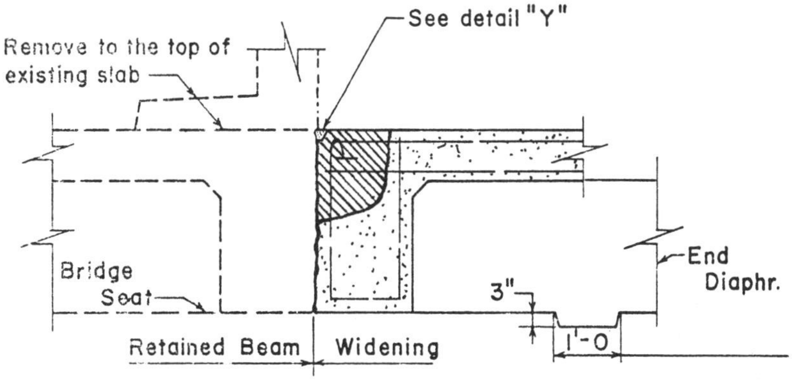


(B) OUTER PORTION SUITABLE FOR WIDENING
(FOR BITUMINOUS WEARING SURFACES ONLY)

## II. R. C. T-BEAM BRIDGES

NOTE: Hatched areas to be cast after supports under widening are removed.

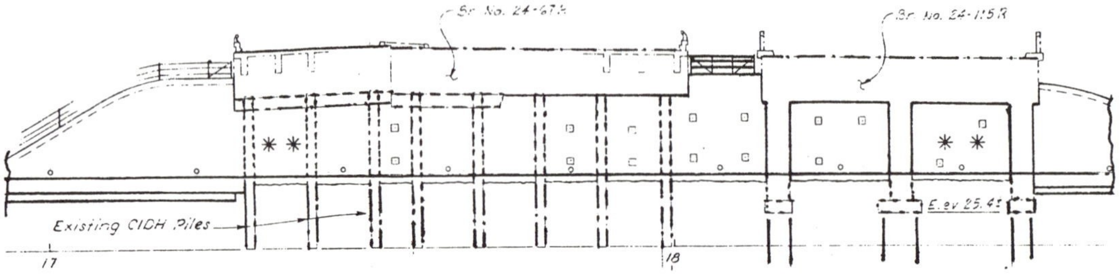


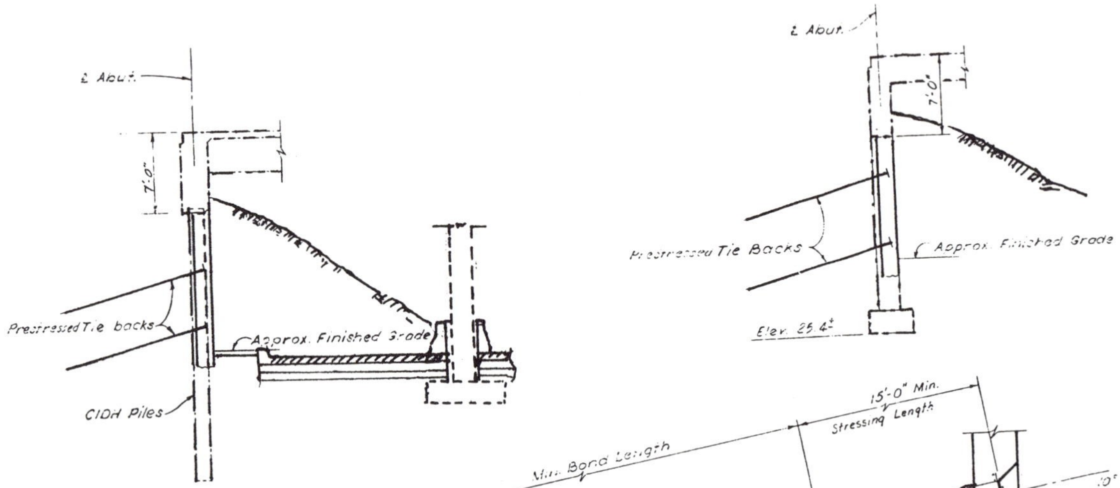


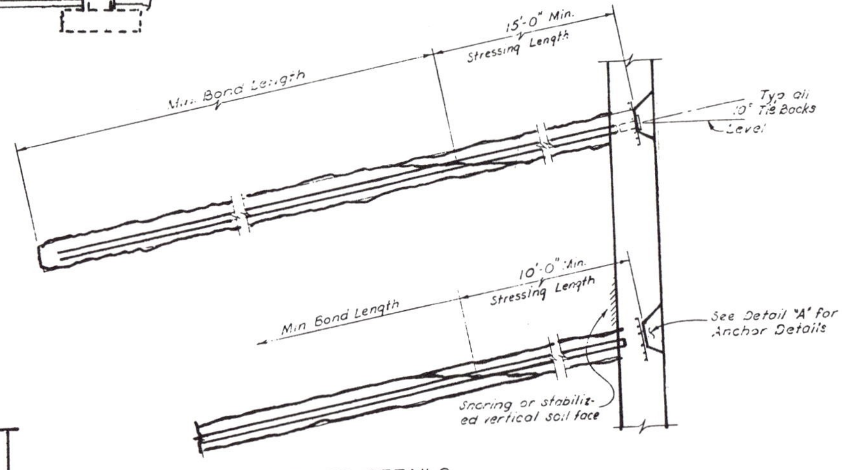


ANCHOR DETAILS

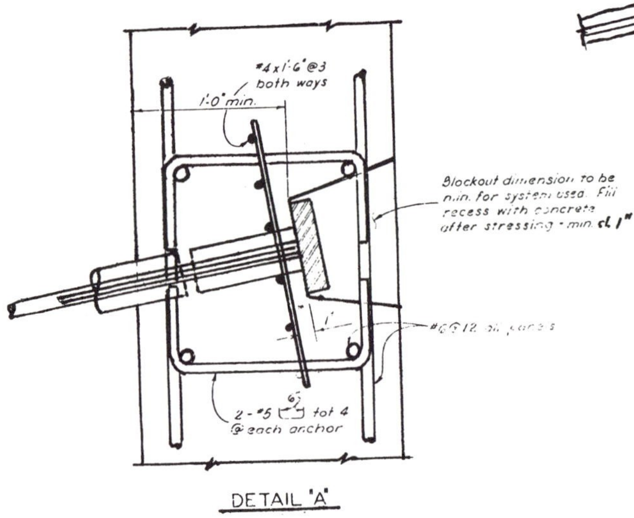


DETAIL "A"

| PANEL | TOTAL NO. TIE BACKS | FINAL WORKING FORCE/TIEBACK |
|---|---|---|
| A | 40 | 40,000 lbs. |
| B | 16 | 35,000 lbs. |
| C | 4 | 35,000 lbs. |
| D | 12 | 44,000 lbs. |
| E | 2 | 26,000 lbs |

$$A_s = \frac{2 \times \text{Final Working Force}}{0.75\ f_s}$$

Panel Concrete: $f_c$ = 3500 psi min. @ 28 days

Anchor hole diameter and bond length to be determined by Contractor based on soil shear resistance of 0.50 Tons per square foot

State of
CALIFORNIA
DEPARTMENT OF TRANSPORTATION

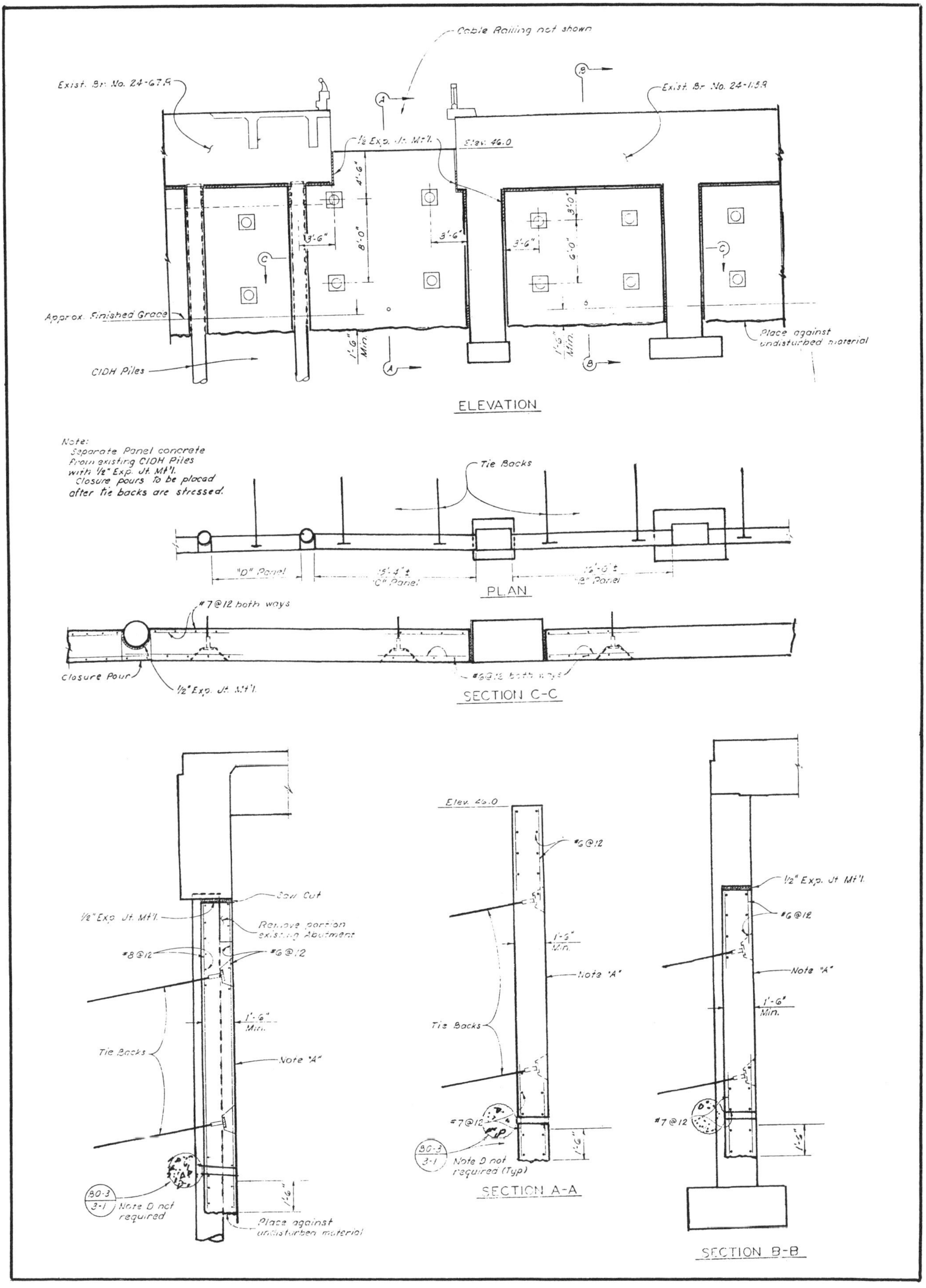
Cable Railing not shown
Exist. Br. No. 24-67.R
Exist. Br. No. 24-158
½ Exp. Jt. Mt'l.
Elev. 46.0
3'-6"
8'-0"
3'-6"
4'-6"
3'-0"
6'-0"
1'-6" Min.
Approx. Finished Grade
CIDH Piles
Place against undisturbed material
ELEVATION
Note:
Separate Panel concrete from existing CIDH Piles with ½" Exp. Jt. Mt'l.
Closure pours to be placed after tie backs are stressed.
Tie Backs
"D" Panel
15'-4"± "C" Panel
12'-0"± "B" Panel
PLAN
#7@12 both ways
Closure Pour
½" Exp. Jt. Mt'l.
#6@12 both ways
SECTION C-C
Saw Cut
½" Exp. Jt. Mt'l.
Remove portion existing Abutment
#8@12
#6@12
1'-6" Min.
Tie Backs
Note "A"
80-3
3-1
Note D not required
Place against undisturbed material
Elev. 46.0
#6@12
1'-6" Min.
Note "A"
Tie Backs
#7@12
Note D not required (Typ)
SECTION A-A
½" Exp. Jt. Mt'l.
#6@12
Note "A"
1'-6" Min.
#7@12
SECTION B-B

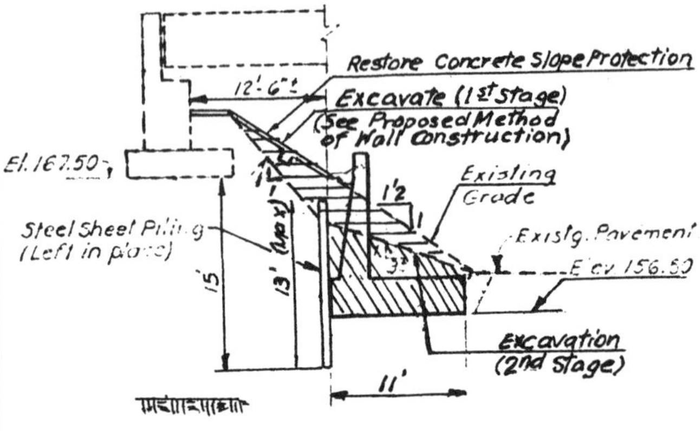


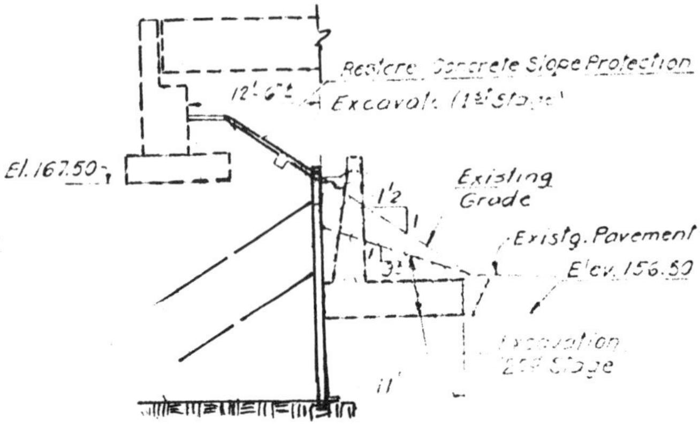


PROPOSED METHOD OF WALL CONSTRUCTION UNDER EXISTING BRIDGE

1. Demolish existing slope protection and excavate area marked thus for an 11 foot section.
2. Install steel sheet piling as shown. Suggested methods are: Pre-auger and jack against Bridge or vibratory driver.
3. Install bulkheads with timber sheathing.
4. Excavate to Elevation 156.50
5. Construct 10 foot section of retaining wall.
6. Backfill, install backdrain and compact
7. Repeat above procedure for additional 10 foot sections
8. Restore concrete slope protection as shown.

Sheet, excavate and construct new pier and retaining wall. Retaining wall and pier shall be complete in place and backfilled before proceding with remainder of work.

Construct bearing pedestals at west abutment, Pier 1 and new pier constructed in Stage I.
Install web splice plates and jacking beam at Pier 2.
Transfer load from jacks beneath beam flanges at Pier 2 to jacks on pier beneath jacking beam.
Remove railing and chain link fence at Pier 2 and install top and bottom flange splice plates, using holes in these members as a template for drilling holes in existing beam flanges.
Reinstall railing and chain link fence.

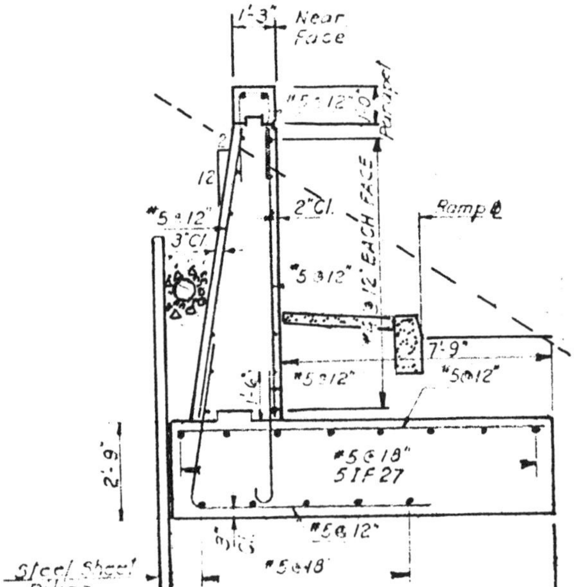


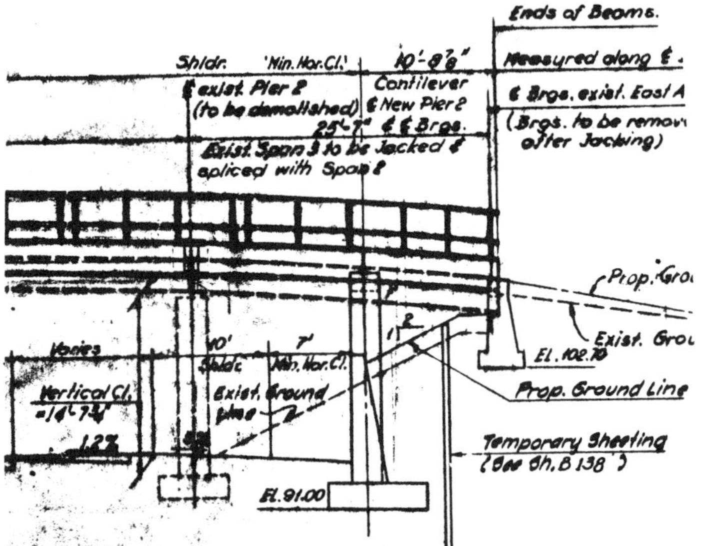

# APPENDIX Q: SPECIFICATION

## Q.1 REPAIR MATERIAL

laterials:

94 lbs. Cement Type I
182 lbs. Fine Aggregate
297 lbs. Coarse Aggregate
5.7 gals. Water
3 Tablespoons Air Entraining Agent
Slump 3" - 5"

Note:

1. Accelerators shall not be used.
2. Coarse aggregate is Grade 57 (3/4" and smaller).
3. For a wheelbarrow load (approx. 2 cu. ft.), use 1 & 1/2 tablespoons air entraining agent.
4. Volume of water will vary slightly.
5. Chloride free materials shall be used.
6. Use only air entraining agents on approved list.

**MATERIAL NAME:** Class III Concrete (Seal)

laterials:

94 lbs. Cement Type I
166 lbs. Fine Aggregate
263 lbs. Coarse Aggregate
5.6 gals. Water
3 Tablespoons Air Entraining Agent
Slump 7" to 9"

1. This is a one-bag tremie mix and is air-entrained, non-vibrated, with water-reducing, retardant admixture.
2. A one-bag mix is approximately two wheelbarrow loads. For one wheelbarrow load (approx. 2 cu. ft.), use 1 & 1/2 tablespoons air entraining agent. Water-reducing retardant used per manufacturer's instructions.
3. Rock shall be coarse aggregate, Grade 57 (#9), 3/4" or smaller.
4. Volume of water will vary slightly.
5. Chloride free materials shall be used.
6. Use only admixtures on the approved list.
7. The above mix may be used for pumped concrete. Variations may be required. Consult with the engineer.

**MATERIAL NAME:** Class IV Concrete (Vibrated)

aterials:

94 lbs. Cement Type I
155 lbs. Fine Aggregate
251 lbs. Coarse Aggregate
4.6 gals. Water
3 Tablespoons Air Entraining Agent
Slump 0" - 3 & 1/2"

Note:

1. Rock shall be coarse aggregate, Grade 57 (3/4" or smaller).
2. Volume of water will vary slightly.
3. Chloride free materials shall be used.
4. For a wheelbarrow load (approx. 2 cu. ft.), use 1 & 1/2 tablespoons of air entraining agent.
5. Water-reducing retardant as per manufacturer's recommendations.
6. Use only admixtures on the approved list.

**TERIAL NAME:** Latex Modified Portland Cement Concrete

laterials:

| | By Weight | By Volume |
|---|---|---|
| Portland Cement | 94 lbs. | 1 cu. ft. |
| Sand | 235 lbs. (dry wt.) | 2.25 cu. ft. |
| Coarse Aggregate* | 178.3 lbs. (dry wt.) | 2.25 cu. ft. |
| Latex Additive | 37.1 lbs. | 4½ gallons |
| Water** | 4 - 25 lbs. | ½ - 3 gallons |

*Coarse aggregate shall be No. 14, No. 16 or No. 16S
**The amount of water required per batch will depend on the moisture content of the sand.

STATE OF FLORIDA DEPARTMENT OF TRANSPORTATION
BRIDGE REPAIR MANUAL

MATERIAL NAME: Portland Cement Grout Filler

Materials:

94 lbs. Cement Type I
436 lbs. Fine Aggregate
5.4 gals. Water
3 tablespoons Air Entraining Agent

Note:

1. Accelerators whall not be used.
2. Volume of water will vary slightly.
3. Chloride free materials shall be used.
4. Use only air entraining agents on approved list.

MATERIAL NAME: Non-Shrink Grout

Non-shrink grout shall be mixed in the following portions by volume:

1) One part Portland Cement
2) One part fine aggregate
3) Non-shrink admixture proportioned as recommended by the manufacturer (for aluminum powder, 1/2 to 3/4 ounce by weight to one sack of cement)

The dry mix shall be first thoroughly mixed to form a uniform mixture. Only enough water shall be added to give a mealy-appearing adhesive mix.

The grout shall be hammered, tamped and rammed into place to produce complete contact between the elements. The material shall be generally placed near the center and built-out toward the edges by compacting as required above.

MATERIAL NAME: Epoxy Grout Filler

Epoxy bonding compound: A two component material that is moisture insensitive for application both above and below water. It must be capable of adhering to wet concrete, steel and fiberglass forms.

Sand: Kiln dried Silica sand (see Material No. 902-2).

Machine mix the two component epoxy bonding compounds in strict accordance with the manufacturer's instructions. Mix in the ratio of 1 part epoxy bonding compound with a maximum of three parts sand. No more epoxy grout filler shall be mixed than can be placed in 20 minutes. Any excess on hand after this time shall be discarded.

MATERIAL NAME: Sand for Epoxy Grout Filler

Kiln dried Silica sand meeting the following gradation requirements:

Standard Sand

| U.S. Sieve | Passing |
|---|---|
| No. 4 | 100% |
| No. 16 | 90-100% |
| No. 30 | 30-50% |
| No. 50 | 0-10% |
| No. 100 | 0-5% |

Alternate Sand

| U.S. Sieve | Passing |
|---|---|
| No. 4 | 100% |
| No. 10 | 90-100% |
| No. 20 | 0-5% |
| No. 40 | 0% |

MATERIAL NAME: Copper Sheeting

Copper sheeting shall weigh a minimum of 10 oz. per square foot or greater, conforming to ASTM B 370. The copper sheets shall be 18" square.

**ITEM NAME:** Painting Structural Steel - Inorganic Zinc

**DESCRIPTION:** Sandblast cleaning and repainting spot areas of deteriorated inorganic zinc paint system.

**APPLICATION:** Spot paint deteriorated areas on structures painted with inorganic zinc system. Applicable to both coastal and inland bridges.

**MATERIALS:** Sandblast Material - M.S. #561-2
Zinc-Rich Primer --- APL
Intermediate Coat -- APL
Finish Coat --------- APL

**CONSTRUCTION METHOD:**

General: Personnel performing this activity must be thoroughly trained with respect to required surface preparation and in the proper techniques for application of coatings. Florida D.O.T. offers a training course on surface preparation and application of protective coatings.

1. All rusted areas are to be spot blasted to a "near-white" condition.
2. Coat the blast-cleaned areas by brush or spray with a single coat (3.0 to 5.0 dry mils) of zinc-rich primer.
3. Twelve to twenty-four hours after application of the primer coat, apply finish coat to a minimum dry film thickness of 3.0 mils. Bridges located in coastal areas require an intermediate coat prior to application of the finish coat

NOTE: Precautions must be taken not to damage the paint system outside the deteriorated area being repaired. This may be accomplished by shielding if necessary, however, any damaged areas from over-blast must be repaired. All mechanical components and electrical equipment must be protected. Blast material in gears and any coating on contactors must be removed.

**ITEM NAME:** Painting Structural Steel - Oil Base Paint

**DESCRIPTION:** Sandblast cleaning and repainting spot areas of deteriorated oil base paint system.

**APPLICATION:** Spot repair of structural steel which has an existing oil base paint coating.

**CONSTRUCTION METHOD:**

General: Same as Item No. 817.

1. All rusted areas are to be spot blasted to a "commercial" blast or wire brushed.
2. The cleaned areas shall then be painted by brush or spray with two coats (min. 2.5 mils dry film thickness per coat) of Florida Specification Code Z-C Primer.
3. A minimum of 48 hours shall elapse between first and second coat of primer.
4. After second coat of primer has cured 48 hours, apply by brush or spray one coat (0.5 mils min. dry film thickness) of Florida Specification Code B-A Aluminum Topcoat.

**MATERIAL NAME:** Sandblast Material

Sand: Silica sand of a maximum particle size no larger than that passing through a 16 mesh screen, U.S. sieve series.

Grit: Crushed grit made of cast iron, malleable iron, steel, or synthetic grits other than sand, of a maximum particle size no larger than that passing through a 16 mesh screen, U.S. sieve series. The largest commercial grade of metal grit permitted by this specification is SAE No. G-25 abrasive material.

Shot: Shot made of cast iron, malleable iron, steel, or synthetic shot of a maximum size no larger than that passing through a 16 mesh screen, U.S. sieve series. The largest commercial grade permitted by this specification is SAE No. S-330.

NOTE: Type of sandblasting material to be furnished shall be as specified on the purchase order.

# Q.2 ADMIXTURE

## CONCRETE ADMIXTURES

**437.1. Description.** This item shall govern the materials used, methods of tests, and construction methods for the use of admixtures in concrete.

**437.2 General.**

An "air-entraining admixture" is defined as a material which, when added to a concrete mixture in the correct quantity, will entrain uniformly dispersed microscopic air.

A "water-reducing, retarding admixture" is defined as a material which, when added to a concrete mixture in the correct quantity, will reduce the quantity of mixing water required to produce concrete of a given consistency and will retard the initial set of the concrete.

A "water-reducing admixture" is defined as a material which, when added to a concrete mixture in the correct quantity, will reduce the quantity of mixing water required to produce concrete of a given consistency.

An "accelerating admixture" is defined as an admixture that accelerates the setting time and the early strength development of concrete.

A "high-range water reducing admixture" is defined as a material which when added to a concrete mix in the correct quantity will produce the results required herein.

• **437.3. Retarding and Water Reducing Admixtures.** The admixture shall meet the requirements of Type A, Type D or Type F admixture as specified in ASTM Designation: C494, modified as follows:

(1) The water-reducing retarder shall retard the initial set of the concrete a minimum of 2 hours and a maximum of 4 hours, at a specified dosage rate, at an ambient temperature of 90 F.

(2) The cement used in any series of tests shall be either the cement proposed for the specific work or a "reference" Type I cement from one mill.

(3) Unless otherwise noted on the plans, the minimum relative durability factor shall be 80.

(4) All concrete being tested will contain entrained air. The air entraining admixture used in the reference and test concrete shall be Neutralized Vinsol Resin.

High-range water reducing admixtures when compared to a standard mix design in accordance with ATSM Designation: C494, in addition to (2) (3) and (4) above shall produce the following:

(1) It shall reduce the required water by a minimum of 15%.

(2) It shall increase the seven day compressive strength of the concrete by a minimum of 25%.

(3) It shall not retard the initial set of concrete by more than one and one-half hours, when tested in accordance with ASTM Designation: C403.

(4) It shall contain no chlorides, air entraining agents or urea.

• **437.4. Air Entraining Admixture.** The admixture shall meet the requirements of ASTM Designation: C260 modified as follows:

(1) The cement used in any series of tests shall be either the cement proposed for specific work or a "reference" Type I cement from one mill.

(2) Unless otherwise noted on the plans, the minimum relative durability factor shall be 80.

The air entraining admixture used in the reference concrete shall be Neutralized Vinsol Resin.

**437.5. Accelerating Admixture.** The admixture shall meet the requirements for a Type C admixture as specified in ASTM C 494 modified as follows:

The accelerating admixture will contain no chlorides, and shall be used in the liquid form only.

Accelerators will be used only to meet special requirements and will require the approval of the Engineer on each specific project.

**437.6. Approval of Admixtures.** The manufacturer shall certify that the material to be furnished meets the requirements of this Item, and of ASTM Designation: C260 or C494, as modified herein, and shall furnish test reports from an approved laboratory, as defined by the item, "Definition of Terms" having prior approval of the Materials and Tests Engineer. At the time of original request for approval of admixtures, the manufacturer shall state in writing the chloride content of the admixtures. No admixture to which chlorides have been added during manufacture will be permitted to be used

In addition to the above, the manufacturer shall furnish the following:

(1) At the time of original request for approval of the admixture, the manufacturer shall supply a five gallon sample of the material to the Materials and Tests Engineer, 38th and Jackson Streets, Austin, Texas 78703.

(2) Each six months after approval of the material, the manufacturer shall furnish a notarized certification indicating that the material originally approved has not been changed or altered in any way. Any change in formulation of an admixture shall require retesting, and shall be approved by Materials and Tests Engineer prior to use.

(3) Prior to approval, an Infrared Spectrophotometry scan, solids content, pH value, and unit weight will be submitted for further identification. A range of values will be given to cover the normal manufacturing range. A change in formulation discovered by any of the tests prescribed herein, or other means, and not reported and retested, may be cause to permanently bar the manufacturer from furnishing admixtures for Departmental projects.

A list of pretested and approved admixtures will be maintained by the Materials and Tests Engineer. This list shall specify the approved dosage rates to be used except as modified herein.

The Department reserves the right to perform any or all of the tests required by ASTM Designation: C 260 and C 494 as a check on the tests reported by the manufacturer. In case of any variance the Departmental tests will govern.

**437.7. Construction Use of Admixtures.** When used in construction in conformance with the basic reference specifications and this specification, the Contractor will be allowed to use any admixture which has been approved. The Contractor shall submit to the Engineer one copy of the invoice showing the admixture or admixtures to be used on the project. Prior to using an admixture in the work, trial mixes shall be made and tested in the field using the materials and equipment to be used on the project.

Mix designs from previous or concurrent jobs may be used without trial batches if it is shown that no substantial change in any of the proposed ingredients has been made.

For air entraining admixtures the dosage shown on the approved list is the dosage utilized in approval tests of the admixture. This dosage *must* be adjusted as necessary to produce the required air content in the concrete within the specified tolerances.

For water reducing, water reducing-retarding, and high range water reducing admixtures, dosage rates different from those on the list of approved admixtures may be used provided they are established on the basis of trial mixes.

An approved retarding admixture (for normal hot weather concreting) may not perform satisfactorily for extended retardation, in which case its use will not be permitted.

All accelerating admixtures dosages will be based on trial mixes and approved by the Engineer.

All admixtures used shall be in the liquid state with the exception of high-range water reducers which may be in the powder form. All liquid admixtures shall be dispensed separately but at the same time as the mixing water except that liquid or powdered high-range water reducers shall be introduced into the concrete at the job site. No admixture shall be dispensed on dry aggregates. Each admixture shall be dispensed separately.

Admixtures shall be agitated as required to prevent separation or sedimentation of solids. Air agitation of Neutralized Vinsol Resin will not be permitted.

Air entraining agents shall be charged into the mixer at the beginning of the batch. Retarding or water reducing admixtures, except for high-range water reducers, shall be charged into the mixer during the last one-third of the batch.

Accelerating admixtures will be used only on the express approval of the Engineer. Accelerating admixtures will not be permitted in bridge decks, direct traffic culvert slabs at any time, nor when Type II cement is specified.

The high-range water reducing admixtures (liquid or powder form) shall be incorporated in the mix entirely at the job site as determined by adequate trial batches, and approved by the Engineer. Normally powdered admixtures will be measured by the weight basis and dispensed at the job site by method approved by the Engineer. The rotation of the mixer shall be sufficient to thoroughly mix the admixture into the concrete. The volume of water in the liquid high-range water reducer shall be taken into account in determining the water/cement ratio of the mix. If during the placement of concrete, an appreciable slump loss is noted, additional high-range water reducing admixture will be required to bring the concrete back into the desired consistency. The addition of water other than in the water reducer will not be permitted at the job site.

When high-range water reducing admixtures are used with Ready-Mixed concrete, the capacity of the truck mixer may be reduced for each batch by 25% of the rated capacity to assure proper mixing of the admixture.

When deemed necessary by the Engineer, the Contractor shall furnish additonal quantities of the admixture being used for further testing. Further use of the admixture will not be allowed until the results of such tests are known and the material meets the requirements of this item.

For individual placements of concrete of 25 cubic yards or more and for all Ready-Mix concrete, the admixture shall be measured and dispensed by a readily adjustable dispenser. When set to a predetermined volume the dispenser shall fill to the preset amount and hold it positively without leakage until the operator releases the content into the mixing water, by some positive means. Unless otherwise shown on the plans, completely automatic dispensing will not be required, except for use with a fully automatic plant.

The calibrated container shall be a measuring reservoir of the type where the level of the admixture is visible at all times. A strip gauge with one ounce increments for air-entraining admixtures and ten ounce increments for water reducing and/or retarding admixtures, shall be attached securely to the measuring apparatus. This strip shall be a material possessing weather-resistant qualities. The accuracy required for these systems shall be plus or minus three percent. The equipment shall visibly show the total amount to be dispensed for ready check by the Engineer.

For contract work, with individual placements of less than 25 cubic yards and with the concrete batched on the job site, the Engineer may waive the requirements for mechanical dispensing equipment.

437.8. **Measurement and Payment.** No additional compensation will be made for the materials, equipment or methods required by this item, but shall be considered subsidiary to the various items included in the contract.

Texas

## Q.3 SHOTCRETE

### PNEUMATICALLY PLACED CONCRETE

431.1. **Description.** This item shall govern for furnishing and placing of "Pneumatically Placed Concrete" for riprap, concrete channel or canal lining, encasement of designated structural steel members, the repair of deteriorated or damaged concrete and for other miscellaneous work as shown on the plans.

431.2. **Materials.** The cement, water, and aggregate shall conform to the requirements of the Item, "Concrete for Structures". Fine aggregate shall conform to the requirements of Table 2, Grade 1, and coarse aggregate shall conform to the requirments of Table 1, Grade 7, unless noted otherwise on the plans.

Air entraining admixtures, retarders and water reducing admixtures, if used, shall comply with the Item, "Concrete Admixtures".

Bar reinforcement and wire fabric reinforcement shall conform to the requirements of the Item, "Reinforcing Steel".

Expansion joint material shall conform to the requirements of the Item, "Concrete Structures".

Steel drive pins, studs or expansion bolts used for the attachment of reinforcing for repair of deteriorated or damaged concrete with pneumatically placed concrete, shall have a minimum diameter of one-eighth of an inch and a minimum length of 2 inches. Size and location of drive pins or studs and method of attachment of reinforcing shall be as specified herein or as detailed on the plans.

The equipment used for driving the pins or studs shall be of the type which uses an explosive for the driving force, and shall be capable of inserting the stud or pin to the required depth without damage to the surrounding concrete.

Expansion hook bolts, (1/4 inch diameter), shall be placed in a drilled hole of the size and depth recommended by the manufacturer. The Engineer may require that a test be made of the driving equipment for steel drive pins and check the pull out quality of the expansion bolts, prior to approving their use.

431.3. **Proportioning and Mixing.** The Contractor shall submit a mix design for approval of the Engineer. The basic mix design shall conform to the following:

| Type | *Minimum of One Part Cement To | 28 Day Compressive Strength Cores |
|---|---|---|
| I | 4 Parts Aggregate | 3600 psi Min. |
| II | 5 Parts Aggregate | 3000 psi Min. |
| III | 7 Parts Aggregate | 2000 psi Min. |

*The Contractor may use a design containing more cement than required by this specification, when approved by the Engineer.

TABLE I

The cement and aggregates shall be measured by volume with enough water added to bring the materials to the desired consistency. Test panels will be required prior to approval of the mix design. The concrete will be applied to a plywood panel and shall be a minimum size of 18" x 18" x 3" in depth. The panel will be shot with the same air pressure and nozzle tip to be used for the production work. The panel will be cured in the same manner required for the particular usage required by the contract.

Three 2 inch diameter cores will be taken from each panel and tested in compression at 7 days. The average strength of the cores shall be a minimum of 70 percent of the strengths required in Table I, herein. Testing of cores shall be in accordance with Test Method Tex-424-A.

The Engineer may require additional test panels during the progress of the work if there is any change in materials, equipment or nozzle operator.

Mixing and application may be done by either by the dry mix or wet mix process. The materials shall be thoroughly and uniformly mixed using a mixer designed for use with pneumatic application. It may be either a paddle type or drum type mixer. Transit mix concrete may be used for the wet process.

All mixing and placing equipment shall be cleaned at regular intervals and be kept in acceptable working condition. The nozzle liner, water and air injection system should be inspected daily and replaced when the parts are worn.

431.4. **Construction Methods.**

**(1) Reinforcement.** All reinforcement to be embedded in pneumatically placed concrete shall be clean and free from loose mill scale, rust, oil, or other coatings which might prevent adequate bond.

Reinforcement shall be secured rigidly in the position indicated on the drawings. The clear distance between reinforcing bars shall be at least 2-1/2 inches.

Minimum clear distance between forms and reinforcement and for cover shall be as shown on the plans. Space shall be provided for splicing bars in the approved manner.

For repair of structures, welded wire fabric shall be held securely approximately three-fourths of an inch out from the surface to be covered. Adjacent sheets shall lap at least 6 inches and sheets shall be fastened together securely by tying at intervals not to exceed 18 inches. In placing the wire fabric, steel drive pins shall be driven to a penetration of not less than 1 inch or 1/4 inch hook bolts installed in accordance with the manufacturer's recommendation into the face of the designated portions to be covered or repaired. The wire fabric shall be fastened securely to each pin or bolt. Any pin that does not reach the desired depth or hook bolt that does not anchor properly in its hole may remain in place but must be supplemented by an additional pin or bolt installation. The welded wire fabric shall have a minimum of 1 inch cover to the finished concrete surface.

For the encasement of designated portions of steel structures, the welded wire fabric shall be bent to a template to conform as nearly as possible to the outlines of the steel members to be encased. Holes not less than 1/2 inch nor more than one inch in diameter, shall be provided in the webs of the members as near as practicable to the flanges for the purpose of attaching the reinforcing fabric. These holes shall be spaced approximately 3 feet on centers. The welded wire fabric shall be held securely approximately three-fourths of an inch out from the surfaces of the members to be encased. Adjacent sheets shall lap at least 6 inches and sheets shall be fastened together securely by tying at intervals not to exceed 18 inches. In placing the wire fabric, three-eighths of an inch round rods shall be fastened to the structural steel through the holes provided in the webs of the members to be encased and the fabric shall be tied securely outside to rods. Ties shall be spaced approximately 12 inches on centers. The formed fabric shall conform, in so far as possible, to the shape of the structural member.

(2) **Existing Structural Steel.** Pneumatically placed concrete for encasing structural steel shall be Type II unless otherwise designated on the plans. All steel and concrete surfaces shall be cleaned thoroughly of all paint, rust, loose mill scale, grease, and such other foreign materials which are likely to prevent adequate bond between the surface to be encased or covered. Concrete surfaces to be covered with pneumatically placed concrete shall be kept damp a day in advance of application and shall be thoroughly cleaned and washed with water and filtered compressed air just prior to the application.

The encasement concrete shall be given a wood float finish and water cured for 4 days.

(3) **Repair of Existing Concrete.** A Type I mixture shall be used for structure repair. All deteriorated or loose concrete shall be removed from the areas designated to be repaired or restored within the limits specified on the plans or designated by the Engineer. Concrete adjacent to a crack shall be removed in such a manner as to leave the existing reinforcing steel throughout the repair area as intact as possible. Concrete and reinforcing steel surfaces which will be in contact with pneumatically placed concrete shall be sandblasted clean, then the surface cleaned of loose material with filtered compressed air.

Exposed areas are to be sprayed with water, followed with another spraying after ten minutes, then not later than ten minutes after the second water spray, the repair area or the cavity will be filled with pneumatically placed concrete, tying in the wire mesh as required. A steel edged screed shall be used to cut surface to original lines. A steel trowel shall be used for final finish.

For curing, the repair area shall have a piece of wet cotton mat taped into place over the repaired area followed with a covering of 4 mil min. sheet plastic also taped into place. The sheet plastic shall be larger than the mat and shall be continuously taped at the edges with 3 inch min. width tape (air duct tape or better) to completely enclose the mat and hold in the moisture. After four days or longer, the mat and cover may be removed.

After the curing period the patches will be tested by striking with a hammer to check for soundness and bond to existing concrete.

(4) **Riprap and Ditch Lining.** Pneumatically placed concrete for riprap and for channel or canal linings shall be the type designated on the plans. The concrete shall be placed within the limits specified on the plans or as designated by the Engineer. The surface shall be given a wood float finish or a gun finish as directed by the Engineer. Curing of riprap and/or ditch lining shall be by either Type I or II membrane in accordance with the Item, "Membrane Curing".

(5) **Operating Requirements for the Dry Mix Process.** The compressor or blower used to supply air shall be capable of delivering a sufficient volume of oil free air, at a pressure range of 30 to 85 psi as required by the size of the nozzle employed. Required capacity of compressor and operating pressures are shown in Table 2 for the various nozzle sizes. Steady pressure must be maintained throughout the placing process. The water pump shall be of sufficient size and capacity to deliver the water to the nozzle at a pressure of not less than 15 psi in excess of the required air pressure.

| Compressor Capacity cu. ft. per Min. | Hose Diameter, inches | Maximum Size of Nozzle Tip, inches | Operating Air Pressure Available, psi |
|---|---|---|---|
| 250 | 1 | 3/4 | 40 |
| 315 | 1 1/4 | 1 | 45 |
| 365 | 1 1/2 | 1 1/4 | 55 |
| 500 | 1 5/8 | 1 1/2 | 65 |
| 600 | 1 3/4 | 1 5/8 | 75 |
| 750 | 2 | 1 3/4 | 85 |

Table 2
COMPRESSOR CAPACITIES

The values shown in Table 2 are based on a hose length of 150 ft. with the nozzle not more than 25 ft. above the delivery equipment. Operating pressures shall be increased approximately 5 psi for each additional 50 ft. of hose and approximately 5 psi for each 25 ft. the nozzle is raised.

(6) **Operating Requirements for the Wet Mix Process.** The pump shall operate so that the line pressure is between 100 psi and 300 psi for delivery hoses with 1½" to 3" diameters. The mixing equipment shall be capable of thoroughly mixing the materials in sufficient quantity to maintain continuous placement. When transit mix concrete is used, this equipment shall conform to the Item, "Ready-Mix Plants".

The use of the wet process will not be permitted for the repair of deteriorated or damaged concrete.

(7) **Rebound.** Rebound material shall not be used.

(8) **Construction Joints.** Particular care shall be given to the formation of construction joints. Unless otherwise noted on the plans, all joints subject to compressive stress or over existing construction joints shall be square butt joints. Tapered joints will be permitted at other locations except the outside 1 inch shall be perpendicular to the surface.

(9) **Placing of Pneumatically Placed Concrete.** Proper consistency shall be controlled at the nozzle valve by the operator for the dry mix process and a low water-cement ratio must be maintained. The consistency of the mix and the water shall be controlled by the mixer pump or by the transit mix truck when used for the wet mix process. The mix shall be sufficiently wet to adhere properly and sufficiently dry so that it will not sag or fall from vertical or inclined surfaces or separate in horizontal work.

When encasing structural steel members or covering portions of structures the concrete may be applied in one coat; however, if the concrete, after being placed, shows any tendency to sag, it shall be applied in two or more coats. Pneumatically placed concrete for overhead work shall be placed in two or more coats as may be necessary to insure proper bond and to eliminate sag. In covering vertical surfaces, placing of the concrete shall begin at the bottom and be completed at the top.

The nozzle shall be held at such distance (2 or 4 feet) and position that the stream of flowing concrete shall impinge as nearly as possible at right angles to the surface being covered. Any deposit of loose sand shall be removed prior to placing any original or succeeding layers of pneumatically placed concrete. Should any deposit of loose sand be covered with pneumatically placed concrete, the concrete shall be removed and replaced with a new coat of pneumatically placed concrete after the receiving surface has been properly cleaned.

Before channel or canal lining or riprap is placed, the earth canal or channel slopes shall have been compacted uniformly and thoroughly and brought to a uniform moist condition. The subgrade for lining shall be excavated and fine graded to the required section. The use of forms for lining will not be required. The surfaces of pneumatically placed concrete for both channel lining and riprap shall be finished accurately by hand floating methods before the concrete has attained its initial set.

The original surface and each surface which is permitted to harden before applying succeeding layers shall be washed with water and air blast, or a stiff hose stream, and loosened material removed. Sand which rebounds and does not fall clear of the work or which collects on horizontal surfaces shall be blown off from time to time to avoid leaving sand pockets. Concrete shall not be applied to a surface containing frost or ice. Where standing or running water is encountered it shall be removed before applying the concrete. No work shall be done without the permission of the Engineer when the temperature is lower than 35 F. After placing, the concrete shall be protected from freezing or quick drying.

**(10) Workmen.** Only experienced foremen, gunmen, nozzlemen, and rodmen shall be employed and satisfactory written evidence of such experience shall be furnished the Engineer or his representative upon demand.

**431.5. Measurement.** Measurement of pneumatically placed concrete for encasement of structural steel members will be by the square foot, in place, of the actual contact area.

Measurement of pneumatically placed concrete for repair and restoration of concrete structures, will be by the cubic foot, in place, using the surface area times the average depth of the patch.



# Q. 4 EXTENSION OF STRUCTURE

## EXTENDING CONCRETE STRUCTURES

**430.1. Description.** This item shall govern for concrete for extending structures and for preparation of the existing structures for extending or widening, including the materials used; the removal of portions of the existing structure, preparation of exposed surfaces of steel and concrete for bonding new construction to old; and the construction of the proposed extensions, all as indicated on the plans.

**430.2. Materials.** All materials shall conform to the requirements of the pertinent items as follows:

| Materials | Item No. |
|---|---|
| Concrete | 420, 421 |
| Expansion Joint Material | 420 and/or 436 |
| Reinforcing Steel | 440 |

**430.3. Construction Methods.** The work shall be performed in accordance with the provisions of the Item, "Concrete Structures", and in conformance with the requirements herein.

The Contractor shall verify all pertinent dimensions of the existing structure, prior to ordering materials required for the extensions.

Portions of the old structure shall be removed to the lines and dimensions shown on the plans, and these materials shall be disposed of as shown on the plans or as directed by the Engineer. Unless otherwise noted on the plans, metal railing shall be removed in such a manner that it will not be damaged, stacked neatly on the right-of-way at convenient loading points which will not interfere with traffic or construction and will remain the property of the Department. Any portion of the existing structure, outside of the limits designated for removal, damaged during the operations of the Contractor shall be restored to its original condition at his entire expense. Explosives shall not be used in the removal of portions of the existing structure unless approved by the Engineer, in writing.

When the headwalls, wingwalls and apron are specified on the plans to be reused in the extended structure, the portion to be reused shall be severed from the old structure to the lines and details shown on the plans. The headwall unit shall be moved to the new location specified, by methods approved by the Engineer, and the extension concrete and reinforcement placed according to the plan details. Any portion of the headwall unit damaged by the moving operation shall be restored to its original condition at the expense of the Contractor.

Unless otherwise noted on the plans, a demolition ball, other swinging weight or impact tool, will be permitted on those portions of the structure not immediately adjacent to the 'break' line of the concrete.

The concrete shall be severed at the 'break' line by pneumatic tools and may be followed by the use of the demolition ball, or other methods acceptable to the Engineer. The final removal of concrete at the 'break' line shall be with pneumatic tools. Damaged concrete shall be treated as specifed above. Bridge slabs (other than multiple box culverts) shall first be sawed along the 'break line one-half of an inch deep prior to beginning the removal of concrete.

Except when otherwise provided on the plans, new reinforcing bars shall be spliced to exposed bars in the old structure by lap splices in accordance with Table 1 of the Item, "Reinforcing Steel". When welded splices are permitted by the plans, they shall conform with the Item, "Structural Welding". For lap splices (not welded), new reinforcing steel need not be tied to existing steel where spacing and/or elevation does not match that of existing steel provided the proper lap length is attained.

Dowels, if required by the plans, shall be installed by grouting reinforcing bars to a minimum length of 12 inches into the old structure. Holes for dowel bars shall be cleaned of all loose material, wetted and filled with a 1:3 mix grout or other approved material, as may be specified on the plans, immediately prior to placing of dowel bars.

Concrete surfaces which will be in contact with new construction shall be roughened and cleaned prior to placing of forms. These surfaces shall be dampened and coated with mortar just prior to placing fresh concrete.

Roadway slabs shall be finished in accordance with Article 420.20 except that when an overlay is required, the slabs shall be given a reasonably smooth surface finish by longitudinal or transverse screeding without any straight edge requirements.

The widened portion of bridges and direct traffic culverts shall not be opened to construction traffic or to the traveling public until authorized by the Engineer in accordance with the following:

> Authorization may be given after the last slab concrete has been in place at least 14 days for light construction traffic not to exceed a three-quarter ton vehicle.
>
> Authorization for normal construction traffic, and when necessary to the traveling public, may be given after the last slab concrete has been in place 30 days. Construction vehicles which exceed the legal load limit will not be authorized to cross structures. Vehicles used exclusively to transport ready-mixed concrete, and which have three axles, will be permitted only if the gross weight does not exceed 51,000 pounds. Speed of such vehicles shall be limited to no more than 10 miles per hour.
> Where a detour is not readily available or economically feasible to use, an occasional crossing of a structure with overweight equipment may be permitted for relocating equipment only but not for hauling material, providing a structural analysis of the structure using the exact equipment in question, indicates no damage will result. The structural analysis must be approved by the Engineer. Temporary matting and/or other requirements may be imposed by the Engineer if an occasional crossing is permitted.

**430.4. Measurement.** The quantities of concrete of the various classifications which will constitute the completed and accepted structure or structures in place will be measured by the cubic yard, each, square foot, square yard, or linear foot as the case may be. Measurement will be as follows:

**(1) General.**

(a) All concrete quantities will be based on the dimensions shown on the plans or those established in writing by the Engineer. Diafram concrete, when required, will be included in the slab measurement.

(b) In determining quantities, no deductions will be made for chamfers less than two inches, embedded portions of structural steel or prestressed concrete beams, piling, anchor bolts, reinforcing steel, drains, weep holes, junction boxes, electrical or telephone conduit, conduit and/or voids for prestressed tendons or for embedded portions of light fixtures.

(c) For Pan Girder Spans, a quantity will be included for the screed setting required to provide proper camber in the roadway surface after form removal.

(d) For Slabs on Steel and Prestressed Beams, a quantity for the haunch between the slab and beams will be included when required. No measurement will be made during construction for variation in the amount of haunch concrete due to deviation from design camber in the beams.

(e) For Slabs on Panels or Tee-Beams, the combination of span length, theoretical camber in beams, computed deflections, and plan vertical curve will be taken into account in determining the quantity for the slab.

Additional concrete which may be required by an adjustment of the profile grade line during construction, to insure proper slab thickness, will not be measured for payment.

## Q.5 WELDING REINFORCING BARS

### REINFORCING STEEL

448.9. **General.** Provisions are made herein for the welding of reinforcing steel by the manual shielded metal-arc process. Other processes may be permitted with the specific approval of the Engineer or may be specified on the plans. When the Cadwell process is permitted, a "C" series splice shall be used with grade 40 reinforcing steel, and a "T" series splice shall be used with grade 60 reinforcing steel, unless specified otherwise on the plans.

Splicing of reinforcing steel by welding shall be done only at locations shown on the plans.

448.10. **Base Metal.** Reinforcing steel to be welded shall be new billet steel conforming to ASTM Designation: A615, and shall also conform to the following chemical composition.

| | |
|---|---|
| Maximum Carbon | 0.40 Percent |
| Maximum Manganese | 1.30 Percent |

Mill test reports will be required in accordance with the Item, "Reinforcing Steel".

448.11. **Filler Metal.** Low hydrogen electrodes as specified in Table 1 will be required for all welding of reinforcing steel. Drying of electrodes shall be as specified in Article 448.3.

• 448.12. **Preheat and Interpass Temperature.** Minimum preheat and interpass temperatures shall be as shown in Table 4.

| Carbon Range | No. 7 & Smaller | No. 8 & Larger |
|---|---|---|
| Up to and including 0.30 | None | 100 |
| 0.31 to 0.35 inclusive | None | 150 |
| 0.36 to 0.40 inclusive | 100 | 250 |
| Unknown | 250 | 400 |

TABLE 4

Preheat and Interpass Temperature
for Reinforcing Steel

For widening projects, use carbon content and bar size of new steel to determine preheat required.

448.13. **Joint Types.** For all bars No. 8 and larger, butt splices will be required. For No. 7 bars and smaller, lap splices will be required.

Fillet welds in lap splices shall be a minimum of 4 inches in length and shall be welded on each side of the lap joint. For bars No. 5 and smaller, welding from one side of the lap will be permitted by the Engineer, when it is impractical to weld from both sides of the joint, but in this case the weld shall be a minimum of 6 inches in length.

Lap welds shall meet the requirements specified in Table 5.

Where possible, all butt splices shall be made in the flat position. All welds for butt splices, except horizontal welds on vertical bars shall be as shown in Figures 4 and 5. The back-up strip will be required when access to the splice is from the top only. When bars may be rotated or access to the splice is available from two sides the double bevel splice may be made and this type weld requires gouging out the root pass similar to a flange splice on structural steel. The root pass may be made using E7010 electrodes for all double beveled splices and the root pass shall be completely removed prior to welding the opposite side. The steel shall be preheated to 400 F if E7010 electrodes are used. Horizontal splices, on vertical bars, shall be as shown in Figure 6.

| Bar Size | "a" | "b" (Max.) | "t" (Min.) | "c" (Max.) | Electrode Size |
|---|---|---|---|---|---|
| No. 4 | .04 in. | 1/8 in. | 1/8 in. | 1/16 in. | 1/8 in. |
| No. 5 | .05 in. | 1/8 in. | 3/16 in. | 1/16 in. | 5/32 in. |
| No. 6 | .06 in. | 1/8 in. | 1/4 in. | 1/16 in. | 5/32 in. |
| No. 7 | .07 in. | 3/16 in. | 5/16 in. | 1/16 in. | 5/32 in. |

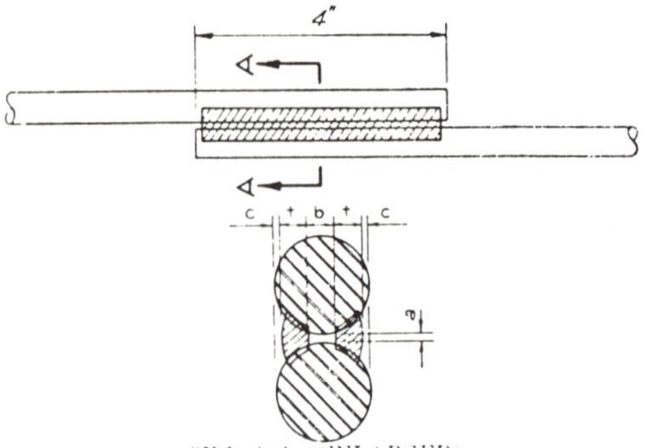


SEC. A-A (ENLARGED)

TABLE 5

Required Dimensions for Lap Splices

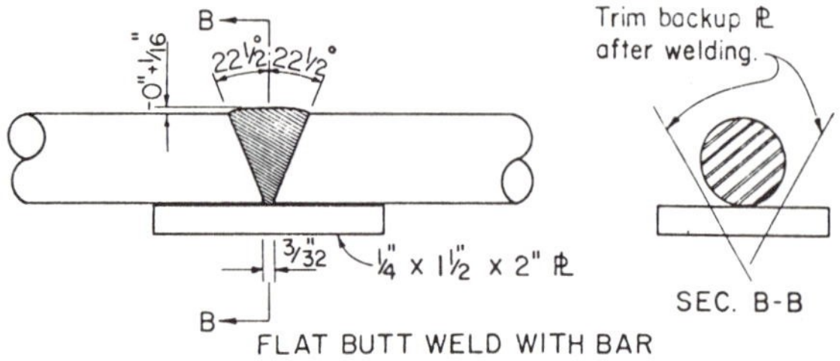


FLAT BUTT WELD WITH BAR
IN HORIZONTAL POSITION
FIGURE 4

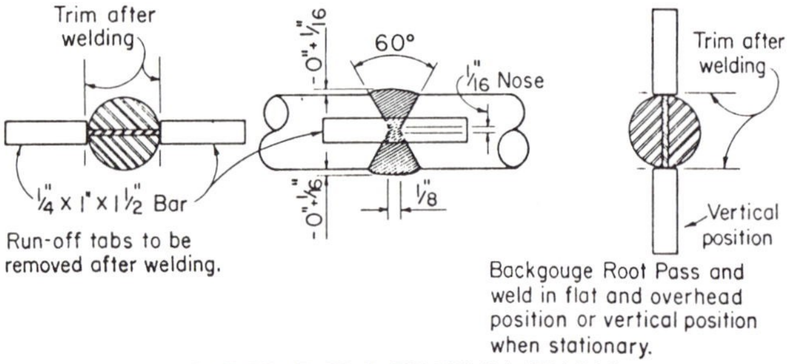


DOUBLE BEVELED BUTT WELD WITH BAR
IN HORIZONTAL POSITION
FIGURE 5

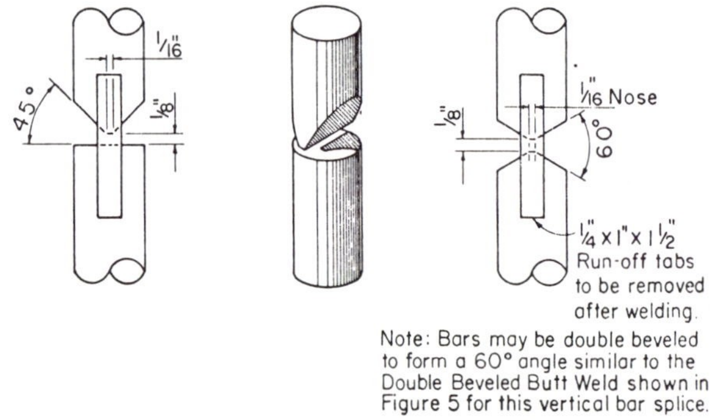


BUTT WELD WITH BAR IN VERTICAL POSITION

448.14. **Widening Projects.** In general, the new reinforcing steel shall be either lap or butt spliced directly to the bar to be extended. When the reinforcement in the old portion of a structure is found to be of the wrong spacing, dowel bars long enough to develop the welded lap or butt splice and also develop the bar in bond, as required in the Item, "Reinforcing Steel", shall be welded to the old steel, and the new reinforcement placed at the correct spacing without welding to the old steel. No measurement or payment will be made for the dowels but will be subsidiary to the other items in the contract.

Both old and new reinforcement shall be cleaned thoroughly prior to the preparation of the joint.

**Texas**

# Q.6 PAINTING EXISTING STEEL

## SECTION 1070—PAINTING EXISTING STRUCTURAL STEEL

**1070.1 DESCRIPTION**—This work is the painting of existing bridges, including painted and unpainted surfaces.

**1070.2 MATERIAL**—Paint. Bulletin 26 and as follows:

- When paint of the same color is applied in two coats, tint the paint for the first coat, as specified in the proposal or as directed.
- Except as specified above, use factory-mixed tint, packed in airtight containers at the place of manufacture, and delivered to the location where it is to be used in the original containers with the seals unbroken. Label or stencil each container to show the manufacturer's name, kind of paint, lot number, and the month and year the paint was packaged.

**1070.3 CONSTRUCTION**—

**(a) Surfaces to Be Painted.** Unless otherwise specified, paint all surfaces previously painted and all surfaces of new metal. Do not paint aluminum and galvanized metal surfaces, unless otherwise directed.

**(b) Cleaning Surfaces.**

**1. General.** Satisfactorily clean surfaces to be painted before applying paint.

The required method of cleaning metal surfaces will be specified in the proposal.

Clean wood surfaces of loose paint, chalking, dirt, and other foreign matter, using hand-operated wire brushes and stiff fiber brushes. Do not damage the surface of the wood.

**2. Solvent Cleaning.** In accordance with the SSPC Surface Preparation Specification, use No. 1 Solvent Cleaning.

Remove soil, cement spatter, drawing compounds, salts, or other foreign matter by brushing with stiff fiber or wire brushes; by cleaning with solutions of alkaline cleaners, provided such cleaners are followed by a fresh water rinse; or by a combination of these methods.

Remove oil, grease, and related substances by wiping or scrubbing the surface with rags or brushes saturated with solvent, by spraying the surface with solvent, or by using vapor degreasing containing stabilized chlorinated hydrocarbon solvents.

Steam cleaning and emulsion cleaners may be used if, after treatment, the surface is washed with hot water or steam to remove any detrimental residue.

To remove detrimental deposits, scrub new weld areas with clean water containing a small amount of corrosion inhibitor or containing a 5% phosphoric acid solution.

**3. Blast Cleaning, Method 1.** Remove heavy deposits of oil, grease, and related substances, as specified in Section 1070.3(b)2. Then, clean metal surfaces to a near-white condition, as specified in Section 1060.3(b)1, using abrasives propelled through nozzles. Completely remove oil, grease, dirt, foreign matter, rust, mill scale, coatings, and old paint, except for slight streaks or discolorations.

If the surface is pitted, slight residues of primer may remain in the bottom of pits.

Clean dry-blasted surfaces with clean bristle, fiber, or hair brushes; with compressed air; or with vacuum. Remove traces of blast residue from the surface and from pockets and corners. Equip compressors with separators or traps to remove water and oil.

To prevent rusting, clean wet-blasted surfaces by rinsing with fresh water with an acceptable corrosion inhibitor added, or with fresh water followed immediately by an inhibitive treatment. Brush to remove residue.

**4. Blast Cleaning, Method 2.** Remove heavy deposits of oil, grease, and related substances, as specified in Section 1070.3(a)2. Then, clean all metal surfaces with abrasives propelled through nozzles or with power tools, completely removing oil, grease, dirt, foreign matter, rust, mill scale, and loose paint, except for slight streaks or discoloration.

Remove stratified rust-scale, using power impact tools. If only minor quantities of stratified rust are present, remove using hand impact tools.

Old, cleaned paint may remain if it cannot be lifted as a layer by inserting the blade of a dull putty knife under it. Verify old paint adhesion by PTM No. 206, minimum Classification 3.

Operate tools to insure that no burrs or sharp ridges are left on the surface and that no sharp cuts are made into the steel.

Use hand tools to clean areas inaccessible to power tools.

After the cleaning operations are completed, remove dust and other loose matter from the surface. If grease or oil is still present, spot clean with solvent.

**5. Other Acceptable Methods of Cleaning.**

**5.a** Dry sandblasting, using compressed-air blast nozzles and dry sand of a maximum particle size no larger than that passing through a size 16-mesh screen, U.S. Standard sieve series.

**5.b** Wet or water-vapor sandblasting, using compressed-air blast nozzles, water, and sand of a maximum particle size no larger than that passing through a size 16-mesh screen, U.S. Standard sieve series.

**5.c** Grit blasting, using compressed-air blast nozzles and crushed grit made of cast iron, malleable iron, steel, or synthetic grits other than sand, of a maximum particle size no larger than that passing through a size 16-mesh screen, U. S. standard sieve series. The largest commercial grade of metal grit permitted is SAE No. G-25 abrasive material.

**5.d** Shot blasting, using compressed-air nozzles and cast iron, malleable iron, steel, or synthetic shot of a maximum size no larger than that passing through a size 16-mesh screen, U.S. Standard sieve series. The largest commercial grade permitted is SAE No. S-330.

**5.e** Other methods of cleaning may be proposed for acceptance.

**6. Safety Precautions.** Fire and explosion hazards are inherent in solvent cleaning operations. Do not start work until safe conditions have been established, in accordance with Subpart D, Sections 1926.55, .56, and .67 and Support F, Section 1926.152 of Title 29 of Code of Federal Regulations. Comply with the Manufacturing Chemists Association procedures for handling and storing dangerous chemical supplements, where the Federal Regulations are not applicable.

Use either petroleum-based mineral spirits (aliphatics), with a minimum flash point of 100F or "Stoddard Solvent," in accordance with ASTM-D484, as the general purpose solvent for cleaning under normal conditions. When the ambient temperature exceeds 80F, use aliphatic mineral spirits with a minimum flash point of 120F. When the ambient temperature exceeds 95F, use heavy mineral spirits with a flash point over 140F. Do not use gasoline or VM. & P. Naptha because of the fire and explosion hazard. Do not use benezene (Benzol) because of its toxicity and flammability.

Take precautions to prevent workers from getting burns from alkali or acid cleaners and to prevent dermatitis. Wear safety eye-protection and protective clothing when handling or using these cleaners. Protect ropes from contact with acid or alkali, to prevent weakening and deterioration.

Solvents are potentially hazardous to health. Use solvents in such a manner so that their concentration in the air being breathed does not constitute a threat to workers' health or life. When used in areas of limited ventilation, where the safe concentration could be exceeded, wear auxiliary breathing equipment, as specified in Table E-4 of Subpart E, Section 1926.103 of Title 29 of Code of Federal Regulations. Clear the fresh air intakes of carbon monoxide or other contaminants from engine exhausts. Do not allow the concentration of solvent in the air to exceed the lower limit of flammability; fire or explosion may result.

**(c) Painting Galvanized Surfaces.** Treat new galvanized surfaces and other galvanized surfaces made bare by the cleaning operations, in accordance with ASTM-D2092, Method A or Method D, before painting.

**(d) Conditions for Painting.** Apply paint only on clean and thoroughly dry surfaces and only during periods of favorable weather. Do not paint when the temperature of the air, paint, or metal is below 40F; when the air is misty; or when conditions are otherwise unsatisfactory. Do not paint damp or frosted surfaces, or surfaces hot enough to cause the paint to blister, to produce a porous paint film, or to cause the vehicle to separate from the pigment.

Do not paint when the steel surface is below the dew point, at a temperature resulting in condensation of moisture. The dew point requirement is satisfied if a thin, clearly defined film of water, applied to the cleaned vertical surface with a damp cloth, evaporates within 15 minutes.

Allow each coat of paint to dry before applying the succeeding coat. Follow the manufacturer's recommendations for drying time, unless a longer time is required. Do not apply a succeeding coat in less than 24 hours.

Apply the prime coat of paint within 24 hours of cleaning or before rusting occurs, but not until the cleaning operations are far enough ahead to eliminate danger of dirt or other material from falling or blowing on the fresh paint. If a protective wash primer is used, the 24-hour period may be extended.

In the event the interval between successive coats exceeds 90 days, clean a test panel, a minimum of 25 square feet, as directed and as specified in Section 1070.3(b)2. Apply the specified paint and thickness to the prepared test panel surface. Test intercoat adhesion by PTM No. 206, minimum Classification 3, within 30 days of application.

If adhesion is developed, clean exposed areas, as specified in Section 1070.3(b)2. If adhesion is not developed within 30 days, reclean, according to the methods orginally used in the contract, then repaint, using the number of coats and types of paint in the proposal.

**(e) Number and Thickness of Coats.** The number of coats and kind of paint will be specified in the proposal.

Thickness of coats:

- **Prime.** Minimum dry film thickness of 2 mils.
- **Second Coat.** Dry film thickness of 1-1/2 mils.

• **Additional Coat.** Minimum dry film thickness of 1-1/4 mils.

A minus tolerance of 1/4-mil total dry film thickness is permitted. If the dry film thickness is less than specified, apply additional coats until the thickness is satisfactory. Measure the dry film thickness by any acceptable method.

**(f) Mixing Paint.** Before use, mix paints thoroughly in original containers. Keep the pigments in suspension by frequent stirring during application. Mix paint by hand or by mechanical power mixers. To insure complete mixing, pour the paint several times from one container to another until the composition is uniform. Strain to remove skins and dried particles. Excessive skinning or partial hardening, due to improper or prolonged storage, will be cause for rejection.

For containers, tools, and equipment used for storing, mixing, handling and applying paint, keep free from accumulation of hardened paint.

**(g) Thinning Paint.** Thoroughly-mixed paint, delivered in containers, is ready for application, and no thinning will be permitted. In cool weather, the paint may be warmed before painting, to reduce the viscosity. Warm the paint by any suitable method, except by the use of an open flame.

**(h) Application.**

**1. General.** Do not paint until the surface has been inspected and accepted.

For general painting and paint striping, apply each coat in two stages, to ensure a maximum coating on parts subject to severe wear and attack.

General painting of the whole surface includes the edges of plates and rolled shapes; corners; crevices; welds; rivet heads; and exposed parts of bolts, nuts and washers. Use acceptable brushes, rollers, spray equipment, or any combination of equipment that gives satisfactory results.

Paint striping, done by brush only, consists of painting the edges of plates and rolled shapes; corners; welds; rivet heads; and exposed parts of bolts, nuts, and washers.

Tint in accordance with Bulletin 26.

Apply the primer by specified painting methods, followed by striping, no sooner than 24 hours later.

Apply the intermediate and finish coats by striping, followed by general painting no sooner than 24 hours later.

Spread paint smoothly and uniformly so every part of the surface is completely covered and so no excess paint collects at any point.

If the paint coating is too thin or if portions of the surface are not completely coated, repaint such portions of the work.

For small cracks, joints, and cavities which were not sealed in a watertight manner by the first coat of paint, fill with dull red primer and linseed oil paste, conforming to the requirements of Federal Specification TT-R-191, Type II; or plastic-calking compound conforming to Federal Specification TT-C-598. Allow to dry before applying the second coat of paint.

**2. Brush Painting.** Use flat brushes, not wider than 4-1/2 inches, or round or oval brushes.

Manipulate the paint under the brush to produce a uniform, even coat in close contact with the metal or with previously applied paint. Work the paint into corners and crevices. Move the brush in a series of small circles to thoroughly fill irregularities in the surface, then brush out and smooth by a series of parallel strokes, until the paint film has an even thickness.

**3. Roller Painting.** Use rollers that do not leave a stippled texture in the paint film. Use rollers only on flat, even surfaces to produce a paint film of even thickness with no skips, runs, sags, or thin areas.

**4. Spray Painting.** Do not use driers or solvents. Keep the paint thoroughly mixed by continuous mechanical agitation, or by frequent intermittent agitation. If air spray is used, provide suitable traps or separators to exclude oil and water from the air.

**5. Surfaces Inaccessible to Normal Painting Methods.** Apply paint by sheepskin daubers, spray, or by other means, if necessary, to ensure thorough coverage of the proper thickness of paint.

**(i) Removal of Unsatisfactory Paint.** Remove the paint, clean the metal as specified and repaint, if rusting occurs; if the paint coat lifts, blisters, wrinkles, or shows evidence of application under unfavorable conditions; if the workmanship is poor; if impure or unauthorized paint has been used; or if for other reasons, the painting is unsatisfactory.

**(j) Stenciling.** After the final coat of paint has dried, stencil the month and year of completion with black paint, at locations directed, in uniform 2- to 3-inch block letters.

**(k) Protection of Structure, Persons, and Property.** For pedestrians, vehicular, and other traffic upon or under the bridge, and for the bridge superstructure and substructure protect against damage or disfigurement by spatters, splashes, and smirches of paint or paint materials. Where necessary for protection, furnish adequate canvas or other suitable cover material.

As directed, remove paint dropped on concrete surfaces, debris from the cleaning operations, paint containers, and refuse.

Penn DOT

# Q.7 STEEL STRUCTURES

441.1. **Description.** This item shall govern for the fabrication and erection of structural steel, and other metals used for steel structures or steel portions of structures, except reinforcing and prestressing steel.

441.2. **Materials.** The metal used for the various portions of the structures shall be as specified and shall conform to the requirements of the Item, "Metal for Structures".

## • FABRICATION

### General

441.3. **Shop Drawings.** Unless otherwise provided on the plans, the Contractor shall prepare and submit detailed shop drawings for each detail of the general plans requiring the use of Structural Steel, Forgings, Wrought Iron, Castings, or Bearings. Camber and erection diagrams will be required. The drawings shall be prepared on sheets 22 by 36 inches. The margin at the left end shall be 1½ inches wide and the others one-half inch. Each sheet shall have a title in the lower right-hand corner. The title shall include the sheet index data shown on the lower right-hand corner of the project plans, sheet numbering for the shop drawings, name of structure or stream, name of Fabricator and name of Contractor.

441.6. **Workmanship.** For railroad underpass structures, shop workmanship shall be in accordance with the latest American Railway Engineering Association Specification for Steel Railway Bridges for Fixed Spans.

All structural materials, before and after fabrication, shall be stored above the ground upon platforms, skids, blocking, or other supports approved by the Engineer. The material shall be kept free from dirt, grease, and other foreign matter and shall be protected as nearly as practicable from corrosion.

Structural steel shall be protected from salt water and salt water spray during transit.

Fabrication and rolling tolerances for rolled shapes, plates, bars, wide flange sections and miscellaneous steel shall be in accordance with ASTM Designation: A 6. Tolerances for fabricated girders shall be in accordance with SDHPT Bulletin C-5.

Rolled sections or fabricated sections of slightly differenct dimensions and weight than the standard sections shown will be acceptable, provided equal or greater Moment of Inertia and Section Modulus for the completed section are provided.

Maximum deviation from flatness for webs of wide flange sections shall be the same as for built-up girders.

Shoes shall be fabricated with a tolerance not greater than the following:

The top bolster shall have the center 75 percent of the long dimension true to one thirty-second inch, with the remainder true to one-sixteenth inch, and shall be true to one thirty-second inch across its entire width in the short dimension.

For a pin and rocker type expansion shoe, the axis of rotation shall coincide with the central axis of the pin.

When the shoe is completely assembled and the top bolster is moved horizontally simulating the movement of the shoe in the finished structure, no point in the plane of the top bolster shall change elevation by more than one-sixteenth inch for the full possible travel of the rocker both ways from the neutral position, nor shall the top bolster change inclination with respect to the horizontal by more than one degree during this same travel.

I-beams and girders shall be fabricated with a tolerance not greater than the following:

The plane of the bearing area of beams and girders shall be perpendicular to the vertical axis of the beam within one-sixteenth inch.

Correction of bearing areas of shoes, beams and girders to the above tolerances, shall be with heat and/or external pressure. Grinding or milling will be permitted if reduction of required thickness of member is not reduced by more than one-sixteenth inch.

Rolled material must be straight before being laid off or worked.

If straightening is necessary, it shall be done by procedures submitted to and approved by the Engineer. Sharp kinks and bends will be cause for rejection of the material, unless corrected to the satisfaction of the Engineer.

Texas

The handling of material, fabrication, blocking of partially completed members and movement of completed members shall be done in such a manner that the safety of workmen and inspection personnel will not be impaired at any time.

All weathering structural steel and other designated steels requiring weathering characteristics shall be maintained at all levels of fabrication and construction as nearly as possible in the condition received from the mill. All shop welds shall be cleaned by power grinding or by blast cleaning to remove welding flux, slag and splatter prior to shipment from the plant.

After all erection, welding and slab concrete placement has been completed, the Contractor shall restore the surfaces of all weathering structural steel as nearly as possible to the above described condition by solvent cleaning, hand cleaning, power brush cleaning or brush-off-blast cleaning, as deemed necessary by the Engineer, to restore the steel to its original clean condition.

All outside surfaces of weathering steel fascia beams, including the underside of the bottom flange, the sides and bottom surfaces of steel bent caps or floor beams and all surfaces of out-rigger-type bent caps extending outside the fascia beams shall receive a Class "B" blast cleaning in addition to the above requirements. No marking will be permitted on the outside face of any fascia beam.

### • Bolted and Welded Structures

441.7. **Cutting, Planing, Facing and Fit of Members.** Sheared edges of plates of more than five-eighths inch thickness and carrying calculated stress shall be planed to a depth of one-fourth inch. Re-entrant cuts shall be filleted to a minimum radius of three-fourths inch, except for the corners of welding access cope holes adjacent to a flange. The fillet and its contiguous cuts shall meet without offset or cutting past the point of tangency (no gouges, etc.).

Unless otherwise permitted by the plans, steel plates for main members shall be cut and fabricated so that the primary direction of rolling is parallel to the direction of the main tensile and/or compressive stresses.

Steel and weld metal may be oxygen cut, provided a true profile, and a smooth and regular surface free from cracks and notches is secured, by the use of a mechanical guide. Hand cutting shall be done only where approved by the Engineer. (Mill scale and extraneous material shall be removed from the cutting side of A514/A517 steel plates along the lines to be cut.)

Edges of all main members which are sheared or oxygen cut, and all other exposed edges to be painted, shall be rounded or chamfered to an approximate one-sixteenth inch dimension by grinding.

In all oxygen cutting, the flame shall be adjusted and manipulated to avoid cutting inside the prescribed lines. Roughness of the cut surfaces shall not be greater than ANSI surface roughness value of 1000 for material up to 4 inches thick and 1600 for material 4 inches to 8 inches thick, except that member ends not subject to calculated stress shall meet ANSI surface roughness value of 2000. Roughness exceeding these values and occasional notches or gouges not more than three-sixteenths of an inch deep, on otherwise satisfactory surfaces, shall be removed by machining or grinding. Cut edges shall be left free of slag. Correction of defects shall be faired to the oxygen cut edges with a slope not exceeding one in 10.

Air carbon-arc or oxygen gouging, oxygen cutting, chipping, or grinding may be used for joint preparation or the removal of defective work or material. Oxygen gouging shall not be used on A514/A517 or A588 weathering steel.

Oxygen cut edges of ASTM A440 steel one-half of an inch or greater in thickness shall be removed to a depth of at least one-eighth of an inch by machining or grinding, except that machine flame cut edges need not be removed, if they are softened after cutting by heating the cut edge uniformly and progressively to a red heat to a depth of at least one-sixteenth of an inch from the edge (1150 F to 1250 F), or post heated by a torch attached to and following the cutting torch with the tips, gas pressure, speed of travel and the distance of the post heating torch from the cutting torch regulated to the thickness of the steel to accomplish the annealing effect described above.

The top and bottom surfaces of steel slabs, base plates and cap plates of columns and pedestals shall be planed or else the steel slabs and base plates hot-straightened. Parts of members in contact with plates shall be faced to fit. For plates over 4 inches thick, top surfaces shall be planed to a ANSI surface roughness value of 250.

Surfaces of bearing plates intended for sliding contact shall have an ANSI surface roughness value of 125.

In planing the surfaces of expansion bearings, the cut of the tool shall be in the direction of expansion.

When stiffeners are required by the plans to be milled, the milled surface shall meet ANSI surface roughness value of 500, and shall provide an even bearing against the flange. Tight-fit, when specified on the plans, shall have at least one point bearing on the flange surface and the remainder with a maximum clearance of one-sixteenth of an inch at any point. Where stiffeners are to be welded to the flange, the opening prior to welding shall not exceed three-sixteenths of an inch, with the fillet weld size increased by the amount of the opening.

Unwelded, cold-bent, load-carrying, rolled-steel plates shall conform to the following:

(1) The bend line shall be at right angles to the direction of rolling.

(2) Bending shall be done so that no cracking of the plate occurs. Minimum bend radii, measured in the concave face of the metal, are shown in Table 1.

| Thickness (t) in inches | Up to 1/2 | Over 1/2 to 1 | Over 1 to 1/2 | Over 1 1/2 to 2 1/2 | Over 2 1/2 to 4 |
|---|---|---|---|---|---|
| Minimum Bend Radii | 2t | 2 1/2 t | 3t | 3 1/2 t | 4t |

High strength steel over 1/2" thick may require hot bending for small radii. For short radius on High Strength (HS) or High Yield Carbon (HYC) Steels, the plates shall be hot bent at a temperature not greater than 1200F.

Allowance for springback of A514/A517 steel should be about 3 times that for carbon steel. For break press forming, the lower die span should be at least 16 times the plate thickness. Multiple hits are advisable. If A514/A517 steel plates to be bent are heated to a temperature greater than 1125 F, they must be requenched and tempered in accordance with the producing mill's practice.

Hot bent plates shall conform to the requirements of (1) above.

**TABLE 1**

**BEND RADII FOR ALL GRADES**

**OF STRUCTURAL STEEL**

(3) Before bending, the corners of the plate shall be rounded to a radius of one-sixteenth of an inch throughout the portion of the plate to be bent.

Finished machining, boring and straightening shall be subsequent to annealing or normalizing structural members. Normalizing and annealing (full annealing) shall be as defined in ASTM E44. The temperatures shall be maintained uniformly throughout the furnace during the heating and cooling so that the temperature at any points on the member will not differ by more than 100 F.

• A514 and/or A517 steel shall not be annealed or normalized and shall be stress relieved only with the approval of the Engineer.

The holding temperature for stress relieving A514 and/or A517 steel shall not exceed 1100 F, except that 950 F shall be maximum for welds and 6 inches surrounding welds.

When required by the plans, bridge shoes, pedestals or other parts which are built up by welding sections of plate together shall be stress relieved in accordance with approved procedures.

**• 441.8. Repair of Defects.**

**(1) Correction of Defects on Edges to be Welded.**

(a) Correction of cutting defects and of occasional notches or gouges less than seven-sixteenths of an inch deep for material up to 4 inches thick, and less than five-eighths of an inch for material over 4 inches thick may be made on steel with yield strengths up through 65,000 psi by welding.

Discontinuities or nonmetallic stringers, opening in plate edges which form the faces of groove welds, which will subsequently be fused with the weld, shall be removed to a depth of five-eighths of an inch and repaired by welding. Laminations opening to these edges shall be removed. Weld repairs shall be made after suitably preparing the defect, welding with low hydrogen electrodes not exceeding five-thirty-second inch in diameter, observing the applicable requirements of SDHPT Bulletin C-5, and grinding the completed weld smooth and flush with the adjacent surfaces.

(b) Occasional notches, gouges or defects in oxygen cut edges of A514/A517 steel may be repaired by welding, when approved by the Engineer under the following conditions:

Cutting defects not more than three-sixteenths of an inch deep in plate edges which will form the faces of a groove weld joint, and which will subsequently be completely fused with the weld may be repaired by welding. Discontinuities or nonmetallic stringers opening to these edges shall be removed to a depth of one-fourth of an inch below the surface by grinding or chipping and the gouge repaired by welding. Laminations opening to these edges shall be removed.

Cutting defects not more than three-sixteenths of an inch deep in plate edges which will form a fillet-welded corner joint shall be repaired by welding only on the part of the edge which will become the faying surface for the joint and the fusion zone of the fillet weld. The part of the defect outside the toe of the completed fillet weld shall be removed by machining or grinding and faired to the oxygen cut surface with a slope not exceeding one in 10. If the actual net cross sectional area which would remain after removal of the discontinuity is 98% or greater than the area of the plate based on nominal dimensions, weld repairs shall be made as specified above using E-11018-M electrodes and grinding the completed weld smooth and flush with the adjacent surface to produce a workman like finish.

**(2) Correction of Defects in Free Running Edges of All Steel.**

(a) Roughness exceeding ANSI surface roughness value of 2000 in oxygen cut surfaces and occasional notches or gouges not more than three-sixteenths of an inch deep on otherwise satisfactory surfaces, shall be removed by machining or grinding to a slope not exceeding one in 10.

(b) In the repair and determination of limits of internal discontinuities visually observed on rolled, sheared or oxygen cut edges and caused by entrapped slag or refractory, deoxidation products, gas pocket or blow holes, the metal removed shall be the minimum necessary to remove the defect or to determine that the permissible limit is not exceeded. All repairs made by welding shall be approved by the Engineer and shall conform to the applicable provisions of SDHPT Bulletin C-5.

The limits of acceptability and the repair of visually observed edge discontinuities in plates 4 inches or less in thickness shall be in accordance with Table 2 where the length of defect is the visible long dimension on the plate edge and the depth, the distance the defect extends into the plate from the edge.

VISIBLE EDGE IMPERFECTIONS FOR PLATE THICKNESS OF 4 INCHES AND UNDER

| Description of discontinuity | Repair required |
|---|---|
| Any discontinuity 1" in length or less. | None - need not be explored. |
| Any discontinuity over 1" in length and 1/8" maximum depth. | None - depth should be explored. |
| Any discontinuity over 1" in length with depth over 1/8" but not greater than 1/4". | Remove - need not weld. |
| Any discontinuity over 1" in length with depth over 1/4" but not greater than 1". | Completely remove and weld. Aggregate length of welding not over 20% of plate edge length being repaired. |
| Any discontinuity over 1" in length with depth greater than 1" | Subject to approval by the Engineer. Gouge out to 1" and block off by welding. Aggregate length of welding not over 20% of plate edge length being repaired unless approved by the Engineer. |

Removal of metal by gouging shall be done in a manner assuring adequate width and slope for welding.

Multiple discontinuities should be considered continuous when located in the same plane within 5 percent of the plate thickness and separated by a distance less than the length of the smaller of two adjacent discontinuities.

TABLE 2

REPAIR OF DISCONTINUITIES

**(3) Repair of Surface Defects.**

The limits of acceptability and repair of surface imperfections for all steels shall be in accordance with ASTM A6.

● **441.9. Heat Curving.** Structural steel with a specified minimum yield point not greater than 50,000 psi may be heat curved. Rolled beams and fabricated girders may not be heat curved if the horizontal radius of curvature measured to the centerline of the web is less than 150 feet or less than the larger of the values calculated (at any and all cross sections throughout the length of the girder or beam) from the following two equations:

$$R = \frac{14\, b\, D}{F_y \, X\, t}$$

$$R = \frac{7500\, b}{F_y \, X}$$

In these equations, $F_y$ is the specified minimum yield point in ksi of steel in the member web, X is the ratio of the total cross-sectional area to the combined cross-sectional area of both flanges, b is the widest flange width in inches, D is the clear distance between flanges in inches, t is the web thickness in inches, and R is the radius in inches.

If the flange thickness exceeds three inches or its width exceeds 30 inches the radius shall not be less than 1000 feet.

Heat curving of individual flange plates may be permitted subject to prior approval of the Engineer.

Camber for rolled beams may be obtained by heating, but girders shall be cambered prior to heat curving by cutting the web to the prescribed arc with proper allowance for shrinkage due to cutting, welding and/or heat curving.

The Contractor shall submit a detailed procedure to the Engineer and obtain his written approval thereof prior to heat curving any beam or girder.

**441.10. Color Coding.** For each approved steel used, a distinct color code shall be required. The color code shall be as specified in ASTM A6. In addition, white shall be required for A36 steel.

The color code used for other steels not specified in ASTM A6 must be submitted to and approved by the Engineer.

The appropriate color(s) shall be placed on the material upon entry into the shop and shall be carried on all pieces to final fabrication. Loss of code marking on any piece and with no other positive identification, shall require testing thereof prior to its use, to re-establish positive identity of the material to the satisfaction of the Engineer.

● **441.11. Straightening Bent Material.** The straightening of plates, angles, other shapes, and built-up members, when permitted by the Engineer, shall be done by methods that will not produce fracture or other injury. Straightening of individual pieces shall be done prior to assembly into a built-up member. Distorted built-up members shall be straightened by mechanical means or, if approved by the Engineer, by the carefully planned and supervised application of a limited amount of localized heat, except that heat straightening of A514/A517 steel members shall be done only under rigidly controlled procedures, each application subject to the approval of the Engineer. In no case shall the maximum temperature of the A514/A517 steel exceed 1100F, nor shall the temperature exceed 950F at the weld metal or within 6 inches thereof. Heat shall not be applied directly on weld metal. In all other steels the temperature of the heated area shall not exceed 1200F (a dull red). In all cases the temperature of the steel shall be controlled by approved temperature indicating devices, such as crayons, liquids, or bimetal thermometers.

Following the straightening of a bend or buckle, the surface of the metal shall be carefully inspected for evidence of fracture.

Correction of errors in camber in welded beams and girders of A514/A517 material shall be done only under rigidly controlled procedures, each application subject to approval of the Engineer.

**441.12. Pins, Pinholes and Rockers.** Pinholes shall be bored true to the specified diameter, smooth and straight, at right angles with the axis of the member, and parallel with each other, unless otherwise required. Pins and pinholes shall be finished to ANSI surface roughness value of 125.

The diameter of the pinhole shall not exceed that of the pin by more than one-fiftieth of an inch for pins, five inches or less in diameter, or one thirty-second of an inch for larger pins. Rockers shall be finished to ANSI surface roughness value of 250.

441.13. **Shop Painting.** The application of shop paint, and the cleaning and coating of sliding surfaces, pins and pinholes, with a mixture of tallow, shall be in accordance with the Item, "Cleaning, Paint and Painting".

441.14. **Marking and Shipping.** All structural members shall be marked in accordance with the erection diagram. The markings shall be over the painted surface. In no case shall shop paint be left off in order to preserve original markings on steel to be painted. Match-marks shall be made with paint in addition to the requirements of Articles 441.19 and 441.25.

Members weighing more than 3 tons shall have the weight marked thereon. The loading, transporting, unloading and storing of material shall be conducted so it will be kept clean and free from injury. Bolts of each length and diameter, and loose nuts or washers of each size, shall be packed separately and shipped in boxes, crates, kegs or barrels. A list and description of the contents shall be plainly marked on the outside of each package.

## ● Bolted Structures

441.15. **Pitch and Edge Distance of Bolts.** Pitch and edge distance not shown on the plans shall be in accordance with the latest edition of AASHTO Standard and Interim Specifications for Highway Bridges.

441.16. **Bolt Holes.** All holes for bolts shall be either punched or drilled. Material forming parts of a member composed of not more than five thicknesses of metal may be punched one-sixteenth inch larger than the nominal diameter of the bolts, if the thickness of the metal is not greater than three-fourths inch for carbon steel, five-eighths inch for HS or one-half inch for XHS steel. For more than five thicknesses or when any of the main material is thicker than shown herein or when required under Article 441.17, all the holes shall be subpunched or subdrilled three-sixteenths inch smaller, and after assembling, reamed one-sixteenth inch larger, or drilled from the solid to one-sixteenth inch larger than the nominal diameter of the bolts.

For punched holes, the diameter of the die shall not exceed that of the punch by more than one-sixteenth inch. If any holes must be enlarged to admit the bolts, they shall be reamed. Holes shall be clean cut without torn or ragged edges. Poor matching of holes will be cause for rejection.

Reamed, punched and drilled holes shall be cylindrical, perpendicular to the member and one-sixteenth inch larger than the nominal diameter of the bolts. Reamers and drills shall be guided by mechanical means. Only for holes not accessible to mechanically guided equipment shall the guiding be by hand. Reaming and drilling shall be done with twist drills, except that for poorly aligned holes, tapered reamers shall be used in conjunction with a template so placed, and held, to force the reaming to the best center of holes for that group. Connecting parts shall be assembled and held securely during reaming or drilling operations and match-marked before disassembling.

441.17. **Preparation of Holes for Field Bolting.** Holes in all field splices of main truss members, box girders, continuous I-beams, and plate girders shall be subpunched (or subdrilled, if required by Article 441.16) and reamed while assembled, or drilled full size with all parts assembled, taking into account their relative position in the finished structure due to grade, camber and curvature. The assembly, including camber, alignment, accuracy of holes and milled joints, shall be approved by the Engineer before reaming or drilling full size is started.

All holes for floor beams and stringer end connections shall be subpunched and reamed to a steel template of not less than one inch thickness or reamed while assembled.

Holes for secondary members such as diaframs, laterals, sway bracing, etc., may be punched full size unless subpunching or subdrilling and reaming is required by Article 441.16.

441.18. **Accuracy of Holes.** Accuracy of all holes punched full size, subpunched or subdrilled, shall be such that a cylindrical pin one-eighth inch smaller in diameter than the nominal size of the punched hole may be entered perpendicular to the face of the member, without drifting, in at least 75 percent of the adjoining holes in the same plane after assembling and prior to any reaming. Pieces not meeting this requirement will be rejected. Any hole which will not pass a pin three-sixteenths inch smaller in diameter than the nominal size of the punched hole will be cause for rejection.

After reaming or drilling, 85 percent of the holes in any adjoining group shall show no offset greater than one thirty-second inch between adjacent thicknesses of metal.

Layout of shop work shall be done so that gage lines for bolts shall not vary from plan dimensions more than one-sixteenth inch. Full size holes in any adjoining group or line shall not vary more than the following:

At least 80 percent of the holes shall be within one-sixteenth inch of plan gage.

Not more than 10 percent of the holes may vary as much as one-eighth inch from plan gage.

Holes varying more than one-eighth inch from plan gage will not be accepted.

441.19. **Shop Assembling.** Each truss or box girder section shall be assembled in its relative position in the shop before reaming is started. Match-marks shall be stamped in the metal at all field connections, in accordance with erection diagrams, at the time reaming is done.

Rigid bolting or clamping shall hold surfaces of assembled parts or members in close contact before reaming is started.

Adjacent sections of beam and girder units shall be assembled in the proper relative positions each will take in the structure before reaming is started. Match-marks shall be stamped in the metal in accordance with erection diagrams at the time reaming is done.

Surfaces of metal in contact shall be cleaned before assembling.

Disassembling after reaming will be required to remove shavings, burrs, etc.

When bolting is required, shop or field, faying surfaces of all joints, including splice plates, shall be cleaned in accordance with the Item "Structural Bolting".

The members shall be free from twists, bends and other deformations. In no case shall tack welding be used in assembly for bolting.

If necessary, the bolt holes shall be spear-reamed for the admission of bolts preparatory to the shop bolting of full-sized punched material. The spear reamer used for this purpose shall be not more than one-sixteenth inch larger than the nominal diameter of the bolts.

Parts not completely bolted in the shop shall be secured by temporary bolts, insofar as practicable, to prevent damage in shipment and handling.

The drifting done during assembling shall be only that required to bring the parts into position, and not sufficient to enlarge the holes or distort the metal. If any holes must be enlarged to admit bolts, they shall be reamed.

441.20. **Bolted Connections.** High-Strength Bolts and Bolting shall be in conformance with the Item, "Metal for Structures", and the Item, "Structural Bolting". The same criteria shall govern for all types of high-strength bolts where the plans or specifications refer to pitch, edge distance, preparation of holes, etc.

441.21. **Preparation and Fit of Members.** When specified on the plans, abutting joints shall be milled and brought to an even bearing. Where joints are not milled the openings shall not exceed one-fourth of an inch.

Floor beams and girders with end connection angles shall be built to exact length back to back of connection angles. If end connections are faced, the finished thickness of the angles shall be not less than that shown on the detail plans.

## ● Welded Structures

441.22. **Welding.** All welding operations, processes, equipment, materials, qualifications of welders, workmanship and inspection shall be in accordance with SDHPT Bulletin C-5, except as provided in Article 441.32.

Nondestructive testing and inspection as herein described will be required for welding performed in the fabricating shop.

Unless otherwise shown on the plans, nondestructive testing (magnetic particle and radiographic) required in the shop will be done by the Contractor and at his expense. This will include furnishing all materials, equipment, tools, labor and incidentals necessary to perform the required testing. The Department may require further tests as necessary in accordance with Article 5.9. In addition, the Department may perform additional testing, including other types.

All magnetic particle inspection and all radiographic inspection shall be done in the presence of and at the locations selected by the Engineer or his authorized representative. The Engineer shall examine and interpret all tests made.

Magnetic particle inspection shall be in accordance with ASTM Designation E109, "Dry Powder Magnetic Particle Inspection".

Unless otherwise shown on the plans for built-up members (plates girders, floor beams and stringers), magnetic particle inspection will be required on 100 percent of the web to flange and bearing stiffener fillet welds on not less than one fabricated piece (as designated by individual piece marks) for each 15 pieces or fraction thereof when the maximum flange thickness is less than two and one half inches and on not less than one piece for each 10 pieces or fraction thereof when the maximum flange thickness is two and one half inches and over. No magnetic particle inspection will be required for rolled sections. Welds requiring repairs shall be retested by magnetic particle inspection after the repairs are made.

Radiographic equipment, procedures, radiographs, identification marks, penetrameters, examination, report, disposition of radiograph and weld surface preparation shall be in accordance with SDHPT Bulletin C-5.

For shop welds of material 65,000 psi yield strength and less, radiographic inspection will be made of the full flange width of 35 percent of all flange splices where the plate thickness at the weld is 2 inches or less and of 50 percent of all flange splices where it is thicker than 2 inches and of 1/5 the depth of the web of 50 percent of the web splices on each structure. If unacceptable work is found, additional radiographs will be made on sections welded by the same equipment and/or operator just prior to and just after the section containing the defect.

For shop welds on steel that has a specified yield strength over 65,000 psi, radiographic inspection shall be made of all groove welds. These welds shall be inspected not less than 48 hours after they are completed.

Welds requiring repairs shall be retested by radiography after repairs are made. All radiographic inspection and necessary repairs shall be done prior to assembly.

When radiographic inspection of particular welds is required by the plans, this shall be in addition to the radiographic inspection required herein.

The quality of all welds shall comply with Article 203, "Quality of Welds" and repairs will be in accordance with Article 204, "Corrections", of SDHPT Bulletin C-5.

**441.23. Preparation of Material for Welding.** Dimensional tolerances, straightness and flatness of structural shapes and plates shall be within the limits prescribed by Article 441.6.

Surfaces to be welded shall be smooth, uniform and free from fins, tears and other defects which would adversely affect the quality of the weld. Surfaces to be welded shall be free from loose scale, slag, rust, grease, or other material. Mill scale that withstands vigorous wire brushing or a light film of drying oil or rust inhibitive coating may remain. Finish of bevels of groove welds shall be milled or ground. Oxygen cut bevels without grinding will not be allowed.

When a zinc rich paint, Protection System II, is specified, surfaces within 4 inches of a groove weld joining main stress carrying members and within 2 inches of fillet welds joining diaframs or lateral bracing to stiffeners or gusset plates shall be sandblast cleaned and coated with linseed oil. After welding is completed, the areas shall be sandblast cleaned and painted as required for the specified paint system.

For other paint systems, surfaces within 2 inches of any weld joining main stress carrying members shall be free from any paint or other material that would prevent proper welding.

Sheared plates for webs of built-up members shall be wide enough to allow for trimming of edges where built-in camber is required. Plates with rolled edges used for webs, shall be trimmed by oxygen cutting.

Preparation of edges by oxygen cutting shall be in accordance with Article 441.7.

The faying surfaces of the web and flange plates and the adjacent surfaces that are to be fillet welded shall be cleaned by grinding prior to assembly and welding of web to flange.

**441.24. Assembly of Parts.** The parts to be joined by fillet welds shall be brought into as close contact as possible, with a maximum separation of three-sixteenths inch. If the separaton is one-sixteenth inch or greater, the leg of the fillet weld shall be increased by an equivalent amount. The separation between faying surfaces of lap joints and of butt joints landing on a backing strip shall not exceed one-sixteenth inch. The fit of joints not sealed by welds throughout their length shall be close enough to exclude water after painting. Where irregularities in rolled shapes or plates, after straightening, prevents this, the procedure necessary to bring them within the above limits shall be subject to the approval of the Engineer. The use of fillers is prohibited, except as specified on the drawings or as approved by the Engineer.

Members to be welded shall be brought into correct alignment and held in position by clamping, welding, or tacking until the joint has been welded. Adequate clamps must be provided to prevent cupping or warping of the flanges when welding them to the web. The clamping devices must be designed to not interfere with the operation or guiding of the automatic welding equipment.

Temporary stiffeners used for jigs and/or warpage control shall not be tack welded to the flange material. Tacking to the web is permissible if the welds are at least d/6 distance away from the flange where "d" is the web depth. The tack weld shall be removed by grinding flush with the parent metal prior to acceptance.

Suitable allowance shall be made for shrinkage, and the joint shall never be restrained on both sides when welding.

Abutting parts to be joined by groove welds shall be aligned carefully. Regardless of the range of stress, all shop groove welds in flange plates shall be ground smooth and flush with the base metal on all surfaces. This shall apply to both parts of equal thickness and parts of unequal thickness.

The surfaces shall be ground so that the radii at the points of transition will be 4 inches minimum.

When groove welds are used to join materials of different thickness or width, there shall be a smooth transition between offset surfaces with a slope of not greater than 1 in 4 in thickness transition, and to the proper radii in the case of width transition (See Figures 1 and 2).

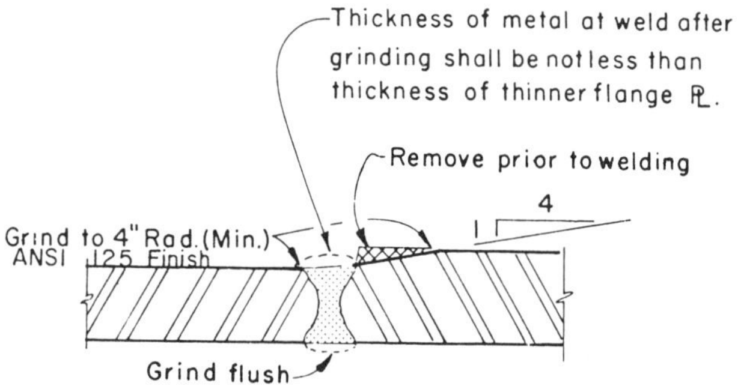


SECTION THRU FLANGE

FIGURE 1

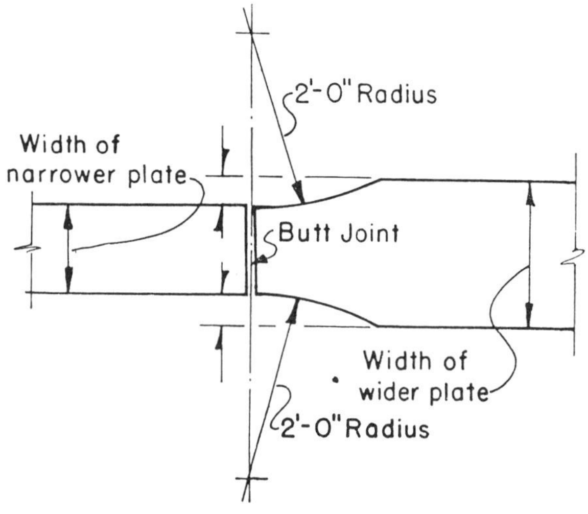


PLAN OF FLANGE

FIGURE 2

Groove welds in web plates need not be ground unless shown on the plans.

Grinding shall be done in the direction of stress, and in a manner that keeps the metal below the blue brittle range. (Below 350F).

Intermediate stiffeners within 12 inches of a splice point shall be shipped tack welded in place. The welding shall be done in the field after the splice is made.

**441.25. End Preparation and Shop Assembly for Field Welds.** Ends of beams and girders shall be prepared in accordance with Figure 3 and the requirements herein or as shown on the plans. The centerline of the land of opposing web and flange bevels shall not deviate from each other by more than one-sixteenth of an inch.

For Shop Assembly, members should be brought into abutting contact in accordance with the blocking diagram. Root faces shall not vary in excess of one-sixteenth of an inch from contact. Corrections by additional cutting and/or grinding shall be made to bring the splice within this tolerance. Finish of bevels for groove welds shall be milled or ground. Oxygen cut bevels without grinding will not be allowed.

Care shall be taken when cutting the bevel adjacent to the web to prevent cutting into the web.

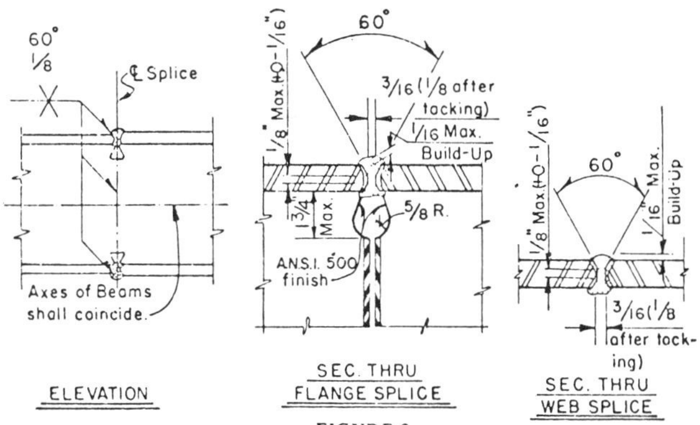


FIGURE 3

End preparation, backing and tolerances for single V groove welds for framing beams or girders shall conform to the details shown on the plans or as permitted by SDHPT Bulletin C-5.

Ends of beams or girders to be welded shall be prepared in the shop taking into account their relative positions in the finished structure due to grade, camber and curvature. Each splice shall be completely shop assembled, checked and while assembled, match-marked.

## ● FIELD ERECTION

**441.26. Methods and Equipment.** Before starting work, the Contractor shall inform the Engineer fully of the method of erection he proposes to follow and the amount and character of the equipment he proposes to use; the adequacy of which shall be subject to the approval of the Engineer. Such approval shall not relieve the Contractor of the responsibility for the safety or adequacy of his methods or equipment or from carrying out the work in full accordance with the plans and specifications. No work shall be done without the sanction of the Engineer.

The Contractor shall prepare and submit detailed erection plans for plate girders (bolted or welded), trusses, and for all railroad underpass structures showing procedures, sequence of work, equipment to be used, falsework, etc., so that a check can be made of the adequacy of the proposed procedure. This detailed procedure shall follow the preliminary procedure previously submitted. Field erection plans for I-beam units will not be required unless specified on the plans.

Spot welding for the purpose of eliminating field erection bolts or for holding steel parts together while bolting will not be permitted.

The Contractor shall provide the falsework and all tools, machinery and appliances, including drift pins and fitting-up bolts, necessary for the expeditious handling of the work. Drift pins sufficient to fill at least one-fourth of the field holes for main connections shall be provided.

All steel beams or girders placed over a traveled roadway or a railroad, shall be securely tied and/or braced in accordance with Standard Plan Sheet SS-MEBR to prevent overturning immediately after erection, and until diaframs are permanently in place. The methods to be used shall be submitted on the erection drawings. When railroad or roadway traffic must be maintained beneath girders or beams already placed, traffic shall be protected against falling objects during the erection of diaframs and other structural members, during the placing of cast-in-place concrete and during the erection and dismantling of forms therefor. The protection shall consist of safety nets (1 inch mesh maximum) or a flooring with openings not larger than 1 inch.

**441.27. Storing Materials.** All material shall be handled in a manner that prevents injury to it. Stored material shall be placed on skids above the ground and kept clean and properly drained. Girders and beams shall be placed upright and shored. Long members, such as columns, shall be supported on skids placed closely enough to prevent excessive deflection.

**441.28. Falsework.** The falsework shall be properly designed for the loads to be supported and shall be substantially constructed and properly maintained. The Contractor shall prepare and submit to the Engineer falsework plans, including design calculations.

The falsework plans shall be complete in all details of members, connections, equipment, etc., so that a structural check can be made of them.

Approval of the Contractor's plans shall not be considered as relieving him of any responsibility.

**441.29. Straightening Bent Material.** Straightening of structural members shall be in accordance with Article 441.11.

**441.30. Handling and Assembling Material.** The parts shall be assembled accurately as shown on the plans. Match-marks shown on the erection drawing shall be shown on each member. The material shall be handled carefully so that no parts will be bent or otherwise damaged. Hammering which will injure or distort the members shall not be done. All bearing and faying surfaces of structural steel in bolted connections shall be cleaned before the connection members are assembled. When A588 steel is used these surfaces shall receive a Class "B" blast cleaning in accordance with the Item, "Cleaning, Paint and Painting" prior to assembly of the connection members. The areas of the outside ply under washers, nuts or bolt heads shall be cleaned prior to installation of the bolts.

Unless erected by the cantilever method, truss spans shall be erected on blocking so located as to provide proper camber. The blocking shall be left in place until the tension chord splices are fully connected and all other truss connections pinned and bolted. Main connections shall have one-half of the holes filled with bolts and erection pins (half bolts and half pins) before swinging the span. Splices and connections carrying traffic during erection shall have three-fourths of the holes so filled.

Fit-up bolts shall be of the same nominal diameter as the connection bolts. Erection pins shall be one thirty-second inch larger.

There shall be no temporary welds for transportation, erection or other purposes on main members, except at approved locations more than one-sixth the depth of the web from the flanges of beams and girders, or in accordance with Standard Plan Sheet SS-MEBR, unless otherwise approved by the Engineer.

**441.32. Field Welding.** Welding and nondestructive testing shall be in accordance with the Item, "Structural Welding".

**441.33. Bearing and Anchorage.** Castings, bearing plates, or shoes shall not be placed upon improperly finished bridge seat bearing areas, but shall have a full and even bearing upon the concrete. Castings, bearing plates, or shoes shall be placed on preformed fabric pads manufactured of all new (unused) materials and composed of multiple layers of prestressed cotton duck, 64 plies per inch of finished pad thickness, impregnated and bound with a high quality rubber compound, containing rot and mildew inhibitors and antioxidants, compounded into resilient pads of uniform thickness. The Shore Durometer hardness of the pad shall be not less than 85 nor more than 95. Pads shall be capable of withstanding 10,000 psi compressive stress without breakdown. A tolerance of ±5% will be allowed from the required pad thickness. Unless otherwise shown on the plans the required thickness of pads will be one-fourth of an inch. Anchor bolt holes shall be not more than one-fourth of an inch larger than the bolt diameter.

Upon completion of welding, all splices, distorted bearing pads and/or expansion shoes shall be restored to an equivalent vertical position at 70F. This shall be done by any approved method of temporarily relieving the load on them.

When the concrete is placed below grade, castings, or bearing plates or shoes may be raised to grade on beds of Portland cement mortar consisting of one part cement, two parts sand and a minimum amount of water. The minimum thickness of mortar bed shall be one-eighth inch and the maximum thickness shall be three-eighths inch. Adequate curing shall be provided.

When it is necessary to raise the castings or bearing plates higher above the concrete than provided by the allowable thickness of the above described materials, the area under the built-up portion shall be prepared in accordance with the specification requirements for construction joints and the build-up placed using an approved latex-based grout mixed in accordance with the manufacturer's recommendations. Steel shims or other approved material shall be used in conjunction with one of the specified materials.

Beams and girders shall have practically full bearing on shoes and base plates. Field corrections shall be made to provide a minimum of 75 percent contact of flange to shoe with no separation greater than one thirty-second of an inch.

Corrections by heat and pressure or through the use of tapered steel shims or layers of steel sheets shall be made to provide bearings to the tolerances given above. Small irregularities may be ground off, but care must be taken not to destroy the finish. The shims or steel sheets used shall be galvanized.

Care shall be taken to set expansion shoes normal to the direction of movement of the unit. Full line bearing of the rocker plate to the base plate shall be provided, and shall be interpreted to mean full bearing on a minimum of 85 percent of the contact surfaces.

All foreign matter shall be removed from machine-finished surfaces immediately prior to placing in the structure. Surfaces designed for sliding movement, one upon the other, shall be given a field coat of graphite spring lubricant when placed in the structure. The rolling surface of rocker shoes shall be painted.

Anchor bolts may be set to the exact location in the concrete when it is placed. Accurately cut and placed templates will be required for positioning the bolts.

In lieu of the above, anchor bolt holes shall be drilled in the exact location or may be formed by insertion of oiled and tapered wooden plugs or tapered metal sleeves into the plastic concrete. The bolts shall be set accurately and fixed in position with non-shrink grout completely filling the holes.

The location of the slotted holes in the expansion shoes in relation to the anchor bolts shall be varied with the prevailing temperature. The nuts on anchor bolts at the expansion ends of spans shall be adjusted to permit the free movement of the span and either lock nuts shall be provided or the threads of the anchor bolts burred.

In setting shoes or bearing plates for steel truss spans, proper allowance shall be made for bottom chord elongation due to dead load.

All shoes and rockers shall be set to be vertical at 70F.

441.34. **Grading Deck on Continuous Units.** Forms shall not be erected until after all welding and/or bolting is complete, the unit positioned, and bearings properly set.

An accurate measurement shall be made of the elevations of girder or beams flanges at all grading control points as shown on the plans.

Subsequent grading of forms and placing and finishing of concrete shall be governed by these measurements only, taking into account the dead load deflections of the slab and rail as shown on the dead load deflection diagram.

441.35. **Misfits.** Corrections of minor misfits and a reasonable amount of reaming will be considered a legitimate part of the operation. Any error in shop work which prevents the proper assembling and fitting-up of parts by the moderate use of drift pins or a moderate amount of reaming and slight chipping or cutting shall be reported immediately to the Engineer for his approval of the method of correction.

The correction shall be made in the presence of the Engineer, who will check the material. Such work is to be done at the entire expense of the Contractor.

441.36. **Paint and Painting.** Unless otherwise provided, the application of field paints shall conform to the requirements of the Item, "Cleaning, Paint and Painting".

For railroad structures, which require that the steel be erected and assembled on falsework and moved into place as a unit, all field paint, except the final coat, shall be applied to the steel while on the falsework and prior to moving into final position. Surfaces inaccessible for painting in the final position shall have the final paint coat applied prior to move-in. Touch up of paint damaged due to move-in to final position will be done, and the final field coat applied.

441.37. **Measurement and Payment.** No direct compensation will be made for "Steel Structures". Measurement and payment for quantities of structural metal, concrete, reinforcement, railing and other proposal items which constitute the completed and accepted structures will be made in accordance with the provisions of pertinent specifications.

# Q.8 STRUCTURAL METAL

442.1. **Description.** This item shall govern for all structural steel, high strength bolts, forgings, steel castings, iron castings, wrought iron, bronze, steel pipe and tubing, aluminum castings and tubing, and other miscellaneous metals used in structures, except reinforcing steel and metal culvert pipe.

442.2. **Process.** All ferrous metals furnished for use under these specifications shall be made by one or more of the following processes only:

Open-hearth, basic oxygen, or electric furnace.

442.3. **Structural Steel.**

(1) **Steel for Main Members.**

Unless otherwise shown on the plans, structural steels listed herein shall conform to the longitudinal Charpy V-notch requirements of Group 1 in accordance with Table A. Sampling and testing shall be in accordance with ASTM A673. The (H) frequency of heat testing shall be used for all material except that (P) frequency shall be used for ASTM A514-517 steels.

- (a) **Carbon Steel.** When shown on the plans and in the proposal as Structural Steel-HYC, the material shall conform to the requirements of ASTM Designation: A36.

(b) **High Strength Steel (HS).** High strength steel when specified on the plans shall have a minimum required yield point of 50 ksi, and will be designated on the plans and in the proposal as Structural Steel-HS.

When so specified, the steel shall conform to one of the following ASTM Steels subject to thickness and physical requirements of the pertinent ASTM Specification:

- 1. ASTM Designation: A572, High-Strength Low-Alloy Columbium-Vanadium Steels of Structural Quality, Grade 50. For welded structures the material will be limited to shapes in ASTM Groups 1, 2 and 3: plates and bars limited to a maximum thickness of 2 inches. For mechanically fastened structures, plates and bars will be limited to a maximum thickness of 4 inches.
- 2. ASTM Designation: A588, High-Strength Low-Alloy Structural Steel With 50,000 psi Minimum Yield Point To 4 Inch Thickness.

(c) Extra High Strength Steel (XHS). Extra high strength steel when specified on the plans shall have a minimum required yield point of 90 ksi and will be designated on the plans and in the proposal as Structural Steel-XHS. When so specified, the steel shall conform to the following:

- 1. ASTM Designation: A514, High-Yield Strength, Quenched and Tempered Alloy Steel Plate.

  The steel furnished shall be suitable for welding. Structural shapes and seamless tubing, meeting the other requirements of A514 steels, will be permitted with a maximum tensile strength of 140 ksi for structural shapes and 145 ksi for seamless tubing.

- 2. ASTM Designation: A517, Pressure Vessel Plates, Alloy Steel, High-Strength, Quenched and Tempered.

The above listed steels are considered weldable. Other steels will require qualification in accordance with SDHPT Bulletin C-5 prior to their use.

(2) Miscellaneous Steel.

(a) Unless otherwise shown on the plans, structural steel for members such as shoes, diaframs, stiffeners (including bearing stiffeners), lateral bracing, diagonals, armor joints and finger joints shall conform to ASTM Designation: A36 or A500, Grade B.

| Material | Thickness Inches | Group 1 See Notes | Group 2 See Notes | Group 3 See Notes |
|---|---|---|---|---|
| A-36, A-440 and A-242 | Up to 4" | 15 ft. lb. at 70F | 15 ft. lb. at 40F | 15 ft. lb. at 10F |
| A-572* | Up to 4" mechanically fastened | 15 ft. lb. at 70F | 15 ft. lb. at 40F | 15 ft. lb. at 10F |
| | Up to 2" welded | 15 ft. lb. at 70F | 15 ft. lb. at 40F | 15 ft. lb. at 10F |
| A-588* | Up to 4" mechanically fastened | 15 ft. lb. at 70F | 15 ft. lb. at 40F | 15 ft. lb. at 10F |
| | Up to 2" welded | 15 ft. lb. at 70F | 15 ft. lb. at 40F | 15 ft. lb. at 10F |
| | Over 2" to 4" welded | 20 ft. lb. at 70F | 20 ft. lb. at 40F | 20 ft. lb. at 10F |
| A-514-517 | Up to 4" mechanically fastened | 25 ft. lb. at 30F | 25 ft. lb. at 0 F | 25 ft. lb. at -30F |
| | Up to 2 1/2" welded | 25 ft. lb. at 30F | 25 ft. lb. at 0 F | 25 ft. lb. at -30F |
| | Over 2 1/2" to 4" welded | 35 ft. lb. at 30F | 35 ft. lb. at 0 F | 35 ft. lb. at -30F |

Notes:

Group 1: Minimum Service Temperature 0 F and above.
Group 2: Minimum Service Temperature from -1F to -30F.
Group 3: Minimum Service Temperature from -31F to -60F.

*If the yield point of the material exceeds 65 ksi, the temperature for the CVN value for acceptability shall be reduced by 15° F for each increment of 10 ksi above 65 ksi.

**TABLE A**
**LONGITUDINAL CHARPY V-NOTCH REQUIREMENTS**

Structural steels used for secondary or nonstress-carrying members will not be subject to impact requirements.

All steels greater than 1/2 inch in thickness used for structural supports for highway signs, luminaires and traffic signals shall conform to the longitudinal Charpy V-notch requirements of Group 1 in accordance with Table A.

(b) Stud shear connectors, slab anchors and anchors on armor joints and finger joints shall conform to the requirements of ASTM Designation: A108, cold drawn bars, Grades 1015, 1018, or 1020, either semi- or fully-killed.

Tensile properties as determined by tests of bar stock after drawing or finishing shall conform to the following:

| | |
|---|---|
| Tensile Strength (Min.) | 60,000 psi |
| Yield Strength (Min.) | 50,000 psi |
| Elongation (Min.) | 20% in 2 inches |
| Reduction of Area (Min.) | 50% |

Tensile properties shall be determined in accordance with the applicable section of ASTM Designation: A370.

The manufacturer shall certify that the studs or anchors as delivered conform with the material requirements of this section.

(c) High strength bolts shall conform to the requirements of ASTM Designation: A325 or A490, unless otherwise shown on the plans.

A mill test report or certification will be required indicating that the bolts conform to these requirements.

(d) Steel piling shall conform to the following:

| Type | ASTM Designation |
|---|---|
| Steel H Piling | A36 |
| Metal Shell Piling (heavier than 10 gauge) | A252* Grade 2 or A36 |
| Sheet Piling (Rolled) | A328* Grade A |
| Sheet Piling (Formed) | A570* Grade A |

*A mill certificate shall be furnished by the manufacturer certifying to the results of tests required by the governing specifications.

Sheet piling of a different configuration than that required by the plans may be used provided the section modulus and weight per foot furnished is equal to or greater than required by the plans.

(e) **Deck Plates.** Material for deck plates shall conform to the following:

Corrosive resistant structural steel conforming to ASTM Designation: A242 or A440. The material must be of weldable quality, and shall contain alloying elements that furnishes corrosion resistance at least twice that of copper bearing structural steel. The type of material to be used with its trade name, shall be stated on the shop plans.

(f) **Rail Posts.** Material for rail posts shall conform to the requirements of ASTM Designation: A36 unless otherwise noted on the plans.

**442.4. Steel Forgings.** Steel forgings from which pins, rollers, trunnions, or other forged parts are to be fabricated, shall conform to the requirements of the Standard Specifications for Carbon-Steel Forgings for General Industrial Use, ASTM Designation: A668, Class C, D, F or G. The above will govern for all Railroad underpass structures. The class of material shall be as designated on the plans.

For other structures, for pins 4 inches in diameter or less, material conforming to ASTM Designation: A108, Grades 1016 to 1030 may be used. The material shall have a minimum yield point of 36 ksi. All forgings shall be thoroughly annealed prior to being machined to form finished parts. Material for pins over 4 inches in diameter shall be in accordance with ASTM Designation: A668, Class C, D, F or G. The class of material shall be as designated on the plans.

**442.5. Steel Castings.** Steel castings shall conform to the specifications for Mild to Medium Strength Carbon Steel Castings for General Application, ASTM Designation: A27, Grade 70-36.

When specified on the plans as Class 70, Class 90, or Class 120, the steel castings shall conform to the specified class of ASTM Designation: A486.

**442.6. Iron Castings.** All iron castings shall be true to pattern in form and dimensions, free from pouring faults, sponginess, cracks, blow holes, and other defects in positions affecting their strength and value for the service intended.

The castings specified shall conform to one of the following ASTM Designations:

| Material | ASTM Designation | Grade |
|---|---|---|
| Gray-Iron Castings | A48 | Class 30 |
| Malleable Castings | A47 | 35018 |
| Ductile Iron Castings | A536 | 60-40-18 |

**442.7. Lead.** Sheet lead shall conform to the requirements of the Specifications for Pig Lead, ASTM Designation: B29.

**442.8. Anchor Bolts.** Unless shown otherwise on the plans, plain and threaded bars used for anchorage purposes shall conform to requirements of ASTM Designation: A36. Headed bolts and nuts shall conform to requirements of ASTM Designation: A307, Grade A. When high strength anchor bolts are designated on the plans, they shall conform to ASTM Designation: A193-B-7. Nuts for high strength anchor bolts shall conform to ASTM Designation: A194-2H.

Threads for anchor bolts and nuts shall be UNC Series, Class 2 fit for 1 inch diameter and smaller. Threads for anchor bolts and nuts over 1 inch diameter shall be 8UN Series, Class 2 fit.

All anchor bolts and nuts, when galvanized, shall be tapped or chased after galvanizing.

Anchor bolts shall not be galvanized unless otherwise noted on the plans.

A mill test report or certification will be required indicating that the material conforms to these requirements. When heat treated material is specified or required, the test report for certification shall include the necessary certification relative to the heat treating process.

**442.9. Steel Pipe.** Steel pipe shall conform to the material requirements shown on the plans, or in other designated specifications.

**442.10. Steel Tubing.** Steel tubing shall conform to ASTM Designation: A500, Grade B, unless otherwise shown on the plans or herein. Tubing conforming to API Standard 5LX, Grade 52, except as noted herein, may be used if produced by a mill recognized as "authorized to produce pipe with the API monogram" and listed as such in the standard API specifications.

The following exceptions to the requirements of API 5LX, Grade 52, will be allowed:

Hydrostatic tests will not be required.

In lieu of a mill test report, a certificate from the manufacturer will be required for each lot or shipment, certifying that the tubing meets the requirements stated above.

442.11. **Pipe Rail.** Pipe shall be construed to include special extruded and bent shapes and shall be of the section indicated on the plans. Pipe may be rolled or extruded to the shape shown on the plans or may be cold pressed from a round pipe or flat plate.

If cold pressed, the design of the press and dies shall result in a pipe of uniform section and free from die marks. After the pipe has been formed to the required section, it shall be cut to the lengths required. The end cuts and notches shall be made at such angles with the axis of the pipe as required to produce vertical end faces and plumb posts when required by the plans. Cutting and notching of pipe shall be done with a saw or machine guided torch or other means that will insure a neat and workmanlike finish.

442.12. **Steel Deep Beam Rail.** The rail element shall be either 10 gage (0.1345 ± 0.008 inches nominal thickness) or 12 gage (0.1046 ± 0.008 inches nominal thickness), exclusive of protective coating, as required by the plans. Rail element shall meet all requirements of AASHTO M-180 except as modified on the plans.

The terminal connector shall be of the same material, but shall not be less than 10 gage.

Unless otherwise noted on the plans the rail element shall be galvanized.

442.13. **Aluminum.**

(1) Unless otherwise shown on the plans, aluminum materials shall conform to the following:

| | |
|---|---|
| Castings | ASTM B108, Alloy A444-T4 |
| Extrusion | ASTM B221, Alloy 6061-T6 |
| Sheet & Plate | ASTM B209, Alloy 2024-T3 |

Test specimens from castings shall be cut, either vertically or horizontally, from the lower 14 inches of the tension flange but not at the junction of the rib or base. The curved surfaces shall be flattened prior to machining. Test specimens shall conform to ASTM Designation: E8, Figure 8.

(2) Aluminum Deep Beam Rail.

| Material | ASTM Designation | Alloy |
|---|---|---|
| Beam Section | B209 | 2024-T3 |
| Terminal Section | B209 | 2024-T3 |

The minimum thickness of the rail element shall be 0.156 inches (nominal) or 0.105 inches (nominal) as required by the plans.

442.14. **Fabrication, Erection, and Painting.** Fabrication, welding, and erection of structural metal shall be in accordance with the Item, "Steel Structures", the Item, "Structural Welding", and SDHPT Bulletin C-5. Paint and painting shall be in accordance with the Item, "Cleaning, Paint and Painting". Aluminum or galvanized steel members shall not require painting.

442.15. **Galvanizing.** Galvanizing of metal fabricated from rolled, pressed, or forged steel shapes, plates, pipe and bars, shall conform to ASTM Designation: A123. Galvanizing of steel or iron castings shall conform to ASTM Designation: A153, Class A. Galvanizing of bolts, nuts, screws, washers, and other miscellaneous hardware, shall conform to ASTM Designation: A153, Class C or D, or B-454, Class 40. The weight of galvanized coating shall be determined according to Test Method Tex-728-I.

Galvanizing will not be required for any material unless specified on the plans or specifications.

442.16. **Measurement.** Measurement of the quantity of structural metal furnished and placed will be based on the weight of metal in the fabricated structure. The weight of erection bolts, paint, or weld metal shall be excluded.

When increases in size or weights of members have been made which were not ordered but approved by the Engineer, the measurement will be made on the sizes or weights shown on the contract plans.

In determining the weight of structural metal in steel or concrete structures, such items as castings, bearing plates, anchor bolts, drains, deck plates, armor joints, finger joints, and all other metal for which no separate measurement is specified, will be considered as Structural Steel.

The weights of rolled shapes and plates shall be computed on the basis of their nominal weights and dimensions using English measurements as shown on the approved shop plans. The weight of castings will be computed from the dimensions shown on the approved shop drawings. Shoes will be measured by the weights shown on the contract plans, or as specified for castings, if weights are not shown.

Deductions will be made for all cuts, copes, perforations, and all holes except bolt holes.

When computed, the weight of metal will be based on Table I.

Splices will be measured as follows:

(1) No additional weight will be allowed for weld metal in a welded splice.

(2) Where a bolted splice is permitted as an alternate for a welded splice, measurement will be made on the basis of a welded splice.

(3) Where a bolted splice is required, the weight of splice material, bolt heads, washers, and nuts, with no deduction for holes, will be measured.

A change in design concept which either increases or reduces the quantity of metal going into the structure or structures, will be measured by actual computed weights of the metal, and the quantity as shown on the plans and in the contract, will be increased or decreased by the revised weights, as the case may be.

| | Pay Weights-Lbs. per Cu. In. | | | | |
|---|---|---|---|---|---|
| Material | Steel | Cast Iron | Wrought Iron | Bronze | Lead |
| Weight | 0.2833 | 0.2604 | 0.2777 | 0.3150 | 0.4085 |

| High Strength Bolts-Weight per Hundred Units | | | | | | | |
|---|---|---|---|---|---|---|---|
| Diameter | 1/2" | 5/8" | 3/4" | 7/8" | 1" | 1-1/8" | 1-1/4" |
| Bolt Heads | | | 15 | 25 | 36 | 52 | 73 |
| Nuts | | | 19 | 28 | 41 | 56 | 73 |
| Washers | | | 12 | 12 | 18 | 21 | 27 |

TABLE I

# Q.9 STRUCTURAL BOLTING

447.1. **Description.** This item shall govern for the materials to be used, and for the method of installation of high strength bolts used in structural connections.

447.2. **Material.**

(1) Bolts of diameters of 1/2 inch to 1-½ inch inclusive, nuts and washers shall conform to the requirements of ASTM Designation: A325 or A490 except as provided herein.

(2) Bolts shall conform to the requirements for heavy hex structural bolts or heavy hex bolts of the ANSI Standard B18.2.1. The length of the bolts shall be such that the point of the bolt will be flush with or outside the face of the nut when completely installed. Dimensions shall conform to Table 1.

(3) When specified on the plans or approved by design change, high strength steel lock-pin and collar fasteners may be used as an alternate for high strength bolts. The shank and head of the high strength steel lock-pin and collar fasteners shall meet the chemical composition and mechanical property requirements of ASTM Designation: A325. The fasteners shall provide a solid shank body of the same nominal diameter and have tensile and shear strength equivalent to or greater than the bolt specified; shall have a cold formed head on one end, of the type and dimensions as approved by the Engineer; a shank length suitable for material thickness fastened, locking grooves, breakneck groove and pull groove on the opposite end. Each fastener shall provide a steel locking collar of proper size for the shank diameter used which by means of suitable installation tools, is cold sweged into the locking grooves forming a head for the grooved end of the fastener after the pull groove section has been removed. The steel locking collar shall be a standard product of an established manufacturer of lock-pin and collar fasteners as approved by the Engineer. Flat washers will not be required when lock-pin and collar fasteners are used.

(4) Unless otherwise specified, threads for structural bolts and nuts up through 1 inch diameter shall be the UNC Series and for over 1 inch diameter and larger shall be the 8UN Series as specified in ANSI B1.1 and shall have Class 2A tolerance for Bolts and Class 2B tolerance for Nuts. Galvanized nuts shall be tapped or chased after galvanizing.

(5) Circular, beveled and clipped washers shall conform to the requirements of ASTM Designation: F436. Their nominal dimensions shall be in accordance with Table 2 of this specification, with washer tolerances as given in ASTM Designation: F436.

Where necessary, washers may be clipped on one side to the dimension given in Table 2.

(6) ASTM Designations: A325 and A490 provide for three types of high-strength bolts. Any of these types may be used when the structural steel is to be painted. Type 3 bolts, nuts and washers shall be used when weathering steel (A-588) is used. Type 2 bolts may not be used when hot-dip galvanized bolts are required. Bolts and nuts of only one type shall be used for any one structure.

**447.3. General.**

(1) Surfaces of bolted parts in contact with the bolt head and nut shall not have a slope of more than 1:20 with respect to a plane normal to the bolt axis. Bolted parts shall fit solidly together when assembled and shall not be separated by gaskets or by any other interposed compressible material. Holes shall be punched, sub-punched and reamed, or drilled as required by the Item, "Steel Structures". Holes shall be of a nominal diameter not more than one- sixteenth of an inch in excess of the nominal bolt diameter. Field erection and fit-up of joints and splices shall conform to the requirements of the Item, "Steel Structures".

(2) When assembled, all joint surfaces, including those adjacent to the bolt head, nuts or washers , shall be free of dirt, rust, loose scale, burrs and other matter that will prevent solid seating of the parts.

(3) Prior to final bolting, shop or field, the faying surface of all joints, including splice plates, shall be given a Class "A" Blast Cleaning, in accordance with the Item "Cleaning, Painting and Painting."

(4) Galvanized surfaces shall be lightly scored by wire brushing or sandblasting.

(5) Bolts and nuts are normally shipped with a thin film of oil or wax on the threads as a protection against rust. This coating should not be removed. If the threads are exposed and corrosion has begun, the threads shall be cleaned and a thin coat of oil applied prior to installation. Oil shall be removed from the exposed portions of the bolts and nuts by use of a solvent when the steel is to be painted.

**447.4. Identification and Dimensions.** Heavy hex structural bolts manufactured to ASTM Designation: A325, Types 1,2 and 3, the dimen-

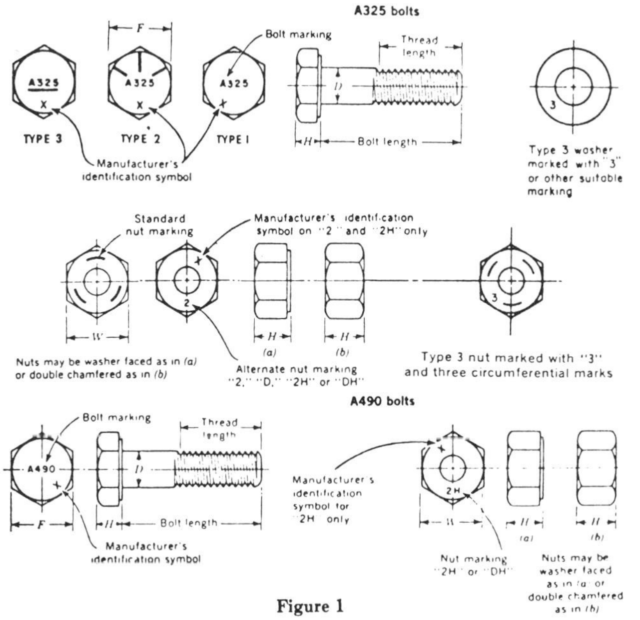


**Figure 1**

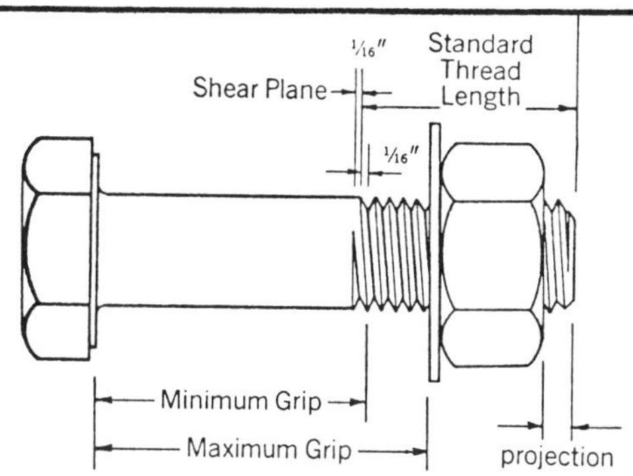


| Thread Length and Projection for A325 and A490 Bolts | | | |
|---|---|---|---|
| Bolt Diam. in. | Standard Thread Length, in. | Available Projection, in. | |
| | | Min. | Max. |
| 1/2 | 1 | 3/32 | 9/32 |
| 5/8 | 1 1/4 | 7/32 | 13/32 |
| 3/4 | 1 3/8 | 7/32 | 13/32 |
| 7/8 | 1 1/2 | 7/32 | 13/32 |
| 1 | 1 3/4 | 11/32 | 17/32 |
| 1 1/8 | 2 | 15/32 | 21/32 |
| 1 1/4 | 2 | 11/32 | 17/32 |
| 1 3/8 | 2 1/4 | 15/32 | 21/32 |
| 1 1/2 | 2 1/4 | 11/32 | 17/32 |

| Nominal Bolt Size, Inches $D$ | Bolt Dimensions, Inches [a] Heavy Hex Structural Bolts | | | Nut Dimensions, Inches Heavy Hex Nuts | |
|---|---|---|---|---|---|
| | Width across flats $F$ | Height, $H$ | Thread length | Width across flats $W$ | Height, $H$ |
| 1/2 | 7/8 | 5/16 | 1 | 7/8 | 31/64 |
| 5/8 | 1 1/16 | 25/64 | 1 1/4 | 1 1/16 | 39/64 |
| 3/4 | 1 1/4 | 15/32 | 1 3/8 | 1 1/4 | 47/64 |
| 7/8 | 1 7/16 | 35/64 | 1 1/2 | 1 7/16 | 55/64 |
| 1 | 1 5/8 | 39/64 | 1 3/4 | 1 5/8 | 63/64 |
| 1 1/8 | 1 13/16 | 11/16 | 2 | 1 13/16 | 1 7/64 |
| 1 1/4 | 2 | 25/32 | 2 | 2 | 1 7/32 |
| 1 3/8 | 2 3/16 | 27/32 | 2 1/4 | 2 3/16 | 1 11/32 |
| 1 1/2 | 2 3/8 | 15/16 | 2 1/4 | 2 3/8 | 1 15/32 |

[a]Refer to Figure 1

Table 1

NUT, BOLT AND THREAD DIMENSIONS

**Hardened Circular and Clipped Circular Washers**

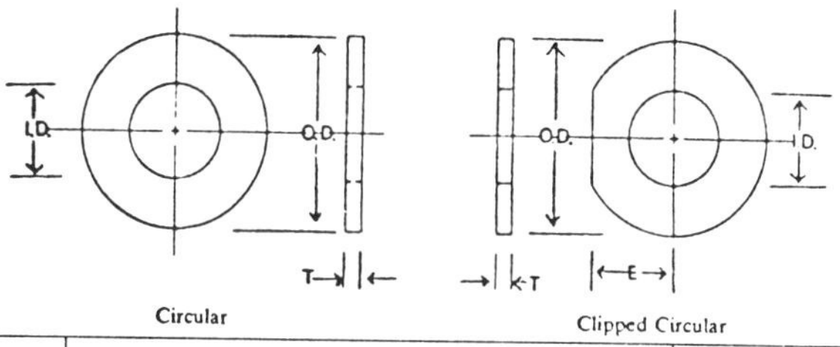


| Bolt Size | Circular and Clipped Circular | | | | Clipped |
|---|---|---|---|---|---|
| | Nominal Outside Diameter (O.D.), in. | Nominal Inside Diameter (I.D.), in. | Thickness (T), in. Min. | Thickness (T), in. Max. | Nominal Edge Distance (E)[a], in. |
| 1/2 | 1 1/16 | 17/32 | 0.097 | 0.177 | 7/16 |
| 5/8 | 1 5/16 | 11/16 | 0.122 | 0.177 | 9/16 |
| 3/4 | 1 15/32 | 13/16 | 0.122 | 0.177 | 21/32 |
| 7/8 | 1 3/4 | 15/16 | 0.136 | 0.177 | 25/32 |
| 1 | 2 | 1 1/8 | 0.136 | 0.177 | 7/8 |
| 1 1/8 | 2 1/4 | 1 1/4 | 0.136 | 0.177 | 1 |
| 1 1/4 | 2 1/2 | 1 3/8 | 0.136 | 0.177 | 1 3/32 |
| 1 3/8 | 2 3/4 | 1 1/2 | 0.136 | 0.177 | 1 7/32 |
| 1 1/2 | 3 | 1 5/8 | 0.136 | 0.177 | 1 5/16 |
| 1 3/4 | 3 3/8 | 1 7/8 | 0.178[b] | 0.28[b] | 1 7/16 |
| 2 | 3 3/4 | 2 1/8 | 0.178[b] | 0.28[b] | 1 3/4 |
| 2 1/4 | 4 | 2 3/8 | 0.24[c] | 0.34[c] | |
| 2 1/2 | 4 1/2 | 2 5/8 | 0.24[c] | 0.34[c] | |
| 2 3/4 | 5 | 2 7/8 | 0.24[c] | 0.34[c] | |
| 3 | 5 1/2 | 3 1/8 | 0.24[c] | 0.34[c] | |
| 3 1/4 | 6 | 3 3/8 | 0.24[c] | 0.34[c] | |
| 3 1/2 | 6 1/2 | 3 5/8 | 0.24[c] | 0.34[c] | |
| 3 3/4 | 7 | 3 7/8 | 0.24[c] | 0.34[c] | |
| 4 | 7 1/2 | 4 1/8 | 0.24[c] | 0.34[c] | |

[a] Clipped edge E shall not be closer than 7/8 of the bolt diameter from the center of the washer.

[b] 3/16 in. nominal.

[c] 1/4 in. nominal.

Table 2-A

**WASHER DIMENSIONS**

sions for which are shown in Table 1 and Fig. 1, are identified on the top of the head by the legend "A325", and the manufacturer's symbol. In addition, Type 1 bolts, at the option of the manufacturer, may be marked with three radial lines 120 degrees apart. Type 2 bolts shall be marked with three radial lines 60 degrees apart. Type 3 bolts shall have the "A325" underlined and the manufacturer may add other distinguishing marks indicating that the bolt is of a weathering type. Bolts manufactured to ASTM Designation: A490 are marked with the legend "A490" and the manufacturer's symbol. Heavy hex nuts for A325 bolts are identified on at least one face by the manufacturer's mark and the number "2" or "2H", by three equally spaced circumferential lines, or by the legend "D" or "DH". Heavy hex nuts for A325 Type 3 bolts shall be marked on one face with three circumferential marks and the numeral "3", and with other distinguishing marks the manufacturer may elect to use. Heavy hex nuts for use on A490 bolts shall be marked with the legend "2H" and the manufacturer's mark or by the legend "DH". Washers for A325 Type 3 bolts shall be marked on one face near the outer edge with the numeral "3", or other distinguishing marks indicating that the washer is of a weathering type. The marking on bearing surfaces of nuts and washers shall be depressed.

**Hardened Beveled Washers**

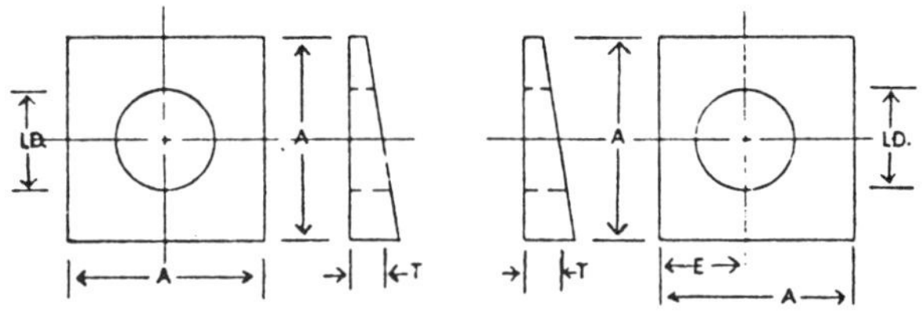


Square                Square Clipped

| Bolt Size, in. | Square Beveled and Clipped Square Beveled[a] | | | | Clipped |
|---|---|---|---|---|---|
| | Minimum Side Dimension (A), in. | Nominal Inside Diameter (I.D.), in. | Mean Thickness (T), in. | Slope or Taper in Thickness | Nominal Edge Distance (E),[b] in. |
| 1/2 | 1 3/4 | 17/32 | 5/16 | 1:6 | 7/16 |
| 5/8 | 1 3/4 | 11/16 | 5/16 | 1:6 | 9/16 |
| 3/4 | 1 3/4 | 13/16 | 5/16 | 1:6 | 21/32 |
| 7/8 | 1 3/4 | 15/16 | 5/16 | 1:6 | 25/32 |
| 1 | 1 3/4 | 1 1/8 | 5/16 | 1:6 | 7/8 |
| 1 1/8 | 2 1/4 | 1 1/4 | 5/16 | 1:6 | 1 |
| 1 1/4 | 2 1/4 | 1 3/8 | 5/16 | 1:6 | 1 3/32 |
| 1 3/8 | 2 1/4 | 1 1/2 | 5/16 | 1:6 | 1 7/32 |
| 1 1/2 | 2 1/4 | 1 5/8 | 5/16 | 1:6 | 1 5/16 |

[a] Rectangular beveled washers shall conform to the dimensions shown above, except that one side may be longer than that shown for the A dimension.

[b] Clipped edge E shall not be closer than 7/8 of the bolt diameter from the center of the washer.

**Table 2-B**
**WASHER DIMENSIONS**

**447.5. Determination of Bolt Lengths.** The grip range of A325 and A490 bolts are shown in Table 3. These bolt lengths are adequate for one circular hardened washer.

Grip Lengths for A325 and A490 Bolts (Includes Allowance for one hardened washer)

| Bolt Length* In. | Bolt Diameter, In. | | | | | | |
|---|---|---|---|---|---|---|---|
| | 1/2 | 5/8 | 3/4 | 7/8 | 1 | 1 1/8 & 1 1/4 | 1 3/8 & 1 1/2 |
| | Range of Grip Lengths, In. | | | | | | |
| 1 1/4 | 5/16-1/2 | -- | -- | -- | -- | -- | -- |
| 1 1/2 | 9/16-3/4 | 5/16-1/2 | -- | -- | -- | -- | -- |
| 1 3/4 | 13/16-1 | 9/16-3/4 | 7/16-5/8 | -- | -- | -- | -- |
| 2 | 1 1/16-1 1/4 | 13/16-1 | 11/16-7/8 | 9/16-3/4 | -- | -- | -- |
| 2 1/4 | 1 5/16-1 1/2 | 1 1/16-1 1/4 | 15/16-1 1/8 | 13/16-1 | 9/16-3/4 | -- | -- |
| 2 1/2 | 1 9/16-1 3/4 | 1 5/16-1 1/2 | 1 3/16-1 3/8 | 1 1/16-1 1/4 | 13/16-1 | 9/16-3/4 | -- |
| 2 3/4 | 1 13/16-2 | 1 9/16-1 3/4 | 1 7/16-1 5/8 | 1 5/16-1 1/2 | 1 1/16-1 1/4 | 13/16-1 | 9/16-3/4 |
| 3 | 2 1/16-2 1/4 | 1 13/16-2 | 1 11/16-1 7/8 | 1 9/16-1 3/4 | 1 5/16-1 1/2 | 1 1/16-1 1/4 | 13/16-1 |
| 3 1/4 | 2 5/16-2 1/2 | 2 1/16-2 1/4 | 1 15/16-2 1/8 | 1 13/16-2 | 1 9/16-1 3/4 | 1 5/16-1 1/2 | 1 1/16-1 1/4 |
| 3 1/2 | 2 9/16-2 3/4 | 2 5/16-2 1/2 | 2 3/16-2 3/8 | 2 1/16-2 1/4 | 1 13/16-2 | 1 9/16-1 3/4 | 1 5/16-1 1/2 |
| 3 3/4 | 2 13/16-3 | 2 9/16-2 3/4 | 2 7/16-2 5/8 | 2 5/16-2 1/2 | 2 1/16-2 1/4 | 1 13/16-2 | 1 9/16-1 3/4 |
| 4 | 3 1/16-3 1/4 | 2 13/16-3 | 2 11/16-2 7/8 | 2 9/16-2 3/4 | 2 5/16-2 1/2 | 2 1/16-2 1/4 | 1 13/16-2 |
| 4 1/4 | 3 5/16-3 1/2 | 3 1/16-3 1/4 | 2 15/16-3 1/8 | 2 13/16-3 | 2 9/16-2 3/4 | 2 5/16-2 1/2 | 2 1/16-2 1/4 |
| 4 1/2 | 3 9/16-3 3/4 | 3 5/16-3 1/2 | 3 3/16-3 3/8 | 3 1/16-3 1/4 | 2 13/16-3 | 2 9/16-2 3/4 | 2 5/16-2 1/2 |
| 4 3/4 | 3 13/16-4 | 3 9/16-3 3/4 | 3 7/16-3 5/8 | 3 5/16-3 1/2 | 3 1/16-3 1/4 | 2 13/16-3 | 2 9/16-2 3/4 |
| 5 | 4 1/16-4 1/4 | 3 13/16-4 | 3 11/16-3 7/8 | 3 9/16-3 3/4 | 3 5/16-3 1/2 | 3 1/16-3 1/4 | 2 13/16-3 |
| 5 1/4 | 4 5/16-4 1/2 | 4 1/16-4 1/4 | 3 15/16-4 1/8 | 3 13/16-4 | 3 9/16-3 3/4 | 3 5/16-3 1/2 | 3 1/16-3 1/4 |
| 5 1/2 | 4 9/16-4 3/4 | 4 5/16-4 1/2 | 4 3/16-4 3/8 | 4 1/16-4 1/4 | 3 13/16-4 | 3 9/16-3 3/4 | 3 5/16-3 1/2 |
| 5 3/4 | 4 13/16-5 | 4 9/16-4 3/4 | 4 7/16-4 5/8 | 4 5/16-4 1/2 | 4 1/16-4 1/4 | 3 13/16-4 | 3 9/16-3 3/4 |
| 6 | 5 1/16-5 1/4 | 4 13/16-5 | 4 11/16-4 7/8 | 4 9/16-4 3/4 | 4 5/16-4 1/2 | 4 1/16-4 1/4 | 3 13/16-4 |
| 6 1/4 | 5 5/16-5 1/2 | 5 1/16-5 1/4 | 4 15/16-5 1/8 | 4 13/16-5 | 4 9/16-4 3/4 | 4 5/16-4 1/2 | 4 1/16-4 1/4 |
| 6 1/2 | 5 9/16-5 3/4 | 5 5/16-5 1/2 | 5 3/16-5 3/8 | 5 1/16-5 1/4 | 4 13/16-5 | 4 9/16-4 3/4 | 4 5/16-4 1/2 |
| 6 3/4 | 5 13/16-6 | 5 9/16-5 3/4 | 5 7/16-5 5/8 | 5 5/16-5 1/2 | 5 1/16-5 1/4 | 4 13/16-5 | 4 9/16-4 3/4 |
| 7 | 6 1/16-6 1/4 | 5 13/16-6 | 5 11/16-5 7/8 | 5 9/16-5 3/4 | 5 5/16-5 1/2 | 5 1/16-5 1/4 | 4 13/16-5 |
| 7 1/4 | 6 5/16-6 1/2 | 6 1/16-6 1/4 | 5 15/16-6 1/8 | 5 13/16-6 | 5 9/16-5 3/4 | 5 5/16-5 1/2 | 5 1/16-5 1/4 |
| 7 1/2 | 6 9/16-6 3/4 | 6 5/16-6 1/2 | 6 3/16-6 3/8 | 6 1/16-6 1/4 | 5 13/16-6 | 5 9/16-5 3/4 | 5 5/16-5 1/2 |
| 7 3/4 | 6 13/16-7 | 6 9/16-6 3/4 | 6 7/16-6 5/8 | 6 5/16-6 1/2 | 6 1/16-6 1/4 | 5 13/16-6 | 5 9/16-5 3/4 |
| 8 | 7 1/16-7 1/4 | 6 13/16-7 | 6 11/16-6 7/8 | 6 9/16-6 3/4 | 6 5/16-6 1/2 | 6 1/16-6 1/4 | 5 13/16-6 |
| 8 1/4 | 7 5/16-7 1/2 | 7 1/16-7 1/4 | 6 15/16-7 1/8 | 6 13/16-7 | 6 9/16-6 3/4 | 6 5/16-6 1/2 | 6 1/16-6 1/4 |
| 8 1/2 | 7 9/16-7 3/4 | 7 5/16-7 1/2 | 7 3/16-7 3/8 | 7 1/16-7 1/4 | 6 13/16-7 | 6 9/16-6 3/4 | 6 5/16-6 1/2 |
| 8 3/4 | 7 13/16-8 | 7 9/16-7 3/4 | 7 7/16-7 5/8 | 7 5/16-7 1/2 | 7 1/16-7 1/4 | 6 13/16-7 | 6 9/16-6 3/4 |
| 9 | 8 1/16-8 1/4 | 7 13/16-8 | 7 11/16-7 7/8 | 7 9/16-7 3/4 | 7 5/16-7 1/2 | 7 1/16-7 1/4 | 6 13/16-7 |

*Lengths above 9 inches are also available. Consult supplier for specific information and data.

**Table 3. Length of Bolts.**

**447.6. Construction Methods.**

(1) Threaded bolts shall be installed with a hardened washer under either the nut or bolt head, whichever is turned in tightening. When the outer plies being fastened have a yield strength of less than 40 KSI, a hardened washer shall be installed under both the head and nut when A490 bolts are used.

(2) All fasteners shall be tightened to give at least the required minimum bolt tension values shown in Table 4 on completion of the joint for the grade and size fastener used.

When lock-pin and collar fasteners are used, the installation tools, methods and procedure shall be as approved by the Engineer.

| Nominal Bolt Size, Inches | Minimum Fastener Tension[a] in Thousands of Pounds (kips) | |
|---|---|---|
| | A325 Bolts | A490 Bolts |
| 1/2 | 12 | 15 |
| 5/8 | 19 | 24 |
| 3/4 | 28 | 35 |
| 7/8 | 39 | 49 |
| 1 | 51 | 64 |
| 1 1/8 | 56 | 80 |
| 1 1/4 | 71 | 102 |
| 1 3/8 | 85 | 121 |
| 1 1/2 | 103 | 148 |

[a] Equal to 70% of specified minimum tensile strengths of bolts, rounded off to the nearest kip.

Table 4
FASTENER TENSION

(3) The following procedure shall apply for bolt installation of a complete connection, regardless of the method of tightening.

(a) A minimum of 20 percent of the holes at a bolted connection shall be filled with erection pins to "fair-up" all holes (each flange and web).

(b) Install bolts in all remaining holes.

(c) Tighten a minimum of 25 percent of the bolts, following a pattern of progression from the center, or most rigid part of the joint toward the free edges, making sure that all plies of the metal in the connection are properly fitted and in contact. Mark these bolts as fit-up bolts.

(d) Tighten all the remaining bolts by the required amount. Remove the erection pins, fill these remaining holes with bolts, and loosen all A325 bolts used for fit-up, and replace all A490 bolts used for fit-up. Tighten these bolts by the required amount.

(e) The tightening crew shall mark finished work with identifying symbol and the Inspector will mark accepted work.

(4) Threaded bolts shall be tightened by power wrenches of adequate capacity and sufficiently supplied with air to perform the required tightening in approximately 10 seconds, by one of the following methods:

**(a)** The Turn-of-the-Nut Method.

**(b)** The Calibrated-Power Wrench Method.

(5) Turn-of-the-Nut Method of tightening.

**(a)** Tighten all bolts not used as fit-up bolts to a "snug-tight" condition. (A snug-tight condition defined as the tightness attained by a few impacts of the power wrench or the full effort of a man using an ordinary spud wrench.)

**(b)** Tighten the nuts or bolts by the amount specified in Table 5. During this operation there shall be no rotation of the part not turned by the wrench. The non-turning element must be held by an adequate wrench.

**(c)** Follow steps (d) and (e) shown in Subsection (3).

When lock-pin and collar fasteners are used, the installation tools, methods and procedure shall be as approved by the Engineer.

| Bolt Length (as measured from underside of head to extreme end of point) | Disposition of Outer Faces of Bolted Parts | | |
|---|---|---|---|
| | Both faces normal to bolt axis | One face normal to bolt axis and other face sloped not more than 1:20 (bevel washer not used) | Both faces sloped not more than 1:20 from normal to bolt axis (bevel washers not used) |
| Up to and including 4 diameters | 1/3 turn | 1/2 turn | 2/3 turn |
| Over 4 diameters but not exceeding 8 diameters | 1/2 turn | 2/3 turn | 5/6 turn |
| Over 8 diameters but not exceeding 12 diameters[b] | 2/3 turn | 5/6 turn | 1 turn |

Nut rotation is relative to bolt, regardless of the element (nut or bolt) being turned. For bolts installed by 1/2 turn and less, the tolerance should be plus or minus 30°; for bolts installed by 2/3 turn and more, the tolerance should be plus or minus 45°.

TABLE 5

**Turn-of-the-Nut Rotation**

(6) Calibrated-Power-Wrench Method of tightening. When calibrating power wrenches, they shall be adjusted to stall-out, or cut-out completely, when a bolt tension of 5% in excess of the value shown in Table 4 is reached. The wrenches shall be calibrated by tightening not less than three typical bolts of each size from the lot to be installed, in a device capable of indicating actual bolt tension and shall be furnished by the Contractor (Skidmore-Wilhelm Bolt Calibrator or equal). The wrenches shall be calibrated at least twice each working day for each bolt diameter being installed. Wrenches shall be recalibrated when significant changes are made in the equipment or hose length, or when a significant difference is noted in the surface condition of the bolts, nuts or washers.

After proper calibration, the following procedures shall be followed:

(a) Tighten all bolts not used as fit-up bolts to the tension required in Table 4.

(b) Follow steps (d) and (e) shown in Subsection (3).

A490 bolts and galvanized A325 bolts shall not be reused. Other A325 bolts may be reused if approved by the Engineer. Retightening previously tightened bolts which may have been loosened by the tightening of adjacent bolts will not be considered as reuse.

447.7. **Inspection.** Procedures for calibration of wrenches and installation of bolts shall be as approved by the Engineer. The inspector will observe the bolt installation to determine that these procedures are followed.

Bolts tightened by the turn-of-the-nut method shall have the outer face of the nut match-marked with the protruding bolt point before final tightening, thus affording the inspector visual means of noting the actual rotation. Such marks shall be made by the operator with a keel, crayon, or spot of paint, after the bolts have been brought up to snug-tight.

The following procedure shall be used to inspect the calibrated power-wrench method of bolt tightening:

Calibrated power-wrench shall be recalibrated as necessary to assure that they are tightening the bolts to the required minimum. During actual bolt tightening, every connection shall have a minimum of 5 percent of the bolts, but not less than one bolt, checked. If any bolt check indicates less than the rotation required by Table 5, the wrench shall be recalibrated and the bolts in the connection shall be re-impacted with the recalibrated wrench as necessary to assure proper tension in all bolts.

The wrench shall be returned to spot check (about 1 in 15) previously tightened bolts which may have been loosened by the tightening of subsequent bolts. Retightening shall continue until all bolts are tightened to the prescribed amount.

When lock-pin and collar fasteners are used or other procedures for tightening threaded bolts are approved by the Engineer, an inspection procedure shall be established to provide a method to assure compliance with the tension requirements of Table 4.

## Q.10 STRUCTURAL WELDING

448.2. **General.** All welds including tack welds to be incorporated therein shall be made by a certified welder. Tack welds shall be cleaned and fused thoroughly with the final weld. Defective, cracked or broken tack welds shall be removed.

Certification for welders shall be in conformance with the requirements of SDHPT Bulletin C-6. Miscellaneous welds may be made by a welder qualified in accordance with Article 448.16.

Welds shall be as required by the contract or erection drawings. The location or size shall not be changed without approval of the Engineer.

The welder shall place his identification mark with crayon or paint near the groove welds made by him.

No welding will be allowed when the air temperature is lower than 20 F, when surfaces are wet or exposed to rain, snow, or wind, or when operators are exposed to inclement conditions that will hamper good workmanship.

Any moisture present at the point of welding shall be driven off in accordance with Table III, before welding commences. Wind breaks shall be required for the protection of all welding operations.

There shall be no temporary welds for transportation, erection or other purposes on main members, except at location more than one-sixth the depth of the web from the flanges of beams and girders or as shown on SS-MEBR Plan Sheet or as approved by the Engineer.

On A514/A517 steel, all groove welds in main members and in flanges of beams and girders subject to tensile stress or reversals of stress shall be finished smooth and flush on all surfaces, including edges, by grinding in the direction of applied stress leaving the surface free from depressions. Chipping may be used provided it is followed by such grinding. Parts joined by groove welds connecting plates of unequal thickness or width shall have a smooth transition between offset surfaces at a slope not greater than one in four with the surface of either part. The surfaces shall be ground so that the radii at the points of transition will be 4 inches minimum.

All groove welds, except when produced with the aid of backing, shall have the root of the initial weld gouged, chipped or otherwise removed to sound metal before welding is started from the second side, except that back gouging will not be required when welding steel piling or armor joints with E 6010 electrodes. The back side shall be thoroughly cleaned before placing back-up pass.

When backing for welds is left in place to become a part of the structure, it shall be a single length insofar as possible. Where more than a single length is needed, they shall be joined by full penetration butt welds. The surfaces of this butt weld shall be ground flush as necessary to obtain proper fit-up in the weld joint.

Before welding over previously deposited metal, all slag shall be removed, and the weld and adjacent base metal shall be cleaned. This requirement shall apply equally to successive layers, successive beads and the crater area.

Arc strikes outside the area of permanent welds must be avoided on all steels. Where they do occur, resulting cracks and blemishes shall be ground out to a smooth contour and checked to insure soundness.

Stringer bead technique shall be used where possible for groove welds on all types of steel. Weaving will not be permitted for A514/A517 steel except in vertical welding, where a weave not exceeding two electrode diameters is permissible for manual shielded metal-arc process.

In all welding processes, the progression for all passes in vertical welding shall be upward using a back step sequence.

Groove welds shall begin and terminate at the ends of a joint on extension bars. Edge preparation and thickness of extension bars shall be the same as that of the member being welded and shall extend a minimum of three-fourths inch beyond the joint. Extension bars shall be removed with a cutting torch or arc-air gouging upon completion and cooling of the weld, and the flange edges shall be ground smooth.

Any defects exposed by the grinding shall be cleaned, filled with weld metal, and reground to a uniform finish. All grinding shall be parallel to the flange. Excess grinding of the parent metal shall be avoided.

448.3. **Filler Metal.** Electrodes for manual shielded metal-arc welding shall conform to the requirments of the latest edition of "Specifications for Mild Steel Covered Arc-Welding Electrodes", AWS A5.1, or to the requirements of the latest edition of "Specifications for Low Alloy Steel Covered Arc-Welding Electrodes", AWS A5.5.

All electrodes and combination of electrode and shielding for gas metal-arc welding for producing weld metal with a minimum specified yield point not exceeding 60,000 psi shall conform to the requirements in the latest edition, "Specification for Mild Steel Electrodes for Gas Metal-Arc Welding", AWS A5.18, AWS A5.28, or "Specification for Mild Steel Electrodes for Flux Cored Arc Welding", AWS A5.20, applicable for the classifications producing weld metal having a minimum impact strength of 20 ft-lb, Charpy V-notch, at a temperature of 0 F or below.

For weld metal with a minimum specified yield strength exceeding 60,000 psi, the Contractor shall demonstrate that each electrode and flux or combination of electrode and shielding medium proposed for use will produce low alloy weld metal having the mechanical properties listed in Table 1 in the as welded condition.

The mechanical properties shall be determined from a multiple pass weld made in accordance with the test requirements of the latest edition of AWS A5.18, AWS A5.28, or AWS A5.20 as applicable.

| GMAW Grade | FCAW Grade | Tensile Strength psi - Min | Yield Strength psi - Min | Elongation, % in 2 inches Min | Impact Strength ft-lb at 0F-Min |
|---|---|---|---|---|---|
| ER80S | E80T | 80,000 | 65,000 | 18 | 20 |
| ER90S | E90T | 90,000 | 78,000 | 17 | 20 |
| ER100S | E100T | 100,000 | 90,000 | 16 | 20 |
| ER110S | E110T | 110,000 | 98,000 | 15 | 20 |

**TABLE 1**

**Required Mechanical Properties for GMAW and FCAW Electrodes**

The mechanical property tests for Grades ER100S, ER110S, E100T and E110T shall be made using ASTM A514/A517 base metal.

A list of approved electrodes and flux electrode combinations will be maintained by the Department. All electrodes used on Department projects must be on the approved list. For approval of electrodes, the manufacturer shall submit to the Materials and Tests Engineer, 38th and Jackson Streets, Austin, Texas, 78703, certified copies of all tests required by the pertinent AWS Specifications. Tests shall be made on electrodes of the same class, size and brand which were manufactured by the same process and with the same materials as the electrodes to be furnished. Tests must be made and approval renewed every 12 months.

For sizes of electrodes not requiring tests by AWS Specifications, test reports shall be furnished for electrodes of the nearest size and of the same classification. The request for approval shall include the manufacturer's certification that the process and material requirements were the same for manufacturing the tested electrodes and those to be furnished, and new test reports, shall be submitted if any changes are made in process or materials during the effective period.

Class of electrodes required will be as shown in Table 2. Electrodes shall be used with the type of current, the polarity and in the positions permitted by AWS A5.1 and A5.5 for manual shielded metal-arc welding. AWS A5.18 or A5.28 and A5.20 Specifications shall govern for gas metal-arc welding and flux cored arc welding.

| Type of Steel | Main Members | | Secondary Members | | |
|---|---|---|---|---|---|
| | Groove & Fillet Welds | | Groove & Fillet Welds | | |
| Steel Piling,<br>A53 Pipe,<br>A500,<br>A501,<br>Armor Joints | E6010<br>E6011<br>E7016<br>E7018 | E60T-8<br>E70S-1B<br>ER70S-2<br>ER70S-3<br>ER70S-6<br>ER70S-7 | E60XX<br>E70XX<br>E70S-1B<br>E70S-2<br>E70S-3<br>E70S-6<br>E70U-1 | E60T-8<br>E7XT-1<br>E7XT-5<br>E7XT-6<br>E7XT-8 | ER70S-2<br>ER70S-3<br>ER70S-6<br>ER70S-7 |
| A36,<br>A441,<br>A572 Grade 50<br>A588,<br>A242 Deck Plates<br>API Pipe | E7016<br>E7018<br>E7XT-1<br>E7XT-5<br>E7XT-6<br>E7XT-8 | ER70S-2<br>ER70S-3<br>ER70S-6<br>ER70S-7 | E7016<br>E7018<br>E7XT-1<br>E7XT-5<br>E7XT-6<br>E7XT-8 | ER70S-2<br>ER70S-3<br>ER70S-6<br>ER70S-7 | |
| A514/A517<br>2½" Thick or Less | E11018M<br>E110T | ER110S | E11018M<br>E110T | ER110S | |
| A514/A517<br>Over 2½" Thick | E10018M<br>E100T | ER100S | E10018M<br>E100T | ER100S | |
| A588, A242,<br>A618 Weathering Steel | E8018,C-3<br>E80T(3) | ER80S(3) | E8018,C-3<br>E80T(3)<br>ER80S(3) | | |
| Reinforcing Steel | E7016 | E7018 | | | |
| A572 Grades 60 and 65 for Light Towers | E8016<br>E8018 | E80T<br>ER80S | | | |

(1) Use of the same type electrode with the next higher mechanical properties, in accordance with AWS A5.1 or A5.5, than those listed will be permitted.

(2) In joints involving base metals of different yield points or strengths, low hydrogen electrodes applicable to the lower strength base metal may be used.

(3) Deposited weld metal for weathering steel shall have the following chemical composition: C, max. %, 0.12; Mn, %, 0.51/1.30; P, max. %, 0.03; S, max %, 0.04; Si, % 0.35/0.80; Cu, %, 0.30/0.75; Ni, %, 0.40/0.80; Cr, %, 0.45/0.70.

TABLE 2

Classifications of Electrodes Permitted

Before use, all electrodes with low hydrogen coverings conforming to AWS A5.1 shall be dried for not less than two hours between 450 and 500 F and electrodes with low hydrogen coverings conforming to AWS A5.5 for not less than one hour at a temperature between 700 and 800 F. Immediately after drying, electrodes shall be stored in ovens held at a temperature of at least 250 F. E70 electrodes not used within four hours, E80 within two hours, E90 within one hour, E100 and E110 within one-half hour after removal from the storage oven shall be redried before use. Electrodes with flux which has been wet, cracked or otherwise damaged, shall not be used. When used for welding A514/A517 steel, electrodes shall be dried at least one hour at temperatures between 700 and 800 F before being used. Electrodes may be redried only once.

Suitable facilities for drying and storage of electrodes shall be furnished at the job site, along with thermometers for checking and controlling the oven temperature.

In humid atmospheres, the times allowed for use without redrying may be reduced.

When a gas or gas mixture is used for gas metal-arc or flux cored arc welding, it shall be of a welding grade having a dew point of -40F or lower. The gas manufacturer shall furnish certification to the Engineer that the gas or gas mixture is suitable for the intended application and will meet the dew point requirements.

Welding wire coils removed from the original package, shall be protected or stored to keep their characteristics or welding properties intact. Rusty coils, or portions of coils that are rusty shall not be used.

Any deviation from the above electrode designation, shall be approved by the Engineer.

● (3) Preheat.

Preheat ahead of welding both groove and fillet welds (including tack welding) will be required as shown in Table 3.

Preheat and interpass temperatures must be sufficient to prevent crack formation. The preheat temperatures shown in Table 3 are minimums and higher preheats may be necessary in highly restrained welds.

When the base metal is below the required temperature, it shall be preheated so the parts being welded are not less than the specified temperature within 3 inches of the point of welding.

For all groove welds, preheat temperature shall be measured on the side opposite to which the heat is applied at points about three inches away from the joint.

Preheating equipment shall be adequate to maintain the entire joint at or above the specified temperature. When possible, a joint shall be completely welded before it is allowed to cool below the specified temperature but shall always be welded sufficiently to prevent cracking before cooling is permitted.

Usually preheat and interpass temperatures shall not exceed 400 F for thickness up to 1 1/2 inches and 450 F for greater thickness. These temperatures shall never be exceeded on A514/A517 steels.

The welder shall have and use approved equipment for checking preheat and interpass temperatures at all times while welding is in progress.

| Thickness of Thickest Part at Point of Welding | Manual or Semi-Automatic Gas Metal-Arc Welding, Flux Cored Arc Welding or Manual Shielded Metal-Arc Welding With Low Hydrogen Electrodes | |
|---|---|---|
| | ASTM A 36; A 242; A 441 A 572 Grades 42, 45, and 50; A 588 | ● ASTM A 514/517 |
| To 3/4, incl. | 50F | 50F |
| Over 3/4 to 1½, incl. | 70F | 125F |
| Over 1½ to 2½, incl. | 150F | 175F |
| Over 2½ | 225F | 225F |

(1) These temperatures are the minimum required for the thinner material shown for each increment, and higher preheat on a step basis will be required for the thicker material within each increment. Preheat and interpass temperatures must be sufficient to prevent crack formation and welding shall be carried continuously to completion or to a point that will assure freedom from cracking before the joint is allowed to cool below the minimum specified preheat and interpass temperature. Temperature above those shown may be required for highly restrained welds.

(2) When E7010 electrodes are permitted for tacking or temporary root pass, the material shall be preheated according to the following:

| Thickness of Thickest Part | Preheat for Tacking or Temporary Root Pass |
|---|---|
| 1/2" and Less | 150 F |
| 9/16" through 3/4" | 200 F |
| 13/16" through 1 1/2" | 300 F |
| Over 1 1/2" | 400 F |

(3) When joining steels of different strengths or thickness with groove welds, the preheat and interpass temperatures for the higher strength steel and the average plate thickness shall be used. For fillet welds, the preheat shall be used for the higher strength steel and the thickest plate being welded.

(4) When the base metal temperature is below 32 F, preheat to at least 70 F and maintain this minimum temperature during welding.

(5) Heat input when welding A514/A517 steel shall not exceed the steel producer's recommendations.

(6) When moisture is present on the base metal it shall be preheated to 200 F before welding is started.

TABLE 3

MINIMUM PREHEAT AND INTERPASS TEMPERATURE FOR MANUAL SHIELDED METAL-ARC WELDING, FLUX CORED ARC WELDING OR GAS METAL-ARC WELDING

**448.6. Quality of Welds.** Weld metal shall be sound throughout.

There shall be no cracks in any weld or weld pass.

There shall be complete fusion between the weld metal and the base metal and between successive passes throughout the joint.

Welds shall be free from overlap and the base metal free from undercut more than one one-hundredth inch deep when its direction is transverse to the primary stress in the part that is undercut. Undercut shall not be more than one thirty-second inch deep when its direction is parallel to the primary stress in the part that is undercut.

All craters shall be filled to the full cross section of the welds.

All welds on A514/517 steel shall be visually examined for longitudinal or transverse cracks not less than 48 hours after completion of welding.

● 448.7. **Corrections.** When welding is unsatisfactory or indicates inferior workmanship, the following corrective measures will be required by the Engineer whose specific approval shall be obtained for making each correction.

When requirements prescribe the removal of part of the weld or a portion of the base metal, removal shall be by oxygen gouging or arc-air gouging.

Oxygen gouging shall not be used on A514/517 steel or for A588 weathering steel.

Backgouging of splices in beams and girders or cutouts of defective welds shall be done by arc-air gouging by a welder qualified to make beam and girder splices.

Where corrections require the deposition of additional weld metal, the sides of the area to be welded shall have sufficient slope to permit depositing new metal.

Defective or unsound welds shall be corrected either by removing and replacing the entire weld, or as follows:

- **Excessive convexity.** Reduce to size by grinding off the excess weld metal.
- **Shrinkage cracks.** Cracks in base metal, craters and excessive porosity. Remove defective portions of base and weld metal down to sound metal and replace with additional sound weld metal.
- **Undercutting, undersize, and excessive concavity.** Clean and deposit additional weld metal.
- **Overlapping and incomplete fusion.** Remove and replace the defective portion of weld.
- **Slag inclusions.** Remove the parts of the weld containing slag and replace with sound weld metal.
- **Removal of adjacent base metal during welding.** Clean and form full size by depositing additional weld metal.

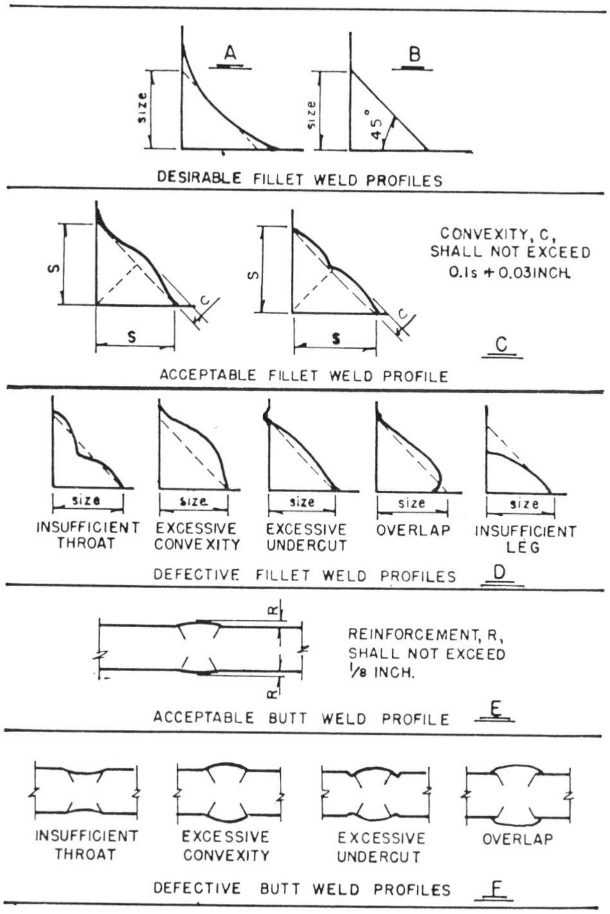


FIGURE 2
Weld Profiles

Where corrections require the deposition of additional weld metal, the electrode used shall be smaller than that used for making the original weld. Surfaces shall be cleaned thoroughly before rewelding.

A cracked weld shall be removed throughout its length, unless the extent of the crack can be ascertained to be limited, in which case the weld metal shall be removed 2 inches beyond each end of the crack and repairs made.

Where work performed after the making of a deficient weld has made the weld inaccessible or has caused new conditions making the correction of the deficiency dangerous or ineffectual, the original conditions shall be restored by removal of welds or members, or both, before making the necessary corrections, or else the deficiency shall be compensated by additional work according to a revised design approved by the Engineer.

Improperly fitted and misaligned parts shall be cut apart and rewelded.

Members distorted by the heat of welding shall be straightened by mechanical means or by the carefully supervised application of a limited amount of localized heat. Heated areas shall not exceed 1200 F as measured by Tempil-sticks or other approved methods for steel up to 65,000 psi yield strength. Parts to be heat straightened shall be substantially free of stress from external forces, except when mechanical means are used in conjunction with the application of heat.

Heat straightening of A514/517 steel shall be done only under rigidly controlled procedures, subject to the approval of the Engineer. In no case shall the maximum temperature of the steel exceed 1100 F. Sharp kinks and bends shall be cause for rejection of the material.

## Q.11 BANK PROTECTION

913-1 **Description:**

The work under this section shall consist of furnishing all materials and constructing bank protection in accordance with the details shown on the plans and the requirements of these specifications. Bank protection shall be dumped riprap, grouted riprap, wire tied riprap, riprap in wire baskets or gabions, and other types of bank protection and shall be constructed at the locations and as shown on the project plans.

913-2 **Materials:**

.01 **Rock:**

(A) **General:**

Rock shall be sound and durable, free from clay or shale seams, cracks or other structural defects and shall have a specific gravity of at least 2.40. Rock used to construct dumped riprap shall be angular in shape. Rock used to construct other types of bank protection may be rounded stones or boulders. Rock shall have a least dimension not less than one-third of its greatest dimension and a gradation in reasonable conformity with that shown herein for the various types of bank protection. Control of the gradation will be by visual inspection.

No source of rock is designated. It shall be the contractor's responsibility to negotiate for the material, obtain the right-of-way and pay all royalties and damages.

The source from which the stone will be obtained shall be selected well in advance of the time when the stone will be required in the work. The acceptability of the stone will be determined by the Engineer. If testing is required, suitable samples of stone shall be taken in the presence of the Engineer at least 25 days in advance of the time when the use of the stone is expected to begin. The approval of some rock fragments from a particular quarry site shall not be construed as constituting the approval of all rock fragments taken from that quarry.

(B) **Grouted Riprap:**

Gradation of the rock for grouted riprap shall be as specified in the special provisions.

(C) **Wire Tied Riprap:**

Rock for wire tied riprap shall be well graded with at least 95 percent, by weight, exceeding the least dimension of the wire mesh opening. The maximum size stone, measured normal to the mat, shall not exceed the mat thickness.

(D) **Dumped Riprap:**

Gradation of the rock for dumped riprap shall be as specified in the special provisions.

The contractor shall provide two samples of rock of at least five tons each, meeting the gradation specified above. The sample at the construction site may be a part of the finished riprap covering. The other sample shall be provided at the quarry. These samples shall be used as a frequent reference for judging the gradation of the riprap supplied. Any difference of opinion between the Engineer and the contractor shall be resolved by dumping and checking the gradation of two random truck loads of stone. Mechanical equipment, a sorting site, and labor needed to assist in checking gradation shall be provided by the contractor at no additional cost to the Department.

(E) **Gabions:**

Rock for gabions shall be well graded, varying in size from four to eight inches.

(F) **Riprap (Slope Mattress):**

Rock for slope mattress shall be well graded with 70 percent, by weight, exceeding four inches. The maximum dimension of a single stone shall not exceed the least dimension of the gabion.

Broken concrete may be used upon approval of the Engineer.

(G) **Rail Bank Protection:**

Rock used to construct rail bank protection (Type 1-Type 6) shall be as specified in the plans.

.02 **Metal Items:**

(A) **Wire Fabric:**

Welded wire fabric shall be galvanized and shall conform to the requirements of AASHTO M 55, except that the minimum weight of the zinc coating shall be 0.15 of an ounce per square foot of actual surface.

Woven wire fabric shall conform to the requirements of ASTM A 116, except that the minimum weight of zinc coating shall be 0.3 of an ounce per square foot or ASTM A 584, Class 1.

Wire fabric shall be of the gauge, spacing, pattern, and dimensions shown on the plans. The selvedge on each sheet of mesh shall be galvanized steel wire two gauges heavier than that used in the body of the mesh.

(B) **Miscellaneous Fittings and Hardware:**

Miscellaneous fittings and hardware shall be of the type and size provided by the manufacturer of the major item to which they apply for the use indicated on the project plans and shall be galvanized in accordance with the requirements of AASHTO M 232.

(C) **Tie Wires:**

Tie wires shall be of good commercial quality and the gauge shall be as shown on the project plans, except that the minimum weight of the zinc coating shall be 0.30 of an ounce per square foot. At the option of the contractor, hog rings may be used on gabions in lieu of tie wires.

(D) **Steel Cable:**

Steel cable shall be zinc-coated steel structural wire rope conforming to the requirements of ASTM A 475, Utilities Grade, Type 1 for the diameter shown on the plans.

(E) **Railroad Rail:**

Railroad rails may be new or used. If used rails are furnished, they shall be free from rust and equal to at least 95 percent of the original section.

(F) **Soil Anchor Stakes:**

Soil anchor stakes shall be steel and of the length called for on the plans. When not specified to be railroad rails, the following items may be used: crane rails with a weight of at least 40 pounds per yard; two inch diameter steel pipe conforming to the requirements of ASTM A 120, or 3" x 3" x 3/8" structural steel angles conforming to the requirements of ASTM A 36 or better. Used rails, pipe or angles may be used provided the material is not rusted or damaged to the extent that the strength of the item is reduced to less than 90 percent of a new item of the same type and nominal size.

.03 **Bedding Material:**

Bedding material shall consist of granular material having a maximum dimension of two inches and shall be free of clay or organic material.

.04 **Grout:**

Grout shall consist of one part portland cement, three parts fine aggregate and one-fifth part hydrated lime, by volume. These materials shall be thoroughly dry mixed and sufficient water shall be added to provide a mixture of thick workable consistency.

Portland cement, fine aggregate and water shall conform to the requirements of Section 1006. Hydrated lime shall conform to the requirements of ASTM C 207, Type N.

Grout that has been mixed more than one hour shall not be used. Retempering of grout will not be permitted.

.05 **Filter Fabric:**

Filter fabric shall conform to the requirements of AASHTO M 287.

.06 **Sacked Concrete:**

Sacked concrete shall be utility concrete conforming to the requirements of Section 922, except that the minimum cement content shall be 376 pounds per cubic yard; the slump shall be from three to five inches; and the aggregate shall conform to the following gradation when tested in accordance with the requirements of Arizona Test Method 201:

| Sieve Size | Percent Passing |
|---|---|
| 2 inch | 100 |
| 1/4 inch | 45- 89 |
| No. 200 | 0- 12 |

Sacks for sacked concrete riprap shall be made of at least ten ounce burlap and shall be approximately 19 1/2 inches x 36 inches measured inside the seams when the sack is laid flat, with an approximate capacity of 1.25 cubic feet. Sound, reclaimed sacks may be used.

913-3 **Construction Requirements:**

.01 **General:**

Areas on which bank protection is to be constructed shall be cleared, grubbed, and excavated or backfilled in accordance with the requirements of the appropriate sections of Division II to produce a ground surface in reasonable conformance with the lines and grades shown on the project plans or established by the Engineer.

Placement through water will not be permitted unless otherwise approved by the Engineer.

.02 **Filter Fabric:**

When filter fabric is required, it shall be placed in the manner and at the locations shown on the project plans. The surface to receive the fabric shall be free of obstructions, depressions and debris. The filter shall be loosely laid and not placed in a stretched condition. The strips shall be placed to provide a minimum 24 inches of overlap for each joint. On horizontal joints, the uphill strip shall overlap the downhill strip. On vertical joints, the upstream strip shall overlap the downstream strip. The fabric shall be protected at all times during construction from extensive exposure to sunlight.

When the maximum size of the rock to be placed on filter fabric exceeds 18 inches, the fabric shall be protected during the placement of the rock by a layer of bedding material. The bedding material shall be spread uniformly on the fabric to a depth of four inches and shall be free of mounds, dips or windrows. Compaction of the bedding material will not be required.

Rock shall be carefully placed on the bedding material and filter fabric in such a manner as not to damage the fabric. If, in the opinion of the Engineer, the fabric is damaged or displaced to the extent that it cannot function as intended, he will order the contractor to remove the rock, regrade the area if necessary, and replace the filter fabric.

.03 **Dumped Riprap:**

The rock shall be placed for its specified thickness in one operation and in a manner which will produce a reasonably well graded mass with a minimum practicable amount of voids with the larger rock evenly distributed throughout the mass.

No method of placing the rock that will cause segregation will be allowed. Hand placing or rearranging of individual rock by mechanical equipment may be necessary to obtain the specified results.

.04 **Wire-tied Riprap:**

After installation of the lower portion of the wire mesh, rock shall be placed in accordance with the requirements of Subsection 913-3.03. After placement of the rock, the upper portion of the wire mesh shall be placed, laced, and tied in accordance with the details shown on the project plans.

.05 **Grouted Riprap:**

Rock for grouted riprap shall be placed in accordance with the requirements of Subsection 913-3.03. The stones shall be thoroughly moistened and any excess of fines shall be sluiced to the underside of the stone blanket before grouting.

The grout may be delivered to the place of final deposit by any means that will insure uniformity and prevent segregation of the grout. If penetration of grout is not obtained by gravity flow into the interstices, the grout shall be spaded or rodded into the interstices to completely fill the voids in the stone blanket. Pressure grouting shall not unseat the stones; and during placing by this method, the grout shall be spaded or rodded into the voids. Penetration of the grout shall be to the depth specified on the project plans. When a rough surface is specified, stone shall be brushed until from one-fourth to one-half of the depth of the maximum size stone is exposed. For a smooth surface, grout shall fill the interstices to within a 1/2 inch of the surface.

Grout shall not be placed when the descending air temperature falls below 40 degrees F. nor until the ascending air temperature rises above 35 degrees F. Temperatures shall be taken in the shade away from artificial heat.

Curing of the grout shall be in accordance with the requirements of Subsection 912-3.09.

At the option of the contractor, shotcrete conforming to the requirements of Section 912 may be furnished in lieu of grout.

.06 **Slope Mattress Riprap:**

The mattress bed shall be excavated to the width, line and grade as staked by the Engineer. The mattress shall be founded on this bed and laid to the lines and dimensions required.

Excavation for toe or cut-off walls shall be made to the neat lines of the wall.

Mattresses shall be fabricated in such a manner that the sides, ends, lid and diaphragms can be assembled at the construction site into rectangular units of the specified sizes. Mattresses are to be of single unit construction, the base, ends and sides either to be woven into a single unit or one edge of these members connected to the base section of the unit in such a manner that strength and flexibility at the point of connection is at least equal to that of the mesh.

All perimeter edges of the mattresses are to be securely selvedged or bound so that the joints formed by tying the selvedges have at least the same strength as the body of the mesh.

Mattresses shall be placed to conform with the details shown on the project plans. Stone shall be placed in close contact within the unit so that maximum fill is obtained. The units may be filled by machine with sufficient hand work to accomplish the requirements of this specification.

Slope mattresses shall be filled with at least two layers of stone. Broken concrete may be placed in the bottom layer with approval of the Engineer. Before the mattress units are filled the longitudinal and lateral edge surfaces of adjoining units shall be tightly connected by means of wire ties placed every four inches or by a spiral tie having a complete loop every four inches. The lid edges of each unit shall be connected in a similar manner to adjacent units. The slope mattress shall be anchored as shown on the project plans. Each anchor stake shall be fastened to the cover mesh with a tie wire.

.07 **Gabions:**

The gabion bed shall be excavated to the width, line and grade as staked by the Engineer. The gabions shall be founded on this bed and laid to the lines and dimensions required.

Excavation for toe or cut-off walls shall be made to the neat lines of the wall.

Gabions shall be fabricated in such a manner that the sides, ends, lid and diaphragms can be assembled at the construction site into rectangular units of the specified sizes. Gabions are to be of single unit construction, the base, ends and sides either to be woven into a single unit or one edge of these members connected to the base section of the unit in such a manner that strength and flexibility at the point of connection is at least equal to that of the mesh.

Where the length of the gabion exceeds its horizontal width the gabion is to be equally divided by diaphragms, of the same mesh and gauge as the body of the gabions, into cells whose length does not exceed the horizontal width. The gabion shall be furnished with the necessary diaphragms secured in proper position on the base section in such a manner that no additional tying at this juncture will be necessary.

All perimeter edges of gabions are to be securely selvedged or bound so that the joints formed by tying the selvedges have at least the same strength as the body of the mesh.

Gabions shall be placed to conform with the project plan details. Stone shall be placed in close contact in the unit so that maximum fill is obtained. The units may be filled by machine with sufficient hand work to accomplish requirements of this specification. The exposed face or faces shall be hand-placed using selected stones to prevent bulging of the gabion cell and to improve appearance. Each cell shall be filled in three lifts. Two connecting tie wires shall be placed as shown on the project plans between each lift in each cell. Care shall be taken to protect the vertical panels and diaphragms from being bent during filling operations.

The last lift of stone in each cell shall be level with the top of the gabion in order to properly close the lid and provide an even surface for the next course.

All gabion units shall be tied together each to its neighbor along all contacting edges in order to form a continuous connecting structure.

Empty gabions stacked on filled gabions shall be laced to the filled gabion at the front, side and back.

**.08 Sacked Concrete Riprap:**

The sacks shall be filled with concrete, loosely packed so as to leave room for folding or tying at the top. Approximately one cubic foot of concrete shall be placed in each sack. Immediately after filling, the sacks shall be placed according to the details shown on the project plans and lightly trampled to cause them to conform with the earth face and with adjacent sacks in place.

The first two courses shall provide a foundation of double thickness. The first foundation course shall consist of a double row of stretchers laid level and adjacent to each other in a neatly trimmed trench. The trench shall be cut back into the slope a sufficient distance to enable proper subsequent placement of the riprap. The second course shall consist of a row of headers placed directly above the double row of stretchers. The third and remaining courses shall consist of stretchers and shall be placed in such a manner that joints in succeeding courses are staggered.

All dirt and debris shall be removed from the top of the sacks before the next course is laid thereon. Stretchers shall be placed so that the folded ends will not be adjacent. Headers shall be placed with the folds toward the earth face. Not more than four vertical courses of sacks shall be placed in any tier until initial set has taken place in the first course of any such tier.

When there will not be proper bearing or bond for the concrete because of delays in placing succeeding layers of sacks or because of the work having been hampered by storms, or mud, or for any cause, a small trench shall be excavated back of the row of sacks already in place, and the trench shall be filled with fresh concrete before the next layer of sacks is laid. The size of the trench and the concrete used for this purpose shall be approved by the Engineer. The Engineer may require header courses at any level to provide additional stability to the riprap.

Sacked concrete riprap shall be cured by being covered with a blanket of wet earth or by being sprinkled with a fine spray of water every two hours during the daytime for a period of four days.

**.09 Rail Bank Protection (Type 1- Type 6):**

Excavation, where required for rock fill, shall be performed in reasonably close conformity to the lines and grades established or shown on the plans.

Rails shall be driven at the locations and to the minimum penetrations shown on the plans. Driving equipment shall be capable of developing sufficient energy to drive the rails to the specified minimum penetration and be approved by the Engineer.

If hard material is encountered during driving before minimum penetration is reached and it has been demonstrated to the satisfaction of the Engineer that additional attempts at driving would result in damage to the rails, the Engineer may order additional work to be performed, such as jetting or drilling, in order that minimum penetration may be obtained or he may order the minimum penetration to be reduced as required by the conditions encountered.

Wire fabric shall be securely fastened to the rails, placed in the trenches and laid on the slopes. The rock backfill shall then be carefully placed so as not to displace the wire fabric or rails. The wire fabric shall entirely enclose the rock backfill.

The completed rock fill shall be backfilled as necessary and the waste material disposed of as directed.

**913-4 Method of Measurement:**

Riprap, except gabions and sacked concrete will be measured by the cubic yard of protection constructed by computing the surface area measured parallel to the protection surface and the total thickness of the riprap measured normal to the protection surface.

Riprap (gabions) will be measured by the cubic yard by computing the volume of the rock filled wire baskets used.

Riprap (sacked concrete) will be measured by the cubic yard of concrete placed in the completed work. The measurement will be based on mixer volumes.

Rail bank protection (Type 1 - Type 6) will be measured by the linear foot. Measurement will be made from top of rail to top of rail (longest rail where rails of two or more lengths are used) and the distance measured will be from end rail to end rail. Where two parallel rows of vertical rails are used, the measurement for payment will be the average of the distance along the two rows.

| Design Size | *Span Inches | *Rise Inches | Minimum Cover Inches | Minimum Gage Required | Coated Thickness Inches | Equivalent Diameter Full Circle Pipe Inches |
|---|---|---|---|---|---|---|
| 1 | 17 | 13 | 12 | 16 | 0.064 | 15 |
| 2 | 21 | 15 | 12 | 16 | 0.064 | 18 |
| 3 | 28 | 20 | 12 | 16 | 0.064 | 24 |
| 4 | 35 | 24 | 12 | 16 | 0.064 | 30 |
| 5 | 42 | 29 | 12 | 14 | 0.079 | 36 |
| 6 | 49 | 33 | 12 | 14 | 0.079 | 42 |
| 7 | 57 | 38 | 12 | 12 | 0.109 | 48 |
| 8 | 64 | 43 | 12 | 12 | 0.109 | 54 |
| 9 | 71 | 47 | 12 | 10 | 0.138 | 60 |

*The fill heights for all sizes of pipe arch are limited to a maximum of 7 feet.

A tolerance of plus or minus one inch or 2 percent of equivalent circular diameter, whichever is greater, will be permissible in span and rise, with all dimensions measured from the inside crests of the corrugations.

Table A
STEEL PIPE ARCH
2 2/3-INCH BY 1/2-INCH CORRUGATIONS

| Design Size | *Span Inches | *Rise Inches | Minimum Cover Inches | Minimum Gage Required | Coated Thickness Inches | Equivalent Diameter Full Circle Pipe Inches |
|---|---|---|---|---|---|---|
| 7 | 53 | 41 | 12 | **14 | 0.064 | 48 |
| 8 | 60 | 46 | 12 | **14 | 0.064 | 54 |
| 9 | 66 | 51 | 12 | **14 | 0.064 | 60 |
| 10 | 73 | 55 | 12 | **14 | 0.064 | 66 |
| 11 | 81 | 59 | 12 | 14 | 0.079 | 72 |
| 12 | 87 | 63 | 12 | 14 | 0.079 | 78 |
| 13 | 95 | 67 | 12 | 12 | 0.109 | 84 |
| 14 | 103 | 71 | 18 | 12 | 0.109 | 90 |
| 15 | 112 | 75 | 18 | 12 | 0.109 | 96 |
| 16 | 117 | 79 | 18 | 12 | 0.109 | 102 |
| 17 | 128 | 83 | 24 | 10 | 0.138 | 108 |
| 18 | 137 | 87 | 24 | 10 | 0.138 | 114 |
| 19 | 142 | 91 | 24 | 10 | 0.138 | 120 |

*The fill heights for all sizes of pipe arch are limited to a maximum of 10 feet.

A tolerance of plus or minus 1 inch or 2 percent of equivalent circular diameter, whichever is greater, will be permissible in span and rise, with all dimensions measured from the inside crests of the corrugations.

**16 gage may be used if the required pipe arch shape can be produced to the satisfaction of the Engineer.

Table B
STEEL PIPE ARCH
3-INCH BY 1-INCH CORRUGATIONS

| Design Size | *Span Inches | *Rise Inches | Minimum Cover Inches | Minimum Gage Required | Coated Thickness Inches | Equivalent Diameter Full Circle Pipe Inches |
|---|---|---|---|---|---|---|
| 11 | 81 | 59 | 12 | 12 | 0.109 | 72 |
| 12 | 87 | 63 | 12 | 12 | 0.109 | 78 |
| 13 | 95 | 67 | 12 | 12 | 0.109 | 84 |
| 14 | 103 | 71 | 18 | 12 | 0.109 | 90 |
| 15 | 112 | 75 | 18 | 12 | 0.109 | 96 |
| 16 | 117 | 79 | 18 | 12 | 0.109 | 102 |
| 17 | 128 | 83 | 24 | 10 | 0.138 | 108 |
| 18 | 137 | 87 | 24 | 10 | 0.138 | 114 |
| 19 | 142 | 91 | 24 | 10 | 0.138 | 120 |

*The fill heights for all sizes of pipe arch are limited to a maximum of 10 feet.

A tolerance of plus or minus one inch or 2 percent of equivalent circular diameter, whichever is greater, will be permissible in span and rise, with all dimensions measured from the inside crests of the corrugations.

Table C
STEEL PIPE ARCH
5-INCH BY 1-INCH CORRUGATIONS

| Design Size | *Span Inches | *Rise Inches | Minimum Cover Inches | Minimum Gage Required | Coated Thickness Inches | Equivalent Diameter Full Circle Pipe Inches |
|---|---|---|---|---|---|---|
| 1 | 17 | 13 | 12 | 16 | 0.060 | 15 |
| 2 | 21 | 15 | 12 | 16 | 0.060 | 18 |
| 3 | 28 | 20 | 12 | 14 | 0.075 | 24 |
| 4 | 35 | 24 | 12 | 14 | 0.075 | 30 |
| 5 | 42 | 29 | 18 | 12 | 0.105 | 36 |
| 6 | 49 | 33 | 18 | 12 | 0.105 | 42 |
| 7 | 57 | 38 | 18 | 10 | 0.135 | 48 |
| 8 | 64 | 43 | 18 | 10 | 0.135 | 54 |
| 9 | 71 | 47 | 18 | 8 | 0.164 | 60 |

*All dimensions are measured from the inside crests of the corrugations. A tolerance of plus or minus one inch or 2 percent of the equivalent circular diameter, whichever is greater will be permissible in span and rise.

*The fill heights for all sizes of pipe arch are limited to a maximum of 7 feet.

Table D
ALUMINUM PIPE ARCH
2 2/3-INCH BY 1/2-INCH CORRUGATIONS

# Q.12 EPOXY BONDING COMPOUND

EPOXY BONDING COMPOUND

4.1G.1. DESCRIPTION.

EPOXY BONDING COMPOUND SHALL INCLUDE THE FURNISHING AND PLACING OF AN EPOXY BONDING COMPOUND ON EXISTING CONCRETE AND/OR EXISTING REINFORCEMENT STEEL BEFORE PLACING NEW CONCRETE OR MORTAR AT THE LOCATIONS PRESCRIBED ON THE PLANS OR DETERMINED BY THE ENGINEER.

4.1G.2. MATERIALS.

EPOXY BONDING COMPOUND SHALL BE A TWO COMPONENT EPOXY-RESIN BONDING SYSTEM FOR APPLICATION TO PORTLAND CEMENT CONCRETE AND SHALL CONFORM TO THE REQUIREMENTS OF ASTM C-881. THE SYSTEM TYPE, GRADE AND CLASS SHALL DEPEND ON THE CONDITIONS OF THE INTENDED USE. THE COLOR SHALL BE CLEAR OR GRAY TO MATCH THE COLOR OF THE ADJACENT CONCRETE.

THE EPOXY BONDING COMPOUND SHALL BE FURNISHED AS TWO SEPARATE COMPONENTS. EACH CONTAINER SHALL BE CLEARLY LABELED AND THE FOLLOWING INFORMATION SHOWN:

(1) SPECIFICATION NUMBER AND TYPE

(2) COMPONENT DESIGNATION (A OR B)

(3) MANUFACTURER'S BATCH NUMBER (A BATCH SHALL CONSIST OF A SINGLE CHARGE OF ALL COMPONENTS IN A MIXING CHAMBER)

(4) EXPIRATION DATE (SHELF LIFE FOR SEPARATE COMPONENTS IN ORIGINAL CONTAINERS)

(5) MIXING RATIO AND DIRECTIONS (BY VOLUME OR WEIGHT AS DESIGNATED BY THE MANUFACTURER)

(6) POTENTIAL HAZARDS AND PRECAUTIONS SHALL BE DISPLAYED IN ACCORDANCE WITH THE FEDERAL HAZARDOUS PRODUCTS LABELING ACT.

4.1G.3. METHODS OF CONSTRUCTION.

ALL SURFACES TO BE BONDED SHALL BE THOROUGHLY BLAST CLEANED WITH NO. 40 BOILER SLAG GRIT OR NO. 2 SANDBLAST SAND OR ABRADED BY APPROVED MECHANICAL MEANS.

AFTER BLAST CLEANING AND JUST PRIOR TO THE APPLICATION OF EPOXY, THE SURFACE SHALL BE VACUUM CLEANED.

THE TWO COMPONENTS OF THE EPOXY BONDING COMPOUND ARE FURNISHED IN SEPARATE CONTAINERS. IT IS ESSENTIAL THAT THE ENTIRE CONTENTS OF BOTH CONTAINERS BE THOROUGHLY BLENDED TOGETHER.

THE EPOXY BONDING COMPOUND SHALL BE MIXED AND APPLIED IN STRICT ACCORDANCE WITH THE MANUFACTURER'S RECOMMENDATIONS.

4.1G.4. QUANTITY AND PAYMENT.

NO SPECIFIC PAYMENT WILL BE MADE FOR EPOXY BONDING COMPOUND AND ALL COSTS THEREOF SHALL BE INCLUDED IN THE UNIT PRICE BID FOR THE CLASS B CONCRETE OF PATCHING CONCRETE DECK OR OTHER APPROPRIATE ITEM.

# Q.13 CRACK REPAIRS

PRESSURE INJECTION

4.1H.1. DESCRIPTION.

PRESSURE INJECTION SHALL INCLUDE THE FURNISHING AND PLACING OF A SUITABLE EPOXY TO SEAL THE CRACKS AT THE APPROXIMATE CRACK LOCATION SHOWN ON THE PLANS OR AS DIRECTED BY THE ENGINEER AT THE TIME OF CONSTRUCTION.

4.1H.2. MATERIALS.

THE EPOXY RESIN SHALL MEET THE REQUIREMENTS OF ASTM C-881 AND THE TYPE AND GRADE SHALL BE SUBMITTED FOR THE ENGINEER'S APPROVAL AFTER THE CONTRACTOR'S CAREFUL ANALYSIS OF THE AREA TO BE INJECTED.

THE SYSTEM SHALL BE 100% SOLIDS WITH COMPONENT 'A' BEING COMPLETELY REACTIVE WITH COMPONENT 'B'. THE SYSTEM SHALL BE SUITABLE FOR USE ON DAMP SURFACES BUT SHALL NOT BE USED WHERE THE TEMPERATURE IS 50 DEGREES OR BELOW.

ALL MATERIALS FURNISHED SHALL BE SHIPPED IN STRONG SUBSTANTIAL CONTAINERS. THE CONTAINERS SHALL BE CLEARLY IDENTIFIED AS "PART A - CONTAINS EPOXY RESIN" AND "PART B - CONTAINS CURING AGENT". THEY SHALL ALSO BE PLAINLY MARKED WITH THE FOLLOWING INFORMATION:

1. NAME OF PRODUCT
2. MIXING PROPORTIONS AND INSTRUCTIONS
3. NAME AND ADDRESS OF MANUFACTURER
4. LOT NUMBER AND BATCH NUMBER
5. DATE OF MANUFACTURE
6. QUANTITY

THE EXPIRATION DATE OF ACCEPTANCE OF THIS MATERIAL SHALL BE ONE YEAR AFTER THE DATE OF MANUFACTURE. ANY UNAUTHORIZED TAMPERING OR BREAKING OF THE SEALS ON THE CONTAINERS BETWEEN THE TIME OF SAMPLING AND DELIVERY TO THE JOB SITE WILL BE CAUSE FOR REJECTION OF THE MATERIAL.

COMPONENT 'A' SHALL BE THE CONDENSATION PRODUCT OF BISPHENOL 'A' AND EPICHLOROHYDRIN. IT SHALL CONTAIN NO NON-REACTIVE DILUTENTS AND SHALL BE DESIGNATED AS A LOW VISCOSITY RESIN SUITABLE FOR INJECTION.

4.1H.3. METHODS OF CONSTRUCTION.

THE ACTUAL LOCATION AND TOTAL QUANTITY OF THE PRESSURE INJECTION OF CRACKS MAY BE ADJUSTED BY FIELD INSPECTION BY THE CONTRACTOR AT THE TIME OF CONSTRUCTION SUBJECT TO THE APPROVAL OF THE ENGINEER.

THE CONTRACTOR SHALL BE REQUIRED TO HAVE PRESENT A MANUFACTURER'S TECHNICAL REPRESENTATIVE FOR THE DURATION OF THE INJECTION PROCESS. ALSO, THE DETAILED METHODS OF REPAIRS AND THE INJECTION PROCEDURE SHOULD BE SUBMITTED FOR THE ENGINEER'S APPROVAL.

THE EPOXY INJECTION EQUIPMENT SHALL BE A POSITIVE DISPLACEMENT PUMP SYSTEM. THE SYSTEM SHALL HAVE A SUITABLE MIXING CHAMBER WHERE THE EPOXY COMPONENTS ARE ACCURATELY METERED AND THOROUGHLY MIXED IMMEDIATELY PRIOR TO INJECTION. A CLEAR, LEGIBLE, AND ACCURATE PRESSURE GAUGE SHALL BE LOCATED IN THE SUPPLY LINE ADJACENT TO THE MIXING CHAMBER.

THE EQUIPMENT SHALL ALSO BE CAPABLE OF PROVIDING A CONTINUOUS AND UNINTERRUPTED PRESSURE HEAD TO CONTINUALLY FORCE THE INJECTION EPOXY INTO THE CRACKS. EPOXY FLOW SHALL BE CAPABLE OF BEING FULLY CONTROLLED BY OPERATOR CONTROLS AT THE MIXING CHAMBER.

ALL WORKING PERSONNEL SHALL BE FAMILIAR WITH THE EQUIPMENT, MATERIALS AND PROCEDURES TO BE USED DURING THE OPERATION. EXTRA (BACKUP) EQUIPMENT TO ASSURE THE CONTINUOUS INJECTION OF EPOXY, IN THE EVENT OF PRIMARY EQUIPMENT FAILURE, SHALL BE REQUIRED.

ALL MATERIALS AND EQUIPMENT, INCLUDING BACKUP EQUIPMENT, SHALL BE AT THE WORK SITE BEFORE INJECTION IS BEGUN. ALL EQUIPMENT SHALL BE IN PROPER CALIBRATION AND IN GOOD WORKING ORDER AS DETERMINED BY, AND TO THE FULL SATISFACTION OF, THE ENGINEER. EPOXY SHALL BE INJECTED ONLY BY THE USE OF AUTOMATIC MECHANICAL PUMPING, METERING, AND MIXING EQUIPMENT AS DESCRIBED ABOVE. PRESSURE POT SYSTEMS AND HAND HELD CAULKING GUNS, OR GREASE GUNS, SHALL NOT BE ALLOWED.

THE TWO COMPONENTS SHALL BE MIXED IN ACCORDANCE WITH THE MANUFACTURER'S RECOMMENDATIONS. THE RATIO OF THE COMPONENTS SHALL BE MAINTAINED WITHIN A TOLERANCE OF FIVE PERCENT.

BEFORE APPLYING THE SURFACE SEAL, ALL LOOSE MATTER SHALL BE REMOVED. SURFACES SHALL BE CLEANED AND PREPARED AS PER THE SEAL MANUFACTURER'S SPECIFICATIONS.

ANY SOLVENT USED FOR CLEANING SHALL BE NON-CHLORINATED. ACCEPTABLE SOLVENTS ARE MINERAL SPIRITS, METHYL ETHYL KETONE, ACETONE, LOW BOILING NAPTHA, XYLENE, OR ANY OTHER NON-CHLORINATED SOLVENT.

PRIOR TO INJECTION OF THE EPOXY IN THE CRACK, A SURFACE SEAL MATERIAL SHALL BE APPLIED TO THE FACE OF THE CRACK. SURFACE SEAL MATERIAL SHOULD BE CAPABLE OF USE ON VERTICAL AS WELL AS HORIZONTAL SURFACES.

OPENINGS IN THE SURFACE SEAL SHALL BE ESTABLISHED ALONG THE CRACK. THE DISTANCE BETWEEN ENTRY PORTS SHALL NOT BE LESS THAN THE THICKNESS OF THE CONCRETE MEMBER BEING REPAIRED.

PORT SPACING AND INSTALLATION

SITES FOR PORTS SHALL BE DRILLED TO A WIDTH AND DEPTH SUFFICIENT TO ASSURE A SNUG FIT OF THE PORT. THESE HOLES SHALL BE CLEANED TO REMOVE ANY DUST OR DEBRIS LEFT BY THE DRILLING OPERATION. CARE SHALL BE EXERCISED TO ASSURE THAT OIL OR OTHER CONTAMINANTS SHALL NOT BE INTRODUCED INTO THE AIR FEED HOSES, OR DEPOSITED ON ANY AIR BLOWN SURFACES.

INJECTION PROCEDURE

NO EPOXY INJECTION, OR SURFACE SEALING, SHALL BE DONE WHEN THE CONCRETE TEMPERATURE OR AMBIENT TEMPERATURE IS, OR IS EXPECTED TO FALL, BELOW 50 DEGREES F DURING THE 24 HOURS FOLLOWING THE TIME OF EPOXY INJECTION. APPLICATION OR INJECTION OF EPOXY AT TEMPERATURES LOWER THAN THAT ALLOWED BY THIS SPECIFICATION, SHALL BE DONE WITH THE WRITTEN APPROVAL OF THE DIRECTOR, MATERIALS BUREAU.

THE INJECTION OF THE ADHESIVE INTO EACH CRACK SHALL BEGIN AT THE ENTRY PORT AT THE LOWEST ELEVATION. INJECTION SHALL CONTINUE AT THE FIRST PORT UNTIL THE EPOXY ADHESIVE BEGINS TO FLOW OUT OF THE PORT AT THE NEXT HIGHEST ELEVATION. THE FIRST PORT SHALL BE PLUGGED AND INJECTION STARTED AT THE SECOND PORT UNTIL THE ADHESIVE FLOWS FROM THE NEXT PORT. THIS SEQUENCE SHALL BE FOLLOWED UNTIL THE ENTIRE CRACK IS REPAIRED.

BEFORE INJECTION OF ANY CRACK, THE AUTOMATIC MIXING AND METERING PUMP SHALL BE ACTIVATED AND A SMALL AMOUNT (ABOUT ONE PINT) OF INJECTION EPOXY SHALL BE MIXED AND WASTED INTO A DISPOSABLE CONTAINER. THE ENGINEER SHALL OBSERVE THIS TRIAL OPERATION, AND BE FULLY SATISFIED THAT THE EQUIPMENT IS WORKING PROPERLY. IF THE EQUIPMENT IS NOT WORKING PROPERLY IT SHALL BE IMMEDIATELY REPAIRED TO FULL WORKING CONDITION OR REPLACED WITH THE BACKUP EQUIPMENT. IF THE BACKUP EQUIPMENT IS USED, ADDITIONAL AND FULLY OPERABLE EQUIPMENT SHALL BE PROVIDED TO REPLACE IT IN CASE OF MALFUNCTIONS.

THE FEED LINE FROM THE MIXING EQUIPMENT SHALL BE SECURELY HELD OR PROPERLY ATTACHED TO THE END MOST PORT (BOTTOM-MOST FOR VERTICAL CRACKS). THE OPERATOR SHALL THEN INITIATE EPOXY INJECTION AND FLOW SHALL BE ALLOWED AT TEN (10) TO FORTY (40) PSI. THE INJECTION PROCEDURE SHALL BE MONITORED TO ASSURE THE EPOXY FLOW DOES NOT CEASE BEFORE THE INJECTED EPOXY EXUDES FROM THE ADJACENT PORT. WHEN EPOXY FLOWS FROM THE ADJACENT PORT, INJECTION SHALL BE STOPPED, THE FEED LINE REMOVED FROM THE PORT, AND THE PORT SEALED. THE FEED LINE SHALL THEN BE ATTACHED TO THE NEXT PORT AND THE PROCEDURE REPEATED UNTIL THE LAST PORT IS SEALED. IF THE EPOXY FLOW STOPS BEFORE EPOXY APPEARS AT THE ADJACENT PORT, THE FEED LINE SHALL BE MOVED TO THE ADJACENT PORT AND THE PORT JUST USED SHALL BE SEALED.

WHEN THE EPOXY SUPPLY IN THE MIXING EQUIPMENT IS ABOUT TO BE EXHAUSTED, IT SHALL BE REPLENISHED. EACH EPOXY COMPONENT SHALL BE THOROUGHLY STIRRED BEFORE ADDING IT TO ITS RESPECTIVE STORAGE TANK IN THE MIXING EQUIPMENT. NO DISCONTINUITY OF EPOXY FLOW THROUGH THE FEED LINES OF EACH OR BOTH COMPONENTS SHALL BE ALLOWED. IN THIS MANNER, A CONTINUOUS INJECTION OPERATION SHALL RESULT.

IN THE EVENT OF LEAKAGE FROM THE CRACK, INJECTION WILL BE STOPPED UNTIL THE LEAK IS SEALED. ANY WORK STOPPAGE OF FIFTEEN (15) MINUTES, OR LONGER, WILL NECESSITATE CLEANING OF THE MIXING CHAMBER AND ANY EQUIPMENT IN CONTACT WITH MIXED EPOXY.

AFTER THE INJECTION PROCESS HAS BEEN COMPLETED AND THE EPOXY ALLOWED TO FULLY CURE, THE INJECTION PORTS, SURFACE SEAL AND ANY SPILLAGE SHALL BE REMOVED FROM ALL SURFACES EASILY VISIBLE TO THE PUBLIC. PORTS MAY BE CUT, OR KNOCKED OFF, WHILE THE SURFACE SEAL AND ANY SPILLAGE SHALL BE GROUND OFF FLUSH WITH THE ORIGINAL SURFACE, USING A HAND GRINDER. ANY DAMAGE TO THE CONCRETE DURING THE CLEANUP PROCEDURE SHALL BE REPAIRED IN A MANNER SATISFACTORY TO THE ENGINEER, AT NO ADDITIONAL COST TO THE STATE.

PAYMENT FOR PRESSURE INJECTION WILL BE MADE FOR THE QUANTITY AS ABOVE DETERMINED, MEASURED IN LINEAR FEET, AT THE UNIT PRICE BID IN THE PROPOSAL FOR THE ITEM PRESSURE INJECTION, WHICH PRICE SHALL BE FULL COMPENSATION FOR THE SATISFACTORY COMPLETE REPAIR OF THE CRACKS, INCLUDING REMOVAL AND DISPOSAL OF UNSOUND CONCRETE, ALL SAW CUTTING, CHIPPING, CLEANING, SAND-BLASTING, DRILLING, FURNISHING AND INSTALLING THE PORTS, FURNISHING EQUIPMENT, FURNISHING DETAILED METHODS OF EPOXY PRESSURE INJECTIONS, TESTS AND RETESTS REQUIRED AS A RESULT OF SUBSTANDARD WORK OR MATERIAL, FURNISHING, INSTALLING, FINISHING, AND CURING EPOXY MATERIALS AND ALL ELSE NECESSARY THEREFOR AND INCIDENTAL THERETO.

THE CONTRACTOR SHALL SUBMIT TO THE ENGINEER A CERTIFICATION FROM THE MANUFACTURER THAT THE EPOXY FURNISHED IS THE MANUFACTURER'S RECOMMENDED EPOXY FOR PRESSURE INJECTION OF CRACKS AND THAT THE METHOD OF APPLICATION PROPOSED BY THE CONTRACTOR IS ACCEPTABLE.

THE CORED INJECTED EPOXY SHALL HAVE PENETRATED A MINIMUM OF 90 PERCENT OF THE VISIBLE CRACK. THE STATE MAY TAKE CORES OF THE REPAIRED CONCRETE TO DETERMINE THE LENGTH OF PENETRATION. IF THE PENETRATION IS LESS THAN 90 PERCENT OF THE VISIBLE CRACK, THE CRACK FROM WHICH THE CORE WAS TAKEN WILL BE DEEMED NOT TO HAVE BEEN REPAIRED.

4.1H.4. QUANTITY AND PAYMENT.

THE QUANTITY OF PRESSURE INJECTION FOR WHICH PAYMENT WILL BE MADE WILL BE THE TOTAL LENGTH OF CRACKS, MEASURED ALONG THE SURFACE OF THE CONCRETE, ACTUALLY REPAIRED IN ACCORDANCE WITH THESE SPECIFICATIONS AT THE LOCATIONS SHOWN ON THE PLANS OR AS DIRECTED BY THE ENGINEER.

## Q.14 REPAIRING SPALLED CONCRETE

REPAIRING SPALLED CONCRETE

4.1P.1. DESCRIPTION

REPAIRING SPALLED CONCRETE SHALL INCLUDE ALL WORK REQUIRED TO REPAIR DETERIORATED CONCRETE SURFACES OF ABUTMENTS, WINGWALLS AND WHERE INDICATED, ALL IN ACCORDANCE WITH THE PLANS AND SPECIFICATIONS OR AS DIRECTED BY THE ENGINEER. THIS WORK CONSISTS OF THE REMOVAL AND DISPOSAL OF ALL LOOSE AND DISINTEGRATED CONCRETE, SAW-CUTTING, THE PREPARATION OF THE SURFACE, CLEANING OR REPLACEMENT OF EXISTING REINFORCEMENT STEEL, FIELD EPOXY COATING EXPOSED EXISTING REINFORCEMENT, THE APPLICATION OF AN EPOXY BONDING COMPOUND AND PLACING OF CLASS B CONCRETE

4.1P.2. MATERIALS

CONCRETE

CONCRETE FOR REPAIR SHALL BE CLASS B, AIR-ENTRAINED, CONFORMING WITH THE APPLICABLE REQUIREMENTS OF DIVISION 4, SECTION 1 OF THE SPECIFICATIONS EXCEPT THAT THE SLUMP SHALL NOT EXCEED 2" + 1/2". THE MAXIMUM SIZE OF COURSE AGGREGATE SHALL BE 3/8 INCH.

THE CONTRACTOR SHALL BE RESPONSIBLE FOR FURNISHING THE DESIGN PROPORTIONS OF CEMENT, FINE AGGREGATE, COARSE AGGREGATE, WATER, AIR-ENTRAINING ADMIXTURE AND WATER-REDUCING ADMIXTURE WHICH WILL PRODUCE A WORKABLE CONCRETE MIX MEETING THE ABOVE REQUIREMENTS FOR CLASS B CONCRETE.

AT THE CONTRACTOR'S OPTION, SUPERPLASTICIZER MAY BE ADDED TO THE CONCRETE MIX AT THE JOB SITE. THE SLUMP OF THE CONCRETE MIX, AS DELIVERED, SHALL CONFORM TO THE REQUIREMENTS ABOVE AND SHALL NOT EXCEED 8 INCHES AFTER THE ADDITION OF THE SUPERPLASTICIZER.

4.1P.3. METHODS OF CONSTRUCTION

THE LATERAL LIMITS OF EACH AREA TO BE REMOVED AND REPLACED WILL BE DELINEATED BY THE ENGINEER AND SUITABLY MARKED. THE OUTLINE OF EACH SUCH AREA SHALL FIRST BE SAW-CUT WHERE REQUIRED TO A DEPTH OF ONE (1) INCH WITH A POWER SAW CAPABLE OF MAKING STRAIGHT CUTS.

ALL DETERIORATED OR UNSOUND CONCRETE SHALL BE REMOVED TO SOUND CONCRETE OR AS DIRECTED BY THE ENGINEER. MINIMUM DEPTH OF REMOVAL SHALL BE TWO (2) INCHES. CONCRETE MAY BE REMOVED BY MEANS OF APPROVED HAND HELD PNEUMATIC CHIPPING HAMMERS, NOT EXCEEDING 30 POUNDS. PNEUMATIC TOOLS SHOULD NOT BE PLACED IN DIRECT CONTACT WITH REINFORCEMENT STEEL. EXTREME CARE SHALL BE TAKEN WHEN REINFORCEMENT STEEL OR ANCHOR BOLTS ARE UNCOVERED SO AS NOT TO DAMAGE THE STEEL OR THE ANCHOR BOLT OR THEIR BOND IN THE SURROUNDING SOUND CONCRETE. REINFORCEMENT STEEL OR OTHER EMBEDDED ITEMS DAMAGED DURING CONCRETE REMOVAL SHALL BE REPAIRED BY THE CONTRACTOR, AS DIRECTED BY THE ENGINEER, AT NO COST TO THE STATE. IF REINFORCEMENT BARS ARE EXPOSED, THE REMOVAL SHALL CONTINUE UNTIL AT LEAST 3/4 OF THE BAR'S CIRCUMFERENCE IS EXPOSED. IF UNSOUND CONCRETE IS ENCOUNTERED AT OR BELOW THE MID-DEPTH OF REINFORCEMENT BARS, REMOVAL SHALL EXTEND TO AT LEAST 3/4 INCH BEYOND THE BARS. THE LIMITS OF THE CAVITY FOR DEEP, NARROW REPAIRS (DEPTH EQUALS OR EXCEEDS TWICE THE WIDTH) SHALL BE UNDERCUT TO LOCK THE REPAIR IN PLACE.

AFTER REMOVAL OF UNSOUND CONCRETE THE SURFACES OF THE REMAINING CONCRETE SHALL BE CLEANED OF ALL LOOSE CONCRETE, DUST, AND OTHER FOREIGN MATERIAL. EMBEDDED ITEMS SHALL BE CLEANED OF ALL LOOSE ADHERING CONCRETE, RUST, AND SCALE. ADHERING CONCRETE SHALL BE REMOVED FROM EXPOSED REINFORCEMENT STEEL AND THE SURFACES OF THE EXPOSED BARS SHALL BE CLEANED AND FIELD EPOXY COATED AS PROVIDED FOR IN DIVISION 4, SECTION 1B OF THE SUPPLEMENTARY SPECIFICATI[illegible]

DETERIORATED REINFORCEMENT SHALL BE REPLACED WITH NEW REINFORCEMENT AS DIRECTED BY THE ENGINEER.

ADEQUATE PROVISIONS MUST BE MADE TO PREVENT PIECES OF BROKEN CONCRETE FROM FALLING BELOW.

DETAILS OF PLATFORMS, CATCHES OR OTHER METHODS OF COLLECTING MATERIALS RESULTING FROM THE PREPARATION OF SPALL AREAS FOR REPAIR SHALL BE OF THE CONTRACTOR'S DESIGN AND SHALL BE SUBMITTED TO THE ENGINEER FOR REVIEW BEFORE COMMENCING CONCRETE REMOVAL. MATERIALS COLLECTED SHALL NOT BE ALLOWED TO ACCUMULATE BUT SHALL BE PROMPTLY REMOVED AND DISPOSED OF AWAY FROM THE SITE.

AFTER COMPLETION OF THE REMOVAL AND CLEANING OPERATIONS AND IMMEDIATELY PRIOR TO THE PLACEMENT OF THE CONCRETE REPAIR MATERIAL, THE SURFACES SHALL RECEIVE A COATING OF THE BONDING COMPOUND APPLIED IN ACCORDANCE WITH THE MANUFACTURER'S RECOMMENDATIONS AND AS DIRECTED BY THE ENGINEER.

THE POT-LIFE OF THE EPOXY BONDING COMPOUND, MIXING PERIOD, MAXIMUM TIME LAPSE BETWEEN MIXING COMPOUND AND PLACING OF THE NEW CONCRETE, ARE ALL DEPENDENT ON THE TEMPERATURE, HUMIDITY, AND WIND CONDITIONS. THE CONTRACTOR SHALL ACQUIANT HIMSELF WITH SUCH INFORMATION AS RECOMMENDED BY THE MANUFACTURER AND SHALL SCHEDULE HIS OPERATIONS ACCORDINGLY.

CONCRETE SHALL BE PLACED IMMEDIATELY WHILE THE COMPOUND IS STILL TACKY. IF THE BONDING COMPOUND DRIES BEFORE THE NEW CONCRETE IS PLACED, ANOTHER COATING OF EPOXY BONDING COMPOUND SHALL BE APPLIED, AS HEREIN DESCRIBED, AT NO ADDITIONAL COST. AFTER THE NEW CONCRETE IS IN PLACE, NORMAL FINISHING OPERATIONS SHALL BE COMPLETED.

ALL DETAILS OF DESIGN, SUCH AS V-GROOVES, CHAMFERS, JOINTS, ETC., IN THE EXISTING ABUTMENTS AND PIER SHALL BE DUPLICATED IN THE REPAIR WORK UNDER THIS SECTION.

PLACING OF CONCRETE, FORMING, REMOVAL OF FORMS, FINISHING AND CURING OF THE NEW CONCRETE SHALL CONFORM WITH THE APPLICABLE REQUIREMENTS OF ARTICLE 4.1.3 OF THE STANDARD AND SUPPLEMENTARY SPECIFICATIONS.

WHATEVER MEANS OF PLACING THE CONCRETE IS EMPLOYED, IT SHALL BE SUCH THAT THE MATERIAL WILL COMPLETELY FILL THE SPACE TO BE REPLACED, BE THOROUGHLY COMPACTED, AND FREE OF AIR POCKETS.

WHEN REPAIRS ARE TO BE MADE OF FAST-SETTING MORTAR, THE MANUFACTURER'S RECOMMENDATIONS SHALL BE STRICTLY FOLLOWED IN ALL PROCEDURES RELATING TO PREPARATION AND PLACEMENT OF THE FAST-SETTING MORTAR, BASED ON PREVAILING CLIMATIC AND JOB CONDITIONS. THE CONTRACTOR SHALL ARRANGE TO HAVE A TECHNICAL REPRESENTATIVE OF THE MANUFACTURER AT THE SITE DURING THE INITIAL PHASE OF THE SPALL REPAIR WORK TO ACQUAINT THE CONTRACTOR'S FORCES WITH THE PROPER METHODS OF PREPARATION AND APPLICATION.

4.1P.4. QUANTITY AND PAYMENT

THE QUANTITY OF REPAIRING SPALLED CONCRETE FOR WHICH PAYMENT WILL BE MADE, WILL BE THE TOTAL AREA OF REPAIR ACTUALLY PERFORMED AS PRESCRIBED REGARDLESS OF DEPTH.

INDIVIDUAL AREA MEASUREMENTS WILL BE COMPUTED TO THE NEAREST HALF SQUARE FOOT. INDIVIDUAL AREAS MEASURING LESS THAN ONE SQUARE FOOT WILL BE CONSIDERED AS ONE (1.0) SQUARE FOOT FOR THE PURPOSE OF PAYMENT. THE ENGINEER'S MEASUREMENTS WILL BE FINAL AND CONCLUSIVE.

PAYMENT FOR REPAIRING SPALLED CONCRETE WILL BE MADE FOR THE QUANTITY AS ABOVE DETERMINED, MEASURED IN SQUARE FEET, AT THE PRICE PER SQUARE FOOT BID FOR THE ITEM REPAIRING SPALLED CONCRETE IN THE PROPOSAL, WHICH PRICES SHALL INCLUDE THE COST OF SAWCUTTING, CATCHING, REMOVING AND DISPOSING OF CONCRETE; CLEANING REMAINING CONCRETE SURFACES; PROTECTING AND REPAIRING UNCOVERED REINFORCEMENT STEEL; ADDITIONAL REINFORCEMENT AS REQUIRED; FURNISHING, MIXING AND APPLYING EPOXY BONDING COMPOUND; FURNISHING, FORMING, PLACING, FINISHING AND CURING CONCRETE; ALL MATERIALS, LABOR, EQUIPMENT AND ALL ELSE NECESSARY THEREFOR AND INCIDENTAL THERETO.

# Q.15 CORING AND GROUTING

4.1M.1. DESCRIPTION.

THE WORK SHALL CONSIST OF CORING HOLES INTO CONCRETE, GROUTING THE HOLES AND PROPERLY POSITIONING REINFORCING BARS, OR ANCHOR BOLTS, INTO THE GROUTED HOLES WHERE SHOWN ON THE PLANS.

FOR PURPOSES OF THIS SPECIFICATION, THE TERMS REINFORCING BARS AND ANCHOR BOLTS SHALL BE CONSIDERED IDENTICAL.

4.1M.2. MATERIALS.

THE GROUT MATERIAL SHALL BE A NON-METALLIC, NON-SHRINK GROUT CONFORMING TO THE REQUIREMENTS OF ARTICLE 4.1.2 OF THE STANDARD SPECIFICATIONS.

4.1M.3. METHODS OF CONSTRUCTION.

EQUIPMENT USED FOR CORING HOLES SHALL BE APPROVED BY THE ENGINEER PRIOR TO USE.

CORING HOLES INTO A STRUCTURAL SLAB OR INTO ANY STRUCTURAL CONCRETE ELEMENT SHALL BE DONE BY MEANS OF A CORE DRILL.

CORING HOLES INTO ANY OTHER STRUCTURAL ELEMENT SHALL BE DONE BY METHODS SATISFACTORY TO THE ENGINEER, UNLESS A SPECIFIC METHOD IS NOTED ELSEWHERE IN THE CONTRACT DOCUMENTS.

CORING WITH A LUBRICANT SHALL NOT BE PERMITTED. WATER WILL NOT BE CONSIDERED A LUBRICANT. CORING METHODS SHALL NOT CAUSE SPALLING OR OTHER DAMAGE TO THE CONCRETE. CONCRETE SPALLED OR OTHERWISE DAMAGED BY THE CONTRACTOR'S OPERATIONS SHALL BE REPAIRED IN A MANNER APPROVED BY, AND TO THE SATISFACTION OF THE ENGINEER. SUCH REPAIR SHALL BE DONE AT THE EXPENSE OF THE CONTRACTOR. HOLES SHALL BE SURFACE DRY AND SHALL HAVE HAD ALL FOREIGN AND LOOSE MATERIAL REMOVED IMMEDIATELY PRIOR TO GROUT PLACEMENT. CORING THROUGH STEEL PARTS OR REINFORCING STEEL SHALL BE PERMITTED ONLY WHERE SHOWN ON THE PLANS AND AS APPROVED BY THE ENGINEER.

GROUT SHALL BE MIXED AND PLACED IN STRICT ACCORDANCE WITH THE MANUFACTURER'S INSTRUCTIONS UNLESS OTHERWISE MODIFIED HEREIN.

NO GROUT PLACEMENT SHALL BE PERMITTED WHEN THE AMBIENT AIR TEMPERATURE IS 50 DEGREES F. OR BELOW DURING THE WORKING DAY. GROUT SHALL BE INSERTED TO A DEPTH SUFFICIENT TO INSURE COMPLETE FILLING OF THE HOLE AFTER INSERTION OF THE REINFORCING BAR.

GROUT MATERIAL SHALL BE THOROUGHLY BRUSHED INTO ALL SURFACES OF THE HOLE IMMEDIATELY PRIOR TO THE ACTUAL PLACEMENT OF THE GROUT.

REINFORCING BARS SHALL BE CLEAN AND DRY PRIOR TO INSERTION INTO THE GROUTED HOLE. THEY SHALL BE INSERTED FULL DEPTH INTO THE HOLE AND SHALL BE MANIPULATED TO INSURE COMPLETE COVERAGE BY THE GROUT. AFTER INSERTION OF THE REINFORCING BAR, ALL EXCESS GROUT SHALL BE STRUCK OFF FLUSH WITH THE CONCRETE FACE. SHOULD THE GROUT FAIL TO FILL THE HOLE AFTER REINFORCING BAR INSERTION, ADDITIONAL GROUT SHALL BE ADDED TO THE HOLE TO ALLOW A FLUSH STRIKE-OFF.

IF THE REINFORCING BAR IS INSERTED IN A HOLE WHICH HAS AN AXIS PREDOMINANTLY HORIZONTAL TO THE GROUND SURFACE, CARE SHALL BE TAKEN TO PREVENT THE REINFORCING BAR FROM CHANGING POSITION PRIOR TO THE SETTING OF THE GROUT, AND TO PREVENT GROUT FROM RUNNING DOWN THE FACE OF THE CONCRETE. THESE PRECAUTIONS SHALL BE DONE IN A MANNER SATISFACTORY TO THE ENGINEER.

4.1M.4. QUANTITY AND PAYMENT.

THE QUANTITY OF CORING AND GROUTING FOR WHICH PAYMENT WILL BE MADE WILL BE THE LENGTH OF HOLES, MEASURED IN LINEAR FEET, INTO WHICH GROUT AND REINFORCING BARS HAVE BEEN INSERTED WHERE SHOWN ON THE PLANS.

PAYMENT FOR CORING AND GROUTING WILL BE MADE FOR THE QUANTITY AS ABOVE DETERMINED, MEASURED IN LINEAR FEET, AT THE UNIT PRICE BID IN THE PROPOSAL FOR THE ITEM CORING AND GROUTING, WHICH PRICE SHALL INCLUDE THE COST OF ALL LABOR, MATERIALS AND EQUIPMENT NECESSARY TO DO THE WORK, EXCEPT THAT REINFORCING BARS AND ANCHOR BOLTS SHALL BE PAID FOR UNDER THEIR APPROPRIATE ITEMS.

## Q.16 PIER REPAIR ABOVE AND BELOW WATER

DESCRIPTION

REPAIRS TO PIER BASES ABOVE EL.+2.9 SHALL CONSIST OF THE FOLLOWING:

1. REMOVAL AND DISPOSAL OF UNSOUND AND DETERIORATED CONCRETE FROM THE PIER BASES OF PIERS 2 AND 4, AS SHOWN ON THE PLANS, AND THE REPLACEMENT THEREOF WITH CLASS B CONCRETE.

2. REMOVAL AND DISPOSAL OF UNSOUND AND DETERIORATED CONCRETE FROM THE NOSE AREA OF ALL PIERS AND THE REPLACEMENT THEREOF WITH CLASS B CONCRETE.

3. THE APPLICATION OF EPOXY WATERPROOFING TO THE TOPS OF ALL PIERS TO THE LIMITS SHOWN ON THE PLANS.

ALL OF THE ABOVE WORK TAKES PLACE ABOVE ELEVATION +2.9 AND SHALL BE PERFORMED IN ACCORDANCE WITH THE PLANS AND AS DIRECTED BY THE ENGINEER.

MATERIALS.

CLASS B CONCRETE MATERIALS SHALL MEET THE REQUIREMENTS SPECIFIED UNDER ARTICLE 4.1.2. CLASS B CONCRETE SHALL BE AIR-ENTRAINED. THE REQUIREMENTS FOR AIR-ENTRAINED CONCRETE, AIR-ENTRAINING ADDITIVES, AND AIR-ENTRAINING ADMIXTURES SHALL BE AS SPECIFIED UNDER ARTICLE 3.12.2.

WELDED STEEL WIRE FABRIC SHALL MEET THE REQUIREMENTS OF A.S.T.M. A185.

EXPANSION TYPE ANCHOR BOLTS AND SHIELDS SHALL MEET THE REQUIREMENTS SPECIFIED UNDER FEDERAL SPECIFICATIONS FF-S-325, GROUP VIII, TYPE 1.

EPOXY BONDING COMPOUND SHALL MEET THE REQUIREMENTS SPECIFIED IN A.A.S.H.T.O. M235-CLASS 1.

EPOXY WATERPROOFING SHALL MEET THE REQUIREMENTS SPECIFIED UNDER ARTICLE 8.5.41. EPOXY WATERPROOFING SHALL BE GREY TO MATCH THE COLOR OF THE ADJACENT CONCRETE.

METHODS OF CONSTRUCTION.

METHODS OF CONSTRUCTION FOR REMOVAL AND DISPOSAL OF UNSOUND AND DETERIORATED CONCRETE SHALL MEET THE REQUIREMENTS SPECIFIED UNDER ARTICLE 2.1.3. AS AMENDED ELSEWHERE HEREIN, AND AS HEREINAFTER SPECIFIED.

WHERE THE TERM "CHIP TO SOUND MATRIX" OR WORDS OF SIMILAR INTENT ARE INDICATED ON THE PLANS, THE LEVEL OR LEVELS OF SOUND MATRIX WILL BE DETERMINED BY THE ENGINEER, BASED ON OBSERVATION OF ACTUAL IN PLACE CONDITIONS, BUT GENERALLY ON THE BASIS THAT CONCRETE WHICH IS DIFFICULT TO REMOVE BY USE OF A PNEUMATIC HAMMER EQUIPPED WITH A SPADE IS SOUND CONCRETE. REMOVAL OF EXISTING CONCRETE SHALL BE PERFORMED IN A MANNER AS NOT TO DAMAGE SOUND MATRIX. EXISTING REINFORCEMENT BARS UNCOVERED BY CHIPPING SHALL BE PRESERVED, IN-PLACE AND IN CORRECT POSITION. SUCH BARS SHALL BE THOROUGHLY CLEANED WITH WIRE BRUSHES BEFORE BEING EMBEDDED IN NEW CONCRETE.

ALL SURFACES TO BE BONDED SHALL BE THOROUGHLY BLAST CLEANED WITH NO. 40 BOILER SLAG GRIT OR NO. 2 SANDBLAST SAND OR ABRADED BY APPROVED MECHANICAL MEANS.

AFTER BLAST CLEANING AND JUST PRIOR TO THE APPLICATION OF EPOXY, THE SURFACE SHALL BE VACUUM CLEANED.

THE TWO COMPONENTS OF THE EPOXY BONDING COMPOUND ARE FURNISHED IN SEPARATE CONTAINERS. IT IS ESSENTIAL THAT THE ENTIRE CONTENTS OF BOTH CONTAINERS BE THOROUGHLY BLENDED TOGETHER. A PADDLE ATTACHED TO A SLOW SPEED ELECTRIC DRILL IS RECOMMENDED FOR MIXING. FOR SMALL BATCHES OF ONE GALLON OR LESS, THOROUGH HAND STIRRING MAY BE SATISFACTORY.

THE CONTRACTOR SHALL SCHEDULE HIS CONCRETING OPERATIONS SO THAT CONCRETE SHALL BE PLACED WHILE THE EPOXY BONDING COMPOUND HAS NOT SET AND IS STILL TACKY. IF IN THE OPINION OF THE ENGINEER, THE BONDING COMPOUND HAS BEGUN TO SET, NO CONCRETE SHALL BE PLACED UNTIL A NEW FILM OF BONDING COMPOUND HAS BEEN APPLIED TO THE REQUIRED AREAS, AT NO COST TO THE STATE

QUANTITY AND PAYMENT

THE QUANTITY OF REPAIRS TO PIER BASES ABOVE EL.+2.9 FOR WHICH PAYMENT WILL BE MADE WILL BE THE VOLUME OF CONCRETE ACTUALLY PLACED WITHIN THE LIMITS SHOWN ON THE PLANS AND AS DIRECTED BY THE ENGINEER.

PAYMENT FOR REPAIRS TO PIER BASES ABOVE EL.+2.9 WILL BE MADE FOR THE QUANTITY AS DETERMINED ABOVE, MEASURED IN CUBIC FEET, AT THE PRICE PER CUBIC FOOT BID FOR THE ITEM REPAIRS TO PIER BASES ABOVE EL.+2.9 IN THE PROPOSAL, WHICH PRICE SHALL INCLUDE ALL THE WORK SPECIFIED HEREIN AND AS SHOWN ON THE PLANS AND THE COST OF FURNISHING ALL LABOR, MATERIALS, EQUIPMENT, INCLUDING REMOVAL AND DISPOSAL OF UNSOUND AND DETERIORATED CONCRETE, WELDED STEEL WIRE FABRIC, ANCHOR BOLTS AND SHIELDS, EPOXY BONDING COMPOUND, EPOXY WATERPROOFING AND ALL ELSE NECESSARY AND INCIDENTAL THERETO.

### REPAIRS TO PIER BASES AT EL. + 2.9 AND BELOW - CONCRETE

DESCRIPTION.

REPAIRS TO PIER BASES AT EL. + 2.9 AND BELOW - CONCRETE APPLIES TO ALL PIERS AND SHALL INCLUDE REMOVAL AND DISPOSAL OF UNSOUND AND DETERIORATED CONCRETE OF PIER BASES AT EL. + 2.9 AND BELOW AND THE REPLACEMENT THEREOF WITH PUMPED CONCRETE FILL, AS INDICATED ON THE PLANS AND IN ACCORDANCE WITH THE SPECIFICATIONS.

MATERIALS.

PUMPED CONCRETE FILL.

PUMPED CONCRETE FILL SHALL HAVE A MINIMUM, ULTIMATE COMPRESSIVE STRENGTH OF 3500 PSI AT THE AGE OF 28 DAYS AND SHALL NOT BE LEANER THAN ONE PART CEMENT TO FIVE PARTS OF AGGREGATE. THE CONTRACTOR SHALL BE RESPONSIBLE FOR, AT HIS OWN COST, PREPARATION AND TESTING OF TRIAL MIXES, WHICH MIXES AND TEST RESULTS SHALL BE APPROVED BY THE ENGINEER BEFORE PLACING PUMPED CONCRETE FILL.

CONCRETE MATERIALS SHALL MEET THE REQUIREMENTS SPECIFIED UNDER ARTICLE 4.1.2. CEMENT SHALL BE TYPE II. THE MAXIMUM SIZE OF COARSE AGGREGATE SHALL BE 3/4 INCH. THE MINIMUM SIZE OF COARSE AGGREGATE SHALL BE 3/8 INCH. THE RATIO OF WEIGHT OF COARSE TO FINE AGGREGATE IN THE MIX SHALL NOT BE LESS THAN 1.0.

METHODS OF CONSTRUCTION.

REMOVALS.

METHODS OF CONSTRUCTION FOR REMOVAL AND DISPOSAL OF UNSOUND AND DETERIORATED CONCRETE ABOVE WATER LINE SHALL MEET THE REQUIREMENTS SPECIFIED UNDER "REPAIRS TO PIER BASES ABOVE EL. + 2.9" AND AS HEREINAFTER SPECIFIED. UNDERWATER SECTIONS SHALL BE CHIPPED WITH A PNEUMATIC HAMMER EQUIPPED WITH A SPADE, USING AN EXTENDED SHAFT ON THE SPADE. WHERE THE WORKMEN OPERATE FROM FLOATS, THE FLOATS SHALL BE SECURED IN POSITION SO THAT THE WORKMEN CAN APPLY PRESSURE BETWEEN THE SPADE AND THE SURFACE BEING CHIPPED BY TAKING LEVERAGE AGAINST THE FLOATS.

PUMPED CONCRETE FILL.

METHODS OF CONSTRUCTION FOR PUMPING CONCRETE FILL SHALL MEET THE REQUIREMENTS SPECIFIED UNDER ARTICLE 4.1.3 AND AS SPECIFIED HEREIN.

THE CONCRETE SHALL BE PLACED UNDER AN AIR PRESSURE SUFFICIENT TO ASSURE A SMOOTH AND CONTINUOUS FLOW. THE DISCHARGE END OF THE HOSE SHALL BE INSERTED INTO THE FORM UNTIL TOP OF CONCRETE FOOTING IS REACHED BEFORE ANY CONCRETE IS PUMPED INTO PLACE. AS THE LEVEL OF THE CONCRETE RISES IN THE FORM, THE HOSE SHALL BE SLOWLY WITHDRAWN, CARE BEING TAKEN THAT THE END OF THE PIPE OR HOSE BE CONTINUOUSLY IMMERSED IN THE NEWLY PLACED CONCRETE. KINKING THE HOSE OR OTHER SOURCES OF SUDDEN SURGES OF FLOW SHALL BE AVOIDED INSOFAR AS POSSIBLE. IN GENERAL, CONCRETE SHALL BE PLACED IN A SMOOTH, CONTINUOUS OPERATION, AT A RATE WHICH WILL AVOID DAMAGE TO THE FORMS AND PERMIT THE CONCRETE TO FLOW ALONG THE LENGTH OF THE FORM TO THE ADJACENT HOSE, AND IN SUCH MANNER AS TO AVOID SEGREGATION OF THE SAND AND CEMENT PARTICLES. THE FORM SHALL BE FILLED AND THE CONCRETE ALLOWED TO SETTLE FOR ONE HOUR, AFTER WHICH THE FORM SHALL BE REFILLED. THE FORM SHALL BE COMPLETELY FILLED WHEN THE CONCRETING OPERATION IS COMPLETE. ALL CONCRETE IN PLACE SHALL BE FREE OF VOIDS, AND STREAKS, HONEYCOMBS, OR OTHER EVIDENCE OF SEGREGATION OR POROUS CONSTRUCTION. A MANIFOLD SHALL BE PROVIDED ON THE HOSE LEADING FROM THE CONCRETE PUMP AND A SUFFICIENT NUMBER OF HOSES SHALL LEAD FROM THE MANIFOLD INTO THE FORM SO THAT ANY ONE SECTION OF FORM BETWEEN BAFFLES WILL BE UNIFORMLY FILLED FROM HOSES SPACED AT INTERVALS NOT TO EXCEED 8 FEET. RUNNING THE CONCRETE ALONG THE FORM A DISTANCE IN EXCESS OF 5 FEET WILL NOT BE PERMITTED, NOR SHALL THE DIFFERENTAIAL HEAD OF THE CONCRETE FILL IN ANY FORM SECTION BETWEEN BAFFLES EXCEED 2 FEET AT ANY TIME DURING THE CONCRETING OPERATION. ADDITIONAL CONCRETE PUMPS MAY BE USED IN LIEU OF A MANIFOLD.

WHEN APPROPRIATE DURING THE PLACING OF THE CONCRETE, THE FORMS SHALL BE VIBRATED LIGHTLY ON THE OUTSIDE TO ASSURE THE COMPLETE FILLING OF THE FORM. THE VIBRATING SHALL BE DONE USING APPROVED EQUIPMENT.

ONLY APPROVED MIXING AND PUMPING EQUIPMENT SHALL BE USED IN THE PREPARATION AND HANDLING OF PUMPED CONCRETE. ALL OIL OR OTHER RUST INHIBITORS SHALL BE REMOVED FROM THE MIXING DRUMS, STIRRING MECHANISMS, AND OTHER PORTIONS OF THE EQUIPMENT IN CONTACT WITH THE CONCRETE BEFORE THE MIXERS ARE USED. ALL MATERIALS SHALL BE ACCURATELY MEASURED BY VOLUME OR WEIGHT AS THEY ARE FED TO THE MIXER.

TOP OF CONCRETE FILL SHALL BE FINISHED WITH TROWELED MORTAR, SO PITCHED, AND TO SUCH LEVEL THAT WATER WILL NOT POND ATOP THE PROTECTION FACING.

QUANTITY AND PAYMENT.

THE QUANTITY OF REPAIRS TO PIER BASES AT EL. + 2.9 AND BELOW - CONCRETE FOR WHICH PAYMENT WILL BE MADE WILL BE THE VOLUME OF PUMPED CONCRETE FILL ACTUALLY PLACED WITHIN THE LIMITS INDICATED ON THE PLANS OR AS REVISED BY AUTHORITY OF THE ENGINEER.

PAYMENT FOR REPAIRS TO PIER BASES AT EL. + 2.9 AND BELOW - CONCRETE WILL BE MADE FOR THE QUANTITY AS ABOVE DETERMINED, MEASURED IN CUBIC FEET, AT THE PRICE PER CUBIC FOOT BID IN THE PROPOSAL FOR THE ITEM REPAIRS TO PIER BASES AT EL. + 2.9 AND BELOW - CONCRETE, WHICH PRICE SHALL INCLUDE THE PERFORMANCE OF ALL THE WORK SPECIFIED HEREIN AND INDICATED ON THE PLANS, AND THE COST OF FURNISHING ALL LABOR, MATERIALS AND EQUIPMENT AND ALL ELSE NECESSARY THEREFOR AND INCIDENTAL THERETO.

# Q.17 REPOINTING STONE MASONRY

RECAULKING STONE MASONRY JOINTS - RIVER PIERS

4.1L.1. DESCRIPTION.

THE WORK UNDER THE ITEM RECAULKING STONE MASONRY JOINTS - RIVER PIERS SHALL CONSIST OF CLEANING OUT AND RECAULKING DETERIORATED JOINTS IN THE ACCESSIBLE PORTIONS OF THE STONE MASONRY FACING OF THE BASES OF THE RIVER PIERS WITH AN EPOXY MATERIAL WHERE SHOWN ON THE PLANS AND AS DELINEATED BY THE ENGINEER AT THE TIME OF CONSTRUCTION.

THE WORK UNDER THE ITEM CLEANING MORTAR JOINTS - RIVER PIERS SHALL CONSIST OF CLEANING ALL MORTAR JOINTS BETWEEN GRANITE BLOCKS PRIOR TO COMMENCING THE REPAIRS IN A MANNER WHICH WILL ENABLE THE ENGINEER TO INSPECT AND DELINEATE THE EXTENT OF RECAULKING TO BE DONE.

4.1L.2. MATERIALS.

FROM TWO FEET ABOVE MEAN HIGH WATER AND ABOVE THE JOINT MATERIAL SHALL BE A 1 TO 3 EPOXY MORTAR.

THE EPOXY COMPOUND SHALL BE FURNISHED AS A TWO-COMPONENT SYSTEM WHOSE COMPONENTS CONFORM TO THE FOLLOWING REQUIREMENTS:

| | |
|---|---|
| COMPONENT A | SHALL BE AN EPOXY RESIN OF THE BISPHENOL-A EPICHLOROHYDRIN TYPE WITH AN EPOXIDE EQUIVALENT OF 180-95 AND SHALL HAVE A VISCOSITY OF 500 TO 700 CPS @ 75 DEGREES F. THE POLYMERIC PORTION SHALL BE A MINIMUM OF 85% ASTM C-1652. |
| COMPONENT B | SHALL CONSIST OF A MODIFIED AMIDO-AMINE WITH A VISCOSITY OF 450-900 CPS @ 75 DEGREES. |
| ADDITIVES: | NO FILLERS, SOLVENTS, PIGMENTS NOR NON-REACTIVE DILUTENTS MAY BE INCORPORATED IN COMPONENT A OR B BY VENDOR OF EPOXY MATERIAL. THE COMPONENT RATIO OF A TO B SHALL BE 2:1 BY VOLUME. |

PROPERTIES OF MIXED EPOXY:

| | |
|---|---|
| COLOR | -STRAW (TRANSPARENT) - VISUAL |
| SOLIDS CONTENT | -98% MINIMUM ASTM D-1078 (ANALYZE COMPS. SEPARATELY) |
| POT LIFE (1 GAL.) | -10-20 MIN. @ 75 DEGREES F. - AASHTO M-200 ADDENDUM 3 |
| INITIAL VISCOSITY | -300-150 CPS @ 75 DEGREES F. - ASTM D-445 (300 GRAM SAMPLE USING BROOKFIELD VISCOMETER MODEL HBT, SPINDLE 2, SPEED 50) |
| INITIAL CURE | -2 1/2 HOURS @ 75 DEGREES F. |
| FINAL CURE | -48 HOURS @ 75 DEGREES F. |
| LOW TEMPERATURE CURE | -MUST HAVE THE ABILITY TO CURE AT TEMPERATURES DOWN TO 30 DEGREES F. |

PROPERTIES OF CURED EPOXY:

NEAT BINDER:

| | |
|---|---|
| HARDNESS (THIN FILM) | -SHORE D 80-90 |
| TENSILE STRENGTH, ASTM D-638 7 DAYS @ 75 DEGREES F. CURE | -7,000 PSI MINIMUM |
| TENSILE MODULUS 7 DAYS @ 75 DEGREES F. CURE | -200,000 - 400,000 PSI |
| FLEXURAL STRENGTH, ASTM D-790 7 DAYS @ 75 DEGREES F. CURE | -12,000 PSI MINIMUM |
| FLEXURAL MODULUS, ASTM D-790 7 DAYS @ 75 DEGREES F. CURE | -300,000 - 400,000 PSI |

PROPERTIES OF MORTAR:

MORTAR:
1 PART BINDER TO 3 PARTS OF SAND BY VOLUME. SAND SHALL BE OTTAWA SILICA SAND NO. 15.

| | |
|---|---|
| ULTIMATE COMPRESSIVE STRENGTH ASTM C-109 (2" CUBES) | -10,000 PSI MINIMUM @ 72 HOURS, 75 DEGREES F. CURE |

FROM TWO FEET ABOVE MEAN HIGH WATER TO TWO FEET BELOW MEAN LOW WATER THE JOINT MATERIAL SHALL BE A 1 TO 1 EPOXY SPLASH ZONE MASTIC, CONFORMING TO THE FOLLOWING SPECIFICATION:

| | |
|---|---|
| GENERIC TYPE: | EPOXY-POLYAMIDE |
| SOLIDS BY VOLUME: | 100% |
| FILM THICKNESS PER COAT: | 1/8 - 1/4 INCH |
| SAG RESISTANCE (@ 75 DEGREES F.): | 1/4 INCH (MINIMUM) |
| DRYING TIME (1/4 INCH FILM): | IN AIR: 4 HOURS MAX. IN SALT WATER: 6-8 HOURS |
| POT LIFE (@ 75 DEGREES F.): | 30 - 45 MINUTES |
| COLOR: | BROWN |
| FLASH POINT (CLEVELAND OPEN CUP): | OVER 300 DEGREES F. |
| MIXING RATIO: | 1 PART CATALYST TO 1 PART BASE (VOLUME) |

GENERAL: THIS EPOXY SPLASH ZONE MASTIC MATERIAL MUST BE CAPABLE OF BONDING TO WATER IMMERSED STEEL, GALVANIZED STEEL, ALUMINUM, CONCRETE, MASONRY OR WOOD SURFACES DURING APPLICATION AND CURING WITH THE MAXIMUM WATER DAMAGE TO THE UNCURED FILM DURING CURING LIMITED TO THE OUTER 20 MILS OF FILM THICKNESS. IN ADDITION, THIS MATERIAL MUST RESIST NORMAL TIDAL ACTION OF WAVES AT THE RIVERS DURING CURING WHEN APPLIED TO AREAS IN THIS SPLASH ZONE.

THE FOLLOWING IS A LIST OF FORMULATORS THAT PRODUCE A SATISFACTORY SPLASH ZONE MASTIC:

ADMIRAL PAINT COMPANY
328 HANLEY INDUSTRIAL COURT
ST. LOUIS, MISSOURI 63144

PITTSBURGH PLATE GLASS
JAY AND WATER STREETS
BROOKLYN, NEW YORK

CHEMICAL DEVELOPMENT COMPANY
ENDICOTT STREET
DANVERS, MASSACHUSETTS

H. V. HARDMAN
571 CORTLAND STREET
BELLEVILLE, NEW JERSEY

MASS AND WALDSTEIN COMPANY
2121 MCCARTER HIGHWAY
NEWARK, NEW JERSEY

4.1L.3. METHODS OF CONSTRUCTION.

THE CONTRACTOR SHALL HAVE SKILLED REPRESENTATIVES OF THE MANUFACTURER OF THE EPOXY COMPOUND AND OF THE SPLASH ZONE MASTIC AT THE SITE DURING THIS REPAIR AND SHALL FOLLOW THEIR DIRECTIONS SUBJECT TO APPROVAL BY THE ENGINEER.

PRIOR TO COMMENCING REPAIRS THE CONTRACTOR SHALL CLEAN ALL MASONRY JOINTS BETWEEN GRANITE BLOCKS TO ENABLE THE ENGINEER TO INSPECT AND DELINEATE THE EXTENT OF THE JOINTS TO BE RECAULKED. THE SURFACES OF ALL MORTAR JOINTS SHALL BE CLEANED USING WIRE BRUSHES AND MURIATIC ACID AS NECESSARY. THE JOINTS SHALL BE KEPT CLEAN UNTIL THEY HAVE BEEN INSPECTED BY THE ENGINEER. THE CONTRACTOR SHALL MAKE ALL NECESSARY PROVISIONS FOR ACCESS BY THE ENGINEER FOR THE PURPOSE OF INSPECTION AND DELINEATION OF THE EXTENT OF THE WORK.

JOINTS IN NEED OF RECAULKING AS DELINEATED BY THE ENGINEER SHALL BE CLEANED OUT TO THE DEPTH NECESSARY TO REACH SOUND MORTAR BUT NOT LESS THAN FOUR INCHES.

THE JOINTS ARE TO BE CLEANED OUT WITH WIRE BRUSHING AND SANDBLASTING. REMOVE DEPOSITS OF OIL AND GREASE WITH A SOLVENT OR AN ALKALINE DETERGENT CLEANER. FLUSH REPEATEDLY AND ALLOW TO AIR DRY.

MIX THE TWO COMPONENTS OF THE MORTAR THOROUGHLY BY HAND OR WITH A MECHANICAL MIXER. USE A DRILL WITH 600 RPM WHEN USING PAINT PADDLE. PRIME SURFACES OF JOINTS WITH THE NEAT TWO COMPONENT EPOXY. FILL JOINTS WITH MORTAR TAMPING IT TIGHTLY. JOINTS SHALL BE FINISHED WITH AN APPROVED POINTING TOOL TO MATCH THE TYPE OF JOINT ORIGINALLY CONSTRUCTED.

THE EPOXY SPLASH ZONE MASTIC IS TO BE PLACED IN ACCORDANCE WITH DIRECTIONS OF THE MANUFACTURER.

NO RECAULKING BELOW TWO (2) FEET BELOW MEAN LOW WATER WILL BE REQUIRED UNDER THIS ITEM. RECAULKING BELOW MEAN LOW WATER WILL BE DONE ONLY AT LOW TIDAL CONDITIONS AND IS NOT INTENDED TO REQUIRE DEWATERING OR ANY OTHER SPECIAL WORK. RECAULKING IS NOT INTENDED TO REQUIRE ANY DISMANTLING OF THE EXISTING FENDER SYSTEM.

4.1L.4. QUANTITY AND PAYMENT

THE QUANTITY OF RECAULKING STONE MASONRY JOINTS FOR WHICH PAYMENT WILL BE MADE WILL BE THE NUMBER OF LINEAR FEET OF STONE MASONRY JOINTS ON THE RIVER PIERS ACTUALLY RECAULKED WHERE SHOWN ON THE PLANS AND AS DELINEATED BY THE ENGINEER AT THE TIME OF CONSTRUCTION.

PAYMENT FOR RECAULKING STONE MASONRY JOINTS WILL BE MADE FOR THE QUANTITY AS DETERMINED ABOVE, AT THE PRICE PER LINEAR FOOT BID FOR THE ITEM RECAULKING STONE MASONRY JOINTS - RIVER PIERS IN THE PROPOSAL, WHICH PRICE SHALL INCLUDE ALL MATERIALS, BOTH EPOXY MORTAR AND EPOXY SPLASH ZONE MASTIC, EQUIPMENT, LABOR AND ALL INCIDENTALS NECESSARY TO COMPLETE THE WORK.

THE QUANTITY OF CLEANING MORTAR JOINTS - RIVER PIERS FOR WHICH PAYMENT WILL BE MADE WILL BE A LUMP SUM COVERING ALL THE WORK REQUIRED TO CLEAN THE MORTAR JOINTS PRIOR TO COMMENCING THE REPAIRS TO ENABLE INSPECTION AND THE DELINEATION OF THE EXTENT OF WORK BY THE ENGINEER, ALL AS SPECIFIED ABOVE.

PAYMENT WILL BE MADE FOR THE QUANTITY AS ABOVE DETERMINED AT THE LUMP SUM PRICE BID FOR THE ITEM CLEANING MORTAR JOINTS - RIVER PIERS IN THE PROPOSAL, WHICH SHALL INCLUDE THE COST OF ALL LABOR, MATERIAL, EQUIPMENT AND INCIDENTALS REQUIRED TO COMPLETE THE WORK AS DESCRIBED IN THESE SPECIFICATIONS.

THE NO. 6 COMMERCIAL BLAST CLEANING PROCEDURE SHALL BE USED ON ALL METAL SURFACES TO REMOVE ALL PAINT, RUST, DIRT AND FOREIGN MATTER IN THE FOLLOWING DETERIORATION-PRONE AREAS:

ALL SURFACES AT AND BENEATH ROADWAY EXPANSION JOINTS

INCLUDING EXPOSED PORTIONS OF EXPANSION JOINTS, CROSS-BEAMS, STRINGERS, FLOORBEAMS, WIND TONGUES, PIN CONNECTIONS, BRACING AND TRUSS MEMBERS WITHIN THE LIMIT OF 5 FEET EITHER SIDE OF THE EXPANSION JOINT.

IN ALL OTHER AREAS THE SURFACES WHICH ARE RUSTED OR EXHIBIT DETERIORATED PAINT SHALL BE SPOT CLEANED BY NO. 6 COMMERCIAL BLAST CLEANING PROCEDURE. SURFACES EXHIBITING PAINT IN GOOD CONDITION SHALL BE CLEANED BY EITHER NO. 1 SOLVENT CLEANING, NO. 2 HAND TOOL CLEANING OR POWER TOOL CLEANING.

PAINT IN GOOD CONDITION IS DEFINED AS BEING IN CONFORMANCE WITH SSPC-VIS2 SHOWING LESS THAN 0.01% SURFACE RUST SO THAT IT IS ONLY NECESSARY TO REMOVE FOREIGN RESIDUE. EXISTING PAINT MUST BE TIGHTLY ADHERED TO THE SURFACE AFTER CLEANING BY SOLVENT OR HAND OR POWER TOOLS, I.E. PAINT CANNOT BE LIFTED AS A LAYER WHEN A DULL PUTTY KNIFE IS INSERTED BENEATH IT. IF THE ENGINEER DETERMINES THAT THE PAINT REMAINING ON THE STEEL SURFACES AFTER CLEANING IS NOT TIGHTLY ADHERED THEN THE SURFACE SHALL BE CLEANED BY THE NO. 6 COMMERICAL BLAST CLEANING PROCEDURE AND SPOT PRIMED.

WHERE IT IS IMPRACTICAL TO USE BLASTING EQUIPMENT, POWER HAND TOOLS MAY BE USED SUBJECT TO THE APPROVAL OF THE ENGINEER EXCEPT THAT THE SURFACE SHALL MEET THE SAME DEGREE OF CLEANNESS AS THAT REQUIRED UNDER NO. 6 COMMERCIAL BLAST CLEANING.

THE SURFACE AFTER BLAST CLEANING SHALL BE DEFINED BY SSPC - VIS. 1, PICTORIAL STANDARDS BSA2, CSA2, OR DSA2 AS APPLICABLE. THE SURFACES SHALL BE THOROUGHLY CLEAN OF PAINT, RUST, MILL SCALE, SLAG, DIRT, OIL, GREASE AND FOREIGN SUBSTANCES.

ALL CONCRETE-STEEL INTERFACES REMAINING AFTER THE REMOVAL OF EXISTING CONCRETE ENCASEMENT SHALL BE CAULKED AFTER CLEANING AND PAINTING. CAULKING SHALL BE IN ACCORDANCE WITH SECTION 4.IN OF THESE SUPPLEMENTARY SPECIFICATIONS, EXCEPT THAT PAYMENT FOR CAULKING WILL BE INCLUDED AS PART OF PAINTING AS DESCRIBED HEREINAFTER.

## Q.18 STRENGTHENING STRUCTURAL MEMBERS

STRUCTURAL STEEL STRENGTHENING REPAIRS

4.3A.1. DESCRIPTION.

THE WORK UNDER THIS SECTION SHALL CONSIST OF REMOVING PARTS OF EXISTING STEEL COMPONENTS AND FURNISHING, FABRICATING AND INSTALLING NEW STRUCTURAL STEEL SHAPES, PLATES AND FASTENERS NECESSARY TO STRENGTHEN EXISTING FLOORSYSTEM MEMBERS AS SHOWN ON THE PLANS AND IN ACCORDANCE WITH THESE SPECIFICATIONS.

INCLUDED SHALL BE THE CUTTING, DRILLING, REMOVAL AND DISPOSAL OF EXISTING STEEL..

ALSO, ALL REQUIRED REPAIRS AND RECONSTRUCTION OR MODIFICATION OF EXISTING STEEL WHICH IS TO REMAIN, ALL WORK REQUIRED FOR STRUCTURAL LIFTING OPERATIONS, REPLACEMENT OF RIVETS WITH HIGH-STRENGTH BOLTS TO THE EXTENT SHOWN ON THE PLANS, AND SHOP PRIMING OF ALL NEW STRUCTURAL STEEL SHALL BE INCLUDED IN THE WORK SPECIFIED IN THIS SECTION.

4.3A.2. MATERIALS.

MATERIALS FOR THIS WORK SHALL MEET THE REQUIREMENTS OF THE FOLLOWING SPECIFICATIONS:

| | | |
|---|---|---|
| STRUCTURAL STEEL | ASTM | A36 |
| HIGH-STRENGTH BOLTS, NUTS AND WASHERS | ASTM | A325 |
| PAINT (SHOP PRIMER) | | AS SPECIFIED UNDER STRUCTURAL STEEL |

4.3A.3. METHODS OF CONSTRUCTION.

GENERAL

THE DETAILS OF THE EXISTING STRUCTURE AND PROPOSED REPAIRS SHOWN ON THE PLANS ARE BASED ON AVAILABLE SHOP DRAWINGS AND/OR THE ORIGINAL CONTRACT DRAWINGS. PRIOR TO FABRICATION THE CONTRACTOR SHALL DETERMINE AND VERIFY ALL NECESSARY DIMENSIONS IN THE FIELD AND SHALL PREPARE DETAILED SHOP DRAWINGS FOR APPROVAL BY THE ENGINEER.

THE TYPES, DETAILS AND DIMENSIONS OF REPAIRS SHOWN ON THE PLANS REPRESENT THE ANTICIPATED NATURE OF THE WORK KNOWN AT THE LETTING OF THE CONTRACT. HOWEVER, THE CONTRACTOR SHALL CALL TO THE IMMEDIATE ATTENTION OF THE ENGINEER IF, DURING THE EXECUTION OF THE WORK, HE DISCOVERS AREAS OF DETERIORATED STRUCTURE WHICH WERE NOT UNCOVERED BY THE ENGINEER DURING THE INSPECTION OF THE STRUCTURE, OR THE TYPE OF REPAIR IS DIFFERENT THAN THAT SHOWN ON THE PLANS. THE ENGINEER SHALL THEN DETERMINE THE REQUIREMENT AND NATURE OF THE REPAIR WORK IN ACCORDANCE WITH THE PROVISIONS OF THE CONTRACT.

JACKING OPERATIONS.

BEFORE STARTING ANY REPAIR, THE CONTRACTOR SHALL RELIEVE CERTAIN SPECIFIED LOADS FROM THE MEMBER TO BE STRENGTHENED, AS REQUIRED, BY MEANS OF JACKING.

FOR PURPOSES OF THIS SPECIFICATION, THE TERM DEFLECTION POINT IS DEFINED AS THE POINT ON THE STRUCTURE, DESIGNATED ON THE PLANS, TO DISPLACED A DISTANCE AS SPECIFIED ON THE PLANS IN ORDER TO RELIEVE THE LOAD IN THE MEMBER BEING STRENGTHENED.

FOR PURPOSES OF THIS SPECIFICATION, THE TERM JACKING POINT IS DEFINED AS A POINT ON THE STRUCTURE, DESIGNATED ON THE PLANS, WHERE THE JACKING FORCE IS APPLIED.

THE MINIMUM JACK CAPACITY REQUIRED FOR EACH JACKING POINT IS SHOWN ON THE PLANS. THE CONTRACTOR SHALL BE RESPONSIBLE FOR PROVIDING SUFFICIENT JACKING CAPACITY AT THE JACKING POINTS TO ACHIEVE THE SPECIFIED DISPLACEMENT AT THE DEFLECTION POINTS.

EACH JACK USED FOR THE LIFTING OPERATION SHALL HAVE THE MINIMUM RATED CAPACITY CLEARLY SHOWN ON THE MANUFACTURER'S NAME PLATE ATTACHED TO EACH JACK. JACKS SHALL HAVE A RATED CAPACITY AT LEAST EQUAL TO THE MINIMUM DESIGNATED ON THE PLANS FOR EACH JACKING POINT. JACKS SHALL BE EQUIPPED WITH PRESSURE GUAGES THAT WILL ENABLE THE APPLIED LIFTING FORCE TO BE MONITORED AT ALL TIMES. THE ENGINEER MAY REQUIRE THAT ANY LIFTING EQUIPMENT WHICH HE DEEMS TO BE INADEQUATE OR FAULTY BE REMOVED FROM THE PROJECT SITE.

THE PLANS DESIGNATE THE DEFLECTION POINTS, REQUIRED DEFLECTIONS, THE JACKING POINTS AND MINIMUM JACK CAPACITIES. THE CONTRACTOR SHALL ENGAGE THE SERVICES OF A NEW JERSEY STATE LICENSED PROFESSIONAL ENGINEER TO DESIGN AND DETAIL THE JACKING SYSTEM. HE SHALL BE AVAILABLE FOR CONSULTATION IN INTERPRETING HIS PLANS AND IN THE RESOLUTION OF PROBLEMS WHICH MAY ARISE DURING THE PERFORMANCE OF THE WORK.

THE CONTRACTOR SHALL FURNISH WORKING DRAWINGS PREPARED, STAMPED AND SIGNED BY A NEW JERSEY STATE LICENSED PROFESSIONAL ENGINEER FOR EACH JACKING SYSTEM REQUIRED. THE WORKING DRAWINGS SHALL NOT ALTER THE NUMBER AND LOCATION OF DEFLECTION AND JACKING POINTS, NOR THE SPECIFIED DEFLECTIONS AND MINIMUM JACK CAPACITIES EXCEPT WHERE APPROVAL IS GIVEN BY THE ENGINEER. THE DRAWINGS SHALL INCLUDE, BUT NEED NOT BE LIMITED TO, THE FOLLOWING INFORMATION:

1. JACKING POINT LOCATIONS.
2. JACK CAPACITIES PROVIDED.
3. DETAILS FOR ALL JACKING EQUIPMENT AND SUPPORT SYSTEMS INCLUDING DISTRIBUTION OF LOADS TO EXISTING GROUND.
4. DETAILS OF REQUIRED MODIFICATIONS TO EXISTING STRUCTURE, IF ANY.
5. TYPE AND GRADE OF ALL MATERIALS.
6. DEFLECTION REQUIRED AT EACH DEFLECTION POINT.
7. METHOD FOR MONITORING DEFLECTIONS.
8. SCHEMATIC HYDRAULIC LAYOUT.

ALL WORK SHALL BE DONE IN ACCORDANCE WITH WORKING DRAWINGS APPROVED BY THE ENGINEER. THE CONTRACTOR MUST HAVE APPROVED WORKING DRAWINGS PRIOR TO THE START OF ANY JACKING OPERATIONS.

THE CONTRACTOR SHALL BEAR ALL COSTS AND/OR DAMAGES WHICH MAY RESULT FROM THE ORDERING OF ANY MATERIALS, OR EQUIPMENT; OR THE USE OF ANY PREPARATORY LABOR PRIOR TO THE APPROVAL OF THE WORKING DRAWINGS.

THE CONTRACTOR SHALL APPLY SUFFICIENT FORCE AT EACH JACKING LOCATION TO ACHIEVE THE REQUIRED DISPLACEMENTS AT THE DEFLECTION POINTS. IF APPLICATION OF A LIFTING FORCE EQUAL TO THE MINIMUM JACK CAPACITY SPECIFIED ON THE PLANS DOES NOT DEVELOP THE REQUIRED DEFLECTION, THE CONTRACTOR SHALL PROVIDE TWICE THE JACKING CAPACITY AND SHALL PROCEED AS APPROVED BY THE ENGINEER.

DURING ALL JACKING OPERATIONS THE CONTRACTOR SHALL ABIDE BY THE LIVE LOAD LIMITATIONS AS SHOWN ON THE PLANS AND IN ARTICLE 1.4.3. OF THE STANDARD SPECIFICATIONS AS AMENDED.

ANY REMOVAL OF CONCRETE ENCASEMENT, MODIFICATION OF EXISTING STRUCTURE, DEVELOPMENT OF SUPPORT CONDITIONS, AND RESTORATION REQUIRED TO PERFORM THE JACKING OPERATION SHALL BE CONSIDERED PART OF THIS WORK AND SHALL BE PERFORMED IN ACCORDANCE WITH THESE SPECIFICATIONS. ENCASEMENT SHALL BE RESTORED AT ALL JOINTS, FASCIAS AND COLUMN MEMBERS.

STRENGTHENING MODIFICATIONS.

THE REMOVAL OF EXISTING CONCRETE ENCASEMENT WHERE REQUIRED TO PERMIT STRUCTURAL STRENGTHENING REPAIR SHALL BE PERFORMED UNDER THE PROVISIONS OF THE ITEM CONCRETE ENCASEMENT REMOVAL.

THE REMOVAL OF EXISTING STEEL AS DESIGNATED ON THE PLANS SHALL BE INCLUDED AS PART OF THE STRUCTURAL STEEL STRENGTHENING REPAIR.

EXISTING CONCRETE ENCASEMENT AND STRUCTURAL STEEL DESIGNATED FOR REMOVAL SHALL BE DISPOSED OF BY METHODS AS THE CONTRACTOR MAY PROPOSE, SUBJECT TO THE APPROVAL OF THE ENGINEER.

EXISTING STEEL EXPOSED AFTER ENCASEMENT REMOVAL, DESIGNATED TO REMAIN, SHALL BE CLEANED AND PRIMED PRIOR TO THE INSTALLATION OF NEW STEEL IN ACCORDANCE WITH THE PROVISIONS OF THE ITEM OF CLEANING AND PAINTING STEEL EXPOSED BY ENCASEMENT REMOVAL OR THE ITEM OF PNEUMATICALLY APPLIED CEMENT MORTAR - ENCASEMENT, AS APPLICABLE.

ALL NEW STRUCTURAL STEEL SHALL BE FURNISHED, FABRICATED, CLEANED AND SHOP PRIMED, AND ERECTED IN ACCORDANCE WITH THE PROVISIONS OF THE ITEM STRUCTURAL STEEL AND IN ACCORDANCE WITH THE CONSTRUCTION SEQUENCE AND DETAILS AS SHOWN ON THE PLANS.

ALL BOLTS SHALL BE 7/8-INCH DIAMETER ASTM A325 HIGH-STRENGTH BOLTS UNLESS OTHERWISE NOTED. HARDENED WASHERS SHALL BE USED UNDER BOTH THE HEAD AND NUT OF ALL HIGH-STRENGTH BOLTS. ALL HOLES FOR 7/8-INCH A325 BOLTS SHALL BE SUB-PUNCHED OR DRILLED 11/16-INCH DIAMETER AND FIELD REAMED TO 15/16-INCH DIAMETER, OR DRILLED FROM SOLID IN THE FIELD FOR ALL COMPONENTS.

EVERY EFFORT SHALL BE MADE TO FIT THE WORK WITH A MINIMUM OF MISMATCH. HOWEVER, IF A MISMATCH OF HOLES SHOULD OCCUR, THE CONTRACTOR SHALL REAM THE MISMATCHED HOLE AND INSTALL THE NEXT LARGER-SIZE BOLT, SUBJECT TO THE APPROVAL OF THE ENGINEER. NO BENDING OR TWISTING OF STEEL WILL BE PERMITTED.

NO WELDING WILL BE PERMITTED EXCEPT WITH THE APPROVAL OF THE ENGINEER.

FLAME CUTTING OF EXISTING STEEL SHALL BE DONE ONLY WHERE APPROVED BY THE ENGINEER. FLAME CUTTING EXISTING STEEL SHALL BE DONE BY EXPERIENCED OPERATORS AND IN THE PRESENCE OF THE ENGINEER OR HIS DESIGNATED REPRESENTATIVE. THE FLAME SHALL BE ADJUSTED AND MANIPULATED TO AVOID OVER-HEATING ADJACENT METAL THAT IS TO REMAIN. ANY EXISTING STEEL THAT HAS BEEN MUTILATED OR DAMAGED BY IMPROPER CUTTING SHALL BE REPLACED BY THE CONTRACTOR AT HIS OWN EXPENSE.

4.3A.4. QUANTITY AND PAYMENT.

THE QUANTITY OF STRUCTURAL STEEL STRENGTHENING REPAIRS FOR WHICH PAYMENT WILL BE MADE WILL BE THE COMPUTED NET WEIGHT OF TYPE DESIGNATED ON THE PLANS, MEASURED IN POUNDS, ACTUALLY FURNISHED, ERECTED AND ACCEPTED BY THE ENGINEER IN ACCORDANCE WITH THESE SPECIFICATIONS.

PAYMENT FOR STRUCTURAL STEEL SRENGTHENING REPAIRS WILL BE MADE FOR THE QUANTITY AS ABOVE DETERMINED, MEASURED IN POUNDS, AT THE PRICE BID PER POUND FOR THE ITEM STRUCTURAL STEEL STRENGTHENING REPAIR FOR EACH TYPE IN THE PROPOSAL, WHICH PRICE SHALL INCLUDE THE COST OF ALL EQUIPMENT, TOOLS, SHOP PRIMING, HANDLING, FALSEWORK, SCAFFOLDING, PLATFORMS, TRANSPORTATION, LABOR, MATERIALS, HIGH STRENGTH BOLTS AND ALL ELSE NECESSARY TO COMPLETE THE WORK AS SHOWN ON THE PLANS IN ACCORDANCE WITH THESE SPECIFICATIONS, EXCEPT AS STATED HEREINAFTER.

ENCASEMENT REMOVAL AND RESTORATION, AND THE CLEANING AND PAINTING OF EXISTING STEEL AT THE STRENGTHENING REPAIR LOCATION SHALL BE PAID FOR UNDER THEIR RESPECTIVE ITEMS IN THE CONTRACT AS SPECIFIED HEREINBEFORE.

IF APPLICATION OF A LIFTING FORCE EQUAL TO THE MINIMUM JACK CAPACITY SPECIFIED ON THE PLANS DOES NOT DEVELOP THE REQUIRED DEFLECTION AND THE CONTRACTOR IS REQUIRED TO REPLACE THE JACK WITH A JACK OF TWICE THE JACKING CAPACITY ORIGINALLY SPECIFIED, A SEPARATE PAYMENT IN THE AMOUNT OF 20 PERCENT OF THE PAYMENT SPECIFIED HEREINBEFORE FOR THE INDIVIDUAL REPAIR LOCATION WILL BE MADE AS FULL COMPENSATION FOR PROVIDING THE ADDITIONAL JACKING CAPACITY.

## Q.19 STRAIGHTENING BENT STEEL MEMBERS

4.3D.1. DESCRIPTION.

THE WORK UNDER THIS ITEM SHALL CONSIST OF STRAIGHTENING TWO (2) BENT BRACING MEMBERS AND ONE (1) BENT STRINGER BY MEANS OF HEATING AND JACKING AT THE LOCATIONS SHOWN ON THE PLANS AND IN ACCORDANCE WITH THESE SPECIFICATIONS.

4.3D.3. METHODS OF CONSTRUCTION.

THE WORK SHALL BE DONE BY METHODS NOT LIKELY TO PRODUCE FRACTURE OR OTHER INJURY TO THE STEEL MEMBERS.

CARE SHALL BE EXERCISED IN THE STRAIGHTENING OPERATIONS. THE CONTRACTOR WILL BE REQUIRED TO EMPLOY CRAFTSMEN SKILLED IN THE ART OF STRAIGHTENING BENT MEMBERS.

THE CONTRACTOR SHALL USE A CONTROLLED METHOD TO STRAIGHTEN THE BENT MEMBERS SUCH AS JACKING, THE USE OF BLOCK AND TACKLE, AND OR THE PROPER APPLICATION OF HEAT. HEATING STRUCTURAL STEEL WILL NOT BE ALLOWED TO EXCEED 1200 DEGREES F, AS INDICATED BY A "CHERRY-RED COLOR" AND MONITORED BY THE ENGINEER BY THE USE OF TEMPERATURE STICKS MADE AVAILABLE BY THE CONTRACTOR. AFTER BEING HEATED, THE METAL SHALL BE COOLED AS SLOWLY AS POSSIBLE. UPON COMPLETION OF STRAIGHTENING, THE STEEL SURFACES SHALL BE CAREFULLY INSPECTED FOR THE PRESENCE OF CRACKS OR OTHER SIGNS OF DISTRESS. CARE SHALL BE EXERCISED IN THE STRAIGHTENING OPERATIONS SO AS NOT TO CAUSE ANY ADDITIONAL DAMAGE TO THE MEMBERS.

IF, IN THE OPINION OF THE ENGINEER, THE CONTRACTOR'S OPERATIONS DAMAGE THE MEMBERS HE IS STRAIGHTENING, HE SHALL BE REQUIRED TO CHANGE HIS METHODS OF OPERATIONS AND MAKE ALL NECESSARY REPAIRS AS ORDERED BY THE ENGINEER AT NO COST TO THE STATE.

THE METHOD THE CONTRACTOR WILL EMPLOY TO STRAIGHTEN THE BENT MEMBERS MAY NECESSITATE STRENGHENING OR TEMPORARY SUPPORTS FOR RELATED MEMBERS TO INSURE ADEQUATE STRENGTH TO RESIST ALL EXISTING AND APPLIED LOADS.

THE CONTRACTOR SHALL OBSERVE ALL MAINTENANCE OF TRAFFIC REQUIREMENTS AS SPECIFIED IN ARTICLE 1.4.3.

THE METHODS OF REPAIR, PROCEDURES, MATERIALS, EQUIPMENT, EQUIPMENT SUPPORTS OR ANCHORAGES, AND STRENGTHENINGS PROPOSED BY THE CONTRACTOR SHALL BE SUBMITTED ON A STANDARD SHOP DRAWING FOR APPROVAL BY THE ENGINEER PRIOR TO BEGINNING OF WORK. APPROVAL SHALL NOT RELIEVE THE CONTRACTOR OF HIS RESPONSIBILITY FOR THE SUCCESSFUL COMPLETION OF THE WORK.

ANY REPLACEMENT, REPAIR OR ADJUSTMENT TO THE SUPERSTRUCTURE STEEL SHALL BE PERFORMED IN CONFORMANCE WITH THE CURRENT REQUIREMENTS OF SECTION 4.3. OF THE STANDARD SPECIFICATIONS.

DURING THE COURSE OF THE WORK, IT MAY BE NECESSARY TO DISCONNECT, SUPPORT, OR ADJUST STEEL, WHICH IS TO REMAIN IN THE STRUCTURE. FOR THESE SITUATIONS, THE CONTRACTOR SHALL SUBMIT HIS PROPOSALS FOR DISCONNECTING, SUPPORTING OR ADJUSTING THE STEEL, AS NECESSARY, TO THE ENGINEER FOR APPROVAL. THE CONTRACTOR'S PROPOSALS SHALL BE APPROVED PRIOR TO THE ACTUAL COMMENCEMENT OF ANY DISCONNECTING, SUPPORTING OR ADJUSTING OF STEEL.

4.3D.4. QUANTITY AND PAYMENT.

THE QUANTITY OF STRAIGHTENING BENT STEEL MEMBERS FOR WHICH PAYMENT WILL BE MADE WILL BE A LUMP SUM COVERING ALL THE WORK SPECIFIED ABOVE.

PAYMENT FOR STRAIGHTENING BENT STEEL MEMBERS WILL BE MADE FOR THE QUANTITY AS ABOVE DETERMINED AT THE LUMP SUM PRICE BID FOR THE ITEM STRAIGHTENING BENT STEEL MEMBERS IN THE PROPOSAL, WHICH PRICE SHALL INCLUDE THE COST OF ALL MATERIALS, LABOR, EQUIPMENT AND ALL ELSE NECESSARY TO COMPLETE THE WORK AT THE LOCATIONS SHOWN ON THE PLANS AND IN ACCORDANCE WITH THESE SPECIFICATIONS.

# Q.20 RIVET REPLACEMENT

RIVET REPLACEMENT

4.3G.1. DESCRIPTION.

RIVET REPLACEMENT INCLUDES THE REMOVAL OF EXISTING DETERIORATED OR REDUCED RIVETS AND PROVIDING HIGH-STENGTH BOLTS FOR THEIR REPLACEMENT AT LOCATIONS AS DELINEATED BY THE ENGINEER DURING CONSTRUCTION OR AS SHOWN ON THE DRAWINGS.

RIVETS REMOVED AND REPLACED TO INSTALL NEW MATERIAL OR CALLED FOR REPLACEMENT SEPARATELY IN CONNECTION WITH A REPAIR OF STRUCTURAL MEMBERS OR TO INSTALL TEMPORARY SUPPORTS ARE NOT INCLUDED UNDER THIS ITEM. ALL SUCH RIVET REPLACEMENTS SHALL BE INCLUDED WITH THE APPLICABLE ITEMS.

4.3G.2. MATERIALS.

HIGH STRENGTH BOLTS AND NUTS: HEAVY HEXAGONAL STRUCTURAL BOLTS, HEAVY HEXAGON NUTS AND HARDENED WASHERS COMPLYING WITH ASTM A325.

4.3G.3. METHODS OF CONSTRUCTION.

REMOVAL OF RIVETS

RIVETS SHALL BE REMOVED BY ONE OF THE FOLLOWING METHODS:

1. SHEAR RIVET HEAD USING A PNEUMATIC RIVET BREAKER (HELLDOG) AND DRIVE OUT RIVET SHANK WITH A PNEUMATIC PUNCH.

2. FLAME CUT RIVET AHEAD 1/16 OF AN INCH ABOVE THE BASE METAL USING A RIVET SCARFING TIP AND DRIVE OUT SHANK USING A PNEMATIC PUNCH.

IF, IN THE OPINION OF THE ENGINEER, RIVET SHANKS CANNOT BE REMOVED BY PUNCHING WITHOUT DAMAGING THE BASE METAL, THE RIVET SHANK SHALL BE REMOVED BY DRILLING.

REMOVE ONLY ONE RIVET AT A TIME FROM ANY ONE MEMBER AND REPLACE IT WITH A HIGH STRENGTH BOLT, UNLESS AUTHORIZED BY THE ENGINEER. THE REPLACEMENT BOLTS SHALL BE PROPERLY TENSIONED PRIOR TO THE REMOVAL OF A SUBSEQUENT RIVET.

INSPECTION

UPON REMOVAL OF EACH RIVET, THE BASE METAL AROUND THE HOLE SHAL BE EXAMINED FOR SURFACE IRREGULARITIES AND DETERIORATION. ALL OXIDIZED MATERIAL SHALL BE REMOVED. LIGHT GRINDING MAY BE ORDERED IF NECESSARY.

HIGH-STRENGTH BOLTS SHALL BE INSTALLED IN ACCORDANCE WITH THE TURN-OF-THE-NUT METHOD OF THE AISC "SPECIFICATIONS FOR STRUCTURAL JOINTS USING ASTM A325 OR A490 BOLTS." HARDENED WASHERS SHALL BE USED UNDER BOTH HEADS AND NUTS OF HIGH-STRENGTH BOLTS.

FIELD QUALITY CONTROL

CHECK THE TENSION OF INSTALLED BOLTS USING A CALIBRATED TORQUE WRENCH TO ROTATE THE NUTS SLOWLY A SMALL AMOUNT IN THE TIGHTENING DIRECTION. BOLTS GIVING READINGS BELOW THE CALIBRATED TORQUE VALUE SHALL BE FURTHER TIGHTENED; IF HIGHER READINGS CANNOT BE OBTAINED, THE BOLTS SHALL BE REMOVED AND DISCARDED.

A READING HIGHER THAN THE CALIBRATED MINIMUM TENSION EQUIVALENT SHALL NOT BE CAUSE FOR REJECTION.

TEN PERCENT OF THE BOLTS IN ANY ONE CONNECTION OR FLANGE ANGLE LEG SHALL BE CHECKED WITH A MINIMUM OF TWO BOLTS BEING REQUIRED AT ALL TIMES FOR EACH CONNECTION.

IF ONE OR MORE BOLTS IN A CONNECTION IS BELOW THE MINIMUM TENSION, AS INDICATED BY THE TORQUE WRENCH, ALL OF THE BOLTS IN THAT CONNECTION SHALL BE CHECKED. THE ENGINEER MAY BE ALLOWED VERIFICATION OF ALL READINGS AT HIS DISCRETION.

4.3G.4. QUANTITY AND PAYMENT.

THE QUANTITY OF RIVET REPLACEMENT FOR WHICH PAYMENT WILL BE MADE WILL BE THE ACTUAL NUMBER OF RIVETS REMOVED AND REPLACED BY THE INSTALLATION OF HIGH-STRENGTH BOLTS AS DELINEATED BY THE ENGINEER AT THE TIME OF CONSTRUCTION OR AS SHOWN IN THE PLANS. THE TENSION TESTING WILL NOT BE SEPARATELY MEASURED FOR PAYMENT BUT ALL COSTS FOR TENSION TESTING WILL BE INCLUDED IN THE RIVET REPLACEMENT.

PAYMENT FOR RIVET REPLACEMENT WILL BE MADE FOR THE QUANTITY OF EACH AS ABOVE DETERMINED AT THE UNIT PRICE BID FOR THE ITEM RIVET REPLACEMENT IN THE PROPOSAL, WHICH PRICE SHALL INCLUDE ALL MATERIALS, LABOR, EQUIPMENT AND INCIDENTALS NECESSARY TO COMPLETE THE WORK.

# Q.21 CLEANING ROLLER NEST BEARINGS

REPAIRING, CLEANING AND PAINTING TRUSS EXPANSION BEARINGS.

ALL TRUSS EXPANSION BEARINGS SHALL BE COMPLETELY CLEANED TO BARE METAL IN ACCORDANCE WITH SSPC NO. 5 WHITE METAL BLAST CLEANING. THE SURFACE AFTER CLEANING SHALL BE DEFINED BY SSPC - VIS. 1, PICTORIAL STANDARDS BSA3, CSA3 OR DSA3 AS APPLICABLE. THE SURFACES SHALL BE THOROUGHLY CLEAN OF PAINT, RUST, MILL SCALE, SLAG, DIRT, OIL, GREASE AND FOREIGN SUBSTANCES.

EXISTING KEEPERS, PLATES AND BOLTS SHALL BE REMOVED AND NEW STAINLESS STEEL KEEPER BARS FABRICATED AND INSTALLED WITH STAINLESS STEEL BOLTS IN THE SEQUENCE OUTLINED ON THE PLANS. SEGMENTAL ROLLERS, ONCE CLEANED, SHALL BE LUBRICATED AT THE BEARING LINES OF SEGMENTAL ROLLERS WITH A PENETRATING OIL AND SHALL BE ALLOWED TO UNDERGO 60 DEGREES F TEMPERATURE CHANGE OVER THE WINTER BEFORE FINAL CLEANING, PAINTING AND LUBRICATING.

EXISTING BOLTS SHALL BE REMOVED BY APPROVED MEANS WHICH WILL NOT DAMAGE THE THREADS. A SUGGESTED METHOD IS TO APPLY A PENETRATING OIL TO THE BOLT HEADS, ALLOW THE OIL TO PENETRATE AND REMOVE BY UNSCREWING THE BOLT WITH A WRENCH OR CLAMPING PLIERS. THIS METHOD WAS USED AT PIERS 54S AND 60 WHERE THE BEARINGS WERE CLEANED AND PRIMED WITH RED LEAD PAINT.

THE KEEPER BARS SHALL BE FABRICATED TO MATCH EXISTING HOLES AND SHALL BE INSTALLED IN ACCORDANCE WITH THE REQUIREMENTS FOR STEEL ERECTION IN ARTICLE 4.3.3.

AFTER ALL THE PENETRATING OIL HAS BEEN REMOVED AND THE SURFACES HAVE BEEN RECLEANED TO A WHITE METAL CONDITION, PAINT SHALL BE APPLIED IN ACCORDANCE WITH THE REQUIREMENTS FOR THE ZONE 3B SYSTEM.

FINAL LUBRICATION SHALL CONSIST OF BRUSHING A THIN COAT OF THE SERVICE GREASE ON THE BEARING SURFACES TOP AND BOTTOM.

WHERE WATER ENTRAPMENT REMAINS BECAUSE OF MORTAR BUILD-UP ADJACENT TO THE MASONRY PLATE AND THE PORTIONS OF THE BURNT-OFF DUST COVERS (PREVIOUSLY REMOVED FROM ALL PIERS) ARE EVIDENT, A CHASE SHALL BE PROVIDED AT THE MIDDLE OF EACH SIDE BY

PAINT, EQUIPMENT, TOOLS, SCAFFOLDING, PLATFORMS, TARPS, AND DROP-CLOTHS LUBRICANTS, STAINLESS STEEL BARS AND BOLTS, MORTAR, AND ALL ELSE NECESSARY TO COMPLETE THE WORK AS SPECIFIED HEREIN AND AS DELINEATED IN THE PLANS. PARTIAL PAYMENT OF 75% OF THE AMOUNT BID WILL BE MADE WHEN THE INITIAL REPAIRING, CLEANING AND LUBRICATING IS COMPLETE.

NO SEPARATE PAYMENT WILL BE MADE FOR CHIPPING A CHASE AND REMOVING A PORTION OF THE DUST COVERS AT THE MIDDLE OF EACH SIDE OF THE BEARINGS. THE COST OF THIS WORK SHALL BE INCLUDED ON THE LUMP SUM PRICE BID FOR THE ITEM REPAIRING, CLEANING, LUBRICATING AND PAINTING TRUSS EXPANSION BEARINGS.

# Q.22 BEARING REPLACEMENT

## NEW EXPANSION BEARINGS

### DESCRIPTION

NEW EXPANSION BEARINGS SHALL INCLUDE THE REMOVAL AND DISPOSAL OF PORTIONS OF THE EXISTING EXPANSION BEARINGS, THE CLEANING AND PAINTING OF ALL FIXED BEARINGS AND EXPANSION BEARINGS THAT ARE TO REMAIN, THE FURNISHING AND INSTALLING OF THE NEW EXPANSION BEARINGS, INCLUDING THE JACKING OF EXISTING GIRDERS, ALL AS INDICATED ON THE PLANS AND IN ACCORDANCE WITH THE SPECIFICATIONS.

### MATERIALS.

STRUCTURAL STEEL COMPONENTS FOR NEW EXPANSION BEARINGS SHALL MEET THE REQUIREMENTS SPECIFIED UNDER CURRENT A.S.T.M. DESIGNATION A588 AND SHALL BE HOT-DIP GALVANIZED AFTER FABRICATION IN ACCORDANCE WITH A.S.T.M. DESIGNATION A123.

BRONZE WASHERS. BRONZE WASHERS SHALL MEET THE REQUIREMENTS OF CURRENT A.S.T.M. DESIGNATION B10C, ALLOY NO. 510 OR FEDERAL SPECIFICATION QQ-B-637, ALLOY 464.

BEDDING MATERIALS. BEDDING MATERIALS FOR MASONRY PLATES SHALL BE FURNISHED AND PLACED ACCORDING TO THE REQUIREMENTS OF 1973 A.A.S.H.O. SPECIFICATIONS FOR HIGHWAY BRIDGES, ART. 2.10.56.

JACKS USED FOR THE JACKING OPERATIONS SHALL HAVE THE RATED CAPACITY SHOWN CLEARLY ON THE MANUFACTURER'S NAME PLATE ATTACHED TO EACH JACK.

### METHODS OF CONSTRUCTION.

METHODS OF CONSTRUCTION FOR REMOVAL AND DISPOSAL OF PORTIONS OF EXISTING BEARINGS SHALL MEET THE REQUIREMENTS SPECIFIED UNDER ARTICLE 2.1.3 AS AMENDED ELSEWHERE HEREIN AND AS INDICATED ON THE PLANS.

METHODS OF CONSTRUCTION INCLUDING FABRICATION, ERECTION AND WELDING OF NEW EXPANSION BEARINGS SHALL MEET THE APPLICABLE REQUIREMENTS SPECIFIED UNDER ARTICLE 4.3.3 AND AS SPECIFIED HEREIN.

THE PLANS INDICATE A SUGGESTED SCHEME FOR JACKING OPERATIONS AND SUPPORTING THE EXISTING STRUCTURE. THE CONTRACTOR MAY USE THE METHOD INDICATED OR USE ALTERNATE METHODS. THE METHOD OF JACKING OPERATIONS AND THE METHOD FOR TEMPORARILY SUPPORTING THE EXISTING STRUCTURE SHALL BE OF THE CONTRACTOR'S CHOOSING BUT WILL BE SUBJECT TO APPROVAL BY THE ENGINEER. THE CONTRACTOR SHALL SUBMIT DETAILED DRAWINGS SHOWING ALL ELEMENTS OF THE PROPOSED JACKING OPERATIONS AND TEMPORARY SUPPORTS HE PROPOSES TO USE, INCLUDING DESIGN CALCULATIONS THEREFOR, FOR APPROVAL BY THE ENGINEER. NO MATERIALS SHALL BE ORDERED AND NO WORK SHALL BE PERFORMED UNTIL WRITTEN APPROVAL BY THE ENGINEER HAS BEEN OBTAINED. APPROVAL BY THE ENGINEER WILL NOT IN ANY WAY RELIEVE THE CONTRACTOR OF HIS RESPONSIBILITY FOR THE SAFETY AND ADEQUACY OF THE JACKING AND SUPPORTS SYSTEMS AND OPERATIONS.

THE CONTRACTOR WILL NOT BE PERMITTED TO JACK THE ENDS OF A GIRDER AFTER THE DECK JOINT AT THAT LOCATION HAS BEEN SEALED.

JACKING OF THE GIRDERS WILL ONLY BE PERMITTED BETWEEN THE HOURS OF TWELVE MID-NIGHT AND FIVE A.M.

THE CONTRACTOR SHALL SUPPLY ALL LIGHTING AND EQUIPMENT THAT IS REQUIRED TO SAFELY PERFORM THE JACKING OPERATIONS AND ALL RELATED WORK AS A NIGHT OPERATION.

NO VEHICULAR TRAFFIC OR PEDESTRIANS WILL BE PERMITTED ON THE STRUCTURE DURING THE ACTUAL JACKING OPERATIONS OR WHILE THE DEAD LOAD OF THE SPAN IS STILL BEING SUPPORTED BY THE JACKS. AS A MEANS TO ACCOMPLISH THIS, THE CONTRACTOR SHALL SET-UP AND MAINTAIN THE DETOUR SHOWN ON THE PLANS. THE DETOUR WILL ONLY BE PERMITTED BETWEEN THE HOURS OF TWELVE MID-NIGHT AND FIVE A.M.

HOWEVER ON DETOUR WILL BE PERMITTED FROM THE END OF THE WORKING DAY ON FRIDAY UNTIL TWELVE MID-NIGHT OF THE FOLLOWING MONDAY.

THE DETOUR NEED ONLY BE IN EFFECT DURING THE ACTUAL JACKING OPERATIONS AND UNTIL THE JACKS HAVE BEEN RELIEVED OF ALL LOAD BY TRANSFERING THE FULL DEAD LOAD OF THAT PARTICULAR SPAN TO THE TEMPORARY SUPPORTS THAT MAKE UP PART OF THE JACKING FRAME.

THE DETOUR SHALL BE COMPLETELY REMOVED BY FIVE A.M., OR SONNER IF NO LONGER REQUIRED, AND ALL SIGNS COVERED OR REMOVED SO THAT NORMAL TRAFFIC FLOW MAY RESUME.

IT SHALL BE UNDERSTOOD BY THE CONTRACTOR THAT THE DETOUR WILL ONLY BE PERMITTED DURING THE JACKING OPERATIONS OF THE GIRDERS. THE TIME LENGTH OF EACH DETOUR SHALL BE A SHORT AS POSSIBLE SO AS TO MINIMIZE ANY INCONVENIENCE TO THE DRIVING PUBLIC.

THE AMOUNT OF JACKING MOVEMENT REQUIRED TO RELEASE AND REMOVE EACH BEARING SHALL BE THE VERY MINIMUM REQUIRED. THE CONTRACTOR SHALL SUBMIT TO THE ENGINEER FOR APPROVAL, AT LEAST 30 DAYS BEFORE THE START OF JACKING OPERATIONS, THE METHOD HE INTENDS TO USE AND THE PROCEDURE HE INTENDS TO FOLLOW IN REMOVING THE BEARING.

THE REMOVAL OF THE BEARING SHALL NOT START UNTIL THE LOAD FROM THE JACKS HAS BEEN TRANSFERRED TO THE TEMPORARY SUPPORTS AND THE TEMPORARY SUPPORTS SECURED.

ALL FIXED BEARINGS, ALL EXPANSION BEARINGS THAT ARE TO REMAIN AND THE UPPER PORTION OF EXPANSION BEARINGS THAT IS TO REMAIN, SHALL BE CLEANED AND PAINTED TWO COATS OF ZINC DUST-ZINC OXIDE PAINT PRIMER. SURFACE PREPARATION SHALL BE IN ACCORDANCE WITH STEEL STRUCTURES PAINTING COUNCIL SSPC-SP6 "COMMERCIAL BLAST CLEANING". THE FIRST COAT OF ZINC DUST-ZINC OXIDE PAINT PRIMER MUST BE APPLIED WITHIN FOUR HOURS AFTER THE SURFACE PREPARATION HAS BEEN COMPLETED.

THE CONTRACTOR'S ATTENTION IS DIRECTED TO THE NECESSITY OF PROVIDING MEANS TO MEASURE CONSTANTLY THE POSSIBLE YIELDING OR DEFLECTION OF THE TEMPORARY SUPPORTS AND SHALL CONTINUOUSLY OBSERVE THIS AT ALL TIMES. THIS PRECAUTION IS OF IMPORTANCE SINCE THE YIELDING OF THE SUPPORT WILL INDUCE PERMANENT RESIDUAL STRESSES IN THE STRUCTURE DURING CONSTRUCTION. ANY YIELDING OR DEFLECTION OF THE TEMPORARY SUPPORTS SHALL BE IMMEDIATELY CORRECTED BY THE CONTRACTOR. THE CONTRACTOR SHALL IMMEDIATELY NOTIFY THE ENGINEER OF SUCH CONDITION. THE CONTRACTOR'S ATTENTION IS DIRECTED TO THE FACT THAT IT IS OF EXTREME IMPORTANCE TO SAFEGUARD THE TEMPORARY SUPPORTS. IT WILL BE THE SOLE RESPONSIBILITY OF THE CONTRACTOR TO PROVIDE AND MAINTAIN ADEQUATE PROTECTION FOR THE TEMPORARY SUPPORTS FOR THE DURATION OF THE CONTRACT. IN THE EVENT THE TEMPORARY SUPPORTS FAIL AND ANY DAMAGE IS DONE TO THE STRUCTURE, THE CONTRACTOR SHALL CORRECT THIS DAMAGE AT HIS OWN EXPENSE. WHEN NO LONGER REQUIRED, AS DETERMINED BY THE ENGINEER, ALL MATERIALS USED FOR THE TEMPORARY SUPPORTS SHALL BECOME THE PROPERTY OF THE CONTRACTOR AND SHALL BE DISPOSED OF BY HIM CLEAR OF THE SITE.

IN THE PERFORMANCE OF THE WORK UNDER THIS ITEM, THE CONTRACTOR WILL BE HELD RESPONSIBLE FOR DAMAGE TO EXISTING CONSTRUCTION TO REMAIN IN PLACE. THE CONTRACTOR SHALL USE EXTREME CARE AND EXERCISE EVERY PRECAUTION NECESSARY TO PREVENT DAMAGE OR INJURY THERETO. ALL DAMAGE AS A RESULT OF THE CONTRACTOR'S OPERATIONS SHALL BE REPAIRED OR REPLACED BY HIM AT HIS OWN EXPENSE TO THE SATISFACTION OF THE ENGINEER AT NO COST TO THE STATE.

QUANTITY AND PAYMENT

THE QUANTITY OF NEW EXPANSION BEARINGS FOR WHICH PAYMENT WILL BE MADE WILL BE A LUMP SUM COVERING ALL THE WORK INDICATED ON THE PLANS AND IN ACCORDANCE WITH THE SPECIFICATIONS.

PAYMENT FOR NEW EXPANSION BEARINGS WILL BE MADE AT THE LUMP SUM PRICE BID IN THE PROPOSAL FOR THE ITEM NEW EXPANSION BEARINGS, WHICH PRICE SHALL INCLUDE THE PERFORMANCE OF ALL THE WORK SPECIFIED HEREIN AND AS SHOWN ON THE PLANS AND THE COST OF FURNISHING ALL LABOR, MATERIALS, EQUIPMENT INCLUDING STRUCTURAL STEEL, BOLTS, NUTS AND WASHERS, JACKS AND JACKING OPERATIONS, TEMPORARY SUPPORTS, REMOVAL AND DISPOSAL OF PORTIONS OF EXISTING EXPANSION BEARINGS, SHIMS, WELDING, GALVANIZING, CLEANING AND PAINING AND ALL ELSE NECESSARY AND INCIDENTAL THERETO.

## Q.23 BEARING REPAIR

DESCRIPTION.

STRAPPING OF BEARINGS SHALL INCLUDE THE FURNISHING AND INSTALLING GALVANIZED STEEL ANGLES TO STRAP BEARINGS TOGETHER, AS INDICATED ON THE PLANS AND IN ACCORDANCE WITH THE SPECIFICATIONS.

MATERIALS.

STRUCTURAL STEEL FOR ANGLES SHALL MEET THE REQUIREMENTS SPECIFIED UNDER ARTICLE 4.3.2 FOR A.S.T.M. DESIGNATION A36 STEEL AND SHALL BE HOT-DIP GALVANIZED IN ACCORDANCE WITH A.S.T.M. DESIGNATION A123.

METHODS OF CONSTRUCTION.

METHODS OF CONSTRUCTION INCLUDING FABRICATION, ERECTION AND WELDING OF STEEL ANGLES TO BASE PLATES SHALL MEET THE APPLICABLE REQUIREMENTS SPECIFIED UNDER ARTICLE 4.3.3 AND AS SPECIFIED HEREIN.

NO PAINTING OF GALVANIZED STEEL ANGLES WILL BE REQUIRED. SURFACES OF GALVANIZED STEEL ANGLES DAMAGED AFTER WELDING OPERATIONS SHALL BE REGALVANIZED IN THE FIELD IN ACCORDANCE WITH THE REQUIREMENTS SPECIFIED UNDER ARTICLE 4.3.3.

QUANTITY AND PAYMENT.

THE QUANTITY OF STRAPPING OF BEARINGS FOR WHICH PAYMENT WILL BE MADE WILL BE A LUMP SUM COVERING ALL THE WORK INDICATED ON THE PLANS AND IN ACCORDANCE WITH THE SPECIFICATIONS.

PAYMENT FOR STRAPPING OF BEARINGS WILL BE MADE AT THE LUMP SUM PRICE BID IN THE PROPOSAL FOR THE ITEM STRAPPING OF BEARINGS, WHICH PRICE SHALL INCLUDE THE PERFORMANCE OF ALL THE WORK SPECIFIED HEREIN AND INDICATED ON THE PLANS, AND THE COST OF FURNISHING ALL LABOR, MATERIALS AND EQUIPMENT INCLUDING STRUCTURAL STEEL, GALVANIZING, WELDING, AND ALL ELSE NECESSARY THEREFOR AND INCIDENTAL THERETO.

## Q.24 TEMPORARY SHIELD OVER RAILROAD

ARTICLE 4.14.1. DESCRIPTION

TEMPORARY SHIELD SHALL INCLUDE THE FURNISHING, ERECTING AND SUBSEQUENT REMOVAL OF A TEMPORARY PROTECTIVE SHIELD OVER THE RAILROAD TRACKS CONSTRUCTED TO PROVIDE A SAFE WORKING PLATFORM DURING CONSTRUCTION OF THE BRIDGE AND TO PRECLUDE THE DROPPING OF ANY MATERIALS THAT MAY ENDANGER THE SAFETY OF THE RAILROAD AND ITS OPERATING TRAIN SERVICES.

ARTICLE 4.14.2. MATERIALS

STRUCTURAL STEEL, UNLESS OTHERWISE INDICATED, OF ALL TYPES, SIZES, SHAPES AND CONFIGURATIONS SHOWN ON THE PLANS, SHALL CONFORM TO THE REQUIREMENTS OF ASTM A36.

PLYWOOD SHALL CONFORM TO THE REQUIREMENTS OF COMMERCIAL STANDARD CS35 TYPE HARDWOOD EXTERIOR TECHNICAL TYPE 1.

TIMBER AND LUMBER TO INCLUDE POSTS, BRIDGING FILLER, NAILERS AND ALL OTHER INDICATED USES AND TERMS SHALL BE SELECT STRUCTURAL GRADE 1800 PSI, SOUTHERN PINE OR EQUAL.

FASTENERS SHALL CONFORM TO THE FOLLOWING: STRUCTURAL STEEL FASTENED WITH NUT, BOLTS, WASHERS OR BENT U-BOLTS SHALL CONFORM TO THE REQUIREMENTS OF ASTM A307 OR AS SPECIFIED IN ARTICLE 4.11A.2 EXCEPT AS OTHERWISE INDICATED ON THE PLANS.

FASTENERS FOR ATTACHMENT OF WOOD TO WOOD MEMBERS AND PLYWOOD TO WOOD CONSTRUCTION SHALL BE OF REQUIRED SIZE COMMERCIALLY AVAILABLE STANDARD COMMON GALVANIZED NAILS; WHERE STRUCTURAL STEEL IS ATTACHED TO WOOD CONSTRUCTION FASTENERS SHALL CONFORM TO THE REQUIREMENTS OF ASTM A307 AS SPECIFIED ABOVE FOR STRUCTURAL STEEL.

ALL FASTENERS SHALL BE GALVANIZED IN CONFORMANCE WITH THE APPLICABLE REQUIREMENTS OF ASTM A123 OR ASTM A153.

ROOFING PAPER SHALL BE 15 POUND FELT CONFORMING TO THE REQUIREMENTS OF ASTM D3158; ADHESIVE AND SEALANT SHALL BE OF COLD OR HOT APPLIED AND OF THE
TYPE RECOMMENDED BY THE PRODUCT MANUFACTURER DESIGNED FOR RAPID AND FULL ADHESION BETWEEN ROOFING PAPER LAYERS AND WOOD SUBSTRATES.

ARTICLE 4.14.3. METHODS OF CONSTRUCTION

THE CONTRACTOR SHALL SUBMIT DETAILED DRAWINGS SHOWING ALL ELEMENTS OF THE SYSTEM HE PROPOSES TO USE AND DESIGN CALCULATIONS THEREFOR, FOR APPROVAL BY THE ENGINEER AND SHALL CONSTRUCT THE SHIELD IN ACCORDANCE WITH THE APPROVED WORKING DRAWINGS.

THE CONTRACTOR SHALL CONSTRUCT THE TEMPORARY SHIELD DURING PERIODS OF TRACK OUTAGES AS SPECIFIED IN DIVISION 1 OF THESE SPECIFICATIONS. THE CONTRACTOR SHALL NOT COMMENCE FURTHER PROGRESSIVE SCHEDULES OF WORK, UNTIL THE PROTECTIVE SHIELD HAS BEEN COMPLETED AND TEMPORARY GROUNDING MEASURES HAVE BEEN IMPLEMENTED.

ALL FASTENERS SHALL BE DRAWN UP TIGHT OR SET FLUSH AND SHOULD ANY TYPE FASTENER BECOME LOOSENED FOR ANY REASON IT OR THEY SHALL BE MAINTAINED AT ALL TIMES DURING CONSTRUCTION.

WALKWAYS OF PLYWOOD CONSTRUCTION SHALL BE PROTECTED AND SEALED, WITHOUT THE USE OF NAILS, AND ALL ROOFING SHALL BE FASTENED IN PLACE USING ASPHALT ADHESIVES.

ALL WORK SHALL PROVIDE RIGIDLY SECURED IN PLACE CONSTRUCTION, FITTED TO THE BRIDGE SKEW AND TO PROVIDE SAFE WORKING CONDITIONS WITHOUT ANY ENDANGERMENT TO THE RAILROAD OPERATIONS, PROPERTY OR OTHER FACILITIES.

UPON COMPLETION OF THE BRIDGE CONSTRUCTION AND PERMANENT BARRICADE AND WHEN DIRECTED BY THE ENGINEER, ALL TEMPORARY CONSTRUCTION ATTACHED TO OR SUSPENDED OR CANTILEVERED FROM THE BRIDGE, SHALL BE REMOVED AND BE DISPOSED OF OFF SITE. REMOVAL SHALL BE CAREFULLY CONDUCTED IN CONSIDERATION OF ALL FACTORS OF TRACK OUTAGES AND SAFETY AS SPECIFIED HEREIN FOR CONSTRUCTION OF THE SHIELD.

ARTICLE 4.14.4. QUANTITY AND PAYMENT

THE QUANTITY OF TEMPORARY SHIELD FOR WHICH PAYMENT WILL BE MADE WILL NOT BE MEASURED BUT ALL COSTS FOR THE WORK OF THIS SECTION SHALL BE INCLUDED IN THE LUMP SUM PRICE FOR THE WORK SHOWN ON THE PLANS.

PAYMENT FOR FURNISHING, ERECTING AND REMOVING OF THE TEMPORARY SHIELD WILL BE MADE FOR THE QUANTITY AS ABOVE DETERMINED, AT THE LUMP SUM PRICE BID FOR THE ITEM TEMPORARY SHIELD IN THE PROPOSAL, WHICH PRICE SHALL INCLUDE ALL COSTS FOR LABOR, MATERIALS, TOOLS, EQUIPMENT FOR THE TEMPORARY PROTECTIVE SHIELD, COMPLETE AS SHOWN ON THE PLANS AND SPECIFIED HEREIN, AND THE SUBSEQUENT REMOVAL AND DISPOSAL WHEN DIRECTED BY THE ENGINEER.

# Q.25 TIEBACKS

## TIEBACKS

### 4.10.1. DESCRIPTION

THE WORK SHALL CONSIST OF DESIGNING, FURNISHING AND INSTALLING A PERMANENT, PRESTRESSED ANCHOR TIEBACK SYSTEM AT THE LOCATIONS INDICATED ON THE PLANS, OR WHERE ORDERED BY THE ENGINEER.

THE WORK WILL REQUIRE THE IN-PLACE TESTING OF THE TIEBACK SYSTEM.

### 4.10.1. MATERIALS

TIEBACK TENDONS SHALL BE FABRICATED FROM SINGLE OR MULTIPLE ELEMENTS OF THE FOLLOWING:

A) STEEL BARS CONFORMING TO ASTM DESIGNATION A-722, "UNCOATED HIGH-STRENGTH STEEL BARS FOR PRESTRESSED CONCRETE."

B) SEVEN-WIRE STRAND CONFORMING TO ASTM DESIGNATION A-416, "UNCOATED SEVEN-WIRE STRESS-RELIEVED STRAND FOR PRESTRESSED CONCRETE."

C) WIRES CONFORMING TO ASTM DESIGNATION A-421, "UNCOATED STRESS-RELIEVED WIRE FOR PRESTRESSED CONCRETE."

D) COMPACT SEVEN-WIRE STRANDS CONFORMING TO ASTM DESIGNATION A-779-80, "UNCOATED SEVEN-WIRE COMPACTED, STRESS-RELIEVED STEEL STRAND FOR PRESTRESSED CONCRETE."

THE TENSION MEMBERS SHALL BE OF SUCH SIZE THAT THE DESIGN LOAD DOES NOT EXCEES 60% OF THE GUARANTEED ULTIMATE TENSILE STRENGTH OF THE TENSION MEMBER.

COUPLERS FOR TENSION MEMBERS SHALL BE CAPABLE OF DEVELOPING 100 PERCENT OF THE GUARANTEED ULTIMATE TENSILE STRENGTH OF THE TENSION MEMBER.

ANCHORAGES SHALL BE CAPABLE OF DEVELOPING 95 PERCENT OF THE GUARANTEED ULTIMATE TENSILE STRENGTH OF THE ANCHOR MATERIAL WHEN TESTED IN AN UNBONDED STATE.

BEARING PLATES SHALL BE FABRICATED FROM MILD STEEL AND BE CAPABLE OF DEVELOPING 95 PERCENT OF THE GUARANTEED MINIMUM ULTIMATE TENSILE STRENGTH OF THE PRESTRESSING STEEL.

ALL END HARDWARE SHALL MEET THE REQUIREMENTS OF ACI 318.

GREASE USED TO COAT THE STRESSING LENGTH SHALL BE COMPOUNDED TO PROVIDE CORROSION INHIBITING AND LUBRICATION PROPERTIES. ACCEPTABLE GREASES FOR THE STRESSING LENGTH SHALL BE:

1. EXXON RUST BAN 326

2. CHEVRON POLYUREA EP GREASE, #2 GRADE OR EQUAL

GREASE OTHER THAN 1 OR 2 ABOVE SHALL BE SUBMITTED TO THE ENGINEER FOR HIS APPROVAL AND SHALL NOT BE ALLOWED TO EXCEED THE MAXIMUM ALLOWABLE QUANTITY OF CERTAIN SUBSTANCES AS SHOWN IN THE FOLLOWING TABLE.

| SUBSTANCE | MAXIMUM ALLOWABLE QUANTITY - PPM | TEST METHOD |
|---|---|---|
| CHLORIDES | 10 | ASTM D-512 |
| NITRATES | 10 | ASTM D-992 |
| SULFIDES | 10 | APHA - "TEST METHODS; SULFIDES IN WATER." |

CENTRALIZERS SHALL CONSIST OF PLASTIC STEEL OR ANY MATERIAL NOT DETRIMENTAL TO THE PRESTRESSING STEEL. WOOD CENTRALIZERS SHALL NOT BE USED. CENTRALIZERS SHALL BE CAPABLE OF POSITIONING THE TENDON IN THE MIDDLE OF THE DRILLED HOLES AND OF PROVIDING NO LESS THAN 0.5 INCHES OF GROUT COVER ALONG THE BONDED LENGTH.

SPACERS SHALL BE USED TO SEPARATE ELEMENTS OF MULTIELEMENT TENDONS. THEY SHALL BE FABRICATED FROM MATERIAL WHICH IS NONDETRIMENTAL TO THE PRESTRESSING STEEL.

A COMBINATION CENTRALIZER--SPACER CAN BE USED.

SHEATHING SHALL BE A SMOOTH PLASTIC TUBE HAVING A MINIMUM WALL THICKNESS OF 0.02 INCHES AND SHALL ENCAPSULATE THE TOTAL STRESSING LENGTH OF THE ANCHOR.

THE GROUT TO BE USED FOR ANCHORAGE SHALL CONSIST OF A PUMPABLE MIXTURE OF TYPES I, II, OR III PORTLAND CEMENT, SAND AND WATER MEETING THE REQUIREMENTS OF ASTM C 150.

CHEMICAL ADDITIVES TO CONTROL BLEED OR RETARD SET MAY BE USED WITH THE ANCHOR GROUT. EXPANSIVE ADDITIVES WILL NOT BE ALLOWED. ADDITIVES IF USED, SHALL BE MIXED IN ACCORDANCE WITH THE MANUFACTURER'S RECOMMENDATIONS. EPOXY RESIN WILL NOT BE ALLOWED AS A SUBSTITUTE FOR CEMENT GROUT.

SHOP DRAWINGS SHALL BE SUBMITTED TO THE ENGINEER FOR WRITTEN APPROVAL.

EACH TIEBACK SHALL HAVE A MINIMUM UNBONDED LENGTH OF 15 FEET (4.58M). THE CONTRACT DRAWINGS INDICATE THE UNBONDED LENGTH REQUIRED FOR EACH TIER OF TIEBACKS. THE TIEBACK SHALL BE INSTALLED AT AN ANGLE VARYING BETWEEN 10 DEGREE AND 30 DEGREE FROM THE HORIZONTAL.

THE MINIMUM TOTAL TIEBACK LENGTHS ARE INDICATED ON THE CONTRACT DRAWINGS. IN NO CASE SHALL THE ANCHOR LENGTH BE LESS THAN 15 FEET.

## 4.10.3. METHODS OF CONSTRUCTION

THE HOLES FOR THE ANCHORS MAY BE EITHER DRIVEN OR DRILLED. CORE DRILLING, ROTARY DRILLING, AUGER DRILLING OR PERCUSSION DRILLING MAY BE USED. IF THE HOLE WILL NOT STAND OPEN, CASING SHALL BE INSTALLED AS REQUIRED TO MAINTAIN A CLEAN AND OPEN HOLE. THE HOLE DIAMETER SHALL NOT BE LESS THAN THREE INCHES IF PRESSURE GROUTING IS USED IN THE BOND LENGTH AND FOUR INCHES IF PRESSURE GROUTING IS NOT USED. (PRESSURE GROUTING IS DEFINED AS GROUTING WITH A PRESSURE GREATER THAN 60 PSI). THE DIAMETER OF THE DRILL BIT SHALL NOT BE SMALLER THAN THE SPECIFIED HOLE DIAMETER MINUS 1/8 INCH. THE HOLE SHALL EXTEND A MINIMUM OF TWO FEET BEYOND THE SPECIFIED BAR LENGTH. THE HOLES SHALL BE DRILLED TO THE INCLINATION SPECIFIED ON THE CONTRACT PLANS WITHIN A PLUS OR MINUS THREE-DEGREE TOLERANCE.

THE TENDON SHALL BE INSTALLED IN THE CASING OR HOLE DRILLED FOR THE ANCHOR. CARE SHALL BE TAKEN TO ENSURE THAT THE TENDON'S CORROSION PROTECTION IS NOT DAMAGED DURING HANDLING OR INSTALLATION. THE TENDON IN THE BOND LENGTH SHALL BE INSTALLED IN SUCH A WAY AS TO ENSURE THAT IT HAS A MINIMUM OF ONE-HALF INCH GROUT COVER. THE BOND LENGTH OF THE TENDON SHALL BE DEGREASED PRIOR TO INSTALLATION BY USING ACETONE, MEK, OR MIBK. NO RESIDUE SHALL BE LEFT ON THE TENDON. OTHER SUBSTANCES MAY BE USED SUBJECT TO APPROVAL BY THE ENGINEER.

IF ARTESIAN OR FLOWING WATER IS ENCOUNTERED IN THE DRILLED HOLE, PRESSURE SHALL BE MAINTAINED ON THE CONSOLIDATION GROUT UNTIL THE GROUT HAS INITIALLY SET.

AFTER A HOLE IS DRILLED TO THE FINAL DEPTH AND WATERTIGHTNESS, IF REQUIRED, IS ATTAINED, THE TENDON SHALL BE INSERTED. ANCHOR TENDONS SHALL NOT BE SUBJECTED TO SHARP BENDS. CENTRALIZERS PROVIDED AT A MAXIMUM OF 10 FEET CENTER TO CENTER SPACING THROUGHOUT THE BOND LENGTH SHALL BE USED TO ENSURE THAT THE TENDONS DO NOT CONTACT THE WALL OF THE DRILL HOLE AND NO LESS THAN 0.5 INCHES OF GROUT COVER ALONG THE BOND LENGTH IS ACHIEVED.

THE GROUTING OPERATION SHALL BE PERFORMED AFTER THE TENDON IS INSERTED. GROUT SHALL ALWAYS BE INJECTED AT THE LOWEST POINT OF THE ANCHOR.

THE GROUTING EQUIPMENT SHALL BE CAPABLE OF CONTINUOUS MIXING AND SHALL PRODUCE GROUT FREE OF LUMPS. THE GROUT PUMP SHALL BE EQUIPPED WITH A GROUT PRESSURE GAGE CAPABLE OF MEASURING 150 PSI.

THE ANNULAR SPACE BETWEEN THE SHEATHING AND THE DRILLED HOLE SHALL BE FILLED WITH GROUT FOR ITS ENTIRE LENGTH.

A PIPE OR TRUMPET INTEGRAL WITH THE BEARING PLATE, SHALL EXTEND FROM THE ANCHOR PLATE A SUFFICIENT DISTANCE TO ENCAPSULATE THE FRONT PORTION OF THE SHEATH. THE TRUMPET VOID SHALL BE FILLED WITH GREASE OR GROUT. ON COMPLETION OF THE WORK, THE ANCHORAGE SHALL BE ENCASED AS SHOWN ON THE CONTRACT DRAWINGS.

### TESTS

ANCHOR TESTING AND STRESSING: EACH ANCHOR SHALL BE TESTED. THE MAXIMUM TEST LOAD SHALL NOT EXCEED 80 PERCENT OF THE GUARANTEED ULTIMATE TENSILE STRENGTH OF THE TENDON. THE FIRST TWO ANCHORS INSTALLED AT EACH SPECIFIED DESIGN LOAD CAPACITY AND 15 PERCENT OF THE REMAINING ANCHORS (LOCATIONS TO BE CHOSEN BY THE ENGINEER) SHALL BE PERFORMANCE TESTED. ALL REMAINING ANCHORS SHALL BE PROOF TESTED.

1. PERFORMANCE TEST

PERFORMANCE TESTS SHALL BE MADE BY INCREMENTALLY LOADING AND UNLOADING THE ANCHOR IN ACCORDANCE WITH THE FOLLOWING SCHEDULE. AT EACH LOAD INCREMENT THE MOVEMENT OF THE TENDON SHALL BE RECORDED TO THE NEAREST .001 INCHES WITH RESPECT TO AN INDEPENDENT FIXED REFERENCE POINT. THE JACK LOAD SHALL BE MONITORED WITH A CALIBRATED LOAD CELL. THE LOAD CELL SHALL BE CALIBRATED BY AN INDEPENDENT TESTING LABORATORY WITHIN 14 DAYS OF START OF TESTING THE ANCHORS. THE CONTRACTOR SHALL PROVIDE THE ENGINEER WITH THE CALIBRATION CURVE BEFORE START OF TESTING.

| CYCLE | LOAD |
|---|---|
| 1 | 0. |
| | 0.25 P |
| 2 | AL |
| | 0.25 P |
| | 0.50 P |
| | 0.25 P |
| 3 | AL |
| | 0.25 P |
| | 0.50 P |
| | 0.75 P |
| | 0.50 P |
| | 0.25 P |

| Cycle | Load |
|---|---|
| 4 | AL |
| | 0.25 P |
| | 0.50 P |
| | 0.75 P |
| | 1.00 P |
| | 0.75 P |
| | 0.50 P |
| | 0.25 P |
| 5 | AL |
| | 0.25 P |
| | 0.50 P |
| | 0.75 P |
| | 1.00 P |
| | 1.25 P |
| | 1.33 P (HOLD 50 MINUTES FOR CREEP TEST) |
| | ADJUST TO TRANSFER LOAD OF .8 P |

P = DESIGN LOAD FOR THE ANCHOR
AL = ALIGNMENT LOAD

AT THE COMPLETION OF THE TEST THE ANCHOR LOAD SHALL BE REDUCED TO 0.80 P AND TRANSFERRED TO THE PERMANENT STRESSING ANCHORAGE.

EACH LOAD INCREMENT SHALL BE HELD UNTIL MOVEMENT CEASES OR FOR TWO MINUTES, WHICHEVER IS LONGER. LOADING AND UNLOADING RATES (TONS PER MINUTE) SHALL BE SUBMITTED BY THE CONTRACTOR FOR APPROVAL.

THE CREEP TEST SHALL CONSIST OF HOLDING THE 1.33 P LOAD FOR 50 MINUTES. WHILE THE LOAD IS MAINTAINED CONSTANT, ANCHOR MOVEMENT (TOTAL MOVEMENT) REFERENCED TO A FIXED POINT, SHALL BE RECORDED AT 0, 1/2, 1, 5, 10, 30 AND 50 MINUTES.

TWO COPIES OF ALL TEST DATA SHALL BE SUBMITTED TO THE ENGINEER.

THE ENGINEER WILL REVIEW ALL PERFORMANCE TESTS TO DETERMINE IF THE ANCHOR IS ACCEPTABLE. AN ANCHOR WILL BE ACCEPTED IF:

A. THE TOTAL ELASTIC MOVEMENT OBTAINED AT THE DESIGN LOAD EXCEEDS 80% OF THE THEORETICAL ELASTIC ELONGATION OF THE STRESSING LENGTH AND IS LESS THAN THE THEORETICAL ELASTIC ELONGATION OF THE STRESSING LENGTH PLUS 50% OF THE BOND LENGTH.

B. THE CREEP MOVEMENT DOES NOT EXCEED 0.080 INCHES DURING THE TIME INCREMENT BETWEEN 5 AND 50 MINUTES REGARDLESS OF TENDON LENGTH AND LOAD.

FOR ANCHORS THAT THE ENGINEER FINDS UNACCEPTABLE, THE CONTRACTOR SHALL SUBMIT A WRITTEN PROPOSAL CONTAINING A SUGGESTED COURSE OF ACTION. THE ACTION TO BE TAKEN WILL BE SUBJECT TO WRITTEN APPROVAL BY THE ENGINEER.

1. PROOF TESTS

THE PROOF TESTS SHALL BE PERFORMED BY INCREMENTALLY LOADING AND UNLOADING THE ANCHOR IN ACCORDANCE WITH THE FOLLOWING SCHEDULE. AT EACH INCREMENT, THE MOVEMENT OF THE TENDON SHALL BE RECORDED TO THE NEAREST .001 INCHES WITH RESPECT TO AN INDEPENDENT FIXED REFERENCE POINT. THE JACK LOAD SHALL BE MONITORED WITH A LOAD CELL.

| LOAD |
|---|
| 0 |
| 0.25 P |
| 0.50 P |
| 0.75 P |
| 1.00 P |
| 1.25 P |
| 1.33 P (HOLD FOR CREEP TEST) |
| ADJUST TO TRANSFER LOAD OF .8 P |

TWO COPIES OF ALL TEST DATA SHALL BE SUBMITTED TO THE ENGINEER.

ACCEPTANCE CRITERIA FOR AN ANCHOR WHICH HAS BEEN PROOF TESTED SHALL BE THE SAME AS THE PERFORMANCE TEST.

THE CREEP TEST SHALL HOLD THE 1.33 P LOAD FOR 5 MINUTES. WITH THE LOAD HELD CONSTANT, ANCHOR MOVEMENT (TOTAL MOVEMENT) SHALL BE RECORDED AT 0 SECONDS, 30 SECONDS, 2 MINUTES, AND 5 MINUTE INTERVALS. IF THE MOVEMENT BETWEEN THE 30 SECONDS AND THE 5 MINUTES READING IS 0.080 INCHES OR MORE, THE LOAD SHALL BE MAINTAINED FOR AN ADDITIONAL 45 MINUTES AND THE MOVEMENT MEASURED. IF THE ADDITIONAL MOVEMENT EXCEEDS 0.080 INCHES, THE ANCHOR SHALL BE REJECTED. ALL MOVEMENTS SHALL BE MEASURED IN RELATION TO A FIXED REFERENCE POINT.

## 4.10.4. QUANTITY AND PAYMENT

THE UNIT PRICE BID SHALL INCLUDE THE COST OF FURNISHING ALL LABOR, EQUIPMENT, AND MATERIAL REQUIRED TO COMPLETE THE WORK.

THE QUANTITY TO BE PAID FOR UNDER THIS ITEM SHALL BE THE NUMBER OF ANCHORS SUCCESSFULLY INSTALLED, TESTED AND ACCEPTED.

# R.1 UTILITY SUPPORT

(a) Gas Lines:

Gas lines or other lines carrying volatile materials shall be of steel pipe, Schedule 40 or heavier, and shall be encased in a steel encasement pipe as noted under "Criteria". Gas lines are normally protected from corrosion by use of imposed low voltage electric currents. Normally, the utility will make provision to electrically insulate the gas line from its support. Lines carrying other volatile materials shall be supported, as required by the utility, with due care taken to protect the structure and traffic.

Access and ventilation shall always be provided in box girder cells containing gas lines.

(b) Water Lines:

Water lines shall preferably be galvanized steel pipe. Where PVC is used, the dead load of water and pipe shall be such that deflections in the pipe shall not exceed 1/360 of the span between supports. As an alternate for PVC pipe, continuous support may be provided.

Some PVC pipe is susceptible to failure due to freezing. For this reason, where a water line is subject to freezing, PVC pipe should be used only with caution. Where freezing may be encountered, consideration should be given to the use of insulation on the pipe.

Care shall be taken that all inertia loads due to dynamic action (water hammer, etc.) can be properly carried. Transverse supports shall be provided for all water lines.

Fire control piping is a special case where unusual care must be taken to handle the inertial loads and associated deflections. Normally the Hydraulic Section of the Bridge Operations Branch will also be involved in this case.

In box girders, care shall be taken to insure that a failure of the water line would not flood the cell with consequential structural failure. Additional weep holes shall be used if necessary.

(c) Sewer Lines:

Where sewer lines are carried on bridges, normally an appropriate encasement pipe must be used. For sewer pipe material, see the utility agreement or the Bridge Division Hydraulic Section.

(d) Telephone and Power Conduit:

Generally, telephone, television cable and power conduit shall be of PVC pipe of a UL approved type and be Schedule 40 or heavier. Where such conduit is buried in concrete curbs or has continuous support, such support is considered to be adequate. Where conduit is supported by hangers or brackets at intervals, the distance between supports shall be small enough to avoid excessive sag between supports. (See PVC pipe, below.) Generally, the conduit shall be designed to support the cable in bending without exceeding working stresses for the conduit material. When the conduit is intended to encase Highway Department electrical wiring, only galvanizing steel conduit shall be used.

(e) Rigid Electrical Conduit for Highway Circuits:

In the case of all new bridge construction where roadway shoulders have not yet been paved and where usable shoulder width is 4' or greater in width, electrical conduit shall be stubbed-out and capped 1'-6" below grade and 3'-0" horizontally toward roadway centerline from the face of the traffic barrier. Longitudinally, this stub-out location should be near the back of pavement seat. The conduit in this location should clear any foreseeable obstructions. The location of the stubbed-out conduit at bridge ends should be clearly shown on the plans. The usual location of the conduit throughout the remainder of the bridge should be in the traffic barrier. Pull boxes shall be provided at a maximum spacing of 200 ft. Their size shall be 6" x 6" x 24" to facilitate pulling of wires.

In the case of existing bridges, an area 2' in width shall be reserved for conduit beginning at a point either 4' or 6' outside the face of usable shoulder. (See attached sketch 8.3.5 D (e) 1.) The fastening for and location of attaching the conduit to the existing bridge should be worked out on a job by job basis.

Types of Supports

The following types of supports have been used. Selection of a particular support type should be based on the needs of the installation and the best economy.

(a) Concrete Embedment - This is the best structural support condition and offers maximum protection to the utility. Its cost may be high for larger conduit and the conduit cannot be replaced. Special care must be taken to handle expansion joints.

(b) Continuous Support - This support condition may be achieved by providing a ledge of concrete to support the conduit. In addition, some type of clamping will be required. The support condition here is very good but the cost may be very high.

(c) Concrete Pedestals - This consists of concrete supports formed at suitable intervals and provided with some type of clamping device.

(d) Pipe Hangers - This is the most usual type of support for utilities to be supported under the bridge deck. It allows the use of standard ordered parts (usually "Grinell") and is very flexible in terms of expansion requirements. It will not normally provide longitudinal support, and transverse support must be provided by a second hanger extending from a girder.

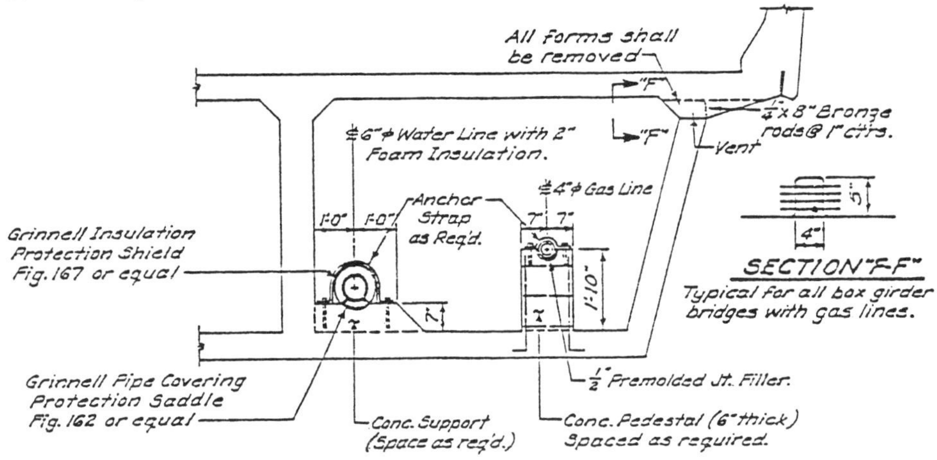


CONC. UTILITY SUPPORTS

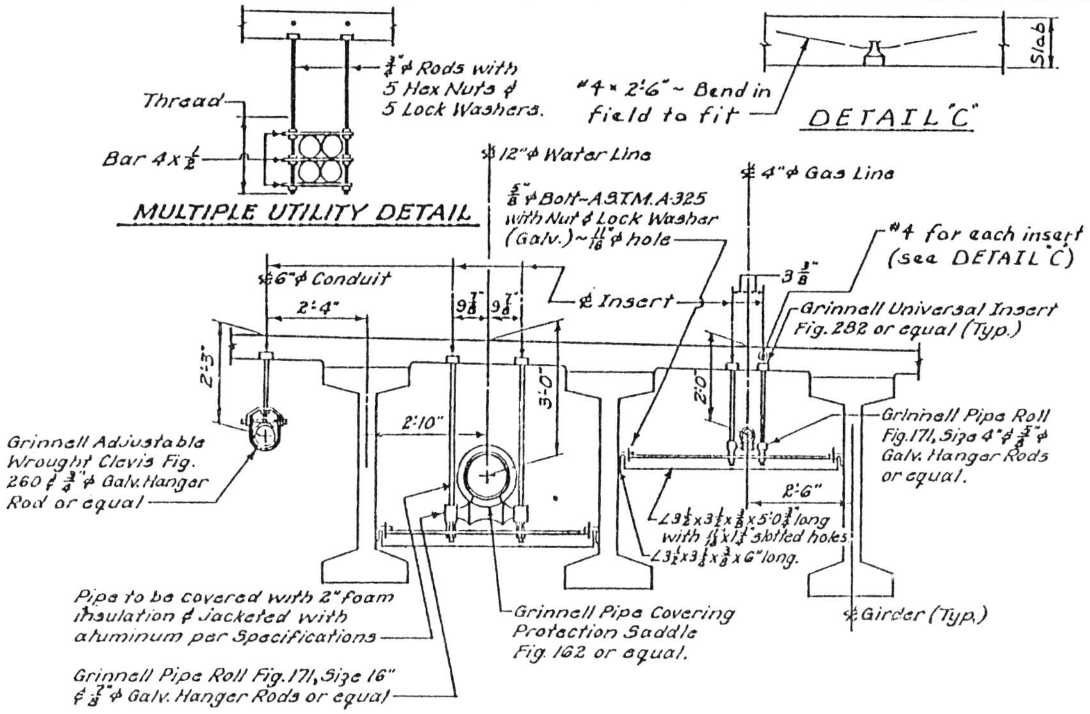


## HANGER TYPE UTILITY SUPPORTS

See Grinnell Catalog No. ph-65.
Details shown are typical. Each specific installation shall be detailed individually.

The following are recommended normal support spacings for PVC pipe. (Ref: Western Plastics Corporation.)

**PVC SCHEDULE 40**

| Nominal Size | Ø O.D. | Ø I.D. | Ø Wall Thickness | Weight Per 100 Ft. | Recommended Support Spacing* |
|---|---|---|---|---|---|
| 3/4 | 1.050 | .824 | .113 | 22.0 | 4 |
| 1 | 1.315 | 1.049 | .133 | 32.0 | 4 1/2 |
| 1 1/4 | 1.660 | 1.380 | .140 | 43.5 | 5 |
| 1 1/2 | 1.900 | 1.610 | .145 | 52.0 | 5 |
| 2 | 2.375 | 2.067 | .154 | 69.5 | 5 |
| 2 1/2 | 2.875 | 2.469 | .203 | 109.5 | 6 |
| 3 | 3.500 | 3.068 | .216 | 143.0 | 6 |
| 4 | 4.500 | 4.026 | .237 | 203.5 | 6 1/2 |
| 5 | (Use 5" PVC Class 200) | | | 276.0 | 7 |
| 6 | 6.625 | 6.065 | .280 | 359.0 | 7 1/2 |
| 8 | (Use 8" PVC Class 160) | | | | 8 |
| 10 | | | | | 8 1/2 |
| 12 | | | | | 9 1/2 |

**PVC SCHEDULE 80**

| Nominal Size | Ø O.D. | Ø I.D. | Ø Wall Thickness | Weight Per 100 Ft. | Recommended Support Spacing* |
|---|---|---|---|---|---|
| 1/2 | .840 | .546 | .147 | 20.5 | 4 1/2 |
| 3/4 | 1.050 | .742 | .154 | 28.0 | 4 1/2 |
| 1 | 1.315 | .957 | .179 | 41.0 | 5 |
| 1 1/4 | 1.660 | 1.278 | .191 | 56.5 | 5 1/2 |
| 1 1/2 | 1.900 | 1.500 | .200 | 68.5 | 5 1/2 |
| 2 | 2.375 | 1.939 | .218 | 94.5 | 6 |
| 2 1/2 | 2.875 | 2.323 | .276 | 144.5 | 6 1/2 |
| 3 | 3.500 | 2.900 | .300 | 193.0 | 7 |
| 4 | 4.500 | 3.826 | .337 | 282.0 | 7 1/2 |
| 5 | 5.563 | 4.813 | .375 | 392.0 | 8 |
| 6 | 6.625 | 5.761 | .432 | 539.0 | 9 |
| 8 | | | | | 9 1/2 |

* Spacings shown are set for 100° maximum temperatures, spacing in feet.

Ø Dimensions in inches.

The physical properties of PVC material are:

E = 410,000 psi
Tensile Strength = 7,300 psi at 78° F.
Working Stress in Bending 4.0 k/in.$^2$
Temperature Coefficient - .035" per 100° F. per foot

**PROPERTY OF PVC PIPE**

# R.2 FOUNDATION REPORT

## FOUNDATION INVESTIGATION REPORT

Throughout the foundation design and construction process it is imperative that good communication and interaction exist between the foundation engineer, structural engineer, and construction engineer. For example, even before the exploration program can be planned, the structural engineer must provide the foundation engineer with the proposed span arrangement and approximate structural loads. This interaction must be continued throughout the design and construction process to insure that the foundation design developed is compatible with the proposed structure type and is cost-effective. The importance of this communication and interaction cannot be overemphasized. Many design and construction problems are prevented through good communication.

The Foundation Investigation Report is the tool used to "communicate" the site conditons and design and construction recommendations to the bridge, roadway design, and construction engineers. The data from subsurface investigations usually are referred to continuously and for many different purposes during the design period, construction period, and, frequently, after completion of the project (resolving claims). Therefore, the foundation report should be clear, concise, accurate. It is an extremely important document whose preparation deserves special care and effort.

### Guidelines for Writing a Good Report

The following guidelines apply to writing a good foundation report:

1. The soils engineer responsible for the report preparation should have a broad enough background in engineering to have some knowledge of the foundation requirements and limitations for various types of structures.

2. The report should contain an interpretation of subsurface conditions.

3. The report should contain specific engineering recommendations for design.

4. The materials and conditions which may be encountered during construction should be discussed.

5. The soils engineer should attempt to anticipate possible design and construction problems and make recommendations for their solution.

6. Recommendations given should be brief, concise, and, where possible, definite; don't be "wishy-washy."

7. Reasons and supporting data for recommendations should be included.

8. Extraneous data which are no use to the designer or project engineer should be omitted.

9. The report should include any special notes which should be placed on the plans or in the special provisions.

### Foundation Report Outline

The following outline may be used as a general guide for presenting data in the foundation investigation report. The outline includes key items for which specific recommendations should be made, if pertinent to the given project.

### Outline

A. Text

1. Introduction

2. Scope of Investigations
   - a. Field explorations
   - b. Lab testing

3. Interpretation of subsurface conditions.

4. Approach Embankment Considerations (Primarily for fills over soft weak subsoils).
   - a. Stability
     1. Excavation and replacement of unsuitable materials
     2. Counterberm
     3. Stage construction - time delay
     4. Other treatment methods - change alignment, lower grade, lightweight fill, etc.
     5. Estimated factors of safety with and without treatment - estimated costs for treatment alternates - recommended treatment
   - b. Settlement of subsoils
     1. Estimated settlement amount
     2. Estimated settlement time
     3. Surcharge height
     4. Special foundation treatment - sand drains, wick drains, soil densification, etc.
     5. Waiting periods
     6. Downdrag on piles
     7. Lateral squeeze of soft subsoils
   - c. Construction considerations
     1. Select fill material - gradation and compaction requirements
     2. Construction monitoring (instrumentation)
   - d. Special notes

5. Spread footing support
   - a. Elevation of bottom of footing - based on frost depth, scour depth, or depth to competent bearing material
   - b. Allowable soil pressure - based on settlement or bearing capacity: considering soil, adjacent foundations, water table, etc.
   - c. Special notes

6. Pile support
   - a. Friction or end bearing or both
   - b. Suitable pile types - reasons for choice and/or exclusion of types
   - c. Pile tip elevations
     1. Estimated tip elevation
     2. Specified tip elevation - explain reasons, such as underlying soft layers, negative skin friction, scour, piles uneconomically long, etc.
   - d. Estimated pile lengths
   - e. Allowable pile loads (design load)
   - f. Estimate of pile group settlement - only of practical significance for pile groups in cohesive soils and large groups in cohesionless soil
   - g. Test piles to establish order lengths - location for maximum utility
   - h. Static pile load tests
   - i. Dynamic pile load tests (pile analyzer)
   - j. Driving criteria - based on dynamic pile formula or wave equation analysis
   - k. Corrosion effects - of particular concern in marine environments
   - l. Special notes
   - m. Actual driving resistance to reach estimated pile length

7. Drilled shaft support
   - a. Shaft diameter
   - b. Shaft length
   - c. Allowable load
   - d. Estimated settlement
   - e. Load tests
   - f. Special notes

8. Special design considerations
   a. Pile or drilled shaft lateral load capacity
   b. Seismic design - design earthquake ground acceleration, liquefaction potential (loose saturated sands and silts)
   c. Lateral earth pressures against retaining walls and high bridge abutments

9. Construction considerations
   a. Water table - fluctuations, control in excavation, pumping, tremie seals, etc.
   c. Excavations - safe slopes for open excavations, need for sheeting, shoring, etc.
   d. Drilled shafts - water table location, artesian water, boulders or obstructions, likely construction method (dry, casing, or slurry)
   e. Adjacent structures - protection against damage from excavation, pile driving, drainage, etc.
   f. Special notes.

Graphic Presentations

1. Map showing project location
2. Detailed plan of the site showing proposed structure(s), borehole locations, and existing structures
3. Laboratory test data
4. Finished boring logs and/or interpreted soil profile

Typical Special Contract Notes

The foundation engineer should include in the Foundation Investigation Report any special notes which he feels should be placed in the contract plans or special provisions. The purpose of such special notes is to bring the contractor's and/or project engineer's attention to certain requirements of the design or construction. Typical special notes relating to pile driving and embankment construction are as follows:

"It is possible that difficult driving of piles may be encountered and it may be necessary to utilize mechanical equipment for removing consolidated material or boulders from the location of piles. This may be accomplished by various types of earth augers, well drilling equipment, or other devices to remove the consolidated material to permit piles to be driven to the desired depth or rated resistance without damage."

"If any obstructions to pile driving are encountered ten (10) feet or less from the bottom of the footing, the contractor shall, if so ordered by the engineer, pull the partially driven pile or piles and remove the obstruction, backfilling the hole with approved suitable material which shall be thoroughly compacted to the satisfaction of the engineer. However, no partially driven pile shall be removed until the engineer is satisfied that the contractor has made every effort to drive the pile through the obstruction. Payment for excavation will be made at the unit price bid for the Structure Excavation Item and for the temporary sheeting under Item ________ when sheeting is used. No other extra payment will be made for this work."

"The ordered length of pile shall be measured below the cut-off elevation shown on the plans. Any additional lengths of pile or splices above the cut-off elevation necessary to facilitate the contractor's operation shall be at his own expense."

"Piles for ________ are driven because of possible future scour of stream bed and shall be driven to the minimum lengths shown on the plans regardless of the resistance to driving. The actual driving resistance is estimated to be _______ tons."

"Piles will be acceptable only when driven to pile driving criteria established by the Chief Bridge Engineer. Prerequisite to establishing those criteria, the contractor shall submit, to the Chief Bridge Engineer, and others as required, Form ____________, 'Pile and Driving Equipment Data'. All information listed on Form ________ shall be provided within fourteen (14) days after the award of the contract. Each separate combination of pile and pile driving equipment proposed by the contractor will require the submission of a corresponding Form ______________."

"Piles for the existing structure should be removed where they interfere with the pile driving for the new structure."

"It shall be the contractor's responsibility to place the cofferdams for __________ so that they will not interfere with the driving of batter piles. Pay lines for the cofferdams shall be as shown on the plans."

"The general subsurface conditions at the site of this structure are as shown on Drawing No. __________."

"Pile driving will not be allowed at the abutments until fill settlement is complete. Estimated settlement time is _____ months after placement of the _____ foot surcharge."

"The contractor shall coordinate his construction schedule to allow installation of embankment monitoring instrumentation by the State forces."

"Instrumentation damaged by contractor personnel shall be repaired or replaced at the contractor's expense. All construction activity in the area of any damaged instrument shall cease until the damage has been corrected."

Foundation Investigation Report July 4, 1982

To: Mr. A. J. Jones
Chief Engineer

From: Mr. A. B. Smith
Chief Foundation Engineer

Subject: Interstate 0 Structure over the Apple Freeway

The Geotechnical section has completed an analysis of the foundation conditions at the site of the subject structure. Our analysis is based on the following information:

1. A 1-inch equals 20 feet plan and profile prepared by the Bridge Division and received in this office April 1, 1982.
2. An interpretation of the boring logs and analysis of soil samples from three drill holes numbered BAF-1 thru 3, 9 auger holes numbered EA-1 thru 9, and one drill hole numbered BAF-4 from which undisturbed samples were taken.
3. Laboratory testing on undisturbed samples from BAF-4.

Subsurface Conditions

The general subsurface conditions are shown on Drawing No. 5 GS 331.

Foundation Recommendations

1. Elevation Assumptions

The foundation recommendations are based on the following bottom of footing elevations:

| | |
|---|---|
| West Abutment | 1011 |
| Pier | 992 |
| East Abutment | 1012 |

Changes to footing elevations may affect the foundation recommendations and should be discussed with this office.

2. Embankment Construction

A. Unsuitable Subexcavation

An approximate 1 to 3-foot thick organic layer exists between approximate stations 92+70 to 94+00 in the area of the east approach embankment. This organic layer should be removed and replaced with granular embankment material.

B. Embankment Material and Placement

The approach embankment shall be constructed of materials placed in accordance with Bridge Design Data Sheet 80-1

C. Embankment Settlement

An estimated 12 inches of fill settlement will occur due to consolidation of the 35-foot thick clay layer underlying the proposed 30-foot high east approach embankment. Estimated settlement time for 90 percent primary settlement is 14 months. Settlement time can be reduced to, (1) 6 months by use of a 10-foot surcharge fill or (2) 2 months through use of either 12-inch diameter sand drains at 9 foot center to center spacing or wick drains at 7.5 foot center to center spacing. Estimated cost for each of these treatments is:

| Treatment | Estimated Settlement Time | Estimated Extra Cost |
|---|---|---|
| Fill only | 14 months | $ --- |
| Fill w/10 foot surcharge | 6 months | 120, 000 |
| Fill w/wick drains | 2 months | 172,000 |
| Fill w/sand drains | 2 months | 385,000 |

It is understood the construction schedule will not allow a 14-month waiting period but will allow up to an 8-month waiting period, therefore, the 10-foot surcharge treatment is recommended as the most cost-effective method to reduce settlement time. The surcharge should be placed for a length of 500 feet back of the bridge ends on both the east and west approach (limits of clay subsoil).

D. Embankment Stability

The estimated immediate end of constructon factor of safety for the proposed 30' high east approach embankment is 1.63. The estimated immediate end-of-construction factor of safety for the proposed 30-foot fill plus 10 foot temporary surcharge is 1.33. Both factors of safety are adequate and no special approach embankment treatment is necessary. Long-term factor of safety will increase as consolidation of the foundation soils occur. The factor of safety for the west approach embankment will be higher as the fill height is 10' less.

E. Embankment Monitoring

It is recommended that fill settlement be monitored with settlement plates and piezometers. Settlement plates should be installed at existing ground elevation at centerline stations 90+00, 93+50, and 96+50. Piezometers to monitor excess pore pressure buildup and dissipation in the clay subsoil are recommended at centerline stations 93+50 and 96+50. A total of three piezometers should be installed at each location - one each at 20, 28, and 36 foot depths. Instrumentation will be installed by State forces.

Abutment Foundation

A. Spread Footings

The abutments may be supported on spread footings placed on compacted select material with a maximum allowable bearing capacity of 3 tons per square foot. The total settlement of the east and west abutments respectively will be 2.6 and 1.9 inches. About 60 percent of the settlement will occur in 2 months after structure construction. This settlement may be reduced by extending the surcharge period. For 90 percent consolidation the surcharge should remain in place a total of 8 months; 2 months longer than required for embankment considerations.

B. Piles

Two pile types were analysed at the abutment; a displacement pile (12" diameter closed end pipe) and a non-displacement pile (12 x 74 H-pile). Displacement type piles are not recommended due to their inability to be driven through the fill and the uncertainity of obtaining penetration in the dense gravel stratum. Non-displacement H-piles are recommended. A 12 x 74 H-pile driven to rock may be designed for a maximum load of 120 tons. Tip reinforcement, such as APF 75500, should be used to prevent tip damage by boulders in the gravel stratum and to insure penetration to rock. Estimated pile lengths are 60 feet at the west abutment and 75 feet at the east abutment.

At the abutments negative skin friction may be expected if the piles are installed before fill settlement is complete. In addition lateral squeeze of the clay subsoil will occur as the clay consolidates. Therefore, to prevent increased vertical downdrag pile loads, bending of abutment piles and rotation of the abutment toward the fill, the abutment piling should not be installed until embankment settlement is complete.

The actual driving resistance estimated to develop the design load for the H-pile at the estimated length is 284 tons. The contractor should size his equipment to achieve this resistance without damaging the pile.

Pier Foundation

A. Spread Footings

The pier may be supported on spread footings placed 4 feet below ground on natural undisturbed soil and designed for a maximum allowable bearing capacity of 3 tons per square foot. Approximately 2.8 inches of settlement is expected at this location with 1 inch occurring immediately and 2 inches occurring in less than 2 months. If a spread footing foundation is chosen, consideration should be given to increasing the structure clearance over the Apple Freeway to account for these settlements. Settlement along the footing axes will be uniform.

B. Piles

A 12" diameter closed end pipe pile and a 12 x 74 H-pile were analysed at the pier. The closed end pipe may be designed for 70 tons with a safety factor of 2 if driven into the dense gravel layer. The estimated length is 40 feet. However a minimum wall thickness of 0.375 inches should be used to prevent overstress during driving. A driving resistance of 192 tons is estimated to reach the estimated length. A conical reinforced point should be used to prevent tip damage due to boulders. The cost per ton on a per foot basis equals $11.

A 12 x 74 H-pile may be designed for 120 tons with a safety factor of 2 if driven to rock. The estimated length is 46 feet. A driving resistance of 203 tons is estimated to reach rock. A reinforced tip similar to APF 75500 should be used to prevent tip damage due to boulders. The cost per ton on a per foot basis equals $8.

We recommend that H-piles be chosen if piles are used because of cost advantages and installation advantages.

5. Special Notes

The following special notes are recommended to be included in the contract documents.

1. The general subsurface conditions at this site are shown on Drawing No 5, GS 331.

2. A 6-month waiting period will be imposed between completion of the 10-foot surcharge on the east embankment. The actual length of the waiting period may be reduced by the Engineer based on an analysis of settlement platform and piezometer readings.

3. The contractor shall coordinate his construction schedule to allow installation of instrumentation by State forces.

4. Instrumentation damaged by contractor personnel shall be repaired or replaced at the contractor's expense. All construction activity in the area of any damaged instrument shall cease until the damage has been corrected.

If piles are used the additional special notes should be provided.

5. Pile driving will not be allowed at the abutments until fill settlement is complete. Estimated maximum settlement time is 6 months after placement of the 10 foot surcharge. This time may be reduced based on interpretation by the State of settlement plate readings.

7. The actual driving resistance to install the 12 x 74 H-piles to the estimated lengths shown on the plans is estimated to be 203 tons at the pier, 284 tons at the east abutment and 250 tons at the west abutment. The contractors equipment shall be capable of overcoming these resistances without inflicting pile damage.

SITE INVESTIGATION REPORT
Project I.D. 1210-00-00
I.H. 43 over Kittycrest Drive
Structure B-5-265/266
Brown County

1. General

This site investigation report is for the twin structures which will carry I.H. 43, station 569± N.B. $R_L$, over Kittycrest Drive as part of the Syene Interchange. Approach fills are expected to be some 25 feet in height, and constructed of silty clays found throughout the area.

Abutment footing grades are anticipated to be at elevation 780±, and the pier footings are expected to be at elevation 755±.

The surface geology is that of glacio-lacustrine clay and silt deposits overlying limestone bedrock.

2. Subsurface Conditions

Six borings were made on the site. Standard Penetration Tests (AASHTO T-206) were made to estimate relative densities, maximum bearing capacities, and pile support values. Soil textures noted are driller's field identification with subsequent verification in the Central Soils Office. Unconfined strength values shown are estimated from pocket penetrometer readings made on split spoon samples. Undisturbed Shelby tube samples were taken of the underlying soft clays to evaluate settlement properties and strength characteristics for an embankment study.

Soils were generally logged as soft to stiff silty clay, with occasional silt layers. A dense sand and gravel layer was encountered just above bedrock at elevation 660±. Approximately a 15 foot surface crust of very stiff silty clay was found throughout the site. Limestone lies fairly uniformly across the site at elevation 650±. Rock cores were taken with results as follows:

| Boring Number | Core Size | % Recovery | Rod |
|---|---|---|---|
| 1 | AXL | 70 | 40% |
| 3 | NXL | 80/100 | 85% |
| 5 | NXL | 100 | 95% |

Water levels observed over a month period from the time of drilling indicate the water table to be at elevation 755±.

3. Bearing Capacity

Footings placed in the upper portion of the surface crust could be designed for a bearing capacity of 2-1/2 TSF. Footings placed below elevation 750±, however, should be designed for a maximum bearing capacity of 1 TSF.

4. Piles

The following values can be used for estimating pile lengths:

| Elevation | Pile Skin Friction (psf) (SF=2) | % End Bearing (Displ. Piles) | % End Bearing (H-Piles) |
|---|---|---|---|
| Surface-745 | 800 | 25-30 | 15-20 |
| 745-695 | 400 | 10-15 | - |
| 695-660 | 600 | 15-20 | 10-15 |
| 660-650 | 800 | 50-60 | 25-30 |

H-piles driven to bedrock should be stressed to a maximum of 12,000 psi in the steel.

5. Caissons

While soil conditions are such that drilled shafts or caissons could be used, they do not appear to offer any advantages at this location.

6. Construction Problems

No particularly unusual construction problems are foreseen. Water table, however, will be near pier footing grades. Should spread footings be used, a minimum amount of time should elapse between excavation and footing construction. Excess moisture and too much reworking of this type soil can cause drastic soil strength reductions.

7. Embankments

Bridge end-slopes were analyzed by a computerized slope stability program utilizing Bishop's method, and the assumption of a circular failure arc. Results from this analysis show a factor of safety against failure of 1.1 for the proposed 2:1 slopes.

Settlement calculations indicate that total settlements on the order of .1 to .2 of a foot can be anticipated over a period of several years. Settlements of this magnitude occurring during this length of time will not cause excessive drag loads to develop on H-piles.

8. Recommendations

Spread footings appear to be the most suitable foundation type for the pier units.

A pile type foundation is suggested for the abutments. Economics should be the deciding factor as to whether to use a greater number of lightly loaded short displacement piles, or fewer, but longer, H-piles driven to bedrock.

Embankment settlement and end-slope stability are not anticipated to be problems. The actual factor of safety will be larger than the calculated value due to three dimensional effects. Settlements will be minimal, and should not adversely affect the structure.

(WIS DOT)

## R.3 SOIL SAMPLE IDENTIFICATION

Definitions of Soil Components and Fractions

1. Grain Size

| Material | Fraction | Sieve Size |
|---|---|---|
| Boulders | | 12"+ |
| Cobbles | | 3" - 12" |
| Gravel | coarse | 3/4" - 3" |
| | fine | No. 4 to 3/4" |
| Sand | coarse | No. 10 to No. 4 |
| | medium | No. 40 to No. 10 |
| | fine | No. 200 to No. 40 |
| Fines (Silt & Clay) | | Passing No. 200 |

2. Coarse- and Fine-Grained Soils

| Descriptive Adjective | Percentage Requirement |
|---|---|
| trace | 1 - 10% |
| little | 10 - 20% |
| some | 20 - 35% |
| and | 35 - 50% |

3. Fine-Grained Soils. Identify in accordance with plasticity characteristics, dry strength, and toughness as described in Table 3.

| | Descriptive Term | | Thickness |
|---|---|---|---|
| Stratified Soils | alternating | | |
| | thick | | |
| | thin | | |
| | with | | |
| | parting | - | 0 to 1/16" thickness |
| | seam | - | 1/16 to 1/2" thickness |
| | layer | - | 1/2 to 12" thickness |
| | stratum | - | greater than 12" thickness |
| | varved Clay | - | alternating seams or layers of sand, silt and clay |
| | pocket | - | small, erratic deposit, usually less than 1 foot |
| | lens | - | lenticular deposit |
| | occasional | - | one or less per foot of thickness |
| | frequent | - | more than one per foot of thickness |

(NAVFAC DM-7)

# R.4 EROSION CONTROL PRACTICE

| Treatment Practice | Advantages | Problems |
|---|---|---|
| **FILL SLOPES** | | |
| BERMS AT TOP OF EMBANKMENT  | Prevent runoff from embankment surface from flowing over face of fill<br>Collect runoff for slope drains or protected ditch<br>Can be placed as a part of the normal construction operation and incorporated into fill or shoulders | Cooperation of construction operators to place final lifts at edge or shaping into berm<br>Difficult to compact outside lift when work is resumed<br>Sediment buildup and berm and slope failure |
| SLOPE DRAINS  | Prevent fill slope erosion caused by embankment surface runoff<br>Can be constructed of full or half section pipe, bituminous, metal, concrete, plastic, or other waterproof material<br>Can be extended as construction progresses<br>May be either temporary or permanent | Permanent construction as needed may not be considered desirable by contractor<br>Removal of temporary drains may disturb growing vegetation<br>Energy dissipation devices are required at the outlets |
| FILL BERMS OR BENCHES  | Slows velocity of slope runoff<br>Collects sediments<br>Provides access for maintenance<br>Collects water, for slope drains<br>May utilize waste | Requires additional fill material if waste is not available<br>May cause sloughing<br>Additional construction area may be needed |
| SEEDING / MULCHING  | Timely application of mulch and seeding decreases the period a slope is subject to severe erosion<br>Mulch that is cut in or otherwise anchored will collect sediment. The furrows made will also hold water and sediment | Seeding season may not be favorable<br>Not 100 percent effective in preventing erosion<br>Watering may be necessary<br>Steep slopes or locations with high velocities may require supplemental treatment. |
| **PROTECTION OF ADJACENT PROPERTY** | | |
| BRUSH BARRIERS  | Use slashing and logs from clearing operation<br>Can be covered and seeded rather than removed<br>Eliminates need for burning or disposal off right-of-way | May be considered unsightly in urban areas |
| STRAW BALE BARRIERS  | Straw is readily available in many areas<br>When properly installed, they filter sediment and some turbidity from runoff | Requires removal<br>Subject to vandal damage<br>Flow is slow through straw requiring considerable area |
| SEDIMENT TRAPS  | Collect much of the sediment spill from fill slopes and storm drain ditches<br>Inexpensive<br>Can be cleaned and expanded to meet need | Does not eliminate all sediment and turbidity<br>Space is not always available |
| SEDIMENT POOLS  | Can be designed to handle large volumes of flow<br>Both sediment and turbidity are removed<br>May be incorporated into permanent erosion control plan | Requires prior planning, additional construction area and/or flow easement<br>If removal is necessary, can present a major effort during final construction stage<br>Clean-out volumes can be large<br>Access for clean-out not always convenient; Anti-seepage baffles required for permanent construction |
| FENCE TRAP 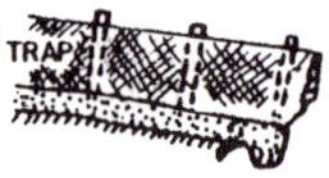 | Low cost<br>Temporary measure can be erected with minimum supervision | Some maintenance needed depending on length of time in place |

| Treatment Practice | Advantages | Problems |
|---|---|---|
| PROTECTION OF ADJACENT PROPERTY (continued) | | |
| ENERGY DISSIPATORS 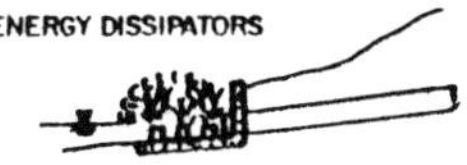 | Slow velocity to permit sediment collection and to minimize channel erosion off project | Collects debris and requires cleaning<br>Requires special design and construction of large shot rock or other suitable material from project |
| LEVEL SPREADERS  | Spreads channel or pipe flow to sheet flow<br>Avoids channel easements and construction off project<br>Simple to construct | Adequate spreader length may not be available<br>Sodding of overflow berm is usually required<br>Must be a part of the permanent erosion control effort<br>Maintenance forces must maintain spreader until no longer required |
| PROTECTION OF STREAM | | |
| CONSTRUCTION DIKE  | Permits work to continue during normal stream stages<br>Controlled flooding can be accomplished during periods of inactivity | Usually requires pumping of work site water into sediment pond<br>Subject to erosion from stream and from direct rainfall on dike |
| COFFERDAM  | Work can be continued during most anticipated stream conditions<br>Clear water can be pumped directly back into stream<br>No material deposited in stream | Expensive |
| TEMPORARY STREAM CHANNEL CHANGE | Prepared channel keeps normal flows away from construction | New channel usually will require protection<br>Stream must be returned to old channel and temporary channel refilled |
| RIP RAP  | Sacked sand with cement or stone easy to stockpile and place<br>Can be installed in increments as needed | Expensive |
| TEMPORARY CULVERTS FOR HAUL ROADS  | Eliminates stream turbulence and turbidity<br>Provides unobstructed passage for fish and other aquatic life<br>Capacity for normal flow can be provided with storm water flowing over the roadway | Space not always available without conflicting with permanent structure work<br>May be expensive, especially for larger sizes of pipe<br>Subject to washout |
| ROCK-LINED LOW-LEVEL CROSSING  | Minimizes stream turbidity<br>Inexpensive<br>May also serve as ditch check or sediment trap | May not be fordable during rainstorms<br>During periods of low flow, passage of fish may be blocked |
| DITCHES | | |
| CHECK DAMS  | Maintains low velocities<br>Catches sediment<br>Can be constructed of logs, shot rock, lumber, masonry or concrete, gabions, sand bags | Close spacing on steep grades<br>Require clean-out<br>Unless keyed at sides and bottom, erosion may occur |
| SEDIMENT TRAPS/ STRAW BALE FILTERS  | Can be located as necessary to collect sediment during construction<br>Clean-out often can be done with on-the-job equipment<br>Simple to construct | Little direction on spacing and size<br>Sediment disposal may be difficult<br>Specification must include provisions for periodic clean-out<br>May require seeding, sodding or pavement when removed during final cleanup |
| SODDING  | Easy to place with a minimum of preparation<br>Can be repaired during construction<br>Immediate protection<br>May be used on sides of paved ditches to provide increased capacity | Requires water during first few weeks<br>Sod not always available<br>Will not withstand high velocity or severe abrasion from sediment load |

| Treatment Practice | Advantages | Problems |
|---|---|---|
| SEEDING WITH MULCH AND MATTING  | Usually least expensive<br>Effective for ditches with low velocity<br>Easily placed in small quantities with inexperienced personnel | Will not withstand medium to high velocity<br>Requires anchoring |
| PAVING, RIPRAP, RUBBLE | Effective for high velocities<br>May be part of the permanent erosion control effort | Cannot always be placed when needed because of construction traffic and final grading and dressing<br>Initial cost is high |
| ROADWAY SURFACE | | |
| CROWNING TO DITCH OR SLOPING TO SINGLE BERM  | Directing the surface water to a prepared or protected ditch minimizes erosion | Requires good construction procedures<br>Can cause local stability problems (sloughing) |
| COMPACTION | The final lift of each day's work should be well compacted and bladed to drain to ditch or berm section<br>Loose or uncompacted material is more subject to erosion | Requires good construction procedures |
| AGGREGATE COVER  | Minimizes surface erosion<br>Permits construction traffic during adverse weather<br>May be used as part of permanent base construction | Requires reworking and compaction if exposed for long periods of time<br>Loss of surface aggregates can be anticipated |
| SEED/MULCH  | Minimizes surface erosion | Must be removed or is lost when construction of pavement is commenced |
| STONE FILLED GABION WALL  | Permits steeper slope<br>No special backfill required<br>Self draining | High cost<br>Requires special techniques to install properly |
| CUT SLOPES | | |
| BERM AT TOP OF CUT  | Diverts water from cut<br>Collects water for slope drains/paved ditches<br>May be constructed before grading is started | Access to top of cut<br>Difficult to build on steep natural slope or rock surface<br>Concentrates water and may require channel protection or energy dissipation devices<br>Can cause water to enter ground, resulting in sloughing of the cut slope |
| DIVERSION DIKE  | Collects and diverts water at a location selected to reduce erosion potential<br>May be incorporated in the permanent project drainage | Access for construction<br>May be continuing maintenance problem if not paved or protected<br>Disturbed material or berm is easily eroded |
| SLOPE BENCHES 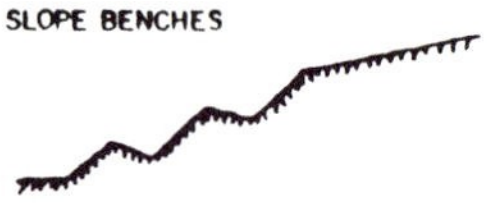 | Slows velocity of surface runoff<br>Collects sediment<br>Provides access to slope for seeding, mulching, and maintenance<br>Collects water for slope drains or may divert water to natural ground | May cause sloughing of slopes if water infiltrates<br>Requires additional construction area<br>Not always possible due to poor material, etc.<br>Requires maintenance to be effective<br>Increases excavation quantities |
| SLOPE DRAINS (PIPE, PAVED, ETC)  | Prevents erosion on the slope<br>Can be temporary or part of permanent construction<br>Can be constructed or extended as grading progresses | Requires supporting effort to collect water<br>Permanent construction is not always compatible with other project work<br>Usually requires some type of energy dissipation |

| Treatment Practice | Advantages | Problems |
|---|---|---|
| CUT SLOPES (continued) | | |
| SEEDING/MULCHING  | The end objective is to have a completely grassed slope. Early placement is a step in this direction. The mulch provides temporary erosion protection until grass is rooted. Temporary or permanent seeding may be used. Mulch should be anchored. Larger slopes can be seeded and mulched with smaller equipment if stage techniques are used. | Difficult to schedule high production units for small increments<br>Time of year may be less desirable<br>May require supplemental water<br>Contractor may perform this operation with untrained or unexperienced personnel and inadequate equipment if stage seeding is required |
| SODDING  | Provides immediate protection<br>Can be used to protect adjacent property from sediment and turbidity | Difficult to place until cut is complete<br>Sod not always available<br>May be expensive |
| SLOPE PAVEMENT, RIPRAP  | Provides immediate protection for high risk areas and under structures<br>May be cast in place or off site | Expensive<br>Difficult to place on high slopes<br>May be difficult to maintain |
| TEMPORARY COVER  | Plastics are available in wide rolls and large sheets that may be used to provide temporary protection for cut or fill slopes<br>Easy to place and remove<br>Useful to protect high risk areas from temporary erosion | Provides only temporary protection<br>Original surface usually requires additional treatment when plastic is removed<br>Must be anchored to prevent wind damage |
| SERRATED SLOPE  | Lowers velocity of surface runoff<br>Collects sediment<br>Holds moisture<br>Minimizes amount of sediment reaching roadside ditches | May cause minor sloughing if water infiltrates<br>Construction compliance |
| FABRIC MATS  | Effective for moderate to high embankment when crown vetch plantings are used<br>Has lower cost features over other methods | Requires anchoring time to promote plant growth.<br>May require periodic maintenance |
| BORROW AREAS | | |
| SELECTIVE GRADING AND SHAPING | Water can be directed to minimize off-site damage<br>Flatter slopes enable mulch to be cut into soil | May not be most economical work method for contractor |
| STRIPPING AND REPLACING OF TOPSOIL | Provides better seed bed<br>Conventional equipment can be used to stockpile and spread topsoil | May restrict volume of material that can be obtained for a site<br>Topsoil stockpiles must be located to minimize sediment damage<br>Cost of rehandling material |

## LIMITING FLOW VELOCITY TO MINIMIZE EROSION

| PERMISSIBLE VELOCITY | | | | |
|---|---|---|---|---|
| | | With Channel Vegetation | | |
| Soil Type | Bare Channel | 6" to 10" in height | 11" to 24" in height | Over 30" in height |
| Sand, Silt, Sandy loam, Silty loam | 1.5 | 2.0 to 3.0 | 2.5 to 3.5 | 3.0 to 4.0 |
| Silty clay loam, Silty clay | 2.0 | 3.0 to 4.0 | 3.5 to 4.5 | 4.0 to 5.0 |
| Clay | 2.5 | 3.0 to 5.0 | 3.0 to 5.5 | 3.0 to 6.0 |

(NAVFAC DM-7)

# R.5 SEEPAGE CONTROL METHODS

| Method | Applicability | Characteristics and Requirements |
|---|---|---|
| Sheet pile cut off wall | Suited especially for stratified soils with high horizontal and low vertical permeability or pervious hydraulic fill materials. May be easily damaged by boulders or buried obstructions. Tongue and groove wood sheeting utilized for shallow excavation in soft to medium soils. Interlocking steel sheetpiling is utilized for deeper cutoff. | Steel sheeting must be carefully driven to maintain interlocks tight. Steel H-pile soldier beams may be used to minimize deviation of sheeting in driving. Some deviation of sheeting from plumb toward the side with least horizontal pressure should be expected. Seepage through interlocks is minimized where tensile force acts across interlocks. For straight wall sheeting an appreciable flow may pass through interlocks. Decrease interlock leakage by filling interlocks with sawdust, bentonite, cement grout, or similar material. |
| Compacted barrier of impervious soil | Formed by compacted backfill in a cutoff trench carried down to impervious material or as a core section in earth dams. | Layers or streaks of pervious material in the impervious zone must be avoided by careful selection and mixing of borrow materials, scarifying lifts, aided by sheepsfoot rolling. A drainage zone downstream of an impervious section of the embankment is necessary in most instances. |
| Grouted or injected cutoff | Applicable where depth or character of foundation materials make sheetpile wall or cutoff trench impractical. Utilized extensively in major hydraulic structures. May be used as a supplement below cutoff sheeting or trenches. | A complete positive grouted cutoff is often difficult and costly to attain, requiring a pattern of holes staggered in rows with carefully planned injection sequence and pressure control. See DM-7.3, Chapter 2 for materials and methods. |
| Slurry trench method | Suited for construction of impervious cutoff trench below groundwater or for stabilizing trench excavation. Applicable whenever cutoff walls in earth are required. Is replacing sheetpile cutoff walls. | Vertical sided trench is excavated below groundwater as slurry with specific gravity generally between 1.2 and 1.8 is pumped back into the trench. Slurry may be formed by mixture of powdered bentonite with fine-grained material removed from the excavation. For a permanent cutoff trench, such as a foundation wall or other diaphragm wall, concrete is tremied to bottom of trench, displacing slurry upward. Alternatively, well graded backfill material is dropped through the slurry in the trench to form a dense mixture that is essentially an incompressible mixture; in working with coarser gravels (which may settle out), to obtain a more reliable key into rock, and a narrower trench, use a cement-bentonite mix. |
| Impervious wall of mixed in-place piles. | Method may be suitable to form cofferdam wall where sheet pile cofferdam is expensive or cannot be driven to suitable depths, or has insufficient rigidity, or requires excessive bracing. | For a cofferdam surrounding an excavation, a line of overlapping mixed in-place piles are formed by a hollow shaft auger or mixing head rotated into the soil while cement grout is pumped through the shaft. Where piles cannot be advanced because of obstructions or boulders, supplementary grouting or injection may be necassary. |
| Freezing - ammonium brine or liquid nitrogen | All types of saturated soils and rock. Forms ice in voids to stop water. Ammonium brine is better for large applications of long duration. Liquid nitrogen is better for small applications of short duration where quick freezing is needed. | Gives temporary mechanical strength to soil. Installation costs are high and refrigeration plant is expensive. Some ground heave occurs. |

# INDEX

## A

**A36**, 42, 184, 778
**A514**, 42, 184, 778
**A517**, 778
**A572**, 42, 184, 778
**A588**, 42, 184, 778
**AASHTO**, 41, 42, 284
**AASHTO-PCI Standard Section**, 120
**AREA**, 58
**Abutment**, 50, 221, 634
**Abutment backwall, details**, 649
**Abutment:** Gravity, 226; Integral, 226; Spill-through, 226; Cellular, 50; Closed end, 50; Counterfort, 226; Diaphragm, 50; Full height, 48, 226; Gravity, 226; Integral, 226; Open or spill-through, 50; Pile bent, 50, 226; Selection, 48; Semi-stub, 49, 226; Stub, 50, 226; Type, 49.
**Acrylic resin**, 146
**Adding members**, 32, 191
**Adding pier**, 500 to 502
**Adding stringers**, 504
**Admixture**, 766
**Admixture, Corrosion inhibiting**, 77
**Aesthetics**, 24
**Aggradatioin**, 262, 269
**Air Carbon Arc Cutting (AAC)**
**Air entraining admixture**, 766
**Alignment**, 281
**Alkyd resin**, 146
**Allowable loads for fillet welds**, 565
**Alternate design**, 29, 30, 31, 35
**Anchor bars**, 99
**Anchor**, 681
**Anchored bulkhead, design criteria**, 672
**Anchored bulkhead**, 239
**Appendix**, 301
**Approach roadway**, 32, 281
**Approach slab**, 57
**Armor, bank**, 266
**Armored joint**, 644
**Asbestos bearing pad**, 631
**Asphalt overlay**, 366

## B

**Back gouging**, 566
**Backwater program**, 722
**Bank protection**, 740, 786
**Bar placement details**, 394
**Barrier, anchorage**, 343
**Basic flood**, 725
**Basic lead silico chromate**, 146
**Bearing capacity**, 701
**Bearing cleaning, roller**, 798
**Bearing hold down device**, 627
**Bearing repair**, 212, 800
**Bearing replacement**, 799
**Bearing retrofit**, 610
**Bearing:** 204, 608; Alignment, 210; Bronze plate, 631; Elastomeric, 207, 616, 628, 630; Forces, 204; Guideline, 212; Hanger assembly, 175, 210, 632; Looseness, 211; Movements, 204; Pad, 631; Pot, 208; Problems, 209; Rocker, 206, 612, 626; Roller, 206; Seat, 212; Sliding plate, 206; Spherical, 208; Steel reinforced elastomeric, 217; Stiffener, 514, 539; Surface, 211; TFE, 207,628.
**Bituminous coating**, 95
**Bolted connection**, 183
**Bolster**, 633
**Bolted and welded structure**, 773
**Bolted structure**, 775
**Bolting**, 561
**Box culvert, details**, 720
**Box girder**, 521
**Box girder, pier cap**, 167
**Braced wall design consideration**, 680
**Braced wall**, 239
**Bracing**, 506
**Breakaway cable terminal**, 288, 748
**Breastwalls**, 634
**Bridge Replacement Program**, 318
**Bridge failure**, 1
**Bridge flooring**, 358
**Bridge foundation**, 268
**Bridge inspection**, 3, 15
**Bridge modification**, 32, 265, 269,
**Bridge railing**, 23, 285
**Bridge replacement**, 6, 14, 33, 42, 266
**Bridge seat elevation**, 666
**Bridge seat**, 222
**Bridge:** Details, 19; Geometry, 21; Hydraulics, 21, 261, 722; Lateral clearance, 21, 282; Vertical clearance, 21, 32; Width, 20, 283;
**Brittle fracture**, 153
**Broom finishing**, 82
**Buttressed wall**, 256

## C

**Caisson**, 245
**Calcium sulfate**, 95
**Cantilever Wall**, 674
**Capacity of pile**, 690,696-699
**Capital cost**, 34, 272
**Carbon Arc Weld (CAW)**, 571
**Catch basin**, 56, 607
**Categorical Exclusioin**, 24, 324
**Cathodic protection**, 75, 144, 150, 152, 260, 702; Impressed system, 75, 152; Non-overlay, 76; Sacrificial, 75, 152.
**Caulking**, 102
**Cement:** High early, 70, 95; High-alumina, 65, 95; Portland, 65, 100.
**Chain drag**, 4
**Channelization**, 269
**Charpy V-notch impact**, 511, 779
**Check dam**, 268
**Chemical grouting**, 254
**Chloride analysis**, 5
**Chloride removal**, 76
**Classification of electrodes**, 784
**Cleaning:** Blast, 149; Hand, 149; Power, 149; Solvent, 149.
**Clear zone width**, 287
**Closed abutment**, 650
**Closure pour**, 298, 755
**Coal tar epoxy**, 88, 259
**Coating**, 144
**Cofferdam**, 240, 267
**Cohesive material**, 690
**Collision damages**, 92, 176, 194
**Commercial blast cleaning**, 149
**Composite action**, 32,118, 186, 512
**Comprehensive evaluation**, 37
**Concrete Bridge Deck**, 62, 81, 83, 319
**Concrete T beam**, 45
**Concrete box girder**, 45
**Concrete cap extension**, 640
**Concrete deck repairs**, 62, 362, 363
**Concrete deck slab**, 44, 332, 334, 337
**Concrete pavement**, 267
**Concrete removal**, 63
**Concrete repair methods**, 94
**Concrete repairs**, 93
**Concrete rigid frame**, 47
**Concrete structure**, 44, 91, 369
**Connection plate location**, 552
**Connection steel:** Cantilever Bracket, 168; Floor beam to girder, 163; Floor beam, 167; Gusset plate, 170; Through girder and beam, 164
**Construction barrier**, 749
**Construction cost**, 35
**Continuous footing**, 56
**Coring and grouting**, 792
**Coring**, 5
**Corrosion action**, 141
**Corrosion control**, 142
**Corrosion protection**, 145, 683
**Corrosion:** 140; Chemical, 142; Differential environment, 142; Electrolyte, 140; Extensive, 6; Fretting, 142; Galvanic, 141; Moderate, 6; Stray current, 142; Stress, 142
**Corrugate metal pipe**, 788
**Corrugated metal deck**, 86, 358
**Cost**, 12, 34, 35
**Countermeasure, Hydraulic**, 263
**Cover plate details**, 542
**Cover plates**, 158, 538
**Crack gage**, 249
**Crack repairs**, 97, 221, 789
**Cracked welds**, 573
**Cracks**, 66, 91, 221
**Creep in skew bridge**, 638
**Creep**, 79
**Crib and metal wall**, 667
**Cross frames**, 541, 547, 554
**Cross frames, end**, 548
**Crutch bent**, 640
**Cultural resources**, 27
**Culvert:**, 274, 706; Details, 716; Design, 709; Extension, 279; Grade, 278; Hydraulic, 277, 706; Metal, 275; Protection, 275
**Curved girder layout**, 560

## D

**Dead loads**, 32
**Deadman anchorage, design criteria**, 673
**Decay**, 197
**Deck alternates**, 61
**Deck condition survey**, 4
**Deck drain**, 368
**Deck drainage**, 56, 600
**Deck joints**, 214, 582
**Deck joint repairs**, 582
**Deck joint retrofit**, 585
**Deck protective system**, 70
**Deck reconstruction procedure**, 7
**Deck reconstruction**, 61
**Deck repair**, 62, 362
**Deck replacement**, 78, 88
**Deck sealants**, 77
**Deck slab**, 82, 332

Deck slab, pouring sequence, 342
Deficiencies, 33
Deficient bridge, 9
Deflection, 42, 505
Deformation, 248
Degradation, 262, 267
Delamination, 4, 70
Design criteria, Bridge, 41
Design defects, 91
Design details, concrete, 106
Design graph, Composite A36, 530
Design method, 42
Design simple span I-beam, 534
Design three span continuous I-beam, 537
Design two span continuous I-beam, 536
Design Example: Concrete box culvert, 709; Concrete slab bridge, 369; Deck drainage, 602; Elastomeric bearing, 616; Pier, 651; Pile, 694; Precast panel, 349; Prestressed box beam, 444; Prestressed girder, 436, 480; Rocker bearing, 612; Steel plate girder, 522
Details, Reinforcements, 107
Diaphragm connection plate, 541
Diaphragms, 506, 546, 554
Differential settlement, 246
Discount Rate, 326
Discretionary fund, 10
Dissimilar metal, 143
Distortion, 573
Double protective system, 79
Doublewal, 236
Dowels, 99, 104
Drainage, 27, 33
Drainage control, 252
Drainage inlet, 607
Drainage studies, Checklist, 726
Drainage, deck, 219
Drilling and dowel, 300
Dry pack, 95, 98
Dumped riprap, 266, 786

## E

Earth pressure table, 668
Earth pressure, computation, 670
Earth pressure, formula, 669
Economic Analysis, 34
Economic life, 35
Economical type of structure, 44
Economical type of concrete bridge, 45
Effects of alloy, 568
Elastomeric bearing, 207, 616, 628, 630
Electrodes, 784
Electrogas Weld (EGW), 571
Electroslag Weld (ESW), 571
Embankment, 686
Embedment length, 300
End cross frames, 548, 549
End diaphragms, 507, 555
Energy conservation, 23
Environmental Assessment, 24, 325
Environmental Categories, 24, 324
Environmental Impact Statement, 24, 324
Epoxy asphalt, 88
Epoxy bonding compound, 789
Epoxy coated bars, 78
Epoxy grout, 69, 119, 404
Epoxy injection, 67, 789
Epoxy mortar, 65, 95, 100
Erosion, 270, 810
Erosion control, 273
Expansion anchor bolts, 99, 391
Expansive soil, 247
Extension of structure, 769
Eye bars, 175

## F

FHWA Rules, 62, 316
FHWA standard details, 538
FHWA, Flood plain process, 728
FHWA, Hydraulic design, 316
FONSI, 25, 26, 33
Fabric reinforced wall, 237
Fabrication, 772
Failure of concrete structure, 96
Fast-setting mortar, 95
Fastener tension, 782
Fatigue, 152, 510
Fatigue categories, 154
Fatigue cracks, 156
Field erection, steel, 777
Field splice, 515
Fillet weld details, 565
Finger joint 597
Fire damages, 92
Flange, splice, 516
Flood hazard evaluation guideline, 725
Flood, 21, 725
Floor system, 521
Flow control structure, 268
Flow in triangular channel, 605
Flux Cored Arc Weld (FCAW), 570
Footing shear key, 671
Footing, design consideration, 646
Foundation: 22, 242, 686; Failure, 245; Pedestal, 671; Repairs, 249; Report, 806; Seals, 56; Selection, 51
Four(4) R program, 2
Fracture, 152
Fracture control plan, 177
Functional requirements, 16
Functionally obsolete, 9
Fund and resource, 16

## G

GREAT cushioin, 290
Gabion, 237, 267, 675, 739, 786
Galvanizing, 145
Gas Metal Arc Weld (GMAW), 570
Gas Tungsten Arc Weld (GTAW), 570
Gas tungsten remelt, 159
General scour, 262
Glue-laminated floor, 86
Gore area, 288
Gradation, 277
Granular material, 691
Grid deck, 81
Grillage footing, 245
Groove weld, 170
Grout, 764
Grouted riprap, 786
Grouting, 253
Guard rail, 285, 287, 745; Bridge related, 747; Connection to bridge, 746; Details, 745; End anchorage, 748; Transition, 287
Guideline, Restoration, 6
Gunite, 101
Gusset plates, 170, 171, 172

## H

HS-25 loading, 41
Half-cell, 5, 70
Hanger assembly, 175, 632
Header, reconstruction, 584
Heat affected zone, 174
Heat curving, 774
Heat straightening, 195
Hi-dri cushion, 289
Honeycombing, 91
Hot asphalt sealer, 95
Hydraulics, 277, 722; Analysis, 264; Design procedure, 271; Report, 729; Study, checklist, 726; Bridge site report, 731;
Hydrocell crush cushion, 289
Hydrogen induced crack, 174

## I

I-beam (A572) bridge, 534
I-beam details, 535
Impact attenuater, 288
Impregnation, 74
Inclinometer, 248
Inclusion, 573
Increase vertical clearance, 503
Inertia crush cushion, 289
Inorganic coating, 150
Insect damage, 197
Instrumentation, 248
Integral abutment, 219, 226
Interest Table, 326
Intersecting weld, 538

## J

Jacketing, 257, 266
Jacking, 182, 225, 796
Joints: 214, 582; Preformed sealer, 217, 586; Finger, 215, 597; Hot poured sealer, 216; Open, 215; Slding plate, 215, 596; Strip seal, 591
Joosten's method, 254

## L

Lack of fusioin, 573
Lamellar tearing, 173
Laminated deck, 198
Lamination, 173
Lap splice table, 389, 390
Lateral bracing, 170, 507, 539, 550, 551, 556
Latex modified concrete, 65, 71, 78, 95
Length of bolts, 782
Life cycle cost, 12, 29
Live load capacity, 32, 80
Live loads, 41
Local scour, 262, 723
Longitudianl weld, 169
Longitudinal stiffener, 169
Low cost bridge, 18
Low slump high density concrete, 65, 70

## M

Magnesia phosphate concrete, 65, 95
Magnetic arc blow, 573
Maintenance of Traffic, 15, 80
Maintenance, 27
Manning's Equation, 604, 707
Marine borer, 197
Metal culvert, 275

Metallic coating, 145
Modifying Structural system, 32, 192
Moisture control, 202
Monitoring, 248

## N

Narrow bridge, 290
National Bridge Inspection Standard, 3, 318
Near white blast cleaning, 149
Needle beam, 251
Non-redundant, 152, 511
Non-shrink grout, 119, 404
Nondestructive testing, 157
Nut, bolt and thread dimension, 781

## O

Oleoresinous, 146
Organic coating, 146
Orthotropic deck, 87, 360
Overflow section, 723
Overlapping, 573
Overturning, 241

## P

PH value, 275
Pachometer, 5
Painting A588, 150
Painting cost, 149
Painting existing steel, 148, 771
Painting, 146, 521, 765
Patching, 62, 64, 100
Peening, 159
Permit, 28, 58
Pier: 232, 651; Cap design, 656; Column, 51, 657; Column bent, 232; Details, 54; Forces, 655; Footing, 51, 658; Hammerhead, 51, 232; Integral, 232; Loads, 652, 653; Repair at water, 793; Repairs, 232; Saddle beam, 642; Seals, 660; Selection, 51; Shape factor, 724; Types, 52
Piezometer, 249
Pigment, 146
Pile bent abutment, 369, 647
Pile design, 690
Pile drivers, 699
Pile driving hammer, 692, 698
Pile encased abutment, 648
Pile foundation, 55, 243, 270,
Pile group action, 692
Pile investigation, 693
Pile jacketing, 258, 704
Pile protection, 259
Pile repairs, 257, 704
Pile, Example problem, 694
Pile, design, 700
Pile: Cast-in-placed, 245; Concrete, 55; Pipe, 244; Precast concrete, 55, 245; Precast concrete, 55; Steel H, 55, 245; Timber, 55, 244
Plank floor, 86
Plasma Arc Cutting (PAC)
Plasma Arc Weld (PAW), 571
Plate girder details, 553
Plate girder, 508
Pneumatically placed concrete, 767
Polymer concrete 66, 74, 95
Polyurethane tile, 88
Poor penetration, 573
Popouts, 91
Porous welds, 573
Post tensioning, 115, 127
Post tensioning, steel, 190
Pouring sequence, 80, 342
Precast concrete box culvert, 718
Precast deck, 83, 348, 350, 401
Precast panel, 84, 349
Precast wing, header, 719
Preformed joint sealer, 586
Preheat, 784, 573
Preliminary design, 40
Preloading, 124
Preplaced aggregate, 98
Present worth method, 36
Preservative, 202
Pressure grouting, 102, 228, 230
Pressure injection, 101, 789
Prestress, 409
Prestress camber, 414, 432
Prestress loss, 421
Prestress transfer, 420
Prestressed Concrete: Alternate, 30; Box beam, 46, 444, 458; Channel girder, 46, 472; Continuity, 46, 416, 477; Design and details, 411, 420, 436, 444; Girder, 45, 409, 411, 414, 418, 436, 449, 453; Slab, 45, 119, 474, 460,; Section property, 449; Selection chart, girder, 418; T girder, 46; 17" voided slab, 460; 21" voided slab, 461; 24" channel beam, 472; 27" box beam, 462; 32" channel beam, 473; 33" box beam, 463; 36" & 45" girder design data, 454; 36" girder detail, 450; 45" girder details, 451; 54" girder design data, 455; 542 girder details, 452; Bearing details, 457; Box beam details, 464 to 470; Box beams, 119; Bridge replacement, 117; Channel beam, 471; Damages, 109 to 114; Deck slab, 119; Defects and countermeasure, 484; Design continuous span, 46, 431, 477; Plank beam, 474; Preloading, 114; Repair design, 125 to 128; Repairs, 109, 110; Restoration of Prestress, 115; Shear keys, 119, 122, 404; Shear transfer, 413; Slab & substructure details, 456; Tie rods, 122, 406
Prestressing, steel, 123, 189, 190, 191
Preventive measures, 269
Primer, 147
Project selection, 10
Protection angle for slab, 644
Protective coating, 76, 103
Protective system, 321

## R

Rail bank protectiion, 741, 786
Railing details, 742
Railing, 283, 742
Railroad structure, 57, 58,
Raising bridge seat, 224, 503
Raking, stone joint, 228
Ranking, 38
Rating alternative, 37
Recaulking, 794
Red lead primer, 147
Rehabilitatiion study, 14
Rehabilitation Procedure, 6, 180, 322
Rehabilitation report, 303
Rehabilitation scheme, 32
Rehabilitation, 42, 180
Reinfoced concrete slab, 369, 374
Reinforced concrete wearing surface, 339
Reinforced earth, 235, 675
Reinforcement over piers, 83
Reinforcement steel, 43, 64, 83, 85, 100, 104, 185, 388, 389
Reinforcing bar table, 388
Reinforcing, 103
Repainting, 148
Repair design consideration, 181
Repair design, stresses, 181
Repair material, 95, 763
Repair methods, concrete, 97
Repair of defects, steel, 773
Repair schemes, 31
Repair spalled concrete, 791
Repair, steel, 186
Repair: Bearing seat, 107; Jacketing, 105; Post tensioning, 106; Reinforcing steel plate, 105; Shear cracks, 105; Spalled concrete, 105, 791; Timber structure, 200
Replacement of rivets, 160, 798
Replacement schemes, 33
Replacement, 42
Repointing, 227, 794
Report, 38
Resetting bearing, 608
Residual stresses, 183
Retaining wall, 234, 667
Retard, 269
Riparian, 60
Riprap, 266, 786
Risk analysis, 270
Rivet replacement, 160, 798
Rock anchor, 255
Rocker bearing design, 612
Rocker bearing details, 626
Roll-in steel structure, 490 to 499
Rolled beam diaphragms, 557
Rolled beam splice details, 558
Rolled beam, 48, 508
Routine maintenance, 269

## S

SSPC, 149
Sacked concrete, 268, 786
Saddle beam, 641
Safety, 23, 33
Sand blasting, 149
Sand cement grout, 404
Sand drain, 689
Scaling, 91
Scour depth, Basic design curve, 724
Scour repairs, 256
Scour, 91, 261, 266, 277, 723
Scupper details, 606
Scupper spacing, 600
Scupper spacing, Design example, 602
Section 4(f), 27
Section 106, 27
Section 404, 28
Seepage control, 252, 810
Seepage, 247
Segmental concrete box girder: 128 to 139; Balanced cantilever, 131; Cast-in-placed, 130; Casting methods, 129; Design consideration, 132; Economical

types of bridge, 132; Erection methods, 130; Incremental launching, 132; Longline system, 130; Problems, 133; Progressive placement, 131; Shortline sytem, 129; Span by span, 131
Selection of rehabilitation, 181
Selection process, 31
Sensitivity, 36
Service level, 284
Service life of steel culvert, 276
Settlement, 91, 241, 246, 687
Settlement, design solution, 689
Shear block, 224
Shear connectors, 518, 545
Shear keys, 119, 404, 407,
Shear lock, 545
Shear strength, 252
Shear, horizontal, 429
Sheathing, 259
Sheet pile, 239, 695
Sheeting, 762
Shielded Metal Arc Weld (SMAW), 570
Shoring, 104
Shotcrete, 101, 767
Shrinkage, 79, 298
Silicate grouting, 254
Skew angle, 43
Skew effects, 223, 415
Skewed Bridge, 83
Skid resistance, 81
Slab haunch details, 559
Slab overhang design, 340
Sliding plate joint, 596
Sliding, 241
Slope and wall, 252
Slope failure, 241
Soil anchor, 255
Soil sample, Visual identification, 809
Sounding, 4
Spalling, 91
Spatter, 573
Spread footing, 55, 242, 270,
Spur dike, 267, 269, 724
Stability, 686
Stabilization, 33, 674
Stabilizing soil, 252
Stage construction, 80, 123, 343, 345
Stay-in-placed form, 89, 346
Steel alternate, 31
Steel I-beam, Design graphs, 530
Steel bolster, 226
Steel bridge, design consideration, 505
Steel bridge, design approach, 511
Steel details, 187
Steel girder selection, 508
Steel girder, 47
Steel girder, Design example, 522
Steel grid deck, 85, 356
Steel structure, 772
Stiffener: 161, 166, 169, 539, 543, 544; Bearing, 162, 539; Longitudinal, 169, 520, 539; Transverse, 161, 166, 171, 519, 538
Stone arch culvert, 276
Stone masonry, 227, 794
Straightening bent material, 774
Straightening bent members, 195, 797
Strand drape, 426
Strands, 124, 409
Stream encroachment, 59
Stream properties, 738
Strengthening steel member, 184, 796
Strengthening, concrete,103
Strengthening, 32
Strengthening, foundation, 250
Stress corrosion crack, 153
Stress riser, 173
Strip seal details, 594
Strip seal joint, 215, 591
Structural bolting, 780
Structural Inventory and Appraisal, 2, 8, 14
Structural metal, 778
Structural welding, 783
Structurally deficient, 9
Structure type selection, 43
Structure type, 44
Structure type, span and loading, 17
Stuctural steel, 184
Stud Weld (SW), 571
Submerged Arc Weld (SAW), 570
Substructure rehabilitation, 231
Substructure retrofit, 640
Substructure, 221, 634
Sufficiency Rating, 8, 330
Superplasticizer, 66
Superstructure, 43
Surcharge, 689
Surface preparation, concrete deck, 63
Surface preparation, concrete, 97
Surface preparation, steel coating, 151

## T

TFE Elastomeric bearing, 628
Tapered culvert inlet, 278
Temporary bridge railing, 750
Temporary shield over railroad, 800
Tendon, 681
Test welding: Dye penetration, 158; Eddy current, 158; Magnetic particle, 158; Radiographic, 158; Ultrasonic, 156; Testing methods, 156
Thermal, 79
TS & L, 40
Thrie beam guard rail, 285
Thrie beam railing, 744
Thrie beam railing, retrofit, 743
Tie rod, 402
Tiebacks, 240, 255, 681, 760, 801
Tied arch action, 103
Timber Sturcture, 197, 574
Timber cap replacement, 578
Timber cap scab, 579
Timber cap strengthening, 578
Timber crib wall, details, 678
Timber deck details, 354
Timber deck, 81, 86
Timber helper bent, 577
Timber pile, Replacement, 581
Timber pile, Shimming, 579
Timber pile, Splice, 581
Timber pile, Sway bracing, 580
Timber stringer repairs, 576
Timber stringer replacement, 576
Torch Brazing (TB)
Torsi onal reinforcement, 396
Toughness, 510
Traffic barrier, 286
Traffic capacity, 294
Traffic control plan, 292, 751
Traffic control, 291, 294
Traffic safety, 33, 281, 742
Traffic, 20, 23,
Transverse tie design, 406
Truck mounted attenuater, 290
Truss portal modification, 194
Truss: 187, 486; Adding or shifting supports, 488; Compression member, 188; Strengthening, 487; Subdivision, 489; Tension member, 187; Upgrading, 487
Turn of the nut method, 782
Type of welds, 565
Types of Abutment, 226
Types of pile, 244
Typical steel coating, 151

## U

US Army Corp of Engineer, 29, 59
US Coast Guard, 28, 58
Undercutting, 573
Underpinning, 245, 250, 266
Uniform annual cost, 36
Uplift, 505
Utilities, 22, 804
Utility support, 804

## V

Value engineering, 28, 29
Vertical alignment, 282
Vertical clearance, 31, 32
Vertical stiffener retrofit, 166
Vibration, 298
Vinyl resin, 146
Visual observation, 4

## W

Wall design criteria, 667
Wall details, 676
Washer dimension, 781
Water quality, 59
Water reducer, 66, 766
Water surface profile, 272
Waterproofing membrane, 73, 79, 88, 367
Waterway opening, 22, 261
Wax bead concrete, 75
Web splice, 516
Weld: nondestructive testing, 157
Weldability, 568
Welded connection, 183
Welded cover plate, 158, 542
Welded splice, reinforcing bars, 392
Welded strucutre, 775
Welded wire fabric, 398
Welding correction, 785
Welding processes, 570
Welding reinforcing bars, 64, 392, 770
Welding symbols, 562 to 563
Welding terms, 566
Welding troubles, 572
Welding, 509
Wetland, 27, 59
Widening details, 755
Widening schemes, 299, 753
Widening under the bridge, 760
Widening, 32, 296, 753
Widening, Design consideration, 296
Wingwalls, 635
Wire tied riprap, 786
Work zone safety, 283
Work zone types, 293
Working stress design, 386, 387

## Z

Zinc-rich, coating, 147

ONE-PIECE CONTINUOUS CONSTRUCTION
AND STURDY ANCHORAGE MEAN NO LEAKS AND NO LOOSENING.
Unidam elastometric bridge expansion joints are also a snap to install and require no maintenance. And that means you save money up front and keep on saving for the life of your bridge. Unidam expansion joints are custom-designed for your project. Each unit comes with design and installation drawings as well as Royston field supervision during installation to make sure everything goes right. To learn more about Unidam bridge expansion joints for new or retrofit installations, write for our new Unidam brochure. Royston Laboratories, Inc., 128 First Street, Pittsburgh, PA 15238. Or phone, 412/828-1500 collect.
a subsidiary of
COLUMBIA CHASE
all about energy
Royston

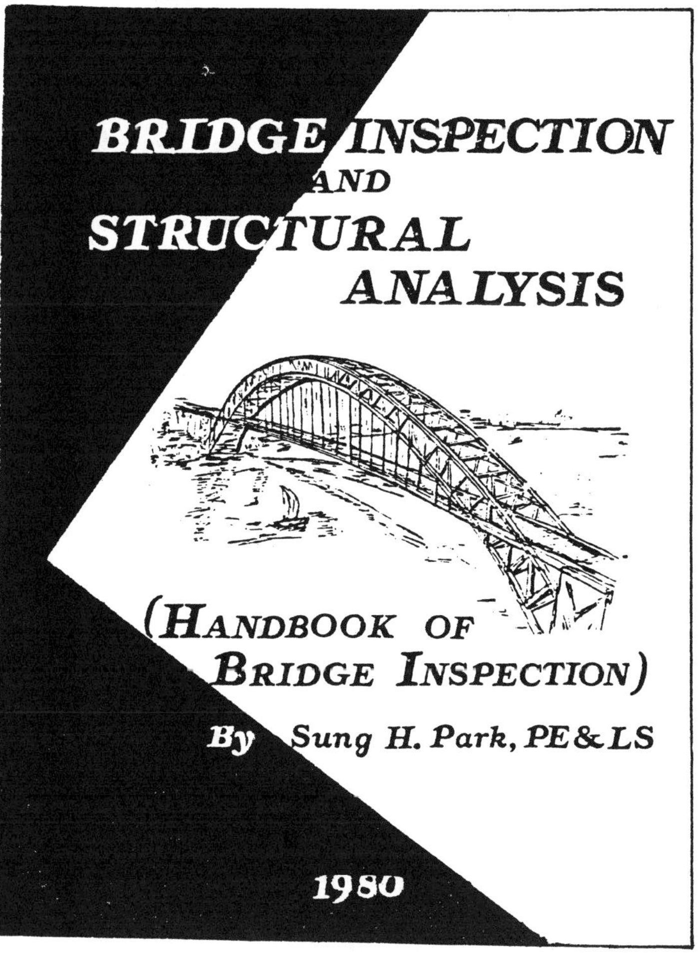

BRIDGE INSPECTION AND STRUCTURAL ANALYSIS
(HANDBOOK OF BRIDGE INSPECTION)
By Sung H. Park, PE&LS
1980